CAMBRIDGE LIBRARY COLLECTION

Books of enduring scholarly value

Astronomy

From ancient times, humans have tried to understand the workings of the world around them. The roots of modern physical science go back to the very earliest mechanical devices such as levers and rollers, the mixing of paints and dyes, and the importance of the heavenly bodies in early religious observance and navigation. The physical sciences as we know them today began to emerge as independent academic subjects during the early modern period, in the work of Newton and other 'natural philosophers', and numerous sub-disciplines developed during the centuries that followed. This part of the Cambridge Library Collection is devoted to landmark publications in this area which will be of interest to historians of science concerned with individual scientists, particular discoveries, and advances in scientific method, or with the establishment and development of scientific institutions around the world.

An Introduction to Practical Astronomy

Although astronomical guides were available in the early nineteenth century, they tended to come from continental presses and were rarely in English. This two-volume work by the clergyman and astronomer William Pearson (1767–1847) aimed, with brilliant success, to compile data from extant sources into one of the first English practical guides to astronomy. Most of the tables were updated and improved versions, and some were wholly reconstructed to streamline the calculation processes. Sir John Herschel dubbed it 'one of the most important and extensive works on that subject which has ever issued from the press', and for his efforts Pearson was awarded the gold medal of the Astronomical Society. First published in 1824, Volume 1 chiefly comprises extensive tables to facilitate the reduction of a range of astronomical observations, including solar and sidereal movements, alongside thorough instructions. In the history of science, Pearson's work reflects the contemporary challenges of celestial study.

An Introduction to Practical Astronomy

Containing Tables For Facilitating the Reduction of Celestial Observations, and a Popular Explanation of Their Construction and Use

Volume 1

William Pearson

Cambridge University Press

CAMBRIDGE
UNIVERSITY PRESS

University Printing House, Cambridge, CB2 8BS, United Kingdom

Published in the United States of America by Cambridge University Press, New York

Cambridge University Press is part of the University of Cambridge.
It furthers the University's mission by disseminating knowledge in the pursuit of
education, learning and research at the highest international levels of excellence.

www.cambridge.org
Information on this title: www.cambridge.org/9781108064057

This edition first published 1824
This digitally printed version 2013

ISBN 978-1-108-06405-7 Paperback

AN

INTRODUCTION

TO

PRACTICAL ASTRONOMY:

CONTAINING

TABLES,

RECENTLY COMPUTED,

FOR FACILITATING THE REDUCTION OF CELESTIAL OBSERVATIONS;

AND

A POPULAR EXPLANATION OF THEIR CONSTRUCTION AND USE.

BY THE REV. W. PEARSON, LL.D. F.R.S. ETC.

RECTOR OF SOUTH KILWORTH, LEICESTERSHIRE, AND TREASURER TO THE ASTRONOMICAL SOCIETY OF LONDON.

"The labour of an Astronomer, in reducing his observations, is so great, that the construction of *convenient* Tables is a matter of considerable importance."

WOODHOUSE. [ASTRON. VOL. I. P. 285.]

VOL. I.

London:

PRINTED FOR THE AUTHOR,

AND SOLD BY MESSRS. LONGMAN, HURST, REES, ORME, BROWN, AND GREEN, PATERNOSTER ROW.

1824.

G. Woodfall, Printer, Angel Court, Skinner Street, London.

TO THE PRESIDENT, VICE-PRESIDENTS, AND MEMBERS AT LARGE,

OF THE

ASTRONOMICAL SOCIETY OF LONDON,

This Volume of Tables,

UNDERTAKEN WITH A VIEW TO PROMOTE THE OBJECTS OF THE SOCIETY,

AND, AT THE SAME TIME, TO SPARE ITS FUNDS,

IS MOST RESPECTFULLY DEDICATED,

BY THEIR OBEDIENT AND VERY HUMBLE SERVANT,

THE TREASURER.

PREFACE.

WHOEVER undertakes to write a work on PRACTICAL ASTRONOMY must suppose his readers acquainted with the general principles of the science, from having previously read some elementary work on the subject; otherwise he would be under the necessity of demonstrating a variety of mathematical formulæ, on which his calculations are founded, and of explaining even the terms, which he must necessarily use, in correcting operations that are purely practical. Fortunately several works are before the public, which give all the information that can be wanted on the THEORY of ASTRONOMY; and many of them are written with that method and perspicuity, which are calculated to render the subject not only intelligible, but highly interesting. The names of VINCE, WOODHOUSE, BRINKLEY, GREGORY, ZACH, DELAMBRE, BESSEL, BIOT, and LITTROW are too well known, from their labours, to require any recommendation; and their works will amply supply what may be considered, by some readers, as omissions in the plan that has been prescribed to the present volume. The Author professes not indeed an intimate knowledge of the more abstruse branches of mathematical science, but flatters himself that he possesses such information as enables him to supply that link, which will connect the theoretical with the practical astronomer, and which was the principal object of his undertaking.

It will probably be expected, that, in a work of this nature, the instruments at present used in practical astronomy be described in the first place, for this appears to be the most natural order to adopt; but the connexion between the observations to be made on the celestial bodies, and the necessary corrections to be applied to them is such, that, before any useful inference can be drawn from instrumental operations, reference must be had to computations depending on suitable formulæ, or to Tables already computed, when they are to be obtained: for until the observer is enabled to convert the *mean* places of the heavenly bodies into the *apparent*, and conversely, he can neither assign a rate to his clock with any degree of accuracy, nor determine the latitude of his observatory. Hence the Author has been induced to divide his work into two volumes; the first, which may be considered in some measure as a compilation, contains a selection of Tables calculated to facilitate the application of all the requisite corrections, and to abridge the labour of computation in a variety of cases that occur in practice; and it is proposed that the second should comprise descriptions of the most useful instruments, illustrated by a series of copper-plate engravings executed in the best style of the art; an account of the most approved methods of making the necessary adjustments; appropriate directions as to the different modes of using the instruments respectively; the exemplification of those various modes in actual practice; and the application of all the requisite corrections, derived from the first volume, for determining the final results of the different operations.

This plan has been adopted, partly with a view of leading the young astronomer by degrees, from one step to another of his progress, lest he should be discouraged by being surrounded at once by difficulties, that might conspire to deter him from persevering in his purpose; and partly from an understanding, that a wish has been expressed by several members of the Astronomical Society of London, to procure a collection of Tables, that may be generally useful, and that may afford the ready means of comparison between the results deduced from the observations of different astronomers. While the co-efficients in the different formulæ, on which Tables of Corrections are founded, vary from one another, it cannot be ascertained whether the discrepancies, in a variety of instances, are principally derivable from the observations themselves, or from the reductions of those observations: but should the same Tables of Corrections be adopted by different astronomers, an estimate may be made of the comparative merits of different instruments, that are used for the same purpose; and it would then appear what confidence may be placed on the powers of an individual instrument. And finally, when the deductions of several astronomers concur in pointing out defects in the corrections thus adopted, the co-efficients themselves may be varied, until it is found that the Tables, arising out of such variations, may constitute a general and permanent code of corrections.

The materials, which the practical astronomer had formerly to consult, lay scattered about in different books; and the explanations were frequently misapprehended by being published in foreign languages; besides, it was no easy matter to select out of the various Tables, that different countries have produced, those particular ones which are most entitled to credit. How far the compiler of the present volume has obviated these difficulties, or any of them, is left to the public to determine. In converting the productions of other more eminent authors to his own purpose, he has found it expedient, frequently to alter the form of a Table when it appeared capable of improvement, either by an enlargement of its subdivisions, or by an alteration in the arrangement of its arguments, or both. In conformity with the modern plan of constructing Tables on the continent, he has avoided, as much as possible, Tables of *double entry*, that require several proportions to be worked; and where such Tables were unavoidably requisite, he has generally extended the subdivisions, so that the quantities may be taken out by simple inspection, except in cases where extraordinary accuracy is required. Into several of the old Tables modern co efficients have been introduced, so as to require re-computation, and *time* has frequently been substituted for *space* in the arguments; but in these cases, and indeed generally, the formulæ have been either annexed to the Tables founded on them, or form a portion of the subsequent explanation. The reader will not be obliged to draw on the credit of the Editor, with respect to the Tables that bear their own marks of authenticity, except so far as the alterations extend; but for several new Tables, and some of them long ones, computed by himself or his assistants, he has to request the indulgence of his readers, who may discover inaccuracies that have escaped detection. He must claim also the same indulgence on behalf of several new Tables, computed and supplied by his friends, whose names they bear, and to whom he begs to express his obligation. The volume has been prepared in a comparatively short space of time, while the Editor resided above eighty miles from the press, so that final revision of single faults has been sometimes confided to the superintendent of the press. But under all these disadvantages it is hoped that few typographical errors will be found beyond what are contained in the pages of *Errata*.

In arranging the plan of tabulating the various quantities that constitute the series of general Tables for finding the precession, and constants of aberration, lunar nutation, and solar nutation, from I. to XIV. inclusive, and in selecting the co-efficients most likely to be depended on, in the present state of practical astronomy, the Compiler could not but distrust his own competence, and therefore gladly availed himself of the advice and labours of the Cape Astronomer, whose wants had directed his attention to this subject, and whose talents are too well known and appreciated to require eulogium.

With such valuable assistance, the Author is encouraged to hope, that his labours may at least tend to promote the objects, for encouraging which that Society was established, of which he has the honour to be Treasurer; though he can hardly expect that the limited circulation of such a work will repay more than one-fourth of the expense that has been incurred in its execution.

The Use of the Tables has been rendered as familiar as the nature of the subject would admit, by popular explanations prefixed to the examples, and by deducing the results at full length, both from the tabulated quantities, and also by logarithmic computations, agreeably to the formulæ on which the Tables are respectively constructed; thus the Reader is enabled to judge of the accuracy of the Tables, and, at the same time, to see in what cases logarithmic computation is preferable to the tabular method of gaining the corrections. The variety of TABLES of REFRACTION introduced, will afford the means of ascertaining what formulæ are the best adapted for particular situations, in which observatories may be placed, with respect to local refractions; and also, of detecting what differences are produced in observations reduced by different Tables.

Some opinion may be formed of the extent of the Author's labours, when it is stated, that, of the 457 pages constituting the volume, 325 contain new Tables, or explanatory matter; 46 are filled with Tables that have been enlarged, or otherwise improved; and 86 only comprise Tables that have been copied in their original state.

During the printing of this volume various improvements have been made in the tabulation of astronomical corrections; many of which have been introduced towards the end of the volume, which addition, while it in some measure deranges the intended order of the Tables, renders the collection more complete, and will not therefore, it is hoped, be considered as an imperfection.

CONTENTS.

ERRATA.

Page	Argument.	Column.	Instead of	Insert
17	Therm. 16 .0	12	0 .2292	0 .2992
18	Fahr. 0°	8. 9. 11. 12	14″. 17″. 5″. 7″.	14°. 17°. 5°. 7°.
19	at the head	12	Therm. 30°	Therm. 50°
			—	—
24	at the head	2	XII	XII
			—	+
28	at the bottom	2	XII	XI
30	III. 16°	8	13 .249	13 .149
35	at the head	12	0—	6—
36	head	......	N′	L′
	II. 0	8	0 .89907	0 .89607
	II. 1	8	0 .89962	0 .89662
44	Dec. 31	1	41	31
45	head	......	Time.	Space.
48	head	......	47	48
49	V 26° 0′	19	1 .335	1 .325
56	VI 50	17	5 .868	5 .768
	VI 19° 52′	17	5 .878	5 .778
	VI 19 54	17	5 .887	5 .787
	VI 19 56	17	5 .896	5 .796
	VI 19 58	17	5 .905	5 .805
	VI 20 0	17	5 .914	5 .814
104	Tab. 1.	head	$\frac{320}{321}$	$\frac{1}{321}$
	Tab. 2.	head	$\frac{320}{321}$	$\frac{1}{321}$
106	Lat. 50°	9	.9350	.6350
107	Lat. 28	8	100162	100172
108	....	head	Lampton.	Lambton.
111	0 46	11^s	3 .367	0 .367
113	γ Pegasi	R. A.	0^h 3^m $58^s.8$	0^h 3^m $58^s.84$
	α Cassiopeæ	R. A.	0 30 21 .2	0 30 20 .99
	α Ursæ Min.	R. A.	0 57 2 .4	0 57 1 .51
115	α Arietis	R. A.	1 57 3 .0	1 57 3 .01
		N. P. D.	67° 23′ 37″	67 23 34 .69
	α Ceti	R. A.	2^h 52^m $52^s.9$	2^h 52^m $52^s.89$
		N. P. D.	86° 37′ 18″.0	86 37 18 .19
	α Persei	R. A.	3^h 11^m $31^s.7$	3 11 31 .25
		N. P. D.	40° 47′ 18″.	40° 47′ 17″.87
117	α Tauri	R. A.	4^h 25^m $36^s.3$	4 25 36 .28
		N. P. D.	73° 51′ 43″	73 51 39 .67
	α Aurigæ	R. A.	5^h 3^m $24^s.6$	5 3 24 .59
		N. P. D.	44° 11′ 51″	44 11 48 .49
	β Orionis	R. A.	5^h 5^m $53^s.6$	5 5 53 .56
		N. P. D.	98° 25′ 0″	98 24 59 .85
	α Tauri	Aber.	3′.748	3″.748
119	β Tauri	R. A.	5^h 14^m $55^s.4$	5 14 55 .41
		N. P. D.	61° 33′ 19″	61 33 16 .89
	α Orionis	R. A.	5^h 45^m $25^s.9$	5 45 25 .92
		N. P. D.	82° 38′ 8″	82 38 6 .04
	α Canis Maj.	N. P. D.	106 28 36	106 28 30 .67
121	β Geminorum	R. A.	7^h 33^m	7 34
123	θ Ursæ Maj.	An. Var.	+ 15″.34	+16 .01
125	β Leonis	An. Var.	— 20 .04	+20 .04
131	α Cor. Bor.	⊙ Nut.	8″.421	0 .421
133	α Scorpii	R. A.	16^h 18^m $22^s.23$	16 18 23 .23
134	head	......	β Herculis	α Herculis
135	ε Sagittarii	An. Var.	+ 0″.95	—1 .14
136	Com. Arg.	head	S S IV 15° VII	I 15 VII
138	γ Aquilæ	An. Var.	— 8″.32	—8″.38
	Com. Arg.	head	S S IV 15° VII	I 15 VII
144	α Piscis Aust.	head	1830	1820

Page	Argument.	Column.	Instead of	Insert
144	α Pegasi	An. Var.	+19″.43	—19 .43
	Common Arg.	head	S S III 0° IX	S S I 15° VII &c.
145	α Ceti	Bessel	$53^s.92$	52 .93
	α Tauri	Maskelyne	35 .00	35 .95
		Bessel	36 .02	36 .12
	β Tauri	An. Var.	$2^s.785$	3 .785
		An. Var.	3″.747	3 .717
	α Virginis	An. Var.	$3^s.444$	3 .144
	α Scorpii	Pond.	$23^s.32$	23 .23
	α Herculis	Bessel	26 .57	26 .67
	γ Draconis	An. Var.	9″.153	0 .714
	ε Sagittarii	An. Var.	$3^s.974$	3 .983
		An. Var.	1″.068	1 .138
	α Aquarii	Piazzi	$33^s.06$	31 .96
146	α Cassiopeæ	Prop. Mot.	$-0^s.005$ Piazzi	—0 .003
			—0 .004 Bessel	—0 .003
		An. Var.	3 .312 Piazzi	3 .314
			3 .317 Bessel	3 .316
		Prop. Mot.	+0″.05 Piazzi	+0 .07
		An. Var.	19 .85 Piazzi	19 .83
			13 .853 Bessel	19 .853
	β Orionis	An. Prec.	$5^s.878$ Mask.	2 .878
	α Hydræ	An. Prec.	3 .952 Mask.	2 .952
		An. Var.	3 .946 Mask.	2 .946
	β Virginis	Prop. Mot.	+0 .041 Bessel	0 .049
		An. Var.	$3^s.114$ Bessel	3 .122
	γ Ursæ Maj.	Prop. Mot.	+0 .019 Bessel	+0 .013
		An. Var.	3 .225 Bessel	3 .219
	α Bootis	An. Var.	18″.007 Bessel	19 .007
	β Ursæ Min.	An. Prec.	$0^s.315$ Bessel	0 .319
		An. Var.	0 .325 Bessel	0 .329
	α Scorpii	An. Prec.	$2^s.657$ Piazzi	3 .657
		An. Prec.	8″.698 Bessel	8 .695
		An. Var.	8 .688 Bessel	8 .685
	β Draconis	An. Prec.	$1^s.344$ Bessel	1 .348
		An. Var.	8″.330 Piazzi	1 .330
			1 .337 Bessel	1 .341
	α Ophiuchi	Prop. Mot.	+0 .18 Piazzi	—0 .18
		An. Var.	3 .18 Piazzi	2 .82
	α Lyræ	An. Prec.	$3^s.013$ Mask.	2 .013
	$α^2$ Capricorni	Prop. Mot.	+0 .008 Bessel	+0 .005
		An. Var.	3 .341 Bessel	3 .338
	α Andromedæ	Prop. Mot.	—0 .009 Piazzi	+0 .009
			—0 .008 Bessel	+0 .008
		An. Var.	3 .054 Piazzi	3 .072
			3 .053 Bessel	3 .069
153	⊙ 1823	5	S 0 10° 19′ 39″.9	9 10 19 39 .9
158	19° 20′	4	$0^s.15$	0 .85
162	24 40	2	9° 33′ 57″. 00	9 33 57 .90
	27 20	4	43 .85	42 . 85
166	9 50	10	15 .96	14 . 96
	10 0	9	38 .88	39 . 88
	10 0	10	15 .86	14 . 86
168	head	......	Table 5.	Table 8.
180	Tab. 25. Alt. 0.	18	1″.6	8 .7
182	56′ 6″	......	15′ 17″. 50	15 17 . 58
	57 12	......	15 35 . 26	15 35 . 56
	58 16	......	15 52 . 91	15 53 . 01
183	26	14	2 . 33	2 . 23
	38	16	4 . 58	4 . 48

Page	Argument.	Column.	Instead of	Insert
184	55°	3	9″.81	6 .84
	49	13	6 .32	6 .34
185	Table 6. 85°	3	2′ 53 .5	1 53 .5
186	29°	6	6 .16	7 .16
	30	6	6 .39	7 .39
189	15	7	57 59 .38	56 59 .38
191	36 30	10	49 30 .35	49 50 .35
192	47 0	6	39 23 .35	39 33 .35
	50	4	35 50 .77	35 59 .77
193	50	4	35 50 .77	35 59 .77
	51 50	6	35 58 .47	35 50 .47
194	63 50	4	24 51 .71	24 41 .71
	65 50	7	24 9 .65	24 9 .25
	67 50	2	20 32 .46	20 22 .46
	74 50	3	14 23 .67	14 23 .37
	84 0	7	6 10 .07	6 10 .03
	89 10	8	0 52 .30	0 52 .36
204	8 0	5	2° 25 18 .14	2 25 18 .74
	9 50	8	2 32 53 .87	1 32 53 .87
205	10 20 }	5	2 27 15 .88	2 27 15 .86
	10 30 }	9	89 .64	39 .64
	10 40	8	1 29 16 .33	1 29 36 .53
	10 50	8	1 28 57 .56	1 28 56 .56
	12 30	5	2 8 11 .29	2 28 11 .29
	13 10 } 13 20 }	9	42 .28	42 .27
	14 50	8	1 12 11 .42	1 12 11 .43
	18 20	2	1 31 42 .14	1 31 42 .15
206	20 0	4	15 .86	14 .86
	20 10	4	15 .96	14 .96
207	Table 2.	head	"head or foot"	"head" only.
228	220°	6	23° 50′ 53″	28 50 53
	221	6	38 27 11	28 27 11
	226	2	198 11 37	193 11 37
232	11 20′	5	18 2 27 .89	18 2 27 .99
	16 40	5	16 34 36 .97	16 34 36 .95
	18 40	2	22 28 26 .00	22 20 26 .00
238	Table 13.	head	Appoximate	Approximate
	27°	11	4′ 33″.80	4 43 .80
	44	4	2 8 .88	2 10 .88
239	25	11	8 43 .36	8 40 .36
	31	2	5 33 .79	5 53 .79
	37	8	10 13 .43	10 53 .43
	39	2	7 11 .29	7 12 .29
240	22	11	10 14 .29	11 14 .29
	51	10	22 26 .37	22 36 .37
	52	4	18 23 .44	18 28 .44
	61	4	20 20 .27	20 30 .27
241	30	9	28 28 .19	18 28 .19
	45	5	23 23 47	23 43 .47
242	22	3	15 2 .58	15 2 .38
	28	2	13 28 .30	18 28 .80
	35	3	22 1 .67	23 1 .67
	35	11	26 41 .79	26 21 .79
243	57	8	42 12 .15	42 12 .12
244	17	11	16 39 .06	16 29 .06
	18	11	17 35 .37	17 25 .37
	19	11	18 31 .36	18 21 .36
	21	8	19 42 .57	19 47 .57
245	11	11	10 16 .47	11 16 .47
	16	6	15 48 .48	15 58 .48
	21	9	21 11 .93	21 1 .93
	23	5	22 33 .14	22 32 .14
	32	7	31 51 .04	30 51 .04
	33	7	32 42 .46	31 42 .46
	45	11	41 26 .91	41 46 .91
246	15	7	15 29 .40	15 29 .48
	16	5	16 28 .86	16 26 .86
	33	3	32 81 .61	32 21 .61
	48	3	44 49 .29	44 9 .29
	48	6	44 27 .14	44 25 .14
	49	3	45 30 .51	44 50 .51
250 251	bottom } bottom }	......	☾'s co-latitude	the co-latitude
252	bottom } 9′ }	8	☾'s co-latitude 16′ 6″.06	the co-latitude 10 6 .06

Page	Argument.	Column.	Instead of	Insert
253	bottom	98	her latitude	the reduced lat.
	bottom	....	☾'s co-lat.	the co-lat.
254	8′	11	3′ 44″.17	2 44 .17
255	38	6	15 3 .57	16 3 .57
257	8	17°	2 26 .42	2 26 .49
259	84°	65	53 54 .83	54 4 .83
260	77	84	58 8 .42	58 8 .52
263	39	20″	16 .54	15 .54
268	.0039	2	0 1 .99	0 1 .90
	.0040	2	0 1 .99	0 1 .95
270	bottom	....	Latitude	co-latitude
	bottom	after	0 .02 M. add......	"and 0 .03 N′.
278	24 Pisc. Aust.	4	22^h 46^m 43^s.53	22 46 34 .53
288	Table 2.	head	Diameter	Breadth
330 } 331 }	....	......:...	Lampton	Lambton
336	Tab. XV. 07	N. P. D.	2	19
	True N P. D.	....	82° 38′ 1″.44	82 38 1 .52
337	from bottom	four lines	Mr. Bates	Mr. Bate
338	from bottom	five lines	dele { "except in one solitary instance of α Tauri."	
343	α Pisc. Aust.	Column 5	0 .37	0 .27
347	Example 1.	Table 1.	−0 .92	+0 .08
		Table 2.	+0 .09	+1 .09
			+0 .002	+2 .07
			−0 .012	−0 .014
	App. R. A.		5^h 45^m 40^s.007	5 45 40 .210
	from bottom	twelve lines	one-third	one-tenth
376	Table 7.	Title	LATITUDE	ALTITUDE
389	first line		latter	reduced paral.
	second line		= hor. par.	= reduced paral.
397	Az. 1° 40	Lat. 50° 0′	10′ 48″.85	10 48 .55
	Az. 11 40	Lat. 56 40	9 4 .24	9 9 .24
	Az. 85 0	Lat. 58 20	0 46 .90	0 46 .60
398	Az. 33 20	Lat. 38 20	1 .2948	1 .2947
401	from bottom	line ten	45° to 90°	45° to 60°

TABLES

FOR FACILITATING

THE REDUCTION OF CELESTIAL OBSERVATIONS.

TABLE I.

A TABLE of MEAN REFRACTIONS to every Ten Minutes of Zenith Distance.

Barometer 29 . 60 Inches. Thermometer 50.

App. Z. D.	Refract.	Diff.	App. Z. D.	Refraction.	Diff.	App. Z. D.	Refraction.	Diff.	App. Z. D.	Refraction.	Diff.	App. Z. D.	Refraction.	Diff.
0° 0	0″.00	″.17	9° 00	9″.00	″.17	18° 00′	18″.50	″.18	27° 00′	29″.00	″.22	36° 00′	41″.40	″.25
0 10	0 .17	.16	9 10	9 .17	.16	18 10	18 .68	.19	27 10	29 .22	.21	36 10	41 .65	.25
0 20	0 .33	.17	9 20	9 .33	.17	18 20	18 .87	.18	27 20	29 .43	.22	36 20	41 .90	.25
0 30	0 .50	.17	9 30	9 .50	.17	18 30	19 .05	.18	27 30	29 .65	.22	36 30	42 .15	.25
0 40	0 .67	.16	9 40	9 .67	.16	18 40	19 .23	.19	27 40	29 .87	.21	36 40	42 .40	.25
0 50	0 .83	.17	9 50	9 .83	.17	18 50	19 .42	.18	27 50	30 .08	.22	36 50	42 .65	.25
1 0	1 .00	.17	10 00	10 .00	.18	19 00	19 .60	.18	28 00	30 .30	.22	37 00	42 .90	.27
1 10	1 .17	.16	10 10	10 .18	.19	19 10	19 .78	.19	28 10	30 .52	.21	37 10	43 .17	.26
1 20	1 .33	.17	10 20	10 .37	.18	19 20	19 .97	.18	28 20	30 .73	.22	37 20	43 .43	.27
1 30	1 .50	.17	10 30	10 .55	.18	19 30	20 .15	.18	28 30	30 .95	.22	37 30	43 .70	.27
1 40	1 .67	.16	10 40	10 .73	.19	19 40	20 .33	.19	28 40	31 .17	.21	37 40	43 .97	.26
1 50	1 .83	.17	10 50	10 .92	.18	19 50	20 .52	.18	28 50	31 .38	.22	37 50	44 .23	.27
2 0	2 .00	.17	11 00	11 .10	.17	20 00	20 .70	.20	29 00	31 .60	.22	38 00	44 .50	.27
2 10	2 .17	.16	11 10	11 .27	.16	20 10	20 .90	.20	29 10	31 .82	.21	38 10	44 .77	.26
2 20	2 .33	.17	11 20	11 .43	.17	20 20	21 .10	.20	29 20	32 .03	.22	38 20	45 .03	.27
2 30	2 .50	.17	11 30	11 .60	.17	20 30	21 .30	.20	29 30	32 .25	.22	38 30	45 .30	.27
2 40	2 .67	.16	11 40	11 .77	.16	20 40	21 .50	.20	29 40	32 .47	.21	38 40	45 .57	.26
2 50	2 .83	.17	11 50	11 .93	.17	20 50	21 .70	.20	29 50	32 .68	.22	38 50	45 .83	.27
3 0	3 .00	.17	12 00	12 .10	.17	21 00	21 .90	.18	30 00	32 .90	.22	39 00	46 .10	.28
3 10	3 .17	.16	12 10	12 .27	.16	21 10	22 .08	.19	30 10	33 .12	.21	39 10	46 .38	.29
3 20	3 .33	.17	12 20	12 .43	.17	21 20	22 .27	.18	30 20	33 .33	.22	39 20	46 .67	.28
3 30	3 .50	.17	12 30	12 .60	.17	21 30	22 .45	.18	30 30	33 .55	.22	39 30	46 .95	.28
3 40	3 .67	.16	12 40	12 .77	.16	21 40	22 .63	.19	30 40	33 .77	.21	39 40	47 .23	.29
3 50	3 .83	.17	12 50	12 .93	.17	21 50	22 .82	.18	30 50	33 .98	.22	39 50	47 .52	.28
4 0	4 .00	.17	13 00	13 .10	.18	22 00	23 .00	.20	31 00	34 .20	.23	40 00	47 .80	.28
4 10	4 .17	.16	13 10	13 .28	.19	22 10	23 .20	.20	31 10	34 .43	.24	40 10	48 .08	.29
4 20	4 .33	.17	13 20	13 .47	.18	22 20	23 .40	.20	31 20	34 .67	.23	40 20	48 .37	.28
4 30	4 .50	.17	13 30	13 .65	.18	22 30	23 .60	.20	31 30	34 .90	.23	40 30	48 .65	.28
4 40	4 .67	.16	13 40	13 .83	.19	22 40	23 .80	.20	31 40	35 .13	.24	40 40	48 .93	.29
4 50	4 .83	.17	13 50	14 .02	.18	22 50	24 .00	.20	31 50	35 .37	.23	40 50	49 .22	.28
5 0	5 .00	.17	14 00	14 .20	.18	23 00	24 .20	.20	32 00	35 .60	.23	41 00	49 .50	.30
5 10	5 .17	.16	14 10	14 .38	.19	23 10	24 .40	.20	32 10	35 .83	.24	41 10	49 .80	.30
5 20	5 .33	.17	14 20	14 .57	.18	23 20	24 .60	.20	32 20	36 .07	.23	41 20	50 .10	.30
5 30	5 .50	.17	14 30	14 .75	.18	23 30	24 .80	.20	32 30	36 .30	.23	41 30	50 .40	.30
5 40	5 .67	.16	14 40	14 .93	.19	23 40	25 .00	.20	32 40	36 .53	.24	41 40	50 .70	.30
5 50	5 .83	.17	14 50	15 .12	.18	23 50	25 .20	.20	32 50	36 .77	.23	41 50	51 .00	.30
6 0	6 .00	.17	15 00	15 .30	.17	24 00	25 .40	.20	33 00	37 .00	.23	42 00	51 .30	.30
6 10	6 .17	.16	15 10	15 .47	.16	24 10	25 .60	.20	33 10	37 .23	.24	42 10	51 .60	.30
6 20	6 .33	.17	15 20	15 .63	.17	24 20	25 .80	.20	33 20	37 .47	.23	42 20	51 .90	.30
6 30	6 .50	.17	15 30	15 .80	.17	24 30	26 .00	.20	33 30	37 .70	.23	42 30	52 .20	.30
6 40	6 .67	.16	15 40	15 .97	.16	24 40	26 .20	.20	33 40	37 .93	.23	42 40	52 .50	.30
6 50	6 .83	.17	15 50	16 .13	.17	24 50	26 .40	.20	33 50	38 .16	.24	42 50	52 .80	.30
7 0	7 .00	.17	16 00	16 .30	.18	25 00	26 .60	.20	34 00	38 .40	.25	43 00	53 .10	.32
7 10	7 .17	.16	16 10	16 .48	.19	25 10	26 .80	.20	34 10	38 .65	.25	43 10	53 .42	.31
7 20	7 .33	.17	16 20	16 .67	.18	25 20	27 .00	.20	34 20	38 .90	.25	43 20	53 .73	.32
7 30	7 .50	17	16 30	16 .85	.18	25 30	27 .20	.20	34 30	39 .15	.25	43 30	54 .05	.32
7 40	7 .67	.16	16 40	17 .03	.19	25 40	27 .40	.20	34 40	39 .40	.25	43 40	54 .37	.31
7 50	7 .83	.17	16 50	17 .22	.18	25 50	27 .60	.20	34 50	39 .65	.25	43 50	54 .68	.32
8 0	8 .00	.17	17 00	17 .40	.18	26 00	27 .80	.20	35 00	39 .90	.25	44 00	55 .00	.32
8 10	8 .17	.16	17 10	17 .58	.19	26 10	28 .00	.20	35 10	40 .15	.25	44 10	55 .32	.31
8 20	8 .33	.17	17 20	17 .77	.18	26 20	28 .20	.20	35 20	40 .40	.25	44 20	55 .63	.32
8 30	8 .50	.17	17 30	17 .95	.18	26 30	28 .40	.20	35 30	40 .65	.25	44 30	55 .95	.32
8 40	8 .67	.16	17 40	18 .13	.19	26 40	28 .60	.20	35 40	40 .90	.25	44 40	56 .27	.31
8 50	8 .83	.17	17 50	18 .32	.18	26 50	28 .80	.20	35 50	41 .15	.25	44 50	56 .58	.32

Dr. BRADLEY'S REFRACTIONS concluded.

TABLE 1.

A TABLE of MEAN REFRACTIONS to every Ten Minutes of Zenith Distance.

Barometer 29 .60 Inches. Thermometer 50.

App. Z.D.	Refraction.	Diff.
45° 00'	0' 56".90	".33
45 10	0 57 .23	.33
45 20	0 57 .56	.33
45 30	0 57 .89	.33
45 40	0 58 .22	.34
45 50	0 58 .56	.34
46 00	0 58 .90	.35
46 10	0 59 .25	.34
46 20	0 59 .59	.35
46 30	0 59 .94	.35
46 40	1 0 .29	.36
46 50	1 0 .65	.35
47 00	1 1 .00	.36
47 10	1 1 .36	.37
47 20	1 1 .73	.36
47 30	1 2 .09	.37
47 40	1 2 .46	.37
47 50	1 2 .83	.37
48 00	1 3 .20	.36
48 10	1 3 .56	.36
48 20	1 3 .92	.37
48 30	1 4 .29	.37
48 40	1 4 .66	.37
48 50	1 5 .03	.37
49 00	1 5 .40	.39
49 10	1 5 .79	.40
49 20	1 6 .19	.40
49 30	1 6 .59	.40
49 40	1 6 .99	.40
49 50	1 7 .39	.41
50 00	1 7 .80	.39
50 10	1 8 .19	.40
50 20	1 8 .59	.40
50 30	1 8 .99	.40
50 40	1 9 .39	.40
50 50	1 9 .79	.41
51 00	1 10 .20	.42
51 10	1 10 .62	.43
51 20	1 11 .05	.43
51 30	1 11 .48	.44
51 40	1 11 .92	.44
51 50	1 12 .36	.44
52 00	1 12 .80	.44
52 10	1 13 .24	.45
52 20	1 13 .69	.45
52 30	1 14 .14	.45
52 40	1 14 .59	.45
52 50	1 15 .04	.46
53 00	1 15 .50	.46
53 10	1 15 .96	.46
53 20	1 16 .42	.47

App. Z.D.	Refraction.	Diff.
53° 30'	1' 16".89	".47
53 40	1 17 .36	.47
53 50	1 17 .83	.47
54 00	1 18 .30	.47
54 10	1 18 .77	.48
54 20	1 19 .25	.48
54 30	1 19 .73	.49
54 40	1 20 .22	.49
54 50	1 20 .71	.49
55 00	1 21 .20	.51
55 10	1 21 .71	.51
55 20	1 22 .22	.51
55 30	1 22 .73	.52
55 40	1 23 .25	.52
55 50	1 23 .77	.53
56 00	1 24 .30	.52
56 10	1 24 .82	.52
56 20	1 25 .34	.53
56 30	1 25 .87	.54
56 40	1 26 .41	.54
56 50	1 26 .95	.55
57 00	1 27 .50	.57
57 10	1 28 .07	.57
57 20	1 28 .64	.58
57 30	1 29 .22	.59
57 40	1 29 .81	.59
57 50	1 30 .40	.60
58 00	1 31 .00	.59
58 10	1 31 .59	.59
58 20	1 32 .18	.60
58 30	1 32 .78	.60
58 40	1 33 .38	.61
58 50	1 33 .99	.61
59 00	1 34 .60	.62
59 10	1 35 .22	.62
59 20	1 35 .84	.63
59 30	1 36 .47	.64
59 40	1 37 .11	.64
59 50	1 37 .75	.65
60 00	1 38 .40	.67
60 10	1 39 .07	.67
60 20	1 39 .74	.68
60 30	1 40 .42	.69
60 40	1 41 .11	.69
60 50	1 41 .80	.70
61 00	1 42 .50	.70
61 10	1 43 .20	.70
61 20	1 43 .90	.71
61 30	1 44 .61	.72
61 40	1 45 .33	.73
61 50	1 46 .06	.74

App. Z.D.	Refraction.	Diff.
62° 00'	1' 46".80	0".76
62 10	1 47 .46	0 .77
62 20	1 48 .33	0 .78
62 30	1 49 .11	0 .79
62 40	1 49 .90	0 .80
62 50	1 50 .70	0 .80
63 00	1 51 .50	0 .80
63 10	1 52 .30	0 .80
63 20	1 53 .10	0 .81
63 30	1 53 .91	0 .82
63 40	1 54 .73	0 .83
63 50	1 55 .56	0 .84
64 00	1 56 .40	0 .85
64 10	1 57 .25	0 .87
64 20	1 58 .12	0 .88
64 30	1 59 .00	0 .89
64 40	1 59 .89	0 .90
64 50	2 0 .79	0 .91
65 00	2 1 .70	0 .92
65 10	2 2 .62	0 .93
65 20	2 3 .55	0 .94
65 30	2 4 .49	0 .96
65 40	2 5 .44	0 .97
65 50	2 6 .42	0 .98
66 00	2 7 .40	1 .00
66 10	2 8 .40	1 .01
66 20	2 9 .41	1 .03
66 30	2 10 .44	1 .04
66 40	2 11 .48	1 .05
66 50	2 12 .53	1 .07
67 00	2 13 .60	1 .08
67 10	2 14 .68	1 .10
67 20	2 15 .78	1 .11
67 30	2 16 .89	1 .12
67 40	2 18 .01	1 .14
67 50	2 19 .15	1 .15
68 00	2 20 .30	1 .16
68 10	2 21 .46	1 .17
68 20	2 22 .63	1 .19
68 30	2 23 .82	1 .21
68 40	2 25 .03	1 .22
68 50	2 26 .25	1 .25
69 00	2 27 .50	1 .28
69 10	2 28 .78	1 .30
69 20	2 30 .08	1 .32
69 30	2 31 .40	1 .34
69 40	2 32 .74	1 .37
69 50	2 34 .11	1 .39
70 00	2 35 .50	1 .41
70 10	2 36 .91	1 .43
70 20	2 38 .34	1 .46

App. Z.D.	Refraction.	Diff.
70° 30'	2' 39".80	1".48
70 40	2 41 .28	1 .50
70 50	2 42 .78	1 .52
71 00	2 44 .30	1 .53
71 10	2 45 .83	1 .56
71 20	2 47 .39	1 .58
71 30	2 48 .97	1 .62
71 40	2 50 .59	1 .64
71 50	2 52 .23	1 .67
72 00	2 53 .90	1 .72
72 10	2 55 .62	1 .75
72 20	2 57 .37	1 .78
72 30	2 59 .15	1 .82
72 40	3 0 .97	1 .85
72 50	3 2 .82	1 .88
73 00	3 4 .70	1 .91
73 10	3 6 .61	1 .94
73 20	3 8 .55	1 .98
73 30	3 10 .53	2 .02
73 40	3 12 .55	2 .06
73 50	3 14 .61	2 .09
74 00	3 16 .70	2 .13
74 10	3 18 .83	2 .18
74 20	3 21 .01	2 .23
74 30	3 23 .24	2 .27
74 40	3 25 .51	2 .32
74 50	3 27 .83	2 .37
75 00	3 30 .20	2 .41
75 10	3 32 .61	2 .47
75 20	3 35 .08	2 .52
75 30	3 37 .60	2 .58
75 40	3 40 .18	2 .63
75 50	3 42 .81	2 .69
76 00	3 45 .50	2 .75
76 10	3 48 .25	2 .81
76 20	3 51 .06	2 .88
76 30	3 53 .94	2 .95
76 40	3 56 .89	3 .02
76 50	3 59 .91	3 .09
77 00	4 3 .00	3 .15
77 10	4 6 .15	3 .25
77 20	4 9 .40	3 .31
77 30	4 12 .71	3 .42
77 40	4 16 .13	3 .49
7 50	4 19 .62	3 .56
78 00	4 23 .18	3 .69
78 10	4 26 .87	3 .79
78 20	4 30 .66	3 .88
78 30	4 34 .54	3 .99
78 40	4 38 .53	4 .11
78 50	4 42 .64	4 .21

App. Z.D.	Refraction.	Diff.
79° 00'	4' 46".85	4".33
79 10	4 51 .18	4 .45
79 20	4 55 .63	4 .58
79 30	5 0 .21	4 .72
79 40	5 4 .93	4 .88
79 50	5 9 .81	5 .02
80 00	5 14 .83	5 .16
80 10	5 19 .99	5 .32
80 20	5 25 .31	5 .50
80 30	5 30 .81	5 .68
80 40	5 36 .49	5 .89
80 50	5 42 .38	6 .07
81 00	5 48 .45	6 .25
81 10	5 54 .70	6 .47
81 20	6 1 .17	6 .71
81 30	6 7 .88	6 .95
81 40	6 14 .83	7 .22
81 50	6 22 .05	7 .50
82 00	6 29 .55	7 .72
82 10	6 37 .27	8 .02
82 20	6 45 .29	8 .35
82 30	6 53 .64	8 .69
82 40	7 2 .33	9 .06
82 50	7 11 .39	9 .45
83 00	7 20 .84	9 .74
83 10	7 30 .58	10 .17
83 20	7 40 .75	10 .64
83 30	7 51 .39	11 .12
83 40	8 2 .51	11 .67
83 50	8 14 .18	12 .23
84 00	8 26 .41	12 .60
84 10	8 39 .01	13 .24
84 20	8 52 .25	13 .94
84 30	9 6 .19	14 .66
84 40	9 20 .85	15 .46
84 50	9 36 .31	16 .19
85 00	9 52 .50	
90 00	32 59 .43	

DECIMAL FACTORS for computing Dr. BRADLEY'S CORRECTIONS of the mean Astronomical REFRACTIONS, depending on the Barometer and Thermometer.

Table 2.

Bar.	29.0	29.1	29.2	29.3	29.4	29.5	29.6	29.7	29.8	29.9	30.0	Bar.
Fahr. Ther.												Fahr. Ther.
20	+ .059	+ .063	+ .067	+ .070	+ .074	+ .077	+ .081	+ .085	+ .088	+ .092	+ .096	20°
21	+ .056	+ .060	+ .064	+ .067	+ .071	+ .075	+ .078	+ .082	+ .085	+ .089	+ .093	21
22	+ .054	+ .057	+ .061	+ .064	+ .068	+ .072	+ .075	+ .079	+ .083	+ .086	+ .090	22
23	+ .051	+ .054	+ .058	+ .062	+ .065	+ .069	+ .072	+ .076	+ .080	+ .083	+ .087	23
24	+ .048	+ .051	+ .055	+ .059	+ .062	+ .066	+ .070	+ .073	+ .077	+ .080	+ .084	24
25	+ .045	+ .049	+ .052	+ .056	+ .060	+ .063	+ .067	+ .070	+ .074	+ .078	+ .081	25
26	+ .042	+ .046	+ .049	+ .053	+ .057	+ .060	+ .064	+ .067	+ .071	+ .075	+ .078	26
27	+ .040	+ .043	+ .047	+ .050	+ .054	+ .057	+ .061	+ .065	+ .068	+ .072	+ .075	27
28	+ .037	+ .040	+ .044	+ .048	+ .051	+ .055	+ .058	+ .062	+ .065	+ .069	+ .073	28
29	+ .034	+ .038	+ .041	+ .045	+ .048	+ .052	+ .055	+ .059	+ .063	+ .066	+ .070	29
30	+ .031	+ .035	+ .038	+ .042	+ .046	+ .049	+ .053	+ .056	+ .060	+ .063	+ .067	30
31	+ .029	+ .032	+ .036	+ .039	+ .043	+ .046	+ .050	+ .053	+ .05[illegible]	+ .061	+ .064	31
32	+ .026	+ .029	+ .033	+ .037	+ .040	+ .044	+ .047	+ .051	+ .054	+ .058	+ .061	32
33	+ .023	+ .027	+ .030	+ .034	+ .037	+ .041	+ .044	+ .048	+ .051	+ .055	+ .059	33
34	+ .021	+ .024	+ .028	+ .031	+ .035	+ .038	+ .042	+ .045	+ .049	+ .052	+ .056	34
35	+ .018	+ .021	+ .025	+ .028	+ .032	+ .036	+ .039	+ .043	+ .046	+ .050	+ .053	35
36	+ .015	+ .019	+ .022	+ .026	+ .029	+ .033	+ .036	+ .040	+ .043	+ .047	+ .050	36
37	+ .013	+ .016	+ .020	+ .023	+ .027	+ .030	+ .034	+ .037	+ .041	+ .044	+ .048	37
38	+ .010	+ .014	+ .017	+ .020	+ .024	+ .027	+ .031	+ .034	+ .038	+ .041	+ .045	38
39	+ .007	+ .011	+ .014	+ .018	+ .021	+ .025	+ .028	+ .032	+ .035	+ .039	+ .042	39
40	+ .005	+ .008	+ .012	+ .015	+ .019	+ .022	+ .026	+ .029	+ .033	+ .036	+ .040	40
41	+ .002	+ .006	+ .009	+ .013	+ .016	+ .020	+ .023	+ .026	+ .030	+ .033	+ .037	41
42	.000	+ .003	+ .007	+ .010	+ .014	+ .017	+ .020	+ .024	+ .027	+ .031	+ .034	42
43	− .003	+ .001	+ .004	+ .008	+ .011	+ .014	+ .018	+ .021	+ .025	+ .028	+ .032	43
44	− .005	− .002	+ .002	+ .005	+ .008	+ .012	+ .015	+ .019	+ .022	+ .026	+ .029	44
45	− .008	− .004	− .001	+ .002	+ .006	+ .009	+ .013	+ .016	+ .020	+ .023	+ .026	45
46	− .010	− .007	− .004	.000	+ .003	+ .007	+ .010	+ .014	+ .017	+ .020	+ .024	46
47	− .013	− .009	− .006	− .003	+ .001	+ .004	+ .008	+ .011	+ .014	+ .018	+ .021	47
48	− .015	− .012	− .009	− .005	− .002	+ .002	+ .005	+ .008	+ .012	+ .015	+ .019	48
49	− .018	− .015	− .011	− .008	− .004	− .001	+ .003	+ .006	+ .009	+ .013	+ .016	49
50	− .020	− .017	− .014	− .010	− .007	− .003	.000	+ .003	+ .007	+ .010	+ .014	50
51	− .023	− .019	− .016	− .013	− .009	− .006	− .003	+ .001	+ .004	+ .008	+ .011	51
52	− .025	− .022	− .019	− .015	− .012	− .008	− .005	− 002	+ .002	+ .005	+ .008	52
53	− .028	− .024	− .021	− .018	− .014	− .011	− .008	− .004	− .001	+ .003	+ .006	53
54	− .030	− .027	− .023	− .020	− .017	− .013	− .010	− .007	− .003	.000	+ .004	54
55	− .032	− .029	− .026	− .022	− .019	− .016	− .012	− .009	− .006	− .002	+ .001	55
56	− .035	− .032	− .028	− .025	− .022	− .018	− .015	− .012	− .008	− .005	− .002	56
57	− .037	− .034	− .031	− .027	− .024	− .021	− .017	− .014	− .011	− .007	− .004	57
58	− .040	− .036	− .033	− .030	− .026	− .023	− .020	− .016	− .013	− .010	− .006	58
59	− .042	− .039	− .035	− .032	− .029	− .025	− .022	− .019	− .015	− .012	− .009	59
60	− .044	− .041	− .038	− .034	− .031	− .028	− .024	− .021	− .018	− .015	− .011	60
61	− .047	− .043	− .040	− .037	− .033	− .030	− .027	− .024	− .020	− .017	− .014	61
62	− .049	− .046	− .042	− .039	− .036	− .032	− .029	− .026	− .023	− .019	− .016	62
63	− .051	− .048	− .045	− .041	− .038	− .035	− .032	− .028	− .025	− .022	− .018	63
64	− .053	− .050	− .047	− .044	− .040	− .037	− .034	− .031	− .027	− .024	− .021	64
65	− .056	− .052	− .049	− .046	− .043	− .039	− .036	− .033	− .030	− .026	− .023	65
66	− .058	− .055	− .052	− .048	− .045	− .042	− .039	− .035	− .032	− .029	− .026	66
67	− .060	− .057	− .054	− .051	− .047	− .044	− .041	− .038	− .034	− .031	− .028	67
68	− .063	− .059	− .056	− .053	− .050	− .046	− .043	− .040	− .037	− .033	− .030	68
69	− .065	− .062	− .058	− .055	− .052	− .049	− .045	− .042	− .039	− .036	− .032	69
70	− .067	− .064	− .061	− .057	− .054	− .051	− .048	− .044	− .041	− .038	− .035	70
71	− .069	− .066	− .063	− .060	− .056	− .053	− .050	− .047	− .044	− .040	− .037	71
72	− .071	− .068	− .065	− .062	− .059	− .055	− .052	− .049	− .046	− .043	− .039	72
73	− .074	− .070	− .067	− .064	− .061	− .058	− .054	− .051	− .048	− .045	− .042	73
74	− .076	− .073	− .069	− .066	− .063	− .060	− .057	− .053	− .050	− .047	− .044	74
75	− .078	− .075	− .072	− .068	− .065	− .062	− .059	− .056	− .053	− .049	− .046	75
76	− .080	− .077	− .074	− .071	− .067	− .064	− .061	− .058	− .055	− .052	− .048	76
77	− .082	− .079	− .076	− .073	− .070	− .066	− .063	− .060	− .057	− .054	− .051	77
78	− .084	− .081	− .078	− .075	− .072	− .069	− .065	− .062	− .059	− .056	− .053	78
79	− .087	− .083	− .080	− .077	− .074	− .071	− .068	− .065	− .061	− .058	− .055	79
80	− .089	− .086	− .082	− .079	− .076	− .073	− .070	− .067	− .064	− .060	− .057	80

Multiply the mean Refraction by the Tabular correcting decimal Number under Bar. at top, and Ther. at side, and the product, applied with the proper Sign, + or −, to the mean Refraction, will give the corrected Refraction.

TABLE 2.

Bar.	30 . 1	30 . 2	30 . 3	30 . 4	30 . 5	30 . 6	30 . 7	30 . 8	30 . 9	31 . 0	Bar.
Fahr. Ther.											Fahr. Ther.
20°	+ .099	+ .103	+ .107	+ .110	+ .114	+ .118	+ .121	+ .125	+ .129	+ .132	20°
21	+ .096	+ .100	+ .104	+ .107	+ .111	+ .115	+ .118	+ .122	+ .126	+ .129	21
22	+ .094	+ .097	+ .101	+ .104	+ .108	+ .112	+ .115	+ .119	+ .123	+ .126	22
23	+ .091	+ .094	+ .098	+ .101	+ .105	+ .109	+ .112	+ .116	+ .120	+ .123	23
24	+ .088	+ .091	+ .095	+ .098	+ .102	+ .106	+ .109	+ .113	+ .117	+ .120	24
25	+ .085	+ .088	+ .092	+ .096	+ .099	+ .103	+ .106	+ .110	+ .114	+ .117	25
26	+ .082	+ .085	+ .089	+ .093	+ .096	+ .100	+ .103	+ .107	+ .111	+ .114	26
27	+ .079	+ .083	+ .086	+ .090	+ .093	+ .097	+ .100	+ .104	+ .108	+ .111	27
28	+ .076	+ .080	+ .083	+ .087	+ .090	+ .094	+ .098	+ .101	+ .105	+ .108	28
29	+ .073	+ .077	+ .080	+ .084	+ .088	+ .091	+ .095	+ .098	+ .102	+ .105	29
30	+ .070	+ .074	+ .078	+ .081	+ .085	+ .088	+ .092	+ .095	+ .099	+ .102	30
31	+ .068	+ .071	+ .075	+ .078	+ .082	+ .085	+ .089	+ .093	+ .096	+ .100	31
32	+ .065	+ .068	+ .072	+ .075	+ .079	+ .083	+ .086	+ .090	+ .093	+ .097	32
33	+ .062	+ .066	+ .069	+ .073	+ .076	+ .080	+ .083	+ .087	+ .090	+ .094	33
34	+ .059	+ .063	+ .066	+ .070	+ .073	+ .077	+ .080	+ .084	+ .088	+ .091	34
35	+ .057	+ .060	+ .064	+ .067	+ .071	+ .074	+ .078	+ .081	+ .085	+ .088	35
36	+ .054	+ .057	+ .061	+ .064	+ .068	+ .071	+ .075	+ .078	+ .082	+ .085	36
37	+ .051	+ .055	+ .058	+ .062	+ .065	+ .069	+ .072	+ .076	+ .079	+ .083	37
38	+ .048	+ .052	+ .055	+ .059	+ .062	+ .066	+ .069	+ .073	+ .076	+ .080	38
39	+ .046	+ .049	+ .053	+ .056	+ .060	+ .063	+ .067	+ .070	+ .074	+ .077	39
40	+ .043	+ .046	+ .050	+ .053	+ .057	+ .060	+ .064	+ .067	+ .071	+ .074	40
41	+ .040	+ .044	+ .047	+ .051	+ .054	+ .058	+ .061	+ .065	+ .068	+ .071	41
42	+ .038	+ .041	+ .045	+ .048	+ .051	+ .055	+ .058	+ .062	+ .065	+ .069	42
43	+ .035	+ .038	+ .042	+ .045	+ .049	+ .052	+ .056	+ .059	+ .063	+ .066	43
44	+ .032	+ .036	+ .039	+ .043	+ .046	+ .050	+ .053	+ .056	+ .060	+ .063	44
45	+ .030	+ .033	+ .037	+ .040	+ .044	+ .047	+ .050	+ .054	+ .057	+ .061	45
46	+ .027	+ .031	+ .034	+ .037	+ .041	+ .044	+ .048	+ .051	+ .054	+ .058	46
47	+ .025	+ .028	+ .031	+ .035	+ .038	+ .042	+ .045	+ .049	+ .052	+ .055	47
48	+ .022	+ .025	+ .029	+ .032	+ .036	+ .039	+ .042	+ .046	+ .049	+ .053	48
49	+ .019	+ .023	+ .026	+ .030	+ .033	+ .036	+ .040	+ .043	+ .047	+ .050	49
50	+ .017	+ .020	+ .024	+ .027	+ .030	+ .034	+ .037	+ .041	+ .044	+ .047	50
51	+ .014	+ .018	+ .021	+ .025	+ .028	+ .031	+ .035	+ .038	+ .041	+ .045	51
52	+ .012	+ .015	+ .019	+ .022	+ .025	+ .029	+ .032	+ .035	+ .039	+ .042	52
53	+ .009	+ .013	+ .016	+ .019	+ .023	+ .026	+ .030	+ .033	+ .036	+ .040	53
54	+ .007	+ .010	+ .014	+ .017	+ .020	+ .024	+ .027	+ .030	+ .034	+ .037	54
55	+ .004	+ .008	+ .011	+ .014	+ .018	+ .021	+ .024	+ .028	+ .031	+ .034	55
56	+ .002	+ .005	+ .009	+ .012	+ .015	+ .019	+ .022	+ .025	+ .029	+ .032	56
57	− .001	+ .003	+ .006	+ .009	+ .013	+ .016	+ .019	+ .023	+ .026	+ .029	57
58	− .003	.000	+ .004	+ .007	+ .010	+ .014	+ .017	+ .020	+ .024	+ .027	58
59	− .006	− .002	+ .001	+ .004	+ .008	+ .011	+ .014	+ .018	+ .021	+ .024	59
60	− .008	− .005	− .001	+ .002	+ .005	+ .009	+ .012	+ .015	+ .019	+ .022	60
61	− .010	− .007	− .004	− .001	+ .003	+ .006	+ .009	+ .013	+ .016	+ .019	61
62	− .013	− .009	− .006	− .003	.000	+ .004	+ .007	+ .010	+ .014	+ .017	62
63	− .015	− .012	− .009	− .005	− .002	+ .001	+ .005	+ .008	+ .011	+ .014	63
64	− .018	− .014	− .011	− .008	− .004	− .001	+ .002	+ .005	+ .009	+ .012	64
65	− .020	− .017	− .013	− .010	− .007	− .004	.000	+ .003	+ .006	+ .010	65
66	− .022	− .019	− .016	− .013	− .009	− .006	− .003	+ .001	+ .004	+ .007	66
67	− .025	− .021	− .018	− .015	− .012	− .008	− .005	− .002	+ .001	+ .005	67
68	− .027	− .024	− .020	− .017	− .014	− .011	− .008	− .004	− .001	+ .002	68
69	− .029	− .026	− .023	− .020	− .016	− .013	− .010	− .007	− .004	.000	69
70	− .032	− .028	− .025	− .022	− .019	− .015	− .012	− .009	− .006	− .003	70
71	− .034	− .031	− .027	− .024	− .021	− .018	− .015	− .011	− .008	− .005	71
72	− .036	− .033	− .030	− .027	− .023	− .020	− .017	− .014	− .011	− .007	72
73	− .038	− .035	− .032	− .029	− .026	− .022	− .019	− .016	− .013	− .010	73
74	− .041	− .038	− .034	− .031	− .028	− .025	− .022	− .018	− .015	− .012	74
75	− .043	− .040	− .037	− .033	− .030	− .027	− .024	− .021	− .017	− .014	75
76	− .045	− .042	− .039	− .036	− .032	− .029	− .026	− .023	− .020	− .017	76
77	− .047	− .044	− .041	− .038	− .035	− .032	− .028	− .025	− .022	− .019	77
78	− .050	− .047	− .043	− .040	− .037	− .034	− .031	− .028	− .024	− .021	78
79	− .052	− .049	− .046	− .042	− .039	− .036	− .033	− .030	− .027	− .023	79
80	− .054	− .051	− .048	− .045	− .042	− .038	− .035	− .032	− .029	− .026	80

Multiply the mean Refraction by the Tabular correcting decimal Number under Bar. at top, and Ther. at side, and the product, applied with the proper Signs, + or —, to the mean Refraction, will give the corrected Refraction.

Table 3.

LOGARITHMS of Dr. BRADLEY'S MEAN REFRACTIONS for every Ten Minutes in Zenith Distance.

Apparent Zen. Dist.	Log. Refract.	Diff.	Apparent Zen. Dist.	Log. Refract.	Diff.	Apparent Zen. Dist.	Log. Refract.	Diff.
0° 0′	$-\infty$		10° 0	1.00000		20° 0′	1.31597	
10	9.23045		10	1.00775	775	10	1.32015	418
20	9.51851	28806	20	1.01578	803	20	1.32428	413
30	9.69897	18046	30	1.02325	747	30	1.32838	410
40	9.82607	12710	40	1.03060	735	40	1.33244	406
50	9.91908	9301	50	1.03822	762	50	1.33646	402
1 0	0.00000	8092	11 0	1.04532	710	21 0	1.34044	398
10	0.06819	6819	10	1.05192	660	10	1.34400	360
20	0.12385	5566	20	1.05805	613	20	1.34772	372
30	0.17609	5224	30	1.06446	641	30	1.35122	350
40	0.22272	4663	40	1.07078	632	40	1.35468	346
50	0.26245	3973	50	1.07664	586	50	1.35832	364
2 0	0.30103	3858	12 0	1.08279	615	22 0	1.36173	341
10	0.33646	3543	10	1.08884	605	10	1.36549	376
20	0.36736	3090	20	1.09447	563	20	1.36922	373
30	0.39794	3058	30	1.10037	590	30	1.37291	369
40	0.42651	2857	40	1.10619	582	40	1.37658	367
50	0.45179	2528	50	1.11160	541	50	1.38021	363
3 0	0.47712	2533	13 0	1.11727	567	23 0	1.38382	361
10	0.50106	2394	10	1.12320	593	10	1.38739	357
20	0.52244	2138	20	1.12937	617	20	1.39094	355
30	0.54407	2163	30	1.13513	576	30	1.39445	351
40	0.56467	2060	40	1.14082	569	40	1.39794	349
50	0.58320	1853	50	1.14675	593	50	1.40140	346
4 0	0.60206	1886	14 0	1.15229	554	24 0	1.40483	343
10	0.62014	1808	10	1.15776	547	10	1.40824	341
20	0.63649	1635	20	1.16346	570	20	1.41462	338
30	0.65321	1672	30	1.16879	533	30	1.41497	335
40	0.66932	1611	40	1.17406	527	40	1.41830	333
50	0.68395	1463	50	1.17955	549	50	1.42160	330
5 0	0.69897	1502	15 0	1.18469	514	25 0	1.42488	328
10	0.71349	1452	10	1.18949	480	10	1.42813	325
20	0.72673	1324	20	1.19396	447	20	1.43136	323
30	0.74033	1360	30	1.19866	470	30	1.43457	321
40	0.75358	1325	40	1.20330	464	40	1.43775	318
50	0.76567	1209	50	1.20763	433	50	1.44091	316
6 0	0.77815	1248	16 0	1.21219	456	26 0	1.44404	313
10	0.79029	1214	10	1.21696	477	10	1.44716	312
20	0.80140	1111	20	1.22194	498	20	1.45025	309
30	0.81291	1151	30	1.22660	466	30	1.45332	307
40	0.82413	1122	40	1.23121	461	40	1.45637	305
50	0.83442	1029	50	1.23603	482	50	1.45939	302
7 0	0.84510	1068	17 0	1.24055	452	27 0	1.46204	301
10	0.85552	1042	10	1.24502	447	10	1.46568	328
20	0.86510	958	20	1.24969	467	20	1.46879	311
30	0.87506	996	30	1.25406	437	30	1.47202	323
40	0.88480	974	40	1.25840	434	40	1.47524	322
50	0.89376	896	50	1.26293	453	50	1.47828	304
8 0	0.90309	933	18 0	1.26717	424	28 0	1.48144	316
10	0.91222	913	10	1.27138	421	10	1.48458	314
20	0.92065	843	20	1.27577	439	20	1.48756	298
30	0.92942	877	30	1.27989	412	30	1.49066	310
40	0.93802	860	40	1.28398	409	40	1.49374	308
50	0.94569	794	50	1.28825	427	50	1.49665	291
9 0	0.95424	828	19 0	1.29226	401	29 0	1.49969	304
10	0.96237	813	10	1.29623	397	10	1.50270	301
20	0.96988	751	20	1.30038	415	20	1.50556	286
30	0.97772	784	30	1.30428	390	30	1.50853	297
40	0.98543	771	40	1.30814	386	40	1.51148	295
50	0.99255	712	50	1.31218	404	50	1.51428	280
		745			379			292

Table 3.

LOGARITHMS of Dr. BRADLEY'S MEAN REFRACTIONS for every Ten Minutes in Zenith Distance.

Apparent Zen. Dist.	Log. Refract.	Diff.	Apparent Zen. Dist.	Log. Refract.	Diff.	Apparent Zen. Dist.	Log. Refract.	Diff.
30° 0′	1 . 51720	289	40° 0′	1 . 67943	253	50° 0	1 . 83123	249
10	1 . 52009	275	10	1 . 68196	262	10	1 . 83372	254
20	1 . 52284	285	20	1 . 68458	250	20	1 . 83626	253
30	1 . 52569	284	30	1 . 68708	250	30	1 . 83879	251
40	1 . 52853	269	40	1 . 68958	256	40	1 . 84130	249
50	1 . 53122	281	50	1 . 69214	247	50	1 . 84379	255
31 0	1 . 53403	291	41 0	1 . 69461	262	51 0	1 . 84634	259
10	1 . 53694	301	10	1 . 69723	261	10	1 . 84893	263
20	1 . 53995	288	20	1 . 69984	259	20	1 . 85156	262
30	1 . 54283	285	30	1 . 70243	258	30	1 . 85418	267
40	1 . 54568	296	40	1 . 70501	256	40	1 . 85685	265
50	1 . 54864	281	50	1 . 70757	255	50	1 . 85950	263
32 0	1 . 55145	280	42 0	1 . 71012	253	52 0	1 . 86213	262
10	1 . 55425	290	10	1 . 71265	252	10	1 . 86475	266
20	1 . 55715	276	20	1 . 71517	250	20	1 . 86741	264
30	1 . 55991	274	30	1 . 71769	249	30	1 . 87005	263
40	1 . 56265	284	40	1 . 72016	247	40	1 . 87268	261
50	1 . 56549	271	50	1 . 72263	246	50	1 . 87529	266
33 0	1 . 56820	269	43 0	1 . 72509	261	53 0	1 . 87795	263
10	1 . 57089	279	10	1 . 72770	252	10	1 . 88058	263
20	1 . 57368	266	20	1 . 73022	258	20	1 . 88321	266
30	1 . 57634	264	30	1 . 73280	256	30	1 . 88587	265
40	1 . 57898	263	40	1 . 73536	247	40	1 . 88852	263
50	1 . 58161	272	50	1 . 73783	253	50	1 . 89115	261
34 0	1 . 58433	282	44 0	1 . 74036	252	54 0	1 . 89376	260
10	1 . 58715	280	10	1 . 74288	245	10	1 . 89636	267
20	1 . 58995	278	20	1 . 74531	249	20	1 . 89903	259
30	1 . 59273	277	30	1 . 74780	248	30	1 . 90162	266
40	1 . 59550	274	40	1 . 75028	238	40	1 . 90428	265
50	1 . 59824	273	50	1 . 75266	245	50	1 . 90693	263
35 0	1 . 60097	272	45 0	1 . 75511	251	55 0	1 . 90956	272
10	1 . 60369	269	10	1 . 75762	250	10	1 . 91228	270
20	1 . 60638	268	20	1 . 76012	248	20	1 . 91498	268
30	1 . 60906	266	30	1 . 76260	247	30	1 . 91766	272
40	1 . 61172	265	40	1 . 76507	253	40	1 . 92038	271
50	1 . 61437	263	50	1 . 76760	252	50	1 . 92309	274
36 0	1 . 61700	262	46 0	1 . 77012	257	56 0	1 . 92583	267
10	1 . 61962	259	10	1 . 77269	248	10	1 . 92850	265
20	1 . 62221	259	20	1 . 77517	255	20	1 . 93115	269
30	1 . 62480	257	30	1 . 77772	253	30	1 . 93384	272
40	1 . 62737	255	40	1 . 78025	258	40	1 . 93656	271
50	1 . 62992	256	50	1 . 78283	250	50	1 . 93927	274
37 0	1 . 63246	272	47 0	1 . 78533	256	57 0	1 . 94201	282
10	1 . 63518	261	10	1 . 78789	261	10	1 . 94483	280
20	1 . 63779	269	20	1 . 79050	252	20	1 . 94763	283
30	1 . 64048	268	30	1 . 79302	258	30	1 . 95046	286
40	1 . 64316	256	40	1 . 79560	257	40	1 . 95332	285
50	1 . 64572	264	50	1 . 79817	255	50	1 . 95617	287
38 0	1 . 64836	263	48 0	1 . 80072	246	58 0	1 . 95904	281
10	1 . 65099	251	10	1 . 80318	246	10	1 . 96185	279
20	1 . 65350	260	20	1 . 80564	250	20	1 . 96464	280
30	1 . 65610	258	30	1 . 80814	250	30	1 . 96744	281
40	1 . 65868	247	40	1 . 81064	247	40	1 . 97025	283
50	1 . 66115	255	50	1 . 81311	246	50	1 . 97308	281
39 0	1 . 66370	263	49 0	1 . 81558	258	59 0	1 . 97589	284
10	1 . 66633	271	10	1 . 81816	263	10	1 . 97873	282
20	1 . 66904	260	20	1 . 82079	262	20	1 . 98155	284
30	1 . 67164	258	30	1 . 82341	260	30	1 . 98439	287
40	1 . 67422	266	40	1 . 82601	259	40	1 . 98726	286
50	1 . 67688	255	50	1 . 82860	263	50	1 . 99012	288

TABLE 3.

LOGARITHMS of Dr. BRADLEY'S MEAN REFRACTIONS for every Ten Minutes in Zenith Distance.

Apparent Zen. Dist.	Log. Refract.	Diff.
60° 0	1 . 99300	294
10	1 . 99594	293
20	1 . 99887	295
30	2 . 00182	297
40	2 . 00479	296
50	2 . 00775	297
61 0	2 . 01072	296
10	2 . 01368	293
20	2 . 01661	296
30	2 . 01957	298
40	2 . 02255	300
50	2 . 02555	302
62 0	2 . 02857	308
10	2 . 03165	310
20	2 . 03475	311
30	2 . 03786	313
40	2 . 04099	316
50	2 . 04415	322
63 0	2 . 04727	311
10	2 . 05038	308
20	2 . 05346	310
30	2 . 05656	312
40	2 . 05968	313
50	2 . 06281	314
64 0	2 . 06595	316
10	2 . 06911	321
20	2 . 07232	323
30	2 . 07555	323
40	2 . 07878	325
50	2 . 08203	326
65 0	2 . 08529	327
10	2 . 08856	328
20	2 . 09184	329
30	2 . 09513	334
40	2 . 09847	334
50	2 . 00181	336
66 0	2 . 10517	339
10	2 . 10856	341
20	2 . 11197	344
30	2 . 11541	345
40	2 . 11886	345
50	2 . 12231	350
67 0	2 . 12581	349
10	2 . 12930	354
20	2 . 13284	353
30	2 . 13637	354
40	2 . 13991	357
50	2 . 14348	358
68 0	2 . 14706	357
10	2 . 15063	358
20	2 . 15421	361
30	2 . 15782	364
40	2 . 16146	364
50	2 . 16510	369
69 0	2 . 16879	375
10	2 . 17254	378
20	2 . 17632	380
30	2 . 18012	383
40	2 . 18395	388
50	2 . 18783	390

Apparent Zen. Dist.	Log. Refract.	Diff.
70° 0'	2 . 19173	392
10	2 . 19565	394
20	2 . 19959	399
30	2 . 20358	400
40	2 . 20758	402
50	2 . 21160	404
71 0	2 . 21564	402
10	2 . 21966	407
20	2 . 22373	408
30	2 . 22781	414
40	2 . 23195	416
50	2 . 23611	419
72 0	2 . 24030	427
10	2 . 24457	431
20	2 . 24888	434
30	2 . 25322	439
40	2 . 25761	441
50	2 . 26202	445
73 0	2 . 26647	446
10	2 . 27093	450
20	2 . 27543	453
30	2 . 27996	458
40	2 . 28454	462
50	2 . 28916	464
74 0	2 . 29380	468
10	2 . 29848	473
20	2 . 30321	480
30	2 . 30801	482
40	2 . 31283	488
50	2 . 31771	492
75 0	2 . 32263	495
10	2 . 32758	502
20	2 . 33260	506
30	2 . 33766	512
40	2 . 34278	515
50	2 . 34793	522
76 0	2 . 35315	526
10	2 . 35841	531
20	2 . 36372	538
30	2 . 36910	545
40	2 . 37455	550
50	2 . 38005	556
77 0	2 . 38561	559
10	2 . 39120	570
20	2 . 39690	572
30	2 . 40262	584
40	2 . 40846	588
50	2 . 41434	591
78 0	2 . 42025	605
10	2 . 42630	612
20	2 . 43242	619
30	2 . 43861	626
40	2 . 44487	636
50	2 . 45123	642
79 0	2 . 45765	651
10	2 . 46416	659
20	2 . 47075	667
30	2 . 47742	678
40	2 . 48420	689
50	2 . 49109	699

Apparent Zen. Dist.	Log. Refract.	Diff.
80° 0	2 . 49808	706
10	2 . 50514	716
20	2 . 51230	728
30	2 . 51958	739
40	2 . 52697	754
50	2 . 53451	763
81 0	2 . 54214	772
10	2 . 54986	785
20	2 . 55771	800
30	2 . 56571	812
40	2 . 57383	829
50	2 . 58212	844
82 0	2 . 59056	852
10	2 . 59908	868
20	2 . 60776	886
30	2 . 61662	903
40	2 . 62565	922
50	2 . 63487	941
83 0	2 . 64428	949
10	2 . 65377	969
20	2 . 66346	992
30	2 . 67338	1013
40	2 . 68351	1037
50	2 . 69388	1062
84 0	2 . 70450	1067
10	2 . 71517	1095
20	2 . 72612	1122
30	2 . 73734	1151
40	2 . 74885	1181
50	2 . 76066	1203
85 0	2 . 77269	

To the Logarithms found in this Table, add the Logarithms for the Barometer and Thermometer taken out of the following Table, and the Sums will be the Logarithms of the corrected Refractions.

LOGARITHMS for correcting Dr. BRADLEY'S MEAN REFRACTIONS.

Table 4.

Fahr. Therm.	Height of the Barometer in English Inches.								Fahr. Therm.
0	29 . 0	29 . 1	29 . 2	29 . 3	29 . 4	29 . 5	29 . 6	29 . 7	0
20	10 . 02497	10 . 02646	10 . 02795	10 . 02944	10 . 03092	10 . 03239	10 . 03386	10 . 05533	20
21	02380	. 02529	. 02678	. 02827	. 02975	. 03122	. 03269	. 03416	21
22	02263	. 02412	. 02561	. 02710	. 02858	. 03005	. 03152	. 03299	22
23	02146	. 02295	. 02444	. 02593	. 02741	. 02888	. 03035	. 03182	23
24	02030	. 02179	. 02328	. 02477	. 02625	. 02772	. 02919	. 03066	24
25	10 . 01914	10 . 02063	10 . 02212	10 . 02361	10 . 02509	10 . 02656	10 . 02803	10 . 02950	25
26	01798	. 01947	. 02096	. 02245	. 02393	. 02540	. 02687	. 02834	26
27	01683	. 01832	. 01981	. 02130	. 02278	. 02425	. 02572	. 02719	27
28	01568	. 01717	. 01866	. 02015	. 02163	. 02310	. 02457	. 02604	28
29	01453	. 01602	. 01751	. 01900	. 02048	. 02195	. 02342	. 02489	29
30	10 . 01339	10 . 01488	10 . 01637	10 . 01786	10 . 01934	10 . 02081	10 . 02228	10 . 02375	30
31	01225	. 01374	. 01523	. 01672	. 01820	. 01967	. 02114	. 02261	31
32	01111	. 01260	. 01409	. 01558	. 01706	. 01853	. 02000	. 02147	32
33	00997	. 01146	. 01295	. 01444	. 01592	. 01739	. 01886	. 02033	33
34	00884	. 01033	. 01182	. 01331	. 01479	. 01626	. 01773	. 01920	34
35	10 . 00771	10 . 00920	10 . 01069	10 . 01218	10 . 01366	10 . 01513	10 . 01661	10 . 01807	35
36	00658	. 00807	. 00956	. 01105	. 01253	. 01400	. 01547	. 01694	36
37	00546	. 00695	. 00844	. 00993	. 01141	. 01288	. 01435	. 01582	37
38	00434	. 00583	. 00732	. 00881	. 01029	. 01176	. 01323	. 01470	38
39	00322	. 00471	. 00620	. 00769	. 00917	. 01064	. 01211	. 01358	39
40	10 . 00211	10 . 00360	10 . 00509	10 . 00658	10 . 00806	10 . 00953	10 . 01100	10 . 01247	40
41	. 00099	. 00248	. 00397	. 00546	. 00694	. 00841	. 00988	. 01135	41
42	9 . 99988	. 00137	. 00286	. 00435	. 00583	. 00730	. 00877	. 01024	42
43	. 99878	. 00027	. 00176	. 00325	. 00473	. 00620	. 00767	. 00914	43
44	. 99767	9 . 99916	. 00065	. 00214	. 00362	. 00509	. 00656	. 00803	44
45	9 . 99657	9 . 99806	9 . 99955	10 . 00104	10 . 00252	10 . 00399	10 . 00546	10 . 00693	45
46	. 99547	. 99696	. 99845	9 . 99904	. 00142	. 00289	. 00436	. 00583	46
47	. 99438	. 99587	. 99736	. 99885	. 00033	. 00180	. 00327	. 00474	47
48	. 99329	. 99478	. 99627	. 99776	9 . 99924	. 00071	. 00218	. 00365	48
49	. 99220	. 99369	. 99518	. 99667	. 99815	9 . 99962	. 00109	. 00256	49
50	9 . 99111	9 . 99260	9 . 99409	9 . 99558	9 . 99706	9 . 99853	10 . 00000	10 . 00147	50
51	. 99003	. 99152	. 99301	. 99450	. 99598	. 99745	9 . 99892	. 00039	51
52	. 98894	. 99043	. 99192	. 99341	. 99489	. 99636	. 99783	9 . 99930	52
53	. 98786	. 98935	. 99084	. 99233	. 99381	. 99528	. 99675	. 99822	53
54	. 98679	. 98828	. 98977	. 99126	. 99274	. 99421	. 99568	. 99715	54
55	9 . 98571	9 . 98720	9 . 98869	9 . 99018	9 . 99166	9 . 99313	9 . 99460	9 . 99607	55
56	. 98474	. 98613	. 98762	. 98911	. 99059	. 99206	. 99353	. 99500	56
57	. 98358	. 98507	. 98656	. 98805	. 98953	. 99100	. 99247	. 99394	57
58	. 98251	. 98400	. 98549	. 98698	. 98846	. 98993	. 99140	. 99287	58
59	. 98145	. 98294	. 98443	. 98592	. 98740	. 98887	. 99034	. 99181	59
60	9 . 98039	9 . 98188	9 . 98337	9 . 98486	9 . 98634	9 . 98781	9 . 98928	9 . 99075	60
61	. 97933	. 98082	. 98231	. 98380	. 98528	. 98675	. 98822	. 98969	61
62	. 97827	. 97976	. 98125	. 98274	. 98422	. 98569	. 98716	. 98863	62
63	. 97722	. 97871	. 98020	. 98169	. 98317	. 98464	. 98611	. 98758	63
64	. 97617	. 97766	. 97915	. 98064	. 98212	. 98359	. 98506	. 98653	64
65	9 . 97512	9 . 97661	9 . 97810	9 . 97959	9 . 98107	9 . 98254	9 . 98401	9 . 98548	65
66	. 97408	. 97557	. 97706	. 97855	. 98003	. 98150	. 98297	. 98444	66
67	. 97303	. 97452	. 97601	. 97750	. 97898	. 98045	. 98192	. 98339	67
68	. 97199	. 97348	. 97497	. 97646	. 97794	. 97941	. 98082	. 98235	68
69	. 97096	. 97245	. 97394	. 97543	. 97691	. 97838	. 97985	. 98132	69
70	9 . 96992	9 . 97141	9 . 97290	9 . 97439	9 . 97587	9 . 97734	9 . 97881	9 . 98028	70
71	. 96889	. 97038	. 97187	. 97336	. 97484	. 97631	. 97778	. 97925	71
72	. 96786	. 96935	. 97084	. 97233	. 97381	. 97528	. 97657	. 97822	72
73	. 96683	. 96832	. 96981	. 97130	. 97278	. 97425	. 97572	. 97719	73
74	. 96580	. 96729	. 96878	. 97027	. 97175	. 97322	. 97469	. 97616	74

Proportional Parts for Hundredths of an Inch of the Barometer.

. 01	. 02	. 03	. 04	. 05	. 06	. 07	. 08	. 09
+ 14	+ 29	+ 43	+ 58	+ 72	+ 87	+ 101	+ 116	+ 131

TABLE 4 concluded.

Fahr. Therm.	Height of the Barometer in English Inches.								Fahr. Therm.
0	29 .8	29 .9	30 .0	30 .1	30 .2	30 .3	30 .4	30 .5	0
20°	10 .03679	10 .03824	10 .03969	10 .04114	10 .04258	10 .04401	10 .04544	10 .04687	20°
21	03569	.03707	.03852	.03997	.04141	.04284	.04427	.04570	21
22	03445	.03590	.03735	.03880	.04024	.04167	.04310	.04453	22
23	03328	.03473	.03618	.03763	.03907	.04050	.04193	.04336	23
24	03212	.03357	.03502	.03647	.03791	.03934	.04077	.04220	24
25	10 .03096	10 .03241	10 .03386	10 .03531	10 .03675	10 .03818	10 .03961	10 .04104	25
26	.02980	.03125	.03270	.03415	.03559	.03702	.03845	.03988	26
27	.02865	.03010	.03155	.03300	.03444	.03587	.03730	.03875	27
28	.02750	.02895	.03040	.03185	.03329	.03472	.03615	.03758	28
29	.02635	.02780	.02925	.03070	.03214	.03357	.03500	.03643	29
30	10 .02521	10 .02666	10 .02811	10 .02956	10 .03100	10 .03243	10 .03386	10 .03529	30
31	.02407	.02552	.02697	.02842	.02986	.03129	.03272	.03415	31
32	.02293	.02438	.02583	.02728	.02872	.03015	.03158	.03301	32
33	.02179	.02324	.02469	.02614	.02758	.02901	.03044	.03187	33
34	.02066	.02211	.02356	.02501	.02645	.02788	.02931	.03074	34
35	10 .01953	10 .02098	10 .02243	10 .02388	10 .02532	10 .02675	10 .02818	10 .02961	35
36	.01840	.01985	.02130	.02275	.02419	.02562	.02705	.02848	36
37	.01728	.01873	.02018	.02163	.02307	.02450	.02593	.02736	37
38	.01616	.01761	.01906	.02051	.02195	.02338	.92481	.02624	38
39	.01504	.01649	.01794	.01939	.02083	.02226	.02369	.02512	39
40	10 .01393	10 .01538	10 .01683	10 .01828	10 .01972	10 .02115	10 .02258	10 .02401	40
41	.01281	.01426	.01571	.01716	.01860	.02003	.02146	.02289	41
42	.01170	.01315	.01460	.01605	.01749	.01892	.02035	.02176	42
43	.01060	.01205	.01350	.01495	.01639	.01782	.01925	.02068	43
44	.00949	.01094	.01239	.01384	.01528	.01671	.01814	.01957	44
45	10 .00839	10 .00984	10 .01129	10 .01274	10 .01418	10 .01561	10 .01704	10 .01847	45
46	.00729	.00874	.01019	.01164	.01308	.01451	.01594	.01737	46
47	.00620	.00765	.00910	.01055	.01199	.01342	.01485	.01628	47
48	.00511	.00656	.00801	.00946	.01090	.01233	·01376	.01519	48
49	.00402	.00547	.00692	.00837	.00981	.01124	.01267	.01410	49
50	10 .00293	10 .00438	10 .00583	10 .00728	10 .00872	10 .01015	10 .01158	10 .01301	50
51	.00185	.00330	.00475	.00620	.00764	.00907	.01050	.01193	51
52	.00076	.00221	.00366	.00511	.00655	.00798	.00941	.01084	52
53	9 .99968	.00113	.00258	.00403	.00547	.00690	.00833	.00976	53
54	.99861	.00006	.00151	.00296	.00440	.00583	.00726	.00869	54
55	9 .99753	9 .99898	10 .00043	10 .00188	10 .00332	10 .00475	10 .00618	10 .00761	55
56	.99646	.99791	9 .99936	.00081	.00225	.00368	.00511	.00654	56
57	.99540	.99685	.99830	9 .99975	.00119	.00262	.00405	.00548	57
58	.99433	.99578	.99723	.99868	.00012	.00155	.00298	.00441	58
59	.99327	.99472	.99617	.99762	9 .99906	.00049	.00192	.00335	59
60	9 .99221	9 .99366	9 .99511	9 .99656	9 .99800	9 .99943	10 .00086	10 .00229	60
61	.99115	.99260	.99405	.99550	.99694	.99837	9 .99980	.00123	61
62	.99009	.99154	.99299	.99444	.99588	.99731	.99874	.00017	62
63	.98904	.99049	.99194	.99339	.99483	.99626	.99769	9 .99912	63
64	.98799	.98944	.99089	.99234	.99378	.99521	.99664	.99807	64
65	9 .98694	9 .98839	9 .98984	9 .99129	9 .99273	9 .99416	9 .99559	9 .99702	65
66	.98590	.98735	.98880	.99025	.99169	.99319	.99455	.99598	66
67	.98485	.98630	.98775	.98920	.99064	.99207	.99350	.99493	67
68	.98381	.98526	.98671	.98816	.98960	.99103	.99246	.99389	68
69	.98278	.98423	.98568	.98713	.98857	.99000	.99143	.99286	69
70	9 .98174	9 .98319	9 .98464	9 .98609	9 .98753	9 .98896	9 .99039	9 .99182	70
71	.98071	.98216	.98361	.98506	.98650	.98793	.98936	.99079	71
72	.97968	.98113	.98258	.98403	.98547	.98690	.98833	.98976	72
73	.97865	.98010	.98155	.98300	.98444	.98587	.98730	.98873	73
74	.97762	.97907	.98052	.98197	.98341	.98484	.98627	.98770	74

Proportional Parts for Hundredths of an Inch of the Barometer.

.01	.02	.03	.04	.05	.06	.07	.08	.09
+ 14	+ 29	+ 43	+ 58	+ 72	+ 87	+ 101	+ 116	+ 131

TABLE 1.

0.760 Bar. Metrique. +10° Therm. Centigrade; or English Barometer 29.94. Fahrenheit's Thermometer 50°.

Zen. Dist.	Refractions.	Logarithms.	Diff. for 1 Min.	Zen. Dist.	Refractions.	Logarithms.	Diff. for 1 Min.	Zen. Dist.	Refractions.	Logarithms.	Diff. for 1 Min.
0° 0′				10° 0′	10″.3	1.0118	7.0	20° 0′	21″.2	1.3264	3.9
10				10	10.5	1.0185		10	21.4	1.3303	
20				20	10.7	1.0259		20	21.6	1.3342	
30				30	10.8	1.0349		30	21.8	1.3380	
40				40	11.0	1.0440		40	22.0	1.3419	
50				50	11.2	1.0471		50	22.2	1.3458	
1 0	1″.0	0.0000	51.4	11 0	11.3	1.0541	6.4	21 0	22.4	1.3496	3.7
10	1.1	0.0514		10	11.5	1.0605		10	22.6	1.3533	
20	1.3	0.1029		20	11.7	1.0669		20	22.8	1.3570	
30	1.5	0.1543		30	11.9	1.0733		30	23.0	1.3607	
40	1.7	0.2058		40	12.0	1.0798		40	23.2	1.3644	
50	1.9	0.2573		50	12.2	1.0862		50	23.4	1.3681	
2 0	2.0	0.3087	29.2	12 0	12.4	1.0929	6.0	22 0	23.5	1.3718	3.6
10	2.2	0.3379		10	12.6	1.0989		10	23.7	1.3754	
20	2.4	0.3671		20	12.8	1.1049		20	23.9	1.3790	
30	2.6	0.3963		30	13.0	1.1109		30	24.1	1.3826	
40	2.8	0.4255		40	13.2	1.1169		40	24.3	1.3862	
50	2.9	0.4549		50	13.4	1.1229		50	24.5	1.3898	
3 0	3.1	0.4841	21.0	13 0	13.5	1.1289	5.5	23 0	24.7	1.3934	3.4
10	3.3	0.5051		10	13.7	1.1344		10	24.9	1.3968	
20	3.4	0.5262		20	13.8	1.1400		20	25.1	1.4002	
30	3.6	0.5472		30	14.0	1.1455		30	25.3	1.4036	
40	3.7	0.5683		40	14.2	1.1511		40	25.5	1.4071	
50	3.9	0.5893		50	14.3	1.1566		50	25.7	1.4105	
4 0	4.1	0.6104	16.1	14 0	14.5	1.1622	5.2	24 0	25.9	1.4140	3.3
10	4.3	0.6265		10	14.7	1.1674		10	26.1	1.4173	
20	4.5	0.6427		20	14.8	1.1726		20	26.3	1.4206	
30	4.6	0.6588		30	15.0	1.1779		30	26.5	1.4240	
40	4.8	0.6750		40	15.2	1.1831		40	26.7	1.4273	
50	4.9	0.6912		50	15.4	1.1883		50	26.9	1.4306	
5 0	5.1	0.7073	13.3	15 0	15.6	1.1936	4.9	25 0	27.2	1.4340	3.3
10	5.3	0.7206		10	15.8	1.1985		10	27.4	1.4373	
20	5.5	0.7339		20	16.0	1.2034		20	27.6	1.4406	
30	5.6	0.7472		30	16.1	1.2083		30	27.8	1.4438	
40	5.8	0.7606		40	16.3	1.2132		40	28.0	1.4471	
50	5.9	0.7739		50	16.5	1.2181		50	28.2	1.4504	
6 0	6.1	0.7872	11.2	16 0	16.7	1.2230	4.6	26 0	28.4	1.4536	3.1
10	6.3	0.7984		10	16.8	1.2276		10	28.6	1.4567	
20	6.5	0.8096		20	17.0	1.2322		20	28.8	1.4599	
30	6.7	0.8208		30	17.2	1.2369		30	29.0	1.4630	
40	6.9	0.8320		40	17.4	1.2415		40	29.2	1.4662	
50	7.0	0.8432		50	17.6	1.2461		50	29.4	1.4693	
7 0	7.2	0.8544	9.8	17 0	17.8	1.2508	4.4	27 0	29.7	1.4725	3.1
10	7.4	0.8642		10	18.0	1.2553		10	30.0	1.4756	
20	7.7	0.8740		20	18.2	1.2597		20	30.2	1.4787	
30	7.8	0.8838		30	18.4	1.2641		30	30.4	1.4818	
40	7.9	0.8936		40	18.6	1.2685		40	30.6	1.4849	
50	8.1	0.9034		50	18.8	1.2729		50	30.8	1.4880	
8 0	8.2	0.9133	8.7	18 0	18.9	1.2773	4.2	28 0	31.0	1.4911	3.0
10	8.4	0.9220		10	19.0	1.2815		10	31.2	1.4941	
20	8.6	0.9307		20	19.2	1.2857		20	31.4	1.4971	
30	8.7	0.9394		30	19.4	1.2899		30	31.6	1.5001	
40	8.9	0.9480		40	19.6	1.2941		40	31.8	1.5031	
50	9.0	0.9567		50	19.8	1.2983		50	32.0	1.5061	
9 0	9.2	0.9652	7.7	19 0	20.0	1.3024	4.0	29 0	32.3	1.5091	3.0
10	9.4	0.9729		10	20.2	1.3064		10	32.5	1.5121	
20	9.6	0.9807		20	20.4	1.3104		20	32.7	1.5150	
30	9.8	0.9884		30	20.6	1.3144		30	32.9	1.5180	
40	10.0	0.9962		40	20.8	1.3184		40	33.1	1.5209	
50	10.1	1.0039		50	21.0	1.3224		50	33.4	1.5239	

TABLE 1. continued.

0.760 Bar. Metrique. +10° Therm. Centigrade; or English Barometer 29.94. Fahrenheit's Thermometer 50°.

Zen. Dist.	Refractions.	Logarithms.	Diff. for 1 Min.	Zen. Dist.	Refractions.	Logarithms.	Diff. for 1 Min.	Zen. Dist.	Refractions.	Logarithms.	Diff. for 1 Min.
30° 00′	33″.6	1 .5268	2.9	40° 00′	0′ 48″.9	1 .6890	2.5	50° 00′	1′ 9″.3	1 .8410	2.6
10	33 .8	1 .5297		10	0 49 .2	1 .6915		10	1 9 .7	1 .8436	
20	34 .1	1 .5326		20	0 49 .5	1 .6941		20	1 10 .1	1 .8462	
30	34 .3	1 .5355		30	0 49 .8	1 .6966		30	1 10 .5	1 .8487	
40	34 .6	1 .5384		40	0 50 .1	1 .6992		40	1 10 .9	1 .8513	
50	34 .8	1 .5413		50	0 50 .4	1 .7017		50	1 11 .3	1 .8539	
31 00	35 .0	1 .5441	2.8	41 00	0 50 .6	1 .7043	2.5	51 00	1 11 .8	1 .8564	2.6
10	35 .2	1 .5469		10	0 50 .9	1 .7068		10	1 12 .2	1 .8590	
20	35 .5	1 .5497		20	0 51 .2	1 .7093		20	1 12 .7	1 .8616	
30	35 .7	1 .5526		30	0 51 .5	1 .7119		30	1 13 .1	1 .8642	
40	36 .0	1 .5554		40	0 51 .8	1 .7144		40	1 13 .5	1 .8668	
50	36 .2	1 .5582		50	0 52 .1	1 .7169		50	1 14 .0	1 .8694	
32 00	36 .4	1 .5611	2.8	42 00	0 52 .4	1 .7195	2.5	52 00	1 14 .4	1 .8719	2.6
10	36 .6	1 .5639		10	0 52 .7	1 .7220		10	1 14 .8	1 .8745	
20	36 .9	1 .5667		20	0 53 .0	1 .7245		20	1 15 .3	1 .8771	
30	37 .1	1 .5695		30	0 53 .3	1 .7271		30	1 15 .8	1 .8797	
40	37 .3	1 .5723		40	0 53 .6	1 .7296		40	1 16 .2	1 .8823	
50	37 .6	1 .5751		50	0 54 .0	1 .7321		50	1 16 .7	1 .8849	
33 00	37 .8	1 .5778	2.7	43 00	0 54 .3	1 .7347	2.5	53 00	1 17 .2	1 .8875	2.6
10	38 .0	1 .5805		10	0 54 .6	1 .7372		10	1 17 .6	1 .8901	
20	38 .3	1 .5835		20	0 54 .9	1 .7397		20	1 18 .1	1 .8927	
30	38 .5	1 .5862		30	0 55 .2	1 .7423		30	1 18 .6	1 .8954	
40	38 .8	1 .5890		40	0 55 .5	1 .7448		40	1 19 .1	1 .8980	
50	39 .0	1 .5917		50	0 55 .8	1 .7473		50	1 19 .6	1 .9006	
34 00	39 .3	1 .5943	2.7	44 00	0 56 .2	1 .7499	2.5	54 00	1 20 .1	1 .9033	2.7
10	39 .5	1 .5970		10	0 56 .5	1 .7524		10	1 20 .6	1 .9060	
20	39 .8	1 .5997		20	0 56 .9	1 .7549		20	1 21 .1	1 .9087	
30	40 .0	1 .6024		30	0 57 .2	1 .7574		30	1 21 .6	1 .9113	
40	40 .3	1 .6051		40	0 57 .5	1 .7599		40	1 22 .1	1 .9140	
50	40 .5	1 .6078		50	0 57 .8	1 .7624		50	1 22 .6	1 .9167	
35 00	40 .8	1 .6105	2.7	45 00	0 58 .2	1 .7650	2.5	55 00	1 23 .1	1 .9193	2.7
10	41 .0	1 .6132		10	0 58 .5	1 .7675		10	1 23 .6	1 .9220	
20	41 .3	1 .6159		20	0 58 .9	1 .7700		20	1 24 .1	1 .9247	
30	41 .5	1 .6185		30	0 59 .2	1 .7725		30	1 24 .6	1 .9274	
40	41 .8	1 .6212		40	0 59 .6	1 .7750		40	1 25 .1	1 .9301	
50	42 .0	1 .6239		50	1 0 .0	1 .7775		50	1 25 .6	1 .9328	
36 00	42 .3	1 .6265	2.6	46 00	1 0 .3	1 .7801	2.5	56 00	1 26 .2	1 .9354	2.7
10	42 .6	1 .6291		10	1 0 .6	1 .7826		10	1 26 .7	1 .9381	
20	42 .8	1 .6318		20	1 1 .0	1 .7851		20	1 27 .3	1 .9408	
30	43 .1	1 .6344		30	1 1 .3	1 .7876		30	1 27 .8	1 .9436	
40	43 .3	1 .6371		40	1 1 .7	1 .7902		40	1 28 .4	1 .9463	
50	43 .6	1 .6397		50	1 2 .0	1 .7927		50	1 29 .0	1 .9490	
37 00	43 .9	1 .6424	2.6	47 00	1 2 .4	1 .7953	2.5	57 00	1 29 .6	1 .9518	2.8
10	44 .1	1 .6450		10	1 2 .8	1 .7978		10	1 30 .2	1 .9546	
20	44 .4	1 .6476		20	1 3 .1	1 .8003		20	1 30 .7	1 .9574	
30	44 .6	1 .6502		30	1 3 .5	1 .8028		30	1 31 .3	1 .9602	
40	45 .0	1 .6528		40	1 3 .8	1 .8054		40	1 31 .9	1 .9629	
50	45 .2	1 .6554		50	1 4 .2	1 .8079		50	1 32 .5	1 .9657	
38 00	45 .6	1 .6580	2.6	48 00	1 4 .6	1 .8105	2.5	58 00	1 33 .1	1 .9685	2.8
10	45 .8	1 .6606		10	1 4 .9	1 .8130		10	1 33 .7	1 .9713	
20	46 .1	1 .6632		20	1 5 .3	1 .8155		20	1 34 .3	1 .9741	
30	46 .3	1 .6658		30	1 5 .7	1 .8181		30	1 34 .9	1 .9769	
40	46 .6	1 .6684		40	1 6 .1	1 .8206		40	1 35 .5	1 .9797	
50	46 .9	1 .6710		50	1 6 .5	1 .8231		50	1 36 .1	1 .9825	
39 00	47 .2	1 .6735	2.6	49 00	1 6 .9	1 .8257	2.5	59 00	1 36 .7	1 .9854	2.9
10	47 .5	1 .6761		10	1 7 .3	1 .8282		10	1 37 .3	1 .9883	
20	47 .8	1 .6787		20	1 7 .7	1 .8308		20	1 38 .0	1 .9912	
30	48 .1	1 .6813		30	1 8 .1	1 .8333		30	1 38 .6	1 .9940	
40	48 .3	1 .6839		40	1 8 .5	1 .8359		40	1 39 .3	1 .9969	
50	48 .6	1 .6865		50	1 8 .9	1 .8385		50	1 39 .9	1 .9998	

TABLE 1. concluded.

0.760 Bar. Metrique. +10° Therm. Centigrade; or English Barometer 29.94. Fahrenheit's Thermometer 50°.

Zen. Dist.	Refractions.	Logarithms	Diff. for 1 Min.
60° 00'	1' 40".6	2.0026	
10	1 41.3	2.0055	
20	1 42.0	2.0084	
30	1 42.7	2.0113	2.9
40	1 43.4	2.0142	
50	1 44.1	2.0172	
61 00	1 44.8	2.0202	
10	1 45.5	2.0232	
20	1 46.2	2.0262	
30	1 46.9	2.0292	3.0
40	1 47.6	2.0322	
50	1 48.4	2.0352	
62 00	1 49.2	2.0381	
10	1 50.0	2.0412	
20	1 50.8	2.0443	
30	1 51.6	2.0474	3.1
40	1 52.4	2.0504	
50	1 53.2	2.0535	
63 00	1 53.9	2.0565	
10	1 54.7	2.0596	
20	1 55.6	2.0627	
30	1 56.4	2.0658	3.1
40	1 57.2	2.0689	
50	1 58.0	2.0721	
64 00	1 58.9	2.0753	
10	1 59.8	2.0785	
20	2 0.7	2.0817	
30	2 1.6	2.0849	3.2
40	2 2.5	2.0881	
50	2 3.4	2.0914	
65 00	2 4.3	2.0946	
10	2 5.2	2.0979	
20	2 6.2	2.1012	
30	2 7.2	2.1046	3.3
40	2 8.2	2.1079	
50	2 9.2	2.1112	
66 00	2 10.2	2.1145	
10	2 11.2	2.1179	
20	2 12.3	2.1213	
30	2 13.8	2.1248	3.4
40	2 14.4	2.1282	
50	2 15.4	2.1316	
67 00	2 16.5	2.1350	
10	2 17.6	2.1385	
20	2 18.7	2.1420	
30	2 19.8	2.1456	3.5
40	2 20.9	2.1491	
50	2 22.1	2.1526	
68 00	2 23.2	2.1561	
10	2 24.4	2.1598	
20	2 25.6	2.1634	
30	2 26.9	2.1691	3.7
40	2 28.1	2.1707	
50	2 29.3	2.1744	
69 00	2 30.6	2.1781	
10	2 31.9	2.1819	
20	2 33.2	2.1857	
30	2 34.6	2.1895	3.8
40	2 36.0	2.1933	
50	2 37.4	2.1971	
70 00	2 38.8	2.2009	
10	2 40.3	2.2048	
20	2 41.8	2.2088	3.9

Zen. Dist.	Refractions.	Logarithms.	Diff. for 1 Min.
70° 30'	2' 43".3	2.2127	3.9
40	2 44.8	2.2167	
50	2 46.3	2.2206	
71 00	2 47.7	2.2246	
10	2 49.3	2.2287	
20	2 51.0	2.2328	
30	2 52.6	2.2369	4.1
40	2 54.3	2.2410	
50	2 55.9	2.2451	
72 00	2 57.6	2.2493	
10	2 59.4	2.2536	
20	3 1.2	2.2579	
30	3 3.0	2.2622	4.3
40	3 4.8	2.2665	
50	3 6.6	2.2708	
73 00	3 8.5	2.2752	
10	3 10.5	2.2797	
20	3 12.5	2.2842	
30	3 14.5	2.2888	4.5
40	3 16.5	2.2933	
50	3 18.5	2.2978	
74 00	3 20.7	2.3024	
10	3 22.9	2.3072	
20	3 25.2	2.3119	
30	3 27.5	2.3166	4.8
40	3 29.8	2.3214	
50	3 32.1	2.3262	
75 00	3 34.3	2.3310	
10	3 36.8	2.3360	
20	3 39.4	2.3411	
30	3 42.0	2.3461	5.0
40	3 44.6	2.3512	
50	3 47.2	2.3562	
76 00	3 49.8	2.3613	
10	3 52.7	2.3667	
20	3 55.7	2.3720	
30	3 58.6	2.3774	5.4
40	4 1.5	2.3828	
50	4 4.5	2.3882	
77 00	4 7.5	2.3936	
10	4 10.5	2.3993	
20	4 14.3	2.4050	
30	4 17.7	2.4107	5.7
40	4 21.1	2.4164	
50	4 24.5	2.4221	
78 00	4 27.9	2.4280	
10	4 31.8	2.4341	
20	4 35.7	2.4403	
30	4 39.7	2.4464	6.1
40	4 43.7	2.4526	
50	4 47.7	2.4587	
79 00	4 51.7	2.4649	
10	4 56.3	2.4715	
20	5 1.0	2.4782	
30	5 5.7	2.4849	6.7
40	5 10.4	2.4916	
50	5 15.1	2.4983	
80 00	5 19.8	2.5049	
10	5 25.1	2.5120	
20	5 30.4	2.5190	
30	5 35.9	2.5262	7.3
40	5 41.5	2.5334	
50	5 47.4	2.5408	

Zen. Dist.	Refractions.	Logarithms.	Diff. for 1 Min.
81° 00'	5' 53".5	2.5483	7.5
10	5 59.9	2.5558	7.7
20	6 6.4	2.5635	7.9
30	6 13.1	2.5714	8.1
40	6 20.0	2.5795	8.2
50	6 27.1	2.5877	8.2
82 00	6 34.4	2.5959	8.2
10	6 41.9	2.6041	8.3
20	6 49.6	2.6124	8.5
30	6 57.7	2.6209	8.8
40	7 6.3	2.6297	9.1
50	7 15.3	2.6388	9.3
83 00	7 24.7	2.6481	9.6
10	7 34.7	2.6577	9.7
20	7 44.9	2.6674	9.8
30	7 55.6	2.6772	10.0
40	8 6.6	2.6872	10.1
50	8 18.2	2.6973	10.2
84 00	8 29.9	2.7075	10.4
10	8 42.3	2.7179	10.7
20	8 55.3	2.7286	11.0
30	9 4.0	2.7396	11.2
40	9 23.4	2.7508	11.4
50	9 38.4	2.7622	11.8
85 00	9 54.3	2.7740	12.0
10	10 10.9	2.7860	12.2
20	10 28.3	2.7982	12.5
30	10 46.7	2.8107	12.8
40	11 6.1	2.8235	13.2
50	11 26.6	2.8367	13.5
86 00	11 48.3	2.8502	13.8
10	12 11.3	2.8640	14.2
20	12 35.6	2.8782	14.5
30	13 1.3	2.8927	14.9
40	13 28.5	2.9076	15.3
50	13 57.3	2.9229	15.7
87 00	14 28.1	2.9386	16.1
10	15 0.9	2.9547	16.6
20	15 36.0	2.9713	17.0
30	16 13.4	2.9883	17.4
40	16 53.2	3.0057	18.1
50	17 36.3	3.0238	18.5
88 00	18 22.2	3.0423	19.0
10	19 11.5	3.0613	19.6
20	20 4.8	3.0809	20.1
30	21 1.9	3.1010	20.7
40	22 3.4	3.1217	21.2
50	23 9.6	3.1429	21.8
89 00	24 21.2	3.1647	22.4
10	25 38.6	3.1871	23.0
20	27 2.2	3.2101	23.4
30	28 32.0	3.2335	24.0
40	30 9.3	3.2575	24.5
50	31 54.3	3.2820	24.7
90 00	33 46.3	3.3067	24.9
10	35 45.9	3.3316	25.1
20	37 53.6	3.3567	25.3
30	40 10.0	3.3820	

TABLE 2.

LOGARITHMS of the FACTORS for the ENGLISH BAROMETER.								For FAHRENHEIT'S THERMOMETER.			
Barom.	Log.	Barom.	Log.	Barom.	Log.	Barom.	Log.	Therm.	Log.	Therm.	Log.
27.50	9.9633	29.00	9.9862	29.60	9.9950	30.20	0.0038	90°	9.9647	30°	0.0187
60	9.9648	01	9.9863	61	9.9952	21	0.0040	89	9.9655	29	0.0196
70	9.9663	02	9.9865	62	9.9953	22	0.0041	88	9.9664	28	0.0206
80	9.9679	03	9.9866	63	9.9955	23	0.0043	87	9.9672	27	0.0215
90	9.9694	04	9.9868	64	9.9956	24	0.0044	86	9.9681	26	0.0225
28.00	9.9710	05	9.9869	65	9.9958	25	0.0046	85	9.9689	25	0.0235
10	9.9725	06	9.9871	66	9.9959	26	0.0047	84	9.9698	24	0.0244
20	9.9741	07	9.9872	67	9.9961	27	0.0049	83	9.9706	23	0.0254
30	9.9757	08	9.9874	68	9.9962	28	0.0050	82	9.9715	22	0.0264
40	9.9772	09	9.9876	69	9.9964	29	0.0052	81	9.9723	21	0.0273
28.50	9.9787	29.10	9.9877	29.70	9.9965	30.30	0.0053	80	9.9732	20	0.0282
51	9.9788	11	9.9878	71	9.9967	31	0.0055	79	9.9741	19	0.0293
52	9.9790	12	9.9879	72	9.9968	32	0.0056	78	9.9749	18	0.0302
53	9.9792	13	9.9881	73	9.9970	33	0.0058	77	9.9758	17	0.0312
54	9.9793	14	9.9882	74	9.9971	34	0.0059	76	9.9766	16	0.0322
55	9.9795	15	9.9884	75	9.6973	35	0.0060	75	9.9775	15	0.0332
56	9.9796	16	9.9885	76	9.9974	36	0.0062	74	9.9784	14	0.0341
57	9.9798	17	9.9887	77	9.9976	37	0.0063	73	9.9793	13	0.0351
58	9.9799	18	9.9888	78	9.9977	38	0.0065	72	9.9802	12	0.0361
59	9.9800	19	8.9890	79	9.9979	39	0.0066	71	9.9811	11	0.0371
28.60	9.9802	29.20	9.9891	29.80	9.9980	30.40	0.0067	70	9.9820	10	0.0381
61	9.9803	21	9.9893	81	9.9981	41	0.0069	69	9.9829	9	0.0391
62	9.9805	22	9.9894	82	9.9983	42	0.0070	68	9.9839	8	0.0401
63	9.9806	23	9.9896	83	9.9984	43	0.0072	67	9.9846	7	0.0411
64	9.9808	24	9.9897	84	9.9986	44	0.0073	66	9.9855	6	0.0421
65	9.9809	25	9.9899	85	9.9987	45	0.0075	65	9.9864	5	0.0431
66	9.9811	26	9.9900	86	9.9989	46	0.0076	64	9.9873	4	0.0441
67	9.9812	27	9.9902	87	9.9990	47	0.0078	63	9.9882	3	0.0451
68	9.9814	28	9.9903	88	9.9992	48	0.0079	62	9.9891	2	0.0461
69	9.9815	29	9.9905	89	9.9993	49	0.0080	61	9.9900	1	0.0471
28.70	9.9817	29.30	9.9906	29.90	9.9995	30.50	0.0081	60	9.9909	0	0.0482
71	9.9819	31	9.9908	91	9.9996	51	0.0083	59	9.9918		
72	9.9820	32	9.9909	92	9.9998	52	0.0084	58	9.9927		
73	9.9822	33	9.9911	93	9.9999	53	0.0086	57	9.9936		
74	9.9823	34	9.9912	94	0.0000	54	0.0087	56	9.9945		
75	9.9825	35	9.9914	95	0.0002	55	0.0089	55	9.9954		
76	9.9826	36	9.9915	96	0.0003	56	0.0090	54	9.9963		
77	9.9828	37	9.9917	97	0.0005	57	0.0092	53	9.9972		
78	9.9829	38	9.9918	98	0.0006	58	0.0093	52	9.9981		
79	9.9831	39	9.9920	99	0.0008	59	0.0095	51	9.9990		
28.80	9.9832	29.40	9.9921	30.00	0.0009	30.60	0.0096	50	0.0000		
81	9.9834	41	9.9923	01	0.0010	70	0.0110	49	0.0009		
82	9.9835	42	9.9924	02	0.0012	80	0.0124	48	0.0018		
83	9.9837	43	9.9926	03	0.0013	90	0.0138	47	0.0027		
84	9.9838	44	9.9927	04	0.0015	31.00	0.0152	46	0.0036		
85	9.9840	45	9.9929	05	0.0016			45	0.0046		
86	9.9841	46	9.9930	06	0.0018			44	0.0055		
87	9.9843	47	9.9931	07	0.0019			43	0.0064		
88	9.9844	48	9.9933	08	0.0021			42	0.0073		
89	9.9845	49	9.9934	09	0.0023			41	0.0083		
28.90	9.9847	29.50	9.9935	30.10	0.0024			40	0.0092		
91	9.9848	51	9.9937	11	0.0025			39	0.0102		
92	9.9850	52	9.9938	12	0.0027			38	0.0111		
93	9.9851	53	9.9940	13	0.0028			37	0.0120		
94	9.9853	54	9.9941	14	0.0030			36	0.0130		
95	9.9854	55	9.9943	15	0.0031			35	0.0139		
96	9.9856	56	9.9944	16	0.0032			34	0.0148		
97	9.9857	57	9.9946	17	0.0034			33	0.0158		
98	9.9859	58	9.9947	18	0.0035			32	0.0168		
99	9.9860	59	9.9949	19	0.0037			31	0.0177		

Log. of Mean Refraction + log. Bar. + log. Ther. = log. of the true Refraction in Seconds.

TABLE 1.

Mean Refraction calculated at 58''.132967 for 45 Degrees of Apparent Zenith Distance × Tang. $\overline{Z-3.6342956r}$ so far as 87° Zen. Dist. below which r is reduced .00462 for each Minute.

Barometer 29.60. Fahrenheit's Thermometer 49° within; 45° without.

Zen. Dist.	Refract.	Diff.
0° 0'	0' 0''.00	.17
10	0 0.17	.17
20	0 0.34	.17
30	0 0.51	.17
40	0 0.68	.16
50	0 0.84	.17
1 0	0 1.01	.17
10	0 1.18	.17
20	0 1.35	.17
30	0 1.52	.17
40	0 1.69	.17
50	0 1.86	.17
2 0	0 2.03	.17
10	0 2.20	.17
20	0 2.37	.16
30	0 2.53	.17
40	0 2.70	.17
50	0 2.87	.17
3 0	0 3.04	.17
10	0 3.21	.17
20	0 3.38	.17
30	0 3.55	.17
40	0 3.72	.17
50	0 3.89	.17
4 0	0 4.06	.17
10	0 4.23	.17
20	0 4.40	.17
30	0 4.57	.17
40	0 4.74	.17
50	0 4.91	.17
5 0	0 5.08	.17
10	0 5.25	.17
20	0 5.42	.17
30	0 5.59	.17
40	0 5.76	.17
50	0 5.93	.17
6 0	0 6.10	.17
10	0 6.27	.17
20	0 6.44	.18
30	0 6.62	.17
40	0 6.79	.17
50	0 6.96	.17
7 0	0 7.13	.17
10	0 7.30	.17
20	0 7.47	.17
30	0 7.64	.18
40	0 7.82	.17
50	0 7.99	.17
8 0	0 8.16	.17
10	0 8.33	.17
20	0 8.50	.18
30	0 8.68	.17
40	0 8.85	.17
50	0 9.02	.18
9 0	0 9.20	.17
10	0 9.37	.17
20	0 9.54	.18
30	0 9.72	.17
40	0 9.89	.17
50	0 10.06	.18

Zen. Dist.	Refract.	Diff.
10° 0'	0' 10''.24	''.17
10	0 10.41	.18
20	0 10.59	.17
30	0 10.76	.18
40	0 10.94	.17
50	0 11.11	.18
11 0	0 11.29	.17
10	0 11.46	.18
20	0 11.64	.17
30	0 11.81	.18
40	0 11.99	.18
50	0 12.17	.17
12 0	0 12.34	.18
10	0 12.52	.18
20	0 12.70	.17
30	0 12.87	.18
40	0 13.05	.17
50	0 13.22	.18
13 0	0 13.40	.18
10	0 13.58	.18
20	0 13.76	.18
30	0 13.94	.18
40	0 14.12	.18
50	0 14.30	.18
14 0	0 14.48	.18
10	0 14.66	.18
20	0 14.84	.18
30	0 15.02	.18
40	0 15.20	.18
50	0 15.38	.18
15 0	0 15.56	.18
10	0 15.74	.18
20	0 15.92	.18
30	0 16.10	.18
40	0 16.28	.19
50	0 16.47	.18
16 0	0 16.65	.18
10	0 16.83	.18
20	0 17.01	.19
30	0 17.20	.18
40	0 17.38	.19
50	0 17.57	.18
17 0	0 17.75	.19
10	0 17.94	.18
20	0 18.12	.19
30	0 18.31	.18
40	0 18.49	.19
50	0 18.68	.18
18 0	0 18.86	.19
10	0 19.05	.19
20	0 19.24	.19
30	0 19.43	.18
40	0 19.61	.19
50	0 19.80	.19
19 0	0 19.99	.19
10	0 20.18	.19
20	0 20.37	.19
30	0 20.56	.19
40	0 20.75	.19
50	0 20.94	.19

Zen. Dist.	Refract.	Diff.
20° 0'	0' 21''.13	.19
10	0 21.32	.19
20	0 21.51	.20
30	0 21.71	.19
40	0 21.90	.19
50	0 22.09	.20
21 0	0 22.29	.19
10	0 22.48	.19
20	0 22.67	.20
30	0 22.87	.19
40	0 23.06	.20
50	0 23.26	.20
22 0	0 23.46	.19
10	0 23.65	.20
20	0 23.85	.20
30	0 24.05	.20
40	0 24.25	.20
50	0 24.45	.20
23 0	0 24.65	.20
10	0 24.85	.20
20	0 25.05	.20
30	0 25.25	.20
40	0 25.45	.20
50	0 25.65	.20
24 0	0 25.85	.20
10	0 26.05	.21
20	0 26.26	.20
30	0 26.46	.20
40	0 26.66	.21
50	0 26.87	.20
25 0	0 27.07	.21
10	0 27.28	.21
20	0 27.49	.20
30	0 27.69	.21
40	0 27.90	.21
50	0 28.11	.21
26 0	0 28.32	.21
10	0 28.53	.21
20	0 28.74	.21
30	0 28.95	.21
40	0 29.16	.21
50	0 29.37	.21
27 0	0 29.58	.22
10	0 29.80	.20
20	0 30.00	.22
30	0 30.22	.22
40	0 30.44	.21
50	0 30.65	.22
28 0	0 30.87	.22
10	0 31.09	.21
20	0 31.30	.22
30	0 31.52	.22
40	0 31.74	.22
50	0 31.96	.22
29 0	0 32.18	.22
10	0 32.40	.22
20	0 32.62	.23
30	0 32.85	.22
40	0 33.07	.22
50	0 33.29	.23

Zen. Dist.	Refract.	Diff.
30° 0'	0' 33''.52	.22
10	0 33.74	.23
20	0 33.97	.23
30	0 34.20	.22
40	0 34.42	.23
50	0 34.65	.23
31 0	0 34.88	.23
10	0 35.11	.23
20	0 35.34	.23
30	0 35.57	.24
40	0 35.81	.23
50	0 36.04	.23
32 0	0 36.27	.24
10	0 36.51	.23
20	0 36.74	.24
30	0 36.98	.24
40	0 37.22	.24
50	0 37.46	.24
33 0	0 37.70	.24
10	0 37.94	.24
20	0 38.18	.24
30	0 38.42	.24
40	0 38.66	.25
50	0 38.91	.24
34 0	0 39.15	.25
10	0 39.40	.24
20	0 39.64	.25
30	0 39.89	.25
40	0 40.14	.25
50	0 40.39	.25
35 0	0 40.64	.25
10	0 40.89	.26
20	0 41.15	.25
30	0 41.40	.26
40	0 41.66	.25
50	0 41.91	.26
36 0	0 42.17	.26
10	0 42.43	.26
20	0 42.69	.26
30	0 42.95	.26
40	0 43.21	.26
50	0 43.47	.27
37 0	0 43.74	.26
10	0 44.00	.27
20	0 44.27	.26
30	0 44.53	.27
40	0 44.80	.27
50	0 45.07	.27
38 0	0 45.34	.28
10	0 45.62	.27
20	0 45.89	.27
30	0 46.16	.28
40	0 46.44	.28
50	0 46.72	.28
39 0	0 47.00	.28
10	0 47.28	.28
20	0 47.56	.28
30	0 47.84	.28
40	0 48.12	.29
50	0 48.41	.28

Zen. Dist.	Refract.	Diff.
40° 0'	0' 48''.69	.29
10	0 48.98	.29
20	0 49.27	.29
30	0 49.56	.29
40	0 49.85	.30
50	0 50.15	.29
41 0	0 50.44	.30
10	0 50.74	.30
20	0 51.04	.30
30	0 51.34	.30
40	0 51.64	.30
50	0 51.94	.31
42 0	0 52.25	.30
10	0 52.55	.31
20	0 52.86	.31
30	0 53.17	.31
40	0 53.48	.31
50	0 53.79	.32
43 0	0 54.11	.31
10	0 54.42	.32
20	0 54.74	.32
30	0 55.06	.32
40	0 55.38	.32
50	0 55.70	.33
44 0	0 56.03	.32
10	0 56.35	.33
20	0 56.68	.33
30	0 57.01	.33
40	0 57.34	.34
50	0 57.68	.33
45 0	0 58.01	.34
10	0 58.35	.34
20	0 58.69	.34
30	0 59.03	.35
40	0 59.38	.34
50	0 59.72	.35
46 0	1 0.07	.35
10	1 0.42	.35
20	1 0.77	.36
30	1 1.13	.35
40	1 1.48	.36
50	1 1.84	.36
47 0	1 2.20	.37
10	1 2.57	.36
20	1 2.93	.37
30	1 3.30	.37
40	1 3.67	.37
50	1 4.04	.37
48 0	1 4.41	.38
10	1 4.79	.38
20	1 5.17	.38
30	1 5.55	.39
40	1 5.94	.39
50	1 6.33	.39
49 0	1 6.72	.39
10	1 7.11	.39
20	1 7.50	.40
30	1 7.90	.40
40	1 8.30	.40
50	1 8.70	.41

Zen. Dist.	Refract.	Diff.
50° 0'	1' 9''.11	.41
10	1 9.52	.41
20	1 9.93	.41
30	1 10.34	.42
40	1 10.76	.42
50	1 11.18	.42
51 0	1 11.60	.43
10	1 12.03	.43
20	1 12.46	.43
30	1 12.89	.44
40	1 13.33	.44
50	1 13.77	.44
52 0	1 14.21	.44
10	1 14.65	.45
20	1 15.10	.45
30	1 15.55	.46
40	1 16.01	.46
50	1 16.47	.46
53 0	1 16.93	.46
10	1 17.39	.47
20	1 17.86	.47
30	1 18.33	.48
40	1 18.81	.48
50	1 19.29	.49
54 0	1 19.78	.49
10	1 20.27	.49
20	1 20.76	.49
30	1 21.25	.50
40	1 21.75	.51
50	1 22.26	.51
55 0	1 22.77	.51
10	1 23.28	.51
20	1 23.79	.52
30	1 24.31	.53
40	1 24.84	.53
50	1 25.37	.54
56 0	1 25.91	.54
10	1 26.45	.54
20	1 26.99	.55
30	1 27.54	.55
40	1 28.09	.56
50	1 28.65	.56
57 0	1 29.21	.57
10	1 29.78	.57
20	1 30.35	.58
30	1 30.93	.58
40	1 31.51	.59
50	1 32.10	.59
58 0	1 32.69	.60
10	1 33.29	.61
20	1 33.90	.61
30	1 34.51	.62
40	1 35.13	.62
50	1 35.75	.63
59 0	1 36.38	.63
10	1 37.01	.64
20	1 37.65	.65
30	1 38.30	.65
40	1 38.95	.66
50	1 39.61	.67

Table 1.

Mean Refraction calculated at 58″. 132967 for 45 Degrees of apparent Zenith Distance × Tang. $\overline{Z-3\ .6342956r}$ so far as 87° Zen. Dist. below which r is reduced .00462 for each Minute.

Barometer 29.60. Fahrenheit's Thermometer 49° within; 45° without.

Zen. Dist.	Refract.	Diff.	Zen. Dist.	Refract.	Diff.	Zen. Dist.	Refract.	Diff.	Zen. Dist.	Refract.	Diff.	Zen. Dist.	Refract.	Diff.	Zen. Dist.	Refract.	Diff.
60° 0′	1′ 40″.28	0″.67	70° 0′	2′ 38″.34	0″.71	75° 0′	3′ 33″.73	1″.22	80° 0′	5′ 19″.18	2″.57	85° 0′	9′ 53″.03	8″.12	88° 54′	23′ 47″.40	22″.50
10	1 40 .95	0 .68	5	2 39 .05	0 .71	5	3 34 .95	1 .23	5	5 21 .75	2 .61	5	10 1 .15	8 .31	57	24 9 .90	23 .04
20	1 41 .63	0 .69	10	2 39 .76	0 .72	10	3 36 .18	1 .24	10	5 24 .36	2 .65	10	10 9 .46	8 .53	89 0	24 32 .94	15 .66
30	1 42 .32	0 .69	15	2 40 .48	0 .73	15	3 37 .42	1 .25	15	5 27 .01	2 .70	15	10 17 99	8 .74	2	24 48 .60	15 .91
40	1 43 .01	0 .70	20	2 41 .21	0 .73	20	3 38 .67	1 .27	20	5 29 .71	2 .73	20	10 26 .73	8 .96	4	25 4 .51	16 .16
50	1 43 .71	0 .71	25	2 41 .94	0 .74	25	3 39 .94	1 .28	25	5 32 .44	2 .77	25	10 35 .69	9 .19	6	25 20 .67	16 .42
61 0	1 44 .42	0 .72	30	2 42 .68	0 .74	30	3 41 .22	1 .30	30	5 35 .21	2 .82	30	10 44 .88	9 .43	8	25 37 .09	16 .68
10	1 45 .14	0 .72	35	2 43 .42	0 .75	35	3 42 .52	1 .31	35	5 38 .03	2 .87	35	10 54 .31	9 .67	10	25 53 .77	16 .94
20	1 45 .86	0 .73	40	2 44 .17	0 .76	40	3 43 .83	1 .32	40	5 40 .90	2 .91	40	11 3 .98	9 .92	12	26 10 .71	17 .20
30	1 46 .59	0 .74	45	2 44 .93	0 .76	45	3 45 .15	1 .34	45	5 43 .81	2 .96	45	11 13 .90	10 .19	14	26 27 .91	17 .49
40	1 47 .33	0 .75	50	2 45 .69	0 .77	50	3 46 .49	1 .35	50	5 46 .77	3 .01	50	11 24 .09	10 .45	16	26 45 .40	17 .76
50	1 48 .08	0 .75	55	2 46 .46	0 .77	55	3 47 .84	1 .37	55	5 49 .78	3 .05	55	11 34 .54	10 .73	18	27 3 .16	18 .04
62 0	1 48 .83	0 .76	71 0	2 47 .23	0 .78	76 0	3 49 .21	1 .38	81 0	5 52 .83	3 .10	86 0	11 45 .27	8 .80	20	27 21 .20	18 .32
10	1 49 .59	0 .77	5	2 48 .01	0 .79	5	3 50 .59	1 .40	5	5 55 .93	3 .16	4	11 54 .07	8 .99	22	27 39 .52	18 .62
20	1 50 .36	0 .78	10	2 48 .80	0 .79	10	3 51 .99	1 .42	10	5 59 .09	3 .21	8	12 3 .06	9 .18	24	27 58 .14	18 .91
30	1 51 .14	0 .79	15	2 49 .59	0 .80	15	3 53 .41	1 .43	15	6 2 .30	3 .26	12	12 12 .24	9 .39	26	28 17 .05	19 .21
40	1 51 .93	0 .80	20	2 50 .39	0 .81	20	3 54 .84	1 .45	20	6 5 .56	3 .32	16	12 21 .63	9 .59	28	28 36 .26	19 .52
50	1 52 .73	0 .80	25	2 51 .20	0 .81	25	3 56 .29	1 .46	25	6 8 .88	3 .38	20	12 31 .22	9 .80	30	28 55 .78	19 .82
63 0	1 53 .53	0 .82	30	2 52 .01	0 .82	30	3 57 .75	1 .48	30	6 12 .26	3 .43	24	12 41 .02	10 .02	32	29 15 .60	20 .14
10	1 54 .35	0 .82	35	2 52 .83	0 .82	35	3 59 .23	1 .50	35	6 15 .69	3 .50	28	12 51 .04	10 .24	34	29 35 .74	20 .45
20	1 55 .17	0 .83	40	2 53 .65	0 .84	40	4 0 .73	1 .52	40	6 19 .19	3 .55	32	13 1 .28	10 .47	36	29 56 .19	20 .78
30	1 56 .00	0 .85	45	2 54 .49	0 .84	45	4 2 .25	1 .53	45	6 22 .74	3 .62	36	13 11 .75	10 .71	38	30 16 .97	21 .10
40	1 56 .85	0 .85	50	2 55 .33	0 .85	50	4 3 .78	1 .55	50	6 26 .36	3 .68	40	13 22 .46	10 .96	40	30 38 .07	21 .44
50	1 57 .70	0 .86	55	2 56 .18	0 .85	55	4 5 .33	1 .56	55	6 30 .04	3 .75	44	13 33 .42	11 .20	42	30 59 .51	21 .77
64 0	1 58 .56	0 .87	72 0	2 57 .03	0 .86	77 0	4 6 .89	1 .59	82 0	6 33 .79	3 .82	48	13 44 .62	11 .46	44	31 21 .28	22 .11
10	1 59 .43	0 .89	5	2 57 .89	0 .87	5	4 8 .48	1 .61	5	6 37 .61	3 .89	52	13 56 .08	11 .73	46	31 43 .39	22 .46
20	2 0 .32	0 89	10	2 58 .76	0 .88	10	4 10 .09	1 .63	10	6 41 .50	3 .96	56	14 7 .81	11 .99	48	32 5 .85	22 .81
30	2 1 .21	0 .91	15	2 59 .64	0 .89	15	4 11 .72	1 .65	15	6 45 .46	4 .03	87 0	14 19 .80	13 .11	50	32 28 .66	23 .16
40	2 2 .12	0 .91	20	3 0 .53	0 .90	20	4 13 .37	1 .66	20	6 49 .49	4 .10	4	14 32 .91	13 .46	52	32 51 .82	23 .53
50	2 3 .03	0 .93	25	3 1 .43	0 .90	25	4 15 .03	1 .69	25	6 53 .59	4 .19	8	14 46 .37	13 .83	54	33 15 .35	23 .89
65 0	2 3 .96	0 .94	30	3 2 .33	0 .91	30	4 16 .72	1 .71	30	6 57 .78	4 .26	12	15 0 .20	14 .20	56	33 39 .24	24 .26
10	2 4 .90	0 .95	35	3 3 .24	0 .92	35	4 18 .43	1 .73	35	7 2 .04	4 .35	16	15 14 .40	14 .60	58	34 3 .50	24 .63
20	2 5 .85	0 .96	40	3 4 .16	0 .93	40	4 20 .16	1 .75	40	7 6 .39	4 .43	20	15 29 .00	15 .00	90 0	34 28 .13	25 .02
30	2 6 .81	0 .97	45	3 5 .09	0 .93	45	4 21 .91	1 .77	45	7 10 .82	4 .51	24	15 44 .00	15 .42	2	34 53 .15	25 .40
40	2 7 .78	0 .99	50	3 6 .02	0 .95	50	4 23 .68	1 .80	50	7 15 .33	4 .60	28	15 59 .42	15 .85	4	35 18 .55	25 .79
50	2 8 .77	1 .00	55	3 6 .97	0 .95	55	4 25 .48	1 .82	55	7 19 .93	4 .70	32	16 15 .27	16 .31	6	35 44 .34	26 .18
66 0	2 9 .77	1 .01	73 0	3 7 .92	0 .96	78 0	4 27 .30	1 .84	83 0	7 24 .63	4 .79	36	16 31 .58	16 .77	8	36 10 .52	26 .58
10	2 10 .78	1 .03	5	3 8 .88	0 .97	5	4 29 .14	1 .87	5	7 29 .42	4 .88	40	16 48 .35	17 .26	10	36 .37 .10	26 .98
20	2 11 .81	1 .04	10	3 9 .85	0 .98	10	4 31 .01	1 .89	10	7 34 .30	4 .99	44	17 5 .61	17 .76	12	37 4 .08	27 .39
30	2 12 .85	1 .05	15	3 10 .83	0 .99	15	4 32 .90	1 .91	15	7 39 .29	5 .09	48	17 23 .37	18 .27	14	37 31 .47	27 .81
40	2 13 .90	1 .07	20	3 11 .82	1 .00	20	4 34 .81	1 .94	20	7 44 .38	5 .19	52	17 41 .64	18 .82	16	37 59 .28	28 .22
50	2 14 .97	1 .08	25	3 12 .82	1 .00	25	4 36 .75	1 .97	25	7 49 .57	5 .30	56	18 0 .46	19 .37	18	38 27 .50	
67 0	2 16 .05	1 .09	30	3 13 .82	1 .02	30	4 38 .72	1 .99	30	7 54 .87	5 .41	88 0	18 19 .83	14 .91			
10	2 17 .14	1 .11	35	3 14 .84	1 .02	35	4 40 .71	2 .02	35	8 0 .28	5 .52	3	18 34 .74	15 .24			
20	2 18 .25	1 .13	40	3 15 .86	1 .04	40	4 42 .73	2 .05	40	8 5 .80	5 .65	6	18 49 .98	15 .59			
30	2 19 .38	1 .14	45	3 16 .90	1 .05	45	4 44 .78	2 .07	45	8 11 .45	5 .76	9	19 5 .57	15 .94			
40	2 20 .52	1 .16	50	3 17 .95	1 .06	50	4 46 .85	2 .11	50	8 17 .21	5 .89	12	19 21 .51	16 .29			
50	2 21 .68	1 .17	55	3 19 .01	1 .06	55	4 48 .96	2 .13	55	8 23 .10	6 .03	15	19 37 .80	16 .67			
68 0	2 22 .85	1 .19	74 0	3 20 .07	1 .08	79 0	4 51 .09	2 .16	84 0	8 29 .13	6 .15	18	19 54 .47	17 .06			
10	2 24 .04	1 .21	5	3 21 .15	1 .08	5	4 53 .25	2 .19	5	8 35 .28	6 .29	21	20 11 .53	17 .45			
20	2 25 .25	1 .22	10	3 22 .23	1 .10	10	4 55 .44	2 .23	10	8 41 .57	6 .43	24	20 28 .98	17 .84			
30	2 26 .47	1 .24	15	3 23 .33	1 .11	15	4 57 .67	2 .25	15	8 48 .00	6 .57	27	20 46 .82	18 .26			
40	2 27 .71	1 .26	20	3 24 .44	1 .12	20	4 59 .92	2 .29	20	8 54 .57	6 .73	30	21 5 .08	18 .68			
50	2 28 .97	1 .28	25	3 25 .56	1 .13	25	5 2 .21	2 .32	25	9 1 .30	6 .88	33	21 23 .76	19 .12			
69 0	2 30 .25	1 .30	30	3 26 .69	1 .15	30	5 4 .53	2 .35	30	9 8 .18	7 .05	36	21 42 .88	19 .57			
10	2 31 .55	1 .32	35	3 27 .84	1 .15	35	5 6 .88	2 .39	35	9 15 .23	7 .20	39	22 2 .45	20 .02			
20	2 32 .87	1 .34	40	3 28 .99	1 .17	40	5 9 .27	2 .42	40	9 22 .43	7 .38	42	22 22 .47	20 .49			
30	2 34 .21	1 .35	45	3 30 .16	1 .18	45	5 11 .69	2 .46	45	9 29 .81	7 .56	45	22 42 .96	20 .99			
40	2 35 .56	1 .38	50	3 31 .34	1 .19	50	5 14 .15	2 .50	50	9 37 .37	7 .73	48	23 3 .95	21 .47			
50	2 36 .94	1 .40	55	3 32 .53	1 .20	55	5 16 .65	2 .53	55	9 45 .10	7 .93	51	23 25 .42	21 .98			

Multiply the Mean Refraction found in this Table by the Sum of the Factors found in Table 2, and the product will be the proper correction to be applied, according to its Algebraic Sign, to the Mean Refraction.

FACTORS for the CORRECTION of the MEAN REFRACTION, by MR. GROOMBRIDGE. TABLE 2.

BAROMETER IN INCHES.

	−		+
28 . 60	. 0350	29 . 60	. 0000
62	. 0342	62	. 0007
64	. 0335	64	. 0014
66	. 0328	66	. 0020
68	. 0321	68	. 0027
28 . 70	. 0314	29 . 70	. 0034
72	. 0306	72	. 0041
74	. 0299	74	. 0047
76	. 0292	76	. 0054
78	. 0285	78	. 0061
28 . 80	. 0278	29 . 80	. 0068
82	. 0271	82	. 0074
84	. 0264	84	. 0081
86	. 0256	86	. 0088
88	. 0249	88	. 0095
28 . 90	. 0242	29 . 90	. 0101
92	. 0235	92	. 0108
94	. 0228	94	. 0115
96	. 0221	96	. 0122
98	. 0214	98	. 0128
29 . 0	. 0207	30 . 0	. 0135
2	. 0200	2	. 0142
4	. 0193	4	. 0149
6	. 0186	6	. 0155
8	. 0179	8	. 0162
29 . 10	. 0172	30 . 10	. 0169
12	. 0165	12	. 0176
14	. 0158	14	. 0182
16	. 0151	16	. 0189
18	. 0144	18	. 0196
29 . 20	. 0137	30 . 20	. 0203
22	. 0130	22	. 0210
24	. 0123	24	. 0216
26	. 0116	26	. 0223
28	. 0109	28	. 0230
29 . 30	. 0102	30 . 30	. 0237
32	. 0096	32	. 0243
34	. 0089	34	. 0250
36	. 0082	36	. 0257
38	. 0075	38	. 0264
29 . 40	. 0068	30 . 40	. 0270
42	. 0061	42	. 0277
44	. 0054	44	. 0284
46	. 0048	46	. 0291
48	. 0041	48	. 0297
29 . 50	. 0034	30 . 50	. 0304
52	. 0027	52	. 0311
54	. 0020	54	. 0318
56	. 0014	56	. 0324
58	. 0007	58	. 0331

FAHRENHEIT'S THERMOMETER.

	Within. +	Without +		Within. −	Without −
21°. 5	. 0632	. 0470	49°. 0	. 0000	. 0080
22 . 0	. 0621	. 0460	49 . 5	. 0011	. 0090
22 . 5	. 0609	. 0450	50 . 0	. 0022	. 0100
23 . 0	. 0598	. 0440	50 . 5	. 0033	. 0110
23 . 5	. 0586	. 0430	51 . 0	. 0044	. 0120
24 . 0	. 0575	. 0420	51 . 5	. 0055	. 0130
24 . 5	. 0563	. 0410	52 . 0	. 0066	. 0140
25 . 0	. 0552	. 0400	52 . 5	. 0077	. 0150
25 . 5	. 0540	. 0390	53 . 0	. 0088	. 0160
26 . 0	. 0529	. 0380	53 . 5	. 0099	. 0170
26 . 5	. 0517	. 0370	54 . 0	. 0110	. 0180
27 . 0	. 0506	. 0360	54 . 5	. 0121	. 0190
27 . 5	. 0494	. 0350	55 . 0	. 0132	. 0200
28 . 0	. 0483	. 0340	55 . 5	. 0143	. 0210
28 . 5	. 0471	. 0330	56 . 0	. 0154	. 0220
29 . 0	. 0460	. 0320	56 . 5	. 0165	. 0230
29 . 5	. 0448	. 0310	57 . 0	. 0176	. 0240
30 . 0	. 0437	. 0300	57 . 5	. 0187	. 0250
30 . 5	. 0425	. 0290	58 . 0	. 0198	. 0260
31 . 0	. 0414	. 0280	58 . 5	. 0209	. 0270
31 . 5	. 0402	. 0270	59 . 0	. 0220	. 0280
32 . 0	. 0391	. 0260	59 . 5	. 0231	. 0290
32 . 5	. 0379	. 0250	60 . 0	. 0242	. 0300
33 . 0	. 0368	. 0240	60 . 5	. 0253	. 0310
33 . 5	. 0356	. 0230	61 . 0	. 0264	. 0320
34 . 0	. 0345	. 0220	61 . 5	. 0275	. 0330
34 . 5	. 0333	. 0210	62 . 0	. 0286	. 0340
35 . 0	. 0322	. 0200	62 . 5	. 0297	. 0350
35 . 5	. 0310	. 0190	63 . 0	. 0308	. 0360
36 . 0	. 0299	. 0180	63 . 5	. 0319	. 0370
36 . 5	. 0287	. 0170	64 . 0	. 0330	. 0380
37 . 0	. 0276	. 0160	64 . 5	. 0341	. 0390
37 . 5	. 0264	. 0150	65 . 0	. 0352	. 0400
38 . 0	. 0253	. 0140	65 . 5	. 0363	. 0410
38 . 5	. 0241	. 0130	66 . 0	. 0374	. 0420
39 . 0	. 0230	. 0120	66 . 5	. 0385	. 0430
39 . 5	. 0218	. 0110	67 . 0	. 0396	. 0440
40 . 0	. 0207	. 0100	67 . 5	. 0407	. 0450
40 . 5	. 0195	. 0090	68 . 0	. 0418	. 0460
41 . 0	. 0184	. 0080	68 . 5	. 0429	. 0470
41 . 5	. 0172	. 0070	69 . 0	. 0440	. 0480
42 . 0	. 0161	. 0060	69 . 5	. 0451	. 0490
42 . 5	. 0149	. 0050	70 . 0	. 0462	. 0500
43 . 0	. 0138	. 0040	70 . 5	. 0473	. 0510
43 . 5	. 0126	. 0030	71 . 0	. 0484	. 0520
44 . 0	. 0115	. 0020	71 . 5	. 0495	. 0530
44 . 5	. 0103	. 0010	72 . 0	. 0506	. 0540
45 . 0	. 0092	—	72 . 5	. 0517	. 0550
45 . 5	. 0080	. 0010	73 . 0	. 0528	. 0560
46 . 0	. 0069	. 0020	73 . 5	. 0539	. 0570
46 . 5	. 0057	. 0030	74 . 0	. 0550	. 0580
47 . 0	. 0046	. 0040	74 . 5	. 0561	. 0590
47 . 5	. 0034	. 0050	75 . 0	. 0572	. 0600
48 . 0	. 0023	. 0060	75 . 5	. 0583	. 0610
48 . 5	. 0011	. 0070	76 . 0	. 0594	. 0620

MEAN REFRACTIONS by GI. PIAZZI.

Zen. Dist.	Mean Refractions.	Diff. in 1°.	Zen. Dist.	Mean Refractions.	Diff. in 1°.
1°	0′ 1″. 0	1″. 0	46	0′ 59″. 2	2″. 2
2	0 2 . 0	1 . 0	47	1 1 . 4	2 . 2
3	0 3 . 0	1 . 0	48	1 3 . 6	2 . 2
4	0 4 . 0	1 . 0	49	1 5 . 8	2 . 4
5	0 5 . 0	1 . 0	50	1 8 . 2	2 . 4
6	0 6 . 0	1 . 0	51	1 10 . 6	2 . 7
7	0 7 . 0	1 . 1	52	1 13 . 3	2 . 7
8	0 8 . 1	1 . 0	53	1 16 . 0	2 . 8
9	0 9 . 1	1 . 1	54	1 18 . 8	3 . 6
10	0 10 . 2	1 . 0	55	1 22 . 4	3 . 2
11	0 11 . 2	1 . 0	56	1 25 . 6	3 . 2
12	0 12 . 2	1 . 0	57	1 28 . 8	3 . 5
13	0 13 . 2	1 . 1	58	1 32 . 3	3 . 7
14	0 14 . 3	1 . 1	59	1 36 . 0	3 . 8
15	0 15 . 4	1 . 0	60	1 39 . 8	4 . 2
16	0 16 . 4	1 . 1	61	1 44 . 0	4 . 4
17	0 17 . 5	1 . 1	62	1 48 . 4	4 . 8
18	0 18 . 6	1 . 1	63	1 53 . 2	4 . 9
19	0 19 . 7	1 . 1	64	1 58 . 1	5 . 4
20	0 20 . 8	1 . 2	65	2 3 . 5	5 . 8
21	0 22 . 0	1 . 1	66	2 9 . 3	6 . 1
22	0 23 . 1	1 . 2	67	2 15 . 4	7 . 0
23	0 24 . 3	1 . 2	68	2 22 . 4	7 . 3
24	0 25 . 5	1 . 2	69	2 29 . 7	8 . 1
25	0 26 . 7	1 . 3	70	2 37 . 8	9 . 0
26	0 28 . 0	1 . 3	71	2 46 . 8	9 . 7
27	0 29 . 3	1 . 3	72	2 56 . 5	9 . 7
28	0 30 . 6	1 . 3	73	3 6 . 2	12 . 1
29	0 31 . 9	1 . 3	74	3 18 . 3	13 . 6
30	0 33 . 2	1 . 3	75	3 31 . 9	15 . 4
31	0 34 . 5	1 . 3	76	3 47 . 3	17 . 6
32	0 35 . 8	1 . 4	77	4 4 . 9	19 . 4
33	0 37 . 2	1 . 4	78	4 24 . 3	23 . 6
34	0 38 . 6	1 . 5	79	4 47 . 9	28 . 2
35	0 40 . 1	1 . 5	80	5 16 . 1	31 . 3
36	0 41 . 6	1 . 6	81	5 47 . 4	40 . 9
37	0 43 . 2	1 . 6	82	6 28 . 3	51 . 2
38	0 44 . 8	1 . 6	83	7 19 . 5	1′ 5 . 4
39	0 46 . 4	1 . 7	84	8 24 . 9	1 20 . 5
40	0 48 . 1	1 . 7	85	9 45 . 4	1 57 . 2
41	0 49 . 8	1 . 8	86	11 42 . 6	2 42 . 5
42	0 51 . 6	1 . 8	87	14 25 . 1	3 37 . 6
43	0 53 . 4	1 . 9	88	18 2 . 7	5 43 . 4
44	0 55 . 3	1 . 9	89	23 46 . 1	8 16 . 9
45	0 57 . 2	2 . 0	90	32 3 . 0	

These Mean Refractions were deduced from actual observations, without reference to any particular theory.

[Log. tan. Zen. Dis. + log. Bar. + log. Ther.—Tab. 3. = Refraction, when the Zenith Distance does not exceed 80°.]

TABLE 1.

Log. Barometer.

Eng. Bar.	Log.	Eng. Bar.	Log.	Eng. Bar.	Log.
29°.00	1.4623	29°.60	1.4713	30°.20	1.4800
29.01	1.4624	29.61	1.4715	30.21	1.4801
29.02	1.4626	29.62	1.4716	30.22	1.4803
29.03	1.4627	29.63	1.4718	30.23	1.4804
29.04	1.4629	29.64	1.4719	30.24	1.4806
29.05	1.4630	29.65	1.4721	30.25	1.4807
29.06	1.4632	29.66	1.4722	30.26	1.4808
29.07	1.4633	29.67	1.4724	30.27	1.4810
29.08	1.4635	29.68	1.4725	30.28	1.4811
29.09	1.4637	29.69	1.4727	30.29	1.4813
29.10	1.4639	29.70	1.4728	30.30	1.4814
29.11	1.4640	29.71	1.4729	30.31	1.4816
29.12	1.4642	29.72	1.4731	30.32	1.4817
29.13	1.4643	29.73	1.4732	30.33	1.4819
29.14	1.4645	29.74	1.4734	30.34	1.4820
29.15	1.4646	29.75	1.4735	30.35	1.4822
29.16	1.4648	29.76	1.4736	30.36	1.4823
29.17	1.4649	29.77	1.4738	30.37	1.4825
29.18	1.4651	29.78	1.4739	30.38	1.4826
29.19	1.4652	29.79	1.4741	30.39	1.4828
29.20	1.4654	29.80	1.4742	30.40	1.4829
29.21	1.4655	29.81	1.4744	30.41	1.4830
29.22	1.4657	29.82	1.4745	30.42	1.4832
29.23	1.4658	29.83	1.4747	30.43	1.4833
29.24	1.4660	29.84	1.4748	30.44	1.4835
29.25	1.4661	29.85	1.4750	30.45	1.4836
29.26	1.4663	29.86	1.4751	30.46	1.4837
29.27	1.4664	29.87	1.4753	30.47	1.4839
29.28	1.4666	29.88	1.4754	30.48	1.4840
29.29	1.4667	29.89	1.4756	30.49	1.4842
29.30	1.4669	29.90	1.4757	30.50	1.4843
29.31	1.4670	29.91	1.4758	30.51	1.4844
29.32	1.4672	29.92	1.4760	30.52	1.4846
29.33	1.4673	29.93	1.4761	30.53	1.4847
29.34	1.4675	29.94	1.4763	30.54	1.4849
29.35	1.4676	29.95	1.4764	30.55	1.4850
29.36	1.4678	29.96	1.4765	30.56	1.4851
29.37	1.4679	29.97	1.4767	30.57	1.4853
29.38	1.4681	29.98	1.4768	30.58	1.4854
29.39	1.4682	29.99	1.4770	30.59	1.4856
29.40	1.4684	30.00	1.4771		
29.41	1.4685	30.01	1.4773		
29.42	1.4687	30.02	1.4774		
29.43	1.4688	30.03	1.4776		
29.44	1.4690	30.04	1.4777		
29.45	1.4691	30.05	1.4779		
29.46	1.4692	30.06	1.4780		
29.47	1.4694	30.07	1.4782		
29.48	1.4695	30.08	1.4783		
29.49	1.4697	30.09	1.4785		
29.50	1.4698	30.10	1.4786		
29.51	1.4700	30.11	1.4787		
29.52	1.4701	30.12	1.4789		
29.53	1.4703	30.13	1.4790		
29.54	1.4704	30.14	1.4792		
29.55	1.4706	30.15	1.4793		
29.56	1.4707	30.16	1.4794		
29.57	1.4709	30.17	1.4796		
29.58	1.4710	30.18	1.4797		
29.59	1.4712	30.19	1.4799		

TABLE 5.

$1-,0001\,(t-50)-\frac{57'',72}{29\ ,60}$

Ther. near Bar.

Ther.	Log.
20°	0.2913
30	0.2909
40	0.2904
50	0.2960
60	0.2896
70	0.2891
80	0.2887

This Table is used for Stars below 10° of app. Altitude.

TABLE 2. Log. Thermom.

$$\left[=\log.\frac{1,0375}{1+,002083(t-32)}\times 1-,0001\,(t-50)\times\frac{57'',72}{29\ .60}\right].$$

Fahr. Ther.	Log.	Fahr. Ther.	Log.	Fahr. Ther.	Log.	Fahr. Ther.	Log.	Fahr. Ther.	Log.	Fahr. Ther.	Log.
10°.0	0.3283	22°.0	0.3163	34°.0	0.3048	46°.0	0.2937	58°.0	0.2827	70°.0	0.2720
10.2	0.3281	22.2	0.3161	34.2	0.3046	46.2	0.2935	58.2	0.2825	70.2	0.2718
10.4	0.3279	22.4	0.3159	34.4	0.3044	46.4	0.2933	58.4	0.2823	70.4	0.2716
10.6	0.3277	22.6	0.3158	34.6	0.3043	46.6	0.2932	58.6	0.2822	70.6	0.2715
10.8	0.3275	22.8	0.3156	34.8	0.3041	46.8	0.2930	58.8	0.2820	70.8	0.2713
11.0	0.3273	23.0	0.3154	35.0	0.3039	47.0	0.2928	59.0	0.2818	71.0	0.2711
11.2	0.3271	23.2	0.3152	35.2	0.3037	47.2	0.2926	59.2	0.2816	71.2	9.2709
11.4	0.3269	23.4	0.3150	35.4	0.3035	47.4	0.2924	59.4	0.2814	71.4	0.2708
11.6	0.3267	23.6	0.3148	35.6	0.3034	47.6	0.2923	59.6	0.2813	71.6	0.2706
11.8	0.3265	23.8	0.3146	35.8	0.3032	47.8	0.2921	59.8	0.2811	71.8	0.2705
12.0	0.3263	24.0	0.3144	36.0	0.3030	48.0	0.2919	60.0	0.2809	72.0	0.2703
12.2	0.3261	24.2	0.3142	36.2	0.3028	48.2	0.2917	60.2	0.2807	72.2	0.2701
12.4	0.3259	24.4	0.3140	36.4	0.3026	48.4	0.2915	60.4	0.2805	72.4	0.2699
12.6	0.3257	24.6	0.3138	36.6	0.3024	48.6	0.2914	60.6	0.2804	72.6	0.2698
12.8	0.3255	24.8	0.3136	36.8	0.3022	48.8	0.9912	60.8	0.2802	72.8	0.2696
13.0	0.3253	25.0	0.3134	37.0	0.3020	49.0	0.2910	61.0	0.2800	73.0	0.2694
13.2	0.3251	25.2	0.3132	37.2	0.3018	49.2	0.2908	61.2	0.2798	73.2	0.2692
13.4	0.3249	25.4	0.3130	37.4	0.3016	49.4	0.2906	61.4	0.2796	73.4	0.2690
13.6	0.3247	25.6	0.3128	37.6	0.3015	49.6	0.2904	61.6	0.2795	73.6	0.2689
13.8	0.3245	25.8	0.3126	37.8	0.3013	49.8	0.2902	61.8	0.2793	73.8	0.2687
14.0	0.3243	26.0	0.3124	38.0	0.3011	50.0	0.2900	62.0	0.2791	74.0	0.2685
14.2	0.3241	26.2	0.3122	38.2	0.3009	50.2	0.2898	62.2	0.2789	74.2	0.2683
14.4	0.3239	26.4	0.3120	38.4	0.3007	50.4	0.2896	62.4	0.2787	74.4	0.2682
14.6	0.3237	26.6	0.3118	38.6	0.3005	50.6	0.2895	62.6	0.2786	74.6	0.2680
14.8	0.3235	26.8	0.3116	38.8	0.3003	50.8	0.2893	62.8	0.2784	74.8	0.2679
15.0	0.3233	27.0	0.3114	39.0	0.3001	51.0	0.2891	63.0	0.2782	75.0	0.2677
15.2	0.3231	27.2	0.3112	39.2	0.2999	51.2	0.2889	63.2	0.2780	75.2	0.2675
15.4	0.3229	27.4	0.3110	39.4	0.2997	51.4	0.2887	63.4	0.2778	75.4	0.2673
15.6	0.3227	27.6	0.3109	39.6	0.2996	51.6	0.2885	63.6	0.2777	75.6	0.2672
15.8	0.3225	27.8	0.3107	39.8	0.2994	51.8	0.2883	63.8	0.2775	75.8	0.2670
16.0	0.3223	28.0	0.3105	40.0	0.2292	52.0	0.2881	64.0	0.2773	76.0	0.2668
16.2	0.3221	28.2	0.3103	40.2	0.2990	52.2	0.2879	64.2	0.2771	76.2	0.2666
16.4	0.3219	28.4	0.3101	40.4	0.2988	52.4	0.2877	64.4	0.2769	76.4	0.2665
16.6	0.3217	28.6	0.3099	40.6	0.2987	52.6	0.2876	64.6	0.2768	76.6	0.2663
16.8	0.3215	28.8	0.3097	40.8	0.2985	52.8	0.2874	64.8	0.2766	76.8	0.2662
17.0	0.3213	29.0	0.3095	41.0	0.2983	53.0	0.2872	65.0	0.2764	77.0	0.2660
17.2	0.3211	29.2	0.3093	41.2	0.2981	53.2	0.2870	65.2	0.2762	77.2	0.2658
17.4	0.3209	29.4	0.3091	41.4	0.2979	53.4	0.2868	65.4	0.2760	77.4	0.2657
17.6	0.3207	29.6	0.3090	41.6	0.2978	53.6	0.2867	65.6	0.2759	77.6	0.2655
17.8	0.3205	29.8	0.3088	41.8	0.2976	53.8	0.2865	65.8	0.2757	77.8	0.2654
18.0	0.3203	30.0	0.3086	42.0	0.2974	54.0	0.2863	66.0	0.2755	78.0	0.2652
18.2	0.3201	30.2	0.3084	42.2	0.2972	54.2	0.2861	66.2	0.2753	78.2	0.2650
18.4	0.3199	30.4	0.3082	42.4	0.2970	54.4	0.2859	66.4	0.2751	78.4	0.2649
18.6	0.3197	30.6	0.3080	42.6	0.2969	54.6	0.2858	66.6	0.2750	78.6	0.2647
18.8	0.3195	30.8	0.3078	42.8	0.2967	54.8	0.2856	66.8	0.2748	78.8	0.2646
19.0	0.3193	31.0	0.3076	43.0	0.2965	55.0	0.2854	67.0	0.2746	79.0	0.2644
19.2	0.3191	31.2	0.3074	43.2	0.2963	55.2	0.2852	67.2	0.2744	79.2	0.2642
19.4	0.3189	31.4	0.3072	43.4	0.2961	55.4	0.2850	67.4	0.2742	79.4	0.2641
19.6	0.3187	31.6	0.3071	43.6	0.2960	55.6	0.2849	67.6	0.2741	79.6	0.2639
19.8	0.3185	31.8	0.3069	43.8	0.2958	55.8	0.2847	67.8	0.2739	79.8	0.2638
20.0	0.3183	32.0	0.3067	44.0	0.2956	56.0	0.2845	68.0	0.2737	80.0	0.2636
20.2	0.3181	32.2	0.3065	44.2	0.2954	56.2	0.2843	68.2	0.2735	80.2	0.2634
20.4	0.3179	32.4	0.3063	44.4	0.2952	56.4	0.2841	68.4	0.2733	80.4	0.2632
20.6	0.3177	32.6	0.3062	44.6	0.2950	56.6	0.2840	68.6	0.2732	80.6	0.2631
20.8	0.3175	32.8	0.3060	44.8	0.2948	56.8	0.2838	68.8	0.2730	80.8	0.2629
21.0	0.3173	33.0	0.3058	45.0	0.2946	57.0	0.2836	69.0	0.2728	81.0	0.2627
21.2	0.3171	33.2	0.3056	45.2	0.2944	57.2	0.2834	69.2	0.2726	81.2	0.2625
21.4	0.3169	33.4	0.3054	45.4	0.2942	57.4	0.2832	69.4	0.2725	81.4	0.2624
21.6	0.3167	33.6	0.3052	45.6	0.2941	57.6	0.2831	69.6	0.2723	81.6	0.2622
21.8	0.3165	33.8	0.3050	45.8	0.2939	57.8	0.2829	69.8	0.2722	81.8	0.2020

TABLE 3.

Zen. Dis.	28°. 5	29°. 0	29°. 5	30°. 0	30°. 5
80°	10″. 5	10″. 7	10″. 9	11″. 1	11″. 4
79	8. 1	8. 3	8. 5	8. 7	8. 9
78	6. 3	6. 4	6. 6	6. 7	6. 9
77	5. 1	5. 2	5. 3	5. 4	5. 6
76	4. 1	4. 2	4. 3	4. 4	4. 5
75	3. 4	3. 4	3. 5	3. 6	3. 7
74	3. 0	3. 0	3. 1	3. 1	3. 2
73	2. 5	2. 5	2. 6	2. 6	2. 6
72	2. 1	2. 1	2. 2	2. 2	2. 2
71	1. 8	1. 8	1. 9	1. 9	1. 9
70	1. 5	1. 5	1. 5	1. 6	1. 6
69	1. 3	1. 3	1. 3	1. 4	1. 4
68	1. 2	1. 2	1. 2	1. 2	1. 2
67	1. 0	1. 0	1. 0	1. 0	1. 0
66	0. 9	0. 9	0. 9	0. 9	0. 9
65	0. 8	0. 8	0. 8	0. 8	0. 8
64	0. 7	0. 7	0. 7	0. 7	0. 7
63	0. 6	0. 6	0. 6	0. 6	0. 6
62	0. 6	0. 6	0. 6	0. 6	0. 6
61	0. 5	0. 5	0. 5	0. 5	0. 5
60	0. 5	0. 5	0. 5	0. 5	0. 5
59	0. 4	0. 4	0. 4	0. 4	0. 4
58	0. 4	0. 4	0. 4	0. 4	0. 4
57	0. 4	0. 4	0. 4	0. 4	0. 4
56	0. 3	0. 3	0. 3	0. 3	0. 3
55	0. 3	0. 3	0. 3	0. 3	0. 3
54	0. 3	0. 3	0. 3	0. 3	0. 3
53	0. 3	0. 3	0. 3	0. 3	0. 3
52	0. 2	0. 2	0. 2	0. 2	0. 2
51	0. 2	0. 2	0. 2	0. 2	0. 2
50	0. 2	0. 2	0. 2	0. 2	0. 2
45	0. 2	0. 2	0. 2	0. 2	0. 2
40	0. 1	0. 1	0. 1	0. 1	0. 1
30	0. 0	0. 0	0. 0	0. 0	0. 0
20	0. 0	0. 0	0. 0	0. 0	0. 0
10	0. 0	0. 0	0. 0	0. 0	0. 0

$$\text{Log. } \kappa\left(\frac{l'}{a \sin. 1''}\right). \tan. \theta'.$$

TABLE 4.

Zen. Dis.		Log.	Diff. in 1′.
80°	0′	1. 3605	6. 90
81	0	1. 4031	7. 62
82	0	1. 4488	8. 21
83	0	1. 4981	9. 07
84	0	1. 5524	10. 13
85	0	1. 6132	11. 0
85	30	1. 6462	11. 37
86	0	1. 6803	12. 07
86	30	1. 7165	13. 23
87	0	1. 7562	13. 75
87	20	1. 7837	14. 55
87	40	1. 8128	14. 65
88	0	1. 8421	15. 50
88	20	1. 8731	16. 05
88	40	1. 9052	16. 75
89	0	1. 9387	17. 25
89	20	1. 9732	18. 0
89	40	2. 0092	18. 60
90	0	2. 0464	

This Table is used for Stars below 10° of apparent Altitude.

Comparative Scales of three different Thermometers.

Fahr.	Réaum.	Centes.	Fahr.	Réaum.	Centes.
	−	−		+	+
0°	14″. 2	17″. 8	45°	5″. 8	7″. 2
1	13. 8	17. 2	46	6. 2	7. 8
2	13. 3	16. 7	47	6. 7	8. 3
3	12. 9	16. 1	48	7. 1	8. 9
4	12. 5	15. 6	49	7. 6	9. 4
5	12. 0	15. 0	50	8. 0	10. 0
6	11. 6	14. 4	51	8. 4	10. 6
7	11. 1	13. 9	52	8. 9	11. 1
8	10. 7	13. 3	53	9. 3	11. 7
9	10. 2	12. 8	54	9. 8	12. 2
10	9. 8	12. 2	55	10. 2	12. 8
11	9. 3	11. 7	56	10. 7	13. 3
12	8. 9	11. 1	57	11. 1	13. 9
13	8. 4	10. 6	58	11. 6	14. 4
14	8. 0	10. 0	59	12. 0	15. 0
15	7. 6	9. 4	60	12. 4	15. 6
16	7. 1	8. 9	61	12. 9	16. 1
17	6. 7	8. 3	62	13. 3	16. 7
18	6. 2	7. 8	63	13. 8	17. 2
19	5. 8	7. 2	64	14. 2	17. 8
20	5. 3	6. 7	65	14. 7	18. 3
21	4. 9	6. 1	66	15. 1	18. 9
22	4. 4	5. 6	67	15. 6	19. 4
23	4. 0	5. 0	68	16. 0	20. 0
24	3. 6	4. 4	69	16. 4	20. 6
25	3. 1	3. 9	70	16. 9	21. 1
26	2. 7	3. 3	71	17. 3	21. 7
27	2. 2	2. 8	72	17. 8	22. 2
28	1. 8	2. 2	73	18. 2	22. 8
29	1. 3	1. 7	74	18. 7	23. 3
30	0. 9	1. 1	75	19. 1	23. 9
31	0. 4	0. 6	76	19. 6	24. 4
32	0. 0	0. 0	77	20. 0	25. 0
	+	+	78	20. 4	25. 6
33	0. 4	0. 6	79	20. 9	26. 1
34	0. 9	1. 1	80	21. 3	26. 7
35	1. 3	1. 7	81	21. 8	27. 2
36	1. 8	2. 2	82	22. 2	27. 8
37	2. 2	2. 8	83	22. 7	28. 3
38	2. 7	3. 3	84	23. 1	28. 9
39	3. 1	3. 9	85	23. 6	29. 4
40	3. 6	4. 4	86	24. 0	30. 0
41	4. 0	5. 0	87	24. 4	30. 6
42	4. 4	5. 6	88	24. 9	31. 1
43	4. 9	6. 1	89	25. 3	31. 7
44	5. 3	6. 7	90	25. 8	32. 2

Use of Tables 4. and 5. for Stars at low Altitudes.

Log. A. in min. of a degree = Table 4. + Ar. Com. Tab. 2. + Tab. 5.

Log. B. = Tab. 4. + Tab. 2. + Log. Bar. + 7.2773.

Log. Refrac. in Seconds = Tab. 2. + Log. Bar. + tan. (Z. D. − A + B.)

Comparative Scales of three different Barometers.

Millim.	French. Inches.	French. Lines.	English. Inches.
736	27	2. 27	28. 977
737	27	2. 71	29. 016
738	27	3. 15	29. 056
739	27	3. 60	29. 095
740	27	4. 04	29. 134
741	27	4. 48	29. 174
742	27	4. 93	29. 213
743	27	5. 37	29. 253
744	27	5. 81	29. 292
745	27	6. 26	29. 331
746	27	6. 70	29. 371
747	27	7. 14	29. 410
748	27	7. 59	29. 449
749	27	8. 03	29. 489
750	27	8. 47	29. 528
751	27	8. 92	29. 568
752	27	9. 36	29. 607
753	27	9. 80	29. 646
754	27	10. 25	29. 686
755	27	10. 69	29. 725
756	27	11. 13	29. 764
757	27	11. 58	29. 804
758	28	0. 02	29. 843
759	28	0. 46	29. 882
760	28	0. 91	29. 922
761	28	1. 35	29. 961
762	28	1. 79	30. 001
763	28	2. 24	30. 040
764	28	2. 68	30. 079
765	28	3. 12	30. 119
766	28	3. 57	30. 158
767	28	4. 01	30. 197
768	28	4. 45	30. 237
769	28	4. 90	30. 276
770	28	5. 34	30. 316
771	28	5. 78	30. 355
772	28	6. 23	30. 394
773	28	6. 67	30. 434
774	28	7. 11	30. 473
775	28	7. 55	30. 513
776	28	8. 00	30. 552
777	28	8. 44	30. 591
778	28	8. 88	30. 631
779	28	9. 33	30. 670

English. Inches.	French. Inches.	French. Lines.	Millim.
29. 0	27	2. 53	736. 6
29. 1	27	3. 65	739. 1
29. 2	27	4. 78	741. 7
29. 3	27	5. 90	744. 2
29. 4	27	7. 03	746. 8
29. 5	27	8. 16	749. 3
29. 6	27	9. 28	751. 8
29. 7	27	10. 41	754. 4
29. 8	27	11. 53	756. 9
29. 9	28	0. 66	759. 5
30. 0	28	1. 79	762. 0
30. 1	28	2. 91	764. 5
30. 2	28	4. 04	767. 1
30. 3	28	5. 16	769. 6
30. 4	28	6. 29	772. 2
30. 5	28	7. 42	774. 7
30. 6	28	8. 55	777. 3
30. 7	28	9. 67	779. 8

DR. YOUNG'S REFRACTIONS.

App. Alt.	Refraction Barom. 30 Therm. 50°	Diff. for 1′ Alt.	Diff. for +1 B.	Diff. for −1° Fa.
0° 0′	33′ 51″	11″. 7	74″	8″. 1
5	32 53	11. 3	71	7. 6
10	31 58	10. 9	69	7. 3
15	31 5	10. 5	67	7. 0
20	30 13	10. 1	65	6. 7
25	29 24	9. 7	63	6. 4
30	28 37	9. 4	61	6. 1
35	27 51	9. 0	59	5. 9
40	27 6	8. 7	58	5. 6
45	26 24	8. 4	56	5. 4
50	25 43	8. 0	55	5. 1
55	25 3	7. 7	53	4. 9
1 0	24 25	7. 4	52	4. 7
5	23 48	7. 1	50	4. 6
10	23 13	6. 9	49	4. 5
15	22 40	6. 6	48	4. 4
20	22 8	6. 3	46	4. 2
25	21 37	6. 1	45	4. 0
30	21 7	5. 9	44	3. 9
35	20 38	5. 7	43	3. 8
40	20 10	5. 5	42	3. 6
45	19 43	5. 3	40	3. 5
50	19 17	5. 1	39	3. 4
55	18 52	4. 9	39	3. 3
2 0	18 29	4. 8	38	3. 2
5	18 5	4. 6	37	3. 1
10	17 43	4. 4	36	3. 0
15	17 21	4. 3	36	2. 9
20	17 0	4. 1	35	2. 8
25	16 40	4. 0	34	2. 8
30	16 21	3. 9	33	2. 7
35	16 2	3. 7	33	2. 7
40	15 43	3. 6	32	2. 6
45	15 25	3. 5	32	2. 5
50	15 8	3. 4	31	2. 4
55	14 51	3. 3	30	2. 3
3 0	14 35	3. 2	30	2. 3
5	14 19	3. 1	29	2. 2
10	14 4	3. 0	29	2. 2
15	13 50	2. 9	28	2. 1
20	13 35	2. 8	28	2. 1
25	13 21	2. 7	27	2. 0
30	13 7	2. 7	27	2. 0
35	12 53	2. 6	26	2. 0
40	12 41	2. 5	26	1. 9
45	12 28	2. 4	25	1. 9
50	12 16	2. 4	25	1. 9
55	12 3	2. 3	25	1. 8
4 0	11 52	2. 2	24. 1	1. 70
10	11 30	2. 1	23. 4	1. 64
20	11 10	2. 0	22. 7	1. 58
30	10 50	1. 9	22. 0	1. 53
40	10 32	1. 8	21. 3	1. 48
50	10 15	1. 7	20. 7	1. 43
5 0	9 58	1. 6	20. 1	1. 38
10	9 42	1. 5	19. 6	1. 34
20	9 27	1. 5	19. 1	1. 30
30	9 11	1. 4	18. 6	1. 26
40	8 58	1. 3	18. 1	1. 22
50	8 45	1. 3	17. 6	1. 19
6 0	8 32	1. 2	17. 2	1. 15
10	8 20	1. 2	16. 8	1. 11
20	8 9	1. 1	16. 4	1. 09
30	7 58	1. 1	16. 0	1. 06
40	7 47	1. 0	15. 7	1. 03
50	7 37	1. 0	15. 3	1. 00

Apparent Altitude.	Refraction. Barometer 30 Therm. 50°	Diff. for 1′ Alt.	Diff. for +1 B.	Diff. for −1° Fa.
7° 0′	7′ 27″	1″. 0	15″. 0	0″. 98
10	7 17	. 9	14. 6	. 95
20	7 8	. 9	14. 3	. 93
30	6 59	. 8	14. 1	. 91
40	6 51	. 8	13. 8	. 89
50	6 43	. 8	13. 5	. 87
8 0	6 35	. 7	13. 3	. 85
10	6 28	. 7	13. 1	. 83
20	6 21	. 7	12. 8	. 82
30	6 14	. 7	12. 6	. 80
40	6 7	. 7	12. 3	. 79
50	6 0	. 6	12. 1	. 77
9 0	5 54	. 6	11. 9	. 76
10	5 47	. 6	11. 7	. 74
20	5 41	. 6	11. 5	. 73
30	5 36	. 6	11. 3	. 71
40	5 30	. 5	11. 1	. 71
50	5 25	. 5	11. 0	. 70
10 0	5 20	. 5	10. 8	. 69
10	5 15	. 5	10. 6	. 67
20	5 10	. 5	10. 4	. 65
30	5 5	. 5	10. 2	. 64
40	5 0	. 5	10. 1	. 63
50	4 56	. 4	9. 9	. 62
11 0	4 51	. 4	9. 8	. 60
10	4 47	. 4	9. 6	. 59
20	4 43	. 4	9. 5	. 58
30	4 39	. 4	9. 4	. 57
40	4 35	. 4	9. 2	. 56
50	4 31	. 4	9. 1	. 55
12 0	4 28. 1	. 38	9. 00	. 556
10	4 24. 4	. 37	8. 86	. 548
20	4 20. 8	. 36	8. 74	. 541
30	4 17. 3	. 35	8. 63	. 533
40	4 13. 9	. 33	8. 51	. 524
50	4 10. 7	. 32	8. 41	. 517
13 0	4 7. 5	. 31	8. 30	. 509
10	4 4. 4	. 31	8. 20	. 503
20	4 1. 4	. 30	8. 10	. 496
30	3 58. 4	. 30	8. 00	. 490
40	3 55. 5	. 29	7. 89	. 482
50	3 52. 6	. 29	7. 79	. 476
14 0	3 49. 9	. 28	7. 70	. 469
10	3 47. 1	. 28	7. 61	. 464
20	3 44. 4	. 27	7. 52	. 458
30	3 41. 8	. 26	7. 43	. 453
40	3 39. 2	. 26	7. 34	. 448
50	3 36. 7	. 25	7. 26	. 444
15 0	3 34. 3	. 24	7. 18	. 439
30	3 27. 3	. 22	6. 95	. 424
16 0	3 20. 6	. 21	6. 73	. 411
30	3 14. 4	. 20	6. 51	. 399
17 0	3 8. 5	. 19	6. 31	. 386
30	3 2. 9	. 18	6. 12	. 374
18 0	2 57. 6	. 17	5. 98	. 362
19 0	2 47. 7	. 16	5. 61	. 340
20 0	2 38. 7	. 15	5. 31	. 322
21 0	2 30. 5	. 13	5. 04	. 305
22 0	2 23. 2	. 12	4. 79	. 290
23 0	2 16. 5	. 11	4. 57	. 276
24 0	2 10. 1	. 10	4. 35	. 264
25 0	2 4. 2	. 09	4. 16	. 252
26 0	1 58. 8	. 09	3. 97	. 241
27 0	1 53. 8	. 08	3. 81	. 230
28 0	1 49. 1	. 08	3. 65	. 219
29 0	1 44. 7	. 07	3. 50	. 209

App. Alt.	Refraction. Barometer 30 Therm. 30°	Diff. for 1′ Alt.	Diff. for +1 B.	Diff. for −1° Fa.
30°	1′ 40″. 5	0″. 07	3″. 36	0″. 201
31	1 36. 6	. 06	3. 23	. 193
32	1 33. 0	. 06	3. 11	. 186
33	1 29. 5	. 06	2. 99	. 179
34	1 26. 1	. 05	2. 88	. 173
35	1 23. 0	. 05	2. 78	. 167
36	1 20. 0	. 05	2. 68	. 161
37	1 17. 1	. 05	2. 58	. 155
38	1 14. 4	. 05	2. 49	. 149
39	1 11. 8	. 04	2. 40	. 144
40	1 9. 3	. 04	2. 32	. 139
41	1 6. 9	. 04	2. 24	. 134
42	1 4. 6	. 038	2. 16	. 130
43	1 2. 4	. 036	2. 09	. 125
44	1 0. 3	. 034	2. 02	. 120
45	0 58. 1	. 034	1. 94	. 117
46	0 56. 1	. 033	1. 88	. 112
47	0 54. 2	. 032	1. 81	. 108
48	0 52. 3	. 031	1. 75	. 104
49	0 50. 5	. 030	1. 69	. 101
50	0 48. 8	. 029	1. 63	. 097
51	0 47. 1	. 028	1. 58	. 094
52	0 45. 4	. 027	1. 52	. 090
53	0 43. 8	. 026	1. 47	. 088
54	0 42. 2	. 026	1. 41	. 085
55	0 40. 8	. 025	1. 36	. 082
56	0 39. 3	. 025	1. 31	. 079
57	0 37. 8	. 025	1. 26	. 076
58	0 36. 4	. 024	1. 22	. 073
59	0 35. 0	. 024	1. 17	. 070
60	0 33. 6	. 023	1. 12	. 067
61	0 32. 3	. 022	1. 08	. 065
62	0 31. 0	. 022	1. 04	. 062
63	0 29. 7	. 021	. 99	. 060
64	0 28. 4	. 021	. 95	. 057
65	0 27. 2	. 020	. 91	. 055
66	0 25. 9	. 020	. 87	. 052
67	0 24. 7	. 020	. 83	. 050
68	0 23. 5	. 020	. 79	. 047
69	0 22. 4	. 020	. 75	. 045
70	0 21. 2	. 020	. 71	. 043
71	0 19. 9	. 020	. 67	. 040
72	0 18. 8	. 019	. 63	. 038
73	0 17. 7	. 018	. 59	. 036
74	0 16. 6	. 018	. 56	. 033
75	0 15. 5	. 018	. 52	. 031
76	0 14. 4	. 018	. 48	. 029
77	0 13. 4	. 017	. 45	. 027
78	0 12. 3	. 017	. 41	. 025
79	0 11. 2	. 017	. 38	. 023
80	0 10. 2	. 017	. 34	. 021
81	0 9. 2	. 017	. 31	. 018
82	0 8. 2	. 017	. 27	. 016
83	0 7. 1	. 017	. 24	. 014
84	0 6. 1	. 017	. 20	. 012
85	0 5. 1	. 017	. 17	. 010
86	0 4. 1	. 017	. 14	. 008
87	0 3. 1	. 017	. 10	. 006
88	0 2. 0	. 017	. 07	. 004
89	0 1. 0	. 017	. 03	. 002

TABLE I.

Logarithms for computing the ANNUAL PRECESSION of a Star in Right Asc. in Time. Argument=the Right Ascens. of ✱ in time.

[Tab. Log. = log. sin. R. A.+.125884.]

S. N.	− 0 +	+ XII −	− I +	+ XIII −	− II +	+ XIV −	− III +	+ XV −	− IV +	+ XVI −	− V +	+ XVII −	S. N.
Min.	Log.	Diff.	Log.	Diff.	Log.	Diff.	Log.	Diff.	Log.	Diff.	Log.	Diff.	Min.
0	− ∞		9. 53888		9. 82485		9. 97537		0. 06342		0. 11083		60
1	7. 76570		9. 54589	701	9. 82812	327	9. 97726	189	0. 06450	108	0. 11133	50	59
2	8. 06673	30103	9. 55278	689	9. 83135	323	9. 97913	187	0. 06558	108	0. 11183	50	58
3	8. 24281	17608	9. 55956	678	9. 83455	320	9. 98098	185	0. 06665	107	0. 11231	48	57
4	8. 36774	12493	9. 56622	666	9. 83772	317	9. 98282	184	0. 06770	105	0. 11279	47	56
5	8. 46464	9690	9. 57278	656	9. 84086	314	9. 98464	182	0. 06875	105	0. 11326	47	55
6	8. 54380	7916	9. 57923	645	9. 84397	311	9. 98645	181	0. 06978	103	0. 11372	46	54
7	8. 61073	6693	9. 58557	634	9. 84705	308	9. 98824	179	0. 07081	103	0. 11417	45	53
8	8. 66870	5797	9. 59182	625	9. 85009	304	9. 99001	177	0. 07182	101	0. 11461	44	52
9	8. 71983	5113	9. 59797	615	9. 85311	302	9. 99177	176	0. 07282	100	0. 11504	43	51
10	8. 76556	4573	9. 60403	606	9. 85610	299	9. 99352	175	0. 07381	99	0. 11547	43	50
11	8. 80693	4137	9. 60999	596	9. 85906	296	9. 99525	173	0. 07489	98	0. 11588	41	49
12	8. 84468	3775	9. 61587	588	9. 86199	293	9. 99696	171	0. 07577	98	0. 11629	41	48
13	8. 87941	3473	9. 62166	579	9. 86490	291	9. 99866	170	0. 07673	96	0. 11669	40	47
14	8. 91156	3215	9. 62736	570	9. 86777	287	0. 00034	168	0. 07768	95	0. 11708	39	46
15	8. 94148	2992	9. 63298	562	9. 87062	285	0. 00201	167	0. 07862	94	0. 11746	38	45
16	8. 96947	2799	9. 63853	555	9. 87345	283	0. 00366	165	0. 07954	92	0. 11783	37	44
17	8. 99575	2628	9. 64399	546	9. 87624	279	0. 00530	164	0. 08046	92	0. 11820	37	43
18	9. 02053	2478	9. 64938	539	9. 87901	277	0. 00693	163	0. 08137	91	0. 11855	35	42
19	9. 04396	2343	9. 65469	531	9. 88176	275	0. 00854	161	0. 08227	90	0. 11890	35	41
20	9. 06618	2222	9. 65994	525	9. 88448	272	0. 01014	160	0. 08316	89	0. 11924	34	40
21	9. 08731	2113	9. 66511	517	9. 88717	269	0. 01172	158	0. 08404	88	0. 11957	33	39
22	9. 10746	2015	9. 67021	510	9. 88984	267	0. 01329	157	0. 08491	87	0. 11989	32	38
23	9. 12670	1924	9. 67524	503	9. 89248	264	0. 01485	156	0. 08577	86	0. 12020	31	37
24	9. 14512	1842	9. 68021	497	9. 89510	262	0. 01639	154	0. 08661	84	0. 12050	30	36
25	9. 16278	1766	9. 68512	491	9. 89770	260	0. 01792	153	0. 08705	84	0. 12080	30	35
26	9. 17974	1696	9. 68996	484	9. 90027	257	0. 01943	151	0. 08828	83	0. 12109	29	34
27	9. 19606	1632	9. 69474	478	9. 90282	255	0. 02093	150	0. 08910	82	0. 12137	28	33
28	9. 21178	1572	9. 69946	472	9. 90535	253	0. 02242	149	0. 08991	81	0. 12164	27	32
29	9. 22694	1516	9. 70412	466	9. 90785	250	0. 02389	147	0. 09071	80	0. 12190	26	31
30	9. 24158	1464	9. 70872	460	9. 91033	248	0. 02535	146	0. 09150	79	0. 12215	25	30
31	9. 25574	1416	9. 71327	455	9. 91279	246	0. 02680	145	0. 09228	78	0. 12240	25	29
32	9. 26944	1370	9. 71776	449	9. 91523	244	0. 02823	143	0. 09305	77	0. 12264	24	28
33	9. 28271	1327	9. 72220	444	9. 91764	241	0. 02965	142	0. 09381	76	0. 12286	22	27
34	9. 29559	1288	9. 72658	438	9. 92003	239	0. 03106	141	0. 09456	75	0. 12308	22	26
35	9. 30808	1249	9. 73092	434	9. 92240	237	0. 03246	140	0. 09530	74	0. 12330	22	25
36	9. 32022	1214	9. 73520	428	9. 92476	236	0. 03384	138	0. 09604	74	0. 12350	20	24
37	9. 33201	1179	9. 73943	423	9. 92709	233	0. 03521	137	0. 09676	72	0. 12369	19	23
38	9. 34349	1148	9. 74361	418	9. 92939	230	0. 03657	136	0. 09747	71	0. 12388	19	22
39	9. 35467	1118	9. 74774	413	9. 93168	229	0. 03792	135	0. 09818	71	0. 12406	18	21
40	9. 36555	1088	9. 75183	409	9. 93395	227	0. 03925	133	0. 09887	69	0. 12423	17	20
41	9. 37617	1062	9. 75587	404	9. 93620	225	0. 04057	132	0. 09956	69	0. 12439	16	19
42	9. 38652	1035	9. 75987	400	9. 93843	223	0. 04188	131	0. 10023	67	0. 12454	15	18
43	9. 39662	1010	9. 76382	395	9. 94064	221	0. 04317	129	0. 10090	67	0. 12469	15	17
44	9. 40648	986	9. 76773	391	9. 94283	219	0. 04446	129	0. 10155	65	0. 12483	14	16
45	9. 41612	964	9. 77159	386	9. 94500	217	0. 04573	127	0. 10220	65	0. 12485	12	15
46	9. 42554	942	9. 77541	382	9. 94715	215	0. 04699	126	0. 10284	64	0. 12507	12	14
47	9. 43475	921	9. 77919	378	9. 94928	213	0. 04824	125	0. 10347	63	0. 12519	12	13
48	9. 44376	901	9. 78293	374	9. 95139	211	0. 04948	124	0. 10409	62	0. 12529	10	12
49	9. 45258	882	9. 78663	370	9. 95349	210	0. 05070	122	0. 10470	61	0. 12538	9	11
50	9. 46122	864	9. 79029	366	9. 95557	208	0. 05191	121	0. 10530	60	0. 12547	9	10
51	9. 46968	846	9. 79391	362	9. 95763	206	0. 05312	121	0. 10590	60	0. 12555	8	9
52	9. 47797	829	9. 79749	358	9. 95967	204	0. 05430	118	0. 10648	58	0. 12562	7	8
53	9. 48610	813	9. 80104	355	9. 96169	202	0. 05548	118	0. 10706	58	0. 12568	6	7
54	9. 49407	797	9. 80455	351	9. 96370	201	0. 05665	117	0. 10762	56	0. 12574	6	6
55	9. 50189	782	9. 80802	347	9. 96568	198	0. 05781	116	0. 10818	56	0. 12578	4	5
56	9. 50956	767	9. 81145	343	9. 96765	197	0. 05895	114	0. 10873	55	0. 12582	4	4
57	9. 51709	753	9. 81486	341	9. 96961	196	0. 06008	113	0. 10927	54	0. 12585	3	3
58	9. 52448	739	9. 81822	336	9. 97155	194	0. 06120	112	0. 10980	53	0. 12587	2	2
59	9. 53175	727	9. 82155	333	9. 97347	192	0. 06232	112	0. 11032	52	0. 12588	1	1
60	9. 53888	713	9. 82485	330	9. 97537	190	0. 06342	110	0. 11083	51	0. 12588	0	0
N. S.	+ XI −	− XXIII +	+ X −	− XXII +	+ IX −	− XXI +	+ VIII −	− XX +	+ VII −	− XIX +	+ VI −	− XVIII +	N. S.

Add to the Logarithm found in the Table, the Log. Tang. of the Stars' Declination; and to the natural number corresponding to this Sum, with its proper Sign, add 3ˢ. 0678, and the result will give the Precession in Right Ascension, in time.

TABLE I. (Nat. Num.)

The Annual Precession of a Star in Right Asc. in Time. Argument =R. A. of ∗ in Time.
[Tab. No. = 1s. 3362 sin. R. A.]

S. N.	− O +	+ XII −	− I +	+ XIII −	− II +	+ XIV −	− III +	+ XV −	− IV +	+ XVI −	− V +	+ XVII −	S. N.
Min.		Diff.		Diff.		Diff.		Diff.		Diff.		Diff.	Min.
0	0s. 000	6	0s. 346	6	0s. 668	5	0s. 945	4	1s. 157	3	1s. 291	1	60
1	0. 006	6	0. 352	6	0. 673	5	0. 949	4	1. 160	3	1. 292	1	59
2	0. 012	5	0. 358	5	0. 678	5	0. 953	4	1. 163	3	1. 293	2	58
3	0. 017	6	0. 363	6	0. 683	5	0. 957	4	1. 166	3	1. 295	1	57
4	0. 023	6	0. 369	6	0. 688	5	0. 961	4	1. 169	3	1. 296	2	56
5	0. 029	6	0. 375	5	0. 693	5	0. 965	4	1. 172	2	1. 298	1	55
6	0. 035	6	0. 380	6	0. 698	5	0. 969	4	1. 174	3	1. 299	2	54
7	0. 041	6	0. 386	6	0. 703	5	0. 973	4	1. 177	3	1. 301	1	53
8	0. 047	6	0. 392	5	0. 708	5	0. 977	4	1. 180	2	1. 302	2	52
9	0. 053	5	0. 397	5	0. 713	5	0. 981	4	1. 182	3	1. 304	1	51
10	0. 058	6	0. 402	6	0. 718	5	0. 985	4	1. 185	3	1. 305	2	50
11	0. 064	6	0. 408	6	0. 723	5	0. 989	4	1. 188	3	1. 307	1	49
12	0. 070	6	0. 414	5	0. 728	5	0. 993	4	1. 191	2	1. 308	1	48
13	0. 076	6	0. 419	6	0. 733	5	0. 997	4	1. 193	3	1. 309	1	47
14	0. 082	5	0. 425	5	0. 738	5	1. 001	4	1. 196	2	1. 310	1	46
15	0. 087	6	0. 430	6	0. 743	5	1. 005	4	1. 198	3	1. 311	1	45
16	0. 093	6	0. 436	6	0. 748	4	1. 009	4	1. 201	2	1. 312	1	44
17	0. 099	6	0. 442	5	0. 752	5	1. 013	4	1. 203	3	1. 313	1	43
18	0. 105	6	0. 447	5	0. 757	4	1. 017	4	1. 206	2	1. 314	1	42
19	0. 111	6	0. 452	5	0. 761	5	1. 021	3	1. 208	3	1. 315	1	41
20	0. 117	6	0. 457	6	0. 766	5	1. 024	4	1. 211	2	1. 316	1	40
21	0. 123	6	0. 463	5	0. 771	5	1. 028	3	1. 213	3	1. 317	1	39
22	0. 129	5	0. 468	6	0. 776	4	1. 031	4	1. 216	2	1. 318	1	38
23	0. 134	6	0. 474	5	0. 780	5	1. 035	3	1. 218	3	1. 319	1	37
24	0. 140	6	0. 479	6	0. 785	5	1. 038	4	1. 221	2	1. 320	1	36
25	0. 146	6	0. 485	5	0. 790	5	1. 042	4	1. 223	3	1. 321	1	35
26	0. 152	6	0. 490	6	0. 795	4	1. 046	4	1. 226	2	1. 322	1	34
27	0. 158	5	0. 496	5	0. 799	5	1. 050	4	1. 228	2	1. 323	1	33
28	0. 163	6	0. 501	5	0. 804	4	1. 054	3	1. 230	3	1. 324	0	32
29	0. 169	5	0. 506	5	0. 808	5	1. 057	3	1. 233	2	1. 324	1	31
30	0. 174	6	0. 511	6	0. 813	5	1. 060	4	1. 235	3	1. 325	1	30
31	0. 180	6	0. 517	5	0. 818	5	1. 064	3	1. 238	2	1. 326	0	29
32	0. 186	6	0. 522	6	0. 823	4	1. 067	4	1. 240	2	1. 326	1	28
33	0. 192	6	0. 528	5	0. 827	5	1. 071	3	1. 242	2	1. 327	0	27
34	0. 198	5	0. 533	6	0. 832	5	1. 074	4	1. 244	2	1. 327	1	26
35	0. 203	6	0. 539	5	0. 837	4	1. 078	3	1. 246	2	1. 328	0	25
36	0. 209	6	0. 544	6	0. 841	5	1. 081	4	1. 248	2	1. 328	1	24
37	0. 215	6	0. 550	5	0. 846	4	1. 085	3	1. 250	2	1. 329	0	23
38	0. 221	6	0. 555	5	0. 850	5	1. 088	4	1. 252	2	1. 329	1	22
39	0. 227	5	0. 560	5	0. 855	4	1. 092	3	1. 254	2	1. 330	1	21
40	0. 232	6	0. 565	5	0. 859	5	1. 095	4	1. 256	2	1. 331	0	20
41	0. 238	6	0. 570	6	0. 864	4	1. 099	3	1. 258	2	1. 331	1	19
42	0. 244	6	0. 576	5	0. 868	5	1. 102	3	1. 260	2	1. 332	0	18
43	0. 250	5	0. 581	5	0. 873	4	1. 105	3	1. 262	2	1. 332	1	17
44	0. 255	6	0. 586	6	0. 877	4	1. 108	4	1. 264	2	1. 333	0	16
45	0. 261	6	0. 592	5	0. 881	5	1. 112	3	1. 266	2	1. 333	1	15
46	0. 267	5	0. 597	5	0. 886	4	1. 115	3	1. 268	2	1. 334	0	14
47	0. 272	6	0. 602	5	0. 890	4	1. 118	3	1. 270	1	1. 334	0	13
48	0. 278	6	0. 607	5	0. 894	5	1. 121	3	1. 271	2	1. 334	1	12
49	0. 284	5	0. 612	5	0. 899	4	1. 124	3	1. 273	1	1: 335	0	11
50	0. 289	6	0. 617	6	0. 903	4	1. 127	3	1. 274	2	1. 335	1	10
51	0. 295	6	0. 623	5	0. 907	4	1. 130	3	1. 276	2	1. 336	0	9
52	0. 301	5	0. 628	5	0. 911	5	1. 143	3	1. 278	1	1. 336	0	8
53	0. 306	6	0. 633	5	0. 916	4	1. 136	3	1. 279	2	1. 336	0	7
54	0. 312	6	0. 638	5	0. 920	4	1. 139	3	1. 281	2	1. 336	0	6
55	0. 318	6	0. 643	5	0. 924	5	1. 142	3	1. 283	2	1. 336	0	5
56	0. 324	5	0. 648	5	0. 929	4	1. 145	3	1. 285	1	1. 336	0	4
57	0. 329	6	0. 653	5	0. 933	4	1. 148	3	1. 286	2	1. 336	0	3
58	0. 335	6	0. 658	5	0. 937	4	1. 151	3	1. 288	2	1. 336	0	2
59	0. 341	5	0. 663	5	0. 941	4	1. 154	3	1. 290	1	1. 336	0	1
60	0. 346		0. 668		0. 945		1. 157		1. 291		1. 336		0
N. S.	+ XI −	− XXIII +	+ X −	− XXII +	+ IX −	− XXI +	+ VIII −	− XX +	+ VII −	− XIX +	+ VI −	− XVIII +	N. S.

Multiply the number in the Table, with its proper Sign, by the natural Tangent of the Star's Declination, and add the constant quantity 3s. 0678 thereto for the Annual Precession.

TABLE II.

Aberration in Right Ascension. To find the Annual Number. Argument = R. A. of * in time.

[Log. tang. A = log. cotang. R. A. — .0374865.]

	O / XII		I / XIII		II / XIV		III / XV		IV / XVI		V / XVII		
	6 + / 0 +	P. P.	6 + / 0 +	P. P.	6 + / 0 +	P. P.	6 + / 0 +	P. P.	6 + / 0 +	P. P.	6 + / 0 +	P. P.	
Min.	A.		A.		A.		A.		A.		A.		Min.
0	3s 0° 0′	0	2s 13° 43′	0	1s 27° 49′	0	1s 12° 32′	0	0s 27° 54′	0	0s 13° 49′	0	60
1	2 29 44	0	2 13 27	0	1 27 33	0	1 12 17	0	0 27 40	0	0 13 35	0	59
2	2 29 27	1	2 13 11	1	1 27 17	1	1 12 2	1	0 27 26	0	0 13 21	0	58
3	2 29 11	1	2 12 55	1	1 27 1	1	1 11 47	1	0 27 11	1	0 13 7	1	57
4	2 28 55	1	2 12 39	1	1 26 40	1	1 11 32	1	0 26 57	1	0 12 53	1	56
5	2 28 39	1	2 12 23	1	1 26 31	1	1 11 17	1	0 26 43	1	0 12 39	1	55
6	2 28 22	2	2 12 6	2	1 26 15	2	1 11 2	2	0 26 29	1	0 12 25	1	54
7	2 28 6	2	2 11 50	2	1 26 0	2	1 10 47	2	0 26 14	2	0 12 11	2	53
8	2 27 49	2	2 11 34	2	1 25 44	2	1 10 33	2	0 26 0	2	0 11 57	2	52
9	2 27 33	2	2 11 18	2	1 25 29	2	1 10 18	2	0 25 46	2	0 11 43	2	51
10	2 27 17	3	2 11 2	3	1 25 13	3	1 10 3	3	0 25 32	2	0 11 30	2	50
11	2 27 1	3	2 10 46	3	1 24 58	3	1 9 48	3	0 25 17	3	0 11 16	3	49
12	2 26 44	3	2 10 30	3	1 24 42	3	1 9 33	3	0 25 3	3	0 11 2	3	48
13	2 26 28	4	2 10 14	4	1 24 27	3	1 9 18	3	0 24 49	3	0 10 48	3	47
14	2 26 11	4	2 9 58	4	1 24 11	4	1 9 4	4	0 24 35	3	0 10 34	3	46
15	2 25 55	4	2 9 42	4	1 23 56	4	1 8 49	4	0 24 20	4	0 10 20	4	45
16	2 25 38	4	2 9 26	4	1 23 40	4	1 8 34	4	0 24 6	4	0 10 7	4	44
17	2 25 22	5	2 9 10	5	1 23 25	4	1 8 19	4	0 23 52	4	0 9 53	4	43
18	2 25 6	5	2 8 53	5	1 23 9	5	1 8 5	5	0 23 38	4	0 9 39	4	42
19	2 24 50	5	2 8 37	5	1 22 54	5	1 7 50	5	0 23 24	4	0 9 25	4	41
20	2 24 33	5	2 8 21	5	1 22 39	5	1 7 35	5	0 23 10	5	0 9 11	5	40
21	2 24 17	6	2 8 5	6	1 22 24	5	1 7 20	5	0 22 55	5	0 8 57	5	39
22	2 24 0	6	2 7 49	6	1 22 8	6	1 7 6	6	0 22 41	5	0 8 44	5	38
23	2 23 44	6	2 7 33	6	1 21 53	6	1 6 51	6	0 22 27	5	0 8 30	5	37
24	2 23 28	6	2 7 18	6	1 21 37	6	1 6 36	6	0 22 13	6	0 8 16	6	36
25	2 23 12	7	2 7 2	7	1 21 22	6	1 6 21	6	0 21 59	6	0 8 2	6	35
26	2 22 55	7	2 6 46	7	1 21 6	7	1 6 7	7	0 21 45	6	0 7 48	6	34
27	2 22 39	7	2 6 30	7	1 20 51	7	1 5 53	7	0 21 30	6	0 7 35	6	33
28	2 22 23	8	2 6 14	8	1 20 36	7	1 5 38	7	0 21 16	7	0 7 21	7	32
29	2 22 7	8	2 5 58	8	1 20 21	7	1 5 23	7	0 21 2	7	0 7 7	7	31
30	2 21 50	8	2 5 42	8	1 20 5	8	1 5 8	8	0 20 48	7	0 6 53	7	30
31	2 21 34	8	2 5 26	8	1 19 50	8	1 4 53	8	0 20 34	7	0 6 39	7	29
32	2 21 17	9	2 5 10	9	1 19 35	8	1 4 39	8	0 20 20	7	0 6 26	7	28
33	2 21 1	9	2 4 54	9	1 19 20	8	1 4 25	8	0 20 6	8	0 6 12	8	27
34	2 20 45	9	2 4 38	9	1 19 4	9	1 4 10	9	0 19 52	8	0 5 58	8	26
35	2 20 29	9	2 4 22	9	1 18 49	9	1 3 56	9	0 19 38	8	0 5 44	8	25
36	2 20 12	10	2 4 7	10	1 18 34	9	1 3 41	9	0 19 24	8	0 5 30	8	24
37	2 19 56	10	2 3 51	10	1 18 19	9	1 3 27	9	0 19 10	9	0 5 16	9	23
38	2 19 40	10	2 3 35	10	1 18 3	10	1 3 12	10	0 18 56	9	0 5 3	9	22
39	2 19 24	11	2 3 19	11	1 17 48	10	1 2 58	10	0 18 42	9	0 4 49	9	21
40	2 19 7	11	2 3 3	11	1 17 33	10	1 2 43	10	0 18 28	9	0 4 35	9	20
41	2 18 51	11	2 2 47	11	1 17 18	10	1 2 29	10	0 18 14	9	0 4 21	9	19
42	2 18 35	11	2 2 32	11	1 17 3	11	1 2 14	11	0 18 0	10	0 4 8	10	18
43	2 18 19	12	2 2 16	12	1 16 48	11	1 2 0	11	0 17 46	10	0 3 54	10	17
44	2 18 2	12	2 2 0	12	1 16 32	11	1 1 45	11	0 17 32	10	0 3 40	10	16
45	2 17 46	12	2 1 44	12	1 16 17	11	1 1 31	11	0 17 18	10	0 3 26	10	15
46	2 17 30	12	2 1 28	12	1 16 2	12	1 1 16	12	0 17 4	11	0 3 13	11	14
47	2 17 14	13	2 1 12	13	1 15 47	12	1 1 2	12	0 16 50	11	0 2 59	11	13
48	2 16 57	13	2 0 57	13	1 15 32	12	1 0 47	12	0 16 36	11	0 2 45	11	12
49	2 16 41	13	2 0 41	13	1 15 17	12	1 0 33	12	0 16 22	11	0 2 31	11	11
50	2 16 25	14	2 0 26	14	1 15 2	13	1 0 18	13	0 16 8	12	0 2 18	12	10
51	2 16 9	14	2 0 10	14	1 14 47	13	1 0 4	13	0 15 54	12	0 2 4	12	9
52	2 15 52	14	1 29 54	14	1 14 32	13	0 29 49	13	0 15 40	12	0 1 50	12	8
53	2 15 36	14	1 29 38	14	1 14 17	13	0 29 35	13	0 15 26	12	0 1 36	12	7
54	2 15 20	15	1 29 23	15	1 14 2	14	0 29 20	14	0 15 12	13	0 1 23	13	6
55	2 15 4	15	1 29 7	15	1 13 47	14	0 29 6	14	0 14 58	13	0 1 9	13	5
56	2 14 48	15	1 28 51	15	1 13 32	14	0 28 52	14	0 14 44	13	0 0 55	13	4
57	2 14 32	15	1 28 35	15	1 13 17	14	0 28 38	14	0 14 30	13	0 0 41	13	3
58	2 14 15	16	1 28 20	16	1 13 2	15	0 28 23	15	0 14 16	14	0 0 28	14	2
59	2 13 59	16	1 28 4	16	1 12 47	15	0 28 9	15	0 14 2	14	0 0 14	14	1
60	2 13 43	16	1 27 49	16	1 12 32	15	0 27 54	15	0 13 49	14	0 0 0	14	0
	6 — / 12 —		6 — / 12 —		6 — / 12 —		6 — / 12 —		6 — / 12 —		6 — / 12 —		
	XI / XXIII		X / XXII		IX / XXI		VIII / XX		VII / XIX		VI / XVIII		

TABLE III.

Aberration in R. A. To find the log. of the Maximum and Maximum Argument = R. A. of the * in time.

[Tab. log. = log. cos. R. A. + log. cosec. A. + .092954. Or = log. sine R. A. + log. sec. A + .130441.]

M	O XII Log.	Diff.	Num.	I XIII Log.	Diff.	Num.	II XIV Log.	Diff.	Num.	III XV Log.	Diff.	Num.	IV XVI Log.	Diff.	Num.	V XVII Log.	Diff.	Num.	M
0	.09295	0	1s. 239	.09569	9	1s. 247	.10296	15	1s. 268	.11252	16	1s. 296	.12167	13	1s. 323	.12814	7	1s. 343	60
1	.09295	1	1. 239	.09578	9	1. 247	.10311	15	1. 268	.11268	16	1. 296	.12180	14	1. 324	.12821	7	1. 344	59
2	.09296	0	1. 239	.09587	9	1. 247	.10326	15	1. 269	.11284	16	1. 297	.12194	13	1. 324	.12828	7	1. 344	58
3	.09296	1	1. 239	.09596	9	1. 247	.10341	15	1. 269	.11300	17	1. 297	.12207	14	1. 325	.12835	8	1. 344	57
4	.09297	1	1. 239	.09605	9	1. 248	.10356	15	1. 269	.11317	16	1. 298	.12221	13	1. 325	.12843	7	1. 344	56
5	.09298	0	1. 239	.09614	10	1. 248	.10371	15	1. 270	.11333	16	1. 298	.12234	13	1. 325	.12850	7	1. 344	55
6	.09298	1	1. 239	.09624	9	1. 248	.10386	15	1. 270	.11349	17	1. 299	.12247	13	1. 326	.12857	6	1. 345	54
7	.09299	1	1. 239	.09633	10	1. 248	.10401	15	1. 271	.11356	15	1. 299	.12260	13	1. 326	.12863	7	1. 345	53
8	.09300	1	1. 239	.09643	10	1. 249	.10416	15	1. 271	.11381	16	1. 300	.12273	13	1. 327	.12870	7	1. 345	52
9	.09301	1	1. 239	.09653	10	1. 249	.10431	15	1. 272	.11397	17	1. 300	.12286	12	1. 327	.12877	7	1. 345	51
10	.09302	3	1. 239	.09663	10	1. 249	.10446	15	1. 272	.11414	16	1. 301	.12298	13	1. 327	.12884	6	1. 345	50
11	.09305	2	1. 239	.09673	10	1. 250	.10461	16	1. 272	.11430	16	1. 301	.12311	12	1. 328	.12890	6	1. 346	49
12	.09307	2	1. 239	.09684	10	1. 250	.10477	16	1. 273	.11446	16	1. 302	.12323	13	1. 328	.12896	6	1. 346	48
13	.09309	3	1. 239	.09694	10	1. 250	.10493	15	1. 273	.11462	16	1. 302	.12336	13	1. 329	.12902	6	1. 346	47
14	.09312	2	1. 239	.09704	11	1. 250	.10508	16	1. 274	.11478	16	1. 303	.12349	12	1. 329	.12908	6	1. 346	46
15	.09314	3	1. 239	.09715	11	1. 251	.10524	16	1. 274	.11494	16	1. 303	.12361	12	1. 329	.12914	5	1. 346	45
16	.09317	2	1. 239	.09726	11	1. 251	.10540	15	1. 275	.11510	16	1. 304	.12373	12	1. 330	.12919	6	1. 347	44
17	.09319	3	1. 239	.09737	11	1. 251	.10555	16	1. 275	.11526	16	1. 304	.12385	12	1. 330	.12925	6	1. 347	43
18	.09322	2	1. 239	.09748	11	1. 252	.10571	15	1. 276	.11542	16	1. 304	.12397	12	1. 330	.12931	5	1. 347	42
19	.09324	3	1. 240	.09759	11	1. 252	.10586	16	1. 276	.11558	16	1. 305	.12409	12	1. 331	.12936	5	1. 347	41
20	.09327	3	1. 240	.09770	11	1. 252	.10602	16	1. 277	.11574	16	1. 305	.12421	12	1. 331	.12941	5	1. 347	40
21	.09330	3	1. 240	.09781	11	1. 253	.10618	16	1. 277	.11590	15	1. 306	.12433	12	1. 332	.12946	5	1. 347	39
22	.09333	4	1. 240	.09792	12	1. 253	.10634	16	1. 278	.11605	16	1. 306	.12445	12	1. 332	.12951	5	1. 348	38
23	.09337	4	1. 240	.09804	11	1. 253	.10650	16	1. 278	.11621	16	1. 307	.12457	11	1. 332	.12956	5	1. 348	37
24	.09341	4	1. 240	.09815	12	1. 254	.10666	16	1. 278	.11637	16	1. 307	.12468	12	1. 333	.12961	5	1. 348	36
25	.09345	4	1. 240	.09827	12	1. 254	.10682	16	1. 279	.11653	15	1. 308	.12480	11	1. 333	.12966	4	1. 348	35
26	.09349	4	1. 240	.09839	12	1. 254	.10698	16	1. 279	.11668	16	1. 308	.12491	11	1. 333	.12970	5	1. 348	34
27	.09353	4	1. 240	.09851	12	1. 255	.10714	16	1. 280	.11684	15	1. 309	.12502	11	1. 334	.12975	4	1. 348	33
28	.09357	4	1. 240	.09863	12	1. 255	.10730	16	1. 280	.11699	16	1. 309	.12513	11	1. 334	.12979	3	1. 348	32
29	.09361	5	1. 241	.09875	12	1. 255	.10746	16	1. 281	.11715	15	1. 310	.12524	11	1. 334	.12982	4	1. 348	31
30	.09366	4	1. 241	.09887	12	1. 256	.10762	16	1. 281	.11730	16	1. 310	.12535	11	1. 335	.12986	3	1. 349	30
31	.09370	5	1. 241	.09899	13	1. 256	.10778	16	1. 282	.11746	15	1. 311	.12546	11	1. 335	.12989	4	1. 349	29
32	.09375	5	1. 241	.09912	12	1. 256	.10794	16	1. 282	.11761	15	1. 311	.12557	10	1. 335	.12993	4	1. 349	28
33	.09380	5	1. 241	.09924	13	1. 257	.10810	17	1. 283	.11776	15	1. 312	.12567	11	1. 336	.12997	4	1. 349	27
34	.09385	6	1. 241	.09937	13	1. 257	.10827	16	1. 283	.11791	15	1. 312	.12578	10	1. 336	.13001	4	1. 349	26
35	.09391	5	1. 241	.09950	13	1. 258	.10843	16	1. 284	.11806	16	1. 312	.12588	10	1. 336	.13005	3	1. 349	25
36	.09396	6	1. 242	.09963	13	1. 258	.10859	16	1. 284	.11822	15	1. 313	.12598	11	1. 337	.13008	3	1. 349	24
37	.09402	5	1. 242	.09976	13	1. 258	.10975	17	1. 285	.11837	15	1. 313	.12609	10	1. 337	.13011	3	1. 349	23
38	.09407	6	1. 242	.09989	13	1. 259	.10892	16	1. 285	.11852	15	1. 314	.12619	10	1. 337	.13014	3	1. 349	22
39	.09413	6	1. 242	.10002	13	1. 259	.10908	16	1. 286	.11867	15	1. 314	.12629	10	1. 338	.13017	2	1. 350	21
40	.09419	6	1. 242	.10015	13	1. 259	.10924	17	1. 286	.11882	15	1. 315	.12639	10	1. 338	.13019	3	1. 350	20
41	.09425	7	1. 242	.10028	14	1. 260	.10941	16	1. 287	.11897	15	1. 315	.12649	9	1. 338	.13022	2	1. 350	19
42	.09432	6	1. 243	.10042	13	1. 260	.10957	17	1. 287	.11912	15	1. 316	.12658	10	1. 338	.13024	3	1. 350	18
43	.09438	7	1. 243	.10055	14	1. 261	.10974	16	1. 288	.11927	14	1. 316	.12668	9	1. 339	.13027	2	1. 350	17
44	.09445	6	1. 243	.10069	13	1. 261	.10990	16	1. 288	.11941	15	1. 317	.12677	10	1. 339	.13029	2	1. 350	16
45	.09451	7	1. 243	.10082	14	1. 261	.11006	16	1. 289	.11956	14	1. 317	.12687	9	1. 339	.13031	2	1. 350	15
46	.09458	7	1. 243	.10096	14	1. 262	.11022	16	1. 289	.11970	15	1. 317	.12696	9	1. 340	.13033	2	1. 350	14
47	.09465	7	1. 244	.10110	14	1. 262	.11038	17	1. 289	.11985	14	1. 318	.12705	9	1. 340	.13035	2	1. 350	13
48	.09472	7	1. 244	.10124	14	1. 263	.11055	17	1. 290	.11999	15	1. 318	.12714	9	1. 340	.13037	2	1. 350	12
49	.09479	8	1. 244	.10138	14	1. 263	.11072	16	1. 290	.12014	14	1. 319	.12723	9	1. 340	.13039	2	1. 350	11
50	.09487	7	1. 244	.10152	14	1. 263	.11088	17	1. 291	.12028	14	1. 319	.12732	9	1. 341	.13041	1	1. 350	10
51	.09494	8	1. 244	.10166	14	1. 264	.11105	16	1. 291	.12042	14	1. 320	.12741	8	1. 341	.13042	1	1. 350	9
52	.09502	8	1. 245	.10180	14	1. 264	.11121	16	1. 292	.12056	14	1. 320	.12740	9	1. 341	.13043	0	1. 350	8
53	.09510	8	1. 245	.10194	14	1. 265	.11137	16	1. 292	.12070	14	1. 320	.12758	8	1. 342	.13043	1	1. 350	7
54	.09518	8	1. 245	.10208	14	1. 265	.11153	17	1. 293	.12084	14	1. 321	.12766	8	1. 342	.13044	0	1. 350	6
55	.09526	8	1. 245	.10222	15	1. 265	.11170	16	1. 293	.12098	14	1. 321	.12774	8	1. 342	.13044	0	1. 350	5
56	.09534	9	1. 246	.10237	14	1. 266	.11186	17	1. 294	.12112	14	1. 322	.12782	8	1. 342	.13044	1	1. 350	4
57	.09543	9	1. 246	.10251	15	1. 266	.11203	16	1. 294	.12126	14	1. 322	.12790	8	1. 343	.13045	0	1. 350	3
58	.09552	9	1. 246	.10266	15	1. 267	.11219	17	1. 295	.12140	14	1. 323	.12798	8	1. 343	.13045	0	1. 350	2
59	.09561	8	1. 246	.10281	15	1. 267	.11236	16	1. 295	.12154	13	1. 323	.12806	8	1. 343	.13045	0	1. 350	1
60	.09569		1. 247	.10296		1. 268	.11252		1. 296	.12167		1. 323	.12814		1. 343	.13045		1. 350	0
	XI XXIII			X XXII			IX XXI			VIII XX			VII XIX			VI XVIII			

Diff. for Columns of Logs.

	6	10	12	15
10″	1	2	2	3
20	2	3	4	5
30	3	5	6	7
45	4	7	8	10
50	5	8	10	12

Add to the Logarithm in the Col. the log. sec. of the Star's Declination, and the Sum will give the log. of the Maximum Aberration in Right Ascension.

Or Multiply the Number in the Col. by the natural Secant of the Star's Declination.

TABLE IV.

Lunar Nutation in R. A. to find log. q. Argument =R. A. of ✱ in time.

[Tab. log. = log. sin. R. A. + .8562514.]

S. N. Min.	+ O − Log.	− XII − Diff.	+ I − Log.	− XIII + Diff.	+ II − Log.	− XIV + Diff.	+ III − Log.	− XV + Diff.	+ IV − Log.	− XVI + Diff.	+ V − Log.	− XVII + Diff.	S. N. Min.
0	− ∞		0. 26925		0. 55522		0. 70574		0. 79378		0. 84120		60
1	8. 49607		0. 27626	701	0. 55849	327	0. 70763	189	0. 79487	109	0. 84171	51	59
2	8. 79709	30102	0. 28315	689	0. 56172	323	0. 70950	187	0. 79595	108	0. 84221	50	58
3	8. 97318	17609	0. 28993	678	0. 56492	320	0. 71135	185	0. 79702	107	0. 84269	48	57
4	9. 09811	12493	0. 29659	666	0. 56809	317	0. 71319	184	0. 79807	105	0. 84316	47	56
5	9. 19500	9689	0. 30314	655	0. 57123	314	0. 71501	182	0. 79911	104	0. 84363	47	55
6	9. 27417	7917	0. 30959	645	0. 57434	311	0. 71682	181	0. 80014	103	0. 84409	46	54
7	9. 34110	6693	0. 31594	635	0. 57741	307	0. 71861	179	0. 80117	103	0. 84454	45	53
8	9. 40907	6797	0. 32219	625	0. 58046	305	0. 72038	177	0. 80219	102	0. 84498	44	52
9	9. 45020	4113	0. 32834	615	0. 58348	302	0. 72213	175	0. 80319	100	0. 84541	43	51
10	9. 49593	4573	0. 33439	605	0. 58647	299	0. 72387	174	0. 80418	99	0. 84583	42	50
11	9. 53728	4135	0. 34036	597	0. 58943	296	0. 72560	173	0. 80516	98	0. 84625	42	49
12	9. 57505	3777	0. 34623	587	0. 59236	293	0. 72732	172	0. 80613	97	0. 84666	41	48
13	9. 60978	3473	0. 35202	579	0. 59526	290	0. 72902	170	0. 80709	96	0. 84706	40	47
14	9. 64193	3215	0. 35773	571	0. 59814	288	0. 73070	168	0. 80804	95	0. 84745	39	46
15	9. 67285	3092	0. 36335	562	0. 60099	285	0. 73237	167	0. 80898	94	0. 84783	38	45
16	9. 69984	2699	0. 36889	554	0. 60381	282	0. 73403	166	0. 80991	93	0. 84820	37	44
17	9. 72612	2628	0. 37436	547	0. 60661	280	0. 73567	164	0. 81083	92	0. 84856	36	43
18	9. 75089	2477	0. 37975	539	0. 60938	277	0. 73730	163	0. 81174	91	0. 84891	35	42
19	9. 77432	2343	0. 38506	531	0. 61212	274	0. 73891	161	0. 81264	90	0. 84926	35	41
20	9. 79655	2223	0. 39030	524	0. 61484	272	0. 74051	160	0. 81353	89	0. 84960	34	40
21	9. 81768	2113	0. 39547	517	0. 61754	270	0. 74210	159	0. 81441	88	0. 84993	33	39
22	9. 83782	2014	0. 40058	511	0. 62021	267	0. 74367	157	0. 81528	87	0. 85025	32	38
23	9. 85707	1925	0. 40561	503	0. 62285	264	0. 74522	155	0. 81614	86	0. 85056	31	37
24	9. 87549	1842	0. 41058	497	0. 62547	262	0. 74675	153	0. 81698	84	0. 85087	31	36
25	9. 89315	1766	0. 41549	491	0. 62807	260	0. 74827	152	0. 81782	84	0. 85117	30	35
26	9. 91011	1696	0. 42033	484	0. 63064	257	0. 74978	151	0. 81865	83	0. 85146	29	34
27	9. 92643	1632	0. 42511	478	0. 63319	255	0. 75128	150	0. 81947	82	0. 85174	28	33
28	9. 94215	1572	0. 42983	472	0. 63571	252	0. 75278	150	0. 82028	81	0. 85200	26	32
29	9. 95731	1516	0. 43449	466	0. 63822	251	0. 75426	148	0. 82108	80	0. 85226	26	31
30	9. 97195	1464	0. 43909	460	0. 64070	248	0. 75572	146	0. 82187	79	0. 85251	25	30
31	9. 98611	1416	0. 44364	455	0. 64316	246	0. 75717	145	0. 82265	78	0. 85276	25	29
32	9. 99981	1370	0. 44813	449	0. 64559	243	0. 75860	143	0. 82342	77	0. 85300	24	28
33	0. 01308	1327	0. 45257	444	0. 64801	242	0. 76002	142	0. 82418	76	0. 85323	23	27
34	0. 02595	1287	0. 45695	438	0. 65040	239	0. 76143	141	0. 82493	75	0. 85245	22	26
35	0. 03845	1250	0. 46128	433	0. 65277	237	0. 76283	140	0. 82567	74	0. 85266	21	25
36	0. 05058	1213	0. 46556	428	0. 65512	235	0. 76421	138	0. 82640	73	0. 85387	21	24
37	0. 06238	1180	0. 46980	424	0. 65745	233	0. 76558	137	0. 82712	72	0. 85407	20	23
38	0. 07386	1148	0. 47398	418	0. 65976	231	0. 76694	136	0. 82783	71	0. 85426	19	22
39	0. 08504	1118	0. 47811	413	0. 66205	229	0. 76829	135	0. 82854	71	0. 85444	18	21
40	0. 09592	1088	0. 48220	409	0. 66432	227	0. 76962	133	0. 82924	70	0. 85460	16	20
41	0. 10653	1061	0. 48624	404	0. 66657	225	0. 77094	132	0. 82993	69	0. 85476	16	19
42	0. 11688	1035	0. 59024	400	0. 66880	223	0. 77225	131	0. 83061	68	0. 85491	15	18
43	0. 12699	1011	0. 49419	395	0. 67100	220	0. 77354	129	0. 83127	66	0. 85505	14	17
44	0. 13685	986	0. 49809	390	0. 67319	219	0. 77483	129	0. 83192	65	0. 85519	14	16
45	0. 14649	964	0. 50196	387	0. 67536	217	0. 77611	128	0. 83257	65	0. 85532	13	15
46	0. 15591	942	0. 50578	382	0. 67752	216	0. 77737	126	0. 83321	64	0. 85544	12	14
47	0. 16512	921	0. 50956	378	0. 67965	213	0. 77861	124	0. 83384	63	0. 85555	11	13
48	0. 17413	901	0. 51330	374	0. 68176	211	0. 77984	123	0. 83446	62	0. 85566	11	12
49	0. 18295	882	0. 51700	370	0. 68386	210	0. 78106	122	0. 83507	61	0. 85576	10	11
50	0. 19159	864	0. 52066	366	0. 68593	207	0. 78227	121	0. 83567	60	0. 85585	9	10
51	0. 20005	846	0. 52428	362	0. 68799	206	0. 78347	120	0. 83626	59	0. 85593	8	9
52	0. 20834	829	0. 52786	258	0. 69003	204	0. 78467	120	0. 83685	59	0. 85599	6	8
53	0. 21647	813	0. 53141	355	0. 69206	203	0. 78586	119	0. 83743	58	0. 85605	6	7
54	0. 22444	797	0. 53491	350	0. 69406	200	0. 78703	117	0. 83800	57	0. 85610	5	6
55	0. 23225	781	0. 53839	348	0. 69605	199	0. 78818	115	0. 83855	55	0. 85615	5	5
56	0. 23993	768	0. 54182	343	0. 69802	197	0. 78932	114	0. 83909	54	0. 85619	4	4
57	0. 24746	753	0. 54522	340	0. 69998	196	0. 79045	113	0. 83963	54	0. 85622	3	3
58	0. 25485	739	0. 54859	337	0. 70191	193	0. 79157	112	0. 84016	53	0. 85624	2	2
59	0. 26211	726	0. 55192	333	0. 70383	192	0. 79268	111	0. 84068	52	0. 85625	1	1
60	0. 26925	714	0. 55522	330	0. 70574	191	0. 79378	110	0. 84120	52	0. 85625	0	0
N. S.	− XI +	+ XXIII −	− X +	+ XXII −	− IX +	+ XXI −	− VIII +	+ XX −	− VII +	+ XIX −	− VI +	+ XVIII −	N. S.

Add the Logarithm found in this Table, to the log. tang. of the Star's Declination, and to the natural number corresponding to the result, with its proper sign, apply −16ˢ. 544, and take out the Logarithm, which call q.

TABLE IV. (Nat. Num.)

Lunar Nutation in R. A. to find q. Argument = R. A. of the * in time.

[Tab. No. = 7″.1822 × sin. R. A.]

S. N.	+ 0 −	− XII +	+ I −	− XIII +	+ II −	− XIV +	+ III −	− XV +	+ IV −	− XVI +	+ V −	− XVII +	S. N.
Min.		Diff.		Diff.		Diff.		Diff.		Diff.		Diff.	Min.
0	0″.000		1″.859		3″.591		5″.079		6″.219		6″.938		60
1	0.032	32	1.890	31	3.618	27	5.101	22	6.235	16	6.946	8	59
2	0.063	31	1.920	30	3.645	27	5.123	22	6.250	15	6.954	8	58
3	0.095	32	1.950	30	3.672	27	5.145	22	6.266	16	6.962	8	57
4	0.126	31	1.980	30	3.699	27	5.167	22	6.281	15	6.969	7	56
5	0.158	32	2.010	30	3.726	27	5.188	21	6.296	15	6.977	8	55
6	0.189	31	2.040	30	3.753	27	5.210	22	6.311	15	6.984	7	54
7	0.220	31	2.070	30	3.780	27	5.231	21	6.326	15	6.991	7	53
8	0.251	31	2.100	30	3.806	26	5.253	22	6.341	15	6.998	7	52
9	0.283	32	2.130	30	3.833	27	5.274	21	6.355	14	7.005	7	51
10	0.314	31	2.160	30	3.859	26	5.295	21	6.370	15	7.012	7	50
11	0.345	31	2.190	30	3.886	27	5.317	22	6.384	14	7.019	7	49
12	0.376	31	2.219	29	3.912	26	5.338	21	6.399	15	7.025	6	48
13	0.408	32	2.249	30	3.938	26	5.359	21	6.413	14	7.032	7	47
14	0.439	31	2.279	30	3.964	26	5.380	21	6.427	14	7.038	6	46
15	0.470	31	2.309	30	3.990	26	5.401	21	6.441	14	7.044	6	45
16	0.501	31	2.338	29	4.016	26	5.421	20	6.455	14	7.050	6	44
17	0.533	32	2.368	30	4.042	26	5.442	21	6.468	13	7.056	6	43
18	0.564	31	2.397	29	4.068	26	5.462	20	6.482	14	7.062	6	42
19	0.595	31	2.427	30	4.094	26	5.482	20	6.495	13	7.068	6	41
20	0.626	31	2.456	29	4.119	25	5.502	20	6.509	14	7.073	5	40
21	0.658	32	2.486	30	4.145	26	5.522	20	6.522	13	7.079	6	39
22	0.689	31	2.515	29	4.170	25	5.542	20	6.535	13	7.084	5	38
23	0.720	31	2.545	30	4.196	26	5.562	20	6.548	13	7.089	5	37
24	0.751	31	2.574	29	4.221	25	5.582	20	6.560	12	7.094	5	36
25	0.783	32	2.603	29	4.247	26	5.601	19	6.573	13	7.099	5	35
26	0.814	31	2.632	29	4.272	25	5.621	20	6.585	12	7.104	5	34
27	0.845	31	2.661	29	4.297	25	5.640	19	6.598	13	7.109	5	33
28	0.876	31	2.690	29	4.322	25	5.660	20	6.610	12	7.113	4	32
29	0.907	31	2.719	29	4.347	25	5.679	19	6.622	12	7.117	4	31
30	0.938	31	2.748	29	4.372	25	5.698	19	6.634	12	7.121	4	30
31	0.969	31	2.777	29	4.397	25	5.717	19	6.646	12	7.125	4	29
32	1.000	31	2.806	29	4.422	25	5.736	19	6.658	12	7.129	4	28
33	1.031	31	2.835	29	4.446	24	5.754	18	6.669	11	7.132	3	27
34	1.062	31	2.864	29	4.471	25	5.773	19	6.681	12	7.136	4	26
35	1.093	31	2.893	29	4.496	25	5.791	18	6.692	11	7.139	3	25
36	1.124	31	2.921	28	4.520	24	5.810	19	6.704	12	7.143	4	24
37	1.155	31	2.950	29	4.545	25	5.828	18	6.715	11	7.146	3	23
38	1.186	31	2.978	28	4.569	24	5.847	19	6.726	11	7.149	3	22
39	1.217	31	3.007	29	4.593	24	5.865	18	6.737	11	7.152	3	21
40	1.247	30	3.035	28	4.617	24	5.883	18	6.748	11	7.155	3	20
41	1.278	31	3.064	29	4.641	24	5.901	18	6.758	10	7.157	2	19
42	1.309	31	3.092	28	4.665	24	5.919	18	6.769	11	7.160	3	18
43	1.340	31	3.120	28	4.689	24	5.937	18	6.779	10	7.162	2	17
44	1.370	30	3.148	28	4.712	23	5.955	18	6.790	11	7.165	3	16
45	1.401	31	3.177	29	4.736	24	5.973	18	6.800	10	7.167	2	15
46	1.432	31	3.205	28	4.759	23	5.990	17	6.810	10	7.169	2	14
47	1.463	31	3.233	28	4.783	24	6.007	17	6.820	10	7.171	2	13
48	1.493	30	3.261	28	4.806	23	6.024	17	6.830	10	7.172	1	12
49	1.524	31	3.289	28	4.829	23	6.041	17	6.839	9	7.174	2	11
50	1.555	31	3.317	28	4.852	23	6.058	17	6.849	10	7.175	1	10
51	1.586	31	3.345	28	4.875	23	6.075	17	6.858	9	7.177	2	9
52	1.616	30	3.372	27	4.898	23	6.091	16	6.868	10	7.178	1	8
53	1.647	31	3.400	28	4.921	23	6.108	17	6.877	9	7.179	1	7
54	1.677	30	3.427	27	4.944	23	6.124	16	6.886	9	7.180	1	6
55	1.708	31	3.455	28	4.967	23	6.140	16	6.895	9	7.181	1	5
56	1.738	30	3.482	27	4.989	22	6.156	16	6.904	9	7.181	0	4
57	1.769	31	3.510	28	5.012	23	6.172	16	6.912	8	7.182	1	3
58	1.799	30	3.537	27	5.034	22	6.188	16	6.921	9	7.182	0	2
59	1.829	30	3.564	27	5.057	23	6.204	16	6.929	8	7.182	0	1
60	1.859	30	3.591	27	5.079	22	6.219	15	6.938	9	7.182	0	0
N. S.	− XI +	+ XXIII −	− X +	+ XXII −	− IX +	+ XXI −	− VIII +	+ XX −	− VII +	+ XIX −	− VI +	+ XVIII −	N. S.

Multiply the Numbers found in the Table by the natural Tangent of the Star's Declination, and apply − 16″.544, and the result gives q.

TABLE V.

Lunar Nutation in R. A. to find log. p. Argument = R. A. of the ✱ in time.

[Tab log. = Cos. R. A. +.9844373.]

S. N.	+ O − / − XII + Log.	Diff.	+ I − / − XIII + Log.	Diff.	+ II − / − XIV + Log.	Diff.	+ III − / − XV + Log.	Diff.	+ IV − / − XVI + Log.	Diff.	+ V − / − XVII + Log.	Diff.	S. N.
Min.													Min.
0	0.98444	1	0.96938	51	0.92197	110	0.83392	190	0.68341	330	0.39743	713	60
1	0.98443	1	0.96887	52	0.92087	111	0.83202	192	0.68011	334	0.39030	726	59
2	0.98442	2	0.96835	53	0.91976	113	0.83010	194	0.67677	336	0.38304	740	58
3	0.98440	3	0.96782	54	0.91863	113	0.82816	195	0.67341	340	0.37564	753	57
4	0.98437	4	0.96728	55	0.91750	114	0.82621	197	0.67001	344	0.36811	767	56
5	0.98433	4	0.96673	56	0.91636	116	0.82424	199	0.66657	347	0.36044	782	55
6	0.98429	6	0.96617	56	0.91520	117	0.82225	201	0.66310	351	0.35262	797	54
7	0.98423	6	0.96561	58	0.91403	117	0.82024	202	0.65959	355	0.34465	813	53
8	0.98417	7	0.96503	58	0.91286	119	0.81822	204	0.65604	358	0.33652	829	52
9	0.98410	8	0.96445	59	0.91167	121	0.81618	206	0.65246	362	0.32823	846	51
10	0.98402	8	0.96386	61	0.91046	121	0.81412	208	0.64884	366	0.31977	863	50
11	0.98394	10	0.96325	61	0.90925	122	0.81204	209	0.64518	370	0.31114	882	49
12	0.98384	10	0.96264	62	0.90803	124	0.80995	212	0.64148	374	0.30232	902	48
13	0.98374	12	0.96202	63	0.90679	125	0.80783	213	0.63774	378	0.29330	921	47
14	0.98362	12	0.96139	64	0.90554	126	0.80570	215	0.63396	382	0.28409	942	46
15	0.98350	12	0.96075	64	0.90428	127	0.80355	217	0.63014	386	0.27467	963	45
16	0.98338	14	0.96011	66	0.90301	128	0.80138	219	0.62628	391	0.26504	987	44
17	0.98324	15	0.95945	67	0.90173	130	0.79919	221	0.62237	395	0.25517	1010	43
18	0.98309	15	0.95878	67	0.90043	131	0.79698	223	0.61842	399	0.24507	1035	42
19	0.98294	16	0.95811	69	0.89912	132	0.79475	225	0.61443	405	0.23472	1061	41
20	0.98278	17	0.95742	69	0.89780	133	0.79250	226	0.61038	408	0.22411	1089	40
21	0.98261	18	0.95673	71	0.89647	135	0.79024	229	0.60630	414	0.21322	1118	39
22	0.98243	19	0.95602	71	0.89512	136	0.78795	231	0.60216	418	0.20204	1147	38
23	0.98224	19	0.95531	72	0.89376	137	0.78564	233	0.59798	423	0.19057	1180	37
24	0.98205	20	0.95459	73	0.89239	138	0.78331	235	0.59375	428	0.17877	1214	36
25	0.98185	22	0.95386	74	0.89101	140	0.78096	237	0.58947	433	0.16663	1249	35
26	0.98163	22	0.95311	75	0.88961	140	0.77859	240	0.58514	439	0.15414	1287	34
27	0.98141	22	0.95236	76	0.88821	143	0.77619	241	0.58075	444	0.14127	1328	33
28	0.98119	24	0.95160	77	0.88678	143	0.77378	244	0.57631	449	0.12799	1370	32
29	0.98095	25	0.95083	78	0.88535	145	0.77134	246	0.57182	454	0.11429	1416	31
30	0.98070	25	0.95005	79	0.88390	146	0.76888	248	0.56728	461	0.10013	1464	30
31	0.98045	26	0.94926	80	0.88244	147	0.76540	250	0.56267	466	0.08549	1516	29
32	0.98019	27	0.94846	81	0.88097	149	0.76390	253	0.55801	472	0.07033	1572	28
33	0.97992	28	0.94765	82	0.87948	150	0.76137	255	0.55329	478	0.05461	1631	27
34	0.97964	29	0.94683	83	0.87798	151	0.75882	257	0.54851	484	0.03830	1697	26
35	0.97935	29	0.94600	83	0.87647	153	0.75625	260	0.54367	490	0.02133	1766	25
36	0.97906	31	0.94517	85	0.87494	154	0.75365	262	0.53877	497	0.00367	1842	24
37	0.97875	31	0.94432	86	0.87340	156	0.75103	264	0.53380	504	9.98525	1924	23
38	0.97844	32	0.94346	87	0.87184	157	0.74839	267	0.52876	510	9.96601	2014	22
39	0.97812	33	0.94259	88	0.87027	158	0.74572	269	0.52366	517	9.94587	2114	21
40	0.97779	34	0.94171	89	0.86869	160	0.74303	272	0.51849	524	9.92473	2222	20
41	0.97745	35	0.94082	90	0.86709	161	0.74031	275	0.51325	532	9.90251	2343	19
42	0.97710	36	0.93992	90	0.86548	162	0.73756	277	0.50793	539	9.87908	2478	18
43	0.97674	36	0.93902	92	0.86386	164	0.73479	279	0.50254	546	9.85430	2628	17
44	0.97638	37	0.93810	92	0.86222	166	0.73200	283	0.49708	554	9.82802	2798	16
45	0.97601	38	0.93718	95	0.86056	167	0.72917	285	0.49154	563	9.80004	2993	15
46	0.97563	39	0.93623	95	0.85889	168	0.72632	287	0.48591	570	9.77011	3215	14
47	0.97524	40	0.93528	96	0.85721	170	0.72345	291	0.48021	579	9.73795	3472	13
48	0.97484	41	0.93432	97	0.85551	171	0.72054	293	0.47442	588	9.70324	3776	12
49	0.97443	41	0.93335	99	0.85380	173	0.71761	296	0.46854	596	9.66548	4136	11
50	0.97402	43	0.93236	99	0.85207	175	0.71465	299	0.46258	606	9.62412	4574	10
51	0.97359	43	0.93137	100	0.85032	176	0.71166	301	0.45652	615	9.57838	5112	9
52	0.97316	44	0.93037	101	0.84856	177	0.70865	305	0.45037	625	9.52726	5798	8
53	0.97272	45	0.92936	103	0.84679	179	0.70560	308	0.44412	634	9.46928	6692	7
54	0.97227	46	0.92833	103	0.84500	181	0.70252	311	0.43778	645	9.40236	7917	6
55	0.97181	47	0.92730	104	0.84319	182	0.69941	314	0.43133	656	9.32319	9690	5
56	0.97134	48	0.92626	106	0.84137	184	0.69627	316	0.42477	666	9.22629	12493	4
57	0.97086	48	0.92520	107	0.83953	185	0.69311	320	0.41811	677	9.10136	17608	3
58	0.97038	50	0.92413	107	0.83768	187	0.68991	324	0.41134	690	8.92528	30103	2
59	0.96988	50	0.92306	109	0.83581	189	0.68667	326	0.40444	701	8.62425		1
60	0.96938		0.92197		0.83392		0.68341		0.39743		− ∞		0
N. S.	+ XI − / − XXIII +		+ X − / − XXII +		+ IX − / − XXI +		+ VIII − / − XX +		+ VII − / − XIX +		+ VI − / − XVIII +		N. S.

Positive Argument.

Cases		
1	$\begin{cases} p+ \\ q+ \end{cases}$	= L
2	$^{p}_{q}\begin{cases} + \\ - \end{cases}$	$= \overset{s}{6} - L$
3	$\begin{cases} p- \\ q- \end{cases}$	$= \overset{s}{6} + L$
4	$\begin{cases} p- \\ q+ \end{cases}$	$= \overset{s}{12} - L$

Add the Logarithm found in the Table to the log. tangent of the ✱'s Declination, and the result gives log. p.

"Log. tangent L = log. p − log. q; and log. Max. Nutation = log. p + log. cosec. L in space.

or = log. p + 8.82391 + log. cosec. L in time.

TABLE V. (Nat. Num.)

Lunar Nutation in R. A. to find p. Argument = R. A. of the * in time.

[Tab. No. = 9″.648 × Cos. R. A.]

S. N. Min.	+ 0 − / − XII +	Diff.	+ I − / − XIII +	Diff.	+ II − / − XIV +	Diff.	+ III − / − XV +	Diff.	+ IV − / − XVI +	Diff.	+ V − / − XVII +	Diff.	S. N. Min.
0	9″.648	0	9″.319	11	8″.355	21	6″.823	30	4″.824	36	2″.497	40	60
1	9 .648	1	9 .308	11	8 .334	21	6 .793	30	4 .788	37	2 .457	41	59
2	9 .647	0	9 .297	11	8 .313	21	6 .763	30	4 .751	37	2 .416	41	58
3	9 .647	1	9 .286	12	8 .292	22	6 .733	31	4 .714	37	2 .375	41	57
4	9 .646	1	9 .274	12	8 .270	22	6 .702	30	4 .677	37	2 .334	41	56
5	9 .645	1	9 .262	12	8 .248	22	6 .672	30	4 .640	37	2 .293	41	55
6	9 .644	1	9 .250	12	8 .226	22	6 .642	31	4 .603	37	2 .252	41	54
7	9 .643	1	9 .238	12	8 .204	22	6 .611	31	4 .566	37	2 .211	41	53
8	9 .642	1	9 .226	12	8 .182	22	6 .580	31	4 .529	37	2 .170	41	52
9	9 .641	2	9 .214	12	8 .160	23	6 .549	31	4 .492	37	2 .129	41	51
10	9 .639	2	9 .202	13	8 .137	23	6 .518	31	4 .455	37	2 .088	41	50
11	9 .637	2	9 .189	13	8 .114	23	6 .487	31	4 .418	38	2 .047	41	49
12	9 .635	2	9 .176	13	8 .091	23	6 .456	31	4 .380	37	2 .006	41	48
13	9 .633	2	9 .163	13	8 .068	23	6 .425	31	4 .343	38	1 .965	41	47
14	9 .631	3	9 .150	13	8 .045	23	6 .394	32	4 .305	38	1 .924	41	46
15	9 .628	3	9 .137	14	8 .022	24	6 .362	32	4 .267	38	1 .883	42	45
16	9 .625	3	9 .123	14	7 .998	23	6 .330	32	4 .229	38	1 .841	41	44
17	9 .622	3	9 .109	14	7 .975	24	6 .298	32	4 .191	38	1 .800	42	43
18	9 .619	4	9 .095	14	7 .951	24	6 .266	32	4 .153	38	1 .758	41	42
19	9 .615	4	9 .081	15	7 .927	24	6 .234	32	4 .115	38	1 .717	42	41
20	9 .611	4	9 .066	14	7 .903	24	6 .202	32	4 .077	38	1 .675	41	40
21	9 .607	4	9 .052	15	7 .879	24	6 .170	32	4 .039	38	1 .634	42	39
22	9 .603	4	9 .037	15	7 .855	25	6 .138	33	4 .001	38	1 .592	41	38
23	9 .599	4	9 .022	15	7 .830	25	6 .105	33	3 .963	39	1 .551	42	37
24	9 .595	4	9 .007	15	7 .805	25	6 .072	33	3 .924	38	1 .509	41	36
25	9 .591	5	8 .992	15	7 .780	25	6 .039	33	3 .886	39	1 .468	42	35
26	9 .586	5	8 .977	15	7 .755	25	6 .006	33	3 .847	38	1 .426	41	34
27	9 .581	5	8 .962	16	7 .730	25	5 .973	33	3 .809	39	1 .385	42	33
28	9 .576	5	8 .946	16	7 .705	25	5 .940	33	3 .770	39	1 .343	42	32
29	9 .571	5	8 .930	16	7 .680	25	5 .907	33	3 .731	39	1 .301	41	31
30	9 .566	6	8 .914	16	7 .655	26	5 .874	34	3 .692	39	1 .260	42	30
31	9 .560	6	8 .898	17	7 .629	26	5 .840	34	3 .653	39	1 .218	42	29
32	9 .554	6	8 .881	16	7 .603	26	5 .806	33	3 .614	39	1 .176	41	28
33	9 .548	6	8 .865	17	7 .577	26	5 .773	34	3 .575	39	1 .135	42	27
34	9 .542	6	8 .848	17	7 .551	26	5 .739	34	3 .536	39	1 .093	42	26
35	9 .536	7	8 .831	17	7 .525	27	5 .705	34	3 .497	39	1 .051	42	25
36	9 .529	7	8 .814	17	7 .498	26	5 .671	34	3 .458	39	1 .009	42	24
37	9 .522	7	8 .797	17	7 .472	27	5 .637	34	3 .419	39	0 .967	42	23
38	9 .515	7	8 .780	18	7 .445	27	5 .603	34	3 .380	40	0 .925	42	22
39	9 .508	7	8 .762	18	7 .418	27	5 .569	35	3 .340	40	0 .883	42	21
40	9 .501	7	8 .744	18	7 .391	27	5 .534	34	3 .300	39	0 .841	42	20
41	9 .494	8	8 .726	18	7 .364	27	5 .500	35	3 .261	40	0 .799	42	19
42	9 .486	8	8 .708	18	7 .337	28	5 .465	35	3 .221	40	0 .757	42	18
43	9 .478	8	8 .690	19	7 .309	28	5 .430	35	3 .181	40	0 .715	42	17
44	9 .470	8	8 .671	18	7 .281	28	5 .395	35	3 .141	40	0 .673	42	16
45	9 .462	8	8 .653	19	7 .253	28	5 .360	35	3 .101	40	0 .631	42	15
46	9 .454	8	8 .634	19	7 .225	28	5 .325	35	3 .061	40	0 .589	42	14
47	9 .446	9	8 .615	19	7 .197	28	5 .290	35	3 .021	40	0 .547	42	13
48	9 .437	9	8 .596	19	7 .169	28	5 .255	35	2 .981	40	0 .505	42	12
49	9 .428	9	8 .577	19	7 .141	28	5 .220	35	2 .941	40	0 .463	42	11
50	9 .419	9	8 .558	19	7 .113	28	5 .185	36	2 .901	40	0 .421	42	10
51	9 .410	9	8 .539	20	7 .085	29	5 .149	36	2 .861	40	0 .379	42	9
52	9 .401	10	8 .519	20	7 .056	28	5 .113	36	2 .821	40	0 .337	42	8
53	9 .391	10	8 .499	20	7 .028	29	5 .077	36	2 .781	41	0 .295	42	7
54	9 .381	10	8 .479	20	6 .999	29	5 .041	36	2 .740	40	0 .253	42	6
55	9 .371	10	8 .459	21	6 .970	29	5 .005	36	2 .700	41	0 .211	42	5
56	9 .361	10	8 .438	20	6 .941	29	4 .969	36	2 .659	40	0 .169	42	4
57	9 .351	10	8 .418	21	6 .912	29	4 .933	36	2 .619	41	0 .127	42	3
58	9 .341	11	8 .397	21	6 .883	30	4 .897	36	2 .578	40	0 .085	43	2
59	9 .330	11	8 .376	21	6 .853	30	4 .861	37	2 .538	41	0 .042	42	1
60	9 .319		8 .355		6 .823		4 .824		2 .497		0 .000		0
N. S.	+ XI − / − XXIII +		+ X − / − XXII +		+ IX − / − XXI +		+ VIII − / − XX +		+ VII − / − XIX +		+ VI − / − XVIII +		N. S.

Positive Argument.

Cases	
1 $\begin{cases} p+ \\ q+ \end{cases}$	= L
2 $\begin{cases} p+ \\ q- \end{cases}$	$= 6^s - L$
3 $\begin{cases} p- \\ q- \end{cases}$	$= 6^s + L$
4 $\begin{cases} p- \\ q+ \end{cases}$	$= 12^s - L$

Multiply the Numbers found in this Table by the natural Tangent of the *'s Declination, and the result gives p.

Tangent L $= \frac{p}{q}$, and Max. Nut. $= \frac{p}{\text{sine L}}$, or $= \frac{q}{\text{cosine L}}$ in space; or $= \frac{1}{15} \times \frac{p}{\text{sine L}}$, or $= \frac{1}{15} \times \frac{q}{\text{cosine L}}$ in time.

TABLE VI.

Solar Nutation in R. A. to find q. and log. q. Argument = R. A. of the * in time.
[Tab. log. = log. sin. R. A. +9 .6001. Tab. Num. = .3982 × sin. R. A.]

S. N.	+ O − XII − +			+ I − XIII − +			+ II − XIV − +			+ III − XV − +			+ IV − XVI − +			+ V − XVII − +			S. N.
M	Log.	Diff	Num.	Log.	Diff	Num.	Log.	Diff.	Num.	Log.	Diff	Num.	Log.	Diff.	Num.	Log.	Diff.	Num.	M.
0	−∞		0″.000	9.0131		0″.103	9.2991		0″.199	9.4496		0″.281	9.5376		0″.345	9.5850		0″.384	60
1	7.2399		0.002	9.0200	69	0.105	9.3023	32	0.200	9.4514	18	0.283	9.5387	11	0.346	9.5855	5	0.385	59
2	7.5409	3010	0.004	9.0269	69	0.106	9.3055	32	0.202	9.4533	19	0.284	9.5398	11	0.346	9.5860	5	0.385	58
3	7.7170	1761	0.006	9.0337	68	0.108	9.3087	32	0.203	9.4551	18	0.285	9.5409	11	0.347	9.5865	5	0.386	57
4	7.8420	1250	0.007	9.0404	67	0.110	9.3119	32	0.205	9.4570	19	0.286	9.5419	10	0.348	9.5870	5	0.386	56
5	7.9389	969	0.009	9.0470	66	0.111	9.3150	31	0.206	9.4588	18	0.288	9.5430	11	0.349	9.5875	5	0.387	55
6	8.0180	691	0.010	9.0535	65	0.113	9.3181	31	0.208	9.4606	18	0.289	9.5440	10	0.350	9.5879	4	0.387	54
7	8.0849	669	0.012	9.0598	63	0.115	9.3212	31	0.209	9.4624	18	0.290	9.5450	10	0.351	9.5884	5	0.387	53
8	8.1429	580	0.014	9.0660	62	0.116	9.3243	31	0.211	9.4642	18	0.291	9.5460	10	0.351	9.5888	4	0.388	52
9	8.1940	511	0.016	9.0721	61	0.118	9.3273	30	0.212	9.4660	18	0.292	9.5470	10	0.352	9.5893	5	0.388	51
10	8.2398	468	0.017	9.0782	61	0.120	9.3303	30	0.214	9.4677	17	0.293	9.5480	10	0.353	9.5897	4	0.389	50
11	8.2811	413	0.019	9.0842	60	0.121	9.3332	29	0.215	9.4695	18	0.295	9.5490	10	0.354	9.5901	4	0.389	49
12	8.3189	378	0.021	9.0901	59	0.123	9.3362	30	0.217	9.4712	17	0.296	9.5500	10	0.355	9.5905	4	0.389	48
13	8.3536	347	0.023	9.0959	58	0.125	9.3391	29	0.218	9.4729	17	0.297	9.5509	9	0.355	9.5909	4	0.390	47
14	8.3858	322	0.024	9.1016	57	0.126	9.3420	29	0.220	9.4746	17	0.298	9.5519	10	0.356	9.5913	4	0.390	46
15	8.4157	299	0.026	9.1072	56	0.128	9.3449	29	0.221	9.4762	16	0.299	9.5528	9	0.357	9.5917	4	0.390	45
16	8.4437	280	0.028	9.1127	55	0.130	9.3477	28	0.223	9.4779	17	0.300	9.5537	9	0.358	9.5920	3	0.391	44
17	8.4700	263	0.029	9.1182	55	0.131	9.3505	28	0.224	9.4795	16	0.302	9.5547	10	0.359	9.5924	4	0.391	43
18	8.4947	247	0.031	9.1236	54	0.133	9.3532	27	0.226	9.4811	16	0.303	9.5556	9	0.359	9.5927	3	0.391	42
19	8.5182	235	0.033	9.1289	53	0.135	9.3560	28	0.227	9.4827	16	0.304	9.5565	9	0.360	9.5931	4	0.392	41
20	8.5404	222	0.035	9.1341	52	0.136	9.3587	27	0.228	9.4843	16	0.305	9.5574	9	0.361	9.5934	3	0.392	40
21	8.5615	211	0.037	9.1393	52	0.138	9.3614	27	0.230	9.4859	16	0.306	9.5583	9	0.362	9.5938	4	0.392	39
22	8.5817	202	0.038	9.1444	51	0.140	9.3640	26	0.231	9.4875	16	0.307	9.5591	8	0.362	9.5941	3	0.393	38
23	8.6009	192	0.040	9.1494	50	0.141	9.3667	27	0.233	9.4890	15	0.308	9.5600	9	0.363	9.5944	3	0.393	37
24	8.6193	184	0.042	9.1544	50	0.143	9.3693	26	0.234	9.4906	16	0.309	9.5608	8	0.364	9.5947	3	0.393	36
25	8.6370	177	0.043	9.1593	49	0.145	9.3719	26	0.235	9.4921	15	0.310	9.5617	9	0.364	9.5950	3	0.393	35
26	8.6540	170	0.045	9.1642	49	0.146	9.3745	26	0.237	9.4936	15	0.311	9.5625	8	0.365	9.5953	3	0.394	34
27	8.6703	163	0.047	9.1690	48	0.148	9.3771	26	0.238	9.4951	15	0.313	9.5633	8	0.366	9.5955	2	0.394	33
28	8.6860	157	0.048	9.1737	47	0.150	9.3796	25	0.240	9.4966	15	0.314	9.5641	8	0.366	9.5958	3	0.394	32
29	8.7012	152	0.050	9.1783	46	0.151	9.3820	24	0.241	9.4980	14	0.315	9.5649	8	0.367	9.5961	3	0.394	31
30	8.7158	146	0.052	9.1829	46	0.153	9.3845	25	0.242	9.4995	15	0.316	9.5657	8	0.368	9.5963	2	0.395	30
31	8.7300	142	0.054	9.1875	46	0.154	9.3869	24	0.243	9.5009	14	0.317	9.5665	8	0.368	9.5966	3	0.395	29
32	8.7437	137	0.055	9.1920	45	0.156	9.3894	25	0.245	9.5024	15	0.318	9.5673	8	0.369	9.5968	2	0.395	28
33	8.7569	132	0.057	9.1964	44	0.158	9.3918	24	0.246	9.5038	14	0.319	9.5680	7	0.370	9.5971	3	0.395	27
34	8.7698	129	0.059	9.2007	43	0.159	9.3942	24	0.248	9.5052	14	0.320	9.5688	8	0.370	9.5973	2	0.396	26
35	8.7823	125	0.060	9.2051	44	0.161	9.3966	24	0.249	9.5066	14	0.321	9.5695	7	0.371	9.5975	2	0.396	25
36	8.7944	121	0.062	9.2094	43	0.163	9.3990	24	0.251	9.5080	14	0.322	9.5702	7	0.372	9.5977	2	0.396	24
37	8.8062	118	0.064	9.2136	42	0.164	9.4013	23	0.252	9.5094	14	0.323	9.5710	8	0.372	9.5979	2	0.396	23
38	8.8177	115	0.066	9.2177	41	0.165	9.4036	23	0.253	9.5108	14	0.324	9.5717	7	0.373	9.5981	2	0.396	22
39	8.8289	112	0.067	9.2219	42	0.167	9.4059	23	0.254	9.5121	13	0.325	9.5724	7	0.374	9.5983	2	0.396	21
40	8.8398	109	0.069	9.2260	41	0.169	9.4082	23	0.256	9.5135	14	0.326	9.5731	7	0.374	9.5984	1	0.397	20
41	8.8504	106	0.071	9.2300	40	0.170	9.4104	22	0.257	9.5148	13	0.327	9.5738	7	0.375	9.5986	2	0.397	19
42	8.8607	103	0.073	9.2340	40	0.172	9.4126	22	0.259	9.5161	13	0.328	9.5745	7	0.375	9.5987	1	0.397	18
43	8.8708	101	0.074	9.2380	40	0.173	9.4148	22	0.260	9.5174	13	0.329	9.5752	7	0.376	9.5989	2	0.397	17
44	8.8807	99	0.076	9.2419	39	0.175	9.4170	22	0.261	9.5187	13	0.330	9.5758	6	0.376	9.5990	1	0.397	16
45	8.8904	97	0.078	9.2457	38	0.176	9.4191	21	0.262	9.5199	12	0.331	9.5765	7	0.377	9.5992	2	0.397	15
46	8.8998	94	0.079	9.2495	38	0.178	9.4213	22	0.264	9.5212	13	0.332	9.5771	6	0.378	9.5993	1	0.397	14
47	8.9090	92	0.081	9.2533	38	0.179	9.4234	21	0.265	9.5224	12	0.333	9.5777	6	0.378	9.5994	1	0.397	13
48	8.9180	90	0.083	9.2571	38	0.181	9.4256	22	0.266	9.5237	13	0.334	9.5783	6	0.379	9.5995	1	0.398	12
49	8.9268	88	0.084	9.2608	37	0.182	9.4277	21	0.268	9.5249	12	0.335	9.5789	6	0.379	9.5996	1	0.398	11
50	8.9354	86	0.086	9.2644	36	0.184	9.4298	21	0.269	9.5261	12	0.336	9.5795	6	0.380	9.5997	1	0.398	10
51	8.9439	85	0.088	9.2681	37	0.185	9.4318	20	0.270	9.5273	12	0.337	9.5801	6	0.380	9.5997	0	0.398	9
52	8.9522	83	0.090	9.2717	36	0.187	9.4339	21	0.271	9.5285	12	0.338	9.5807	6	0.381	9.5998	1	0.398	8
53	8.9603	81	0.091	9.2752	35	0.188	9.4359	20	0.273	9.5297	12	0.339	9.5813	6	0.381	9.5998	0	0.398	7
54	8.9683	80	0.093	9.2787	35	0.190	9.4379	20	0.274	9.5309	12	0.339	9.5818	5	0.382	9.5999	1	0.398	6
55	8.9761	78	0.095	9.2822	35	0.191	9.4399	20	0.275	9.5320	11	0.340	9.5824	6	0.382	9.5999	0	0.398	5
56	8.9838	77	0.096	9.2857	35	0.193	9.4418	19	0.276	9.5332	12	0.341	9.5829	5	0.383	9.6000	1	0.398	4
57	8.9913	75	0.098	9.2891	34	0.194	9.4438	20	0.278	9.5343	11	0.342	9.5835	6	0.383	9.6000	0	0.398	3
58	8.9987	74	0.100	9.2924	33	0.196	9.4457	19	0.279	9.5354	11	0.343	9.5840	5	0.384	9.6001	1	0.398	2
59	9.0060	73	0.101	9.2958	34	0.198	9.4477	20	0.280	9.5365	11	0.344	9.5845	5	0.384	9.6001	0	0.398	1
60	9.0131	71	0.103	9.2991	33	0.199	9.4496	19	0.281	9.5376	11	0.345	9.5850	5	0.384	9.6001	0	0.398	0
N. S.	− + XII XXIII + −			− + X XXII + −			− + IX XXI + −			− + VIII XX + −			− + VII XIX + −			− + VI XVIII + −			N. S.

Add to the log. in the Table the log. tangent of the Star's Declination, and to the natural number corresponding to this sum apply the constant quantity −0″.917, and the log. of the result gives log. q. Or, multiply the number in the Table by the natural tang. of the Star's Declination, and apply −0″.917, and the result gives q.

TABLE VII.

Solar Nutation in R. A. to find p, and log. p. Argument = R. A. of the ∗ in time.

[Tab. log. = log. cos. R. A. +9 .6374897. Tab. Num. = .434 × cos. R. A.]

S. N.	+ O −		− XII +	+ I −		− XIII +	+ II −		− XIV +	+ III −		− XV +	+ IV −		− XVI +	+ V −		− XVII +	S. N.
M.	Log.	Diff.	Num.	Log.	Diff.	Num.	Log.	Diff.	Num.	Log.	Diff.	Num.	Log.	Diff.	Num.	Log.	Diff.	Num.	M
0	9.6375	0	0″.434	9.6224	5	0″.419	9.5750	11	0″.376	9.4870	19	0″.307	9.3365	33	0″.217	9.0505	71	0″.112	60
1	9.6375	0	.434	9.6219	5	.419	9.5739	11	.375	9.4851	20	.305	9.3332	34	.215	9.0434	73	.110	59
2	9.6375	1	.434	9.6214	5	.418	9.5728	11	.374	9.4831	19	.304	9.3298	33	.214	9.0361	74	.109	58
3	9.6374	0	.434	9.6209	6	.418	9.5717	11	.373	9.4812	20	.302	9.3265	34	.212	9.0287	75	.107	57
4	9.6374	1	.434	9.6203	5	.417	9.5706	12	.372	9.4792	19	.301	9.3231	35	.210	9.0212	77	.105	56
5	9.6373	0	.434	9.6198	6	.417	9.5694	11	.371	9.4773	20	.300	9.3196	35	.209	9.0135	78	.103	55
6	9.6373	1	.434	9.6192	5	.416	9.5683	12	.370	9.4753	20	.299	9.3161	35	.207	9.0057	80	.101	54
7	9.6372	0	.434	9.6187	6	.416	9.5671	12	.369	9.4733	20	.297	9.3126	35	.205	8.9977	81	.099	53
8	9.6372	1	.434	9.6181	6	.415	9.5659	12	.368	9.4713	21	.296	9.3091	36	.204	8.9896	83	.097	52
9	9.6371	0	.434	9.6175	6	.415	9.5647	12	.367	9.4692	20	.295	9.3055	37	.202	8.9813	85	.096	51
10	9.6371	1	.433	9.6169	6	.414	9.5635	12	.366	9.4672	21	.293	9.3018	36	.200	8.9728	86	.094	50
11	9.6370	1	.433	9.6163	6	.414	9.5623	12	.365	9.4651	21	.292	9.2982	37	.199	8.9642	88	.092	49
12	9.6369	1	.433	9.6157	6	.413	9.5611	13	.364	9.4630	22	.290	9.2945	38	.197	8.9554	90	.090	48
13	9.6368	1	.433	9.6151	6	.413	9.5598	12	.363	9.4608	21	.289	9.2907	38	.195	8.9464	92	.088	47
14	9.6367	1	.433	9.6145	6	.412	9.5586	13	.362	9.4587	22	.288	9.2869	38	.194	8.9372	94	.087	46
15	9.6366	2	.433	9.6139	7	.411	9.5573	12	.361	9.4565	21	.286	9.2831	38	.192	8.9278	97	.085	45
16	9.6364	1	.433	9.6132	6	.410	9.5561	13	.360	9.4544	22	.285	9.2793	39	.190	8.9181	99	.083	44
17	9.6363	2	.433	9.6126	7	.410	9.5548	13	.359	9.4522	22	.283	9.2754	40	.189	8.9082	101	.081	43
18	9.6361	1	.432	9.6119	7	.409	9.5535	13	.358	9.4500	22	.282	9.2714	40	.187	8.8981	103	.079	42
19	9.6360	2	.432	9.6112	7	.409	9.5522	13	.357	9.4478	22	.280	9.2674	40	.185	8.8878	106	.077	41
20	9.6358	1	.432	9.6105	7	.408	9.5509	14	.356	9.4456	23	.279	9.2634	41	.183	8.8772	109	.075	40
21	9.6357	2	.432	9.6098	7	.407	9.5495	13	.355	9.4433	23	.278	9.2593	42	.182	8.8663	112	.074	39
22	9.6355	2	.432	9.6091	7	.407	9.5482	14	.354	9.4410	23	.276	9.2551	41	.180	8.8551	115	.072	38
23	9.6353	2	.432	9.6084	8	.406	9.5468	14	.352	9.4387	23	.275	9.2510	42	.178	8.8436	118	.070	37
24	9.6351	2	.432	9.6076	7	.405	9.5454	14	.351	9.4364	24	.273	9.2468	43	.177	8.8318	121	.068	36
25	9.6349	2	.432	9.6069	7	.405	9.5440	14	.350	9.4340	24	.272	9.2425	44	.175	8.8197	125	.066	35
26	9.6347	2	.431	9.6062	8	.404	9.5426	14	.349	9.4316	24	.270	9.2381	43	.173	8.8072	129	.064	34
27	9.6345	3	.431	9.6054	7	.403	9.5412	14	.348	9.4292	24	.269	9.2338	44	.171	8.7943	132	.062	33
28	9.6342	2	.431	9.6047	8	.402	9.5398	15	.347	9.4268	25	.267	9.2294	45	.170	8.7811	137	.060	32
29	9.6340	3	.431	9.6039	8	.402	9.5383	14	.346	9.4243	24	.266	9.2249	46	.168	8.7674	141	.059	31
30	9.6337	2	.430	9.6031	8	.401	9.5369	15	.345	9.4219	25	.264	9.2203	46	.166	8.7532	146	.057	30
31	9.6335	3	.430	9.6023	8	.400	9.5354	14	.343	9.4194	24	.263	9.2157	46	.164	8.7386	152	.055	29
32	9.6332	3	.430	9.6015	8	.400	9.5340	15	.342	9.4170	25	.261	9.2111	47	.163	8.7234	157	.053	28
33	9.6329	2	.430	9.6007	8	.399	9.5325	15	.341	9.4145	26	.260	9.2064	48	.161	8.7077	165	.051	27
34	9.6327	3	.429	9.5999	8	.398	9.5310	15	.340	9.4119	26	.258	9.2016	49	.159	8.6914	170	.049	26
35	9.6324	3	.429	9.5991	9	.397	9.5295	15	.339	9.4093	26	.257	9.1967	49	.157	8.6744	177	.047	25
36	9.6321	3	.429	9.5982	8	.397	9.5280	16	.337	9.4067	26	.255	9.1918	50	.156	8.6567	184	.045	24
37	9.6318	3	.428	9.5974	9	.396	9.5264	15	.336	9.4041	27	.254	9.1868	50	.154	8.6383	192	.043	23
38	9.6315	3	.428	9.5965	8	.395	9.5249	16	.335	9.4014	26	.252	9.1818	51	.152	8.6195	202	.041	22
39	9.6312	4	.428	9.5957	9	.394	9.5233	16	.334	9.3988	27	.251	9.1767	52	.150	8.5989	211	.040	21
40	9.6308	3	.427	9.5948	9	.393	9.5217	16	.333	9.3961	27	.249	9.1715	52	.148	8.5778	222	.038	20
41	9.6305	4	.427	9.5939	9	.393	9.5201	16	.332	9.3934	28	.247	9.1663	53	.147	8.5556	235	.036	19
42	9.6301	3	.427	9.5930	9	.392	9.5185	16	.330	9.3906	27	.246	9.1610	54	.145	8.5321	247	.034	18
43	9.6298	4	.426	9.5921	10	.391	9.5169	16	.329	9.3879	28	.244	9.1556	55	.143	8.5074	263	.032	17
44	9.6294	3	.426	9.5911	9	.390	9.5153	17	.327	9.3851	28	.243	9.1501	55	.141	8.4811	280	.030	16
45	9.6291	4	.426	9.5902	9	.390	9.5136	16	.326	9.3823	29	.241	9.1446	56	.140	8.4531	299	.028	15
46	9.6287	4	.425	9.5893	10	.389	9.5120	17	.325	9.3794	29	.240	9.1390	57	.138	8.4232	322	.027	14
47	9.6283	4	.425	9.5883	9	.388	9.5103	17	.324	9.3765	29	.238	9.1333	58	.136	8.3910	347	.025	13
48	9.6279	4	.425	9.5874	10	.387	9.5086	17	.322	9.3736	30	.236	9.1275	59	.134	8.3563	378	.023	12
49	9.6275	4	.424	9.5864	10	.386	9.5069	18	.321	9.3706	29	.235	9.1216	60	.132	8.3185	413	.021	11
50	9.6271	4	.424	9.5854	10	.385	9.5051	17	.320	9.3677	30	.233	9.1156	61	.13[illegible]	8.2772	468	.019	10
51	9.6267	5	.423	9.5844	10	.384	9.5034	18	.319	9.3647	30	.232	9.1095	61	.129	8.2314	511	.017	9
52	9.6262	4	.423	9.5834	10	.383	9.5016	18	.317	9.3617	31	.230	9.1034	62	.127	8.1803	580	.015	8
53	9.6258	5	.422	9.5824	10	.383	9.4998	18	.316	9.3586	31	.228	9.0972	63	.125	8.1223	669	.013	7
54	9.6253	4	.422	9.5814	10	.382	9.4980	18	.315	9.3555	31	.227	9.0909	65	.123	8.0554	791	.011	6
55	9.6249	5	.421	9.5804	11	.381	9.4962	18	.313	9.3524	31	.225	9.0844	66	.121	7.9763	969	.009	5
56	9.6244	5	.421	9.5793	10	.380	9.4944	19	.312	9.3493	32	.224	9.0778	67	.120	7.8794	1250	.008	4
57	9.6239	5	.420	9.5783	11	.379	9.4925	18	.311	9.3461	32	.222	9.0711	68	.118	7.7544	1761	.006	3
58	9.6234	5	.420	9.5772	11	.378	9.4907	19	.310	9.3429	32	.220	9.0643	69	.116	7.5783	3000	.004	2
59	9.6229	5	.419	9.5761	11	.377	9.4888	18	.308	9.3397	32	.219	9.0574	69	.114	7.2783		.002	1
60	9.6224		.419	9.5750		.376	9.4870		.307	9.3365		.217	9.0505		.112	− ∞		.000	0
N. S.	+ XI −		− XXIII +	+ X −		− XXII +	+ IX −		− XXI +	+ VIII −		− XX +	+ VII −		− XIX +	+ VI −		− XVIII +	N. S.

Positive Argument.

Cases		
1	$\left\{ \begin{matrix} p+ \\ q+ \end{matrix} \right.$	= S
2	$\left\{ \begin{matrix} p+ \\ q- \end{matrix} \right.$	= $\overset{s}{6}$ − S
3	$\left\{ \begin{matrix} p- \\ q- \end{matrix} \right.$	= $\overset{s}{6}$ + S
4	$\left\{ \begin{matrix} p- \\ q+ \end{matrix} \right.$	= $\overset{s}{12}$ − S

Add the log. in the Table to the log. tang. of the ∗'s Declin. and the result is log. p.

Or multiply the numbers by the natural tang. of the ∗'s Declin. and the result is p.

Tangent $S = \frac{p}{q}$, and Max. Sol. Nut. $= \frac{p}{\text{sine S}}$, or $= \frac{q}{\text{cosine S}}$. Or log. tang. S = log. p − log. q, and log. max. in time = log. p + 8.82391 + log. cosec. S.
= log. q + 8.82391 + log. sec. S.

TABLE VIII. (By Inspection.)

The Annual Precession of a Star in North Polar Distance. Argument = R. A. of the * in time.

[Tab. No. = Cos. R. A. 20″. 0436.]

Min.	O − XII +	Diff.	I − XII +	Diff.	II − XIV +	Diff.	III − XV +	Diff.	IV − XVI +	Diff.	V − XVII +	Diff.	Min.
0	20″.044	0	19″.361	23	17″.358	44	14″.173	62	10″.022	76	5″.188	85	60
1	20 .044	1	19 .338	23	17 .314	44	14 .111	62	9 .946	76	5 .103	84	59
2	20 .043	1	19 .315	24	17 .270	44	14 .049	63	9 .870	76	5 .019	85	58
3	20 .042	1	19 .291	24	17 .226	45	13 .986	63	9 .794	77	4 .934	85	57
4	20 .041	2	19 .267	24	17 .181	45	13 .923	63	9 .717	76	4 .849	85	56
5	20 .039	2	19 .243	25	17 .136	46	13 .860	63	9 .641	77	4 .764	85	55
6	20 .037	3	19 .218	25	17 .090	46	13 .797	63	9 .564	77	4 .679	85	54
7	20 .034	3	19 .193	25	17 .044	46	13 .734	64	9 .487	78	4 .594	85	53
8	20 .031	3	19 .168	26	16 .998	47	13 .670	64	9 .409	77	4 .509	85	52
9	20 .028	3	19 .142	26	16 .951	47	13 .606	64	9 .322	77	4 .424	86	51
10	20 .025	4	19 .116	27	16 .904	47	13 .542	65	9 .255	78	4 .338	85	50
11	20 .021	5	19 .089	27	16 .857	47	13 .477	65	9 .177	77	4 .253	86	49
12	20 .016	5	19 .062	27	16 .810	48	13 .412	65	9 .100	78	4 .167	85	48
13	20 .011	5	19 .035	27	16 .762	48	13 .347	66	9 .022	79	4 .082	86	47
14	20 .006	5	19 .008	28	16 .714	48	13 .281	66	8 .943	78	3 .996	86	46
15	20 .001	6	18 .980	28	16 .666	49	13 .215	66	8 .865	79	3 .910	85	45
16	19 .995	6	18 .952	29	16 .617	49	13 .249	66	8 .787	79	3 .825	86	44
17	19 .989	7	18 .923	29	16 .568	49	13 .083	66	8 .708	79	3 .739	86	43
18	19 .982	7	18 .894	29	16 .519	50	13 .017	66	8 .629	79	3 .643	86	42
19	19 .975	7	18 .865	30	16 .469	50	12 .951	67	8 .550	79	3 .567	86	41
20	19 .968	8	18 .835	30	16 .419	50	12 .884	67	8 .471	80	3 .481	87	40
21	19 .960	8	18 .805	31	16 .369	51	12 .817	68	8 .391	79	3 .394	86	39
22	19 .952	9	18 .774	31	16 .318	51	12 .749	68	8 .312	80	3 .308	86	38
23	19 .943	9	18 .743	31	16 .267	51	12 .681	68	8 .232	80	3 .222	86	37
24	19 .934	9	18 .712	31	16 .216	52	12 .613	68	8 .152	80	3 .136	87	36
25	19 .925	10	18 .681	32	16 .164	52	12 .545	68	8 .072	80	3 .049	86	35
26	19 .915	10	18 .649	32	16 .112	52	12 .477	68	7 .992	80	2 .963	87	34
27	19 .905	11	18 .617	33	16 .060	52	12 .409	69	7 .912	81	2 .876	86	33
28	19 .894	11	18 .574	33	16 .008	53	12 .340	69	7 .831	81	2 .790	87	32
29	19 .883	11	18 .551	33	15 .955	53	12 .271	69	7 .751	81	2 .703	87	31
30	19 .872	12	18 .518	33	15 .902	54	12 .202	69	7 .670	81	2 .616	86	30
31	19 .860	12	18 .485	34	15 .848	54	12 .133	70	7 .589	81	2 .530	87	29
32	19 .848	12	18 .451	35	15 .794	54	12 .063	70	7 .508	81	2 .443	87	28
33	19 .836	13	18 .416	35	15 .740	55	11 .993	70	7 .427	81	2 .356	87	27
34	19 .823	13	18 .381	35	15 .687	55	11 .923	71	7 .346	82	2 .269	87	26
35	19 .810	13	18 .346	35	15 .632	55	11 .852	71	7 .264	81	2 .182	87	25
36	19 .797	14	18 .311	36	15 .577	55	11 .781	71	7 .183	82	2 .095	87	24
37	19 .783	14	18 .275	36	15 .522	56	11 .710	71	7 .101	82	2 .008	87	23
38	19 .769	15	18 .239	37	15 .466	56	11 .639	71	7 .019	82	1 .921	87	22
39	19 .754	15	18 .202	37	15 .410	56	11 .568	71	6 .937	82	1 .834	87	21
40	19 .739	15	18 .165	37	15 .354	56	11 .497	72	6 .855	82	1 .747	87	20
41	19 .724	16	18 .128	37	15 .298	57	11 .425	72	6 .773	82	1 .660	87	19
42	19 .708	16	18 .091	38	15 .241	57	11 .353	72	6 .691	83	1 .573	88	18
43	19 .692	17	18 .053	38	15 .184	57	11 .281	73	6 .608	82	1 .485	87	17
44	19 .675	17	18 .015	38	15 .127	57	11 .208	73	6 .526	83	1 .398	87	16
45	19 .658	17	17 .977	39	15 .070	58	11 .135	73	6 .443	83	1 .311	87	15
46	19 .641	17	17 .938	39	15 .012	58	11 .062	73	6 .360	83	1 .224	88	14
47	19 .624	18	17 .899	40	14 .954	58	10 .989	73	6 .277	83	1 .136	87	13
48	19 .606	18	17 .859	40	14 .896	59	10 .916	73	6 .194	83	1 .049	89	12
49	19 .588	19	17 .819	40	14 .837	59	10 .843	74	6 .111	84	0 .960	86	11
50	19 .569	19	17 .779	41	14 .778	59	10 .769	74	6 .027	83	0 .874	86	10
51	19 .550	20	17 .738	41	14 .719	60	10 .695	74	5 .944	84	0 .788	88	9
52	19 .530	20	17 .697	41	14 .659	60	10 .621	74	5 .860	83	0 .700	88	8
53	19 .510	20	17 .656	41	14 .599	60	10 .547	74	5 .777	84	0 .612	87	7
54	19 .490	21	17 .615	42	14 .539	60	10 .473	75	5 .693	84	0 .525	88	6
55	19 .469	21	17 .573	42	14 .479	61	10 .398	75	5 .609	84	0 .437	87	5
56	19 .448	21	17 .531	43	14 .418	61	10 .323	75	5 .525	84	0 .350	88	4
57	19 .427	22	17 .488	43	14 .357	61	10 .248	75	5 .441	85	0 .262	87	3
58	19 .405	22	17 .445	43	14 .296	61	10 .173	75	5 .356	84	0 .175	88	2
59	19 .383	22	17 .402	44	14 .235	62	10 .098	76	5 .272	84	0 .087	87	1
60	19 .361		17 .358		14 .173		10 .022		5 188		0 .000		0
	+ XI − XXIII		+ X − XXII		+ IX − XXI		+ VIII − XX		+ VII − XIX		+ VI − XVIII		

TABLE IX.

Aberration in N. P. D. to find log. q. Argument = R. A. of the * in time.
[Tab. log. = log. Cos. R. A. +1 .3065322.]

S. N.	− O + / + XII −		− I + / + XIII −		− II + / + XIV −		− III + / + XV −		− IV + / + XVI −		− V + / + XVII −		S. N.
Min.	Log.	Diff.	Log.	Diff.	Log.	Diff.	Log.	Diff.	Log.	Diff.	Log.	Diff.	Min.
0	1 .30653	1	1 .29148	51	1 .24406	110	1 .15602	191	1 .00550	330	0 .71953	714	60
1	1 .30652	1	1 .29097	53	1 .24296	111	1 .15411	192	1 .00220	333	0 .71239	726	59
2	1 .30651	1	1 .29044	53	1 .24185	112	1 .15219	194	0 .99887	337	0 .70513	739	58
3	1 .30650	3	1 .28991	54	1 .24073	113	1 .15025	195	0 .99550	340	0 .69774	753	57
4	1 .30647	4	1 .28937	55	1 .23960	114	1 .14830	197	0 .99210	343	0 .69021	768	56
5	1 .30643	5	1 .28882	56	1 .23846	116	1 .14633	199	0 .98867	347	0 .68253	781	55
6	1 .30638	5	1 .28826	56	1 .23730	117	1 .14434	200	0 .98520	351	0 .67472	797	54
7	1 .30633	6	1 .28770	57	1 .23613	118	1 .14234	202	0 .98169	355	0 .66675	813	53
8	1 .30627	7	1 .28713	58	1 .23495	119	1 .14032	204	0 .97814	358	0 .65862	829	52
9	1 .30620	8	1 .28655	60	1 .23376	120	1 .13828	206	0 .97456	362	0 .65033	846	51
10	1 .30612	9	1 .28595	60	1 .23256	121	1 .13622	208	0 .97094	366	0 .64187	864	50
11	1 .30603	9	1 .28535	61	1 .23135	122	1 .13414	210	0 .96728	370	0 .63323	882	49
12	1 .30594	10	1 .28474	62	1 .23013	124	1 .13204	211	0 .96358	374	0 .62441	901	48
13	1 .30584	12	1 .28412	63	1 .22889	125	1 .12993	213	0 .95984	378	0 .61540	921	47
14	1 .30572	12	1 .28349	64	1 .22764	126	1 .12780	215	0 .95606	382	0 .60619	942	46
15	1 .30560	13	1 .28285	65	1 .22638	127	1 .12565	218	0 .95224	387	0 .59677	964	45
16	1 .30547	14	1 .28220	66	1 .22511	128	1 .12347	219	0 .94837	390	0 .58713	986	44
17	1 .30533	14	1 .28154	67	1 .22383	130	1 .12128	220	0 .94447	395	0 .57727	1011	43
18	1 .30519	15	1 .28087	67	1 .22253	131	1 .11908	223	0 .94052	400	0 .56716	1035	42
19	1 .30504	16	1 .28020	68	1 .22122	132	1 .11685	225	0 .93652	404	0 .55681	1061	41
20	1 .30488	17	1 .27952	69	1 .21990	133	1 .11460	227	0 .93248	409	0 .54620	1098	40
21	1 .30471	18	1 .27883	71	1 .21857	135	1 .11233	229	0 .92839	413	0 .53532	1118	39
22	1 .30453	19	1 .27812	71	1 .21722	136	1 .11004	231	0 .92426	418	0 .52414	1148	38
23	1 .30434	19	1 .27741	72	1 .21586	137	1 .10773	233	0 .92008	423	0 .51266	1180	37
24	1 .30415	21	1 .27669	74	1 .21449	138	1 .10540	236	0 .91585	429	0 .50086	1213	36
25	1 .30394	21	1 .27595	74	1 .21311	140	1 .10304	237	0 .91156	433	0 .48873	1250	35
26	1 .30373	22	1 .27521	75	1 .21171	141	1 .10067	239	0 .90723	438	0 .47623	1287	34
27	1 .30351	22	1 .27446	76	1 .21030	142	1 .09828	240	0 .90285	444	0 .46336	1327	33
28	1 .30329	24	1 .27370	77	1 .20888	143	1 .09588	244	0 .89841	449	0 .45009	1370	32
29	1 .30305	25	1 .27293	78	1 .20745	145	1 .09344	246	0 .89392	455	0 .43639	1416	31
30	1 .30280	25	1 .27215	79	1 .20600	146	1 .09098	248	0 .88937	460	0 .42223	1464	30
31	1 .30255	26	1 .27136	80	1 .20454	148	1 .08850	250	0 .88477	466	0 .40759	1516	29
32	1 .30229	27	1 .27056	81	1 .20306	149	1 .08600	253	0 .88011	472	0 .39243	1572	28
33	1 .30202	28	1 .26975	82	1 .20157	150	1 .08347	255	0 .87539	478	0 .37671	1632	27
34	1 .30174	29	1 .26893	83	1 .20007	151	1 .08092	257	0 .87061	484	0 .36039	1696	26
35	1 .30145	30	1 .26810	84	1 .19856	153	1 .07835	260	0 .86577	491	0 .34343	1766	25
36	1 .30115	31	1 .26726	85	1 .19703	154	1 .07575	262	0 .86086	497	0 .32577	1842	24
37	1 .30084	31	1 .26641	85	1 .19549	155	1 .07313	264	0 .85589	503	0 .30735	1924	23
38	1 .30053	32	1 .26556	87	1 .19394	157	1 .07049	267	0 .85086	511	0 .28811	2015	22
39	1 .30021	32	1 .26469	88	1 .19237	158	1 .06782	270	0 .84575	517	0 .26796	2113	21
40	1 .29989	34	1 .26381	89	1 .19079	160	1 .06512	272	0 .84058	524	0 .24683	2222	20
41	1 .29955	35	1 .26292	90	1 .18919	161	1 .06240	274	0 .83534	531	0 .22461	2343	19
42	1 .29920	36	1 .26202	91	1 .18758	163	1 .05966	277	0 .83003	539	0 .20118	2478	18
43	1 .29884	36	1 .26111	92	1 .18595	164	1 .05689	280	0 .82464	547	0 .17640	2628	17
44	1 .29848	37	1 .26019	93	1 .18431	165	1 .05409	282	0 .81917	554	0 .15012	2799	16
45	1 .29811	38	1 .25926	94	1 .18266	167	1 .05127	285	0 .81363	562	0 .12213	2992	15
46	1 .29773	39	1 .25832	95	1 .18099	168	1 .04842	287	0 .80801	571	0 .09221	3215	14
47	1 .29734	40	1 .25737	96	1 .17931	170	1 .04555	291	0 .80230	579	0 .06006	3473	13
48	1 .29694	41	1 .25641	97	1 .17761	172	1 .04264	293	0 .79651	587	0 .02533	3775	12
49	1 .29653	42	1 .25544	98	1 .17589	173	1 .03971	296	0 .79064	597	9 .98758	4137	11
50	1 .29611	42	1 .25446	99	1 .17416	174	1 .03675	299	0 .78467	605	9 .94621	4573	10
51	1 .29569	43	1 .25347	100	1 .17242	176	1 .03376	302	0 .77862	615	9 .90048	5113	9
52	1 .29526	44	1 .25247	102	1 .17066	178	1 .03074	305	0 .77247	625	9 .84035	5797	8
53	1 .29482	45	1 .25145	102	1 .16888	179	1 .02769	307	0 .76622	635	9 .79138	6693	7
54	1 .29437	46	1 .25043	104	1 .16709	180	1 .02462	311	0 .75987	644	9 .72445	7916	6
55	1 .29391	47	1 .24939	104	1 .16529	182	1 .02151	314	0 .75343	656	9 .64529	9690	5
56	1 .29344	48	1 .24835	106	1 .16347	184	1 .01837	317	0 .74687	666	9 .54839	12493	4
57	1 .29296	49	1 .24729	106	1 .16163	186	1 .01520	320	0 .74021	678	9 .42346	17609	3
58	1 .29247	49	1 .24623	108	1 .15977	187	1 .01200	323	0 .73343	689	9 .24737	30102	2
59	1 .29198	50	1 .24515	109	1 .15790	188	1 .00877	327	0 .72654	701	8 .94635		1
60	1 .29148		1 .24406		1 .15602		1 .00550		0 .71953		− ∞		0
N. S.	− XI + / + XXIII −		− X + / + XXII −		− IX + / + XXI −		− VIII + / + XX −		− VII + / + XIX −		− VI + / + XVIII −		N. S.

Positive Argument.

Cases		
1	$\{ p+ , q+ \}$	$= A'$
2	$\{ p+ , q- \}$	$= 6^s - A'$
3	$\{ p- , q- \}$	$= 6^s + A'$
4	$\{ p- , q+ \}$	$= 12^s - A'$

To the Logarithms found in this Table add the log. tang. of the *'s Declination, and the result gives q.

Let A' be the Annual Number, then tangent $A' = \log. p. - \log. q.$ and Max. Aber. $\begin{cases} = \log. p. + \log. \text{Cos. Dec.} - \log. \text{sine } A'. \\ \text{or} \\ = \log. q. + \log. \text{Cos. Dec.} - \log. \text{Cos. } A'. \end{cases}$

TABLE IX. (Nat. Num.)

Aberration in North Polar Distance, to find q. Argument = R. A. of the * in time.

[Tab. No. =20″. 255 × Cos. R. A.]

S. N. Min.	− O + / + XII −	Diff.	− I + / + XIII −	Diff.	− II + / + XIV −	Diff.	− III + / + XV −	Diff.	− IV + / + XVI −	Diff.	− V + / + XVII −	Diff.	S. N. Min.
0	20″. 255	0	19″. 565	23	17″. 541	44	14″. 322	63	10″. 128	77	5″. 243	85	60
1	20. 255	1	19. 542	24	17. 497	45	14. 259	63	10. 051	77	5. 158	86	59
2	20. 254	0	19. 518	24	17. 452	45	14. 196	63	9. 974	77	5. 072	86	58
3	20. 254	1	19. 494	24	17. 407	45	14. 133	63	9. 897	77	4. 986	86	57
4	20. 252	2	19. 470	25	17. 362	46	14. 070	64	9. 820	77	4. 900	86	56
5	20. 250	2	19. 445	25	17. 316	46	14. 006	64	9. 743	78	4. 814	86	55
6	20. 248	2	19. 420	25	17. 270	46	13. 942	64	9. 665	78	4. 728	86	54
7	20. 246	3	19. 395	25	17. 224	47	13. 878	64	9. 587	78	4. 642	86	53
8	20. 243	3	19. 370	26	17. 177	47	13. 814	65	9. 509	78	4. 556	86	52
9	20. 240	4	19. 344	26	17. 130	47	13. 749	65	9. 431	78	4. 470	86	51
10	20. 236	4	19. 318	27	17. 083	48	13. 684	65	9. 353	78	4. 384	86	50
11	20. 232	5	19. 291	27	17. 035	48	13. 619	66	9. 275	79	4. 298	86	49
12	20. 227	5	19. 264	28	16. 987	48	13. 553	66	9. 196	79	4. 212	86	48
13	20. 222	5	19. 236	28	16. 939	49	13. 487	66	9. 117	79	4. 126	87	47
14	20. 217	6	19. 208	28	16. 890	49	13. 421	66	9. 038	79	4. 039	87	46
15	20. 211	5	19. 180	28	16. 841	49	13. 355	67	8. 959	79	3. 952	87	45
16	20. 206	7	19. 152	29	16. 792	50	13. 288	67	8. 880	80	3. 865	87	44
17	20. 199	6	19. 123	30	16. 742	50	13. 221	67	8. 800	80	3. 778	87	43
18	20. 193	7	19. 093	30	16. 692	50	13. 154	67	8. 720	80	3. 691	87	42
19	20. 186	8	19. 063	30	16. 642	50	13. 087	68	8. 640	80	3. 604	87	41
20	20. 178	8	19. 033	31	16. 592	51	13. 019	68	8. 560	80	3. 517	87	40
21	20. 170	8	19. 002	31	16. 541	51	12. 951	68	8. 480	80	3. 430	87	39
22	20. 162	9	18. 971	31	16. 490	51	12. 883	68	8. 400	80	3. 343	87	38
23	20. 153	9	18. 940	31	16. 439	52	12. 815	69	8. 320	81	3. 256	87	37
24	20. 144	10	18. 909	32	16. 387	52	12. 746	69	8. 239	81	3. 169	87	36
25	20. 134	9	18. 877	32	16. 335	52	12. 677	69	8. 158	81	3. 082	87	35
26	20. 125	10	18. 845	32	16. 283	53	12. 608	69	8. 077	81	2. 995	87	34
27	20. 115	10	18. 813	33	16. 230	53	12. 539	69	7. 996	81	2. 908	88	33
28	20. 105	11	18. 780	33	16. 177	54	12. 470	70	7. 915	82	2. 820	88	32
29	20. 094	12	18. 747	34	16. 123	54	12. 400	70	7. 833	82	2. 732	88	31
30	20. 082	12	18. 713	34	16. 069	54	12. 330	70	7. 751	82	2. 644	88	30
31	20. 070	12	18. 679	34	16. 015	54	12. 260	70	7. 669	82	2. 556	88	29
32	20. 058	12	18. 645	35	15. 961	55	12. 190	71	7. 587	82	2. 468	88	28
33	20. 046	13	18. 610	35	15. 906	55	12. 119	71	7. 505	82	2. 380	88	27
34	20. 033	13	18. 575	35	15. 851	55	12. 048	71	7. 423	82	2. 292	88	26
35	20. 020	14	18. 540	36	15. 796	55	11. 977	71	7. 341	82	2. 204	88	25
36	20. 006	14	18. 504	36	15. 741	56	11. 906	72	7. 259	82	2. 116	88	24
37	19. 992	15	18. 468	37	15. 685	56	11. 834	72	7. 177	83	2. 028	88	23
38	19. 977	15	18. 431	37	15. 629	56	11. 762	72	7. 094	83	1. 940	88	22
39	19. 962	15	18. 394	37	15. 573	57	11. 690	72	7. 011	83	1. 852	88	21
40	19. 947	15	18. 357	38	15. 516	57	11. 618	73	6. 928	83	1. 764	88	20
41	19. 932	16	18. 319	38	15. 459	57	11. 545	73	6. 845	83	1. 676	88	19
42	19. 916	17	18. 281	38	15. 402	57	11. 472	73	6. 762	83	1. 588	88	18
43	19. 899	16	18. 243	38	15. 345	58	11. 399	73	6. 679	84	1. 500	88	17
44	19. 883	17	18. 205	39	15. 287	58	11. 326	73	6. 595	84	1. 412	88	16
45	19. 866	17	18. 166	39	15. 229	58	11. 253	73	6. 511	84	1. 324	88	15
46	19. 849	18	18. 127	40	15. 171	59	11. 180	74	6. 427	84	1. 236	88	14
47	19. 831	18	18. 087	40	15. 112	59	11. 106	74	6. 343	84	1. 148	88	13
48	19. 813	19	18. 047	40	15. 053	59	11. 032	74	6. 259	84	1. 060	89	12
49	19. 794	19	18. 007	41	14. 994	60	10. 958	74	6. 175	84	0. 971	88	11
50	19. 775	19	17. 966	41	14. 934	60	10. 884	75	6. 091	84	0. 883	88	10
51	19. 756	20	17. 925	41	14. 874	60	10. 809	75	6. 007	84	0. 795	88	9
52	19. 736	20	17. 884	42	14. 814	61	10. 734	75	5. 923	85	0. 707	89	8
53	19. 716	21	17. 842	42	14. 753	61	10. 659	75	5. 838	85	0. 618	88	7
54	19. 696	21	17. 800	42	14. 692	61	10. 584	76	5. 753	85	0. 530	88	6
55	19. 675	21	17. 758	43	14. 631	61	10. 508	76	5. 668	85	0. 442	88	5
56	19. 654	22	17. 715	43	14. 570	62	10. 432	76	5. 583	85	0. 354	89	4
57	19. 632	22	17. 672	43	14. 508	62	10. 356	76	5. 498	85	0. 265	88	3
58	19. 610	22	17. 629	44	14. 446	62	10. 280	76	5. 413	85	0. 177	88	2
59	19. 588	23	17. 585	44	14. 384	62	10. 204	76	5. 328	85	0. 089	89	1
60	19. 565		17. 541		14. 322		10. 128		5. 243		0. 000		0
N. S.	− XI + / + XXIII −		− X + / + XXII −		− IX + / + XXI −		− VIII + / + XX −		− VII + / + XIX −		− VI + / + XVIII −		N. S.

Multiply the numbers found in the Table by the Natural Tangent of the Star's Declination and the result will give q. Tang. $A' = \frac{p}{q}$, and Maximum $= \frac{p \text{ cos. Dec.}}{\text{sine } A'}$, or $= \frac{q \text{ cos. Dec.}}{\text{cosine } A'}$.

TABLE X.

Aberration in North Polar Distance to find log. p. Argument = R. A. of the * in time.

[Tab. log. = log. sin. R. A. + 1.2690457.]

S. N.	+ 0 −	− XII +	+ I −	− XIII +	+ II −	− XIV +	+ III −	− XV +	+ IV −	− XVI +	+ V −	− XVII +	S. N.
Min.	Log.	Diff.	Log.	Diff.	Log.	Diff.	Log.	Diff.	Log.	Diff.	Log.	Diff.	Min.
0	− ∞		0.68204	701	0.96802	326	1.11853	189	1.20657	109	1.25399	51	60
1	8.90886	30103	0.68905	689	0.97128	323	1.12042	187	1.20766	108	1.25450	50	59
2	9.20989	17608	0.69594	678	0.97451	321	1.12229	185	1.20874	107	1.25500	48	58
3	9.38597	12493	0.70272	666	0.97772	316	1.12414	184	1.20981	105	1.25548	47	57
4	9.51090	9690	0.70938	656	0.98088	314	1.12598	182	1.21086	104	1.25595	47	56
5	9.60780	7916	0.71594	645	0.98402	311	1.12780	181	1.21190	103	1.25642	46	55
6	9.68696	6693	0.72239	634	0.98713	308	1.12961	179	1.21293	103	1.25688	45	54
7	9.75389	5799	0.72873	625	0.99021	305	1.13140	177	1.21396	102	1.25733	44	53
8	9.81186	5113	0.73498	615	0.99326	301	1.13317	176	1.21498	100	1.25777	43	52
9	9.86299	4574	0.74113	606	0.99627	299	1.13493	174	1.21598	99	1.25820	42	51
10	9.90873	4136	0.74719	596	0.99926	296	1.13667	173	1.21697	98	1.25862	42	50
11	9.95009	3776	0.75315	588	1.00222	293	1.13840	172	1.21795	98	1.25904	41	49
12	9.98785	3472	0.75903	579	1.00515	291	1.14012	170	1.21893	96	1.25945	40	48
13	0.02257	3215	0.76482	570	1.00806	288	1.14182	169	1.21989	95	1.25985	39	47
14	0.05472	2992	0.77052	563	1.01094	284	1.14351	167	1.22084	94	1.26024	38	46
15	0.08464	2799	0.77615	554	1.01378	283	1.14518	165	1.22178	93	1.26062	37	45
16	0.11263	2628	0.78169	546	1.01661	279	1.14683	164	1.22271	92	1.26099	37	44
17	0.13891	2478	0.78715	539	1.01940	277	1.14847	162	1.22363	91	1.26136	36	43
18	0.16369	2343	0.79254	532	1.02217	275	1.15009	161	1.22454	90	1.26172	35	42
19	0.18712	2222	0.79786	524	1.02492	272	1.15170	160	1.22544	88	1.26207	33	41
20	0.20934	2113	0.80310	517	1.02764	269	1.15330	159	1.22632	88	1.26240	33	40
21	0.23047	2015	0.80827	510	1.03033	267	1.15489	157	1.22720	87	1.26273	32	39
22	0.25062	1924	0.81337	504	1.03300	264	1.15646	155	1.22807	86	1.26305	31	38
23	0.26986	1842	0.81841	496	1.03564	262	1.15801	154	1.22893	85	1.26336	31	37
24	0.28828	1766	0.82337	491	1.03826	260	1.15955	153	1.22978	84	1.26367	30	36
25	0.30594	1696	0.82828	484	1.04086	257	1.16108	151	1.23062	83	1.26397	29	35
26	0.32290	1632	0.83312	478	1.04343	255	1.16259	150	1.23145	82	1.26426	28	34
27	0.33922	1572	0.83790	472	1.04598	253	1.16409	149	1.23227	80	1.26454	26	33
28	0.35494	1516	0.84262	466	1.04851	250	1.16558	147	1.23307	80	1.26480	26	32
29	0.37010	1464	0.84728	461	1.05101	248	1.16705	146	1.23387	79	1.26506	25	31
30	0.38474	1416	0.85189	454	1.05349	246	1.16851	145	1.23466	78	1.26531	25	30
31	0.39890	1370	0.85643	449	1.05595	244	1.16996	143	1.23544	77	1.26556	24	29
32	0.41260	1328	0.86092	444	1.05839	241	1.17139	142	1.23621	76	1.26580	23	28
33	0.42588	1287	0.86536	439	1.06080	240	1.17281	141	1.23697	75	1.26603	22	27
34	0.43875	1247	0.86975	433	1.06320	237	1.17422	140	1.23772	74	1.26625	21	26
35	0.45124	1214	0.87408	428	1.06557	235	1.17562	138	1.23846	74	1.26646	20	25
36	0.46338	1180	0.87836	423	1.06792	233	1.17700	137	1.23920	73	1.26666	20	24
37	0.47518	1147	0.88259	418	1.07025	231	1.17837	136	1.23993	71	1.26686	19	23
38	0.48665	1118	0.88677	414	1.07256	228	1.17973	135	1.24064	70	1.26705	18	22
39	0.49783	1089	0.89091	408	1.07484	227	1.18108	133	1.24134	69	1.26723	16	21
40	0.50872	1061	0.89499	404	1.07711	225	1.18241	132	1.24203	68	1.26739	16	20
41	0.51933	1035	0.89903	400	1.07936	223	1.18373	131	1.24271	68	1.26755	15	19
42	0.52968	1010	0.90303	395	1.08159	221	1.18504	130	1.24339	67	1.26770	15	18
43	0.53978	996	0.90698	391	1.08380	219	1.18634	128	1.24406	66	1.26785	14	17
44	0.54964	964	0.91089	386	1.08599	217	1.18762	127	1.24472	65	1.26799	13	16
45	0.55928	942	0.91475	382	1.08816	215	1.18889	126	1.24537	64	1.26812	12	15
46	0.56870	921	0.91857	378	1.09031	213	1.19015	125	1.24601	63	1.26824	11	14
47	0.57791	901	0.92235	374	1.09244	212	1.19140	124	1.24664	61	1.26835	10	13
48	0.58692	883	0.92609	370	1.09456	209	1.19264	123	1.24725	61	1.26845	10	12
49	0.59575	863	0.92979	366	1.09665	208	1.19387	121	1.24786	60	1.26855	9	11
50	0.60438	846	0.93345	362	1.09873	206	1.19508	120	1.24846	59	1.26864	8	10
51	0.61284	829	0.93707	359	1.10079	204	1.19628	119	1.24905	59	1.26872	6	9
52	0.62113	823	0.94066	354	1.10283	202	1.19747	118	1.24964	58	1.26878	6	8
53	0.62936	787	0.94420	351	1.10485	201	1.19865	117	1.25022	57	1.26884	5	7
54	0.63723	782	0.94771	347	1.10686	199	1.19982	115	1.25079	56	1.26889	5	6
55	0.64505	767	0.95118	344	1.10885	197	1.20097	114	1.25135	54	1.26894	4	5
56	0.65272	753	0.95462	340	1.11082	195	1.20211	113	1.25189	53	1.26898	3	4
57	0.66025	740	0.95802	336	1.11277	194	1.20324	112	1.25242	53	1.26901	2	3
58	0.66765	726	0.96138	334	1.11471	192	1.20436	111	1.25295	52	1.26903	2	2
59	0.67491	713	0.96472	330	1.11663	190	1.20547	110	1.25347	52	1.26905	0	1
60	0.68204		0.96802		1.11853		1.20657		1.25399		1.26905		0
N. S.	− XI +	+ XXIII −	− X +	+ XXII −	− IX +	+ XXI −	− VIII +	+ XX −	− VII +	+ XIX −	− VI +	+ XVIII −	N. S.

Positive Argument.

Cases	
1 $\left\{ \begin{matrix} p+ \\ q+ \end{matrix} \right.$	$= A'$
2 $\left\{ \begin{matrix} p+ \\ q- \end{matrix} \right.$	$= 6^s - A'$
3 $\left\{ \begin{matrix} p- \\ q- \end{matrix} \right.$	$= 6^s + A'$
4 $\left\{ \begin{matrix} p- \\ q+ \end{matrix} \right.$	$= 12^s - A'$

To the Logarithm found in this Table, add the Log. Tang. of the *'s Declination; and to the natural number, with the proper Sign belonging to the Sum, add 8″.066, and take out the log. of the result, which call p.

Log. Tangent $A' =$ log. $p -$ log. q; and Log. Max. of Aberration = log. $p -$ log. sine $A' +$ log. Cos. Dec.; or = log $q +$ log. sec. $A' +$ log. Cos. Dec.

TABLE X. (Nat. Num.)

Aberration in North Polar Distance to find p. Argument $=$ R. A. of the $*$ in Time.

[Tab. No. $= 18''.580 \times$ sin. R. A.]

S. N. Min.	+ 0 − / − XII +	Diff.	+ I − / − XIII +	Diff.	+ II − / − XIV +	Diff.	+ III − / − XV +	Diff.	+ IV − / − XVI +	Diff.	+ V − / − XVII +	Diff.	S. N. Min.
0	0″.000	81	4″.809	78	9″.290	71	13″.138	57	16″.090	40	17″.947	21	60
1	0 .081	81	4 .887	78	9 .361	70	13 .195	57	16 .130	40	17 .968	21	59
2	0 .162	81	4 .965	78	9 .431	69	13 .252	57	16 .170	40	17 .989	20	58
3	0 .243	81	5 .043	78	9 .500	69	13 .309	56	16 .210	40	18 .009	19	57
4	0 .324	81	5 .121	78	9 .569	70	13 .365	57	16 .250	39	18 .028	20	56
5	0 .405	81	5 .199	78	9 .639	69	13 .422	56	16 .289	39	18 .048	19	55
6	0 .486	81	5 .277	78	9 .708	69	13 .478	56	16 .328	39	18 .067	19	54
7	0 .567	81	5 .355	77	9 .777	69	13 .534	55	16 .367	38	18 .086	18	53
8	0 .648	81	5 .432	78	9 .846	69	13 .589	56	16 .405	38	18 .104	18	52
9	0 .729	81	5 .510	78	9 .915	69	13 .645	55	16 .443	37	18 .122	18	51
10	0 .810	81	5 .588	77	9 .984	68	13 .700	54	16 .480	38	18 .140	17	50
11	0 .891	81	5 .665	77	10 .052	67	13 .754	54	16 .518	37	18 .157	17	49
12	0 .972	81	5 .742	78	10 .119	68	13 .808	54	16 .555	37	18 .174	17	48
13	1 .053	81	5 .820	77	10 .187	67	13 .852	54	16 .592	36	18 .191	16	47
14	1 .134	81	5 .897	77	10 .254	68	13 .916	53	16 .628	36	18 .207	16	46
15	1 .215	81	5 .974	76	10 .322	67	13 .969	53	16 .664	36	18 .223	15	45
16	1 .296	81	6 .050	77	10 .389	68	14 .022	53	16 .700	35	18 .238	15	44
17	1 .377	81	6 .127	76	10 .457	67	14 .075	53	16 .735	35	18 .253	15	43
18	1 .458	81	6 .203	76	10 .524	67	14 .128	53	16 .770	35	18 .268	15	42
19	1 .539	80	6 .279	76	10 .591	66	14 .181	52	16 .805	34	18 .283	15	41
20	1 .619	81	6 .355	76	10 .657	66	14 .233	53	16 .839	34	18 .298	14	40
21	1 .700	81	6 .431	76	10 .723	66	14 .286	51	16 .873	34	18 .312	13	39
22	1 .781	81	6 .507	76	10 .789	66	14 .337	51	16 .907	33	18 .325	14	38
23	1 .862	80	6 .583	75	10 .855	66	14 .388	51	16 .940	33	18 .339	13	37
24	1 .942	81	6 .658	76	10 .921	66	14 .439	51	16 .973	33	18 .352	12	36
25	2 .023	81	6 .734	76	10 .987	65	14 .490	51	17 .006	33	18 .364	12	35
26	2 .104	80	6 .810	75	11 .052	65	14 .541	50	17 .039	32	18 .376	12	34
27	2 .184	80	6 .885	75	11 .117	65	14 .591	50	17 .071	32	18 .388	11	33
28	2 .264	81	6 .960	75	11 .182	65	14 .641	50	17 .103	32	18 .399	11	32
29	2 .345	81	7 .035	75	11 .247	64	14 .691	49	17 .135	31	18 .410	10	31
30	2 .426	80	7 .110	75	11 .311	64	14 .740	50	17 .166	31	18 .420	11	30
31	2 .506	80	7 .185	75	11 .375	64	14 .790	49	17 .197	30	18 .431	10	29
32	2 .586	81	7 .260	75	11 .439	64	14 .839	49	17 .227	30	18 .441	9	28
33	2 .667	80	7 .335	75	11 .503	64	14 .888	48	17 .257	30	18 .450	9	27
34	2 .747	80	7 .410	74	11 .567	63	14 .936	48	17 .287	30	18 .459	9	26
35	2 .827	80	7 .484	73	11 .630	63	14 .984	48	17 .317	29	18 .468	9	25
36	2 .907	80	7 .557	74	11 .693	63	15 .032	48	17 .346	29	18 .477	9	24
37	2 .987	80	7 .631	74	11 .756	63	15 .080	47	17 .375	28	18 .486	8	23
38	3 .067	80	7 .705	74	11 .819	62	15 .127	47	17 .403	29	18 .494	8	22
39	3 .147	79	7 .779	73	11 .881	62	15 .174	46	17 .432	28	18 .502	7	21
40	3 .226	80	7 .852	74	11 .943	62	15 .220	47	17 .460	27	18 .509	7	20
41	3 .306	80	7 .926	73	12 .005	61	15 .267	46	17 .487	27	18 .516	7	19
42	3 .386	79	7 .999	74	12 .066	62	15 .313	46	17 .514	27	18 .523	6	18
43	3 .465	80	8 .073	73	12 .128	61	15 .359	45	17 .541	27	18 .529	6	17
44	3 .545	79	8 .146	72	12 .189	62	15 .404	45	17 .568	26	18 .535	5	16
45	3 .624	80	8 .218	73	12 .251	61	15 .449	45	17 .594	26	18 .540	5	15
46	3 .704	79	8 .291	72	12 .312	60	15 .494	44	17 .620	26	18 .545	5	14
47	3 .783	80	8 .363	72	12 .372	60	15 .538	44	17 .646	25	18 .550	4	13
48	3 .863	79	8 .435	72	12 .432	60	15 .582	44	17 .671	25	18 .554	4	12
49	3 .942	80	8 .507	72	12 .492	60	15 .626	43	17 .696	24	18 .558	4	11
50	4 .022	79	8 .579	72	12 .552	59	15 .669	44	17 .720	24	18 .562	4	10
51	4 .101	80	8 .651	72	12 .611	60	15 .713	43	17 .744	24	18 .566	3	9
52	4 .181	79	8 .723	72	12 .671	59	15 .756	43	17 .768	24	18 .569	3	8
53	4 .260	79	8 .795	72	12 .730	59	15 .799	43	17 .792	23	18 .572	2	7
54	4 .339	78	8 .867	71	12 .789	59	15 .842	42	17 .815	23	18 .574	2	6
55	4 .417	78	8 .938	70	12 .848	58	15 .884	42	17 .838	22	18 .576	2	5
56	4 .495	79	9 .008	71	12 .906	59	15 .926	42	17 .860	23	18 .578	1	4
57	4 .574	79	9 .079	70	12 .965	58	15 .968	42	17 .883	22	18 .579	0	3
58	4 .653	78	9 .149	71	13 .023	58	16 .010	40	17 .905	21	18 .579	1	2
59	4 .731	78	9 .220	70	13 .081	57	16 .050	40	17 .926	21	18 .580	0	1
60	4 .809		9 .290		13 .138		16 .090		17 .947		18 .580		0
N. S.	− XI + / + XXIII −		− X + / + XXII −		− IX + / + XXI −		− VIII + / + XX −		− VII + / + XIX −		− VI + / + XVIII −		N. S.

Positive Argument.

Cases	
1 $\begin{cases} p+ \\ q+ \end{cases}$	$=$ A′
2 $\begin{cases} p+ \\ q- \end{cases}$	$= \overset{s}{6} -$ A′
3 $\begin{cases} p- \\ q- \end{cases}$	$= \overset{s}{6} +$ A′
4 $\begin{cases} p- \\ q+ \end{cases}$	$= \overset{s}{12} -$ A′

Multiply the numbers in this Table by the natural Tangent of the $*$'s Declination, and apply to the products $+8''.066$, and the results will give p.

Tangent A′ $= \frac{p}{q}$; and Max. Aber. $= \frac{p \text{ cos. Dec.}}{\text{sine A}'} = \frac{q \text{ cos. Dec.}}{\text{cos. A}'}$; or $= p$ cosec. A′ cos. Dec. $= q$. sec. A′. cos. Dec.

TABLE XI. (By Inspection.)

Lunar Nutation in North Polar Distance. To find the Annual Number L′. Argument = R. A. of the ✱ in time.

[Log. tang. L′ = log. tang. R. A. + .1281798.]

	O XII		I XIII		II XIV		III XV		IV XVI		V XVII		
	12 − 6−	P. P.	12 − 6−	P. P.	12 − 6−	P. P.	12 − 6−	P. P.	12 − 6 −	P. P.	12 − 0 −	P. P.	
Min.	L′.		L′.		L′.		L′.		L′.		L′.		Min.
0	0s 0° 0′	0	0s 19° 48′	0	1s 7° 48′	0	1s 23° 20′	0	2s 6° 45′	0	2s 18° 43′	0	60
1	0 0 20	0	0 20 7	0	1 8 5	0	1 23 35	0	2 6 58	0	2 18 55	0	59
2	0 0 40	1	0 20 26	1	1 8 21	1	1 23 49	0	2 7 10	0	2 19 6	0	58
3	0 1 0	1	0 20 45	1	1 8 38	1	1 24 3	1	2 7 23	0	2 19 18	0	57
4	0 1 21	1	0 21 4	1	1 8 55	1	1 24 17	1	2 7 35	1	2 19 29	1	56
5	0 1 41	2	0 21 23	1	1 9 12	1	1 24 32	1	2 7 48	1	2 19 41	1	55
6	0 2 1	2	0 21 42	2	1 9 28	2	1 24 46	1	2 8 0	1	2 19 52	1	54
7	0 2 21	2	0 22 1	2	1 9 45	2	1 25 0	2	2 8 12	1	2 20 4	1	53
8	0 2 41	3	0 22 20	2	1 10 1	2	1 25 14	2	2 8 24	2	2 20 15	1	52
9	0 3 1	3	0 22 39	3	1 10 17	2	1 25 28	2	2 8 37	2	2 20 37	2	51
10	0 3 21	3	0 22 57	3	1 10 33	3	1 25 42	2	2 8 49	2	2 20 38	2	50
11	0 3 41	4	0 23 16	3	1 10 50	3	1 25 56	3	2 9 2	2	2 20 50	2	49
12	0 4 2	4	0 23 35	4	1 11 6	3	1 26 10	3	2 9 14	2	2 21 1	2	48
13	0 4 22	4	0 23 54	4	1 11 23	4	1 26 24	3	2 9 26	3	2 21 12	2	47
14	0 4 42	5	0 24 12	4	1 11 39	4	1 26 38	3	2 9 38	3	2 21 23	2	46
15	0 5 2	5	0 24 31	5	1 11 55	4	1 28 52	4	2 9 51	3	2 21 35	3	45
16	0 5 22	5	0 24 49	5	1 12 11	4	1 27 6	4	2 10 3	3	2 21 46	3	44
17	0 5 42	6	0 25 8	5	1 12 27	5	1 27 20	4	2 10 15	3	2 21 58	3	43
18	0 6 2	6	0 25 26	5	1 12 43	5	1 27 33	4	2 10 27	4	2 22 9	3	42
19	0 6 22	6	0 25 45	6	1 12 59	5	1 27 47	5	2 10 39	4	2 22 20	3	41
20	0 6 42	7	0 26 3	6	1 13 15	5	1 28 1	5	2 10 51	4	2 22 31	4	40
21	0 7 2	7	0 26 22	6	1 13 31	6	1 28 15	5	2 11 4	4	2 22 43	4	39
22	0 7 22	7	0 26 40	7	1 13 47	6	1 28 28	5	2 11 16	4	2 22 54	4	38
23	0 7 42	8	0 26 59	7	1 14 3	6	1 28 42	6	2 11 28	5	2 23 6	4	37
24	0 8 2	8	0 27 17	7	1 14 18	6	1 28 55	6	2 11 40	5	2 23 17	4	36
25	0 8 22	8	0 27 35	8	1 14 34	7	1 29 9	6	2 11 52	5	2 23 28	4	35
26	0 8 42	9	0 27 53	8	1 14 50	7	1 29 22	6	2 12 4	5	2 23 39	5	34
27	0 9 2	9	0 28 11	8	1 15 6	7	1 29 36	6	2 12 16	5	2 23 51	5	33
28	0 9 22	9	0 28 29	8	1 15 21	8	1 29 49	7	2 12 28	6	2 24 2	5	32
29	0 9 42	10	0 28 48	9	1 15 37	8	2 0 3	7	2 12 40	6	2 24 13	5	31
30	0 10 2	10	0 29 6	9	1 15 52	8	2 0 16	7	2 12 52	6	2 24 24	5	30
31	0 10 22	10	0 29 24	9	1 16 8	8	2 0 30	7	2 13 4	6	2 24 36	5	29
32	0 10 42	11	0 29 42	10	1 16 23	9	2 0 43	8	2 13 16	6	2 24 47	5	28
33	0 11 1	11	1 0 0	10	1 16 39	9	2 0 56	8	2 13 28	7	2 24 58	6	27
34	0 11 21	11	1 0 17	10	1 16 54	9	2 1 9	8	2 13 39	7	2 25 9	6	26
35	0 11 41	12	1 0 35	11	1 17 10	9	2 1 23	9	2 13 51	7	2 25 21	6	25
36	0 12 1	12	1 0 53	11	1 17 25	10	2 1 36	9	2 14 3	7	2 25 32	6	24
37	0 12 21	12	1 1 11	11	1 17 40	10	2 1 49	9	2 14 15	7	2 25 43	6	23
38	0 12 40	13	1 1 29	11	1 17 55	10	2 2 2	9	2 14 27	8	2 25 54	7	22
39	0 13 0	13	1 1 47	12	1 18 10	10	2 2 15	10	2 14 39	8	2 26 5	7	21
40	0 13 20	13	1 2 4	12	1 18 25	11	2 2 28	10	2 14 50	8	2 26 16	7	20
41	0 13 40	14	1 2 22	12	1 18 41	11	2 2 41	10	2 15 2	8	2 26 28	7	19
42	0 13 59	14	1 2 39	13	1 18 56	11	2 2 54	10	2 15 14	8	2 26 39	7	18
43	0 14 19	14	1 2 57	13	1 19 11	12	2 3 7	11	2 15 26	9	2 26 50	8	17
44	0 14 38	15	1 3 14	13	1 19 26	12	2 3 20	11	2 15 37	9	2 27 1	8	16
45	0 14 58	15	1 3 32	14	1 19 41	12	2 3 33	11	2 15 49	9	2 27 13	8	15
46	0 15 17	15	1 3 49	14	1 19 55	12	2 3 46	11	2 16 1	9	2 27 24	8	14
47	0 15 37	16	1 4 6	14	1 20 10	13	2 3 59	12	2 16 13	9	2 27 35	8	13
48	0 15 56	16	1 4 23	14	1 20 25	13	2 4 12	12	2 16 24	10	2 27 46	9	12
49	0 16 16	16	1 4 41	15	1 20 40	13	2 4 25	12	2 16 36	10	2 27 57	9	11
50	0 16 35	17	1 4 58	15	1 20 55	13	2 4 38	12	2 16 47	10	2 28 8	9	10
51	0 16 55	17	1 5 15	15	1 21 10	14	2 4 51	13	2 16 59	10	2 28 20	9	9
52	0 17 14	17	1 5 32	16	1 21 24	14	2 5 3	13	2 17 11	10	2 28 31	9	8
53	0 17 34	18	1 5 49	16	1 21 39	14	2 5 16	13	2 17 23	11	2 28 42	10	7
54	0 17 53	18	1 6 6	16	1 21 53	14	2 5 28	13	2 17 34	11	2 28 53	10	6
55	0 18 12	18	1 6 23	17	1 22 8	15	2 5 41	13	2 17 46	11	2 29 4	10	5
56	0 18 31	19	1 6 40	17	1 22 22	15	2 5 54	13	2 17 57	11	2 29 15	10	4
57	0 18 50	19	1 6 57	17	1 22 37	15	2 6 7	13	2 18 9	11	2 29 27	10	3
58	0 19 9	19	1 7 14	17	1 22 51	15	2 6 19	13	2 18 20	12	2 29 38	11	2
59	0 19 29	20	1 7 31	18	1 23 6	15	2 6 32	13	2 18 32	12	2 29 49	11	1
60	0 19 48	20	1 7 48	18	1 23 20	15	2 6 45	13	2 18 43	12	3 0 0	11	0
	6 + 0+		6 + 0+		6 + 0+		6 + 0+		6 + 0+		6 + 0+		
	XI XXIII		X XXII		IX XXI		VIII XX		VII XIX		VI XVIII		

TABLE XII. (By Inspection.)

Lunar Nutation in North Polar Distance. To find Log. Max. and Max. Argument = R. A. of the * in time.

[Log. Max. = log. sin. R. A. + .9844373 — log. sin. N′.]

	O		XII	I		XIII	II		XIV	III		XV	IV		XVI	V		XVII	
Min.	Log. Max.	Diff.	Max.	Log. Max.	Diff.	Max.	Log. Max.	Diff.	Max.	Log. Max.	Diff.	Max.	Log. Max.	Diff.	Max.	Log. Max.	Diff.	Max.	Min.
0	0.85627	0	7″.183	0.86768	36	7″.374	0.89907	55	7″.872	0.92968	55	8″.505	0.95879	41	9″.095	0.97788	21	9″.504	60
1	0.85627	0	7.183	0.86804	36	7.380	0.89962	54	7.882	0.93023	54	8.516	0.95920	41	9.104	0.97809	21	9.508	59
2	0.85627	1	7.183	0.86840	36	7.386	0.89716	55	7.892	0.93077	54	8.527	0.95961	40	9.112	0.97830	21	9.513	58
3	0.85628	2	7.183	0.86876	38	7.392	0.89771	56	7.902	0.93131	54	8.537	0.96001	41	9.121	0.97851	21	9.517	57
4	0.85630	3	7.183	0.86914	38	7.399	0.89827	55	7.912	0.93185	54	8.548	0.96042	40	9.129	0.97872	20	9.522	56
5	0.85633	4	7.183	0.86952	39	7.405	0.89882	56	7.922	0.93239	53	8.559	0.96082	39	9.138	0.97892	19	9.526	55
6	0.85637	4	7.184	0.86991	39	7.412	0.89938	55	7.932	0.93292	54	8.569	0.96121	39	9.146	0.97911	19	9.531	54
7	0.85641	6	7.184	0.87030	40	7.418	0.89993	56	7.942	0.93346	53	8.580	0.96160	39	9.154	0.97930	19	9.535	53
8	0.85647	6	7.185	0.87070	40	7.425	0.90049	56	7.952	0.93399	53	8.590	0.96199	39	9.162	0.97949	19	9.539	52
9	0.85653	6	7.187	0.87110	41	7.432	0.90105	57	7.963	0.93452	53	8.601	0.96238	38	9.170	0.97968	19	9.543	51
10	0.85659	7	7.188	0.87151	40	7.439	0.90162	56	7.973	0.93505	52	8.611	0.96276	38	9.178	0.97987	18	9.547	50
11	0.85666	7	7.189	0.87191	42	7.446	0.90218	56	7.983	0.93557	53	8.621	0.96314	39	9.186	0.98005	18	9.551	49
12	0.85673	7	7.190	0.87233	42	7.453	0.90274	56	7.994	0.93610	52	8.632	0.96353	37	9.194	0.98023	17	9.555	48
13	0.85680	9	7.191	0.87275	42	7.460	0.90330	56	8.004	0.93662	53	8.642	0.96390	37	9.203	0.98040	17	9.559	47
14	0.85689	9	7.193	0.87317	42	7.468	0.90386	56	8.014	0.93715	52	8.653	0.96427	37	9.210	0.98057	16	9.563	46
15	0.85698	10	7.194	0.87359	44	7.475	0.90442	57	8.024	0.93767	52	8.663	0.96464	36	9.218	0.98073	17	9.566	45
16	0.85708	11	7.195	0.87403	44	7.482	0.90499	57	8.035	0.93819	52	8.673	0.96500	37	9.226	0.98090	16	9.570	44
17	0.85719	11	7.197	0.87447	43	7.490	0.90556	56	8.046	0.93871	51	8.684	0.96537	36	9.234	0.98106	15	9.573	43
18	0.85730	12	7.199	0.87490	45	7.497	0.90612	57	8.056	0.93922	51	8.694	0.96573	36	9.241	0.98121	15	9.577	42
19	0.85742	13	7.202	0.87535	45	7.505	0.90669	57	8.067	0.93973	51	8.704	0.96609	35	9.249	0.98136	15	9.580	41
20	0.85755	13	7.204	0.87580	45	7.513	0.90726	56	8.077	0.94024	51	8.715	0.96644	34	9.257	0.98151	15	9.583	40
21	0.85768	15	7.206	0.87625	45	7.521	0.90782	57	8.088	0.94075	51	8.725	0.96678	35	9.264	0.98166	14	9.587	39
22	0.85783	15	7.208	0.87670	45	7.529	0.90839	56	8.098	0.94126	50	8.735	0.96713	34	9.271	0.98180	13	9.590	38
23	0.85798	16	7.211	0.87715	47	7.536	0.90895	57	8.109	0.94176	50	8.745	0.96747	35	9.279	0.98193	14	9.593	37
24	0.85814	16	7.214	0.87762	47	7.544	0.90952	57	8.120	0.94226	50	8.755	0.96782	34	9.286	0.98207	13	9.596	36
25	0.85830	18	7.216	0.87809	46	7.553	0.91009	57	8.130	0.94276	50	8.765	0.96816	33	9.293	0.98220	13	9.599	35
26	0.85848	18	7.219	0.87855	46	7.561	0.91066	56	8.141	0.94326	50	8.775	0.96849	33	9.300	0.98233	12	9.602	34
27	0.85866	18	7.222	0.87901	48	7.569	0.91122	57	8.151	0.94376	50	8.786	0.96882	33	9.307	0.98245	12	9.604	33
28	0.85884	19	7.225	0.87949	48	7.577	0.91179	57	8.162	0.94426	49	8.796	0.96915	32	9.314	0.98257	12	9.607	32
29	0.85903	19	7.228	0.87997	48	7.585	0.91236	57	8.173	0.94475	50	8.806	0.96947	32	9.321	0.98269	11	9.609	31
30	0.85922	20	7.232	0.88045	48	7.594	0.91293	56	8.184	0.94525	49	8.816	0.96979	31	9.328	0.98280	11	9.612	30
31	0.85942	21	7.235	0.88093	48	7.602	0.91349	57	8.194	0.94574	48	8.826	0.97010	32	9.335	0.98291	10	9.614	29
32	0.85963	21	7.238	0.88141	49	7.611	0.91406	57	8.205	0.94622	48	8.835	0.97042	32	9.342	0.98301	10	9.617	28
33	0.85984	22	7.242	0.88190	50	7.619	0.91463	57	8.216	0.94670	48	8.845	0.97074	30	9.349	0.98311	10	9.619	27
34	0.86006	22	7.245	0.88240	49	7.628	0.91520	56	8.226	0.94718	48	8.855	0.97104	31	9.355	0.98321	10	9.621	26
35	0.86028	23	7.249	0.88289	50	7.637	0.91576	57	8.237	0.94766	47	8.865	0.97135	30	9.362	0.98331	9	9.623	25
36	0.86051	23	7.253	0.88339	50	7.646	0.91633	56	8.248	0.94813	48	8.874	0.97165	30	9.368	0.98340	9	9.625	24
37	0.86074	24	7.257	0.88389	51	7.654	0.91689	57	8.258	0.94861	47	8.884	0.97195	29	9.375	0.98349	8	9.627	23
38	0.86098	24	7.261	0.88440	51	7.663	0.91746	56	8.269	0.94908	46	8.894	0.97224	29	9.381	0.98357	7	9.629	22
39	0.86122	25	7.265	0.88491	51	7.672	0.91802	57	8.280	0.94955	46	8.903	0.97253	29	9.387	0.98364	8	9.631	21
40	0.86147	25	7.269	0.88542	51	7.681	0.91859	56	8.290	0.95001	47	8.913	0.97282	28	9.393	0.98372	7	9.632	20
41	0.86172	26	7.273	0.88593	51	7.690	0.91915	56	8.302	0.95047	46	8.923	0.97310	29	9.399	0.98379	7	9.634	19
42	0.86198	27	7.278	0.88644	51	7.699	0.91971	56	8.313	0.95094	45	8.932	0.97339	28	9.406	0.98386	7	9.635	18
43	0.86225	27	7.282	0.88695	53	7.708	0.92027	57	8.324	0.95140	45	8.941	0.97367	29	9.412	0.98393	6	9.637	17
44	0.86252	28	7.287	0.88748	52	7.718	0.92084	56	8.334	0.95185	46	8.951	0.97396	27	9.418	0.98399	6	9.638	16
45	0.86280	28	7.291	0.88800	53	7.727	0.92140	56	8.345	0.95230	45	8.960	0.97323	27	9.424	0.98405	5	9.640	15
46	0.86308	29	7.296	0.88853	52	7.736	0.92196	55	8.355	0.95276	45	8.970	0.97450	26	9.430	0.98410	4	9.641	14
47	0.86337	29	7.301	0.88905	53	7.746	0.92251	56	8.366	0.95321	45	8.979	0.97476	25	9.436	0.98414	5	9.642	13
48	0.86366	30	7.306	0.88958	52	7.755	0.92307	55	8.377	0.95366	44	8.988	0.97501	25	9.441	0.98419	4	9.643	12
49	0.86396	31	7.311	0.89010	53	7.764	0.92362	56	8.387	0.95410	44	8.997	0.97526	26	9.446	0.98423	4	9.644	11
50	0.86427	31	7.316	0.89063	54	7.774	0.92418	55	8.398	0.95454	43	9.006	0.97552	25	9.452	0.98427	4	9.645	10
51	0.86458	33	7.321	0.89117	54	7.784	0.92473	56	8.409	0.95497	44	9.015	0.97577	25	9.458	0.98431	3	9.645	9
52	0.86491	33	7.327	0.89171	54	7.793	0.92529	55	8.420	0.95541	43	9.024	0.97602	24	9.463	0.98434	3	9.646	8
53	0.86524	34	7.332	0.89225	54	7.803	0.92584	56	8.430	0.95584	43	9.033	0.97626	24	9.468	0.98437	3	9.647	7
54	0.86558	34	7.338	0.89279	54	7.813	0.92640	55	8.441	0.95627	43	9.042	0.97650	24	9.473	0.98440	1	9.647	6
55	0.86592	35	7.344	0.89333	54	7.822	0.92695	55	8.452	0.95670	42	9.051	0.97674	23	9.479	0.98441	1	9.648	5
56	0.86627	35	7.349	0.89387	54	7.832	0.92750	55	8.463	0.95712	42	9.060	0.97679	23	9.484	0.98442	1	9.648	4
57	0.86652	35	7.355	0.89441	55	7.842	0.92805	54	8.473	0.95754	42	9.069	0.97720	23	9.489	0.98443	0	9.648	3
58	0.86697	35	7.362	0.89496	55	7.852	0.92859	55	8.484	0.95796	42	9.078	0.97743	22	9.494	0.98443	1	9.648	2
59	0.86732	36	7.368	0.89551	56	7.862	0.92914	54	8.495	0.95838	41	9.086	0.97765	23	9.499	0.98444	0	9.648	1
60	0.86768		7.374	0.89607		7.872	0.92968		8.505	0.95879		9.095	0.97788		9.504	0.98444			0
	XI		XXIII	X		XXII	IX		XXI	VIII		XX	VII		XIX	VI		XVIII	

TABLE XIII. (By Inspection.)

Solar Nutation in North Polar Distance to find the annual number S′. Argument = R. A. of * in time.

[Log. tang. S′ = log. tang. R. A. + .0373884.]

	O XII		I XIII		II XIV		III XV		IV XVI		V XVII		
	12 − 6 −	P. P.	12 − 6 −	P. P.	12 − 6 −	P. P.	12 − 6 −	P. P.	12 − 6 −	P. P.	12 − 6 −	P. P.	
Min.	S′.		S′.		S′.		S′.		S′.		S′.		Min.
0	0s 0° 0′	0	0s 16° 17′	0	1s 2° 11′	0	1s 17° 28′	0	2s 2° 5′	0	2s 16° 11′	0	60
1	0 0 17	0	0 16 34	0	1 2 26	0	1 17 43	0	2 2 20	0	2 16 25	0	59
2	0 0 33	1	0 16 50	1	1 2 42	1	1 17 58	1	2 2 34	0	2 16 39	0	58
3	0 0 49	1	0 17 6	1	1 2 57	1	1 18 13	1	2 2 49	1	2 16 53	1	57
4	0 1 5	1	0 17 22	1	1 3 13	1	1 18 28	1	2 3 3	1	2 17 7	1	56
5	0 1 22	1	0 17 38	1	1 3 28	1	1 18 43	1	2 3 17	1	2 17 21	1	55
6	0 1 38	2	0 17 54	2	1 3 44	2	1 18 58	2	2 3 32	1	2 17 35	1	54
7	0 1 54	2	0 18 10	2	1 4 0	2	1 19 13	2	2 3 46	2	2 17 49	2	53
8	0 2 11	2	0 18 26	2	1 4 16	2	1 19 27	2	2 4 0	2	2 18 3	2	52
9	0 2 27	2	0 18 42	2	1 4 31	2	1 19 42	2	2 4 15	2	2 18 17	2	51
10	0 2 43	3	0 18 58	3	1 4 47	3	1 19 57	3	2 4 29	2	2 18 31	2	50
11	0 2 59	3	0 19 14	3	1 5 2	3	1 20 12	3	2 4 43	3	2 18 45	3	49
12	0 3 16	3	0 19 30	3	1 5 18	3	1 20 27	3	2 4 57	3	2 18 58	3	48
13	0 3 33	4	0 19 46	4	1 5 33	3	1 20 42	3	2 5 12	3	2 19 12	3	47
14	0 3 49	4	0 20 2	4	1 5 49	4	1 20 57	3	2 5 26	3	2 19 26	3	46
15	0 4 6	4	0 20 18	4	1 6 4	4	1 21 12	4	2 5 40	4	2 19 40	4	45
16	0 4 22	4	0 20 34	4	1 6 20	4	1 21 26	4	2 5 54	4	2 19 53	4	44
17	0 4 38	5	0 20 50	5	1 6 35	4	1 21 41	4	2 6 9	4	2 20 7	4	43
18	0 4 55	5	0 21 6	5	1 6 50	5	1 21 56	4	2 6 23	4	2 20 21	4	42
19	0 5 11	5	0 21 22	5	1 7 6	5	1 22 11	5	2 6 37	4	2 20 35	4	41
20	0 5 27	5	0 21 38	5	1 7 21	5	1 22 25	5	2 6 51	5	2 20 49	5	40
21	0 5 43	6	0 21 54	6	1 7 37	5	1 22 40	5	2 7 5	5	2 21 3	5	39
22	0 6 0	6	0 22 10	6	1 7 52	6	1 22 54	5	2 7 19	5	2 21 17	5	38
23	0 6 16	6	0 22 26	6	1 8 8	6	1 23 9	6	2 7 23	5	2 21 31	5	37
24	0 6 32	6	0 22 42	6	1 8 23	6	1 23 23	6	2 7 47	6	2 21 44	6	36
25	0 6 48	7	0 22 58	7	1 8 38	6	1 23 38	6	2 8 1	6	2 21 58	6	35
26	0 7 5	7	0 23 14	7	1 8 54	7	1 23 53	6	2 8 15	6	2 22 12	6	34
27	0 7 21	7	0 23 30	7	1 9 9	7	1 24 7	7	2 8 29	6	2 22 26	6	33
28	0 7 37	8	0 23 46	8	1 9 24	7	1 24 22	7	2 8 43	7	2 22 39	7	32
29	0 7 54	8	0 24 2	8	1 9 40	7	1 24 36	7	2 8 58	7	2 22 53	7	31
30	0 8 10	8	0 24 18	8	1 9 55	8	1 24 51	7	2 9 12	7	2 23 7	7	30
31	0 8 27	8	0 24 34	8	1 10 10	8	1 25 5	8	2 9 26	7	2 23 21	7	29
32	0 8 43	9	0 24 50	9	1 10 25	8	1 25 20	8	2 9 40	7	2 23 35	7	28
33	0 8 59	9	0 25 6	9	1 10 41	8	1 25 35	8	2 9 54	8	2 23 49	8	27
34	0 9 16	9	0 25 22	9	1 10 56	9	1 25 49	8	2 10 8	8	2 24 3	8	26
35	0 9 32	9	0 25 38	9	1 11 11	9	1 26 4	9	2 10 22	8	2 24 17	8	25
36	0 9 48	10	0 25 53	10	1 11 26	9	1 26 19	9	2 10 36	8	2 24 30	8	24
37	0 10 5	10	0 26 9	10	1 11 42	9	1 26 33	9	2 10 50	9	2 24 44	9	23
38	0 10 21	10	0 26 25	10	1 11 57	10	1 26 48	9	2 11 4	9	2 24 58	9	22
39	0 10 37	11	0 26 41	10	1 12 12	10	1 27 2	10	2 11 18	9	2 25 12	9	21
40	0 10 53	11	0 26 56	11	1 12 27	10	1 27 17	10	2 11 32	9	2 25 25	9	20
41	0 11 10	11	0 27 12	11	1 12 42	10	1 27 31	10	2 11 46	10	2 25 39	10	19
42	0 11 26	11	0 27 28	11	1 12 57	11	1 27 46	10	2 12 0	10	2 25 53	10	18
43	0 11 42	12	0 27 44	11	1 13 12	11	1 28 0	11	2 12 14	10	2 26 7	10	17
44	0 11 58	12	0 28 0	12	1 13 27	11	1 28 15	11	4 12 28	10	2 26 20	10	16
45	0 12 15	12	0 28 15	12	1 13 43	11	1 28 29	11	2 12 42	11	2 26 34	10	15
46	0 12 31	12	0 28 31	12	1 13 58	12	1 28 44	11	2 12 56	11	2 26 48	11	14
47	0 12 47	12	0 28 47	12	1 14 13	12	1 28 58	12	2 13 10	11	2 27 2	11	13
48	0 13 3	13	0 29 3	13	1 14 28	12	1 29 13	12	2 13 24	11	2 27 15	11	12
49	0 13 20	13	0 29 18	13	1 14 43	12	1 29 27	12	2 13 38	12	2 27 29	11	11
50	0 13 36	13	0 29 34	13	1 14 58	13	1 29 42	12	2 13 52	12	2 27 43	12	10
51	0 13 52	14	0 29 50	13	1 15 13	13	1 29 56	13	2 14 6	12	2 27 57	12	9
52	0 14 8	14	1 0 6	14	1 15 28	13	2 0 11	13	2 14 20	12	2 28 10	12	8
53	0 14 25	14	1 0 21	14	1 15 43	13	2 0 25	13	2 14 34	13	2 28 24	12	7
54	0 14 41	14	1 0 37	14	1 15 58	14	2 0 39	13	2 14 48	13	2 28 38	12	6
55	0 14 57	15	1 0 53	14	1 16 13	14	2 0 54	14	2 15 2	13	2 28 52	13	5
56	0 15 13	15	1 1 9	14	1 16 28	14	2 1 8	14	2 15 16	13	2 29 5	13	4
57	0 15 29	15	1 1 24	15	1 16 43	14	2 1 23	14	2 15 30	14	2 29 19	13	3
58	0 15 45	15	1 1 40	15	1 16 58	15	2 1 37	14	2 15 44	14	2 29 33	13	2
59	0 16 1	16	1 1 55	15	1 17 13	15	2 1 51	14	2 15 58	14	2 29 47	13	1
60	0 16 17	16	1 2 11	15	1 17 28	15	2 2 5	14	2 16 11	14	3 0 0	13	0
	6 + 0 +		6 + 0 +		6 + 0 +		6 + 0 +		6 + 0 +		6 + 0 +		
	XI XXIII		X XXII		IX XXI		VIII XX		VII XIX		VI XVIII		

TABLE XIV. (By Inspection.)

Solar Nutation in North Polar Distance. To find Log. Max. and Max. Argument = R. A. of the * in time.

[Log. Max. = log. sin. R. A. +9.6374897 − log. sin, S′.]

	O	XII	I	XIII	II	XIV	III	XV	IV	XVI	V	XVII	
Min.	Log. Max.	Max.	Log. Max.	Max.	Log. Max.	Max.	Log. Max.	Max.	Log. Max.	Max.	Log. Max.	Max.	Min.
0	9.600	0″.398	9.603	0″.401	9.611	0″.408	9.620	0″.417	9.629	0.426	9.635	0″.432	60
1	9.600	0.398	9.603	0.401	9.611	0.408	9.620	0.417	9.629	0.426	9.635	0.432	59
2	9.600	0.398	9.603	0.401	9.641	0.408	9.620	0.417	9.629	0.426	9.635	0.432	58
3	9.600	0.398	9.603	0.401	9.611	0.408	9.620	0.417	9.629	0.426	9.635	0.432	57
4	9.600	0.398	9.603	0.401	9.611	0.408	9.620	0.417	9.629	0.426	9.635	0.432	56
5	9.600	0.398	9.603	0.401	9.611	0.408	9.620	0.417	9.629	0.426	9.636	0.433	55
6	9.600	0.398	9.603	0.401	9.611	0.408	9.621	0.418	9.629	0.426	9.636	0.433	54
7	9.600	0.398	9.603	0.401	9.611	0.408	9.621	0.418	9.630	0.427	9.636	0.433	53
8	9.600	0.398	9.603	0.401	9.611	0.409	9.621	0.418	9.630	0.427	9.636	0.433	52
9	9.600	0.398	9.603	0.401	9.611	0.409	9.621	0.418	9.630	0.427	9.636	0.433	51
10	9.600	0.398	9.604	0.402	9.612	0.409	9.621	0.418	9.630	0.427	9.636	0.433	50
11	9.600	0.398	9.604	0.402	9.612	0.409	9.621	0.418	9.630	0.427	9.636	0.433	49
12	9.600	0.398	9.604	0.402	9.612	0.409	9.621	0.418	9.630	0.427	9.636	0.433	48
13	9.600	0.398	9.604	0.402	9.612	0.409	9.622	0.419	9.630	0.427	9.636	0.433	47
14	9.600	0.398	9.604	0.402	9.612	0.409	9.622	0.419	9.630	0.427	9.636	0.433	46
15	9.600	0.398	9.604	0.402	9.612	0.409	9.622	0.419	9.631	0.428	9.636	0.433	45
16	9.600	0.398	9.604	0.402	9.612	0.409	9.622	0.419	9.631	0.428	9.636	0.433	44
17	9.600	0.398	9.604	0.402	9.612	0.409	9.622	0.419	9.631	0.428	9.636	0.433	43
18	9.600	0.398	9.604	0.402	9.612	0.409	9.622	0.419	9.631	0.428	9.636	0.433	42
19	9.600	0.398	9.604	0.402	9.612	0.410	9.623	0.420	9.631	0.428	9.636	0.433	41
20	9.600	0.398	9.605	0.403	9.612	0.410	9.623	0.420	9.631	0.428	9.636	0.433	40
21	9.600	0.398	9.605	0.403	9.612	0.410	9.623	0.420	9.631	0.428	9.636	0.433	39
22	9.600	0.398	9.605	0.403	9.613	0.410	9.623	0.420	9.631	0.428	9.636	0.433	38
23	9.600	0.398	9.605	0.403	9.613	0.410	9.623	0.420	9.632	0.429	9.637	0.434	37
24	9.601	0.399	9.605	0.403	9.613	0.410	9.623	0.420	9.632	0.429	9.637	0.434	36
25	9.601	0.399	9.605	0.403	9.613	0.410	9.624	0.421	9.632	0.429	9.637	0.434	35
26	9.601	0.399	9.605	0.403	9.613	0.410	9.624	0.421	9.632	0.429	9.637	0.434	34
27	9.601	0.399	9.606	0.404	9.613	0.410	9.624	0.421	9.632	0.429	9.637	0.434	33
28	9.601	0.399	9.606	0.404	9.613	0.410	9.624	0.421	9.632	0.429	9.637	0.434	32
29	9.601	0.399	9.606	0.404	9.614	0.411	9.624	0.421	9.632	0.429	9.637	0.434	31
30	9.601	0.399	9.606	0.404	9.614	0.411	9.624	0.421	9.632	0.429	9.637	0.434	30
31	9.601	0.399	9.606	0.404	9.614	0.411	9.625	0.422	9.632	0.429	9.637	0.434	29
32	9.601	0.399	9.606	0.404	9.615	0.412	9.635	0.422	9.633	0.430	9.637	0.434	28
33	9.601	0.399	9.606	0.404	9.615	0.412	9.625	0.422	9.633	0.430	9.637	0.434	27
34	9.601	0.399	9.606	0.404	9.615	0.412	9.625	0.422	9.633	0.430	9.637	0.434	26
35	9.601	0.399	9.607	0.405	9.615	0.412	9.625	0.422	9.633	0.430	9.637	0.434	25
36	9.601	0.399	9.607	0.405	9.616	0.413	9.625	0.422	9.633	0.430	9.637	0.434	24
37	9.601	0.399	9.607	0.405	9.616	0.413	9.626	0.423	9.633	0.430	9.637	0.434	23
38	9.601	0.399	9.607	0.405	9.616	0.413	9.626	0.423	9.633	0.430	9.637	0.434	22
39	9.601	0.399	9.607	0.405	9.616	0.413	9.626	0.423	9.633	0.430	9.637	0.434	21
40	9.601	0.399	9.607	0.405	9.616	0.413	9.626	0.423	9.634	0.431	9.637	0.434	20
41	9.601	0.399	9.607	0.405	9.616	0.414	9.626	0.423	9.634	0.431	9.637	0.434	19
42	9.601	0.399	9.608	0.406	9.617	0.414	9.626	0.423	9.634	0.431	9.637	0.434	18
43	9.601	0.399	9.608	0.406	9.617	0.414	9.626	0.423	9.634	0.431	9.637	0.434	17
44	9.601	0.399	9.608	0.406	9.617	0.414	9.626	0.423	9.634	0.431	9.637	0.434	16
45	9.601	0.399	9.608	0.406	9.617	0.414	9.627	0.424	9.634	0.431	9.637	0.434	15
46	9.601	0.399	9.608	0.406	9.617	0.415	9.627	0.424	9.634	0.431	9.637	0.434	14
47	9.601	0.399	9.608	0.406	9.617	0.415	9.627	0.424	9.634	0.431	9.637	0.434	13
48	9.602	0.400	9.609	0.407	9.618	0.415	9.627	0.424	9.634	0.431	9.637	0.434	12
49	9.602	0.400	9.609	0.407	9.618	0.415	9.627	0.424	9.634	0.431	9.637	0.434	11
50	9.602	0.400	9.609	0.407	9.618	0.415	9.627	0.424	9.634	0.431	9.637	0.434	10
51	9.602	0.400	9.609	0.407	9.618	0.415	9.627	0.424	9.634	0.431	9.637	0.434	9
52	9.602	0.400	9.609	0.407	9.618	0.416	9.628	0.425	9.634	0.431	9.637	0.434	8
53	9.602	0.400	9.609	0.407	9.619	0.416	9.628	0.425	9.635	0.432	9.637	0.434	7
54	9.602	0.400	9.609	0.407	9.619	0.416	9.628	0.425	9.635	0.432	9.637	0.434	6
55	9.602	0.400	9.609	0.407	9.619	0.416	9.628	0.425	9.635	0.432	9.637	0.434	5
56	9.602	0.400	9.609	0.407	9.619	0.416	9.628	0.425	9.635	0.432	9.637	0.434	4
57	9.602	0.400	9.610	0.408	9.619	0.416	9.628	0.425	9.635	0.432	9.637	0.434	3
58	9.602	0.400	9.610	0.408	9.619	0.417	9.628	0.425	9.635	0.432	9.637	0.434	2
59	9.602	0.400	9.610	0.408	9.620	0.417	9.629	0.426	9.635	0.432	9.637	0.434	1
60	9.602	0.400	9.610	0.408	9.620	0.417	9.629	0.436	9.635	0.432	9.637	0.434	0
	XI	XXIII	X	XXII	IX	XXI	VIII	XX	VII	XIX	VI	XVIII	

TABLE XV.

GENERAL TABLE of daily precessions in Right Ascension and North Polar Distance, when the annual quantities are known.

Jan.	1	2	3	4	5	6	7	8	9	10	11	12	13	14	15	16	17	18	19	20	Jan.
1	0.003	0.005	0.008	0.011	0.014	0.016	0.019	0.022	0.025	0.027	0.030	0.032	0.035	0.038	0.042	0.044	0.047	0.050	0.052	0.054	1
2	0.005	0.011	0.016	0.022	0.027	0.033	0.038	0.044	0.049	0.055	0.060	0.066	0.071	0.076	0.081	0.088	0.093	0.098	0.104	0.110	2
3	0.008	0.016	0.025	0.033	0.041	0.049	0.058	0.066	0.074	0.082	0.090	0.098	0.107	0.116	0.123	0.132	0.140	0.148	0.156	0.164	3
4	0.011	0.022	0.033	0.044	0.055	0.066	0.077	0.088	0.099	0.110	0.121	0.132	0.143	0.154	0.165	0.176	0.187	0.198	0.208	0.219	4
5	0.014	0.027	0.041	0.055	0.069	0.082	0.096	0.110	0.123	0.137	0.151	0.164	0.178	0.192	0.207	0.220	0.233	0.226	0.260	0.274	5
6	0.016	0.033	0.049	0.066	0.082	0.099	0.115	0.132	0.148	0.164	0.181	0.198	0.214	0.230	0.246	0.264	0.280	0.296	0.312	0.329	6
7	0.019	0.038	0.058	0.077	0.096	0.115	0.134	0.153	0.173	0.192	0.211	0.230	0.249	0.278	0.288	0.306	0.326	0.346	0.364	0.384	7
8	0.022	0.044	0.066	0.088	0.110	0.132	0.153	0.175	0.197	0.219	0.242	0.264	0.285	0.306	0.330	0.350	0.372	0.394	0.416	0.438	8
9	0.025	0.049	0.074	0.099	0.123	0.148	0.173	0.197	0.222	0.247	0.271	0.296	0.321	0.346	0.369	0.394	0.419	0.444	0.468	0.493	9
10	0.027	0.055	0.082	0.110	0.137	0.164	0.192	0.219	0.247	0.274	0.301	0.328	0.356	0.384	0.411	0.438	0.466	0.494	0.521	0.548	10
11	0.030	0.060	0.090	0.121	0.151	0.181	0.211	0.241	0.271	0.301	0.332	0.362	0.392	0.422	0.453	0.482	0.512	0.542	0.573	0.603	11
12	0.033	0.066	0.099	0.132	0.164	0.197	0.230	0.263	0.296	0.329	0.361	0.394	0.427	0.460	0.492	0.526	0.559	0.592	0.625	0.658	12
13	0.036	0.071	0.107	0.142	0.178	0.214	0.249	0.285	0.321	0.356	0.292	0.428	0.463	0.498	0.534	0.570	0.606	0.642	0.677	0.712	13
14	0.038	0.077	0.115	0.153	0.192	0.230	0.269	0.307	0.345	0.384	0.322	0.460	0.499	0.538	0.576	0.614	0.652	0.690	0.729	0.767	14
15	0.041	0.082	0.123	0.164	0.206	0.247	0.288	0.329	0.370	0.411	0.353	0.494	0.535	0.576	0.618	0.658	0.699	0.740	0.781	0.822	15
16	0.044	0.088	0.132	0.175	0.219	0.263	0.307	0.351	0.395	0.438	0.482	0.526	0.570	0.614	0.657	0.702	0.746	0.790	0.833	0.877	16
17	0.047	0.093	0.140	0.186	0.233	0.279	0.326	0.373	0.419	0.466	0.512	0.558	0.605	0.652	0.699	0.746	0.792	0.838	0.885	0.932	17
18	0.049	0.099	0.148	0.197	0.247	0.296	0.345	0.395	0.444	0.493	0.543	0.592	0.641	0.690	0.741	0.790	0.839	0.888	0.937	0.986	18
19	0.052	0.104	0.156	0.208	0.260	0.312	0.364	0.416	0.469	0.521	0.572	0.624	0.676	0.728	0.780	0.832	0.885	0.933	0.999	1.041	19
20	0.055	0.110	0.164	0.219	0.274	0.329	0.384	0.438	0.493	0.548	0.603	0.658	0.713	0.768	0.832	0.876	0.931	0.986	1.041	1.096	20
21	0.058	0.115	0.173	0.230	0.288	0.345	0.403	0.460	0.518	0.575	0.633	0.690	0.748	0.806	0.864	0.920	0.978	1.036	1.093	1.151	21
22	0.060	0.121	0.181	0.241	0.301	0.362	0.422	0.482	0.543	0.603	0.663	0.724	0.784	0.844	0.903	0.964	1.025	1.086	1.145	1.206	22
23	0.063	0.126	0.189	0.252	0.315	0.378	0.441	0.504	0.567	0.630	0.693	0.756	0.819	0.882	0.945	1.008	1.071	1.134	1.197	1.260	23
24	0.066	0.132	0.197	0.263	0.329	0.395	0.460	0.526	0.592	0.658	0.724	0.790	0.855	0.920	0.987	1.052	1.118	1.184	1.250	1.315	24
25	0.069	0.137	0.206	0.274	0.343	0.411	0.480	0.548	0.617	0.685	0.754	0.822	0.891	0.960	1.029	1.096	1.165	1.234	1.302	1.370	25
26	0.071	0.142	0.214	0.285	0.356	0.427	0.499	0.570	0.641	0.712	0.783	0.854	0.926	0.998	1.068	1.140	1.211	1.282	1.353	1.425	26
27	0.074	0.148	0.222	0.296	0.370	0.444	0.518	0.592	0.666	0.740	0.814	0.888	0.962	1.036	1.110	1.184	1.258	1.332	1.405	1.480	27
28	0.077	0.153	0.230	0.307	0.384	0.460	0.537	0.614	0.690	0.767	0.844	0.920	0.997	1.074	1.152	1.228	1.304	1.380	1.457	1.534	28
29	0.079	0.159	0.238	0.318	0.397	0.477	0.556	0.636	0.715	0.795	0.874	0.954	1.033	1.112	1.191	1.272	1.351	1.430	1.510	1.590	29
30	0.082	0.164	0.247	0.329	0.411	0.493	0.575	0.658	0.740	0.822	0.904	0.986	1.068	1.150	1.233	1.316	1.398	1.480	1.562	1.644	30
31	0.085	0.170	0.255	0.340	0.425	0.510	0.595	0.680	0.764	0.849	0.935	1.020	1.105	1.190	1.275	1.360	1.444	1.528	1.614	1.699	31
Feb.																					Feb.
1	0.088	0.175	0.263	0.351	0.438	0.526	0.614	0.701	0.789	0.877	0.964	1.052	1.140	1.228	1.314	1.402	1.490	1.578	1.666	1.754	1
2	0.090	0.181	0.271	0.362	0.452	0.543	0.633	0.723	0.814	0.904	0.995	1.086	1.176	1.266	1.356	1.446	1.537	1.628	1.718	1.808	2
3	0.093	0.186	0.279	0.373	0.466	0.559	0.652	0.745	0.838	0.932	1.025	1.118	1.211	1.304	1.398	1.490	1.583	1.676	1.770	1.863	3
4	0.096	0.192	0.288	0.384	0.480	0.575	0.671	0.767	0.863	0.959	1.055	1.150	1.246	1.342	1.440	1.534	1.630	1.726	1.822	1.918	4
5	0.099	0.197	0.296	0.399	0.493	0.592	0.690	0.797	0.888	0.986	1.085	1.184	1.282	1.380	1.479	1.594	1.685	1.778	1.874	1.973	5
6	0.101	0.203	0.304	0.406	0.507	0.608	0.710	0.811	0.912	1.014	1.115	1.216	1.318	1.420	1.521	1.622	1.723	1.824	1.926	2.028	6
7	0.104	0.208	0.312	0.416	0.521	0.625	0.729	0.833	0.937	1.041	1.146	1.250	1.354	1.458	1.563	1.666	1.770	1.874	1.978	2.082	7
8	0.107	0.214	0.321	0.427	0.534	0.641	0.748	0.855	0.962	1.069	1.175	1.282	1.389	1.496	1.602	1.710	1.817	1.924	2.030	2.137	8
9	0.110	0.219	0.329	0.438	0.548	0.658	0.767	0.877	0.986	1.096	1.206	1.316	1.425	1.534	1.644	1.754	1.863	1.972	2.082	2.192	9
10	0.112	0.225	0.337	0.449	0.562	0.674	0.786	0.899	1.011	1.123	1.236	1.348	1.462	1.572	1.686	1.798	1.910	2.022	2.134	2.247	10
11	0.115	0.230	0.345	0.460	0.575	0.690	0.806	0.921	1.036	1.151	1.265	1.380	1.496	1.612	1.725	1.842	1.957	2.072	2.186	2.302	11
12	0.118	0.236	0.353	0.471	0.589	0.707	0.825	0.943	1.060	1.178	1.296	1.414	1.532	1.650	1.767	1.886	2.003	2.120	2.238	2.536	12
13	0.121	0.241	0.362	0.482	0.603	0.723	0.844	0.964	1.085	1.206	1.326	1.446	1.567	1.688	1.809	1.928	2.049	2.170	2.291	2.411	13
14	0.123	0.247	0.370	0.493	0.617	0.740	0.863	0.986	1.110	1.233	1.357	1.480	1.603	1.726	1.851	1.972	2.096	2.220	2.343	2.466	14
15	0.126	0.252	0.378	0.504	0.630	0.756	0.882	1.008	1.134	1.261	1.386	1.512	1.631	1.764	1.890	2.016	2.142	2.268	2.395	2.521	15
16	0.129	0.258	0.386	0.515	0.644	0.773	0.901	1.030	1.159	1.288	1.417	1.546	1.674	1.802	1.932	2.060	2.189	2.318	2.447	2.576	16
17	0.132	0.263	0.395	0.526	0.658	0.789	0.921	1.052	1.184	1.315	1.447	1.578	1.710	1.842	1.974	2.104	2.236	2.368	2.499	2.630	17
18	0.134	0.269	0.403	0.537	0.671	0.806	0.940	1.074	1.208	1.343	1.477	1.612	1.746	1.880	2.013	2.148	2.282	2.416	2.551	2.685	18
19	0.137	0.274	0.411	0.548	0.685	0.822	0.959	1.096	1.233	1.370	1.507	1.644	1.781	1.918	2.055	2.192	2.329	2.466	2.603	2.740	19
20	0.140	0.279	0.419	0.559	0.699	0.838	0.978	1.118	1.258	1.397	1.537	1.676	1.816	1.956	2.097	2.236	2.376	2.516	2.655	2.795	20
21	0.142	0.285	0.427	0.570	0.712	0.855	0.997	1.140	1.282	1.425	1.567	1.710	1.852	1.994	2.136	2.280	2.422	2.564	2.707	2.850	21
22	0.145	0.290	0.436	0.581	0.726	0.871	1.017	1.162	1.307	1.452	1.597	1.742	1.888	2.034	2.178	2.324	2.469	2.614	2.759	2.904	22
23	0.148	0.296	0.444	0.592	0.740	0.888	1.036	1.184	1.332	1.480	1.628	1.776	1.924	2.072	2.220	2.368	2.516	2.664	2.811	2.960	23
24	0.151	0.301	0.452	0.603	0.754	0.904	1.055	1.206	1.356	1.507	1.658	1.808	1.959	2.110	2.262	2.412	2.562	2.712	2.863	3.014	24
25	0.153	0.307	0.460	0.614	0.767	0.921	1.074	1.228	1.381	1.534	1.688	1.842	1.995	2.148	2.301	2.456	2.609	2.762	2.915	3.069	25
26	0.156	0.312	0.469	0.625	0.781	0.937	1.093	1.249	1.406	1.562	1.718	1.874	2.030	2.186	2.343	2.498	2.655	2.812	2.967	3.124	26
27	0.159	0.318	0.477	0.636	0.795	0.954	1.112	1.271	1.430	1.589	1.749	1.908	2.066	2.224	2.385	2.542	2.701	2.860	3.019	3.178	27
28	0.162	0.323	0.485	0.647	0.808	0.970	1.132	1.293	1.455	1.617	1.778	1.940	2.102	2.264	2.424	2.586	2.748	2.910	3.072	3.233	28
	1	2	3	4	5	6	7	8	9	10	11	12	13	14	15	16	17	18	19	20	

In the months of January and February of every Leap Year, take out the precession for the preceding day.

TABLE XV.

GENERAL TABLE of daily Precessions, &c. continued.

Mar.	1	2	3	4	5	6	7	8	9	10	11	12	13	14	15	16	17	18	19	20	Mar.
1	0.164	0.329	0.493	0.658	0.822	0.986	1.151	1.315	1.480	1.644	1.808	1.972	2.137	2.302	2.466	2.630	2.795	2.960	3.124	3.288	1
2	0.167	0.334	0.501	0.669	0.836	1.003	1.170	1.337	1.504	1.671	1.839	2.006	2.173	2.340	2.508	2.674	2.841	3.008	3.176	3.343	2
3	0.170	0.340	0.510	0.680	0.849	1.019	1.189	1.359	1.529	1.699	1.868	2.038	2.208	2.378	2.547	2.718	2.888	3.058	3.228	3.398	3
4	0.173	0.345	0.518	0.690	0.863	1.036	1.208	1.381	1.554	1.726	1.899	2.072	2.244	2.416	2.589	2.762	2.935	3.108	3.280	3.452	4
5	0.175	0.351	0.526	0.701	0.877	1.052	1.228	1.403	1.578	1.754	1.929	2.104	2.280	2.456	2.631	2.806	2.981	3.156	3.332	3.507	5
6	0.178	0.356	0.534	0.712	0.891	1.069	1.247	1.425	1.603	1.781	1.960	2.138	2.316	2.494	2.673	2.850	3.028	3.206	3.384	3.562	6
7	0.181	0.362	0.543	0.723	0.904	1.085	1.266	1.447	1.628	1.808	1.989	2.170	2.351	2.532	2.712	2.894	3.075	3.256	3.436	3.617	7
8	0.184	0.367	0.551	0.734	0.918	1.101	1.285	1.469	1.652	1.836	2.019	2.202	2.386	2.570	2.754	2.938	3.121	3.304	3.488	3.672	8
9	0.186	0.373	0.559	0.745	0.932	1.118	1.304	1.491	1.677	1.863	2.050	2.236	2.422	2.608	2.796	2.982	3.168	3.354	3.540	3.726	9
10	0.189	0.378	0.567	0.756	0.945	1.134	1.323	1.512	1.702	1.891	2.079	2.268	2.457	2.646	2.835	3.024	3.214	3.404	3.592	3.781	10
11	0.192	0.384	0.575	0.767	0.959	1.151	1.343	1.534	1.726	1.918	2.110	2.302	2.494	2.686	2.877	3.068	3.260	3.452	3.644	3.836	11
12	0.195	0.389	0.584	0.778	0.973	1.167	1.362	1.556	1.751	1.945	2.140	2.334	2.529	2.724	2.919	3.112	3.307	3.502	3.696	3.891	12
13	0.197	0.395	0.592	0.789	0.986	1.184	1.381	1.578	1.776	1.973	2.170	2.368	2.565	2.762	2.958	3.156	3.354	3.552	3.748	3.946	13
14	0.200	0.400	0.600	0.800	1.000	1.200	1.400	1.600	1.800	2.000	2.200	2.400	2.600	2.800	3.000	3.200	3.400	3.600	3.800	4.000	14
15	0.203	0.406	0.608	0.811	1.014	1.217	1.419	1.622	1.825	2.028	2.231	2.434	2.636	2.838	3.042	3.244	3.447	3.650	3.853	4.055	15
16	0.206	0.411	0.617	0.822	1.028	1.233	1.439	1.644	1.850	2.055	2.261	2.466	2.672	2.878	3.084	3.288	3.494	3.700	3.905	4.110	16
17	0.208	0.416	0.625	0.833	1.041	1.249	1.458	1.666	1.874	2.082	2.290	2.498	2.707	2.916	3.123	3.332	3.540	3.748	3.957	4.165	17
18	0.211	0.422	0.633	0.844	1.055	1.266	1.477	1.688	1.899	2.110	2.321	2.532	2.743	2.954	3.165	3.376	3.587	3.798	4.009	4.220	18
19	0.214	0.427	0.641	0.855	1.069	1.282	1.496	1.710	1.923	2.137	2.351	2.564	2.778	2.992	3.207	3.420	3.623	3.846	4.061	4.274	19
20	0.216	0.433	0.649	0.866	1.082	1.299	1.515	1.732	1.948	2.165	2.381	2.598	2.814	3.030	3.246	3.464	3.680	3.896	4.113	4.329	20
21	0.219	0.438	0.658	0.877	1.096	1.315	1.534	1.754	1.973	2.192	2.411	2.630	2.849	3.068	3.288	3.508	3.727	3.946	4.165	4.384	21
22	0.222	0.444	0.666	0.888	1.110	1.332	1.554	1.776	1.997	2.219	2.442	2.664	2.886	3.108	3.330	3.552	3.773	3.994	4.217	4.439	22
23	0.225	0.449	0.674	0.899	1.123	1.348	1.573	1.797	2.022	2.247	2.471	2.696	2.921	3.146	3.369	3.594	3.819	4.044	4.269	4.494	23
24	0.227	0.455	0.682	0.910	1.137	1.365	1.592	1.820	2.047	2.274	2.502	2.730	2.957	3.184	3.411	3.640	3.867	4.094	4.321	4.548	24
25	0.230	0.460	0.690	0.921	1.151	1.381	1.611	1.841	2.071	2.302	2.532	2.762	2.992	3.222	3.453	3.682	3.912	4.142	4.373	4.603	25
26	0.233	0.466	0.699	0.932	1.165	1.397	1.630	1.863	2.096	2.329	2.562	2.794	3.027	3.260	3.495	3.726	3.959	4.192	4.425	4.658	26
27	0.236	0.471	0.707	0.943	1.178	1.414	1.649	1.885	2.121	2.356	2.592	2.828	3.063	3.298	3.534	3.770	4.006	4.242	4.477	4.713	27
28	0.238	0.477	0.715	0.954	1.192	1.430	1.669	1.907	2.145	2.384	2.622	2.860	3.099	3.338	3.576	3.814	4.052	4.290	4.529	4.768	28
29	0.241	0.482	0.723	0.964	1.206	1.447	1.688	1.929	2.170	2.411	2.653	2.894	3.135	3.376	3.618	3.858	4.099	4.340	4.581	4.822	29
30	0.244	0.488	0.732	0.975	1.219	1.463	1.707	1.951	2.195	2.439	2.682	2.926	3.170	3.414	3.657	3.902	4.146	4.390	4.633	4.877	30
31	0.247	0.493	0.740	0.986	1.233	1.480	1.726	1.973	2.219	2.466	2.713	2.960	3.206	3.452	3.699	3.946	4.192	4.438	4.685	4.932	31
Apr.																					Apr.
1	0.249	0.499	0.748	0.997	1.247	1.496	1.745	1.995	2.244	2.493	2.743	2.992	3.241	3.490	3.741	3.990	4.239	4.488	4.737	4.987	1
2	0.252	0.504	0.756	1.008	1.260	1.512	1.765	2.017	2.269	2.521	2.772	3.024	3.277	3.530	3.780	4.034	4.286	4.538	4.799	5.042	2
3	0.255	0.510	0.764	1.019	1.274	1.529	1.784	2.039	2.293	2.548	2.803	3.058	3.313	3.568	3.822	4.078	4.332	4.586	4.841	5.096	3
4	0.258	0.515	0.773	1.030	1.288	1.545	1.803	2.060	2.318	2.576	2.833	3.090	3.348	3.606	3.864	4.120	4.378	4.636	4.894	5.151	4
5	0.260	0.521	0.781	1.041	1.302	1.562	1.822	2.082	2.343	2.603	2.864	3.124	3.384	3.644	3.906	4.164	4.425	4.686	4.946	5.206	5
6	0.263	0.526	0.789	1.052	1.315	1.578	1.841	2.104	2.367	2.630	2.893	3.156	3.419	3.682	3.945	4.208	4.471	4.734	4.998	5.261	6
7	0.266	0.532	0.797	1.063	1.329	1.595	1.860	2.126	2.392	2.658	2.924	3.190	3.455	3.720	3.987	4.252	4.518	4.784	5.050	5.316	7
8	0.269	0.537	0.806	1.074	1.343	1.611	1.880	2.148	2.417	2.685	2.954	3.222	3.491	3.760	4.029	4.296	4.565	4.834	5.102	5.370	8
9	0.271	0.543	0.814	1.085	1.356	1.628	1.899	2.170	2.441	2.713	2.984	3.256	3.527	3.798	4.068	4.340	4.611	4.882	5.154	5.425	9
10	0.274	0.548	0.822	1.096	1.370	1.644	1.918	2.192	2.466	2.740	3.014	3.288	3.562	3.836	4.110	4.384	4.658	4.932	5.206	5.480	10
11	0.277	0.553	0.830	1.107	1.384	1.660	1.937	2.214	2.491	2.767	3.044	3.320	3.597	3.874	4.152	4.428	4.705	4.982	5.258	5.535	11
12	0.279	0.559	0.838	1.118	1.397	1.677	1.956	2.236	2.515	2.794	3.074	3.354	3.633	3.912	4.191	4.472	4.751	5.030	5.310	5.590	12
13	0.282	0.564	0.847	1.129	1.411	1.693	1.976	2.258	2.540	2.822	3.104	3.386	3.669	3.952	4.233	4.516	4.798	5.080	5.362	5.644	13
14	0.285	0.570	0.855	1.140	1.425	1.710	1.995	2.280	2.565	2.850	3.135	3.420	3.705	3.990	4.275	4.560	4.845	5.130	5.414	5.699	14
15	0.288	0.575	0.863	1.151	1.439	1.726	2.014	2.302	2.589	2.877	3.165	3.452	3.740	4.028	4.317	4.604	4.891	5.178	5.466	5.754	15
16	0.290	0.581	0.871	1.162	1.452	1.743	2.033	2.324	2.614	2.904	3.195	3.486	3.776	4.066	4.356	4.648	4.938	5.228	5.518	5.809	16
17	0.293	0.586	0.880	1.173	1.466	1.759	2.052	2.345	2.639	2.931	3.225	3.518	3.811	4.104	4.398	4.690	4.984	5.278	5.570	5.864	17
18	0.296	0.592	0.888	1.184	1.480	1.776	2.071	2.367	2.663	2.959	3.256	3.552	3.847	4.142	4.440	4.734	5.030	5.326	5.622	5.918	18
19	0.299	0.597	0.896	1.195	1.493	1.792	2.091	2.389	2.688	2.987	3.285	3.584	3.883	4.182	4.479	4.778	5.077	5.376	5.675	5.973	19
20	0.301	0.603	0.904	1.206	1.507	1.808	2.110	2.411	2.713	3.014	3.315	3.616	3.918	4.220	4.521	4.822	5.124	5.426	5.727	6.028	20
21	0.304	0.608	0.912	1.217	1.521	1.825	2.129	2.433	2.737	3.041	3.346	3.650	3.954	4.258	4.563	4.866	5.170	5.474	5.779	6.083	21
22	0.307	0.614	0.921	1.228	1.534	1.841	2.148	2.455	2.762	3.069	3.375	3.682	3.989	4.296	4.602	4.910	5.217	5.524	5.831	6.138	22
23	0.310	0.619	0.929	1.238	1.548	1.858	2.167	2.477	2.787	3.096	3.406	3.716	4.025	4.334	4.644	4.954	5.264	5.574	5.883	6.192	23
24	0.312	0.625	0.937	1.249	1.562	1.874	2.187	2.499	2.811	3.124	3.436	3.748	4.061	4.374	4.686	4.998	5.310	5.622	5.935	6.247	24
25	0.315	0.630	0.945	1.260	1.576	1.891	2.206	2.521	2.836	3.151	3.467	3.782	4.097	4.412	4.728	5.042	5.357	5.672	5.987	6.302	25
26	0.318	0.636	0.954	1.271	1.589	1.907	2.225	2.543	2.861	3.178	3.496	3.814	4.132	4.450	4.767	5.086	5.404	5.722	6.039	6.357	26
27	0.321	0.641	0.962	1.282	1.603	1.923	2.244	2.565	2.885	3.206	3.526	3.846	4.167	4.488	4.809	5.130	5.450	5.770	6.091	6.412	27
28	0.323	0.647	0.970	1.293	1.617	1.940	2.263	2.587	2.910	3.233	3.557	3.880	4.203	4.526	4.851	5.174	5.497	5.820	6.143	6.466	28
29	0.326	0.652	0.978	1.304	1.630	1.956	2.282	2.608	2.935	3.261	3.586	3.912	4.238	4.564	4.890	5.216	5.543	5.870	6.195	6.521	29
30	0.329	0.658	0.986	1.315	1.644	1.973	2.302	2.630	2.959	3.288	3.617	3.946	4.275	4.604	4.932	5.260	5.589	5.918	6.247	6.576	30
	1	2	3	4	5	6	7	8	9	10	11	12	13	14	15	16	17	18	19	20	

TABLE XV.

GENERAL TABLE of daily Precessions, &c. continued.

May.	1	2	3	4	5	6	7	8	9	10	11	12	13	14	15	16	17	18	19	20	May.
1	0.332	0.663	0.995	1.326	1.658	1.989	2.321	2.652	2.984	3.315	3.647	3.978	4.310	4.642	4.972	5.304	5.636	5.968	6.299	6.631	1
2	0.334	0.669	1.003	1.337	1.671	2.006	2.340	2.674	3.009	3.343	3.677	4.012	4.346	4.680	5.013	5.348	5.683	6.018	6.352	6.686	2
3	0.337	0.674	1.011	1.348	1.685	2.022	2.360	2.696	3.033	3.370	3.707	4.044	4.382	4.720	5.055	5.392	5.729	6.066	6.403	6.740	3
4	0.340	0.680	1.019	1.359	1.699	2.039	2.378	2.718	3.058	3.398	3.738	4.078	4.417	4.756	5.097	5.436	5.796	6.116	6.456	6.795	4
5	0.343	0.685	1.028	1.370	1.713	2.055	2.398	2.740	3.083	3.425	3.768	4.110	4.453	4.796	5.138	5.480	5.823	6.165	6.508	6.850	5
6	0.345	0.690	1.036	1.381	1.726	2.071	2.417	2.762	3.107	3.452	3.797	4.142	4.488	4.834	5.178	5.524	5.869	6.214	6.559	6.905	6
7	0.348	0.696	1.044	1.392	1.740	2.088	2.436	2.784	3.132	3.480	3.828	4.176	4.524	4.872	5.220	5.568	5.916	6.264	6.612	6.960	7
8	0.351	0.701	1.052	1.403	1.754	2.104	2.455	2.806	3.156	3.507	3.858	4.208	4.557	4.910	5.261	5.612	5.962	6.312	6.663	7.014	8
9	0.353	0.707	1.060	1.414	1.767	2.121	2.474	2.828	3.181	3.535	3.888	4.242	4.595	4.948	5.302	5.656	6.009	6.362	6.716	7.069	9
10	0.356	0.712	1.069	1.425	1.781	2.137	2.493	2.850	3.206	3.562	3.918	4.274	4.630	4.986	5.343	5.700	6.056	6.411	6.768	7.124	10
11	0.359	0.718	1.077	1.436	1.795	2.154	2.513	2.872	3.230	3.590	3.949	4.308	4.667	5.026	5.384	5.744	6.102	6.461	6.820	7.179	11
12	0.362	9.723	1.085	1.447	1.808	2.170	2.532	2.893	3.255	3.617	3.978	4.340	4.702	5.064	5.425	5.785	6.148	6.510	6.872	7.234	12
13	0.364	0.729	1.093	1.458	1.822	2.186	2.551	2.915	3.280	3.644	4.008	4.372	4.738	5.102	5.466	5.830	6.195	6.558	6.924	7.288	13
14	0.367	0.734	1.101	1.469	1.836	2.203	2.570	2.937	3.304	3.672	4.039	4.406	4.773	5.140	5.508	5.874	6.241	6.608	6.976	7.343	14
15	0.370	0.740	1.110	1.480	1.850	2.219	2.589	2.959	3.329	3.699	4.069	4.438	4.808	5.178	5.549	5.918	6.288	6.657	7.028	7.398	15
16	0.373	0.745	1.118	1.491	1.863	2.236	2.608	2.981	3.354	3.726	4.099	4.472	4.844	5.216	5.589	5.962	6.335	6.707	7.080	7.453	16
17	0.375	0.751	1.126	1.502	1.877	2.252	2.628	3.003	3.378	3.754	4.129	4.504	4.880	5.256	5.631	6.006	6.381	6.756	7.132	7.508	17
18	0.378	0.756	1.134	1.512	1.891	2.269	2.647	3.025	3.403	3.781	4.160	4.538	4.916	5.294	5.672	6.050	6.428	6.806	7.184	7.562	18
19	0.381	0.762	1.143	1.524	1.904	2.285	2.660	3.047	3.428	3.809	4.189	4.570	4.951	5.332	5.713	6.094	6.475	6.855	7.237	7.618	19
20	0.384	0.767	1.151	1.534	1.918	2.302	2.684	3.069	3.452	3.836	4.220	4.603	4.986	5.368	5.754	6.138	6.521	6.904	7.288	7.672	20
21	9.386	0.773	1.159	1.545	1.932	2.318	2.704	3.091	3.477	3.863	4.250	4.636	5.022	5.408	5.795	6.182	6.568	6.954	7.340	7.727	21
22	0.389	0.778	1.167	1.556	1.945	2.334	2.724	3.113	3.502	3.891	4.279	4.668	5.058	5.448	5.836	6.226	6.615	7.003	7.393	7.782	22
23	0.392	0.784	1.175	1.563	1.959	2.351	2.743	3.135	3.526	3.918	4.310	4.702	5.094	5.486	5.877	6.270	6.661	7.052	7.444	7.836	23
24	0.395	0.789	1.184	1.578	1.973	2.367	2.762	3.156	3.551	3.946	4.340	4.735	5.129	5.524	5.919	6.312	6.707	7.102	7.497	7.891	24
25	0.397	0.795	1.192	1.589	1.987	2.384	2.781	3.178	3.576	3.973	4.371	4.768	5.165	5.562	5.960	6.356	6.754	7.151	7.549	7.946	25
26	0.400	0.800	1.200	1.600	2.000	2.400	2.800	3.200	3.600	4.000	4.400	4.800	5.200	5.600	6.000	6.400	6.800	7.200	7.600	8.001	26
27	0.403	0.806	1.208	1.611	2.014	2.417	2.819	3.222	3.625	4.028	4.431	4.834	5.236	5.638	6.042	6.444	6.847	7.250	7.653	8.056	27
28	0.406	0.811	1.217	1.622	2.028	2.433	2.839	3.244	3.650	4.055	4.461	4.866	5.272	5.678	6.083	6.488	6.894	7.299	7.705	8.110	28
29	0.408	0.817	1.225	1.633	2.041	2.450	2.858	3.266	3.674	4.083	4.491	4.900	5.308	5.716	6.124	6.532	6.940	7.349	7.757	8.165	29
30	0.411	0.822	1.233	1.644	2.055	2.466	2.877	3.288	3.699	4.110	4.521	4.932	5.343	5.754	6.165	6.576	6.987	7.398	7.809	8.220	30
31	0.414	0.827	1.241	1.655	2.069	2.482	2.896	3.310	3.724	4.137	4.551	4.964	5.378	5.792	6.206	6.620	7.034	7.447	7.861	8.275	31
June.																					June.
1	0.416	0.833	1.249	1.666	2.082	2.499	2.915	3.332	3.748	4.165	4.581	4.998	5.414	5.830	6.247	6.664	7.080	7.497	7.913	8.330	1
2	0.419	0.838	1.258	1.677	2.096	2.515	2.935	3.354	3.773	4.192	4.611	5.030	5.450	5.870	6.288	6.708	7.127	7.546	7.965	8.384	2
3	0.422	0.844	1.266	1.688	2.110	2.532	2.954	3.376	3.798	4.220	4.642	5.064	5.486	5.908	6.330	6.752	7.174	7.596	8.018	8.439	3
4	0.425	0.849	1.274	1.699	2.124	2.548	2.973	3.398	3.822	4.247	4.672	5.096	5.521	5.946	6.371	6.796	7.220	7.644	8.069	8.494	4
5	0.427	0.855	1.282	1.710	2.137	2.564	2.992	3.420	3.847	4.274	4.702	5.129	5.557	5.984	6.412	6.840	7.267	7.693	8.121	8.549	5
6	0.430	0.860	1.291	1.721	2.151	2.581	3.011	3.441	3.872	4.302	4.732	5.162	5.592	6.022	6.453	6.882	7.313	7.743	8.174	8.604	6
7	0.433	0.866	1.299	1.732	2.165	2.593	3.030	3.463	3.896	4.329	4.763	5.196	5.628	6.060	6.494	6.926	7.359	7.792	8.225	8.658	7
8	0.436	0.871	1.307	1.743	2.178	2.614	3.050	3.485	3.921	4.357	4.792	5.228	5.664	6.100	6.535	6.970	7.406	7.842	8.278	8.713	8
9	0.438	0.877	1.315	1.754	2.192	2.630	3.069	3.507	3.946	4.384	4.822	5.260	5.699	6.138	6.576	7.014	7.453	7.891	8.330	8.768	9
10	0.441	0.882	1.323	1.765	2.206	2.647	3.088	3.529	3.970	4.411	4.853	5.294	5.735	6.176	6.617	7.058	7.499	7.941	8.381	8.823	10
11	0.444	0.888	1.332	1.776	2.219	2.663	3.107	3.551	3.995	4.439	4.882	5.326	5.770	6.214	6.658	7.102	7.546	7.990	8.434	8.878	11
12	0.447	0.893	1.340	1.786	2.233	2.680	3.126	3.573	4.020	4.466	4.913	5.360	5.806	6.252	6.699	7.146	7.593	8.040	8.486	8.932	12
13	0.449	0.899	1.348	1.797	2.247	2.696	3.146	3.595	4.044	4.494	4.943	5.392	5.842	6.292	6.741	7.190	7.639	8.089	8.538	8.987	13
14	0.452	0.904	1.356	1.808	2.261	2.713	3.165	3.617	4.069	4.521	4.974	5.426	5.878	6.330	6.782	7.234	7.686	8.138	8.590	9.042	14
15	0.455	0.910	1.365	1.819	2.274	2.729	3.184	3.639	4.094	4.548	5.003	5.458	5.913	6.368	6.823	7.278	7.733	8.197	8.642	9.097	15
16	0.458	0.915	1.373	1.830	2.288	2.745	3.203	3.661	4.118	4.576	5.033	5.491	5.948	6.406	6.864	7.322	7.779	8.236	8.694	9.152	16
17	0.460	0.921	1.381	1.841	2.302	2.762	3.222	3.683	4.143	4.603	5.064	5.524	5.984	6.444	6.905	7.366	7.826	8.286	8.746	9.206	17
18	0.463	0.926	1.389	1.852	2.315	2.778	3.241	3.704	4.168	4.631	5.093	5.556	6.019	6.482	6.946	7.408	7.872	8.335	8.799	9.261	18
19	0.466	0.932	1.397	1.863	2.329	2.795	3.261	3.726	4.192	4.658	5.124	5.590	6.056	6.522	6.987	7.452	7.918	8.384	8.850	9.316	19
20	0.469	0.937	1.406	1.874	2.343	2.811	3.280	3.748	4.217	4.685	5.154	5.622	6.091	6.560	7.028	7.496	7.965	8.434	8.902	9.371	20
21	0.471	0.943	1.414	1.885	2.356	2.828	3.299	3.770	4.242	4.713	5.184	5.656	6.127	6.598	7.069	7.540	8.012	8.483	8.955	9.426	21
22	0.474	0.948	1.422	1.896	2.370	2.844	3.318	3.792	4.266	4.740	5.214	5.688	6.162	6.636	7.110	7.584	8.058	8.532	9.006	9.480	22
23	0.477	0.953	1.430	1.907	2.384	2.861	3.337	3.814	4.291	4.768	5.245	5.722	6.198	6.674	7.162	7.628	8.105	8.582	9.059	9.535	23
24	0.480	0.959	1.439	1.918	2.398	2.877	3.357	3.836	4.316	4.795	5.275	5.754	6.234	6.714	7.193	7.672	8.152	8.631	9.111	9.590	24
25	0.482	0.964	1.447	1.929	2.411	2.893	3.376	3.858	4.340	4.822	5.304	5.786	6.269	6.752	7.234	7.716	8.198	8.680	9.162	9.645	25
26	0.485	0.970	1.455	1.940	2.425	2.910	3.395	3.880	4.365	4.850	5.335	5.820	6.305	6.790	7.275	7.760	8.245	8.730	9.215	9.700	26
27	0.488	0.975	1.463	1.951	2.439	2.926	3.414	3.902	4.389	4.877	5.365	5.852	6.340	6.828	7.316	7.804	8.291	8.779	9.266	9.754	27
28	0.490	0.981	1.471	1.962	2.452	2.943	3.433	3.924	4.414	4.905	5.395	5.886	6.376	6.866	7.357	7.848	8.338	8.828	9.319	9.809	28
29	0.493	0.986	1.480	1.973	2.466	2.959	3.452	3.946	4.439	4.932	5.425	5.918	6.411	6.904	7.398	7.892	8.385	8.877	9.371	9.864	29
30	0.496	0.992	1.488	1.984	2.480	2.976	3.472	3.968	4.463	4.959	5.456	5.952	6.448	6.944	7.440	7.936	8.431	8.926	9.422	9.919	30
	1	2	3	4	5	6	7	8	9	10	11	12	13	14	15	16	17	18	19	20	

TABLE XV.

GENERAL TABLE of daily Precessions, &c. continued.

JULY.	1	2	3	4	5	6	7	8	9	10	11	12	13	14	15	16	17	18	19	20	JULY.
1	0.499	0.997	1.496	1.995	2.493	2.992	3.491	3.989	4.488	4.987	5.485	5.984	6.483	6.982	7.480	7.978	8.477	8.976	9.475	9.974	1
2	0.501	1.003	1.504	2.006	2.507	3.009	3.510	4.011	4.513	5.014	5.513	6.018	6.519	7.020	7.521	8.022	8.524	9.026	9.527	10.028	2
3	0.504	1.008	1.512	2.017	2.521	3.025	3.529	4.033	4.537	5.042	5.546	6.050	6.554	7.058	7.562	8.066	8.590	9.075	9.579	10.083	3
4	0.507	1.014	1.521	2.028	2.535	3.041	3.548	4.055	4.562	5.069	5.576	6.082	6.589	7.096	7.603	8.110	8.617	9.124	9.631	10.138	4
5	0.510	1.019	1.529	2.039	2.548	3.058	3.567	4.077	4.587	5.096	5.606	6.116	6.625	7.134	7.644	8.154	8.664	9.174	9.683	10.193	5
6	0.512	1.025	1.537	2.050	2.562	3.074	3.587	4.099	4.611	5.124	5.636	6.148	6.661	7.174	7.686	8.198	8.710	9.223	9.735	10.248	6
7	0.515	1.030	1.545	2.060	2.576	3.091	3.606	4.121	4.636	5.151	5.667	6.182	6.697	7.212	7.727	8.242	8.757	9.272	9.787	10.302	7
8	0.518	1.036	1.554	2.071	2.589	3.107	3.625	4.143	4.661	5.179	5.696	6.214	6.732	7.250	7.768	8.286	8.804	9.322	9.840	10.357	8
9	0.521	1.041	1.562	2.082	2.603	3.124	3.644	4.165	4.685	5.206	5.727	6.248	6.768	7.288	7.809	8.330	8.850	9.371	9.891	10.412	9
10	0.523	1.047	1.570	2.093	2.617	3.140	3.663	4.187	4.710	5.233	5.757	6.280	6.803	7.326	7.850	8.374	8.897	9.420	9.943	10.467	10
11	0.526	1.052	1.578	2.104	2.630	3.156	3.683	4.209	4.735	5.261	5.786	6.312	6.839	7.366	7.892	8.418	8.944	9.470	9.995	10.522	11
12	0.529	1.058	1.586	2.115	2.644	3.173	3.702	4.231	4.759	5.288	5.817	6.346	6.875	7.404	7.933	8.462	8.990	9.519	10.047	10.576	12
13	0.532	1.063	1.595	2.126	2.658	3.189	3.721	4.252	4.784	5.316	5.847	6.378	6.910	7.442	7.973	8.504	9.036	9.568	10.100	10.631	13
14	0.534	1.069	1.603	2.137	2.672	3.206	3.740	4.274	4.809	5.343	5.878	6.412	6.946	7.480	8.014	8.548	9.083	9.618	10.152	10.686	14
15	0.537	1.074	1.611	2.148	2.685	3.222	3.759	4.296	4.833	5.370	5.907	6.444	6.981	7.518	8.055	8.592	9.129	9.667	10.203	10.741	15
16	0.540	1.080	1.619	2.159	2.699	3.239	3.778	4.318	4.858	5.398	5.938	6.478	7.017	7.556	8.096	8.636	9.176	9.716	10.256	10.796	16
17	0.543	1.085	1.628	2.170	2.713	3.255	3.798	4.340	4.883	5.425	5.968	6.510	7.053	7.596	8.138	8.680	9.223	9.766	10.308	10.850	17
18	0.545	1.091	1.636	2.181	2.726	3.272	3.817	4.362	4.907	5.453	5.998	6.544	7.089	7.634	8.179	8.724	9.269	9.815	10.360	10.905	18
19	0.548	1.096	1.644	2.192	2.740	3.288	3.836	4.384	4.932	5.480	6.028	6.576	7.124	7.672	8.220	8.768	9.316	9.864	10.412	10.960	19
20	0.551	1.101	1.652	2.203	2.754	3.304	3.855	4.406	4.957	5.507	6.058	6.608	7.159	7.710	8.261	8.812	9.363	9.914	10.464	11.015	20
21	0.553	1.107	1.660	2.214	2.767	3.321	3.874	4.428	4.981	5.535	6.088	6.642	7.195	7.748	8.302	8.856	9.409	9.963	10.516	11.070	21
22	0.556	1.112	1.669	2.225	2.781	3.337	3.894	4.450	5.006	5.562	6.118	6.674	7.231	7.788	8.344	8.900	9.456	10.012	10.568	11.124	22
23	0.559	1.118	1.677	2.236	2.795	3.354	3.913	4.472	5.031	5.590	6.149	6.708	7.267	7.826	8.385	8.944	9.503	10.062	10.621	11.179	23
24	0.562	1.123	1.685	2.247	2.809	3.370	3.932	4.494	5.055	5.617	6.179	6.740	7.302	7.864	8.426	8.988	9.549	10.111	10.672	11.234	24
25	0.564	1.129	1.693	2.258	2.822	3.387	3.951	4.516	5.080	5.644	6.209	6.774	7.338	7.902	8.467	9.032	9.596	10.160	10.724	11.289	25
26	0.567	1.134	1.702	2.269	2.836	3.403	3.970	4.537	5.105	5.672	6.239	6.806	7.373	7.940	9.507	9.074	9.642	10.210	10.777	11.344	26
27	0.570	1.140	1.710	2.280	2.850	3.419	3.989	4.559	5.129	5.699	6.270	6.838	7.409	7.978	8.548	9.118	9.688	10.259	10.828	11.398	27
28	0.573	1.145	1.718	2.291	2.863	3.436	4.008	4.581	5.154	5.727	6.299	6.872	7.444	8.016	8.589	9.162	9.735	10.308	10.881	11.453	28
29	0.575	1.151	1.726	2.302	2.877	3.452	4.028	4.603	5.179	5.754	6.329	6.904	7.480	8.056	8.631	9.206	9.782	10.358	10.933	11.508	29
30	0.578	1.156	1.734	2.313	2.891	3.469	4.047	4.625	5.203	5.781	6.360	6.938	7.516	8.094	8.672	9.250	9.828	10.407	10.984	11.563	30
31	0.581	1.162	1.743	2.324	2.904	3.485	4.066	4.647	5.228	5.809	6.389	6.970	7.551	8.132	8.713	9.294	9.875	10.456	11.037	11.618	31
AUG.																					AUG.
1	0.584	1.167	1.751	2.334	2.918	3.502	4.085	4.669	5.253	5.836	6.420	7.004	7.587	8.170	8.754	9.338	9.922	10.506	11.089	11.672	1
2	0.586	1.173	1.759	2.345	2.932	3.518	4.105	4.691	5.277	5.864	6.450	7.036	7.623	8.210	8.796	9.392	9.968	10.555	11.141	11.727	2
3	0.589	1.178	1.767	2.356	2.946	3.535	4.124	4.713	5.302	5.891	6.481	7.070	7.659	8.248	8.837	9.426	10.015	10.604	11.193	11.782	3
4	0.592	1.184	1.776	2.367	2.959	3.551	4.143	4.735	5.327	5.918	6.510	7.102	7.694	8.286	8.878	9.470	10.062	10.654	11.245	11.836	4
5	0.595	1.189	1.784	2.378	2.973	3.567	4.162	4.757	5.351	5.946	6.540	7.134	7.729	8.324	8.919	9.514	10.108	10.703	11.297	11.891	5
6	0.597	1.195	1.792	2.389	2.987	3.584	4.181	4.779	5.376	5.973	6.571	7.168	7.765	8.362	8.960	9.558	10.155	10.752	11.349	11.946	6
7	0.600	1.200	1.800	2.400	3.000	3.600	4.200	4.800	5.401	6.001	6.600	7.200	7.800	8.400	9.000	9.600	10.201	10.802	11.402	12.001	7
8	0.603	1.206	1.808	2.411	3.014	3.617	4.220	4.822	5.425	6.028	6.631	7.234	7.837	8.440	9.042	9.644	10.247	10.851	11.453	12.056	8
9	0.606	1.211	1.817	2.422	3.028	3.633	4.239	4.844	5.450	6.055	6.661	7.266	7.872	8.478	9.083	9.688	10.294	10.900	11.505	12.110	9
10	0.608	1.217	1.825	2.433	3.041	3.650	4.258	4.866	5.475	6.083	6.691	7.300	7.908	8.516	9.124	9.732	10.341	10.950	11.558	12.165	10
11	0.611	1.222	1.833	2.444	3.055	3.666	4.277	4.888	5.499	6.110	6.721	7.332	7.943	8.554	9.165	9.776	10.387	10.999	11.609	12.220	11
12	0.614	1.228	1.841	2.455	3.069	3.683	4.296	4.910	5.524	6.138	6.752	7.366	7.979	8.592	9.206	9.820	10.434	11.048	11.662	12.275	12
13	0.617	1.233	1.850	2.466	3.083	3.699	4.316	4.932	5.549	6.165	6.782	7.398	8.015	8.632	9.248	9.864	10.481	11.098	11.714	12.330	13
14	0.519	1.238	1.858	2.477	3.096	3.715	4.335	4.954	5.573	6.192	6.811	7.430	8.050	8.670	9.289	9.908	10.537	11.147	11.765	12.384	14
15	0.622	1.244	1.866	2.488	3.110	3.732	4.354	4.976	5.598	6.220	6.842	7.464	8.086	8.708	9.330	9.952	10.574	11.196	11.818	12.439	15
16	0.625	1.249	1.874	2.499	3.124	3.748	4.373	4.998	5.622	6.247	6.872	7.496	8.121	8.746	9.371	9.996	10.620	11.245	11.869	12.494	16
17	0.627	1.255	1.882	2.510	3.137	3.765	4.392	5.020	5.647	6.275	6.902	7.530	8.167	8.784	9.412	10.040	10.667	11.294	11.922	12.549	17
18	0.630	1.260	1.890	2.521	3.151	3.781	4.411	5.042	5.672	6.302	6.932	7.562	8.192	8.822	9.453	10.084	10.714	11.344	11.974	12.604	18
19	0.633	1.266	1.899	2.532	3.165	3.798	4.431	5.064	5.696	6.329	6.963	7.596	8.229	8.862	9.494	10.128	10.760	11.393	12.025	12.658	19
20	0.636	1.271	1.907	2.543	3.178	3.814	4.450	5.085	5.721	6.357	6.992	7.628	8.264	8.900	9.535	10.170	10.806	11.442	12.078	12.713	20
21	0.638	1.277	1.915	2.554	3.192	3.831	4.469	5.107	5.746	6.384	7.023	7.662	8.300	8.938	9.576	10.214	10.853	11.492	12.130	12.768	21
22	0.641	1.282	1.923	2.565	3.206	3.847	4.488	5.129	5.770	6.412	7.053	7.694	8.335	8.976	9.617	10.258	10.899	11.541	12.182	12.823	22
23	0.644	1.288	1.932	2.576	3.220	3.863	4.537	5.151	5.795	6.439	7.083	7.726	8.370	9.014	9.658	10.312	10.946	11.590	12.234	12.878	23
24	0.647	1.293	1.940	2.587	3.233	3.880	4.526	5.173	5.820	6.466	7.113	7.760	8.406	9.052	9.699	10.346	10.993	11.640	12.286	12.932	24
25	0.649	1.299	1.948	2.598	3.247	3.896	4.546	5.195	5.844	6.494	7.143	7.792	8.442	9.082	9.741	10.390	11.038	11.689	12.338	12.987	25
26	0.652	1.304	1.956	2.608	3.261	3.913	4.565	5.217	5.869	6.521	7.174	7.826	8.478	9.130	9.782	10.434	11.086	11.738	12.390	13.042	26
27	0.655	1.310	1.965	2.619	3.274	3.929	4.584	5.239	5.894	6.547	7.203	7.858	8.513	9.168	9.823	10.478	11.133	11.788	12.441	13.097	27
28	0.658	1.315	1.973	2.630	3.288	3.946	4.603	5.261	5.918	6.576	7.234	7.892	8.549	9.206	9.864	10.522	11.179	11.837	12.494	13.152	28
29	0.660	1.321	1.981	2.641	3.302	3.962	4.622	5.283	5.943	6.603	7.264	7.924	8.584	9.244	9.905	10.566	11.226	11.886	12.546	13.206	29
30	0.663	1.327	1.989	2.652	3.315	3.978	4.642	5.305	5.968	6.631	7.293	7.956	8.620	9.284	9.947	10.610	11.273	11.935	12.599	13.261	30
31	0.666	1.332	1.997	2.663	3.329	3.995	4.661	5.327	5.992	6.658	7.324	7.990	8.656	9.322	9.988	10.654	11.319	11.984	12.650	13.315	31
	1	2	3	4	5	6	7	8	9	10	11	12	13	14	15	16	17	18	19	20	

TABLE XV.

GENERAL TABLE of daily Precessions, &c. continued.

Sep.	1	2	3	4	5	6	7	8	9	10	11	12	13	14	15	16	17	18	19	20	Sep.
1	0.669	1.337	2.006	2.674	3.343	4.011	4.680	5.348	6.017	6.686	7.354	8.022	8.691	9.360	10.029	10.696	11.365	12.034	12.703	13.371	1
2	0.671	1.343	2.014	2.685	3.357	4.028	4.699	5.370	6.042	6.713	7.385	8.055	8.727	9.398	10.070	10.740	11.412	12.084	12.755	13.426	2
3	0.674	1.348	2.022	2.696	3.370	4.044	4.718	5.392	6.066	6.740	7.414	8.088	8.762	9.436	10.110	10.784	11.458	12.133	12.806	13.481	3
4	0.677	1.354	2.030	2.707	3.384	4.061	4.737	5.414	6.091	6.768	7.445	8.122	8.798	9.474	10.152	10.828	11.505	12.182	12.859	13.536	4
5	0.680	1.360	2.039	2.718	3.398	4.077	4.757	5.436	6.116	6.795	7.475	8.155	8.834	9.513	10.194	10.872	11.552	12.232	12.911	13.590	5
6	0.682	1.365	2.047	2.729	3.411	4.094	4.776	5.458	6.140	6.823	7.505	8.188	8.870	9.552	10.234	10.916	11.598	12.281	12.963	13.645	6
7	0.685	1.370	2.055	2.740	3.425	4.110	4.795	5.480	6.165	6.850	7.535	8.220	8.905	9.590	10.275	10.960	11.645	12.330	13.015	13.700	7
8	6.688	1.375	2.063	2.751	3.439	4.126	4.814	5.502	6.190	6.877	7.565	8.253	8.940	9.628	10.316	11.004	11.692	12.380	13.067	13.755	8
9	0.690	1.381	2.071	2.762	3.452	4.143	4.833	5.524	6.214	6.905	7.595	8.286	8.976	9.666	10.357	11.048	11.738	12.429	13.119	13.810	9
10	0.693	1.386	2.080	2.773	3.466	4.159	4.853	5.546	6.239	6.933	7.625	8.318	9.012	9.705	10.398	11.092	11.785	12.478	13.172	13.864	10
11	0.696	1.392	2.088	2.784	3.480	4.176	4.872	5.568	6.264	6.960	7.656	8.351	9.048	9.744	10.440	11.136	11.832	12.528	13.224	13.920	11
12	0.699	1.397	2.096	2.795	3.494	4.192	4.891	5.590	6.288	6.987	7.686	8.384	9.083	9.782	10.481	11.180	11.878	12.577	13.275	13.974	12
13	0.701	1.403	2.104	2.806	3.507	4.209	4.910	5.612	6.313	7.014	7.716	8.417	9.119	9.820	10.522	11.224	11.925	12.626	13.327	14.029	13
14	0.704	1.408	2.113	2.817	3.521	4.225	4.929	5.633	6.338	7.042	7.746	8.450	9.154	9.858	10.563	11.266	11.971	12.676	13.380	14.084	14
15	0.707	1.414	2.121	2.828	3.535	4.242	4.948	5.655	6.362	7.069	7.777	8.483	9.190	9.896	10.604	11.310	12.017	12.725	13.431	14.138	15
16	0.710	1.420	2.129	2.839	3.548	4.258	4.968	5.677	6.387	7.097	7.806	8.516	9.226	9.935	10.645	11.354	12.064	12.774	13.484	14.193	16
17	0.712	1.425	2.137	2.850	3.562	4.274	4.987	5.699	6.412	7.124	7.836	8.549	9.261	9.974	10.686	11.398	12.111	12.824	13.536	14.248	17
18	0.715	1.430	2.145	2.861	3.576	4.291	5.006	5.721	6.436	7.151	7.867	8.582	9.297	10.012	10.727	11.442	12.157	12.873	13.587	14.303	18
19	0.718	1.436	2.154	2.872	3.589	4.307	5.025	5.743	6.461	7.179	7.896	8.615	9.332	10.050	10.768	11.486	12.204	12.922	13.640	14.358	19
20	0.721	1.441	2.162	2.882	3.603	4.324	5.044	5.765	6.486	7.206	7.927	8.648	9.368	10.088	10.809	11.530	12.251	12.972	13.692	14.412	20
21	0.723	1.447	2.170	2.893	3.617	4.340	5.064	5.787	6.510	7.234	7.957	8.681	9.404	10.127	10.851	11.574	12.297	13.021	13.744	14.467	21
22	0.726	1.452	2.178	2.904	3.631	4.357	5.083	5.809	6.535	7.261	7.988	8.714	9.440	10.166	10.892	11.618	12.344	13.070	13.796	14.522	22
23	0.729	1.458	2.187	2.915	3.644	4.373	5.102	5.931	6.560	7.288	8.017	8.746	9.475	10.204	10.933	11.662	12.391	13.120	13.848	14.577	23
24	0.732	1.463	2.195	2.926	3.658	4.389	5.121	5.853	6.584	7.316	8.047	8.779	9.510	10.242	10.974	11.706	12.437	13.169	13.900	14.632	24
25	0.734	1.469	2.203	2.937	3.672	4.406	5.140	5.875	6.609	7.343	8.078	8.812	9.546	10.280	11.015	11.750	12.484	13.218	13.952	14.686	25
26	0.737	1.474	2.211	2.948	3.685	4.422	5.159	5.896	6.634	7.371	8.107	8.845	9.581	10.318	11.056	11.792	12.530	13.268	14.005	14.741	26
27	0.740	1.480	2.219	2.959	3.699	4.439	5.179	5.918	6.658	7.398	8.138	8.878	9.618	10.357	11.097	11.836	12.576	13.317	14.056	14.796	27
28	0.743	1.485	2.228	2.970	3.713	4.455	5.198	5.940	6.683	7.425	8.168	8.911	9.653	10.396	11.138	11.880	12.623	13.366	14.108	14.851	28
29	0.745	1.491	2.236	2.981	3.726	4.472	5.217	5.962	6.708	7.453	8.198	8.944	9.689	10.434	11.179	11.924	12.670	13.406	14.171	14.906	29
30	0.748	1.496	2.244	2.992	3.740	4.488	5.236	5.984	6.732	7.480	8.228	8.977	9.724	10.472	11.220	11.968	12.716	13.465	14.212	14.960	30
Oct.																					Oct.
1	0.751	1.502	2.252	3.003	3.754	4.505	5.255	6.006	6.757	7.508	8.259	9.010	9.760	10.511	11.262	12.012	12.763	13.514	14.265	15.051	1
2	0.754	1.507	2.260	3.014	3.768	4.521	5.275	6.028	6.782	7.535	8.289	9.042	9.796	10.550	11.303	12.056	12.810	13.564	14.317	15.070	2
3	0.756	1.512	2.269	3.025	3.781	4.537	5.294	6.050	6.806	7.562	8.318	9.075	9.831	10.588	11.344	12.100	12.856	13.613	14.368	15.125	3
4	0.760	1.518	2.277	3.036	3.795	4.554	5.313	6.072	6.831	7.590	8.349	9.108	9.867	10.626	11.385	12.144	12.903	13.662	14.421	15.180	4
5	0.762	1.523	2.285	3.047	3.809	4.570	5.332	6.094	6.855	7.617	8.379	9.141	9.902	10.664	11.426	12.188	12.949	13.711	14.462	15.234	5
6	0.764	1.529	2.293	3.058	3.822	4.587	5.351	6.116	6.880	7.645	8.409	9.174	9.938	10.702	11.467	12.232	12.996	13.760	14.525	15.290	6
7	0.767	1.534	2.302	3.069	3.836	4.603	5.370	6.138	6.905	7.672	8.439	9.207	9.973	10.741	11.508	12.276	13.043	13.810	14.577	15.344	7
8	0.770	1.540	2.310	3.080	3.850	4.620	5.390	6.160	6.929	7.699	8.470	9.240	10.010	10.780	11.559	12.320	13.089	13.859	14.628	15.399	8
9	0.773	1.543	2.318	3.091	3.863	4.636	5.409	6.181	6.954	7.727	8.499	9.273	10.045	10.818	11.580	12.362	13.135	13.908	14.681	15.453	9
10	0.775	1.551	2.326	3.102	3.877	4.653	5.428	6.203	6.979	7.754	8.530	9.306	10.081	10.856	11.631	12.406	13.182	13.958	14.733	15.508	10
11	0.778	1.556	2.334	3.113	3.891	4.669	5.447	6.225	7.003	7.782	8.560	9.338	10.116	10.894	11.673	12.450	13.228	14.007	14.785	15.563	11
12	0.781	1.562	2.343	3.124	3.905	4.685	5.466	6.247	7.028	7.809	8.590	9.371	10.151	10.932	11.714	12.494	13.275	14.056	14.837	15.618	12
13	0.784	1.567	2.351	3.135	3.918	4.702	5.485	6.269	7.053	7.836	8.620	9.404	10.187	10.971	11.755	12.538	13.322	14.106	14.889	15.673	13
14	0.786	1.573	2.359	3.146	3.932	4.718	5.505	6.291	7.077	7.864	8.650	9.437	10.223	11.010	11.796	12.582	13.368	14.155	14.941	15.728	14
15	0.789	1.578	2.367	3.156	3.946	4.735	5.524	6.313	7.102	7.891	8.681	9.470	10.259	11.048	11.837	12.626	13.415	14.204	14.993	15.782	15
16	0.792	1.584	2.376	3.167	3.959	4.751	5.543	6.335	7.127	7.919	8.710	9.503	10.294	11.086	11.878	12.670	13.462	14.254	15.046	15.837	16
17	0.795	1.589	2.384	3.178	3.973	4.768	5.562	6.357	7.151	7.946	8.741	9.536	10.330	11.124	11.919	12.714	13.508	14.303	15.097	15.892	17
18	0.797	1.595	2.392	3.189	3.987	4.784	5.581	6.379	7.176	7.973	8.771	9.568	10.365	11.162	11.960	12.758	13.555	14.352	15.149	15.947	18
19	0.800	1.600	2.400	3.200	4.000	4.800	5.601	6.401	7.201	8.001	8.800	9.601	10.401	11.201	12.001	12.802	13.602	14.402	15.202	16.002	19
20	0.803	1.606	2.408	3.211	4.014	4.817	5.620	6.423	7.225	8.028	8.831	9.634	10.437	11.240	12.042	12.845	13.648	14.451	15.253	16.056	20
21	0.806	1.611	2.417	3.222	4.028	4.833	5.639	6.444	7.250	8.056	8.861	9.667	10.472	11.278	12.084	12.888	13.694	14.500	15.306	16.111	21
22	0.808	1.617	2.425	3.233	4.042	4.850	5.658	6.466	7.275	8.083	8.892	9.700	10.508	11.316	12.125	12.932	13.741	14.550	15.358	16.166	22
23	0.811	1.622	2.433	3.244	4.055	4.866	5.677	6.488	7.299	8.110	8.921	9.733	10.543	11.354	12.166	12.976	13.787	14.599	15.409	16.221	23
24	0.814	1.628	2.441	3.255	4.069	4.883	5.696	6.510	7.324	8.138	8.952	9.766	10.579	11.392	12.207	13.020	13.834	14.648	15.462	16.276	24
25	0.817	1.633	2.450	3.266	4.083	4.899	5.716	6.532	7.349	8.165	8.982	9.799	10.615	11.431	12.248	13.064	13.881	14.698	15.514	16.330	25
26	0.819	1.639	2.458	3.277	4.096	4.916	5.735	6.554	7.373	8.193	9.012	9.832	10.651	11.470	12.289	13.108	13.927	14.747	15.566	16.385	26
27	0.822	1.644	2.466	3.288	4.110	4.932	5.754	6.576	7.398	8.220	9.042	9.864	10.686	11.508	12.330	13.152	13.974	14.796	15.618	16.440	27
28	0.825	1.649	2.474	3.299	4.124	4.948	5.773	6.598	7.423	8.247	9.072	9.897	10.721	11.546	12.371	13.196	14.021	14.846	15.670	16.495	28
29	0.827	1.655	2.482	3.310	4.137	4.965	5.792	6.620	7.447	8.275	9.102	9.930	10.757	11.585	12.412	13.240	14.067	14.895	15.722	16.550	29
30	0.830	1.660	2.491	3.321	4.151	4.981	5.812	6.642	7.472	8.302	9.132	9.963	10.793	11.624	12.453	13.284	14.114	14.944	15.774	16.604	30
31	0.833	1.666	2.499	3.332	4.165	4.998	5.831	6.664	7.497	8.330	9.163	9.996	10.829	11.662	12.495	13.328	14.161	14.994	15.827	16.659	31
	1	2	3	4	5	6	7	8	9	10	11	12	13	14	15	16	17	18	19	20	

TABLE XV.

GENERAL TABLE of daily Precessions, &c. concluded.

Nov.	1	2	3	4	5	6	7	8	9	10	11	12	13	14	15	16	17	18	19	20	Nov.
1	0.836	1.671	2.507	3.343	4.179	5.014	5.850	6.686	7.521	8.357	9.193	10.028	10.864	11.700	12.535	13.372	14.207	15.042	15.878	16.714	1
2	0.838	1.677	2.515	3.354	4.192	5.031	5.870	6.708	7.546	8.384	9.223	10.062	10.901	11.740	12.576	13.416	14.254	15.092	15.930	16.769	2
3	0.841	1.682	2.524	3.365	4.206	5.047	5.889	6.729	7.571	8.412	9.253	10.094	10.936	11.778	12.618	13.458	14.300	15.141	15.983	16.824	3
4	0.844	1.688	2.532	3.376	4.219	5.064	5.907	6.751	7.595	8.439	9.280	10.128	10.971	11.814	12.659	13.502	14.346	15.190	16.034	16.878	4
5	0.847	1.693	2.540	3.387	4.233	5.080	5.927	6.773	7.620	8.467	9.313	10.160	11.007	11.854	12.700	13.546	14.393	15.240	16.087	16.933	5
6	0.849	1.699	2.548	3.398	4.247	5.096	5.946	6.795	7.645	8.494	9.343	10.192	11.042	11.892	12.741	13.590	14.440	15.289	16.139	16.988	6
7	0.852	1.704	2.556	3.409	4.261	5.113	5.965	6.817	7.669	8.521	9.374	10.226	11.078	11.930	12.782	13.634	14.486	15.338	16.190	17.043	7
8	0.855	1.710	2.565	3.420	4.274	5.129	5.984	6.839	7.694	8.549	9.403	10.258	11.113	11.968	12.823	13.678	14.533	15.388	16.243	17.098	8
9	0.858	1.715	2.573	3.430	4.288	5.146	6.003	6.861	7.719	8.576	9.434	10.292	11.149	12.006	12.864	13.722	14.580	15.438	16.295	17.152	9
10	0.860	1.721	2.581	3.441	4.302	5.162	6.023	6.883	7.743	8.604	9.464	10.324	11.185	12.046	12.906	13.766	14.626	15.487	16.347	17.207	10
11	0.863	1.726	2.589	3.452	4.316	5.179	6.042	6.905	7.768	8.631	9.495	10.358	11.221	12.084	12.947	13.810	14.673	15.536	16.399	17.262	11
12	0.866	1.732	2.598	3.463	4.329	5.195	6.061	6.927	7.793	8.658	9.524	10.390	11.256	12.122	12.987	13.854	14.720	15.586	16.451	17.317	12
13	0.869	1.737	2.606	3.474	4.343	5.211	6.080	6.949	7.817	8.686	9.554	10.422	11.291	12.160	13.029	13.898	14.766	15.635	16.503	17.372	13
14	0.871	1.743	2.614	3.485	4.357	5.228	6.099	6.971	7.842	8.713	9.585	10.456	11.327	12.198	13.070	13.942	14.813	15.684	16.555	17.426	14
15	0.874	1.748	2.622	3.496	4.370	5.244	6.118	6.992	7.867	8.741	9.614	10.488	11.362	12.236	13.111	13.984	14.859	15.734	16.608	17.481	15
16	0.877	1.754	2.630	3.507	4.384	5.261	6.138	7.014	7.891	8.768	9.645	10.522	11.399	18.276	13.152	14.028	14.905	15.783	16.659	17.536	16
17	0.880	1.759	2.639	3.518	4.398	5.277	6.157	7.036	7.916	8.795	9.675	10.554	11.434	12.314	13.193	14.072	14.952	15.832	16.711	17.591	17
18	0.882	1.765	2.647	3.529	4.411	5.294	6.176	7.058	7.941	8.823	9.705	10.588	11.470	12.352	13.234	14.116	14.999	15.882	16.764	17.646	18
19	0.885	1.770	2.655	3.540	4.425	5.310	6.195	7.080	7.965	8.850	9.735	10.620	11.505	12.390	13.275	14.160	15.045	15.930	16.813	17.700	19
20	0.888	1.776	2.663	3.551	4.439	5.326	6.214	7.102	7.990	8.878	9.766	10.652	11.540	12.428	13.317	14.204	15.092	15.979	16.868	17.755	20
21	0.891	1.781	2.672	3.562	4.453	5.343	6.234	7.124	8.015	8.905	9.796	10.686	11.577	12.468	13.358	14.248	15.139	16.029	16.920	17.810	21
22	0.893	1.786	2.680	3.573	4.466	5.359	6.253	7.146	8.039	8.932	9.825	10.718	11.612	12.506	13.398	14.292	15.185	16.078	16.971	17.865	22
23	0.896	1.792	2.688	3.584	4.480	5.376	6.272	7.168	8.064	8.960	9.856	10.752	11.648	12.544	13.440	14.336	15.232	16.128	17.024	17.920	23
24	0.899	1.797	2.696	3.595	4.494	5.392	6.291	7.190	8.088	8.987	9.886	10.784	11.683	12.582	13.481	14.380	15.278	16.177	17.075	17.974	24
25	0.901	1.803	2.704	3.606	4.507	5.409	6.310	7.212	8.113	9.014	9.916	10.818	11.719	12.620	13.521	14.424	15.325	16.226	17.127	18.029	25
26	0.904	1.808	2.713	3.617	4.521	5.425	6.329	7.234	8.138	9.042	9.946	10.850	11.754	12.658	13.563	14.468	15.372	16.276	17.180	18.084	26
27	0.907	1.814	2.721	3.628	4.535	5.441	6.349	7.256	8.162	9.069	9.977	10.882	11.790	12.698	13.604	14.512	15.418	16.325	17.231	18.139	27
28	0.910	1.819	2.729	3.639	4.548	5.458	6.368	7.277	8.187	9.097	10.006	10.916	11.826	12.736	13.645	14.554	15.464	16.374	17.284	18.194	28
29	0.912	1.825	2.737	3.650	4.562	5.474	6.387	7.299	8.212	9.124	10.037	10.948	11.861	12.774	13.686	14.598	15.511	16.423	17.336	18.248	19
30	0.915	1.830	2.745	3.660	4.576	5.491	6.406	7.321	8.236	9.152	10.067	10.982	11.897	12.812	13.728	14.642	15.557	16.473	17.388	18.303	30
Dec.																					Dec.
1	0.918	1.836	2.754	3.672	4.589	5.507	6.425	7.343	8.261	9.179	10.097	11.014	11.932	12.850	13.768	14.686	15.604	16.522	17.440	18.358	1
2	0.921	1.841	2.762	3.683	4.603	5.524	6.444	7.365	9.286	9.206	10.127	11.048	11.968	12.888	13.809	14.730	15.651	16.572	17.492	18.413	2
3	0.923	1.847	2.770	3.694	4.617	5.540	6.464	7.387	8.310	9.234	10.157	11.080	12.004	12.928	13.851	14.774	15.697	16.621	17.544	18.468	3
4	0.926	1.852	2.778	3.704	4.631	5.557	6.483	7.409	8.335	9.261	10.188	11.113	12.040	12.966	13.892	14.818	15.744	16.670	17.596	18.522	4
5	0.929	1.858	2.787	3.715	4.644	5.573	6.502	7.431	8.360	9.289	10.217	11.146	12.075	13.004	13.933	14.862	15.791	16.720	17.649	18.577	5
6	0.932	1.863	2.794	3.726	4.658	5.589	6.521	7.453	8.384	9.316	10.246	11.178	12.110	13.042	13.974	14.906	15.834	16.769	17.700	18.632	6
7	0.934	1.869	2.803	3.737	4.672	5.606	6.540	7.475	8.409	9.343	10.278	11.212	12.146	13.080	14.016	14.950	15.884	16.818	17.752	18.687	7
8	0.937	1.874	2.811	3.748	4.685	5.622	6.560	7.497	8.440	9.371	10.311	11.244	12.182	13.120	14.056	14.994	15.937	16.887	17.814	18.742	8
9	0.940	1.880	2.819	3.759	4.699	5.639	6.579	7.519	8.458	9.398	10.338	11.278	12.218	13.158	14.097	15.038	15.977	16.917	17.856	18.796	9
10	0.943	1.885	2.828	3.770	4.713	5.655	6.598	7.540	8.483	9.426	10.368	11.310	12.253	13.196	14.139	15.080	16.023	16.966	17.909	18.851	10
11	0.945	1.891	2.836	3.781	4.727	5.672	6.617	7.562	8.508	9.453	10.399	11.344	12.289	13.234	14.180	15.124	16.070	17.016	17.961	18.906	11
12	0.948	1.896	2.844	3.792	4.740	5.688	6.636	7.584	8.532	9.480	10.428	11.376	12.324	13.272	14.221	15.168	16.116	17.065	18.012	18.961	12
13	0.951	1.902	2.852	3.803	4.754	5.705	6.655	7.606	8.557	9.508	10.459	11.409	12.360	13.310	14.262	15.212	16.163	17.114	18.065	19.016	13
14	0.954	1.907	2.861	3.814	4.768	5.721	6.675	7.628	8.582	9.535	10.489	11.442	12.396	13.350	14.303	15.256	16.210	17.164	18.117	19.070	14
15	0.956	1.913	2.869	3.825	4.781	5.738	6.694	7.650	8.606	9.563	10.519	11.475	12.432	13.388	14.344	15.300	16.256	17.213	18.169	19.125	15
16	0.959	1.918	2.877	3.836	4.795	5.754	6.713	7.672	8.631	9.590	10.549	11.508	12.467	13.426	14.385	15.344	16.303	17.262	18.221	19.180	16
17	0.962	1.923	2.885	3.847	4.809	5.770	6.732	7.694	8.656	9.617	10.579	11.540	12.502	13.464	14.426	15.388	16.350	17.312	18.273	19.235	17
18	0.964	1.929	2.893	3.858	4.822	5.787	6.751	7.716	8.680	9.645	10.609	11.574	12.538	13.502	14.467	15.432	16.396	17.361	18.325	19.290	18
19	0.967	1.934	2.902	3.869	4.836	5.803	6.771	7.738	8.705	9.672	10.639	11.606	12.574	13.542	14.508	15.476	16.443	17.410	18.377	19.344	19
20	0.970	1.940	2.910	3.880	4.850	5.820	6.790	7.760	8.730	9.700	10.670	11.640	12.610	13.580	14.550	15.520	16.490	17.460	18.430	19.399	20
21	0.973	1.945	2.918	3.890	4.864	5.836	6.809	7.782	8.754	9.727	10.700	11.672	12.645	13.618	14.591	15.564	16.536	17.509	18.481	19.454	21
22	0.975	1.951	2.926	3.902	4.877	5.853	6.828	7.804	8.780	9.754	10.730	11.706	12.681	13.656	14.632	15.608	16.584	17.559	18.534	19.509	22
23	0.978	1.956	2.935	3.913	4.891	5.869	6.847	7.825	8.804	9.782	10.760	11.738	12.716	13.694	14.673	15.650	15.629	17.608	18.586	19.564	23
24	0.981	1.962	2.943	3.924	4.905	5.885	6.866	7.847	8.828	9.809	10.791	11.770	12.751	13.732	14.714	15.694	16.675	17.657	18.637	19.618	24
25	0.984	1.967	2.951	3.935	4.918	5.902	6.886	7.869	8.853	9.837	10.820	11.804	12.788	13.772	14.755	15.738	15.722	17.706	18.690	19.673	25
26	0.986	1.973	2.959	3.946	4.932	5.918	6.905	7.891	8.878	9.864	10.850	11.836	12.823	13.810	14.796	15.782	16.769	17.755	18.742	19.728	26
27	0.989	1.978	2.967	3.957	4.946	5.935	6.924	7.913	8.902	9.891	10.881	11.870	12.859	13.848	14.837	15.826	16.815	17.805	18.793	19.783	27
28	0.992	1.984	2.976	3.968	4.960	5.951	6.943	7.935	8.927	9.919	10.911	11.902	12.894	13.886	14.878	15.870	16.862	17.854	18.846	19.838	28
29	0.995	1.989	2.984	3.978	4.973	5.968	6.962	7.957	8.952	9.946	10.941	11.936	12.930	13.924	14.919	15.914	16.909	17.904	18.898	19.892	29
30	0.997	1.995	2.992	3.989	4.987	5.984	6.982	7.979	8.976	9.974	10.971	11.968	12.966	13.964	14.961	15.958	16.955	17.952	18.950	19.947	30
41	1.000	2.000	3.000	4.000	5.000	6.000	7.000	8.000	9.000	10.001	11.000	12.000	13.000	14.000	15.000	16.000	17.000	18.000	19.000	20.000	31
	1	2	3	4	5	6	7	8	9	10	11	12	13	14	15	16	17	18	19	20	

TABLE XVI.

Minutes of Time to be added to the Sun's Longitude at Noon on any Day, to give the proper Argument for the Time of a Star's Meridian Passage. Arguments =Right Ascension of the Star in Time at the top; and Day of the Month at the side.

Day.	O	I	II	III	IV	V	VI	VII	VIII	IX	X	XI	XII	XIII	XIV	XV	XVI	XVII	XVIII	XIX	XX	XXI	XXII	XXIII
	Min.	Min.	Min.	Min.	Min.	Min.	Min.	Min.	Min.	Min.	Min	Min.	Min.	Min.	Min.	Min.	Min.	Min.	Min.	Min.	Min.	Min.	Min.	Min.
January 0	13	15	18	20	23	25	28	30	33	36	38	41	43	46	48	51	53	56	59	1	2	5	8	10
10	11	14	17	19	21	24	26	29	32	34	37	39	42	44	47	49	52	55	57	60	1	4	7	9
20	10	13	15	18	20	23	25	28	30	33	35	38	40	43	46	48	51	53	56	59	0	2	5	7
30	8	11	13	16	18	20	23	26	28	31	33	36	38	41	43	46	49	51	53	57	59	0	3	5
February 9	7	9	11	14	17	19	21	24	26	29	32	34	37	39	42	44	47	49	52	55	57	60	1	4
19	5	7	10	12	15	17	20	22	25	27	30	32	35	38	41	42	45	47	50	52	55	58	0	2
March 1	2	5	8	10	12	15	18	20	23	25	28	30	33	35	38	40	43	45	47	50	52	55	58	0
11	1	4	7	9	11	13	16	19	21	23	26	28	31	33	36	38	41	43	46	48	51	53	56	59
21	0	2	5	8	10	12	14	17	20	22	25	27	29	32	34	37	39	41	44	46	49	51	54	57
31	58	1	4	7	9	11	13	16	19	21	23	26	28	31	33	36	38	40	43	45	47	50	53	55
April 10	57	0	2	5	8	10	12	14	17	20	22	25	27	29	32	34	36	39	41	44	46	49	51	54
20	55	57	0	3	5	8	10	12	15	18	20	23	25	27	30	32	35	37	39	42	44	47	49	52
30	53	55	57	1	4	7	9	11	13	16	19	21	23	26	28	31	33	36	38	40	43	45	47	50
May 10	51	54	56	0	2	5	7	10	12	14	17	19	21	24	26	29	31	34	36	38	41	43	45	48
20	49	52	54	57	0	3	5	8	10	12	15	17	19	22	24	27	29	32	34	36	39	41	43	46
30	47	49	52	54	57	2	4	6	8	11	13	16	18	20	22	25	27	29	32	34	37	39	41	44
June 9	45	48	50	52	55	0	2	5	7	9	12	14	16	19	21	23	26	28	31	33	36	38	40	43
19	43	46	48	50	53	55	0	3	5	7	10	12	14	17	19	21	24	26	29	31	33	36	38	41
29	41	44	46	49	51	53	56	1	4	6	8	11	13	16	18	20	23	25	27	29	32	34	37	39
July 9	40	43	45	47	50	52	55	0	2	5	7	9	12	14	17	19	21	23	26	28	31	33	35	38
19	38	41	43	45	48	50	52	55	0	3	5	7	10	12	14	17	19	21	24	26	28	31	33	36
29	37	39	41	44	46	49	51	54	56	1	4	6	8	11	13	16	18	20	23	25	27	29	32	34
August 8	36	38	40	43	45	48	50	52	55	0	2	5	7	10	12	14	17	19	21	24	26	28	31	33
18	34	36	38	41	43	45	48	50	53	55	0	3	5	8	10	12	14	17	19	22	24	26	29	31
28	32	35	37	39	42	44	46	49	51	54	57	1	4	6	8	11	13	16	18	20	22	25	27	30
Septemb. 7	31	33	36	38	40	43	45	48	50	53	56	0	2	5	7	10	12	14	17	19	21	24	26	28
17	29	32	34	37	39	42	44	46	49	52	54	57	1	4	6	8	11	13	16	18	20	23	25	27
27	28	31	33	36	38	41	43	45	48	51	53	56	0	2	5	7	10	12	14	17	19	22	24	26
October 7	27	29	32	34	37	39	41	44	46	49	51	54	57	0	3	5	8	10	12	14	18	20	22	25
17	26	28	31	33	36	38	40	43	45	47	50	53	55	58	1	4	7	9	11	13	16	19	21	23
27	25	27	29	32	35	37	39	42	44	46	49	52	54	57	0	2	5	7	10	12	14	17	20	22
Novemb. 6	23	25	28	30	33	35	37	40	42	45	47	50	52	55	58	0	3	5	8	10	12	15	18	20
16	21	23	26	28	31	33	36	38	41	43	46	48	51	54	56	59	1	4	7	9	11	13	16	19
26	20	22	25	27	29	32	35	37	40	42	45	47	50	52	55	58	0	2	5	8	10	12	14	17
Decemb. 6	18	20	23	25	28	30	33	36	38	41	43	46	48	51	53	56	59	0	3	5	8	10	13	15
16	17	19	21	24	26	29	32	34	37	39	41	44	47	49	52	55	57	60	1	4	7	9	11	14
26	15	17	19	22	24	27	30	32	35	37	39	42	45	47	50	53	55	58	0	2	5	8	10	13
31	13	15	18	21	23	26	29	31	34	36	38	41	44	46	49	52	54	57	60	0	3	6	8	11

This Table will give the Argument for Table XVIII. to the accuracy of One Minute; but in cases where the Seconds are regarded, the following Table must be substituted.

TABLE XVII.

To adapt the Sun's Longitude, as calculated for any given Observatory, to the Meridian of any other Observatory.
Argument = the Difference of Longitude of the two places.

E. W.	− O +	− I +	− II +	− III +	− IV +	− V +	− VI +	− VII +	− VIII +	− IX +	− X +	− XI +	E. W.
Min.													Min.
0	0′ 0″.0	2′ 27″.8	4′ 55″.7	7′ 23″.5	9′ 51″.4	12′ 19″.2	14′ 47″.1	17′ 14″.9	19′ 42″.8	22′ 10″.6	24′ 38″.5	27′ 6″.3	0
1	0 2.5	2 30.3	4 58.2	7 26.0	9 53.8	12 21.7	14 49.5	17 17.4	19 45.2	22 13.0	24 40.9	27 8.7	1
2	0 4.9	2 32.7	5 0.6	7 28.4	9 56.2	12 24.1	14 51.9	17 19.8	19 47.6	22 15.4	24 43.3	27 11.1	2
3	0 7.4	2 35.2	5 3.1	7 30.9	9 58.7	12 26.6	14 54.4	17 22.3	19 50.1	22 17.9	24 45.8	27 13.6	3
4	0 9.9	2 37.7	5 5.6	7 33.4	10 1.2	12 29.1	14 56.9	17 24.8	19 52.6	22 20.4	24 48.3	27 16.1	4
5	0 12.3	2 40.1	5 8.0	7 35.8	10 3.6	12 31.5	14 59.3	17 27.2	19 55.0	22 22.8	24 50.7	27 18.5	5
6	0 14.8	2 42.6	5 10.5	7 38.3	10 6.1	12 34.0	15 1.8	17 29.7	19 57.5	22 25.3	24 53.2	27 21.0	6
7	0 17.2	2 45.1	5 12.9	7 40.7	10 8.5	12 36.4	15 4.2	17 32.1	19 59.9	22 27.7	24 55.6	27 23.4	7
8	0 19.7	2 47.5	5 15.4	7 43.2	10 11.0	12 38.9	15 6.7	17 34.6	20 2.4	22 30.2	24 58.1	27 25.9	8
9	0 22.2	2 50.0	5 17.9	7 45.7	10 13.5	12 41.4	15 9.2	17 37.1	20 4.9	22 32.7	25 0.6	27 28.4	9
10	0 24.6	2 52.4	5 20.3	7 48.1	10 15.9	12 43.8	15 11.6	17 39.5	20 7.3	22 35.1	25 3.0	27 30.8	10
11	0 27.1	2 54.9	5 22.8	7 50.6	10 18.4	12 46.3	15 14.1	17 42.0	20 9.8	22 37.6	25 5.5	27 33.3	11
12	0 29.6	2 57.4	5 25.3	7 53.1	10 20.9	12 48.8	15 16.6	17 44.5	20 12.3	22 40.1	25 8.0	27 35.8	12
13	0 32.0	2 59.8	5 27.7	7 55.5	10 23.3	12 51.2	15 19.0	17 46.9	20 14.7	22 42.5	25 10.4	27 38.2	13
14	0 34.5	3 2.3	5 30.2	7 58.0	10 25.8	12 53.7	15 21.5	17 49.4	20 17.2	22 45.0	25 12.9	27 40.7	14
15	0 37.0	3 4.8	5 32.7	8 0.5	10 28.3	12 56.2	15 24.0	17 51.9	20 19.7	22 47.5	25 15.4	27 43.2	15
16	0 39.4	3 7.2	5 35.1	8 2.9	10 30.7	12 58.6	15 26.4	17 54.3	20 22.1	22 49.9	25 17.8	27 45.6	16
17	0 41.9	3 9.7	5 37.6	8 5.4	10 33.2	13 1.1	15 28.9	17 56.8	20 24.6	22 52.4	25 20.3	27 48.1	17
18	0 44.4	3 12.2	5 40.1	8 7.9	10 35.7	13 3.6	15 31.4	17 59.3	20 27.1	22 54.9	25 22.8	27 50.6	18
19	0 46.8	3 14.6	5 42.5	8 10.3	10 38.1	13 6.0	15 33.8	18 1.7	20 29.5	22 57.3	25 25.2	27 53.0	19
20	0 49.3	3 17.1	5 45.0	8 12.8	10 40.6	13 8.5	15 36.3	18 4.2	20 32.0	22 59.8	25 27.7	27 55.5	20
21	0 51.7	3 19.5	5 47.4	8 15.2	10 43.0	13 10.9	15 38.7	18 6.6	20 34.4	23 2.2	25 30.1	27 57.9	21
22	0 54.2	3 22.0	5 49.9	8 17.7	10 45.5	13 13.4	15 41.2	18 9.1	20 36.9	23 4.7	25 32.6	28 0.4	22
23	0 56.7	3 24.5	5 52.4	8 20.2	10 48.0	13 15.9	15 43.7	18 11.6	20 39.4	23 7.2	25 35.1	28 2.9	23
24	0 59.1	3 26.9	5 54.8	8 22.6	10 50.4	13 18.3	15 46.1	18 14.0	20 41.8	23 9.6	25 37.5	28 5.3	24
25	1 1.6	3 29.4	5 57.3	8 25.1	10 52.9	13 20.8	15 48.6	18 16.5	20 44.3	23 12.1	25 40.0	28 7.8	25
26	1 4.1	3 31.9	5 59.8	8 27.6	10 55.4	13 23.3	15 51.1	18 19.0	20 46.8	23 14.6	25 42.5	28 10.3	26
27	1 6.5	3 34.3	6 2.2	8 30.0	10 57.8	13 25.7	15 53.5	18 21.4	20 49.2	23 17.0	25 44.9	28 12.7	27
28	1 9.0	3 36.8	6 4.7	8 32.5	11 0.3	13 28.2	15 56.0	18 23.9	20 51.7	23 19.5	25 47.4	28 15.2	28
29	1 11.5	3 39.3	6 7.2	8 35.0	11 2.8	13 30.7	15 58.5	18 26.4	20 54.2	23 22.0	25 49.9	28 17.7	29
30	1 13.9	3 41.7	6 9.6	8 37.4	11 5.2	13 33.1	16 0.9	18 28.8	20 56.6	23 24.4	25 52.3	28 20.1	30
31	1 16.4	3 44.2	6 12.1	8 39.9	11 7.7	13 35.6	16 3.4	18 31.3	20 59.1	23 26.9	25 54.8	28 22.6	31
32	1 18.8	3 46.6	6 14.5	8 42.3	11 10.1	13 38.0	16 5.8	18 33.7	21 1.5	23 29.3	25 57.2	28 25.0	32
33	1 21.3	3 49.1	6 17.0	8 44.8	11 12.6	13 40.5	16 8.3	18 36.2	21 4.0	23 31.8	25 59.7	28 27.5	33
34	1 23.8	3 51.6	6 19.5	8 47.3	11 15.1	13 43.0	16 10.8	18 38.7	21 6.5	23 34.3	26 2.2	28 30.0	34
35	1 26.2	3 54.0	6 21.9	8 49.7	11 17.5	13 45.4	16 13.2	18 41.1	21 8.9	23 36.7	26 4.6	28 32.4	35
36	1 28.7	3 56.5	6 24.4	8 52.2	11 20.0	13 47.9	16 15.7	18 43.6	21 11.4	23 39.2	26 7.1	28 34.9	36
37	1 31.2	3 59.0	6 26.9	8 54.7	11 22.5	13 50.4	16 18.2	18 46.1	21 13.9	23 41.7	26 9.6	28 37.4	37
38	1 33.6	4 1.4	6 29.3	8 57.1	11 24.9	13 52.8	16 20.6	18 48.5	21 16.3	23 44.1	26 12.0	28 39.8	38
39	1 36.1	4 3.9	6 31.8	8 59.6	11 27.4	13 55.3	16 23.1	18 51.0	21 18.8	23 46.6	26 14.5	28 42.3	39
40	1 38.6	4 6.4	6 34.3	9 2.1	11 29.9	13 57.8	16 25.6	18 53.5	21 21.3	23 49.1	26 17.0	28 44.8	40
41	1 41.0	4 8.8	6 36.7	9 4.5	11 32.3	14 0.2	16 28.0	18 55.9	21 23.7	23 51.5	26 19.4	28 47.2	41
42	1 43.5	4 11.3	6 39.2	9 7.0	11 34.8	14 2.7	16 30.5	18 58.4	21 26.2	23 54.0	26 21.9	28 49.7	42
43	1 46.0	4 13.8	6 41.7	9 9.5	11 37.3	14 5.2	16 33.0	19 0.9	21 28.7	23 56.5	26 24.4	28 52.2	43
44	1 48.4	4 16.2	6 44.1	9 11.9	11 39.7	14 7.6	16 35.4	19 3.3	21 31.1	23 58.9	26 26.8	28 54.6	44
45	1 50.9	4 18.7	6 46.6	9 14.4	11 42.2	14 10.1	16 37.9	19 5.8	21 33.6	24 1.4	26 29.3	28 57.1	45
46	1 53.3	4 21.1	6 49.0	9 16.8	11 44.6	14 12.5	16 40.3	19 8.2	21 36.0	24 3.8	26 31.7	28 59.5	46
47	1 55.8	4 23.6	6 51.5	9 19.3	11 47.1	14 15.0	16 42.8	19 10.7	21 38.5	24 6.3	26 34.2	29 2.0	47
48	1 58.3	4 26.1	6 54.0	9 21.8	11 49.6	14 17.5	16 45.3	19 13.2	21 41.0	24 8.8	26 36.7	29 4.5	48
49	2 0.7	4 28.5	6 56.4	9 24.2	11 52.0	14 19.9	16 47.7	19 15.6	21 43.4	24 11.2	26 39.1	29 6.9	49
50	2 3.2	4 31.0	6 58.9	9 26.7	11 54.5	14 22.4	16 50.2	19 18.1	21 45.9	24 13.7	26 41.6	29 9.4	50
51	2 5.7	4 33.5	7 1.4	9 29.2	11 57.0	14 24.9	16 52.7	19 20.6	21 48.4	24 16.2	26 44.1	29 11.9	51
52	2 8.1	4 35.9	7 3.8	9 31.6	11 59.4	14 27.3	16 55.1	19 23.0	21 50.8	24 18.6	26 46.5	29 14.3	52
53	2 10.6	4 38.4	7 6.3	9 34.1	12 1.9	14 29.8	16 57.6	19 25.5	21 53.3	24 21.1	26 49.0	29 16.8	53
54	2 13.1	4 40.9	7 8.8	9 36.6	12 4.4	14 32.3	17 0.1	19 28.0	21 55.8	24 23.6	26 51.5	29 19.3	54
55	2 15.5	4 43.3	7 11.2	9 39.0	12 6.8	14 34.7	17 2.5	19 30.4	21 58.2	24 26.0	26 53.9	29 21.7	55
56	2 18.0	4 45.8	7 13.7	9 41.5	12 9.3	14 37.2	17 5.0	19 32.9	22 0.7	24 28.5	26 56.4	29 24.2	56
57	2 20.5	4 48.3	7 16.2	9 44.0	12 11.8	14 39.7	17 7.5	19 35.4	22 3.2	24 31.0	26 58.9	29 26.7	57
58	2 22.9	4 50.7	7 18.6	9 46.4	12 14.2	14 42.1	17 9.9	19 37.8	22 5.6	24 33.4	27 1.3	29 29.1	58
59	2 25.4	4 53.2	7 21.1	9 48.9	12 16.7	14 44.6	17 12.4	19 40.3	22 8.1	24 35.9	27 3.8	29 31.6	59
60	2 27.8	4 55.7	7 23.5	9 51.4	12 19.2	14 47.1	17 14.9	19 42.8	22 10.6	24 38.5	27 6.3	29 34.2	60

Also, to find the additive Corrections, contained in the preceding Table, with greater precision.
Argument = the Star's R. A. — the Sun's R. A.

TABLE XVIII.

GENERAL TABLE of Aberration, Lunar Nutation, and Solar Nutation, when the Maximum is known.

− VI + O	1	2	3	4	5	6	7	8	9	10	11	12	13	14	15	16	17	18	19	20	
0° 0′	.000	.000	.000	.000	.000	.000	.000	.000	.000	.000	.000	.000	.000	.000	.000	.000	.000	.000	.000	.000	60′
2	.001	.001	.001	.002	.003	.004	.004	.005	.006	.006	.007	.007	.008	.009	.009	.010	.010	.011	.011	.012	58
4	.002	.002	.003	.004	.006	.008	.008	.010	.011	.012	.014	.014	.016	.017	.018	.020	.020	.022	.022	.024	56
6	.002	.003	.005	.006	.009	.011	.012	.015	.016	.018	.020	.021	.024	.025	.027	.029	.030	.032	.033	.036	54
8	.003	.005	.007	.009	.012	.014	.016	.019	.021	.024	.026	.028	.031	.033	.035	.038	.040	.042	.044	.047	52
10	.003	.006	.009	.012	.015	.017	.020	.023	.026	.029	.032	.035	.038	.041	.043	.047	.049	.052	.055	.058	50
12	.004	.007	.011	.014	.018	.021	.025	.028	.032	.035	.039	.042	.046	.050	.052	.057	.059	.063	.066	.070	48
14	.004	.008	.013	.017	.021	.025	.029	.033	.037	.041	.046	.049	.054	.058	.061	.066	.069	.074	.077	.082	46
16	.005	.009	.014	.019	.024	.029	.033	.038	.042	.047	.052	.056	.062	.066	.070	.075	.079	.085	.088	.094	44
18	.005	.010	.015	.021	.027	.032	.037	.043	.047	.053	.058	.063	.069	.074	.079	.084	.089	.095	.099	.105	42
20	.006	.012	.017	.023	.029	.035	.041	.047	.052	.058	.064	.070	.076	.082	.087	.093	.099	.105	.110	.116	40
22	.006	.013	.019	.025	.032	.039	.045	.052	.058	.064	.071	.077	.084	.090	.096	.103	.109	.116	.122	.128	38
24	.007	.014	.021	.028	.035	.043	.049	.057	.064	.070	.078	.084	.092	.098	.105	.113	.119	.127	.133	.140	36
26	.007	.015	.023	.030	.038	.046	.053	.062	.069	.076	.084	.091	.099	.106	.114	.122	.129	.137	.144	.152	34
28	.008	.016	.025	.032	.041	.049	.057	.066	.074	.082	.090	.098	.106	.114	.123	.131	.139	.147	.155	.164	32
30	.009	.017	.026	.035	.044	.052	.061	.070	.079	.087	.096	.105	.113	.122	.131	.140	.148	.157	.166	.175	30
32	.009	.018	.028	.037	.047	.056	.065	.075	.085	.093	.103	.112	.121	.131	.140	.150	.158	.168	.177	.187	28
34	.010	.020	.029	.040	.050	.060	.069	.080	.090	.099	.110	.119	.129	.139	.149	.159	.168	.179	.188	.199	26
36	.010	.021	.031	.042	.053	.064	.073	.085	.095	.105	.116	.126	.137	.147	.158	.168	.178	.189	.199	.211	24
38	.011	.022	.033	.044	.056	.067	.077	.089	.100	.111	.122	.133	.144	.155	.166	.177	.188	.199	.210	.222	22
40	.012	.023	.035	.047	.058	.070	.081	.093	.105	.116	.128	.140	.151	.163	.174	.186	.197	.209	.221	.233	20
42	.012	.024	.037	.049	.061	.074	.086	.098	.111	.122	.135	.147	.159	.172	.183	.196	.207	.220	.232	.245	18
44	.013	.025	.039	.051	.064	.078	.090	.103	.116	.128	.142	.154	.167	.180	.192	.205	.217	.231	.243	.257	16
46	.014	.027	.041	.054	.067	.081	.094	.108	.121	.134	.148	.161	.175	.188	.201	.214	.227	.242	.254	.269	14
48	.014	.028	.042	.056	.070	.084	.098	.112	.126	.140	.154	.168	.182	.196	.210	.223	.237	.252	.265	.280	12
50	.015	.029	.044	.058	.073	.087	.102	.116	.131	.145	.160	.175	.189	.204	.218	.232	.247	.262	.276	.291	10
52	.015	.030	.046	.060	.076	.091	.106	.121	.137	.151	.167	.182	.197	.212	.227	.242	.257	.273	.288	.303	8
54	.016	.032	.047	.063	.079	.095	.110	.126	.142	.157	.174	.189	.205	.220	.236	.252	.267	.284	.299	.315	6
56	.016	.033	.049	.065	.082	.099	.114	.131	.147	.163	.180	.196	.213	.228	.245	.261	.277	.294	.310	.327	4
58	.017	.034	.051	.068	.085	.102	.118	.136	.152	.169	.186	.203	.220	.236	.254	.270	.287	.304	.321	.338	2
1° 0′	.017	.035	.052	.070	.087	.105	.122	.140	.157	.175	.192	.209	.227	.244	.262	.279	.297	.314	.332	.349	29° 0′
2	.018	.036	.054	.072	.090	.109	.127	.145	.163	.181	.199	.216	.235	.253	.271	.289	.307	.325	.343	.361	58
4	.018	.037	.055	.075	.093	.113	.131	.150	.168	.187	.206	.223	.243	.261	.280	.299	.317	.336	.354	.373	56
6	.019	.039	.057	.077	.096	.117	.135	.155	.173	.193	.212	.230	.251	.269	.289	.308	.327	.346	.365	.385	54
8	.019	.040	.059	.079	.099	.119	.139	.159	.178	.199	.218	.237	.258	.277	.298	.317	.337	.356	.376	.396	52
10	.020	.041	.061	.081	.102	.122	.143	.163	.183	.204	.224	.244	.265	.285	.306	.326	.347	.366	.387	.407	50
12	.020	.042	.063	.084	.105	.126	.147	.168	.189	.210	.231	.251	.273	.294	.315	.336	.357	.377	.399	.419	48
14	.021	.043	.064	.086	.108	.130	.151	.173	.195	.216	.238	.258	.281	.302	.324	.345	.367	.388	.410	.431	46
16	.022	.045	.066	.088	.111	.134	.155	.178	.200	.222	.244	.265	.289	.310	.333	.354	.377	.399	.421	.443	44
18	.022	.046	.068	.090	.114	.137	.159	.182	.205	.228	.250	.272	.296	.318	.341	.363	.387	.409	.432	.454	42
20	.023	.047	.070	.093	.116	.140	.163	.186	.210	.233	.256	.279	.303	.326	.349	.372	.396	.419	.443	.465	40
22	.023	.048	.071	.095	.119	.144	.167	.191	.216	.239	.263	.286	.311	.334	.358	.382	.406	.430	.454	.477	38
24	.024	.049	.073	.098	.122	.148	.171	.196	.221	.245	.270	.293	.319	.342	.367	.392	.416	.441	.465	.489	36
26	.025	.050	.075	.100	.125	.151	.175	.201	.226	.251	.276	.300	.326	.350	.376	.401	.426	.451	.476	.501	34
28	.025	.051	.077	.102	.128	.154	.179	.205	.231	.257	.282	.307	.333	.358	.385	.410	.436	.461	.487	.513	32
30	.026	.052	.079	.105	.131	.157	.183	.209	.236	.262	.288	.314	.340	.366	.393	.419	.445	.471	.498	.524	30
32	.026	.053	.080	.107	.134	.161	.188	.214	.242	.268	.295	.321	.348	.375	.402	.429	.455	.482	.509	.536	28
34	.027	.054	.082	.109	.137	.165	.192	.219	.247	.274	.302	.328	.356	.383	.411	.438	.465	.493	.520	.548	26
36	.028	.056	.084	.112	.140	.169	.196	.224	.252	.280	.308	.335	.364	.391	.420	.447	.475	.503	.531	.560	24
38	.028	.057	.085	.114	.143	.172	.200	.229	.257	.286	.314	.342	.372	.399	.428	.456	.485	.513	.542	.571	22
40	.029	.058	.087	.116	.145	.175	.204	.233	.262	.291	.320	.349	.379	.407	.436	.465	.495	.523	.553	.582	20
42	.030	.059	.089	.118	.148	.179	.208	.238	.268	.297	.327	.356	.387	.416	.445	.475	.505	.534	.564	.594	18
44	.030	.060	.091	.121	.151	.183	.212	.243	.273	.303	.334	.363	.395	.424	.454	.485	.515	.545	.575	.606	16
46	.031	.062	.092	.123	.154	.186	.216	.248	.278	.309	.340	.370	.402	.432	.463	.494	.525	.556	.586	.618	14
48	.031	.063	.094	.126	.157	.189	.220	.252	.283	.315	.346	.377	.409	.440	.472	.503	.535	.566	.597	.629	12
50	.032	.064	.096	.128	.160	.192	.224	.256	.288	.320	.352	.384	.416	.448	.480	.512	.544	.576	.608	.640	10
52	.033	.065	.098	.130	.163	.196	.228	.261	.294	.326	.359	.391	.424	.457	.489	.522	.554	.587	.619	.652	8
54	.033	.066	.100	.133	.166	.200	.232	.266	.299	.332	.366	.398	.432	.465	.498	.531	.564	.598	.630	.664	6
56	.034	.068	.102	.135	.169	.203	.236	.271	.304	.338	.372	.405	.439	.473	.507	.540	.574	.608	.641	.676	4
58	.035	.069	.103	.138	.172	.206	.240	.275	.309	.344	.378	.412	.446	.481	.516	.549	.584	.618	.652	.687	2
2° 0′	.035	.070	.105	.140	.175	.209	.244	.279	.314	.349	.384	.419	.453	.489	.524	.558	.593	.628	.663	.698	28° 0′
	1	2	3	4	5	6	7	8	9	10	11	12	13	14	15	16	17	18	19	20	V + XI −

TABLE XVIII.

GENERAL TABLE of Aberration, &c. continued.

− VI + O	1	2	3	4	5	6	7	8	9	10	11	12	13	14	15	16	17	18	19	20	
2° 0′	.035	.070	.105	.140	.175	.209	.244	.279	.314	.349	.384	.419	.454	.489	.524	.558	.593	.628	.663	.698	28° 0′
2	.036	.072	.107	.143	.178	.213	.249	.284	.320	.355	.391	.426	.462	.497	.533	.568	.603	.639	.674	.710	58
4	.037	.073	.109	.145	.181	.217	.253	.289	.325	.361	.398	.433	.470	.505	.542	.578	.613	.650	.685	.722	56
6	.037	.074	.111	.147	.184	.221	.257	.294	.330	.367	.404	.440	.477	.513	.551	.587	.623	.660	.696	.734	54
8	.038	.075	.112	.149	.187	.224	.261	.298	.335	.373	.410	.447	.484	.521	.559	.596	.633	.670	.707	.745	52
10	.038	.076	.113	.151	.189	.227	.265	.302	.340	.378	.416	.454	.491	.529	.567	.605	.643	.680	.718	.756	50
12	.039	.077	.115	.154	.192	.231	.269	.307	.346	.384	.423	.461	.499	.538	.576	.615	.653	.691	.729	.768	48
14	.039	.078	.117	.157	.195	.235	.273	.312	.351	.390	.430	.468	.507	.546	.585	.624	.663	.702	.740	.780	46
16	.040	.079	.119	.159	.198	.238	.277	.317	.356	.396	.436	.475	.515	.554	.594	.633	.673	.713	.751	.792	44
18	.040	.080	.121	.161	.201	.241	.281	.322	.361	.402	.442	.482	.522	.562	.603	.642	.683	.723	.762	.803	42
20	.040	.081	.122	.163	.204	.244	.285	.326	.366	.407	.448	.488	.529	.570	.611	.651	.692	.733	.773	.814	40
22	.041	.083	.124	.166	.207	.248	.289	.331	.372	.413	.455	.495	.537	.579	.620	.661	.702	.744	.785	.826	38
24	.042	.084	.126	.169	.210	.252	.293	.336	.377	.419	.462	.502	.545	.587	.629	.671	.712	.755	.796	.838	36
26	.043	.085	.128	.171	.213	.256	.297	.341	.382	.425	.468	.509	.553	.595	.638	.680	.722	.765	.807	.850	34
28	.044	.086	.130	.173	.216	.259	.301	.345	.387	.431	.474	.516	.560	.603	.646	.689	.732	.775	.818	.861	32
30	.044	.087	.131	.175	.218	.262	.305	.349	.392	.436	.480	.523	.567	.611	.654	.698	.741	.785	.829	.872	30
32	.045	.089	.133	.178	.221	.266	.309	.354	.398	.442	.487	.530	.575	.619	.663	.708	.751	.796	.840	.884	28
34	.046	.090	.135	.180	.224	.270	.313	.359	.404	.448	.494	.537	.583	.627	.672	.718	.761	.807	.851	.896	26
36	.046	.091	.137	.182	.227	.273	.317	.364	.409	.454	.500	.544	.591	.635	.681	.727	.771	.817	.862	.908	24
38	.047	.092	.139	.184	.230	.276	.321	.368	.415	.460	.506	.551	.598	.643	.690	.736	.781	.827	.873	.920	22
40	.047	.093	.140	.186	.233	.279	.325	.372	.419	.465	.512	.558	.605	.651	.698	.745	.791	.837	.884	.931	20
42	.048	.095	.142	.189	.236	.283	.330	.377	.425	.471	.519	.565	.613	.660	.707	.755	.801	.848	.895	.943	18
44	.048	.096	.144	.192	.239	.287	.334	.382	.430	.477	.526	.572	.621	.668	.716	.764	.811	.859	.906	.955	16
46	.049	.097	.146	.194	.242	.291	.338	.387	.435	.483	.532	.579	.629	.676	.725	.773	.821	.869	.917	.967	14
48	.049	.098	.147	.196	.245	.294	.342	.392	.440	.489	.538	.586	.636	.684	.733	.782	.831	.879	.928	.978	12
50	.049	.099	.148	.198	.247	.297	.346	.396	.445	.494	.544	.593	.643	.692	.741	.791	.840	.889	.939	.989	10
52	.050	.101	.150	.201	.250	.301	.350	.401	.451	.500	.551	.600	.651	.701	.750	.801	.850	.900	.950	1.001	8
54	.051	.102	.152	.203	.253	.305	.354	.406	.456	.506	.557	.607	.659	.709	.759	.811	.860	.911	.961	1.013	6
56	.051	.103	.154	.205	.256	.308	.358	.411	.461	.512	.563	.614	.666	.717	.768	.820	.870	.921	.972	1.025	4
58	.052	.104	.156	.207	.259	.311	.362	.415	.466	.518	.569	.621	.673	.725	.777	.829	.880	.931	.983	1.036	2
3° 0′	.052	.105	.157	.209	.262	.314	.366	.419	.471	.523	.575	.628	.680	.733	.785	.838	.890	.941	.994	1.047	27° 0′
2	.053	.107	.159	.212	.265	.318	.371	.424	.477	.529	.582	.635	.688	.741	.794	.848	.900	.952	1.005	1.059	58
4	.054	.108	.161	.215	.268	.322	.375	.429	.482	.535	.589	.642	.696	.749	.803	.857	.910	.963	1.016	1.071	56
6	.054	.109	.163	.217	.271	.325	.379	.434	.487	.541	.596	.649	.704	.757	.812	.866	.920	.974	1.027	1.083	54
8	.055	.110	.165	.219	.274	.338	.383	.438	.492	.547	.602	.656	.711	.765	.821	.875	.930	.984	1.038	1.094	52
10	.055	.111	.166	.221	.276	.331	.387	.442	.497	.552	.608	.663	.718	.773	.829	.884	.939	.994	1.049	1.105	50
12	.056	.112	.168	.224	.279	.335	.391	.447	.503	.558	.615	.670	.726	.782	.838	.894	.949	1.005	1.060	1.117	48
14	.057	.113	.170	.227	.282	.339	.395	.452	.508	.564	.622	.677	.734	.790	.847	.903	.959	1.016	1.071	1.129	46
16	.057	.114	.172	.229	.285	.343	.399	.457	.513	.570	.628	.684	.742	.798	.856	.912	.969	1.026	1.082	1.141	44
18	.058	.115	.173	.231	.288	.346	.403	.461	.518	.576	.634	.691	.749	.806	.864	.921	.979	1.036	1.093	1.152	42
20	.058	.116	.174	.233	.291	.349	.407	.465	.523	.581	.640	.698	.756	.814	.872	.930	.988	1.046	1.104	1.163	40
22	.059	.118	.176	.236	.294	.353	.411	.470	.529	.587	.647	.705	.764	.822	.881	.940	.998	1.057	1.116	1.175	38
24	.060	.119	.178	.238	.397	.357	.415	.475	.534	.593	.653	.712	.772	.830	.890	.950	1.008	1.068	1.127	1.187	36
26	.060	.120	.180	.240	.300	.360	.419	.480	.539	.599	.659	.719	.779	.838	.899	.959	1.018	1.078	1.138	1.199	34
28	.061	.121	.182	.242	.303	.363	.423	.484	.544	.605	.665	.726	.786	.846	.908	.968	1.028	1.088	1.149	1.210	22
30	.061	.122	.183	.244	.305	.366	.427	.488	.549	.611	.671	.732	.793	.854	.916	.977	1.037	1.098	1.160	1.221	30
32	.062	.124	.185	.247	.308	.370	.432	.493	.555	.617	.678	.739	.801	.863	.925	.987	1.047	1.109	1.171	1.233	28
34	.063	.125	.187	.250	.311	.374	.436	.498	.560	.623	.685	.746	.809	.871	.934	.996	1.057	1.120	1.182	1.245	26
36	.063	.126	.189	.252	.314	.378	.440	.503	.565	.629	.691	.753	.817	.879	.943	1.005	1.067	1.130	1.193	1.257	24
38	.064	.127	.191	.254	.317	.381	.444	.508	.570	.635	.697	.760	.824	.887	.951	1.014	1.077	1.140	1.204	1.268	22
40	.064	.128	.192	.256	.320	.384	.448	.512	.575	.640	.703	.767	.831	.895	.959	1.023	1.087	1.150	1.215	1.279	20
42	.065	.130	.194	.259	.323	.388	.452	.517	.581	.646	.710	.774	.839	.904	.968	1.033	1.097	1.161	1.226	1.291	18
44	.066	.131	.196	.261	.326	.392	.456	.522	.586	.652	.717	.781	.847	.912	.977	1.043	1.107	1.172	1.237	1.303	16
46	.066	.132	.198	.263	.329	.395	.460	.527	.591	.658	.723	.788	.855	.920	.986	1.052	1.117	1.182	1.248	1.315	14
48	.067	.133	.200	.265	.332	.398	.464	.531	.596	.664	.729	.795	.862	.928	.995	1.061	1.127	1.192	1.259	1.326	12
50	.067	.134	.201	.267	.334	.401	.468	.535	.601	.669	.735	.802	.869	.936	1.003	1.070	1.136	1.202	1.270	1.337	10
52	.068	.136	.203	.270	.337	.405	.472	.540	.607	.675	.742	.809	.877	.944	1.012	1.080	1.146	1.213	1.281	1.349	8
54	.069	.137	.205	.273	.340	.409	.476	.545	.612	.681	.749	.816	.885	.952	1.021	1.089	1.156	1.224	1.292	1.361	6
56	.069	.138	.207	.275	.343	.412	.480	.550	.617	.687	.755	.823	.893	.960	1.030	1.098	1.166	1.235	1.303	1.373	4
58	.070	.139	.208	.277	.346	.415	.484	.554	.622	.693	.761	.830	.900	.968	1.038	1.107	1.176	1.245	1.314	1.384	2
4° 0	.070	.140	.209	.279	.349	.418	.488	.558	.627	.698	.767	.837	.907	.976	1.046	1.116	1.185	1.255	1.325	1.395	26° 0′
	1	2	3	4	5	6	7	8	9	10	11	12	13	14	15	16	17	18	19	20	V + XI −

GENERAL TABLE of Aberration, &c. continued.

S. −VI +O	1	2	3	4	5	6	7	8	9	10	11	12	13	14	15	16	17	18	19	20	
4° 0	.070	.140	.209	.280	.349	.418	.489	.558	.627	.698	.767	.837	.907	.976	1.046	1.116	1.185	1.255	1.335	1.395	26° 0′
2	.071	.141	.211	.283	.352	.422	.493	.563	.633	.704	.774	.844	.915	.985	1.055	1.126	1.195	1.266	1.336	1.407	58
4	.072	.142	.213	.285	.355	.426	.497	.568	.634	.710	.781	.851	.923	.993	1.064	1.135	1.205	1.277	1.347	1.419	56
6	.072	.143	.215	.287	.358	.430	.501	.573	.644	.716	.787	.858	.930	1.001	1.073	1.144	1.215	1.287	1.358	1.431	54
8	.073	.144	.217	.289	.361	.433	.505	.577	.649	.722	.793	.865	.937	1.009	1.082	1.153	1.225	1.297	1.369	1.442	52
10	.073	.145	.218	.291	.363	.436	.509	.581	.654	.727	.799	.872	.944	1.017	1.090	1.162	1.235	1.307	1.380	1.453	50
12	.074	.147	.220	.294	.366	.440	.513	.586	.660	.733	.806	.879	.952	1.026	1.099	1.172	1.245	1.318	1.392	1.465	48
14	.075	.148	.222	.296	.369	.444	.517	.591	.665	.739	.813	.886	.960	1.034	1.108	1.182	1.255	1.329	1.403	1.477	46
16	.075	.149	.224	.298	.372	.447	.521	.596	.670	.745	.819	.893	.968	1.042	1.117	1.191	1.265	1.340	1.414	1.489	44
18	.076	.150	.226	.300	.375	.450	.525	.600	.675	.751	.825	.900	.975	1.050	1.125	1.200	1.275	1.350	1.425	1.500	42
20	.076	.151	.227	.302	.378	.453	.529	.604	.680	.756	.831	.907	.982	1.058	1.133	1.209	1.284	1.360	1.436	1.511	40
22	.077	.153	.229	.305	.381	.457	.533	.609	.686	.762	.838	.914	.990	1.066	1.142	1.219	1.294	1.371	1.447	1.523	38
24	.078	.154	.231	.308	.384	.461	.537	.614	.691	.768	.845	.921	.998	1.074	1.151	1.228	1.304	1.382	1.458	1.535	36
26	.078	.155	.233	.310	.387	.465	.541	.619	.696	.774	.851	.928	1.006	1.082	1.160	1.237	1.314	1.392	1.469	1.547	34
28	.079	.156	.234	.312	.390	.468	.545	.624	.701	.780	.857	.935	1.013	1.090	1.169	1.246	1.324	1.402	1.480	1.558	32
30	.079	.157	.235	.314	.392	.471	.549	.628	.706	.785	.863	.942	1.020	1.098	1.177	1.255	1.334	1.412	1.491	1.569	30
32	.080	.159	.237	.317	.395	.475	.554	.633	.712	.791	.870	.949	1.028	1.107	1.186	1.265	1.344	1.423	1.502	1.581	28
34	.080	.160	.239	.319	.398	.479	.558	.638	.717	.797	.877	.956	1.036	1.115	1.195	1.275	1.354	1.434	1.513	1.593	26
36	.081	.161	.241	.321	.401	.482	.562	.643	.722	.803	.883	.963	1.044	1.123	1.204	1.284	1.364	1.445	1.524	1.605	24
38	.081	.162	.243	.323	.404	.485	.566	.647	.727	.809	.889	.970	1.051	1.131	1.212	1.293	1.374	1.455	1.535	1.616	22
40	.081	.163	.244	.325	.407	.488	.570	.651	.732	.814	.895	.976	1.058	1.139	1.220	1.302	1.383	1.465	1.546	1.627	20
42	.082	.165	.246	.328	.410	.492	.574	.656	.738	.820	.902	.983	1.066	1.148	1.229	1.312	1.393	1.476	1.557	1.639	18
44	.083	.166	.248	.331	.413	.496	.578	.661	.743	.826	.909	.990	1.074	1.156	1.238	1.321	1.403	1.487	1.568	1.651	16
46	.083	.167	.250	.333	.416	.500	.582	.666	.748	.832	.915	.997	1.081	1.164	1.247	1.330	1.413	1.497	1.579	1.663	14
48	.084	.168	.252	.335	.419	.503	.586	.670	.753	.838	.921	1.004	1.088	1.172	1.256	1.339	1.423	1.507	1.590	1.674	12
50	.084	.169	.253	.337	.421	.506	.590	.674	.758	.843	.927	1.011	1.095	1.180	1.264	1.348	1.432	1.517	1.601	1.685	10
52	.085	.170	.255	.340	.424	.510	.594	.679	.764	.849	.934	1.018	1.103	1.188	1.273	1.358	1.442	1.528	1.612	1.697	8
54	.086	.171	.257	.343	.427	.514	.598	.684	.770	.855	.941	1.025	1.111	1.196	1.282	1.367	1.452	1.539	1.623	1.709	6
56	.086	.172	.259	.345	.430	.517	.602	.689	.775	.861	.947	1.032	1.119	1.204	1.291	1.376	1.462	1.550	1.634	1.721	4
58	.087	.173	.261	.347	.433	.520	.606	.693	.780	.867	.953	1.039	1.126	1.212	1.299	1.385	1.472	1.560	1.645	1.732	2
5° 0	.087	.174	.262	.349	.436	.523	.610	.697	.785	.872	.959	1.046	1.133	1.220	1.307	1.394	1.482	1.570	1.656	1.743	25° 0′
2	.088	.176	.264	.352	.439	.527	.614	.702	.790	.878	.966	1.053	1.141	1.229	1.316	1.404	1.492	1.581	1.667	1.755	58
4	.089	.177	.266	.354	.442	.531	.618	.707	.795	.884	.973	1.060	1.149	1.237	1.325	1.414	1.502	1.591	1.678	1.767	56
6	.089	.178	.268	.356	.445	.534	.622	.712	.800	.890	.979	1.067	1.157	1.245	1.334	1.423	1.512	1.601	1.689	1.779	54
8	.090	.179	.269	.358	.448	.537	.626	.716	.805	.896	.985	1.074	1.164	1.253	1.343	1.432	1.522	1.611	1.700	1.790	52
10	.090	.180	.270	.360	.450	.540	.630	.720	.810	.901	.991	1.080	1.171	1.261	1.351	1.441	1.531	1.621	1.711	1.801	50
12	.091	.182	.272	.363	.453	.544	.635	.725	.816	.907	.998	1.087	1.179	1.269	1.360	1.451	1.541	1.632	4.722	1.813	48
14	.092	.183	.274	.366	.456	.548	.639	.730	.821	.913	1.004	1.094	1.187	1.277	1.369	1.460	1.551	1.643	1.733	1.825	46
16	.092	.184	.276	.368	.459	.552	.643	.735	.826	.919	1.010	1.101	1.194	1.285	1.378	1.469	1.561	1.653	1.744	1.837	44
18	.093	.185	.278	.370	.462	.555	.647	.740	.831	.925	1.016	1.108	1.201	1.293	1.386	1.478	1.571	1.663	1.755	1.848	42
20	.093	.186	.279	.372	.465	.558	.651	.744	.836	.930	1.022	1.115	1.208	1.301	1.394	1.487	1.580	1.673	1.766	1.859	40
22	.094	.188	.281	.375	.468	.562	.655	.749	.842	.936	1.029	1.122	1.216	1.310	1.403	1.497	1.590	1.684	1.777	1.871	38
24	.095	.189	.283	.377	.471	.566	.659	.754	.848	.942	1.036	1.129	1.224	1.318	1.412	1.507	1.600	1.695	1.788	1.883	36
26	.095	.190	.285	.379	.474	.569	.663	.759	.853	.948	1.042	1.136	1.232	1.326	1.421	1.516	1.610	1.705	1.799	1.895	34
28	.096	.191	.287	.381	.477	.572	.667	.763	.858	.954	1.048	1.143	1.239	1.334	1.430	1.525	1.620	1.715	1.810	1.906	32
30	.096	.192	.288	.383	.479	.575	.671	.767	.863	.959	1.054	1.149	1.246	1.342	1.438	1.534	1.629	1.725	1.821	1.917	30
32	.097	.194	.290	.386	.482	.579	.675	.772	.868	.965	1.061	1.156	1.254	1.350	1.447	1.544	1.639	1.736	1.832	1.929	28
34	.098	.195	.292	.389	.485	.583	.679	.777	.873	.971	1.068	1.163	1.262	1.358	1.456	1.553	1.649	1.747	1.843	1.941	26
36	.098	.196	.294	.391	.488	.586	.683	.782	.878	.977	1.074	1.170	1.269	1.366	1.465	1.562	1.659	1.757	1.854	1.953	24
38	.099	.197	.295	.393	.491	.589	.687	.786	.883	.982	1.080	1.177	1.276	1.374	1.473	1.571	1.669	1.767	1.865	1.964	22
40	.099	.198	.296	.395	.494	.592	.691	.790	.888	.987	1.086	1.184	1.283	1.382	1.481	1.580	1.678	1.777	1.876	1.975	20
42	.100	.199	.298	.398	.497	.596	.695	.795	.894	.993	1.093	1.191	1.291	1.391	1.490	1.590	1.688	1.788	1.887	1.987	18
44	.101	.200	.300	.401	.500	.600	.699	.800	.899	.999	1.100	1.198	1.299	1.399	1.499	1.599	1.698	1.799	1.898	1.999	16
46	.101	.201	.302	.403	.503	.604	.703	.805	.904	1.005	1.106	1.205	1.307	1.407	1.508	1.608	1.708	1.809	1.909	2.011	14
48	.102	.202	.304	.405	.506	.607	.707	.809	.909	1.011	1.112	1.212	1.314	1.415	1.517	1.617	1.718	1.819	1.920	2.022	12
50	.102	.203	.305	.407	.508	.610	.711	.813	.914	1.016	1.118	1.219	1.321	1.423	1.525	1.626	1.728	1.829	1.931	2.033	10
52	.103	.205	.307	.410	.511	.614	.716	.818	.920	1.022	1.125	1.226	1.329	1.431	1.534	1.636	1.738	1.840	1.942	2.045	8
54	.104	.206	.309	.412	.514	.618	.720	.823	.926	1.028	1.132	1.233	1.337	1.439	1.543	1.646	1.748	1.851	1.953	2.057	6
56	.104	.207	.311	.414	.517	.621	.724	.828	.931	1.034	1.138	1.240	1.345	1.447	1.552	1.655	1.758	1.861	1.964	2.069	4
58	.105	.208	.313	.416	.520	.624	.728	.832	.936	1.040	1.144	1.247	1.352	1.455	1.560	1.664	1.768	1.871	1.975	2.080	2
6° 0	.105	.209	.314	.418	.523	.627	.732	.836	.941	1.045	1.150	1.254	1.359	1.463	1.568	1.673	1.777	1.881	1.986	2.091	24° 0′
	1	2	3	4	5	6	7	8	9	10	11	12	13	14	15	16	17	18	19	20	V + XI−

TABLE XVIII.

GENERAL TABLE of Aberration, &c. continued.

S. − VI + O	1	2	3	4	5	6	7	8	9	10	11	12	13	14	15	16	17	18	19	20	
6° 0	.105	.209	.314	.418	.523	.627	.732	.836	.941	1.045	1.150	1.254	1.359	1.463	1.568	1.673	1.777	1.881	1.986	2.091	24° 0′
2	.106	.211	.316	.421	.526	.631	.736	.841	.947	1.051	1.157	1.261	1.367	1.472	1.577	1.683	1.787	1.892	1.997	2.103	58
4	.106	.212	.318	.424	.529	.635	.740	.846	.952	1.057	1.164	1.268	1.375	1.480	1.586	1.692	1.797	1.903	2.008	2.115	56
6	.107	.213	.320	.426	.532	.638	.744	.851	.957	1.063	1.170	1.275	1.382	1.488	1.595	1.701	1.807	1.913	2.019	2.126	54
8	.107	.214	.321	.428	.535	.641	.748	.855	.962	1.069	1.176	1.282	1.389	1.496	1.603	1.710	1.817	1.923	2.030	2.137	52
10	.107	.215	.322	.430	.537	.644	.752	.859	.967	1.074	1.182	1.289	1.396	1.504	1.611	1.719	1.826	1.933	2.041	2.148	50
12	.108	.217	.324	.433	.540	.648	.756	.864	.973	1.080	1.189	1.296	1.404	1.512	1.620	1.729	1.836	1.944	2.052	2.160	48
14	.109	.218	.326	.435	.543	.652	.760	.869	.978	1.086	1.195	1.303	1.412	1.520	1.629	1.738	1.846	1.955	2.063	2.172	46
16	.109	.219	.328	.437	.546	.656	.764	.874	.983	1.092	1.201	1.310	1.420	1.528	1.638	1.747	1.856	1.965	2.074	2.184	44
18	.110	.220	.330	.439	.549	.659	.768	.878	.988	1.098	1.207	1.317	1.427	1.536	1.647	1.756	1.866	1.975	2.085	2.195	42
20	.110	.221	.331	.441	.552	.662	.772	.882	.993	1.103	1.213	1.324	1.434	1.544	1.655	1.765	1.875	1.985	2.096	2.206	40
22	.111	.222	.333	.444	.555	.666	.776	.887	.999	1.109	1.220	1.331	1.442	1.553	1.664	1.775	1.885	1.996	2.107	2.218	38
24	.112	.223	.335	.447	.558	.670	.780	.892	1.004	1.115	1.227	1.338	1.450	1.561	1.673	1.784	1.895	2.007	2.118	2.230	36
26	.112	.224	.337	.449	.561	.673	.784	.897	1.009	1.121	1.233	1.345	1.458	1.569	1.682	1.793	1.905	2.018	2.129	2.242	34
28	.113	.225	.339	.451	.564	.676	.788	.902	1.014	1.127	1.239	1.351	1.465	1.577	1.690	1.802	1.215	2.028	2.140	2.251	32
30	.113	.226	.340	.453	.566	.679	.792	.906	1.019	1.132	1.245	1.358	1.472	1.585	1.698	1.811	1.924	2.038	2.151	2.264	30
32	.114	.228	.342	.456	.569	.683	.797	.911	1.025	1.138	1.252	1.365	1.480	1.593	1.707	1.821	1.934	2.049	2.162	2.276	28
34	.115	.229	.344	.458	.572	.687	.801	.916	1.030	1.144	1.259	1.372	1.488	1.601	1.716	1.831	1.944	2.060	2.173	2.288	26
36	.115	.230	.346	.460	.575	.691	.805	.921	1.035	1.150	1.265	1.379	1.495	1.609	1.725	1.840	1.954	2.070	2.184	2.300	24
38	.116	.231	.347	.462	.578	.694	.809	.925	1.040	1.156	1.271	1.386	1.502	1.617	1.734	1.849	1.964	2.080	2.195	2.311	22
40	.116	.232	.348	.464	.581	.697	.813	.929	1.045	1.161	1.277	1.393	1.509	1.625	1.742	1.858	1.974	2.090	2.206	2.322	20
42	.117	.234	.350	.467	.584	.701	.817	.934	1.051	1.167	1.284	1.400	1.517	1.634	1.751	1.868	1.984	2.101	2.217	2.334	18
44	.118	.235	.352	.470	.587	.705	.821	.939	1.056	1.173	1.291	1.407	1.525	1.642	1.760	1.877	1.994	2.112	2.228	2.346	16
46	.118	.236	.354	.472	.590	.708	.825	.944	1.061	1.179	1.297	1.414	1.533	1.650	1.769	1.886	2.004	2.122	2.239	2.358	14
48	.119	.237	.356	.474	.593	.711	.829	.948	1.066	1.185	1.303	1.421	1.540	1.658	1.777	1.895	2.014	2.132	2.250	2.369	12
50	.119	.238	.357	.476	.595	.714	.833	.952	1.071	1.190	1.309	1.428	1.547	1.666	1.785	1.904	2.023	2.142	2.261	2.380	10
52	.120	.240	.359	.479	.598	.718	.837	.957	1.077	1.196	1.316	1.435	1.555	1.674	1.794	1.914	2.033	2.153	2.272	2.392	8
54	.121	.241	.361	.482	.601	.722	.841	.962	1.082	1.202	1.323	1.442	1.563	1.682	1.803	1.923	2.043	2.164	2.283	2.404	6
56	.121	.242	.363	.484	.604	.725	.845	.967	1.087	1.208	1.329	1.449	1.571	1.690	1.812	1.932	2.053	2.174	2.294	2.415	4
58	.122	.243	.365	.486	.607	.728	.849	.971	1.092	1.214	1.335	1.456	1.578	1.698	1.820	1.941	2.063	2.184	2.305	2.426	2
7° 0′	.122	.244	.366	.488	.609	.731	.853	.975	1.097	1.219	1.341	1.463	1.585	1.706	1.828	1.950	2.072	2.194	2.316	2.437	23° 0′
2	.123	.246	.368	.491	.612	.735	.857	.980	1.103	1.225	1.348	1.470	1.593	1.715	1.837	1.960	2.082	2.205	2.328	2.449	58
4	.124	.247	.370	.493	.615	.739	.861	.985	1.108	1.231	1.354	1.477	1.601	1.723	1.846	1.969	2.092	2.216	2.339	2.461	56
6	.124	.248	.372	.495	.618	.743	.865	.990	1.113	1.237	1.360	1.484	1.608	1.731	1.855	1.978	2.102	2.226	2.350	2.473	54
8	.125	.249	.373	.497	.621	.746	.869	.994	1.118	1.243	1.366	1.491	1.615	1.739	1.863	1.987	2.112	2.236	2.361	2.484	52
10	.125	.250	.374	.499	.624	.749	.873	.998	1.124	1.248	1.372	1.497	1.622	1.747	1.871	1.996	2.122	2.246	2.372	2.495	50
12	.126	.251	.376	.502	.627	.753	.878	1.003	1.129	1.254	1.379	1.504	1.630	1.755	1.880	2.006	2.132	2.257	2.383	2.507	48
14	.127	.252	.378	.505	.630	.757	.882	1.008	1.134	1.260	1.386	1.511	1.638	1.763	1.889	2.015	2.142	2.267	2.394	2.519	46
16	.127	.253	.380	.507	.633	.760	.886	1.013	1.139	1.266	1.392	1.518	1.645	1.771	1.890	2.024	2.152	2.277	2.405	2.531	44
18	.128	.254	.382	.509	.636	.763	.890	1.017	1.144	1.271	1.398	1.525	1.652	1.779	1.907	2.033	2.161	2.287	2.415	2.542	42
20	.128	.255	.383	.511	.638	.766	.894	1.021	1.149	1.276	1.404	1.532	1.659	1.787	1.915	2.042	2.170	2.297	2.425	2.553	40
22	.129	.257	.385	.514	.641	.770	.898	1.026	1.155	1.282	1.411	1.539	1.667	1.795	1.924	2.052	2.180	2.308	2.436	2.565	38
24	.130	.258	.387	.516	.644	.774	.902	1.031	1.160	1.288	1.418	1.546	1.675	1.803	1.933	2.061	2.190	2.319	2.447	2.577	36
26	.130	.259	.389	.518	.647	.777	.906	1.036	1.165	1.294	1.424	1.553	1.683	1.811	1.942	2.070	2.200	2.330	2.458	2.589	34
28	.131	.260	.391	.520	.650	.780	.910	1.040	1.170	1.300	1.430	1.560	1.690	1.819	1.950	2.079	2.210	2.340	2.468	2.600	32
30	.131	.261	.392	.522	.653	.783	.914	1.044	1.175	1.305	1.436	1.566	1.697	1.827	1.958	2.088	2.219	2.350	2.480	2.611	30
32	.132	.263	.394	.525	.656	.787	.918	1.049	1.181	1.311	1.443	1.573	1.705	1.836	1.967	2.098	2.229	2.361	2.491	2.623	28
34	.132	.264	.396	528	.659	.791	.922	1.054	1.186	1.317	1.449	1.580	1.713	1.844	1.976	2.107	2.239	2.371	2.502	2.635	26
36	.132	.265	.398	.530	.662	.794	.926	1.059	1.191	1.323	1.456	1.587	1.720	1.852	1.985	2.116	2.249	2.381	2.513	2.646	24
38	.133	.266	.399	.532	.665	.797	.930	1.063	1.196	1.329	1.461	1.594	1.727	1.860	1.993	2.125	2.259	2.391	2.524	2.657	22
40	.133	.267	.400	.534	.667	.800	.934	1.067	1.201	1.334	1.467	1.601	1.734	1.868	2.001	2.134	2.268	2.401	2.535	2.668	20
42	.134	.269	.402	.537	.670	.804	.938	1.072	1.207	1.340	1.474	1.608	1.742	1.876	2.010	2.144	2.278	2.412	2.546	2.680	18
44	.135	.270	.404	.539	.673	.808	.942	1.077	1.212	1.346	1.481	1.615	1.750	1.884	2.019	2.153	2.288	2.423	2.557	2.692	16
46	.135	.271	.406	.541	.676	.812	.946	1.082	1.217	1.352	1.487	1.622	1.758	1.892	2.028	2.162	2.298	2.433	2.568	2.704	14
48	.136	.272	.408	.543	.679	.815	.950	1.086	1.222	1.358	1.493	1.629	1.765	1.900	2.036	2.171	2.308	2.443	2.679	2.715	12
50	.136	.273	.409	.545	681	.818	.954	1.090	1.227	1.363	1.499	1.635	1.772	1.908	2.044	2.180	2.317	2.453	2.590	2.726	10
52	.137	.274	.411	.548	.684	.822	.958	1.095	1.333	1.369	1.506	1.642	1.780	1.916	2.053	2.190	2.327	2.454	2.601	2.738	8
54	.138	.275	.413	.551	.687	.826	.962	1.100	1.238	1.375	1.513	1.649	1.788	1.924	2.062	2.199	2.337	2.475	2.612	2.750	6
56	.138	.276	.415	.553	.690	.829	.966	1.105	1.243	1.381	1.519	1.656	1.795	1.932	2.071	2.208	2.347	2.485	2.623	2.761	4
58	.139	.277	.417	.555	.693	.832	.970	1.109	1.248	1.387	1.525	1.663	1.802	1.940	2.079	2.217	2.357	2.495	2.634	2.772	2
8° 0′	.139	.278	.418	.557	.696	.835	.974	1.113	1.253	1.392	1.531	1.670	1.809	1.948	2.087	2.226	2.363	2.505	2.644	2.783	22° 0′
	1	2	3	4	5	6	7	8	9	10	11	12	13	14	15	16	17	18	19	20	V + XI −

TABLE XVIII.

GENERAL TABLE of Aberration, &c. continued.

S. − VI + O	1	2	3	4	5	6	7	8	9	10	11	12	13	14	15	16	17	18	19	20	
8° 0′	.139	.278	.418	.557	.696	.835	.974	1.113	1.253	1.392	1.531	1.670	1.809	1.948	2.087	2.226	2.366	2.504	2.644	2.783	22° 0′
2	.140	.280	.420	.560	.699	.839	.978	1.118	1.259	1.398	1.538	1.677	1.817	1.957	2.096	2.236	2.376	2.515	2.655	2.795	58
4	.140	.281	.422	.562	.702	.843	.982	1.123	1.264	1.404	1.545	1.684	1.825	1.965	2.105	2.246	2.386	2.526	2.666	2.807	56
6	.141	.282	.424	.564	.705	.846	.986	1.128	1.269	1.410	1.551	1.691	1.833	1.973	2.114	2.255	2.396	2.537	2.677	2.819	54
8	.142	.283	.425	.566	.708	.849	.990	1.132	1.274	1.416	1.557	1.698	1.840	1.981	2.123	2.264	2.406	2.547	2.688	2.830	52
10	.142	.284	.426	.568	.710	.852	.994	1.136	1.279	1.421	1.563	1.705	1.847	1.989	2.131	2.273	2.415	2.557	2.699	2.841	50
12	.143	.286	.428	.571	.713	.856	.999	1.141	1.284	1.427	1.570	1.712	1.855	1.997	2.139	2.283	2.425	2.568	2.700	2.853	48
14	.143	.287	.430	.574	.716	.860	1.003	1.146	1.289	1.433	1.576	1.719	1.863	2.005	2.148	2.292	2.435	2.579	2.721	2.865	46
16	.144	.288	.432	.576	.719	.864	1.007	1.151	1.294	1.439	1.582	1.726	1.870	2.013	2.157	2.301	2.445	2.489	2.732	2.877	44
18	.145	.289	.434	.578	.722	.867	1.011	1.155	1.299	1.444	1.588	1.733	1.877	2.021	2.166	2.310	2.455	2.599	2.733	2.888	42
20	.145	.290	.435	.580	.725	.870	1.015	1.159	1.304	1.449	1.591	1.739	1.884	2.029	2.174	2.319	2.464	2.609	2.754	2.899	40
22	.146	.292	.437	.583	.728	.874	1.019	1.164	1.310	1.455	1.601	1.746	1.892	2.037	2.183	2.329	2.474	2.620	2.765	2.911	38
24	.146	.293	.439	.585	.731	.878	1.023	1.169	1.315	1.461	1.608	1.753	1.900	2.045	2.192	2.338	2.484	2.630	2.776	2.923	36
26	.147	.294	.441	.587	.734	.881	1.027	1.174	1.320	1.467	1.614	1.760	1.907	2.053	2.201	2.347	2.494	2.640	2.787	2.934	34
28	.148	.295	.442	.589	.737	.884	1.031	1.178	1.325	1.473	1.620	1.767	1.914	2.061	2.209	2.356	2.504	2.650	2.798	2.945	32
30	.148	.296	.443	.591	.739	.887	1.035	1.182	1.330	1.478	1.626	1.773	1.921	2.069	2.217	2.365	2.513	2.660	2.808	2.956	30
32	.149	.297	.445	.594	.742	.891	1.039	1.187	1.336	1.484	1.633	1.780	1.929	2.077	2.226	2.375	2.523	2.671	2.819	2.968	28
34	.149	.298	.447	.597	.745	.895	1.043	1.192	1.341	1.490	1.640	1.787	1.937	2.085	2.235	2.384	2.533	2.682	2.830	2.980	26
36	.150	.299	.449	.599	.748	.898	1.047	1.197	1.346	1.496	1.646	1.794	1.945	2.093	2.244	2.393	2.543	2.692	2.841	2.991	24
38	.151	.300	.451	.601	.751	.901	1.051	1.201	1.351	1.502	1.652	1.801	1.952	2.101	2.252	2.402	2.552	2.702	2.852	3.002	22
40	.151	.301	.452	.603	.753	.904	1.055	1.205	1.356	1.507	1.658	1.808	1.959	2.109	2.260	2.411	2.561	2.712	2.863	3.013	20
42	.152	.303	.454	.606	.756	.908	1.059	1.210	1.362	1.513	1.665	1.815	1.967	2.117	2.269	2.421	2.571	2.723	2.874	3.025	18
44	.152	.304	.456	.608	.759	.912	1.063	1.215	1.367	1.519	1.671	1.822	1.975	2.125	2.278	2.430	2.581	2.734	2.885	3.037	16
46	.153	.305	.458	.610	.762	.915	1.067	1.220	1.372	1.525	1.677	1.829	1.982	2.133	2.287	2.439	2.591	2.744	2.896	3.049	14
48	.154	.306	.460	.612	.765	.918	1.071	1.224	1.377	1.531	1.683	1.836	1.989	2.141	2.295	2.448	2.601	2.754	2.907	3.060	12
50	.154	.307	.461	.614	.768	.921	1.075	1.228	1.382	1.536	1.689	1.842	1.996	2.149	2.303	2.457	2.610	2.764	2.918	3.071	10
52	.155	.309	.463	.617	.771	.925	1.079	1.233	1.388	1.542	1.696	1.849	2.004	2.158	2.312	2.467	2.620	2.775	2.929	3.083	8
54	.155	.310	.465	.620	.774	.929	1.083	1.238	1.393	1.548	1.703	1.856	2.012	2.166	2.321	2.476	2.630	2.786	2.940	3.095	6
56	.156	.311	.467	.622	.777	.933	1.087	1.243	1.398	1.554	1.709	1.863	2.020	2.174	2.330	2.485	2.640	2.796	2.951	3.107	4
58	.156	.312	.468	.624	.780	.936	1.091	1.247	1.403	1.559	1.715	1.870	2.027	2.182	2.338	2.494	2.650	2.806	2.962	3.118	2
9° 0′	.156	.313	.469	.626	.782	.939	1.095	1.251	1.408	1.564	1.721	1.877	2.034	2.190	2.347	2.503	2.659	2.816	2.972	3.129	21° 0′
2	.157	.315	.471	.629	.785	.943	1.099	1.256	1.414	1.570	1.728	1.884	2.042	2.198	2.356	2.513	2.669	2.827	2.983	3.141	58
4	.157	.316	.473	.631	.788	.947	1.103	1.261	1.419	1.576	1.734	1.891	2.050	2.206	2.365	2.522	2.679	2.837	2.994	3.153	56
6	.158	.317	.475	.633	.791	.950	1.107	1.266	1.424	1.582	1.740	1.898	2.057	2.214	2.374	2.531	2.689	2.847	3.005	3.164	54
8	.159	.318	.477	.635	.794	.953	1.111	1.270	1.429	1.588	1.746	1.905	2.064	2.222	2.382	2.540	2.699	2.857	3.016	3.175	52
10	.159	.319	.478	.637	.797	.956	1.115	1.274	1.434	1.593	1.752	1.911	2.071	2.230	2.390	2.549	2.708	2.867	3.027	3.186	50
12	.160	.320	.480	.640	.800	.960	1.119	1.279	1.440	1.599	1.759	1.918	2.079	2.238	2.399	2.559	2.718	2.878	3.038	3.198	48
14	.160	.321	.482	.643	.803	.964	1.123	1.284	1.445	1.605	1.765	1.925	2.087	2.246	2.408	2.568	2.728	2.889	3.049	3.210	46
16	.161	.322	.484	.645	.806	.967	1.127	1.289	1.450	1.611	1.771	1.932	2.094	2.254	2.417	2.577	2.738	2.899	3.060	3.222	44
18	.162	.323	.486	.647	.809	.970	1.131	1.293	1.455	1.617	1.777	1.939	2.101	2.262	2.425	2.586	2.748	2.909	3.071	3.233	42
20	.162	.324	.487	.649	.811	.973	1.135	1.297	1.460	1.622	1.783	1.946	2.108	2.270	2.433	2.595	2.757	2.919	3.081	3.244	40
22	.163	.326	.489	.652	.814	.977	1.139	1.302	1.465	1.628	1.790	1.953	2.116	2.278	2.442	2.605	2.767	2.930	3.092	3.256	38
24	.163	.327	.491	.654	.817	.981	1.143	1.307	1.470	1.634	1.797	1.960	2.124	2.286	2.451	2.614	2.777	2.941	3.103	3.268	36
26	.164	.328	.493	.656	.820	.984	1.147	1.312	1.475	1.640	1.803	1.967	2.132	2.294	2.460	2.623	2.787	2.951	3.114	3.279	34
28	.165	.329	.494	.658	.823	.987	1.151	1.316	1.480	1.646	1.809	1.974	2.139	2.302	2.468	2.632	2.797	2.961	3.125	3.290	32
30	.165	.330	.495	.660	.825	.990	1.155	1.320	1.485	1.651	1.815	1.980	2.146	2.310	2.476	2.641	2.806	2.971	3.136	3.301	30
32	.166	.332	.497	.663	.828	.994	1.159	1.325	1.491	1.659	1.822	1.987	2.154	2.319	2.485	2.651	2.816	2.982	3.147	3.313	28
34	.166	.333	.499	.666	.831	.998	1.163	1.330	1.496	1.663	1.829	1.994	2.162	2.327	2.494	2.660	2.826	2.992	3.158	3.325	26
36	.167	.334	.501	.668	.834	1.001	1.167	1.335	1.501	1.669	1.835	2.001	2.169	2.335	2.503	2.669	2.836	3.002	3.169	3.336	24
38	.168	.335	.503	.670	.837	1.004	1.171	1.339	1.506	1.674	1.841	2.008	2.176	2.343	2.511	2.678	2.848	3.012	3.180	3.347	22
40	.168	.336	.504	.672	.840	1.007	1.175	1.343	1.511	1.679	1.817	2.015	2.183	2.351	2.519	2.687	2.855	3.022	3.190	3.358	20
42	.169	.338	.506	.675	.843	1.011	1.179	1.348	1.517	1.685	1.854	2.022	2.191	2.359	2.528	2.696	2.865	3.033	3.201	3.370	18
44	.169	.339	.508	.677	.846	1.015	1.183	1.353	1.522	1.691	1.860	2.029	2.199	2.367	2.537	2.705	2.875	3.044	3.212	3.382	16
46	.170	.340	.510	.679	.849	1.019	1.187	1.358	1.527	1.697	1.866	2.036	2.206	2.375	2.546	2.714	2.885	3.054	3.223	3.393	14
48	.171	.341	.511	.681	.852	1.021	1.191	1.362	1.532	1.703	1.872	2.043	2.213	2.383	2.554	2.723	2.894	3.064	3.234	3.404	12
50	.171	.342	.512	.683	.854	1.025	1.195	1.366	1.537	1.708	1.878	2.049	2.220	2.391	2.562	2.732	2.903	3.074	3.245	3.415	10
52	.172	.343	.514	.686	.857	1.029	1.199	1.371	1.543	1.714	1.885	2.055	2.228	2.399	2.571	2.742	2.913	3.085	3.256	3.427	8
54	.172	.344	.516	.689	.860	1.033	1.203	1.376	1.548	1.720	1.892	2.062	2.236	2.407	2.580	2.751	2.923	3.095	3.267	3.439	6
56	.173	.345	.518	.691	.863	1.036	1.207	1.381	1.553	1.726	1.898	2.069	2.243	2.415	2.589	2.760	2.933	3.105	3.278	3.451	4
58	.174	.346	.520	.693	.866	1.039	1.211	1.385	1.558	1.732	1.906	2.076	2.250	2.423	2.597	2.769	2.943	3.115	3.289	3.462	2
10° 0	.174	.347	.521	.695	.868	1.042	1.215	1.389	1.563	1.737	1.910	2.083	2.257	2.431	2.605	2.778	2.952	3.125	3.299	3.473	20° 0′
	1	2	3	4	5	6	7	8	9	10	11	12	13	14	15	16	17	18	19	20	V + XI −

TABLE XVIII.

GENERAL TABLE of Aberration, &c. continued.

S. − VI + O	1	2	3	4	5	6	7	8	9	10	11	12	13	14	15	16	17	18	19	20	
10° 0′	.174	.347	.521	.695	.868	1.042	1.216	1.389	1.563	1.737	1.910	2.084	2.257	2.431	2.605	2.778	2.952	3.126	3.299	3.473	20° 0′
2	.175	.349	.523	.698	.872	1.046	1.220	1.394	1.569	1.743	1.917	2.091	2.265	2.439	2.614	2.788	2.962	3.137	3.310	3.485	58
4	.176	.350	.525	.700	.874	1.050	1.224	1.399	1.574	1.749	1.924	2.098	2.273	2.447	2.623	2.797	2.972	3.147	3.321	3.497	56
6	.176	.351	.527	.702	.877	1.053	1.228	1.404	1.579	1.755	1.930	2.105	2.281	2.455	2.632	2.806	2.982	3.157	3.332	3.508	54
8	.177	.352	.529	.704	.880	1.056	1.232	1.408	1.584	1.760	1.936	2.112	2.288	2.463	2.640	2.815	2.992	3.167	3.343	3.519	52
10	.177	.353	.530	.706	.883	1.059	1.236	1.412	1.589	1.765	1.942	2.118	2.295	2.471	2.648	2.824	3.001	3.177	3.354	3.530	50
12	.178	.355	.532	.709	.886	1.063	1.240	1.417	1.594	1.771	1.949	2.125	2.303	2.479	2.657	2.833	3.011	3.188	3.365	3.542	48
14	.178	.356	.534	.711	.889	1.067	1.244	1.422	1.599	1.777	1.955	2.132	2.311	2.487	2.666	2.842	3.021	3.199	3.376	3.554	46
16	.179	.357	.536	.713	.892	1.070	1.248	1.427	1.604	1.783	1.961	2.139	2.318	2.495	2.675	2.851	3.031	3.209	3.387	3.565	44
18	.179	.358	.537	.715	895	1.073	1.252	1.431	1.609	1.789	1.967	2.146	2.325	2.503	2.683	2.860	3.040	3.219	3.398	3.578	42
20	.179	.359	.538	.717	.897	1.076	1.256	1.435	1.614	1.794	1.973	2.152	2.332	2.511	2.691	2.869	3.049	3.229	3.408	3.587	40
22	.180	.361	.540	.720	.900	1.080	1.260	1.440	1.620	1.800	1.980	2.159	2.340	2.519	2.700	2.879	3.059	3.240	3.419	3.599	38
24	.181	.362	.542	.723	.903	1.084	1.264	1.445	1.625	1.806	1.987	2.166	2.348	2.527	2.709	2.888	3.069	3.250	3.430	3.611	36
26	.181	.363	.544	.725	.906	1.087	1 268	1.450	1.630	1.812	1.993	2.173	2.355	2.535	2.717	2.897	3.079	3.260	3.441	3.623	34
28	.182	.364	.546	.727	.909	1.090	1.272	1.454	1.635	1.817	1.999	2.180	2.362	2.543	2.725	2.906	3.089	3.270	3.452	3.634	32
30	.182	.365	.547	.729	.911	1.093	1.276	1.458	1.640	1.822	2.005	2.187	2.369	2.551	2.733	2.915	3.098	3.280	3.462	3.645	30
32	.183	.366	.549	.732	.914	1.097	1.280	1.463	1.646	1.828	2.012	2.194	2.377	2.559	2.742	2.925	3.108	3.291	3.473	3.657	28
34	.184	.367	.551	.734	.917	1.101	1.284	1.468	1.651	1.834	2.019	2.201	2.385	2.567	2.751	2.934	3.118	3.302	3.484	3.669	26
36	.184	.368	.553	.736	.920	1.105	1.288	1.473	1.656	1.840	2.024	2.208	2.392	2.575	2.760	2.943	3.128	3.312	3.495	3.680	24
38	.185	.369	.554	.738	.923	1.108	1.292	1.477	1.661	1.846	2.030	2.215	2.399	2.583	2.768	2.952	3.138	3.322	3.506	3.691	22
40	.185	.370	.555	.740	.926	1.111	1.296	1.481	1.666	1.851	2.036	2.221	2.406	2.591	2.776	2.961	3.147	3.332	3.517	3.702	20
42	.186	.372	.557	.743	.929	1.115	1.300	1.486	1.672	1.857	2.043	2 228	2.414	2.599	2.785	2.971	3.157	3.343	3.528	3.714	18
44	.187	.373	.559	.746	.932	1.119	1.304	1.491	1.677	1.863	2.050	2.235	2.422	2.607	2.794	2.980	3.167	3.353	3.539	3.726	16
46	.187	.374	.561	.748	.935	1.122	1.308	1.496	1.682	1.869	2.056	2.242	2.429	2.615	2.803	2.989	3.177	3.363	3.550	3.737	14
48	.188	.375	.563	.750	.938	1.125	1.312	1.500	1.687	1.875	2.062	2.249	2.436	2.623	2.811	2.998	3.186	3.373	3.561	3.748	12
50	.188	.376	.564	.752	.940	1.128	1.316	1.504	1.692	1.880	2.068	2 255	2 443	2.631	2.819	3.007	3.195	3.383	3.571	3.759	10
52	.189	.378	.566	.755	.943	1.132	1.320	1.509	1.697	1.886	2.075	2.262	2.451	2.639	2.828	3.017	3.205	3.394	3.582	3.771	8
54	.190	.379	.568	.757	.946	1.136	1.324	1.514	1.702	1.892	2.081	2.269	2.459	2.647	2.837	3.026	3.215	3.405	3.593	3.783	6
56	.190	.380	.570	.759	.949	1.139	1.328	1.518	1.707	1.898	2.087	2.276	2.466	2.655	2.846	3.035	3.225	3.415	3.604	3.796	4
58	.191	.381	.571	.761	.952	1.142	1.332	1.522	1.712	1.903	2.093	2.283	2.473	2.663	2.854	3.044	3.235	3.425	3.615	3.806	2
11° 0′	.191	.382	.572	.763	.954	1.145	1.336	1.526	1.717	1.908	2.099	2.290	2.480	2.671	2.862	3.053	3.244	3.435	3.625	3.816	19° 0′
2	.192	.383	.574	.766	.957	1.149	1.340	1.531	1.723	1.914	2.106	2.297	2.488	2.679	2.871	3.062	3.254	3.446	3.636	3.828	58
4	.193	.384	.576	.769	.960	1.153	1.344	1.536	1.728	1.920	2.112	2.304	2.496	2.687	2.880	3.071	3.264	3.456	3.647	3.840	56
6	.193	.385	.578	.771	.963	1.156	1.348	1.541	1.733	1.926	2.118	2.311	2.504	2.695	2.889	3.080	3.274	3.466	3.658	3.851	54
8	.194	.386	.580	.773	.966	1.159	1.352	1.545	1.738	1.932	2.124	2.318	2*.511	2.703	2.898	3.089	3.283	3.476	3.669	3.862	52
10	.194	.387	.581	.775	.968	1.162	1.356	1.549	1.743	1.937	2.130	2.324	2.518	2.711	2.905	3.098	3.292	3.486	3.680	3.873	50
12	.195	.389	.583	.778	.971	1.166	1.360	1.554	1.749	1.943	2.137	2.331	2.526	2.719	2.914	3.108	3.302	3.497	3.691	3.885	48
14	.196	.390	.585	.780	.974	1.170	1 364	1.559	1.754	1.949	2.144	2.338	2.534	2.727	2.923	3.117	3.312	3.507	3.702	3.897	46
16	.196	.291	.587	.782	.977	1.173	1.368	1.564	1.759	1.955	2.150	2.345	2.541	2.735	2.932	3.126	3.322	3.517	3.713	3.908	44
18	.197	.392	.589	.784	.980	1.176	1.372	1.568	1.764	1.960	2.156	2.352	2.548	2.743	2.940	3.135	3.332	3.527	3.724	3.919	42
20	.197	.393	590	.786	.983	1.179	1.376	1.572	1.769	1.965	2.162	2.358	2.555	2.751	2.948	3.144	3.341	3.537	3.734	3.930	40
22	.198	.395	·592	.789	.986	1.183	1.380	1.577	1.774	1.971	2.169	2.365	2.563	2.759	2.957	3.153	3.351	3.548	3.745	3.942	38
24	.198	.396	594	.791	.989	1.187	1.384	1.582	1.779	1.977	2.175	2.372	2.571	2.767	2.966	3.162	3.361	3.559	3.756	3.954	36
26	.199	.397	.596	.793	.992	1.190	1.388	1.587	1.784	1.983	2.181	2.379	2.578	2.775	2.975	3.171	3.371	3.569	3.767	3.965	34
28	.199	.398	.597	.795	.995	1.193	1 392	1.591	1.789	1.989	2.187	2.386	2.585	2.783	2.983	3.180	3.380	3.579	3.778	3.976	32
30	.199	.399	.598	.797	.997	1.196	1.396	1.595	1.794	1.994	2.193	2.392	2.592	2.791	2.991	3.189	3.389	3.589	3.788	3.987	30
32	.200	.400	.600	.800	1 000	1.200	1.400	1.600	1.800	2.000	2.200	2.399	2.600	2 799	3.000	3.199	3.499	3.600	3.799	3.999	28
34	.201	.401	.602	.802	1 003	1.204	1.404	1.605	1.805	2.006	2.207	2.406	2.608	2.807	3.009	3.208	3.409	3.610	3.810	4.011	26
36	.201	.402	.604	.804	1.006	1.207	1.408	1.610	1.810	2.012	2.213	2.413	2.615	2.815	3.017	3.217	3.419	3.620	3.821	4.022	24
38	.202	.403	.606	.806	1.009	1.210	1.412	1.614	1.815	2.017	2.219	2.420	2.622	2.823	3.025	3.226	3.428	3.630	3.832	4.033	22
40	202	.404	.607	.808	1 011	1.213	1.416	1.618	1.820	2.022	2.225	2.427	2.629	2.831	3.033	3.235	3.438	3.640	3.842	4.044	20
42	.203	.406	.609	.811	1.014	1.217	1.420	1.623	1.825	2.028	2.232	2.434	2.637	2.839	3.042	3.245	3.448	3.651	3.853	4.056	18
44	.204	.407	.611	.814	1.017	1.221	1.424	1.628	1.831	2.034	2.238	2.441	2.645	2.847	3.051	3.254	3.458	3.661	3.864	4.068	16
46	.204	.408	.613	.816	1.020	1.224	1.428	1.632	1.836	2.040	2.244	2.448	2.652	2.855	3.050	3.263	3.468	3.671	3.875	4.079	14
48	.205	.409	.614	.818	1.023	1.227	1.432	1.636	1.841	2.046	2.250	2.455	2.659	2.863	3.068	3.272	3.477	3.681	3.886	4.090	12
50	.205	.410	.615	.820	1.025	1.230	1 436	1.640	1.846	2.051	2.256	2.461	2.666	2.871	3.076	3.281	3.486	3.691	3.896	4.101	10
52	.206	.412	.617	.823	1.028	1.234	1.440	1.645	1.851	2.057	2.263	2.468	2.674	2.879	3.085	3.290	3.496	3.702	3.907	4.113	8
54	.207	.413	.619	.826	1.031	1.238	1.444	1.650	1.856	2.063	2.269	2.475	2.682	2.887	3.094	3.299	3.506	3.712	3.918	4.125	6
56	.207	.414	.621	.828	1.034	1.242	1.448	1.655	1.861	2.069	2.275	2.482	2.689	2.895	3.103	3.308	3.516	3.723	3.929	4.137	4
58	.208	.415	.623	.830	1.037	1.245	1.452	1.659	1.866	2.074	2.281	2.489	2.696	2.903	3.111	3.317	3.525	3.733	3.940	4.148	2
12° 0′	.208	.416	.624	.832	1.040	1.248	1.456	1.664	1.871	2.079	2.287	2.495	2.703	2.911	3.119	3.326	3.534	3.743	3.950	4.159	18° 0
	1	2	3	4	5	6	7	8	9	10	11	12	13	14	15	16	17	18	19	20	V + XI −

TABLE XVIII.

GENERAL TABLE of Aberration, &c. continued.

S. − VI + O	1	2	3	4	5	6	7	8	9	10	11	12	13	14	15	16	17	18	19	20	
12° 0′	.208	.416	.624	.832	1.040	1.248	1.455	1.663	1.871	2.079	2.287	2.495	2.703	2.911	3.119	3.326	3.534	3.742	3.950	4.158	18° 0′
2	.209	.418	.626	.835	1.043	1.252	1.459	1.668	1.877	2.085	2.294	2.502	2.711	2.919	3.128	3.336	3.544	3.753	3.961	4.170	58
4	.209	.419	.628	.837	1.046	1.256	1.463	1.673	1.882	2.091	2.300	2.509	2.719	2.927	3.137	3.345	3.55-	3.764	3.972	4.182	56
6	.210	.420	.630	.839	1.049	1.259	1.467	1.678	1.887	2.097	2.306	2.516	2.726	2.935	3.145	3.354	3.56-	3.774	3.983	4.193	54
8	.211	.421	.631	.841	1.052	1.262	1.471	1.682	1.892	2.103	2.312	2.523	2.733	2.943	3.153	3.363	3.574	3.784	3.994	4.204	52
10	.211	.422	.632	.843	1.054	1.265	1.475	1.686	1.897	2.108	2.318	2.529	2.740	2.951	3.161	3.372	3.583	3.794	4.004	4.215	50
12	.212	.423	.634	.846	1.057	1.269	1.479	1.691	1.902	2.114	2.325	2.536	2.748	2.959	3.170	3.382	3.593	3.805	4.015	4.227	48
14	.212	.424	.636	.848	1.060	1.273	1.483	1.696	1.907	2.120	2.332	2.543	2.756	2.967	3.179	3.391	3.603	3.815	4.026	4.239	46
16	.213	.425	.638	.850	1.063	1.276	1.487	1.701	1.912	2.126	2.338	2.550	2.763	2.975	3.188	3.400	3.613	3.825	4.037	4.250	44
18	.214	.426	.640	.852	1.066	1.279	1.491	1.705	1.917	2.131	2.344	2.557	2.770	2.983	3.196	3.409	3.62[illegible]	3.835	4.048	4.261	42
20	.214	.427	.641	.854	1.068	1.282	1.495	1.709	1.922	2.136	2.350	2.563	2.777	2.990	3.204	3.418	3.63[illegible]	3.845	4.058	4.272	40
22	.215	.429	.643	.857	1.071	1.286	1.499	1.714	1.928	2.142	2.357	2.570	2.785	2.998	3.213	3.427	3.641	3.856	4.069	4.284	38
24	.215	.430	.645	.860	1.074	1.290	1.503	1.719	1.933	2.148	2.363	2.577	2.792	3.006	3.222	3.436	3.651	3.866	4.080	4.296	36
26	.216	.431	.647	.862	1.077	1.293	1.507	1.724	1.938	2.154	2.369	2.584	2.799	3.014	3.231	3.445	3.661	3.876	4.091	4.307	34
28	.216	.432	.648	.864	1.080	1.296	1.511	1.728	1.943	2.159	2.375	2.591	2.806	3.022	3.239	3.454	3.671	3.886	4.102	4.318	32
30	.216	.433	.649	.866	1.082	1.299	1.515	1.732	1.948	2.164	2.381	2.597	2.813	3.030	3.247	3.463	3.680	3.896	4.112	4.329	30
32	.217	.435	.651	.869	1.085	1.303	1.519	1.737	1.954	2.170	2.388	2.604	2.821	3.038	3.256	3.472	3.690	3.907	4.123	4.341	28
34	.217	.436	.653	.871	1.088	1.307	1.523	1.742	1.959	2.176	2.394	2.611	2.829	3.046	3.265	3.481	3.700	3.917	4.134	4.353	26
36	.218	.437	.655	.873	1.091	1.310	1.527	1.746	1.964	2.182	2.400	2.618	2.837	3.054	3.273	3.490	3.710	3.927	4.145	4.364	24
38	.219	.438	.657	.875	1.094	1.313	1.531	1.750	1.969	2.188	2.406	2.625	2.844	3.062	3.281	3.499	3.719	3.937	4.156	4.375	22
40	.219	.439	.658	.877	1.096	1.316	1.535	1.754	1.974	2.193	2.412	2.631	[illegible].851	3.070	3.289	3.508	3.728	3.947	4.166	4.386	20
42	.220	.440	.660	.879	1.099	1.320	1.539	1.759	1.979	2.199	2.419	2.638	2.859	3.078	3.298	3.518	3.738	3.958	4.177	4.398	18
44	.220	.441	.662	.882	1.102	1.324	1.543	1.764	1.984	2.205	2.425	2.645	2.867	3.086	3.307	3.527	3.748	3.968	4.188	4.409	16
46	.221	.442	.664	.884	1.105	1.327	1.547	1.769	1.989	2.211	2.431	2.652	2.874	3.094	3.316	3.536	3.758	3.978	4.199	4.420	14
48	.222	.443	.665	.886	1.108	1.330	1.551	1.773	1.994	2.216	2.437	2.659	2.881	3.102	3.324	3.545	3.767	3.988	4.210	4.431	12
50	.222	.444	.666	.888	1.110	1.333	1.555	1.777	1.999	2.221	2.443	2.665	2.888	3.110	3.332	3.554	3.776	3.998	4.220	4.442	10
52	.223	.446	.668	.891	1.113	1.337	1.559	1.782	2.005	2.227	2.450	2.672	2.896	3.118	3.341	3.563	3.786	4.009	4.231	4.454	8
54	.223	.447	.670	.894	1.116	1.341	1.563	1.787	2.010	2.233	2.457	2.679	2.903	3.126	3.350	3.572	3.796	4.019	4.242	4.466	6
56	.224	.448	.672	.896	1.119	1.344	1.567	1.792	2.015	2.239	2.463	2.686	2.910	3.134	3.358	3.581	3.806	4.029	4.253	4.477	4
58	.225	.449	.674	.898	1.122	1.347	1.571	1.796	2.020	2.245	2.469	2.693	2.917	3.142	3.366	3.590	3.815	4.039	4.264	4.488	2
13° 0′	.225	.450	.675	.900	1.125	1.350	1.575	1.800	2.025	2.250	2.475	2.700	2.924	3.149	3.374	3.599	3.824	4.049	4.274	4.499	17° 0′
2	.226	.452	.677	.903	1.128	1.354	1.579	1.805	2.030	2.256	2.482	2.707	2.932	3.157	3.383	3.608	3.834	4.060	4.285	4.511	58
4	.226	.453	.679	.905	1.131	1.358	1.583	1.810	2.035	2.262	2.488	2.714	2.940	3.165	3.392	3.617	3.844	4.070	4.296	4.523	56
6	.227	.454	.681	.907	1.134	1.361	1.587	1.814	2.040	2.268	2.494	2.721	2.947	3.173	3.401	3.626	3.854	4.080	4.307	4.534	54
8	.228	.455	.682	.909	1.137	1.364	1.591	1.818	2.045	2.273	2.500	2.727	2.954	3.181	3.409	3.635	3.863	4.090	4.318	4.545	52
10	.228	.456	.683	.911	1.139	1.367	1.594	1.822	2.050	2.278	2.506	2.733	2.961	3.189	3.417	3.644	3.872	4.100	4.328	4.556	50
12	.229	.457	.685	.914	1.142	1.371	1.598	1.827	2.056	2.284	2.513	2.740	2.969	3.197	3.426	3.654	3.882	4.111	4.339	4.568	48
14	.229	.458	.687	.916	1.145	1.375	1.602	1.832	2.031	2.290	2.519	2.747	2.977	3.205	3.435	3.663	3.892	4.121	4.350	4.579	46
16	.230	.459	.689	.918	1.148	1.378	1.606	1.837	2.066	2.296	2.525	2.754	2.984	3.213	3.443	3.672	3.902	4.131	4.361	4.590	44
18	.231	.460	.691	.920	1.151	1.381	1.610	1.841	2.071	2.301	2.531	2.761	2.991	3.221	3.451	3.681	3.911	4.141	4.372	4.601	42
20	.231	.461	.692	.922	1.153	1.384	1.614	1.845	2.076	2.306	2.537	2.767	2.998	3.229	3.459	3.690	3.920	4.151	4.382	4.612	40
22	.232	.463	.694	.925	1.156	1.388	1.618	1.850	2.081	2.312	2.544	2.774	3.006	3.237	3.468	3.699	3.930	4.162	4.393	4.624	38
24	.232	.464	.696	.928	1.159	1.392	1.622	1.855	2.086	2.318	2.550	2.781	3.014	3.24	3.477	3.708	3.940	4.172	4.404	4.636	36
26	.233	.465	.698	.930	1.162	1.395	1.626	1.860	2.091	2.324	2.556	2.788	3.021	3.253	3.486	3.717	3.950	4.182	4.415	4.647	34
28	.233	.466	.699	.932	1.165	1.398	1.630	1.864	2.096	2.330	2.562	2.795	3.028	3.261	3.494	3.726	3.959	4.192	4.426	4.658	32
30	.233	.467	.700	.934	1.167	1.401	1.634	1.868	2.101	2.335	2.568	2.801	3.035	3.268	3.502	3.735	3.968	4.202	4.436	4.669	30
32	.234	.469	.702	.937	1.170	1.405	1.638	1.873	2.106	2.341	2.575	2.808	3.043	3.276	3.511	3.744	3.978	4.213	4.447	4.681	28
34	.234	.470	.704	.939	1.173	1.409	1.642	1.878	2.111	2.347	2.581	2.815	3.051	3.284	3.520	3.753	3.988	4.223	4.458	4.692	26
36	.235	.471	.706	.941	1.176	1.412	1.646	1.882	2.116	2.353	2.587	2.822	3.058	3.292	3.528	3.762	3.998	4.233	4.469	4.703	24
38	.236	.472	.708	.943	1.179	1.415	1.650	1.886	2.121	2.358	2.593	2.829	3.065	3.300	3.536	3.771	4.008	4.243	4.479	4.714	22
40	.236	.473	.709	.945	1.181	1.418	1.654	1.890	2.126	2.363	2.599	2.835	3.072	3.308	3.544	3.780	4.017	4.253	4.489	4.725	20
42	.237	.474	.711	.948	1.184	1.422	1.658	1.895	2.132	2.369	2.606	2.842	3.080	3.316	3.553	3.790	4.027	4.264	4.500	4.737	18
44	.237	.475	.713	.950	1.187	1.426	1.662	1.900	2.137	2.375	2.612	2.849	3.087	3.324	3.562	3.799	4.037	4.274	4.511	4.749	16
46	.238	.476	.715	.952	1.190	1.429	1.666	1.905	2.142	2.381	2.618	2.856	3.094	3.332	3.571	3.808	4.047	4.284	4.522	4.760	14
48	.239	.477	.716	.954	1.193	1.432	1.670	1.909	2.147	2.386	2.624	2.863	3.101	3.340	3.579	3.817	4.056	4.294	4.533	4.771	12
50	.239	.478	.717	.956	1.196	1.435	1.674	1.913	2.152	2.391	2.630	2.869	3.108	3.347	3.587	3.826	4.065	4.304	4.543	4.782	10
52	.240	.480	.719	.959	1.199	1.439	1.678	1.918	2.157	2.397	2.637	2.876	3.116	3.355	3.596	3.835	4.075	4.314	4.554	4.794	8
54	.240	.481	.721	.962	1.202	1.443	1.682	1.923	2.162	2.403	2.543	2.883	3.124	3.363	3.605	3.844	4.085	4.324	4.565	4.805	6
56	.241	.482	.723	.964	1.205	1.446	1.686	1.927	2.167	2.409	2.649	2.890	3.131	3.371	3.613	3.853	4.095	4.334	4.576	4.816	4
58	.242	.483	.725	.966	1.208	1.449	1.690	1.931	2.172	2.414	2.655	2.897	3.138	3.379	3.621	3.862	4.104	4.344	4.586	4.827	2
14° 0′	.242	.484	.726	.968	1.210	1.452	1.693	1.935	2.177	2.419	2.661	2.903	3.145	3.387	3.629	3.871	4.113	4.354	4.596	4.838	16° 0′
	1	2	3	4	5	6	7	8	9	10	11	12	13	14	15	16	17	18	19	20	V + XI −

TABLE XVIII.

GENERAL TABLE of Aberration, &c. continued.

S. – VI + O	1	2	3	4	5	6	7	8	9	10	11	12	13	14	15	16	17	18	19	20	
14° 0'	.242	.484	.726	.968	1.210	1.452	1.693	1.935	2.177	2.419	2.661	2.903	3.145	3.337	3.629	3.871	4.113	4.354	4.596	4.838	16° 0'
2	.243	.486	.728	.971	1.213	1.456	1.697	1.940	2.183	2.425	2.668	2.910	3.153	3.395	3.638	3.880	4.123	4.365	4.607	4.850	58
4	.243	.487	.730	.973	1.216	1.459	1.701	1.945	2.188	2.431	2.674	2.917	3.161	3.403	3.647	3.889	4.133	4.375	4.618	4.862	56
6	.244	.488	.732	.975	1.219	1.462	1.705	1.950	2.193	2.437	2.680	2.924	3.168	3.411	3.655	3.898	4.142	4.085	4.629	4.873	54
8	.245	.489	.733	.977	1.222	1.465	1.709	1.954	2.198	2.442	2.686	2.931	3.175	3.419	3.663	3.907	4.151	4.395	4.640	4.884	52
10	.245	.490	.734	.979	1.224	1.468	1.713	1.958	2.203	2.447	2.692	2.937	3.182	3.426	3.671	3.916	4.160	4.405	4.650	4.895	50
12	.246	.491	.736	.982	1.227	1.472	1.717	1.963	2.208	2.453	2.699	2.944	3.190	3.434	3.680	3.925	4.170	4.416	4.661	4.907	48
14	.246	.492	.738	.984	1.230	1.476	1.721	1.968	2.213	2.459	2.705	2.951	3.197	3.442	3.689	3.934	4.180	4.426	4.672	4.918	46
16	.247	.493	.740	.986	1.233	1.479	1.725	1.972	2.218	2.465	2.711	2.958	3.204	3.450	3.697	3.943	4.190	4.436	4.683	4.929	44
18	.248	.494	.742	.988	1.236	1.482	1.729	1.976	2.223	2.471	2.717	2.965	3.211	3.458	3.705	3.952	4.199	4.446	4.694	4.940	42
20	.248	.495	.743	.990	1.238	1.485	1.733	1.980	2.228	2.476	2.723	2.971	3.218	3.466	3.713	3.961	4.208	4.456	4.704	4.951	40
22	.249	.497	.745	.993	1.241	1.489	1.737	1.985	2.234	2.482	2.730	2.978	3.226	3.474	3.722	3.970	4.218	4.467	4.715	4.963	38
24	.249	.498	.747	.996	1.244	1.493	1.741	1.990	2.239	2.488	2.736	2.985	3.234	3.482	3.731	3.979	4.228	4.477	4.726	4.975	36
26	.250	.499	.749	.998	1.247	1.496	1.745	1.995	2.244	2.494	2.742	2.992	3.241	3.490	3.740	3.988	4.238	4.487	4.737	4.986	34
28	.250	.500	.750	1.000	1.250	1.499	1.749	1.999	2.249	2.499	2.748	2.999	3.248	3.498	3.748	3.997	4.248	4.497	4.747	4.997	32
30	.250	.501	.751	1.002	1.252	1.502	1.753	2.003	2.254	2.504	2.754	3.005	3.255	3.505	3.756	4.006	4.257	4.507	4.757	5.008	30
32	.251	.502	.753	1.005	1.255	1.506	1.757	2.008	2.259	2.510	2.761	3.012	3.262	3.513	3.765	4.015	4.267	4.518	4.768	5.020	28
34	.251	.503	.755	1.007	1.258	1.510	1.761	2.013	2.264	2.516	2.767	3.019	3.270	3.521	3.774	4.024	4.277	4.528	4.779	5.031	26
36	.252	.504	.757	1.009	1.261	1.513	1.765	2.018	2.269	2.522	2.773	3.026	3.277	3.529	3.782	4.033	4.287	4.538	4.790	5.042	24
38	.253	.505	.759	1.011	1.264	1.516	1.769	2.022	2.274	2.527	2.779	3.032	3.284	3.537	3.790	4.042	4.296	4.548	4.801	5.053	22
40	.253	.506	.760	1.013	1.266	1.519	1.772	2.026	2.279	2.532	2.785	3.038	3.292	3.545	3.798	4.051	4.305	4.558	4.811	5.064	20
42	.254	.508	.762	1.016	1.269	1.523	1.776	2.031	2.284	2.538	2.792	3.045	3.300	3.553	3.807	4.060	4.315	4.568	4.822	5.076	18
44	.254	.509	.764	1.018	1.272	1.527	1.780	2.036	2.289	2.544	2.798	3.052	3.307	3.561	3.816	4.069	4.325	4.578	4.833	5.087	16
46	.255	.510	.766	1.020	1.275	1.529	1.784	2.040	2.294	2.550	2.804	3.059	3.314	3.569	3.824	4.078	4.334	4.588	4.844	5.098	14
48	.256	.511	.767	1.022	1.278	1.533	1.788	2.044	2.299	2.555	2.810	3.066	3.321	3.577	3.832	4.087	4.343	4.598	4.854	5.109	12
50	.256	.512	.768	1.024	1.280	1.536	1.792	2.048	2.304	2.560	2.816	3.072	3.328	3.584	3.840	4.096	4.352	4.608	4.864	5.120	10
52	.257	.514	.770	1.027	1.283	1.540	1.796	2.053	2.309	2.566	2.823	3.079	3.336	3.592	3.849	4.105	4.362	4.619	4.875	5.132	8
54	.257	.515	.772	1.029	1.286	1.544	1.800	2.058	2.314	2.572	2.829	3.086	3.344	3.600	3.858	4.114	4.372	4.629	4.886	5.143	6
56	.258	.516	.774	1.031	1.289	1.547	1.804	2.063	2.319	2.578	2.835	3.093	3.351	3.608	3.866	4.123	4.382	4.639	4.897	5.154	4
58	.259	.517	.776	1.033	1.292	1.550	1.808	2.067	2.324	2.583	2.841	3.100	3.358	3.616	3.874	4.132	4.392	4.649	4.908	5.165	2
15° 0'	.259	.518	.777	1.035	1.294	1.553	1.812	2.071	2.329	2.588	2.847	3.106	3.365	3.623	3.882	4.141	4.400	4.659	4.918	5.176	15° 0'
2	.260	.519	.779	1.038	1.297	1.557	1.816	2.076	2.335	2.594	2.854	3.113	3.373	3.631	3.891	4.150	4.410	4.669	4.929	5.188	58
4	.260	.520	.781	1.040	1.300	1.561	1.820	2.081	2.340	2.600	2.860	3.120	3.380	3.639	3.900	4.159	4.420	4.679	4.940	5.200	56
6	.261	.521	.783	1.042	1.303	1.564	1.824	2.085	2.345	2.606	2.866	3.127	3.387	3.647	3.909	4.168	4.430	4.689	4.951	5.211	54
8	.262	.522	.784	1.044	1.306	1.567	1.828	2.089	2.350	2.611	2.872	3.134	3.394	3.655	3.917	4.177	4.439	4.699	4.961	5.222	52
10	.262	.523	.785	1.046	1.308	1.570	1.831	2.093	2.355	2.616	2.878	3.140	3.401	3.663	3.925	4.186	4.448	4.709	4.971	5.233	50
12	.263	.525	.787	1.049	1.311	1.574	1.835	2.098	2.360	2.622	2.885	3.147	3.409	3.671	3.934	4.195	4.458	4.720	4.982	5.245	48
14	.263	.526	.789	1.052	1.314	1.578	1.839	2.103	2.365	2.628	2.891	3.154	3.417	3.679	3.943	4.204	4.468	4.730	4.993	5.256	46
16	.264	.527	.791	1.054	1.317	1.581	1.843	2.107	2.370	2.634	2.897	3.161	3.424	3.687	3.951	4.213	4.477	4.740	5.004	5.267	44
18	.264	.528	.792	1.056	1.320	1.584	1.847	2.111	2.375	2.639	2.903	3.167	3.431	3.695	3.959	4.222	4.486	4.750	5.014	5.278	42
20	.264	.529	.793	1.058	1.322	1.587	1.851	2.115	2.380	2.644	2.909	3.173	3.438	3.702	3.967	4.231	4.495	4.760	5.024	5.289	40
22	.265	.531	.795	1.061	1.325	1.591	1.855	2.120	2.385	2.650	2.916	3.180	3.446	3.710	3.976	4.240	4.505	4.770	5.035	5.301	38
24	.265	.532	.797	1.063	1.328	1.594	1.859	2.125	2.390	2.656	2.922	3.187	3.453	3.718	3.985	4.249	4.515	4.780	5.046	5.312	36
26	.266	.533	.799	1.065	1.331	1.597	1.863	2.130	2.395	2.662	2.928	3.194	3.460	3.726	3.993	4.258	4.525	4.790	5.057	5.323	34
28	.267	.534	.800	1.067	1.334	1.600	1.867	2.134	2.400	2.667	2.934	3.201	3.467	3.734	4.001	4.267	4.534	4.800	5.068	5.334	32
30	.267	.535	.801	1.069	1.336	1.603	1.871	2.138	2.405	2.672	2.940	3.207	3.474	3.741	4.009	4.276	4.543	4.810	5.078	5.345	30
32	.268	.536	.803	1.072	1.339	1.607	1.875	2.143	2.410	2.678	2.946	3.214	3.482	3.749	4.018	4.285	4.553	4.821	5.089	5.357	28
34	.268	.537	.805	1.074	1.342	1.611	1.879	2.148	2.415	2.684	2.952	3.221	3.489	3.757	4.027	4.294	4.563	4.831	5.100	5.368	26
36	.269	.538	.807	1.076	1.345	1.614	1.883	2.152	2.420	2.690	2.958	3.228	3.496	3.765	4.035	4.303	4.573	4.841	5.111	5.379	24
38	.270	.539	.809	1.078	1.348	1.617	1.887	2.156	2.425	2.695	2.964	3.234	3.503	3.773	4.043	4.312	4.582	4.851	5.121	5.380	22
40	.270	.540	.810	1.080	1.350	1.620	1.890	2.160	2.430	2.700	2.970	3.240	3.510	3.780	4.051	4.321	4.591	4.861	5.131	5.401	20
42	.271	.542	.812	1.083	1.353	1.624	1.894	2.165	2.436	2.706	2.977	3.247	3.518	3.788	4.060	4.330	4.601	4.871	5.142	5.413	18
44	.271	.543	.814	1.085	1.356	1.628	1.898	2.170	2.441	2.712	2.983	3.254	3.526	3.796	4.069	4.339	4.611	4.881	5.153	5.424	16
46	.272	.544	.816	1.087	1.359	1.631	1.902	2.175	2.446	2.718	2.989	3.261	3.533	3.804	4.077	4.348	4.620	4.891	5.164	5.435	14
48	.273	.545	.818	1.089	1.362	1.634	1.906	2.179	2.451	2.723	2.995	3.268	3.540	3.812	4.085	4.357	4.629	4.901	5.174	5.446	12
50	.273	.546	.819	1.091	1.364	1.637	1.910	2.183	2.456	2.728	3.001	3.274	3.547	3.820	4.093	4.366	4.638	4.911	5.184	5.457	10
52	.274	.547	.821	1.094	1.367	1.641	1.914	2.188	2.461	2.734	3.008	3.281	3.555	3.828	4.102	4.375	4.648	4.922	5.195	5.469	8
54	.274	.548	.823	1.097	1.370	1.645	1.918	2.193	2.466	2.740	3.014	3.288	3.562	3.836	4.111	4.384	4.658	4.932	5.206	5.480	6
56	.275	.549	.825	1.099	1.373	1.648	1.922	2.197	2.471	2.746	3.020	3.295	3.569	3.844	4.119	4.393	4.668	4.942	5.217	5.491	4
58	.276	.550	.826	1.101	1.376	1.651	1.926	2.201	2.476	2.751	3.026	3.302	3.576	3.852	4.127	4.402	4.677	4.952	5.227	5.502	2
16° 0	.276	.551	.827	1.103	1.378	1.654	1.929	2.205	2.481	2.756	3.032	3.308	3.583	3.859	4.135	4.410	4.686	4.962	5.237	5.513	14° 0'
	1	2	3	4	5	6	7	8	9	10	11	12	13	14	15	16	17	18	19	20	V + XI –

TABLE XVIII.

GENERAL TABLE of Aberration, &c. continued.

S. − VI + O	1	2	3	4	5	6	7	8	9	10	11	12	13	14	15	16	17	18	19	20	
16° 0′	.276	.551	.827	1.103	1.378	1.654	1.929	2.205	2.481	2.756	3.032	3.308	3.583	3.859	4.135	4.410	4.686	4.962	5.237	5.513	14° 0′
2	.277	.553	.829	1.105	1.381	1.658	1.933	2.210	2.486	2.762	3.039	3.315	3.591	3.867	4.144	4.419	4.696	4.972	5.248	5.525	58
4	.277	.554	.831	1.107	1.384	1.661	1.937	2.215	2.491	2.768	3.045	3.322	3.599	3.875	4.153	4.428	4.706	4.982	5.259	5.536	56
6	.278	.555	.833	1.110	1.387	1.665	1.941	2.220	2.496	2.774	3.051	3.329	3.606	3.883	4.161	4.437	4.716	4.992	5.270	5.547	54
8	.278	.556	.834	1.112	1.390	1.668	1.945	2.224	2.501	2.779	3.057	3.335	3.613	3.891	4.169	4.446	4.725	5.002	5.280	5.558	52
10	.278	.557	.835	1.114	1.392	1.671	1.949	2.228	2.506	2.784	3.063	3.341	3.620	3.898	4.177	4.455	4.734	5.012	5.290	5.569	50
12	.279	.558	.837	1.117	1.395	1.675	1.953	2.233	2.511	2.790	3.070	3.348	3.628	3.906	4.186	4.464	4.744	5.022	5.301	5.581	48
14	.279	.559	.839	1.119	1.398	1.678	1.957	2.238	2.516	2.796	3.076	3.355	3.635	3.914	4.195	4.473	4.754	5.032	5.312	5.592	46
16	.280	.560	.841	1.121	1.401	1.681	1.961	2.242	2.521	2.802	3.082	3.362	3.642	3.922	4.203	4.482	4.763	5.042	5.323	5.603	44
18	.281	.561	.843	1.123	1.404	1.684	1.965	2.246	2.526	2.807	3.088	3.369	3.649	3.930	4.211	4.491	4.772	5.052	5.333	5.614	42
20	.281	.563	.844	1.125	1.406	1.687	1.969	2.250	2.531	2.812	3.094	3.375	3.656	3.937	4.219	4.500	4.781	5.062	5.343	5.625	40
22	.282	.564	.846	1.128	1.409	1.691	1.973	2.255	2.536	2.818	3.100	3.382	3.664	3.945	4.228	4.509	4.791	5.072	5.354	5.636	38
24	.282	.565	.848	1.130	1.412	1.695	1.977	2.260	2.541	2.824	3.106	3.389	3.671	3.953	4.236	4.518	4.801	5.082	5.365	5.647	36
26	.283	.566	.850	1.132	1.415	1.698	1.981	2.264	2.546	2.830	3.112	3.396	3.678	3.961	4.244	4.527	4.810	5.092	5.376	5.658	34
28	.284	.567	.851	1.134	1.418	1.701	1.985	2.268	2.551	2.835	3.118	3.402	3.685	3.969	4.252	4.536	4.819	5.102	5.386	5.669	32
30	.284	.568	.852	1.136	1.420	1.704	1.988	2.272	2.556	2.840	3.124	3.408	3.692	3.976	4.260	4.544	4.828	5.112	5.396	5.680	30
32	.285	.570	.854	1.139	1.423	1.708	1.992	2.277	2.561	2.846	3.131	3.415	3.700	3.984	4.269	4.553	4.838	5.122	5.407	5.692	28
34	.285	.571	.856	1.141	1.426	1.712	1.996	2.282	2.566	2.852	3.137	3.422	3.707	3.992	4.278	4.562	4.848	5.132	5.418	5.703	26
36	.286	.572	.858	1.143	1.429	1.715	2.000	2.286	2.571	2.858	3.143	3.429	3.714	4.000	4.286	4.571	4.858	5.142	5.429	5.714	24
38	.287	.573	.859	1.145	1.432	1.718	2.004	2.290	2.576	2.863	3.149	3.436	3.721	4.008	4.294	4.580	4.867	5.152	5.439	5.725	22
40	.287	.574	.860	1.147	1.434	1.721	2.008	2.294	2.581	2.868	3.155	3.442	3.728	4.015	4.302	4.589	4.876	5.162	5.449	5.736	20
42	.288	.575	.862	1.150	1.437	1.725	2.012	2.299	2.586	2.874	3.162	3.449	3.736	4.023	4.311	4.598	4.886	5.173	5.460	5.748	18
44	.288	.576	.864	1.152	1.440	1.729	2.016	2.304	2.591	2.880	3.168	3.456	3.744	4.031	4.320	4.607	4.895	5.183	5.471	5.759	16
46	.289	.577	.866	1.154	1.443	1.732	2.020	2.309	2.596	2.886	3.174	3.463	3.751	4.039	4.328	4.616	4.904	5.193	5.482	5.770	14
48	.290	.578	.867	1.156	1.446	1.735	2.024	2.313	2.601	2.891	3.180	3.469	3.758	4.047	4.336	4.625	4.913	5.203	5.492	5.781	12
50	.290	.579	.868	1.158	1.448	1.738	2.027	2.317	2.606	2.896	3.186	3.475	3.765	4.054	4.344	4.633	4.922	5.213	5.502	5.792	10
52	.291	.581	.870	1.161	1.451	1.742	2.031	2.322	2.611	2.902	3.192	3.482	3.773	4.062	4.353	4.642	4.932	5.223	5.513	5.803	8
54	.291	.582	.872	1.163	1.454	1.745	2.035	2.327	2.616	2.908	3.198	3.488	3.780	4.070	4.362	4.651	4.942	5.233	5.524	5.814	6
56	.292	.583	.874	1.165	1.457	1.748	2.039	2.331	2.621	2.914	3.204	3.495	3.787	4.078	4.370	4.660	4.952	5.243	5.535	5.825	4
58	.292	.584	.876	1.167	1.460	1.751	2.043	2.335	2.626	2.919	3.210	3.502	3.794	4.086	4.378	4.669	4.961	5.253	5.545	5.836	2
17° 0′	.292	.585	.877	1.169	1.462	1.754	2.047	2.339	2.631	2.924	3.216	3.509	3.801	4.093	4.386	4.678	4.970	5.263	5.555	5.847	13° 0′
2	.293	.586	.879	1.172	1.465	1.756	2.051	2.344	2.636	2.930	3.223	3.516	3.809	4.101	4.395	4.687	4.980	5.273	5.566	5.859	58
4	.293	.587	.881	1.175	1.468	1.762	2.055	2.349	2.641	2.936	3.229	3.523	3.816	4.109	4.403	4.696	4.990	5.283	5.577	5.870	56
6	.294	.588	.883	1.177	1.471	1.765	2.059	2.353	2.646	2.942	3.235	3.530	3.823	4.117	4.411	4.705	5.000	5.293	5.588	5.881	54
8	.295	.589	.885	1.179	1.474	1.768	2.063	2.357	2.651	2.947	3.241	3.536	3.830	4.125	4.419	4.714	5.009	5.303	5.598	5.892	52
10	.295	.590	.886	1.181	1.476	1.771	2.066	2.361	2.656	2.952	3.247	3.542	3.837	4.132	4.427	4.722	5.018	5.313	5.608	5.903	50
12	.296	.592	.888	1.184	1.479	1.775	2.070	2.366	2.662	2.958	3.253	3.549	3.845	4.140	4.436	4.731	5.028	5.323	5.619	5.915	48
14	.296	.593	.890	1.186	1.482	1.779	2.074	2.371	2.667	2.964	3.259	3.556	3.852	4.148	4.445	4.740	5.038	5.333	5.630	5.926	46
16	.297	.594	.892	1.188	1.485	1.782	2.078	2.376	2.672	2.969	3.265	3.563	3.859	4.156	4.453	4.749	5.047	5.343	5.641	5.937	44
18	.298	.595	.893	1.190	1.488	1.785	2.082	2.380	2.677	2.974	3.271	3.569	3.866	4.164	4.461	4.758	5.056	5.353	5.651	5.948	42
20	.298	.596	.894	1.192	1.490	1.788	2.086	2.384	2.682	2.979	3.277	3.575	3.873	4.171	4.469	4.767	5.065	5.363	5.661	5.959	40
22	.299	.597	.896	1.195	1.493	1.792	2.090	2.389	2.687	2.985	3.284	3.582	3.881	4.179	4.478	4.776	5.075	5.373	5.672	5.970	38
24	.299	.598	.898	1.197	1.496	1.795	2.094	2.394	2.692	2.991	3.290	3.589	3.889	4.187	4.487	4.785	5.085	5.383	5.683	5.981	36
26	.300	.599	.900	1.199	1.499	1.798	2.098	2.398	2.697	2.997	3.296	3.596	3.896	4.195	4.495	4.794	5.095	5.393	5.694	5.992	34
28	.301	.600	.901	1.201	1.502	1.801	2.102	2.402	2.702	3.002	3.302	3.603	3.903	4.203	4.503	4.803	5.104	5.403	5.704	6.003	32
30	.301	.601	.902	1.203	1.504	1.804	2.105	2.406	2.706	3.007	3.308	3.609	3.909	4.210	4.511	4.811	5.112	5.413	5.714	6.014	30
32	.302	.603	.904	1.206	1.507	1.808	2.109	2.411	2.711	3.013	3.314	3.616	3.917	4.218	4.520	4.820	5.122	5.423	5.725	6.026	28
34	.302	.604	.906	1.208	1.510	1.812	2.113	2.416	2.716	3.019	3.320	3.623	3.924	4.226	4.528	4.829	5.132	5.433	5.736	6.037	26
36	.303	.605	.908	1.210	1.513	1.815	2.117	2.420	2.721	3.025	3.326	3.630	3.931	4.234	4.536	4.838	5.141	5.443	5.746	6.048	24
38	.304	.606	.909	1.212	1.515	1.818	2.121	2.424	2.726	3.030	3.332	3.636	3.938	4.242	4.544	4.847	5.150	5.453	5.756	6.059	22
40	.304	.607	.910	1.214	1.517	1.821	2.124	2.428	2.731	3.035	3.338	3.642	3.945	4.249	4.552	4.856	5.159	5.463	5.766	6.070	20
42	.305	.609	.912	1.217	1.520	1.825	2.128	2.433	2.736	3.041	3.345	3.649	3.953	4.257	4.561	4.865	5.169	5.473	5.777	6.081	18
44	.305	.610	.014	1.219	1.523	1.828	2.132	2.438	2.741	3.047	3.351	3.656	3.960	4.265	4.570	4.874	5.179	5.483	5.788	6.092	16
46	.306	.611	.916	1.221	1.526	1.831	2.136	2.442	2.746	3.053	3.357	3.663	3.967	4.273	4.578	4.883	5.188	5.493	5.799	6.103	14
48	.306	.612	.918	1.223	1.529	1.834	2.140	2.446	2.751	3.058	3.363	3.669	3.974	4.280	4.586	4.892	5.197	5.503	5.809	6.114	12
50	.306	.613	.919	1.225	1.531	1.837	2.144	2.450	2.756	3.063	3.369	3.675	3.981	4.287	4.594	4.900	5.206	5.513	5.819	6.125	10
52	.307	.614	.921	1.228	1.534	1.841	2.148	2.455	2.761	3.069	3.375	3.682	3.989	4.295	4.603	4.909	5.216	5.523	5.830	6.136	8
54	.307	.615	.923	1.230	1.537	1.845	2.152	2.460	2.766	3.075	3.381	3.689	3.996	4.303	4.611	4.918	5.226	5.533	5.841	6.147	6
56	.308	.616	.925	1.232	1.540	1.848	2.156	2.464	2.771	3.080	3.387	3.696	4.003	4.311	4.619	4.927	5.235	5.543	5.851	6.158	4
58	.309	.617	.926	1.234	1.543	1.851	2.160	2.468	2.776	3.085	3.393	3.702	4.010	4.319	4.627	4.936	5.244	5.553	5.861	6.169	2
18° 0′	.309	.618	.927	1.236	1.545	1.854	2.163	2.472	2.781	3.090	3.399	3.708	4.017	4.326	4.635	4.944	5.253	5.562	5.871	6.180	12° 0′
	1	2	3	4	5	6	7	8	9	10	11	12	13	14	15	16	17	18	19	20	V + XI −

TABLE XVIII.

GENERAL TABLE of Aberration, &c. continued.

S. — VI + O	1	2	3	4	5	6	7	8	9	10	11	12	13	14	15	16	17	18	19	20	
18° 0′	.309	.618	.927	1.236	1.545	1.854	2.163	2.472	2.781	3.090	3.399	3.708	4.017	4.326	4.635	4.944	5.253	5.562	5.871	6.180	12° 0′
2	.310	.620	.929	1.239	1.548	1.858	2.167	2.477	2.786	3.096	3.406	3.715	4.025	4.334	4.644	4.953	5.263	5 572	5.882	6.192	58
4	.310	.621	.931	1.241	1.551	1.862	2.171	2.482	2.791	3.102	3.412	3.722	4.032	4.342	4.653	4.962	5.273	5.582	5.893	6.203	56
6	.311	.622	.933	1.243	1.554	1.865	2.175	2.486	2.796	3.108	3 418	3.729	4.039	4.350	4.661	4.971	5.282	5.592	5.904	6.214	54
8	.312	.623	.934	1.245	1.557	1.868	2.179	2.490	2.801	3.113	3.424	3.735	4 046	4.358	4.669	4.980	5.291	5.602	5.914	6.225	52
10	.312	.624	.935	1.247	1.559	1.871	2.182	2.494	2.806	3.118	3.430	3.741	4.053	4.365	4 677	4.988	5 300	5.612	5.924	6.236	50
12	.313	.625	.937	1.250	1.562	1.875	2.186	2.499	2.811	3.124	3.436	3.748	4.061	4.373	4 686	4.997	5 310	5.622	5.935	6 247	48
14	.313	.626	.939	1.252	1.565	1.878	2.190	2.504	2.816	3.130	3.442	3.755	4.068	4.381	4.694	5.006	5 320	5.632	5.946	6.258	46
16	.314	.627	.941	1.254	1.568	1.881	2.194	2.508	2.821	3.135	3.448	3.762	4.075	4.389	4.702	5.015	5.329	5.642	5.956	6.269	44
18	.315	.628	.943	1.256	1.571	1.884	2.198	2.512	2.826	3 140	3.454	3.768	4.082	4.396	4.710	5.024	5.338	5.652	5.966	6.280	42
20	.315	.629	.944	1.258	1.573	1.887	2.202	2.516	2.831	3.145	3.460	3.774	4.089	4.403	4.718	5.033	5.347	5.662	5.976	6.291	40
22	.316	.631	.946	1.261	1.576	1 891	2.206	2.521	2.836	3.151	3.466	3.781	4.097	4.411	4.727	5.042	5.357	5.672	5.987	6.302	38
24	.316	.632	.948	1.263	1.579	1.895	2.210	2.526	2.841	3.157	3.472	3.788	4.104	4.419	4.736	5.051	5.367	5.682	5.998	6.313	36
26	.317	.633	.950	1.265	1.582	1.898	2.214	2.530	2.846	3.163	3 478	3.795	4.111	4.427	4.744	5.060	5.376	5.692	6 009	6.324	34
28	.317	.634	.951	1.267	1.585	1.901	2.218	2.534	2.851	3.168	3 484	3.802	4.118	4.435	4.752	5.069	5.385	5.702	6.019	6.335	32
30	.317	.635	.952	1.269	1.587	1.904	2.221	2.538	2 856	3.173	3.490	3.808	4.125	4.442	4.760	5.077	5.394	5.711	6.029	6.346	30
32	.318	.636	.954	1.272	1.590	1.908	2.225	2.543	2.861	3.179	3.497	3.815	4.133	4.450	4.769	5.086	5.404	5.721	6.040	6.357	28
34	.318	.637	.956	1.274	1.593	1.911	2.229	2.548	2.866	3.185	3.503	3.822	4.140	4.458	4.777	5.095	5 414	5.731	6.051	6.368	26
36	.319	.638	.958	1.276	1.596	1.914	2.233	2.552	2.871	3.191	3.509	3.829	4.147	4.466	4.785	5.104	5 423	5.741	6.061	6.379	24
38	.320	.639	.959	1.278	1.598	1.917	2.237	2.556	2.876	3.196	3.515	3.835	4.154	4.474	4.793	5.113	5.432	5.751	6.071	6.390	22
40	.320	.640	.960	1.280	1.600	1.920	2.240	2.560	2.881	3.201	3.521	3.841	4.161	4.481	4.801	5.121	5.441	5.761	6.081	6.401	20
42	.321	.642	.962	1.283	1.603	1.924	2.244	2.565	2.886	3.207	3.527	3.848	4.169	4.489	4.810	5.130	5 451	5.771	6.092	6.412	18
44	.321	.643	.964	1.285	1.606	1.928	2.248	2.570	2.891	3.213	3.533	3.855	4.176	4.497	4.818	5.139	5.461	5.781	6.103	6.423	16
46	.322	.644	.966	1.287	1.609	1.931	2.252	2.574	2 896	3.218	3 539	3.862	4.183	4.505	4.826	5.148	5.470	5.791	6.114	6.434	14
48	.323	.645	.968	1.289	1.612	1.934	2.256	2.578	2.901	3.223	3.545	3.868	4.190	4.512	4.834	5.157	5.479	5.801	6 124	6.445	12
50	.323	.646	.969	1.291	1.614	1.937	2.260	2.582	2.905	3.228	3.551	3.874	4.197	4.519	4.842	5.165	5.488	5.811	6.134	6.456	10
52	.324	.647	.971	1.294	1.617	1.941	2.264	2.587	2 910	3.234	3.557	3.881	4.205	4.527	4.851	5.174	5.498	5.821	6.145	6.467	8
54	.324	.648	.973	1.296	1.620	1.945	2.268	2.592	2.915	3.240	3.563	3 888	4.212	4.535	4.860	5.183	5.507	5.831	6.156	6.478	6
56	.325	.649	.975	1.298	1.623	1.948	2.272	2.596	2.920	3 246	3.569	3.895	4.219	4.543	4.868	5.192	5.516	5.841	6.166	6.489	4
58	.326	.650	.976	1.300	1.626	1.951	2.276	2.600	2.925	3.251	3.575	3.901	4.226	4.551	4.876	5.201	5.525	5.851	6.176	6.500	2
19° 0′	.326	.651	.977	1.302	1.628	1.954	2.279	2.604	2.930	3.256	3.581	3.907	4.233	4.558	4.884	5.209	5.534	5.860	6.186	6.511	11° 0′
2	.327	.653	.979	1.305	1.631	1.958	2.283	2.609	2.935	3.262	3.587	3.914	4.240	4 566	4.893	5.218	5 544	5.370	6.197	6.522	58
4	.327	.654	.981	1.307	1.634	1.961	2.287	2.614	2.940	3.268	3.593	3 921	4.247	4.574	4.901	5.227	5.554	5.880	6.208	6.533	56
6	.328	.655	.983	1.309	1.637	1.964	2.291	2.618	2.945	3.273	3.599	3.928	4.254	4.582	4.909	5.236	5 563	5.890	6.218	6.544	54
8	.328	.656	.984	1.311	1.640	1.967	2.295	2.622	2.950	3.278	3.605	3.934	4.261	4.589	4.917	5.245	5.572	5.900	6.228	6.555	52
10	.328	.657	.985	1.313	1.642	1.970	2.298	2.626	2.955	3.283	3.611	3.940	4.268	4.596	4.925	5.253	5.581	5.910	6.238	6.566	50
12	.329	.658	.987	1.316	1.645	1.974	2.302	2.631	2.960	3.289	3.618	3.947	4 276	4.604	4.934	5.262	5.591	5.920	6.249	6.577	48
14	.329	.659	.989	1.318	1.648	1.977	2.306	2.636	2.965	3.295	3.624	3.954	4.283	4.612	4.942	5.271	5.601	5.930	6.260	6.588	46
16	.330	.660	.991	1.320	1.651	1.980	2.310	2.640	2.970	3.301	3.630	3.961	4.290	4.620	4.950	5.280	5.610	5.940	6.270	6.599	44
18	.331	.661	.992	1.322	1.653	1.983	2.314	2.644	2.975	3.306	3.636	3.967	4.297	4.628	4.958	5.289	5 619	5.950	6 280	6.610	42
20	.331	.662	.993	1.324	1.655	1.986	2.317	2.648	2.980	3.311	3.642	3.973	4.304	4.635	4.966	5.297	5.628	5.959	6.290	6.621	40
22	.332	.664	.995	1.327	1.658	1.990	2.321	2.653	2.985	3.317	3.648	3 980	4.311	4.643	4.975	5 306	5.638	5.969	6.301	6.632	38
24	.332	.665	.997	1.329	1.661	1.994	2.325	2.658	2.990	3.323	3.654	3.987	4.318	4.651	4.983	5.315	5.648	5.979	6 312	6.643	36
26	.333	.666	.999	1.331	1.664	1.997	2.329	2.662	2.995	3.328	3.660	3.994	4.325	4.659	4.991	5.324	5.657	5.989	6.322	6.654	34
28	.334	.667	1.000	1.333	1.667	2.000	2.333	2.666	3.000	3.333	3.666	4.000	4.332	4.666	4.999	5.333	5.666	5.999	6.332	6.665	32
30	.334	.668	1.001	1.335	1.669	2.003	2.337	2.670	3.004	3.338	3.672	4.006	4.339	4.673	5.007	5.341	5.675	6.009	6 342	6.676	30
32	.335	.669	1.003	1.338	1.672	2.007	2.341	2.675	3.009	3.344	3.678	4.013	4.347	4.681	5.016	5.350	5.685	6.019	6.353	6 687	28
34	.335	.670	1.005	1.340	1.675	2.010	2.345	2.680	3 014	3.350	3.684	4.020	4 354	4.689	5 024	5.359	5 694	6.029	6.364	6.698	26
36	.336	.671	1.007	1.342	1.678	2.013	2.349	2.684	3.019	3.356	3.690	4.027	4.361	4.697	5.032	5.368	5.703	6.039	6.375	6.709	24
38	.337	.672	1.009	1.344	1.681	2.016	2.353	2.688	3.024	3.361	3.696	4.033	4.368	4.705	5.040	5.377	5.712	6.049	6.385	6.720	22
40	.337	.673	1.010	1.346	1.683	2.019	2.356	2.692	3.029	3.366	3.702	4.039	4.375	4.712	5.048	5 385	5.721	6.058	6.395	6.731	20
42	.338	.675	1.012	1.349	1.686	2.023	2.360	2.697	3 034	3.372	3.703	4.046	4.383	4.720	5.057	5.394	5.731	6.068	6.406	6.742	18
44	.338	.676	1.014	1.351	1.689	2.027	2.364	2.702	3 039	3.378	3.714	4 053	4.390	4.728	5.065	5.403	5.741	6 078	6.416	6.753	16
46	.339	.677	1.015	1.353	1.692	2.030	2.368	2.706	3.044	3.383	3.720	4.059	4.397	4.736	5.073	5 412	5 750	6.088	6.426	6.764	14
48	.339	.678	1.017	1.355	1.694	2.033	2.372	2.710	3.049	3.388	3.726	4.055	4.404	4.743	5.081	5.420	5.759	6.098	6.436	6.775	12
50	.339	.679	1.018	1.357	1.696	2.036	2.375	2.714	3.054	3.393	3.732	4.071	4.411	4.750	5.089	5.428	5.868	6.107	6.446	6.786	10
52	.340	.680	1.020	1.360	1.699	2.040	2.379	2.719	3.059	3.399	3.738	4.078	4.418	4.758	5.098	5.437	5.878	6.117	6.457	6.797	8
54	.340	.681	1.022	1.362	1.702	2.043	2.383	2.724	3.064	3.405	3.744	4.085	4.425	4.766	5.106	5.446	5.887	6.127	6.468	6.808	6
56	.341	.682	1.024	1.364	1.705	2.046	2.387	2.728	3.069	3.410	3 750	4.092	4.432	4 774	5.114	5.455	5.896	6.137	6.478	6.819	4
58	.342	.683	1.025	1.366	1.708	2.049	2.391	2 732	3.074	3.415	3.756	4.098	4.439	4.781	5.122	5 464	5 905	6.147	6.488	6 830	2
20° 0′	.342	.684	1.026	1.368	1.710	2.052	2.394	2.736	3.078	3.420	3.762	4.104	4.446	4.788	5.130	5.472	5.914	6.156	6.498	6.840	10° 0′
	1	2	3	4	5	6	7	8	9	10	11	12	13	14	15	16	17	18	19	20	V + XI —

GENERAL TABLE of Aberration, &c. continued.

S. − VI + O	1	2	3	4	5	6	7	8	9	10	11	12	13	14	15	16	17	18	19	20	
20° 0′	.342	.684	1.026	1.368	1.710	2.052	2.394	2.736	3.078	3.420	3.762	4.104	4.446	4.788	5.130	5.472	5.814	6.156	6.498	6.840	10° 0′
2	.343	.686	1.028	1.371	1.713	2.056	2.398	2.741	3.083	3.426	3.768	4.111	4.454	4.796	5.139	5.481	5.824	6.166	6.509	6.851	58
4	.343	.687	1.030	1.373	1.716	2.060	2.402	2.746	3.088	3.432	3.774	4.118	4.461	4.804	5.147	5.490	5.834	6.176	6.520	6.862	56
6	.344	.688	1.032	1.375	1.719	2.063	2.406	2.750	3.093	3.438	3.780	4.125	4.468	4.812	5.155	5.499	5.843	6.186	6.530	6.873	54
8	.345	.689	1.033	1.377	1.722	2.066	2.410	2.754	3.098	3.443	3.786	4.131	4.475	4.820	5.163	5.508	5.852	6.196	6.540	6.884	52
10	.345	.690	1.034	1.379	1.724	2.069	2.413	2.758	3.103	3.448	3.792	4.137	4.482	4.827	5.171	5.516	5.861	6.206	6.550	6.895	50
12	.346	.691	1.036	1.382	1.727	2.073	2.417	2.763	3.108	3.454	3.798	4.144	4.489	4.835	5.180	5.525	5.871	6.216	6.561	6.906	48
14	.346	.692	1.038	1.384	1.730	2.076	2.421	2.768	3.113	3.460	3.804	4.151	4.496	4.843	5.188	5.534	5.880	6.226	6.572	6.917	46
16	.347	.693	1.040	1.386	1.733	2.079	2.425	2.772	3.118	3.465	3.810	4.158	4.503	4.851	5.196	5.543	5.889	6.236	6.582	6.928	44
18	.348	.694	1.042	1.388	1.735	2.082	2.429	2.776	3.123	3.470	3.816	4.164	4.510	4.858	5.204	5.552	5.898	6.246	6.592	6.939	42
20	.348	.695	1.043	1.390	1.737	2.085	2.432	2.780	3.127	3.475	3.822	4.170	4.517	4.865	5.212	5.560	5.907	6.255	6.602	6.950	40
22	.349	.696	1.045	1.393	1.740	2.089	2.436	2.785	3.132	3.481	3.828	4.177	4.525	4.873	5.221	5.569	5.917	6.265	6.613	6.961	38
24	.349	.697	1.047	1.395	1.743	2.092	2.440	2.790	3.137	3.487	3.834	4.184	4.532	4.881	5.229	5.578	5.927	6.275	6.624	6.972	36
26	.350	.698	1.049	1.397	1.746	2.095	2.444	2.794	3.142	3.492	3.840	4.191	4.539	4.889	5.237	5.587	5.936	6.285	6.634	6.983	34
28	.350	.699	1.050	1.399	1.749	2.098	2.448	2.798	3.147	3.497	3.846	4.197	4.546	4.896	5.245	5.595	5.945	6.295	6.644	6.994	32
30	.350	.700	1.051	1.401	1.751	2.101	2.451	2.802	3.152	3.502	3.852	4.203	4.553	4.903	5.253	5.603	5.954	6.304	6.654	7.004	30
32	.351	.702	1.053	1.404	1.754	2.105	2.455	2.807	3.157	3.508	3.858	4.210	4.560	4.911	5.262	5.612	5.963	6.314	6.665	7.015	28
34	.351	.703	1.055	1.406	1.757	2.109	2.459	2.811	3.162	3.514	3.864	4.217	4.567	4.919	5.270	5.621	5.972	6.324	6.676	7.026	26
36	.352	.704	1.057	1.408	1.760	2.112	2.463	2.815	3.167	3.519	3.870	4.223	4.574	4.927	5.278	5.630	5.981	6.334	6.686	7.037	24
38	.353	.705	1.058	1.410	1.763	2.115	2.467	2.819	3.172	3.524	3.876	4.229	4.581	4.934	5.286	5.639	5.990	6.344	6.696	7.048	22
40	.353	.706	1.059	1.412	1.765	2.118	2.471	2.823	3.176	3.529	3.882	4.235	4.588	4.941	5.294	5.647	5.999	6.353	6.706	7.059	20
42	.354	.707	1.061	1.415	1.768	2.122	2.475	2.828	3.181	3.535	3.888	4.242	4.595	4.949	5.303	5.656	6.009	6.363	6.717	7.070	18
44	.354	.708	1.063	1.417	1.771	2.125	2.479	2.833	3.186	3.541	3.894	4.249	4.602	4.957	5.311	5.665	6.319	6.373	6.727	7.081	16
46	.355	.709	1.065	1.419	1.774	2.128	2.483	2.837	3.191	3.547	3.900	4.256	4.609	4.965	5.319	5.674	6.028	6.383	6.737	7.092	14
48	.356	.710	1.067	1.421	1.776	2.131	2.487	2.841	3.196	3.552	3.906	4.262	4.616	4.971	5.327	5.682	6.037	6.393	6.747	7.103	12
50	.356	.711	1.068	1.423	1.778	2.134	2.490	2.845	3.201	3.557	3.912	4.268	4.623	4.979	5.335	5.690	6.046	6.402	6.757	7.113	10
52	.357	.713	1.070	1.426	1.782	2.138	2.494	2.850	3.206	3.563	3.918	4.275	4.631	4.987	5.344	5.699	6.056	6.412	6.768	7.124	8
54	.357	.714	1.072	1.428	1.785	2.141	2.498	2.855	3.211	3.569	3.924	4.282	4.638	4.995	5.352	5.708	6.065	6.422	6.779	7.135	6
56	.358	.715	1.073	1.430	1.788	2.144	2.502	2.859	3.216	3.574	3.930	4.288	4.645	5.003	5.360	5.717	6.074	6.432	6.789	7.146	4
58	.358	.716	1.074	1.432	1.790	2.147	2.506	2.863	3.221	3.579	3.936	4.294	4.652	5.010	5.368	5.726	6.083	6.442	6.799	7.157	2
21° 0′	.358	.717	1.075	1.434	1.792	2.150	2.509	2.867	3.225	3.584	3.942	4.300	4.659	5.017	5.376	5.734	6.092	6.451	6.809	7.167	9° 0′
2	.259	.718	1.077	1.436	1.795	2.154	2.513	2.872	3.230	3.590	3.948	4.307	4.666	5.025	5.384	5.743	6.102	6.461	6.820	7.178	58
4	.359	.719	1.079	1.438	1.798	2.158	2.517	2.877	3.235	3.596	3.954	4.314	4.673	5.033	5.392	5.752	6.111	6.471	6.831	7.189	56
6	.360	.720	1.081	1.440	1.801	2.161	2.521	2.881	3.240	3.601	3.960	4.321	4.680	5.041	5.400	5.761	6.120	6.481	6.841	7.200	54
8	.361	.721	1.082	1.442	1.803	2.164	2.525	2.885	3.245	3.606	3.966	4.327	4.687	5.048	5.408	5.769	6.129	6.490	6.851	7.211	52
10	.361	.722	1.083	1.444	1.805	2.167	2.528	2.889	3.250	3.611	3.972	4.333	4.694	5.055	5.416	5.777	6.138	6.499	6.861	7.222	50
12	.362	.724	1.085	1.447	1.808	2.171	2.532	2.894	3.255	3.617	3.978	4.340	4.701	5.063	5.425	5.786	6.148	6.509	6.872	7.233	48
14	.362	.725	1.087	1.449	1.811	2.174	2.536	2.898	3.260	3.623	3.984	4.347	4.708	5.071	5.433	5.795	6.157	6.519	6.882	7.244	46
16	.363	.726	1.089	1.451	1.814	2.177	2.540	2.902	3.265	3.628	3.990	4.353	4.715	5.079	5.441	5.804	6.166	6.529	6.892	7.255	44
18	.364	.727	1.090	1.453	1.817	2.180	2.544	2.906	3.270	3.633	3.996	4.359	4.722	5.086	5.449	5.813	6.175	6.539	6.902	7.266	42
20	.364	.728	1.091	1.455	1.819	2.183	2.547	2.910	3.274	3.638	4.002	4.365	4.729	5.093	5.457	5.821	6.184	6.548	6.912	7.276	40
22	.365	.729	1.093	1.458	1.822	2.187	2.551	2.915	3.279	3.644	4.008	4.372	4.737	5.101	5.466	5.830	6.194	6.558	6.923	7.287	38
24	.365	.730	1.095	1.460	1.825	2.190	2.555	2.920	3.284	3.650	4.014	4.379	4.744	5.109	5.474	5.839	6.204	6.568	6.934	7.298	36
26	.366	.731	1.097	1.462	1.828	2.193	2.559	2.924	3.289	3.655	4.020	4.386	4.751	5.117	5.482	5.848	6.213	6.578	6.944	7.309	34
28	.367	.732	1.099	1.464	1.831	2.196	2.563	2.928	3.294	3.660	4.026	4.392	4.758	5.124	5.490	5.856	6.222	6.588	6.954	7.320	32
30	.367	.733	1.100	1.466	1.833	2.199	2.566	2.932	3.299	3.665	4.032	4.398	4.765	5.131	5.498	5.864	6.231	6.597	6.964	7.330	30
32	.368	.734	1.102	1.469	1.836	2.203	2.570	2.937	3.304	3.671	4.038	4.405	4.772	5.139	5.506	5.873	6.241	6.607	6.975	7.341	28
34	.368	.735	1.104	1.471	1.839	2.206	2.574	2.942	3.309	3.677	4.044	4.412	4.779	5.147	5.514	5.882	6.250	6.617	6.985	7.352	26
36	.369	.736	1.106	1.473	1.842	2.209	2.578	2.946	3.314	3.682	4.050	4.418	4.786	5.155	5.522	5.891	6.259	6.627	6.995	7.363	24
38	.369	.737	1.107	1.475	1.844	2.212	2.582	2.950	3.319	3.687	4.056	4.424	4.793	5.162	5.530	5.899	6.268	6.637	7.005	7.374	22
40	.369	.738	1.108	1.477	1.846	2.215	2.585	2.954	3.323	3.692	4.061	4.430	4.800	5.169	5.538	5.907	6.277	6.646	7.015	7.384	20
42	.370	.730	1.110	1.480	1.849	2.219	2.589	2.959	3.328	3.698	4.067	4.437	4.807	5.177	5.547	5.916	6.287	6.656	7.026	7.395	18
44	.370	.741	1.112	1.482	1.852	2.222	2.593	2.963	3.333	3.704	4.073	4.444	4.814	5.185	5.555	5.925	6.296	6.666	7.036	7.406	16
46	.371	.742	1.114	1.484	1.855	2.225	2.597	2.967	3.338	3.709	4.079	4.451	4.821	5.193	5.563	5.934	6.305	6.676	7.046	7.417	14
48	.372	.743	1.115	1.486	1.858	2.228	2.600	2.971	3.343	3.714	4.085	4.457	4.828	5.200	5.571	5.943	6.314	6.685	7.056	7.428	12
50	.372	.744	1.116	1.488	1.860	2.231	2.603	2.975	3.347	3.719	4.091	4.463	4.835	5.207	5.579	5.951	6.323	6.694	7.066	7.438	10
52	.373	.745	1.118	1.490	1.863	2.235	2.607	2.980	3.352	3.725	4.097	4.470	4.842	5.215	5.587	5.960	6.332	6.704	7.077	7.449	8
54	.373	.746	1.120	1.492	1.866	2.239	2.611	2.985	3.357	3.731	4.103	4.477	4.849	5.223	5.595	5.969	6.341	6.714	7.088	7.460	6
56	.374	.747	1.122	1.494	1.869	2.242	2.615	2.989	3.362	3.736	4.109	4.483	4.856	5.231	5.603	5.978	6.350	6.724	7.098	7.471	4
58	.375	.748	1.123	1.496	1.871	2.245	2.619	2.993	3.367	3.741	4.115	4.489	4.863	5.238	5.611	5.986	6.359	6.734	7.108	7.482	2
22° 0	.375	.749	1.124	1.498	1.873	2.248	2.622	2.097	3.372	3.746	4.121	4.495	4.870	5.245	5.619	5.994	6.368	6.743	7.118	7.492	8° 0
	1	2	3	4	5	6	7	8	9	10	11	12	13	14	15	16	17	18	19	20	V + XI −

TABLE XVIII.

GENERAL TABLE of Aberration, &c. continued.

S. − VI + O	1	2	3	4	5	6	7	8	9	10	11	12	13	14	15	16	17	18	19	20	
22° 0	.375	.749	1.124	1.498	1.873	2.248	2.622	2.997	3.372	3.746	4.121	4.495	4.870	5.245	5.619	5.994	6.368	6.743	7.118	7.492	8° 0′
2	.376	.751	1.126	1.501	1.876	2.252	2.626	3.002	3.377	3.752	4.127	4.502	4.877	5.253	5.628	6.003	6.378	6.753	7.129	7.503	58
4	.376	.752	1.128	1.503	1.879	2.255	2.630	3.006	3.382	3.758	4.133	4.509	4.884	5.261	5.636	6.012	6.387	6.763	7.139	7.514	56
6	.377	.753	1.130	1.505	1.882	2.258	2.634	3.010	3.387	3.763	4.139	4.516	4.891	5.268	5.644	6.021	6.396	6.773	7.149	7.525	54
8	.377	.754	1.131	1.507	1.885	2.261	2.638	3.014	3.392	3.768	4.145	4.522	4.898	5.275	5.652	6.029	6.405	6.782	7.159	7.536	52
10	.377	.755	1.132	1.509	1.887	2.264	2.641	3.018	3.396	3.773	4.150	4.528	4.905	5.282	5.660	6.037	6.414	6.791	7.169	7.546	50
12	.378	.756	1.134	1.512	1.890	2.268	2.645	3.023	3.401	3.779	4.156	4.535	4.912	5.290	5.668	6.046	6.424	6.801	7.179	7.557	48
14	.378	.757	1.136	1.514	1.893	2.271	2.649	3.028	3.406	3.785	4.162	4.542	4.919	5.298	5.676	6.055	6.433	6.811	7.189	7.568	46
16	.379	.758	1.138	1.516	1.896	2.274	2.653	3.032	3.411	3.790	4.168	4.548	4.926	5.306	5.684	6.064	6.442	6.821	7.199	7.579	44
18	.379	.759	1.139	1.518	1.898	2.277	2.657	3.036	3.416	3.795	4.174	4.554	4.933	5.313	5.692	6.072	6.451	6.831	7.209	7.590	42
20	.380	.760	1.140	1.520	1.900	2.280	2.660	3.040	3.420	3.800	4.180	4.560	4.940	5.320	5.700	6.080	6.460	6.840	7.219	7.600	40
22	.381	.761	1.142	1.523	1.903	2.284	2.664	3.045	3.425	3.806	4.186	4.567	4.947	5.328	5.708	6.089	6.470	6.850	7.230	7.611	38
24	.381	.762	1.144	1.525	1.906	2.287	2.668	3.049	3.430	3.812	4.192	4.574	4.954	5.336	5.716	6.098	6.479	6.860	7.241	7.622	36
26	.382	.763	1.146	1.527	1.909	2.290	2.672	3.053	3.435	3.817	4.198	4.580	4.961	5.344	5.724	6.107	6.488	6.870	7.251	7.633	34
28	.383	.764	1.147	1.529	1.911	2.293	2.676	3.057	3.440	3.822	4.204	4.586	4.968	5.351	5.732	6.115	6.497	6.879	7.261	7.644	32
30	.383	.765	1.148	1.531	1.913	2.296	2.679	3.061	3.444	3.827	4.210	4.592	4.975	5.358	5.740	6.123	6.506	6.888	7.271	7.654	30
32	.384	.767	1.150	1.534	1.916	2.300	2.683	3.066	3.449	3.833	4.216	4.599	4.982	5.366	5.749	6.132	6.515	6.898	7.282	7.665	28
34	.384	.768	1.152	1.536	1.919	2.303	2.687	3.071	3.454	3.839	4.222	4.606	4.989	5.374	5.757	6.141	6.524	6.908	7.292	7.676	26
36	.385	.769	1.154	1.538	1.922	2.306	2.691	3.075	3.459	3.844	4.228	4.612	4.996	5.381	5.765	6.150	6.533	6.918	7.302	7.687	24
38	.385	.770	1.155	1.540	1.925	2.309	2.695	3.079	3.464	3.849	4.234	4.618	5.003	5.388	5.773	6.158	6.542	6.928	7.312	7.697	22
40	.385	.771	1.156	1.542	1.927	2.312	2.698	3.083	3.468	3.854	4.239	4.624	5.010	5.395	5.781	6.166	6.551	6.937	7.322	7.707	20
42	.386	.772	1.158	1.544	1.930	2.316	2.702	3.088	3.473	3.860	4.245	4.631	5.017	5.403	5.789	6.175	6.560	6.947	7.333	7.718	18
44	.386	.773	1.160	1.546	1.933	2.319	2.706	3.092	3.478	3.866	4.251	4.638	5.024	5.411	5.797	6.184	6.569	6.957	7.343	7.729	16
46	.387	.774	1.162	1.548	1.936	2.322	2.710	3.096	3.483	3.871	4.257	4.645	5.031	5.419	5.805	6.193	6.578	6.966	7.353	7.740	14
48	.388	.775	1.163	1.550	1.938	2.325	2.713	3.100	3.488	3.876	4.263	4.651	5.038	5.426	5.813	6.201	6.587	6.975	7.363	7.751	12
50	.388	.776	1.164	1.552	1.940	2.328	2.716	3.104	3.493	3.881	4.269	4.657	5.045	5.433	5.821	6.209	6.596	6.984	7.373	7.761	10
52	.389	.778	1.166	1.555	1.943	2.332	2.720	3.109	3.498	3.887	4.275	4.664	5.052	5.441	5.829	6.218	6.606	6.994	7.384	7.772	8
54	.389	.779	1.168	1.557	1.946	2.335	2.724	3.114	3.503	3.892	4.281	4.671	5.059	5.449	5.837	6.227	6.615	7.004	7.394	7.783	6
56	.390	.780	1.170	1.559	1.949	2.338	2.728	3.118	3.508	2.897	4.287	4.677	5.066	5.456	5.845	6.235	6.624	7.014	7.404	7.794	4
58	.391	.781	1.171	1.561	1.952	2.341	2.732	3.122	3.513	3.902	4.293	4.683	5.073	5.463	5.853	6.243	6.633	7.024	7.414	7.805	2
23° 0′	.391	.782	1.172	1.563	1.954	2.344	2.735	3.126	3.517	3.907	4.298	4.689	5.079	5.470	5.861	6.252	6.642	7.033	7.424	7.815	7° 0′
2	.392	.783	1.174	1.566	1.957	2.348	2.739	3.131	3.522	3.913	4.304	4.696	5.086	5.478	5.869	6.261	6.652	7.043	7.435	7.826	58
4	.392	.784	1.176	1.568	1.960	2.351	2.743	3.135	3.527	3.919	4.310	4.703	5.093	5.486	5.877	6.270	6.661	7.053	7.445	7.837	56
6	.393	.785	1.178	1.570	1.963	2.354	2.747	3.139	3.532	3.924	4.316	4.709	5.100	5.494	5.885	6.279	6.670	7.063	7.455	7.848	54
8	.393	.786	1.179	1.572	1.965	2.357	2.751	3.143	3.537	3.929	4.322	4.716	5.107	5.501	5.893	6.287	6.679	7.072	7.465	7.858	52
10	.393	.787	1.180	1.574	1.967	2.360	2.754	3.147	3.541	3.934	4.328	4.721	5.114	5.508	5.901	6.295	6.688	7.081	7.475	7.868	50
12	.394	.788	1.182	1.576	1.970	2.364	2.758	3.152	3.546	3.940	4.334	4.728	5.121	5.516	5.909	6.304	6.697	7.091	7.485	7.879	48
14	.394	.789	1.184	1.578	1.973	2.368	2.762	3.157	3.551	3.946	4.340	4.735	5.128	5.524	5.916	6.313	6.706	7.101	7.495	7.890	46
16	.395	.790	1.186	1.580	1.976	2.371	2.766	3.161	3.556	3.951	4.346	4.741	5.135	5.531	5.925	6.321	6.715	7.111	7.505	7.901	44
18	.396	.791	1.187	1.582	1.978	2.374	2.770	3.165	3.561	3.956	4.352	4.747	5.142	5.538	5.932	6.329	6.724	7.120	7.515	7.912	42
20	.396	.792	1.188	1.584	1.980	2.377	2.773	3.169	3.565	3.961	4.357	4.753	5.149	5.545	5.941	6.337	6.733	7.129	7.525	7.922	40
22	.397	.794	1.190	1.587	1.983	2.381	2.777	3.174	3.570	3.967	4.363	4.760	5.156	5.553	5.949	6.346	6.743	7.139	7.536	7.933	38
24	.397	.795	1.192	1.589	1.986	2.384	2.781	3.178	3.575	3.973	4.369	4.767	5.163	5.561	5.957	6.355	6.752	7.149	7.546	7.944	36
26	.398	.796	1.194	1.591	1.989	2.387	2.785	3.182	3.580	3.978	4.375	4.773	5.170	5.569	5.965	6.364	6.761	7.159	7.556	7.955	34
28	.399	.797	1.195	1.593	1.991	2.390	2.788	3.186	3.585	3.983	4.381	4.779	5.177	5.576	5.973	6.372	6.770	7.169	7.566	7.965	32
30	.399	.798	1.196	1.595	1.994	2.393	2.791	3.190	3.589	3.988	4.386	4.785	5.184	5.583	5.981	6.380	6.779	7.178	7.576	7.975	30
32	.400	.799	1.198	1.598	1.997	2.397	2.795	3.195	3.594	3.994	4.392	4.792	5.191	5.591	5.989	6.389	6.788	7.188	7.587	7.986	28
34	.400	.800	1.200	1.600	2.000	2.400	2.799	3.199	3.599	3.999	4.398	4.799	5.198	5.599	5.997	6.398	6.797	7.198	7.597	7.997	26
36	.401	.801	1.202	1.602	2.003	2.403	2.803	3.203	3.604	4.004	4.404	4.805	5.205	5.606	6.005	6.407	6.806	7.208	7.607	8.008	24
38	.401	.802	1.203	1.604	2.005	2.406	2.807	3.207	3.609	4.009	4.410	4.811	5.212	5.613	6.013	6.415	6.815	7.217	7.617	8.018	22
40	.401	.803	1.204	1.606	2.007	2.409	2.810	3.211	3.613	4.014	4.416	4.817	5.218	5.620	6.021	6.423	6.824	7.226	7.627	8.028	20
42	.402	.804	1.206	1.608	2.010	2.413	2.814	3.216	3.618	4.020	4.422	4.824	5.225	5.628	6.029	6.432	6.833	7.236	7.638	8.039	18
44	.402	.805	1.208	1.610	2.013	2.417	2.818	3.221	3.623	4.026	4.428	4.831	5.232	5.636	6.037	6.441	6.842	7.246	7.648	8.050	16
46	.403	.806	1.210	1.612	2.016	2.420	2.822	3.225	3.628	4.031	4.434	4.837	5.239	5.643	6.045	6.449	6.851	7.255	7.658	8.061	14
48	.404	.807	1.211	1.614	2.018	2.423	2.826	3.229	3.633	4.036	4.440	4.843	5.246	5.650	6.053	6.457	6.860	7.264	7.668	8.072	12
50	.404	.808	1.212	1.616	2.020	2.426	2.829	3.233	3.637	4.041	4.445	4.849	5.253	5.657	6.061	6.465	6.869	7.273	7.678	8.082	10
52	.405	.810	1.214	1.619	2.023	2.429	2.833	3.238	3.642	4.047	4.451	4.856	5.260	5.665	6.069	6.474	6.878	7.283	7.688	8.093	8
54	.405	.811	1.216	1.621	2.026	2.432	2.837	3.242	3.647	4.052	4.457	4.863	5.267	5.673	6.077	6.483	6.887	7.293	7.698	8.104	6
56	.406	.812	1.218	1.623	2.029	2.435	2.841	3.246	3.652	4.057	4.463	4.869	5.274	5.680	6.085	6.492	6.896	7.303	7.708	8.115	4
58	.407	.813	1.219	1.625	2.032	2.438	2.844	3.250	3.657	4.062	4.469	4.875	5.281	5.687	6.093	6.500	6.905	7.312	7.718	8.125	2
24° 0	.407	.814	1.220	1.627	2.034	2.441	2.847	3.254	3.661	4.067	4.474	4.881	5.288	5.694	6.101	6.508	6.914	7.321	7.728	8.135	6° 0
	1	2	3	4	5	6	7	8	9	10	11	12	13	14	15	16	17	18	19	20	V + XI −

TABLE XVIII.

GENERAL TABLE of Aberration, &c. continued.

S. − VI + O	1	2	3	4	5	6	7	8	9	10	11	12	13	14	15	16	17	18	19	20	
24° 0′	.407	.814	1.220	1.627	2.034	2.441	2.847	3.254	3.661	4.067	4.474	4.881	5.288	5.694	6.101	6.508	6.914	7.321	7.728	8.135	6° 0′
2	.407	.815	1.222	1.630	2.037	2.444	2.851	3.259	3.666	4.073	4.480	4.888	5.295	5.702	6.109	6.517	6.924	7.331	7.739	8.146	58
4	.408	.816	1.224	1.632	2.040	2.447	2.855	3.263	3.671	4.079	4.486	4.895	5.302	5.710	6.117	6.526	6.933	7.341	7.749	8.157	56
6	.409	.817	1.226	1.634	2.043	2.450	2.859	3.267	3.676	4.084	4.492	4.901	5.309	5.717	6.125	6.534	6.942	7.351	7.759	8.168	54
8	.409	.818	1.227	1.636	2.045	2.453	2.863	3.271	3.681	4.089	4.498	4.907	5.316	5.724	6.133	6.542	6.951	7.360	7.769	8.178	52
10	.409	.819	1.228	1.638	2.047	2.456	2.866	3.275	3.685	4.094	4.503	4.913	5.322	5.731	6.141	6.550	6.960	7.369	7.779	8.188	50
12	.410	.820	1.230	1.640	2.050	2.460	2.870	3.280	3.690	4.100	4.509	4.920	5.329	5.739	6.149	6.559	6.969	7.379	7.789	8.199	48
14	.410	.821	1.232	1.642	2.053	2.463	2.874	3.284	3.695	4.105	4.515	4.926	5.336	5.747	6.157	6.568	6.978	7.389	7.799	8.210	46
16	.411	.822	1.234	1.644	2.056	2.466	2.878	3.288	3.700	4.110	4.521	4.932	5.343	5.755	6.165	6.577	6.987	7.399	7.809	8.221	44
18	.412	.823	1.235	1.646	2.058	2.469	2.881	3.292	3.704	4.115	4.527	4.938	5.350	5.762	6.173	6.585	6.996	7.408	7.819	8.231	42
20	.412	.824	1.236	1.648	2.060	2.472	2.884	3.296	3.708	4.120	4.532	4.944	5.357	5.769	6.181	6.593	7.005	7.417	7.829	8.241	40
22	.413	.825	1.238	1.651	2.063	2.476	2.888	3.301	3.713	4.126	4.538	4.951	5.364	5.777	6.189	6.602	7.014	7.427	7.839	8.252	38
24	.413	.826	1.240	1.653	2.066	2.479	2.892	3.306	3.718	4.132	4.544	4.958	5.371	5.785	6.197	6.611	7.023	7.437	7.849	8.263	36
26	.414	.827	1.242	1.655	2.069	2.482	2.896	3.310	3.723	4.137	4.550	4.964	5.378	5.792	6.205	6.619	7.032	7.446	7.859	8.274	34
28	.415	.828	1.243	1.657	2.072	2.485	2.900	3.314	3.728	4.142	4.556	4.970	5.385	5.799	6.213	6.627	7.041	7.455	7.869	8.284	32
30	.415	.829	1.244	1.659	2.074	2.488	2.903	3.318	3.732	4.147	4 562	4.976	5.391	5.806	6.221	6.635	7.050	7.464	7.879	8.294	30
32	.416	.831	1.246	1.661	2.077	2.492	2.907	3.323	3.737	4.153	4.568	4.983	5.398	5.814	6.229	6.644	7.059	7.474	7.890	8.305	28
34	.416	.832	1.248	1.663	2.080	2.495	2.911	3.327	3.742	4.158	4.574	4.990	5.405	5.822	6.237	6.653	7.068	7.484	7.900	8.316	26
36	.417	.833	1.250	1.665	2.083	2.498	2.915	3.331	3.747	4.163	4.580	4.996	5.412	5.829	6.245	6.661	7.077	7.494	7.910	8.327	24
38	.417	.834	1.251	1.667	2.085	2.501	2.918	3.335	3.752	4.168	4.586	5.002	5.419	5.836	6.253	6.669	7.086	7.503	7.920	8.337	22
40	.417	.835	1.252	1.669	2.087	2.504	2.921	3.339	3.756	4.173	4.591	5.008	5.425	5.843	6.260	6.677	7.095	7.512	7.930	8.347	20
42	.418	.836	1.254	1.672	2.090	2.508	2.925	3.344	3.761	4.179	4.597	5.015	5.432	5.851	6.268	6.686	7.104	7.522	7.940	8.358	18
44	.418	.837	1.256	1.674	2.093	2.511	2.929	3.348	3.766	4.185	4.603	5.022	5.439	5.859	6.276	6.695	7.113	7.532	7.950	8.369	16
46	.419	.838	1.258	1.676	2.096	2.514	2.933	3.352	3.771	4.190	4.609	5.028	5.446	5.866	6.284	6.704	7.122	7.542	7.960	8.380	14
48	.420	.839	1.259	1.678	2.098	2.517	2.937	3.356	3.776	4.195	4.615	5.034	5.453	5.873	6.292	6.712	7.131	7.551	7.970	8.390	12
50	.420	.840	1.260	1.680	2.100	2.520	2.940	3.360	3.780	4.200	4.620	5.040	5.460	5.880	6.300	6.720	7.140	7.560	7.980	8.400	10
52	.421	.841	1.262	1.682	2.103	2.524	2.944	3.365	3.785	4.206	4.626	5.047	5.467	5.888	6.308	6.729	7.149	7.570	7.990	8.411	8
54	.421	.842	1.264	1.684	2.106	2.527	2.948	3.369	3.790	4.211	4.632	5.053	5.474	5.896	6.316	6.738	7.158	7.580	8.000	8.422	6
56	.422	.843	1.266	1.686	2.109	2.530	2.952	3.373	3.795	4.216	4.638	5.059	5.481	5.903	6.324	6.746	7.167	7.589	8.010	8.432	4
58	.423	.844	1.267	1.688	2.111	2.533	2.955	3.377	3.800	4.221	4.644	5.065	5.488	5.910	6.332	6.754	7.176	7.598	8.020	8.442	2
25° 0′	.423	.845	1.268	1.690	2.113	2.536	2.958	3.381	3.804	4.226	4.649	5.071	5.494	5.917	6.339	6.762	7.185	7.607	8.030	8.452	5° 0′
2	.424	.847	1.270	1.693	2.116	2.539	2.962	3.386	3.809	4.232	4.655	5.078	5.501	5.925	6.347	6.771	7.194	7.617	8.040	8.463	58
4	.424	.848	1.272	1.695	2.119	2.542	2.966	3.390	3.814	4.238	4.661	5.085	5.508	5.933	6.355	6.780	7.203	7.627	8.050	8.474	56
6	.424	.849	1.274	1.697	2.122	2.545	2.970	3.394	3.819	4.243	4.667	5.091	5.515	5.940	6.363	6.788	7.212	7.637	8.060	8.485	54
8	.425	.850	1.275	1.699	2.124	2.548	2.974	3.398	3.823	4.248	4.673	5.097	5.522	5.947	6.371	6.796	7.221	7.646	8.070	8.495	52
10	.425	.851	1.276	1.701	2.126	2.551	2.977	3.402	3.827	4.253	4.678	5.103	5.528	5.954	6.379	6.804	7.229	7.655	8.080	8.505	50
12	.426	.852	1.278	1.704	2.129	2.555	2.981	3.407	3.832	4.259	4.684	5.110	5.535	5.962	6.387	6 813	7.238	7.665	8.090	8.516	48
14	.426	.853	1.280	1.706	2.132	2.558	2.985	3.411	3.837	4.264	4.690	5.117	5.542	5.969	6.395	6.822	7.247	7.675	8.100	8.527	46
16	.427	.854	1.282	1.708	2.135	2.561	2.989	3.415	3.842	4.269	4.696	5.123	5.549	5.976	6.403	6.830	7.256	7.684	8.110	8.538	44
18	.428	.855	1.283	1.710	2.137	2.564	2.992	3.419	3.847	4.274	4.702	5.129	5.556	5.983	6.410	6.838	7.265	7.693	8.120	8.548	42
20	.428	.856	1.284	1.712	2.139	2.567	2.995	3.423	3.851	4.279	4.707	5.135	5.563	5.990	6.418	6.846	7.274	7.702	8.130	8.558	40
22	.429	.857	1.286	1.714	2.142	2.571	2.999	3.428	3.856	4.285	4.713	5.142	5 570	5.998	6.426	6.855	7.283	7.712	8.140	8.569	38
24	.429	.858	1.288	1.716	2.145	2.574	3.003	3.432	3.861	4.290	4.719	5.149	5.577	6.006	6.434	6.864	7.292	7.722	8.150	8.580	36
26	.430	.859	1.290	1.718	2.148	2.577	3.007	3.436	3.866	4.295	4.725	5.155	5.584	6.013	6.442	6.872	7.301	7.731	8.160	8.590	34
28	.431	.860	1.291	1.720	2.151	2.580	3.011	3.440	3.871	4.300	4.731	5.161	5.591	6.020	6.450	6.880	7.310	7.740	8.170	8.600	32
30	.431	.861	1.292	1.722	2.153	2.583	3.014	3.444	3.875	4.305	4.736	5.166	5.597	6.027	6.458	6.888	7.319	7.749	8.180	8.610	30
32	.432	.862	1.294	1.725	2.156	2.587	3.018	3.449	3.880	4.311	4.742	5.173	5.604	6.035	6.466	6.897	7.328	7.759	8.190	8.621	28
34	.432	.863	1.296	1.727	2.159	2.590	3.022	3.453	3.885	4.316	4.748	5.180	5.611	6.043	6.474	6.906	7.337	7.769	8.200	8.632	26
36	.432	.864	1.297	1.729	2.162	2.593	3.026	3.457	3.890	4.321	4.754	5.186	5.618	6.050	6.482	6.914	7.346	7.778	8.210	8.643	24
38	.433	.865	1.298	1.731	2.164	2.596	3.029	3.461	3.894	4.326	4.760	5.192	5.625	6.057	6.490	6.922	7.355	7.787	8.220	8.653	22
40	.433	.866	1.299	1.733	2.166	2.599	3.032	3.465	3.898	4.331	4.765	5.198	5.631	6.064	6.497	6.930	7.363	7.796	8.230	8.663	20
42	.434	.868	1.301	1.735	2.169	2.603	3.036	3.470	3.903	4.337	4.771	5.205	5.638	6.072	6.505	6.939	7.372	7.806	8.240	8.674	18
44	.434	.869	1.303	1.737	2.172	2.606	3.040	3.474	3.908	4.343	4.777	5.212	5.645	6.080	6.513	6.948	7.381	7.816	8.250	8.685	16
46	.435	.870	1.305	1.739	2.175	2.609	3.044	3.478	3.913	4.348	4.783	5.217	5.652	6.087	6.521	6.956	7.390	7.826	8.260	8.695	14
48	.436	.871	1.306	1.741	2.177	2.612	3.047	3.482	3.918	4.353	4.788	5.223	5.659	6.094	6.529	6.964	7.399	7.835	8.270	8.705	12
50	.436	.872	1.307	1.743	2.179	2.615	3.050	3.486	3.922	4.358	4.793	5.229	5.665	6.101	6.536	6.972	7.408	7.844	8.279	8.715	10
52	.437	.873	1.309	1.746	2.182	2.618	3.054	3.491	3.927	4.364	4.799	5.236	5.672	6.109	6.544	6.981	7.417	7.854	8.289	8.726	8
54	.437	.874	1.311	1.748	2.185	2.621	3.058	3.495	3.932	4.369	4.805	5.242	5.679	6.116	6.552	6.990	7.426	7.864	8.299	8.737	6
56	.438	.875	1.313	1.750	2.188	2.624	3.062	3.499	3.937	4.374	4.811	5.248	5.686	6.123	6.560	6.998	7.435	7.873	8.309	8.747	4
58	.438	.876	1.314	1.752	2.190	2.627	3.066	3.503	3.941	4.379	4.817	5.254	5.693	6.130	6.568	7.006	7.444	7.882	8.319	8.757	2
26° 0	.438	.877	1.315	1.754	2.192	2.630	3.069	3.507	3.945	4.384	4.822	5.260	5.699	6.137	6.575	7.014	7.452	7.891	8.329	8.767	4° 0
	1	2	3	4	5	6	7	8	9	10	11	12	13	14	15	16	17	18	19	20	V + XI −

TABLE XVIII.

GENERAL TABLE of Aberration, &c. continued.

S. − VI + O	1	2	3	4	5	6	7	8	9	10	11	12	13	14	15	16	17	18	19	20	
26° 0′	.438	.877	1.315	1.754	2.192	2.630	3.069	3.507	3.945	4.384	4.822	5.260	5.699	6.137	6.575	7.014	7.452	7.891	8.329	8.767	4° 0′
2	.439	.878	1.317	1.756	2.195	2.634	3.073	3.512	3.950	4.390	4.828	5.267	5.706	6.145	6.583	7.023	7.461	7.901	8.339	8.778	58
4	.439	.879	1.319	1.758	2.198	2.637	3.077	3.516	3.955	4.395	4.834	5.274	5.713	6.153	6.591	7.032	7.470	7.911	8.349	8.789	56
6	.440	.880	1.321	1.760	2.201	2.640	3.081	3.520	3.960	4.400	4.840	5.280	5.720	6.160	6.599	7.040	7.479	7.920	8.359	8.799	54
8	.441	.881	1.322	1.762	2.203	2.643	3.084	3.524	3.965	4.405	4.846	5.286	5.727	6.167	6.607	7.048	7.488	7.929	8.369	8.809	52
10	.441	.882	1.323	1.764	2.205	2.646	3.087	3.528	3.969	4.410	4.851	5.292	5.733	6.174	6.615	7.056	7.497	7.938	8.379	8.819	50
12	.442	.883	1.325	1.766	2.208	2.650	3.091	3.533	3.974	4.416	4.857	5.299	5.740	6.182	6.623	7.065	7.506	7.948	8.389	8.830	48
14	.442	.884	1.327	1.768	2.211	2.653	3.095	3.537	3.979	4.421	4.863	5.305	5.747	6.189	6.631	7.073	7.515	7.958	8.399	8.841	46
16	.443	.885	1.329	1.770	2.214	2.656	3.099	3.541	3.984	4.426	4.869	5.311	5.754	6.196	6.639	7.081	7.524	7.967	8.409	8.852	44
18	.444	.886	1.330	1.772	2.216	2.659	3.102	3.545	3.988	4.431	4.875	5.317	5.761	6.203	6.647	7.089	7.533	7.976	8.419	8.862	42
20	.444	.887	1.331	1.774	2.218	2.662	3.105	3.549	3.992	4.436	4.880	5.323	5.767	6.210	6.654	7.097	7.541	7.985	8.428	8.872	40
22	.445	.888	1.333	1.777	2.221	2.665	3.109	3.554	3.997	4.442	4.886	5.330	5.774	6.218	6.662	7.106	7.550	7.995	8.438	8.883	38
24	.445	.889	1.335	1.779	2.224	2.668	3.113	3.558	4.002	4.447	4.892	5.336	5.781	6.226	6.670	7.115	7.559	8.005	8.448	8.894	36
26	.446	.890	1.337	1.781	2.227	2.671	3.117	3.562	4.007	4.452	4.898	5.342	5.788	6.233	6.678	7.123	7.568	8.014	8.458	8.904	34
28	.446	.891	1.338	1.783	2.229	2.674	3.120	3.566	4.012	4.457	4.903	5.348	5.795	6.240	6.686	7.131	7.577	8.023	8.468	8.914	32
30	.446	.892	1.339	1.785	2.231	2.677	3.123	3.570	4.016	4.462	4.908	5.354	5.801	6.247	6.693	7.139	7.585	8.032	8.478	8.924	30
32	.447	.894	1.341	1.787	2.234	2.681	3.127	3.574	4.021	4.468	4.914	5.361	5.808	6.255	6.701	7.148	7.594	8.042	8.488	8.935	28
34	.447	.895	1.343	1.789	2.237	2.684	3.131	3.578	4.026	4.473	4.920	5.367	5.815	6.262	6.709	7.157	7.603	8.051	8.498	8.946	26
36	.448	.896	1.344	1.791	2.240	2.687	3.135	3.582	4.031	4.478	4.926	5.373	5.822	6.269	6.717	7.165	7.612	8.060	8.508	8.956	24
38	.449	.897	1.345	1.793	2.242	2.690	3.139	3.586	4.035	4.483	4.932	5.380	5.828	6.276	6.725	7.173	7.621	8.069	8.518	8.966	22
40	.449	.898	1.346	1.795	2.244	2.693	3.142	3.590	4.039	4.488	4.937	5.385	5.834	6.283	6.732	7.181	7.630	8.078	8.527	8.976	20
42	.450	.899	1.348	1.798	2.247	2.696	3.146	3.595	4.044	4.494	4.943	5.392	5.841	6.291	6.740	7.190	7.639	8.088	8.537	8.987	18
44	.450	.900	1.350	1.800	2.250	2.699	3.150	3.599	4.049	4.499	4.949	5.398	5.848	6.299	6.748	7.198	7.648	8.098	8.547	8.998	16
46	.451	.901	1.352	1.802	2.253	2.702	3.154	3.603	4.054	4.504	4.955	5.405	5.855	6.306	6.756	7.206	7.657	8.107	8.557	9.008	14
48	.451	.902	1.353	1.804	2.255	2.705	3.157	3.607	4.059	4.509	4.960	5.411	5.862	6.314	6.764	7.214	7.666	8.116	8.567	9.018	12
50	.451	.903	1.354	1.806	2.257	2.708	3.160	3.611	4.063	4.514	4.965	5.417	5.868	6.320	6.771	7.222	7.674	8.125	8.577	9.028	10
52	.452	.904	1.356	1.808	2.260	2.712	3.164	3.616	4.068	4.520	4.971	5.424	5.875	6.328	6.779	7.231	7.683	8.135	8.587	9.039	8
54	.452	.905	1.358	1.810	2.263	2.715	3.168	3.620	4.073	4.525	4.977	5.430	5.882	6.335	6.787	7.240	7.692	8.145	8.597	9.050	6
56	.453	.906	1.360	1.812	2.266	2.718	3.172	3.624	4.078	4.530	4.983	5.436	5.889	6.342	6.795	7.248	7.701	8.154	8.607	9.060	4
58	.454	.907	1.361	1.814	2.268	2.721	3.175	3.628	4.082	4.535	4.989	5.442	5.896	6.349	6.803	7.256	7.710	8.163	8.617	9.070	2
27° 0′	.454	.908	1.362	1.816	2.270	2.724	3.178	3.632	4.086	4.540	4.994	5.448	5.902	6.356	6.810	7.264	7.718	8.172	8.626	9.080	3° 0′
2	.455	.909	1.364	1.818	2.273	2.728	3.182	3.637	4.091	4.546	5.000	5.455	5.909	6.364	6.818	7.273	7.727	8.182	8.636	9.091	58
4	.455	.910	1.366	1.820	2.276	2.731	3.186	3.641	4.096	4.551	5.006	5.461	5.916	6.371	6.826	7.281	7.736	8.192	8.646	9.102	56
6	.456	.911	1.368	1.822	2.279	2.734	3.190	3.645	4.101	4.556	5.012	5.467	5.923	6.378	6.834	7.289	7.745	8.202	8.656	9.112	54
8	.457	.912	1.369	1.824	2.281	2.737	3.193	3.649	4.105	4.561	5.017	5.473	5.930	6.385	6.842	7.297	7.754	8.210	8.666	9.122	52
10	.457	.913	1.370	1.826	2.283	2.740	3.196	3.653	4.109	4.566	5.022	5.479	5.936	6.392	6.849	7.305	7.762	8.218	8.675	9.132	50
12	.458	.914	1.372	1.829	2.286	2.743	3.200	3.657	4.114	4.572	5.028	5.486	5.943	6.400	6.857	7.314	7.771	8.228	8.685	9.143	48
14	.458	.915	1.374	1.831	2.289	2.746	3.204	3.661	4.119	4.577	5.034	5.492	5.950	6.407	6.865	7.323	7.780	8.238	8.695	9.153	46
16	.459	.916	1.376	1.833	2.292	2.749	3.208	3.665	4.124	4.582	5.040	5.498	5.957	6.414	6.873	7.331	7.789	8.247	8.705	9.163	44
18	.459	.917	1.377	1.835	2.294	2.752	3.211	3.669	4.129	4.587	5.046	5.504	5.963	6.421	6.881	7.339	7.798	8.256	8.715	9.173	42
20	.459	.918	1.378	1.837	2.296	2.755	3.214	3.673	4.133	4.592	5.051	5.510	5.969	6.428	6.888	7.347	7.806	8.265	8.724	9.183	40
22	.460	.919	1.380	1.839	2.299	2.759	3.218	3.678	4.138	4.598	5.057	5.517	5.976	6.436	6.896	7.356	7.815	8.275	8.734	9.194	38
24	.460	.921	1.382	1.841	2.302	2.762	3.222	3.682	4.143	4.603	5.063	5.523	5.983	6.444	6.904	7.364	7.824	8.285	8.744	9.205	36
26	.461	.922	1.383	1.843	2.305	2.765	3.226	3.686	4.148	4.608	5.069	5.529	5.990	6.451	6.912	7.372	7.833	8.294	8.754	9.215	34
28	.462	.923	1.384	1.845	2.307	2.768	3.229	3.690	4.152	4.613	5.074	5.535	5.997	6.458	6.919	7.380	7.842	8.303	8.764	9.225	32
30	.462	.924	1.385	1.847	2.309	2.771	3.232	3.694	4.156	4.618	5.079	5.541	6.003	6.465	6.926	7.388	7.850	8.312	8.773	9.235	30
32	.463	.925	1.387	1.849	2.312	2.774	3.236	3.699	4.161	4.623	5.085	5.548	6.010	6.473	6.934	7.397	7.859	8.322	8.783	9.246	28
34	.463	.926	1.389	1.851	2.315	2.777	3.240	3.703	4.166	4.628	5.091	5.554	6.017	6.480	6.942	7.405	7.868	8.331	8.793	9.257	26
36	.464	.927	1.391	1.853	2.318	2.780	3.244	3.707	4.171	4.633	5.097	5.560	6.024	6.487	6.950	7.413	7.877	8.340	8.803	9.267	24
38	.464	.928	1.392	1.855	2.320	2.783	3.247	3.711	4.175	4.638	5.103	5.566	6.030	6.494	6.958	7.421	7.886	8.349	8.813	9.277	22
40	.464	.929	1.393	1.857	2.322	2.786	3.250	3.715	4.179	4.643	5.108	5.572	6.036	6.501	6.965	7.429	7.894	8.358	8.823	9.287	20
42	.465	.930	1.395	1.860	2.325	2.789	3.254	3.719	4.184	4.649	5.114	5.578	6.043	6.509	6.973	7.438	7.903	8.368	8.833	9.298	18
44	.465	.931	1.397	1.862	2.328	2.792	3.258	3.723	4.189	4.654	5.120	5.584	6.050	6.516	6.981	7.446	7.912	8.377	8.842	9.308	16
46	.466	.932	1.399	1.864	2.331	2.795	3.262	3.727	4.194	4.659	5.126	5.590	6.057	6.523	6.989	7.454	7.921	8.386	8.852	9.318	14
48	.467	.933	1.400	1.866	2.333	2.798	3.265	3.731	4.198	4.664	5.131	5.596	6.064	6.530	6.997	7.462	7.929	8.395	8.862	9.328	12
50	.467	.934	1.401	1.868	2.335	2.801	3.268	3.735	4.202	4.669	5.136	5.602	6.070	6.537	7.004	7.470	7.937	8.404	8.871	9.338	10
52	.468	.935	1.403	1.870	2.338	2.805	3.272	3.740	4.207	4.675	5.142	5.609	6.077	6.545	7.012	7.479	7.946	8.414	8.881	9.349	8
54	.468	.936	1.405	1.872	2.341	2.808	3.276	3.744	4.212	4.680	5.148	5.615	6.084	6.552	7.020	7.487	7.955	8.423	8.891	9.359	6
56	.469	.937	1.406	1.874	2.344	2.811	3.280	3.748	4.217	4.685	5.154	5.621	6.091	6.559	7.028	7.495	7.964	8.432	8.901	9.369	4
58	.470	.938	1.407	1.876	2.346	2.814	3.283	3.752	4.221	4.690	5.159	5.627	6.097	6.566	7.035	7.503	7.973	8.441	8.911	9.379	2
28° 0	.470	.939	1.408	1.878	2.348	2.817	3.286	3.756	4.225	4.695	5.164	5.633	6.103	6.573	7.042	7.511	7.981	8.450	8.920	9.389	2° 0
	1	2	3	4	5	6	7	8	9	10	11	12	13	14	15	16	17	18	19	20	V + XI −

GENERAL TABLE of Aberration, &c. continued.

S. − V + O	1	2	3	4	5	6	7	8	9	10	11	12	13	14	15	16	17	18	19	20	
28° 0′	.470	.939	1.408	1.878	2.348	2.817	3.286	3.756	4.225	4.695	5.164	5.634	6.103	6.573	7.042	7.511	7.981	8.250	8.920	9.390	2° 0′
2	.471	.940	1.410	1.880	2.350	2.820	3.290	3.760	4.230	4.700	5.170	5.640	6.110	6.581	7.050	7.520	7.990	8.261	8.930	9.400	58
4	.471	.941	1.412	1.882	2.353	2.823	3.294	3.764	4.235	4.705	5.176	5.646	6.117	6.588	7.058	7.529	7.999	8.269	8.940	9.410	56
6	.472	.942	1.414	1.884	2.356	2.826	3.298	3.768	4.240	4.710	5.182	5.652	6.124	6.595	7.066	7.537	8.008	8.279	8.950	9.420	54
8	.472	.943	1.415	1.886	2.358	2.829	3.301	3.772	4.244	4.715	5.187	5.658	6.131	6.602	7.074	7.545	8.016	8.288	8.959	9.430	52
10	.472	.944	1.416	1.888	2.360	2.832	3.304	3.776	4.248	4.720	5.192	5.664	6.137	6.609	7.081	7.553	8.024	8.297	8.968	9.440	50
12	.473	.945	1.418	1.890	2.363	2.836	3.308	3.781	4.253	4.726	5.198	5.671	6.144	6.616	7.089	7.562	8.033	8.307	8.978	9.451	48
14	.473	.946	1.420	1.892	2.366	2.839	3.312	3.785	4.258	4.731	5.204	5.677	6.151	6.623	7.097	7.570	8.042	8.316	8.988	9.462	46
16	.474	.947	1.422	1.894	2.369	2.842	3.316	3.789	4.263	4.736	5.210	5.683	6.158	6.630	7.105	7.578	8.051	8.325	8.998	9.472	44
18	.475	.948	1.423	1.896	2.371	2.845	3.319	3.793	4.267	4.741	5.216	5.689	6.164	6.637	7.112	7.586	8.060	8.334	9.008	9.482	42
20	.475	.949	1.424	1.898	2.373	2.848	3.322	3.797	4.271	4.746	5.221	5.695	6.170	6.644	7.119	7.594	8.068	8.343	9.017	9.492	40
22	.476	.950	1.426	1.901	2.376	2.851	3.326	3.801	4.276	4.752	5.227	5.702	6.177	6.652	7.127	7.603	8.077	8.353	9.027	9.503	38
24	.476	.951	1.428	1.903	2.379	2.854	3.330	3.805	4.281	4.757	5.233	5.708	6.184	6.659	7.135	7.611	8.086	8.362	9.037	9.514	36
26	.477	.952	1.430	1.905	2.382	2.857	3.334	3.809	4.286	4.762	5.239	5.714	6.191	6.666	7.143	7.619	8.095	8.371	9.047	9.524	34
28	.477	.953	1.431	1.907	2.384	2.860	3.337	3.813	4.290	4.767	5.244	5.720	6.197	6.673	7.150	7.627	8.103	8.380	9.057	9.534	32
30	.477	.954	1.432	1.909	2.386	2.863	3.340	3.817	4.294	4.772	5.249	5.726	6.203	6.680	7.157	7.635	8.111	8.389	9.066	9.544	30
32	.478	.955	1.434	1.911	2.389	2.866	3.344	3.822	4.299	4.777	5.255	5.732	6.210	6.688	7.165	7.643	8.120	8.399	9.076	9.554	28
34	.478	.956	1.436	1.913	2.392	2.869	3.348	3.826	4.304	4.782	5.261	5.738	6.217	6.695	7.173	7.651	8.129	8.408	9.086	9.564	26
36	.479	.957	1.437	1.915	2.395	2.872	3.352	3.830	4.309	4.787	5.267	5.744	6.224	6.702	7.181	7.659	8.138	8.417	9.096	9.574	24
38	.480	.958	1.438	1.917	2.397	2.875	3.355	3.834	4.313	4.792	5.272	5.750	6.230	6.709	7.189	7.667	8.147	8.426	9.105	9.584	22
40	.480	.959	1.439	1.919	2.399	2.878	3.358	3.838	4.317	4.797	5.277	5.756	6.236	6.716	7.196	7.675	8.155	8.435	9.114	9.594	20
42	.481	.961	1.441	1.921	2.402	2.882	3.362	3.842	4.322	4.803	5.283	5.763	6.243	6.724	7.204	7.684	8.164	8.445	9.124	9.605	18
44	.481	.962	1.443	1.923	2.405	2.885	3.366	3.846	4.327	4.808	5.289	5.769	6.250	6.731	7.212	7.692	8.173	8.454	9.134	9.615	16
46	.482	.963	1.445	1.925	2.407	2.888	3.370	3.850	4.332	4.813	5.295	5.775	6.257	6.738	7.220	7.700	8.182	8.463	9.144	9.625	14
48	.482	.964	1.446	1.927	2.409	2.891	3.373	3.854	4.336	4.818	5.300	5.781	6.263	6.745	7.227	7.708	8.190	8.472	9.154	9.635	12
50	.482	.965	1.447	1.929	2.411	2.894	3.376	3.858	4.340	4.823	5.305	5.787	6.269	6.752	7.234	7.716	8.198	8.481	9.163	9.645	10
52	.483	.966	1.449	1.931	2.414	2.897	3.380	3.863	4.345	4.828	5.311	5.794	6.276	6.759	7.242	7.725	8.207	8.491	9.173	9.656	8
54	.483	.967	1.451	1.933	2.417	2.900	3.384	3.867	4.350	4.833	5.317	5.800	6.283	6.766	7.250	7.733	8.216	8.500	9.183	9.666	6
56	.484	.968	1.452	1.935	2.420	2.903	3.388	3.871	4.355	4.838	5.323	5.806	6.290	6.773	7.258	7.741	8.225	8.509	9.193	9.676	4
58	.485	.969	1.453	1.937	2.422	2.906	3.391	3.875	4.359	4.843	5.328	5.812	6.297	6.780	7.265	7.749	8.234	8.518	9.202	9.686	2
29° 0′	.485	.970	1.454	1.939	2.424	2.909	3.394	3.879	4.363	4.848	5.333	5.818	6.303	6.787	7.272	7.757	8.242	8.527	9.211	9.696	1° 0′
2	.486	.971	1.456	1.941	2.427	2.912	3.398	3.883	4.368	4.854	5.339	5.824	6.310	6.795	7.280	7.766	8.251	8.537	9.221	9.707	58
4	.486	.972	1.458	1.943	2.430	2.915	3.402	3.887	4.373	4.859	5.345	5.830	6.317	6.802	7.288	7.774	8.260	8.546	9.231	9.717	56
6	.487	.973	1.460	1.945	2.433	2.918	3.406	3.891	4.378	4.864	5.351	5.836	6.324	6.809	7.296	7.782	8.269	8.555	9.241	9.727	54
8	.487	.974	1.461	1.947	2.435	2.921	3.409	3.895	4.382	4.869	5.356	5.842	6.330	6.816	7.303	7.790	8.277	8.564	9.251	9.737	52
10	.487	.975	1.462	1.949	2.437	2.924	3.412	3.899	4.386	4.874	5.361	5.848	6.336	6.823	7.310	7.798	8.285	8.573	9.260	9.747	50
12	.488	.976	1.464	1.952	2.440	2.927	3.416	3.903	4.391	4.879	5.367	5.855	6.343	6.830	7.318	7.806	8.294	8.582	9.270	9.758	48
14	.488	.977	1.466	1.954	2.443	2.930	3.420	3.907	4.396	4.884	5.373	5.861	6.350	6.837	7.326	7.814	8.303	8.591	9.280	9.768	46
16	.489	.978	1.468	1.956	2.446	2.933	3.423	3.911	4.401	4.889	5.379	5.867	6.357	6.844	7.334	7.822	8.312	8.600	9.290	9.778	44
18	.490	.979	1.469	1.958	2.448	2.936	3.426	3.915	4.405	4.894	5.384	5.873	6.363	6.851	7.341	7.830	8.320	8.609	9.299	9.788	42
20	.490	.980	1.470	1.960	2.450	2.939	3.429	3.919	4.409	4.899	5.389	5.879	6.369	6.858	7.349	7.838	8.328	8.618	9.308	9.798	40
22	.491	.981	1.472	1.962	2.453	2.943	3.433	3.923	4.414	4.904	5.395	5.885	6.376	6.866	7.357	7.847	8.337	8.627	9.318	9.808	38
24	.491	.982	1.474	1.964	2.456	2.946	3.437	3.927	4.419	4.909	5.401	5.891	6.383	6.873	7.365	7.855	8.346	8.636	9.328	9.818	36
26	.492	.983	1.475	1.966	2.458	2.949	3.441	3.931	4.424	4.914	5.407	5.897	6.389	6.880	7.372	7.863	8.355	8.645	9.338	9.828	34
28	.492	.984	1.476	1.968	2.460	2.952	3.444	3.935	4.428	4.919	5.412	5.903	6.395	6.887	7.379	7.871	8.363	8.654	9.347	9.838	32
30	.492	.985	1.477	1.970	2.462	2.955	3.447	3.939	4.432	4.924	5.417	5.909	6.401	6.894	7.386	7.879	8.371	8.663	9.356	9.848	30
32	.493	.986	1.479	1.972	2.465	2.958	3.451	3.944	4.437	4.930	5.423	5.915	6.408	6.901	7.394	7.887	8.380	8.673	9.366	9.859	28
34	.493	.987	1.481	1.974	2.468	2.961	3.455	3.948	4.442	4.935	5.429	5.921	6.415	6.908	7.402	7.895	8.389	8.682	9.376	9.869	26
36	.494	.988	1.483	1.976	2.471	2.964	3.459	3.952	4.447	4.940	5.435	5.927	6.422	6.915	7.410	7.903	8.398	8.691	9.386	9.879	24
38	.495	.989	1.484	1.978	2.473	2.967	3.462	3.956	4.451	4.945	5.440	5.933	6.428	6.922	7.417	7.911	8.407	8.700	9.396	9.889	22
40	.495	.990	1.485	1.980	2.475	2.970	3.465	3.960	4.455	4.950	5.445	5.939	6.434	6.929	7.424	7.919	8.415	8.709	9.405	9.899	20
42	.496	.991	1.487	1.982	2.478	2.973	3.469	3.964	4.460	4.955	5.451	5.946	6.441	6.937	7.432	7.928	8.424	8.719	9.415	9.910	18
44	.496	.992	1.489	1.984	2.481	2.976	3.473	3.968	4.465	4.960	5.457	5.952	6.448	6.944	7.440	7.936	8.433	8.728	9.425	9.920	16
46	.497	.993	1.490	1.986	2.483	2.979	3.476	3.972	4.469	4.965	5.462	5.958	6.455	6.951	7.448	7.944	8.441	8.737	9.434	9.930	14
48	.498	.994	1.491	1.988	2.485	2.982	3.479	3.976	4.473	4.970	5.467	5.964	6.461	6.958	7.455	7.952	8.449	8.746	9.443	9.940	12
50	.498	.995	1.492	1.990	2.487	2.985	3.482	3.980	4.477	4.975	5.472	5.970	6.467	6.965	7.462	7.960	8.457	8.755	9.452	9.950	10
52	.499	.996	1.494	1.992	2.490	2.988	3.486	3.984	4.482	4.980	5.478	5.976	6.474	6.972	7.470	7.968	8.466	8.764	9.462	9.960	8
54	.499	.997	1.496	1.994	2.493	2.991	3.490	3.988	4.487	4.985	5.484	5.982	6.481	6.979	7.478	7.976	8.475	8.773	9.472	9.970	6
56	.500	.998	1.498	1.996	2.496	2.994	3.494	3.992	4.492	4.990	5.490	5.988	6.488	6.986	7.486	7.984	8.484	8.782	9.482	9.980	4
58	.500	.999	1.499	1.998	2.498	2.997	3.497	3.996	4.496	4.995	5.495	5.994	6.494	6.993	7.493	7.092	8.492	8.791	9.491	9.990	2
30° 0	.500	1.000	1.500	2.000	2.500	3.000	3.500	4.000	4.500	5.000	5.500	6.000	6.500	7.000	7.500	8.000	8.500	9.000	9.500	10.000	0° 0
	1	2	3	4	5	6	7	8	9	10	11	12	13	14	15	16	17	18	19	20	V + XI −

TABLE XVIII.

GENERAL TABLE of Aberration, &c. continued.

S. −VII + I	1	2	3	4	5	6	7	8	9	10	11	12	13	14	15	16	17	18	19	20	
0° 0′	.500	1.000	1.500	2.000	2.500	3.000	3.500	4.000	4.500	5.000	5.500	6.000	6.500	7.000	7.500	8.000	8.500	9.000	9.500	10.000	30° 0′
2	.501	1.001	1.502	2.002	2.503	3.003	3.504	4.004	4.505	5.005	5.506	6.006	6.507	7.008	7.508	8.008	8.509	9.010	9.510	10.010	58
4	.501	1.002	1.503	2.004	2.505	3.006	3.507	4.008	4.509	5.010	5.512	6.012	6.514	7.015	7.516	8.016	8.518	9.019	9.520	10.020	56
6	.502	1.003	1.505	2.006	2.508	3.009	3.511	4.012	4.514	5.015	5.517	6.018	6.520	7.022	7.523	8.024	8.526	9.028	9.529	10.030	54
8	.502	1.004	1.507	2.008	2.510	3.012	3.515	4.016	4.519	5.020	5.522	6.024	6.527	7.029	7.531	8.032	8.535	9.037	9.539	10.040	52
10	.503	1.005	1.508	2.010	2.513	3.015	3.518	4.020	4.523	5.025	5.528	6.030	6.533	7.036	7.538	8.040	8.543	9.046	9.548	10.050	50
12	.503	1.006	1.510	2.012	2.516	3.018	3.522	4.024	4.528	5.030	5.534	6.036	6.540	7.043	7.546	8.048	8.552	9.055	9.558	10.060	48
14	.504	1.007	1.511	2.014	2.518	3.021	3.525	4.028	4.532	5.035	5.539	6.042	6.546	7.050	7.553	8.056	8.560	9.064	9.567	10.070	46
16	.504	1.008	1.513	2.016	2.520	3.024	3.529	4.032	4.537	5.040	5.545	6.048	6.553	7.057	7.561	8.064	8.569	9.073	9.577	10.080	44
18	.505	1.009	1.514	2.018	2.523	3.027	3.532	4.036	4.541	5.045	5.550	6.054	6.559	7.064	7.568	8.072	8.577	9.082	9.586	10.090	42
20	.505	1.010	1.515	2.020	2.525	3.030	3.535	4.040	4.545	5.050	5.555	6.060	6.565	7.070	7.575	8.080	8.585	9.091	9.595	10.100	40
22	.506	1.011	1.517	2.022	2.528	3.033	3.539	4.044	4.550	5.055	5.561	6.066	6.572	7.077	7.583	8.088	8.594	9.100	9.605	10.110	38
24	.506	1.012	1.519	2.024	2.531	3.036	3.543	4.048	4.555	5.060	5.567	6.072	6.579	7.085	7.591	8.096	8.603	9.109	9.615	10.120	36
26	.507	1.013	1.520	2.026	2.533	3.039	3.546	4.052	4.559	5.065	5.572	6.078	6.585	7.092	7.598	8.104	8.611	9.118	9.624	10.130	34
28	.507	1.014	1.522	2.028	2.536	3.042	3.550	4.056	4.564	5.070	5.578	6.084	6.592	7.099	7.605	8.112	8.620	9.127	9.634	10.140	32
30	.508	1.015	1.523	2.030	2.538	3.045	3.553	4.060	4.568	5.075	5.583	6.090	6.598	7.106	7.613	8.120	8.628	9.136	9.643	10.150	30
32	.508	1.016	1.525	2.032	2.541	3.048	3.557	4.064	4.573	5.080	5.589	6.096	6.605	7.113	7.620	8.128	8.637	9.145	9.653	10.160	28
34	.509	1.017	1.526	2.034	2.543	3.051	3.560	4.068	4.577	5.085	5.594	6.102	6.611	7.120	7.628	8.136	8.645	9.154	9.662	10.170	26
36	.509	1.018	1.528	2.036	2.546	3.054	3.564	4.072	4.582	5.090	5.600	6.108	6.618	7.127	7.636	8.144	8.653	9.163	9.672	10.180	24
38	.510	1.019	1.529	2.038	2.548	3.057	3.567	4.076	4.586	5.095	5.605	6.114	6.624	7.134	7.643	8.152	8.662	9.172	9.681	10.190	22
40	.510	1.020	1.530	2.040	2.550	3.060	3.570	4.080	4.590	5.100	5.610	6.120	6.630	7.140	7.650	8.160	8.670	9.181	9.690	10.200	20
42	.511	1.021	1.532	2.042	2.553	3.063	3.574	4.084	4.595	5.105	5.616	6.126	6.637	7.148	7.658	8.168	8.679	9.190	9.700	10.210	18
44	.511	1.022	1.534	2.044	2.556	3.066	3.578	4.088	4.599	5.110	5.622	6.132	6.644	7.155	7.666	8.176	8.688	9.199	9.710	10.220	16
46	.512	1.023	1.535	2.046	2.558	3.069	3.581	4.092	4.603	5.115	5.627	6.138	6.650	7.162	7.673	8.184	8.696	9.208	9.719	10.230	14
48	.512	1.024	1.537	2.048	2.561	3.072	3.585	4.096	4.608	5.120	5.633	6.144	6.657	7.169	7.681	8.192	8.705	9.217	9.729	10.240	12
50	.513	1.025	1.538	2.050	2.563	3.075	3.588	4.100	4.613	5.125	5.638	6.150	6.663	7.176	7.688	8.200	8.713	9.226	9.738	10.250	10
52	.513	1.026	1.540	2.052	2.566	3.078	3.592	4.104	4.618	5.130	5.644	6.156	6.670	7.183	7.696	8.208	8.722	9.235	9.748	10.260	8
54	.514	1.027	1.541	2.054	2.568	3.081	3.595	4.108	4.622	5.135	5.649	6.162	6.676	7.190	7.704	8.216	8.730	9.244	9.757	10.270	6
56	.514	1.028	1.543	2.056	2.571	3.084	3.599	4.112	4.627	5.140	5.655	6.168	6.683	7.197	7.711	8.224	8.739	9.253	9.767	10.280	4
58	.515	1.029	1.544	2.058	2.573	3.087	3.602	4.116	4.631	5.145	5.660	6.174	6.688	7.204	7.718	8.232	8.747	9.262	9.776	10.290	2
1° 0′	.515	1.030	1.545	2.060	2.575	3.090	3.605	4.120	4.635	5.150	5.665	6.180	6.695	7.211	7.725	8.240	8.755	9.271	9.785	10.300	29° 0′
2	.516	1.031	1.547	2.062	2.578	3.093	3.609	4.124	4.640	5.155	5.671	6.186	6.702	7.218	7.733	8.248	8.764	9.280	9.795	10.310	58
4	.516	1.032	1.549	2.064	2.581	3.096	3.613	4.128	4.645	5.160	5.677	6.192	6.709	7.225	7.741	8.256	8.773	9.289	9.805	10.320	56
6	.517	1.033	1.550	2.066	2.583	3.099	3.616	4.132	4.649	5.165	5.682	6.198	6.715	7.232	7.748	8.264	8.781	9.298	9.814	10.330	54
8	.517	1.034	1.552	2.068	2.586	3.102	3.620	4.136	4.654	5.170	5.688	6.204	6.722	7.239	7.756	8.272	8.790	9.307	9.824	10 340	52
10	.518	1.035	1.553	2.070	2.588	3.105	3.623	4.140	4.658	5.175	5.693	6.210	6.728	7.246	7.763	8.280	8.798	9.316	9.833	10.350	50
12	.518	1.036	1.555	2.072	2.591	3.108	3.627	4.144	4.663	5.180	5.699	6.216	6.735	7.253	7.771	8.288	8.807	9.325	9.843	10.360	48
14	.519	1.037	1.556	2.074	2.593	3.111	3.630	4.148	4.667	5.185	5.704	6.222	6.741	7.260	7.778	8.296	8.815	9.334	9.852	10.370	46
16	.519	1.038	1.558	2.076	2.596	3.114	3.634	4.152	4.672	5.190	5.710	6.228	6.748	7.267	7.786	8.304	8.824	9.343	9.862	10.380	44
18	.520	1.039	1.559	2.078	2.598	3.117	3.637	4.156	4.676	5.195	5.715	6.234	6.754	7.274	7.793	8.312	8.832	9.352	9.871	10.390	42
20	.520	1.040	1.560	2.080	2.600	3.120	3.640	4.160	4.680	5.200	5.720	6.240	6.760	7.281	7.800	8.320	8.840	9.361	9.880	10.400	40
22	.521	1.041	1.562	2.082	2.603	3.123	3.644	4.164	4.685	5.205	5.726	6.246	6.767	7.288	7.808	8.328	8.849	9.370	9.890	10.410	38
24	.521	1.042	1.564	2.084	2.605	3.126	3.648	4.168	4.690	5.210	5.732	6.252	6.774	7.295	7.816	8.336	8.858	9.379	9.900	10.420	36
26	.522	1.043	1.565	2.086	2.608	3.129	3.651	4.172	4.694	5.215	5.737	6.258	6.780	7.302	7.823	8.344	8.866	9.388	9.909	10.430	34
28	.522	1.044	1.567	2.088	2.611	3.132	3.655	4.176	4.699	5.220	5.743	6.264	6.787	7.309	7.831	8.352	8.875	9.397	9.919	10.440	32
30	.523	1.045	1.568	2.090	2.613	3.135	3.658	4.180	4.703	5.225	5.748	6.270	6.793	7.316	7.838	8.360	8.883	9.406	9.928	10.450	30
32	.523	1.046	1.570	2.092	2.616	3.138	3.662	4.184	4.708	5.230	5.754	6.276	6.800	7.323	7.846	8.368	8.892	9.415	9.938	10.460	28
34	.524	1.047	1.571	2.094	2.618	3.141	3.665	4.188	4.713	5.235	5.759	6.282	6.807	7.330	7.853	8.376	8.900	9.424	9.947	10.470	26
36	.524	1.048	1.573	2.096	2.621	3.144	3.669	4.192	4.717	5.240	5.765	6.288	6.813	7.337	7.861	8.384	8.909	9.433	9.957	10.480	24
38	.525	1.049	1.574	2.098	2.623	3.147	3.672	4.196	4.721	5.245	5.770	6.294	6.820	7.344	7.868	8.392	8.917	9.442	9.966	10.490	22
40	.525	1.050	1.575	2.100	2.625	3.150	3.675	4.200	4.725	5.250	5.775	6.300	6.825	7.350	7.875	8.400	8.925	9.450	9.975	10.500	20
42	.526	1.051	1.577	2.102	2.628	3.153	3.679	4.204	4.730	5.255	5.781	6.306	6.832	7.357	7.883	8.408	8.934	9.459	9.985	10.510	18
44	.526	1.052	1.578	2.104	2.631	3.156	3.682	4.208	4.734	5.260	5.787	6.312	6.838	7.364	7.890	8.416	8.942	9.468	9.994	10.520	16
46	.527	1.053	1.580	2.106	2.633	3.159	3.686	4.212	4.739	5.265	5.792	6.318	6.845	7.371	7.898	8.424	8.951	9.477	10.004	10.530	14
48	.527	1.054	1.581	2.108	2.635	3.162	3.689	4.216	4.743	5.270	5.798	6.324	6.851	7.378	7.905	8.432	8.959	9.486	10.013	10.540	12
50	.528	1.055	1.582	2.110	2.637	3.165	3.692	4.220	4.747	5.275	5.803	6.330	6.857	7.384	7.912	8.440	8.967	9.494	10.022	10.550	10
52	.528	1.056	1.584	2.112	2.640	3.168	3.696	4.224	4.752	5.280	5.809	6.336	6.864	7.391	7.920	8.448	8.976	9.503	10.032	10.560	8
54	.529	1.057	1.586	2.114	2.643	3.171	3.700	4.228	4.756	5.285	5.814	6.342	6.871	7.398	7.927	8.456	8.984	9.512	10.041	10.570	6
56	.529	1.058	1.587	2.116	2.645	3.174	3.703	4.232	4.761	5.290	5.819	6.348	6.877	7.405	7.934	8.464	8.992	9.521	10.050	10.580	4
58	.530	1.059	1.589	2.118	2.648	3.177	3.706	4.236	4.765	5.295	5.824	6.354	6.883	7.412	7.941	8.471	9.000	9.530	10.059	10.590	2
2° 0	.530	1.060	1.590	2.120	2.650	3.180	3.709	4.239	4.769	5.299	5.829	6.360	6.889	7.418	7.948	8.478	9.008	9.538	10.068	10.599	28° 0
	1	2	3	4	5	6	7	8	9	10	11	12	13	14	15	16	17	18	19	20	IV + X −

GENERAL TABLE of Aberration, &c. continued.

S. − VII + I	1	2	3	4	5	6	7	8	9	10	11	12	13	14	15	16	17	18	19	20	
2° 0′	.530	1 .060	1 .590	2 .120	2 .650	3 .180	3 .709	4 .239	4 .769	5 .299	5 .829	6 .360	6 .889	7 .418	7 .948	8 .478	9 .008	9 .538	10 .068	10 .598	28° 0′
2	.530	1 .061	1 .592	2 .122	2 .653	3 .183	3 .713	4 .243	4 .774	5 .304	5 .835	6 .366	6 .896	7 .426	7 .956	8 .486	9 .017	9 .547	10 .078	10 .608	58
4	.531	1 .062	1 .593	2 .124	2 .655	3 .186	3 .717	4 .247	4 .778	5 .309	5 .840	6 .371	6 .902	7 .433	7 .964	8 .494	9 .025	9 .556	10 .087	10 .618	56
6	.531	1 .063	1 .594	2 .126	2 .658	3 .188	3 .720	4 .251	4 .783	5 .314	5 .846	6 .376	6 .908	7 .440	7 .971	8 .502	9 .034	9 .565	10 .097	10 .628	54
8	.532	1 .064	1 .596	2 .128	2 .660	3 .191	3 .724	4 .255	4 .787	5 .319	5 .851	6 .382	6 .914	7 .447	7 .979	8 .510	9 .042	9 .574	10 .106	10 .638	52
10	.532	1 .065	1 .597	2 .130	2 .662	3 .194	3 .727	4 .259	4 .791	5 .324	5 .856	6 .388	6 .921	7 .454	7 .986	8 .518	9 .050	9 .582	10 .115	10 .648	50
12	.533	1 .066	1 .599	2 .132	2 .664	3 .197	3 .731	4 .263	4 .796	5 .329	5 .862	6 .394	6 .928	7 .461	7 .994	8 .526	9 .059	9 .591	10 .125	10 .658	48
14	.533	1 .067	1 .601	2 .134	2 .667	3 .200	3 .734	4 .267	4 .801	5 .334	5 .867	6 .400	6 .934	7 .468	8 .001	8 .534	9 .068	9 .600	10 .134	10 .668	46
16	.534	1 .068	1 .602	2 .136	2 .670	3 .203	3 .738	4 .271	4 .805	5 .339	5 .873	6 .406	6 .941	7 .475	8 .009	8 .542	9 .076	9 .609	10 .144	10 .678	44
18	.534	1 .069	1 .604	2 .138	2 .672	3 .206	3 .741	4 .275	4 .810	5 .344	5 .878	6 .412	6 .947	7 .482	8 .016	8 .550	9 .085	9 .618	10 .153	10 .688	42
20	.535	1 .070	1 .605	2 .139	2 .674	3 .209	3 .744	4 .279	4 .814	5 .348	5 .883	6 .418	6 .953	7 .488	8 .023	8 .558	9 .093	9 .627	10 .162	10 .697	40
22	.535	1 .071	1 .607	2 .141	2 .677	3 .212	3 .748	4 .283	4 .819	5 .353	5 .889	6 .424	6 .960	7 .495	8 .031	8 .566	9 .101	9 .636	10 .172	10 .707	38
24	.536	1 .072	1 .608	2 .143	2 .680	3 .215	3 .751	4 .287	4 .823	5 .358	5 .894	6 .430	6 .966	7 .502	8 .038	8 .574	9 .109	9 .645	10 .181	10 .717	36
26	.536	1 .073	1 .610	2 .145	2 .682	3 .218	3 .755	4 .291	4 .828	5 .363	5 .900	6 .436	6 .973	7 .509	8 .045	8 .582	9 .117	9 .654	10 .191	10 .727	34
28	.537	1 .074	1 .611	2 .147	2 .685	3 .221	3 .758	4 .295	4 .832	5 .368	5 .905	6 .442	6 .979	7 .516	8 .052	8 .590	9 .125	9 .663	10 .200	10 .737	32
30	.537	1 .075	1 .612	2 .149	2 .687	3 .224	3 .761	4 .298	4 .836	5 .373	5 .910	6 .448	6 .985	7 .522	8 .059	8 .597	9 .134	9 .672	10 .209	10 .746	30
32	.538	1 .076	1 .614	2 .151	2 .690	3 .227	3 .765	4 .302	4 .841	5 .378	5 .916	6 .454	6 .992	7 .529	8 .067	8 .606	9 .143	9 .681	10 .219	10 .756	28
34	.538	1 .077	1 .615	2 .153	2 .692	3 .230	3 .768	4 .306	4 .845	5 .383	5 .922	6 .460	6 .998	7 .536	8 .074	8 .614	9 .151	9 .690	10 .228	10 .766	26
36	.539	1 .078	1 .617	2 .155	2 .695	3 .233	3 .772	4 .310	4 .850	5 .388	5 .927	6 .466	7 .005	7 .543	8 .082	8 .622	9 .160	9 .699	10 .238	10 .776	24
38	.539	1 .079	1 .618	2 .157	2 .697	3 .236	3 .775	4 .314	4 .854	5 .393	5 .933	6 .472	7 .011	7 .550	8 .089	8 .629	9 .168	9 .708	10 .247	10 .786	22
40	.540	1 .080	1 .619	2 .159	2 .699	3 .239	3 .778	4 .318	4 .858	5 .398	5 .938	6 .478	7 .017	7 .556	8 .096	8 .636	9 .176	9 .716	10 .256	10 .796	20
42	.540	1 .081	1 .621	2 .161	2 .702	3 .242	3 .782	4 .322	4 .863	5 .403	5 .944	6 .484	7 .024	7 .563	8 .104	8 .644	9 .184	9 .725	10 .266	10 .806	18
44	.541	1 .082	1 .623	2 .163	2 .704	3 .245	3 .785	4 .326	4 .867	5 .408	5 .949	6 .490	7 .030	7 .570	8 .111	8 .652	9 .192	9 .734	10 .275	10 .816	16
46	.541	1 .083	1 .624	2 .165	2 .707	3 .248	3 .789	4 .330	4 .872	5 .413	5 .955	6 .496	7 .036	7 .577	8 .119	8 .660	9 .201	9 .743	10 .284	10 .825	14
48	.542	1 .084	1 .626	2 .167	2 .709	3 .251	3 .792	4 .334	4 .876	5 .418	5 .960	6 .502	7 .042	7 .584	8 .126	8 .668	9 .209	9 .752	10 .293	10 .835	12
50	.542	1 .085	1 .627	2 .169	2 .711	3 .253	3 .795	4 .338	4 .880	5 .422	5 .964	6 .507	7 .048	7 .590	8 .133	8 .676	9 .218	9 .760	10 .302	10 .844	10
52	.543	1 .086	1 .629	2 .171	2 .714	3 .256	3 .799	4 .342	4 .885	5 .427	5 .970	6 .513	7 .055	7 .597	8 .141	8 .684	9 .227	9 .769	10 .312	10 .854	8
54	.543	1 .087	1 .630	2 .173	2 .716	3 .259	3 .803	4 .346	4 .889	5 .432	5 .975	6 .519	7 .062	7 .604	8 .148	8 .692	9 .235	9 .778	10 .321	10 .863	6
56	.544	1 .088	1 .632	2 .175	2 .719	3 .262	3 .807	4 .350	4 .893	5 .437	5 .981	6 .525	7 .068	7 .611	8 .156	8 .700	9 .243	9 .787	10 .330	10 .873	4
58	.544	1 .089	1 .633	2 .177	2 .721	3 .265	3 .810	4 .354	4 .897	5 .441	5 .986	6 .531	7 .074	7 .618	8 .163	8 .708	9 .251	9 .796	10 .339	10 .882	2
3° 0′	.545	1 .089	1 .634	2 .179	2 .723	3 .268	3 .813	4 .357	4 .902	5 .446	5 .991	6 .536	7 .081	7 .625	8 .170	8 .715	9 .259	9 .804	10 .348	10 .892	27° 0′
2	.545	1 .090	1 .636	2 .181	2 .726	3 .271	3 .817	4 .361	4 .907	5 .451	5 .997	6 .542	7 .088	7 .632	8 .178	8 .723	9 .268	9 .813	10 .358	10 .902	58
4	.546	1 .091	1 .637	2 .183	2 .728	3 .274	3 .821	4 .365	4 .911	5 .456	6 .002	6 .548	7 .094	7 .639	8 .185	8 .731	9 .276	9 .822	10 .367	10 .912	56
6	.546	1 .092	1 .639	2 .185	2 .731	3 .277	3 .824	4 .369	4 .916	5 .461	6 .008	6 .554	7 .101	7 .646	8 .193	8 .739	9 .285	9 .831	10 .377	10 .922	54
8	.547	1 .093	1 .640	2 .187	2 .733	3 .280	3 .827	4 .373	4 .920	5 .466	6 .013	6 .560	7 .107	7 .653	8 .200	8 .747	9 .294	9 .840	10 .386	10 .932	52
10	.547	1 .094	1 .641	2 .188	2 .735	3 .283	3 .830	4 .377	4 .924	5 .471	6 .018	6 .566	7 .113	7 .660	8 .207	8 .754	9 .301	9 .848	10 .395	10 .942	50
12	.548	1 .095	1 .643	2 .190	2 .738	3 .286	3 .834	4 .381	4 .929	5 .476	6 .024	6 .572	7 .120	7 .667	8 .215	8 .762	9 .310	9 .857	10 .405	10 .952	48
14	.548	1 .096	1 .644	2 .192	2 .741	3 .289	3 .837	4 .385	4 .933	5 .481	6 .029	6 .578	7 .126	7 .674	8 .222	8 .770	9 .318	9 .866	10 .414	10 .961	46
16	.549	1 .097	1 .646	2 .194	2 .743	3 .292	3 .841	4 .389	4 .937	5 .486	6 .035	6 .584	7 .133	7 .681	8 .229	8 .778	9 .326	9 .875	10 .423	10 .971	44
18	.549	1 .098	1 .648	2 .196	2 .746	3 .294	3 .844	4 .393	4 .941	5 .491	6 .040	6 .590	7 .139	7 .688	8 .236	8 .786	9 .334	9 .884	10 .432	10 .980	42
20	.550	1 .099	1 .649	2 .198	2 .748	3 .297	3 .847	4 .396	4 .946	5 .495	6 .045	6 .595	7 .145	7 .694	8 .243	8 .793	9 .342	9 .892	10 .441	10 .990	40
22	.550	1 .100	1 .651	2 .200	2 .751	3 .300	3 .851	4 .400	4 .951	5 .500	6 .051	6 .601	7 .152	7 .701	8 .251	8 .801	9 .351	9 .901	10 .451	10 .999	38
24	.551	1 .101	1 .652	2 .202	2 .753	3 .303	3 .854	4 .404	4 .955	5 .505	6 .056	6 .607	7 .158	7 .708	8 .258	8 .809	9 .359	9 .910	10 .460	11 .009	36
26	.551	1 .102	1 .654	2 .204	2 .756	3 .306	3 .858	4 .408	4 .960	5 .510	6 .061	6 .613	7 .164	7 .715	8 .266	8 .817	9 .368	9 .919	10 .469	11 .019	34
28	.552	1 .103	1 .655	2 .206	2 .758	3 .309	3 .861	4 .412	4 .964	5 .514	6 .066	6 .619	7 .170	7 .722	8 .273	8 .825	9 .376	9 .928	10 .478	11 .029	32
30	.552	1 .104	1 .656	2 .208	2 .760	3 .312	3 .864	4 .416	4 .968	5 .519	6 .071	6 .624	7 .176	7 .728	8 .280	8 .832	9 .384	9 .936	10 .487	11 .038	30
32	.552	1 .105	1 .658	2 .210	2 .763	3 .315	3 .868	4 .420	4 .972	5 .524	6 .077	6 .630	7 .183	7 .735	8 .288	8 .840	9 .392	9 .945	10 .497	11 .048	28
34	.553	1 .106	1 .659	2 .212	2 .765	3 .318	3 .871	4 .424	4 .977	5 .529	6 .082	6 .636	7 .189	7 .742	8 .295	8 .848	9 .400	9 .954	10 .506	11 .058	26
36	.553	1 .107	1 .661	2 .214	2 .768	3 .321	3 .875	4 .428	4 .981	5 .534	6 .088	6 .642	7 .195	7 .749	8 .302	8 .855	9 .408	9 .962	10 .515	11 .068	24
38	.554	1 .108	1 .662	2 .216	2 .770	3 .324	3 .878	4 .432	4 .985	5 .539	6 .093	6 .648	7 .201	7 .756	8 .309	8 .863	9 .416	9 .971	10 .524	11 .078	22
40	.554	1 .109	1 .663	2 .217	2 .772	3 .326	3 .881	4 .435	4 .989	5 .544	6 .098	6 .653	7 .207	7 .762	8 .316	8 .870	9 .424	9 .979	10 .533	11 .087	20
42	.555	1 .110	1 .665	2 .219	2 .775	3 .329	3 .885	4 .439	4 .994	5 .549	6 .104	6 .659	7 .214	7 .769	8 .323	8 .878	9 .433	9 .988	10 .543	11 .097	18
44	.555	1 .111	1 .666	2 .221	2 .777	3 .332	3 .888	4 .443	4 .998	5 .554	6 .109	6 .665	7 .220	7 .776	8 .330	8 .886	9 .441	9 .996	10 .552	11 .107	16
46	.556	1 .112	1 .668	2 .223	2 .780	3 .335	3 .891	4 .447	5 .003	5 .559	6 .114	6 .671	7 .226	7 .782	8 .337	8 .893	9 .449	10 .005	10 .561	11 .117	14
48	.556	1 .113	1 .669	2 .225	2 .782	3 .338	3 .894	4 .450	5 .007	5 .564	6 .119	6 .677	7 .232	7 .789	8 .344	8 .901	9 .457	10 .014	10 .570	11 .127	12
50	.557	1 .114	1 .670	2 .227	2 .784	3 .341	3 .897	4 .454	5 .011	5 .568	6 .125	6 .682	7 .238	7 .795	8 .351	8 .908	9 .465	10 .022	10 .579	11 .136	10
52	.557	1 .115	1 .672	2 .229	2 .787	3 .345	3 .901	4 .458	5 .016	5 .573	6 .131	6 .688	7 .245	7 .802	8 .359	8 .916	9 .473	10 .031	10 .589	11 .146	8
54	.558	1 .116	1 .674	2 .231	2 .789	3 .347	3 .904	4 .462	5 .020	5 .578	6 .136	6 .694	7 .251	7 .808	8 .367	8 .924	9 .482	10 .040	10 .598	11 .156	6
56	.558	1 .117	1 .675	2 .233	2 .792	3 .350	3 .908	4 .466	5 .025	5 .583	6 .141	6 .700	7 .257	7 .815	8 .374	8 .932	9 .491	10 .049	10 .607	11 .165	4
58	.559	1 .118	1 .677	2 .235	2 .794	3 .353	3 .911	4 .470	5 .029	5 .588	6 .146	6 .705	7 .263	7 .821	8 .381	8 .940	9 .499	10 .058	10 .616	11 .175	2
4° 0	.559	1 .118	1 .678	2 .237	2 .796	3 .355	3 .914	4 .474	5 .033	5 .592	6 .151	6 .710	7 .269	7 .828	8 .388	8 .948	9 .507	10 .066	10 .625	11 .184	26° 0
	1	2	3	4	5	6	7	8	9	10	11	12	13	14	15	16	17	18	19	20	IV + X −

TABLE XVIII.

GENERAL TABLE of Aberration, &c. continued.

S. −VII + I	1	2	3	4	5	6	7	8	9	10	11	12	13	14	15	16	17	18	19	20	
4° 0′	.559	1.118	1.678	2.237	2.796	3.355	3.914	4.474	5.033	5.592	6.151	6.710	7.269	7.828	8.388	8.948	9.507	10.066	10.625	11.184	26° 0′
2	.560	1.119	1.680	2.239	2.799	3.358	3.918	4.478	5.038	5.597	6.157	6.716	7.276	7.835	8.396	8.956	9.515	10.075	10.634	11.194	58
4	.560	1.120	1.681	2.241	2.801	3.361	3.921	4.482	5.042	5.602	6.162	6.722	7.282	7.842	8.403	8.964	9.523	10.083	10.643	11.204	56
6	.561	1.121	1.683	2.243	2.804	3.364	3.925	4.486	5.046	5.607	6.168	6.728	7.289	7.849	8.410	8.971	9.531	10.092	10.652	11.213	54
8	.561	1.122	1.684	2.245	2.806	3.367	3.928	4.490	5.050	5.612	6.173	6.734	7.295	7.856	8.417	8.979	9.539	10.100	10.661	11.223	52
10	.562	1.123	1.685	2.246	2.808	3.370	3.931	4.493	5.054	5.616	6.178	6.740	7.301	7.862	8.424	8.986	9.547	10.108	10.670	11.222	50
12	.562	1.124	1.687	2.248	2.811	3.373	3.935	4.497	5.059	5.621	6.184	6.746	7.308	7.869	8.432	8.994	9.556	10.117	10.680	11.242	48
14	.563	1.125	1.688	2.250	2.813	3.376	3.938	4.501	5.063	5.626	6.189	6.752	7.314	7.876	8.439	9.002	9.564	10.126	10.689	11.252	46
16	.563	1.126	1.690	2.252	2.816	3.379	3.942	4.505	5.068	5.631	6.194	6.758	7.320	7.883	8.446	9.009	9.572	10.135	10.698	11.261	44
18	.564	1.127	1.691	2.254	2.818	3.382	3.945	4.509	5.072	5.636	6.199	6.764	7.326	7.890	8.453	9.017	9.580	10.143	10.707	11.271	42
20	.564	1.128	1.692	2.256	2.820	3.384	3.948	4.512	5.076	5.640	6.204	6.769	7.332	7.896	8.460	9.024	9.588	10.152	10.716	11.280	40
22	.565	1.129	1.694	2.258	2.823	3.387	3.952	4.516	5.081	5.645	6.210	6.775	7.339	7.903	8.468	9.032	9.597	10.161	10.726	11.290	38
24	.565	1.130	1.695	2.260	2.825	3.390	3.955	4.520	5.085	5.650	6.215	6.781	7.345	7.910	8.475	9.039	9.605	10.170	10.735	11.300	36
26	.565	1.131	1.697	2.262	2.828	3.393	3.959	4.524	5.090	5.655	6.220	6.787	7.352	7.917	8.482	9.047	9.613	10.179	10.744	11.309	34
28	.566	1.132	1.698	2.264	2.830	3.396	3.962	4.528	5.094	5.660	6.225	6.793	7.358	7.924	8.489	9.054	9.621	10.187	10.753	11.319	32
30	.566	1.133	1.699	2.266	2.832	3.399	3.965	4.531	5.098	5.664	6.230	6.798	7.364	7.930	8.496	9.062	9.629	10.196	10.762	11.328	30
32	.567	1.134	1.701	2.268	2.835	3.402	3.969	4.535	5.103	5.669	6.236	6.804	7.371	7.937	8.504	9.070	9.637	10.205	10.771	11.338	28
34	.567	1.135	1.702	2.270	2.837	3.405	3.972	4.539	5.107	5.674	6.241	6.810	7.377	7.944	8.511	9.078	9.645	10.213	10.780	11.348	26
36	.568	1.136	1.704	2.272	2.840	3.408	3.976	4.543	5.111	5.679	6.247	6.815	7.383	7.951	8.518	9.085	9.653	10.222	10.789	11.357	24
38	.568	1.137	1.705	2.274	2.842	3.411	3.979	4.547	5.115	5.684	6.252	6.821	7.389	7.958	8.525	9.093	9.661	10.231	10.798	11.367	22
40	.569	1.138	1.706	2.275	2.844	3.413	3.982	4.550	5.119	5.688	6.257	6.826	7.395	7.964	8.532	9.100	9.669	10.239	10.807	11.376	20
42	.569	1.139	1.708	2.277	2.847	3.416	3.986	4.554	5.124	5.693	6.263	6.832	7.401	7.971	8.540	9.108	9.678	10.248	10.817	11.386	18
44	.570	1.140	1.710	2.279	2.849	3.419	3.989	4.558	5.128	5.698	6.268	6.838	7.407	7.978	8.547	9.116	9.686	10.257	10.826	11.396	16
46	.570	1.141	1.711	2.281	2.852	3.422	3.992	4.562	5.133	5.703	6.273	6.843	7.413	7.984	8.554	9.124	9.695	10.265	10.835	11.405	14
48	.571	1.142	1.713	2.283	2.854	3.425	3.995	4.566	5.137	5.708	6.278	6.849	7.419	7.991	8.561	9.132	9.703	10.274	10.844	11.415	12
50	.571	1.142	1.714	2.285	2.856	3.427	3.998	4.570	5.141	5.712	6.283	6.854	7.425	7.997	8.568	9.140	9.711	10.282	10.853	11.424	10
52	.572	1.143	1.716	2.287	2.859	3.430	4.002	4.574	5.146	5.717	6.289	6.860	7.432	8.004	8.576	9.148	9.719	10.291	10.862	11.434	8
54	.572	1.144	1.717	2.289	2.861	3.433	4.005	4.578	5.150	5.722	6.294	6.866	7.438	8.011	8.583	9.156	9.727	10.300	10.871	11.444	6
56	.573	1.145	1.719	2.291	2.864	3.436	4.009	4.582	5.154	5.727	6.300	6.872	7.445	8.017	8.590	9.163	9.735	10.308	10.880	11.453	4
58	.573	1.146	1.720	2.293	2.866	3.439	4.012	4.586	5.158	5.732	6.305	6.878	7.451	8.024	8.597	9.171	9.743	10.317	10.889	11.463	2
5° 0′	.574	1.147	1.721	2.294	2.868	3.442	4.015	4.589	5.162	5.736	6.310	6.883	7.457	8.030	8.604	9.178	9.751	10.325	10.898	11.472	25° 0′
2	.574	1.148	1.723	2.296	2.871	3.445	4.019	4.593	5.167	5.741	6.316	6.889	7.464	8.037	8.612	9.186	9.760	10.334	10.908	11.482	58
4	.575	1.149	1.724	2.298	2.873	3.448	4.022	4.597	5.171	5.746	6.321	6.895	7.470	8.044	8.619	9.194	9.768	10.343	10.917	11.492	56
6	.575	1.150	1.726	2.300	2.876	3.451	4.026	4.601	5.176	5.751	6.326	6.901	7.476	8.051	8.626	9.202	9.776	10.351	10.926	11.501	54
8	.576	1.151	1.727	2.302	2.878	3.454	4.029	4.605	5.180	5.756	6.331	6.907	7.482	8.058	8.633	9.210	9.784	10.360	10.935	11.511	52
10	.576	1.152	1.728	2.304	2.880	3.456	4.032	4.608	5.184	5.760	6.336	6.912	7.488	8.064	8.640	9.216	9.792	10.368	10.944	11.520	50
12	.577	1.153	1.730	2.306	2.883	3.459	4.036	4.612	5.189	5.765	6.341	6.918	7.494	8.071	8.647	9.224	9.800	10.377	10.953	11.530	48
14	.577	1.154	1.731	2.308	2.885	3.462	4.039	4.616	5.193	5.770	6.346	6.924	7.500	8.078	8.654	9.232	9.808	10.385	10.962	11.539	46
16	.577	1.155	1.733	2.310	2.888	3.465	4.043	4.620	5.197	5.774	6.351	6.929	7.506	8.084	8.661	9.239	9.816	10.394	10.971	11.549	44
18	.578	1.156	1.734	2.312	2.890	3.468	4.046	4.624	5.201	5.779	6.356	6.935	7.512	8.091	8.668	9.247	9.824	10.402	10.980	11.558	42
20	.578	1.157	1.735	2.313	2.892	3.470	4.048	4.627	5.205	5.783	6.361	6.940	7.518	8.097	8.675	9.254	9.832	10.410	10.989	11.567	40
22	.579	1.158	1.737	2.315	2.895	3.473	4.052	4.631	5.210	5.788	6.367	6.946	7.525	8.104	8.683	9.262	9.840	10.419	10.998	11.577	38
24	.579	1.159	1.738	2.317	2.897	3.476	4.055	4.635	5.214	5.793	6.372	6.952	7.531	8.111	8.690	9.270	9.848	10.427	11.007	11.586	36
26	.580	1.160	1.740	2.319	2.900	3.479	4.059	4.639	5.218	5.798	6.378	6.957	7.537	8.117	8.697	9.277	9.856	10.436	11.016	11.596	34
28	.580	1.161	1.741	2.321	2.902	3.482	4.062	4.643	5.222	5.803	6.383	6.963	7.543	8.124	8.704	9.285	9.864	10.444	11.025	11.605	32
30	.581	1.161	1.742	2.323	2.904	3.484	4.065	4.646	5.226	5.807	6.388	6.968	7.549	8.130	8.711	9.292	9.872	10.452	11.034	11.614	30
32	.581	1.162	1.744	2.325	2.907	3.487	4.069	4.650	5.231	5.812	6.394	6.974	7.556	8.137	8.719	9.300	9.881	10.461	11.043	11.624	28
34	.582	1.163	1.745	2.327	2.909	3.490	4.072	4.654	5.235	5.817	6.399	6.980	7.562	8.144	8.726	9.308	9.889	10.470	11.052	11.634	26
36	.582	1.164	1.747	2.329	2.911	3.493	4.076	4.658	5.239	5.822	6.404	6.985	7.568	8.151	8.733	9.315	9.897	10.478	11.061	11.643	24
38	.583	1.165	1.748	2.331	2.913	3.496	4.079	4.662	5.243	5.827	6.409	6.991	7.574	8.157	8.740	9.323	9.905	10.487	11.070	11.653	22
40	.583	1.166	1.749	2.332	2.915	3.498	4.082	4.665	5.248	5.831	6.414	6.996	7.580	8.164	8.747	9.330	9.913	10.495	11.079	11.662	20
42	.583	1.167	1.751	2.334	2.918	3.501	4.086	4.669	5.253	5.836	6.419	7.002	7.587	8.171	8.754	9.338	9.921	10.504	11.088	11.672	18
44	.584	1.168	1.752	2.336	2.920	3.504	4.089	4.673	5.257	5.840	6.424	7.008	7.593	8.178	8.761	9.346	9.929	10.513	11.097	11.681	16
46	.584	1.169	1.754	2.338	2.923	3.507	4.092	4.676	5.261	5.845	6.429	7.013	7.600	8.184	8.768	9.353	9.937	10.521	11.106	11.691	14
48	.585	1.170	1.755	2.340	2.925	3.510	4.095	4.680	5.265	5.850	6.434	7.019	7.606	8.191	8.775	9.360	9.945	10.530	11.115	11.700	12
50	.585	1.171	1.756	2.342	2.927	3.513	4.098	4.683	5.269	5.854	6.439	7.025	7.611	8.197	8.782	9.367	9.953	10.538	11.124	11.709	10
52	.586	1.172	1.758	2.344	2.930	3.516	4.102	4.687	5.374	5.859	6.445	7.031	7.618	8.204	8.789	9.375	9.961	10.547	11.133	11.719	8
54	.586	1.173	1.759	2.346	2.932	3.519	4.105	4.691	5.278	5.864	6.450	7.037	7.624	8.211	8.796	9.383	9.969	10.555	11.142	11.728	6
56	.587	1.174	1.761	2.348	2.935	3.522	4.109	4.695	5.282	5.869	6.456	7.043	7.630	8.217	8.803	9.390	9.977	10.564	11.151	11.738	4
58	.587	1.175	1.762	2.350	2.937	3.525	4.112	4.699	5.286	5.874	6.461	7.049	7.636	8.224	8.810	9.397	9.985	10.572	11.160	11.747	2
6° 0	.588	1.176	1.763	2.352	2.939	3.527	4.115	4.702	5.290	5.878	6.466	7.054	7.642	8.230	8.817	9.404	9.992	10.580	11.168	11.756	24° 0
	1	2	3	4	5	6	7	8	9	10	11	12	13	14	15	16	17	18	19	20	IV + X −

TABLE XVIII.

GENERAL TABLE of Aberration, &c. continued.

S. −VII + I	1	2	3	4	5	6	7	8	9	10	11	12	13	14	15	16	17	18	19	20	
6° 0′	.588	1.176	1.763	2.352	2.939	3.527	4.115	4.702	5.290	5.878	6.466	7.054	7.642	8.230	8.817	9.404	9.992	10.580	11.168	11.756	24° 0′
2	.588	1.177	1.765	2.354	2.942	3.530	4.119	4.706	5.295	5.883	6.471	7.060	7.648	8.237	8.824	9.412	10.000	10.589	11.177	11.766	58
4	.589	1.178	1.766	2.356	2.944	3.533	4.122	4.710	5.299	5.888	6.476	7.066	7.654	8.243	8.831	9.420	10.008	10.597	11.186	11.775	56
6	.589	1.179	1.768	2.358	2.947	3.536	4.125	4.714	5.303	5.892	6.481	7.071	7.660	8.250	8.838	9.427	10.016	10.606	11.195	11.784	54
8	.589	1.179	1.769	2.360	2.949	3.539	4.128	4.718	5.307	5.897	6.486	7.077	7.666	8.256	8.845	9.435	10.024	10.614	11.204	11.793	52
10	.590	1.180	1.770	2.361	2.951	3.541	4.131	4.721	5.311	5.901	6.491	7.082	7.672	8.262	8.852	9.442	10.032	10.622	11.212	11.802	50
12	.591	1.181	1.772	2.363	2.954	3.544	4.135	4.725	5.316	5.906	6.497	7.088	7.678	8.269	8.859	9.450	10.040	10.631	11.221	11.812	48
14	.591	1.182	1.773	2.365	2.956	3.547	4.138	4.729	5.320	5.911	6.502	7.094	7.684	8.275	8.866	9.458	10.048	10.639	11.230	11.822	46
16	.592	1.183	1.775	2.367	2.958	3.550	4.141	4.733	5.324	5.916	6.507	7.099	7.690	8.282	8.873	9.465	10.056	10.648	11.239	11.831	44
18	.592	1.184	1.776	2.369	2.960	3.553	4.144	4.737	5.328	5.921	6.512	7.105	7.696	8.288	8.880	9.473	10.064	10.656	11.248	11.840	42
20	.593	1.185	1.777	2.370	2.962	3.555	4.147	4.740	5.332	5.925	6.517	7.110	7.702	8.294	8.887	9.480	10 072	10.664	11.257	11.850	40
22	.593	1.186	1.779	2.372	2.965	3.558	4.151	4.744	5.337	5.930	6.523	7.116	7.709	8.301	8.894	9.488	10.080	10.673	11.266	11.860	38
24	.594	1.187	1.780	2.374	2.967	3.561	4.154	4.748	5.341	5.935	6.528	7.122	7.715	8.307	8.901	9.496	10.088	10.681	11.275	11.869	36
26	.594	1.188	1.782	2.376	2.970	3.564	4.158	4.752	5.345	5.939	6.533	7.127	7.721	8.314	8.908	9.503	10.096	10.690	11.284	11.878	34
28	.595	1.189	1.783	2.378	2.972	3.567	4.161	4.756	5.349	5.944	6.538	7.133	7.727	8.320	8.915	9.511	10.104	10.698	11.293	11.887	32
30	.595	1 190	1.784	2.379	2.974	3.569	4.164	4.759	5.353	5.948	6.543	7.138	7.733	8.327	8.922	9.518	10.112	10.706	11.301	11.896	30
32	.595	1.191	1.786	2.381	2.977	3.572	4.168	4.763	5.358	5.953	6.549	7.144	7.739	8.334	8.929	9.526	10.120	10.715	11.310	11.906	28
34	.596	1.192	1.787	2.383	2.979	3.575	4.171	4.767	5.362	5.958	6.554	7.150	7.745	8.341	8.936	9.533	10.128	10.723	11.319	11.915	26
36	.596	1.193	1.789	2.385	2.982	3.578	4.174	4.770	5.366	5.963	6.559	7.155	7.751	8.347	8.943	9.541	10.136	10.731	11.328	11.924	24
38	.597	1.194	1.790	2.387	2.984	3.581	4.177	4.774	5.370	5.968	6.564	7.161	7.757	8.354	8.950	9.548	10.144	10.740	11.337	11.934	22
40	.597	1.194	1.791	2.389	2.986	3.583	4.180	4.777	5.374	5.972	6.569	7.166	7.763	8.360	8.957	9.555	10.151	10.748	11.346	11.944	20
42	.598	1.195	1.793	2.391	2.989	3.586	4.184	4.781	5.379	5.977	6.575	7.172	7.769	8.367	8.964	9.563	10.159	10.757	11.355	11.954	18
44	.598	1.196	1.794	2.393	2.991	3.589	4.187	4.785	5.383	5.982	6.580	7.178	7.775	8.373	8.971	9.570	10.167	10.765	11.364	11.963	16
46	.599	1.197	1.796	2.395	2.994	3.592	4.190	4.789	5.387	5.987	6.585	7.183	7.781	8.380	8.978	9.578	10.175	10.774	11.373	11.972	14
48	.599	1.198	1.797	2.397	2.996	3.595	4.193	4.793	5.391	5.991	6.590	7.189	7.787	8.387	8.985	9.585	10.183	10.782	11.382	11.981	12
50	.600	1.199	1.798	2.398	2.998	3.597	4.196	4.796	5.395	5.995	6.595	7.194	7.793	8.392	8.992	9.592	10.191	10.790	11.390	11.990	10
52	.600	1.200	1.800	2.400	3.000	3.600	4.200	4.800	5.400	6.000	6.600	7.200	7.799	8.399	9.000	9.600	10.199	10.799	11.399	12.000	8
54	.601	1.201	1.801	2.402	3.003	3.603	4.203	4.804	5.404	6.005	6.605	7.206	7.806	8.406	9.007	9.608	10.207	10.807	11.408	12.010	6
56	.601	1.202	1.803	2.404	3.005	3.606	4.207	4.808	5.408	6.009	6.610	7.211	7.812	8.412	9.014	9.615	10.215	10.816	11.417	12.019	4
58	.602	1.203	1.804	2.406	3.007	3.609	4.210	4.812	5.412	6.014	6.615	7.217	7.818	8.419	9.021	9.622	10.223	10.824	11.426	12.028	2
7° 0′	.602	1.204	1.805	2.407	3.009	3.611	4.213	4.815	5.416	6.018	6.620	7.222	7.824	8.425	9.028	9.630	10.231	10.832	11.434	12.037	23° 0′
2	.602	1.205	1.807	2.409	3.012	3.614	4.217	4.819	5.421	6.023	6.625	7.228	7.830	8.432	9.035	9.638	10.239	10.841	11.443	12.046	58
4	.603	1.206	1.808	2.411	3.014	3.617	4.220	4.823	5.425	6.028	6.630	7.234	7.836	8.439	9.042	9.645	10.247	10.849	11.452	12.055	56
6	.603	1.207	1.810	2.413	3.017	3.620	4.223	4.826	5.429	6.032	6.635	7.240	7.842	8.445	9.049	9.653	10.255	10.858	11.461	12.064	54
8	.604	1.208	1.811	2.415	3.019	3.623	4.226	4.830	5.433	6.037	6.640	7.245	7.848	8.452	9.056	9.660	10.263	10.866	11.470	12.073	52
10	.604	1.208	1.812	2.417	3.021	3.625	4.229	4.833	5.437	6.041	6.645	7.250	7.854	8.458	9.062	9.667	10.270	10.874	11.478	12.082	50
12	.605	1.209	1.814	2.419	3.024	3.628	4.233	4.837	5.442	6.046	6.651	7.256	7.860	8.465	9.069	9.675	10.278	10.883	11.487	12.092	48
14	.605	1.210	1.815	2.421	3.026	3.631	4.236	4.841	5.446	6.051	6.656	7.262	7.866	8.471	9.076	9.682	10.286	10.891	11.496	12.102	46
16	.606	1.211	1.817	2.423	3.028	3.634	4.239	4.845	5.450	6.056	6.661	7.267	7.872	8.478	9.083	9.690	10.294	10.900	11.505	12.111	44
18	.606	1.212	1.818	2.425	3.030	3.637	4.242	4.848	5.454	6.061	6.666	7.273	7.878	8.484	9.090	9.697	10.302	10.908	11.514	12.111	42
20	.607	1.213	1.819	2.426	3.032	3.639	4.245	4.852	5.458	6.065	6.671	7.278	7.884	8.490	9.097	9.704	10.310	10.916	11.523	12.130	40
22	.607	1.214	1.821	2.428	3.035	3.642	4.249	4.856	5.463	6.070	6.677	7.284	7.890	8.497	9.104	9.712	10.318	10.925	11.532	12.140	38
24	.608	1.215	1.822	2.430	3.037	3.645	4.252	4.860	5.467	6.075	6.682	7.290	7.896	8.503	9.111	9.719	10.326	10.933	11.541	12.149	36
26	.608	1.216	1.824	2.432	3.040	3.648	4.255	4.863	5.471	6.079	6.687	7.295	7.902	8.510	9.118	9.727	10.334	10.942	11.550	12.159	34
28	.609	1.217	1.825	2.434	3.042	3.651	4.258	4.867	5.475	6.084	6.692	7.301	7.908	8.516	9.125	9.734	10.342	10.950	11.559	12.168	32
30	.609	1.218	1.826	2.435	3.044	3.653	4.261	4.870	5.479	6.088	6.697	7.306	7.914	8.522	9.132	9.741	10.349	10.958	11.567	12.177	30
32	.609	1.219	1.828	2.437	3.047	3.656	4.265	4.874	5.484	6.093	6.702	7.312	7.920	8.529	9.139	9.749	10.357	10.967	11.576	12.187	28
34	.610	1.220	1.829	2.439	3.049	3.659	4.268	4.878	5.488	6.098	6.707	7.317	7.926	8.536	9.146	9.756	10.365	10.975	11.585	12.196	26
36	.610	1.221	1.831	2.441	3.051	3.662	4.272	4.882	5.492	6.102	6.712	7.322	7.932	8.542	9.153	9.764	10.373	10.984	11.594	12.205	24
38	.611	1.222	1.832	2.442	3.053	3.664	4.275	4.885	5.496	6.107	6.717	7.327	7.938	8.548	9.160	9.771	10.381	10.992	11.603	12.214	22
40	.611	1.222	1.833	2.444	3.055	3.666	4.278	4.889	5.500	6.111	6.722	7.332	7.944	8.555	9.167	9.778	10.389	11.000	11.611	12.223	20
42	.611	1.223	1.835	2.446	3.058	3.669	4.282	4.893	5.504	6.116	6.727	7.338	7.950	8.562	9.174	9.786	10.397	11.008	11.620	12.233	18
44	.612	1.224	1.836	2.448	3.060	3.672	4.285	4.897	5.508	6.121	6.732	7.344	7.956	8.569	9.181	9.793	10.405	11.016	11.629	12.242	16
46	.612	1.225	1.838	2.450	3.063	3.675	4.288	4.901	5.512	6.125	6.737	7.349	7.962	8.575	9.188	9.800	10.412	11.024	11.638	12.251	14
48	.613	1.226	1.839	2.452	3.065	3.678	4.291	4.904	5.516	6.130	6.742	7.355	7.968	8.582	9.195	9.807	10.420	11.032	11.647	12.260	12
50	.613	1.227	1.840	2.454	3.067	3.680	4.294	4.907	5.520	6.134	6.747	7.360	7.974	8.588	9.201	9.814	10.427	11.040	11.655	12.269	10
52	.614	1.228	1.842	2.456	3.070	3.683	4.298	4.911	5.525	6.139	6.752	7.366	7.980	8.595	9.208	9.821	10.435	11.049	11.664	12.278	8
54	.614	1.229	1.843	2.458	3.072	3.686	4.301	4.915	5.529	6.144	6.757	7.372	7.986	8.601	9.215	9.828	10.443	11.057	11.673	12.287	6
56	.615	1.230	1.845	2.460	3.074	3.689	4.304	4.918	5.533	6.148	6.762	7.377	7.992	8.608	9.222	9.835	10.451	11.066	11.681	12.296	4
58	.615	1.231	1.846	2.462	3.076	3.692	4.307	4.922	5.537	6.153	6.767	7.383	7.998	8.614	9.229	9.842	10.459	11.074	11.690	12.305	2
8° 0	.616	1.231	1.847	2.463	3.078	3.694	4.310	4.925	5.541	6.157	6.773	7.388	8.004	8.620	9.235	9.850	10.466	11.082	11.698	12.314	22° 0
	1	2	3	4	5	6	7	8	9	10	11	12	3	14	15	16	17	18	19	20	IV + X −

TABLE XVIII.

GENERAL TABLE of Aberration, &c. continued.

S. −VII +I	1	2	3	4	5	6	7	8	9	10	11	12	13	14	15	16	17	18	19	20	
8° 0′	.616	1.231	1.847	2.463	3.078	3.694	4.310	4.925	5.541	6.157	6.773	7.388	8.004	8.620	9.235	9.851	10.466	11.082	11.698	12.314	22° 0′
2	.616	1.232	1.849	2.465	3.081	3.697	4.314	4.929	5.546	6.162	6.778	7.394	8.010	8.626	9.242	9.859	10.474	11.091	11.707	12.323	58
4	.617	1.233	1.850	2.467	3.083	3.700	4.317	4.933	5.550	6.167	6.783	7.399	8.016	8.633	9.249	9.866	10.482	11.099	11.716	12.332	56
6	.617	1.234	1.852	2.469	3.085	3.702	4.320	4.937	5.554	6.171	6.788	7.405	8.022	8.639	9.256	9.874	10.490	11.108	11.725	12.341	54
8	.617	1.235	1.853	2.471	3.087	3.705	4.323	4.940	5.558	6.176	6.793	7.410	8.028	8.646	9.263	9.881	10.498	11.116	11.734	12.350	52
10	.618	1.236	1.854	2.472	3.090	3.708	4.326	4.944	5.562	6.180	6.798	7.415	8.034	8.652	9.270	9.888	10.506	11.124	11.742	12.359	50
12	.618	1.237	1.856	2.474	3.093	3.711	4.330	4.948	5.566	6.185	6.803	7.421	8.040	8.659	9.277	9.896	10.514	11.133	11.751	12.369	48
14	.619	1.238	1.857	2.476	3.095	3.714	4.333	4.952	5.570	6.189	6.808	7.426	8.046	8.665	9.284	9.903	10.522	11.141	11.760	12.378	46
16	.619	1.239	1.859	2.478	3.097	3.716	4.336	4.955	5.574	6.194	6.813	7.431	8.052	8.672	9.291	9.910	10.530	11.149	11.768	12.387	44
18	.620	1.240	1.860	2.480	3.099	3.719	4.339	4.959	5.578	6.198	6.818	7.436	8.058	8.678	9.298	9.917	10.538	11.157	11.777	12.396	42
20	.620	1.240	1.861	2.481	3.101	3.721	4.342	4.962	5.583	6.202	6.823	7.442	8.063	8.684	9.304	9.924	10.545	11.165	11.785	12.405	40
22	.621	1.241	1.863	2.483	3.104	3.724	4.346	4.966	5.588	6.207	6.828	7.448	8.069	8.691	9.311	9.932	10.553	11.174	11.794	12.414	38
24	.621	1.242	1.864	2.485	3.106	3.727	4.349	4.970	5.592	6.212	6.833	7.454	8.075	8.697	9.318	9.939	10.561	11.182	11.803	12.423	36
26	.622	1.243	1.866	2.487	3.109	3.730	4.352	4.973	5.595	6.216	6.838	7.459	8.081	8.704	9.325	9.946	10.568	11.190	11.811	12.432	34
28	.622	1.244	1.867	2.489	3.111	3.733	4.355	4.977	5.599	6.221	6.843	7.464	8.087	8.710	9.332	9.953	10.576	11.198	11.820	12.441	32
30	.623	1.245	1.868	2.490	3.113	3.735	4.358	4.980	5.603	6.225	6.848	7.470	8.093	8.716	9.338	9.960	10.583	11.206	11.828	12.450	30
32	.623	1.246	1.870	2.492	3.116	3.738	4.361	4.984	5.607	6.230	6.853	7.476	8.099	8.723	9.345	9.968	10.591	11.215	11.837	12.460	28
34	.624	1.247	1.871	2.494	3.118	3.741	4.365	4.988	5.611	6.235	6.858	7.482	8.105	8.729	9.352	9.975	10.599	11.223	11.846	12.469	26
36	.624	1.248	1.872	2.496	3.120	3.744	4.368	4.991	5.615	6.239	6.863	7.487	8.111	8.736	9.359	9.982	10.606	11.231	11.854	12.478	24
38	.624	1.249	1.873	2.498	3.122	3.747	4.371	4.994	5.619	6.243	6.868	7.493	8.117	8.742	9.366	9.989	10.614	11.239	11.863	12.487	22
40	.625	1.250	1.874	2.499	3.124	3.749	4.374	4.998	5.623	6.248	6.873	7.498	8.123	8.748	9.372	9.996	10.621	11.247	11.871	12.496	20
42	.625	1.251	1.876	2.501	3.127	3.752	4.377	5.002	5.628	6.253	6.878	7.504	8.129	8.755	9.379	10.004	10.629	11.256	11.880	12.505	18
44	.626	1.252	1.877	2.503	3.129	3.755	4.380	5.006	5.632	6.258	6.883	7.509	8.135	8.761	9.386	10.011	10.637	11.264	11.889	12.514	16
46	.626	1.253	1.879	2.505	3.131	3.758	4.383	5.010	5.636	6.262	6.888	7.515	8.141	8.767	9.393	10.019	10.645	11.272	11.897	12.523	14
48	.627	1.254	1.880	2.507	3.133	3.760	4.386	5.013	5.640	6.267	6.893	7.520	8.147	8.773	9.400	10.026	10.653	11.280	11.906	12.532	12
50	.627	1.254	1.881	2.508	3.135	3.762	4.389	5.017	5.644	6.271	6.898	7.525	8.152	8.779	9.406	10.033	10.660	11.288	11.914	12.541	10
52	.627	1.255	1.883	2.510	3.138	3.765	4.392	5.021	5.648	6.276	6.903	7.531	8.158	8.785	9.413	10.041	10.668	11.296	11.923	12.550	8
54	.628	1.256	1.884	2.512	3.140	3.768	4.396	5.025	5.652	6.280	6.908	7.536	8.164	8.791	9.420	10.048	10.676	11.304	11.932	12.559	6
56	.628	1.257	1.886	2.514	3.143	3.771	4.399	5.028	5.656	6.285	6.913	7.542	8.170	8.797	9.427	10.056	10.684	11.312	11.940	12.568	4
58	.629	1.258	1.887	2.516	3.145	3.774	4.402	5.031	5.660	6.289	6.918	7.547	8.176	8.803	9.434	10.063	10.692	11.320	11.949	12.577	2
9° 0′	.629	1.259	1.888	2.517	3.147	3.776	4.405	5.035	5.664	6.293	6.923	7.552	8.181	8.810	9.440	10.070	10.699	11.328	11 957	12.586	21° 0′
2	.630	1.260	1.890	2.519	3.150	3.779	4.409	5.039	5.668	6.298	6.928	7.558	8.187	8.817	9.447	10.078	10.707	11.336	11.966	12.596	58
4	.630	1.261	1.891	2.521	3.152	3.782	4.412	5.043	5.672	6.302	6.933	7.563	8.193	8.823	9.454	10.085	10.715	11.344	11.975	12.605	56
6	.631	1.262	1.893	2.523	3.154	3.785	4.415	5.046	5.676	6.307	6.938	7.569	8.199	8.830	9.461	10.092	10.723	11.352	11.983	12.614	54
8	.631	1.263	1.894	2.525	3.156	3.788	4.418	5.050	5.680	6.312	6.943	7.574	8.205	8.836	9.468	10.099	10.731	11.360	11.992	12.623	52
10	.632	1.263	1.895	2.526	3.158	3.790	4.421	5.053	5.684	6.316	6.948	7.579	8.211	8.842	9.474	10.106	10.738	11.368	12.000	12.632	50
12	.632	1.264	1.897	2.528	3.161	3.793	4.425	5.057	5.689	6.321	6.953	7.585	8.217	8.849	9.481	10.114	10.746	11.377	12.009	12.641	48
14	.633	1.265	1.898	2.530	3.163	3.796	4.428	5.061	5.693	6.325	6.958	7.590	8.223	8.855	9.488	10.121	10.754	11.385	12.018	12.650	46
16	.633	1.266	1.900	2.532	3.165	3.798	4.431	5.064	5.697	6.330	6.963	7.596	8.229	8.862	9.495	10.128	10.761	11.393	12.026	12.659	44
18	.634	1.267	1.901	2.534	3.167	3.801	4.434	5.068	5.701	6.334	6.968	7.601	8.235	8.868	9.502	10.135	10.769	11.401	12.035	12.668	42
20	.634	1.268	1.902	2.535	3.169	3.803	4.437	5.071	5.705	6.338	6.973	7.606	8.240	8.874	9.508	10.142	10.776	11.409	12.043	12.677	40
22	.634	1.269	1.904	2.537	3.172	3.806	4.441	5.075	5.709	6.343	6.978	7.612	8.246	8.881	9.515	10.150	10.784	11.418	12.052	12.686	38
24	.635	1.270	1.905	2.539	3.174	3.809	4.444	5.079	5.713	6.347	6.983	7.617	8.252	8.887	9.522	10.157	10.792	11.426	12.061	12.695	36
26	.635	1.271	1.906	2.541	3.176	3.811	4.447	5.082	5.717	6.352	6.988	7.623	8.258	8.894	9.529	10.164	10.800	11.434	12.070	12.704	34
28	.636	1.272	1.907	2.543	3.178	3.814	4.450	5.086	5.721	6.357	6.993	7.628	8.264	8.900	9.536	10.171	10.807	11.442	12.078	12.713	32
30	.636	1.272	1.908	2.544	3.180	3.817	4.453	5.089	5.725	6.361	6.997	7.633	8.270	8.906	9.542	10.178	10.814	11.450	12.086	12.722	30
32	.636	1.273	1.910	2.546	3.182	3.820	4.456	5.093	5.729	6.366	7.002	7.639	8.276	8.913	9.549	10.186	10.822	11.458	12.095	12.731	28
34	.637	1.274	1.911	2.548	3.185	3.823	4.459	5.097	5.733	6.370	7.007	7.644	8.282	8.919	9.556	10.193	10.830	11.466	12.104	12.740	26
36	.637	1.275	1.913	2.550	3.187	3.825	4.462	5.100	5.737	6.375	7.012	7.650	8.288	8.925	9.563	10.200	10.837	11.474	12.112	12.749	24
38	.637	1.276	1.914	2.552	3.190	3.828	4.465	5.104	5.741	6.379	7.017	7.655	8.294	8.931	9.570	10,.207	10.845	11.482	12.121	12.758	22
40	.638	1.277	1.915	2.553	3.192	3.830	4.468	5.107	5.745	6.383	7.022	7.660	8.299	8.937	9.576	10.214	10.852	11.490	12.129	12.766	20
42	.638	1.278	1.917	2.555	3.195	3.833	4.472	5.111	5.749	6.388	7.027	7.666	8.305	8.944	9.583	10.222	10.860	11.498	12.138	12.775	18
44	.639	1.279	1.918	2.557	3.197	3.836	4.475	5.115	5.753	6.393	7.032	7.671	8.311	8.950	9.590	10.229	10.868	11.506	12.147	12.784	16
46	.640	1.280	1.920	2.559	3.199	3.839	4.478	5.118	5.757	6.397	7.037	7.677	8.317	8.956	9.596	10.236	10.875	11.514	12.155	12.793	14
48	.640	1.281	1.921	2.561	3.201	3.841	4.481	5.122	5.761	6.402	7.042	7.682	8.323	8.962	9.603	10.243	10.883	11.522	12.163	12.802	12
50	.641	1.281	1.922	2.562	3.203	3.843	4.484	5.125	5.765	6.406	7.047	7.687	8.328	8.968	9.609	10.250	10.890	11.530	12.171	12.811	10
52	.641	1.282	1.924	2.564	3.206	3.846	4.488	5.129	5.769	6.411	7.052	7.693	8.334	8.975	9.616	10.257	10.898	11.538	12.180	12.820	8
54	.642	1.283	1.925	2.566	3.208	3.849	4.491	5.133	5.773	6.415	7.057	7.698	8.340	8.981	9.623	10.264	10.906	11.546	12.188	12.829	6
56	.642	1.284	1.926	2.568	3.210	3.852	4.494	5.136	5.777	6.420	7.062	7.704	8.346	8.988	9.630	10.271	10.914	11.554	12.197	12.838	4
58	.643	1.285	1.927	2.570	3.212	3.855	4.497	5.140	5.781	6.424	7.067	7.709	8.352	8.994	9.636	10.278	10.921	11.562	12.205	12.847	2
10° 0	.643	1.286	1.928	2.571	3.214	3.857	4.500	5.142	5.785	6.428	7.071	7.714	8.357	9.000	9.642	10.285	10.927	11.570	12.213	12.856	20° 0
	1	2	3	4	5	6	7	8	9	10	11	12	13	14	15	16	17	18	19	20	IV+ X−

TABLE XVIII.

GENERAL TABLE of Aberration, &c. continued.

S. – VII + I	1	2	3	4	5	6	7	8	9	10	11	12	13	14	15	16	17	18	19	20	
10° 0′	.643	1.286	1.928	2.571	3.214	3.857	4.500	5.142	5.785	6.428	7.071	7.714	8.357	9.000	9.642	10.284	10.927	11.570	12.213	12.856	20° 0′
2	.643	1.287	1.930	2.573	3.217	3.860	4.503	5.146	5.789	6.433	7.076	7.720	8.363	9.006	9.649	10.292	10.935	11.578	12.222	12.865	58
4	.644	1.288	1.931	2.575	3.219	3.863	4.506	5.150	5.793	6.438	7.081	7.725	8.369	9.012	9.656	10.299	10.943	11.586	12.230	12.874	56
6	.644	1.289	1.933	2.577	3.221	3.865	4.509	5.153	5.797	6.442	7.086	7.730	8.374	9.018	9.662	10.306	10.950	11.594	12.239	12.883	54
8	.645	1.290	1.934	2.579	3.223	3.868	4.512	5.157	5.801	6.446	7.090	7.735	8.379	9.024	9.669	10.313	10.958	11.602	12.247	12.891	52
10	.645	1.290	1.935	2.580	3.225	3.870	4.515	5.160	5.805	6.450	7.095	7.740	8.385	9.030	9.675	10.320	10.965	11.610	12.255	12.900	50
12	.645	1.291	1.937	2.582	3.228	3.873	4.519	5.164	5.809	6.455	7.100	7.746	8.391	9.037	9.682	10.328	10.973	11.618	12.264	12.909	48
14	.646	1.292	1.938	2.584	3.230	3.876	4.522	5.168	5.813	6.459	7.105	7.751	8.397	9.043	9.689	10.335	10.981	11.626	12.272	12.918	46
16	.646	1.293	1.940	2.586	3.232	3.878	4.525	5.171	5.817	6.464	7.110	7.756	8.403	9.049	9.696	10.342	10.988	11.634	12.281	12.927	44
18	.647	1.294	1.941	2.588	3.234	3.881	4.528	5.175	5.821	6.468	7.114	7.761	8.409	9.055	9.702	10.349	10.996	11.642	12.289	12.936	42
20	.647	1.295	1.942	2.589	3.236	3.883	4.531	5.178	5.825	6.472	7.119	7.766	8.414	9.061	9.709	10.356	11.003	11.650	12.297	12.944	40
22	.647	1.296	1.943	2.591	3.239	3.886	4.534	5.182	5.829	6.477	7.124	7.772	8.420	9.068	9.716	10.364	11.011	11.658	12.306	12.953	38
24	.648	1.297	1.945	2.593	3.241	3.889	4.537	5.186	5.833	6.482	7.129	7.777	8.426	9.074	9.723	10.371	11.019	11.666	12.315	12.962	36
26	.648	1.298	1.946	2.595	3.243	3.892	4.540	5.189	5.837	6.486	7.134	7.783	8.432	9.080	9.729	10.378	11.026	11.674	12.323	12.971	34
28	.649	1.299	1.948	2.597	3.245	3.894	4.543	5.193	5.841	6.490	7.139	7.788	8.438	9.086	9.736	10.385	11.034	11.682	12.332	12.980	32
30	.649	1.299	1.949	2.598	3.247	3.897	4.546	5.196	5.845	6.495	7.144	7.793	8.443	9.092	9.742	10.392	11.041	11.690	12.340	12.989	30
32	.650	1.300	1.950	2.600	3.250	3.900	4.550	5.200	5.849	6.500	7.149	7.799	8.449	9.099	9.749	10.399	11.049	11.698	12.349	12.998	28
34	.650	1.301	1.951	2.601	3.252	3.903	4.553	5.203	5.853	6.504	7.154	7.804	8.455	9.105	9.756	10.406	11.057	11.706	12.357	13.007	26
36	.651	1.302	1.953	2.603	3.254	3.905	4.556	5.207	5.857	6.509	7.159	7.810	8.461	9.111	9.762	10.413	11.064	11.714	12.366	13.016	24
38	.651	1.303	1.954	2.605	3.256	3.908	4.559	5.210	5.861	6.513	7.164	7.815	8.467	9.117	9.769	10.420	11.071	11.722	12.374	13.025	22
40	.652	1.303	1.955	2.607	3.258	3.910	4.562	5.213	5.865	6.517	7.169	7.820	8.472	9.123	9.775	10.427	11.078	11.730	12.382	13.034	20
42	.652	1.304	1.957	2.609	3.261	3.913	4.565	5.217	5.869	6.522	7.174	7.826	8.478	9.130	9.782	10.434	11.086	11.738	12.391	13.043	18
44	.653	1.305	1.958	2.611	3.263	3.916	4.568	5.220	5.873	6.526	7.179	7.831	8.484	9.136	9.789	10.441	11.094	11.746	12.399	13.052	16
46	.653	1.306	1.960	2.612	3.265	3.918	4.571	5.224	5.877	6.531	7.184	7.836	8.490	9.142	9.795	10.448	11.101	11.754	12.408	13.061	14
48	.654	1.307	1.961	2.614	3.267	3.921	4.574	5.228	5.881	6.535	7.189	7.841	8.495	9.148	9.802	10.455	11.109	11.762	12.416	13.070	12
50	.654	1.308	1.962	2.615	3.269	3.923	4.577	5.231	5.885	6.539	7.193	7.846	8.500	9.154	9.808	10.462	11.116	11.770	12.424	13.078	10
52	.654	1.309	1.964	2.617	3.272	3.926	4.580	5.235	5.889	6.544	7.198	7.852	8.506	9.160	9.815	10.469	11.124	11.778	12.433	13.087	8
54	.655	1.310	1.965	2.619	3.274	3.928	4.583	5.239	5.893	6.548	7.203	7.857	8.512	9.166	9.822	10.477	11.132	11.786	12.441	13.096	6
56	.655	1.311	1.966	2.621	3.276	3.931	4.586	5.242	5.897	6.553	7.208	7.862	8.517	9.172	9.828	10.484	11.139	11.794	12.450	13.105	4
58	.656	1.312	1.967	2.623	3.278	3.934	4.589	5.246	5.901	6.557	7.213	7.867	8.523	9.178	9.835	10.491	11.147	11.802	12.458	13.114	2
11° 0′	.656	1.312	1.968	2.624	3.280	3.936	4.592	5.249	5.905	6.561	7.217	7.872	8.528	9.184	9.841	10.498	11.154	11.810	12.466	13.122	19° 0′
2	.656	1.313	1.970	2.626	3.283	3.939	4.596	5.253	5.909	6.566	7.222	7.878	8.534	9.191	9.848	10.505	11.162	11.818	12.475	13.131	58
4	.657	1.314	1.971	2.628	3.285	3.942	4.599	5.256	5.913	6.570	7.227	7.883	8.540	9.197	9.855	10.512	11.170	11.826	12.483	13.140	56
6	.657	1.315	1.973	2.629	3.287	3.945	4.602	5.260	5.917	6.574	7.232	7.888	8.546	9.203	9.861	10.519	11.177	11.834	12.491	13.149	54
8	.658	1.316	1.974	2.631	3.289	3.947	4.605	5.263	5.921	6.579	7.237	7.893	8.552	9.209	9.867	10.526	11.185	11.842	12.499	13.158	52
10	.658	1.317	1.975	2.632	3.291	3.950	4.608	5.266	5.924	6.583	7.241	7.899	8.557	9.215	9.874	10.533	11.191	11.849	12.507	13.166	50
12	.658	1.318	1.977	2.634	3.294	3.953	4.611	5.270	5.928	6.588	7.246	7.905	8.563	9.222	9.881	10.540	11.199	11.857	12.516	13.175	48
14	.659	1.319	1.978	2.636	3.296	3.956	4.614	5.274	5.932	6.592	7.251	7.910	8.569	9.228	9.888	10.547	11.206	11.865	12.524	13.184	46
16	.659	1.320	1.979	2.638	3.298	3.958	4.617	5.277	5.936	6.596	7.256	7.916	8.575	9.234	9.894	10.554	11.214	11.873	12.532	13.192	44
18	.660	1.321	1.980	2.640	3.300	3.961	4.620	5.281	5.940	6.600	7.260	7.921	8.581	9.240	9.901	10.561	11.221	11.881	12.540	13.201	42
20	.660	1.321	1.981	2.642	3.302	3.963	4.623	5.284	5.944	6.604	7.265	7.926	8.586	9.246	9.907	10.568	11.228	11.888	12.548	13.209	40
22	.661	1.322	1.983	2.644	3.305	3.966	4.626	5.288	5.948	6.609	7.270	7.932	8.592	9.252	9.914	10.575	11.236	11.897	12.557	13.218	38
24	.661	1.323	1.984	2.646	3.307	3.969	4.629	5.291	5.952	6.613	7.275	7.937	8.598	9.258	9.920	10.582	11.243	11.905	12.565	13.227	36
26	.662	1.324	1.986	2.648	3.309	3.972	4.632	5.295	5.256	6.618	7.280	7.942	8.604	9.264	9.927	10.589	11.251	11.913	12.574	13.235	34
28	.662	1.324	1.987	2.650	3.311	3.974	4.635	5.298	5.960	6.622	7.285	7.947	8.610	9.270	9.933	10.596	11.258	11.921	12.582	13.244	32
30	.663	1.325	1.988	2.652	3.313	3.976	4.638	5.301	5.964	6.626	7.289	7.952	8.615	9.276	9.939	10.602	11.265	11.928	12.590	13.252	30
32	.663	1.326	1.990	2.654	3.316	3.979	4.642	5.305	5.968	6.631	7.294	7.958	8.621	9.283	9.946	10.609	11.273	11.937	12.599	13.261	28
34	.664	1.327	1.991	2.655	3.318	3.982	4.645	5.309	5.972	6.635	7.299	7.963	8.627	9.289	9.953	10.616	11.280	11.945	12.607	13.270	26
36	.664	1.328	1.992	2.657	3.320	3.984	4.648	5.312	5.976	6.640	7.304	7.968	8.633	9.295	9.959	10.623	11.288	11.953	12.615	13.279	24
38	.665	1.329	1.993	2.658	3.322	3.987	4.651	5.315	5.980	6.644	7.309	7.973	8.639	9.301	9.966	10.630	11.295	11.960	12.623	13.288	22
40	.665	1.330	1.994	2.659	3.324	3.989	4.654	5.318	5.983	6.648	7.313	7.978	8.643	9.307	9.972	10.637	11.302	11.967	12.631	13.296	20
42	.665	1.331	1.996	2.661	3.327	3.992	4.657	5.322	5.987	6.653	7.318	7.984	8.649	9.314	9.979	10.644	11.310	11.975	12.640	13.305	18
44	.666	1.332	1.997	2.663	3.329	3.995	4.660	5.326	5.991	6.657	7.323	7.989	8.655	9.320	9.986	10.651	11.317	11.983	12.648	13.314	16
46	.666	1.333	1.999	2.665	3.331	3.997	4.663	5.329	5.995	6.662	7.328	7.994	8.661	9.326	9.992	10.658	11.325	11.991	12.657	13.323	14
48	.667	1.334	2.000	2.666	3.333	4.000	4.666	5.333	5.999	6.666	7.333	7.999	8.666	9.332	9.999	10.665	11.332	11.999	12.665	13.332	12
50	.667	1.334	2.001	2.668	3.335	4.002	4.669	5.336	6.003	6.670	7.337	8.004	8.671	9.338	10.005	10.672	11.339	12.006	12.673	13.340	10
52	.667	1.335	2.003	2.670	3.338	4.005	4.672	5.340	6.007	6.675	7.342	8.010	8.677	9.344	10.012	10.680	11.347	12.014	12.681	13.349	8
54	.668	1.336	2.004	2.672	3.340	4.008	4.675	5.344	6.011	6.679	7.347	8.015	8.683	9.350	10.019	10.687	11.354	12.022	12.689	13.357	6
56	.668	1.337	2.005	2.674	3.342	4.010	4.678	5.347	6.015	6.683	7.351	8.020	8.690	9.356	10.024	10.694	11.361	12.030	12.697	13.365	4
58	.669	1.337	2.006	2.676	3.344	4.013	4.681	5.351	6.019	6.687	7.356	8.025	8.695	9.362	10.031	10.701	11.368	12.037	12.705	13.374	2
12° 0	.669	1.338	2.007	2.677	3.346	4.015	4.684	5.354	6.022	6.691	7.360	8.030	8.699	9.368	10.038	10.708	11.376	12.014	12.713	13.382	18° 0
	1	2	3	4	5	6	7	8	9	10	11	12	13	14	15	16	17	18	19	20	IV + X –

TABLE XVIII.

GENERAL TABLE of Aberration, &c. continued.

S. − VII + I	1	2	3	4	5	6	7	8	9	10	11	12	13	14	15	16	17	18	19	20	
12° 0′	.669	1.338	2.007	2.677	3.346	4.015	4.684	5.354	6.022	6.691	7.361	8.030	8.699	9.369	10.038	10.706	11.376	12.045	12.713	13.383	18° 0′
2	.670	1.339	2.009	2.679	3.349	4.018	4.687	5.358	6.026	6.696	7.366	8.036	8.705	9.375	10.045	10.713	11.384	12.053	12.722	13.392	58
4	.670	1.340	2.011	2.681	3.351	4.021	4.690	5.362	6.030	6.701	7.371	8.041	8.711	9.381	10.052	10.720	11.391	12.061	12.731	13.401	56
6	.671	1.341	2.012	2.683	3.353	4.024	4.693	5.365	6.034	6.705	7.376	8.046	8.717	9.387	10.058	10.727	11.398	12.069	12.739	13.410	54
8	.671	1.342	2.013	2.684	3.355	4.026	4.696	5.368	6.038	6.709	7.381	8.051	8.722	9.393	10.064	10.734	11.405	12.076	12.747	13.418	52
10	.671	1.343	2.014	2.685	3.357	4.028	4.699	5.371	6.042	6.713	7.385	8.056	8.727	9.398	10.070	10.741	11.412	12.083	12.755	13.426	50
12	.672	1.344	2.016	2.687	3.359	4.031	4.702	5.375	6.046	6.718	7.390	8.061	8.733	9.404	10.077	10.748	11.420	12.091	12.763	13.435	48
14	.672	1.345	2.017	2.689	3.361	4.034	4.705	5.379	6.050	6.722	7.395	8.066	8.739	9.410	10.084	10.755	11.428	12.099	12.771	13.444	46
16	.673	1.346	2.018	2.691	3.363	4.036	4.708	5.382	6.054	6.726	7.400	8.071	8.744	9.416	10.090	10.762	11.435	12.107	12.779	13.453	44
18	.673	1.347	2.019	2.693	3.365	4.038	4.711	5.385	6.058	6.730	7.404	8.076	8.749	9.422	10.096	10.769	11.442	12.115	12.787	13.461	42
20	.673	1.347	2.020	2.694	3.367	4.040	4.714	5.388	6.061	6.734	7.408	8.081	8.754	9.428	10.102	10.775	11.449	12.122	12.795	13.469	40
22	.674	1.348	2.022	2.696	3.370	4.043	4.717	5.392	6.065	6.739	7.413	8.087	8.760	9.434	10.109	10.782	11.457	12.130	12.804	13.478	38
24	.674	1.349	2.024	2.698	3.372	4.046	4.720	5.396	6.069	6.744	7.418	8.092	8.766	9.440	10.116	10.789	11.464	12.138	12.812	13.487	36
26	.675	1.350	2.025	2.700	3.374	4.049	4.723	5.399	6.073	6.748	7.423	8.097	8.772	9.446	10.122	10.796	11.471	12.146	12.820	13.496	34
28	.676	1.351	2.026	2.701	3.376	4.052	4.726	5.402	6.077	6.752	7.428	8.102	8.777	9.452	10.128	10.803	11.478	12.154	12.828	13.504	32
30	.676	1.351	2.027	2.702	3.378	4.054	4.729	5.405	6.080	6.756	7.432	8.107	8.782	9.458	10.134	10.809	11.485	12.161	12.836	13.512	30
32	.677	1.352	2.029	2.704	3.381	4.057	4.732	5.409	6.084	6.761	7.437	8.113	8.788	9.464	10.141	10.816	11.493	12.169	12.845	13.521	28
34	.677	1.353	2.030	2.706	3.383	4.060	4.735	5.413	6.088	6.765	7.442	8.118	8.794	9.470	10.148	10.823	11.501	12.177	12.853	13.530	26
36	.678	1.354	2.031	2.708	3.385	4.062	4.738	5.416	6.092	6.769	7.447	8.123	8.800	9.476	10.154	10.830	11.508	12.185	12.861	13.539	24
38	.678	1.355	2.032	2.710	3.387	4.064	4.741	5.419	6.096	6.773	7.451	8.128	8.805	9.482	10.160	10.837	11.515	12.192	12.869	13.547	22
40	.678	1.356	2.033	2.711	3.389	4.066	4.744	5.422	6.100	6.777	7.455	8.133	8.810	9.488	10.166	10.844	11.522	12.199	12.877	13.555	20
42	.679	1.357	2.035	2.713	3.391	4.069	4.747	5.426	6.104	6.782	7.460	8.138	8.816	9.494	10.173	10.851	11.530	12.207	12.886	13.564	18
44	.679	1.358	2.037	2.715	3.393	4.072	4.750	5.430	6.108	6.787	7.465	8.143	8.822	9.500	10.180	10.858	11.537	12.215	12.894	13.573	16
46	.680	1.359	2.038	2.717	3.395	4.075	4.753	5.433	6.112	6.791	7.470	8.148	8.828	9.506	10.186	10.865	11.544	12.223	12.902	13.582	14
48	.680	1.360	2.039	2.719	3.397	4.077	4.756	5.436	6.116	6.795	7.474	8.153	8.833	9.512	10.192	10.872	11.551	12.231	12.910	13.590	12
50	.680	1.360	2.040	2.720	3.399	4.079	4.759	5.439	6.119	6.799	7.478	8.158	8.838	9.518	10.198	10.878	11.558	12.238	12.918	13.598	10
52	.681	1.361	2.042	2.722	3.402	4.082	4.762	5.443	6.123	6.804	7.483	8.164	8.844	9.524	10.205	10.885	11.566	12.246	12.926	13.607	8
54	.681	1.362	2.043	2.724	3.404	4.085	4.765	5.447	6.127	6.808	7.488	8.169	8.850	9.530	10.212	10.892	11.573	12.254	12.934	13.616	6
56	.682	1.363	2.044	2.726	3.406	4.088	4.768	5.450	6.131	6.812	7.493	8.174	8.856	9.536	10.218	10.899	11.580	12.262	12.942	13.624	4
58	.682	1.364	2.045	2.727	3.408	4.090	4.771	5.453	6.135	6.816	7.498	8.179	8.861	9.542	10.224	10.906	11.587	12.269	12.950	13.632	2
13° 0′	.682	1.364	2.046	2.728	3.410	4.092	4.774	5.456	6.138	6.820	7.502	8.184	8.866	9.548	10.230	10.912	11.594	12.276	12.958	13.640	17° 0′
2	.683	1.365	2.048	2.730	3.413	4.095	4.777	5.460	6.142	6.825	7.507	8.190	8.872	9.554	10.237	10.919	11.602	12.284	12.966	13.649	58
4	.683	1.366	2.049	2.732	3.415	4.098	4.780	5.464	6.146	6.829	7.512	8.195	8.878	9.560	10.244	10.926	11.609	12.292	12.974	13.658	56
6	.684	1.367	2.050	2.734	3.417	4.101	4.783	5.467	6.150	6.833	7.517	8.200	8.884	9.566	10.250	10.933	11.616	12.300	12.982	13.667	54
8	.684	1.368	2.051	2.736	3.419	4.103	4.786	5.470	6.154	6.837	7.522	8.205	8.889	9.572	10.256	10.940	11.623	12.307	12.990	13.674	52
10	.684	1.368	2.052	2.737	3.421	4.105	4.789	5.473	6.157	6.841	7.526	8.210	8.894	9.578	10.262	10.946	11.630	12.314	12.998	13.682	50
12	.685	1.369	2.054	2.739	3.423	4.108	4.792	5.477	6.161	6.846	7.531	8.215	8.900	9.584	10.269	10.953	11.638	12.322	13.006	13.691	48
14	.685	1.370	2.056	2.741	3.425	4.111	4.795	5.481	6.165	6.850	7.536	8.220	8.906	9.590	10.276	10.960	11.646	12.330	13.014	13.700	46
16	.686	1.371	2.057	2.743	3.427	4.113	4.798	5.484	6.169	6.854	7.540	8.225	8.911	9.596	10.282	10.967	11.653	12.338	13.022	13.709	44
18	.686	1.372	2.058	2.744	3.429	4.115	4.801	5.487	6.173	6.858	7.544	8.230	8.916	9.602	10.288	10.974	11.660	12.345	13.030	13.717	42
20	.686	1.373	2.059	2.745	3.431	4.117	4.804	5.490	6.176	6.862	7.548	8.235	8.921	9.607	10.294	10.980	11.667	12.352	13.038	13.725	40
22	.687	1.374	2.061	2.747	3.434	4.120	4.807	5.494	6.180	6.867	7.553	8.241	8.927	9.613	10.301	10.987	11.674	12.360	13.047	13.734	38
24	.687	1.375	2.062	2.749	3.436	4.123	4.810	5.498	6.184	6.872	7.558	8.246	8.933	9.619	10.308	10.994	11.681	12.368	13.055	13.743	36
26	.688	1.376	2.063	2.751	3.438	4.126	4.813	5.501	6.188	6.876	7.563	8.251	8.939	9.625	10.314	11.001	11.688	12.376	13.063	13.751	34
28	.688	1.377	2.064	2.752	3.440	4.128	4.816	5.504	6.192	6.880	7.568	8.256	8.944	9.631	10.320	11.008	11.695	12.384	13.071	13.759	32
30	.688	1.377	2.065	2.753	3.442	4.130	4.819	5.507	6.195	6.884	7.572	8.261	8.949	9.637	10.326	11.014	11.702	12.391	13.079	13.767	30
32	.689	1.378	2.067	2.755	3.444	4.133	4.822	5.511	6.199	6.889	7.577	8.266	8.955	9.643	10.333	11.021	11.710	12.399	13.087	13.776	28
34	.689	1.379	2.068	2.757	3.446	4.136	4.825	5.515	6.203	6.893	7.582	8.271	8.961	9.649	10.339	11.028	11.717	12.407	13.095	13.785	26
36	.690	1.380	2.069	2.759	3.448	4.139	4.828	5.518	6.207	6.897	7.587	8.276	8.966	9.655	10.345	11.035	11.724	12.415	13.103	13.793	24
38	.691	1.381	2.070	2.761	3.450	4.141	4.831	5.521	6.211	6.901	7.591	8.281	8.971	9.661	10.351	11.042	11.731	12.422	13.111	13.801	22
40	.691	1.381	2.071	2.762	3.452	4.143	4.833	5.524	6.214	6.905	7.595	8.286	8.976	9.667	10.357	11.047	11.738	12.429	13.119	13.809	20
42	.692	1.382	2.073	2.764	3.455	4.146	4.836	5.528	6.218	6.910	7.600	8.291	8.982	9.673	10.364	11.054	11.746	12.437	13.127	13.818	18
44	.692	1.383	2.075	2.766	3.457	4.149	4.839	5.532	6.222	6.914	7.605	8.296	8.988	9.679	10.371	11.061	11.753	12.445	13.135	13.827	16
46	.693	1.384	2.076	2.768	3.459	4.151	4.842	5.535	6.226	6.918	7.610	8.301	8.993	9.685	10.377	11.068	11.760	12.453	13.143	13.835	14
48	.693	1.385	2.077	2.769	3.461	4.153	4.845	5.538	6.230	6.922	7.614	8.306	8.998	9.691	10.383	11.075	11.767	12.460	13.151	13.843	12
50	.693	1.385	2.078	2.770	3.463	4.155	4.848	5.541	6.233	6.926	7.618	8.311	9.003	9.696	10.389	11.081	11.774	12.467	13.159	13.851	10
52	.694	1.386	2.080	2.772	3.465	4.158	4.851	5.545	6.237	6.931	7.623	8.316	9.009	9.702	10.396	11.088	11.781	12.475	13.167	13.860	8
54	.694	1.387	2.081	2.774	3.467	4.161	4.854	5.548	6.241	6.935	7.628	8.321	9.015	9.708	10.402	11.095	11.788	12.483	13.175	13.869	6
56	.695	1.388	2.082	2.776	3.469	4.164	4.857	5.551	6.245	6.939	7.633	8.326	9.021	9.714	10.408	11.102	11.795	12.490	13.183	13.878	4
58	.695	1.389	2.083	2.778	3.471	4.166	4.860	5.554	6.249	6.943	7.637	8.331	9.026	9.720	10.414	11.109	11.802	12.497	13.191	13.886	2
14° 0	.695	1.389	2.084	2.779	3.473	4.168	4.863	5.557	6.252	6.947	7.641	8.336	9.031	9.725	10.420	11.115	11.809	12.504	13.199	13.894	16° 0
	1	2	3	4	5	6	7	8	9	10	11	12	13	14	15	16	17	18	19	20	IV + X −

TABLE XVIII.

GENERAL TABLE of Aberration, &c. continued.

S. − VII + I	1	2	3	4	5	6	7	8	9	10	11	12	13	14	15	16	17	18	19	20	
14° 0′	.695	1.389	2.084	2.779	3.473	4.168	4.863	5.557	6.252	6.947	7.641	8.236	9.031	9.725	10.420	11.115	11.809	12.504	13.199	13.894	16° 0′
2	.696	1.390	2.086	2.781	3.476	4.171	4.866	5.561	6.256	6.952	7.646	8.241	9.037	9.731	10.427	11.122	11.817	12.512	13.207	13.903	58
4	.696	1.391	2.087	2.783	3.478	4.174	4.869	5.565	6.260	6.956	7.651	8.246	9.043	9.737	10.433	11.129	11.824	12.520	13.215	13.911	56
6	.697	1.392	2.088	2.785	3.480	4.177	4.872	5.568	6.264	6.960	7.656	8.251	9.048	9.743	10.439	11.136	11.831	12.528	13.223	13.919	54
8	.697	1.393	2.089	2.786	3.482	4.179	4.875	5.571	6.268	6.964	7.661	8.256	9.053	9.749	10.445	11.142	11.838	12.535	13.231	13.927	52
10	.697	1.394	2.090	2.787	3.484	4.181	4.877	5 574	6.271	6.968	7.665	8.361	9.058	9.755	10.451	11.148	11.845	12.542	13.239	13.935	50
12	.698	1.395	2.092	2.789	3.486	4.184	4.880	5.578	6.275	6.972	7.670	8.366	9.064	9.761	10.458	11.155	11.853	12.550	13.247	13.944	48
14	.698	1.396	2.094	2.791	3.488	4.187	4.883	5.582	6.279	6.976	7.675	8.371	9.070	9.767	10.465	11.162	11.860	12.558	13.255	13.953	46
16	.699	1.397	2.095	2.793	3.490	4.189	4.886	5.585	6.283	6.980	7.679	8.376	9.075	9.773	10.471	11.169	11.867	12.565	13.263	13.961	44
18	.699	1.398	2.096	2.794	3.492	4.191	4.889	5.588	6.287	6.984	7.683	8.381	9.080	9.779	10.477	11.176	11.874	12.572	13.271	13.969	42
20	.699	1.398	2.097	2.795	3.494	4.193	4.892	5.591	6.290	6.988	7.687	8.386	9.085	9.784	10.483	11.182	11.881	12.579	13.278	13.977	40
22	.700	1.399	2.099	2.797	3.497	4.196	4.895	5.595	6.294	6.993	7.692	8.391	9.091	9.790	10.489	11.189	11.888	12.587	13.286	13.986	38
24	.700	1.400	2.100	2.799	3.499	4.199	4.898	5.598	6.298	6.997	7.697	8.396	9.097	9.796	10.495	11.196	11.895	12.595	13.294	13.994	36
26	.701	1.401	2.101	2.801	3.501	4.202	4.901	5.601	6.302	7.001	7.702	8.401	9.102	9.802	10.501	11.202	11.902	12.602	13.302	14.002	34
28	.701	1.402	2.102	2.803	3.503	4.204	4.904	5.604	6.305	7.005	7.707	8.406	9.107	9.808	10.507	11.208	11.909	12.609	13.310	14.010	32
30	.701	1.402	2.103	2.804	3.505	4.206	4.906	5.607	6.308	7.009	7.711	8.411	9.112	9.813	10.513	11.214	11.915	12.616	13.317	14.018	30
32	.702	1.403	2.105	2.806	3.507	4.209	4.909	5.611	6.312	7.014	7.716	8.416	9.118	9.819	10.520	11.221	11.923	12.624	13.325	14.027	28
34	.702	1.404	2.106	2.808	3.509	4.212	4.912	5.615	6.316	7.018	7.721	8.421	9.124	9.825	10.527	11.228	11.930	12.632	13.333	14.036	26
36	.703	1.405	2.107	2.810	3.511	4.214	4.915	5.618	6.320	7.022	7.725	8.426	9.129	9.831	10.533	11.235	11.937	12.640	13.341	14.044	24
38	.703	1.406	2.108	2.811	3.513	4.216	4.918	5.621	6.324	7.026	7.729	8.431	9.134	9.837	10.539	11.242	11.944	12.647	13.349	14.052	22
40	.703	1.406	2.109	2.812	3.515	4.218	4.921	5.624	6.327	7.030	7.733	8.436	9.139	9.842	10.545	11.248	11.951	12.654	13.357	14.060	20
42	.704	1.407	2.111	2.814	3.517	4.221	4.924	5.628	6.331	7.035	7.738	8.441	9.145	9.848	10.551	11.255	11.958	12.662	13.365	14.069	18
44	.704	1.408	2.112	2.816	3.519	4.224	4.927	5 631	6.335	7.039	7.743	8.446	9.151	9.854	10.557	11.262	11.965	12.670	13.373	14.077	16
46	.705	1.409	2.113	2.818	3.521	4.326	4.930	5.634	6.339	7.043	7.747	8.451	9.156	9.860	10.563	11.268	11.972	12.677	13.381	14.085	14
48	.705	1.410	2.114	2.819	3.523	4.228	4.933	5.637	6.342	7.047	7.751	8.456	9.161	9.865	10.569	11.274	11.979	12.684	13.389	14.093	12
50	.705	1.410	2.115	2.820	3.525	4.230	4.935	5.640	6.345	7.051	7.755	8.461	9.166	9.870	10.575	11.280	11.985	12.691	13.396	14.101	10
52	.706	1.411	2.117	2.821	3.528	4.233	4.938	5.644	6.349	7.055	7.760	8.466	9.172	9.876	10.581	11.287	11.993	12.699	13.404	14.110	8
54	.706	1.412	2.118	2.823	3.530	4.236	4.941	5.648	6.353	7.059	7.765	8.471	9.178	9.882	10.587	11.294	12.000	12.707	13.412	14.118	6
56	.707	1.413	2.119	2.825	3.532	4.239	4.944	5.651	6.357	7.063	7.770	8.476	9.183	9.888	10.594	11.301	12.007	12.714	13.420	14.126	4
58	.707	1.414	2.120	2.827	3.534	4.241	4.947	5.654	6.361	7.067	7.775	8.481	9.188	9.894	10.600	11.308	12.014	12.721	13.428	14.134	2
15° 0′	.707	1.414	2.121	2.828	3.536	4.243	4.950	5.657	6.364	7.071	7.779	8.486	9.193	9.900	10.607	11.314	12.021	12.728	13.435	14.142	15° 0′
2	.708	1.415	2.123	2.830	3.538	4.246	4.953	5.661	6.368	7.076	7.784	8.491	9.199	9.906	10.613	11.321	12.028	12.736	13.443	14.151	58
4	.708	1.416	2.124	2.832	3.540	4.249	4.956	5.664	6.372	7.080	7.789	8.496	9.204	9.912	10.619	11.328	12.035	12.744	13.451	14.160	56
6	.709	1.417	2.125	2.834	3.542	4.251	4.959	5.667	6.376	7.084	7.793	8.501	9.209	9.918	10.625	11.334	12.042	12.751	13.459	14.168	54
8	.709	1.418	2.126	2.836	3.544	4.253	4.962	5.670	6.379	7.088	7.797	8.506	9.214	9.923	10.631	11.340	12.049	12.758	13.467	14.176	52
10	.709	1.418	2.127	2.837	3.546	4.255	4.964	5.673	6.382	7.092	7.801	8.510	9.219	9.928	10.637	11.346	12.055	12.765	13.474	14.184	50
12	.710	1.419	2.129	2.839	3.548	4.258	4.967	5.677	6.386	7.096	7.806	8.515	9.225	9.934	10.644	11.353	12.063	12.773	13.482	14.192	48
14	.710	1.420	2.131	2.841	3.550	4.261	4.970	5.681	6.390	7.100	7.811	8.520	9.231	9.940	10.651	11.360	12.070	12.781	13.490	14.200	46
16	.711	1.421	2.132	2.843	3.552	4.263	4.973	5.684	6.394	7.104	7.815	8.525	9.236	9.946	10.657	11.367	12.077	12.788	13.498	14.208	44
18	.711	1.422	2.133	2.844	3.554	4.265	4.976	5.687	6.398	7.108	7.819	8.530	9.241	9.952	10.663	11.374	12.084	12.795	13.506	14.216	42
20	.711	1.422	2.134	2.845	3.556	4.267	4.979	5.690	6.401	7.112	7.823	8.535	9.246	9.957	10.669	11.380	12.091	12.802	13.513	14.224	40
22	.712	1.423	2.136	2.847	3.558	4.270	4 982	5.694	6.405	7.117	7.828	8.540	9.252	9.963	10.675	11.387	12.098	12.810	13.521	14.233	38
24	.712	1.424	2.137	2.849	3.560	4.273	4.985	5.697	6.409	7.121	7.833	8.545	9.258	9.969	10.681	11.394	12.105	12.818	13.529	14.241	36
26	.713	1.425	2.138	2.851	3.562	4.276	4.988	5.700	6.413	7.125	7.838	8.550	9.263	9.975	10.687	11.400	12.112	12.825	13.537	14.249	34
28	.713	1.426	2.139	2.852	3.564	4.278	4.991	5.703	6.416	7.129	7.842	8.555	9.268	9.981	10.693	11.406	12.119	12.832	13.545	14.257	32
30	.713	1.427	2.140	2.853	3.566	4.280	4.993	5.706	6.419	7.133	7.846	8.559	9.273	9.986	10.699	11.412	12.125	12.839	13.552	14.265	30
32	.714	1.428	2.142	2.855	3.568	4.283	4.996	5.710	6.423	7.137	7.851	8.564	9.279	9.992	10.705	11.419	12.132	12.847	13.560	14.273	28
34	.714	1.429	2.143	2.857	3.570	4.286	4.999	5.713	6.427	7.141	7.856	8.569	9.284	9.998	10.711	11.426	12.139	12.854	13.568	14.281	26
36	.715	1.430	2.144	2.859	3.572	4.288	5.002	5.716	6.431	7.145	7.860	8.574	9.289	10.004	10.717	11.433	12.146	12.861	13.576	14.289	24
38	.715	1.431	2.145	2.860	3.574	4.290	5.005	5.719	6.435	7.149	7.864	8.579	9.294	10.009	10.723	11.439	12.153	12.868	13.584	14.297	22
40	.715	1.431	2.146	2.861	3.576	4.292	5.007	5.722	6.438	7.153	7.868	8.583	9.299	10.014	10.729	11.445	12.160	12.875	13.591	14.305	20
42	.716	1.432	2.148	2.863	3.579	4.295	5.010	5.726	6.442	7.157	7.873	8.588	9.305	10.020	10.736	11.452	12.167	12.883	13.599	14.314	18
44	.716	1.433	2.149	2.865	3.581	4.298	5.013	5.730	6.446	7.161	7.878	8.593	9.311	10.026	10.742	11.459	12.174	12.891	13.607	14.322	16
46	.717	1.434	2.150	2.867	3.583	4.300	5.016	5.733	6.450	7.165	7.883	8.598	9.316	10.032	10.748	11.465	12.181	12.898	13.615	14.330	14
48	.717	1.435	2.151	2.868	3.585	4.303	5.019	5.736	6.453	7.169	7.888	8.603	9.321	10.038	10.754	11.471	12.188	12.905	13.622	14.338	12
50	.717	1.435	2.152	2.869	3.587	4.305	5.021	5.739	6.456	7.173	7.892	8.608	9.326	10.044	10.760	11.477	12.195	12.912	13.629	14.346	10
52	.718	1.436	2.154	2.871	3.589	4.308	5.024	5.743	6.460	7.177	7.897	8.613	9.331	10.050	10.766	11.484	12.202	12.920	13.637	14.355	8
54	.718	1.437	2.155	2.873	3.591	4.310	5.027	5.746	6.464	7.181	7.901	8.618	9.336	10.056	10.772	11.491	12.209	12.927	13.645	14.363	6
56	.719	1.438	2.156	2.875	3.593	4.312	5.030	5.749	6.468	7.185	7.905	8.623	9.341	10.061	10.778	11.497	12.216	12.934	13.653	14.371	4
58	.719	1.439	2.157	2.876	3.595	4.314	5.033	5.752	6.471	7.189	7.909	8.628	9.346	10.066	10.784	11.503	12.223	12.941	13.660	14.379	2
16° 0	.719	1.439	2.158	2.877	3.597	4.316	5.035	5.755	6.474	7.193	7.913	8.632	9.351	10.071	10.790	11.509	12.229	12.948	13.667	14.387	14° 0
	1	2	3	4	5	6	7	8	9	10	11	12	13	14	15	16	17	18	19	20	IV + X −

TABLE XVIII.

GENERAL TABLE of Aberration, &c. continued.

S. − VII + I	1	2	3	4	5	6	7	8	9	10	11	12	13	14	15	16	17	18	19	20	
16° 0′	.719	1.439	2.158	2.877	3.597	4.316	5.035	5.755	6.474	7.193	7.912	8.632	9.346	10.067	10.790	11.509	12.228	12.948	13.667	14.386	14° 0′
2	.719	1.439	2.159	2.879	3.599	4.319	5.038	5.759	6.478	7.198	7.917	8.637	9.353	10.074	10.796	11.516	12.236	12.956	13.675	14.395	58
4	.720	1.440	2.160	2.881	3.601	4.322	5.041	5.762	6.482	7.202	7.922	8.642	9.360	10.081	10.802	11.523	12.243	12.964	13.683	14.404	56
6	.720	1.441	2.161	2.883	3.603	4.324	5.044	5.765	6.486	7.206	7.927	8.647	9.366	10.087	10.808	11.530	12.250	12.971	13.691	14.412	54
8	.721	1.442	2.162	2.884	3.605	4.326	5.047	5.768	6.489	7.210	7.931	8.652	9.372	10.093	10.814	11.536	12.257	12.978	13.699	14.420	52
10	.721	1.443	2.164	2.885	3.607	4.328	5.050	5.771	6.492	7.214	7.935	8.657	9.378	10.099	10.820	11.542	12.264	12.985	13.706	14.428	50
12	.721	1.443	2.165	2.887	3.609	4.331	5.053	5.775	6.496	7.218	7.940	8.662	9.384	10.105	10.827	11.549	12.271	12.993	13.714	14.436	48
14	.722	1.444	2.166	2.889	3.611	4.334	5.056	5.778	6.500	7.222	7.945	8.667	9.389	10.111	10.833	11.556	12.278	13.001	13.722	14.444	46
16	.722	1.445	2.167	2.891	3.613	4.336	5.059	5.781	6.504	7.226	7.949	8.672	9.394	10.117	10.839	11.562	12.285	13.008	13.730	14.452	44
18	.723	1.446	2.168	2.893	3.615	4.338	5.062	5.784	6.507	7.230	7.953	8.677	9.399	10.123	10.845	11.568	12.292	13.014	13.737	14.460	42
20	.723	1.447	2.170	2.894	3.617	4.340	5.064	5.787	6.510	7.234	7.957	8.681	9.404	10.128	10.851	11.574	12.298	13.021	13.744	14.468	40
22	.723	1.447	2.171	2.896	3.619	4.343	5.067	5.791	6.514	7.238	7.962	8.686	9.410	10.134	10.857	11.581	12.305	13.028	13.752	14.476	38
24	.724	1.448	2.172	2.898	3.621	4.346	5.070	5.794	6.518	7.242	7.967	8.691	9.415	10.140	10.863	11.588	12.312	13.035	13.760	14.484	36
26	.724	1.449	2.173	2.900	3.623	4.348	5.073	5.797	6.522	7.246	7.971	8.696	9.420	10.146	10.869	11.594	12.319	13.042	13.768	14.492	34
28	.725	1.450	2.174	2.901	3.625	4.350	5.076	5.800	6.525	7.250	7.975	8.700	9.425	10.151	10.875	11.600	12.325	13.049	13.775	14.500	32
30	.725	1.451	2.176	2.902	3.627	4.352	5.078	5.803	6.528	7.254	7.979	8.704	9.430	10.156	10.881	11.606	12.331	13.056	13.782	14.508	30
32	.725	1.451	2.177	2.904	3.629	4.355	5.081	5.807	6.532	7.258	7.984	8.709	9.436	10.162	10.887	11.613	12.338	13.064	13.790	14.516	28
34	.726	1.452	2.178	2.906	3.631	4.358	5.084	5.810	6.536	7.262	7.989	8.714	9.441	10.168	10.893	11.620	12.345	13.071	13.798	14.524	26
36	.726	1.453	2.179	2.908	3.633	4.360	5.087	5.813	6.540	7.266	7.993	8.719	9.446	10.174	10.899	11.626	12.352	13.078	13.806	14.532	24
38	.727	1.454	2.180	2.909	3.635	4.362	5.090	5.816	6.543	7.270	7.997	8.724	9.451	10.179	10.905	11.632	12.359	13.085	13.813	14.540	22
40	.727	1.455	2.182	2.910	3.637	4.364	5.092	5.819	6.546	7.274	8.001	8.728	9.456	10.184	10.911	11.638	12.365	13.092	13.820	14.548	20
42	.727	1.455	2.183	2.912	3.639	4.367	5.095	5.823	6.550	7.278	8.006	8.733	9.462	10.190	10.917	11.645	12.372	13.100	13.828	14.556	18
44	.728	1.456	2.184	2.914	3.641	4.370	5.098	5.826	6.554	7.282	8.011	8.738	9.467	10.196	10.923	11.652	12.379	13.107	13.836	14.564	16
46	.728	1.457	2.185	2.916	3.643	4.372	5.101	5.829	6.558	7.286	8.015	8.743	9.472	10.202	10.929	11.658	12.385	13.114	13.844	14.572	14
48	.729	1.458	2.186	2.917	3.645	4.374	5.104	5.832	6.561	7.290	8.019	8.748	9.477	10.207	10.935	11.664	12.392	13.121	13.851	14.580	12
50	.729	1.459	2.188	2.918	3.647	4.376	5.106	5.835	6.564	7.294	8.023	8.752	9.482	10.212	10.941	11.670	12.399	13.128	13.858	14.588	10
52	.729	1.459	2.189	2.920	3.649	4.379	5.109	5.839	6.568	7.298	8.028	8.757	9.488	10.218	10.947	11.677	12.406	13.136	13.866	14.596	8
54	.730	1.460	2.190	2.922	3.651	4.382	5.112	5.842	6.572	7.302	8.033	8.762	9.493	10.224	10.953	11.684	12.413	13.143	13.874	14.604	6
56	.730	1.461	2.191	2.923	3.653	4.384	5.115	5.845	6.576	7.306	8.037	8.767	9.498	10.230	10.959	11.690	12.420	13.150	13.882	14.612	4
58	.731	1.462	2.192	2.924	3.655	4.386	5.118	5.848	6.579	7.310	8.041	8.772	9.503	10.235	10.964	11.696	12.427	13.157	13.889	14.620	2
17° 0′	.731	1.463	2.194	2.925	3.657	4.388	5.120	5.851	6.582	7.314	8.045	8.776	9.508	10.240	10.971	11.702	12.433	13.164	13.896	14.628	13° 0′
2	.731	1.463	2.195	2.927	3.659	4.391	5.123	5.855	6.586	7.318	8.050	8.781	9.513	10.246	10.977	11.709	12.440	13.172	13.904	14.636	58
4	.732	1.464	2.196	2.929	3.661	4.394	5.126	5.858	6.590	7.322	8.054	8.786	9.518	10.251	10.983	11.716	12.447	13.179	13.912	14.644	56
6	.732	1.465	2.197	2.931	3.663	4.396	5.129	5.861	6.594	7.326	8.058	8.791	9.523	10.256	10.989	11.722	12.454	13.186	13.919	14.652	54
8	.733	1.466	2.198	2.932	3.665	4.398	5.131	5.864	6.597	7.330	8.062	8.796	9.528	10.261	10.994	11.728	12.461	13.193	13.926	14.659	52
10	.733	1.467	2.200	2.933	3.667	4.400	5.133	5.867	6.600	7.333	8.066	8.800	9.533	10.266	11.000	11.734	12.467	13.200	13.933	14.666	50
12	.733	1.467	2.201	2.935	3.669	4.403	5.136	5.871	6.604	7.337	8.071	8.805	9.539	10.272	11.006	11.741	12.474	13.208	13.941	14.674	48
14	.734	1.468	2.202	2.937	3.671	4.406	5.139	5.874	6.608	7.341	8.076	8.810	9.544	10.278	11.012	11.748	12.481	13.215	13.949	14.682	46
16	.734	1.469	2.203	2.939	3.673	4.408	5.142	5.877	6.612	7.345	8.080	8.815	9.549	10.284	11.018	11.754	12.488	13.222	13.957	14.690	44
18	.735	1.470	2.204	2.940	3.675	4.410	5.145	5.880	6.615	7.349	8.084	8.820	9.554	10.289	11.024	11.760	12.495	13.229	13.964	14.698	42
20	.735	1.471	2.206	2.941	3.677	4.412	5.147	5.883	6.618	7.353	8.088	8.824	9.559	10.294	11.030	11.766	12.501	13.236	13.971	14.706	40
22	.735	1.471	2.207	2.943	3.679	4.415	5.150	5.886	6.622	7.357	8.093	8.829	9.565	10.300	11.036	11.772	12.508	13.244	13.979	14.714	38
24	.736	1.472	2.208	2.945	3.681	4.418	5.153	5.889	6.626	7.361	8.098	8.834	9.570	10.306	11.042	11.778	12.515	13.251	13.987	14.722	36
26	.736	1.473	2.209	2.947	3.683	4.420	5.156	5.892	6.630	7.365	8.102	8.839	9.575	10.312	11.048	11.784	12.522	13.258	13.995	14.730	34
28	.737	1.474	2.210	2.949	3.685	4.422	5.159	5.895	6.633	7.369	8.106	8.844	9.580	10.317	11.054	11.790	12.528	13.265	14.002	14.738	32
30	.737	1.475	2.212	2.950	3.686	4.424	5.161	5.898	6.636	7.373	8.110	8.848	9.585	10.322	11.059	11.796	12.534	13.272	14.009	14.746	30
32	.737	1.475	2.213	2.952	3.688	4.427	5.164	5.902	6.640	7.377	8.115	8.853	9.590	10.328	11.065	11.803	12.541	13.279	14.017	14.754	28
34	.738	1.476	2.214	2.954	3.690	4.429	5.167	5.905	6.644	7.381	8.119	8.858	9.595	10.334	11.071	11.810	12.548	13.286	14.024	14.762	26
36	.738	1.477	2.215	2.955	3.692	4.431	5.170	5.908	6.647	7.385	8.123	8.863	9.600	10.340	11.077	11.816	12.555	13.293	14.031	14.770	24
38	.739	1.478	2.216	2.956	3.694	4.433	5.173	5.911	6.650	7.389	8.127	8.867	9.605	10.345	11.083	11.822	12.561	13.300	14.038	14.777	22
40	.739	1.479	2.218	2.957	3.696	4.435	5.175	5.914	6.653	7.392	8.131	8.871	9.610	10.350	11.089	11.828	12.567	13.306	14.045	14.784	20
42	.739	1.479	2.219	2.959	3.698	4.438	5.178	5.918	6.657	7.396	8.136	8.876	9.615	10.356	11.095	11.835	12.574	13.314	14.053	14.792	18
44	.740	1.480	2.220	2.961	3.700	4.441	5.181	5.921	6.661	7.400	8.141	8.881	9.620	10.361	11.101	11.842	12.581	13.321	14.061	14.800	16
46	.740	1.480	2.221	2.963	3.702	4.443	5.184	5.924	6.665	7.404	8.145	8.886	9.625	10.366	11.107	11.848	12.588	13.328	14.069	14.808	14
48	.741	1.481	2.222	2.964	3.704	4.445	5.186	5.927	6.668	7.408	8.149	8.890	9.630	10.371	11.113	11.854	12.595	13.335	14.076	14.816	12
50	.741	1.482	2.224	2.965	3.706	4.447	5.188	5.930	6.671	7.412	8.153	8.894	9.635	10.376	11.118	11.860	12.601	13.342	14.083	14.824	10
52	.741	1.482	2.225	2.967	3.708	4.450	5.191	5.933	6.675	7.416	8.158	8.899	9.640	10.382	11.124	11.866	12.608	13.349	14.090	14.832	8
54	.742	1.483	2.226	2.969	3.710	4.453	5.194	5.936	6.679	7.420	8.162	8.904	9.645	10.388	11.130	11.872	12.614	13.356	14.097	14.840	6
56	.742	1.484	2.227	2.971	3.712	4.455	5.197	5.939	6.682	7.424	8.166	8.909	9.650	10.393	11.136	11.878	12.620	13.363	14.104	14.847	4
58	.743	1.485	2.228	2.972	3.714	4.457	5.200	5.942	6.685	7.428	8.170	8.913	9.655	10.398	11.141	11.884	12.626	13.369	14.111	14.854	2
18° 0′	.743	1.486	2.229	2.973	3.716	4.459	5.202	5.945	6.688	7.431	8.174	8.917	9.660	10.404	11.147	11.890	12.633	13.376	14.118	14.862	12° 0
	1	2	3	4	5	6	7	8	9	10	11	12	13	14	15	16	17	18	19	20	IV + X −

TABLE XVIII.

GENERAL TABLE of Aberration, &c. continued.

S. −VII + I	1	2	3	4	5	6	7	8	9	10	11	12	13	14	15	16	17	18	19	20	
18° 0′	.743	1.486	2.229	2.973	3.716	4.459	5.202	5.945	6.688	7.431	8.174	8.917	9.660	10.404	11.147	11.890	12.633	13.376	14.119	14.862	12° 0′
2	.743	1.486	2.231	2.975	3.718	4.462	5.205	5.949	6.692	7.435	8.179	8.922	9.666	10.410	11.153	11.897	12.640	13.384	14.127	14.870	58
4	.744	1.487	2.232	2.977	3.720	4.465	5.208	5.952	6.696	7.439	8.184	8.927	9.671	10.416	11.159	11.904	12.647	13.391	14.135	14.878	56
6	.744	1.488	2.233	2.978	3.722	4.467	5.211	5.955	6.700	7.443	8.188	8.932	9.676	10.422	11.165	11.910	12.654	13.398	14.142	14.886	54
8	.745	1.489	2.234	2.979	3.724	4.469	5.214	5.958	6.703	7.447	8.192	8.937	9.681	10.427	11.171	11.916	12.661	13.405	14.149	14.894	52
10	.745	1.490	2.235	2.980	3.726	4.471	5.216	5.961	6.706	7.451	8.196	8.941	9.686	10.432	11.177	11.922	12.667	13.412	14.157	14.902	50
12	.745	1.490	2.237	2.982	3.728	4.474	5.219	5.964	6.710	7.455	8.201	8.946	9.691	10.438	11.183	11.928	12.674	13.419	14.165	14.910	48
14	.746	1.491	2.238	2.984	3.730	4.476	5.222	5.967	6.714	7.459	8.205	8.951	9.696	10.443	11.189	11.934	12.681	13.426	14.172	14.918	46
16	.746	1.492	2.239	2.986	3.732	4.478	5.225	5.970	6.717	7.463	8.209	8.956	9.701	10.448	11.195	11.940	12.687	13.433	14.179	14.926	44
18	.747	1.493	2.240	2.987	3.734	4.480	5.227	5.973	6.720	7.467	8.213	8.960	9.706	10.453	11.200	11.946	12.693	13.440	14.186	14.933	42
20	.747	1.494	2.241	2.988	3.735	4.482	5.229	5.976	6.723	7.470	8.217	8.964	9.711	10.458	11.205	11.952	12.699	13.446	14.193	14.940	40
22	.747	1.494	2.242	2.990	3.737	4.485	5.232	5.980	6.727	7.474	8.222	8.969	9.717	10.464	11.211	11.959	12.706	13.454	14.201	14.948	38
24	.748	1.495	2.243	2.992	3.739	4.488	5.235	5.983	6.731	7.478	8.227	8.973	9.722	10.470	11.217	11.966	12.713	13.461	14.209	14.956	36
26	.748	1.496	2.244	2.994	3.741	4.490	5.238	5.986	6.735	7.482	8.231	8.979	9.727	10.476	11.223	11.972	12.720	13.468	14.217	14.964	34
28	.749	1.497	2.245	2.995	3.743	4.492	5.241	5.989	6.738	7.486	8.235	8.983	9.732	10.481	11.229	11.978	12.727	13.475	14.224	14.972	32
30	.749	1.498	2.247	2.996	3.745	4.494	5.243	5.992	6.741	7.490	8.239	8.988	9.737	10.486	11.235	11.984	12.733	13.482	14.231	14.980	30
32	.749	1.499	2.249	2.998	3.747	4.497	5.246	5.995	6.745	7.494	8.244	8.993	9.742	10.492	11.241	11.990	12.740	13.489	14.239	14.988	28
34	.750	1.500	2.250	3.000	3.749	4.499	5.249	5.998	6.749	7.498	8.248	8.998	9.747	10.497	11.247	11.996	12.747	13.496	14.246	14.996	26
36	.750	1.501	2.251	3.002	3.751	4.501	5.252	6.001	6.752	7.502	8.252	9.003	9.752	10.502	11.253	12.002	12.753	13.503	14.253	15.004	24
38	.751	1.502	2.252	3.003	3.753	4.503	5.254	6.004	6.755	7.506	8.256	9.008	9.757	10.507	11.258	12.008	12.758	13.510	14.260	15.011	22
40	.751	1.503	2.253	3.004	3.754	4.505	5.256	6.007	6.758	7.509	8.260	9.012	9.762	10.512	11.263	12.014	12.765	13.516	14.267	15.018	20
42	.751	1.503	2.254	3.006	3.756	4.508	5.259	6.010	6.762	7.513	8.265	9.017	9.767	10.518	11.269	12.020	12.772	13.523	14.275	15.026	18
44	.752	1.504	2.255	3.008	3.758	4.511	5.262	6.013	6.766	7.517	8.269	9.022	9.772	10.524	11.275	12.026	12.778	13.530	14.282	15.034	16
46	.752	1.505	2.256	3.009	3.760	4.514	5.265	6.016	6.769	7.521	8.273	9.026	9.777	10.530	11.281	12.032	12.784	13.537	14.289	15.042	14
48	.753	1.505	2.257	3.010	3.762	4.516	5.268	6.019	6.772	7.525	8.277	9.030	9.782	10.535	11.287	12.038	12.790	13.544	14.296	15.049	12
50	.753	1.506	2.258	3.011	3.764	4.518	5.270	6.022	6.775	7.528	8.281	9.034	9.786	10.540	11.292	12.044	12.797	13.550	14.303	15.056	10
52	.753	1.506	2.260	3.013	3.766	4.520	5.273	6.026	6.779	7.532	8.286	9.039	9.791	10.546	11.298	12.051	12.804	13.557	14.311	15.064	8
54	.754	1.507	2.261	3.015	3.768	4.522	5.276	6.029	6.783	7.536	8.290	9.044	9.796	10.551	11.304	12.058	12.811	13.564	14.318	15.072	6
56	.754	1.507	2.262	3.017	3.770	4.524	5.279	6.032	6.786	7.540	8.294	9.048	9.801	10.556	11.310	12.064	12.818	13.571	14.325	15.080	4
58	.755	1.508	2.263	3.018	3.772	4.526	5.281	6.035	6.789	7.544	8.298	9.052	9.806	10.561	11.316	12.070	12.824	13.578	14.332	15.087	2
19° 0′	.755	1.509	2.264	3.019	3.774	4.528	5.283	6.038	6.792	7.547	8.302	9.056	9.811	10.566	11.321	12.076	12.830	13.584	14.339	15.094	11° 0′
2	.755	1.509	2.266	3.021	3.776	4.531	5.286	6.041	6.796	7.551	8.307	9.061	9.816	10.572	11.327	12.082	12.837	13.592	14.347	15.102	58
4	.756	1.510	2.267	3.023	3.778	4.534	5.289	6.044	6.800	7.555	8.311	9.066	9.821	10.577	11.333	12.088	12.844	13.599	14.355	15.110	56
6	.756	1.511	2.268	3.024	3.780	4.537	5.292	6.047	6.804	7.559	8.315	9.071	9.826	10.582	11.339	12.094	12.851	13.606	14.362	15.118	54
8	.757	1.512	2.269	3.025	3.782	4.539	5.294	6.050	6.807	7.563	8.319	9.075	9.831	10.587	11.344	12.100	12.857	13.613	14.369	15.125	52
10	.757	1.513	2.270	3.026	3.783	4.541	5.296	6.053	6.810	7.566	8.323	9.079	9.836	10.592	11.349	12.106	12.863	13.620	14.376	15.132	50
12	.757	1.513	2.272	3.028	3.785	4.543	5.299	6.056	6.814	7.570	8.328	9.084	9.841	10.598	11.355	12.112	12.870	13.627	14.384	15.140	48
14	.758	1.514	2.273	3.030	3.787	4.545	5.302	6.059	6.818	7.574	8.332	9.089	9.846	10.604	11.361	12.118	12.877	13.634	14.391	15.148	46
16	.758	1.515	2.274	3.032	3.789	4.547	5.305	6.062	6.821	7.578	8.336	9.094	9.851	10.610	11.367	12.124	12.883	13.641	14.398	15.156	44
18	.759	1.516	2.275	3.033	3.792	4.549	5.308	6.065	6.824	7.582	8.340	9.098	9.856	10.615	11.373	12.130	12.889	13.648	14.405	15.163	42
20	.759	1.517	2.276	3.034	3.793	4.551	5.310	6.068	6.827	7.585	8.344	9.102	9.861	10.620	11.378	12.136	12.895	13.654	14.412	15.170	40
22	.759	1.517	2.277	3.036	3.795	4.554	5.313	6.071	6.831	7.589	8.348	9.107	9.866	10.626	11.384	12.142	12.902	13.661	14.420	15.178	38
24	.760	1.518	2.278	3.038	3.797	4.557	5.316	6.074	6.835	7.593	8.352	9.112	9.871	10.631	11.390	12.148	12.909	13.668	14.427	15.186	36
26	.760	1.519	2.279	3.040	3.799	4.559	5.319	6.077	6.838	7.597	8.356	9.117	9.876	10.636	11.396	12.154	12.915	13.675	14.434	15.194	34
28	.760	1.520	2.280	3.041	3.801	4.561	5.321	6.080	6.841	7.601	8.360	9.121	9.881	10.641	11.401	12.160	12.921	13.682	14.441	15.201	32
30	.760	1.521	2.281	3.042	3.802	4.563	5.323	6.083	6.844	7.604	8.364	9.125	9.885	10.646	11.406	12.166	12.927	13.688	14.448	15.208	30
32	.760	1.521	2.283	3.044	3.804	4.566	5.326	6.086	6.848	7.608	8.369	9.130	9.890	10.651	11.412	12.172	12.934	13.695	14.456	15.216	28
34	.761	1.522	2.284	3.046	3.806	4.568	5.329	6.089	6.852	7.612	8.373	9.135	9.895	10.656	11.418	12.178	12.941	13.702	14.463	15.224	26
36	.761	1.523	2.285	3.047	3.808	4.570	5.332	6.092	6.855	7.616	8.377	9.140	9.900	10.661	11.424	12.184	12.947	13.709	14.470	15.232	24
38	.762	1.524	2.286	3.048	3.810	4.572	5.334	6.095	6.858	7.620	8.381	9.144	9.905	10.666	11.429	12.190	12.953	13.716	14.477	15.239	22
40	.762	1.525	2.287	3.049	3.812	4.574	5.336	6.098	6.861	7.623	8.385	9.148	9.910	10.672	11.434	12.196	12.959	13.722	14.484	15.246	20
42	.762	1.525	2.289	3.051	3.814	4.577	5.339	6.101	6.865	7.627	8.390	9.153	9.915	10.678	11.440	12.202	12.966	13.729	14.492	15.254	18
44	.763	1.526	2.290	3.053	3.816	4.579	5.342	6.104	6.869	7.631	8.394	9.158	9.920	10.683	11.446	12.208	12.973	13.736	14.499	15.262	16
46	.763	1.526	2.291	3.055	3.818	4.581	5.345	6.107	6.872	7.635	8.398	9.162	9.925	10.688	11.452	12.214	12.979	13.743	14.506	15.270	14
48	.764	1.527	2.292	3.056	3.820	4.583	5.347	6.110	6.875	7.639	8.402	9.166	9.930	10.693	11.457	12.220	12.985	13.750	14.513	15.277	12
50	.764	1.528	2.293	3.057	3.821	4.585	5.349	6.113	6.878	7.642	8.406	9.170	9.935	10.698	11.462	12.226	12.991	13.756	14.520	15.284	10
52	.764	1.528	2.294	3.059	3.823	4.588	5.352	6.116	6.882	7.646	8.410	9.175	9.940	10.704	11.468	12.232	12.998	13.763	14.527	15.292	8
54	.765	1.529	2.295	3.061	3.825	4.590	5.355	6.119	6.885	7.650	8.414	9.180	9.945	10.709	11.474	12.238	13.004	13.770	14.534	15.299	6
56	.705	1.530	2.296	3.062	3.827	4.592	5.358	6.122	6.888	7.654	8.418	9.184	9.950	10.714	11.480	12.244	13.010	13.776	14.541	15.306	4
58	.766	1.531	2.297	3.063	3.829	4.594	5.360	6.125	6.891	7.657	8.422	9.188	9.955	10.719	11.485	12.250	13.016	13.782	14.548	15.313	2
20° 0	.766	1.532	2.298	3.064	3.830	4.596	5.362	6.128	6.894	7.660	8.426	9.182	9.959	10.724	11.490	12.256	13.022	13.788	14.554	15.320	10 0
	1	2	3	4	5	6	7	8	9	10	11	12	13	14	15	16	17	18	19	20	IV + X −

TABLE XVIII.

GENERAL TABLE of Aberration, &c. continued.

S. − VII + I	1	2	3	4	5	6	7	8	9	10	11	12	13	14	15	16	17	18	19	20	
20° 0′	.766	1.532	2.298	3.064	3.830	4.596	5.362	6.128	6.894	7.660	8.426	9.192	9.959	10.724	11.490	12.256	13.022	13.788	14.554	15.320	10° 0′
2	.766	1.532	2.300	3.066	3.832	4.599	5.365	6.131	6.898	7.664	8.431	9.197	9.964	10.730	11.496	12.262	13.029	13.795	14.562	15.328	58
4	.767	1.533	2.301	3.068	3.834	4.602	5.368	6.134	6.902	7.668	8.435	9.202	9.969	10.735	11.502	12.268	13.036	13.802	14.569	15.336	56
6	.767	1.534	2.302	3.070	3.836	4.604	5.371	6.137	6.905	7.672	8.439	9.207	9.974	10.740	11.508	12.274	13.042	13.809	14.576	15.344	54
8	.768	1.535	2.303	3.071	3.838	4.606	5.373	6.140	6.908	7.676	8.443	9.212	9.979	10.745	11.513	12.280	13.048	13.815	14.583	15.351	52
10	.768	1.536	2.304	3.072	3.840	4.608	5.375	6.143	6.911	7.679	8.447	9.216	9.983	10.750	11.518	12.286	13.054	13.822	14.590	15.358	50
12	.768	1.536	2.305	3.074	3.842	4.611	5.378	6.146	6.915	7.683	8.452	9.221	9.988	10.756	11.524	12.292	13.061	13.829	14.598	15.366	48
14	.769	1.537	2.306	3.076	3.844	4.613	5.381	6.149	6.919	7.687	8.456	9.226	9.993	10.761	11.530	12.298	13.068	13.836	14.605	15.374	46
16	.769	1.538	2.307	3.077	3.846	4.615	5.384	6.152	6.922	7.691	8.460	9.230	9.998	10.766	11.536	12.304	13.074	13.843	14.612	15.382	44
18	.770	1.539	2.308	3.078	3.848	4.617	5.386	6.155	6.925	7.695	8.464	9.234	10.003	10.771	11.541	12.310	13.080	13.850	14.619	15.389	42
20	.770	1.540	2.309	3.079	3.849	4.619	5.388	6.158	6.928	7.698	8.468	9.238	10.007	10.776	11.546	12.316	13.086	13.856	14.626	15.396	40
22	.770	1.540	2.310	3.081	3.851	4.622	5.391	6.161	6.932	7.702	8.472	9.243	10.012	10.782	11.552	12.322	13.093	13.864	14.634	15.404	38
24	.771	1.541	2.311	3.083	3.853	4.624	5.394	6.164	6.936	7.706	8.476	9.248	10.017	10.787	11.558	12.328	13.100	13.871	14.641	15.411	36
26	.771	1.541	2.312	3.085	3.855	4.626	5.397	6.167	6.940	7.710	8.480	9.252	10.022	10.792	11.564	12.334	13.107	13.878	14.648	15.418	34
28	.772	1.542	2.313	3.086	3.857	4.628	5.399	6.170	6.943	7.713	8.484	9.256	10.027	10.797	11.569	12.340	13.113	13.885	14.655	15.425	32
30	.772	1.543	2.315	3.087	3.858	4.630	5.401	6.173	6.946	7.716	8.488	9.260	10.031	10.802	11.574	12.346	13.119	13.892	14.662	15.432	30
32	.772	1.543	2.316	3.089	3.860	4.633	5.404	6.176	6.949	7.720	8.493	9.265	10.036	10.808	11.580	12.352	13.125	13.898	14.669	15.440	28
34	.773	1.544	2.317	3.091	3.862	4.635	5.407	6.179	6.952	7.724	8.497	9.270	10.041	10.813	11.586	12.358	13.131	13.904	14.676	15.448	26
36	.773	1.545	2.318	3.092	3.864	4.637	5.410	6.182	6.855	7.728	8.501	9.274	10.046	10.818	11.592	12.364	13.137	13.910	14.683	15.456	24
38	.774	1.546	2.319	3.093	3.866	4.639	5.412	6.185	6.958	7.732	8.505	9.278	10.051	10.823	11.597	12.370	13.143	13.916	14.690	15.463	22
40	.774	1.547	2.320	3.094	3.867	4.641	5.414	6.188	6.961	7.735	8.509	9.282	10.055	10.828	11.602	12.376	13.149	13.922	14.696	15.470	20
42	.774	1.547	2.322	3.096	3.869	4.644	5.417	6.191	6.965	7.739	8.513	9.287	10.060	10.834	11.608	12.382	13.156	13.929	14.703	15.478	18
44	.774	1.548	2.323	3.098	3.871	4.646	5.420	6.194	6.969	7.743	8.517	9.292	10.065	10.839	11.614	12.388	13.163	13.936	14.710	15.485	16
46	.774	1.549	2.324	3.099	3.873	4.648	5.423	6.197	6.972	7.747	8.521	9.296	10.070	10.844	11.620	12.394	13.169	13.943	14.717	15.492	14
48	.775	1.550	2.325	3.100	3.875	4.650	5.425	6.200	6.975	7.750	8.525	9.300	10.075	10.849	11.625	12.400	13.175	13.950	14.724	15.499	12
50	.775	1.551	2.326	3.101	3.877	4.652	5.427	6.203	6.978	7.753	8.529	9.304	10.079	10.854	11.630	12.406	13.181	13.956	14.731	15.506	10
52	.775	1.551	2.328	3.103	3.879	4.655	5.430	6.206	6.982	7.757	8.533	9.309	10.084	10.860	11.636	12.412	13.187	13.963	14.738	15.514	8
54	.776	1.552	2.329	3.105	3.881	4.657	5.433	6.209	6.985	7.761	8.537	9.314	10.089	10.865	11.642	12.418	13.193	13.970	14.745	15.522	6
56	.776	1.552	2.330	3.107	3.883	4.659	5.436	6.212	6.988	7.765	8.541	9.318	10.094	10.870	11.647	12.424	13.199	13.976	14.752	15.530	4
58	.777	1.553	2.331	3.108	3.885	4.661	5.438	6.215	6.991	7.769	8.545	9.322	10.099	10.875	11.652	12.429	13.205	13.982	14.759	15.537	2
21° 0′	.777	1.554	2.332	3.109	3.886	4.663	5.440	6.217	6.994	7.772	8.549	9.326	10.103	10.880	11.657	12.434	13.211	13.988	14.766	15.544	9° 0′
2	.777	1.554	2.333	3.111	3.888	4.666	5.443	6.220	6.998	7.776	8.553	9.331	10.108	10.886	11.663	12.440	13.218	13.995	14.773	15.552	58
4	.778	1.555	2.334	3.113	3.890	4.668	5.446	6.223	7.002	7.780	8.557	9.336	10.113	10.891	11.669	12.446	13.225	14.002	14.780	15.559	56
6	.778	1.556	2.335	3.114	3.892	4.670	5.449	6.226	7.005	7.784	8.561	9.340	10.118	10.896	11.675	12.452	13.231	14.009	14.787	15.566	54
8	.779	1.557	2.336	3.115	3.894	4.672	5.451	6.229	7.008	7.787	8.565	9.344	10.123	10.901	11.680	12.458	13.237	14.016	14.794	15.573	52
10	.779	1.558	2.337	3.116	3.895	4.674	5.453	6.232	7.011	7.790	8.569	9.348	10.127	10.906	11.685	12.464	13.243	14.022	14.801	15.580	50
12	.779	1.558	2.338	3.118	3.897	4.677	5.456	6.235	7.015	7.794	8.573	9.353	10.132	10.912	11.691	12.470	13.249	14.029	14.808	15.588	48
14	.780	1.559	2.339	3.120	3.899	4.679	5.459	6.238	7.018	7.798	8.577	9.358	10.137	10.917	11.697	12.476	13.255	14.036	14.815	15.595	46
16	.780	1.560	2.340	3.121	3.901	4.681	5.462	6.241	7.021	7.802	8.581	9.362	10.142	10.922	11.702	12.482	13.261	14.042	14.822	15.602	44
18	.781	1.561	2.341	3.122	3.903	4.683	5.464	6.244	7.024	7.805	8.585	9.366	10.147	10.927	11.707	12.487	13.267	14.048	14.829	15.609	42
20	.781	1.562	2.342	3.123	3.904	4.685	5.466	6.246	7.027	7.808	8.589	9.370	10.151	10.932	11.712	12.492	13.273	14.054	14.835	15.616	40
22	.781	1.562	2.344	3.125	3.906	4.688	5.469	6.249	7.031	7.812	8.593	9.375	10.156	10.937	11.718	12.498	13.280	14.061	14.842	15.624	38
24	.782	1.563	2.345	3.127	3.908	4.690	5.472	6.252	7.035	7.816	8.597	9.380	10.161	10.942	11.724	12.504	13.287	14.068	14.849	15.631	36
26	.782	1.563	2.346	3.128	3.910	4.692	5.474	6.255	7.038	7.820	8.601	9.384	10.166	10.947	11.729	12.510	13.293	14.075	14.856	15.638	34
28	.783	1.564	2.347	3.129	3.912	4.694	5.476	6.258	7.041	7.823	8.605	9.388	10.170	10.952	11.734	12.516	13.299	14.082	14.863	15.645	32
30	.783	1.565	2.348	3.130	3.913	4.696	5.478	6.261	7.044	7.826	8.609	9.392	10.174	10.956	11.739	12.522	13.305	14.088	14.870	15.652	30
32	.783	1.565	2.349	3.132	3.915	4.699	5.481	6.264	7.048	7.830	8.613	9.397	10.179	10.962	11.745	12.528	13.311	14.095	14.877	15.660	28
34	.783	1.566	2.350	3.134	3.917	4.701	5.484	6.267	7.051	7.834	8.617	9.402	10.184	10.967	11.751	12.534	13.317	14.102	14.884	15.667	26
36	.783	1.567	2.351	3.136	3.919	4.703	5.487	6.270	7.054	7.838	8.621	9.406	10.189	10.972	11.756	12.540	13.323	14.108	14.891	15.674	24
38	.784	1.568	2.352	3.137	3.921	4.705	5.489	6.273	7.057	7.841	8.625	9.410	10.194	10.977	11.761	12.545	13.329	14.114	14.898	15.681	22
40	.784	1.569	2.353	3.138	3.922	4.707	5.491	6.275	7.060	7.844	8.628	9.414	10.198	10.982	11.766	12.550	13.335	14.120	14.904	15.688	20
42	.784	1.569	2.355	3.140	3.924	4.709	5.494	6.278	7.064	7.848	8.632	9.418	10.203	10.988	11.772	12.556	13.342	14.127	14.911	15.696	18
44	.785	1.570	2.356	3.142	3.926	4.711	5.497	6.281	7.067	7.852	8.636	9.422	10.208	10.993	11.778	12.562	13.348	14.134	14.918	15.703	16
46	.785	1.570	2.357	3.143	3.928	4.713	5.500	6.284	7.070	7.856	8.640	9.426	10.213	10.998	11.784	12.568	13.354	14.140	14.925	15.710	14
48	.786	1.571	2.358	3.144	3.930	4.715	5.502	6.287	7.073	7.859	8.644	9.430	10.217	11.003	11.789	12.574	13.360	14.146	14.932	15.717	12
50	.786	1.572	2.359	3.145	3.931	4.717	5.504	6.290	7.076	7.862	8.648	9.434	10.221	11.008	11.794	12.580	13.366	14.152	14.938	15.724	10
52	.786	1.572	2.360	3.147	3.933	4.720	5.507	6.293	7.080	7.866	8.652	9.439	10.226	11.013	11.800	12.586	13.372	14.159	14.945	15.732	8
54	.787	1.573	2.361	3.149	3.935	4.722	5.510	6.296	7.083	7.870	8.656	9.444	10.231	11.018	11.805	12.592	13.378	14.166	14.952	15.739	6
56	.787	1.574	2.362	3.150	3.937	4.724	5.512	6.299	7.086	7.874	8.660	9.448	10.236	11.023	11.810	12.698	13.384	14.172	14.959	15.746	4
58	.788	1.575	2.363	3.151	3.939	4.726	5.514	6.302	7.089	7.877	8.664	9.452	10.240	11.028	11.815	12.603	13.390	14.178	14.966	15.753	2
22° 0	.788	1.576	2.364	3.152	3.940	4.728	5.516	6.304	7.092	7.880	8.668	9.456	10.244	11.032	11.820	12.608	13.396	14.184	14.972	15.760	8° 0
	1	2	3	4	5	6	7	8	9	10	11	12	13	14	15	16	17	18	19	20	IV + X −

TABLE XVIII.

GENERAL TABLE of Aberration, &c. continued.

S. − VII + I	1	2	3	4	5	6	7	8	9	10	11	12	13	14	15	16	17	18	19	20	
22° 0′	.788	1 .576	2 .364	3 .152	3 .940	4 .728	5 .516	6 .304	7 .092	7 .880	8 .668	9 .456	10 .244	11 .032	11 .820	12 .608	13 .396	14 .184	14 .972	15 .760	8° 0′
2	.788	1 .576	2 .365	3 .154	3 .942	4 .731	5 .519	6 .307	7 .096	7 .884	8 .672	9 .461	10 .249	11 .038	11 .826	12 .614	13 .402	14 .191	14 .979	15 .768	58
4	.789	1 .577	2 .366	3 .156	3 .944	4 .733	5 .522	6 .310	7 .099	7 .888	8 .676	9 .466	10 .254	11 .043	11 .832	12 .620	13 .408	14 .198	14 .986	15 .775	56
6	.789	1 .578	2 .367	3 .158	3 .946	4 .735	5 .525	6 .313	7 .102	7 .892	8 .680	9 .470	10 .259	11 .048	11 .837	12 .626	13 .414	14 .204	14 .993	15 .782	54
8	.790	1 .579	2 .368	3 .159	3 .948	4 .737	5 .527	6 .316	7 .105	7 .895	8 .684	9 .474	10 .264	11 .053	11 .842	12 .631	13 .420	14 .210	15 .000	15 .789	52
10	.790	1 .580	2 .369	3 .160	3 .949	4 .739	5 .529	6 .318	7 .108	7 .898	8 .688	9 .478	10 .268	11 .058	11 .847	12 .636	13 .426	14 .216	15 .006	15 .796	50
12	.790	1 .580	2 .371	3 .162	3 .951	4 .741	5 .532	6 .321	7 .112	7 .902	8 .692	9 .483	10 .273	11 .063	11 .853	12 .642	13 .433	14 .223	15 .013	15 .804	48
14	.791	1 .581	2 .372	3 .163	3 .953	4 .743	5 .535	6 .324	7 .115	7 .906	8 .696	9 .488	10 .278	11 .068	11 .859	12 .648	13 .439	14 .230	15 .020	15 .811	46
16	.791	1 .581	2 .373	3 .164	3 .955	4 .745	5 .537	6 .327	7 .118	7 .910	8 .700	9 .492	10 .283	11 .073	11 .864	12 .654	13 .445	14 .236	15 .027	15 .818	44
18	.792	1 .582	2 .374	3 .165	3 .957	4 .747	5 .539	6 .330	7 .121	7 .913	8 .704	9 .496	10 .287	11 .078	11 .869	12 .660	13 .451	14 .242	15 .034	15 .825	42
20	.792	1 .583	2 .375	3 .166	3 .958	4 .750	5 .541	6 .333	7 .124	7 .916	8 .708	9 .500	10 .291	11 .082	11 .874	12 .666	13 .457	14 .248	15 .040	15 .832	40
22	.792	1 .583	2 .376	3 .168	3 .960	4 .752	5 .544	6 .336	7 .128	7 .920	8 .712	9 .504	10 .296	11 .088	11 .880	12 .672	13 .463	14 .255	15 .047	15 .840	38
24	.792	1 .584	2 .377	3 .170	3 .962	4 .754	5 .547	6 .339	7 .131	7 .924	8 .716	9 .508	10 .301	11 .093	11 .886	12 .678	13 .469	14 .262	15 .054	15 .847	36
26	.792	1 .585	2 .378	3 .171	3 .964	4 .756	5 .550	6 .342	7 .134	7 .928	8 .720	9 .512	10 .306	11 .098	11 .891	12 .684	13 .475	14 .268	15 .061	15 .854	34
28	.793	1 .586	2 .379	3 .172	3 .966	4 .758	5 .552	6 .345	7 .137	7 .931	8 .724	9 .516	10 .310	11 .103	11 .896	12 .689	13 .481	14 .274	15 .068	15 .861	32
30	.793	1 .587	2 .380	3 .173	3 .967	4 .760	5 .554	6 .347	7 .140	7 .934	8 .727	9 .520	10 .314	11 .108	11 .901	12 .694	13 .487	14 .280	15 .074	15 .868	30
32	.793	1 .587	2 .381	3 .175	3 .969	4 .763	5 .557	6 .350	7 .144	7 .938	8 .731	9 .525	10 .319	11 .113	11 .907	12 .700	13 .493	14 .287	15 .081	.15 .875	28
34	.794	1 .588	2 .382	3 .177	3 .971	4 .765	5 .560	6 .353	7 .147	7 .942	8 .735	9 .530	10 .324	11 .118	11 .912	12 .706	13 .499	14 .294	15 .088	15 .882	26
36	.794	1 .588	2 .383	3 .179	3 .973	4 .767	5 .562	6 .356	7 .150	7 .945	8 .739	9 .534	10 .329	11 .123	11 .917	12 .712	13 .505	14 .300	15 .095	15 .890	24
38	.795	1 .589	2 .384	3 .180	3 .975	4 .769	5 .564	6 .359	7 .153	7 .948	8 .743	9 .538	10 .333	11 .128	11 .922	12 .717	13 .511	14 .306	15 .101	15 .896	22
40	.795	1 .590	2 .385	3 .181	3 .976	4 .771	5 .566	6 .361	7 .156	7 .951	8 .746	9 .542	10 .337	11 .132	11 .927	12 .722	13 .517	14 .312	15 .107	15 .903	20
42	.795	1 .590	2 .387	3 .183	3 .978	4 .773	5 .569	6 .364	7 .160	7 .955	8 .750	9 .546	10 .342	11 .137	11 .933	12 .728	13 .523	14 .319	15 .114	15 .910	18
44	.796	1 .591	2 .388	3 .185	3 .980	4 .775	5 .572	6 .367	7 .163	7 .959	8 .754	9 .550	10 .347	11 .142	11 .938	12 .734	13 .529	14 .326	15 .121	15 .917	16
46	.796	1 .592	2 .389	3 .186	3 .982	4 .777	5 574	6 .370	7 .166	7 .963	8 .758	9 .554	10 .351	11 .147	11 .943	12 .740	13 .535	14 .332	15 .128	15 .924	14
48	.797	1 .593	2 .390	3 .187	3 .983	4 .779	5 .576	6 .373	7 .169	7 .966	8 .762	9 .558	10 .355	11 .152	11 .948	12 .745	13 .541	14 .338	15 .135	15 .931	12
50	.797	1 .594	2 .391	3 .188	3 .984	4 .781	5 .578	6 .375	7 .172	7 .969	8 .766	9 .562	10 .359	11 .156	11 .953	12 .750	13 .547	14 .344	15 .141	15 .938	10
52	.797	1 .594	2 .392	3 .190	3 .986	4 .784	5 .581	6 .378	7 .176	7 .973	8 .770	9 .567	10 .364	11 .162	11 .959	12 .756	13 .553	14 .351	15 .148	15 .945	8
54	.798	1 .595	2 .393	3 .192	3 .988	4 .786	5 .584	6 .381	7 .179	7 .977	8 .774	9 .572	10 .369	11 .167	11 .965	12 .762	13 .559	14 .358	15 .155	15 .952	6
56	.798	1 .595	2 .394	3 .193	3 .990	4 .788	5 .587	6 .384	7 .182	7 .980	8 .778	9 .576	10 .374	11 .172	11 .970	12 .768	13 .565	14 .364	15 .162	15 .959	4
58	.798	1 .596	2 .395	3 .194	3 .992	4 .790	5 .589	6 .387	7 .185	7 .983	8 .782	9 .580	10 .379	11 .177	11 .975	12 .773	13 .571	14 .370	15 .168	15 .966	2
23° 0′	.799	1 .597	2 .396	3 .195	3 .993	4 .792	5 .591	6 .389	7 .188	7 .986	8 .785	9 .584	10 .383	11 .182	11 .980	12 .778	13 .577	14 .376	15 .174	15 .972	7° 0′
2	.799	1 .597	2 .397	3 .197	3 .995	4 .794	5 .594	6 .392	7 .191	7 .990	8 .789	9 .588	10 .388	11 .187	11 .986	12 .784	13 .583	14 .382	15 .181	15 .980	58
4	.799	1 .598	2 .398	3 .199	3 .997	4 .796	5 .597	6 .395	7 .194	7 .994	8 .793	9 .592	10 .393	11 .192	11 .991	12 .790	13 .589	14 .388	15 .188	15 .987	56
6	.799	1 .599	2 .399	3 .200	3 .999	4 .798	5 .599	6 .398	7 .197	7 .998	8 .797	9 .596	10 .398	11 .197	11 .996	12 .796	13 .595	14 .394	15 .195	15 .994	54
8	.800	1 .600	2 .400	3 .201	4 .001	4 .800	5 .601	6 .401	7 .200	8 .001	8 .801	9 .600	10 .402	11 .202	12 .001	12 .801	13 .601	14 .400	15 .201	16 .001	52
10	.800	1 .601	2 .401	3 .202	4 .002	4 .802	5 .603	6 .403	7 .203	8 .004	8 .804	9 .604	10 .405	11 .206	12 .006	12 .806	13 .606	14 .406	15 .207	16 .008	50
12	.800	1 .601	2 .402	3 .204	4 .004	4 .805	5 .606	6 .406	7 .207	8 .008	8 .808	9 .609	10 .410	11 .211	12 .011	12 .812	13 .612	14 .413	15 .214	16 .015	48
14	.801	1 .602	2 .403	3 .206	4 .006	4 .807	5 .609	6 .409	7 .211	8 .012	8 .812	9 .614	10 .415	11 .216	12 .016	12 .817	13 .618	14 .420	15 .221	16 022	46
16	.801	1 .602	2 .404	3 .207	4 .008	4 .809	5 .611	6 .412	7 .214	8 .015	8 .816	9 .618	10 .420	11 .222	12 .021	12 .822	13 .624	14 .427	15 .228	16 .029	44
18	.802	1 .603	2 .405	3 .208	4 .010	4 .811	5 .613	6 .414	7 .217	8 .018	8 .820	9 .622	10 .424	11 .226	12 .026	12 .827	13 .630	14 .434	15 .235	16 .036	42
20	.802	1 .604	2 .406	3 .209	4 .011	4 .813	5 .615	6 .416	7 .220	8 .021	8 .823	9 .626	10 .428	11 .230	12 .031	12 .832	13 .636	14 .440	15 .241	16 .043	40
22	.802	1 .604	2 .408	3 .211	4 .013	4 .815	5 .618	6 .419	7 .223	8 .025	8 .827	9 .630	10 .433	11 .235	12 .037	12 .838	13 .642	14 .446	15 .248	16 .050	38
24	.803	1 .605	2 .409	3 .212	4 .015	4 .817	5 .621	6 .422	7 .227	8 .029	8 .831	9 .634	10 .438	11 .240	12 .043	12 .844	13 .648	14 .452	15 .255	16 .057	36
26	.803	1 .606	2 .410	3 .213	4 .017	4 .819	5 .624	6 .425	7 .230	8 .033	8 .835	9 .638	10 .442	11 .245	12 048	12 .850	13 .654	14 458	15 .262	16 .064	34
28	.804	1 .607	2 .411	3 .214	4 .018	4 .821	5 .626	6 .428	7 .233	8 .036	8 .839	9 .642	10 .446	11 .250	12 .053	12 .856	13 660	14 .464	15 .268	16 .071	32
30	.804	1 .608	2 .412	3 .215	4 .019	4 .823	5 .628	6 .431	7 .235	8 .039	8 .843	9 .646	10 .450	11 .254	12 .058	12 .862	13 .666	14 .470	15 .274	16 .078	30
32	.804	1 .608	2 .413	3 .217	4 .021	4 .826	5 .630	6 .434	7 .238	8 .043	8 .847	9 .651	10 .455	11 .259	12 .064	12 .868	13 .672	14 .476	15 .281	16 .085	28
34	.805	1 .609	2 .414	3 .219	4 .023	4 .828	5 .633	6 .437	7 .241	8 .047	8 .851	9 .656	10 .460	11 .264	12 .069	12 .874	13 .678	14 .482	15 .288	16 .092	26
36	.805	1 .609	2 .415	3 .220	4 .025	4 .830	5 .635	6 .440	7 .244	8 .050	8 .855	9 .660	10 .465	11 .269	12 .074	12 .880	13 .684	14 .488	15 .294	16 .099	24
38	.805	1 .610	2 .416	3 .221	4 .027	4 .832	5 .637	6 .443	7 .247	8 .053	8 .859	9 .664	10 .469	11 .274	12 .079	12 .885	13 .690	14 .494	15 .300	16 .106	22
40	.806	1 .611	2 .417	3 .222	4 .028	4 .834	5 .639	6 .445	7 .250	8 .056	8 .862	9 .668	10 .473	11 .278	12 .084	12 .890	13 .695	14 .500	15 .306	16 .112	20
42	.806	1 .611	2 .418	3 .224	4 .030	4 .836	5 .642	6 .448	7 .254	8 .060	8 866	9 .672	10 .478	11 .283	12 .089	12 .896	13 .701	14 .507	15 .313	16 .119	18
44	.806	1 .612	2 .419	3 .226	4 .032	4 .838	5 .645	6 .451	7 .257	8 .064	8 .870	9 .676	10 .483	11 .288	12 .094	12 .901	13 .707	14 .514	15 .320	16 .126	16
46	.806	1 .613	2 .420	3 .228	4 .034	4 .840	5 .647	6 .454	7 .260	8 .067	8 .874	9 .680	10 .487	11 .293	12 .099	12 .906	13 .713	14 .520	15 .327	16 .133	14
48	.807	1 .614	2 .421	3 .229	4 .036	4 .842	5 .649	6 .456	7 .263	8 .070	8 .877	9 .684	10 .491	11 .298	12 .104	12 .911	13 .719	14 .526	15 .333	16 .140	12
50	.807	1 .615	2 .422	3 .230	4 .037	4 .844	5 .651	6 .458	7 .266	8 .073	8 .880	9 .688	10 .495	11 .302	12 .109	12 .916	13 .724	14 .532	15 .339	16 .146	10
52	.807	1 .615	2 .423	3 .232	4 .039	4 .846	5 .654	6 .461	7 .269	8 .077	8 .884	9 .692	10 .500	11 .307	12 .115	12 .922	13 .730	14 .538	15 .346	16 .153	8
54	.808	1 .616	2 .424	3 .233	4 .041	4 .848	5 .657	6 .464	7 .272	8 .081	8 .888	9 .696	10 .505	11 .312	12 .120	12 .928	13 .736	14 .544	15 .353	16 .160	6
56	.808	1 .616	2 .425	3 .234	4 .043	4 .850	5 .659	6 .467	7 .275	8 .084	8 .892	9 .700	10 .509	11 .317	12 .125	12 .934	13 .742	14 .550	15 .359	16 .167	4
58	.809	1 .617	2 .426	3 .235	4 .044	4 .852	5 .661	6 .470	7 .278	8 .087	8 .896	9 .704	10 .513	11 .322	12 .130	12 .939	13 .748	14 .556	15 .365	16 .174	2
24° 0	.809	1 .618	2 .427	3 .236	4 .045	4 .854	5 .663	6 .472	7 .281	8 .090	8 .899	9 .708	10 .517	11 .326	12 .135	12 .944	13 .753	14 .562	15 .371	16 .180	6° 0
	1	2	3	4	5	6	7	8	9	10	11	12	13	14	15	16	17	18	19	20	IV + X −

TABLE XVIII.

GENERAL TABLE of Aberration, &c. continued.

S. −VII + I	1	2	3	4	5	6	7	8	9	10	11	12	13	14	15	16	17	18	19	20	
24° 0'	.809	1.618	2.427	3.236	4.045	4.854	5.663	6.472	7.281	8.090	8.899	9.708	10.517	11.326	12.135	12.944	13.753	14.562	15.371	16.180	6° 0'
2	.809	1.618	2.428	3.238	4.047	4.856	5.666	6.475	7.285	8.094	8.903	9.712	10.522	11.331	12.141	12.950	13.759	14.569	15.378	16.187	58
4	.810	1.619	2.429	3.240	4.049	4.858	5.669	6.478	7.288	8.098	8.907	9.716	10.527	11.336	12.147	12.956	13.765	14.575	15.385	16.194	56
6	.810	1.619	2.430	3.241	4.051	4.860	5.671	6.481	7.291	8.101	8.911	9.720	10.531	11.341	12.152	12.962	13.771	14.581	15.392	16.201	54
8	.810	1.620	2.431	3.242	4.053	4.862	5.673	6.484	7.294	8.104	8.915	9.724	10.535	11.346	12.157	12.967	13.777	14.587	15.398	16.208	52
10	.811	1.621	2.432	3.243	4.054	4.864	5.675	6.486	7.297	8.107	8.918	9.728	10.539	11.350	12.161	12.972	13.783	14.593	15.404	16.214	50
12	.811	1.621	2.433	3.245	4.056	4.867	5.678	6.489	7.300	8.111	8.922	9.733	10.544	11.355	12.166	12.978	13.789	14.599	15.411	16.221	48
14	.811	1.622	2.434	3.247	4.058	4.869	5.681	6.492	7.303	8.115	8.926	9.737	10.549	11.360	12.171	12.983	13.796	14.605	15.418	16.228	46
16	.811	1.623	2.435	3.248	4.060	4.871	5.683	6.495	7.306	8.118	8.930	9.741	10.554	11.365	12.176	12.988	13.802	14.611	15.424	16.235	44
18	.812	1.624	2.436	3.249	4.061	4.873	5.685	6.497	7.309	8.121	8.933	9.745	10.558	11.370	12.181	12.993	13.807	14.617	15.430	16.242	42
20	.812	1.625	2.437	3.250	4.062	4.875	5.687	6.499	7.312	8.124	8.936	9.749	10.562	11.374	12.186	12.998	13.812	14.623	15.436	16.248	40
22	.812	1.625	2.438	3.252	4.064	4.877	5.690	6.502	7.315	8.128	8.940	9.753	10.567	11.379	12.192	13.004	13.817	14.630	15.443	16.255	38
24	.813	1.626	2.439	3.253	4.066	4.879	5.693	6.505	7.318	8.132	8.944	9.757	10.572	11.384	12.197	13.010	13.823	14.636	15.450	16.262	36
26	.813	1.626	2.440	3.254	4.068	4.881	5.695	6.508	7.321	8.135	8.948	9.761	10.576	11.389	12.202	13.016	13.829	14.642	15.456	16.269	34
28	.814	1.627	2.441	3.255	4.070	4.883	5.697	6.511	7.324	8.138	8.952	9.765	10.580	11.394	12.207	13.021	13.835	14.648	15.462	16.276	32
30	.814	1.628	2.442	3.256	4.071	4.885	5.699	6.513	7.327	8.141	8.955	9.769	10.584	11.398	12.212	13.026	13.840	14.654	15.468	16.282	30
32	.814	1.628	2.443	3.258	4.073	4.887	5.702	6.516	7.330	8.145	8.959	9.774	10.589	11.403	12.217	13 032	13.846	14.660	15.475	16.289	28
34	.815	1.629	2.444	3.260	4.075	4.889	5.705	6.519	7.333	8.149	8.963	9.778	10.594	11.408	12.222	13.037	13.852	14.666	15.482	16.296	26
36	.815	1.630	2.445	3.261	4.077	4.891	5.707	6.522	7.336	8.152	8.967	9.782	10.598	11.413	12.227	13.042	13.858	14.672	15.488	16.303	24
38	.816	1.631	2.446	3.262	4.078	4.893	5.709	6.524	7.339	8.155	8.971	9.786	10.602	11.418	12.232	13.047	13.863	14.678	15.494	16.310	22
40	.816	1.632	2.447	3.263	4.079	4.895	5.711	6.526	7.342	8.158	8.974	9.790	10.606	11.422	12.237	13.052	13.868	14.684	15.500	16.316	20
42	.816	1.632	2.448	3.264	4.081	4.897	5.714	6.529	7.345	8.162	8.978	9.794	10.611	11.427	12.242	13.058	13.874	14.691	15.507	16.323	18
44	.817	1.633	2.449	3.265	4.083	4.899	5.716	6.532	7.348	8.166	8.982	9.798	10.615	11.432	12.247	13.064	13.880	14.697	15.514	16.330	16
46	.817	1.633	2.450	3.266	4.085	4.901	5.718	6.535	7.351	8.169	8.986	9.802	10.619	11.436	12.252	13.070	13.886	14.703	15.520	16.337	14
48	.817	1.634	2.451	3.267	4.086	4.903	5.720	6.538	7.354	8.172	8.990	9.806	10.623	11.440	12.257	13.075	13.892	14.709	15.526	16.343	12
50	.818	1.635	2.452	3.269	4.087	4.905	5.722	6.540	7.357	8.175	8.993	9.810	10.627	11.444	12.262	13.080	13.897	14.715	15.532	16.350	10
52	.818	1.635	2.453	3.270	4.089	4.907	5.725	6.543	7.360	8.179	8.997	9.814	10.632	11.449	12.267	13.086	13.903	14.721	15.539	16.357	8
54	.818	1.636	2.455	3.271	4.091	4.909	5.728	6.546	7.363	8.183	9.001	9.818	10.637	11.454	12.272	13.091	13.909	14.727	15.546	16.364	6
56	.818	1.636	2.456	3.273	4.093	4.911	5.730	6.549	7.366	8.186	9.005	9.822	10.641	11.459	12.277	13.096	13.915	14.733	15.552	16.371	4
58	.819	1.637	2.457	3.275	4.095	4.913	5.732	6.551	7.369	8.189	9.008	9.826	10.645	11.464	12.282	13.101	13.920	14.739	15.558	16.378	2
25° 0'	.819	1.638	2.458	3.277	4.096	4.915	5.734	6.553	7.372	8.192	9.011	9.830	10.649	11.468	12.287	13.106	13.925	14.745	15.564	16.384	5° 0'
2	.819	1.638	2.459	3.279	4.098	4.917	5.737	6.556	7.375	8.196	9.015	9.834	10.654	11.473	12.293	13.112	13.931	14.751	15.571	16.391	58
4	.820	1.639	2.460	3.280	4.100	4.919	5.740	6.559	7.378	8.199	9.019	9.838	10.659	11.478	12.298	13.118	13.937	14.757	15.577	16.398	56
6	.820	1.640	2.461	3.281	4.102	4.921	5.742	6.562	7.381	8.202	9.023	9.842	10.663	11.483	12.303	13.124	13.943	14.763	15.583	16.404	54
8	.821	1.641	2.462	3.282	4.103	4.923	5.744	6.565	7.384	8.205	9.026	9.846	10.667	11.488	12.308	13.129	13.949	14.769	15.589	16.410	52
10	.821	1.642	2.463	3.283	4.104	4.925	5.746	6.567	7.387	8.208	9.029	9.850	10.671	11.492	12.313	13.134	13.954	14.775	15.595	16.416	50
12	.821	1.642	2.463	3.285	4.106	4.927	5.749	6.570	7.390	8.212	9.033	9.854	10.676	11.497	12.318	13.140	13.960	14.781	15.602	16.423	48
14	.822	1.643	2.464	3.287	4.108	4.929	5.751	6.573	7.393	8.216	9.037	9.858	10.680	11.502	12.323	13.145	13.966	14.787	15.609	16.430	46
16	.822	1.644	2.465	3.288	4.110	4.931	5.753	6.576	7.396	8.219	9.041	9.862	10.684	11.506	12.328	13.150	13.972	14.793	15.615	16.437	44
18	.822	1.645	2.466	3.289	4.111	4.933	5.755	6.578	7.399	8.222	9.044	9.866	10.688	11.510	12.333	13.155	13.977	14.799	15.621	16.444	42
20	.823	1.645	2.467	3.290	4.112	4.935	5.757	6.580	7.402	8.225	9.048	9.870	10.692	11.514	12.337	13.160	13.982	14.805	15.627	16.450	40
22	.823	1.645	2.468	3.292	4.114	4.937	5.760	6.583	7.405	8.229	9.052	9.874	10.697	11.519	12.342	13.165	13.988	14.811	15.634	16.457	38
24	.823	1.646	2.469	3.294	4.116	4.939	5.763	6.586	7.408	8.232	9.056	9.878	10.702	11.524	12.347	13.170	13.994	14.817	15.640	16.464	36
26	.823	1.646	2.470	3.295	4.118	4.941	5.765	6.588	7.411	8.235	9.059	9.882	10.706	11.529	12.352	13.175	13.999	14.823	15.646	16.470	34
28	.824	1.647	2.471	3.296	4.120	4.943	5.767	6.590	7.414	8.238	9.062	9.886	10.710	11.534	12.357	13.180	14.004	14.828	15.652	16.476	32
30	.824	1.648	2.472	3.297	4.121	4.945	5.769	6.592	7.417	8.241	9.065	9.889	10.714	11.538	12.361	13.185	14.009	14.833	15.658	16.482	30
32	.824	1.648	2.473	3.299	4.123	4.947	5.772	6.595	7.420	8.245	9.069	9.894	10.719	11.543	12.366	13.190	14.015	14.840	15.665	16.489	28
34	.825	1.649	2.474	3.300	4.125	4.949	5.774	6.598	7.423	8.249	9.073	9.898	10.723	11.548	12.371	13.196	14.021	14.846	15.672	16.496	26
36	.825	1.650	2.475	3.301	4.127	4.951	5.776	6.601	7.426	8.252	9.077	9.902	10.727	11.552	12.376	13.202	14.027	14.851	15.678	16.503	24
38	.826	1.651	2.476	3.302	4.128	4.953	5.778	6.603	7.429	8.255	9.081	9.906	10.731	11.556	12.381	13.207	14.033	14.857	15.684	16.510	22
40	.826	1.652	2.477	3.304	4.129	4.955	5.780	6.606	7.432	8.258	9.084	9.910	10.735	11.560	12.386	13.212	14.038	14.864	15.690	16.516	20
42	.826	1.652	2.478	3.305	4.131	4.957	5.783	6.609	7.435	8.262	9.088	9.914	10.740	11.565	12.391	13.218	14.044	14.870	15.697	16.523	18
44	.826	1.653	2.479	3.307	4.133	4.959	5.786	6.612	7.438	8.265	9.092	9.918	10.745	11.570	12.396	13.223	14.050	14.876	15.703	16.530	16
46	.827	1.653	2.480	3.308	4.135	4.961	5.788	6.615	7.441	8.268	9.095	9.922	10.749	11.575	12.401	13.228	14.056	14.882	15.709	16.536	14
48	.827	1.654	2.481	3.309	4.136	4.963	5.790	6.617	7.444	8.271	9.098	9.926	10.753	11.580	12.406	13.233	14.061	14.888	15.715	16.542	12
50	.827	1.655	2.482	3.310	4.137	4.965	5.792	6.619	7.447	8.274	9.101	9.930	10.757	11.584	12.411	13.238	14.066	14.893	15.721	16.548	10
52	.827	1.655	2.483	3.312	4.139	4.967	5.795	6.622	7.450	8.278	9.105	9.934	10.761	11.589	12.416	13.244	14.072	14.899	15.727	16.555	8
54	.828	1.656	2.484	3.313	4.141	4.969	5.797	6.625	7.453	8.281	9.109	9.938	10.765	11.594	12.421	13.249	14.078	14.905	15.733	16.562	6
56	.828	1.656	2.485	3.314	4.143	4.971	5.799	6.628	7.456	8.284	9.113	9.942	10.769	11.598	12.426	13.254	14.083	14.911	15.739	16.568	4
58	.829	1.657	2.486	3.315	4.144	4.973	5.801	6.630	7.459	8.287	9.116	9.945	10.773	11.602	12.431	13.259	14.088	14.917	15.745	16.574	2
26° 0	.829	1.658	2.487	3.316	4.145	4.974	5.803	6.632	7.461	8.290	9.119	9.948	10.777	11.606	12.435	13.264	14.093	14.922	15.751	16.580	4° 0
	1	2	3	4	5	6	7	8	9	10	11	12	13	14	15	16	17	18	19	20	IV + X −

TABLE XVIII.

GENERAL TABLE of Aberration, &c. continued.

S. −VII +I	1	2	3	4	5	6	7	8	9	10	11	12	13	14	15	16	17	18	19	20	
26° 0′	.829	1.658	2.487	3.316	4.145	4.974	5.803	6.632	7.461	8.290	9.119	9.948	10.777	11.606	12.435	13.264	14.093	14.922	15.751	16.580	4° 0′
2	.829	1.658	2.488	3.318	4.147	4.976	5.806	6.635	7.464	8.294	9.123	9.952	10.782	11.611	12.440	13.270	14.099	14.928	15.758	16.587	58
4	.830	1.659	2.489	3.319	4.149	4.978	5.809	6.638	7.467	8.297	9.126	9.956	10.786	11.615	12.445	13.275	14.105	14.934	15.764	16.593	56
6	.830	1.659	2.490	3.321	4.151	4.980	5.811	6.641	7.470	8.301	9.130	9.960	10.791	11.620	12.450	13.281	14.111	14.940	15.771	16.600	54
8	.830	1.660	2.491	3.322	4.152	4.982	5.813	6.643	7.473	8.304	9.134	9.964	10.795	11.625	12.455	13.286	14.117	14.946	15.777	16.607	52
10	.831	1.661	2.492	3.323	4.153	4.984	5.815	6.645	7.476	8.307	9.138	9.968	10.799	11.630	12.460	13.291	14.122	14.952	15.783	16.614	50
12	.831	1.661	2.493	3.325	4.155	4.986	5.818	6.648	7.479	8.310	9.142	9.972	10.804	11.635	12.465	13.296	14.128	14.958	15.790	16.621	48
14	.831	1.662	2.494	3.326	4.156	4.988	5.820	6.650	7.482	8.314	9.145	9.976	10.808	11.640	12.470	13.301	14.134	14.964	15.796	16.627	46
16	.832	1.663	2.495	3.327	4.158	4.990	5.822	6.653	7.485	8.317	9.149	9.980	10.812	11.644	12.475	13.306	14.139	14.970	15.802	16.634	44
18	.832	1.664	2.496	3.328	4.159	4.992	5.824	6.655	7.488	8.320	9.152	9.984	10.816	11.648	12.480	13.311	14.144	14.976	15.808	16.640	42
20	.832	1.665	2.497	3.329	4.161	4.994	5.826	6.658	7.491	8.323	9.155	9.988	10.820	11.652	12.484	13.316	14.149	14.982	15.814	16.646	40
22	.833	1.665	2.498	3.331	4.163	4.996	5.829	6.661	7.494	8.326	9.159	9.992	10.824	11.657	12.489	13.322	14.155	14.988	15.820	16.652	38
24	.833	1.666	2.499	3.332	4.165	4.998	5.831	6.663	7.497	8.330	9.162	9.995	10.828	11.661	12.494	13.327	14.161	14.994	15.826	16.658	36
26	.833	1.666	2.500	3.334	4.167	5.000	5.833	6.666	7.499	8.333	9.166	9.999	10.832	11.666	12.499	13.332	14.166	15.000	15.832	16.665	34
28	.834	1.667	2.501	3.335	4.169	5.002	5.835	6.668	7.502	8.336	9.169	10.003	10.836	11.670	12.504	13.337	14.171	15.005	15.838	16.671	32
30	.834	1.668	2.502	3.336	4.170	5.003	5.837	6.671	7.505	8.339	9.173	10.006	10.840	11.674	12.508	13.342	14.176	15.010	15.844	16.678	30
32	.834	1.668	2.503	3.338	4.172	5.005	5.840	6.674	7.508	8.343	9.177	10.010	10.845	11.679	12.513	13.348	14.182	15.016	15.850	16.685	28
34	.835	1.669	2.504	3.339	4.173	5.007	5.842	6.676	7.511	8.346	9.180	10.014	10.849	11.683	12.518	13.353	14.188	15.021	15.856	16.691	26
36	.835	1.669	2.505	3.340	4.175	5.009	5.844	6.679	7.513	8.349	9.184	10.018	10.853	11.688	12.523	13.358	14.893	15.027	15.862	16.698	24
38	.835	1.670	2.506	3.341	4.176	5.011	5.846	6.681	7.516	8.352	9.187	10.022	10.857	11.692	12.528	13.363	14.198	15.033	15.868	16.704	22
40	.836	1.671	2.507	3.342	4.178	5.013	5.848	6.684	7.519	8.355	9.191	10.026	10.861	11.696	12.532	13.368	14.203	15.038	15.874	16.710	20
42	.836	1.671	2.507	3.344	4.180	5.015	5.851	6.687	7.522	8.359	9.195	10.030	10.866	11.701	12.537	13.374	14.209	15.044	15.881	16.717	18
44	.836	1.672	2.508	3.345	4.181	5.017	5.854	6.689	7.525	8.362	9.198	10.034	10.871	11.706	12.542	13.379	14.214	15.050	15.887	16.723	16
46	.836	1.672	2.509	3.346	4.183	5.019	5.856	6.692	7.528	8.365	9.202	10.038	10.875	11.711	12.547	13.384	14.220	15.056	15.893	16.730	14
48	.837	1.673	2.510	3.347	4.184	5.021	5.858	6.694	7.531	8.368	9.205	10.042	10.879	11.716	12.552	13.389	14.226	15.062	15.899	16.736	12
50	.837	1.674	2.511	3.348	4.185	5.023	5.860	6.697	7.534	8.371	9.208	10.046	10.883	11.720	12.557	13.394	14.231	15.068	15.905	16.742	10
52	.837	1.674	2.512	3.350	4.187	5.025	5.863	6.699	7.537	8.375	9.212	10.050	10.887	11.725	12.562	13.399	14.237	15.074	15.911	16.749	8
54	.838	1.675	2.513	3.351	4.188	5.027	5.865	6.702	7.540	8.378	9.215	10.053	10.891	11.729	12.566	13.404	14.242	15.079	15.917	16.755	6
56	.838	1.675	2.514	3.353	4.190	5.029	5.867	6.704	7.542	8.381	9.219	10.057	10.895	11.734	12.571	13.409	14.247	15.085	15.923	16.762	4
58	.838	1.676	2.515	3.354	4.192	5.031	5.869	6.707	7.545	8.384	9.222	10.061	10.899	11.738	12.576	13.414	14.252	15.091	15.929	16.768	2
27° 0′	.839	1.677	2.516	3.355	4.193	5.032	5.871	6.709	7.548	8.387	9.226	10.064	10.903	11.742	12.580	13.418	14.257	15.096	15.935	16.774	3° 0′
2	.839	1.677	2.517	3.357	4.195	5.034	5.874	6.712	7.551	8.391	9.230	10.068	10.908	11.747	12.585	13.424	14.263	15.102	15.941	16.781	58
4	.839	1.678	2.518	3.358	4.196	5.036	5.876	6.714	7.554	8.394	9.233	10.072	10.912	11.751	12.590	13.429	14.268	15.108	15.947	16.787	56
6	.839	1.679	2.519	3.359	4.198	5.038	5.878	6.717	7.556	8.397	9.237	10.076	10.916	11.756	12.595	13.434	14.274	15.113	15.953	16.794	54
8	.840	1.680	2.520	3.360	4.199	5.040	5.880	6.719	7.559	8.400	9.240	10.080	10.920	11.760	12.600	13.439	14.279	15.119	15.959	16.800	52
10	.840	1.681	2.521	3.361	4.201	5.042	5.882	6.722	7.562	8.403	9.243	10.084	10.924	11.764	12.604	13.444	14.284	15.124	15.965	16.806	50
12	.840	1.681	2.522	3.363	4.203	5.044	5.885	6.725	7.565	8.406	9.247	10.088	10.928	11.769	12.609	13.450	14.290	15.130	15.971	16.812	48
14	.841	1.682	2.523	3.364	4.204	5.046	5.887	6.727	7.568	8.409	9.250	10.092	10.932	11.773	12.614	13.455	14.295	15.135	15.977	16.818	46
16	.841	1.682	2.524	3.365	4.206	5.048	5.889	6.730	7.570	8.412	9.254	10.096	10.936	11.778	12.619	13.460	14.301	15.141	15.983	16.824	44
18	.842	1.683	2.525	3.366	4.208	5.050	5.891	6.732	7.573	8.415	9.257	10.099	10.940	11.782	12.624	13.465	14.306	15.147	15.989	16.830	42
20	.842	1.684	2.526	3.367	4.209	5.051	5.893	6.735	7.576	8.418	9.260	10.102	10.944	11.786	12.628	13.470	14.311	15.152	15.994	16.836	40
22	.842	1.684	2.527	3.369	4.211	5.053	5.896	6.737	7.579	8.422	9.264	10.106	10.948	11.791	12.633	13.475	14.316	15.158	16.001	16.843	38
24	.842	1.685	2.527	3.370	4.212	5.055	5.898	6.740	7.582	8.425	9.267	10.109	10.952	11.795	12.637	13.480	14.323	15.164	16.007	16.849	36
26	.843	1.685	2.528	3.372	4.214	5.057	5.900	6.743	7.585	8.428	9.271	10.112	10.956	11.800	12.642	13.485	14.328	15.170	16.013	16.856	34
28	.843	1.686	2.529	3.373	4.215	5.059	5.902	6.745	7.588	8.431	9.274	10.116	10.960	11.804	12.646	13.490	14.333	15.176	16.019	16.862	32
30	.843	1.687	2.530	3.374	4.217	5.060	5.904	6.747	7.591	8.434	9.277	10.120	10.964	11.808	12.651	13.495	14.338	15.182	16.025	16.868	30
32	.843	1.687	2.531	3.876	4.219	5.062	5.907	6.749	7.594	8.438	9.281	10.124	10.969	11.813	12.656	13.500	14.344	15.188	16.031	16.875	28
34	.844	1.688	2.532	3.377	4.220	5.064	5.909	6.752	7.597	8.441	9.284	10.128	10.973	11.817	12.661	13.505	14.349	15.194	16.037	16.881	26
36	.344	1.688	2.533	3.378	4.222	5.066	5.811	6.754	7.599	8.444	9.288	10.132	10.977	11.822	12.666	13.510	14.355	15.199	16.043	16.888	24
38	.845	1.689	2.534	3.379	4.223	5.068	5.913	6.757	7.602	8.447	9.292	10.136	10.981	11.826	12.671	13.515	14.260	15.205	16.049	16.894	22
40	.845	1.690	2.535	3.380	4.225	5.070	5.915	6.760	7.605	8.450	9.295	10.140	10.985	11.830	12.675	13.520	14.365	15.210	16.055	16.900	20
42	.845	1.690	2.536	3.382	4.227	5.072	5.918	6.762	7.608	8.453	9.299	10.144	10.989	11.835	12.680	13.525	14.370	15.216	16.061	16.906	18
44	.846	1.691	2.537	3.383	4.228	5.074	5.920	6.765	7.611	8.456	9.302	10.147	10.993	11.839	12.684	13.530	14.376	15.221	16.067	16.912	16
46	.846	1.691	2.538	3.384	4.230	5.076	5.922	6.767	7.613	8.459	9.306	10.151	10.997	11.844	12.689	13.535	14.381	15.227	16.073	16.918	14
48	.846	1.692	2.539	3.385	4.231	5.078	5.924	6.770	7.616	8.462	9.309	10.154	11.001	11.848	12.694	13.540	14.386	15.232	16.079	16.924	12
50	.847	1.693	2.540	3.386	4.233	5.079	5.926	6.772	7.619	8.465	9.312	10.158	11.005	11.852	12.698	13.544	14.391	15.238	16.084	16.930	10
52	.847	1.693	2.540	3.388	4.235	5.081	5.928	6.774	7.622	8.469	9.316	10.162	11.009	11.856	12.703	13.549	14.397	15.244	16.090	16.937	8
54	.847	1.694	2.541	3.389	4.236	5.083	5.930	6.777	7.624	8.472	9.319	10.165	11.013	11.860	12.707	13.554	14.402	15.249	16.096	16.943	6
56	.847	1.694	2.542	3.390	4.238	5.085	5.932	6.779	7.627	8.475	9.322	10.169	11.017	11.864	12.712	13.559	14.407	15.255	16.102	16.950	4
58	.848	1.695	2.543	3.391	4.239	5.087	5.934	6.782	7.630	8.478	9.326	10.172	11.021	11.868	12.716	13.564	14.412	15.261	16.108	16.956	2
28° 0	.848	1.696	2.544	3.392	4.240	5.088	5.936	6.784	7.633	8.481	9.329	10.176	11.024	11.872	12.720	13.568	14.417	15.266	16.114	16.962	2° 0
	1	2	3	4	5	6	7	8	9	10	11	12	13	14	15	16	17	18	19	20	IV + X −

TABLE XVIII.

GENERAL TABLE of Aberration, &c. continued.

S − VII + I	1	2	3	4	5	6	7	8	9	10	11	12	13	14	15	16	17	18	19	20	
28° 0′	.848	1.696	2.544	3.392	4.240	5.088	5.936	6.784	7.633	8.481	9.329	10.177	11.025	11.873	12.720	13.569	14.417	15.265	16.114	16.962	2° 0′
2	.848	1.696	2.545	3.394	4.242	5.090	5.939	6.787	7.636	8.484	9.333	10.181	11.029	11.877	12.725	13.574	14.423	15.271	16.120	16.968	58
4	.849	1.697	2.546	3.395	4.243	5.092	5.941	6.789	7.638	8.487	9.336	10.185	11.033	11.882	12.730	13.579	14.428	15.276	16.126	16.974	56
6	.849	1.697	2.547	3.396	4.245	5.094	5.943	6.792	7.641	8.490	9.339	10.189	11.037	11.886	12.734	13.584	14.433	15.282	16.131	16.980	54
8	.849	1.698	2.548	3.397	4.247	5.096	5.945	6.794	7.643	8.493	9.343	10.193	11.041	11.890	12.739	13.589	14.438	15.287	16.136	16.986	52
10	.850	1.699	2.549	3.398	4.248	5.098	5.947	6.797	7.646	8.496	9.346	10.196	11.045	11.894	12.744	13.594	14.443	15.292	16.142	16.992	50
12	.850	1.699	2.550	3.400	4.250	5.100	5.949	6.799	7.649	8.499	9.350	10.200	11.049	11.899	12.749	13.599	14.449	15.298	16.148	16.998	48
14	.850	1.700	2.551	3.401	4.252	5.102	5.952	6.802	7.652	8.502	9.353	10.203	11.053	11.903	12.754	13.603	14.454	15.304	16.154	17.004	46
16	.850	1.701	2.552	3.403	4.253	5.103	5.954	6.805	7.654	8.505	9.356	10.207	11.057	11.908	12.758	13.608	14.459	15.309	16.160	17.010	44
18	.851	1.701	2.552	3.404	4.255	5.105	5.956	6.807	7.657	8.508	9.359	10.211	11.061	11.912	12.763	13.613	14.464	15.315	16.165	17.016	42
20	.851	1.702	2.553	3.405	4.256	5.107	5.958	6.809	7.660	8.511	9.362	10.214	11.065	11.916	12.767	13.618	14.469	15.320	16.171	17.022	40
22	.851	1.702	2.554	3.406	4.258	5.109	5.960	6.812	7.663	8.514	9.366	10.218	11.069	11.921	12.772	13.623	14.475	15.326	16.177	17.028	38
24	.852	1.703	2.555	3.408	4.259	5.111	5.963	6.814	7.666	8.517	9.369	10.221	11.073	11.925	12.776	13.628	14.480	15.332	16.183	17.034	36
26	.852	1.703	2.556	3.409	4.261	5.112	5.965	6.817	7.668	8.520	9.373	10.225	11.077	11.930	12.781	13.632	14.485	15.337	16.189	17.040	34
28	.852	1.704	2.557	3.410	4.262	5.114	5.967	6.819	7.671	8.523	9.376	10.228	11.081	11.934	12.786	13.637	14.490	15.343	16.194	17.046	32
30	.853	1.705	2.558	3.411	4.263	5.116	5.969	6.821	7.674	8.526	9.379	10.232	11.085	11.938	12.790	13.642	14.495	15.348	16.200	17.052	30
32	.853	1.705	2.559	3.412	4.265	5.118	5.971	6.824	7.677	8.529	9.383	10.236	11.089	11.942	12.795	13.647	14.500	15.354	16.206	17.059	28
34	.853	1.706	2.560	3.414	4.266	5.119	5.973	6.826	7.679	8.532	9.386	10.239	11.093	11.946	12.800	13.652	14.505	15.359	16.212	17.065	26
36	.853	1.707	2.561	3.415	4.268	5.121	5.975	6.829	7.682	8.535	9.390	10.242	11.096	11.950	12.804	13.656	14.510	15.364	16.217	17.072	24
38	.854	1.707	2.562	3.416	4.269	5.123	5.977	6.831	7.684	8.538	9.393	10.246	11.100	11.954	12.808	13.661	14.515	15.369	16.223	17.078	22
40	.854	1.708	2.563	3.417	4.271	5.125	5.979	6.833	7.687	8.542	9.396	10.250	11.104	11.958	12.812	13.666	14.520	15.374	16.229	17.084	20
42	.854	1.709	2.564	3.419	4.273	5.127	5.982	6.836	7.690	8.545	9.400	10.254	11.108	11.963	12.817	13.671	14.526	15.380	16.235	17.090	18
44	.855	1.709	2.565	3.420	4.274	5.129	5.984	6.838	7.693	8.548	9.403	10.257	11.112	11.967	12.821	13.676	14.531	15.386	16.241	17.096	16
46	.855	1.710	2.565	3.421	4.276	5.130	5.986	6.841	7.696	8.551	9.407	10.261	11.116	11.972	12.826	13.680	14.537	15.392	16.247	17.102	14
48	.855	1.710	2.566	3.422	4.277	5.132	5.988	6.843	7.699	8.554	9.410	10.265	11.120	11.976	12.831	13.685	14.542	15.398	16.253	17.108	12
50	.856	1.711	2.567	3.423	4.278	5.134	5.990	6.845	7.702	8.557	9.413	10.268	11.124	11.980	12.835	13.690	14.547	15.404	16.259	17.114	10
52	.856	1.712	2.568	3.425	4.280	5.136	5.992	6.848	7.705	8.560	9.417	10.272	11.128	11.984	12.840	13.695	14.552	15.410	16.264	17.120	8
54	.856	1.712	2.569	3.426	4.281	5.138	5.994	6.850	7.707	8.563	9.420	10.276	11.132	11.988	12.844	13.700	14.557	15.415	16.270	17.126	6
56	.856	1.713	2.570	3.427	4.283	5.139	5.996	6.853	7.710	8.566	9.423	10.279	11.135	11.992	12.849	13.705	14.562	15.420	16.275	17.132	4
58	.857	1.713	2.571	3.428	4.284	5.141	5.998	6.855	7.713	8.569	9.426	10.283	11.139	11.996	12.853	13.709	14.567	15.425	16.281	17.138	2
29° 0′	.857	1.714	2.572	3.429	4.286	5.143	6.000	6.857	7.715	8.572	9.429	10.286	11.143	12.000	12.857	13.714	14.572	15.430	16.287	17.144	1° 0
2	.857	1.715	2.573	3.431	4.288	5.145	6.003	6.860	7.718	8.575	9.433	10.290	11.147	12.005	12.862	13.719	14.577	15.436	16.293	17.150	58
4	.858	1.715	2.574	3.432	4.289	5.147	6.005	6.862	7.720	8.578	9.436	10.293	11.151	12.009	12.866	13.724	14.582	15.441	16.299	17.156	56
6	.858	1.716	2.575	3.433	4.291	5.148	6.007	6.865	7.723	8.581	9.440	10.297	11.155	12.014	12.871	13.729	14.587	15.446	16.304	17.162	54
8	.858	1.716	2.575	3.434	4.292	5.150	6.009	6.867	7.726	8.584	9.443	10.301	11.159	12.018	12.876	13.733	14.592	15.451	16.310	17.168	52
10	.859	1.717	2.576	3.435	4.293	5.152	6.011	6.869	7.728	8.587	9.446	10.304	11.163	12.022	12.880	13.738	14.597	15.456	16.315	17.174	50
12	.859	1.718	2.577	3.437	4.295	5.154	6.013	6.872	7.731	8.590	9.450	10.308	11.167	12.026	12.885	13.743	14.602	15.462	16.321	17.180	48
14	.859	1.718	2.577	3.438	4.296	5.156	6.015	6.874	7.733	8.593	9.453	10.312	11.171	12.030	12.889	13.748	14.607	15.467	16.327	17.186	46
16	.859	1.719	2.578	3.439	4.298	5.157	6.017	6.877	7.735	8.596	9.456	10.315	11.174	12.034	12.893	13.753	14.612	15.472	16.332	17.192	44
18	.860	1.719	2.579	3.440	4.300	5.159	6.019	6.879	7.738	8.599	9.459	10.319	11.178	12.038	12.898	13.757	14.617	15.477	16.338	17.198	42
20	.860	1.720	2.580	3.441	4.301	5.161	6.021	6.881	7.741	8.602	9.462	10.322	11.182	12.042	12.902	13.762	14.622	15.482	16.343	17.204	40
22	.860	1.721	2.581	3.443	4.303	5.163	6.023	6.883	7.744	8.605	9.466	10.326	11.186	12.046	12.906	13.767	14.627	15.488	16.349	17.210	38
24	.861	1.721	2.582	3.444	4.304	5.164	6.025	6.886	7.747	8.608	9.469	10.330	11.190	12.050	12.811	13.771	14.632	15.494	16.355	17.215	36
26	.861	1.722	2.583	3.445	4.306	5.166	6.027	6.888	7.749	8.610	9.472	10.333	11.193	12.054	12.915	13.776	14.637	15.499	16.360	17.221	34
28	.861	1.722	2.584	3.446	4.307	5.168	6.029	6.890	7.752	8.613	9.475	10.337	11.197	12.058	12.919	13.780	14.642	15.505	16.366	17.226	32
30	.862	1.723	2.585	3.447	4.308	5.170	6.031	6.892	7.755	8.616	9.478	10.340	11.201	12.062	12.923	13.785	14.647	15.510	16.371	17.232	30
32	.862	1.724	2.586	3.448	4.310	5.172	6.034	6.895	7.758	8.619	9.482	10.344	11.205	12.067	12.928	13.790	14.653	15.516	16.377	17.238	28
34	.862	1.724	2.587	3.449	4.311	5.173	6.036	6.897	7.760	8.622	9.485	10.348	11.209	12.071	12.933	13.795	14.658	15.521	16.383	17.244	26
36	.862	1.725	2.587	3.450	4.313	5.175	6.038	6.899	7.763	8.625	9.488	10.351	11.213	12.076	12.937	13.800	14.663	15.526	16.388	17.250	24
38	.863	1.725	2.588	3.451	4.314	5.177	6.040	6.902	7.766	8.628	9.491	10.355	11.217	12.080	12.942	13.805	14.668	15.531	16.394	17.256	22
40	.863	1.726	2.589	3.452	4.316	5.179	6.042	6.905	7.768	8.631	9.494	10.358	11.221	12.084	12.947	13.810	14.673	15.536	16.399	17.262	20
42	.863	1.727	2.590	3.454	4.318	5.181	6.044	6.908	7.771	8.634	9.498	10.362	11.225	12.088	12.952	13.815	14.678	15.542	16.405	17.268	18
44	.864	1.727	2.591	3.455	4.319	5.182	6.046	6.911	7.774	8.637	9.501	10.365	11.228	12.092	12.956	13.820	14.683	15.547	16.411	17.274	16
46	.864	1.728	2.592	3.456	4.321	5.184	6.048	6.913	7.776	8.640	9.505	10.368	11.232	12.096	12.960	13.824	14.688	15.552	16.416	17.280	14
48	.864	1.728	2.593	3.457	4.322	5.185	6.050	6.915	7.779	8.643	9.508	10.371	11.236	12.100	12.965	13.829	14.693	15.557	16.422	17.286	12
50	.865	1.729	2.594	3.458	4.323	5.187	6.052	6.917	7.781	8.646	9.511	10.374	11.239	12.104	12.969	13.834	14.698	15.562	16.427	17.292	10
52	.865	1.730	2.595	3.460	4.325	5.189	6.054	6.920	7.784	8.649	9.514	10.378	11.243	12.108	12.973	13.839	14.703	15.568	16.433	17.298	8
54	.865	1.730	2.595	3.461	4.326	5.190	6.056	6.922	7.786	8.652	9.517	10.382	11.247	12.112	12.978	13.843	14.708	15.573	16.438	17.303	6
56	.865	1.731	2.596	3.462	4.328	5.192	6.058	6.924	7.789	8.655	9.520	10.385	11.250	12.116	12.982	13.848	14.712	15.578	16.444	17.309	4
58	.866	1.731	2.597	3.463	4.329	5.194	6.060	6.926	7.792	8.658	9.523	10.389	11.254	12.120	12.986	13.852	14.717	15.583	16.449	17.315	2
30° 0	.866	1.732	2.598	3.464	4.330	5.196	6.062	6.928	7.794	8.660	9.526	10.392	11.258	12.124	12.990	13.856	14.722	15.588	16.454	17.320	0
	1	2	3	4	5	6	7	8	9	10	11	12	13	14	15	16	17	18	19	20	IV + X −

GENERAL TABLE of Aberration, &c. continued.

S −VIII +II	1	2	3	4	5	6	7	8	9	10	11	12	13	14	15	16	17	18	19	20	
0° 0′	.866	1.732	2.598	3.464	4.330	5.196	6.062	6.928	7.794	8.660	9.526	10.392	11.258	12.124	12.990	13.856	14.722	15.588	16.454	17.320	30° 0′
2	.866	1.733	2.599	3.466	4.332	5.198	6.064	6.931	7.797	8.663	9.530	10.396	11.262	12.128	12.995	13.861	14.727	15.594	16.460	17.326	58
4	.867	1.733	2.599	3.467	4.333	5.200	6.066	6.933	7.799	8.666	9.533	10.399	11.266	12.132	13.000	13.866	14.732	15.599	16.466	17.332	56
6	.867	1.734	2.600	3.468	4.335	5.201	6.068	6.936	7.802	8.669	9.537	10.403	11.270	12.136	13.004	13.870	14.737	15.604	16.471	17.338	54
8	.867	1.734	2.601	3.469	4.336	5.203	6.070	6.938	7.805	8.672	9.540	10.407	11.273	12.140	13.008	13.875	14.742	15.609	16.477	17.344	52
10	.868	1.735	2.602	3.470	4.337	5.205	6.072	6.940	7.807	8.675	9.543	10.410	11.277	12.144	13.012	13.880	14.747	15.614	16.482	17.350	50
12	.868	1.736	2.603	3.472	4.339	5.207	6.074	6.943	7.810	8.678	9.546	10.414	11.281	12.148	13.017	13.885	14.752	15.620	16.488	17.356	48
14	.868	1.736	2.604	3.473	4.340	5.209	6.076	6.945	7.813	8.681	9.549	10.417	11.285	12.152	13.021	13.889	14.757	15.625	16.494	17.362	46
16	.868	1.737	2.605	3.474	4.342	5.210	6.078	6.947	7.815	8.684	9.552	10.421	11.289	12.156	13.025	13.894	14.762	15.630	16.499	17.367	44
18	.869	1.737	2.606	3.475	4.344	5.212	6.080	6.949	7.818	8.686	9.555	10.425	11.292	12.160	13.029	13.898	14.766	15.635	16.504	17.373	42
20	.869	1.738	2.607	3.476	4.345	5.214	6.082	6.951	7.820	8.689	9.558	10.428	11.296	12.164	13.033	13.902	14.771	15.640	16.509	17.378	40
22	.869	1.739	2.608	3.477	4.347	5.216	6.085	6.954	7.823	8.692	9.562	10.432	11.300	12.169	13.038	13.907	14.776	15.646	16.514	17.384	38
24	.869	1.739	2.608	3.478	4.348	5.218	6.087	6.956	7.825	8.695	9.565	10.435	11.304	12.173	13.042	13.912	14.781	15.651	16.519	17.390	36
26	.870	1.740	2.609	3.479	4.350	5.219	6.089	6.959	7.828	8.698	9.568	10.438	11.308	12.178	13.047	13.916	14.786	15.656	16.524	17.396	34
28	.870	1.740	2.610	3.480	4.351	5.221	6.091	6.961	7.830	8.701	9.571	10.441	11.311	12.182	13.052	13.921	14.791	15.661	16.529	17.402	32
30	.870	1.741	2.611	3.481	4.352	5.222	6.093	6.963	7.833	8.704	9.574	10.444	11.315	12.186	13.056	13.926	14.796	15.666	16.534	17.408	30
32	.870	1.741	2.612	3.482	4.354	5.224	6.095	6.966	7.836	8.707	9.578	10.448	11.319	12.190	13.061	13.931	14.801	15.672	16.540	17.414	28
34	.871	1.742	2.613	3.483	4.355	5.226	6.097	6.968	7.839	8.710	9.581	10.451	11.323	12.194	13.065	13.935	14.806	15.677	16.546	17.420	26
36	.871	1.743	2.613	3.484	4.357	5.227	6.099	6.970	7.841	8.713	9.584	10.455	11.327	12.198	13.069	13.940	14.810	15.682	16.552	17.425	24
38	.871	1.743	2.614	3.485	4.358	5.229	6.101	6.972	7.844	8.716	9.587	10.459	11.330	12.202	13.073	13.944	14.815	15.687	16.558	17.431	22
40	.872	1.744	2.615	3.487	4.359	5.231	6.103	6.974	7.846	8.718	9.590	10.462	11.334	12.206	13.077	13.948	14.820	15.692	16.564	17.436	20
42	.872	1.744	2.616	3.489	4.361	5.233	6.105	6.977	7.849	8.721	9.593	10.466	11.338	12.210	13.082	13.953	14.825	15.697	16.570	17.442	18
44	.872	1.745	2.617	3.490	4.362	5.235	6.107	6.979	7.852	8.724	9.596	10.469	11.342	12.214	13.086	13.958	14.830	15.703	16.576	17.448	16
46	.872	1.745	2.618	3.491	4.364	5.237	6.109	6.982	7.854	8.726	9.599	10.472	11.345	12.218	13.091	13.963	14.835	15.708	16.581	17.453	14
48	.873	1.746	2.619	3.492	4.365	5.238	6.111	6.984	7.857	8.729	9.602	10.475	11.349	12.222	13.095	13.967	14.840	15.713	16.587	17.459	12
50	.873	1.746	2.620	3.493	4.366	5.239	6.113	6.986	7.859	8.732	9.605	10.478	11.352	12.226	13.099	13.972	14.845	15.718	16.591	17.464	10
52	.873	1.747	2.621	3.495	4.368	5.241	6.115	6.989	7.862	8.735	9.609	10.482	11.356	12.230	13.103	13.977	14.850	15.724	16.597	17.470	8
54	.874	1.747	2.622	3.496	4.269	5.243	6.117	6.991	7.864	8.738	9.612	10.485	11.359	12.233	13.107	13.981	14.855	15.729	16.603	17.476	6
56	.874	1.748	2.622	3.497	4.371	5.245	6.119	6.993	7.867	8.740	9.615	10.489	11.363	12.237	13.111	13.986	14.860	15.734	16.608	17.481	4
58	.874	1.748	2.623	3.498	4.372	5.246	6.120	6.995	7.869	8.743	9.618	10.493	11.367	12.241	13.115	13.990	14.864	15.739	16.613	17.487	2
1° 0′	.875	1.749	2.624	3.499	4.373	5.248	6.122	6.997	7.872	8.746	9.621	10.496	11.370	12.244	13.119	13.994	14.869	15.744	16.618	17.492	29° 0′
2	.875	1.750	2.625	3.500	4.375	5.250	6.124	7.000	7.875	8.749	9.624	10.400	11.374	12.248	13.124	13.999	14.874	15.749	16.623	17.498	58
4	.875	1.750	2.626	3.501	4.376	5.252	6.126	7.002	7.877	8.752	9.627	10.503	11.378	12.252	13.128	14.003	14.878	15.754	16.628	17.504	56
6	.875	1.751	2.626	3.502	4.378	5.253	6.128	7.004	7.880	8.754	9.630	10.506	11.381	12.256	13.132	14.008	14.883	15.759	16.633	17.509	54
8	.876	1.751	2.627	3.503	4.379	5.255	6.130	7.006	7.882	8.757	9.633	10.509	11.385	12.260	13.136	14.012	14.887	15.763	16.638	17.515	52
10	.876	1.752	2.628	3.504	4.380	5.256	6.132	7.008	7.884	8.760	9.636	10.512	11.388	12.264	13.140	14.016	14.892	15.768	16.644	17.520	50
12	.876	1.752	2.629	3.506	4.382	5.258	6.134	7.011	7.887	8.763	9.639	10.516	11.392	12.268	13.145	14.021	14.897	15.774	16.650	17.526	48
14	.876	1.753	2.630	3.507	4.383	5.260	6.136	7.013	7.890	8.766	9.642	10.519	11.396	12.272	13.149	14.025	14.902	15.779	16.655	17.532	46
16	.877	1.754	2.630	3.508	4.385	5.261	6.138	7.015	7.892	8.768	9.645	10.523	11.399	12.275	13.153	14.030	14.907	15.784	16.661	17.538	44
18	.877	1.754	2.631	3.509	4.386	5.263	6.140	7.017	7.895	8.771	9.648	10.527	11.403	12.279	13.157	14.034	14.911	15.789	16.666	17.543	42
20	.877	1.755	2.632	3.510	4.387	5.265	6.142	7.019	7.897	8.774	9.651	10.530	11.407	12.284	13.161	14.038	14.916	15.794	16.671	17.548	40
22	.877	1.756	2.633	3.511	4.389	5.267	6.144	7.022	7.900	8.777	9.655	10.534	11.411	12.288	13.166	14.043	14.921	15.799	16.676	17.554	38
24	.878	1.756	2.634	3.512	4.290	5.269	6.146	7.024	7.902	8.780	9.658	10.537	11.415	12.292	13.170	14.048	14.926	15.804	16.681	17.559	36
26	.878	1.757	2 635	3.513	4.392	5.270	6.148	7.027	7.905	8.782	9.661	10.540	11.418	12.296	13.175	14.053	14.931	15.808	16.687	17.565	34
28	.878	1.757	2.636	3.514	4.393	5.272	6.150	7.029	7.907	8.785	9.664	10.543	11.422	12.300	13.179	14.057	14.936	15.814	16.692	17.571	32
30	.879	1.758	2.637	3.515	4.394	5.273	6.152	7.031	7.909	8.788	9.637	10.546	11.425	12.304	13.183	14.062	14.940	15.818	16.697	17.576	30
32	.879	1.758	2.638	3.517	4.396	5.275	6.154	7.034	7.912	8.791	9.670	10.550	11.429	12.308	13.187	14.067	14.945	15.823	16.703	17.582	28
34	.879	1.759	2.638	3.518	4.397	5.277	6.156	7.036	7.915	8.794	9.673	10.553	11.432	12.311	13.191	14.071	14.950	15.829	16.708	17.588	26
36	.879	1.759	2.639	3.519	4.399	5.278	6.158	7.038	7.917	8.796	9.676	10.556	11.436	12.315	13.195	14.076	14.955	15.834	16.714	17.593	24
38	.880	1.760	2.640	3.520	4.400	5.280	6.159	7.040	7.920	8.799	9.679	10.559	11.439	12.319	13.199	14.080	14.959	15.839	16.719	17.599	22
40	.880	1.760	2.641	3.521	4.401	5.281	6.161	7.042	7.922	8.802	9.682	10.562	11.442	12.322	13.203	14.084	14.964	15.844	16.724	17.604	20
42	.880	1.761	2.642	3.522	4.403	5.283	6.163	7.045	7.925	8.805	9.686	10.566	11.446	12.326	13.207	14.089	14.969	15.849	16.730	17.610	18
44	.881	1.761	2.643	3.523	4.404	5.285	6.165	7.047	7.928	8.808	9.689	10.569	11.450	12.330	13.212	14.093	14.974	15.854	16.735	17.616	16
46	.881	1.762	2.643	3.524	4.406	5.287	6.167	7.049	7.930	8.810	9.692	10.573	11.453	12.334	13.216	14.098	14.978	15.858	16.740	17.621	14
48	.881	1.762	2.644	3.525	4.407	5.288	6.169	7.051	7.932	8.813	9.695	10.577	11.457	12.338	13.220	14.102	14.983	15.863	16.745	17.627	12
50	.882	1.763	2.645	3.526	4.408	5.290	6.171	7.053	7.934	8.816	9.698	10.580	11.461	12.342	13.224	14.106	14.987	15.868	16.750	17.632	10
52	.882	1.764	2.646	3.528	4.410	5.292	6.173	7.056	7.937	8.819	9.701	10.584	11.465	12.346	13.229	14.111	14.992	15.874	16.756	17.638	8
54	.882	1.764	2.646	3.529	4.411	5.284	6.175	7.058	7.940	8.822	9.704	10.587	11.469	12.350	13.233	14.115	14.997	15.879	16.761	17.644	6
56	.882	1.765	2.647	3.530	3.413	5.295	6.177	7.060	7.942	8.824	9.707	10.590	11.472	12.354	13.237	14.120	15.001	15.884	16.767	17.649	4
58	.883	1.765	2.648	3.531	4.414	5.297	6.179	7.062	7.045	8.827	9.710	10.593	11.476	12.358	13.241	14.124	15.006	15.889	16.772	17.655	2
2° 0′	.883	1.766	2.649	3.532	4.415	5.298	6.181	7.064	7.947	8.830	9.713	10.596	11.479	12.362	13.245	14.128	15.011	15.894	16.777	17.660	28° 0′
	1	2	3	4	5	6	7	8	9	10	11	12	13	14	15	16	17	18	19	20	III + IX −

TABLE XVIII.

GENERAL TABLE of Aberration, &c. continued.

S. −VIII +II	1	2	3	4	5	6	7	8	9	10	11	12	13	14	15	16	17	18	19	20	
2° 0′	.883	1.766	2.649	3.532	4.415	5.298	6.181	7.064	7.947	8.830	9.713	10.596	11.479	12.362	13.245	14.128	15.011	15.894	16.777	17.660	28° 0′
2	.883	1.767	2.650	3.533	4.417	5.300	6.183	7.066	7.950	8.833	9.716	10.600	11.483	12.366	13.249	14.133	15.016	15.898	16.782	17.666	58
4	.883	1.767	2.651	3.534	4.418	5.301	6.185	7.069	7.952	8.835	9.719	10.603	11.486	12.370	13.253	14.137	15.020	15.904	16.787	17.671	56
6	.884	1.768	2.651	3.535	4.420	5.303	6.187	7.071	7.955	8.838	9.722	10.606	11.490	12.373	13.257	14.142	15.025	15.908	16.792	17.676	54
8	.884	1.768	2.652	3.536	4.421	5.304	6.188	7.073	7.957	8.841	9.725	10.609	11.493	12.377	13.261	14.146	15.030	15.913	16.797	17.681	52
10	.884	1.769	2.653	3.537	4.422	5.306	6.190	7.075	7.059	8.843	9.727	10.612	11.496	12.380	13.265	14.150	15.034	15.918	16.802	17.686	50
12	.884	1.769	2.654	3.538	4.423	5.308	6.192	7.077	7.962	8.846	9.731	10.616	11.500	12.384	13.269	14.154	15.039	15.923	16.808	17.692	48
14	.885	1.770	2.655	3.539	4.425	5.310	6.194	7.079	7.964	8.849	9.734	10.619	11.503	12.388	13.273	14.158	15.043	15.928	16.813	17.698	46
16	.885	1.770	2.655	3.541	4.426	5.311	6.196	7.081	7.967	8.851	9.737	10.622	11.507	12.392	13.277	14.162	15.048	15.933	16.818	17.703	44
18	.885	1.771	2.656	3.542	4.427	5.313	6.198	7.083	7.969	8.854	9.740	10.625	11.511	12.396	13.281	14.166	15.052	15.937	16.823	17.709	42
20	.886	1.771	2.657	3.543	4.428	5.314	6.200	7.085	7.971	8.857	9.743	10.628	11.514	12.400	13.285	14.170	15.056	15.942	16.828	17.714	40
22	.886	1.772	2.658	3.544	4.430	5.316	6.202	7.087	7.974	8.860	9.746	10.632	11.518	12.404	13.289	14.175	15.061	15.947	16.833	17.720	38
24	.886	1.772	2.659	3.545	4.431	5.318	6.204	7.090	7.976	8.863	9.749	10.635	11.521	12.408	13.293	14.179	15.066	15.952	16.838	17.725	36
26	.886	1.773	2.660	3.546	4.432	5.319	6.206	7.092	7.979	8.865	9.752	10.638	11.525	12.411	13.297	14.183	15.070	15.956	16.843	17.730	34
28	.887	1.773	2.660	3.547	4.434	5.321	6.207	7.094	7.981	8.868	9.754	10.641	11.528	12.415	13.301	14.187	15.075	15.961	16.848	17.735	32
30	.887	1.774	2.661	3.548	4.435	5.322	6.209	7.096	7.983	8.870	9.757	10.644	11.531	12.418	13.305	14.192	15.079	15.966	16.853	17.740	30
32	.887	1.775	2.662	3.549	4.437	5.324	6.211	7.099	7.986	8.873	9.760	10.648	11.535	12.422	13.309	14.197	15.084	15.971	16.859	17.746	28
34	.887	1.775	2.663	3.550	4.438	5.326	6.213	7.101	7.988	8.876	9.763	10.651	11.539	12.426	13.314	14.201	15.089	15.976	16.864	17.752	26
36	.888	1.776	2.664	3.551	4.440	5.327	6.215	7.103	7.991	8.879	9.766	10.654	11.543	12.430	13.318	14.206	15.093	15.980	16.869	17.757	24
38	.888	1.776	2.664	3.552	4.441	5.329	6.217	7.105	7.993	8.881	9.769	10.657	11.546	12.434	13.322	14.210	15.098	15.985	16.874	17.763	22
40	.888	1.777	2.665	3.553	4.442	5.330	6.219	7.107	7.995	8.884	9.772	10.660	11.549	12.438	13.326	14.214	15.102	15.990	16.879	17.768	20
42	.888	1.777	2.666	3.555	4.443	5.332	6.221	7.109	7.998	8.887	9.775	10.664	11.553	12.442	13.330	14.218	15.107	15.995	16.884	17.774	18
44	.889	1.778	2.667	3.556	4.445	5.333	6.223	7.111	8.000	8.890	9.778	10.667	11.556	12.446	13.334	14.222	15.111	15.999	16.889	17.779	16
46	.889	1.778	2.668	3.557	4.446	5.335	6.225	7.113	8.003	8.893	9.781	10.670	11.560	12.450	13.338	14.226	15.116	16.004	16.894	17.784	14
48	.889	1.779	2.668	3.558	4.447	5.337	6.227	7.115	8.005	8.895	9.784	10.673	11.563	12.453	13.342	14.230	15.120	16.009	16.899	17.789	12
50	.890	1.779	2.669	3.559	4.448	5.338	6.228	7.117	8.007	8.897	9.787	10.676	11.566	12.456	13.345	14.234	15.124	16.014	16.904	17.794	10
52	.890	1.780	2.670	3.560	4.450	5.340	6.230	7.120	8.010	8.900	9.790	10.680	11.570	12.460	13.349	14.239	15.129	16.019	16.909	17.800	8
54	.890	1.780	2.671	3.561	4.451	5.342	6.232	7.122	8.012	8.902	9.793	10.683	11.573	12.464	13.353	14.243	15.134	16.024	16.914	17.805	6
56	.890	1.781	2.671	3.562	4.452	5.343	6.234	7.124	8.015	8.905	9.795	10.686	11.577	12.467	13.357	14.248	15.138	16.028	16.919	17.810	4
58	.891	1.781	2.672	3.563	4.454	5.345	6.236	7.126	8.017	8.908	9.798	10.689	11.580	12.471	13.361	14.252	15.143	16.033	16.924	17.815	2
3° 0′	.891	1.782	2.673	3.564	4.455	5.346	6.237	7.128	8.019	8.910	9.801	10.692	11.583	12.474	13.365	14.256	15.147	16.038	16.929	17.820	27° 0′
2	.891	1.782	2.674	3.565	4.457	5.348	6.239	7.130	8.022	8.913	9.804	10.696	11.587	12.478	13.369	14.261	15.152	16.043	16.934	17.826	58
4	.891	1.783	2.675	3.566	4.458	5.350	6.241	7.133	8.024	8.916	9.807	10.699	11.590	12.482	13.373	14.265	15.156	16.048	16.939	17.831	56
6	.892	1.784	2.676	3.567	4.459	5.352	6.243	7.135	8.027	8.919	9.810	10.702	11.594	12.485	13.377	14.269	15.161	16.053	16.944	17.836	54
8	.892	1.784	2.676	3.568	4.461	5.353	6.245	7.137	8.029	8.921	9.812	10.705	11.597	12.489	13.381	14.274	15.165	16.057	16.949	17.841	52
10	.892	1.785	2.677	3.569	4.462	5.354	6.246	7.139	8.031	8.923	9.815	10.708	11.600	12.492	13.385	14.278	15.170	16.062	16.954	17.846	50
12	.892	1.785	2.678	3.570	4.463	5.356	6.248	7.141	8.034	8.926	9.818	10.712	11.604	12.496	13.389	14.282	15.175	16.067	16.959	17.852	48
14	.893	1.786	2.679	3.571	4.465	5.357	6.250	7.143	8.036	8.928	9.821	10.715	11.607	12.500	13.393	14.286	15.179	16.072	16.964	17.857	46
16	.893	1.786	2.680	3.572	4.466	5.359	6.252	7.145	8.039	8.931	9.824	10.718	11.611	12.503	13.397	14.290	15.183	16.076	16.969	17.862	44
18	.893	1.787	2.680	3.573	4.467	5.361	6.254	7.147	8.041	8.934	9.827	10.721	11.614	12.507	13.401	14.294	15.187	16.081	16.974	17.867	42
20	.894	1.787	2.681	3.575	4.468	5.362	6.255	7.149	8.043	8.936	9.830	10.724	11.617	12.510	13.404	14.298	15.192	16.086	16.979	17.872	40
22	.894	1.788	2.682	3.576	4.469	5.364	6.257	7.151	8.045	8.939	9.833	10.728	11.621	12.514	13.408	14.302	15.197	16.091	16.984	17.877	38
24	.894	1.788	2.683	3.577	4.471	5.365	6.259	7.153	8.048	8.942	9.836	10.731	11.625	12.518	13.412	14.306	15.201	16.096	16.989	17.883	36
26	.894	1.789	2.684	3.578	4.472	5.367	6.261	7.155	8.050	8.944	9.839	10.734	11.629	12.522	13.416	14.310	15.205	16.100	16.904	17.888	34
28	.895	1.789	2.684	3.579	4.474	5.368	6.263	7.157	8.052	8.947	9.841	10.737	11.632	12.526	13.420	14.314	15.209	16.104	16.999	17.893	32
30	.895	1.790	2.685	3.580	4.475	5.370	6.265	7.159	8.054	8.949	9.844	10.740	11.635	12.530	13.424	14.318	15.213	16.108	17.003	17.898	30
32	.895	1.790	2.686	3.581	4.476	5.372	6.267	7.161	8.057	8.952	9.847	10.743	11.639	12.534	13.428	14.323	15.218	16.113	17.008	17.904	28
34	.895	1.791	2.686	3.582	4.478	5.373	6.269	7.164	8.059	8.955	9.850	10.746	11.642	12.538	13.432	14.327	15.223	16.118	17.013	17.909	26
36	.896	1.791	2.687	3.583	4.479	5.375	6.271	7.166	8.061	8.957	9.852	10.749	11.645	12.542	13.436	14.332	15.227	16.122	17.018	17.914	24
38	.896	1.792	2.688	3.584	4.480	5.376	6.272	7.168	8.064	8.960	9.855	10.752	11.648	12.545	13.440	14.336	15.232	16.127	17.023	17.919	22
40	.896	1.792	2.689	3.585	4.481	5.377	6.274	7.170	8.066	8.962	9.858	10.754	11.651	12.548	13.444	14.340	15.236	16.132	17.028	17.924	20
42	.896	1.793	2.690	3.586	4.483	5.379	6.276	7.172	8.069	8.965	9.861	10.758	11.655	12.552	13.448	14.344	15.241	16.137	17.033	17.930	18
44	.897	1.793	2.691	3.587	4.484	5.380	6.278	7.174	8.071	8.068	9.864	10.761	11.659	12.556	13.452	14.348	15.245	16.142	17.038	17.935	16
46	.897	1.794	2.692	3.588	4.486	5.382	6.280	7.176	8.074	8.970	9.867	10.764	11.662	12.559	13.456	14.352	15.250	16.147	17.043	17.940	14
48	.897	1.794	2.692	3.589	4.487	5.384	6.281	7.178	8.076	8.973	9.870	10.767	11.665	12.562	13.460	14.356	15.254	16.151	17.048	17.945	12
50	.898	1.795	2.693	3.590	4.488	5.385	6.283	7.180	8.078	8.975	9.873	10.770	11.668	12.566	13.463	14.360	15.258	16.156	17.053	17.950	10
52	.898	1.796	2.694	3.591	4.489	5.387	6.285	7.182	8.081	8.978	9.876	10.774	11.672	12.570	13.467	14.364	15.263	16.161	17.058	17.955	8
54	.898	1.796	2.694	3.592	4.491	5.388	6.287	7.184	8.083	8.981	9.879	10.777	11.675	12.574	13.471	14.368	15.267	16.165	17.063	17.961	6
56	.898	1.797	2.695	3.593	4.492	5.390	6.289	7.186	8.085	8.984	9.882	10.780	11.679	12.577	13.475	14.372	15.271	16.170	17.068	17.966	4
58	.899	1.797	2.695	3.594	4.493	5.391	6.290	7.188	8.087	8.986	9.885	10.783	11.682	12.581	13.479	14.376	15.275	16.174	17.072	17.971	2
4° 0	.899	1.798	2.696	3.595	4.494	5.393	6.292	7.190	8.089	8.988	9.887	10.786	11.685	12.584	13.482	14.380	15.279	16.178	17.077	17.976	26° 0
	1	2	3	4	5	6	7	8	9	10	11	12	13	14	15	16	17	18	19	20	III + IX −

GENERAL TABLE of Aberration, &c. continued.

S. −VIII +II	1	2	3	4	5	6	7	8	9	10	11	12	13	14	15	16	17	18	19	20	
4° 0′	.899	1.798	2.696	3.595	4.494	5.393	6.292	7.190	8.089	8.988	9.887	10.786	11.684	12.584	13.482	14.380	15.279	16.178	17.077	17.976	26° 0′
2	.899	1.798	2.697	3.596	4.496	5.295	6.294	7.193	8.092	8.991	9.890	10.789	11.688	12.588	13.486	14.385	15.284	16.183	17.082	17.982	58
4	.899	1.799	2.698	3.597	4.497	5.396	6.296	7.195	8.094	8.993	9.893	10.792	11.691	12.592	13.490	14.389	15.289	16.188	17.087	17.987	56
6	.899	1.799	2.699	3.598	4.498	5.398	6.298	7.197	8.097	8.996	9.895	10.795	11.695	12.595	13.494	14.394	15.293	16.193	17.092	17.992	54
8	.900	1.800	2.699	3.599	4.499	5.399	6.299	7.199	8.099	8.999	9.898	10.797	11.698	12.599	13.498	14.398	15.298	16.197	17.097	17.997	52
10	.900	1.800	2.700	3.600	4.500	5.400	6.301	7.201	8.101	9.001	9.901	10.800	11.701	12.602	13.502	14.402	15.302	16.202	17.102	18.002	50
12	.900	1.801	2.701	3.601	4.502	5.402	6.303	7.203	8.104	9.004	9.904	10.803	11.705	12.606	13.506	14.406	15.307	16.207	17.107	18.007	48
14	.900	1.801	2.702	3.602	4.503	5.403	6.305	7.205	8.106	9.006	9.907	10.807	11.708	12.609	13.510	14.410	15.311	16.211	17.112	18.012	46
16	.901	1.802	2.703	3.603	4.505	5.405	6.306	7.207	8.108	9.009	9.909	10.810	11.711	12.612	13.513	14.414	15.315	16.216	17.116	18.017	44
18	.901	1.802	2.704	3.604	4.506	5.407	6.308	7.209	8.110	9.011	9.912	10.813	11.714	12.615	13.517	14.418	15.319	16.220	17.121	18.022	42
20	.901	1.803	2.704	3.605	4.507	5.408	6.309	7.211	8.112	9.013	9.914	10.816	11.717	12.618	13.520	14.422	15.323	16.224	17.125	18.026	40
22	.901	1.803	2.705	3.606	4.509	5.410	6.311	7.213	8.115	9.016	9.917	10.820	11.721	12.622	13.524	14.426	15.328	16.229	17.130	18.032	38
24	.902	1.804	2.706	3.607	4.510	5.412	6.313	7.215	8.117	9.018	9.920	10.823	11.724	12.626	13.528	14.430	15.332	16.233	17.135	18.037	36
26	.902	1.804	2.707	3.608	4.511	5.413	6.315	7.217	8.119	9.021	9.923	10.826	11.728	12.629	13.532	14.434	15.336	16.238	17.139	18.042	34
28	.902	1.805	2.707	3.609	4.512	5.415	6.316	7.219	8.121	9.024	9.926	10.829	11.731	12.632	13.536	14.438	15.340	16.242	17.144	18.047	32
30	.903	1.805	2.708	3.610	4.513	5.416	6.318	7.221	8.123	9.026	9.929	10.832	11.734	12.636	13.539	14.442	15.344	16.246	17.149	18.052	30
32	.903	1.806	2.709	3.611	4.515	5.418	6.320	7.223	8.126	9.029	9.932	10.835	11.738	12.640	13.543	14.446	15.349	16.251	17.154	18.057	28
34	.903	1.806	2.710	3.612	4.516	5.419	6.322	7.225	8.128	9.031	9.935	10.838	11.741	12.644	13.547	14.450	15.353	16.256	17.159	18.062	26
36	.903	1.807	2.710	3.613	4.517	5.421	6.324	7.227	8.131	9.034	9.937	10.841	11.744	12.647	13.551	14.454	15.358	16.260	17.164	18.067	24
38	.904	1.807	2.711	3.614	4.518	5.422	6.325	7.229	8.133	9.036	9.940	10.844	11.747	12.651	13.555	14.458	15.362	16.265	17.169	18.072	22
40	.904	1.808	2.712	3.615	4.519	5.423	6.327	7.231	8.135	9.038	9.942	10.846	11.750	12.654	13.558	14.462	15.366	16.270	17.173	18.077	20
42	.904	1.808	2.713	3.616	4.521	5.425	6.329	7.233	8.138	9.041	9.945	10.849	11.754	12.658	13.562	14.466	15.371	16.275	17.178	18.082	18
44	.904	1.809	2.713	3.617	4.522	5.426	6.331	7.235	8.140	9.043	9.948	10.852	11.757	12.662	13.566	14.470	15.375	16.279	17.183	18.087	16
46	.904	1.809	2.714	3.618	4.523	5.428	6.333	7.237	8.142	9.046	9.951	10.855	11.760	12.665	13.570	14.474	15.379	16.284	17.188	18.092	14
48	.905	1.810	2.714	3.619	4.524	5.429	6.335	7.239	8.144	9.049	9.954	10.858	11.763	12.669	13.574	14.478	15.383	16.288	17.193	18·097	12
50	.905	1.910	2.715	3.620	4.525	5.430	6.336	7.241	8.146	9.051	9.956	10.860	11.766	12.672	13.577	14.482	15.387	16.292	17.197	18.102	10
52	.905	1.811	2.716	3.621	4.527	5.432	6.338	7.243	8.149	9.054	9.959	10.864	11.770	12.676	13.581	14.486	15.391	16.297	17.202	18.107	8
54	.905	1.811	2.717	3.622	4.528	5.434	6.340	7.245	8.151	9.056	9.962	10.867	11.773	12.679	13.585	14.490	15.395	16.301	17.207	18.112	6
56	.905	1.812	2.717	3.623	4.530	5.435	6.342	7.247	8.153	9.059	9.964	10.870	11.776	12.682	13.588	14.493	15.399	16.306	17.211	18.117	4
58	.906	1.812	2.718	3.624	4.531	5.437	6.343	7.249	8.155	9.061	9.967	10.873	11.779	12.685	13.592	14.497	15.403	16.310	17.216	18.122	2
5° 0′	.906	1.813	2.719	3.625	4.532	5.438	6.344	7.250	8.157	9.063	9.969	10.876	11.782	12.688	13.594	14.500	15.407	16.314	17.220	18.126	25° 0′
2	.906	1.813	2.720	3.626	4.534	5.440	6.346	7.252	8.160	9.066	9.973	10.879	11.786	12.692	13.598	14.504	15.412	16.319	17.225	18.131	58
4	.907	1.814	2.721	3.627	4.535	5.441	6.348	7.254	8.162	9.068	9.976	10.882	11.789	12.696	13.602	14.508	15.416	16.323	17.229	18.136	56
6	.907	1.814	2.722	3.628	4.536	5.443	6.350	7.256	8.164	9.071	9.978	10.885	11.792	12.699	13.606	14.512	15.420	16.328	17.234	18.141	54
8	.907	1.815	2.722	3.629	4.537	5.444	6.352	7.258	8.166	9.073	9.981	10.888	11.795	12.703	13.610	14.516	15.424	16.332	17.239	18.146	52
10	.908	1.815	2.723	3.630	4.538	5.445	6.353	7.260	8.168	9.075	9.983	10.890	11.798	12.706	13.613	14.520	15.428	16.336	17.243	18.151	50
12	.908	1.816	2.724	3.631	4.540	5.447	6.355	7.262	8.171	9.078	9.986	10.894	11.802	12.710	13.617	14.524	15.433	16.341	17.248	18.156	48
14	.908	1.816	2.724	3.632	4.541	5.449	6.357	7.264	8.173	9.080	9.989	10.897	11.805	12.713	13.621	14.528	15.437	16.345	17.253	18.161	46
16	.908	1.817	2.725	3.633	4.542	5.450	6.358	7.266	8.175	9.083	9.992	10.900	11.808	12.716	13.624	14.532	15.441	16.350	17.258	18.166	44
18	.909	1.817	2.726	3.634	4.543	5.452	6.359	7.268	8.177	9.085	9.995	10.903	11 811	12.719	13.628	14.536	15.445	16.354	17.263	18.171	42
20	.909	1.818	2.726	3.635	4.544	5.453	6.361	7.270	8.179	9.088	9.997	10.906	11.814	12.722	13.631	14.540	15.449	16.358	17.267	18.176	40
22	.909	1.818	2.727	3.636	4.546	5.455	6.363	7.272	8.182	9.091	10.000	10.909	11.818	12.726	13.635	14.544	15.454	16.363	17.272	18.181	38
24	.909	1.819	2.728	3.637	4.547	5.456	6.365	7.274	8.184	9.093	10.003	10.912	11.821	12.730	13.639	14.548	15.458	16.367	17.277	18.186	36
26	.909	1.819	2.729	3.638	4.548	5.458	6.367	7.276	8.186	9.096	10.005	10.915	11.824	12.733	13.643	14.552	15.462	16.371	17.281	18.191	34
28	.910	1.820	2.729	3.639	4.549	5.459	6.368	7.278	8.188	9.098	10.008	10.918	11.827	12.737	13.646	14.556	15.466	16.375	17.286	18.196	32
30	.910	1.820	2.730	3.640	4.550	5.460	6.370	7.280	8.190	9.100	10.010	10.920	11.830	12.740	13.650	14.560	15.470	16.380	17.290	18.200	30
32	.910	1.821	2.731	3.641	4.552	5.462	6.372	7.282	8.192	9.103	10.013	10.923	11.833	12.744	13.654	14.564	15.474	16.384	17.295	18.205	28
34	.910	1.821	2.732	3.642	4.553	5.463	6.374	7.284	8.194	9.105	10.016	10.926	11.836	12.747	13.658	14.568	15.478	16.388	17.299	18.210	26
36	.910	1.821	2.733	3.643	4.554	5.465	6.376	7.286	8.196	9.108	10.018	10.929	11.839	12.750	13.661	14.571	15.482	16.392	17.303	18.215	24
38	.911	1.822	2.733	3.644	4.555	5.466	6.377	7.288	8.198	9.110	10.021	10.931	11.842	12.753	13.664	14.575	15.486	16.396	17.308	18.220	22
40	.911	1.822	2.734	3.645	4.556	5.467	6.378	7.289	8.200	9.112	10.023	10.934	11.845	12.756	13.667	14.578	15.490	16.400	17.312	18.224	20
42	.911	1.823	2.735	3.646	4.558	5.469	6.380	7.291	8.203	9.115	10.026	10.937	11.849	12.760	13.671	14.582	15.494	16.405	17.317	18.229	18
44	.912	1.823	2.735	3.647	4.559	5.470	6.382	7.293	8.205	9.117	10.029	10.940	11.852	12.764	13.675	14.586	15.498	16.409	17.322	18.234	16
46	.912	1.824	2.736	3.648	4.560	5.472	6.384	7.295	8.207	9.120	10.031	10.942	11.855	12.767	13.679	14.590	15.502	16.414	17.326	18.239	14
48	.912	1.824	2.736	3.649	4.561	5.473	6.385	7.297	8.209	9.122	10.034	10.945	11.858	12.771	13.682	14.594	15.506	16.418	17.331	18.244	12
50	.912	1.825	2.737	3.649	4.562	5.474	6.387	7.299	8.211	9.124	10.036	10.948	11.861	12.774	13.686	14.598	15.510	16.422	17.335	18.248	10
52	.913	1.825	2.738	3.650	4.564	5.476	6.389	7.301	8.214	9.127	10.039	10.951	11.864	12.778	13.690	14.602	15.514	16.427	17.340	18.253	8
54	.913	1.826	2.739	3.651	4.565	5.477	6.391	7.303	8.216	9.129	10.042	10.954	11.867	12.781	13.694	14.606	15.518	16.431	17.345	18.258	6
56	.913	1.826	2.740	3.652	4.566	5.479	6.392	7.305	8.218	9.132	10.045	10.957	11.870	12.784	13.697	14.609	15.522	16.436	17.349	18.263	4
58	.914	1.827	2.741	3.653	4.507	5.400	6.304	7.307	8.220	9.134	10.047	10.959	11.873	12.787	13.700	14.613	15.526	16.440	17.254	18.268	2
6° 0	.914	1.827	2.741	3.654	4.568	5.481	6.395	7.308	8.222	9.136	10.050	10.962	11.876	12.790	13.703	14.616	15.530	16.444	17.358	18.272	24° 0
	1	2	3	4	5	6	7	8	9	10	11	12	13	14	15	16	17	18	19	20	III+ IX−

TABLE XVIII.

GENERAL TABLE of Aberration, &c. continued.

S. −VIII +II	1	2	3	4	5	6	7	8	9	10	11	12	13	14	15	16	17	18	19	20	
6° 0′	.914	1.827	2.741	3.654	4.568	5.481	6.395	7.308	8.222	9.136	10.050	10.962	11.876	12.790	13.703	14.616	15.530	16.444	17.358	18.272	24° 0′
2	.914	1.827	2.742	3.655	4.570	5.483	6.397	7.310	8.225	9.139	10.053	10.965	11.879	12.794	13.707	14.620	15.535	16.449	17.363	18.277	58
4	.914	1.828	2.742	3.656	4.571	5.484	6.399	7.312	8.227	9.141	10.055	10.968	11.882	12.797	13.710	14.624	15.539	16.453	17.367	18.281	56
6	.914	1.828	2.743	3.657	4.572	5.486	6.400	7.314	8.229	9.143	10.058	10.971	11.885	12.800	13.714	14.628	15.543	16.458	17.372	18.286	54
8	.915	1.829	2.743	3.658	4.573	5.487	6.402	7.316	8.231	9.145	10.060	10.974	11.888	12.803	13.718	14.632	15.547	16.462	17.376	18.290	52
10	.915	1.829	2.744	3.659	4.574	5.488	6.403	7.318	8.233	9.147	10.062	10.976	11.891	12.806	13.721	14.636	15.551	16.466	17.380	18.294	50
12	.915	1.830	2.745	3.660	4.576	5.490	6.405	7.320	8.235	9.150	10.065	10.979	11.894	12.810	13.725	14.640	15.555	16.470	17.385	18.299	48
14	.915	1.830	2.746	3.661	4.577	5.491	6.407	7.322	8.237	9.152	10.067	10.982	11.897	12.813	13.728	14.644	15.559	16.474	17.389	18.304	46
16	.915	1.831	2.746	3.662	4.578	5.493	6.408	7.324	8.239	9.155	10.070	10.985	11.900	12.816	13.732	14.647	15.563	16.478	17.394	18.308	44
18	.916	1.831	2.747	3.663	4.579	5.494	6.410	7.326	8.241	9.157	10.073	10.988	11.903	12.819	13.735	14.651	15.567	16.482	17.398	18.313	42
20	.916	1.832	2.748	3.664	4.580	5.495	6.411	7.327	8.243	9.159	10.075	10.990	11.906	12.822	13.738	14.654	15.570	16.486	17.402	18.318	40
22	.916	1.832	2.749	3.665	4.581	5.497	6.413	7.329	8.246	9.161	10.078	10.993	11.909	12.826	13.742	14.658	15.575	16.491	17.407	18.323	38
24	.916	1.833	2.749	3.666	4.582	5.498	6.415	7.331	8.248	9.164	10.080	10.996	11.912	12.829	13.746	14.662	15.579	16.495	17.412	18.328	36
26	.916	1.833	2.750	3.667	4.583	5.500	6.416	7.333	8.250	9.167	10.083	10.999	11.915	12.832	13.749	14.666	15.583	16.500	17.416	18.333	34
28	.917	1.834	2.750	3.668	4.584	5.501	6.418	7.335	8.252	9.169	10.086	11.002	11.918	12.835	13.753	14.670	15.587	16.504	17.421	18.337	32
30	.917	1.834	2.751	3.668	4.585	5.502	6.419	7.337	8.254	9.171	10.088	11.004	11.921	12.838	13.756	14.674	15.591	16.508	17.425	18.342	30
32	.917	1.834	2.752	3.669	4.587	5.504	6.421	7.339	8.256	9.174	10.091	11.007	11.925	12.842	13.760	14.678	15.595	16.512	17.430	18.347	28
34	.917	1.835	2.753	3.670	4.588	5.505	6.423	7.341	8.258	9.176	10.093	11.010	11.928	12.846	13.764	14.682	15.599	16.516	17.434	18.351	26
36	.917	1.835	2.754	3.671	4.589	5.507	6.424	7.343	8.260	9.178	10.096	11.013	11.931	12.849	13.767	14.685	15.603	16.520	17.438	18.356	24
38	.918	1.836	9.754	3.672	4.590	5.508	6.426	7.345	8.262	9.180	10.098	11.016	11.934	12.853	13.771	14.689	15.607	16.524	17.442	18.360	22
40	.918	1.836	2.755	3.673	4.591	5.509	6.428	7.346	8.264	9.182	10.100	11.018	11.937	12.856	13.774	14.692	15.610	16.528	17.446	18.364	20
42	.918	1.837	2.756	3.674	4.593	5.511	6.430	7.348	8.266	9.185	10.103	11.021	11.940	12.860	13.778	14.696	15.614	16.532	17.451	18.369	18
44	.918	1.837	2.756	3.675	4.594	5.512	6.432	7.350	8.268	9.187	10.105	11.024	11.943	12.863	13.781	14.700	15.618	16.536	17.455	18.373	16
46	.919	1.838	2.757	3.676	4.595	5.514	6.433	7.352	8.270	9.190	10.108	11.027	11.946	12.866	13.785	14.703	15.622	16.540	17.460	18.378	14
48	.919	1.838	2.757	3.677	4.596	5.515	6.435	7.354	8.272	9.192	10.111	11.030	11.949	12.869	13.788	14.707	15.626	16.544	17.464	18.383	12
50	.919	1.839	2.758	3.678	4.597	5.516	6.436	7.355	8.274	9.194	10.113	11.032	11.952	12.872	13.791	14.710	15.629	16.548	17.468	18.388	10
52	.920	1.839	2.759	3.679	4.599	5.518	6.438	7.357	8.277	9.197	10.116	11.035	11.955	12.876	13.795	14.714	15.633	16.553	17.473	18.393	8
54	.920	1.840	2.760	3.680	4.600	5.519	6.440	7.359	8.279	9.199	10.119	11.038	11.958	12.879	13.798	14.718	15.637	16.557	17.477	18.397	6
56	.920	1.840	2.761	3.681	4.601	5.121	6.441	7.361	8.281	9.201	10.121	11.041	11.961	12.882	13.802	14.722	15.641	16.562	17.482	18.402	4
58	.920	1.841	2.761	3.682	4.602	5.522	6.443	7.363	8.283	9.203	10.124	11.044	11.964	12.885	13.805	14.725	15.645	16.566	17.486	18.406	2
7° 0′	.921	1.841	2.762	3.682	4.603	5.523	6.444	7.364	8.285	9.205	10.126	11.046	11.967	12.888	13.808	14.728	15.649	16.570	17.490	18.410	23° 0′
2	.921	1.841	2.763	3.683	4.604	5.525	6.446	7.366	8.287	9.208	10.129	11.049	11.970	12.892	13.812	14.732	15.653	16.574	17.495	18.415	58
4	.921	1.842	2.763	3.684	4.605	5.526	6.448	7.368	8.289	9.210	10.131	11.052	11.973	12.895	13.816	14.736	15.657	16.578	17.499	18.419	56
6	.921	1.842	2.764	3.685	4.606	5.528	6.449	7.370	8.291	9.212	10.134	11.055	11.976	12.898	13.819	14.739	15.661	16.582	17.503	18.424	54
8	.922	1.843	2.764	3.686	4.607	5.529	6.451	7.372	8.293	9.214	10.136	11.058	11.979	12.901	13.822	14.743	15.664	16.586	17.507	18.428	52
10	.922	1.843	2.765	3.687	4.608	5.530	6.452	7.373	8.295	9.216	10.138	11.060	11.982	12.904	13.825	14.746	15.668	16.590	17.511	18.432	50
12	.922	1.844	2.766	3.688	4.610	5.532	6.454	7.375	8.297	9.219	10.141	11.063	11.985	12.907	13.829	14.750	15.672	16.594	17.516	18.437	48
14	.922	1.844	2.766	3.689	4.611	5.533	6.455	7.377	8.299	9.222	10.144	11.066	11.988	12.910	13.832	14.754	15.676	16.598	17.520	18.442	46
16	.922	1.845	2.767	3.690	4.612	5.534	6.457	7.379	8.301	9.224	10.146	11.069	11.991	12.913	13.835	14.757	15.680	16.602	17.525	18.447	44
18	.923	1.845	2.767	3.691	4.613	5.536	6.458	7.381	8.303	9.226	10.149	11.072	11.994	12.916	13.838	14.761	15.683	16.606	17.529	18.451	42
20	.923	1.846	2.768	3.691	4.614	5.537	6.459	7.382	8.305	9.228	10.151	11.074	11.996	12.918	13.841	14.764	15.687	16.610	17.533	18.456	40
22	.923	1.846	2.769	3.692	4.615	5.539	6.461	7.384	8.307	9.231	10.154	11.077	11.999	12.922	13.845	14.768	15.691	16.614	17.538	18.461	38
24	.923	1.847	2.770	3.693	4.616	5.540	6.463	7.386	8.309	9.233	10.156	11.079	12.002	12.925	13.848	14.772	15.695	16.618	17.542	18.465	36
26	.923	1.847	2.771	3.694	4.617	5.541	6.464	7.388	8.311	9.235	10.159	11.082	12.005	12.928	13.851	14.775	15.699	16.622	17.546	18.470	34
28	.924	1.848	2.771	3.695	4.618	5.542	6.466	7.390	8.313	9.237	10.161	11.084	12.008	12.931	13.855	14.779	15.702	16.626	17.550	18.474	32
30	.924	1.848	2.772	3.696	4.619	5.543	6.467	7.391	8.315	9.239	10.163	11.086	12.010	12.934	13.858	14.782	15.706	16.630	17.554	18.478	30
32	.924	1.848	2.773	3.697	4.620	5.545	6.469	7.393	8.317	9.242	10.166	11.089	12.013	12.938	13.862	14.786	15.710	16.634	17.559	18.483	28
34	.924	1.849	2.773	3.698	4.621	5.546	6.471	7.395	8.319	9.244	10.168	11.092	12.016	12.941	13.865	14.790	15.714	16.638	17.563	18.487	26
36	.925	1.849	2.774	3.699	4.622	5.547	6.472	7.397	8.321	9.246	10.171	11.095	12.019	12.944	13.869	14.793	15.718	16.642	17.567	18.492	24
38	.925	1.850	2.774	3.700	4.624	5.548	6.474	7.399	8.323	9.248	10.173	11.098	12.022	12.947	13.872	14.797	15.722	16.646	17.571	18.496	22
40	.925	1.850	2.775	3.700	4.625	5.550	6.475	7.400	8.325	9.250	10.175	11.100	12.025	12.950	13.875	14.800	15.725	16.650	17.575	18.500	20
42	.925	1.850	2.776	3.701	4.627	5.552	6.477	7.402	8.327	9.253	10.178	11.103	12.028	12.954	13.879	14.804	15.729	16.654	17.580	18.505	18
44	.925	1.851	2.776	3.702	4.628	5.553	6.479	7.404	8.329	9.255	10.180	11.106	12.031	12.957	13.882	14.808	15.733	16.658	17.584	18.509	16
46	.926	1.851	2.777	3.703	4.629	5.555	6.480	7.406	8.331	9.257	10.183	11.109	12.034	12.960	13.886	14.811	15.737	16.662	17.588	18.513	14
48	.926	1.852	2.777	3.704	4.630	5.556	6.482	7.408	8.333	9.259	10.185	11.112	12.037	12.963	13.889	14.815	15.740	16.666	17.592	18.518	12
50	.926	1.852	2.778	3.704	4.631	5.557	6.483	7.409	8.335	9.261	10.187	11.114	12.040	12.966	13.892	14.818	15.744	16.670	17.596	18.522	10
52	.926	1.852	2.779	3.705	4.632	5.559	6.485	7.411	8.337	9.264	10.190	11.117	12.043	12.969	13.895	14.822	15.748	16.674	17.601	18.527	8
54	.927	1.853	2.780	3.706	4.633	5.560	6.486	7.413	8.339	9.266	10.192	11.119	12.046	12.972	13.898	14.825	15.752	16.678	17.605	18.531	6
56	.927	1.853	2.781	3.707	4.634	5.561	6.488	7.415	8.341	9.268	10.195	11.122	12.049	12.975	13.902	14.829	15.755	16.682	17.609	18.536	4
58	.927	1.854	2.781	3.708	4.635	5.562	6.489	7.416	8.343	9.270	10.197	11.124	12.052	12.978	13.905	14.832	15.759	16.686	17.613	18.540	2
8° 0	.927	1.854	2.782	3.709	4.636	5.563	6.490	7.417	8.345	9.272	10.199	11.126	12.054	12.980	13.907	14.835	15.762	16.690	17.617	18.544	22° 0
	1	2	3	4	5	6	7	8	9	10	11	12	13	14	15	16	17	18	19	20	III+ IX−

TABLE XVIII.

GENERAL TABLE of Aberration, &c. continued.

S. −VIII + II	1	2	3	4	5	6	7	8	9	10	11	12	13	14	15	16	17	18	19	20	
8° 0′	.927	1.854	2.782	3.709	4.636	5.563	6.490	7.417	8.345	9.272	10.199	11.126	12.054	12.980	13.907	14.835	15.762	16.690	17.617	18.544	22° 0′
2	.927	1.855	2.783	3.710	4.637	5.565	6.492	7.419	8.347	9.275	10.202	11.129	12.056	12.984	13.911	14.838	15.766	16.694	17.621	18.549	58
4	.927	1.855	2.783	3.711	4.638	5.566	6.494	7.421	8.349	9.277	10.204	11.132	12.059	12.987	13.914	14.842	15.770	16.698	17.625	18.553	56
6	.928	1.856	2.784	3.712	4.639	5.568	6.496	7.428	8.351	9.279	10.207	11.134	12.062	12.990	13.918	14.845	15.773	16.702	17.629	18.558	54
8	.928	1.856	2.784	3.713	4.640	5.569	6.497	7.424	8.353	9.281	10.209	11.137	12.065	12.993	13.921	14.849	15.777	16.705	17.633	18.562	52
10	.928	1.857	2.785	3.713	4.641	5.570	6.498	7.426	8.354	9.283	10.211	11.140	12.068	12.996	13.924	14.852	15.780	16.708	17.637	18.566	50
12	.928	1.857	2.786	3.714	4.642	5.572	6.500	7.428	8.356	9.286	10.214	11.143	12.071	12.999	13.928	14.856	15.784	16.712	17.642	18.571	48
14	.929	1.858	2.786	3.715	4.643	5.573	6.502	7.430	8.358	9.288	10.216	11.145	12.074	13.003	13.931	14.860	15.788	16.716	17.646	18.575	46
16	.929	1.858	2.787	3.716	4.644	5.574	6.503	7.432	8.360	9.290	10.219	11.148	12.077	13.006	13.935	14.863	15.792	16.720	17.650	18.580	44
18	.929	1.859	2.787	3.716	4.645	5.575	6.505	7.433	8.362	9.292	10.221	11.150	12.079	13.009	13.938	14.867	15.796	16.724	17.654	18.584	42
20	.929	1.859	2.788	3.717	4.647	5.576	6.506	7.435	8.364	9.294	10.223	11.152	12.082	13.012	13.941	14.870	15.799	16.728	17.658	18.588	40
22	.929	1.859	2.789	3.718	4.648	5.578	6.508	7.437	8.366	9.296	10.226	11.155	12.085	13.015	13.944	14.874	15.803	16.732	17.662	18.592	38
24	.930	1.860	2.789	3.719	4.649	5.579	6.509	7.439	8.368	9.298	10.228	11.158	12.088	13.018	13.947	14.877	15.807	16.736	17.666	18.596	36
26	.930	1.860	2.790	3.720	4.650	5.581	6.511	7.440	8.370	9.300	10.230	11.161	12.091	13.021	13.950	14.880	15.811	16.740	17.670	18.600	34
28	.930	1.861	2.790	3.721	4.651	5.582	6.512	7.442	8.372	9.302	10.232	11.163	12.093	13.024	13.953	14.883	15.814	16.744	17.674	18.604	32
30	.930	1.861	2.791	3.722	4.652	5.583	6.513	7.443	8.374	9.304	10.234	11.166	12.096	13.026	13.956	14.886	15.817	16.748	17.678	18.608	30
32	.931	1.861	2.792	3.723	4.653	5.585	6.515	7.445	8.376	9.307	10.237	11.169	12.099	13.029	13.960	14.890	15.821	16.752	17.682	18.613	28
34	.931	1.862	2.792	3.724	4.654	5.586	6.516	7.447	8.378	9.309	10.239	11.171	12.102	13.032	13.963	14.894	15.825	16.756	17.686	18.617	26
36	.931	1.862	2.793	3.725	4.655	5.587	6.518	7.449	8.380	9.311	10.242	11.174	12.104	13.035	13.966	14.897	15.828	16.759	17.690	18.622	24
38	.931	1.863	2.793	3.725	4.656	5.588	6.519	7.450	8.382	9.313	10.244	11.176	12.107	13.038	13.969	14.901	15.832	16.763	17.694	18.626	22
40	.932	1.863	2.794	3.726	4.657	5.589	6.520	7.452	8.383	9.315	10.247	11.178	12.109	13.040	13.972	14.904	15.835	16.766	17.698	18.630	20
42	.932	1.863	2.795	3.727	4.659	5.591	6.522	7.454	8.385	9.317	10.250	11.181	12.112	13.044	13.976	14.908	15.839	16.770	17.702	18.634	18
44	.932	1.864	2.796	3.728	4.660	5.592	6.523	7.455	8.387	9.319	10.252	11.183	12.115	13.047	13.979	14.911	15.843	16.774	17.706	18.638	16
46	.932	1.864	2.796	3.729	4.661	5.593	6.524	7.457	8.389	9.321	10.254	11.186	12.117	13.050	13.982	14.914	15.846	16.778	17.710	18.642	14
48	.932	1.865	2.797	3.730	4.662	5.594	6.526	7.459	8.391	9.323	10.256	11.188	12.120	13.053	13.985	14.917	15.850	16.782	17.714	18.646	12
50	.933	1.865	2.798	3.730	4.663	5.595	6.528	7.460	8.393	9.325	10.258	11.190	12.123	13.056	13.988	14.920	15.853	16.786	17.718	18.650	10
52	.933	1.865	2.799	3.730	4.664	5.597	6.530	7.462	8.395	9.328	10.261	11.193	12.126	13.059	13.992	14.924	15.857	16.790	17.722	18.655	8
54	.933	1.866	2.799	3.731	4.665	5.598	6.531	7.464	8.397	9.330	10.263	11.196	12.129	13.062	13.995	14.928	15.861	16.794	17.726	18.659	6
56	.933	1.866	2.800	3.732	4.666	5.600	6.532	7.465	8.399	9.332	10.266	11.198	12.132	13.065	13.998	14.931	15.864	16.797	17.730	18.664	4
58	.933	1.867	2.800	3.733	4.667	5.601	6.534	7.467	8.400	9.334	10.268	11.201	12.135	13.068	14.001	14.935	15.868	16.801	17.734	18.668	2
9° 0′	.934	1.867	2.801	3.734	4.668	5.602	6.535	7.469	8.402	9.336	10.270	11.204	12.137	13.070	14.004	14.938	15.871	16.804	17.738	18.672	21° 0′
2	.934	1.867	2.802	3.735	4.669	5.604	6.537	7.471	8.404	9.338	10.273	11.207	12.140	13.073	14.007	14.942	15.875	16.808	17.742	18.676	58
4	.934	1.868	2.802	3.736	4.670	5.605	6.538	7.472	8.406	9.340	10.275	11.210	12.143	13.076	14.010	14.945	15.879	16.812	17.746	18.680	56
6	.934	1.868	2.803	3.737	4.671	5.606	6.540	7.474	8.408	9.342	10.277	11.212	12.145	13.079	14.013	14.948	15.882	16.816	17.750	18.684	54
8	.934	1.868	2.803	3.738	4.672	5.607	6.541	7.475	8.410	9.344	10.279	11.214	12.148	13.082	14.016	14.951	15.886	16.820	17.754	18.688	52
10	.935	1.869	2.804	3.739	4.673	5.608	6.542	7.477	8.412	9.346	10.281	11.216	12.150	13.085	14.019	14.954	15.889	16.824	17.758	18.692	50
12	.935	1.869	2.805	3.740	4.674	5.610	6.544	7.479	8.414	9.349	10.284	11.219	12.153	13.088	14.023	14.958	15.893	16.828	17.762	18.697	48
14	.935	1.870	2.805	3.741	4.675	5.611	6.545	7.480	8.416	9.351	10.286	11.221	12.156	13.091	14.026	14.961	15.896	16.831	17.766	18.701	46
16	.935	1.870	2.806	3.742	4.676	5.612	6.547	7.482	8.418	9.353	10.289	11.224	12.159	13.094	14.029	14.964	15.900	16.835	17.770	18.706	44
18	.936	1.871	2.806	3.742	4.677	5.613	6.549	7.483	8.420	9.355	10.291	11.226	12.161	13.097	14.032	14.967	15.903	16.839	17.774	18.710	42
20	.936	1.871	2.807	3.743	4.678	5.614	6.550	7.485	8.421	9.357	10.293	11.228	12.164	13.100	14.035	14.970	15.906	16.842	17.778	18.714	40
22	.936	1.871	2.808	3.744	4.679	5.616	6.552	7.487	8.423	9.359	10.296	11.231	12.167	13.103	14.038	14.974	15.910	16.846	17.782	18.718	38
24	.936	1.872	2.808	3.745	4.680	5.617	6.553	7.489	8.425	9.361	10.298	11.233	12.170	13.106	14.041	14.977	15.913	16.850	17.786	18.722	36
26	.936	1.872	2.809	3.746	4.681	5.618	6.555	7.490	8.427	9.363	10.300	11.236	12.172	13.109	14.044	14.980	15.917	16.853	17.790	18.726	34
28	.937	1.873	2.809	3.747	4.682	5.619	6.556	7.492	8.429	9.365	10.302	11.238	12.175	13.112	14.047	14.983	15.920	16.857	17.793	18.730	32
30	.937	1.873	2.810	3.747	4.683	5.620	6.557	7.493	8.430	9.367	10.304	11.240	12.177	13.114	14.050	14.986	15.923	16.860	17.797	18.734	30
32	.937	1.873	2.811	3.748	4.685	5.622	6.559	7.495	8.432	9.369	10.307	11.243	12.180	13.117	14.054	14.990	15.927	16.864	17.801	18.738	28
34	.937	1.874	2.811	3.749	4.686	5.623	6.560	7.497	8.434	9.371	10.309	11.245	12.182	13.120	14.057	14.994	15.931	16.868	17.805	18.742	26
36	.937	1.874	2.812	3.750	4.687	5.624	6.562	7.499	8.436	9.373	10.311	11.248	12.185	13.123	14.060	14.998	15.934	16.872	17.809	18.746	24
38	.938	1.875	2.812	3.750	4.688	5.625	6.563	7.500	8.438	9.375	10.313	11.250	12.188	13.126	14.063	15.001	15.938	16.875	17.813	18.750	22
40	.938	1.875	2.813	3.751	4.689	5.626	6.564	7.502	8.439	9.377	10.315	11.252	12.190	13.128	14.066	15.004	15.941	16.878	17.816	18.754	20
42	.938	1.875	2.814	3.752	4.690	5.628	6.566	7.504	8.441	9.379	10.318	11.255	12.193	13.131	14.069	15.008	15.945	16.882	17.820	18.758	18
44	.938	1.876	2.814	3.753	4.691	5.629	6.567	7.506	8.443	9.381	10.320	11.257	12.195	13.134	14.072	15.011	15.948	16.886	17.824	18.762	16
46	.938	1.876	2.815	3.754	4.692	5.630	6.569	7.507	8.445	9.383	10.322	11.260	12.198	13.137	14.075	15.014	15.952	16.890	17.828	18.766	14
48	.939	1.877	2.815	3.755	4.693	5.631	6.570	7.509	8.447	9.385	10.324	11.262	12.200	13.140	14.078	15.017	15.955	16.893	17.832	18.770	12
50	.939	1.877	2.816	3.755	4.694	5.632	6.571	7.510	8.448	9.387	10.326	11.264	12.203	13.142	14.081	15.020	15.958	16.896	17.835	18.774	10
52	.939	1.177	2.817	3.756	4.695	5.634	6.573	7.512	8.450	9.389	10.329	11.267	12.206	13.145	14.084	15.024	15.962	16.900	17.839	18.778	8
54	.939	1.878	2.817	3.757	4.696	5.635	6.574	7.514	8.452	9.391	10.331	11.269	12.209	13.148	14.087	15.027	15.965	16.903	17.843	18.782	6
56	.939	1.878	2.818	3.758	4.697	5.636	6.576	7.515	8.453	9.393	10.333	11.272	12.211	13.151	14.090	15.030	15.969	16.907	17.847	18.786	4
58	.940	1.879	2.818	3.758	4.698	5.637	6.577	7.517	8.455	9.395	10.335	11.274	12.214	13.154	14.093	15.033	15.972	16.911	17.850	18.790	2
10° 0	.940	1.879	2.819	3.759	4.699	5.638	6.578	7.518	8.457	9.397	10.337	11.276	12.216	13.156	14.096	15.036	15.975	16.914	17.854	18.794	20° 0
	1	2	3	4	5	6	7	8	9	10	11	12	13	14	15	16	17	18	19	20	III + IX −

TABLE XVIII.

GENERAL TABLE of Aberration, &c. continued.

S. −VIII + II	1	2	3	4	5	6	7	8	9	10	11	12	13	14	15	16	17	18	19	20	
10° 0′	.940	1 .879	2 .819	3 .759	4 .699	5 .638	6 .578	7 .518	8 .457	9 .397	10 .337	11 .276	12 .216	13 .156	14 .096	15 .036	15 .975	16 .914	17 .854	18 .794	20° 0′
2	.940	1 .879	2 .820	3 .760	4 .700	5 .639	6 .580	7 .520	8 .459	9 .399	10 .340	11 .279	12 .219	13 .159	14 .099	15 .039	15 .979	16 .918	17 .858	18 .798	58
4	.940	1 .880	2 .820	3 .761	4 .701	5 .641	6 .581	7 .521	8 .461	9 .401	10 .342	11 .281	12 .222	13 .162	14 .102	15 .042	15 .982	16 .922	17 .862	18 .802	56
6	.940	1 .880	2 .821	3 .762	4 .702	5 .642	6 .583	7 .523	8 .463	9 .403	10 .344	11 .284	12 .224	13 .165	14 .105	15 .045	15 .985	16 .925	17 .866	18 .806	54
8	.941	1 .881	2 .821	3 .762	4 .703	5 .643	6 .584	7 .524	8 .465	9 .405	10 .346	11 .286	12 .227	13 .168	14 .108	15 .048	15 .988	16 .929	17 .870	18 .810	52
10	.941	1 .881	2 .822	3 .763	4 .703	5 .644	6 .585	7 .526	8 .466	9 .407	10 .348	11 .288	12 .229	13 .170	14 .110	15 .050	15 .991	16 .932	17 .873	18 .814	50
12	.941	1 .881	2 .823	3 .764	4 .704	5 .646	6 .587	7 .527	8 .468	9 .409	10 .351	11 .291	12 .232	13 .173	14 .113	15 .054	15 .995	16 .936	17 .877	18 .818	48
14	.941	1 .882	2 .823	3 .765	4 .705	5 .647	6 .588	7 .529	8 .470	9 .411	10 .353	11 .293	12 .235	13 .176	14 .116	15 .057	15 .998	16 .940	17 .881	18 .822	46
16	.941	1 .882	2 .824	3 .766	4 .706	5 .648	6 .590	7 .530	8 .472	9 .413	10 .355	11 .296	12 .237	13 .179	14 .119	15 .060	16 .002	16 .944	17 .885	18 .826	44
18	.942	1 .883	2 .824	3 .766	4 .707	5 .649	6 .591	7 .532	8 .474	9 .415	10 .357	11 .298	12 .240	13 .182	14 .122	15 .063	16 .005	16 .947	17 .889	18 .830	42
20	.942	1 .883	2 .825	3 .767	4 .708	5 .650	6 .592	7 .533	8 .475	9 .417	10 .359	11 .300	12 .242	13 .184	14 .125	15 .066	16 .008	16 .950	17 .892	18 .834	40
22	.942	1 .883	2 .826	3 .768	4 .709	5 .652	6 .594	7 .535	8 .477	9 .419	10 .361	11 .303	12 .245	13 .187	14 .128	15 .070	16 .012	16 .954	17 .896	18 .838	38
24	.942	1 .884	2 .826	3 .769	4 .710	5 .653	6 .595	7 .537	8 .479	9 .421	10 .363	11 .305	12 .247	13 .190	14 .131	15 .073	16 .015	16 .958	17 .900	18 .842	36
26	.942	1 .884	2 .827	3 .770	4 .711	5 .654	6 .597	7 .538	8 .481	9 .423	10 .365	11 .308	12 .250	13 .193	14 .134	15 .076	16 .019	16 .961	17 .903	18 .845	34
28	.943	1 .885	2 .827	3 .770	4 .712	5 .655	6 .598	7 .540	8 .483	9 .425	10 .367	11 .310	12 .253	13 .196	14 .137	15 .079	16 .022	16 .965	17 .907	18 .849	32
30	.943	1 .885	2 .828	3 .771	4 .713	5 .656	6 .599	7 .541	8 .484	9 .426	10 .369	11 .312	12 .255	13 .198	14 .140	15 .082	16 .025	16 .968	17 .910	18 .852	30
32	.943	1 .885	2 .829	3 .772	4 .714	5 .658	6 .601	7 .543	8 .486	9 .428	10 .372	11 .315	12 .258	13 .201	14 .143	15 .086	16 .029	16 .972	17 .914	18 .856	28
34	.943	1 .886	2 .829	3 .772	4 .715	5 .659	6 .602	7 .545	8 .488	9 .430	10 .374	11 .317	12 .260	13 .203	14 .146	15 .089	16 .032	16 .976	17 .918	18 .860	26
36	.943	1 .886	2 .830	3 .773	4 .716	5 .660	6 .603	7 .546	8 .490	9 .432	10 .376	11 .320	12 .263	13 .206	14 .149	15 .092	16 .036	16 .979	17 .922	18 .864	24
38	.943	1 .887	2 .830	3 .774	4 .717	5 .661	6 .604	7 .548	8 .492	9 .434	10 .378	11 .322	12 .265	13 .208	14 .152	15 .095	16 .039	16 .983	17 .926	18 .868	22
40	.944	1 .887	2 .831	3 .774	4 .718	5 .662	6 .605	7 .549	8 .493	9 .436	10 .380	11 .324	12 .267	13 .210	14 .154	15 .098	16 .042	16 .986	17 .929	18 .872	20
42	.944	1 .887	2 .832	3 .775	4 .719	5 .663	6 .607	7 .551	8 .495	9 .438	10 .383	11 .326	12 .270	13 .213	14 .157	15 .102	16 .046	16 .990	17 .933	18 .876	18
44	.944	1 .888	2 .832	3 .776	4 .720	5 .664	6 .608	7 .552	8 .497	9 .440	10 .385	11 .328	12 .272	13 .216	14 .160	15 .106	16 .049	16 .993	17 .937	18 .880	16
46	.944	1 .888	2 .833	3 .777	4 .721	5 .665	6 .610	7 .554	8 .498	9 .442	10 .387	11 .330	12 .275	13 .219	14 .163	15 .109	16 .052	16 .996	17 .940	18 .884	14
48	.944	1 .889	2 .834	3 .778	4 .722	5 .666	6 .611	7 .555	8 .500	9 .444	10 .389	11 .332	12 .277	13 .222	14 .166	15 .112	16 .055	16 .999	17 .944	18 .888	12
50	.945	1 .889	2 .834	3 .778	4 .723	5 .667	6 .612	7 .557	8 .501	9 .446	10 .391	11 .334	12 .279	13 .224	14 .169	15 .114	16 .058	17 .002	17 .947	18 .892	10
52	.945	1 .889	2 .835	3 .779	4 .724	5 .669	6 .614	7 .559	8 .503	9 .448	10 .393	11 .337	12 .282	13 .227	14 .172	15 .117	16 .062	17 .006	17 .951	18 .896	8
54	.945	1 .890	2 .835	3 .780	4 .725	5 .670	6 .615	7 .560	8 .505	9 .450	10 .395	11 .339	12 .285	13 .230	14 .175	15 .120	16 .065	17 .010	17 .955	18 .900	6
56	.945	1 .890	2 .836	3 .781	4 .726	5 .671	6 .617	7 .562	8 .507	9 .452	10 .397	11 .342	12 .287	13 .233	14 .178	15 .123	16 .068	17 .013	17 .958	18 .903	4
58	.945	1 .891	2 .836	3 .781	4 .727	5 .672	6 .618	7 .563	8 .508	9 .454	10 .399	11 .344	12 .290	13 .236	14 .181	15 .125	16 .071	17 .017	17 .962	18 .907	2
11° 0′	.946	1 .891	2 .837	3 .782	4 .728	5 .673	6 .619	7 .564	8 .510	9 .455	10 .401	11 .346	12 .292	13 .238	14 .183	15 .128	16 .074	17 .020	17 .965	18 .910	19° 0′
2	.946	1 .891	2 .837	3 .783	4 .729	5 .675	6 .621	7 .566	8 .512	9 .457	10 .404	11 .349	12 .295	13 .241	14 .186	15 .132	16 .078	17 .024	17 .969	18 .914	58
4	.946	1 .892	2 .838	3 .784	4 .730	5 .676	6 .622	7 .567	8 .514	9 .459	10 .406	11 .351	12 .297	13 .244	14 .189	15 .135	16 .081	17 .027	17 .973	18 .918	56
6	.946	1 .892	2 .838	3 .785	4 .731	5 .677	6 .623	7 .569	8 .515	9 .461	10 .408	11 .354	12 .300	13 .246	14 .192	15 .138	16 .084	17 .030	17 .976	18 .922	54
8	.946	1 .893	2 .839	3 .786	4 .732	5 .678	6 .624	7 .571	8 .517	9 .463	10 .410	11 .356	12 .302	13 .248	14 .195	15 .141	16 .087	17 .033	17 .980	18 .926	52
10	.947	1 .893	2 .839	3 .786	4 .732	5 .679	6 .625	7 .572	8 .518	9 .465	10 .412	11 .358	12 .304	13 .250	14 .197	15 .144	16 .090	17 .036	17 .983	18 .930	50
12	.947	1 .893	2 .840	3 .787	4 .733	5 .680	6 .627	7 .574	8 .520	9 .467	10 .414	11 .360	12 .307	13 .253	14 .200	15 .147	16 .094	17 .040	17 .987	18 .934	48
14	.947	1 .894	2 .840	3 .788	4 .734	5 .681	6 .628	7 .575	8 .522	9 .469	10 .416	11 .362	12 .309	13 .256	14 .203	15 .150	16 .097	17 .044	17 .991	18 .938	46
16	.947	1 .894	2 .841	3 .789	4 .735	5 .682	6 .629	7 .577	8 .523	9 .471	10 .418	11 .364	12 .312	13 .259	14 .206	15 .153	16 .100	17 .047	17 .994	18 .942	44
18	.947	1 .895	2 .841	3 .790	4 .736	5 .683	6 .631	7 .578	8 .525	9 .473	10 .420	11 .366	12 .314	13 .262	14 .209	15 .156	16 .103	17 .051	17 .998	18 .945	42
20	.947	1 .895	2 .842	3 .790	4 .737	5 .684	6 .632	7 .579	8 .527	9 .474	10 .421	11 .368	12 .316	13 .264	14 .211	15 .158	16 .106	17 .054	18 .001	18 .948	40
22	.948	1 .895	2 .843	3 .791	4 .738	5 .686	6 .634	7 .581	8 .529	9 .476	10 .423	11 .371	12 .319	13 .267	14 .214	15 .162	16 .110	17 .058	18 .005	18 .952	38
24	.948	1 .896	2 .843	3 .792	4 .739	5 .687	6 .635	7 .582	8 .531	9 .478	10 .425	11 .373	12 .321	13 .269	14 .217	15 .165	16 .113	17 .061	18 .008	18 .956	36
26	.948	1 .896	2 .844	3 .792	4 .740	5 .688	6 .636	7 .584	8 .533	9 .480	10 .427	11 .376	12 .324	13 .272	14 .220	15 .168	16 .116	17 .064	18 .012	18 .960	34
28	.948	1 .897	2 .844	3 .793	4 .741	5 .689	6 .637	7 .585	8 .534	9 .482	10 .429	11 .378	12 .326	13 .274	14 .223	15 .171	16 .119	17 .067	18 .015	18 .963	32
30	.948	1 .897	2 .845	3 .793	4 .742	5 .690	6 .638	7 .587	8 .535	9 .483	10 .431	11 .380	12 .328	13 .276	14 .225	15 .174	16 .122	17 .070	18 .018	18 .966	30
32	.948	1 .897	2 .846	3 .794	4 .743	5 .691	6 .640	7 .589	8 .537	9 .485	10 .433	11 .382	12 .331	13 .279	14 .228	15 .177	16 .125	17 .074	18 .022	18 .970	28
34	.949	1 .898	2 .846	3 .795	4 .744	5 .692	6 .641	7 .591	8 .539	9 .487	10 .435	11 .384	12 .333	13 .282	14 .231	15 .180	16 .128	17 .077	18 .025	18 .974	26
36	.949	1 .898	2 .847	3 .796	4 .745	5 .693	6 .643	7 .592	8 .540	9 .489	10 .437	11 .386	12 .336	13 .285	14 .234	15 .183	16 .131	17 .080	18 .029	18 .977	24
38	.949	1 .899	2 .847	3 .796	4 .746	5 .694	6 .644	7 .593	8 .542	9 .491	10 .439	11 .388	12 .338	13 .288	14 .237	15 .186	16 .134	17 .083	18 .032	18 .981	22
40	.949	1 .899	2 .848	3 .797	4 .746	5 .695	6 .645	7 .594	8 .543	9 .492	10 .441	11 .390	12 .340	13 .290	14 .239	15 .188	16 .137	17 .086	18 .035	18 .984	20
42	.949	1 .899	2 .848	3 .798	4 .747	5 .697	6 .647	7 .596	8 .545	9 .494	10 .444	11 .393	12 .343	13 .293	14 .242	15 .191	16 .140	17 .090	18 .039	18 .988	18
44	.950	1 .900	2 .849	3 .799	4 .748	5 .698	6 .648	7 .597	8 .547	9 .496	10 .446	11 .395	12 .345	13 .295	14 .245	15 .194	16 .143	17 .093	18 .043	18 .992	16
46	.950	1 .900	2 .850	3 .800	4 .749	5 .699	6 .649	7 .599	8 .548	9 .498	10 .448	11 .398	12 .348	13 .298	14 .247	15 .197	16 .146	17 .096	18 .047	18 .996	14
48	.950	1 .900	2 .850	3 .800	4 .750	5 .700	6 .650	7 .600	8 .550	9 .500	10 .450	11 .400	12 .350	13 .300	14 .250	15 .200	16 .149	17 .099	18 .050	19 .000	12
50	.950	1 .900	2 .851	3 .801	4 .751	5 .701	6 .651	7 .601	8 .551	9 .502	10 .452	11 .402	12 .352	13 .302	14 .252	15 .202	16 .152	17 .102	18 .053	19 .004	10
52	.950	1 .901	2 .851	3 .802	4 .751	5 .702	6 .653	7 .603	8 .553	9 .504	10 .454	11 .404	12 .355	13 .305	14 .255	15 .206	16 .156	17 .106	18 .056	19 .008	8
54	.951	1 .901	2 .852	3 .802	4 .752	5 .703	6 .654	7 .604	8 .555	9 .506	10 .456	11 .406	12 .357	13 .307	14 .258	15 .209	16 .159	17 .110	18 .060	19 .012	6
56	.951	1 .901	2 .852	3 .803	4 .753	5 .704	6 .655	7 .606	8 .557	9 .508	10 .458	11 .408	12 .359	13 .310	14 .261	15 .212	16 .163	17 .113	18 .063	19 .015	4
58	.951	1 .902	2 .853	3 .803	4 .754	5 .705	6 .656	7 .608	8 .559	9 .510	10 .460	11 .410	12 .361	13 .312	14 .264	15 .215	16 .166	17 .117	18 .067	19 .019	2
12° 0	.951	1 .902	2 .853	3 .804	4 .755	5 .706	6 .657	7 .609	8 .560	9 .511	10 .462	11 .412	12 .363	13 .314	14 .266	15 .218	16 .169	17 .120	18 .071	19 .022	18° 0
	1	2	3	4	5	6	7	8	9	10	11	12	13	14	15	16	17	18	19	20	III + IX −

TABLE XVIII.

GENERAL TABLE of Aberration, &c. continued.

S. −VIII +II	1	2	3	4	5	6	7	8	9	10	11	12	13	14	15	16	17	18	19	20	
12° 0′	.951	1 .902	2 .853	3 .804	4 .755	5 .706	6 .657	7 .609	8 .560	9 .511	10 .462	11 .412	12 .363	13 .314	14 .266	15 .218	16 .169	17 .120	18 .071	19 .022	18° 0′
2	.951	1 .902	2 .854	3 .805	4 .756	5 .708	6 .659	7 .611	8 .562	9 .513	10 .464	11 .415	12 .366	13 .317	14 .269	15 .221	16 .172	17 .124	18 .075	19 .026	58
4	.951	1 .903	2 .855	3 .806	4 .757	5 .709	6 .660	7 .612	8 .564	9 .515	10 .466	11 .417	12 .368	13 .320	14 .272	15 .224	16 .175	17 .127	18 .078	19 .030	56
6	.952	1 .903	2 .855	3 .807	4 .758	5 .710	6 .662	7 .614	8 .565	9 .517	10 .468	11 .420	12 .371	13 .322	14 .275	15 .227	16 .178	17 .130	18 .082	19 .033	54
8	.952	1 .904	2 .856	3 .808	4 .759	5 .711	6 .663	7 .615	8 .567	9 .519	10 .470	11 .422	12 .373	13 .325	14 .277	15 .230	16 .181	17 .133	18 .085	19 .037	52
10	.952	1 .904	2 .856	3 .808	4 .760	5 .712	6 .664	7 .616	8 .568	9 .520	10 .472	11 .424	12 .376	13 .328	14 .280	15 .232	16 .184	17 .136	18 .088	19 .040	50
12	.952	1 .904	2 .857	3 .809	4 .761	5 .713	6 .666	7 .618	8 .570	9 .522	10 .474	11 .426	12 .379	13 .331	14 .283	15 .235	16 .187	17 .140	18 .092	19 .044	48
14	.952	1 .905	2 .857	3 .809	4 .762	5 .714	6 .667	7 .619	8 .572	9 .524	10 .476	11 .428	12 .381	13 .333	14 .286	15 .238	16 .190	17 .143	18 .095	19 .047	46
16	.953	1 .905	2 .858	3 .810	4 .763	5 .715	6 .668	7 .621	8 .573	9 .526	10 .478	11 .430	12 .383	13 .336	14 .288	15 .241	16 .193	17 .146	18 .098	19 .050	44
18	.953	1 .905	2 .858	3 .811	4 .764	5 .716	6 .669	7 .622	8 .575	9 .527	10 .480	11 .432	12 .385	13 .338	14 .291	15 .244	16 .196	17 .149	18 .101	19 .053	42
20	.953	1 .906	2 .859	3 .811	4 .764	5 .717	6 .670	7 .623	8 .576	9 .528	10 .481	11 .434	12 .387	13 .340	14 .293	15 .246	16 .199	17 .152	18 .104	19 .056	40
22	.953	1 .906	2 .859	3 .812	4 .765	5 .718	6 .672	7 .625	8 .578	9 .530	10 .483	11 .436	12 .390	13 .343	14 .296	15 .249	16 .202	17 .156	18 .108	19 .060	38
24	.953	1 .906	2 .860	3 .813	4 .766	5 .719	6 .673	7 .626	8 .580	9 .532	10 .485	11 .438	12 .392	13 .345	14 .299	15 .252	16 .205	17 .159	18 .112	19 .064	36
26	.953	1 .906	2 .860	3 .814	4 .767	5 .720	6 .674	7 .628	8 .581	9 .534	10 .487	11 .440	12 .394	13 .348	14 .301	15 .255	16 .208	17 .162	18 .115	19 .067	34
28	.954	1 .907	2 .861	3 .814	4 .768	5 .721	6 .675	7 .629	8 .583	9 .536	10 .489	11 .442	12 .396	13 .350	14 .304	15 .258	16 .211	17 .165	18 .118	19 .071	32
30	.954	1 .907	2 .861	3 .815	4 .769	5 .722	6 .676	7 .630	8 .584	9 .537	10 .491	11 .444	12 .398	13 .352	14 .306	15 .260	16 .214	17 .168	18 .121	19 .074	30
32	.954	1 .907	2 .862	3 .816	4 .770	5 .724	6 .678	7 .632	8 .586	9 .539	10 .493	11 .447	12 .401	13 .355	14 .309	15 .263	16 .217	17 .171	18 .125	19 .078	28
34	.954	1 .908	2 .862	3 .817	4 .771	5 .725	6 .679	7 .633	8 .587	9 .541	10 .495	11 .449	12 .403	13 .357	14 .312	15 .266	16 .220	17 .174	18 .128	19 .082	26
36	.954	1 .908	2 .863	3 .817	4 .772	5 .726	6 .680	7 .634	8 .589	9 .543	10 .497	11 .451	12 .406	13 .360	14 .314	15 .269	16 .223	17 .177	18 .131	19 .085	24
38	.955	1 .909	2 .863	3 .818	4 .773	5 .727	6 .681	7 .636	8 .590	9 .545	10 .499	11 .454	12 .408	13 .362	14 .317	15 .272	16 .226	17 .180	18 .134	19 .089	22
40	.955	1 .909	2 .864	3 .818	4 .773	5 .728	6 .682	7 .637	8 .591	9 .546	10 .501	11 .456	12 .410	13 .364	14 .319	15 .274	16 .228	17 .182	18 .137	19 .092	20
42	.955	1 .909	2 .864	3 .819	4 .774	5 .729	6 .684	7 .639	8 .593	9 .548	10 .503	11 .458	12 .413	13 .367	14 .322	15 .277	16 .231	17 .186	18 .141	19 .096	18
44	.955	1 .910	2 .865	3 .820	4 .775	5 .730	6 .685	7 .640	8 .594	9 .550	10 .505	11 .460	12 .415	13 .369	14 .325	15 .280	16 .234	17 .189	18 .144	19 .100	16
46	.955	1 .910	2 .865	3 .821	4 .776	5 .731	6 .686	7 .642	8 .595	9 .552	10 .507	11 .462	12 .417	13 .372	14 .327	15 .283	16 .237	17 .192	18 .148	19 .104	14
48	.955	1 .911	2 .866	3 .821	4 .777	5 .732	6 .687	7 .643	8 .597	9 .554	10 .509	11 .464	12 .419	13 .374	14 .330	15 .286	16 .240	17 .195	18 .151	19 .107	12
50	.956	1 .911	2 .866	3 .822	4 .777	5 .733	6 .688	7 .644	8 .599	9 .555	10 .511	11 .466	12 .421	13 .376	14 .332	15 .288	16 .243	17 .198	18 .154	19 .110	10
52	.956	1 .911	2 .867	3 .823	4 .778	5 .734	6 .690	7 .646	8 .601	9 .557	10 .513	11 .468	12 .424	13 .379	14 .335	15 .291	16 .246	17 .201	18 .158	19 .114	8
54	.956	1 .912	2 .867	3 .823	4 .779	5 .735	6 .691	7 .647	8 .602	9 .559	10 .515	11 .470	12 .426	13 .381	14 .337	15 .293	16 .249	17 .205	18 .161	19 .117	6
56	.956	1 .912	2 .868	3 .824	4 .780	5 .736	6 .692	7 .648	8 .604	9 .560	10 .517	11 .472	12 .428	13 .384	14 .340	15 .296	16 .252	17 .208	18 .164	19 .120	4
58	.956	1 912	2 .868	3 .824	4 .781	5 .737	6 .693	7 .649	8 .606	9 .562	10 .518	11 .474	12 .430	13 .386	14 .342	15 .298	16 .255	17 .211	18 .167	19 .123	2
13° 0′	.956	1 .913	2 .869	3 .825	4 .782	5 .738	6 .694	7 .650	8 .607	9 .563	10 .519	11 .476	12 .432	13 .388	14 .344	15 .300	16 .257	17 .214	18 .170	19 .126	17° 0′
2	.956	1 .913	2 .870	3 .826	4 .783	5 .739	6 .696	7 .652	8 .609	9 .565	10 .521	11 .478	12 .435	13 .391	14 .347	15 .303	16 .260	17 .217	18 .174	19 .130	58
4	.957	1 .913	2 .870	3 .827	4 .784	5 .740	6 .697	7 .653	8 .611	9 .567	10 .523	11 .480	12 .437	13 .393	14 .350	15 .306	16 .263	17 .220	18 .177	19 .134	56
6	.957	1 .914	2 .871	3 .827	4 .785	5 .741	6 .698	7 .655	8 .612	9 .569	10 .525	11 .482	12 .439	13 .396	14 352	15 .309	16 .266	17 .223	18 .180	19 .137	54
8	.957	1 .914	2 .871	3 .828	4 .786	5 .742	6 .699	7 .656	8 .613	9 .571	10 .527	11 .484	12 .441	13 .398	14 355	15 .312	16 .269	17 .226	18 .183	19 .141	52
10	.957	1 .914	2 .872	3 .829	4 .786	5 .743	6 .700	7 .657	8 .614	9 .572	10 .529	11 .486	12 .443	13 .400	14 .357	15 .314	16 .271	17 .228	18 .186	19 .144	50
12	.957	1 .914	2 .872	3 .830	4 .787	5 .744	6 .702	7 .659	8 .616	9 .574	10 .531	11 .488	12 .446	13 .403	14 .360	15 .317	16 .274	17 .232	18 .190	19 .148	48
14	.958	1 .915	2 .873	3 .830	4 .787	5 .745	6 .703	7 .660	8 .618	9 .576	10 .533	11 .490	12 .448	13 .405	14 .363	15 .320	16 .277	17 .235	18 .193	19 .151	46
16	.958	1 .915	2 .873	3 .831	4 .788	5 .746	6 .704	7 .662	8 .619	9 .577	10 .535	11 .492	12 .450	13 .408	14 .365	15 .323	16 .280	17 .238	18 .196	19 .154	44
18	.958	1 .915	2 .874	3 .831	4 .789	5 .747	6 .705	7 .663	8 .621	9 .579	10 .537	11 .494	12 .452	13 .410	14 .368	15 .326	16 .283	17 .241	18 .199	19 .157	42
20	.958	1 .916	2 .874	3 .832	4 .790	5 .748	6 .706	7 .664	8 .622	9 .580	10 .538	11 .496	12 .454	13 .412	14 .370	15 .328	16 .286	17 .244	18 .202	19 .160	40
22	.958	1 .916	2 .875	3 .833	4 .791	5 .749	6 .708	7 .666	8 .624	9 .582	10 .540	11 .498	12 .456	13 .415	14 .373	15 .331	16 .289	17 .247	18 .205	19 .164	38
24	.958	1 .916	2 .875	3 .833	4 .792	5 .750	6 .709	7 .667	8 .625	9 .584	10 .542	11 .500	12 .459	13 .417	14 .375	15 .334	16 .292	17 .250	18 .208	19 .167	36
26	.959	1 .917	2 .876	3 .834	4 .793	5 .751	6 .710	7 .669	8 .627	9 .585	10 .544	11 .502	12 .461	13 .420	14 .378	15 .337	16 .295	17 .253	18 .211	19 .170	34
28	.959	1 .917	2 .876	3 .834	4 .794	5 .752	6 .711	7 .670	8 .628	9 .587	10 .546	11 .504	12 .463	13 .422	14 .381	15 .340	16 .298	17 .256	18 .214	19 .173	32
30	.959	1 .918	2 .877	3 .835	4 .794	5 .753	6 .712	7 .671	8 .629	9 .588	10 .547	11 .506	12 .465	13 .424	14 .383	15 .342	16 .300	17 .258	18 .217	19 .176	30
32	.959	1 .918	2 .877	3 .836	4 .795	5 .754	6 .714	7 .673	8 .631	9 .590	10 .549	11 .508	12 .468	13 .427	14 .386	15 .345	16 .303	17 .262	18 .221	19 .180	28
34	.959	1 .918	2 .878	3 .837	4 .796	5 .755	6 .715	7 .674	8 .633	9 .592	10 .551	11 .510	12 .470	13 .429	14 .388	15 .347	16 .306	17 .265	18 .224	19 .183	26
36	.959	1 .918	2 .878	3 .837	4 .797	5 .756	6 .716	7 .675	8 .634	9 .593	10 .553	11 .512	12 .472	13 .432	14 .391	15 .350	16 .309	17 .268	18 .227	19 .186	24
38	.960	1 .919	2 .879	3 .838	4 .797	5 .757	6 .717	7 .676	8 .636	9 .595	10 .555	11 .514	12 .474	13 .434	14 .393	15 .352	16 .312	17 .271	18 .230	19 .189	22
40	.960	1 .919	2 .879	3 .839	4 .798	5 .758	6 .718	7 .677	8 .637	9 .596	10 .556	11 .516	12 .476	13 .436	14 .395	15 .354	16 .314	17 .274	18 .233	19 .192	20
42	.960	1 .920	2 .879	3 .840	4 .799	5 .759	6 .719	7 .679	8 .639	9 .598	10 .558	11 .518	12 .478	13 .438	14 .398	15 .357	16 .317	17 .277	18 .237	19 .196	18
44	.960	1 .920	2 .880	3 .841	4 .800	5 .760	6 .720	7 .680	8 .641	9 .600	10 .560	11 .520	12 .480	13 .440	14 .400	15 .360	16 .320	17 .280	18 .240	19 .200	16
46	.960	1 .920	2 .880	3 .841	4 .800	5 .761	6 .721	7 .682	8 .642	9 .602	10 .562	11 .522	12 .482	13 .442	14 .403	15 .363	16 .323	17 .283	18 .243	19 .204	14
48	.960	1 .921	2 .881	3 .842	4 .801	5 .762	6 .722	7 .683	8 .643	9 .603	10 .564	11 .524	12 .484	13 .444	14 .405	15 .366	16 .326	17 .286	18 .246	19 .207	12
50	.961	1 .921	2 .881	3 .842	4 .802	5 .763	6 .723	7 .684	8 .644	9 .605	10 .566	11 .526	12 .486	13 .446	14 .407	15 .368	16 .328	17 .288	18 .249	19 .210	10
52	.961	1 .921	2 .882	3 .843	4 .803	5 .764	6 .725	7 .686	8 .646	9 .607	10 .568	11 .528	12 .489	13 .449	14 .410	15 .371	16 .331	17 .292	18 .252	19 .214	8
54	.961	1 .922	2 .882	3 .843	4 .804	5 .765	6 .726	7 .687	8 .647	9 .609	10 .570	11 .530	12 .491	13 .452	14 .412	15 .373	16 .334	17 .295	18 .255	19 .217	6
56	.961	1 .922	2 .883	3 .844	4 .805	5 .766	6 .727	7 .688	8 .649	9 .610	10 .571	11 .532	12 .493	13 .454	14 .415	15 .376	16 .337	17 .298	18 .258	19 .220	4
58	.961	1 .923	2 .883	3 .844	4 .806	5 .767	6 .728	7 .689	8 .650	9 .612	10 .573	11 .534	12 .495	13 .456	14 .417	15 .378	16 .339	17 .300	18 .261	19 .223	2
14° 0	.961	1 .923	2 .884	3 .845	4 .806	5 .768	6 .729	7 .690	8 .651	9 .613	10 .574	11 .536	12 .497	13 .458	14 .419	15 .380	16 .341	17 .302	18 .264	19 .226	16° 0
	1	2	3	4	5	6	7	8	9	10	11	12	13	14	15	16	17	18	19	20	III + IX −

TABLE XVIII.

GENERAL TABLE of Aberration, &c. continued.

S. −VIII +II	1	2	3	4	5	6	7	8	9	10	11	12	13	14	15	16	17	18	19	20	
14° 0′	.961	1.923	2.884	3.845	4.806	5.768	6.729	7.690	8.651	9.613	10.574	11.536	12.497	13.458	14.419	15.380	16.341	17.302	18.264	19.226	16° 0′
2	.961	1.923	2.884	3.846	4.807	5.769	6.730	7.691	8.653	9.615	10.576	11.538	12.499	13.460	14.422	15.383	16.344	17.305	18.267	19.230	58
4	.962	1.923	2.885	3.846	4.808	5.770	6.731	7.693	8.654	9.617	10.578	11.540	12.501	13.462	14.424	15.386	16.347	17.308	18.270	19.233	56
6	.962	1.924	2.885	3.847	4.809	5.771	6.732	7.695	8.656	9.618	10.580	11.541	12.503	13.464	14.427	15.388	16.349	17.311	18.273	19.236	54
8	.962	1.924	2.886	3.847	4.809	5.772	6.733	7.696	8.657	9.620	10.582	11.543	12.505	13.466	14.429	15.391	16.352	17.314	18.276	19.239	52
10	.962	1.924	2.886	3.848	4.810	5.772	6.734	7.697	8.658	9.621	10.583	11.544	12.506	13.468	14.431	15.394	16.355	17.316	18.279	19.242	50
12	.962	1.924	2.887	3.849	4.811	5.773	6.736	7.699	8.660	9.623	10.585	11.546	12.509	13.471	14.434	15.397	16.358	17.320	18.283	19.246	48
14	.963	1.925	2.887	3.849	4.812	5.774	6.737	7.700	8.661	9.625	10.587	11.548	12.511	13.473	14.436	15.399	16.361	17.323	18.286	19.249	46
16	.963	1.925	2.888	3.850	4.813	5.775	6.738	7.701	8.663	9.626	10.589	11.550	12.513	13.476	14.439	15.402	16.364	17.326	18.289	19.252	44
18	.963	1.926	2.888	3.850	4.813	5.776	6.739	7.702	8.664	9.628	10.591	11.552	12.515	13.478	14.441	15.404	16.366	17.329	18.292	19.255	42
20	.963	1.926	2.889	3.851	4.814	5.777	6.740	7.703	8.666	9.629	10.592	11.554	12.517	13.480	14.443	15.406	16.369	17.332	18.295	19.258	40
22	.963	1.926	2.889	3.852	4.815	5.778	6.741	7.705	8.668	9.631	10.594	11.556	12.519	13.482	14.446	15.409	16.372	17.335	18.298	19.261	38
24	.963	1.926	2.890	3.853	4.816	5.779	6.742	7.706	8.669	9.632	10.596	11.558	12.521	13.484	14.448	15.411	16.375	17.338	18.301	19.264	36
26	.963	1.926	2.890	3.853	4.817	5.780	6.743	7.707	8.671	9.634	10.597	11.560	12.523	13.486	14.450	15.414	16.377	17.341	18.304	19.267	34
28	.964	1.927	2.891	3.854	4.817	5.781	6.744	7.708	8.672	9.635	10.599	11.562	12.525	13.488	14.452	15.416	16.380	17.344	18.307	19.270	32
30	.964	1.927	2.891	3.855	4.818	5.782	6.745	7.709	8.673	9.636	10.600	11.564	12.527	13.490	14.454	15.418	16.382	17.346	18.309	19.272	30
32	.964	1.927	2.891	3.856	4.819	5.783	6.747	7.711	8.675	9.638	10.602	11.566	12.529	13.493	14.457	15.421	16.385	17.349	18.312	19.276	28
34	.964	1.928	2.892	3.856	4.820	5.784	6.748	7.712	8.676	9.640	10.604	11.568	12.531	13.495	14.459	15.423	16.388	17.352	18.315	19.279	26
36	.964	1.928	2.892	3.857	4.821	5.785	6.749	7.713	8.678	9.641	10.605	11.570	12.533	13.498	14.462	15.426	16.390	17.355	18.318	19.282	24
38	.964	1.928	2.893	3.857	4.821	5.786	6.750	7.714	8.679	9.643	10.607	11.571	12.535	13.500	14.464	15.428	16.393	17.358	18.321	19.285	22
40	.964	1.929	2.893	3.858	4.822	5.786	6.751	7.715	8.680	9.644	10.608	11.572	12.537	13.502	14.466	15.430	16.395	17.360	18.324	19.288	20
42	.965	1.929	2.894	3.859	4.823	5.787	6.752	7.717	8.682	9.646	10.610	11.574	12.539	13.504	14.469	15.433	16.398	17.363	18.327	19.292	18
44	.965	1.929	2.894	3.859	4.824	5.788	6.753	7.718	8.683	9.647	10.612	11.576	12.541	13.506	14.471	15.435	16.400	17.366	18.330	19.295	16
46	.965	1.929	2.895	3.860	4.825	5.789	6.754	7.719	8.684	9.649	10.614	11.578	12.543	13.508	14.473	15.438	16.403	17.369	18.333	19.298	14
48	.965	1.930	2.895	3.860	4.825	5.790	6.755	7.720	8.686	9.650	10.615	11.580	12.545	13.510	14.475	15.440	16.405	17.372	18.336	19.301	12
50	.965	1.930	2.896	3.861	4.826	5.791	6.756	7.721	8.687	9.652	10.617	11.582	12.547	13.512	14.477	15.442	16.408	17.374	18.339	19.304	10
52	.965	1.930	2.896	3.862	4.827	5.792	6.758	7.723	8.689	9.654	10.619	11.584	12.550	13.515	14.480	15.445	16.411	17.377	18.342	19.307	8
54	.965	1.931	2.897	3.862	4.828	5.793	6.759	7.724	8.690	9.655	10.621	11.586	12.552	13.517	14.482	15.447	16.413	17.379	18.345	19.310	6
56	.966	1.931	2.897	3.863	4.829	5.794	6.760	7.725	8.691	9.657	10.622	11.588	12.554	13.520	14.485	15.450	16.415	17.382	18.347	19.313	4
58	.966	1.931	2.898	3.863	4.829	5.795	6.761	7.726	8.692	9.658	10.624	11.590	12.556	13.522	14.487	15.452	16.418	17.384	18.350	19.316	2
15° 0′	.966	1.932	2.898	3.864	4.830	5.796	6.762	7.727	8.693	9.659	10.625	11.592	12.558	13.524	14.489	15.454	16.420	17.386	18.352	19.318	15° 0′
2	.966	1.932	2.898	3.865	4.831	5.797	6.763	7.729	8.695	9.661	10.627	11.594	12.560	13.526	14.492	15.457	16.423	17.389	18.355	19.322	58
4	.966	1.932	2.899	3.865	4.831	5.798	6.764	7.730	8.696	9.662	10.629	11.596	12.562	13.528	14.494	15.459	16.426	17.391	18.358	19.325	56
6	.966	1.932	2.899	3.866	4.832	5.799	6.765	7.731	8.697	9.664	10.631	11.598	12.564	13.530	14.496	15.462	16.428	17.394	18.361	19.328	54
8	.967	1.933	2.900	3.866	4.833	5.799	6.766	7.732	8.699	9.666	10.632	11.599	12.566	13.532	14.498	15.464	16.430	17.396	18.364	19.331	52
10	.967	1.933	2.900	3.867	4.833	5.800	6.767	7.733	8.700	9.667	10.634	11.600	12.567	13.534	14.500	15.466	16.433	17.398	18.367	19.334	50
12	.967	1.933	2.900	3.868	4.834	5.801	6.768	7.735	8.702	9.669	10.636	11.602	12.569	13.536	14.503	15.469	16.436	17.402	18.370	19.337	48
14	.967	1.934	2.901	3.868	4.835	5.802	6.769	7.736	8.703	9.670	10.637	11.604	12.571	13.538	14.505	15.471	16.439	17.405	18.373	19.340	46
16	.967	1.934	2.901	3.869	4.835	5.803	6.770	7.737	8.705	9.672	10.639	11.606	12.573	13.540	14.507	15.474	16.441	17.408	18.376	19.343	44
18	.967	1.934	2.902	3.869	4.836	5.804	6.771	7.738	8.706	9.673	10.640	11.608	12.575	13.542	14.509	15.476	16.444	17.411	18.379	19.346	42
20	.967	1.935	2.902	3.870	4.837	5.805	6.772	7.739	8.707	9.674	10.641	11.610	12.577	13.544	14.511	15.478	16.446	17.414	18.381	19.348	40
22	.968	1.935	2.903	3.871	4.838	5.806	6.773	7.741	8.709	9.676	10.643	11.612	12.579	13.546	14.514	15.481	16.449	17.417	18.384	19.352	38
24	.968	1.935	2.903	3.871	4.839	5.807	6.774	7.742	8.710	9.677	10.645	11.614	12.581	13.548	14.516	15.483	16.451	17.420	18.387	19.355	36
26	.968	1.935	2.904	3.872	4.840	5.808	6.775	7.743	8.711	9.679	10.647	11.616	12.583	13.550	14.518	15.486	16.454	17.422	18.390	19.358	34
28	.968	1.936	2.904	3.872	4.840	5.809	6.776	7.744	8.712	9.681	10.648	11.617	12.585	13.552	14.520	15.488	16.456	17.424	18.393	19.361	32
30	.968	1.936	2.905	3.873	4.841	5.809	6.777	7.745	8.713	9.682	10.650	11.618	12.586	13.554	14.522	15.490	16.458	17.426	18.395	19.364	30
32	.968	1.936	2.905	3.874	4.842	5.810	6.778	7.747	8.715	9.684	10.652	11.620	12.588	13.556	14.525	15.493	16.461	17.429	18.398	19.367	28
34	.969	1.937	2.906	3.874	4.842	5.811	6.779	7.748	8.716	9.685	10.654	11.622	12.590	13.558	14.527	15.495	16.464	17.432	18.401	19.370	26
36	.969	1.937	2.906	3.875	4.843	5.812	6.780	7.749	8.718	9.687	10.655	11.623	12.592	13.560	14.529	15.498	16.466	17.435	18.404	19.373	24
38	.969	1.937	2.907	3.875	4.843	5.813	6.781	7.750	8.719	9.688	10.657	11.625	12.594	13.562	14.531	15.500	16.469	17.438	18.407	19.376	22
40	.969	1.938	2.907	3.876	4.844	5.813	6.782	7.751	8.720	9.689	10.658	11.626	12.595	13.564	14.533	15.502	16.471	17.440	18.409	19.378	20
42	.969	1.938	2.907	3.876	4.845	5.814	6.783	7.753	8.722	9.691	10.660	11.628	12.597	13.566	14.536	15.505	16.474	17.443	18.412	19.381	18
44	.969	1.938	2.908	3.877	4.846	5.815	6.784	7.754	8.723	9.692	10.662	11.630	12.599	13.568	14.538	15.507	16.476	17.445	18.415	19.384	16
46	.969	1.938	2.908	3.877	4.847	5.816	6.785	7.755	8.725	9.694	10.663	11.632	12.601	13.570	14.540	15.510	16.479	17.447	18.417	19.387	14
48	.970	1.939	2.909	3.878	4.847	5.817	6.786	7.756	8.726	9.695	10.665	11.634	12.603	13.572	14.542	15.512	16.481	17.450	18.420	19.390	12
50	.970	1.939	2.909	3.878	4.848	5.818	6.787	7.757	8.726	9.696	10.666	11.636	12.605	13.574	14.544	15.514	16.483	17.452	18.422	19.392	10
52	.970	1.939	2.909	3.879	4.849	5.819	6.788	7.758	8.728	9.698	10.668	11.638	12.607	13.576	14.546	15.516	16.486	17.455	18.425	19.395	8
54	.970	1.940	2.910	3.879	4.850	5.820	6.789	7.759	8.729	9.699	10.669	11.640	12.609	13.578	14.548	15.518	16.488	17.458	18.428	19.398	6
56	.970	1.940	2.910	3.880	4.851	5.821	6.790	7.760	8.731	9.701	10.671	11.642	12.611	13.580	14.550	15.520	16.491	17.460	18.430	19.401	4
58	.970	1.941	2.911	3.880	4.851	5.822	6.791	7.761	8.732	9.702	10.672	11.643	12.613	13.582	14.552	15.522	16.493	17.463	18.433	19.404	2
16° 0	.970	1.941	2.911	3.881	4.852	5.822	6.792	7.762	8.733	9.703	10.673	11.644	12.614	13.584	14.554	15.524	16.495	17.466	18.436	19.406	14° 0
	1	2	3	4	5	6	7	8	9	10	11	12	13	14	15	16	17	18	19	20	III+ IX−

TABLE XVIII.

GENERAL TABLE of Aberration, &c. continued.

S. −VIII +II	1	2	3	4	5	6	7	8	9	10	11	12	13	14	15	16	17	18	19	20	
16° 0′	.970	1.941	2.911	3.881	4.852	5.822	6.792	7.762	8.733	9.703	10.673	11.644	12.614	13.584	14.554	15.524	16.495	17.466	18.436	19.406	14° 0′
2	.970	1.941	2.912	3.882	4.853	5.823	6.793	7.764	8.735	9.705	10.675	11.646	12.616	13.586	14.557	15.527	16.498	17.469	18.439	19.409	58
4	.970	1.941	2.912	3.883	4.853	5.824	6.794	7.765	8.736	9.706	10.677	11.648	12.618	13.588	14.559	15.529	16.500	17.471	18.442	19.412	56
6	.971	1.941	2.912	3.883	4.854	5.825	6.795	7.766	8.737	9.708	10.678	11.649	12.620	13.590	14.561	15.532	16.503	17.474	18.444	19.414	54
8	.971	1.942	2.913	3.884	4.854	5.826	6.796	7.767	8.738	9.709	10.679	11.651	12.621	13.592	14.563	15.534	16.505	17.476	18.447	19.417	52
10	.971	1.942	2.913	3.884	4.855	5.826	6.797	7.768	8.739	9.710	10.681	11.652	12.623	13.594	14.565	15.536	16.507	17.478	18.449	19.420	50
12	.971	1.942	2.913	3.885	4.856	5.827	6.798	7.770	8.741	9.712	10.683	11.654	12.625	13.596	14.568	15.539	16.510	17.481	18.452	19.423	48
14	.971	1.942	2.914	3.885	4.857	5.828	6.799	7.771	8.742	9.713	10.685	11.656	12.627	13.598	14.570	15.541	16.512	17.483	18.455	19.426	46
16	.971	1.942	2.914	3.886	4.857	5.829	6.800	7.772	8.743	9.715	10.686	11.657	12.629	13.600	14.572	15.544	16.515	17.486	18.457	19.428	44
18	.972	1.943	2.914	3.886	4.858	5.829	6.801	7.773	8.744	9.716	10.688	11.659	12.630	13.602	14.574	15.546	16.517	17.488	18.460	19.431	42
20	.972	1.943	2.915	3.887	4.859	5.830	6.802	7.774	8.745	9.717	10.689	11.660	12.632	13.604	14.576	15.548	16.519	17.490	18.462	19.434	40
22	.972	1.943	2.915	3.888	4.860	5.831	6.803	7.775	8.747	9.719	10.691	11.662	12.634	13.606	14.578	15.550	16.522	17.493	18.465	19.437	38
24	.972	1.944	2.916	3.888	4.861	5.832	6.804	7.776	8.748	9.720	10.692	11.664	12.636	13.608	14.580	15.552	16.524	17.495	18.468	19.440	36
26	.972	1.944	2.916	3.889	4.861	5.833	6.805	7.777	8.749	9.722	10.694	11.665	12.638	13.610	14.582	15.554	16.526	17.498	18.470	19.442	34
28	.972	1.944	2.916	3.889	4.862	5.834	6.806	7.778	8.750	9.723	10.695	11.667	12.639	13.612	14.584	15.556	16.528	17.500	18.473	19.445	32
30	.972	1.945	2.917	3.890	4.862	5.834	6.807	7.779	8.751	9.724	10.696	11.668	12.641	13.614	14.586	15.558	16.530	17.502	18.475	19.448	30
32	.973	1.945	2.917	3.890	4.863	5.835	6.808	7.780	8.753	9.726	10.698	11.670	12.643	13.616	14.588	15.560	16.533	17.505	18.478	19.451	28
34	.973	1.945	2.918	3.891	4.863	5.836	6.809	7.781	8.754	9.727	10.699	11.672	12.645	13.618	14.590	15.562	16.535	17.507	18.480	19.453	26
36	.973	1.945	2.918	3.891	4.864	5.837	6.810	7.782	8.755	9.728	10.701	11.673	12.646	13.619	14.592	15.564	16.537	17.510	18.483	19.456	24
38	.973	1.946	2.919	3.892	4.864	5.837	6.811	7.783	8.756	9.729	10.702	11.675	12.648	13.621	14.594	15.566	16.539	17.512	18.485	19.458	22
40	.973	1.946	2.919	3.892	4.865	5.838	6.811	7.784	8.757	9.730	10.703	11.676	12.649	13.622	14.596	15.568	16.541	17.514	18.487	19.460	20
42	.973	1.946	2.919	3.893	4.866	5.839	6.812	7.786	8.759	9.732	10.705	11.678	12.651	13.624	14.598	15.571	16.544	17.517	18.490	19.463	18
44	.973	1.946	2.920	3.893	4.867	5.840	6.813	7.787	8.760	9.733	10.707	11.680	12.653	13.626	14.600	15.573	16.546	17.519	18.493	19.466	16
46	.973	1.946	2.920	3.894	4.867	5.841	6.814	7.788	8.761	9.735	10.708	11.681	12.655	13.628	14.602	15.576	16.549	17.522	18.495	19.469	14
48	.974	1.947	2.921	3.894	4.868	5.841	6.815	7.789	8.762	9.736	10.709	11.683	12.656	13.630	14.604	15.578	16.551	17.524	18.498	19.472	12
50	.974	1.947	2.921	3.895	4.869	5.842	6.816	7.790	8.763	9.737	10.711	11.684	12.658	13.632	14.606	15.580	16.553	17.526	18.500	19.475	10
52	.974	1.947	2.921	3.896	4.870	5.843	6.817	7.791	8.765	9.739	10.713	11.686	12.660	13.634	14.608	15.582	16.556	17.529	18.503	19.478	8
54	.974	1.948	2.922	3.896	4.870	5.844	6.818	7.792	8.766	9.740	10.714	11.688	12.662	13.636	14.610	15.584	16.558	17.531	18.506	19.480	6
56	.974	1.948	2.922	3.897	4.871	5.845	6.819	7.793	8.767	9.742	10.716	11.689	12.664	13.638	14.612	15.586	16.560	17.534	18.508	19.483	4
58	.974	1.948	2.923	3.897	4.871	5.845	6.820	7.794	8.768	9.743	10.717	11.690	12.665	13.640	14.614	15.588	16.562	17.536	18.511	19.485	2
17° 0′	.974	1.949	2.923	3.898	4.872	5.846	6.821	7.795	8.769	9.744	10.718	11.692	12.667	13.642	14.616	15.590	16.564	17.538	18.513	19.488	13° 0
2	.974	1.949	2.923	3.898	4.873	5.847	6.822	7.796	8.771	9.746	10.720	11.694	12.669	13.644	14.618	15.592	16.567	17.541	18.516	19.491	58
4	.975	1.949	2.924	3.899	4.873	5.848	6.823	7.797	8.772	9.747	10.721	11.696	12.671	13.646	14.620	15.594	16.569	17.543	18.518	19.493	56
6	.975	1.949	2.924	3.899	4.874	5.849	6.824	7.798	8.773	9.748	10.723	11.698	12.672	13.647	14.621	15.596	16.572	17.546	18.521	19.496	54
8	.975	1.950	2.925	3.900	4.874	5.849	6.825	7.799	8.774	9.749	10.724	11.699	12.674	13.649	14.623	15.598	16.574	17.548	18.523	19.498	52
10	.975	1.950	2.925	3.900	4.875	5.850	6.825	7.800	8.775	9.750	10.725	11.700	12.675	13.650	14.625	15.600	16.576	17.550	18.525	19.500	50
12	.975	1.950	2.925	3.901	4.876	5.851	6.826	7.801	8.777	9.752	10.727	11.702	12.677	13.652	14.627	15.602	16.578	17.553	18.528	19.503	48
14	.975	1.950	2.926	3.901	4.876	5.852	6.827	7.802	8.778	9.753	10.729	11.704	12.679	13.654	14.629	15.604	16.580	17.555	18.531	19.506	46
16	.975	1.950	2.926	3.902	4.877	5.853	6.828	7.803	8.779	9.754	10.730	11.705	12.681	13.656	14.631	15.606	16.582	17.558	18.534	19.509	44
18	.976	1.951	2.926	3.902	4.877	5.854	6.829	7.804	8.780	9.756	10.732	11.707	12.682	13.658	14.633	15.608	16.584	17.560	18.536	19.511	42
20	.976	1.951	2.927	3.903	4.878	5.854	6.830	7.805	8.781	9.757	10.733	11.708	12.684	13.660	14.635	15.610	16.586	17.562	18.538	19.514	40
22	.976	1.951	2.927	3.903	4.879	5.855	6.831	7.806	8.783	9.759	10.735	11.710	12.686	13.662	14.637	15.612	16.589	17.565	18.541	19.517	38
24	.976	1.952	2.927	3.904	4.880	5.856	6.832	7.807	8.784	9.760	10.736	11.712	12.688	13.664	14.639	15.614	16.591	17.567	18.543	19.519	36
26	.976	1.952	2.928	3.904	4.880	5.857	6.833	7.808	8.785	9.761	10.737	11.713	12.689	13.665	14.641	15.616	16.593	17.570	18.546	19.522	34
28	.976	1.952	2.928	3.905	4.881	5.858	6.834	7.809	8.786	9.762	10.738	11.715	12.691	13.667	14.642	15.618	16.595	17.572	18.548	19.524	32
30	.976	1.953	2.929	3.905	4.882	5.858	6.834	7.810	8.787	9.763	10.739	11.716	12.692	13.668	14.644	15.620	16.597	17.574	18.550	19.526	30
32	.976	1.953	2.929	3.906	4.883	5.859	6.835	7.811	8.788	9.765	10.741	11.718	12.694	13.670	14.646	15.622	16.599	17.576	18.553	19.529	28
34	.977	1.953	2.930	3.906	4.883	5.860	6.836	7.812	8.789	9.766	10.742	11.720	12.696	13.672	14.648	15.624	16.601	17.578	18.555	19.531	26
36	.977	1.953	2.930	3.907	4.884	5.861	6.837	7.813	8.790	9.767	10.744	11.721	12.697	13.673	14.650	15.626	16.603	17.580	18.557	19.534	24
38	.977	1.954	2.930	3.907	4.884	5.861	6.838	7.814	8.791	9.768	10.745	11.723	12.699	13.675	14.651	15.628	16.605	17.582	18.559	19.536	22
40	.977	1.954	2.931	3.908	4.885	5.862	6.838	7.815	8.792	9.769	10.746	11.724	12.700	13.676	14.653	15.630	16.607	17.584	18.561	19.538	20
42	.977	1.954	2.931	3.908	4.886	5.863	6.839	7.816	8.794	9.771	10.748	11.726	12.702	13.678	14.655	15.632	16.610	17.587	18.564	19.541	18
44	.977	1.954	2.932	3.909	4.886	5.863	6.840	7.817	8.795	9.772	10.749	11.727	12.704	13.680	14.657	15.634	16.612	17.589	18.566	19.543	16
46	.977	1.954	2.932	3.909	4.887	5.864	6.841	7.818	8.796	9.773	10.751	11.728	12.705	13.682	14.659	15.636	16.614	17.591	18.569	19.546	14
48	.977	1.955	2.932	3.910	4.887	5.864	6.842	7.819	8.797	9.774	10.752	11.729	12.707	13.684	14.661	15.638	16.616	17.593	18.571	19.548	12
50	.978	1.955	2.933	3.910	4.888	5.865	6.843	7.820	8.798	9.775	10.753	11.730	12.708	13.686	14.663	15.640	16.618	17.596	18.573	19.550	10
52	.978	1.955	2.933	3.911	4.889	5.866	6.844	7.821	8.799	9.777	10.755	11.732	12.710	13.688	14.665	15.642	16.620	17.598	18.576	19.553	8
54	.978	1.955	2.934	3.911	4.889	5.867	6.845	7.822	8.800	9.778	10.756	11.734	12.712	13.690	14.667	15.644	16.622	17.600	18.578	19.555	6
56	.978	1.955	2.934	3.912	4.890	5.868	6.846	7.823	8.801	9.780	10.758	11.735	12.713	13.692	14.669	15.646	16.624	17.602	18.581	19.558	4
58	.978	1.956	2.934	3.912	4.890	5.868	6.847	7.824	8.802	9.781	10.759	11.737	12.715	13.693	14.671	15.648	16.626	17.604	18.583	19.561	2
18° 0	.978	1.956	2.935	3.913	4.891	5.869	6.847	7.825	8.803	9.782	10.760	11.738	12.716	13.694	14.672	15.650	16.628	17.606	18.585	19.564	12° 0
	1	2	3	4	5	6	7	8	9	10	11	12	13	14	15	16	17	18	19	20	III+ IX−

TABLE XVIII.

GENERAL TABLE of Aberration, &c. continued.

S. −VIII +II	1	2	3	4	5	6	7	8	9	10	11	12	13	14	15	16	17	18	19	20	
18° 0′	.978	1.956	2.935	3.913	4.891	5.869	6.847	7.825	8.803	9.782	10.760	11.738	12.716	13.694	14.672	15.650	16.628	17.606	18.585	19.564	12° 0′
2	.978	1.956	2.935	3.913	4.892	5.870	6.848	7.826	8.805	9.784	10.762	11.740	12.718	13.696	14.674	15.652	16.631	17.609	18.588	19.567	58
4	.978	1.957	2.935	3.914	4.892	5.871	6.849	7.827	8.806	9.785	10.763	11.742	12.720	13.698	14.676	15.654	16.633	17.611	18.590	19.569	56
6	.979	1.957	2.935	3.914	4.893	5.872	6.850	7.828	8.807	9.786	10.765	11.743	12.721	13.699	14.678	15.656	16.635	17.614	18.593	19.572	54
8	.979	1.958	2.936	3.914	4.893	5.872	6.850	7.829	8.808	9.787	10.766	11.745	12.723	13.701	14.680	15.658	16.637	17.616	18.595	19.574	52
10	.979	1.958	2.936	3.915	4.894	5.873	6.851	7.830	8.809	9.788	10.767	11.746	12.724	13.702	14.681	15.660	16.639	17.618	18.597	19.576	50
12	.979	1.958	2.936	3.915	4.895	5.874	6.852	7.831	8.810	9.789	10.768	11.748	12.726	13.704	14.683	15.662	16.641	17.620	18.600	19.578	48
14	.979	1.958	2.937	3.916	4.895	5.874	6.853	7.832	8.811	9.790	10.769	11.749	12.727	13.706	14.685	15.664	16.643	17.622	18.602	19.580	46
16	.979	1.958	2.937	3.916	4.896	5.875	6.854	7.833	8.812	9.791	10.770	11.750	12.729	13.707	14.687	15.666	16.645	17.624	18.604	19.582	44
18	.979	1.959	2.937	3.917	4.896	5.875	6.854	7.834	8.813	9.792	10.771	11.751	12.730	13.709	14.689	15.668	16.647	17.626	18.606	19.584	42
20	.979	1.959	2.938	3.917	4.897	5.876	6.855	7.835	8.814	9.793	10.772	11.752	12.731	13.710	14.690	15.670	16.649	17.628	18.608	19.586	40
22	.979	1.959	2.938	3.918	4.898	5.877	6.856	7.836	8.815	9.795	10.774	11.754	12.733	13.712	14.692	15.672	16.651	17.630	18.610	19.589	38
24	.980	1.959	2.939	3.918	4.898	5.878	6.857	7.837	8.816	9.796	10.775	11.756	12.735	13.714	14.694	15.674	16.653	17.632	18.612	19.591	36
26	.980	1.959	2.939	3.919	4.899	5.879	6.858	7.838	8.817	9.797	10.777	11.757	12.736	13.715	14.696	15.676	16.655	17.634	18.614	19.594	34
28	.980	1.960	2.939	3.919	4.899	5.879	6.858	7.839	8.818	9.798	10.778	11.759	12.738	13.717	14.697	15.678	16.657	17.636	18.616	19.596	32
30	.980	1.960	2.940	3.920	4.900	5.880	6.859	7.839	8.819	9.799	10.779	11.760	12.739	13.718	14.698	15.679	16.659	17.638	18.618	19.598	30
32	.980	1.960	2.940	3.920	4.901	5.881	6.860	7.840	8.821	9.801	10.781	11.762	12.741	13.720	14.700	15.681	16.661	17.641	18.621	19.601	28
34	.980	1.960	2.941	3.921	4.901	5.881	6.861	7.841	8.822	9.802	10.782	11.763	12.743	13.722	14.702	15.683	16.663	17.643	18.623	19.603	26
36	.980	1.960	2.941	3.921	4.902	5.882	6.862	7.842	8.823	9.803	10.784	11.764	12.744	13.724	14.704	15.685	16.665	17.646	18.626	19.606	24
38	.980	1.961	2.941	3.921	4.902	5.882	6.863	7.843	8.824	9.804	10.785	11.765	12.746	13.726	14.706	15.687	16.667	17.648	18.628	19.608	22
40	.981	1.961	2.942	3.922	4.903	5.883	6.864	7.844	8.825	9.805	10.786	11.766	12.747	13.728	14.708	15.688	16.669	17.650	18.630	19.610	20
42	.981	1.961	2.942	3.922	4.904	5.884	6.865	7.845	8.826	9.807	10.788	11.768	12.749	13.730	14.710	15.690	16.671	17.652	18.633	19.613	18
44	.981	1.961	2.942	3.923	4.904	5.884	6.866	7.846	8.827	9.808	10.789	11.769	12.750	13.732	14.712	15.692	16.673	17.654	18.635	19.615	16
46	.981	1.961	2.942	3.923	4.904	5.885	6.867	7.847	8.828	9.809	10.790	11.770	12.752	13.733	14.714	15.694	16.675	17.656	18.637	19.617	14
48	.981	1.962	2.943	3.924	4.905	5.885	6.867	7.848	8.829	9.810	10.791	11.771	12.753	13.735	14.716	15.696	16.677	17.658	18.639	19.620	12
50	.981	1.962	2.943	3.924	4.905	5.886	6.868	7.849	8.830	9.811	10.792	11.772	12.754	13.736	14.717	15.698	16.679	17.660	18.641	19.622	10
52	.981	1.962	2.943	3.925	4.906	5.887	6.869	7.850	8.831	9.812	10.794	11.774	12.756	13.738	14.719	15.700	16.681	17.662	18.643	19.624	8
54	.981	1.962	2.944	3.925	4.906	5.888	6.869	7.851	8.832	9.813	10.795	11.776	12.757	13.739	14.721	15.702	16.683	17.664	18.645	19.626	6
56	.981	1.962	2.944	3.926	4.907	5.889	6.870	7.852	8.833	9.814	10.796	11.777	12.759	13.740	14.722	15.703	16.685	17.666	18.647	19.628	4
58	.982	1.963	2.944	3.926	4.907	5.889	6.871	7.853	8.834	9.815	10.797	11.779	12.760	13.741	14.723	15.705	16.687	17.668	18.649	19.630	2
19° 0′	.982	1.963	2.945	3.927	4.908	5.890	6.871	7.853	8.835	9.816	10.798	11.780	12.761	13.742	14.724	15.706	16.688	17.670	18.651	19.632	11° 0′
2	.982	1.963	2.945	3.927	4.909	5.891	6.872	7.854	8.836	9.818	10.800	11.782	12.763	13.744	14.726	15.708	16.690	17.672	18.654	19.635	58
4	.982	1.963	2.946	3.928	4.909	5.891	6.873	7.855	8.837	9.819	10.801	11.783	12.764	13.746	14.728	15.710	16.692	17.674	18.656	19.637	56
6	.982	1.963	2.946	3.928	4.910	5.892	6.874	7.856	8.838	9.820	10.802	11.784	12.766	13.747	14.729	15.711	16.694	17.676	18.658	19.639	54
8	.982	1.964	2.946	3.928	4.910	5.892	6.874	7.857	8.839	9.821	10.803	11.785	12.767	13.749	14.731	15.713	16.696	17.678	18.660	19.642	52
10	.982	1.964	2.947	3.929	4.911	5.893	6.875	7.857	8.840	9.822	10.804	11.786	12.768	13.750	14.732	15.714	16.697	17.680	18.662	19.644	50
12	.982	1.964	2.947	3.929	4.912	5.894	6.876	7.858	8.841	9.823	10.806	11.788	12.770	13.752	14.735	15.716	16.699	17.682	18.664	19.646	48
14	.982	1.964	2.947	3.930	4.912	5.894	6.877	7.859	8.842	9.824	10.807	11.789	12.771	13.754	14.737	15.718	16.701	17.684	18.666	19.648	46
16	.982	1.964	2.947	3.930	4.913	5.895	6.878	7.860	8.843	9.825	10.808	11.790	12.773	13.755	14.739	15.720	16.703	17.686	18.668	19.650	44
18	.983	1.965	2.948	3.930	4.913	5.895	6.878	7.861	8.844	9.826	10.809	11.791	12.774	13.757	14.740	15.722	16.705	17.688	18.670	19.652	42
20	.983	1.965	2.948	3.931	4.914	5.896	6.879	7.862	8.845	9.827	10.810	11.792	12.775	13.758	14.741	15.724	16.707	17.690	18.672	19.654	40
22	.983	1.965	2.948	3.931	4.915	5.897	6.880	7.863	8.846	9.829	10.812	11.794	12.777	13.760	14.743	15.726	16.709	17.692	18.674	19.657	38
24	.983	1.966	2.949	3.932	4.915	5.898	6.881	7.864	8.847	9.830	10.813	11.795	12.779	13.762	14.745	15.728	16.711	17.694	18.676	19.659	36
26	.983	1.966	2.949	3.932	4.915	5.899	6.882	7.865	8.848	9.831	10.814	11.797	12.780	13.763	14.746	15.729	16.712	17.696	18.678	19.662	34
28	.983	1.966	2.949	3.932	4.916	5.899	6.882	7.866	8.849	9.832	10.815	11.799	12.782	13.765	14.748	15.731	16.714	17.698	18.680	19.664	32
30	.983	1.967	2.950	3.933	4.916	5.900	6.883	7.866	8.849	9.833	10.816	11.800	12.783	13.766	14.749	15.732	16.715	17.699	18.682	19.666	30
32	.983	1.967	2.950	3.933	4.917	5.901	6.884	7.867	8.850	9.834	10.818	11.802	12.785	13.768	14.751	15.734	16.717	17.701	18.684	19.668	28
34	.984	1.967	2.950	3.934	4.917	5.901	6.885	7.868	8.851	9.835	10.819	11.803	12.786	13.770	14.753	15.736	16.719	17.703	18.686	19.670	26
36	.984	1.967	2.950	3.934	4.918	5.902	6.886	7.869	8.852	9.836	10.820	11.804	12.788	13.771	14.754	15.737	16.721	17.705	18.688	19.672	24
38	.984	1.968	2.951	3.934	4.918	5.902	6.886	7.870	8.853	9.837	10.821	11.805	12.789	13.773	14.756	15.739	16.723	17.707	18.690	19.674	22
40	.984	1.968	2.951	3.935	4.919	5.903	6.887	7.870	8.854	9.838	10.822	11.806	12.790	13.774	14.757	15.740	16.724	17.708	18.692	19.676	20
42	.984	1.968	2.951	3.935	4.920	5.904	6.888	7.871	8.855	9.839	10.823	11.808	12.792	13.776	14.759	15.742	16.726	17.710	18.694	19.678	18
44	.984	1.968	2.952	3.936	4.920	5.904	6.888	7.872	8.856	9.840	10.824	11.809	12.793	13.777	14.760	15.744	16.728	17.712	18.696	19.680	16
46	.984	1.968	2.952	3.936	4.921	5.905	6.889	7.873	8.857	9.841	10.825	11.810	12.794	13.778	14.762	15.745	16.730	17.714	18.698	19.682	14
48	.984	1.969	2.952	3.936	4.921	5.905	6.890	7.873	8.858	9.842	10.826	11.811	12.795	13.779	14.763	15.747	16.732	17.716	18.700	19.684	12
50	.984	1.969	2.953	3.937	4.922	5.906	6.890	7.874	8.859	9.843	10.827	11.812	12.796	13.780	14.764	15.748	16.733	17.718	18.702	19.686	10
52	.984	1.969	2.953	3.937	4.922	5.907	6.891	7.875	8.860	9.844	10.829	11.814	12.798	13.782	14.766	15.750	16.735	17.720	18.704	19.688	8
54	.984	1.969	2.953	3.938	4.923	5.907	6.892	7.876	8.861	9.845	10.830	11.815	12.799	13.784	14.768	15.752	16.737	17.722	18.706	19.690	6
56	.985	1.969	2.953	3.938	4.923	5.908	6.893	7.877	8.862	9.846	10.831	11.816	12.801	13.785	14.770	15.754	16.739	17.724	18.708	19.692	4
58	.985	1.970	2.954	3.939	4.924	5.908	6.894	7.878	8.863	9.847	10.832	11.817	12.802	13.787	14.772	15.756	16.741	17.726	18.710	19.694	2
20° 0	.985	1.970	2.954	3.939	4.924	5.909	6.894	7.879	8.863	9.848	10.833	11.818	12.803	13.788	14.773	15.758	16.742	17.727	18.711	19.696	10° 0
	1	2	3	4	5	6	7	8	9	10	11	12	13	14	15	16	17	18	19	20	III + IX −

TABLE XVIII.

GENERAL TABLE of Aberration, &c. continued.

S. −VIII +II	1	2	3	4	5	6	7	8	9	10	11	12	13	14	15	16	17	18	19	20	
20° 0′	.985	1.970	2.954	3.939	4.924	5.909	6.894	7.879	8.863	9.848	10.833	11.818	12.803	13.788	14.773	15.758	16.742	17.726	18.711	19.696	10° 0′
2	.985	1.970	2.954	3.939	4.925	5.910	6.895	7.880	8.864	9.849	10.834	11.820	12.805	13.790	14.775	15.760	16.744	17.728	18.713	19.698	58
4	.985	1.970	2.955	3.940	4.925	5.910	6.896	7.881	8.865	9.850	10.835	11.821	12.806	13.791	14.776	15.761	16.745	17.730	18.715	19.700	56
6	.985	1.970	2.955	3.940	4.926	5.911	6.896	7.881	8.866	9.851	10.836	11.822	12.807	13.792	14.778	15.763	16.747	17.732	18.717	19.702	54
8	.985	1.970	2.955	3.940	4.926	5.911	6.897	7.882	8.867	9.852	10.837	11.823	12.808	13.793	14.779	15.764	16.749	17.734	18.719	19.704	52
10	.985	1.971	2.956	3.941	4.927	5.912	6.897	7.883	8.868	9.853	10.838	11.824	12.809	13.794	14.780	15.765	16.750	17.736	18.721	19.706	50
12	.985	1.971	2.956	3.941	4.927	5.913	6.898	7.884	8.869	9.854	10.840	11.826	12.811	13.796	14.782	15.767	16.752	17.738	18.723	19.708	48
14	.985	1.971	2.956	3.942	4.927	5.913	6.899	7.884	8.870	9.855	10.841	11.827	12.812	13.797	14.783	15.768	16.754	17.739	18.725	19.710	46
16	.986	1.971	2.956	3.942	4.928	5.914	6.900	7.885	8.871	9.856	10.842	11.828	12.814	13.799	14.785	15.770	16.755	17.741	18.726	19.712	44
18	.986	1.972	2.957	3.943	4.928	5.914	6.900	7.885	8.871	9.857	10.843	11.829	12.815	13.801	14.786	15.771	16.757	17.743	18.728	19.714	42
20	.986	1.972	2.957	3.943	4.929	5.915	6.901	7.886	8.872	9.858	10.844	11.830	12.816	13.802	14.787	15.772	16.758	17.744	18.730	19.716	40
22	.986	1.972	2.957	3.943	4.929	5.916	6.902	7.887	8.873	9.859	10.845	11.832	12.818	13.804	14.789	15.774	16.760	17.746	18.732	19.718	38
24	.986	1.972	2.958	3.944	4.930	5.916	6.902	7.888	8.874	9.860	10.846	11.833	12.819	13.805	14.790	15.775	16.762	17.748	18.734	19.720	36
26	.986	1.972	2.958	3.944	4.930	5.917	6.903	7.889	8.875	9.861	10.847	11.834	12.820	13.806	14.792	15.777	16.763	17.750	18.736	19.722	34
28	.986	1.973	2.958	3.945	4.931	5.917	6.903	7.889	8.876	9.862	10.848	11.835	12.821	13.807	14.793	15.778	16.765	17.752	18.738	19.724	32
30	.986	1.973	2.959	3.945	4.931	5.918	6.904	7.890	8.877	9.863	10.849	11.836	12.822	13.808	14.794	15.780	16.767	17.754	18.740	19.726	30
32	.986	1.973	2.959	3.945	4.932	5.919	6.905	7.891	8.878	9.864	10.851	11.838	12.824	13.810	14.796	15.782	16.769	17.756	18.742	19.728	28
34	.986	1.973	2.959	3.946	4.932	5.919	6.905	7.892	8.879	9.865	10.852	11.839	12.825	13.811	14.797	15.783	16.770	17.757	18.744	19.730	26
36	.987	1.973	2.959	3.946	4.933	5.920	6.906	7.892	8.880	9.866	10.853	11.840	12.826	13.812	14.799	15.785	16.772	17.759	18.746	19.732	24
38	.987	1.974	2.960	3.947	4.933	5.920	6.906	7.893	8.880	9.867	10.854	11.841	12.827	13.813	14.800	15.786	16.773	17.761	18.748	19.734	22
40	.987	1.974	2.960	3.947	4.934	5.921	6.907	7.894	8.881	9.868	10.855	11.842	12.828	13.814	14.801	15.788	16.775	17.762	18.749	19.736	20
42	.987	1.974	2.960	3.947	4.934	5.921	6.908	7.895	8.882	9.869	10.856	11.843	12.830	13.816	14.803	15.790	16.777	17.764	18.751	19.738	18
44	.987	1.974	2.961	3.948	4.935	5.922	6.909	7.896	8.883	9.870	10.857	11.844	12.831	13.817	14.805	15.791	16.778	17.765	18.753	19.740	16
46	.987	1.974	2.961	3.948	4.935	5.922	6.909	7.896	8.884	9.871	10.858	11.845	12.832	13.819	14.806	15.793	16.780	17.767	18.754	19.742	14
48	.987	1.974	2.961	3.949	4.936	5.923	6.910	7.897	8.884	9.872	10.859	11.846	12.833	13.821	14.808	15.794	16.782	17.769	18.756	19.744	12
50	.987	1.974	2.962	3.949	4.936	5.923	6.911	7.898	8.885	9.872	10.860	11.847	12.834	13.822	14.809	15.796	16.783	17.770	18.757	19.745	10
52	.987	1.974	2.962	3.949	4.936	5.924	6.912	7.899	8.886	9.873	10.861	11.848	12.836	13.824	14.811	15.798	16.785	17.772	18.759	19.747	8
54	.987	1.974	2.962	3.950	4.937	5.924	6.912	7.900	8.887	9.874	10.862	11.849	12.837	13.825	14.812	15.799	16.786	17.773	18.761	19.749	6
56	.987	1.974	2.962	3.950	4.937	5.925	6.913	7.900	8.888	9.875	10.863	11.850	12.838	13.826	14.814	15.801	16.788	17.775	18.762	19.751	4
58	.988	1.975	2.963	3.951	4.938	5.925	6.913	7.901	8.889	9.876	10.864	11.851	12.839	13.827	14.815	15.802	16.790	17.777	18.764	19.753	2
21° 0′	.988	1.975	2.963	3.951	4.938	5.926	6.914	7.902	8.889	9.877	10.865	11.852	12.840	13.828	14.816	15.803	16.791	17.778	18.766	19.754	9° 0′
2	.988	1.975	2.963	3.951	4.939	5.927	6.915	7.903	8.890	9.878	10.866	11.854	12.842	13.830	14.818	15.805	16.793	17.780	18.768	19.756	58
4	.988	1.975	2.963	3.952	4.939	5.927	6.915	7.904	8.891	9.879	10.867	11.855	12.843	13.831	14.819	15.806	16.794	17.781	18.770	19.758	56
6	.988	1.975	2.963	3.952	4.940	5.928	6.916	7.904	8.892	9.880	10.868	11.856	12.844	13.832	14.820	15.808	16.796	17.783	18.771	19.760	54
8	.988	1.975	2.964	3.952	4.940	5.928	6.916	7.905	8.892	9.881	10.869	11.857	12.845	13.833	14.821	15.809	16.797	17.785	18.773	19.762	52
10	.988	1.976	2.964	3.953	4.941	5.929	6.917	7.905	8.893	9.881	10.870	11.858	12.846	13.834	14.822	15.810	16.798	17.786	18.774	19.763	50
12	.988	1.976	2.964	3.953	4.941	5.930	6.918	7.906	8.894	9.882	10.871	11.860	12.848	13.836	14.824	15.812	16.800	17.788	18.776	19.765	48
14	.988	1.976	2.965	3.953	4.942	5.930	6.918	7.907	8.895	9.883	10.872	11.861	12.849	13.837	14.825	15.813	16.802	17.790	18.778	19.767	46
16	.988	1.976	2.965	3.953	4.942	5.931	6.919	7.907	8.896	9.884	10.873	11.862	12.850	13.838	14.827	15.815	16.803	17.791	18.780	19.769	44
18	.988	1.977	2.965	3.954	4.942	5.931	6.919	7.908	8.896	9.885	10.874	11.863	12.851	13.839	14.828	15.816	16.805	17.793	18.782	19.771	42
20	.989	1.977	2.966	3.954	4.943	5.932	6.920	7.909	8.897	9.886	10.875	11.864	12.852	13.840	14.829	15.818	16.806	17.794	18.783	19.772	40
22	.989	1.977	2.966	3.954	4.943	5.932	6.920	7.910	8.898	9.887	10.876	11.865	12.853	13.842	14.831	15.820	16.808	17.796	18.785	19.774	38
24	.989	1.977	2.966	3.955	4.944	5.933	6.921	7.910	8.899	9.888	10.877	11.866	12.854	13.843	14.832	15.821	16.809	17.798	18.787	19.776	36
26	.989	1.977	2.966	3.955	4.944	5.933	6.921	7.911	8.900	9.889	10.878	11.867	12.855	13.844	14.833	15.822	16.811	17.799	18.788	19.777	34
28	.989	1.978	2.966	3.956	4.944	5.934	6.922	7.911	8.900	9.889	10.879	11.868	12.856	13.845	14.834	15.823	16.812	17.801	18.789	19.779	32
30	.989	1.978	2.967	3.956	4.945	5.934	6.922	7.912	8.901	9.890	10.879	11.869	12.857	13.846	14.835	15.824	16.813	17.802	18.791	19.780	30
32	.989	1.978	2.967	3.956	4.945	5.935	6.923	7.913	8.902	9.891	10.880	11.870	12.859	13.848	14.837	15.826	16.815	17.804	18.793	19.782	28
34	.989	1.978	2.967	3.957	4.946	5.935	6.924	7.914	8.903	9.892	10.881	11.871	12.860	13.849	14.838	15.828	16.817	17.806	18.795	19.784	26
36	.989	1.978	2.967	3.957	4.946	5.936	6.924	7.914	8.904	9.893	10.882	11.872	12.861	13.850	14.840	15.829	16.818	17.807	18.796	19.785	24
38	.989	1.979	2.968	3.958	4.946	5.936	6.925	7.915	8.904	9.893	10.883	11.873	12.862	13.851	14.841	15.831	16.820	17.809	18.798	19.787	22
40	.989	1.979	2.968	3.958	4.947	5.937	6.926	7.916	8.905	9.894	10.884	11.874	12.863	13.852	14.842	15.832	16.821	17.810	18.799	19.788	20
42	.989	1.979	2.968	3.958	4.947	5.937	6.927	7.917	8.906	9.895	10.885	11.875	12.864	13.854	14.844	15.834	16.823	17.812	18.801	19.790	18
44	.990	1.979	2.969	3.958	4.948	5.938	6.927	7.917	8.907	9.896	10.886	11.876	12.865	13.855	14.845	15.835	16.824	17.814	18.803	19.792	16
46	.990	1.979	2.969	3.958	4.948	5.938	6.928	7.918	8.908	9.897	10.887	11.877	12.866	13.856	14.846	15.836	16.826	17.815	18.804	19.794	14
48	.990	1.980	2.969	3.959	4.948	5.939	6.928	7.918	8.909	9.898	10.888	11.878	12.867	13.857	14.847	15.837	16.827	17.817	18.806	19.796	12
50	.990	1.980	2.970	3.959	4.949	5.939	6.929	7.919	8.909	9.899	10.889	11.879	12.868	13.858	14.848	15.838	16.828	17.818	18.808	19.798	10
52	.990	1.980	2.970	3.959	4.949	5.940	6.930	7.920	8.910	9.900	10.890	11.880	12.870	13.860	14.850	15.840	16.830	17.820	18.810	19.800	8
54	.990	1.980	2.970	3.960	4.950	5.940	6.930	7.920	8.910	9.901	10.891	11.881	12.871	13.861	14.851	15.841	16.831	17.821	18.811	19.802	6
56	.990	1.980	2.970	3.960	4.950	5.941	6.931	7.921	8.911	9.902	10.892	11.882	12.872	13.862	14.852	15.842	16.832	17.822	18.813	19.803	4
58	.990	1.981	2.971	3.961	4.951	5.941	6.931	7.921	8.911	9.903	10.893	11.883	12.873	13.863	14.853	15.843	16.833	17.823	18.814	19.805	2
22° 0	.990	1.981	2.971	3.961	4.951	5.942	6.932	7.922	8.912	9.903	10.893	11.884	12.874	13.864	14.854	15.844	16.834	17.824	18.815	19.806	8° 0
	1	2	3	4	5	6	7	8	9	10	11	12	13	14	15	16	17	18	19	20	III + IX −

TABLE XVIII.

GENERAL TABLE of Aberration, &c. continued.

S. −VIII +II	1	2	3	4	5	6	7	8	9	10	11	12	13	14	15	16	17	18	19	20	
22° 0′	.990	1.981	2.971	3.961	4.951	5.942	6.932	7.922	8.912	9.903	10.893	11.884	12.874	13.864	14.854	15.844	16.834	17.824	18.815	19.806	8° 0′
2	.990	1.981	2.971	3.961	4.951	5.942	6.933	7.923	8.913	9.904	10.894	11.885	12.875	13.865	14.856	15.846	16.836	17.826	18.816	19.808	58
4	.990	1.981	2.971	3.962	4.952	5.943	6.933	7.923	8.914	9.905	10.895	11.886	12.876	13.866	14.857	15.847	16.837	17.827	18.818	19.810	56
6	.990	1.981	2.971	3.962	4.952	5.943	6.934	7.924	8.915	9.906	10.896	11.887	12.877	13.867	14.858	15.848	16.839	17.829	18.819	19.811	54
8	.991	1.981	2.972	3.962	4.953	5.944	6.934	7.924	8.915	9.907	10.897	11.887	12.878	13.868	14.859	15.849	16.840	17.830	18.821	19.813	52
10	.991	1.981	2.972	3.963	4.953	5.944	6.935	7.925	8.916	9.907	10.898	11.888	12.879	13.869	14.860	15.850	16.841	17.832	18.823	19.814	50
12	.991	1.981	2.972	3.963	4.953	5.944	6.935	7.926	8.917	9.908	10.898	11.889	12.880	13.870	14.861	15.852	16.843	17.834	18.825	19.816	48
14	.991	1.981	2.972	3.963	4.954	5.945	6.936	7.927	8.918	9.909	10.899	11.890	12.881	13.871	14.862	15.853	16.845	17.835	18.827	19.818	46
16	.991	1.981	2.972	3.963	4.954	5.945	6.936	7.927	8.919	9.910	10.900	11.891	12.882	13.872	14.863	15.855	16.846	17.837	18.828	19.819	44
18	.991	1.982	2.973	3.964	4.954	5.946	6.937	7.928	8.919	9.910	10.901	11.891	12.882	13.873	14.864	15.856	16.848	17.838	18.830	19.821	42
20	.991	1.982	2.973	3.964	4.955	5.946	6.937	7.929	8.920	9.911	10.902	11.892	12.883	13.874	14.866	15.857	16.849	17.840	18.831	19.822	40
22	.991	1.982	2.973	3.964	4.955	5.947	6.938	7.930	8.921	9.912	10.903	11.893	12.884	13.876	14.868	15.859	16.851	17.842	18.833	19.824	38
24	.991	1.982	2.973	3.965	4.956	5.947	6.938	7.930	8.921	9.912	10.904	11.894	12.885	13.877	14.869	15.860	16.852	17.843	18.834	19.825	36
26	.991	1.982	2.973	3.965	4.956	5.948	6.939	7.931	8.922	9.913	10.904	11.895	12.886	13.878	14.870	15.862	16.853	17.844	18.836	19.827	34
28	.991	1.983	2.974	3.965	4.957	5.948	6.939	7.931	8.922	9.913	10.905	11.896	12.887	13.879	14.871	15.863	16.854	17.845	18.837	19.828	32
30	.991	1.983	2.974	3.966	4.957	5.949	6.940	7.932	8.923	9.914	10.906	11.897	12.889	13.880	14.872	15.864	16.855	17.846	18.838	19.829	30
32	.991	1.983	2.974	3.966	4.957	5.949	6.941	7.933	8.924	9.915	10.907	11.898	12.890	13.881	14.874	15.866	16.857	17.848	18.840	19.831	28
34	.992	1.983	2.974	3.966	4.958	5.950	6.941	7.933	8.924	9.916	10.908	11.899	12.891	13.882	14.875	15.867	16.858	17.849	18.841	19.832	26
36	.992	1.983	2.974	3.966	4.958	5.950	6.942	7.934	8.925	9.916	10.909	11.900	12.892	13.883	14.876	15.868	16.859	17.850	18.843	19.834	24
38	.992	1.984	2.975	3.967	4.959	5.951	6.942	7.934	8.925	9.917	10.909	11.901	12.893	13.884	14.877	15.969	16.860	17.851	18.844	19.835	22
40	.992	1.984	2.975	3.967	4.959	5.951	6.943	7.935	8.926	9.918	10.910	11.902	12.894	13.885	14.878	15.870	16.861	17.852	18.845	19.836	20
42	.992	1.984	2.975	3.967	4.959	5.951	6.943	7.936	8.927	9 919	10.911	11.903	12.895	13.886	14.879	15.872	16.863	17.854	18.847	19.838	18
44	.992	1.984	2.976	3.968	4.960	5.952	6.944	7.936	8.928	9.920	10.912	11.904	12.896	13.887	14.880	15.873	16.864	17.855	18.848	19.840	16
46	.992	1.984	2.976	3.968	4.960	5.952	6.944	7.937	8.928	9.921	10.913	11.905	12.897	13.888	14.881	15.874	16.866	17.857	18.850	19.842	14
48	.992	1.984	2.976	3.968	4.961	5.953	6.945	7.937	8.929	9.921	10.913	11.905	12.897	13.889	14.882	15.875	16.867	17.858	18.851	19.843	12
50	.992	1.984	2.977	3.969	4.961	5.953	6.945	7.938	8.930	9.922	10.914	11.906	12.898	13.890	14.883	15.876	16.868	17.859	18.852	19.844	10
52	.992	1.984	2.977	3.969	4.961	5.953	6.946	7.938	8.931	9.923	10.915	11.907	12.899	13.891	14.884	15.877	16.869	17.861	18.854	19.846	8
54	.992	1.984	2.977	3.969	4.962	5.954	6.946	7.939	8.931	9.924	10.916	11.908	12.900	13.892	14.885	15.878	16.871	17.862	18.855	19.848	6
56	.992	1.984	2.977	3.969	4.962	5.954	6.947	7.939	8.932	9.924	10.917	11.908	12.901	13.893	14.886	15.879	16.872	17.863	18.857	19.849	4
58	.992	1.985	2.978	3.970	4.963	5.955	6.947	7.940	8.932	9.925	10.917	11.909	12.902	13.894	14.887	15.880	16.873	17.865	18.858	19.851	2
23° 0′	.993	1.985	2.978	3.970	4.963	5.955	6.948	7.940	8.933	9.926	10.918	11.910	12.903	13.895	14.888	15.881	16.874	17.866	18.859	19.852	7° 0′
2	.993	1.985	2.978	3.970	4.963	5.955	6.948	7.941	8.934	9.927	10.919	11.911	12.904	13.896	14.889	15.882	16.875	17.868	18.861	19.854	58
4	.993	1.985	2.978	3.971	4.964	5.956	6.949	7.941	8.934	9.927	10.920	11.912	12.905	13.897	14.890	15.883	16.876	17.869	18.862	19.855	56
6	.993	1.985	2.978	3.971	4.964	5.956	6.949	7.942	8.935	9.928	10.921	11.912	12.906	13.898	14.891	15.884	16.877	17.870	18.863	19.857	54
8	.993	1.986	2.979	3.971	4.964	5.957	6.950	7.942	8.935	9.928	10.921	11.913	12.906	13.899	14.892	15.885	16.878	17.871	18.864	19.858	52
10	.993	1.986	2.979	3.972	4.965	5.957	6.950	7.943	8.936	9.929	10.922	11.914	12.907	13.900	14.893	15.886	16.879	17.872	18.865	19.859	50
12	.993	1.986	2.979	3.972	4.965	5.957	6.951	7.944	8.937	9.930	10.923	11.915	12.908	13.901	14.894	15.888	16.881	17.874	18.867	19.861	48
14	.993	1.986	2.979	3.972	4.965	5.958	6.951	7.944	8.937	9.930	10.924	11.916	12.909	13.902	14.895	15.889	16.882	17.875	18.868	19.862	46
16	.993	1.986	2.979	3.972	4.965	5.958	6.952	7.945	8.938	9.931	10.924	11.916	12.910	13.903	14.896	15.890	16.883	17.876	18.869	19.863	44
18	.993	1.986	2.980	3.973	4.966	5.959	6.952	7.945	8.938	9.931	10.925	11.917	12.911	13.904	14.897	15.891	16.884	17.877	18.870	19.864	42
20	.993	1.987	2.980	3.973	4.966	5.959	6.953	7.946	8.939	9.932	10.925	11.918	12.912	13 905	14.899	15.892	16.885	17.878	18.871	19.865	40
22	.993	1.987	2.980	3.973	4.966	5.959	6.953	7.947	8.940	9.933	10.926	11.919	12.913	13.906	14.900	15.894	16.887	17.880	18.873	19.867	38
24	.993	1.987	2.980	3.973	4.967	5.960	6.954	7.947	8.940	9.934	10.927	11.920	12.914	13.907	14.901	15.895	16.888	17.881	18.874	19.868	36
26	.993	1.987	2.980	3.973	4.967	5.960	6.954	7.948	8.941	9.934	10.927	11.920	12.915	13.908	14.902	15.896	16.889	17.882	18.876	19.870	34
28	.994	1.987	2.981	3.974	4.968	5 961	6.955	7.948	8.941	9.935	10.928	11.921	12.915	13.909	14.903	15.897	16.890	17.883	18.877	19.871	32
30	.994	1.987	2.981	3.974	4.968	5.961	6.955	7.949	8.942	9.936	10.929	11.922	12.916	13.910	14.904	15.898	16.891	17.884	18.878	19.872	30
32	.994	1.987	2.981	3.974	4.968	5.961	6.955	7.949	8.943	9.937	10.930	11.923	12.917	13.911	14.905	15.899	16.893	17.886	18.880	19.874	28
34	.994	1.987	2.981	3.975	4.969	5.962	6.956	7.950	8.943	9.937	10.931	11.924	12.918	13.912	14.906	15.900	16.894	17.887	18.881	19.875	26
36	.994	1.987	2.981	3.975	4.969	5.962	6.956	7.950	8.944	9.938	10.932	11.924	12 919	13.913	14.907	15.901	16.895	17.888	18.882	19.876	24
38	.994	1.988	2.982	3.975	4.969	5.963	6.957	7.951	8.944	9.938	10.932	11.925	12 919	13.914	14.907	15.902	16.896	17.889	18.883	19.877	22
40	.994	1.988	2.982	3.976	4.970	5.963	6.957	7.951	8.945	9.939	10.933	11.926	12.920	13.915	14.908	15.903	16.897	17.890	18.884	19.878	20
42	.994	1.988	2.982	3.976	4.970	5.963	6.958	7.952	8.946	9 940	10.934	11.927	12.921	13.916	14.909	15.904	16.893	17.892	18.886	19.880	18
44	.994	1.988	2.982	3.976	4.970	5.964	6.958	7.952	8.946	9.940	10.935	11.928	12.922	13.917	14.910	15.905	16.899	17.893	18.887	19.881	16
46	.994	1.988	2.982	3.976	4.970	5.964	6.959	7.953	8.947	9.941	10.936	11.928	12.923	13.918	14.911	15.906	16.900	17.891	18.888	19.882	14
48	.994	1.988	2.983	3.977	4.971	5.965	6.959	7.953	8.947	9.941	10.936	11.929	12.924	13.919	14.912	15.907	16.901	17.895	18.889	19.883	12
50	.994	1.988	2.983	3.977	4.971	5.965	6.960	7.954	8 948	9.942	10.937	11.930	12.925	13.920	14.913	15.908	16.902	17.896	18.890	19.884	10
52	.994	1.988	2.983	3.977	4.971	5.965	6.960	7.954	8.949	9.943	10.938	11.931	12.926	13.921	14.914	15.909	16.903	17.898	18.892	19.886	8
54	.994	1.988	2.983	3.977	4.972	5.966	6.961	7.955	8.949	9.943	10.939	11.932	12.927	13.522	14.915	15.910	16.904	17.899	18.893	19.887	6
56	.994	1.988	2.983	3.977	4.972	5.966	6.961	7.955	8.950	9.944	10.939	11.933	12.928	13.923	14.916	15.911	16.905	17.900	18.894	19.888	4
58	.994	1.989	2.934	3.978	4.973	5.967	6.962	7.956	8.950	9.944	10.940	11.933	12.928	13.923	14.917	15.912	16.906	17.901	18.895	19.889	2
24° 0	.995	1.989	2.984	3.978	4.973	5.967	6.962	7.956	8.951	9.945	10.940	11.934	12.929	13.924	14.918	15.912	16.907	17.902	18.896	19.890	6° 0
	1	2	3	4	5	6	7	8	9	10	11	12	13	14	15	16	17	18	19	20	III + IX −

TABLE XVIII.

GENERAL TABLE of Aberration, &c. continued.

S. −VIII +II	1	2	3	4	5	6	7	8	9	10	11	12	13	14	15	16	17	18	19	20	
24° 0	.995	1.989	2.984	3.978	4.973	5.967	6.962	7.956	8.951	9.945	10.940	11.934	12.929	13.924	14.918	15.912	16.907	17.902	18.896	19.890	6° 0'
2	.995	1.989	2.984	3.978	4.973	5.967	6.962	7.957	8.952	9.946	10.941	11.935	12.930	13.925	14.919	15.913	16.908	17.903	18.897	19.892	58
4	.995	1.989	2.984	3.978	4.973	5.968	6.963	7.957	8.952	9.946	10.941	11.936	12.931	13.926	14.920	15.914	16.909	17.904	18.898	19.893	56
6	.995	1.989	2.984	3.978	4.973	5.968	6.963	7.958	8.952	9.947	10.942	11.937	12.932	13.927	14.921	15.915	16.910	17.905	18.899	19.894	54
8	.995	1.989	2.984	3.979	4.974	5.968	6.963	7.958	8.953	9.947	10.942	11.937	12.933	13.927	14.922	15.916	16.911	17.905	18.900	19.895	52
10	.995	1.990	2.984	3.979	4.974	5.969	6.964	7.959	8.953	9.948	10.943	11.938	12.933	13.928	14.923	15.917	16.912	17.906	18.901	19.896	50
12	.995	1.990	2.984	3.979	4.974	5.969	6.064	7.959	8.954	9.949	10.944	11.939	12.934	13.929	14.924	15.918	16.913	17.907	18.903	19.898	48
14	.995	1.990	2.984	3.979	4.975	5.970	6.965	7.960	8.954	9.949	10.944	11.940	12.935	13.930	14.925	15.919	16.914	17.908	18.904	19.899	46
16	.995	1.990	2.984	3.979	4.975	5.970	6.965	7.960	8.955	9.950	10.945	11.941	12.936	13.931	14.926	15.920	16.915	17.910	18.905	19.900	44
18	.995	1.990	2.985	3.980	4.975	5.970	6.965	7.960	8.955	9.950	10.945	11.941	12.936	13.931	14.927	15.921	16.916	17.911	18.906	19.901	42
20	.995	1.990	2.985	3.980	4.976	5.971	6.966	7.961	8.956	9.951	10.946	11.942	12.937	13.932	14.927	15.922	16.917	17.912	18.907	19.902	40
22	.995	1.990	2.985	3.980	4.976	5.971	6.966	7.961	8.957	9.952	10.947	11.943	12.938	13.933	14.928	15.923	16.918	17.914	18.909	19.904	38
24	.995	1.990	2.985	3.981	4.976	5.971	6.967	7.962	8.957	9.952	10.947	11.943	12.939	13.934	14.929	15.924	16.919	17.915	18.910	19.905	36
26	.995	1.990	2.985	3.981	4.976	5.971	6.967	7.962	8.958	9.953	10.948	11.944	12.939	13.935	14.930	15.925	16.920	17.916	18.911	19.906	34
28	.995	1.991	2.986	3.981	4.977	5.972	6.968	7.962	8.958	9.953	10.948	11.944	12.940	13.935	14.930	15.926	16.921	17.917	18.912	19.907	32
30	.995	1.991	2.986	3.982	4.977	5.972	6.968	7.963	8.959	9.954	10.949	11.945	12.940	13.936	14.931	15.926	16.922	17.918	18.913	19.908	30
32	.995	1.991	2.986	3.982	4.977	5.972	6.968	7.963	8.959	9.955	10.950	11.946	12.941	13.937	14.932	15.927	16.923	17.919	18.914	19.910	28
34	.995	1.991	2.986	3.982	4.977	5.973	6.969	7.964	8.960	9.955	10.950	11.946	12.942	13.938	14.933	15.928	16.924	17.920	18.915	19.911	26
36	.995	1.991	2.986	3.982	4.977	5.973	6.969	7.964	8.960	9.956	10.951	11.947	12.943	13.939	14.934	15.929	16.925	17.921	18.916	19.912	24
38	.996	1.991	2.987	3.982	4.978	5.973	6.969	7.965	8.960	9.956	10.951	11.947	12.943	13.939	14.934	15.930	16.926	17.922	18.917	19.913	22
40	.996	1.991	2.987	3.983	4.978	5.974	6.970	7.965	8.961	9.957	10.952	11.948	12.944	13.940	14.935	15.930	16.926	17.922	18.918	19.914	20
42	.996	1.991	2.987	3.983	4.979	5.974	6.970	7.966	8.961	9.957	10.953	11.949	12.945	13.941	14.936	15.931	16.927	17.923	18.919	19.915	18
44	.996	1.991	2.987	3.983	4.979	5.975	6.971	7.966	8.962	9.958	10.953	11.950	12.946	13.942	14.937	15.932	16.928	17.924	18.920	19.916	16
46	.996	1.991	2.987	3.983	4.979	5.975	6.971	7.967	8.962	9.958	10.954	11.951	12.947	13.943	14.938	15.933	16.929	17.925	18.921	19.917	14
48	.996	1.992	2.988	3.984	4.980	5.975	6.971	7.967	8.962	9.958	10.954	11.951	12.948	13.943	14.939	15.934	16.930	17.926	18.922	19.918	12
50	.996	1.992	2.988	3.984	4.980	5.976	6.972	7.968	8.963	9.959	10.955	11.952	12.948	13.944	14.940	15.935	16.931	17.927	18.923	19.919	10
52	.996	1.992	2.988	3.984	4.980	5.976	6.972	7.968	8.964	9.960	10.956	11.953	12.949	13.945	14.941	15.936	16.932	17.928	18.924	19.920	8
54	.996	1.992	2.988	3.984	4.980	5.976	6.972	7.969	8.964	9.960	10.956	11.953	12.949	13.945	14.941	15.937	16.933	17.929	18.925	19.921	6
56	.996	1.992	2.988	3.984	4.980	5.976	6.972	7.969	8.965	9.961	10.957	11.954	12.950	13.946	14.942	15.938	16.934	17.930	18.926	19.922	4
58	.996	1.992	2.988	3.985	4.981	5.977	6.973	7.969	8.965	9.961	10.957	11.954	12.950	13.946	14.942	15.939	16.935	17.931	18.927	19.923	2
25° 0'	.996	1.992	2.989	3.985	4.981	5.977	6.973	7.970	8.966	9.962	10.958	11.955	12.951	13.947	14.943	15.940	16.936	17.932	18.928	19.924	5° 0'
2	.996	1.992	2.989	3.985	4.981	5.977	6.973	7.970	8.966	9.962	10.959	11.956	12.952	13.948	14.944	15.941	16.937	17.933	18.929	19.925	58
4	.996	1.992	2.989	3.985	4.981	5.978	6.974	7.970	8.967	9.963	10.959	11.956	12.952	13.948	14.945	15.942	16.938	17.934	18.930	19.926	56
6	.996	1.992	2.989	3.985	4.981	5.978	6.974	7.971	8.967	9.963	10.960	11.957	12.953	13.949	14.946	15.943	16.939	17.935	18.931	19.927	54
8	.996	1.993	2.989	3.985	4.982	5.978	6.974	7.971	8.967	9.964	10.960	11.957	12.953	13.949	14.946	15.944	16.939	17.935	18.932	19.928	52
10	.996	1.993	2.989	3.986	4.982	5.979	6.975	7.972	8.968	9.964	10.961	11.958	12.954	13.950	14.947	15.944	16.940	17.936	18.932	19.929	50
12	.996	1.993	2.989	3.986	4.982	5.979	6.975	7.972	8.968	9.965	10.962	11.959	12.955	13.951	14.948	15.945	16.941	17.937	18.933	19.930	48
14	.996	1.993	2.989	3.986	4.982	5.979	6.976	7.972	8.969	9.965	10.962	11.959	12.955	13.952	14.948	15.945	16.941	17.938	18.934	19.931	46
16	.997	1.993	2.989	3.986	4.982	5.979	6.976	7.972	8.969	9.966	10.963	11.960	12.956	13.953	14.949	15.946	16.942	17.939	18.935	19.932	44
18	.997	1.993	2.990	3.987	4.983	5.980	6.976	7.973	8.969	9.966	10.963	11.960	12.956	13.954	14.949	15.946	16.943	17.939	18.936	19.933	42
20	.997	1.993	2.990	3.987	4.983	5.980	6.977	7.973	8.970	9.967	10.964	11.961	12.957	13.954	14.950	15.947	16.943	17.940	18.937	19.934	40
22	.997	1.993	2.990	3.987	4.983	5.980	6.977	7.973	8.970	9.967	10.965	11.962	12.958	13.955	14.951	15.948	16.944	17.941	18.938	19.935	38
24	.997	1.993	2.990	3.987	4.984	5.981	6.977	7.974	8.971	9.968	10.965	11.962	12.958	13.955	14.951	15.948	16.945	17.942	18.939	19.936	36
26	.997	1.993	2.990	3.987	4.984	5.981	6.977	7.974	8.971	9.968	10.965	11.963	12.959	13.956	14.952	15.949	16.946	17.943	18.940	19.937	34
28	.997	1.994	2.991	3.988	4.984	5.981	6.978	7.974	8.971	9.968	10.966	11.963	12.959	13.956	14.952	15.949	16.946	17.944	18.940	19.938	32
30	.997	1.994	2.991	3.988	4.985	5.982	6.978	7.975	8.972	9.969	10.966	11.964	12.960	13.957	14.953	15.950	16.947	17.944	18.941	19.938	30
32	.997	1.994	2.991	3.988	4.985	5.982	6.978	7.975	8.972	9.969	10.966	11.964	12.961	13.958	14.954	15.951	16.948	17.945	18.942	19.939	28
34	.997	1.994	2.991	3.988	4.985	5.982	6.979	7.976	8.973	9.970	10.967	11.965	12.961	13.958	14.954	15.952	16.949	17.946	18.943	19.940	26
36	.997	1.994	2.991	3.988	4.985	5.982	6.979	7.976	8.973	9.970	10.967	11.965	12.962	13.959	14.955	15.952	16.950	17.947	18.944	19.941	24
38	.997	1.994	2.991	3.988	4.986	5.982	6.979	7.977	8.973	9.971	10.968	11.966	12.962	13.959	14.956	15.953	16.950	17.948	18.944	19.942	22
40	.997	1.994	2.991	3.989	4.986	5.983	6.980	7.977	8.974	9.971	10.968	11.966	12.963	13.960	14.957	15.954	16.951	17.948	18.945	19.942	20
42	.997	1.994	2.991	3.989	4.986	5.983	6.980	7.977	8.974	9.972	10.969	11.967	12.964	13.961	14.958	15.955	16.952	17.949	18.946	19.943	18
44	.997	1.994	2.991	3.989	4.986	5.983	6.981	7.978	8.975	9.972	10.969	11.967	12.964	13.961	14.959	15.956	16.953	17.950	18.947	19.944	16
46	.997	1.994	2.991	3.989	4.986	5.983	6.981	7.978	8.975	9.973	10.970	11.967	12.965	13.962	14.960	15.957	16.954	17.951	18.948	19.945	14
48	.997	1.995	2.992	3.989	4.987	5.983	6.981	7.978	8.975	9.973	10.970	11.968	12.965	13.962	14.960	15.957	16.954	17.951	18.949	19.946	12
50	.997	1.995	2.992	3.989	4.987	5.984	6.982	7.979	8.976	9.974	10.971	11.968	12.966	13.963	14.961	15.958	16.955	17.952	18.950	19.947	10
52	.997	1.995	2.992	3.989	4.987	5.984	6.982	7.979	8.976	9.974	10.972	11.969	12.966	13.964	14.962	15.959	16.956	17.953	18.951	19.948	8
54	.997	1.995	2.992	3.989	4.987	5.984	6.982	7.980	8.977	9.975	10.972	11.969	12.967	13.964	14.962	15.960	16.957	17.954	18.952	19.949	6
56	.997	1.995	2.992	3.989	4.987	5.984	6.982	7.980	8.977	9.975	10.973	11.969	12.967	13.965	14.963	15.961	16.957	17.955	18.953	19.950	4
58	.998	1.995	2.993	3.990	4.988	5.984	6.983	7.980	8.978	9.976	10.973	11.970	12.968	13.965	14.964	15.961	16.958	17.955	18.954	19.951	2
26° 0	.998	1.995	2.993	3.990	4.988	5.985	6.983	7.981	8.978	9.976	10.974	11.970	12.968	13.966	14.964	15.962	16.959	17.956	18.954	19.952	4° 0
	1	2	3	4	5	6	7	8	9	10	11	12	13	14	15	16	17	18	19	20	III + IX −

TABLE XVIII.

GENERAL TABLE of Aberration, &c. continued.

S. −VIII +II	1	2	3	4	5	6	7	8	9	10	11	12	13	14	15	16	17	18	19	20	
26° 0	.998	1.995	2.993	3.990	4.988	5.985	6.983	7.981	8.978	9.976	10.974	11.970	12.968	13.966	14.964	15.962	16.959	17.956	18.954	19.952	4° 0′
2	.998	1.995	2.993	3.990	4.988	5.985	6.983	7.981	8.978	9.976	10.974	11.971	12.969	13.967	14.965	15.963	16.960	17.957	18.955	19.953	58
4	.998	1.995	2.993	3.990	4.988	5.986	6.983	7.981	8.979	9.977	10.975	11.971	12.969	13.967	14.965	15.993	16.960	17.958	18.956	19.954	56
6	.998	1.995	2.993	3.990	4.988	5.986	6.983	7.981	8.979	9.977	10.975	11.972	12.970	13.968	14.966	15.964	16.961	17.959	18.957	19.955	54
8	.998	1.996	2.993	3.990	4.988	5.986	6.983	7.982	8.979	9.977	10.975	11.972	12.970	13.968	14.966	15.964	16.961	17.959	18.957	19.955	52
10	.998	1.996	2.993	3.991	4.989	5.987	6.984	7.982	8.980	9.978	10.976	11.973	12.971	13.969	14.967	15.965	16.962	17.960	18.958	19.956	50
12	.998	1.996	2.993	3.991	4.989	5.987	6.984	7.982	8.980	9.978	10.976	11.974	12.972	13.970	14.968	15.966	16.963	17.961	18.959	19.957	48
14	.998	1.996	2.993	3.991	4.989	5.987	6.985	7.983	8.981	9.979	10.977	11.974	12.972	13.970	14.968	15.966	16.963	17.962	18.960	19.958	46
16	.998	1.996	2.993	3.991	4.989	5.987	6.985	7.983	8.981	9.979	10.977	11.975	12.973	13.971	14.969	15.967	16.964	17.963	18.961	19.959	44
18	.998	1.996	2.993	3.991	4.989	5.987	6.985	7.983	8.981	9.979	10.977	11.975	12.973	13.971	14.969	15.967	16.964	17.963	18.961	19.959	42
20	.998	1.996	2.994	3.992	4.990	5.988	6.986	7.984	8.982	9.980	10.978	11.976	12.974	13.972	14.970	15.968	16.965	17.964	18.962	19.960	40
22	.998	1.996	2.994	3.992	4.990	5.988	6.986	7.984	8.982	9.980	10.978	11.976	12.974	13.972	14.970	15.969	16.966	17.965	18.963	19.961	38
24	.998	1.996	2.994	3.992	4.990	5.988	6.986	7.984	8.982	9.980	10.978	11.977	12.975	13.973	14.971	15.969	16.966	17.965	18.963	19.961	36
26	.998	1.996	2.994	3.992	4.990	5.988	6.986	7.984	8.982	9.980	10.978	11.977	12.975	13.973	14.971	15.970	16.967	17.966	18.964	19.962	34
28	.998	1.996	2.994	3.992	4.990	5.988	6.986	7.985	8.983	9.980	10.979	11.977	12.975	13.973	14.972	15.970	16.967	17.966	18.964	19.962	32
30	.998	1.996	2.994	3.993	4.991	5.989	6.987	7.985	8.983	9.981	10.979	11.978	12.976	13.974	14.972	15.971	16.968	17.967	18.965	19.963	30
32	.998	1.996	2.994	3.993	4.991	5.989	6.987	7.985	8.983	9.981	10.979	11.978	12.976	13.974	14.973	15.972	16.969	17.968	18.966	19.964	28
34	.998	1.996	2.994	3.993	4.991	5.989	6.987	7.986	8.984	9.982	10.980	11.979	12.977	13.975	14.973	15.972	16.969	17.968	18.966	19.964	26
36	.998	1.996	2.994	3.993	4.991	5.989	6.987	7.986	8.984	9.982	10.980	11.979	12.977	13.975	14.974	15.973	16.970	17.969	18.967	19.965	24
38	.998	1.997	2.994	3.993	4.991	5.989	6.987	7.986	8.984	9.982	10.980	11.979	12.977	13.975	14.974	15.973	16.970	17.969	18.967	19.965	22
40	.998	1.997	2.995	3.993	4.992	5.990	6.988	7.987	8.985	9.983	10.981	11.980	12.978	13.976	14.975	15.974	16.971	17.970	18.968	19.966	20
42	.998	1.997	2.995	3.993	4.992	5.990	6.988	7.987	8.985	9.983	10.981	11.980	12.978	13.976	14.975	15.974	16.972	17.971	18.969	19.967	18
44	.998	1.997	2.995	3.993	4.992	5.990	6.988	7.987	8.985	9.984	10.982	11.981	12.979	13.977	14.976	15.975	16.972	17.971	18.969	19.967	16
46	.998	1.997	2.995	3.993	4.992	5.990	6.988	7.987	8.985	9.984	10.982	11.981	12.979	13.977	14.976	15.975	16.973	17.972	18.970	19.968	14
48	.998	1.997	2.995	3.994	4.992	5.990	6.988	7.988	8.986	9.984	10.982	11.981	12.979	13.977	14.977	15.975	16.973	17.972	18.970	19.968	12
50	.999	1.997	2.995	3.994	4.992	5.991	6.989	7.988	8.986	9.985	10.983	11.982	12.980	13.978	14.977	15.976	16.974	17.973	18.971	19.969	10
52	.999	1.997	2.995	3.994	4.992	5.991	6.989	7.988	8.986	9.985	10.983	11.982	12.980	13.978	14.978	15.976	16.975	17.974	18.972	19.970	8
54	.999	1.997	2.995	3.994	4.992	5.991	6.989	7.989	8.987	9.985	10.984	11.983	12.981	13.979	14.978	15.977	16.975	17.974	18.972	19.970	6
56	.999	1.997	2.995	3.994	4.992	5.991	6.989	7.989	8.987	9.985	10.984	11.983	12.981	13.979	14.979	15.977	16.976	17.975	18.973	19.971	4
58	.999	1.997	2.995	3.994	4.993	5.991	6.989	7.989	8.987	9.986	10.984	11.983	12.982	13.979	14.979	15.977	16.976	17.975	18.973	19.971	2
27° 0′	.999	1.997	2.996	3.995	4.993	5.992	6.990	7.989	8.988	9.986	10.985	11.984	12.982	13.980	14.980	15.978	16.977	17.276	18.974	19.972	3° 0′
2	.999	1.997	2.996	3.995	4.993	5.992	6.990	7.990	8.988	9.986	10.985	11.984	12.983	13.981	14.980	15.978	16.977	17.976	18.975	19.973	58
4	.999	1.997	2.996	3.995	4.993	5.992	6.990	7.990	8.988	9.987	10.986	11.985	12.983	13.981	14.981	15.979	16.978	17.977	18.975	19.973	56
6	.999	1.997	2.996	3.995	4.993	5.992	6.991	7.990	8.988	9.987	10.986	11.985	12.984	13.982	14.981	15.979	16.978	17.977	18.976	19.974	54
8	.999	1.998	2.996	3.995	4.993	5.992	6.991	7.990	8.988	9.987	10.986	11.985	12.984	13.982	14.981	15.979	16.978	17.977	18.976	19.974	52
10	.999	1.998	2.996	3.995	4.994	5.993	6.991	7.990	8.989	9.988	10.987	11.986	12.985	13.983	14.982	15.980	16.979	17.278	18.977	19.975	50
12	.999	1.998	2.996	3.995	4.994	5.993	6.992	7.990	8.989	9.988	10.987	11.986	12.985	13.984	14.982	15.980	16.979	17.978	18.977	19.976	48
14	.999	1.998	2.996	3.995	4.994	5.993	6.992	7.990	8.989	9.989	10.988	11.987	12.986	13.984	14.983	15.981	16.980	17.979	18.978	19.976	46
16	.999	1.998	2.996	3.995	4.994	5.993	6.992	7.990	8.989	9.989	10.988	11.987	12.986	13.985	14.983	15.981	16.980	17.979	18.978	19.977	44
18	.999	1.998	2.996	3.996	4.994	5.993	6.992	7.991	8.990	9.989	10.988	11.987	12.986	13.985	14.983	15.981	16.980	17.979	18.979	19.977	42
20	.999	1.998	2.997	3.996	4.995	5.994	6.993	7.991	8.990	9.989	10.989	11.988	12.987	13.986	14.984	15.982	16.981	17.980	18.979	19.978	40
22	.999	1.998	2.997	3.996	4.995	5.994	6.993	7.991	8.990	9.989	10.989	11.988	12.987	13.986	14.984	15.982	16.982	17.981	18.980	19.979	38
24	.999	1.998	2.997	3.996	4.995	5.994	6.993	7.991	8.991	9.990	10.989	11.988	12.987	13.986	14.985	15.983	16.982	17.981	18.980	19.979	36
26	.999	1.998	2.997	3.996	4.995	5.994	6.993	7.991	8.991	9.990	10.989	11.988	12.987	13.986	14.985	15.983	16.983	17.982	18.981	19.980	34
28	.999	1.998	2.997	3.996	4.995	5.994	6.993	7.991	8.991	9.990	10.989	11.989	12.988	13.987	14.985	15.983	16.983	17.982	18.981	19.980	32
30	.999	1.998	2.997	3.996	4.995	5.994	6.993	7.992	8.991	9.991	10.990	11.989	12.988	13.987	14.986	15.984	16.984	17.983	18.982	19.981	30
32	.999	1.998	2.997	3.996	4.995	5.994	6.993	7.992	8.992	9.991	10.990	11.989	12.988	13.987	14.986	15.984	16.984	17.984	18.983	19.982	28
34	.999	1.998	2.997	3.996	4.995	5.994	6.993	7.992	8.992	9.991	10.090	11.989	12.989	13.987	14.986	15.985	16.985	17.984	18.983	19.982	26
36	.999	1.998	2.997	3.996	4.995	5.994	6.993	7.992	8.992	9.991	10.990	11.989	12.989	13.987	14.986	15.985	16.985	17.985	18.984	19.983	24
38	.999	1.998	2.997	3.997	4.996	5.995	6.994	7.992	8.992	9.991	10.990	11.990	12.989	13.988	14.987	15.985	16.985	17.985	18.984	19.983	22
40	.999	1.998	2.998	3.997	4.996	5.995	6.994	7.993	8.993	9.992	10.991	11.990	12.990	13.988	14.987	15.986	16.986	17.986	18.985	19.984	20
42	.999	1.998	2.998	3.997	4.996	5.995	6.994	7.993	8.993	9.992	10.991	11.990	12.990	13.988	14.987	15.986	16.986	17.986	18.985	19.984	18
44	.999	1.998	2.998	3.997	4.996	5.995	6.994	7.993	8.993	9.992	10.991	11.991	12.990	13.989	14.988	15.987	16.987	17.987	18.986	19.985	16
46	.999	1.998	2.998	3.997	4.996	5.995	6.994	7.993	8.993	9.992	10.991	11.991	12.990	13.989	14.988	15.987	16.987	17.987	18.986	19.985	14
48	.999	1.998	2.998	3.997	4.996	5.995	6.994	7.993	8.993	9.992	10.991	11.991	12.990	13.989	14.988	15.987	16.987	17.987	18.986	19.985	12
50	.999	1.999	2.998	3.997	4.996	5.996	6.995	7.994	8.994	9.993	10.992	11.992	12.991	13.990	14.989	15.988	16.988	17.988	18.987	19.986	10
52	.999	1.999	2.998	3.997	4.996	5.996	6.995	7.994	8.994	9.993	10.992	11.992	12.991	13.990	14.989	15.988	16.988	17.988	18.987	19.986	8
54	.999	1.999	2.998	3.997	4.996	5.996	6.995	7.994	8.994	9.993	10.992	11.992	12.991	13.991	14.990	15.989	16.989	17.989	18.988	19.987	6
56	.999	1.999	2.998	3.097	4.996	5.996	6.995	7.994	8.994	9.993	10.992	11.992	12.991	13.991	14.990	15.989	16.989	17.989	18.988	19.987	4
58	.999	1.999	2.998	3.998	4.997	5.996	6.995	7.994	8.994	9.994	10.992	11.993	12.991	13.991	14.990	15.989	16.990	17.989	18.988	19.987	2
28° 0	.999	1.999	2.998	3.998	4.997	5.996	6.996	7.995	8.995	9.994	10.993	11.993	12.992	13.992	14.991	15.990	16.990	17.990	18.989	19.988	2° 0
	1	2	3	4	5	6	7	8	9	10	11	12	13	14	15	16	17	18	19	20	III + IX −

GENERAL TABLE of Aberration, &c. concluded.

S. −VIII + II	1	2	3	4	5	6	7	8	9	10	11	12	13	14	15	16	17	18	19	20	
28° 0	.999	1.999	2.998	3.998	4.997	5.996	6.996	7.995	8.995	9.994	10.993	11.993	12.992	13.992	14.991	15.990	16.990	17.989	18.989	19.988	2° 0'
2	.999	1.990	2.998	3.998	4.997	5.996	6.996	7.995	8.995	9.994	10.993	11.993	12.992	13.992	14.991	15.990	16.990	17.989	18.989	19.988	58
4	.999	1.999	2.998	3.998	4.997	5.996	6.996	7.995	8.995	9.994	10.993	11.993	12.992	13.992	14.991	15.991	16.991	17.990	18.989	19.989	56
6	.999	1.999	2.998	3.998	4.997	5.996	6.996	7.995	8.995	9.994	10.993	11.993	12.992	13.992	14.991	15.991	16.991	17.990	18.989	19.989	54'
8	.999	1.999	2.998	3.998	4.997	5.996	6.996	7.995	8.995	9.994	10.993	11.993	12.992	13.992	14.992	15.991	16.991	17.990	18.990	19.989	52
10	.999	1.999	2.999	3.998	4.997	5.997	6.996	7.996	8.995	9.995	10.994	11.994	12.993	13.993	14.992	15.992	16.992	17.991	18.990	19.990	50
12	.999	1.998	2.999	3.998	4.997	5.997	6.996	7.996	8.995	9.995	10.994	11.994	12.993	13.993	14.992	15.992	16.992	17.991	18.990	19.990	48
14	.999	1.999	2.999	3.998	4.997	5.997	6.996	7.996	8.995	9.995	10.994	11.994	12.993	13.993	14.993	15.992	16.992	17.992	18.991	19.991	46
16	.999	1.999	2.999	3.998	4.997	5.997	6.996	7.996	8.995	9.995	10.994	11.994	12.993	13.993	14.993	15.992	16.992	17.992	18.991	19.991	44
18	.999	1.999	2.999	3.998	4.997	5.997	6.996	7.996	8.995	9.995	10.994	11.994	12.993	13.993	14.993	15.992	16.992	17.992	18.991	19.991	42
20	.999	1.999	2.999	3.998	4.998	5.998	6.997	7.997	8.996	9.996	10.995	11.995	12.994	13.994	14.993	15.993	16.993	17.993	18.991	19.992	40
22	.999	1.999	2.999	3.998	4.998	5.998	6.997	7.997	8.996	9.996	10.995	11.995	12.994	13.994	14.994	15.993	16.993	17.993	18.992	19.992	38
24	.999	1.999	2.999	3.998	4.998	5.998	6.997	7.997	8.996	9.996	10.995	11.995	12.994	13.994	14.994	15.993	16.993	17.993	18.992	19.992	36
26	.999	1.999	2.999	3.998	4.998	5.998	6.997	7.997	8.996	9.996	10.995	11.995	12.994	13.994	14.994	15.993	16.993	17.993	18.992	19.992	34
28	.999	1.999	2.999	3.998	4.998	5.998	6.997	7.997	8.996	9.996	10.995	11.995	12.994	13.994	14.994	15.993	16.993	17.993	18.992	19.992	32
30	.999	1.999	2.999	3.999	4.998	5.998	6.998	7.997	8.997	9.997	10.996	11.996	12.995	13.995	14.995	15.994	16.994	17.994	18.993	19.993	30
32	.999	1.999	2.999	3.999	4.998	5.998	6.998	7.997	8.997	9.997	10.996	11.996	12.995	13.995	14.995	15.994	16.994	17.994	18.993	19.993	28
34	.999	1.999	2.999	3.999	4.998	5.998	6.998	7.997	8.997	9.997	10.996	11.996	12.995	13.995	14.995	15.994	16.994	17.994	18.993	19.993	26
36	.999	1.999	2.999	3.999	4.998	5.998	6.998	7.997	8.997	9.997	10.996	11.996	12.995	13.995	14.995	15.994	16.994	17.994	18.993	19.993	24
38	.999	1.999	2.999	3.999	4.998	5.998	6.998	7.997	8.997	9.997	10.996	11.996	12.995	13.995	14.995	15.994	16.994	17.994	18.994	19.994	22
40	.999	1.999	2.999	3.999	4.999	5.998	6.998	7.998	8.997	9.997	10.997	11.997	12.996	13.996	14.996	15.995	16.995	17.995	18.994	19.994	20
42	.999	1.999	2.999	3.999	4.999	5.998	6.998	7.998	8.998	9.997	10.997	11.997	12.996	13.996	14.996	15.995	16.995	17.995	18.994	19.994	18
44	.999	1.999	2.999	3.999	4.999	5.998	6.998	7.998	8.998	9.997	10.997	11.997	12.996	13.996	14.996	15.995	16.995	17.995	18.995	19.995	16
46	.999	1.999	2.999	3.999	4.999	5.998	6.998	7.998	8.998	9.997	10.997	11.997	12.996	13.996	14.996	15.995	16.995	17.995	18.995	19.995	14
48	.999	1.999	2.999	3.999	4.999	5.999	6.998	7.998	8.998	9.998	10.997	11.997	12.996	13.996	14.996	15.995	16.995	17.995	18.995	19.995	12
50	1.000	1.999	2.999	3.999	4.999	5.999	6.999	7.998	8.998	9.998	10.998	11.997	12.997	13.997	14.997	15.996	16.996	17.996	18.996	19.996	10
52	1.000	1.999	2.999	3.999	4.999	5.999	6.999	7.998	8.998	9.998	10.998	11.997	12.997	13.997	14.997	15.996	16.996	17.996	18.996	19.996	8
54	1.000	1.999	2.999	3.999	4.999	5.999	6.999	7.998	8.998	9.998	10.998	11.997	12.997	13.997	14.997	15.996	16.996	17.996	18.996	19.996	6
56	1.000	1.999	2.999	3.999	4.999	5.999	6.999	7.998	8.998	9.998	10.998	11.998	12.997	13.997	14.997	15.996	16.996	17.996	18.996	19.996	4
58	1.000	1.999	2.999	3.999	4.999	5.999	6.999	7.998	8.998	9.998	10.998	11.998	12.997	13.997	14.997	15.996	16.996	17.996	18.996	19.996	2
29° 0	1.000	2.000	2.999	3.999	4.999	5.999	6.999	7.999	8.999	9.998	10.999	11.998	12.998	13.998	14.998	15.997	16.997	17.997	18.997	19.997	1° 0'
2	1.000	2.000	2.999	3.999	4.999	5.999	6.999	7.999	8.999	9.999	10.999	11.998	12.998	13.998	14.998	15.997	16.997	17.997	18.997	19.997	58
4	1.000	2.000	2.999	3.999	4.999	5.999	6.999	7.999	8.999	9.999	10.999	11.998	12.998	13.998	14.998	15.997	16.997	17.997	18.997	19.997	56
6	1.000	2.000	2.999	3.999	4.999	5.999	6.999	7.999	8.999	9.999	10.999	11.998	12.998	13.998	14.998	15.997	16.997	17.997	18.997	19.997	54
8	1.000	2.000	2.999	3.999	4.999	5.999	6.999	7.999	8.999	9.999	10.999	11.998	12.998	13.998	14.998	15.997	16.997	17.997	18.997	19.997	52
10	1.000	2.000	3.000	3.999	4.999	5.999	6.999	7.999	8.999	9.999	10.999	11.999	12.999	13.998	14.998	15.998	16.998	17.998	18.998	19.998	50
12	1.000	2.000	3.000	3.999	4.999	5.999	6.999	7.999	8.999	9.999	10.999	11.999	12.999	13.998	14.998	15.998	16.998	17.998	18.998	19.998	48
14	1.000	2.000	3.000	3.999	4.999	5.999	6.999	7.999	8.999	9.999	10.999	11.999	12.999	13.998	14.998	15.998	16.998	17.998	18.998	19.998	46
16	1.000	2.000	3.000	3.999	4.999	5.999	6.999	7.999	8.999	9.999	10.999	11.999	12.999	13.998	14.998	15.998	16.998	17.998	18.998	19.998	44
18	1.000	2.000	3.000	3.999	4.999	5.999	6.999	7.999	8.999	9.999	10.999	11.999	12.999	13.998	14.998	15.998	16.998	17.998	18.998	19.998	42
20	1.000	2.000	3.000	4.000	5.000	5.999	6.999	7.999	8.999	9.999	10.999	11.999	12.999	13.999	14.999	15.999	16.999	17.999	18.998	19.998	40
22	1.000	2.000	3.000	4.000	5.000	5.999	6.999	7.999	8.999	9.999	10.999	11.999	12.999	13.999	14.999	15.999	16.999	17.999	18.998	19.998	38
24	1.000	2.000	3.000	4.000	5.000	5.999	6.999	7.999	8.999	9.999	10.999	11.999	12.999	13.999	14.999	15.999	16.999	17.999	18.998	19.998	36
26	1.000	2.000	3.000	4.000	5.000	5.999	6.999	7.999	8.999	9.999	10.999	11.999	12.999	13.999	14.999	15.999	16.999	17.999	18.998	19.998	34
28	1.000	2.000	3.000	4.000	5.000	5.999	6.999	7.999	8.999	9.999	10.999	11.999	12.999	13.999	14.999	15.999	16.999	17.999	18.998	19.998	32
30	1.000	2.000	3.000	4.000	5.000	6.000	7.000	8.000	8.999	9.999	10.999	11.999	12.999	13.999	14.999	15.999	16.999	17.999	18.999	19.999	30
32	1.000	2.000	3.000	4.000	5.000	6.000	7.000	8.000	8.999	9.999	10.999	11.999	12.999	13.999	14.999	15.999	16.999	17.999	18.999	19.999	28
34	1.000	2.000	3.000	4.000	5.000	6.000	7.000	8.000	8.999	9.999	10.999	11.999	12.999	13.999	14.999	15.999	16.999	17.999	18.999	19.999	26
36	1.000	2.000	3.000	4.000	5.000	6.000	7.000	8.000	8.999	9.999	10.999	11.999	12.999	13.999	14.999	15.999	16.999	17.999	18.999	19.999	24
38	1.000	2.000	3.000	4.000	5.000	6.000	7.000	8.000	8.999	9.999	10.999	11.999	12.999	13.999	14.999	15.999	16.999	17.999	18.999	19.999	22
40	1.000	2.000	3.000	4.000	5.000	6.000	7.000	8.000	9.000	10.000	11.000	12.000	13.000	13.999	14.999	15.999	16.999	17.999	18.999	19.999	20
42	1.000	2.000	3.000	4.000	5.000	6.000	7.000	8.000	9.000	10.000	11.000	12.000	13.000	13.999	14.999	15.999	16.999	17.999	18.999	19.999	18
44	1.000	2.000	3.000	4.000	5.000	6.000	7.000	8.000	9.000	10.000	11.000	12.000	13.000	13.999	14.999	15.999	16.999	17.999	18.999	19.999	16
46	1.000	2.000	3.000	4.000	5.000	6.000	7.000	8.000	9.000	10.000	11.000	12.000	13.000	13.999	14.999	15.999	16.999	17.999	18.999	19.999	14
48	1.000	2.000	3.000	4.000	5.000	6.000	7.000	8.000	9.000	10.000	11.000	12.000	13.000	13.999	14.999	15.999	16.999	17.999	18.999	19.999	12
50	1.000	2.000	3.000	4.000	5.000	6.000	7.000	8.000	9.000	10.000	11.000	12.000	13.000	14.000	15.000	16.000	17.000	18.000	19.000	20.000	10
52	1.000	2.000	3.000	4.000	5.000	6.000	7.000	8.000	9.000	10.000	11.000	12.000	13.000	14.000	15.000	16.000	17.000	18.000	19.000	20.000	8
54	1.000	2.000	3.000	4.000	5.000	6.000	7.000	8.000	9.000	10.00	11.000	12.000	13.000	14.000	15.000	16.000	17.000	18.000	19.000	20.000	6
56	1.000	2.000	3.000	4.000	5.000	6.000	7.000	8.000	9.000	10.000	11.000	12.000	13.000	14.000	15.000	16.000	17.000	18.000	19.000	20.000	4
58	1.000	2.000	3.000	4.000	5.000	6.000	7.000	8.000	9.000	10.000	11.000	12.000	13.000	14.000	15.000	16.000	17.000	18.000	19.000	20.000	2
0° 0	1.000	2.000	3.000	4.000	5.000	6.000	7.000	8.000	9.000	10.000	11.000	12.000	13.000	14.000	15.000	16.000	17.000	18.000	19.000	20.000	0° 0
	1	2	3	4	5	6	7	8	9	10	11	12	13	14	15	16	17	18	19	20	III+ IX−

(The greatest Aberration, according to Delambre, $=20''.255$.)

Table A.

Argument = R. A. ∗ − long. ⊙.

	S. O / S. VI	S. I / S. VII	S. II / S. VIII	
	− −	− +	− +	
0° 0′	19″.42	16″.82	9″.71	30° 0′
0 30	19 .42	16 .73	9 .56	29 30
1 0	19 .41	16 .64	9 .41	29 0
1 30	19 .41	16 .56	9 .27	28 30
2 0	19 .41	16 .47	9 .12	28 0
2 30	19 .40	16 .38	8 .97	27 30
3 0	19 .39	16 .29	8 .82	27 0
3 30	19 .38	16 .20	8 .67	26 30
4 0	19 .37	16 .10	8 .51	26 0
4 30	19 .36	16 .01	8 .36	25 30
5 0	19 .34	15 .91	8 .21	25 0
5 30	19 .33	15 .81	8 .06	24 30
6 0	19 .31	15 .71	7 .90	24 0
6 30	19 .29	15 .61	7 .75	23 30
7 0	19 .27	15 .51	7 .59	23 0
7 30	19 .25	15 .41	7 .43	22 30
8 0	19 .23	15 .30	7 .27	22 0
8 30	19 .21	15 .20	7 .12	21 30
9 0	19 .18	15 .09	6 .96	21 0
9 30	19 .15	14 .99	6 .80	20 30
10 0	19 .12	14 .88	6 .64	20 0
10 30	19 .09	14 .77	6 .48	19 30
11 0	19 .06	14 .66	6 .32	19 0
11 30	19 .03	14 .55	6 .16	18 30
12 0	18 .99	14 .43	6 .00	18 0
12 30	18 .96	14 .32	5 .84	17 30
13 0	18 .92	14 .20	5 .68	17 0
13 30	18 .88	14 .09	5 .52	16 30
14 0	18 .84	13 .97	5 .35	16 0
14 30	18 .80	13 .85	5 .19	15 30
15 0	18 .76	13 .73	5 .03	15 0
15 30	18 .72	13 .61	4 .87	14 30
16 0	18 .67	13 .49	4 .70	14 0
16 30	18 .62	13 .37	4 .54	13 30
17 0	18 .57	13 .24	4 .37	13 0
17 30	18 .52	13 .12	4 .21	12 30
18 0	18 .47	12 .99	4 .04	12 0
18 30	18 .42	12 .87	3 .88	11 30
19 0	18 .36	12 .74	3 .71	11 0
19 30	18 .31	12 .61	3 .54	10 30
20 0	18 .25	12 .48	3 .37	10 0
20 30	18 .19	12 .35	3 .21	9 30
21 0	18 .13	12 .22	3 .04	9 0
21 30	18 .07	12 .09	2 .87	8 30
22 0	18 .00	11 .96	2 .70	8 0
22 30	17 .94	11 .83	2 .54	7 30
23 0	17 .87	11 .69	2 .37	7 0
23 30	17 .81	11 .55	2 .20	6 30
24 0	17 .74	11 .41	2 .03	6 0
24 30	17 .67	11 .28	1 .86	5 30
25 0	17 .60	11 .14	1 .69	5 0
25 30	17 .53	11 .00	1 .52	4 30
26 0	17 .45	10 .86	1 .35	4 0
26 30	17 .38	10 .72	1 .19	3 30
27 0	17 .30	10 .58	1 .02	3 0
27 30	17 .23	10 .44	0 .85	2 30
28 0	17 .15	10 .29	0 .68	2 0
28 30	17 .07	10 .15	0 .51	1 30
29 0	16 .98	10 .00	0 .34	1 0
29 30	16 .90	9 .86	0 .17	0 30
30 0	16 .82	9 .71	0 .00	0 0
	− +	− +	− +	
	S. XI / S. V	S. X / S. IV	S. IX / S. III	

Table B.

Argument = R. A. ∗ + long. ⊙.

	S. O / S. VI	S. I / S. VII	S. II / S. VIII	
	+ −	+ −	+ −	
0° 0′	0″.84	0″.73	0″.42	30 0
0 30	0 .84	0 .72	0 .42	29 30
1 0	0 .84	0 .72	0 .41	29 0
1 30	0 .84	0 .72	0 .40	28 30
2 0	0 .84	0 .71	0 .39	28 0
2 30	0 .84	0 .71	0 .39	27 30
3 0	0 .84	0 .70	0 .38	27 0
3 30	0 .84	0 .70	0 .38	26 30
4 0	0 .84	0 .69	0 .37	26 0
4 30	0 .84	0 .69	0 .36	25 30
5 0	0 .83	0 .69	0 .35	25 0
5 30	0 .83	0 .68	0 .35	24 30
6 0	0 .83	0 .68	0 .34	24 0
6 30	0 .83	0 .68	0 .33	23 30
7 0	0 .83	0 .67	0 .33	23 0
7 30	0 .83	0 .67	0 .32	22 30
8 0	0 .83	0 .66	0 .31	22 0
8 30	0 .83	0 .66	0 .31	21 30
9 0	0 .83	0 .65	0 .30	21 0
9 30	0 .83	0 .65	0 .29	20 30
10 0	0 .83	0 .64	0 .29	20 0
10 30	0 .82	0 .64	0 .28	19 30
11 0	0 .82	0 .63	0 .27	19 0
11 30	0 .82	0 .63	0 .27	18 30
12 0	0 .82	0 .62	0 .26	18 0
12 30	0 .82	0 .62	0 .25	17 30
13 0	0 .82	0 .61	0 .25	17 0
13 30	0 .82	0 .61	0 .24	16 30
14 0	0 .81	0 .60	0 .23	16 0
14 30	0 .81	0 .60	0 .23	15 30
15 0	0 .81	0 .59	0 .22	15 0
15 30	0 .81	0 .59	0 .21	14 30
16 0	0 .81	0 .58	0 .20	14 0
16 30	0 .80	0 .58	0 .20	13 30
17 0	0 .80	0 .57	0 .19	13 0
17 30	0 .80	0 .57	0 .18	12 30
18 0	0 .80	0 .56	0 .17	12 0
18 30	0 .80	0 .56	0 .17	11 30
19 0	0 .79	0 .55	0 .16	11 0
19 30	0 .79	0 .55	0 .15	10 30
20 0	0 .79	0 .54	0 .15	10 0
20 30	0 .79	0 .54	0 .14	9 30
21 0	0 .78	0 .53	0 .13	9 0
21 30	0 .78	0 .53	0 .13	8 30
22 0	0 .78	0 .52	0 .12	8 0
22 30	0 .78	0 .51	0 .11	7 30
23 0	0 .77	0 .50	0 .10	7 0
23 30	0 .77	0 .50	0 .10	6 30
24 0	0 .77	0 .49	0 .09	6 0
24 30	0 .76	0 .49	0 .08	5 30
25 0	0 .76	0 .48	0 .07	5 0
25 30	0 .76	0 .47	0 .07	4 30
26 0	0 .75	0 .47	0 .06	4 0
26 30	0 .75	0 .46	0 .05	3 30
27 0	0 .75	0 .46	0 .04	3 0
27 30	0 .74	0 .45	0 .04	2 30
28 0	0 .74	0 .44	0 .03	2 0
28 30	0 .74	0 .44	0 .03	1 30
29 0	0 .73	0 .43	0 .02	1 0
29 30	0 .73	0 .43	0 .01	0 30
30 0	0 .73	0 .42	0 .00	0 0
	+ −	+ −	+ +	
	S. XI / S. V	S. X / S. IV	S. IX / S. III	

Take the two quantities out of their respective Tables, and Multiply their Sum, taken algebraically, by the Secant of the Star's Declination, and the product will be the Aberration in Space; which product, divided by 15, will be the Aberration in Time.

(The greatest Abberration, according to DELAMBRE, =20″. 255.)

TABLE C.

Argument = R. A. ✱ − Long. ☉.

		s. O − / s. VI +	s. I − / s. VII +	s. II − / s. VIII +		
0°	0′	0″.00	9″.71	16 .82	30°	0′
0	30	0 .17	9 .86	16 .90	29	30
1	0	0 .34	10 .00	16 .98	29	0
1	30	0 .51	10 .15	17 .07	28	30
2	0	0 .68	10 .29	17 .15	28	0
2	30	0 .85	10 .44	17 .23	27	30
3	0	1 .02	10 .58	17 .30	27	0
3	30	1 .19	10 .72	17 .38	26	30
4	0	1 .35	10 .86	17 .45	26	0
4	30	1 .52	11 .00	17 .53	25	30
5	0	1 .69	11 .14	17 .60	25	0
5	30	1 .86	11 .28	17 .67	24	30
6	0	2 .03	11 .41	17 .74	24	0
6	30	2 .20	11 .55	17 .81	23	30
7	0	2 .37	11 .69	17 .87	23	0
7	30	2 .54	11 .83	17 .94	22	30
8	0	2 .70	11 .96	18 .00	22	0
8	30	2 .87	12 .09	18 .07	21	30
9	0	3 .04	12 .22	18 .13	21	0
9	30	3 .21	12 .35	18 .19	20	30
10	0	3 .37	12 .48	18 .25	20	0
10	30	3 .54	12 .61	18 .31	19	30
11	0	3 .71	12 .74	18 .36	19	0
11	30	3 .88	12 .87	18 .42	18	30
12	0	4 .04	12 .99	18 .47	18	0
12	30	4 .21	13 .12	18 .52	17	30
13	0	4 .37	13 .24	18 .57	17	0
13	30	4 .54	13 .37	18 .62	16	30
14	0	4 .70	13 .49	18 .67	16	0
14	30	4 .87	13 .61	18 .72	15	30
15	0	5 .03	13 .73	18 .76	15	0
15	30	5 .19	13 .85	18 .80	14	30
16	0	5 .35	13 .97	18 .84	14	0
16	30	5 .52	14 .09	18 .88	13	30
17	0	5 .68	14 .20	18 .92	13	0
17	30	5 .84	14 .32	18 .96	12	30
18	0	6 .00	14 .43	18 .99	12	0
18	30	6 .16	14 .55	19 .03	11	30
19	0	6 .32	14 .66	19 .06	11	0
19	30	6 .48	14 .77	19 .09	10	30
20	0	6 .64	14 .88	19 .12	10	0
20	30	6 .80	14 .99	19 .15	9	30
21	0	6 .96	15 .09	19 .18	9	0
21	30	7 .12	15 .20	19 .21	8	30
22	0	7 .27	15 .30	19 .23	8	0
22	30	7 .43	15 .41	19 .25	7	30
23	0	7 .59	15 .51	19 .27	7	0
23	30	7 .75	15 .61	19 .29	6	30
24	0	7 .90	15 .71	19 .31	6	0
24	30	8 .06	15 .81	19 .33	5	30
25	0	8 .21	15 .91	19 .34	5	0
25	30	8 .36	16 .01	19 .36	4	30
26	0	8 .51	16 .10	19 .37	4	0
26	30	8 .67	16 .20	19 .38	3	30
27	0	8 .82	16 .29	19 .39	3	0
27	30	8 .97	16 .38	19 .40	2	30
28	0	9 .12	16 .47	19 .41	2	0
28	30	9 .27	16 .56	19 .41	1	30
29	0	9 .41	16 .64	19 .41	1	0
29	30	9 .56	16 .73	19 .42	0	30
30	0	9 .71	16 .82	19 .42	0	0
		+ − / s. XI s. V	+ − / s. X s. IV	+ − / s. IX s. III		

TABLE D.

Argument = R. A. ✱ + Long. ☉.

		s. O + / s. VI −	s. I + / s. VII −	s. II + / s. VIII −		
0°	0′	0″.00	0″.42	0″.73	30°	0′
0	30	0 .01	0 .43	0 .73	29	30
1	0	0 .02	0 .43	0 .73	29	0
1	30	0 .03	0 .44	0 .74	28	30
2	0	0 .03	0 .44	0 .74	28	0
2	30	0 .04	0 .45	0 .74	27	30
3	0	0 .04	0 .46	0 .75	27	0
3	30	0 .05	0 .46	0 .75	26	30
4	0	0 .06	0 .47	0 .75	26	0
4	30	0 .07	0 .47	0 .76	25	30
5	0	0 .07	0 .48	0 .76	25	0
5	30	0 .08	0 .49	0 .76	24	30
6	0	0 .09	0 .49	0 .77	24	0
6	30	0 .10	0 .50	0 .77	23	30
7	0	0 .10	0 .50	0 .77	23	0
7	30	0 .11	0 .51	0 .78	22	30
8	0	0 .12	0 .52	0 .78	22	0
8	30	0 .13	0 .53	0 .78	21	30
9	0	0 .13	0 .53	0 .78	21	0
9	30	0 .14	0 .54	0 .79	20	30
10	0	0 .15	0 .54	0 .79	20	0
10	30	0 .15	0 .55	0 .79	19	30
11	0	0 .16	0 .55	0 .79	19	0
11	30	0 .17	0 .56	0 .80	18	30
12	0	0 .17	0 .56	0 .80	18	0
12	30	0 .18	0 .57	0 .80	17	30
13	0	0 .19	0 .57	0 .80	17	0
13	30	0 .20	0 .58	0 .80	16	30
14	0	0 .20	0 .58	0 .81	16	0
14	30	0 .21	0 .59	0 .81	15	30
15	0	0 .22	0 .59	0 .81	15	0
15	30	0 .23	0 .60	0 .81	14	30
16	0	0 .23	0 .60	0 .81	14	0
16	30	0 .24	0 .61	0 .82	13	30
17	0	0 .25	0 .61	0 .82	13	0
17	30	0 .25	0 .62	0 .82	12	30
18	0	0 .26	0 .62	0 .82	12	0
18	30	0 .27	0 .63	0 .82	11	30
19	0	0 .27	0 .63	0 .82	11	0
19	30	0 .28	0 .64	0 .82	10	30
20	0	0 .29	0 .64	0 .83	10	0
20	30	0 .29	0 .65	0 .83	9	30
21	0	0 .30	0 .65	0 .83	9	0
21	30	0 .31	0 .66	0 .83	8	30
22	0	0 .31	0 .66	0 .83	8	0
22	30	0 .32	0 .67	0 .83	7	30
23	0	0 .33	0 .67	0 .83	7	0
23	30	0 .33	0 .68	0 .83	6	30
24	0	0 .34	0 .68	0 .83	6	0
24	30	0 .35	0 .68	0 .83	5	30
25	0	0 .35	0 .69	0 .83	5	0
25	30	0 .36	0 .69	0 .84	4	30
26	0	0 .37	0 .69	0 .84	4	0
26	30	0 .38	0 .70	0 .84	3	30
27	0	0 .38	0 .70	0 .84	3	0
27	30	0 .39	0 .71	0 .84	2	30
28	0	0 .39	0 .71	0 .84	2	0
28	30	0 .40	0 .72	0 .84	1	30
29	0	0 .41	0 .72	0 .84	1	0
29	30	0 .42	0 .72	0 .84	0	30
30	0	0 .42	0 .73	0 .84	0	0
		− + / s. XI s. V	− + / s. X s. IV	− + / s. IX s. III		

TABLE E.

Arguments.

(Long. ☉ + Dec. ✱) and (Long. ☉ − Dec. ✱.)

		s. O + / s. VI −	s. I + / s. VII −	s. II + / s. VIII −		
0°	0′	4″.03	3″.49	2″.02	30°	0′
0	30	4 .03	3 .48	1 .99	29	30
1	0	4 .03	3 .46	1 .96	29	0
1	30	4 .03	3 .44	1 .93	28	30
2	0	4 .03	3 .42	1 .89	28	0
2	30	4 .03	3 .40	1 .86	27	30
3	0	4 .03	3 .38	1 .83	27	0
3	30	4 .02	3 .36	1 .80	26	30
4	0	4 .02	3 .34	1 .77	26	0
4	30	4 .02	3 .32	1 .74	25	30
5	0	4 .02	3 .30	1 .70	25	0
5	30	4 .01	3 .28	1 .67	24	30
6	0	4 .01	3 .26	1 .64	24	0
6	30	4 .00	3 .24	1 .61	23	30
7	0	4 .00	3 .22	1 .58	23	0
7	30	3 .99	3 .20	1 .55	22	30
8	0	3 .99	3 .18	1 .51	22	0
8	30	3 .98	3 .16	1 .48	21	30
9	0	3 .98	3 .13	1 .45	21	0
9	30	3 .97	3 .11	1 .42	20	30
10	0	3 .97	3 .09	1 .38	20	0
10	30	3 .96	3 .07	1 .35	19	30
11	0	3 .96	3 .04	1 .31	19	0
11	30	3 .95	3 .02	1 .28	18	30
12	0	3 .95	3 .00	1 .25	18	0
12	30	3 .94	2 .98	1 .22	17	30
13	0	3 .93	2 .95	1 .18	17	0
13	30	3 .92	2 .93	1 .15	16	30
14	0	3 .91	2 .90	1 .11	16	0
14	30	3 .91	2 .88	1 .08	15	30
15	0	3 .90	2 .85	1 .04	15	0
15	30	3 .89	2 .83	1 .01	14	30
16	0	3 .88	2 .80	0 .98	14	0
16	30	3 .87	2 .78	0 .95	13	30
17	0	3 .86	2 .75	0 .91	13	0
17	30	3 .85	2 .77	0 .88	12	30
18	0	3 .84	2 .70	0 .84	12	0
18	30	3 .83	2 .68	0 .81	11	30
19	0	3 .81	2 .65	0 .77	11	0
19	30	3 .80	2 .62	0 .74	10	30
20	0	3 .79	2 .59	0 .70	10	0
20	30	3 .78	2 .57	0 .67	9	30
21	0	3 .77	2 .54	0 .63	9	0
21	30	3 .76	2 .51	0 .60	8	30
22	0	3 .74	2 .48	0 .56	8	0
22	30	3 .73	2 .46	0 .53	7	30
23	0	3 .71	2 .43	0 .49	7	0
23	30	3 .70	2 .40	0 .46	6	30
24	0	3 .68	2 .37	0 .42	6	0
24	30	3 .67	2 .34	0 .39	5	30
25	0	3 .66	2 .31	0 .35	5	0
25	30	3 .64	2 .29	0 .32	4	30
26	0	3 .63	2 .26	0 .28	4	0
26	30	3 .61	2 .23	0 .25	3	30
27	0	3 .59	2 .20	0 .21	3	0
27	30	3 .58	2 .17	0 .18	2	30
28	0	3 .56	2 .14	0 .14	2	0
28	30	3 .55	2 .11	0 .11	1	30
29	0	3 .53	2 .08	0 .07	1	0
29	30	3 .51	2 .05	0 .04	0	30
30	0	3 .49	2 .02	0 .00	0	0
		+ − / s. XI s. V	+ − / s. X s. IV	+ − / s. IX s. III		

Multiply the sum of the numbers taken from Tables C. and D. with their algebraic signs, by the sine of the Star's declination, and the product will be the first part of the correction; to which add algebraically the second and third parts, taken successively from Table E. and the amount will be the whole correction in North Polar Distance. If the correction should be required in *North Declination*, all the signs must be changed.

(The Ratio of the Elliptic Axes, according to LA PLACE, =20″.111 : 14″.971.)

TABLE F.
Argument = R. A. ✱ − Long. ☽'s ☊.

	S. O − / S. VI +	S. I − / S. VII +	S. II − / S. VIII +	
0° 0′	8″.77	7″.60	4″.39	30° 0′
0 30	8.77	7.56	4.32	29 30
1 0	8.77	7.52	4.25	29 0
1 30	8.77	7.48	4.19	28 30
2 0	8.77	7.44	4.12	28 0
2 30	8.76	7.40	4.05	27 30
3 0	8.76	7.36	3.98	27 0
3 30	8.75	7.32	3.91	26 30
4 0	8.75	7.27	3.84	26 0
4 30	8.74	7.23	3.78	25 30
5 0	8.74	7.18	3.71	25 0
5 30	8.73	7.14	3.64	24 30
6 0	8.72	7.10	3.57	24 0
6 30	8.72	7.05	3.50	23 30
7 0	8.71	7.00	3.43	23 0
7 30	8.70	6.96	3.36	22 30
8 0	8.69	6.91	3.29	22 0
8 30	8.67	6.87	3.22	21 30
9 0	8.66	6.82	3.14	21 0
9 30	8.65	6.77	3.07	20 30
10 0	8.64	6.72	3.00	20 0
10 30	8.62	6.67	2.93	19 30
11 0	8.61	6.62	2.86	19 0
11 30	8.60	6.57	2.79	18 30
12 0	8.58	6.52	2.71	18 0
12 30	8.57	6.47	2.64	17 30
13 0	8.55	6.41	2.56	17 0
13 30	8.53	6.36	2.49	16 30
14 0	8.51	6.31	2.42	16 0
14 30	8.49	6.26	2.35	15 30
15 0	8.47	6.20	2.27	15 0
15 30	8.45	6.15	2.20	14 30
16 0	8.43	6.09	2.12	14 0
16 30	8.41	6.04	2.05	13 30
17 0	8.39	5.98	1.97	13 0
17 30	8.37	5.93	1.90	12 30
18 0	8.34	5.87	1.82	12 0
18 30	8.32	5.81	1.75	11 30
19 0	8.29	5.75	1.67	11 0
19 30	8.27	5.70	1.60	10 30
20 0	8.24	5.64	1.52	10 0
20 30	8.21	5.58	1.45	9 30
21 0	8.19	5.52	1.37	9 0
21 30	8.16	5.46	1.30	8 30
22 0	8.13	5.39	1.22	8 0
22 30	8.10	5.34	1.15	7 30
23 0	8.07	5.28	1.07	7 0
23 30	8.04	5.22	1.00	6 30
24 0	8.01	5.16	0.92	6 0
24 30	7.98	5.10	0.84	5 30
25 0	7.94	5.03	0.76	5 0
25 30	7.91	4.97	0.69	4 30
26 0	7.88	4.90	0.61	4 0
26 30	7.84	4.84	0.54	3 30
27 0	7.81	4.78	0.46	3 0
27 30	7.78	4.72	0.39	2 30
28 0	7.74	4.65	0.31	2 0
28 30	7.71	4.59	0.23	1 30
29 0	7.67	4.52	0.15	1 0
29 30	7.61	4.46	0.08	0 30
30 0	7.60	4.39	0.00	0 0
	− S. XI / + S. V	− S. X / + S. IV	− S. IX / + S. III	

TABLE G.
Argument = R. A. ✱ + Long. ☽'s ☊.

	S. O − / S. VI +	S. I − / S. VII +	S. II − / S. VIII +	
0° 0′	1″.29	1″.11	0.64	30° 0′
0 30	1.29	1.10	0.63	29 30
1 0	1.28	1.10	0.62	29 0
1 30	1.28	1.09	0.61	28 30
2 0	1.28	1.09	0.60	28 0
2 30	1.28	1.08	0.59	27 30
3 0	1.28	1.08	0.58	27 0
3 30	1.28	1.07	0.57	26 30
4 0	1.28	1.07	0.56	26 0
4 30	1.28	1.06	0.55	25 30
5 0	1.28	1.06	0.54	25 0
5 30	1.28	1.05	0.53	24 30
6 0	1.28	1.04	0.52	24 0
6 30	1.28	1.04	0.51	23 30
7 0	1.28	1.03	0.50	23 0
7 30	1.27	1.02	0.49	22 30
8 0	1.27	1.01	0.48	22 0
8 30	1.27	1.01	0.47	21 30
9 0	1.27	1.00	0.46	21 0
9 30	1.27	0.99	0.45	20 30
10 0	1.27	0.98	0.44	20 0
10 30	1.27	0.98	0.43	19 30
11 0	1.26	0.97	0.42	19 0
11 30	1.26	0.96	0.41	18 30
12 0	1.26	0.95	0.40	18 0
12 30	1.26	0.95	0.39	17 30
13 0	1.25	0.94	0.38	17 0
13 30	1.25	0.93	0.37	16 30
14 0	1.25	0.92	0.35	16 0
14 30	1.25	0.92	0.34	15 30
15 0	1.24	0.91	0.33	15 0
15 30	1.24	0.90	0.32	14 30
16 0	1.24	0.89	0.31	14 0
16 30	1.23	0.89	0.30	13 30
17 0	1.23	0.88	0.29	13 0
17 30	1.23	0.87	0.28	12 30
18 0	1.22	0.86	0.27	12 0
18 30	1.22	0.85	0.26	11 30
19 0	1.22	0.84	0.25	11 0
19 30	1.21	0.84	0.24	10 30
20 0	1.21	0.83	0.22	10 0
20 30	1.20	0.82	0.21	9 30
21 0	1.20	0.81	0.20	9 0
21 30	1.20	0.80	0.19	8 30
22 0	1.19	0.79	0.18	8 0
22 30	1.19	0.78	0.17	7 30
23 0	1.18	0.77	0.16	7 0
23 30	1.18	0.77	0.15	6 30
24 0	1.17	0.76	0.13	6 0
24 30	1.17	0.75	0.12	5 30
25 0	1.16	0.74	0.11	5 0
25 30	1.16	0.73	0.10	4 30
26 0	1.15	0.72	0.09	4 0
26 30	1.15	0.71	0.08	3 30
27 0	1.14	0.70	0.07	3 0
27 30	1.14	0.69	0.06	2 30
28 0	1.13	0.68	0.04	2 0
28 30	1.12	0.67	0.03	1 30
29 0	1.12	0.66	0.02	1 0
29 30	1.11	0.65	0.01	0 30
30 0	1.11	0.64	0.00	0 0
	− S. XI / + S. V	− S. X / + S. IV	− S. IX / + S. III	

TABLE H.
Argument = Long. of ☽'s ☊.

	S. O − / S. VI +	S. I − / S. VII +	S. II − / S. VIII +	
0° 0′	0″.00	8″.62	14″.93	30° 0′
0 30	0.15	8.75	15.01	29 30
1 0	0.30	8.88	15.08	29 0
1 30	0.45	9.01	15.16	28 30
2 0	0.60	9.14	15.23	28 0
2 30	0.75	9.27	15.30	27 30
3 0	0.90	9.39	15.36	27 0
3 30	1.05	9.52	15.43	26 30
4 0	1.20	9.64	15.50	26 0
4 30	1.35	9.77	15.57	25 30
5 0	1.50	9.89	15.63	25 0
5 30	1.65	10.02	15.69	24 30
6 0	1.80	10.14	15.75	24 0
6 30	1.95	10.26	15.81	23 30
7 0	2.10	10.38	15.87	23 0
7 30	2.25	10.50	15.93	22 30
8 0	2.40	10.62	15.99	22 0
8 30	2.55	10.74	16.05	21 30
9 0	2.70	10.85	16.10	21 0
9 30	2.85	10.97	16.15	20 30
10 0	2.99	11.08	16.20	20 0
10 30	3.14	11.20	16.25	19 30
11 0	3.29	11.31	16.30	19 0
11 30	3.44	11.43	16.35	18 30
12 0	3.59	11.54	16.40	18 0
12 30	3.74	11.65	16.45	17 30
13 0	3.88	11.76	16.49	17 0
13 30	4.03	11.87	16.54	16 30
14 0	4.17	11.98	16.58	16 0
14 30	4.32	12.09	16.62	15 30
15 0	4.46	12.19	16.66	15 0
15 30	4.61	12.30	16.70	14 30
16 0	4.75	12.40	16.73	14 0
16 30	4.90	12.51	16.77	13 30
17 0	5.04	12.61	16.88	13 0
17 30	5.19	12.71	16.84	12 30
18 0	5.33	12.81	16.87	12 0
18 30	5.47	12.91	16.90	11 30
19 0	5.61	13.01	16.93	11 0
19 30	5.76	13.11	16.96	10 30
20 0	5.90	13.21	16.98	10 0
20 30	6.04	13.31	17.01	9 30
21 0	6.18	13.40	17.03	9 0
21 30	6.32	13.50	17.06	8 30
22 0	6.46	13.59	17.08	8 0
22 30	6.60	13.68	17.10	7 30
23 0	6.74	13.77	17.12	7 0
23 30	6.88	13.86	17.14	6 30
24 0	7.01	13.95	17.15	6 0
24 30	7.15	14.04	17.17	5 30
25 0	7.29	14.13	17.18	5 0
25 30	7.42	14.22	17.19	4 30
26 0	7.56	14.30	17.20	4 0
26 30	7.70	14.38	17.21	3 30
27 0	7.83	14.46	17.22	3 0
27 30	7.97	14.54	17.22	2 30
28 0	8.10	14.62	17.23	2 0
28 30	8.23	14.70	17.23	1 30
29 0	8.36	14.78	17.24	1 0
29 30	8.49	14.86	17.24	0 30
30 0	8.62	14.93	17.24	0 0
	+ S. XI / − S. V	+ S. X / − S. IV	+ S. IX / − S. III	

Multiply the Sum of the Corrections taken from Tables F. and G. by the Tangent of the Star's Declination; and if the Declination be South, change the Signs of these two Tables; to the Product apply the quantity taken out of Table H. to obtain the whole Correction.

(The Ratio of the Elliptic Axes, according to La Place, =20″. 111 : 14″.971.)

Table I.

Argument = R. A. ✱ − Long. ☽'s ☊.

	S. 0 − / S. VI +	S. I − / S. VII +	S. II − / S. VIII +	
0° 0′	0″.00	4″.39	7″.60	30° 0′
0 30	0 .08	4 .46	7 .64	29 30
1 0	0 .15	4 .52	7 .67	29 0
1 30	0 .23	4 .59	7 .71	28 30
2 0	0 .31	4 .65	7 .74	28 0
2 30	0 .39	4 .72	7 .78	27 30
3 0	0 .46	4 .78	7 .81	27 0
3 30	0 .54	4 .84	7 .84	26 30
4 0	0 .61	4 .90	7 .88	26 0
4 30	0 .69	4 .97	7 .91	25 30
5 0	0 .76	5 .03	7 .94	25 0
5 30	0 .84	5 .10	7 .98	24 30
6 0	0 .92	5 .16	8 .01	24 0
6 30	1 .00	5 .22	8 .04	23 30
7 0	1 .07	5 .28	8 .07	23 0
7 30	1 .15	5 .34	8 .10	22 30
8 0	1 .22	5 .39	8 .13	22 0
8 30	1 .30	5 .46	8 .16	21 30
9 0	1 .37	5 .52	8 .19	21 0
9 30	1 .45	5 .58	8 .21	20 30
10 0	1 .52	5 .64	8 .24	20 0
10 30	1 .60	5 .70	8 .27	19 30
11 0	1 .67	5 .75	8 .29	19 0
11 30	1 .75	5 .81	8 .32	18 30
12 0	1 .82	5 .87	8 .34	18 0
12 30	1 .90	5 .93	8 .37	17 30
13 0	1 .97	5 .98	8 .39	17 0
13 30	2 .05	6 .04	8 .41	16 30
14 0	2 .12	6 .09	8 .43	16 0
14 30	2 .20	6 .15	8 .45	15 30
15 0	2 .27	6 .20	8 .47	15 0
15 30	2 .35	6 .26	8 .49	14 30
16 0	2 .42	6 .31	8 .51	14 0
16 30	2 .49	6 .36	8 .53	13 30
17 0	2 .56	6 .41	8 .55	13 0
17 30	2 .64	6 .47	8 .57	12 30
18 0	2 .71	6 .52	8 .58	12 0
18 30	2 .79	6 .57	8 .60	11 30
19 0	2 .86	6 .62	8 .61	11 0
19 30	2 .93	6 .67	8 .62	10 30
20 0	3 .00	6 .72	8 .64	10 0
20 30	3 .07	6 .77	8 .65	9 30
21 0	3 .14	6 .82	8 .66	9 0
21 30	3 .22	6 .87	8 .67	8 30
22 0	3 .29	6 .91	8 .69	8 0
22 30	3 .36	6 .96	8 .70	7 30
23 0	3 .43	7 .00	8 .71	7 0
23 30	3 .50	7 .05	8 .72	6 30
24 0	3 .57	7 .10	8 .72	6 0
24 30	3 .64	7 .14	8 .73	5 30
25 0	3 .71	7 .18	8 .74	5 0
25 30	3 .78	7 .23	8 .74	4 30
26 0	3 .84	7 .27	8 .75	4 0
26 30	3 .91	7 .32	8 .75	3 30
27 0	3 .98	7 .36	8 .76	3 0
27 30	4 .05	7 .40	8 .76	2 30
28 0	4 .12	7 .44	8 .77	2 0
28 30	4 .19	7 .48	8 .77	1 30
29 0	4 .25	7 .52	8 .77	1 0
29 30	4 .32	7 .56	8 .77	0 30
30 0	4 .39	7 .60	8 .77	0 0
	+ S. XI / − S. V	+ S. X / − S. IV	+ S. IX / − S. III	

Table K.

Argument = R. A. ✱ + Long. ☽'s ☊.

	S. 0 − / S. VI +	S. I − / S. VII +	S. II − / S. VIII +	
0° 0′	0″.00	0″.64	1″.11	30° 0′
0 30	0 .01	0 .65	1 .11	29 30
1 0	0 .02	0 .66	1 .12	29 0
1 30	0 .03	0 .67	1 .12	28 30
2 0	0 .04	0 .68	1 .13	28 0
2 30	0 .06	0 .69	1 .14	27 30
3 0	0 .07	0 .70	1 .14	27 0
3 30	0 .08	0 .71	1 .15	26 30
4 0	0 .09	0 .72	1 .15	26 0
4 30	0 .10	0 .73	1 .16	25 30
5 0	0 .11	0 .74	1 .16	25 0
5 30	0 .12	0 .75	1 .17	24 30
6 0	0 .13	0 .76	1 .17	24 0
6 30	0 .15	0 .77	1 .18	23 30
7 0	0 .16	0 .77	1 .18	23 0
7 30	0 .17	0 .78	1 .19	22 30
8 0	0 .18	0 .79	1 .19	22 0
8 30	0 .19	0 .80	1 .20	21 30
9 0	0 .20	0 .81	1 .20	21 0
9 30	0 .21	0 .82	1 .20	20 30
10 0	0 .22	0 .83	1 .21	20 0
10 30	0 .24	0 .84	1 .21	19 30
11 0	0 .25	0 .84	1 .22	19 0
11 30	0 .26	0 .85	1 .22	18 30
12 0	0 .27	0 .86	1 .22	18 0
12 30	0 .28	0 .87	1 .23	17 30
13 0	0 .29	0 .88	1 .23	17 0
13 30	0 .30	0 .89	1 .23	16 30
14 0	0 .31	0 .89	1 .24	16 0
14 30	0 .32	0 .90	1 .24	15 30
15 0	0 .33	0 .91	1 .24	15 0
15 30	0 .34	0 .92	1 .25	14 30
16 0	0 .35	0 .92	1 .25	14 0
16 30	0 .37	0 .93	1 .25	13 30
17 0	0 .38	0 .94	1 .25	13 0
17 30	0 .39	0 .95	1 .26	12 30
18 0	0 .40	0 .95	1 .26	12 0
18 30	0 .41	0 .96	1 .26	11 30
19 0	0 .42	0 .97	1 .26	11 0
19 30	0 .43	0 .98	1 .27	10 30
20 0	0 .44	0 .98	1 .27	10 0
20 30	0 .45	0 .99	1 .27	9 30
21 0	0 .46	1 .00	1 .27	9 0
21 30	0 .47	1 .01	1 .27	8 30
22 0	0 .48	1 .01	1 .27	8 0
22 30	0 .49	1 .02	1 .27	7 30
23 0	0 .50	1 .03	1 .28	7 0
23 30	0 .51	1 .04	1 .28	6 30
24 0	0 .52	1 .04	1 .28	6 0
24 30	0 .53	1 .05	1 .28	5 30
25 0	0 .54	1 .06	1 .28	5 0
25 30	0 .55	1 .06	1 .28	4 30
26 0	0 .56	1 .07	1 .28	4 0
26 30	0 .57	1 .07	1 .28	3 30
27 0	0 .58	1 .08	1 .28	3 0
27 30	0 .59	1 .08	1 .28	2 30
28 0	0 .60	1 .09	1 .28	2 0
28 30	0 .61	1 .09	1 .28	1 30
29 0	0 .62	1 .10	1 .28	1 0
29 30	0 .63	1 .10	1 .29	0 30
30 0	0 .64	1 .11	1 .29	0 0
	+ S. XI / − S. V	+ S. X / − S. IV	+ S. IX / − S. III	

The Sum of the Corrections taken out of these two Tables, according to their Algebraic Signs, will give the whole Correction, in North Polar Distance; but if it be required for North *Declination*, the Signs must be changed.

The Tables F, G, H, I, and K, will also give the *Solar Nutation*, provided 2 ⊙ Longitude be substituted for ☽'s ☊ in the several Arguments; and provided the resulting corrections be multiplied by 0.06221.

(In Space.)

TABLE 1.

The Change of Aberration in Right Ascension, corresponding to a change of one Degree in the Right Ascension of a Star's Place.

Arguments = the ∗'s Declination at top, and at the side = R. A. ∗ — ☉'s Longitude.

+		−	0°	10°	20°	30°	40°	50°	60°	70°	80°	+		−
S.		S.										S.		S.
O	0°	VI	0″.000	0″.000	0″.000	0″.000	0″.000	0″.000	0″.000	0″.000	0″.000	VI	0°	O
	10		0 .059	0 .060	0 .063	0 .068	0 .077	0 .092	0 .118	0 .172	0 .339		10	
	20		0 .116	0 .118	0 .124	0 .134	0 .151	0 .180	0 .232	0 .339	0 .668		20	
I	0	VII	0 .169	0 .172	0 .180	0 .196	0 .221	0 .264	0 .339	0 .495	0 .976	V	0	XI
	10		0 .218	0 .221	0 .232	0 .252	0 .284	0 .339	0 .436	0 .637	1 .254		10	
	20		0 .260	0 .264	0 .276	0 .300	0 .339	0 .404	0 .519	0 .759	1 .495		20	
II	0	VIII	0 .294	0 .298	0 .312	0 .339	0 .383	0 .457	0 .587	0 .858	1 .690	IV	0	X
	10		0 .318	0 .323	0 .339	0 .368	0 .416	0 .495	0 .636	0 .931	1 .833		10	
	20		0 .334	0 .339	0 .355	0 .385	0 .436	0 .519	0 .668	0 .976	1 .922		20	
III	0	IX	0 .339	0 .344	0 .361	0 .391	0 .442	0 .527	0 .678	0 .991	1 .952	III	0	IX

TABLE 2.

The Change of Aberration in Right Ascension corresponding to a change of one Degree in the Declination of a Star's Place.

Arguments = the ∗'s declination at top, and at the side = R. A. ∗ — ☉'s Longitude.

−		+	0°	10°	20°	30°	40°	50°	60°	70°	80°	+		−
S.		S.										S.		S.
O	0°	VI	0″.000	0″.061	0″.131	0″.226	0″.371	0″.629	1″.174	2″.723	11″.069	VI	0°	O
	10		0 .000	0 .060	0 .129	0 .222	0 .366	0 .619	1 .156	2 .681	10 .901		10	
	20		0 .000	0 .057	0 .123	0 .212	0 .349	0 .591	1 .103	2 .558	10 .402		20	
I	0	VII	0 .000	0 .053	0 .114	0 .196	0 .322	0 .544	1 .017	2 .363	9 .586	V	0	XI
	10		0 .000	0 .047	0 .101	0 .173	0 .284	0 .481	0 .899	2 .086	8 .480		10	
	20		0 .000	0 .039	0 .084	0 .145	0 .239	0 .404	0 .755	1 .750	7 .115		20	
II	0	VIII	0 .000	0 .030	0 .066	0 .113	0 .186	0 .314	0 .587	1 .361	5 .535	IV	0	X
	10		0 .000	0 .021	0 .045	0 .077	0 .127	0 .215	0 .402	0 .931	3 .786		10	
	20		0 .000	0 .011	0 .023	0 .039	0 .065	0 .109	0 .204	0 .473	1 .922		20	
III	0	IX	0 .000	0 .000	0 .000	0 .000	0 .000	0 .000	0 .000	0 .000	0 .000	III	0	IX

TABLE 3.

The Change of Aberration in Declination, corresponding to a change of one Degree in the Right Ascension of a Star's Place.

Arguments = the ∗'s declination at the top, and at the side = R. A. ∗ — ☉'s Longitude.

+		−	10°	20°	30°	40°	50°	60°	70°	80°	90°	−		+
S.		S.										S.		S.
O	0°	VI	0″.059	0″.116	0″.170	0″.218	0″.260	0″.293	0″.318	0″.334	0″.339	VI	0°	O
	10		0 .058	0 .114	0 .167	0 .214	0 .256	0 .289	0 .314	0 .329	0 .334		10	
	20		0 .055	0 .109	0 .159	0 .205	0 .244	0 .276	0 .299	0 .314	0 .318		20	
I	0	VII	0 .051	0 .100	0 .147	0 .189	0 .225	0 .254	0 .276	0 .289	0 .293	V	0	XI
	10		0 .045	0 .089	0 .130	0 .167	0 .199	0 .225	0 .244	0 .256	0 .260		10	
	20		0 .038	0 .075	0 .109	0 .140	0 .167	0 .189	0 .205	0 .214	0 .218		20	
II	0	VIII	0 .029	0 .058	0 .085	0 .109	0 .130	0 .147	0 .159	0 .167	0 .170	IV	0	X
	10		0 .020	0 .040	0 .058	0 .075	0 .089	0 .100	0 .109	0 .114	0 .116		10	
	20		0 .010	0 .020	0 .029	0 .038	0 .045	0 .051	0 .055	0 .059	0 .059		20	
III	0	IX	0 .000	0 .000	0 .000	0 .000	0 .000	0 .000	0 .000	0 .000	0 .000	III	0	IX

TABLE 4.

The Change of Aberration in Declination corresponding to a change of one Degree in the Declination of a Star's Place.

Arguments = the ∗'s declination at the top, and at the side = R. A. ∗ — ☉'s Longitude.

+		−	0°	10°	20°	30°	40°	50°	60°	70°	80°	+		−
S.		S.										S.		S.
O	0°	VI	0″.000	0″.000	0″.000	0″.000	0″.000	0″.000	0″.000	0″.000	0″.000	VI	0°	O
	10		0 .059	0 .058	0 .055	0 .051	0 .045	0 .038	0 .029	0 .020	0 .010		10	
	20		0 .116	0 .114	0 .109	0 .100	0 .089	0 .075	0 .058	0 .040	0 .020		20	
I	0	VII	0 .170	0 .167	0 .159	0 .147	0 .130	0 .109	0 .085	0 .058	0 .029	V	0	XI
	10		0 .218	0 .214	0 .205	0 .189	0 .167	0 .140	0 .109	0 .075	0 .038		10	
	20		0 .260	0 .256	0 .244	0 .225	0 .199	0 .167	0 .130	0 .089	0 .045		20	
II	0	VIII	0 .293	0 .289	0 .276	0 .254	0 .225	0 .189	0 .147	0 .100	0 .051	IV	0	X
	10		0 .318	0 .314	0 .299	0 .276	0 .244	0 .205	0 .159	0 .109	0 .055		10	
	20		0 .334	0 .329	0 .314	0 .289	0 .256	0 .214	0 .167	0 .114	0 .058		20	
III	0	IX	0 .339	0 .334	0 .318	0 .293	0 .260	0 .218	0 .170	0 .116	0 .059	III	0	IX

TABLE 5.

The CHANGE of LUNAR NUTATION in RIGHT ASCENSION, corresponding to a change of ONE DEGREE in the RIGHT ASCENSION of a STAR'S PLACE.

Arguments = the ∗'s declination at the top, and at the side = R. A. ∗ − ☊ ☾.

+ S.		− S.	0°	10°	20°	30°	40°	50°	60°	70°	80°	+ S.		− S.
O	0°	VI	0″.000	0″.000	0″.000	0″.000	0″.000	0″.000	0″.000	0″.000	0″.000	VI	0°	O
	10		0 .000	0 .005	0 .009	0 .015	0 .021	0 .030	0 .044	0 .070	0 .145		20	
	20		0 .000	0 .009	0 .018	0 .029	0 .042	0 .060	0 .087	0 .138	0 .285		10	
I	0	VII	0 .000	0 .013	0 .027	0 .042	0 .062	0 .088	0 .127	0 .202	0 .417	V	0	XI
	10		0 .000	0 .017	0 .034	0 .055	0 .079	0 .113	0 .164	0 .259	0 .535		20	
	20		0 .000	0 .020	0 .041	0 .065	0 .094	0 .134	0 .195	0 .302	0 .624		10	
II	0	VIII	0 .000	0 .022	0 .046	0 .073	0 .107	0 .152	0 .220	0 .350	0 .721	IV	0	X
	10		0 .000	0 .024	0 .050	0 .080	0 .116	0 .164	0 .239	0 .379	0 .783		20	
	20		0 .000	0 .025	0 .053	0 .084	0 .121	0 .172	0 .251	0 .397	0 .820		10	
III	0	IX	0 .000	0 .026	0 .054	0 .085	0 .123	0 .175	0 .254	0 .403	0 .833	III	0	IX

TABLE 6.

The CHANGE of LUNAR NUTATION in RIGHT ASCENSION, corresponding to a change of ONE DEGREE in the DECLINATION of a STAR'S PLACE.

Arguments = the ∗'s declination at the top, and at the side = R. A. ∗ − ☊ ☾.

− S.		+ S.	0°	10°	20°	30°	40°	50°	60°	70°	80°	+ S.		− S.
O	0°	VI	0″.147	0″.151	0″.166	0″.196	0″.250	0″.356	0″.588	1″.256	4″.871	VI	0°	O
	10		0 .145	0 .149	0 .164	0 .193	0 .246	0 .350	0 .579	1 .237	4 .797		20	
	20		0 .138	0 .142	0 .156	0 .184	0 .235	0 .334	0 .552	1 .180	4 .577		10	
I	0	VII	0 .127	0 .131	0 .144	0 .170	0 .217	0 .308	0 .509	1 .087	4 .219	V	0	XI
	10		0 .113	0 .116	0 .127	0 .150	0 .192	0 .272	0 .450	0 .962	3 .732		20	
	20		0 .094	0 .098	0 .107	0 .126	0 .161	0 .229	0 .378	0 .807	3 .131		10	
II	0	VIII	0 .073	0 .076	0 .083	0 .098	0 .125	0 .178	0 .294	0 .628	2 .436	IV	0	X
	10		0 .050	0 .052	0 .057	0 .067	0 .085	0 .121	0 .200	0 .428	1 .662		20	
	20		0 .025	0 .027	0 .029	0 .034	0 .043	0 .062	0 .102	0 .218	0 .846		10	
III	0	IX	0 .000	0 .000	0 .000	0 .000	0 .000	0 .000	0 .000	0 .000	0 .000	III	0	IX

TABLE 7.

The CHANGE of LUNAR NUTATION in DECLINATION.

Argument = R. A. ∗ − ☊ ☾.

+ S.		− S.		− S.		+ S.
O	0°	VI	0″.166	VI	0°	O
	10		0 .164		20	
	20		0 .156		10	
I	0	VII	0 .144	V	0	XI
	10		0 .127		20	
	20		0 .107		10	
II	0	VIII	0 .083	IV	0	X
	10		0 .057		20	
	20		0 .029		10	
III	0	IX	0 .000	III	0	IX

TABLE 8.

The DIFFERENCES that may be allowed in the R. A. and Dec. of a STAR'S PLACE, before the ABERRATION in R. A. will change *one Second*.

Argument = Declination.	Differences allowable. In R. A.	Differences allowable. In Dec.
0°	3° 0′	4° 0′
10	3 0 2 0 1 0	1 0 7 0 9 0
20	2 0 1 0 0 0	6 0 7 0 8 0
30	2 0 1 0 0 0	3 0 4 0 4 0
40	2 0 1 0 0 0	1 0 1 30 3 0
50	2 0 1 0 0 0	0 0 1 20 1 30
60	1 30 1 0 0 0	0 0 0 30 1 0
70	1 0 0 30 0 0	0 0 0 15 0 20
80	0 20 0 0	0 0 0 10

TABLE 9.

The DIFFERENCES that may be allowed in the R. A. and Dec. of a STAR'S PLACE, before the ABERRATION in DECLINATION will change *one Second*.

Argument = Declination.	Differences allowable. In R. A.	Differences allowable. In Dec.
0°	8° 0′ 0 0	2° 0′ 3 0
10	8 0 3 0	2 0 3 0
20	7 0 2 0	1 30 3 0
30	5 0 3 0	2 0 3 0
40	3 0 2 0	2 0 3 30
50	3 30 2 0	2 0 3 0
60	3 0 2 0	3 0 5 0
70	3 0 1 30	2 0 7 0
80	3 0 2 0	0 0 8 0

TABLE 10.

The DIFFERENCES that may be allowed in the R. A. and Dec. of a STAR'S PLACE, before the NUTATION in RIGHT ASCENSION will change *one Second*.

Argument = Declination.	Differences allowable. In R. A.	Differences allowable. In Dec.
0°	20° 0′	6° 0′
10	10 0 5 0	6 0 6 30
20	10 0 6 0	5 0 6 0
30	10 0 6 0 3 0	2 0 4 0 5 0
40	7 0 4 0 2 0	1 0 3 0 4 0
50	6 0 3 0 1 0	0 0 2 30 3 0
60	4 0 2 0 1 0	0 45 1 30 1 45
70	2 0 1 0 0 0	0 30 0 45 0 54
80	1 0 0 30	0 10 0 12

TABLE 11.

The DIFFERENCES that may be allowed in the RIGHT ASCENSION and DECLINATION of a STAR'S PLACE, before the LUNAR NUTATION in DECLINATION will amount to *one Second*.

Argument = Declination.	Differences allowable. In R. A.	Differences allowable. In Dec.
0° up to 90°	7°	90°

DIURNAL ABERRATION.

A Table of the Diurnal Aberration in Right Ascension of Stars, in Space, at the time of their Culmination.

Arguments=the Latitude of the Observatory at the Top, and the Declination of the Star at the Side.

Dec.	0°	10°	20°	30°	40°	50°	60°	70°	80°	90°	Dec.
0°	0".30	0".30	0".29	0".26	0".23	0".20	0".15	0".10	0".05	0".00	0°
10	0 .31	0 .30	0 .29	0 .27	0 .24	0 .20	0 .15	0 .11	0 .05	0 .00	10
20	0 .32	0 .32	0 .30	0 .28	0 .25	0 .21	0 .16	0 .11	0 .06	0 .00	20
30	0 .35	0 .35	0 .33	0 .30	0 .27	0 .23	0 .18	0 .12	0 .06	0 .00	30
35	0 .37	0 .37	0 .35	0 .32	0 .28	0 .24	0 .19	0 .13	0 .06	0 .00	35
40	0 .40	0 .39	0 .37	0 .34	0 .30	0 .26	0 .20	0 .14	0 .07	0 .00	40
45	0 .43	0 .42	0 .40	0 .37	0 .33	0 .28	0 .22	0 .15	0 .07	0 .00	45
50	0 .47	0 .47	0 .44	0 .41	0 .36	0 .30	0 .24	0 .16	0 .08	0 .00	50
55	0 .53	0 .52	0 .50	0 .46	0 .41	0 .34	0 .27	0 .18	0 .09	0 .00	55
60	0 .60	0 .60	0 .57	0 .53	0 .47	0 .39	0 .30	0 .21	0 .11	0 .00	60
63	0 .67	0 .66	0 .63	0 .58	0 .51	0 .43	0 .34	0 .23	0 .12	0 .00	63
66	0 .75	0 .74	0 .70	0 .65	0 .57	0 .48	0 .37	0 .26	0 .13	0 .00	66
69	0 .85	0 .84	0 .80	0 .74	0 .65	0 .54	0 .42	0 .29	0 .15	0 .00	69
72	0 .98	0 .97	0 .92	0 .85	0 .75	0 .63	0 .49	0 .34	0 .17	0 .00	72
75	1 .18	1 .16	1 .10	1 .02	0 .90	0 .76	0 .59	0 .40	0 .20	0 .00	75
78	1 .46	1 .44	1 .37	1 .27	1 .12	0 .94	0 .73	0 .50	0 .25	0 .00	78
81	1 .94	1 .91	1 .83	1 .68	1 .49	1 .25	0 .97	0 .67	0 .34	0 .00	81
84	2 .91	2 .86	2 .73	2 .52	2 .23	1 .87	1 .45	1 .00	0 .51	0 .00	84
87	5 .81	5 .72	5 .46	5 .03	4 .45	3 .73	2 .90	1 .99	1 .01	0 .00	87
88	8 .71	8 .58	8 .19	7 .54	6 .67	5 .60	4 .36	2 .98	1 .52	0 .00	88
Polaris.	9 .96	9 .80	9 .35	8 .62	7 .63	6 .40	4 .98	3 .41	1 .73	0 .00	Polaris.
89	17 .42	17 .15	16 .37	15 .09	13 .34	11 .20	8 .71	5 .96	3 .03	0 .00	89

A GENERAL TABLE of ABERRATION of LIGHT, adapted for the PLANETS and COMETS.

Arguments D—g D+g	Aberration.	Arguments D—g D+g	Aberration.	Arguments D—g D+g	Aberration.	Arguments D—g D+g	Aberration.	Arg. D—g D+g	Aberration.	Arg. D—g D+g	Aberration.
0° 0'	0".00	0° 27'	6".24	0° 54'	24".97	1° 21'	56".18	1° 48'	9[illegible]".87	2° 15'	156".05
0 1	0 .01	0 28	6 .71	0 55	25 .90	1 22	57 .58	1 49	101 .73	2 16	158 .37
0 2	0 .03	0 29	7 .20	0 56	26 .85	1 23	58 .99	1 50	103 .61	2 17	160 .71
0 3	0 .08	0 30	7 .71	0 57	27 .82	1 24	60 .42	1 51	105 .50	2 18	163 .06
0 4	0 .14	0 31	8 .23	0 58	28 .80	1 25	61 .86	1 52	107 .41	2 19	165 .43
0 5	0 .22	0 32	8 .77	0 59	29 .80	1 26	63 .33	1 53	109 .33	2 20	167 .82
0 6	0 .31	0 33	9 .32	1 0	30 .82	1 27	64 .81	1 54	111 .28	2 21	170 .23
0 7	0 .42	0 34	9 .90	1 1	31 .86	1 28	66 .31	1 55	113 .24	2 22	172 .65
0 8	0 .55	0 35	10 .49	1 2	32 .91	1 29	67 .82	1 56	115 .22	2 23	175 .10
0 9	0 .70	0 36	11 .10	1 3	33 .98	1 30	69 .36	1 57	117 .21	2 24	177 .56
0 10	0 .86	0 37	11 .72	1 4	35 .07	1 31	70 .91	1 58	119 .22	2 25	180 .03
0 11	1 .03	0 38	12 .36	1 5	36 .18	1 32	72 .47	1 59	121 .25	2 26	182 .52
0 12	1 .23	0 39	13 .02	1 6	37 .30	1 33	74 .05	2 0	123 .30	2 27	185 .03
0 13	1 .45	0 40	13 .70	1 7	38 .44	1 34	75 .66	2 1	125 .36	2 28	187 .55
0 14	1 .68	0 41	14 .39	1 8	39 .59	1 35	77 .28	2 2	127 .44	2 29	190 .10
0 15	1 .93	0 42	15 .10	1 9	40 .77	1 36	78 .91	2 3	129 .54	2 30	192 .66
0 16	2 .19	0 43	15 .83	1 10	41 .96	1 37	80 .56	2 4	131 .66	2 31	195 .24
0 17	2 .47	0 44	16 .58	1 11	43 .16	1 38	82 .23	2 5	133 .79	2 32	197 .83
0 18	2 .77	0 45	17 .34	1 12	44 .39	1 39	83 .92	2 6	135 .94	2 33	200 .44
0 19	3 .09	0 46	18 .12	1 13	45 .63	1 40	85 .62	2 7	138 .11	2 34	203 .07
0 20	3 .42	0 47	18 .91	1 14	46 .89	1 41	87 .35	2 8	140 .29	2 35	205 .71
0 21	3 .77	0 48	19 .72	1 15	48 .16	1 42	89 .09	2 9	142 .49	2 36	208 .38
0 22	4 .14	0 49	20 .56	1 16	49 .46	1 43	90 .84	2 10	144 .71	2 37	211 .06
0 23	4 .53	0 50	21 .41	1 17	50 .77	1 44	92 .61	2 11	146 .94	2 38	213 .76
0 24	4 .93	0 51	22 .27	1 18	52 .09	1 45	94 .40	2 12	149 .19	2 39	216 .47
0 25	5 .35	0 52	23 .15	1 19	53 .44	1 46	96 .21	2 13	151 .46	2 40	219 .20
0 26	5 .79	0 53	24 .05	1 20	54 .80	1 47	98 .03	2 14	153 .75		

ARGUMENT = HOUR ANGLE, counted from the Meridian.

S.	M. 0	M. 1	M. 2	M. 3	M. 4	M. 5	M. 6	M. 7	M. 8	M. 9	M. 10	M. 11	M. 12	M. 13	M. 14	S.
0	0″.0	2″.0	7″.8	17″.7	31″.4	49″.1	70″.7	96″.2	125.7	159″.0	196″.3	237″.5	282″.7	331″.8	384″.7	0
1	0.0	2.0	8.0	17.9	31.7	49.4	71.1	96.6	126.2	159.6	197.0	238.3	283.5	332.6	385.6	1
2	0.0	2.1	8.1	18.1	31.9	49.7	71.5	97.1	126.7	160.2	197.6	239.0	284.2	333.4	386.5	2
3	0.0	2.2	8.2	18.3	32.2	50.1	71.9	97.6	127.2	160.8	198.3	239.7	285.0	334.3	387.5	3
4	0.0	2.2	8.4	18.5	32.5	50.4	72.3	98.1	127.8	161.4	198.9	240.4	285.8	335.2	388.4	4
5	0.0	2.3	8.5	18.7	32.7	50.7	72.7	98.5	128.3	162.0	199.6	241.2	286.6	336.0	389.3	5
6	0.0	2.4	8.7	18.9	33.0	51.1	73.1	99.0	128.8	162.6	200.3	241.9	287.4	336.9	390.2	6
7	0.0	2.4	8.8	19.1	33.3	51.4	73.5	99.4	129.4	163.2	200.9	242.6	288.2	337.7	391.1	7
8	0.0	2.5	8.9	19.3	33.5	51.7	73.9	99.9	129.9	163.8	201.6	243.3	289.0	338.6	392.1	8
9	0.0	2.6	9.1	19.5	33.8	52.1	74.3	100.4	130.4	164.4	202.2	244.1	289.8	339.4	393.0	9
10	0.0	2.7	9.2	19.7	34.1	52.4	74.7	100.8	131.0	165.0	202.9	244.8	290.6	340.3	393.9	10
11	0.1	2.7	9.4	19.9	34.4	52.7	75.1	101.3	131.5	165.6	203.6	245.5	291.4	341.2	394.8	11
12	0.1	2.8	9.5	20.1	34.6	53.1	75.5	101.8	132.0	166.2	204.2	246.2	292.2	342.0	395.8	12
13	0.1	2.9	9.6	20.3	34.9	53.4	75.9	102.3	132.6	166.8	204.9	247.0	293.0	342.9	396.7	13
14	0.1	3.0	9.8	20.5	35.2	53.8	76.3	102.7	133.1	167.4	205.6	247.7	293.8	343.7	397.6	14
15	0.1	3.1	9.9	20.7	35.5	54.1	76.7	103.2	133.6	168.0	206.3	248.5	294.6	344.6	398.6	15
16	0.1	3.1	10.1	20.9	35.7	54.5	77.1	103.7	134.2	168.6	206.9	249.2	295.4	345.5	399.5	16
17	0.2	3.2	10.2	21.2	36.0	54.8	77.5	104.2	134.7	169.2	207.6	249.9	296.2	346.3	400.5	17
18	0.2	3.3	10.4	21.4	36.3	55.1	77.9	104.6	135.3	169.8	208.3	250.7	297.0	347.2	401.4	18
19	0.2	3.4	10.5	21.6	36.6	55.5	78.3	105.1	135.8	170.4	208.9	251.4	297.8	348.1	402.3	19
20	0.2	3.5	10.7	21.8	36.9	55.8	78.8	105.6	136.4	171.0	209.6	252.2	298.6	349.0	403.3	20
21	0.3	3.6	10.8	22.0	37.2	56.2	79.2	106.1	136.9	171.6	210.3	252.9	299.4	349.8	404.2	21
22	0.3	3.7	11.0	22.3	37.4	56.5	79.6	106.6	137.4	172.2	211.0	253.6	300.2	350.7	405.1	22
23	0.3	3.8	11.1	22.5	37.7	56.9	80.0	107.0	138.0	172.9	211.6	254.4	301.0	351.6	406.0	23
24	0.3	3.8	11.3	22.7	38.0	57.3	80.4	107.5	138.5	173.5	212.3	255.1	301.8	352.5	407.0	24
25	0.3	3.9	11.5	22.9	38.3	57.6	80.8	108.0	139.1	174.1	213.0	255.9	302.6	353.3	408.0	25
26	0.4	4.0	11.6	23.1	38.6	58.0	81.3	108.5	139.6	174.7	213.7	256.6	303.5	354.2	408.9	26
27	0.4	4.1	11.8	23.4	38.9	58.3	81.7	109.0	140.2	175.3	214.4	257.4	304.3	355.1	409.9	27
28	0.4	4.2	11.9	23.6	39.2	58.7	82.1	109.5	140.7	175.9	215.1	258.1	305.1	356.0	410.8	28
29	0.5	4.3	12.1	23.8	39.5	59.0	82.5	110.0	141.3	176.6	215.8	258.9	305.9	356.9	411.7	29
30	0.5	4.4	12.3	24.0	39.8	59.4	83.0	110.4	141.8	177.2	216.4	259.6	306.7	357.7	412.7	30
31	0.5	4.5	12.4	24.3	40.1	59.8	83.4	110.9	142.4	177.8	217.1	260.4	307.5	358.6	413.6	31
32	0.6	4.6	12.6	24.5	40.3	60.1	83.8	111.4	143.0	178.4	217.8	261.1	308.4	359.5	414.6	32
33	0.6	4.7	12.8	24.7	40.6	60.5	84.2	111.9	143.5	179.0	218.5	261.9	309.2	360.3	415.6	33
34	0.6	4.8	12.9	25.0	40.9	60.8	84.7	112.4	144.1	179.7	219.2	262.6	310.0	361.2	416.6	34
35	0.7	4.9	13.1	25.2	41.2	61.2	85.1	112.9	144.6	180.3	219.9	263.4	310.8	362.1	417.5	35
36	0.7	5.0	13.3	25.4	41.5	61.6	85.5	113.4	145.2	180.9	220.6	264.1	311.6	363.0	418.4	36
37	0.7	5.1	13.4	25.7	41.8	61.9	86.0	113.9	145.8	181.6	221.3	264.9	312.5	363.9	419.4	37
38	0.8	5.2	13.6	25.9	42.1	62.3	86.4	114.4	146.3	182.2	222.0	265.7	313.3	364.8	420.3	38
39	0.8	5.3	13.8	26.2	42.5	62.7	86.8	114.9	146.9	182.8	222.7	266.4	314.2	365.7	421.3	39
40	0.9	5.4	14.0	26.4	42.8	63.0	87.3	115.4	147.5	183.4	223.4	267.2	315.0	366.5	422.2	40
41	0.9	5.6	14.1	26.6	43.1	63.4	87.7	115 9	148.0	184.1	224.1	267.9	315.8	367.5	423.2	41
42	1.0	5.7	14.3	26.9	43.4	63.8	88.1	116.4	148.6	184.7	224.8	268.7	316.6	368.4	424.2	42
43	1.0	5.8	14.5	27.1	43.7	64.2	88.6	116.9	149.2	185.4	225.5	269.5	317.4	369.3	425.1	43
44	1.1	5.9	14.7	27.4	44.0	64.5	89.0	117.4	149.7	186.0	226.2	270.2	318.3	370.2	426.1	44
45	1.1	6.0	14.8	27.6	44.3	64.9	89.5	117.9	150.3	186.6	226.9	271.0	319.1	371.1	427.0	45
46	1.2	6.1	15.0	27.9	44.6	65.3	89.9	118.4	150.9	187.3	227.6	271.8	319.9	372.0	428.0	46
47	1.2	6.2	15.2	28.1	44.9	65.7	90.3	118.9	151.5	187.9	228.3	272.6	320.8	372.9	429.0	47
48	1.3	6.4	15.4	28.3	45.2	66.0	90.8	119.5	152.0	188.5	229.0	273.3	321.6	373.8	430.0	48
49	1.3	6.5	15.6	28.6	45.5	66.4	91.2	120.0	152.6	189.2	229.7	274.1	322.4	374.7	430.9	49
50	1.4	6.6	15.8	28.8	45.9	66.8	91.7	120.5	153.2	189.8	230.4	274.9	323.3	375.6	431.9	50
51	1.4	6.7	15.9	29.1	46.2	67.2	92.1	121.0	153.8	190.5	231.1	275.6	324.1	376.5	432.8	51
52	1.5	6.8	16.1	29.4	46.5	67.6	92.6	121.5	154.4	191.1	231.8	276.4	325.0	377.4	433.8	52
53	1.5	7.0	16.3	29.6	46.8	68.0	93.0	122.0	154.9	191.8	232.5	277.2	325.8	378.3	434.8	53
54	1.6	7.1	16.5	29.9	47.1	68.3	93.5	122.5	155.5	192.4	233.3	278.0	326.7	379.2	435.7	54
55	1.6	7.2	16.7	30.1	47.5	68.7	93.9	123.1	156.1	193.1	234.0	278.9	327.5	380.2	436.7	55
56	1.7	7.3	16.9	30.4	47.8	69.1	94.4	123.6	156.7	193.7	234.7	279.5	328.4	381.1	437.7	56
57	1.8	7.5	17.1	30.6	48.1	69.5	94.8	124.1	157.3	194.4	235.4	280.3	329.2	382.0	438.7	57
58	1.8	7.6	17.3	30.9	48.4	69.9	95.3	124.6	157.8	195.0	236.1	281.1	330.0	382.9	439.6	58
59	1.9	7.7	17.5	31.1	48.8	70.3	95.7	125.1	158.4	195.7	236.8	281.9	330.9	383.8	440.6	59
60	2.0	7.8	17.7	31.4	49.1	70.7	96.2	125.7	159.0	196.3	237.5	282.7	331.8	384.7	441.6	60

REDUCTION to the MERIDIAN.

PART I.—According to DELAMBRE.

ARGUMENT = HOUR ANGLE, counted from the Meridian.

S.	M. 15	M. 16	M. 17	M. 18	M. 19	M. 20	M. 21	M. 22	M. 23	M. 24	M. 25	M. 26	M. 27	S.
0	441″.6	502″.5	567″.1	635″.8	708″.3	784″.9	865″.3	949″.6	1037″.8	1129″.9	1225″.9	1325″.9	1429″.7	0
1	442.6	503.5	568.2	636.9	709.5	786.2	866.6	951.0	1039.3	1131.4	1227.5	1327.6	1431.4	1
2	443.6	504.6	569.3	638.1	710.8	787.5	868.0	952.4	1040.8	1133.0	1229.2	1329.3	1433.2	2
3	444.6	505.6	570.4	639.3	712.1	788.8	869.4	953.8	1042.3	1134.6	1230.8	1331.0	1434.9	3
4	445.6	506.7	571.6	640.5	713.4	790.1	870.8	955.3	1043.8	1136.2	1232.5	1332.7	1436.7	4
5	446.5	507.7	572.7	641.7	714.6	791.4	872.1	956.7	1045.3	1137.8	1234.1	1334.4	1438.5	5
6	447.5	508.8	573.8	642.9	715.9	792.7	873.5	958.2	1046.8	1139.3	1235.7	1336.1	1440.3	6
7	448.5	509.8	574.9	644.1	717.1	794.0	874.9	959.6	1048.3	1140.9	1237.3	1337.8	1442.1	7
8	449.5	510.9	576.1	645.3	718.4	795.4	876.3	961.1	1049.8	1142.5	1239.0	1339.5	1443.9	8
9	450.5	511.9	577.2	646.4	719.6	796.7	877.6	962.5	1051.3	1144.0	1240.6	1341.2	1445.6	9
10	451.5	513.0	578.3	647.6	720.9	798.0	879.0	963.9	1052.8	1145.6	1242.3	1342.9	1447.4	10
11	452.5	514.0	579.4	648.8	722.1	799.3	880.4	965.4	1054.3	1147.2	1243.9	1344.6	1449.2	11
12	453.5	515.1	580.6	650.0	723.4	800.7	881.8	966.9	1055.9	1148.8	1245.6	1346.3	1451.0	12
13	454.5	516.1	581.7	651.2	724.6	802.0	883.2	968.3	1057.4	1150.4	1247.2	1348.0	1452.8	13
14	455.5	517.2	582.8	652.4	725.9	803.3	884.6	969.8	1058.9	1152.0	1248.9	1349.7	1454.5	14
15	456.5	518.3	583.9	653.6	727.1	804.6	886.0	971.2	1060.4	1153.6	1250.5	1351.4	1456.3	15
16	457.5	519.4	585.1	654.8	728.4	806.0	887.4	972.7	1062.0	1155.2	1252.2	1353.2	1458.1	16
17	458.5	520.4	586.2	656.0	729.6	807.3	888.8	974.1	1063.5	1156.8	1253.8	1354.9	1459.9	17
18	459.5	521.4	587.3	657.2	730.9	808.6	890.2	975.5	1065.0	1158.3	1255.5	1356.6	1461.6	18
19	460.5	522.5	588.4	658.4	732.2	809.9	891.6	977.0	1066.5	1159.9	1257.1	1358.3	1463.4	19
20	461.5	523.6	589.6	659.6	733.5	811.3	893.0	978.5	1068.1	1161.5	1258.8	1360.1	1465.2	20
21	462.5	524.6	590.7	660.8	734.7	812.6	894.4	979.9	1069.6	1163.1	1260.4	1361.8	1466.9	21
22	463.5	525.7	591.9	662.0	736.0	813.9	895.8	981.4	1071.1	1164.7	1262.1	1363.5	1468.7	22
23	464.5	526.8	593.0	663.2	737.2	815.2	897.2	982.9	1072.6	1166.3	1263.7	1365.2	1470.5	23
24	465.5	527.9	594.1	664.4	738.5	816.6	898.6	984.4	1074.2	1167.9	1265.4	1367.0	1472.3	24
25	466.5	528.9	595.2	665.6	739.7	817.9	900.0	985.8	1075.7	1169.5	1267.0	1368.7	1474.0	25
26	467.5	530.0	596.4	666.8	741.0	819.2	901.4	987.3	1077.2	1171.1	1268.7	1370.4	1475.9	26
27	468.5	531.1	597.5	668.0	742.3	820.5	902.8	988.8	1078.7	1172.7	1270.3	1372.1	1477.7	27
28	469.5	532.2	598.7	669.2	743.6	821.9	904.2	990.3	1080.3	1174.3	1272.1	1373.9	1479.5	28
29	470.5	533.2	599.8	670.4	744.8	823.2	905.6	991.8	1081.8	1175.9	1273.7	1375.6	1481.3	29
30	471.5	534.3	601.0	671.6	746.1	824.6	907.0	993.2	1083.3	1177.5	1275.4	1377.4	1483.1	30
31	472.6	535.4	602.1	672.8	747.4	825.9	908.4	994.7	1084.8	1179.1	1277.1	1379.0	1484.9	31
32	473.6	536.5	603.3	674.1	748.7	827.3	909.8	996.2	1086.4	1180.7	1278.8	1380.8	1486.7	32
33	474.6	537.5	604.4	675.3	749.9	828.6	911.2	997.6	1087.9	1182.3	1280.4	1382.5	1488.5	33
34	475.6	538.6	605.6	676.5	751.2	829.9	912.6	999.1	1089.5	1183.9	1282.1	1384.2	1490.3	34
35	476.6	539.7	606.7	677.7	752.5	831.2	914.0	1000.6	1091.0	1185.5	1283.8	1385.9	1492.1	35
36	477.6	540.8	607.9	678.9	753.8	832.6	915.5	1002.1	1092.6	1187.1	1285.5	1387.7	1493.9	36
37	478.7	541.9	609.0	680.1	755.0	833.9	916.9	1003.5	1094.1	1188.7	1287.1	1389.4	1495.7	37
38	479.7	543.0	610.2	681.3	756.3	835.3	918.3	1005.0	1095.7	1190.3	1288.8	1391.2	1497.5	38
39	480.7	544.1	611.3	682.5	757.6	836.6	919.7	1006.5	1097.2	1191.9	1290.5	1392.9	1499.3	39
40	481.7	545.2	612.5	683.8	758.9	838.0	921.1	1008.0	1098.8	1193.5	1292.2	1394.7	1501.1	40
41	482.8	546.2	613.6	685.0	760.2	839.3	922.5	1009.4	1100.3	1195.1	1293.8	1396.4	1502.9	41
42	483.8	547.3	614.8	686.2	761.5	840.7	923.9	1010.9	1101.9	1196.7	1295.5	1398.2	1504.7	42
43	484.8	548.4	615.9	687.4	762.8	842.0	925.3	1012.4	1103.4	1198.3	1297.2	1399.9	1506.5	43
44	485.8	549.5	617.1	688.7	764.0	843.4	926.8	1013.9	1105.0	1199.9	1298.9	1401.7	1508.4	44
45	486.9	550.6	618.2	689.9	765.3	844.7	928.2	1015.4	1106.5	1201.5	1300.5	1403.4	1510.2	45
46	487.9	551.7	619.4	691.1	766.6	846.1	929.6	1016.9	1108.1	1203.1	1302.2	1405.2	1512.0	46
47	488.9	552.8	620.5	692.3	767.9	847.5	931.0	1018.4	1109.6	1204.7	1303.9	1406.9	1513.8	47
48	490.0	553.9	621.7	693.6	769.2	848.9	932.4	1019.9	1111.2	1206.4	1305.6	1408.7	1515.6	48
49	491.0	555.0	622.8	694.8	770.5	850.2	933.8	1021.4	1112.7	1208.0	1307.3	1410.4	1517.4	49
50	492.0	556.1	624.0	696.0	771.8	851.6	935.2	1022.8	1114.3	1209.6	1309.0	1412.2	1519.2	50
51	493.1	557.2	625.2	697.2	773.1	852.9	936.6	1024.3	1115.8	1211.2	1310.7	1413.9	1521.0	51
52	494.1	558.3	626.4	698.4	774.5	854.3	938.1	1025.8	1117.4	1212.9	1312.4	1415.7	1522.9	52
53	495.2	559.4	627.5	699.6	775.8	855.6	939.5	1027.3	1118.9	1214.5	1314.1	1417.4	1524.7	53
54	496.2	560.5	628.7	700.9	777.1	857.1	940.9	1028.8	1120.5	1216.1	1315.7	1419.2	1526.5	54
55	497.2	561.6	629.9	702.2	778.4	858.4	942.3	1030.3	1122.0	1217.7	1317.4	1420.9	1528.3	55
56	498.2	562.7	631.1	703.5	779.7	859.8	943.8	1031.8	1123.6	1219.4	1319.1	1422.7	1530.2	56
57	499.2	563.8	632.2	704.7	781.0	861.1	945.2	1033.3	1125.1	1221.0	1320.8	1424.4	1532.0	57
58	500.3	564.9	633.4	705.9	782.3	862.5	946.6	1034.8	1126.7	1222.6	1322.5	1426.2	1533.8	58
59	501.4	566.0	634.6	707.1	783.6	863.9	948.1	1036.3	1128.3	1224.2	1324.2	1427.9	1535.6	59
60	502.5	567.1	635.8	708.3	784.9	865.3	949.6	1037.8	1129.9	1225.9	1325.9	1429.7	1537.5	60

Part I.—According to DELAMBRE.

Argument = Hour Angle, counted from the Meridian.

S.	M. 28	M. 29	M. 30	M. 31	M. 32	M. 33	M. 34	M. 35	S.
0	1537″.5	1649″.0	1764″.6	1884″.0	2007″.4	2134″.6	2265″.6	2400″.6	0
1	1539 .3	1650 .9	1766 .6	1886 .0	2009 .4	2136 .8	2267 .8	2402 .9	1
2	1541 .1	1652 .8	1768 .5	1888 .0	2011 .5	2138 .9	2270 .0	2405 .2	2
3	1542 .9	1654 .7	1770 .5	1890 .0	2013 .6	2141 .1	2272 .2	2407 .5	3
4	1544 .8	1656 .6	1772 .4	1892 .1	2015 .7	2143 .2	2274 .5	2409 .8	4
5	1546 .6	1658 .5	1774 .4	1894 .1	2017 .8	2145 .3	2276 .7	2412 .0	5
6	1548 .4	1660 .4	1776 .3	1896 .1	2019 .9	2147 .5	2278 .9	2414 .3	6
7	1550 .2	1662 .3	1778 .3	1898 .1	2022 .0	2149 .7	2281 .2	2416 .6	7
8	1552 .1	1664 .2	1780 .3	1900 .2	2024 .1	2151 .8	2283 .4	2418 .9	8
9	1553 .9	1666 .1	1782 .3	1902 .2	2026 .2	2153 .9	2285 .6	2421 .2	9
10	1555 .8	1668 .0	1784 .2	1904 .3	2028 .3	2156 .1	2287 .8	2423 .5	10
11	1557 .6	1669 .9	1786 .2	1906 .3	2030 .5	2158 .3	2290 .0	2425 .8	11
12	1559 .5	1671 .9	1788 .2	1908 .4	2032 .5	2160 .5	2292 .3	2428 .1	12
13	1561 .3	1673 .8	1790 .1	1910 .4	2034 .6	2162 .6	2294 .5	2430 .4	13
14	1563 .2	1675 .7	1792 .1	1912 .4	2036 .7	2164 .8	2296 .8	2432 .7	14
15	1565 .0	1677 .6	1794 .1	1914 .4	2038 .8	2166 .9	2299 .0	2435 .0	15
16	1566 .9	1679 .5	1796 .1	1916 .5	2040 .9	2169 .1	2301 .3	2437 .3	16
17	1568 .7	1681 .4	1798 .1	1918 .5	2043 .0	2171 .2	2303 .6	2439 .6	17
18	1570 .5	1683 .3	1800 .0	1920 .6	2045 .1	2173 .4	2305 .8	2441 .9	18
19	1572 .4	1685 .2	1802 .0	1922 .6	2047 .2	2175 .6	2308 .0	2444 .2	19
20	1574 .3	1687 .2	1804 .0	1924 .7	2049 .3	2177 .8	2310 .2	2446 .5	20
21	1576 .1	1689 .1	1805 .9	1926 .7	2051 .4	2179 .9	2312 .4	2448 .8	21
22	1578 .0	1691 .0	1807 .9	1928 .8	2053 .5	2182 .1	2314 .7	2451 .1	22
23	1579 .8	1692 .9	1809 .9	1930 .8	2055 .7	2184 .3	2316 .9	2453 .4	23
24	1581 .7	1694 .8	1811 .9	1932 .9	2057 .8	2186 .5	2319 .2	2455 .7	24
25	1583 .5	1696 .7	1813 .9	1935 .0	2059 .9	2188 .6	2321 .5	2458 .0	25
26	1585 .3	1698 .6	1815 .8	1937 .0	2062 .0	2190 .8	2323 .7	2460 .3	26
27	1587 .2	1700 .5	1817 .8	1939 .0	2064 .1	2193 .0	2325 .9	2462 .6	27
28	1589 .1	1702 .5	1819 .8	1941 .1	2066 .2	2195 .2	2328 .2	2464 .9	28
29	1590 .9	1704 .4	1821 .8	1943 .1	2068 .3	2197 .3	2330 .4	2467 .2	29
30	1592 .7	1706 .3	1823 .8	1945 .2	2070 .4	2199 .5	2332 .7	2469 .5	30
31	1594 .6	1708 .2	1825 .8	1947 .2	2072 .6	2201 .7	2334 .9	2471 .8	31
32	1596 .5	1710 .2	1827 .8	1949 .3	2074 .7	2203 .9	2337 .2	2474 .2	32
33	1598 .3	1712 .1	1829 .8	1951 .3	2076 .8	2206 .1	2339 .4	2476 .5	33
34	1600 .2	1714 .0	1831 .8	1953 .4	2078 .9	2208 .3	2341 .7	2478 .8	34
35	1602 .1	1715 .9	1833 .8	1955 .5	2081 .0	2210 .5	2343 .9	2481 .1	35
36	1604 .0	1717 .9	1835 .8	1957 .6	2083 .2	2212 .7	2346 .2	2483 .5	36
37	1605 .9	1719 .8	1837 .8	1959 .6	2085 .3	2214 .9	2348 .5	2485 .8	37
38	1607 .7	1721 .7	1839 .8	1961 .7	2087 .4	2217 .1	2350 .7	2488 .1	38
39	1609 .6	1723 .6	1841 .8	1963 .7	2089 .6	2219 .3	2353 .0	2490 .4	39
40	1611 .5	1725 .6	1843 .8	1965 .8	2091 .7	2221 .5	2355 .2	2492 .8	40
41	1613 .3	1727 .5	1845 .8	1967 .8	2093 .8	2223 .7	2357 .5	2495 .1	41
42	1615 .2	1729 .5	1847 .8	1969 .9	2095 .9	2225 .9	2359 .7	2497 .4	42
43	1617 .1	1731 .5	1849 .8	1972 .0	2098 .0	2228 .1	2361 .9	2499 .7	43
44	1619 .0	1733 .4	1851 .8	1974 .1	2100 .2	2230 .3	2364 .2	2502 .1	44
45	1620 .8	1735 .3	1853 .8	1976 .1	2102 .3	2232 .5	2366 .4	2504 .4	45
46	1622 .7	1737 .2	1855 .8	1978 .2	2104 .5	2234 .7	2368 .7	2506 .7	46
47	1624 .6	1739 .2	1857 .8	1980 .3	2106 .6	2236 .9	2371 .0	2509 .0	47
48	1626 .5	1741 .2	1859 .8	1982 .4	2108 .8	2239 .1	2373 .3	2511 .4	48
49	1628 .3	1743 .1	1861 .8	1984 .4	2110 .9	2241 .3	2375 .5	2513 .7	49
50	1630 .2	1745 .1	1863 .8	1986 .5	2113 .1	2243 .5	2377 .8	2516 .1	50
51	1632 .1	1747 .0	1865 .8	1988 .6	2115 .2	2245 .7	2380 .1	2518 .4	51
52	1634 .0	1749 .0	1867 .8	1990 .7	2117 .4	2247 .9	2382 .4	2520 .8	52
53	1635 .9	1750 .9	1869 .8	1992 .7	2119 .6	2250 .1	2384 .6	2523 .1	53
54	1637 .7	1752 .9	1871 .8	1994 .8	2121 .7	2252 .3	2386 .9	2525 .4	54
55	1639 .6	1754 .8	1873 .8	1996 .9	2123 .8	2254 .5	2389 .2	2527 .7	55
56	1641 .5	1756 .8	1875 .9	1999 .0	2126 .0	2256 .7	2391 .5	2530 .1	56
57	1643 .3	1758 .7	1877 .9	2001 .0	2128 .1	2258 .9	2393 .7	2532 .4	57
58	1645 .2	1760 .7	1879 .9	2003 .1	2130 .3	2261 .1	2396 .0	2534 .8	58
59	1647 .1	1762 .6	1882 .0	2005 .3	2132 .4	2263 .4	2398 .3	2537 .1	59
60	1649 .0	1764 .6	1884 .0	2007 .4	2134 .6	2265 .6	2400 .6	2539 .5	60

Part II.—From Schumacher.

Argument = Hour Angle, counted from the Meridian.

M.	S.	+	M.	S.	+	M.	S.	+
0	0	0″.00	14	30	0″.41	24	50	3″.55
1	0	0 .00		40	0 .43	25	0	3 .64
2	0	0 .00		50	0 .45		10	3 .74
3	0	0 .00	15	0	0 .47		20	3 .84
4	0	0 .00		10	0 .49		30	3 .94
5	0	0 .01		20	0 .52		40	4 .05
	10	0 .01		30	0 .54		50	4 .15
	20	0 .01		40	0 .56	26	0	4 .26
	30	0 .01		50	0 .59		10	4 .37
	40	0 .01	16	0	0 .61		20	4 .48
	50	0 .01		10	0 .64		30	4 .60
6	0	0 .01		20	0 .67		40	4 .72
	10	0 .01		30	0 .69		50	4 .83
	20	0 .01		40	0 .72	27	0	4 .96
	30	0 .02		50	0 .75		10	5 .08
	40	0 .02	17	0	0 .78		20	5 .20
	50	0 .02		10	0 .81		30	5 .33
7	0	0 .02		20	0 .84		40	5 .46
	10	0 .02		30	0 .88		50	5 .60
	20	0 .03		40	0 .91	28	0	5 .73
	30	0 .03		50	0 .95		10	5 .87
	40	0 .03	18	0	0 .98		20	6 .01
	50	0 .04		10	1 .02		30	6 .15
8	0	0 .04		20	1 .06		40	6 .30
	10	0 .04		30	1 .09		50	6 .44
	20	0 .05		40	1 .13	29	0	6 .59
	30	0 .05		50	1 .18		10	6 .75
	40	0 .05	19	0	1 .22		20	6 .90
	50	0 .06		10	1 .26		30	7 .06
9	0	0 .06		20	1 .30		40	7 .22
	10	0 .07		30	1 .35		50	7 .38
	20	0 .07		40	1 .40	30	0	7 .55
	30	0 .08		50	1 .44		10	7 .62
	40	0 .08	20	0	1 .49		20	7 .89
	50	0 .09		10	1 .54		30	8 .06
10	0	0 .09		20	1 .60		40	8 .24
	10	0 .10		30	1 .65		50	8 .42
	20	0 .11		40	1 .70	31	0	8 .61
	30	0 .11		50	1 .76		10	8 .79
	40	0 .12	21	0	1 .82		20	8 .98
	50	0 .13		10	1 .87		30	9 .17
11	0	0 .14		20	1 .93		40	9 .37
	10	0 .15		30	1 .99		50	9 .57
	20	0 .15		40	2 .06	32	0	9 .77
	30	0 .16		50	2 .12		10	9 .97
	40	0 .17	22	0	2 .19		20	10 .18
	50	0 .18		10	2 .25		30	10 .39
12	0	0 .19		20	2 .32		40	10 .61
	10	0 .21		30	2 .39		50	10 .82
	20	0 .22		40	2 .46	33	0	11 .04
	30	0 .23		50	2 .54		10	11 .27
	40	0 .24	23	0	2 .61		20	11 .50
	50	0 .25		10	2 .69		30	11 .73
13	0	0 .27		20	2 .77		40	11 .96
	10	0 .28		30	2 .85		50	12 .20
	20	0 .30		40	2 .93	34	0	12 .44
	30	0 .31		50	3 .01		20	12 .94
	40	0 .33	24	0	3 .10		40	13 .45
	50	0 .34		10	3 .18	35	0	13 .97
14	0	0 .36		20	3 .27		20	14 .51
	10	0 .38		30	3 .36		40	15 .06
	20	0 .39		40	3 .45	36	0	15 .63

A TABLE of NATURAL VERSED SINES adapted to SECONDS of TIME.

Argument = the Horary Angle counted from the Meridian.

S.	M. 0	M. 1	M. 2	M. 3	M. 4	M. 5	M. 6	M. 7	M. 8	M. 9	M. 10	M. 11	M. 12	M. 13	M. 14	S.
0	0	9	38	86	152	238	343	466	609	771	952	1152	1370	1608	1865	0
1	0	10	39	87	153	240	345	469	612	774	955	1155	1374	1612	1870	1
2	0	10	39	88	155	241	346	471	614	777	958	1159	1378	1617	1874	2
3	0	10	40	89	156	243	348	473	617	780	961	1162	1382	1621	1879	3
4	0	11	41	90	157	244	350	475	619	782	964	1166	1386	1625	1883	4
5	0	11	41	91	159	246	352	478	622	785	968	1169	1390	1629	1887	5
6	0	12	42	92	160	248	354	480	624	788	971	1173	1393	1633	1892	6
7	0	12	43	93	161	249	356	482	627	791	974	1176	1397	1637	1896	7
8	0	12	43	93	163	251	358	484	630	794	977	1180	1401	1641	1901	8
9	0	13	44	94	164	253	360	487	632	797	981	1183	1405	1646	1905	9
10	0	13	45	95	165	254	362	489	635	800	984	1187	1409	1650	1910	10
11	0	13	45	96	167	256	364	491	637	803	987	1190	1413	1654	1914	11
12	0	14	46	97	168	257	366	493	640	806	990	1194	1416	1658	1919	12
13	0	14	47	98	169	259	368	496	643	809	993	1197	1420	1662	1923	13
14	1	14	47	99	171	261	370	498	645	811	997	1201	1424	1667	1928	14
15	1	15	48	100	172	262	372	500	648	814	1000	1205	1428	1671	1932	15
16	1	15	49	102	173	264	374	503	650	817	1003	1208	1432	1675	1937	16
17	1	16	50	103	175	266	376	505	653	820	1006	1212	1436	1679	1941	17
18	1	16	50	104	176	267	378	507	656	823	1010	1215	1440	1683	1946	18
19	1	17	51	105	177	269	380	510	658	826	1013	1219	1444	1688	1950	19
20	1	17	52	106	179	271	382	512	661	829	1016	1222	1448	1692	1955	20
21	1	17	53	107	180	272	384	514	664	832	1020	1226	1451	1696	1960	21
22	1	18	53	108	182	274	386	517	666	835	1023	1230	1455	1700	1964	22
23	1	18	54	109	183	276	388	519	669	838	1026	1233	1459	1705	1968	23
24	2	19	55	110	184	278	390	521	672	841	1029	1237	1463	1709	1973	24
25	2	19	56	111	185	279	392	524	674	844	1033	1241	1467	1713	1978	25
26	2	20	56	112	187	281	394	526	677	847	1036	1244	1471	1717	1982	26
27	2	20	57	113	188	283	396	528	680	850	1039	1248	1475	1722	1987	27
28	2	20	58	114	190	284	398	531	682	853	1043	1251	1479	1726	1992	28
29	2	21	59	116	191	286	400	533	685	856	1046	1255	1483	1730	1997	29
30	3	21	59	117	193	288	402	535	688	859	1049	1259	1487	1734	2001	30
31	3	22	60	118	194	289	404	538	690	862	1053	1262	1491	1739	2005	31
32	3	22	61	119	196	291	406	540	693	865	1056	1266	1495	1743	2010	32
33	3	23	62	120	197	293	408	543	696	868	1059	1270	1499	1747	2014	33
34	3	23	63	121	198	295	410	545	699	871	1062	1273	1503	1751	2019	34
35	3	24	64	122	200	297	412	547	701	874	1066	1277	1507	1756	2024	35
36	3	24	64	123	201	299	415	550	704	877	1069	1281	1511	1760	2028	36
37	4	25	65	124	203	300	417	552	707	880	1073	1284	1515	1764	2033	37
38	4	25	66	126	204	302	419	555	709	883	1076	1288	1519	1769	2038	38
39	4	26	67	127	206	304	421	557	712	886	1079	1292	1523	1773	2042	39
40	4	26	68	128	207	306	423	559	715	889	1083	1295	1527	1777	2047	40
41	4	27	68	129	209	307	425	562	718	892	1086	1299	1531	1782	2052	41
42	5	28	69	130	210	309	427	564	720	896	1090	1303	1535	1786	2056	42
43	5	28	70	131	212	311	429	567	723	899	1093	1307	1539	1790	2061	43
44	5	29	71	133	213	313	432	569	726	902	1096	1310	1543	1795	2066	44
45	5	29	72	134	215	315	434	572	729	905	1100	1314	1547	1799	2070	45
46	6	30	73	135	216	316	436	574	732	908	1103	1318	1551	1804	2075	46
47	6	30	74	136	218	318	438	577	734	911	1107	1321	1555	1808	2080	47
48	6	31	75	137	219	320	440	579	737	914	1110	1325	1559	1812	2084	48
49	6	31	75	139	221	322	442	582	740	917	1114	1329	1563	1817	2089	49
50	7	32	76	140	222	324	444	584	743	920	1117	1333	1567	1821	2094	50
51	7	33	77	141	225	326	447	587	745	923	1120	1336	1571	1825	2099	51
52	7	33	78	142	226	328	449	589	748	927	1124	1340	1575	1830	2103	52
53	7	34	79	144	227	329	451	592	751	930	1127	1344	1580	1834	2108	53
54	8	34	80	145	229	331	453	594	754	933	1131	1348	1584	1839	2113	54
55	8	35	81	146	230	333	455	597	757	936	1134	1352	1588	1843	2117	55
56	8	36	82	147	232	335	458	599	760	939	1138	1355	1592	1847	2122	56
57	8	36	83	149	233	337	460	602	763	942	1141	1359	1596	1852	2127	57
58	9	37	84	150	235	339	462	604	765	945	1145	1363	1600	1856	2132	58
59	9	37	85	151	236	341	464	607	768	949	1148	1367	1604	1861	2136	59
60	9	38	86	152	238	343	466	609	771	952	1152	1370	1608	1865	2141	60

A TABLE of NATURAL VERSED SINES adapted to SECONDS of TIME.

Argument = the Horary Angle counted from the Meridian.

S.	M. 15	M. 16	M. 17	M. 18	M. 19	M. 20	M. 21	M. 22	M. 23	M. 24	M. 25	M. 26	M. 27	M. 28	M. 29	S.
0	2141	2436	2750	3083	3434	3805	4195	4604	5031	5478	5944	6428	6931	7454	7995	0
1	2146	2441	2755	3088	3441	3812	4202	4611	5039	5486	5952	6436	6940	7463	8004	1
2	2151	2446	2761	3093	3447	3818	4208	4618	5046	5494	5960	6445	6949	7472	8013	2
3	2155	2451	2766	3099	3453	3824	4215	4625	5053	5501	5967	6453	6957	7481	8023	3
4	2160	2456	2771	3106	3459	3831	4222	4632	5061	5509	5975	6461	6966	7489	8032	4
5	2165	2461	2777	3111	3465	3837	4228	4639	5068	5516	5983	6469	6974	7498	8041	5
6	2170	2466	2782	3117	3471	3843	4235	4646	5075	5524	5991	6478	6983	7507	8050	6
7	2175	2472	2788	3123	3477	3850	4242	4653	5083	5531	5999	6486	6991	7516	8059	7
8	2179	2477	2793	3128	3483	3856	4248	4660	5090	5539	6007	6494	7000	7525	8069	8
9	2184	2482	2799	3134	3489	3863	4255	4667	5097	5547	6015	6502	7009	7534	8078	9
10	2189	2487	2804	3140	3495	3869	4262	4674	5105	5554	6023	6511	7017	7543	8087	10
11	2194	2492	2809	3146	3501	3875	4269	4681	5112	6562	6031	6519	7026	7552	8096	11
12	2198	2497	2815	3151	3507	3882	4275	4688	5119	5570	6039	6527	7034	7561	8106	12
13	2203	2502	2820	3157	3513	3888	4282	4695	5127	5577	6047	6536	7043	7570	8115	13
14	2208	2507	2826	3163	3519	3895	4289	4702	5134	5585	6055	6544	7052	7579	8124	14
15	2213	2513	2831	3169	3525	3901	4295	4709	5141	5593	6063	6551	7060	7587	8133	15
16	2218	2518	2837	3175	3531	3907	4302	4716	5149	5600	6071	6560	7069	7596	8143	16
17	2223	2523	2842	3180	3538	3914	4309	4723	5156	5608	6079	6569	7078	7605	8152	17
18	2227	2528	2848	3186	3544	3920	4316	4730	5163	5616	6087	6577	7086	7614	8161	18
19	2232	2533	2853	3192	3550	3927	4322	4737	5171	5623	6095	6585	7095	7623	8170	19
20	2237	2538	2859	3198	3556	3933	4329	4744	5178	5631	6103	6594	7103	7632	8180	20
21	2242	2544	2864	3204	3562	3940	4336	4751	5186	5639	6111	6602	7112	7641	8189	21
22	2247	2549	2870	3209	3568	3946	4343	4758	5193	5647	6119	6610	7121	7650	8198	22
23	2252	2554	2875	3215	3574	3952	4350	4766	5200	5654	6127	6619	7130	7659	8207	23
24	2257	2559	2881	3221	3581	3959	4356	4773	5208	5662	6135	6627	7138	7668	8217	24
25	2262	2564	2886	3227	3587	3965	4363	4780	5215	5670	6143	6636	7147	7677	8226	25
26	2266	2570	2892	3233	3593	3972	4370	4787	5223	5678	6151	6644	7156	7686	8235	26
27	2272	2575	2897	3239	3599	3978	4377	4794	5230	5685	6159	6652	7164	7695	8245	27
28	2276	2580	2903	3244	3605	3985	4383	4801	5237	5693	6167	6661	7173	7704	8254	28
29	2281	2585	2908	3250	3611	3991	4390	4808	5245	5701	6175	6669	7182	7713	8263	29
30	2286	2590	2914	3256	3618	3998	4397	4815	5252	5709	6183	6678	7190	7722	8273	30
31	2291	2596	2919	3262	3624	4004	4404	4822	5260	5716	6192	6686	7199	7731	8282	31
32	2296	2601	2925	3268	3630	4011	4411	4829	5267	5724	6200	6694	7208	7740	8291	32
33	2301	2606	2930	3274	3636	4017	4418	4837	5275	5732	6208	6703	7217	7749	8301	33
34	2306	2611	2936	3280	3642	4024	4424	4844	5282	5740	6216	6711	7226	7758	8210	34
35	2311	2617	2942	3286	3648	4030	4431	4851	5290	5747	6224	6719	7234	7767	8319	35
36	2316	2622	2947	3291	3655	4037	4438	4858	5297	5755	6232	6728	7243	7776	8329	36
37	2321	2627	2952	3297	3661	4043	4445	4865	5305	5763	6240	6736	7251	7785	8338	37
38	2326	2632	2958	3303	4667	4050	4452	4872	5312	5771	6248	6745	7260	7794	8848	38
39	2331	2638	2964	3309	3673	4056	4459	4880	5320	5779	6256	6753	7269	7804	8357	39
40	2335	2643	2970	3315	3680	4063	4465	4887	5327	5786	6264	6762	7278	7813	8366	40
41	2340	2648	2975	3321	3686	4070	4472	4894	5335	5794	6273	6770	7286	7822	8376	41
42	2345	2654	2981	3327	3692	4076	4479	4901	5342	5802	6281	6778	7295	7831	8385	42
43	2350	2659	2986	3333	3698	4083	4486	4908	5350	5810	6289	6787	7304	7840	8394	43
44	2355	2664	2992	3339	3705	4089	4493	4916	5357	5818	6297	6795	7313	7849	8404	44
45	2360	2669	2998	3345	3711	4096	4500	4923	5365	5826	6305	6804	7321	7858	8413	45
46	2365	2675	3003	3351	3717	4102	4507	4930	5372	5833	6313	6812	7330	7867	8423	46
47	2270	2680	3009	3357	3723	4109	4514	4937	5380	5841	6322	6821	7339	7876	8432	47
48	2375	2685	3015	3363	3730	4116	4520	4944	5387	5849	6330	6829	7348	7885	8442	48
49	2380	2691	3020	3369	3736	4122	4527	4952	5395	5857	6338	6838	7357	7894	8451	49
50	2385	2696	3026	3375	3742	4129	4534	4959	5402	5865	6346	6846	7365	7903	8460	50
51	2390	2701	3031	3380	3749	4135	4541	4966	5410	5873	6354	6355	7374	7913	8470	51
52	2395	2707	3037	3386	3755	4142	4548	4973	5417	5881	6362	6363	7383	7922	8479	52
53	2401	2712	3043	3392	3761	4149	4555	4981	5425	5888	6371	6872	7392	7931	8489	53
54	2406	2718	3049	3398	3767	4155	4562	4988	5433	5896	6379	6880	7401	7940	8498	54
55	2411	2723	3054	3404	3774	4162	4569	4995	5440	5904	6387	6889	7410	7949	8508	55
56	2416	2728	3060	3410	3780	4168	4576	5002	5448	5912	6395	6897	7418	7958	8517	56
57	2421	2734	3066	3416	3786	4175	4583	5010	5455	5920	6403	6906	7427	7968	8527	57
58	2426	2739	3071	3422	3793	4182	4590	5017	5463	5928	6411	6914	7436	7977	8536	58
59	2431	2744	3077	3428	3799	4188	4597	5024	5470	5936	6420	6923	7445	7986	8546	59
60	2436	2750	3083	3434	3805	4195	4604	5031	5478	5944	6428	6931	7454	7995	8555	60

[From MENDOZA.]

TABLE 1. (The Compression $= \frac{320}{321}$.)

The MEASURE of a DEGREE, in each parallel of Latitude, both for the SPHERE, and SPHEROID, calculated in Minutes of the Equator.

Par. of Lat.	Degree on the Sphere.	Difference.	Degree on the Spheroid.	Difference.
0°	60'.000	0.009	60.000	0'.009
1	59.991	0.028	59.991	0.027
2	59.963	0.045	59.964	0.046
3	59.918	0.064	59.918	0.063
4	59.854	0.082	59.855	0.082
5	59.772	0.101	59.773	0.100
6	59.671	0.118	59.673	0.117
7	59.553	0.137	59.556	0.136
8	59.416	0.155	59.420	0.154
9	59.261	0.173	59.266	0.172
10	59.088	0.190	59.094	0.190
11	58.898	0.209	58.904	0.207
12	58.689	0.227	58.697	0.226
13	58.462	0.244	58.471	0.243
14	58.218	0.262	58.228	0.260
15	57.956	0.280	57.968	0.279
16	57.676	0.298	57.689	0.295
17	57.378	0.315	57.394	0.314
18	57.063	0.332	57.080	0.330
19	56.731	0.349	56.750	0.348
20	56.382	0.367	56.402	0.365
21	56.015	0.384	56.037	0.382
22	55.631	0.401	55.655	0.398
23	55.230	0.417	55.257	0.416
24	54.813	0.435	54.841	0.432
25	54.378	0.450	54.409	0.449
26	53.928	0.468	53.960	0.465
27	53.460	0.483	53.495	0.482
28	52.977	0.500	53.013	0.497
29	52.477	0.515	52.516	0.514
30	51.962		52.002	

Par. of Lat.	Degree on the Sphere.	Difference.	Degree on the Spheroid.	Difference.
30°	51.962	0'.532	52'.002	0'.529
31	51.430	0.547	51.473	0.546
32	50.883	0.563	50.927	0.560
33	50.320	0.578	50.367	0.576
34	49.742	0.593	49.791	0.592
35	49.149	0.608	49.199	0.606
36	48.541	0.623	48.593	0.621
37	47.918	0.637	47.972	0.636
38	47.281	0.652	47.336	0.650
39	46.629	0.666	46.686	0.664
40	45.963	0.680	46.022	0.679
41	45.283	0.694	45.343	0.692
42	44.589	0.708	44.651	0.706
43	43.881	0.721	43.945	0.720
44	43.160	0.734	43.225	0.732
45	42.426	0.746	42.493	0.746
46	41.680	0.760	41.747	0.759
47	40.920	0.772	40.988	0.771
48	40.148	0.784	40.217	0.783
49	39.364	0.797	39.434	0.796
50	38.567	0.808	38.638	0.807
51	37.759	0.819	37.831	0.820
52	36.940	0.831	37.011	0.830
53	36.109	0.842	36.181	0.842
54	35.267	0.852	35.339	0.852
55	34.415	0.863	34.487	0.863
56	33.552	0.874	33.624	0.874
57	32.678	0.883	32.750	0.884
58	31.795	0.893	31.866	0.893
59	30.[illegible]	0.902	30.973	0.903
60	30.000		30.070	

Par. of Lat.	Degree on the Sphere.	Difference.	Degree on the Spheroid.	Difference.
60°	30'.000	0'.911	30.070	0'.912
61	29.089	0.921	29.158	0.921
62	28.168	0.929	28.237	0.930
63	27.239	0.937	27.307	0.938
64	26.302	0.945	26.369	0.947
65	25.357	0.953	25.422	0.954
66	24.404	0.960	24.468	0.962
67	23.444	0.968	23.506	0.969
68	22.476	0.974	22.537	0.976
69	21.502	0.981	21.561	0.983
70	20.521	0.987	20.578	0.989
71	19.534	0.993	19.589	0.996
72	18.541	0.999	18.593	1.001
73	17.542	1.004	17.592	1.006
74	16.538	1.009	16.586	1.012
75	15.529	1.014	15.574	1.016
76	14.515	1.018	14.558	1.021
77	13.497	1.022	13.537	1.025
78	12.475	1.026	12.512	1.029
79	11.449	1.030	11.483	1.033
80	10.419	1.033	10.450	1.036
81	9.386	1.036	9.414	1.038
82	8.350	1.038	8.376	1.041
83	7.312	1.040	7.335	1.044
84	6.272	1.043	6.292	1.046
85	5.229	1.044	5.246	1.048
86	4.185	1.045	4.198	1.049
87	3.140	1.046	3.149	1.049
88	2.094	1.047	2.100	1.050
89	1.047	1.047	1.050	1.050
90	0.000		0.000	

TABLE 2. (The Compression $= \frac{320}{321}$.)

The MEASURES of different ARCS of the MERIDIAN, in the SPHEROID, from the EQUATOR to the POLE, and also of the respective DEGREES of LATITUDE, calculated in Minutes of the Equator.

Lat.	Arc. of Mer.	Deg. of Lat.	Difference.
0	0'.000	59'.628	
1	59.628	59.629	0'.001
2	119.257	59.630	.001
3	178.887	59.631	.001
4	238.518	59.632	.001
5	298.150	59.634	.002
6	357.784	59.636	.002
7	417.420	59.638	.002
8	477.058	59.641	.003
9	536.699	59.644	.003
10	596.343	59.647	.003
11	655.990	59.651	.004
12	715.641	59.655	.004
13	775.296	59.659	.004
14	834.955	59.663	.004
15	894.618	59.668	.005
16	954.286	59.673	.005
17	1013.959	59.679	.006
18	1073.638	59.685	.006
19	1133.323	59.691	.006
20	1193.014	59.697	.006
21	1252.711	59.703	.006
22	1312.414	59.710	.007
23	1372.124	59.717	.007
24	1431.841	59.724	.007
25	1491.565	59.732	.008
26	1551.297	59.739	.007
27	1611.036	59.747	.008
28	1670.783	59.755	.008
29	1730.538	59.764	.009
30	1790.302		.008

Lat.	Arc. of Mer.	Deg. of Lat.	Difference.
30	1790.302	59'.772	0'.008
31	1850.074	59.781	.009
32	1909.855	59.789	.008
33	1969.644	59.798	.009
34	2029.442	59.807	.009
35	2089.249	59.816	.009
36	2149.065	59.826	.010
37	2208.891	59.835	.009
38	2268.726	59.844	.009
39	2328.570	59.854	.010
40	2388.424	59.863	.009
41	2448.287	59.873	.010
42	2508.160	59.883	.010
43	2568.043	59.893	.010
44	2627.936	59.902	.009
45	2687.838	59.912	.010
46	2747.750	59.922	.010
47	2807.672	59.931	.009
48	2867.603	59.941	.010
49	2927.544	59.951	.010
50	2987.495	59.960	.009
51	3047.455	59.970	.010
52	3107.425	59.979	.009
53	3167.404	59.989	.010
54	3227.393	59.998	.009
55	3287.391	60.007	.009
56	3347.398	60.016	.009
57	3407.414	60.025	.009
58	3467.439	60.034	.009
59	3527.473	60.042	.008
60	3587.515		.009

Lat.	Arc. of Mer.	Deg. of Lat.	Difference.
60	3587.515	60'.051	0'.009
61	3647.566	60.059	.008
62	3707.625	60.067	.008
63	3767.692	60.075	.008
64	3827.767	60.082	.007
65	3887.849	60.090	.008
66	3947.939	60.097	.007
67	4008.036	60.104	.007
68	4068.140	60.111	.007
69	4128.251	60.117	.006
70	4188.368	60.124	.007
71	4248.492	60.130	.006
72	4308.622	60.135	.005
73	4368.757	60.141	.006
74	4428.898	60.146	.005
75	4489.044	60.151	.005
76	4549.195	60.155	.004
77	4609.350	60.160	.005
78	4669.510	60.164	.004
79	4729.674	60.167	.003
80	4789.841	60.171	.004
81	4850.012	60.174	.003
82	4910.186	60.176	002
83	4970.362	60.179	.003
84	5030.541	60.181	.002
85	5090.722	60.182	.001
86	5150.904	60.184	.002
87	5211.088	60.185	.001
88	5271.273	60.185	.000
89	5331.458	60.186	.001
90	5391.644		

TABLE 3.

A TABLE of the COMPARATIVE DEGREES of the MERIDIAN for eight different SPHEROIDS, in MINUTES of the EQUATOR; calculated for 62 parallels of Latitude, by the REV. MICHAEL WARD, L.L.B.

Lat.	Compression. $\frac{1}{301}$	Compression. $\frac{1}{306}$	Compression. $\frac{1}{311}$	Compression. $\frac{1}{316}$	Compression. $\frac{1}{321}$	Compression. $\frac{1}{326}$	Compression. $\frac{1}{331}$	Compression. $\frac{1}{334}$	Lat.
0°	59′ .60199	59′ .60848	59′ .61476	59′ .62085	59′ .62675	59′ .63246	59′ .63801	59′ .64126	0°
1	.60218	.60866	.61494	.62101	.62691	.63263	.63818	.64142	1
2	.60273	.60920	.61546	.62153	.62742	.63314	.63867	.64191	2
3	.60364	.61009	.61634	.62240	.62827	.63398	.63950	.64272	3
4	.60491	.61134	.61757	.62361	.62946	.63516	.64065	.64387	4
5	.60653	.61295	.61915	.62516	.63099	.63667	.64213	.64534	5
6	.60852	.61490	.62107	.62706	.63286	.63849	.64394	.64713	6
7	.61087	.61721	.62334	.62931	.63506	.64065	.64607	.64924	7
8	.61356	.61986	.62595	.63189	.63759	.64314	.64853	.65167	8
9	.61661	.62285	.62890	.63479	.64045	.64596	.65130	.65438	9
10	.62000	.62619	.63218	.63801	.64362	.64909	.65438	.65748	10
11	.62374	.62987	.63580	.64155	.64712	.65254	.65778	.66084	11
12	.62781	.63387	.63974	.64543	.65094	.65630	.66148	.66451	12
13	.63221	.63820	.64400	.64962	.65507	.66036	.66549	.66848	13
14	.63694	.64286	.64858	.65413	.65951	.66473	.66979	.67274	14
15	.64199	.64782	.65347	.65895	.66425	.66940	.67438	.67730	15
16	.64736	.65310	.65866	.66406	.66928	.67435	.67926	.68213	16
17	.65303	.65869	.66416	.66946	.67460	.67959	.68445	.68725	17
18	.65901	.66456	.66994	.67516	.68020	.68511	.68986	.69263	18
19	.66528	.67072	.67601	.68113	.68609	.69090	.69556	.69829	19
20	.67184	.67718	.68236	.68739	.69224	.69696	.70153	.70420	20
21	.67867	.68391	.68897	.69388	.69865	.70327	.70774	.71036	21
22	.68578	.69089	.69585	.70065	.70531	.70983	.71421	.71676	22
23	.69315	.69814	.70298	.70767	.71222	.71663	.72091	.72340	23
24	.70077	.70564	.71036	.71493	.71937	.72367	.72784	.73027	24
25	.70863	.71337	.71798	.72243	.72674	.73093	.73499	.73736	25
26	.71672	.72133	.72582	.73015	.73433	.73841	.74235	.74465	26
27	.72504	.72952	.73386	.73808	.74214	.74609	.74992	.75215	27
28	.73358	.73791	.74212	.74621	.75014	.75397	.75768	.75985	28
29	.74232	.74651	.75058	.75453	.75834	.76204	.76563	.76772	29
30	.75125	.75590	.75923	.76302	.76671	.77029	.77375	.77577	30
31	.76036	.76426	.76805	.77170	.77526	.77871	.78204	.78399	31
32	.76964	.77339	.77703	.78054	.78396	.78728	.79048	.79236	32
33	.77909	.78268	.78617	.78954	.79282	.79600	.79907	.80087	33
34	.78868	.79212	.79545	.79868	.80181	.80486	.80780	.80951	34
35	.79841	.80169	.80486	.80794	.81093	.81384	.81665	.81828	35
36	.80826	.81137	.81440	.81732	.82017	.82294	.82560	.82716	36
37	.81822	.82118	.82405	.82682	.82952	.83214	.83467	.83614	37
38	.82829	.83107	.83379	.83641	.83896	.84142	.84382	.84521	38
39	.83844	.84106	.84361	.84608	.84848	.85079	.85305	.85436	39
40	.84867	.85112	.85351	.85582	.85807	.86025	.86235	.86358	40
41	.85896	.86125	.86347	.86563	.86772	.86975	.87171	.87286	41
42	.86930	.87142	.87348	.87548	.87742	.87930	.88112	.88218	42
43	.87968	.88163	.88353	.88537	.88715	.88888	.89056	.89153	43
44	.89009	.89186	.89360	.89528	.89691	.89895	.90002	.90091	44
45	.90050	.90211	.90369	.90521	.90667	.90812	.91150	.91030	45
46	.91092	.91236	.91377	.91513	.91644	.91773	.91897	.91967	46
47	.92133	.92260	.92384	.92504	.92621	.92734	.92844	.92007	47
48	.93170	.93281	.93389	.93493	.93594	.93693	.93789	.93843	48
49	.94205	.94298	.94390	.94479	.94564	.94648	.94730	.94775	49
50	.95234	.95310	.95386	.95459	.95529	.95595	.95665	.95702	50
51	.96256	.96317	.96376	.96433	.96488	.96544	.96594	.96624	51
52	.97272	.97315	.97358	.97400	.97440	.97481	.97518	.97539	52
53	.98278	.98305	.98332	.98359	.98384	.98409	.98433	.98447	53
54	.99275	.99285	.99297	.99308	.99319	.99327	.99339	.99346	54
55	60 .00260	60 .00255	60 .00251	60 .00247	60 .00243	60 .00239	60 .00235	60 .00233	55
56	.01232	.01211	.01192	.01173	.01155	.01136	.01121	.01110	56
57	.02191	.02155	.02120	.02087	.02054	.02021	.01992	.01974	57
58	.03135	.03084	.03035	.02986	.02939	.02894	.02851	.02825	58
59	.04063	.03998	.03934	.03871	.03810	.03752	.03696	.03662	59
60	.04975	.04894	.04815	.04739	.04665	.04594	.04524	.04483	60
61	.05868	.05772	.05680	.05590	.05503	.05418	.05337	.05289	61
62	.06528	.06631	.06524	.06422	.06322	.06220	.06130	.06077	62
	$\frac{1}{301}$	$\frac{1}{306}$	$\frac{1}{311}$	$\frac{1}{316}$	$\frac{1}{321}$	$\frac{1}{326}$	$\frac{1}{331}$	$\frac{1}{334}$	

N.B. In the two preceding Tables, the *Compression* must be read $\frac{1}{321}$, instead of $\frac{320}{321}$.

TABLE 4.

Lengths, in MINUTES of the EQUATOR, of one DEGREE of LONGITUDE in each of 62 successive PARALLELS of LATITUDE, calculated for EIGHT different SPHEROIDS, by the REV. MICHAEL WARD, L.L.B.

Lat.	$\frac{1}{301}$	$\frac{1}{306}$	$\frac{1}{311}$	$\frac{1}{316}$	$\frac{1}{321}$	$\frac{1}{326}$	$\frac{1}{331}$	$\frac{1}{334}$	Lat.
0°	60 .0000	60 .0000	60 .0000	60 .0000	60 .0000	60 .0000	60 .0000	60 .0000	0°
1	59 .9909	59 .9909	59 .9909	59 .9909	59 .9909	59 .9909	59 .9909	59 .9909	1
2	59 .9637	59 .9636	59 .9636	59 .9636	59 .9636	59 .9636	59 .9636	59 .9636	2
3	59 .9183	59 .9183	59 .9182	59 .9182	59 .9182	59 .9182	59 .9182	59 .9182	3
4	59 .8548	59 .8548	59 .8547	59 .8547	59 .8547	59 .8547	59 .8547	59 .8547	4
5	59 .7732	59 .7732	59 .7731	59 .7731	59 .7731	59 .7730	59 .7730	59 .7730	5
6	59 .6734	59 .6734	59 .6734	59 .6733	59 .6733	59 .6733	59 .6732	59 .6732	6
7	59 .5556	59 .5556	59 .5556	59 .5555	59 .5555	59 .5554	59 .5554	59 .5554	7
8	59 .4199	59 .4198	59 .4198	59 .4197	59 .4196	59 .4196	59 .4195	59 .4195	8
9	59 .2661	59 .2660	59 .2659	59 .2658	59 .2657	59 .2656	59 .2656	59 .2655	9
10	59 .0943	59 .0942	59 .0941	59 .0940	59 .0939	59 .0938	59 .0937	59 .0936	10
11	58 .9047	58 .9046	58 .9045	58 .9044	58 .9043	58 .9041	58 .9040	58 .9039	11
12	58 .6972	58 .6971	58 .6970	58 .6969	58 .6967	58 .6966	58 .6965	58 .6964	12
13	58 .4720	58 .4718	58 .4717	58 .4715	58 .4714	58 .4712	58 .4711	58 .4709	13
14	58 .2290	58 .2288	58 .2287	58 .2285	58 .2283	58 .2281	58 .2280	58 .2279	14
15	57 .9684	57 .9682	57 .9680	57 .9678	57 .9676	57 .9674	57 .9672	57 .9671	15
16	57 .6902	57 .6899	57 .6897	57 .6895	57 .6893	57 .6891	57 .6889	57 .6888	16
17	57 .3945	57 .3942	57 .3939	57 .3937	57 .3935	57 .3933	57 .3931	57 .3930	17
18	57 .0814	57 .0811	57 .0808	57 .0806	57 .0804	57 .0801	57 .0798	57 .0796	18
19	56 .7510	56 .7507	56 .7504	56 .7501	56 .7498	56 .7495	56 .7492	56 .7490	19
20	56 .4034	56 .4030	56 .4027	56 .4023	56 .4020	56 .4017	56 .4014	56 .4012	20
21	56 .0387	56 .0383	56 .0379	56 .0375	56 .0371	56 .0368	56 .0365	56 .0362	21
22	55 .6569	55 .6565	55 .6561	55 .6557	55 .6553	55 .6549	55 .6545	55 .6543	22
23	55 .2582	55 .2577	55 .2573	55 .2569	55 .2565	55 .2561	55 .2557	55 .2555	23
24	54 .8428	54 .8423	54 .8418	54 .8413	54 .8409	54 .8405	54 .8400	54 .8398	24
25	54 .4107	54 .4101	54 .4096	54 .4091	54 .4086	54 .4081	54 .4076	54 .4073	25
26	53 .9620	53 .9615	53 .9609	53 .9604	53 .9599	53 .9594	53 .9589	53 .9386	26
27	53 .4969	53 .4963	53 .4958	53 .4952	53 .4946	53 .4941	53 .4936	53 .4933	27
28	53 .0156	53 .0149	53 .0143	53 .0137	53 .0131	53 .0126	53 .0121	53 .0118	28
29	52 .5181	52 .5174	52 .5168	52 .5162	52 .5155	52 .5149	52 .5143	52 .5140	29
30	52 .0046	52 .0039	52 .0032	52 .0026	52 .0019	52 .0013	52 .0007	52 .0004	30
31	51 .4752	51 .4745	51 .4739	51 .4732	51 .4725	51 .4718	51 .4712	51 .4708	31
32	50 .9303	50 .9295	50 .9288	50 .9280	50 .9273	50 .9266	50 .9260	50 .9256	32
33	50 .3698	50 .3690	50 .3682	50 .3674	50 .3667	50 .3660	50 .3653	50 .3649	33
34	49 .7939	49 .7930	49 .7922	49 .7914	49 .7907	49 .7899	49 .7892	49 .7887	34
35	49 .2028	49 .2019	49 .2011	49 .2002	49 .1994	49 .1987	49 .1680	49 .1975	35
36	48 .5968	48 .5959	48 .5950	48 .5941	48 .5933	48 .5925	48 .5917	48 .5914	36
37	47 .9758	47 .9748	47 .9739	47 .9730	47 .9722	47 .9714	47 .9706	47 .9701	37
38	47 .3401	47 .3391	47 .3381	47 .3371	47 .3362	47 .3355	47 .3348	47 .3342	38
39	46 .6901	46 .6891	46 .6881	46 .6871	46 .6862	46 .6853	46 .6845	46 .6840	39
40	46 .0257	46 .0247	46 .0237	46 .0227	46 .0218	46 .0209	46 .0199	46 .0195	40
41	45 .3473	45 .3463	45 .3453	45 .3443	45 .3434	45 .3424	45 .3414	45 .3409	41
42	44 .6550	44 .6539	44 .6529	44 .6519	44 .6508	44 .6499	44 .6490	44 .6485	42
43	43 .9490	43 .9479	43 .9469	43 .9459	43 .9450	43 .9440	43 .9429	43 .9424	43
44	43 .2298	43 .2286	43 .2275	43 .2264	43 .2253	43 .2243	43 .2233	43 .2227	44
45	42 .4968	42 .4957	42 .4946	42 .4936	42 .4926	42 .4915	42 .4905	42 .4899	45
46	41 .7512	41 .7500	41 .7489	41 .7478	41 .7467	41 .7457	41 .7447	41 .7440	46
47	40 .9926	40 .9914	40 .9903	40 .9892	40 .9881	40 .9870	40 .9859	40 .9854	47
48	40 .2215	40 .2203	40 .2191	40 .2180	40 .2169	40 .2159	40 .2148	40 .2142	48
49	39 .4381	39 .4369	39 .4357	39 .4346	39 .4335	39 .4324	39 .4313	39 .4307	49
50	38 .6425	38 .6413	38 .6402	38 .6391	38 .6380	38 .6368	38 .6357	38 .9350	50
51	37 .8350	37 .8338	37 .8326	37 .8315	37 .8304	37 .8293	37 .8282	37 .8275	51
52	37 .0160	37 .0148	37 .0136	37 .0124	37 .0112	37 .0101	37 .0090	37 .0083	52
53	36 .1854	36 .1842	36 .1830	36 .1818	36 .1807	36 .1796	36 .1785	36 .1778	53
54	35 .3438	35 .3426	35 .3414	35 .3402	35 .3391	35 .3380	35 .3369	35 .3363	54
55	34 .4914	34 .4902	34 .4890	34 .4978	34 .4866	34 .4955	34 .4844	34 .4838	55
56	33 .6283	33 .6270	33 .6258	33 .6246	33 .6235	33 .6224	33 .6213	33 .6206	56
57	32 .7548	32 .7536	32 .7524	32 .7512	32 .7500	32 .7489	32 .7478	32 .7470	57
58	31 .8712	31 .8700	31 .8688	31 .8676	31 .8665	31 .8654	31 .8643	31 .8636	58
59	30 .9778	30 .9766	30 .9754	30 .9743	30 .9732	30 .9721	30 .9710	30 .9704	59
60	30 .0748	30 .0736	30 .0724	30 .0713	30 .0702	30 .0691	30 .0680	30 .0675	60
61	29 .1626	29 .1614	29 .1602	29 .1591	29 .1580	29 .1569	29 .1558	29 .1551	61
62	28 .2416	28 .2404	28 .2392	28 .2381	28 .2370	28 .2359	28 .2348	28 .2341	62
	$\frac{1}{301}$	$\frac{1}{306}$	$\frac{1}{311}$	$\frac{1}{316}$	$\frac{1}{321}$	$\frac{1}{326}$	$\frac{1}{331}$	$\frac{1}{334}$	

TABLE 5.

Decimal Degrees of Longitude and Latitude in the Spheroid by Puissant, and of Latitude by Svanberg, in French Metres, in each parallel Degree of Latitude from the Equator to the Pole.

Par. of Lat.	Puissant. $\frac{1}{335}$ Longitude.	Puissant. $\frac{1}{335}$ Latitude.	Svanberg. $\frac{1}{323}$ Latitude.	Par. of Lat.	Puissant. $\frac{1}{335}$ Longitude.	Puissant. $\frac{1}{335}$ Latitude.	Svanberg. $\frac{1}{323}$ Latitude.	Par. of Lat.	Puissant. $\frac{1}{335}$ Longitude.	Puissant. $\frac{1}{335}$ Latitude.	Svanberg. $\frac{1}{323}$ Latitude.
0	100149.4	99552.5	99537.45	34	86269.5	99789.7	99777.06	67	49731.8	100233.9	100237.84
1	100137.1	99552.9	99537.68	35	85461.0	99802.2	99789.92	68	48358.3	100245.9	100250.34
2	100100.3	99553.8	99538.36	36	84631.4	99814.9	99803.00	69	46972.4	100257.5	100262.60
3	100038.9	99555.1	99539.49	37	83780.9	99827.8	99816.26	70	45574.8	100269.0	100274.61
4	99953.0	99556.9	99541.09	38	82909.7	99840.9	99829.73	71	44165.9	100280.2	100286.35
5	99842.5	99559.0	99543.14	39	82018.1	99854.1	99843.35	72	42746.0	100291.1	100297.80
6	99707.6	99561.8	99545.63	40	81106.2	99867.5	99857.13	73	41315.3	100301.7	100308.97
7	99548.2	99564.7	99548.56	41	80174.1	99881.0	99871.07	74	39874.4	100312.0	100319.82
8	99364.3	99568.2	99551.95	42	79222.3	99894.6	99885.13	75	38423.4	100322.0	100330.37
9	99156.2	99572.1	99555.77	43	78250.9	99908.3	99899.31	76	36962.8	100331.7	100340.46
10	98923.6	99576.4	99560.03	44	77260.1	99922.1	99913.59	77	35493.0	100341.1	100350.48
11	98666.8	99581.2	99564.73	45	76250.1	99936.0	99927.96	78	34014.2	100350.1	100360.03
12	98385.8	99586.3	99569.85	46	75221.3	99950.0	99942.41	79	32527.0	100358.8	100369.21
13	98080.6	99591.8	99575.41	47	74173.8	99964.0	99956.92	80	31031.6	100367.2	100378.04
14	97751.3	99597.8	99581.38	48	73108.0	99978.0	99971.47	81	29528.5	100375.1	100386.49
15	97398.1	99604.2	99587.76	49	72024.0	99992.1	99986.05	82	28017.9	100382.7	100394.56
16	97020.9	99610.9	99594.55	50	70922.1	100006.2	100000.66	83	26500.3	100389.9	100402.24
17	96616.9	99618.0	99601.74	51	69802.6	100020.3	100015.27	84	24976.1	100396.8	100409.53
18	96195.1	99625.4	99609.32	52	68665.8	100034.4	100029.87	85	23445.6	100403.2	100416.42
19	95746.8	99633.4	99617.29	53	67512.0	100048.4	100044.44	86	21909.2	100409.3	100422.88
20	95274.9	99641.6	99625.64	54	66341.3	100062.4	100058.97	87	20367.3	100414.9	100428.95
21	94779.6	99650.2	99634.35	55	65154.2	100076.3	100073.44	88	18820.3	100420.1	100434.57
22	94260.9	99659.1	99643.43	56	63950.9	100090.2	100087.85	89	17268.6	100424.9	100439.77
23	93719.1	99668.4	99652.86	57	62731.7	100103.9	100102.19	90	15712.6	100429.3	100444.54
24	93154.2	99678.0	99662.64	58	61496.8	100117.6	100116.41	91	14152.6	100433.2	100448.87
25	92566.4	99687.9	99672.75	59	60246.7	100131.2	100130.53	92	12589.0	100436.8	100452.75
26	91955.8	99698.1	99683.18	60	58981.5	100144.6	100144.53	93	11022.3	100439.9	100456.18
27	91322.6	99708.6	99693.93	61	57701.6	100157.9	100158.39	94	9452.9	100442.5	100459.17
28	90666.9	99719.4	99704.98	62	56407.4	100171.0	100162.09	95	7881.0	100444.7	100461.70
29	89988.9	99730.5	99716.33	63	55099.1	100184.0	100185.62	96	6307.2	100446.5	100463.77
30	89288.6	99741.9	99727.27	64	53777.1	100196.8	100198.98	97	4731.8	100447.8	100465.39
31	88566.4	99753.5	99739.86	65	52441.7	100209.4	100212.14	98	3155.7	100448.7	100466.54
32	87822.4	99765.3	99152.02	66	51093.1	100221.7	100225.10	99	1577.8	100449.2	100467.24
33	87056.7	99777.4	99764.42	67	49731.8	100233.9	100237.84	100	0000.0	100449.4	100467.47

TABLE 6.

Degrees of Latitude and Longitude in Toises by Cagnoli. Compression = $\frac{1}{[illegible]}$.

Par. of lat.	Latitude.	Diff.	Longit.	Diff.	Par. of lat.	Latitude.	Diff.	Longit.	Diff.	Par. of lat.	Latitude.	Diff.	Longit.	Diff.	Par. of lat.	Latitude.	Diff.	Longit.	Diff.
0°	56756		57136		23	56843	7	52622	380	46°	57050	10	39759	710	69	57253	7	20536	930
1	56756	0	57128	8	24	56850	7	52226	396	47	57060	10	39036	723	70	57260	7	19599	937
2	56757	1	57102	26	25	56857	7	51814	412	48	57070	10	38302	734	71	57266	6	18657	942
3	56758	1	57058	44	26	56865	8	51387	427	49	57080	10	37556	746	72	57272	6	17709	948
4	56759	1	56998	60	27	56873	8	50944	443	50	57090	10	36799	757	73	57278	6	16756	953
5	56760	1	56920	78	28	56881	8	50485	459	51	57100	10	36030	769	74	57284	6	15798	958
6	56762	2	56825	95	29	56889	8	50012	473	52	57110	10	35250	780	75	57289	5	14835	963
7	56764	2	56713	112	30	56898	9	49523	489	53	57119	9	34459	791	76	57294	5	13866	969
8	56767	3	56584	129	31	56906	8	49019	504	54	57129	10	33657	801	77	57298	4	12893	973
9	56770	3	56437	147	32	56915	9	48500	519	55	57138	9	32845	812	78	57302	4	11917	976
10	56773	3	56274	163	33	56924	9	47966	534	56	57148	10	32024	821	79	57306	4	10937	980
11	56777	4	56093	181	34	56934	10	47418	548	57	57157	9	31192	832	80	57310	4	9953	984
12	56781	4	55896	197	35	56943	9	46855	563	58	57166	9	30350	842	81	57313	3	8967	986
13	56785	4	55681	215	36	56952	9	46277	578	59	57175	9	29499	851	82	57316	3	7978	989
14	56789	4	55450	231	37	56962	10	45686	591	60	57184	9	28640	859	83	57319	3	6986	992
15	56794	5	55202	248	38	56971	9	45081	605	61	57192	8	27772	868	84	57321	2	5992	994
16	56789	5	54937	265	39	56981	10	44462	619	62	57201	9	26894	878	85	57323	2	4996	996
17	56804	5	54655	282	40	56991	10	43829	633	63	57209	8	26009	885	86	57324	1	3999	997
18	56810	6	54357	298	41	57001	10	43183	646	64	57217	8	25115	894	87	57325	1	3000	999
19	56816	6	54053	314	42	57011	10	42524	659	65	57225	8	24213	902	88	57326	1	2000	1000
20	56822	6	53712	331	43	57021	10	41852	672	66	57232	7	23305	908	89	57327	1	1000	1000
21	56829	7	53365	347	44	57030	9	41167	685	67	57239	7	22389	916	90	57327	0	0000	1000
22	56836	7	53002	363	45	57040	10	40469	698	68	57246	7	21466	923					

The Metre is to the English Yard as 39 .3702 : 36 ; or as 1 .09362 : 1. The Toise is to the English Fathom as 76 .736 : 72 ; or as 1 .06577 : 1.

TABLE 7.

The MEASURE of each DEGREE, of three different denominations, in each PARALLEL of LATITUDE from the EQUATOR to either POLE, in English Fathoms, by LIEUT.-COL. W. LAMPTON. Compression $= \frac{1}{304}$.

Paral. of Lat.	Meridional in Fathoms.	Perpendicular in Fath.	Longitudinal in Fathoms.	Paral. of Lat.	Meridional in Fathoms.	Perpendicular in Fath.	Longitudinal in Fathoms.	Paral. of Lat.	Meridional in Fathoms.	Perpendicular in Fath.	Longitudinal in Fathoms.
0°	60458 .64	60857 .05	60857 .05	30°	60607 .4	60906 .9	52746 .9	60°	60906 .7	61007 .0	30503 .5
1	60458 .8	60857 .1	60847 .8	31	60616 .5	60910 .0	52210 .0	61	60915 .7	61010 .0	29578 .2
2	60459 .8	60857 .3	60820 .2	32	60625 .8	60913 .1	51657 .2	62	60924 .5	61012 .9	28643 .8
3	60460 .3	60857 .6	60774 .2	33	60635 .2	60916 .2	51088 .6	63	60933 .1	61015 .8	27700 .6
4	60461 .5	60858 .0	60709 .8	34	60644 .8	60919 .4	50504 .5	64	60941 .4	61018 .6	26748 .8
5	60463 .2	60858 .6	60627 .0	35	60654 .5	60922 .7	49904 .9	65	60949 .6	61021 .3	25788 .7
6	60465 .1	60859 .2	60525 .8	36	60664 .4	60926 .0	49290 .2	66	60957 .5	61024 .0	24820 .7
7	60467 .5	60860 .0	60406 .4	37	60674 .3	60929 .3	48660 .3	67	60965 .3	61026 .6	23845 .0
8	60470 .1	60860 .9	60268 .6	38	60684 .4	60932 .7	48015 .6	68	60972 .7	61029 .0	22861 .9
9	60473 .2	60861 .9	60112 .6	39	60694 .6	60936 .1	47355 .2	69	60979 .8	61031 .4	21871 .7
10	60476 .5	60863 .0	59938 .4	40	60704 .8	60939 .5	46682 .4	70	60986 .7	61033 .7	20874 .8
11	60480 .3	60864 .3	59746 .1	41	60715 .1	60943 .0	45994 .2	71	60993 .4	61035 .9	19871 .4
12	60484 .3	60865 .7	59535 .6	42	60725 .4	60946 .4	45292 .0	72	60999 .7	61038 .0	18861 .8
13	60488 .7	60867 .1	59307 .1	43	60735 .8	60949 .9	44576 .0	73	61005 .7	61040 .1	17846 .4
14	60493 .4	60868 .7	59060 .6	44	60746 .3	60953 .4	43846 .2	74	61011 .5	61042 .0	16825 .4
15	60498 .4	60870 .4	58796 .3	45	60756 .7	60956 .9	43103 .0	75	61016 .8	61043 .8	15799 .3
16	60503 .8	60872 .2	58514 .1	46	60767 .2	60960 .4	42346 .6	76	61022 .0	61045 .5	14768 .2
17	60509 .4	60874 .1	58214 .2	47	60777 .6	60963 .9	41577 .3	77	61026 .7	61047 .1	13732 .6
18	60515 .4	60876 .1	57896 .6	48	60788 .0	60967 .4	40795 .1	78	61031 .2	61048 .5	12692 .7
19	60521 .6	60878 .2	57561 .4	49	60798 .4	60970 .8	40000 .5	79	61035 .3	61049 .9	11648 .9
20	60528 .2	60880 .4	57208 .8	50	60808 .7	60974 .3	39193 .5	80	61039 .1	61051 .2	10601 .4
21	60535 .0	60882 .7	56838 .9	51	60819 .0	60977 .7	38374 .5	81	61042 .5	61052 .3	9550 .7
22	60542 .0	60885 .0	56451 .6	52	60829 .2	60981 .1	37543 .7	82	61045 .6	61053 .4	8497 .0
23	60549 .4	60887 .5	56047 .2	53	60839 .3	60984 .5	36701 .4	83	61048 .3	61054 .3	7440 .6
24	60557 .0	60890 .0	55625 .8	54	60849 .3	60987 .9	35847 .8	84	61050 .7	61055 .1	6382 .0
25	60564 .8	60892 .7	55187 .5	55	60859 .3	60991 .2	34983 .1	85	61052 .7	61055 .7	5321 .4
26	60572 .9	60895 .4	54732 .4	56	60869 .0	60994 .4	34107 .6	86	61054 .3	61056 .3	4259 .1
27	60581 .2	60898 .2	54260 .6	57	60878 .7	60997 .6	33221 .7	87	61055 .6	61056 .7	3195 .5
28	60589 .7	60901 .0	53772 .4	58	60888 .2	61000 .8	32325 .5	88	61056 .5	61057 .0	2130 .9
29	60598 .4	60903 .0	53267 .8	59	60897 .5	61004 .0	31419 .4	89	61057 .1	61057 .2	1065 .6
30	60607 .4	60906 .9	52746 .9	60	60906 .6	61007 .0	30503 .5	90	61057 .25	61057 .25	0000 .0

TABLE 8.

The same in English Feet. Compression $= \frac{1}{334}$. [From Dr. Rees' Cyclopædia.]

Lat.	Meridional.	Perpendic.	Longitudinal.	Lat.	Meridional.	Perpendic.	Longitudinal.	Lat.	Meridional.	Perpendic.	Longitudinal.
0°	362909	365094	365094	30°	363724	365360	316404	60°	365368	365920	182958
1	362910	365095	365040	31	363773	365377	313188	61	365416	365937	177408
2	362913	365097	364872	32	363823	365395	309876	62	365463	365953	171804
3	362918	365099	364596	33	363874	365413	306468	63	365509	365969	166146
4	362925	365102	364212	34	363926	365431	302964	64	365554	365984	160428
5	362934	365106	363714	35	363979	365449	299364	65	365598	365998	154668
6	362945	365110	363108	36	364033	365468	295674	66	365641	366011	148860
7	362958	365115	362394	37	364088	365487	291894	67	365682	366023	143010
8	362973	365120	361566	38	364144	365506	288024	68	365722	366035	137112
9	362989	365125	360630	39	364201	365525	284064	69	365761	366047	131178
10	363007	365131	359586	40	364258	365544	280020	70	365799	366058	125196
11	363027	365138	358428	41	364315	365563	275892	71	365836	366069	119168
12	363049	365146	357162	42	364372	365582	271680	72	365871	366079	113124
13	363073	365156	355794	43	364430	365601	267384	73	365904	366089	107034
14	363099	365163	354318	44	364488	365620	263004	74	365935	366099	100908
15	363127	365172	352734	45	364546	365640	258546	75	365965	366108	94752
16	363157	365181	351042	46	364604	365660	253410	76	365993	366117	88572
17	363189	365191	349242	47	364662	365679	249396	77	366019	366126	82362
18	363223	365201	347340	48	364720	365698	244704	78	366043	366134	76122
19	363258	365211	345324	49	364777	365717	239934	79	366065	366141	69858
20	363295	365222	343254	50	364834	365736	235086	80	366084	366148	63576
21	363333	365233	340968	51	364891	365755	230166	81	366102	366154	57276
22	363370	365245	338628	52	364948	365774	225180	82	366118	366159	50958
23	363410	365257	336186	53	365004	365793	220128	83	366133	366164	44622
24	363451	365269	333648	54	365059	365812	215010	84	366146	366168	38274
25	363494	365282	331014	55	365113	365831	209826	85	366159	366172	31914
26	363538	365296	328284	56	365166	365849	204576	86	366168	366176	25542
27	363583	365311	325458	57	365218	365867	199266	87	366175	366179	19164
28	363629	365327	322536	58	365269	365885	193890	88	366180	366181	12780
29	363676	365343	319518	59	365319	365903	188454	89	366183	366183	6390
30	363724	365360	316404	60	365368	365920	182958	90	366184	366184	0000

SPACE INTO TIME.

To convert Degrees and parts of the Equator into Sidereal Time.

	H.	M.		M.	S.		S.
1°	0	4	1′	0	4	1″	0.066
2	0	8	2	0	8	2	0.133
3	0	12	3	0	12	3	0.200
4	0	16	4	0	16	4	0.266
5	0	20	5	0	20	5	0.333
6	0	24	6	0	24	6	0.400
7	0	28	7	0	28	7	0.466
8	0	32	8	0	32	8	0.533
9	0	36	9	0	36	9	0.600
10	0	40	10	0	40	10	0.666
11	0	44	11	0	44	11	0.733
12	0	48	12	0	48	12	0.799
13	0	52	13	0	52	13	0.866
14	0	56	14	0	56	14	0.933
15	1	0	15	1	0	15	1.000
16	1	4	16	1	4	16	1.066
17	1	8	17	1	8	17	1.133
18	1	12	18	1	12	18	1.200
19	1	16	19	1	16	19	1.266
20	1	20	20	1	20	20	1.333
25	1	40	21	1	24	21	1.400
30	2	0	22	1	28	22	1.466
35	2	20	23	1	32	23	1.533
40	2	40	24	1	36	24	1.600
45	3	0	25	1	40	25	1.666
50	3	20	26	1	44	26	1.733
55	3	40	27	1	48	27	1.799
60	4	0	28	1	52	28	1.866
65	4	20	29	1	56	29	1.933
70	4	40	30	2	0	30	2.000
75	5	0	31	2	4	31	2.066
80	5	20	32	2	8	32	2.133
90	6	0	33	2	12	33	2.200
100	6	40	34	2	16	34	2.266
110	7	20	35	2	20	35	2.333
120	8	0	36	2	24	36	2.400
130	8	40	37	2	28	37	2.466
140	9	20	38	2	32	38	2.533
150	10	0	39	2	36	39	2.600
160	10	40	40	2	40	40	2.666
170	11	20	41	2	44	41	2.733
180	12	0	42	2	48	42	2.799
190	12	40	43	2	52	43	2.866
200	13	20	44	2	56	44	2.933
210	14	0	45	3	0	45	3.000
220	14	40	46	3	4	46	3.066
230	15	20	47	3	8	47	3.133
240	16	0	48	3	12	48	3.200
250	16	40	49	3	16	49	3.266
260	17	20	50	3	20	50	3.333
270	18	0	51	3	24	51	3.400
280	18	40	52	3	28	52	3.466
290	19	20	53	3	32	53	3.533
300	20	0	54	3	36	54	3.600
310	20	40	55	3	40	55	3.666
320	21	20	56	3	44	56	3.733
330	22	0	57	3	48	57	3.799
340	22	40	58	3	52	58	3.866
350	23	20	59	3	56	59	3.933
360	24	0	60	4	0	60	4.000

Or to convert Degrees and parts of Terrestrial Longitude into Time.

TIME INTO SPACE.

To convert Sidereal Time into Degrees and Parts of the Equator.

H.		M.			S.		
1	15°	1	0°	15′	1	0′	15″
2	30	2	0	30	2	0	30
3	45	3	0	45	3	0	45
4	60	4	1	0	4	1	0
5	75	5	1	15	5	1	15
6	90	6	1	30	6	1	30
7	105	7	1	45	7	1	45
8	120	8	2	0	8	2	0
9	135	9	2	15	9	2	15
10	150	10	2	30	10	2	30
11	165	11	2	45	11	2	45
12	180	12	3	0	12	3	0
13	195	13	3	15	13	3	15
14	210	14	3	30	14	3	30
15	225	15	3	45	15	3	45
16	240	16	4	0	16	4	0
17	255	17	4	15	17	4	15
18	270	18	4	30	18	4	30
19	285	19	4	45	19	4	45
20	300	20	5	0	20	5	0
21	315	21	5	15	21	5	15
22	330	22	5	30	22	5	30
23	345	23	5	45	23	5	45
24	360	24	6	0	24	6	0
		25	6	15	25	6	15
TENTHS.		26	6	30	26	6	30
0s.1	1″.5	27	6	45	27	6	45
0.2	3.0	28	7	0	28	7	0
0.3	4.5	29	7	15	29	7	15
0.4	6.0	30	7	30	30	7	30
0.5	7.5	31	7	45	31	7	45
0.6	9.0	32	8	0	32	8	0
0.7	10.5	33	8	15	33	8	15
0.8	12.0	34	8	30	34	8	30
0.9	13.5	35	8	45	35	8	45
1.0	15.0	36	9	0	36	9	0
HUNDREDTHS.		37	9	15	37	9	15
		38	9	30	38	9	30
0s.01	0″.15	39	9	45	39	9	45
.02	0.30	40	10	0	40	10	0
.03	0.45	41	10	15	41	10	15
.04	0.60	42	10	30	42	10	30
.05	0.75	43	10	45	43	10	45
.06	0.90	44	11	0	44	11	0
.07	1.05	45	11	15	45	11	15
.08	1.20	46	11	30	46	11	30
.09	1.35	47	11	45	47	11	45
.10	1.50	48	12	0	48	12	0
		49	12	15	49	12	15
THOUSANDTHS.		50	12	30	50	12	30
0s.001	0″.015	51	12	45	51	12	45
.002	.030	52	13	0	52	13	0
.003	.045	53	13	15	53	13	15
.004	.060	54	13	30	54	13	30
.005	.075	55	13	45	55	13	45
.006	.090	56	14	0	56	14	0
.007	.105	57	14	15	57	14	15
.008	.120	58	14	30	58	14	30
.009	.155	59	14	45	59	14	45
.010	.150	60	15	0	60	15	0

Or to convert Time into Degrees and Parts of Terrestrial Longitude.

To change mean Solar into Sidereal Time.

Solar Days.	Add	Solar Minutes	Add Seconds.	Solar Seconds	Add Parts of Sec.
	h. m. s.				
1	0 3 56 .556	1	0ˢ.164	1	0ˢ.003
2	0 7 53 .112	2	0 .329	2	0 .006
3	0 11 49 .668	3	0 .493	3	0 .008
4	0 15 46 .224	4	0 .658	4	0 .011
5	0 19 42 .780	5	0 .822	5	0 .014
6	0 23 39 .336	6	0 .986	6	0 .017
7	0 27 35 .892	7	1 .150	7	0 .019
8	0 31 32 .448	8	1 .315	8	0 .022
9	0 35 29 .004	9	1 .479	9	0 .025
10	0 39 25 .560	10	1 .643	10	0 .027
11	0 43 22 .116	11	1 .807	11	0 .030
12	0 47 18 .672	12	1 .972	12	0 .033
13	0 51 15 .228	13	2 .136	13	0 .036
14	0 55 11 .784	14	2 .300	14	0 .038
15	0 59 8 .340	15	2 .464	15	0 .041
16	1 3 4 .896	16	2 .629	16	0 .044
17	1 7 1 .452	17	2 .893	17	0 .047
18	1 10 58 .008	18	3 .057	18	0 .050
19	1 14 54 .564	19	3 .221	19	0 .053
20	1 18 51 .120	20	3 .286	20	0 .055
21	1 22 47 .676	21	3 .450	21	0 .058
22	1 26 44 .232	22	3 .614	22	0 .061
23	1 30 40 .788	23	3 .779	23	0 .064
24	1 34 37 .344	24	3 .943	24	0 .066
25	1 38 33 .900	25	4 .108	25	0 .069
26	1 42 30 .456	26	4 .272	26	0 .072
27	1 46 27 .012	27	4 .436	27	0 .075
28	1 50 23 .568	28	4 .600	28	0 .077
29	1 54 20 .124	29	4 .764	29	0 .080
30	1 58 16 .680	30	4 .928	30	0 .082
31	2 2 13 .236	31	5 .092	31	0 .085
32	2 6 9 .792	32	5 .257	32	0 .088
33	2 10 6 .348	33	5 .421	33	0 .091
34	2 14 2 .904	34	5 .585	34	0 .094
35	2 17 59 .460	35	5 .750	35	0 .097
Sol. hrs.	m. s.	36	5 .914	36	0 .100
1	0 9 .8565	37	6 .078	37	0 .103
2	0 19 .713	38	6 .242	38	0 .106
3	0 29 .569	39	6 .407	39	0 .108
4	0 39 .426	40	6 .571	40	0 .111
5	0 49 .282	41	6 .735	41	0 .114
6	0 59 .139	42	6 .900	42	0 .116
7	1 8 .995	43	7 .064	43	0 .119
8	1 18 .852	44	7 .228	44	0 .122
9	1 28 .708	45	7 .393	45	0 .125
10	1 38 .565	46	7 .557	46	0 .128
11	1 48 .421	47	7 .722	47	0 .131
12	1 58 .278	48	7 .886	48	0 .133
13	2 8 .134	49	8 .050	49	0 .136
14	2 17 .991	50	8 .214	50	0 .138
15	2 27 .847	51	8 .378	51	0 .141
16	2 37 .704	52	8 .543	52	0 .144
17	2 47 .560	53	8 .707	53	0 .147
18	2 57 .417	54	8 .872	54	0 .150
19	3 7 .273	55	9 .036	55	0 .152
20	3 17 .130	56	9 .200	56	0 .155
21	3 26 .987	57	9 .364	57	0 .157
22	3 36 .844	58	9 .528	58	0 .159
23	3 46 .700	59	9 .692	59	0 .162
24	3 56 .556	60	9 .856	60	0 .164

To change Sidereal into mean Solar Time.

Sidereal Days.	Subtract	Sidereal Min.	Subtract Seconds.	Sidereal Seconds	Subtract Parts of a Sec.
	h. m. s.				
1	0 3 55 .908	1	0ˢ.164	1	0ˢ.003
2	0 7 51 .816	2	0 .328	2	0 .005
3	0 11 47 .724	3	0 .491	3	0 .008
4	0 15 43 .632	4	0 .655	4	0 .011
5	0 19 39 .540	5	0 .819	5	0 .014
6	0 23 35 .448	6	0 .983	6	0 .016
7	0 27 31 .356	7	1 .147	7	0 .019
8	0 31 27 .264	8	1 .311	8	0 .022
9	0 35 23 .172	9	1 .474	9	0 .025
10	0 39 19 .080	10	1 .636	10	0 .027
11	0 43 14 .988	11	1 .802	11	0 .030
12	0 47 10 .896	12	1 .966	12	0 .032
13	0 51 6 .804	13	2 .130	13	0 .035
14	0 55 2 .712	14	2 .294	14	0 .038
15	0 58 58 .620	15	2 .457	15	0 .041
16	1 2 54 .528	16	2 .621	16	0 .044
17	1 6 50 .436	17	2 .785	17	0 .046
18	1 10 46 .344	18	2 .949	18	0 .049
19	1 14 42 .252	19	3 .113	19	0 .052
20	1 18 38 .160	20	3 .277	20	0 .055
21	1 22 34 .068	21	3 .440	21	0 .057
22	1 26 29 .976	22	3 .604	22	0 .060
23	1 30 25 .884	23	3 .768	23	0 .063
24	1 34 21 .792	24	3 .932	24	0 .066
25	1 38 17 .700	25	4 .096	25	0 .068
26	1 42 13 .608	26	4 .259	26	0 .071
27	1 46 9 .516	27	4 .423	27	0 .074
28	1 50 5 .424	28	4 .587	28	0 .076
29	1 54 1 .332	29	4 .751	29	0 .079
30	1 57 57 .240	30	4 .915	30	0 .082
31	2 1 53 .148	31	5 .079	31	0 .085
32	2 5 49 .056	32	5 .242	32	0 .087
33	2 9 44 .964	33	5 .406	33	0 .090
34	2 13 40 .872	34	5 .570	34	0 .093
35	2 17 36 .780	35	5 .734	35	0 .096
Sid. hrs.	m. s.	36	5 .898	36	0 .098
1	0 9 .829	37	6 .062	37	0 .101
2	0 19 .659	38	6 .225	38	0 .104
3	0 29 .483	39	6 .389	39	0 .106
4	0 39 .318	40	6 .553	40	0 .109
5	0 49 .147	41	6 .717	41	0 .112
6	0 58 .977	42	6 .881	42	0 .115
7	1 8 .806	43	7 .044	43	0 .117
8	1 18 .636	44	7 .208	44	0 .120
9	1 28 .465	45	7 .372	45	0 .123
10	1 38 .295	46	7 .536	46	0 .126
11	1 48 .124	47	7 .699	47	0 .128
12	1 57 .954	48	7 .864	48	0 .131
13	2 7 .783	49	8 .027	49	0 .134
14	2 17 .613	50	8 .191	50	0 .137
15	2 27 .442	51	8 .355	51	0 .139
16	2 37 .272	52	8 .519	52	0 .142
17	2 47 .101	53	8 .683	53	0 .145
18	2 56 .931	54	8 .646	54	0 .147
19	3 6 .760	55	9 .010	55	0 .150
20	3 16 .590	56	9 .174	56	0 .153
21	3 26 .419	57	9 .338	57	0 .156
22	3 36 .249	58	9 .502	58	0 .158
23	3 46 .078	59	9 .666	59	0 .161
24	3 55 .908	60	9 .829	60	0 .164

This Table may be used to show the Sun's Right Ascension also, in Sidereal Time.

A Table of the Proportional Parts of a Clock's Daily Rate.

The Arguments are, the Daily Rate at the top and bottom, and the interval at either side.

h.	m.	1s	2s	3s	4s	5s	6s	7s	8s	9s	10s	11s	12s	13s	14s	15s	16s	h.	m.
0	12	0.008	0.017	0.025	0.033	0.042	0.050	0.058	0.067	0.075	0.083	0.092	0.100	0.108	0.117	0.125	0.133	0	12
	24	0.017	0.033	0.050	0.067	0.083	0.100	0.117	0.133	0.150	0.167	0.183	0.200	0.217	0.233	0.250	0.267		24
	36	0.025	0.050	0.075	0.100	0.125	0.150	0.175	0.200	0.225	0.250	0.275	0.300	0.325	0.350	0.375	0.400		36
	48	0.033	0.067	0.100	0.133	0.167	0.200	0.233	0.267	0.300	0.333	3.367	0.400	0.433	0.467	0.500	0.533		48
1	0	0.042	0.083	0.125	0.167	0.208	0.250	0.292	0.333	0.375	0.417	0.458	0.500	0.542	0.583	0.625	0.667	1	0
	12	0.050	0.100	0.150	0.200	0.250	0.300	0.350	0.400	0.450	0.500	0.550	0.600	0.650	0.700	0.750	0.800		12
	24	0.058	0.117	0.175	0.233	0.292	0.350	0.408	0.467	0.525	0.583	0.642	0.700	0.758	0.817	0.875	0.933		24
	36	0.067	0.133	0.200	0.267	0.333	0.400	0.467	0.533	0.600	0.667	0.733	0.800	0.867	0.933	1.000	1.067		36
	48	0.075	0.150	0.225	0.300	0.375	0.450	0.525	0.600	0.675	0.750	0.825	0.900	0.975	1.050	1.125	1.200		48
2	0	0.083	0.167	0.250	0.333	0.417	0.500	0.583	0.667	0.750	0.833	0.917	1.000	1.083	1.167	1.250	1.333	2	0
	12	0.092	0.183	0.275	0.367	0.458	0.550	0.642	0.783	0.825	0.917	1.008	1.100	1.192	1.283	1.375	1.467		12
	24	0.100	0.200	0.300	0.400	0.500	0.600	0.700	0.800	0.900	1.000	1.100	1.200	1.300	1.400	1.500	1.600		24
	36	0.108	0.217	0.325	0.433	0.542	0.650	0.758	0.867	0.975	1.083	1.192	1.300	1.408	1.517	1.625	1.733		36
	48	0.117	0.233	0.350	0.467	0.583	0.700	0.817	0.933	1.050	1.167	1.283	1.400	1.517	1.633	1.750	1.867		48
3	0	0.125	0.250	0.375	0.500	0.625	0.750	0.875	1.000	1.125	1.250	1.375	1.500	1.625	1.750	1.875	2.000	3	0
	12	0.133	0.267	0.400	0.533	0.667	0.800	0.933	1.067	1.200	1.333	1.467	1.600	1.733	1.867	2.000	2.133		12
	24	0.142	0.283	0.425	0.567	0.708	0.850	0.992	1.133	1.275	1.417	1.558	1.700	1.842	1.983	2.125	2.267		24
	36	0.150	0.300	0.450	0.600	0.750	0.900	1.050	1.200	1.350	1.500	1.650	1.800	1.950	2.100	2.250	2.400		36
	48	0.158	0.317	0.475	0.633	0.792	0.950	1.108	1.267	1.425	1.583	1.742	1.900	2.058	2.217	2.375	2.533		48
4	0	0.167	0.333	0.500	0.667	0.833	1.000	1.167	1.333	1.500	1.667	1.833	2.000	2.167	2.333	2.500	2.667	4	0
	12	0.175	0.350	0.525	0.700	0.875	1.050	1.225	1.400	1.575	1.750	1.925	2.100	2.275	2.450	2.625	2.800		12
	24	0.183	0.367	0.550	0.733	0.917	1.100	1.283	1.467	1.650	1.833	2.017	2.200	2.383	2.567	2.750	2.933		24
	36	0.192	0.383	0.575	0.767	0.958	1.150	1.342	1.533	1.725	1.917	2.108	2.300	2.492	2.683	2.875	3.067		36
	48	0.200	0.400	0.600	0.800	1.000	1.200	1.400	1.600	1.800	2.000	2.200	2.400	2.600	2.800	3.000	3.200		48
5	0	0.208	0.417	0.625	0.833	1.042	1.250	1.458	1.667	1.875	2.083	2.292	2.500	2.708	2.917	3.125	3.333	5	0
	12	0.217	0.433	0.650	0.867	1.083	1.300	1.517	1.733	1.950	2.167	2.383	2.600	2.817	3.033	3.250	3.467		12
	24	0.225	0.450	0.675	0.900	1.125	1.350	1.575	1.800	2.025	2.250	2.475	2.700	2.925	3.150	3.375	3.600		24
	36	0.233	0.467	0.700	0.933	1.167	1.400	1.633	1.867	2.100	2.333	2.567	2.800	3.033	3.267	3.500	3.733		36
	48	0.242	0.483	0.725	0.967	1.208	1.450	1.692	1.933	2.175	2.417	2.658	2.900	3.142	3.383	3.625	3.867		48
6	0	0.250	0.500	0.750	1.000	1.250	1.500	1.750	2.000	2.250	2.500	2.750	3.000	3.250	3.500	3.750	4.000	6	0
	12	0.258	0.517	0.775	1.033	1.292	1.550	1.808	2.067	2.325	2.583	2.842	3.100	3.358	3.617	3.875	4.133		12
	24	0.267	0.533	0.800	1.067	1.333	1.600	1.867	2.133	2.400	2.667	2.933	3.200	3.467	3.733	4.000	4.267		24
	36	0.275	0.550	0.825	1.100	1.375	1.650	1.925	2.200	2.475	2.750	3.025	3.300	3.575	3.850	4.125	4.400		36
	48	0.283	0.567	0.850	1.133	1.417	1.700	1.983	2.267	3.550	2.833	3.117	3.400	3.683	3.967	4.250	4.533		48
7	0	0.292	0.583	0.875	1.167	1.458	1.750	2.042	2.333	2.625	2.917	3.208	3.500	3.792	4.083	4.375	4.667	7	0
	12	0.300	0.600	0.900	1.200	1.500	1.800	2.100	2.400	2.700	3.000	3.300	3.600	3.900	4.200	4.500	4.800		12
	24	0.308	0.617	0.925	1.233	1.542	1.850	2.158	2.467	2.775	3.083	3.392	3.700	4.008	4.317	4.625	4.933		24
	36	0.317	0.633	0.950	1.267	1.583	1.900	2.217	2.533	2.850	3.167	3.483	3.800	4.117	4.433	4.750	5.067		36
	48	0.325	0.650	0.975	1.300	1.625	1.950	2.275	2.600	2.925	3.250	3.575	3.900	4.225	4.550	4.875	5.200		48
8	0	0.333	0.667	1.000	1.333	1.667	2.000	2.333	2.667	3.000	3.333	3.667	4.000	4.333	4.667	5.000	5.333	8	0
	12	0.342	0.683	1.025	1.367	1.708	2.050	2.392	2.733	3.075	3.417	3.758	4.100	4.442	4.783	5.125	5.467		12
	24	0.350	0.700	1.050	1.400	1.750	2.100	2.450	2.800	3.150	3.500	3.850	4.200	4.550	4.900	5.250	5.600		24
	36	0.358	0.717	1.075	1.433	1.792	2.150	2.508	2.867	3.225	3.583	3.942	4.300	4.658	5.017	5.375	5.733		36
	48	0.367	0.733	1.100	1.467	1.833	2.200	2.567	2.933	3.300	3.667	4.033	4.400	4.767	5.133	5.500	5.867		48
9	0	0.375	0.750	1.125	1.500	1.875	2.250	2.625	3.000	3.375	3.750	4.125	4.500	4.875	5.250	5.625	6.000	9	0
	12	0.383	0.767	1.150	1.533	1.917	2.300	2.683	3.067	3.450	3.833	4.217	4.600	4.983	5.367	5.750	6.133		12
	24	0.392	0.783	1.175	1.567	1.958	2.350	2.742	3.133	3.525	3.917	4.308	4.700	5.092	5.483	5.875	6.267		24
	36	0.400	0.800	1.200	1.600	2.000	2.400	2.800	3.200	3.600	4.000	4.400	4.800	5.200	5.600	6.000	6.400		36
	48	0.408	0.817	1.225	1.633	2.042	2.450	2.858	3.267	3.675	4.083	4.492	4.900	5.308	5.717	6.125	6.533		48
10	0	0.417	0.833	1.250	1.667	2.083	2.500	2.917	3.333	3.750	4.167	4.583	5.000	5.417	5.833	6.250	6.667	10	0
	12	0.425	0.850	1.275	1.700	2.125	2.550	2.975	3.400	3.825	4.250	4.675	5.100	5.525	5.950	6.275	6.800		12
	24	0.433	0.867	1.300	1.733	2.167	2.600	3.033	3.467	3.900	4.333	4.767	5.200	5.633	6.037	6.500	6.933		24
	36	0.442	0.883	1.325	1.767	2.208	2.650	3.092	3.533	3.975	4.417	4.858	5.300	5.742	6.183	6.625	7.067		36
	48	0.450	0.900	1.350	1.800	2.250	2.700	3.150	3.600	4.050	4.500	4.950	5.400	5.850	6.300	6.750	7.200		48
11	0	0.458	0.917	1.375	1.833	2.292	2.750	3.208	3.667	4.125	4.583	5.042	5.500	5.958	6.417	6.875	7.333	11	0
	12	0.467	0.933	1.400	1.867	2.333	2.800	3.267	3.733	4.200	4.667	5.133	5.600	6.067	6.533	7.000	7.467		12
	24	0.475	0.950	1.425	1.900	2.375	2.850	3.325	3.800	4.275	4.750	5.225	5.700	6.175	6.650	7.125	7.600		24
	36	0.483	0.967	1.450	1.933	2.417	2.900	3.383	3.867	4.350	4.833	5.317	5.800	6.283	6.767	7.250	7.733		36
	48	0.492	0.983	1.475	1.967	2.458	2.950	3.442	3.933	4.425	4.917	5.408	5.900	6.392	6.883	7.275	7.867		48
12	0	0.500	1.000	1.500	2.000	2.500	3.000	3.500	4.000	4.500	5.000	5.500	6.000	6.500	7.000	7.500	8.000	12	0
		.1 .01	.2 .02	.3 .03	.4 .04	.5 .05	.6 .06	.7 .07	.8 .08	.9 .09	1.0 .1	1.1 .11	1.2 .12	1.3 .13	1.4 .14	1.5 .15	1.6 .16		

TIME.

The Table of the Proportional Parts of a Clock's Daily Rate concluded.

The Arguments are, the Daily Rate at the top and bottom, and the Interval at either side.

h.	m.	1s	2s	3s	4s	5s	6s	7s	8s	9s	10s	11s	12s	13s	14s	15s	16s	h.	m.
12	0	0.500	1.000	1.500	2.000	2.500	3.000	3.500	4.000	4.500	5.000	5.500	6.000	6.500	7.000	7.500	8.000	12	0
	12	0.508	1.017	1.525	2.033	2.542	3.050	3.558	4.067	4.575	5.083	5.592	6.100	6.608	7.117	7.625	8.133		12
	24	0.517	1.033	1.550	2.067	2.583	3.100	3.617	4.133	4.650	5.167	5.683	6.200	6.717	7.233	7.750	8.267		24
	36	0.525	1.050	1.575	2.100	2.625	3.150	3.675	4.200	4.725	5.250	5.775	6.300	6.825	7.350	7.875	8.400		36
	48	0.533	1.067	1.600	2.133	2.667	3.200	3.733	4.267	4.800	5.333	5.867	6.400	6.933	7.467	8.000	8.533		48
13	0	0.542	1.083	1.625	2.167	2.708	3.250	3.792	4.333	4.875	5.417	5.958	6.500	7.042	7.583	8.125	8.667	13	0
	12	0.550	1.100	1.650	2.200	2.750	3.300	3.850	4.400	4.950	5.500	6.050	6.600	7.150	7.700	8.250	8.800		12
	24	0.558	1.117	1.675	2.233	2.792	3.350	3.908	4.467	5.025	5.583	6.142	6.700	7.258	7.817	8.375	8.933		24
	33	0.567	1.133	1.700	2.267	2.833	3.400	3.967	4.533	5.100	5.666	6.233	6.800	7.367	7.933	8.500	9.067		36
	48	0.575	1.150	1.725	2.300	2.875	3.450	4.025	4.600	5.175	5.750	6.325	6.900	7.475	8.050	8.625	9.200		48
14	0	0.583	1.167	1.750	2.333	2.917	3.500	4.083	4.667	5.250	5.833	6.417	7.000	7.583	8.167	8.750	9.333	14	0
	12	0.592	1.183	1.775	2.367	2.958	3.550	4.142	4.733	5.325	5.917	6.508	7.100	7.692	8.283	8.875	9.467		12
	24	0.600	1.200	1.800	2.400	3.000	3.600	4.200	4.800	5.400	6.000	6.600	7.200	7.800	8.400	9.000	9.600		24
	36	0.608	1.217	1.825	2.433	3.042	3.650	4.258	4.867	5.475	6.083	6.692	7.300	7.908	8.517	9.125	9.733		36
	48	0.617	1.233	1.850	2.467	3.083	3.700	4.317	4.933	5.550	6.167	6.783	7.400	8.017	8.633	9.250	9.867		48
15	0	0.625	1.250	1.875	2.500	3.125	3.750	4.375	5.000	5.625	6.250	6.875	7.500	8.125	8.750	9.375	10.000	15	0
	12	0.633	1.267	1.900	2.533	3.167	3.800	4.433	5.067	5.700	6.333	6.967	7.600	8.233	8.867	9.500	10.133		12
	24	0.642	1.283	1.925	2.567	3.208	3.850	4.492	5.133	5.775	6.417	7.058	7.700	8.342	8.983	9.625	10.267		24
	36	0.650	1.300	1.950	2.600	3.250	3.900	4.550	5.200	5.850	6.500	7.150	7.800	8.450	9.100	9.750	10.400		36
	48	0.658	1.317	1.975	2.633	3.292	3.950	4.608	5.267	5.925	6.583	7.242	7.900	8.558	9.217	9.875	10.533		48
16	0	0.667	1.333	2.000	2.667	3.333	4.000	4.667	5.333	6.000	6.667	7.333	8.000	8.667	9.333	10.000	10.667	16	0
	12	0.675	1.350	2.025	2.700	3.375	4.050	4.725	5.400	6.075	6.750	7.425	8.100	8.775	9.450	10.125	10.800		12
	24	0.683	1.367	2.050	2.733	3.417	4.100	4.783	5.467	6.150	6.833	7.517	8.200	8.883	9.567	10.250	10.933		24
	36	0.692	1.383	2.075	2.767	3.458	4.150	4.842	5.533	6.225	6.917	7.608	8.300	8.992	9.683	10.375	11.067		36
	48	0.700	1.400	2.100	2.800	3.500	4.200	4.900	5.600	6.300	7.000	7.700	8.400	9.100	9.800	10.500	11.200		48
17	0	0.708	1.417	2.125	2.833	3.542	4.250	4.958	5.667	6.375	7.083	7.792	8.500	9.208	9.917	10.625	11.333	17	0
	12	0.717	1.433	2.150	2.867	3.583	4.300	5.017	5.733	6.450	7.167	7.883	8.600	9.317	10.033	10.750	11.467		12
	24	0.725	1.450	2.175	2.900	3.625	4.350	5.075	5.800	6.525	7.250	7.975	8.700	9.425	10.150	10.875	11.600		24
	36	0.733	1.467	2.200	2.933	3.667	4.400	5.133	5.867	6.600	7.333	8.067	8.800	9.533	10.267	11.000	11.733		36
	48	0.742	1.483	2.225	2.967	3.708	4.450	5.192	5.933	6.675	7.417	8.158	8.900	9.642	10.383	11.125	11.867		48
18	0	0.750	1.500	2.250	3.000	3.750	4.500	5.250	6.000	6.750	7.500	8.250	9.000	9.750	10.500	11.250	12.000	18	0
	12	0.758	1.517	2.275	3.033	3.792	4.550	5.308	6.067	6.825	7.583	8.342	9.100	9.858	10.617	11.375	12.133		12
	24	0.767	1.533	2.300	3.067	3.833	4.600	5.367	6.133	6.900	7.667	8.433	9.200	9.967	10.733	11.500	12.267		24
	36	0.775	1.550	2.325	3.100	3.875	4.650	5.425	6.200	6.975	7.750	8.525	9.300	10.075	10.850	11.625	12.400		36
	48	0.783	1.567	2.350	3.133	3.917	4.700	5.483	6.267	7.050	7.833	8.617	9.400	10.183	10.967	11.750	12.533		48
19	0	0.792	1.583	2.375	3.167	3.958	4.750	5.542	6.333	7.125	7.917	8.708	9.500	10.292	11.083	11.875	12.667	19	0
	12	0.800	1.600	2.400	3.200	4.000	4.800	5.600	6.400	7.200	8.000	8.800	9.600	10.400	11.200	12.000	12.800		12
	24	0.808	1.617	2.425	3.233	4.042	4.850	5.658	6.467	7.275	8.083	8.892	9.700	10.508	11.317	12.125	12.933		24
	36	0.817	1.633	2.450	3.267	4.083	4.900	5.717	6.533	7.350	8.167	8.983	9.800	10.617	11.433	12.250	13.067		36
	48	0.825	1.650	2.475	3.300	4.125	4.950	5.775	6.600	7.425	8.250	9.075	9.900	10.725	11.550	12.375	13.200		48
20	0	0.833	1.667	2.500	3.333	4.167	5.000	5.833	6.667	7.500	8.333	9.167	10.000	10.833	11.667	12.500	13.333	20	0
	12	0.842	1.683	2.525	3.367	4.208	5.050	5.892	6.733	7.575	8.417	9.258	10.100	10.942	11.783	12.625	13.467		12
	24	0.850	1.700	2.550	3.400	4.250	5.100	5.950	6.800	7.650	8.500	9.350	10.200	11.050	11.900	12.750	13.600		24
	36	0.858	1.717	2.575	3.433	4.292	5.150	6.008	6.867	7.725	8.583	9.442	10.300	11.158	12.017	12.875	13.733		36
	48	0.867	1.733	2.600	3.467	4.333	5.200	6.067	6.933	7.800	8.667	9.533	10.400	11.267	12.133	13.000	13.867		48
21	0	0.875	1.750	2.625	3.500	4.375	5.250	6.125	7.000	7.875	8.750	9.625	10.500	11.375	12.250	13.125	14.000	21	0
	12	0.883	1.767	2.650	3.533	4.417	5.300	6.183	7.067	7.950	8.833	9.717	10.600	11.483	12.367	13.250	14.133		12
	24	0.892	1.783	2.675	3.567	4.458	5.350	6.242	7.133	8.025	8.917	9.808	10.700	11.592	12.483	13.375	14.267		24
	36	0.900	1.800	2.700	3.600	4.500	5.400	6.300	7.200	8.100	9.000	9.900	10.800	11.700	12.600	13.500	14.400		36
	48	0.908	1.817	2.725	3.633	4.542	5.450	6.358	7.267	8.175	9.083	9.992	10.900	11.808	12.717	13.625	14.533		48
22	0	0.917	1.833	2.750	3.667	4.583	5.500	6.417	7.333	8.250	9.167	10.083	11.000	11.917	12.833	13.750	14.667	22	0
	12	0.925	1.850	2.775	3.700	4.625	5.550	6.475	7.400	8.325	9.250	10.175	11.100	12.025	12.950	13.875	14.800		12
	24	0.933	1.867	2.800	3.733	4.667	5.600	6.533	7.467	8.400	9.333	10.267	11.200	12.133	13.067	14.000	14.933		24
	36	0.942	1.883	2.825	3.767	4.708	5.650	6.592	7.533	8.475	9.417	10.358	11.300	12.242	13.183	14.125	15.067		36
	48	0.950	1.900	2.850	3.800	4.750	5.700	6.650	7.600	8.550	9.500	10.450	11.400	12.350	13.300	14.250	15.200		48
23	0	0.958	1.917	2.875	3.833	4.792	5.750	6.708	7.667	8.625	9.583	10.542	11.500	12.458	13.417	14.375	15.333	23	0
	12	0.967	1.933	2.900	3.867	4.833	5.800	6.767	7.733	8.700	9.667	10.633	11.600	12.567	13.533	14.500	15.467		12
	24	0.975	1.950	2.925	3.900	4.875	5.850	6.825	7.800	8.775	9.750	10.725	11.700	12.675	13.650	14.625	15.600		24
	36	0.983	1.967	2.950	3.933	4.917	5.900	6.883	7.867	8.850	9.833	10.817	11.800	12.783	13.767	14.750	15.733		36
	48	0.992	1.983	2.975	3.967	4.958	5.950	6.942	7.933	8.925	9.917	10.908	11.900	12.892	13.883	14.875	15.867		48
24	0	1.000	2.000	3.000	4.000	5.000	6.000	7.000	8.000	9.000	10.000	11.000	12.000	13.000	14.000	15.000	16.000	24	0
		.1 .01	.2 .02	.3 .03	.4 .04	.5 .05	.6 .06	.7 .07	.8 .08	.9 .09	1 .0 .1	1 .1 .11	1 .2 .12	1 .3 .13	1 .4 .14	1 .5 .15	1 .6 .16		

The *first* Sheet of the *following* Table was printed before it was discovered that the North Polar Distances of the Greenwich Stars for 1820 are many of them *erroneous*, as given in the Nautical Almanac of 1824; the Reader is therefore requested to substitute the mean places given in the Nautical Almanac of 1823, which may be found in a subsequent Table, before the respective corrections are applied to those *twelve* Stars. Or any mean places hereafter determined may be adopted for all the 48 Stars.

The CORRECTIONS of 48 principal Stars, both in RIGHT ASCENSION in time, and in NORTH POLAR DISTANCE, adapted for every Degree of the SUN'S LONGITUDE, and of the MOON'S ASCENDING NODE.

1830. Common Argument. + S.	γ PEGASI 1820. Right Ascension. 0h 3m 58s.8 Ann. Var. + 3s.08 Aber.	☾ Nut.	☉ Nut.	N. P. Distance. 75° 49′ 0″ Ann. Var. — 20″.20 Aber.	☾ Nut.	☉ Nut.	1830. Common Argument. — S.
III 0° IX	1s.277	1s.117	0s.062	9″.196	7″.183	0″.398	III 0° IX
1	1.277	1.117	.062	9.19	7.18	.39	29
2	1.276	1.116	.062	9.18	7.17	.39	28
3	1.275	1.115	.062	9.18	7.17	.39	27
4	1.274	1.114	.062	9.17	7.16	.39	26
5	1.273	1.113	.062	9.16	7.15	.39	25
6	1.271	1.111	.062	9.15	7.14	.39	24
7	1.269	1.109	.062	9.13	7.13	.39	23
8	1.267	1.107	.061	9.11	7.12	.39	22
9	1.264	1.104	.061	9.09	7.10	.39	21
10	1.260	1.101	.061	9.07	7.08	.39	20
11	1.255	1.097	.061	9.04	7.06	.39	19
12	1.250	1.092	.061	9.00	7.03	.39	18
13	1.245	1.088	.060	8.96	7.00	.39	17
14	1.240	1.084	.060	8.92	6.97	.39	16
15	1.234	1.079	.060	8.87	6.94	.38	15
16	1.229	1.074	.060	8.83	6.91	.38	14
17	1.222	1.069	.059	8.79	6.87	.38	13
18	1.215	1.063	.059	8.74	6.83	.38	12
19	1.209	1.057	.058	8.69	6.79	.38	11
20	1.201	1.050	.058	8.63	6.75	.37	10
21	1.193	1.043	.058	8.58	6.70	.37	9
22	1.184	1.036	.057	8.52	6.65	.37	8
23	1.176	1.029	.057	8.45	6.61	.37	7
24	1.167	1.021	.057	8.39	6.56	.36	6
25	1.158	1.013	.056	8.32	6.51	.36	5
26	1.149	1.004	.056	8.25	6.46	.36	4
27	1.139	0.995	.055	8.18	6.40	.35	3
28	1.128	0.986	.055	8.11	6.34	.35	2
29	1.118	0.977	.054	8.03	6.28	.35	1
IV 0 X	1.106	0.967	.054	7.95	6.22	.34	II 0 VIII
1	1.094	0.957	.053	7.88	6.16	.34	29
2	1.082	0.947	.053	7.80	6.09	.34	28
3	1.071	0.937	.052	7.71	6.02	.33	27
4	1.059	0.926	.051	7.62	5.95	.33	26
5	1.047	0.915	.051	7.53	5.88	.33	25
6	1.034	0.903	.050	7.43	5.81	.32	24
7	1.020	0.892	.050	7.33	5.73	.32	23
8	1.006	0.880	.049	7.24	5.66	.31	22
9	0.992	0.867	.048	7.14	5.58	.31	21
10	0.977	0.855	.047	7.03	5.50	.30	20
11	0.963	0.842	.047	6.93	5.42	.30	19
12	0.948	0.829	.046	6.82	5.34	.30	18
13	0.934	0.816	.045	6.72	5.25	.29	17
14	0.918	0.803	.045	6.61	5.17	.29	16
IV 15 X	0.903	0.790	.044	6.50	5.08	.28	I 15 VII
	Arg. = ☉Long. + 8s 28° 47′	Arg. = ☾☊ + 6s 8° 24′	Arg. = 2☉Long. + 6s 6° 50′	Arg. = ☉Long. + 1s 27° 12′	Arg. = ☾☊ + 11s 28° 29′	Arg. = 2☉Long. + 11s 28° 47′	
	Aber.	☾ Nut.	☉ Nut.	Aber.	☾ Nut.	☉ Nut.	
	R. A.			N. P. D.			
	ALGENIB 1830.						

1830. Common Argument. + S.	α CASSIOPEÆ 1820. Right Ascension. 0h 30m 21s.2 Ann. Var. + 3s.31 Aber.	☾ Nut.	☉ Nut.	N. P. Distance. 34° 27′ 4″ Ann. Var. — 19″.80 Aber.	☾ Nut.	☉ Nut.	1830. Common Argument. — S.
III 0° IX	2s.197	1s.516	0s.078	16″.743	7″.235	0″.399	III 0° IX
1	2.196	1.515	.078	16.73	7.23	.39	29
2	2.195	1.514	.078	16.72	7.22	.39	28
3	2.193	1.513	.078	16.71	7.22	.39	27
4	2.191	1.512	.078	16.70	7.21	.39	26
5	2.188	1.510	.078	16.68	7.21	.39	25
6	2.185	1.508	.078	16.65	7.20	.39	24
7	2.181	1.505	.077	16.62	7.19	.39	23
8	2.176	1.502	.077	16.58	7.18	.39	22
9	2.171	1.498	.077	16.54	7.16	.39	21
10	2.165	1.494	.077	16.49	7.14	.39	20
11	2.158	1.489	.077	16.44	7.11	.39	19
12	2.150	1.484	.076	16.38	7.08	.39	18
13	2.141	1.478	.076	16.31	7.05	.39	17
14	2.131	1.471	.076	16.24	7.02	.39	16
15	2.120	1.464	.075	16.16	6.98	.39	15
16	2.109	1.457	.075	16.09	6.95	.38	14
17	2.098	1.449	.075	16.01	6.92	.38	13
18	2.087	1.441	.074	15.91	6.88	.38	12
19	2.076	1.433	.074	15.82	6.84	.38	11
20	2.064	1.425	.073	15.72	6.80	.37	10
21	2.051	1.415	.073	15.62	6.75	.37	9
22	2.037	1.405	.072	15.52	6.71	.37	8
23	2.022	1.395	.072	15.41	6.66	.37	7
24	2.006	1.385	.071	15.29	6.61	.36	6
25	1.990	1.374	.071	15.17	6.56	.36	5
26	1.974	1.362	.070	15.04	6.50	.36	4
27	1.957	1.350	.070	14.91	6.45	.36	3
28	1.939	1.337	.069	14.78	6.39	.35	2
29	1.921	1.324	.068	14.64	6.33	.35	1
IV 0 X	1.900	1.311	.067	14.49	6.27	.35	II 0 VIII
1	1.880	1.300	.067	14.33	6.20	.34	29
2	1.860	1.285	.066	14.19	6.14	.34	28
3	1.841	1.271	.065	14.03	6.07	.34	27
4	1.821	1.257	.064	13.88	6.00	.33	26
5	1.800	1.241	.063	13.70	5.92	.33	25
6	1.777	1.226	.062	13.52	5.85	.32	24
7	1.753	1.210	.061	13.36	5.78	.32	23
8	1.730	1.194	.061	13.18	5.70	.31	22
9	1.706	1.177	.060	13.00	5.62	.31	21
10	1.681	1.160	.060	12.80	5.54	.30	20
11	1.656	1.143	.059	12.62	5.46	.30	19
12	1.631	1.125	.058	12.43	5.38	.30	18
13	1.605	1.108	.057	12.23	5.29	.29	17
14	1.580	1.090	.056	12.03	5.21	.29	16
IV 15 X	1.554	1.071	.055	11.83	5.12	.28	I 15 VII
	Arg. = ☉Long. + 8s 21° 35′	Arg. = ☾☊ + 7s 7° 53′	Arg. = 2☉Long. + 7s 2° 52′	Arg. = ☉Long. + 0s 8° 34′	Arg. = ☾☊ + 1s 19° 40′	Arg. = 2☉Long. + 11s 21° 36′	
	Aber.	☾ Nut.	☉ Nut.	Aber.	☾ Nut.	☉ Nut.	
	R. A.			N. P. D.			
	SCHEDIR 1830.						

1830. Common Argument. + S.	α URSÆ MINORIS 1820. Right Ascension. 0h 57m 2s.4 Ann. Var. + 14s.74 Aber.	☾ Nut.	☉ Nut.	N. P. Distance. 1° 39′ 5″.5 Ann. Var. — 19″.45 Aber.	☾ Nut.	☉ Nut.	1830. Common Argument. — S.
III 0° IX	44s.71	22s.96	1s.048	20″.10	7″.37	0″.40	III 0° IX
1	44.69	22.95	1.048	20.09	7.37	.40	29
2	44.66	22.93	1.047	20.08	7.36	.40	28
3	44.63	22.91	1.046	20.06	7.36	.40	27
4	44.59	22.89	1.045	20.04	7.35	.40	26
5	44.54	22.86	1.044	20.02	7.34	.40	25
6	44.47	22.83	1.042	19.99	7.33	.40	24
7	44.38	22.79	1.040	19.95	7.32	.40	23
8	44.27	22.74	1.038	19.90	7.30	.40	22
9	44.15	22.68	1.035	19.85	7.28	.39	21
10	44.01	22.62	1.032	19.79	7.26	.39	20
11	43.86	22.55	1.029	19.72	7.24	.39	19
12	43.70	22.47	1.025	19.65	7.21	.39	18
13	43.54	22.38	1.021	19.58	7.18	.39	17
14	43.37	22.28	1.017	19.50	7.15	.39	16
15	43.19	22.17	1.012	19.42	7.12	.39	15
16	42.99	22.06	1.007	19.33	7.09	.38	14
17	42.77	21.94	1.002	19.23	7.05	.38	13
18	42.54	21.82	0.996	19.12	7.01	.38	12
19	42.28	21.70	0.990	19.01	6.97	.38	11
20	42.01	21.57	0.984	18.89	6.92	.38	10
21	41.73	21.43	0.978	18.76	6.88	.37	9
22	41.44	21.29	0.971	18.63	6.83	.37	8
23	41.14	21.14	0.964	18.50	6.78	.37	7
24	40.83	20.98	0.957	18.36	6.73	.37	6
25	40.51	20.81	0.949	18.21	6.68	.36	5
26	40.17	20.64	0.941	18.04	6.62	.36	4
27	39.82	20.46	0.933	17.90	6.57	.36	3
28	39.46	20.27	0.924	17.73	6.51	.35	2
29	39.09	20.08	0.916	17.57	6.45	.35	1
IV 0 X	38.71	19.88	0.907	17.41	6.38	.35	II 0 VIII
1	38.32	19.69	0.898	17.22	6.32	.34	29
2	37.92	19.49	0.888	17.04	6.25	.34	28
3	37.50	19.27	0.878	16.86	6.18	.34	27
4	37.07	19.04	0.868	16.65	6.12	.33	26
5	36.63	18.81	0.858	16.46	6.04	.33	25
6	36.18	18.57	0.847	16.25	5.96	.32	24
7	35.71	18.33	0.836	16.04	5.88	.32	23
8	35.23	18.09	0.824	15.83	5.81	.32	22
9	34.74	17.84	0.813	15.61	5.73	.31	21
10	34.24	17.58	0.802	15.38	5.64	.31	20
11	33.73	17.33	0.790	15.15	5.56	.30	19
12	33.21	17.08	0.778	14.93	5.48	.30	18
13	32.68	16.80	0.766	14.70	5.39	.29	17
14	32.15	16.51	0.753	14.45	5.30	.29	16
IV 15 X	31.61	16.23	0.741	14.21	5.22	.28	I 15 VII
	Arg. = ☉Long. + 8s 13° 51′	Arg. = ☾☊ + 8s 16° 7′	Arg. = 2☉Long. + 8s 13° 3′	Arg. = ☉Long. + 11s 16° 58′	Arg. = ☾☊ + 11s 10° 21′	Arg. = 2☉Long. + 11s 13° 51′	
	Aber.	☾ Nut.	☉ Nut.	Aber.	☾ Nut.	☉ Nut.	
	R. A.			N. P. D.			
	POLARIS 1830.						

The CORRECTIONS of 48 principal Stars, &c. continued.

1830. Common Argument.	γ PEGASI 1820. Right Ascension. h. m. s. 0 3 58 .8 Ann. Var. +3s.08			γ PEGASI 1820. N. P. Distance. 75° 49′ 0″. Ann. Var. −20″.20			α CASSIOPEÆ 1820. Right Ascension. h. m. s. 0 30 21 .2 Ann. Var. +3s.31			α CASSIOPEÆ 1820. N. P. Distance. 34° 27′ 4″. Ann. Var. −19″.80			α URSÆ MINORIS 1820. Right Ascension. h. m. s. 0 57 2 .4 Ann. Var. +14s.74			α URSÆ MINORIS 1820. N. P. Distance. 1° 39′ 5″.5 Ann. Var. −19″.45			1830. Common Argument.
+ s. − s.	Aber.	☾ Nut.	☉ Nut.	Aber.	☾ Nut.	☉ Nut.	Aber.	☾ Nut.	☉ Nut.	Aber.	☾ Nut.	☉ Nut.	Aber.	☾ Nut.	☉ Nut.	Aber.	☾ Nut.	☉ Nut.	+ s. − s.
IV 15° X	0s.903	0s.790	0s.044	6″.50	5″.08	0″.28	1s.554	1s.071	0s.055	11″.83	5″.12	0″.28	31s.61	16s.23	0s.741	14″.21	5″.22	0″.28	I 15° VII
16	.886	.776	.043	6.38	4.99	.28	1.528	1.053	.054	11.62	5.03	.28	31.06	15.96	.728	13.96	5.12	.28	14
17	.870	.762	.042	6.27	4.90	.27	1.500	1.033	.053	11.41	4.93	.27	30.50	15.68	.714	13.71	5.03	.27	13
18	.854	.747	.041	6.15	4.81	.27	1.471	1.014	.052	11.19	4.84	.27	29.93	15.37	.701	13.44	4.93	.27	12
19	.837	.732	.041	6.02	4.71	.26	1.441	0.994	.051	10.97	4.74	.26	29.34	15.06	.687	13.17	4.83	.26	11
20	.820	.717	.040	5.90	4.62	.26	1.412	0.973	.050	10.75	4.65	.26	28.74	14.75	.673	12.91	4.73	.26	10
21	.804	.703	.039	5.78	4.52	.25	1.383	0.954	.049	10.52	4.56	.25	28.14	14.45	.659	12.65	4.64	.25	9
22	.786	.687	.038	5.66	4.42	.24	1.352	0.933	.048	10.29	4.46	.25	27.53	14.14	.645	12.36	4.54	.25	8
23	.768	.672	.037	5.53	4.32	.24	1.322	0.912	.047	10.06	4.35	.24	26.91	13.82	.631	12.09	4.44	.24	7
24	.750	.657	.036	5.40	4.22	.23	1.292	0.891	.046	9.83	4.25	.23	26.28	13.50	.616	11.81	4.33	.23	6
25	.732	.641	.036	5.27	4.12	.23	1.261	0.869	.045	9.59	4.15	.23	25.64	13.17	.601	11.52	4.23	.23	5
26	.714	.625	.035	5.14	4.02	.22	1.230	0.847	.044	9.35	4.04	.22	25.00	12.84	.586	11.25	4.12	.22	4
27	.695	.608	.034	5.00	3.91	.22	1.198	0.825	.042	9.11	3.94	.22	24.35	12.50	.570	10.93	4.02	.22	3
28	.677	.592	.033	4.87	3.81	.21	1.166	0.803	.041	8.86	3.83	.21	23.69	12.17	.555	10.65	3.91	.21	2
29	.658	.576	.032	4.73	3.70	.21	1.133	0.781	.040	8.62	3.73	.21	23.03	11.83	.540	10.35	3.80	.21	1
V 0 XI	.639	.559	.031	4.60	3.59	.20	1.100	0.758	.039	8.37	3.62	.20	22.36	11.49	.524	10.05	3.69	.20	I 0 VII
1	.619	.542	.030	4.46	3.48	.19	1.066	0.735	.038	8.11	3.51	.19	21.68	11.14	.508	9.74	3.57	.19	29
2	.599	.524	.029	4.31	3.37	.19	1.032	0.712	.037	7.86	3.39	.19	21.00	10.78	.492	9.43	3.46	.19	28
3	.580	.507	.028	4.17	3.26	.18	0.998	0.688	.035	7.59	3.28	.18	20.31	10.42	.476	9.12	3.34	.18	27
4	.559	.490	.027	4.03	3.15	.17	0.963	0.664	.034	7.35	3.17	.17	19.61	10.06	.459	8.81	3.23	.17	26
5	.540	.472	.026	3.88	3.04	.17	0.929	0.641	.033	7.07	3.06	.17	18.89	9.70	.443	8.49	3.12	.17	25
6	.519	.454	.025	3.73	2.92	.16	0.893	0.616	.032	6.80	2.94	.16	18.16	9.33	.426	8.17	2.99	.16	24
7	.499	.436	.024	3.59	2.81	.16	0.858	0.592	.030	6.53	2.82	.16	17.45	8.96	.409	7.85	2.88	.16	23
8	.478	.418	.023	3.44	2.69	.15	0.822	0.567	.029	6.26	2.71	.15	16.73	8.59	.392	7.53	2.76	.15	22
9	.457	.400	.022	3.29	2.57	.14	0.787	0.543	.028	5.99	2.59	.14	16.01	8.21	.375	7.20	2.64	.14	21
10	.437	.382	.021	3.14	2.45	.14	0.751	0.518	.027	5.72	2.47	.14	15.28	7.85	.358	6.87	2.52	.14	20
11	.416	.363	.020	2.99	2.34	.13	0.715	0.493	.025	5.44	2.35	.13	14.54	7.47	.341	6.54	2.40	.13	19
12	.394	.345	.019	2.84	2.22	.12	0.679	0.468	.024	5.17	2.23	.12	13.80	7.09	.323	6.21	2.27	.12	18
13	.373	.326	.018	2.68	2.10	.12	0.642	0.443	.023	4.89	2.11	.12	13.06	6.71	.306	5.88	2.15	.12	17
14	.352	.308	.017	2.53	1.98	.11	0.606	0.417	.022	4.61	1.99	.11	12.32	6.33	.289	5.54	2.03	.11	16
15	.330	.289	.016	2.38	1.86	.10	0.568	0.392	.020	4.33	1.87	.10	11.56	5.94	.271	5.20	1.91	.10	15
16	.309	.270	.015	2.22	1.74	.10	0.532	0.366	.019	4.04	1.75	.10	10.80	5.55	.253	4.86	1.78	.10	14
17	.287	.251	.014	2.07	1.61	.09	0.494	0.341	.018	3.76	1.62	.09	10.04	5.16	.235	4.52	1.65	.09	13
18	.265	.232	.013	1.92	1.49	.08	0.457	0.315	.016	3.47	1.50	.08	9.27	4.77	.217	4.18	1.53	.08	12
19	.243	.213	.012	1.75	1.37	.08	0.419	0.289	.015	3.19	1.38	.08	8.51	4.37	.200	3.84	1.40	.08	11
20	.221	.194	.011	1.59	1.25	.07	0.382	0.263	.014	2.93	1.25	.07	7.75	3.98	.182	3.49	1.28	.07	10
21	.200	.175	.010	1.44	1.12	.06	0.344	0.237	.012	2.61	1.13	.06	6.98	3.59	.164	3.14	1.15	.06	9
22	.178	.155	.009	1.28	1.00	.06	0.306	0.211	.011	2.33	1.01	.06	6.22	3.19	.146	2.79	1.03	.06	8
23	.156	.136	.008	1.12	0.88	.05	0.268	0.185	.009	2.04	0.88	.05	5.45	2.80	.128	2.45	0.90	.05	7
24	.134	.117	.006	0.96	0.75	.04	0.230	0.159	.008	1.75	0.76	.04	4.68	2.40	.110	2.10	0.77	.04	6
25	.111	.097	.005	0.80	0.63	.03	0.192	0.132	.007	1.46	0.63	.03	3.90	2.00	.091	1.75	0.64	.03	5
26	.089	.078	.004	0.64	0.50	.03	0.154	0.106	.005	1.17	0.51	.03	3.12	1.60	.073	1.40	0.52	.03	4
27	.067	.059	.003	0.48	0.38	.02	0.115	0.079	.004	0.78	0.38	.02	2.34	1.20	.055	1.05	0.39	.02	3
28	.045	.039	.002	0.32	0.25	.01	0.077	0.053	.003	0.48	0.25	.01	1.56	0.80	.037	0.70	0.26	.01	2
29	.022	.020	.001	0.16	0.13	.01	0.038	0.026	.001	0.23	0.13	.01	0.78	0.40	.018	0.35	0.13	.01	1
VI 0 O	.000	.000	.000	0.00	0.00	.00	0.000	0.000	.000	0.00	0.00	.00	0.00	0.00	.000	0.00	0.00	.00	O 0 VI
	Arg. = ☉ Long. + 8s 28° 47′	Arg. = ☾ ☊ + 6s 8° 24′	Arg. = 2 ☉ Long. + 6s 6° 50′	Arg. = ☉ Long. + 1s 27° 12′	Arg. = ☾ ☊ + 11s 28° 9′	Arg. = 2 ☉ Long. + 11s 28° 47′	Arg. = ☉ Long. + 8s 21° 35′	Arg. = ☾ ☊ + 7s 7° 53′	Arg. = 2 ☉ Long. + 7s 2° 52′	Arg. = ☉ Long. + 0s 8° 34′	Arg. = ☾ ☊ + 11s 19° 40′	Arg. = 2 ☉ Long. + 11s 21° 36′	Arg. = ☉ Long. + 8s 13° 51′	Arg. = ☾ ☊ + 8s 16° 7′	Arg. = 2 ☉ Long. + 8s 13° 3′	Arg. = ☉ Long. + 11s 16° 58′	Arg. = ☾ ☊ + 11s 10° 21′	Arg. = 2 ☉ Long. + 11s 13° 51′	
	Aber.	☽ Nut	☉ Nut.	Aber.	☾ Nut.	☉ Nut	Aber.	☾ Nut.	☉ Nut.	Aber.	☾ Nut.	☉ Nut.	Aber.	☾ Nut.	☉ Nut.	Aber.	☾ Nut.	☉ Nut	
	R. A.			N. P. D.			R. A.			N. P. D.			R. A.			N. P. D.			
	ALGENIB 1830.						SCHEDIR 1830.						POLARIS 1830.						

The CORRECTIONS of 48 principal Stars, &c. continued.

1830. Common Argument.	α ARIETIS 1820.						α CETI 1820.						α PERSEI 1820.						1830. Common Argument.
	Right Ascension. h. m. s. 1 57 3 .0			N. P. Distance. 67° 23′ 37″.			Right Ascension. h. m. s. 2 52 52 .9			N. P. Distance. 86° 37′ 18″.			Right Ascension. h. m. s. 3 11 31 .7			N. P. Distance. 40° 47′ 18″.			
	Ann. Var. +3^s.35			Ann. Var. −17″.40			Ann. Var. +3^s.12			Ann. Var. −14″.75			Ann. Var. +4^s.20			Ann. Var. −13″.50			
+ − s. s.	Aber.	☾ Nut.	⊙Nut.	Aber.	☾ Nut.	⊙Nut.	Aber.	☾ Nut.	⊙Nut.	Aber.	☾ Nut.	⊙Nut.	Aber.	☾ Nut.	⊙Nut.	Aber.	☾ Nut.	⊙Nut.	+ − s. s.
III 0° IX	1^s.372	1^s.224	0^s.067	7″.853	7″.848	0″.408	1^s.295	1°.117	0^s.063	7″.344	8″.434	0″.416	1^s.993	1^s.596	0^s.087	11″.50	8″.634	0″.418	III 0° IX
1	1.371	1.223	.067	7.85	7.84	.40	1.294	1.116	.063	7.34	8.43	.41	1.992	1.595	.087	11.49	8.63	.41	29
2	1.370	1.222	.067	7.84	7.84	.40	1.293	1.115	.063	7.33	8.42	.41	1.991	1.594	.087	11.48	8.62	.41	28
3	1.369	1.221	.067	7.84	7.83	.40	1.292	1.114	.063	7.33	8.42	.41	1.989	1.593	.086	11.47	8.62	.41	27
4	1.368	1.220	.067	7.83	7.82	.40	1.291	1.113	.063	7.32	8.41	.41	1.987	1.591	.086	11.46	8.61	.41	26
5	1.366	1.218	.067	7.82	7.81	.40	1.290	1.112	.063	7.32	8.40	.41	1.985	1.589	.086	11.45	8.60	.41	25
6	1.364	1.216	.067	7.81	7.80	.40	1.288	1.110	.063	7.31	8.39	.41	1.982	1.586	.086	11.43	8.59	.41	24
7	1.362	1.214	.066	7.80	7.79	.40	1.286	1.108	.063	7.30	8.37	.41	1.978	1.583	.085	11.41	8.57	.41	23
8	1.359	1.212	.066	7.78	7.77	.40	1.283	1.106	.062	7.28	8.35	.41	1.974	1.580	.085	11.39	8.55	.41	22
9	1.355	1.209	.066	7.76	7.75	.40	1.280	1.103	.062	7.26	8.33	.41	1.969	1.576	.085	11.36	8.53	.41	21
10	1.351	1.206	.066	7.74	7.73	.40	1.276	1.100	.062	7.24	8.31	.41	1.964	1.572	.085	11.33	8.50	.41	20
11	1.347	1.202	.066	7.71	7.70	.40	1.272	1.096	.062	7.21	8.28	.41	1.958	1.567	.084	11.29	8.47	.41	19
12	1.342	1.198	.065	7.68	7.67	.40	1.268	1.092	.062	7.18	8.25	.41	1.951	1.561	.084	11.25	8.44	.41	18
13	1.337	1.193	.065	7.65	7.64	.40	1.263	1.088	.061	7.15	8.22	.41	1.943	1.555	.084	11.20	8.41	.41	17
14	1.331	1.188	.065	7.62	7.61	.40	1.257	1.083	.061	7.12	8.18	.40	1.934	1.548	.083	11.15	8.37	.40	16
15	1.325	1.182	.065	7.58	7.57	.39	1.251	1.078	.061	7.09	8.14	.40	1.925	1.541	.083	11.10	8.33	.40	15
16	1.319	1.176	.064	7.55	7.54	.39	1.245	1.073	.061	7.05	8.10	.40	1.916	1.534	.083	11.00	8.29	.40	14
17	1.312	1.170	.064	7.51	7.50	.39	1.239	1.067	.060	7.02	8.06	.40	1.906	1.526	.082	10.98	8.25	.40	13
18	1.305	1.164	.064	7.47	7.46	.39	1.232	1.061	.060	6.98	8.02	.40	1.896	1.518	.082	10.92	8.21	.40	12
19	1.297	1.158	.063	7.42	7.41	.39	1.225	1.055	.059	6.94	7.98	.39	1.885	1.509	.081	10.86	8.16	.40	11
20	1.289	1.150	.063	7.37	7.36	.38	1.218	1.049	.059	6.90	7.93	.39	1.873	1.500	.081	10.80	8.11	.39	10
21	1.281	1.142	.063	7.33	7.32	.38	1.209	1.042	.059	6.85	7.87	.39	1.860	1.489	.080	10.73	8.06	.39	9
22	1.272	1.134	.062	7.28	7.27	.38	1.200	1.034	.058	6.81	7.82	.39	1.846	1.478	.080	10.66	8.01	.39	8
23	1.262	1.126	.062	7.23	7.22	.37	1.192	1.027	.058	6.76	7.76	.38	1.833	1.468	.079	10.59	7.95	.39	7
24	1.252	1.118	.061	7.17	7.16	.37	1.184	1.020	.058	6.71	7.71	.38	1.821	1.457	.079	10.50	7.89	.38	6
25	1.242	1.108	.061	7.12	7.11	.37	1.174	1.011	.057	6.65	7.65	.38	1.807	1.446	.078	10.42	7.83	.38	5
26	1.232	1.099	.060	7.06	7.05	.37	1.163	1.002	.057	6.60	7.58	.37	1.792	1.434	.077	10.33	7.76	.38	4
27	1.222	1.089	.060	7.69	7.98	.36	1.153	0.994	.056	6.54	7.52	.37	1.776	1.421	.077	10.24	7.69	.37	3
28	1.211	1.080	.059	6.93	7.92	.36	1.142	0.985	.056	6.48	7.45	.37	1.760	1.408	.076	10.14	7.62	.37	2
29	1.199	1.069	.059	6.87	7.86	.36	1.132	0.976	.055	6.42	7.38	.36	1.743	1.394	.075	10.04	7.55	.37	1
IV 0 X	1.187	1.059	.058	6.80	6.79	.35	1.121	0.966	.055	6.36	7.31	.36	1.726	1.380	.075	9.95	7.48	.36	II 0 VIII
1	1.175	1.049	.057	6.73	6.72	.35	1.110	0.956	.054	6.30	7.24	.36	1.709	1.366	.074	9.85	7.41	.36	29
2	1.164	1.038	.057	6.66	6.65	.35	1.098	0.946	.054	6.23	7.16	.35	1.691	1.352	.073	9.75	7.34	.35	28
3	1.151	1.026	.056	6.58	6.57	.34	1.085	0.936	.053	6.16	7.08	.35	1.672	1.338	.072	9.64	7.25	.35	27
4	1.136	1.014	.055	6.51	6.50	.34	1.072	0.925	.052	6.09	7.00	.34	1.653	1.322	.071	9.53	7.16	.35	26
5	1.123	1.001	.055	6.43	6.42	.33	1.060	0.914	.052	6.02	6.91	.34	1.633	1.306	.070	9.41	7.07	.34	25
6	1.109	0.989	.054	6.35	6.34	.33	1.046	0.902	.051	5.94	6.82	.34	1.612	1.289	.070	9.29	6.98	.34	24
7	1.094	0.976	.053	6.27	6.26	.33	1.033	0.891	.051	5.86	6.73	.33	1.592	1.273	.069	9.17	6.89	.33	23
8	1.080	0.963	.053	6.18	6.17	.32	1.020	0.879	.050	5.79	6.64	.33	1.572	1.256	.068	9.05	6.80	.33	22
9	1.065	0.950	.052	6.10	6.09	.32	1.006	0.867	.049	5.71	6.56	.32	1.550	1.239	.067	8.92	6.71	.32	21
10	1.050	0.936	.051	6.01	6.00	.31	0.991	0.854	.048	5.62	6.46	.32	1.527	1.221	.066	8.79	6.61	.32	20
11	1.034	0.922	.051	5.92	5.91	.31	0.976	0.841	.048	5.54	6.37	.81	1.504	1.202	.065	8.66	6.52	.32	19
12	1.019	0.908	.050	5.83	5.82	.30	0.962	0.828	.047	5.46	6.27	.31	1.481	1.184	.064	8.53	6.42	.31	18
13	1.002	0.894	.049	5.74	5.73	.30	0.946	0.816	.046	5.37	6.17	.30	1.458	1.136	.063	8.39	6.32	.31	17
14	0.986	0.879	.048	5.65	5.64	.29	0.931	0.803	.045	5.28	6.07	.30	1.434	1.148	.062	8.25	6.22	.30	16
IV 15 X	0.969	0.864	.047	5.55	5.54	.29	0.915	0.789	.045	5.19	5.97	.29	1.409	1.128	.061	8.11	6.11	.30	I 15 VII
	Arg. = ⊙ Long. + 7^s 28° 26′	Arg. = ☾ ☊ + 6^s 11° 1′	Arg. = 2 ⊙ Long. + 6^s 8° 59′	Arg. = ⊙ Long. + 1^s 0° 2′	Arg. = ☾ ☊ + 10^s 22° 52′	Arg. = 2 ⊙ Long. + 10^s 28° 26′	Arg. = ⊙ Long. + 7^s 14° 11′	Arg. = ☾ ☊ + 6^s 1° 26′	Arg. = 2 ⊙ Long. + 6^s 1° 9′	Arg. = ⊙ Long. + 2^s 23° 8′	Arg. = ☾ ☊ + 10^s 8° 15′	Arg. = 2 ⊙ Long. + 10^s 14° 56′	Arg. = ⊙ Long. + 7^s 9° 30′	Arg. = ☾ ☊ + 6^s 18° 13′	Arg. = 2 ⊙ Long. + 6^s 14° 57′	Arg. = ⊙ Long. + 11^s 3° 7′	Arg. = ☾ ☊ + 10^s 3° 47′	Arg. = 2 ⊙ Long. + 10^s 9° 30′	
	Aber.	☾ Nut.	⊙Nut.	Aber.	☾ Nut.	⊙Nut.	Aber.	☾ Nut.	⊙Nut.	Aber.	☾ Nut.	⊙Nut.	Aber.	☾ Nut.	⊙Nut.	Aber.	☾ Nut.	⊙Nut.	
	R. A.			N. P. D.			R. A.			N. P. D.			R. A.			N. P. D.			
	HAMAL 1830.						MENKAB 1830.						MIRFAK 1830.						

The CORRECTIONS of 48 principal Stars, &c. continued.

1830. Common Argument.	α ARIETIS 1820.						α CETI 1820.						α PERSEI 1820.						1830. Common Argument.
	Right Ascension. $1^h\ 57^m\ 3^s.0$			N. P. Distance. $67^\circ\ 23'\ 37''$.			Right Ascension. $2^h\ 52^m\ 52^s.9$			N. P. Distance. $86^\circ\ 37'\ 18''$.			Right Ascension. $3^h\ 11^m\ 31^s.7$			N. P. Distance. $40^\circ\ 47'\ 18''$.			
	Ann. Var. $+3^s.35$			Ann. Var. $-17''.40$			Ann. Var. $+3^s.12$			Ann. Var. $-14''.75$			Ann. Var. $+4^s.20$			Ann. Var. $-13''.50$			
+ S. − S.	Aber.	☾ Nut.	☉ Nut.	Aber.	☾ Nut.	☉ Nut.	Aber.	☾ Nut.	☉ Nut.	Aber.	☾ Nut.	☉ Nut.	Aber.	☾ Nut.	☉ Nut.	Aber.	☾ Nut.	☉ Nut.	+ S. − S.
IV 15° X	$0^s.969$	$0^s.864$	$0^s.047$	5″.55	5″.54	0″.29	$0^s.915$	$0^s.789$	$0^s.045$	5″.19	5″.97	0″.29	$1^s.409$	$1^s.128$	$0^s.061$	8″.11	6″.11	0″.30	I 15° VII
16	.952	.849	.047	5.46	5.45	.28	.899	.775	.044	5.10	5.86	.29	1.384	1.108	.060	7.97	6.00	.29	14
17	.935	.834	.046	5.36	5.35	.28	.882	.761	.043	5.01	5.75	.28	1.359	1.088	.059	7.82	5.88	.28	13
18	.917	.818	.045	5.26	5.25	.27	.866	.746	.042	4.92	5.64	.28	1.334	1.067	.058	7.67	5.77	.28	12
19	.899	.802	.044	5.15	5.14	.27	.848	.732	.041	4.82	5.53	.27	1.307	1.045	.056	7.52	5.66	.27	11
20	.881	.786	.043	5.04	5.03	.26	.831	.717	.040	4.72	5.42	.27	1.280	1.024	.055	7.37	5.55	.27	10
21	.863	.770	.042	4.94	4.93	.26	.815	.702	.040	4.62	5.31	.26	1.254	1.003	.054	7.22	5.44	.26	9
22	.844	.753	.041	4.83	4.82	.25	.797	.687	.039	4.52	5.19	.26	1.227	0.981	.053	7.07	5.32	.26	8
23	.825	.736	.040	4.72	4.71	.25	.779	.671	.038	4.42	5.07	.25	1.200	0.958	.052	6.91	5.20	.25	7
24	.806	.718	.039	4.62	4.61	.24	.761	.656	.037	4.32	4.96	.24	1.172	0.936	.051	6.74	5.08	.25	6
25	.787	.701	.038	4.51	4.50	.23	.742	.640	.036	4.21	4.84	.24	1.143	0.914	.049	6.58	4.96	.24	5
26	.767	.684	.037	4.39	4.38	.23	.724	.624	.035	4.11	4.72	.23	1.114	0.892	.048	6.42	4.83	.23	4
27	.747	.666	.036	4.27	4.26	.22	.705	.607	.034	4.00	4.59	.23	1.085	0.868	.047	6.25	4.71	.23	3
28	.727	.648	.035	4.16	4.15	.22	.686	.591	.033	3.89	4.47	.22	1.056	0.845	.046	6.08	4.58	.22	2
29	.707	.630	.034	4.04	4.03	.21	.667	.575	.032	3.78	4.34	.21	1.027	0.821	.044	5.91	4.45	.22	1
V 0 XI	.686	.612	.033	3.93	3.92	.20	.648	.558	.031	3.67	4.22	.21	0.997	0.798	.043	5.74	4.32	.21	I 0 VII
1	.665	.593	.032	3.81	3.80	.20	.628	.542	.030	3.56	4.09	.20	0.966	0.773	.042	5.57	4.18	.20	29
2	.644	.574	.031	3.68	3.67	.19	.608	.524	.029	3.45	3.96	.20	0.935	0.748	.040	5.38	4.06	.20	28
3	.622	.555	.030	3.56	3.55	.19	.588	.507	.028	3.33	3.83	.19	0.904	0.724	.039	5.21	3.92	.19	27
4	.601	.536	.029	3.44	3.43	.18	.567	.489	.027	3.22	3.69	.18	0.873	0.699	.038	5.03	3.78	.18	26
5	.580	.517	.028	3.32	3.31	.17	.547	.472	.026	3.11	3.57	.18	0.842	0.673	.036	4.85	3.65	.18	25
6	.557	.498	.027	3.19	3.18	.17	.527	.454	.025	2.98	3.43	.17	0.810	0.648	.035	4.66	3.51	.17	24
7	.535	.478	.026	3.07	3.06	.16	.516	.436	.024	2.87	3.29	.16	0.778	0.623	.034	4.48	3.37	.16	23
8	.513	.458	.025	2.94	2.93	.15	.485	.418	.023	2.75	3.16	.16	0.746	0.597	.032	4.30	3.23	.16	22
9	.491	.438	.024	2.81	2.81	.15	.464	.400	.022	2.63	3.02	.15	0.714	0.571	.031	4.11	3.09	.15	21
10	.468	.418	.023	2.68	2.68	.14	.442	.382	.021	2.51	2.88	.14	0.682	0.545	.029	3.92	2.95	.14	20
11	.446	.398	.022	2.55	2.55	.13	.421	.363	.020	2.39	2.75	.14	0.649	0.519	.028	3.73	2.81	.14	19
12	.424	.378	.021	2.43	2.43	.13	.400	.345	.019	2.27	2.61	.13	0.616	0.492	.027	3.54	2.67	.13	18
13	.401	.357	.020	2.29	2.29	.12	.378	.326	.018	2.15	2.47	.12	0.583	0.466	.025	3.35	2.52	.12	17
14	.378	.337	.018	2.16	2.16	.11	.357	.307	.017	2.02	2.33	.11	0.549	0.439	.024	3.16	2.38	.12	16
15	.355	.316	.017	2.03	2.03	.11	.335	.289	.016	1.90	2.18	.11	0.516	0.412	.022	2.97	2.23	.11	15
16	.331	.296	.016	1.90	1.90	.10	.313	.270	.015	1.77	2.04	.10	0.482	0.385	.021	2.77	2.09	.10	14
17	.308	.275	.015	1.76	1.76	.09	.291	.251	.014	1.65	1.89	.09	0.448	0.358	.019	2.58	1.94	.09	13
18	.285	.254	.014	1.63	1.63	.08	.269	.232	.013	1.53	1.75	.09	0.414	0.331	.018	2.38	1.79	.09	12
19	.261	.233	.013	1.50	1.50	.08	.247	.213	.012	1.40	1.61	.08	0.380	0.304	.016	2.19	1.64	.08	11
20	.238	.212	.012	1.36	1.36	.07	.225	.194	.011	1.27	1.46	.07	0.346	0.277	.015	1.99	1.50	.07	10
21	.214	.191	.010	1.23	1.23	.06	.202	.174	.010	1.15	1.32	.07	0.312	0.249	.013	1.79	1.35	.07	9
22	.191	.170	.009	1.09	1.09	.06	.180	.155	.009	1.02	1.17	.06	0.277	0.222	.012	1.60	1.20	.06	8
23	.167	.149	.008	0.96	0.96	.05	.158	.136	.008	0.89	1.03	.05	0.243	0.194	.010	1.40	1.05	.05	7
24	.143	.128	.007	0.82	0.82	.04	.136	.117	.007	0.77	0.88	.04	0.209	0.167	.009	1.20	0.90	.04	6
25	.120	.107	.006	0.69	0.69	.04	.113	.097	.006	0.64	0.74	.04	0.174	0.139	.008	1.00	0.75	.04	5
26	.096	.086	.005	0.55	0.55	.03	.090	.078	.004	0.51	0.59	.03	0.139	0.111	.006	0.80	0.60	.03	4
27	.072	.064	.003	0.41	0.41	.03	.068	.058	.003	0.38	0.44	.02	0.104	0.084	.004	0.60	0.45	.02	3
28	.048	.043	.002	0.27	0.27	.02	.045	.039	.002	0.26	0.29	.01	0.069	0.056	.003	0.40	0.30	.01	2
29	.024	.021	.001	0.14	0.14	.01	.023	.019	.001	0.13	0.15	.01	0.035	0.028	.001	0.20	0.15	.01	1
VI 0 O	.000	.000	.000	0.00	0.00	.00	.000	.000	.000	0.00	0.00	.00	0.000	0.000	.000	0.00	0.00	.00	O 0 VI
	Arg. = ☉ Long. + $7^s\ 28^\circ\ 26'$	Arg. = ☾ ☊ + $6^s\ 11^\circ\ 1'$	Arg. = 2☉ Long. + $6^s\ 8^\circ\ 59'$	Arg. = ☉ Long. + $1^s\ 0^\circ\ 2'$	Arg. = ☾ ☊ + $10^s\ 22^\circ\ 52'$	Arg. = 2☉ Long. + $10^s\ 28^\circ\ 26'$	Arg. = ☉ Long. + $7^s\ 14^\circ\ 11'$	Arg. = ☾ ☊ + $6^s\ 1^\circ\ 26'$	Arg. = 2☉ Long. + $6^s\ 1^\circ\ 9'$	Arg. = ☉ Long. + $2^s\ 23^\circ\ 8'$	Arg. = ☾ ☊ + $10^s\ 8^\circ\ 15'$	Arg. = 2☉ Long. + $10^s\ 14^\circ\ 56'$	Arg. = ☉ Long. + $7^s\ 9^\circ\ 30'$	Arg. = ☾ ☊ + $6^s\ 18^\circ\ 13'$	Arg. = 2☉ Long. + $6^s\ 14^\circ\ 57'$	Arg. = ☉ Long. + $11^s\ 3^\circ\ 7'$	Arg. = ☾ ☊ + $10^s\ 3^\circ\ 47'$	Arg. = 2☉ Long. + $10^s\ 9^\circ\ 30'$	
	Aber.	☾ Nut.	☉ Nut.	Aber.	☾ Nut.	☉ Nut.	Aber.	☾ Nut.	☉ Nut.	Aber.	☾ Nut.	☉ Nut.	Aber.	☾ Nut.	☉ Nut.	Aber.	☾ Nut.	☉ Nut.	
	R. A.			N. P. D.			R. A.			N. P. D.			R. A.			N. P. D.			
	HAMAL 1830.						MENKAB 1830.						MIRFAK 1830.						

The CORRECTIONS of 48 principal Stars, &c. continued.

1830. Common Argument. + s. / − s.	α TAURI 1820. Right Ascension. 4h. 25m. 36s.3 Ann. Var. + 3s.43 Aber.	☾ Nut.	☉ Nut.	N. P. Distance. 73° 51′ 43″. Ann. Var. − 7″.95 Aber.	☾ Nut.	☉ Nut.	1830. Common Argument. + s. / − s.
III 0° IX	1s.388	1s.231	0s.068	3″.748	9″.301	0″.429	III 0° IX
1	1.387	1.230	.068	3.74	9.30	.42	29
2	1.386	1.229	.068	3.74	9.29	.42	28
3	1.385	1.228	.068	3.74	9.28	.42	27
4	1.383	1.227	.068	3.74	9.27	.42	26
5	1.381	1.227	.068	3.73	9.26	.42	25
6	1.379	1.225	.068	3.73	9.25	.42	24
7	1.377	1.223	.067	3.72	9.23	.42	23
8	1.374	1.220	.067	3.71	9.21	.42	22
9	1.371	1.217	.067	3.70	9.19	.42	21
10	1.367	1.214	.067	3.70	9.17	.42	20
11	1.362	1.210	.067	3.69	9.14	.42	19
12	1.357	1.206	.067	3.67	9.11	.42	18
13	1.352	1.201	.066	3.65	9.07	.42	17
14	1.346	1.196	.066	3.64	9.03	.42	16
15	1.340	1.190	.066	3.62	8.98	.41	15
16	1.334	1.184	.065	3.60	8.93	.41	14
17	1.327	1.178	.065	3.59	8.88	.41	13
18	1.319	1.172	.065	3.57	8.83	.41	12
19	1.311	1.165	.064	3.55	8.78	.40	11
20	1.302	1.158	.064	3.52	8.73	.40	10
21	1.294	1.150	.064	3.50	8.68	.40	9
22	1.285	1.142	.063	3.47	8.62	.40	8
23	1.276	1.134	.063	3.45	8.56	.40	7
24	1.267	1.125	.062	3.43	8.49	.39	6
25	1.257	1.116	.062	3.40	8.42	.39	5
26	1.246	1.107	.061	3.37	8.35	.39	4
27	1.235	1.097	.061	3.34	8.27	.38	3
28	1.224	1.087	.060	3.31	8.20	.38	2
29	1.213	1.077	.059	3.28	8.13	.37	1
IV 0 X	1.201	1.067	.059	3.25	8.05	.37	II 0 VIII
1	1.189	1.056	.058	3.21	7.97	.37	29
2	1.176	1.044	.058	3.18	7.88	.36	28
3	1.163	1.032	.057	3.15	7.80	.36	27
4	1.150	1.021	.056	3.11	7.71	.36	26
5	1.136	1.009	.056	3.07	7.62	.35	25
6	1.122	0.997	.055	3.03	7.52	.35	24
7	1.107	0.984	.054	2.99	7.42	.34	23
8	1.092	0.970	.054	2.95	7.32	.34	22
9	1.077	0.956	.053	2.91	7.22	.33	21
10	1.061	0.942	.052	2.87	7.12	.33	20
11	1.046	0.928	.051	2.83	7.01	.32	19
12	1.030	0.914	.051	2.79	6.91	.32	18
13	1.014	0.900	.050	2.74	6.80	.31	17
14	0.997	0.886	.049	2.70	6.69	.31	16
IV 15 X	0.980	0.871	.048	2.65	6.57	.30	I 15 VII
	Arg. = ☉ Long. + 6s 21° 42′	Arg. = ☾ ☊ + 6s 3° 33′	Arg. = 2 ☉ Long. + 6s 2° 49′	Arg. = ☉ Long. + 1s 23° 15′	Arg. = ☾ ☊ + 9s 17° 54′	Arg. = 2 ☉ Long. + 9s 21° 42′	
	Aber.	☾ Nut.	☉ Nut.	Aber.	☾ Nut.	☉ Nut.	
	R. A.			N. P. D.			
	ALDEBARAM 1830.						

1830. Common Argument. + s. / − s.	α AURIGÆ 1820. Right Ascension. 5h. 3m. 24s.6 Ann. Var. + 4s.41 Aber.	☾ Nut.	☉ Nut.	N. P. Distance. 44° 11′ 51″. Ann. Var. − 4″.57 Aber.	☾ Nut.	☉ Nut.	1830. Common Argument. + s. / − s.
III 0° IX	1s.928	1s.589	0s.088	8″.105	9″.523	0″.432	III 0° IX
1	1.927	1.588	.088	8.10	9.52	.43	29
2	1.926	1.587	.088	8.09	9.51	.43	28
3	1.924	1.586	.088	8.09	9.50	.43	27
4	1.922	1.584	.088	8.08	9.49	.43	26
5	1.920	1.582	.088	8.07	9.48	.43	25
6	1.917	1.580	.087	8.06	9.46	.43	24
7	1.914	1.577	.087	8.05	9.44	.43	23
8	1.910	1.574	.087	8.03	9.42	.43	22
9	1.906	1.570	.087	8.01	9.40	.43	21
10	1.901	1.566	.087	7.99	9.38	.42	20
11	1.895	1.561	.086	7.96	9.35	.42	19
12	1.888	1.555	.086	7.93	9.32	.42	18
13	1.880	1.548	.086	7.90	9.28	.42	17
14	1.871	1.541	.085	7.87	9.24	.42	16
15	1.861	1.534	.085	7.83	9.20	.42	15
16	1.852	1.527	.084	7.79	9.15	.42	14
17	1.842	1.520	.084	7.75	9.10	.41	13
18	1.832	1.511	.084	7.71	9.05	.41	12
19	1.822	1.502	.083	7.67	9.00	.41	11
20	1.812	1.492	.083	7.62	8.95	.41	10
21	1.800	1.482	.082	7.57	8.89	.40	9
22	1.787	1.472	.082	7.52	8.82	.40	8
23	1.774	1.462	.081	7.46	8.76	.40	7
24	1.761	1.451	.080	7.41	8.69	.39	6
25	1.748	1.439	.080	7.35	8.62	.39	5
26	1.733	1.427	.079	7.29	8.55	.39	4
27	1.718	1.414	.078	7.23	8.47	.38	3
28	1.702	1.402	.078	7.16	8.40	.38	2
29	1.686	1.389	.077	7.09	8.31	.38	1
IV 0 X	1.670	1.375	.076	7.02	8.23	.37	II 0 VIII
1	1.652	1.361	.075	6.95	8.15	.37	29
2	1.634	1.346	.075	6.88	8.07	.37	28
3	1.617	1.332	.074	6.80	7.98	.36	27
4	1.599	1.316	.073	6.72	7.89	.36	26
5	1.580	1.300	.072	6.64	7.80	.35	25
6	1.560	1.284	.071	6.56	7.70	.35	24
7	1.540	1.278	.070	6.47	7.60	.34	23
8	1.520	1.251	.069	6.38	7.50	.34	22
9	1.499	1.234	.068	6.30	7.40	.34	21
10	1.487	1.216	.067	6.21	7.29	.33	20
11	1.454	1.198	.066	6.12	7.18	.33	19
12	1.432	1.179	.065	6.02	7.07	.32	18
13	1.410	1.161	.064	5.93	6.96	.32	17
14	1.387	1.141	.063	5.83	6.85	.31	16
IV 15 X	1.364	1.122	.062	5.73	6.73	.31	I 15 VII
	Arg. = ☉ Long. + 6s 12° 56′	Arg. = ☾ ☊ + 6s 5° 46′	Arg. = 2 ☉ Long. + 6s 4° 41′	Arg. = ☉ Long. + 9s 25° 37′	Arg. = ☾ ☊ + 9s 10° 29′	Arg. = 2 ☉ Long. + 9s 12° 51′	
	Aber.	☾ Nut.	☉ Nut.	Aber.	☾ Nut.	☉ Nut.	
	R. A.			N. P. D.			
	CAPELLA 1830.						

1830. Common Argument. + s. / − s.	β ORIONIS 1820. Right Ascension. 5h. 5m. 53s.6 Ann. Var. + 2s.88 Aber.	☾ Nut.	☉ Nut.	N. P. Distance. 98° 25′ 0″. Ann. Var. − 4″.92 Aber.	☾ Nut.	☉ Nut.	1830. Common Argument. + s. / − s.
III 0° IX	1s.359	1s.038	0s.057	10″.643	9″.532	0″.433	III 0° IX
1	1.358	1.037	.057	10.64	9.53	.43	29
2	1.357	1.036	.057	10.63	9.52	.43	28
3	1.356	1.035	.057	10.62	9.51	.43	27
4	1.354	1.034	.057	10.61	9.50	.43	26
5	1.352	1.033	.057	10.60	9.49	.43	25
6	1.350	1.032	.057	10.59	9.47	.43	24
7	1.348	1.030	.057	10.56	9.45	.43	23
8	1.346	1.028	.056	10.54	9.43	.43	22
9	1.343	1.026	.056	10.52	9.41	.43	21
10	1.339	1.023	.056	10.49	9.39	.43	20
11	1.334	1.020	.056	10.46	9.36	.43	19
12	1.329	1.016	.056	10.42	9.33	.42	18
13	1.323	1.012	.056	10.38	9.29	.42	17
14	1.317	1.007	.055	10.33	9.25	.42	16
15	1.311	1.002	.055	10.28	9.21	.42	15
16	1.306	0.997	.055	10.23	9.16	.42	14
17	1.300	0.992	.055	10.18	9.11	.41	13
18	1.293	0.987	.054	10.12	9.06	.41	12
19	1.285	0.982	.054	10.06	9.01	.41	11
20	1.276	0.975	.054	9.99	8.96	.41	10
21	1.268	0.968	.053	9.93	8.90	.41	9
22	1.259	0.962	.053	9.86	8.83	.40	8
23	1.250	0.955	.053	9.80	8.77	.40	7
24	1.241	0.948	.052	9.72	8.70	.40	6
25	1.231	0.940	.052	9.64	8.63	.39	5
26	1.222	0.933	.051	9.56	8.56	.39	4
27	1.211	0.925	.051	9.47	8.48	.39	3
28	1.199	0.916	.050	9.39	8.41	.38	2
29	1.188	0.907	.050	9.30	8.33	.38	1
IV 0 X	1.176	0.898	.049	9.21	8.24	.37	II 0 VIII
1	1.164	0.889	.049	9.12	8.16	.37	29
2	1.152	0.880	.048	9.02	8.08	.37	28
3	1.140	0.871	.048	8.93	7.99	.36	27
4	1.126	0.861	.047	8.82	7.90	.36	26
5	1.112	0.850	.047	8.71	7.81	.35	25
6	1.098	0.838	.046	8.60	7.71	.35	24
7	1.084	0.827	.046	8.49	7.61	.35	23
8	1.070	0.816	.045	8.37	7.51	.34	22
9	1.055	0.806	.044	8.26	7.40	.34	21
10	1.040	0.794	.044	8.14	7.29	.33	20
11	1.024	0.782	.043	8.02	7.18	.33	19
12	1.009	0.771	.042	7.90	7.07	.32	18
13	0.993	0.759	.042	7.78	6.96	.32	17
14	0.977	0.747	.041	7.65	6.85	.31	16
IV 15 X	0.960	0.734	.040	7.52	6.74	.31	I 15 VII
	Arg. = ☉ Long. + 6s 12° 20′	Arg. = ☾ ☊ + 5s 28° 47′	Arg. = 2 ☉ Long. + 5s 29° 0′	Arg. = ☉ Long. + 3s 3° 42′	Arg. = ☾ ☊ + 9s 10° 4′	Arg. = 2 ☉ Long. + 9s 12° 20	
	Aber.	☾ Nut.	☉ Nut.	Aber.	☾ Nut.	☉ Nut.	
	R. A.			N. P. D.			
	RIGEL 1830.						

The CORRECTIONS of 48 principal Stars, &c. continued.

1830. Common Argument.	α TAURI 1820.						α AURIGÆ 1820.						β ORIONIS 1820.						1830. Common Argument.
	Right Ascension. 4h 25m 36s.3			N. P. Distance. 73° 51′ 43″.			Right Ascension. 5h 3m 24s.6			N. P. Distance. 44° 11′ 51″.			Right Ascension. 5h 5m 53s.6			N. P. Distance. 98° 25′ 0″.			
	Ann. Var. +3s.43			Ann. Var. −7″.95			Ann. Var. +4s.41			Ann. Var. −4″.57			Ann. Var. +2s.88			Ann. Var. −4″.92			
+ − S. S.	Aber.	☾ Nut.	⊙ Nut.	Aber.	☾ Nut.	⊙ Nut.	Aber.	☾ Nut.	⊙ Nut.	Aber.	☾ Nut.	⊙ Nut.	Aber.	☾ Nut.	⊙ Nut.	Aber.	☾ Nut.	⊙ Nut.	+ − S. S.
IV 15° X	0s.980	0s.871	0s.048	2″.65	6″.57	0″.30	1s.364	1s.122	0s.062	5″.73	6″.73	0″.31	0s.960	0s.734	0s.040	7″.52	6″.74	0″.31	I 15° VII
16	.963	.855	.047	2.61	6.46	.30	1.349	1.102	.061	5.63	6.62	.30	.943	.721	.040	7.39	6.62	.30	14
17	.945	.839	.046	2.56	6.34	.29	1.314	1.082	.060	5.53	6.49	.29	.926	.707	.039	7.25	6.50	.30	13
18	.927	.823	.045	2.51	6.22	.29	1.290	1.062	.059	5.43	6.37	.29	.908	.694	.038	7.11	6.37	.29	12
19	.908	.807	.045	2.46	6.09	.28	1.264	1.041	.058	5.32	6.24	.28	.890	.680	.037	6.96	6.25	.28	11
20	.890	.791	.044	2.41	5.97	.28	1.239	1.020	.057	5.21	6.12	.28	.872	.666	.037	6.83	6.12	.28	10
21	.872	.775	.043	2.36	5.85	.27	1.214	0.999	.055	5.10	6.00	.27	.854	.653	.036	6.69	6.00	.27	9
22	.853	.758	.042	2.31	5.72	.26	1.188	0.977	.054	4.99	5.86	.27	.836	.639	.035	6.55	5.87	.27	8
23	.833	.741	.041	2.26	5.59	.26	1.160	0.955	.053	4.88	5.73	.26	.816	.624	.034	6.40	5.73	.26	7
24	.814	.724	.040	2.21	5.47	.25	1.132	0.933	.052	4.76	5.59	.25	.798	.610	.033	6.25	5.60	.25	6
25	.795	.707	.039	2.15	5.33	.25	1.105	0.910	.050	4.65	5.46	.25	.779	.595	.033	6.10	5.47	.25	5
26	.776	.689	.038	2.10	5.20	.24	1.078	0.888	.049	4.54	5.33	.24	.760	.580	.032	5.96	5.33	.24	4
27	.755	.671	.037	2.04	5.06	.23	1.050	0.864	.048	4.42	5.19	.23	.740	.565	.031	5.79	5.19	.24	3
28	.735	.653	.036	1.99	4.93	.23	1.022	0.842	.047	4.30	5.05	.23	.720	.550	.030	5.64	5.06	.23	2
29	.714	.634	.035	1.93	4.79	.22	0.993	0.817	.045	4.18	4.91	.22	.700	.535	.029	5.48	4.91	.22	1
V 0 XI	.693	.616	.034	1.87	4.65	.21	0.964	0.794	.044	4.06	4.77	.22	.679	.519	.029	5.32	4.77	.22	I 0 VII
1	.672	.597	.033	1.82	4.51	.21	0.935	0.769	.043	3.93	4.62	.21	.658	.503	.028	5.16	4.62	.21	29
2	.651	.578	.032	1.76	4.36	.20	0.905	0.745	.041	3.81	4.47	.28	.637	.487	.027	5.00	4.47	.20	28
3	.629	.559	.031	1.70	4.22	.19	0.875	0.721	.040	3.68	4.32	.20	.616	.471	.026	4.83	4.32	.20	27
4	.607	.540	.030	1.64	4.07	.19	0.845	0.696	.039	3.55	4.17	.19	.595	.455	.025	4.66	4.17	.19	26
5	.586	.521	.029	1.58	3.93	.18	0.814	0.671	.037	3.42	4.02	.18	.574	.438	.024	4.50	4.03	.18	25
6	.563	.501	.028	1.52	3.78	.17	0.784	0.645	.036	3.29	3.87	.18	.552	.422	.023	4.32	3.87	.18	24
7	.542	.481	.027	1.46	3.63	.17	0.753	0.620	.034	3.17	3.72	.17	.531	.415	.022	4.16	3.72	.17	23
8	.519	.461	.025	1.40	3.48	.16	0.722	0.594	.033	3.04	3.57	.16	.508	.388	.021	3.98	3.57	.16	22
9	.497	.441	.024	1.34	3.33	.15	0.691	0.568	.031	2.90	3.41	.15	.487	.372	.020	3.81	3.41	.16	21
10	.474	.421	.023	1.28	3.18	.15	0.659	0.542	.030	2.77	3.25	.15	.464	.355	.019	3.64	3.25	.15	20
11	.451	.401	.022	1.22	3.03	.14	0.628	0.517	.028	2.64	3.10	.14	.442	.338	.019	3.46	3.10	.14	19
12	.428	.380	.021	1.16	2.87	.13	0.596	0.491	.027	2.50	2.94	.13	.419	.321	.018	3.28	2.94	13	18
13	.405	.360	.020	1.10	2.72	.13	0.564	0.464	.026	2.37	2.78	.13	.397	.303	.017	3.11	2.78	.13	17
14	.382	.339	.019	1.03	2.56	.12	0.532	0.437	.024	2.23	2.63	.12	.374	.286	.016	2.93	2.62	.12	16
15	.359	.318	.018	0.97	2.41	.11	0.499	0.411	.023	2.10	2.46	.11	.351	.268	.015	2.75	2.46	.11	15
16	.335	.297	.016	0.91	2.25	.10	0.466	0.384	.022	1.96	2.30	.10	.328	.251	.014	2.57	2.30	.10	14
17	.311	.277	.015	0.84	2.09	.10	0.433	0.356	.020	1.82	2.14	.10	.305	.233	.013	2.39	2.14	.10	13
18	.288	.256	.014	0.78	1.93	.09	0.401	0.329	.018	1.68	1.98	.09	.282	.216	.012	2.21	1.98	.09	12
19	.264	.235	.013	0.71	1.77	.08	0.367	0.303	.017	1.54	1.81	.08	.259	.198	.011	2.03	1.82	.08	11
20	.241	.212	.012	0.65	1.61	.07	0.334	0.275	.015	1.41	1.65	.07	.235	.180	.010	1.85	1.65	.08	10
21	.217	.193	.011	0.59	1.45	.07	0.302	0.248	.014	1.27	1.49	.07	.212	.162	.009	1.66	1.49	.07	9
22	.193	.171	.009	0.52	1.29	.06	0.268	0.221	.012	1.13	1.32	.06	.189	.144	.008	1.48	1.32	.06	8
23	.169	.150	.008	0.46	1.13	.05	0.235	0.193	.011	0.99	1.16	.05	.165	.126	.007	1.30	1.16	.05	7
24	.145	.129	.007	0.39	0.97	.04	0.202	0.166	.009	0.85	1.00	.05	.142	.109	.006	1.11	1.00	.05	6
25	.121	.108	.006	0.33	0.81	.04	0.168	0.139	.008	0.71	0.83	.04	.118	.091	.005	0.93	0.83	.04	5
26	.097	.086	.005	0.26	0.65	.03	0.135	0.111	.006	0.57	0.66	.03	.095	.073	.004	0.74	0.66	.03	4
27	.073	.065	.004	0.20	0.49	.02	0.101	0.073	.005	0.42	0.50	.02	.071	.054	.003	0.56	0.50	.02	3
28	.048	.043	.002	0.13	0.32	.02	0.067	0.055	.003	0.28	0.33	.02	.047	.036	.002	0.37	0.33	.02	2
29	.024	.021	.001	0.06	0.16	.01	0.034	0.028	.001	0.14	0.17	.01	.024	.018	.001	0.19	0.16	.01	1
VI 0 O	.000	.000	.000	0.00	0.00	.00	0.000	0.000	.000	0.00	0.00	.00	.000	.000	.000	0.00	0.00	.00	O 0 VI
	Arg. = ⊙ Long. + 6s 21° 42′	Arg. = ☾ ☊ + 6s 3° 33′	Arg. = 2 ⊙ Long. + 6s 2° 49′	Arg. = ⊙ Long. + 1s 23° 15′	Arg. = ☾ ☊ + 9s 17° 54′	Arg. = 2 ⊙ Long. + 9s 21° 42′	Arg. = ⊙ Long. + 6s 12° 56′	Arg. = ☾ ☊ + 6s 5° 46′	Arg. = 2 ⊙ Long. + 6s 4° 41′	Arg. = ⊙ Long. + 9s 25° 37′	Arg. = ☾ ☊ + 9s 10° 29′	Arg. = 2 ⊙ Long. + 9s 12° 51′	Arg. = ⊙ Long. + 6s 12° 20′	Arg. = ☾ ☊ + 5s 28° 47′	Arg. = 2 ⊙ Long. + 5s 29° 0′	Arg. = ⊙ Long. + 3s 3° 42′	Arg. = ☾ ☊ + 9s 10° 4′	Arg = 2 ⊙ Long. + 9s 12° 22′	
	Aber.	☾ Nut.	⊙ Nut.	Aber.	☾ Nut.	⊙ Nut.	Aber.	☾ Nut.	⊙ Nut.	Aber.	☾ Nut.	⊙ Nut.	Aber.	☾ Nut.	⊙ Nut.	Aber.	☾ Nut	⊙ Nut.	
	R. A.			N. P. D.			R. A.			N. P. D.			R. A.			N. P. D.			
	ALDEBARAM 1830.						CAPELLA 1830.						RIGEL 1830.						

The CORRECTIONS of 48 principal Stars, &c. continued.

1830. Common Argument.	β TAURI 1820.						α ORIONIS 1820.						α CANIS MAJORIS 1820.						1830. Common Argument.
	Right Ascension. h. m. s. 5 14 55 .4			N. P. Distance. 61° 33′ 19″.			Right Ascension. h. m. s. 5 45 25 .9			N. P. Distance. 82° 38′ 8″.			Right Ascension. h. m. s. 6 37 12 .9			N. P. Distance. 106° 28′ 36″.			
	Ann. Var. + 3s. 78			Ann. Var. − 3″. 83			Ann. Var. + 3s. 25			Ann. Var. − 1″. 37			Ann. Var. + 2s. 61			Ann. Var. + 4″. 36			
+ s. − s.	Aber.	☾ Nut.	☉ Nut.	Aber.	☾ Nut.	☉ Nut.	Aber.	☾ Nut.	☉ Nut.	Aber.	☾ Nut.	☉ Nut.	Aber.	☾ Nut.	☉ Nut.	Aber.	☾ Nut.	☉ Nut.	+ s. − s.
III 0° IX	1s.531	1s.360	0s.075	2″. 451	9″. 568	0″. 433	1s.362	1s.166	0s.066	5″. 619	9″. 641	0″. 434	1s.405	0s.964	0s.054	12″. 971	9″. 591	0″. 433	III 0° IX
1	1 .530	1 .359	.075	2 . 45	9 . 56	. 43	1 .361	1 .165	.066	5 . 61	9 . 64	. 43	1 .404	.964	.054	12 . 96	9 . 59	. 43	29
2	1 .529	1 .358	.075	2 . 45	9 . 55	. 43	1 .360	1 .164	.066	5 . 60	9 . 63	. 43	1 .403	.963	.054	12 . 95	9 . 58	. 43	28
3	1 .528	1 .357	.075	2 . 45	9 . 54	. 43	1 .359	1 .163	.066	5 . 60	9 . 62	. 43	1 .402	.962	.054	12 . 94	9 . 57	. 43	27
4	1 .527	1 .356	.075	2 . 44	9 . 53	. 43	1 .358	1 .162	.066	5 . 59	9 . 61	. 43	1 .401	.961	.054	12 . 93	9 . 56	. 43	26
5	1 .525	1 .354	.075	2 . 44	9 . 51	. 43	1 .357	1 .161	.066	5 . 59	9 . 60	. 43	1 .400	.960	.054	12 . 91	9 . 55	. 43	25
6	1 .523	1 .352	.075	2 . 44	9 . 50	. 43	1 .355	1 .160	.066	5 . 58	9 . 58	. 43	1 .398	.959	.954	12 . 89	9 . 53	. 43	24
7	1 .520	1 .350	.074	2 . 43	9 . 49	. 43	1 .353	1 .158	.066	5 . 57	9 . 56	. 43	1 .396	.957	.054	12 . 87	9 . 51	. 43	23
8	1 .517	1 .347	.074	2 . 43	9 . 47	. 43	1 .350	1 .156	.065	5 . 56	9 . 54	. 43	1 .393	.955	.053	12 . 84	9 . 49	. 43	22
9	1 .514	1 .344	.074	2 . 42	9 . 45	. 43	1 .347	1 .153	.065	5 . 55	9 . 52	. 43	1 .390	.953	.053	12 . 81	9 . 47	. 43	21
10	1 .510	1 .340	.074	2 . 42	9 . 43	. 43	1 .343	1 .150	.065	5 . 54	9 . 50	. 43	1 .386	.950	.053	12 . 78	9 . 44	. 43	20
11	1 .505	1 .336	.074	2 . 41	9 . 40	. 43	1 .339	1 .146	.065	5 . 52	9 . 47	. 43	1 .382	.947	.053	12 . 74	9 . 41	. 43	19
12	1 .499	1 .331	.073	2 . 40	9 . 37	. 42	1 .334	1 .142	.065	5 . 50	9 . 44	. 42	1 .377	.943	.053	12 . 69	9 . 38	. 42	18
13	1 .493	1 .325	.073	2 . 39	9 . 33	. 42	1 .329	1 .137	.064	5 . 48	9 . 40	. 42	1 .371	.939	.053	12 . 64	9 . 34	. 42	17
14	1 .486	1 .319	.073	2 . 38	9 . 29	. 42	1 .323	1 .132	.064	5 . 46	9 . 36	. 42	1 .364	.935	.052	12 . 58	9 . 30	. 42	16
15	1 .479	1 .313	.072	2 . 37	9 . 24	. 42	1 .316	1 .126	.064	5 . 43	9 . 31	. 42	1 .357	.931	.052	12 . 52	9 . 26	. 42	15
16	1 .472	1 .307	.072	2 . 36	9 . 19	. 42	1 .309	1 .120	.064	5 . 40	9 . 26	. 42	1 .351	.926	.052	12 . 46	9 . 21	. 42	14
17	1 .465	1 .300	.072	2 . 34	9 . 15	. 42	1 .302	1 .114	.063	5 . 37	9 . 21	. 42	1 .344	.922	.052	12 . 40	9 . 16	. 41	13
18	1 .457	1 .293	.071	2 . 33	9 . 10	. 41	1 .295	1 .108	.063	5 . 34	9 . 16	. 41	1 .337	.917	.051	12 . 33	9 . 11	. 41	12
19	1 .449	1 .286	.071	2 . 32	9 . 05	. 41	1 .288	1 .102	.062	5 . 31	9 . 11	. 41	1 .329	.912	.051	12 . 27	9 . 06	. 41	11
20	1 .440	1 .278	.070	2 . 30	8 . 99	. 41	1 .281	1 .095	.062	5 . 28	9 . 06	. 41	1 .320	.906	.051	12 . 19	9 . 01	. 41	10
21	1 .431	1 .270	.070	2 . 29	8 . 92	. 41	1 .272	1 .088	.062	5 . 25	9 . 00	. 41	1 .311	.900	.050	12 . 11	8 . 95	. 41	9
22	1 .421	1 .261	.069	2 . 27	8 . 86	. 40	1 .263	1 .081	.061	5 . 22	8 . 94	. 40	1 .301	.893	.050	12 . 02	8 . 88	. 40	8
23	1 .411	1 .252	.069	2 . 25	8 . 80	. 40	1 .254	1 .073	.061	5 . 18	8 . 87	. 40	1 .292	.886	.050	11 . 93	8 . 82	. 40	7
24	1 .400	1 .242	.068	2 . 24	8 . 73	. 40	1 .245	1 .065	.060	5 . 14	8 . 80	. 40	1 .283	.880	.049	11 . 83	8 . 76	. 40	6
25	1 .388	1 .232	.068	2 . 22	8 . 66	. 39	1 .235	1 .057	.060	5 . 09	8 . 73	. 39	1 .273	.872	.049	11 . 74	8 . 69	. 39	5
26	1 .376	1 .222	.067	2 20	8 . 59	. 39	1 .225	1 .048	.059	5 . 06	8 . 66	. 39	1 .262	.865	.049	11 . 65	8 . 62	. 39	4
27	1 .364	1 .211	.067	2 . 18	8 . 52	. 39	1 .214	1 .038	.059	5 . 01	8 . 58	. 39	1 .251	.858	.048	11 . 54	8 . 54	. 39	3
28	1 .352	1 .200	.066	2 . 16	8 . 43	. 38	1 .203	1 .028	.058	4 . 97	8 . 50	. 38	1 .249	.850	.048	11 . 43	8 . 45	. 38	2
29	1 .339	1 .190	.066	2 . 14	8 . 36	. 38	1 .192	1 .019	.058	4 . 92	8 . 42	. 38	1 .228	.842	.047	11 . 32	8 . 37	. 38	1
IV 0 X	1 .326	1 .178	.065	2 . 12	8 . 27	. 38	1 .180	1 .010	.057	4 . 87	8 . 33	. 38	1 .217	.834	.047	11 . 21	8 . 30	. 37	II 0 VIII
1	1 .313	1 .165	.064	2 . 10	8 . 18	. 37	1 .178	0 .999	.057	4 . 82	8 . 25	. 37	1 .204	.825	.046	11 . 10	8 . 21	. 37	29
2	1 .299	1 .152	.064	2 . 08	8 . 10	. 37	1 .156	0 .988	.056	4 . 77	8 . 16	. 37	1 .191	.816	.046	10 . 99	8 . 12	. 37	28
3	1 .284	1 .140	.063	2 . 05	8 . 02	. 36	1 .143	0 .977	.055	4 . 72	8 . 07	. 36	1 .179	.807	.045	10 . 87	8 . 03	. 36	27
4	1 .269	1 .127	.062	2 . 03	7 . 93	. 36	1 .130	0 .966	.055	4 . 66	7 . 98	. 36	1 .165	.798	.045	10 . 74	7 . 94	. 36	26
5	1 .254	1 .113	.061	2 . 01	7 . 83	. 35	1 .116	0 .955	.054	4 . 61	7 . 89	. 36	1 .151	.789	.044	10 . 62	7 . 85	. 35	25
6	1 .238	1 .099	.061	1 . 98	7 . 73	. 35	1 .102	0 .942	.053	4 . 54	7 . 79	. 35	1 .136	.779	.044	10 . 49	7 . 75	. 35	24
7	1 .222	1 .085	.060	1 . 96	7 . 63	. 35	1 .087	0 .930	.053	4 . 48	7 . 69	. 35	1 .121	.769	.043	10 . 34	7 . 65	. 35	23
8	1 .206	1 .071	.059	1 . 93	7 . 53	. 34	1 .073	0 .918	.052	4 . 43	7 . 59	. 34	1 .107	.759	.043	10 . 21	7 . 55	. 34	22
9	1 .190	1 .056	.058	1 . 90	7 . 43	. 34	1 .059	0 .905	.051	4 . 37	7 . 49	. 34	1 .091	.749	.042	10 . 08	7 . 45	. 34	21
10	1 .172	1 .041	.057	1 . 87	7 . 32	. 33	1 .043	0 .892	.051	4 . 30	7 . 38	. 33	1 .075	.738	.041	9 . 92	7 . 34	. 33	20
11	1 .155	1 .025	.057	1 . 85	7 . 21	. 33	1 .027	0 .879	.050	4 . 24	7 . 27	. 33	1 .060	.727	.041	9 . 77	7 . 23	. 33	19
12	1 .138	1 .010	.056	1 . 82	7 . 10	. 32	1 .012	0 .865	.049	4 . 17	7 . 16	. 32	1 .044	.716	.040	9 . 62	7 . 12	. 32	18
13	1 .119	0 .994	.055	1 . 79	6 . 99	. 32	0 .996	0 .852	.048	4 . 11	7 . 04	. 32	1 .028	.704	.039	9 . 47	7 . 01	. 32	17
14	1 .101	0 .977	.054	1 . 76	6 . 87	. 31	0 .980	0 .838	.047	4 . 04	6 . 93	. 31	1 .011	.693	.03[illegible]	9 . 32	6 . 89	. 31	16
IV 15 X	1 .082	0 .961	.053	1 . 73	6 . 76	. 31	0 .962	0 .823	.047	3 . 97	6 . 81	. 31	0 .992	.681	.038	9 . 16	6 . 77	. 31	I 15 VII
	Arg. = ☉ Long. + 6s 10° 13′	Arg. = ☾ ☊ + 6s 2° 50′	Arg. = 2 ☉ Long. + 6s 2° 18′	Arg. = ☉ Long. + 10s 19° 23′	Arg. = ☾ ☊ + 9s 8° 20′	Arg. = 2 ☉ Long. + 9s 10° 13′	Arg. = ☉ Long. + 6s 13° 12′	Arg. = ☾ ☊ + 6s 0° 15′	Arg. = 2 ☉ Long. + 6s 0° 12′	Arg. = ☉ Long. + 2s 29° 50′	Arg. = ☾ ☊ + 9s 2° 36′	Arg. = 2 ☉ Long. + 9s 3° 13′	Arg. = ☉ Long. + 5s 21° 21′	Arg. = ☾ ☊ + 6s 1° 51′	Arg. = 2 ☉ Long. + 6s 1° 30′	Arg. = ☉ Long. + 2s 25° 50′	Arg. = ☾ ☊ + 5s 22° 58′	Arg. = 2 ☉ Long. + 8s 21° 22′	
	Aber.	☾ Nut.	☉ Nut.	Aber.	☾ Nut.	☉ Nut.	Aber.	☾ Nut.	☉ Nut.	Aber.	☾ Nut.	☉ Nut.	Aber.	☾ Nut.	☉ Nut.	Aber.	☾ Nut.	☉ Nut.	
	R. A.			N. P. D.			R. A.			N. P. D.			R. A.			N. P. D.			
	NATH 1830.						BETEIGEUZE 1830.						SIRIUS 1830.						

The CORRECTIONS of 48 principal Stars, &c. continued.

1830.	β TAURI 1820.						α ORIONIS 1820.						α CANIS MAJORIS 1820.						1830.
Common Argument.	Right Ascension. h. m. s. 5 14 55 . 4			N. P. Distance. 61° 33′ 19″			Right Ascension. h. m. s. 5 45 25s. 9			N. P. Distance. 82° 38′ 8″			Right Ascension. h. m. s. 6 37 12 . 9			N. P. Distance. 106° 28′ 36″			Common Argument.
	Ann. Var. + 3s. 78			Ann. Var. − 3″. 83			Ann. Var. + 3s . 25			Ann. Var. − 1″. 37			Ann. Var. + 2s . 61			Ann. Var. + 4″. 37			
+ − S. S.	Aber.	☾ Nut.	☉ Nut.	Aber.	☾ Nut.	☉ Nut.	Aber.	☾ Nut.	☉ Nut.	Aber.	☾ Nut.	☉ Nut.	Aber.	☾ Nut.	☉ Nut.	Aber.	☾ Nut.	☉ Nut.	+ − S. S.
IV 15° X	1s. 082	0s. 961	0s. 053	1″. 73	6″. 76	0″. 31	0s. 962	0s. 823	0s. 047	3″. 97	6″. 81	0″. 31	0s. 992	0s. 681	0s. 038	9″. 16	6″. 77	0″. 31	I 15° VII
16	1 . 063	. 944	. 052	1 . 70	6 . 64	. 30	. 946	. 809	. 046	3 . 90	6 . 69	. 30	. 975	. 669	. 038	9 . 00	6 . 65	. 30	14
17	1 . 044	. 927	. 051	1 . 67	6 . 52	. 30	. 928	. 795	. 045	3 . 83	6 . 57	. 30	. 957	. 657	. 037	8 . 83	6 . 53	. 30	13
18	1 . 024	. 909	. 050	1 . 64	6 . 39	. 29	. 911	. 780	. 044	3 . 76	6 . 45	. 29	. 939	. 644	. 036	8 . 66	6 . 41	. 29	12
19	1 . 004	. 891	. 049	1 . 61	6 . 27	. 28	. 893	. 764	. 043	3 . 68	6 . 32	. 28	. 921	. 632	. 035	8 . 49	6 . 28	. 28	11
20	0 . 983	. 872	. 048	1 . 57	6 . 14	. 28	. 875	. 749	. 042	3 . 61	6 . 19	. 28	. 902	. 618	. 035	8 . 32	6 . 16	. 28	10
21	0 . 963	. 855	. 047	1 . 54	6 . 02	. 27	. 857	. 733	. 042	3 . 53	6 . 07	. 27	. 883	. 606	. 034	8 . 15	6 . 03	. 27	9
22	0 . 942	. 837	. 046	1 . 51	5 . 88	. 27	. 838	. 717	. 041	3 . 46	5 . 93	. 27	. 864	. 593	. 033	7 . 98	5 . 90	. 27	8
23	0 . 921	. 817	. 045	1 . 47	5 . 75	. 26	. 819	. 701	. 040	3 . 38	5 . 80	. 26	. 845	. 580	. 032	7 . 80	5 . 77	. 26	7
24	0 . 900	. 798	. 044	1 . 44	5 . 62	. 25	. 801	. 685	. 039	3 . 30	5 . 67	. 25	. 825	. 566	. 032	7 . 62	5 . 63	. 25	6
25	0 . 878	. 780	. 043	1 . 41	5 . 48	. 25	. 781	. 668	. 038	3 . 22	5 . 53	. 25	. 815	. 552	. 031	7 . 43	5 . 50	. 25	5
26	0 . 856	. 761	. 042	1 . 37	5 . 35	. 24	. 762	. 652	. 037	3 . 15	5 . 39	. 24	. 786	. 539	. 030	7 . 25	5 . 36	. 24	4
27	0 . 833	. 741	. 041	1 . 33	5 . 21	. 24	. 742	. 635	. 036	3 . 06	5 . 25	. 24	. 764	. 525	. 029	7 . 05	5 . 22	. 24	3
28	0 . 811	. 721	. 040	1 . 30	5 . 07	. 23	. 722	. 618	. 035	2 . 98	5 . 11	. 23	. 744	. 511	. 029	6 . 87	5 . 08	. 23	2
29	0 . 788	. 700	. 039	1 . 26	4 . 92	. 22	. 702	. 600	. 034	2 . 89	4 . 97	. 22	. 723	. 497	. 028	6 . 67	4 . 94	. 22	1
V 0 XI	0 . 766	. 680	. 038	1 . 23	4 . 78	. 22	. 682	. 583	. 033	2 . 81	4 . 82	. 22	. 702	. 482	. 027	6 . 48	4 . 80	. 22	I 0 VII
1	0 . 742	. 659	. 036	1 . 19	4 . 64	. 21	. 661	. 565	. 032	2 . 72	4 . 67	. 21	. 681	. 467	. 026	6 . 28	4 . 65	. 21	29
2	0 . 718	. 638	. 035	1 . 15	4 . 49	. 20	. 639	. 547	. 031	2 . 64	4 . 52	. 20	. 659	. 452	. 025	6 . 18	4 . 50	. 20	28
3	0 . 695	. 617	. 034	1 . 11	4 . 34	. 20	. 618	. 529	. 030	2 . 55	4 . 37	. 20	. 637	. 437	. 025	5 . 88	4 . 35	. 20	27
4	0 . 671	. 596	. 033	1 . 07	4 . 18	. 19	. 597	. 511	. 029	2 . 46	4 . 22	. 19	. 616	. 422	. 024	5 . 68	4 . 20	. 19	26
5	0 . 647	. 575	. 032	1 . 04	4 . 04	. 18	. 576	. 493	. 028	2 . 37	4 . 07	. 18	. 594	. 407	. 023	5 . 48	4 . 05	. 18	25
6	0 . 622	. 552	. 030	1 . 00	3 . 88	. 18	. 554	. 474	. 027	2 . 28	3 . 91	. 18	. 571	. 391	. 022	5 . 27	3 . 89	. 18	24
7	0 . 598	. 531	. 029	0 . 96	3 . 73	. 17	. 532	. 455	. 026	2 . 19	3 . 76	. 17	. 548	. 376	. 021	5 . 16	3 . 74	. 17	23
8	0 . 573	. 509	. 028	0 . 92	3 . 58	. 16	. 510	. 436	. 025	2 . 10	3 . 61	. 16	. 526	. 361	. 020	4 . 85	3 . 59	. 16	22
9	0 . 548	. 487	. 027	0 . 88	3 . 42	. 16	. 488	. 417	. 024	2 . 01	3 . 45	. 16	. 504	. 345	. 019	4 . 64	3 . 43	. 16	21
10	0 . 523	. 465	. 026	0 . 84	3 . 26	. 15	. 466	. 398	. 023	1 . 92	3 . 29	. 15	. 480	. 329	. 018	4 . 43	3 . 27	. 15	20
11	0 . 498	. 442	. 024	0 . 80	3 . 11	. 14	. 443	. 379	. 021	1 . 83	3 . 13	. 14	. 457	. 313	. 018	4 . 22	3 . 12	. 14	19
12	0 . 473	. 419	. 023	0 . 76	2 . 95	. 13	. 421	. 360	. 020	1 . 74	2 . 97	. 13	. 433	. 297	. 017	4 . 00	2 . 96	. 13	18
13	0 . 447	. 397	. 022	0 . 72	2 . 79	. 13	. 398	. 341	. 019	1 . 64	2 . 81	. 13	. 410	. 281	. 016	3 . 79	2 . 80	. 13	17
14	0 . 422	. 374	. 021	0 . 67	2 . 63	. 12	. 375	. 321	. 018	1 . 55	2 . 65	. 12	. 387	. 265	. 015	3 . 57	2 . 64	. 12	16
15	0 . 396	. 351	. 019	0 . 63	2 . 47	. 11	. 352	. 301	. 017	1 . 45	2 . 49	. 11	. 363	. 249	. 014	3 . 35	2 . 48	. 11	15
16	0 . 370	. 328	. 018	0 . 59	2 . 31	. 10	. 329	. 281	. 016	1 . 36	2 . 33	. 10	. 339	. 233	. 013	3 . 13	2 . 32	. 10	14
17	0 . 344	. 305	. 017	0 . 55	2 . 15	. 10	. 306	. 262	. 015	1 . 26	2 . 16	. 10	. 315	. 216	. 012	2 . 91	2 . 15	. 10	13
18	0 . 318	. 282	. 016	0 . 51	1 . 98	. 09	. 283	. 242	. 014	1 . 17	2 . 00	. 09	. 291	. 200	. 011	2 . 69	1 . 99	. 09	12
19	0 . 292	. 259	. 014	0 . 47	1 . 82	. 08	. 260	. 222	. 013	1 . 07	1 . 83	. 08	. 267	. 183	. 010	2 . 47	1 . 82	. 08	11
20	0 . 266	. 236	. 013	0 . 43	1 . 66	. 08	. 236	. 202	. 011	0 . 97	1 . 67	. 08	. 243	. 167	. 009	2 . 25	1 . 66	. 08	10
21	0 . 239	. 212	. 012	0 . 38	1 . 49	. 07	. 213	. 182	. 010	0 . 88	1 . 51	. 07	. 219	. 151	. 008	2 . 03	1 . 50	. 07	9
22	0 . 213	. 189	. 010	0 . 34	1 . 33	. 06	. 190	. 162	. 009	0 . 78	1 . 34	. 06	. 195	. 134	. 008	1 . 80	1 . 33	. 06	8
23	0 . 187	. 166	. 009	0 . 30	1 . 16	. 05	. 166	. 142	. 008	0 . 68	1 . 17	. 05	. 171	. 117	. 007	1 . 58	1 . 17	. 05	7
24	0 . 160	. 142	. 008	0 . 26	1 . 00	. 05	. 143	. 122	. 007	0 . 59	1 . 01	. 05	. 147	. 101	. 006	1 . 36	1 . 00	. 05	6
25	0 . 134	. 119	. 007	0 . 21	0 . 83	. 04	. 119	. 102	. 006	0 . 49	0 . 84	. 04	. 123	. 084	. 005	1 . 13	0 . 84	. 04	5
26	0 . 107	. 095	. 005	0 . 17	0 . 67	. 03	. 095	. 081	. 005	0 . 39	0 . 67	. 03	. 098	. 067	. 004	0 . 91	0 . 67	. 03	4
27	0 . 080	. 071	. 004	0 . 13	0 . 50	. 02	. 071	. 061	. 003	0 . 29	0 . 51	. 02	. 074	. 051	. 003	0 . 68	0 . 50	. 02	3
28	0 . 054	. 048	. 003	0 . 08	0 . 33	. 02	. 048	. 041	. 002	0 . 20	0 . 34	. 02	. 049	. 034	. 002	0 . 45	0 . 33	. 02	2
29	0 . 027	. 024	. 001	0 . 04	0 . 17	. 01	. 024	. 020	. 001	0 . 10	0 . 17	. 01	. 025	. 017	. 001	0 . 23	0 . 17	. 01	1
VI 0 O	0 . 000	. 000	. 000	0 . 00	0 . 00	. 00	. 000	. 000	. 000	0 . 00	0 . 00	. 00	. 000	. 000	. 000	0 . 00	0 . 00	. 00	O 0 VI
	Arg. = ☉ Long. + 6s 10° 13′	Arg. = ☾ ☊ + 6s 2° 50′	Arg. = 2☉ Long. + 6s 2° 18′	Arg. = ☉ Long. + 10s 19° 23′	Arg. = ☾ ☊ + 9s 8° 20′	Arg. = 2☉ Long. + 9s 10° 13′	Arg. = ☉ Long. + 6s 13° 12′	Arg. = ☾ ☊ + 6s 0° 15′	Arg. = 2☉ Long. + 6s 0° 12′	Arg. = ☉ Long. + 2s 29° 50′	Arg. = ☾ ☊ + 9s 2° 36′	Arg. = 2☉ Long. + 9s 3° 13′	Arg. = ☉ Long. + 5s 21° 21′	Arg. = ☾ ☊ + 6s 1° 51′	Arg. = 2☉ Long. + 6s 1° 30′	Arg. = ☉ Long. + 2s 25° 50′	Arg. = ☾ ☊ + 8s 22° 58′	Arg. = 2☉ Long. + 8s 21° 22′	
	Aber.	☾ Nut.	☉ Nut.	Aber.	☾ Nut.	☉ Nut.	Aber.	☾ Nut.	☉ Nut.	Aber.	☾ Nut.	☉ Nut.	Aber.	☾ Nut.	☉ Nut.	Aber.	☾ Nut.	☉ Nut.	
	R. A.			N. P. D.			R. A.			N. P. D.			R. A.			N. P. D.			
	NATH 1830.						BETEIGEUZE 1830.						SIRIUS 1830.						

The CORRECTIONS of 48 principal Stars, &c. continued.

1830. Common Argument.	α GEMINORUM 1820.						α CANIS MINORIS 1820.						β GEMINORUM 1820.						1830. Common Argument.
	Right Ascension. h. m. s. 7 23 6 .11			N. P. Distance. 57° 43′ 36″.31			Right Ascension. h. m. s. 7 29 52 .67			N. P. Distance. 84° 19′ 15″.69			Right Ascension. h. m. s. 7 33 17 .47			N. P. Distance. 61° 32′ 52″.45			
	Ann. Var. +3s.85			Ann. Var. +7″.06			Ann. Var. +3s.15			Ann. Var. +8″.54			Ann. Var. +3s.69			Ann. Var. +8″.00			
+ S.	Aber.	☾ Nut.	⊙ Nut.	Aber.	☾ Nut.	⊙ Nut.	Aber.	☾ Nut.	⊙ Nut.	Aber.	☾ Nut.	⊙ Nut.	Aber.	☾ Nut.	⊙ Nut.	Aber.	☾ Nut.	⊙ Nut.	− S.
III 0° IX	1s.581	1.392	0s.077	4″.550	9″.370	0″.43	1s.341	1.149	0s.063	6″.383	9″.326	0″.429	1s.516	1.346	0s.075	4″.006	9″.292	0″.429	III 0° IX
1	1.580	1.391	.077	4.55	9.37	.43	1.340	1.148	.063	6.38	9.32	.42	1.515	1.345	.075	4.00	9.29	.42	29
2	1.579	1.390	.077	4.55	9.36	.43	1.339	1.147	.063	6.38	9.31	.42	1.514	1.344	.075	4.00	9.28	.42	28
3	1.578	1.389	.077	4.54	9.35	.43	1.338	1.146	.063	6.37	9.31	.42	1.513	1.343	.075	4.00	9.27	.42	27
4	1.577	1.388	.077	4.54	9.34	.43	1.337	1.145	.063	6.37	9.30	.42	1.512	1.342	.075	4.00	9.26	.42	26
5	1.575	1.386	.077	4.53	9.33	.43	1.335	1.143	.063	6.36	9.29	.42	1.510	1.340	.075	3.99	9.25	.42	25
6	1.573	1.384	.077	4.52	9.32	.43	1.333	1.141	.063	6.35	9.28	.42	1.508	1.338	.075	3.99	9.23	.42	24
7	1.570	1.382	.077	4.51	9.30	.43	1.331	1.139	.062	6.34	9.26	.42	1.505	1.336	.074	3.98	9.21	.42	23
8	1.567	1.379	.076	4.50	9.28	.43	1.328	1.137	.062	6.33	9.24	.42	1.502	1.333	.074	3.97	9.19	.42	22
9	1.563	1.376	.076	4.49	9.26	.42	1.325	1.134	.062	6.31	9.22	.42	1.498	1.330	.074	3.96	9.17	.42	21
10	1.559	1.372	.076	4.48	9.23	.42	1.321	1.131	.062	6.29	9.20	.42	1.494	1.327	.074	3.95	9.15	.42	20
11	1.554	1.368	.076	4.47	9.20	.42	1.317	1.127	.062	6.27	9.17	.42	1.489	1.323	.074	3.94	9.12	.42	19
12	1.548	1.363	.075	4.45	9.17	.42	1.312	1.123	.062	6.25	9.14	.42	1.484	1.318	.073	3.92	9.09	.42	18
13	1.542	1.357	.075	4.43	9.13	.42	1.307	1.119	.061	6.23	9.10	.42	1.478	1.312	.073	3.90	9.05	.42	17
14	1.535	1.351	.075	4.41	9.09	.42	1.301	1.114	.061	6.20	9.06	.42	1.471	1.306	.073	3.89	9.01	.42	16
15	1.527	1.344	.074	4.39	9.05	.42	1.295	1.109	.061	6.17	9.02	.41	1.464	1.300	.072	3.87	8.97	.41	15
16	1.520	1.338	.074	4.37	9.01	.41	1.289	1.104	.061	6.14	8.97	.41	1.457	1.294	.072	3.85	8.93	.41	14
17	1.512	1.331	.074	4.35	8.96	.41	1.282	1.098	.060	6.11	8.92	.41	1.450	1.287	.072	3.83	8.88	.41	13
18	1.503	1.324	.073	4.33	8.91	.41	1.275	1.092	.060	6.08	8.87	.41	1.442	1.280	.071	3.81	8.83	.41	12
19	1.495	1.316	.073	4.30	8.86	.41	1.268	1.086	.060	6.04	8.82	.41	1.434	1.272	.071	3.79	8.78	.41	11
20	1.486	1.308	.072	4.27	8.80	.40	1.260	1.079	.059	6.00	8.76	.40	1.424	1.264	.071	3.76	8.72	.40	10
21	1.476	1.299	.072	4.24	8.74	.40	1.251	1.072	.059	5.96	8.70	.40	1.414	1.256	.070	3.74	8.67	.40	9
22	1.465	1.290	.071	4.21	8.68	.40	1.242	1.064	.058	5.92	8.64	.40	1.404	1.248	.070	3.71	8.61	.40	8
23	1.455	1.281	.071	4.18	8.62	.40	1.234	1.057	.058	5.87	8.58	.40	1.395	1.239	.069	3.69	8.54	.40	7
24	1.444	1.271	.070	4.15	8.56	.39	1.225	1.049	.058	5.83	8.52	.39	1.385	1.230	.069	3.66	8.47	.39	6
25	1.432	1.261	.070	4.12	8.48	.39	1.215	1.041	.057	5.79	8.45	.39	1.373	1.220	.068	3.63	8.41	.39	5
26	1.421	1.251	.069	4.09	8.41	.39	1.206	1.032	.057	5.74	8.38	.39	1.362	1.210	.067	3.60	8.34	.39	4
27	1.408	1.240	.069	4.06	8.34	.38	1.195	1.023	.056	5.69	8.30	.38	1.350	1.199	.067	3.57	8.26	.38	3
28	1.395	1.228	.068	4.02	8.26	.38	1.183	1.013	.056	5.64	8.22	.38	1.337	1.188	.066	3.53	8.18	.38	2
29	1.382	1.217	.067	3.98	8.18	.38	1.172	1.004	.055	5.58	8.15	.38	1.325	1.178	.066	3.50	8.12	.38	1
IV 0 X	1.369	1.205	.067	3.94	8.10	.37	1.162	0.994	.055	5.53	8.07	.37	1.312	1.166	.065	3.47	8.04	.37	II 0 VIII
1	1.355	1.192	.066	3.90	8.02	.37	1.150	0.984	.054	5.47	7.99	.37	1.298	1.154	.064	3.43	7.96	.37	29
2	1.340	1.180	.065	3.86	7.94	.36	1.137	0.974	.053	5.42	7.91	.36	1.284	1.141	.064	3.40	7.87	.36	28
3	1.326	1.168	.065	3.82	7.85	.36	1.124	0.963	.953	5.36	7.82	.36	1.270	1.129	.063	3.36	7.78	.36	27
4	1.310	1.154	.064	3.77	7.76	.36	1.111	0.952	.052	5.29	7.78	.36	1.256	1.116	.062	3.32	7.70	.36	26
5	1.294	1.140	.063	3.72	7.67	.35	1.098	0.940	.052	5.23	7.64	.35	1.241	1.102	.061	3.28	7.61	.35	25
6	1.278	1.124	.062	3.68	7.57	.35	1.084	0.928	.051	5.17	7.54	.35	1.226	1.088	.061	3.24	7.52	.35	24
7	1.262	1.111	.061	3.63	7.48	.34	1.070	0.916	.050	5.10	7.44	.34	1.210	1.074	.060	3.20	7.42	.34	23
8	1.245	1.096	.061	3.58	7.38	.34	1.056	0.904	.050	5.03	7.35	.34	1.194	1.060	.059	3.16	7.32	.34	22
9	1.228	1.081	.060	3.53	7.28	.33	1.041	0.892	.049	4.96	7.25	.33	1.178	1.046	.058	3.12	7.22	.33	21
10	1.210	1.065	.059	3.48	7.17	.33	1.026	0.878	.048	4.89	7.14	.33	1.160	1.030	.057	3.07	7.11	.33	20
11	1.192	1.050	.058	3.43	7.06	.32	1.011	0.866	.048	4.82	6.73	.32	1.143	1.015	.057	3.02	7.00	.32	19
12	1.174	1.033	.057	3.38	6.96	.32	0.996	0.853	.047	4.75	6.93	.32	1.126	1.000	.056	2.98	6.90	.32	18
13	1.156	1.018	.056	3.32	6.85	.31	0.980	0.839	.046	4.67	6.82	.31	1.108	0.984	.055	2.93	6.79	.31	17
14	1.137	1.001	.055	3.27	6.74	.31	0.964	0.825	.045	4.59	6.71	.31	1.090	0.968	.054	2.88	6.68	.31	16
IV 15 X	1.118	0.983	.054	3.21	6.62	.30	0.947	0.811	.045	4.52	6.59	.30	1.071	0.951	.053	2.83	6.57	.30	I 15 VII
	Arg. = ⊙ Long. + 5s 10° 40′	Arg. = ☾ ☊ + 5s 21° 2′	Arg. = 2⊙ Long. + 5s 25° 9′	Arg. = ⊙ Long. + 7s 2° 7′	Arg. = ☾ ☊ + 8s 14° 20′	Arg. = 2⊙ Long. + 8s 10° 39′	Arg. = ⊙ Long. + 5s 9° 6′	Arg. = ☾ ☊ + 5s 28° 48′	Arg. = 2⊙ Long. + 5s 29° 0′	Arg. = ⊙ Long. + 3s 6° 54′	Arg. = ☾ ☊ + 8s 12° 47′	Arg. = 2⊙ Long. + 8s 9° 6′	Arg. = ⊙ Long. + 5s 8° 2′	Arg. = ☾ ☊ + 5s 24° 2′	Arg. = 2⊙ Long. + 5s 25° 9′	Arg. = ⊙ Long. + 6s 14° 32′	Arg. = ☾ ☊ + 8s 11° 53′	Arg. = 2⊙ Long. + 8s 8° 2′	
	Aber.	☾ Nut.	⊙ Nut.	Aber.	☾ Nut.	⊙ Nut.	Aber.	☾ Nut.	⊙ Nut.	Aber.	☾ Nut.	⊙ Nut.	Aber.	☾ Nut.	⊙ Nut.	Aber.	☾ Nut.	⊙ Nut.	
	R. A.			N. P. D.			R. A.			N. P. D.			R. A.			N. P. D.			
	CASTOR 1830.						PROCYON 1830.						POLLUX 1830.						

The CORRECTIONS of 48 principal Stars, &c. continued.

1830. Common Argument.	α GEMINORUM 1820.						α CANIS MINORIS 1820.						β GEMINORUM 1820.						1830. Common Argument.
	Right Ascension. h. m. s. 7 23 6 .11			N. P. Distance. 57° 43′ 36″. 31			Right Ascension. h. m. s. 7 29 52 .67			N. P. Distance. 84° 19′ 15″. 69			Right Ascension. h. m. s. 7 34 17 .47			N. P. Distance. 61° 32′ 52″. 45			
	Ann. Var. + 3s. 85			Ann. Var. + 7″. 06			Ann. Var. + 3s. 15			Ann. Var. + 8″.54			Ann. Var. + 3s. 69			Ann. Var. + 8″. 00			
+ S. — S.	Aber.	☾ Nut.	⊙ Nut.	Aber.	☾ Nut.	⊙ Nut.	Aber.	☾ Nut.	⊙ Nut.	Aber.	☾ Nut.	⊙ Nut.	Aber.	☾ Nut.	⊙ Nut.	Aber.	☾ Nut.	⊙ Nut.	+ S. — S.
IV 15° X	1s.118	0s.983	9s.054	3″.21	6″.62	0″.30	0s.947	0s.811	0s.045	4″.52	6″.59	0″.30	1s.071	0s.951	0s.053	2″.83	6″.57	0″.30	I 15° VII
16	1 .097	.966	.053	3 .18	6 .50	.30	.931	.797	.044	4 .43	6 .48	.30	1 .052	.935	.052	2 .78	6 .45	.30	14
17	1 .077	.948	.053	3 .10	6 .38	.29	.914	.783	.043	4 .36	6 .36	.29	1 .032	.917	.051	2 .73	6 .33	.29	13
18	1 .057	.930	.052	3 .04	6 .26	.29	.897	.768	.042	4 .27	6 .24	.29	1 .013	.900	.050	2 .68	6 .21	.29	12
19	1 .036	.912	.050	2 .98	6 .14	.28	.879	.752	.041	4 .18	6 .11	.28	0 .993	.882	.049	2 .62	6 .08	.28	11
20	1 .015	.893	.049	2 .92	6 .02	.28	.861	.737	.040	4 .10	5 .99	.28	0 .973	.865	.048	2 .57	5 .96	.28	10
21	0 .994	.875	.048	2 .86	5 .89	.27	.843	.722	.040	4 .02	5 .87	.27	0 .953	.847	.047	2 .52	5 .84	.27	9
22	0 .973	.857	.047	2 .80	5 .77	.26	.824	.716	.039	3 .93	5 .74	.26	0 .933	.828	.046	2 .47	5 .72	.26	8
23	0 .951	.837	.046	2 .74	5 .64	.26	.806	.691	.038	3 .84	5 .61	.26	0 .912	.810	.045	2 .41	5 .59	.26	7
24	0 .928	.817	.045	2 .67	5 .50	.25	.788	.674	.037	3 .75	5 .48	.25	0 .890	.791	.044	2 .35	5 .46	.25	6
25	0 .906	.798	.044	2 .61	5 .37	.25	.769	.658	.036	3 .66	5 .35	.25	0 .868	.772	.043	2 .30	5 .33	.25	5
26	0 .884	.778	.043	2 .54	5 .24	.24	.750	.642	.035	3 .57	5 .22	.24	0 .847	.753	.042	2 .24	5 .20	.24	4
27	0 .860	.758	.042	2 .47	5 .10	.23	.730	.625	.034	3 .48	5 .08	.23	0 .825	.733	.041	2 .18	5 .06	.23	3
28	0 .837	.738	.041	2 .41	4 .97	.23	.711	.608	.033	3 .39	4 .94	.23	0 .803	.714	.040	2 .12	4 .92	.23	2
29	0 .814	.717	.040	2 .34	4 .83	.22	.691	.592	.032	3 .29	4 .81	.22	0 .781	.693	.039	2 .06	4 .78	.22	1
V 0 XI	0 .791	.696	.038	2 .27	4 .69	.21	.671	.574	.032	3 .20	4 .67	.21	0 .758	.673	.038	2 .00	4 .65	.21	I 0 VII
1	0 .767	.674	.037	2 .20	4 .54	.21	.650	.557	.031	3 .10	4 .52	.21	0 .734	.653	.036	1 .94	4 .50	.21	29
2	0 .742	.653	.036	2 .13	4 .39	.20	.629	.539	.030	3 .00	4 .37	.20	0 .711	.632	.035	1 .88	4 .36	.20	28
3	0 .717	.631	.035	2 .06	4 .25	.20	.608	.521	.029	2 .90	4 .23	.19	0 .687	.611	.034	1 .82	4 .22	.19	27
4	0 .792	.609	.034	1 .99	4 .11	.19	.587	.503	.028	2 .80	4 .08	.19	0 .664	.590	.033	1 .76	4 .07	.19	26
5	0 .667	.588	.033	1 .92	3 .96	.18	.566	.485	.027	2 .70	3 .94	.18	0 .640	.569	.032	1 .69	3 .92	.18	25
6	0 .642	.566	.031	1 .85	3 .81	.17	.545	.467	.026	2 .59	3 .79	.17	0 .616	.547	.030	1 .63	3 .77	.17	24
7	0 .617	.543	.030	1 .77	3 .66	.17	.523	.448	.025	2 .49	3 .64	.17	0 .592	.526	.029	1 .57	3 .62	.17	23
8	0 .592	.521	.029	1 .70	3 .50	.16	.502	.429	.024	2 .39	3 .49	.16	0 .567	.504	.028	1 .50	3 .47	.16	22
9	0 .566	.498	.028	1 .63	3 .35	.15	.480	.411	.023	2 .29	3 .34	.15	0 .543	.482	.027	1 .43	3 .32	.15	21
10	0 .540	.476	.026	1 .55	3 .20	.15	.458	.392	.022	2 .18	3 .19	.15	0 .518	. 60	.026	1 .37	3 .17	.15	20
11	0 .514	.453	.025	1 .48	3 .05	.14	.436	.373	.020	2 .08	3 .03	.14	0 .493	.438	.024	1 .30	3 .02	.14	19
12	0 .488	.430	.024	1 .40	2 .89	.13	.414	.354	.019	1 .97	2 .88	.13	0 .468	.416	.023	1 .24	2 .87	.13	18
13	0 .462	.407	.023	1 .33	2 .74	.13	.392	.335	.018	1 .87	2 .73	.13	0 .443	.393	.022	1 .16	2 .71	.13	17
14	0 .435	.384	.021	1 .25	2 .58	.12	.369	.316	.017	1 .76	2 .57	.12	0 .417	.371	.021	1 .10	2 .56	.12	16
15	0 .408	.360	.020	1 .18	2 .42	.11	.347	.297	.016	1 .65	2 .41	.11	0 .392	.348	.019	1 .04	2 .40	.11	15
16	0 .382	.336	.019	1 .10	2 .26	.10	.324	.277	.015	1 .54	2 .25	.10	0 .366	.325	.018	0 .97	2 .24	.10	14
17	0 .355	.312	.017	1 .02	2 .10	.10	.301	.258	.014	1 .43	2 .09	.10	0 .340	.302	.017	0 .90	2 .08	.10	13
18	0 .328	.289	.016	0 .94	1 .94	.09	.278	.238	.013	1 .33	1 .94	.09	0 .314	.279	.016	0 .83	1 .93	.09	12
19	0 .301	.265	.015	0 .87	1 .78	.08	.255	.219	.012	1 .22	1 .78	.08	0 .288	.256	.014	0 .76	1 .77	.08	11
20	0 .274	.241	.013	0 .79	1 .62	.07	.233	.199	.011	1 .11	1 .64	.07	0 .263	.234	.013	0 .69	1 .61	.07	10
21	0 .247	.217	.012	0 .71	1 .46	.07	.209	.179	.010	1 .00	1 .46	.07	0 .237	.210	.012	0 .63	1 .45	.07	9
22	0 .220	.194	.011	0 .63	1 .30	.06	.187	.160	.009	0 .89	1 .30	.06	0 .211	.187	.010	0 .56	1 .29	.06	8
23	0 .193	.170	.009	0 .55	1 .14	.05	.163	.140	.008	0 .78	1 .14	.05	0 .185	.164	.009	0 .49	1 .13	.05	7
24	0 .165	.146	.008	0 .48	0 .98	.05	.140	.120	.007	0 .67	0 .98	.04	0 .159	.141	.008	0 .42	0 .97	.04	6
25	0 .138	.121	.007	0 .40	0 .82	.04	.117	.100	.006	0 .56	0 .81	.04	0 .132	.118	.007	0 .35	0 .81	.04	5
26	0 .111	.097	.005	0 .32	0 .66	.03	.094	.080	.004	0 .45	0 .65	.03	0 .106	.094	.005	0 .28	0 .65	.03	4
27	0 .083	.073	.004	0 .24	0 .49	.02	.070	.060	.003	0 .33	0 .49	.02	0 .079	.071	.004	0 .21	0 .49	.02	3
28	0 .055	.048	.003	0 .16	0 .33	.02	.047	.040	.002	0 .22	0 .33	.02	0 .053	.047	.003	0 .14	0 .32	.02	2
29	0 .028	.024	.001	0 .08	0 .16	.01	.023	.020	.001	0 .11	0 .16	.01	0 .026	.023	.001	0 .07	0 .16	.01	1
VI 0 O	0 .000	.000	.000	9 .00	0 .00	.00	.000	.000	.000	0 .00	0 .00	.00	0 .000	.000	.000	0 .00	0 .00	.00	O 0 VI
	Arg. = ⊙ Long. + 5s 10° 40′	Arg. = ☾ ☊ + 5s 24° 2′	Arg. = 2 ⊙ Long. + 5s 25° 9′	Arg. = ⊙ Long. + 7s 2° 7′	Arg. = ☾ ☊ + 8s 14° 20′	Arg. = 2 ⊙ Long. + 8s 10° 39′	Arg. = ⊙ Long. + 5s 9° 6′	Arg. = ☾ ☊ + 5s 28° 48′	Arg. = 2 ⊙ Long. + 5s 29° 0′	Arg. = ⊙ Long. + 3s 6° 54′	Arg. = ☾ ☊ + 8s 12° 47′	Arg. = 2 ⊙ Long. + 8s 9° 6′	Arg. = ⊙ Long. + 5s 8° 2′	Arg. = ☾ ☊ + 5s 24° 2′	Arg. = 2 ⊙ Long. + 5s 25° 9′	Arg. = ⊙ Long. + 6s 14° 32′	Arg. = ☾ ☊ + 8s 11° 53′	Arg. = 2 ⊙ Long. + 8s 8° 2′	
	Aber.	☾ Nut.	⊙ Nut.	Aber.	☾ Nut.	⊙ Nut.	Aber.	☾ Nut.	⊙ Nut.	Aber.	☾ Nut.	⊙ Nut.	Aber.	☾ Nut.	⊙ Nut.	Aber.	☾ Nut.	⊙ Nut.	
	R. A.			N. P. D.			R. A.			N. P. D.			R. A.			N. P. D.			
	CASTOR 1830.						PROCYON 1830.						POLLUX 1830.						

The CORRECTIONS of 48 principal Stars, &c. continued.

1830. Common Argument.	α HYDRÆ 1820.						θ URSÆ MAJORIS 1820.						α LEONIS 1820.						1830. Common Argument.
	Right Ascension. h. m. s. 9 18 44 . 61			N. P. Distance. 97° 52′ 57″. 19			Right Ascension. h. m. s. 9 20 45 . 21			N. P. Distance. 37° 30′ 30″. 40			Right Ascension. h. m. s. 9 58 46 . 74			N. P. Distance. 77° 9′ 23″. 62			
	Ann. Var. + 2s. 95			Ann. Var. + 15″. 19			Ann. Var. + 4s. 07			Ann. Var. + 15″. 34			Ann. Var. + 3s. 21			Ann. Var. + 17″. 33			
+ − s. s.	Aber.	☾ Nut.	⊙ Nut.	Aber.	☾ Nut.	⊙ Nut.	Aber.	☾ Nut.	⊙ Nut.	Aber.	☾ Nut.	⊙ Nut.	Aber.	☾ Nut.	⊙ Nut.	Aber.	☾ Nut.	⊙ Nut.	+ − s. s.
III 0° IX	1s.299	1s.059	0s.059	9″.875	8″.299	0″.414	2s.109	1s.632	0s.088	13″.153	8″.275	0″.413	1s.300	1s.166	0s.065	6″.974	7″.882	0″.408	III 0° IX
1	1 .298	1 .059	.059	9 .87	8 .29	.41	2 .108	1 .631	.088	13 .14	8 .27	.413	1 .299	1 .165	.065	6 .97	7 .88	.40	29
2	1 .297	1 .058	.059	9 .86	8 .28	.41	2 .107	1 .630	.088	13 .13	8 .26	.413	1 .298	1 .164	.065	6 .96	7 .87	.40	28
3	1 .296	1 .057	.059	9 .86	8 .28	.41	2 .105	1 .629	.088	13 .12	8 .26	.413	1 .297	1 .163	.065	6 .96	7 .87	.40	27
4	1 .295	1 .056	.059	9 .85	8 .27	.41	2 .103	1 .628	.088	13 .11	8 .25	.412	1 .296	1 .162	.065	6 .95	7 .86	.40	26
5	1 .293	1 .055	.059	9 .84	8 .26	.41	2 .101	1 .626	.088	13 .10	8 .24	.412	1 .295	1 .161	.065	6 .94	7 .85	.40	25
6	1 .291	1 .053	.059	9 .83	8 .25	.41	2 .098	1 .624	.088	13 .08	8 .23	.411	1 .293	1 .159	.065	6 .93	7 .84	.40	24
7	1 .289	1 .051	.059	9 .81	8 .24	.41	2 .094	1 .621	.087	13 .06	8 .22	.410	1 .291	1 .157	.065	6 .92	7 .83	.40	23
8	1 .286	1 .049	.058	9 .79	8 .22	.41	2 .090	1 .618	.087	13 .03	8 .20	.409	1 .288	1 .155	.064	6 .91	7 .81	.40	22
9	1 .283	1 .046	.058	9 .76	8 .20	.41	2 .085	1 .614	.087	13 .00	8 .18	.408	1 .284	1 .153	.064	6 .89	7 .79	.40	21
10	1 .280	1 .043	.058	9 .73	8 .18	.41	2 .080	1 .610	.087	12 .97	8 .16	.407	1 .280	1 .150	.064	6 .87	7 .77	.40	20
11	1 .276	1 .040	.058	9 .70	8 .15	.41	2 .074	1 .605	.086	12 .93	8 .14	.406	1 .276	1 .147	.064	6 .85	7 .75	.40	19
12	1 .272	1 .036	.058	9 .66	8 .12	.40	2 .066	1 .599	.086	12 .88	8 .11	.405	1 .271	1 .143	.064	6 .82	7 .72	.40	18
13	1 .267	1 .032	.057	9 .62	8 .09	.40	2 .057	1 .592	.086	12 .82	8 .08	.403	1 .266	1 .138	.063	6 .79	7 .69	.40	17
14	1 .261	1 .027	.057	9 .58	8 .06	.40	2 .048	1 .585	.085	12 .76	8 .04	.401	1 .261	1 .132	.063	6 .76	7 .65	.40	16
15	1 .254	1 .022	.057	9 .53	8 .02	.40	2 .038	1 .578	.085	12 .70	8 .00	.399	1 .255	1 .126	.063	6 .73	7 .61	.39	15
16	1 .248	1 .018	.057	9 .49	7 .98	.40	2 .028	1 .570	.085	12 .64	7 .96	.397	1 .249	1 .120	.063	6 .70	7 .57	.39	14
17	1 .241	1 .013	.056	9 .44	7 .94	.40	2 .017	1 .562	.084	12 .58	7 .92	.395	1 .243	1 .114	.062	6 .67	7 .53	.39	13
18	1 .234	1 .007	.056	9 .38	7 .90	.39	2 .006	1 .553	.084	12 .52	7 .88	.393	1 .237	1 .108	.062	6 .64	7 .49	.39	12
19	1 .227	1 .001	.056	9 .33	7 .85	.39	1 .995	1 .544	.083	12 .44	7 .83	.391	1 .230	1 .102	.061	6 .60	7 .45	.39	11
20	1 .220	0 .995	.055	9 .28	7 .80	.39	1 .983	1 .535	.083	12 .36	7 .78	.388	1 .222	1 .095	.061	6 .55	7 .41	.38	10
21	1 .212	0 .988	.055	9 .22	7 .75	.39	1 .971	1 .525	.082	12 .28	7 .73	.386	1 .213	1 .088	.061	6 .51	7 .36	.38	9
22	1 .203	0 .981	.055	9 .15	7 .70	.38	1 .958	1 .515	.082	12 .20	7 .68	.383	1 .204	1 .081	.060	6 .46	7 .31	.38	8
23	1 .195	0 .974	.054	9 .08	7 .65	.38	1 .944	1 .504	.081	12 .11	7 .63	.381	1 .196	1 .073	.060	6 .42	7 .26	.38	7
24	1 .186	0 .967	.054	9 .02	7 .59	.38	1 .929	1 .492	.080	12 .02	7 .57	.378	1 .186	1 .065	.059	6 .37	7 .20	.37	6
25	1 .177	0 .960	.054	8 .95	7 .53	.38	1 .913	1 .480	.080	11 .93	7 .51	.375	1 .177	1 .057	.059	6 .32	7 .14	.37	5
26	1 .167	0 .952	.053	8 .87	7 .47	.37	1 .897	1 .468	.079	11 .82	7 .45	.372	1 .167	1 .048	.058	6 .27	7 .08	.37	4
27	1 .157	0 .943	.053	8 .79	7 .40	.37	1 .880	1 .455	.078	11 .72	7 .38	.368	1 .157	1 .039	.058	6 .22	7 .02	.36	3
28	1 .146	0 .934	.052	8 .71	7 .33	.37	1 .862	1 .442	.078	11 .62	7 .32	.365	1 .147	1 .029	.057	6 .16	6 .96	.36	2
29	1 .135	0 .926	.052	8 .63	7 .27	.36	1 .845	1 .428	.077	11 .51	7 .25	.362	1 .136	1 .020	.057	6 .10	6 .89	.36	1
IV 0 X	1 .124	0 .917	.051	8 .55	7 .19	.36	1 .827	1 .414	.076	11 .39	7 .17	.358	1 .125	1 .010	.056	6 .04	6 .82	.35	II 0 VIII
1	1 .112	0 .907	.051	8 .45	7 .12	.36	1 .809	1 .400	.075	11 .27	7 .10	.354	1 .113	0 .999	.056	5 .98	6 .75	.35	29
2	1 .100	0 .898	.050	8 .36	7 .04	.35	1 .789	1 .375	.075	11 .15	7 .02	.350	1 .102	0 .988	.055	5 .91	6 .68	.35	28
3	1 .088	0 .887	.050	8 .27	6 .96	.35	1 .770	1 .370	.074	11 .02	6 .95	.347	1 .090	0 .977	.055	5 .85	6 .61	.34	27
4	1 .076	0 .877	.049	8 .17	6 .88	.34	1 .750	1 .354	.073	10 .90	6 .86	.342	1 .077	0 .966	.054	5 .78	6 .54	.34	26
5	1 .063	0 .867	.048	8 .08	6 .80	.34	1 .729	1 .338	.072	10 .78	6 .78	.339	1 .064	0 .955	.053	5 .71	6 .46	.33	25
6	1 .050	0 .856	.048	7 .98	6 72	.33	1 .707	1 .321	.071	10 .64	6 .70	.334	1 .051	0 .943	.053	5 .64	6 .37	.33	24
7	1 .036	0 .845	.047	7 .88	6 .63	.33	1 .684	1 .304	.070	10 .50	6 .61	.330	1 .037	0 .931	.052	5 .57	6 .29	.33	23
8	1 .023	0 .833	.046	7 .77	6 .54	.33	1 .661	1 .286	.069	10 .36	6 .52	.325	1 .023	0 .918	.051	5 .49	6 .21	.32	22
9	1 .008	0 .822	.046	7 .67	6 .45	.32	1 .638	1 .269	.068	10 .21	6 .43	.321	1 .010	0 .905	.050	5 .42	6 .12	.32	21
10	0 .994	0 .810	.045	7 .56	6 .36	.32	1 .625	1 .250	.067	10 .07	6 .34	.316	0 .994	0 .892	.050	5 .34	6 .04	.31	20
11	0 .979	0 .799	.045	7 .44	6 .27	.31	1 .592	1 .231	.066	9 .92	6 .25	.312	0 .979	0 .879	.049	5 .26	5 .95	.31	19
12	0 .964	0 .787	.044	7 .33	6 .17	.31	1 .569	1 .213	.065	9 .76	6 .16	.307	0 .965	0 .866	.048	5 .18	5 .86	.30	18
13	0 .948	0 .775	.043	7 .22	6 .07	.30	1 .544	1 .194	.064	9 .62	6 .06	.302	0 .950	0 .852	.048	5 .10	5 .77	.30	17
14	0 .933	0 .762	.042	7 .10	5 .97	.30	1 .518	1 .174	.063	9 .46	5 .96	.297	0 .934	0 .838	.047	5 .02	5 .67	.29	16
IV 15 X	0 .917	0 .748	.042	6 .97	5 .87	.29	1 .492	1 .154	.062	9 .30	5 .86	.292	0 .918	0 .823	.046	4 .93	5 .57	.29	I 15 VII
	Arg. = ⊙ Long. + 4s 12° 38′	Arg. = ☾ ☊ + 6s 3° 42′	Arg. = 2⊙ Long. + 6s 3° 0′	Arg. = ⊙ Long. + 2s 17° 32′	Arg. = ☾ ☊ + 7s 18° 37′	Arg. = 2⊙ Long. + 7s 12° 38′	Arg. = ⊙ Long. + 4s 12° 5′	Arg. = ☾ ☊ + 5s 6° 45′	Arg. = 2⊙ Long. + 5s 10° 47′	Arg. = ⊙ Long. + 6s 19° 55′	Arg. = ☾ ☊ + 7s 18° 3′	Arg. = 2⊙ Long. + 7s 11° 50′	Arg. = ⊙ Long. + 4s 2° 22′	Arg. = ☾ ☊ + 5s 23° 47′	Arg. = 2⊙ Long. + 5s 24° 56′	Arg. = ⊙ Long. + 4s 3° 46′	Arg. = ☾ ☊ + 7s 8° 0′	Arg. = 2⊙ Long. + 7s 2° 21′	
	Aber.	☾ Nut.	⊙ Nut.	Aber.	☾ Nut.	⊙ Nut.	Aber.	☾ Nut.	⊙ Nut.	Aber.	☾ Nut.	⊙ Nut.	Aber.	☾ Nut.	⊙ Nut.	Aber.	☾ Nut.	⊙ Nut.	
	R. A.			N. P. D.			R. A.			N. P. D.			R. A.			N. P. D.			
	ALPHARD 1830.						θ URSÆ MAJORIS 1830.						REGULUS 1830.						

The CORRECTIONS of 48 principal Stars, &c. continued.

1830. Common Argument.			α HYDRÆ 1820.						θ URSÆ MAJORIS 1820.						α LEONIS 1820.						1830. Common Argument.		
			Right Ascension. h. m. s. 9 18 44 . 61			N. P. Distance. 97° 52′ 57″. 19			Right Ascension. h. m. s. 9 20 45 . 21			N. P. Distance. 37° 30′ 30″. 40			Right Ascension. h. m. s. 9 58 46 . 74			N. P. Distance. 79° 9′ 23″. 62					
			Ann. Var. $+2^{s}.95$			Ann. Var. +15″. 19			Ann. Var. $+4^{s}.07$			Ann. Var. +15″. 34			Ann. Var. $+3^{s}.21$			Ann. Var. + 17″. 33					
+ S.		− S.	Aber.	☾ Nut.	☉ Nut.	Aber.	☾ Nut	☉ Nut.	Aber.	☾ Nut.	☉ Nut.	Aber.	☾ Nut.	☉ Nut.	Aber.	☾ Nut.	☉ Nut.	Aber.	☾ Nut.	☉ Nut.	+ S.		− S.
IV	15°	X	$0^{s}.917$	$0^{s}.748$	$0^{s}.042$	6″.97	5″.87	0″.29	$1^{s}.492$	$1^{s}.154$	$0^{s}.062$	9″.30	5″.86	0″.292	$0^{s}.918$	$0^{s}.823$	$0^{s}.046$	4″.93	5″.57	0″.29	I	15°	VII
	16		.891	.735	.041	6.85	5.77	.29	1.466	1.133	.061	9.14	5.75	.287	.892	.809	.045	4.84	5.47	.28		14	
	17		.885	.722	.040	6.73	5.66	.28	1.439	1.112	.060	8.97	5.65	.282	.886	.794	.044	4.76	5.37	.28		13	
	18		.868	.708	.039	6.60	5.55	.28	1.411	1.091	.059	8.80	5.54	.276	.868	.780	.043	4.67	5.27	.27		12	
	19		.851	.694	.039	6.47	5.44	.27	1.383	1.070	.058	8.62	5.43	.271	.851	.764	.043	4.57	5.17	.27		11	
	20		.833	.680	.038	6.34	5.33	.27	1.355	1.099	.057	8.45	5.32	.265	.834	.749	.042	4.48	5.06	.26		10	
	21		.816	.666	.037	6.21	5.22	.26	1.327	1.027	.055	8.28	5.21	.260	.817	.733	.041	4.39	4.96	.26		9	
	22		.799	.652	.036	6.08	5.11	.26	1.299	1.005	.054	8.10	5.10	.255	.800	.717	.040	4.29	4.86	.25		8	
	23		.781	.637	.036	5.94	5.00	.25	1.270	0.982	.053	7.92	4.98	.249	.781	.701	.039	4.19	4.75	.25		7	
	24		.763	.622	.035	5.80	4.88	.24	1.240	0.959	.052	7.73	4.87	.243	.763	.685	.038	4.09	4.64	.24		6	
	25		.744	.607	.034	5.66	4.77	.24	1.210	0.936	.050	7.54	4.75	.237	.745	.668	.937	4.00	4.52	.23		5	
	26		.726	.592	.033	5.52	4.65	.23	1.180	0.913	.049	7.36	4.63	.231	.726	.651	.036	3.90	4.41	.23		4	
	27		.706	.577	.032	5.37	4.52	.23	1.149	0.879	.048	7.16	4.51	.225	.707	.634	.035	3.80	4.29	.22		3	
	28		.688	.561	.031	5.23	4.40	.22	1.118	0.865	.047	6.97	4.39	.219	.688	.618	.034	3.70	4.18	.22		2	
	29		.668	.546	.030	5.08	4.28	.21	1.086	0.841	.045	6.77	4.27	.213	.669	.601	.033	3.59	4.06	.21		1	
V	0	XI	.649	.530	.030	4.94	4.15	.21	1.056	0.816	.044	6.58	4.14	.207	.650	.583	.032	3.49	3.94	.20	I	0	VII
	1		.629	.513	.029	4.78	4.02	.20	1.023	0.791	.043	6.37	4.02	.200	.630	.565	.031	3.38	3.82	.20		29	
	2		.609	.497	.028	4.63	3.89	.19	0.990	0.766	.041	6.17	3.88	.194	.609	.547	.030	3.27	3.70	.19		28	
	3		.589	.481	.027	4.48	3.77	.19	0.957	0.741	.040	5.97	3.76	.188	.589	.529	.029	3.17	3.58	.19		27	
	4		.568	.464	.026	4.32	3.64	.18	0.924	0.716	.039	5.76	3.63	.181	.569	.511	.028	3.06	3.46	.18		26	
	5		.548	.447	.025	4.17	3.51	.18	0.891	0.690	.037	5.56	3.50	.175	.549	.493	.027	2.95	3.33	.17		25	
	6		.528	.430	.024	4.01	3.37	.17	0.857	0.663	.036	5.35	3.37	.168	.528	.474	.026	2.83	3.20	.17		24	
	7		.507	.413	.023	3.85	3.24	.16	0.823	0.637	.034	5.14	3.23	.161	.507	.455	.025	2.72	3.08	.16		23	
	8		.486	.396	.022	3.69	3.11	.16	0.790	0.611	.033	4.92	3.10	.155	.487	.436	.024	2.61	2.95	.15		22	
	9		.465	.379	.021	3.53	2.97	.15	0.756	0.585	.031	4.71	2.97	.148	.466	.417	.023	2.50	2.82	.15		21	
	10		.444	.362	.020	3.37	2.84	.14	0.721	0.558	.030	4.50	2.83	.141	.444	.398	.022	2.38	2.69	.14		20	
	11		.422	.344	.019	3.21	2.70	.13	0.686	0.532	.029	4.28	2.69	.135	.423	.379	.021	2.27	2.56	.13		19	
	12		.400	.327	.018	3.05	2.56	.13	0.652	0.504	.027	4.06	2.56	.128	.401	.360	.020	2.15	2.43	.13		18	
	13		.379	.309	.017	2.88	2.43	.12	0.617	0.477	.026	3.84	2.42	.121	.379	.341	.019	2.04	2.30	.12		17	
	14		.357	.292	.016	2.72	2.29	.11	0.582	0.450	.024	3.62	2.28	.114	.358	.321	.018	1.92	2.17	.11		16	
	15		.335	.274	.015	2.55	2.15	.11	0.546	0.422	.023	3.40	2.14	.107	.336	.301	.017	1.80	2.04	.11		15	
	16		.313	.256	.014	2.39	2.01	.10	0.510	0.394	.021	3.18	2.00	.100	.314	.282	.016	1.69	1.91	.10		14	
	17		.292	.237	.013	2.22	1.86	.09	0.474	0.367	.020	2.95	1.86	.093	.292	.262	.015	1.57	1.77	.09		13	
	18		.269	.220	.012	2.05	1.72	.09	0.438	0.339	.018	2.73	1.72	.086	.270	.242	.013	1.45	1.64	.08		12	
	19		.247	.202	.011	1.88	1.58	.08	0.402	0.311	.017	2.50	1.58	.079	.247	.222	.012	1.33	1.50	.08		11	
	20		.225	.184	.010	1.71	1.44	.07	0.366	0.283	.015	2.28	1.44	.072	.225	.202	.011	1.21	1.37	.07		10	
	21		.203	.165	.009	1.54	1.30	.06	0.330	0.255	.014	2.06	1.29	.065	.203	.181	.010	1.09	1.23	.06		9	
	22		.181	.147	.008	1.37	1.15	.06	0.294	0.227	.012	1.83	1.15	.058	.181	.162	.009	0.97	1.10	.06		8	
	23		.158	.129	.007	1.20	1.01	.05	0.257	0.199	.011	1.60	1.01	.050	.158	.142	.008	0.85	0.96	.05		7	
	24		.136	.111	.006	1.03	0.87	.04	0.221	0.171	.009	1.38	0.87	.043	.136	.122	.007	0.73	0.82	.04		6	
	25		.113	.093	.005	0.86	0.72	.04	0.185	0.142	.008	1.15	0.72	.036	.113	.102	.006	0.61	0.69	.04		5	
	26		.091	.074	.004	0.69	0.58	.03	0.148	0.114	.006	0.92	0.58	.029	.091	.082	.005	0.49	0.55	.03		4	
	27		.068	.056	.003	0.52	0.44	.02	0.111	0.086	.005	0.69	0.43	.022	.068	.061	.003	0.37	0.41	.02		3	
	28		.045	.037	.002	0.34	0.29	.01	0.074	0.057	.003	0.46	0.29	.014	.046	.041	.002	0.24	0.27	.02		2	
	29		.023	.018	.001	0.17	0.14	.01	0.037	0.029	.002	0.23	0.14	.007	.023	.020	.001	0.12	0.14	.01		1	
VI	0	O	.000	.000	.000	0.00	0.00	.00	0.000	0.000	.000	0.00	0.00	.000	.000	.000	.000	0.00	0.00	.00	O	0	VI
			Arg. = ☉ Long. + $4^{s}\ 12^{\circ}\ 38'$	Arg. = ☾ ☊ + $6^{s}\ 3^{\circ}\ 42'$	Arg. = 2 ☉ Long. + $6^{s}\ 3^{\circ}\ 0'$	Arg. = ☉ Long. + $2^{s}\ 17^{\circ}\ 32'$	Arg. = ☾ ☊ + $7^{s}\ 18^{\circ}\ 37'$	Arg. = 2 ☉ Long. + $7^{s}\ 12^{\circ}\ 38'$	Arg. = ☉ Long. + $4^{s}\ 12^{\circ}\ 5'$	Arg. = ☾ ☊ + $5^{s}\ 6^{\circ}\ 45'$	Arg. = 2 ☉ Long. + $5^{s}\ 10^{\circ}\ 47'$	Arg. = ☉ Long. + $6^{s}\ 19^{\circ}\ 55'$	Arg. = ☾ ☊ + $7^{s}\ 18^{\circ}\ 3'$	Arg. = 2 ☉ Long. + $7^{s}\ 11^{\circ}\ 50'$	Arg. = ☉ Long. + $4^{s}\ 2^{\circ}\ 22'$	Arg. = ☾ ☊ + $5^{s}\ 23^{\circ}\ 47'$	Arg. = 2 ☉ Long. + $5^{s}\ 24^{\circ}\ 56'$	Arg. = ☉ Long. + $4^{s}\ 3^{\circ}\ 46'$	Arg. = ☾ ☊ + $7^{s}\ 8^{\circ}\ 0'$	Arg = 2 ☉ Long. + $7^{s}\ 2^{\circ}\ 21'$			
			Aber.	☾ Nut.	☉ Nut.	Aber.	☾ Nut.	☉ Nut.	Aber.	☾ Nut.	☉ Nut.	Aber.	☾ Nut.	☉ Nut.	Aber.	☾ Nut.	☉ Nut.	Aber.	☾ Nut	☉ Nut.			
			R. A.			N. P. D.			R. A.			N. P. D.			R. A.			N. P. D.					
			ALPHARD 1830.						θ URSÆ MAJORIS 1830.						REGULUS 1830.								

The CORRECTIONS of 48 principal Stars, &c. continued.

1830. Common Argument.	α URSÆ MAJORIS 1820. Right Ascension. 10h 52m 31s.99 Ann. Var. + 3s.83			N. P. Distance. 27° 16′ 46″.60 Ann. Var. + 19″.30			β LEONIS 1820. Right Ascension. 11h 39m 52s.45 Ann. Var. + 3s.07			N. P. Distance. 74° 25′ 17″.93 Ann. Var. − 20″.04			β VIRGINIS 1820. Right Ascension. 11h 41m 19s.35 Ann. Var. + 3s.12			N. P. Distance. 87° 13′ 14″.66 Ann. Var. + 20″.00			1830. Common Argument.
+ s. − s.	Aber.	☾ Nut.	⊙ Nut.	Aber.	☾ Nut.	⊙ Nut.	Aber.	☾ Nut.	⊙ Nut.	Aber.	☾ Nut.	⊙ Nut.	Aber.	☾ Nut.	⊙ Nut.	Aber.	☾ Nut.	⊙ Nut.	+ s. − s.
III 0° IX	2s.719	1s.858	0s.093	17″.284	7″.417	0″.401	1s.287	1.128	0s.062	9″.118	7.204	0″.398	1s.241	1.104	0s.061	8″.042	7″.200	0″.398	III 0° IX
1	2.718	1.857	.093	17.27	7.41	.40	1.286	1.127	.062	9.11	7.20	.39	1.241	1.104	.061	8.04	7.20	.39	29
2	2.716	1.856	.093	17.26	7.40	.40	1.285	1.126	.062	9.10	7.20	.39	1.240	1.103	.061	8.03	7.19	.39	28
3	2.713	1.854	.093	17.25	7.40	.40	1.284	1.125	.062	9.10	7.19	.39	1.240	1.102	.061	8.02	7.19	.39	27
4	2.709	1.852	.093	17.23	7.39	.40	1.283	1.124	.062	9.09	7.19	.39	1.239	1.101	.061	8.01	7.18	.39	26
5	2.705	1.850	.093	17.21	7.38	.40	1.281	1.123	.062	9.08	7.18	.39	1.238	1.100	.061	8.00	7.17	.39	25
6	2.701	1.847	.093	17.18	7.37	.40	1.279	1.121	.062	9.06	7.17	.39	1.236	1.098	.061	7.99	7.16	.39	24
7	2.696	1.844	.092	17.15	7.36	.40	1.277	1.119	.062	9.04	7.16	.39	1.234	1.096	.061	7.98	7.15	.39	23
8	2.691	1.840	.092	17.11	7.35	.40	1.274	1.117	.061	9.02	7.14	.30	1.231	1.094	.060	7.96	7.13	.39	22
9	2.685	1.835	.092	17.07	7.33	.40	1.271	1.114	.061	9.00	7.12	.39	1.228	1.092	.060	7.94	7.11	.39	21
10	2.678	1.830	.092	17.03	7.31	.40	1.268	1.111	.061	8.98	7.10	.39	1.224	1.089	.060	7.92	7.09	.39	20
11	2.670	1.824	.091	16.98	7.29	.39	1.264	1.107	.061	8.95	7.08	.39	1.220	1.086	.060	7.89	7.07	.39	19
12	2.660	1.817	.091	16.92	7.26	.39	1.259	1.103	.061	8.92	7.05	.39	1.216	1.082	.060	7.86	7.04	.39	18
13	2.650	1.810	.091	16.85	7.23	.39	1.254	1.099	.060	8.88	7.02	.39	1.211	1.077	.059	7.83	7.01	.39	17
14	2.638	1.802	.090	16.77	7.20	.39	1.248	1.094	.060	8.84	6.99	.39	1.206	1.072	.059	7.80	6.98	.39	16
15	2.625	1.794	.090	16.69	7.17	.39	1.242	1.089	.060	8.80	6.96	.38	1.200	1.067	.059	7.76	6.95	.38	15
16	2.611	1.785	.089	16.60	7.13	.39	1.236	1.084	.060	8.76	6.93	.38	1.194	1.062	.059	7.72	6.92	.38	14
17	2.597	1.776	.089	16.51	7.09	.38	1.230	1.078	.059	8.72	6.89	.38	1.188	1.057	.058	7.68	6.88	.38	13
18	2.583	1.766	.088	16.42	7.05	.38	1.224	1.072	.059	8.67	6.85	.38	1.181	1.051	.058	7.64	6.84	.38	12
19	2.568	1.756	.088	16.33	7.01	.38	1.217	1.066	.059	8.62	6.81	.38	1.174	1.045	.058	7.60	6.80	.38	11
20	2.553	1.745	.087	16.23	6.97	.38	1.210	1.060	.058	8.56	6.77	.37	1.167	1.038	.057	7.56	6.76	.37	10
21	2.537	1.734	.087	16.13	6.92	.37	1.201	1.053	.058	8.50	6.72	.37	1.154	1.031	.057	7.51	6.72	.37	9
22	2.520	1.722	.086	16.02	6.87	.37	1.193	1.046	.058	8.44	6.67	.37	1.151	1.024	.057	7.45	6.67	.37	8
23	2.501	1.710	.086	15.91	6.83	.37	1.184	1.039	.057	8.38	6.63	.37	1.143	1.018	.056	7.40	6.62	.37	7
24	2.481	1.697	.085	15.79	6.77	.37	1.175	1.031	.057	8.32	6.58	.36	1.134	1.010	.056	7.35	6.57	.36	6
25	2.461	1.683	.084	15.66	6.72	.36	1.166	1.022	.056	8.25	6.53	.36	1.125	1.001	.055	7.28	6.52	.36	5
26	2.440	1.669	.084	15.53	6.67	.36	1.156	1.013	.056	8.19	6.48	.36	1.116	0.993	.055	7.22	6.47	.36	4
27	2.419	1.654	.083	15.40	6.61	.36	1.146	1.004	.055	8.12	6.42	.35	1.106	0.984	.054	7.16	6.41	.35	3
28	2.397	1.639	.082	15.26	6.55	.35	1.135	0.995	.055	8.04	6.36	.35	1.096	0.975	.054	7.10	6.35	.35	2
29	2.375	1.625	.081	15.12	6.48	.35	1.125	0.986	.054	7.97	6.30	.35	1.086	0.966	.053	7.03	6.29	.35	1
IV 0 X	2.352	1.610	.080	14.97	6.42	.35	1.114	0.976	.054	7.89	6.24	.34	1.075	0.956	.053	6.96	6.23	.34	II 0 VIII
1	2.328	1.593	.080	14.81	6.36	.34	1.102	0.966	.053	7.82	6.18	.34	1.064	0.947	.052	6.89	6.17	.34	29
2	2.303	1.577	.079	14.65	6.29	.34	1.091	0.956	.053	7.73	6.12	.34	1.053	0.937	.052	6.82	6.10	.34	28
3	2.277	1.559	.078	14.50	6.22	.34	1.079	0.946	.052	7.65	6.05	.33	1.042	0.927	.051	6.74	6.04	.33	27
4	2.251	1.540	.077	14.32	6.15	.33	1.067	0.935	.051	7.56	5.97	.33	1.029	0.916	.051	6.66	5.96	.33	26
5	2.224	1.522	.076	14.15	6.08	.33	1.053	0.923	.051	7.47	5.90	.33	1.018	0.905	.050	6.58	5.89	.33	25
6	2.197	1.502	.075	13.97	6.00	.32	1.040	0.912	.050	7.37	5.83	.32	1.004	0.893	.049	6.50	5.82	.32	24
7	2.169	1.482	.074	13.79	5.92	.32	1.027	0.900	.050	7.28	5.75	.32	0.991	0.882	.049	6.42	5.74	.32	23
8	2.140	1.462	.073	13.61	5.85	.32	1.013	0.888	.049	7.18	5.67	.31	0.978	0.870	.048	6.33	5.67	.31	22
9	2.110	1.442	.072	13.42	5.77	.31	0.999	0.876	.048	7.08	5.60	.31	0.964	0.858	.047	6.24	5.58	.31	21
10	2.080	1.421	.071	13.22	5.68	.31	0.984	0.863	.047	6.98	5.52	.30	0.950	0.845	.047	6.15	5.51	.30	20
11	2.049	1.401	.070	13.02	5.60	.30	0.970	0.850	.047	6.87	5.43	.30	0.936	0.832	.046	6.06	5.43	.30	19
12	2.018	1.380	.069	12.82	5.51	.30	0.955	0.837	.046	6.77	5.36	.30	0.922	0.820	.045	5.97	5.35	.30	18
13	1.986	1.358	.068	12.62	5.43	.29	0.940	0.824	.045	6.67	5.27	.29	0.907	0.807	.045	5.88	5.27	.29	17
14	1.954	1.336	.067	12.42	5.34	.29	0.925	0.811	.045	6.56	5.18	.29	0.892	0.794	.044	5.78	5.18	.29	16
IV 15 X	1.920	1.313	.066	12.22	5.25	.28	0.909	0.797	.044	6.44	5.09	.28	0.877	0.781	.043	5.68	5.08	.28	I 15 VII
	Arg. = ⊙ Long. + 3s 18° 7′	Arg. = ☾ ☊ + 4s 18° 58′	Arg. = 2 ⊙ Long. + 4s 21° 45′	Arg. = ⊙ Long. + 6s 3° 27′	Arg. = ☾ ☊ + 6s 21° 58′	Arg. = 2 ⊙ Long. + 6s 18° 7′	Arg. = ⊙ Long. + 3s 5° 21′	Arg. = ☾ ☊ + 5s 20° 55′	Arg. = 2 ⊙ Long. + 5s 22° 38′	Arg. = ⊙ Long. + 4s 6° 19′	Arg. = ☾ ☊ + 6s 6° 34′	Arg. = 2 ⊙ Long. + 6s 5° 5′	Arg. = ⊙ Long. + 3s 4° 56′	Arg. = ☾ ☊ + 5s 28° 25′	Arg. = 2 ⊙ Long. + 5s 28° 43′	Arg. = ⊙ Long. + 3s 6° 51′	Arg. = ☾ ☊ + 6s 6° 5′	Arg. = 2 ⊙ Long. + 6s 5° 58′	
	Aber.	☾ Nut.	⊙ Nut.	Aber.	☾ Nut.	⊙ Nut.	Aber.	☾ Nut.	⊙ Nut.	Aber.	☾ Nut.	⊙ Nut.	Aber.	☾ Nut.	⊙ Nut.	Aber.	☾ Nut.	⊙ Nut.	
	R. A.			N. P. D.			R. A.			N. P. D.			R. A.			N. P. D.			
	DUBHE 1830.						DENEBOLA 1830.						ZAVIJAVA 1830.						

The CORRECTIONS of 48 principal Stars, &c. continued.

1830. Common Argument.	α URSÆ MAJORIS 1820.						β LEONIS 1820.						β VIRGINIS 1820.						1830. Common Argument.
	Right Ascension. h. 10 m. 52 s. 31 .99			N. P. Distance. 27° 16′ 46″. 60			Right Ascension. h. 11 m. 39 s. 52 . 45			N. P. Distance. 74° 25′ 17″. 93			Right Ascension. h. 11 m. 41 s. 19 . 35			N. P. Distance. 87° 13′ 14″. 66			
	Ann. Var. + 3s . 83			Ann. Var. + 19″. 30			Ann. Var. + 3s . 07			Ann. Var. + 20″. 04			Ann. Var. + 3s . 12			Ann. Var. + 20″.00			
+ s. — s.	Aber.	☾ Nut.	⊙ Nut.	Aber.	☾ Nut.	⊙ Nut.	Aber.	☾ Nut.	⊙ Nut.	Aber.	☾ Nut.	⊙ Nut.	Aber.	☾ Nut.	⊙ Nut.	Aber.	☾ Nut.	⊙ Nut.	+ s. — s.
IV 15° X	1s.920	1s.313	0s.066	12″. 22	5″. 25	0″. 28	0s . 909	0s . 797	0s . 044	6″. 44	5″. 09	0″. 28	0s . 877	0s . 781	0s . 043	5″. 68	5″. 08	0″. 28	I 15° VII
16	1 .886	1 .290	.065	12 . 00	5 . 16	. 28	. 893	. 783	. 043	6 . 23	5 . 01	. 28	. 862	. 767	. 042	5 . 58	5 . 00	. 28	14
17	1 .851	1 .266	.063	11 . 78	5 . 06	. 27	. 876	. 768	. 042	6 . 21	4 . 92	. 27	. 846	. 753	. 042	5 . 48	4 . 91	. 27	13
18	1 .816	1 .242	.062	11 . 55	4 . 97	. 27	. 860	. 754	. 041	6 . 09	4 . 82	. 27	. 830	. 739	. 041	5 . 38	4 . 82	. 27	12
19	1 .780	1 .218	.061	11 . 32	4 . 87	. 26	. 842	. 738	. 041	5 . 98	4 . 72	. 26	. 813	. 724	. 040	5 . 27	4 . 72	. 26	11
20	1 .745	1 .194	.060	11 . 10	4 . 77	. 26	. 826	. 724	. 040	5 . 86	4 . 63	. 26	. 797	. 710	. 039	5 . 16	4 . 62	. 26	10
21	1 .710	1 .170	.058	10 . 87	4 . 67	. 25	. 809	. 709	. 039	5 . 74	4 . 53	. 25	. 781	. 695	. 038	5 . 06	4 . 53	. 25	9
22	1 .672	1 .144	.057	10 . 63	4 . 57	. 25	. 792	. 694	. 038	5 . 61	4 . 43	. 24	. 764	. 680	. 038	4 . 95	4 . 43	. 24	8
23	1 .633	1 .118	.056	10 . 40	4 . 47	. 24	. 764	. 678	. 037	5 . 48	4 . 33	. 24	. 747	. 665	. 037	4 . 84	4 . 33	. 24	7
24	1 .596	1 .092	.055	10 . 16	4 . 36	. 24	. 756	. 662	. 036	5 . 36	4 . 23	. 23	. 730	. 649	. 036	4 . 73	4 . 23	. 23	6
25	1 .559	1 .066	.053	9 . 91	4 . 26	. 23	. 738	. 646	. 036	5 . 23	4 . 13	. 23	. 712	. 633	. 035	4 . 61	4 . 12	. 23	5
26	1 .520	1 .040	.052	9 . 66	4 . 15	. 22	. 719	. 631	. 035	5 . 10	4 . 03	. 22	. 695	. 617	. 034	4 . 50	4 . 02	. 22	4
27	1 .479	1 .012	.051	9 . 40	4 . 04	. 22	. 700	. 613	. 034	4 . 97	3 . 92	. 22	. 676	. 601	. 033	4 . 38	3 . 91	. 22	3
28	1 .440	0 .985	.049	9 . 15	3 . 93	. 21	. 682	. 597	. 033	4 . 83	3 . 82	. 21	. 658	. 585	. 032	4 . 26	3 . 81	. 21	2
29	1 .400	0 .957	.048	8 . 90	3 . 82	. 21	. 663	. 581	. 032	4 . 69	3 . 71	. 20	. 640	. 569	. 031	4 . 14	3 . 70	. 20	1
V 0. XI	1 .359	0 .930	.047	8 . 64	3 . 71	. 20	. 644	. 564	. 031	4 . 56	3 . 60	. 20	. 622	. 553	. 030	4 . 02	3 . 60	. 20	I 0 VII
1	1 .316	0 .901	.045	8 . 37	3 . 60	. 19	. 624	. 547	. 030	4 . 42	3 . 49	. 19	. 602	. 536	. 030	3 . 89	3 . 49	. 19	29
2	1 .274	0 .871	.044	8 . 10	3 . 48	. 19	. 604	. 529	. 029	4 . 28	3 . 38	. 19	. 583	. 518	. 029	3 . 77	3 . 37	. 19	28
3	1 .232	0 .842	.042	7 . 84	3 . 37	. 18	. 584	. 512	. 028	4 . 14	3 . 27	. 18	. 563	. 502	. 028	3 . 65	3 . 26	. 18	27
4	1 .190	0 .813	.041	7 . 57	3 . 25	. 18	. 563	. 494	. 027	3 . 99	3 . 16	. 17	. 544	. 484	. 027	3 . 52	3 . 15	. 17	26
5	1 .148	0 .785	.039	7 . 30	3 . 14	. 17	. 543	. 477	. 026	3 . 85	3 . 04	. 17	. 525	. 467	. 026	3 . 40	3 . 04	. 17	25
6	1 .104	0 .755	.038	7 . 02	3 . 02	. 16	. 523	. 458	. 025	3 . 70	2 . 93	. 16	. 505	. 449	. 025	3 . 27	2 . 92	. 16	24
7	1 .061	0 .725	.036	6 . 74	2 . 90	. 16	. 503	. 440	. 024	3 . 56	2 . 81	. 16	. 485	. 432	. 024	3 . 14	2 . 81	. 16	23
8	1 .017	0 .695	.035	6 . 47	2 . 78	. 15	. 482	. 422	. 023	3 . 41	2 . 70	. 15	. 465	. 413	. 023	3 . 01	2 . 69	. 15	22
9	0 .972	0 .665	.033	6 . 19	2 . 66	. 14	. 461	. 403	. 022	3 . 26	2 . 58	. 14	. 445	. 395	. 022	2 . 86	2 . 58	. 14	21
10	0 .928	0 .635	.032	5 . 91	2 . 54	. 14	. 440	. 385	. 021	3 . 12	2 . 46	. 14	. 424	. 377	. 021	2 . 75	2 . 46	. 14	20
11	0 .883	0 .605	.030	5 . 62	2 . 41	. 13	. 418	. 367	. 020	2 . 96	2 . 34	. 13	. 404	. 359	. 020	2 . 62	2 . 34	. 13	19
12	0 .838	0 .574	.029	5 . 34	2 . 29	. 12	. 397	. 348	. 019	2 . 81	2 . 22	. 12	. 383	. 341	. 019	2 . 48	2 . 22	. 12	18
13	0 .794	0 .543	.027	5 . 05	2 . 17	. 12	. 376	. 329	. 018	2 . 66	2 . 10	. 12	. 363	. 323	. 018	2 . 35	2 . 10	. 12	17
14	0 .748	0 .512	.026	4 . 76	2 . 04	. 11	. 354	. 311	. 017	2 . 51	1 . 98	. 11	. 342	. 304	. 017	2 . 21	1 . 98	. 11	16
15	0 .702	0 .480	.024	4 . 47	1 . 92	. 10	. 333	. 291	. 016	2 . 35	1 . 86	. 10	. 321	. 285	. 016	2 . 08	1 . 86	. 10	15
16	0 .656	0 .449	.022	4 . 17	1 . 79	. 10	. 311	. 272	. 015	2 . 20	1 . 74	. 10	. 300	. 267	. 015	1 . 94	1 . 74	. 10	14
17	0 .610	0 .417	.021	3 . 88	1 . 67	. 09	. 289	. 253	. 014	2 . 04	1 . 62	. 09	. 279	. 248	. 014	1 . 81	1 . 62	. 09	13
18	0 .564	0 .386	.019	3 . 59	1 . 54	. 08	. 267	. 234	. 013	1 . 89	1 . 49	. 08	. 258	. 229	. 013	1 . 67	1 . 49	. 08	12
19	0 .518	0 .354	.018	3 . 29	1 . 41	. 08	. 245	. 215	. 012	1 . 74	1 . 37	. 08	. 236	. 210	. 012	1 . 53	1 . 37	. 08	11
20	0 .471	0 .322	.016	3 . 00	1 . 29	. 07	. 223	. 195	. 011	1 . 58	1 . 25	. 07	. 215	. 192	. 011	1 . 39	1 . 25	. 07	10
21	0 .425	0 .290	.015	2 . 70	1 . 16	. 06	. 201	. 176	. 010	1 . 42	1 . 13	. 06	. 194	. 173	. 010	1 . 26	1 . 12	. 06	9
22	0 .378	0 .258	.013	2 . 40	1 . 03	. 06	. 179	. 157	. 009	1 . 27	1 . 00	. 06	. 173	. 154	. 008	1 . 12	1 . 00	. 06	8
23	0 .331	0 .226	.011	2 . 11	0 . 90	. 05	. 157	. 137	. 008	1 . 11	0 . 88	. 05	. 151	. 135	. 007	0 . 98	0 . 87	. 05	7
24	0 .284	0 .194	.010	1 . 81	0 . 78	. 04	. 135	. 118	. 006	0 . 95	0 . 75	. 04	. 130	. 116	. 006	0 . 84	0 . 75	. 04	6
25	0 .237	0 .162	.008	1 . 51	0 . 65	. 04	. 112	. 095	. 005	0 . 80	0 . 63	. 03	. 108	. 096	. 005	0 . 70	0 . 63	. 03	5
26	0 .190	0 .130	.006	1 . 21	0 . 52	. 03	. 090	. 079	. 004	0 . 64	0 . 50	. 03	. 086	. 077	. 004	0 . 56	0 . 50	. 03	4
27	0 .142	0 .097	.005	0 . 91	0 . 39	. 02	. 067	. 059	. 003	0 . 48	0 . 38	. 02	. 065	. 058	. 003	0 . 42	0 . 38	. 02	3
28	0 .094	0 .065	.003	0 . 60	0 . 26	. 01	. 045	. 039	. 002	0 . 32	0 . 25	. 01	. 043	. 039	. 002	0 . 28	0 . 25	. 01	2
29	0 .047	0 .033	.002	0 . 30	0 . 13	. 01	. 022	. 020	. 001	0 . 16	0 . 13	. 01	. 022	. 019	. 001	0 . 14	0 . 13	. 01	1
VI 0 O	0 .000	0 .000	.000	0 . 00	0 . 00	. 00	. 000	. 000	. 000	0 . 00	0 . 00	. 00	. 000	. 000	. 000	0 . 00	0 . 00	. 00	O 0 VI
	Arg.= ⊙ Long.+ 3s 18° 7′	Arg. = ☾ ☊ + 4s 18° 58′	Arg.=2⊙Long.+ 4s 24° 45′	Arg.=⊙Long. + 6s 3° 27′	Arg. = ☾ ☊ + 6s 21° 58′	Arg.=2⊙Long.+ 6s 18° 7′	Arg.= ⊙ Long.+ 3s 5° 21′	Arg. = ☾ ☊ + 5s 20° 55′	Arg.=2⊙Long.+ 5s 22° 38′	Arg.=⊙Long.+ 4s 6° 19′	Arg. = ☾ ☊ + 6s 6° 34′	Arg.=2⊙Long.+ 6s 5° 5′	Arg.= ⊙ Long.+ 3s 4° 56′	Arg. = ☾ ☊ + 5s 28° 25′	Arg.=2⊙Long.+ 5s 28° 43′	Arg.=⊙Long. + 3s 6° 51′	Arg. = ☾ ☊ + 6s 6° 5′	Arg.=2⊙Long.+ 6s 5° 58′	
	Aber.	☾ Nut.	⊙ Nut.	Aber.	☾ Nut.	⊙ Nut.	Aber.	☾ Nut.	⊙ Nut.	Aber.	☾ Nut.	⊙ Nut.	Aber.	☾ Nut.	⊙ Nut.	Aber.	☾ Nut.	⊙ Nut.	
	R. A.			N. P. D.			R. A.			N. P. D.			R. A.			N. P. D.			
	DUBHE 1830.						DENEBOLA 1830.						ZAVIJAVA 1830.						

The CORRECTIONS of 48 principal Stars, &c. continued.

1830. Common Argument.	γ URSÆ MAJORIS 1820. Right Ascension. h. m. s. 11 44 19 .13 Ann. Var. + 3s. 20			N. P. Distance. 35° 18′ 15″. 25 Ann. Var. + 20″. 00			α VIRGINIS 1820. Right Ascension. h. m. s. 13 15 43 .46 Ann. Var. + 3s. 14			N. P. Distance. 100° 13′ 3″. 28 Ann. Var. + 18″. 95			ζ URSÆ MAJORIS 1820. Right Ascension. h. m. s. 13 16 39 .51 Ann. Var. + 2s. 42			N. P. Distance. 34° 7′ 53″. 72 Ann. Var. + 18″. 98			1830. Common Argument.
+ − s. s.	Aber.	☾ Nut.	⊙ Nut.	Aber.	☾ Nut.	⊙ Nut.	Aber.	☾ Nut.	⊙ Nut.	Aber.	☾ Nut.	⊙ Nut.	Aber.	☾ Nut.	⊙ Nut.	Aber.	☾ Nut.	⊙ Nut.	+ − s. s.
III 0° IX	2s.141	1s.463	0s.076	16″.933	7″.194	0″.398	1s.272	1s.168	0s.063	7″.658	7″.477	0″.402	2s.227	1s.248	0s.070	18″.503	7″.490	0″.402	III 0° IX
1	2.140	1.462	.076	16.92	7.19	.39	1.272	1.167	.063	7.65	7.47	.40	2.226	1.247	.070	18.49	7.49	.40	29
2	2.139	1.461	.076	16.91	7.18	.39	1.271	1.166	.063	7.65	7.47	.40	2.225	1.246	.070	18.48	7.48	.40	28
3	2.137	1.460	.076	16.90	7.18	.39	1.270	1.165	.063	7.64	7.46	.40	2.224	1.245	.070	18.47	7.47	.40	27
4	2.135	1.459	.076	16.89	7.17	.39	1.269	1.164	.063	7.64	7.46	.40	2.222	1.243	.070	18.45	7.46	.40	26
5	2.132	1.458	.076	16.87	7.16	.39	1.268	1.163	.063	7.63	7.45	.40	2.220	1.241	.070	18.43	7.45	.40	25
6	2.129	1.456	.076	16.85	7.15	.39	1.266	1.161	.063	7.62	7.44	.40	2.217	1.239	.069	18.40	7.44	.40	24
7	2.125	1.453	.075	16.82	7.14	.39	1.264	1.159	.062	7.61	7.43	.40	2.213	1.237	.069	18.37	7.43	.40	23
8	2.121	1.450	.075	16.78	7.12	.39	1.261	1.156	.062	7.59	7.41	.40	2.208	1.235	.069	18.33	7.42	.40	22
9	2.116	1.447	.075	16.74	7.10	.39	1.257	1.153	.062	7.57	7.39	.40	2.202	1.233	.069	18.29	7.40	.40	21
10	2.110	1.443	.075	16.69	7.08	.39	1.253	1.150	.062	7.55	7.37	.40	2.195	1.230	.069	18.24	7.38	.40	20
11	2.103	1.438	.075	16.63	7.06	.39	1.249	1.146	.062	7.52	7.34	.39	2.187	1.226	.069	18.19	7.35	.39	19
12	2.095	1.433	.074	16.57	7.03	.39	1.244	1.142	.062	7.49	7.31	.39	2.179	1.221	.068	18.13	7.32	.39	18
13	2.086	1.427	.074	16.50	7.00	.39	1.239	1.137	.061	7.46	7.28	.39	2.170	1.216	.068	18.06	7.29	.39	17
14	2.077	1.420	.074	16.43	6.97	.39	1.234	1.132	.061	7.43	7.25	.39	2.161	1.210	.068	17.98	7.26	.39	16
15	2.067	1.413	.073	16.35	6.94	.38	1.229	1.127	.061	7.40	7.22	.39	2.151	1.204	.068	17.89	7.23	.39	15
16	2.057	1.406	.073	16.27	6.91	.38	1.223	1.122	.061	7.36	7.19	.39	2.141	1.199	.067	17.80	7.19	.39	14
17	2.047	1.399	.073	16.19	6.88	.38	1.217	1.116	.060	7.32	7.15	.38	2.130	1.193	.067	17.70	7.15	.38	13
18	2.036	1.391	.072	16.11	6.84	.38	1.210	1.110	.060	7.28	7.11	.38	2.118	1.187	.067	17.60	7.11	.38	12
19	2.025	1.383	.072	16.01	6.80	.38	1.203	1.104	.060	7.24	7.07	.38	2.105	1.180	.066	17.50	7.07	.38	11
20	2.013	1.375	.072	15.91	6.76	.37	1.196	1.097	.059	7.20	7.03	.38	2.092	1.172	.066	17.40	7.03	.38	10
21	2.000	1.366	.071	15.81	6.71	.37	1.188	1.090	.059	7.15	6.98	.37	2.079	1.164	.065	17.29	6.98	.37	9
22	1.986	1.357	.071	15.70	6.67	.37	1.180	1.082	.058	7.10	6.93	.37	2.065	1.157	.065	17.17	6.93	.37	8
23	1.972	1.347	.070	15.59	6.62	.37	1.171	1.074	.058	7.05	6.88	.37	2.051	1.148	.064	17.05	6.88	.37	7
24	1.957	1.337	.069	15.47	6.57	.36	1.162	1.067	.058	7.00	6.84	.37	2.036	1.139	.064	16.93	6.83	.37	6
25	1.941	1.326	.069	15.34	6.52	.36	1.152	1.058	.057	6.94	6.78	.36	2.020	1.130	.063	16.79	6.78	.36	5
26	1.925	1.315	.068	15.22	6.47	.36	1.143	1.049	.057	6.88	6.72	.36	2.002	1.121	.063	16.65	6.73	.36	4
27	1.908	1.303	.068	15.10	6.41	.35	1.133	1.040	.056	6.82	6.66	.36	1.985	1.111	.062	16.49	6.67	.36	3
28	1.890	1.292	.067	14.95	6.35	.35	1.122	1.031	.056	6.76	6.60	.35	1.967	1.100	.062	16.34	6.61	.35	2
29	1.872	1.279	.066	14.81	6.29	.35	1.112	1.021	.055	6.70	6.54	.35	1.949	1.091	.061	16.20	6.55	.35	1
IV 0 X	1.853	1.267	.066	14.67	6.23	.34	1.101	1.011	.055	6.63	6.48	.35	1.930	1.080	.061	16.04	6.48	.35	II 0 VIII
1	1.834	1.253	.065	14.52	6.17	.34	1.090	1.010	.054	6.56	6.41	.34	1.910	1.069	.060	15.88	6.42	.34	29
2	1.815	1.240	.064	14.36	6.10	.34	1.079	0.990	.053	6.49	6.34	.34	1.890	1.058	.059	15.72	6.35	.34	28
3	1.796	1.227	.064	14.20	6.03	.33	1.067	0.979	.053	6.42	6.27	.34	1.869	1.046	.059	15.54	6.28	.34	27
4	1.775	1.213	.063	14.03	5.96	.33	1.054	0.967	.052	6.35	6.20	.33	1.847	1.033	.058	15.36	6.21	.33	26
5	1.754	1.199	.062	13.87	5.89	.33	1.042	0.956	.052	6.27	6.13	.33	1.825	1.021	.057	15.18	6.13	.33	25
6	1.732	1.184	.061	13.69	5.82	.32	1.029	0.944	.051	6.19	6.05	.32	1.802	1.008	.057	14.99	6.06	.32	24
7	1.710	1.168	.061	13.51	5.74	.32	1.016	0.932	.050	6.12	6.97	.32	1.779	0.995	.056	14.79	6.98	.32	23
8	1.687	1.152	.060	13.33	5.67	.31	1.002	0.919	.050	6.03	5.89	.32	1.754	0.982	.055	14.60	5.90	.32	22
9	1.663	1.136	.059	13.15	5.58	.31	0.988	0.907	.049	5.95	5.82	.31	1.730	0.968	.054	14.40	5.82	.31	21
10	1.639	1.120	.058	12.96	5.51	.30	0.974	0.894	.048	5.86	5.73	.31	1.706	0.954	.054	14.19	5.73	.31	20
11	1.615	1.104	.057	12.76	5.43	.30	0.959	0.880	.048	5.78	5.64	.30	1.680	0.940	.053	13.88	5.65	.30	19
12	1.591	1.088	.057	12.57	5.35	.30	0.945	0.867	.047	5.69	5.56	.30	1.654	0.926	.052	13.76	5.56	.30	18
13	1.567	1.070	.056	12.38	5.26	.29	0.930	0.854	.046	5.60	5.47	.29	1.629	0.912	.051	13.54	5.46	.29	17
14	1.541	1.052	.055	12.18	5.17	.29	0.915	0.839	.045	5.51	5.38	.29	1.603	0.897	.050	13.32	5.39	.29	16
IV 15 X	1.514	1.034	.054	11.97	5.08	.28	0.899	0.825	.045	5.42	5.29	.28	1.576	0.881	.050	13.08	5.29	.28	I 15 VII
	Arg. = ⊙ Long. + 3s 4° 8′	Arg. = ☾ ☊ + 4s 21° 42′	Arg. = 2 ⊙ Long. + 4s 27° 21′	Arg. = ⊙ Long. + 5s 17° 30′	Arg. = ☾ ☊ + 6s 5° 5′	Arg. = 2 ⊙ Long. + 6s 4° 7′	Arg. = ⊙ Long. + 2s 9° 22′	Arg. = ☾ ☊ + 6s 5° 24′	Arg. = 2 ⊙ Long. + 6s 4° 30′	Arg. = ⊙ Long. + 2s 3° 35′	Arg. = ☾ ☊ + 5s 5° 16′	Arg. = 2 ⊙ Long. + 5s 9° 22′	Arg. = ⊙ Long. + 2s 9° 9′	Arg. = ☾ ☊ + 4s 14° 14′	Arg. = 2 ⊙ Long. + 4s 13° 40′	Arg. = ⊙ Long. + 4s 28° 44′	Arg. = ☾ ☊ + 5s 4° 41′	Arg. = 2 ⊙ Long. + 5s 9° 9′	
	Aber.	☾ Nut.	⊙ Nut.	Aber.	☾ Nut.	⊙ Nut.	Aber.	☾ Nut.	⊙ Nut.	Aber.	☾ Nut.	⊙ Nut.	Aber.	☾ Nut.	⊙ Nut.	Aber.	☾ Nut.	⊙ Nut.	
	R. A.			N. P. D.			R. A.			N. P. D.			R. A.			N. P. D.			
	PHECDA 1830.						SPICA VIRGINIS 1830.						MIZAR 1830.						

The CORRECTIONS of 48 principal Stars, &c. continued.

1830. Common Argument.	γ URSÆ MAJORIS 1820. Right Ascension. h. m. s. 11 44 19 . 13 Ann. Var. +3ˢ. 20			γ URSÆ MAJORIS 1820. N. P. Distance. 35° 18′ 15.″25 Ann. Var. +20″. 00			α VIRGINIS 1820. Right Ascension. h. m. s. 13 15 43 .46 Ann. Var. +3ˢ. 14			α VIRGINIS 1820. N. P. Distance. 100° 13′ 3″. 28 Ann. Var. +18″. 95			ζ URSÆ MAJORIS 1820. Right Ascension. h. m. s. 13 16 39 . 51 Ann. Var. +2ˢ. 42			ζ URSÆ MAJORIS 1820. N. P. Distance. 34° 7′ 53″. 72 Ann. Var. + 18″. 98			1830. Common Argument.
+ s. — s.	Aber.	☾ Nut.	☉Nut.	Aber.	☾ Nut.	☉Nut.	Aber.	☾ Nut.	☉Nut.	Aber.	☾ Nut.	☉Nut.	Aber.	☾ Nut.	☉Nut.	Aber.	☾ Nut.	☉Nut.	+ s. — s.
IV 15° X	1ˢ. 514	1ˢ. 034	0ˢ. 054	11″.97	5″. 08	3″. 28	0ˢ. 899	0ˢ. 825	0ˢ. 045	5″. 42	5″. 29	0″. 28	1ˢ. 576	0ˢ. 881	0ˢ. 050	13″. 08	5″. 29	0″. 28	I 15° VII
16	1 . 487	1 . 016	. 053	11 .75	4 . 99	. 28	. 883	. 810	. 044	5 . 32	5 . 19	. 28	1 . 539	. 865	. 049	12 . 65	5 . 20	. 28	14
17	1 . 460	0 . 997	. 052	11 .54	4 . 91	. 27	. 867	. 796	. 043	5 . 22	5 . 10	. 27	1 . 521	. 850	. 048	12 . 62	5 . 11	. 27	13
18	1 . 431	0 . 978	. 051	11 .32	4 . 81	. 27	. 851	. 781	. 042	5 . 12	5 . 01	. 27	1 . 491	. 833	. 047	12 . 38	5 . 01	. 27	12
19	1 . 402	0 . 958	. 050	11 .10	4 . 72	. 26	. 833	. 765	. 041	5 . 02	4 . 91	. 26	1 . 461	. 816	. 046	12 . 14	4 . 91	. 26	11
20	1 . 375	0 . 939	. 049	10 .88	4 . 62	. 26	. 816	. 750	. 040	4 . 92	4 . 81	. 26	1 . 432	. 801	. 045	11 . 90	4 . 81	. 26	10
21	1 . 347	0 . 920	. 048	10 .65	4 . 53	. 25	. 800	. 735	. 040	4 . 82	4 . 71	. 25	1 . 402	. 785	. 044	11 . 65	4 . 71	. 25	9
22	1 . 318	0 . 900	. 047	10 .42	4 . 43	. 24	. 783	. 719	. 039	4 . 72	4 . 61	. 25	1 . 372	. 767	. 043	11 . 39	4 . 61	. 25	8
23	1 . 289	0 . 880	. 046	10 .19	4 . 33	. 24	. 765	. 702	. 038	4 . 61	4 . 50	. 24	1 . 341	. 750	. 042	11 . 13	4 . 51	. 24	7
24	1 . 259	0 . 860	. 045	9 .95	4 . 22	. 23	. 747	. 686	. 037	4 . 50	4 . 39	. 24	1 . 310	. 733	. 041	10 . 88	4 . 40	. 24	6
25	1 . 229	0 . 838	. 044	9 .70	4 . 12	. 23	. 729	. 669	. 036	4 . 39	4 . 28	. 23	1 . 279	. 716	. 040	10 . 62	4 . 29	. 23	5
26	1 . 198	0 . 817	. 043	9 .46	4 . 01	. 22	. 711	. 653	. 035	4 . 28	4 . 18	. 22	1 . 247	. 697	. 039	10 . 36	4 . 18	. 22	4
27	1 . 166	0 . 796	. 041	9 .21	3 . 91	. 22	. 692	. 635	. 034	4 . 17	4 . 07	. 22	1 . 214	. 679	. 038	10 . 08	4 . 07	. 22	3
28	1 . 135	0 . 775	. 040	8 .97	3 . 81	. 21	. 674	. 608	. 033	4 . 06	3 . 96	. 21	1 . 182	. 661	. 037	9 . 80	3 . 96	. 21	2
29	1 . 103	0 . 754	. 039	8 .72	3 . 70	. 20	. 655	. 601	. 032	3 . 94	3 . 85	. 21	1 . 149	. 642	. 036	9 . 53	3 . 85	. 21	1
V 0 XI	1 . 071	0 . 732	. 038	8 .46	3 . 60	. 20	. 636	. 584	. 032	3 . 83	3 . 74	. 20	1 . 115	. 624	. 035	9 . 26	3 . 74	. 20	I 0 VII
1	1 . 038	0 . 709	. 037	8 .20	3 . 49	. 19	. 517	. 566	. 031	3 . 71	3 . 62	. 19	1 . 080	. 565	. 034	8 . 97	3 . 63	. 19	29
2	1 . 005	0 . 686	. 036	7 .94	3 . 37	. 19	. 597	. 548	. 030	3 . 59	3 . 51	. 19	1 . 046	. 585	. 033	8 . 68	3 . 51	. 19	28
3	0 . 972	0 . 664	. 035	7 .68	3 . 26	. 18	. 577	. 530	. 029	3 . 48	3 . 40	. 18	1 . 012	. 566	. 032	8 . 40	3 . 40	. 18	27
4	0 . 938	0 . 641	. 033	7 .41	3 . 15	. 17	. 557	. 511	. 028	3 . 36	3 . 28	. 18	0 . 976	. 547	. 031	8 . 11	3 . 28	. 18	26
5	0 . 905	0 . 618	. 032	7 .15	3 . 04	. 17	. 538	. 493	. 027	3 . 24	3 . 16	. 17	0 . 941	. 527	. 030	7 . 82	3 . 16	. 17	25
6	0 . 870	0 . 595	. 031	6 .88	2 . 92	. 16	. 517	. 474	. 026	3 . 11	3 . 04	. 16	0 . 906	. 507	. 028	7 . 52	3 . 04	. 16	24
7	0 . 835	0 . 572	. 030	6 .61	2 . 81	. 16	. 497	. 456	. 025	2 . 99	2 . 92	. 16	0 . 870	. 487	. 027	7 . 23	2 . 92	. 16	23
8	0 . 802	0 . 547	. 028	6 .33	2 . 69	. 15	. 476	. 437	. 024	2 . 86	2 . 80	. 15	0 . 835	. 467	. 026	6 . 93	2 . 80	. 15	22
9	0 . 767	0 . 524	. 027	6 .06	2 . 57	. 14	. 456	. 418	. 023	2 . 74	2 . 68	. 14	0 . 799	. 447	. 025	6 . 63	2 . 68	. 14	21
10	0 . 732	0 . 500	. 026	5 .78	2 . 46	. 14	. 435	. 399	. 022	2 . 62	2 . 56	. 14	0 . 762	. 427	. 024	6 . 33	2 . 56	. 14	20
11	0 . 697	0 . 476	. 025	5 .51	2 . 34	. 13	. 414	. 380	. 020	2 . 49	2 . 43	. 13	0 . 725	. 406	. 023	6 . 02	2 . 44	. 13	19
12	0 . 661	0 . 452	. 024	5 .23	2 . 22	. 12	. 393	. 360	. 019	2 . 37	2 . 31	. 12	0 . 688	. 386	. 022	5 . 71	2 . 31	. 12	18
13	0 . 626	0 . 428	. 022	4 .95	2 . 10	. 12	. 372	. 341	. 018	2 . 24	2 . 18	. 12	0 . 651	. 365	. 020	5 . 41	2 . 19	. 12	17
14	0 . 590	0 . 403	. 021	4 .66	1 . 98	. 11	. 350	. 322	. 017	2 . 11	2 . 06	. 11	0 . 614	. 344	. 019	5 . 10	2 . 06	. 11	16
15	0 . 554	0 . 378	. 020	4 .38	1 . 86	. 10	. 329	. 302	. 016	1 . 98	1 . 93	. 10	0 . 576	. 323	. 018	4 . 79	1 . 94	. 10	15
16	0 . 518	0 . 353	. 018	4 .08	1 . 74	. 10	. 307	. 282	. 015	1 . 85	1 . 81	. 10	0 . 538	. 302	. 017	4 . 47	1 . 81	. 10	14
17	0 . 481	0 . 328	. 017	3 .80	1 . 61	. 09	. 286	. 262	. 014	1 . 72	1 . 68	. 09	0 . 501	. 280	. 016	4 . 16	1 . 68	. 09	13
18	0 . 445	0 . 304	. 016	3 .51	1 . 49	. 08	. 264	. 242	. 013	1 . 59	1 . 55	. 08	0 . 463	. 259	. 015	3 . 84	1 . 55	. 08	12
19	0 . 408	0 . 279	. 015	3 .22	1 . 37	. 08	. 242	. 222	. 012	1 . 46	1 . 43	. 08	0 . 425	. 238	. 013	3 . 52	1 . 43	. 08	11
20	0 . 371	0 . 254	. 013	2 .94	1 . 25	. 07	. 221	. 202	. 011	1 . 33	1 . 30	. 07	0 . 386	. 216	. 012	3 . 21	1 . 30	. 07	10
21	0 . 335	0 . 228	. 012	2 .64	1 . 12	. 06	. 199	. 183	. 010	1 . 19	1 . 17	. 06	0 . 348	. 195	. 011	2 . 89	1 . 17	. 06	9
22	0 . 298	0 . 203	. 011	2 .35	1 . 00	. 06	. 177	. 162	. 009	1 . 07	1 . 04	. 06	0 . 310	. 174	. 010	2 . 57	1 . 04	. 06	8
23	0 . 261	0 . 178	. 009	2 .06	0 . 87	. 05	. 155	. 142	. 008	0 . 93	0 . 91	. 05	0 . 272	. 152	. 009	2 . 25	0 . 91	. 05	7
24	0 . 224	0 . 153	. 008	1 .77	0 . 75	. 04	. 133	. 122	. 007	0 . 80	0 . 78	. 04	0 . 233	. 131	. 007	1 . 94	0 . 78	. 04	6
25	0 . 187	0 . 128	. 007	1 .48	0 . 63	. 03	. 111	. 002	. 006	0 . 67	0 . 65	. 04	0 . 195	. 119	. 006	1 . 62	0 . 65	. 04	5
26	0 . 150	0 . 102	. 005	1 .18	0 . 50	. 03	. 089	. 082	. 004	0 . 53	0 . 52	. 03	0 . 156	. 087	. 005	1 . 29	0 . 52	. 03	4
27	0 . 112	0 . 077	. 004	0 .89	0 . 38	. 02	. 067	. 061	. 003	0 . 40	0 . 39	. 02	0 . 117	. 065	. 004	0 . 97	0 . 39	. 02	3
28	0 . 075	0 . 051	. 003	0 .59	0 . 25	. 01	. 044	. 041	. 002	0 . 27	0 . 26	. 01	0 . 078	. 044	. 002	0 . 65	0 . 26	. 01	2
29	0 . 037	0 . 026	. 001	0 .30	0 . 13	. 01	. 022	. 020	. 001	0 . 13	0 . 13	. 01	0 . 039	. 022	. 001	0 . 32	0 . 13	. 01	1
VI 0 O	0 . 000	0 . 000	. 000	0 .00	0 . 00	. 00	. 000	. 000	. 000	0 . 00	0 . 00	. 00	0 . 000	. 000	. 000	0 . 00	0 . 00	. 00	O 0 VI
	Arg. = ☉ Long. + 3ˢ 4° 8′	Arg. = ☾ ☊ + 4ˢ 21° 42′	Arg. = 2☉ Long. + 4ˢ 27° 21′	Arg. = ☉ Long. + 5ˢ 17° 30′	Arg. = ☾ ☊ + 6ˢ 5° 5′	Arg. = 2☉ Long. + 6ˢ 4° 7′	Arg. = ☉ Long. + 2ˢ 9° 22′	Arg. = ☾ ☊ + 6ˢ 5° 24′	Arg. = 2☉ Long. + 6ˢ 4° 30′	Arg. = ☉ Long. + 2ˢ 3° 35′	Arg. = ☾ ☊ + 5ˢ 5° 16′	Arg. = 2☉ Long. + 5ˢ 9° 22′	Arg. = ☉ Long. + 2ˢ 9° 9′	Arg. = ☾ ☊ + 4ˢ 14° 14′	Arg. = 2☉ Long. + 4ˢ 13° 40′	Arg. = ☉ Long. + 4ˢ 28° 44′	Arg. = ☾ ☊ + 5ˢ 4° 41′	Arg. = 2☉ Long. + 5ˢ 9° 9′	
	Aber.	☾ Nut.	☉Nut.	Aber.	☽ Nut.	☉Nut.	Aber.	☾ Nut.	☉Nut.	Aber.	☾ Nut.	☉Nut	Aber.	☾ Nut.	☉Nut.	Aber.	☾ Nut.	☉Nut.	
	R. A.			N. P. D.			R. A.			N. P. D.			R. A.			N. P. D.			
	PHECDA 1830.						SPICA VIRGINIS 1830.						MIZAR 1830.						

The CORRECTIONS of 48 principal Stars, &c. continued.

η URSÆ MAJORIS 1820. Right Ascension h. m. s. 13 40 26 . 43; Ann. Var. + 2s. 38. N. P. Distance 39° 47′ 5″. 21; Ann. Var. + 18″. 20.

1830. Common Argument. + S.	R. A. Aber.	R. A. ☾ Nut.	R. A. ⊙ Nut.	N. P. D. Aber.	N. P. D. ☾ Nut.	N. P. D. ⊙ Nut.	1830. Common Argument. − S.
III 0° IX	1s.966	1s.111	0s.057	18″. 00	7″. 688	0″. 405	III 0° IX
1	1 .965	1 .111	.057	17 . 99	7 . 68	. 40	29
2	1 .964	1 .110	.057	17 . 98	7 . 68	. 40	28
3	1 .963	1 .109	.057	17 . 96	7 . 67	. 40	27
4	1 .961	1 .108	.057	17 . 94	7 . 67	. 40	26
5	1 .959	1 .107	.057	17 . 92	7 . 66	. 40	25
6	1 .956	1 .105	.057	17 . 89	7 . 65	. 40	24
7	1 .952	1 .103	.057	17 . 86	7 . 64	. 40	23
8	1 .948	1 .100	.057	17 . 82	7 . 62	. 40	22
9	1 .943	1 .097	.056	17 . 77	7 . 60	. 40	21
10	1 .938	1 .094	.056	17 . 72	7 . 58	. 40	20
11	1 .932	1 .090	.056	17 . 66	7 . 55	. 40	19
12	1 .925	1 .086	.056	17 . 60	7 . 52	. 40	18
13	1 .917	1 .082	.056	17 . 53	7 . 49	. 39	17
14	1 .909	1 .078	.055	17 . 46	7 . 46	. 39	16
15	1 .900	1 .073	.055	17 . 38	7 . 42	. 39	15
16	1 .891	1 .068	.055	17 . 30	7 . 39	. 39	14
17	1 .881	1 .062	.055	17 . 21	7 . 35	. 39	13
18	1 .871	1 .056	.064	17 . 11	7 . 31	. 38	12
19	1 .860	1 .050	.054	17 . 01	7 . 27	. 38	11
20	1 .848	1 .044	.054	16 . 91	7 . 23	. 38	10
21	1 .835	1 .037	.053	16 . 80	7 . 18	. 38	9
22	1 .823	1 .030	.053	16 . 68	7 . 13	. 38	8
23	1 .810	1 .023	.053	16 . 55	7 . 07	. 37	7
24	1 .796	1 .015	.052	16 . 43	7 . 02	. 37	6
25	1 .781	1 .007	.052	16 . 30	6 . 97	. 37	5
26	1 .766	0 .998	.051	16 . 17	6 . 91	. 36	4
27	1 .751	0 .990	.051	16 . 03	6 . 85	. 36	3
28	1 .735	0 .980	.050	15 . 89	6 . 78	. 36	2
29	1 .720	0 .971	.050	15 . 75	6 . 72	. 35	1
IV 0 X	1 .703	0 .962	.049	15 . 60	6 . 66	. 35	II 0 VIII
1	1 .685	0 .952	.049	15 . 43	6 . 59	. 35	29
2	1 .668	0 .942	.048	15 . 26	6 . 52	. 34	28
3	1 .649	0 .932	.048	15 . 10	6 . 45	. 34	27
4	1 .629	0 .921	.047	14 . 92	6 . 37	. 34	26
5	1 .611	0 .910	.047	14 . 74	6 . 30	. 33	25
6	1 .590	0 .898	.046	14 . 56	6 . 22	. 33	24
7	1 .570	0 .887	.046	14 . 37	6 . 14	. 32	23
8	1 .550	0 .875	.045	14 . 18	6 . 06	. 32	22
9	1 .528	0 .863	.044	13 . 98	5 . 97	. 31	21
10	1 .505	0 .850	.044	13 . 77	5 . 88	. 31	20
11	1 .483	0 .837	.043	13 . 57	5 . 80	. 31	19
12	1 .461	0 .825	.042	13 . 37	5 . 71	. 30	18
13	1 .438	0 .812	.042	13 . 15	5 . 62	. 30	17
14	1 .414	0 .799	.041	12 . 94	5 . 53	. 29	16
IV 15 X	1 .390	0 .786	.040	12 . 72	5 . 44	. 29	I 15 VII
	Arg. = ⊙ Long. + 2s 3° 49′	Arg. = ☾ ☊ + 4s 20° 54′	Arg. = 2⊙ Long. + 4s 26° 36′	Arg. = ⊙ Long. + 4s 21° 27′	Arg. = ☾ ☊ + 4s 27° 39′	Arg. = 2⊙ Long. + 5s 2° 51′	

ALKAID 1830.

α BOOTIS 1820. Right Ascension h. m. s. 14 7 27 . 44; Ann. Var. + 2s. 73. N. P. Distance 69° 52′ 31″. 89; Ann. Var. + 18″. 99.

1830. Common Argument. + S.	R. A. Aber.	R. A. ☾ Nut.	R. A. ⊙ Nut.	N. P. D. Aber.	N. P. D. ☾ Nut.	N. P. D. ⊙ Nut.	1830. Common Argument. − S.
III 0° IX	1s.353	1s.030	0s.056	12″. 441	7″. 951	0″. 408	III 0° IX
1	1 .352	1 .030	.056	12 . 43	7 . 95	. 40	29
2	1 .351	1 .029	.056	12 . 42	7 . 95	. 40	28
3	1 .350	1 .028	.056	12 . 41	7 . 94	. 40	27
4	1 .349	1 .027	.056	12 . 40	7 . 94	. 40	26
5	1 .348	1 .026	.056	12 . 39	7 . 93	. 40	25
6	1 .346	1 .024	.056	12 . 37	7 . 92	. 40	24
7	1 .344	1 .022	.056	12 . 35	7 . 91	. 40	23
8	1 .341	1 .020	.055	12 . 33	7 . 89	. 40	22
9	1 .337	1 .018	.055	12 . 30	7 . 87	. 40	21
10	1 .333	1 .015	.055	12 . 27	7 . 85	. 40	20
11	1 .328	1 .012	.055	12 . 23	7 . 82	. 40	19
12	1 .323	1 .008	.055	12 . 18	7 . 79	. 40	18
13	1 .318	1 .004	.054	12 . 13	7 . 76	. 40	17
14	1 .312	1 .000	.054	12 . 08	7 . 73	. 40	16
15	1 .306	0 .995	.054	12 . 02	7 . 69	. 39	15
16	1 .300	0 .990	.054	11 . 96	7 . 65	. 39	14
17	1 .293	0 .983	.054	11 . 90	7 . 61	. 39	13
18	1 .286	0 .979	.053	11 . 84	7 . 57	. 39	12
19	1 .279	0 .973	.053	11 . 77	7 . 53	. 39	11
20	1 .271	0 .967	.053	11 . 70	7 . 48	. 38	10
21	1 .263	0 .961	.052	11 . 62	7 . 43	. 38	9
22	1 .255	0 .955	.052	11 . 54	7 . 38	. 38	8
23	1 .246	0 .948	.052	11 . 46	7 . 33	. 38	7
24	1 .237	0 .941	.051	11 . 37	7 . 27	. 37	6
25	1 .226	0 .933	.051	11 . 28	7 . 21	. 37	5
26	1 .217	0 .925	.050	11 . 19	7 . 15	. 37	4
27	1 .205	0 .917	.050	11 . 09	7 . 08	. 36	3
28	1 .194	0 .909	.049	10 . 99	7 . 02	. 36	2
29	1 .183	0 .901	.049	10 . 89	6 . 96	. 36	1
IV 0 X	1 .171	0 .892	.049	10 . 78	6 . 89	. 35	II 0 VIII
1	1 .160	0 .883	.048	10 . 67	6 . 82	. 35	29
2	1 .148	0 .874	.048	10 . 56	6 . 75	. 35	28
3	1 .135	0 .864	.047	10 . 44	6 . 67	. 34	27
4	1 .121	0 .854	.046	10 . 32	6 . 60	. 34	26
5	1 .108	0 .843	.046	10 . 20	6 . 52	. 33	25
6	1 .094	0 .832	.045	10 . 07	5 . 44	. 33	24
7	1 .080	0 .822	.045	9 . 93	6 . 35	. 33	23
8	1 .066	0 .811	.044	9 . 80	6 . 27	. 32	22
9	1 .051	0 .800	.044	9 . 66	6 . 18	. 32	21
10	1 .036	0 .789	.043	9 . 52	6 . 09	. 31	20
11	1 .020	0 .777	.042	9 . 38	6 . 00	. 31	19
12	1 .005	0 .765	.042	9 . 24	5 . 91	. 30	18
13	0 .989	0 .753	.041	9 . 10	5 . 82	. 30	17
14	0 .973	0 .741	.040	8 . 95	5 . 73	. 29	16
IV 15 X	0 .956	0 .728	.040	8 . 79	5 . 62	. 29	I 15 VII
	Arg. = ⊙ Long. + 1s 25° 46′	Arg. = ☾ ☊ + 5s 18° 50′	Arg. = 2⊙ Long. + 5s 20° 54′	Arg. = ⊙ Long. + 3s 28° 18′	Arg. = ☾ ☊ + 4s 19° 59′	Arg. = 2⊙ Long. + 4s 25° 45′	

ARCTURUS 1830.

α² LIBRÆ 1820. Right Ascension h. m. s. 14 40 56 . 47; Ann. Var. + 3s. 29. N. P. Distance 105° 17′ 9″. 73; Ann. Var. + 15″. 20.

1830. Common Argument. + S.	R. A. Aber.	R. A. ☾ Nut.	R. A. ⊙ Nut.	N. P. D. Aber.	N. P. D. ☾ Nut.	N. P. D. ⊙ Nut.	1830. Common Argument. − S.
III 0° IX	1s.334	1s.196	0s.066	6″. 146	8″. 307	0″. 414	III 0° IX
1	1 .333	1 .195	.066	6 . 14	8 . 30	. 41	29
2	1 .332	1 .194	.066	6 . 13	8 . 29	. 41	28
3	1 .331	1 .193	.066	6 . 13	8 . 29	. 41	27
4	1 .330	1 .192	.066	6 . 12	8 . 28	. 41	26
5	1 .329	1 .191	.066	6 . 12	8 . 27	. 41	25
6	1 .327	1 .189	.066	6 . 11	8 . 26	. 41	24
7	1 .325	1 .187	.066	6 . 10	8 . 24	. 41	23
8	1 .322	1 .184	.065	6 . 09	8 . 22	. 41	22
9	1 .318	1 .181	.065	6 . 08	8 . 20	. 41	21
10	1 .314	1 .178	.065	6 . 06	8 . 18	. 41	20
11	1 .310	1 .174	.065	6 . 04	8 . 15	. 41	19
12	1 .305	1 .170	.065	6 . 02	8 . 12	. 41	18
13	1 .300	1 .165	.064	5 . 99	8 . 09	. 40	17
14	1 .294	1 .160	.064	5 . 96	8 . 06	. 40	16
15	1 .288	1 .154	.064	5 . 93	8 . 02	. 40	15
16	1 .282	1 .148	.064	5 . 90	7 . 98	. 40	14
17	1 .275	1 .142	.063	5 . 87	7 . 94	. 40	13
18	1 .268	1 .136	.063	5 . 84	7 . 90	. 39	12
19	1 .261	1 .130	.062	5 . 81	7 . 86	. 39	11
20	1 .253	1 .123	.062	5 . 77	7 . 81	. 39	10
21	1 .245	1 .116	.062	5 . 73	7 . 76	. 39	9
22	1 .237	1 .108	.061	5 . 69	7 . 70	. 39	8
23	1 .228	1 .100	.061	5 . 65	7 . 65	. 38	7
24	1 .219	1 .092	.060	5 . 61	7 . 59	. 38	6
25	1 .210	1 .083	.060	5 . 57	7 . 53	. 38	5
26	1 .200	1 .074	.059	5 . 52	7 . 47	. 37	4
27	1 .189	1 .065	.059	5 . 47	7 . 40	. 37	3
28	1 .178	1 .055	.058	5 . 42	7 . 34	. 37	2
29	1 .167	1 .045	.058	5 . 37	7 . 27	. 36	1
IV 0 X	1 .155	1 .034	.057	5 . 32	7 . 20	. 36	II 0 VIII
1	1 .143	1 .024	.057	5 . 27	7 . 12	. 35	29
2	1 .131	1 .014	.056	5 . 21	7 . 05	. 35	28
3	1 .119	1 .003	.055	5 . 15	6 . 97	. 35	27
4	1 .106	0 .991	.055	5 . 09	6 . 89	. 34	26
5	1 .192	0 .979	.054	5 . 03	6 . 81	. 34	25
6	1 .078	0 .967	.053	4 . 97	6 . 72	. 33	24
7	1 .064	0 .954	.053	4 . 91	6 . 64	. 33	23
8	1 .050	0 .941	.052	4 . 84	6 . 55	. 33	22
9	1 .036	0 .929	.051	4 . 77	6 . 46	. 32	21
10	1 .021	0 .915	.051	4 . 70	6 . 36	. 32	20
11	1 .006	0 .902	.050	4 . 63	6 . 27	. 31	19
12	0 .991	0 .888	.049	4 . 56	6 . 17	. 31	18
13	0 .975	0 .874	.048	4 . 49	6 . 08	. 30	17
14	0 .959	0 .860	.047	4 . 42	5 . 98	. 30	16
IV 15 X	0 .942	0 .845	.047	4 . 34	5 . 88	. 29	I 15 VII
	Arg. = ⊙ Long. + 1s 17° 11′	Arg. = ☾ ☊ + 6s 6° 27′	Arg. = 2⊙ Long. + 6s 5° 15′	Arg. = ⊙ Long. + 1s 18° 25′	Arg. = ☾ ☊ + 4s 11° 11′	Arg. = 2⊙ Long. + 4s 17° 11′	

KIFFA AUSTRALIS 1830.

The CORRECTIONS of 48 principal Stars, &c. continued.

η URSÆ MAJORIS 1820. — ALKAID 1830.

1830. Common Argument.	Right Ascension. 13h 40m 26s.43			N. P. Distance. 39° 47′ 5″.21		
	Ann. Var. +2s.38			Ann Var. + 13″.20		
+ S. — S.	Aber.	☾ Nut.	⊙ Nut.	Aber.	☾ Nut.	⊙ Nut.
IV 15° X	1s.390	0s.786	0s.040	12″.72	5″.44	0″.29
16	1 .365	.772	.039	12 .50	5 .34	.28
17	1 .340	.757	.039	12 .27	5 .24	.28
18	1 .315	.743	.038	12 .03	5 .14	.27
19	1 .289	.728	.037	11 .80	5 .04	.27
20	1 .263	.713	.037	11 .56	4 .94	.26
21	1 .237	.699	.036	11 .32	4 .84	.25
22	1 .210	.684	.035	11 .08	4 .73	.25
23	1 .182	.668	.034	10 .83	4 .63	.24
24	1 .155	.653	.033	10 .58	4 .52	.24
25	1 .128	.637	.033	10 .32	4 .41	.23
26	1 .099	.622	.032	10 .07	4 .30	.23
27	1 .070	.605	.031	9 .80	4 .18	.22
28	1 .042	.588	.030	9 .53	4 .07	.21
29	1 .012	.572	.029	9 .26	3 .96	.21
V 0 XI	0 .983	.555	.028	9 .00	3 .85	.20
1	0 .953	.538	.028	8 .72	3 .73	.20
2	0 .922	.521	.027	8 .44	3 .61	.19
3	0 .892	.504	.026	8 .16	3 .49	.18
4	0 .861	.487	.025	7 .88	3 .37	.18
5	0 .830	.469	.024	7 .60	3 .25	.17
6	0 .799	.452	.023	7 .31	3 .13	.16
7	0 .767	.434	.022	7 .02	3 .00	.16
8	0 .736	.416	.021	6 .73	2 .88	.15
9	0 .704	.398	.020	6 .44	2 .75	.15
10	0 .672	.380	.019	6 .15	2 .63	.14
11	0 .640	.361	.019	5 .85	2 .50	.13
12	0 .607	.343	.018	5 .56	2 .37	.13
13	0 .575	.324	.017	5 .26	2 .25	.12
14	0 .542	.306	.016	4 .96	2 .12	.11
15	0 .508	.287	.015	4 .66	1 .99	.10
16	0 .475	.268	.014	4 .35	1 .86	.10
17	0 .442	.249	.013	4 .04	1 .73	.09
18	0 .408	.231	.012	3 .73	1 .60	.08
19	0 .374	.212	.011	3 .43	1 .46	.08
20	0 .341	.193	.010	3 .12	1 .33	.07
21	0 .307	.174	.009	2 .81	1 .20	.06
22	0 .274	.155	.008	2 .50	1 .07	.06
23	0 .239	.135	.007	2 .19	0 .94	.05
24	0 .206	.116	.006	1 .88	0 .80	.04
25	0 .172	.097	.005	1 .57	0 .67	.04
26	0 .137	.077	.004	1 .26	0 .54	.03
27	0 .103	.058	.003	0 .94	0 .40	.02
28	0 .069	.039	.002	0 .63	0 .27	.01
29	0 .034	.019	.001	0 .31	0 .13	.01
VI 0 O	0 .000	.000	.000	0 .00	0 .00	.00
	Arg. = ⊙ Long. + 2s 3° 49′	Arg. = ☾ ☊ + 4s 20° 54′	Arg. = 2⊙ Long. + 4s 26° 36′	Arg. = ⊙ Long. + 4s 21° 27′	Arg. = ☾ ☊ + 4s 27° 39′	Arg. = 2⊙ Long. + 5s 2° 51′
	Aber.	☾ Nut.	⊙ Nut.	Aber.	☾ Nut.	⊙ Nut.
	R. A.			N. P. D.		

α BOOTIS 1820. — ARCTURUS 1830.

1830. Common Argument.	Right Ascension. 14h 7m 27s.44			N. P. Distance. 69° 52′ 31″.89		
	Ann. Var. +2s.73			Ann. Var. +18″.99		
+ S. — S.	Aber.	☾ Nut.	⊙ Nut.	Aber.	☾ Nut.	⊙ Nut.
IV 15° X	0s.956	0s.728	0s.040	8″.79	5″.62	0″.29
16	.939	.715	.039	8 .64	5 .53	.28
17	.922	.702	.038	8 .48	5 .43	.28
18	.905	.689	.037	8 .32	5 .32	.27
19	.887	.675	.037	8 .15	5 .22	.27
20	.869	.661	.036	7 .99	5 .12	.26
21	.851	.648	.035	7 .83	5 .01	.26
22	.832	.634	.035	7 .66	4 .90	.25
23	.813	.620	.034	7 .48	4 .79	.25
24	.795	.605	.033	7 .31	4 .68	.24
25	.776	.590	.032	7 .14	4 .57	.23
26	.757	.576	.031	6 .96	4 .45	.23
27	.737	.561	.030	6 .78	4 .33	.22
28	.717	.546	.030	6 .60	4 .22	.22
29	.697	.531	.029	6 .41	4 .10	.21
V 0 XI	.677	.516	.028	6 .22	3 .98	.20
1	.656	.500	.027	6 .03	3 .86	.20
2	.635	.484	.026	5 .84	3 .74	.19
3	.614	.467	.025	5 .65	3 .61	.19
4	.593	.451	.025	5 .46	3 .49	.18
5	.572	.435	.024	5 .26	3 .36	.17
6	.550	.418	.023	5 .16	3 .23	.17
7	.528	.402	.022	4 .86	3 .11	.16
8	.517	.385	.021	4 .66	2 .98	.15
9	.485	.369	.020	4 .46	2 .85	.15
10	.463	.352	.019	4 .26	2 .72	.14
11	.440	.335	.018	4 .15	2 .59	.13
12	.417	.318	.017	3 .84	2 .46	.13
13	.395	.301	.016	3 .64	2 .33	.12
14	.373	.284	.015	3 .43	2 .19	.11
15	.350	.266	.014	3 .22	2 .06	.11
16	.327	.249	.014	3 .01	1 .92	.10
17	.304	.231	.013	2 .79	1 .79	.09
18	.281	.214	.012	2 .58	1 .65	.08
19	.258	.196	.011	2 .37	1 .52	.08
20	.234	.179	.010	2 .16	1 .38	.07
21	.211	.161	.009	1 .95	1 .24	.06
22	.188	.143	.008	1 .73	1 .11	.06
23	.165	.125	.007	1 .52	0 .97	.05
24	.142	.108	.006	1 .30	0 .83	.04
25	.118	.090	.005	1 .09	0 .70	.04
26	.095	.072	.004	0 .87	0 .56	.03
27	.071	.054	.003	0 .65	0 .42	.02
28	.047	.036	.002	0 .43	0 .28	.01
29	.024	.018	.001	0 .22	0 .14	.01
VI 0 O	.000	.000	.000	0 .00	0 .00	.00
	Arg. = ⊙ Long. + 1s 25° 46′	Arg. = ☾ ☊ + 5s 18° 50′	Arg. = 2⊙ Long. + 5s 20° 54′	Arg. = ⊙ Long. + 3s 28° 18′	Arg. = ☾ ☊ + 4s 19° 59′	Arg. = 2 ⊙ Long. + 4s 25° 45′
	Aber.	☾ Nut.	⊙ Nut.	Aber.	☾ Nut.	⊙ Nut.
	R. A.			N. P. D.		

α² LIBRÆ 1820. — KIFFA AUSTRALIS 1830.

1830. Common Argument.	Right Ascension. 14h 40m 56s.47			N. P. Distance. 105° 17′ 9″.73			1830. Common Argument.
	Ann. Var. +3s.29			Ann. Var. +15″.20			
+ S. — S.	Aber.	☾ Nut.	⊙ Nut	Aber.	☾ Nut.	⊙ Nut.	+ S. — S.
IV 15° X	0s.942	0s.845	0s.047	4″.34	5″.88	0″.29	I 15° VII
16	.926	.829	.046	4 .26	5 .77	.29	14
17	.909	.814	.045	4 .18	5 .67	.28	13
18	.892	.799	.044	4 .10	5 .56	.28	12
19	.874	.783	.043	4 .02	5 .45	.27	11
20	.857	.767	.042	3 .94	5 .34	.27	10
21	.839	.752	.041	3 .86	5 .23	.26	9
22	.820	.736	.041	3 .78	5 .12	.25	8
23	.802	.719	.040	3 .69	5 .00	.25	7
24	.784	.702	.039	3 .61	4 .88	.24	6
25	.765	.685	.038	3 .52	4 .77	.24	5
26	.746	.668	.037	3 .43	4 .65	.23	4
27	.726	.651	.036	3 .34	4 .53	.23	3
28	.706	.633	.035	3 .25	4 .41	.22	2
29	.687	.616	.034	3 .16	4 .28	.21	1
V 0 XI	.667	.598	.033	3 .07	4 .16	.21	I 0 VII
1	.647	.579	.032	2 .98	4 .03	.20	29
2	.626	.561	.031	2 .88	3 .90	.19	28
3	.606	.543	.030	2 .79	3 .77	.19	27
4	.585	.524	.029	2 .69	3 .64	.18	26
5	.564	.505	.028	2 .59	3 .51	.18	25
6	.542	.486	.027	2 .50	3 .38	.17	24
7	.521	.467	.026	2 .40	3 .25	.16	23
8	.499	.447	.025	2 .30	3 .11	.16	22
9	.478	.428	.024	2 .20	2 .98	.15	21
10	.456	.408	.023	2 .10	2 .84	.14	20
11	.434	.389	.021	2 .00	2 .70	.13	19
12	.412	.369	.020	1 .90	2 .57	.13	18
13	.390	.349	.019	1 .79	2 .43	.12	17
14	.367	.329	.018	1 .69	2 .29	.11	16
15	.345	.309	.017	1 .59	2 .15	.11	15
16	.322	.289	.016	1 .48	2 .01	.10	14
17	.300	.268	.015	1 .38	1 .87	.09	13
18	.277	.248	.014	1 .27	1 .73	.09	12
19	.254	.228	.013	1 .17	1 .58	.08	11
20	.231	.207	.011	1 .06	1 .44	.07	10
21	.208	.186	.010	0 .96	1 .30	.06	9
22	.185	.166	.009	0 .85	1 .16	.06	8
23	.163	.145	.008	0 .75	1 .01	.05	7
24	.140	.125	.007	0 .64	0 .87	.04	6
25	.116	.104	.006	0 .54	0 .73	.04	5
26	.093	.083	.005	0 .43	0 .58	.03	4
27	.070	.062	.004	0 .32	0 .44	.02	3
28	.047	.042	.002	0 .21	0 .29	.01	2
29	.023	.021	.001	0 .11	0 .15	.01	1
VI 0 O	.000	.000	.000	0 .00	0 .00	.00	O 0 VI
	Arg. = ⊙ Long. + 1s 17° 11′	Arg. = ☾ ☊ + 6s 6° 27′	Arg. = 2⊙ Long. + 6s 5° 15′	Arg. = ⊙ Long. + 1s 18° 25′	Arg. = ☾ ☊ + 4s 11° 11′	Arg. = 2⊙ Long. + 4s 17° 11′	
	Aber.	☾ Nut.	⊙ Nut.	Aber.	☾ Nut.	⊙ Nut.	
	R. A.			N. P. D.			

The CORRECTIONS of 48 principal Stars, &c. continued.

1830. Common Argument.	β URSÆ MINORIS 1820.						α CORONÆ BOREALIS 1820.						α SERPENTIS 1820.						1830. Common Argument.
	Right Ascension. h. m. s. 14 51 20.36			N. P. Distance. 15° 6′ 31″.83			Right Ascension. h. m. s. 15 27 4.34			N. P. Distance. 62° 40′ 22″.78			Right Ascension. h. m. s. 15 35 24.67			N. P. Distance. 83° 0′ 1″.33			
	Ann. Var.—0s.32			Ann. Var.+14″.70			Ann. Var.+2s.53			Ann. Var.+12″.49			Ann. Var.+2s.94			Ann. Var.+11″.73			
+ — S. S.	Aber.	☾ Nut.	⊙ Nut.	Aber.	☾ Nut.	⊙ Nut.	Aber.	☾ Nut.	⊙ Nut.	Aber.	☾ Nut.	⊙ Nut.	Aber.	☾ Nut.	⊙ Nut.	Aber.	☾ Nut.	⊙ Nut.	+ — S. S.
III 0° IX	4s.942	1s.745	0s.079	20″.249	8″.412	0″.415	1s.473	0s.931	0s.051	15″.011	8″.791	8″.421	1s.323	1s.060	0s.059	9″.935	8″.873	0″.422	III 0° IX
1	4.940	1.744	.079	20.23	8.41	.41	1.472	.931	.051	15.00	8.79	.42	1.322	1.060	.059	9.93	8.87	.42	29
2	4.937	1.743	.079	20.21	8.40	.41	1.471	.930	.051	14.99	8.78	.42	1.321	1.059	.059	9.92	8.86	.42	28
3	4.933	1.742	.079	20.19	8.40	.41	1.470	.929	.051	14.98	8.77	.42	1.320	1.058	.059	9.91	8.85	.42	27
4	4.928	1.740	.079	20.17	8.39	.41	1.469	.928	.051	14.97	8.76	.42	1.319	1.057	.059	9.90	8.84	.42	26
5	4.922	1.738	.079	20.15	8.38	.41	1.467	.927	.051	14.96	8.75	.42	1.318	1.056	.059	9.89	8.83	.42	25
6	4.915	1.736	.079	20.12	8.37	.41	1.465	.926	.051	14.94	8.74	.42	1.316	1.055	.059	9.88	8.82	.42	24
7	4.906	1.733	.078	20.09	8.35	.41	1.462	.924	.051	14.92	8.72	.42	1.314	1.053	.059	9.86	8.80	.42	23
8	4.895	1.729	.078	20.05	8.33	.41	1.459	.922	.051	14.89	8.70	.42	1.311	1.051	.058	9.84	8.78	.42	22
9	4.882	1.725	.078	20.00	8.31	.41	1.456	.920	.050	14.85	8.68	.42	1.307	1.048	.058	9.82	8.76	.42	21
10	4.867	1.720	.078	19.95	8.29	.41	1.452	.918	.050	14.80	8.66	.42	1.303	1.045	.058	9.79	8.74	.42	20
11	4.851	1.714	.078	19.89	8.26	.41	1.447	.915	.050	14.75	8.63	.41	1.299	1.041	.058	9.75	8.71	.41	19
12	4.834	1.708	.077	19.82	8.23	.41	1.442	.912	.050	14.69	8.60	.41	1.294	1.037	.058	9.71	8.68	.41	18
13	4.816	1.701	.077	19.74	8.20	.40	1.436	.908	.050	14.63	8.56	.41	1.289	1.033	.057	9.67	8.65	.41	17
14	4.796	1.693	.077	19.65	8.16	.40	1.429	.904	.050	14.57	8.52	.41	1.284	1.028	.057	9.63	8.61	.41	16
15	4.774	1.685	.076	19.55	8.12	.40	1.422	.900	.049	14.50	8.48	.41	1.278	1.023	.057	9.59	8.57	.41	15
16	4.751	1.677	.076	19.45	8.08	.40	1.415	.895	.049	14.44	8.44	.41	1.272	1.018	.057	9.54	8.53	.41	14
17	4.726	1.669	.076	19.35	8.04	.40	1.408	.891	.049	14.37	8.40	.40	1.265	1.013	.056	9.49	8.48	.40	13
18	4.700	1.660	.075	19.24	8.00	.39	1.401	.886	.049	14.29	8.36	.40	1.258	1.008	.056	9.44	8.43	.40	12
19	4.672	1.650	.075	19.13	7.96	.39	1.393	.881	.048	14.20	8.31	.40	1.251	1.002	.056	9.39	8.38	.40	11
20	4.643	1.639	.074	19.02	7.91	.39	1.384	.875	.048	14.11	8.25	.40	1.243	0.996	.056	9.33	8.33	.40	10
21	4.613	1.629	.074	18.90	7.86	.39	1.374	.869	.048	14.01	8.20	.39	1.235	0.989	.055	9.27	8.28	.39	9
22	4.581	1.618	.073	18.77	7.80	.39	1.364	.863	.047	13.92	8.14	.39	1.227	0.982	.055	9.21	8.22	.39	8
23	4.548	1.607	.073	18.64	7.55	.38	1.355	.857	.047	13.82	8.09	.39	1.219	0.975	.054	9.14	8.16	.39	7
24	4.514	1.595	.072	18.50	7.69	.38	1.345	.850	.047	13.72	8.03	.38	1.210	0.967	.054	9.07	8.10	.39	6
25	4.479	1.582	.072	18.35	7.63	.38	1.335	.843	.046	13.61	7.97	.38	1.200	0.960	.054	9.00	8.04	.38	5
26	4.442	1.569	.071	18.20	7.57	.37	1.323	.836	.046	13.50	7.90	.38	1.190	0.952	.053	8.92	7.97	.38	4
27	4.403	1.555	.070	18.04	7.50	.37	1.311	.828	.045	13.38	7.83	.38	1.180	0.944	.053	8.84	7.90	.38	3
28	4.363	1.541	.070	17.87	7.43	.37	1.300	.821	.045	13.25	7.76	.37	1.169	0.935	.052	8.76	7.83	.37	2
29	4.322	1.527	.069	17.70	7.36	.36	1.288	.814	.045	13.13	7.69	.37	1.158	0.927	.052	8.68	7.76	.37	1
IV 0 X	4.279	1.512	.068	17.52	7.29	.36	1.275	.806	.044	13.00	7.61	.36	1.146	0.917	.051	8.60	7.68	.37	II 0 VIII
1	4.234	1.496	.068	17.35	7.22	.36	1.262	.798	.044	12.87	7.53	.36	1.134	0.908	.051	8.51	7.61	.36	29
2	4.189	1.480	.067	17.17	7.14	.35	1.249	.790	.043	12.73	7.45	.35	1.122	0.898	.050	8.41	7.52	.36	28
3	4.144	1.464	.066	16.99	7.06	.35	1.235	.781	.043	12.60	7.37	.35	1.110	0.888	.050	8.32	7.44	.35	27
4	4.097	1.447	.065	16.79	6.98	.34	1.221	.772	.042	12.45	7.28	.35	1.097	0.878	.049	8.22	7.35	.35	26
5	4.048	1.430	.065	16.58	6.89	.34	1.206	.763	.042	12.30	7.20	.34	1.083	0.867	.048	8.12	7.26	.35	25
6	3.998	1.411	.064	16.37	6.81	.34	1.191	.753	.041	12.14	7.11	.34	1.070	0.857	.048	8.02	7.17	.34	24
7	3.946	1.393	.063	16.17	6.72	.33	1.176	.743	.041	11.99	7.02	.34	1.056	0.846	.047	7.92	7.08	.34	23
8	3.893	1.374	.062	15.95	6.63	.33	1.160	.733	.040	11.83	6.92	.33	1.042	0.835	.046	7.82	6.99	.33	22
9	3.839	1.356	.061	15.73	6.54	.32	1.144	.723	.040	11.67	6.83	.33	1.028	0.823	.046	7.71	6.89	.33	21
10	3.785	1.336	.060	15.50	6.45	.32	1.127	.713	.039	11.50	6.73	.32	1.013	0.811	.045	7.60	6.79	.32	20
11	3.729	1.316	.060	15.27	6.35	.31	1.111	.702	.038	11.32	6.63	.32	0.998	0.799	.045	7.48	6.69	.32	19
12	3.672	1.296	.059	15.04	6.26	.31	1.094	.692	.038	11.15	6.53	.31	0.982	0.787	.044	7.37	6.59	.31	18
13	3.613	1.275	.058	14.80	6.16	.30	1.076	.681	.037	10.98	6.43	.31	0.967	0.775	.043	7.26	6.49	.31	17
14	3.553	1.255	.057	14.56	6.06	.30	1.059	.669	.037	10.80	6.32	.30	0.951	0.762	.042	7.14	6.38	.30	16
IV 15 X	3.493	1.234	.056	14.32	5.95	.29	1.041	.658	.036	10.61	6.22	.30	0.935	0.749	.042	7.02	6.27	.30	I 15 VII
	Arg.=⊙ Long.+ 1s 14° 43′	Arg.=☾☊ + 2s 26° 45′	Arg.=2⊙ Long.+ 2s 25° 59′	Arg.=⊙ Long.+ 4s 15° 5′	Arg.=☾☊ + 4s 8° 45′	Arg.=2⊙ Long.+ 4s 14° 43′	Arg.=⊙ Long. + 1s 5° 45′	Arg.=☾☊ + 5s 17° 17′	Arg.=2⊙ Long.+ 5s 19° 38′	Arg.=⊙ Long.+ 3s 22° 28′	Arg.=☾☊ + 4s 0° 17′	Arg.=2⊙ Long.+ 4s 5° 46′	Arg.=⊙ Long.+ 1s 3° 43′	Arg.=☾☊ + 5s 27° 30′	Arg.=2⊙ Long.+ 5s 27° 58′	Arg.=⊙ Long.+ 3s 8° 22′	Arg.=☾☊ + 3s 28° 24′	Arg.=2⊙ Long.+ 4s 3° 43′	
	Aber.	☾ Nut.	⊙ Nut.	Aber.	☾ Nut.	⊙ Nut.	Aber.	☾ Nut.	⊙ Nut.	Aber.	☾ Nut.	⊙ Nut.	Aber.	☾ Nut.	⊙ Nut.	Aber.	☾ Nut.	⊙ Nut.	
	R. A.			N. P. D.			R. A.			N. P. D.			R. A.			N. P. D.			
	KOCAB 1830.						GEMMA 1830.						UNUKALHAY 1830.						

The CORRECTIONS of 48 principal Stars, &c. continued.

1830. Common Argument.	β URSÆ MINORIS 1820.						α CORONÆ BOREALIS 1820.						α SERPENTIS 1820.						1830. Common Argument.
	Right Ascension. h. m. s. 14 51 20 . 36			N. P. Distance. 15° 6′ 31″. 83			Right Ascension. h. m. s. 15 27 4 . 34			N. P. Distance. 62° 40′ 22″. 78			Right Ascension. h. m. s. 15 35 24 . 67			N. P. Distance. 83° 0′ 1″. 33			
	Ann. Var —0s. 32			Ann. Var. + 14′. 70			Ann. Var. +2s. 53			Ann. Var. + 12″. 49			Ann. Var. +2s. 94			Ann. Var. +11″. 73			
+ S. — S.	Aber.	☾ Nut.	⊙ Nut.	Aber.	☾ Nut.	⊙ Nut.	Aber.	☾ Nut.	⊙ Nut.	Aber.	☾ Nut.	⊙ Nut.	Aber.	☾ Nut.	⊙ Nut.	Aber.	☾ Nut.	⊙ Nut.	+ S. — S.
IV 15° X	3s. 493	1s. 234	0s. 056	14″. 32	5″. 95	0″. 29	1s. 041	0s. 658	0s. 036	10″. 61	6″. 22	0″. 30	0s. 935	0s. 749	0s. 042	7″. 02	6″. 27	0″. 30	I 15° VII
16	3 . 432	1 . 212	. 055	14 . 06	5 . 85	. 29	1 . 022	. 646	. 035	10 . 42	6 . 11	. 29	. 918	. 736	. 041	6 . 89	6 . 16	. 29	14
17	3 . 370	1 . 190	. 054	13 . 80	5 . 74	. 28	1 . 003	. 635	. 035	10 . 23	6 . 00	. 28	. 902	. 722	. 040	6 . 77	6 . 05	. 29	13
18	3 . 307	1 . 168	. 053	13 . 54	5 . 63	. 28	0 . 985	. 623	. 034	10 . 03	5 . 88	. 28	. 885	. 709	. 039	6 . 64	5 . 93	. 28	12
19	3 . 242	1 . 144	. 052	13 . 27	5 . 52	. 27	0 . 965	. 610	. 033	9 . 84	5 . 76	. 27	. 867	. 695	. 039	6 . 51	5 . 82	. 28	11
20	3 . 176	1 . 110	. 051	13 . 00	5 . 41	. 27	0 . 945	. 598	. 033	9 . 64	5 . 65	. 27	. 850	. 680	. 038	6 . 38	5 . 70	. 27	10
21	3 . 109	1 . 097	. 050	12 . 74	5 . 30	. 26	0 . 926	. 586	. 032	9 . 44	5 . 53	. 26	. 832	. 666	. 037	6 . 25	5 . 58	. 27	9
22	3 . 042	1 . 073	. 049	12 . 46	5 . 18	. 26	0 . 906	. 573	. 031	9 . 24	5 . 42	. 26	. 814	. 652	. 036	6 . 11	5 . 46	. 26	8
23	2 . 974	1 . 050	. 048	12 . 18	5 . 07	. 25	0 . 886	. 560	. 031	9 . 03	5 . 29	. 25	. 796	. 638	. 035	5 . 97	5 . 34	. 25	7
24	2 . 905	1 . 025	. 046	11 . 90	4 . 95	. 24	0 . 865	. 547	. 030	8 . 82	5 . 17	. 25	. 777	. 623	. 035	5 . 84	5 . 22	. 25	6
25	2 . 835	1 . 000	. 045	11 . 61	4 . 83	. 24	0 . 844	. 534	. 029	8 . 61	5 . 04	. 24	. 759	. 608	. 034	5 . 70	5 . 09	. 24	5
26	2 . 763	0 . 976	. 044	11 . 32	4 . 71	. 23	0 . 823	. 521	. 029	8 . 39	4 . 92	. 24	. 740	. 593	. 033	5 . 55	4 . 97	. 24	4
27	2 . 690	0 . 950	. 043	11 . 02	4 . 58	. 23	0 . 802	. 507	. 028	8 . 17	4 . 79	. 23	. 720	. 577	. 032	5 . 41	4 . 83	. 23	3
28	2 . 617	0 . 925	. 042	10 . 72	4 . 46	. 22	0 . 780	. 494	. 027	7 . 95	4 . 66	. 22	. 701	. 562	. 031	5 . 26	4 . 70	. 22	2
29	2 . 544	0 . 899	. 041	10 . 42	4 . 34	. 21	0 . 758	. 480	. 026	7 . 73	4 . 53	. 22	. 682	. 546	. 030	5 . 12	4 . 57	. 22	1
V 0 XI	2 . 471	0 . 872	. 040	10 . 12	4 . 21	. 21	0 . 736	. 466	. 026	7 . 51	4 . 40	. 21	. 662	. 530	. 030	4 . 97	4 . 44	. 21	I 0 VII
1	2 . 397	0 . 845	. 038	9 . 81	4 . 08	. 20	0 . 714	. 452	. 025	7 . 28	4 . 27	. 20	. 641	. 514	. 029	4 . 82	4 . 30	. 20	29
2	2 . 321	0 . 818	. 037	9 . 50	3 . 95	. 19	0 . 691	. 437	. 024	7 . 05	4 . 13	. 20	. 621	. 498	. 028	4 . 66	4 . 17	. 20	28
3	2 . 244	0 . 792	. 036	9 . 18	3 . 82	. 19	0 . 668	. 422	. 023	6 . 81	3 . 99	. 19	. 601	. 481	. 027	4 . 51	4 . 03	. 19	27
4	2 . 167	0 . 765	. 035	8 . 87	3 . 69	. 18	0 . 645	. 408	. 022	6 . 58	3 . 85	. 18	. 580	. 464	. 026	4 . 35	3 . 89	. 18	26
5	2 . 089	0 . 737	. 033	8 . 55	3 . 56	. 18	0 . 622	. 393	. 022	6 . 34	3 . 71	. 18	. 559	. 448	. 025	4 . 19	3 . 75	. 18	25
6	2 . 010	0 . 709	. 032	8 . 23	3 . 42	. 17	0 . 598	. 378	. 021	6 . 10	3 . 57	. 17	. 538	. 431	. 024	4 . 03	3 . 61	. 17	24
7	1 . 929	0 . 681	. 031	7 . 91	3 . 28	. 16	0 . 575	. 363	. 020	5 . 86	3 . 43	. 16	. 517	. 414	. 023	3 . 87	3 . 46	. 16	23
8	1 . 848	0 . 653	. 030	7 . 58	3 . 15	. 16	0 . 552	. 348	. 019	5 . 62	3 . 29	. 16	. 496	. 397	. 022	3 . 71	3 . 32	. 16	22
9	1 . 768	0 . 625	. 028	7 . 25	3 . 01	. 15	0 . 528	. 333	. 018	5 . 38	3 . 15	. 15	. 474	. 380	. 021	3 . 55	3 . 18	. 15	21
10	1 . 688	0 . 596	. 027	6 . 92	2 . 88	. 14	0 . 503	. 318	. 017	5 . 13	3 . 01	. 14	. 452	. 362	. 020	3 . 39	3 . 03	. 14	20
11	1 . 607	0 . 567	. 026	6 . 58	2 . 74	. 14	0 . 479	. 303	. 017	4 . 88	2 . 86	. 14	. 431	. 345	. 019	3 . 23	2 . 88	. 14	19
12	1 . 525	0 . 538	. 024	6 . 25	2 . 60	. 13	0 . 455	. 287	. 016	4 . 64	2 . 72	. 13	. 408	. 327	. 018	3 . 07	2 . 74	. 13	18
13	1 . 443	0 . 510	. 023	5 . 91	2 . 46	. 12	0 . 430	. 272	. 015	4 . 39	2 . 57	. 12	. 387	. 310	. 017	2 . 90	2 . 59	. 12	17
14	1 . 360	0 . 481	. 022	5 . 58	2 . 32	. 11	0 . 406	. 257	. 014	4 . 14	2 . 42	. 12	. 365	. 292	. 016	2 . 74	2 . 44	. 12	16
15	1 . 277	0 . 451	. 020	5 . 24	2 . 17	. 11	0 . 381	. 241	. 013	3 . 88	2 . 27	. 11	. 342	. 274	. 015	2 . 57	2 . 30	. 11	15
16	1 . 193	0 . 422	. 019	4 . 89	2 . 03	. 10	0 . 356	. 225	. 012	3 . 63	2 . 12	. 10	. 320	. 256	. 014	2 . 40	2 . 14	. 10	14
17	1 . 109	0 . 392	. 018	4 . 55	1 . 89	. 09	0 . 331	. 209	. 011	3 . 37	1 . 97	. 09	. 297	. 238	. 013	2 . 23	1 . 99	. 09	13
18	1 . 025	0 . 362	. 016	4 . 20	1 . 75	. 09	0 . 306	. 193	. 011	3 . 12	1 . 82	. 09	. 275	. 220	. 012	2 . 06	1 . 84	. 09	12
19	0 . 940	0 . 332	. 015	3 . 86	1 . 60	. 08	0 . 280	. 177	. 010	2 . 86	1 . 68	. 08	. 252	. 202	. 011	1 . 89	1 . 69	. 08	11
20	0 . 856	0 . 312	. 014	3 . 51	1 . 46	. 07	0 . 255	. 161	. 009	2 . 60	1 . 53	. 07	. 230	. 184	. 010	1 . 72	1 . 54	. 07	10
21	0 . 772	0 . 273	. 012	3 . 16	1 . 32	. 06	0 . 230	. 146	. 008	2 . 35	1 . 37	. 07	. 207	. 166	. 009	1 . 55	1 . 39	. 07	9
22	0 . 687	0 . 243	. 011	2 . 82	1 . 17	. 06	0 . 205	. 130	. 007	2 . 09	1 . 22	. 06	. 184	. 147	. 008	1 . 38	1 . 23	. 06	8
23	0 . 602	0 . 213	. 010	2 . 47	1 . 03	. 05	0 . 180	. 113	. 006	1 . 83	1 . 07	. 05	. 161	. 129	. 007	1 . 21	1 . 08	. 05	7
24	0 . 517	0 . 183	. 008	2 . 12	0 . 88	. 04	0 . 154	. 097	. 005	1 . 57	0 . 92	. 04	. 139	. 111	. 006	1 . 04	0 . 93	. 04	6
25	0 . 431	0 . 152	. 007	1 . 77	0 . 73	. 04	0 . 129	. 081	. 004	1 . 31	0 . 77	. 04	. 116	. 092	. 005	0 . 87	0 . 78	. 04	5
26	0 . 345	0 . 122	. 006	1 . 41	0 . 59	. 03	0 . 103	. 065	. 004	1 . 05	0 . 61	. 03	. 092	. 074	. 004	0 . 69	0 . 62	. 03	4
27	0 . 259	0 . 092	. 004	1 . 06	0 . 44	. 02	0 . 077	. 049	. 003	0 . 79	0 . 46	. 02	. 069	. 056	. 003	0 . 52	0 . 46	. 02	3
28	0 . 172	0 . 061	. 003	0 . 71	0 . 29	. 01	0 . 051	. 032	. 002	0 . 43	0 . 31	. 01	. 046	. 037	. 002	0 . 35	0 . 31	. 01	2
29	0 . 086	0 . 031	. 001	0 . 35	0 . 15	. 01	0 . 026	. 016	. 001	0 . 26	0 . 15	. 01	. 023	. 019	. 001	0 . 17	0 . 15	. 01	1
VI 0 O	0 . 000	0 . 000	. 000	0 . 00	0 . 00	. 00	0 . 000	. 000	. 000	0 . 00	0 . 00	. 00	. 000	. 000	. 000	0 . 00	0 . 00	. 00	O 0 VI
	Arg. = ⊙ Long. + 1s 14° 43′	Arg. = ☾ ☊ + 2s 26° 45′	Arg. = 2 ⊙ Long. + 2s 25° 59′	Arg. = ⊙ Long. + 4s 15° 5′	Arg. = ☾ ☊ + 4s 8° 45′	Arg. = 2 ⊙ Long. + 4s 14° 43′	Arg. = ⊙ Long. + 1s 5° 45′	Arg. = ☾ ☊ + 5s 17° 17′	Arg. = 2 ⊙ Long. + 5s 19° 38′	Arg. = ⊙ Long. + 3s 22° 28′	Arg. = ☾ ☊ + 4s 0° 17′	Arg. = 2 ⊙ Long. + 4s 5° 46′	Arg. = ⊙ Long. + 1s 3° 43′	Arg. = ☾ ☊ + 5s 27° 30′	Arg. = 2 ⊙ Long. + 5s 27° 58′	Arg. = ⊙ Long. + 3s 8° 22′	Arg. = ☾ ☊ + 3s 28° 24′	Arg. = 2 ⊙ Long. + 4s 3° 43′	
	Aber.	☾ Nut.	⊙ Nut.	Aber.	☾ Nut.	⊙ Nut.	Aber.	☾ Nut.	⊙ Nut.	Aber.	☾ Nut.	⊙ Nut.	Aber.	☾ Nut.	⊙ Nut.	Aber.	☾ Nut.	⊙ Nut.	
	R. A.			N. P. D.			R. A.			N. P. D.			R. A.			N. P. D.			
	KOCAB 1830.						GEMMA 1830.						UNUKALHAY 1830.						

The CORRECTIONS of 48 principal Stars, &c. continued.

1830. Common Argument.	α SCORPII 1820.						α HERCULIS 1820.						β DRACONIS 1820.						1830. Common Argument.
	Right Ascension. h. 16 m. 18 s. 22 .23			N. P. Distance. 116° 1′ 16″.10			Right Ascension. h. 17 m. 6 s. 26 .83			N. P. Distance. 75° 23′ 45″.59			Right Ascension. h. 17 m. 26 s. 21 .9			N. P. Distance. 37° 33 37″.6			
	Ann. Var. + 3s.66			Ann. Var. + 8″.62			Ann. Var. + 2s.73			Ann. Var. + 4″.48			Ann. Var. + 1s.34			Ann. Var. + 2″.97			
+ S.	Aber.	☾ Nut.	⊙ Nut.	Aber.	☾ Nut.	⊙ Nut.	Aber.	☾ Nut.	⊙ Nut.	Aber.	☾ Nut.	⊙ Nut.	Aber.	☾ Nut.	⊙ Nut.	Aber.	☾ Nut.	⊙ Nut.	− S.
III 0° IX	1s.481	1s.320	0s.073	3″.784	9″.249	0″.428	1s.389	0s.979	0s.054	12″.418	9″.533	0″.433	2s.211	0s.502	s.034	19″.626	9″.603	0″.434	III 0° IX
1	1.480	1.319	.073	3.78	9.24	.42	1.388	.979	.054	12.41	9.53	.43	2.210	.502	.034	19.61	9.60	.43	29
2	1.479	1.318	.073	3.78	9.24	.42	1.387	.978	.054	12.40	9.52	.43	2.208	.502	.034	19.60	9.59	.43	28
3	1.478	1.317	.073	3.78	9.23	.42	1.386	.977	.054	12.39	9.52	.43	2.206	.501	.034	19.59	9.58	.43	27
4	1.477	1.316	.073	3.77	9.22	.42	1.385	.976	.054	12.38	9.51	.43	2.204	.501	.034	19.57	9.57	.43	26
5	1.475	1.314	.073	3.77	9.21	.42	1.383	.975	.054	12.37	9.50	.43	2.201	.500	.034	19.55	9.56	.43	25
6	1.473	1.312	.073	3.77	9.20	.42	1.381	.974	.054	12.35	9.48	.43	2.197	.499	.034	19.53	9.54	.43	24
7	1.470	1.310	.072	3.76	9.18	.42	1.379	.972	.054	12.33	9.46	.43	2.193	.498	.034	19.50	9.52	.43	23
8	1.467	1.307	.072	3.75	9.16	.42	1.376	.970	.054	12.30	9.44	.43	2.188	.497	.034	19.47	9.50	.43	22
9	1.464	1.304	.072	3.74	9.14	.42	1.373	.968	.053	12.27	9.42	.43	2.183	.496	.034	19.42	9.48	.43	21
10	1.460	1.301	.072	3.73	9.11	.42	1.369	.965	.053	12.24	9.39	.43	2.177	.495	.034	19.36	9.46	.43	20
11	1.456	1.297	.072	3.72	9.08	.42	1.364	.962	.053	12.20	9.36	.43	2.170	.493	.033	19.29	9.43	.43	19
12	1.451	1.292	.071	3.71	9.04	.42	1.359	.958	.053	12.15	9.33	.42	2.162	.391	.033	19.22	9.40	.43	18
13	1.445	1 287	.071	3.70	9.00	.42	1.353	.954	.053	12.10	9.29	.42	2.154	.489	.033	19.14	9.36	.42	17
14	1.438	1.281	.071	3.68	8.96	.42	1.347	.950	.052	12.05	9.25	.42	2.144	.487	.033	19.05	9.32	.42	16
15	1.431	1.275	.071	3.66	8.92	.41	1.341	.945	.052	12.00	9.21	.42	2.134	.485	.033	18.96	9.27	.42	15
16	1.424	1.269	.070	3.64	8.88	.41	1.334	.940	.052	11.94	9.17	.42	2.123	.483	.033	18.88	9.22	.42	14
17	1.416	1.262	.070	3.62	8.83	.41	1.327	.935	.052	11.88	9.12	.41	2.112	.481	.033	18.79	9.17	.42	13
18	1.408	1.255	.070	3.60	8.78	.41	1.320	.930	.051	11.82	9.07	.41	2.101	.478	.032	18.68	9.12	.41	12
19	1.400	1.248	.069	3.58	8.73	.41	1.312	.925	.051	11.75	9.02	.41	2.089	.475	.032	18.56	9.07	.41	11
20	1.392	1.240	.069	3.56	8.68	.40	1.304	.920	.051	11.68	8.96	.41	2.077	.472	.032	18.44	9.02	.41	10
21	1.382	1.232	.068	3.53	8.63	.40	1.296	.914	.050	11.60	8.90	.40	2.064	.469	.032	18.32	8.96	.41	9
22	1.372	1.223	.068	3.51	8.57	.40	1.288	.907	.050	11.51	8.83	.40	2.050	.466	.032	18.20	8.90	.40	8
23	1.363	1.215	.067	3.49	8.51	.39	1.279	.901	.050	11.43	8.76	.40	2.035	.463	.031	18.08	8.83	.40	7
24	1.353	1.206	.067	3.46	8.44	.39	1.269	.894	.049	11.35	8.70	.40	2.020	.459	.031	17.94	8.76	.40	6
25	1.342	1.197	.066	3.43	8.37	.39	1.259	.887	.049	11.26	8.63	.39	2.004	.455	.031	17.80	8.69	.39	5
26	1.331	1.187	.066	3.40	8.30	.38	1.248	.879	.049	11.17	8.56	.39	1.988	.451	.031	17.65	8.62	.39	4
27	1.329	1.176	.065	3.37	8.23	.38	1.237	.871	.048	11.07	8.48	.39	1.971	.447	.030	17.50	8.55	.39	3
28	1.308	1.165	.064	3.34	8.15	.38	1.226	.863	.048	10.97	8.40	.38	1.953	.443	.030	17.34	8.47	.38	2
29	1.295	1.154	.064	3.31	8.08	.37	1.214	.856	.047	10.87	8.32	.38	1.934	.439	.030	17.17	8.39	.38	1
IV 0 X	1.282	1.143	.063	3.28	8.00	.37	1.202	.847	.047	10.76	8.24	.38	1.914	.434	.029	17.00	8.31	.38	II 0 VIII
1	1.269	1.131	.063	3.24	7.92	.37	1.190	.838	.046	10.65	8.16	.37	1.895	.430	.029	16.82	8.22	.37	29
2	1.256	1.119	.062	3.21	7.84	.36	1.177	.829	.046	10.53	8.07	.37	1.874	.426	.029	16.64	8.14	.37	28
3	1.242	1.106	.061	3.17	7.75	.36	1.164	.820	.045	10.42	7.99	.36	1.853	.421	.029	16.46	8.05	.36	27
4	1.228	1.093	.061	3.14	7.66	.35	1.151	.811	.045	10.30	7.90	.36	1.832	.416	.028	16.27	7.96	.36	26
5	1.213	1.081	.060	3.10	7.57	.35	1.137	.801	.044	10.18	7.80	.35	1.811	.411	.028	16.09	7.86	.35	25
6	1.198	1.068	.059	3.06	7.47	.35	1.123	.791	.044	10.05	7.71	.35	1.788	.406	.028	15.89	7.76	.35	24
7	1.183	1.054	.058	3.02	7.38	.34	1.108	.781	.043	9.92	7.61	.35	1.765	.401	.027	15.68	7.66	.35	23
8	1.167	1.039	.058	2.98	7.28	.34	1.093	.771	.043	9.78	7.51	.34	1.742	.395	.027	15.47	7.56	.34	22
9	1.151	1.025	.057	2.94	7.18	.33	1.078	.760	.042	9.64	7.40	.34	1.718	.390	.027	15.25	7.46	.34	21
10	1.134	1.010	.056	2.90	7.07	.33	1.062	.749	.041	9.50	7.29	.33	1.693	.384	.026	15.03	7.35	.33	20
11	1.117	0.995	.055	2.85	6.97	.32	1.047	.738	.041	9.36	7.18	.33	1.667	.378	.026	14.81	7.24	.33	19
12	1.100	0.980	.054	2.81	6.87	.32	1.031	.727	.040	9.22	7.07	.32	1.641	.372	.025	14.59	7.13	.32	18
13	1.082	0.965	.053	2.77	6.76	.31	1.015	.716	.039	9.08	6.97	.32	1.616	.367	.025	14.36	7.02	.32	17
14	1.065	0.949	.053	2.72	6.65	.31	0.998	.704	.039	9.93	6.85	.31	1.590	.361	.024	14.12	6.90	.31	16
IV 15 X	1.047	0.933	.052	2.67	6.53	.30	0.982	.692	.038	9.78	6.73	.31	1.563	.355	.024	13.88	6.78	.31	I 15 VII
	Arg. = ⊙ Long. + 0s 23° 24′	Arg. = ☾ ☊ + 6s 5° 50′	Arg. = 2 ⊙ Long. + 6s 4° 44′	Arg. = ⊙ Long. + 11s 28° 0′	Arg. = ☾ ☊ + 3s 19° 21′	Arg. = 2 ⊙ Long. + 3s 24° 23′	Arg. = ⊙ Long. + 0s 12° 12′	Arg. = ☾ ☊ + 5s 27° 45′	Arg. = 2 ⊙ Long. + 5s 28° 10′	Arg. = ⊙ Long. + 3s 5° 25′	Arg. = ☾ ☊ + 3s 9° 58′	Arg. = 2 ⊙ Long. + 3s 12° 26′	Arg. = ⊙ Long. + 0s 7° 40′	Arg. = ☾ ☊ + 5s 15° 59′	Arg. = 2 ⊙ Long. + 5s 20° 52′	Arg. = ⊙ Long. + 3s 6° 48′	Arg. = ☾ ☊ + 3s 6° 15′	Arg. = 2 ⊙ Long. + 3s 7° 40′	
	Aber.	☾ Nut.	⊙ Nut.	Aber.	☾ Nut.	⊙ Nut.	Aber.	☾ Nut.	⊙ Nut.	Aber.	☾ Nut.	⊙ Nut.	Aber.	☾ Nut.	⊙ Nut.	Aber.	☾ Nut.	⊙ Nut.	
	R. A.			N. P. D.			R. A.			N. P. D.			R. A.			N. P. D.			
	ANTARES 1830.						RASALGETI 1830.						ALWAID 1830.						

The CORRECTIONS of 48 principal Stars, &c. continued.

1830. Common Argument.	α SCORPII 1820. Right Ascension. h. m. s. 16 18 22 . 23 Ann. Var. + 3ˢ. 66			N. P. Distance. 116° 1′ 16″. 10 Ann. Var. + 8″. 62			β HERCULIS 1820. Right Ascension. h. m. s. 17 6 26 . 83 Ann. Var. + 2ˢ. 73			N. P. Distance. 75° 23′ 45″. 59 Ann. Var. + 4″. 48			β DRACONIS 1820. Right Ascension. h. m. s. 17 26 21 . 9 Ann. Var. + 1ˢ. 34			N. P. Distance. 37° 33′ 37″. 6 Ann. Var. + 2″. 97			1830. Common Argument.
+ S. − S.	Aber.	☾ Nut.	☉ Nut.	Aber.	☾ Nut.	☉ Nut.	Aber.	☾ Nut.	☉ Nut.	Aber.	☾ Nut.	☉ Nut.	Aber.	☾ Nut.	☉ Nut.	Aber.	☾ Nut.	☉ Nut.	+ S. − S.
IV 15° X	1^s.047	0^s.933	0^s.052	2″.67	6″.53	0″.30	0^s.982	0^s.692	0^s.038	8″.78	6″.73	0″.31	1^s.563	0^s.355	0^s.024	13″.88	6″.78	0″.31	I 15° VII
16	1.028	.916	.051	2.63	6.42	.30	.964	.679	.037	8.62	6.61	.30	1.535	.348	.024	13.63	6.66	.30	14
17	1.009	.900	.050	2.58	6.30	.29	.946	.667	.037	8.46	6.49	.30	1.508	.342	.023	13.38	6.54	.30	13
18	0.990	.882	.049	2.53	6.18	.29	.928	.654	.036	8.30	6.37	.29	1.479	.336	.023	13.13	6.42	.29	12
19	0.970	.865	.048	2.48	6.06	.28	.910	.641	.035	8.14	6.25	.28	1.449	.329	.022	12.87	6.29	.28	11
20	0.951	.848	.047	2.43	5.94	.28	.892	.629	.035	7.98	6.12	.28	1.420	.322	.022	12.61	6.16	.28	10
21	0.931	.830	.046	2.38	5.82	.27	.873	.616	.034	7.82	6.00	.27	1.391	.316	.021	12.35	6.04	.27	9
22	0.911	.812	.045	2.33	5.69	.26	.854	.602	.033	7.65	5.86	.27	1.361	.309	.021	12.08	5.91	.27	8
23	0.891	.794	.044	2.28	5.56	.26	.835	.588	.032	7.47	5.73	.26	1.330	.302	.020	11.81	5.78	.26	7
24	0.870	.770	.043	2.22	5.43	.25	.815	.575	.032	7.30	5.60	.25	1.299	.295	.020	11.54	5.64	.25	6
25	0.849	.757	.042	2.17	5.30	.25	.796	.562	.031	7.13	5.46	.25	1.268	.288	.019	11.26	5.51	.25	5
26	0.827	.738	.041	2.12	5.17	.24	.776	.547	.030	6.95	5.33	.24	1.236	.281	.019	10.97	5.37	.24	4
27	0.806	.718	.040	2.06	5.03	.23	.756	.533	.029	6.76	5.19	.24	1.203	.273	.018	10.68	5.23	.24	3
28	0.785	.699	.039	2.01	4.90	.23	.736	.519	.029	6.58	5.05	.23	1.171	.266	.018	10.40	5.09	.23	2
29	0.762	.680	.038	1.95	4.76	.22	.715	.504	.028	6.40	4.91	.22	1.139	.258	.017	10.12	4.95	.22	1
V 0 XI	0.741	.660	.037	1.89	4.63	.21	.695	.490	.027	6.21	4.77	.22	1.106	.251	.017	9.82	4.81	.22	I 0 VII
1	0.718	.640	.035	1.83	4.48	.21	.673	.474	.026	6.02	4.52	.21	1.071	.243	.016	9.51	4.66	.21	29
2	0.695	.619	.034	1.78	4.33	.20	.652	.459	.025	5.83	4.47	.20	1.038	.235	.016	9.21	4.51	.20	28
3	0.672	.599	.033	1.72	4.19	.19	.630	.444	.024	5.64	4.33	.20	1.003	.228	.015	8.91	4.36	.20	27
4	0.649	.579	.032	1.66	4.05	.19	.608	.428	.024	5.44	4.18	.19	0.968	.220	.015	8.60	4.21	.19	26
5	0.626	.558	.031	1.60	3.90	.18	.586	.413	.023	5.25	4.03	.18	0.934	.212	.014	8.29	4.06	.18	25
6	0.602	.537	.030	1.54	3.75	.17	.564	.397	.022	5.05	3.87	.18	0.898	.204	.014	7.98	3.90	.18	24
7	0.578	.516	.029	1.48	3.61	.17	.542	.382	.021	4.85	3.72	.17	0.863	.196	.013	7.67	3.75	.17	23
8	0.554	.494	.027	1.42	3.46	.16	.520	.366	.020	4.65	3.57	.16	0.826	.188	.013	7.35	3.59	.16	22
9	0.530	.473	.026	1.36	3.31	.15	.497	.350	.019	4.45	3.41	.16	0.792	.180	.012	7.03	3.44	.16	21
10	0.506	.451	.025	1.29	3.16	.15	.475	.334	.018	4.25	3.25	.15	0.756	.171	.012	6.71	3.28	.15	20
11	0.482	.429	.024	1.23	3.01	.14	.452	.318	.018	4.04	3.10	.14	0.719	.163	.011	6.38	3.12	.14	19
12	0.457	.407	.023	1.17	2.85	.13	.428	.302	.017	3.84	2.94	.13	0.682	.155	.010	6.06	2.96	.13	18
13	0.433	.386	.021	1.11	2.70	.13	.406	.286	.016	3.63	2.78	.13	0.646	.147	.010	5.74	2.81	.13	17
14	0.408	.364	.020	1.04	2.55	.12	.382	.270	.015	3.42	2.62	.12	0.609	.138	.009	5.41	2.65	.12	16
15	0.383	.341	.019	0.98	2.39	.11	.359	.253	.014	3.21	2.46	.11	0.572	.130	.009	5.07	2.48	.11	15
16	0.358	.319	.018	0.91	2.23	.10	.335	.236	.013	3.00	2.30	.10	0.534	.121	.008	4.74	2.32	.10	14
17	0.332	.297	.016	0.85	2.07	.10	.312	.220	.012	2.79	2.14	.10	0.497	.113	.008	4.41	2.15	.10	13
18	0.307	.274	.015	0.79	1.92	.09	.288	.203	.011	2.58	1.98	.09	0.459	.104	.007	4.07	1.99	.09	12
19	0.282	.251	.014	0.72	1.76	.08	.264	.186	.010	2.37	1.82	.08	0.421	.095	.006	3.74	1.83	.08	11
20	0.257	.229	.013	0.66	1.60	.07	.241	.170	.009	2.15	1.65	.08	0.384	.087	.006	3.40	1.66	.08	10
21	0.231	.206	.011	0.59	1.44	.07	.217	.153	.008	1.94	1.49	.07	0.345	.078	.005	3.07	1.50	.07	9
22	0.206	.184	.010	0.53	1.29	.06	.193	.136	.008	1.73	1.32	.06	0.308	.070	.005	2.73	1.34	.06	8
23	0.180	.161	.009	0.46	1.13	.05	.169	.119	.007	1.51	1.16	.05	0.269	.061	.004	2.39	1.17	.05	7
24	0.155	.138	.008	0.40	0.97	.04	.145	.102	.006	1.30	1.00	.05	0.231	.053	.004	2.05	1.00	.05	6
25	0.129	.115	.006	0.33	0.81	.04	.121	.085	.005	1.09	0.83	.04	0.193	.044	.003	1.71	0.84	.04	5
26	0.103	.092	.005	0.26	0.65	.03	.097	.068	.004	0.87	0.67	.03	0.155	.035	.002	1.37	0.67	.03	4
27	0.078	.069	.004	0.20	0.48	.02	.073	.051	.003	0.65	0.50	.02	0.116	.026	.002	1.03	0.50	.02	3
28	0.052	.046	.003	0.13	0.32	.01	.049	.034	.002	0.43	0.33	.02	0.077	.018	.001	0.68	0.34	.02	2
29	0.026	.023	.001	0.07	0.16	.01	.024	.017	.001	0.22	0.17	.01	0.039	.009	.001	0.34	0.17	.01	1
VI 0 O	0.000	.000	.000	0.00	0.00	.00	.000	.000	.000	0.00	0.00	.00	0.000	.000	.000	0.00	0.00	.00	O 0 VI
	Arg. = ☉ Long. + 0ˢ 23° 24′	Arg. = ☾ ☊ + 6ˢ 5° 50′	Arg. = 2☉ Long. + 6ˢ 4° 44′	Arg. = ☉ Long. + 11ˢ 28° 0	Arg. = ☾ ☊ + 3ˢ 19° 21′	Arg. = 2☉ Long. + 3ˢ 24° 23′	Arg. = ☉ Long. + 0ˢ 12° 12′	Arg. = ☾ ☊ + 5ˢ 27° 45′	Arg. = 2☉ Long. + 5ˢ 28° 10′	Arg. = ☉ Long. + 3ˢ 5° 25′	Arg. = ☾ ☊ + 3ˢ 9° 58′	Arg. = 2☉ Long. + 3ˢ 12° 26′	Arg. = ☉ Long. + 0ˢ 7° 40′	Arg. = ☾ ☊ + 5ˢ 15° 59′	Arg. = 2☉ Long. + 5ˢ 20° 52′	Arg. = ☉ Long. + 3ˢ 6° 48′	Arg. = ☾ ☊ + 3ˢ 6° 15′	Arg. = 2☉ Long. + 3ˢ 7° 40′	
	Aber.	☾ Nut.	☉ Nut.	Aber.	☾ Nut.	☉ Nut.	Aber.	☾ Nut.	☉ Nut.	Aber.	☾ Nut.	☉ Nut.	Aber.	☾ Nut.	☉ Nut.	Aber.	☾ Nut.	☉ Nut.	
	R. A.			N. P. D			R. A.			N. P. D.			R. A.			N. P. D.			
	ANTARES 1830.						RASALGETI 1830.						ALWAID 1830.						

The CORRECTIONS of 48 principal Stars, &c. continued.

1830. Common Argument.	α OPHIUCHI 1820. Right Ascension. 17h 26m 35s.16 Ann. Var. +2^{s}.77			N. P. Distance. 77° 18′ 0″.55 Ann. Var. +3″.10			γ DRACONIS 1820. Right Ascension. 17h 52m 25s.89 Ann. Var. +1^{s}.38			N. P. Distance. 38° 29′ 8″.45 Ann. Var. +0″.70			ε SAGITTARII 1820. Right Ascension. 18h 12m 13s.1 Ann. Var. +3^{s}.97			N. P. Distance. 124° 27′ 32″.6 Ann. Var. +0″.95			1830. Common Argument.
+ S. − S.	Aber.	☾ Nut.	⊙ Nut.	Aber.	☾ Nut.	⊙ Nut.	Aber.	☾ Nut.	⊙ Nut.	Aber.	☾ Nut.	⊙ Nut.	Aber.	☾ Nut.	⊙ Nut.	Aber.	☾ Nut.	⊙ Nut.	+ S. − S.
III 0° IX	1^{s}.382	1^{s}.005	0^{s}.055	11″.926	9″.604	0″.434	2^{s}.169	0^{s}.502	0^{s}.070	19″.562	9″.647	0″.434	1^{s}.637	1^{s}.430	0^{s}.085	3″.896	9″.642	0″.434	III 0° IX
1	1.381	1.004	.055	11.91	9.60	.43	2.168	.502	.070	19.55	9.64	.43	1.636	1.429	.085	3.89	9.64	.43	29
2	1.380	1.002	.055	11.90	9.59	.43	2.166	.502	.070	19.54	9.63	.43	1.635	1.428	.085	3.89	9.64	.43	28
3	1.379	1.000	.055	11.89	9.59	.43	2.164	.501	.070	19.53	9.62	.43	1.634	1.427	.085	3.89	9.63	.43	27
4	1.378	0.998	.055	11.88	9.58	.43	2.162	.501	.070	19.51	9.61	.43	1.632	1.426	.085	3.89	9.62	.43	26
5	1.377	0.996	.055	11.87	9.57	.43	2.159	.500	.070	19.49	9.60	.43	1.630	1.424	.085	3.88	9.61	.43	25
6	1.375	0.994	.055	11.85	9.55	.43	2.156	.499	.070	19.46	9.59	.43	1.628	1.422	.084	3.88	9.59	.43	24
7	1.373	0.992	.055	11.83	9.53	.43	2.152	.498	.069	19.43	9.57	.43	1.625	1.419	.084	3.87	9.57	.43	23
8	1.370	0.990	.055	11.81	9.51	.43	2.147	.497	.069	19.39	9.55	.43	1.622	1.416	.084	3.86	9.55	.43	22
9	1.367	0.988	.055	11.78	9.49	.43	2.142	.496	.069	19.34	9.53	.43	1.618	1.413	.084	3.85	9.53	.43	21
10	1.363	0.986	.055	11.75	9.46	.43	2.136	.494	.069	19.29	9.50	.43	1.613	1.409	.084	3.84	9.50	.43	20
11	1.358	0.983	.054	11.71	9.43	.43	2.129	.493	.069	19.23	9.47	.43	1.607	1.405	.083	3.83	9.47	.43	19
12	1.353	0.980	.054	11.67	9.40	.43	2.121	.491	.068	19.16	9.44	.43	1.601	1.400	.083	3.82	9.44	.43	18
13	1.347	0.976	.054	11.62	9.36	.42	2.113	.489	.068	19.08	9.40	.42	1.595	1.394	.083	3.81	9.40	.42	17
14	1.341	0.971	.054	11.57	9.32	.42	2.104	.487	.068	18.99	9.36	.42	1.588	1.387	.082	3.80	9.36	.42	16
15	1.335	0.966	.054	11.51	9.28	.42	2.094	.485	.068	18.90	9.32	.42	1.581	1.380	.082	3.78	9.32	.42	15
16	1.328	0.961	.053	11.45	9.24	.42	2.084	.483	.067	18.80	9.27	.42	1.573	1.373	.081	3.76	9.28	.42	14
17	1.321	0.956	.053	11.39	9.19	.42	2.073	.480	.067	18.70	9.22	.42	1.565	1.366	.081	3.74	9.23	.42	13
18	1.314	0.951	.053	11.33	9.14	.41	2.062	.478	.067	18.60	9.17	.41	1.557	1.359	.081	3.72	9.18	.41	12
19	1.307	0.945	.052	11.27	9.09	.41	2.050	.475	.066	18.49	9.12	.41	1.548	1.351	.080	3.69	9.13	.41	11
20	1.299	0.939	.052	11.20	9.03	.41	2.037	.472	.066	18.38	9.06	.41	1.539	1.343	.080	3.66	9.07	.41	10
21	1.290	0.933	.052	11.12	8.97	.41	2.024	.469	.065	18.26	9.00	.41	1.529	1.334	.079	3.64	9.01	.41	9
22	1.281	0.927	.051	11.04	8.91	.40	2.010	.466	.065	18.13	8.94	.40	1.519	1.325	.079	3.61	8.94	.40	8
23	1.272	0.920	.051	10.97	8.84	.40	1.996	.463	.064	18.00	8.87	.40	1.508	1.315	.078	3.59	8.87	.40	7
24	1.262	0.913	.051	10.89	8.77	.40	1.981	.459	.064	17.86	8.80	.40	1.496	1.306	.078	3.56	8.81	.40	6
25	1.252	0.906	.050	10.80	8.70	.39	1.966	.455	.063	17.72	8.74	.39	1.484	1.295	.077	3.53	8.74	.39	5
26	1.241	0.898	.050	10.71	8.63	.39	1.950	.451	.063	17.57	8.67	.39	1.472	1.283	.076	3.50	8.67	.39	4
27	1.231	0.890	.049	10.62	8.55	.39	1.932	.447	.062	17.42	8.59	.39	1.459	1.272	.076	3.47	8.59	.39	3
28	1.220	0.882	.049	10.52	8.47	.38	1.914	.443	.062	17.27	8.51	.38	1.446	1.261	.075	3.44	8.51	.38	2
29	1.209	0.874	.048	10.42	8.39	.38	1.897	.439	.061	17.11	8.42	.38	1.432	1.250	.074	3.41	8.42	.38	1
IV 0 X	1.197	0.865	.048	10.32	8.31	.38	1.879	.434	.061	16.94	8.34	.38	1.418	1.237	.074	3.37	8.34	.38	II 0 VIII
1	1.184	0.856	.047	10.22	8.22	.37	1.859	.430	.060	16.77	8.25	.37	1.403	1.224	.073	3.34	8.26	.37	29
2	1.172	0.847	.047	10.11	8.14	.37	1.839	.426	.059	16.59	8.17	.37	1.388	1.212	.072	3.30	8.17	.37	28
3	1.159	0.838	.046	10.00	8.05	.36	1.819	.421	.059	16.40	8.08	.36	1.373	1.199	.071	3.27	8.08	.36	27
4	1.145	0.828	.046	9.88	7.96	.36	1.798	.416	.058	16.21	7.99	.36	1.357	1.185	.070	3.23	7.99	.36	26
5	1.131	0.818	.045	9.76	7.87	.35	1.776	.411	.057	16.02	7.90	.35	1.341	1.171	.070	3.19	7.90	.35	25
6	1.117	0.808	.045	9.63	7.77	.35	1.753	.406	.057	15.82	7.79	.35	1.324	1.156	.069	3.15	7.80	.35	24
7	1.103	0.798	.044	9.51	7.67	.35	1.731	.401	.056	15.61	7.69	.35	1.307	1.141	.068	3.11	7.70	.35	23
8	1.088	0.787	.044	9.38	7.57	.34	1.709	.395	.055	15.41	7.60	.34	1.290	1.125	.067	3.07	7.60	.34	22
9	1.073	0.776	.043	9.25	7.46	.34	1.685	.390	.054	15.20	7.49	.34	1.271	1.109	.066	3.03	7.50	.34	21
10	1.058	0.765	.042	9.12	7.35	.33	1.660	.384	.054	14.98	7.38	.33	1.253	1.093	.065	2.98	7.38	.33	20
11	1.042	0.754	.042	8.99	7.24	.33	1.635	.379	.053	14.75	7.27	.33	1.234	1.077	.064	2.94	7.27	.33	19
12	1.026	0.743	.041	8.85	7.13	.32	1.611	.373	.052	14.53	7.16	.32	1.216	1.061	.063	2.89	7.16	.32	18
13	1.010	0.731	.040	8.71	7.02	.32	1.586	.367	.051	14.30	7.05	.32	1.197	1.044	.062	2.85	7.05	.32	17
14	0.993	0.719	.040	8.57	6.91	.31	1.560	.361	.050	14.06	6.93	.31	1.178	1.027	.061	2.80	6.94	.31	16
IV 15 X	0.976	0.706	.039	8.42	6.79	.31	1.533	.355	.050	13.82	6.81	.31	1.158	1.010	.060	2.75	6.82	.31	I 15 VII
	Arg. = ⊙ Long. + 0ˢ 7° 34′	Arg. = ☾ ☊ + 5ˢ 28° 49′	Arg. = 2⊙ Long. + 5ˢ 29° 2′	Arg. = ⊙ Long. + 3ˢ 3° 4′	Arg. = ☾ ☊ + 3ˢ 6° 9′	Arg. = 2⊙ Long. + 3ˢ 7° 34′	Arg. = ⊙ Long. + 0ˢ 1° 41′	Arg. = ☾ ☊ + 5ˢ 27° 8′	Arg. = 2 ⊙ Long. + 5ˢ 27° 36′	Arg. = ⊙ Long. + 3ˢ 1° 29′	Arg. = ☾ ☊ + 3ˢ 1° 21′	Arg. = 2⊙ Long. + 3ˢ 1° 41′	Arg. = ⊙ Long. + 11ˢ 27° 2′	Arg. = ☾ ☊ + 5ˢ 29° 0′	Arg. = 2⊙ Long. + 5ˢ 29° 15′	Arg. = ⊙ Long. + 8ˢ 20° 29′	Arg. = ☾ ☊ + 2ˢ 27° 36′	Arg. = 2⊙ Long. + 2ˢ 27° 5′	
	Aber.	☾ Nut.	⊙ Nut.	Aber.	☾ Nut.	⊙ Nut.	Aber.	☾ Nut.	⊙ Nut.	Aber.	☾ Nut.	⊙ Nut.	Aber.	☾ Nut.	⊙ Nut.	Aber.	☾ Nut.	⊙ Nut.	
	R. A.			N. P. D.			R. A.			N. P. D.			R. A.			N. P. D.			
	RASALAGUE 1830.						ETAMIN 1830.						KAUS AUSTRALIS 1830.						

The CORRECTIONS of 48 principal Stars, &c. continued.

1830. Common Argument. + − S. S.	α OPHIUCHI 1820. Right Ascension. h. m. s. 17 26 35 .16 Ann. Var. +2s.77 Aber.	☾ Nut.	☉ Nut.	N. P. Distance. 77° 18′ 0″.55 Ann. Var. +3″.10 Aber.	☾ Nut.	☉ Nut.
IV 15° X	0s.976	0s.706	0s.039	8″.42	6″.79	0″.30
16	.959	.694	.038	8 .27	6 .67	.30
17	.941	.681	.038	8 .12	6 .55	.29
18	.923	.668	.037	7 .97	6 .42	.29
19	.905	.655	.036	7 .81	6 .29	.28
20	.887	.642	.036	7 .66	6 .17	.28
21	.869	.629	.035	7 .50	6 .04	.27
22	.850	.616	.034	7 .34	5 .91	.26
23	.831	.602	.033	7 .17	5 .78	.26
24	.812	.588	.032	7 .00	5 .64	.25
25	.792	.574	.032	6 .83	5 .51	.25
26	.772	.559	.031	6 .66	5 .37	.24
27	.752	.544	.030	6 .48	5 .23	.23
28	.732	.530	.029	6 .31	5 .09	.23
29	.712	.516	.028	6 .14	4 .95	.22
V 0 XI	.691	.501	.028	5 .96	4 .81	.21
1	.669	.485	.027	5 .78	4 .66	.21
2	.648	.469	.026	5 .59	4 .51	.20
3	.627	.454	.025	5 .41	4 .36	.19
4	.605	.438	.024	5 .22	4 .21	.19
5	.583	.422	.023	5 .04	4 .06	.18
6	.561	.406	.022	4 .85	3 .90	.17
7	.539	.390	.022	4 .66	3 .75	.17
8	.517	.374	.021	4 .47	3 .59	.16
9	.495	.358	.020	4 .27	3 .44	.15
10	.472	.342	.019	4 .07	3 .28	.15
11	.449	.325	.018	3 .83	3 .13	.14
12	.427	.309	.017	3 .68	2 .97	.13
13	.403	.292	.016	3 .48	2 .81	.13
14	.381	.275	.015	3 .28	2 .65	.12
15	.357	.258	.014	3 .08	2 .48	.11
16	.334	.241	.013	2 .88	2 .32	.10
17	.310	.224	.012	2 .68	2 .16	.10
18	.287	.207	.011	2 .47	1 .99	.09
19	.263	.190	.011	2 .27	1 .83	.08
20	.239	.173	.010	2 .07	1 .67	.07
21	.216	.156	.009	1 .86	1 .50	.07
22	.193	.139	.008	1 .66	1 .34	.06
23	.169	.122	.007	1 .45	1 .17	.05
24	.145	.105	.006	1 .25	1 .01	.04
25	.121	.087	.005	1 .04	0 .84	.04
26	.097	.070	.004	0 .83	0 .67	.03
27	.072	.052	.003	0 .62	0 .50	.02
28	.048	.035	.002	0 .42	0 .33	.01
29	.024	.017	.001	0 .21	0 .17	.01
VI 0 O	.000	.000	.000	0 .00	0 .00	.00
	Arg. = ☉ Long. + 0s 7° 34′	Arg. = ☾ ☊ + 5s 23° 49′	Arg. = 2☉ Long. + 5s 29° 2′	Arg. = ☉ Long. + 3s 3° 4′	Arg. = ☾ ☊ + 3s 6° 9′	Arg. = 2☉ Long. + 3s 7° 31′
	Aber.	☾ Nut.	☉ Nut.	Aber.	☾ Nut.	☉ Nut.
	R. A.			N. P. D.		
	RASALAGUE 1830.					

1830. Common Argument. + − S. S.	γ DRACONIS 1820. Right Ascension. h m. s. 17 52 25 .89 Ann. Var. +1s.38 Aber.	☾ Nut.	☉ Nut.	N. P. Distance. 38° 29′ 8″.45 Ann. Var. +0″.70 Aber.	☾ Nut.	☉ Nut.
IV 15° X	1s.533	0s.355	0s.050	13″.82	6″.81	0″.30
16	1 .506	.348	.049	13 .58	6 .69	.30
17	1 .479	.342	.048	13 .33	6 .57	.29
18	1 .451	.335	.047	13 .08	6 .45	.29
19	1 .422	.329	.046	12 .82	6 .32	.28
20	1 .393	.322	.045	12 .57	6 .19	.28
21	1 .364	.316	.044	12 .31	6 .06	.27
22	1 .335	.309	.043	12 .04	5 .93	.26
23	1 .305	.302	.042	11 .77	5 .80	.26
24	1 .274	.295	.041	11 .50	5 .67	.25
25	1 .243	.288	.040	11 .21	5 .53	.25
26	1 .213	.280	.039	10 .93	5 .39	.24
27	1 .182	.273	.038	10 .65	5 .25	.23
28	1 .150	.266	.037	10 .37	5 .11	.23
29	1 .117	.258	.036	10 .08	4 .97	.22
V 0 XI	1 .084	.251	.035	9 .78	4 .82	.21
1	1 .051	.243	.034	9 .47	4 .67	.21
2	1 .008	.235	.033	9 .17	4 .52	.20
3	0 .984	.228	.032	8 .87	4 .37	.19
4	0 .950	.220	.031	8 .57	4 .22	.19
5	0 .916	.212	.030	8 .26	4 .07	.18
6	0 .881	.204	.028	7 .95	3 .91	.17
7	0 .846	.196	.027	7 .64	3 .76	.17
8	0 .811	.188	.026	7 .32	3 .61	.16
9	0 .776	.180	.025	7 .00	3 .45	.15
10	0 .741	.171	.024	6 .68	3 .29	.15
11	0 .705	.163	.023	6 .36	3 .14	.14
12	0 .669	.155	.022	6 .04	2 .98	.13
13	0 .633	.147	.020	5 .72	2 .82	.13
14	0 .597	.138	.019	5 .39	2 .66	.12
15	0 .561	.130	.018	5 .06	2 .49	.11
16	0 .524	.121	.017	4 .72	2 .33	.10
17	0 .487	.113	.016	4 .39	2 .16	.10
18	0 .450	.104	.015	4 .06	2 .00	.09
19	0 .413	.095	.013	3 .72	1 .83	.08
20	0 .376	.087	.012	3 .39	1 .67	.07
21	0 .339	.078	.011	3 .06	1 .51	.07
22	0 .302	.070	.010	2 .72	1 .34	.06
23	0 .264	.061	.009	2 .38	1 .18	.05
24	0 .227	.052	.007	2 .05	1 .01	.04
25	0 .189	.044	.006	1 .71	0 .84	.04
26	0 .152	.035	.005	1 .37	0 .67	.03
27	0 .114	.026	.004	1 .02	0 .51	.02
28	0 .076	.017	.002	0 .68	0 .34	.01
29	0 .038	.009	.001	0 .34	0 .17	.01
VI 0 O	0 .000	.000	.000	0 .00	0 .00	.00
	Arg. = ☉ Long. + 0s 1° 41′	Arg. = ☾ ☊ + 5s 27° 8′	Arg. = 2☉ Long. + 5s 27° 36′	Arg. = ☉ Long. + 3s 1° 29′	Arg. = ☾ ☊ + 3s 1° 21′	Arg. = 2☉ Long. + 3s 1° 41′
	Aber.	☾ Nut.	☉ Nut.	Aber.	☾ Nut.	☉ Nut.
	R. A.			N. P. D.		
	ETAMIN 1830.					

1830. Common Argument. + − S. S.	ε SAGITTARII 1820. Right Ascension. h. m. s. 18 12 13 .1 Ann. Var. +3s.97 Aber.	☾ Nut.	☉ Nut.	N. P. Distance. 124° 27′ 32″.6 Ann. Var. +0″.95 Aber.	☾ Nut.	☉ Nut.	1830. Common Argument. + − S. S.
IV 15° X	1s.158	1s.010	0s.060	2″.75	6″.82	0″.30	IV 15° VII
16	1 .136	0 .992	.059	2 .70	6 .69	.30	14
17	1 .115	0 .974	.058	2 .65	6 .57	.29	13
18	1 .094	0 .955	.057	2 .61	6 .45	.29	12
19	1 .072	0 .936	.056	2 .55	6 .32	.28	11
20	1 .051	0 .917	.055	2 .50	6 .19	.28	10
21	1 .029	0 .899	.054	2 .45	6 .06	.27	9
22	1 .007	0 .879	.052	2 .40	5 .93	.26	8
23	0 .984	0 .859	.051	2 .34	5 .80	.26	7
24	0 .961	0 .839	.050	2 .29	5 .66	.25	6
25	0 .938	0 .819	.049	2 .23	5 .53	.25	5
26	0 .915	0 .799	.048	2 .18	5 .39	.24	4
27	0 .891	0 .778	.046	2 .12	5 .25	.23	3
28	0 .867	0 .757	.045	2 .06	5 .11	.23	2
29	0 .843	0 .736	.044	2 .01	4 .97	.22	1
V 0 XI	0 .818	0 .715	.043	1 .95	4 .83	.21	I 0 VII
1	0 .793	0 .692	.041	1 .89	4 .68	.21	29
2	0 .768	0 .670	.040	1 .83	4 .53	.20	28
3	0 .743	0 .648	.039	1 .77	4 .37	.19	27
4	0 .717	0 .626	.037	1 .71	4 .22	.19	26
5	0 .691	0 .604	.036	1 .65	4 .07	.18	25
6	0 .665	0 .581	.035	1 .58	3 .92	.17	24
7	0 .639	0 .558	.033	1 .52	3 .77	.17	23
8	0 .612	0 .535	.032	1 .46	3 .61	.16	32
9	0 .586	0 .512	.030	1 .39	3 .45	.15	21
10	0 .559	0 .488	.029	1 .33	3 .29	.15	20
11	0 .533	0 .465	.028	1 .27	3 .14	.14	19
12	0 .506	0 .441	.026	1 .20	2 .98	.13	18
13	0 .478	0 .417	.025	1 .14	2 .82	.13	17
14	0 .451	0 .393	.023	1 .07	2 .66	.12	16
15	0 .423	0 .369	.022	1 .01	2 .49	.11	15
16	0 .396	0 .345	.020	0 .94	2 .33	.10	14
17	0 .368	0 .321	.019	0 .88	2 .17	.10	13
18	0 .340	0 .297	.018	0 .81	2 .00	.09	12
19	0 .312	0 .272	.016	0 .74	1 .84	.08	11
20	0 .284	0 .248	.015	0 .68	1 .67	.07	10
21	0 .256	0 .223	.013	0 .61	1 .51	.07	9
22	0 .228	0 .199	.012	0 .54	1 .34	.06	8
23	0 .200	0 .174	.010	0 .47	1 .17	.05	7
24	0 .171	0 .150	.009	0 .41	1 .01	.04	6
25	0 .143	0 .125	.007	0 .34	0 .84	.04	5
26	0 .114	0 .100	.006	0 .27	0 .67	.03	4
27	0 .086	0 .075	.004	0 .20	0 .51	.02	3
28	0 .057	0 .050	.003	0 .14	0 .34	.01	2
29	0 .029	0 .025	.001	0 .07	0 .17	.01	1
VI 0 O	0 .000	0 .000	.000	0 .00	0 .00	.00	O 0 VI
	Arg. = ☉ Long. + 11s 27° 2′	Arg. = ☾ ☊ + 5s 29° 0′	Arg. = 2☉ Long. + 5s 29° 15′	Arg. = ☉ Long. + 8s 20° 29′	Arg. = ☾ ☊ + 2s 27° 36′	Arg. = 2☉ Long. + 2s 27° 5′	
	Aber.	☾ Nut.	☉ Nut.	Aber.	☾ Nut.	☉ Nut.	
	R. A.			N. P. D.			
	KAUS AUSTRALIS 1830.						

The CORRECTIONS of 48 principal Stars, &c. continued.

α LYRÆ 1820. (VEGA 1830.)

Right Ascension: 18h 30m 50s.89. Ann Var. +2ˢ.03.

N. P. Distance: 51° 22′ 39″.45. Ann. Var. −3″.00.

1830. Common Argument. +	R. A. Aber.	R. A. ☾ Nut.	R. A. ⊙ Nut.	N. P. D. Aber.	N. P. D. ☾ Nut.	N. P. D. ⊙ Nut.	1830. Common Argument. −
S. III 0°	1ˢ.727	0ˢ.969	0ˢ.040	17″.871	9″.608	0″.434	S. IX 0°
1	1.726	.968	.040	17.86	9.60	.434	29
2	1.725	.967	.040	17.85	9.59	.434	28
3	1.723	.966	.040	17.83	9.59	.433	27
4	1.721	.965	.040	17.81	9.58	.433	26
5	1.719	.964	.040	17.79	9.57	.432	25
6	1.716	.963	.040	17.76	9.55	.432	24
7	1.713	.961	.040	17.73	9.53	.431	23
8	1.709	.959	.040	17.69	9.51	.430	22
9	1.705	.957	.039	17.65	9.49	.429	21
10	1.701	.955	.039	17.60	9.47	.428	20
11	1.696	.952	.039	17.54	9.44	.427	19
12	1.690	.949	.039	17.48	9.41	.425	18
13	1.683	.945	.039	17.41	9.37	.423	17
14	1.676	.941	.039	17.33	9.33	.421	16
15	1.668	.936	.039	17.25	9.28	.419	15
16	1.660	.931	.039	17.16	9.23	.417	14
17	1.651	.926	.038	17.07	9.18	.415	13
18	1.642	.921	.038	16.98	9.13	.413	12
19	1.632	.916	.038	16.89	9.08	.411	11
20	1.622	.911	.038	16.80	9.03	.408	10
21	1.611	.905	.037	16.70	8.97	.405	9
22	1.600	.898	.037	16.59	8.91	.402	8
23	1.589	.892	.037	16.46	8.84	.399	7
24	1.577	.885	.037	16.33	8.77	.396	6
25	1.565	.878	.036	16.19	8.70	.393	5
26	1.552	.870	.036	16.06	8.63	.390	4
27	1.538	.863	.036	15.92	8.55	.387	3
28	1.524	.855	.035	15.77	8.47	.383	2
29	1.510	.847	.035	15.62	8.39	.379	1
IV 0	1.495	.838	.035	15.47	8.31	.376	VIII 0
1	1.480	.829	.034	15.31	8.22	.372	29
2	1.464	.821	.034	15.15	8.14	.368	28
3	1.448	.812	.034	14.99	8.05	.364	27
4	1.431	.803	.033	14.82	7.96	.360	26
5	1.413	.793	.033	14.63	7.86	.355	25
6	1.395	.783	.032	14.45	7.76	.351	24
7	1.378	.773	.032	14.26	7.66	.346	23
8	1.360	.763	.032	14.07	7.56	.342	22
9	1.341	.753	.031	13.88	7.46	.337	21
10	1.321	.742	.031	13.68	7.35	.332	20
11	1.301	.731	.030	13.48	7.24	.327	19
12	1.281	.720	.030	13.27	7.13	.322	18
13	1.261	.708	.029	13.06	7.02	.317	17
14	1.241	.697	.029	12.84	6.91	.312	16
IV 15	1.220	.685	.028	12.62	6.78	.307	VII 15

Arguments (R. A.): Aber.: Arg. = ⊙ Long. + 11ˢ 22° 50′; ☾ Nut.: Arg. = ☾ ☊ + 6ˢ 5° 30′; ⊙ Nut.: Arg. = 2⊙ Long. + 6ˢ 4° 8′.

Arguments (N. P. D.): Aber.: Arg. = ⊙ Long. + 2ˢ 24° 30′; ☾ Nut.: Arg. = ☾ ☊ + 2ˢ 24° 11′; ⊙ Nut.: Arg. = 2⊙ Long. + 2ˢ 22° 5′.

γ AQUILÆ 1830. (TARAZED 1830.)

Right Ascension: 19h 37m 42s.28. Ann. Var. +2″.85.

N. P. Distance: 79° 49′ 2″.22. Ann. Var. −8″.38.

1830. Common Argument. +	R. A. Aber.	R. A. ☾ Nut.	R. A. ⊙ Nut.	N. P. D. Aber.	N. P. D. ☾ Nut.	N. P. D. ⊙ Nut.	1830. Common Argument. −
S. III 0°	1ˢ.353	1ˢ.021	0ˢ.057	11″.112	9″.270	0″.428	S. IX 0°
1	1.352	1.021	.057	11.10	9.27	.428	29
2	1.351	1.020	.057	11.09	9.26	.428	28
3	1.350	1.019	.057	11.08	9.25	.428	27
4	1.349	1.018	.057	11.07	9.24	.427	26
5	1.348	1.017	.057	11.06	9.23	.427	25
6	1.346	1.015	.057	11.04	9.22	.426	24
7	1.344	1.013	.057	11.02	9.20	.425	23
8	1.341	1.011	.056	11.00	9.18	.424	22
9	1.338	1.009	.056	10.98	9.16	.423	21
10	1.334	1.006	.056	10.95	9.13	.422	20
11	1.329	1.003	.056	10.92	9.10	.420	19
12	1.324	0.999	.056	10.88	9.07	.419	18
13	1.318	0.995	.056	10.84	9.03	.417	17
14	1.312	0.991	.055	10.79	8.99	.415	16
15	1.306	0.986	.055	10.74	8.95	.413	15
16	1.300	0.981	.055	10.69	8.91	.411	14
17	1.294	0.976	.055	10.63	8.86	.409	13
18	1.287	0.971	.054	10.57	8.81	.407	12
19	1.280	0.965	.054	10.51	8.76	.405	11
20	1.271	0.959	.054	10.45	8.70	.402	10
21	1.263	0.953	.053	10.38	8.64	.400	9
22	1.254	0.946	.053	10.31	8.58	.397	8
23	1.245	0.939	.053	10.24	8.52	.394	7
24	1.236	0.932	.052	10.16	8.46	.391	6
25	1.226	0.925	.052	10.07	8.39	.388	5
26	1.217	0.917	.051	9.99	8.32	.385	4
27	1.206	0.909	.051	9.90	8.25	.381	3
28	1.195	0.901	.050	9.81	8.17	.378	2
29	1.184	0.892	.050	9.72	8.10	.374	1
IV 0	1.172	0.884	.049	9.62	8.02	.371	VIII 0
1	1.160	0.875	.049	9.52	7.94	.367	29
2	1.148	0.865	.048	9.42	7.86	.363	28
3	1.135	0.856	.048	9.32	7.77	.359	27
4	1.121	0.846	.047	9.21	7.68	.355	26
5	1.108	0.836	.047	9.10	7.58	.351	25
6	1.094	0.825	.046	8.99	7.49	.346	24
7	1.080	0.815	.046	8.87	7.39	.342	23
8	1.066	0.804	.045	8.75	7.30	.337	22
9	1.051	0.793	.044	8.63	7.20	.333	21
10	1.035	0.782	.044	8.51	7.09	.328	20
11	1.020	0.770	.043	8.38	6.99	.323	19
12	1.005	0.758	.042	8.25	6.88	.318	18
13	0.988	0.746	.042	8.12	6.77	.313	17
14	0.972	0.734	.041	7.99	6.66	.308	16
IV 15	0.956	0.722	.040	7.86	6.55	.303	VII 15

Arguments (R. A.): Aber.: Arg. = ⊙ Long. + 11ˢ 7° 16′; ☾ Nut.: Arg. = ☽ ☊ + 6ˢ 2° 42′; ⊙ Nut.: Arg. = 2⊙ Long. + 6ˢ 2° 11′.

Arguments (N. P. D.): Aber.: Arg. = ⊙ Long. + 2ˢ 22° 17′; ☾ Nut.: Arg. = ☾ ☊ + 2ˢ 11° 14′; ⊙ Nut.: Arg. = 2⊙ Long. + 2ˢ 7° 16′.

α AQUILÆ 1830. (ALTAIR 1830.)

Right Ascension: 19h 42m 0s.14. Ann. Var. +2ˢ.93.

N. P. Distance: 81° 35′ 55″.39. Ann. Var. −9″.06.

1830. Common Argument. +	R. A. Aber.	R. A. ☾ Nut.	R. A. ⊙ Nut.	N. P. D. Aber.	N. P. D. ☾ Nut.	N. P. D. ⊙ Nut.	1830. Common Argument. −
S. III 0°	1ˢ.345	1ˢ.034	0ˢ.059	10″.489	9″.237	0″.428	S. IX 0°
1	1.344	1.034	.059	10.48	9.23	.428	29
2	1.343	1.033	.059	10.47	9.22	.428	28
3	1.342	1.032	.059	10.46	9.22	.428	27
4	1.341	1.031	.059	10.45	9.21	.427	26
5	1.340	1.030	.059	10.44	9.20	.427	25
6	1.338	1.028	.059	10.42	9.19	.426	24
7	1.336	1.026	.059	10.40	9.17	.425	23
8	1.333	1.024	.058	10.38	9.15	.424	22
9	1.330	1.022	.058	10.35	9.13	.423	21
10	1.326	1.019	.058	10.32	9.10	.422	20
11	1.322	1.016	.058	10.29	9.07	.420	19
12	1.317	1.012	.058	10.25	9.04	.419	18
13	1.312	1.008	.058	10.21	9.00	.417	17
14	1.306	1.004	.057	10.17	8.96	.415	16
15	1.300	0.999	.057	10.12	8.92	.413	15
16	1.294	0.994	.057	10.07	8.87	.411	14
17	1.287	0.989	.057	10.02	8.82	.409	13
18	1.280	0.984	.056	9.97	8.77	.407	12
19	1.272	0.978	.056	9.91	8.72	.405	11
20	1.264	0.972	.055	9.85	8.67	.402	10
21	1.256	0.965	.055	9.78	8.61	.400	9
22	1.247	0.958	.055	9.71	8.55	.397	8
23	1.238	0.952	.054	9.65	8.49	.394	7
24	1.229	0.945	.054	9.57	8.43	.391	6
25	1.220	0.937	.054	9.50	8.36	.388	5
26	1.210	0.929	.053	9.42	8.29	.385	4
27	1.199	0.921	.053	9.34	8.21	.381	3
28	1.188	0.912	.052	9.25	8.14	.378	2
29	1.177	0.904	.052	9.16	8.07	.374	1
IV 0	1.165	0.895	.051	9.07	7.99	.371	VIII 0
1	1.153	0.886	.051	8.98	7.91	.367	29
2	1.141	0.877	.050	8.89	7.83	.363	28
3	1.128	0.867	.050	8.79	7.74	.359	27
4	1.115	0.857	.049	8.68	7.65	.355	26
5	1.101	0.847	.048	8.58	7.56	.351	25
6	1.087	0.836	.048	8.47	7.46	.346	24
7	1.063	0.825	.047	8.36	7.37	.342	23
8	1.059	0.814	.047	8.25	7.27	.337	22
9	1.045	0.803	.046	8.14	7.17	.333	21
10	1.029	0.792	.045	8.02	7.06	.328	20
11	1.014	0.780	.045	7.90	6.96	.323	19
12	0.999	0.768	.044	7.78	6.86	.318	18
13	0.983	0.756	.043	7.66	6.75	.313	17
14	0.967	0.744	.042	7.54	6.64	.308	16
IV 15	0.951	0.732	.042	7.41	6.52	.303	VII 15

Arguments (R. A.): Aber.: Arg. = ⊙ Long. + 11ˢ 6° 15′; ☾ Nut.: Arg. = ☾ ☊ + 6ˢ 2° 17′; ⊙ Nut.: Arg. = 2⊙ Long. + 6ˢ 1° 51′.

Arguments (N. P. D.): Aber.: Arg. = ⊙ Long. + 2ˢ 22° 58′; ☾ Nut.: Arg. = ☾ ☊ + 2ˢ 10° 21′; ⊙ Nut.: Arg. = 2⊙ Long. + 2ˢ 6° 16′.

The CORRECTIONS of 48 principal Stars, &c. continued.

1830. Common Argument.	α LYRÆ 1820. Right Ascension.			α LYRÆ 1820. N. P. Distance.			γ AQUILÆ 1820. Right Ascension.			γ AQUILÆ 1820. N. P. Distance.			α AQUILÆ 1820. Right Ascension.			α AQUILÆ 1820. N. P. Distance.			1830. Common Argument.
	h. m. s. 18 30 50 .89			51° 22′ 39″. 45			h. m. s. 19 37 42 .28			79° 49′ 2″. 22			h. m. s. 19 42 0 .14			81° 35′ 55″. 39			
	Ann. Var. +2s. 03			Ann. Var.—3″. 00			Ann. Var. +2s. 85			Ann. Var.—8″. 32			Ann. Var. +2s. 93			Ann. Var.—9″. 06			
+ S. − S.	Aber.	☾ Nut.	⊙Nut.	Aber.	☾ Nut.	⊙Nut.	Aber.	☾ Nut.	⊙Nut.	Aber.	☾ Nut.	⊙Nut.	Aber.	☾ Nut.	⊙Nut.	Aber.	☾ Nut.	⊙Nut.	+ S. − S.
IV 15° X	1s. 220	0s. 685	0s. 028	12″.62	6″. 78	0″. 307	0s. 956	0s. 722	0s. 040	7″. 86	6″. 55	0″. 303	0s. 951	0s. 732	0s. 042	7″. 41	6″. 52	0″. 303	IV 15° VII
16	1 . 198	. 673	. 028	12 .40	6 . 67	. 302	. 939	. 708	. 040	7 . 72	6 . 43	. 297	. 934	. 718	. 041	7 . 28	6 . 41	. 297	14
17	1 . 176	. 661	. 027	12 .18	6 . 55	. 296	. 922	. 696	. 039	7 . 58	6 . 32	. 292	. 917	. 705	. 040	7 . 15	6 . 30	. 292	13
18	1 . 154	. 648	. 027	11 .95	6 . 42	. 290	. 905	. 683	. 038	7 . 43	6 . 20	. 286	. 900	. 692	. 039	7 . 01	6 . 18	. 286	12
19	1 . 131	. 635	. 026	11 .71	6 . 29	. 285	. 887	. 669	. 037	7 . 28	6 . 07	. 281	. 882	. 678	. 039	6 . 87	6 . 06	. 281	11
20	1 . 108	. 622	. 026	11 .47	6 . 17	. 279	. 869	. 656	. 037	7 . 14	5 . 95	. 275	. 864	. 664	. 038	6 . 73	5 . 93	. 275	10
21	1 . 086	. 609	. 025	11 .23	6 . 05	. 273	. 851	. 642	. 036	6 . 99	5 . 83	. 270	. 846	. 651	. 037	6 . 59	5 . 81	. 270	9
22	1 . 062	. 596	. 025	10 .99	5 . 92	. 267	. 832	. 628	. 035	6 . 84	5 . 70	. 264	. 828	. 637	. 036	6 . 45	5 . 68	. 264	8
23	1 . 038	. 583	. 024	10 .74	5 . 78	. 261	. 814	. 614	. 034	6 . 68	5 . 57	. 258	. 809	. 622	. 035	6 . 31	5 . 56	. 258	7
24	1 . 014	. 569	. 023	10 .50	5 . 64	. 255	. 795	. 600	. 033	6 . 53	5 . 45	. 252	. 791	. 608	. 035	6 . 16	5 . 43	. 252	6
25	0 . 989	. 556	. 023	10 .24	5 . 51	. 249	. 776	. 585	. 033	6 . 37	5 . 32	. 246	. 772	. 593	. 034	6 . 01	5 . 30	. 246	5
26	0 . 965	. 542	. 022	9 .99	5 . 37	. 243	. 757	. 571	. 032	6 . 22	5 . 18	. 240	. 753	. 579	. 033	5 . 86	5 . 17	. 240	4
27	0 . 939	. 527	. 022	9 .72	5 . 23	. 236	. 737	. 556	. 031	6 . 05	5 . 04	. 233	. 732	. 563	. 032	5 . 71	5 . 03	. 233	3
28	0 . 914	. 513	. 021	9 .46	5 . 09	. 230	. 717	. 541	. 030	5 . 88	4 . 91	. 227	. 713	. 548	. 031	5 . 55	4 . 89	. 227	2
29	0 . 888	. 499	. 021	9 .20	4 . 95	. 223	. 697	. 526	. 029	5 . 72	4 . 77	. 220	. 693	. 533	. 030	5 . 40	4 . 76	. 220	1
V 0 XI	0 . 863	. 485	. 020	8 .93	4 . 81	. 217	. 677	. 511	. 028	5 . 56	4 . 64	. 214	. 673	. 518	. 030	5 . 24	4 . 62	. 214	I 0 VII
1	0 . 836	. 469	. 019	8 .65	4 . 66	. 210	. 656	. 495	. 028	5 . 39	4 . 49	. 207	. 652	. 502	. 029	5 . 08	4 . 48	. 207	29
2	0 . 809	. 454	. 019	8 .37	4 . 51	. 203	. 635	. 478	. 027	5 . 22	4 . 35	. 201	. 632	. 486	. 028	4 . 92	4 . 33	. 201	28
3	0 . 783	. 439	. 018	8 .10	4 . 36	. 197	. 614	. 463	. 026	5 . 04	4 . 20	. 194	. 611	. 469	. 027	4 . 76	4 . 18	. 194	27
4	0 . 756	. 424	. 018	7 .82	4 . 21	. 190	. 593	. 447	. 025	4 . 87	4 . 06	. 188	. 589	. 453	. 026	4 . 59	4 . 04	188	26
5	0 . 729	. 409	. 017	7 .54	4 . 06	. 183	. 572	. 432	. 024	4 . 69	3 . 91	. 181	. 568	. 437	. 025	4 . 43	3 . . 90	. 181	25
6	0 . 701	. 393	. 016	7 .26	3 . 90	. 176	. 550	. 415	. 023	4 . 52	3 . 76	. 174	. 547	. 420	. 024	4 . 26	3 . 75	. 174	24
7	0 . 673	. 378	. 016	6 .97	3 . 75	. 169	. 528	. 398	. 022	4 . 34	3 . 62	. 167	. 526	. 403	. 023	4 . 09	3 . 60	. 167	23
8	0 . 646	. 362	. 015	6 .68	3 . 59	. 162	. 517	. 382	. 021	4 . 16	3 . 47	. 160	. 503	. 387	. 022	3 . 92	3 . 45	. 160	32
9	0 . 618	. 347	. 014	6 .39	3 . 44	. 155	. 485	. 365	. 020	3 . 98	3 . 32	. 153	. 482	. 370	. 021	3 . 75	3 . 30	. 153	21
10	0 . 589	. 331	. 014	6 .10	3 . 28	. 148	. 462	. 349	. 019	3 . 80	3 . 17	. 146	. 460	. 353	. 020	3 . 58	3 . 15	. 146	20
11	0 . 561	. 315	. 013	5 .81	3 . 13	. 141	. 440	. 332	. 019	3 . 61	3 . 01	. 139	. 438	. 336	. 019	3 . 41	3 . 00	. 139	19
12	0 . 533	. 299	. 012	5 .52	2 . 97	. 134	. 417	. 315	. 018	3 . 43	2 . 86	. 132	. 415	. 319	. 018	3 . 23	2 . 85	. 132	18
13	0 . 504	. 283	. 012	5 .22	2 . 81	. 127	. 395	. 298	. 017	3 . 24	2 . 71	. 125	. 393	. 302	. 017	3 . 06	2 . 70	. 125	17
14	0 . 476	. 267	. 011	4 .92	2 . 64	. 120	. 371	. 281	. 016	3 . 06	2 . 55	. 118	. 371	. 285	. 016	2 . 89	2 . 54	. 118	16
15	0 . 446	. 250	. 010	4 .62	2 . 48	. 112	. 350	. 264	. 015	2 . 87	2 . 39	. 111	. 348	. 267	. 015	2 . 71	2 . 39	. 111	15
16	0 . 417	. 234	. 010	4 .32	2 . 32	. 105	. 327	. 247	. 014	2 . 68	2 . 24	. 103	. 325	. 250	. 014	2 . 53	2 . 23	. 103	14
17	0 . 388	. 217	. 009	4 .01	2 . 18	. 097	. 304	. 229	. 013	2 . 49	2 . 04	. 096	. 302	. 232	. 013	2 . 35	2 . 07	. 096	13
18	0 . 358	. 201	. 008	3 .71	1 . 99	. 090	. 281	. 212	. 012	2 . 31	1 . 92	. 089	. 279	. 215	. 012	2 . 18	1 . 92	. 089	12
19	0 . 329	. 184	. 008	3 .40	1 . 83	. 083	. 258	. 195	. 011	2 . 12	1 . 76	. 081	. 256	. 197	. 011	2 . 00	1 . 76	. 081	11
20	0 . 299	. 168	. 007	3 .10	1 . 67	. 075	. 235	. 177	. 010	1 . 93	1 . 61	.074	. 233	. 179	. 010	1 . 82	1 . 60	. 074	10
21	0 . 269	. 152	. 006	2 .79	1 . 50	. 068	. 211	. 160	. 009	1 . 74	1 . 45	. 067	. 210	. 162	. 009	1 . 64	1 . 44	. 067	9
22	0 . 240	. 135	. 006	2 .48	1 . 34	. 060	. 188	. 142	. 008	1 . 55	1 . 29	. 060	. 187	. 144	. 008	1 . 46	1 . 28	. 060	8
23	0 . 210	. 118	. 005	2 .18	1 . 17	. 053	. 165	. 124	. 007	1 . 35	1 . 13	. 052	. 164	. 126	. 007	1 . 28	1 . 12	. 052	7
24	0 . 181	. 101	. 004	1 .87	1 . 00	. 045	. 142	. 107	. 006	1 . 16	0 . 97	. 045	. 141	. 108	. 006	1 . 10	0 . 96	. 045	6
25	0 . 151	. 085	. 003	1 .56	0 . 81	. 038	. 118	. 089	. 005	0 . 97	0 . 81	. 037	. 118	. 090	. 005	0 . 92	0 . 81	. 037	5
26	0 . 121	. 068	. 003	1 .25	0 . 67	. 030	. 095	. 071	. 004	0 . 78	0 . 65	. 030	. 094	. 072	. 004	0 . 73	0 . 64	. 030	4
27	0 . 090	. 051	. 002	0 .94	0 . 50	. 023	. 071	. 053	. 003	0 . 58	0 . 48	. 022	. 071	. 054	. 003	0 . 55	0 . 48	. 022	3
28	0 . 060	. 034	. 001	0 .62	0 . 34	. 015	. 047	. 036	. 002	0 . 39	0 . 32	. 015	. 047	. 036	. 002	0 . 37	0 . 32	. 015	2
29	0 . 030	. 017	. 001	0 .31	0 . 17	. 008	. 024	. 018	. 001	0 . 19	0 . 16	. 007	. 024	. 018	. 001	0 . 18	0 . 16	. 007	1
VI 0 O	0 . 000	. 000	. 000	0 .00	0 . 00	. 000	. 000	. 000	. 000	0 . 00	0 . 00	. 000	. 000	. 000	. 000	0 . 00	0 . 00	. 000	O 0 VI
	Arg.=⊙ Long.+11s 22° 50′	Arg.=☾☊ + 6s 5° 30′	Arg.=2⊙Long.+6s 4° 8′	Arg.=⊙ Long.+2s 24° 30′	Arg.=☾☊ + 2s 24° 11′	Arg.=2⊙Long.−2s 22° 5′	Arg.=⊙ Long.+11s 7° 16′	Arg.=☾☊ + 6s 2° 42′	Arg.=2⊙Long.+6s 2° 11′	Arg.=⊙ Long.+2s 22° 17′	Arg.=☾☊ + 2s 11° 14′	Arg.=2⊙ Long.+2s 7° 16′	Arg.=⊙ Long.+11s 6° 15′	Arg.=☾☊ + 6s 2° 17′	Arg.=2⊙Long.+6s 1° 51′	Arg.=⊙ Long.+2s 22° 58′	Arg.=☾☊ + 2s 10° 21′	Arg.=2⊙Long.+2s 6° 16′	
	Aber.	☾ Nut.	⊙Nut.	Aber.	☾ Nut.	⊙Nut.	Aber.	☾ Nut.	⊙Nut.	Aber.	☾ Nut.	⊙Nut.	Aber.	☾ Nut.	⊙Nut.	Aber.	☾ Nut.	⊙Nut.	
	R. A.			N. P. D.			R. A.			N. P. D.			R. A.			N. P. D.			
	VEGA 1830.						TARAZED 1830.						ALTAIR 1830.						

The CORRECTIONS of 48 principal Stars, &c. continued.

1830. Common Argument.	β AQUILÆ 1820. Right Ascension. h. m. s. 19 46 28 .41 Ann. Var. +2s.95			N. P. Distance. 84° 2' 4''.13 Ann. Var. −8''.57			α² CAPRICORNI 1820. Right Ascension. h. m. s. 20 8 3 .65 Ann. Var. +3s.33			N. P. Distance. 103° 5' 37''.35 Ann. Var. −10''.80			α CYGNI 1820. Right Ascension. h. m. s. 20 35 18 .06 Ann. Var. +2s.04			N. P. Distance. 45° 21' 28''.59 Ann. Var. −12''.63			1830. Common Argument.
+ − S. S.	Aber.	☾ Nut.	☉ Nut.	Aber.	☾ Nut.	☉ Nut.	Aber.	☾ Nut.	☉ Nut.	Aber.	☾ Nut.	☉ Nut.	Aber.	☾ Nut.	☉ Nut.	Aber.	☾ Nut.	☉ Nut.	+ − S. S.
III 0° IX	1s.336	1s.060	0s.057	9''.800	9''.203	0''.427	1s.335	1s.199	0s.066	4''.944	9''.018	0''.424	1s.839	0s.836	0s.045	18''.215	8''.758	0''.420	III 0° IX
1	1.335	1.060	.057	9.79	9.20	.427	1.334	1.198	.066	4.94	9.01	.424	1.838	.836	.045	18.20	8.75	.420	29
2	1.334	1.059	.057	9.78	9.19	.427	1.333	1.197	.066	4.93	9.00	.424	1.837	.835	.045	18.19	8.74	.420	28
3	1.333	1.058	.057	9.77	9.18	.426	1.332	1.196	.066	4.93	9.00	.423	1.835	.834	.045	18.18	8.74	.419	27
4	1.332	1.057	.057	9.76	9.17	.426	1.331	1.195	.066	4.92	8.99	.423	1.833	.833	.045	18.16	8.73	.419	26
5	1.330	1.056	.057	9.75	9.16	.425	1.329	1.194	.066	4.92	8.98	.422	1.831	.832	.045	18.14	8.72	.418	25
6	1.328	1.054	.057	9.74	9.15	.425	1.327	1.192	.066	4.91	8.97	.422	1.829	.831	.045	18.12	8.71	.418	24
7	1.326	1.052	.057	9.72	9.13	.424	1.324	1.190	.066	4.90	8.95	.421	1.826	.829	.045	18.09	8.69	.417	23
8	1.323	1.050	.056	9.70	9.11	.423	1.321	1.187	.066	4.89	8.93	.420	1.822	.827	.045	18.05	8.67	.416	22
9	1.320	1.047	.056	9.68	9.09	.422	1.318	1.184	.065	4.88	8.91	.419	1.818	.825	.044	18.00	8.65	.415	21
10	1.316	1.044	.056	9.65	9.07	.421	1.314	1.181	.065	4.87	8.88	.418	1.813	.823	.044	17.95	8.63	.414	20
11	1.312	1.040	.056	9.62	9.04	.420	1.310	1.177	.065	4.85	8.85	.417	1.807	.820	.044	17.89	8.60	.413	19
12	1.307	1.036	.056	9.58	9.01	.418	1.305	1.173	.065	4.83	8.82	.416	1.800	.817	.044	17.82	8.57	.412	18
13	1.302	1.032	.056	9.54	8.97	.416	1.300	1.168	.064	4.81	8.78	.414	1.793	.814	.044	17.75	8.53	.410	17
14	1.296	1.028	.055	9.50	8.93	.414	1.295	1.163	.064	4.79	8.74	.412	1.785	.811	.044	17.67	8.49	.408	16
15	1.290	1.023	.055	9.46	8.88	.412	1.289	1.158	.064	4.77	8.70	.410	1.776	.807	.044	17.59	8.45	.406	15
16	1.284	1.018	.055	9.41	8.84	.410	1.283	1.152	.064	4.75	8.66	.408	1.768	.803	.043	17.50	8.41	.404	14
17	1.277	1.013	.055	9.37	8.79	.408	1.276	1.146	.063	4.73	8.61	.406	1.759	.799	.043	17.41	8.37	.402	13
18	1.270	1.008	.054	9.32	8.74	.406	1.269	1.140	.063	4.71	8.56	.404	1.750	.795	.043	17.31	8.32	.399	12
19	1.263	1.002	.054	9.27	8.69	.404	1.262	1.133	.062	4.68	8.51	.402	1.740	.791	.043	17.21	8.27	.397	11
20	1.255	0.996	.054	9.21	8.64	.401	1.254	1.126	.062	4.65	8.46	.399	1.729	.786	.042	17.11	8.22	.395	10
21	1.247	0.989	.053	9.14	8.58	.399	1.246	1.119	.062	4.62	8.41	.396	1.717	.781	.042	17.00	8.17	.392	9
22	1.238	0.982	.053	9.08	8.52	.396	1.237	1.111	.061	4.58	8.35	.393	1.705	.775	.042	16.89	8.12	.390	8
23	1.229	0.975	.053	9.02	8.46	.393	1.229	1.103	.061	4.55	8.28	.390	1.693	.770	.041	16.78	8.06	.387	7
24	1.220	0.967	.052	8.95	8.40	.390	1.220	1.094	.060	4.52	8.22	.387	1.680	.766	.041	16.65	8.00	.384	6
25	1.210	0.960	.052	8.87	8.33	.387	1.210	1.086	.060	4.48	8.16	.384	1.667	.758	.041	16.51	7.93	.381	5
26	1.200	0.952	.051	8.80	8.26	.384	1.190	1.078	.059	4.44	8.10	.381	1.653	.752	.040	16.37	7.87	.377	4
27	1.190	0.944	.051	8.72	8.18	.380	1.189	1.068	.059	4.40	8.03	.378	1.638	.745	.040	16.23	7.80	.374	3
28	1.180	0.935	.050	8.64	8.12	.377	1.178	1.058	.058	4.36	7.95	.374	1.624	.738	.040	16.08	7.73	.371	2
29	1.168	0.927	.050	8.56	8.04	.373	1.167	1.048	.058	4.32	7.88	.471	1.609	.732	.039	15.94	7.66	.367	1
IV 0 X	1.156	0.918	.049	8.47	7.96	.370	1.156	1.038	.057	4.28	7.80	.367	1.593	.724	.039	15.78	7.58	.364	II 0 VIII
1	1.144	0.908	.049	8.39	7.88	.366	1.143	1.028	.057	4.24	7.72	.364	1.577	.717	.039	15.62	7.51	.360	29
2	1.132	0.898	.048	9.30	7.80	.362	1.131	1.017	.056	4.19	7.64	.360	1.560	.710	.038	15.45	7.43	.356	28
3	1.120	0.888	.048	8.21	7.72	.358	1.119	1.005	.055	4.14	7.56	.356	1.542	.701	.038	15.28	7.35	.352	27
4	1.108	0.878	.047	8.12	7.63	.354	1.106	0.993	.055	4.10	7.47	.352	1.525	.693	.037	15.10	7.26	.348	26
5	1.094	0.868	.047	8.02	7.53	.350	1.093	0.982	.054	4.05	7.38	.347	1.507	.685	.037	14.92	7.17	.344	25
6	1.080	0.857	.046	7.92	7.44	.345	1.079	0.968	.053	4.00	7.28	.343	1.488	.677	.036	14.73	7.08	.340	24
7	1.066	0.845	.046	7.82	7.34	.341	1.065	0.956	.053	3.94	7.19	.339	1.469	.668	.036	14.54	6.99	.335	23
8	1.052	0.834	.045	7.72	7.25	.336	1.051	0.943	.052	3.89	7.10	.334	1.450	.659	.035	14.35	6.90	.331	22
9	1.037	0.823	.044	7.61	7.15	.332	1.036	0.931	.051	3.84	7.00	.329	1.430	.650	.035	14.15	6.80	.326	21
10	1.022	0.811	.044	7.50	7.04	.327	1.021	0.617	.051	3.78	6.90	.325	1.408	.640	.034	13.94	6.70	.322	20
11	1.007	0.799	.043	7.38	6.94	.322	1.006	0.904	.050	3.73	6.80	.320	1.387	.631	.034	13.74	6.61	.317	19
12	0.992	0.787	.042	7.27	6.83	.317	0.991	0.890	.049	3.67	6.70	.315	1.366	.622	.033	13.53	6.51	.312	18
13	0.976	0.775	.042	7.16	6.73	.312	0.975	0.876	.048	3.62	6.59	.310	1.345	.612	.033	13.32	6.41	.307	17
14	0.960	0.762	.041	7.04	6.62	.307	0.960	0.862	.048	3.56	6.48	.305	1.323	.602	.032	13.10	6.30	.303	16
IV 15 X	0.944	0.749	.040	6.92	6.50	.302	0.943	0.847	.047	3.49	6.37	.300	1.300	.591	.032	12.88	6.19	.397	I 15 VII
	Arg. = ☉Long. + 11s 5° 11'	Arg. = ☾☊ + 6s 1° 39'	Arg. = 2☉Long. + 6s 1° 23'	Arg. = ☉Long. + 2s 24° 26'	Arg. = ☾☊ + 2s 9° 26'	Arg. = 2☉Long. + 2s 5° 12'	Arg. = ☉Long. + 11s 0° 12'	Arg. = ☾☊ + 5s 26° 12'	Arg. = 2☉Long. + 5s 26° 55'	Arg. = ☉Long. + 3s 29° 32'	Arg. = ☾☊ + 2s 4° 54'	Arg. = 2☉Long. + 2s 0° 2'	Arg. = ☉Long. + 10s 23° 29'	Arg. = ☾☊ + 6s 28° 32'	Arg. = 2☉Long. + 6s 23° 47'	Arg. = ☉Long. + 2s 0° 38'	Arg. = ☾☊ + 1s 29° 0'	Arg. = 2☉Long. + 1s 23° 14'	
	Aber.	☾ Nut.	☉ Nut.	Aber.	☾ Nut.	☉ Nut.	Aber.	☾ Nut.	☉ Nut.	Aber.	☾ Nut.	☉ Nut.	Aber.	☾ Nut.	☉ Nut.	Aber.	☾ Nut.	☉ Nut.	
	R. A.			N. P. D.			R. A.			N. P. D.			R. A.			N. P. D.			
	ALSHAIRN 1830.						SECUNDA GIEDI 1830.						DENEB 1830.						

The CORRECTIONS of 48 principal Stars, &c. continued.

1830. Common Argument.	β AQUILÆ 1820.						α² CAPRICORNI 1820.						α CYGNI 1820.						1830. Common Argument.
	Right Ascension. h. 19 m. 46 s. 28.41			N. P. Distance. 84° 2′ 4″.13			Right Ascension. h. 20 m. 8 s. 3.65			N. P. Distance. 103° 5′ 37″.35			Right Ascension. h. 20 m. 35 s. 18.06			N. P. Distance. 45° 21′ 28″.59			
	Ann. Var. + 2s.95			Ann. Var. − 8″.57			Ann. Var. + 3s.33			Ann. Var. − 10″.80			Ann. Var. + 2s.04			Ann. Var. − 12″.63			
+ S. − S.	Aber.	☾ Nut.	☉ Nut.	Aber.	☾ Nut.	☉ Nut.	Aber.	☾ Nut.	☉ Nut.	Aber.	☾ Nut.	☉ Nut.	Aber.	☾ Nut.	☉ Nut.	Aber.	☾ Nut.	☉ Nut.	+ S. − S.
IV 15° X	0s.944	0s.749	0s.040	6″.92	6″.50	0″.302	0s.943	0s.847	0s.047	3″.49	6″.37	0″.300	1s.300	0s.591	0s.032	12″.88	6″.19	0″.297	I 15° VII
16	.927	.736	.040	6.80	6.38	.297	.927	.832	.046	3.43	6.26	.295	1.277	.581	.031	12.65	6.08	.292	14
17	.910	.723	.039	6.68	6.27	.291	.910	.817	.045	3.37	6.14	.289	1.254	.570	.031	12.41	5.97	.286	13
18	.893	.709	.038	6.55	6.15	.285	.893	.801	.044	3.30	6.03	.284	1.230	.559	.030	12.19	5.86	.281	12
19	.875	.695	.037	6.42	6.03	.280	.875	.785	.043	3.24	5.91	.278	1.205	.548	.030	11.94	5.74	.275	11
20	.858	.681	.037	6.29	5.91	.274	.857	.769	.042	3.17	5.79	.273	1.181	.538	.029	11.70	5.63	.270	10
21	.840	.667	.036	6.16	5.79	.269	.840	.754	.042	3.11	5.67	.267	1.157	.527	.028	11.46	5.51	.264	9
22	.821	.652	.035	6.03	5.66	.263	.821	.738	.041	3.04	5.55	.261	1.132	.515	.028	11.21	5.39	.259	8
23	.804	.638	.034	5.89	5.53	.257	.803	.721	.040	2.97	5.42	.255	1.106	.503	.027	10.86	5.27	.253	7
24	.785	.623	.033	5.76	5.41	.251	.785	.704	.039	2.90	5.30	.249	1.080	.492	.026	10.70	5.15	.247	6
25	.766	.608	.033	5.62	5.28	.245	.765	.687	.038	2.83	5.17	.243	1.054	.480	.026	10.45	5.02	.241	5
26	747	.593	.032	5.48	5.15	.239	.746	.670	.037	2.77	5.04	.237	1.028	.468	.025	10.19	4.90	.235	4
27	.727	.577	.031	5.33	5.01	.232	.726	.652	.036	2.69	4.91	.231	1.000	.456	.024	9.91	4.77	.229	3
28	.717	.562	.030	5.19	4.87	.226	.717	.635	.035	2.62	4.78	.225	0.974	.443	.024	9.65	4.64	.223	2
29	.687	.546	.029	5.05	4.74	.220	.687	.617	.034	2.55	4.64	.218	0.946	.431	.023	9.38	4.51	.216	1
V 0 XI	.668	.531	.028	4.90	4.60	.214	.668	.600	.033	2.47	4.51	.212	0.920	.418	.023	9.11	4.38	.210	I 0 VII
1	.647	.514	.028	4.75	4.46	.207	.647	.581	.032	2.40	4.37	.206	0.891	.405	.022	8.83	4.24	.204	29
2	.627	.497	.027	4.60	4.32	.200	.627	.562	.031	2.32	4.23	.199	0.862	.392	.021	8.55	4.11	.197	28
3	.606	.481	.026	4.45	4.17	.194	.606	.544	.030	2.24	4.09	.193	0.834	.380	.020	8.27	3.97	.191	27
4	.585	.464	.025	4.29	4.03	.188	.585	.525	.029	2.16	3.95	.186	0.805	.367	.020	7.98	3.84	.184	26
5	.564	.448	.024	4.14	3.89	.181	.564	.516	.028	2.09	3.81	.179	0.777	.354	.019	7.70	3.70	.177	25
6	.543	.431	.023	3.98	3.74	.174	.543	.487	.027	2.01	3.66	.172	0.747	.340	.018	7.41	3.56	.171	24
7	.522	.414	.022	3.82	3.59	.167	.522	.468	.026	1.93	3.52	.166	0.718	.326	.018	7.11	3.42	.164	23
8	.500	.398	.021	3.66	3.44	.160	.500	.448	.025	1.85	3.37	.159	0.688	.313	.017	6.81	3.28	.157	22
9	.478	.380	.020	3.50	3.29	.153	.478	.429	.024	1.77	3.22	.152	0.658	.300	.016	6.52	3.14	.151	21
10	.457	.362	.019	3.34	3.14	.146	.456	.410	.023	1.69	3.08	.145	0.628	.286	.015	6.22	2.99	.144	20
11	.434	.345	.019	3.19	2.95	.139	.434	.390	.021	1.61	2.93	.138	0.598	.272	.015	5.92	2.85	.137	19
12	.412	.327	.018	3.03	2.84	.132	.412	.370	.020	1.53	2.78	.132	0.568	.258	.014	5.62	2.70	.130	18
13	.390	.310	.017	2.86	2.69	.125	.390	.350	.019	1.44	2.63	.124	0.538	.244	.013	5.32	2.56	.123	17
14	.368	.292	.016	2.70	2.53	.118	.368	.330	.018	1.36	2.48	.117	0.507	.230	.612	5.02	2.42	.116	16
15	.345	.274	.015	2.53	2.38	.110	.345	.310	.017	1.28	2.33	.110	0.476	.216	.012	4.71	2.27	.109	15
16	.322	.256	.014	2.37	2.24	.103	.322	.289	.016	1.19	2.17	.103	0.444	.202	.011	4.40	2.12	.101	14
17	.300	.238	.013	2.20	2.07	.096	.300	.269	.015	1.11	2.02	.095	0.413	.188	.010	4.08	1.97	.094	13
18	.277	.220	.012	2.03	1.91	.089	.277	.249	.014	1.03	1.87	.088	0.382	.174	.009	3.78	1.82	.087	12
19	.254	.202	.011	1.87	1.75	.081	.254	.228	.013	0.94	1.72	.081	0.350	.159	.009	3.47	1.67	.080	11
20	.232	.184	.010	1.70	1.60	.074	.231	.208	.011	0.86	1.56	.074	0.319	.145	.008	3.16	1.52	.073	10
21	.209	.166	.009	1.53	1.44	.067	.209	.187	.010	0.77	1.41	.066	0.287	.131	.007	2.85	1.37	.066	9
22	.186	.147	.008	1.36	1.28	.059	.186	.167	.009	0.69	1.25	.059	0.256	.116	.006	2.52	1.22	.078	8
23	.163	.129	.007	1.19	1.12	.052	.163	.146	.008	0.60	1.10	.052	0.224	.102	.005	2.21	1.07	.051	7
24	.140	.111	.006	1.03	0.96	.045	.140	.125	.007	0.52	0.94	.044	0.193	.088	.005	1.91	0.92	.044	6
25	.117	.093	.005	0.86	0.80	.037	.117	.105	.006	0.43	0.77	.037	0.161	.073	.004	1.59	0.77	.037	5
26	.093	.074	.004	0.68	0.64	.030	.093	.084	.005	0.35	0.63	.030	0.129	.058	.003	1.27	0.61	.029	4
27	.070	.056	.003	0.51	0.48	.022	.070	.063	.003	0.26	9.47	.022	0.096	.044	.002	0.95	0.46	.022	3
28	.047	.037	.002	0.34	0.32	.015	.047	.042	.002	0.17	0.31	.015	0.064	.029	.002	0.64	0.31	.015	2
29	.023	.019	.001	0.17	0.16	.007	.023	.021	.001	0.09	0.16	.007	0.032	.015	.001	0.32	0.15	.007	1
VI 0 O	.000	.000	.000	0.00	0.00	.000	.000	.000	.000	0.00	0.00	.000	0.000	.000	.000	0.00	0.00	.000	O 0 VI
	Arg. = ☉ Long. + 11s 5° 11′	Arg. = ☾ ☊ + 6s 1° 39′	Arg. = 2 ☉ Long. + 6s 1° 23′	Arg. = ☉ Long. + 2s 24° 26′	Arg. = ☾ ☊ + 2s 9° 26′	Arg. = 2 ☉ Long. + 2s 5° 12′	Arg. = ☉ Long. + 11s 0° 12′	Arg. = ☾ ☊ + 5s 26° 12′	Arg. = 2 ☉ Long. + 5s 26° 55′	Arg. = ☉ Long. + 3s 29° 32′	Arg. = ☾ ☊ + 2s 4° 54′	Arg. = 2 ☉ Long. + 2s 0° 2′	Arg. = ☉ Long. + 10s 23° 29′	Arg. = ☾ ☊ + 6s 28° 32	Arg. = 2 ☉ Long. + 6s 23° 47′	Arg. = ☉ Long. + 2s 0° 38′	Arg. = ☾ ☊ + 1s 29° 0′	Arg. = 2 ☉ Long. + 1s 23° 14′	
	Aber.	☾ Nut.	☉ Nut.	Aber.	☾ Nut.	☉ Nut.	Aber.	☾ Nut.	☉ Nut.	Aber.	☾ Nut.	☉ Nut.	Aber.	☾ Nut.	☉ Nut.	Aber.	☾ Nut.	☉ Nut.	
	R. A.			N. P. D.			R. A.			N. P. D.			R. A.			N. P. D.			
	ALSHAIN 1830.						SECUNDA GIEDI 1830.						DENEB 1830.						

The CORRECTIONS of 48 principal Stars, &c. continued.

1830. Common Argument.	α CEPHEI 1820. Right Ascension. 21h 14m 16s.71 Ann Var. +1s.42			N. P. Distance. 28° 10′ 28″.25 Ann. Var. −14″.90			β CEPHEI 1820. Right Ascension. 21h 26m 17s.96 Ann. Var. +0s.81			N. P. Distance. 20° 13′ 40″.64 Ann. Var. −15″.70			α AQUARII 1820. Right Ascension. 21h 56m 32s.26 Ann. Var. +3s.09			N. P. Distance. 91° 11′ 20″.55 Ann. Var. −17″.37			1830. Common Argument.
+ − S. S.	Aber.	☾ Nut.	⊙ Nut.	Aber.	☾ Nut.	⊙ Nut.	Aber.	☾ Nut.	⊙ Nut.	Aber.	☾ Nut.	⊙ Nut.	Aber.	☾ Nut.	⊙ Nut.	Aber.	☾ Nut.	⊙ Nut.	+ − S. S.
III 0° IX	2s.733	1s.037	0s.050	19″.847	8″.350	0″.415	3s.718	1s.404	0s.064	20″.181	8″.221	0″.412	1s.269	1s.108	0s.063	7″.883	7″.902	0″.408	III 0° IX
1	2.732	1.036	.050	19.83	8.35	.415	3.717	1.403	.064	20.17	8.22	.412	1.268	1.107	.063	7.88	7.90	.408	29
2	2.730	1.035	.050	19.82	8.34	.415	3.715	1.402	.064	20.16	8.21	.412	1.267	1.106	.063	7.87	7.89	.408	28
3	2.728	1.034	.050	19.81	8.33	.414	3.712	1.401	.064	20.15	8.20	.411	1.266	1.105	.063	7.87	7.89	.407	27
4	2.725	1.033	.050	19.79	8.32	.414	3.708	1.399	.064	20.13	8.19	.411	1.265	1.104	.063	7.86	7.88	.407	26
5	2.722	1.032	.050	19.77	8.31	.413	3.704	1.397	.064	20.10	8.18	.410	1.264	1.102	.063	7.85	7.87	.406	25
6	2.718	1.031	.050	19.74	8.30	.413	3.699	1.395	.064	20.07	8.17	.410	1.262	1.101	.063	7.84	7.86	.406	24
7	2.713	1.029	.050	19.71	8.28	.412	3.692	1.393	.064	20.04	8.16	.409	1.260	1.099	.063	7.83	7.85	.405	23
8	2.707	1.027	.050	19.67	8.26	.411	3.683	1.391	.063	20.00	8.14	.408	1.257	1.097	.062	7.81	7.83	.404	22
9	2.700	1.025	.049	19.62	8.24	.410	3.672	1.388	.063	19.95	8.12	.407	1.254	1.094	.062	7.79	7.81	.403	21
10	2.692	1.022	.049	19.56	8.22	.409	3.661	1.384	.063	19.89	8.10	.406	1.251	1.091	.062	7.77	7.79	.402	20
11	2.683	1.019	.049	19.50	8.19	.408	3.648	1.379	.063	19.82	8.07	.405	1.247	1.087	.062	7.74	7.76	.401	19
12	2.673	1.015	.049	19.43	8.16	.407	3.634	1.373	.063	19.75	8.04	.404	1.242	1.083	.062	7.71	7.73	.400	18
13	2.663	1.011	.049	19.35	8.13	.405	3.619	1.367	.062	19.67	8.01	.402	1.237	1.079	.061	7.68	7.70	.398	17
14	2.652	1.006	.049	19.27	8.10	.403	3.603	1.361	.062	19.58	7.98	.400	1.231	1.074	.061	7.65	7.67	.396	16
15	2.640	1.001	.048	19.18	8.06	.401	3.586	1.355	.062	19.49	7.94	.398	1.225	1.069	.061	7.61	7.63	.394	15
16	2.627	0.997	.048	19.09	8.02	.399	3.569	1.349	.062	19.40	7.90	.396	1.219	1.064	.061	7.57	7.60	.392	14
17	2.613	0.992	.048	18.99	7.98	.397	3.551	1.342	.061	19.30	7.86	.394	1.213	1.059	.060	7.53	7.56	.390	13
18	2.599	0.986	.048	18.88	7.94	.395	3.533	1.335	.061	19.19	7.82	.392	1.206	1.053	.060	7.49	7.51	.388	12
19	2.584	0.980	.047	18.77	7.90	.392	3.514	1.327	.061	19.08	7.78	.390	1.199	1.047	.060	7.45	7.47	.386	11
20	2.568	0.973	.047	18.65	7.85	.390	3.494	1.319	.060	18.96	7.73	.387	1.192	1.041	.059	7.41	7.42	.384	10
21	2.551	0.967	.047	18.53	7.80	.387	3.473	1.310	.060	18.83	7.68	.385	1.184	1.034	.059	7.36	7.38	.381	9
22	2.534	0.961	.046	18.40	7.75	.385	3.451	1.301	.059	18.70	7.63	.382	1.176	1.026	.058	7.31	7.32	.379	8
23	2.516	0.954	.046	18.27	7.69	.382	3.426	1.292	.059	18.57	7.57	.379	1.168	1.019	.058	7.26	7.27	.376	7
24	2.497	0.947	.046	18.14	7.63	.379	3.399	1.282	.059	18.44	7.51	.376	1.159	1.012	.057	7.20	7.22	.373	6
25	2.477	0.939	.045	18.00	7.57	.376	3.370	1.272	.058	18.29	7.45	.373	1.150	1.003	.057	7.14	7.16	.370	5
26	2.457	0.932	.045	17.85	7.51	.373	3.341	1.261	.058	18.14	7.39	.370	1.140	0.995	.057	7.08	7.10	.367	4
27	2.436	0.924	.045	17.69	7.44	.370	3.311	1.250	.057	17.98	7.32	.367	1.130	0.986	.056	7.02	7.04	.364	3
28	2.414	0.915	.044	17.53	7.37	.366	3.281	1.239	.057	17.81	7.26	.363	1.120	0.977	.056	6.96	6.98	.361	2
29	2.391	0.906	.044	17.37	7.31	.363	3.251	1.227	.056	17.64	7.19	.360	1.109	0.967	.055	6.89	6.91	.357	1
IV 0 X	2.367	0.897	.043	17.20	7.23	.359	3.220	1.215	.055	17.47	7.12	.357	1.099	0.958	.055	6.82	6.84	.354	II 0 VIII
1	2.343	0.888	.043	17.02	7.16	.356	3.189	1.203	.055	17.30	7.05	.353	1.087	0.948	.054	6.75	6.77	.350	29
2	2.318	0.879	.042	17.83	7.08	.352	3.156	1.190	.054	17.12	6.97	.349	1.075	0.938	.053	6.68	6.70	.346	28
3	2.292	0.869	.042	16.64	7.01	.348	3.121	1.178	.054	16.93	6.90	.345	1.064	0.928	.053	6.61	6.63	.343	27
4	2.266	0.859	.041	16.45	6.93	.344	3.084	1.164	.053	16.74	6.82	.341	1.052	0.917	.052	6.53	6.55	.339	26
5	2.239	0.849	.041	16.26	6.84	.340	3.046	1.150	.052	16.53	6.73	.337	1.039	0.906	.052	6.46	6.47	.334	25
6	2.212	0.838	.040	16.06	6.75	.336	3.008	1.135	.052	16.32	6.65	.333	1.026	0.895	.051	6.37	6.39	.330	24
7	2.184	0.827	.040	15.85	6.67	.331	2.969	1.120	.051	16.11	6.56	.329	1.012	0.884	.050	6.29	6.31	.326	23
8	2.155	0.816	.039	15.65	6.58	.327	2.929	1.105	.050	15.91	6.47	.324	0.999	0.872	.050	6.21	6.22	.322	22
9	2.125	0.805	.039	15.43	6.48	.322	2.889	1.090	.050	15.69	6.39	.320	0.985	0.860	.049	6.12	6.14	.317	21
10	2.094	0.793	.038	15.21	6.39	.318	2.848	1.074	.049	15.45	6.30	.315	0.971	0.847	.048	6.03	6.05	.313	20
11	2.063	0.782	.038	14.98	6.30	.313	2.806	1.058	.048	15.23	6.20	.311	0.956	0.835	.048	5.94	5.96	.308	19
12	2.031	0.770	.037	14.75	6.21	.309	2.763	1.043	.048	15.00	6.11	.306	0.942	0.822	.047	5.86	5.87	.304	18
13	1.999	0.758	.037	14.52	6.11	.304	2.719	1.027	.047	14.76	6.02	.301	0.927	0.809	.046	5.77	5.78	.299	17
14	1.966	0.746	.036	14.28	6.01	.299	2.674	1.010	.046	14.52	5.92	.296	0.912	0.796	.045	5.67	5.68	.294	16
IV 15 X	1.932	0.733	.035	14.04	5.91	.293	2.629	0.992	.045	14.27	5.82	.291	0.897	0.783	.045	5.57	5.58	.289	I 15 VII
	Arg. = ⊙ Long. + 10s 13° 50′	Arg. = ☾ ☊ + 8s 0° 29′	Arg. = 2 ⊙ Long. + 7s 25° 6′	Arg. = ⊙ Long. + 1s 17° 30′	Arg. = ☾ ☊ + 1s 19° 33′	Arg. = 2 ⊙ Long. + 1s 13° 50′	Arg. = ⊙ Long. + 10s 10° 49′	Arg. = ☾ ☊ + 8s 17° 42′	Arg. = 2 ⊙ Long. + 8s 15° 13′	Arg. = ⊙ Long. + 1s 12° 25′	Arg. = ☾ ☊ + 1s 16° 47′	Arg. = 2 ⊙ Long. + 1s 10° 49′	Arg. = ⊙ Long. + 10s 2° 58′	Arg. = ☾ ☊ + 5s 29° 26′	Arg. = 2 ⊙ Long. + 5s 29° 33′	Arg. = ⊙ Long. + 3s 2° 31′	Arg. = ☾ ☊ + 1s 8° 37′	Arg. = 2 ⊙ Long. + 1s 2° 56′	
	Aber.	☾ Nut.	⊙ Nut.	Aber.	☾ Nut.	⊙ Nut.	Aber.	☾ Nut.	⊙ Nut.	Aber.	☾ Nut.	⊙ Nut.	Aber.	☾ Nut.	⊙ Nut.	Aber.	☾ Nut.	⊙ Nut.	
	R. A.			N. P. D.			R. A.			N. P. D.			R. A.			N. P. D.			
	ALDERAMIN 1830.						ALPHIRK 1830.						SADALMELIK 1830.						

The CORRECTIONS of 48 principal Stars, &c. continued.

1830. Common Argument.	α CEPHEI 1820. Right Ascension. 21h 14m 16s.71 Ann. Var. +1s.42			α CEPHEI 1820. N. P. Distance. 28° 10′ 28″.25 Ann. Var. −14″.90			β CEPHEI 1820. Right Ascension. 21h 26m 17s.96 Ann. Var. +0s.81			β CEPHEI 1820. N. P. Distance. 20° 13′ 40″.64 Ann. Var. −15″.70			α AQUARII 1820. Right Ascension. 21h 56m 32s.26 Ann. Var. +3s.09			α AQUARII 1820. N. P. Distance. 91° 11′ 20″.55 Ann. Var. −17″.37			1830. Common Argument.
+ s.	Aber.	☾ Nut.	⊙ Nut.	Aber.	☾ Nut.	⊙ Nut.	Aber.	☾ Nut.	⊙ Nut.	Aber.	☾ Nut.	⊙ Nut.	Aber.	☾ Nut.	⊙ Nut.	Aber.	☾ Nut.	⊙ Nut.	− s.
IV 15° X	1s.932	0s.733	0s.035	14″.04	5″.91	0″.293	2s.629	0s.992	0s.045	14″.27	5″.82	0″.291	0s.897	0s.783	0s.045	5″.57	5″.58	0″.289	I 15° VII
16	1.899	.720	.035	13.79	5.80	.288	2.583	.974	.044	14.02	5.71	.286	.881	.769	.044	5.48	5.49	.284	14
17	1.863	.707	.034	13.54	5.69	.283	2.536	.956	.044	13.77	5.61	.281	.865	.755	.043	5.38	5.39	.278	13
18	1.828	.693	.033	13.28	5.58	.278	2.488	.938	.043	13.51	5.50	.275	.848	.741	.042	5.27	5.28	.273	12
19	1.792	.679	.033	13.02	5.47	.272	2.439	.920	.042	13.24	5.39	.270	.831	.726	.041	5.17	5.18	.268	11
20	1.756	.666	.032	12.75	5.37	.267	2.390	.901	.041	12.97	5.28	.264	.814	.711	.040	5.07	5.07	.262	10
21	1.720	.652	.031	12.49	5.26	.261	2.340	.883	.040	12.70	5.18	.259	.797	.697	.040	4.97	4.97	.257	9
22	1.683	.638	.031	12.23	5.14	.255	2.289	.864	.039	12.43	5.07	.253	.780	.682	.039	4.86	4.86	.251	8
23	1.644	.623	.030	11.95	5.03	.250	2.238	.844	.038	12.15	4.95	.248	.763	.666	.038	4.74	4.75	.246	7
24	1.606	.609	.029	11.67	4.91	.244	2.186	.824	.038	11.87	4.83	.242	.745	.651	.037	4.63	4.64	.240	6
25	1.569	.594	.029	11.39	4.79	.238	2.133	.805	.037	11.58	4.72	.236	.727	.635	.036	4.52	4.53	.234	5
26	1.529	.580	.028	11.11	4.67	.232	2.080	.785	.036	11.30	4.60	.230	.710	.619	.035	4.41	4.42	.228	4
27	1.489	.564	.027	10.81	4.55	.226	2.026	.764	.035	11.00	4.48	.224	.691	.603	.034	4.29	4.30	.222	3
28	1.449	.549	.026	10.52	4.43	.220	1.971	.744	.034	10.70	4.36	.218	.672	.587	.033	4.18	4.19	.216	2
29	1.408	.534	.026	10.23	4.30	.214	1.915	.723	.033	10.40	4.23	.212	.655	.571	.032	4.06	4.07	.210	1
V 0 XI	1.367	.519	.025	9.93	4.18	.207	1.859	.702	.032	10.09	4.11	.206	.635	.554	.031	3.94	3.95	.204	I 0 VII
1	1.325	.503	.024	9.62	4.05	.201	1.802	.680	.031	9.78	3.98	.200	.615	.537	.030	3.82	3.83	.198	29
2	1.283	.487	.023	9.31	3.92	.195	1.744	.658	.030	9.47	3.86	.193	.596	.520	.030	3.70	3.71	.192	28
3	1.240	.471	.023	9.01	3.79	.188	1.687	.637	.029	9.16	3.73	.187	.576	.503	.029	3.58	3.59	.185	27
4	1.198	.454	.022	8.69	3.66	.182	1.628	.615	.028	8.85	3.60	.180	.556	.486	.028	3.45	3.46	.179	26
5	1.155	.438	.021	8.38	3.53	.175	1.570	.593	.027	8.53	3.47	.174	.536	.468	.027	3.33	3.34	.173	25
6	1.111	.421	.020	8.06	3.39	.169	1.510	.571	.026	8.20	3.34	.167	.516	.450	.026	3.20	3.21	.166	24
7	1.167	.405	.020	7.75	3.26	.162	1.450	.548	.025	7.88	3.21	.161	.496	.432	.025	3.08	3.09	.160	23
8	1.123	.388	.019	7.43	3.12	.155	1.390	.526	.024	7.56	3.08	.154	.475	.414	.024	2.95	2.96	.153	22
9	0.978	.371	.018	7.11	2.99	.149	1.330	.503	.023	7.23	2.94	.148	.454	.396	.023	2.82	2.83	.146	21
10	0.933	.354	.017	6.78	2.85	.142	1.269	.480	.022	6.90	2.81	.141	.433	.378	.022	2.69	2.70	.140	20
11	0.889	.337	.016	6.46	2.72	.135	1.209	.457	.021	6.57	2.68	.134	.413	.360	.020	2.57	2.57	.133	19
12	0.844	.320	.015	6.14	2.58	.128	1.147	.433	.020	6.13	2.54	.127	.392	.342	.019	2.44	2.44	.126	18
13	0.799	.303	.015	5.81	2.44	.121	1.085	.410	.019	5.90	2.40	.120	.371	.323	.018	2.30	2.31	.119	17
14	0.753	.286	.014	5.47	2.30	.114	1.023	.387	.018	5.57	2.27	.113	.349	.305	.017	2.17	2.18	.113	16
15	0.706	.268	.013	5.14	2.16	.107	0.960	.363	.017	5.22	2.13	.106	.328	.286	.016	2.04	2.04	.106	15
16	0.660	.250	.012	4.80	2.02	.100	0.897	.339	.015	4.88	1.99	.099	.306	.268	.015	1.90	1.91	.099	14
17	0.614	.233	.011	4.46	1.88	.093	0.834	.315	.014	4.53	1.85	.092	.285	.249	.014	1.77	1.77	.092	13
18	0.567	.215	.010	4.12	1.73	.086	0.771	.291	.013	4.19	1.71	.085	.263	.230	.013	1.64	1.64	.085	12
19	0.521	.198	.010	3.78	1.59	.079	0.707	.267	.012	3.85	1.57	.078	.242	.211	.012	1.50	1.50	.078	11
20	0.474	.180	.009	3.44	1.45	.072	0.644	.244	.011	3.50	1.43	.071	.220	.192	.011	1.37	1.37	.071	10
21	0.427	.162	.008	3.10	1.30	.065	0.581	.219	.010	3.16	1.29	.064	.198	.173	.010	1.23	1.23	.064	9
22	0.380	.144	.007	2.76	1.16	.058	0.517	.195	.009	2.81	1.14	.057	.176	.154	.009	1.10	1.10	.057	8
23	0.333	.126	.006	2.42	1.02	.051	0.453	.171	.008	2.46	1.00	.050	.155	.135	.008	0.96	0.96	.050	7
24	0.286	.109	.005	2.08	0.87	.043	0.389	.147	.007	2.11	0.86	.043	.133	.116	.007	0.82	0.82	.043	6
25	0.239	.091	.004	1.73	0.73	.036	0.324	.123	.006	1.76	0.72	.036	.111	.098	.006	0.69	0.69	.036	5
26	0.191	.072	.003	1.39	0.58	.029	0.260	.098	.005	1.41	0.57	.029	.089	.078	.004	0.55	0.55	.029	4
27	0.143	.054	.003	1.04	0.44	.022	0.195	.074	.004	1.06	0.43	.022	.066	.059	.003	0.41	0.41	.021	3
28	0.095	.036	.002	0.69	0.29	.014	0.130	.049	.002	0.70	0.29	.014	.044	.039	.002	0.28	0.28	.014	2
29	0.048	.018	.001	0.35	0.15	.007	0.065	.025	.001	0.35	0.14	.007	.022	.020	.001	0.14	0.14	.007	1
VI 0 O	0.000	.000	.000	0.00	0.00	.000	0.000	.000	.000	0.00	0.00	.000	.000	.000	.000	0.00	0.00	.000	O 0 VI
	Arg.=⊙Long.+10s 13° 50′	Arg.=☾☊+8s 0° 29′	Arg.=2⊙Long.+7s 25° 6′	Arg.=⊙Long.+1s 17° 30′	Arg.=☾☊+1s 19° 33′	Arg.=2⊙Long.+1s 13° 50′	Arg.=⊙Long.+10s 10° 49′	Arg.=☾☊+8s 17° 42′	Arg.=2⊙Long.+8s 15° 13′	Arg.=⊙Long.+1s 12° 25′	Arg.=☾☊+1s 16° 47′	Arg.=2⊙Long.+1s 10° 49′	Arg.=⊙Long.+10s 2° 58′	Arg.=☾☊+5s 29° 26′	Arg.=2⊙Long.+5s 29° 33′	Arg.=⊙Long.+3s 2° 31′	Arg.=☾☊+1s 8° 37′	Arg.=2⊙Long.+1s 2° 56′	
	Aber.	☾ Nut.	⊙ Nut.	Aber.	☾ Nut.	⊙ Nut.	Aber.	☾ Nut.	⊙ Nut.	Aber.	☾ Nut.	⊙ Nut.	Aber.	☾ Nut.	⊙ Nut.	Aber.	☾ Nut.	⊙ Nut.	
	R. A.			N. P. D.			R. A.			N. P. D.			R. A.			N. P. D.			
	ALDERAMIN 1830.						ALPHIRK 1830.						SADALMELIK 1830.						

The CORRECTIONS of 48 principal Stars, &c. continued.

α PISCIS AUSTRINI 1820. — FOMALHAUT 1830.

Right Ascension: 22h 47m 40s.96 — Ann. Var. + 3s.34. N. P. Distance: 120° 34′ 25″.80 — Ann. Var. − 19″.10

1830. Common Argument +	R. A. Aber.	R. A. ☾ Nut.	R. A. ☉ Nut.	N. P. D. Aber.	N. P. D. ☾ Nut.	N. P. D. ☉ Nut.	1830. Common Argument −
s. III 0° IX	1″.451	1″.243	0″.071	10″.600	7″.451	0″.402	s. III 0° IX
1	1.450	1.242	.071	10.59	7.45	.40	29
2	1.449	1.241	.071	10.58	7.44	.40	28
3	1.448	1.240	.071	10.57	7.44	.40	27
4	1.447	1.239	.071	10.56	7.43	.40	26
5	1.445	1.238	.071	10.55	7.42	.40	25
6	1.443	1.236	.071	10.53	7.41	.40	24
7	1.441	1.234	.070	10.51	7.40	.40	23
8	1.438	1.231	.070	10.49	7.38	.40	22
9	1.434	1.228	.070	10.47	7.36	.40	21
10	1.430	1.225	.070	10.44	7.34	.40	20
11	1.425	1.221	.070	10.41	7.31	.39	19
12	1.420	1.217	.070	10.37	7.28	.39	18
13	1.414	1.212	.069	10.33	7.25	.39	17
14	1.408	1.207	.069	10.28	7.22	.39	16
15	1.401	1.201	.069	10.23	7.19	.39	15
16	1.394	1.195	.068	10.18	7.16	.39	14
17	1.387	1.189	.068	10.13	7.12	.38	13
18	1.379	1.182	.068	10.07	7.08	.38	12
19	1.371	1.175	.067	10.01	7.04	.38	11
20	1.363	1.168	.067	9.95	7.00	.38	10
21	1.354	1.161	.066	9.89	6.96	.38	9
22	1.344	1.152	.066	9.82	6.91	.37	8
23	1.335	1.144	.065	9.75	6.86	.37	7
24	1.325	1.135	.065	9.67	6.81	.37	6
25	1.314	1.126	.064	9.60	6.75	.36	5
26	1.303	1.117	.064	9.52	6.69	.36	4
27	1.292	1.108	.063	9.44	6.63	.36	3
28	1.280	1.098	.063	9.35	6.57	.35	2
29	1.278	1.087	.062	9.26	6.51	.35	1
IV 0 X	1.256	1.076	.062	9.17	6.45	.35	II 0 VIII
1	1.243	1.065	.061	8.07	6.38	.34	29
2	1.230	1.054	.060	8.98	6.32	.34	28
3	1.217	1.042	.060	8.88	6.25	.34	27
4	1.202	1.030	.059	8.78	6.17	.33	26
5	1.188	1.018	.058	8.67	6.10	.33	25
6	1.173	0.105	.057	8.56	6.02	.32	24
7	1.158	0.992	.057	8.45	5.94	.32	23
8	1.142	0.979	.056	8.34	5.86	.32	22
9	1.126	0.965	.055	8.22	5.78	.31	21
10	1.110	0.952	.054	8.10	5.70	.31	20
11	1.093	0.937	.054	7.99	5.62	.30	19
12	1.077	0.923	.053	7.87	5.53	.30	18
13	1.060	0.908	.052	7.75	5.45	.29	17
14	1.043	0.893	.051	7.62	5.36	.29	16
IV 15 X	1.025	0.878	.050	7.49	5.27	.28	I 15 VII
	Arg. = ☉ Long. + 9s 19° 26′	Arg. = ☾ ☊ + 5s 13° 8′	Arg. = 2☉ Long. + 5s 8° 41′	Arg. = ☉ Long. + 5s 7° 34′	Arg. = ☾ ☊ + 0s 23° 34′	Arg. = 2☉ Long. + 0s 19° 26′	

α PEGASI 1820. — MARKAB 1830.

Right Ascension: 22h 55m 48s.24 — Ann. Var. + 2s.98. N. P. Distance: 75° 45′ 35″.96 — Ann. Var. − 19″.43

1830. Common Argument +	R. A. Aber.	R. A. ☾ Nut.	R. A. ☉ Nut.	N. P. D. Aber.	N. P. D. ☾ Nut.	N. P. D. ☉ Nut.	1830. Common Argument −
s. III 0° IX	1″.287	1″.080	0″.057	10″.285	7″.397	0″.401	s. III 0° IX
1	1.286	1.080	.057	10.28	7.39	.40	29
2	1.285	1.079	.057	10.27	7.39	.40	28
3	1.284	1.078	.057	10.26	7.38	.40	27
4	1.283	1.077	.057	10.25	7.38	.40	26
5	1.282	1.076	.057	10.24	7.37	.40	25
6	1.281	1.074	.057	10.23	7.36	.40	24
7	1.279	1.072	.057	10.21	7.35	.40	23
8	1.276	1.070	.056	10.19	7.33	.40	22
9	1.273	1.067	.056	10.17	7.31	.40	21
10	1.269	1.064	.056	10.14	7.29	.40	20
11	1.265	1.061	.056	10.11	7.27	.39	19
12	1.260	1.057	.056	10.07	7.24	.39	18
13	1.255	1.053	.056	10.03	7.21	.39	17
14	1.249	1.048	.055	9.98	7.18	.39	16
15	1.243	1.043	.055	9.93	7.15	.39	15
16	1.237	1.018	.055	9.88	7.11	.39	14
17	1.231	1.013	.055	9.83	7.07	.38	13
18	1.224	1.027	.054	9.78	7.03	.38	12
19	1.217	1.021	.054	9.72	6.99	.38	11
20	1.210	1.015	.054	9.66	6.95	.38	10
21	1.202	1.008	.053	9.60	6.90	.38	9
22	1.193	1.001	.053	9.53	6.86	.37	8
23	1.185	1.094	.053	9.46	6.81	.37	7
24	1.176	1.087	.052	9.39	6.76	.37	6
25	1.166	0.979	.052	9.32	6.71	.36	5
26	1.157	0.971	.051	9.24	6.65	.36	4
27	1.147	0.962	.051	9.16	6.59	.36	3
28	1.136	0.953	.050	9.07	6.53	.35	2
29	1.125	0.944	.050	8.99	6.47	.35	1
IV 0 X	1.114	0.935	.049	8.90	6.41	.35	II 0 VIII
1	1.102	0.925	.049	8.81	6.34	.34	29
2	1.091	9.915	.048	8.72	6.27	.34	28
3	1.080	0.905	.048	8.62	6.20	.34	27
4	1.067	0.895	.047	8.52	6.13	.33	26
5	1.054	0.884	.047	8.42	6.06	.33	25
6	1.041	0.873	.046	8.31	5.98	.32	24
7	1.028	0.862	.046	8.20	5.91	.32	23
8	1.014	0.850	.045	8.10	5.83	.32	22
9	1.000	0.838	.044	7.98	5.75	.31	21
10	0.985	0.826	.044	7.87	5.66	.31	20
11	0.970	0.814	.043	7.76	5.58	.30	19
12	0.956	0.802	.042	7.64	5.50	.30	18
13	0.941	0.790	.042	7.52	5.41	.29	17
14	0.926	0.777	.041	7.40	5.32	.29	16
IV 15 X	0.910	0.763	.040	7.27	5.23	.28	I 15 VII
	Arg. = ☉ Long. + 9s 17° 16′	Arg. = ☾ ☊ + 6s 8° 23′	Arg. = 2☉ Long. + 6s 7° 9′	Arg. = ☉ Long. + 2s 1° 49′	Arg. = ☾ ☊ + 0s 20′ 59′	Arg. = 2☉ Long. + 0s 17° 17′	

α ANDROMEDÆ 1820. — SIRRAH 1830.

Right Ascension: 23h 59m 6s.36 — Ann. Var. + 3s.08. N. P. Distance: 61° 54′ 10″.31 — Ann. Var. − 19″.99

1830. Common Argument +	R. A. Aber.	R. A. ☾ Nut.	R. A. ☉ Nut.	N. P. D. Aber.	N. P. D. ☾ Nut.	N. P. D. ☉ Nut.	1830. Common Argument −
s. III 0° IX	1″.405	1″.156	0″.063	11″.936	7″.183	0″.398	s. III 0° IX
1	1.404	1.155	.063	11.93	7.18	.398	29
2	1.403	1.154	.063	11.92	7.17	.398	28
3	1.402	1.153	.063	11.91	7.17	.397	27
4	1.401	1.152	.063	11.90	7.16	.397	26
5	1.400	1.151	.063	11.89	7.15	.396	25
6	1.398	1.150	.063	11.87	7.14	.396	24
7	1.396	1.148	.063	11.85	7.13	.395	23
8	1.393	1.146	.062	11.82	7.12	.394	22
9	1.389	1.143	.062	11.79	7.10	.393	21
10	1.385	1.140	.062	11.76	7.08	.392	20
11	1.380	1.136	.062	11.72	7.06	.391	19
12	1.375	1.132	.062	11.68	7.03	.390	18
13	1.369	1.127	.061	11.63	7.00	.388	17
14	1.363	1.122	.061	11.58	6.97	.386	16
15	1.356	1.116	.061	11.52	6.94	.384	15
16	1.349	1.111	.061	11.46	6.91	.382	14
17	1.342	1.105	.060	11.40	6.87	.380	13
18	1.335	1.099	.060	11.34	6.83	.378	12
19	1.328	1.093	.060	11.28	6.79	.376	11
20	1.320	1.086	.059	11.21	6.75	.374	10
21	1.311	1.079	.059	11.14	6.71	.372	9
22	1.302	1.071	.058	11.06	6.66	.370	8
23	1.293	1.064	.058	10.98	6.61	.367	7
24	1.283	1.056	.058	10.90	6.56	.364	6
25	1.273	1.048	.057	10.81	6.51	.361	5
26	1.262	1.039	.057	10.72	6.46	.358	4
27	1.251	1.029	.056	10.63	6.40	.355	3
28	1.240	1.020	.056	10.53	6.34	.351	2
29	1.229	1.010	.055	10.43	6.28	.348	1
IV 0 X	1.216	1.000	.055	10.33	6.22	.345	II 0 VIII
1	1.204	0.990	.054	10.22	6.16	.341	29
2	1.191	0.980	.053	10.11	6.09	.338	28
3	1.178	0.969	.053	10.00	6.03	.334	27
4	1.164	0.957	.052	9.89	5.96	.330	26
5	1.150	0.946	.052	9.77	5.88	.326	25
6	1.136	0.934	.051	9.65	5.81	.322	24
7	1.121	0.922	.050	9.52	5.74	.318	23
8	1.106	0.910	.050	9.40	5.66	.314	22
9	1.091	0.898	.049	9.27	5.58	.309	21
10	1.075	0.885	.048	9.13	5.50	.305	20
11	1.060	0.871	.048	9.00	5.42	.300	19
12	1.044	0.858	.047	8.86	5.34	.296	18
13	1.027	0.845	.046	8.72	5.26	.291	17
14	1.010	0.831	.045	8.58	5.17	.286	16
IV 15 X	0.992	0.817	.045	8.43	5.08	.281	I 15 VII
	Arg. = ☉ Long. + 9s 0° 6′	Arg. = ☾ ☊ + 6s 17° 21′	Arg. = 2☉ Long. + 6s 14° 13′	Arg. = ☉ Long. + 1s 6° 48′	Arg. = ☾ ☊ + 0s 0° 8′	Arg. = 2☉ Long. + 0s 0° 7′	

The CORRECTIONS of 48 principal Stars, &c. continued.

α PISCIS AUSTRINI 1830. — Right Ascension 22h. 47m. 40s.96; Ann. Var. + 3s.34. N. P. Distance 120° 34′ 25″.80; Ann. Var. + 19″.10.

1830. Common Argument. + S.	– S.	R. A. Aber.	R. A. ☾ Nut.	R. A. ⊙ Nut.	N. P. D. Aber.	N. P. D. ☾ Nut.	N. P. D. ⊙ Nut.	1830. Common Argument. + S.	– S.
IV 15°	X	1s.025	0s.878	0s.050	7″.49	5″.27	0″.28	III 0°	IX
16		1.007	.863	.049	7.36	5.13	.28	29	
17		0.988	.847	.048	7.22	5.08	.27	28	
18		0.970	.831	.048	7.18	4.98	.27	27	
19		0.951	.814	.047	6.94	4.88	.26	26	
20		0.932	.798	.046	6.80	4.78	.26	25	
21		0.912	.782	.045	6.66	4.68	.25	24	
22		0.892	.765	.044	6.52	4.58	.25	23	
23		0.872	.748	.043	6.37	4.48	.24	22	
24		0.852	.731	.042	6.22	4.38	.24	21	
25		0.831	.713	.041	6.07	4.27	.23	20	
26		0.811	.685	.040	5.92	4.17	.22	19	
27		0.790	.676	.039	5.77	4.06	.22	18	
28		0.768	.658	.038	5.62	3.95	.21	17	
29		0.746	.640	.037	5.46	3.84	.21	16	
V 0	XI	0.725	.622	.036	5.30	3.73	.20	15	
1		0.703	.603	.034	5.14	3.61	.19	14	
2		0.680	.583	.033	4.97	3.49	.19	13	
3		0.658	.564	.032	4.81	3.38	.18	12	
4		0.635	.545	.031	4.64	3.26	.18	11	
5		0.612	.526	.030	4.48	3.15	.17	10	
6		0.589	.506	.029	4.31	3.03	.16	9	
7		0.566	.486	.028	4.14	2.91	.16	8	
8		0.543	.465	.027	3.96	2.79	.15	7	
9		0.519	.445	.025	3.79	2.67	.14	6	
10		0.496	.425	.024	3.62	2.55	.14	5	
11		0.472	.404	.023	3.45	2.42	.13	4	
12		0.448	.384	.022	3.27	2.30	.12	3	
13		0.424	.363	.021	3.10	2.18	.12	2	
14		0.400	.342	.020	2.92	2.05	.11	1	
15		0.375	.321	.018	2.74	1.93	.10	II 0	VIII
16		0.350	.300	.017	2.56	1.80	.10	29	
17		0.326	.279	.016	2.38	1.67	.09	28	
18		0.301	.258	.015	2.20	1.55	.08	27	
19		0.276	.237	.014	2.02	1.42	.08	26	
20		0.252	.216	.012	1.84	1.29	.07	25	
21		0.227	.194	.011	1.65	1.16	.06	24	
22		0.202	.173	.010	1.47	1.04	.06	23	
23		0.177	.152	.009	1.29	0.91	.05	22	
24		0.152	.130	.007	1.11	0.78	.04	21	
25		0.127	.109	.006	0.93	0.65	.04	20	
26		0.101	.087	.005	0.74	0.52	.03	19	
27		0.076	.065	.004	0.56	0.39	.02	18	
28		0.051	.043	.003	0.37	0.26	.01	17	
29		0.025	.022	.001	0.19	0.13	.01	16	
VI 0	0	0.000	.000	.000	0.00	0.00	.00	I 15	VII
		Arg. = ⊙ Long. + 9s 19° 26′	Arg. = ☾ ☊ + 5s 13° 8′	Arg. = 2⊙ Long. + 5s 8° 41′	Arg. = ⊙ Long. + 5s 7° 31′	Arg. = ☾ ☊ + 0s 23° 31′	Arg. = 2⊙ Long. + 0s 19° 26′		
		Aber.	☾ Nut.	⊙ Nut.	Aber.	☾ Nut.	⊙ Nut.		
		R. A.			N. P. D.				
		FOMALHAUT 1830.							

α PEGASI 1820. — Right Ascension 22h. 55m. 48s.24; Ann. Var. + 2s.98. N. P. Distance 75° 45′ 35″.96; Ann. Var. + 19″.43.

1830. Common Argument. + S.	– S.	R. A. Aber.	R. A. ☾ Nut.	R. A. ⊙ Nut.	N. P. D. Aber.	N. P. D. ☾ Nut.	N. P. D. ⊙ Nut.	1830. Common Argument. + S.	– S.
IV 15°	X	0s.910	0s.763	0s.040	7″.27	5″.23	0″.28	III 0°	IX
16		.894	.750	.040	7.14	5.14	.28	29	
17		.877	.736	.039	7.01	5.05	.27	28	
18		.861	.722	.038	6.88	4.95	.27	27	
19		.843	.708	.037	6.74	4.85	.26	26	
20		.826	.694	.037	6.61	4.76	.26	25	
21		.810	.680	.036	6.47	4.66	.25	24	
22		.792	.665	.035	6.33	4.56	.25	23	
23		.775	.650	.034	6.18	4.45	.24	22	
24		.757	.635	.033	6.04	4.34	.24	21	
25		.738	.619	.033	5.90	4.24	.23	20	
26		.720	.604	.032	5.75	4.14	.22	19	
27		.701	.588	.031	5.60	4.03	.22	18	
28		.682	.572	.030	5.45	3.92	.21	17	
29		.663	.557	.029	5.30	3.81	.21	16	
V 0	XI	.644	.541	.029	5.14	3.70	.20	15	
1		.624	.524	.028	4.98	3.59	.19	14	
2		.604	.507	.027	4.82	3.47	.19	13	
3		.584	.490	.026	4.67	3.36	.18	12	
4		.564	.473	.025	4.51	3.24	.18	11	
5		.544	.456	.024	4.34	3.13	.17	10	
6		.523	.439	.023	4.17	3.01	.16	9	
7		.503	.422	.022	4.01	2.89	.16	8	
8		.482	.404	.021	3.85	2.77	.15	7	
9		.461	.387	.020	3.68	2.65	.14	6	
10		.440	.369	.019	3.51	2.53	.14	5	
11		.419	.351	.019	3.34	2.41	.13	4	
12		.398	.333	.018	3.17	2.28	.12	3	
13		.376	.315	.017	3.30	2.16	.12	2	
14		.355	.297	.016	2.83	2.04	.11	1	
15		.333	.279	.015	2.66	1.91	.10	II 0	VIII
16		.311	.261	.014	2.48	1.79	.10	29	
17		.289	.242	.013	2.31	1.66	.09	28	
18		.267	.224	.012	2.13	1.54	.08	27	
19		.245	.206	.011	1.96	1.41	.08	26	
20		.223	.187	.010	1.78	1.28	.07	25	
21		.201	.169	.009	1.61	1.16	.06	24	
22		.179	.151	.008	1.43	1.03	.06	23	
23		.157	.132	.007	1.25	0.90	.05	22	
24		.135	.113	.006	1.08	0.77	.04	21	
25		.112	.094	.005	0.90	0.65	.04	20	
26		.089	.076	.004	0.72	0.52	.03	19	
27		.067	.057	.003	0.54	0.39	.02	18	
28		.045	.038	.002	0.36	0.26	.01	17	
29		.022	.019	.001	0.18	0.13	.01	16	
VI 0	0	.000	.000	.000	0.00	0.00	.00	I 15	VII
		Arg. = ⊙ Long. + 9s 17° 16′	Arg. = ☾ ☊ + 6s 8° 23′	Arg. = 2 ⊙ Long. + 6s 7° 9′	Arg. = ⊙ Long. + 2s 1° 49′	Arg. = ☾ ☊ + 0s 20° 59′	Arg. = 2⊙ Long. + 0s 17° 17′		
		Aber.	☾ Nut.	⊙ Nut.	Aber.	☾ Nut.	⊙ Nut.		
		R. A.			N. P. D.				
		MARKAB 1830.							

α ANDROMEDÆ 1820. — Right Ascension 23h. 59m. 6s.36; Ann. Var. + 3s.08. N. P. Distance 61° 54′ 10″.31; Ann. Var. – 19″.99.

1830. Common Argument. + S.	– S.	R. A. Aber.	R. A. ☾ Nut.	R. A. ⊙ Nut.	N. P. D. Aber.	N. P. D. ☾ Nut.	N. P. D. ⊙ Nut.	1830. Common Argument. + S.	– S.
IV 15°	X	0s.992	0s.817	0s.045	8″.43	5″.08	0″.281	III 0°	IX
16		.975	.803	.044	8.28	4.99	.276	29	
17		.957	.788	.043	8.13	4.90	.271	28	
18		.939	.773	.042	7.98	4.81	.266	27	
19		.921	.757	.041	7.82	4.71	.261	26	
20		.902	.742	.040	7.67	4.62	.256	25	
21		.884	.727	.040	7.51	4.52	.250	24	
22		.865	.712	.039	7.35	4.42	.245	23	
23		.845	.696	.038	7.18	4.32	.239	22	
24		.825	.679	.037	7.01	4.22	.234	21	
25		.805	.663	.036	6.84	4.12	.228	20	
26		.786	.646	.035	6.67	4.02	.223	19	
27		.765	.629	.034	6.49	3.91	.217	18	
28		.745	.613	.033	6.32	3.81	.211	17	
29		.724	.596	.032	6.15	3.70	.205	16	
V 0	XI	.703	.578	.032	5.97	3.59	.199	15	
1		.681	.560	.031	5.78	3.48	.193	14	
2		.659	.542	.030	5.60	3.37	.187	13	
3		.638	.525	.029	5.42	3.26	.181	12	
4		.616	.507	.028	5.23	3.15	.174	11	
5		.593	.488	.027	5.04	3.04	.168	10	
6		.571	.470	.026	4.85	2.92	.162	9	
7		.548	.452	.025	4.66	2.81	.156	8	
8		.526	.432	.024	4.47	2.69	.149	7	
9		.503	.413	.023	4.27	2.57	.143	6	
10		.480	.395	.022	4.08	2.45	.136	5	
11		.457	.376	.020	3.88	2.33	.130	4	
12		.433	.357	.019	3.68	2.21	.123	3	
13		.410	.338	.018	3.48	2.10	.116	2	
14		.387	.318	.017	3.28	1.98	.110	1	
15		.363	.299	.016	3.08	1.86	.103	II 0	VIII
16		.339	.279	.015	2.88	1.74	.096	29	
17		.316	.259	.014	2.68	1.62	.089	28	
18		.292	.240	.013	2.48	1.49	.083	27	
19		.268	.220	.012	2.27	1.37	.076	26	
20		.244	.200	.011	2.07	1.25	.069	25	
21		.220	.181	.010	1.87	1.12	.062	24	
22		.196	.161	.009	1.66	1.00	.055	23	
23		.171	.141	.008	1.45	0.87	.049	22	
24		.147	.121	.007	1.25	0.75	.042	21	
25		.123	.101	.006	1.04	0.63	.035	20	
26		.098	.081	.004	0.73	0.50	.028	19	
27		.074	.061	.003	0.63	0.38	.021	18	
28		.049	.040	.002	0.42	0.25	.014	17	
29		.025	.020	.001	0.21	0.13	.007	16	
VI 0	0	.000	.000	.000	0.00	0.00	.000	I 15	VII
		Arg. = ⊙ Long. + 9s 0° 6′	Arg. = ☾ ☊ + 6s 17° 21′	Arg. = 2 ⊙ Long. + 6s 14° 13′	Arg. = ⊙ Long. + 1s 6° 48′	Arg. = ☾ ☊ + 0s 0° 8′	Arg. = 2 ⊙ Long. + 0s 0° 7′		
		Aber.	☾ Nut.	⊙ Nut.	Aber.	☾ Nut.	⊙ Nut.		
		R. A.			N. P. D.				
		SIRRAH 1830.							

A TABLE of the COMPARATIVE MEAN PLACES of the preceding 48 principal Stars, as determined by different eminent Astronomers.

OBSERVATIONS REDUCED TO THE EPOCH 1820.

Stars.	RIGHT ASCENSION.									NORTH POLAR DISTANCE.								
	Common.		Seconds of Time.					Ann. Var. by Schumacher.	Secular Variation.	Common.		Seconds of Space.					Ann. Var. by Schumacher.	Secular Variation.
			Pond.	Maskelyne.	Bessel.	Piazzi.	Zach.					Pond.	Piazzi.	Zach.	Brinkley.	Oriani.		
	h.	m.						+									−	+
γ Pegasi	0	3	58s.84	58s.52	58s.80	58s.59	58s.36	3s.077	+0s.0096	75°	48′	60″.15	62″.38	52″.30	60″.82	″	20″.050	0″.0133
α Cassiop.	0	30	20.99			20.80		3.327	+0.0508	34	27	4.08	4.55		2.67		19.836	.0644
α Ursæ Min.	0	57	1.51		0.54	0.66		14.657	+5.8040	1	39	5.57	5.67		5.70	5.25	19.470	.4044
α Arietis	1	57	3.01	2.54	2.94	2.60	2.50	3.352	+0.0200	67	23	34.69	35.28	30.74	35.00		17.383	.2400
α Ceti	2	52	52.89	52.48	53.92	51.69	52.39	3.121	+0.0096	86	37	18.19	20.05	11.10	20.26		14.524	.3155
α Persei	3	11	31.25			31.11		4.220	+0.0489	40	47	17.87	17.31		17.18		13.435	.4555
α Tauri	4	25	36.28	35.00	36.02	35.83	35.95	3.425	+0.0108	73	51	39.67	38.67	36.78	40.43	39.45	7.916	.4577
α Aurigæ	5	3	24.59	24.30	24.42	24.21	24.43	4.409	+0.0185	44	11	48.49	48.01	50.14	45.31	47.16	4.457	.4622
β Orionis	5	5	53.56	53.29	53.42	53.36	53.38	2.877	+0.0043	98	24	59.85	61.52	50.44	61.80	62.00	4.699	.4111
β Tauri	5	14	55.41	55.10	55.22	55.12	55.11	2.785	+0.0093	61	33	16.89	19.02	12.54	18.06	17.83	3.747	.5400
α Orionis	5	45	25.92	25.62	25.70	25.72	25.65	3.245	+0.0033	82	38	6.04	8.42	0.98	6.89	6.42	1.275	.4733
																	+	
α Canis Maj.	6	37	12.90	12.70	12.78	12.83	13.20	2.644	+0.0004	106	28	30.67	34.18	20.48	33.13	37.01	4.440	.3800
α Gemin.	7	23	6.11	5.74	5.70	5.55	5.86	3.842	−0.0121	57	43	36.31	37.22	34.48	37.56		7.170	.5266
α Canis Min.	7	29	52.67	52.34	52.45	52.38	52.16	3.146	−0.0043	84	19	15.69	13.48	8.56	15.72	15.76	8.647	.4222
β Gemin.	7	34	17.47	17.07	17.20	16.98	17.06	3.682	−0.0124	61	32	52.45	53.72	48.44	51.43		8.078	.4911
α Hydræ	9	18	44.61	44.34	44.33	44.24	44.33	2.945	−0.0015	97	52	57.19	58.30	51.10	59.22		15.226	.2733
θ Ursæ Maj.	9	20	45.21			45.09		4.072	−0.0567	37	30	30.40	33.97			34.31	16.011	.3444
α Leonis	9	58	46.74	46.40	46.49	46.28	46.62	3.202	−0.0102	77	9	23.62	22.54	18.92	23.75	23.28	17.252	.2333
α Ursæ Maj.	10	52	31.99			31.76		3.804	−0.0862	27	16	46.60	47.09		46.08	46.74	19.291	.1600
β Leonis	11	39	52.45	52.17	52.27	52.11	52.08	3.065	−0.0079	74	25	17.93	15.54	15.32	16.54		20.015	.0355
β Virginis	11	41	19.35	18.85	19.20	18.96	19.02	3.122	−0.0007	87	13	14.66	15.06	10.66			20.257	.0333
γ Ursæ Maj.	11	44	19.13			18.76		3.209	−0.0456	35	18	15.25	17.81		14.92	16.26	20.045	.0288
																		−
α Virginis	13	15	43.46	43.25	43.38	43.27	43.35	3.444	+0.0111	100	12	63.28	63.58	53.00	63.04	63.03	18.963	.1311
ζ Ursæ Maj.	13	16	39.51			38.28		2.422	−0.0132	34	7	53.72	53.62		53.23		18.959	.1222
η Ursæ Maj.	13	40	26.43			25.67		2.370	−0.0109	39	47	5.21	4.16		5.46		18.151	.1533
α Bootis	14	7	27.44	27.14	27.27	27.01	27.02	2.728	+0.0012	69	52	31.89	31.22	23.62	32.05	31.67	18.957	.2155
α^2 Libræ	14	40	56.47	56.09	56.40	56.28	56.15	3.302	+0.0156	105	16	69.73	70.81	59.78	71.79		15.306	.3133
								−										+
β Ursæ Min.	14	51	20.36			20.31		0.305	−0.1114	15	6	31.83	35.92		33.18	33.47	14.820	.0288
								+										−
α Cor. Bor.	15	27	4.34	3.96	4.22	3.79	4.26	2.531	+0.0024	62	40	22.78	20.79	20.32	22.46	22.78	12.408	.2955
α Serpentis	15	35	24.67	24.27	24.57	24.42	24.38	2.948	+0.0064	82	59	61.33	61.78	56.40	61.01	62.59	11.734	.3488
α Scorpii	16	18	23.32	22 95	23.35	23.39	23.27	3.664	+0.0157	116	1	16.10	16.77	2.94	18.41	15.60	8.575	.4844
α Herculis	17	6	26.83	26.46	26.57	26.31	26.57	2.725	+0.0037	75	23	45.59	43.47	36.24	46.38	46.42	4.517	.3866
β Draconis	17	26	21.92			21.88		1.343	+0.0053	37	33	37.58	39.88				2.909	.1911
α Ophiuchi	17	26	35.16	34.73	34.95	34.70	34.78	2.773	+0.0034	77	17	60.55	58.49	54.72	61.84	61.09	3.031	.4000
γ Draconis	17	52	25.89			25.51		1.388	+0.0035	38	29	8.45	10.24		8.77		9.153	.2022
																	−	
ε Sagittarii	18	12	13.10			13.32		3.974	−0.0003	124	27	32.48	23.30				(1.068)	.5800
α Lyræ	18	30	50.89	50.62	50.72	50.50	50.47	2.027	+0.0016	51	22	39.45	40.80	42.82	40.91	41.54	3.003	.2911
γ Aquilæ	19	37	42.28	41.94	42.16	42.05	41.76	2.854	−0.0009	79	48	62.22	62.56	55.96	63.06		8.343	.3755
α Aquilæ	19	41	60.14	59.82	60.03	59.82	59.78	2.927	−0.0015	81	35	55.39	54.86	50.00	57.35	57.07	9.089	.3955
β Aquilæ	19	46	28.41	28.12	28.34	28.18	28.08	2.949	−0.0015	84	1	64.13	64.27	58.92	65.37		8.567	.3688
α^2 Capric.	20	8	3.70	3.45	3.69	3.61	3.58	3.337	−0.0081	103	5	37.35	36.78	26.84	36.46	34.65	10.716	.4111
α Cygni	20	35	18.06	17.67	17.90	17.52	17.58	2.036	+0.0022	45	21	28.59	28.46	32.32	30.13	30.11	12.611	.2266
α Cephei	21	14	16.71			15.94		1.431	−0.0062	28	10	28.25	31.95		29.10	29.37	14.977	.1311
β Cephei	21	26	17.06			17.06		0.802	−0.0315	20	13	40.64	45.49		42.28	43.39	15.593	.0644
α Aquarii	21	56	32.26	31.98	32.23	33.06	31.89	3.082	−0.0043	91	11	20.55	20.52	12.62	20.91		17.261	.2266
α Pis. Austr.	22	47	40.96	40.78	41.10	40.96	40.98	3.339	−0.0218	120	34	25.80	23.18	8.90		22.13	18.922	.1488
α Pegasi	22	55	48.24	47.92	48.17	47.92	47.73	2.979	+0.0052	75	45	35.96	36.64	28.46	37.55	37.84	19.327	.1155
α Androm.	23	59	6.36	5.99	6.23	5.85	5.87	3.073	+0.0176	61	54	10.31	13.63	5.93	11.90		19.917	.0044
	Common.		Pond.	Maskelyne.	Bessel.	Piazzi.	Zach.	Ann. Var. by Schumacher.	Secular Variation.	Common.		Pond.	Piazzi.	Zach.	Brinkley.	Oriani.	Ann. Var. by Schumacher.	Secular Variation.
	RIGHT ASCENSION.									NORTH POLAR DISTANCE.								

By the assistance of this Table and of the next following one, the practical Astronomer may adopt the Mean Places and Annual Variations of any of the different Observers, and may use them *ad libitum* with the preceding Table of Corrections. The Mean Places are brought up with the Annual Variations of the respective Observers; Mr. Pond's being chiefly taken from the Nautical Almanac of 1823.

A TABLE of the COMPARATIVE ANNUAL PRECESSIONS, SECULAR CHANGES of PRECESSION, PROPER MOTIONS, and ANNUAL VARIATIONS, of the 48 principal Sta both in RIGHT ASCENSION in TIME, and in NORTH POLAR DISTANCE, for the *beginning of the Century*, according to different eminent Astronomers.

1800.	RIGHT ASCENSION IN TIME.									NORTH POLAR DISTANCE.								
	Annual Precession.			Proper Motions.			Annual Variations.			Annual Precession.			Proper Motions.			Annual Variations.		
STARS.	Mas-kelyne.	Piazzi.	Bessel.	Mas-kelyne.	Piazzi.	Bessel.	Mas-kelyne.	Piazzi.	Bessel.	Mas-kelyne.	Piazzi.	Bessel.	Mas-kelyne.	Piazzi.	Bessel.	Mas-kelyne.	Piazzi.	Bessel.
	+	+	+				+	+	+	−	−	−				−	−	−
γ Pegasi	3ˢ.075	3ˢ.073	3ˢ.072	−0ˢ.006	−0ˢ.002	+0ˢ.003	3ˢ.069	3ˢ.071	3ˢ.075	20″.05	20″.06	20″.044	−0″.15	+0″.09	−0″.01	20″.20	19″.97	20″.054
α Cassiop.		3.317	3.313		−0.005	+0.004		3.312	3.317		19.90	19.883		+0.05	+0.03		19.85	13.853
α Ursæ Min.		12.940	12.929			+0.093		12.940	13.022		19.54	19.524		0.00	−0.02		19.54	19.544
α Arietis	3.339	3.338	3.336	+0.007	+0.013	+0.012	3.346	3.351	3.348	17.54	17.55	17.536	+0.07	+0.20	+0.10	17.47	17.35	17.436
α Ceti	3.123	3.122	3.120	−0.008	−0.005	−0.001	3.115	3.117	3.119	14.67	14.68	14.670	−0.08	+0.15	+0.08	14.75	14.53	14.590
α Persei		4.210	4.207		−0.014	+0.003		4.196	4.210		13.55	13.535		+0.02	0.00		13.53	13.535
α Tauri	3.423	3.422	3.420	+0.002	+0.003	+0.003	3.425	3.425	3.423	8.11	8.12	8.117	+0.12	+0.21	+0.10	7.99	7.91	8.017
α Aurigæ	4.401	4.400	4.397	+0.014	+0.008	+0.009	4.415	4.408	4.406	5.01	5.03	5.025	+0.44	+0.44	+0.44	4.57	4.59	4.585
β Orionis	5.878	2.877	2.875	−0.002	−0.003	+0.001	2.876	2.874	2.876	4.76	4.78	4.770	−0.16	+0.02	−0.02	4.92	4.76	4.790
β Tauri	3.780	3.779	3.776	+0.001	−0.002	+0.007	3.781	3.777	3.783	4.02	4.00	4.026	+0.10	+0.17	+0.20	3.92	3.83	3.826
α Orionis	3.243	3.241	3.239	+0.001	−0.002	+0.005	3.244	3.239	3.244	1.36	1.39	1.368	−0.13	−0.03	−0.01	1.49	1.42	1.378
										+	+	+				+	+	+
α Canis Maj.	2.681	2.679	2.678	−0.028	−0.034	−0.034	2.653	2.645	2.644	3.17	3.17	3.165	+1.04	+1.14	+1.20	4.21	4.31	4.365
α Geminor.	3.863	3.862	3.859	−0.010	−0.011	−0.015	3.853	3.851	3.844	7.02	7.01	7.006	+0.04	+0.10	+0.05	7.06	7.11	7.056
α Canis Min.	3.195	3.193	3.191	−0.053	−0.047	−0.044	3.142	3.146	3.147	7.59	7.58	7.576	+0.95	+0.98	+0.98	8.54	8.56	8.556
β Geminor.	3.737	3.737	3.734	−0.049	−0.048	−0.049	3.688	3.689	3.685	7.93	7.92	7.917	0.00	+0.11	+0.06	7.93	8.03	7.977
α Hydræ	3.952	2.950	2.948	−0.006	−0.010	−0.003	3.946	2.940	2.945	15.24	15.24	15.229	−0.14	+0.05	+0.01	15.10	15.29	15.239
θ Ursæ Maj.		4.197	4.193		−0.120	−0.109		4.077	4.084		15.34	15.322		+0.60	+0.61		15.94	15.932
α Leonis	3.227	3.225	3.223	−0.015	−0.019	−0.020	3.212	3.206	3.203	17.25	17.27	17.259	−0.08	+0.01	−0.06	17.17	17.28	17.199
α Ursæ Maj.		3.839	3.836		−0.016	−0.015		3.823	3.821		19.17	19.151		0.00	+0.11			19.261
β Leonis	3.105	3.104	3.102	−0.038	−0.035	−0.035	3.067	3.069	3.067	19.97	19.98	19.961	+0.07	+0.08	+0.05	20.04	20.06	20.011
β Virginis	3.076	3.075	3.073	+0.049	+0.051	+0.041	3.125	3.126	3.114	19.98	19.99	19.972	+0.24	+0.30	+0.28	20.22	20.29	20.252
γ Ursæ Maj.		3.208	3.206		+0.004	+0.019		3.212	3.225		20.01	19.992		+0.03	+0.05		20.04	20.042
α Virginis	3.148	3.145	3.144	−0.001	−0.006	−0.003	3.147	3.139	3.141	18 99	19.01	18.991	−0.19	+0.03	0.00	18.80	19.04	18.991
ζ Ursæ Maj.		2.425	2.424		−0.005	+0.001		2.420	2.425		18.98	18.958		+0.01	+0.03		18.99	18.988
η Ursæ Maj.		2.390	2.389		−0.033	−0.017		2.357	2.372		18.20	18.181		0.00	0.00		18.20	18.181
α Bootis	2.812	2.811	2.809	−0.084	−0.078	−0.081	2.728	2.733	2.728	17.07	17.08	17.067	+1.72	+1.96	+1.94	18.79	19.04	18.007
α² Libræ	3.304	3.303	3.301	−0.007	−0.013	−0.002	3.297	3.290	3.299	15.36	15.38	15.365	−0.15	+0.08	+0.01	15.21	15.46	15.375
		−	−					−	−									
β Ursæ Min.		0.320	0.315		−0.020	−0.010		0.340	0.325		14.71	14.694		+0.18	+0.12		14.89	14.814
		+	+					+	+									
α Cor. Bor.	2.528	2.527	2.525	+0.017	−0.007	+0.005	2.545	2.520	2.530	12.46	12.48	12.463	+0.03	+0.10	+0.01	12.49	12.58	12.473
α Serpentis	2.937	2.936	2.934	+0.007	−0.007	+0.012	2.944	2.929	2.946	11.89	11.91	11.894	−0.19	−0.05	−0.09	11.70	11.86	11.804
α Scorpii	3.658	2.657	3.654	0.000	−0.003	+0.007	3.658	3.654	3.661	8.69	8.70	8.698	−0.26	+0.10	−0.01	8.43	8.80	8.688
α Herculis	2.731	2.730	2.728	0.000	−0.007	−0.004	2.731	2.723	2.724	4.71	4.72	4.720	−0.23	−0.12	−0.12	4.48	4.60	4.600
β Draconis		1.348	1.344		−0.018	−0.007		8.330	1.337		2.97	2.971		0.00	−0.02		2.97	2.951
α Ophiuchi	2.772	2.771	2.769	+0.004	+0.006	+0.003	2.776	2.777	2.772	2.99	3.00	2.993	+0.05	+0.18	+0.12	3.04	3.18	3.113
γ Draconis		1.388	1.387		−0.021	0.000		1.367	1.387		0.70	0.703		+0.07	+0.05		0.77	0.753
										−	−	−				−	−	−
ε Sagittarii		3.986	3.983		−0.010	0.000		3.976	3.983		0.95	0.953		+0.08	−0.07		0.87	1.023
α Lyræ	3.013	2.011	2.010	+0.015	+0.015	+0.017	2.028	2.026	2.027	2.64	2.63	2.631	−0.27	−0.25	−0.31	2.91	2.88	2.941
γ Aquilæ	2.853	2.851	2.849	−0.007	+0.005	+0.005	2.846	2.856	2.854	8.22	8.22	8.213	−0.16	−0.04	−0.05	8.38	8.26	8.263
α Aquilæ	2.893	2.892	2.890	+0.032	+0.034	+0.038	2.925	2.926	2.928	8.56	8.56	8.553	−0.54	−0.38	−0.46	9.10	8.94	9.013
β Aquilæ	2.947	2.945	2.943	−0.002	−0.002	+0.005	2.945	2.943	2.948	8.91	8.91	8.898	+0.35	+0.54	+0.41	8.56	8.37	8.488
α² Capric.	3.336	3.335	3.333	+0.003	+0.003	+0.008	3.339	3.338	3.341	10.56	10.55	10.545	−0.26	−0.25	−0.08	10.82	10.80	10.625
α Cygni	2.042	2.040	2.039	−0.005	−0.005	−0.004	2.037	2.035	2.035	12.53	12.53	12.521	−0.03	0.00	−0.04	12.56	12.53	12.561
α Cephei		1.419	1.418		+0.018	+0.013		1.437	1.431		15.01	15.001		+0.07	+0.05		14.94	14.951
β Cephei		0.821	0.821		−0.011	−0.012		0.810	0.809		15.70	15.688		+0.03	+0.11		15.67	15.578
α Aquarii	3.086	3.085	3.083	−0.005	−0.008	0.000	3.081	3.077	3.083	17.17	17.18	17.160	−0.19	+0.05	−0.05	17.36	17.13	17.210
α Pisc. Aust.	3.320	3.319	3.317	+0.023	+0.022	+0.027	3.343	3.341	3.344	19.04	19.04	19.026	−0.06	+0.26	+0.14	19.10	18.78	18.886
α Pegasi	2.976	2.975	2.973	−0.004	+0.001	+0.005	2.972	2.976	2.978	19.25	19.28	19.240	−0.18	+0.07	−0.06	19.43	19.19	19.300
α Androm.	3.065	3.063	3.061	+0.005	−0.009	−0.008	3.070	3.054	3.053	20.05	20.06	20.045	+0.06	+0.21	+0.13	19.99	19.85	19.915
	Mas-kelyne.	Piazzi.	Bessel.	Mas-kelyne.	Piazzi.	Bessel.	Mas-kelyne.	Piazzi.	Bessel.	Mas-kelyne.	Piazzi.	Bessel.	Mas-kelyne.	Piazzi.	Bessel.	Mas-kelyne.	Piazzi.	Bessel.
	Annual Precession.			Proper Motions.			Annual Variations.			Annual Precession.			Proper Motions.			Annual Variations.		
	RIGHT ASCENSION IN TIME.									NORTH POLAR DISTANCE.								

A TABLE of the CHANGES that take place in the ARGUMENTS of the preceding 48 principal STARS, in the course of 30 Years.

STARS.	RIGHT ASCENSION.									NORTH POLAR DISTANCE.								
	A.		Cha. in 30 Yrs.	L.		Cha. in 30 Yrs.	S.		Cha. in 30 Yrs.	A′.		Cha. in 30 Yrs.	L′.		Cha. in 30 Yrs.	S′.		Cha. in 30 Yrs.
	1800	1830		1800	1830		1800	1830		1800	1830		1800	1830		1800	1830	
	s ° ′	s ° ′	′	s ° ′	s ° ′	′	s ° ′	s ° ′	′	s ° ′	s ° ′	′	s ° ′	s ° ′	′	s ° ′	s ° ′	′
γ Pegasi	8 29 13	8 28 47	−26	6 8 18	6 8 24	+ 6	6 6 46	6 6 50	+ 4	1 27 37	1 27 12	−25	11 29 1	11 28 29	−32	11 29 12	11 28 47	−25
α Cassio.	8 22 3	8 21 35	−28	7 7 52	7 7 53	+ 1	7 2 14	7 2 16	+ 2	0 9 2	0 8 34	−28	11 20 13	11 19 40	−33	11 22 2	11 21 36	−26
α Ur.Mi.	8 15 45	8 13 51	−114	8 17 11	8 16 7	−64	8 14 22	8 13 3	−79	11 18 38	11 16 55	−103	11 12 38	11 10 21	−137	11 15 45	11 13 51	−114
α Arietis	7 28 52	7 28 26	−26	6 11 1	6 11 1	0	6 8 59	6 8 59	0	1 0 33	1 0 2	−31	10 23 21	10 22 52	−29	10 28 53	10 28 26	−27
α Ceti	7 14 35	7 14 11	−24	6 1 23	6 1 26	+ 3	6 1 7	6 1 9	+ 2	2 23 22	2 23 8	−14	10 8 37	10 8 15	−22	10 14 35	10 14 11	−24
α Persei	7 10 1	7 9 30	−31	6 18 23	6 18 13	−10	6 15 6	6 14 57	− 9	11 3 48	11 3 7	−41	10 4 16	10 3 47	−29	10 10 1	10 9 30	−31
α Tauri	6 22 7	6 21 42	−25	6 3 30	6 3 28	− 2	6 2 50	6 2 49	− 1	1 23 12	1 23 15	+ 3	9 18 15	9 17 54	−21	9 22 7	9 21 42	−25
α Aurigæ	6 13 22	6 12 56	−26	6 5 59	6 5 46	−13	6 4 52	6 4 41	−11	9 26 36	9 25 37	−59	9 10 55	9 10 29	−26	9 13 22	9 12 51	−31
β Orion.	6 12 40	6 12 20	−20	5 28 45	5 28 47	+ 2	5 28 59	5 29 0	+ 1	3 3 49	3 3 42	− 7	9 10 21	9 10 4	−17	9 12 40	9 12 20	−20
β Tauri	6 10 39	6 10 13	−26	6 2 57	6 2 50	− 7	6 2 24	6 2 18	− 6	10 20 58	10 19 23	−95	9 8 41	9 8 20	−21	9 10 39	9 10 13	−26
α Orion.	6 3 35	6 3 13	−22	6 0 16	6 0 15	− 1	6 0 14	6 0 12	− 2	2 28 12	2 28 23	+11	9 2 55	9 2 36	−19	9 3 35	9 3 13	−22
α Ca.Ma.	5 21 39	5 21 21	−18	6 1 47	6 1 51	+ 4	6 1 27	6 1 30	+ 3	2 16 0	2 25 50	−10	8 23 13	8 22 58	−15	8 21 39	8 21 22	−17
α Gem.	5 11 7	5 10 40	−27	5 24 9	5 24 2	− 7	11 25 15	5 25 9	− 6	7 3 20	7 2 6	−74	8 14 29	8 14 6	−23	8 11 6	8 10 39	−27
α Ca.Mi.	5 9 29	5 9 6	−23	5 28 47	5 28 48	+ 1	5 29 1	5 29 0	− 1	3 6 54	3 6 54	0	8 13 6	8 12 47	−19	8 9 28	8 9 6	−22
β Gem.	5 8 29	5 8 2	−27	5 24 7	5 24 2	− 5	5 25 14	5 25 9	− 5	6 15 27	6 14 32	−55	8 12 15	8 11 53	−22	8 8 28	8 8 2	−26
α Hydræ	4 13 1	4 12 38	−23	6 3 37	6 3 42	+ 5	6 2 56	6 3 0	+ 4	2 17 46	2 17 32	−14	7 18 59	7 18 37	−22	7 13 1	7 12 38	−23
θ Ur.Ma.	4 12 36	4 12 5	−31	5 6 54	5 6 45	− 9	5 10 55	5 10 47	− 8	6 20 34	6 19 55	−39	7 18 35	7 18 3	−32	7 12 36	7 12 5	−31
α Leonis	4 2 48	4 2 22	−26	5 23 44	5 23 47	+ 3	5 24 54	5 24 56	+ 2	4 4 12	4 3 46	−26	7 8 27	7 8 0	−27	7 2 45	7 2 21	−24
α Ur.Ma.	3 18 38	3 18 7	−31	4 19 1	4 18 58	− 3	4 24 50	4 24 45	− 5	6 3 59	6 3 27	−32	6 22 34	6 21 58	−36	6 18 38	6 18 7	−31
β Leonis	3 5 46	3 5 21	−25	5 20 50	5 20 55	+ 5	5 22 32	5 22 38	+ 6	4 6 45	4 6 19	−26	6 7 7	6 6 34	−33	6 5 47	6 5 21	−26
β Virg.	3 5 23	3 4 56	−27	5 28 19	5 28 25	+ 6	5 28 38	5 28 43	+ 5	3 7 16	3 6 51	−25	6 6 37	6 6 5	−32	6 5 23	6 4 58	−25
γ Ur.Ma.	3 4 34	3 4 8	−26	4 21 44	4 21 42	− 2	4 27 21	4 27 21	0	5 17 54	5 17 30	−24	6 5 37	6 5 5	−32	6 4 34	6 4 7	−27
α Virg.	2 9 47	2 9 22	−25	6 5 29	6 5 33	+ 4	6 4 28	6 4 30	+ 2	2 3 58	2 3 32	−26	5 5 36	5 5 6	−30	5 9 47	5 9 22	−25
ζ Ur.Ma.	2 9 28	2 9 9	−19	4 14 5	4 14 14	+ 9	4 20 3	4 20 12	+ 9	4 29 4	4 28 44	−20	5 5 14	5 4 51	−23	5 9 28	5 9 9	−19
η Ur.Ma.	2 3 8	2 2 49	−19	4 20 44	4 20 54	+10	4 26 26	4 26 36	+10	4 21 41	4 21 27	−14	4 28 1	4 27 39	−22	5 3 9	5 2 51	−18
α Bootis	1 26 7	1 25 46	−21	5 18 42	5 18 50	+ 8	5 20 48	5 20 54	+ 6	3 28 37	3 28 18	−19	4 20 23	4 19 59	−24	4 26 8	4 25 45	−23
α² Libræ	1 17 35	1 17 11	−24	6 6 27	6 6 27	0	6 5 15	6 5 15	0	1 18 50	1 18 25	−25	4 11 37	4 11 11	−26	4 17 35	4 17 11	−24
β Ur.Mi.	1 14 40	1 14 43	+ 3	2 26 24	2 26 45	+21	2 25 36	2 25 59	+23	4 14 46	4 15 5	+19	4 8 42	4 8 45	+ 3	4 14 40	4 14 43	+ 3
α Cor.Bo	1 6 3	1 5 45	−18	5 17 9	5 17 17	+ 8	5 19 31	5 19 38	+ 7	3 22 41	5 22 28	−13	4 0 34	4 0 17	−17	4 6 3	4 5 46	−17
α Serp.	1 4 4	1 3 43	−21	5 27 26	5 27 30	+ 4	5 27 55	5 27 58	+ 3	3 8 32	3 8 22	−10	3 28 45	3 28 24	−21	4 4 5	4 3 43	−22
α Scorpii	0 23 50	0 23 24	−26	6 5 54	6 5 50	− 4	6 4 48	6 4 44	− 4	11 28 49	11 28 0	−49	3 19 43	3 19 21	−22	3 23 49	3 23 23	−26
α Herc.	0 12 28	0 12 12	−16	5 27 42	5 27 45	− 2	5 28 8	5 28 10	+ 2	3 5 34	3 5 25	− 9	3 10 14	3 9 58	−16	3 12 32	3 12 13	−19
β Draco.	0 7 53	0 7 40	−13	5 15 43	5 15 59	+16	5 18 18	5 18 33	+15	3 6 58	3 6 48	−10	3 6 22	3 6 15	− 7	3 7 49	3 7 40	− 9
α Ophiu.	0 7 49	0 7 34	−15	5 28 45	5 28 49	+ 4	5 28 59	5 29 2	+ 3	3 3 12	3 3 4	− 8	3 6 25	3 6 9	−16	3 7 53	3 7 34	−19
γ Draco.	0 1 50	0 1 41	− 9	5 26 46	5 27 2	+16	5 27 22	5 27 36	+14	3 1 37	3 1 29	− 8	3 1 29	3 1 21	− 8	3 1 50	3 1 41	− 9
ε Sagitt.	11 27 29	11 27 2	−27	5 29 9	5 29 0	− 9	5 29 19	5 29 12	− 7	8 21 57	8 20 29	−88	2 27 58	2 27 36	−22	2 27 29	2 27 5	−24
α Lyræ	11 23 5	11 22 50	−15	6 5 19	6 5 30	+11	6 4 18	6 4 28	+10	2 24 40	2 24 30	−10	2 24 22	2 24 11	−11	2 23 5	2 22 50	−15
γ Aquilæ	11 7 37	11 7 16	−21	6 2 38	6 2 42	+ 4	6 2 8	6 2 11	+ 3	2 22 22	2 22 17	− 5	2 11 31	2 11 14	−17	2 7 37	2 7 16	−21
α Aquilæ	11 6 36	11 6 15	−21	6 2 13	6 2 17	+ 4	6 1 48	6 1 51	+ 3	2 23 8	2 22 58	−10	2 10 39	2 10 21	−18	2 6 37	2 6 16	−21
β Aquilæ	11 5 33	11 5 11	−22	6 1 36	6 1 39	+ 3	6 1 18	6 1 20	+ 2	2 24 34	2 24 26	− 8	2 9 45	2 9 26	−19	2 5 33	2 5 12	−21
α² Capri.	11 0 26	11 0 2	−24	5 26 14	5 26 12	− 2	5 26 56	5 26 55	− 1	3 29 39	3 29 32	− 7	2 5 16	2 4 54	−22	2 0 25	2 0 2	−23
α Cygni	10 23 44	10 23 29	−15	6 28 19	6 28 32	+13	6 23 37	6 23 47	+10	2 0 52	2 0 38	−14	1 29 13	1 29 0	−13	1 23 29	1 23 14	−15
α Cephei	10 14 1	10 13 50	−11	8 0 14	8 0 29	+15	7 24 48	7 25 6	+18	1 17 32	1 17 30	− 2	1 19 57	1 19 48	− 9	1 14 1	1 13 50	−11
β Cephei	10 10 56	10 10 49	− 7	8 17 38	8 17 54	+16	8 14 52	8 15 11	+19	1 12 33	1 12 25	− 8	1 16 54	1 16 47	− 7	1 10 56	1 10 49	− 7
α Aquar.	10 3 21	10 2 58	−23	5 29 22	5 29 26	+ 4	5 29 29	5 29 33	+ 4	3 2 50	3 2 31	−19	1 9 4	1 8 37	−27	1 3 21	1 2 56	−25
α Pis.Au.	9 19 53	9 19 26	−27	5 13 6	5 13 8	+ 2	5 16 9	5 16 10	+ 1	5 8 4	5 7 34	−30	0 24 3	0 23 34	−29	0 19 53	0 19 26	−27
α Pegasi	9 17 41	9 17 16	−25	6 8 17	6 8 23	+ 6	6 6 44	6 6 49	+ 5	2 2 27	2 2 5	−22	0 21 26	0 20 59	−27	0 17 41	0 17 17	−24
α Andro.	9 0 32	9 0 6	−26	6 17 16	6 17 21	+ 5	6 14 8	6 14 13	+ 5	1 7 8	1 6 43	−25	0 0 39	0 0 8	−31	0 0 34	0 0 6	−23
	1800	1830		1800	1830		1800	1830		1800	1830		1800	1830		1800	1830	
	A.			L.			S.			A′.			L′.			S′.		
	RIGHT ASCENSION.									NORTH POLAR DISTANCE.								

The CHANGES of the SUBSIDIARY NUMBERS (and consequently of the ARGUMENTS), as given in this Table, will render the CORRECTIONS of the 48 principal Stars (from page 113 to page 144 inclusive) suitable for *any year*, before or after 1830 provided that due proportions be taken for the elapsed periods, and that the signs be changed for any time antecedent to the year 1830, when this epoch is used. The constants given in this Table, being recomputed, should be used in preference to those subjoined to the Table of " Corrections of 48 principal Stars, unless the *Corrigenda* be duly attended to, as given in page 339 hereafter.

A TABLE of the Annual Precessions and Maxima of Aberration, Lunar Nutation, and Solar Nutation of the 48 principal Stars, both in Right Ascension in Time, and in North Polar Distance, for the beginning of the Years 1800 and 1830.

Bayer's Names of the Stars.	RIGHT ASCENSION IN TIME. Precession. 1800	Precession. 1830	Aberration. 1800 Max.	Aberration. 1830 Max.	☾ Nutation. 1800 Max.	☾ Nutation. 1830 Max.	⊙ Nutation. 1800 Max.	⊙ Nutation. 1830 Max.	NORTH POLAR DISTANCE. Precession. 1800	Precession. 1830	Aberration. Max. 1800	Aberration. Max. 1830	☾ Nutation. 1800 Max.	☾ Nutation. 1830 Max.	⊙ Nutation. 1800 Max.	⊙ Nutation. 1830 Max.	Ancient Names of the Stars.
	+	+							—	—							
γ Pegasi	3ˢ.072	3ˢ.074	1ˢ.277	1ˢ.278	1ˢ.117	1ˢ.117	0ˢ.061	0ˢ.062	20″.042	20″.040	9″.196	9″.196	7″.183	7″.183	0″.398	0″.398	Algenib
α Cassiopeæ	3.315	3.330	2.137	2.197	1.509	1.516	0.078	0.078	19.880	19.867	16.747	16.743	7.229	7.235	0.399	0.399	Schedir
α Ursæ Min.	12.932	15.367	40.916	44.708	20.917	22.956	0.953	1.048	19.522	19.372	19.920	20.096	7.329	7.371	0.400	0.400	Polaris
α Arietis	3.336	3.342	1.370	1.372	1.220	1.224	0.067	0.067	17.535	17.462	7.881	7.853	7.830	7.848	0.407	0.408	Hamal
α Ceti	3.120	3.123	1.294	1.295	1.123	1.117	0.062	0.063	14.670	14.640	7.374	7.344	8.418	8.434	0.416	0.416	Menkab
α Persei	4.207	4.221	1.988	1.994	1.592	1.596	0.087	0.087	13.535	13.398	11.527	11.501	8.612	8.634	0.418	0.418	Mirfak
α Tauri	3.421	3.423	1.387	1.388	1.231	1.231	0.068	0.068	8.116	7.978	3.795	3.748	9.289	9.301	0.429	0.429	Aldebaran
α Aurigæ	4.397	4.402	1.926	1.928	1.589	1.589	0.088	0.088	5.025	4.837	8.124	8.105	9.517	9.523	0.432	0.432	Capella
β Orionis	2.875	2.876	1.359	1.359	1.034	1.038	0.058	0.057	4.771	4.651	10.659	10.643	9.526	9.532	0.433	0.433	Rigel
β Tauri	3.777	3.769	1.530	1.531	1.359	1.360	0.075	0.075	4.026	3.865	2.493	2.451	9.562	9.568	0.433	0.433	Nath
α Orionis	3.240	3.241	1.361	1.362	1.179	1.166	0.063	0.066	1.369	1.224	5.629	5 619	9.639	9.641	0.434	0.434	Beteigeuze
									+	+							
α Canis Maj.	2.678	2.678	1.405	1.405	0.970	0.964	0.053	0.053	3.165	3.281	12.963	12.971	9.595	9.591	0.434	0.433	Sirius
α Gemin.	3.860	3.854	1.582	1.581	1.394	1.394	0.077	0.077	7.004	7.162	4.529	4.550	9.382	9.370	0.430	0.430	Castor
α Canis Min.	3.192	3.190	1.342	1.341	1.148	1.149	0.064	0.063	7.564	7.702	6.355	6.383	9.337	9.326	0.429	0.429	Procyon
β Gemin.	3.734	3.731	1.517	1.516	1.345	1.348	0.075	0.075	7.916	8.064	3.962	4.006	9.307	9.294	0.429	0.429	Pollux
α Hydræ	2.949	2.948	1.299	1.299	1.061	1.059	0.059	0.059	15.227	15.314	9.854	9.875	8.316	8.299	0.414	0.414	Alphard
ϑ Ursæ Maj.	4.194	4.175	2.117	2.109	1.638	1.632	0.088	0.088	15.319	15.434	13.134	13.153	8.309	8.275	0.414	0.413	
α Leonis	3.224	3.220	1.300	1.300	1.166	1.166	0.064	0.065	17.257	17.326	6.946	6.974	7.895	7.882	0.408	0.408	Regulus
α Ursæ Maj.	3.819	3.811	2.735	2.719	1.825	1.815	0.093	0.093	19.148	19.197	17.256	17.284	7.430	7.417	0.401	0.401	Dubhe
β Leonis	3.103	3.100	1.287	1.287	1.128	1.128	0.062	0.062	19.959	19.973	9.117	9.118	7.206	7.204	0.398	0.398	Benebola
β Virginis	3.074	3.076	1.241	1.241	1.102	1.104	0.061	0.061	19.970	19.981	8.039	8.042	7.204	7.290	0.398	0.398	Zavijava
γ Ursæ Maj.	3.207	3.192	2.150	2.141	1.468	1.463	0.076	0.076	19.990	20.000	16.882	16.933	7.197	7.194	0.398	0.398	Phecda
α Virginis	3.144	3.147	1.270	1.272	1.135	1.137	0.063	0.063	18.989	18.945	7.676	7.658	7.466	7.477	0.402	0.402	Spica Virg.
ζ Ursæ Maj.	2.425	2.420	2.235	2.227	1.254	1.248	0.063	0.063	18.956	18.922	18.514	18.503	7.481	7.490	0.402	0.402	Mizar
η Ursæ Maj.	2.390	2.386	1.972	1.966	1.111	1.111	0.057	0.057	18.184	18.134	18.022	18.000	7.679	7.688	0.404	0.405	Alkaid
α Bootis	2.810	2.809	1.354	1.353	1.030	1.030	0.057	0.056	17.066	17.002	12.452	12.441	7.937	7.951	0.408	0.408	Arcturus
α² Libræ	3.301	3.305	1.332	1.334	1.193	1.196	0.066	0.066	15.363	15.272	6.184	6.146	8.288	8.397	0.413	0.414	Kiffa Aust.
	—	—															
β Ursæ Min.	0.318	0.287	4.981	4.942	1.759	1.745	0.079	0.079	14.691	14.712	20.251	20.249	8.414	8.412	0.415	0.415	Kocab
	+	+															
α Cor. Bor.	2.526	2.526	1.474	1.473	0.932	0.931	0.051	0.051	12.460	12.375	15.020	15.011	8.778	8.791	0.421	0.421	Gemma
α Serpentis	2.935	2.938	1.322	1.323	1.057	1.060	0.058	0.059	11.893	11.788	9.973	9 935	8.859	8.873	0.422	0.422	Unukalhay
α Scorpii	3.655	3.659	1.480	1.481	1.322	1.320	0.073	0.073	8.695	8.550	3.849	3.784	9.235	9.249	0.428	0.428	Antares
α Herculis	2.729	2.729	1.389	1.389	0 985	0.979	0.055	0.054	4.719	4.603	12.424	12.418	9.529	9.533	0.433	0.433	Rasalgeti
β Draconis	1.349	1.349	2.212	2.211	0.503	0.502	0.028	0.026	2.970	2.912	19.630	19.626	9.598	9.603	0.434	0.434	Alwaid
α Ophiuchi	2.770	2.770	1.382	1.382	0.996	0.996	0.055	0.055	2.993	2.873	11.930	11.926	9.620	9.604	0.434	0.434	Rasalgue
γ Draconis	1.388	1.389	2.170	2.169	0.502	0.502	0.028	0.028	0.703	0.639	19.562	19.562	9.646	9.647	0.434	0.434	Etamin
									—	—							
ε Sagittarii	3.984	3.983	1.638	1.637	1.433	1.430	0.079	0.079	0.952	1.126	3.890	3.896	9.644	9.642	0.343	0.343	Kaus Austr.
α Lyræ	2.010	2.010	1.726	1.727	0.727	0.727	0.040	0.040	2.630	2.717	17.864	17.871	9.612	9.608	0.434	0.434	Vega
γ Aquilæ	2.850	2.850	1.353	1.353	1.025	1.021	0.055	0.057	8.211	8.328	10.995	11.112	9.281	9.270	0.429	0.428	Tarazed
α Aquilæ	2.891	2.889	1.345	1.345	1.040	1.040	0.058	0.059	8.551	8.667	10.497	10.489	9.249	9.237	0.428	0.428	Altair
β Aquilæ	2.944	2.943	1.336	1.336	1.057	1.060	0.059	0.059	8.903	9.021	9.782	9.800	9.214	9.203	0.427	0.427	Alshairn
α² Capricor.	3.334	3.331	1.356	1.355	1.200	1.199	0.066	0.066	10.542	10.666	4.900	4.944	9.033	9.018	0.425	0.424	Sec. Giedi
α Cygni	2.039	2.040	1.836	1.839	0.834	0.836	0.044	0.045	12.519	12.588	18.236	18.215	8.769	8.758	0.421	0.420	Deneb
α Cephei	1.419	1.416	2.723	2.733	1.031	1.037	0.049	0.050	14.999	15.042	19.816	19.847	8.359	8.350	0.415	0.415	Alderamin
β Cephei	0.822	0.811	3.695	3.718	1.392	1.404	0.063	0.064	15.687	15.711	20.182	20.181	8.226	8.221	0.412	0.412	Alphirk
α Aquarii	3.084	3.082	1.270	1.269	1.102	1.108	0.062	0.063	17.158	17.228	7.858	7.883	7.918	7.902	0.408	0.408	Sadalmelik
α Pis. Austr.	3.318	3.311	1.454	1.451	1.246	1.243	0.068	0.068	19.024	19.068	10.573	10.600	7.463	7.451	0.402	0.402	Fomalhaut
α Pegasi	2.974	2.989	1.287	1.287	1.080	1.080	0.060	0.060	19.238	19.274	10.259	10.270	7.407	7.397	0.401	0.401	Markab
α Androm.	3.059	3.066	1.403	1.405	1.152	1.156	0.063	0.063	20.043	20.044	11.925	11.936	7.183	7.183	0.398	0.398	Sirrah
	1800	1830	Max. 1800	Max. 1830	Max. 1800	Max. 1830	Max. 1800	Max. 1830	1800	1830	Max. 1800	Max. 1830	Max. 1800	Max. 1830	Max. 1800	Max. 1830	
	Precession.		Aberration.		☾ Nutation.		⊙ Nutation.		Precession.		Aberration.		☾ Nutation.		⊙ Nutation.		
	RIGHT ASCENSION IN TIME.								NORTH POLAR DISTANCE.								

The Numbers contained in this Table were computed by means of the first Series of General Tables, numbered I. to XIV., near the beginning of this Volume, as were also the Arguments contained in the preceding Tables.

BESSEL'S TABLE OF CORRECTIONS,

IN RIGHT ASCENSION,

OF

DR. MASKELYNE'S 36 PRINCIPAL STARS,

FOR THE MOMENT OF CULMINATION.

(1818.)

[To precede Page 149.]

TABLE 1.

Years.	Reduction for Time. In Jan. and Feb.	k
1800	....	−0 . 10
1810	....	−0 . 53
1813	....	−0 . 25
1814	....	−0 . 49
1815	....	−0 . 74
1816 B	−0 . 98	+0 . 02
1817	....	−0 . 22
1818	....	−0 . 46
1819	....	−0 . 70
1820 B	−0 . 95	+0 . 05
1821	....	−0 . 19
1822	....	−0 . 43
1823	....	−0 . 67
1824 B	−0 . 92	+0 . 08
1825	....	−0 . 16
1826	....	−0 . 40
1827	....	−0 . 64
1828 B	−0 . 89	+0 . 11
1829	....	−0 . 13
1830	....	−0 . 37
1831	....	−0 . 61
1832	−0 . 86	+0 . 14

TABLE 2.

Places.	Reduction for Long. of Observatory.
Berlin	−0 . 03
Copenhagen	−0 . 03
Dorpat	−0 . 07
Göttingen	−0 . 02
Gotha	−0 . 02
Greenwich	+0 . 01
Königsberg	−0 . 06
Manheim	−0 . 02
Mayland	−0 . 02
Munich	−0 . 03
Palermo	−0 . 03
Paris	0 . 00
Petersburg	−0 . 08
Vienna	−0 . 04

TABLE 3.

Stars.	Day of Culmination with the Sun.	Reduction for Date. Before.	After.
γ Pegasi	March 21	+ 1	+ 2
α Arietis	April 21	+ 1	+ 2
α Ceti	May 6	+ 1	+ 2
α Tauri	May 29	+ 1	+ 2
α Aurigæ supra	June 7	+ 1	+ 2
——— infra	December 8	+ 1 . 5	+ 2 . 5
β Orionis	June 8	+ 1	+ 2
β Tauri	June 10	+ 1	+ 2
α Orionis	June 17	+ 1	+ 2
α Canis Majoris	June 30	+ 1	+ 2
α Geminorum	July 11	+ 1	+ 2
α Canis Minoris	July 13	+ 1	+ 2
β Geminorum	July 14	+ 1	+ 2
α Hydræ	August 9	+ 1	+ 2
α Leonis	August 20	+ 1	+ 2
β Leonis	September 17	+ 1	+ 2
β Virginis	September 17	+ 1	+ 2
α Virginis	October 13	+ 1	+ 2
α Bootis	October 27	+ 1	+ 2
α^1 Libræ	November 4	+ 1	+ 2
α^2 Libræ	November 4	+ 1	+ 2
α Coronæ	November 16	+ 1	+ 2
α Serpentis	November 18	+ 1	+ 2
α Scorpii	November 28	+ 1	+ 2
α Herculis	December 9	+ 1	+ 2
α Ophiuchi	December 13	+ 1	+ 2
α Lyræ supra	December 28	+ 1	+ 2
——— infra	June 28	+ 0 . 5	+ 1 . 5
γ Aquilæ	January 12	0	+ 1
α Aquilæ	January 13	0	+ 1
β Aquilæ	January 14	0	+ 1
α^1 Capricorni	January 19	0	+ 1
α^2 Capricorni	January 19	0	+ 1
α Cygni supra	January 26	0	+ 1
——— infra	July 29	+ 0 . 5	+ 1 . 5
α Aquarii	February 15	0	+ 1
α Piscis Austrini	February 28	0	+ 1
α Pegasi	March 2	0	+ 1
α Andromedæ	March 20	0	+ 1

Jan. 0 means Dec. 31 of the preceding Year, when the Sun's Longitude is supposed to be $\overset{s}{9}$ 10° 0′; *k* is equal to the Complement of the Sun's daily Motion in Longitude from $\overset{s}{9}$ 10° to the Sun's Longitude on Dec. 31, and is applied as a Reduction to that Time.

The amount of the three Quantities taken from the three preceding Tables, according to the Time, Place, and Star in question, must be applied to the Day of the Month proposed, in order to gain the Star's Corrections in Right Ascension for the moment of Culmination, from the *succeeding* Table, which is adapted for Sidereal Days.

☞ In the *principal Table*, the respective Columns of the *upper* Portion contain the Mean Right Ascension of each Star, for the beginning of each Year, combined with the Lunar Nutation for each 100 Days successively; and the wide Columns of the lower Portion, contain the aggregate of the Precession, Aberration, and Solar Nutation, in one Sum, to be applied to the former combined Sum. The narrow Columns, improperly headed "☾ Nut." and "☉ Nut." being only the *Columns* of *Differences* of the contiguous Columns.

BESSEL'S TABLE

of 1818,

for determining the true apparent Right Ascensions of Dr. Maskelyne's 36 Stars.

	γ Pegasi.		α Arietis.		α Ceti.		α Tauri.		α Aurigæ.		β Orionis.		β Tauri.		α Orionis.		α Can. Maj.	
	h. m. 0 4	☾ Nut.	h. m. 1 57	☾ Nut.	h. m. 2 52	☾ Nut.	h. m. 4 25	☾ Nut.	h. m. 5 3	☾ Nut.	h. m. 5 5	☾ Nut.	h. m. 5 15	☾ Nut.	h. m. 5 45	☾ Nut.	h. m. 6 37	☾ Nut.
1822. Jan. 0	5s. 399	+ 85	10s. 098	+ 96	59s. 720	+ 79	43s. 549	+ 90	33s. 925	+118	59s. 718	+ 71	3s. 428	+ 98	32s. 752	+ 81	18s. 518	+ 69
April 10	6 . 484	81	10 . 194	90	59 . 799	75	43 . 639	84	34 . 043	112	59 . 789	66	3 . 526	92	32 . 833	76	18 . 587	64
July 19	5 . 565	76	10 . 284	87	59 . 874	69	43 . 727	78	34 . 155	105	59 . 855	61	3 . 618	86	32 . 909	71	18 . 651	60
Oct. 27	5 . 641	72	10 . 371	82	59 . 943	64	43 . 801	73	34 . 260	98	59 . 916	55	3 . 704	79	32 . 980	65	18 . 711	55
Dec. 66	5 . 713		10 . 453		60 . 007		43 . 874		34 . 358		59 . 971		3 . 783		32 . 045		18 . 766	
1823. Jan. 0	8 . 771	66	13 . 784	77	63 . 112	58	47 . 281	67	38 . 740	90	62 . 832	50	7 . 544	73	36 . 268	59	21 . 392	50
April 10	8 . 837	60	13 . 861	96	63 . 170	51	47 . 348	59	38 . 830	81	62 . 882	44	7 . 617	64	36 . 327	51	21 . 442	44
July 19	8 . 897	53	13 . 930	62	63 . 221	43	47 . 407	50	38 . 911	70	62 . 926	36	7 . 681	54	36 . 378	43	21 . 486	38
Oct. 27	8 . 950	46	13 . 992	54	63 . 264	35	47 . 457	43	38 . 981	59	62 . 962	28	7 . 735	46	36 . 421	35	21 . 524	31
Dec. 66	8 . 996		14 . 046		63 . 299		47 . 500		39 . 040		62 . 990		7 . 781		36 . 456		21 . 555	
1824. Jan. 0	12 . 063	39	17 . 387	48	66 . 413	30	50 . 916	36	43 . 435	52	65 . 860	23	11 . 553	39	39 . 690	29	24 . 189	26
April 10	12 . 102	32	17 . 435	40	66 . 443	21	50 . 952	27	43 . 487	40	65 . 883	15	11 . 592	28	39 . 719	19	24 . 215	18
July 19	12 . 134	24	17 . 475	30	66 . 464	12	50 . 979	17	43 . 527	27	65 . 898	8	11 . 620	18	39 . 738	11	24 . 233	11
Oct. 27	12 . 158	16	17 . 505	21	66 . 476	3	50 . 996	8	43 . 554	16	65 . 906	— 1	11 . 638	7	39 . 749	1	24 . 244	4
Dec. 66	12 . 174		17 . 526		66 . 479		51 . 004		43 . 570		65 . 905		11 . 645		39 . 750		24 . 248	
1825. Jan. 0	15 . 251	9	20 . 879	14	69 . 604	— 2	54 . 432	1	47 . 980	5	68 . 784	6	15 . 430	0	42 . 996	— 5	26 . 891	— 2
April 10	15 . 260	0	20 . 893	6	69 . 602	11	54 . 433	— 8	47 . 985	— 4	68 . 778	14	15 . 430	— 11	42 . 991	14	26 . 889	9
July 19	15 . 260	— 8	20 . 899	— 5	69 . 591	20	54 . 425	19	47 . 981	14	68 . 764	23	15 . 419	21	42 . 977	22	26 . 880	16
Oct. 27	15 . 252	18	20 . 894	14	69 . 571	29	54 . 406	27	47 . 967	24	68 . 741	30	15 . 398	32	42 . 955	32	26 . 864	25
Dec. 66	15 . 234		20 . 880		69 . 542		54 . 379		47 . 943		68 . 711		15 . 366		42 . 923		26 . 839	

	Aber.	⊙Nut.	Aber.	⊙Nut.	Aber.	⊙Nut.	Aber.	⊙Nut.	Aber.	⊙Nut.	Aber.	⊙Nut.	Aber.	⊙Nut.	Aber.	⊙Nut.	Aber.	⊙Nut.
Jan. 0	—0s. 156	—113	+0s. 550	—118	+0s. 788	— 87	+1s. 212	— 27	+1s. 819	— 7	+1s. 278	— 9	+1s. 469	+ 17	+1s. 349	+ 38	+1s. 425	+ 65
10	0 . 269	107	0 . 432	133	0 . 701	111	1 . 185	65	1 . 812	64	1 . 269	49	1 . 486	— 31	1 . 387	— 5	1 . 490	19
20	0 . 376	97	0 . 299	145	0 . 590	131	1 . 120	103	1 . 748	123	1 . 220	89	1 . 455	80	1 . 382	51	1 . 509	— 32
30	0 . 473	81	0 . 154	148	0 . 459	143	1 . 017	132	1 . 625	170	1 . 131	122	1 . 375	119	1 . 331	90	1 . 477	75
Feb. 9	0 . 554	61	0 . 006	144	0 . 316	148	0 . 885	155	1 . 455	216	1 . 009	150	1 . 256	154	1 . 241	125	1 . 402	116
19	0 . 615	34	—0 . 138	130	0 . 168	146	0 . 730	168	1 . 245	235	0 . 859	169	1 . 102	177	1 . 116	150	1 . 286	147
Mar. 1	0 . 649	2	0 . 268	108	0 . 022	134	0 . 562	173	1 . 010	249	0 . 690	178	0 . 925	190	0 . 966	168	1 . 139	173
11	0 . 651	+ 33	0 . 376	78	—0 . 112	115	0 . 389	165	0 . 761	247	0 . 512	179	0 . 735	193	0 . 798	174	0 . 966	186
21	0 . 618	73	0 . 454	41	0 . 227	85	0 . 224	147	0 . 514	231	0 . 333	166	0 . 542	181	0 . 624	169	0 . 780	191
31	0 . 545	111	0 . 495	0	0 . 312	53	0 . 077	122	0 . 283	202	0 . 167	152	0 . 361	161	0 . 455	157	0 . 589	186
Apr. 10	0 . 434	156	0 . 495	+ 49	0 . 365	10	—0 . 045	86	0 . 031	160	0 . 015	112	0 . 200	129	0 . 298	133	0 . 403	171
20	0 . 278	193	0 . 446	96	0 . 375	+ 30	0 . 131	47	—0 . 079	112	—0 . 097	98	0 . 071	92	0 . 165	104	0 . 232	150
30	0 . 085	232	0 . 350	145	0 . 345	80	0 . 178	1	0 . 191	53	0 . 195	49	—0 . 021	46	0 . 061	67	0 . 082	120
May 10	+0 . 147	262	0 . 205	190	0 . 265	122	0 . 179	+ 43	0 . 244	+ 5	0 . 244	8	0 . 067	0	—0 . 006	29	—0 . 038	88
20	0 . 409	290	0 . 015	237	0 . 143	165	0 . 136	92	0 . 239	68	0 . 252	+ 36	0 . 067	+ 48	0 . 035	+ 15	0 . 126	49
30	0 . 699	308	+0 . 222	261	+0 . 022	204	0 . 044	135	0 . 171	126	0 . 216	78	+0 . 019	96	0 . 020	55	0 . 175	12
June 9	1 . 007	319	0 . 483	296	0 . 226	238	+0 . 091	177	0 . 045	186	0 . 138	120	0 . 077	148	+0 . 035	97	0 . 187	+ 29
19	1 . 326	322	0 . 779	316	0 . 464	265	0 . 268	213	+0 . 141	237	0 . 018	157	0 . 225	189	0 . 132	135	0 . 158	66
29	1 . 648	315	1 . 095	328	0 . 729	285	0 . 481	245	0 . 378	284	+0 . 139	191	0 . 414	226	0 . 267	170	0 . 092	104
July 9	1 . 963	302	1 . 423	333	1 . 014	299	0 . 726	269	0 . 662	323	0 . 330	220	0 . 640	256	0 . 437	201	0 . 012	137
19	2 . 265	280	1 . 756	329	1 . 313	305	0 . 995	288	0 . 985	355	0 . 550	244	0 . 896	283	0 . 638	227	0 . 149	169
29	2 . 545	254	2 . 085	318	1 . 618	304	1 . 283	301	1 . 340	380	0 . 794	262	1 . 179	304	0 . 865	248	0 . 318	197
Aug. 8	2 . 799	220	2 . 403	301	1 . 922	297	1 . 584	307	1 . 720	397	1 . 056	275	1 . 483	318	1 . 113	265	0 . 515	222
18	3 . 019	186	2 . 704	279	2 . 219	285	1 . 891	309	2 . 117	408	1 . 331	284	1 . 801	328	1 . 378	278	0 . 737	243
28	3 . 205	145	2 . 983	252	2 . 504	266	2 . 200	305	2 . 525	412	1 . 615	288	2 . 129	332	1 . 656	287	0 . 980	261
Sep. 7	3 . 350	107	3 . 235	224	2 . 770	246	2 . 505	298	2 . 937	412	1 . 903	287	2 . 461	333	1 . 948	292	1 . 241	274
17	3 . 457	69	3 . 459	192	3 . 016	221	2 . 803	288	3 . 349	406	2 . 190	282	2 . 794	330	2 . 235	293	1 . 515	286
27	3 . 526	35	3 . 651	160	3 . 237	197	3 . 091	273	3 . 755	394	2 . 472	273	3 . 124	322	2 . 528	291	1 . 801	291
Oct. 7	3 . 561	1	3 . 811	128	3 . 434	169	3 . 364	255	4 . 149	377	2 . 745	261	3 . 446	311	2 . 819	285	2 . 092	293
17	3 . 562	— 27	3 . 939	96	3 . 603	140	3 . 619	234	4 . 526	356	3 . 006	244	3 . 757	295	3 . 104	275	2 . 385	289
27	3 . 535	52	4 . 035	64	3 . 743	112	3 . 853	211	4 . 882	327	3 . 250	224	4 . 052	273	3 . 373	260	2 . 674	281
Nov. 6	3 . 483	73	4 . 099	32	3 . 855	81	4 . 064	183	5 . 209	292	3 . 474	199	4 . 327	250	3 . 639	242	2 . 955	266
16	3 . 410	88	4 . 131	2	3 . 936	50	4 . 247	152	5 . 501	253	3 . 673	170	4 . 577	219	3 . 881	217	3 . 221	244
26	3 . 322	101	4 . 133	— 27	3 . 986	20	4 . 399	118	5 . 754	205	3 . 843	137	4 . 796	184	4 . 098	186	3 . 465	216
Dec. 6	3 . 221	110	4 . 106	56	4 . 006	— 11	4 . 517	80	5 . 959	150	3 . 980	100	4 . 980	142	4 . 284	151	3 . 681	181
16	3 . 111	114	4 . 050	80	3 . 995	42	4 . 597	41	6 . 109	95	4 . 080	60	5 . 122	97	4 . 435	111	3 . 862	141
26	2 . 997	115	3 . 970	104	3 . 953	70	4 . 638	— 1	6 . 204	33	4 . 140	18	5 . 219	49	4 . 540	63	4 . 003	96
36	2 . 882		3 . 866		3 . 883		4 . 637		6 . 237		4 . 158		5 . 268		4 . 614		4 . 099	
	γ Pegasi.		α Arietis.		α Ceti.		α Tauri.		α Aurigæ.		β Orionis.		β Tauri.		α Orionis.		α Can. Maj.	

BESSEL'S TABLE

of 1818,

for determining the true apparent Right Ascensions of Dr. Maskelyne's 36 Stars, continued.

	α Gemin.		α Can. Min.		β Gemin.		α Hydræ.		α Leonis.		β Leonis.		β Virginis.		α Virginis.		α Bootis.	
	h. m. 7 23	☾ Nut.	h. m. 7 29	☾ Nut.	h. m. 7 34	☾ Nut.	h. m. 9 18	☾ Nut.	h. m. 9 58	☾ Nut.	h. m. 11 39	☾ Nut.	h. m. 11 41	☾ Nut.	h. m. 13 15	☾ Nut.	h. m. 14 7	☾ Nut.
1822. Jan. 0	14ˢ.177	+ 90	59ˢ.342	+ 78	25ˢ.358	+ 87	50ˢ.696	+ 78	53ˢ.593	+ 74	59ˢ.108	+ 69	26ˢ.042	+ 75	50ˢ.147	+ 84	33ˢ.416	+ 61
April 10	14 .267	82	59 .420	74	25 .445	79	50 .774	73	53 .667	69	59 .177	63	26 .117	71	50 .231	80	33 .477	55
July 19	14 .349	75	59 .494	68	25 .524	72	50 .847	68	53 .736	62	59 .240	56	26 .188	64	50 .311	75	33 .532	49
Oct. 27	14 .424	67	59 .562	61	25 .596	65	50 .915	63	53 .798	55	59 .296	49	26 .252	59	50 .386	69	33 .581	42
Dec. 66	14 .491		59 .623		25 .661		50 .978		53 .853		59 .345		26 .311		50 .455		33 .623	
1823. Jan. 0	18 .314	60	62 .751	56	29 .326	57	53 .904	58	57 .041	49	62 .397	44	29 .418	54	53 .578	65	36 .343	37
April 10	18 .374	49	62 .807	48	29 .383	48	53 .962	51	57 .090	41	62 .441	35	29 .472	45	53 .643	58	36 .380	29
July 19	18 .423	40	62 .855	40	29 .431	38	54 .013	43	57 .131	32	62 .476	27	29 .517	38	53 .701	49	36 .409	21
Oct. 27	18 .463	29	62 .895	32	29 .469	29	54 .056	37	57 .163	24	62 .503	18	29 .555	30	53 .750	42	36 .430	14
Dec. 66	18 .492		62 .927		29 .498		54 .093		57 .187		62 .521		29 .585		53 .792		36 .444	
1824. Jan. 0	22 .327	22	66 .065	26	33 .174	21	57 .028	31	60 .385	18	65 .584	12	32 .702	24	56 .924	37	39 .173	8
April 10	22 .349	11	66 .091	16	33 .195	11	57 .059	23	60 .403	8	65 .596	3	32 .726	16	56 .961	28	39 .181	0
July 19	22 .360	0	66 .107	8	33 .206	0	57 082	15	60 .411	− 1	65 .599	− 5	32 .742	7	56 .989	19	39 .181	− 8
Oct. 27	22 .360	− 11	66 .115	− 1	33 .206	− 11	57 .097	7	60 .410	9	65 .594	14	32 .749	− 2	57 .008	10	39 .173	16
Dec. 66	22 .349		66 .114		33 .195		57 .104		60 .401		65 .580		32 .747		57 .018		39 .157	
1825. Jan. 0	26 .198	19	69 .263	7	36 .885	18	60 .049	1	63 .610	16	68 .653	21	35 .875	8	60 .161	5	41 .896	22
April 10	26 .179	29	69 .256	16	36 .867	28	60 .050	− 7	63 .594	25	68 .632	28	35 .867	16	60 .166	− 4	41 .874	29
July 19	26 .150	39	69 .240	25	36 .839	38	60 .043	15	63 .569	33	68 .604	37	35 .851	24	60 .162	14	41 .845	36
Oct. 27	26 .111	50	69 .215	33	36 .801	49	60 .028	24	63 .536	42	68 .567	46	35 .827	33	60 .148	22	41 .809	45
Dec. 66	26 .061		69 .182		36 .752		60 .004		63 .494		68 .521		35 .794		60 .126		41 .764	

	Aber.	⊙ Nut.	Aber.	⊙ Nut.	Aber.	⊙ Nut.	Aber.	⊙ Nut.	Aber.	⊙ Nut.	Aber.	⊙ Nut.	Aber.	⊙ Nut.	Aber.	⊙ Nut.	Aber.	⊙ Nut.
Jan. 0	+1ˢ.577	+157	+1ˢ.337	+135	+1ˢ.499	+164	+1ˢ.055	+226	+1ˢ.889	+269	+0ˢ.351	+322	+0ˢ.338	+316	−0ˢ.206	+330	−0ˢ.546	+324
10	1 .734	103	1 .472	88	1 .663	111	1 .281	184	1 .158	233	0 .673	302	0 .654	294	+0 .124	325	0 .222	331
20	1 .837	43	1 .560	37	1 .774	53	1 .465	135	1 .391	186	0 .973	268	0 .948	262	0 .449	311	+0 .109	325
30	1 .880	− 13	1 .597	− 10	1 .827	− 1	1 .600	89	1 .577	140	1 .243	231	1 .210	225	0 .760	289	0 .434	313
Feb. 9	1 .867	67	1 .587	56	1 .826	55	1 .689	37	1 .717	88	1 .474	187	1 .435	184	1 .049	257	0 .747	290
19	1 .800	112	1 .531	95	1 .771	99	1 .726	− 6	1 .805	42	1 .661	144	1 .619	142	1 .306	224	1 .037	265
Mar. 1	1 .688	151	1 .436	127	1 .672	138	1 .720	49	1 .847	− 5	1 .805	97	1 .761	98	1 .530	188	1 .302	231
11	1 .537	176	1 .309	150	1 .554	164	1 .671	83	1 .842	43	1 .902	56	1 .859	58	1 .718	153	1 .533	199
21	1 .361	191	1 .159	163	1 .370	180	1 .588	109	1 .799	77	1 .958	16	1 .917	22	1 .871	117	1 .732	162
31	1 .170	194	0 .996	165	1 .190	186	1 .479	128	1 .722	101	1 .974	− 17	1 .939	− 11	1 .988	85	1 .894	128
Apr. 10	0 .976	186	0 .831	159	1 .004	179	1 .351	138	1 .621	117	1 .957	46	1 .928	37	2 .073	53	2 .022	93
20	0 .790	168	0 .672	145	0 .825	164	1 .213	142	1 .504	127	1 .911	69	1 .891	59	2 .126	25	2 .115	60
30	0 .622	139	0 .527	122	0 .661	138	1 .071	137	1 .377	128	1 .842	85	1 .832	73	2 .151	2	2 .175	30
May 10	0 .483	108	0 .405	96	0 .523	110	0 .934	128	1 .249	125	1 .757	98	1 .759	85	2 .153	− 20	2 .205	2
20	0 .375	67	0 .309	64	0 .413	72	0 .806	112	1 .124	114	1 .659	103	1 .674	92	2 .133	41	2 .207	− 26
30	0 .308	27	0 .245	30	0 .341	35	0 .694	93	1 .010	101	1 .556	107	1 .582	97	2 .092	59	2 .181	50
June 9	0 .281	+ 16	0 .215	+ 5	0 .306	+ 6	0 .601	72	0 .909	83	1 .449	106	1 .485	96	2 .033	75	2 .131	72
19	0 .297	58	0 .220	39	0 .312	45	0 .529	49	0 .826	64	1 .343	103	1 .389	94	1 .958	88	2 .059	92
29	0 .355	100	0 .259	74	0 .357	84	0 .480	24	0 .762	43	1 .240	95	1 .295	88	1 .870	100	1 .967	110
July 9	0 .455	137	0 .333	106	0 .441	121	0 .456	+ 1	0 .719	19	1 .145	86	1 .207	80	1 .770	107	1 .857	124
19	0 .592	174	0 .439	137	0 .562	155	0 .457	28	0 .700	+ 4	1 .059	74	1 .127	70	1 .663	111	1 .733	135
29	0 .766	207	0 .576	164	0 .717	188	0 .485	56	0 .704	29	0 .985	58	1 .057	55	1 .552	111	1 .598	142
Aug. 8	0 .973	236	0 .740	190	0 .905	217	0 .541	84	0 .733	56	0 .927	40	1 .002	38	1 .441	106	1 .456	143
18	1 .209	262	0 .930	213	1 .121	241	0 .625	112	0 .789	84	0 .887	17	0 .964	16	1 .335	95	1 .313	138
28	1 .471	286	1 .143	234	1 .363	266	0 .737	141	0 .873	112	0 .870	+ 8	0 .948	+ 9	1 .240	77	1 .175	128
Sep. 7	1 .757	307	1 .377	254	1 .629	288	0 .878	171	0 .985	143	0 .878	39	0 .957	38	1 .163	53	1 .047	108
17	2 .064	324	1 .631	270	1 .917	306	1 .049	201	1 .128	174	0 .917	72	0 .995	73	1 .110	22	0 .939	84
27	2 .388	338	1 .901	283	2 .223	321	1 .250	229	1 .302	205	0 .989	109	1 .068	108	1 .088	+ 14	0 .855	51
Oct. 7	2 .726	349	2 .184	295	2 .544	335	1 .479	257	1 .507	237	1 .098	150	1 .176	148	1 .102	57	0 .804	11
17	3 .075	355	2 .479	300	2 .879	340	1 .736	279	1 .744	265	1 .248	187	1 .324	187	1 .159	100	0 .793	+ 32
27	3 .430	355	2 .779	303	3 .219	345	2 .015	298	2 .009	292	1 .435	228	1 .511	226	1 .259	149	0 .825	82
Nov. 6	3 .785	348	3 .082	296	3 .564	339	2 .313	312	2 .301	311	1 .663	263	1 .737	260	1 .408	192	0 .907	128
16	4 .133	334	3 .378	286	3 .903	327	2 .625	318	2 .612	326	1 .926	295	1 .997	291	1 .600	237	1 .035	179
26	4 .467	311	3 .664	267	4 .230	308	2 .943	316	2 .938	331	2 .221	318	2 .288	313	1 .837	272	1 .214	222
Dec. 6	4 .778	279	3 .931	239	4 .538	278	3 .259	302	3 .269	327	2 .539	333	2 .601	326	2 .109	302	1 .436	263
16	5 .057	239	4 .170	206	4 .816	241	3 .561	280	3 .596	314	2 .872	337	2 .927	333	2 .411	322	1 .699	293
26	5 .296	192	4 .376	164	5 .057	196	3 .841	250	3 .910	290	3 .209	330	3 .260	322	2 .733	331	1 .992	315
36	5 .488		4 .540		5 .253		4 .091		4 .200		3 .539		3 .582		3 .064		2 .307	
	α Gemin.		α Can. Min.		β Gemin.		α Hydræ.		α Leonis.		β Leonis.		β Virginis.		α Virginis.		α Bootis.	

BESSEL'S TABLE

of 1818,

for determining the true apparent Right Ascensions of Dr. Maskelyne's 36 Stars, continued.

	α¹ Libræ.		α² Libræ.		α Coronæ.		α Serpentis.		α Scorpii.	
	h. m. 14 40	☾ Nut.	h. m. 14 41	☾ Nut.	h. m. 15 27	☾ Nut.	h. m. 15 35	☾ Nut.	h. m. 16 18	☾ Nut.
1822. Jan. 0	52ˢ.081	+ 90	3ˢ.502	+ 89	9ˢ.941	+ 53	31ˢ.051	+ 72	31ˢ.245	+ 98
April 10	52 .171	85	3 .591	85	9 .994	48	31 .123	66	31 .313	93
July 19	52 .256	79	3 .676	80	10 .042	42	31 .189	61	31 .406	87
Oct. 27	52 .335	74	3 .756	74	10 .084	36	31 .250	55	31 .493	82
Dec. 66	52 .409		3 .830		10 .120		31 .305		31 .575	
1823. Jan. 0	55 .687	69	7 .111	69	12 .647	32	34 .238	49	35 .212	75
April 10	55 .756	61	7 .180	61	12 .679	24	34 .287	43	35 .287	67
July 19	55 .817	54	7 .241	54	12 .703	17	34 .330	35	35 .354	58
Oct. 27	55 .871	46	7 .295	45	12 .720	10	34 .365	27	35 .412	50
Dec. 66	55 .917		7 .340		12 .730		34 .392		35 .462	
1824. Jan. 0	59 .205	39	10 .631	39	15 .266	5	37 .334	21	39 .110	43
April 10	59 .244	31	10 .670	31	15 .271	− 2	37 .355	14	39 .153	33
July 19	59 .275	22	10 .701	22	15 .269	10	37 .369	5	39 .186	23
Oct. 27	59 .297	12	10 .723	13	15 .259	17	37 .374	− 2	39 .209	13
Dec. 66	59 .309		10 .736		15 .242		37 .372		39 .222	
1825. Jan. 0	62 .608	6	14 .037	7	17 .787	22	40 .324	8	42 .882	6
April 10	62 .614	− 3	14 .044	− 3	17 .765	28	40 .316	17	42 .888	− 4
July 19	62 .611	12	14 .041	13	17 .737	35	40 .299	25	42 .884	15
Oct. 27	62 .599	22	14 .028	22	17 .702	42	40 .274	33	42 .869	26
Dec. 66	62 .577		14 .006		17 .660		40 .241		42 .843	

	α Herculis.		α Ophiuchi.		α Lyræ.		γ Aquilæ.	
	h. m. 17 6	☾ Nut.	h. m. 17 26	☾ Nut.	h. m. 18 30	☾ Nut.	h. m. 19 37	☾ Nut.
1822. Jan. 0	32ˢ.671	+ 67	41ˢ.028	+ 69	55ˢ.108	+ 54	48ˢ.350	+ 76
April 10	32 .738	62	41 .097	63	55 .162	51	48 .424	70
July 19	32 .800	56	41 .160	59	55 .213	48	48 .494	64
Oct. 27	32 .856	52	41 .219	54	55 .261	44	48 .558	60
Dec. 66	32 .908		41 .273		55 .305		48 .618	
1823. Jan. 0	35 .623	47	44 .033	49	57 .322	41	51 .455	54
April 10	35 .670	39	44 .082	41	57 .363	37	51 .509	48
July 19	35 .709	34	44 .123	35	57 .400	31	51 .557	41
Oct. 27	35 .743	25	44 .158	26	57 .431	27	51 .598	35
Dec. 66	35 .768		44 .186		57 .458		51 .633	
1824. Jan. 0	38 .492	21	46 .955	22	59 .481	23	54 .478	29
April 10	38 .513	13	46 .977	15	59 .504	18	54 .507	21
July 19	38 .526	5	46 .992	7	59 .522	12	54 .528	13
Oct. 27	38 .531	− 3	46 .999	− 1	59 .534	6	54 .541	5
Dec. 66	38 .528		46 .998		59 .540		54 .546	
1825. Jan. 0	41 .261	7	49 .777	7	61 .570	3	57 .401	0
April 10	41 .254	16	49 .770	13	61 .573	− 3	57 .401	− 9
July 19	41 .238	22	49 .757	22	61 .570	9	57 .392	16
Oct. 27	41 .216	31	49 .735	29	61 .561	14	57 .376	24
Dec. 66	41 .185		49 .706		61 .547		57 .352	

	α¹ Libræ.		α² Libræ.		α Coronæ.		α Serpentis.		α Scorpii.	
	Aber.	⊙Nut.	Aber.	⊙Nut.	Aber.	⊙Nut.	Aber.	⊙Nut.	Aber.	⊙Nut.
Jan. 0	−0ˢ.688	+322	−0ˢ.690	+323	−1ˢ.020	+294	−0ˢ.935	+280	−1ˢ.198	+290
10	0 .366	332	0 .367	332	0 .726	315	0 .655	298	0 .908	315
20	0 .034	330	0 .035	330	0 .411	326	0 .357	308	0 .593	332
30	+0 .296	322	0 .295	322	0 .085	330	0 .049	310	0 .261	340
Feb. 9	0 .618	305	0 .617	305	+0 .245	323	+0 .261	304	+0 .079	340
19	0 .923	285	0 .922	285	0 .568	311	0 .565	293	0 .419	337
Mar. 1	1 .208	258	1 .207	259	0 .879	291	0 .858	275	0 .756	325
11	1 .466	231	1 .466	231	1 .170	266	1 .133	256	1 .081	310
21	1 .697	202	1 .697	202	1 .436	238	1 .389	232	1 .391	293
31	1 .899	173	1 .899	173	1 674	207	1 .621	207	1 .684	272
Apr. 10	2 .072	143	2 .072	144	1 .881	175	1 .828	180	1 .956	249
20	2 .215	114	2 .216	114	1 .056	141	2 .008	154	2 .205	224
30	2 .329	86	2 .330	86	1 .197	107	2 .162	124	2 .429	196
May 10	2 .415	58	2 .416	58	2 .304	72	2 .286	96	2 .625	165
20	2 .473	29	2 .474	30	2 .376	39	2 .382	66	2 .790	133
30	2 .502	2	2 .504	2	2 .415	4	2 .448	36	2 .923	97
June 9	2 .504	− 26	2 .506	− 25	2 .419	− 28	2 .484	6	3 .020	61
19	2 .478	50	2 .481	51	2 .391	60	2 .490	− 24	3 .081	24
29	2 .428	76	2 .430	75	2 .331	90	2 .466	54	3 .105	− 16
July 9	2 .352	97	2 .355	97	2 .241	117	2 .412	80	3 .089	52
19	2 .255	116	2 .258	116	2 .124	142	2 .332	106	3 .037	88
29	2 .139	129	2 .142	130	1 .982	161	2 .226	126	2 .949	118
Aug. 8	2 .010	139	2 .012	138	1 .821	176	2 .100	142	2 .831	145
18	1 .871	139	1 .874	139	1 .645	183	1 .958	152	2 .686	163
28	1 .732	134	1 .735	134	1 .462	184	1 .806	154	2 .523	174
Sep. 7	1 .598	118	1 .601	119	1 .278	175	1 .652	149	2 .349	174
17	1 .480	96	1 .482	96	1 .103	159	1 .503	135	2 .175	165
27	1 .384	66	1 .386	65	0 .944	134	1 .368	111	2 .010	143
Oct. 7	1 .318	25	1 .321	26	0 .810	99	1 .257	80	1 .867	114
17	1 .293	+ 17	1 .295	+ 17	0 .711	59	1 .177	43	1 .753	75
27	1 .310	68	1 .312	68	0 .652	11	1 .134	+ 1	1 .678	27
Nov. 6	1 .378	117	1 .380	117	0 .641	+ 39	1 .135	50	1 .651	+ 23
16	1 .495	168	1 .497	168	0 .680	94	1 .185	98	1 .674	80
26	1 .663	213	1 .665	213	0 .774	144	1 .283	148	1 .754	132
Dec. 6	1 .876	256	1 .878	256	0 .918	195	1 .431	192	1 .886	185
16	2 .132	288	2 .134	287	1 .113	238	1 .623	231	2 .071	230
26	2 .420	312	2 .421	313	1 .351	273	1 .854	263	2 .301	268
36	2 .732		2 .734		1 .624		2 .117		2 .569	
	α¹ Libræ.		α² Libræ.		α Coronæ.		α Serpentis.		α Scorpii.	

	α Herculis.		α Ophiuchi.		α Lyræ.		γ Aquilæ.	
	Aber.	⊙Nut.	Aber.	⊙Nut.	Aber.	⊙Nut.	Aber.	⊙Nut.
Jan. 0	−1ˢ.265	+209	−1ˢ.295	+192	−1ˢ.703	+114	−1ˢ.294	+ 69
10	1 .056	240	1 .103	225	1 .589	160	1 .225	104
20	0 .816	265	0 .878	250	1 .429	203	1 .121	138
30	0 .551	282	0 .628	271	1 .226	240	0 .983	169
Feb. 9	0 .269	292	0 .357	284	0 .986	277	0 .814	198
19	+0 .023	298	0 .073	292	0 .709	301	0 .616	222
Mar. 1	0 .321	295	+0 .219	293	0 .408	319	0 .394	244
11	0 .616	29[illegible]	0 .513	291	0 .089	331	0 .150	262
21	0 .906	279	0 .803	284	+0 .242	335	+0 .112	277
31	1 .185	265	1 .087	273	0 .577	334	0 .389	287
Apr. 10	1 .450	247	1 .360	258	0 .911	324	0 .676	293
20	1 .697	225	1 .618	240	1 .235	309	0 .969	294
30	1 .922	202	1 .858	218	1 .544	287	1 .263	290
May 10	2 .124	174	2 .076	192	1 .831	259	1 .553	279
20	2 .298	144	2 .268	163	2 .090	225	1 .832	263
30	2 .442	111	2 .431	131	2 .315	187	2 .095	241
June 9	2 .553	75	2 .562	96	2 .502	142	2 .336	213
19	2 .628	40	2 .658	59	2 .644	96	2 .549	179
29	2 .668	0	2 .717	20	2 .740	46	2 .728	140
July 9	2 .668	− 35	2 .737	− 18	2 .786	− 3	2 .868	100
19	2 .633	72	2 .719	50	2 .78[illegible]	54	2 .968	56
29	2 .561	105	2 .663	90	2 .729	100	3 .024	13
Aug. 8	2 .456	134	2 .573	125	2 .629	145	3 .037	− 32
18	2 .322	157	2 .450	148	2 .484	183	3 .005	70
28	2 .165	174	2 .302	169	2 .301	215	2 .935	107
Sep. 7	1 .991	181	2 .133	178	2 .086	238	2 .828	134
17	1 .810	181	1 .955	182	1 .848	253	2 .694	157
27	1 .629	170	1 .773	174	1 .595	256	2 .537	170
Oct. 7	1 .459	151	1 .599	157	1 .339	249	2 .367	173
17	1 .308	122	1 .442	132	1 .090	232	2 .194	168
27	1 .186	86	1 .310	98	0 .858	204	2 .026	153
Nov. 6	1 .100	44	1 .212	59	0 .654	170	1 .873	133
16	1 .056	+ 4	1 .153	12	0 .484	125	1 .740	102
26	1 .060	50	1 .141	+ 33	0 .359	78	1 .638	71
Dec. 6	1 .110	100	1 .174	82	0 .281	24	1 .567	32
16	1 .210	144	1 .256	126	0 .257	+ 27	1 .535	+ 5
26	1 .354	185	1 .382	167	0 .284	80	1 .540	43
36	1 .539		1 .549		0 .364		1 .583	
	α Herculis.		α Ophiuchi.		α Lyræ.		γ Aquilæ.	

BESSEL'S TABLE

of 1818,

for determining the true apparent Right Ascensions of Dr. Maskelyne's 36 Stars, concluded.

	α Aquilæ.		β Aquilæ.		α¹ Capric.		α² Capric.		α Cygni.		α Aquarii.		α Pisc. Aust.		α Pegasi.		α Androm.	
	h. m. 19 42	☾ Nut.	h. m. 19 46	☾ Nut.	h. m. 20 7	☾ Nut.	h. m. 20 8	☾ Nut.	h. m. 20 35	☾ Nut.	h. m. 21 56	☾ Nut.	h. m. 22 47	☾ Nut.	h. m. 22 55	☾ Nut.	h. m. 23 59	☾ Nut.
1822. Jan. 0	6ˢ.387	+ 75	34ˢ.747	+ 76	47ˢ.210	+ 80	11ˢ.030	+ 79	22ˢ.068	+ 72	38ˢ.959	+ 77	48ˢ.672	+ 65	54ˢ.562	+ 82	12ˢ.697	+ 94
April 10	6 .462	70	34 .823	70	47 .290	73	11 .109	74	22 .140	69	39 .036	72	48 .737	58	54 .644	79	12 .791	90
July 19	6 .532	65	34 .893	66	47 .363	67	11 .183	67	22 .209	69	39 .108	66	48 .795	49	54 .723	74	12 .881	88
Oct. 27	6 .597	60	34 .959	60	47 .430	61	11 .250	61	22 .278	67	39 .174	60	48 .844	42	54 .797	69	12 .969	83
Dec. 66	6 .657		35 .019		47 .491		11 .311		22 .345		39 .234		48 .886		54 .866		12 .052	
1823. Jan. 0	9 .567	55	37 .951	50	50 .806	55	14 .630	55	24 .364	65	42 .300	55	52 .214	35	57 .826	65	16 .104	80
April 10	9 .622	49	38 .007	48	50 .861	46	14 .685	46	24 .429	61	42 .355	47	52 .219	25	57 .891	58	16 .184	73
July 19	9 .671	41	38 .055	41	50 .907	38	14 .731	38	24 .490	58	42 .402	39	52 .274	15	57 .949	51	16 .257	66
Oct. 27	9 .712	33	38 .096	33	50 .945	28	14 .769	29	24 .548	54	42 .441	32	52 .289	6	58 .000	44	16 .323	61
Dec. 66	9 .745		38 .129		51 .973		14 .798		24 .602		42 .473		52 .295		58 .044		16 .384	
1824. Jan. 0	12 .664	29	41 .070	29	54 .299	22	18 .128	22	26 .626	51	45 .547	27	55 .635	− 1	61 .013	39	19 .444	55
April 10	12 .693	21	41 .099	20	54 .321	14	18 .150	13	26 .677	45	45 .574	18	55 .634	10	61 .052	31	19 .499	47
July 19	12 714	12	41 .119	11	54 .335	3	18 .163	4	26 722	41	45 .592	9	55 .624	21	61 .083	23	19 .546	30
Oct. 27	12 .726	5	41 .130	4	54 .338	− 5	18 .167	− 6	26 763	34	45 .601	− 1	55 .603	30	61 .106	14	19 .585	30
Dec. 66	12 731		41 .134		54 .333		18 .161		26 .797		45 .600		55 .573		61 .120		19 .615	
1825. Jan. 0	15 .659	− 1	44 .085	− 2	57 .670	12	21 .503	12	38 .828	30	48 .686	6	58 .925	36	64 .099	9	22 .686	24
April 10	15 .658	9	44 .083	10	57 .658	22	21 .491	21	38 .858	26	48 .680	14	58 .889	45	64 .108	1	22 .710	16
July 19	15 .649	17	44 .073	19	57 .636	30	21 .470	31	38 .884	19	48 .666	23	58 .844	54	64 .109	− 9	22 .726	7
Oct. 27	15 .632	25	44 .054	26	57 .606	39	21 .439	39	38 .903	12	48 .643	32	58 790	62	64 .100	16	22 .733	− 3
Dec. 66	15 .607		44 .028		57 .567		21 .400		38 .915		48 .611		58 728		64 .084		22 .730	

	Aber.	☉Nut.	Aber.	☉Nut.	Aber.	☉Nut.	Aber.	☉Nut.	Aber.	☉Nut.	Aber.	☉Nut.	Aber.	☉Nut.	Aber.	☉Nut.	Aber.	☉Nut.
Jan. 0	−1ˢ.279	+ 66	−1ˢ.265	+ 63	−1ˢ.252	+ 58	−1ˢ.251	+ 57	−1ˢ.608	− 52	−0ˢ.843	− 41	−0ˢ 711	− 97	−0ˢ.560	− 90	−0ˢ.212	−139
10	1 .213	102	1 .202	99	1 .194	92	1 .194	92	1 .666	3	0 .884	16	0 .802	66	0 .650	71	0 .351	131
20	1 .111	136	1 .103	135	1 .102	126	1 .102	126	1 .663	+ 48	0 .900	+ 11	0 .868	39	0 .721	50	0 .482	116
30	0 .975	167	0 .970	163	0 .976	157	0 .976	156	1 .615	98	0 .889	40	0 .907	9	0 .771	26	0 .598	96
Feb. 9	0 .808	195	0 .807	191	0 .819	186	0 .820	186	1 .517	148	0 .849	70	0 .916	+ 24	0 797	+ 4	0 .694	68
19	0 .613	220	0 .616	216	0 .633	212	0 .634	211	1 .369	196	0 779	101	0 .892	57	0 793	35	0 .762	36
Mar. 1	0 .393	242	0 .400	239	0 .421	236	0 423	236	1 .173	240	0 .678	133	0 .835	94	0 758	80	0 .798	+ 2
11	0 .151	261	0 .161	257	0 .105	357	0 .187	258	0 .933	279	0 .545	163	0 741	130	0 .678	97	0 .796	44
21	+0 .110	275	+0 .096	273	+0 .072	375	+0 .071	276	0 .654	313	0 .382	195	0 .611	169	0 .581	144	0 752	90
31	0 .385	287	0 .369	285	0 .347	29	0 .347	290	0 .341	339	0 .187	224	0 .442	206	0 .437	180	0 .662	134
Apr. 10	0 .672	294	0 .654	292	0 .638	301	0 .637	301	0 .002	358	0 .037	252	0 .236	242	0 .257	217	0 .528	198
20	0 .966	295	0 .946	295	1 .939	308	0 .938	308	+0 .356	369	0 .289	273	+0 .006	274	0 .040	248	0 .350	219
30	1 .261	292	1 .241	292	1 .247	309	1 .246	310	0 .725	369	0 .502	293	0 .280	304	+0 .208	275	0 .131	258
May 10	1 .553	282	1 .533	284	1 .556	305	1 .556	305	1 .094	361	0 .855	304	0 .584	326	0 .483	296	+0 .127	289
20	1 .835	267	1 .817	269	1 .861	294	1 .861	294	1 .455	344	1 .159	310	0 .910	343	0 779	312	0 .416	313
30	2 .102	245	2 .086	248	2 .155	276	2 .155	277	1 799	318	1 .469	308	1 .253	351	1 .091	318	0 .729	330
June 9	2 .347	217	2 .334	221	2 .431	253	2 .432	252	2 .117	282	1 .777	298	1 .604	351	1 .409	316	1 .059	338
19	2 .564	185	2 .555	199	2 .684	222	2 .684	223	2 .399	242	1 .075	281	1 .955	342	1 725	306	1 .397	338
29	2 .749	147	2 .744	151	2 .906	185	2 .907	186	2 .641	192	1 .356	155	2 .297	324	2 .031	288	1 735	328
July 9	2 .896	106	2 .895	112	3 .091	145	3 .093	146	2 .833	142	2 .611	255	2 .621	298	2 .319	264	2 .063	311
19	3 .002	61	3 .067	67	3 .236	101	3 .239	100	2 .975	85	2 .836	188	2 .919	265	2 .583	232	2 .374	287
29	3 .063	18	3 .074	24	3 .337	55	3 .339	57	3 .060	29	3 .024	149	3 .182	224	2 .815	196	2 .661	258
Aug. 8	3 .081	− 25	3 .098	− 20	3 .392	9	3 .396	9	3 .089	− 27	3 .173	105	3 406	177	3 .011	156	2 .919	223
18	3 .056	63	3 .078	60	3 .401	− 33	3 405	− 33	3 .062	79	3 .278	63	3 .583	131	3 .167	115	3 .142	185
28	2 .993	100	3 .018	96	3 .[illegible]	75	3 .372	74	2 .983	130	3 .341	18	3 714	78	3 .282	73	3 .327	145
Sep. 7	2 .893	130	2 .922	126	3 .293	107	3 .298	107	2 .853	172	3 .359	− 20	3 792	32	3 .355	33	3 .472	106
17	2 763	152	2 796	148	3 .186	135	3 .191	134	2 .681	208	3 .339	56	3 .824	− 15	3 .398	− 5	3 .578	65
27	2 .611	165	2 .648	163	3 .051	152	3 .057	153	2 .473	234	3 .283	84	3 .809	54	3 .383	38	3 .643	29
Oct. 7	2 .446	170	2 .485	168	2 .899	162	2 .904	161	2 .239	253	3 .199	108	3 .755	91	3 .345	67	3 .672	− 8
17	2 .276	166	2 .317	164	2 737	161	2 743	161	1 .986	262	3 .091	124	3 .664	116	3 .278	90	3 .664	38
27	2 .110	151	2 .153	150	2 .576	150	2 .582	150	1 .724	259	2 .967	131	3 .548	137	3 .188	106	3 .626	67
Nov. 6	1 .959	131	2 .003	131	2 .426	133	2 .432	133	1 .465	250	2 .836	133	3 411	149	3 .082	118	3 .559	90
16	1 .828	102	1 .872	103	2 .293	106	2 .299	106	1 .215	228	2 .703	127	3 .262	151	2 .964	123	3 .469	110
26	1 726	70	1 769	71	2 .187	76	2 .193	76	0 .987	202	2 .576	116	3 .111	149	2 .841	124	3 .359	125
Dec. 6	1 .656	33	1 .698	34	2 .111	40	2 .117	41	0 .785	167	2 .460	101	2 .962	139	2 .717	119	3 .234	136
16	1 .623	+ 4	1 .664	+ 1	2 .071	5	2 .076	4	0 .618	127	2 .359	78	2 .823	125	2 .598	112	3 .698	141
26	1 .627	43	1 .665	40	2 .066	+ 33	2 .072	+ 32	0 .491	83	2 .281	58	2 .698	105	2 .486	99	2 .957	142
36	1 .670		1 .705		2 .099		2 .104		0 .408		2 .223		2 .593		2 .387		2 .815	
	α Aquilæ.		β Aquilæ.		α¹ Capric.		α² Capric.		α Cygni.		α Aquarii.		α Pisc. Aust.		α Pegasi.		α Androm.	

TABLE 1.

EPOCHS of the MEAN LONGITUDES of the SUN, and of his PERIGEE.

Years.	Mean Longitude of the Sun.	Mean Longitude of his Perigee.	Years.	Mean Longitude of the Sun.	Mean Longitude of his Perigee.	Years.	Mean Longitude of the Sun.	Mean Longitude of his Perigee.
1750	9s 11° 0′ 11″.5	9s 8° 37′ 28″	1800C.	9s 10° 53′ 29″.8	9s 9° 29′ 3″	1850	9s 10° 46′ 48″.2	9s 10° 20′ 38″
1751	9 10 45 51 .8	9 8 38 30	1801	9 10 39 10 .2	9 9 30 5	1851	9 10 32 28 .5	9 10 21 40
1752B.	9 10 31 32 .2	9 8 39 31	1802	9 10 24 50 .6	9 9 31 7	1852B.	9 10 18 8 .9	9 10 22 42
1753	9 11 16 20 .9	9 8 40 33	1803	9 10 10 30 .9	9 9 32 9	1853	9 11 2 57 .6	9 10 23 44
1754	9 11 2 1 .3	9 8 41 35	1804B.	9 9 56 11 .3	9 9 33 11	1854	9 10 48 38 .0	9 10 24 46
1755	9 10 47 41 .6	9 8 42 37	1805	9 10 41 0 .0	9 9 34 13	1855	9 10 34 18 .3	9 10 25 48
1756B.	9 10 33 22 .0	9 8 43 39	1806	9 10 26 40 .4	9 9 35 15	1856B.	9 10 19 58 .7	9 10 26 50
1757	9 11 18 10 .7	9 8 44 41	1807	9 10 12 20 .7	9 9 36 17	1857	9 11 4 47 .4	9 10 27 52
1758	9 11 3 51 .1	9 8 45 43	1808B.	9 9 58 1 .1	9 9 37 18	1858	9 10 50 27 .8	9 10 28 54
1759	9 10 49 31 .5	9 8 46 45	1809	9 10 42 49 .8	9 9 38 20	1859	9 10 36 8 .2	9 10 29 56
1760B.	9 10 35 11 .8	9 8 47 47	1810	9 10 28 30 .2	9 9 39 22	1860B.	9 10 21 48 .5	9 10 30 57
1761	9 11 20 0 .5	9 8 48 49	1811	9 10 14 10 .5	9 9 40 24	1861	9 11 6 37 .2	9 10 31 59
1762	9 11 5 40 .9	9 8 49 51	1812B.	9 9 59 50 .9	9 9 41 26	1862	9 10 52 17 .6	9 10 33 1
1763	9 10 51 21 .2	9 8 50 52	1813	9 10 44 39 .6	9 9 42 28	1863	9 10 37 58 .0	9 10 34 3
1764B.	9 10 37 1 .6	9 8 51 54	1814	9 10 30 20 .0	9 9 43 30	1864B.	9 10 23 38 .3	9 10 35 5
1765	9 11 21 50 .3	9 8 52 56	1815	9 10 16 0 .3	9 9 44 32	1865	9 11 8 27 .0	9 10 36 7
1766	9 11 7 30 .7	9 8 53 58	1816B.	9 10 14 0 .7	9 9 45 34	1866	9 10 54 7 .4	9 10 37 9
1767	9 10 53 11 .0	9 8 55 0	1817	9 10 46 29 .4	9 9 46 36	1867	8 10 39 47 .8	9 10 38 11
1768B.	9 10 38 51 .4	9 8 56 2	1818	9 10 32 9 .8	9 9 47 38	1868B.	9 10 25 28 .1	9 10 39 13
1769	9 11 23 40 .1	9 8 57 4	1819	9 10 17 50 .1	9 9 48 39	1869	9 11 10 16 .8	9 10 40 15
1770	9 11 9 20 .5	9 8 58 6	1820B.	9 10 3 30 .5	9 9 49 41	1870	9 10 55 57 .2	9 10 41 17
1771	9 10 55 0 .8	9 8 59 8	1821	9 10 48 19 .2	9 9 50 43	1871	9 10 41 37 .6	9 10 42 18
1772B.	9 10 40 41 .2	9 9 0 10	1822	9 10 33 59 .6	9 9 51 45	1872B.	9 10 27 17 .9	9 10 43 20
1773	9 11 25 29 .9	9 9 1 12	1823	0 10 19 39 .9	9 9 52 47	1873	9 11 12 6 .6	9 10 44 22
1774	9 11 11 10 .3	9 9 2 13	1824B.	9 10 5 20 .3	9 9 53 49	1874	9 10 57 47 .0	9 10 45 24
1775	9 10 56 50 .6	9 9 3 15	1825	9 10 50 9 .0	9 9 54 51	1875	9 10 43 27 .4	9 10 46 26
1776B.	9 10 42 31 .0	9 9 4 17	1826	9 10 35 49 .4	9 9 55 52	1876B.	9 10 29 7 .8	9 10 47 28
1777	9 11 27 19 .7	9 9 5 19	1827	9 10 21 29 .7	9 9 56 54	1877	9 11 13 56 .4	9 10 48 30
1778	9 11 13 0 .1	9 9 6 21	1828B.	9 10 7 10 .1	9 9 57 56	1878	9 10 59 36 .8	9 10 49 32
1779	9 10 58 40 .4	9 9 7 23	1829	9 10 51 58 .8	9 9 58 58	1879	9 10 45 17 .2	9 10 50 34
1780B.	9 10 44 20 .8	9 9 8 25	1830	9 10 37 39 .2	9 10 0 0	1880B.	9 10 30 57 .6	9 10 51 36
1781	9 11 29 9 .5	9 9 9 27	1831	9 10 23 19 .5	9 10 1 2	1881	9 11 15 46 .3	9 10 52 38
1782	9 11 14 49 .9	9 9 10 29	1832B.	9 10 8 59 .9	9 10 2 4	1882	9 11 1 26 .6	9 10 53 39
1783	9 10 0 30 .2	9 9 11 31	1833	9 10 53 48 .6	9 10 3 6	1883	9 10 47 7 .0	9 10 54 41
1784B.	9 10 46 10 .6	9 9 12 33	1834	9 10 39 29 .0	9 10 4 8	1884B.	9 10 32 47 .4	9 10 55 43
1785	9 11 30 59 .3	9 9 13 34	1835	9 10 25 9 .3	9 10 5 10	1885	9 11 17 36 .1	9 10 56 45
1786	9 11 16 39 .7	9 9 14 36	1836B.	9 10 10 49 .7	9 10 6 12	1886	9 11 3 16 .4	9 10 57 47
1787	9 10 2 20 .0	9 9 15 38	1837	9 10 55 38 .4	9 10 7 13	1887	9 10 48 56 .8	9 10 58 49
1788B.	9 10 48 0 .4	9 9 16 40	1838	9 10 41 18 .8	9 10 8 15	1888B.	9 10 34 37 .2	9 10 59 51
1789	9 11 32 49 .1	9 9 17 42	1839	9 10 26 59 .1	9 10 9 17	1889	9 11 19 25 .9	9 11 0 53
1790	9 11 18 29 .5	9 9 18 44	1840B.	9 10 12 39 .5	9 10 10 19	1890	9 11 5 6 .2	9 11 1 55
1791	9 11 4 9 .8	9 9 19 46	1841	9 10 57 28 .2	9 10 11 21	1891	9 10 50 46 .6	9 11 2 57
1792B.	9 10 49 50 .2	9 9 20 48	1842	9 10 43 8 .6	9 10 12 23	1892B.	9 10 36 27 .0	9 11 3 59
1793	9 11 34 38 .9	9 9 21 50	1843	9 10 28 48 .9	9 10 13 25	1893	9 11 21 15 .7	9 11 5 0
1794	9 11 20 19 .3	9 9 22 52	1844B.	9 10 14 29 .3	9 10 14 27	1894	9 11 6 56 .0	9 11 6 2
1795	9 11 5 59 .6	9 9 23 54	1845	9 10 59 18 .0	9 10 15 29	1895	9 10 52 36 .4	9 11 7 4
1796B.	9 10 51 40 .0	9 9 24 55	1846	9 10 44 58 .4	9 10 16 31	1896B.	9 10 38 16 .8	9 11 8 6
1797	9 11 36 28 .7	9 9 25 57	1847	9 10 30 38 .7	9 10 17 33	1897	9 11 23 5 .5	9 11 9 8
1798	9 11 22 9 .1	9 9 26 59	1848B.	9 10 16 19 .1	9 10 18 35	1898	9 11 8 45 .8	9 11 10 10
1799	9 11 7 49 .4	9 9 28 1	1849	9 11 1 7 .8	9 10 19 37	1899	9 10 54 26 .2	9 11 11 12
1800C.	9 10 53 29 .8	9 9 29 3	1850	9 10 46 48 .2	9 10 20 38	1900B.	9 10 40 6 .6	9 11 12 14
Years.	Mean Longitude of the Sun.	Mean Longitude of his Perigee.	Years.	Mean Longitude of the Sun.	Mean Longitude of his Perigee.	Years.	Mean Longitude of the Sun.	Mean Longitude of his Perigee.

These EPOCHS are calculated for the NOON of January 1, mean Time at GREENWICH, and are useful when the Sun's mean Anomaly is used as an *Argument*.

TABLE 2. The Mean Longitude of the Sun and of his Perigee for each Day in the Year, and the number of each Day counted from the beginning of the Year, with its fractional part of the Year.

JANUARY.

Years Com.	Years Bis.	☉'s Long. s	°	′	″	Perigee.	Day.	Frac.
1	1	0ˢ	0°	0′	0″.0	0″.0	1	.000
2	2	0	0	59	08.3	0.2	2	.003
3	3	0	1	58	16.7	0.3	3	.006
4	4	0	2	57	25.0	0.5	4	.008
5	5	0	3	56	33.3	0.7	5	.011
6	6	0	4	55	41.6	0.8	6	.014
7	7	0	5	54	50.0	1.0	7	.017
8	8	0	6	53	58.3	1.2	8	.019
9	9	0	7	53	6.6	1.4	9	.022
10	10	0	8	52	15.0	1.5	10	.025
11	11	0	9	51	23.3	1.7	11	.028
12	12	0	10	50	31.6	1.9	12	.030
13	13	0	11	49	40.0	2.0	13	.033
14	14	0	12	48	48.3	2.2	14	.036
15	15	0	13	47	56.6	2.3	15	.039
16	16	0	14	47	4.9	2.5	16	.041
17	17	0	15	46	13.3	2.7	17	.044
18	18	0	16	45	21.6	2.9	18	.046
19	19	0	17	44	29.9	3.0	19	.049
20	20	0	18	43	38.3	3.2	20	.052
21	21	0	19	42	46.6	3.4	21	.055
22	22	0	20	41	54.9	3.6	22	.058
23	23	0	21	41	3.3	3.7	23	.060
24	24	0	22	40	11.6	3.9	24	.063
25	25	0	23	39	19.9	4.0	25	.066
26	26	0	24	38	28.2	4.2	26	.068
27	27	0	25	37	36.6	4.4	27	.071
28	28	0	26	36	44.9	4.6	28	.074
29	29	0	27	35	53.2	4.7	29	.077
30	30	0	28	35	1.6	4.9	30	.079
31	31	0	29	34	9.9	5.1	31	.082

FEBRUARY.

Years Com.	Years Bis.	☉'s Long. s	°	′	″	Perigee.	Day.	Frac.
1	1	1ˢ	0°	33′	18″.2	5″.3	32	.085
2	2	1	1	32	26.6	5.4	33	.088
3	3	1	2	31	34.9	5.6	34	.091
4	4	1	3	30	43.2	5.8	35	.093
5	5	1	4	29	51.5	5.9	36	.096
6	6	1	5	28	59.9	6.1	37	.099
7	7	1	6	28	8.2	6.3	38	.102
8	8	1	7	27	16.5	6.4	39	.104
9	9	1	8	26	24.9	6.6	40	.107
10	10	1	9	25	33.2	6.8	41	.109
11	11	1	10	24	41.5	7.0	42	.112
12	12	1	11	23	49.8	7.1	43	.115
13	13	1	12	22	58.2	7.3	44	.118
14	14	1	13	22	6.5	7.5	45	.120
15	15	1	14	21	14.8	7.6	46	.123
16	16	1	15	20	23.2	7.8	47	.126
17	17	1	16	19	31.5	8.0	48	.129
18	18	1	17	18	39.8	8.1	49	.131
19	19	1	18	17	48.2	8.3	50	.134
20	20	1	19	16	56.5	8.5	51	.137
21	21	1	20	16	4.8	8.7	52	.140
22	22	1	21	15	13.1	8.8	53	.142
23	23	1	22	14	21.5	9.0	54	.145
24	24	1	23	13	29.8	9.2	55	.148
25	25	1	24	12	38.1	9.3	56	.151
26	26	1	25	11	46.5	9.5	57	.154
27	27	1	26	10	54.8	9.7	58	.157
28	28	1	27	10	3.1	9.8	59	.159
	29	1	28	9	11.4	10.0	60	.162

MARCH.

Years Com.	Years Bis.	☉'s Long. s	°	′	″	Perigee.	Day.	Frac.
1	0	1ˢ	28°	9′	11″.4	10″.0	60	.162
2	1	1	29	8	19.8	10.2	61	.164
3	2	2	0	7	28.1	10.3	62	.167
4	3	2	1	6	36.4	10.5	63	.170
5	4	2	2	5	44.8	10.7	64	.173
6	5	2	3	4	53.1	10.9	65	.175
7	6	2	4	4	1.4	11.0	66	.178
8	7	2	5	3	9.8	11.2	67	.181
9	8	2	6	2	18.1	11.4	68	.184
10	9	2	7	1	26.4	11.5	69	.186
11	10	2	8	0	34.8	11.7	70	.189
12	11	2	8	59	43.1	11.9	71	.192
13	12	2	9	58	51.4	12.0	72	.195
14	13	2	10	57	59.7	12.2	73	.197
15	14	2	11	57	8.1	12.4	74	.200
16	15	2	12	56	16.4	12.6	75	.203
17	16	2	13	55	24.7	12.7	76	.206
18	17	2	14	54	33.1	12.9	77	208
19	18	2	15	53	41.4	13.1	78	.211
20	19	2	16	52	49.7	13.2	79	.214
21	20	2	17	51	58.1	13.4	80	.217
22	21	2	18	51	6.4	13.6	81	.219
23	22	2	19	50	14.7	13.7	82	.222
24	23	2	20	49	23.0	13.9	83	.225
25	24	2	21	48	31.4	14.1	84	.227
26	25	2	22	47	39.7	14.3	85	.230
27	26	2	23	46	48.0	14.4	86	.233
28	27	2	24	45	56.4	14.6	87	.236
29	28	2	25	45	4.7	14.8	88	.239
30	29	2	26	44	13.0	14.9	89	.241
31	30	2	27	43	21.4	15.1	90	.244
	31	2	28	42	29.7	15.3	91	.246

APRIL.

Years Com.	Years Bis.	☉'s Long. s	°	′	″	Perigee.	Day.	Frac.
1	0	2ˢ	28°	42′	29″.7	15″.3	91	.246
2	1	2	29	41	38.0	15.4	92	.249
3	2	3	0	40	46.3	15.6	93	.252
4	3	3	1	39	54.7	15.8	94	.255
5	4	3	2	39	3.0	15.9	95	.258
6	5	3	3	38	11.3	16.1	96	.260
7	6	3	4	37	19.7	16.3	97	.263
8	7	3	5	36	28.0	16.5	98	.266
9	8	3	6	35	36.3	16.6	99	.269
10	9	3	7	34	44.7	16.8	100	.271
11	10	3	8	33	53.0	17.0	101	.274
12	11	3	9	33	1.3	17.1	102	.277
13	12	3	10	32	9.6	17.3	103	.280
14	13	3	11	31	18.0	17.5	104	.282
15	14	3	12	30	26.3	17.6	105	.285
16	15	3	13	29	34.6	17.8	106	.288
17	16	3	14	28	43.0	18.0	107	.291
18	17	3	15	27	51.3	18.2	108	.293
19	18	3	16	26	59.6	18.3	109	.296
20	19	3	17	26	8.0	18.5	110	.299
21	20	3	18	25	16.3	18.7	111	.302
22	21	3	19	24	24.6	18.8	112	.304
23	22	3	20	23	32.9	19.0	113	.307
24	23	3	21	22	41.3	19.2	114	.309
25	24	3	22	21	49.6	19.3	115	.312
26	25	3	23	20	57.9	19.5	116	.315
27	26	3	24	20	6.3	19.7	117	.318
28	27	3	25	19	14.6	19.9	118	.320
29	28	3	26	18	22.9	20.0	119	.323
30	29	3	27	17	31.2	20.2	120	.326
	30	3	28	16	39.6	20.4	121	.329

MAY.

Years Com.	Years Bis.	☉'s Long. s	°	′	″	Perigee.	Day.	Frac.
1	0	3ˢ	28°	16′	39″.6	20″.4	121	329
2	1	3	29	15	47.9	20.5	122	.331
3	2	4	0	14	56.2	20.7	123	.334
4	3	4	1	14	4.6	20.9	124	.337
5	4	4	2	13	12.9	21.0	125	.340
6	5	4	3	12	21.2	21.2	126	.342
7	6	4	4	11	29.6	21.4	127	.345
8	7	4	5	10	37.9	21.5	128	.348
9	8	4	6	9	46.2	21.7	129	.351
10	9	4	7	8	54.5	21.9	130	.353
11	10	4	8	8	2.9	22.1	131	.356
12	11	4	9	7	11.2	22.2	132	.359
13	12	4	10	6	19.5	22.4	133	.362
14	13	4	11	5	27.9	22.6	134	.364
15	14	4	12	4	36.2	22.7	135	.367
16	15	4	13	3	44.5	22.9	136	.370
17	16	4	14	2	52.9	23.1	137	.373
18	17	4	15	2	1.2	23.2	138	.375
19	18	4	16	1	9.5	23.4	139	.378
20	19	4	17	0	17.8	23.6	140	.381
21	20	4	17	59	26.2	23.8	141	.383
22	21	4	18	58	34.5	23.9	142	.386
23	22	4	19	57	42.8	24.1	143	.389
24	23	4	20	56	51.2	24.3	144	.392
25	24	4	21	55	59.5	24.4	145	.395
26	25	4	22	55	7.8	24.6	146	.397
27	26	4	23	54	16.2	24.8	147	.400
28	27	4	24	53	24.5	24.9	148	.403
29	28	4	25	52	32.8	25.1	149	.406
30	29	4	26	51	41.1	25.3	150	.408
31	30	4	27	50	49.5	25.5	151	.411
	31	4	28	49	57.8	25.6	152	.414

JUNE.

Years Com.	Years Bis.	☉'s Long. s	°	′	″	Perigee.	Day.	Frac.
1	0	4ˢ	28°	49′	57″.8	25″.6	152	.414
2	1	4	29	49	6.1	25.8	153	.416
3	2	5	0	48	14.5	26.0	154	.419
4	3	5	1	47	22.8	26.1	155	.422
5	4	5	2	46	31.1	26.3	156	.425
6	5	5	3	45	39.5	26.5	157	.427
7	6	5	4	44	47.8	26.6	158	.430
8	7	5	5	43	56.1	26.8	159	.433
9	8	5	6	43	4.4	27.0	160	.436
10	9	5	7	42	12.8	27.2	161	.438
11	10	5	8	41	21.1	27.3	162	.441
12	11	5	9	40	29.4	27.5	163	.444
13	12	5	10	39	37.8	27.7	164	.447
14	13	5	11	38	46.1	27.8	165	.449
15	14	5	12	37	54.4	28.0	166	.452
16	15	5	13	37	2.8	28.2	167	.455
17	16	5	14	36	11.1	28.3	168	.458
18	17	5	15	35	19.4	28.5	169	.460
19	18	5	16	34	27.7	28.7	170	.463
20	19	5	17	33	36.1	28.8	171	.465
21	20	5	18	32	44.4	29.0	172	.468
22	21	5	19	31	52.7	29.2	173	.471
23	22	5	20	31	1.1	29.4	174	.473
24	23	5	21	30	9.4	29.5	175	.476
25	24	5	22	29	17.7	29.7	176	.479
26	25	5	23	28	26.0	29.9	177	.482
27	26	5	24	27	34.4	30.0	178	.485
28	27	5	25	26	42.7	30.2	179	.487
29	28	5	26	25	51.0	30.4	180	.490
30	29	5	27	24	59.4	30.5	181	.493
	30	5	28	24	7.7	30.7	182	.496

TABLE 2.

The Mean Longitude of the Sun, &c. concluded.

JULY.

Years. Com.	Bis.	☉'s Long.	Perigee.	Day.	Frac.
1	0	5ˢ 28° 24′ 7″.7	30″.7	182	.496
2	1	5 29 23 16.0	30.9	183	.499
3	2	6 0 22 24.4	31.1	184	.502
4	3	6 1 21 32.7	31.2	185	.504
5	4	6 2 20 41.0	31.4	186	.507
6	5	6 3 19 49.3	31.6	187	.509
7	6	6 4 18 57.7	31.7	188	.512
8	7	6 5 18 6.0	31.9	189	.515
9	8	6 6 17 14.3	32.1	190	.518
10	9	6 7 16 22.7	32.2	191	.520
11	10	6 8 15 31.0	32.4	192	.523
12	11	6 9 14 39.3	32.6	193	.526
13	12	6 10 13 47.7	32.7	194	.529
14	13	6 11 12 56.0	32.9	195	.531
15	14	6 12 12 4.3	33.1	196	.534
16	15	6 13 11 12.6	33.3	197	.537
17	16	6 14 10 21.0	33.4	198	.540
18	17	6 15 9 29.3	33.6	199	.542
19	18	6 16 8 37.6	33.8	200	.545
20	19	6 17 7 46.0	33.9	201	.548
21	20	6 18 6 54.3	34.1	202	.551
22	21	6 19 6 2.6	34.3	203	.553
23	22	6 20 5 11.0	34.4	204	.556
24	23	6 21 4 19.3	34.6	205	.559
25	24	6 22 3 27.6	34.8	206	.562
26	25	6 23 2 36.0	35.0	207	.564
27	26	6 24 1 44.3	35.1	208	.567
28	27	6 25 0 52.6	35.3	209	.570
29	28	6 26 0 1.0	35.5	210	.573
30	29	6 26 59 9.3	35.6	211	.575
31	30	6 27 58 17.6	35.8	212	.578
	31	6 28 57 26.0	36.0	213	.581

AUGUST.

Years. Com.	Bis.	☉'s Long.	Perigee.	Day.	Frac.
1	0	6ˢ 28° 57′ 26″.0	36″.0	213	.581
2	1	6 29 56 34.3	36.1	214	.583
3	2	7 0 55 42.6	36.3	215	.586
4	3	7 1 54 51.0	36.5	216	.589
5	4	7 2 53 59.3	36.7	217	.592
6	5	7 3 53 7.6	36.8	218	.594
7	6	7 4 52 16.0	37.0	219	.597
8	7	7 5 51 24.3	37.2	220	.600
9	8	7 6 50 32.6	37.3	221	.602
10	9	7 7 49 41.0	37.5	222	.605
11	10	7 8 48 49.3	37.7	223	.608
12	11	7 9 47 57.6	37.8	224	.610
13	12	7 10 47 5.9	38.0	225	.613
14	13	7 11 46 14.2	38.2	226	.616
15	14	7 12 45 22.5	38.3	227	.619
16	15	7 13 44 30.9	38.5	228	.622
17	16	7 14 43 39.2	38.7	229	.625
18	17	7 15 42 47.5	38.9	230	.627
19	18	7 16 41 55.9	39.0	231	.630
20	19	7 17 41 4.2	39.2	232	.633
21	20	7 18 40 12.5	39.4	233	.636
22	21	7 19 39 29.9	39.5	234	.638
23	22	7 20 38 29.2	39.7	235	.641
24	23	7 21 37 37.5	39.9	236	.644
25	24	7 22 36 45.8	40.0	237	.647
26	25	7 23 35 54.2	40.2	238	.649
27	26	7 24 35 2.5	40.4	239	.052
28	27	7 25 34 10.8	40.6	240	.655
29	28	7 26 33 19.2	40.7	241	.657
30	29	7 27 32 27.6	40.9	242	.660
31	30	7 28 31 35.9	41.1	243	.663
	31	7 29 30 44.2	41.2	244	.666

SEPTEMBER.

Years. Com.	Bis.	☉'s Long.	Perigee.	Day.	Frac.
1	0	7ˢ 29° 30′ 44″.2	41″.2	244	.666
2	1	8 0 29 51.5	41.4	245	.668
3	2	8 1 29 0.8	41.6	246	.671
4	3	8 2 28 9.1	41.7	247	.673
5	4	8 3 27 17.4	41.9	248	.675
6	5	8 4 26 25.8	42.1	249	.678
7	6	8 5 25 34.1	42.2	250	.681
8	7	8 6 24 42.5	42.4	251	.684
9	8	8 7 23 50.8	42.6	252	.687
10	9	8 8 22 59.1	42.8	253	.689
11	10	8 9 22 7.4	42.9	254	.692
12	11	8 10 21 15.8	43.1	255	.695
13	12	8 11 20 24.1	43.3	256	.698
14	13	8 12 19 32.4	43.4	257	.701
15	14	8 13 18 40.8	43.6	258	.703
16	15	8 14 17 49.1	43.8	259	.706
17	16	8 15 16 57.4	43.9	260	.709
18	17	8 16 16 5.8	44.1	261	.711
19	18	8 17 15 14.1	44.3	262	.714
20	19	8 18 14 22.4	44.5	263	.717
21	20	8 19 13 30.7	44.6	264	.720
22	21	8 20 12 39.1	44.8	265	.722
23	22	8 21 11 47.4	45.0	266	.725
24	23	8 22 10 55.7	45.1	267	.728
25	24	8 23 10 4.1	45.3	268	.731
26	25	8 24 9 12.4	45.5	269	.733
27	26	8 25 8 20.7	45.6	270	.736
28	27	8 26 7 29.1	45.8	271	.739
29	28	8 27 6 37.4	46.0	272	.742
30	29	8 28 5 45.7	46.2	273	.744
	30	8 29 4 54.0	46.3	274	.747

OCTOBER.

Years. Com.	Bis.	☉'s Long.	Perigee.	Day.	Frac.
1	0	8ˢ 29° 4′ 54″.0	46″.3	274	.747
2	1	9 0 4 2.4	46.5	275	.750
3	2	9 1 3 10.7	46.7	276	.753
4	3	9 2 2 19.0	46.9	277	.755
5	4	9 3 1 27.4	47.0	278	.758
6	5	9 4 0 35.7	47.2	279	.761
7	6	9 4 59 44.0	47.4	280	.763
8	7	9 5 58 52.4	47.5	281	.766
9	8	9 6 58 0.7	47.7	282	.769
10	9	9 7 57 9.0	47.9	283	.772
11	10	9 8 56 17.3	48.0	284	.775
12	11	9 9 55 25.7	48.2	285	.777
13	12	9 10 54 34.0	48.4	286	.780
14	13	9 11 53 42.4	48.5	287	.782
15	14	9 12 52 50.7	48.7	288	.785
16	15	9 13 51 59.0	48.9	289	.788
17	16	9 14 51 7.3	49.0	290	.791
18	17	9 15 50 15.7	49.2	291	.793
19	18	9 16 49 24.0	49.4	292	.796
20	19	9 17 48 32.3	49.6	293	.799
21	20	9 18 47 40.6	49.7	294	.802
22	21	9 19 46 49.0	49.9	295	.804
23	22	9 20 45 57.3	50.1	296	.807
24	23	9 21 45 5.6	50.2	297	.810
25	24	9 22 44 14.0	50.4	298	.813
26	25	9 23 43 22.3	50.6	299	.815
27	26	9 24 42 30.6	50.7	300	.818
28	27	9 25 41 39.0	50.9	301	.821
29	28	9 26 40 47.3	51.1	302	.824
30	29	9 27 39 55.6	51.2	303	.826
31	30	9 28 39 4.0	51.4	304	.829
	31	9 29 38 12.4	51.6	305	.832

NOVEMBER.

Years. Com.	Bis.	☉'s Long.	Perigee.	Day.	Frac.
1	0	9ˢ 29° 38′ 12″.4	51″.6	305	.832
2	1	10 0 37 20.6	51.8	306	.835
3	2	10 1 36 29.0	51.9	307	.838
4	3	10 2 35 37.3	52.1	308	.840
5	4	10 3 34 45.6	52.3	309	.843
6	5	10 4 33 54.3	52.4	310	.846
7	6	10 5 33 2.3	52.6	311	.848
8	7	10 6 32 10.6	52.8	312	.851
9	8	10 7 31 19.0	52.9	313	.854
10	9	10 8 30 27.3	53.0	314	.856
11	10	10 9 29 35.6	53.3	315	.859
12	11	10 10 28 44.0	53.5	316	.862
13	12	10 11 27 52.3	53.6	317	.865
14	13	10 12 27 0.7	53.8	318	.867
15	14	10 13 26 8.9	54.0	319	.870
16	15	10 14 25 17.2	54.1	320	.873
17	16	10 15 24 25.5	54.3	321	.876
18	17	10 16 23 33.9	54.5	322	.878
19	18	10 17 22 42.2	54.6	323	.881
20	19	10 18 21 50.5	54.8	324	.884
21	20	10 19 20 58.9	55.0	325	.887
22	21	10 20 20 7.2	55.2	326	.890
23	22	10 21 19 15.5	55.3	327	.893
24	23	10 22 18 23.9	55.5	328	.895
25	24	10 23 17 32.2	55.7	329	.898
26	25	10 24 16 40.5	55.8	330	.900
27	26	10 25 15 48.8	56.0	331	.903
28	27	10 26 14 57.2	56.2	332	.906
29	28	10 27 14 5.5	56.3	333	.909
30	29	10 28 13 13.8	56.5	334	.911
	30	10 29 12 22.2	56.7	335	.914

DECEMBER.

Years. Com.	Bis.	☉'s Long.	Perigee.	Day.	Frac.
1	0	10ˢ 29° 12′ 22″.2	56″.7	335	.914
2	1	11 0 11 30.5	56.8	336	.917
3	2	11 1 10 38.8	57.0	337	.920
4	3	11 2 9 47.2	57.2	338	.922
5	4	11 3 8 55.5	57.4	339	.925
6	5	11 4 8 3.8	57.5	340	.928
7	6	11 5 7 12.1	57.7	341	.931
8	7	11 6 6 20.5	57.9	342	.933
9	8	11 7 5 28.8	58.0	343	.936
10	9	11 8 4 37.1	58.2	344	.939
11	10	11 9 3 45.5	58.4	345	.942
12	11	11 10 2 53.8	58.5	346	.944
13	12	11 11 2 2.2	58.7	347	.947
14	13	11 12 1 10.5	58.9	348	.950
15	14	11 13 0 18.8	59.0	349	.953
16	15	11 13 59 27.1	59.2	350	.955
17	16	11 14 58 35.4	59.4	351	.958
18	17	11 15 57 43.8	59.6	352	.961
19	18	11 16 56 52.1	59.7	353	.964
20	19	11 17 56 0.4	59.9	354	.966
21	20	11 18 55 8.8	60.1	355	.969
22	21	11 19 54 17.1	60.2	356	.972
23	22	11 20 53 25.4	60.4	357	.974
24	23	11 21 52 33.8	60.6	358	.977
25	24	11 22 51 42.1	60.7	359	.980
26	25	11 23 50 50.4	60.9	360	.983
27	26	11 24 49 58.7	61.1	361	.985
28	27	11 25 49 7.1	61.3	362	.988
29	28	11 26 48 15.4	61.4	363	.991
30	29	11 27 47 23.7	61.6	364	.994
31	30	11 28 46 32.1	61.8	365	.997
	31	11 29 45 40.4	61.9	366	0

SOLAR TABLES.

TABLE 3. COMPARATIVE EPOCHS of the SUN'S TRUE LONGITUDE, and of his MEAN RIGHT ASCENSION.

(The former to be used with the Nautical Almanac of any Year, in order to obtain the *Arguments* for the Corrections due to any other Year, past or future, and the latter to be used with the subjoined Table.)

Years.	Long.	Right Ascens.	Years.	Long.	Right Ascens.	Years.	Long.	Right Ascens.	Years.	Long.	Right Ascens.	Years.	Long.	Right Ascens.
		m. s.			m. s.			m. s.			m. s.			m. s.
1750	60′ .2	3 18 .60	1780B.	44′ .3	6 11 .84	1810	28′ .5	1 11 .96	1840B.	12′ .7	4 5 .21	1870	55′ .9	3 1 .91
1751	45 .9	2 21 .29	1781	89 .2	5 14 .53	1811	14 .2	0 14 .66	1841	57 .5	3 7 .90	1871	41 .6	2 4 .60
1752B.	31 .5	5 20 .54	1782	74 .8	4 17 .22	1812B.	59 .8	3 13 .91	1842	43 .1	2 10 .60	1872B.	27 .3	5 3 .85
1753	76 .3	4 23 .23	1783	60 .5	3 19 .92	1813	44 .6	2 16 .60	1843	28 .8	1 13 .29	1873	72 .1	4 6 .54
1754	62 .0	3 25 .93	1784B.	46 .2	6 19 .16	1814	30 .3	1 19 .29	1844B.	14 .5	4 12 .54	1874	57 .8	3 9 .24
1755	47 .7	2 29 .62	1785	91 .0	5 21 .86	1815	16 .0	0 21 .99	1845	59 .3	3 15 .23	1875	43 .5	2 11 .93
1756B.	33 .4	5 27 .87	1786	76 .7	4 24 .55	1816B.	1 .7	3 21 .24	1846	45 .0	2 17 .93	1876B.	29 .1	5 11 .18
1757	78 .2	4 30 .56	1787	62 .3	3 27 .24	1817	46 .5	2 23 .93	1847	30 .6	1 20 .62	1877	73 .9	4 13 .87
1758	63 .9	3 33 .26	1788B.	48 .0	6 26 .49	1818	32 .2	1 26 .62	1848B.	16 .3	4 19 .87	1878	59 .6	3 16 .57
1759	49 .5	2 35 .95	1789	92 .8	5 29 .19	1819	17 .8	0 29 .32	1849	61 .1	3 22 .56	1879	45 .3	2 19 .26
1760B.	35 .2	5 35 .19	1790	78 .5	4 31 .88	1820B.	3 .5	3 28 .56	1850	46 .8	2 25 .26	1880B.	31 .0	5 18 .51
1761	80 .0	4 37 .89	1791	64 .2	3 34 .57	1821	48 .3	2 31 .25	1851	32 .5	1 27 .95	1881	75 .8	4 21 .20
1762	65 .7	3 40 .58	1792B.	49 .8	6 33 .82	1822	34 .0	1 33 .95	1852B.	18 .1	4 27 .20	1882	61 .4	3 23 .90
1763	51 .3	2 43 .27	1793	94 .6	5 36 .51	1823	19 .7	0 36 .64	1853	62 .9	3 29 .89	1883	47 .1	2 26 .59
1764B.	37 .0	5 42 .52	1794	80 .3	4 39 .21	1824B.	5 .3	3 35 .89	1854	48 .6	2 32 .59	1884B.	32 .8	5 25 .84
1765	81 .8	4 45 .22	1795	66 .0	3 41 .90	1825	50 .2	2 38 .58	1855	34 .3	1 35 .28	1885	77 .6	4 28 .53
1766	67 .5	3 47 .91	1796B.	51 .6	6 41 .15	1826	35 .8	1 41 .28	1856B.	20 .0	4 34 .53	1886	63 .3	3 31 .23
1767	53 .2	2 50 .60	1797	96 .5	5 43 .84	1827	21 .5	0 43 .97	1857	64 .8	3 37 .22	1887	48 .9	2 33 .92
1768B.	38 .8	5 49 .85	1798	82 .2	4 46 .54	1828B.	7 .2	3 43 .22	1858	50 .5	2 39 .92	1888B.	34 .6	5 33 .17
1769	83 .7	4 52 .55	1799	67 .8	3 49 .23	1829	52 .0	2 45 .91	1859	36 .1	1 42 .61	1889	79 .4	4 35 .86
1770	69 .3	3 55 .24	1800C.	53 .5	2 51 .92	1830	37 .6	1 48 .61	1860B.	21 .8	4 41 .86	1890	65 .1	3 38 .56
1771	55 .0	2 57 .93	1801	39 .2	1 54 .61	1831	23 .3	0 51 .30	1861	66 .6	3 44 .55	1891	50 .8	2 41 .25
1772B.	40 .7	5 57 .18	1802	24 .8	0 57 .31	1832B.	9 .0	3 50 .55	1862	52 .3	2 47 .25	1892B.	36 .4	5 40 .50
1773	85 .5	4 59 .87	1803	10 .5	0 0 .00	1833	53 .8	2 53 .24	1863	38 .0	1 49 .94	1893	81 .3	4 43 .19
1774	71 .2	4 2 .57	1804B.	56 .2	2 59 .25	1834	39 .5	1 55 .94	1864B.	23 .6	4 49 .19	1894	66 .9	3 45 .89
1775	56 .8	3 5 .26	1805	41 .0	2 1 .93	1835	25 .2	0 58 .63	1865	68 .4	3 51 .88	1895	52 .6	2 48 .58
1776B.	42 .5	6 4 .51	1806	26 .7	1 4 .64	1836B.	10 .8	3 57 .88	1866	54 .1	2 54 .58	1896B.	38 .3	5 47 .83
1777	87 .3	5 7 .20	1807	12 .3	0 7 .33	1837	55 .6	3 0 .57	1867	39 .8	1 57 .27	1897	83 .1	4 50 .52
1778	73 .0	4 9 .90	1808B.	58 .0	3 6 .58	1838	41 .3	2 3 .27	1868B	25 .5	4 56 .52	1898	68 .8	3 53 .22
1779	58 .7	3 12 .59	1809	42 .8	2 9 .27	1839	27 .0	1 5 .96	1869	70 .3	3 59 .21	1899	54 .4	2 55 .91

TABLE 4. The SUN'S Mean RIGHT ASCENSION for the Year 1803 at Greenwich.

	JANUARY.	FEBRUARY.	MARCH.	APRIL.	MAY.	JUNE.	JULY.	AUGUST.	SEPTEMBER.	OCTOBER.	NOVEMBER.	DECEMBER.	
	h. m. s.	h. m. s.	h. m. s.	h. m. s.	h. m. s.	h. m. s.	h. m. s.	h. m. s.	h. m. s.	h. m. s.	h. m. s.	h. m. s.	
0	18 36 45.6	20 38 58.8	22 29 22.3										0
1	40 42.1	42 55.3	33 18.9	0 35 32.1	2 33 48.7	4 36 2.0	6 34 18.6	8 36 31.8	10 38 45.1	12 37 1.7	14 39 14.9	16 37 31.6	1
2	44 38.7	46 51.9	37 15.5	39 28.6	37 45.3	39 57.5	38 15.2	40 28.4	42 41.6	40 58.3	43 11.5	41 28.1	2
3	48 35.2	50 48.4	41 12.0	43 25.2	41 41.8	43 55.1	42 11.7	44 25.0	46 38.2	44 54.8	47 8.0	45 24.7	3
4	52 31.8	54 45.0	45 8.5	47 21.8	45 38.4	47 51.6	46 8.3	48 21.5	50 34.7	48 51.4	51 4.6	49 21.3	4
5	56 28.3	58 41.5	49 5.1	51 18.3	49 35.0	51 48.2	50 4.8	52 18.1	54 31.3	52 47.9	55 1.2	53 17.8	5
6	19 0 24.9	21 2 38.1	53 1.7	55 14.9	53 31.5	55 44.7	54 1.4	56 14.6	58 27.8	56 44.5	58 57.7	57 14.4	6
7	4 21.4	6 34.7	56 58.2	59 11.4	57 28.1	59 41.3	57 57.9	9 0 11.2	11 2 24.4	13 0 41.0	15 2 54.3	17 1 10.9	7
8	8 18.0	10 31.2	23 0 54.8	1 3 8.0	3 1 24.6	5 3 37.9	7 1 54.5	4 7.7	6 21.0	4 37.6	6 50.8	5 7.5	8
9	12 14.6	14 27.8	4 51.3	7 4.5	5 21.2	7 34.4	5 51.1	8 4.3	10 17.5	8 34.2	10 47.4	9 4.0	9
10	16 11.1	18 24.3	8 47.9	11 1.1	9 17.7	11 31.0	9 47.6	12 0.8	14 14.1	12 30.7	14 43.9	13 0.6	10
11	20 7.7	22 20.9	12 44.4	14 57.6	13 14.3	15 27.5	13 44.2	15 57.4	18 10.6	16 27.3	18 40.5	16 57.1	11
12	24 4.2	26 17.4	16 41.0	18 54.2	17 10.9	19 24.1	17 40.7	19 53.5	22 7.2	20 23.8	22 37.0	20 53.7	12
13	28 0.8	30 14.0	20 36.5	22 50.7	21 7.4	23 20.6	21 37.3	23 50.5	26 3.7	24 20.4	26 33.6	24 50.2	13
14	31 57.3	34 10.5	24 33.1	26 47.3	25 4.0	27 17.2	25 33.8	27 47.1	30 0.3	28 16.9	30 30.1	28 46.8	14
15	35 53.9	38 7.1	28 30.6	30 43.9	29 1.5	31 13.7	29 30.4	31 43.6	33 56.8	32 13.5	34 26.7	32 43.4	15
16	39 50.4	42 3.7	32 27.2	34 40.4	32 57.1	35 10.3	33 27.0	35 40.1	37 53.4	36 10.0	38 23.2	36 39.9	16
17	43 47.0	46 0.2	36 23.8	38 37.0	36 53.6	39 6.8	37 23.5	39 36.7	41 49.9	40 6.6	42 19.8	40 36.5	17
18	47 43.5	49 56.8	40 20.3	42 33.5	40 50.2	43 3.4	41 20.1	43 33.3	45 46.5	44 3.1	46 16.3	44 33.0	18
19	51 40.1	53 53.3	44 16.9	46 30.1	44 46.7	46 59.9	45 16.6	47 29.8	49 43.1	47 59.7	50 12.9	48 29.6	19
20	55 36.7	57 49.9	48 13.4	50 26.6	48 43.3	50 56.5	49 13.2	51 26.4	53 39.6	51 56.2	54 9.4	52 26.1	20
21	59 33.2	22 1 46.4	52 10.0	54 23.2	52 39.8	54 53.1	53 9.7	55 22.9	57 36.2	55 52.8	16 0 6.0	56 22.7	21
22	20 3 29.8	5 43.0	56 6.5	58 19.8	56 36.4	58 49.6	57 6.3	59 19.5	12 1 32.7	59 49.3	2 2.6	18 0 19.3	22
23	7 26.3	9 39.5	0 0 3.1	2 2 16.3	4 0 33.0	6 2 46.2	8 1 2.8	10 3 16.1	5 29.3	14 3 45.9	5 59.1	4 15.8	23
24	11 22.9	13 36.1	3 59.7	6 12.9	4 29.5	6 42.7	4 59.4	7 12.6	9 25.8	7 42.5	9 55.7	8 12.4	24
25	15 19.4	17 32.6	7 56 2	10 9.4	8 26.1	10 39.3	8 56.0	11 9.2	13 22.4	11 39.0	13 52.3	12 8.9	25
26	19 16.0	21 29.2	11 52.8	14 6.0	12 22.0	14 35.8	12 53.5	15 5.7	17 18.9	15 35.6	17 48.8	16 5.5	26
27	23 12.5	25 25.8	15 49.3	18 1.5	16 19.2	18 32.4	16 49.1	19 2.3	21 15.5	19 32.2	21 45.4	20 2.0	27
28	27 9.1	29 22.3	19 45.9	21 59.1	20 15.7	22 29.0	20 45.6	22 58.8	25 12.0	23 28.7	25 41.9	23 58.6	28
29	31 5.7		23 42.4	25 55.6	24 12.3	26 25.5	24 42.2	26 55.4	29 8.6	27 25.3	29 38.5	27 55.1	29
30	35 2.2		27 39.0	29 52.2	28 8.8	30 22.1	28 38.7	30 51.9	33 5.1	31 21.8	33 34.0	31 51.7	30
31	38 58.8		31 38.5		32 5.4		32 35.3	34 48.5		35 18.4		35 48.2	31

When the Numbers in this Table are used with a Leap Year, the Right Ascension of the *preceding* Day must be taken in the Months of JANUARY and FEBRUARY.
By means of these two Tables, the Sidereal and Solar Clocks may at any Noon be compared with each other.

TABLE 5.

Conversion of the SUN'S LONGITUDE into RIGHT ASCENSION in Time, by MR. J. UTTING.

Argument. ☉'s Long.	0ˢ VIˢ h. h. 0+ 12+	Diff.	Variation.	Iˢ VIIˢ h. h. 0+ 12+	Diff.	Variation.	IIˢ VIIIˢ h. h. 0+ 12+	Diff.	Variation.	Argument. ☉'s Long.
	h. m. s.			h. m. s.			h. m. s.			
0° 0′	0 0 0 .00	36ˢ.69	0ˢ.00	1 51 37 .39	38ˢ.21	1ˢ.20	3 51 15 .25	41ˢ.65	1ˢ.30	30° 0′
10	0 0 36 .69	36 .69	0 .00	1 52 15 .60	38 .23	1 .20	3 51 56 .90	41 .68	1 .30	50
20	0 1 13 .38	36 .70	0 .01	1 52 53 .83	38 .25	1 .21	3 52 38 .58	41 .69	1 .29	40
30	0 1 50 .08	36 .69	0 .01	1 53 32 .08	38 .26	1 .21	3 53 20 .27	41 .71	1 .29	30
40	0 2 26 .77	36 .69	0 .02	1 54 10 .34	38 .28	1 .21	3 54 1 .98	41 .73	1 .29	20
50	0 3 3 .46	36 .70	0 .03	1 54 48 .62	38 .30	1 .22	3 54 43 .71	41 .75	1 .28	10
1 0	0 3 40 .16	36 .69	0 .04	1 55 26 .92	38 .31	1 .22	3 55 25 .46	41 .76	1 .28	29 0
10	0 4 16 .85	36 .69	0 .04	1 56 5 .23	38 .32	1 .23	3 56 7 .22	41 .79	1 .27	50
20	0 4 53 .54	36 .70	0 .05	1 56 43 .55	38 .35	1 .23	3 56 49 .01	41 .80	1 .27	40
30	0 5 30 .24	36 .70	0 .06	1 57 21 .90	38 .36	1 .23	3 57 30 .81	41 .82	1 .26	30
40	0 6 6 .94	36 .69	0 .07	1 58 0 .26	38 .38	1 .24	3 58 12 .63	41 .84	1 .26	20
50	0 6 43 .63	36 .70	0 .08	1 58 38 .64	38 .39	1 .24	3 58 54 .47	41 .86	1 .25	10
2 0	0 7 20 .33	36 .70	0 .09	1 59 17 .03	38 .41	1 .25	3 59 36 .33	41 .88	1 .25	28 0
10	0 7 57 .03	36 .70	0 .10	1 59 55 .44	38 .43	1 .25	4 0 18 .21	41 .89	1 .25	50
20	0 8 33 .73	36 .70	0 .10	2 0 33 .87	38 .44	1 .25	4 1 0 .10	41 .91	1 .24	40
30	0 9 10 .43	36 .71	0 .11	2 1 12 .31	38 .46	1 .26	4 1 42 .01	41 .94	1 .24	30
40	0 9 47 .14	36 .70	0 .12	2 1 50 .77	38 .48	1 .26	4 2 23 .95	41 .94	1 .24	20
50	0 10 23 .84	36 .71	0 .13	2 2 29 .25	38 .49	1 .26	4 3 5 .89	41 .97	1 .23	10
3 0	0 11 0 .55	36 .71	0 .14	2 3 7 .74	38 .51	1 .27	4 3 47 .86	41 .99	1 .23	27 0
10	0 11 37 .26	36 .71	0 .14	2 3 46 .25	38 .52	1 .27	4 4 29 .85	42 .00	1 .22	50
20	0 12 13 .97	36 .71	0 .15	2 4 24 .77	38 .56	1 .28	4 5 11 .85	42 .02	1 .22	40
30	0 12 50 .68	36 .72	0 .16	2 5 3 .33	38 .56	1 .28	4 5 53 .87	42 .04	1 .21	30
40	0 13 27 .40	36 .71	0 .17	2 5 41 .89	38 .58	1 .28	4 6 35 .91	42 .05	1 .21	20
50	0 14 4 .11	36 .72	0 .17	2 6 20 .47	38 .60	1 .29	4 7 17 .96	42 .08	1 .20	10
4 0	0 14 40 .83	36 .72	0 .18	2 6 59 .07	38 .61	1 .29	4 8 0 .04	42 .09	1 .20	26 0
10	0 15 17 .55	36 .73	0 .19	2 7 37 .68	38 .64	1 .30	4 8 42 .13	42 .11	1 .19	50
20	0 15 54 .28	36 .72	0 .20	2 8 16 .32	38 .64	1 .30	4 9 24 .24	42 .12	1 .19	40
30	0 16 31 .00	36 .73	0 .20	2 8 54 .96	38 .67	1 .30	4 10 6 .36	42 .15	1 .18	30
40	0 17 7 .73	36 .74	0 .21	2 9 33 .63	38 .69	1 .31	4 10 48 .51	42 .16	1 .18	20
50	0 17 44 .47	36 .73	0 .22	2 10 12 .32	38 .70	1 .31	4 11 30 .67	42 .18	1 .17	10
5 0	0 18 21 .20	36 .74	0 .23	2 10 51 .02	38 .72	1 .32	4 12 12 .85	42 .19	1 .17	25 0
10	0 18 57 .94	36 .74	0 .24	2 11 29 .74	38 .74	1 .32	4 12 55 .04	42 .21	1 .16	50
20	0 19 34 .68	36 .74	0 .25	2 12 8 .48	38 .75	1 .32	4 13 37 .25	42 .23	1 .16	40
30	0 20 11 .42	36 .75	0 .25	2 12 47 .23	38 .78	1 .32	4 14 19 .48	42 .25	1 .15	30
40	0 20 48 .17	36 .75	0 .26	2 13 26 .01	38 .79	1 .33	4 15 1 .73	42 .26	1 .15	20
50	0 21 24 .92	36 .75	0 .27	2 14 4 .80	38 .81	1 .33	4 15 43 .99	42 .28	1 .14	10
6 0	0 22 1 .67	36 .76	0 .28	2 14 43 .61	38 .83	1 .33	4 16 26 .27	42 .31	1 .14	24 0
10	0 22 38 .43	36 .76	0 .29	2 15 22 .44	38 .84	1 .33	4 17 8 .58	42 .30	1 .13	50
20	0 23 15 .19	36 .77	0 .30	2 16 1 .28	38 .87	1 .33	4 17 50 .88	42 .33	1 .13	40
30	0 23 51 .96	36 .77	0 .30	2 16 40 .15	38 .88	1 .34	4 18 33 .21	42 .34	1 .12	30
40	0 24 28 .73	36 .77	0 .31	2 17 19 .03	38 .90	1 .34	4 19 15 .55	42 .37	1 .11	20
50	0 25 5 .50	36 .77	0 .32	2 17 57 .93	38 .92	1 .34	4 19 57 .92	42 .38	1 .10	10
7 0	0 25 42 .27	36 .79	0 .32	2 18 36 .85	38 .93	1 .35	4 20 40 .30	42 .39	1 .10	23 0
10	0 26 19 .06	36 .78	0 .33	2 19 15 .78	38 .96	1 .35	4 21 22 .69	42 .42	1 .09	50
20	0 26 55 .84	36 .79	0 .34	2 19 54 .74	38 .97	1 .35	4 22 5 .11	42 .42	1 .09	40
30	0 27 32 .63	36 .79	0 .35	2 20 33 .71	38 .99	1 .36	4 22 47 .53	42 .44	1 .08	30
40	0 28 9 .42	36 .80	0 .35	2 21 12 .70	39 .01	1 .36	4 23 29 .97	42 .46	1 .08	20
50	0 28 46 .22	36 .80	0 .36	2 21 51 .71	39 .03	1 .36	4 24 12 .43	42 .48	1 .07	10
8 0	0 29 23 .02	36 .81	0 .37	2 22 30 .74	39 .05	1 .37	4 24 54 .91	42 .49	1 .07	22 0
10	0 29 59 .83	36 .81	0 .37	2 23 9 .79	39 .07	1 .37	4 25 37 .40	42 .50	1 .06	50
20	0 30 36 .64	36 .82	0 .38	2 23 48 .86	39 .08	1 .37	4 26 19 .90	42 .53	1 .05	40
30	0 31 13 .46	36 .82	0 .39	2 24 27 .94	39 .10	1 .37	4 27 2 .43	42 .53	1 .04	30
40	0 31 50 .28	36 .83	0 .40	2 25 7 .04	39 .13	1 .38	4 27 44 .96	42 .56	1 .04	20
50	0 32 27 .11	36 .83	0 .40	2 25 46 .17	39 .14	1 .38	4 28 27 .52	42 .57	1 .03	10
9 0	0 33 3 .94	36 .84	0 .41	2 26 25 .31	39 .16	1 .38	4 29 10 .09	42 .58	1 .03	21 0
10	0 33 40 .78	36 .84	0 .42	2 27 4 .47	39 .18	1 .38	4 29 52 .67	42 .60	1 .02	50
20	0 34 17 .62	36 .85	0 .43	2 27 43 .65	39 .19	1 .39	4 30 35 .27	42 .61	1 .01	40
30	0 34 54 .47	36 .85	0 .44	2 28 22 .84	39 .22	1 .39	4 31 17 .83	42 .63	1 .01	30
40	0 35 31 .32	36 .86	0 .45	2 29 2 .06	39 .24	1 .39	4 32 0 .51	42 .64	1 .00	20
50	0 36 8 .18	36 .87	0 .45	2 29 41 .30	39 .25	1 .40	4 32 43 .15	42 .66	1 .00	10
10 0	0 36 45 .05		0 .46	2 30 20 .55		1 .40	4 33 25 .81		0 .99	20 0
☉'s Long. Argument.	h. h. 12− 24− Vˢ XIˢ	Diff.	Variation.	h. h. 12− 24− IVˢ Xˢ	Diff.	Variation.	h. h. 12− 24− IIIˢ IXˢ	Diff.	Variation.	☉'s Long. Argument.

The Obliquity of the Ecliptic is taken = 23° 27′ 57″, and the Variation of Right Ascension corresponds to 100 of diminution in the Obliquity.

TABLE 5.

Conversion of the Sun's Longitude into Right Ascension in Time, continued.

Argument. ⊙'s Long.	0s VIs h. h. 0+ 12+	Diff.	Variation.	Is VIIs h. h. 0+ 12+	Diff.	Variation.	IIs VIIIs h. h. 0+ 12+	Diff.	Variation.	Argument. ⊙'s Long.
	h. m. s.			h. m. s.			h. m. s.			
10° 0′	0 36 45.05	36s.87	0s.46	2 30 20.55	39s.27	1s.40	4 33 25.81	42s.68	0s.99	20° 0′
10	0 37 21.92	36.88	0.47	2 30 59.82	39.30	1.40	4 34 8.49	42.68	0.98	50
20	0 37 58.80	36.88	0.48	2 31 39.12	39.31	1.40	4 34 51.17	42.71	0.98	40
30	0 38 35.68	36.89	0.48	2 32 18.43	39.33	1.40	4 35 33.88	42.71	0.97	30
40	0 39 12.57	36.89	0.49	2 32 57.76	39.35	1.40	4 36 16.59	42.73	0.96	20
50	0 39 49.46	36.91	0.50	2 33 37.11	39.37	1.41	4 36 59.32	42.75	0.95	10
11 0	0 40 26.37	36.90	0.50	2 34 16.48	39.39	1.41	4 37 42.07	42.76	0.95	19 0
10	0 41 3.27	36.92	0.51	2 34 55.87	39.41	1.41	4 38 24.83	42.77	0.94	50
20	0 41 40.19	36.92	0.52	2 35 35.28	39.43	1.41	4 39 7.60	42.79	0.93	40
30	0 42 17.11	36.93	0.52	2 36 14.71	39.44	1.41	4 39 50.39	42.80	0.93	30
40	0 42 54.04	36.93	0.53	2 36 54.15	39.47	1.42	4 40 33.19	42.82	0.92	20
50	0 43 30.97	36.94	0.54	2 37 33.62	39.49	1.42	4 41 16.01	42.82	0.91	10
12 0	0 44 7.91	36.95	0.54	2 38 13.11	39.50	1.42	4 41 58.83	42.85	0.91	18 0
10	0 44 44.86	36.96	0.55	2 38 52.61	39.53	1.42	4 42 41.68	42.85	0.90	50
20	0 45 21.82	36.96	0.55	2 39 32.14	39.54	1.42	4 43 24.53	42.87	0.90	40
30	0 45 58.78	36.97	0.56	2 40 11.68	39.57	1.43	4 44 7.40	42.88	0.89	30
40	0 46 35.75	36.98	0.57	2 40 51.25	39.58	1.43	4 44 50.28	42.90	0.88	20
50	0 47 12.73	36.98	0.58	2 41 30.83	39.61	1.43	4 45 33.18	42.91	0.88	10
13 0	0 47 49.71	37.00	0.59	2 42 10.44	39.62	1.43	4 46 16.09	42.92	0.87	17 0
10	0 48 26.71	37.00	0.59	2 42 50.06	39.64	1.43	4 46 59.01	42.93	0.86	50
20	0 49 3.71	37.00	0.60	2 43 29.70	39.66	1.43	4 47 41.94	42.95	0.85	40
30	0 49 40.71	37.02	0.61	2 44 9.36	39.69	1.44	4 48 24.89	42.96	0.85	30
40	0 50 17.73	37.02	0.62	2 44 49.05	39.70	1.44	4 49 7.85	42.97	0.84	20
50	0 50 54.75	37.04	0.62	2 45 28.75	39.72	1.44	4 49 50.82	42.98	0.83	10
14 0	0 51 31.79	37.04	0.63	2 46 8.47	39.74	1.44	4 50 33.80	43.00	0.82	16 0
10	0 52 8.83	37.04	0.64	2 46 48.21	39.76	1.44	4 51 16.80	43.01	0.81	50
20	0 52 45.87	37.06	0.64	2 47 27.97	39.79	1.44	4 51 59.81	43.02	0.81	40
30	0 53 22.93	37.06	0.65	2 48 7.76	39.80	1.44	4 52 42.83	43.04	0.80	30
40	0 53 59.99	37.08	0.66	2 48 47.56	30.82	1.44	4 53 25.87	43.04	0.79	20
50	0 54 37.07	37.08	0.66	2 49 27.38	39.84	1 44	4 54 8.91	43.06	0.78	10
15 0	0 55 14.15	37.09	0.67	2 50 7.22	39.86	1.44	4 54 51.97	43.07	0.78	15 0
10	0 55 51.24	37.10	0.68	2 50 47.08	39.88	1.44	4 55 35.04	43.08	0.77	50
20	0 56 28.34	37.10	0.68	2 51 26.96	39.91	1.44	4 56 18.12	43.09	0.76	40
30	0 57 5.44	37.12	0.69	2 52 6.87	39.92	1.44	4 57 1.21	43.10	0.75	30
40	0 57 42.56	37.13	0.70	2 52 46.79	39.94	1.44	4 57 44.31	43.12	0.75	20
50	0 58 19.69	37.13	0.70	2 53 26.73	39.96	1.44	4 58 27.43	43.12	0.74	10
16 0	0 58 56.82	37.14	0.71	2 54 6.69	39.98	1.44	4 59 10.55	43.14	0.73	14 0
10	0 59 33.96	37.16	0.72	2 54 46.67	40.00	1.44	4 59 53.69	43.15	0.72	50
20	1 0 11.12	37.16	0.72	2 55 26.67	40.02	1.44	5 0 36.84	43.15	0.71	40
30	1 0 48.28	37.17	0.73	2 56 6.69	40.05	1.44	5 1 19.99	43.17	0.70	30
40	1 1 25.45	37.19	0.74	2 56 46.74	40.06	1.44	5 2 3.16	43.18	0.70	20
50	1 2 2.64	37.19	0.74	2 57 26.80	40.08	1.44	5 2 46.34	43.19	0.69	10
17 0	1 2 39.83	37.20	0.75	2 58 6.88	40.10	1.44	5 3 29.53	43.20	0.68	13 0
10	1 3 17.03	37.21	0.76	2 58 46.98	40.13	1.44	5 4 12.73	43.21	0.67	50
20	1 3 54.24	37.22	0.76	2 59 27.11	40.14	1.44	5 4 55.94	43.22	0.67	40
30	1 4 31.46	37.23	0.77	3 0 7.25	40.16	1.44	5 5 39.16	43.23	0.66	30
40	1 5 8.69	37.24	0.78	3 0 47.41	40.18	1.44	5 6 22.39	43.24	0.65	20
50	1 5 45.93	37.25	0.78	3 1 27.59	40.21	1.44	5 7 5.63	43.25	0.65	10
18 0	1 6 23.18	37.26	0.79	3 2 7.80	40.22	1.44	5 7 48.88	43.26	0.64	12 0
10	1 7 0.44	37.27	0.80	3 2 48.02	40.25	1.44	5 8 32.14	43.26	0.63	50
20	1 7 37.71	37.29	0.80	3 3 28.27	40.26	1.44	5 9 15.40	43.28	0.62	40
30	1 8 15.00	37.29	0.81	3 4 8.53	40.28	1.44	5 9 58.68	43.29	0.62	30
40	1 8 52.29	37.30	0.82	3 4 48.81	40.31	1.44	5 10 41.97	43.29	0.61	20
50	1 9 29.59	37.32	0.83	3 5 29.12	40.32	1.44	5 11 25.26	43.31	0.60	10
19 0	1 10 6.91	37.32	0.83	3 6 9.44	40.35	1.44	5 12 8.57	43.31	0.59	11 0
10	1 10 44.23	37.33	0.84	3 6 49.79	40.36	1.44	5 12 51.88	43.32	0.58	50
20	1 11 21.56	37.35	0.15	3 7 30.15	40.39	1.44	5 13 35.20	43.34	0.57	40
30	1 11 58.91	37.36	0.85	3 8 10.54	40.40	1.44	5 14 18.54	43.34	0.57	30
40	1 12 36.27	37.37	0.86	3 8 50.94	40.43	1.44	5 15 1.88	43.34	0.56	20
50	1 13 13.64	37.38	0.87	3 9 31.37	40.45	1.44	5 15 45.22	43.36	0.55	10
20 0	1 13 51.02		0.87	3 10 11.82		1.44	5 16 28.58		0.54	10 0
⊙'s Long. Argument.	h. h. 12− 24− Vs XIs	Diff.	Variation.	h. h. 12− 24− IVs Xs	Diff.	Variation.	h. h. 12− 24− IIIs IXs	Diff.	Variation.	⊙'s Long. Argument.

The obliquity of the Ecliptic is taken = 23° 27′ 57″, and the variation of Right Ascension corresponds to 100″ of diminution in the obliquity.

TABLE 5.

Conversion of the SUN'S LONGITUDE into RIGHT ASCENSION in Time, concluded.

Argument. ⊙'s Long.	0s VIs h. 0+ h. 12+	Diff.	Variation.	Is VIIs h. 0+ h. 12+	Diff.	Variation.	IIs VIIIs h. 0+ h. 12+	Diff.	Variation.	Argument. ⊙'s Long.
	h. m. s.			h. m. s.			h. m. s.			
20° 0′	1 13 51.02	37s.39	0s.87	3 10 11.82	40s.46	1s.44	5 16 28.58	43s.36	0s.54	10° 0′
10	1 14 28.41	37.40	0.88	3 10 52.28	40.49	1.44	5 17 11.94	43.37	0.53	50
20	1 15 5.81	37.41	0.88	3 11 32.77	40.50	1.44	5 17 55.31	43.38	0.52	40
30	1 15 43.22	37.43	0.89	3 12 13.27	40.53	1.44	5 18 38.69	43.39	0.52	30
40	1 16 20.65	37.44	0.90	3 12 53.80	40.55	1.44	5 19 22.08	43.40	0.51	20
50	1 16 58.09	37.45	0.91	3 13 34.35	40.57	1.43	5 20 5.48	43.40	0.50	10
21 0	1 17 35.54	37.46	0.91	3 14 14.92	40.59	1.43	5 20 48.88	43.41	0.49	9 0
10	1 18 13.00	37.47	0.92	3 14 55.51	40.60	1.43	5 21 32.29	43.42	0.48	50
20	1 18 50.47	37.48	0.92	3 15 36.11	40.63	1.43	5 22 15.71	43.42	0.47	40
30	1 19 27.95	37.50	0.93	3 16 16.74	40.65	1.43	5 22 59.13	43.43	0.46	30
40	1 20 5.45	37.51	0.93	3 16 57.39	40.67	1.43	5 23 42.56	43.44	0.45	20
50	1 20 42.96	37.52	0.94	3 17 38.06	40.69	1.43	5 24 26.00	43.44	0.44	10
22 0	1 21 20.48	37.53	0.94	3 18 18.75	40.71	1.43	5 25 9.44	43.45	0.43	8 0
10	1 21 58.01	37.55	0.95	3 18 59.46	40.73	1.43	5 25 52.89	43.46	0.42	50
20	1 22 35.56	37.55	0.96	3 19 40.19	40.75	1.43	5 26 36.35	43.46	0.41	40
30	1 23 13.11	37.58	0.96	3 20 20.94	40.77	1.42	5 27 19.81	43.47	0.41	30
40	1 23 50.69	37.58	0.97	3 21 1.71	40.79	1.42	5 28 3.28	43.48	0.40	20
50	1 24 28.27	37.59	0.98	3 21 42.50	40.81	1.42	5 28 46.76	43.48	0.39	10
23 0	1 25 5.86	37.61	0.98	3 22 23.31	40.83	1.42	5 29 30.24	43.49	0.38	7 0
10	1 25 43.47	37.63	0.99	3 23 4.14	40.85	1.42	5 30 13.73	43.49	0.37	50
20	1 26 21.10	37.63	0.99	3 23 44.99	40.87	1.42	5 30 57.22	43.50	0.36	40
30	1 26 58.73	37.65	1.00	3 24 25.86	40.89	1.41	5 31 40.72	43.51	0.35	30
40	1 27 36.38	37.66	1.00	3 25 6.75	40.91	1.41	5 32 24.23	43.51	0.35	20
50	1 28 14.04	37.67	1.01	3 25 47.66	40.93	1.41	5 33 7.74	43.51	0.34	10
24 0	1 28 51.71	37.69	1.01	3 26 28.59	40.95	1.41	5 33 51.25	43.52	0.33	6 0
10	1 29 29.40	37.70	1.02	3 27 9.54	40.97	1.41	5 34 34.77	43.52	0.32	50
20	1 30 7.10	37.71	1.02	3 27 50.51	41.00	1.41	5 35 18.29	43.53	0.31	40
30	1 30 44.81	37.73	1.03	3 28 31.51	41.01	1.40	5 36 1.82	43.53	0.30	30
40	1 31 22.54	37.74	1.04	3 29 12.52	41.03	1.40	5 36 45.35	43.54	0.29	20
50	1 32 0.28	37.75	1.04	3 29 53.55	41.05	1.40	5 37 28.89	43.54	0.28	10
25 0	1 32 38.03	37.77	1.05	3 30 34.60	41.07	1.39	5 38 12.43	43.55	0.27	5 0
10	1 33 15.80	37.78	1.05	3 31 15.67	41.09	1.39	5 38 55.98	43.55	0.26	50
20	1 33 53.58	37.80	1.06	3 31 56.76	41.11	1.39	5 39 39.53	43.55	0.25	40
30	1 34 31.38	37.81	1.06	3 32 37.87	41.13	1.39	5 40 23.08	43.56	0.25	30
40	1 35 9.19	37.82	1.07	3 33 19.00	41.15	1.38	5 41 6.64	43.56	0.24	20
50	1 35 47.01	37.84	1.07	3 34 0.15	41.17	1.38	5 41 50.20	43.57	0.23	10
26 0	1 36 24.85	37.85	1.08	3 34 41.32	41.19	1.38	5 42 33.77	43.57	0.22	4 0
10	1 37 2.70	37.87	1.08	3 35 22.51	41.21	1.38	5 43 17.34	43.57	0.21	50
20	1 37 40.57	37.88	1.09	3 36 3 72	41.23	1.37	5 44 0.91	43.57	0.20	40
30	1 38 18.45	37.90	1.09	3 36 44.95	41.25	1.37	5 44 44.48	43.58	0.19	30
40	1 38 56.35	37.91	1.10	3 37 26.20	41.27	1.37	5 45 28.06	43.58	0.18	20
50	1 39 34.26	37.92	1.10	3 38 7.47	41.28	1.36	5 46 11.64	43.58	0.17	10
27 0	1 40 12.18	37.94	1.11	3 38 48.75	41.31	1.36	5 46 55.22	43.59	0.16	3 0
10	1 40 50.12	37.95	1.11	3 39 30.06	41.33	1.36	5 47 38.81	43.58	0.15	50
20	1 41 28.07	37.97	1.12	3 40 11.39	41.34	1.35	5 48 22.39	43.59	0.14	40
30	1 42 6.04	37.98	1.12	3 40 52.73	41.37	1.35	5 49 5.98	43.59	0.14	30
40	1 42 44.02	38.00	1.13	3 41 34.10	41.38	1.35	5 49 49.57	43.60	0.13	20
50	1 43 22.02	38.02	1.13	3 42 15.48	41.41	1.34	5 50 33.17	43.59	0.12	10
28 0	1 44 0.04	38.02	1.14	3 42 56.89	41.43	1.34	5 51 16.76	43.60	0.11	2 0
10	1 44 38.06	38.05	1.14	3 43 38.32	41.44	1.34	5 52 0.36	43.60	0.10	50
20	1 45 16.11	38.06	1.15	3 44 19.76	41.46	1.33	5 52 43.96	43.60	0.10	40
30	1 45 54.17	38.07	1.15	3 45 1.22	41.48	1.33	5 53 27.56	43.60	0.09	30
40	1 46 32.24	38.09	1.16	3 45 42.70	41.51	1.33	5 54 11.16	43.60	0.08	20
50	1 47 10.33	38.11	1.16	3 46 24.21	41.52	1.32	5 54 54.76	43.61	0.07	10
29 0	1 47 48.44	38.11	1.17	3 47 5.73	41.54	1.32	5 55 38.37	43.60	0.06	1 0
10	1 48 26.55	38.14	1.17	3 47 47.27	41.56	1.32	5 56 21.97	43.61	0.05	50
20	1 49 4.69	38.15	1.18	3 48 28.83	41.57	1.31	5 57 5.58	43.60	0.04	40
30	1 49 42.84	38.17	1.18	3 49 10.40	41.60	1.31	5 57 49.18	43.61	0.03	30
40	1 50 21.01	38.18	1.19	3 49 52.00	41.62	1.31	5 58 32.79	43.60	0.02	20
50	1 50 59.19	38.20	1.19	3 50 33.62	41.63	1.30	5 59 16.39	43.61	0.01	10
30 0	1 51 37.39		1.20	3 51 15.25		1.30	6 0 0.00		0.00	0 0
⊙'s Long. Argument.	h. 12− h. 24− Vs XIs	Diff.	Variation.	h. 12− h. 24− IVs Xs	Diff.	Variation.	h. 12− h. 24− IIIs IXs	Diff.	Variation.	⊙'s Long. Argument.

The obliquity of the Ecliptic is taken =23° 27′ 57″, and the Variation of Right Ascension corresponds to 100″ of diminution in the Obliquity.

TABLE 6.

The Sun's Declination to every 10′ of his Longitude, with the Differences, and Variations for 100″ change of the Obliquity, by Mr. J. Utting.

Argument. ⊙'s Long.	O North. VI South. Declination.	Difference.	Variation.	I North. VII South. Declination.	Difference.	Variation.	II North. VIII South. Declination.	Difference.	Variation.	Argument. ⊙'s Long.
0° 0′	0° 0′ 0″.00	238″.92	0″.00	11° 29′ 3″.83	210″.99	46″.81	20° 10′ 22″.19	126″.98	84″.64	30° 0′
10	0 3 58.92	238.92	0.27	11 32 34.82	210.66	47.06	20 12 29.17	126.34	84.80	50
20	0 7 57.84	238.92	0.53	11 36 5.48	210.36	47.30	20 14 35.51	125.74	84.96	40
30	0 11 56.76	238.91	0.80	11 39 35.84	210.04	47.55	20 16 41.25	125.10	85.12	30
40	0 15 55.67	238.90	1.07	11 43 5.88	209.72	47.79	20 18 46.35	124.51	85.28	20
50	0 19 54.57	238.90	1.33	11 46 35.60	209.41	48.05	20 20 50.86	123.88	85.44	10
1 0	0 23 53.47	238.88	1.60	11 50 5.01	209.08	48.29	20 22 54.74	123.24	85.60	29 0
10	0 27 52.35	238.88	1.87	11 53 34.09	208.76	48.53	20 24 57.98	122.64	85.76	50
20	0 31 51.23	238.85	2.13	11 57 2.85	208.43	48.77	20 27 0.62	122.01	85.91	40
30	0 35 50.08	238.85	2.40	12 0 31.28	208.11	49.01	20 29 2.63	121.41	86.07	30
40	0 39 48.93	238.83	2.67	12 3 59.39	207.78	49.25	20 31 4.04	120.75	86.23	20
50	0 43 47.76	238.81	2.93	12 7 27.17	207.45	49.49	20 33 4.79	120.12	86.38	10
2 0	0 47 46.57	238.78	3.20	12 10 54.62	207.12	49.73	20 35 4.91	119.50	86.54	28 0
10	0 51 45.35	238.77	3.47	12 14 21.74	206.78	49.98	20 37 4.41	118.86	86.69	50
20	0 55 44.12	238.74	3.73	12 17 48.52	206.45	50.22	20 39 3.27	118.25	86.84	40
30	0 59 42.86	238.72	4.00	12 21 14.97	206.12	50.46	20 41 1.52	117.61	86.99	30
40	1 3 41.58	238.69	4.27	12 24 41.09	205.76	50.70	20 42 59.13	116.98	87.14	20
50	1 7 40.27	238.66	4.53	12 28 6.85	205.44	50.94	20 44 56.11	116.33	87.29	10
3 0	1 11 38.93	238.63	4.80	12 31 32.29	205.09	51.18	20 46 52.44	115.70	87.44	27 0
10	1 15 37.56	238.60	5.07	12 34 57.38	204.75	51.42	20 48 48.14	115.04	87.59	50
20	1 19 36.16	238.56	5.33	12 38 22.13	204.40	51.66	20 50 43.18	114.43	87.73	40
30	1 23 34.72	238.53	5.60	12 41 46.53	204.05	51.90	20 52 37.61	113.77	87.88	30
40	1 27 33.25	238.49	5.87	12 45 10.58	203.70	52.14	20 54 31.38	113.15	88.02	20
50	1 31 31.74	238.45	6.13	12 48 34.28	203.36	52.38	20 56 24.53	112.47	88.17	10
4 0	1 35 30.19	238.41	6.40	12 51 57.64	202.99	52.62	20 58 17.00	111.85	88.31	26 0
10	1 39 28.60	238.37	6.67	12 55 20.63	202.66	52.86	21 0 8.85	111.19	88.45	50
20	1 43 26.97	238.32	6.93	12 58 43.29	202.27	53.10	21 2 0.04	110.54	88.59	40
30	1 47 25.29	238.28	7.20	13 2 5.56	201.93	53.34	21 3 50.58	109.91	88.73	30
40	1 51 23.57	238.23	7.47	13 5 27.49	201.58	53.58	21 5 40.49	109.25	88.87	20
50	1 55 21.80	238.18	7.73	13 8 49.07	201.21	53.82	21 7 29.74	108.59	89.01	10
5 0	1 59 19.98	238.13	8.00	13 12 10.28	200.83	54.06	21 9 18.33	107.95	89.15	25 0
10	2 3 18.11	238.08	8.27	13 15 31.11	200.50	54.30	21 11 6.28	107.29	89.29	50
20	2 7 16.19	238.02	8.53	13 18 51.61	200.10	54.53	21 12 53.57	106.62	89.42	40
30	2 11 14.21	237.97	8.80	13 22 11.71	199.74	54.77	21 14 40.19	105.98	89.56	30
40	2 15 12.18	237.91	9.07	13 25 31.45	199.38	55.00	21 16 26.17	105.31	89.69	20
50	2 19 10.09	237.85	9.33	13 28 50.83	199.00	55.24	21 18 11.48	104.67	89.83	10
6 0	2 23 7.94	287.78	9.60	13 32 9.83	198.63	55.47	21 19 56.15	103.98	89.97	24 0
10	2 27 5.72	237.73	9.87	13 35 28.46	198.24	55.70	21 21 40.13	103.33	90.10	50
20	2 31 3.45	237.66	10.13	13 38 46.70	197.88	55.94	21 23 23.46	102.67	90.24	40
30	2 35 1.11	237.59	10.40	13 42 4.58	197.49	56.18	21 25 6.13	102.00	90.37	30
40	2 38 58.70	237.53	10.67	13 45 22.07	197.12	56.41	21 26 48.13	101.35	90.50	20
50	2 42 56.23	237.45	10.93	13 48 39.19	196.73	56.65	21 28 29.48	100.67	90.63	10
7 0	2 46 53.68	237.39	11.19	13 51 55.92	196.34	56.87	21 30 10.15	100.02	90.76	23 0
10	2 50 51.07	237.31	11.45	13 55 12.26	195.96	57.10	21 31 50.17	99.32	90.89	50
20	2 54 48.38	237.23	11.72	13 58 28.22	195.57	57.34	21 33 29.49	98.66	91.02	40
30	2 58 45.61	237.17	11.98	14 1 43.79	195.19	57.57	21 35 8.15	98.00	91.15	30
40	3 2 42.77	237.08	12.25	14 4 58.98	194.79	57.80	21 36 46.15	97.30	91.27	20
50	3 6 39.85	237.00	12.51	14 8 13.77	194.39	58.03	21 38 23.45	96.66	91.40	10
8 0	3 10 36.85	236.92	12.78	14 11 28.16	194.00	58.26	21 40 0.11	95.95	91.53	22 0
10	3 14 33.77	236.84	13.04	14 14 42.16	193.60	58.49	21 41 36.06	95.30	91.66	50
20	3 18 30.61	236.75	13.31	14 17 55.76	193.22	58.72	21 43 11.36	94.62	91.78	40
30	3 22 27.36	236.66	13.58	14 21 8.98	192.79	58.95	21 44 45.98	93.93	91.90	30
40	3 26 24.02	236.58	13.84	14 24 21.77	192.40	59.17	21 46 19.91	93.26	92.02	20
50	3 30 20.60	236.48	14.11	14 27 34.17	192.00	59.40	21 47 53.17	92.56	92.14	10
9 0	3 34 17.08	236.40	14.38	14 30 46.17	191.60	59.63	21 49 25.73	91.90	92.26	21 0
10	3 38 13.48	236.29	14.64	14 33 57.77	191.19	59.86	21 50 57.63	91.20	92.38	50
20	3 42 9.77	236.21	14.91	14 37 8.96	190.77	60.09	21 52 28.83	90.52	92.50	40
30	3 46 5.98	236.10	15.17	14 40 19.73	190.37	60.32	21 53 59.35	89.84	92.62	30
40	3 50 2.08	236.01	15.44	14 43 30.10	189.95	60.54	21 55 29.19	89.15	92.74	20
50	3 53 58.09	235.91	15.70	14 46 40.05	189.54	60.77	21 56 58.34	88.47	92.85	10
10 0	3 57 54.00		15.97	14 49 49.59		61.00	21 58 26.81		92.97	20 0
	V North. XI South.			IV North. X South.			III North. IX South.			

The obliquity of the Ecliptic is taken =23° 27′ 57″.

TABLE 6.

The Sun's Declination to every 10′ of his Longitude, &c. continued.

Argument. ☉'s Long.	O^s North. VI^s South. Declination.	Difference.	Variation.	I^s North. VII^s South. Declination.	Difference.	Variation.	II^s North. VIII^s South. Declination.	Difference.	Variation.	Argument. ☉'s Long.
10° 0′	3° 57′ 54″.00	235″.91	15″.97	14° 49′ 49″.59	189″.54	61″.00	21° 58′ 26″.81	88″.47	92″.97	20° 0′
10	4 1 49.80	235.80	16.23	14 52 58.71	189.12	61.23	21 59 54.57	87.76	93.08	50
20	4 5 45.50	235.70	16.50	14 56 7.42	188.71	61.45	22 1 21.65	87.08	93.20	40
30	4 9 41.10	235.60	16.76	14 59 15.71	188.29	61.68	22 2 48.06	86.41	93.31	30
40	4 13 36.59	235.49	17.03	15 2 23.56	187.85	61.90	22 4 13.73	85.67	93.42	20
50	4 17 31.96	235.37	17.29	15 5 31.01	187.45	62.13	22 5 38.75	85.02	93.53	10
11 0	4 21 27.23	235.27	17.56	15 8 38.03	187.02	62.35	22 7 3.08	84.33	93.64	19 0
10	4 25 22.39	235.16	17.82	15 11 44.61	186.58	62.58	22 8 26.67	83.59	93.75	50
20	4 29 17.43	235.04	18.09	15 14 50.78	186.17	62.80	22 9 49.60	82.93	93.85	40
30	4 33 12.35	234.92	18.35	15 17 56.52	185.74	63.02	22 11 11.83	82.23	93.96	30
40	4 37 7.16	234.81	18.61	15 21 1.81	185.29	63.25	22 12 33.33	81.50	94.06	20
50	4 41 1.85	234.69	18.88	15 24 6.66	184.85	63.47	22 13 54.17	80.84	94.17	10
12 0	4 44 56.41	234.56	19.14	15 27 11.11	184.45	63.69	22 15 14.29	80.12	94.27	18 0
10	4 48 50.86	234.45	19.40	15 30 15.09	183.98	63.91	22 16 33.73	79.44	94.37	50
20	4 52 45.18	234.32	19.67	15 33 18.64	183.55	64.13	22 17 52.44	78.71	94.47	40
30	4 56 39.37	234.19	19.93	15 36 21.76	183.12	64.35	22 19 10.45	78.01	94.57	30
40	5 0 33.43	234.06	20.19	15 39 24.44	182.68	64.57	22 20 27.76	77.31	94.67	20
50	5 4 27.37	233.94	20.46	15 42 26.67	182.23	64.79	22 21 44.37	76.61	94.77	10
13 0	5 8 21.17	233.80	20.72	15 45 28.45	181.78	65.01	22 23 0.25	75.88	94.87	17 0
10	5 12 14.84	233.67	20.98	15 48 29.78	181.33	65.23	22 24 15.45	75.20	94.97	50
20	5 16 8.38	233.54	21.25	15 51 30.68	180.90	65.45	22 25 29.94	74.49	95.06	40
30	5 20 1.78	233.40	21.51	15 54 31.11	180.43	65.67	22 26 43.73	73.79	95.16	30
40	5 23 55.04	233.26	21.77	15 57 31.09	179.98	65.89	22 27 56.78	73.05	95.26	20
50	5 27 48.16	233.12	22.03	16 0 30.63	179.54	66.10	22 29 9.14	72.36	95.36	10
14 0	5 31 41.14	232.98	22.30	16 3 29.70	179.07	66.32	22 30 20.76	71.62	95.45	16 0
10	5 35 33.99	232.85	22.56	16 6 28.32	178.62	66.54	22 31 31.71	70.95	95.54	50
20	5 39 26.66	232.67	22.83	16 9 26.47	178.15	66.75	22 32 41.92	70.21	95.64	40
30	5 43 19.20	232.54	23.09	16 12 24.15	177.68	66.97	22 33 51.41	69.49	95.73	30
40	5 47 11.60	232.40	23.36	16 15 21.39	177.24	67.18	22 35 0.20	68.79	95.82	20
50	5 51 3.85	232.25	23.62	16 18 18.15	176.76	67.40	22 36 8.26	68.06	95.91	10
15 0	5 54 55.94	232.09	23.88	16 21 14.46	176.31	67.61	22 37 15.62	67.36	96.00	15 0
10	5 58 47.88	231.94	24.15	16 24 10.28	175.82	67.83	22 38 22.24	66.62	96.09	50
20	6 2 39.66	231.78	24.41	16 27 5.66	175.38	68.04	22 39 28.16	65.92	96.17	40
30	6 6 31.29	231.63	24.67	16 30 0.55	174.89	68.25	22 40 33.35	65.19	96.26	30
40	6 10 22.75	231.46	24.93	16 32 54.96	174.41	68.46	22 41 37.82	64.47	96.35	20
50	6 14 14.06	231.31	25.19	16 35 48.90	173.94	68.67	22 42 41.58	63.76	96.43	10
16 0	6 18 5.20	231.14	25.45	16 38 42.37	173.47	68.88	22 43 44.60	63.02	96.52	14 0
10	6 21 56.18	230.98	25.71	16 41 35.36	172.99	69.09	22 44 46.90	62.30	96.60	50
20	6 25 46.99	230.81	25.97	16 44 27.87	172.51	69.30	22 45 48.50	61.60	96.68	40
30	6 29 37.64	230.65	26.23	16 47 19.89	172.02	69.51	22 46 49.34	60.84	96.76	30
40	6 33 28.12	230.48	26.49	16 50 11.44	171.55	69.72	22 47 49.48	60.14	96.84	20
50	6 37 18.43	230.31	26.75	16 53 2.49	171.05	69.93	22 48 48.88	59.40	96.92	10
17 0	6 41 8.56	230.13	27.01	16 55 53.07	170.58	70.14	22 49 47.56	58.68	97.00	13 0
10	6 44 58.52	229.96	27.27	16 58 43.14	170.07	70.35	22 50 45.52	57.96	97.08	50
20	6 48 48.30	229.78	27.53	17 1 32.74	169.60	70.55	22 51 42.72	57.20	97.15	40
30	6 52 37.90	229.60	27.79	17 4 21.84	169.10	70.76	22 52 39.22	56.50	97.22	30
40	6 56 27.33	229.43	28.05	17 7 10.44	168.60	70.97	22 53 35.00	55.78	97.30	20
50	7 0 16.57	229.24	28.31	17 9 58.54	168.10	71.17	22 54 30.00	55.00	97.37	10
18 0	7 4 5.63	229.06	28.57	17 12 46.18	167.64	71.38	22 55 24.32	54.32	97.44	12 0
10	7 7 54.51	228.88	28.83	17 15 33.28	167.10	71.59	22 56 17.88	53.56	97.51	50
20	7 11 43.19	228.68	29.09	17 18 19.91	166.63	71.79	22 57 10.72	52.84	97.57	40
30	7 15 31.69	228.50	29.35	17 21 6.01	166.10	72.00	22 58 2.82	52.10	97.64	30
40	7 19 20.00	228.31	29.61	17 23 51.63	165.62	72.20	22 58 54.18	51.36	97.70	20
50	7 23 8.12	228.12	29.86	17 26 36.73	165.10	72.40	22 59 44.82	50.64	97.77	10
19 0	7 26 56.04	227.92	30.12	17 29 21.33	164.60	72.60	23 0 34.72	49.90	97.84	11 0
10	7 30 43.78	227.74	30.38	17 32 5.40	164.07	72.80	23 1 23.88	49.16	97.90	50
20	7 34 31.31	227.53	30.64	17 34 49.00	163.60	73.00	23 2 12.31	48.43	97.97	40
30	7 38 18.64	227.33	30.90	17 37 32.06	163.06	73.20	23 2 59.98	47.67	98.03	30
40	7 42 5.78	227.14	31.15	17 40 14.61	162.55	73.40	23 3 46.94	46.96	98.09	20
50	7 45 52.71	226.93	31.41	17 42 56.65	162.04	73.60	23 4 33.14	46.20	98.15	10
20 0	7 49 39.43	226.72	31.67	17 45 38.18	161.53	73.80	23 5 18.62	45.48	98.21	10 0
	V^s North. XI^s South.			IV^s North. X^s South.			III^s North. IX^s South.			

The obliquity of the Ecliptic is taken =23° 27′ 57″.

TABLE 6.

The Sun's Declination to every 10′ of his Longitude, &c. concluded.

Argument. ☉'s Long.		O North. VI South. Declination.	Difference.	Variation.	I North. VII South. Declination.	Difference.	Variation.	II North. VIII South. Declination.	Difference.	Variation.	Argument. ☉'s Long.	
			226′.72			161″.53			45″.48			
20°	0′	7° 49′ 39″.43	226 .52	31″.67	17° 45′ 38″.18	161 .00	73″.80	23° 5′ 18″.62	44 .73	98″.21	10°	0′
	10	7 53 25 .95	226 .32	31 .93	17 48 19 .18	160 .48	74 .00	23 6 3 .35	43 .99	98 .27		50
	20	7 57 12 .27	226 .10	32 .19	17 50 59 .66	159 .97	74 .19	23 6 47 .34	43 .25	98 .32		40
	30	8 0 58 .37	225 .90	32 .45	17 53 39 .63	159 .43	74 .39	23 7 30 .59	42 .51	98 .38		30
	40	8 4 44 .27	225 .68	32 .70	17 56 19 .06	158 .89	74 .58	23 8 13 .10	41 .78	98 .44		20
	50	8 8 29 .95	225 .46	32 .96	17 58 57 .95	158 .39	74 .78	23 8 54 .88	41 .00	98 .49		10
21	0	8 12 15 .41	225 .25	33 .22	18 1 36 .34	157 .86	74 .97	23 9 35 .88	40 .28	98 .55	9	0
	10	8 16 0 .66	225 .04	33 .48	18 4 14 .20	157 .33	75 .17	23 10 16 .16	39 .55	98 .60		50
	20	8 19 45 .70	224 .80	33 .73	18 6 51 .53	156 .80	75 .36	23 10 55 .71	38 .79	98 .65		40
	30	8 23 30 .50	224 .60	33 .99	18 9 28 .33	156 .25	75 .56	23 11 34 .50	38 .05	98 .70		30
	40	8 27 15 .10	224 .37	34 .24	18 12 4 .58	155 .73	75 .75	23 12 12 .55	37 .31	98 .75		20
	50	8 30 59 .47	224 .14	34 .50	18 14 40 .31	155 .17	75 .95	23 12 49 .86	36 .55	98 .80		10
22	0	8 34 43 .61	223 .92	34 .76	18 17 15 .48	154 .66	76 .14	23 13 26 .41	35 .79	98 .85	8	0
	10	8 38 27 .53	223 .69	35 .01	18 19 50 .14	154 .11	76 .33	23 14 2 .20	35 .09	98 .90		50
	20	8 42 11 .22	223 .46	35 .27	18 22 24 .25	153 .58	76 .52	23 14 37 .29	34 .30	98 .94		40
	30	8 45 54 .68	223 .22	35 .52	18 24 57 .83	153 .01	76 .71	23 15 11 .59	33 .57	98 .99		30
	40	8 49 37 .90	222 .99	35 .78	18 27 30 .84	152 .49	76 .90	23 15 45 .16	32 .82	99 .03		20
	50	8 53 20 .89	222 .76	36 .03	18 30 3 .33	151 .92	77 .09	23 16 17 .98	32 .06	99 .08		10
23	0	8 57 3 .65	222 .52	36 .29	18 32 35 .25	151 .38	77 .28	23 16 50 .04	31 .35	99 .12	7	0
	10	9 0 46 .17	222 .29	36 .54	18 35 6 .63	150 .85	77 .47	23 17 21 .39	30 .57	99 .16		50
	20	9 4 28 .46	222 .04	36 .80	18 37 37 .48	150 .28	77 .65	23 17 51 .96	29 .82	99 .20		40
	30	9 8 10 .50	221 .79	37 .05	18 40 7 .76	149 .74	77 .84	23 18 21 .78	29 .08	99 .24		30
	40	9 11 52 .29	221 .57	37 .31	18 42 37 .50	149 .18	78 .02	23 18 50 .86	28 .30	99 .27		20
	50	9 15 33 .86	221 .30	37 .56	18 45 6 .68	148 .61	78 .21	23 19 19 .16	27 .57	99 .31		10
24	0	9 19 15 .16	221 .06	37 .82	18 47 35 .29	148 .06	78 .40	23 19 46 .73	26 .84	99 .35	6	0
	10	9 22 56 .22	220 .81	38 .07	18 50 3 .35	147 .50	78 .59	23 20 13 .57	26 .06	99 .39		50
	20	9 26 37 .03	220 .57	38 .33	18 52 30 .85	146 .95	78 .77	23 20 39 .63	25 .33	99 .42		40
	30	9 30 17 .60	220 .30	38 .58	18 54 57 .80	146 .39	78 .96	23 21 4 .96	24 .57	99 .45		30
	40	9 33 57 .00	220 .05	38 .84	18 57 24 .19	145 .81	79 .14	23 21 29 .53	23 .80	99 .49		20
	50	9 37 37 .95	219 .80	39 .09	18 59 50 .00	145 .26	79 .33	23 21 53 .33	23 .06	99 .52		10
25	0	9 41 17 .75	219 .54	39 .34	19 2 15 .26	144 .69	79 .51	23 22 16 .39	22 .32	99 .55	5	0
	10	9 44 57 .29	219 .29	39 .59	19 4 39 .95	144 .12	79 .69	23 22 38 .71	21 .56	99 .58		50
	20	9 48 36 .58	219 .02	39 .85	19 7 4 .07	143 .53	79 .87	23 23 0 .27	20 .77	99 .61		40
	30	9 52 15 .60	218 .75	40 .10	19 9 27 .60	142 .98	80 .05	23 23 21 .04	20 .06	99 .64		30
	40	9 55 54 .35	218 .49	40 .35	19 11 50 .58	142 .40	80 .23	23 23 41 .10	19 .30	99 .67		20
	50	9 59 32 .84	218 .23	40 .60	19 14 12 .98	141 .82	80 .41	23 24 0 .40	18 .54	99 .70		10
26	0	10 3 11 .07	217 .95	40 .85	19 16 34 .80	141 .27	80 .59	33 24 18 .94	17 .79	99 .72	4	0
	10	10 6 49 .02	217 .69	41 .10	19 18 56 .07	140 .66	80 .77	23 24 36 .73	17 .03	99 .74		50
	20	10 10 26 .71	217 .40	41 .35	19 21 16 .73	140 .09	80 .94	23 24 53 .76	16 .26	99 .77		40
	30	10 14 4 .11	217 .15	41 .60	19 23 36 .82	139 .51	81 .12	23 25 10 .02	15 .51	99 .79		30
	40	10 17 41 .26	216 .85	41 .85	19 25 56 .33	138 .93	81 .29	23 25 25 .53	14 .78	99 .81		20
	50	10 21 18 .11	216 .59	42 .10	19 28 15 .26	138 .35	81 .47	23 25 40 .31	14 .00	99 .83		10
27	0	10 24 54 .70	216 .31	42 .35	19 30 33 .61	137 .75	81 .64	23 25 54 .31	13 .25	99 .85	3	0
	10	10 28 31 .01	216 .03	42 .60	19 32 51 .36	137 .18	81 .81	23 26 7 .56	12 .50	99 .87		50
	20	10 32 7 .04	215 .74	43 .85	19 35 8 .54	136 .58	81 .98	23 26 20 .06	11 .74	99 .88		40
	30	10 35 42 .78	215 .46	43 .10	19 37 25 .12	135 .98	82 .15	23 26 31 .80	10 .98	99 .90		30
	40	10 39 18 .24	215 .17	43 .35	19 39 41 .10	135 .41	82 .32	23 26 42 .78	10 .22	99 .91		20
	50	10 42 53 .41	214 .89	43 .60	19 41 56 .51	134 .81	83 .49	23 26 53 .00	9 .47	99 .93		10
28	0	10 46 28 .30	214 .59	43 .85	19 44 11 .32	134 .21	82 .66	23 27 2 .47	8 .69	99 .94	2	0
	10	10 50 2 .89	214 .31	44 .10	19 46 25 .53	133 .62	82 .83	23 27 11 .16	7 .96	99 .95		50
	20	10 53 37 .20	214 .01	44 .35	19 48 39 .15	133 .02	83 .00	23 27 19 .12	7 .21	99 .95		40
	30	10 57 11 .21	213 .72	44 .60	19 50 52 .17	132 .42	83 .16	23 27 26 .33	6 .43	99 .96		30
	40	11 0 44 .93	213 .41	44 .85	19 53 4 .59	131 .81	83 .33	23 27 32 .76	5 .69	99 .97		20
	50	11 4 18 .34	213 .13	45 .10	19 55 16 .40	131 .22	83 .50	23 27 38 .45	4 .90	99 .98		10
29	0	11 7 51 .47	212 .81	45 .34	19 57 27 .62	130 .62	83 .66	23 27 43 .35	4 .19	99 .98	1	0
	10	11 11 24 .28	212 .52	45 .59	19 59 38 .24	130 .02	83 .83	23 27 47 .54	3 .40	99 .99		50
	20	11 14 56 .80	212 .22	45 .83	20 1 48 .26	129 .38	83 .99	23 27 50 .94	2 .66	99 .99		40
	30	11 18 29 .02	211 .91	46 .08	20 3 57 .64	128 .81	84 .15	23 27 53 .60	1 .90	99 .99		30
	40	11 22 0 .93	211 .61	46 .32	20 6 6 .45	128 .18	84 .32	23 27 55 .50	1 .13	99 .99		20
	50	11 25 32 .54	211 .29	46 .57	20 8 14 .63	127 .56	84 .48	23 27 56 .63	0 .37	100 .00		10
30	0	11 29 3 .83		46 .81	20 10 22 .19		84 .64	23 27 57 .00		100 .00	0	0
		V North. XI South.			IV North. X South.			III North. IX South.				

The Obliquity of the Ecliptic is taken = 23° 27′ 57″.

TABLE 7. Proportional Parts of the Sun's diurnal Changes of Right Ascension, and of Declination. Arguments = the Diurnal Change at the Top or Bottom, and the Hour Angle from Noon, or the Longitude, as the case may be, on the Side.

Arg. Hr. Angle	1	2	3	4	5	6	7	8	9	10	11	12	13	14	15	16	17	18	19	20	Arg. Long.
h. m.																					
0 4	.003	.006	.008	.011	.014	.017	.019	.022	.025	.028	.031	.033	.036	.039	.042	.044	.047	.050	.053	.056	1°
8	.005	.011	.017	.022	.028	.033	.039	.044	.050	.056	.062	.067	.072	.078	.083	.688	.094	.100	.106	.112	2
12	.008	.017	.025	.033	.042	.050	.058	.067	.075	.083	.092	.100	.108	.117	.125	.133	.142	.150	.158	.167	3
16	.011	.023	.033	.044	.056	.067	.077	.089	.100	.111	.123	.133	.144	.156	.167	.177	.189	.200	.211	.223	4
20	.014	.028	.042	.055	.070	.083	.097	.111	.125	.139	.153	.167	.180	.195	.208	.222	.236	.250	.264	.278	5
24	.017	.033	.050	.067	.083	.100	.117	.133	.150	.167	.184	.200	.217	.233	.250	.267	.283	.300	.317	.333	6
28	.020	.039	.058	.078	.097	.117	.136	.155	.175	.195	.214	.233	.253	.272	.292	.311	.330	.350	.370	.389	7
32	.022	.044	.067	.089	.111	.133	.156	.177	.200	.223	.245	.267	.289	.311	.333	.355	.377	.400	.423	.445	8
36	.025	.050	.075	.100	.125	.150	.175	.200	.225	.250	.275	.300	.325	.350	.375	.400	.425	.450	.475	.500	9
40	.028	.056	.083	.111	.139	.167	.194	.222	.250	.278	.306	.333	.361	.389	.417	.444	.472	.500	.528	.556	10
44	.030	.061	.092	.122	.153	.183	.214	.244	.275	.306	.337	.367	.397	.428	.458	.488	.519	.550	.581	.612	11
48	.033	.067	.100	.133	.167	.200	.233	.267	.300	.333	.367	.400	.433	.467	.500	.533	.567	.600	.633	.667	12
52	.036	.073	.108	.144	.181	.217	.252	.289	.325	.361	.398	.433	.469	.506	.542	577	.614	.650	.686	.723	13
56	.039	.078	.117	.155	.195	.233	.272	.311	.350	.389	.428	.467	.505	.545	.583	.622	.661	.700	.739	.778	14
1 0	.042	.083	.125	.167	.208	.250	.292	.333	.375	.417	.459	.500	.542	.583	.625	.667	.708	.750	.792	.833	15
4	.045	.089	.133	.178	.222	.267	.311	.355	.400	.445	.489	.533	.578	.622	.667	.711	.755	.800	.845	.889	16
8	.047	.090	.142	.189	.236	.283	.331	.377	.425	.473	.520	.567	.614	.661	.708	.755	.802	.850	.898	.945	17
12	.050	.100	.150	.200	.250	.300	.350	.400	.450	.500	.550	.600	.650	.700	.750	.800	.850	.900	.950	1.000	18
16	.053	.106	.158	.211	.264	.317	.369	.422	.475	.528	.581	.633	.686	.739	.792	.844	.897	.950	1.003	1.056	19
20	.055	.111	.167	.222	.278	.333	.389	.444	.500	.556	.612	.667	.722	.778	.833	.888	.944	1.000	1.056	1.012	20
24	.058	.117	.175	.233	.292	.350	.408	.467	.525	.583	.642	.700	.758	.817	.875	.933	.992	1.050	1.108	1.167	21
28	.061	.123	.183	.244	.306	.367	.427	.489	.550	.611	.673	.733	.794	.856	.917	.977	1.039	1.100	1.161	1.223	22
32	.064	.128	.192	.255	.320	.383	.447	.511	.575	.639	.703	.767	.830	.895	.958	1.022	1.086	1.150	1.214	1.278	23
36	.067	.133	.200	.267	.333	.400	.467	.533	.600	.667	.733	.800	.867	.933	1.000	1.067	1.133	1.200	1.267	1.333	24
40	.070	.139	.208	.278	.347	.417	.486	.555	.625	.695	.764	.833	.903	.972	1.042	1.111	1.180	1.250	1.320	1.389	25
44	.072	.144	.217	.289	.361	.433	.506	.577	.650	.723	.794	.867	.939	1.011	1.083	1.155	1.227	1.300	1.373	1.445	26
48	.075	.150	.225	.300	.375	.450	.525	.600	.675	.750	.825	.900	.975	1.050	1.125	1.200	1.275	1.350	1.425	1.500	27
52	.078	.156	.233	.311	.389	.467	.544	.622	.700	.778	.856	.933	1.011	1.089	1.167	1.244	1.322	1.400	1.478	1.556	28
56	.080	.161	.242	.322	.403	.483	.564	.644	.725	.806	.886	.967	1.047	1.128	1.208	1.288	1.369	1.450	1.531	1.612	29
2 0	.083	.167	.250	.333	.417	.500	.583	.667	.750	.833	.917	1.000	1.083	1.167	1.250	1.333	1.417	1.500	1.583	1.667	30
4	.086	.173	.258	.344	.431	.517	.602	.689	.775	.861	.948	1.033	1.119	1.206	1.292	1.377	1.464	1.550	1.636	1.723	31
8	.089	.178	.267	.355	.445	.533	.622	.711	.800	.889	.978	1.067	1.155	1.245	1.333	1.422	1.511	1.600	1.689	1.778	32
12	.092	.183	.275	.367	.458	.550	.642	.733	.825	.917	1.009	1.100	1.192	1.283	1.375	1.467	1.558	1.650	1.742	1.833	33
16	.095	.189	.283	.378	.472	.567	.661	.755	.850	.945	1.039	1.133	1.228	1.322	1.417	1.511	1.605	1.700	1.795	1.889	34
20	.097	.194	.292	.389	.486	.583	.681	.777	.875	.973	1.070	1.167	1.264	1.361	1.458	1.555	1.652	1.750	1.848	1.945	35
24	.100	.200	.300	.400	.500	.600	.700	.800	.900	1.000	1.100	1.200	1.300	1.400	1.500	1.600	1.700	1.800	1.900	2.000	36
28	.103	.206	.308	.411	.514	.617	.719	.822	.925	1.028	1.131	1.233	1.336	1.439	1.542	1.644	1.747	1.850	1.953	2.056	37
32	.105	.211	.317	.422	.528	.633	.739	.844	.950	1.056	1.162	1.267	1.372	1.478	1.583	1.688	1.794	1.900	2.006	2.112	38
36	.108	.217	.325	.433	.542	.650	.758	.867	.975	1.083	1.192	1.300	1.408	1.517	1.625	1.733	1.842	1.950	2.058	2.167	39
40	.111	.223	.333	.444	.556	.667	.777	.889	1.000	1.111	1.223	1.333	1.444	1.556	1.667	1.777	1.889	2.000	2.111	2.223	40
44	.114	.228	.342	.455	.570	.683	.797	.911	1.025	1.139	1.253	1.367	1.480	1.595	1.708	1.822	1.936	2.050	2.164	2.278	41
48	.117	.233	.350	.467	.583	.700	.817	.933	1.050	1.167	1.283	1.400	1.517	1.633	1.750	1.867	1.983	2.100	2.217	2.333	42
52	.120	.239	.358	.478	.597	.717	.836	.955	1.075	1.195	1.314	1.433	1.553	1.672	1.792	1.911	2.030	2.150	2.270	2.389	43
56	.122	.244	.367	.489	.611	.733	.856	.977	1.100	1.223	1.345	1.467	1.589	1.711	1.833	1.955	2.077	2.200	2.323	2.445	44
3 0	.125	.250	.375	.500	.625	.750	.875	1.000	1.125	1.250	1.375	1.500	1.625	1.750	1.875	2.000	2.125	2.250	2.375	2.500	45
4	.128	.256	.383	.511	.639	.767	.894	1.022	1.150	1.278	1.406	1.533	1.661	1.789	1.917	2.044	2.172	2.300	2.428	2.556	46
8	.130	.261	.392	.522	.653	.783	.914	1.044	1.175	1.306	1.437	1.567	1.697	1.828	1.958	2.088	2.219	2.350	2.481	2.612	47
12	.133	.267	.400	.533	.667	.800	.933	1.067	1.200	1.333	1.467	1.600	1.733	1.867	2.000	2.133	2.267	2.400	2.533	2.667	48
16	.136	.273	.408	.544	.681	.817	.952	1.089	1.225	1.361	1.498	1.633	1.769	1.906	2.042	2.177	2.314	2.450	2.586	2.723	49
20	.139	.278	.417	.555	.695	.833	.972	1.111	1.250	1.389	1.528	1.667	1.805	1.945	2.083	2.222	2.361	2.500	2.639	2.778	50
24	.142	.283	.425	.567	.708	.850	.992	1.133	1.275	1.417	1.559	1.700	1.842	1.983	2.125	2.267	2.408	2.550	2.692	2.833	51
28	.145	.289	.433	.578	.722	.867	1.011	1.155	1.300	1.445	1.589	1.733	1.878	2.022	2.167	2.311	2.455	2.600	2.745	2.889	52
32	.147	.294	.442	.589	.736	.883	1.031	1.177	1.325	1.473	1.620	1.767	1.914	2.061	2.208	2.355	2.502	2.650	2.798	2.945	53
36	.150	.300	.450	.600	.750	.900	1.050	1.200	1.350	1.500	1.650	1.800	1.950	2.100	2.250	2.400	2.550	2.700	2.850	3.000	54
40	.153	.306	.458	.611	.764	.917	1.069	1.222	1.375	1.528	1.681	1.833	1.986	2.139	2.292	2.444	2.597	2.750	2.903	3.056	55
44	.155	.311	.467	.622	.778	.933	1.089	1.244	1.400	1.556	1.712	1.867	2.022	2.278	2.333	2.488	2.644	2.800	2.956	3.112	56
48	.158	.317	.475	.633	.792	.950	1.108	1.267	1.425	1.583	1.742	1.900	2.058	2.217	2.375	2.533	2.692	2.850	3.008	3.167	57
52	.161	.323	.483	.644	.806	.967	1.127	1.289	1.450	1.611	1.773	1.933	2.094	2.256	2.417	2.577	2.739	2.900	3.061	3.223	58
56	.164	.328	.492	.655	.820	.983	1.147	1.311	1.475	1.639	1.803	1.967	2.130	2.295	2.458	2.622	2.786	2.950	3.114	3.278	59
4 0	.167	.333	.500	.667	.833	1.000	1.167	1.333	1.500	1.667	1.833	2.000	2.167	2.333	2.500	2.667	2.833	3.000	3.167	3.333	60
Hour Angle from Noon.	.1	.2	.3	.4	.5	.6	.7	.8	.9	.10	.11	.12	.13	.14	.15	.16	.17	.18	.19	.20	Long. from Greenwich.

This Table is useful for adapting the Right Ascension or Declination of the Sun, as given in the Nautical Almanac, for Noon at Greenwich, to any other Time, either in the same or in any different Longitude.

TABLE 7. Proportional Parts of the Sun's diurnal Changes, &c. continued.

Arg. Hr.Angle h. m.	1	2	3	4	5	6	7	8	9	10	11	12	13	14	15	16	17	18	19	20	Arg. Long.
4 0	.167	.333	.500	.667	0.833	1.000	1.167	1.333	1.500	1.667	1.833	2.000	2.167	2.333	2.500	2.667	2.833	3.000	3.167	3.333	60°
4	.170	.339	.508	.678	0.847	1.017	1.186	1.355	1.525	1.695	1.864	2.033	2.203	2.372	2.542	2.711	2.880	3.050	3.220	3.389	61
8	.173	.344	.517	.689	0.861	1.033	1.205	1.377	1.550	1.723	1.895	2.067	2.249	2.411	2.583	2.755	2.927	3.100	3.273	3.445	62
12	.175	.350	.525	.700	0.875	1.050	1.225	1.400	1.575	1.750	1.925	2.100	2.275	2.450	2.625	2.800	2.975	3.150	3.325	3.500	63
16	.178	.356	.533	.711	0.889	1.067	1.244	1.422	1.600	1.778	1.956	2.133	2.311	2.489	2.667	2.844	3.022	3.200	3.378	3.556	64
20	.181	.361	.542	.722	0.903	1.083	1.263	1.444	1.625	1.806	1.987	2.167	2.347	2.528	2.708	2.888	3.069	3.250	3.431	3.612	65
24	.183	.367	.550	.733	0.917	1.100	1.283	1.467	1.650	1.833	2.017	2.200	2.383	2.567	2.750	2.933	3.117	3.300	3.483	3.667	66
28	.186	.373	.558	.744	0.931	1.117	1.302	1.489	1.675	1.861	2.048	2.233	2.419	2.606	2.792	2.977	3.164	3.350	3.536	3.723	67
32	.189	.378	.567	.755	0.945	1.133	1.322	1.511	1.700	1.889	2.078	2.267	2.455	2.645	2.833	3.022	3.211	3.400	3.589	3.778	68
36	.192	.383	.575	.767	0.958	1.150	1.342	1.533	1.725	1.917	2.109	2.300	2.492	2.683	2.875	3.067	3.258	3.450	3.642	3.833	69
40	.195	.389	.583	.778	0.972	1.167	1.361	1.555	1.750	1.945	2.140	2.333	2.528	2.722	2.917	3.111	3.305	3.500	3.695	3.889	70
44	.198	.394	.592	.789	0.986	1.183	1.380	1.577	1.775	1.973	2.170	2.367	2.564	2.761	2.958	3.155	8.352	3.550	3.748	3.945	71
48	.200	.400	.600	.800	1.000	1.200	1.400	1.600	1.800	2.000	2.200	2.400	2.600	2.800	3.000	3.200	3.400	3.600	3.800	4.000	72
52	.203	.406	.608	.811	1.014	1.217	1.419	1.622	1.825	2.028	2.231	2.433	2.636	2.839	3.042	3.244	3.447	3.650	3.853	4.056	73
56	.206	.411	.617	.822	1.028	1.233	1.438	1.644	1.850	2.056	2.262	2.467	2.672	2.878	3.083	3.288	3.494	3.700	3.906	4.112	74
5 0	.208	.417	.625	.833	1.042	1.250	1.458	1.667	1.875	2.083	2.292	2.500	2.708	2.917	3.125	3.333	3.542	3.750	3.958	4.167	75
4	.211	.423	.633	.844	1.056	1.267	1.477	1.689	1.900	2.111	2.323	2.533	2.744	2.956	3.167	3.377	3.589	3.800	4.011	4.223	76
8	.214	.428	.642	.855	1.070	1.283	1.497	1.711	1.925	2.139	2.353	2.567	2.780	2.995	3.208	3.422	3.636	3.850	4.064	4.278	77
12	.217	.433	.650	.867	1.083	1.300	1.517	1.733	1.950	2.167	2.384	2.600	2.817	3.033	3.250	3.467	3.683	3.900	4.117	4.333	78
16	.220	.439	.658	.878	1.097	1.317	1.536	1.755	1.975	2.195	2.414	2.633	2.853	3.072	3.292	3.511	3.730	3.950	4.170	4.389	79
20	.223	.444	.667	.889	1.111	1.333	1.555	1.777	2.000	2.223	2.445	2.667	2.889	3.111	3.333	3.555	3.777	4.000	4.223	4.445	80
24	.225	.450	.675	.900	1.125	1.350	1.575	1.800	2.025	2.250	2.475	2.700	2.925	3.150	3.375	3.600	3.825	4.050	4.275	4.500	81
28	.228	.456	.683	.911	1.139	1.367	1.594	1.822	2.050	2.278	2.506	2.733	2.961	3.189	3.417	3.644	3.872	4.100	4.328	4.556	82
32	.231	.461	.692	.922	1.153	1.383	1.613	1.844	2.075	2.306	2.537	2.767	2.997	3.228	3.458	3.688	3.919	4.150	4.381	4.612	83
36	.233	.467	.700	.933	1.167	1.400	1.633	1.867	2.100	2.333	2.567	2.800	3.033	3.267	3.500	3.733	3.967	4.200	4.433	4.667	84
40	.236	.473	.708	.944	1.181	1.417	1.652	1.889	2.125	2.361	2.598	2.833	3.069	3.306	3.542	3.777	4.014	4.250	4.486	4.723	85
44	.239	.478	.717	.955	1.195	1.433	1.672	1.911	2.150	2.389	2.628	2.867	3.105	3.345	3.583	3.822	4.061	4.300	4.539	4.778	86
48	.242	.483	.725	.967	1.208	1.450	1.692	1.933	2.175	2.417	2.659	2.900	3.142	3.383	3.625	3.867	4.108	4.350	4.592	4.833	87
52	.245	.489	.733	.978	1.222	1.467	1.711	1.955	2.200	2.445	2.689	2.933	3.178	3.422	3.667	3.911	4.155	4.400	4.645	4.889	88
56	.248	.494	.742	.989	1.236	1.483	1.730	1.977	2.225	2.473	2.720	2.967	3.214	3.461	3.708	3.955	4.202	4.450	4.698	4.945	89
6 0	.250	.500	.750	1.000	1.250	1.500	1.750	2.000	2.250	2.500	2.750	3.000	3.250	3.500	3.750	4.000	4.250	4.500	4.750	5.000	90
4	.253	.506	.758	1.011	1.264	1.517	1.769	2.022	2.275	2.528	2.781	3.033	3.286	3.539	3.792	4.044	4.297	4.550	4.803	5.056	91
8	.256	.511	.767	1.022	1.278	1.533	1.788	2.044	2.300	2.556	2.812	3.067	3.322	3.578	3.833	4.088	4.344	4.600	4.856	5.112	92
12	.258	.517	.775	1.033	1.292	1.550	1.808	2.067	2.325	2.583	2.842	3.100	3.358	3.617	3.875	4.133	4.392	4.650	4.908	5.167	93
16	.261	.523	.783	1.044	1.306	1.567	1.827	2.089	2.350	2.611	2.873	3.133	3.394	3.656	3.917	4.177	4.439	4.700	4.961	5.223	94
20	.264	.528	.792	1.055	1.320	1.583	1.847	2.111	2.375	2.639	2.903	3.167	3.430	3.695	3.958	4.222	4.486	4.750	5.014	5.278	95
24	.267	.533	.800	1.067	1.333	1.600	1.867	2.133	2.400	2.667	2.934	3.200	3.467	3.733	4.000	4.267	4.533	4.800	5.067	5.333	96
28	.270	.539	.808	1.078	1.347	1.617	1.886	2.155	2.425	2.695	2.964	3.233	3.503	3.772	4.042	4.311	4.580	4.850	5.120	5.389	97
32	.273	.544	.817	1.089	1.361	1.633	1.905	2.177	2.450	2.723	2.995	3.267	3.539	3.811	4.083	4.355	4.627	4.900	5.173	5.445	98
36	.275	.550	.825	1.100	1.375	1.650	1.925	2.200	2.475	2.750	3.025	3.300	3.575	3.850	4.125	4.400	4.675	4.950	5.225	5.500	99
40	.278	.556	.833	1.111	1.389	1.667	1.944	2.222	2.500	2.778	3.056	3.333	3.611	3.889	4.167	4.444	4.722	5.000	5.278	5.556	100
44	.281	.561	.842	1.122	1.403	1.683	1.963	2.244	2.525	2.806	3.087	3.367	3.647	3.928	4.208	4.488	4.769	5.050	5.331	5.612	101
48	.283	.567	.850	1.133	1.417	1.700	1.983	2.267	2.550	2.833	3.117	3.400	3.683	3.967	4.250	4.533	4.817	5.100	5.383	5.667	102
52	.286	.573	.858	1.144	1.431	1.717	2.002	2.289	2.575	2.861	3.148	3.433	3.719	4.006	4.292	4.577	4.864	5.150	5.436	5.723	103
56	.289	.578	.867	1.155	1.445	1.733	2.022	2.311	2.600	2.889	3.178	3.467	3.755	4.045	4.333	4.622	4.911	5.200	5.489	5.778	104
7 0	.292	.583	.875	1.167	1.458	1.750	2.042	2.333	2.625	2.917	3.209	3.500	3.792	4.083	4.375	4.667	4.958	5.250	5.542	5.833	105
4	.295	.589	.883	1.178	1.472	1.767	2.061	2.355	2.650	2.945	3.239	3.533	3.828	4.122	4.417	4.711	5.005	5.300	5.595	5.889	106
8	.298	.594	.892	1.189	1.486	1.783	2.080	2.377	2.675	2.973	3.270	3.567	3.864	4.161	4.458	4.755	5.052	5.350	5.648	5.945	107
12	.300	.600	.900	1.200	1.500	1.800	2.100	2.400	2.700	3.000	3.300	3.600	3.900	4.200	4.500	4.800	5.100	5.400	5.700	6.000	108
16	.303	.606	.908	1.211	1.514	1.817	2.119	2.422	2.725	3.028	3.331	3.633	3.936	4.239	4.542	4.844	5.147	5.450	5.753	6.056	109
20	.306	.611	.917	1.222	1.528	1.833	2.138	2.444	2.750	3.056	3.362	3.667	3.972	4.278	4.583	4.888	5.194	5.500	5.806	6.112	110
24	.308	.617	.925	1.233	1.542	1.850	2.158	2.467	2.775	3.083	3.392	3.700	4.008	4.317	4.625	4.933	5.242	5.550	5.858	6.167	111
28	.311	.623	.933	1.244	1.556	1.867	2.177	2.489	2.800	3.111	3.423	3.733	4.044	4.356	4.667	4.977	5.289	5.600	5.911	6.223	112
32	.314	.628	.942	1.255	1.570	1.883	2.197	2.511	2.825	3.139	3.453	3.767	4.080	4.395	4.708	5.022	5.336	5.650	5.964	6.278	113
36	.317	.633	.950	1.267	1.583	1.900	2.217	2.533	2.850	3.167	3.484	3.800	4.117	4.433	4.750	5.067	5.383	5.700	6.017	6.333	114
40	.320	.639	.958	1.278	1.597	1.917	2.236	2.555	2.875	3.195	3.514	3.833	4.153	4.472	4.792	5.111	5.430	5.750	6.070	6.389	115
44	.323	.644	.967	1.289	1.611	1.933	2.255	2.577	2.900	3.223	3.545	3.867	4.189	4.511	4.833	5.155	5.477	5.800	6.123	6.445	116
48	.325	.650	.975	1.300	1.625	1.950	2.275	2.600	2.925	3.250	3.575	3.900	4.225	4.550	4.875	5.200	5.525	5.850	6.175	6.500	117
52	.328	.656	.983	1.311	1.639	1.967	2.294	2.622	2.950	3.278	3.606	3.933	4.261	4.589	4.917	5.244	5.572	5.900	6.228	6.556	118
56	.331	.661	.992	1.322	1.653	1.983	2.313	2.644	2.975	3.306	3.637	3.967	4.297	4.628	4.958	5.288	5.619	5.950	6.281	6.612	119
8 0	.333	.667	1.000	1.333	1.667	2.000	2.333	2.667	3.000	3.333	3.667	4.000	4.333	4.667	5.000	5.333	5.667	6.000	6.333	6.667	120
Hr.Angle from Noon.	.1	.2	.3	.4	.5	.6	.7	.8	.9	.10	.11	.12	.13	.14	.15	.16	.17	.18	.19	.20	Lon.fr. Greenwich.

TABLE 7. Proportional Parts of the Sun's diurnal Changes, &c. concluded.

Arg. Hr.Angle	1	2	3	4	5	6	7	8	9	10	11	12	13	14	15	16	17	18	19	20	Arg. Long.
h. m.																					
8 0	.333	0.667	1.000	1.333	1.667	2.000	2.333	2.667	3.000	3.333	3.667	4.000	4.333	4.667	5.000	5.333	5.667	6.000	6.333	6.667	120°
4	.336	0.673	1.008	1.344	1.681	2.017	2.352	2.689	3.025	3.361	3.698	4.033	4.369	4.706	5.042	5.377	5.714	6.050	6.386	6.723	121
8	.339	0.678	1.017	1.355	1.695	2.033	2.372	2.711	3.050	3.389	3.728	4.067	4.405	4.745	5.083	5.422	5.761	6.100	6.439	6.778	122
12	.342	0.683	1.025	1.367	1.708	2.050	2.392	2.733	3.075	3.417	3.759	4.100	4.442	4.783	5.125	5.467	5.808	6.150	6.492	6.833	123
16	.345	0.689	1.033	1.378	1.722	2.067	2.411	2.755	3.100	3.445	3.789	4.133	4.478	4.822	5.167	5.511	5.855	6.200	6.545	6.889	124
20	.348	0.694	1.042	1.389	1.736	2.083	2.430	2.777	3.125	3.473	3.820	4.167	4.514	4.861	5.208	5.555	5.902	6.250	6.598	6.945	125
24	.350	0.700	1.050	1.400	1.750	2.100	2.450	2.800	3.150	3.500	3.850	4.200	4.550	4.900	5.250	5.600	5.950	6.300	6.650	7.000	126
28	.353	0.706	1.058	1.411	1.764	2.117	2.469	2.822	3.175	3.528	3.881	4.233	4.586	4.939	5.292	5.644	5.997	6.350	6.703	7.056	127
32	.356	0.711	1.067	1.422	1.778	2.133	2.488	2.844	3.200	3.556	3.912	4.267	4.622	4.978	5.333	5.688	6.044	6.400	6.756	7.112	128
36	.358	0.717	1.075	1.433	1.792	2.150	2.508	2.867	3.225	3.583	3.942	4.300	4.658	5.017	5.375	5.733	6.092	6.450	6.808	7.167	129
40	.361	0.723	1.083	1.444	1.806	2.167	2.527	2.889	3.250	3.611	3.973	4.333	4.694	5.056	5.417	5.777	6.139	6.500	6.861	7.223	130
44	.364	0.728	1.092	1.455	1.820	2.183	2.547	2.911	3.275	3.639	4.003	4.367	4.730	5.095	5.458	5.822	6.186	6.550	6.914	7.278	131
48	.367	0.733	1.100	1.467	1.833	2.200	2.567	2.933	3.300	3.667	4.034	4.400	4.767	5.133	5.500	5.867	6.233	6.600	6.967	7.333	132
52	.370	0.739	1.108	1.478	1.847	2.217	2.586	2.955	3.325	3.695	4.064	4.433	4.803	5.172	5.542	5.911	6.280	6.650	7.020	7.389	133
56	.373	0.744	1.117	1.489	1.861	2.233	2.605	2.977	3.350	3.723	4.095	4.467	4.839	5.211	5.583	5.955	6.327	6.700	7.073	7.445	134
9 0	.375	0.750	1.125	1.500	1.875	2.250	2.625	3.000	3.375	3.750	4.125	4.500	4.875	5.250	5.625	6.000	6.375	6.750	7.125	7.500	135
4	.378	0.756	1.133	1.511	1.889	2.267	2.644	3.022	3.400	3.778	4.156	4.533	4.911	5.289	5.667	6.044	6.422	6.800	7.178	7.556	136
8	.381	0.761	1.142	1.522	1.903	2.283	2.663	3.044	3.425	3.806	4.187	4.567	4.947	5.328	5.708	6.088	6.469	6.850	7.231	7.612	137
12	.383	0.767	1.150	1.533	1.917	2.300	2.683	3.067	3.450	3.833	4.217	4.600	4.983	5.367	5.750	6.133	6.517	6.900	7.283	7.667	138
16	.386	0.773	1.158	1.544	1.931	2.317	2.702	3.089	3.475	3.861	4.248	4.633	5.019	5.406	5.792	6.177	6.564	6.950	7.336	7.723	139
20	.389	0.778	1.167	1.555	1.945	2.333	2.722	3.111	3.500	3.889	4.278	4.667	5.055	5.445	5.833	6.222	6.611	7.000	7.389	7.778	140
24	.392	0.783	1.175	1.567	1.958	2.350	2.742	3.133	3.525	3.917	4.309	4.700	5.092	5.483	5.875	6.267	6.658	7.050	7.442	7.833	141
28	.395	0.789	1.183	1.578	1.972	2.367	2.761	3.155	3.550	3.955	4.339	4.733	5.128	5.522	5.917	6.311	6.705	7.100	7.495	7.889	142
32	.398	0.794	1.192	1.589	1.986	2.383	2.780	3.177	3.575	3.973	4.370	4.767	5.164	5.561	5.958	6.355	6.752	7.150	7.548	7.945	143
36	.400	0.800	1.200	1.600	2.000	2.400	2.800	3.200	3.600	4.000	4.400	4.800	5.200	5.600	6.000	6.400	6.800	7.200	7.600	8.000	144
40	.403	0.806	1.208	1.611	2.014	2.417	2.819	3.222	3.625	4.028	4.431	4.833	5.236	5.639	6.042	6.444	6.847	7.250	7.653	8.056	145
44	.406	0.811	1.217	1.622	2.028	2.433	2.838	3.244	3.650	4.056	4.462	4.867	5.272	5.678	6.083	6.488	6.894	7.300	7.706	8.112	146
48	.408	0.817	1.225	1.633	2.042	2.450	2.858	3.267	3.675	4.083	4.492	4.900	5.308	5.717	6.125	6.533	6.942	7.350	7.758	8.167	147
52	.411	0.823	1.233	1.644	2.056	2.467	2.877	3.289	3.700	4.111	4.523	4.933	5.344	5.756	6.167	6.577	6.989	7.400	7.811	8.223	148
56	.414	0.829	1.242	1.655	2.070	2.483	2.897	3.311	3.725	4.139	4.553	4.967	5.380	5.795	6.208	6.622	7.036	7.450	7.864	8.278	149
10 0	.417	0.833	1.250	1.667	2.083	2.500	2.917	3.333	3.750	4.167	4.584	5.000	5.417	5.833	6.250	6.667	7.083	7.500	7.917	8.333	150
4	.420	0.839	1.258	1.678	2.097	2.517	2.936	3.355	3.775	4.195	4.624	5.033	5.453	5.872	6.292	6.711	7.130	7.550	7.970	8.389	151
8	.423	0.844	1.267	1.689	2.111	2.533	2.955	3.377	3.800	4.223	4.655	5.067	5.489	5.911	6.333	6.755	7.177	7.600	8.023	8.445	152
12	.425	0.850	1.275	1.700	2.125	2.550	2.975	3.400	3.825	4.250	4.675	5.100	5.525	5.950	6.375	6.800	7.225	7.650	8.075	8.500	153
16	.428	0.856	1.283	1.711	2.139	2.567	2.994	3.422	3.850	4.278	4.706	5.133	5.561	5.989	6.417	6.844	7.272	7.700	8.128	8.556	154
20	.431	0.861	1.292	1.722	2.153	2.583	3.013	3.444	3.875	4.306	4.737	5.167	5.597	6.028	6.458	6.888	7.319	7.750	8.181	8.612	155
24	.433	0.867	1.300	1.733	2.167	2.600	3.033	3.467	3.900	4.333	4.767	5.200	5.633	6.067	6.500	6.933	7.367	7.800	8.233	8.667	156
28	.436	0.873	1.308	1.744	2.181	2.617	3.052	3.489	3.925	4.361	4.798	5.233	5.669	6.106	6.542	6.977	7.414	7.850	8.286	8.723	157
32	.439	0.878	1.317	1.755	2.195	2.633	3.072	3.511	3.950	4.389	4.828	5.267	5.705	6.145	6.583	7.022	7.461	7.900	8.339	8.778	158
36	.442	0.883	1.325	1.767	2.208	2.650	3.092	3.533	3.975	4.417	4.859	5.300	5.742	6.183	6.625	7.067	7.508	7.950	8.392	8.833	159
40	.445	0.889	1.333	1.778	2.222	2.667	3.111	3.555	4.000	4.445	4.889	5.333	5.778	6.222	6.667	7.111	7.555	8.000	8.445	8.889	160
44	.448	0.894	1.342	1.789	2.236	2.683	3.130	3.577	4.025	4.473	4.920	5.367	5.814	6.261	6.708	7.155	7.602	8.050	8.498	8.945	161
48	.450	0.900	1.350	1.800	2.250	2.700	3.150	3.600	4.050	4.500	4.950	5.400	5.850	6.300	6.750	7.200	7.650	8.100	8.550	9.000	162
52	.453	0.906	1.358	1.811	2.264	2.717	3.169	3.622	4.075	4.528	4.981	5.433	5.886	6.339	6.792	7.244	7.697	8.150	8.603	9.056	163
56	.456	0.911	1.367	1.822	2.278	2.733	3.188	3.644	4.100	4.556	5.012	5.467	5.922	6.378	6.833	7.288	7.744	8.200	8.656	9.112	164
11 0	.458	0.917	1.375	1.833	2.292	2.750	3.208	3.667	4.125	4.583	5.042	5.500	5.958	6.417	6.875	7.333	7.792	8.250	8.708	9.167	165
4	.461	0.923	1.383	1.844	2.306	2.767	3.227	3.689	4.150	4.611	5.073	5.533	5.994	6.456	6.917	7.377	7.839	8.300	8.761	9.223	166
8	.464	0.928	1.392	1.855	2.320	2.783	3.247	3.711	4.175	4.639	5.103	5.567	6.030	6.495	6.958	7.422	7.886	8.350	8.814	9.278	167
12	.467	0.933	1.400	1.867	2.333	2.800	3.267	3.733	4.200	4.667	5.134	5.600	6.067	6.533	7.000	7.467	7.933	8.400	8.867	9.333	168
16	.470	0.939	1.408	1.878	2.347	2.817	3.286	3.755	4.225	4.695	5.164	5.633	6.103	6.572	7.042	7.511	7.980	8.450	8.920	9.389	169
20	.473	0.944	1.417	1.889	2.361	2.833	3.305	3.777	4.250	4.723	5.195	5.667	6.139	6.611	7.083	7.555	8.027	8.500	8.973	9.445	170
24	.475	0.950	1.425	1.900	2.375	2.850	3.325	3.800	4.275	4.750	5.225	5.700	6.175	6.650	7.125	7.600	8.075	8.550	9.025	9.500	171
28	.478	0.956	1.433	1.911	2.389	2.867	3.344	3.822	4.300	4.778	5.256	5.733	6.211	6.689	7.167	7.644	8.122	8.600	9.078	9.556	172
32	.481	0.961	1.442	1.922	2.403	2.883	3.363	3.844	4.325	4.806	5.287	5.767	6.247	6.728	7.208	7.688	8.169	8.650	9.131	9.012	173
36	.483	0.967	1.450	1.933	2.417	2.900	3.383	3.867	4.350	4.833	5.317	5.800	6.283	6.767	7.250	7.733	8.217	8.700	9.183	9.667	174
40	.486	0.973	1.458	1.944	2.431	2.917	3.402	3.889	4.375	4.861	5.348	5.833	6.319	6.806	7.292	7.777	8.264	8.750	9.236	9.723	175
44	.489	0.978	1.467	1.955	2.445	2.933	3.422	3.911	4.400	4.889	5.378	5.867	6.355	6.845	7.333	7.822	8.311	8.800	9.289	9.778	176
48	.492	0.983	1.475	1.967	2.458	2.950	3.442	3.933	4.425	4.917	5.409	5.900	6.392	6.883	7.375	7.867	8.358	8.850	9.342	9.833	177
52	.495	0.989	1.483	1.978	2.472	2.967	3.461	3.955	4.450	4.945	5.439	5.933	6.428	6.922	7.417	7.911	8.406	8.900	9.395	9.889	178
56	.498	0.994	1.492	1.989	2.486	2.983	3.480	3.977	4.475	4.973	5.470	5.967	6.464	6.961	7.458	7.955	8.453	8.950	9.448	9.945	179
12 0	.500	1.000	1.500	2.000	2.500	3.000	3.500	4.000	4.500	5.000	5.500	6.000	6.500	7.000	7.500	8.000	8.500	9.000	9.500	10.000	180
Hr.Angle from Noon.	.1	.2	.3	.4	.5	.6	.7	.8	.9	.10	.11	.12	.13	.14	.15	.16	.17	.18	.19	.20	Long. fr Greenwich.

TABLE 8. The Reduction of the Ecliptic to the Equator, calculated for Obliquity 23° 27′ 57″ to every 10′ of Longitude of the points of the Ecliptic, with the Diff. and Var. for 100″ Var. of the Obliquity. Arg. = the Long. of the point of the Ecliptic. By Mr. J. Utting.

Argument.	0s – Reduction.	VIs – Difference	Variation.	Is – Reduction.	VIIs – Difference.	Variation.	IIs – Reduction.	VIIIs – Difference.	Variation.	Arg.
0° 0′	0° 0′ 0″.00	49″.62	0″.00	2° 5′ 39″.16	26″.78	17″.93	2° 11′ 11″.24	24″.81	19″.55	30° 0′
10	0 0 49.62	49.62	0.12	2 6 5.94	26.55	18.00	2 10 46.43	25.11	19.50	50
20	0 1 39.24	49.62	0.24	2 6 32.49	26.31	18.07	2 10 21.32	25.38	19.44	40
30	0 2 28.86	49.61	0.35	2 6 58.80	26.06	18.13	2 9 55.94	25.64	19.38	30
40	0 3 18.47	49.61	0.47	2 7 24.86	25.82	18.20	2 9 30.30	25.94	19.32	20
50	0 4 8.08	49.60	0.59	2 7 50.68	25.60	18.27	2 9 4.36	26.21	19.26	10
1 0	0 4 57.68	49.59	0.70	2 8 16 28	25.32	18.33	2 8 38.15	26.48	19.20	29 0
10	0 5 47.27	49.57	0.82	2 8 41 60	25.08	18.40	2 8 11.67	26.77	19.14	50
20	0 6 36.84	49.57	0.93	2 9 6 68	24.85	18.46	2 7 44.90	27.07	19.08	40
30	0 7 26.41	49.56	1.05	2 9 31 53	24.59	18.52	2 7 17.83	27.31	19.01	30
40	0 8 15.97	49.54	1.16	2 9 56 12	24.34	18.59	2 6 50.52	27.60	18.95	20
50	0 9 5.51	49.52	1.28	2 10 20 46	24.11	18.65	2 6 20.92	27.86	18.88	10
2 0	0 9 55.03	49.51	1.39	2 10 44.57	23.84	18.71	2 5 55.06	28.16	18.82	28 0
10	0 10 44.54	49.49	1.51	2 11 8.41	23.59	18.77	2 5 26.90	28.41	18.75	50
20	0 11 34.03	49.46	1.62	2 11 32.00	23.35	18.83	2 4 58.49	28.69	18.68	40
30	0 12 23.49	49.45	1.74	2 11 55.35	23.09	18.89	2 4 29.80	28.98	18.61	30
40	0 13 12.94	49.42	1.85	2 12 18.44	22.85	18.95	2 4 0.82	29.24	18.54	20
50	0 14 2.36	49.39	1.97	2 12 41.29	22.57	19.01	2 3 31.58	29.50	18.47	10
3 0	0 14 51.75	49.37	2.08	2 13 3.86	22.33	19.07	2 3 2.08	29.77	18.40	27 0
10	0 15 41.12	49.34	2.20	2 13 26.19	22.28	19.13	2 2 32.31	30.05	18.33	50
20	0 16 30.46	49.31	2.31	2 13 48.47	21.61	19.18	2 2 2.26	30.30	18.26	40
30	0 17 19.77	49.28	2.43	2 14 10.08	21.54	19.24	2 1 31.96	30.59	18.19	30
40	0 18 9.05	49.25	2.54	2 14 31.62	21.30	19.30	2 1 1.37	30.83	18.12	20
50	0 18 58.30	49.22	2.66	2 14 52.92	21.06	19.35	2 0 30.54	31.11	18.04	10
4 0	0 19 47.52	49.17	2.77	2 15 13.98	20.76	19.41	1 59 59.43	31.37	17.97	26 0
10	0 20 36.69	49.14	2.89	2 15 34.74	20.52	19.47	1 59 28.06	31.63	17.90	50
20	0 21 25.83	49.11	3.00	2 15 55.26	20.27	19.52	1 58 56.43	31.88	17.82	40
30	0 22 14.94	49.06	3.12	2 16 15.53	19.98	19.57	1 58 24.55	32.17	17.75	30
40	0 23 4.00	49.02	3.23	2 16 35.51	19.75	19.63	1 57 52.38	32.40	17.67	20
50	0 23 53.02	48.98	3.35	2 16 55.26	19.44	19.68	1 57 19.98	32.67	17.60	10
5 0	0 24 42.00	48.94	3.46	2 17 14.70	19.21	19.73	1 56 47.31	32.91	17.52	25 0
10	0 25 30.94	48.89	3.58	2 17 33.91	18.92	19.78	1 56 14.40	33.19	17.44	50
20	0 26 19.83	48.84	3.69	2 17 52.83	18.67	19.83	1 55 41.21	33.44	17.36	40
30	0 27 8.67	48.79	3.81	2 18 11.50	18.37	19.88	1 55 7.77	33.69	17.28	30
40	0 27 57.46	48.74	3.92	2 18 29.87	18.13	19.93	1 54 34.08	33.93	17.20	20
50	0 28 46.20	48.70	4.04	2 18 48.00	17.89	19.97	1 54 0.15	34.21	17.12	10
6 0	0 29 34.90	48.64	4.15	2 19 5.89	17.57	20.02	1 53 25.94	34.44	17.04	24 0
10	0 30 23.54	48.58	4.26	2 19 23.46	17.32	20.07	1 52 51.50	34.69	16.96	50
20	0 31 12.12	48.53	4.38	2 19 40.78	17.04	20.11	1 52 16.81	34.94	16.87	40
30	0 32 0.65	48.48	4.49	2 19 57.82	16.79	20.16	1 51 41.87	35.18	16.79	30
40	0 32 49.13	48.41	4.60	2 20 14.61	16.48	20.20	1 51 6.69	35.45	16.70	20
50	0 33 37.54	48.35	4.72	2 20 31.09	16.22	20.25	1 50 31.24	35.69	16.62	10
7 0	0 34 25.89	48.29	4.83	2 20 47.31	15.96	20.29	1 49 55.55	35.91	16.54	23 0
10	0 35 14.18	48.23	4.94	2 21 3.27	15.66	20.33	1 49 19.64	36.18	16.45	50
20	0 36 2.41	48.16	5.06	2 21 18.93	15.40	20.38	1 48 43.46	36.41	16.36	40
30	0 36 50.57	48.10	5.17	2 21 34.33	15.14	20.42	1 48 7.05	36.64	16.27	30
40	0 37 38.67	48.03	5.28	2 21 49.47	14.83	20.46	1 47 30.41	36.88	16.18	20
50	0 38 26.70	47.96	5.39	2 22 4.30	14.59	20.50	1 46 53.53	37.12	16.09	10
8 0	0 39 14.66	47.89	5.50	2 22 18.89	14.29	20.54	1 46 16.41	37.38	16.00	22 0
10	0 40 2.55	47.82	5.61	2 22 33.18	13.98	20.58	1 45 39.03	37.59	15.91	50
20	0 40 50.37	47.75	5.73	2 22 47.16	13.73	20.62	1 45 1.44	37.84	15.82	40
30	0 41 38.12	47.66	5.84	2 23 0.89	13.45	20.65	1 44 23.60	38.05	15.72	30
40	0 42 25.78	47.60	5.95	2 23 14.34	13.18	20.69	1 43 45.55	38.30	15.63	20
50	0 43 13.38	47.52	6.06	2 23 27.52	12.87	20.73	1 43 7.25	38.54	15.53	10
9 0	0 44 0.90	47.44	6.17	2 23 40.39	12.61	20.76	1 42 28.71	38.73	15.44	21 0
10	0 44 48.34	47.35	6.28	2 23 53.00	12.32	20.80	1 41 49.98	38.98	15.34	50
20	0 45 35.69	47.28	6.39	2 24 5.32	12.02	20.83	1 41 11.00	39.22	15.25	40
30	0 46 22.97	47.19	6.50	2 24 17.34	11.77	20.86	1 40 31.78	39.43	15.15	30
40	0 47 10.16	47.10	6.61	2 24 29.11	11.45	20.90	1 39 52.35	39.65	15.06	20
50	0 47 57.26	47.02	6.72	2 24 40.56	11.19	20.93	1 39 12.70	38.88	15.96	10
10 0	0 48 44.28		6.83	2 24 51.75		20.96	1 38 32.82		15.86	20 0
	Vs +	XIs +		IVs +	Xs +		IIIs +	Xs +		

TABLE 8.

The Reduction of the Ecliptic to the Equator, &c. continued.

Arg.	0s − VIs − Reduction.	Difference	Variation.	Is − VIIs − Reduction.	Difference.	Variation.	IIs − VIIIs − Reduction.	Difference.	Variation.	Arg.
10° 0′	0° 48′ 44″.28	47″.02	6″.83	2° 24′ 51″.75	11″.19	20″.96	1° 38′ 32″.82	39″.88	14″.86	20° 0
10	0 49 31 .22	46 .94	6 .94	2 25 2 .64	10 .89	20 .99	1 37 52 .72	40 .10	14 .76	50
20	0 50 18 .07	46 .85	7 .06	2 25 13 .23	10 .59	21 .02	1 37 12 .40	40 .32	14 .67	40
30	0 51 4 .82	46 .75	7 .17	2 25 23 .56	10 .33	21 .05	1 36 31 .87	40 .53	14 .57	30
40	0 51 51 .48	46 .66	7 .28	2 25 33 .57	10 .01	21 .08	1 35 51 .11	40 .76	14 .47	20
50	0 52 38 .04	46 .56	7 .39	2 25 43 .33	9 .76	21 .11	1 35 10 .16	40 .95	14 .37	10
11 0	0 53 24 .52	46 .48	7 .50	2 25 52 .77	9 .44	21 .14	1 34 28 .99	41 .17	14 .27	19 0
10	0 54 10 .90	46 .38	7 .61	2 26 1 .93	9 .16	21 .17	1 33 47 .59	41 .40	14 .17	50
20	0 54 57 .18	46 .28	7 .72	2 26 10 .81	8 .88	21 .19	1 33 5 .98	41 .61	14 .07	40
30	0 55 43 .36	46 .18	7 .83	2 26 19 .40	8 .59	21 .22	1 32 24 .17	41 .81	13 .97	30
40	0 56 29 .44	46 .08	7 .94	2 26 27 .70	8 .30	21 .24	1 31 42 .15	42 .02	13 .87	20
50	0 57 15 .42	45 .98	8 .05	2 26 35 .67	7 .97	21 .27	1 30 59 .92	42 .23	13 .77	10
12 0	0 58 1 .30	45 .88	8 .16	2 26 43 .39	7 .72	21 .29	1 30 17 .50	42 .42	13 .66	18 0
10	0 58 47 .05	45 .75	8 .27	2 26 50 .81	7 .42	21 .32	1 29 34 .87	42 .63	13 .56	50
20	0 59 32 .72	45 .67	8 .38	2 26 57 .93	7 .12	21 .34	1 28 52 .03	42 .84	13 .45	40
30	1 0 18 .29	45 .57	8 .48	2 27 4 .74	6 .81	21 .36	1 28 8 .99	43 .04	13 .34	30
40	1 1 3 .73	45 .44	8 .59	2 27 11 .29	6 .55	21 .38	1 27 25 .74	43 .25	13 .24	20
50	1 1 49 .07	45 .34	8 .70	2 27 17 .53	6 .24	21 .40	1 26 42 .31	43 .43	13 .13	10
13 0	1 2 34 .29	45 .22	8 .80	2 27 23 .46	5 .93	21 .42	1 25 58 .68	43 .63	13 .02	17 0
10	1 3 19 .40	45 .11	8 .91	2 27 29 .12	5 .66	21 .44	1 25 14 .87	43 .81	12 .91	50
20	1 4 4 .41	45 .01	9 .01	2 27 34 .49	5 .37	21 .46	1 24 30 .86	44 .01	12 .80	40
30	1 4 49 .29	44 .88	9 .12	2 27 39 .53	5 .04	21 .48	1 23 46 .63	44 .23	12 .69	30
40	1 5 34 .05	44 .76	9 .22	2 27 44 .30	4 .77	21 .50	1 23 2 .25	44 .38	12 .58	20
50	1 6 18 .70	44 .65	9 .33	2 27 48 .79	4 .49	21 .51	1 22 17 .66	44 .59	12 .47	10
14 θ	1 7 3 .21	44 .51	9 .43	2 27 52 .93	4 .14	21 .53	1 21 32 .89	44 .77	12 .35	16 0
10	1 7 47 .62	44 .41	9 .54	2 27 56 .81	3 .88	21 .55	1 20 47 .93	44 .96	12 .25	50
20	1 8 31 .91	44 .29	9 .64	2 28 0 .38	3 .57	21 .56	1 20 2 .80	45 .13	12 .13	40
30	1 9 16 .06	44 .15	9 .75	2 28 3 .62	3 .24	21 .58	1 19 17 .49	45 .31	12 .01	30
40	1 10 0 .10	44 .04	9 .85	2 28 6 .64	3 .02	21 .59	1 18 31 .99	45 .50	11 .90	20
50	1 10 44 .01	43 .91	9 .96	2 28 9 .30	2 .66	21 .61	1 17 46 .31	45 .68	11 .79	10
15 0	1 11 27 .78	43 .77	10 .06	3 28 11 .70	2 .40	21 .62	1 17 0 .47	45 .84	11 .67	15 0
10	1 12 11 .43	43 .65	10 .17	2 28 13 .77	2 .07	21 .63	1 16 14 .42	46 .05	11 .55	50
20	1 12 54 .94	43 .51	10 .27	2 28 15 .53	1 .76	21 .63	1 15 28 .22	46 .20	11 .43	40
30	1 13 38 .33	43 .39	10 .37	2 28 17 .00	1 .47	21 .64	1 14 41 .86	46 .36	11 .32	30
40	1 14 21 .59	43 .26	10 .48	2 28 18 .21	1 .21	21 .65	1 13 55 .31	46 .55	11 .20	20
50	1 15 4 .70	43 .11	10 .58	2 28 19 .07	0 .86	21 .65	1 13 8 .60	46 .71	11 .09	10
16 0	1 15 47 .69	42 .99	10 .68	2 28 19 .67	0 .60	21 .66	1 12 21 .72	46 .88	10 .97	14 0
10	1 16 30 .53	42 .84	10 .78	2 28 19 .93	0 .26	21 .66	1 11 34 .68	47 .04	10 .85	50
20	1 17 13 .23	42 .70	10 .88	2 28 19 .90	0 .03	21 .66	1 10 47 .47	47 .21	10 .73	40
30	1 17 55 .78	42 .55	10 .98	2 28 19 .59	0 .31	21 .67	1 10 0 .10	47 .37	10 .62	30
40	1 18 38 .20	42 .42	11 .08	2 28 18 .95	0 .64	21 .67	1 9 12 .59	47 .51	10 .50	20
50	1 19 20 .47	42 .27	11 .18	2 28 18 .02	0 .93	21 .67	1 8 24 .90	47 .69	10 .38	10
17 0	1 20 2 .61	42 .14	11 .28	2 28 16 .79	1 .23	21 .67	1 7 37 .06	47 .84	10 .26	13 0
10	1 20 44 .59	41 .98	11 .38	2 28 15 .24	1 .55	21 .67	1 6 49 .08	47 .98	10 .14	50
20	1 21 26 .44	41 .85	11 .48	2 28 13 .42	1 .82	21 .67	1 6 0 .92	48 .16	10 .02	40
30	1 22 8 .13	41 .69	11 .58	2 28 11 .29	2 .13	21 .67	1 5 12 .62	48 .30	9 .90	30
40	1 22 49 .65	41 .52	11 .67	2 28 8 .83	2 .46	21 .67	1 4 24 .18	48 .44	9 .78	20
50	1 23 31 .05	41 .40	11 .77	2 28 6 .09	2 .74	21 .67	1 3 35 .59	48 .59	9 .66	10
18 0	1 24 12 .28	41 .23	11 .87	2 28 3 .03	3 .06	21 .66	1 2 46 .84	48 .75	9 .53	12 0
10	1 24 53 .37	41 .09	11 .97	2 27 59 .69	3 .34	21 .66	1 1 57 .97	48 .87	9 .41	50
20	1 25 34 .30	40 .93	12 .06	2 27 56 .02	3 .67	21 .65	1 1 8 .94	49 .03	9 .29	40
30	1 26 15 .07	40 .77	12 .16	2 27 52 .07	3 .95	21 .65	1 0 19 .78	49 .16	9 .16	30
40	1 26 55 .68	40 .61	12 .26	2 27 47 .81	4 .26	21 .65	0 59 30 .48	49 .30	9 .04	20
50	1 27 36 .14	40 .46	12 .36	2 27 43 .24	4 .57	21 .64	0 58 41 .04	49 .44	8 .92	10
19 0	1 28 16 .42	40 .28	12 .45	2 27 38 .38	4 .86	21 .64	0 57 51 .47	49 .57	8 .79	11 0
10	1 28 56 .56	40 .14	12 .55	2 27 33 .19	5 .19	21 .63	0 57 1 .77	49 .70	8 .67	50
20	1 29 36 .53	39 .97	12 .64	2 27 27 .74	5 .45	21 .63	0 56 11 .94	49 .83	8 .54	40
30	1 30 16 .33	39 .80	12 .74	2 27 21 .93	5 .81	21 .62	0 55 21 .97	49 .97	8 .42	30
40	1 30 55 .97	39 .64	12 .83	2 27 15 .86	6 .07	21 .61	0 54 31 .87	50 .10	8 .29	20
50	1 31 35 .45	39 .48	12 .93	2 27 9 .45	6 .41	21 .60	0 53 41 .66	50 .21	8 .17	10
20 0	1 32 14 .74	39 .29	13 .02	2 27 2 .76	6 .69	21 .59	0 52 51 .33	50 .33	8 .04	10 0
	Vs + XIs +			IVs + Xs +			IIIs + IXs +			

TABLE 5.

The Reduction of the Ecliptic to the Equator, &c. concluded.

Arg.	0ˢ – VIˢ – Reduction.	Difference.	Variation.	Iˢ – VIIˢ – Reduction.	Difference.	Variation.	IIˢ – VIIIˢ – Reduction.	Difference.	Variation	Arg.
		39″.29			6″.69			50″.33		
20° 0	1° 32′ 14″.74		13″.02	2° 27′ 2″.76		21″.59	0° 52′ 51″.33		8″.04	10° 0′
		39.13			6.99			50.46		
10	1 32 53.87		13.12	2 26 55.77		21.58	0 52 0.87		7.92	50
		38.97			7.32			50.58		
20	1 33 32.84		13.21	2 26 48.45		21.57	0 51 10.29		7.79	40
		38.80			7.57			50.70		
30	1 34 11.64		13.30	2 26 40.88		21.56	0 50 19.59		7.66	30
		38.61			7.93			50.81		
40	1 34 50.25		13.40	2 26 32.95		21.55	0 49 28.78		7.53	20
		38.45			8.21			50.93		
50	1 35 28.70		13.49	2 26 24.74		21.53	0 48 37.85		7.41	10
		38.27			8.51			51.03		
21 0	1 36 6.97		13.58	2 26 16.23		21.52	0 47 46.82		7.28	9 0
		38.09			8.83			51.15		
10	1 36 45.06		13.67	2 26 17.40		21.51	0 46 55.67		7.15	50
		37.91			9.11			51.25		
20	1 37 22.97		13.77	2 25 58.29		21.49	0 46 4.42		7.02	40
		37.74			9.43			51.37		
30	1 38 0.71		13.86	2 25 48.86		21.47	0 45 13.05		6.89	30
		37.55			9.72			51.46		
40	1 38 38.26		13.95	2 25 39.14		21.45	0 44 21.59		6.76	20
		37.39			10.07			51.57		
50	1 39 15.65		14.04	2 25 29.07		21.43	0 43 30.02		6.63	10
		37.18			10.33			51.67		
22 0	1 39 52.83		14.13	2 25 18.74		21.41	0 42 38.35		6.49	8 0
		37.00			10.62			51.76		
10	1 40 29.83		14.22	2 25 8.12		21.39	0 41 46.59		6.36	50
		36.81			10.96			51.86		
20	1 41 6.64		14.31	2 24 57.16		21.37	0 40 54.73		6.23	40
		36.64			11.23			51.96		
30	1 41 43.28		14.40	2 24 45.93		21.35	0 40 2.77		6.10	30
		36.43			11.55			52.04		
40	1 42 19.71		14.49	2 24 34.38		21.33	0 39 10.73		5.97	20
		36.26			11.85			52.14		
50	1 42 55.97		14.58	2 24 22.53		21.30	0 38 18.59		5.84	10
		36.08			12.16			52.23		
23 0	1 43 32.05		14.67	2 24 10.37		21.28	0 37 26.36		5.70	7 0
		35.87			12.44			52.31		
10	1 44 7.92		14.76	2 23 57.93		21.25	0 36 34.05		5.57	50
		35.65			12.77			52.40		
20	1 44 43.57		14.84	2 23 45.16		21.23	0 35 41.65		5.44	40
		35.50			13.07			52.48		
30	1 45 19.07		14.93	2 23 32.09		21.20	0 34 49.17		5.31	30
		35.29			13.37			52.56		
40	1 45 54.36		15.02	2 23 18.72		21.18	0 33 56.61		5.18	20
		35.08			13.65			52.64		
50	1 46 29.44		15.10	2 23 5.07		21.15	0 33 3.97		5.05	10
		34.91			13.95			52.70		
24 0	1 47 4.35		15.19	2 22 51.12		21.12	0 32 11.27		4.92	6 0
		34.68			14.26			52.79		
10	1 47 39.03		15.28	2 22 36.86		21.08	0 31 18.48		4.78	50
		34.51			14.56			52.87		
20	1 48 13.54		15.36	2 22 22.30		21.05	0 30 25.61		4.65	40
		34.27			14.88			52.93		
30	1 48 47.81		15.44	2 22 7.42		21.02	0 29 32.68		4.51	30
		34.12			15.16			53.00		
40	1 49 21.93		15.53	2 21 52.26		20.98	0 28 39.68		4.37	20
		33.87			15.46			53.06		
50	1 49 55.80		15.61	2 21 36.80		20.95	0 27 46.62		4.24	10
		33.70			15.75			53.13		
25 0	1 50 29.50		15.69	2 21 21.05		20.92	0 26 53.49		4.10	5 0
		33.47			16.07			53.19		
10	1 51 2.97		15.77	2 21 4.98		20.88	0 26 0.30		3.96	50
		33.27			16.37			53.25		
20	1 51 36.24		15.85	2 20 48.61		20.85	0 25 7.05		3.83	40
		33.06			16.66			53.31		
30	1 52 9.30		15.93	2 20 31.95		20.81	0 24 13.74		3.69	30
		32.86			16.93			53.37		
40	1 52 42.16		16.01	2 20 15.02		20.77	0 23 20.37		3.55	20
		32.63			17.27			53.41		
50	1 53 14.79		16.09	2 19 57.75		20.73	0 22 26.96		3.42	10
		32.43			17.54			53.48		
26 0	1 53 47.22		16.17	2 19 40.21		20.69	0 21 33.48		3.28	4 0
		32.21			17.83			53.51		
10	1 54 19.43		16.25	2 19 22.38		20.65	0 20 39.97		3.15	50
		32.02			18.15			53.57		
20	1 54 51.45		16.33	2 19 4.23		20.61	0 19 46.40		3.01	40
		31.77			18.43			53.62		
30	1 55 23.22		16.41	2 18 45.80		20.57	0 18 52.78		2.88	30
		31.58			18.73			53.65		
40	1 55 54.80		16.48	2 18 27.07		20.52	0 17 59.13		2.74	20
		31.35			19.05			53.70		
50	1 56 26.15		16.55	2 18 8.02		20.48	0 17 5.43		2.61	10
		31.14			19.31			53.74		
27 0	1 56 57.29		16.64	2 17 48.71		20.44	0 16 11.69		2.47	3 0
		30.91			19.62			53.78		
10	1 57 28.20		16.72	2 17 29.09		20.39	0 15 17.91		2.33	50
		30.71			19.89			53.81		
20	1 57 58.91		16.79	2 17 9.20		20.35	0 14 24.10		2.20	40
		30.46			20.22			53.84		
30	1 58 29.37		16.87	2 16 48.98		20.30	0 13 30.26		2.06	30
		30.26			20.49			53.88		
40	1 58 59.63		16.94	2 16 28.49		20.26	0 12 36.38		1.92	20
		30.02			20.78			53.91		
50	1 59 29.65		17.02	2 16 7.71		20.21	0 11 42.47		1.79	10
		29.80			21.09			53.93		
28 0	1 59 59.45		17.09	2 15 46.62		20.17	0 10 48.54		1.65	2 0
		29.59			21.35			53.95		
10	2 0 29.04		17.16	2 15 25.27		20.12	0 9 54.59		1.51	50
		29.36			21.67			53.99		
20	2 0 58.40		17.24	2 15 3.60		20.07	0 9 0.60		1.38	40
		29.11			21.93			54.00		
30	2 1 27.51		17.31	2 14 41.67		20.02	0 8 6.60		1.24	30
		28.89			22.24			54.02		
40	2 1 56.40		17.38	2 14 19.43		19.97	0 7 12.58		1.10	20
		28.66			22.52			54.01		
50	2 2 25.06		17.45	2 13 56.91		19.92	0 6 18.54		0.97	10
		28.42			22.83			54.04		
29 0	2 2 53.48		17.52	2 13 34.08		19.87	0 5 24.50		0.83	1 0
		28.21			23.08			54.07		
10	2 3 21.69		17.59	2 13 11.00		19.82	0 4 30.43		0.69	50
		27.98			23.39			54.07		
20	2 3 49.67		17.65	2 12 47.61		19.76	0 3 36.36		0.56	40
		27.72			23.67			54.08		
30	2 4 17.39		17.72	2 12 23.94		19.71	0 2 42.28		0.42	30
		27.49			23.94			54.09		
40	2 4 44.88		17.79	2 12 0.00		19.66	0 1 48.19		0.28	20
		27.26			24.26			54.10		
50	2 5 12.14		17.86	2 11 35.74		19.60	0 0 54.09		0.14	10
		27.02			24.50			54.09		
30 0	2 5 39.16		17.93	2 11 11.24		19.55	0 0 0.00		0.00	0 0
	Vˢ + XIˢ +			IVˢ + Xˢ +			IIIˢ + IXˢ +			

TABLE 9.

The SUN's Semi-diameter, Horary Motion in Long. and Horizontal Parallax. Argument = the Sun's mean Anomaly.

Deg.	Signs 0. Semi-diameter.	Signs 0. Horary motion.	Signs 0. Horizon. Parallax.	I. Semi-diameter.	I. Horary motion.	I. Horizon. Parallax.	II. Semi-diameter.	II. Horary motion.	II. Horizon. Parallax.	Deg.
0°	16′ 17″.79	2′ 32″.92	8″.95	16′ 15″.49	2′ 32″.20	8″.93	16′ 9″.30	2′ 30″.28	8″.87	30°
1	16 17.79	2 32.92	8.95	16 15.34	2 32.15	8.93	16 9.05	2 30.20	8.87	29
2	16 17.78	2 32.92	8.95	16 15.18	2 32.10	8.93	16 8.79	2 30.12	8.87	28
3	16 17.77	2 32.91	8.95	16 15.02	2 32.05	8.92	16 8.54	2 30.04	8.87	27
4	16 17.75	2 32.91	8.95	16 14.85	2 32.00	8.92	16 8.28	2 29.96	8.86	26
5	16 17.73	2 32.90	8.95	16 14.69	2 31.95	8.92	16 8.02	2 29.88	8.86	25
6	16 17.70	2 32.89	8.95	16 14.51	2 31.90	8.92	16 7.75	2 29.80	8.86	24
7	16 17.66	2 32.88	8.95	16 14.34	2 31.84	8.92	16 7.49	2 29.71	8.86	23
8	16 17.62	2 32.87	8.95	16 14.16	2 31.78	8.92	16 7.22	2 29.63	8.85	22
9	16 17.58	2 32.86	8.95	16 13.97	2 31.73	8.91	16 6.95	2 29.54	8.85	21
10	16 17.53	2 32.84	8.95	16 13.78	2 31.67	8.91	16 6.68	2 29.46	8.85	20
11	16 17.47	2 32.82	8.95	16 13.59	2 31.61	8.91	16 6.41	2 29.38	8.85	19
12	16 17.41	2 32.80	8.95	16 13.39	2 31.55	8.91	16 6.14	2 29.30	8.84	18
13	16 17.35	2 32.78	8.95	16 13.20	2 31.48	8.91	16 5.86	2 29.21	8.84	17
14	16 17.28	2 32.76	8.95	16 12.99	2 31.42	8.91	16 5.59	2 29.13	8.84	16
15	16 17.20	2 32.73	8.95	16 12.78	2 31.36	8.90	16 5.31	2 29.04	8.84	15
16	16 17.12	2 32.71	8.94	16 12.57	2 31.29	8.90	16 5.04	2 28.96	8.83	14
17	16 17.04	2 32.68	8.94	16 12.36	2 31.22	8.90	16 4.76	2 28.87	8.83	13
18	16 16.95	2 32.65	8.94	16 12.14	2 31.16	8.90	16 4.48	2 28.78	8.83	12
19	16 16.85	2 32.62	8.94	16 11.92	2 31.09	8.90	16 4.20	2 28.70	8.83	11
20	16 16.75	2 32.59	8.94	16 11.69	2 31.02	8.89	16 3.92	2 28.61	8.82	10
21	16 16.65	2 32.56	8.94	16 11.46	2 30.95	8.89	16 3.64	2 28.52	8.82	9
22	16 16.54	2 32.53	8.94	16 11.24	2 30.88	8.89	16 3.36	2 28.44	8.82	8
23	16 16.42	2 32.49	8.94	16 11.01	2 30.80	8.89	16 3.07	2 28.35	8.82	7
24	16 16.30	2 32.45	8.94	16 10.77	2 30.73	8.89	16 2.79	2 28.26	8.81	6
25	16 16.18	2 32.41	8.94	16 10.54	2 30.66	8.88	16 2.51	2 28.18	8.81	5
26	16 16.05	2 32.37	8.93	16 10.29	2 30.58	8.88	16 2.23	2 28.09	8.81	4
27	16 15.92	2 32.33	8.93	16 10.05	2 30.51	8.88	16 1.95	2 28.00	8.80	3
28	16 15.78	2 32.29	8.93	16 9.80	2 30.43	8.88	16 1.67	2 27.92	8.80	2
29	16 15.63	2 32.24	8.93	16 9.55	2 30.35	8.87	16 1.38	2 27.83	8.80	1
30	16 15.49	2 32.20	8.93	16 9.30	2 30.28	8.87	16 1.10	2 27.74	8.80	0
	XI.			X.			IX.			

Deg.	III. Semi-diameter.	III. Horary motion.	III. Horizon. Parallax.	IV. Semi-diameter.	IV. Horary motion.	IV. Horizon. Parallax.	V. Semi-diameter.	V. Horary motion.	V. Horizon. Parallax.	Deg.
0°	16′ 1″.10	2′ 27″.74	8″.80	15′ 53″.17	2′ 25″.32	8″.72	15′ 47″.53	2′ 23″.60	8″.67	30°
1	16 0.82	2 27.66	8.79	15 52.94	2 25.24	8.72	15 47.40	2 23.56	8.67	29
2	16 0.54	2 27.57	8.79	15 52.70	2 25.17	8.72	15 47.27	2 23.52	8.67	28
3	16 0.26	2 27.48	8.79	15 52.48	2 25.10	8.72	15 47.15	2 23.48	8.67	27
4	15 59.98	2 27.40	8.78	15 52.25	2 25.03	8.72	15 47.03	2 23.45	8.67	26
5	15 59.70	2 27.31	8.78	15 52.03	2 24.96	8.71	15 46.92	2 23.41	8.67	25
6	15 59.42	2 27.23	8.78	15 51.81	2 24.90	8.71	15 46.81	2 23.38	8.67	24
7	15 59.14	2 27.14	8.78	15 51.59	2 24.84	8.71	15 46.70	2 23.35	8.67	23
8	15 58.87	2 27.06	8.78	15 51.37	2 24.77	8.71	15 46.60	2 23.32	8.66	22
9	15 58.59	2 26.97	8.77	15 51.16	2 24.70	8.71	15 46.50	2 23.29	8.66	21
10	15 58.32	2 26.89	8.77	15 50.96	2 24.64	8.70	15 46.41	2 23.26	8.66	20
11	15 58.05	2 26.80	8.77	15 50.75	2 24.58	8.70	15 46.32	2 23.23	8.66	19
12	15 57.77	2 26.72	8.77	15 50.55	2 24.52	8.70	15 46.24	2 23.20	8.66	18
13	15 57.50	2 26.64	8.77	15 50.35	2 24.46	8.70	15 46.16	2 23.18	8.66	17
14	15 57.23	2 26.56	8.76	15 50.15	2 24.40	8.70	15 46.08	2 23.16	8.66	16
15	15 56.96	2 26.47	8.76	15 49.96	2 24.34	8.69	15 46.01	2 23.14	8.66	15
16	15 56.70	2 26.39	8.76	15 49.77	2 24.28	8.69	15 45.95	2 23.12	8.66	14
17	15 56.44	2 26.31	8.76	15 49.59	2 24.23	8.69	15 45.88	2 23.10	8.66	13
18	15 56.17	2 26.23	8.75	15 49.41	2 24.17	8.69	15 45.83	2 23.09	8.66	12
19	15 55.91	2 26.15	8.75	15 49.23	2 24.12	8.69	15 45.78	2 23.07	8.66	11
20	15 55.65	2 26.07	8.75	15 49.06	2 24.06	8.69	15 45.73	2 23.05	8.66	10
21	15 55.39	2 25.99	8.75	15 48.89	2 24.01	8.69	15 45.68	2 23.04	8.66	9
22	15 55.13	2 25.91	8.74	15 48.72	2 23.96	8.68	15 45.64	2 23.03	8.66	8
23	15 54.88	2 25.84	8.74	15 48.56	2 23.91	8.68	15 45.61	2 23.02	8.66	7
24	15 54.63	2 25.76	8.74	15 48.40	2 23.86	8.68	15 45.58	2 23.01	8.65	6
25	15 54.38	2 25.68	8.74	15 48.24	2 23.82	8.68	15 45.55	2 23.00	8.65	5
26	15 54.14	2 25.61	8.73	15 48.09	2 23.77	8.68	15 45.53	2 23.00	8.65	4
27	15 53.89	2 25.53	8.73	15 47.94	2 23.73	8.68	15 45.52	2 22.99	8.65	3
28	15 53.65	2 25.46	8.73	15 47.80	2 23.68	8.68	15 45.51	2 22.99	8.65	2
29	15 53.41	2 25.39	8.73	15 47.66	2 23.64	8.67	15 45.50	2 22.99	8.65	1
30	15 53.17	2 25.32	8.72	15 47.53	2 23.60	8.67	15 45.50	2 22.99	8.65	0
	VIII.			VII.			VI.			

The Sun's mean Anomaly, in this Table, is the Longitude of the Sun's Perigee subtracted from his Mean Longitude, agreeably to the French Tables.

If the Sun's Semi-diameter be assumed, or be hereafter determined to be too great or too small by one second, as given in this Table, for the mean distance from the Earth, the Variation due to any other Semi-diameter may be obtained from the following Formula, viz. 1″ + 0″.017 cos. (☉'s Longitude + 2ˢ 20½°).

TABLE 10. Reduction to either Solstice.

Argument = the Sun's distance in Right Ascension in Time from the Solstice.

Arg. m. s.	Reduction.	Diff.	Var. in 100″ change of obliquity.
0 0	0′ 0″.00	0″.02	0 .0000
10	0 0 .02	0 .06	0 .0000
20	0 0 .08	0 .10	0 .0001
30	0 0 .18	0 .14	0 .0002
40	0 0 .32	0 .18	0 .0003
50	0 0 .50	0 .22	0 .0005
1 0	0 0 .72	0 .26	0 .0007
10	0 0 .98	0 .29	0 .0009
20	0 1 .27	0 .34	0 .0011
30	0 1 .61	0 .38	0 .0014
40	0 1 .99	0 .42	0 .0018
50	0 2 .41	0 .56	0 0022
2 0	0 2 .87	0 .50	0 .0026
10	0 3 .37	0 .54	0 .0031
20	0 3 .91	0 .57	0 .0036
30	0 4 .48	0 .62	0 .0041
40	0 5 .10	0 .66	0 .0046
50	0 5 .76	0 .69	0 .0053
3 0	0 6 .45	0 .74	0 .0059
10	0 7 .19	0 .78	0 .0066
20	0 7 .97	0 .81	0 .0073
30	0 8 .78	0 .86	0 .0080
40	0 9 .64	0 .89	0 .0088
50	0 10 .53	0 .94	0 .0096
4 0	0 11 .47	0 .97	0 .0104
10	0 12 .44	1 .02	0 .0113
20	0 13 .46	1 .06	0 .0122
30	0 14 .52	1 .09	0 .0132
40	0 15 .61	1 .14	0 .0142
50	0 16 .75	1 .18	0 .0152
5 0	0 17 .93	1 .21	0 .0163
10	0 19 .14	1 .26	0 .0174
20	0 20 .40	1 .29	0 .0185
30	0 21 .69	1 .34	0 .0197
40	0 23 .03	1 .37	0 .0209
50	0 24 .40	1 .42	0 .0221
6 0	0 25 .82	1 .45	0 .0234
10	0 27 .27	1 .49	0 .0247
20	0 28 .76	1 .54	0 .0261
30	0 30 .30	1 .57	0 .0274
40	0 31 .37	1 .62	0 .0289
50	0 33 .49	1 .65	0 .0303
7 0	0 35 .14	1 .69	0 .0318
10	0 36 .83	1 .73	0 .0333
20	0 38 .56	1 .78	0 .0349
30	0 40 .34	1 .81	0 .0366
40	0 42 .15	1 .85	0 .0381
50	0 44 .00	1 .89	0 .0399
8 0	0 45 .89	1 .94	0 .0416
10	0 47 .83	1 .97	0 .0434
20	0 49 .80	2 .01	0 .0452
30	0 51 .81	2 .05	0 .0470
40	0 53 .86	2 .09	0 .0489
50	0 55 .95	2 .13	0 .0508
9 0	0 58 .08	2 .18	0 .0527
10	1 0 .26	2 .21	0 .0547
20	1 2 .47	2 .25	0 .0567
30	1 4 .72	2 .29	0 .0587
40	1 7 .01	2 .33	0 .0608
50	1 9 .34	2 .37	0 .0629
10 0	1 11 .71		0 .0651

Arg. m. s.	Reduction.	Diff.	Var. in 100″ change of obliquity.
		2″.37	
10 0	1′ 11″.71	2 .41	0 .0651
10	1 14 .12	2 .45	0 .0673
20	1 16 .57	2 .49	0 .0695
30	1 19 .06	2 .53	0 .0717
40	1 21 .59	2 .57	0 .0740
50	1 24 .16	2 .61	0 .0763
11 0	1 26 .77	2 .65	0 .0787
10	1 29 .42	2 .69	0 .0811
20	1 32 .11	2 .72	0 .0836
30	1 34 .83	2 .77	0 .0860
40	1 37 .60	2 .81	0 .0886
50	1 40 .41	2 .85	0 .0911
12 0	1 43 .26	2 .89	0 .0937
10	1 46 .15	2 .93	0 .0963
20	1 49 .08	2 .96	0 .0990
30	1 52 .04	2 .01	0 .1017
40	1 55 .05	3 .05	0 .1044
50	1 58 .10	3 .09	0 .1072
13 0	2 1 .19	3 .12	0 .1100
10	2 4 .31	3 .17	0 .1128
20	2 7 .48	3 .21	0 .1157
30	2 10 .69	3 .25	0 .1186
40	2 13 .94	3 .28	0 .1216
50	2 17 .22	3 .33	0 .1246
14 0	2 20 .55	3 .36	0 .1276
10	2 23 .91	3 .41	0 .1307
20	2 27 .32	3 .45	0 .1338
30	2 30 .77	3 .48	0 .1369
40	2 34 .25	3 .53	0 .1401
50	2 37 .78	3 .56	0 .1433
15 0	2 41 .34	3 .61	0 .1465
10	2 44 .95	3 .64	0 .1498
20	2 48 .59	3 .69	0 .1531
30	2 52 .28	3 .72	0 .1564
40	2 56 .00	3 .77	0 .1598
50	2 59 .77	3 .80	0 .1632
16 0	3 3 .57	3 .85	0 .1667
10	3 7 .42	3 .88	0 .1702
20	3 11 .30	3 .92	0 .1737
30	3 15 .22	3 .97	0 .1773
40	3 19 .19	4 .00	0 .1809
50	3 23 .19	4 .05	0 .1845
17 0	3 27 .24	4 .08	0 .1882
10	3 31 .32	4 .12	0 .1919
20	3 35 .44	4 .16	0 .1957
30	3 39 .60	4 .21	0 .1994
40	3 43 .81	4 .24	0 .2033
50	3 48 .05	4 .28	0 .2071
18 0	3 52 .33	4 .33	0 .2110
10	3 56 .66	4 .36	0 .2149
20	4 1 .02	4 .40	0 .2189
30	4 5 .42	4 .44	0 .2229
40	4 9 .86	4 .48	0 .2269
50	4 14 .34	4 .53	0 .2310
19 0	4 18 .87	4 .56	0 .2351
10	4 23 .43	4 .60	0 .2393
20	4 28 .03	4 .64	0 .2435
30	4 32 .67	4 .68	0 .2477
40	4 37 .35	4 .72	0 .2520
50	4 42 .07	4 .76	0 .2563
20 0	4 46 .83		0 .2606

Arg. m. s.	Reduction.	Diff.	Var. in 100″ change of obliquity.
		4″.76	
20 0	4′ 46′ .83	4 .80	0 .2606
10	4 51 .63	4 .84	0 .2650
20	4 56 .47	4 .88	0 .2694
30	5 1 .35	4 .92	0 .2738
40	5 6 .27	4 .96	0 .2783
50	5 11 .23	5 .00	0 .2728
21 0	5 16 .23	5 .04	0 .2873
10	5 21 .27	5 .07	0 .2919
20	5 26 .34	5 .12	0 .2965
30	5 31 .46	5 .16	0 .3012
40	5 36 .62	5 .20	0 .3059
50	5 41 .82	5 .24	0 .3106
22 0	5 47 .06	5 .28	0 .3154
10	5 52 .34	5 .32	0 .3102
20	5 57 .66	5 .35	0 .3250
30	6 3 .01	5 .40	0 .3299
40	6 8 .41	5 .44	0 .3349
50	6 13 .85	5 .48	0 .3398
23 0	6 19 .33	5 .51	0 .3448
10	6 24 .84	5 .56	0 .3498
20	6 30 .40	5 .60	0 .3549
30	6 36 .00	5 .64	0 .3600
40	6 41 .64	5 .67	0 .3651
50	6 47 .31	5 .72	0 .3703
24 0	6 53 .03	5 .75	0 .3755
10	6 58 .78	5 .80	0 .3807
20	7 4 .58	5 .84	0 .3860
30	7 10 .42	5 .87	0 .3913
40	7 16 .29	5 .92	0 .3967
50	7 22 .21	5 .95	0 .4021
25 0	7 28 .16	6 .00	0 .4075
10	7 34 .16	6 .03	0 .4130
20	7 40 .19	6 .08	0 .4185
30	7 46 .27	6 .11	0 .4240
40	7 52 .38	6 .15	0 .4296
50	7 58 .53	6 .20	0 .4352
26 0	8 4 .73	6 .23	0 .4408
10	8 10 .96	6 .28	0 .4465
20	8 17 .24	6 .31	0 .4522
30	8 23 .55	6 .35	0 .4579
40	8 29 .90	6 .40	0 .4637
50	8 36 .30	6 .43	0 .4695
27 0	8 42 .73	6 .47	0 .4754
10	8 49 .20	6 .52	0 .4813
20	8 55 .72	6 .55	0 .4873
30	9 2 .27	6 .59	0 .4932
40	9 8 .86	6 .63	0 .4993
50	9 15 .49	6 .68	0 .5053
28 0	9 22 .17	6 .71	0 .5114
10	9 28 .88	6 .75	0 .5175
20	9 35 .63	6 .79	0 .5237
30	9 42 .42	6 .83	0 .5299
40	9 49 .25	6 .87	0 .5361
50	9 56 .12	6 .91	0 .5424
29 0	10 3 .03	6 .95	0 .5487
10	10 9 .98	6 .99	0 .5550
20	10 16 .97	7 .04	0 .5614
30	10 24 .01	7 .07	0 .5678
40	10 31 .08	7 .11	0 .5743
50	10 38 .19	7 .15	0 .5808
30 0	10 43 .34		0 .5873

This Table gives the Differences between the Obliquity 23° 27′ 40″ and the Sun's Declination, at various points of the Ecliptic near the Solstices, and the Column of Variations in 100″ change of the Obliquity, will give the further correction for any given diminution of the Obliquity.

TABLE 10. Reduction to either Solstice, concluded.

Arg. m.	s.	Reduction.	Diff.	Var. in 100″ change of obliquity.
30	0	10′ 45″.84	7″.15	0.5873
	10	10 52.53	7.19	0.5939
	20	10 59.76	7.23	0.6005
	30	11 7.03	7.27	0.6071
	40	11 14.34	7.31	0.6138
	50	11 21.69	7.35	0.6205
31	0	11 29.08	7.39	0.6273
	10	11 36.51	7.43	0.6341
	20	11 43.97	7.46	0.6409
	30	11 51.48	7.51	0.6478
	40	11 59.03	7.55	0.6547
	50	12 6.61	7.58	0.6616
32	0	12 14.24	7.63	0.6686
	10	12 21.91	7.67	0.6756
	20	12 29.62	7.71	0.6827
	30	12 37.56	7.74	0.6897
	40	12 45.15	7.79	0.6969
	50	12 52.98	7.83	0.7040
33	0	13 0.84	7.86	0.7112
	10	13 8.75	7.91	0.7184
	20	13 16.70	7.95	0.7257
	30	13 24.68	7.98	0.7330
	40	13 32.71	8.03	0.7403
	50	13 40.77	8.06	0.7477
34	0	13 48.88	8.11	0.7551
	10	13 57.03	8.15	0.7626
	20	14 5.21	8.18	0.7701
	30	14 13.44	8.23	0.7776
	40	14 21.70	8.26	0.7852
	50	14 30.01	8.31	0.7928
35	0	14 38.35	8.34	0.8004
	10	14 46.73	8.38	0.8081
	20	14 55.16	8.43	0.8158
	30	15 3.62	8.46	0.8235
	40	15 12.13	8.51	0.8313
	50	15 20.67	8.54	0.8391
36	0	15 29.25	8.58	0.8470
	10	15 37.88	8.63	0.8549
	20	15 46.54	8.66	0.8628
	30	15 55.24	8.70	0.8708
	40	16 3.98	8.74	0.8788
	50	16 12.76	8.78	0.8868
37	0	16 21.59	8.83	0.8949
	10	16 30.45	8.86	0.9050
	20	16 39.35	8.90	0.9112
	30	16 48.29	8.94	0.9194
	40	16 57.27	8.98	0.9276
	50	17 6.29	9.02	0.9359
38	0	17 15.36	9.07	0.9442
	10	17 24.46	9.10	0.9025
	20	17 33.60	9.14	0.9609
	30	17 42.78	9.18	0.9693
	40	17 52.00	9.22	0.9778
	50	18 1.26	9.26	0.9863
39	0	18 10.56	9.30	0.9948
	10	18 19.90	9.34	1.0034
	20	18 29.28	9.38	1.0120
	30	18 38.70	9.42	1.0206
	40	18 48.15	9.45	1.0293
	50	18 57.65	9.50	1.0380
40	0	19 7.19	9.54	1.0468

Arg. m.	s.	Reduction.	Diff.	Var. in 100″ change of obliquity.
40	0	19′ 7″.19	9″.54	1.0468
	10	19 16.77	9.58	1.0556
	20	19 26.39	9.62	1.0644
	30	19 36.05	9.66	1.0732
	40	19 45.75	9.70	1.0821
	50	19 55.48	9.73	1.0910
41	0	20 5.26	9.78	1.1000
	10	20 15.08	9.82	1.1090
	20	20 24.94	9.86	1.1181
	30	20 34.83	9.89	1.1272
	40	20 44.77	9.94	1.1363
	50	20 54.74	9.97	1.1455
42	0	21 4.76	10.02	1.1547
	10	21 14.81	10.05	1.1639
	20	21 24.91	10.10	1.1732
	30	21 35.04	10.13	1.1825
	40	21 45.22	10.18	1.1919
	50	21 55.44	10.22	1.2013
43	0	22 5.69	10.25	1.2107
	10	22 15.99	10.30	1.2202
	20	22 26.32	10.33	1.2297
	30	22 36.70	10.38	1.2392
	40	22 47.11	10.41	1.2488
	50	22 57.56	10.46	1.2584
44	0	23 8.06	10.50	1.2681
	10	23 18.59	10.53	1.2778
	20	23 29.16	10.57	1.2875
	30	23 39.78	10.62	1.2973
	40	23 50.43	10.65	1.3071
	50	24 1.12	10.69	1.3169
45	0	24 11.86	10.74	1.3268
	10	24 22.63	10.77	1.3367
	20	24 33.44	10.81	1.3467
	30	24 44.29	10.85	1.3567
	40	24 55.18	10.89	1.3667
	50	25 6.12	10.94	1.3768
46	0	25 17.09	10.97	1.3869
	10	25 28.10	11.01	1.3971
	20	25 39.15	11.05	1.4073
	30	25 50.24	11.09	1.4175
	40	26 1.37	11.13	1.4278
	50	26 12.54	11.17	1.4381
47	0	26 23.75	11.21	1.4484
	10	26 35.00	11.25	1.4587
	20	26 46.29	11.29	1.4692
	30	26 57.62	11.33	1.4796
	40	27 8.98	11.36	1.4901
	50	27 20.39	11.41	1.5006
48	0	27 31.84	11.45	1.5112
	10	27 43.33	11.49	1.5218
	20	27 54.86	11.53	1.5324
	30	28 6.43	11.57	1.5431
	40	28 18.04	11.61	1.5538
	50	28 29.68	11.64	1.5645
49	0	28 41.37	11.09	1.5753
	10	28 53.10	11.73	1.5861
	20	29 4.86	11.76	1.5970
	30	29 16.67	11.81	1.6079
	40	29 28.52	11.85	1.6188
	50	29 40.40	11.88	1.6298
50	0	29 52.33	11.93	1.6408

Arg. m.	s.	Reduction.	Diff.	Var. in 100″ change of obliquity.
50	0	29′ 52″.33	11′.93	1.6408
	10	30 4.29	11.96	1.6519
	20	30 16.30	12.01	1.6630
	30	30 28.34	12.04	1.6741
	40	30 40.43	12.09	1.6853
	50	30 52.55	12.12	1.6965
51	0	31 4.72	12.17	1.7078
	10	31 16.92	12.20	1.7191
	20	31 29.17	12.25	1.7304
	30	31 41.45	12.28	1.7417
	40	31 53.77	12.32	1.7531
	50	32 6.14	12.37	1.7645
52	0	32 18.54	12.40	1.7760
	10	32 30.98	12.44	1.7875
	20	32 43.47	12.49	1.7991
	30	32 55.99	12.52	1.8106
	40	33 8.55	12.56	1.8223
	50	33 20.15	12.60	1.8339
53	0	33 33.79	12.64	1.8456
	10	33 46.47	12.68	1.8574
	20	33 59.20	12.73	1.8692
	30	34 11.96	12.76	1.8810
	40	34 24.76	12.80	1.8929
	50	34 37.60	12.84	1.9048
54	0	34 50.48	12.88	1.9167
	10	35 3.40	12.92	1.9287
	20	35 16.36	12.96	1.9407
	30	35 29.36	13.00	1.9527
	40	35 42.39	13.03	1.9648
	50	35 55.47	13.08	1.9769
55	0	36 8.59	13.12	1.9891
	10	36 21.75	13.16	2.0013
	20	36 34.95	13.20	2.0136
	30	36 48.19	13.24	2.0258
	40	37 1.47	13.28	2.0382
	50	37 14.78	13.31	2.0505
56	0	37 28.14	13.36	2.0629
	10	37 41.54	13.40	2.0753
	20	37 54.07	13.43	2.0878
	30	38 8.45	13.48	2.1003
	40	38 21.96	13.51	2.1228
	50	38 35.52	13.56	2.1254
57	0	38 49.12	13.60	2.1380
	10	39 2.75	13.63	2.1507
	20	39 16.43	13.68	2.1634
	30	39 30.14	13.71	2.1762
	40	39 43.90	13.76	2.1889
	50	39 57.69	13.79	2.2017
58	0	40 11.53	13.84	2.2146
	10	40 25.40	13.87	2.2275
	20	40 39.31	13.91	2.2405
	30	40 53.27	13.96	2.2534
	40	41 7.26	13.99	2.2665
	50	41 21.29	13.03	2.2795
59	0	41 35.36	14.07	2.2926
	10	41 49.48	14.12	2.3057
	20	42 3.63	14.15	2.3189
	30	42 17.82	14.19	2.3321
	40	42 32.05	14.23	2.3454
	50	42 46.32	14.27	2.3587
60	0	43 0.64	14.32	2.3720

This Table gives the Differences between the Obliquity 23° 27′ 40″ and the Sun's Declination at various points of the Ecliptic near the Solstices, and the Column of Variations in 100″ change of the Obliquity, will give the further Correction for any given diminution of the Obliquity.

SOLAR TABLES.

TABLE 11.

The Sun's Latitude, or Deviation from the Ecliptic.

Arg.	A+B+N	2B−C	3C−4B	B−2E	Arg.	A+B+N	2B−C	3C−4B	B−2E	Arg.	A+B+N	2B−C	3C−4B	B−2E
0	−0″.00	+0″.10	+0″.24	+0″.16	340	−0″.58	−0″.08	−0″.07	−0″.11	670	+0″.60	−0″.02	−0″.17	−0″.06
10	0.04	0.10	0.25	0.16	350	0.56	0.08	0.09	0.11	680	0.62	0.02	0.16	0.05
20	0.08	0.10	0.25	0.16	360	0.53	0.09	0.10	0.12	690	0.64	0.01	0.15	0.04
30	0.12	0.09	0.25	0.16	370	0.50	0.09	0.12	0.13	700	0.65	0.01	0.14	0.03
40	0.16	0.09	0.25	0.15	380	0.47	0.09	0.13	0.14	710	0.66	0.00	0.12	0.02
50	0.20	0.09	0.25	0.15	390	0.44	0.10	0.15	0.14	720	0.66	+0.01	0.11	0.01
60	0.24	0.08	0.25	0.14	400	0.40	0.10	0.16	0.15	730	0.67	0.01	0.09	0.00
70	0.28	0.08	0.25	0.14	410	0.36	0.10	0.17	0.15	740	0.67	0.02	0.08	+0.01
80	0.32	0.08	0.24	0.14	420	0.32	0.10	0.18	0.16	750	0.68	0.03	0.06	0.02
90	0.36	0.07	0.24	0.13	430	0.28	0.10	0.19	0.16	760	0.68	0.03	0.05	0.03
100	0.40	0.07	0.23	0.13	440	0.24	0.10	0.20	0.16	770	0.67	0.04	0.03	0.04
110	0.44	0.06	0.23	0.12	450	0.20	0.11	0.21	0.16	780	0.67	0.04	0.02	0.05
120	0.47	0.06	0.22	0.11	460	0.16	0.11	0.22	0.16	790	0.66	0.05	0.00	0.06
130	0.50	0.05	0.21	0.10	470	0.12	0.11	0.23	0.16	800	0.65	0.06	+0.01	0.07
140	0.53	0.04	0.20	0.09	480	0.08	0.10	0.23	0.16	810	0.64	0.06	0.03	0.08
150	0.56	0.04	0.19	0.08	490	0.04	0.10	0.24	0.16	820	0.62	0.07	0.04	0.09
160	0.58	0.03	0.18	0.07	500	0.00	0.10	0.24	0.16	830	0.60	0.07	0.06	0.10
170	0.60	0.02	0.17	0.06	510	+0.04	0.10	0.25	0.16	840	0.58	0.08	0.07	0.11
180	0.62	0.02	0.16	0.05	520	0.08	0.10	0.25	0.16	850	0.56	0.08	0.09	0.11
190	0.64	0.01	0.15	0.04	530	0.12	0.09	0.25	0.16	860	0.53	0.09	0.10	0.12
200	0.65	0.01	0.14	0.03	540	0.16	0.09	0.25	0.16	870	0.50	0.09	0.12	0.13
210	0.66	−0.00	0.12	0.02	550	0.20	0.09	0.25	0.15	880	0.47	0.09	0.13	0.14
220	0.66	0.01	0.11	0.01	560	0.24	0.08	0.25	0.15	890	0.44	0.10	0.15	0.14
230	0.67	0.01	0.09	0.00	570	0.28	0.08	0.25	0.14	900	0.40	0.10	0.16	0.15
240	0.67	0.02	0.08	−0.01	580	0.32	0.08	0.24	0.14	910	0.36	0.10	0.17	0.15
250	0.68	0.03	0.06	0.02	590	0.36	0.07	0.24	0.13	920	0.32	0.10	0.18	0.16
260	0.68	0.03	0.05	0.03	600	0.40	0.07	0.23	0.13	930	0.28	0.10	0.19	0.16
270	0.67	0.04	0.03	0.04	610	0.44	0.06	0.23	0.13	940	0.24	0.10	0.20	0.16
280	0.67	0.04	0.02	0.05	620	0.47	0.06	0.22	0.12	950	0.20	0.11	0.21	0.16
290	0.66	0.05	0.00	0.06	630	0.50	0.05	0.21	0.11	960	0.16	0.11	0.22	0.16
300	0.65	0.06	0.01	0.07	640	0.53	0.04	0.20	0.10	970	0.12	0.11	0.23	0.16
310	0.64	0.06	0.03	0.08	650	0.56	0.04	0.19	0.09	980	0.08	0.10	0.23	0.16
320	0.62	0.07	0.04	0.09	660	0.58	0.03	0.18	0.08	990	0.04	0.10	0.24	0.16
330	0.60	0.07	0.06	0.10	670	0.60			0.07	1000	0.00	0.10	0.24	0.16
Arg.	A+B+N	2B−C	2C−4B	B−2E	Arg.	A+B+N	2B−C	3C−4B	B−2E	Arg.	A+B+N	2B−C	3C−4B	B−2E

TABLE 12.

The Corrections of the Sun's Longitude, Right Ascension, and Declination, depending on his Latitude.

Argument = Sun's Long.	Corrections of Long.	Corrections of R. A.	Corrections of Declin.	Argument = Sun's Long.	Corrections of Long.	Corrections of R. A.	Corrections of Declin.	Argument = Sun's Long.	Corrections of Long.	Corrections of R. A.	Corrections of Declin.
s s	+	+	−	s s	+	+	−	s s	−	−	−
0 0° VI	0″.43	0″.40	0″.92	II 0° VIII	0″.22	0″.23	0″.97	IV 0° X	0″.22	0″.23	0″.97
5	0.43	0.40	0.92	5	0.18	0.19	0.98	5	0.25	0.26	0.97
10	0.43	0.39	0.92	10	0.15	0.16	0.99	10	0.28	0.28	0.96
15	0.42	0.39	0.93	15	0.11	0.12	0.99	15	0.31	0.31	0.96
20	0.41	0.38	0.93	20	0.08	0.08	1.00	20	0.33	0.33	0.95
25	0.39	0.37	0.93	25	0.04	0.04	1.00	25	0.36	0.34	0.94
I 0 VII	0.38	0.36	0.94	III 0 IX	0.00	0.00	1.00	V 0 XI	0.38	0.36	0.94
					−	−			+	+	
5	0.36	0.34	0.94	5	0.04	0.04	1.00	5	0.39	0.37	0.93
10	0.33	0.33	0.95	10	0.08	0.08	0.99	10	0.41	0.38	0.93
15	0.31	0.31	0.96	15	0.11	0.12	0.99	15	0.42	0.39	0.93
20	0.28	0.28	0.96	20	0.15	0.16	0.99	20	0.43	0.39	0.92
25	0.25	0.26	0.97	25	0.18	0.19	0.98	25	0.43	0.40	0.92
II 0 VIII	0.22	0.23	0.97	IV 0 X	0.22	0.23	0.97	VI 0 XII	0.43	0.40	0.92

In Table 11, the Argument supposes the Circle divided into 1000 parts, and the amount of the four quantities, taken with their Algebraic Signs, gives the Sun's Latitude. In the formation of the respective Arguments A is put = ☉'s Mean Longitude — ☾'s do.; B = Heliocentric Longitude of the Earth; C = that of Venus; E = that of Jupiter; and N = the Supplement of the ☾ ☊'s Longitude, all taken millesimally.

In Table 12, the Sun's Latitude is supposed to be = + 1″; and the Tabular Numbers taken therefrom, multiplied into the Latitude (found in Table 11.) will give the respective Corrections; but if the Latitude be negative, the Signs must be changed.

TABLE 13.

The Sun's Horary Motion in Longitude, Right Ascension, and Declination, with his Semi-Diameter in Sidereal Time.

Argument = the Sun's true Longitude, or corresponding Day of the Month.

Day of the Month.	☉'s true Long.	Motion in Long.	Diff. in 5°.	Motion in R.A.	Diff. in 5°.	Motion in Decl.	Diff. in 5°.	Semi-diameter in Sidereal Time.	Diff. in 5°.
	s							m. s.	
Mar. 21	O 0°	148″.77		136″.46		59″.23		1 4.4	
			0.43		−0.23		0.36		.0
26	5	148.34		136.23		58.87		4.4	
			0.43		+0.09		0.74		.0
31	10	147.91		136.32		58.13		4.4	
			0.43		0.41		1.11		.1
April 5	15	147.48		136.73		57.02		4.5	
			0.43		0.71		1.49		.2
10	20	147.05		137.44		55.53		4.7	
			0.42		0.98		1.85		.2
15	25	146.63		138.42		53.68		4.9	
			0.41		1.23		2.23		.3
20	I 0	146.22		139.65		51.45		5.2	
			0.40		1.47		2.60		.4
25	5	145.82		141.12		48.85		5.6	
			0.38		1.65		2.96		.4
May 1	10	145.44		142.77		45.89		6.0	
			0.36		1.76		3.33		.4
6	15	145.08		144.53		42.56		6.4	
			0.34		1.85		3.66		.4
11	20	144.74		146.38		38.89		6.8	
			0.32		1.87		4.00		.4
16	25	144.42		148.25		34.89		7.2	
			0.29		1.81		4.32		.4
21	II 0	144.13		150.06		30.57		7.6	
			0.26		1.67		4.61		.4
27	5	143.87		151.73		25.96		8.0	
			0.23		1.48		4.87		.3
June 1	10	143.64		153.21		21.09		8.3	
			0.20		1.21		5.08		.2
6	15	143.44		154.42		16.01		8.5	
			0.16		0.88		5.24		.2
11	20	143.28		155.30		10.77		8.7	
			0.13		0.53		5.36		.1
16	25	143.15		155.83		5.41		8.8	
			0.09		0.12		5.41		.1
22	III 0	143.06		155.95		0.00		8.9	
			0.06		0.28		5.41		.1
27	5	143.00		155.67		5.41		8.8	
			0.02		0.68		5.34		.1
July 2	10	142.98		154.99		10.75		8.7	
			0.02		1.04		5.21		.2
7	15	143.00		153.95		15.96		8.5	
			0.06		1.34		5.05		.3
13	20	143.06		152.61		21.01		8.2	
			0.09		1.64		4.82		.4
18	25	143.15		150.97		25.83		7.8	
			0.13		1.80		4.56		.4
23	IV 0	143.28		149.17		30.39		7.4	
			0.16		1.93		4.25		.4
28	5	143.44		147.24		34.64		7.0	
			0.20		1.97		3.96		.5
Aug. 3	10	143.64		145.27		38.60		6.5	
			0.23		1.92		3.61		.4
8	15	143.87		143.35		42.21		6.1	
			0.26		1.87		3.26		.4
13	20	144.13		141.48		45.47		5.7	
			0.29		1.71		2.91		.4
18	25	144.42		139.77		48.38		5.3	
			0.32		1.55		2.54		.4
23	V 0	144.74		138.22		50.92		4.9	
			0.34		1.26		2.19		.3
29	5	145.08		136.96		53.11		4.6	
			0.36		1.02		1.81		.3
Sept. 3	10	145.44		135.94		54.92		4.3	
			0.38		0.74		1.46		.2
8	15	145.82		135.20		56.38		4.1	
			0.40		0.43		1.09		.1
13	20	146.22		134.77		57.47		4.0	
			0.41		0.10		0.72		.0
18	25	146.63		134.87		58.19		4.0	
			0.42		0.02		0.36		.1
23	VI 0	147.05		134.89		58.55		4.1	
Day of the Month.	☉'s true Long.	Motion in Long.	Diff. in 5°.	Motion in R.A.	Diff. in 5°.	Motion in Decl.	Diff. in 5°.	Semi-diameter in Sidereal Time.	Diff. in 5°.

Day of the Month.	☉'s true Long.	Motion in Long.	Diff. in 5°	Motion in R. A.	Diff. in 5°.	Motion in Decl.	Diff. in 5°.	Semi-diameter in Sidereal Time.	Diff. in 5°.
	s							m. s.	
Sept. 23	VI 0°	147″.05		134″.89		58″.55		1 4.1	
			0.43		0.76		0.00		.1
28	5	147.48		135.65		58.53		4.2	
			0.43		0.68		0.40		.2
Oct. 3	10	147.91		136.33		58.13		4.4	
			0.43		1.21		0.78		.3
9	15	148.34		137.54		57.35		4.7	
			0.43		1.51		1.17		.4
14	20	148.77		139.05		56.18		5.1	
			0.43		1.81		1.56		.4
19	25	149.20		140.86		54.62		5.5	
			0.41		2.02		1.94		.5
24	VII 0	149.61		142.88		52.64		6.0	
			0.41		2.31		2.38		.5
29	5	150.02		145.19		50.26		6.5	
			0.38		2.45		2.81		.6
Nov. 3	10	150.40		147.64		47.45		7.1	
			0.37		2.57		3.22		.6
8	15	150.77		150.21		44.23		7.7	
			0.35		2.63		3.62		.6
13	20	151.12		152.84		40.61		8.3	
			0.32		2.61		4.03		.5
18	25	151.44		155.45		36.58		8.8	
			0.30		2.53		4.40		.6
22	VIII 0	151.74		157.98		32.18		9.4	
			0.27		2.33		4.75		.5
27	5	152.01		160.31		27.43		9.9	
			0.23		2.08		5.08		.4
Dec. 2	10	152.24		162.39		22.35		10.3	
			0.22		1.73		5.32		.4
7	15	152.46		164.12		17.03		10.7	
			0.16		1.35		5.56		.3
12	20	152.62		165.43		11.47		11.0	
			0.13		0.85		5.70		.1
17	25	152.75		166.28		5.77		11.1	
			0.09		0.33		5.77		.1
22	IX 0	152.84		166.61		0.00		11.2	
			0.06		0.17		5.78		.1
27	5	152.90		166.44		5.78		11.1	
			0.02		0.68		5.71		.1
Jan. 1	10	152.92		165.76		11.49		11.0	
			0.02		1.16		5.58		.2
6	15	152.90		164.60		17.07		10.8	
			0.06		1.57		5.37		.3
11	20	152.84		163.03		22.44		10.5	
			0.10		1.95		5.12		.4
16	25	152.74		161.08		27.56		10.1	
			0.13		2.19		4.80		.5
20	X 0	152.61		158.89		32.36		9.6	
			0.16		2.40		4.47		.5
25	5	152.45		156.49		36.83		9.1	
			0.21		2.51		4.08		.6
30	10	152.24		153.98		40.91		8.5	
			0.23		2.54		3.69		.6
Feb. 4	15	152.01		151.44		44.60		7.9	
			0.27		2.49		3.28		.5
9	20	151.74		143.95		47.88		7.4	
			0.30		2.39		2.85		.6
14	25	151.44		146.56		50.73		6.8	
			0.32		2.24		2.44		.5
19	XI 0	151.12		144.32		53.17		6.3	
			0.34		1.98		2.02		.5
24	5	150.78		142.34		55.19		5.8	
			0.36		1.77		1.61		.4
March 1	10	150.42		140.57		56.80		5.4	
			0.39		1.48		1.20		.3
6	15	150.03		139.09		58.00		5.1	
			0.41		1.18		0.81		.3
11	20	149.62		137.91		58.81		4.8	
			0.42		0.88		0.40		.2
16	25	149.20		137.03		59.21		4.6	
			0.43		0.57		0.02		.1
21	XII 0	148.77		136.46		59.23		4.5	
Day of the Month.	☉'s true Long.	Motion in Long.	Diff. in 5°.	Motion in R. A.	Diff. in 5°.	Motion in Decl.	Diff. in 5°.	Semi-diameter in Sidereal Time.	Diff. in 5°.

If a Clock indicating *Solar* Time be used, two-tenths of a Second must be subtracted from the Tabular value of the Semi-diameter's Time of Meridian passage, and the remainder will be the measure in Solar Time.

TABLE 14.

THE MEAN OBLIQUITY of the ECLIPTIC for the beginning of each of 150 Years.

Years.	Ann Dim. 0″.48 Delambre's Obliquity.	Ann. Dim. 0″.43 Brinkley's Obliquity.	Years.	Ann. Dim. 0″.48 Delambre's Obliquity.	Ann. Dim. 0″.43 Brinkley's Obliquity.	Years.	Ann. Dim. 0″.48 Delambre's Obliquity.	Ann. Dim. 0″.43 Brinkley's Obliquity.
1750	23° 28′ 21″.00	23° 28′ 17″.54	1800	23° 27′ 57″.00	23° 27′ 56″.04	1850	23° 27′ 33″.00	23° 27′ 34″.54
1751	23 28 20.52	23 28 17.11	1801	23 27 56.52	23 27 55.61	1851	23 27 32.52	23 27 34.11
1752	23 28 20.04	23 28 16.68	1802	23 27 56.04	23 27 55.18	1852	23 27 32.04	23 27 33.68
1753	23 28 19.56	23 28 16.25	1803	23 27 55.56	23 27 54.75	1853	23 27 31.56	23 27 33.25
1754	23 28 19.08	23 28 15.82	1804	23 27 55.08	23 27 54.32	1854	23 27 31.08	23 27 32.82
1755	23 28 18.60	23 28 15.39	1805	23 27 54.60	23 27 53.89	1855	23 27 30.60	23 27 32.39
1756	23 28 18.12	23 28 14.96	1806	23 27 54.12	23 27 53.46	1856	23 27 30.12	23 27 31.96
1757	23 28 17.64	23 28 14.53	1807	23 27 53.64	23 27 53.03	1857	23 27 29.64	23 27 31.53
1758	23 28 17.16	23 28 14.10	1808	23 27 53.16	23 27 52.60	1858	23 27 29.16	23 27 31.10
1759	23 28 16.68	23 28 13.67	1809	23 27 52.68	23 27 52.17	1859	23 27 28.68	23 27 30.67
1760	23 28 16.20	23 28 13.24	1810	23 27 52.20	23 27 51.74	1860	23 27 28.20	23 27 30.24
1761	23 28 15.72	23 28 12.81	1811	23 27 51.72	23 27 51.31	1861	23 27 27.72	23 27 29.31
1762	23 28 15.24	22 28 12.38	1812	23 27 51.24	23 27 50.88	1862	23 27 27.24	23 27 29.38
1763	23 28 14.76	23 28 11.95	1813	23 27 50.76	23 27 50.45	1863	23 27 26.76	23 27 28.95
1764	23 28 14.28	23 28 11.52	1814	23 27 50.28	23 27 50.02	1864	23 27 26.28	23 27 28.52
1765	23 28 13.80	23 28 11.09	1815	23 27 49.80	23 27 49.59	1865	23 27 25.80	23 27 28.09
1766	23 28 13.32	23 28 10.66	1816	23 27 49.32	23 27 49.16	1866	23 27 25.32	23 27 27.66
1767	23 28 12.84	23 28 10.23	1817	23 27 48.84	23 27 48.73	1867	23 27 24.84	23 27 27.23
1768	23 28 12.36	23 28 9.80	1818	23 27 48.36	23 27 48.30	1868	23 27 24.36	23 27 26.80
1769	23 28 11.88	23 28 9.37	1819	23 27 47.88	23 27 47.87	1869	23 27 23.88	23 27 26.37
1770	23 28 11.40	23 28 8.94	1820	23 27 47.40	23 27 47.44	1870	23 27 23.40	23 27 25.94
1771	23 28 10.92	23 28 8.51	1821	23 27 46.92	23 27 47.01	1871	23 27 22.92	23 27 25.51
1772	23 28 10.44	23 28 8.08	1822	23 27 46.44	23 27 46.58	1872	23 27 22.44	23 27 25.08
1773	23 28 9.96	23 28 7.65	1823	23 27 45.96	23 27 46.15	1873	23 27 21.96	23 27 24.65
1774	23 28 9.48	23 28 7.22	1824	23 27 45.48	23 27 45.72	1874	23 27 21.48	23 27 24.22
1775	23 28 9.00	23 28 6.79	1825	23 27 45.00	23 27 45.29	1875	23 27 21.00	23 27 23.79
1776	23 28 8.52	23 28 6.36	1826	23 27 44.52	23 27 44.86	1876	23 27 20.52	23 27 23.36
1777	23 28 8.04	23 28 5.93	1827	23 27 44.04	23 27 44.43	1877	23 27 20.04	23 27 22.93
1778	23 28 7.56	23 28 5.50	1828	23 27 43.56	23 27 44.00	1878	23 27 19.56	23 27 22.50
1779	23 28 7.08	23 28 5.07	1829	23 27 43.08	23 27 43.57	1879	23 27 19.08	23 27 22.07
1780	23 28 6.60	23 28 4.64	1830	23 27 42.60	23 27 43.14	1880	23 27 18.60	23 27 21.64
1781	23 28 6.12	23 28 4.21	1831	23 27 42.12	23 27 42.71	1881	23 27 18.12	23 27 21.21
1782	23 28 5.64	23 28 3.78	1832	23 27 41.64	23 27 42.28	1882	23 27 17.64	23 27 20.78
1783	23 28 5.16	23 28 3.35	1833	23 27 41.16	23 27 41.85	1883	23 27 17.16	23 27 20.35
1784	23 28 4.68	23 28 2.92	1834	23 27 40.68	23 27 41.42	1884	23 27 16.68	23 27 19.92
1785	23 28 4.20	23 28 2.49	1835	23 27 40.20	23 27 40.99	1885	23 27 16.20	23 27 19.49
1786	23 28 3.72	23 28 2.06	1836	23 27 39.72	23 27 40.56	1886	23 27 15.72	23 27 19.06
1787	23 28 3.24	23 28 1.63	1837	23 27 39.24	23 27 40.13	1887	23 27 15.24	23 27 18.63
1788	23 28 2.76	23 28 1.20	1838	23 27 38.76	23 27 39.70	1888	23 27 14.76	23 27 18.20
1789	23 28 2.28	23 28 0.77	1839	23 27 38.28	23 27 39.27	1889	23 27 14.28	23 27 17.77
1790	23 28 1.80	23 28 0.34	1840	23 27 37.80	23 27 38.84	1890	23 27 13.80	23 27 17.34
1791	23 28 1.32	23 27 59.91	1841	23 27 37.32	23 27 38.41	1891	23 27 13.32	23 27 16.91
1792	23 28 0.84	23 27 59.48	1842	23 27 36.84	23 27 37.98	1892	23 27 12.84	23 27 16.48
1793	23 28 0.36	23 27 59.05	1843	23 27 36.36	23 27 37.55	1893	23 27 12.36	23 27 16.05
1794	23 27 59.88	23 27 58.62	1844	23 27 35.88	23 27 37.12	1894	23 27 11.88	23 27 15.62
1795	23 27 59.40	23 27 58.19	1845	23 27 35.40	23 27 36.69	1895	23 27 11.40	23 27 15.19
1796	23 27 58.92	23 27 57.76	1846	23 27 34.92	23 27 36.26	1896	23 27 10.92	23 27 14.76
1797	23 27 58.44	23 27 57.33	1847	23 27 34.44	23 27 35.83	1897	23 27 10.44	23 27 14.33
1798	23 27 57.96	23 27 56.90	1848	23 27 33.96	23 27 35.40	1898	23 27 9.96	23 27 13.90
1799	23 27 57.48	23 27 56.47	1849	23 27 33.48	23 27 34.97	1899	23 27 9.48	23 27 13.47
1800	23 27 57.00	23 27 56.04	1850	23 27 33.00	23 27 34.54	1900	23 27 9.00	23 27 13.04

The Annual Diminution determined by Bessel, from Bradley's Observations, is very nearly a *Mean* between Delambre's and Brinkley's, viz. — 0″.457, and by a recent determination — 0″.46.

TABLE 15. Monthly Obliq. of the Ecliptic.

	Delambre.	Brinkley.
January 1	0″.000	0″.000
February 1........	0 .042	0 .038
March 1	0 .079	0 .071
April 1	0 .120	0 .107
May 1...............	0 .159	0 .142
June 1...............	0 .200	0 .179
July 1...............	0 .239	0 .214
August 1...........	0 .280	0 .251
September 1	0 .321	0 .287
October 1	0 .360	0 .323
November 1	0 .401	0 .359
December 1	0 .441	0 .395

TABLE 17. Sol. Equat. of the Obliquity.

Arg. ⊙'s true Long. s	s	Equation.	Corresponding Days of the Month.	
O 0°	VI	+0″.43	Mar. 21	Sept. 23
5		0 .43	26	28
10		0 .41	31	Oct. 3
15		0 .37	April 5	9
20		0 .33	10	14
25		0 .29	15	19
I 0	VII	0 .22	20	24
5		0 .15	25	29
10		0 .08	May 1	Nov. 3
15		0 .00	6	8
20		−0 .08	11	13
25		0 .22	16	18
II 0	VIII	0 .29	21	22
5		0 .33	27	27
10		0 .37	June 1	Dec. 2
15		0 .41	6	7
20		0 .43	11	12
25		0 .43	16	17
III 0	IX	0 .43	22	22
5		0 .41	27	27
10		0 .37	July 2	Jan. 1
15		0 .33	7	6
20		0 .29	13	11
25		0 .22	18	16
IV 0	X	0 .15	23	20
5		0 .08	28	25
10		0 .00	Aug. 3	30
15		+0 .08	8	Feb. 4
20		0 .15	13	9
25		0 .22	18	14
V 0	XI	0 .29	23	19
5		0 .33	29	24
10		0 .38	Sept. 3	Mar. 1
15		0 .41	8	6
20		0 .43	13	11
25		0 .43	18	16
VI 0	XII	0 .43	23	21

TAB. 19. Sol. Nut. of the Equinoxes in R.A.

Arg. ⊙'s true Long.	O IX + / VI III +	I X + / VII IV +	Arg. ⊙'s true Long.
0°	0ˢ.00	0ˢ.06	30°
3	0 .01	0 .06	27
6	0 .02	0 .07	24
9	0 .03	0 .07	21
12	0 .03	0 .07	18
15	0 .03	0 .07	15
18	0 .04	0 .07	12
21	0 .05	0 .07	9
24	0 .05	0 .07	6
27	0 .06	0 .06	3
30	0 .06	0 .06	0
	II XI + / VIII V +		

TABLE 16. Lunar Equation of the Obliquity. [= 9″.648 cosin Long. ☾'s ☊.]

Arg. Long. ☾'s ☊.	O VI + −	I VII + −	II VIII + −	Arg. Long. ☾'s ☊.
0°	9″.65	8″.34	4″.82	30°
1	9 .64	8 .25	4 .67	29
2	9 .63	8 .16	4 .53	28
3	9 .62	8 .08	4 .37	27
4	9 .61	7 .94	4 .22	26
5	9 .60	7 .89	4 .07	25
6	9 .58	7 .79	3 .92	24
7	9 .56	7 .69	3 .76	23
8	9 .54	7 .59	3 .61	22
9	9 .52	7 .49	3 .45	21
10	9 .50	7 .38	3 .39	20
11	9 .46	7 .27	3 .14	19
12	9 .42	7 .16	2 .98	18
13	9 .39	7 .04	2 .82	17
14	9 .35	6 .93	2 .66	16
15	9 .31	6 .81	2 .49	15
16	9 .26	6 .69	2 .33	14
17	9 .21	6 .57	2 .17	13
18	9 .16	6 .45	2 .00	12
19	9 .11	6 .32	1 .84	11
20	9 .06	6 .19	1 .67	10
21	8 .99	6 .06	1 .51	9
22	8 .93	5 .93	1 .34	8
23	8 .87	5 .80	1 .17	7
24	8 .80	5 .66	1 .01	6
25	8 .73	5 .53	0 .84	5
26	8 .65	5 .38	0 .67	4
27	8 .57	5 .25	0 .50	3
28	8 .50	5 .11	0 .33	2
29	8 .43	4 .96	0 .17	1
30	8 .34	4 .82	0 .00	0
	+ − XI V	X IV + −	IX III + −	

TABLE 18. Lun. Nut. of the Equinoxes in R.A.

Arg. Long. ☾'s ☊.	− + O VI	− + I VII	− + II VIII	Arg. Long. ☾'s ☊.
0°	0ˢ.00	0ˢ.55	0ˢ.97	30°
1	0 .02	0 .57	0 .97	29
2	0 .04	0 .59	0 .98	28
3	0 .06	0 .60	0 .99	27
4	0 .08	0 .62	1 .00	26
5	0 .10	0 .64	1 .00	25
6	0 .11	0 .65	1 .01	24
7	0 .13	0 .67	1 .02	23
8	0 .15	0 .69	1 .02	22
9	0 .17	0 .70	1 .03	21
10	0 .19	0 .71	1 .04	20
11	0 .21	0 .73	1 .05	19
12	0 .23	0 .75	1 .05	18
13	0 .25	0 .76	1 .06	17
14	0 .26	0 .78	1 .06	16
15	0 .28	0 .79	1 .07	15
16	0 .30	0 .81	1 .07	14
17	0 .32	0 .82	1 .08	13
18	0 .34	0 .83	1 .08	12
19	0 .36	0 .84	1 .08	11
20	0 .37	0 .86	1 .09	10
21	0 .39	0 .87	1 .09	9
22	0 .41	0 .88	1 .09	8
23	0 .43	0 .89	1 .09	7
24	0 .44	0 .90	1 .10	6
25	0 .46	0 .91	1 .10	5
26	0 .48	0 .92	1 .10	4
27	0 .50	0 .94	1 .10	3
28	0 .51	0 .95	1 .10	2
29	0 .53	0 .96	1 .10	1
30	0 .56	0 .97	1 .10	0
	XI V + −	X IV + −	IX III + −	

TABLE 20. Greenwich Observations of the Obliquity.

Years.	Summer Solstices.	Winter Solstices.	Observers.
1753	....	23° 28′ 14″.66	Bradley.
1754	23° 28′ 17″.64	23 28 16 .27	do.
1755	23 28 15 .82	23 28 14 .53	do.
1756	23 28 12 .93	23 28 12 .72	do.
1757	23 28 13 .37	23 28 12 .35	do.
1758	23 28 11 .56	23 28 13 .60	do.
1759	23 28 9 .11	23 28 14 .41	do.
1760	23 28 10 .47	....	do.
1765	23 28 9 .50	23 28 6 .40	Maskelyne
1766	23 28 5 .90	23 28 9 .15	do.
1767	23 28 7 .60	23 28 2 .21	do.
1768	23 28 10 .07	23 28 5 .87	do.
1769	23 28 7 .84	23 28 6 .99	do.
1770	23 28 9 .01	23 28 4 .64	do.
1771	23 28 8 .24	23 28 2 .62	do.
1772	23 28 8 .22	....	do.
1775	....	23 28 3 .79	do.
1776	23 28 5 .90	23 27 58 .00	do.
1777	23 28 7 .10	23 28 0 .70	do.
1778	23 28 5 .30	23 27 58 .20	do.
1779	23 28 25 .30	23 27 58 .70	do.
1780	23 28 5 .40	23 27 58 .30	do.
1781	23 28 2 .60	....	do.
1785	23 27 59 .00	23 27 53 .90	do.
1786	23 27 58 .50	23 27 56 .75	do.
1787	23 27 59 .80	....	do.
1788	23 28 0 .75	23 27 57 .10	do.
1789	23 27 57 .80	33 27 56 .90	do.
1790	23 28 1 .15	23 27 52 .15	do.
1791	23 27 59 .17	23 27 49 .00	do.
1798	23 27 55 .09	23 27 43 .31	do.
1799	23 27 55 .44	23 27 44 .86	do.
1800	23 27 55 .91	23 27 44 .67	do.
1801	23 27 56 .06	23 27 41 .04	do.
1802	23 27 54 .10	23 27 44 .25	do.
1803	23 27 55 .80	....	do.
1804	23 27 53 .63	23 27 43 .51	do.
1805	23 27 50 .89	23 27 43 .05	do.
1806	23 27 52 .64	23 27 42 .78	do.
1807	23 27 52 .95	23 27 47 .28	do.
1808	23 27 51 .65	....	do.
1809	23 27 49 .54	23 27 44 .80	do.
1810	23 27 52 .72	23 27 40 .32	do.
1811	23 27 53 .90	23 27 39 .18	Pond.
1812	23 27 51 .90	23 27 46 .8	do.
1813	23 27 51 .3	23 27 46 .2	do.
1814	23 27 49 .7	23 27 44 .5	do.
1815	23 27 49 .4	23 27 44 .8	do.
1816	23 27 49 .4	23 27 43 .8	do.
1817	23 27 47 .9	23 27 42 .6	do.
1818	23 27 49 .4	23 27 42 .9	do.
1819	23 27 47 .2	23 27 45 .7	do.
1820	23 27 46 .5		do.
1821			
1822			
1823			
1824			
1825			
1826			
1827			
1828			
1829			
1830			

TABLE 21.

General Table of the Corrections of Noon, by corresponding equal Altitudes. PART I. (From DELAMBRE.)

Arg. ☉'s Long.	Log. d D. / 360° −	Diff.	Log. d D. tan. D. / 360° +	Diff.
0°	0.5968	−2	0.0000	+8.4384
1	0.5966	3	8.4384	3007
2	0.5963	5	8.7391	1756
3	0.5958	5	8.9147	1243
4	0.5953	7	9.0390	961
5	0.5946	8	9.1351	783
6	0.5938	9	9.2134	659
7	0.5929	10	9.2793	568
8	0.5919	11	9.3361	498
9	0.5908	13	9.3859	443
10	0.5895	14	9.4302	398
11	0.5881	15	9.4700	360
12	0.5866	16	9.5060	328
13	0.5850	17	9.5388	301
14	0.5833	19	9.5689	278
15	0.5814	20	9.5967	256
16	0.5794	21	9.6223	238
17	0.5773	23	9.6461	222
18	0.5750	24	9.6683	206
19	0.5726	25	9.6889	192
20	0.5701	27	9.7081	180
21	0.5674	28	9.7261	169
22	0.5646	30	9.7430	158
23	0.5616	31	9.7588	147
24	0.5585	33	9.7735	139
25	0.5552	34	9.7874	130
26	0.5518	35	9.8004	121
27	0.5483	37	9.8125	114
28	0.5446	39	9.8239	106
29	0.5407	41	9.8345	99
30	0.5366	42	9.8444	91
31	0.5324	44	9.8535	86
32	0.5280	46	9.8621	79
33	0.5234	47	9.8700	73
34	0.5187	50	9.8773	66
35	0.5137	51	9.8839	61
36	0.5086	53	9.8900	55
37	0.5033	55	9.8955	50
38	0.4978	58	9.9005	44
39	0.4920	60	9.9049	39
40	0.4860	62	9.9088	33
41	0.4798	64	9.9121	28
42	0.4734	66	9.9149	22
43	0.4668	69	9.9171	18
44	0.4599	72	9.9189	12
45	0.4527	74	9.9201	7
46	0.4453	77	9.9208	+1
47	0.4376	80	9.9209	−4
48	0.4296	83	9.9205	9
49	0.4213	86	9.9196	14
50	0.4127	89	9.9182	20
51	0.4038	92	9.9162	26
52	0.3946	96	9.9136	32
53	0.3850	100	9.9104	37
54	0.3750	103	9.9067	43
55	0.3647	108	9.9024	49
56	0.3539	112	9.8975	55
57	0.3427	116	9.8920	62
58	0.3311	121	9.8858	69
59	0.3190	126	9.8789	76
60	0.3064		9.8713	

Arg. ☉'s Long.	Log. d D. / 360° −	Diff.	Log. d D. tan. D. / 360° +	Diff.
60°	0.3064	−132	9.8713	−82
61	0.2932	137	9.8631	91
62	0.2795	143	9.8540	98
63	0.2652	150	9.8442	106
64	0.2502	157	9.8336	116
65	0.2345	164	9.8220	124
66	0.2181	172	9.8096	134
67	0.2009	181	9.7962	145
68	0.1828	191	9.7817	156
69	0.1637	201	9.7661	168
70	0.1436	212	9.7493	181
71	0.1224	225	9.7312	196
72	0.0999	239	9.7116	210
73	0.0760	255	9.6906	229
74	0.0505	272	9.6677	247
75	0.0233	292	9.6430	269
76	9.9941	314	9.6161	293
77	9.9627	341	9.5868	322
78	9.9286	372	9.5546	353
79	9.8914	409	9.5193	392
80	9.8505	452	9.4801	437
81	9.8053	507	9.4364	494
82	9.7546	576	9.3870	564
83	9.6970	666	9.3306	656
84	9.6304	789	9.2650	780
85	9.5515	966	9.1870	959
86	9.4549	1246	9.0911	1241
87	9.3303	1758	8.9670	1754
88	9.1545	3003	8.7916	3001
89	8.8542	8.8542	8.4915	8.4915
90	+	+	−	+
91	8.8512	3016	8.4886	3014
92	9.1528	1772	8.7900	1757
93	9.3290	1247	8.9657	1241
94	9.4537	968	9.0898	962
95	9.5505	789	9.1860	780
96	9.6294	666	9.2640	655
97	9.6960	575	9.3295	564
98	9.7535	507	9.3859	493
99	9.8042	452	9.4352	437
100	9.8494	407	9.4789	391
101	9.8901	372	9.5180	353
102	9.9273	340	9.5533	321
103	9.9613	314	9.5854	293
104	9.9927	291	9.6147	268
105	0.0218	271	9.6415	246
106	0.0489	254	9.6661	228
107	0.0743	238	9.6889	210
108	0.0981	225	9.7099	195
109	0.1206	211	9.7294	180
110	0.1417	201	9.7474	168
111	0.1618	189	9.7642	155
112	0.1807	181	9.7797	144
113	0.1988	171	9.7941	133
114	0.2159	164	9.8074	124
115	0.2323	155	9.8198	114
116	0.2478	149	9.8312	106
117	0.2627	143	9.8418	97
118	0.2770	136	9.8515	90
119	0.2906	131	9.8605	82
120	0.3037		9.8687	

Arg. ☉'s Long.	Log. d D. / 360° +	Diff.	Log. d D. tan. D. / 360° −	Diff.
120°	0.3037	+126	9.8687	+75
121	0.3163	120	9.8762	68
122	0.3283	116	9.8830	61
123	0.3399	111	9.8891	54
124	0.3510	106	9.8945	49
125	0.3616	103	9.8994	42
126	0.3719	99	9.9036	37
127	0.3818	95	9.9073	30
128	0.3913	92	9.9103	25
129	0.4005	89	9.9128	20
130	0.4094	85	9.9148	14
131	0.4179	82	9.9162	8
132	0.4261	79	9.9170	+3
133	0.4340	76	9.9173	−2
134	0.4416	74	9.9171	7
135	0.4490	71	9.9164	13
136	0.4561	68	9.9151	18
137	0.4629	66	9.9133	23
138	0.4695	64	9.9110	29
139	0.4759	61	9.9081	34
140	0.4820	59	9.9047	39
141	0.4879	57	9.9008	44
142	0.4936	55	9.8964	50
143	0.4991	52	9.8914	56
144	0.5043	51	9.8858	62
145	0.5094	49	9.8796	67
146	0.5143	47	9.8729	73
147	0.5190	45	9.8656	80
148	0.5235	44	9.8576	86
149	0.5279	42	9.8490	92
150	0.5321	40	9.8398	100
151	0.5361	38	9.8298	106
152	0.5399	37	9.8192	114
153	0.5436	35	9.8078	122
154	0.5471	34	9.7956	130
155	0.5505	32	9.7826	139
156	0.5537	30	9.7687	148
157	0.5567	29	9.7539	158
158	0.5596	28	9.7381	169
159	0.5624	27	9.7212	180
160	0.5651	25	9.7032	193
161	0.5676	24	9.6839	207
162	0.5700	22	9.6632	221
163	0.5722	21	9.6411	239
164	0.5743	20	9.6172	256
165	0.5763	18	9.5916	278
166	0.5781	17	9.5638	301
167	0.5798	16	0.5337	329
168	0.5814	15	9.5008	360
169	0.5829	14	9.4648	398
170	0.5843	12	9.4250	443
171	0.5855	11	9.3807	498
172	0.5866	10	9.3309	569
173	0.5876	9	9.2740	659
174	0.5885	8	9.2081	783
175	0.5893	7	9.1298	961
176	0.5900	6	9.0337	1243
177	0.5906	4	8.9094	1756
178	0.5910	3	8.7338	3007
179	0.5913	2	8.4331	8.4331
180	0.5915		0.0000	

In this Table, D = the Sun's Declination at Noon, and dD. = his Horary Motion in Declination.

TABLE 21.

General Table of the Corrections of Noon, &c. PART I. concluded.

Arg. ⊙'s Long.	Log dn. / 360° +	Diff.	Log dn. tan n / 360° —	Diff.
180°	0 .5915	+ 1	0 .0000	+8.4334
181	0 .5916	0	8 .4334	3010
182	0 .5916	— 1	8 .7344	1760
183	0 .5915	2	8 .9104	1201
184	0 .5913	3	9 .0305	1010
185	0 .5910	4	9 .1315	786
186	0 .5906	6	9 .2101	662
187	0 .5900	7	9 .2763	572
188	0 .5893	8	9 .3335	501
189	0 .5885	9	9 .3836	447
190	0 .5876	10	9 .4283	401
191	0 .5866	11	9 .4684	304
192	0 .5855	12	9 .5048	332
193	0 .5843	14	9 .5380	306
194	0 .5829	15	9 .5686	281
195	0 .5814	16	9 .5967	261
196	0 .5798	17	9 .6228	242
197	0 .5781	18	9 .6470	226
198	0 .5763	19	9 .6696	210
199	0 .5744	21	9 .6906	198
200	0 .5723	22	9 .7104	185
201	0 .5701	23	9 .7289	173
202	0 .5678	25	9 .7462	163
203	0 .5653	26	9 .7625	153
204	0 .5627	27	9 .7778	144
205	0 .5600	29	9 .7922	135
206	0 .5571	30	9 .8057	127
207	0 .5541	32	9 .8184	119
208	0 .5509	33	9 .8303	111
209	0 .5476	35	9 .8414	104
210	0 .5441	36	9 .8518	98
211	0 .5405	38	9 .8616	92
212	0 .5367	40	9 .8708	85
213	0 .5327	41	9 .8793	79
214	0 .5286	43	9 .8872	73
215	0 .5243	45	9 .8945	67
216	0 .5198	47	9 .9012	62
217	0 .5151	49	9 .9074	56
218	0 .5102	51	9 .9130	51
219	0 .5051	53	9 .9181	45
220	0 .4998	55	9 .9226	40
221	0 .4943	57	9 .9266	35
222	0 .4886	60	9 .9301	30
223	0 .4826	62	9 .9331	24
224	0 .4764	64	9 .9355	19
225	0 .4700	67	+9 .9374	14
226	0 .4633	70	9 .9388	8
227	0 .4563	73	9 .9396	+ 3
228	0 .4490	76	9 .9399	— 3
229	0 .4414	79	9 .9396	8
230	0 .4335	82	9 .9388	13
231	0 .4253	85	9 .9375	18
232	0 .4168	89	9 .9357	24
233	0 .4079	93	9 .9333	30
234	0 .3986	97	9 .9303	36
235	0 .3889	101	9 .9267	42
236	0 .3788	105	9 .9225	49
237	0 .3683	109	9 .9176	56
238	0 .3574	114	9 .9120	62
239	0 .3460	120	9 .9058	69
240	0 .3340		9 .8989	

Arg. ⊙'s Long.	Log dn. / 360° +	Diff.	Log dn. tan n / 360° +	Diff.
240°	0 .3340	— 125	9 .8989	— 76
241	0 .3215	131	9 .8913	84
242	0 .3084	137	9 .8829	92
243	0 .2947	144	9 .8737	101
244	0 .2803	151	9 .8636	110
245	0 .2652	158	9 .8526	118
246	0 .2494	166	9 .8408	128
247	0 .2328	175	8 .8280	139
248	0 .2153	186	9 .8141	151
249	0 .1967	196	9 .7990	163
250	0 .1771	208	9 .7827	176
251	0 .1563	220	9 .7651	191
252	0 .1343	234	9 .7460	206
253	0 .1109	250	9 .7254	224
254	0 .0859	268	9 .7030	243
255	0 .0591	288	9 .6787	264
256	0 .0303	311	9 .6523	290
257	9 .9992	338	9 .6233	318
258	9 .9654	368	9 .5915	350
259	9 .9286	406	9 .5565	389
260	9 .8880	449	9 .5176	435
261	9 .8431	505	9 .4741	491
262	9 .7926	574	9 .4250	562
263	9 .7352	664	9 .3688	654
264	9 .6688	787	9 .3034	779
265	9 .5901	966	9 .2255	958
266	9 .4935	1247	9 .1297	1242
267	9 .3688	1759	9 .0055	1757
268	9 .1929	3017	8 .8298	3013
269	8 .8912		8 .5285	
270	....		....	
271	— 8 .8939	+3003	— 8 .5312	+3001
272	9 .1942	1757	8 .8313	1753
273	9 .3699	1246	9 .0066	1241
274	9 .4945	965	9 .1307	958
275	9 .5910	787	9 .2265	779
276	9 .6697	665	9 .3044	654
277	9 .7362	574	9 .3698	562
278	9 .7936	505	9 .4260	492
279	9 .8441	450	9 .4752	435
280	9 .8891	406	9 .5187	389
281	9 .9297	369	9 .5576	351
282	9 .9666	338	9 .5927	318
283	0 .0004	312	9 .6245	291
284	0 .0316	289	9 .6536	265
285	0 .0605	268	9 .6801	244
286	0 .0873	251	9 .7045	224
287	0 .1124	235	9 .7269	208
288	0 .1359	221	9 .7477	191
289	0 .1580	208	9 .7668	177
290	0 .1788	196	9 .7845	163
291	0 .1984	186	9 .8008	152
292	0 .2170	177	9 .8160	140
293	0 .2347	167	9 .8300	129
294	0 .2514	159	9 .8429	120
295	0 .2673	152	9 .8549	109
296	0 .2825	144	9 .8658	102
297	0 .2969	138	9 .8760	92
298	0 .3107	132	9 .8852	85
299	0 .3239	126	9 .8937	77
300	0 .3365		9 .9014	

Arg. ⊙'s Long.	Log dn. / 360° —	Diff.	Log dn. tan n / 360° —	Diff.
300°	0 .3365	+120	9 .9014	+ 70
301	0 .3485	115	9 .9084	63
302	0 .3600	110	9 .9147	56
303	0 .3710	106	9 .9203	49
304	0 .3816	102	9 .9252	43
305	0 .3918	97	9 .9295	37
306	0 .4015	94	9 .9332	31
307	0 .4109	89	9 .9363	25
308	0 .4198	86	9 .9388	19
309	0 .4284	83	9 .9407	14
310	0 .4367	80	9 .9421	8
311	0 .4447	76	9 .9429	+ 3
312	0 .4523	74	9 .9432	— 2
313	0 .4597	70	9 .9430	8
314	0 .4667	68	9 .9422	13
315	0 .4735	65	9 .9409	18
316	0 .4800	63	9 .9391	24
317	0 .4863	60	9 .9367	29
218	0 .4923	58	9 .9338	34
319	0 .4981	56	9 .9304	39
320	0 .5037	54	9 .9265	45
321	0 .5091	52	9 .9220	50
322	0 .5143	49	9 .9170	55
323	0 .5192	47	9 .9115	61
324	0 .5239	46	9 .9054	67
325	0 .5285	44	9 .8987	73
326	0 .5329	42	9 .8914	79
327	0 .5371	40	9 .8835	85
328	0 .5411	39	9 .8750	91
329	0 .5450	37	9 .8659	97
330	0 .5487	35	9 .8562	104
331	0 .5522	33	9 .8458	111
332	0 .5555	32	9 .8347	118
333	0 .5587	30	9 .8229	126
334	0 .5617	29	9 .8103	134
335	0 .5646	28	9 .7969	143
336	0 .5674	27	9 .7826	153
337	0 .5701	25	9 .7673	163
338	0 .5726	24	9 .7510	173
339	0 .5750	22	9 .7337	184
340	0 .5772	21	9 .7153	197
341	0 .5793	20	9 .6956	211
342	0 .5813	19	9 .6745	226
343	0 .5832	17	9 .6519	242
344	0 .5849	16	9 .6277	260
345	0 .5865	15	9 .6017	281
346	0 .5880	14	9 .5736	305
347	0 .5894	12	9 .5431	332
248	0 .5906	11	9 .5099	363
349	0 .5917	10	9 .4736	402
350	0 .5927	9	9 .4334	446
351	0 .5936	8	9 .3888	501
352	0 .5944	7	9 .3387	572
353	0 .5951	6	9 .2815	662
354	0 .5957	5	9 .2153	786
355	0 .5962	4	9 .1367	965
356	0 .5966	2	9 .0402	1246
357	0 .5968	1	8 .9156	1759
358	0 .5969	0	8 .7397	3010
359	0 .5969	— 1	8 .4387	8.4387
360	0 .5968		0 .0000	

TABLE 22.

General Table of the Corrections of Noon by corresponding Equal Altitudes. Part II. (From Delambre.)

Argument. Half the Interval = t.	Log t Sin. (15°. t.) +	Diff.	Log t Tan. (15°. t.) +	Diff.	Argument. = t.	Log t Sin. (15°. t.) +	Diff.	Log t Tan. (15°. t.) +	Diff.	Argument. = t.	Log t Sin. (15°. t.) +	Diff.	Log t Tan. (15°. t.) −	Diff.
h. m.					h. m.					h. m.				
O 0	…………		…………		IV 0	0.6645		0.3634		VIII 0	0.9656		0.6645	
							+36		−132			+101		+261
5	0.5821		0.5820		5	0.6681		0.3502		5	0.9757		0.6906	
		+1		−2			37		140			103		256
10	0.5822		0.5818		10	0.6718		0.3362		10	0.9860		0.7162	
		2		4			38		149			106		251
15	0.5824		0.5814		15	0.6756		0.3213		15	0.9966		0.7413	
		2		5			39		159			109		247
20	0.5826		0.5809		20	0.6795		0.3054		20	1.0075		0.7460	
		3		6			40		169			111		244
25	0.5829		0.5803		25	0.6835		0.2885		25	1.0186		0.7904	
		4		8			41		181			114		240
30	0.5833		0.5795		30	0.6876		0.2704		30	1.0300		0.8144	
		4		9			42		194			116		237
35	0.5837		0.5786		35	0.6918		0.2510		35	1.0416		0.8381	
		5		10			43		209			120		235
40	0.5842		0.5776		40	0.6961		0.2301		40	1.0536		0.8616	
		6		12			43		226			123		234
45	0.5848		0.5764		45	0.7004		0.2075		45	1.0659		0.8850	
		7		13			44		245			126		232
50	0.5855		0.5751		50	0.7048		0.1830		50	1.0785		0.9082	
		7		15			45		267			129		231
55	0.5862		0.5736		55	0.7093		0.1563		55	1.0914		0.9313	
		8		16			46		293			133		229
I 0	0.5870		0.5720		V 0	0.7139		0.1270		IX 0	1.1047		0.9542	
		9		18			47		322			137		230
5	0.5879		0.5702		5	0.7186		0.0948		5	1.1184		0.9772	
		9		20			49		358			141		230
10	0.5888		0.5682		10	0.7235		0.0590		10	1.1325		1.0002	
		10		21			50		402			145		230
15	0.5898		0.5661		15	0.7285		0.0188		15	1.1470		1.0232	
		11		22			51		455			150		230
20	0.5909		0.5639		20	0.7336		9.9733		20	1.1620		1.0462	
		11		24			52		523			154		232
25	0.5920		0.5615		25	0.7388		9.9210		25	1.1774		1.0694	
		12		26			53		612			159		233
30	0.5932		0.5589		30	0.7441		9.8598		30	1.1933		1.0927	
		13		28			54		734			164		236
35	0.5945		0.5561		35	0.7495		9.7864		35	1.2097		1.1163	
		14		29			55		911			170		238
40	0.5959		0.5532		40	0.7550		9.6953		40	1.2267		1.1401	
		14		31			56		1191			176		240
45	0.5973		0.5501		45	0.7606		9.5762		45	1.2443		1.1641	
		15		33			57		1702			182		244
50	0.5988		0.5468		50	0.7663		9.4060		50	1.2625		1.1885	
		16		35			59		−2951			189		248
55	0.6004		0.5433		55	0.7722		9.1109		55	1.2814		1.2133	
		17		37			60					196		253
II 0	0.6021		0.5396		VI 0	0.7782		…………		X 0	1.3010		1.2386	
		17		39			61					205		257
5	0.6038		0.5357		5	0.7843		9.1230		5	1.3215		1.2643	
		18		41			62		+3071			213		263
10	0.6056		0.5316		10	0.7905		9.4301		10	1.3428		1.2906	
		19		43			63		1823			222		271
15	0.6075		0.5273		15	0.7968		9.6124		15	1.3650		1.3177	
		19		45			65		1312			233		279
20	0.6094		0.5228		20	0.8033		9.7436		20	1.3883		1.3456	
		20		48			66		1032			244		287
25	0.6114		0.5180		25	0.8099		9.8468		25	1.4127		1.3743	
		21		50			67		855			257		297
30	0.6135		0.5130		30	0.8166		9.9323		30	1.4384		1.4040	
		22		53			69		734			270		308
35	0.6157		0.5077		35	0.8235		0.0057		35	1.4654		1.4348	
		22		56			70		645			286		322
40	0.6179		0.5021		40	0.8305		0.0702		40	1.4940		1.4670	
		23		58			72		578			303		336
45	0.6202		0.4963		45	0.8377		0.1280		45	1.5243		1.5006	
		24		61			74		524			323		354
50	0.6226		0.4902		50	0.8451		0.1804		50	1.5566		1.5360	
		25		64			75		481			346		375
55	0.6251		0.4838		55	0.8526		0.2285		55	1.5912		1.5735	
		26		67			76		447			372		398
III 0	0.6277		0.4771		VII 0	0.8602		0.2732		XI 0	1.6284		1.6133	
		26		70			78		416			403		427
5	0.6303		0.4701		5	0.8680		0.3148		5	1.6687		1.6560	
		27		74			79		392			439		462
10	0.6330		0.4627		10	0.8759		0.3540		10	1.7126		1.7022	
		28		78			81		371			483		503
15	0.6358		0.4549		15	0.8840		0.3911		15	1.7609		1.7525	
		29		82			83		353			538		555
20	0.6387		0.4467		20	0.8923		0.4264		20	1.8147		1.8080	
		29		86			85		336			606		623
25	0.6416		0.4381		25	0.9008		0.4600		25	1.8753		1.8703	
		30		90			87		323			697		710
30	0.6446		0.4291		30	0.9095		0.4923		30	1.9450		1.9413	
		31		95			88		310			819		831
35	0.6477		0.4196		35	0.9183		0.5233		35	2.0269		2.0244	
		32		101			90		300			998		1006
40	0.6509		0.4095		40	0.9273		0.5533		40	2.1267		2.1250	
		33		106			92		290			1277		1285
45	0.6542		0.3989		45	0.9365		0.5823		45	2.2544		2.2535	
		34		112			95		281			1790		1795
50	0.6576		0.3877		50	0.9460		0.6104		50	2.4334		2.4330	
		34		118			97		274			3040		3043
55	0.6610		0.3759		55	0.9557		0.6378		55	2.7374		2.7373	
		35		125			99		267					
IV 0	0.6645		0.3634		VIII 0	0.9656		0.6645		XII 0	…………		…………	

TABLE 23.

CORRECTIONS for EQUAL ALTITUDES, in determining the included Noon or Midnight. (From ZACH.)

Argument = half the interval between the corresponding Observations = t.

Arg. = t. h. m.	Angle α. +	Diff. for 1′.	Angle β. −	Diff. for 1′.
0 0	44° 40′ 16″		12° 26′ 9″	
		0′ 2″		0′ 1″
5	44 40 27		12 26 3	
		0 5		0 4
10	44 40 52		12 25 43	
		0 8		0 7
15	44 41 33		12 25 9	
		0 11		0 10
20	44 42 30		12 24 21	
		0 15		0 12
25	44 43 43		12 23 19	
		0 18		0 15
30	44 45 13		12 22 3	
		0 21		0 18
35	44 47 0		12 20 33	
		0 25		0 21
40	44 49 3		12 18 49	
		0 28		0 24
45	44 51 22		12 16 51	
		0 31		0 26
50	44 53 38		12 14 39	
		0 35		0 29
55	44 56 51		12 12 14	
		0 37		0 32
1 0	45 0 0		12 9 34	
		0 41		0 35
5	45 3 25		12 6 40	
		0 44		0 38
10	45 7 7		12 3 31	
		0 48		0 40
15	45 11 6		12 0 8	
		0 51		0 43
20	45 15 22		11 56 31	
		0 54		0 46
25	45 19 54		11 52 39	
		0 58		0 49
30	45 24 43		11 48 32	
		1 1		0 52
35	45 29 49		11 44 11	
		1 5		0 55
40	45 35 12		11 39 34	
		1 8		0 58
45	45 40 52		11 34 42	
		1 11		1 1
50	45 46 49		11 29 35	
		1 15		1 4
55	45 53 3		11 24 13	
		1 18		1 8
2 0	45 59 34		11 18 35	
		1 22		1 11
5	46 6 23		11 12 42	
		1 25		1 14
10	46 13 29		11 6 32	
		1 29		1 17
15	46 20 52		11 0 5	
		1 32		1 20
20	46 28 33		10 53 24	
		1 36		1 23
25	46 36 31		10 46 26	
		1 39		1 27
30	46 44 46		10 39 10	
		1 43		1 30
35	46 53 19		10 31 39	
		1 46		1 34
40	47 2 11		10 23 49	
		1 50		1 38
45	47 11 20		10 15 42	
		1 53		1 41
50	47 20 47		10 7 18	
		1 57		1 45
55	47 30 32		9 58 35	
		2 0		1 48
3 0	47 40 35		9 49 35	
		2 4		1 52
5	47 50 56		9 40 15	
		2 8		1 56
10	48 1 35		9 30 37	
		2 12		1 59
15	48 12 33		9 20 40	
		2 15		2 3
20	48 23 49		9 10 23	
		2 19		2 7
25	48 35 24		8 59 46	
		2 23		2 11
30	48 47 17		8 48 49	
		2 26		2 15
35	48 59 29		8 37 32	
		2 30		2 19
40	49 12 0		8 25 53	

Arg. = t. h. m.	Angle α. +	Diff. for 1′.	Angle β. −	Diff. for 1′.
		2′ 34″		2′ 24″
3 45	49° 24′ 50″		8° 13′ 52″	
		2 38		2 28
50	49 37 59		8 1 31	
		2 42		2 33
55	49 51 27		7 48 47	
		2 45		2 37
4 0	50 5 13		7 35 40	
		2 49		2 42
5	50 19 18		7 22 10	
		2 53		2 47
10	50 33 43		7 8 16	
		2 57		2 51
15	50 48 28		6 53 58	
		3 1		2 57
20	51 3 32		6 39 15	
		3 5		3 2
25	51 18 55		6 24 6	
		3 8		3 7
30	51 34 37		6 8 32	
		3 12		3 12
35	51 50 39		5 52 31	
		3 16		3 18
40	52 7 1		5 36 2	
		3 20		3 23
45	52 23 42		5 19 6	
		3 24		3 29
50	52 40 43		5 1 41	
		3 28		3 35
55	52 58 3		4 43 47	
		3 32		3 41
5 0	53 15 43		4 25 23	
		3 36		3 47
5	53 33 43		4 6 28	
		3 40		3 53
10	53 52 2		3 47 1	
		3 44		4 0
15	54 10 41		3 27 1	
		3 48		4 7
20	54 29 40		3 6 28	
		3 52		4 13
25	54 48 59		2 45 21	
		3 56		4 20
30	55 8 37		2 23 38	
		3 59		4 28
35	55 28 34		2 1 19	
		4 3		4 36
40	55 48 51		1 38 22	
		4 7		4 43
45	56 9 27		1 14 47	
		4 11		4 51
50	56 30 23		0 50 33	
		4 15		4 59
55	56 51 40		0 25 37	
		4 19		5 8
6 0	57 13 14		0 0 0	
		4 23	+	5 16
5	57 35 7		0 26 21	
		4 27		5 25
10	57 57 20		0 53 26	
		4 30		5 35
15	58 19 51		1 21 17	
		4 34		5 44
20	58 42 41		1 49 56	
		4 38		5 53
25	59 5 50		2 19 22	
		4 41		6 4
30	59 29 16		2 49 41	
		4 45		6 14
35	59 53 1		3 20 53	
		4 48		6 25
40	60 17 4		3 52 57	
		4 52		6 36
45	60 41 24		4 25 57	
		4 56		6 48
50	61 6 2		4 59 55	
		4 59		6 59
55	61 30 57		5 34 52	
		5 2		7 11
7 0	61 56 9		6 10 49	
		5 6		7 24
5	62 21 37		6 47 51	
		5 9		7 37
10	62 47 21		7 25 58	
		5 12		7 51
15	63 13 21		8 5 12	
		5 15		8 5
20	63 39 37		8 45 37	

Arg. = t. h. m.	Angle α. +	Diff. for 1′.	Angle β. +	Diff. for 1′.
		5′ 18″		8′ 19″
7 25	64° 6′ 8″		9° 27′ 14″	
		5 21		8 38
30	64 32 53		10 10 6	
		5 24		8 50
35	64 59 52		10 54 14	
		5 27		9 6
40	65 27 6		11 39 44	
		5 30		9 22
45	65 54 34		12 26 35	
		5 32		9 39
50	66 22 14		13 14 52	
		5 34		9 57
55	66 50 5		14 4 37	
		5 37		10 15
8 0	67 18 8		14 55 53	
		5 39		10 34
5	67 46 24		15 48 42	
		5 41		10 53
10	68 14 50		16 43 9	
		5 43		11 13
15	68 43 26		17 39 15	
		5 45		11 34
20	69 12 12		18 37 4	
		5 47		11 55
25	69 41 7		19 36 40	
		5 49		12 17
30	70 10 10		20 38 4	
		5 50		12 39
35	70 39 22		21 41 21	
		5 52		13 2
40	71 8 40		22 46 32	
		5 53		13 26
45	71 38 4		23 53 41	
		5 54		13 50
50	72 7 34		25 2 51	
		5 55		14 15
55	72 37 10		26 14 5	
		5 56		14 40
9 0	73 6 50		27 27 25	
		5 57		15 6
5	73 36 33		28 42 54	
		5 57		15 32
10	74 6 19		30 0 33	
		5 58		15 58
15	74 36 8		31 20 24	
		5 58		16 25
20	75 5 58		32 42 30	
		5 58		16 52
25	75 35 48		34 6 51	
		5 58		17 19
30	76 5 38		35 33 26	
		5 58		17 46
35	76 35 28		37 2 17	
		5 58		18 13
40	77 5 16		38 33 25	
		5 58		18 40
45	77 35 2		40 6 47	
		5 57		19 7
50	78 4 45		41 42 21	
		5 56		19 33
55	78 34 24		43 20 7	
		5 55		19 59
10 0	79 3 57		45 0 0	
		5 54		20 23
5	79 33 27		46 41 56	
		5 53		20 47
10	80 2 50		48 25 58	
		5 51		21 9
15	80 32 5		50 11 41	
		5 50		21 31
20	81 1 13		51 59 18	
		5 48		21 51
25	81 30 13		53 48 34	
		5 46		22 10
30	81 59 4		55 39 22	
		5 44		22 26
35	82 27 45		57 31 34	
		5 42		22 41
40	82 56 15		59 24 59	
		5 40		22 54
45	83 24 34		61 19 27	
		5 38		23 4
50	83 52 42		63 14 50	
		5 35		23 12
55	84 20 38		65 10 58	
		5 33		23 19
11 0	84 48 21		67 7 9	

TABLE 24.

Corrections for Equal Altitudes, in determining the included Noon or Midnight. (From Zach.)

Argument = the Sun's apparent Longitude.

Long. ⊙	a	Diff. for 1°.	b	Diff. for 1°.	Long ⊙	a	Diff. for 1°.	b	Diff. for 1°.	Long. ⊙	a	Diff. for 1°.	b	Diff. for 1°.
s O 0°	−15″.253	0″.018	−0″.000	0′.472	s IV 0°	+ 7″.828	0″.219	+12″.893	0″.182	s VIII 0°	+ 8″.289	0″.245	−13″.653	0″.279
5	15.164	0.038	2.360	0.459	5	8.923	0.204	13.803	0.095	5	7.066	0.262	12.257	0.369
10	14.973	0.057	4.653	0.433	10	9.943	0.185	14.277	0.005	10	5.757	0.274	10.414	0.444
15	14.687	0.077	6.822	0.399	15	10.873	0.170	14.303	0.080	15	4.387	0.287	8.194	0.510
20	14.304	0.095	8.815	0.353	20	11.712	0.150	13.902	0.159	20	2.954	0.293	5.646	0.553
25	13.827	0.115	10.582	0.279	25	12.462	0.131	13.106	0.232	25	1.486	0.297	2.879	0.576
I 0	13.252	0.134	12.071	0.232	V 0	13.116	0.113	11.946	0.295	IX 0	− 0.000	0.297	+ 0.000	0.577
5	12.583	0.152	13.233	0.159	5	13.680	0.093	10.470	0.350	5	1.489	0.294	2.884	0.554
10	11.821	0.172	14.031	0.078	10	14.146	0.075	8.718	0.394	10	2.960	0.287	5.656	0.511
15	10.963	0.189	14.421	0.007	15	14.522	0.056	6.746	0.429	15	4.397	0.277	8.214	0.448
20	10.017	0.206	14.384	0.096	20	14.803	0.037	4.600	0.453	20	5.780	0.263	10.456	0.372
25	8.987	0.223	13.902	0.187	25	14.989	0.018	2.333	0.467	25	7.099	0.247	12.315	0.283
II 0	7.874	0.237	12.969	0.274	VI 0	15.081	0.001	− 0.000	0.469	X 0	8.335	0.230	13.729	0.199
5	6.687	0.251	11.600	0.355	5	15.076	0.021	2.347	0.461	5	9.487	0.210	14.675	0.092
10	5.432	0.262	9.827	0.425	10	14.973	0.040	4.653	0.442	10	10.538	0.190	15.135	0.005
15	4.124	0.270	7.704	0.481	15	14.772	0.060	6.862	0.411	15	11.488	0.179	15.112	0.095
20	2.774	0.276	5.301	0.521	20	14.471	0.080	8.918	0.370	20	12.333	0.147	14.639	0.179
25	1.394	0.279	2.700	0.540	25	14.069	0.101	10.768	0.317	25	13.066	0.126	13.743	0.254
III 0	+ 0.000	0.279	+0.000	0.540	VII 0	13.564	0.123	12.354	0.233	XI 0	13.695	0.104	12.474	0.319
5	1.394	0.275	2.700	0.518	5	12.951	0.145	13.621	0.178	5	14.216	0.093	10.880	0.373
10	2.767	0.268	5.292	0.478	10	12.225	0.166	14.511	0.095	10	14.631	0.062	9.017	0.416
15	4.111	0.260	7.680	0.422	15	11.393	0.187	14.987	0.007	15	14.940	0.042	6.940	0.447
20	5.412	0.248	9.789	0.351	20	10.460	0.207	15.021	0.082	20	15.148	0.021	4.707	0.467
25	6.653	0.235	11.542	0.270	25	9.422	0.227	14.609	0.191	25	15.251	0.000	2.374	0.475
IV 0	7.828		12.893		VIII 0	8.289		13.653		XII 0	−15.253		0.000	

TABLE 25.

The Sun's Parallax in Altitude. (By Mr. J. Utting.)

Arguments = the Sun's Altitude, and Horizontal Parallax.

Alt.	8″.5	8″.6	8″.7	8″.8	8″.9	9″.0	Alt.	8″.5	8″.6	8″.7	8″8	8″.9	9″.0	Alt.	8″.5	8″.6	1″.6	8″.8	8″.9	9″.0
1°	8″.50	8″.60	8″.70	8″.80	8″.90	9″.00	31°	7″.29	7″.37	7″.46	7″.54	7″.63	7″.71	61°	4″.12	4″.17	4″.22	4″.27	4″.31	4″.36
2	8.49	8.59	8.69	8.79	8.89	8.99	32	7.21	7.29	7.38	7.46	7.55	7.63	62	3.99	4.04	4.08	4.13	4.18	4.23
3	8.49	8.59	8.69	8.79	8.89	8.99	33	7.13	7.21	7.30	7.38	7.46	7.55	63	3.86	3.90	3.95	4.00	4.04	4.09
4	8.48	8.58	8.68	8.78	8.88	8.98	34	7.05	7.13	7.21	7.30	7.38	7.46	64	3.73	3.77	3.81	3.86	3.90	3.95
5	8.47	8.57	8.67	8.77	8.87	8.97	35	6.96	7.04	7.13	7.21	7.29	7.37	65	3.59	3.63	3.68	3.72	3.76	3.80
6	8.45	8.55	8.65	8.75	8.85	8.95	36	6.88	6.96	7.04	7.12	7.20	7.28	66	3.45	3.50	3.54	3.58	3.62	3.66
7	8.44	8.54	8.64	8.73	8.83	8.93	37	6.79	6.87	6.95	7.03	7.11	7.19	67	3.32	3.36	3.40	3.44	3.48	3.52
8	8.42	8.52	8.62	8.71	8.81	8.91	38	6.70	6.78	6.86	6.93	7.01	7.09	68	3.18	3.22	3.26	3.30	3.33	3.37
9	8.40	8.50	8.59	8.69	8.79	8.89	39	6.61	6.68	6.76	6.84	6.92	6.99	69	3.05	3.08	3.12	3.15	3.19	3.23
10	8.37	8.47	8.57	8.67	8.76	8.86	40	6.51	6.59	6.66	6.74	6.82	6.89	70	2.91	2.94	2.98	3.01	3.04	3.08
11	8.34	8.44	8.54	8.64	8.74	8.83	41	6.41	6.49	6.57	6.64	6.72	6.79	71	2.77	2.80	2.83	2.86	2.90	2.93
12	8.31	8.41	8.51	8.61	8.70	8.80	42	6.32	6.39	6.47	6.54	6.61	6.69	72	2.63	2.66	2.69	2.72	2.75	2.78
13	8.28	8.38	8.48	8.57	8.67	8.77	43	6.22	6.29	6.36	6.44	6.51	6.58	73	2.49	2.51	2.54	2.57	2.60	2.63
14	8.25	8.35	8.44	8.54	8.64	8.73	44	6.11	6.19	6.26	6.33	6.40	6.47	74	2.34	2.37	2.40	2.43	2.45	2.48
15	8.21	8.31	8.40	8.50	8.60	8.69	45	6.01	6.08	6.15	6.22	6.29	6.36	75	2.20	2.23	2.25	2.28	2.30	2.33
16	8.17	8.27	8.36	8.46	8.56	8.65	46	5.90	5.97	6.04	6.11	6.18	6.25	76	2.06	2.08	2.10	2.13	2.15	2.18
17	8.13	8.22	8.32	8.42	8.51	8.61	47	5.80	5.87	5.93	6.00	6.07	6.14	77	1.91	1.93	1.96	1.98	2.00	2.02
18	8.08	8.18	8.27	8.37	8.46	8.56	48	5.69	5.75	5.82	5.89	5.96	6.02	78	1.77	1.79	1.81	1.83	1.85	1.87
19	8.04	8.13	8.23	8.32	8.42	8.51	49	5.58	5.64	5.71	5.77	5.84	5.90	79	1.62	1.64	1.66	1.68	1.70	1.72
20	7.99	8.08	8.18	8.27	8.36	8.46	50	5.46	5.53	5.59	5.66	5.72	5.79	80	1.48	1.49	1.51	1.53	1.55	1.56
21	7.94	8.03	8.12	8.22	8.31	8.40	51	5.35	5.41	5.48	5.54	5.60	5.66	81	1.33	1.35	1.36	1.38	1.39	1.41
22	7.88	7.97	8.07	8.16	8.25	8.34	52	5.23	5.29	5.36	5.42	5.48	5.54	82	1.18	1.20	1.21	1.22	1.24	1.25
23	7.82	7.92	8.00	8.10	8.19	8.28	53	5.12	5.18	5.24	5.30	5.36	5.42	83	1.04	1.05	1.06	1.07	1.08	1.10
24	7.76	7.86	7.95	8.04	8.13	8.22	54	5.00	5.05	5.11	5.17	5.23	5.29	84	0.89	0.90	0.91	0.92	0.93	0.94
25	7.70	7.79	7.88	7.98	8.07	8.16	55	4.88	4.93	4.99	5.05	5.10	5.16	85	0.74	0.75	0.76	0.77	0.78	0.78
26	7.64	7.73	7.82	7.91	8.00	8.09	56	4.75	4.81	4.87	4.92	4.98	5.03	86	0.59	0.60	0.61	0.61	0.62	0.63
27	7.57	7.66	7.75	7.84	7.93	8.02	57	4.63	4.68	4.74	4.79	4.85	4.90	87	0.44	0.45	0.46	0.46	0.47	0.47
28	7.51	7.59	7.68	7.77	7.86	7.95	58	4.50	4.56	4.61	4.66	4.72	4.77	88	0.30	0.30	0.30	0.31	0.31	0.31
29	7.43	7.52	7.61	7.70	7.78	7.87	59	4.38	4.43	4.48	4.53	4.58	4.64	89	0.15	0.15	0.15	0.15	0.16	0.16
30	7.36	7.45	7.53	7.62	7.71	7.79	60	4.25	4.30	4.35	4.40	4.45	4.50	90	0.00	0.00	0.00	0.00	0.00	0.00

TABLE 1.

Epochs of the Mean Longitude of the Moon's Ascending Node.

	Years.	Mean Long. of ☾'s ☊.				Years.	Mean Long. of ☾'s ☊.			
C.	1800	1s	3°	14′	5″.6	1860	10s	12°	46′	15″.1
	1801	0	13	54	22.3	1861	9	23	23	21.1
	1802	11	24	34	38.9	1862	9	4	3	37.7
	1803	11	5	14	55.5	1863	8	14	43	54.3
B.	1804	10	15	55	12.2	1864	7	25	24	11.0
	1805	9	26	32	18.2	1865	7	6	0	57.0
	1806	9	7	12	34.8	1866	6	16	41	33.6
	1807	8	17	52	51.5	1867	5	27	21	50.3
B.	1808	7	28	33	8.1	1868	5	8	2	6.9
	1809	7	9	10	14.1	1869	4	18	39	12.9
	1810	6	19	50	30.7	1870	3	29	19	29.5
	1811	6	0	30	47.4	1871	3	9	59	46.2
B.	1812	5	11	11	4.0	1872	2	20	40	2.8
	1813	4	21	48	10.0	1873	2	1	17	8.8
	1814	4	2	28	26.7	1874	1	11	57	25.5
	1815	3	13	8	43.3	1875	0	22	37	42.1
B.	1816	2	23	48	59.9	1876	0	3	16	58.7
	1817	2	4	26	5.9	1877	11	13	55	4.7
	1818	1	15	6	22.6	1878	10	24	35	21.4
	1819	0	25	46	39.2	1879	10	5	15	37.0
B.	1820	0	6	25	55.9	1880	9	15	55	54.7
	1821	11	17	4	1.9	1881	8	26	33	0.7
	1822	10	27	44	18.5	1882	8	7	12	57.3
	1823	10	8	24	35.1	1883	7	17	53	33.9
B.	1824	9	19	4	51.8	1884	6	28	33	50.6
	1825	8	29	41	57.8	1885	6	9	10	56.6
	1826	8	10	22	14.4	1886	5	19	51	13.2
	1827	7	21	2	31.1	1887	5	0	31	29.9
B.	1828	7	1	42	47.7	1888	4	11	11	46.5
	1829	6	12	19	53.7	1889	3	21	48	52.5
	1830	5	23	0	10.3	1890	3	2	29	9.1
	1831	5	3	40	26.0	1891	2	13	9	25.7
B.	1832	4	14	20	43.6	1892	1	23	49	42.4
	1833	3	24	57	49.6	1893	1	4	26	48.4
	1834	3	5	38	6.3	1894	0	15	7	5.1
	1835	2	16	18	22.9	1895	11	25	47	21.7
B.	1836	1	26	58	39.5	1896	11	6	27	38.3
	1837	1	7	35	45.5	1897	10	17	3	44.3
	1838	0	18	16	2.2	1898	9	27	45	1.0
	1839	11	28	56	18.8	1899	9	8	25	17.6
B.	1840	11	9	36	35.5	1900	8	19	4	34.3
	1841	10	20	13	41.5	1901	7	29	41	40.3
	1842	10	0	54	8.1	1902	7	10	21	57.7
	1843	9	11	34	14.7	1903	6	21	2	14.4
B.	1844	8	22	14	31.4	1904	6	1	42	31.0
	1845	8	2	51	37.4	1905	5	12	19	37.0
	1846	7	13	31	54.0	1906	4	22	59	53.6
	1847	6	24	12	10.7	1907	4	3	40	10.5
B.	1848	6	4	52	27.3	1908	3	14	20	26.9
	1849	5	15	29	33.3	1909	2	24	57	32.9
	1850	4	26	9	59.9	1910	2	5	37	49.5
	1851	4	6	50	6.6	1911	1	16	18	6.2
B.	1852	3	17	30	23.1	1912	0	26	58	22.8
	1853	2	28	7	29.2	1913	0	7	35	38.8
	1854	2	8	47	45.9	1914	11	18	15	45.4
	1855	1	19	28	2.5	1915	10	28	56	2.1
B.	1856	1	0	8	19.1	1916	10	9	36	18.7
	1857	0	10	45	25.1	1917	9	20	13	24.7
	1858	11	21	25	41.0	1918	9	0	53	41.3
	1859	11	2	5	58.4	1919	8	11	34	8.0
B.	1860	10	12	46	15.1	1920	7	22	14	14.6

TABLE 2.

Retrograde Motion of the Moon's Ascending Node for Each Day in the Year.
To be subtracted from the Longitude of the Epoch.
In Jan. and Feb. the Column marked Com. must only be used.

Com.	Bis.	JANUARY.			FEBRUARY.			MARCH.			APRIL.			MAY.			JUNE.		
1	0	0°	0′	0″	1°	38′	30″	3°	7′	28″	4°	45′	58″	6°	21′	17″	7°	59′	47″
2	1	0	3	11	1	41	40	3	10	38	4	49	8	6	24	27	8	2	57
3	2	0	6	21	1	44	51	3	13	49	4	52	19	6	27	38	8	6	8
4	3	0	9	32	1	48	2	3	17	0	4	55	30	6	30	49	8	9	19
5	4	0	12	43	1	51	12	3	20	10	4	58	40	6	33	59	8	12	29
6	5	0	15	53	1	54	23	3	23	21	5	1	51	6	37	10	8	15	40
7	6	0	19	4	1	57	34	3	26	32	5	5	1	6	40	21	8	18	50
8	7	0	22	15	2	0	44	3	29	42	5	8	12	6	43	31	8	22	1
9	8	0	25	25	2	3	55	3	32	53	5	11	23	6	46	42	8	26	12
10	9	0	28	36	2	7	6	3	36	4	5	14	33	6	49	53	8	28	22
11	10	0	31	46	2	10	16	3	39	14	5	17	44	6	53	3	8	31	33
12	11	0	34	57	2	13	27	3	42	25	5	20	55	6	56	14	8	34	44
13	12	0	38	8	2	16	38	3	45	35	5	24	5	6	59	24	8	37	54
14	13	0	41	18	2	19	48	3	48	46	5	27	16	7	2	35	8	41	5
15	14	0	44	29	2	22	59	3	51	57	5	30	27	7	5	46	8	44	16
16	15	0	47	40	2	26	9	3	55	7	5	33	37	7	8	56	8	47	26
17	16	0	50	50	2	29	20	3	58	18	5	36	48	7	12	7	8	50	37
18	17	0	54	1	2	32	31	4	1	29	5	39	58	7	15	18	8	53	47
19	18	0	57	12	2	35	41	4	4	39	5	43	9	7	18	28	8	56	58
20	19	1	0	22	2	38	52	4	7	50	5	46	20	7	21	39	9	0	9
21	20	1	3	33	2	42	3	4	11	0	5	49	30	7	24	50	9	3	19
22	21	1	6	43	2	45	13	4	14	11	5	52	41	7	28	0	9	6	30
23	22	1	9	54	2	48	24	4	17	22	5	55	52	7	31	11	9	9	41
24	23	1	13	5	2	51	35	4	20	32	5	59	2	7	34	21	9	12	51
25	24	1	16	15	2	54	45	4	23	43	6	2	13	7	37	32	9	16	2
26	25	1	19	26	2	57	56	4	26	54	6	5	24	7	40	43	9	19	13
27	26	1	22	37	3	1	6	4	30	4	6	8	34	7	43	53	9	22	23
28	27	1	25	47	3	4	17	4	33	15	6	11	45	7	47	4	9	25	34
29	28	1	28	58	3	7	28	4	36	26	6	14	55	7	50	15	9	28	44
30	29	1	32	9				4	39	36	6	18	6	7	53	25	9	31	55
31	30	1	35	19				4	42	47	6	21	17	7	56	36	9	35	6
	31							4	45	58				7	59	47			

Com.	Bis.	JULY.			AUGUST.			SEPTEMB.			OCTOBER.			NOVEMBER.			DECEMBER.		
1	0	9°	35′	6″	11°	13′	36″	12°	52′	5″	14°	27′	25″	16°	5′	54″	17°	41′	14″
2	1	9	38	16	11	16	46	12	55	26	14	30	35	16	9	5	17	44	24
3	2	9	41	27	11	19	57	12	58	27	14	33	46	16	12	16	17	47	35
4	3	9	44	38	11	23	8	13	1	37	14	36	56	16	15	26	17	50	46
5	4	9	47	48	11	26	18	13	4	48	14	40	7	16	18	37	17	53	56
6	5	9	50	59	11	29	29	13	7	59	14	43	18	16	21	48	17	57	7
7	6	9	54	10	11	32	39	13	11	9	14	46	28	16	24	58	18	0	17
8	7	9	57	20	11	35	50	13	14	20	14	49	39	16	28	9	18	3	28
9	8	10	0	31	11	39	1	13	17	31	14	52	50	16	31	20	18	6	39
10	9	10	3	42	11	42	11	13	20	41	14	56	0	16	34	30	18	9	49
11	10	10	6	52	11	45	22	13	23	52	14	59	11	16	37	41	18	13	0
12	11	10	10	3	11	48	33	13	27	2	15	2	22	16	40	51	18	16	11
13	12	10	13	13	11	51	43	13	30	13	15	5	32	16	44	2	18	19	21
14	13	10	16	24	11	54	54	13	33	24	15	8	43	16	47	13	18	22	32
15	14	10	19	35	11	58	5	13	36	34	15	11	54	16	50	23	18	25	43
16	15	10	22	45	12	1	15	13	39	45	15	15	4	16	53	34	18	28	53
17	16	10	25	56	12	4	26	13	42	56	15	18	15	16	56	45	18	32	4
18	17	10	29	7	12	7	36	13	46	6	15	21	25	16	59	55	18	35	14
19	18	10	32	17	12	10	47	13	49	17	15	24	36	17	3	6	18	38	25
20	19	10	35	28	12	13	58	13	52	28	15	27	47	17	6	17	18	41	36
21	20	10	38	39	12	17	8	13	55	38	15	30	57	17	9	27	18	44	46
22	21	10	41	49	12	20	19	13	58	49	15	34	8	17	12	38	18	47	57
23	22	10	45	0	12	23	30	14	1	59	15	37	19	17	15	48	18	51	8
24	23	10	48	10	12	26	40	14	5	10	15	40	29	17	18	59	18	54	18
25	24	10	51	21	12	29	51	14	8	21	15	43	40	17	22	10	18	57	29
26	25	10	54	32	12	33	2	14	11	31	15	46	51	17	25	20	19	0	40
27	26	10	57	42	12	36	12	14	14	42	15	50	1	17	28	31	19	3	50
28	27	11	0	53	12	39	23	14	17	53	15	53	12	17	31	42	19	7	1
29	28	11	4	4	12	42	33	14	21	3	15	56	22	17	34	52	19	10	11
30	29	11	7	14	12	45	44	14	24	14	15	59	33	17	38	3	19	13	22
31	30	11	10	25	12	48	55	14	27	25	16	2	41	17	41	14	19	16	33
	31	11	13	36	12	52	5				16	5	54				19	19	43.4

The Epochs in Table 1. are for Midnight preceding Jan. 1, at Greenwich, but may be used for Noon of Jan. 1, if 1′ 35″.3 be subtracted from each Tabular Number.

TABLE 3.

The Moon's Semi-diameter for every Second of her Equatorial Horizontal Parallax. (By Mr. J. Utting.)

	53′	54′	55′	56′	57′	58′	59′	60′	61′
0″	14′ 26″.87	14′ 43″.22	14′ 59″.58	15′ 15″.94	15′ 32″.29	15′ 48″.65	16′ 5″.00	16′ 21″.36	16′ 37″.72
1	14 27.14	14 43.50	14 59.85	15 16.21	15 32.56	15 48.92	16 5.28	16 21.63	16 37.99
2	14 27.41	14 43.77	15 0.13	15 16.48	15 32.84	15 49.19	16 5.55	16 21.91	16 38.26
3	14 27.69	14 44.04	15 0.40	15 16.75	15 33.11	15 49.47	16 5.82	16 22.18	16 38.53
4	14 27.96	14 44.31	15 0.67	15 17.03	15 33.38	15 49.74	16 6.09	16 22.45	16 38.81
5	14 28.23	14 44.59	15 0.94	15 17.30	15 33.66	15 50.01	16 6.37	16 22.72	16 39.08
6	14 28.50	14 44.86	15 1.22	15 17.50	15 33.93	15 50.28	16 6.64	16 23.00	16 39.36
7	14 28.78	14 45.13	15 1.49	15 17.84	15 34.20	15 50.56	16 6.91	16 23.27	16 39.62
8	14 29.05	14 45.40	15 1.76	15 18.12	15 34.47	15 50.82	16 7.18	16 23.54	16 39.90
9	14 29.32	14 45.68	15 2.08	15 18.39	15 34.75	15 51.10	16 7.46	16 23.81	16 40.17
10	14 29.59	14 45.95	15 2.31	15 18.66	15 35.02	15 51.37	16 7.73	16 24.09	16 40.44
11	14 29.87	14 46.22	15 2.58	15 18.93	15 35.29	15 51.65	16 8.00	16 24.36	16 40.71
12	14 30.14	14 46.50	15 2.85	15 19.21	15 35.26	15 51.92	16 8.28	16 24.63	16 40.99
13	14 30.41	14 46.77	15 3.12	15 19.48	15 35.84	15 52.19	16 8.55	16 24.90	16 41.26
14	14 30.68	14 47.04	15 3.40	15 19.75	15 36.11	15 52.46	16 8.82	16 25.18	16 41.53
15	14 30.96	14 47.31	15 3.67	15 20.03	15 36.38	15 52.74	16 9.09	16 25.45	16 41.81
16	14 31.23	14 47.59	15 3.94	15 20.30	15 36.65	15 52.91	16 9.37	16 25.72	16 42.08
17	14 31.50	14 47.86	15 4.21	15 20.57	15 36.93	15 53.28	16 9.64	16 25.99	16 42.35
18	14 31.77	14 48.13	15 4.49	15 20.84	15 37.20	15 53.55	16 9.91	16 26.26	16 42.62
19	14 32.05	14 48.40	15 4.76	15 21.12	15 37.47	15 53.83	16 10.18	16 26.54	16 42.90
20	14 32.32	14 48.68	15 5.03	15 21.39	15 37.74	15 54.10	16 10.46	16 26.81	16 43.17
21	14 32.59	14 48.95	15 5.30	15 21.66	15 38.02	15 54.37	16 10.73	16 27.08	16 43.44
22	14 32.87	14 49.22	15 5.58	15 21.93	15 38.29	15 54.65	16 11.00	16 27.36	16 43.71
23	14 33.14	14 49.49	15 5.85	15 22.21	15 38.56	15 54.92	16 11.27	16 27.63	16 43.99
24	14 33.41	14 49.77	15 6.12	15 22.48	15 38.83	15 55.19	16 11.55	16 27.90	16 44.26
25	14 33.68	14 50.04	15 6.40	15 22.75	15 39.11	15 55.46	16 11.82	16 28.18	16 44.53
26	14 33.96	14 50.31	15 6.67	15 23.02	15 39.38	15 55.74	16 12.09	16 28.45	16 44.80
27	14 34.23	14 50.58	15 6.94	15 23.30	15 39.65	15 56.01	16 12.36	16 28.72	16 45.08
28	14 34.50	14 50.86	15 7.21	15 23.57	15 39.92	15 56.28	16 12.64	16 28.99	16 45.35
29	14 34.77	14 51.13	15 7.49	15 23.84	15 40.20	15 56.55	16 12.91	16 29.27	16 45.62
30	14 35.05	14 51.40	15 7.76	15 24.11	15 40.47	15 56.83	16 13.18	16 29.54	16 45.89
31	14 35.32	14 51.67	15 8.03	15 24.39	15 40.74	15 57.10	16 13.45	16 29.81	16 46.17
32	14 35.59	14 51.95	15 8.30	15 24.66	15 41.02	15 57.37	16 13.73	16 30.08	16 46.44
33	14 35.86	14 52.22	15 8.58	15 24.93	15 41.29	15 57.64	16 14.00	16 30.36	16 46.71
34	14 36.14	14 52.49	15 8.85	15 25.20	15 41.56	15 57.92	16 14.27	16 30.63	16 46.98
35	14 36.41	14 52.76	15 9.12	15 25.48	15 41.83	15 58.19	16 14.55	16 30.90	16 47.26
36	14 36.68	14 53.04	15 9.39	15 25.75	15 42.11	15 58.46	16 14.82	16 31.17	16 47.53
37	14 36.95	14 53.31	15 9.67	15 26.02	15 42.38	15 58.73	16 15.09	16 31.45	16 47.80
38	14 37.23	14 53.58	15 9.94	15 26.29	15 42.65	15 59.01	16 15.36	16 31.72	16 48.07
39	14 37.50	14 53.86	15 10.21	15 26.57	15 42.92	15 59.28	16 15.64	16 31.99	16 48.35
40	14 37.77	14 54.13	15 10.48	15 26.84	15 43.20	15 59.55	16 15.91	16 32.26	16 48.62
41	14 38.04	14′ 54.40	15 10.76	15 27.11	15 43.47	15 59.82	16 16.18	16 32.54	16 48.89
42	14 38.32	14 54.67	15 11.03	15 27.39	15 43.74	16 0.10	16 16.45	16 32.81	16 49.17
43	14 38.59	14 54.95	15 11.30	15 27.66	15 44.01	16 0.37	16 16.73	16 33.08	16 49.44
44	14 38.86	14 55.22	15 11.57	15 27.93	15 44.29	16 0.64	16 17.00	16 33.35	16 49.71
45	14 39.14	14 55.49	15 11.85	15 28.20	15 44.56	16 0.92	16 17.27	16 33.63	16 49.98
46	14 39.41	14 55.76	15 12.12	15 28.48	15 44.83	16 1.19	16 17.54	16 33.90	16 50.26
47	14 39.68	14 56.04	15 12.39	15 28.75	15 45.10	16 1.46	16 17.82	16 34.17	16 50.53
48	14 39.95	14 56.31	15 12.66	15 29.02	15 45.38	16 1.73	16 18.09	16 34.44	16 50.80
49	14 40.23	14 56.58	15 12.94	15 29.29	15 45.65	16 2.01	16 18.36	16 34.72	16 51.07
50	14 40.50	14 56.85	15 13.21	15 29.57	15 45.92	16 2.28	16 18.63	16 34.99	16 51.35
51	14 40.77	14 57.13	15 13.48	15 29.84	15 46.19	16 2.55	16 18.91	16 35.26	16 51.62
52	14 41.04	14 57.40	15 13.76	15 30.11	15 46.47	16 2.82	16 19.18	16 35.54	16 51.89
53	14 41.32	14 57.67	15 14.03	15 30.39	15 46.74	16 3.10	16 19.45	16 35.81	16 52.16
54	14 41.59	14 57.94	15 14.30	15 30.66	15 47.01	16 3.37	16 19.72	16 36.08	16 52.44
55	14 41.86	14 58.22	15 14.57	15 30.93	15 47.29	16 3.64	16 20.00	16 36.35	16 52.71
56	14 42.13	14 58.49	15 14.85	15 31.20	15 47.56	16 3.91	16 20.27	16 36.63	16 52.98
57	14 42.41	14 58.76	15 15.12	15 31.47	15 47.83	16 4.19	16 20.54	16 36.90	16 53.25
58	14 42.68	14 59.03	15 15.39	15 31.75	15 48.10	16 4.46	16 20.81	16 37.17	16 53.53
59	14 42.95	14 59.31	15 15.66	15 32.02	15 48.38	16 4.73	16 21.09	16 37.44	16 53.80
60	14 43.22	14 59.58	15 15.94	15 32.29	15 48.65	16 5.00	16 21.36	16 37.72	16 54.07

The Semi-diameters given by Burg, exceed these by about 1″. 15 each.

TABLE 4.

Reduction of the Moon's Parallax in the Spheroid, to every Degree of Latitude, and for every Minute of her Equatorial Parallax.

Argument = the Moon's Equatorial Horizontal Parallax. (By Mr. J. Utting.)

Compression = $\frac{1}{309}$ and $\frac{1}{317}$.

Arg. = Lat.	54′		55′		56′		57′		58′		59′		60′		61′		62′	
	$\frac{1}{309}$	$\frac{1}{317}$	$\frac{1}{309}$	$\frac{1}{317}$	$\frac{1}{309}$	$\frac{1}{317}$	$\frac{1}{309}$	$\frac{1}{317}$	$\frac{1}{309}$	$\frac{1}{317}$	$\frac{1}{309}$	$\frac{1}{317}$	$\frac{1}{309}$	$\frac{1}{317}$	$\frac{1}{309}$	$\frac{1}{317}$	$\frac{1}{309}$	$\frac{1}{317}$
1°	0″.00	0″.00	0″.00	0″.00	0″.00	0″.00	0″.00	0″.00	0″.00	0″.00	0″.00	0″.00	0″.00	0″.00	0″.00	0″.00	0″.00	0″.00
2	0.01	0.01	0.01	0.01	0.01	0.01	0.01	0.01	0.01	0.01	0.01	0.01	0.01	0.01	0.01	0.01	0.01	0.01
3	0.03	0.03	0.03	0.03	0.03	0.03	0.03	0.03	0.03	0.03	0.03	0.03	0.03	0.03	0.03	0.03	0.03	0.03
4	0.05	0.05	0.05	0.05	0.05	0.05	0.05	0.05	0.06	0.05	0.06	0.05	0.06	0.06	0.06	0.06	0.06	0.06
5	0.08	0.08	0.08	0.08	0.08	0.08	0.08	0.08	0.09	0.08	0.09	0.08	0.09	0.09	0.09	0.09	0.09	0.09
6	0.12	0.12	0.12	0.11	0.12	0.12	0.12	0.12	0.12	0.12	0.13	0.12	0.13	0.12	0.13	0.13	0.13	0.13
7	0.16	0.15	0.17	0.15	0.17	0.16	0.17	0.16	0.17	0.16	0.17	0.16	0.18	0.17	0.18	0.17	0.18	0.17
8	0.21	0.20	0.21	0.20	0.21	0.20	0.22	0.21	0.22	0.21	0.23	0.21	0.23	0.22	0.23	0.22	0.24	0.23
9	0.26	0.25	0.27	0.25	0.27	0.26	0.28	0.26	0.28	0.27	0.29	0.27	0.29	0.28	0.30	0.28	0.30	0.28
10	0.32	0.31	0.33	0.31	0.33	0.32	0.34	0.32	0.34	0.33	0.35	0.33	0.35	0.34	0.36	0.35	0.37	0.35
11	0.38	0.37	0.39	0.38	0.39	0.38	0.40	0.39	0.41	0.40	0.42	0.40	0.42	0.41	0.43	0.42	0.44	0.42
12	0.45	0.44	0.46	0.45	0.47	0.45	0.48	0.46	0.48	0.47	0.49	0.48	0.50	0.49	0.51	0.50	0.52	0.50
13	0.53	0.51	0.54	0.52	0.55	0.53	0.56	0.54	0.57	0.55	0.58	0.56	0.59	0.57	0.60	0.58	0.61	0.59
14	0.61	0.59	0.62	0.60	0.63	0.62	0.64	0.63	0.65	0.64	0.67	0.65	0.68	0.66	0.69	0.67	0.70	0.68
15	0.70	0.68	0.71	0.69	0.72	0.71	0.74	0.72	0.75	0.73	0.76	0.74	0.78	0.76	0.79	0.77	0.80	0.78
16	0.79	0.77	0.81	0.79	0.82	0.80	0.84	0.81	0.85	0.83	0.87	0.84	0.88	0.86	0.89	0.87	0.91	0.89
17	0.89	0.87	0.91	0.88	0.92	0.90	0.94	0.92	0.96	0.93	0.97	0.95	0.99	0.96	1.01	0.98	1.02	1.00
18	1.00	0.97	1.02	0.99	1.04	1.00	1.05	1.02	1.06	1.04	1.08	1.06	1.10	1.08	1.12	1.09	1.14	1.11
19	1.10	1.08	1.12	1.10	1.14	1.12	1.16	1.14	1.18	1.16	1.20	1.18	1.23	1.20	1.25	1.22	1.27	1.24
20	1.22	1.19	1.24	1.21	1.26	1.23	1.29	1.25	1.31	1.27	1.33	1.30	1.36	1.32	1.38	1.34	1.40	1.36
21	1.34	1.30	1.36	1.33	1.39	1.35	1.41	1.37	1.44	1.40	1.46	1.42	1.49	1.45	1.51	1.47	1.54	1.50
22	1.46	1.42	1.49	1.45	1.52	1.48	1.55	1.50	1.57	1.53	1.60	1.56	1.63	1.58	1.66	1.61	1.68	1.63
23	1.60	1.55	1.62	1.58	1.65	1.61	1.68	1.64	1.71	1.66	1.74	1.69	1.76	1.72	1.79	1.75	1.82	1.78
24	1.73	1.68	1.75	1.71	1.79	1.74	1.82	1.77	1.85	1.80	1.88	1.83	1.92	1.87	1.95	1.90	1.98	1.93
25	1.86	1.81	1.90	1.85	1.93	1.88	1.97	1.91	2.00	1.95	2.04	1.98	2.08	2.01	2.11	2.05	2.14	2.08
26	2.00	1.95	2.04	1.99	2.08	2.02	2.11	2.06	2.15	2.10	2.19	2.13	2.33	2.17	2.26	2.20	2.30	2.24
27	2.15	2.09	2.19	2.13	2.23	2.17	2.27	2.21	2.31	2.25	2.35	2.29	2.39	2.33	2.42	2.36	2.46	2.40
28	2.30	2.24	2.35	2.28	2.39	2.32	2.42	2.36	2.46	2.41	2.51	2.45	2.55	2.49	2.59	2.53	2.64	2.57
29	2.45	2.39	2.49	2.43	2.54	2.48	2.59	2.52	2.63	2.56	2.68	2.61	2.72	2.65	2.77	2.70	2.82	2.74
30	2.61	2.54	2.65	2.59	2.70	2.63	2.75	2.68	2.80	2.73	2.85	2.78	2.90	2.82	2.95	2.87	3.00	2.92
31	2.77	2.70	2.82	2.74	2.87	2.79	2.92	2.84	2.98	2.89	3.03	2.94	3.08	2.99	3.12	3.04	3.17	3.09
32	2.93	2.86	2.99	2.91	3.04	2.96	3.10	3.01	3.15	3.07	3.20	3.12	3.25	3.17	3.31	3.22	3.36	3.28
33	3.10	3.02	3.15	3.07	3.21	3.13	3.26	3.18	3.32	3.24	3.38	3.29	3.44	3.35	3.50	3.41	3.55	3.46
34	3.26	3.18	3.32	3.24	3.38	3.30	3.44	3.36	3.51	3.41	3.57	3.47	3.63	3.53	3.69	3.59	3.75	3.65
35	3.44	3.34	3.50	3.41	3.56	3.47	3.63	3.53	3.69	3.59	3.76	3.65	3.82	3.72	3.89	3.78	3.95	3.84
36	3.61	3.51	3.68	3.58	3.75	3.64	3.80	3.71	3.87	3.77	3.94	3.84	4.01	3.90	4.07	3.97	4.14	4.03
37	3.79	3.68	3.85	3.75	3.92	3.82	3.99	3.89	4.06	3.96	4.13	4.02	4.20	4.09	4.27	4.16	4.34	4.23
38	3.95	3.86	4.03	3.93	4.10	4.00	4.18	4.07	4.25	4.14	4.33	4.21	4.40	4.28	4.58	4.36	1.55	4.43
39	4.14	4.03	4.21	4.10	4.29	4.18	4.37	4.25	4.45	4.32	4.51	4.40	4.59	4.48	4.67	4.55	4.75	4.63
40	4.32	4.20	4.40	4.28	4.48	4.36	4.56	4.44	4.64	4.51	4.72	4.59	4.80	4.67	4.88	4.75	4.96	4.83
41	4.50	4.38	4.58	4.46	4.66	4.54	4.75	4.62	4.83	4.70	4.92	4.78	5.00	4.87	5.09	4.95	5.17	5.03
42	4.68	4.56	4.77	4.64	4.85	4.73	4.94	4.81	5.03	4.89	5.12	5.98	5.20	5.06	5.28	5.15	5.37	5.23
43	4.86	4.73	4.95	4.82	5.04	4.91	5.14	5.00	5.22	5.09	5.31	5.17	5.40	5.26	5.49	5.35	5.58	5.44
44	5.05	4.91	5.14	5.00	5.23	5.09	5.32	5.18	5.42	5.28	5.51	5.37	5.61	5.46	5.70	5.55	5.80	5.64
45	5.22	5.09	5.32	5.18	5.42	5.28	5.52	5.37	5.62	5.47	5.71	5.56	5.81	5.66	5.91	5.75	6.01	5.84

TABLE 4.

REDUCTION of the MOON'S PARALLAX in the SPHEROID, &c. concluded.

Compression $= \frac{1}{309}$ and $\frac{1}{317}$.

Arg. = Lat.	54′		55′		56′		57′		58′		59′		60′		61′		62′	
	$\frac{1}{309}$	$\frac{1}{317}$	$\frac{1}{309}$	$\frac{1}{317}$	$\frac{1}{309}$	$\frac{1}{317}$	$\frac{1}{309}$	$\frac{1}{317}$	$\frac{1}{309}$	$\frac{1}{317}$	$\frac{1}{309}$	$\frac{1}{317}$	$\frac{1}{309}$	$\frac{1}{317}$	$\frac{1}{309}$	$\frac{1}{317}$	$\frac{1}{309}$	$\frac{1}{317}$
46°	5″.41	5″.27	5″.51	5″.37	5″.61	5″.46	5″.72	5″.56	5″.82	5″.66	5″.91	5″.76	6″.01	5″.85	6″.11	5″.95	6″.21	6″.05
47	5.60	5.45	5.70	5.55	5.80	5.65	5.90	5.75	6.01	5.85	6.11	5.95	6.22	6.05	6.32	6.15	6.43	6.25
48	5.78	5.62	5.88	5.73	5.99	5.83	6.10	5.94	6.21	6.04	6.32	6.16	6.42	6.25	6.53	6.35	6.64	6.46
49	5.96	5.80	6.07	5.91	6.18	6.02	6.30	6.12	6.41	6.23	6.51	6.32	6.62	6.45	6.73	6.55	6.85	6.66
50	6.14	5.98	6.26	6.09	6.37	6.20	6.49	6.31	6.59	6.42	6.71	6.53	6.82	6.64	6.94	6.75	7.05	6.86
51	6.32	6.15	6.44	6.27	6.56	6.38	6.69	6.50	6.79	6.61	6.91	6.72	7.02	6.83	7.15	6.95	7.27	7.07
52	6.50	6.33	6.62	6.44	6.74	6.56	6.86	6.68	6.98	6.80	7.11	6.91	7.22	7.03	7.35	7.12	7.46	7.24
53	6.68	6.50	6.80	6.62	6.93	6.74	7.05	6.86	7.18	6.98	7.29	7.10	7.42	7.22	7.55	7.34	7.67	7.46
54	6.86	6.67	6.99	6.80	7.12	6.92	7.24	7.04	7.36	7.17	7.49	7.29	7.62	7.41	7.75	7.54	7.88	7.66
55	7.04	9.84	7.17	6.97	7.29	7.09	7.42	7.22	7.55	7.35	7.69	7.47	7.82	7.60	7.95	7.73	8.08	7.85
56	7.21	7.01	7.33	7.14	7.47	7.27	7.60	7.40	7.74	7.53	7.88	7.66	8.01	7.79	8.15	7.92	8.28	8.05
57	7.37	7.17	7.51	7.31	7.65	7.44	7.79	7.57	7.92	7.70	8.05	7.84	8.19	7.97	8.33	8.10	8.47	8.24
58	7.54	7.33	7.68	7.47	7.82	7.61	7.95	7.74	8.10	7.88	8.24	8.01	8.38	8.15	8.52	8.29	8.66	8.42
59	7.71	7.49	7.85	7.63	7.99	7.77	8.13	7.91	8.28	8.05	8.42	8.19	8.57	8.33	8.70	8.47	8.85	8.60
60	7.87	7.65	8.01	7.79	8.16	7.93	8.30	8.08	8.45	8.22	8.60	8.36	8.74	8.50	8.89	8.64	9.03	8.78
61	8.02	7.80	8.17	7.95	8.32	8.09	8.48	8.24	8.63	8.38	8.77	8.53	8.92	8.67	9.07	8.82	9.22	8.96
62	8.18	7.95	8.33	8.10	8.49	8.25	8.63	8.40	8.79	8.54	8.94	8.69	9.09	8.84	9.24	8.99	9.39	9.13
63	8.33	8.10	8.49	8.25	8.64	8.40	8.79	8.55	8.95	8.70	9.11	8.85	9.26	9.00	9.41	9.15	9.56	9.30
64	8.48	8.24	8.63	8.40	8.79	8.55	8.95	8.70	9.11	8.85	9.27	9.01	9.42	9.16	9.58	9.31	9.74	9.47
65	8.63	8.38	8.78	8.54	8.94	8.69	9.10	8.85	9.26	9.00	9.42	9.16	9.58	9.32	9.74	9.47	9.90	9.63
66	8.76	8.52	8.92	8.68	9.09	8.84	9.25	8.99	9.41	9.15	9.57	9.31	9.74	9.47	9.90	9.62	10.06	9.78
67	8.89	8.65	9.06	8.81	9.23	8.97	9.39	9.13	9.55	9.29	9.72	9.45	9.89	9.61	10.04	9.77	10.21	9.93
68	9.03	8.78	9.20	8.94	9.36	9.10	9.53	9.26	9.70	9.43	9.87	9.59	10.02	9.75	10.19	9.91	10.36	10.08
69	9.16	8.90	9.32	9.06	9.49	9.23	9.66	9.39	9.83	9.56	10.01	9.72	10.17	9.89	10.34	10.05	10.51	10.22
70	9.28	9.02	9.45	9.18	9.62	9.35	9.80	9.52	9.97	9.69	10.13	9.85	10.31	10.02	10.48	10.19	10.66	10.35
71	9.39	9.13	9.57	9.30	9.74	9.47	9.92	9.64	10.08	9.81	10.26	9.98	10.44	10.14	10.61	10.31	10.78	10.48
72	9.50	9.24	9.68	9.41	9.86	9.59	10.03	9.75	10.21	9.92	10.38	10.09	10.56	10.26	10.73	10.44	10.91	10.61
73	9.61	9.34	9.79	9.51	9.97	9.69	10.14	9.86	10.32	10.03	10.50	10.21	10.68	10.38	10.86	10.55	11.04	10.72
74	9.71	9.44	9.90	9.61	10.07	9.79	10.25	9.96	10.43	10.14	10.62	10.31	10.79	10.49	10.98	10.66	11.15	10.84
75	9.81	9.53	10.00	9.71	10.17	9.88	10.35	10.06	10.54	10.24	10.71	10.41	10.90	10.59	11.08	10.77	11.27	10.94
76	9.90	9.62	10.08	9.80	10.26	9.97	10.45	10.15	10.64	10.33	10.81	10.51	11.00	10.69	11.18	10.86	11.37	11.04
77	9.99	9.70	10.16	9.88	10.35	10.06	10.54	10.24	10.72	10.42	10.90	10.60	11.09	10.78	11.28	10.96	11.46	11.14
78	10.05	9.78	10.24	9.96	10.43	10.14	10.62	10.32	10.80	10.50	10.99	10.68	11.18	10.86	11.37	11.04	11.55	11.22
79	10.12	9.85	10.32	10.03	10.51	10.21	10.70	10.39	10.88	10.58	11.07	10.76	11.26	10.94	11.44	11.12	11.63	11.30
80	10.19	9.91	10.38	10.09	10.57	10.28	10.75	10.46	10.94	10.64	11.14	10.83	11.33	11.01	11.51	11.20	11.70	11.38
81	10.25	9.97	10.44	10.15	10.63	10.34	10.81	10.52	11.01	10.71	11.20	10.89	11.39	11.08	11.57	11.26	11.77	11.45
82	10.30	10.02	10.50	10.21	10.69	10.39	10.87	10.58	11.07	10.76	11.26	10.95	11.44	11.14	11.64	11.32	11.83	11.51
83	10.35	10.07	10.55	10.26	10.73	10.44	10.92	10.63	11.12	10.82	11.31	11.00	11.50	11.19	11.69	11.37	11.89	11.56
84	10.39	10.11	10.59	10.30	10.77	10.48	10.97	10.67	11.16	10.86	11.36	11.05	11.54	11.23	11.74	11.42	11.93	11.61
85	10.43	10.14	10.62	10.33	10.81	10.52	11.01	10.71	11.20	10.90	11.39	11.08	11.58	11.27	11.78	11.46	11.98	11.65
86	10.46	10.17	10.65	10.36	10.84	10.55	11.04	10.74	11.23	10.92	11.42	11.11	11.62	11.30	11.81	11.49	12.01	11.68
87	10.48	10.19	10.68	10.38	10.86	10.57	11.06	10.76	11.26	10.95	11.44	11.14	11.64	11.33	11.84	11.51	12.04	11.70
88	10.50	10.21	10.69	10.40	10.88	10.59	11.08	10.77	11.27	10.96	11.46	11.15	11.66	11.34	11.86	11.53	12.05	11.72
89	10.51	10.22	10.70	10.41	10.89	10.60	11.09	10.78	11.29	10.97	11.47	11.16	11.67	11.35	11.87	11.54	12.06	11.73
90	10.51	10.22	10.70	10.41	10.89	10.60	11.09	10.79	11.29	10.98	11.47	11.17	11.67	11.36	11.87	11.55	12.07	11.74

TABLE 5.

LOGARITHMS of the EARTH'S RADII, in each parallel of Latitude, calculated for three different Compressions; the Equatorial Radius being Unity.

Lat.	1/300	1/317	1/330	Lat.	1/300	1/317	1/330	Lat.	1/300	1/317	1/330
0°	0 . 0000000	0 . 0000000	0 . 0000000	30°	9 . 9996402	9 . 9996594	9 . 9996727	60°	9 . 9989151	9 . 9989733	9 . 9990138
1	9 . 9999995	9 . 9999995	9 . 9999995	31	96181	96386	96527	61	88932	89527	89940
2	9982	9983	9983	32	95957	96174	96323	62	88720	89325	89745
3	9960	9962	9964	33	95728	95957	96115	63	88512	89128	89556
4	9930	9934	9936	34	95496	95737	95904	64	88308	88935	89371
5	9890	9896	9900	35	95261	95514	95690	65	88111	88748	89192
6	9843	9851	9857	36	95023	95288	95473	66	87918	88566	89017
7	9786	9798	9806	37	94781	95059	95254	67	87732	88389	88847
8	9721	9737	9747	38	94537	94829	95032	68	87552	88219	88684
9	9648	9668	9680	39	94291	94596	94809	69	87378	88055	88526
10	9566	9590	9606	40	94044	94362	94584	70	87210	87897	88373
11	9477	9505	9524	41	93794	94126	94357	71	87050	87745	88228
12	9379	9412	9435	42	93543	93887	94129	72	86896	87599	88089
13	9273	9312	9338	43	93291	93649	93900	73	86750	87461	87956
14	9158	9204	9235	44	93038	93411	93670	74	86611	87330	87829
15	9037	9088	9124	45	92786	93172	93440	75	86479	87206	87710
16	8908	8966	9007	46	92533	92933	93210	76	86356	87089	87598
17	8771	8837	8882	47	92280	92692	92980	77	86240	86979	87493
18	8627	8701	8751	48	92028	92453	92751	78	86131	86877	87395
19	8476	8558	8614	49	91776	92215	92522	79	86031	86782	87303
20	8318	8409	8470	50	91525	91979	92295	80	85940	86695	87220
21	8153	8253	8320	51	91277	91744	92069	81	85857	86617	87145
22	7983	8091	8164	52	91030	91510	91844	82	85782	86546	87077
23	7805	7922	8003	53	90785	91278	91621	83	85716	86482	87017
24	7621	7748	7836	54	90542	91048	91401	84	85659	86428	86965
25	7431	7569	7663	55	90302	90821	91183	85	85610	86382	86920
26	7236	7383	7485	56	90065	90597	90967	86	85570	86345	86884
27	7035	7193	7303	57	89831	90376	90755	87	85539	86316	86856
28	6829	6997	7115	58	89600	90158	90545	88	85517	86296	86837
29	6618	6798	6923	59	89374	89943	90340	89	85504	86283	86825
30	6402	6594	6727	60	89151	89733	90138	90	85499	86278	86820

TABLE 6.

ANGLES of the VERTICAL with the RADIUS; or REDUCTION of the LATITUDE, in each Parallel.

Lat.	1/300	1/317	1/330	Lat.	1/300	1/317	1/330	Lat.	1/300	1/317	1/330
	—	—	—		—	—	—		—	—	—
0°	0′ 0″.0	0′ 0″.0	0′ 0″.0	30°	9′ 55″.4	9′ 23″.5	9′ 1″.3	60°	9′ 57″.4	9′ 25″.3	9′ 2″.8
1	0 24 .0	0 22 .6	0 21 .8	31	10 7 .2	9 34 .6	9 11 .9	61	9 45 .1	9 13 .6	8 51 .6
2	0 47 .9	0 45 .3	0 43 .5	32	10 18 .1	9 45 .0	9 21 .9	62	9 32 .0	9 1 .3	8 39 .8
3	1 11 .8	1 7 .9	1 5 .2	33	10 28 .3	9 54 .6	9 31 .1	63	9 18 .3	8 48 .2	8 27 .3
4	1 35 .5	1 30 .4	1 26 .9	34	10 37 .8	10 3 .6	9 39 .7	64	9 3 .8	8 34 .6	8 14 .2
5	1 59 .2	1 52 .8	1 48 .4	35	10 46 .4	10 11 .7	9 47 .6	65	8 48 .7	8 20 .3	8 0 .4
6	2 22 .7	2 15 .1	2 9 .8	36	10 54 .3	10 19 .2	9 54 .8	66	8 32 .9	8 5 .3	7 46 .1
7	2 46 .1	2 37 .2	2 31 .0	37	11 1 .4	10 25 .9	10 1 .2	67	8 16 .6	7 49 .8	7 31 .2
8	3 9 .2	2 59 .1	2 52 .0	38	11 7 .7	10 31 .9	10 6 .9	68	7 59 .6	7 33 .8	7 15 .8
9	3 32 .1	3 20 .8	3 12 .9	39	11 13 .2	10 37 .1	10 11 .9	69	7 42 .0	7 17 .1	6 59 .8
10	3 54 .8	3 42 .2	3 33 .5	40	11 17 .9	10 41 .5	10 16 .1	70	7 23 .8	6 59 .9	6 43 .8
11	4 17 .2	4 3 .4	3 53 .8	41	11 21 .7	10 45 .1	10 19 .6	71	7 5 .1	6 42 .3	6 26 .3
12	4 39 .3	4 24 .3	4 13 .9	42	11 24 .7	10 47 .9	10 22 .4	72	6 45 .9	6 24 .0	6 8 .9
13	5 1 .0	4 44 .9	4 33 .7	43	11 26 .9	10 50 .0	10 24 .3	73	6 26 .2	6 5 .4	5 50 .9
14	5 22 .4	5 5 .1	4 53 .1	44	11 28 .2	10 51 .2	10 25 .5	74	6 6 .0	5 46 .3	5 32 .6
15	5 43 .4	5 25 .0	5 12 .2	45	11 28 .7	10 51 .7	10 26 .0	75	5 45 .4	5 26 .8	5 13 .8
16	6 3 .9	5 44 .4	5 30 .9	46	11 28 .4	10 51 .4	10 25 .7	76	5 24 .3	5 6 .8	4 54 .6
17	6 24 .1	6 3 .5	5 49 .2	47	11 27 .3	10 50 .3	10 24 .6	77	5 2 .8	4 46 .5	4 35 .1
18	6 43 .7	6 22 .1	6 7 .0	48	11 25 .2	10 48 .4	10 22 .7	78	4 41 .0	4 25 .8	4 15 .3
19	7 2 .9	6 40 .2	6 24 .5	49	11 22 .3	10 45 .7	10 20 .1	79	4 18 .8	4 4 .9	3 55 .2
20	7 21 .6	6 57 .9	6 41 .4	50	11 18 .6	10 42 .2	10 16 .8	80	3 56 .3	3 43 .6	3 34 .7
21	7 39 .7	7 15 .1	6 57 .9	51	11 14 .1	10 37 .9	10 12 .7	81	3 33 .5	3 22 .0	3 14 .0
22	7 57 .3	7 31 .7	7 13 .9	52	11 8 .8	10 32 .8	10 7 .8	82	3 10 .4	3 0 .2	2 53 .0
23	8 14 .2	7 47 .8	7 29 .3	53	11 2 .6	10 27 .0	10 2 .2	83	2 47 .2	2 38 .2	2 31 .9
24	8 30 .7	8 3 .3	7 44 .2	54	10 55 .7	10 20 .4	9 55 .9	84	2 23 .7	2 15 .9	2 10 .5
25	8 46 .4	8 18 .2	7 58 .6	55	10 47 .9	10 13 .1	9 48 .8	85	2 0 .0	2 53 .5	1 49 .0
26	9 1 .6	8 32 .6	8 12 .4	56	10 39 .4	10 5 .0	9 41 .0	86	1 36 .2	1 31 .0	1 27 .4
27	9 16 .1	8 46 .3	8 25 .5	57	10 30 .0	9 56 .1	9 32 .6	87	1 12 .3	1 8 .3	1 5 .6
28	9 29 .9	8 59 .3	8 38 .1	58	10 19 .9	9 46 .6	9 23 .4	88	0 48 .2	0 45 .6	0 43 .8
29	9 43 .0	9 11 .8	8 50 .0	59	10 9 .0	9 36 .3	9 13 .4	89	0 24 .1	0 22 .8	0 21 .9
30	9 55 .4	9 23 .5	9 1 .3	60	9 57 .4	9 25 .3	9 2 .8	90	0 0 .0	0 0 .0	0 0 .0

TABLE 7.

The Augmentation of the Moon's Semidiameter in Altitude, for every 10″ of her Horizontal Semidiameter. (By Mr. J. Utting.)

Argument = the Moon's Altitude.

☾'s Alt.	14′ 30″	14′ 40″	14′ 50″	15′ 0″	15′ 10″	15′ 20″	15′ 30″	15′ 40″	15′ 50″	16′ 0″	16′ 10″	16′ 20″	16′ 30″	16′ 40″	16′ 50″	17′ 0″
1°	0″.24	0″.24	0″.25	0″.25	0″.26	0″.26	0″.27	0″.27	0″.28	0″.29	0″.29	0″.30	0″.31	0″.31	0″.32	0″.33
2	0.48	0.48	0.49	0.50	0.51	0.52	0.54	0.55	0.56	0.58	0.59	0.60	0.62	0.63	0.64	0.65
3	0.71	0.72	0.73	0.75	0.76	0.78	0.80	0.82	0.84	0.86	0.88	0.90	0.92	0.93	0.95	0.97
4	0.95	0.96	0.98	1.00	1.02	1.04	1.07	1.09	1.12	1.15	1.17	1.20	1.23	1.25	1.27	1.30
5	1.18	1.20	1.22	1.25	1.28	1.31	1.34	1.37	1.40	1.43	1.46	1.49	1.53	1.56	1.59	1.62
6	1.41	1.44	1.47	1.50	1.53	1.56	1.60	1.63	1.67	1.71	1.75	1.79	1.83	1.86	1.90	1.94
7	1.65	1.68	1.71	1.75	1.79	1.83	1.87	1.91	1.95	2.00	2.04	2.08	2.13	2.17	2.21	2.26
8	1.88	1.92	1.96	2.00	2.04	2.09	2.14	2.18	2.23	2.28	2.33	2.38	2.43	2.48	2.53	2.58
9	2.11	2.15	2.20	2.25	2.30	2.35	2.40	2.45	2.50	2.56	2.61	2.67	2.73	2.78	2.84	2.90
10	2.35	2.40	2.45	2.50	2.55	2.61	2.67	2.73	2.79	2.85	2.91	2.97	3.03	3.09	3.15	3.22
11	2.58	2.63	2.69	2.75	2.81	2.87	2.94	3.00	3.06	3.13	3.19	3.26	3.33	3.40	3.47	3.54
12	2.81	2.87	2.93	3.00	3.06	3.13	3.20	3.27	3.34	3.41	3.48	3.55	3.63	3.70	3.78	3.86
13	3.04	3.11	3.18	3.25	3.32	3.39	3.47	3.54	3.61	3.69	3.77	3.85	3.93	4.01	4.09	4.18
14	3.27	3.34	3.42	3.50	3.57	3.65	3.73	3.81	3.89	3.97	4.05	4.14	4.23	4.31	4.40	4.49
15	3.50	3.58	3.66	3.74	3.82	3.90	3.99	4.07	4.16	4.25	4.34	4.43	4.52	4.61	4.70	4.80
16	3.73	3.81	3.89	3.98	4.07	4.16	4.25	4.34	4.43	4.53	4.62	4.71	4.81	4.91	5.01	5.11
17	3.95	4.04	4.13	4.22	4.31	4.41	4.51	4.60	4.70	4.80	4.90	5.00	5.10	5.20	5.31	5.42
18	4.17	4.26	4.36	4.46	4.56	4.66	4.76	4.86	4.96	5.07	5.17	5.28	5.39	5.50	5.61	5.73
19	4.40	4.50	4.60	4.70	4.80	4.91	5.02	5.13	5.24	5.35	5.46	5.57	5.68	5.80	5.92	6.04
20	4.62	4.72	4.83	4.94	5.05	5.16	5.27	5.38	5.50	5.62	5.73	5.85	5.97	6.09	6.22	6.35
21	4.84	4.95	5.06	5.18	5.29	5.40	5.52	5.64	5.76	5.89	6.01	6.13	6.26	6.39	6.52	6.65
22	5.06	5.18	5.30	5.42	5.53	5.65	5.77	5.90	6.03	6.16	6.29	6.42	6.55	6.68	6.81	6.95
23	5.28	5.40	5.52	5.65	5.77	5.89	6.02	6.15	6.28	6.42	6.55	6.69	6.83	6.97	7.11	7.25
24	5.49	5.62	5.75	5.88	6.01	6.14	6.27	6.40	6.54	6.68	6.82	6.96	7.11	7.25	7.39	7.54
25	5.71	5.84	5.97	6.11	6.24	6.38	6.52	6.66	6.80	6.94	7.09	7.24	7.39	7.54	7.69	7.84
26	5.92	6.06	6.20	6.34	6.48	6.62	6.76	6.90	7.05	7.20	7.35	7.50	7.66	7.81	7.97	8.13
27	6.13	6.27	6.41	6.56	6.70	6.85	7.00	7.15	7.30	7.46	7.61	7.77	7.93	8.09	8.25	8.42
28	6.34	6.49	6.64	6.79	6.94	7.09	7.24	7.40	7.56	7.72	7.88	8.04	8.20	8.37	8.54	8.71
29	6.55	6.70	6.85	7.01	6.16	7.32	7.48	7.64	7.80	7.97	8.13	8.30	8.47	8.64	8.82	9.00
30	6.75	6.91	7.07	7.23	6.39	7.55	7.71	7.88	8.05	8.22	8.39	8.56	8.74	8.92	9.10	9.28
31	6.95	7.11	7.28	7.45	7.61	7.77	7.94	8.11	8.29	8.47	8.64	8.82	9.00	9.18	9.37	9.56
32	7.15	7.32	7.49	7.67	7.83	8.00	8.17	8.35	8.53	8.72	8.90	9.08	9.26	9.45	9.61	9.84
33	7.35	7.52	7.70	7.88	8.05	8.22	8.40	8.58	8.77	8.96	9.14	9.33	9.52	9.72	9.92	10.12
34	7.55	7.73	7.91	8.09	8.27	8.45	8.63	8.82	9.01	9.20	9.39	9.58	9.78	9.98	10.18	10.39
35	7.74	7.92	8.11	8.30	8.48	8.66	8.85	9.04	9.24	9.44	9.63	9.83	10.03	10.24	10.45	10.66
36	7.93	8.12	8.31	8.50	8.69	8.88	9.07	9.27	9.47	9.67	9.87	10.07	10.28	10.49	10.70	10.92
37	8.12	8.31	8.50	8.70	8.89	9.09	9.29	9.49	9.69	9.90	10.11	10.32	10.53	10.74	10.95	11.17
38	8.31	8.50	8.70	8.90	9.10	9.30	9.51	9.71	9.92	10.13	10.34	10.56	10.78	10.99	11.20	11.42
39	8.49	8.69	8.89	9.10	9.30	9.51	9.72	9.93	10.14	10.36	10.58	10.80	11.02	11.23	11.44	11.66
40	8.67	8.88	9.09	9.30	9.51	9.72	9.93	10.14	10.36	10.58	10.80	11.03	11.26	11.48	11.70	11.92
41	8.85	9.06	9.27	9.49	9.70	9.92	10.14	10.36	10.58	10.80	11.03	11.26	11.49	11.72	11.95	12.18
42	9.03	9.24	9.46	9.68	9.90	10.12	10.34	10.56	10.79	11.02	11.25	11.48	11.72	11.96	12.20	12.44
43	9.21	9.43	9.65	9.87	10.09	10.31	10.54	10.77	11.00	11.23	11.47	11.71	11.95	12.19	12.43	12.68
44	9.38	9.60	9.82	10.05	10.28	10.51	10.74	10.97	11.20	11.44	11.68	11.92	12.17	12.42	12.67	12.92
45	9.55	9.77	10.00	10.23	10.46	10.69	10.93	11.17	11.41	11.65	11.89	12.14	12.39	12.64	12.89	13.15

TABLE 7.

The AUGMENTATION of the MOON'S SEMIDIAMETER in ALTITUDE, &c. concluded.

Argument = the Moon's Altitude.

☾'s Alt.	14′ 30″	14′ 40″	14′ 50″	15′ 0″	15′ 10″	15′ 20″	15′ 30″	15′ 40″	15′ 50″	16′ 0″	16′ 10″	16′ 20″	16′ 30″	16′ 40″	16′ 50″	17′ 0″
46°	9″.72	9″.95	10″.18	10″.41	10″.64	10″.88	11″.12	11″.36	11″.60	11″.85	12″.10	12″.35	12″.61	12″.86	13″.12	13″.38
47	9.89	10.12	10.35	10.59	10.83	11.07	11.31	11.55	11.80	12.05	12.30	12.56	12.82	13.08	13.34	13.61
48	10.05	10.28	10.52	10.76	11.00	11.24	11.49	11.74	11.99	12.25	12.50	12.77	13.03	13.29	13.56	13.83
49	10.21	10.45	10.69	10.93	11.17	11.42	11.67	11.92	12.18	12.44	12.70	12.96	13.23	13.50	13.77	14.04
50	10.37	10.61	10.85	11.10	11.34	11.59	11.85	12.11	12.37	12.63	12.89	13.16	13.43	13.70	13.97	14.25
51	10.52	10.76	11.01	11.26	11.51	11.76	12.02	12.28	12.54	12.81	13.08	13.35	13.63	13.90	14.18	14.46
52	10.67	10.92	11.17	11.42	11.67	11.93	12.19	12.45	12.72	12.99	13.26	13.54	13.82	14.10	14.38	14.66
53	10.81	11.06	11.31	11.57	11.83	12.09	12.36	12.63	12.90	13.17	13.45	13.73	14.01	14.29	14.57	14.86
54	10.95	11.20	11.46	11.72	11.98	12.25	12.52	12.79	13.06	13.34	13.62	13.90	14.19	14.48	14.77	15.06
55	11.09	11.35	11.61	11.87	12.13	12.40	12.68	12.95	13.23	13.51	13.79	14.08	14.37	14.66	14.95	15.25
56	11.22	11.48	11.74	12.01	12.28	12.55	12.83	13.11	13.39	13.67	13.96	14.25	14.55	14.84	15.14	15.44
57	11.35	11.61	11.88	12.15	12.42	12.70	12.98	13.26	13.54	13.83	14.12	14.42	14.72	15.02	15.32	15.62
58	11.48	11.75	12.02	12.29	12.56	12.84	13.12	13.41	13.70	13.99	14.28	14.58	14.88	15.18	15.48	15.79
59	11.60	11.87	12.14	12.42	12.70	12.98	13.26	13.55	13.84	14.14	14.44	14.74	15.04	15.34	15.65	15.96
60	11.72	11.99	12.27	12.55	12.83	13.11	13.40	13.69	13.99	14.29	14.59	14.89	15.20	15.51	15.82	16.13
61	11.84	12.11	12.39	12.67	12.95	13.24	13.53	13.83	14.13	14.43	14.73	15.04	15.35	15.66	15.97	16.29
62	11.95	12.23	12.51	12.79	13.08	13.37	13.66	13.96	14.26	14.57	14.88	15.19	15.50	15.81	16.13	16.45
63	12.06	12.34	12.62	12.91	13.20	13.49	13.79	14.09	14.39	14.70	15.01	15.32	15.64	15.96	16.28	16.60
64	12.17	12.45	12.73	13.02	13.31	13.61	13.91	14.21	14.52	14.83	15.14	15.46	15.78	16.10	16.42	16.75
65	12.27	12.55	12.84	13.13	13.43	13.73	14.03	14.34	14.65	14.96	15.27	15.59	15.91	16.23	16.56	16.89
66	12.37	12.66	12.95	13.24	13.54	13.84	14.14	14.45	14.76	15.08	15.40	15.72	16.04	16.37	16.70	17.03
67	12.46	12.75	13.04	13.34	13.64	13.94	14.25	14.56	14.87	15.19	15.51	15.83	16.16	16.49	16.82	17.16
68	12.55	12.84	13.14	13.44	13.74	14.05	14.36	14.67	14.98	15.30	15.62	15.95	16.23	16.61	16.94	17.28
69	12.64	12.93	13.23	13.53	13.84	14.15	14.46	14.77	15.09	15.41	15.73	16.06	16.39	16.72	17.06	17.40
70	12.72	13.02	13.32	13.62	13.93	14.24	14.55	14.87	15.19	15.51	15.84	16.17	16.50	16.83	17.17	17.51
71	12.80	13.10	13.40	13.71	14.02	14.33	14.64	14.96	15.28	15.61	15.94	16.27	16.60	16.94	17.28	17.62
72	12.88	13.18	13.48	13.79	14.10	14.41	14.73	15.05	15.37	15.70	16.03	16.36	16.70	17.04	17.38	17.73
73	12.95	13.25	13.56	13.87	14.18	14.49	14.81	15.13	15.46	15.79	16.12	16.45	16.79	17.13	17.48	17.83
74	13.02	13.32	13.63	13.94	14.25	14.57	14.89	15.21	15.54	15.87	16.20	16.54	16.88	17.22	17.57	17.92
75	13.08	13.39	13.70	14.01	14.32	14.64	14.96	15.29	15.62	15.95	16.28	16.62	16.96	17.31	17.66	18.01
76	13.14	13.45	13.76	14.07	14.39	14.71	15.03	15.36	15.69	16.02	16.36	16.70	17.04	17.39	17.74	18.09
77	13.19	13.50	13.81	14.13	14.45	14.77	15.09	15.42	15.75	16.09	16.43	16.77	17.11	17.46	17.81	18.17
78	13.24	13.55	13.86	14.18	14.50	14.82	15.15	15.48	15.81	16.15	16.49	16.83	17.18	17.53	17.88	18.24
79	13.29	13.60	13.91	14.23	14.55	14.87	15.20	15.53	15.87	16.21	16.55	16.89	17.24	17.59	17.94	18.30
80	13.33	13.64	13.96	14.28	14.60	14.92	15.25	15.58	15.92	16.26	16.60	16.94	17.30	17.65	18.00	18.36
81	13.37	13.68	14.00	14.32	14.64	14.97	15.30	15.63	15.97	16.31	16.65	16.99	17.35	17.70	18.06	18.42
82	13.40	13.72	14.04	14.36	14.68	15.01	15.34	15.67	16.01	16.35	16.69	17.04	17.39	17.75	18.11	18.47
83	13.43	13.75	14.07	14.39	14.72	15.05	15.38	15.71	16.05	16.39	16.73	17.08	17.43	17.79	18.15	18.51
84	13.46	13.78	14.10	14.42	14.75	15.08	15.41	15.74	16.08	16.42	16.77	17.12	17.47	17.83	18.19	18.55
85	13.48	13.80	14.12	14.44	14.77	15.10	15.43	15.77	16.11	16.45	16.80	17.15	17.50	17.86	18.22	18.58
86	13.50	13.82	14.14	14.46	14.79	15.12	15.45	15.79	16.13	16.47	16.82	17.17	17.52	17.88	18.24	18.60
87	13.52	13.84	14.16	14.48	14.81	15.14	15.47	15.81	16.15	16.49	16.84	17.19	17.54	17.90	18.26	18.62
88	13.53	13.85	14.17	14.49	14.82	15.15	15.48	15.82	16.16	16.50	16.85	17.20	17.55	17.91	18.27	18.63
89	13.54	13.86	14.18	14.50	14.83	15.16	15.49	15.83	16.17	16.51	16.86	17.21	17.56	17.92	18.28	18.64
90	13.54	13.86	14.18	14.50	14.83	15.16	15.49	15.83	16.17	16.51	16.86	17.21	17.57	17.93	18.29	18.65

TABLE 8.—The Moon's Parallax to every 10′ of her Altitude, and for every Minute of her Horizontal Parallax, with the difference for 100″, variation of the Horizontal Parallax. Argument = the Moon's Altitude. (By Mr. J. Utting.)

Arg. ☾'s Alt	54′	55′	56′	57′	58′	59′	60′	61′	62′	Diff. for 100″.
0° 0	54′ 0″.00	55′ 0″.00	56′ 0″.00	57′ 0″.00	58′ 0″.00	59′ 0″.00	60′ 0″.00	61′ 0″.00	62′ 0″.00	100″.00
10	53 59.99	54 59.99	55 59.99	56 59.99	57 59.99	58 59.98	59 59.98	60 59.98	61 59.98	100.00
20	53 59.95	54 59.94	55 59.94	56 59.94	57 59.94	58 59.94	59 59.94	60 59.94	61 59.94	100.00
30	53 59.88	54 59.87	55 59.87	56 59.87	57 59.87	58 59.87	59 59.86	60 59.86	61 59.86	100.00
40	53 59.78	54 59.78	55 59.77	56 59.77	57 59.76	58 59.76	59 59.76	60 59.75	61 59.75	99.99
50	53 59.66	54 59.65	55 59.64	56 59.64	57 59.63	58 59.63	59 59.62	60 59.61	61 59.61	99.99
1 0	53 59.51	54 59.50	55 59.49	56 59.48	57 59.47	58 59.46	59 59.45	60 59.44	61 59.43	99.98
10	53 59.33	54 59.32	55 59.30	56 59.29	57 59.28	58 59.27	59 59.25	60 59.24	61 59.23	99.98
20	53 59.12	54 59.11	55 59.09	56 59.07	57 59.06	58 59.04	59 59.03	60 59.01	61 58.99	99.97
30	53 58.89	54 58.87	55 58.85	56 58.83	57 58.81	58 58.79	59 58.77	60 58.75	61 58.73	99.97
40	53 58.63	54 58.60	55 58.58	56 58.55	57 58.53	58 58.50	59 58.48	60 58.45	61 58.43	99.96
50	53 58.34	54 58.31	55 58.28	56 58.25	57 58.22	58 58.19	59 58.16	60 58.13	61 58.10	99.95
2 0	53 58.03	54 57.99	55 57.95	56 57.92	57 57.88	58 57.84	59 57.81	60 57.77	61 57.73	99.94
10	53 57.68	54 57.64	55 57.60	56 57.56	57 57.51	58 57.47	59 57.43	60 57.38	61 57.34	99.93
20	53 57.31	54 57.26	55 57.21	56 57.16	57 57.11	58 57.06	59 57.02	60 56.97	61 56.92	99.92
30	53 56.92	54 56.86	55 56.80	56 56.74	57 56.69	58 56.63	59 56.57	60 56.52	61 56.46	99.90
40	53 56.49	54 56.43	55 56.36	56 56.30	57 56.23	58 56.17	59 56.10	60 56.04	61 55.97	99.89
50	53 56.04	54 55.97	55 55.89	56 55.82	57 55.75	58 55.67	59 55.60	60 55.53	61 55.45	99.88
3 0	53 55.56	54 55.48	55 55.40	56 55.31	57 55.23	58 55.15	59 55.07	60 54.98	61 54.90	99.86
10	53 55.05	54 54.96	55 54.87	56 54.78	57 54.69	58 54.59	59 54.50	60 54.41	61 54.32	99.85
20	53 54.52	54 54.42	55 54.32	56 54.21	57 54.11	58 54.01	59 53.91	60 53.81	61 53.71	99.83
30	53 53.96	54 53.84	55 53.73	56 53.62	57 53.51	58 53.40	59 53.29	60 53.17	61 53.06	99.81
40	53 53.37	54 53.24	55 53.12	56 53.00	57 52.88	58 52.75	59 52.63	60 52.51	61 52.39	99.80
50	53 52.75	54 52.62	55 52.48	56 52.35	57 52.21	58 52.08	59 51.95	60 51.81	61 51.68	99.78
4 0	53 52.11	54 51.96	55 51.82	56 51.67	57 51.52	58 51.38	59 51.23	60 51.08	61 50.94	99.76
10	53 51.44	54 51.28	55 51.12	56 50.96	57 50.80	58 50.64	59 50.48	60 50.33	61 50.17	99.74
20	53 50.74	54 50.57	55 50.39	56 50.22	57 50.05	58 49.88	59 49.71	60 49.54	61 49.37	99.71
30	53 50.01	54 49.83	55 49.64	56 49.46	57 49.27	58 49.06	59 48.90	60 48.72	61 48.53	99.69
40	53 49.26	54 49.06	55 48.86	56 48.66	57 48.46	58 48.26	59 48.07	60 47.87	61 47.67	99.67
50	53 48.48	54 48.27	55 48.05	56 47.84	57 47.62	58 47.41	59 47.20	60 46.98	61 46.77	99.64
5 0	53 47.67	54 47.44	55 47.21	56 46.99	57 46.76	58 46.53	59 46.30	60 46.07	61 45.84	99.62
10	53 46.84	54 46.59	55 46.35	56 46.10	57 45.86	58 45.62	59 45.37	60 45.13	61 44.89	99.59
20	53 45.97	54 45.71	55 45.45	56 45.19	57 44.93	58 44.67	59 44.41	60 44.15	61 43.90	99.57
30	53 45.08	54 44.81	55 45.53	56 44.26	57 43.98	58 43.70	59 43.43	60 43.15	61 42.87	99.54
40	53 44.17	54 43.87	55 43.58	56 43.29	57 42.99	58 42.70	59 42.41	60 42.11	61 41.82	99.51
50	53 43.22	54 42.91	55 42.60	56 42.29	57 41.98	58 41.67	59 41.36	60 41.05	61 40.74	99.48
6 0	53 42.25	54 41.92	55 41.59	56 41.26	57 40.94	58 40.61	59 40.28	60 39.95	61 39.62	99.45
10	53 41.25	54 40.90	55 40.56	56 40.21	57 39.86	58 39.52	59 39.17	60 38.82	61 38.47	99.42
20	53 40.23	54 39.86	55 39.49	56 39.13	57 38.76	58 38.40	59 38.03	60 37.66	61 37.30	99.39
30	53 39.17	54 38.78	55 38.40	56 38.02	57 37.63	58 37.24	59 36.86	60 36.47	61 36.09	99.36
40	53 38.09	54 37.69	55 37.28	56 36.88	57 36.47	58 36.06	59 35.66	60 35.25	61 34.85	99.32
50	53 36.98	54 36.56	55 36.13	56 35.71	57 35.28	58 34.85	59 34.43	60 34.00	61 33.58	99.29
7 0	53 35.85	54 35.40	55 34.96	56 34.51	57 34.06	58 33.61	59 33.17	60 32.72	61 32.27	99.25
10	53 34.69	54 34.22	55 33.75	56 33.28	57 32.81	58 32.34	59 31.87	60 31.41	61 30.94	99.22
20	53 33.50	54 33.01	55 32.52	56 32.03	57 31.53	58 31.04	59 30.55	60 30.06	61 29.57	99.18
30	53 32.28	54 31.77	55 31.25	56 30.74	57 30.23	58 29.71	59 29.20	60 28.69	61 28.17	99.14
40	53 31.04	54 30.50	55 29.96	56 29.43	57 28.89	58 28.36	59 27.82	60 27.28	61 26.75	99.11
50	53 29.77	54 29.21	55 28.65	56 28.09	57 27.53	58 26.97	59 26.41	60 25.85	61 25.29	99.07
8 0	53 28.47	54 27.88	55 27.30	56 26.72	57 26.13	58 25.55	59 24.97	60 24.38	61 23.80	99.03
10	53 27.14	54 26.53	55 25.93	56 25.32	57 24.71	58 24.10	59 23.49	60 22.88	61 22.28	98.99
20	53 25.79	54 25.16	55 24.52	56 23.89	57 23.26	58 22.62	59 21.99	60 21.36	61 20.72	98.94
30	53 24.41	54 23.75	55 23.09	56 22.43	57 21.78	58 21.12	59 20.46	60 19.80	61 19.14	98.90
40	53 23.00	54 22.32	55 21.63	56 20.95	57 20.26	58 19.58	59 18.89	60 18.21	61 17.52	98.86
50	53 21.57	54 20.86	55 20.15	56 19.44	57 18.72	58 18.01	59 17.30	60 16.59	61 15.88	98.81
9 0	53 20.11	54 19.37	55 18.63	56 17.89	57 17.16	58 16.42	59 15.68	60 14.94	61 14.20	98.77
10	53 18.62	54 17.86	55 17.09	56 16.32	57 15.56	58 14.79	59 14.02	60 13.26	61 12.49	98.72
20	53 17.11	54 16.31	55 15.52	56 14.72	57 13.93	58 13.14	59 12.34	60 11.55	61 10.75	98.68
30	53 15.57	54 14.74	55 13.92	56 13.10	57 12.27	58 11.45	59 10.63	60 9.81	61 8.98	98.63
40	53 14.00	54 13.14	55 12.29	56 11.44	57 10.59	58 9.74	59 8.88	60 8.03	61 7.18	98.58
50	53 12.40	54 11.52	55 10.64	56 9.76	57 8.87	58 7.99	59 7.11	60 6.23	61 5.35	98.53
10 0	53 10.78	54 9.87	55 8.95	56 8.04	57 7.13	58 6.22	59 5.31	60 4.40	61 3.49	98.48

TABLE 8.

The Moon's Parallax to every 10′ of her Altitude, &c. continued.

Arg. ☾'s Alt.	54′	55′	56′	57′	58′	59′	60′	61′	62′	Diff. for 100″.
10° 0	53 10″.78	54′ 9″.87	55′ 8″.95	56′ 8″.04	57′ 7″.13	58′ 6″.22	59′ 5″.31	60′ 4″.40	61′ 3″.49	98″.48
10	53 9 .13	54 8 .19	55 7 .24	56 6 .30	57 5 .36	58 4 .42	59 3 .47	60 2 .53	61 1 .59	98 .43
20	53 7 .45	54 6 .48	55 5 .50	56 4 .53	57 3 .56	58 2 .58	59 1 .61	60 0 .64	60 59 .66	98 .38
30	53 5 .75	54 4 .74	55 3 .74	56 2 .73	57 1 .73	58 0 .72	58 59 .71	59 58 .71	60 57 .71	98 .33
40	53 4 .01	54 2 .98	55 1 .94	56 0 .90	56 59 .87	57 58 .83	58 57 .79	59 56 .76	60 55 .72	98 .27
50	53 2 .26	54 1 .19	55 0 .12	55 59 .05	56 57 .98	57 56 .91	58 55 .84	59 54 .77	60 53 .70	98 .22
11 0	53 0 .47	53 59 .37	54 58 .27	55 57 .17	56 56 .06	57 54 .96	58 53 .86	59 52 .76	60 51 .65	98 .16
10	52 58 .66	53 57 .52	54 56 .39	55 55 .25	56 54 .12	57 52 .98	58 51 .84	59 50 .71	60 49 .57	98 .11
20	52 56 .82	53 55 .65	54 54 .48	55 53 .31	56 52 .14	57 50 .97	58 49 .80	59 48 .63	60 47 .46	98 .05
30	52 54 .96	53 53 .75	54 52 .55	55 51 .34	56 50 .14	57 48 .93	58 47 .73	59 46 .52	60 45 .32	97 .99
40	52 53 .06	53 51 .82	54 50 .58	55 49 .34	56 48 .11	57 46 .87	58 45 .63	59 44 .39	60 43 .15	97 .93
50	52 51 .14	53 49 .87	54 48 .59	55 47 .32	56 46 .04	57 44 .77	58 43 .49	59 42 .22	60 40 .94	97 .87
12 0	52 49 .20	53 47 .89	54 46 .58	55 45 .26	56 43 .95	57 42 .64	58 41 .33	59 40 .02	60 38 .71	97 .81
10	52 47 .23	53 45 .88	54 44 .53	55 43 .18	56 41 .83	57 40 .49	58 39 .14	59 37 .79	60 36 .44	97 .75
20	52 45 .23	53 43 .84	54 42 .46	55 41 .07	56 39 .69	57 38 .30	58 36 .92	59 35 .53	60 34 .15	97 .69
30	52 43 .20	53 41 .78	54 40 .35	55 38 .93	56 37 .51	57 36 .09	58 34 .67	59 33 .24	60 31 .82	97 .63
40	52 41 .15	53 39 .69	54 38 .23	55 36 .77	56 35 .30	57 33 .84	58 32 .38	59 30 .92	60 29 .46	97 .57
50	52 39 .07	53 37 .57	54 36 .07	55 34 .57	56 33 .07	57 31 .57	58 30 .07	59 28 .57	60 27 .08	97 .50
13 0	52 36 .96	53 35 .42	54 33 .88	55 32 .35	56 30 .81	57 29 .27	58 27 .72	59 26 .19	60 24 .66	97 .44
10	52 34 .83	53 33 .25	54 31 .67	55 30 .09	56 28 .52	57 26 .94	58 25 .36	59 23 .78	60 22 .21	97 .37
20	52 32 .67	53 31 .05	54 29 .43	55 27 .81	56 26 .20	57 24 .58	58 22 .96	59 21 .34	60 19 .73	97 .30
30	52 30 .48	53 28 .82	54 27 .16	55 25 .51	56 23 .85	57 22 .19	58 20 .53	59 18 .87	60 17 .22	97 .24
40	52 28 .26	53 26 .57	54 24 .87	55 23 .17	56 21 .47	57 19 .77	58 18 .07	59 16 .37	60 14 .67	97 .17
50	52 26 .02	53 24 .28	54 22 .54	55 20 .80	56 19 .06	57 17 .32	58 15 .58	59 13 .84	60 12 .10	97 .10
14 0	52 23 .76	53 21 .98	54 20 .19	55 18 .41	56 16 .63	57 14 .85	58 13 .06	59 11 .28	60 9 .50	97 .03
10	52 21 .46	53 19 .64	54 17 .82	55 15 .99	56 14 .17	57 12 .34	58 10 .52	59 8 .69	60 6 .87	96 .96
20	52 19 .14	53 17 .28	54 15 .41	55 13 .54	56 11 .67	57 9 .81	58 7 .94	59 6 .07	60 4 .20	96 .89
30	52 16 .80	53 14 .89	54 12 .98	55 11 .06	56 9 .15	57 7 .24	58 5 .33	59 3 .42	60 1 .51	96 .81
40	52 14 .43	53 12 .47	54 10 .52	55 8 .56	56 6 .60	57 4 .65	58 2 .69	59 0 .74	59 58 .78	96 .74
50	52 12 .03	53 10 .03	54 8 .03	55 6 .03	56 4 .03	57 2 .03	58 0 .03	58 58 .03	59 56 .03	96 .67
15 0	52 9 .60	53 7 .56	54 5 .51	55 3 .47	56 1 .42	57 59 .38	57 57 .33	58 55 .29	59 53 .24	96 .59
10	52 7 .15	53 5 .06	54 2 .97	55 0 .88	55 58 .79	56 56 .70	57 54 .61	58 52 .52	59 50 .43	96 .52
20	52 4 .67	53 2 .53	54 0 .40	54 58 .26	55 56 .12	56 53 .99	57 51 .85	58 49 .72	59 47 .58	96 .44
30	52 2 .16	52 59 .98	53 57 .80	54 55 .62	55 53 .43	56 51 .25	57 49 .07	58 46 .89	59 44 .71	96 .36
40	51 59 .63	52 57 .40	53 55 .17	54 52 .94	55 50 .71	56 48 .49	57 46 .26	58 44 .03	59 41 .80	96 .28
50	51 57 .07	52 54 .80	53 52 .52	54 50 .24	55 47 .97	56 45 .69	57 43 .41	58 41 .14	59 38 .86	96 .21
16 0	51 54 .49	52 52 .16	53 49 .84	54 47 .52	55 45 .19	56 42 .87	57 40 .54	58 38 .22	59 35 .89	96 .13
10	51 51 .88	52 49 .50	53 47 .13	54 44 .76	55 42 .39	56 40 .01	57 37 .64	58 35 .27	59 32 .90	96 .05
20	51 49 .24	52 46 .82	53 44 .40	54 41 .98	55 39 .55	56 37 .13	57 34 .71	58 32 .29	59 29 .87	95 .96
30	51 46 .58	52 44 .11	53 41 .63	54 39 .16	55 36 .69	56 34 .22	57 31 .75	58 29 .28	59 26 .81	95 .88
40	51 43 .89	52 41 .37	53 38 .84	54 36 .32	55 33 .80	56 31 .28	57 28 .76	58 26 .74	59 23 .72	95 .80
50	51 41 .17	52 38 .60	53 36 .03	54 33 .46	55 30 .89	56 28 .32	57 25 .74	58 23 .17	59 20 .60	95 .72
17 0	51 38 .43	52 35 .81	53 33 .18	54 30 .56	55 27 .94	56 25 .32	57 22 .70	58 20 .08	59 17 .45	95 .63
10	51 35 .66	52 32 .99	53 30 .31	54 27 .64	55 24 .97	56 22 .29	57 19 .62	58 16 .95	59 14 .28	95 .55
20	51 32 .86	52 30 .14	53 27 .41	54 24 .69	55 21 .96	56 19 .24	57 16 .52	58 13 .79	59 11 .07	95 .46
30	51 30 .04	52 27 .27	53 24 .49	54 21 .72	55 18 .94	56 16 .16	57 13 .38	58 10 .60	59 7 .83	95 .37
40	51 27 .20	52 24 .37	53 21 .54	54 18 .71	55 15 .88	56 13 .05	57 10 .22	58 7 .39	59 4 .56	95 .28
50	51 24 .32	52 21 .44	53 18 .56	54 15 .67	55 12 .79	56 9 .91	57 7 .03	58 4 .14	59 1 .26	95 .20
18 0	51 21 .42	52 18 .49	53 15 .55	54 12 .61	55 9 .68	56 6 .74	57 3 .80	58 0 .87	58 57 .93	95 .11
10	51 18 .50	52 15 .51	53 12 .52	54 9 .53	55 6 .53	56 3 .54	57 0 .55	57 57 .56	58 54 .57	95 .02
20	51 15 .55	52 12 .50	53 9 .46	54 6 .41	55 3 .36	56 0 .32	56 57 .27	57 54 .23	58 51 .18	94 .92
30	51 12 .57	52 9 .47	53 6 .37	54 3 .27	55 0 .17	55 57 .07	56 53 .97	57 50 .86	58 47 .76	94 .83
40	51 9 .56	52 6 .41	53 3 .25	54 0 .10	54 56 .94	55 53 .78	56 50 .63	57 47 .47	58 44 .32	94 .74
50	51 6 .54	52 3 .32	53 0 .11	53 56 .90	54 53 .69	55 50 .47	56 47 .26	57 44 .05	58 40 .84	94 .65
19 0	51 3 .48	52 0 .21	52 56 .94	53 53 .67	54 50 .40	55 47 .14	56 43 .87	57 40 .60	58 37 .33	94 .55
10	51 0 .40	51 57 .07	52 53 .75	53 50 .42	54 47 .10	55 43 .77	56 40 .44	57 37 .12	58 33 .79	94 .46
20	50 57 .29	51 53 .91	52 50 .52	53 47 .14	54 43 .76	55 40 .37	56 36 .99	57 33 .61	58 30 .22	94 .36
30	50 54 .16	51 50 .72	52 47 .28	53 43 .83	54 40 .39	55 36 .95	56 33 .51	57 30 .07	58 26 .63	94 .26
40	50 51 .00	51 47 .50	52 44 .00	53 40 .50	54 37 .00	55 33 .50	56 30 .00	57 26 .50	58 23 .00	94 .17
50	50 47 .81	51 44 .26	52 40 .70	53 37 .14	54 33 .58	55 30 .02	56 26 .46	57 22 .90	58 19 .34	94 .07
20 0	50 44 .60	51 40 .99	52 37 .37	53 33 .75	54 30 .13	55 26 .51	56 22 .89	57 19 .27	58 15 .66	93 .97

TABLE 8.

The Moon's Parallax to every 10′ of her Altitude, &c. continued.

Arg. ☾'s Alt.	54′	55′	56′	57′	58′	59′	60′	61′	62′	Diff. for 100″.
20° 0	50′ 44″.60	51′ 40″.99	52′ 37″.37	53′ 33″.75	54′ 30″.13	55′ 26″.51	56′ 22″.89	57′ 19″.27	58′ 15″.66	93″.97
10	50 41.37	51 37.69	52 34.01	53 30.33	54 26.65	55 22.98	56 19.30	57 15.62	58 11.94	93.87
20	50 38.11	51 34.37	52 30.64	53 26.90	54 23.16	55 19.42	56 15.68	57 11.94	58 8.20	93.77
30	50 34.82	51 31.02	52 27.22	53 23.42	54 19.62	55 15.82	56 12.02	57 8.22	58 4.42	93.67
40	50 31.50	51 27.64	52 23.78	53 19.92	54 16.06	55 12.20	56 8.34	57 4.48	58 0.62	93.56
50	50 28.17	51 24.24	52 20.32	53 16.40	54 12.47	55 8.55	56 4.63	57 0.70	57 56.78	93.46
21 0	50 24.80	51 20.82	52 16.83	53 12.85	54 8.86	55 4.87	56 0.89	56 56.90	57 52.92	93.36
10	50 21.41	51 17.36	52 13.31	53 9.27	54 5.22	55 1.17	55 57.12	56 53.07	57 49.03	93.25
20	50 17.99	51 13.88	52 9.77	53 5.66	54 1.55	54 57.44	55 53.33	56 49.22	57 45.10	93.15
30	50 14.55	51 10.38	52 6.20	53 2.03	53 57.86	54 53.68	55 49.51	56 45.33	57 41.15	93.04
40	50 11.09	51 6.85	52 2.61	52 38.37	53 54.13	54 49.89	55 45.65	56 41.41	57 37.17	92.93
50	50 7.59	51 3.29	51 28.99	52 54.69	53 50.38	54 46.07	55 41.77	56 37.47	57 33.16	92.83
22 0	50 4.08	50 59.71	51 55.34	52 50.97	53 46.60	54 42.23	55 37.86	53 33.49	57 29.12	92.72
10	50 0.53	50 56.10	51 51.66	52 47.23	53 42.79	54 38.36	55 33.92	56 29.49	57 25.06	92.61
20	49 56.96	50 52.46	51 47.96	52 43.46	53 38.96	54 34.46	55 29.96	56 25.46	57 20.96	92.50
30	49 53.37	50 48.80	51 44.24	52 39.67	53 35.10	54 30.53	55 25.97	56 21.40	57 16.83	92.39
40	49 49.75	50 45.12	51 40.48	52 35.85	53 31.21	54 26.58	55 21.94	56 17.31	57 12.68	92.27
50	49 46.11	50 41.40	51 36.70	52 32.00	53 27.30	54 22.60	55 17.90	56 13.19	57 8.49	92.16
23 0	49 42.44	50 37.67	51 32.90	52 28.13	53 23.36	54 18.59	55 13.82	56 9.05	57 4.28	92.05
10	49 38.74	50 33.90	51 29.06	52 24.23	53 19.39	54 14.55	55 9.71	56 4.87	57 0.04	91.94
20	49 35.02	50 30.11	51 25.21	52 20.30	53 15.39	54 10.49	55 5.58	56 0.67	56 55.76	91.82
30	49 31.27	50 26.30	51 21.32	52 16.35	53 11.37	54 6.39	55 1.42	55 56.44	56 51.46	91.71
40	49 27.50	50 22.46	51 17.41	52 12.37	53 7.32	54 2.28	54 57.23	55 52.18	56 47.13	91.59
50	49 23.71	50 18.59	51 13.48	52 8.36	53 3.24	53 58.13	54 53.01	55 47.89	56 42.78	91.47
24 0	49 19.89	50 14.70	51 9.51	52 4.33	52 59.14	53 53.95	54 48.76	55 43.58	56 38.39	91.35
10	49 16.04	50 10.78	51 5.52	52 0.27	52 55.01	53 49.75	54 44.49	55 39.23	56 33.97	91.24
20	49 12.17	50 6.84	51 1.51	51 56.18	52 50.85	53 45.52	54 40.19	55 34.86	56 29.53	91.12
30	49 8.27	50 2.87	50 57.47	51 52.07	52 46.67	53 41.26	54 35.86	55 30.46	56 25.06	91.00
40	49 4.35	49 59.88	50 53.40	51 47.93	52 42.45	53 36.98	54 31.50	55 26.03	56 20.55	90.88
50	49 0.41	49 54.86	50 49.31	51 43.76	52 38.22	53 32.67	54 27.12	55 21.57	56 16.02	90.75
25 0	48 56.44	49 50.82	50 45.19	51 39.57	52 33.95	53 28.33	54 22.71	55 17.09	56 11.47	90.63
10	48 52.44	49 46.75	50 41.05	51 35.36	52 29.66	53 23.96	54 18.27	55 12.57	56 6.88	90.51
20	48 48.42	49 42.65	50 36.88	51 31.11	52 25.34	53 19.57	54 13.80	55 8.03	56 2.26	90.38
30	48 44.38	49 38.53	50 32.69	51 26.84	52 21.00	53 15.15	54 9.31	55 3.46	55 57.62	90.26
40	48 40.31	49 34.39	50 28.47	51 22.55	52 16.63	53 10.71	54 4.79	54 58.87	55 52.95	90.13
50	48 36.21	49 30.22	50 24.22	51 18.22	52 12.23	53 6.23	54 0.24	54 54.24	55 48.24	90.01
26 0	48 32.09	49 26.02	50 19.95	51 13.88	52 7.80	53 1.73	53 55.66	54 49.59	55 43.51	89.88
10	48 27.95	49 21.80	56 15.65	51 9.50	52 3.35	52 57.20	53 51.05	54 44.91	55 38.76	89.75
20	48 23.78	49 17.55	50 11.33	51 5.10	51 58.88	52 52.65	53 46.42	54 40.20	55 33.97	89.62
30	48 19.59	49 13.28	50 6.98	51 0.68	51 54.37	52 48.07	53 41.76	54 35.46	55 29.16	89.49
40	48 15.37	49 8.99	50 2.61	50 56.22	51 49.84	52 43.46	53 37.08	54 30.70	55 24.31	89.36
50	48 11.13	49 4.67	49 58.21	50 51.75	51 45.29	52 38.82	53 32.36	54 25.90	55 19.44	89.23
27 0	48 6.86	49 0.32	49 53.78	50 47.24	51 40.70	52 34.16	53 27.62	54 21.08	55 14.54	89.10
10	48 2.57	48 55.95	49 49.33	50 42.71	51 36.09	52 29.47	53 22.86	54 16.24	55 9.62	88.97
20	47 58.25	48 51.56	49 44.86	50 38.16	51 31.46	52 24.76	53 18.06	54 11.36	55 4.66	88.84
30	47 53.91	48 47.14	49 40.36	50 33.58	51 26.80	52 20.02	53 13.24	54 6.46	54 59.68	88.70
40	47 49.55	48 42.69	49 35.83	50 28.97	51 22.11	52 15.25	53 8.39	54 1.53	54 54.67	88.57
50	47 45.16	48 38.22	49 31.28	50 24.34	51 17.40	52 10.46	53 3.51	53 56.57	54 49.63	88.43
28 0	47 40.75	48 33.73	49 26.70	50 19.68	51 12.66	52 5.63	52 58.61	53 51.59	54 44.57	88.29
10	47 36.31	48 29.21	49 22.10	50 15.00	51 7.89	52 0.79	52 53.68	53 46.58	54 39.47	88.16
20	47 31.85	48 24.66	49 17.48	50 10.29	51 3.10	51 55.91	52 48.73	53 41.54	54 34.35	88.02
30	47 27.37	48 20.10	49 12.83	50 5.56	50 58.28	51 51.01	52 43.74	53 36.47	54 29.20	87.88
40	47 22.86	48 15.50	49 8.15	50 0.79	50 53.44	51 46.09	52 38.73	53 31.38	54 24.02	87.74
50	47 18.33	48 10.89	49 3.45	49 56.01	50 48.57	51 41.13	52 33.69	53 26.26	54 18.82	87.60
29 0	47 13.77	48 6.25	48 58.72	49 51.20	50 43.68	51 36.15	52 28.63	53 21.11	54 13.59	87.46
10	47 9.19	48 1.58	48 53.97	49 46.36	50 38.76	51 31.15	52 23.54	53 15.93	54 8.33	87.32
20	47 4.58	47 56.89	48 49.20	49 41.50	50 33.81	51 26.12	52 18.42	53 10.73	54 3.04	87.18
30	46 59.95	47 52.17	48 44.40	49 36.62	50 28.84	51 21.06	52 13.28	53 5.50	53 57.72	87.04
40	46 55.30	47 47.43	48 39.57	49 31.71	50 23.84	51 15.98	52 8.11	53 0.25	53 52.38	86.89
50	46 50.62	47 42.67	48 34.72	49 26.77	50 18.82	51 10.87	52 2.91	52 54.96	53 47.01	86.75
30 0	46 45.92	47 37.88	48 29.85	49 21.81	50 13.77	51 5.73	51 57.69	52 49.65	53 41.61	86.60

TABLE 8.

The MOON'S PARALLAX to every 10′ of her Altitude, &c. continued.

Arg. ☾'s Alt.	54′	55′	56′	57′	58′	59′	60′	61′	62′	Diff. for 100″.
30° 0′	46′ 45″.92	47′ 37″.88	48′ 29″.85	49′ 21″.81	50′ 13″.77	51′ 5″.73	51′ 57″.69	52′ 49″.65	53′ 41″.61	86″.60
10	46 41 .20	47 33 .07	48 24 .95	49 16 .82	50 8 .69	51 0 .57	51 52 .44	52 44 .32	53 36 .19	86 .46
20	46 36 .45	47 28 .24	48 20 .02	49 11 .81	50 3 .59	50 55 .38	51 47 .17	52 38 .95	53 30 .74	86 .31
30	46 31 .68	47 23 .38	48 15 .07	49 6 .77	49 58 .47	50 50 .17	51 41 .87	52 33 .56	59 25 .26	86 .16
40	46 26 .88	47 18 .49	48 10 .10	49 1 .71	49 53 .32	50 44 .93	51 36 .54	52 28 .15	53 19 .75	86 .01
50	46 22 .06	47 13 .58	48 5 .10	48 56 .62	49 48 .14	50 39 .66	51 31 .18	52 22 .70	53 14 .22	85 .87
31 0	46 17 .22	47 8 .65	48 0 .08	48 51 .51	49 42 .94	50 34 .37	51 25 .80	52 17 .23	53 8 .66	85 .72
10	46 12 .36	47 3 .70	47 55 .04	48 46 .38	49 37 .72	50 29 .06	51 20 .40	52 11 .74	53 3 .08	85 .57
20	46 7 .47	46 58 .72	47 49 .97	48 41 .21	49 32 .46	50 23 .71	51 14 .96	52 6 .21	52 57 .46	85 .42
30	46 2 .55	46 53 .72	47 44 .87	48 36 .03	49 27 .19	50 18 .35	51 9 .50	52 0 .66	52 51 .82	85 .26
40	45 57 .62	46 48 .69	47 39 .75	48 30 .82	49 21 .89	50 12 .95	51 4 .02	51 55 .09	52 46 .15	85 .11
50	45 52 .66	46 43 .63	47 34 .61	48 25 .58	49 16 .56	50 7 .53	50 58 .51	51 49 .49	52 40 .46	84 .96
32 0	45 47 .68	46 38 .56	47 29 .44	48 20 .32	49 11 .21	50 2 .09	50 52 .97	51 43 .86	52 34 .74	84 .80
10	45 42 .67	46 33 .46	47 24 .25	48 15 .04	49 5 .83	49 56 .62	50 47 .41	51 38 .20	52 28 .99	84 .65
20	45 37 .64	46 28 .34	47 19 .04	48 9 .73	49 0 .43	49 51 .13	50 41 .82	51 32 .52	52 23 .22	84 .50
30	45 32 .59	46 23 .19	47 13 .80	48 4 .40	48 55 .00	49 45 .61	50 36 .21	51 26 .81	52 17 .42	84 .34
40	45 27 .51	46 18 .02	47 8 .53	47 59 .04	48 49 .55	49 40 .06	50 30 .57	51 21 .08	52 11 .59	84 .18
50	45 22 .41	46 12 .83	47 3 .24	47 53 .66	48 44 .07	49 34 .49	50 24 .90	51 15 .32	52 5 .74	84 .03
33 0	45 17 .29	46 7 .61	46 57 .93	47 48 .25	48 38 .57	49 28 .89	50 19 .21	51 9 .53	51 59 .85	83 .87
10	45 12 .15	46 2 .37	46 52 .60	47 42 .82	48 33 .05	49 23 .27	50 13 .50	51 3 .72	51 53 .95	83 .71
20	45 6 .98	45 57 .11	46 47 .24	47 37 .37	48 27 .50	49 17 .63	50 7 .76	50 57 .89	51 48 .01	83 .55
30	45 1 .79	45 51 .82	46 41 .86	47 31 .89	48 21 .92	49 11 .96	50 1 .99	50 52 .02	51 42 .05	83 .39
40	44 56 .58	45 46 .51	46 36 .45	47 26 .39	48 16 .32	49 6 .26	49 56 .20	50 46 .13	51 36 .07	83 .23
50	44 51 .34	45 41 .18	46 31 .02	47 20 .86	48 10 .70	49 0 .54	49 50 .38	50 40 .22	51 30 .06	83 .07
34 0	44 46 .08	45 35 .82	46 25 .57	47 15 .31	48 5 .05	48 54 .79	49 44 .54	50 34 .28	51 24 .02	82 .90
10	44 40 .80	45 30 .44	46 20 .09	47 9 .73	47 59 .38	48 49 .02	49 38 .67	50 28 .31	51 17 .96	82 .74
20	44 35 .50	45 25 .04	46 14 .59	47 4 .13	47 53 .68	48 43 .23	49 32 .77	50 22 .32	51 11 .87	82 .58
30	44 30 .17	45 19 .62	46 9 .06	46 58 .51	47 47 .96	48 37 .41	49 26 .85	50 16 .30	51 5 .75	82 .41
40	44 24 .82	45 14 .17	46 3 .52	46 52 .86	47 42 .21	48 31 .56	49 20 .91	50 10 .26	50 59 .61	82 .25
50	44 19 .45	45 8 .70	45 57 .95	46 47 .19	47 36 .44	48 25 .69	49 14 .94	50 4 .19	50 53 .44	82 .08
35 0	44 14 .05	45 3 .20	45 52 .35	46 41 .50	47 30 .65	48 19 .80	49 8 .95	49 58 .10	50 47 .25	81 .92
10	44 8 .64	44 57 .68	45 46 .73	46 35 .78	47 24 .83	48 13 .88	49 2 .93	49 51 .98	50 41 .02	81 .75
20	44 3 .20	44 52 .14	45 41 .09	46 30 .04	47 18 .99	48 7 .94	48 56 .88	49 45 .83	50 34 .78	81 .58
30	43 57 .73	44 46 .58	45 35 .43	46 24 .27	47 13 .12	48 1 .97	48 50 .82	49 39 .66	50 28 .51	81 .41
40	43 52 .25	44 41 .00	45 29 .74	46 18 .49	47 7 .23	47 55 .98	48 44 .72	49 33 .47	50 22 .21	81 .24
50	43 46 .74	44 35 .39	45 24 .03	46 12 .67	47 1 .32	47 49 .96	48 38 .60	49 27 .25	50 15 .89	81 .07
36 0	43 41 .22	44 29 .76	45 18 .30	46 6 .84	46 55 .38	47 43 .93	48 32 .46	49 21 .00	50 9 .54	80 .90
10	43 35 .66	44 24 .10	45 12 .54	46 0 .98	46 49 .42	47 37 .86	48 26 .29	49 14 .73	50 3 .17	80 .73
20	43 30 .09	44 18 .43	45 6 .76	45 55 .10	46 43 .43	47 31 .77	48 20 .10	49 8 .44	49 56 .77	80 .56
30	43 24 .50	44 12 .73	45 0 .96	45 49 .19	46 37 .42	47 25 .65	48 13 .88	49 2 .12	49 50 .35	80 .39
40	43 18 .88	44 7 .01	44 55 .13	45 43 .26	46 31 .39	47 19 .52	48 7 .64	48 55 .77	49 43 .90	80 .21
50	43 13 .24	44 1 .26	44 49 .29	45 37 .31	46 25 .33	47 13 .36	48 1 .38	48 49 .40	49 37 .42	80 .04
37 0	43 7 .58	43 55 .50	44 43 .42	45 31 .33	46 19 .25	47 7 .17	47 55 .09	48 43 .01	49 30 .92	79 .86
10	43 1 .90	43 49 .71	44 37 .53	45 25 .34	46 13 .15	47 0 .97	47 48 .78	48 36 .59	49 24 .40	79 .69
20	42 56 .19	43 43 .90	44 31 .61	45 19 .31	46 7 .02	46 54 .73	47 42 .43	48 30 .14	49 17 .85	79 .51
30	42 50 .46	43 38 .07	44 25 .67	45 13 .27	46 0 .87	46 48 .47	47 36 .07	48 23 .67	49 11 .27	79 .34
40	42 44 .72	43 32 .21	44 19 .71	45 7 .20	45 54 .70	46 42 .19	47 29 .69	48 17 .18	49 4 .68	79 .16
50	42 38 .95	43 26 .33	44 13 .72	45 1 .11	45 48 .50	46 35 .89	47 23 .27	48 10 .66	48 58 .05	78 .98
38 0	42 33 .15	43 20 .44	44 7 .72	44 55 .00	45 42 .28	46 29 .56	47 16 .84	48 4 .12	48 51 .40	78 .80
10	42 27 .34	43 14 .51	44 1 .69	44 48 .86	45 36 .03	46 23 .21	47 10 .38	47 57 .55	48 44 .73	78 .63
20	42 21 .51	43 8 .57	43 55 .64	44 42 .70	45 29 .77	46 16 .83	47 3 .90	47 50 .96	48 38 .03	78 .44
30	42 15 .65	43 2 .61	43 49 .56	44 36 .52	45 23 .48	46 10 .43	46 57 .39	47 44 .35	48 31 .30	78 .26
40	42 9 .77	42 56 .62	43 43 .47	44 30 .32	45 17 .16	46 4 .01	46 50 .86	47 37 .71	48 24 .55	78 .08
50	42 3 .87	42 50 .61	43 37 .35	44 24 .09	45 10 .83	45 57 .57	46 44 .30	47 31 .04	48 17 .78	77 .90
39 0	41 57 .95	42 44 .58	43 31 .21	44 17 .84	45 4 .47	45 51 .10	46 37 .73	47 24 .35	48 10 .98	77 .71
10	41 52 .01	42 38 .53	43 25 .05	44 11 .57	44 58 .09	45 44 .60	46 31 .12	47 17 .64	48 4 .16	77 .53
20	41 46 .05	42 32 .46	43 18 .86	44 5 .27	44 51 .68	45 38 .09	46 24 .50	47 10 .91	47 57 .31	77 .35
30	41 40 .06	42 26 .36	43 12 .66	43 58 .96	44 45 .25	45 31 .55	46 17 .85	47 4 .15	47 50 .44	77 .16
40	41 34 .06	42 20 .24	43 6 .43	43 52 .62	44 38 .80	45 24 .99	46 11 .18	46 57 .36	47 43 .55	76 .98
50	41 28 .03	42 14 .11	43 0 .18	43 46 .26	44 32 .33	45 18 .41	46 4 .48	46 50 .55	47 36 .63	76 .79
40 0	41 21 .98	42 7 .95	42 53 .91	43 39 .87	44 25 .83	45 11 .80	45 57 .76	46 43 .72	47 29 .68	76 .60

TABLE 8.

The Moon's Parallax to every 10′ of her Altitude, &c. continued.

Arg. ☾'s Alt	54′	55′	56′	57′	58′	59′	60′	61′	62′	Diff. for 100″.
40° 0	41′ 21″.98	42′ 7″.95	42′ 53″.91	43′ 39″.87	44′ 25″.83	45′ 11″.80	45′ 57″.76	46′ 43″.72	47′ 29″.69	76″.60
10	41 15 .92	42 1 .77	42 47 .62	43 33 .47	44 19 .32	45 5 .17	45 51 .02	46 36 .87	47 22 .72	76 .42
20	41 9 .83	41 55 .56	42 41 .30	43 27 .04	44 12 .78	44 58 .51	45 44 .25	46 29 .99	47 15 .73	76 .23
30	41 3 .72	41 49 .34	42 34 .96	43 20 .59	44 6 .21	44 51 .84	45 37 .46	46 23 .09	47 8 .71	76 .04
40	40 57 .58	41 43 .09	42 28 .61	43 14 .12	43 59 .63	44 45 .14	45 30 .65	46 16 .16	47 1 .67	75 .85
50	40 51 .43	41 36 .83	42 22 .23	43 7 .62	43 53 .02	44 38 .42	45 23 .81	46 9 .21	46 54 .61	75 .66
41 0	40 45 .26	41 30 .54	42 15 .82	43 1 .11	43 46 .39	44 31 .67	45 16 .95	46 2 .24	46 47 .52	75 .47
10	40 39 .07	41 24 .23	42 9 .40	42 54 .57	43 39 .74	44 24 .91	45 10 .07	45 55 .24	46 40 .41	75 .28
20	40 32 .85	41 17 .90	42 2 .96	42 48 .01	43 33 .06	44 18 .11	45 3 .17	45 48 .22	46 33 .27	75 .09
30	40 26 .62	41 11 .55	41 56 .49	42 41 .43	43 26 .37	44 11 .30	44 56 .24	45 41 .18	46 26 .11	74 .90
40	40 20 .36	41 5 .18	41 50 .00	42 34 .83	43 19 .65	44 4 .47	44 49 .29	45 34 .11	46 18 .93	74 .70
50	40 14 .09	40 58 .79	41 43 .50	42 28 .20	43 12 .91	43 57 .61	44 42 .32	45 27 .02	46 11 .73	74 .51
42 0	40 7 .79	40 52 .38	41 36 .97	42 21 .56	43 6 .14	43 50 .73	44 35 .32	45 19 .91	45 4 .50	74 .31
10	40 1 .47	40 45 .94	41 30 .42	42 14 .89	42 59 .36	43 43 .83	44 28 .30	45 12 .77	45 57 .25	74 .12
20	39 55 .14	40 39 .49	41 23 .84	42 8 .20	42 52 .55	43 36 .91	44 21 .26	45 5 .62	45 49 .97	73 .92
30	39 48 .78	40 33 .02	41 17 .25	42 1 .49	42 45 .72	43 29 .96	44 14 .20	44 58 .43	45 42 .67	73 .73
40	39 42 .40	40 26 .52	41 10 .64	41 54 .76	42 38 .88	43 22 .99	44 7 .11	44 51 .23	45 35 .35	73 .53
50	39 36 .00	40 20 .00	41 4 .00	41 48 .00	42 32 .00	43 16 .00	44 0 .00	44 44 .00	45 28 .00	73 .33
43 0	39 29 .59	40 13 .47	40 57 .35	41 41 .23	42 25 .11	43 8 .99	43 52 .87	44 36 .75	45 20 .64	73 .14
10	39 23 .15	40 6 .91	40 50 .67	41 34 .43	42 18 .20	43 1 .96	43 45 .72	44 29 .48	45 13 .24	72 .94
20	39 16 .69	40 0 .33	40 43 .98	41 27 .62	42 11 .26	42 54 .90	43 38 .54	44 22 .19	45 5 .83	72 .74
30	39 10 .21	39 53 .74	40 37 .26	41 20 .78	42 4 .30	42 47 .83	43 31 .35	44 14 .87	44 58 .39	72 .54
40	39 3 .72	39 47 .12	40 30 .52	41 13 .92	41 57 .32	42 40 .73	43 24 .13	44 7 .53	44 50 .93	72 .34
50	38 57 .20	39 40 .48	40 23 .76	41 7 .04	41 50 .32	42 33 .61	43 16 .89	44 0 .17	44 43 .45	72 .14
44 0	38 50 .66	39 33 .82	40 16 .98	41 0 14	41 43 .30	42 26 .46	43 9 .62	43 52 .78	44 35 .94	71 .93
10	38 44 .10	39 27 .14	40 10 .18	40 53 .22	41 36 .26	42 19 .30	43 2 .34	43 45 .38	44 28 .42	71 .73
20	38 37 .53	39 20 .44	40 3 .36	40 46 .28	41 29 .20	42 12 .11	42 55 .03	43 37 .95	44 20 .87	71 .53
30	38 30 .93	39 13 .73	39 56 .52	40 39 .32	41 22 .11	42 4 .91	42 47 .70	43 30 .50	44 13 .29	71 .33
40	38 24 .32	39 6 .99	39 49 .66	40 32 .33	41 15 .01	41 57 .68	42 40 .35	43 23 .02	44 5 .70	71 .12
50	38 17 .68	39 0 .23	39 42 .78	40 25 .33	41 7 .88	41 50 .43	42 32 .98	43 15 .53	43 58 .08	70 .92
45 0	38 11 .03	38 53 .45	39 35 .88	40 18 .31	41 0 .73	41 43 .16	42 25 .58	43 8 .01	43 50 .44	70 .71
10	38 4 .35	38 46 .65	39 28 .96	40 11 .26	40 53 .56	41 35 .87	42 18 .17	43 0 .47	43 42 .77	70 .50
20	37 57 .66	38 39 .84	39 22 .02	40 4 .20	40 46 .38	41 28 .55	42 10 .73	42 52 .91	43 35 .09	70 .30
30	37 50 .95	38 33 .00	39 15 .06	39 57 .11	40 39 .16	41 21 .22	42 3 .27	42 45 .33	43 27 .38	70 .09
40	37 44 .21	38 26 .14	39 8 .07	39 50 .00	40 31 .93	41 13 .86	41 55 .79	42 37 .72	43 19 .65	69 .88
50	37 37 .46	38 19 .27	39 1 .07	39 42 .88	40 24 .68	41 6 .49	41 48 .29	42 30 .10	43 11 .90	69 .67
46 0	37 30 .69	38 12 .37	38 54 .05	39 35 .73	40 17 .41	40 59 .09	41 40 .77	42 22 .45	43 4 .13	69 .47
10	37 23 .90	38 5 .46	38 47 .01	39 28 .57	40 10 .12	40 51 .67	41 33 .23	42 14 .78	42 56 .33	69 .26
20	37 17 .10	37 58 .52	38 39 .95	39 21 .38	40 2 .81	40 44 .23	41 25 .66	42 7 .09	42 48 .52	69 .05
30	37 10 .27	37 51 .57	38 32 .87	39 14 .17	39 55 .47	40 36 .78	41 18 .08	41 59 .38	42 40 .68	68 .84
40	37 3 .42	37 44 .60	38 25 .77	39 6 .95	39 48 .12	40 29 .30	41 10 .47	41 51 .64	42 32 .82	68 .62
50	36 56 .56	37 37 .61	38 18 .65	38 59 .70	39 40 .75	40 21 .80	41 2 .84	41 43 .89	42 24 .94	68 .41
47 0	36 49 .67	37 30 .59	38 11 .51	38 52 .43	39 23 .35	40 14 .27	40 55 .19	41 36 .11	42 17 .03	68 .20
10	36 42 .77	37 23 .57	38 4 .36	38 45 .15	39 25 .94	40 6 .73	40 47 .53	41 28 .32	42 9 .11	67 .99
20	36 35 .85	37 16 .52	37 57 .18	38 37 .84	39 18 .51	39 59 .17	40 39 .84	41 20 .50	42 1 .16	67 .77
30	36 28 .91	37 9 .45	37 49 .98	38 30 .52	39 11 .05	39 51 .59	40 32 .12	41 12 .66	41 53 .20	67 .56
40	36 21 .95	37 2 .36	37 42 .77	38 23 .17	39 3 .58	39 43 .99	40 24 .39	41 4 .80	41 45 .21	67 .34
50	36 14 .98	36 55 .26	37 35 .53	38 15 .81	38 56 .09	39 36 .37	40 16 .64	40 56 .92	41 37 .20	67 .13
48 0	36 7 .98	36 48 .13	37 28 .28	38 8 .43	38 48 .57	39 28 .72	40 8 .87	40 49 .02	41 29 .17	66 .91
10	36 0 .97	36 40 .99	37 21 .01	38 1 .02	38 4 .04	39 21 .06	40 1 .08	40 41 .10	41 21 .11	66 .70
20	35 53 .94	36 33 .82	37 13 .71	37 53 .60	38 33 .49	39 13 .38	39 53 .27	40 33 .15	41 13 .04	66 .48
30	35 46 .89	36 26 .65	37 6 .40	37 46 .16	38 25 .92	39 5 .67	39 45 .43	40 25 .19	41 4 .95	66 .26
40	35 39 .82	36 19 .45	36 59 .07	37 38 .70	38 18 .33	38 57 .95	39 37 .58	40 17 .21	40 56 .83	66 .04
50	35 32 .74	36 12 .23	36 51 .73	37 31 .22	38 10 .72	38 50 .21	39 29 .71	40 9 .20	40 48 .70	65 .83
49 0	35 25 .63	36 4 .99	36 44 .36	37 23 .72	38 3 .09	38 42 .45	39 21 .81	40 1 .18	40 40 .54	65 .61
10	35 18 .51	35 57 .74	36 36 .97	37 16 .20	37 55 .44	38 34 .67	39 13 .90	39 53 .13	40 32 .36	65 .39
20	35 11 .37	35 50 .47	36 29 .57	37 8 .67	37 47 .77	38 26 .87	39 5 .96	39 45 .06	40 24 .16	65 .17
30	35 4 .21	34 43 .18	36 22 .15	37 1 .11	37 40 .08	38 19 .05	38 58 .01	39 36 .98	40 15 .95	64 .94
40	34 57 .04	35 35 .87	36 14 .70	36 53 .54	37 32 .37	38 11 .21	38 50 .04	39 28 .87	40 7 .71	64 .72
50	34 49 .84	35 28 .54	36 7 .24	36 45 .95	37 24 .65	38 3 .35	38 42 .05	39 20 .75	39 59 .45	64 .50
50 0	34 42 .63	35 21 .20	35 50 .77	36 38 .33	37 16 .90	37 55 .47	38 34 .04	39 12 .60	39 51 .17	64 .28

TABLE 8.

The Moon's Parallax to every 10′ of her Altitude, &c. continued.

Arg. ☾'s Alt.	54′	55′	56′	57′	58′	59′	60′	61′	62′	Diff. for 100″.
50° 0′	34′ 42″.63	35′ 21″.20	35′ 50″.77	36′ 38″.33	37′ 16″.90	37′ 55″.47	38′ 34″.04	39′ 12″.60	39′ 51″.17	64″.28
10	34 35.40	35 13.84	35 52.27	36 30.70	37 9.14	37 47.57	38 26.00	39 4.44	39 42.87	64.06
20	34 28.16	35 6.46	35 44.76	36 23.05	37 1.35	37 39.65	38 17.95	38 56.25	39 34.55	63.83
30	34 20.89	34 59.06	35 37.22	36 15.39	36 53.55	37 31.72	38 9.88	38 48.05	39 26.21	63.61
40	34 13.61	34 51.64	35 29.67	36 7.70	36 45.73	37 23.76	38 1.79	38 39.82	39 17.85	63.38
50	34 6.31	34 44.21	35 22.10	36 0.00	36 37.89	37 15.79	37 53.68	38 31.58	39 9.47	63.16
51 0	33 59.00	34 36.76	35 14.52	35 52.28	36 30.03	37 7.79	37 45.55	38 23.31	39 1.07	62.93
10	33 51.67	34 29.29	35 6.91	35 44.54	36 22.16	36 59.78	37 37.41	38 15.03	38 52.65	62.70
20	33 44.31	34 21.80	34 59.29	35 36.78	36 14.26	36 51.75	37 29.24	38 6.73	38 44.21	62.48
30	33 36.95	34 14.30	34 51.65	35 29.00	36 6.35	36 43.70	37 21.05	37 58.40	38 35.75	62.25
40	33 29.56	34 6.78	34 43.99	35 21.21	35 58.42	36 35.63	37 12.85	37 50.06	38 27.28	62.02
50	33 22.16	33 59.24	34 36.32	35 13.39	35 58.47	36 27.55	37 4.62	37 41.70	38 18.78	61.80
52 0	33 14.74	33 51.68	34 28.62	35 5.56	35 42.50	36 19.44	36 56.38	37 33.32	38 10.26	61.57
10	33 7.31	33 44.11	34 20.91	34 57.71	35 34.52	36 11.32	36 48.12	37 24.92	38 1.72	61.34
20	32 59.86	33 36.52	34 13.18	34 49.85	35 26.51	36 3.18	36 39.84	37 16.50	37 53.17	61.11
30	32 52.39	33 28.91	34 5.44	34 41.96	35 18.49	35 55.02	36 31.54	37 8.07	37 44.59	60.88
40	32 44.90	33 21.29	33 57.68	34 34.06	35 10.45	35 46.84	36 23.22	36 59.61	37 36.00	60.65
50	32 37.40	33 13.65	33 49.90	34 26.14	35 2.39	35 38.64	36 14.89	36 51.14	37 27.38	60.41
53 0	32 29.88	33 5.99	33 42.10	34 18.21	34 54.32	35 30.43	36 6.53	36 42.64	37 18.75	60.18
10	32 22.35	32 58.31	33 34.28	34 10.25	34 46.22	35 22.19	35 58.16	36 34.13	37 10.10	59.95
20	32 14.79	32 50.62	33 26.45	34 2.28	34 38.11	35 13.94	35 49.77	36 25.60	37 1.43	59.72
30	32 7.23	32 42.92	33 18.60	33 54.29	34 29.98	35 5.67	35 41.36	36 17.05	36 52.74	59.48
40	31 59.64	32 35.19	33 10.74	33 46.29	34 21.84	34 57.39	35 32.93	36 8.48	36 44.03	59.25
50	31 52.04	32 27.45	33 2.86	33 38.27	34 13.67	34 49.08	35 24.49	35 59.90	36 35.31	59.01
54 0	31 44.42	32 19.69	32 54.96	33 30.23	34 5.49	34 40.76	35 16.03	35 51.29	36 26.56	58.78
10	31 36.79	32 11.92	32 47.04	33 22.17	33 57.29	34 32.42	35 7.55	35 42.67	36 17.80	58.54
20	31 29.14	32 4.13	32 39.11	33 14.09	33 49.08	34 24.06	34 59.05	35 34.03	36 9.02	58.31
30	31 21.48	31 56.32	32 31.16	33 6.00	33 40.85	34 15.69	34 50.53	35 25.37	36 0.22	58.07
40	31 13.80	31 48.50	32 23.20	32 57.90	33 32.60	34 7.30	34 42.00	35 16.70	35 51.40	57.83
50	31 6.10	31 40.66	32 15.21	32 49.77	33 24.33	33 58.89	34 33.44	35 8.00	35 42.56	57.60
55 0	30 58.39	31 32.80	32 7.22	32 41.63	33 16.05	33 50.46	34 24.88	34 59.29	35 33.70	57.36
10	30 50.66	31 24.93	31 59.20	32 33.47	33 7.75	33 42.02	34 16.29	34 50.56	35 24.83	57.12
20	30 42.92	31 17.04	31 51.17	32 25.30	32 59.43	33 33.56	34 17.68	34 41.81	35 15.94	56.88
30	30 35.16	31 9.14	31 43.12	32 17.11	32 51.09	33 25.07	33 59.06	34 33.05	35 7.03	56.64
40	30 27.38	31 1.22	31 35.06	32 8.90	32 42.74	33 16.58	33 50.42	34 24.26	34 58.10	56.40
50	30 19.59	30 53.29	31 26.98	32 0.68	32 34.38	33 8.07	33 41.77	34 15.46	34 49.16	56.16
56 0	30 11.78	30 45.34	31 18.89	31 52.44	32 25.99	32 59.54	33 33.09	34 6.65	34 40.20	55.92
10	30 3.96	30 37.37	31 10.78	31 44.18	32 17.59	32 51.00	33 24.40	33 57.81	34 31.22	55.68
20	29 56.13	30 29.39	31 2.65	31 35.91	32 9.17	32 42.44	33 15.70	33 48.96	34 22.22	55.44
30	29 48.28	30 21.39	30 54.51	31 27.62	32 0.74	32 33.86	33 6.97	33 40.09	34 13.21	55.19
40	29 40.41	30 13.38	30 46.35	31 19.32	31 52.29	32 25.26	32 58.23	33 31.20	34 4.17	54.95
50	29 32.53	30 5.35	30 38.18	31 11.00	31 43.83	32 16.65	32 49.47	33 22.30	33 55.12	54.71
57 0	29 24.63	29 57.31	30 29.99	31 2.67	31 35.34	32 8.02	32 40.70	33 13.38	33 46.06	54.46
10	29 16.72	29 49.25	30 21.78	30 54.31	31 26.85	31 59.38	32 31.91	33 4.44	33 36.97	54.22
20	29 8.79	29 41.18	30 13.56	30 45.95	31 18.33	31 50.71	32 23.10	32 55.49	33 27.87	53.98
30	29 0.85	29 33.09	30 5.33	30 37.57	31 9.80	31 42.04	32 14.28	32 46.52	33 18.76	53.73
40	28 52.89	29 24.99	29 57.08	30 29.17	31 1.26	31 33.35	32 5.44	32 37.53	33 9.62	53.48
50	28 44.92	29 16.87	29 48.81	30 20.75	30 52.70	31 24.64	31 56.58	32 28.53	33 0.47	53.24
58 0	28 36.94	29 8.73	29 40.53	30 12.32	30 44.12	31 15.91	31 47.71	32 19.50	32 51.30	52.99
10	28 28.94	29 0.59	29 32.23	30 3.88	30 35.53	31 7.17	31 38.82	32 10.47	32 42.12	52.75
20	28 20.92	28 52.42	29 23.92	29 55.42	30 26.92	30 58.42	31 29.92	32 1.41	32 32.91	52.50
30	28 12.90	28 44.25	29 15.60	29 46.95	30 18.30	30 49.65	31 21.00	31 52.35	32 23.70	52.25
40	28 4.85	28 36.05	29 7.25	29 38.46	30 9.66	30 40.86	31 12.06	31 43.26	32 14.46	52.00
50	27 56.79	28 27.85	28 58.90	29 29.95	30 1.00	30 32.05	31 3.11	31 34.16	32 5.21	51.75
59 0	27 48.72	28 19.63	28 50.53	29 21.43	29 52.33	30 23.23	30 54.14	31 25.04	31 55.94	51.50
10	27 40.64	28 11.39	28 42.14	29 12.90	29 43.65	30 14.40	30 45.15	31 15.91	31 46.66	51.25
20	27 32.54	28 3.14	28 33.74	29 4.35	29 34.95	30 5.55	30 36.15	31 6.76	31 37.36	51.00
30	27 24.42	27 54.88	28 25.33	28 55.78	29 26.23	29 56.69	30 27.14	30 57.59	31 28.04	50.75
40	27 16.30	27 46.60	28 16.90	28 47.20	29 17.50	29 47.81	30 18.11	30 48.41	31 18.71	50.50
50	27 8.16	27 38.31	28 8.46	28 38.61	29 8.76	29 38.91	30 9.06	30 39.21	31 9.36	50.25
60 0	27 0.00	27 30.00	28 0.00	28 30.00	29 0.00	29 30.00	30 0.00	30 30.00	31 0.00	50.00

TABLE 8.

The Moon's Parallax to every 10′ of her Altitude, &c. continued.

Arg. ☾'s Alt	54′	55′	56′	57′	58′	59′	60′	61′	62′	Diff. for 100″.
60° 0′	27′ 0″.00	27′ 30″.00	28′ 0″.00	28′ 30″.00	29′ 0″.00	29′ 30″.00	30′ 0″.00	30′ 30″.00	31′ 0″.00	50″.00
10	26 51.83	27 21.68	27 51.53	28 21.38	28 51.23	29 21.07	29 50.92	30 20.77	30 50.62	49.75
20	26 43.65	27 13.35	27 43.04	28 12.74	28 44.44	29 12.13	29 41.83	30 11.53	30 41.23	49.50
30	26 35.45	27 5.00	27 34.54	28 4.09	28 33.63	29 3.18	29 32.72	30 2.27	30 31.82	49.24
40	26 27.24	26 56.64	27 26.03	27 55.42	28 24.82	28 54.21	29 23.60	29 53.00	30 22.39	48.99
50	26 19.02	26 48.26	27 17.50	27 46.74	28 15.98	28 45.23	29 14.47	29 43.71	30 12.95	48.74
61 0	26 10.78	26 39.87	27 8.96	27 38.05	28 7.14	28 36.23	29 5.31	29 34.40	30 3.49	48.48
10	26 2.53	26 31.47	27 0.41	27 29.34	27 58.28	28 27.21	28 56.15	29 25.08	29 54.02	48.23
20	25 54.27	26 23.05	26 51.84	27 20.62	27 49.40	28 18.18	28 46.97	29 15.75	29 44.53	47.97
30	25 45.99	26 14.62	26 43.25	27 11.88	27 40.51	28 9.14	28 37.77	29 6.40	29 35.03	47.72
40	25 37.71	26 6.18	26 34.66	27 3.13	27 31.61	28 0.09	28 28.56	28 57.04	29 25.51	47.46
50	25 29.40	25 57.73	26 26.05	26 54.37	27 22.69	27 51.01	28 19.34	28 47.66	29 15.98	47.20
62 0	25 21.09	25 49.26	26 17.42	26 45.59	27 13.76	27 41.93	28 10.10	28 38.27	29 6.43	46.95
10	25 12.76	25 40.77	26 8.79	26 36.80	27 4.82	27 32.83	28 0.84	28 28.86	28 56.87	46.69
20	25 4.42	25 32.28	26 0.14	26 28.00	26 55.86	27 23.72	27 51.58	28 19.44	28 47.30	46.43
30	24 56.07	25 23.77	25 51.48	26 19.18	26 46.89	27 14.59	27 42.29	28 10.00	28 37.70	46.17
40	24 47.70	25 15.25	25 42.80	26 10.35	26 37.90	27 5.45	27 33.00	28 0.55	28 28.10	45.92
50	24 39.32	25 6.72	25 34.11	26 1.50	26 28.90	26 56.29	27 23.69	27 51.08	28 18.48	45.66
63 0	24 30.93	24 58.17	25 25.41	25 52.65	26 19.89	26 47.13	27 14.37	27 41.61	28 8.84	45.40
10	24 22.53	24 49.61	25 16.69	25 43.78	26 10.86	26 37.94	27 5.03	27 32.11	27 59.20	45.14
20	24 14.11	24 41.04	25 7.97	25 34.89	26 1.82	26 28.75	26 25.68	27 22.61	27 49.53	44.88
30	24 5.68	24 32.45	24 59.22	25 26.00	25 52.77	26 19.54	26 46.31	27 13.08	27 39.86	44.62
40	23 57.24	24 23.86	24 50.47	25 17.09	25 43.70	26 10.32	26 36.93	27 3.55	27 30.17	44.36
50	23 48.79	24 15.25	24 51.71	25 8.17	25 34.62	26 1.08	26 27.54	26 54.00	27 20.46	44.10
64 0	23 40.32	24 6.62	24 32.93	24 59.23	25 25.53	25 51.83	26 18.14	26 44.44	27 10.74	43.84
10	23 31.85	23 57.99	24 24.14	24 50.28	25 16.43	25 42.57	26 8.72	26 34.86	27 1.01	43.58
20	23 23.36	23 49.34	24 15.33	24 41.32	25 7.31	25 33.30	25 59.29	26 25.27	26 51.26	43.31
30	23 14.86	23 40.69	24 6.52	24 32.35	24 58.18	25 24.01	25 49.84	26 15.67	26 41.50	43.05
40	23 6.34	23 32.02	23 57.69	24 23.36	24 49.04	25 14.71	25 40.38	26 6.06	26 31.73	42.79
50	22 57.82	23 23.33	23 48.85	24 14.36	24 39.88	25 5.40	25 30.91	25 56.43	26 21.94	42.53
65 0	22 49.28	23 14.64	23 40.00	24 5.35	24 30.71	24 56.07	25 21.43	25 46.78	26 12.14	42.26
10	22 40.74	23 5.93	23 31.13	23 56.33	24 21.53	24 46.73	25 11.93	25 37.13	26 2.33	42.00
20	22 32.18	22 57.22	23 22.26	23 47.30	24 12.34	24 37.38	25 2.42	25 47.46	25 52.50	41.73
30	22 23.61	22 48.49	23 13.37	23 38.25	24 3.13	24 28.01	24 52.90	25 17.78	25 42.66	41.47
40	22 15.02	22 39.75	23 4.47	23 29.19	23 53.91	24 18.64	24 43.36	25 8.08	25 32.81	41.20
50	22 6.43	22 30.99	22 55.56	23 20.12	23 44.69	24 9.65	24 33.81	24 58.38	25 22.94	40.94
66 0	21 57.83	22 22.23	22 46.64	23 11.04	23 35.44	23 59.85	24 24.25	24 48.66	25 13.06	40.67
10	21 49.21	22 13.46	22 37.70	23 1.95	23 26.19	23 50.43	24 14.68	24 38.92	25 3.17	40.41
20	21 40.58	22 4.67	22 28.75	22 52.84	23 16.92	23 41.01	24 5.09	24 29.18	24 53.26	40.14
30	21 31.95	21 55.87	22 19.80	22 43.72	23 7.65	23 31.57	23 55.50	24 19.42	24 43.35	39.87
40	21 23.30	21 47.06	22 10.83	22 34.59	22 58.36	23 22.12	23 45.89	24 9.65	24 33.42	39.61
50	21 14.64	21 38.24	22 1.85	22 25.45	22 49.06	23 12.66	23 36.27	23 59.87	24 23.47	39.34
67 0	21 5.97	21 29.41	21 52.86	22 16.30	22 39.71	23 3.19	23 26.63	23 50.08	24 13.52	39.07
10	20 57.29	21 20.57	21 43.85	22 7.14	22 30.42	22 53.70	23 16.99	23 40.27	24 3.55	38.81
20	20 48.60	21 11.72	21 34.84	21 57.96	22 21.09	22 44.21	23 7.33	23 30.45	23 53.57	38.54
30	20 39.89	21 2.86	21 25.82	21 48.78	22 11.74	22 34.70	22 57.66	23 20.62	23 43.58	38.27
40	20 31.18	20 53.98	21 16.78	21 39.58	22 2.38	22 25.18	22 47.98	23 10.78	23 33.58	38.00
50	20 32.46	20 45.10	21 7.74	21 30.37	21 53.01	22 15.65	22 38.29	23 0.93	23 23.56	37.73
68 0	20 13.73	20 36.20	20 58.68	21 21.15	21 43.63	22 6.11	22 28.58	22 51.06	23 13.54	37.46
10	20 4.98	20 27.30	20 49.61	21 11.93	21 34.24	21 56.55	22 18.87	22 41.18	23 3.50	37.19
20	19 56.23	20 18.38	20 40.53	21 2.68	21 24.84	21 46.99	22 9.14	22 31.29	22 53.45	36.92
30	19 47.46	20 9.45	20 31.44	20 53.43	21 15.42	21 37.41	21 59.40	22 21.39	22 43.38	36.65
40	19 38.69	20 0.52	20 22.35	20 44.17	21 6.00	21 27.83	21 49.66	22 11.48	22 33.31	36.38
50	19 29.91	19 51.57	20 13.24	20 34.90	20 56.57	21 18.23	21 39.90	22 1.56	22 23.23	36.11
69 0	19 21.11	19 42.61	20 4.12	20 25.62	20 47.12	21 8.62	21 30.12	21 51.63	22 13.13	35.84
10	19 12.31	19 33.65	19 54.99	20 16.33	20 37.67	20 59.00	21 20.34	21 41.68	22 3.02	35.57
20	19 3.50	19 24.67	19 45.85	20 7.02	20 28.20	20 49.37	21 10.55	21 31.73	21 52.90	35.29
30	18 54.67	19 15.68	19 36.70	19 57.71	20 18.72	20 39.73	21 0.75	21 21.76	21 42.77	35.02
40	18 45.84	19 6.69	19 27.54	19 48.39	20 9.23	20 30.08	20 50.93	21 11.78	21 32.63	34.75
50	18 37.00	18 57.68	19 18.37	19 39.05	19 59.74	20 20.42	20 41.11	21 1.79	21 22.48	34.48
70 0	18 28.15	18 48.67	19 9.19	19 29.71	19 50.23	20 10.75	20 31.27	20 51.79	21 12.31	34.20

TABLE 8.

The MOON'S PARALLAX to every 10′ of her Altitude, &c. continued.

Arg. ☾'s Alt.	54′	55′	56′	57′	58′	59′	60′	61′	62′	Diff. for 100″
70° 0′	18′ 28″.15	18′ 48″.67	19′ 9″.19	19′ 29″.71	19′ 50″.23	20′ 10″.75	20′ 31″.27	20′ 51″.79	21′ 12″.31	34″.20
10	18 19 .28	18 39 .64	19 0 .00	19 20 .36	19 40 .71	20 1 .07	20 21 .43	20 41 .78	21 2 .14	33 .93
20	18 10 .41	18 30 .61	18 50 .80	19 10 .99	19 31 .19	19 51 .38	20 11 .57	20 31 .76	20 51 .96	33 .65
30	18 1 .53	18 21 .56	18 41 .59	19 1 .62	19 21 .65	19 41 .68	20 1 .70	20 21 .73	20 41 .76	33 .38
40	17 52 .65	18 12 .51	18 32 .37	18 52 .24	19 12 .10	19 31 .96	19 51 .83	20 11 .69	20 31 .56	33 .11
50	17 43 .75	18 3 .45	18 23 .15	18 42 .85	19 2 .54	19 22 .24	19 41 .94	20 1 .64	20 21 .34	32 .83
71 0	17 34 .84	17 54 .38	18 13 .91	18 33 .44	18 52 .98	19 12 .51	19 32 .05	19 51 .58	20 11 .11	32 .56
10	17 25 .93	17 45 .29	17 4 .66	18 24 .03	18 43 .40	19 2 .77	19 22 .14	19 41 .51	20 0 .88	32 .28
20	17 17 .00	17 36 .20	17 55 .41	18 14 .61	18 33 .82	18 53 .02	19 12 .22	19 31 .43	19 50 .63	32 .01
30	17 8 .07	17 27 .11	17 46 .14	18 5 .18	18 24 .22	18 43 .26	19 2 .30	19 21 .34	19 40 .37	31 .73
40	16 59 .13	17 18 .00	17 36 .87	17 55 .74	18 14 .62	18 33 .49	18 52 .36	19 11 .23	19 30 .11	31 .45
50	16 50 .17	17 8 .88	17 27 .59	17 46 .30	18 5 .00	18 23 .71	18 42 .42	19 1 .12	19 19 .83	31 .18
72 0	16 41 .22	16 59 .76	17 18 .30	17 36 .84	17 55 .38	18 13 .92	18 32 .46	18 51 .00	19 9 .54	30 .90
10	16 32 .25	16 50 .62	17 9 .00	17 27 .37	17 45 .75	18 4 .12	18 22 .50	18 40 .87	18 59 .25	30 .62
20	16 23 .27	16 41 .48	16 59 .69	17 17 .90	17 36 .11	17 54 .31	18 12 .52	18 30 .73	18 48 .94	30 .35
30	16 14 .29	16 32 .33	16 50 .37	17 8 .41	17 26 .46	17 44 .50	18 2 .54	18 20 .58	18 38 .63	30 .07
40	16 5 .29	16 23 .17	16 41 .05	16 58 .92	17 16 .80	17 34 .67	17 52 .55	18 10 .42	18 28 .30	29 .79
50	15 56 .29	16 14 .00	16 31 .71	16 49 .42	17 7 .13	17 24 .84	17 42 .55	18 0 .26	18 17 .97	29 .52
73 0	15 47 .28	16 4 .83	16 22 .37	16 39 .91	16 57 .45	17 15 .00	17 32 .54	17 50 .08	18 7 .62	29 .24
10	15 38 .27	15 55 .64	16 13 .02	16 30 .39	16 47 .77	17 5 .14	17 22 .52	17 39 .89	17 57 .27	28 .96
20	15 29 .24	15 46 .45	16 3 .66	16 20 .87	16 38 .08	16 55 .28	17 12 .49	17 29 .70	17 46 .91	28 .68
30	15 20 .21	15 37 .25	15 54 .29	16 11 .33	16 28 .37	16 45 .41	17 2 .46	17 19 .50	17 36 .54	28 .40
40	15 11 .17	15 28 .04	15 44 .92	16 1 .79	16 18 .66	16 35 .54	16 52 .41	17 9 .28	17 26 .16	28 .12
50	15 2 .12	15 18 .83	15 35 .53	15 52 .24	16 8 .95	16 25 .65	16 42 .36	16 59 .06	17 15 .77	27 .84
74 0	14 53 .07	15 9 .60	15 26 .14	15 42 .68	15 59 .22	16 15 .76	16 32 .29	16 48 .83	17 5 .37	27 .56
10	14 44 .00	15 0 .37	15 16 .74	15 33 .11	15 49 .48	16 5 .85	16 22 .22	16 38 .59	16 54 .96	27 .28
20	14 34 .93	14 51 .13	15 7 .34	15 23 .54	15 39 .74	15 55 .94	16 12 .15	16 28 .35	16 44 .55	27 .00
30	14 25 .85	14 41 .89	14 57 .92	15 13 .96	15 29 .99	15 46 .02	16 2 .06	16 18 .09	16 34 .13	26 .72
40	14 16 .77	14 32 .63	14 48 .50	15 4 .37	15 20 .23	15 36 .10	15 51 .96	16 7 .83	16 23 .17	26 .44
50	14 7 .67	14 23 .67	14 39 .07	14 54 .77	15 10 .46	15 26 .16	15 41 .86	15 57 .56	16 13 .26	26 .16
75 0	13 58 .57	14 14 .10	14 29 .63	14 45 .16	15 0 .69	15 16 .22	15 31 .75	15 47 .28	16 2 .81	25 .88
10	13 49 .47	14 4 .83	14 20 .19	14 35 .55	14 50 .91	15 6 .27	15 21 .63	15 36 .99	15 52 .35	25 .60
20	13 40 .35	13 55 .54	14 10 .74	14 25 .93	14 41 .12	14 56 .31	15 11 .50	15 26 .69	15 41 .89	25 .32
30	13 31 .23	13 46 .25	14 1 .28	14 16 .30	14 31 .32	14 46 .35	15 1 .37	15 16 .39	15 31 .41	25 .04
40	13 22 .10	13 36 .96	13 51 .81	14 6 .66	14 21 .52	14 36 .37	14 51 .23	15 6 .08	15 20 .93	24 .76
50	13 12 .97	13 27 .65	13 42 .34	13 57 .02	14 11 .71	14 26 .39	14 41 .08	14 55 .76	15 10 .45	24 .47
76 0	13 3 .83	13 18 .34	13 32 .86	13 47 .37	14 1 .89	14 16 .40	14 30 .92	14 45 .43	14 59 .95	24 .19
10	12 54 .68	13 9 .02	13 23 .37	13 37 .72	13 52 .06	14 6 .41	14 20 .75	14 35 .10	14 49 .45	23 .91
20	12 45 .52	12 59 .70	13 13 .88	13 28 .05	13 42 .23	13 56 .41	14 10 .58	14 24 .76	14 38 .94	23 .63
30	12 36 .36	12 50 .37	13 4 .38	13 18 .38	13 32 .39	13 46 .40	14 0 .40	14 14 .41	14 28 .42	23 .34
40	12 27 .20	12 41 .03	12 54 .87	13 8 .71	13 22 .54	13 36 .38	13 50 .22	14 4 .05	14 17 .89	23 .06
50	12 18 .02	12 31 .69	12 45 .36	12 59 .02	13 12 .69	13 26 .36	13 40 .02	13 53 .69	14 7 .36	22 .78
77 0	12 8 .84	12 22 .34	12 35 .84	12 49 .33	13 2 .83	13 16 .33	13 29 .82	13 43 .32	13 56 .82	22 .50
10	11 59 .66	12 12 .98	12 26 .31	12 39 .64	12 52 .96	13 6 .29	13 19 .62	13 32 .94	13 46 .27	22 .21
20	11 50 .46	12 3 .62	12 16 .78	12 29 .93	12 43 .09	12 56 .25	13 9 .40	13 22 .56	13 35 .72	21 .93
30	11 41 .26	11 54 .25	12 7 .24	12 20 .22	12 33 .21	12 46 .20	12 59 .18	13 12 .17	13 25 .16	21 .64
40	11 32 .06	11 44 .88	11 57 .69	12 10 .51	12 23 .32	12 36 .14	12 48 .96	13 1 .77	13 14 .59	21 .36
50	11 22 .85	11 35 .50	11 48 .14	12 0 .79	12 13 .43	12 26 .08	12 38 .72	12 51 .37	13 4 .01	21 .08
78 0	11 13 .63	11 26 .11	11 38 .58	11 51 .06	12 3 .53	12 16 .01	12 28 .48	12 40 .96	12 53 .43	20 .79
10	11 4 .41	11 16 .72	11 29 .02	11 41 .32	11 53 .63	12 5 .93	12 18 .24	12 30 .54	12 42 .81	20 .51
20	10 55 .19	11 7 .32	11 19 .45	11 31 .58	11 43 .72	11 55 .85	12 7 .98	12 20 .12	12 32 .25	20 .22
30	10 45 .95	10 57 .91	11 9 .88	11 21 .84	11 33 .80	11 45 .76	11 57 .72	12 9 .69	12 21 .65	19 .94
40	10 36 .71	10 48 .51	11 0 .30	11 12 .09	11 23 .88	11 35 .67	11 47 .46	11 59 .25	12 11 .04	19 .65
50	10 27 .47	10 39 .09	10 50 .71	11 2 .33	11 13 .95	11 25 .57	11 37 .19	11 48 .81	12 0 .43	19 .37
79 0	10 18 .22	10 29 .67	10 41 .12	10 52 .57	11 4 .02	11 15 .46	11 26 .91	11 38 .36	11 49 .81	19 .08
10	10 8 .97	10 20 .24	10 31 .52	10 42 .80	10 54 .08	11 5 .35	11 16 .63	11 27 .91	11 39 .18	18 .80
20	9 59 .71	10 10 .81	10 21 .92	10 33 .02	10 44 .13	10 55 .24	11 6 .34	11 17 .45	11 28 .55	18 .51
30	9 50 .44	10 1 .38	10 12 .31	10 23 .25	10 34 .18	10 45 .11	10 56 .05	11 6 .98	11 17 .92	18 .22
40	9 41 .17	9 51 .93	10 2 .70	10 13 .46	10 24 .22	10 34 .99	10 45 .75	10 56 .51	11 7 .27	17 .94
50	9 31 .90	9 42 .49	9 53 .08	10 3 .67	10 14 .26	10 24 .85	10 35 .44	10 46 .03	10 56 .62	17 .65
80 0	9 22 .62	9 33 .04	9 43 .46	9 53 .88	10 4 .30	10 14 .71	10 25 .13	10 35 .55	10 45 .97	17 .36

TABLE 8.

The Moon's Parallax to every 10′ of her Altitude, &c. concluded.

Arg. ☾'s Alt.	54′	55′	56′	57′	58′	59′	60′	61′	62′	Diff. for 100″.
80° 0	9′ 22″.62	9′ 33″.04	9′ 43″.46	9′ 53″.88	10′ 4″.30	10′ 14″.71	10′ 25″.13	10′ 35″.55	10′ 45″.97	17″.36
10	9 13.34	9 23.58	9 33.83	9 44.08	9 54.32	10 4.57	10 14.82	10 25.07	10 35.31	17.08
20	9 4.05	9 14.12	9 24.20	9 34.27	9 44.35	9 54.42	10 4.50	10 14.57	10 24.65	16.79
30	8 54.75	9 4.66	9 14.56	9 24.46	9 34.37	9 44.27	9 54.17	10 4.07	10 13.98	16.50
40	8 45.46	8 55.19	9 4.92	9 14.65	9 24.38	9 34.11	9 43.84	9 53.57	10 3.30	16.22
50	8 36.15	8 45.71	8 55.27	9 4.83	9 14.39	9 23.95	9 33.50	9 43.06	9 52.62	15.93
81 0	8 26.85	8 36.23	8 45.62	8 55.01	9 4.39	9 13.78	9 23.16	9 32.53	9 41.94	15.64
10	8 17.54	8 26.75	8 35.96	8 45.18	8 54.39	9 3.60	9 12.82	9 22.03	9 31.25	15.36
20	8 8.22	8 17.26	8 26.30	8 35.35	8 44.39	8 53.43	9 2.47	9 11.51	9 20.55	15.07
30	7 58.90	8 7.77	8 16.64	8 25.51	8 34.38	8 43.25	8 52.11	9 0.98	9 9.85	14.78
40	7 49.58	7 58.28	8 6.97	8 15.67	8 24.36	8 33.06	8 41.75	8 50.45	8 59.15	14.49
50	7 40.25	7 48.78	7 57.30	8 5.82	8 14.34	8 22.87	8 31.39	8 39.91	8 48.44	14.21
82 0	7 30.92	7 39.27	7 47.62	7 55.97	8 4.32	8 12.67	8 21.02	8 29.37	8 37.72	13.92
10	7 21.59	7 29.76	7 37.94	7 46.12	7 54.30	8 2.47	8 10.65	8 18.83	8 27.01	13.63
20	7 12.25	7 20.25	7 28.26	7 36.26	7 44.27	7 52.27	8 0.27	8 8.28	8 16.28	13.34
30	7 2.90	7 10.74	7 18.57	7 26.40	7 34.23	7 42.06	7 49.89	7 57.73	8 5.56	13.05
40	6 53.56	7 1.22	7 8.88	7 16.53	7 24.19	7 31.85	7 39.51	7 47.17	7 54.83	12.76
50	6 44.21	6 51.69	6 59.18	7 6.67	7 14.15	7 21.64	7 29.12	7 36.61	7 44.09	12.48
83 0	6 34.86	6 42.17	6 49.48	6 56.79	7 4.11	7 11.42	7 18.73	7 26.04	7 33.35	12.19
10	6 25.50	6 32.64	6 39.78	6 46.92	6 54.06	7 1.19	7 8.33	7 15.47	7 22.61	11.90
20	6 16.14	6 23.11	6 30.07	6 37.04	6 44.00	6 50.97	6 57.93	7 4.90	7 11.87	11.61
30	6 6.78	6 13.57	6 20.36	6 27.16	6 33.95	6 40.74	6 47.53	6 54.32	7 1.12	11.32
40	5 57.41	6 4.03	6 10.65	6 17.27	6 23.89	6 30.51	6 37.13	6 43.74	6 50.36	11.03
50	5 48.04	5 54.49	6 0.93	6 7.38	6 13.83	6 20.27	6 26.72	6 33.16	6 39.61	10.74
84 0	5 38.67	5 44.94	5 51.22	5 57.49	6 3.76	6 10.07	6 16.30	6 22.57	6 28.85	10.45
10	5 29.30	5 35.40	5 41.49	5 47.59	5 53.69	5 59.79	6 5.89	6 11.98	6 18.08	10.16
20	5 19.92	5 25.84	5 31.77	5 37.69	5 43.62	5 49.54	5 55.47	6 1.39	6 7.32	9.87
30	5 10.54	5 16.29	5 22.04	5 27.79	5 33.54	5 39.29	5 45.04	5 50.80	5 56.55	9.58
40	5 1.16	5 6.74	5 12.31	5 17.89	5 23.47	5 29.04	5 34.62	5 40.20	5 45.77	9.29
50	4 51.77	4 57.18	5 2.58	5 7.98	5 13.39	5 18.79	5 24.19	5 29.59	5 35.00	9.01
85 0	4 42.38	4 47.62	4 52.84	4 58.07	5 3.30	5 8.53	5 13.76	5 18.99	5 24.22	8.72
10	4 32.99	4 38.05	4 43.11	4 48.16	4 53.22	4 58.27	5 3.33	5 8.38	5 13.44	8.43
20	4 23.60	4 28.48	4 33.37	4 38.25	4 43.13	4 48.01	4 52.89	4 57.77	5 2.65	8.14
30	4 14.21	4 18.92	4 23.62	4 28.33	4 33.04	4 37.75	4 42.45	4 47.16	4 51.87	7.85
40	4 4.81	4 9.34	4 13.88	4 18.41	4 22.94	4 27.48	4 32.01	4 36.55	4 41.08	7.56
50	3 55.41	3 59.77	4 4.13	4 8.49	4 12.85	4 17.21	4 21.57	4 25.93	4 30.29	7.27
86 0	3 46.01	3 50.20	3 54.38	3 58.57	4 2.75	4 6.94	4 11.12	4 15.31	4 19.49	6.98
10	3 36.61	3 40.62	3 44.63	3 48.64	3 52.65	3 56.66	4 0.68	4 4.69	4 8.70	6.69
20	3 27.20	3 31.04	3 34.88	3 38.71	3 42.55	3 46.39	3 50.23	3 54.06	3 57.90	6.40
30	3 17.80	3 21.46	3 25.12	3 28.79	3 32.45	3 36.11	3 39.77	3 43.44	3 47.10	6.10
40	3 8.39	3 11.88	3 15.37	3 18.86	3 22.34	3 25.83	3 29.32	3 32.81	3 36.30	5.81
50	2 58.98	3 2.29	3 5.61	3 8.92	3 12.24	3 15.55	3 18.87	3 22.18	3 25.49	5.52
87 0	2 49.57	2 52.71	2 55.85	2 58.99	3 2.13	3 5.27	3 8.41	3 11.55	3 14.69	5.23
10	2 40.16	2 43.12	2 46.09	2 49.05	2 52.02	2 54.99	2 57.95	3 0.92	3 3.88	4.94
20	2 30.74	2 33.53	2 33.33	2 39.12	2 41.91	2 44.70	2 47.49	2 50.28	2 53.07	4.65
30	2 21.33	2 23.94	2 26.56	2 29.18	2 31.80	2 34.41	2 37.03	2 39.65	2 42.26	4.36
40	2 11.91	2 14.35	2 16.80	2 19.24	2 21.68	2 24.12	2 26.57	2 29.01	2 31.45	4.07
50	2 2.49	2 4.76	2 7.03	2 9.30	2 11.57	2 13.84	2 16.10	2 18.37	2 20.64	3.78
88 0	1 53.07	1 55.17	1 57.26	1 59.36	2 1.45	2 3.54	2 5.64	2 7.73	2 9.83	3.49
10	1 43.65	1 45.57	1 47.49	1 49.41	1 51.33	1 53.25	1 55.17	1 57.09	1 59.01	3.20
20	1 34.23	1 35.98	1 37.73	1 39.47	1 41.22	1 42.96	1 44.71	1 46.45	1 48.20	2.91
30	1 24.81	1 26.38	1 27.95	1 29.53	1 31.10	1 32.67	1 34.24	1 35.81	1 37.38	2.62
40	1 15.39	1 16.79	1 18.18	1 19.58	1 20.98	1 22.37	1 23.77	1 25.16	1 26.56	2.33
50	1 5.97	1 7.19	1 8.41	1 9.63	1 10.86	1 12.08	1 13.30	1 14.52	1 15.74	2.04
89 0	0 56.55	0 57.59	0 58.64	0 59.69	1 0.73	1 1.78	1 2.83	1 3.88	1 4.92	1.75
10	0 47.12	0 47.99	0 48.87	0 49.74	0 50.61	0 51.49	0 52.30	0 53.23	0 54.10	1.45
20	0 37.70	0 38.40	0 39.09	0 39.79	0 40.49	0 41.19	0 41.89	0 42.59	0 43.28	1.16
30	0 28.27	0 28.80	0 29.32	0 29.84	0 30.37	0 30.89	0 31.42	0 31.94	0 32.46	0.87
40	0 18.85	0 19.20	0 19.55	0 19.90	0 20.25	0 20.59	0 20.94	0 21.29	0 21.64	0.58
50	0 9.42	0 9.60	0 9.77	0 9.95	0 10.12	0 10.30	0 10.47	0 10.65	0 10.82	0.29
90 0	0 0.00	0 0.00	0 0.00	0 0.00	0 0.00	0 0.00	0 0.00	0 0.00	0 0.00	0.00

Enter the Table with the ☾'s Altitude as the Argument, at the side, and with her Horizontal Parallax at the head of the Table; and take out the ☾'s Parallax in Altitude, making proportion for the odd Minutes, &c., of the ☾'s Altitude, and for Seconds, &c., of the ☾'s Horizontal Parallax. The last column contains the difference for 100″ variation for the ☾'s Horizontal Parallax: but the differences for 10′ of Altitude must be found by subtraction.

TABLE 9.—TIME of the MOON'S SEMI-DIAMETER passing the MERIDIAN. (LAMBERT'S Table I. extended.)

Arguments = The Moon's Motion in Right Ascension in Twelve Solar Hours, at the Head; and the Apparent Semi-diameter at the Meridian Passage, on the Side.

☾'s App. Semi-diameter.	5° 10′	5° 20′	5° 30′	5° 40′	5° 50′	6° 0′	6° 10′	6° 20′	6° 30′	6° 40′	6° 50′	7° 0′	7° 10′	7° 20′	7° 30′	7° 40′	7° 50′	8° 0′	8° 10′	8° 20′
14′ 30″	59″.66	59″.71	59″.77	59″.82	59″.87	59″.93	59″.98	60″.03	60″.09	60″.14	60″.20	60″.26	60″.31	60″.36	60″.42	60″.47	60″.52	60″.58	60″.63	60″.68
31	59.73	59.78	59.84	59.89	59.94	60.00	60.05	60.10	60.16	60.21	60.27	60.33	60.38	60.43	60.49	60.54	60.59	60.65	60.70	60.75
32	59.80	59.85	59.91	59.96	60.01	60.07	60.12	60.17	60.23	60.28	60.34	60.40	60.45	60.50	60.59	60.61	60.66	60.72	60.77	60.82
33	59.87	59.92	59.98	60.03	60.08	60.14	60.19	60.24	60.30	60.35	60.41	60.47	60.52	60.57	60.63	60.68	60.73	60.79	60.84	60.89
34	59.94	59.99	60.05	60.10	60.15	60.21	60.26	60.31	60.37	60.42	60.48	60.54	60.59	60.64	60.70	60.75	60.80	60.86	60.91	60.96
35	60.00	60.06	60.12	60.17	60.22	60.28	60.33	60.38	60.44	60.49	60.54	60.60	60.65	60.70	60.76	60.81	60.87	60.93	60.98	61.03
36	60.07	60.13	60.19	60.24	60.29	60.35	60.40	60.45	60.51	60.56	60.61	60.67	60.72	60.77	60.83	60.88	60.94	61.00	61.05	61.10
37	60.14	60.20	60.26	60.31	60.36	60.42	60.47	60.51	60.58	60.63	60.68	60.74	60.79	60.84	60.90	60.95	61.01	61.07	61.12	61.17
38	60.21	60.27	60.33	60.38	60.43	60.49	60.54	60.58	60.65	60.70	60.75	60.81	60.86	60.91	60.97	61.02	61.08	61.14	61.19	61.24
39	60.28	60.34	60.40	60.45	60.50	60.56	60.61	60.65	60.72	60.77	60.82	60.88	60.93	60.98	61.04	61.09	61.15	61.21	61.26	61.31
40	60.35	60.40	60.46	60.51	60.56	60.62	60.67	60.73	60.79	60.84	60.89	60.95	61.00	61.05	61.11	61.16	61.21	61.27	61.32	61.38
41	60.42	60.47	60.53	60.58	60.63	60.69	60.74	60.80	60.86	60.91	60.96	61.02	61.07	61.12	61.18	61.23	61.28	61.34	61.39	61.45
42	60.49	60.54	60.60	60.65	60.70	60.76	60.81	60.87	60.93	60.98	61.03	61.09	61.14	61.19	61.25	61.30	61.35	61.41	61.46	61.52
43	60.56	60.61	60.67	60.72	60.77	60.83	60.88	60.94	61.00	61.05	61.10	61.16	61.21	61.26	61.32	61.37	61.42	61.48	61.53	61.59
44	60.63	60.68	60.74	60.79	60.84	60.90	60.95	61.01	61.07	61.12	61.17	61.23	61.28	61.33	61.39	61.44	61.49	61.55	61.60	61.66
45	60.69	60.74	60.80	60.85	60.91	60.97	61.02	61.07	61.13	61.18	61.24	61.30	61.35	61.40	61.46	61.51	61.56	61.62	61.67	61.73
46	60.76	60.81	60.87	60.92	60.98	61.04	61.09	61.14	61.20	61.25	61.31	61.37	61.42	61.47	61.53	61.58	61.63	61.69	61.74	61.80
47	60.83	60.88	60.94	60.99	61.05	61.11	61.16	61.21	61.27	61.32	61.38	61.44	61.49	61.54	61.60	61.65	61.70	61.76	61.81	61.87
48	60.90	60.95	61.01	61.06	61.12	61.18	61.23	61.28	61.34	61.39	61.45	61.51	61.56	61.61	61.67	61.72	61.77	61.83	61.88	61.94
49	60.97	61.02	61.08	61.13	61.19	61.25	61.30	61.35	61.41	61.46	61.52	61.58	61.63	61.68	61.74	61.79	61.84	61.90	61.95	62.01
50	61.03	61.09	61.15	61.20	61.25	61.31	61.36	61.42	61.48	61.53	61.58	61.64	61.69	61.75	61.81	61.86	61.91	61.97	62.02	62.08
51	61.10	61.16	61.22	61.27	61.32	61.38	61.43	61.49	61.55	61.60	61.65	61.71	61.76	61.82	61.88	61.93	61.98	62.04	62.09	62.15
52	61.17	61.23	61.29	61.34	61.39	61.45	61.50	61.56	61.62	61.67	61.72	61.78	61.83	61.89	61.95	62.00	62.05	62.11	62.16	62.22
53	61.24	61.30	61.36	61.41	61.46	61.52	61.57	61.63	61.69	61.74	61.79	61.85	61.90	61.96	62.02	62.07	62.12	62.18	62.23	62.29
54	61.31	61.37	61.43	61.48	61.53	61.59	61.64	61.70	61.76	61.81	61.86	61.92	61.97	62.03	62.09	62.14	62.19	62.25	62.30	62.36
55	61.37	61.43	61.49	61.54	61.60	61.66	61.71	61.76	61.82	61.87	61.93	61.99	62.04	62.09	62.15	62.20	62.26	62.32	62.37	62.43
56	61.44	61.50	61.56	61.61	61.67	61.73	61.78	61.83	61.89	61.94	62.00	62.06	62.11	62.16	62.22	62.27	62.33	62.39	62.44	62.50
57	61.51	61.57	61.63	61.68	61.74	61.80	61.85	61.90	61.96	62.01	62.07	62.13	62.18	62.23	62.29	62.34	62.40	62.46	62.51	62.57
58	61.58	61.64	61.70	61.75	61.81	61.87	61.92	61.97	62.03	62.08	62.14	62.20	62.25	62.30	62.36	62.41	62.47	62.53	62.58	62.64
59	61.65	61.71	61.77	61.82	61.88	61.94	61.99	62.04	62.10	62.15	62.21	62.27	62.32	62.37	62.43	62.48	62.54	62.60	62.65	62.71
15′ 0″	61.72	61.77	61.83	61.88	61.94	62.00	62.05	62.11	62.17	62.22	62.27	62.33	62.38	62.44	62.50	62.55	62.61	62.67	62.72	62.78
1	61.79	61.84	61.90	61.95	62.01	62.07	62.12	62.18	62.24	62.29	62.34	62.40	62.45	62.51	62.57	62.62	62.68	62.74	62.79	62.85
2	61.86	61.91	61.97	62.02	62.08	62.14	62.19	62.25	62.31	62.36	62.41	62.47	62.52	62.58	62.64	62.69	62.75	62.81	62.86	62.92
3	61.93	61.98	62.04	62.09	62.15	62.21	62.26	62.32	62.38	62.43	62.48	62.54	62.59	62.65	62.71	62.76	62.82	62.88	62.93	62.99
4	62.00	62.05	62.11	62.16	62.22	62.28	62.33	62.39	62.45	62.50	62.55	62.60	62.66	62.72	62.78	62.83	62.89	62.95	63.00	63.06
5	62.06	62.12	62.18	62.23	62.28	62.34	62.39	62.45	62.51	62.56	62.62	62.68	62.73	62.79	62.85	62.90	62.95	63.01	63.06	63.12
6	62.13	62.19	62.25	62.30	62.35	62.41	62.46	62.52	62.58	62.63	62.69	62.75	62.80	62.86	62.92	62.97	63.02	63.08	63.13	63.19
7	62.20	62.26	62.32	62.37	62.42	62.48	62.53	62.59	62.65	62.70	62.76	62.82	62.87	62.93	62.99	63.04	63.09	63.15	63.20	63.26
8	62.27	62.33	62.39	62.44	62.49	62.55	62.60	62.66	62.72	62.77	62.83	62.89	62.94	63.00	63.06	63.11	63.16	63.22	63.27	63.33
9	62.34	62.40	62.46	62.51	62.56	62.62	62.67	62.73	62.79	62.84	62.90	62.96	63.01	63.07	63.13	63.18	63.23	63.29	63.34	63.40
10	62.40	62.46	62.52	62.57	62.63	62.69	62.74	62.80	62.86	62.91	62.97	63.03	63.08	63.14	63.20	63.25	63.30	63.36	63.41	63.47
11	62.47	62.53	62.59	62.64	62.70	62.76	62.81	62.87	62.93	62.97	63.04	63.10	63.15	63.21	63.27	63.32	63.37	63.43	63.48	63.54
12	62.54	62.60	62.66	62.71	62.77	62.83	62.88	62.94	63.00	63.05	63.11	63.17	63.22	63.28	63.34	63.39	63.44	63.50	63.55	63.61
13	62.61	62.67	62.73	62.78	62.84	62.90	62.95	63.01	63.07	63.11	63.18	63.24	63.29	63.35	63.41	63.46	63.51	63.57	63.62	63.68
14	62.68	62.74	62.80	62.85	62.91	62.97	63.02	63.08	63.14	63.18	63.25	63.31	63.36	63.42	63.48	63.53	63.58	63.64	63.69	63.75
15	62.75	62.80	62.86	62.91	62.97	63.03	63.08	63.14	63.20	63.25	63.31	63.37	63.42	63.48	63.54	63.59	63.65	63.71	63.76	63.82
16	62.82	62.87	62.93	62.98	63.04	63.10	63.15	63.21	63.27	63.32	63.38	63.44	63.49	63.55	63.61	63.66	63.72	63.78	63.83	63.89
17	62.89	62.94	63.00	63.05	63.11	63.17	63.22	63.28	63.34	63.39	63.45	63.51	63.56	63.62	63.68	63.73	63.79	63.85	63.90	63.96
18	62.96	63.01	63.06	63.12	63.18	63.24	63.29	63.35	63.41	63.46	63.52	63.58	63.63	63.69	63.75	63.80	63.86	63.92	63.97	64.03
19	63.03	63.08	63.13	63.19	63.25	63.31	63.36	63.42	63.48	63.53	63.59	63.65	63.70	63.76	63.82	63.87	63.93	63.99	64.04	64.10
20	63.09	63.15	63.21	63.26	63.32	63.38	63.43	63.49	63.55	63.60	63.66	63.72	63.77	63.83	63.89	63.94	64.00	64.06	64.11	64.17
21	63.16	63.22	63.28	63.33	63.39	63.45	63.50	63.56	63.62	63.67	63.73	63.79	63.84	63.90	63.96	64.01	64.07	64.13	64.18	64.24
22	63.23	63.29	63.35	63.40	63.46	63.52	63.57	63.63	63.69	63.74	63.80	63.86	63.91	63.97	64.03	64.08	64.14	64.20	64.25	64.31
23	63.30	63.36	63.42	63.47	63.53	63.59	63.64	63.70	63.76	63.81	63.87	63.93	63.98	64.05	64.10	64.15	64.21	64.27	64.32	64.38
24	63.37	63.43	63.49	63.54	63.60	63.66	63.71	63.77	63.83	63.88	63.94	64.00	64.05	64.11	64.17	64.22	64.28	64.34	64.39	64.45
25	63.43	63.49	63.55	63.60	63.66	63.72	63.77	63.83	63.89	63.95	64.01	64.07	64.12	64.18	64.24	64.29	64.35	64.41	64.46	64.52
26	63.50	63.56	63.62	63.67	63.73	63.79	63.84	63.90	63.96	64.02	64.08	64.14	64.19	64.25	64.31	64.36	64.42	64.48	64.53	64.59
27	63.57	63.63	63.69	63.74	63.80	63.86	63.91	63.97	64.03	64.09	64.15	64.21	64.26	64.32	64.38	64.43	64.49	64.55	64.60	64.66
28	63.64	63.70	63.76	63.81	63.87	63.93	63.97	64.04	64.10	64.16	64.22	64.28	64.33	64.39	64.45	64.50	64.56	64.62	64.67	64.73
29	63.71	63.77	63.83	63.88	63.94	64.00	64.05	64.11	64.17	64.23	64.29	64.35	64.40	64.46	64.52	64.57	64.64	64.69	64.74	64.80
30	63.77	63.83	63.89	63.95	64.01	64.07	64.12	64.18	64.24	64.29	64.35	64.41	64.46	64.52	64.58	64.64	64.70	64.76	64.81	64.87

Multiply the Tabular Number of Seconds by the Natural Secant of the Moon's Declination North or South, and the Product will be the correct Interval required.

TABLE 9.

Time of the Moon's Semi-diameter passing the Meridian, continued.

☾'s App. Semi-diameter.	5° 10′	5° 20′	5° 30′	5° 40′	5° 50′	6° 0′	6° 10′	6° 20′	6° 30′	6° 40′	6° 50′	7° 0′	7° 10′	7° 20′	7° 30′	7° 40′	7° 50′	8° 0′	8° 10′	8° 20′
15′ 30″	63s.77	63s.83	63s.89	63s.95	64s.01	64s.07	64s.12	64s.18	64s.24	64s.29	64s.35	64s.41	64s.46	64s.52	64s.58	64s.64	64s.70	64s.76	64s.81	64s.87
31	63.84	63.90	63.96	64.02	64.08	64.14	64.19	64.25	64.31	64.36	64.42	64.48	64.53	64.59	64.65	64.71	64.77	64.83	64.88	64.94
32	63.91	63.97	64.03	64.09	64.15	64.21	64.26	64.32	64.38	64.43	64.49	64.55	64.60	64.66	64.72	64.78	64.84	64.90	64.95	65.01
33	63.98	64.04	64.10	64.16	64.22	64.28	64.33	64.39	64.45	64.50	64.56	64.62	64.67	64.73	64.79	64.85	64.91	64.97	65.02	65.08
34	64.05	64.11	64.17	64.23	64.29	64.35	64.40	64.46	64.52	64.57	64.63	64.69	64.74	64.80	64.86	64.92	64.98	65.04	65.09	65.15
35	64.12	64.18	64.24	64.29	64.35	64.41	64.46	64.52	64.58	64.64	64.70	64.76	64.81	64.87	64.93	64.98	65.04	65.10	65.16	65.22
36	64.19	64.25	64.31	64.36	64.42	64.48	64.53	64.59	64.65	64.71	64.77	64.83	64.88	64.94	65.00	65.05	65.11	65.17	65.23	65.29
37	64.26	64.32	64.38	64.43	64.49	64.55	64.60	64.66	64.72	64.78	64.84	64.90	64.95	65.01	65.07	65.12	65.18	65.24	65.30	65.36
38	64.33	64.39	64.45	64.50	64.56	64.62	64.67	64.73	64.79	64.85	64.91	64.97	65.02	65.08	65.14	65.19	65.25	65.31	65.37	65.43
39	64.40	64.46	64.52	64.57	64.63	64.69	64.74	64.80	64.86	64.92	64.98	65.04	65.09	65.15	65.21	65.26	65.32	65.38	65.44	65.50
40	64.46	64.52	64.58	64.64	64.70	64.76	64.81	64.87	64.93	64.98	65.04	65.10	65.16	65.22	65.28	65.33	65.39	65.45	65.51	65.57
41	64.53	64.59	64.65	64.71	64.77	64.83	64.88	64.94	65.00	65.05	65.11	65.17	65.23	65.29	65.35	65.40	65.46	65.52	65.58	65.64
42	64.60	64.66	64.72	64.78	64.84	64.90	64.95	65.01	65.07	65.12	65.18	65.24	65.30	65.36	65.42	65.47	65.53	65.59	65.65	65.71
43	64.67	64.73	64.79	64.85	64.91	64.97	65.02	65.08	65.14	65.19	65.25	65.31	65.37	65.43	65.49	65.54	65.60	65.66	65.72	65.78
44	64.74	64.80	64.86	64.92	64.98	65.04	65.09	65.15	65.21	65.26	65.32	65.38	65.44	65.50	65.56	65.61	65.67	65.72	65.79	65.85
45	64.80	64.86	64.92	64.98	65.04	65.10	65.15	65.21	65.27	65.33	65.39	65.45	65.50	65.56	65.62	65.68	65.74	65.80	65.86	65.92
46	64.87	64.93	64.99	65.05	65.11	65.17	65.22	65.28	65.34	65.40	65.46	65.52	65.57	65.63	65.69	65.75	65.81	65.87	65.93	65.99
47	64.94	65.00	65.06	65.12	65.18	65.24	65.29	65.35	65.41	65.47	65.53	65.59	65.64	65.70	65.76	65.82	65.88	66.94	66.00	66.06
48	65.01	65.07	65.13	65.19	65.25	65.31	65.36	65.42	65.48	65.54	65.60	65.66	65.71	65.77	65.83	65.89	65.95	66.01	66.07	66.13
49	65.08	65.14	65.20	65.26	65.32	65.38	65.43	65.49	65.55	65.61	65.67	65.73	65.78	65.54	65.90	65.96	66.02	66.08	66.14	66.20
50	65.15	65.21	65.27	65.32	65.38	65.44	65.50	65.56	65.62	65.68	65.74	65.80	65.85	65.91	65.97	66.03	66.09	66.15	66.20	66.26
51	65.22	65.28	65.34	65.39	65.45	65.51	65.57	65.63	65.69	65.75	65.81	65.87	65.92	65.98	66.04	66.10	66.16	66.22	66.27	66.33
52	65.29	65.35	65.41	65.46	65.52	65.58	65.64	65.70	65.76	65.82	65.88	65.94	65.99	66.05	66.11	66.17	66.23	66.29	66.34	66.40
53	65.36	65.42	64.48	65.53	65.59	65.65	65.71	65.77	65.83	65.89	65.95	66.01	66.06	66.12	66.18	66.24	66.30	66.36	66.41	66.47
54	65.43	65.49	65.55	65.60	65.66	65.72	65.78	65.84	65.90	65.96	66.02	66.08	66.13	66.19	66.25	66.31	66.37	66.43	66.48	66.54
55	65.49	65.55	65.61	65.67	65.73	65.79	65.85	65.91	65.97	66.02	66.08	66.14	66.20	66.26	66.32	66.38	66.44	66.50	66.55	66.61
56	65.56	65.62	65.68	65.74	65.80	65.86	65.92	65.98	66.04	66.09	66.15	66.21	66.27	66.33	66.39	66.45	66.51	66.57	66.62	66.68
57	65.63	65.69	65.75	65.81	65.87	65.93	65.99	66.05	66.10	66.16	66.22	66.28	66.34	66.40	66.46	66.52	66.58	66.64	66.69	66.75
58	65.70	65.76	65.82	65.88	65.94	66.00	66.06	66.12	66.17	66.23	66.29	66.35	66.41	66.47	66.53	66.59	66.65	66.71	66.76	66.82
59	65.77	65.83	65.89	65.95	66.01	66.07	66.13	66.19	66.24	66.30	66.36	66.42	66.48	66.54	66.60	66.66	66.72	66.78	66.83	66.89
16 0	65.84	65.90	65.96	66.01	66.07	66.13	66.19	66.25	66.31	66.37	66.43	66.49	66.55	66.61	66.67	66.72	66.78	66.84	66.90	66.96
1	65.91	65.97	66.03	66.08	66.14	66.20	66.26	66.32	66.38	66.44	66.50	66.56	66.62	66.68	66.74	66.79	66.85	66.91	66.97	67.03
2	65.98	66.04	66.10	66.15	66.21	66.27	66.33	66.39	66.45	66.51	66.57	66.63	66.69	66.75	66.81	66.86	66.92	66.98	67.04	67.10
3	66.05	66.11	66.17	66.22	66.28	66.34	66.40	66.46	66.52	66.58	66.64	66.70	66.76	66.82	66.88	66.93	66.99	67.06	67.11	67.17
4	66.12	66.18	66.24	66.29	66.35	66.41	66.47	66.55	66.59	66.65	66.71	66.77	66.83	66.89	66.95	67.00	67.06	67.13	67.18	67.24
5	66.18	66.24	66.30	66.36	66.42	66.48	66.54	66.60	66.66	66.72	66.78	66.84	66.89	66.95	67.01	67.07	67.13	67.19	67.25	67.31
6	66.25	66.31	66.37	66.43	66.49	66.55	66.61	66.67	66.73	66.79	66.85	66.91	66.96	67.02	67.08	67.14	67.20	67.26	67.32	67.38
7	66.32	66.38	66.44	66.50	66.56	66.62	66.68	66.74	66.80	66.86	66.92	66.98	67.03	67.09	67.15	67.21	67.27	67.33	67.39	67.45
8	66.39	66.45	66.51	66.57	66.63	66.69	66.75	66.81	66.87	66.93	66.99	67.05	67.10	67.16	67.22	67.28	67.34	67.40	67.46	67.52
9	66.46	66.52	66.58	66.64	66.70	66.76	66.82	66.88	66.94	67.00	67.06	67.12	67.17	67.23	67.29	67.35	67.41	67.47	67.53	67.59
10	66.52	66.58	66.64	66.70	66.76	66.82	66.88	66.94	67.00	67.06	67.12	67.18	67.24	67.30	67.36	67.42	67.48	67.54	67.60	67.66
11	66.59	66.65	66.71	66.77	66.83	66.89	66.95	67.01	67.07	67.13	67.19	67.25	67.31	67.37	67.43	67.49	67.55	67.61	67.67	67.73
12	66.66	66.72	66.78	66.84	66.90	66.96	67.02	67.08	67.14	67.20	67.26	67.32	67.38	67.44	67.50	67.56	67.62	67.68	67.74	67.80
13	66.73	66.79	66.85	66.91	66.97	67.03	67.09	67.15	67.21	67.27	67.33	67.39	67.45	67.51	67.57	67.63	67.69	67.75	67.81	67.87
14	66.80	66.86	66.92	66.98	67.04	67.10	67.16	67.22	67.28	67.34	67.40	67.46	67.52	67.58	67.64	67.70	67.76	67.82	67.88	67.94
15	66.86	66.92	66.98	67.04	67.10	67.17	67.23	67.29	67.35	67.41	67.47	67.53	67.59	67.65	67.71	67.77	67.83	67.89	67.95	68.01
16	66.93	66.99	67.05	67.11	67.17	67.24	67.30	67.36	67.42	67.48	67.54	67.60	67.66	67.72	67.78	67.84	67.90	67.96	68.02	68.08
17	67.00	67.06	67.12	67.18	67.24	67.31	67.37	67.43	67.49	67.55	67.61	67.67	67.73	67.79	67.85	67.91	67.97	68.03	68.09	68.15
18	67.07	67.13	67.19	67.25	67.31	67.38	67.44	67.50	67.56	67.62	67.68	67.74	67.80	67.86	67.92	67.98	68.04	68.10	68.16	68.22
19	67.14	67.20	67.26	67.32	67.38	67.45	67.51	67.57	67.63	67.69	67.75	67.81	67.87	67.93	67.99	68.05	68.11	68.17	68.23	68.29
20	67.21	67.27	67.33	67.39	67.45	67.51	67.57	67.63	67.69	67.75	67.81	67.87	67.93	67.99	68.05	68.11	68.17	68.24	68.30	68.36
21	67.28	67.34	67.40	67.46	67.52	67.58	67.64	67.70	67.76	67.82	67.88	67.94	68.00	68.06	68.12	68.18	68.24	68.31	68.37	68.43
22	67.35	67.41	67.47	67.53	67.59	67.65	67.71	67.77	67.83	67.89	67.95	68.01	68.07	68.13	68.19	68.25	68.31	68.38	68.44	68.50
23	67.42	67.48	67.54	67.60	67.66	67.72	67.78	67.84	67.90	67.96	68.02	68.08	68.14	68.20	68.26	68.32	68.38	68.45	68.51	68.57
24	67.49	67.55	67.61	67.66	67.73	67.79	67.85	67.91	67.97	68.03	68.09	68.15	68.21	68.27	68.33	68.39	62.45	68.52	68.58	68.64
25	67.55	67.61	67.67	67.73	67.79	67.86	67.92	67.98	68.04	68.10	68.16	68.22	68.28	68.34	68.40	68.46	68.52	68.58	68.64	68.70
26	67.62	67.68	67.74	67.80	67.86	67.93	67.99	68.05	68.11	68.17	68.23	68.29	68.35	68.41	68.47	68.53	68.59	68.65	68.71	68.77
27	67.69	67.75	67.81	67.87	67.93	68.00	68.06	68.12	68.18	68.24	68.30	68.36	68.42	68.48	68.54	68.60	68.66	68.72	68.78	68.84
28	67.76	67.82	67.88	67.94	68.00	68.07	68.13	68.19	68.25	68.31	68.37	68.43	68.49	68.55	68.61	68.67	68.73	68.79	68.85	68.91
29	67.83	67.89	67.95	68.01	68.07	68.14	68.20	68.26	68.32	68.38	68.44	68.50	68.56	68.62	68.68	68.74	68.80	68.86	68.92	68.98
30	67.89	67.95	68.02	68.08	68.14	68.20	68.26	68.32	68.38	68.44	68.50	68.57	68.63	68.69	68.75	68.81	68.87	68.93	68.99	69.05

Multiply the Tabular Number of Seconds by the Natural Secant of the Moon's Declination North or South, and the Product will be the correct Interval required.

TABLE 9.

TIME of the MOON'S SEMI-DIAMETER passing the MERIDIAN, concluded.

☾'s App. Semi-diameter.	5° 10′	5° 20′	5° 30′	5° 40′	5° 50′	6° 0′	6° 10′	6° 20′	6° 30′	6° 40′	6° 50′	7° 0′	7° 10′	7° 20′	7° 30′	7° 40′	7° 50′	8° 0′	8° 10′	8° 20′
16° 30″	67.89	67.95	68.02	68.08	68.14	68.20	68.26	68.32	68.38	68.44	68.50	68.57	68.63	68.69	68.75	68.81	68.87	68.93	68.99	69.05
31	67.96	68.02	68.09	68.15	68.21	68.27	68.33	68.39	68.45	68.51	68.57	68.64	68.70	68.76	68.82	68.88	68.94	69.00	69.06	69.11
32	68.03	68.09	68.16	68.22	68.28	68.34	68.40	68.46	68.52	68.58	68.64	38.71	68.77	68.83	68.89	68.95	69.01	69.07	69.13	69.18
33	68.10	68.16	68.23	68.29	68.35	68.41	68.47	68.53	68.59	68.65	68.71	68.78	68.84	68.90	68.96	69.02	69.08	69.14	69.20	69.25
34	68.17	68.23	68.30	68.36	68.42	68.48	68.54	68.60	68.66	68.72	68.73	68.85	68.91	68.97	69.03	69.09	69.15	69.21	69.27	69.32
35	68.24	68.30	68.36	68.42	68.48	68.54	68.60	68.66	68.73	68.79	68.85	68.91	68.97	69.03	69.10	69.16	69.22	69.28	69.34	69.39
36	68.31	68.37	68.43	68.49	68.55	68.61	68.67	68.73	68.80	68.86	68.92	68.98	69.04	69.10	69.17	69.23	69.29	69.35	69.41	69.46
37	68.38	68.44	68.50	68.56	68.62	68.68	68.74	68.80	68.87	68.93	68.99	69.05	69.11	69.17	69.24	69.30	69.36	69.42	69.48	69.53
38	68.45	68.51	68.57	68.63	68.69	68.75	68.81	68.87	68.94	69.00	69.06	69.12	69.18	69.24	69.31	69.37	69.43	69.49	69.55	69.60
39	68.52	68.58	68.64	68.70	68.76	68.82	68.88	68.94	69.01	69.07	69.13	69.19	69.25	69.31	69.38	69.44	69.50	69.56	69.62	69.67
40	68.58	68.64	68.70	68.76	68.82	68.89	68.95	69.01	69.07	69.13	69.19	69.26	69.32	69.38	69.44	69.50	69.56	69.63	69.69	69.75
41	68.65	68.71	68.77	68.83	68.89	68.96	69.02	69.08	69.14	69.20	69.26	69.33	69.39	69.45	66.51	69.57	69.63	69.70	69.76	69.82
42	68.72	68.78	68.84	68.90	68.96	69.03	69.09	69.15	69.21	69.27	69.33	69.40	69.46	69.52	69.58	69.64	69.70	69.77	69.83	69.89
43	68.79	68.85	68.91	68.97	69.03	69.10	69.16	69.22	69.28	69.34	69.40	69.47	69.53	69.59	69.65	69.71	69.77	69.84	69.90	69.96
44	68.86	68.92	68.98	69.04	69.10	69.17	69.23	69.29	69.35	69.41	69.47	69.54	69.60	69.66	69.72	69.78	69.84	69.91	69.97	70.03
45	68.92	68.98	69.05	69.11	69.17	69.23	69.29	69.35	69.42	69.48	69.54	69.60	69.66	69.72	69.79	69.85	69.91	69.98	70.04	70.10
46	68.99	69.05	69.12	69.18	69.24	69.30	69.36	69.42	69.49	69.55	69.61	69.67	69.73	69.79	69.86	69.92	69.98	70.05	70.11	70.17
47	69.06	69.12	69.19	69.25	69.31	69.37	69.43	69.49	69.56	69.62	69.68	69.74	69.80	69.86	62.93	69.99	70.05	70.12	70.18	70.24
48	69.13	69.19	69.26	69.32	69.38	69.44	69.50	69.56	69.63	59.69	69.75	69.81	69.87	69.93	70.00	70.06	70.12	70.19	70.25	70.31
49	69.20	69.26	69.33	69.39	69.45	69.51	69.57	69.63	69.70	69.76	69.82	69.88	69.94	70.00	70.07	70.13	70.19	70.26	70.32	70.38
50	69.26	69.32	69.39	69.45	69.51	69.58	69.64	69.70	69.76	69.82	69.88	69.95	70.01	70.07	70.14	70.20	70.26	70.33	70.39	70.45
51	69.33	69.39	69.46	69.52	69.58	69.65	69.71	69.77	69.83	69.89	69.95	70.02	70.08	70.14	70.21	70.27	70.33	70.40	70.46	70.52
52	69.40	69.46	69.53	69.59	69.65	69.72	68.78	69.84	69.90	69.96	70.02	70.09	70.15	70.21	70.28	70.34	70.40	70.47	70.53	70.59
53	69.47	69.53	69.60	69.66	69.72	69.79	69.85	69.91	69.97	70.03	70.09	70.16	70.22	70.28	70.35	70.41	70.47	70.54	70.60	70.66
54	69.54	69.60	69.67	69.73	69.79	69.86	69.92	69.98	70.04	70.10	70.16	70.23	70.29	70.35	70.42	70.48	70.54	70.61	70.67	70.73
55	69.61	69.67	69.73	69.79	69.85	69.92	69.98	70.04	70.11	70.17	70.23	70.30	70.36	70.42	70.49	70.55	70.61	70.67	70.73	70.79
56	69.68	69.74	69.80	69.86	69.92	69.99	70.05	70.11	70.18	70.24	70.30	70.37	70.43	70.49	70.56	70.62	70.68	70.74	70.80	70.86
57	69.75	69.81	69.87	69.93	69.99	70.06	70.12	70.18	70.25	70.31	70.37	70.44	70.50	70.56	70.63	70.69	70.75	70.81	70.87	70.93
58	69.82	69.88	69.94	70.00	70.06	70.13	70.19	70.25	70.32	70.38	70.44	70.51	70.57	70.65	70.70	70.76	70.82	70.88	70.94	71.00
59	69.89	69.95	70.01	70.07	70.13	70.20	70.26	70.32	70.39	70.45	70.51	70.58	70.64	70.72	70.77	70.83	70.89	70.95	71.01	71.07
17 0	69.95	70.01	70.08	70.14	70.20	70.27	70.33	70.39	70.46	70.52	70.58	70.64	70.70	70.76	70.83	70.89	70.95	71.02	71.08	71.14
1	70.02	70.08	70.15	70.21	70.26	70.34	70.40	70.46	70.53	70.59	70.65	70.71	70.77	70.83	70.90	70.96	71.02	71.09	71.15	71.21
2	70.09	70.15	70.22	70.28	70.34	70.41	70.47	70.53	70.60	70.66	70.72	70.78	70.84	70.90	70.97	71.03	71.09	71.16	71.22	71.28
3	70.16	70.22	70.29	70.35	70.41	70.48	70.54	70.60	70.67	70.73	70.79	70.85	70.91	70.97	71.04	71.10	71.16	71.23	71.29	71.35
4	70.23	70.29	70.26	70.42	70.48	70.55	70.61	70.67	70.74	70.80	70.86	70.92	70.98	71.04	71.11	71.17	71.23	71.30	71.36	71.42
5	70.30	70.36	70.42	70.48	70.54	70.61	70.67	70.73	70.80	70.86	70 92	70.99	71.05	71.11	71.18	71.24	71.30	71.36	71.42	71.48
6	70.37	70.43	70.49	70.55	70.61	70.68	70.73	70.80	70.87	70.93	70.99	71.06	71.12	71.18	71.25	71.31	71.37	71.43	71.49	71.55
7	70.44	70.50	70.56	70.62	70.68	70.75	70.81	70.88	70.94	71.00	71.06	71.13	71.19	71.25	71.32	71.38	71.44	71.50	71.56	71.62
8	70.51	70.57	70.63	70.69	70.75	70.82	70.88	70.95	71.01	71.07	71.13	71.20	71.26	71.32	71.39	71.45	71.51	71.57	71.63	71.69
9	70.58	70.64	70.70	70.76	70.82	70.89	71.95	71.02	71.08	71.14	71.20	71.27	71.33	71.39	71.46	71.52	71.58	71.64	71.70	71.76
10	70.64	70.70	70.77	70.83	70.89	70.96	71.02	71.08	71.15	71.21	71.27	71.33	71.39	71.45	71.52	71.58	71.64	71.71	71.77	71.83
11	70.71	70.77	70.84	70.90	70.96	71.03	71.09	71.15	71.22	71.28	71.34	71.40	71.46	71.52	71.59	71.65	71.71	71.78	71.84	71.90
12	70.78	70.84	70.91	70.97	71.03	71.10	71.16	71.22	71.29	71.35	71.41	71.47	71.53	71.59	71.66	71.72	71.78	71.85	71.91	71.97
13	70.85	70.91	70.98	71.04	71.10	71.17	71.23	71.29	71.36	71.42	71.48	71.54	71.60	71.66	71.73	71.79	71.85	71.92	71.98	72.04
14	70.92	70.98	71.05	71.11	71.17	71.24	71.30	71.36	71.43	71.49	71.55	71.61	71.67	71.73	71.80	71.86	71.92	71.99	72.05	72.11
15	70.99	71.05	71.11	71.17	71.23	71.30	71.36	71.42	71.49	71.55	71.61	71.68	71.74	71.80	71.87	71.93	71.99	72.05	72.11	72.17
16	71.06	71.12	71.18	71.24	71.30	71.37	71.43	71.49	71.56	71.62	71.68	71.75	71.81	71.87	71.94	72.00	72.06	72.12	72.18	72.24
17	71.13	71.19	71.25	71.31	71.37	71.44	71.50	71.56	71.63	71.69	71.75	71.82	71.88	71.94	72.01	72.07	72.13	72.19	72.25	72.31
18	71.20	71.26	71.32	71.38	71.44	71.51	71.57	71.63	71.70	71.76	71.82	71.89	71.95	72.01	72.08	72.14	72.20	72.26	72.32	72.38
19	71.27	71.33	71.39	71.45	71.51	71.58	71.64	71.70	71.77	71.83	71.89	71.96	72.02	72.08	72.15	72.21	72.27	72.33	72.39	72.45
20	71.33	71.39	71.46	71.52	71.58	71.65	71.71	71.77	71.84	71.90	71.96	72.02	72.08	72.14	72.21	72.27	72.33	72.40	72.46	72.52
21	71.40	71.46	71.53	71.59	71.65	71.72	71.78	71.84	71.91	71.97	72.03	72.09	72.15	72.21	72.28	72.34	72.40	72.47	72.53	72.59
22	71.47	71.53	71.60	71.66	71.72	71.79	71.85	71.91	71.98	72.04	72.10	72.16	72.22	72.28	72.35	72.41	72.47	72.54	72.60	72.66
23	71.54	71.60	71.67	71.73	71.79	71.86	71.92	71.98	72.05	72.11	72.17	72.23	72.29	72.35	72.42	72.48	72.54	72.61	72.67	72.73
24	71.61	71.67	71.74	71.80	71.86	71.93	71.99	72.05	72.12	72.18	72.24	72.30	72.36	72.42	72.49	72.55	72.61	72.68	72.74	72.80
25	71.68	71.74	71.80	71.86	71.92	71.99	72.05	72.11	72.18	72.24	72.30	72.37	72.43	72.49	72.56	72.62	72.68	72.74	72.80	72.86
26	71.75	71.81	71.87	71.93	71.99	72.06	72.12	72.18	72.25	72.31	72.37	72.44	72.50	72.56	72.63	72.69	72.75	72.81	72.87	72.93
27	71.82	71.88	71.94	72.00	72.06	72.13	72.19	72.25	72.32	72.38	72.44	72.51	72.57	72.63	72.70	72.76	72.82	72.88	72.94	73.00
28	71.89	71.95	72.01	72.07	72.13	72.20	72.26	72.32	72.39	72.45	72.51	72.58	72.64	72.70	72.77	72.83	72.89	72.95	73.01	73.07
29	71.96	72.02	72.08	72.14	72.20	72.27	72.33	72.39	72.46	72.52	72.58	72.65	72.71	72.77	72.84	72.90	72.96	73.02	73.08	73.14
30	72.02	72.08	72.15	72.21	72.27	72.34	72.40	72.46	72.53	72.59	72.65	72.71	72.77	72.83	72.90	72.96	73.02	73.09	73.15	73.21

Multiply the Tabular Number of Seconds by the Natural Secant of the Moon's Declination, North or South, and the Product will be the correct Interval required.

TABLE 10. SECONDS of the LUNAR DISC, that pass the Meridian in ONE SECOND of SIDEREAL TIME. (BURG's Table extended.)

Arguments = Moon's Declination at the Side, and her Motion in Right Ascension in 12 Solar Hours, at the Head.

☾'s Declination.	5° 10′	5° 20′	5° 30′	5° 40′	5° 50′	6° 0′	6° 10′	6° 20′	6° 30′	6° 40′	6° 50′	7° 0′	7° 10′	7° 20′	7° 30′	7° 40′	7° 50′	8° 0′	8° 10′	8° 20′
0° 0	14s.57	14s.55	14s.54	14s.52	14s.51	14s.50	14s.48	14s.47	14s.46	14s.44	14s.43	14s.42	14s.40	14s.39	14s.38	14s.36	14s.35	14s.33	14s.31	14s.30
0 30	14.57	14.55	14.54	14.52	14.51	14.50	14.48	14.47	14.46	14.44	14.43	14.42	14.40	14.39	14.38	14.36	14.35	14.33	14.31	14.30
1 0	14.57	14.55	14.54	14.52	14.51	14.50	14.48	14.47	14.46	14.44	14.43	14.42	14.40	14.39	14.38	14.36	14.35	14.33	14.31	14.30
1 30	14.56	14.54	14.53	14.51	14.50	14.49	14.47	14.46	14.45	14.44	14.42	14.41	14.39	14.38	14.37	14.35	14.34	14.32	14.30	14.29
2 0	14.56	14.54	14.53	14.51	14.50	14.49	14.47	14.46	14.45	14.44	14.42	14.41	14.39	14.38	14.37	14.35	14.34	14.32	14.30	14.29
2 30	14.55	14.53	14.52	14.50	14.49	14.48	14.46	14.45	14.44	14.43	14.41	14.40	14.38	14.37	14.36	14.34	14.33	14.31	14.29	14.28
3 0	14.55	14.53	14.52	14.50	14.49	14.48	14.46	14.45	14.44	14.43	14.41	14.40	14.38	14.37	14.36	14.34	14.33	14.31	14.29	14.28
3 30	14.54	14.52	14.51	14.49	14.48	14.47	14.45	14.44	14.43	14.42	14.40	14.39	14.37	14.36	14.35	14.33	14.32	14.30	14.28	14.27
4 0	14.53	14.51	14.50	14.48	14.47	14.46	14.44	14.43	14.42	14.41	14.39	14.38	14.36	14.35	14.34	14.32	14.31	14.29	14.27	14.26
4 30	14.52	14.50	14.49	14.47	14.46	14.45	14.43	14.42	14.41	14.40	14.38	14.37	14.35	14.34	14.33	14.31	14.30	14.28	14.26	14.25
5 0	14.51	14.49	14.48	14.46	14.45	14.44	14.42	14.41	14.40	14.39	14.37	14.36	14.34	14.33	14.32	14.30	14.29	14.27	14.25	14.24
5 30	14.50	14.48	14.47	14.45	14.44	14.43	14.41	14.40	14.39	14.38	14.36	14.35	14.33	14.32	14.31	14.29	14.28	14.26	14.24	14.23
6 0	14.49	14.47	14.46	14.44	14.43	14.42	14.40	14.39	14.38	14.37	14.35	14.33	14.32	14.31	14.30	14.28	14.27	14.25	14.23	14.22
6 0	14.48	14.46	14.45	14.43	14.42	14.41	14.39	14.38	14.37	14.36	14.34	14.32	14.31	14.30	14.29	14.27	14.26	14.24	14.22	14.21
7 0	14.46	14.44	14.43	14.41	14.40	14.39	14.37	14.36	14.35	14.34	14.32	14.30	14.29	14.28	14.27	14.25	14.24	14.22	14.20	14.19
7 30	14.45	14.43	14.42	14.40	14.39	14.38	14.36	14.35	14.34	14.33	14.31	14.29	14.28	14.27	14.26	14.24	14.23	14.21	14.19	14.18
8 0	14.43	14.41	14.40	14.38	14.37	14.36	14.34	14.33	14.32	14.31	14.29	14.27	14.26	14.25	14.24	14.22	14.21	14.19	14.17	14.16
8 30	14.41	14.39	14.38	14.36	14.35	14.34	14.32	14.31	14.30	14.29	14.27	14.26	14.24	14.23	14.22	14.20	14.19	14.18	14.16	14.14
9 0	14.39	14.37	14.36	14.34	14.33	14.32	14.30	14.29	14.28	14.27	14.25	14.24	14.22	14.21	14.20	14.18	14.17	14.16	14.14	14.12
9 30	14.37	14.35	14.34	14.32	14.31	14.30	14.28	14.27	14.26	14.25	14.23	14.22	14.20	14.19	14.18	14.16	14.15	14.14	14.12	14.10
10 0	14.35	14.33	14.32	14.30	14.29	14.28	14.26	14.25	14.24	14.23	14.21	14.20	14.18	14.17	14.16	14.14	14.13	14.12	14.10	14.08
10 30	14.33	14.31	14.30	14.28	14.27	14.26	14.24	14.23	14.22	14.21	14.19	14.18	14.16	14.15	14.14	14.12	14.11	14.10	14.08	14.06
11 0	14.30	14.28	14.27	14.25	14.24	14.23	14.21	14.20	14.19	14.18	14.16	14.15	14.13	14.12	14.11	14.09	14.08	14.07	14.05	14.03
11 30	14.28	14.26	14.25	14.23	14.22	14.21	14.19	14.18	14.17	14.16	14.14	14.13	14.11	14.10	14.09	14.07	14.06	14.05	14.03	14.01
12 0	14.25	14.23	14.22	14.20	14.19	14.18	14.16	14.15	14.14	14.13	14.11	14.10	14.08	14.07	14.06	14.04	14.03	14.02	14.00	13.99
12 30	14.23	14.21	14.20	14.18	14.17	14.16	14.14	14.13	14.12	14.11	14.09	14.08	14.06	14.05	14.03	14.02	14.01	13.99	13.98	13.97
13 0	14.20	14.18	14.17	14.15	14.14	14.13	14.11	14.10	14.09	14.08	14.06	14.05	14.03	14.02	14.00	13.98	13.97	13.96	13.95	13.94
13 30	14.17	14.15	14.14	14.12	14.11	14.10	14.08	14.07	14.06	14.05	14.03	14.02	14.00	13.99	13.97	13.96	13.94	13.93	13.92	13.91
14 0	14.14	14.12	14.11	14.09	14.08	14.07	14.05	14.04	14.03	14.02	14.00	13.99	13.97	13.96	13.94	13.93	13.91	13.90	13.89	13.88
14 30	14.11	14.09	14.08	14.06	14.05	14.04	14.02	14.01	14.00	13.99	13.97	13.96	13.94	13 93	13.91	13.90	13.88	13.87	13.86	13.85
15 0	14.08	14.06	14.05	14.03	14.02	14.01	13.99	13.98	13.97	13.96	13.94	13.93	13.91	13.90	13.88	13.87	13.85	13.84	13.83	13.82
15 30	14.05	14.03	14.02	14.00	13.99	13.98	13.96	13.95	13.94	13.93	13.91	13.90	13.88	13.87	13.85	13.84	13.82	13.81	13.80	13.79
16 0	14.01	13.99	13.98	13.96	13.95	13.94	13.92	13.91	13.90	13.89	13.88	13.86	13.84	13.83	13.81	13.80	13.78	13.77	13.76	13.75
16 30	13.97	13.96	13.95	13.93	13.92	13.91	13.89	13.88	13.87	13.86	13.85	13.83	13.81	13.80	13.78	13.77	13.75	13.74	13.73	13.72
17 0	13.93	13.92	13.91	13.89	13.88	13.87	13.85	13.84	13.83	13.82	13.81	13.79	13.77	13.76	13.74	13.73	13.71	13.70	13.69	13.68
17 30	13.89	13.88	13.87	13.86	13.84	13.83	13.81	13.80	13.79	13.78	13.77	13.75	13.73	13.72	13.71	13.69	13.68	13.67	13.66	13.64
18 0	13.85	13.84	13.83	13.82	13.80	13.79	13.78	13.76	13.75	13.74	13.73	13.71	13.69	13.68	13.67	13.66	13.64	13.63	13.62	13.60
18 30	13.81	13.80	13.79	13.78	13.77	13.75	13.74	13.73	13.71	13.69	13.68	13.67	13.66	13.64	13.63	13.62	13.61	13.59	13.58	13.56
19 0	13.77	13.76	13.75	13.73	13.72	13.71	13.69	13.68	13.67	13.65	13.64	13.63	13.62	13.60	13.59	13.58	13.57	13.55	13.54	13.52
19 30	13.73	13.72	13.71	13.69	13.68	13.67	13.65	13.64	13.63	13.61	13.60	13.59	13 58	13.56	13.55	13.54	13.52	13.51	13.50	13.48
20 0	13.69	13.68	13.67	13.65	13.64	13.63	13.61	13.60	13.59	13.57	13.56	13.55	13.53	13.52	13.51	13.50	13.48	13.47	13.46	13.44
20 30	13.65	13.64	13.63	13.61	13.60	13.59	13.57	13.56	13.55	13.53	13.52	13.51	13.49	13.48	13.47	13.46	13.44	13.43	13.42	13.40
21 0	13.60	13.59	13.58	13.56	13.55	13.54	13.52	13.51	13.50	13.48	13.47	13.46	13.44	13.43	13.42	13.41	13.39	13.38	13.37	13.35
21 30	13.56	13.55	13.54	13.52	13.51	13.50	13.48	13.47	13.46	13.44	13.43	13.42	13.40	13.39	13.38	13.37	13.35	13.34	13.33	13.31
22 0	13.51	13.50	13.49	13.47	13.46	13.45	13.43	13.42	13.41	13.39	13.38	13.37	13.35	13.34	13.33	13.32	13.30	13.29	13.28	13.26
22 30	13.46	13.45	13.44	13.42	13.41	13.40	13.38	13.37	13.36	13.34	13.33	13.32	13.30	13.29	13.28	13.27	13.25	13.24	13.23	13.21
23 0	13.41	13.40	13.39	13.37	13.36	13.35	13.33	13.32	13.31	13.29	13.28	13.27	13.25	13.24	13.23	13.22	13.20	13.19	13.18	13.16
23 30	13.36	13.35	13.34	13.32	13.31	13.30	13.28	13.27	13.26	13.24	13.23	13.22	13.20	13.19	13.18	13.17	13.15	13.14	13.13	13.11
24 0	13.31	13.30	13.29	13.27	13.26	13.25	13.23	13.22	13.21	13.19	13.18	13.17	13.15	13.14	13.13	13.12	13.10	13.09	13.08	13.06
24 30	13.26	13.25	13.24	13.22	13.21	13.20	13.18	13.17	13.16	13.14	13.13	13.12	13.11	13.09	13.08	13.07	13.05	13.04	13.03	13.01
25 0	13.20	13.19	13.18	13.16	13.15	13.14	13.12	13.11	13.10	13.09	13.08	13.07	13.06	13.04	13.03	13.02	13.00	12.99	12.98	12.96
25 30	13.15	13.14	13.13	13.11	13.10	13.09	13.07	13.06	13.05	13.04	13.03	13.02	13.00	12.99	12.98	12.97	12.95	12.94	12.93	12.91
26 0	13.09	13.08	13.07	13.05	13.04	13.03	13.01	13.00	12.99	12.98	12.97	12.96	12.94	12.93	12.92	12.91	12.90	12.88	12.87	12.85
26 30	13.04	13.03	13.02	13.00	12.99	12.98	12.96	12.95	12.94	12.93	12.92	12.91	12.89	12.88	12.87	12.86	12.85	12.83	12.82	12.80
27 0	12.98	12.97	12.96	12.94	12.93	12.92	12.90	12.89	12.88	12.87	12.86	12.85	12.83	12.82	12.81	12.80	12.79	12.77	12.76	12.74
27 30	12.92	12.91	12.90	12.88	12.87	12.86	12.84	12.83	12.82	12.81	12.80	12.79	12.77	12.76	12.75	12.74	12.73	12.72	12.71	12.69
28 0	12.86	12.85	12.84	12.82	12.81	12.80	12.78	12.77	12.76	12.75	12.74	12.73	12.71	12.70	12.69	12.68	12.67	12.66	12.65	12.63
28 30	12.80	12.79	12.78	12.76	12.75	12.74	12.72	12.71	12.70	12.69	12.68	12.67	12.65	12.64	12.63	12.62	12.61	12.60	12.59	12.57
29 0	12.74	12.73	12.72	12.70	12.69	12.68	12.66	12.65	12.64	12.63	12.62	12.61	12.59	12.58	12.57	12.56	12.55	12.54	12.53	12.51
29 30	12.68	12.67	12.66	12.64	12.63	12.62	12.60	12.59	12.58	12.57	12.56	12.55	12.53	12.52	12.51	12.50	12.49	12.48	12.47	12.45
30 0	12.62	12.60	12.59	12.58	12.57	12.56	12.54	12.53	12.52	12.51	12.50	12.49	12.47	12.46	12.45	12.44	12.43	12.42	12.41	12.39

Divide the true apparent Semi-diameter of the Moon by the proper Tabular Number, and the Quotient will give the Measure of the Semi-diameter's passage in Sidereal Seconds.

TABLE II.—INTERPOLATION by SECOND DIFFERENCES derived from Intervals of 12 Hours. Arguments = the Mean of the 2 nearest Second Differences at the Head; and the Time for which the Correction is required from Noon or Midnight, at the Side.

MINUTES OF SECOND DIFFERENCES.

h.	m.	h.	m.	1′	2′	3′	4′	5′	6′	7′	8′	9′	10′	11′	12′	13′	14′	15′	h.	m.	h.	m.
0	0	12	0	0″.0	0″.0	0.0	0″.0	0″.0	0.0	0″.0	0″.0	0″.0	0.0	0″.0	0.0	0″.0	0.0	0.0	12	0	0	0
0	10	11	50	0.4	0.8	1.2	1.6	2.0	2.4	2.9	3.3	3.7	4.1	4.5	4.9	5.3	5.8	6.2	11	50	0	10
0	20	11	40	0.8	1.6	2.4	3.2	4.1	4.9	5.7	6.5	7.3	8.1	8.9	9.7	11.6	11.4	12.2	11	40	0	20
0	30	11	30	1.2	2.4	3.6	4.8	6.0	7.2	8.4	9.6	10.8	12.0	13.2	14.4	15.6	16.8	18.0	11	30	0	30
0	40	11	20	1.6	3.1	4.7	6.3	7.9	9.4	11.0	12.6	14.2	15.7	17.3	18.9	20.4	22.0	23.6	11	20	0	40
0	50	11	10	1.9	3.9	5.8	7.8	9.7	11.6	13.6	15.5	17.4	19.4	21.3	23.3	25.2	27.2	29.1	11	10	0	50
1	0	11	0	2.3	4.6	6.9	9.2	11.5	13.8	16.0	18.3	20.6	22.9	25.2	27.5	29.8	32.0	34.3	11	0	1	0
1	10	10	50	2.6	5.3	7.9	10.5	13.2	15.8	18.4	21.1	23.7	26.3	29.0	31.6	34.2	36.8	39.5	10	50	1	10
1	20	10	40	3.0	5.9	8.9	11.9	14.8	17.8	20.7	23.7	26.7	29.6	32.6	35.6	38.5	41.4	44.4	10	40	1	20
1	30	10	30	3.3	6.6	9.8	13.1	16.4	19.7	23.0	26.3	29.5	32.8	36.1	39.4	41.7	46.0	49.3	10	30	1	30
1	40	10	20	3.6	7.2	10.8	14.4	17.9	21.5	25.1	28.7	32.3	35.9	39.5	43.1	46.6	50.2	53.8	10	20	1	40
1	50	10	10	3.9	7.8	11.6	15.5	19.4	23.3	27.2	31.1	34.9	38.8	42.7	46.6	50.5	54.4	58.3	10	10	1	50
2	0	10	0	4.2	8.3	12.5	16.7	20.8	25.0	29.2	33.3	37.5	41.7	45.8	50.0	54.2	58.4	62.5	10	0	2	0
2	10	9	50	4.4	8.9	13.3	17.8	22.2	26.6	31.1	35.5	39.9	44.4	48.8	53.3	57.7	62.2	66.6	9	50	2	10
2	20	9	40	4.7	9.4	14.1	18.8	23.5	28.2	32.9	37.6	42.3	47.0	51.7	56.4	61.1	65.8	70.5	9	40	2	20
2	30	9	30	4.9	9.9	14.8	19.8	24.7	29.7	34.6	39.6	44.5	49.5	54.4	59.4	64.3	69.2	74.2	9	30	2	30
2	40	9	20	5.2	10.4	15.6	20.7	25.9	31.1	36.3	41.5	46.7	51.9	57.0	62.2	67.4	72.6	77.8	9	20	2	40
2	50	9	10	5.4	18.8	16.2	21.6	27.1	32.5	37.9	43.3	48.7	54.1	59.5	64.9	70.4	75.8	81.2	9	10	2	50
3	0	9	0	5.6	11.3	16.9	22.5	28.1	33.8	39.4	45.0	50.6	56.3	61.9	67.5	73.2	78.8	84.4	9	0	3	0
3	10	8	50	5.8	11.7	17.5	23.3	29.1	35.0	40.8	46.6	52.4	58.3	64.1	69.9	75.8	81.6	87.4	8	50	3	10
3	20	8	40	6.0	12.0	18.1	24.1	30.1	36.1	42.1	48.1	54.2	60.2	66.2	72.2	78.2	84.2	90.2	8	40	3	20
3	30	8	30	6.2	12.4	18.6	24.8	31.0	37.2	43.4	49.6	55.8	62.0	68.2	74.4	80.6	86.8	93.0	8	30	3	30
3	40	8	20	6.4	12.7	19.1	25.5	31.8	38.2	44.6	50.9	57.3	63.7	70.0	76.4	82.8	89.2	95.5	8	20	3	40
3	50	8	10	6.5	13.0	19.6	26.1	32.6	39.1	45.7	52.2	58.7	65.2	71.7	78.3	84.8	91.4	97.9	8	10	3	50
4	0	8	0	6.7	13.3	20.0	26.7	33.3	40.0	46.7	53.3	60.0	66.7	73.3	80.0	86.7	93.4	100.0	8	0	4	0
4	20	7	40	6.9	13.8	20.8	27.7	34.6	41.5	48.4	55.4	62.3	69.2	76.1	83.1	89.9	96.8	103.8	7	40	4	20
4	40	7	20	7.1	14.3	21.4	28.5	35.6	42.8	49.9	57.0	64.2	71.3	78.4	85.6	92.7	99.8	106.9	7	20	4	40
5	0	7	0	7.3	14.6	21.9	29.2	36.5	43.8	51.0	58.3	65.6	72.9	80.2	87.5	94.8	102.0	109.3	7	0	5	0
5	20	6	40	7.4	14.8	22.2	29.6	37.0	44.4	51.9	59.3	66.7	74.1	81.5	88.9	96.3	103.8	111.2	6	40	5	20
5	40	6	20	7.5	15.0	22.4	29.9	37.4	44.9	52.3	59.8	67.3	74.8	82.2	89.7	97.2	104.6	112.1	6	20	5	40
6	0	6	0	7.5	15.0	22.5	30.0	37.5	45.0	52.5	60.0	67.5	75.0	82.5	90.0	97.5	105.0	112.5	6	0	6	0

SECONDS OF SECOND DIFFERENCES.

h.	m.	h.	m.																h.	m.	h.	m.
0	0	12	0	0″.0	0″.0	0″.0	0″.0	0″.0	0″.0	0″.0	0″.0	0″.0	0″.0	0.0	0″.0	0.0	0″.0	0″.0	12	15	0	0
0	10	11	50	0.0	0.0	0.0	0.0	0.0	0.1	0.1	0.1	0.2	0.2	0.2	0.3	0.3	0.3	0.4	12	50	0	10
0	20	11	40	0.0	0.0	0.0	0.0	0.1	0.1	0.2	0.3	0.3	0.4	0.4	0.5	0.6	0.7	0.8	11	40	0	20
0	30	11	30	0.0	0.0	0.1	0.1	0.1	0.2	0.3	0.4	0.5	0.6	0.7	0.8	0.9	1.0	1.1	11	30	0	30
0	40	11	20	0.0	0.1	0.1	0.1	0.1	0.3	0.4	0.5	0.7	0.8	0.9	1.0	1.1	1.3	1.5	11	20	0	40
0	50	11	10	0.0	0.1	0.1	0.1	0 2	0.3	0.5	0.6	0.8	1.0	1.1	1.3	1.5	1.6	1.8	11	10	0	50
1	0	11	0	0.0	0.1	0.1	0.1	0.2	0.4	0.6	0.8	0.9	1.1	1.3	1.5	1.7	1.9	2.0	11	0	1	0
1	10	10	50	0.0	0.1	0.1	0.2	0.2	0.4	0.7	0.9	1.1	1.3	1.5	1.8	2.0	2.2	2.4	10	50	1	10
1	20	10	40	0.0	0.1	0.1	0.2	0.2	0.5	0.8	1.0	1.2	1.5	1.7	2.0	2.2	2.5	2.7	10	40	1	20
1	30	10	30	0.0	0.1	0.2	0.2	0.3	0.5	0.8	1.1	1.4	1.6	1.9	2.2	2.4	2.7	3.0	10	30	1	30
1	40	10	20	0.1	0.1	0.2	0.2	0.3	0.6	0.9	1.2	1.5	1.8	2.1	2.4	2.7	3.0	3.3	10	20	1	40
1	50	10	10	0.1	0.1	0.2	0.3	0.3	0.6	0.9	1.3	1.6	1.9	2.2	2.6	2.9	3.2	3.5	10	10	1	50
2	0	10	0	0.1	0.1	0.2	0.3	0.3	0.7	1.0	1.4	1.7	2.1	2.4	2.8	3.1	3.5	3.8	10	0	2	0
2	10	9	50	0.1	0.1	0.2	0.3	0.4	0.7	1.1	1.5	1.8	2.2	2.6	3.0	3.3	3.7	4.0	9	50	2	10
2	20	9	40	0.1	0.2	0.2	0.3	0.4	0.8	1.2	1.6	1.9	2.3	2.7	3.1	3.5	3.9	4.2	9	40	2	20
2	30	9	30	0.1	0.2	0.2	0.3	0.4	0.8	1.2	1.6	2.0	2.5	2.9	3.3	3.7	4.1	4.5	9	30	2	30
2	40	9	20	0.1	0.2	0.3	0.3	0.4	0.9	1.3	1.7	2.1	2.6	3.0	3.5	3.9	4.3	4.7	9	20	2	40
2	50	9	10	0.1	0.2	0.3	0.4	0.4	0.9	1.4	1.8	2.2	2.7	3.1	3.6	4.0	4.5	4.9	9	10	2	50
3	0	9	0	0.1	0.2	0.3	0.4	0.5	0.9	1.4	1.9	2.3	2.8	3.3	3.8	4.2	4.7	5.1	9	0	3	0
3	10	8	50	0.1	0.2	0.3	0.4	0.5	1.0	1.5	1.9	2.4	2.9	3.4	3.9	4.4	4.9	5.3	8	50	3	10
3	20	8	40	0.1	0.2	0.3	0.4	0.5	1.0	1.5	2.0	2.5	3.0	3.5	4.0	4.5	5.0	5.5	8	40	3	20
3	30	8	30	0.1	0.2	0.3	0.4	0.5	1.0	1.5	2.1	2.6	3.1	3.6	4.1	4.6	5.2	5.7	8	30	3	30
3	40	8	20	0.1	0.2	0.3	0.4	0.5	1.1	1.6	2.1	2.6	3.2	3 7	4.2	4.7	5.3	5.8	8	20	3	40
3	50	8	10	0.1	0.2	0.3	0.4	0.5	1.1	1.6	2.2	2.7	3.3	3.8	4.3	4.8	5.4	6.0	8	10	3	50
4	0	8	0	0.1	0.2	0.3	0.4	0.6	1.1	1.6	2.2	2.8	3.4	3.8	4.4	5.0	5.6	6.2	8	0	4	0
4	20	7	40	0.1	0.2	0.3	0.5	0.6	1.2	1.7	2.3	2.9	3.5	4.0	4.6	5.2	5.8	6.4	7	40	4	20
4	40	7	20	0.1	0.2	0.4	0.5	0.6	1.2	1.7	2.4	3.0	3.6	4.2	4.8	5.3	5.9	6.6	7	20	4	40
5	0	7	0	0.1	0.2	0.4	0.5	0.6	1.2	1.7	2.4	3.0	3.6	4.3	4.9	5.5	6.1	6.7	7	0	5	0
5	20	6	40	0.1	0.2	0.4	0.5	0.6	1.2	1.7	2.5	3.1	3.7	4.3	4.9	5.6	6.2	6.8	6	40	5	20
5	40	6	20	0.1	0.2	0.4	0.5	0.6	1.3	1.8	2.5	3.1	3.8	4.4	5.0	5.6	6.2	6.9	6	20	5	40
6	0	6	0	0.1	0.2	0.4	0.5	0.6	1.3	1.8	2.5	3.2	3.8	4.4	5.0	5.7	6.3	7.0	6	0	6	0
				1″	2″	3″	4″	5″	10″	15″	20″	25″	30″	35′	40″	45″	50′	55″				

TABLE 12.

FACTORS by which the Equatorial Interval in a Transit Instrument must be multiplied, in order to obtain the Time of the Moon's meridian passage over the space contained between two of its lines or wires.

Arguments = the Horizontal Parallax at the Head; and at the Side, the Tabular Number taken out of Table 10, agreeably to its Arguments.

	50′	Diff.	53′	Diff.	56′	Diff.	59′	Diff.	62′	Diff.	65′	Diff.	68′	Diff.	
12^s.30	1.208	10	1.207	10	1.206	10	1.206	10	1.205	10	1.205	10	1.204	10	12^s.30
12.40	1.198	10	1.197	9	1.196	9	1.196	10	1.195	10	1.195	10	1.194	10	12.40
12.50	1.188	9	1.188	10	1.187	10	1.186	9	1.185	10	1.185	10	1.184	9	12.50
12.60	1.179	9	1.178	9	1.177	9	1.177	9	1.175	8	1.175	8	1.175	9	12.60
12.70	1.170	9	1.169	9	1.168	9	1.168	9	1.167	9	1.167	9	1.166	9	12.70
12.80	1.161	9	1.160	9	1.159	9	1.159	9	1.158	9	1.158	9	1.157	9	12.80
12.90	1.152	9	1.151	9	1.150	8	1.150	9	1.149	9	1.149	9	1.148	9	12.90
13.00	1.143	9	1.142	9	1.142	9	1.141	9	1.140	9	1.140	9	1.139	9	13.00
13.10	1.134	9	1.133	8	1.133	9	1.132	9	1.131	9	1.131	9	1.130	8	13.10
13.20	1.125	8	1.125	9	1.124	8	1.123	8	1.122	7	1.122	8	1.122	9	13.20
13.30	1.117	8	1.116	8	1.116	8	1.115	8	1.114	8	1.114	8	1.113	8	13.30
13.40	1.109	9	1.108	8	1.108	8	1.107	8	1.106	8	1.106	8	1.105	8	13.40
13.50	1.100	8	1.100	8	1.100	9	1.099	9	1.098	8	1.098	9	1.097	8	13.50
13.60	1.092	8	1.092	8	1.091	8	1.090	8	1.090	8	1.089	8	1.089	8	13.60
13.70	1.084	7	1.084	8	1.083	7	1.082	7	1.082	8	1.081	7	1.081	8	13.70
13.80	1.077	8	1.076	8	1.076	8	1.075	8	1.074	8	1.074	8	1.073	8	13.80
13.90	1.069	8	1.068	8	1.068	8	1.067	8	1.066	7	1.066	8	1.065	7	13.90
14.00	1.061	7	1.060	7	1.060	8	1.059	7	1.059	8	1.058	7	1.058	8	14.00
14.10	1.054	8	1.053	8	1.052	7	1.052	8	1.051	7	1.051	8	1.050	7	14.10
14.20	1.046	7	1.045	7	1.045	7	1.044	7	1.044	7	1.043	7	1.043	8	14.20
14.30	1.039	7	1.038	7	1.038	8	1.037	7	1.037	8	1.036	7	1.035	7	14.30
14.40	1.032	7	1.031	7	1.030	7	1.030	7	1.029	7	1.029	7	1.028	7	14.40
14.50	1.025	7	1.024	7	1.023	7	1.023	7	1.022	7	1.022	7	1.021	7	14.50
14.60	1.018		1.017		1.016		1.016		1.015		1.015		1.014		14.60

TABLE 13.

CORRECTION of the MOON's observed ZENITH DISTANCE.

Arguments = the Moon's Declination at the Side; and at the Head, the Moon's Semi-diameter ± the Tabular Number taken from Table 10, multiplied by the Meridian Interval of Time elapsed between the two Observations of Right Ascension and of Zenith Distance, accordingly as the preceding or following Limb was observed.

☾'s Dec.	1000″	950″	900″	850″	800″	750″	700″	650″	600″	550″	500″	450″	400″	350″	300″	250″	200″	150″	100″	☾'s Dec.
0°	0″.0	0″.0	0″.0	0″.0	0″.0	0″.0	0″.0	0″.0	0″.0	0″.0	0″.0	0″.0	0″.0	0″.0	0″.0	0″.0	0″.0	0″.0	0″.0	0°
2	0.1	0.1	0.1	0.1	0.0	0.0	0.0	0.0	0.0	0.0	0.0	0.0	0.0	0.0	0.0	0.0	0.0	0.0	0.0	2
4	0.2	0.2	0.1	0.1	0.1	0.1	0.1	0.0	0.0	0.0	0.0	0.0	0.0	0.0	0.0	0.0	0.0	0.0	0.0	4
6	0.3	0.2	0.2	0.2	0.1	0.1	0.1	0.1	0.1	0.1	0.1	0.0	0.0	0.0	0.0	0.0	0.0	0.0	0.0	6
8	0.3	0.3	0.3	0.3	0.2	0.2	0.2	0.1	0.1	0.1	0.1	0.1	0.1	0.0	0.0	0.0	0.0	0.0	0.0	8
10	0.4	0.3	0.3	0.3	0.3	0.2	0.2	0.2	0.2	0.1	0.1	0.1	0.1	0.1	0.0	0.0	0.0	0.0	0.0	10
12	0.5	0.4	0.4	0.3	0.3	0.3	0.3	0.2	0.2	0.2	0.1	0.1	0.1	0.1	0.0	0.0	0.0	0.0	0.0	12
14	0.6	0.5	0.5	0.4	0.4	0.3	0.3	0.2	0.2	0.2	0.1	0.1	0.1	0.1	0.1	0.0	0.0	0.0	0.0	14
16	0.7	0.6	0.6	0.5	0.5	0.4	0.3	0.3	0.2	0.2	0.2	0.1	0.1	0.1	0.1	0.0	0.0	0.0	0.0	16
18	0.8	0.7	0.6	0.6	0.5	0.5	0.4	0.3	0.3	0.2	0.2	0.1	0.1	0.1	0.1	0.0	0.0	0.0	0.0	18
20	0.9	0.8	0.7	0.6	0.6	0.5	0.4	0.3	0.3	0.3	0.2	0.2	0.1	0.1	0.1	0.0	0.0	0.0	0.0	20
22	1.0	0.9	0.8	0.7	0.6	0.6	0.5	0.4	0.3	0.3	0.3	0.2	0.1	0.1	0.1	0.1	0.0	0.0	0.0	22
24	1.1	1.0	0.9	0.8	0.7	0.6	0.5	0.4	0.4	0.3	0.3	0.2	0.2	0.1	0.1	0.1	0.0	0.0	0.0	24
26	1.2	1.1	1.0	0.9	0.8	0.7	0.6	0.5	0.4	0.4	0.3	0.3	0.2	0.1	0.1	0.1	0.0	0.0	0.0	26
28	1.3	1.2	1.1	1.0	0.8	0.7	0.6	0.5	0.5	0.4	0.3	0.3	0.2	0.2	0.1	0.1	0.1	0.0	0.0	28
30	1.4	1.3	1.2	1.1	0.9	0.8	0.7	0.6	0.5	0.4	0.3	0.3	0.2	0.2	0.1	0.1	0.1	0.1	0.0	30

Let b be the moment of Meridian Passage of the Moon's enlightened limb, at the middle wire, and d the moment when the Zenith Distance was taken; then $b - d$ will express the time elapsed, or Factor involved in the head Argument. The Sign + must be used with the Tabular Number when the first limb is observed; but the Sign − if the second. This Correction will always have a Sign contrary to that of the Declination (which is + for N. and − for S.) and the Declination will always be diminished thereby: the quantity $b - d$ will be negative, if the Observation of the Zenith Distance was taken *after* that of the limb in R. A. and *vice versâ*,

This Table would serve for the Stars, Planets, or Sun, if, instead of the Tabular Number taken from Table 10. the Number ± 15 were used with the Factor $b - d$.

TABLE 14.—DECREASE of the MOON'S (or SUN'S) DIAMETER when inclined to the Horizon, supposing the Horizontal Diameter (of each) to be 30′.
Arguments = Zenith Distance at the Head, or Altitude at the Foot, and Inclination at the Side.

Inclination.	80°	79°	78°	77°	76°	74°	72°	70°	68°	66°	64°	60°	56°	50°	44°	38°	30°	20°	10°	Inclination.
0°	0″.0	0″.0	0″.0	0″.0	0″.0	0″.0	0″.0	0″.0	0″.0	0″.0	0″.0	0″.0	0″.0	0″.0	0″.0	0″.0	0″.0	0″.0	0″.0	0°
3	0.0	0.0	0.0	0.0	0.0	0.0	0.0	0.0	0.0	0.0	0.0	0.0	0.0	0.0	0.0	0.0	0.0	0.0	0.0	3
6	0.2	0.1	0.1	0.1	0.1	0.1	0.0	0.0	0.0	0.0	0.0	0.0	0.0	0.0	0.0	0.0	0.0	0.0	0.0	6
9	0.4	0.3	0.3	0.2	0.2	0.2	0.1	0.1	0.1	0.1	0.1	0.0	0.0	0.0	0.0	0.0	0.0	0.0	0.0	9
12	0.7	0.6	0.5	0.4	0.4	0.3	0.2	0.2	0.1	0.1	0.1	0.1	0.1	0.1	0.0	0.0	0.0	0.0	0.0	12
15	1.1	0.9	0.7	0.6	0.6	0.4	0.3	0.3	0.2	0.2	0.2	0.1	0.1	0.1	0.1	0.0	0.0	0.0	0.0	15
18	1.5	1.2	1.0	0.9	0.8	0.6	0.5	0.4	0.3	0.2	0.2	0.1	0.1	0.1	0.1	0.1	0.0	0.0	0.0	18
21	2.0	1.6	1.4	1.2	1.0	0.8	0.6	0.5	0.4	0.3	0.3	0.2	0.2	0.2	0.1	0.1	0.1	0.1	0.1	21
24	2.5	2.1	1.8	1.5	1.3	1.1	0.8	0.7	0.6	0.5	0.4	0.3	0.3	0.2	0.2	0.1	0.1	0.1	0.1	24
27	3.1	2.6	2.2	1.9	1.6	1.4	1.0	0.9	0.7	0.6	0.5	0.4	0.3	0.2	0.2	0.2	0.1	0.1	0.1	27
30	3.8	3.2	2.7	2.3	2.0	1.7	1.3	1.0	0.9	0.7	0.6	0.5	0.4	0.3	0.2	0.2	0.2	0.1	0.1	30
33	4.5	3.8	3.2	2.7	2.4	2.0	1.5	1.2	1.0	0.9	0.8	0.6	0.5	0.4	0.3	0.2	0.2	0.2	0.1	33
36	5.2	4.4	3.7	3.2	2.8	2.3	1.7	1.4	1.2	1.0	0.9	0.7	0.5	0.4	0.3	0.3	0.2	0.2	0.2	36
39	6.0	5.0	4.3	3.7	3.2	2.6	2.0	1.6	1.4	1.2	1.0	0.8	0.6	0.5	0.4	0.3	0.3	0.2	0.2	39
42	6.8	5.7	4.8	4.1	3.6	2.9	2.3	1.9	1.6	1.3	1.1	0.9	0.7	0.5	0.4	0.4	0.3	0.2	0.2	42
45	7.6	6.3	5.4	4.6	4.0	3.3	2.5	2.1	1.7	1.5	1.3	1.0	0.8	0.6	0.5	0.4	0.3	0.3	0.2	45
48	8.4	7.0	5.9	5.1	4.5	3.6	2.8	2.3	1.9	1.6	1.4	1.1	0.9	0.7	0.5	0.4	0.4	0.3	0.3	48
51	9.2	7.7	6.5	5.6	4.9	4.0	3.0	2.5	2.1	1.8	1.5	1.2	0.9	0.7	0.6	0.5	0.4	0.3	0.3	51
54	9.9	8.3	7.0	6.1	5.3	4.3	3.3	2.7	2.3	1.9	1.6	1.3	1.0	0.8	0.6	0.5	0.5	0.4	0.3	54
57	10.7	8.9	7.6	6.5	5.7	4.6	3.5	2.9	2.4	2.1	1.8	1.4	1.1	0.8	0.7	0.6	0.5	0.4	0.4	57
60	11.4	9.5	8.1	6.9	6.1	4.9	3.8	3.1	2.6	2.2	1.9	1.5	1.2	0.9	0.7	0.6	0.5	0.4	0.4	60
63	12.1	10.1	8.6	7.3	6.4	5.2	4.0	3.3	2.7	2.3	2.0	1.5	1.2	1.0	0.7	0.6	0.5	0.4	0.4	63
66	12.7	10.6	9.0	7.7	6.7	5.5	4.2	3.5	2.9	2.5	2.1	1.6	1.3	1.0	0.8	0.7	0.6	0.5	0.4	66
69	13.2	11.1	9.4	8.1	7.0	5.7	4.4	3.6	3.0	2.6	2.2	1.7	1.3	1.0	0.8	0.7	0.6	0.5	0.4	69
72	13.7	11.5	9.7	8.4	7.3	5.9	4.6	3.7	3.1	2.7	2.3	1.8	1.4	1.1	0.9	0.7	0.6	0.5	0.4	72
75	14.1	11.8	10.0	8.7	7.5	6.1	4.7	3.8	3.2	2.8	2.4	1.8	1.4	1.1	0.9	0.7	0.6	0.5	0.5	75
78	14.5	12.1	10.3	8.9	7.7	6.3	4.8	3.9	3.3	2.8	2.4	1.9	1.5	1.1	0.9	0.8	0.6	0.5	0.5	78
81	14.8	12.3	10.5	9.0	7.9	6.4	4.9	4.0	3.4	2.9	2.5	1.9	1.5	1.2	0.9	0.8	0.7	0.5	0.5	81
84	15.0	12.5	10.6	9.1	8.0	6.5	5.0	4.1	3.4	2.9	2.5	1.9	1.5	1.2	0.9	0.8	0.7	0.5	0.5	84
87	15.1	12.6	10.7	9.2	8.1	6.5	5.0	4.1	3.5	3.0	2.5	1.9	1.5	1.2	0.9	0.8	0.7	0.5	0.5	87
90	15.2	12.7	10.7	9.3	8.1	6.6	5.1	4.1	3.5	3.0	2.5	1.9	1.5	1.2	1.0	0.8	0.7	0.5	0.5	90
	10°	11°	12°	13°	14°	16°	18°	20°	22°	24°	26°	30°	34°	40°	46°	52°	60°	70°	80°	

TABLE 15.—The LUMINOUS PORTION of the MOON in DIGITS, or Twefth Parts of the Diameter.
Argument = the Angular Distance from the Sun, when West of the Moon.

Arg.	0°	10°	20°	30°	40°	50°	60°	70°	80°	90°	100°	110°	120°	130°	140°	150°	160°	170°	Arg.
0°	0.000	0.091	0.362	0.804	1.404	2.144	3.000	3.948	4.959	6.000	7.041	8.053	9.000	9.856	10.596	11.196	11.638	11.909	10
1	0.001	0.110	0.399	0.857	1.472	2.224	3.092	4.047	5.062	6.104	7.145	8.150	9.090	9.936	10.663	11.247	11.673	11.926	9
2	0.004	0.131	0.437	0.912	1.542	2.306	3.184	4.147	5.166	6.209	7.247	8.247	9.179	10.014	10.728	11.297	11.706	11.941	8
3	0.009	0.154	0.477	0.969	1.612	2.389	3.277	4.247	5.269	6.314	7.349	8.344	9.267	10.091	10.791	11.346	11.737	11.955	7
4	0.015	0.179	0.519	1.026	1.684	2.474	3.370	4.347	5.373	6.418	7.451	8.440	9.355	10.167	10.254	11.392	11.767	11.967	6
5	0.023	0.205	0.563	1.085	1.758	2.559	3.465	4.448	5.477	6.523	7.552	8.535	9.441	10.242	10.915	11.437	11.793	11.977	5
6	0.033	0.233	0.608	1.146	1.833	2.645	3.560	4.549	5.582	6.627	7.653	8.630	9.526	10.316	10.974	11.481	11.821	11.985	4
7	0.045	0.263	0.654	1.209	1.909	2.733	3.656	4.651	5.686	6.731	7.753	8.723	9.611	10.388	11.031	11.523	11.846	11.991	3
8	0.059	0.294	0.703	1.272	1.986	2.821	3.753	4.753	5.791	6.834	7.853	8.816	9.694	10.458	11.088	11.563	11.869	11.996	2
9	0.074	0.327	0.753	1.337	2.064	2.910	3.850	4.855	5.896	6.938	7.953	8.908	9.776	10.528	11.143	11.601	11.890	11.999	1
10	0.091	0.362	0.804	1.404	2.144	3.000	3.948	4.959	6.000	7.041	8.053	9.000	9.856	10.596	11.196	11.638	11.909	12.000	0
	170°	160°	150°	140°	130°	120°	110°	100°	90°	80°	70°	60°	50°	40°	30°	20°	10°	0°	

Argument = the Angular Distance from the Sun, when East of the Moon.

This Table may be used also for Venus, provided the Argument be made equal to the Angle formed by two lines drawn from the Planet to the Sun and Earth respectively.

TABLE 1.—Reduction of the Equator to the Ecliptic, calculated for Obliquity 23° 27′ 57″ to every 10′ of Right Ascension in Space, with the Diff. and Var. for 100″ Var. of the Obliquity. Arg. = the Signs and Deg. of the Equator, or the Right Ascen. of the Diff. Points of the Ecliptic in Space.

Arg.	O + VI + (s)			I + VII + (s)			II + VIII + (s)			Arg.
	Reduction.	Difference.	Variation.	Reduction.	Difference.	Variation.	Reduction.	Difference	Variation.	
0° 0′	0° 0′ 0″.00	54″.09	0″.00	2° 11′ 11″.24	24″.50	19″.55	2° 5′ 39″.16	27″.02	17″.93	30° 0′
10	0 0 54 .09	54 .10	0 .14	2 11 35 .74	24 .26	19 .60	2 5 12 .14	27 .26	17 .86	50
20	0 1 48 .19	54 .09	0 .28	2 12 0 .00	23 .94	19 .66	2 4 44 .88	27 .49	17 .79	40
30	0 2 42 .28	54 .08	0 .42	2 12 23 .94	23 .67	19 .71	2 4 17 .39	27 .72	17 .72	30
40	0 3 36 .36	54 .07	0 .56	2 12 47 .61	23 .39	19 .76	2 3 49 .67	27 .98	17 .65	20
50	0 4 30 .43	54 .07	0 .69	2 13 11 .00	23 .08	19 .82	2 3 21 .69	28 .21	17 .59	10
1 0	0 5 24 .50	54 .04	0 .83	2 13 34 .08	22 .83	19 .87	2 2 53 .48	28 .42	17 .52	29 0
10	0 6 18 .54	54 .04	0 .97	2 13 56 .91	22 .52	19 .92	2 2 25 .06	28 .66	17 .45	50
20	0 7 12 .58	54 .02	1 .10	2 14 19 .43	22 .24	19 .97	2 1 56 .40	28 .89	17 .38	40
30	0 8 6 .60	54 .00	1 .24	2 14 41 .67	21 .93	20 .02	2 1 27 .51	29 .11	17 .31	30
40	0 9 0 .60	53 .99	1 .38	2 15 3 .60	21 .67	20 .07	2 0 58 .40	29 .36	17 .24	20
50	0 9 54 .59	53 .95	1 .51	2 15 25 .27	21 .35	20 .12	2 0 29 .04	29 .59	17 .16	10
2 0	0 10 48 .54	53 .93	1 .65	2 15 46 .62	21 .09	20 .17	1 59 59 .45	29 .80	17 .09	28 0
10	0 11 42 .47	53 .91	1 .79	2 16 7 .71	20 .78	20 .21	1 59 29 .65	30 .02	17 .02	50
20	0 12 36 .38	53 .88	1 .92	2 16 28 .49	20 .49	20 .26	1 58 59 .63	30 .26	16 .94	40
30	0 13 30 .26	53 .84	2 .06	2 16 48 .98	20 .22	20 .30	1 58 29 .37	30 .46	16 .87	30
40	0 14 24 .10	53 .81	2 .20	2 17 9 .20	19 .89	20 .35	1 57 58 .91	30 .71	16 .79	20
50	0 15 17 .91	53 .78	2 .33	2 17 29 .09	19 .62	20 .39	1 57 28 .20	30 .91	16 .72	10
3 0	0 16 11 .69	53 .74	2 .47	2 17 48 .71	19 .31	20 .44	1 56 57 .29	31 .14	16 .64	27 0
10	0 17 5 .43	53 .70	2 .61	2 18 8 .02	19 .05	20 .48	1 56 26 .15	31 .35	18 .55	50
20	0 17 59 .13	53 .65	2 .74	2 18 27 .07	18 .73	20 .52	1 55 54 .80	31 .58	16 .48	40
30	0 18 52 .78	53 .62	2 .88	2 18 45 .80	18 .43	20 .57	1 55 23 .22	31 .77	16 .41	30
40	0 19 46 .40	53 .57	3 .01	2 19 4 .23	18 .15	20 .61	1 54 51 .45	32 .02	16 .33	20
50	0 20 39 .97	53 .51	3 .15	2 19 22 .38	17 .83	20 .65	1 54 19 .43	32 .21	16 .25	10
4 0	9 21 33 .48	53 .48	3 .28	2 19 40 .21	17 .54	20 .69	1 53 47 .22	32 .43	16 .17	26 0
10	0 22 26 .96	53 .41	3 .42	2 19 57 .75	17 .27	20 .73	1 53 14 .79	32 .63	16 .09	50
20	0 23 20 .37	53 .37	3 .55	2 20 15 .02	16 .93	20 .77	1 52 42 .16	32 .86	16 .01	40
30	0 24 13 .74	53 .31	3 .69	2 20 31 .95	16 .66	20 .81	1 52 9 .30	33 .06	15 .93	30
40	0 25 7 .05	53 .25	3 .83	2 20 48 .61	16 .37	20 .85	1 51 36 .24	33 .27	15 .85	20
50	0 26 0 .30	53 .19	3 .96	2 21 4 .98	16 .07	20 .88	1 51 2 .97	33 .47	15 .77	10
5 0	0 26 53 .49	53 .13	4 .10	2 21 21 .05	15 .75	20 .92	1 50 29 .50	33 .70	15 .69	25 0
10	0 27 46 .62	53 .06	4 .24	2 21 36 .80	15 .46	20 .95	1 49 55 .80	33 .87	15 .61	50
20	0 28 39 .68	53 .00	4 .37	2 21 52 .26	15 .16	20 .98	1 49 21 .93	34 .12	15 .53	40
30	0 29 32 .68	52 .93	4 .51	2 22 7 .42	14 .88	21 .02	1 48 47 .81	34 .27	15 .44	30
40	0 30 25 .61	52 .87	4 .65	2 22 22 .30	14 .56	21 .05	1 48 13 .54	34 .51	15 .36	20
50	0 31 18 .48	52 .79	4 .78	2 22 36 .86	14 .26	21 .08	1 47 39 .03	34 .68	15 .28	10
6 0	0 32 11 .27	52 .70	4 .92	2 22 51 .12	13 .95	21 .12	1 47 4 .35	34 .91	15 .19	24 0
10	0 33 3 .97	52 .64	5 .05	2 23 5 .07	13 .65	21 .15	1 46 29 .44	35 .08	15 .10	50
20	0 33 56 .61	52 .56	5 .18	2 23 18 .72	13 .37	21 .18	1 45 54 .36	35 .29	15 .02	40
30	0 34 49 .17	52 .48	5 .31	2 23 32 .09	13 .07	21 .20	1 45 19 .07	35 .50	14 .93	30
40	0 35 41 .65	52 .40	5 .44	2 23 45 .16	12 .77	21 .23	1 44 43 .57	35 .65	14 .84	20
50	0 36 34 .05	52 .31	5 .57	2 23 57 .93	12 .44	21 .25	1 44 7 .92	35 .87	14 .76	10
7 0	0 37 26 .36	52 .23	5 .70	2 24 10 .37	12 .16	21 .28	1 43 32 .05	36 .08	14 .67	23 0
10	0 38 18 .59	52 .14	5 .84	2 24 22 .53	11 .85	21 .30	1 42 55 .97	36 .26	14 .58	50
20	0 39 10 .73	52 .04	5 .97	2 24 34 .38	11 .55	21 .33	1 42 19 .71	36 .43	14 .49	40
30	0 40 2 .77	51 .96	6 .10	2 24 45 .93	11 .23	21 .35	1 41 43 .28	36 .64	14 .40	30
40	0 40 54 .73	51 .86	6 .23	2 24 57 .16	10 .96	21 .37	1 41 6 .64	36 .81	14 .31	20
50	0 41 46 .59	51 .76	6 .36	2 25 8 .12	10 .62	21 .39	1 40 29 .83	37 .00	14 .22	10
8 0	0 42 38 .35	51 .67	6 .49	2 25 18 .14	10 .33	21 .41	1 39 52 .83	37 .18	14 .13	22 0
10	0 43 30 .02	51 .57	6 .63	2 25 29 .07	10 .07	21 .43	1 39 15 .65	37 .39	14 .04	50
20	0 44 21 .59	51 .46	6 .76	2 25 39 .14	9 .72	21 .45	1 38 38 .26	37 .55	13 .95	40
30	0 45 13 .05	51 .37	6 .89	2 25 48 .86	9 .43	21 .47	1 38 0 .71	37 .74	13 .86	30
40	0 46 4 .42	51 .25	7 .02	2 25 58 .29	9 .11	21 .49	1 37 22 .97	37 .91	13 .77	20
50	0 46 55 .67	51 .15	7 .15	2 26 17 .40	8 .83	21 .51	1 36 45 .06	38 .09	13 .67	10
9 0	0 47 46 .82	51 .03	7 .28	2 26 16 .23	8 .51	21 .52	1 36 6 .97	38 .27	13 .58	21 0
10	0 48 37 .85	50 .93	7 .41	2 26 24 .74	8 .21	21 .53	1 35 28 .70	38 .45	13 .49	50
20	0 49 28 .78	50 .81	7 .53	2 26 32 .95	7 .93	21 .55	1 34 50 .25	38 .61	13 .40	40
30	0 50 19 .59	50 .70	7 .66	2 26 40 .88	7 .57	21 .56	1 34 11 .64	38 .80	13 .30	30
40	0 51 10 .29	50 .58	7 .79	2 26 48 .45	7 .32	21 .57	1 33 32 .84	38 .97	13 .21	20
50	0 52 0 .87	50 .46	7 .92	2 26 55 .77	6 .99	21 .58	2 32 53 .87	39 .13	13 .12	10
10 0	0 52 51 .33		8 .04	2 27 2 .76		21 .59	1 32 14 .74		13 .02	20 0
	XI − V − (s)			X − IV − (s)			IX − III − (s)			

This Table may be used either for obtaining the approximate Longitude of a Star, Planet, &c. or for giving the Sun's true Longitude from his observed Right Ascension.

It may also be used instead of Table 8. of the Solar Tables, provided III be subtracted from the Argument.

ZODIACAL TABLES.

TABLE 1.

Reduction of the Equator to the Ecliptic, &c. continued.

Arg.	0^s + / VI^s + Reduction.	Difference.	Variation.	I^s + / VII^s + Reduction.	Differ.	Variation.	II^s + / VIII^s + Reduction.	Difference.	Variation.	Arg.
10° 0	0° 52′ 51″.33	50″.33	8″.04	2° 27′ 2″.76	6.69	21.59	1° 32′ 14″.74	39″.29	13″.02	20° 0′
10	0 53 41.66	50.21	8.17	2 27 9.45	6.41	21.60	1 31 35.45	39.48	12.93	50
20	0 54 31.87	50.10	8.29	2 27 15.88	6.07	21.61	1 30 55.97	89.64	12.83	40
30	0 55 21.97	49.97	8.42	2 27 21.93	5.81	21.62	1 30 16.33	39.80	12.74	30
40	0 56 11.94	49.83	8.54	2 27 27.74	5.45	21.63	1 29 16.33	39.97	12.64	20
50	0 57 1.77	49.70	8.67	2 27 33.19	5.19	21.63	1 28 57.56	40.14	12.55	10
11 0	0 57 51.47	49.57	8.79	2 27 38.38	4.86	21.64	1 28 16.42	40.28	12.45	19 0
10	0 58 41.04	49.44	8.92	2 27 43.24	4.57	21.64	1 27 36.14	40.46	12.36	50
20	0 59 30.48	49.30	9.04	2 27 47.81	4.26	21.65	1 26 55.68	40.61	12.26	40
30	1 0 19.78	49.16	9.16	2 27 52.07	3.95	21.65	1 26 15.07	40.77	12.16	30
40	1 1 8.94	49.03	9.29	2 27 56.02	3.67	21.65	1 25 34.30	40.93	12.06	20
50	1 1 57.97	48.87	9.41	2 27 59.69	3.34	21.66	1 24 53.37	41.09	11.97	10
12 0	1 2 46.84	48.75	9.53	2 28 3.03	3.06	21.66	1 24 12.28	41.23	11.87	18 0
10	1 3 35.59	48.59	9.66	2 28 6.09	2.74	21.67	1 23 31.05	41.40	11.77	50
20	1 4 24.18	48.44	9.78	2 28 8.83	2.46	21.67	1 22 49.65	41.52	11.67	40
30	1 5 12.62	48.30	9.90	2 8 11.29	2.13	21.67	1 22 8.13	41.69	11.58	30
40	1 6 0.92	48.16	10.02	2 28 13.42	1.82	21.67	1 21 26.44	41.85	11.48	20
50	1 6 49.08	47.98	10.14	2 28 15.24	1.55	21.67	1 20 44.59	41.98	11.38	10
13 0	1 7 37.06	47.84	10.26	2 28 16.79	1.23	21.67	1 20 2.61	42.14	11.28	17 0
10	1 8 24.90	47.69	10.38	2 28 18.02	0.93	21.67	1 19 20.47	42.26	11.18	50
20	1 9 12.59	47.51	10.50	2 28 18.95	0.64	21.67	1 18 38.20	42.42	11.08	40
30	1 10 0.10	47.37	10.62	2 28 19.59	0.31	21.67	1 17 55.78	42.55	10.98	30
40	1 10 47.47	47.21	10.73	2 28 19.90	0.03	21.66	1 17 13.23	42.70	10.88	20
50	1 11 34.68	47.04	10.85	2 28 19.93	0.26	21.66	1 16 30.53	42.84	10.78	10
14 0	1 12 21.72	46.88	10.97	2 28 19.67	0.60	21.66	1 15 47.69	42.99	10.68	16 0
10	1 13 8.60	46.71	11.09	2 28 19.07	0.86	21.65	1 15 4.70	43.11	10.58	50
20	1 13 55.31	46.55	11.20	2 28 18.21	1.21	21.65	1 14 21.59	43.26	10.48	40
30	1 14 41.86	46.36	11.32	2 28 17.00	1.47	21.64	1 13 38.33	43.39	10.37	30
40	1 15 28.22	46.20	11.43	2 28 15.53	1.76	21.63	1 12 54.94	43.51	10.27	20
50	1 16 14.42	46.05	11.55	2 28 13.77	2.07	21.63	1 12 11.42	43.65	10.17	10
15 0	1 17 0.47	45.84	11.67	2 28 11.70	2.40	21.62	1 11 27.78	43.77	10.06	15 0
10	1 17 46.31	45.68	11.79	2 28 9.30	2.66	21.61	1 10 44.01	43.91	9.96	50
20	1 18 31.99	45.50	11.90	2 28 6.64	3.02	21.59	1 10 0.10	44.04	9.85	40
30	1 19 17.49	45.31	12.01	2 28 3.62	3.24	21.58	1 9 16.06	44.15	9.75	30
40	1 20 2.80	45.13	12.13	2 28 0.38	3.57	21.56	1 8 31.91	44.29	9.64	20
50	1 20 47.93	44.96	12.25	2 27 56.81	3.88	21.55	1 7 47.62	44.41	9.54	10
16 0	1 21 32.89	44.77	12.35	2 27 52.93	4.14	21.53	1 7 3.21	44.51	9.43	14 0
10	1 22 17.66	44.59	12.47	2 27 48.79	4.49	21.51	1 6 18.70	44.65	9.33	50
20	1 23 2.25	44.38	12.58	2 27 44.30	4.77	21.50	1 5 34.05	44.76	9.22	40
30	1 23 46.63	44.23	12.69	2 27 39.53	5.04	21.48	1 4 49.29	44.88	9.12	30
40	1 24 30.86	44.01	12.80	2 27 34.49	5.37	21.46	1 4 4.41	45.01	9.01	20
50	1 25 14.87	43.81	12.91	2 27 29.12	5.66	21.44	1 3 19.40	45.11	8.91	10
17 0	1 25 58.68	43.63	13.02	2 27 23.46	5.93	21.42	1 2 34.29	45.22	8.80	13 0
10	1 26 42.31	43.43	13.13	2 27 17.53	6.24	21.40	1 1 49.07	45.34	8.70	50
20	1 27 25.74	43.25	13.24	2 27 11.29	6.55	21.38	1 1 3.73	45.44	8.59	40
30	1 28 8.99	43.04	13.34	2 27 4.74	6.81	21.36	1 0 18.29	45.57	8.48	30
40	1 28 52.03	42.84	13.45	2 26 57.93	7.12	21.34	1 59 32.72	45.67	8.38	20
50	1 29 34.87	42.63	13.56	2 26 50.81	7.42	21.32	1 58 47.05	45.75	8.27	10
18 0	1 30 17.50	42.42	13.66	2 26 43.39	7.72	21.29	0 58 1.30	45.88	8.16	12 0
10	1 30 59.92	42.23	13.77	2 26 35.67	7.97	21.27	0 57 15.42	45.98	8.05	50
20	1 31 42.14	42.02	13.87	2 26 27.70	8.30	21.24	0 56 29.44	46.08	7.94	40
30	1 32 24.17	41.81	13.97	2 26 19.40	8.59	21.22	0 55 43.36	46.18	7.83	30
40	1 33 5.98	41.61	14.07	2 26 10.81	8.88	21.19	0 54 57.18	46.28	7.72	20
50	1 33 47.59	41.40	14.17	2 26 1.93	9.16	21.17	0 54 10.90	46.38	7.61	10
19 0	1 34 28.99	41.17	14.27	2 25 52.77	9.44	21.14	0 53 24.52	46.48	7.50	11 0
10	1 35 10.16	40.95	14.37	2 25 43.33	9.76	21.11	0 52 38.04	46.56	7.39	50
20	1 35 51.11	40.76	14.47	2 25 33.57	10.01	21.08	0 51 51.48	46.66	7.28	40
30	1 36 31.87	40.53	14.57	2 25 23.56	10.33	21.05	0 51 4.82	46.75	7.17	30
40	1 37 12.40	40.32	14.67	2 25 13.23	10.59	21.02	0 50 18.07	46.85	7.06	20
50	1 37 52.72	40.10	14.76	2 25 2.64	10.89	20.99	0 49 31.22	46.94	6.94	10
20 0	1 38 32.82		14.86	2 24 51.75		20.96	0 48 44.28		6.83	10 0
	XI^s − / V^s −			X^s − / IV^s −			IX^s − / III^s −			

This Table may be used either for obtaining the approximate Longitude of a Star, Planet, &c. or for giving the Sun's true Longitude from his observed Right Ascension.

It may also be used instead of Table 8. of the Solar Tables, provided III^s be subtracted from the Argument.

ZODIACAL TABLES.

TABLE 1.

Reduction of the Equator to the Ecliptic, &c. concluded.

Arg.	0s + VIs + Reduction.	Difference.	Variation.	Is + VIIs + Reduction.	Difference.	Variation.	IIs + VIIIs + Reduction.	Difference.	Variation	Arg.
20° 0'	1° 38' 32".82	39".88	15".86	2° 24' 51".75	11".19	20".96	0° 48' 44".28	47".02	6".83	10° 0'
10	1 39 12.70	39.65	15.96	2 24 40.56	11.45	20.93	0 47 57.26	47.10	6.72	50
20	1 39 52.35	39.43	15.06	2 24 29.11	11.77	20.90	0 47 10.16	47.19	6.61	40
30	1 40 31.78	39.22	15.15	2 24 17.34	12.02	20.86	0 46 22.97	47.28	6.50	30
40	1 41 11.00	38.98	15.25	2 24 5.32	12.32	20.83	0 45 35.69	47.35	6.39	20
50	1 41 49.98	38.73	15.34	2 23 53.00	12.61	20.80	0 44 48.34	47.44	6.28	10
21 0	1 42 28.71	38.54	15.44	2 23 40.39	12.87	20.76	0 44 0.90	47.52	6.17	9 0
10	1 43 7.25	38.30	15.53	2 23 27.52	13.18	20.73	0 43 13.38	47.60	6.06	50
20	1 43 45.55	38.05	15.63	2 23 14.34	13.45	20.69	0 42 25.78	47.66	5.95	40
30	1 44 23.60	37.84	15.72	2 23 0.89	13.73	20.65	0 41 38.12	47.75	5.84	30
40	1 45 1.44	37.59	15.82	2 22 47.16	13.98	20.62	0 40 50.37	47.82	5.73	20
50	1 45 39.03	37.38	15.91	2 22 33.18	14.29	20.58	0 40 2.55	47.89	5.61	10
22 0	1 46 16.41	37.12	16.00	2 22 18.89	14.59	20.54	0 39 14.66	47.96	5.50	8 0
10	1 46 53.53	36.88	16.09	2 22 4.30	14.83	20.50	0 38 26.70	48.03	5.39	50
20	1 47 30.41	36.64	16.18	2 21 49.47	15.14	20.46	0 37 38.67	48.10	5.28	40
30	1 48 7.05	36.41	16.27	2 21 34.33	15.40	20.42	0 36 50.57	48.16	5.17	30
40	1 48 43.46	36.18	16.36	2 21 18.93	15.66	20.38	0 36 2.41	48.23	5.06	20
50	1 49 19.64	35.91	16.45	2 21 3.27	15.96	20.33	0 35 14.18	48.29	4.94	10
23 0	1 49 55.55	35.69	16.54	2 20 47.31	16.22	20.29	0 34 25.89	48.35	4.83	7 0
10	1 50 31.24	35.45	16.62	2 20 31.09	16.48	20.25	0 33 37.54	48.41	4.72	50
20	1 51 6.69	35.18	16.70	2 20 14.61	16.79	20.20	0 32 49.13	48.48	4.60	40
30	1 51 41.87	34.94	16.79	2 19 57.82	17.04	20.16	0 32 0.65	48.53	4.49	30
40	1 52 16.81	34.69	16.87	2 19 40.78	17.32	20.11	0 31 12.12	48.58	4.38	20
50	1 52 51.50	34.44	16.96	2 19 23.46	17.57	20.07	0 30 23.54	48.64	4.26	10
24 0	1 53 25.94	34.21	17.04	2 19 5.89	17.89	20.02	0 29 34.90	48.70	4.15	6 0
10	1 54 0.15	33.93	17.12	2 18 48.00	18.13	19.97	0 28 46.20	48.74	4.04	50
20	1 54 34.08	33.69	17.20	2 18 29.87	18.37	19.93	0 27 57.46	48.79	3.92	40
30	1 55 7.77	33.44	17.28	2 18 11.50	18.67	19.88	0 27 8.67	48.84	3.81	30
40	1 55 41.21	33.19	17.36	2 17 52.83	18.92	19.83	0 26 19.83	48.89	3.69	20
50	1 56 14.40	32.91	17.44	2 17 33.91	19.21	19.78	0 25 30.94	48.94	3.58	10
25 0	1 56 47.31	32.67	17.52	2 17 14.70	19.44	19.73	0 24 42.00	48.98	3.46	5 0
10	1 57 19.98	32.40	17.60	2 16 55.26	19.75	19.68	0 23 53.02	49.02	3.35	50
20	1 57 52.38	32.17	17.67	2 16 35.51	19.98	19.63	0 23 4.00	49.06	3.23	40
30	1 58 24.55	31.88	17.75	2 16 15.53	20.27	19.57	0 22 14.94	49.11	3.12	30
40	1 58 56.43	31.63	17.82	2 15 55.26	20.52	19.52	0 21 25.83	49.14	3.00	20
50	1 59 28.06	31.37	17.90	2 15 34.74	20.76	19.47	0 20 36.69	49.17	2.89	10
26 0	1 59 59.43	31.11	17.97	2 15 13.98	21.06	19.41	0 19 47.52	49.22	2.77	4 0
10	2 0 30.54	30.83	18.04	2 14 52.92	21.30	19.35	0 18 58.30	49.25	2.66	50
20	2 1 1.37	30.59	18.12	2 14 31.62	21.54	19.30	0 18 9.05	49.28	2.54	40
30	2 1 31.96	30.30	18.19	2 14 10.08	21.61	19.24	0 17 19.77	49.31	2.43	30
40	2 2 2.26	30.05	18.26	2 13 48.47	22.28	19.18	0 16 30.46	49.34	2.31	20
50	2 2 32.31	29.77	18.33	2 13 26.19	22.33	19.13	0 15 41.12	49.37	2.20	10
27 0	2 3 2.08	29.50	18.40	2 13 3.86	22.57	19.07	0 14 51.75	49.39	2.08	3 0
10	2 3 31.58	29.24	18.47	2 12 41.29	22.85	19.01	0 14 2.36	49.42	1.97	50
20	2 4 0.82	28.98	18.54	2 12 18.44	23.09	18.95	0 13 12.94	49.45	1.85	40
30	2 4 29.80	28.69	18.61	2 11 55.35	23.35	18.89	0 12 23.49	49.46	1.74	30
40	2 4 58.49	28.41	18.68	2 11 32.00	23.59	18.83	0 11 34.03	49.49	1.62	20
50	2 5 26.90	28.16	18.75	2 11 8.41	23.84	18.77	0 10 44.54	49.51	1.51	10
28 0	2 5 55.06	27.86	18.82	2 10 44.57	24.11	18.71	0 9 55.03	49.52	1.39	2 0
10	2 6 20.92	27.60	18.88	2 10 20.46	24.34	18.65	0 9 5.51	49.54	1.28	50
20	2 6 50.52	27.31	18.95	2 9 56.12	24.59	18.59	0 8 15.97	49.56	1.16	40
30	2 7 17.83	27.07	19.01	2 9 31.53	24.85	18.52	0 7 26.41	49.57	1.05	30
40	2 7 44.90	26.77	19.08	2 9 6.68	25.08	18.46	0 6 36.84	49.57	0.93	20
50	2 8 11.67	26.48	19.14	2 8 41.60	25.32	18.40	0 5 47.27	49.59	0.82	10
29 0	2 8 38.15	26.21	19.20	2 8 16.28	25.60	18.33	0 4 57.68	49.60	0.70	1 0
10	2 9 4.36	25.94	19.26	2 7 50.68	25.82	18.27	0 4 8.08	49.61	0.59	50
20	2 9 30.30	25.64	19.32	2 7 24.86	26.06	18.20	0 3 18.47	49.61	0.47	40
30	2 9 55.94	25.38	19.38	2 6 58.80	26.31	18.13	0 2 28.86	49.62	0.35	30
40	2 10 21.32	25.11	19.44	2 6 32.49	26.55	18.07	0 1 39.24	49.62	0.24	20
50	2 10 46.43	24.81	19.50	2 6 5.94	26.78	18.00	0 0 49.62	49.62	0.12	10
30 0	2 11 11.24		19.55	2 5 39.16		17.93	0 0 0.00		0.00	0 0
	XIs − Vs −			Xs − IVs −			Xs − IIIs −			

This Table may be used either for obtaining the approximate Longitude of a Star, Planet, &c. or for giving the Sun's true Longitude from his observed Right Ascension.

It may also be used instead of Table 8. of the Solar Tables, provided IIIs be subtracted from the Argument.

TABLE 2. CORRECTION of the APPROXIMATE LONGITUDE of a ZODIACAL STAR, or PLANET. (DR. MASKELYNE's enlarged.)

Arguments = the Right Ascension in Space, or Time, at the Side; and approximate Latitude, at the Head, or Foot.

Signs of R. A. 0ˢ with approximate Latitude N. + S. − ; VIˢ with approximate Latitude N. − S. +

Time. N. + S. − h. m. / N. − S. + h.	Space.	0° 10′	0° 20′	0° 30′	0° 40′	0° 50′	1° 0′	1° 10′	1° 20′	1° 30′	Space.	Time. N. − S. + h. m. / N. + S. − h.
0 0 XII	0° 0′	3′ 58″.9	7′ 57″.8	11′ 56″.8	15′ 55″.7	19′ 54″.7	23′ 53″.7	27′ 52″.7	31′ 51″.7	35′ 50″.7	30° 0	XII 0 XXIV
2	0 30	3 58 .9	7 57 .8	11 56 .8	15 55 .7	19 54 .6	23 53 .6	27 52 .5	31 51 .6	35 50 .6	29 30	58
4	1 0	3 58 .9	7 57 .8	11 56 .7	15 55 .6	19 54 .5	23 53 .4	27 52 .3	31 51 .4	35 50 .4	29 0	56
6	1 30	3 58 .8	7 57 .7	11 56 .5	15 55 .4	19 54 .2	23 53 .6	27 52 .0	31 51 .0	35 49 .9	28 30	54
8	2 0	3 58 .8	7 57 .6	11 56 .3	15 55 .1	19 53 .9	23 52 .8	27 51 .7	31 50 .6	35 49 .4	28 0	52
10	2 30	3 58 .7	7 57 .4	11 56 .1	15 54 .8	19 53 .5	23 52 .3	27 51 .1	31 49 .9	35 48 .6	27 30	50
12	3 0	3 58 .6	7 57 .2	11 55 .8	15 54 .5	19 53 .0	23 51 .7	27 50 .4	31 49 .1	35 47 .8	27 0	48
14	3 30	3 58 .5	7 57 .0	11 55 .4	15 53 .9	19 52 .4	23 51 .0	27 49 .5	31 48 .1	35 46 .7	26 30	46
16	4 0	3 58 .3	7 56 .7	11 55 .0	15 53 .4	19 51 .8	23 50 .2	27 48 .6	31 47 .0	35 45 .5	26 0	44
18	4 30	3 58 .2	7 56 .4	11 54 .6	15 52 .8	19 51 .0	23 49 .2	27 47 .5	31 46 .7	35 44 .1	25 30	42
20	5 0	3 58 .0	7 56 .0	11 54 .1	15 52 .1	19 50 .1	23 48 .2	27 46 .3	31 44 .4	35 42 .6	25 0	40
22	5 30	3 57 .8	7 55 .6	11 53 .5	15 51 .3	19 49 .1	23 47 .0	27 45 .0	31 42 .8	35 40 .8	24 30	38
24	6 0	3 57 .6	7 55 .2	11 52 .9	15 50 .5	19 48 .1	23 45 .8	27 43 .6	31 41 .2	35 38 .9	24 0	36
26	6 30	3 57 .4	7 54 .8	11 52 .2	15 49 .6	19 47 .0	23 44 .4	27 41 .9	31 39 .4	35 36 .9	23 30	34
28	7 0	3 57 .2	7 54 .3	11 51 .4	15 48 .6	19 45 .8	23 43 .0	27 40 .2	31 37 .5	35 34 .7	23 0	32
30	7 30	3 56 .9	7 53 .8	11 50 .6	15 47 .6	19 44 .5	23 41 .4	27 38 .3	31 35 .4	35 32 .3	22 30	30
32	8 0	3 56 .6	7 53 .2	11 49 .8	15 46 .5	19 43 .1	23 39 .7	27 36 .4	31 33 .2	35 29 .8	22 0	28
34	8 30	3 56 .3	7 52 .6	11 48 .9	15 45 .2	19 41 .6	23 37 .9	27 34 .3	31 30 .7	35 27 .1	21 30	26
36	9 0	3 56 .0	7 51 .9	11 47 .9	15 43 .9	19 40 .0	23 36 .0	27 32 .1	31 28 .1	35 24 .3	21 0	24
38	9 30	3 55 .7	7 51 .3	11 46 .9	15 42 .6	19 38 .3	23 34 .0	27 29 .7	31 25 .4	35 21 .3	20 30	22
40	10 0	3 55 .3	7 50 .6	11 45 .9	15 41 .2	19 36 .5	23 31 .9	27 27 .2	31 22 .6	35 18 .1	20 0	20
42	10 30	3 54 .9	7 49 .8	11 44 .8	15 39 .7	19 34 .8	23 29 .6	27 24 .6	31 19 .6	35 14 .7	19 30	18
44	11 0	3 54 .5	7 49 .0	11 43 .6	15 38 .2	19 33 .0	23 27 .3	27 21 .9	31 16 .6	35 11 .2	19 0	16
46	11 30	3 54 .1	7 48 .2	11 42 .4	15 36 .6	19 30 .8	23 24 .8	27 19 .1	31 13 .3	35 7 .5	18 30	14
48	12 0	3 53 .7	7 47 .4	11 41 .1	15 34 .8	19 28 .6	23 22 .3	27 16 .2	31 9 .9	35 3 .8	18 0	12
50	12 30	3 53 .3	7 46 .5	11 39 .8	15 33 .0	19 26 .5	23 19 .6	27 13 .0	31 6 .4	34 59 .7	17 30	10
52	13 0	3 52 .8	7 45 .6	11 38 .4	15 31 .2	19 24 .1	23 16 .9	27 9 .8	31 2 .7	34 55 .6	17 0	8
54	13 30	3 52 .3	7 44 .6	11 36 .0	15 29 .3	19 21 .7	23 14 .0	27 6 .4	30 58 .6	34 51 .2	16 30	6
56	14 0	3 51 .8	7 43 .6	11 35 .5	15 27 .4	19 19 .2	23 11 .1	27 3 .0	30 54 .5	34 46 .8	16 0	4
58	14 30	3 51 .3	7 42 .6	11 33 .9	15 25 .3	19 16 .6	23 8 .0	26 59 .4	30 50 .6	34 42 .2	15 30	2
I 0 XIII	15 0	3 50 .8	7 41 .6	11 32 .2	15 23 .2	19 14 .0	23 4 .8	26 55 .7	30 46 .5	34 37 .5	15 0	XI 0 XXIII
2	15 30	3 50 .2	7 40 .4	11 30 .6	15 21 .0	19 11 .2	23 1 .5	26 51 .8	30 42 .1	34 32 .5	14 30	58
4	16 0	3 49 .6	7 39 .3	11 29 .0	15 18 .7	19 8 .4	22 58 .1	26 47 .8	30 37 .6	34 27 .4	14 0	56
6	16 30	3 49 .1	7 38 .1	11 27 .3	15 16 .4	19 5 .5	22 54 .6	26 43 .7	30 32 .9	34 22 .2	13 30	54
8	17 0	3 48 .5	7 36 .8	11 25 .5	15 14 .0	19 2 .6	22 51 .0	26 39 .5	30 28 .1	34 16 .8	13 0	52
10	17 30	3 47 .9	7 35 .7	11 23 .6	15 11 .5	18 59 .4	22 47 .3	26 35 .2	30 23 .1	34 11 .2	12 30	50
12	18 0	3 47 .3	7 34 .5	11 21 .7	15 8 .9	18 56 .2	22 43 .5	26 30 .8	30 17 .9	34 5 .5	12 0	48
14	18 30	3 46 .6	7 33 .2	11 19 .7	15 6 .3	18 52 .9	22 39 .6	26 25 .2	30 12 .7	33 59 .2	11 30	46
16	19 0	3 45 .9	7 31 .8	11 17 .7	15 3 .7	18 49 .6	22 35 .6	26 21 .5	30 7 .4	33 53 .6	11 0	44
18	19 30	3 45 .2	7 30 .4	11 15 .7	15 0 .9	18 46 .2	22 31 .4	26 16 .7	30 2 .0	33 47 .7	10 30	42
20	20 0	3 44 .5	7 29 .0	11 13 .6	14 58 .1	18 42 .7	22 27 .2	26 11 .8	29 56 .4	33 41 .8	10 0	40
22	20 30	3 43 .8	7 27 .6	11 11 .4	14 55 .2	18 38 .0	22 22 .8	26 6 .8	29 50 .5	33 34 .8	9 30	38
24	21 0	3 43 .0	7 26 .1	11 9 .1	14 52 .2	18 35 .3	22 18 .4	26 1 .7	29 44 .7	33 27 .8	9 0	36
26	21 30	3 42 .3	7 24 .6	11 6 .9	14 49 .2	18 31 .5	22 13 .9	25 56 .4	29 38 .7	33 21 .0	8 30	34
28	22 0	3 41 .5	7 23 .0	11 4 .6	14 46 .2	18 27 .7	22 9 .3	25 50 .9	29 32 .5	33 14 .1	8 0	32
30	22 30	3 40 .7	7 21 .4	11 2 .2	14 43 .0	18 23 .7	22 4 .5	25 45 .3	29 26 .2	33 6 .9	7 30	30
32	23 0	3 39 .9	7 19 .8	10 59 .8	14 39 .7	18 19 .7	21 59 .7	25 39 .7	29 19 .7	32 59 .7	7 0	28
34	23 30	3 39 .1	7 18 .2	10 57 .4	14 36 .4	18 15 .6	21 54 .7	25 33 .9	29 13 .1	32 52 .3	6 30	26
36	24 0	3 38 .3	7 16 .5	10 55 .0	14 33 .1	18 11 .4	21 49 .7	25 28 .1	29 6 .4	32 44 .8	6 0	24
38	24 30	3 37 .5	7 14 .8	10 52 .4	14 29 .7	18 7 .1	21 44 .5	25 22 .0	28 59 .5	32 37 .0	5 30	22
40	25 0	3 36 .6	7 13 .1	10 49 .6	14 26 .2	18 2 .7	21 39 .3	25 15 .9	28 52 .6	32 29 .2	5 0	20
42	25 30	3 35 .7	7 11 .3	10 46 .9	14 22 .8	17 58 .3	21 34 .0	25 9 .7	28 45 .4	32 21 .2	4 30	18
44	26 0	3 34 .8	7 9 .5	10 44 .2	14 19 .3	17 53 .8	21 28 .6	25 3 .4	28 38 .2	32 13 .1	4 0	16
46	26 30	3 33 .9	7 7 .7	10 41 .4	14 15 .5	17 49 .2	21 23 .0	24 57 .0	28 30 .8	32 4 .8	3 30	14
48	27 0	3 32 .9	7 5 .8	10 38 .6	14 11 .5	17 44 .5	21 17 .4	24 50 .4	28 23 .3	31 56 .5	3 0	12
50	27 30	3 31 .9	7 3 .9	10 35 .8	14 7 .7	17 39 .7	21 11 .6	24 43 .7	28 15 .6	31 47 .9	2 30	10
52	28 0	3 30 .9	7 1 .9	10 32 .9	14 3 .9	17 34 .9	21 5 .8	24 36 .9	28 7 .9	31 39 .0	2 0	8
54	28 30	3 29 .9	6 59 .9	10 29 .9	13 59 .9	17 29 .9	20 59 .9	24 29 .9	28 0 .4	31 30 .1	1 30	6
56	29 0	3 28 .9	6 57 .9	10 26 .9	13 55 .9	17 24 .9	20 53 .9	24 22 .8	27 52 .9	31 21 .1	1 0	4
58	29 30	3 27 .9	6 55 .9	10 23 .8	13 51 .8	17 19 .8	20 47 .8	24 15 .6	27 44 .3	31 11 .9	0 30	2
II 0 XIV	30 0	3 26 .9	6 53 .8	10 20 .7	13 47 .7	17 14 .7	20 41 .6	24 8 .3	27 35 .6	31 2 .7	0 0	X 0 XXII

Signs of R. A. Vˢ with approximate Latitude N. − S. + ; XIˢ with approximate Latitude N. + S. −

The approximate Latitude = the Ecliptic Declination ± the Declination of the Star or Planet.

The Corrections contained in this Table are calculated for an Obliquity of the Ecliptic = 23° 28′; therefore a further Correction proportionate to the *diminution* must be applied from TABLE 4, which is adapted for this purpose.

ZODIACAL TABLES.

TABLE 2.

Correction of the approximate Longitude, &c. continued.

Signs of R. A. 0ˢ with approximate Latitude N. + S. − ; VIˢ with approximate Latitude N. − S. +

Time. N. + S. − (h. m.)	Time. N. − S. + (h.)	Space.	1° 40′	1° 50′	2° 0′	2° 10′	2° 20′	2° 30′	2° 40′	2° 50′	3° 0′	Space.	Time. N. − S. + (h. m.)	Time. N. + S. − (h.)
0 0	XII	0° 0	39′ 49″.8	43′ 48″.9	47′ 48″.1	51′ 47″.3	55′ 46″.5	59′ 45″.8	63′ 45″.2	67′ 44″.6	71′ 44″.0	30° 0′	XII 0	XXIV
2		0 30	39 49 .7	43 48 .7	47 47 .9	51 47 .1	55 46 .3	59 45 .6	63 45 .0	67 44 .3	71 43 .7	29 30	58	
4		1 0	39 49 .5	43 48 .5	47 47 .7	51 46 .8	55 46 .0	59 45 .3	63 44 .6	67 43 .9	71 43 .3	29 0	56	
6		1 30	39 49 .0	43 47 .9	47 47 .0	51 46 .2	55 45 .3	59 44 .6	63 43 .8	67 43 .2	71 42 .5	28 30	54	
8		2 0	39 48 .4	43 47 .3	47 46 .3	51 45 .4	55 44 .5	59 43 .6	63 42 .8	67 42 .1	71 41 .4	28 0	52	
10		2 30	39 47 .5	43 46 .3	47 45 .3	51 44 .3	55 43 .3	59 41 .3	63 41 .5	67 40 .7	71 40 .0	27 30	50	
12		3 0	39 46 .5	43 45 .3	47 44 .2	51 43 .0	55 41 .9	59 40 .9	63 39 .9	67 39 .0	71 38 .1	27 0	48	
14		3 30	39 45 .3	43 44 .0	47 42 .7	51 41 .5	55 40 .2	59 39 .2	63 38 .0	67 37 .0	71 35 .9	26 30	46	
16		4 0	39 44 .0	43 42 .5	47 41 .1	51 39 .7	55 38 .4	59 37 .1	63 35 .8	67 34 .7	71 33 .5	26 0	44	
18		4 30	39 42 .5	43 40 .7	47 39 .2	51 37 .7	55 36 .2	59 34 .7	63 33 .3	67 32 .1	71 30 .7	25 30	42	
20		5 0	39 40 .8	43 38 .8	47 37 .2	51 35 .5	55 33 .8	59 32 .2	63 30 .6	67 29 .1	71 27 .6	25 0	40	
22		5 30	39 38 .8	43 36 .7	47 34 .9	51 33 .8	55 31 .1	59 29 .3	63 27 .5	67 25 .8	71 24 .2	24 30	38	
24		6 0	39 36 .7	43 34 .6	47 32 .4	51 30 .3	55 28 .2	59 26 .2	63 24 .2	67 22 .2	71 20 .4	24 0	36	
26		6 30	39 34 .4	43 32 .1	47 29 .6	51 27 .3	55 25 .0	59 22 .7	63 20 .5	67 18 .4	71 16 .3	23 30	34	
28		7 0	39 32 .0	43 29 .4	47 26 .7	51 24 .1	55 21 .6	59 19 .1	63 16 .6	67 14 .2	71 11 .9	23 0	32	
30		7 30	39 29 .4	43 26 .4	47 23 .5	51 20 .7	55 18 .9	59 15 .1	63 12 .4	67 9 .8	71 7 .2	22 30	30	
32		8 0	39 26 .6	43 23 .4	47 20 .2	51 17 .1	55 14 .0	59 10 .9	63 7 .9	67 5 .0	71 2 .1	22 0	28	
34		8 30	39 23 .5	43 20 .1	47 16 .6	51 13 .1	55 9 .6	59 6 .4	63 3 .2	66 59 .9	70 56 .7	21 30	26	
36		9 0	39 20 .4	43 16 .6	47 12 .8	51 9 .0	55 5 .3	59 1 .7	62 58 .1	66 54 .5	70 51 .0	21 0	24	
38		9 30	39 17 .3	43 12 .9	47 8 .7	51 4 .6	55 0 .6	58 56 .6	62 52 .7	66 48 .8	70 45 .0	20 30	22	
40		10 0	39 14 .1	43 9 .0	47 4 .5	51 0 .1	54 55 .7	58 51 .3	62 47 .1	66 42 .8	70 38 .6	20 0	20	
42		10 30	39 10 .1	43 4 .8	47 0 .0	50 55 .1	54 50 .4	58 45 .7	62 41 .2	66 36 .5	70 31 .9	19 30	18	
44		11 0	39 5 .9	43 0 .6	46 55 .4	50 50 .2	54 45 .0	58 39 .9	62 34 .9	66 29 .9	70 24 .9	19 0	16	
46		11 30	39 1 .3	42 56 .1	46 50 .4	50 44 .9	54 39 .3	58 33 .8	62 28 .4	66 23 .0	70 17 .6	18 30	14	
48		12 0	38 57 .6	42 51 .5	46 45 .4	50 39 .4	54 33 .4	58 27 .5	62 21 .6	66 15 .7	70 10 .0	18 0	12	
50		12 30	38 53 .2	42 46 .6	46 40 .1	50 33 .6	54 27 .2	58 20 .8	62 14 .5	66 8 .2	70 2 .0	17 30	10	
52		13 0	38 48 .6	42 41 .6	46 34 .6	50 27 .7	54 20 .8	58 13 .9	62 7 .1	66 0 .4	69 53 .7	17 0	8	
54		13 30	38 43 .7	42 36 .3	46 28 .5	50 21 .5	54 14 .1	58 6 .7	61 59 .5	65 52 .3	69 45 .1	16 30	6	
56		14 0	38 38 .8	42 30 .8	46 22 .2	50 15 .0	54 7 .1	57 59 .3	61 51 .6	65 43 .8	69 36 .2	16 0	4	
58		14 30	38 33 .7	42 25 .2	46 16 .3	50 8 .3	53 59 .9	57 51 .6	61 43 .4	65 35 .1	69 27 .0	15 30	2	
I 0	XIII	15 0	38 28 .4	42 19 .4	46 10 .4	50 1 .4	53 52 .5	57 43 .7	61 34 .8	65 26 .1	69 17 .4	15 0	XI 0	XXIII
2		15 30	38 22 .9	42 13 .3	46 3 .8	49 54 .2	53 44 .8	57 35 .4	61 26 .1	65 16 .8	69 7 .5	14 30	58	
4		16 0	38 17 .3	42 7 .2	45 57 .0	49 46 .9	53 36 .9	57 26 .9	61 17 .0	65 7 .2	68 57 .3	14 0	56	
6		16 30	38 11 .4	42 0 .7	45 50 .0	49 39 .3	53 28 .7	57 18 .1	61 7 .6	64 57 .3	68 46 .9	13 30	54	
8		17 0	38 5 .4	41 54 .1	45 42 .8	49 31 .5	53 20 .3	57 9 .2	60 58 .0	64 47 .0	68 36 .1	13 0	52	
10		17 30	37 59 .3	41 47 .2	45 35 .4	49 23 .4	53 11 .6	56 59 .9	60 48 .2	64 36 .5	68 25 .0	12 30	50	
12		18 0	37 53 .1	41 40 .3	45 27 .7	49 15 .2	53 2 .8	56 50 .4	60 38 .0	64 25 .6	68 13 .6	12 0	48	
14		18 30	37 46 .5	41 33 .1	45 19 .9	49 6 .7	52 53 .6	56 40 .5	60 27 .5	64 14 .5	68 1 .8	11 30	46	
16		19 0	37 39 .6	41 25 .8	45 11 .9	48 58 .0	52 44 .2	56 30 .5	60 16 .8	64 3 .1	67 49 .6	11 0	44	
18		19 30	37 32 .7	41 18 .2	45 3 .6	48 49 .0	52 34 .5	56 20 .1	60 5 .7	63 51 .5	67 37 .2	10 30	42	
20		20 0	37 25 .7	41 10 .4	44 55 .1	48 39 .9	52 24 .7	56 9 .6	59 54 .5	63 39 .5	67 24 .5	10 0	40	
22		20 30	37 18 .4	41 2 .4	44 46 .4	48 30 .5	52 14 .5	55 58 .7	59 42 .9	63 27 .3	67 11 .5	9 30	38	
24		21 0	37 11 .1	40 54 .3	44 37 .6	48 20 .9	52 4 .2	55 47 .6	59 31 .1	63 14 .7	66 58 .1	9 0	36	
26		21 30	37 3 .6	40 45 .9	44 28 .5	48 11 .1	51 53 .6	55 36 .3	59 19 .1	63 1 .9	66 54 .5	8 30	34	
28		22 0	36 55 .8	40 37 .5	44 19 .3	48 1 .1	51 42 .9	55 24 .8	59 6 .7	62 48 .7	66 30 .7	8 0	32	
30		22 30	36 47 .8	40 28 .8	44 9 .8	47 45 .8	51 31 .9	55 12 .9	58 54 .0	62 35 .3	66 16 .5	7 30	30	
32		23 0	36 39 .8	40 20 .0	44 0 .1	47 40 .3	51 20 .6	55 0 .8	58 41 .1	62 21 .6	66 1 .9	7 0	28	
34		23 30	36 32 .0	40 10 .9	43 50 .2	47 29 .7	51 9 .0	54 48 .4	58 28 .0	62 7 .6	65 47 .1	6 30	26	
36		24 0	36 23 .2	40 1 .7	43 40 .2	47 19 .0	50 57 .2	54 35 .8	58 14 .5	61 53 .2	65 32 .0	6 0	24	
38		24 30	36 14 .6	39 52 .3	43 29 .9	47 7 .7	50 45 .2	54 22 .0	58 0 .7	61 38 .7	65 16 .6	5 30	22	
40		25 0	36 5 .9	39 42 .6	43 19 .4	46 56 .2	50 33 .0	54 9 .9	57 46 .8	61 23 .8	65 0 .8	5 0	20	
42		25 30	35 56 .8	39 32 .8	43 9 .7	46 44 .6	50 20 .4	53 56 .5	57 32 .6	61 8 .7	64 44 .8	4 30	18	
44		26 0	35 47 .7	39 22 .9	42 57 .9	46 32 .8	50 7 .6	53 43 .0	57 18 .1	60 53 .3	64 28 .5	4 0	16	
46		26 30	35 38 .6	39 12 .7	42 46 .8	46 20 .8	49 54 .3	53 29 .2	57 3 .4	60 37 .7	64 11 .9	3 30	14	
48		27 0	35 29 .4	39 2 .4	42 35 .5	46 8 .6	49 41 .0	53 15 .1	56 48 .3	60 21 .7	63 55 .0	3 0	12	
50		27 30	35 19 .8	38 51 .9	42 24 .0	45 56 .1	49 27 .7	53 0 .7	56 33 .0	60 5 .4	63 37 .8	2 30	10	
52		28 0	35 10 .1	38 41 .3	42 12 .4	45 43 .5	49 14 .2	52 46 .2	56 17 .4	59 48 .9	63 20 .3	2 0	8	
54		28 30	35 0 .2	38 30 .4	42 0 .5	45 30 .7	49 0 .7	52 31 .3	56 1 .7	59 32 .2	63 2 .6	1 30	6	
56		29 0	34 50 .2	38 19 .4	41 48 .5	45 17 .7	48 47 .0	52 16 .3	55 45 .6	59 15 .0	62 44 .5	1 0	4	
58		29 30	34 39 .9	38 8 .1	41 36 .2	45 4 .4	48 32 .7	52 1 .0	55 39 .3	58 57 .7	62 26 .2	0 30	2	
II 0	XIV	30 0	34 29 .6	37 56 .8	41 23 .9	44 51 .0	48 18 .2	51 45 .5	55 12 .8	58 40 .1	62 7 .5	0 0	X 0	XXII

Signs of R. A. Vˢ with approximate Latitude N. − S. + ; XIˢ with approximate Latitude N. + S. −

The approximate Latitude = the Ecliptic Declination ± the Declination of the Star or Planet.

The Corrections contained in this Table are calculated for an Obliquity of the Ecliptic = 23° 28′; therefore a further Correction proportionate to the *diminution* must be applied from Table 4, which is adapted for this purpose.

TABLE 2.

Correction of the approximate Longitude, &c. continued.

Signs of R. A. 0 s with approximate Latitude N. + S. −   VI s with approximate Latitude N. − S. +

Time. N. + S. − / N. − S. + (h. m. h.)	Space.	3° 10′	3° 20′	3° 30′	3° 40′	3° 50′	4° 0′	4° 10′	4° 20′	4° 30′	Space.	Time. N. − S. + / N. + S. − (h. m. h.)
0 0 XII	0° 0′	75′ 43″.5	79′ 43″.1	83′ 42″.7	87′ 42″.3	91′ 42″.2	95′ 42″.1	99′ 42″.1	103′ 42″.1	107′ 42″.2	30° 0′	XII 0 XXIV
2	0 30	75 43.2	79 42.8	83 42.3	87 42.0	91 41.9	95 41.8	99 41.8	103 41.8	107 41.8	29 30	58
4	1 0	75 42.8	79 42.4	83 41.9	87 41.6	91 41.4	95 41.3	99 41.2	103 41.2	107 41.2	29 0	56
6	1 30	75 41.9	79 41.5	83 41.0	87 40.6	91 40.3	95 40.1	99 40.0	103 40.0	107 39.9	28 30	54
8	2 0	75 40.7	79 40.2	83 39.7	87 39.2	91 38.9	95 38.6	99 38.4	103 38.4	107 38.4	28 0	52
10	2 30	75 39.2	79 38.7	83 38.0	87 37.4	91 37.0	95 36.6	99 36.3	103 36.2	107 36.1	27 30	50
12	3 0	75 37.3	79 36.9	83 35.9	87 35.2	91 34.7	95 34.2	99 33.9	103 33.6	107 33.4	27 0	48
14	3 30	75 35.1	79 34.8	83 33.4	87 32.6	91 31.9	95 31.3	99 30.4	103 30.5	107 30.2	26 30	46
16	4 0	75 32.5	79 32.4	83 30.5	87 29.6	91 28.8	95 28.1	99 27.5	103 27.0	107 26.5	26 0	44
18	4 30	75 29.5	79 28.3	83 27.2	87 26.2	91 25.2	95 24.4	99 23.7	103 22.9	107 22.3	25 30	42
20	5 0	75 26.2	79 24.9	83 23.6	87 22.4	91 21.3	95 20.3	99 19.3	103 18.4	107 17.7	25 0	40
22	5 30	75 22.6	79 21.1	83 19.6	87 18.2	91 16.9	95 15.7	99 14.5	103 13.4	107 12.4	24 30	38
24	6 0	75 18.6	79 16.9	83 15.2	87 13.7	91 12.1	95 10.7	99 9.3	103 8.0	107 6.7	24 0	36
26	6 30	75 14.3	79 12.3	83 10.4	87 8.7	91 6.8	95 5.2	99 3.6	103 2.1	107 0.6	23 30	34
28	7 0	75 9.6	79 7.4	83 5.3	87 3.3	91 1.2	94 59.3	98 57.5	102 55.7	106 54.1	23 0	32
30	7 30	75 4.6	79 2.3	82 59.8	86 57.5	90 53.1	94 53.0	98 50.9	102 48.9	106 46.9	22 30	30
32	8 0	74 59.3	78 56.8	82 53.9	86 51.3	90 48.7	94 46.2	98 43.9	102 41.6	106 39.4	22 0	28
34	8 30	74 53.6	78 50.7	82 47.6	86 44.7	90 41.8	94 39.1	98 36.3	102 33.8	106 31.3	21 30	26
36	9 0	74 47.6	78 44.2	82 40.9	86 37.7	90 34.5	94 31.4	98 28.4	102 25.5	106 22.7	21 0	24
38	9 30	74 41.2	78 37.5	82 33.9	86 30.4	90 26.8	94 23.3	98 20.0	102 16.7	106 12.6	20 30	22
40	10 0	74 34.5	78 30.5	82 26.5	86 22.7	90 18.7	94 14.9	38 11.2	102 7.6	106 3.1	20 0	20
42	10 30	74 27.5	78 23.0	82 18.7	86 14.4	90 10.1	94 6.0	98 1.9	101 57.9	105 53.2	19 30	18
44	11 0	74 20.1	78 15.2	82 10.5	86 5.8	90 1.2	93 56.7	97 52.2	101 47.8	105 42.9	19 0	16
46	11 30	74 12.4	78 7.1	82 1.9	85 56.8	89 51.8	93 47.0	97 42.0	101 37.3	105 32.1	18 30	14
48	12 0	74 4.3	77 58.6	81 53.0	85 47.5	89 42.1	93 36.9	97 31.4	101 26.2	105 21.0	18 0	12
50	12 30	74 55.9	77 49.7	81 43.7	85 38.7	89 31.9	93 26.1	97 20.3	101 14.6	105 9.0	17 30	10
52	13 0	73 47.1	77 40.5	81 34.0	85 27.6	89 21.3	93 15.0	97 8.8	101 2.7	104 56.7	17 0	8
54	13 30	73 38.0	77 31.0	81 24.0	85 17.1	89 10.2	93 3.5	96 56.3	100 50.9	104 43.7	16 30	6
56	14 0	73 28.6	77 21.1	81 13.6	85 6.2	88 58.8	92 51.7	96 44.5	100 37.4	104 30.4	16 0	4
58	14 30	73 18.8	77 10.8	81 2.8	84 54.9	88 47.0	92 39.3	96 31.6	100 24.0	104 16.5	15 30	2
I 0 XIII	15 0	73 8.7	77 0.2	80 51.7	84 43.2	88 34.8	92 26.5	96 18.4	100 10.2	104 2.2	15 0	XI 0 XXIII
2	15 30	72 58.3	77 49.2	80 40.1	84 31.1	88 22.2	92 13.3	96 4.6	99 55.9	103 47.4	14 30	58
4	16 0	72 47.6	76 37.9	80 28.2	84 18.7	88 9.2	91 59.8	95 50.5	99 41.2	103 32.1	14 0	56
6	16 30	72 36.5	76 26.2	80 15.9	84 5.8	87 55.7	91 45.7	95 35.8	99 26.1	103 16.3	13 30	54
8	17 0	72 25.0	76 14.2	80 3.3	83 52.6	87 41.9	91 31.3	95 20.8	99 10.4	103 0.0	13 0	52
10	17 30	72 13.3	76 1.8	80 50.3	83 39.0	87 27.7	91 16.4	95 5.3	98 54.3	102 43.3	12 30	50
12	18 0	72 1.2	75 49.1	79 36.9	83 25.0	87 13.1	91 1.2	94 49.4	98 37.7	102 26.2	12 0	48
14	18 30	72 40.8	75 36.0	79 23.2	83 10.6	86 58.1	90 45.5	94 33.0	98 20.7	102 8.5	11 30	46
16	19 0	71 36.0	75 22.6	79 9.2	82 55.9	86 42.6	90 29.4	94 16.3	98 3.3	101 50.4	11 0	44
18	19 30	71 23.0	75 8.8	78 54.7	82 40.7	86 26.8	90 12.9	94 9.0	97 45.4	101 31.7	10 30	42
20	20 0	71 9.6	74 54.2	78 39.9	82 25.2	86 10.6	89 56.0	93 41.5	97 27.1	101 12.7	10 0	40
22	20 30	71 55.8	74 40.7	78 24.7	82 9.3	85 53.9	89 38.7	93 23.2	97 8.2	100 53.1	9 30	38
24	21 0	70 41.7	74 25.4	78 9.2	81 53.0	85 36.8	89 21.0	93 4.6	96 48.9	100 33.2	9 0	36
26	21 30	70 27.4	73 10.3	77 53.3	81 36.4	85 19.5	39 2.9	92 45.8	96 29.3	100 12.7	8 30	34
28	22 0	70 12.8	73 54.8	77 37.0	81 19.4	85 1.8	88 44.4	92 26.7	96 9.3	99 51.9	8 0	32
30	22 30	69 57.8	73 39.1	77 20.5	81 2.1	84 43.6	88 25.3	92 6.9	95 48.7	99 30.5	7 30	30
32	23 0	69 42.4	73 23.0	77 3.6	80 44.4	84 25.0	88 5.8	91 46.8	95 27.7	99 8.8	7 0	28
34	23 30	69 26.8	72 6.5	76 46.3	80 26.2	84 6.0	87 46.0	91 26.2	95 6.3	98 46.6	6 30	26
36	24 0	69 10.8	72 49.7	76 28.6	80 7.7	83 46.7	87 25.9	91 5.3	94 44.5	98 23.9	6 0	24
38	24 30	68 54.5	72 32.6	76 10.6	79 48.8	83 26.5	87 5.3	91 43.8	94 22.1	98 0.7	5 30	22
40	25 0	68 37.9	72 15.1	75 52.3	79 29.6	83 6.9	86 44.4	90 21.9	93 59.4	97 37.1	5 0	20
42	25 30	68 21.0	71 57.3	75 33.6	79 10.0	82 46.5	86 23.0	89 59.6	93 36.7	97 13.1	4 30	18
44	26 0	67 3.8	71 39.2	75 14.6	78 50.0	82 25.6	86 1.2	89 36.9	93 12.7	96 48.7	4 0	16
46	26 30	67 46.3	71 20.4	74 55.2	78 29.7	82 4.4	85 49.0	89 13.9	92 48.7	96 23.6	3 30	14
48	27 0	67 28.4	70 2.0	74 35.5	78 9.1	81 42.8	85 16.5	88 50.4	92 24.3	95 58.1	3 0	12
50	27 30	66 10.3	70 42.9	74 15.4	77 48.1	81 20.8	84 53.6	88 26.5	91 59.5	95 32.3	2 30	10
52	28 0	66 51.8	70 23.4	73 55.0	77 26.7	80 58.4	84 30.3	88 2.2	91 34.2	95 6.2	2 0	8
54	28 30	66 83.1	70 3.2	73 34.1	77 5.0	80 35.7	84 6.6	87 37.5	91 8.4	94 39.5	1 30	6
56	29 0	66 14.0	69 43.6	73 13.2	76 42.9	80 12.6	83 42.5	87 12.4	90 42.3	94 12.4	1 0	4
58	29 30	65 54.8	69 23.2	72 51.8	76 20.5	79 49.2	82 18.0	86 46.9	90 15.8	93 44.8	0 30	2
II 0 XIV	30 0	65 35.1	69 2.5	72 30.0	75 57.7	79 25.4	82 53.1	86 21.0	89 48.9	93 16.9	0 0	X 0 XXII

Signs of R. A. V s with approximate Latitude N. − S. +   XI s with approximate Latitude N. + S. −

The approximate Latitude = the Ecliptic Declination ± the Declination of the Star or Planet.

The Corrections contained in this Table are calculated for an Obliquity of the Ecliptic = 23° 28′; therefore a further Correction proportionate to the *diminution* must be applied from Table 4, which is adapted for this purpose.

TABLE 2.

Correction of the approximate Longitude, &c. continued.

Signs of R. A. 0 (s) with approximate Latitude N. + S. — VI (s) with approximate Latitude N. — S. +

Time. N. + N. — S. — S. +	Space.	4° 40′	4° 50′	5° 0′	5° 10′	5° 20′	5° 30′	5° 40′	5° 50′	6° 0′	Space.	Time. N. — N. + S. + S. —
h. m. h.												h. m. h.
O 0 XII	0° 0′	111′ 42″.4	115′ 42″.8	119′ 43″.2	123′ 43″.7	127′ 44″.4	131′ 45″.1	135′ 45″.9	139′ 46″.9	143′ 47″.9	30° 0′	XII 0 XXIV
2	0 30	111 42.0	115 42.3	119 42.8	123 43.3	127 44.1	131 44.8	135 45.6	139 46.3	143 47.5	29 30	58
4	1 0	111 41.4	115 41.7	119 42.1	123 42.6	127 43.2	131 43.9	135 44.7	139 45.4	143 46.6	29 0	56
6	1 30	111 40.1	115 40.3	119 40.7	123 41.2	127 41.7	131 42.4	135 43.1	139 43.8	143 44.9	28 30	54
8	2 0	111 38.4	115 38.5	119 38.8	123 39.2	127 39.7	131 40.3	135 40.9	139 41.6	143 42.7	28 0	52
10	2 30	111 36.1	115 36.1	119 36.3	123 36.6	127 37.0	131 37.5	135 38.1	139 38.7	143 39.2	27 30	50
12	3 0	111 33.3	115 33.3	119 33.3	123 33.4	127 33.8	131 34.2	135 34.8	139 35.4	143 35.1	27 0	48
14	3 30	111 29.8	115 29.8	119 29.8	123 29.7	127 30.1	131 30.3	135 30.7	139 31.2	143 31.3	26 30	46
16	4 0	111 26.1	115 25.9	119 25.7	123 25.7	127 25.7	131 25.8	125 26.1	139 26.5	143 27.0	26 0	44
18	4 30	111 21.7	115 21.3	119 21.1	123 20.9	127 20.7	131 20.7	135 20.8	139 21.0	143 21.4	25 30	42
20	5 0	111 17.0	115 16.4	119 15.9	123 15.5	127 15.2	131 15.0	135 14.9	139 15.0	143 15.2	25 0	40
22	5 30	111 11.6	115 10.9	119 10.2	123 9.6	127 9.1	131 8.7	135 8.4	139 8.3	143 7.2	24 30	38
24	6 0	111 5.8	115 4.8	119 3.9	123 3.1	127 2.4	131 1.8	135 1.3	139 1.0	143 0.7	24 0	36
26	6 30	110 59.3	114 58.2	118 56.9	122 56.0	126 55.1	130 54.3	134 53.5	138 53.0	142 52.4	23 30	34
28	7 0	110 52.5	114 51.0	118 49.6	122 48.4	126 47.2	130 46.2	134 45.2	138 44.4	142 43.6	23 0	32
30	7 30	110 45.1	114 43.4	118 41.6	122 40.3	126 38.8	130 37.5	134 36.2	138 35.1	142 33.6	22 30	30
32	8 0	110 37.3	114 35.3	118 33.3	122 31.6	126 29.8	130 28.2	134 26.7	138 25.3	142 24.1	22 0	28
34	8 30	110 28.8	114 26.6	118 24.3	122 22.3	126 20.3	130 18.3	134 16.5	138 14.8	142 13.2	21 30	26
36	9 0	110 20.0	114 17.4	118 14.8	122 12.5	126 10.0	130 7.8	134 5.7	138 3.7	142 1.8	21 0	24
38	9 30	110 10.6	114 7.7	118 4.7	122 2.0	125 59.3	129 56.7	133 54.3	137 51.9	141 49.7	20 30	22
40	10 0	110 0.7	113 57.4	117 54.1	121 51.0	125 48.0	129 45.1	133 42.3	137 39.6	141 37.0	20 0	20
42	10 30	109 50.3	113 46.6	117 43.0	121 39.0	125 36.0	129 32.8	133 29.7	137 26.5	141 23.6	19 30	18
44	11 0	109 39.4	113 35.3	117 31.3	121 27.5	125 23.5	129 20.0	133 16.5	137 12.9	141 9.6	19 0	16
46	11 30	109 28.0	113 23.5	117 19.1	121 14.8	125 10.5	129 6.0	133 2.6	136 58.6	140 54.9	18 30	14
48	12 0	109 16.1	113 11.2	117 6.3	121 1.6	124 57.0	128 52.5	132 48.1	136 43.8	140 39.6	18 0	12
50	12 30	109 3.7	112 58.3	116 53.0	120 57.9	124 42.8	128 37.9	132 33.0	136 28.3	140 23.6	17 30	10
52	13 0	108 50.8	112 44.9	116 39.2	120 33.6	124 28.1	128 22.7	132 17.4	136 12.2	140 7.1	17 0	8
54	13 30	108 37.4	112 31.1	116 24.9	120 18.8	124 12.8	128 6.9	132 1.1	135 55.4	139 49.8	16 30	6
56	14 0	108 23.5	112 16.7	116 10.0	120 3.4	123 56.9	127 50.5	131 44.2	135 38.0	139 31.9	16 0	4
58	14 30	108 9.1	112 1.7	115 54.6	119 47.4	123 40.4	127 33.7	131 26.7	135 20.0	139 13.4	15 30	2
I 0 XIII	15 0	107 54.2	111 46.4	115 38.6	119 30.9	123 23.4	127 16.3	131 8.7	135 1.5	138 54.3	15 0	XI 0 XXIII
2	15 30	107 38.8	111 30.5	115 22.1	119 13.9	123 5.8	126 58.0	130 50.0	134 42.2	138 34.5	14 30	58
4	16 0	107 23.0	111 14.0	115 5.1	118 56.3	122 47.7	126 39.1	130 30.7	134 22.3	138 14.1	14 0	56
6	16 30	107 6.6	110 57.1	114 47.6	118 38.2	122 29.0	126 19.8	130 10.8	134 1.8	137 53.5	13 30	54
8	17 0	106 49.8	110 39.6	114 29.6	118 19.6	122 9.7	126 0.0	129 50.3	133 40.8	137 31.4	13 0	52
10	17 30	106 32.4	110 21.7	114 11.0	118 0.4	121 49.9	125 39.5	129 29.2	133 19.1	137 9.1	12 30	50
12	18 0	106 14.6	110 3.2	113 51.9	117 40.7	121 29.5	125 18.4	129 7.6	132 56.8	136 46.2	12 0	48
14	18 30	105 56.2	109 44.3	113 32.3	117 20.4	121 8.6	124 56.9	128 45.3	132 33.9	136 22.5	11 30	46
16	19 0	105 37.5	109 24.8	113 12.1	116 59.6	120 47.1	124 34.8	128 22.5	132 10.4	135 58.2	11 0	44
18	19 30	105 18.2	109 4.9	112 51.5	116 38.3	120 25.1	124 12.1	127 59.1	131 46.3	135 33.5	10 30	42
20	20 0	104 58.5	108 44.4	112 30.3	116 16.4	120 2.5	123 48.8	127 35.1	131 21.6	135 8.2	10 0	40
22	20 30	104 38.2	108 23.4	112 8.6	115 53.9	119 39.3	123 25.3	127 10.4	130 56.2	134 42.6	9 30	38
24	21 0	104 17.4	108 1.8	111 46.3	115 30.9	119 15.6	123 0.3	126 45.2	130 30.2	134 15.5	9 0	36
26	21 30	103 56.3	107 39.9	111 23.7	115 7.5	118 51.4	122 35.4	126 19.4	130 3.8	133 48.2	8 30	34
28	22 0	103 34.7	107 17.4	111 0.5	114 43.6	118 26.7	122 9.9	125 53.3	129 36.8	133 20.4	8 0	32
30	22 30	103 12.6	106 54.4	111 36.8	114 19.1	118 1.4	121 43.8	125 26.4	129 9.1	132 51.9	7 30	30
32	23 0	102 50.0	106 31.1	110 12.6	113 54.0	117 35.6	121 17.2	124 58.9	128 40.8	132 22.8	7 0	28
34	23 30	102 26.9	106 7.3	109 47.9	113 28.5	117 9.2	120 50.0	124 30.7	128 11.9	131 53.1	6 30	26
36	24 0	102 3.3	105 42.9	109 22.6	113 2.4	116 42.2	120 22.2	124 2.3	127 42.5	131 22.8	6 0	24
38	24 30	101 39.4	105 18.1	108 56.9	112 35.8	116 14.8	119 53.9	123 33.1	127 12.4	130 51.9	5 30	22
40	25 0	101 14.9	104 52.7	108 30.6	112 8.7	115 46.8	119 25.0	123 3.4	126 41.8	130 20.4	5 0	20
42	25 30	100 50.0	104 26.9	108 3.9	111 41.1	115 18.3	118 55.6	122 33.1	126 10.6	129 48.3	4 30	18
44	26 0	100 24.5	104 0.6	107 36.7	111 12.9	114 49.2	118 25.7	122 2.2	125 38.9	129 15.6	4 0	16
46	26 30	99 58.7	103 33.8	107 9.0	110 44.3	114 19.7	117 55.2	121 28.8	125 6.0	128 42.3	3 30	14
48	27 0	99 32.4	103 6.5	106 40.8	110 15.2	113 49.6	117 24.2	120 58.8	124 33.6	128 8.5	3 0	12
50	27 30	99 5.7	102 38.8	106 12.1	109 45.6	113 19.0	116 52.6	120 26.3	124 0.2	127 34.0	2 30	10
52	28 0	98 38.4	102 10.6	105 42.9	109 15.4	112 47.9	116 20.5	119 53.2	123 26.1	126 59.0	2 0	8
54	28 30	98 10.8	101 42.0	105 13.3	108 44.8	112 16.3	115 47.9	119 19.6	122 51.5	126 23.4	1 30	6
56	29 0	97 42.6	101 12.8	104 43.2	108 13.6	111 44.1	115 14.7	118 45.4	122 16.3	125 47.2	1 0	4
58	29 30	97 14.1	100 43.3	104 12.6	107 42.0	111 11.5	114 41.0	118 10.6	121 40.5	125 10.4	0 30	2
II 0 XIV	30 0	96 45.0	100 13.2	103 41.5	107 9.8	110 38.3	114 6.8	117 35.3	121 4.2	124 33.1	0 0	X 0 XXII

Signs of R. A. V (s) with approximate Latitude N. — S. + XI (s) with approximate Latitude N. + S. —

The approximate Latitude = the Ecliptic Declination ± the Declination of the Star or Planet.

The Corrections contained in this Table are calculated for an Obliquity of the Ecliptic = 23° 28′; therefore a further Correction proportionate to the *diminution* must be applied from Table 4, which is adapted for this purpose.

ZODIACAL TABLES.

TABLE 2.

Correction of the approximate Longitude, &c. continued.

Signs of R. A. I with approximate Latitude N. + S. − | VII with approximate Latitude N. − S. +

Time. N. + S. − / N. − S. +	Space.	0° 10′	0° 20′	0° 30′	0° 40′	0° 50′	1° 0′	1° 10′	1° 20′	1° 30′	Space.	Time. N. − S. + / N. + S. −
h. m. h.												h. m. h.
II 0 XIV	0° 0′	3′ 26″.9	6′ 53″.8	10′ 20″.7	13′ 47″.7	17′ 14″.7	20′ 41″.6	24′ 8″.3	27′ 35″.6	31′ 2″.7	30° 0′	X 0 XXII
2	0 30	3 25.9	6 51.7	10 17.6	13 43.5	17 9.4	20 35.3	24 1.1	27 27.2	30 53.2	29 30	58
4	1 0	3 24.8	6 49.6	10 14.4	13 39.2	17 4.0	20 28.9	23 53.8	27 18.7	30 43.7	29 0	56
6	1 30	3 23.8	6 47.4	10 11.2	13 34.9	16 58.6	20 22.4	23 46.2	27 9.9	30 33.7	28 30	54
8	2 0	3 22.7	6 45.2	10 7.9	13 30.5	16 53.2	20 15.8	23 38.5	27 1.2	30 23.9	28 0	52
10	2 30	3 21.5	6 43.0	10 4.5	13 26.0	16 47.6	20 9.2	23 30.7	26 52.2	30 13.9	27 30	50
12	3 0	3 20.3	6 40.7	10 1.1	13 21.5	16 41.9	20 2.4	23 22.8	26 43.2	30 3.8	27 0	48
14	3 30	3 19.2	6 38.5	9 57.7	13 16.9	16 36.2	19 55.5	23 14.8	26 34.1	29 53.5	26 30	46
16	4 0	3 18.1	6 36.2	9 54.2	13 12.3	16 30.4	19 48.6	23 6.7	26 24.9	29 43.1	26 0	44
18	4 30	3 17.0	6 33.8	9 50.7	13 7.6	16 24.5	19 41.5	22 58.5	26 15.4	29 32.5	25 30	42
20	5 0	3 15.8	6 31.4	9 47.2	13 2.9	16 18.6	19 34.4	22 50.2	26 6.0	29 21.8	25 0	40
22	5 30	3 14.5	6 29.0	9 43.6	12 58.0	16 12.6	19 27.2	22 41.7	25 56.3	29 10.9	24 30	38
24	6 0	3 13.3	6 26.6	9 39.9	12 53.2	16 6.5	19 19.9	22 33.2	25 46.6	29 0.0	24 0	36
26	6 30	3 12.1	6 24.1	9 36.1	12 48.2	16 0.3	19 12.5	22 24.6	25 36.6	28 48.9	23 30	34
28	7 0	3 10.8	6 21.6	9 32.4	12 43.3	15 54.1	19 5.0	22 15.9	25 26.7	28 37.7	23 0	32
30	7 30	3 9.6	6 19.1	9 28.6	12 38.2	15 47.8	18 57.4	22 7.0	25 16.6	28 26.3	22 30	30
32	8 0	3 8.3	6 16.6	9 24.8	12 33.1	15 41.4	18 49.7	21 58.1	25 6.4	28 14.9	22 0	28
34	8 30	3 7.0	6 13.9	9 21.0	12 27.9	15 35.0	18 42.0	21 49.0	24 56.1	28 3.3	21 30	26
36	9 0	3 5.6	6 11.3	9 17.1	12 22.7	15 28.4	18 34.2	21 39.9	24 45.7	28 51.5	21 0	24
38	9 30	3 4.4	6 8.7	9 13.1	12 17.5	15 21.3	18 26.2	21 30.6	24 35.1	27 39.6	20 30	22
40	10 0	3 3.1	6 6.1	9 9.1	12 12.1	15 15.2	18 18.3	21 21.3	24 24.5	27 27.6	20 0	20
42	10 30	3 1.7	6 3.3	9 5.0	12 6.8	15 8.5	18 10.2	21 11.9	24 13.7	27 15.5	19 30	18
44	11 0	3 0.3	6 0.6	9 0.9	12 1.3	15 1.7	18 2.0	21 2.4	24 2.8	27 3.2	19 0	16
46	11 30	2 59.0	5 57.8	8 56.8	11 55.8	14 54.8	17 53.8	20 52.7	23 51.8	26 50.9	18 30	14
48	12 0	2 57.6	5 55.1	8 52.7	11 50.2	14 47.8	17 45.4	20 43.0	23 40.7	26 38.4	18 0	12
50	12 30	2 56.2	5 52.3	8 48.5	11 44.6	14 40.1	17 37.0	20 33.2	23 29.4	26 25.7	17 30	10
52	13 0	2 54.8	5 49.5	8 44.2	11 39.2	14 33.8	17 28.5	20 23.3	23 18.1	26 13.0	17 0	8
54	13 30	2 53.4	5 46.6	8 39.9	11 33.3	14 26.7	17 20.0	20 13.3	23 6.7	26 0.0	16 30	6
56	14 0	2 51.9	5 43.7	8 35.6	11 27.5	14 19.4	17 11.3	20 3.2	22 55.2	25 47.1	16 0	4
58	14 30	2 50.4	5 39.8	8 31.2	11 21.7	14 12.2	17 2.6	19 53.0	22 43.5	25 34.0	15 30	2
III 0 XV	15 0	2 48.9	5 37.9	8 26.8	11 15.8	14 4.8	16 53.8	19 42.8	22 31.8	25 20.8	15 0	IX 0 XXI
2	15 30	2 47.5	5 34.9	8 22.4	11 9.9	13 57.4	16 44.8	19 32.3	22 19.9	25 7.5	14 30	58
4	16 0	2 46.0	5 31.9	8 17.9	11 3.9	13 49.9	16 35.9	19 21.9	22 8.0	24 54.1	14 0	56
6	16 30	2 44.5	5 28.9	8 13.4	10 57.9	13 42.4	16 26.8	19 11.4	21 55.9	24 40.5	13 30	54
8	17 0	2 43.0	5 25.9	8 8.8	10 51.8	13 34.8	16 17.7	19 0.8	21 43.8	24 26.8	13 0	52
10	17 30	2 41.4	5 22.8	8 4.2	10 45.7	13 27.2	16 8.5	18 50.0	21 31.6	24 13.0	12 30	50
12	18 0	2 39.8	5 19.7	7 59.6	10 39.5	13 19.4	15 59.3	18 39.2	21 19.2	23 59.1	12 0	48
14	18 30	2 38.3	5 16.6	7 54.9	10 33.3	13 11.6	15 50.0	18 28.3	21 6.8	23 45.1	11 30	46
16	19 0	2 36.8	5 13.5	7 50.2	10 27.0	13 3.8	15 40.6	18 17.4	20 54.2	23 31.0	11 0	44
18	19 30	2 35.2	5 10.4	7 45.5	10 20.7	12 55.8	15 31.1	18 6.3	20 41.6	23 16.8	10 30	42
20	20 0	2 33.6	5 7.2	7 40.7	10 14.3	12 47.9	15 21.6	17 55.2	20 28.8	23 2.5	10 0	40
22	20 30	2 32.0	5 4.0	7 35.9	10 7.9	12 37.9	15 11.9	17 43.9	20 16.0	22 48.0	9 30	38
24	21 0	2 30.3	5 0.7	7 31.1	10 1.4	12 31.8	15 2.2	17 32.7	20 3.1	22 33.5	9 0	36
26	21 30	2 28.7	4 57.5	7 26.2	9 54.9	12 23.7	14 52.4	17 21.3	19 50.1	22 18.9	8 30	34
28	22 0	2 27.1	4 54.2	7 21.3	9 48.4	12 15.5	14 42.6	17 9.8	19 37.0	22 4.2	8 0	32
30	22 30	2 25.5	4 50.9	7 16.4	9 41.4	12 7.3	14 32.7	16 58.2	19 23.8	21 49.3	7 30	30
32	23 0	2 23.9	4 47.6	7 11.4	9 35.2	11 59.0	14 22.8	16 46.6	19 10.5	21 34.4	7 0	28
34	23 30	2 22.1	4 44.3	7 6.4	3 28.6	11 50.6	14 12.8	16 34.9	18 57.2	21 19.3	6 30	26
36	24 0	2 20.4	4 40.9	7 1.3	9 21.8	11 42.2	14 2.7	16 23.2	18 43.7	21 4.2	6 0	24
38	24 30	2 18.8	4 37.5	6 56.2	9 15.0	11 33.7	13 47.5	16 11.3	18 30.2	20 49.0	5 30	22
40	25 0	2 17.1	4 34.1	6 51.1	9 8.2	11 25.2	13 42.3	15 59.4	18 16.5	20 33.7	5 0	20
42	25 30	2 15.4	4 30.7	6 46.0	9 0.9	11 16.2	13 32.0	15 47.4	18 2.8	20 18.3	4 30	18
44	26 0	2 13.7	4 27.2	6 40.8	8 54.4	11 8.1	13 21.7	15 35.4	17 49.0	20 2.7	4 0	16
46	26 30	2 11.9	4 23.7	6 35.6	8 47.4	10 59.5	13 11.3	15 23.2	17 35.1	19 47.1	3 30	14
48	27 0	2 10.1	4 20.2	6 30.4	8 40.5	10 50.7	13 0.8	15 11.0	17 21.2	19 31.4	3 0	12
50	27 30	2 8.4	4 16.7	6 25.1	8 33.5	10 41.9	12 50.3	14 58.7	17 7.2	19 15.5	2 30	10
52	28 0	2 6.7	4 13.2	6 19.8	8 26.4	10 33.1	12 39.7	14 46.4	16 53.1	18 59.7	2 0	8
54	28 30	2 4.9	4 9.2	6 14.5	8 19.3	10 24.2	12 29.1	14 34.0	16 38.9	18 43.7	1 30	6
56	29 0	2 3.1	4 6.1	6 9.2	8 12.2	10 15.3	12 18.4	14 21.5	16 24.7	18 27.7	1 0	4
58	29 30	2 1.3	4 2.5	6 3.8	8 5.1	10 6.3	12 7.6	14 8.9	16 10.3	18 11.6	0 30	2
IV 0 XVI	30 0	1 59.5	3 58.9	5 58.4	7 57.9	9 57.3	11 56.8	13 56.3	15 55.9	17 55.4	0 0	VIII 0 XX

Signs of R. A. IV with approximate Longitude N. − S. + | X with approximate Latitude N. + S. −

The approximate Latitude = the Ecliptic Declination ± the Declination of the Star or Planet.

The Corrections contained in this Table are calculated for an Obliquity of the Ecliptic = 23° 28′; therefore a further Correction proportionate to the *diminution* must be applied from Table 4, which is adapted for this purpose.

ZODIACAL TABLES.

TABLE 2.

Correction of the approximate Longitude, &c. continued.

Signs of R. A. I with approximate Latitude N. + / S. − ; VII with approximate Latitude N. − / S. +

Time. N. + S. − / N. − S. + (h. m. h.)	Space.	1° 40′	1° 50′	2° 0′	2° 10′	2° 20′	2° 30′	2° 40′	2° 50′	3° 0′	Space.	Time. N. − S. + / N. + S. − (h. m. h.)
II 0 XIV	0° 0′	34′ 29″.6	37′ 56″.8	41′ 23″.9	44′ 51″.0	48′ 18″.2	51′ 45″.5	55′ 12″.8	58′ 40″.1	62′ 7″.5	30° 0′	X 0 XXII
2	0 30	34 19.1	37 45.2	41 11.3	44 37.3	48 3.5	51 29.7	54 55.8	58 12.2	61 48.5	29 30	58
4	1 0	34 8.5	37 33.5	40 58.5	44 23.5	47 48.6	51 13.7	54 38.9	58 4.1	61 29.4	29 0	56
6	1 30	33 57.7	37 21.6	40 45.6	44 9.4	47 33.4	50 57.4	54 21.5	57 45.6	61 9.8	28 30	54
8	2 0	33 46.7	37 9.5	40 32.4	43 55.2	47 18.1	50 41.0	54 4.0	57 27.0	60 50.1	28 0	52
10	2 30	33 35.6	36 57.3	40 19.0	43 40.7	47 2.5	50 24.3	53 46.2	57 8.2	60 30.0	27 30	50
12	3 0	33 24.3	36 44.9	40 5.4	43 26.0	46 46.7	50 7.4	53 28.2	56 48.9	60 9.8	27 0	48
14	3 30	33 12.9	36 32.3	39 51.7	43 11.1	46 30.7	49 50.2	53 9.8	56 29.4	59 49.1	26 30	46
16	4 0	33 1.3	36 19.6	39 37.8	42 56.1	46 14.5	49 32.9	52 51.3	56 9.8	59 28.3	26 0	44
18	4 30	32 49.6	36 6.6	39 23.7	42 40.8	45 58.0	49 15.3	52 32.5	55 49.8	59 7.1	25 30	42
20	5 0	32 37.7	35 53.5	39 9.5	42 25.4	45 41.4	48 57.4	52 13.6	55 29.6	58 45.8	25 0	40
22	5 30	32 25.6	35 40.3	38 55.0	42 9.7	45 24.5	48 39.3	51 54.2	55 9.1	58 24.1	24 30	38
24	6 0	32 13.4	35 26.9	38 40.4	41 53.9	45 7.5	48 21.1	51 34.7	54 48.4	58 2.2	24 0	36
26	6 30	32 1.1	35 13.3	38 25.6	41 37.9	44 50.2	48 2.6	51 14.9	54 27.4	57 39.9	23 30	34
28	7 0	31 48.6	34 59.6	38 10.6	41 21.7	44 32.7	47 43.9	50 55.0	54 6.2	57 17.5	23 0	32
30	7 30	31 36.0	34 45.7	37 55.4	41 5.3	44 15.0	47 24.9	50 34.8	53 44.8	56 54.7	22 30	30
32	8 0	31 23.2	34 31.7	37 40.1	40 48.7	43 57.2	47 5.8	50 14.4	53 23.1	56 31.8	22 0	28
34	8 30	31 10.3	34 17.5	37 24.6	40 31.9	43 39.1	46 46.4	49 53.7	53 1.1	56 8.5	21 30	26
36	9 0	30 57.3	34 3.1	37 9.0	40 14.9	43 20.8	46 26.8	49 32.9	52 38.9	55 45.0	21 0	24
38	9 30	30 44.1	33 48.6	36 53.1	39 57.7	43 2.3	46 7.0	49 11.7	52 16.5	55 21.2	20 30	22
40	10 0	30 30.8	33 33.9	36 37.1	39 40.4	42 43.7	45 47.0	48 50.3	51 53.8	54 57.2	20 0	20
42	10 30	30 17.3	33 19.0	36 20.9	39 22.9	42 22.8	45 26.8	48 28.7	51 30.9	54 32.9	19 30	18
44	11 0	30 3.7	33 4.1	36 4.6	39 5.2	42 5.8	45 6.4	48 6.9	51 7.7	54 8.5	19 0	16
46	11 30	29 49.9	32 48.9	35 48.1	38 47.3	41 46.5	44 45.7	47 44.9	50 44.3	53 43.7	18 30	14
48	12 0	29 36.0	32 33.7	35 31.5	38 29.2	41 27.0	44 24.9	47 22.8	50 20.7	53 18.7	18 0	12
50	12 30	29 22.0	32 28.3	35 14.7	38 11.0	41 7.3	44 3.8	47 0.3	49 56.8	52 53.4	17 30	10
52	13 0	29 7.8	32 2.8	34 57.7	37 52.6	40 47.5	43 42.6	46 37.7	49 32.8	52 27.9	17 0	8
54	13 30	28 53.5	31 47.0	34 40.5	37 34.0	40 27.5	43 21.2	46 11.8	49 8.4	52 2.1	16 30	6
56	14 0	28 39.1	31 31.1	34 23.2	37 15.3	40 7.4	42 59.6	45 51.8	48 43.9	51 36.2	16 0	4
58	14 30	28 24.6	31 15.1	34 5.8	36 56.4	39 47.0	42 37.7	45 28.4	48 19.1	51 10.0	15 30	2
III 0 XV	15 0	28 9.9	30 59.0	33 48.1	36 37.3	39 26.4	42 15.7	45 4.9	47 54.2	50 43.7	15 0	IX 0 XXI
2	15 30	27 55.0	30 42.8	33 30.3	36 18.0	39 5.6	41 53.4	44 41.2	47 29.0	50 16.9	14 30	58
4	16 0	27 40.1	30 26.4	33 12.4	35 58.6	38 44.7	41 31.0	44 17.3	47 3.6	49 50.0	14 0	56
6	16 30	27 25.1	30 9.8	32 54.3	35 39.0	38 23.6	41 8.4	43 53.2	46 38.0	49 22.8	13 30	54
8	17 0	27 9.9	29 53.0	32 36.1	35 19.2	38 2.4	40 45.7	43 28.9	46 12.2	48 55.5	13 0	52
10	17 30	26 54.6	29 36.1	32 17.7	34 59.3	37 41.0	40 22.7	43 4.4	45 46.1	48 27.9	12 30	50
12	18 0	26 39.1	29 19.1	31 59.2	34 39.3	37 19.4	39 59.5	42 39.7	45 19.9	48 0.1	12 0	48
14	18 30	26 23.6	29 2.0	31 40.5	34 19.0	36 55.6	39 36.1	42 14.8	44 53.4	47 32.1	11 30	46
16	19 0	26 7.9	28 44.8	31 21.7	33 58.6	36 35.6	39 12.6	41 49.7	44 26.8	47 3.9	11 0	44
18	19 30	25 52.1	28 27.4	31 2.7	33 38.1	36 13.5	38 48.9	41 24.4	43 59.9	46 35.4	10 30	42
20	20 0	25 36.2	28 9.9	30 43.6	33 17.4	35 51.2	38 25.0	40 58.9	43 32.8	46 6.8	10 0	40
22	20 30	25 20.2	27 52.4	30 24.4	32 56.6	35 28.8	38 1.0	40 33.2	43 5.5	45 37.9	9 30	38
24	21 0	25 4.0	27 34.6	30 5.1	32 35.6	35 6.2	37 36.8	40 7.4	42 38.1	45 8.8	9 0	36
26	21 30	24 47.8	27 16.7	29 45.6	32 14.4	34 43.4	37 12.4	39 41.5	42 10.4	44 39.5	8 30	34
28	22 0	24 31.4	26 58.6	29 25.9	31 53.1	34 20.4	36 47.8	39 15.2	41 42.6	44 10.0	8 0	32
30	22 30	24 14.9	26 40.5	29 6.0	31 31.7	33 57.3	36 23.0	38 48.8	41 14.5	43 40.3	7 30	30
32	23 0	23 58.3	26 22.2	28 46.1	31 10.1	33 34.1	35 58.1	38 22.2	40 46.3	43 10.4	7 0	28
34	23 30	23 41.6	26 3.8	28 26.1	30 48.4	33 10.7	35 33.0	37 55.4	39 17.9	42 40.3	6 30	26
36	24 0	23 24.7	25 45.3	28 5.9	30 26.5	32 47.1	35 7.8	37 28.5	39 49.3	42 10.1	6 0	24
38	24 30	23 7.8	25 26.7	27 45.6	30 4.5	32 23.4	34 42.4	37 1.4	39 10.5	41 39.6	5 30	22
40	25 0	22 50.8	25 7.9	27 25.2	29 42.4	31 59.6	34 16.9	36 34.2	38 51.5	41 8.9	5 0	20
42	25 30	22 33.7	24 49.2	27 4.6	29 20.1	31 35.7	33 51.2	35 6.7	38 22.3	40 37.9	4 30	18
44	26 0	22 16.4	24 30.2	26 43.9	28 57.7	31 11.5	33 25.3	35 39.1	37 53.0	40 6.9	4 0	16
46	26 30	22 9.1	24 11.1	26 24.1	28 45.2	30 47.3	32 59.3	35 11.4	37 23.5	39 35.7	3 30	14
48	27 0	21 41.7	23 51.9	26 2.1	28 12.5	30 22.8	32 33.1	34 43.5	36 53.9	39 4.3	3 0	12
50	27 30	21 24.2	23 32.6	25 41.1	27 49.7	29 58.2	32 4.8	34 15.4	36 24.1	38 32.7	2 30	10
52	28 0	21 6.5	23 13.2	25 19.9	27 26.7	29 33.5	31 40.3	33 47.2	35 54.1	38 1.0	2 0	8
54	28 30	20 48.8	22 53.7	24 58.7	27 3.7	29 8.7	31 13.7	33 18.8	35 23.9	37 29.1	1 30	6
56	29 0	20 30.9	22 34.0	24 37.3	26 40.5	28 43.7	30 46.9	32 50.2	34 53.6	36 56.9	1 0	4
58	29 30	20 12.0	22 14.3	24 15.8	26 17.2	28 18.6	30 20.0	32 21.5	34 23.1	36 24.6	0 30	2
IV 0 XVI	30 0	19 54.9	21 54.5	23 54.1	25 53.8	27 53.3	29 53.0	31 52.7	33 52.4	35 52.2	0 0	VIII 0 XX

Signs of R. A. IV with approximate Latitude N. − / S. + ; X with approximate Latitude N. + / S. −

The approximate Latitude = the Ecliptic Declination ± the Declination of the Star or Planet.

The Corrections contained in this Table are calculated for an Obliquity of the Ecliptic = 23° 28′; therefore a further Correction proportionate to the *diminution* must be applied from Table 4, which is adapted for this purpose.

ZODIACAL TABLES.

TABLE 2.

CORRECTION of the APPROXIMATE LONGITUDE, &c. continued.

Signs of R. A. $\overset{s}{\text{I}}$ with approximate Latitude N. + S. − $\qquad$ $\overset{s}{\text{VII}}$ with approximate Latitude N. − S. +

Time. N. + N. − / S. − S. +	Space.	3° 10′	3° 20′	3° 30′	3° 40′	3° 50′	4° 0′	4° 10′	4° 20′	4° 30′	Space.	Time. N. − N. + / S. + S. −
h. m. h.												h. m. h.
II 0 XIV	0° 0′	65′ 34″.9	69′ 2″.3	72′ 30″.0	75′ 57″.7	79′ 25″.4	82′ 53″.1	86′ 21″.0	89′ 48″.9	93′ 16″.9	30° 0′	X 0 XXII
2	0 30	65 14 .9	68 41 .3	72 7 .6	75 34 .5	79 1 .1	82 27 .8	85 54 .1	89 21 .5	92 48 .5	29 30	58
4	1 0	64 54 .7	68 20 .2	71 45 .0	75 11 .1	78 36 .6	82 2 .3	85 28 .0	88 53 .8	92 19 .7	29 0	56
6	1 30	64 34 .1	67 58 .6	71 22 .9	74 47 .2	78 1 .7	81 36 .2	85 0 .9	88 25 .6	91 50 .4	28 30	54
8	2 0	64 13 .3	67 36 .8	70 59 .7	74 23 .1	77 46 .5	81 9 .9	84 33 .5	87 57 .1	91 20 .8	28 0	52
10	2 30	63 52 .1	67 14 .5	70 36 .1	73 58 .5	77 20 .8	80 43 .1	84 5 .6	87 28 .1	90 50 .8	27 30	50
12	3 0	63 30 .7	66 52 .0	70 12 .3	73 33 .7	76 54 .9	80 16 .1	83 37 .4	86 58 .8	90 20 .3	27 0	48
14	3 30	63 8 .9	66 28 .9	69 48 .2	73 8 .5	76 28 .5	79 48 .6	83 8 .2	86 29 .0	89 49 .3	26 30	46
16	4 0	62 46 .9	66 5 .6	69 24 .0	72 43 .1	76 1 .9	79 20 .8	82 39 .8	85 58 .8	89 18 .0	26 0	44
18	4 30	62 24 .5	65 41 .9	69 9 .4	72 17 .2	75 34 .8	78 52 .6	81 10 .4	85 28 .2	88 46 .1	25 30	42
20	5 0	62 2 .0	65 18 .1	68 34 .6	71 51 0	75 7 .5	78 24 .1	81 40 .7	84 57 .3	88 14 .1	25 0	40
22	5 30	61 39 .1	64 54 .0	68 9 .2	71 24 .4	74 39 .7	77 55 .1	80 10 .5	84 26 .0	87 41 .5	24 30	38
24	6 0	61 16 .0	64 29 .8	67 43 .7	70 57 .7	74 11 .8	77 25 .9	80 40 .0	83 54 .3	87 8 .6	24 0	36
26	6 30	60 52 .5	64 5 .1	67 17 .8	70 30 .5	73 43 .2	76 56 .1	79 9 .2	83 22 .2	86 35 .3	23 30	34
28	7 0	60 28 .8	63 40 .2	66 51 .6	70 3 .1	73 14 .6	76 26 .1	79 38 .0	82 49 .7	86 1 .6	23 0	32
30	7 30	60 4 .7	63 14 .9	66 25 .0	69 36 .1	72 45 .5	75 55 .8	79 6 .3	82 16 .9	85 27 .4	22 30	30
32	8 0	59 40 .5	62 49 .4	65 58 .3	69 7 .2	72 16 .2	75 25 .3	78 34 .4	81 43 .7	84 52 .9	22 0	28
34	8 30	59 15 .9	62 23 .5	65 31 .1	68 38 .7	71 45 .9	74 54 .2	78 2 .1	81 10 .0	84 17 .9	21 30	26
36	9 0	58 51 .2	61 57 .4	65 3 .7	68 10 .0	71 16 .4	74 22 .9	77 29 .5	80 36 .1	83 42 .7	21 0	24
38	9 30	58 26 .1	61 30 .9	64 35 .9	67 40 .9	70 46 .0	73 51 .1	76 56 .5	80 1 .7	83 6 .9	20 30	22
40	10 0	58 0 .8	61 4 .3	64 7 .9	67 11 .6	70 15 .4	73 19 .2	76 23 .1	79 27 .0	82 31 .0	20 0	20
42	10 30	58 35 .1	60 37 3	63 39 .5	66 41 .4	69 44 .3	72 46 .8	75 49 .3	78 52 .0	81- 54 .6	19 30	18
44	11 0	57 9 .3	60 10 .1	63 11 .0	66 12 .0	69 13 .0	72 14 .1	75 15 .3	78 16 .5	81 17 .8	19 0	16
46	11 30	56 33 .1	59 42 .5	62 42 .0	65 41 .7	68 41 .3	71 41 .0	74 40 .9	77 40 .6	80 40 .6	18 30	14
48	12 0	56 16 .7	59 14 .8	62 12 .9	65 11 .2	68 9 .4	71 7 .7	74 6 .1	77 4 .6	80 3 .1	18 0	12
50	12 30	55 50 .0	58 46 .7	61 43 .4	64 40 .3	67 37 .0	70 33 .9	73 31 .1	76 28 .1	79 25 .2	17 30	10
52	13 0	55 23 .2	58 18 .4	61 13 .7	64 9 .1	67 4 .5	70 0 .0	72 55 .6	75 51 .2	78 46 .9	17 0	8
54	13 30	54 56 .0	57 49 .7	60 43 .7	63 37 .6	66 31 .6	69 25 .6	72 19 .7	75 14 .0	78 8 .3	16 30	6
56	14 0	54 28 .6	57 20 .9	60 13 .4	63 5 .9	65 58 .4	68 51 .0	71 43 .7	74 36 .4	77 29 .3	16 0	4
58	14 30	54 1 .9	56 52 .2	59 42 .8	62 34 .8	65 24 .9	67 16 .0	71 7 .2	74 58 .6	76 49 .9	15 30	2
III 0 XV	15 0	53 33 .0	56 22 .4	59 11 .9	62 1 .5	64 51 .1	67 40 .8	70 30 .5	73 20 .4	76 10 .2	15 0	IX 0 XXI
2	15 30	53 4 .8	55 52 .7	58 40 .8	61 28 .9	64 17 .0	67 5 .2	69 53 .5	72 41 .8	75 30 .2	14 30	58
4	16 0	52 36 .4	55 22 .9	58 9 .4	60 56 .0	63 42 .6	66 29 .3	69 16 .1	72 2 .8	74 49 .8	14 0	56
6	16 30	52 7 .7	54 52 .7	57 37 .7	60 22 .8	63 7 .9	65 53 .1	68 38 .4	71 23 .6	74 9 .1	13 30	54
8	17 0	51 38 .9	54 22 .4	57 5 .9	59 49 .4	62 33 .0	65 16 .6	68 0 .4	70 44 .1	73 28 .0	13 0	52
10	17 30	51 9 .8	53 51 .7	56 33 .6	59 15 .6	61 57 .7	64 49 .8	67 22 .1	70 4 .3	72 46 .6	12 30	50
12	18 0	50 40 .5	53 20 .8	56 1 .2	58 41 .7	61 22 .2	64 2 .8	66 43 .4	69 24 .1	72 4 .8	12 0	48
14	18 30	50 10 .9	52 49 .6	55 28 .5	58 7 .4	60 46 .3	63 25 .3	66 4 .5	68 43 .6	71 22 .8	11 30	46
16	19 0	49 41 .1	52 18 .3	54 55 .6	57 32 .9	60 10 .3	62 47 .7	65 25 .2	68 2 .7	70 40 .4	11 0	44
18	19 30	49 11 .0	51 46 .6	54 22 .3	56 58 .1	59 33 .8	62 9 .7	64 45 .2	67 21 .7	69 57 .7	10 30	42
20	20 0	48 40 .8	51 14 .8	53 48 .8	56 23 .1	58 57 .2	61 31 .5	64 5 .8	66 40 .2	69 14 .6	10 0	40
22	20 30	48 10 .3	50 42 .7	53 15 .2	55 47 .7	58 20 .3	60 53 .0	63 25 .8	65 58 .4	68 31 .3	9 30	38
24	21 0	47 39 .6	50 10 .4	52 41 .3	55 12 .2	57 43 .2	60 14 .2	62 45 .3	65 16 .4	67 47 .6	9 0	36
26	21 30	46 8 .6	49 37 .8	52 7 .1	54 36 .3	57 5 .7	59 35 .0	62 4 .6	64 34 .1	67 3 .6	8 30	34
28	22 0	46 37 .4	49 5 .1	51 32 .7	54 0 .3	56 28 .0	58 55 .8	61 23 .6	63 51 .4	66 19 .3	8 0	32
30	22 30	46 6 .0	48 32 .0	50 58 .0	53 23 .9	55 50 .1	58 15 .6	60 42 .4	63 8 .6	65 34 .7	7 30	30
32	23 0	45 34 .5	47 58 .8	50 23 .1	52 47 .4	55 11 .9	57 36 .2	60 0 .7	62 25 3	64 49 .8	7 0	28
34	23 30	45 2 .8	47 25 .3	49 48 .0	52 10 .6	54 33 .4	56 56 .1	59 18 .8	61 41 .8	64 4 .7	6 30	26
36	24 0	44 30 .9	46 51 .7	49 12 .7	51 33 .6	53 54 .6	56 15 .7	58 36 .8	60 58 .0	63 19 .2	6 0	24
38	24 30	43 58 .7	46 17 .7	48 37 .1	50 56 .3	53 15 .7	55 35 .0	57 54 .5	60 14 .0	62 33 .5	5 30	22
40	25 0	43 26 .3	45 43 .8	48 1 .3	50 18 .8	52 36 .5	54 54 .1	57 11 .8	59 29 .6	61 47 .4	5 0	20
42	25 30	42 53 .7	45 9 .5	47 25 .3	49 41 .1	51 57 .0	54 12 .9	56 28 .8	58 45 .0	61 1 .1	4 30	18
44	26 0	42 21 .0	44 35 .0	46 49 .1	49 3 .2	51 17 .3	53 31 .5	55 45 .5	58 0 .1	60 14 .4	4 0	16
46	26 30	41 48 .1	44 0 .3	46 12 .6	48 25 .0	50 37 .4	52 49 .8	55 2 .0	57 15 .0	59 27 .6	3 30	14
48	27 0	41 15 .0	43 25 .4	45 35 .9	47 46 .6	49 57 .2	52 7 .9	54 18 .2	56 29 .5	58 40 .4	3 0	12
50	27 30	40 41 .5	42 50 .3	44 59 .0	47 8 .0	49 16 .8	51 25 .8	53 34 .3	55 43 .9	57 53 .0	2 30	10
52	28 0	40 7 9	42 15 .0	44 22 .0	46 29 .1	48 36 .2	50 43 .4	52 50 .0	54 57 .9	57 5 .3	2 0	8
54	28 30	39 34 .2	41 39 .5	43 44 .7	45 50 .0	47 55 .3	50 0 .8	52 5 .7	54 11 .7	56 17 .4	1 30	6
56	29 0	39 0 .3	41 3 .8	43 7 .3	45 10 .8	47 14 .3	49 17 .9	51 21 .1	53 25 .2	55 29 .1	1 0	4
58	29 30	38 26 .2	40 27 .9	42 29 .6	44 31 .3	46 33 .1	48 34 .9	50 36 .6	52 38 .7	54 40 .7	0 30	2
IV 0 XVI	30 0	37 52 .0	39 51 .8	41 51 .7	43 51 .6	45 51 .6	47 51 .6	49 51 .7	51 51 .8	53 51 .9	0 0	VIII 0 XX

Signs of R. A. $\overset{s}{\text{IV}}$ with approximate Latitude N. − S. + $\qquad$ $\overset{s}{\text{X}}$ with approximate Latitude N. + S. −

The approximate Latitude = the Ecliptic Declination ± the Declination of the Star or Planet.

The Corrections contained in this Table are calculated for an Obliquity of the Ecliptic = 23° 28′; therefore a further Correction proportionate to the *diminution* must be applied from TABLE 4, which is adapted for this purpose.

TABLE 2.

Correction of the approximate Longitude, &c. continued.

Signs of R. A. I^s with approximate Latitude N. + S. − | VII^s with approximate Latitude N. − S. +

Time. N. + N. − S. − S. +	Space.	4° 40′	4° 50′	5° 0′	5° 10′	5° 20′	5° 30′	5° 40′	5° 50′	6° 0′	Space.	Time. N. − N. + S. + S. −
h. m. h.												h. m. h.
II 0 XIV	0° 0′	96′ 45″.0	100′ 13″.2	103′ 41″.5	107′ 9″.8	110′ 38″.3	114′ 6″.8	117′ 35″.6	121′ 4″.2	124′ 33″.1	30° 0′	X 0 XXII
2	0 30	96 15.5	99 42.6	103 14.7	106 37.1	110 4.5	113 32.0	116 59.7	120 27.3	123 55.1	29 30	58
4	1 0	95 45.7	99 11.7	102 37.8	106 4.1	109 30.4	112 56.8	116 23.4	119 50.0	123 16.8	29 0	56
6	1 30	95 15.3	98 40.3	102 5.3	105 30.4	108 55.7	112 21.0	115 46.4	119 12.0	122 37.6	28 30	54
8	2 0	94 44.6	98 8.4	101 32.4	104 56.4	108 20.6	111 44.8	115 9.1	118 33.6	121 58.1	28 0	52
10	2 30	94 13.4	97 36.1	100 58.9	104 21.8	107 44.8	111 7.9	114 31.1	117 54.5	121 17.7	27 30	50
12	3 0	93 41.7	97 3.4	100 25.1	103 46.8	107 8.7	110 30.7	113 52.8	117 15.0	120 37.3	27 0	48
14	3 30	93 9.7	96 30.2	99 50.7	103 11.2	106 32.0	109 52.8	113 13.8	116 34.8	119 55.9	26 30	46
16	4 0	92 37.2	95 56.5	99 15.9	102 35.3	105 54.9	109 14.6	112 34.4	115 54.2	119 14.2	26 0	44
18	4 30	92 4.2	95 22.3	98 40.6	101 58.8	105 17.2	108 35.7	111 54.3	115 13.0	118 31.7	25 30	42
20	5 0	91 30.9	94 47.8	98 4.9	101 22.0	104 39.2	107 56.5	111 13.9	114 31.4	117 48.9	25 0	40
22	5 30	90 57.0	94 12.9	97 28.7	100 44.6	104 0.5	107 16.6	110 32.8	113 49.1	117 5.4	24 30	38
24	6 0	90 23.0	93 37.7	96 52.1	100 6.8	103 21.5	106 36.4	109 51.4	113 6.4	116 21.6	24 0	36
26	6 30	89 48.4	93 1.8	96 15.1	99 28.5	102 41.9	105 55.6	109 9.4	112 22.9	115 37.0	23 30	34
28	7 0	89 13.5	92 25.5	95 37.6	98 49.8	102 2.0	105 14.4	108 27.0	111 39.1	114 52.0	23 0	32
30	7 30	88 38.1	91 48.3	94 59.6	98 10.5	101 21.5	104 32.6	107 43.8	110 54.7	114 6.9	22 30	30
32	8 0	88 2.3	91 11.8	94 21.3	97 30.9	100 40.6	103 50.4	107 0.3	110 10.0	113 20.4	22 0	28
34	8 30	87 26.1	90 34.2	93 42.5	96 50.8	99 59.2	103 7.7	106 16.5	109 24.8	112 33.7	21 30	26
36	9 0	86 49.5	89 56.3	93 3.3	96 10.3	99 17.4	102 24.6	105 31.9	108 39.2	111 46.7	21 0	24
38	9 30	86 12.5	89 18.0	92 23.6	95 30.3	98 35.0	101 40.9	104 46.9	107 52.9	110 59.0	20 30	22
40	10 0	85 35.1	88 39.3	91 43.6	94 47.9	97 52.3	100 56.8	104 1.5	107 6.2	110 11.0	20 0	20
42	10 30	84 57.2	88 0.1	91 3.0	94 6.0	97 9.1	100 12.2	103 15.5	106 18.9	109 22.3	19 30	18
44	11 0	84 19.2	87 20.6	90 22.1	93 23.7	96 25.5	99 27.3	102 29.2	105 31.2	108 33.2	19 0	16
46	11 30	83 40.6	86 40.7	89 40.7	92 40.9	95 41.4	98 41.8	101 42.3	104 42.9	107 43.4	18 30	14
48	12 0	83 1.7	86 0.4	88 58.9	91 57.7	94 56.9	97 55.9	100 55.0	103 54.2	106 53.3	18 0	12
50	12 30	82 22.3	85 19.6	88 16.9	91 14.2	94 11.9	97 9.5	100 7.2	103 4.9	106 2.6	17 30	10
52	13 0	81 42.6	84 38.5	87 34.4	90 30.4	93 26.5	96 22.7	99 19.0	102 15.3	105 11.5	17 0	8
54	13 30	81 2.6	83 57.0	86 51.4	89 46.3	92 40.7	95 35.5	98 30.2	101 25.1	104 19.9	16 30	6
56	14 0	80 22.1	83 15.1	86 8.1	89 1.8	91 54.5	94 47.9	97 41.1	100 34.6	103 28.0	16 0	4
58	14 30	79 41.2	82 32.9	85 24.4	88 16.8	91 7.8	93 59.7	96 51.5	99 43.5	102 35.5	15 30	2
III 0 XV	15 0	79 0.2	81 50.2	84 40.3	87 31.5	90 20.7	93 11.2	96 1.5	98 52.0	101 42.7	15 0	IX 0 XXI
2	15 30	78 18.6	81 7.1	83 55.8	86 45.4	89 33.3	92 22.1	95 11.1	98 0.0	100 49.2	14 30	58
4	16 0	77 36.6	80 23.7	83 10.9	85 58.9	88 45.5	91 32.7	94 20.4	97 7.7	99 55.3	14 0	56
6	16 30	76 54.4	79 40.0	82 25.6	85 11.8	87 57.2	90 42.8	93 28.9	96 14.8	99 0.9	13 30	54
8	17 0	76 11.9	78 55.9	81 40.0	84 24.4	87 8.5	89 52.6	92 37.0	95 21.5	98 6.1	13 0	52
10	17 30	75 29.0	78 11.5	80 53.8	83 36.7	86 19.3	89 1.9	91 44.8	94 27.8	97 10.8	12 30	50
12	18 0	74 45.7	77 26.6	80 7.3	82 48.6	85 29.7	88 10.9	90 52.3	93 33.7	96 15.1	12 0	48
14	18 30	74 2.1	76 41.4	79 20.5	82 0.3	84 39.8	87 19.4	89 59.2	92 39.0	95 18.9	11 30	46
16	19 0	73 18.1	75 55.8	78 33.3	81 11.6	83 49.6	86 27.6	89 5.8	91 44.0	94 22.3	11 0	44
18	19 30	72 33.6	75 9.9	77 45.8	80 22.5	82 58.9	85 35.3	88 11.9	90 58.5	93 25.2	10 30	42
20	20 0	71 48.8	74 23.7	76 58.0	79 33.1	82 7.9	84 42.7	87 17.7	89 52.7	92 27.8	10 0	40
22	20 30	71 3.7	73 37.1	76 10.0	78 43.3	81 16.4	83 49.7	86 23.0	88 56.4	91 29.9	9 30	38
24	21 0	70 18.4	72 50.2	75 21.6	77 53.1	80 24.6	82 56.3	85 28.0	87 59.8	90 31.7	9 0	36
26	21 30	69 32.9	72 3.0	74 32.7	77 2.6	79 32.5	82 2.5	84 32.5	87 2.7	89 32.9	8 30	34
28	22 0	68 47.0	71 15.4	73 43.5	76 11.7	78 40.0	81 8.3	83 36.7	86 5.2	88 33.8	8 0	32
30	22 30	68 1.0	70 27.5	72 53.9	75 20.4	77 47.5	80 13.7	82 40.5	85 7.3	87 34.3	7 30	30
32	23 0	67 14.6	69 39.2	72 4.0	74 28.8	76 54.4	79 18.8	81 43.9	84 9.1	86 34.4	7 0	28
34	23 30	66 27.9	68 50.7	71 13.8	73 37.0	76 0.8	78 23.5	80 47.0	83 10.4	85 34.0	6 30	26
36	24 0	65 40.9	68 1.8	70 23.3	72 44.8	75 6.8	77 27.9	79 49.7	82 11.4	84 33.3	6 0	24
38	24 30	64 53.4	67 12.7	69 32.4	71 52.2	74 12.3	76 32.0	78 52.0	81 12.0	83 32.2	5 30	22
40	25 0	64 5.5	66 23.2	68 41.2	70 59.3	73 17.4	75 35.6	77 53.9	80 12.2	82 30.7	5 0	20
42	25 30	63 17.4	65 33.4	67 49.7	70 6.0	72 22.4	74 38.9	76 55.5	79 12.0	81 28.8	4 30	18
44	26 0	62 29.0	64 43.3	66 57.9	69 12.4	71 27.0	73 41.9	75 56.7	78 11.7	80 26.6	4 0	16
46	26 30	61 37.3	63 53.1	66 5.8	68 18.6	70 31.4	72 44.5	74 57.6	77 10.8	79 24.0	3 30	14
48	27 0	60 52.3	63 2.5	65 13.3	67 24.5	69 35.5	71 46.8	73 58.2	76 9.6	78 21.0	3 0	12
50	27 30	60 1.2	62 11.5	64 20.6	66 30.1	68 39.3	70 48.7	72 58.4	75 8.0	77 17.7	2 30	10
52	28 0	59 12.7	61 20.1	63 27.6	65 35.2	67 42.8	69 50.5	71 58.3	74 6.1	76 14.0	2 0	8
54	28 30	58 23.0	60 28.6	62 34.3	64 40.1	66 45.9	68 51.8	70 57.9	73 3.9	75 10.0	1 30	6
56	29 0	57 32.9	59 36.8	61 40.7	63 44.7	65 48.7	67 52.8	69 57.0	72 1.3	74 5.6	1 0	4
58	29 30	56 42.7	58 44.8	60 46.9	62 49.0	64 51.3	66 53.6	68 55.9	70 58.4	73 3.9	0 30	2
IV 0 XVI	30 0	55 52.1	57 52.4	59 52.7	61 53.0	63 53.5	65 54.0	67 54.5	69 55.1	71 55.8	0 0	VIII 0 XX

Signs of R. A. IV^s with approximate Latitude N. − S. + | X^s with approximate Latitude N. + S. −

The approximate Latitude = the Ecliptic Declination ± the Declination of the Star or Planet.

The Corrections contained in this Table are calculated for an Obliquity of the Ecliptic = 23° 28′; therefore a further Correction proportionate to the *diminution* must be applied from Table 4, which is adapted for this purpose.

ZODIACAL TABLES.

TABLE 2.

Correction of the approximate Longitude, &c. continued.

Signs of R. A. II with approximate Latitude N. + S. − | VIII with approximate Latitude N. − S. +

Time. N. + S. − / N. − S. +	Space.	0° 10′	0° 20′	0° 30′	0° 40′	0° 50′	1° 0′	1° 10′	1° 20′	1° 30′	Space.	Time. N. − S. + / N. + S. −
h. m. h.												h. m. h.
IV 0 XVI	0° 0′	1′ 59″.5	3′ 58″.9	5′ 58″.4	7′ 57″.9	9′ 57″.3	11′ 56″.8	13′ 56″.3	15′ 55″.9	17′ 55″.4	30° 0′	VIII 0 XX
2	0 30	1 57.7	3 55.3	5 53.0	7 50.7	9 48.3	11 46.0	13 43.6	15 41.4	17 39.0	29 30	58
4	1 0	1 55.9	3 51.7	5 47.5	7 43.4	9 39.2	11 35.1	13 30.9	15 26.8	17 22.5	29 0	56
6	1 30	1 54.1	3 48.1	5 42.0	7 33.1	9 30.1	11 24.1	13 18.1	15 12.2	17 6.2	28 30	54
8	2 0	1 52.2	3 44.4	5 36.5	7 28.7	9 20.9	11 13.1	13 5.3	14 57.5	16 49.7	28 0	52
10	2 30	1 50.4	3 40.7	5 31.0	7 21.3	9 11.7	11 2.0	12 52.4	14 42.7	16 33.1	27 30	50
12	3 0	1 48.5	3 36.9	5 25.4	7 13.9	9 2.4	10 50.8	12 39.4	14 27.9	16 16.4	27 0	48
14	3 30	1 46.7	3 33.1	5 19.9	7 6.4	8 53.1	10 39.7	12 26.4	14 13.0	15 59.7	26 30	46
16	4 0	1 44.8	3 29.4	5 14.3	6 58.9	8 43.7	10 28.5	12 13.3	13 58.0	15 42.9	26 0	44
18	4 30	1 42.9	3 25.7	5 9.6	6 51.4	8 34.3	10 17.2	12 0.1	13 48.0	15 25.9	25 30	42
20	5 0	1 41.0	3 21.9	5 2.9	6 43.9	8 24.9	10 5.9	11 46.9	13 27.9	15 8.9	25 0	40
22	5 30	1 39.1	3 18.2	4 57.3	6 36.3	8 15.4	9 54.5	11 33.6	13 12.8	14 51.9	24 30	38
24	6 0	1 37.2	3 14.4	4 51.6	6 28.7	8 5.9	9 43.1	11 20.3	12 57.6	14 34.8	24 0	36
26	6 30	1 35.3	3 10.6	4 45.8	6 21.1	7 56.4	9 31.7	11 6.9	12 42.3	14 17.6	23 30	34
28	7 0	1 33.4	3 6.8	4 40.0	6 13.4	7 46.8	9 20.2	10 53.5	12 27.0	14 0.4	23 0	32
30	7 30	1 31.4	3 2.9	4 34.3	6 5.7	7 37.2	9 8.7	10 40.1	12 11.6	13 43.1	22 30	30
32	8 0	1 29.5	2 59.0	4 28.5	5 58.0	7 27.5	8 57.1	10 26.6	11 56.1	13 25.7	22 0	28
34	8 30	1 27.6	2 55.1	4 22.7	5 50.2	7 17.8	8 45.5	10 13.0	11 40.6	13 8.3	21 30	26
36	9 0	1 25.7	2 51.2	4 16.9	5 42.4	7 8.1	8 33.8	9 59.4	11 25.1	12 50.8	21 0	24
38	9 30	1 23.7	2 47.4	4 11.1	5 34.6	6 58.4	8 22.1	9 45.8	11 9.5	12 33.2	20 30	22
40	10 0	1 21.8	2 43.5	4 5.2	5 26.8	6 48.6	8 10.3	9 32.1	10 53.9	12 15.6	20 0	20
42	10 30	1 19.8	2 39.6	3 59.3	5 19.0	6 38.8	7 58.5	9 18.4	10 38.2	11 57.9	19 30	18
44	11 0	1 17.9	2 35.6	3 53.4	5 11.1	6 28.9	7 46.7	9 4.6	10 22.4	11 40.2	19 0	16
46	11 30	1 15.9	2 31.7	3 47.5	5 3.3	6 19.1	7 34.9	8 50.8	10 6.6	11 22.4	18 30	14
48	12 0	1 13.9	2 27.7	3 41.5	4 55.4	6 9.2	7 23.0	8 36.9	9 50.7	11 4.6	18 0	12
50	12 30	1 11.9	2 23.7	3 35.5	4 47.4	5 59.3	7 11.1	8 23.0	9 34.8	10 46.7	17 30	10
52	13 0	1 9.8	2 19.7	3 29.5	4 39.4	5 49.3	6 59.1	8 9.0	9 18.9	10 28.8	17 0	8
54	13 30	1 7.9	2 15.7	3 23.6	4 31.4	5 39.3	6 47.2	7 55.0	9 2.9	10 10.8	16 30	6
56	14 0	1 6.0	2 11.7	3 17.6	4 23.4	5 29.3	6 35.2	7 41.0	8 46.9	9 52.8	16 0	4
58	14 30	1 4.0	2 7.7	3 11.6	4 15.4	5 19.4	6 23.1	7 27.0	8 30.9	9 34.7	15 30	2
V 0 XVII	15 0	1 1.9	2 3.7	3 5.5	4 7.4	5 9.2	6 11.0	7 12.9	8 14.8	9 16.6	15 0	VII 0 XIX
2	15 30	0 59.9	1 59.7	2 59.5	3 59.3	4 59.1	5 59.0	6 58.8	7 58.7	8 58.5	14 30	58
4	16 0	0 57.9	1 55.7	2 53.4	3 51.2	4 49.0	5 46.9	6 44.7	7 42.5	8 40.3	14 0	56
6	16 30	0 55.9	1 51.6	2 47.3	3 43.1	4 38.9	5 34.7	6 30.5	7 26.3	8 22.1	13 30	54
8	17 0	0 53.9	1 47.5	2 41.2	3 35.0	4 28.7	5 22.5	6 16.3	7 10.0	8 3.8	13 0	52
10	17 30	0 51.9	1 43.5	2 35.1	3 26.8	4 18.6	5 10.3	6 2.1	6 53.8	7 45.5	12 30	50
12	18 0	0 49.9	1 39.4	2 29.0	3 18.7	4 8.4	4 58.1	5 47.8	6 37.5	7 27.2	12 0	48
14	18 30	0 47.9	1 35.3	2 22.9	3 10.6	3 58.2	4 45.9	5 33.5	6 21.1	7 8.9	11 30	46
16	19 0	0 45.8	1 31.2	2 16.8	3 2.4	3 48.0	4 33.6	5 19.1	6 4.7	6 50.5	11 0	44
18	19 30	0 43.7	1 27.2	2 10.7	2 54.2	3 37.8	4 21.3	5 4.8	5 48.4	6 32.2	10 30	42
20	20 0	0 41.6	1 23.1	2 4.6	2 46.0	3 27.5	4 8.9	4 50.4	5 32.0	6 13.8	10 0	40
22	20 30	0 39.6	1 19.0	1 58.4	2 38.3	3 17.2	3 56.6	4 36.1	5 15.5	5 55.4	9 30	38
24	21 0	0 37.6	1 14.9	1 52.2	2 29.5	3 6.9	3 44.3	4 21.7	4 59.0	5 36.9	9 0	36
26	21 30	0 35.6	1 10.8	1 45.9	2 21.3	2 56.6	3 31.9	4 7.3	4 42.6	5 18.4	8 30	34
28	22 0	0 33.6	1 6.6	1 39.7	2 13.0	2 46.3	3 19.5	3 52.8	4 26.1	4 59.9	8 0	32
30	22 30	0 31.5	1 2.5	1 33.6	2 4.8	2 36.0	3 7.2	3 38.4	4 9.6	4 41.3	7 30	30
32	23 0	0 29.4	0 58.3	1 27.4	1 56.5	2 25.6	2 54.8	3 23.9	3 53.0	4 22.6	7 0	28
34	23 30	0 27.4	0 54.1	1 21.2	1 48.2	2 15.6	2 42.4	3 9.4	3 36.4	4 3.8	6 30	26
36	24 0	0 25.5	0 49.9	1 15.0	1 39.9	2 4.9	2 29.9	2 54.9	3 19.8	3 45.0	6 0	24
38	24 30	0 23.5	0 45.9	1 8.8	1 31.7	1 54.6	2 17.5	2 40.4	3 3.2	3 26.2	5 30	22
40	25 0	0 21.6	0 41.8	1 2.6	1 23.4	1 44.2	2 5.0	2 25.8	2 46.6	3 7.4	5 0	20
42	25 30	0 19.4	0 37.7	0 56.3	1 13.1	1 33.8	1 52.5	2 11.3	2 30.0	2 48.7	4 30	18
44	26 0	0 17.3	0 33.6	0 50.0	1 6.8	1 23.4	1 40.0	1 56.7	2 13.4	2 30.0	4 0	16
46	26 30	0 15.3	0 29.6	0 43.9	0 58.4	1 13.0	1 27.6	1 42.2	1 56.8	2 11.4	3 30	14
48	27 0	0 13.2	0 25.5	0 37.7	0 50.0	1 2.6	1 15.2	1 27.6	1 40.1	1 52.7	3 0	12
50	27 30	0 11.0	0 21.5	0 31.6	0 41.9	0 52.3	1 2.6	1 13.0	1 23.5	1 34.0	2 30	10
52	28 0	0 8.8	0 17.3	0 25.5	0 33.7	0 41.9	0 50.0	0 58.4	1 6.8	1 15.2	2 0	8
54	28 30	0 6.6	0 13.1	0 19.4	0 25.6	0 31.5	0 37.5	0 43.8	0 50.1	0 56.5	1 30	6
56	29 0	0 4.4	0 8.8	0 13.2	0 17.4	0 21.0	0 25.0	0 29.2	0 33.4	0 37.7	1 0	4
58	29 30	0 2.2	0 4.4	0 6.6	0 8.7	0 10.5	0 12.5	0 14.6	0 16.7	0 18.8	0 30	2
VI 0 XVIII	30 0	0 0.0	0 0.0	0 0.0	0 0.0	0 0.0	0 0.0	0 0.0	0 0.0	0 0.0	0 0	VI 0 XVIII

Signs of R. A. III with approximate Latitude N. − S. + | IX with approximate Latitude N. + S. −

The approximate Latitude = the Ecliptic Declination ± the Declination of the Star or Planet.

The Corrections contained in this Table are calculated for an Obliquity of the Ecliptic = 23° 28′; therefore a further Correction proportionate to the *diminution* must be applied from Table 4, which is adapted for this purpose.

TABLE 2.

Correction of the approximate Longitude, &c. continued.

Signs of R. A. $\overset{s}{\text{II}}$ with approximate Latitude N. + S. − ; $\overset{s}{\text{VIII}}$ with approximate Latitude N. − S. +

Time. N. + S. − (h. m.)	Time. N. − S. + (h.)	Space.	1° 40′	1° 50′	2° 0′	2° 10′	2° 20′	2° 30′	2° 40′	2° 50′	3° 0′	Space.	Time. N. − S. + (h. m.)	Time. N. + S. − (h.)
IV 0	XVI	0° 0′	19′ 54″.9	21′ 54″.5	23′ 54″.1	25′ 53″.8	27′ 53″.3	29′ 53″.0	31′ 52″.7	33′ 52″.4	35′ 52″.2	30° 0′	VIII 0	XX
2		0 30	19 36.9	21 34.6	23 32.3	25 30.2	27 28.0	29 25.8	31 23.8	33 21.6	35 19.5	29 30	58	
4		1 0	19 18.7	21 14.6	23 10.5	25 6.5	27 2.6	28 58.6	30 54.6	32 50.7	34 46.8	29 0	56	
6		1 30	19 0.4	20 54.5	22 48.6	24 42.7	26 37.0	28 31.2	30 25.3	32 19.6	34 13.8	28 30	54	
8		2 0	18 42.0	20 34.3	22 26.6	24 18.8	26 11.2	28 3.6	29 55.9	31 48.4	33 40.9	28 0	52	
10		2 30	18 23.6	20 14.0	22 4.3	23 54.8	25 45.4	27 35.8	29 26.6	31 16.9	33 7.6	27 30	50	
12		3 0	18 5.0	19 53.6	21 42.1	23 30.8	25 19.4	27 8.0	28 56.7	30 45.4	32 34.3	27 0	48	
14		3 30	17 46.4	19 33.2	21 19.8	23 6.6	24 53.3	26 40.1	28 26.9	30 13.7	32 0.6	26 30	46	
16		4 0	17 27.6	19 12.6	20 57.4	22 42.3	24 27.1	26 12.0	27 57.0	29 41.9	31 26.9	26 0	44	
18		4 30	17 8.9	18 51.9	20 35.8	22 17.8	24 0.8	25 43.8	27 26.9	29 9.9	30 53.0	25 30	42	
20		5 0	16 50.0	18 31.1	20 12.2	21 53.3	23 34.4	35 15.6	26 56.7	28 37.9	30 19.1	25 0	40	
22		5 30	16 31.0	18 10.2	19 49.4	21 28.6	23 7.8	24 47.2	26 26.4	28 5.7	29 45.0	24 30	38	
24		6 0	16 12.0	17 49.3	19 26.6	21 3.9	22 41.2	24 18.6	25 56.0	27 33.4	29 10.8	24 0	36	
26		6 30	15 52.9	17 28.3	19 3.7	20 39.1	22 14.5	23 50.0	25 25.4	27 0.9	28 36.4	23 30	34	
28		7 0	15 33.8	17 7.2	18 40.7	20 14.2	21 47.7	23 21.2	24 54.7	26 28.4	28 1.9	23 0	32	
30		7 30	15 14.6	16 46.1	18 17.6	19 44.2	21 20.7	22 52.4	24 23.9	25 55.6	27 27.2	22 30	30	
32		8 0	14 55.3	16 24.9	17 54.5	19 24.1	20 53.7	22 23.4	23 53.0	25 22.8	27 52.5	22 0	28	
34		8 30	14 36.9	16 3.6	17 31.2	18 58.9	20 26.6	21 54.3	22 22.0	24 49.8	26 32.6	21 30	26	
36		9 0	14 16.4	15 42.2	17 7.9	18 33.6	19 59.4	21 25.1	22 50.9	24 16.7	25 42.6	21 0	24	
38		9 30	13 56.9	15 20.7	16 44.5	18 8.2	19 32.1	20 55.8	22 19.7	23 43.5	25 7.5	20 30	22	
40		10 0	13 37.4	14 59.2	16 21.0	17 42.8	19 4.7	20 26.5	21 48.4	23 10.3	24 32.4	20 0	20	
42		10 30	13 17.3	14 37.6	15 57.4	17 17.3	18 37.2	19 57.0	21 17.0	22 36.9	23 57.0	19 30	18	
44		11 0	12 58.1	14 15.9	15 33.8	16 51.7	18 9.6	19 27.5	20 45.5	22 3.4	23 21.5	19 0	16	
46		11 30	12 38.4	13 54.2	15 10.1	16 26.0	17 41.9	18 57.9	20 13.9	21 29.8	22 45.9	18 30	14	
48		12 0	12 18.6	13 32.4	14 46.3	16 0.2	17 14.2	18 28.2	19 42.2	20 56.2	22 10.2	18 0	12	
50		12 30	11 58.7	13 10.5	14 22.5	15 34.4	16 46.4	17 58.4	19 10.4	20 22.4	21 34.4	17 30	10	
52		13 0	11 38.7	12 48.6	13 58.6	15 8.5	16 18.5	17 28.5	18 38.5	19 48.5	20 58.5	17 0	8	
54		13 30	11 18.7	12 26.7	13 34.6	14 42.5	15 50.5	16 58.5	18 6.5	19 14.5	20 22.5	16 30	6	
56		14 0	10 58.7	12 4.7	13 10.6	14 16.5	15 22.5	16 28.5	17 34.5	18 40.5	19 46.5	16 0	4	
58		14 30	10 38.6	11 42.6	12 36.5	13 50.4	14 54.4	15 58.4	17 2.3	18 6.3	19 10.3	15 30	2	
V 0	XVII	15 0	10 18.5	11 20.4	12 12.3	13 24.3	14 26.2	15 28.2	16 30.1	17 32.1	18 34.1	15 0	VII 0	XIX
2		15 30	9 58.4	10 58.2	11 58.1	13 58.1	13 57.9	14 57.9	15 57.8	16 57.8	17 57.7	14 30	58	
4		16 0	9 38.2	10 36.0	11 33.9	12 31.8	13 29.6	14 27.5	15 25.5	16 23.4	17 21.3	14 0	56	
6		16 30	9 18.0	10 13.7	11 9.6	12 5.4	13 1.3	13 57.1	14 53.0	15 48.9	16 44.8	13 30	54	
8		17 0	8 57.7	9 51.4	10 45.2	11 39.0	12 32.9	13 26.7	14 20.5	15 14.4	16 8.3	13 0	52	
10		17 30	8 37.3	9 29.0	10 20.8	11 12.6	12 4.4	12 56.1	13 48.0	14 39.8	15 31.6	12 30	50	
12		18 0	8 16.9	9 6.6	9 56.3	10 46.1	11 35.8	12 25.5	13 15.4	14 5.1	14 54.9	12 0	48	
14		18 30	7 56.5	8 44.1	9 31.8	10 19.5	11 7.2	11 54.9	12 42.7	13 30.4	14 18.1	11 30	46	
16		19 0	7 36.0	8 21.6	9 7.3	9 52.9	10 38.6	11 24.3	12 9.9	12 55.6	13 41.3	11 0	44	
18		19 30	7 15.5	7 59.1	8 42.7	9 26.3	10 9.9	10 53.6	11 37.1	12 20.8	13 4.4	10 30	42	
20		20 0	6 55.0	7 36.5	8 18.0	8 59.6	9 41.1	10 22.8	11 4.3	11 45.9	12 27.5	10 0	40	
22		20 30	6 34.5	7 13.9	7 53 4	8 32.9	9 12.4	9 51.9	10 28.4	11 10.9	11 50.5	9 30	38	
24		21 0	6 13.9	6 51.3	7 28.7	8 6.1	8 43.6	9 21.0	9 58.4	10 35.9	11 13.4	9 0	36	
26		21 30	5 53.3	6 28.6	7 4.0	7 39.3	8 14.7	8 50.1	8 22.4	10 0.8	10 36.3	8 30	34	
28		22 0	5 32.7	6 5.9	6 39.2	7 12.5	7 45.8	8 19.1	8 52.4	9 25.7	9 59.1	8 0	32	
30		22 30	5 12.0	5 43.2	6 14.4	6 45.6	7 16.9	7 48.1	8 19.3	8 50.6	9 22.9	7 30	30	
32		23 0	4 51.3	5 20.4	5 49.5	6 18.7	6 47.9	7 17.0	7 46.2	8 15.4	8 44.6	7 0	28	
34		23 30	4 30.6	4 57.6	5 24.7	5 51.7	6 18.9	6 45.9	7 13.0	7 40.2	8 7.2	6 30	26	
36		24 0	4 9.8	4 34.8	4 59.8	5 24.8	5 49.8	6 14.8	6 39.8	7 4.9	7 29.8	6 0	24	
38		24 30	3 49.1	4 12.0	4 34.9	4 57.8	5 20.8	5 43.7	6 6.6	6 29.6	6 52.5	5 30	22	
40		25 0	3 28.3	3 49.1	4 9.9	4 30.8	4 51.7	5 12.5	5 33.4	5 54.3	6 15.1	5 0	20	
42		25 30	3 7.5	3 26.3	3 45.0	4 3.8	4 22.6	4 41.3	5 0.1	5 19.0	5 37.7	4 30	18	
44		26 0	2 46.7	3 3.4	3 20.0	3 36.7	3 53.5	4 10.1	4 26.8	4 43.6	5 0.3	4 0	16	
46		26 30	2 26.0	2 40.5	2 55.1	3 9.7	3 24.4	3 38.9	3 53.5	4 8.2	4 22.8	3 30	14	
48		27 0	2 5.2	2 17.6	2 30.1	2 42.6	2 55.2	3 7.7	3 20.2	3 32.7	3 45.3	3 0	12	
50		27 30	1 44.4	1 54.7	2 0.1	2 15.6	2 26.3	2 36.6	2 46.9	2 57.3	3 7.8	2 30	10	
52		28 0	1 23.5	1 31.8	1 40.1	1 48.5	1 57.0	2 5.4	2 13.6	2 21.9	2 30.2	2 0	8	
54		28 30	1 2.7	1 8.9	1 15.0	1 11.4	1 27.8	1 34.1	1 40.2	1 46.5	1 52.8	1 30	6	
56		29 0	0 41.8	0 46.0	0 50.0	0 54.3	0 58.5	1 2.7	1 6.9	1 11.0	1 15.3	1 0	4	
58		29 30	0 20.9	0 23.0	0 25.0	0 27.2	0 29.3	0 31.4	0 33.3	0 35.5	0 37.7	0 30	2	
VI 0	XVIII	30 0	0 0.0	0 0.0	0 0.0	0 0.0	0 0.0	0 0.0	0 0.0	0 0.0	0 0.0	0 0	VI 0	XVIII

Signs of R. A. $\overset{s}{\text{III}}$ with approximate Latitude N. − S. + ; $\overset{s}{\text{IX}}$ with approximate Latitude N. + S. −

The approximate Latitude = the Ecliptic Declination ± the Declination of the Star or Planet.

The Corrections contained in this Table are calculated for an Obliquity of the Ecliptic = 23° 28′; therefore a further Correction proportionate to the *diminution* must be applied from Table 4, which is adapted for this purpose.

TABLE 2.

Correction of the approximate Longitude, &c. continued.

Signs of R. A. II^s with approximate Latitude N. + / S. − ; VIII^s with approximate Latitude N. − / S. +

Time. N. + N. − / S. − S. +	Space.	3° 10′	3° 20′	3° 30′	3° 40′	3° 50′	4° 0′	4° 10′	4° 20′	4° 30′	Space.	Time. N. − N. + / S. + S. −
h. m. h.												h. m. h.
IV 0 XVI	0° 0′	37′ 52″.0	39′ 51″.8	41′ 51″.7	43′ 51″.6	45′ 51 .6	47′ 51″.6	49′ 51″.7	51′ 51″.8	53′ 51″.9	30° 0′	VIII 0 XX
2	0 30	37 17 .6	39 16 .6	41 13 .6	43 11 .7	45 9 .9	47 8 .1	49 6 .3	51 4 .6	53 2 .9	29 30	58
4	1 0	36 43 .0	38 39 .2	40 35 .4	42 31 .7	44 28 .0	46 24 .4	48 20 .8	50 17 .2	52 13 .7	29 0	56
6	1 30	36 8 .2	38 2 .6	39 57 .0	41 51 .4	43 45 .9	45 40 .4	47 35 .0	49 29 .6	51 24 .3	28 30	54
8	2 0	35 33 .3	37 25 .8	39 18 .4	41 11 .0	43 3 .6	44 56 .3	46 49 .0	48 41 .8	50 34 .6	28 0	52
10	2 30	34 58 .2	36 48 .9	38 39 .6	40 30 .3	42 21 .2	44 11 .9	46 2 .8	47 53 .6	49 44 .6	27 30	50
12	3 0	34 22 .9	36 11 .8	38 0 .6	39 49 .5	41 38 .5	43 27 .4	45 16 .4	47 5 .3	48 54 .5	27 0	48
14	3 30	33 47 .5	35 34 .5	37 21 .5	39 8 .5	40 55 .6	42 42 .6	44 39 .7	46 16 .8	48 4 .1	26 30	46
16	4 0	33 12 .0	34 57 .1	36 42 .2	38 27 .3	40 12 .5	41 57 .7	43 42 .9	45 28 .2	47 13 .6	26 0	44
18	4 30	32 36 .3	34 19 .6	36 2 .7	37 45 .9	39 29 .2	41 12 .5	42 55 .9	44 39 .3	46 22 .8	25 30	42
20	5 0	32 0 .5	33 41 .9	35 23 .1	37 4 .4	38 45 .8	40 27 .2	42 8 .7	43 50 .2	45 31 .8	25 0	40
22	5 30	31 24 .5	33 3 .9	34 43 .2	36 22 .7	38 2 .2	39 41 .7	41 21 .3	43 0 .9	44 40 .5	24 30	38
24	6 0	30 48 .2	32 25 .8	34 3 .2	35 40 .8	37 18 .4	38 56 .0	40 33 .7	42 11 .4	43 49 .1	24 0	36
26	6 30	30 11 .9	31 47 .5	33 23 .1	34 58 .8	36 34 .4	38 10 .1	39 45 .9	41 21 .7	42 57 .5	23 30	34
28	7 0	29 35 .5	31 9 .0	32 42 .9	34 16 .6	35 50 .3	37 24 .1	38 57 .9	40 31 .8	42 5 .7	23 0	32
30	7 30	28 59 .0	30 31 .4	32 2 .5	33 34 .2	35 6 .0	36 37 .9	38 9 .7	39 41 .7	41 13 .7	22 30	30
32	8 0	28 22 .3	29 52 .7	31 21 .9	32 51 .7	34 21 .6	35 51 .5	37 21 .4	38 51 .4	40 21′.5	22 0	28
34	8 30	27 45 .0	29 13 .0	30 41 .2	32 9 .0	33 37 .0	35 5 .0	36 32 .9	38 1 .0	39 29 .1	21 30	26
36	9 0	27 8 .5	28 34 .1	30 0 .3	31 26 .2	32 52 .2	34 18 .3	35 44 .3	37 10 .4	38 36 .5	21 0	24
38	9 30	26 31 .4	27 55 .2	29 19 .9	30 43 .3	32 12 .3	33 31 .4	34 55 .5	36 19 .6	37 43 .7	20 30	22
40	10 0	25 54 .2	27 16 .1	28 38 .2	30 0 .2	31 22 .3	32 44 .4	34 6 .5	35 28 .6	36 50 .8	20 0	20
42	10 30	25 16 .9	26 36 .9	27 56 .9	29 17 .0	30 37 .1	31 57 .2	33 17 .3	34 37 .5	35 57 .7	19 30	18
44	11 0	24 39 .4	25 57 .5	27 15 .5	28 33 .6	29 51 .8	31 9 .9	32 28 .0	33 46 .3	35 4 .5	19 0	16
46	11 30	24 1 .9	25 18 .0	26 34 .0	27 50 .1	29 6 .3	30 22 .4	31 38 .6	32 54 .8	34 11 .1	18 30	14
48	12 0	23 24 .2	24 38 .3	25 52 .4	27 6 .5	28 20 .6	29 34 .8	30 49 .0	32 3 .2	33 17 .5	18 0	12
50	12 30	22 46 .5	23 58 .5	25 15 .6	26 22 .8	27 34 .9	28 47 .1	29 59 .3	31 11 .5	32 23 .8	17 30	10
52	13 0	22 8 .6	23 18 .6	24 28 .7	25 38 .9	26 49 .0	27 59 .2	29 9 .4	30 19 .6	31 29 .9	17 0	8
54	13 30	21 30 .7	22 18 .7	23 46 .8	24 54 .9	26 3 .0	27 11 .2	28 19 .4	29 27 .6	30 35 .9	16 30	6
56	14 0	20 52 .6	21 58 .6	23 4 .7	24 10 .8	25 16 .9	26 23 .1	27 29 .3	28 35 .5	29 41 .7	16 0	4
58	14 30	20 14 .4	21 18 .4	22 22 .5	23 26 .6	24 30 .7	25 34 .9	26 39 .0	27 43 .2	28 47 .4	15 30	2
V 0 XVII	15 0	19 36 .1	20 38 .1	21 40 .2	22 42 .3	23 44 .4	24 46 .5	25 48 .6	26 50 .8	27 53 .0	15 0	VII 0 XIX
2	15 30	18 57 .8	19 57 .8	20 57 .8	21 57 .9	22 58 .0	23 58 .0	24 58 .1	25 58 .3	26 58 .5	14 30	58
4	16 0	18 19 .3	19 17 .3	20 15 .3	21 13 .3	22 11 .4	23 9 .4	24 7 .5	25 5 .7	26 3 .8	14 0	56
6	16 30	17 40 .8	18 36 .8	19 32 .7	20 28 .7	21 24 .8	22 20 .8	23 16 .8	24 12 .9	25 9 .0	13 30	54
8	17 0	17 2 .2	17 56 .1	18 50 .0	19 44 .0	20 38 .0	21 32 .0	22 26 .0	23 20 .0	24 14 .1	13 0	52
10	17 30	16 23 .6	17 15 .4	18 7 .3	18 59 .2	19 51 .2	20 43 .1	21 45 .1	22 27 .1	23 19 .1	12 30	50
12	18 0	15 44 .8	16 34 .6	17 24 .4	18 14 .3	19 4 .2	19 54 .1	20 54 .0	21 34 .1	22 24 .0	12 0	48
14	18 30	15 6 .0	15 53 .8	16 41 .5	17 29 .4	18 17 .2	19 5 .1	19 52 .9	20 40 .9	21 28 .8	11 30	46
16	19 0	14 27 .0	15 12 .9	15 58 .5	16 44 .3	17 30 .1	18 15 .9	19 1 .7	19 47 .6	20 33 .4	11 0	44
18	19 30	13 48 .1	14 31 .9	15 15 .5	15 59 .2	16 43 .0	17 26 .7	18 10 .4	18 54 .2	19 38 .1	10 30	42
20	20 0	13 9 .1	13 50 .7	14 32 .4	15 14 .0	15 55 .7	16 37 .3	17 19 .0	18 0 .7	18 42 .6	10 0	40
22	20 30	12 30 .0	13 9 .6	13 49 .2	15 28 .8	15 8 .4	15 48 .0	16 27 .6	17 7 .2	17 47 .0	9 30	38
24	21 0	11 50 .8	12 28 .3	13 5 .9	13 43 .4	14 20 .9	14 58 .5	15 36 .0	16 13 .6	16 51 .2	9 0	36
26	21 30	11 11 .7	11 47 .0	12 22 .6	12 58 .0	13 33 .5	14 9 .0	14 44 .4	15 20 .0	15 55 .5	8 30	34
28	22 0	10 32 .4	11 5 .7	11 39 .1	12 12 .5	12 45 .9	13 19 .3	13 52 .7	14 26 .2	14 59 .7	8 0	32
30	22 30	9 53 .2	10 24 .4	10 55 .7	11 27 .1	11 58 .4	12 29 .7	13 1 .0	13 32 .4	14 3 .8	7 30	30
32	23 0	9 13 .8	9 42 .9	10 12 .2	10 41 .5	11 10 .7	11 39 .9	12 9 .2	12 38 .5	13 7 .8	7 0	28
34	23 30	8 34 .5	9 1 .5	9 28 .7	9 55 .9	10 23 .1	10 49 .2	11 17 .4	11 44 .6	12 11 .8	6 30	26
36	24 0	7 55 .0	8 20 .0	8 45 .1	9 10 .2	9 35 .3	10 0 .3	10 25 .4	10 50 .5	11 15 .7	6 0	24
38	24 30	7 15 .6	7 48 .5	8 1 .5	8 24 .6	8 47 .6	9 10 .5	9 33 .5	9 56 .5	10 19 .6	5 30	22
40	25 0	6 36 .0	6 56 .9	7 17 .8	7 38 .8	7 59 .7	8 20 .6	8 41 .5	9 2 .4	9 23 .4	5 0	20
42	25 30	5 56 .6	6 14 .3	6 34 .2	6 53 .1	7 11 .9	7 30 .7	7 49 .5	8 8 .4	8 27 .3	4 30	18
44	26 0	5 17 .0	5 33 .7	5 50 .4	6 7 .2	6 23 .9	6 40 .6	6 57 .4	7 14 .2	7 31 .0	4 0	16
46	26 30	4 37 .5	4 52 .1	5 6 .7	5 21 .4	5 36 .0	5 50 .6	6 5 .3	6 20 .0	6 34 .7	3 30	14
48	27 0	3 57 .8	4 10 .3	4 22 .9	4 35 .5	4 48 .0	5 0 .5	5 13 .2	5 25 .7	5 38 .3	3 0	12
50	27 30	3 18 .3	3 28 .7	3 39 .2	3 59 .6	4 0 .1	4 10 .5	4 21 .1	4 31 .5	4 42 .0	2 30	10
52	28 0	2 38 .6	2 47 .0	2 55 .4	3 3 .7	3 12 .1	3 20 .4	3 28 .8	3 37 .2	3 45 .6	2 0	8
54	23 30	1 59 .0	2 5 .3	2 11 .6	2 17 .8	2 24 .1	2 30 .3	2 36 .7	2 42 .9	2 49 .3	1 30	6
56	29 0	1 19 .3	1 23 .6	1 27 .8	1 31 .9	1 36 .1	1 40 .2	1 44 .4	1 48 .6	1 52 .8	1 0	4
58	29 30	0 30 .7	0 41 .8	0 43 .9	0 45 .9	0 48 .0	0 50 .1	0 52 .3	0 54 .3	0 56 .5	0 30	2
VI 0 XVIII	30 0	0 0 .0	0 0 .0	0 0 .0	0 0 .0	0 0 .0	0 0 .0	0 0 .0	0 0 .0	0 0 .0	0 0	VI 0 XVIII

Signs of R. A. III^s with approximate Latitude N. − / S. + ; IX^s with approximate Latitude N. + / S. −

The approximate Latitude = the Ecliptic Declination ± the Declination of the Star or Planet.

The Corrections contained in this Table are calculated for an Obliquity of the Ecliptic = 23° 28′; therefore a further Correction proportionate to the *diminution* must be applied from Table 4, which is adapted for this purpose.

ZODIACAL TABLES.

TABLE 2.

CORRECTION of the APPROXIMATE LONGITUDE, &c. concluded.

Signs of R. A. II^s with approximate Latitude N. + / S. − | VIII^s with approximate Latitude N. − / S. +

Time. N. + S. − / N. − S. +	Space.	4° 40′	4° 50′	5° 0′	5° 10′	5° 20′	5° 30′	5° 40′	5° 50′	6° 0′	Space.	Time. N. − S. + / N. + S. −
h. m. h.												h. m. h.
IV 0 XVI	0° 0′	55′ 52″.1	57′ 52″.4	59′ 52″.7	61′ 53″.0	63′ 53″.5	65′ 54″.0	67′ 54″.5	69′ 55″.1	71′ 55″.8	30° 0′	VIII 0 XX
2	0 30	55 1.3	56 59.7	58 58.2	60 56.7	62 55.3	64 54.1	66 52.7	68 51.5	70 50.4	29 30	58
4	1 0	54 10.3	56 6.9	58 3.5	60 0.2	61 57.0	63 53.9	65 50.8	67 47.7	69 44.8	29 0	56
6	1 30	53 19.0	55 13.7	57 8.5	59 3.4	60 58.3	62 53.4	64 48.4	66 43.5	68 38.7	28 30	54
8	2 0	52 27.5	54 20.4	56 13.3	58 6.4	59 59.5	61 52.6	63 45.8	65 39.1	67 32.4	28 0	52
10	2 30	51 3.7	53 26.7	55 17.8	57 9.0	59 0.2	60 51.5	62 43.0	64 34.9	66 25.7	27 30	50
12	3 0	50 43.7	52 32.9	54 22.1	56 11.4	58 0.8	59 50.2	61 39.7	63 29.2	65 18.8	27 0	48
14	3 30	49 51.4	51 38.7	53 26.7	55 13.4	57 0.5	58 48.6	60 36.2	62 23.8	64 11.5	26 30	46
16	4 0	48 59.0	50 44.4	52 29.9	54 15.5	56 1.0	57 46.7	59 32.4	61 18.2	63 4.0	26 0	44
18	4 30	48 6.3	49 49.8	51 33.4	53 17.1	55 0.7	56 44.6	58 28.4	60 12.2	61 56.1	25 30	42
20	5 0	47 13.4	48 55.0	50 36.7	52 18.5	54 0.3	55 42.2	57 24.1	59 6.0	60 48.1	25 0	40
22	5 30	46 20.3	47 59.9	49 39.7	51 19.6	52 59.5	54 39.5	56 24.5	57 59.5	59 39.6	24 30	38
24	6 0	45 27.1	47 4.7	48 42.6	50 20.6	51 58.6	53 36.6	55 14.7	56 52.8	58 31.0	24 0	36
26	6 30	44 33.4	46 9.2	47 45.2	49 21.3	50 57.4	52 33.4	54 9.6	55 45.7	57 22.0	23 30	34
28	7 0	43 39.6	45 13.6	46 47.6	48 21.8	49 55.9	51 30.0	53 4.2	54 38.5	56 12.9	23 0	32
30	7 30	42 45.6	44 17.7	45 49.8	47 22.0	48 54.2	50 26.4	51 58.7	53 31.0	55 3.4	22 30	30
32	8 0	41 51.5	43 21.6	44 51.8	46 22.1	47 52.2	49 22.5	50 52.9	52 23.2	53 53.7	22 0	28
34	8 30	40 57.2	42 25.3	43 53.5	45 21.8	46 50.0	48 18.4	49 46.9	51 15.2	52 43.7	21 30	26
36	9 0	40 2.7	41 28.8	42 55.1	44 21.4	45 47.7	47 14.1	48 40.6	50 7.0	51 33.5	21 0	24
38	9 30	39 8.0	40 32.1	41 56.5	43 20.8	44 45.1	46 9.6	47 34.1	48 58.5	50 23.0	20 30	22
40	10 0	38 13.1	39 35.3	40 57.7	42 20.0	43 42.4	45 4.9	46 27.3	47 49.8	49 12.4	20 0	20
42	10 30	37 18.0	38 38.3	39 58.6	41 19.0	42 39.4	44 0.0	45 20.4	46 40.9	48 1.5	19 30	18
44	11 0	36 22.8	37 41.1	38 59.4	40 17.8	41 36.3	42 54.8	44 13.3	45 31.9	46 50.4	19 0	16
46	11 30	35 27.4	36 33.7	37 59.9	39 16.4	40 31.9	41 49.4	43 6.0	44 22.5	45 39.1	18 30	14
48	12 0	34 31.8	35 36.1	37 0.5	38 14.9	39 29.4	40 43.8	41 58.4	43 12.9	44 27.6	18 0	12
50	12 30	33 36.1	34 48.4	36 0.8	37 13.2	38 25.6	39 38.2	40 50.6	42 3.1	43 15.8	17 30	10
52	13 0	32 40.2	33 50.6	35 0.9	36 11.3	37 21.7	38 32.3	39 42.7	40 53.2	42 3.9	17 0	8
54	13 30	31 44.2	32 52.5	34 0.9	35 9.2	36 17.6	37 26.2	38 34.6	39 43.1	40 51.7	16 30	6
56	14 0	30 48.1	31 54.3	33 0.7	34 7.0	35 13.4	36 19.9	37 26.3	38 32.9	39 39.4	16 0	4
58	14 30	29 51.8	30 56.0	32 0.3	33 4.6	34 9.0	35 13.5	36 17.9	37 22.4	38 26.9	15 30	2
V 0 XVII	15 0	28 55.3	29 57.5	30 59.8	32 2.1	33 4.5	34 6.9	35 9.3	36 11.8	37 14.3	15 0	VII 0 XIX
2	15 30	27 58.7	28 58.9	29 59.2	30 59.4	31 59.8	33 0.2	34 0.5	35 1.0	36 1.4	14 30	58
4	16 0	27 2.0	28 0.2	28 58.4	29 56.6	30 54.9	31 53.3	32 51.6	33 50.0	34 48.4	14 0	56
6	16 30	26 5.2	27 1.3	27 57.5	28 53.7	29 49.9	30 46.2	31 42.5	32 38.9	33 35.2	13 30	54
8	17 0	25 8.2	26 2.3	26 56.5	27 50.6	28 44.8	29 39.0	30 33.3	31 27.6	32 21.9	13 0	52
10	17 30	24 11.2	25 3.2	25 55.3	26 47.4	27 39.6	28 31.7	29 24.0	30 16.2	31 8.4	12 30	50
12	18 0	23 14.0	24 4.0	24 54.0	25 44.1	26 34.2	27 24.3	28 14.5	29 4.6	29 54.8	12 0	48
14	18 30	22 16.7	23 4.7	23 52.6	24 40.8	25 28.9	26 16.7	27 4.9	27 52.9	28 41.1	11 30	46
16	19 0	21 19.3	22 5.2	22 51.1	23 37.3	24 23.0	25 9.0	25 55.1	26 41.1	27 27.2	11 0	44
18	19 30	20 21.8	21 5.6	21 49.6	22 33.6	23 17.6	24 1.2	24 45.2	25 29.2	26 13.2	10 30	42
20	20 0	19 24.2	20 5.8	20 47.9	21 29.7	22 11.5	22 53.3	23 35.2	24 17.2	24 59.1	10 0	40
22	20 30	18 26.6	19 6.1	19 46.1	20 25.8	21 5.8	21 45.4	22 25.1	23 4.0	23 44.9	9 30	38
24	21 0	17 28.8	18 6.3	18 44.1	19 21.8	19 59.5	20 37.2	21 14.9	21 52.7	22 30.5	9 0	36
26	21 30	16 31.0	17 6.5	17 42.2	18 17.8	18 53.5	19 29.0	20 4.7	20 40.3	21 16.1	8 30	34
28	22 0	15 33.1	16 6.6	16 40.1	17 13.6	17 47.1	18 20.7	18 54.3	19 27.8	20 1.5	8 0	32
30	22 30	14 35.2	15 6.6	15 38.0	16 9.4	16 40.9	17 12.3	17 43.8	18 15.3	18 46.9	7 30	30
32	23 0	13 37.1	14 6.4	14 35.7	15 5.1	15 34.4	16 3.8	16 33.2	17 2.6	17 32.1	7 0	28
34	23 30	12 39.0	13 6.3	13 33.5	14 0.8	14 28.2	14 55.3	15 23.6	15 49.9	16 17.5	6 30	26
36	24 0	11 40.8	12 6.0	12 31.1	12 56.3	13 21.5	13 46.7	14 11.9	14 37.1	15 2.4	6 0	24
38	24 30	10 42.6	11 5.7	11 28.8	11 51.9	12 15.2	12 38.1	13 2.2	13 24.3	13 47.6	5 30	22
40	25 0	9 44.3	10 5.3	10 26.3	10 47.3	11 8.3	11 29.3	11 50.3	12 11.3	12 32.4	5 0	20
42	25 30	8 46.1	9 5.0	9 23.9	9 42.7	10 1.8	10 20.6	10 39.5	10 58.4	11 17.5	4 30	18
44	26 0	7 47.7	8 4.5	8 21.3	8 38.0	8 54.9	9 11.7	9 28.5	9 45.3	10 2.2	4 0	16
46	26 30	6 49.4	7 4.1	7 18.8	7 33.4	7 48.2	8 2.9	8 17.6	8 32.3	8 47.1	3 30	14
48	27 0	5 50.9	6 3.5	6 16.1	6 28.7	6 41.3	6 53.9	7 6.5	7 19.1	7 31.8	3 0	12
50	27 30	4 52.5	5 3.0	5 13.5	5 24.0	5 34.4	5 43.0	5 55.5	6 6.2	6 15.7	2 30	10
52	28 0	3 54.0	4 2.4	4 10.8	4 19.2	4 27.5	4 36.0	4 44.4	4 52.9	5 1.3	2 0	8
54	28 30	2 55.6	3 1.8	3 8.2	3 14.5	3 20.7	3 27.0	3 33.4	3 39.8	3 46.1	1 30	6
56	29 0	1 57.1	2 1.2	2 5.5	2 9.6	2 13.8	2 18.0	2 22.2	2 26.5	2 30.7	1 0	4
58	29 30	0 58.5	1 0.6	1 2.8	1 4.9	1 6.9	1 9.0	1 11.2	1 13.3	1 5.4	0 30	2
VI 0 XVIII	30 0	0 0.0	0 0.0	0 0.0	0 0.0	0 0.0	0 0.0	0 0.0	0 0.0	0 0.0	0 0	VI 0 XVIII

Signs of R. A. III^s with approximate Latitude N. − / S. + | IX^s with approximate Latitude N. + / S. −

The approximate Latitude = the Ecliptic Declination ± the Declination of the Star or Planet.

The Corrections contained in this Table are calculated for an Obliquity of the Ecliptic = 23° 28′; therefore a further Correction proportionate to the *diminution* must be applied from TABLE 4, which is adapted for this purpose.

TABLE 3. CORRECTION of the APPROXIMATE LATITUDE of a ZODIACAL STAR or PLANET.

Arguments = the Ecliptic Declination at the Side; and approximate Latitude at the Head or Foot.

APPROXIMATE LATITUDE.

Ecliptic Declin.	10′	20′	30′	40′	50′	1° 0′	1° 10′	1° 20′	1° 30′	1° 40′	1° 50′	2° 0′	Ecliptic Declin.
0° 0′	0′ 49″.7	1′ 39″.3	2′ 28″.9	3′ 18″.6	4′ 8″.2	4′ 57″.8	5′ 47″.5	6′ 37″.2	7′ 26″.7	8′ 16″.4	9′ 6″.1	9′ 56″.4	0° 0′
0 30	0 49 .7	1 39 .2	2 28 .8	3 18 .5	4 8 .0	4 57 .6	5 47 .3	6 36 .9	7 26 .5	8 16 .1	9 5 .8	9 55 .8	0 30
1 0	0 49 .6	1 39 .1	2 28 .7	3 18 .3	4 7 .8	4 57 .3	5 46 .9	6 36 .4	7 26 .0	8 15 .5	9 5 .1	9 54 .7	1 0
1 30	0 49 .5	1 38 .9	2 28 .4	3 17 .8	4 7 .2	4 56 .6	5 46 .2	6 35 .0	7 25 .0	8 14 .5	9 4 .0	9 53 .4	1 30
2 0	0 49 .3	1 38 .6	2 27 .9	3 17 .2	4 6 .5	4 55 .8	5 45 .2	6 34 .4	7 23 .7	8 13 .1	9 2 .4	9 51 .7	2 0
2 30	0 49 .2	1 38 .2	2 27 .3	3 16 .5	4 5 .6	4 54 .6	5 43 .9	6 32 .8	7 22 .0	8 11 .2	9 0 .3	9 49 .4	2 30
3 0	0 49 .0	1 37 .8	2 26 .6	3 15 .6	4 4 .4	4 53 .3	5 42 .2	6 31 .1	7 20 .0	8 8 .8	8 57 .8	9 46 .7	3 0
3 30	0 48 .7	1 37 .3	2 25 .8	3 14 .4	4 4 .0	4 51 .6	5 40 .4	6 28 .8	7 17 .5	8 6 .1	8 54 .8	9 43 .4	3 30
4 0	0 48 .4	1 36 .6	2 24 .9	3 13 .1	4 1 .5	4 49 .7	5 38 .1	6 26 .4	7 14 .7	8 3 .0	8 51 .3	9 39 .6	4 0
4 30	0 48 .0	1 35 .9	2 23 .8	3 11 .7	3 59 .7	4 47 .5	5 35 .6	6 23 .5	7 11 .5	7 59 .4	8 47 .3	9 35 .3	4 30
5 0	0 47 .6	1 35 .1	2 22 .6	3 10 .1	3 57 .7	4 45 .2	5 32 .8	6 20 .3	7 7 .8	7 55 .4	8 42 .9	9 30 .5	5 0
5 30	0 47 .1	1 34 .2	2 21 .3	3 8 .3	3 55 .4	4 42 .6	5 29 .6	6 16 .7	7 3 .8	7 50 .9	8 38 .0	9 25 .2	5 30
6 0	0 46 .6	1 33 .2	2 19 .8	3 6 .4	3 53 .0	4 39 .8	5 26 .3	6 12 .8	6 59 .6	7 46 .1	8 32 .7	9 19 .4	6 0
6 30	0 46 .1	1 32 .1	2 18 .2	3 4 .3	3 50 .3	4 36 .5	5 22 .6	6 8 .6	6 54 .8	7 40 .8	8 26 .9	9 13 .0	6 30
7 0	0 45 .5	1 31 .0	2 16 .5	3 2 .0	3 47 .5	4 33 .0	5 18 .6	6 4 .0	6 49 .6	7 35 .1	8 20 .6	9 6 .1	7 0
7 30	0 44 .9	1 29 .8	2 14 .6	2 59 .5	3 44 .4	4 29 .3	5 14 .2	5 59 .1	6 44 .0	7 28 .9	8 13 .8	8 58 .7	7 30
8 0	0 44 .2	1 28 .5	2 12 .7	2 56 .9	3 41 .1	4 25 .4	5 9 .6	5 53 .8	6 38 .1	7 22 .3	8 6 .6	8 50 .8	8 0
8 30	0 43 .5	1 27 .0	2 10 .5	2 54 .1	3 37 .6	4 21 .1	5 4 .6	5 48 .3	6 31 .7	7 15 .2	7 58 .8	8 42 .3	8 30
9 0	0 42 .8	1 25 .5	2 8 .2	2 51 .1	3 33 .9	4 16 .6	4 59 .3	5 42 .2	6 25 .0	7 7 .8	7 50 .6	8 33 .4	9 0
9 30	0 42 .0	1 23 .9	2 7 .9	2 47 .9	3 29 .9	4 11 .8	4 53 .8	5 36 .9	6 17 .8	6 59 .8	7 41 .9	8 23 .8	9 30
10 0	0 41 .2	1 22 .3	2 3 .4	2 44 .5	3 25 .7	4 6 .9	4 48 .0	5 29 .2	6 10 .3	6 51 .5	7 32 .7	8 13 .8	10 0
10 30	0 40 .3	1 20 .5	2 0 .8	2 41 .0	3 21 .3	4 1 .5	4 41 .8	5 22 .2	6 2 .3	6 42 .7	7 22 .9	8 3 .2	10 30
11 0	0 39 .4	1 18 .7	1 58 .0	2 37 .3	3 16 .7	3 56 .0	4 35 .4	5 14 .7	5 54 .0	6 33 .4	7 12 .7	7 52 .1	11 0
11 30	0 38 .3	1 16 .7	1 55 .1	2 33 .3	3 11 .8	3 50 .1	4 28 .6	5 6 .8	5 45 .1	6 23 .6	7 1 .9	7 40 .4	11 30
12 0	0 37 .2	1 14 .7	1 52 .0	2 29 .4	3 6 .7	3 44 .0	4 21 .4	4 58 .7	5 36 .1	6 13 .3	6 50 .8	7 28 .3	12 0
12 30	0 36 .3	1 12 .5	1 48 .8	2 25 .1	3 1 .3	3 37 .6	4 13 .8	4 50 .2	5 26 .3	6 2 .6	6 38 .9	7 15 .4	12 30
13 0	0 35 .3	1 10 .3	1 45 .4	2 20 .6	2 55 .8	3 31 .0	4 6 .0	4 41 .3	5 16 .4	5 51 .6	6 26 .6	7 2 .0	13 0
13 30	0 34 .1	1 8 .0	1 41 .0	2 16 .0	2 49 .9	3 23 .9	3 57 .8	4 32 .0	5 5 .9	5 39 .9	6 13 .8	6 48 .0	13 30
14 0	0 32 .8	1 5 .6	1 38 .4	2 11 .2	2 43 .9	3 16 .6	3 49 .5	4 22 .3	4 55 .1	5 27 .9	6 0 .7	6 33 .5	14 0
14 30	0 31 .5	1 3 .1	1 34 .6	2 6 .1	2 37 .6	3 9 .0	3 40 .6	4 12 .2	4 43 .7	5 15 .4	5 46 .8	6 18 .4	14 30
15 0	0 30 .2	1 0 .4	1 30 .7	2 0 .9	2 31 .1	3 1 .3	3 31 .5	4 1 .8	4 32 .0	5 2 .4	5 32 .5	6 2 .7	15 0
15 30	0 28 .9	0 57 .7	1 26 .6	1 55 .4	2 24 .3	2 53 .1	3 21 .9	3 50 .9	4 19 .7	4 48 .7	5 17 .5	5 46 .4	15 30
16 0	0 27 .6	0 54 .9	1 22 .4	1 49 .7	2 17 .3	2 44 .7	3 12 .2	3 39 .6	4 7 .1	4 34 .6	5 2 .0	5 29 .5	16 0
16 30	0 26 .1	0 52 .0	1 18 .0	1 43 .9	2 10 .0	2 35 .9	3 1 .9	3 27 .9	3 54 .0	4 20 .0	4 45 .9	5 2 .0	16 30
17 0	0 24 .5	0 49 .0	1 13 .5	1 37 .9	2 2 .4	2 26 .9	2 51 .4	3 15 .8	3 40 .4	4 4 .9	4 29 .4	4 53 .9	17 0
17 30	0 22 .9	0 45 .8	1 8 .8	1 31 .7	1 54 .6	2 17 .5	2 40 .4	3 3 .4	3 26 .3	3 49 .2	4 12 .1	4 35 .2	17 30
18 0	0 21 .3	0 42 .6	1 3 .9	1 25 .3	1 46 .6	2 7 .9	2 29 .0	2 50 .5	3 11 .8	3 33 .1	3 54 .4	4 15 .8	18 0
18 30	0 20 .6	0 39 .3	0 58 .9	1 18 .6	1 38 .2	1 57 .7	2 17 .3	2 37 .1	2 56 .7	3 16 .4	3 35 .9	3 55 .6	18 30
19 0	0 18 .9	0 35 .8	0 53 .7	1 11 .7	1 29 .6	1 47 .5	2 5 .3	2 23 .4	2 41 .2	2 59 .2	3 16 .9	3 35 .1	19 0
19 30	0 16 .1	0 32 .3	0 48 .4	1 4 .6	1 20 .7	1 36 .8	1 52 .9	2 9 .1	2 25 .2	2 41 .4	2 57 .4	3 13 .6	19 30
20 0	0 14 .3	0 28 .6	0 42 .9	0 57 .3	1 11 .6	1 25 .9	1 40 .2	1 54 .5	2 8 .8	2 23 .1	2 37 .4	2 51 .7	20 0
20 30	0 12 .4	0 24 .8	0 37 .2	0 49 .7	1 2 .1	1 14 .5	1 27 .0	1 39 .3	1 51 .8	2 4 .1	2 16 .5	2 29 .0	20 30
21 0	0 10 .5	0 21 .0	0 31 .4	0 41 .9	0 52 .4	1 2 .9	1 13 .4	1 23 .8	1 34 .3	1 44 .7	1 55 .2	2 5 .7	21 0
21 30	0 8 .4	0 17 .0	0 25 .4	0 33 .8	0 42 .3	0 50 .8	0 59 .2	1 7 .8	1 16 .2	1 24 .6	1 33 .1	1 41 .6	21 30
22 0	0 6 .3	0 12 .8	0 19 .2	0 25 .6	0 32 .1	0 38 .5	0 44 .9	0 51 .3	0 57 .7	1 4 .1	1 10 .5	1 16 .9	22 0
22 30	0 4 .2	0 8 .5	0 12 .8	0 17 .1	0 21 .4	0 25 .5	0 29 .9	0 34 .2	0 38 .5	0 42 .7	0 47 .0	0 51 .3	22 30
23 0	0 2 .1	0 4 .2	0 6 .3	0 8 .4	0 10 .5	0 12 .6	0 14 .7	0 16 .8	0 18 .9	0 21 .0	0 23 .1	0 25 .2	23 0
23 28	0 0 .0	0 0 .0	0 0 .0	0 0 .0	0 0 .0	0 0 .0	0 0 .0	0 0 .0	0 0 .0	0 0 .0	0 0 .0	0 0 .0	23 28
	10′	20′	30′	40′	50′	1° 0′	1° 10′	1° 20′	1° 30′	40′	1° 50′	2° 0′	

APPROXIMATE LATITUDE.

Subtract the Number taken out of this Table from the approximate Latitude (found as before directed) and the remainder will be the Latitude to the Obliquity of the Ecliptic 23° 28′; and the further Correction for the *diminution* of the Obliquity, must be taken from TABLE 5, which is adapted for this purpose.

TABLE 3.

Correction of the approximate Latitude, &c. continued.

APPROXIMATE LATITUDE.

Ecliptic Declin.	2° 10′	2° 20′	2° 30′	2° 40′	2° 50′	3° 0′	3° 10′	3° 20′	3° 30′	3° 40′	3° 50′	4° 0′	Ecliptic Declin.
0° 0′	10′ 45″.4	11′ 35″.1	12′ 24″.8	13′ 14″.5	14′ 4″.2	14′ 53″.7	15′ 43″.8	16′ 33″.5	17′ 23″.3	18′ 13″.1	19′ 2″.9	19′ 52″.7	0° 0′
0 30	10 45.0	11 34.8	12 24.5	13 14.2	14 3.9	14 53.4	15 43.3	16 33.1	17 22.8	18 12.6	19 2.3	19 52.1	0 30
1 0	10 44.3	11 33.9	12 23.6	13 13.2	14 2.8	14 52.4	15 42.1	16 31.8	17 21.5	18 11.3	19 1.0	19 50.7	1 0
1 30	10 42.9	11 32.5	12 22.0	13 11.6	14 1.0	14 50.6	15 40.1	16 29.7	17 19.3	18 9.4	18 59.0	19 48.2	1 30
2 0	10 41.1	11 30.5	12 19.8	13 9.2	13 58.5	14 48.0	15 37.4	16 26.8	17 16.3	18 5.7	18 55.2	19 44.9	2 0
2 30	10 38.6	11 27.8	12 16.9	13 6.2	13 55.3	14 44.5	15 33.8	16 23.0	17 12.3	18 1.4	18 50.7	19 40.3	2 30
3 0	10 35.6	11 24.5	12 13.5	13 2.4	13 51.4	14 40.3	15 29.4	16 18.4	17 7.4	17 56.3	18 45.5	19 34.9	3 0
3 30	10 32.1	11 20.7	12 9.3	12 58.1	13 46.8	14 35.4	15 24.2	16 12.9	17 1.7	17 50.3	18 39.1	19 28.1	3 30
4 0	10 28.0	11 16.3	12 4.6	12 53.0	13 41.4	14 29.8	15 18.2	16 6.6	16 55.1	17 43.5	18 32.0	19 20.4	4 0
4 30	10 23.3	11 11.3	11 59.2	12 47.4	13 35.3	14 23.3	15 11.4	15 59.4	16 47.6	17 35.6	18 23.5	19 11.7	4 30
5 0	10 18.1	11 5.7	11 53.3	12 40.9	13 28.5	14 16.1	15 3.8	15 51.4	16 39.1	17 26.8	18 14.8	19 2.2	5 0
5 30	10 12.4	10 59.5	11 46.6	12 33.8	13 21.0	14 8.1	14 55.3	15 42.5	16 29.8	17 17.0	18 4.3	18 51.5	5 30
6 0	10 6.0	10 52.7	11 39.3	12 26.0	13 12.7	13 59.4	14 46.1	15 32.8	16 19.6	17 6.3	17 53.1	18 39.9	6 0
6 30	9 59.1	10 45.2	11 31.4	12 17.5	13 3.7	13 49.8	14 36.0	15 22.2	16 8.4	16 54.7	17 40.8	18 27.1	6 30
7 0	9 51.7	10 37.2	11 22.8	12 8.3	12 54.0	13 39.5	14 25.1	15 10.8	15 56.4	16 42.4	17 27.7	18 13.4	7 0
7 30	9 43.6	10 28.6	11 13.5	11 58.5	12 43.5	13 28.4	14 13.6	14 58.4	15 43.5	16 28.8	17 13.5	17 58.6	7 30
8 0	9 35.1	10 19.4	11 3.6	11 47.9	12 32.3	13 16.6	14 0.9	14 45.2	15 29.6	16 14.4	16 58.4	17 42.8	8 0
8 30	9 26.8	10 9.5	10 53.0	11 36.7	12 20.3	13 3.9	13 47.5	14 41.1	15 14.8	15 58.7	16 42.1	17 25.8	8 30
9 0	9 17.9	9 59.0	10 41.8	11 24.7	12 7.6	12 50.4	13 33.3	14 16.2	14 59.1	15 42.0	16 24.9	17 7.9	9 0
9 30	9 6.7	9 48.3	10 29.9	11 12.0	11 44.1	12 36.1	13 18.2	14 0.4	14 42.5	15 25.2	16 6.6	16 48.8	9 30
10 0	8 55.0	9 36.2	10 17.4	10 58.6	11 39.8	12 21.1	13 2.3	13 43.8	14 25.2	15 6.4	15 47.4	16 28.7	10 0
10 30	8 43.5	9 23.8	10 4.1	10 44.5	11 24.9	12 5.1	12 45.6	13 26.0	14 6.5	14 46.8	15 27.0	16 7.4	10 30
11 0	8 31.4	9 10.9	9 50.2	10 29.6	11 9.0	11 48.4	12 27.9	13 7.2	13 46.8	14 26.2	15 5.7	15 45.2	11 0
11 30	8 18.8	8 57.2	9 35.6	10 14.5	10 52.7	11 30.9	12 9.4	12 47.7	13 26.3	14 4.7	14 43.2	15 21.7	11 30
12 0	8 5.5	8 42.9	9 20.3	9 58.7	10 35.3	11 12.5	11 50.0	12 27.4	13 4.8	13 42.3	14 19.8	14 57.3	12 0
12 30	7 51.6	8 27.4	9 4.2	9 41.1	10 17.1	10 53.3	11 29.7	12 6.0	12 42.4	13 18.8	13 55.2	14 31.5	12 30
13 0	7 37.2	8 12.4	8 47.5	9 22.8	9 58.2	10 33.2	11 8.5	11 43.7	12 19.0	12 54.3	13 29.5	14 4.8	13 0
13 30	7 22.0	7 56.0	8 30.1	9 4.3	9 38.4	10 12.2	10 46.4	11 20.4	11 54.5	12 28.7	13 2.7	13 36.8	13 30
14 0	7 6.3	7 39.1	8 12.0	8 44.8	9 17.7	9 50.3	10 23.4	10 56.2	11 29.1	12 2.0	12 34.9	13 7.8	14 0
14 30	6 49.8	7 21.5	7 53.1	8 24.8	9 1.3	9 27.9	9 59.5	10 31.0	11 2.6	11 34.2	12 5.9	12 37.6	14 30
15 0	6 32.9	7 3.2	7 33.5	8 3.8	8 34.0	9 4.3	9 34.6	10 4.8	10 35.1	11 5.5	11 35.8	12 6.2	15 0
15 30	6 15.2	6 44.0	7 13.1	7 42.1	8 11.0	8 39.9	9 8.8	9 37.6	10 6.7	10 35.5	11 4.5	11 33.5	15 30
16 0	5 57.0	6 24.5	6 52.0	7 19.5	7 47.0	8 14.5	8 42.0	9 9.5	9 37.1	10 4.6	10 32.1	10 59.7	16 0
16 30	5 38.0	6 4.1	6 30.0	6 56.2	7 22.2	7 48.2	8 14.2	8 40.3	9 6.5	9 32.4	9 58.6	10 24.6	16 30
17 0	5 18.4	5 42.9	6 7.4	6 31.9	6 56.5	7 21.0	7 45.5	8 10.1	8 34.7	8 59.2	9 23.8	9 48.4	17 0
17 30	4 58.0	5 21.0	5 43.9	6 7.0	6 29.9	6 54.9	7 15.8	7 38.8	8 1.9	8 24.7	8 47.8	9 10.9	17 30
18 0	4 37.1	4 58.3	5 19.8	5 41.1	6 2.4	6 23.8	6 45.2	7 6.5	7 27.9	7 49.3	8 10.7	8 32.1	18 0
18 30	4 15.4	4 35.0	4 54.7	5 14.4	5 34.1	5 53.8	6 13.4	6 33.0	6 52.7	7 12.5	7 32.3	7 51.9	18 30
19 0	3 53.0	4 10.9	4 28.9	4 46.8	5 4.8	5 22.8	5 40.6	5 58.6	6 16.6	6 34.6	6 52.6	7 10.6	19 0
19 30	3 29.8	3 46.1	4 2.1	4 18.5	4 34.6	4 50.7	5 6.8	5 23.0	5 39.3	5 55.4	6 11.8	6 27.9	19 30
20 0	3 6.0	3 20.4	3 34.7	3 49.1	4 3.4	4 17.7	4 32.1	4 46.4	5 0.8	5 15.1	5 29.8	5 43.9	20 0
20 30	2 41.4	2 54.0	3 6.3	3 18.9	3 31.2	3 43.6	3 56.1	4 8.5	4 21.0	4 33.4	4 46.2	4 58.4	20 30
21 0	2 16.1	2 26.7	2 37.2	2 47.7	2 58.1	3 8.6	3 19.1	3 29.6	3 40.1	3 50.6	4 1.2	4 11.7	21 0
21 30	1 50.0	1 58.6	2 7.0	2 15.6	2 24.1	2 32.5	2 40.9	2 49.3	2 57.9	3 6.6	3 14.9	3 23.4	21 30
22 0	1 23.2	1 29.7	1 36.1	1 42.5	1 49.1	1 55.4	2 1.8	2 8.0	2 14.6	2 21.1	2 27.5	2 33.9	22 0
22 30	0 55.6	1 0.0	1 4.2	1 8.5	1 12.9	1 17.1	1 21.3	1 25.6	1 29.9	1 34.3	1 38.5	1 42.7	22 30
23 0	0 27.3	0 29.6	0 31.5	0 33.6	0 35.7	0 37.8	0 39.8	0 42.0	0 44.1	0 46.2	0 48.3	0 50.4	23 0
23 28	0 0.0	0 0.0	0 0.0	0 0.0	0 0.0	0 0.0	0 0.0	0 0.0	0 0.0	0 0.0	0 0.0	0 0.0	23 28
	2° 10′	2° 20′	2° 30′	2° 40′	2° 50′	3° 0′	3° 10′	3° 20′	3° 30′	3° 40′	3° 50′	4° 0′	

APPROXIMATE LATITUDE.

Subtract the Number taken out of this Table from the approximate Latitude (found as before directed) and the remainder will be the Latitude to the Obliquity of the Ecliptic 23° 28′; and the further Correction for the *diminution* of the Obliquity, must be taken from Table 5, which is adapted for this purpose.

TABLE 3.

CORRECTION of the APPROXIMATE LATITUDE, &c. concluded.

APPROXIMATE LATITUDE.

Ecliptic Declin.	4° 10′	4° 20′	4° 30′	4° 40′	4° 50′	5° 0′	5° 10′	5° 20′	5° 30′	5° 40′	5° 50′	6° 0′	Ecliptic Declin.
0° 0′	20′ 42″.6	21′ 32″.5	22′ 22″.4	23′ 12″.2	24′ 2″.2	24′ 52″.1	25′ 42″.1	26′ 32″.1	27′ 22″.1	28′ 12″.1	29′ 2″.2	29′ 52″.3	0° 0′
0 30	20 42.1	21 31.9	22 21.7	23 11.6	24 1.5	24 51.4	25 41.4	26 31.4	27 21.4	28 11.4	29 1.5	29 51.5	0 30
1 0	20 40.5	21 30.3	22 20.0	23 9.9	23 59.6	24 49.6	25 39.4	26 29.3	27 19.3	28 9.3	28 59.4	29 49.3	1 0
1 30	20 37.8	21 27.6	22 17.2	23 6.9	23 56.6	24 46.5	25 36.3	26 26.0	27 15.8	28 5.7	28 55.7	29 45.5	1 30
2 0	20 34.2	21 23.7	22 13.3	23 2.8	23 52.4	24 42.1	25 31.8	26 21.3	27 11.0	28 0.7	28 50.5	29 40.2	2 0
2 30	20 29.4	21 18.8	22 8.2	22 57.5	23 46.9	24 36.3	25 26.0	26 15.3	27 4.7	27 54.2	28 43.8	29 33.4	2 30
3 0	20 23.7	21 12.8	22 1.9	22 51.1	23 40.2	24 29.4	25 18.8	26 7.9	26 57.1	27 46.4	28 35.7	29 25.1	3 0
3 30	20 16.8	21 5.6	21 54.5	22 43.4	23 32.2	24 21.2	25 10.2	25 59.1	26 48.0	27 37.1	28 26.1	29 15.2	3 30
4 0	20 9.0	20 57.4	21 46.0	22 34.6	23 23.0	24 11.7	25 0.3	25 49.0	26 37.6	27 26.3	28 15.0	29 3.8	4 0
4 30	20 0.0	20 48.1	21 36.3	22 24.5	23 12.7	24 1.0	24 49.2	25 37.5	26 25.8	27 14.1	28 2.4	28 50.9	4 30
5 0	19 50.0	20 37.7	21 25.5	22 13.3	23 1.1	23 48.9	24 36.8	25 24.6	26 12.6	27 0.5	27 48.4	28 36.4	5 0
5 30	19 38.8	20 26.4	21 13.5	22 0.9	22 48.2	23 35.6	24 23.0	25 10.5	25 57.9	26 45.4	27 32.8	28 20.4	5 30
6 0	19 26.7	20 13.8	21 0.4	21 47.2	22 34.1	23 21.0	24 7.9	24 54.9	25 41.8	26 28.8	27 15.8	28 2.8	6 0
6 30	19 13.4	19 59.9	20 46.1	21 32.3	22 18.7	23 5.1	23 51.5	23 38.0	25 24.3	26 10.9	26 57.2	27 43.8	6 30
7 0	18 59.1	19 44.8	20 30.6	21 16.3	22 2.1	22 47.9	23 33.7	24 19.6	25 5.4	25 51.7	26 37.2	27 23.2	7 0
7 30	18 43.6	19 28.7	20 14.0	20 59.0	21 44.1	22 29.8	23 14.6	23 59.8	24 45.0	25 30.4	26 15.6	27 0.9	7 30
8 0	18 27.1	19 11.6	19 56.1	20 40.5	21 25.0	22 9.6	22 54.1	23 38.6	24 23.2	25 7.8	25 52.4	26 37.0	8 0
8 30	18 9.4	18 53.3	19 37.0	19 20.8	21 4.5	21 48.3	22 32.2	23 16.0	23 59.9	24 43.8	25 27.7	26 11.6	8 30
9 0	17 50.8	18 33.8	19 16.8	19 59.8	20 42.8	21 25.9	22 9.0	22 52.1	23 35.2	24 18.3	25 1.4	25 44.7	9 0
9 30	17 31.0	18 13.1	18 55.4	19 37.5	20 19.8	21 2.1	21 44.3	22 26.7	23 8.9	23 51.3	24 33.6	25 16.2	9 30
10 0	17 10.0	17 51.3	18 32.7	19 14.0	19 55.5	20 36.9	21 18.3	21 59.8	22 41.2	23 22.7	24 4.3	24 45.9	10 0
10 30	16 47.9	17 28.3	18 8.8	18 49.3	19 29.9	20 10.3	20 51.0	21 31.5	22 12.0	22 52.6	23 33.3	24 14.0	10 30
11 0	16 24.8	17 4.2	17 43.8	18 23.3	19 2.9	19 42.4	20 22.2	21 1.7	21 41.3	22 21.0	23 0.7	23 40.5	11 0
11 30	16 0.4	16 38.8	17 17.4	17 56.0	18 34.6	19 13.1	19 51.9	20 30.4	21 9.1	21 47.8	22 26.5	23 5.3	11 30
12 0	15 34.8	16 12.3	16 49.8	17 27.4	18 4.9	18 42.5	19 20.2	19 57.7	20 35.3	21 13.0	21 50.7	22 28.4	12 0
12 30	15 7.9	15 44.5	16 20.9	16 57.4	17 33.9	18 10.4	18 47.0	19 23.5	20 0.1	20 36.6	21 13.2	21 49.8	12 30
13 0	14 40.0	15 15.4	15 50.8	16 26.1	17 1.5	17 36.9	18 12.3	18 47.7	19 23.2	19 58.6	20 34.1	21 9.6	13 0
13 30	14 10.9	14 45.1	14 19.3	15 53.5	16 27.7	17 1.9	17 36.2	18 10.5	18 44.8	19 19.0	19 53.3	20 27.7	13 30
14 0	13 40.8	14 13.6	14 46.5	15 19.5	15 52.6	16 25.6	16 58.6	17 31.7	18 4.7	18 37.7	19 10.9	19 44.0	14 0
14 30	13 9.3	13 40.9	14 12.4	14 44.2	15 16.0	15 47.6	16 19.5	16 51.3	17 23.0	17 54.8	18 26.7	18 58.5	14 30
15 0	12 36.5	13 6.9	13 37.2	14 7.7	14 38.1	15 8.0	15 38.9	16 9.3	16 39.8	17 10.3	17 40.8	18 11.3	15 0
15 30	12 2.5	12 31.5	13 0.6	13 29.8	13 58.6	14 27.7	14 56.7	15 25.8	15 54.9	16 24.0	16 53.1	17 22.3	15 30
16 0	11 27.3	11 54.8	12 22.5	12 50.5	13 17.7	13 45.2	14 13.0	14 40.7	15 8.3	15 36.0	16 3.7	16 31.4	16 0
16 30	10 50.9	11 16.8	11 43.0	12 9.5	12 35.3	13 1.4	13 27.7	13 53.9	14 20.0	14 46.2	15 12.5	15 38.7	16 30
17 0	10 13.0	10 37.6	11 2.2	11 26.8	11 51.5	12 16.1	12 40.8	13 5.4	13 30.1	13 54.8	14 19.5	14 44.2	17 0
17 30	9 33.8	9 56.9	10 19.9	10 42.9	11 6.1	11 29.1	11 52.3	12 15.4	12 38.4	13 1.6	13 24.7	13 47.8	17 30
18 0	8 53.5	9 14.9	9 36.3	9 57.7	10 19.2	10 40.6	11 2.1	11 23.6	11 45.0	12 6.5	12 28.0	12 49.5	18 0
18 30	8 11.6	8 31.4	8 51.2	9 10.8	9 30.7	9 50.5	10 10.2	10 30.1	10 49.9	11 9.7	11 29.4	11 49.2	18 30
19 0	7 28.6	7 46.6	8 4.6	8 22.6	8 40.7	8 58.9	9 16.7	9 34.8	9 52.9	10 11.0	10 29.0	10 47.0	19 0
19 30	6 44.0	7 0.3	7 16.5	7 32.7	7 49.0	8 5.4	8 21.5	8 37.8	8 54.0	9 10.4	9 26.6	9 42.9	19 30
20 0	5 58.2	6 12.6	6 27.0	6 41.3	6 55.8	7 10.2	7 24.6	7 39.0	7 53.4	8 7.9	8 22.3	8 36.8	20 0
20 30	5 10.8	5 23.3	5 36.7	5 48.3	6 0.8	6 13.3	6 25.9	6 38.4	6 51.3	7 3.5	7 15.9	7 28.6	20 30
21 0	4 22.1	4 32.7	4 43.3	4 53.8	5 4.3	5 14.9	5 25.4	5 36.0	5 46.6	5 57.1	6 7.6	6 18.2	21 0
21 30	3 31.8	3 40.4	3 49.1	3 57.4	4 6.0	4 14.6	4 23.0	4 31.6	4 40.0	4 48.7	4 57.3	5 5.7	21 30
22 0	2 40.3	2 46.8	2 53.2	2 59.6	3 6.1	3 12.6	3 18.8	3 25.4	3 31.9	3 38.4	3 45.0	3 51.2	22 0
22 30	1 47.0	1 51.4	1 55.7	2 0.0	2 4.4	2 8.7	2 12.8	2 17.2	2 21.5	2 25.9	2 30.5	2 34.5	22 30
23 0	0 52.5	0 54.6	0 56.7	0 58.8	1 0.9	1 3.0	1 5.1	1 7.2	1 9.3	1 11.5	1 13.6	1 15.7	23 0
23 28	0 0.0	0 0.0	0 0.0	0 0.0	0 0.0	0 0.0	0 0.0	0 0.0	0 0.0	0 0.0	0 0.0	0 0.0	23 28
	4° 10′	4° 20′	4° 30′	4° 40′	4° 50′	5° 0′	5° 10′	5° 20′	5° 30′	5° 40′	5° 50′	6° 0′	

APPROXIMATE LATITUDE.

Subtract the Number taken out of this Table from the approximate Latitude, (found as before directed) and the Remainder will be the Latitude to the Obliquity of the Ecliptic 23° 28′; and the further Correction for the *diminution* of the Obliquity, must be taken from TABLE 5, which is adapted for this purpose.

TABLE 4.—Correction of the Longitude of a Zodiacal Star or Planet, arising out of 100″ Diminution of the Obliquity of the Ecliptic. Arguments = Longitude once corrected, at the Side; and Latitude once corrected, at the Head.

N. + S. — Long.	N. — S. + Long.	LATITUDE. 0° 30′	1° 0′	1° 30′	2° 0′	2° 30′	3° 0′	3° 30′	4° 0′	4° 30′	5° 0′	5° 30′	6° 0′	45°	Long.	Long.
S.	S.														S.	S.
6 0°	0 0°	0″.8	1″.7	2″.6	3″.5	4″.3	5″.2	6″.1	7″.0	7″.8	8″.7	9″.6	10″.5	100″.0	6 0°	12 0°
6 5	0 5	0.8	1.6	2.6	3.5	4.3	5.2	6.1	6.9	7.7	8.6	9.5	10.4	99.5	5 25	11 25
6 10	0 10	0.8	1.6	2.5	3.4	4.2	5.1	6.0	6.8	7.6	8.5	9.4	10.3	98.5	5 20	11 20
6 15	0 15	0.8	1.5	2.4	3.4	4.2	5.0	5.9	6.7	7.5	8.3	9.2	10.1	96.6	5 15	11 15
6 20	0 20	0.8	1.5	2.3	3.3	4.1	4.9	5.7	6.5	7.3	8.1	9.0	9.8	94.1	5 10	11 10
6 25	0 25	0.7	1.4	2.2	3.2	4.0	4.7	5.5	6.3	7.0	7.8	8.6	9.5	90.7	5 5	11 5
7 0	1 0	0.7	1.4	2.2	3.1	3.8	4.5	5.3	6.0	6.7	7.5	8.2	9.0	86.5	5 0	11 0
7 5	1 5	0.7	1.3	2.1	3.0	3.6	4.3	5.0	5.7	6.4	7.1	7.8	8.5	81.8	4 25	10 25
7 10	1 10	0.6	1.3	1.0	2.7	3.4	4.0	4.7	5.3	6.0	6.7	7.3	8.0	76.5	4 20	10 20
7 15	1 15	0.6	1.2	1.8	2.5	3.1	3.7	4.3	5.0	5.6	6.2	6.8	7.4	70.7	4 15	10 15
7 20	1 20	0.6	1.1	1.7	2.2	2.8	3.3	3.9	4.5	5.1	5.7	6.2	6.7	64.2	4 10	10 10
7 25	1 25	0.5	1.0	1.5	2.0	2.5	3.0	3.5	4.0	4.5	5.0	5.5	6.0	57.3	4 5	10 5
8 0	2 0	0.4	0.8	1.2	1.7	2.1	2.6	3.0	3.5	3.8	4.3	4.7	5.2	50.0	4 0	10 0
8 5	2 5	0.3	0.7	1.0	1.4	1.8	2.2	2.6	3.0	3.3	3.7	4.0	4.3	42.2	3 25	9 25
8 10	2 10	0.2	0.5	0.8	1.1	1.4	1.8	2.1	2.3	2.6	3.0	3.2	3.5	33.4	3 20	9 20
8 15	2 15	0.2	0.4	0.6	0.8	1.0	1.4	1.6	1.8	2.0	2.3	2.5	2.7	25.9	3 15	9 15
8 20	2 20	0.1	0.3	0.4	0.6	0.8	0.9	1.0	1.2	1.3	1.5	1.6	1.8	10.1	3 10	9 10
8 25	2 25	0.1	0.2	0.2	0.3	0.4	0.5	0.5	0.6	0.6	0.7	0.8	0.9	8.7	3 5	9 5
9 0	3 0	0.0	0.0	0.0	0.0	0.0	0.0	0.0	0.0	0.0	0.0	0.0	0.0	0.0	3 0	9 0
		LATITUDE.													S. — N. +	S. + N. —

If the Latitude exceed 6°, take the Number out of the Column headed 45°, and multiply it by the natural Tangent of the Latitude once corrected, and the Product will give the Tabular Correction. If the Obliquity be greater than 23° 28′, the Signs must be changed.

TABLE 5.—Correction of the Latitude of a Zodiacal Star or Planet, arising out of 100″ Diminution of the Obliquity of the Ecliptic. Argument = Longitude of the Star or Planet once corrected.

N. — S. + Long.	N. + S. — Long.	Correc. for 100″.	Long.	Long.	N. — S. + Long.	N. + S. — Long.	Correc. for 100″.	Long.	Long.	N. — S. + Long.	N. + S. — Long.	Correc. for 100″.	Long.	Long.
S.	S.		S.	S.	S.	S.		S.	S.	S.	S.		S.	S.
6 0°	0 0°	0″.0	6 0°	12 0°	7 0°	1 0°	50″.0	5 0°	11 0°	8 0°	2 0°	86″.5	4 0°	10 0°
6 1	0 1	1.7	5 29	11 29	7 1	1 1	51.5	4 29	10 29	8 1	2 1	87.4	3 29	9 29
6 2	0 2	3.5	5 28	11 28	7 2	1 2	52.8	4 28	10 28	8 2	2 2	88.1	3 28	9 28
6 3	0 3	5.2	5 27	11 27	7 3	1 3	54.4	4 27	10 27	8 3	2 3	88.9	3 27	9 27
6 4	0 4	7.0	5 26	11 26	7 4	1 4	55.8	4 26	10 26	8 4	2 4	89.7	3 26	9 26
6 5	0 5	8.7	5 25	11 25	7 5	1 5	57.3	4 25	10 25	8 5	2 5	90.6	3 25	9 25
6 6	0 6	10.5	5 24	11 24	7 6	1 6	58.7	4 24	10 24	8 6	2 6	91.4	3 24	9 24
6 7	0 7	12.2	5 23	11 23	7 7	1 7	60.2	4 23	10 23	8 7	2 7	92.0	3 23	9 23
6 8	0 8	13.8	5 22	11 22	7 8	1 8	61.5	4 22	10 22	8 8	2 8	92.6	3 22	9 22
6 9	0 9	15.7	5 21	11 21	7 9	1 9	62.8	4 21	10 21	8 9	2 9	93.2	3 21	9 21
6 10	0 10	17.3	5 20	11 20	7 10	1 10	64.2	4 20	10 20	8 10	2 10	93.8	3 20	9 20
6 11	0 11	19.0	5 19	11 19	7 11	1 11	65.5	4 19	10 19	8 11	2 11	94.4	3 19	9 19
6 12	0 12	20.8	5 18	11 18	7 12	1 12	66.8	4 18	10 18	8 12	2 12	95.0	3 18	9 18
6 13	0 13	22.5	5 17	11 17	7 13	1 13	68.2	4 17	10 17	8 13	2 13	95.6	3 17	9 17
6 14	0 14	24.2	5 16	11 16	7 14	1 14	69.3	4 16	10 16	8 14	2 14	96.0	3 16	9 16
6 15	0 15	25.8	5 15	11 15	7 15	1 15	70.6	4 15	10 15	8 15	2 15	96.4	3 15	9 15
6 16	0 16	27.5	5 14	11 14	7 16	1 16	71.8	4 14	10 14	8 16	2 16	96.9	3 14	9 14
6 17	0 17	29.3	5 13	11 13	7 17	1 17	73.0	4 13	10 13	8 17	2 17	97.3	3 13	9 13
6 18	0 18	30.8	5 12	11 12	7 18	1 18	74.3	4 12	10 12	8 18	2 18	97.8	3 12	9 12
6 19	0 19	32.4	5 11	11 11	7 19	1 19	75.5	4 11	10 11	8 19	2 19	98.2	3 11	9 11
6 20	0 20	34.2	5 10	11 10	7 20	1 20	76.5	4 10	10 10	8 20	2 20	98.5	3 10	9 10
6 21	0 21	35.8	5 9	11 9	7 21	1 21	77.6	4 9	10 9	8 21	2 21	98.7	3 9	9 9
6 22	0 22	37.5	5 8	11 8	7 22	1 22	78.6	4 8	10 8	8 22	2 22	99.0	3 8	9 8
6 23	0 23	39.0	5 7	11 7	7 23	1 23	79.7	4 7	10 7	8 23	2 23	99.2	3 7	9 7
6 24	0 24	40.7	5 6	11 6	7 24	1 24	80.7	4 6	10 6	8 24	2 24	99.4	3 6	9 6
6 25	0 25	42.2	5 5	11 5	7 25	1 25	81.7	4 5	10 5	8 25	2 25	99.6	3 5	9 5
6 26	0 26	43.8	5 4	11 4	7 26	1 26	82.7	4 4	10 4	8 26	2 26	99.7	3 4	9 4
6 27	0 27	45.3	5 3	11 3	7 27	1 27	83.7	4 3	10 3	8 27	2 27	99.8	3 3	9 3
6 28	0 28	46.8	5 2	11 2	7 28	1 28	84.7	4 2	10 2	8 28	2 28	99.9	3 2	9 2
6 29	0 29	48.5	5 1	11 1	7 29	1 29	85.6	4 1	10 1	8 29	2 29	99.9	3 1	9 1
7 0	1 0	50.0	5 0	11 0	8 0	2 0	86.5	4 0	10 0	9 0	3 0	100.0	3 0	9 0
			S. — N. +	S. + N. —				S. — N. +	S. + N. —				S. — N. +	S. + N. —

When the Obliquity of the Ecliptic is taken greater than 23° 28′, the Signs of the Table must be changed.

TABLE 6. The PRINCIPAL ZODIACAL STARS that may be OCCULTED by the MOON, with the corresponding places of the ☾'s ☊, when the OCCULTATIONS may be *expected*.

CONSTELLATIONS.

☾'s ☊	♓	Ceti.	♈	♉	Aurigæ.	♊	♋	♌	♍	♎	♏	Ophiuchi	♐	♑	♒
s															
XI 27°	ε			ηfβ	ϰ	ιυ	γ	στ	βαi		πσα		τ	γδ	λφ
24	ε			ηfβ	ϰ	ιυ	γ	ϱστ	βαi		πσατ		τ	γδ	λφ
21	ε			ηfβ	ϰ	ιυ	γ	ϱστ	βαi		πσατ		τ	γδ	λφ
18	ε			ηfβ	ϰ	υϰ	γ	ϱτ	βαi		πσατ		τ	γδ	λφ
15				ηfβ		υϰ	γ	αϱτ	αi		πσατ		στ	γδ	λφ
12				ηfβ		υϰ	γ	αϱτ	αi		πσατ		στ	γδ	λφ
9				ηfβ		ϰ	γ	αϱτ	i		πσατ		στψ	γδ	λ
6				ηfβ		ϰ	γ	αϱτ	i		πσατ		στψ	γδ	λ
3				ηfβ		ϰ	γ	αϱτ	i		πσατ		στψ		λ
0				ηfβ		ϰ	γ	αϱτυ	i		πσατ		σψ	ϑ	λ
X 27				ηfβ		ϰ		αϱυ	i		πσατ		σψ	ϑ	
24				ηfβ		ϰ		αϱυ	i		πσατ	A	σψ	ϑ	
21				ηfβ		ϰ	δ	αϱυ	i		πσα	A	σψ	ϑ	
18				ηfβ		εϰ	δ	αυ	i		πσα	A	σψ	ϑ	
15				ηfβ		εϰ	δ	υ			πσα	A	λσψ	ϑ	
12				ηf		εϰ	δ	υ			πσα	A	λσψ	ϑ	ϑ
9				ηf		εϰ	δ	υ			πσα	A	λσψ	ϑ	ϑ
6				ηf		εϰ	δ	υ			πσα	A	λσψ	ϑ	ϑ
3				ηf		εϰ	δ	υ			πσα	A	λψ	ϑ	ϑ
0				ηf		εϰ	δ	υ			πσα	A	λ		ϑ
IX 27				ηf		εδϰ	δ	ξπυ			πσα	A	λ		ϑ
24				ηf		εδϰ	δ	ξπυ			σα	Aϑ	λ		ϑ
21				ηf132		εδϰ		ξοπ			σα	Aϑ	λιν		ϑ
18				ηf132		εδϰ		ξοπ			σα	Aϑ	λιν		ϑ
15				ηf132		δϰ		ξοπ			σ	AϑB	λινο		ϑ
12				ηf132		δϰ		ξοπ			σ	ϑB	ινο		ϑ
9				ηf132		δϰ		ξοπ			δσ	ϑB	ινοπ		ϑ
6				ηf132		μδϰ		ξοπ			δ	ϑB	ινοπ		
3	ω			ηf132		ημδϰ		ξοπ			δ	ϑB	ινοπ		
0	ω			fA132		ημδϰ		ξοπ			δ	ϑB	ινοπ		
VIII 27				A132		ημζ		ξοπ			δ	ϑB	ινοπ		
24			δ	A2ϰ132		ημζ		ξοπ	i		δ	B	ινοπ		
21			δ	A2ϰ		ημζ		ξοπ	i		δ	B	οπd		
18			δ	A2ϰ		ημζ		ξοπ	i		δ	B	οπd	β	
15			δ	A2ϰ		ημζ		οπ	i		δ	B	1.2μπd	β	
12			δ	A2ϰι		ημνζ		οπ	i		δ		1.2μπd	β	
9			δ	A2ϰιζ		ημνζ	lα	οπ	i		δ	ϱ	1.2μπd	β	
6			δ	A2ϰιζ		ηνζ	1.2α	οπ	i	ϰλ		ϱ	1.2μd	β	
3			δ	A2ϰιζ		νζ	1.2α	οπ	i	ϰλ		ϱ	1.2μd	β	
0			δ	2ϰιζ		νζ	1.2α	οπ	αi	ϰλ	β	ϱ	1.2μd	β	
VII 27	ζ		δ	2ϰιζ		ν	1.2α	οπ	αi	ϰλ	βν	ϱ	1.2μd	β	
24	ζ			ιζ		ν	1.2α	οπ	αi	ϰλ	βν	ϱ	1.2μd	β	
21	ζ			ιζ		ν	1.2α	οπυ	αi	αϰλ	βν	ϱ	1.2μd	β	
18	ζ			ειζ		ν	1.2α	οπυ	αi	αϰλ	βν	ϱ	1.2μd	β	
15	ζ			ειζ		ν	1.2α	οπυ	α	αϰλ	βν	ϱ	1.2μd	β	
12	εζ			εζ		ν	1.2α	οπυ	αλ	αϰλ	βν	ϱ	2μd	β	
9	εζ			εζ		ν	1.2α	οπυ	αλ	α4ζ	βν	ϱ	2μd	β	
6	εζ			ε		ν	1.2α	οπυ	αλ	α4ζ	βν			β	
3	εζ			ε			1.2α	οπυ	αλ	α4ζ	βν			β	
0	εζ			ιδε			1.2α	οπυ	αλ	α4ζϑ	ν			β	
VI 27	ε			1.2δε			1.2α	οπυ	λ	α4ζϑ	ν			β	
24	ε			1.2δε			1.2α	οπυ	λ	α4ζϑ				β	
21	ε			1.2δε			1.2α	οπυ	λ	4ζϑ				β	ϑ
18	εο			1.2δε			1.2α	ξοπυ	λ	4ζϑ				β	ϑ
15	εο			1.2δε			1.2α	ξοπυ	λ	4ζϑ				β	ϑ
12	ο			1.2δ			1.2α	ξοπ	ϰ	4ζγϑ		φ		β	ϑ
9	ο			1.2δα			1.2α	ξοπτ	ϑϰ	4ζγϑ		φ		β	ϑ
6	ο			1.2δα			1.2α	ξοπτ	ϑϰ	γϑ		φ		β	ϑ
3	ο			1.2δα			12.α	ξοπτ	ϑϰ	γϑ		φ		β	ϑ
0	ο			1.2δα				ξοπτ	βηϑϰ	γϑ		φ		β	ϑ
☾'s ☊	♓	Ceti.	♈	♉	Aurigæ.	♊	♋	♌	♍	♎	♏	Ophiuchi	♐	♑	♒

CONSTELLATIONS.

The Letters in this Table are BAYER'S, and the Numerals FLAMSTEED'S.

ZODIACAL TABLES.

TABLE 6.

The PRINCIPAL ZODIACAL STARS that may be OCCULTED by the MOON, &c. continued.

CONSTELLATIONS.

☾'s ☊	♓	Ceti.	♈	♉	Aurigæ.	♊	♋	♌	♍	♎	♏	Ophiuchi	♐	♑	♒
s															
V 27°	ο			γ1.2δα				ξοτ	βηϑκ	γϑ		φ		β	ϑ
24	ο			γ1.2δα				ξοτ	βηϑκ	γϑ		φ		β	ϑ
21	ο			γ1.2δα				ξοτ	βηϑκ	γϑ		φ	d	β	ϑ
18				γ1.2δα				ξτ	βηϑκ	γϑ		φ	d	β	ϑλ
15				γ1.2δα				ξϱτ	βηγϑκ	γϑ		φ	d	β	λφ
12				γ1.2δα				ξϱτ	βηγϑκ	γ		φ	d		λφ
9	ν	μ		γ1.2δα				ϱσ	βηγκ	γψ		φ	d		λφ
6	ν	μ		γ1.2δα				αϱσ	βηγκ	γψ		φ	d		λφ
3	ν	μ		γ1.2δα		ν		αϱσ	βηγ	γψ		φ	d		λφ
0	ν	μ		γ1.2δα		ν		αϱσ	ηγ	γψ		φ	d		λφ
IV 27	νη	2ξμ		γ1.2δα		ν		αϱσ	γ	γψ		φ	d		λφ
24	νη	2ξμ		γ1.2δα		νζ		αϱσ	γ	γψ		φ	d		λφ
21	νη	2ξμ		γ1.2δα		νζ		αϱσ	γ	γψ		φ	2μd		λφ
18	νη	2ξμ		γ1.2δα		νζ	δ	αϱσ	γ	γψ		φ	2μd	ϑ	φ
15	νη	2ξμ		γ1.2δα		νζ	δ	αϱσ	γ	γ		φ	2μπd	ϑ	
12	νη	2ξμ		γ1.2δα		νζκ	δ	ασ	γ	γ		φ	2μπd	ϑ	
9	νη	2ξμ		γ1.2δα		νζκ	δ	α		γϑ		φ	2μϑ	ϑ	
6	νη	2ξμ		1.2δαζ		νζκ	δ	α		γϑ		φ	2μοπ	ϑ	
3	νη	2ξμ		1.2δζ		νζδκ	δ			γϑ		φ	2μοπ	ϑγ	
0	νη	2ξμ		1.2δζ		νζδκ	δ			γϑ		φ	2μοπ	ϑγ	
III 27	νη	2ξμ		1.2δζ		ηνζδκ	δ			γϑ		ϱ	2μοπ	ϑγ	
24	νη	2ξμ		1.2δεζ		ηζδκ	δ			γϑ		ϱ	2μ1νοπ	ϑγ	
21	νη	2ξμ		1.2δεζ		ηζδκ	δ			γϑ		ϱ	2μ1νοπ	ϑγ	
18	νη	2ξμ		1.2δεζ		ηδκ				γϑ		ϱ	2μ1νοπ	γ	
15	νη	2ξμ		1.2δεζ		ηδκ				γϑ		ϱ	1νο	γ	
12	νη	μ		1.2δειζ		ηδκ				γϑ		ϱ	1νο	γ	
9	νη	μ		1.2δειζ		ηδκ		η		γϑ		ϱ	1νο	γ	
6	νη			ειζ		ηδκ		η		4ζγϑ	κ	ϱ	1ν	γ	
3	νη			ει		ηκ		η		4ζγϑ	ν	ϱ	1ν	γ	
0	νη			ει		ηκ		η		4ζϑ	ν	ϱ	1ν	ε	2τ
II 27	νη			ει132		ηεκ	γ	ηι	κ	4ζϑ	βν		1ν	ε	2τ
24	νη			ει132		εκ	γ	ηι	κ	4ζϑ	βν	λψ		ε	2τ
21	νη			ε132		εκ	γ	ηι	κ	4ζ	βν	B	λψ	ε	2τ
18	ν			ι132		εκ	γ	ηι	κ	4ζ	βν	B	λψ	ε	2τ
15	ν			ι132		εκ	γ	ηι	κ	4ζ	βν	B	λσψ	ε	2τ
12	ν			2κ132		εκ	γ	ηι	κ	4ζλ	βν	ϑB	λσψ	ε	2τ
9	ν			2κ132		εκ	γ	ηι	κ	4ζκλ	βν	ϑB	λσψ	ε	2τ
6				2κ132		εκ	γ	ηι	κ	4ζκλ	β	ϑB	λσψ	ε	2τ
3				2κ132		εκ	γ	ηι	κ	κλ	β	ϑB	λσψ	ε	2τ
0	ο			A2κ132		εκ		η	γκ	ακλ	β	ϑB	λσψ	ε	2τ
I 27	ο			A2κ132		ευκ		η	γκ	ακλ		ϑB	λσψ	ε	2τ
24	ο			A2κ		υκ		η	γκλ	ακλ		ϑB	λσψ	ε	2τ
21	ο			A2κ		υ		η	γϑλ	ακλ	δ	Aϑ	στψ	ε	2τ
18	ο			A2κ		υ		η	γϑλ	ακλ	δ	Aϑ	στ	ε	2τ
12	ο			A		ιυ		η	γϑλ	ακ	δ	Aϑ	στ	ε	2τ
9	ο		δ	A		ιυ		η	γϑλ	α	δ	A	στ	ε	2τ
6	ο		δ	A		ιυ		η	γϑλ	α	δ	A	στ	ε	
3	ζο		δ	A		ιυ		η	ηγϑλ	α	δ	A	στ	ε	
0	ζο		δ			ιυ		η	ηγϑλ		δσ	A	τ	ε	
0 27	ζ		δ			ιυ		η	ηγϑλ		δσ	A	τb	ε	
24	ζ		δ	β		ιυ		η	ηϑ		δσα	A	τb	ε	
21	ζ		δ	ηβ		ιυ		ησ	ηϑ		δσα	A	τb	ε	
18	ζ		δ	ηβ		ιυ		ησ	η		σα	A	τb	ε	
15	εζ		δ	ηβ		ιυ		ησ	βη		σα	A	τ	ε	
12	εζ		δ	ηβ	κ	ιυ		ησ	βηα		σα	A	τ		
9	εζ			ηβ	κ	ιυ		π	βηα		σα	A	τ		
6	εζ			ηβ	κ	ιυ	γ	σ	βηα		σα		δτ	εγδ	
3	ε			ηβ	κ	ιυ	γ	σ	βα		πσα		δτ	γδ	
0	ε			ηβ	κ	ιυ	γ	σ	βα		πσα		δτ	γδ	
☾'s ☊	♓	Ceti.	♈	♉	Aurigæ.	♊	♋	♌	♍	♎	♏	Ophiuchi	♐	♑	♒

CONSTELLATIONS.

The Letters in this Table are BAYER's, and the Numerals FLAMSTEED's.

TABLE 7.—The Longitude and Altitude of the Nonagesimal Degree of the Ecliptic to every Deg. of the R. A. of the Mid-heaven, with the Diff. for 100′ of R. A., the Diff. for 100′ Var. of Lat. North of Greenwich, and the Diff. for 100″ Var. of the Obliq. of the Eclip. The Lat. of Greenwich is taken at 51° 28′ 40″, and reduced to the Earth's centre, viz. 51° 17′ 48″, the Compres. being = $\frac{1}{309}$, and the Obliq. of the Eclip. 23° 28′. (By Mr. J. Utting.) Argument = the R. A. of the Mid-heaven.

Arg. The R. A. of the Mid-heaven.	Long. of Nonag.	Diff. for 100′ of R. A.	Var. for 100′ Var. of Lat.	Var. for 100″ Var. of Obliq.	Alt. of Nonag.	Diff. for 100′ of R. A.	Var. for 100′ Var. of Lat.	Var. for 100″ Var. of Obliq.	Arg. The R. A. of the Mid-heaven.
			+	—			—	—	
0°	26° 25 38″	4405″	5025″	91″.8	44° 17′ 14″	2133″	4899″	44″.5	0°
1	27 9 41	4393	4960	90.1	44 38 34	2118	4915	45.7	1
2	27 53 37	4380	4895	88.5	44 59 45	2105	4931	46.9	2
3	28 37 24	4368	4831	86.8	45 20 48	2090	4948	48.0	3
4	29 21 5	4355	4767	85.2	45 41 42	2075	4964	49.1	4
5	30 4 38	4343	4704	83.5	46 2 27	2058	4980	50.1	5
6	30 48 4	4333	4641	81.9	46 23 2	2045	4997	51.2	6
7	31 31 24	4322	4579	80.4	46 43 29	2030	5014	52.2	7
8	32 14 37	4313	4517	78.8	47 3 47	2012	5031	53.3	8
9	32 57 45	4303	4457	77.3	47 23 54	1997	5048	54.4	9
10	33 40 47	4295	4396	75.7	47 43 52	1980	5065	55.5	10
11	34 23 44	4287	4336	74.2	48 3 40	1963	5082	56.6	11
12	35 6 36	4280	4276	72.8	48 23 18	1947	5100	57.6	12
13	35 49 24	4272	4217	71.3	48 42 46	1927	5117	58.6	13
14	36 32 7	4267	4157	69.9	49 2 2	1912	5134	59.6	14
15	37 14 47	4260	4098	68.4	49 21 9	1893	5151	60.6	15
16	37 57 23	4252	4039	67.0	49 40 5	1875	5169	61.6	16
17	38 39 54	4248	3981	65.6	49 58 50	1855	5187	62.6	17
18	39 22 23	4242	3923	64.3	50 17 23	1838	5204	63.5	18
19	40 4 48	4238	3865	62.9	50 35 46	1818	5222	64.5	19
20	40 47 11	4232	3808	61.5	50 53 57	1798	5239	65.4	20
21	41 29 30	4230	3751	60.2	51 11 56	1780	5256	66.3	21
22	42 11 48	4225	3694	59.0	51 29 44	1762	5274	67.2	22
23	42 54 3	4222	3637	57.7	51 47 21	1740	5292	68.0	23
24	43 36 15	4220	3581	56.5	52 4 45	1720	5309	68.9	24
25	44 18 27	4217	3524	55.2	52 21 57	1700	5326	69.8	25
26	45 0 37	4213	3468	54.0	52 38 57	1678	5344	70.7	26
27	45 42 45	4210	3412	52.8	52 55 44	1658	5361	71.5	27
28	46 24 51	4207	3356	51.7	53 12 19	1637	5378	72.4	28
29	47 6 55	4207	3300	50.5	53 28 41	1615	5395	73.2	29
30	47 48 59	4205	3245	49.3	53 44 50	1593	5411	74.1	30
31	48 31 1	4203	3190	48.2	54 0 46	1572	5428	74.9	31
32	49 13 4	4202	3135	47.1	54 16 29	1550	5445	75.7	32
33	49 55 5	4200	3080	45.9	54 31 59	1527	5462	76.4	33
34	50 37 5	4202	3025	44.8	54 47 15	1505	5478	77.2	34
35	51 19 6	4198	2970	43.7	55 2 18	1482	5494	78.0	35
36	52 1 5	4198	2915	42.7	55 17 7	1458	5511	78.7	36
37	52 43 4	4200	2861	41.6	55 31 42	1435	5526	79.4	37
38	53 25 4	4198	2806	40.6	55 46 3	1413	5542	80.2	38
39	54 7 2	4198	2752	39.5	56 0 11	1388	5558	80.9	39
40	54 49 1	4198	2698	38.5	56 14 4	1365	5573	81.6	40
41	55 31 0	4198	2643	37.5	56 27 43	1340	5588	82.3	41
42	56 12 59	4200	2589	36.6	56 41 7	1317	5603	83.0	42
43	56 54 59	4198	2534	35.6	56 54 17	1292	5618	83.7	43
44	57 36 58	4200	2480	34.7	57 7 12	1268	5633	84.4	44
45	58 18 58	4200	2426	33.7	57 19 53	1242	5648	85.1	45
46	59 0 58	4200	2373	32.8	57 32 18	1218	5662	85.7	46
47	59 42 58	4202	2320	31.9	57 44 29	1192	5676	86.3	47
48	60 24 59	4203	2266	31.0	57 56 24	1167	5690	86.9	48
49	61 7 2	4203	2212	30.1	58 8 4	1142	5704	87.5	49
50	61 49 4	4205	2158	29.2	58 19 29	1115	5717	88.1	50
51	62 31 7	4205	2104	28.3	58 30 38	1090	5730	88.6	51
52	63 13 10	4207	2051	27.5	58 41 32	1063	5743	89.2	52
53	63 55 14	4208	1997	26.6	58 52 10	1037	5756	89.7	53
54	64 37 19	4208	1943	25.8	59 2 32	1012	5768	90.3	54
55	65 19 25	4210	1889	24.9	59 12 39	985	5781	90.8	55
56	66 1 31	4212	1835	24.1	59 22 30	957	5793	91.3	56
57	66 43 38	4212	1782	23.3	59 32 4	930	5804	91.8	57
58	67 25 46	4213	1728	22.5	59 41 22	903	5814	92.3	58
59	68 7 56	4215	1673	21.7	59 50 24	877	5825	92.8	59
60	68 50 5		1620	20.9	59 59 10		5836	93.3	60

Monsieur Biot proposes to discard the terms *Mid-heaven*, or *medium cœli*, and *nonagesimal* degree; and to substitute the R. A., the Long., and the Lat. of the *Zenith*.

TABLE 7.

The LONGITUDE and ALTITUDE of the NONAGESIMAL DEGREE, continued.

Arg. The R. A. of the Mid-heaven.	Long. of Nonag.	Diff. for 100′ of R. A.	Var. for 100′ Var. of Lat.	Var. for 100″ Var. of Obliq.	Alt. of Nonag.	Diff. for 100′ of R. A.	Var. for 100′ Var. of Lat.	Var. for 100″ Var. of Obliq.	Arg. The R. A. of the Mid-heaven.
			+	−			−	−	
60°	68° 50′ 5″	4217″	1620″	20″.9	59° 59′ 10″	850″	5836″	93″.3	60°
61	69 32 15	4220	1566	20 .1	60 7 40	822	5847	93 .7	61
62	70 14 27	4218	1512	19 .4	60 15 53	793	5857	94 .1	62
63	70 56 38	4222	1459	18 .6	60 23 49	765	5867	94 .6	63
64	71 38 51	4223	1405	17 .9	60 31 28	740	5876	95 .0	64
65	72 21 5	4225	1351	17 .1	60 38 52	710	5886	95 .4	65
66	73 3 20	4223	1297	16 .4	60 45 58	682	5894	95 .7	66
67	73 45 34	4227	1243	15 .6	60 52 47	655	5902	96 .0	67
68	74 27 50	4228	1190	14 .9	60 59 20	625	5911	96 .4	68
69	75 10 7	4230	1136	14 .1	61 5 35	597	5918	96 .7	69
70	75 52 25	4228	1080	13 .4	61 11 33	568	5925	97 .0	70
71	76 34 42	4232	1028	12 .7	61 17 14	542	5932	97 .2	71
72	77 17 1	4232	974	12 .0	61 22 39	512	5940	97 .5	72
73	77 59 20	4233	921	11 .3	61 27 46	482	5946	97 .7	73
74	78 41 40	4235	866	10 .6	61 32 35	455	5952	98 .0	74
75	79 24 1	4235	812	9 .9	61 37 8	425	5958	98 .2	75
76	80 6 22	4237	758	9 .3	61 41 23	395	5964	98 .4	76
77	80 48 44	4237	704	8 .6	61 45 20	368	5969	98 .6	77
78	81 31 6	4240	650	7 .9	61 49 1	337	5973	98 .8	78
79	82 13 30	4238	595	7 .3	61 52 23	310	5977	99 .0	79
80	82 55 53	4238	541	6 .6	61 55 29	278	5981	99 .2	80
81	83 38 16	4238	487	5 .9	61 58 16	252	5985	99 .3	81
82	84 20 39	4242	433	5 .3	62 0 47	222	5988	99 .5	82
83	85 3 4	4242	379	4 .6	62 3 0	188	5991	99 .6	83
84	85 45 29	4242	325	4 .0	62 4 53	162	5993	99 .7	84
85	86 27 54	4243	271	3 .3	62 6 30	135	5995	99 .8	85
86	87 10 18	4242	217	2 .6	62 7 51	102	5997	99 .8	86
87	87 52 44	4243	163	2 .0	62 8 52	75	5998	99 .9	87
88	88 35 9	4243	109	1 .3	62 9 37	43	5999	100 .0	88
89	89 17 35	4242	54	0 .7	62 10 3	15	6000	100 .0	89
			+	−					
90	90 0 0	4242	0	0 .0	62 10 12	15	6000	100 .0	90
			−	+					
91	90 42 25	4243	54	0 .7	62 10 3	43	6000	100 .0	91
92	91 24 51	4243	109	1 .3	62 9 37	75	5999	100 .0	92
93	92 7 16	4242	163	2 .0	62 8 52	102	5998	99 .9	93
94	92 49 42	4243	217	2 .6	62 7 51	135	5997	99 .8	94
95	93 32 6	4242	271	3 .3	62 6 30	162	5995	99 .8	95
96	94 14 31	4242	325	4 .0	62 4 53	188	5993	99 .7	96
97	94 56 56	4242	379	4 .6	62 3 0	222	5991	99 .6	97
98	95 39 21	4238	433	5 .3	62 0 47	252	5988	99 .5	98
99	96 21 44	4238	487	5 .9	61 58 16	278	5985	99 .3	99
100	97 4 7	4238	541	6 .6	61 55 29	310	5981	99 .2	100
101	97 46 30	4240	595	7 .3	61 52 23	337	5977	99 .0	101
102	98 28 54	4237	650	7 .9	61 49 1	368	5973	98 .8	102
103	99 11 16	4237	704	8 .6	61 45 20	395	5969	98 .6	103
104	99 53 38	4236	758	9 .3	61 41 23	425	5964	98 .4	104
105	100 35 59	4235	812	9 .9	61 37 8	455	5958	98 .2	105
106	101 18 20	4233	866	10 .6	61 32 35	482	5952	98 .0	106
107	102 0 40	4232	921	11 .3	61 27 46	512	5946	97 .7	107
108	102 42 59	4232	974	12 .0	61 22 39	542	5940	97 .5	108
109	103 25 18	4228	1028	12 .7	61 17 14	568	5932	97 .2	109
110	104 7 35	4230	1080	13 .4	61 11 33	597	5925	97 .0	110
111	104 49 53	4228	1136	14 .1	61 5 35	625	5918	96 .7	111
112	105 32 10	4227	1190	14 .9	60 59 20	655	5911	96 .4	112
113	106 14 26	4223	1243	15 .6	60 52 47	682	5902	96 .0	113
114	106 56 40	4225	1297	16 .4	60 45 58	710	5894	95 .7	114
115	107 38 55	4223	1351	17 .1	60 38 52	740	5886	95 .4	115
116	108 21 9	4222	1405	17 .9	60 31 28	765	5876	95 .0	116
117	109 3 22	4218	1459	18 .6	60 23 49	793	5867	94 .6	117
118	109 45 33	4220	1512	19 .4	60 15 53	822	5857	94 .1	118
119	110 27 45	4217	1566	20 .1	60 7 40	850	5847	93 .7	119
120	111 9 55		1620	20 .9	59 59 10		5836	93 .3	120

TABLE 7.

The Longitude and Altitude of the Nonagesimal Degree, continued.

Arg. The R. A. of the Mid-heaven.	Long. of Nonag.	Diff. for 100' of R. A.	Var. for 100' Var. of Lat.	Var. for 100" Var. of Obliq.	Alt. of Nonag.	Diff. for 100' of R. A.	Var. for 100' Var. of Lat.	Var. for 100" Var. of Obliq.	Arg. The R. A. of the Mid-heaven.
			−	+					
120°	111° 9' 55"	4215"	1620"	20".9	59° 59' 10"	877"	5836"	93".3	120°
121	111 52 4	4213	1673	21 .7	59 50 24	903	5825	92 .8	121
122	112 34 14	4212	1728	22 .5	59 41 22	930	5814	92 .8	122
123	113 16 22	4212	1782	23 .3	59 32 4	957	5804	91 .8	123
124	113 58 29	4210	1835	24 .1	59 22 30	985	5793	91 .3	124
125	114 40 35	4208	1889	24 .9	59 12 39	1012	5781	90 .8	125
126	115 22 41	4208	1943	25 .8	59 2 32	1037	5768	90 .3	126
127	116 4 46	4207	1997	26 .6	58 52 10	1063	5756	89 .7	127
128	116 46 50	4205	2051	27 .5	58 41 32	1090	5743	89 .2	128
129	117 28 53	4205	2104	28 .3	58 30 38	1115	5730	88 .6	129
130	118 10 56	4203	2158	29 .2	58 19 29	1142	5717	88 .1	130
131	118 52 58	4203	2212	30 .1	58 8 4	1167	5704	87 .5	131
132	119 35 1	4202	2266	31 .0	57 56 24	1192	5690	86 .9	132
133	120 17 2	4200	2320	31 .9	57 44 29	1218	5676	86 .3	133
134	120 59 2	4200	2373	32 .8	57 32 18	1242	5662	85 .7	134
135	121 41 2	4200	2426	33 .7	57 19 53	1268	5648	85 .1	135
136	122 23 2	4198	2480	34 .7	57 7 12	1292	5633	84 .4	136
137	123 5 1	4200	2534	35 .6	56 54 17	1317	5618	83 .7	137
138	123 47 1	4198	2589	36 .6	52 41 7	1340	5603	83 .0	138
139	124 29 0	4198	2643	37 .5	56 27 43	1365	5588	82 .3	139
140	125 10 59	4198	2698	38 .5	56 14 4	1388	5573	81 .6	140
141	125 52 58	4198	2752	39 .5	56 0 11	1413	5558	80 .9	141
142	126 34 56	4200	2806	40 .6	55 46 3	1435	5542	80 .2	142
143	127 16 56	4198	2861	41 .6	55 31 42	1458	5526	79 .4	143
144	127 58 55	4198	2915	42 .7	55 17 7	1482	5511	78 .7	144
145	128 40 54	4202	2970	43 .7	55 2 18	1505	5494	78 .0	145
146	129 22 55	4200	3025	44 .8	54 47 15	1527	5478	77 .2	146
147	130 4 55	4202	3080	45 .9	54 31 59	1550	5462	76 .4	147
148	130 46 56	4203	3135	47 .1	54 16 29	1572	5445	75 .7	148
149	131 28 59	4205	3190	48 .2	54 0 46	1593	5428	74 .9	149
150	132 11 1	4207	3245	49 .3	53 44 50	1615	5411	74 .1	150
151	132 53 5	4207	3300	50 .5	53 28 41	1637	5395	73 .2	151
152	133 35 9	4210	3356	51 .7	53 12 19	1658	5378	72 .4	152
153	134 17 15	4213	3412	52 .8	52 55 44	1678	5361	71 .5	153
154	134 59 23	4217	3468	54 .0	52 38 57	1700	5344	70 .7	154
155	135 41 33	4220	3524	55 .2	52 21 57	1720	5326	69 .8	155
156	136 23 45	4222	3581	56 .5	52 4 45	1740	5309	68 .9	156
157	137 5 57	4225	3637	57 .7	51 47 21	1762	5292	68 .0	157
158	137 48 12	4230	3694	59 .0	51 29 44	1780	5274	67 .2	158
159	138 30 30	4232	3751	60 .2	51 11 56	1798	5256	66 .3	159
160	139 12 49	4238	3808	61 .5	50 53 57	1818	5239	65 .4	160
161	139 55 12	4242	3865	62 .9	50 35 46	1838	5222	64 .5	161
162	140 37 37	4248	3923	64 .3	50 17 23	1855	5204	63 .5	162
163	141 20 6	4252	3981	65 .6	49 58 50	1875	5187	62 .6	163
164	142 2 37	4260	4039	67 .0	49 40 5	1893	5169	61 .6	164
165	142 45 13	4267	4098	68 .4	49 21 9	1912	5151	60 .6	165
166	143 27 53	4272	4157	69 .9	49 2 2	1927	5134	59 .6	166
167	144 10 36	4280	4217	71 .3	48 42 46	1947	5117	58 .6	167
168	144 53 24	4287	4276	72 .8	48 23 18	1963	5100	57 .6	168
169	145 36 16	4295	4336	74 .2	48 3 40	1980	5082	56 .6	169
170	146 19 13	4303	4396	75 .7	47 43 52	1997	5065	55 .5	170
171	147 2 15	4313	4457	77 .3	47 23 54	2012	5048	54 .4	171
172	147 5 23	4322	4517	78 .8	47 3 47	2030	5031	53 .3	172
173	148 28 36	4333	4579	80 .4	46 43 29	2045	5014	52 .2	173
174	149 11 56	4343	4641	81 .9	46 23 2	2058	4997	51 .2	174
175	149 55 22	4355	4704	83 .5	46 2 27	2075	4980	50 .1	175
176	150 38 55	4368	4767	85 .2	45 41 42	2090	4964	49 .1	176
177	151 22 36	4380	4831	86 .8	45 20 48	2105	4948	48 .0	177
178	152 6 23	4393	4895	88 .5	44 59 45	2118	4931	46 .9	178
179	152 50 19	4405	4960	90 .1	44 38 34	2133	4915	45 .7	179
180	153 34 22		5025	91 .8	44 17 14		4899	44 .5	180

TABLE 7.

The Longitude and Altitude of the Nonagesimal Degree, continued.

Arg. The R. A. of the Mid-heaven.	Long. of Nonag.	Diff. for 100′ of R. A.	Var. for 100′ Var. of Lat.	Var. for 100″ Var. of Obliq.	Alt. of Nonag.	Diff. for 100′ of R. A.	Var. for 100′ Var. of Lat.	Var. for 100″ Var. of Obliq.	Arg. The R. A. of the Mid-heaven.
			−	+			−	−	
180°	153° 34′ 22″	4422″	5025″	91″.8	44° 17′ 14″	2147″	4899″	44″.5	180°
181	154 18 35	4435	5090	93 .6	43 55 46	2160	4883	43 .3	181
182	155 2 56	4453	5156	95 .4	43 31 10	2172	4867	42 .1	182
183	155 47 28	4470	5222	97 .1	43 12 27	2187	4852	41 .0	183
184	156 32 10	4487	5290	98 .9	42 50 35	2197	4836	39 .8	184
185	157 17 2	4505	5358	100 .7	42 28 37	2212	4822	38 .6	185
186	158 2 5	4525	5426	102 .6	42 6 30	2222	4807	37 .7	186
187	158 47 20	4545	5496	104 .5	41 44 17	2232	4792	36 .8	187
188	159 32 47	4567	5565	106 .5	41 21 58	2245	4778	36 .0	188
189	160 18 27	4588	5636	108 .4	40 59 31	2255	4764	35 .1	189
190	161 4 20	4610	5707	110 .3	40 36 58	2265	4751	34 .2	190
191	161 50 26	4638	5778	112 .3	40 14 19	2275	4738	32 .6	191
192	162 36 49	4662	5851	114 .3	39 51 34	2285	4725	30 .9	192
193	163 23 26	4687	5925	116 .4	39 28 43	2295	4713	29 .3	193
194	164 10 18	4717	5999	118 .4	39 5 46	2302	4700	27 .6	194
195	164 57 28	4745	6074	120 .4	38 42 45	2313	4688	26 .0	195
196	165 44 55	4775	6150	122 .6	38 19 37	2318	4676	24 .6	196
197	166 32 40	4807	6226	124 .8	37 56 26	2328	4666	23 .2	197
198	167 20 44	4838	6304	127 .0	37 33 9	2335	4655	21 .9	198
199	168 9 7	4875	6382	129 .2	37 9 48	2342	4645	20 .5	199
200	168 57 52	4910	6461	131 .4	36 46 23	2347	4636	19 .1	200
201	169 46 58	4947	6541	133 .7	36 22 55	2357	4627	17 .7	201
202	170 36 26	4988	6621	136 .0	35 59 21	2358	4618	16 .2	202
203	171 26 19	5027	6703	138 .2	35 35 46	2365	4610	14 .8	203
204	172 16 35	5070	6785	140 .5	35 12 7	2372	4603	13 .3	204
205	173 7 17	5117	6868	142 .8	34 48 24	2373	4596	11 .9	205
206	173 58 27	5158	6952	145 .2	34 24 40	2377	4590	10 .4	206
207	174 50 2	5212	7036	147 .5	34 0 54	2383	4585	8 .9	207
208	175 42 9	5260	7122	149 .9	33 37 4	2382	4580	7 .4	208
209	176 34 45	5312	7208	152 .2	33 13 15	2387	4576	5 .9	209
210	177 27 52	5370	7294	154 .6	32 49 23	2387	4573	4 .4	210
211	178 21 34	5423	7383	157 .0	32 25 31	2390	4570	2 .8	211
212	179 15 48	5487	7472	159 .4	32 1 37	2388	4568	1 .2	212
								+	
213	180 10 40	5545	7561	161 .7	31 37 45	2390	4566	0 .4	213
214	181 6 7	5615	7651	164 .1	31 13 50	2388	4565	2 .0	214
215	182 2 16	5678	7741	166 .5	30 49 57	2388	4566	3 .6	215
216	182 59 3	5750	7831	169 .1	30 26 4	2383	4567	5 .3	216
217	183 56 33	5827	7921	171 .8	30 2 14	2383	4570	7 .0	217
218	184 54 49	5903	8013	174 .4	29 38 24	2377	4573	8 .7	218
219	185 53 51	5983	8104	177 .1	29 14 38	2375	4577	10 .4	219
220	186 53 41	6068	8196	179 .7	23 50 53	2370	4582	12 .1	220
221	187 54 22	6153	8288	182 .3	38 27 11	2363	4588	13 .9	221
222	188 55 54	6247	8379	184 .9	28 3 33	2357	4596	15 .7	222
223	189 58 22	6340	8469	187 .5	27 39 59	2350	4604	17 .5	223
224	191 1 46	6442	8559	190 .1	27 16 29	2340	4613	19 .3	224
225	192 6 11	6543	8648	192 .7	26 53 5	2332	4624	21 .1	225
226	198 11 37	6652	8736	195 .3	26 29 46	2320	4637	23 .0	226
227	194 18 8	6767	8823	197 .7	26 6 34	2310	4650	24 .8	227
228	195 25 48	6883	8908	199 .9	25 43 28	2297	4664	26 .7	228
229	196 34 38	7003	8992	202 .2	25 20 30	2283	4680	28 .5	229
230	197 44 40	7135	9072	204 .5	24 57 40	2267	4698	30 .4	230
231	198 56 1	7265	9149	206 .5	24 35 0	2252	4717	32 .4	231
232	200 8 40	7402	9223	208 .5	24 12 29	2235	4737	34 .5	232
233	201 22 41	7552	9294	210 .5	23 50 8	2215	4758	36 .5	233
234	202 38 12	7697	9360	212 .5	23 27 59	2195	4781	38 .6	234
235	203 55 10	7853	9421	214 .5	23 6 2	2172	4805	40 .6	235
236	205 13 42	8013	9476	216 .5	22 44 19	2150	4831	42 .7	236
237	206 33 50	8185	9525	218 .5	22 22 49	2125	4858	44 .9	237
238	207 55 41	8360	9567	220 .6	22 1 34	2097	4887	47 .0	238
239	209 19 17	8537	9600	222 .0	21 40 36	2068	4918	49 .0	239
240	210 44 39		9623	221 .4	21 19 55		4950	51 .3	240

TABLE 7.

The Longitude and Altitude of the Nonagesimal Degree, continued.

Arg. The R. A. of the Mid-heaven.	Long. of Nonag.	Diff. for 100′ of R. A.	Var. for 100′ Var. of Lat.	Var. for 100″ Var. of Obliq.	Alt. of Nonag.	Diff. for 100′ of R. A.	Var. for 100′ Var. of Lat.	Var. for 100″ Var. of Obliq.	Arg. The R. A. of the Mid-heaven.
			−	+			−	+	
240°	210° 44′ 39″	8727″	9623″	221″.4	21° 19′ 55″	2038″	4950″	51″.3	240°
241	212 11 55	8920	9638	221.0	20 59 32	2005	4983	53.5	241
242	213 41 7	9118	9638	220.7	20 39 29	1970	5019	55.7	242
243	215 12 18	9327	9627	220.5	20 19 47	1933	5055	57.8	243
244	216 45 34	9542	9603	220.0	20 0 27	1895	5093	60.0	244
245	218 20 59	9757	9566	219.1	19 41 30	1852	5132	62.2	245
246	219 58 33	9973	9507	216.8	19 22 59	1810	5173	64.4	246
247	221 38 17	10212	9429	214.5	19 4 53	1760	5215	66.6	247
248	223 20 24	10442	9333	212.3	18 47 17	1715	5258	68.7	248
249	225 4 49	10678	9214	210.0	18 30 8	1660	5302	70.9	249
250	226 51 36	10915	9075	207.7	18 13 32	1605	5346	73.1	250
251	228 40 45	11165	8904	203.8	17 57 29	1548	5392	75.2	251
252	230 32 24	11405	8712	199.2	17 42 0	1488	5438	77.2	252
253	232 26 27	11817	8488	193.9	17 27 7	1423	5484	79.3	253
254	234 22 57	12060	8234	187.9	17 12 53	1358	5530	81.3	254
255	236 21 53	12305	7949	181.3	16 59 18	1288	5575	83.4	255
256	238 23 16	12368	7633	173.9	16 46 25	1215	5621	85.2	256
257	240 26 57	12605	7282	165.8	16 34 16	1140	5665	87.0	257
258	242 33 0	12830	6897	156.8	16 22 52	1063	5707	88.7	258
259	244 41 18	13043	6477	147.0	16 12 14	978	5748	90.5	259
260	246 51 44	13245	6024	136.7	16 2 27	897	5789	92.3	260
261	249 4 11	13430	5537	125.5	15 53 29	810	5825	93.4	261
262	251 18 29	13610	5015	113.7	15 45 23	720	5860	94.6	262
263	253 34 35	13772	4465	101.1	15 38 11	632	5891	95.7	263
264	255 52 18	13908	3886	87.7	15 31 52	537	5918	96.9	264
265	258 11 23	14023	3281	74.1	15 26 30	440	5942	98.0	265
266	260 31 37	14125	2652	59.9	15 22 6	345	5963	99.0	266
267	262 52 52	14193	2006	45.3	15 18 39	247	5979	99.6	267
268	265 14 48	14248	1344	30.4	15 16 11	148	5990	100.0	268
269	267 37 17	14272	674	15.2	15 14 42	50	5998	100.0	269
270	270 0 0	14272	0	0.0	15 14 12	50	6000	100.0	270
			+	−					
271	272 22 43	14248	674	15.2	15 14 42	148	5998	100.0	271
272	274 45 12	14193	1344	30.4	15 16 11	247	5990	100.0	272
273	277 7 8	14125	2006	45.3	15 18 39	345	5979	99.6	273
274	279 28 23	14023	2652	59.9	15 22 6	440	5963	99.0	274
275	281 48 37	13908	3281	74.1	15 26 30	537	5942	98.0	275
276	284 7 42	13772	3886	87.7	15 51 52	632	5918	96.9	276
277	286 25 25	13610	4465	101.1	15 38 11	720	5891	95.7	277
278	288 41 31	13430	5015	113.7	15 45 23	810	5860	94.6	278
279	290 55 49	13245	5537	125.5	15 53 29	897	5825	93.4	279
280	293 8 16	13043	6024	136.7	16 2 27	978	5789	92.3	280
281	295 18 42	12830	6477	147.0	16 12 14	1063	5748	90.5	281
282	297 27 0	12605	6897	156.8	16 22 52	1140	5707	88.7	282
283	299 33 3	12368	7282	165.8	16 34 16	1215	5665	87.0	283
284	301 36 44	12305	7633	173.9	16 46 25	1288	5621	85.2	284
285	303 38 7	12060	7949	181.3	16 59 18	1358	5575	83.4	285
286	305 37 3	11817	8234	187.9	17 12 53	1423	5530	81.3	286
287	307 33 33	11405	8488	193.9	17 27 7	1488	5484	79.3	287
288	309 27 36	11165	8712	199.2	17 42 0	1548	5438	77.2	288
289	311 19 15	10915	8904	203.8	17 57 29	1605	5392	75.2	289
290	313 8 24	10678	9075	207.7	18 13 32	1660	5346	73.1	290
291	314 55 11	10442	9214	210.0	18 30 8	1715	5302	70.9	291
292	316 39 36	10212	9333	212.3	18 47 17	1760	5258	68.7	292
293	318 21 43	9973	9429	214.5	19 4 53	1810	5215	66.6	293
294	320 1 27	9757	9507	216.8	19 22 59	1852	5173	64.4	294
295	321 39 1	9542	9566	219.1	19 41 30	1895	5132	62.2	295
296	323 14 26	9327	9603	220.0	20 0 27	1933	5093	60.0	296
297	324 47 42	9118	9627	220.5	20 19 47	1970	5055	57.8	297
298	326 18 53	8920	9638	220.7	20 39 29	2005	5019	55.7	298
299	327 48 5	8727	9638	221.0	20 59 32	2038	4983	53.5	299
300	329 15 21		9623	221.4	21 19 55		4950	51.3	300

TABLE 7.

The Longitude and Altitude of the Nonagesimal Degree, concluded.

Arg. The R. A. of the Mid-heaven.	Long. of Nonag.	Diff. for 100′ of R. A.	Var. for 100′ Var. of Lat.	Var. for 100″ Var. of Obliq.	Alt. of Nonag.	Diff. for 100′ of R. A.	Var. for 100′ Var. of Lat.	Var. for 100″ Var. of Obliq.	Arg. The R. A. of the Mid-heaven.
			+	−			−	+	
300°	329° 15′ 21″		9623″	221″.4	21° 19′ 55″		4950″	51″.3	300°
		8537″				2068″			
301	330 40 43		9600	222 .0	21 40 36		4918	49 .0	301
		8360				2097			
302	332 4 19		9567	220 .6	22 1 34		4887	47 .0	302
		8185				2125			
303	333 26 10		9525	218 .5	22 22 49		4858	44 .9	303
		8013				2150			
304	334 46 18		9476	216 .5	22 44 19		4831	42 .7	304
		7853				2172			
305	336 4 50		9421	214 .5	23 6 2		4805	40 .6	305
		7697				2195			
306	337 21 48		9360	212 .5	23 27 59		4781	38 .6	306
		7552				2215			
307	338 37 19		9294	210 .5	23 50 8		4758	36 .5	307
		7402				2235			
308	339 51 20		9223	208 .5	24 12 29		4737	34 .5	308
		7265				2252			
309	341 3 59		9149	206 .5	24 35 0		4713	32 .4	309
		7135				2267			
310	342 15 20		9072	204 .5	24 57 40		4698	30 .4	310
		7003				2283			
311	343 25 22		8992	202 .2	25 20 30		4680	28 .5	311
		6883				2297			
312	344 34 12		8908	199 .9	25 43 28		4664	26 .7	312
		6767				2310			
313	345 41 52		8823	197 .7	26 6 34		4650	24 .8	313
		6652				2320			
314	346 48 23		8736	195 .3	26 29 46		4637	23 .0	314
		6543				2332			
315	347 53 49		8648	192 .7	26 53 5		4624	21 .1	315
		6442				2340			
316	348 58 14		8559	190 .1	27 16 29		4613	19 .3	316
		6340				2350			
317	350 1 38		8469	187 .5	27 39 59		4604	17 .5	317
		6247				2357			
318	351 4 6		8379	184 .9	28 3 33		4596	15 .7	318
		6153				2363			
319	352 5 38		8288	182 .3	28 27 11		4588	13 .9	319
		6068				2370			
320	353 6 19		8196	179 .7	28 50 53		4582	12 .1	320
		5983				2375			
321	354 6 9		8104	177 .1	29 14 38		4577	10 .4	321
		5903				2377			
322	355 5 11		8013	174 .4	29 38 24		4573	8 .7	322
		5827				2383			
323	356 3 27		7921	171 .8	30 2 14		4570	7 .0	323
		5750				2383			
324	357 0 57		7831	169 .1	30 26 4		4567	5 .3	324
		5678				2388			
325	357 57 44		7741	166 .5	30 49 57		4566	3 .6	325
		5615				2388			
326	358 53 53		7651	164 .1	31 13 50		4565	2 .0	326
		5545				2390			
327	359 49 20		7561	161 .7	31 37 44		4566	0 .4	327
		5487				2388			
								−	
328	0 44 12		7472	159 .4	32 1 37		4568	1 .2	328
		5423				2390			
329	1 38 26		7383	157 .0	32 25 31		4570	2 .8	329
		5370				2387			
330	2 32 8		7294	154 .6	32 49 23		4573	4 .4	330
		5312				2387			
331	3 25 15		7208	152 .2	33 13 15		4576	5 .9	331
		5260				2382			
332	4 17 51		7122	149 .9	33 37 4		4580	7 .4	332
		5212				2383			
333	5 9 58		7036	147 .5	34 0 54		4585	8 .9	333
		5158				2377			
334	6 1 33		6952	145 .2	34 24 40		4590	10 .4	334
		5117				2373			
335	6 52 43		6868	142 .8	34 48 24		4596	11 .9	335
		5070				2372			
336	7 43 25		6785	140 .5	35 12 7		4603	13 .3	336
		5027				2365			
337	8 33 41		6703	138 .2	35 35 46		4610	14 .8	337
		4988				2358			
338	9 23 34		6621	136 .0	35 59 21		4618	16 .2	338
		4947				2357			
339	10 13 2		6541	133 .7	36 22 55		4627	17 .7	339
		4910				2347			
340	11 2 8		6461	131 .4	36 46 23		4636	19 .1	340
		4875				2342			
341	11 50 53		6382	129 .2	37 9 48		4645	20 .5	341
		4838				2335			
342	12 39 16		6304	127 .0	37 33 9		4655	21 .9	342
		4807				2328			
343	13 27 20		6226	124 .8	37 56 26		4666	23 .2	343
		4775				2318			
344	14 15 5		6150	122 .6	38 19 37		4676	24 .6	344
		4745				2313			
345	15 2 32		6074	120 .4	38 42 45		4688	26 .0	345
		4717				2302			
346	15 49 42		5999	118 .4	39 5 46		4700	27 .6	346
		4687				2295			
347	16 36 34		5925	116 .4	39 28 43		4713	29 .3	347
		4662				2285			
348	17 23 11		5851	114 .3	39 51 34		4725	30 .9	348
		4638				2275			
349	18 9 34		5778	112 .3	40 14 19		4738	32 .6	349
		4610				2265			
350	18 55 40		5707	110 .3	40 36 58		4751	34 .2	350
		4588				2255			
351	19 41 33		5636	108 .4	40 59 31		4764	35 .1	351
		4567				2245			
352	20 27 13		5565	106 .5	41 21 58		4778	36 .0	352
		4545				2232			
353	21 12 40		5496	104 .5	41 44 17		4792	36 .8	353
		4525				2222			
354	21 57 55		5426	102 .6	42 6 30		4807	37 .7	354
		4505				2212			
355	22 42 58		5358	100 .7	42 28 37		4822	38 .6	355
		4487				2197			
356	23 27 50		5290	98 .9	42 50 35		4836	39 .8	356
		4470				2187			
357	24 12 32		5222	97 .1	43 12 27		4852	41 .0	357
		4453				2172			
358	24 57 4		5156	95 .4	43 34 10		4867	42 .1	358
		4435				2160			
359	25 41 25		5090	93 .6	43 55 46		4883	43 .3	359
		4422				2147			
360	26 25 38		5025	91 .8	44 17 14		4899	44 .5	360

TABLE 8. The ANGLE of POSITION of each point of the ECLIPTIC, to every 10′ of its Longitude, calculated for Obliquity 23° 27′, with the Differences, and Variation for 100″ change of the Obliquity. (By MR. J. UTTING.) Argument = the Longitude of the point of the Ecliptic.

Arg.	0ˢ + / VIˢ − Angle of Position.	Difference.	Variation.	Iˢ + / VIIˢ − Angle of Position.	Difference.	Variation.	IIˢ + / VIIIˢ − Angle of Position.	Difference.	Variation.	Arg.
0° 0′	23° 27′ 0″.00	0″.32	100″.00	20° 35′ 21″.12	114″.34	90″.19	12° 14′ 13″.95	215″.50	56″.76	30° 0′
10	23 26 59.68	0.96	100.00	20 33 26.78	114.97	90.08	12 10 38.45	215.95	56.50	50
20	23 26 58.72	1.58	99.99	20 31 31.81	115.59	89.96	12 7 2.50	216.41	56.23	40
30	23 26 57.14	2.24	99.99	20 29 36.22	116.22	89.84	12 3 26.09	216.87	55.97	30
40	23 26 54.90	2.87	99.99	20 27 40.00	116.83	89.72	11 59 49.22	217.32	55.71	20
50	23 26 52.03	3.51	99.99	20 25 43.17	117.46	89.61	11 56 11.90	217.77	55.45	10
1 0	23 26 48.52	4.13	99.99	20 23 45.71	118.08	89.49	11 52 34.13	218.21	55.18	29 0
10	23 26 44.39	4.79	99.98	20 21 47.63	118.69	89.37	11 48 55.92	218.66	54.91	50
20	23 26 39.60	5.41	99.98	20 19 48.94	119.31	89.25	11 45 17.26	219.11	54.65	40
30	23 26 34.19	6.07	99.98	20 17 49.63	119.94	89.13	11 41 38.15	219.55	54.38	30
40	23 26 28.12	6.68	99.98	20 15 49.69	120.55	89.01	11 37 58.60	219.99	54.11	20
50	23 26 21.44	7.32	99.97	20 13 49.14	121.17	88.89	11 34 18.61	220.43	53.84	10
2 0	23 26 14.12	7.96	99.97	20 11 47.97	121.80	88.77	11 30 38.18	220.87	53.57	28 0
10	23 26 6.16	8.61	99.96	20 9 46.17	122.40	88.64	11 26 57.31	221.30	53.30	50
20	23 25 57.55	9.23	99.95	20 7 43.77	123.02	88.52	11 23 16.01	221.74	53.03	40
30	23 25 48.32	10.87	99.94	20 5 40.75	123.64	88.39	11 19 34.27	222.16	52.75	30
40	23 25 38.45	10.52	99.93	20 3 37.11	124.26	88.26	11 15 52.11	222.61	52.48	20
50	23 25 27.93	11.16	99.92	20 1 32.85	124.88	88.14	11 12 9.50	223.02	52.21	10
3 0	23 25 16.77	11.79	99.91	19 59 27.97	125.49	88.01	11 8 26.48	223.46	51.93	27 0
10	23 25 4.98	12.43	99.90	19 57 22.48	126.11	87.88	11 4 43.02	223.87	51.65	50
20	23 24 52.55	13.05	99.89	19 55 16.37	126.72	87.75	11 0 59.15	224.31	51.38	40
30	23 24 39.50	13.71	99.88	19 53 9.65	127.33	87.62	10 57 14.84	224.71	51.10	30
40	23 24 25.79	14.32	99.86	19 51 2.32	127.96	87.49	10 53 30.13	225.14	50.83	20
50	23 24 11.47	14.98	99.85	19 48 54.36	128.56	87.36	10 49 44.99	225.56	50.55	10
4 0	23 23 56.49	15.61	99.84	19 46 45.80	129.18	87.23	10 45 59.43	225.96	50.27	26 0
10	23 23 40.88	16.26	99.83	19 44 36.62	129.79	87.10	10 42 13.47	226.38	49.99	50
20	23 23 24.62	16.88	99.81	19 42 26.83	130.41	86.96	10 38 27.09	226.79	49.71	40
30	23 23 7.74	17.52	99.80	19 40 16.42	131.01	86.83	10 34 40.30	227.21	49.43	30
40	23 22 50.22	18.15	99.79	19 38 5.41	131.63	86.70	10 30 53.09	227.60	49.15	20
50	23 22 32.07	18.81	99.77	19 35 53.78	132.25	86.57	10 27 5.49	228.00	48.87	10
5 0	23 22 13.26	19.43	99.76	19 33 41.53	132.84	86.43	10 23 17.49	228.42	48.59	25 0
10	23 21 53.83	20.05	99.74	19 31 28.69	133.47	86.29	10 19 29.07	228.81	48.31	50
20	23 21 33.78	20.71	99.72	19 29 15.22	134.09	86.15	10 15 40.26	229.21	48.02	40
30	23 21 13.07	21.35	99.70	19 27 1.13	134.67	86.01	10 11 51.05	229.61	47.73	30
40	23 20 51.72	21.98	99.68	19 24 46.46	135.28	85.87	10 8 1.44	230.00	47.45	20
50	23 20 29.74	22.64	99.66	19 22 31.18	135.92	85.73	10 4 11.44	230.39	47.17	10
6 0	23 20 7.10	23.24	99.64	19 20 15.26	136.50	85.59	10 0 21.05	230.78	46.88	24 0
10	23 19 43.86	23.88	99.62	19 17 58.76	137.12	85.44	9 56 30.27	231.17	46.59	50
20	23 19 19.98	24.53	99.59	19 15 41.64	137.73	85.30	9 52 39.10	231.55	46.30	40
30	23 18 55.45	25.17	99.57	19 13 23.91	138.32	85.15	9 48 47.55	231.93	46.01	30
40	23 18 30.28	25.80	99.55	19 11 5.59	138.94	85.01	9 44 55.62	232.32	45.72	20
50	23 18 4.48	26.45	99.52	19 8 46.65	139.55	84.87	9 41 3.30	232.70	45.43	10
7 0	23 17 38.03	27.08	99.50	19 6 27.10	140.14	84.72	9 37 10.60	233.08	45.14	23 0
10	23 17 10.95	27.71	99.47	19 4 6.96	140.77	84.58	9 33 17.52	233.44	44.85	50
20	23 16 43.24	28.34	99.44	19 1 46.19	141.35	84.44	9 29 24.08	233.82	44.55	40
30	23 16 14.90	28.99	99.42	18 59 24.84	141.97	84.29	9 25 30.26	234.19	44.26	30
40	23 15 45.91	29.63	99.39	18 57 2.87	142.57	84.15	9 21 36.07	234.56	43.96	20
50	23 15 16.28	30.26	99.36	18 54 40.30	143.18	84.00	9 17 41.51	234.92	43.67	10
8 0	23 14 46.02	30.90	99.34	18 52 17.12	143.76	83.85	9 13 46.59	235.28	43.37	22 0
10	23 14 15.12	31.53	99.31	18 49 53.36	144.37	83.70	9 9 51.31	235.66	43.07	50
20	23 13 43.59	32.17	99.28	18 47 28.99	144.98	83.55	9 5 55.65	236.01	42.78	40
30	23 13 11.42	32.82	99.26	18 45 4.01	145.58	83.40	9 1 59.61	236.36	42.48	30
40	23 12 38.60	33.44	99.23	18 42 38.43	146.18	83.25	8 58 3.28	236.71	42.18	20
50	23 12 5.16	34.09	99.20	18 40 12.25	146.77	83.10	8 54 6.57	237.07	41.88	10
9 0	23 11 31.07	34.71	99.17	18 37 45.48	147.38	82.94	8 50 9.50	237.43	41.58	21 0
10	23 10 56.36	35.36	99.14	18 35 18.10	147.97	82.78	8 46 12.07	237.76	41.28	50
20	23 10 21.00	35.98	99.10	18 32 50.13	148.57	82.62	8 42 14.31	238.12	40.98	40
30	23 9 45.02	36.64	99.07	18 30 21.56	149.17	82.46	8 38 16.19	238.45	40.68	30
40	23 9 8.38	37.26	99.04	18 27 52.39	149.77	82.30	8 34 17.74	238.80	40.38	20
50	23 8 31.12	37.90	99.00	18 25 22.62	150.36	82.14	8 30 18.94	239.13	40.08	10
10 0	23 7 53.22		98.97	18 22 52.26		81.98	8 26 19.81		39.77	20 0
	Vˢ − / XIˢ +			IVˢ − / Xˢ +			IIIˢ − / IXˢ +			

TABLE 8.

The Angle of Position of each point of the Ecliptic, &c. continued.

Arg.	0s + / VIs − Angle of Position.	Difference	Variation.	Is + / VIIs − Angle of Position.	Difference.	Variation.	IIs + / VIIIs − Angle of Position.	Difference.	Variation.	Arg.
10° 0'	23° 7' 53".22	38".54	98".97	18° 22' 52".26	150".96	81".98	8° 26' 19".81	239".47	39".77	20° 0'
10	23 7 14.68	39.16	98.93	18 20 21.30	151.55	81.81	8 22 20.34	239.81	39.46	50
20	23 6 35.52	39.83	98.90	18 17 49.75	152.15	81.65	8 18 20.53	240.12	39.16	40
30	23 5 55.69	40.44	98.86	18 15 17.60	152.73	81.49	8 14 20.41	240.47	38.85	30
40	23 5 15.25	41.09	98.82	18 12 44.87	153.35	81.32	8 10 19.94	240.79	38.55	20
50	23 4 34.16	41.71	98.78	18 10 11.52	153.91	81.16	8 6 19.15	241.11	38.24	10
11 0	23 3 52.45	42.37	98.74	18 7 37.61	154.51	81.00	8 2 18.04	241.43	37.93	19 0
10	23 3 10.08	42.99	98.70	18 5 3.10	155.11	80.83	7 58 16.61	241.76	37.62	50
20	23 2 27.09	43.63	98.66	18 2 27.89	155.71	80.66	7 54 14.85	242.07	37.31	40
30	23 1 43.46	44.26	98.62	17 59 52.28	156.28	80.49	7 50 12.78	242.39	37.00	30
40	23 0 59.20	44.91	98.57	17 57 16.00	156.89	80.32	7 46 10.39	242.69	36.69	20
50	23 0 14.29	45.53	98.53	17 54 39.11	157.44	80.15	7 42 7.70	242.99	36.38	10
12 0	22 59 28.76	46.18	98.49	17 52 1.67	158.07	79.98	7 38 4.71	243.32	36.07	18 0
10	22 58 42.58	46.80	98.45	17 49 23.60	158.64	79.81	7 34 1.39	243.62	35.76	50
20	22 57 55.78	47.46	98.40	17 46 44.96	159.22	79.64	7 29 57.77	243.91	35.45	40
30	22 57 8.32	48.08	98.36	17 44 5.74	159.81	79.47	7 25 53.86	244.22	35.14	30
40	22 56 20.24	48.71	98.32	17 41 25.93	160.40	79.30	7 21 49.64	244.51	34.83	20
50	22 55 31.53	49.37	98.27	17 38 45.53	160.97	79.13	7 17 45.13	244.81	34.51	10
13 0	22 54 42.16	49.99	98.23	17 36 4.56	161.53	78.96	7 13 40.32	245.09	34.20	17 0
10	22 53 52.17	50.62	98.18	17 33 23.03	162.17	78.79	7 9 35.23	245.39	33.89	50
20	22 53 1.55	51.27	98.13	17 30 40.86	162.74	78.61	7 5 29.84	245.67	33.58	40
30	22 52 10.28	51.89	98.08	17 27 58.12	163.30	78.43	7 1 24.17	245.96	33.26	30
40	22 51 18.39	52.54	98.03	17 25 14.82	163.89	78.26	6 57 18.21	246.24	32.94	20
50	22 50 25.85	53.17	97.98	17 22 30.93	164.46	78.08	6 53 11.97	246.52	32.62	10
14 0	22 49 32.68	53.80	97.93	17 19 46.47	165.04	77.91	6 49 5.45	246.79	32.30	16 0
10	22 48 38.88	54.46	97.88	17 17 1.43	165.62	77.73	6 44 58.66	247.07	31.98	50
20	22 47 44.42	55.06	97.83	17 14 15.81	166.20	77.55	6 40 51.59	247.34	31.66	40
30	22 46 49.36	55.72	97.79	17 11 29.61	166.77	77.37	6 36 44.25	247.60	31.34	30
40	22 45 53.64	56.35	97.74	17 8 42.84	167.36	77.19	6 32 36.65	247.88	31.02	20
50	22 44 57.29	56.98	97.69	17 5 55.48	167.92	77.01	6 28 28.77	248.14	30.70	10
15 0	22 44 0.31	57.63	97.64	17 3 7.56	168.49	76.82	6 24 20.63	248.40	30.38	15 0
10	22 43 2.68	58.24	97.59	17 0 19.07	169.07	76.63	6 20 12.23	248.65	30.05	50
20	22 42 4.44	58.90	97.53	16 57 30.00	169.64	76.44	6 16 3.58	248.92	29.73	40
30	22 41 5.54	59.52	97.48	16 54 40.36	170.22	76.26	6 11 54.66	249.16	29.41	30
40	22 40 6.02	60.17	97.42	16 51 50.14	170.77	76.07	6 7 45.50	249.41	29.08	20
50	22 39 5.85	60.80	97.37	16 48 59.37	171.36	75.88	6 3 36.09	249.66	28.76	10
16 0	22 38 5.05	61.42	97.31	16 46 8.01	171.91	75.69	5 59 26.43	249.90	28.44	14 0
10	22 37 3.63	62.06	97.25	16 43 16.10	172.49	75.50	5 55 16.53	250.15	28.11	50
20	22 36 1.57	62.70	97.19	16 40 23.61	173.05	75.31	5 51 6.38	250.38	27.79	40
30	22 34 58.87	63.34	97.13	16 37 30.56	173.61	75.12	5 46 56.00	250.63	27.46	30
40	22 33 55.53	63.98	97.07	16 34 36.97	174.20	74.92	5 42 45.37	250.86	27.14	20
50	22 32 51.55	64.60	97.01	16 31 42.75	174.75	74.73	5 38 34.51	251.09	26.81	10
17 0	22 31 46.95	65.23	96.95	16 28 48.00	175.31	74.53	5 34 23.42	251.31	26.48	13 0
10	22 30 41.72	65.89	96.89	16 25 52.69	175.86	74.33	5 30 12.11	251.54	26.16	50
20	22 29 35.83	66.50	96.83	16 22 56.83	176.43	74.14	5 26 0.57	251.77	25.83	40
30	22 28 29.33	67.14	96.77	16 20 0.40	177.00	73.94	5 21 48.80	251.98	25.50	30
40	22 27 22.19	67.77	96.71	16 17 3.40	177.55	73.74	5 17 36.82	252.21	25.18	20
50	22 26 14.42	68.44	96.65	16 14 5.85	178.10	73.54	5 13 24.61	252.42	24.85	10
18 0	22 25 5.98	69.03	96.59	16 11 7.75	178.67	73.34	5 9 12.19	252.63	24.52	12 0
10	22 23 56.95	69.68	96.52	16 8 9.08	179.22	73.14	5 4 59.56	252.83	24.19	50
20	22 22 47.27	70.32	96.45	16 5 9.86	179.77	72.93	5 0 46.73	253.05	23.85	40
30	22 21 36.95	70.95	96.38	16 2 10.09	180.33	72.73	4 56 33.68	253.24	23.52	30
40	22 28 26.00	71.58	96.31	15 59 9.76	180.89	72.53	4 52 20.41	253.46	23.19	20
50	22 19 14.42	72.20	96.24	15 56 8.87	181.42	72.32	4 48 6.98	253.64	22.85	10
19 0	22 18 2.22	72.87	96.17	15 53 7.45	181.98	72.12	4 43 53.34	253.84	22.52	11 0
10	22 16 49.35	73.48	96.10	15 50 5.47	182.54	71.91	4 39 39.50	254.04	22.18	50
20	22 15 35.87	74.10	96.03	15 47 2.93	183.08	71.70	4 35 25.46	254.22	21.85	40
30	22 14 21.77	74.77	95.96	15 43 59.85	183.63	71.50	4 31 11.24	254.42	21.52	30
40	22 13 7.00	75.38	95.89	15 40 56.22	184.17	71.29	4 26 56.82	254.59	21.18	20
50	22 11 51.62	76.01	95.82	15 37 52.05	184.73	71.08	4 22 42.23	254.77	20.85	10
20 0	22 10 35.61		95.75	15 34 47.32		70.88	4 18 27.46		20.52	10 0
	Vs − / XIs +			IVs − / Xs +			IIIs − / IXs +			

TABLE 8.

The Angle of Position of each Point of the Ecliptic, &c. concluded.

Arg.	0s + / VIs −			Is + / VIIs −			IIs + / VIIIs −			Arg.
	Angle of Position.	Difference.	Variation.	Angle of Position.	Difference.	Variation.	Angle of Position.	Difference.	Variation.	
20° 0′	22° 10′ 35″.61	76″.66	95″.75	15° 34′ 47″.32	185″.26	70″.88	4° 18′ 27″.46	254″.96	20″.52	10° 0′
10	22 9 18.95	77.26	95.68	15 31 42.06	185.80	70.67	4 14 12.50	255.12	20.19	50
20	22 8 1.69	77.94	95.60	15 28 36.26	186.36	70.46	4 9 57.38	255.31	19.85	40
30	22 6 43.75	78.54	95.53	15 25 29.90	186.88	70.25	4 5 42.07	255.47	19.52	30
40	22 5 25.21	79.18	95.46	15 22 23.02	187.44	70.04	4 1 26.60	255.64	19.18	20
50	22 4 6.03	79.82	95.38	15 19 15.58	187.97	69.83	3 57 10.96	255.80	18.85	10
21 0	22 2 46.21	80.44	95.31	15 16 7.61	188.50	69.61	3 52 55.16	255.96	18.51	9 0
10	22 1 25.77	81.08	95.23	15 12 59.11	189.04	69.40	3 48 39.20	256.12	18.18	50
20	22 0 4.69	81.71	95.15	15 9 50.07	189.57	69.18	3 44 23.08	256.27	17.84	40
30	21 58 42.98	82.35	95.07	15 6 40.50	190.12	68.96	3 40 6.81	256.43	17.50	30
40	21 57 20.63	82.97	94.99	15 3 30.38	190.64	68.74	3 35 50.38	256.58	17.16	20
50	21 55 57.66	83.59	94.91	15 0 19.74	191.17	68.52	3 31 33.80	256.73	16.83	10
22 0	21 54 34.07	84.25	94.83	14 57 8.57	191.70	68.30	3 27 17.07	256.86	16.49	8 0
10	21 53 9.82	84.87	94.75	14 53 56.87	192.23	68.08	3 23 0.21	257.01	16.15	50
20	21 51 44.95	85.51	94.67	14 50 44.64	192.77	67.85	3 18 43.20	257.15	15.81	40
30	21 50 19.44	86.13	94.59	14 47 31.87	193.28	67.63	3 14 26.05	257.28	15.47	30
40	21 48 53.31	86.75	94.51	14 44 18.59	193.81	67.41	3 11 8.77	257.41	15.13	20
50	21 47 26.56	87.40	94.43	14 41 4.78	194.34	67.18	3 5 51.36	257.54	14.79	10
23 0	21 45 59.16	88.01	94.35	14 37 50.44	194.85	66.96	3 1 33.82	257.67	14.44	7 0
10	21 44 31.15	88.67	94.26	14 34 35.59	195.37	66.73	2 57 16.15	257.79	14.10	50
20	21 43 2.48	89.28	94.17	14 31 20.22	195.91	66.50	2 52 58.36	257.91	13.76	40
30	21 41 33.20	89.92	94.09	14 28 4.31	196.41	66.27	2 48 40.45	258.03	13.42	30
40	21 40 3.28	90.56	94.00	14 24 47.90	196.93	66.04	2 44 22.42	258.14	13.08	20
50	21 38 32.72	91.16	93.91	14 21 30.97	197.45	65.81	2 40 4.28	258.25	12.74	10
24 0	21 37 1.56	91.82	93.82	14 18 13.52	197.96	65.58	2 35 46.03	258.37	12.39	6 0
10	21 35 29.74	92.43	93.73	14 14 55.56	198.48	65.35	2 31 27.66	258.46	12.05	50
20	21 33 57.31	93.07	93.64	14 11 37.08	198.99	65.12	2 27 9.20	258.57	11.71	40
30	21 32 24.24	93.68	93.55	14 8 18.09	199.49	64.89	2 22 50.63	258.67	11.37	30
40	21 30 50.56	94.33	93.46	14 4 58.60	200.02	64.66	2 18 31.96	258.77	11.03	20
50	21 29 16.23	94.96	93.37	14 1 38.58	200.51	64.43	2 14 13.19	258.86	10.69	10
25 0	21 27 41.27	95.58	93.28	13 58 18.07	201.03	64.19	2 9 54.33	258.95	10.35	5 0
10	21 26 5.69	96.21	93.18	13 54 57.04	201.52	63.95	2 5 35.38	259.03	10.00	50
20	21 24 29.48	96.85	93.09	13 51 35.52	202.04	63.72	2 1 16.35	259.12	9.66	40
30	21 22 52.63	97.47	92.99	13 48 13.48	202.51	63.48	1 56 57.23	259.21	9.32	30
40	21 21 15.16	98.08	92.89	13 44 50.97	203.05	63.24	1 52 38.02	259.28	8.98	20
50	21 19 37.08	98.73	92.80	13 41 27.92	203.52	63.00	1 48 18.74	259.36	8.64	10
26 0	21 17 58.35	99.35	92.70	13 38 4.40	204.03	62.76	1 43 59.38	259.43	8.29	4 0
10	21 16 19.00	99.98	92.60	13 34 40.37	204.53	62.52	1 39 39.95	259.49	7.95	50
20	21 14 39.02	100.62	92.50	13 31 15.84	205.01	62.28	1 35 20.46	259.57	7.60	40
30	21 12 58.40	101.24	92.41	13 27 50.83	205.52	62.03	1 31 0.89	259.63	7.26	30
40	21 11 17.16	101.84	92.31	13 24 25.31	206.00	61.79	1 26 41.26	259.69	6.91	20
50	21 9 35.32	102.49	92.21	13 20 59.31	206.49	61.55	1 22 21.57	259.74	6.57	10
27 0	21 7 52.83	103.12	92.11	13 17 32.82	206.99	61.31	1 18 1.83	259.80	6.22	3 0
10	21 6 9.71	103.76	92.00	13 14 5.83	207.46	61.06	1 13 42.03	259.86	5.88	50
20	21 4 25.95	104.36	91.90	13 10 38.37	207.96	60.81	1 9 22.17	259.90	5.53	40
30	21 2 41.59	104.99	91.80	13 7 10.41	208.44	60.56	1 5 2.27	259.94	5.19	30
40	21 0 56.60	105.62	91.70	13 3 41.97	208.92	60.31	1 0 42.33	259.99	4.84	20
50	20 59 10.98	106.23	91.60	13 0 13.05	209.39	60.06	0 56 22.34	260.03	4.50	10
28 0	20 57 24.75	106.86	91.49	12 56 43.66	209.89	59.81	0 52 2.31	260.07	4.15	2 0
10	20 55 37.89	107.50	91.39	12 53 13.77	210.35	59.56	0 47 42.24	260.09	3.80	50
20	20 53 50.39	108.12	91.28	12 49 43.42	210.83	59.31	0 43 22.15	260.13	3.46	40
30	20 52 2.27	108.75	91.17	12 46 12.59	211.31	59.06	0 39 2.02	260.16	3.11	30
40	20 50 13.52	109.38	91.06	12 42 41.28	211.79	58.81	0 34 41.86	260.18	2.77	20
50	20 48 24.14	109.97	90.96	12 39 9.49	212.25	58.56	0 30 21.68	260.20	2.42	10
29 0	20 46 34.17	110.62	90.85	12 35 37.24	212.70	58.31	0 26 1.48	260.22	2.08	1 0
10	20 44 43.55	111.24	90.74	12 32 4.54	213.20	58.05	0 21 41.26	260.23	1.74	50
20	20 42 52.31	111.86	90.63	12 28 31.34	213.65	57.80	0 17 21.03	260.25	1.39	40
30	20 41 0.45	112.48	90.52	12 24 57.69	214.12	57.54	0 13 0.78	260.26	1.05	30
40	20 39 7.97	113.10	90.41	12 21 23.57	214.57	57.28	0 8 40.52	260.26	0.70	20
50	20 37 14.87	113.75	90.30	12 17 49.00	215.05	57.02	0 4 20.26	260.26	0.35	10
30 0	20 35 21.12		90.19	12 14 13.95		56.76	0 0 0.00		0.00	0 0
	Vs − / XIs +			IVs − / Xs +			IIIs − / IXs +			

TABLE 9.

AUGMENTATION of the ECLIPTIC ANGLE of POSITION, for Zodiacal Stars. (Dr. MASKELYNE'S, expanded.)

Arguments = the Ecliptic Angle of Position at the Side, and the Latitude of the Star at the Head.

Eclipt. Ang.	0° 10′	0° 20′	0° 30′	0° 40′	0° 50′	1° 0′	1° 10′	1° 20	1° 30′	1° 40′	1° 50′	2° 0′	2° 10′
0°	0″.0	0″.0	0″.0	0″.0	0″.0	0″.0	0″.0	0″.0	0″.0	0″.0	0″.0	0″.0	0″.0
1	0.0	0.1	0.1	0.2	0.3	0.5	0.7	0.9	1.2	1.5	1.8	2.2	2.6
2	0.0	0.1	0.2	0.4	0.7	1.1	1.5	1.9	2.4	3.0	3.7	4.4	5.2
3	0.1	0.2	0.4	0.7	1.1	1.6	2.2	2.9	3.7	4.5	5.5	6.6	7.8
4	0.1	0.2	0.5	0.9	1.5	2.2	3.0	3.9	4.9	6.0	7.3	8.8	10.4
5	0.1	0.3	0.7	1.2	1.9	2.8	3.7	4.9	6.2	7.6	9.2	11.0	12.9
6	0.2	0.4	0.8	1.4	2.2	3.3	4.5	5.9	7.4	9.1	11.0	13.2	15.5
7	0.2	0.5	1.0	1.7	2.6	3.8	5.2	6.9	8.6	10.6	12.9	15.4	18.1
8	0.2	0.6	1.1	1.9	3.0	4.4	5.9	7.9	9.9	12.2	14.8	17.6	20.6
9	0.3	0.7	1.2	2.1	3.4	5.0	6.7	8.9	11.2	13.8	16.7	19.8	23.3
10	0.3	0.8	1.4	2.4	3.8	5.5	7.5	9.9	12.4	15.3	18.6	22.1	26.0
11	0.4	0.9	1.5	2.6	4.1	6.1	8.4	10.9	13.7	16.9	20.5	24.4	28.7
12	0.4	1.0	1.7	2.9	4.6	6.7	9.2	12.0	15.0	18.5	22.4	26.7	31.3
13	0.5	1.1	1.8	3.2	5.0	7.2	9.9	13.0	16.3	20.2	24.4	29.0	34.0
14	0.5	1.2	1.9	3.4	5.4	7.8	10.7	14.0	17.6	21.7	26.3	31.4	36.8
15	0.6	1.3	2.1	3.7	5.8	8.4	11.4	15.0	18.9	23.3	28.2	33.7	39.6
16	0.6	1.4	2.2	3.9	6.2	9.0	12.3	16.1	20.3	25.0	30.2	36.1	42.3
17	0.7	1.5	2.4	4.2	6.6	9.6	13.1	17.1	21.6	26.6	32.2	38.4	45.1
18	0.7	1.6	2.5	4.4	7.0	10.2	13.9	18.1	22.9	28.3	34.3	40.9	48.0
19	0.8	1.7	2.7	4.7	7.4	10.8	14.7	19.2	24.3	30.0	36.3	43.3	50.9
20	0.8	1.8	2.9	5.0	7.8	11.4	15.6	20.3	25.7	31.7	38.4	45.8	53.8
21	0.9	1.9	3.0	5.3	8.3	12.0	16.4	21.4	27.1	33.4	40.4	48.2	56.6
22	0.9	2.0	3.1	5.6	8.7	12.7	17.2	22.5	28.5	35.2	42.6	50.7	59.5
23	1.0	2.1	3.3	5.9	9.2	13.3	18.1	23.7	30.0	37.0	44.8	53.3	1 2.5
23 27′ 57″	1.0	2.2	3.4	6.1	9.5	13.7	18.6	24.3	30.7	37.8	45.7	54.5	1 4.0

AUGMENTATION of the ECLIPTIC ANGLE of POSITION, continued.

Eclipt. Ang.	2° 20′	2° 30′	2° 40′	2° 50′	3° 0′	3° 10′	3° 20′	3° 30′	3° 40′	3° 50′	4° 0′	4° 10′	4° 20′
0°	0′ 0″.0	0′ 0″.0	0′ 0″.0	0′ 0″.0	0′ 0″.0	0′ 0″.0	0′ 0″.0	0′ 0″.0	0′ 0″.0	0′ 0.0	0′ 0″.0	0′ 0″.0	0 0″.0
1	0 3.0	0 3.4	0 3.9	0 4.4	0 4.9	0 5.4	0 6.0	0 6.6	0 7.3	0 8.0	0 8.8	0 9.5	0 10.3
2	0 6.0	0 6.8	0 7.8	0 8.8	0 9.9	0 11.0	0 12.2	0 13.4	0 14.7	0 16.1	0 17.6	0 19.1	0 20.6
3	0 9.0	0 10.3	0 11.7	0 13.2	0 14.8	0 16.5	0 18.3	0 20.2	0 22.1	0 24.2	0 26.4	0 28.6	0 30.9
4	0 12.0	0 13.7	0 15.6	0 17.7	0 19.8	0 22.1	0 24.5	0 26.9	0 29.5	0 32.3	0 35.2	0 38.2	0 41.2
5	0 15.0	0 17.2	0 19.5	0 22.1	0 24.7	0 27.5	0 30.6	0 33.8	0 37.0	0 40.5	0 44.0	0 47.7	0 51.6
6	0 18.0	0 20.6	0 23.4	0 26.5	0 29.7	0 33.1	0 36.7	0 40.5	0 44.5	0 48.6	0 52.9	0 57.4	1 2.1
7	0 21.0	0 24.1	0 27.4	0 30.9	0 34.7	0 38.7	0 42.9	0 47.3	0 52.0	0 56.7	1 1.8	1 7.1	1 12.6
8	0 24.0	0 27.6	0 31.4	0 35.4	0 39.7	0 44.3	0 49.1	0 54.2	0 59.5	1 5.0	1 10.8	1 16.8	1 23.1
9	0 27.0	0 31.1	0 35.4	0 39.9	0 44.8	0 50.0	0 55.4	1 1.0	1 7.0	1 13.2	1 19.8	1 26.6	1 33.7
10	0 30.1	0 34.6	0 39.4	0 44.4	0 49.9	0 55.7	1 1.7	1 7.9	1 14.5	1 21.5	1 28.8	1 36.4	1 44.3
11	0 33.2	0 38.2	0 43.5	0 49.0	0 55.0	1 1.3	1 8.0	1 14.9	1 22.2	1 29.9	1 37.9	1 46.2	1 54.9
12	0 36.3	0 41.7	0 47.6	0 53.6	1 0.1	1 7.0	1 14.3	1 21.9	1 29.9	1 38.3	1 47.0	1 56.1	2 5.6
13	0 39.4	0 45.3	0 51.7	0 58.3	1 5.3	1 12.7	1 20.6	1 29.0	1 37.7	1 46.8	1 56.3	2 6.1	2 16.5
14	0 42.6	0 48.9	0 55.8	1 2.9	1 10.5	1 18.6	1 27.1	1 36.1	1 45.5	1 55.3	2 5.6	2 16.2	2 27.4
15	0 45.8	0 52.6	0 59.9	1 7.5	1 15.8	1 24.5	1 33.6	1 43.2	1 53.3	2 3.9	2 14.9	2 26.4	2 38.4
16	0 49.0	0 56.3	1 4.1	1 12.2	1 21.1	1 30.4	1 40.1	1 50.5	2 1.2	2 12.6	2 24.4	2 36.7	2 49.5
17	0 52.3	1 0.0	1 8.3	1 17.0	1 26.5	1 36.4	1 46.8	1 57.8	2 9.3	2 21.4	2 34.0	2 47.1	3 0.7
18	0 55.6	1 3.8	1 12.6	1 21.9	1 32.0	1 42.5	1 53.6	2 5.2	2 17.5	2 30.3	2 43.6	2 57.6	3 12.0
19	0 58.9	1 7.6	1 16.9	1 26.8	1 37.5	1 48.6	2 0.4	2 12.7	2 25.7	2 39.2	2 53.4	3 8.2	3 23.5
20	1 2.3	1 11.5	1 21.3	1 31.8	1 43.0	1 54.8	2 7.2	2 20.3	2 34.0	2 48.3	3 3.3	3 18.9	3 35.2
21	1 5.7	1 15.4	1 25.8	1 36.9	1 48.7	2 1.1	2 14.2	2 27.9	2 42.3	2 57.5	3 13.3	3 29.8	3 47.0
22	1 9.1	1 19.3	1 30.3	1 42.0	1 54.3	2 7.4	2 21.2	2 35.7	2 50.8	3 6.7	3 23.5	3 40.8	3 58.9
23	1 12.5	1 23.3	1 34.9	1 47.1	2 0.0	2 13.7	2 28.2	2 43.6	2 59.5	3 16.0	3 33.7	3 51.9	4 10.9
23 27′ 57″	1 14.3	1 25.3	1 37.0	1 49.5	2 2.9	2 17.0	2 31.8	2 47.4	3 3.7	3 20.8	3 38.7	3 57.3	4 16.7

TABLE 9.

Augmentation of the Ecliptic Angle of Position, concluded.

Eclipt. Ang.	4° 30′	4° 40′	4° 50′	5° 0′	5° 10′	5° 20′	5° 30′	5° 40′	5° 50′	6° 0′	6° 10′	6° 20′	6° 30′
0°	0′ 0″.0	0′ 0″.0	0′ 0″.0	0′ 0″.0	0′ 0″.0	0′ 0″.0	0′ 0″.0	0′ 0″.0	0′ 0″.0	0′ 0″.0	0′ 0″.0	0′ 0″.0	0′ 0″.0
1	0 11.1	0 11.9	0 12.8	0 13.7	0 14.7	0 15.6	0 16.6	0 17.6	0 18.7	0 19.8	0 20.9	0 22.1	0 23.3
2	0 22.2	0 23.9	0 25.7	0 27.5	0 29.4	0 31.3	0 33.3	0 35.4	0 37.5	0 39.7	0 41.9	0 44.2	0 46.6
3	0 33.4	0 36.0	0 38.6	0 41.3	0 44.1	0 47.0	0 50.0	0 53.1	0 56.3	0 59.6	1 3.0	1 6.4	1 9.9
4	0 44.6	0 48.0	1 51.5	0 55.1	0 58.9	1 2.7	1 6.7	1 10.8	0 15.1	1 19.5	1 24.0	1 28.6	1 33.3
5	0 55.8	1 0.1	1 4.4	1 9.0	1 13.6	1 18.4	1 23.4	1 28.6	1 33.9	1 39.4	1 45.1	1 50.9	1 56.7
6	1 7.0	1 12.1	1 17.4	1 22.8	1 28.4	1 34.2	1 40.2	1 46.4	1 52.8	1 59.4	2 6.2	2 13.2	2 20.2
7	1 18.3	1 24.2	1 30.4	1 36.7	1 43.3	1 50.1	1 57.1	2 4.3	2 11.8	2 19.5	2 27.4	2 35.6	2 43.8
8	1 29.6	1 36.4	1 43.5	1 50.7	1 58.2	2 6.0	2 14.0	2 22.3	2 30.9	2 39.7	2 48.7	2 58.0	3 7.5
9	1 41.0	1 48.7	1 56.5	2 4.8	2 13.3	2 22.0	2 31.1	2 40.4	2 50.0	3 0.0	3 10.2	3 20.6	3 31.3
10	1 52.4	2 1.0	2 9.7	2 19.0	2 28.4	2 38.1	2 48.2	2 58.6	3 9.3	3 20.4	3 31.7	3 43.3	3 55.3
11	2 3.9	2 13.3	2 23.0	2 33.2	2 43.6	2 54.3	3 5.4	3 16.9	3 28.7	3 40.9	3 53.4	4 6.2	4 19.4
12	2 15.5	2 25.8	2 36.4	2 47.5	2 58.8	3 10.6	3 22.7	3 35.3	3 48.2	4 1.5	3 15.2	3 29.2	4 43.6
13	2 27.2	2 38.3	2 49.9	3 1.9	3 14.2	3 27.0	3 40.2	3 53.8	4 7.8	4 22.3	4 37.2	4 52.4	5 8.1
14	2 39.0	2 51.0	3 3.5	3 16.4	3 29.7	3 43.5	3 57.8	4 12.5	4 27.7	4 43.3	4 59.3	5 15.8	5 32.8
15	2 50.9	3 3.8	3 17.2	3 31.1	3 45.4	4 0.2	4 15.6	4 31.4	4 47.7	5 4.5	5 21.6	5 39.4	5 57.6
16	3 2.9	3 16.7	3 31.0	3 45.9	4 1.3	4 17.1	4 33.6	4 50.5	5 7.9	5 25.9	5 44.2	6 3.1	6 22.7
17	3 15.0	3 29.8	3 45.1	4 0.9	4 17.3	4 34.1	4 51.7	5 9.8	5 28.3	5 47.4	6 7.0	6 27.2	6 48.1
18	3 27.2	3 43.0	3 59.2	4 16.0	4 33.4	4 51.4	5 10.0	5 29.2	5 48.8	6 9.2	6 30.2	6 51.8	7 13.7
19	3 39.6	3 56.3	4 13.4	4 31.3	4 49.7	5 8.9	5 28.5	5 48.8	6 9.6	6 31.3	6 53.5	7 16.3	7 39.6
20	3 52.2	4 9.8	4 27.9	4 46.8	5 6.3	5 26.5	5 47.3	6 8.7	6 30.8	6 53.6	7 17.1	7 41.1	8 5.9
21	4 4.9	4 23.4	4 42.6	5 2.5	5 23.1	5 44.4	6 6.3	6 28.9	6 52.2	7 16.3	7 41.0	8 6.4	8 32.5
22	4 17.7	4 37.2	4 57.5	5 18.4	5 40.1	6 2.5	6 25.6	6 49.4	7 13.9	7 39.2	8 5.2	8 31.9	8 59.5
23	4 30.8	4 51.2	5 12.6	5 34.6	5 57.2	6 20.7	6 45.1	7 10.2	7 36.0	8 2.5	8 29.8	8 57.8	9 26.9
23 27′ 57″	4 36.9	4 57.9	5 19.6	5 42.1	6 5.3	6 29.3	6 54.1	7 19.7	7 46.1	8 13.4	8 41.4	9 10.1	9 39.6

TABLE 10.

The Angle between the Ecliptic and a Line parallel to the Equator, with the Differences and Variations for 100″ Variation of the Obliquity from 23° 27′ 57″.

Argument = the Ecliptic's Declination.

Eclipt. Dec.	Angle of Ecliptic and Paral.	Diff.	Var. for 100″ Var. of Obliq.	Eclip. Dec.	Angle of Ecliptic and Paral.	Diff.	Var. for 100″ Var. of Obliq.	Eclip. Dec.	Angle of Ecliptic and Paral.	Diff.	Var. for 100″ Var. of Obliq.
0° 0′	23° 27′ 57″	—	100″	12° 0′	20° 19′ 0″	16′ 44″	117″	20° 10′	12° 15′ 29″	—	200″
0 30	23 27 39	0′ 18″	100	12 30	20 1 16	17 44	119	20 20	11 58 18	17′ 11″	204
1 0	23 26 44	0 55	100	13 0	19 42 28	18 48	121	20 30	11 40 31	17 47	210
1 30	23 25 14	1 30	100	13 30	19 22 35	19 53	123	20 40	11 22 5	18 26	215
2 0	23 23 7	2 7	101	14 0	19 1 30	21 5	126	20 50	11 2 58	19 7	222
2 30	23 20 23	2 44	101	14 30	18 39 10	22 20	128	21 0	10 43 5	19 53	229
3 0	23 17 2	3 21	101	15 0	18 15 30	23 40	131	21 5	10 32 51	10 14	233
3 30	23 13 5	3 57	101	15 20	17 58 56	16 34	133	21 10	10 22 23	10 28	237
4 0	23 8 29	4 36	101	15 40	17 41 42	17 14	135	21 15	10 11 42	10 41	241
4 30	23 3 15	5 14	102	16 0	17 23 46	17 56	138	21 20	10 0 45	10 57	246
5 0	22 57 23	5 52	102	16 20	17 5 5	18 41	141	21 25	9 49 34	11 11	250
5 30	22 50 52	6 31	103	16 40	16 45 36	19 29	144	21 30	9 38 7	11 27	255
6 0	22 43 40	7 12	103	17 0	16 25 18	20 18	147	21 35	9 26 22	11 45	260
6 30	22 35 48	7 52	104	17 20	16 4 6	21 12	150	21 40	9 14 18	12 4	266
7 0	22 27 14	8 34	105	17 40	15 41 56	22 10	154	21 45	9 1 56	12 22	273
7 30	22 17 58	9 16	106	18 0	15 18 44	23 12	158	21 50	8 49 13	12 43	279
8 0	22 7 59	9 59	107	18 20	14 54 26	24 18	163	21 55	8 36 7	13 6	286
8 30	21 57 14	10 45	108	18 40	14 28 55	25 31	168	22 0	8 22 37	13 30	294
9 0	21 45 43	11 31	109	19 0	14 2 5	26 50	173	22 5	8 8 40	13 57	303
9 30	21 33 26	12 17	110	19 15	13 41 0	21 5	178	22 10	7 54 15	14 25	312
10 0	21 20 19	13 7	111	19 30	13 19 3	21 57	183	22 15	7 39 18	14 57	322
10 30	21 6 21	13 58	112	19 45	12 56 6	22 57	189	22 20	7 23 47	15 31	334
11 0	20 51 30	14 51	114	20 0	12 32 7	23 59	195	22 25	7 7 38	16 9	347
11 30	20 35 44	15 46	115	20 10	12 15 29	16 38	200	22 30	6 50 46	16 52	361

TABLE 11. The Ecliptic's Longitude, corresponding to every 10′ of its Declination; with the Differences and Var. for 100″ Var. of the Obliquity. The Argument = the Ecliptic's Declination. (Mr. J. Utting.)

Eclipt. Dec.	Ecliptic's Longitude.	Diff. in 10′.	Variation.	Ecliptic's Dec.	Ecliptic's Longitude.	Diff. in 10′.	Variation.
0° 0′	0° 0′ 0″.0	1506″.8	+0″.0	12° 0′	31° 28′ 29″.9	1732″.0	+141″.2
10	0 25 6.8	1506.8	1.7	10	31 57 21.9	1740.1	143.9
20	0 50 13.6	1507.0	3.4	20	32 26 22.0	1748.3	146.6
30	1 15 20.6	1507.2	5.1	30	32 55 30.3	1756.8	149.4
40	1 40 27.8	1507.5	6.7	40	33 24 47.2	1765.6	152.2
50	2 5 35.3	1507.8	8.4	50	33 54 12.8	1774.6	155.1
1 0	2 30 43.1	1508.2	10.1	13 0	34 23 47.4	1784.0	158.0
10	2 55 51.3	1508.7	11.8	10	34 53 31.4	1793.6	160.9
20	3 21 0.0	1509.2	13.5	20	35 23 25.0	1803.5	163.9
30	3 46 9.2	1509.8	15.2	30	35 53 28.5	1813.6	167.0
40	4 11 19.0	1510.6	16.9	40	36 23 42.1	1824.3	170.1
50	4 36 29.6	1511.2	18.6	50	36 54 6.4	1835.0	173.3
2 0	5 1 40.8	1512.1	20.3	14 0	37 24 41.4	1846.3	176.5
10	5 26 52.9	1513.0	22.0	10	37 55 27.7	1858.0	179.8
20	5 52 5.9	1514.0	23.7	20	38 26 25.7	1869.9	183.2
30	6 17 19.9	1514.9	25.4	30	38 57 35.6	1882.3	186.8
40	6 42 34.8	1516.1	27.1	40	39 28 57.9	1895.1	190.1
50	7 7 50.9	1517.3	28.9	50	40 0 33.0	1908.3	193.7
3 0	7 33 8.2	1518.5	30.6	15 0	40 32 21.3	1922.1	197.4
10	7 58 26.7	1519.8	32.3	10	41 4 23.4	1936.3	201.1
20	8 23 46.5	1521.3	34.1	20	41 36 39.7	1951.0	205.0
30	8 49 7.8	1522.7	35.8	30	42 9 10.7	1966.3	208.9
40	9 14 30.5	1524.2	37.5	40	42 41 57.0	1982.0	213.0
50	9 39 54.7	1525.9	39.3	50	43 14 59.0	1998.6	217.1
4 0	10 5 20.6	1527.6	41.0	16 0	43 48 17.6	2015.5	221.4
10	10 30 48.2	1529.3	42.8	10	44 21 53.1	2033.3	225.7
20	10 56 17.5	1531.2	44.6	20	44 55 46.4	2051.7	230.2
30	11 21 48.7	1533.2	46.4	30	45 29 58.1	2071.0	234.8
40	11 47 21.9	1535.1	48.2	40	46 4 29.1	2090.9	239.6
50	12 12 57.0	1537.3	50.0	50	46 39 20.0	2111.7	244.5
5 0	12 38 34.3	1539.4	51.8	17 0	47 14 31.7	2133.5	249.6
10	13 4 13.7	1541.7	53.6	10	47 50 5.2	2156.1	254.9
20	13 29 55.4	1544.1	55.4	20	48 26 1.3	2179.9	260.3
30	13 55 39.5	1546.5	57.2	30	49 2 21.2	2204.7	266.0
40	14 21 26.0	1549.0	59.1	40	49 39 5.9	2230.8	271.8
50	14 47 15.0	1551.7	60.9	50	50 16 16.7	2258.1	277.8
6 0	15 13 6.7	1554.3	62.8	18 0	50 53 54.8	2286.7	284.1
10	15 39 1.0	1557.1	64.6	10	51 32 1.5	2316.8	290.6
20	16 4 58.1	1560.1	66.5	20	52 10 38.3	2348.6	297.4
30	16 30 58.2	1563.0	68.4	30	52 49 46.9	2382.0	304.5
40	16 57 1.2	1566.1	70.3	40	53 29 28.9	2417.5	312.0
50	17 23 7.3	1569.3	72.2	50	54 9 46.4	2454.8	319.7
7 0	17 49 16.6	1572.6	74.2	19 0	54 50 41.2	2494.5	327.9
10	18 15 29.2	1576.0	76.1	10	55 32 15.7	2536.7	336.5
20	18 41 45.2	1579.4	78.1	20	56 14 32.4	2581.7	345.6
30	19 8 4.6	1583.1	80.1	30	56 57 34.1	2629.4	355.1
40	19 34 27.7	1586.8	82.0	40	57 41 23.5	2680.9	365.2
50	20 0 54.5	1590.6	84.0	50	58 26 4.4	2735.9	376.0
8 0	20 27 25.1	1594.5	86.1	20 0	59 11 40.3	2795.1	387.5
10	20 53 59.6	1598.6	88.1	10	59 58 15.4	2859.2	399.7
20	21 20 38.2	1602.7	90.2	20	60 45 54.6	2928.6	412.9
30	21 47 20.9	1607.1	92.2	30	61 34 43.2	3004.3	427.0
40	22 14 8.0	1611.4	94.3	40	62 24 47.5	3086.9	442.4
50	22 40 59.4	1616.0	96.4	50	63 16 14.4	3178.1	459.0
9 0	23 7 55.4	1620.8	98.6	21 0	64 9 12.5	3278.7	477.3
10	23 34 56.2	1625.4	100.7	10	65 3 51.2	3391.0	497.4
20	24 2 1.6	1630.5	102.9	20	66 0 22.2	3517.1	519.7
30	24 29 12.1	1635.6	105.1	30	66 58 59.3	3660.0	544.7
40	24 56 27.7	1640.8	107.3	40	67 59 59.3	3824.0	572.9
50	25 23 48.5	1646.1	109.5	50	69 3 43.3	4014.0	605.3
10 0	25 51 14.6	1651.8	111.8	22 0	70 10 37.3	4238.3	642.6
10	26 18 46.4	1657.4	114.1	10	71 21 15.6	4508.4	687.0
20	26 46 23.8	1663.3	116.4	20	72 36 24.0	4841.5	740.6
30	27 14 7.1	1669.3	118.8	30	73 57 5.5	5266.7	807.3
40	27 41 56.4	1675.5	121.1	40	75 24 52.2	5835.8	894.0
50	28 9 51.9	1681.9	123.5	50	77 2 8.0	6653.5	1012.8
11 0	28 37 53.8	1688.5	126.0	23 0	78 53 1.5	7976.5	1191.5
10	29 6 2.3	1695.2	128.4	10	81 5 58.0	10736.5	1508.0
20	29 34 17.5	1702.2	130.9	20	84 4 54.5	21305.5	2355.0
30	30 2 39.7	1709.3	133.4	23 27 57″	90 0 0.0		
40	30 31 9.0	1716.6	136.0				
50	30 59 45.6	1724.3	138.6				
12 0	31 28 29.9		141.2				

In this Table, the Longitudes of all the Points in the first Quadrant only of the Ecliptic are given; in the second Quadrant, the Longitudes are obtained by subtracting the Tabular quantity from 180°; in the third Quadrant, 180° are added; and in the fourth, the Tabular Numbers must be subtracted from 360°.

The Obliquity of the Ecliptic is taken = 23° 27′ 57″.

TABLE 12. The Ecliptic's Right Ascension in Space, corresponding to every 10′ of its Declination, with the Differences, and Var. for 100″ Var. of the Obliquity. The Argument = the Ecliptic's Declination. (Mr. J. Utting.)

Eclipt. Dec.	Ecliptic's R. A.	Diff. in 10′.	Variation.
0° 0′	0° 0′ 0″.0	1382″.2	+0″.0
10	0 23 2.2	1382.3	1.8
20	0 46 4.5	1382.4	3.7
30	1 9 6.9	1382.7	5.5
40	1 32 9.6	1383.0	7.4
50	1 55 12.6	1383.4	9.2
1 0	2 18 16.0	1384.0	11.0
10	2 41 20.0	1384.6	12.9
20	3 4 24.6	1385.2	14.7
30	3 27 29.8	1386.1	16.6
40	3 50 35.9	1386.9	18.4
50	4 13 42.8	1387.8	20.3
2 0	4 36 50.6	1388.9	22.1
10	4 59 59.5	1390.0	24.0
20	5 23 9.5	1391.2	25.8
30	5 46 20.7	1392.5	27.7
40	6 9 33.2	1393.9	29.6
50	6 32 47.1	1395.4	31.5
3 0	6 56 2.5	1397.0	33.3
10	7 19 19.5	1398.6	35.2
20	7 42 38.1	1400.4	37.1
30	8 5 58.5	1402.3	39.0
40	8 29 20.8	1404.2	40.9
50	8 52 45.0	1406.2	42.8
4 0	9 16 11.2	1408.4	44.7
10	9 39 39.6	1410.6	46.7
20	10 3 10.2	1413.0	48.6
30	10 26 43.2	1415.4	50.5
40	10 50 18.6	1417.9	52.5
50	11 13 56.5	1420.6	54.4
5 0	11 37 37.0	1423.3	56.4
10	12 1 20.3	1426.2	58.4
20	12 25 6.5	1429.1	60.4
30	12 48 55.6	1432.1	62.4
40	13 12 47.7	1435.4	64.4
50	13 36 43.1	1438.6	66.4
6 0	14 0 41.7	1442.0	68.4
10	14 24 43.7	1445.5	70.4
20	14 48 49.2	1449.1	72.5
30	15 12 58.3	1452.8	74.6
40	15 37 11.1	1456.8	76.6
50	16 1 27.9	1460.7	78.7
7 0	16 25 48.6	1464.8	80.8
10	16 50 13.4	1469.0	83.0
20	17 14 42.4	1473.4	85.1
30	17 39 15.8	1477.9	87.3
40	18 3 53.7	1482.4	89.4
50	18 28 36.1	1487.3	91.6
8 0	18 53 23.4	1492.2	93.8
10	19 18 15.6	1497.2	96.0
20	19 43 12.8	1502.4	98.3
30	20 8 15.2	1507.7	100.5
40	20 33 22.9	1513.2	102.8
50	20 58 36.1	1518.8	105.1
9 0	21 23 54.9	1524.7	107.4
10	21 49 19.6	1530.7	109.8
20	22 14 50.3	1536.7	112.1
30	22 40 27.0	1543.1	114.5
40	23 6 10.1	1549.5	116.9
50	23 31 59.6	1556.2	119.4
10 0	23 57 55.8	1563.0	121.9
10	24 23 58.8	1570.1	124.4
20	24 50 8.9	1577.3	126.9
30	25 16 26.2	1584.7	129.4
40	25 42 50.9	1592.4	132.0
50	26 9 23.3	1600.2	134.6
11 0	26 36 3.5	1608.2	137.3
10	27 2 51.7	1616.6	140.0
20	27 29 48.3	1625.0	142.7
30	27 56 53.3	1633.9	145.5
40	28 24 7.2	1642.8	148.2
50	28 51 30.0	1652.1	151.1
12 0	29 19 2.1		153.9

Ecliptic's Dec.	Ecliptic's R. A.	Diff. in 10′.	Variation.
12° 0′	29° 19′ 2″.1	1661″.6	+153″.9
10	29 46 43.7	1671.4	156.9
20	30 14 35.1	1681.5	159.8
30	30 42 36.6	1691.8	162.8
40	31 10 48.4	1702.6	165.9
50	31 39 11.0	1713.5	169.0
13 0	32 7 44.5	1724.8	172.2
10	32 36 29.3	1736.4	175.4
20	33 5 25.7	1748.5	178.7
30	33 34 34.2	1760.8	182.0
40	34 3 55.0	1773.5	185.4
50	34 33 28.5	1786.7	188.9
14 0	35 3 15.2	1800.3	192.4
10	35 33 15.5	1814.2	196.0
20	36 3 29.7	1828.6	199.6
30	36 33 58.3	1843.5	203.4
40	37 4 41.8	1858.8	207.2
50	37 35 40.6	1874.8	211.1
15 0	38 6 55.4	1891.2	215.1
10	38 38 26.6	1908.2	219.2
20	39 10 14.8	1925.7	223.4
30	39 42 20.5	1943.9	227.7
40	40 14 44.4	1962.8	232.1
50	40 47 27.2	1982.3	236.7
16 0	41 20 29.5	2002.6	241.3
10	41 53 52.1	2023.6	246.0
20	42 27 35.7	2045.4	250.9
30	43 1 41.1	2068.1	256.0
40	43 36 9.2	2091.8	261.2
50	44 11 1.0	2116.2	266.5
17 0	44 46 17.2	2141.9	272.1
10	45 21 59.1	2168.6	277.8
20	45 58 7.7	2196.4	283.7
30	46 34 44.1	2225.5	289.8
40	47 11 49.6	2255.9	296.2
50	47 49 25.5	2287.8	302.8
18 0	48 27 33.3	2321.3	309.6
10	49 6 14.6	2356.3	316.8
20	49 45 30.9	2393.2	324.2
30	50 25 24.1	2432.1	331.9
40	51 5 56.2	2473.0	340.0
50	51 47 9.2	2516.3	348.5
19 0	52 29 5.5	2562.1	357.4
10	53 11 47.6	2610.7	366.8
20	53 55 18.3	2662.3	376.6
30	54 39 40.6	2717.3	387.1
40	55 24 57.9	2776.1	398.1
50	56 11 14.0	2839.1	409.9
20 0	56 58 33.1	2906.8	422.4
10	57 46 59.9	2979.6	435.7
20	58 36 39.5	3058.7	450.0
30	59 27 38.2	3144.5	465.4
40	60 20 2.7	3238.1	482.2
50	61 14 0.8	3341.0	500.4
21 0	62 9 41.8	3454.8	520.3
10	63 7 16.6	3581.0	542.2
20	64 6 57.6	3722.5	566.4
30	65 9 0.1	3882.9	593.7
40	66 13 43.0	4065.9	624.3
50	67 21 28.9	4278.2	659.6
22 0	68 32 47.1	4527.8	700.4
10	69 48 14.9	4827.7	748.8
20	71 8 42.6	5196.8	807.3
30	72 35 19.4	5667.3	880.0
40	74 9 46.7	6294.7	974.3
50	75 54 41.4	7194.1	1103.8
23 0	77 54 35.5	8446.0	1298.7
10	80 18 41.5	11067.9	1643.8
20	83 33 9.4	23210.6	2566.7
23 27 57″	90 0 0.0		

The Right Ascension of the first Quadrant only of the Ecliptic is given in this Table; that of the second Quadrant is obtained by subtracting the Tabular Numbers from 180°; that of the third Quadrant requires 180° to be added; and that of the fourth Quadrant is the Tabular Quantity subtracted from 360°.

The Obliquity of the Ecliptic is taken = 23° 27′ 57″.

TABLE 13.—The Moon's appoximate Parallax in Longitude, for 60′ of Horizontal Parallax in the Spheroid. (Mr. J. Utting.)

Arguments = the Altitude of the Nonagesimal Degree at the Side; and the Moon's Distance therefrom, at the Head.

Alt. of the Nonag.	The ☾'s Distance from the Nonagesimal Degree of the Ecliptic.									
	1°	2°	3°	4°	5°	6°	7°	8°	9°	10°
10°	0′ 10″.91	0′ 21″.82	0′ 32″.72	0′ 43″.61	0′ 54″.48	1′ 5″.34	1′ 16″ 18	1′ 27″.00	1′ 37″.79	1′ 48″.55
11	0 11 .98	0 23 .97	0 35 .95	0 47 .91	0 59 .87	1 11 .80	1 23 71	1 35 .60	1 47 .46	1 59 .28
12	0 13 .06	0 26 .12	0 39 .17	0 52 .20	1 5 .24	1 18 .24	1 31 22	1 44 .17	1 57 .09	2 9 .97
13	0 14 .13	0 28 .26	0 42 .38	0 56 .49	1 10 .58	1 24 .65	1 38 69	1 52 .71	2 6 .68	2 20 .62
14	0 15 .20	0 30 .39	0 45 .58	1 0 .75	1 15 .91	1 31 .04	1 46 14	2 1 .21	2 16 .24	2 31 .23
15	0 16 .26	0 32 .51	0 48 .76	1 4 .99	1 21 .21	1 37 .39	1 53 .55	2 9 .67	2 25 .76	2 41 .80
16	0 17 .32	0 34 .63	0 51 .93	1 9 .22	1 26 .49	1 43 .72	2 0 .93	2 18 .10	2 35 .23	2 52 .31
17	0 18 .37	0 36 .73	0 55 .09	1 13 .42	1 31 .74	1 50 .02	2 8 .27	2 26 .49	2 44 .65	3 2 .78
18	0 19 .42	0 38 .82	0 58 .23	1 17 .60	1 36 .96	1 56 .28	2 15 .57	2 34 .82	2 54 .03	3 13 .18
19	0 20 .46	0 40 .90	1 1 .34	1 21 .76	1 42 .16	2 2 .51	2 29 .57	2 43 .12	3 3 .35	3 23 .52
20	0 21 .49	0 42 .97	1 4 .44	1 25 .89	1 47 .32	2 8 .70	2 30 .05	2 51 .36	3 12 .61	3 33 .81
21	0 22 .52	0 45 .03	1 7 .52	1 29 .99	1 52 .45	2 14 .85	2 37 .23	2 59 .55	3 21 .82	3 44 .03
22	0 23 .54	0 47 .07	1 10 .58	1 34 .07	1 57 .54	2 20 .97	2 44 .35	3 7 .69	3 30 .96	3 54 .18
23	0 24 .55	0 49 .09	1 13 .62	1 38 .12	2 2 .60	2 27 .03	2 51 .82	3 15 .77	3 40 .04	4 4 .26
24	0 25 .56	0 51 .10	1 16 .64	1 42 .14	2 7 .62	2 33 .06	2 58 .45	3 23 .78	3 49 .05	4 14 .26
25	0 26 .55	0 53 .10	1 19 .63	1 46 .13	2 12 .60	2 39 .03	3 5 .42	3 31 .74	3 58 .00	4 24 .19
26	0 27 .54	0 55 .08	1 22 .60	1 50 .08	2 17 .54	2 44 .96	3 12 .33	3 39 .64	4 6 .87	4 34 .04
27	0 28 .52	0 57 .04	1 25 .54	1 54 .00	2 22 44	2 50 .84	3 19 .18	3 47 .46	4 15 .67	4 33 .80
28	0 29 .49	0 58 .98	1 28 .45	1 57 .90	2 27 .30	2 56 .66	3 26 .46	3 55 .22	4 24 .39	4 53 .48
29	0 30 .46	1 0 .91	1 31 .33	2 1 .75	2 32 .11	3 2 .44	3 32 .70	4 2 .90	4 33 .02	5 3 .07
30	0 31 .41	1 2 .82	1 34 .20	2 5 .56	2 36 .88	3 8 .15	3 39 .36	4 10 .51	4 41 .58	5 12 .57
31	0 32 .36	1 4 .71	1 37 .04	2 9 .34	2 41 .60	3 13 .81	3 45 .96	4 18 .05	4 50 .05	5 21 .97
32	0 33 .29	1 6 .58	1 39 .84	2 13 .08	2 46 .27	3 19 .41	3 52 .49	4 25 .50	4 58 .43	5 31 .27
33	0 34 .22	1 8 .43	1 42 .61	2 16 .78	2 50 .88	3 24 .95	3 58 .95	4 32 .88	5 6 .72	5 40 .47
34	0 35 .13	1 10 .26	1 45 .36	2 20 .43	2 55 .45	3 30 .43	4 5 .33	4 40 .17	5 14 .92	5 49 .57
35	0 36 .03	1 12 .07	1 48 .07	2 24 .04	2 59 .97	3 35 .84	4 11 .64	4 47 .38	5 23 .04	5 58 .56
36	0 36 .93	1 13 .85	1 50 .75	2 27 .61	3 4 .42	3 41 .19	4 17 .88	4 54 .49	5 31 .02	6 7 .44
37	0 37 .81	1 15 .61	1 53 .39	2 31 .13	3 8 .83	3 46 .46	4 24 .03	5 1 .52	5 38 .94	6 12 .21
38	0 38 .68	1 17 .35	1 56 .00	2 34 .61	3 13 .17	3 51 .67	4 30 .11	5 8 .46	5 46 .72	6 24 .87
39	0 39 .54	1 19 .07	1 58 .57	2 38 .04	3 17 .46	3 56 .81	4 36 .10	5 15 .30	5 54 .41	6 33 .41
40	0 40 .38	1 20 .76	2 1 .11	2 41 .42	3 21 .68	4 1 .88	4 42 .01	5 22 .05	6 2 .00	6 41 .83
41	0 41 .22	1 22 .43	2 3 .61	2 44 .75	3 25 .85	4 6 .88	4 47 .83	5 28 .70	6 9 .47	6 50 .12
42	0 42 .04	1 24 .07	2 6 .07	2 48 .03	3 29 .95	4 11 .80	4 53 .57	5 35 .25	6 16 .83	6 58 .30
43	0 42 .85	1 25 .69	2 8 .50	2 51 .27	3 33 .99	4 16 .64	4 59 .21	5 41 .70	6 24 .08	7 6 .34
44	0 43 .65	1 27 .28	2 8 .88	2 54 .41	3 37 .96	4 21 .40	5 4 .77	5 48 .04	6 31 .21	7 14 .25
45	0 44 .43	1 28 .84	2 13 .22	2 57 .56	3 41 .86	4 26 .09	5 10 .23	5 54 .28	6 38 .22	7 22 .04
46	0 45 .20	1 30 .38	2 15 .53	3 0 .64	3 45 .70	4 30 .69	5 15 .60	6 0 .41	6 45 .11	7 29 .68
47	0 45 .95	1 31 .89	2 17 .80	3 3 .66	3 49 .47	4 35 .21	5 20 .87	6 6 .43	6 51 .88	7 37 .30
48	0 46 .69	1 33 .37	2 20 .02	3 6 .62	3 53 .17	4 39 .65	5 26 .04	6 12 .33	6 58 .52	7 44 .56
49	0 47 .42	1 34 .82	2 22 .21	3 9 .53	3 56 .79	4 44 .00	5 31 .11	6 18 .13	7 5 .03	7 51 .79
50	0 48 .13	1 36 .24	2 24 .31	3 12 .38	4 0 .35	4 48 .26	5 36 .09	6 23 .81	7 11 .41	7 58 .88
51	0 48 .83	1 37 .63	2 26 .42	3 15 .17	4 3 .84	4 52 .44	5 40 .96	6 29 .37	7 17 .66	8 5 .82
52	0 49 .51	1 39 .00	2 28 .47	3 17 .89	4 7 .25	4 56 .53	5 45 .72	6 34 .81	7 23 .78	8 12 61
53	0 50 .18	1 40 .34	2 30 .47	3 20 .56	4 10 .58	5 0 .53	5 50 .39	6 40 .13	7 29 .76	8 19 .25
54	0 50 .83	1 41 .64	2 32 .42	3 23 .17	4 13 .84	5 4 .44	5 54 .94	6 45 .34	7 35 .61	8 25 .74
55	0 51 .47	1 42 .93	2 34 .34	3 25 .71	4 17 .02	5 8 .25	5 59 .39	6 50 .41	7 41 .32	8 32 .08
56	0 52 .09	1 44 .20	2 36 .20	3 28 .19	4 20 .12	5 11 .97	6 3 .72	6 55 .37	7 46 .88	8 38 .26
57	0 52 .69	1 45 .39	2 38 .01	3 30 .61	4 23 .14	5 15 .59	6 7 .95	7 0 .19	7 52 .31	8 44 .28
58	0 53 .28	1 46 .55	2 39 .78	3 32 .96	4 26 .08	5 19 .12	6 12 .06	7 4 .89	7 57 .59	8 50 .14
59	0 53 .85	1 47 .69	2 41 .50	3 35 .26	4 28 .95	5 22 .55	6 16 .06	7 9 .46	8 2 .73	8 55 .84
60	0 54 .41	1 48 .81	2 43 .17	3 37 .49	4 31 .73	5 25 .89	6 19 .95	7 13 .90	8 7 .71	9 1 .38
61	0 54 .95	1 49 .89	2 44 .79	3 39 .64	4 34 .43	5 29 .12	6 23 .72	7 18 .20	8 12 .55	9 6 .75
62	0 55 .47	1 50 .93	2 46 .36	3 41 .73	4 37 .05	5 32 .26	6 27 .38	7 22 .38	8 17 .24	9 11 .96
63	0 55 .98	1 51 .94	2 47 .88	3 43 .75	4 39 .58	5 35 .29	6 30 .91	7 26 .42	8 21 .78	9 17 .00
64	0 56 .47	1 52 .92	2 49 .34	3 45 .71	4 42 .01	5 38 .22	6 34 .33	7 30 .32	8 26 .17	9 21 .87
65	0 56 .94	1 53 .86	2 50 .76	3 47 .60	4 44 .37	5 41 .05	6 37 .62	7 34 .08	8 30 .40	9 26 .56
66	0 57 .40	1 54 .78	2 52 .12	3 49 .42	4 46 .63	5 43 .77	6 40 .80	7 37 .71	8 34 .48	9 31 .09
67	0 57 .84	1 55 .65	2 53 .43	3 51 .16	4 48 .81	5 46 .39	6 43 .85	7 41 .19	8 38 .40	9 35 .44
68	0 58 .26	1 56 .49	2 54 .69	3 52 .84	4 50 .91	5 48 .90	6 46 .78	7 44 .54	8 42 .16	9 39 .61
69	0 58 .66	1 57 .29	2 55 .90	3 54 .45	4 52 .92	5 51 .31	6 49 .59	7 47 .75	8 45 .76	9 43 .61
70	0 59 .04	1 58 .06	2 57 .05	3 55 .98	4 54 .84	5 53 .61	6 52 .27	7 50 .81	8 49 .20	9 47 .43

TABLE 13.

The MOON'S APPROXIMATE PARALLAX in LONGITUDE, continued.

Alt. of the Nonag.	The ☾'s Distance from the Nonagesimal Degree of the Ecliptic. 11°	12°	13°	14°	15°	16°	17°	18°	19°	20°
10°	1′ 59″.28	2′ 9″.97	2′ 20″.62	2′ 31″.23	2′ 41″.80	2′ 52″.31	3′ 2″.77	3′ 13″.18	3′ 23″.52	3′ 33″.81
11	2 11.07	2 22.82	2 34.52	2 46.18	2 57.79	3 9.34	3 20.83	3 32.27	3 43.64	3 54.94
12	2 22.82	2 35.62	2 48.37	3 1.07	3 13.72	3 26.31	3 38.83	3 51.29	4 3.68	4 16.00
13	2 34.52	2 48.37	3 2.17	3 15.91	3 29.60	3 43.22	3 56.77	4 10.25	4 23.65	4 36.98
14	2 46.18	3 1.07	3 15.91	3 30.69	3 45.41	4 0.06	4 14.63	4 29.11	4 43.54	4 57.87
15	2 57.79	3 13.72	3 29.60	3 45.41	4 1.15	4 16.82	4 32.42	4 47.93	5 3.35	5 18.68
16	3 9.34	3 26.31	3 43.22	4 0.05	4 16.82	4 33.51	4 50.12	5 6.64	5 23.06	5 39.39
17	3 20.83	3 38.83	3 56.77	4 14.63	4 32.42	4 50.12	5 7.73	5 25.25	5 42.67	5 59.99
18	3 32.27	3 51.29	4 10.25	4 29.11	4 47.93	5 6.64	5 25.25	5 43.77	6 2.18	6 20.48
19	3 43.64	4 3.68	4 23.65	4 43.54	5 3.35	5 23.06	5 42.67	6 2.18	6 21.58	6 40.86
20	3 54.94	4 16.00	4 36.98	4 57.87	5 18.68	5 39.39	5 59.99	6 20.48	6 40.86	7 1.12
21	4 6.17	4 28.23	4 50.21	5 12.11	5 33.91	5 55.61	6 17.20	6 38.67	7 0.02	7 21.25
22	4 17.32	4 40.39	5 3.37	5 26.25	5 49.04	6 11.73	6 34.29	6 56.73	7 19.06	7 41.25
23	4 28.40	4 52.46	5 16.42	5 40.30	6 4.06	6 27.72	6 51.26	7 14.67	7 37.95	8 1.10
24	4 39.39	5 4.44	5 29.39	5 54.24	6 18.97	6 43.60	7 8.11	7 32.48	7 56.71	8 20.81
25	4 50.30	5 16.32	5 42.25	6 8.07	6 33.77	6 59.36	7 24.82	7 50.15	8 15.33	8 43.36
26	5 1.12	5 28.11	5 55.00	6 21.79	6 48.44	7 15.00	7 41.40	8 7.67	8 33.79	8 59.76
27	5 11.85	5 39.80	6 7.65	6 35.39	7 3.00	7 30.49	7 57.84	8 25.05	8 52.10	9 18.99
28	5 22.49	5 51.39	6 20.19	6 48.87	7 17.43	7 45.85	8 14.13	8 42.27	9 10.24	9 38.05
29	5 33.02	6 2.87	6 32.61	7 2.23	7 31.72	8 1.07	8 30.28	8 59.33	9 28.22	9 56.93
30	5 43.46	6 14.24	6 44.91	7 15.46	7 45.88	8 16.15	8 46.27	9 16.23	9 46.02	10 15.63
31	5 33.79	6 25.50	6 57.09	7 28.56	7 59.89	8 31.07	9 2.10	9 32.96	10 3.65	10 34.15
32	6 4.01	6 36.64	7 9.14	7 41.52	8 13.76	8 45.84	9 17.75	9 49.52	10 21.09	10 52.47
33	6 14.12	6 47.65	7 21.06	7 54.34	8 27.47	9 0.44	9 33.25	10 5.89	10 38.34	11 10.60
34	6 24.12	6 58.55	7 32.85	8 7.02	8 41.03	9 14.88	9 48.57	10 22.08	10 55.40	11 28.51
35	6 34.00	7 9.31	7 44.50	8 19.54	8 54.43	9 29.16	10 3.71	10 38.08	11 12.26	11 46.23
36	6 43.76	7 19.95	7 56.00	8 31.92	9 7.66	9 43.26	10 18.66	10 53.89	11 28.91	12 3.72
37	6 53.39	7 30.45	8 7.36	8 44.13	9 20.74	9 57.18	10 13.43	11 9.50	11 45.35	12 21.00
38	7 2.91	7 40.81	8 18.58	8 56.19	9 33.64	10 10.92	10 48.00	11 24.90	12 1.58	12 38.04
39	7 11.29	7 51.04	8 29.64	9 8.09	9 46.37	10 24.47	11 2.38	11 40.09	12 17.59	12 54.87
40	7 21.54	8 1.12	8 40.54	9 19.81	9 58.91	10 37.84	11 16.56	11 55.07	12 33.37	13 10.45
41	7 30.66	8 11.05	8 51.29	9 31.37	10 11.28	10 51.00	11 30.53	12 9.84	12 48.93	13 27.79
42	7 39.63	8 20.83	9 1.88	9 42.76	10 33.46	11 3.97	11 44.28	12 24.38	13 4.25	13 43.89
43	7 48.47	8 30.46	9 12.30	9 53.97	10 35.45	11 16.74	11 57.83	12 38.70	13 19.33	13 59.73
44	7 57.17	8 39.94	9 22.55	10 4.99	10 47.26	11 29.32	12 11.16	12 52.78	13 34.16	14 15.31
45	8 5.72	8 49.26	9 32.63	10 15.83	10 58.86	11 41.67	12 24.27	13 6.63	13 48.76	14 30.64
46	8 14.02	8 58.41	9 42.54	10 26.48	11 10.25	11 53.81	12 37.13	13 20.24	14 3.10	14 45.70
47	8 22.38	9 7.41	9 52.27	10 36.95	11 21.44	12 5.72	12 49.78	13 33.60	14 17.18	15 0.50
48	8 30.48	9 16.23	10 1.82	10 47.22	11 32.42	12 18.42	13 2.19	13 46.72	14 30.99	15 15.01
49	8 38.42	9 24.89	10 11.18	10 57.29	11 43.20	12 28.89	13 14.36	13 59.59	14 44.55	15 29.25
50	8 46.21	9 33.37	10 20.36	11 7.16	11 53.76	12 40.15	13 26.29	14 12.20	14 57.84	15 43.20
51	8 53.83	9 41.68	10 29.35	11 16.83	12 4.10	12 51.16	13 37.98	14 24.55	15 10.85	15 56.88
52	9 1.29	9 49.81	10 38.15	11 26.29	12 14.23	13 1.94	13 49.41	14 36.63	15 23.59	16 10.26
53	9 8.59	9 57.76	10 46.75	11 35.55	12 24.13	13 12.48	14 0.59	14 48.45	15 36.04	16 23.34
54	9 15.72	10 5.53	10 55.16	11 44.60	12 33.80	13 22.78	14 11.52	15 0.00	15 48.20	16 36.12
55	9 22.68	10 13.12	11 3.37	11 53.43	12 43.24	13 32.84	14 22.19	15 11.28	16 0.08	16 48.60
56	9 29.48	10 20.52	11 11.37	12 2.03	12 52.45	13 42.65	14 32.59	15 22.28	16 11.07	17 0.77
57	9 36.09	10 27.73	11 19.18	12 10.41	13 1.43	13 52.21	14 42.73	15 32.99	16 22.96	17 12.63
58	9 42.53	10 34.75	11 26.77	12 18.57	13 10.17	14 1.52	14 52.60	15 43.42	16 33.95	17 24.18
59	9 48.80	10 41.57	11 34.15	12 26.52	13 18.66	14 10.56	15 2.20	15 53.57	16 44.64	17 35.41
60	9 54.88	10 48.20	11 41.33	12 34.24	13 26.92	14 19.35	15 11.52	16 3.42	16 55.02	17 46.31
61	10 0.79	10 54.64	11 48.29	12 41.72	13 34.93	14 27.88	15 20.57	16 12.08	17 5.09	17 56.90
62	10 6.51	11 0.87	11 55.03	12 48.97	13 42.68	14 36.14	15 29.34	16 22.24	17 14.86	18 7.15
63	10 12.04	11 6.90	12 1.56	12 55.99	13 50.19	14 44.14	15 37.82	16 31.21	17 24.30	18 17.07
64	10 17.39	11 12.73	12 7.86	13 2.78	13 57.45	14 51.86	15 46.01	16 39.87	17 33.42	18 26.66
65	10 22.55	11 18.36	12 13.95	13 9.32	14 4.45	14 59.32	15 53.92	16 48.23	17 42.23	18 35.91
66	10 27.54	11 23.77	12 19.81	13 15.62	14 11.19	15 6.50	16 1.54	16 56.28	17 50.73	18 44.82
67	10 32.45	11 28.98	12 25.45	13 21.68	14 17.68	15 13.41	16 8.87	17 4.03	17 58.87	18 53.39
68	10 36.89	11 33.98	12 30.86	13 27.50	14 23.90	15 20.04	16 15.90	17 11.46	18 6.70	19 1.61
69	10 41.29	11 38.77	12 36.04	13 33.07	14 29.86	15 26.39	16 22.63	17 18.57	18 14.20	19 9.49
70	10 45.49	11 43.34	12 40.99	13 38.40	14 35.56	15 32.45	16 29.06	17 25.37	18 21.36	19 17.02

TABLE 13.

The Moon's approximate Parallax in Longitude, continued.

Alt. of the Nonag.	The ☾'s Distance from the Nonagesimal Degree of the Ecliptic.									
	21°	22°	23°	24°	25°	26°	27°	28°	29°	30°
10°	3′ 44″.03	3′ 54″.18	4′ 4″.26	4′ 14″.27	4′ 24″.19	4′ 34″.04	4′ 43″.80	4′ 53″.48	5′ 3″.07	5′ 12″.57
11	4 6.17	4 17.32	4 28.40	4 39.39	4 50.30	5 1.12	5 11.85	5 22.49	5 33.02	5 43.46
12	4 28.23	4 40.39	4 52.46	5 4.44	5 16.32	5 28.11	5 39.80	5 51.39	6 2.87	6 14.24
13	4 50.21	5 3.37	5 16.42	5 29.39	5 42.25	5 55.00	6 7.65	6 20.19	6 32.61	6 44.91
14	5 12.11	5 26.25	5 40.30	5 54.24	6 8.07	6 21.79	6 35.39	6 48.87	7 2.23	7 15.46
15	5 33.91	5 49.04	6 4.06	6 18.98	6 33.77	6 48.44	7 3.00	7 17.43	7 31.72	7 45.88
16	5 55.61	6 11.73	6 27.72	6 43.60	6 59.36	7 15.00	7 30.49	7 45.85	8 1.07	8 16.15
17	6 17.20	6 34.29	6 51.26	7 8.11	7 24.82	7 41.40	7 57.84	8 14.13	8 30.28	8 46.27
18	6 38.67	6 56.73	7 14.67	7 32.48	7 50.15	8 7.67	8 25.05	8 42.27	8 59.33	9 16.23
19	7 0.02	7 19.06	7 37.95	7 56.71	8 15.33	8 33.79	8 52.10	9 10.24	9 28.22	9 46.02
20	7 21.25	7 41.25	8 1.10	8 20.81	8 40.36	8 59.76	9 18.99	9 38.05	9 56.93	10 15.63
21	7 42.34	8 3.29	8 24.09	8 44.74	9 5.23	9 25.03	9 45.70	10 5.67	10 25.46	10 45.06
22	8 3.29	8 25.19	8 46.93	9 8.51	9 29.94	9 51.18	10 12.24	10 33.12	10 53.81	10 14.29
23	8 24.09	8 46.93	9 9.61	9 32.13	9 54.47	10 16.62	10 38.60	11 0.37	11 21.95	11 43.31
24	8 44.74	9 8.51	9 32.13	9 55.56	10 18.82	10 41.89	11 4.76	11 27.42	11 49.88	12 12.12
25	9 5.23	9 29.94	9 54.47	10 18.82	10 42.98	11 6.95	11 30.71	11 54.26	12 17.60	12 40.71
26	9 25.03	9 51.18	10 16.62	10 41.89	11 6.95	11 31.80	11 56.46	12 20.89	12 45.10	13 4.07
27	9 45.70	10 12.24	10 38.60	11 4.76	11 30.71	11 56.46	12 81.98	12 47.29	13 12.36	13 37.18
28	10 5.67	10 33.12	11 0.37	11 27.42	11 54.26	12 20.89	12 47.29	13 13.45	13 39.38	14 5.05
29	10 25.46	10 53.81	11 21.95	11 49.88	12 17.60	12 45.10	13 12.35	13 39.38	14 6.14	14 32.66
30	10 45.06	10 14.29	11 43.31	12 12.12	12 40.71	13 4.07	13 37.18	14 5.05	14 32.66	15 0.00
31	11 4.46	11 34.57	12 4.47	12 34.15	13 3.59	13 32.80	14 1.76	14 30.46	14 58.90	15 27.07
32	11 23.66	11 54.64	12 25.40	12 55.93	13 26.23	13 56.29	14 26.08	14 55.60	15 24.88	15 53.86
33	11 42.65	12 14.49	12 46.11	13 17.49	13 48.63	14 19.52	14 50.14	15 20.49	15 50.57	16 20.35
34	12 1.43	12 34.11	13 6.58	13 38.79	14 10.77	14 42.48	15 13.93	15 45.09	16 15.97	16 46.54
35	12 19.99	12 53.52	13 26.81	13 59.86	14 32.65	15 5.18	15 37.43	16 9.40	16 41.07	17 12.44
36	12 38.32	13 12.68	13 46.79	14 20.66	14 54.27	15 27.60	16 0.66	16 24.67	17 5.87	17 38.01
37	12 56.42	13 31.60	14 6.53	14 41.21	15 15.62	15 49.75	16 23.59	16 57.13	17 30.36	18 3.27
38	13 14.28	13 50.28	14 26.00	15 1.48	15 36.68	16 11.60	16 46.22	17 20.53	17 54.52	18 28.19
39	13 31.91	14 8.69	14 45.22	15 21.48	15 57.46	16 33.15	17 8.54	17 43.61	18 18.36	18 52.77
40	13 49.28	14 26.85	15 4.16	15 41.20	16 17.95	16 54.40	17 30.55	18 6.37	18 41.87	19 17.01
41	14 6.40	14 44.75	15 22.83	16 0.64	16 38.15	17 15.34	17 52.23	18 28.80	19 5.03	19 40.91
42	14 23.26	15 2.38	15 41.21	16 19.78	16 58.03	17 35.97	18 13.60	18 50.89	19 27.84	20 4.43
43	14 39.86	15 19.73	15 59.32	16 38.62	17 17.61	17 56.28	18 34.63	19 12.63	19 50.30	20 27.58
44	14 56.20	15 36.80	16 17.13	16 57.15	17 36.87	18 16.27	18 55.33	19 34.04	20 12.40	20 50.38
45	15 12.26	15 53.59	16 34.64	17 15.38	17 55.81	18 35.91	19 15.67	19 55.07	20 34.11	21 12.78
46	15 28.04	16 10.09	16 51.85	17 33.29	18 14.42	18 55.21	19 35.66	20 15.74	20 55.47	21 34.80
47	15 43.54	16 26.29	17 8.75	17 50.89	18 32.70	19 14.17	19 55.29	20 36.05	21 16.43	21 56.43
48	15 58.75	16 42.19	17 25.33	18 8.15	18 50.63	19 32.78	20 14.57	20 55.98	21 37.02	22 17.65
49	16 13.67	16 57.79	17 41.60	18 25.08	19 8.23	19 51.02	20 33.46	21 15.52	21 57.20	22 38.47
50	16 28.29	17 13.08	17 57.54	18 41.68	19 25.48	20 8.92	20 52.00	21 34.69	22 16.99	22 58.88
51	16 42.62	17 28.05	18 13.16	18 57.94	19 42.37	20 26.44	21 10.14	21 53.47	22 26.37	23 18.86
52	16 56.63	17 42.70	18 23.44	19 13.85	19 58.89	20 43.58	21 27.90	22 11.81	22 55.33	23 38.42
53	17 10.34	17 57.03	18 43.39	19 29.40	20 15.06	21 0.35	21 45.26	22 29.76	23 13.87	23 57.54
54	17 23.49	18 10.53	18 57.99	19 44.60	20 30.86	21 16.73	22 2.23	22 47.32	23 31.99	24 16.23
55	17 36.81	18 24.69	19 12.25	19 59.44	20 46.28	21 32.73	22 18.79	23 4.44	23 49.68	24 34.48
56	17 49.56	18 38.02	19 26.15	20 13.91	21 1.32	21 48.33	22 34.95	23 21.15	24 6.93	24 52.27
57	18 1.99	18 51.02	19 39.70	20 28.02	21 15.98	22 3.53	22 50.69	23 37.42	24 23.74	25 9.60
58	18 14.09	19 3.66	19 52.89	20 41.75	21 30.24	22 18.33	23 6.02	23 53.28	24 40.11	25 26.48
59	18 25.85	19 15.96	20 5.72	20 55.11	21 44.12	22 32.72	23 20.92	24 8.69	24 56.03	25 42.90
60	18 37.28	19 27.91	20 18.18	21 8.08	21 57.59	22 46.70	23 35.40	24 23.66	25 11.49	25 58.84
61	18 48.37	19 39.50	20 20.27	21 20.66	22 10.67	23 0.27	23 49.45	24 38.19	25 26.49	26 14.31
62	18 59.11	19 50.73	20 41.98	21 32.86	22 23.34	23 13.41	24 3.06	24 52.26	25 41.02	26 29.29
63	19 9.51	20 1.60	20 53.32	21 44.66	22 35.60	23 26.13	24 16.23	25 5.88	25 55.08	26 43.80
64	19 19.56	20 12.09	21 4.27	21 56.06	22 47.45	23 38.42	24 28.96	25 19.05	26 8.68	26 57.82
65	19 29.25	20 22.23	21 14.84	22 7.06	22 58.88	23 50.28	24 41.24	25 31.75	26 21.79	27 11.35
66	19 38.59	20 32.00	21 25.02	22 17.66	23 9.89	24 1.70	24 53.07	25 43.97	26 34.42	27 24.38
67	19 47.57	20 41.38	21 34.81	22 27.85	23 20.48	24 12.68	25 4.44	25 55.74	26 46.57	27 36.90
68	19 56.19	20 50.38	21 44.20	22 37.13	23 30.63	24 23.21	25 15.36	26 7.03	26 58.23	27 48.93
69	20 4.44	20 59.01	21 53.20	22 47.00	23 40.37	24 33.32	25 25.81	26 17.84	27 9.39	28 0.45
70	20 12.32	21 7.25	22 1.80	22 55.95	23 49.67	24 42.96	25 35.80	26 28.17	27 20.06	28 11.45

TABLE 13.

The Moon's approximate Parallax in Longitude, continued.

Alt. of the Nonag.	The ☾'s Distance from the Nonagesimal Degree of the Ecliptic.									
	31°	32°	33°	34°	35°	36°	37°	38°	39°	40°
10°	5′ 21″.97	5′ 31″.27	5′ 40″.47	5′ 49″.57	5′ 58″.56	6′ 7″.44	6′ 16″.21	6′ 24″.87	6′ 33″.41	6′ 41″.83
11	5 53.79	6 4.01	6 14.12	6 24.12	6 34.00	6 43.76	6 53.39	7 2.91	7 12.29	7 21.54
12	6 25.50	6 36.64	6 47.65	6 58.55	7 9.31	7 19.95	7 30.45	7 40.81	7 51.04	8 1.12
13	6 57.09	7 9.14	7 21.06	7 32.85	7 44.50	7 56.00	8 7.36	8 18.58	8 29.64	8 40.54
14	7 28.56	7 41.52	7 54.34	8 7.02	8 19.54	8 31.92	8 44.13	8 56.19	9 8.09	9 19.81
15	7 59.89	8 13.76	8 27.47	8 41.03	8 54.43	9 7.66	9 20.74	9 33.64	9 46.37	9 58.91
16	8 31.07	8 45.84	9 0.44	9 14.88	9 29.16	9 43.26	9 57.18	10 10.92	10 24.47	10 37.84
17	9 2.10	9 17.75	9 33.25	9 48.57	10 3.71	10 18.66	10 33.43	10 48.00	11 2.38	11 16.56
18	9 32.96	9 49.52	10 5.89	10 22.08	10 38.08	10 53.89	11 9.50	11 24.90	11 40.09	11 55.07
19	10 3.65	10 21.09	10 38.34	10 55.40	11 12.26	11 28.91	11 45.35	12 1.58	12 17.59	12 33.37
20	10 34.15	10 52.47	11 10.60	11 28.51	11 46.23	12 3.72	12 21.00	12 38.04	12 54.87	13 10.45
21	11 4.46	11 23.66	11 42.65	12 1.43	12 19.99	12 38.32	12 56.42	13 14.28	13 31.91	13 49.28
22	11 34.57	11 54.64	12 14.49	12 34.11	12 53.52	13 12.68	13 31.60	13 50.28	14 8.69	14 26.85
23	12 4.47	12 25.40	12 46.11	13 6.58	13 26.81	13 46.79	14 6.53	14 26.00	14 45.22	15 4.16
24	12 34.15	12 55.93	13 17.49	13 38.79	13 59.86	14 20.66	14 41.21	15 1.48	15 21.48	15 41.20
25	13 3.59	13 26.23	13 48.63	14 10.77	14 32.65	14 54.27	15 15.62	15 36.68	15 57.46	16 17.95
26	13 32.80	13 56.29	14 19.52	14 42.48	15 5.18	15 27.60	15 49.75	16 11.60	16 33.15	16 54.40
27	14 1.76	14 26.08	14 50.14	15 13.93	15 37.43	16 0.66	16 23.59	16 46.22	17 8.54	17 30.55
28	14 30.46	14 55.60	15 20.49	15 45.08	16 9.40	16 24.67	16 57.13	17 20.53	17 43.61	18 6.37
29	14 58.90	15 24.88	15 50.57	16 15.97	16 41.07	17 5.87	17 30.36	17 54.52	18 18.36	18 41.87
30	15 27.07	15 53.86	16 20.35	16 46.54	17 12.44	17 38.01	18 3.27	28 28.19	18 52.77	19 17.01
31	15 54.94	16 22.54	16 49.83	17 16.82	17 43.49	18 9.83	18 35.85	19 1.52	19 26.85	19 51.82
32	16 22.54	16 50.92	17 19.01	17 46.77	18 14.22	18 41.32	19 8.09	19 34.50	20 0.56	20 26.25
33	16 49.83	17 19.01	17 47.87	18 16.41	18 44.61	19 12.47	19 39.99	20 7.13	20 33.91	21 0.31
34	17 16.82	17 46.78	18 16.41	18 45.71	19 14.66	19 43.26	20 11.51	20 39.38	21 6.88	21 33.98
35	17 43.49	18 14.22	18 44.61	19 14.66	19 44.36	20 13.70	20 42.67	21 11.26	21 39.47	22 7.28
36	18 9.83	18 41.32	19 12.47	19 43.26	20 13.70	20 43.80	21 13.46	21 42.74	22 11.66	22 40.15
37	18 35.85	19 8.09	19 39.99	20 11.51	20 42.67	21 13.46	21 43.85	22 13.85	22 43.45	23 12.62
38	19 1.52	19 34.50	20 7.13	20 39.38	21 11.26	21 42.74	22 13.85	22 44.54	23 14.81	23 44.66
39	19 26.85	20 0.56	20 33.91	21 6.88	21 39.47	22 11.66	22 43.45	23 14.81	23 45.75	24 16.27
40	19 51.82	20 26.25	21 0.31	21 33.98	22 7.28	22 40.15	23 12.62	23 44.66	24 16.27	24 47.42
41	20 16.43	20 51.57	21 26.33	22 0.70	22 34.68	23 8.23	23 41.37	24 14.08	24 46.34	25 18.14
42	20 40.66	21 16.50	21 51.96	22 27.01	23 1.67	23 35.89	24 9.69	24 43.05	25 15.95	25 48.38
43	21 4.51	21 41.04	22 17.19	22 52.91	23 28.23	24 3.11	24 37.57	25 11.57	25 45.10	26 18.17
44	21 27.99	22 5.20	22 42.02	23 18.41	23 54.38	24 29.91	25 5.00	25 39.62	26 13.79	26 47.46
45	21 51.07	22 28.95	23 6.42	23 23.47	24 20.08	24 56.25	25 31.26	26 7.22	26 41.98	27 16.27
46	22 13.75	22 52.29	23 30.41	24 8.10	24 45.35	25 22.14	25 58.47	26 34.33	27 9.70	27 44.57
47	22 36.03	23 15.20	23 53.96	24 32.27	25 10.15	25 47.55	26 24.50	27 0.96	27 36.91	28 12.38
48	22 57.89	23 37.70	24 17.08	24 56.02	25 34.50	26 12.52	26 50.05	27 27.09	28 3.63	28 39.66
49	23 19.33	23 59.76	24 39.76	25 19.30	25 58.38	26 36.98	27 15.10	27 52.72	28 29.83	29 6.42
50	23 40.35	24 21.38	25 1.93	25 42.11	26 21.79	27 0.97	27 39.66	28 17.84	28 55.51	29 32.68
51	24 0.93	24 42.55	25 23.74	26 4.45	26 44.70	27 24.46	28 3.72	28 42.45	29 20.66	29 58.34
52	24 21.08	25 3.29	25 45.05	26 26.33	27 7.14	27 47.44	28 27.25	29 6.53	29 45.28	30 23.48
53	24 40.78	25 23.56	26 5.89	26 47.73	27 29.08	28 9.92	28 50.26	29 30.08	30 9.36	30 48.07
54	25 0.03	25 43.36	26 26.24	27 8.63	27 50.52	28 31.90	29 12.76	29 53.08	30 32.87	31 12.09
55	25 18.82	26 2.72	26 46.10	27 29.02	28 11.45	28 53.34	29 34.72	30 15.55	30 55.83	31 35.55
56	25 37.15	26 21.56	27 5.49	27 48.92	28 31.86	29 14.27	29 56.14	30 37.46	31 18.23	31 58.42
57	25 55.01	26 39.93	27 24.38	28 8.31	28 51.74	29 34.64	30 17.02	30 58.81	31 40.05	32 20.71
58	26 12.40	26 58.23	27 42.77	28 27.20	29 11.11	29 54.49	30 37.33	31 19.60	32 1.30	32 42.41
59	26 29.31	27 15.23	28 0.65	28 45.55	29 29.94	30 13.78	30 57.08	31 19.81	32 21.95	33 3.52
60	26 45.73	27 32.12	28 18.02	29 3.88	29 48.23	30 32.53	31 16.27	31 39.44	32 42.03	33 24.01
61	27 1.67	27 48.52	28 34.87	29 20.68	30 5.98	30 50.71	31 34.89	32 18.49	33 1.50	33 43.90
62	27 17.11	28 4.40	28 51.20	29 37.45	30 23.18	31 8.34	31 52.94	32 36.95	33 20.37	34 3.17
63	27 32.05	28 19.78	29 7.00	29 53.68	30 39.82	31 25.39	32 10.39	32 54.81	33 38.63	34 21.82
64	27 46.49	28 34.64	29 22.27	30 9.35	30 55.90	31 41.87	32 27.27	33 12.06	33 56.27	34 39.84
65	28 0.42	28 48.97	29 37.00	30 24.49	31 11.42	31 57.77	32 43.55	33 28.72	34 13.29	34 57.23
66	28 13.84	29 2.77	29 51.19	30 39.05	31 26.36	32 13.08	32 59.23	33 44.76	34 29.69	35 13.97
67	28 26.74	29 16.05	30 4.83	30 53.06	31 40.73	32 27.81	33 14.30	34 0.19	34 45.46	35 30.08
68	28 39.13	29 28.80	30 17.93	31 6.51	31 54.52	32 41.94	33 28.77	34 14.99	35 0.58	35 45.54
69	28 50.99	29 41.00	30 30.47	31 19.37	32 7.72	32 55.47	33 42.62	34 29.17	35 15.07	36 0.34
70	29 2.32	29 52.66	30 42.46	31 31.63	32 20.35	33 8.41	33 55.88	34 42.72	35 28.92	36 14.48

ZODIACAL TABLES.

TABLE 13.

The Moon's approximate Parallax in Longitude, continued.

Alt. of the Nonag.	The ☾'s Distance from the Nonagesimal Degree of the Ecliptic.									
	41°	42°	43°	44°	45°	46°	47°	48°	49°	50°
10°	6' 50".12	6' 58".30	7' 6".34	7' 14".25	7' 22".04	7' 29".68	7' 37".19	7' 44".56	7' 51".79	7' 58".88
11	7 30 .66	7 39 .63	7 48 .47	7 57 .17	8 5 .72	8 14 .02	8 22 .38	8 30 .48	8 38 .42	8 46 .21
12	8 11 .05	8 20 .83	8 30 .46	8 39 .94	8 49 .26	8 58 .41	9 7 .41	9 16 .23	9 24 .89	9 33 .37
13	8 51 .29	9 1 .88	9 12 .30	9 22 .55	9 32 .63	9 42 .54	9 52 .27	10 1 .82	10 11 .18	10 20 .36
14	9 31 .37	9 42 .76	9 53 .97	10 4 .99	10 15 .83	10 16 .48	10 36 .95	10 47 .22	10 57 .29	11 7 .16
15	10 11 .28	10 33 .46	10 35 .45	10 47 .26	10 58 .86	11 10 .25	11 21 .44	11 32 .42	11 43 .20	11 53 .76
16	10 51 .00	11 3 .97	11 16 .74	11 29 .32	11 41 .67	11 53 .81	12 5 .72	12 18 .42	12 28 .89	12 40 .15
17	11 30 .58	11 44 .28	11 57 .83	12 11 .16	12 24 .27	12 37 .13	12 49 .78	13 2 .19	13 14 .36	13 26 .29
18	12 9 .84	12 24 .38	12 38 .70	12 52 .78	13 6 .63	13 20 .24	13 33 .60	13 46 .72	13 59 .59	14 12 .20
19	12 48 .93	13 4 .25	13 19 .33	13 34 .16	13 48 .76	14 3 .10	14 17 .18	14 30 .99	14 44 .55	14 57 .84
20	13 27 .79	13 43 .89	13 59 .73	14 15 .31	14 30 .64	14 45 .70	15 0 .50	15 15 .01	15 29 .25	15 43 .20
21	14 6 .40	14 23 .26	14 39 .86	14 56 .20	15 12 .26	15 28 .04	15 43 .54	15 58 .75	16 13 .67	16 28 .29
22	14 44 .75	15 2 .58	15 19 .73	15 36 .80	15 33 .59	16 10 .09	16 26 .29	16 42 .19	16 57 .79	17 13 .08
23	15 22 .83	15 41 .21	15 59 .32	16 17 .13	16 34 .64	16 51 .85	17 8 .75	17 25 .33	17 41 .60	17 57 .54
24	16 0 .64	16 19 .78	16 38 .62	16 57 .15	17 15 .38	17 33 .29	17 50 .89	18 8 .15	18 25 .08	18 41 .68
25	16 38 .15	16 58 .03	17 17 .61	17 36 .87	17 55 .81	18 14 .42	18 32 .70	18 50 .63	19 8 .23	19 25 .48
26	17 15 .34	17 35 .97	17 56 .28	18 16 .27	18 35 .91	18 55 .21	19 14 .17	19 32 .78	19 51 .02	20 8 .92
27	17 52 .23	18 13 .60	18 34 .63	18 55 .33	19 15 .67	19 35 .66	19 55 .29	20 14 .57	20 33 .46	20 52 .00
28	18 28 .30	18 50 .89	19 12 .63	19 34 .04	19 55 .07	20 15 .74	20 36 .05	20 55 .98	21 15 .52	21 34 .69
29	19 5 .03	19 27 .84	19 50 .30	20 12 .40	20 34 .11	20 55 .47	21 16 .43	21 37 .02	21 57 .20	22 16 .99
30	19 40 .91	20 4 .43	20 27 .58	20 50 .38	21 12 .78	21 34 .80	21 56 .43	22 17 .65	22 38 .47	22 58 .88
31	20 16 .43	20 40 .66	21 4 .51	21 27 .99	21 51 .07	22 13 .75	22 36 .03	22 57 .89	23 19 .33	23 40 .35
32	20 51 .57	21 16 .50	21 41 .04	22 5 .20	22 28 .95	22 52 .29	23 15 .20	23 37 .70	23 59 .76	24 21 .38
33	21 26 .33	21 51 .96	22 17 .19	22 42 .02	23 6 .42	23 30 .41	23 53 .96	24 17 .08	24 39 .76	25 1 .98
34	22 0 .70	22 27 .01	22 52 .91	23 18 .41	23 43 .47	24 8 .10	24 32 .27	24 56 .02	25 19 .30	25 42 .11
35	22 34 .68	22 1 .67	23 28 .23	23 54 .38	24 20 .08	24 45 .35	25 10 .15	25 34 .50	25 58 .38	26 41 .79
36	23 8 .23	23 35 .89	24 3 .11	24 29 .91	24 56 .25	25 22 .14	25 47 .55	26 12 .52	26 36 .98	27 0 .97
37	23 41 .37	24 9 .69	24 37 .57	25 5 .00	25 31 .96	25 58 .47	26 24 .50	26 50 .05	27 15 .10	27 39 .66
38	24 14 .08	24 43 .05	25 11 .57	25 39 .62	26 7 .22	26 34 .33	27 0 .96	27 27 .09	27 52 .72	28 17 .84
39	24 48 .34	25 15 .95	25 45 .10	26 13 .79	26 41 .98	27 9 .70	27 36 .91	28 3 .63	28 29 .83	28 55 .51
40	25 18 .14	25 48 .38	26 18 .17	26 47 .46	27 16 .27	27 44 .57	28 12 .38	28 39 .66	29 6 .42	29 32 .68
41	25 49 .49	26 20 .36	26 50 .75	27 20 .65	27 50 .05	28 18 .95	28 47 .32	29 15 .17	29 42 .48	30 9 .25
42	26 20 .30	26 51 .85	27 22 .85	27 53 .34	28 23 .33	28 52 .79	29 21 .74	29 50 .14	30 18 .00	30 45 .30
43	26 50 .75	27 22 .85	27 54 .43	28 25 .52	28 56 .08	29 56 .12	29 55 .61	30 24 .56	30 52 .95	31 20 .79
44	27 20 .65	27 53 .34	28 25 .52	28 57 .17	29 28 .31	29 58 .90	30 28 .95	30 58 .43	31 27 .36	31 55 .70
45	27 50 .05	28 23 .33	28 56 .08	29 28 .31	30 0 .00	30 31 .14	31 1 .72	31 31 .74	32 1 .17	32 30 .03
46	28 18 .95	28 52 .79	29 26 .12	29 58 .90	30 31 .14	31 2 .82	31 33 .93	32 4 .46	32 34 .41	63 3 .76
47	28 47 .32	29 21 .74	29 55 .61	30 28 .95	31 1 .72	31 33 .93	32 5 .56	32 36 .61	33 7 .05	33 36 .90
48	29 15 .17	29 50 .14	30 24 .56	30 58 .43	31 31 .74	32 4 .46	32 36 .61	33 8 .15	33 39 .09	34 9 .41
49	29 42 .48	30 18 .00	30 52 .95	31 27 .36	32 1 .17	32 34 .41	33 7 .05	33 39 .09	34 10 .51	34 41 .31
50	30 9 .25	30 45 .30	31 20 .79	31 55 .70	32 30 .03	33 3 .76	33 36 .90	34 9 .41	34 41 .31	35 12 .56
51	30 35 .47	31 12 .04	31 48 .04	32 23 .46	32 58 .29	33 32 .52	34 6 .12	34 39 .12	35 11 .47	35 43 .18
52	31 1 .13	31 38 .21	32 14 .72	32 50 .63	33 25 .95	34 0 .65	34 34 .73	35 8 .19	35 40 .99	36 13 .14
53	31 26 .23	32 3 .81	32 40 .80	33 17 .20	33 52 .99	34 28 .16	35 2 .70	35 36 .61	36 9 .85	36 42 .45
54	31 50 .75	32 28 .82	33 6 .29	33 43 .16	34 19 .42	34 55 .05	35 30 .04	36 4 .38	36 38 .06	37 11 .07
55	32 14 .68	32 53 .23	33 31 .18	34 8 .51	34 45 .22	35 21 .30	35 56 .72	36 31 .50	37 5 .60	37 39 .02
56	32 38 .03	33 17 .04	33 55 .45	34 33 .22	35 10 .38	35 46 .90	36 22 .75	36 57 .94	37 32 .46	38 6 .28
57	33 0 .78	33 40 .25	34 19 .09	34 57 .32	35 34 .90	36 11 .84	36 48 .10	37 23 .71	37 58 .62	38 32 .85
58	33 22 .93	34 2 .83	34 42 .12	35 20 .77	35 58 .78	36 36 .12	37 12 .80	37 48 .80	38 24 .11	38 58 .71
59	33 44 .47	34 24 .80	35 4 .51	35 43 .58	36 21 .99	36 59 .74	37 36 .81	38 13 .20	38 48 .88	39 23 .86
60	34 5 .39	34 46 .14	35 26 .26	36 5 .73	36 44 .54	37 22 .68	38 0 .14	38 36 .89	39 12 .95	39 48 .28
61	34 25 .69	35 6 .85	35 47 .37	36 27 .22	37 6 .42	37 44 .94	38 22 .77	38 59 .89	39 36 .30	40 11 .99
62	34 45 .36	35 26 .90	36 7 .82	36 48 .04	37 27 .62	38 6 .50	38 44 .69	39 22 .17	39 58 .93	40 34 .95
63	35 4 .39	35 46 .32	36 27 .59	37 8 .20	37 48 .13	38 27 .37	39 5 .91	39 43 .73	40 20 .82	40 57 .18
64	35 22 .78	36 5 .08	36 46 .71	37 27 .67	38 7 .96	38 47 .53	39 26 .41	40 4 .57	40 41 .98	41 18 .66
65	35 40 .58	36 23 .18	37 5 .16	37 46 .47	38 27 .08	39 7 .00	39 46 .19	40 24 .67	41 2 .39	41 39 .38
66	35 57 .62	36 40 .61	37 22 .93	38 4 .56	38 45 .51	39 25 .74	40 5 .25	40 44 .03	41 22 .06	41 59 .33
67	36 14 .05	36 57 .38	37 40 .01	38 21 .97	39 3 .22	39 43 .76	40 23 .57	41 2 .65	41 40 .97	42 18 .53
68	36 29 .83	37 13 .46	37 56 .42	38 38 .67	39 20 .23	40 1 .06	40 41 .16	41 20 .52	41 59 .12	42 36 .95
69	36 44 .93	37 28 .87	38 12 .12	38 54 .67	39 36 .51	40 17 .62	40 58 .00	41 37 .63	42 16 .50	42 54 .59
70	36 59 .38	37 43 .60	38 27 .13	39 9 .96	39 52 .07	40 33 .45	41 14 .09	41 53 .98	42 33 .10	43 11 .45

TABLE 13.

The Moon's approximate Parallax in Longitude, continued.

Alt. of the Nonag.	The ☾'s Distance from the Nonagesimal Degree of the Ecliptic.									
	51°	52°	53°	54°	55°	56°	57°	58°	59°	60°
10°	8′ 5″.82	8′ 12″.61	8′ 19″.25	8′ 25″.74	8′ 32″.08	8′ 38″.26	8′ 44″.28	8′ 50″.14	8′ 55″.84	9′ 1″.38
11	8 53.83	9 1.29	9 8.59	9 15.72	9 22.68	9 29.48	9 36.09	9 42.53	9 48.80	9 54.88
12	9 41.68	9 49.81	9 57.76	10 5.53	10 13.12	10 20.52	10 27.73	10 34.75	10 41.57	10 48.20
13	10 29.35	10 38.15	10 46.75	10 55.16	11 3.37	11 11.37	11 19.18	11 26.77	11 34.15	11 41.33
14	11 16.83	11 26.29	11 35.55	11 44.60	11 53.43	12 2.03	12 10.41	12 18.57	12 26.52	12 34.24
15	12 4.10	12 14.23	12 24.13	12 33.80	12 43.24	12 52.45	13 1.43	13 10.17	13 18.66	13 26.92
16	12 51.16	13 1.94	13 12.48	13 22.78	13 32.84	13 42.65	13 52.21	14 1.52	14 10.56	14 19.35
17	13 37.98	13 49.41	14 0.59	14 11.52	14 22.19	14 32.59	14 42.73	14 52.60	15 2.20	15 11.52
18	14 24.55	14 36.63	14 48.45	15 0.00	15 11.28	15 22.28	15 32.99	15 43.42	15 53.57	16 3.42
19	15 10.85	15 23.59	15 36.04	15 48.20	16 0.08	16 11.67	16 22.96	16 33.95	16 44.64	16 55.02
20	15 56.88	16 10.26	16 23.34	16 36.12	16 48.60	17 0.77	17 12.63	17 24.18	17 35.41	17 46.31
21	16 42.62	16 56.63	17 10.34	17 23.49	17 36.81	17 49.56	18 1.99	18 14.09	18 25.85	18 37.28
22	17 28.05	17 42.70	17 57.03	18 10.53	18 24.69	18 38.02	18 51.02	19 3.66	19 15.96	19 27.91
23	18 13.16	18 28.44	18 43.39	18 57.99	19 12.25	19 26.15	19 39.70	19 52.89	20 5.72	20 18.18
24	18 57.94	19 13.85	19 29.40	19 44.60	19 59.44	20 13.91	20 28.02	20 41.75	20 55.11	21 8.08
25	19 42.37	19 58.89	20 15.06	20 30.86	20 46.28	21 1.32	21 15.98	21 30.24	21 44.12	21 57.59
26	20 26.44	20 43.58	21 0.35	21 16.73	21 32.73	21 48.33	22 3.53	22 18.33	22 32.72	22 46.70
27	21 10.14	21 27.90	21 45.26	22 2.23	22 18.79	22 34.95	22 50.69	23 6.02	23 20.92	23 35.40
28	21 53.47	22 11.81	22 29.76	22 47.32	23 4.44	23 21.15	23 37.42	23 58.28	24 8.69	24 23.66
29	22 36.37	22 55.33	23 13.87	23 31.99	23 49.68	24 6.93	24 23.74	24 40.11	24 56.03	25 11.49
30	23 18.86	23 38.42	23 57.54	24 16.23	24 34.48	24 52.27	25 9.60	25 26.48	25 42.90	25 58.84
31	24 0.93	24 21.08	24 40.78	25 0.03	25 18.82	25 37.15	25 55.01	26 12.40	26 29.31	26 45.73
32	24 42.55	25 3.29	25 23.56	25 43.36	26 2.72	26 21.56	26 39.93	26 58.23	27 15.23	27 32.12
33	25 23.74	25 45.05	26 5.89	26 26.24	26 46.10	27 5.49	27 24.38	27 42.77	28 0.65	28 18.02
34	26 4.45	26 26.33	26 47.73	27 8.63	27 29.02	27 48.92	28 8.31	28 27.20	28 45.55	29 3.38
35	26 44.70	27 7.14	27 29.08	27 50.52	28 11.45	28 31.86	28 51.74	29 11.11	29 29.94	29 48.23
36	27 24.46	27 47.44	28 9.92	28 31.90	28 53.34	29 14.27	29 34.64	29 54.49	30 13.78	30 32.53
37	28 3.72	28 27.25	28 50.26	29 12.76	29 34.72	29 56.14	30 17.02	30 37.33	30 57.08	31 16.27
38	28 42.45	29 6.53	29 30.08	29 53.08	30 15.55	30 37.46	30 58.81	31 19.60	31 19.81	31 39.44
39	29 20.66	29 45.28	30 9.36	30 32.87	30 55.83	31 18.23	31 40.05	32 1.30	32 31.95	32 42.03
40	29 58.34	30 23.48	30 48.07	31 12.09	31 35.55	31 58.42	32 20.71	32 42.41	33 3.52	33 24.01
41	30 35.47	31 1.13	31 26.23	31 50.75	32 14.68	32 38.03	33 0.78	33 32.93	33 44.47	34 5.39
42	31 12.04	31 38.21	32 3.81	32 28.82	32 53.23	33 17.04	33 40.25	34 2.83	34 24.80	34 46.14
43	31 48.04	32 14.72	32 40.80	33 6.29	33 31.18	33 55.45	34 19.09	34 42.12	35 4.51	35 26.26
44	32 23.46	32 50.63	33 17.20	33 43.16	34 8.51	34 33.22	34 57.32	35 20.77	35 43.58	36 5.73
45	32 58.29	33 25.95	33 52.99	34 19.42	34 45.22	35 10.38	35 34.90	35 58.78	36 21.99	36 44.54
46	33 32.52	34 0.65	34 28.16	34 55.05	35 21.30	35 46.90	36 11.84	36 36.12	36 59.74	37 22.68
47	34 6.12	34 34.73	35 2.70	35 30.04	35 56.72	36 22.75	36 48.10	37 12.80	37 36.81	38 0.14
48	34 39.12	35 8.19	35 36.61	36 4.38	36 31.50	36 57.94	37 23.71	37 48.80	38 13.20	38 36.89
49	35 11.47	35 40.99	36 9.85	36 38.06	37 5.60	37 32.46	37 58.62	38 24.11	38 48.88	39 12.95
50	35 43.18	36 13.14	36 42.45	37 11.07	37 39.02	38 6.28	38 32.85	38 58.71	39 23.86	39 48.28
51	36 15.24	36 44.64	37 14.36	37 43.41	38 11.77	38 39.42	39 6.38	39 32.61	39 58.12	40 22.90
52	36 44.64	37 15.46	37 45.60	38 15.05	38 43.80	39 11.84	39 39.17	40 5.78	40 31.65	40 56.78
53	37 14.36	37 45.60	38 16.15	38 46.00	39 15.14	39 43.56	40 11.25	40 38.21	41 4.43	41 29.90
54	37 43.41	38 15.05	38 46.00	39 16.13	39 45.75	40 14.54	40 42.60	41 9.91	41 36.47	42 2.26
55	38 11.77	38 43.80	39 15.14	39 45.75	40 15.64	40 44.79	41 13.20	41 40.85	42 7.73	42 33.86
56	38 39.42	39 11.84	39 43.56	40 14.54	40 44.79	41 14.31	41 43.04	42 11.02	42 38.25	43 4.68
57	39 6.38	39 39.17	40 11.25	40 42.60	41 13.20	41 43.04	42 12.15	42 40.44	43 7.97	43 34.72
58	39 32.61	40 5.78	40 38.21	41 9.91	41 40.85	42 11.02	42 40.44	43 9.06	43 36.91	44 3.95
59	39 58.12	40 31.65	41 4.43	41 36.47	42 7.73	42 38.25	43 7.97	43 36.91	44 5.04	44 32.38
60	40 22.90	40 56.78	41 29.90	42 2.26	42 33.86	43 4.68	43 34.72	44 3.95	44 32.38	45 0.00
61	40 46.94	41 21.15	41 54.60	42 27.30	42 59.20	43 30.33	44 0.66	44 30.19	44 58.90	45 26.79
62	41 10.25	41 44.77	42 18.55	42 51.55	43 23.77	43 55.18	44 25.81	44 55.61	45 24.60	45 52.75
63	41 32.79	42 7.64	42 41.72	43 15.02	43 47.53	44 19.24	44 50.14	45 20.22	45 49.47	46 17.88
64	41 54.58	42 29.73	43 4.11	43 37.70	44 10.50	44 42.48	45 13.65	45 43.99	46 13.50	46 42.16
65	42 15.59	42 51.05	43 25.71	43 59.59	44 32.66	45 4.91	45 36.34	46 6.93	46 36.69	47 5.59
66	42 35.85	43 11.58	43 46.52	44 20.67	44 54.00	45 26.51	45 58.19	46 29.03	46 59.02	47 28.15
67	42 55.32	43 31.32	44 6.53	44 40.93	45 14.52	45 47.28	46 19.20	46 50.28	47 20.49	47 49.85
68	43 14.01	43 50.27	44 25.74	45 0.38	45 34.22	46 7.21	46 39.37	47 10.67	47 41.11	48 10.67
69	43 31.91	44 8.42	44 44.13	45 19.02	45 53.08	46 26.30	46 58.68	47 30.20	48 0.85	48 30.62
70	43 49.00	44 25.76	45 1.70	45 36.82	46 11.10	46 44.55	47 17.13	47 48.86	48 19.71	48 49.67

TABLE 13.

The MOON'S APPROXIMATE PARALLAX in LONGITUDE, continued.

Alt. of the Nonag.	The ☾'s Distance from the Nonagesimal Degree of the Ecliptic.									
	61°	62°	63°	64°	65°	66°	67°	68°	69°	70°
10°	9′ 6″.75	9′ 11″.96	9′ 17″.00	9′ 21″.87	9′ 26″.56	9′ 31″.09	9′ 35″.44	9′ 39″.62	9′ 43″.61	9′ 47″.43
11	10 0 .79	10 6 .51	10 12 .04	10 17 .39	10 22 .55	10 27 .54	10 32 .45	10 36 .89	10 41 .29	10 45 .49
12	10 54 .64	11 0 .87	11 6 .90	11 12 .73	11 18 .36	11 23 .77	11 28 .98	11 33 .98	11 38 .77	11 43 .34
13	11 48 .29	11 55 .03	12 1 .56	12 7 .86	12 13 .95	12 19 .81	12 25 .45	12 30 .86	12 36 .04	12 40 .99
14	12 41 .72	12 48 .97	12 55 .99	13 2 .78	13 9 .32	13 15 .62	13 21 .68	13 27 .50	13 33 .07	13 38 .40
15	13 34 .93	13 42 .68	13 50 .19	13 57 .45	14 4 .45	14 11 .19	14 17 .68	14 23 .90	14 29 .86	14 35 .56
16	14 27 .88	14 36 .14	14 44 .14	14 51 .86	14 59 .32	15 6 .50	15 13 .41	15 20 .04	15 26 .39	15 32 .45
17	15 20 .57	15 29 .34	15 37 .82	15 46 .01	15 53 .92	16 1 .54	16 8 .87	16 15 .90	16 22 .63	16 39 .06
18	16 12 .98	16 22 .24	16 31 .21	16 39 .87	16 48 .23	16 56 .28	17 4 .03	17 11 .46	17 18 .57	17 35 .37
19	17 5 .09	17 14 .86	17 24 .30	17 33 .42	17 42 .23	17 50 .73	17 58 .87	18 6 .70	18 14 .20	18 31 .36
20	17 56 .90	18 7 .15	18 17 .07	18 26 .66	18 35 .91	18 44 .82	18 53 .39	19 1 .61	19 9 .49	19 17 .02
21	18 48 .37	18 59 .11	19 9 .51	19 19 .56	19 29 .25	19 38 .59	19 42 .57	19 56 .19	20 4 .44	20 12 .32
22	19 39 .50	19 50 .73	20 1 .60	20 12 .09	20 22 .23	20 32 .00	20 41 .38	20 50 .38	20 59 .01	21 7 .25
23	20 30 .27	20 41 .98	20 53 .32	21 4 .27	21 14 .84	21 25 .02	21 34 .81	21 44 .20	21 53 .20	22 1 .80
24	21 20 .66	21 32 .86	21 44 .66	21 56 .06	22 7 .06	22 17 .66	22 27 .85	22 37 .13	22 47 .00	22 55 .95
25	22 10 .67	22 23 .34	22 35 .60	22 47 .45	22 58 .88	23 9 .89	23 20 .48	23 30 .63	23 40 .37	23 49 .67
26	23 0 .27	23 13 .41	23 26 .13	23 38 .42	23 50 .28	24 1 .70	24 12 .68	24 23 .21	24 33 .32	24 42 .96
27	23 49 .45	24 3 .06	24 16 .23	24 28 .96	24 41 .24	24 53 .07	25 4 .44	25 15 .36	25 25 .81	25 35 .80
28	24 38 .19	24 52 .26	25 5 .88	25 19 .05	25 31 .75	25 43 .97	25 55 .74	26 7 .03	26 17 .84	26 28 .17
29	25 26 .49	25 41 .02	25 55 .08	26 8 .63	26 21 .79	26 34 .42	26 46 .57	26 58 .23	27 9 .39	27 20 .06
30	26 14 .31	26 29 .29	26 43 .80	26 57 .82	27 11 .35	27 24 .38	27 36 .90	27 48 .93	28 0 .45	28 11 .45
31	27 1 .67	27 17 .11	27 32 .05	27 46 .49	28 0 .42	28 13 .84	28 26 .74	28 39 .13	28 50 .99	29 2 .32
32	27 48 .52	28 4 .40	28 19 .78	28 34 .64	28 48 .97	29 2 .77	29 16 .05	29 28 .80	29 41 .00	29 52 .66
33	28 34 .87	28 51 .20	29 7 .00	29 22 .27	29 37 .00	29 51 .19	30 4 .83	30 17 .93	30 30 .47	30 42 .46
34	29 20 .68	29 37 .45	29 53 .68	30 9 .35	30 24 .49	30 39 .05	30 53 .06	31 6 .51	31 19 .37	31 31 .68
35	30 5 .98	30 23 .18	30 39 .82	30 55 .90	31 11 .42	31 26 .36	31 40 .73	31 54 .52	32 7 .72	32 20 .35
36	30 50 .71	31 8 .34	31 25 .39	31 41 .87	31 57 .77	32 13 .08	32 27 .81	32 41 .94	32 55 .47	33 8 .41
37	31 34 .89	31 52 .94	32 10 .39	32 27 .27	32 43 .55	32 59 .23	33 14 .30	33 28 .77	33 42 .62	33 55 .88
38	32 18 .49	32 36 .95	32 54 .81	33 12 .06	33 28 .72	33 44 .76	34 0 .19	34 14 .99	34 29 .17	34 42 .72
39	33 1 .50	33 20 .37	33 38 .63	33 56 .27	34 13 .29	34 29 .69	34 45 .46	35 0 .58	35 15 .07	35 28 .92
40	33 43 .90	34 3 .17	34 21 .82	34 39 .84	34 57 .23	35 13 .97	35 30 .08	35 45 .54	36 0 .34	36 14 .48
41	34 25 .69	34 45 .36	35 4 .39	35 22 .78	35 40 .58	35 57 .62	36 14 .05	36 29 .83	36 44 .93	36 59 .38
42	35 6 .85	35 26 .90	35 46 .32	36 5 .08	36 23 .18	36 40 .61	36 57 .38	37 13 .46	37 28 .87	37 43 .60
43	35 47 .37	36 7 .82	36 27 .59	36 46 .71	37 5 .16	37 22 .93	37 40 .01	37 56 .42	38 12 .12	38 27 .13
44	36 27 .22	36 48 .04	37 8 .20	37 27 .67	37 46 .47	38 4 .56	38 21 .97	38 38 .67	38 54 .67	39 9 .96
45	37 6 .42	37 27 .62	37 48 .13	38 7 .96	38 27 .08	38 45 .51	39 3 .22	39 20 .23	39 36 .51	39 52 .07
46	37 44 .94	38 6 .50	38 27 .37	38 47 .53	39 7 .00	39 25 .74	39 43 .76	40 1 .06	40 17 .62	40 33 .45
47	38 22 .77	38 44 .69	39 5 .91	39 26 .41	39 46 .19	40 5 .25	40 23 .57	40 41 .16	40 58 .00	41 14 .09
48	38 59 .89	39 22 .17	39 43 .73	40 4 .57	40 24 .67	40 44 .03	41 2 .65	41 20 .52	41 37 .63	41 53 .98
49	39 36 .30	39 58 .93	40 20 .82	40 41 .98	41 2 .39	41 22 .06	41 40 .97	41 59 .12	42 16 .50	42 33 .10
50	40 11 .99	40 34 .95	40 57 .18	41 18 .66	41 39 .38	41 59 .33	42 18 .53	42 36 .95	42 54 .59	43 11 .45
51	40 46 .94	41 10 .25	41 32 .79	41 54 .58	42 15 .59	42 35 .85	42 55 .32	43 14 .01	43 31 .91	43 49 .00
52	41 21 .15	41 44 .77	42 7 .64	42 29 .73	42 51 .05	43 11 .58	43 31 .32	43 50 .27	44 8 .42	44 25 .76
53	41 54 .60	42 18 .55	42 41 .72	43 4 .11	43 25 .71	43 46 .52	44 6 .53	44 25 .74	44 44 .13	45 1 .70
54	42 27 .30	42 51 .55	43 15 .02	43 37 .70	43 59 .59	44 20 .67	44 40 .93	45 0 .38	45 19 .02	45 36 .82
55	42 59 .20	43 23 .77	43 47 .53	44 10 .50	44 32 .66	44 54 .00	45 14 .52	45 34 .22	45 53 .08	46 11 .10
56	43 30 .33	43 55 .18	44 19 .24	44 42 .48	45 4 .91	45 26 .51	45 47 .28	46 7 .21	46 26 .30	46 44 .55
57	44 0 .66	44 25 .81	44 50 .14	45 13 .65	45 36 .34	45 58 .19	46 19 .20	46 39 .37	46 58 .68	47 17 .13
58	44 30 .19	44 55 .61	45 20 .22	45 43 .99	46 6 .93	46 29 .03	46 50 .28	47 10 .67	47 30 .20	47 48 .86
59	44 58 .90	45 24 .60	45 49 .47	46 13 .50	46 36 .69	46 59 .02	47 20 .49	47 41 .11	48 0 .85	48 19 .71
60	45 26 .79	45 52 .75	46 17 .88	46 42 .16	47 5 .59	47 28 .15	47 49 .85	48 10 .67	48 30 .62	48 49 .67
61	45 53 .86	46 20 .08	46 45 .45	47 9 .97	47 33 .62	47 56 .42	48 18 .33	48 39 .36	48 59 .51	49 18 .75
62	46 20 .08	46 46 .55	47 12 .16	47 36 .91	48 0 .80	48 23 .80	48 45 .93	49 7 .16	49 27 .49	49 46 .92
63	46 45 .45	47 12 .16	47 38 .01	48 2 .99	48 27 .08	48 50 .30	49 12 .62	49 34 .05	49 54 .57	50 14 .18
64	47 9 .97	47 36 .91	48 2 .99	48 28 .18	48 52 .50	49 15 .92	49 38 .44	50 0 .05	50 20 .75	50 40 .52
65	47 33 .62	48 0 .80	48 27 .08	48 52 .50	49 17 .01	49 40 .63	50 3 .34	50 25 .13	50 46 .00	51 5 .94
66	47 56 .42	48 23 .80	48 50 .30	49 15 .92	49 40 .63	50 4 .44	50 27 .33	50 49 .30	51 10 .33	51 30 .43
67	48 18 .33	48 45 .93	49 12 .62	49 38 .44	50 3 .34	50 27 .33	50 50 .38	51 12 .52	51 33 .71	51 53 .96
68	48 39 .36	49 7 .16	49 34 .05	50 0 .05	50 25 .13	50 49 .30	51 12 .52	51 34 .81	51 56 .15	52 16 .56
69	48 59 .51	49 27 .49	49 54 .57	50 20 .75	50 46 .00	51 10 .33	51 33 .71	51 56 .15	52 17 .65	52 38 .20
70	49 18 .75	49 46 .92	50 14 .18	50 40 .52	51 5 .94	51 30 .43	51 53 .96	52 16 .56	52 38 .20	52 58 .88

TABLE 13.

The Moon's approximate Parallax in Longitude, continued.

Alt. of the Nonag.	The ☾'s Distance from the Nonagesimal Degree of the Ecliptic.									
	71°	72°	73°	74°	75°	76°	77°	78°	79°	80°
10°	9′ 51″.08	9′ 54″.54	9′ 57″.82	10′ 0″.92	10′ 3″.83	10′ 6″.56	10′ 9″.11	10′ 11″.45	10′ 13″.65	10′ 15″.64
11	10 49 .49	10 53 .29	10 56 .90	11 0 .30	11 3 .50	11 6 .50	11 9 .31	11 11 .90	11 14 .29	10 16 .47
12	11 47 .70	11 51 .85	11 55 .78	11 59 .49	12 2 .98	12 6 .25	12 9 .30	12 12 .13	12 14 .73	12 17 .11
13	12 45 .70	12 50 .18	12 54 .44	12 58 .45	13 2 .22	13 5 .77	13 9 .07	13 12 .13	13 14 .94	13 17 .52
14	13 43 .47	13 48 .29	13 52 .86	13 57 .18	14 1 .24	14 5 .05	14 8 .60	14 11 .89	14 14 .92	14 17 .69
15	14 40 .98	14 47 .14	14 51 .03	14 55 .65	15 0 .00	15 4 .07	15 7 .87	15 11 .39	15 14 .63	15 17 .59
16	15 38 .23	15 43 .73	15 48 .94	15 53 .85	15 48 .48	16 2 .82	16 6 .86	16 10 .61	16 14 .06	16 17 .22
17	16 35 .18	16 41 .02	16 44 .53	16 51 .76	16 56 .68	17 1 .28	17 5 .56	17 9 .54	17 13 .20	17 16 .55
18	17 31 .85	17 38 .01	17 43 .85	17 49 .37	17 54 .56	17 59 .42	18 3 .95	18 8 .15	18 12 .02	18 15 .56
19	18 28 .19	18 34 .68	18 40 .83	18 46 .64	18 52 .11	18 57 .23	19 2 .00	19 6 .43	19 10 .51	19 14 .24
20	19 24 .19	19 31 .01	19 37 .47	19 43 .57	19 49 .32	19 54 .70	19 59 .71	20 4 .37	20 8 .65	20 12 .57
21	20 19 .83	20 26 .97	20 33 .76	20 40 .14	20 46 .16	20 51 .80	20 57 .05	21 11 .93	21 6 .42	21 10 .52
22	21 15 .11	21 22 .58	21 29 .66	21 36 .34	21 42 .63	21 48 .53	21 54 .02	21 59 .11	22 3 .81	22 8 .10
23	22 9 .99	22 17 .78	22 25 .17	22 33 .14	22 38 .70	22 44 .85	22 50 .58	22 55 .89	23 0 .79	23 5 .26
24	23 4 .48	23 12 .59	23 20 .27	23 27 .53	23 34 .36	23 40 .76	23 46 .72	23 52 .25	23 57 .35	24 2 .01
25	23 58 .54	24 6 .96	24 14 .94	24 22 .48	24 29 .58	24 36 .23	24 42 .43	24 48 .18	24 53 .47	24 58 .31
26	24 52 .16	25 0 .90	25 9 .18	25 17 .00	25 24 .36	25 31 .26	25 37 .69	25 43 .65	25 49 .14	25 54 .16
27	25 45 .32	25 54 .38	26 2 .95	26 11 .05	26 18 .67	26 25 .81	26 32 .47	26 38 .65	26 44 .34	26 49 .54
28	26 38 .02	26 47 .38	26 56 .25	27 4 .63	27 12 .51	27 19 .89	27 26 .78	27 33 .17	27 39 .05	27 44 .42
29	27 30 .22	27 39 .89	27 49 .05	27 57 .70	28 5 .85	28 13 .47	28 20 .58	28 27 .18	28 33 .25	28 38 .79
30	28 21 .93	28 31 .90	28 41 .35	28 50 .27	28 58 .67	29 6 .53	29 13 .87	29 20 .67	29 26 .93	29 32 .65
31	29 13 .12	29 23 .39	29 33 .12	29 42 .31	29 50 .96	29 59 .05	30 6 .62	30 13 .62	30 20 .07	30 25 .97
32	30 3 .77	30 14 .34	30 24 .35	30 33 .81	30 42 .71	31 51 .04	30 58 .81	31 6 .02	31 12 .66	31 18 .73
33	30 53 .87	31 4 .74	31 15 .02	31 24 .75	31 33 .90	32 42 .46	31 50 .44	31 57 .85	32 4 .67	32 10 .91
34	31 43 .42	31 54 .57	32 5 .13	32 15 .11	32 24 .50	32 33 .30	32 41 .50	32 49 .10	32 56 .11	33 2 .51
35	32 32 .38	32 43 .81	32 54 .65	33 4 .88	33 14 .51	33 23 .53	33 31 .95	33 39 .75	33 46 .94	33 53 .50
36	33 20 .74	33 32 .46	33 43 .57	33 54 .06	34 3 .93	34 13 .17	34 21 .79	34 29 .79	34 37 .15	34 43 .88
37	34 8 .49	34 20 .49	34 26 .87	34 42 .61	34 52 .71	35 2 .18	35 11 .00	35 19 .19	35 26 .72	35 33 .61
38	34 55 .63	35 7 .90	35 19 .54	35 30 .52	35 40 .86	35 50 .55	35 59 .58	36 7 .95	36 15 .66	36 22 .71
39	35 42 .12	35 54 .67	36 1 .56	36 17 .78	36 28 .35	36 38 .26	36 47 .49	36 56 .05	37 3 .93	37 11 .13
40	36 27 .96	36 40 .78	36 52 .92	37 4 .38	37 15 .19	37 25 .30	37 54 .73	37 43 .47	37 51 .52	37 58 .88
41	37 13 .13	37 26 .22	37 38 .60	37 50 .31	38 1 .34	38 11 .66	38 21 .28	38 30 .20	38 38 .41	38 45 .92
42	37 57 .63	38 10 .97	38 23 .61	38 35 .55	38 46 .79	38 57 .32	39 7 .13	39 16 .23	39 24 .61	39 32 .27
43	38 41 .43	38 55 .02	39 7 .91	39 20 .08	39 31 .53	39 42 .26	39 52 .26	40 1 .54	40 10 .18	40 17 .89
44	39 24 .53	39 38 .37	39 51 .50	40 3 .89	40 15 .56	40 26 .49	40 36 .68	40 46 .12	40 54 .82	41 2 .78
45	40 6 .90	40 20 .99	40 34 .35	40 46 .97	40 58 .85	41 9 .97	41 20 .34	41 29 .95	41 38 .81	41 26 .91
46	40 48 .54	41 2 .88	41 16 .47	41 29 .31	41 41 .38	41 52 .70	42 3 .25	42 13 .03	42 22 .05	42 30 .28
47	41 29 .43	41 44 .01	41 57 .82	42 10 .88	42 23 .16	42 34 .66	42 45 .39	42 55 .33	43 4 .50	43 12 .87
48	42 9 .57	42 24 .38	42 38 .42	42 51 .68	43 4 .16	43 15 .85	43 26 .75	43 36 .86	43 46 .17	43 54 .68
49	42 48 .93	43 3 .97	43 18 .23	43 31 .70	43 44 .37	43 56 .25	44 7 .31	44 17 .59	44 27 .04	44 35 .68
50	43 27 .51	43 42 .79	43 57 .26	44 10 .93	44 23 .79	44 35 .84	44 47 .08	44 57 .50	45 7 .09	45 15 .86
51	44 5 .29	44 20 .80	44 35 .48	44 49 .34	45 2 .39	45 14 .61	45 26 .02	45 36 .59	45 46 .32	45 55 .22
52	44 42 .28	44 57 .99	45 12 .88	45 26 .94	45 40 .18	45 52 .57	46 4 .13	46 14 .85	46 24 .72	46 33 .74
53	45 18 .44	45 34 .37	45 49 .46	46 3 .71	46 17 .12	46 29 .69	46 41 .40	46 52 .26	47 2 .26	47 11 .40
54	45 53 .79	46 9 .91	46 25 .20	46 39 .64	46 53 .22	47 5 .95	47 17 .81	47 28 .82	47 38 .95	47 48 .21
55	46 28 .28	46 44 .61	47 0 .09	47 14 .71	47 28 .46	47 41 .35	47 53 .36	48 4 .51	48 14 .76	48 24 .14
56	47 1 .93	47 18 .46	47 34 .13	47 48 .92	48 2 .84	48 15 .88	48 28 .04	48 39 .32	48 49 .70	48 59 .19
57	47 34 .71	47 51 .44	48 7 .29	48 22 .25	48 36 .33	48 49 .53	49 1 .83	49 13 .24	49 23 .74	49 33 .34
58	48 6 .64	48 23 .55	48 39 .57	48 54 .71	49 8 .95	49 22 .29	49 34 .73	49 46 .26	49 56 .88	50 6 .59
59	48 37 .69	48 54 .77	49 10 .96	49 26 .26	49 40 .65	49 54 .14	50 6 .71	50 18 .37	50 29 .10	50 38 .92
60	49 7 .84	49 25 .10	49 41 .46	49 56 .92	50 11 .46	50 25 .08	50 37 .78	50 49 .56	51 0 .41	51 10 .33
61	49 37 .09	49 54 .53	50 11 .05	50 26 .66	50 41 .34	50 55 .09	51 7 .93	51 19 .32	51 30 .78	51 40 .80
62	50 5 .44	50 23 .04	50 39 .72	50 55 .48	51 10 .30	51 24 .19	51 37 .14	51 49 .15	52 0 .21	52 10 .32
63	50 32 .87	50 50 .62	51 7 .47	51 23 .36	51 38 .32	51 52 .34	52 5 .41	52 17 .53	52 28 .69	52 38 .89
64	50 59 .38	51 17 .29	51 34 .28	51 50 .31	52 5 .41	52 19 .55	52 32 .73	52 44 .95	52 56 .21	53 6 .50
65	51 24 .95	51 43 .01	52 0 .15	52 16 .31	52 31 .53	52 45 .79	52 59 .09	53 11 .41	53 22 .76	53 33 .14
66	51 49 .59	52 7 .80	52 25 .06	52 41 .36	52 56 .70	53 11 .07	53 24 .47	53 36 .90	53 48 .34	53 58 .80
67	52 13 .28	52 31 .63	52 49 .01	53 5 .44	53 20 .90	53 35 .38	53 48 .88	54 1 .40	54 12 .93	54 23 .47
68	52 36 .01	52 54 .50	53 12 .01	53 28 .56	53 44 .13	53 58 .71	54 12 .31	54 24 .92	54 36 .54	54 47 .15
69	52 57 .78	53 16 .40	53 34 .03	53 50 .69	54 6 .38	54 21 .06	54 34 .75	54 47 .44	54 59 .14	55 9 .82
70	53 18 .59	53 37 .32	53 55 .08	54 11 .85	54 27 .62	54 42 .41	54 56 .10	55 8 .97	55 20 .74	55 31 .50

TABLE 13.

The Moon's approximate Parallax in Longitude, concluded.

Alt. of the Nonag.	The ☽'s Distance from the Nonagesimal Degree of the Ecliptic.									
	81°	82°	83°	84°	85°	86°	87°	88°	89°	90°
10°	10′ 17″.44	10′ 19″.05	10′ 20″.47	10′ 21″.71	10′ 22″.75	10′ 23″.61	10′ 24″.28	10′ 24″.75	10′ 25″.04	10′ 25″.13
11	11 18.45	11 20.22	11 21.78	11 23.14	11 24.39	11 25.23	11 25.97	11 26.49	11 26.80	11 26.91
12	12 19.27	12 21.20	12 22.90	12 24.38	12 25.63	12 26.66	12 27.46	12 28.03	12 28.37	12 28.48
13	13 19.85	13 21.94	13 23.78	13 25.38	13 26.73	13 27.85	13 28.71	13 29.33	13 29.70	13 29.82
14	14 20.20	14 22.44	14 24.43	14 26.15	14 27.60	14 28.80	14 29.73	14 30.39	14 30.79	14 30.92
15	15 20.28	15 22.68	15 24.80	15 26.64	15 28.20	15 29.40	15 30.47	15 31.18	15 31.60	15 31.75
16	16 20.08	16 22.64	16 24.90	16 28.86	16 28.52	16 29.88	16 30.93	16 31.69	16 32.14	16 32.29
17	17 19.58	17 22.30	17 24.69	17 26.77	17 28.53	17 29.97	17 31.09	17 31.90	17 32.37	17 32.54
18	18 18.76	18 21.64	18 24.17	18 26.37	18 28.23	18 29.75	18 30.94	18 31.79	18 32.29	18 32.46
19	19 17.61	19 20.64	19 23.31	19 25.63	19 27.59	19 29.19	19 30.44	19 31.33	19 31.86	19 32.05
20	20 16.11	20 19.29	20 22.10	20 24.53	20 26.59	20 28.27	20 29.59	20 30.52	20 31.08	20 31.27
21	21 14.24	21 17.57	21 20.51	21 23.06	21 25.22	21 26.98	21 28.36	21 29.34	21 29.93	21 30.12
22	22 11.98	22 15.46	22 18.53	22 21.20	22 23.45	22 25.30	22 26.74	22 27.76	22 28.38	22 28.58
23	23 9.31	23 12.94	23 16.14	23 18.92	23 21.27	23 23.20	23 24.70	23 25.77	23 26.41	23 26.63
24	24 6.22	24 10.00	24 13.34	24 16.23	24 18.68	24 20.69	24 22.25	24 23.36	24 24.03	24 24.29
25	25 2.69	25 6.62	25 10.08	25 13.09	25 15.63	25 17.72	25 19.34	25 20.50	25 21.19	25 21.43
26	25 58.71	26 2.78	26 6.37	26 9.49	26 12.13	26 14.29	26 15.97	26 17.18	26 17.90	26 18.14
27	26 54.25	26 58.46	27 2.18	27 5.41	27 8.15	27 10.38	27 12.12	27 13.37	27 14.12	27 14.37
28	27 49.29	27 53.65	27 57.50	28 0.84	28 3.67	28 5.98	28 7.78	28 9.07	28 9.84	28 10.10
29	28 43.82	28 48.32	28 52.30	28 55.75	28 58.68	29 1.07	29 2.92	29 4.25	29 5.05	29 5.31
30	29 37.84	29 42.48	29 46.58	29 50.14	29 53.15	29 55.62	29 57.53	29 58.90	29 59.73	30 0.00
31	30 31.31	30 36.09	30 40.31	30 43.98	30 47.08	30 49.62	30 51.59	30 53.00	30 53.85	30 54.14
32	31 24.22	31 29.14	31 33.49	31 37.26	31 40.45	31 43.06	31 45.09	31 46.55	31 47.42	31 47.71
33	32 16.55	32 81.61	32 26.09	32 29.96	32 33.24	32 35.92	32 38.01	32 39.51	32 40.40	32 40.70
34	33 8.31	33 13.50	33 18.09	33 22.07	33 25.43	33 28.19	33 30.34	33 31.87	33 32.79	33 33.09
35	33 59.45	34 4.77	34 9.48	34 13.56	34 17.01	34 19.84	34 22.04	34 23.61	34 24.55	34 24.87
36	34 49.98	34 55.43	35 0.25	35 4.43	35 7.97	35 10.87	35 13.13	35 14.74	35 15.70	35 16.03
37	35 39.86	35 45.45	35 50.38	35 54.66	35 58.29	36 1.25	36 3.56	36 5.21	36 6.20	36 6.53
38	36 29.09	36 34.81	36 39.86	36 44.24	36 47.95	36 50.98	36 53.34	36 55.03	36 56.04	36 56.38
39	37 17.66	37 23.50	37 28.65	37 33.14	37 36.94	37 40.04	27 42.45	37 44.18	37 45.21	37 45.55
40	38 5.55	38 11.52	38 16.79	38 21.36	38 25.23	38 28.40	38 30.86	38 32.63	38 33.68	38 34.04
41	38 52.74	38 58.83	39 4.21	39 8.87	39 12.82	39 16.06	39 18.57	39 20.37	39 21.45	39 21.81
42	39 39.21	39 45.43	39 50.92	39 55.67	39 59.70	40 3.00	40 5.57	40 7.40	40 8.50	40 8.87
43	40 24.96	40 31.30	40 36.89	40 41.74	40 45.84	40 49.21	40 51.82	40 53.69	40 54.81	40 55.20
44	41 9.98	41 16.43	41 22.13	41 27.07	41 31.25	41 34.68	41 37.34	41 39.25	41 40.39	41 40.77
45	41 54.24	42 0.81	42 6.61	42 11.62	42 15.90	42 19.39	42 22.09	42 24.04	42 25.20	42 25.58
46	42 37.74	42 44.42	42 50.32	42 55.44	42 59.77	43 3.32	43 6.07	43 8.05	43 9.23	43 9.62
47	43 20.46	43 27.25	43 33.25	43 38.45	43 42.85	43 46.46	43 49.26	43 51.27	43 52.47	43 52.87
48	44 2.38	44 49.29	44 15.38	44 20.67	44 27.14	44 28.81	44 31.66	44 33.69	44 34.91	44 35.32
49	44 43.50	45 30.51	44 56.70	45 2.07	45 6.61	45 10.33	45 13.23	45 15.29	45 16.53	45 16.95
50	45 23.81	45 30.92	45 37.20	45 42.65	45 47.27	45 51.04	45 53.98	45 56.08	45 57.34	45 57.76
51	46 3.28	46 10.49	46 16.86	46 22.39	46 27.08	46 30.91	46 33.79	46 36.02	46 37.30	46 37.73
52	46 41.91	46 49.23	46 55.69	47 1.30	47 6.04	47 9.93	47 12.95	47 15.11	47 16.41	47 16.84
53	47 19.68	47 27.10	47 33.66	47 39.34	47 44.15	47 48.09	47 51.15	47 53.34	47 54.65	47 55.09
54	47 56.60	48 4.12	48 10.75	48 16.51	48 21.38	48 25.37	48 28.47	48 30.69	48 32.02	48 32.46
55	48 32.63	48 40.24	48 46.97	48 52.80	48 57.73	49 1.77	49 4.90	49 7.15	49 8.50	49 8.95
56	49 7.79	49 15.49	49 22.29	49 28.19	49 33.18	49 37.27	49 40.44	49 42.72	49 44.08	49 44.54
57	49 42.05	49 49.82	49 56.71	50 2.68	50 7.72	50 11.86	50 15.07	50 17.37	50 18.75	50 19.21
58	50 15.39	50 23.26	50 30.22	50 36.25	50 41.35	50 45.54	50 48.79	50 51.11	50 52.51	50 52.97
59	50 47.82	50 55.76	51 2.80	51 8.90	51 14.06	51 18.29	51 21.57	51 23.92	51 25.33	51 25.80
60	51 19.31	51 27.35	51 34.45	51 40.61	51 45.83	51 50.10	51 53.42	51 55.79	51 57.22	51 57.69
61	51 49.86	51 57.98	52 5.16	52 11.38	52 16.65	52 20.96	52 24.31	52 26.71	52 28.15	52 28.63
62	52 19.48	52 27.68	52 34.92	52 41.20	52 46.52	52 50.87	52 54.25	52 56.68	52 58.13	52 58.61
63	52 48.13	52 56.40	53 3.72	53 10.05	53 15.42	53 19.81	53 23.22	53 25.67	53 27.14	53 27.62
64	53 15.82	53 24.17	53 31.54	53 37.93	53 43.35	53 47.78	53 51.22	53 53.69	53 55.17	53 55.66
65	53 42.53	53 50.95	53 58.38	53 54.83	54 10.30	54 14.76	54 18.24	54 20.72	54 22.21	54 22.71
66	54 8.27	54 16.76	54 24.25	54 30.75	54 36.25	54 40.75	54 44.26	54 46.76	54 48.26	54 48.76
67	54 33.02	54 41.57	54 49.11	54 55.66	55 1.20	55 5.74	55 9.27	55 11.79	55 13.30	55 13.82
68	54 56.77	55 5.38	55 12.98	55 19.58	55 25.16	55 29.73	55 33.29	55 35.83	55 37.35	55 37.86
69	55 19.50	55 28.17	55 35.84	55 42.48	55 48.10	55 52.70	55 56.28	55 58.84	56 0.37	56 0.89
70	55 41.24	55 49.97	55 57.68	55 4.36	56 10.02	56 14.65	56 18.26	56 20.83	56 22.38	56 22.89

TABLE 14. The MOON'S APPROXIMATE PARALLAX in LATITUDE, for every 10′ of the Alt. of the Nonag. Degree, and for 60′ of Hor. Par. in the Spheroid.

Degrees.	Argument = the Altitude of the Nonagesimal Degree of the Ecliptic.						Degrees.
	Minutes.						
	0′	10′	20′	30′	40′	50′	
10°	59′ 5″.31	59′ 3′.47	59′ 1″.61	58′ 59″.71	58′ 57″.79	58′ 55′.84	10°
11	58 53 .86	58 51′.84	58 49 .80	58 47 .73	58 45 .63	58 43 .49	11
12	58 41 .33	58 39 .14	58 36 .92	58 34 .67	58 32 .38	58 30 .07	12
13	58 27 .73	58 25 .36	58 22 .96	58 20 .53	58 18 .07	58 15 .58	13
14	58 13 .06	58 10 .52	58 7 .94	58 5 .33	58 2 .69	58 0 .03	14
15	57 57 .33	57 54 .61	57 51 .85	57 49 .07	57 46 .26	57 43 .41	15
16	57 40 .54	57 37 .64	57 34 .71	57 31 .75	57 28 .76	57 25 .74	16
17	57 22 .70	57 19 .62	57 16 .52	57 13 .38	57 10 .22	57 7 .03	17
18	57 3 .80	57 0 .55	56 57 .27	56 53 .97	56 50 .63	56 47 .26	18
19	56 43 .87	56 40 .44	56 36 .99	56 33 .51	56 30 .00	56 26 .46	19
20	56 22 .89	56 19 .30	56 15 .68	56 12 .02	56 8 .34	56 4 .63	20
21	56 0 .89	55 57 .12	55 53 .33	55 49 .51	55 45 .65	55 41 .77	21
22	55 37 .86	55 33 .92	55 29 .96	55 25 .97	55 21 .94	55 17 .90	22
23	55 13 .82	55 9 .71	55 5 .58	55 1 .42	54 57 .23	54 53 .01	23
24	54 48 .76	54 44 .49	54 40 .19	54 35 .86	54 31 .50	54 27 .12	24
25	54 22 .71	54 18 .27	54 13 .80	54 9 .31	54 4 .79	54 0 .24	25
26	53 55 .66	53 51 .05	53 46 .42	53 41 .76	53 37 .08	53 32 .36	26
27	53 27 .62	53 22 .86	53 18 .06	53 13 .24	53 8 .39	53 3 .51	27
28	52 58 .61	52 53 .68	52 48 .73	52 43 .74	52 38 .73	52 33 .69	28
29	52 28 .63	52 23 .54	52 18 .42	52 13 .28	52 8 .11	52 2 .91	29
30	51 57 .69	51 52 .44	51 47 .17	51 41 .87	51 36 .54	51 31 .18	30
31	51 25 .80	51 20 .40	51 14 .96	51 9 .50	51 4 .02	50 58 .51	31
32	50 52 .97	50 47 .41	50 41 .82	50 36 .21	50 30 .57	50 24 .90	32
33	50 19 .21	50 13 .50	50 7 .76	50 1 .99	49 56 .20	49 50 .38	33
34	49 44 .54	49 38 .67	49 32 .77	49 26 .85	49 20 .91	49 14 .94	34
35	49 8 .95	49 2 .93	48 56 .88	48 50 .82	48 44 .72	48 38 .60	35
36	48 32 .46	48 26 .29	48 20 .10	48 13 .88	48 7 .64	48 1 .38	36
37	47 55 .09	47 48 .78	47 42 .43	47 36 .07	47 29 .69	47 23 .27	37
38	47 16 .84	47 10 .38	47 3 .90	46 57 .39	46 50 .86	46 44 .30	38
39	46 37 .73	46 31 .12	46 24 .50	46 17 .85	46 11 .18	46 4 .48	39
40	45 57 .76	45 51 .02	45 44 .25	45 37 .46	45 30 .65	45 23 .81	40
41	45 16 .95	45 10 .07	45 3 .17	44 56 .24	44 49 .29	44 42 .32	41
42	44 35 .32	44 28 .30	44 21 .26	44 14 .20	44 7 .11	44 0 .00	42
43	43 52 .87	43 45 .72	43 38 .54	43 31 .35	43 24 .13	43 16 .89	43
44	43 9 .62	43 2 .34	42 55 .03	42 47 .70	42 40 .35	42 32 .98	44
45	42 25 .58	42 18 .17	42 10 .73	42 3 .27	41 55 .79	41 48 .29	45
46	41 40 .77	41 33 .23	41 25 .66	41 18 .08	41 10 .47	41 2 .84	46
47	40 55 .19	40 47 .53	40 39 .84	40 32 .12	40 24 .39	40 16 .64	47
48	40 8 .87	40 1 .08	39 53 .27	39 45 .43	39 37 .58	39 29 .71	48
49	39 21 .81	39 13 .90	39 5 .96	38 58 .01	38 50 .04	38 42 .05	49
50	38 34 .04	38 26 .00	38 17 .95	38 9 .88	38 1 .79	37 53 .68	50
51	37 45 .55	37 37 .41	37 29 .24	37 21 .05	37 12 .85	37 4 .62	51
52	36 56 .38	36 48 .12	36 39 .84	36 31 .54	36 23 .22	36 14 .89	52
53	36 6 .53	35 58 .16	35 49 .77	35 41 .36	35 32 .93	35 24 .49	53
54	35 16 .03	35 7 .55	34 59 .05	34 50 .53	34 42 .00	34 33 .44	54
55	34 24 .88	34 16 .29	34 7 .68	33 59 .06	33 50 .42	33 41 .77	55
56	33 33 .09	33 24 .40	33 15 .70	33 6 .97	32 58 .23	32 49 .47	56
57	32 40 .70	32 31 .91	32 23 .10	32 14 .28	32 5 .44	31 56 .58	57
58	31 47 .71	31 38 .82	31 29 .92	31 21 .00	31 12 .06	31 3 .11	58
59	30 54 .14	30 45 .15	30 36 .15	30 27 .14	30 18 .11	30 9 .06	59
60	30 0 .00	29 50 .92	29 41 .83	29 32 .72	29 23 .60	29 14 .47	60
61	29 5 .31	28 56 .15	28 46 .97	28 37 .77	28 28 .56	28 19 .34	61
62	28 10 .10	28 0 .84	27 51 .58	27 42 .29	27 33 .00	27 23 .69	62
63	27 14 .37	27 5 .03	26 25 .68	26 46 .31	26 36 .93	26 27 .54	63
64	26 18 .14	26 8 .72	25 59 .29	25 49 .84	25 40 .38	25 30 .91	64
65	25 21 .43	25 11 .93	25 2 .42	24 52 .90	24 43 .36	24 33 .81	65
66	24 24 .25	24 14 .68	24 5 .09	23 55 .50	23 45 .89	23 36 .27	66
67	23 26 .63	23 16 .99	23 7 .33	22 57 .66	22 47 .98	22 38 .29	67
68	22 28 .58	22 18 .87	22 9 .14	21 59 .40	21 49 .66	21 39 .90	68
69	21 30 .12	21 20 .34	21 10 .55	21 0 .75	20 50 .93	20 41 .11	69
70	20 31 .27	20 21 .43	20 11 .57	20 1 .70	19 51 .83	19 41 .94	70

TABLE 15. Augmentation of the Moon's approximate Parallax in Longitude; or Diminution of the Moon's approximate Parallax in Latitude.

☾'s Lat.	10′	20′	30′	40′	50′	60′	Diff. for 10′ of Parallax.
0° 10′	0″.00	0″.01	0″.01	0″.01	0″.01	0″.02	0″.003
20	0 .01	0 .02	0 .03	0 .04	0 .05	0 .06	0 .010
30	0 .02	0 .05	0 .07	0 .09	0 .11	0 .14	0 .023
40	0 .04	0 .08	0 .12	0 .16	0 .20	0 .24	0 .041
50	0 .06	0 .13	0 .19	0 .25	0 .32	0 .38	0 .063
1 0	0 .09	0 .18	0 .27	0 .37	0 .46	0 .55	0 .091
10	0 .12	0 .25	0 .37	0 .50	0 .62	0 .75	0 .124
20	0 .16	0 .32	0 .49	0 .65	0 .81	0 .97	0 .162
30	0 .21	0 .41	0 .62	0 .82	1 .03	1 .23	0 .206
40	0 .25	0 .51	0 .76	1 .02	1 .27	1 .52	0 .254
50	0 .31	0 .61	0 .92	1 .23	1 .54	1 .84	0 .307
2 0	0 .37	0 .73	1 .10	1 .46	1 .83	2 .19	0 .366
10	0 .43	0 .86	1 .29	1 .72	2 .15	2 .58	0 .429
20	0 .50	1 .00	1 .49	1 .99	2 .49	2 .99	0 .498
30	0 .57	1 .14	1 .71	2 .29	2 .86	3 .43	0 .572
40	0 .65	1 .30	1 .95	2 .60	3 .25	3 .90	0 .650
50	0 .73	1 .47	2 .20	2 .94	3 .67	4 .41	0 .734
3 0	0 .82	1 .65	2 .47	3 .29	4 .12	4 .94	0 .823
10	0 .92	1 .84	2 .75	3 .67	4 .59	5 .51	0 .918
20	1 .02	2 .03	3 .05	4 .07	5 .08	6 .10	1 .017
30	1 .12	2 .24	3 .36	4 .48	5 .61	6 .73	1 .121
40	1 .23	2 .46	3 .69	4 .92	6 .15	7 .38	1 .231
50	1 .35	2 .69	4 .04	5 .38	6 .73	8 .07	1 .345
4 0	1 .47	2 .93	4 .40	5 .86	7 .33	8 .79	1 .465
10	1 .59	3 .18	4 .77	6 .36	7 .95	9 .54	1 .590
20	1 .72	3 .44	5 .16	6 .88	8 .60	10 .32	1 .720
30	1 .86	3 .71	5 .57	7 .42	9 .28	11 .13	1 .855
40	2 .00	3 .99	5 .99	7 .98	9 .98	11 .97	1 .996
50	2 .14	4 .28	6 .42	8 .56	10 .71	12 .85	2 .141
5 0	2 .29	4 .58	6 .88	9 .17	11 .46	13 .75	2 .292
10	2 .45	4 .90	7 .34	9 .79	12 .24	14 .69	2 .448
20	2 .61	5 .22	7 .83	10 .44	13 .04	15 .65	2 .609
30	2 .78	5 .55	8 .33	11 .10	13 .88	16 .65	2 .775

Arguments for the Augmentation of the ☾'s approximate Par. in Long. = the ☾'s Lat. at the Side; and the approx. Par. in Long. once corrected by repetition, at the Head. Arguments for the Diminution of the ☾'s approx. Par. in Lat. = the ☾'s Lat. at the Side; and approx. Par. in Lat. at the Head.

TABLE 16. The Second Correction of the Moon's Parallax in Latitude, subtractive.

☾'s Lat.	10′	20′	30′	40′	50′	60′	Diff. for 10′ Par.
0° 10′	0′ 1″.75	0′ 3″.49	0′ 5″.24	0′ 6″.98	0′ 8″.73	0′ 10″.47	1″.745
20	0 3 .49	0 6 .98	0 10 .47	0 13 .96	0 17 .45	0 20 .94	3 .491
30	0 5 .24	0 10 .47	0 15 .71	0 20 .94	0 26 .18	0 31 .42	5 .236
40	0 6 .98	0 13 .96	0 20 .94	0 27 .92	0 34 .91	0 41 .89	6 .981
50	0 8 .73	0 17 .45	0 26 .18	0 34 .91	0 43 .63	0 52 .36	8 .726
1 0	0 10 .47	0 20 .94	0 31 .41	0 41 .89	0 52 .36	1 2 .83	10 .471
10	0 12 .22	0 24 .43	0 36 .65	0 48 .87	1 1 .08	1 13 .30	12 .216
20	0 13 .96	0 27 .92	0 41 .88	0 55 .85	1 9 .81	1 23 .77	13 .961
30	0 15 .71	0 31 .41	0 47 .12	0 62 .82	1 18 .53	1 34 .24	15 .706
40	0 17 .45	0 34 .90	0 52 .35	1 9 .80	1 27 .25	1 44 .70	17 .451
50	0 19 .20	0 38 .39	0 57 .59	1 16 .78	1 35 .98	1 55 .17	19 .195
2 0	0 20 .94	0 41 .88	1 2 .82	1 23 .76	1 44 .70	2 5 .64	20 .940
10	0 22 .68	0 45 .37	1 8 .05	1 30 .74	1 53 .42	2 16 .10	22 .684
20	0 24 .43	0 48 .86	1 13 .28	1 37 .71	2 2 .14	2 26 .57	24 .428
30	0 26 .17	0 52 .34	1 18 .51	1 44 .69	2 10 .86	2 37 .03	26 .172
40	0 27 .92	0 55 .83	1 23 .75	1 51 .66	2 19 .58	2 47 .49	27 .915
50	0 29 .66	0 59 .32	1 28 .98	1 58 .63	2 28 .29	2 57 .95	29 .659
3 0	0 31 .40	1 2 .80	1 34 .20	2 5 .61	2 37 .01	3 8 .41	31 .402
10	0 33 .14	1 6 .29	1 39 .43	2 12 .58	2 45 .72	3 18 .87	33 .144
20	0 34 .89	1 9 .77	1 44 .66	2 19 .55	2 54 .43	3 29 .32	34 .887
30	0 36 .63	1 13 .26	1 49 .89	2 26 .52	3 3 .15	3 39 .77	36 .629
40	0 38 .37	1 16 .74	1 55 .11	2 33 .48	3 11 .86	3 50 .23	38 .371
50	0 40 .11	1 20 .23	2 0 .34	2 40 .45	3 20 .56	4 0 .68	40 .113
4 0	0 41 .85	1 23 .71	2 5 .56	2 47 .42	3 29 .27	4 11 .12	41 .854
10	0 43 .59	1 27 .19	2 10 .78	2 54 .38	3 37 .97	4 21 .57	43 .595
20	0 45 .34	1 30 .67	2 16 .01	3 1 .34	3 46 .68	4 32 .01	45 .335
30	0 47 .08	1 34 .15	2 21 .23	3 8 .30	3 55 .38	4 42 .45	47 .075
40	0 48 .82	1 37 .63	2 26 .45	3 15 .26	4 4 .08	4 52 .89	48 .815
50	0 50 .55	1 41 .11	2 31 .66	3 22 .22	4 12 .77	5 3 .33	50 .555
5 0	0 52 .29	1 44 .59	2 36 .88	3 29 .17	4 21 .47	5 13 .76	52 .293
10	0 54 .03	1 48 .06	2 42 .10	3 36 .13	4 30 .16	5 24 .19	54 .032
20	0 55 .77	1 51 .54	2 47 .31	3 43 .08	4 38 .85	5 34 .62	55 .770
30	0 57 .51	1 55 .01	2 52 .52	3 50 .03	4 47 .54	5 45 .04	57 .507

Arguments = the ☾'s Lat. at the Side; and the Tabular Quantity taken from the Table of the ☾'s approx. Par. in Long. (or Table 13.) by entering it with the Alt. of the Nonag. Deg. and the Complement of the ☾'s distance from thence.

TABLE 17.—CORRECTIONS of the MOON'S PARALLAX in LONGITUDE and in LATITUDE, accordingly as her Hor. Par. is less or greater than 60′.
Arguments = the Moon's Parallax in Long. or Lat., at the Side; and the Difference between her Horizontal Parallax and 60′, at the Head.

	Difference of the ☽'s Horizontal Parallax from 60′.							Difference of the ☽'s Horizontal Parallax from 60′.					
☽'s Par. in Long. or Lat.	Difference for Minutes of the ☽'s Par. in Long.						☽'s Par. in Long. or Lat.	Difference for Seconds of the ☽'s Par. in Long.					
	1′	2′	3′	4′	5′	6′		1′	2′	3′	4′	5′	6′
1′	0′ 1″	0′ 2″	0′ 3″	0′ 4″	0′ 5″	0′ 6″	1″	0″.02	0″.03	0″.05	0″.07	0″.08	0″.10
2	0 2	0 4	0 6	0 8	0 10	0 12	2	0.03	0.07	0.10	0.13	0.17	0.20
3	0 3	0 6	0 9	0 12	0 15	0 18	3	0.05	0.10	0.15	0.20	0.25	0.30
4	0 4	0 8	0 12	0 16	0 20	0 24	4	0.07	0.13	0.20	0.27	0.33	0.40
5	0 5	0 10	0 15	0 20	0 25	0 30	5	0.08	0.17	0.25	0.33	0.42	0.50
6	0 6	0 12	0 18	0 24	0 30	0 36	6	0.10	0.20	0.30	0.40	0.50	0.60
7	0 7	0 14	0 21	0 28	0 35	0 42	7	0.12	0.23	0.35	0.47	0.58	0.70
8	0 8	0 16	0 24	0 32	0 40	0 48	8	0.13	0.27	0.40	0.53	0.67	0.80
9	0 9	0 18	0 27	0 36	0 45	0 54	9	0.15	0.30	0.45	0.60	0.75	0.90
10	0 10	0 20	0 30	0 40	0 50	1 0	10	0.17	0.33	0.50	0.67	0.83	1.00
11	0 11	0 22	0 33	0 44	0 55	1 6	11	0.18	0.37	0.55	0.73	0.92	1.10
12	0 12	0 24	0 36	0 48	1 0	1 12	12	0.20	0.40	0.60	0.80	1.00	1.20
13	0 13	0 26	0 39	0 52	1 5	1 18	13	0.22	0.43	0.65	0.87	1.08	1.30
14	0 14	0 28	0 42	0 56	1 10	1 24	14	0.23	0.47	0.70	0.93	1.17	1.40
15	0 15	0 30	0 45	1 0	1 15	1 30	15	0.25	0.50	0.75	1.00	1.25	1.50
16	0 16	0 32	0 48	1 4	1 20	1 36	16	0.27	0.53	0.80	1.07	1.33	1.60
17	0 17	0 34	0 51	1 8	1 25	1 42	17	0.28	0.57	0.85	1.13	1.42	1.70
18	0 18	0 36	0 54	1 12	1 30	1 48	18	0.30	0.60	0.90	1.20	1.50	1.80
19	0 19	0 38	0 57	1 16	1 35	1 54	19	0.32	0.63	0.95	1.27	1.58	1.90
20	0 20	0 40	1 0	1 20	1 40	2 0	20	0.33	0.67	1.00	1.33	1.67	2.00
21	0 21	0 42	1 3	1 24	1 45	2 6	21	0.35	0.70	1.05	1.40	1.75	2.10
22	0 22	0 44	1 6	1 28	1 50	2 12	22	0.37	0.73	1.10	1.47	1.83	2.20
23	0 23	0 46	1 9	1 32	1 55	2 18	23	0.38	0.77	1.15	1.53	1.92	2.30
24	0 24	0 48	1 12	1 36	2 0	2 24	24	0.40	0.80	1.20	1.60	2.00	2.40
25	0 25	0 50	1 15	1 40	2 5	2 30	25	0.42	0.83	1.25	1.67	2.08	2.50
26	0 26	0 52	1 18	1 44	2 10	2 36	26	0.43	0.87	1.30	1.73	2.17	2.60
27	0 27	0 54	1 21	1 48	2 15	2 42	27	0.45	0.90	1.35	1.80	2.25	2.70
28	0 28	0 56	1 24	1 52	2 20	2 48	28	0.47	0.93	1.40	1.87	2.33	2.80
29	0 29	0 58	1 27	1 56	2 25	2 54	29	0.48	0.97	1.45	1.93	2.42	2.90
30	0 30	1 0	1 30	2 0	2 30	3 0	30	0.50	1.00	1.50	2.00	2.50	3.00
31	0 31	1 2	1 33	2 4	2 35	3 6	31	0.52	1.03	1.55	2.07	2.58	3.10
32	0 32	1 4	1 36	2 8	2 40	3 12	32	0.53	1.07	1.60	2.13	2.67	3.20
33	0 33	1 6	1 39	2 12	2 45	3 18	33	0.55	1.10	1.65	2.20	2.75	3.30
34	0 34	1 8	1 42	2 16	2 50	3 24	34	0.57	1.13	1.70	2.27	2.83	3.40
35	0 35	1 10	1 45	2 20	2 55	3 30	35	0.58	1.17	1.75	2.33	2.92	3.50
36	0 36	1 12	1 48	2 24	3 0	3 36	36	0.60	1.20	1.80	2.40	3.00	3.60
37	0 37	1 14	1 51	2 28	3 5	3 42	37	0.62	1.23	1.85	2.47	3.08	3.70
38	0 38	1 16	1 54	2 32	3 10	3 48	38	0.63	1.27	1.90	2.53	3.17	3.80
39	0 39	1 18	1 57	2 36	3 15	3 54	39	0.65	1.30	1.95	2.60	3.25	3.90
40	0 40	1 20	2 0	2 40	3 20	4 0	40	0.67	1.33	2.00	2.67	3.33	4.00
41	0 41	1 22	2 3	2 44	3 25	4 6	41	0.68	1.37	2.05	2.73	3.42	4.10
42	0 42	1 24	2 6	2 48	3 30	4 12	42	0.70	1.40	2.10	2.80	3.50	4.20
43	0 43	1 26	2 9	2 52	3 35	4 18	43	0.72	1.43	2.15	2.87	3.58	4.30
44	0 44	1 28	2 12	2 56	3 40	4 24	44	0.73	1.47	2.20	2.93	3.67	4.40
45	0 45	1 30	2 15	3 0	3 45	4 30	45	0.75	1.50	2.25	3.00	3.75	4.50
46	0 46	1 32	2 18	3 4	3 50	4 36	46	0.77	1.53	2.30	3.07	3.83	4.60
47	0 47	1 34	2 21	3 8	3 55	4 42	47	0.78	1.57	2.35	3.13	3.92	4.70
48	0 48	1 36	2 24	3 12	4 0	4 48	48	0.80	1.60	2.40	3.20	4.00	4.80
49	0 49	1 38	2 27	3 16	4 5	4 54	49	0.82	1.63	2.45	3.27	4.08	4.90
50	0 50	1 40	2 30	3 20	4 10	5 0	50	0.83	1.67	2.50	3.33	4.17	5.00
51	0 51	1 42	2 33	3 24	4 15	5 6	51	0.85	1.70	2.55	3.40	4.25	5.10
52	0 52	1 44	2 36	3 28	4 20	5 12	52	0.87	1.73	2.60	3.47	4.33	5.20
53	0 53	1 46	2 39	3 32	4 25	5 18	53	0.88	1.77	2.65	3.53	4.42	5.30
54	0 54	1 48	2 42	3 36	4 30	5 24	54	0.90	1.80	2.70	3.60	4.50	5.40
55	0 55	1 50	2 45	3 40	4 35	5 30	55	0.92	1.83	2.75	3.67	4.58	5.50
56	0 56	1 52	2 48	3 44	4 40	4 36	56	0.93	1.87	2.80	3.73	4.67	5.60
57	0 57	1 54	2 51	3 48	4 45	5 42	57	0.95	1.90	2.85	3.80	4.75	5.70
58	0 58	1 56	2 54	3 52	4 50	5 48	58	0.97	1.93	2.90	3.87	4.83	5.80
59	0 59	1 58	2 57	3 56	4 55	5 54	59	0.98	1.97	2.95	3.93	4.92	5.90
60	1 0	2 0	3 0	4 0	5 0	6 0	60	1.00	2.00	3.00	4.00	5.00	6.00

TABLE 13. The MOON'S PARALLAX in RIGHT ASCENSION, to every Degree of her Declination, and to every Minute of her approximate Parallax in Right Ascension. Arguments = the ☾'s Declin. at the Head; and her approx. Par. in R. A. at the Side. (Mr. J. UTTING.)

Approx. Par. in R. A.	The ☾'s Declination.																			
	1°		2°		3°		4°		5°		6°		7°		8°		9°		10°	
1′	1′	0″.01	1′	0″.04	1′	0″.08	1′	0″.15	1′	0″.23	1′	0″.33	1′	0″.45	1′	0″.59	1′	0″.75	1′	0″.93
2	2	0.02	2	0.07	2	0.16	2	0.29	2	0.46	2	0.66	2	0.90	2	1.18	2	1.50	2	1.85
3	3	0.03	3	0.11	3	0.25	3	0.44	3	0.69	3	0.99	3	1.35	3	1.77	3	2.24	3	2.78
4	4	0.04	4	0.15	4	0.33	4	0.59	4	0.92	4	1.32	4	1.80	4	2.36	4	2.99	4	3.70
5	5	0.05	5	0.18	5	0.41	5	0.73	5	1.15	5	1.65	5	2.25	5	2.95	5	3.74	5	4.63
6	6	0.06	6	0.22	6	0.49	6	0.88	6	1.38	6	1.98	6	2.70	6	3.54	6	4.49	6	5.55
7	7	0.06	7	0.26	7	0.58	7	1.03	7	1.60	7	2.31	7	3.15	7	4.13	7	5.24	7	6.48
8	8	0.07	8	0.29	8	0.66	8	1.17	8	1.83	8	2.64	8	3.60	8	4.72	8	5.98	8	7.41
9	9	0.08	9	0.33	9	0.74	9	1.32	9	2.06	9	2.98	9	4.05	9	5.31	9	6.73	9	8.33
10	10	0.09	10	0.37	10	0.82	10	1.47	10	2.29	10	3.31	10	4.51	10	5.90	10	7.48	10	9.26
11	11	0.10	11	0.40	11	0.91	11	1.61	11	2.52	11	3.63	11	4.96	11	6.49	11	8.23	11	10.18
12	12	0.11	12	0.44	12	0.99	12	1.76	12	2.75	12	3.97	12	5.41	12	7.08	12	8.98	12	11.11
13	13	0.12	13	0.48	13	1.07	13	1.90	13	2.98	13	4.30	13	5.86	13	7.67	13	9.72	13	12.03
14	14	0.13	14	0.51	14	1.15	14	2.05	14	3.21	14	4.63	14	6.31	14	8.26	14	10.47	14	12.96
15	15	0.14	15	0.55	15	1.24	15	2.20	15	3.44	15	4.96	15	6.76	15	8.85	15	11.21	15	13.88
16	16	0.15	16	0.59	16	1.32	16	2.34	16	3.67	16	5.29	16	7.21	16	9.44	16	11.97	16	14.81
17	17	0.16	17	0.62	17	1.40	17	2.49	17	3.90	17	5.62	17	7.66	17	10.03	17	12.72	17	15.74
18	18	0.16	18	0.66	18	1.48	18	2.64	18	4.13	18	5.95	18	8.11	18	10.62	18	13.46	18	16.66
19	19	0.17	19	0.70	19	1.56	19	2.78	19	4.35	19	6.28	19	8.56	19	11.20	19	14.21	19	17.59
20	20	0.18	20	0.73	20	1.65	20	2.93	20	4.58	20	6.61	20	9.01	20	11.79	20	14.96	20	18.51
21	21	0.19	21	0.77	21	1.73	21	3.08	21	4.81	21	6.94	21	9.46	21	12.38	21	15.71	21	19.44
22	22	0.20	22	0.81	22	1.81	22	3.22	22	5.04	22	7.27	22	9.91	22	12.97	22	16.45	22	20.36
23	23	0.21	23	0.84	23	1.89	23	3.37	23	5.27	23	7.60	23	10.36	23	13.56	23	17.20	23	21.29
24	24	0.22	24	0.88	24	1.98	24	3.52	24	5.50	24	7.93	24	10.81	24	14.15	24	17.95	24	22.22
25	25	0.23	25	0.91	25	2.06	25	3.66	25	5.73	25	8.26	25	11.26	25	14.74	25	18.70	25	23.14
26	26	0.24	26	0.95	26	2.14	26	3.81	26	5.96	26	8.59	26	11.71	26	15.33	26	19.45	26	24.07
27	27	0.25	27	0.99	27	2.22	27	3.96	27	6.19	27	8.92	27	12.16	27	15.92	27	20.19	27	24.99
28	28	0.25	28	1.03	28	2.31	28	4.10	28	6.42	28	9.26	28	12.61	28	16.51	28	20.94	28	25.92
29	29	0.26	29	1.06	29	2.39	29	4.25	29	6.65	29	9.59	29	13.06	29	17.10	29	21.69	29	26.85
30	30	0.27	30	1.10	30	2.47	30	4.40	30	6.88	30	9.92	30	13.52	30	17.69	30	22.44	30	27.77
31	31	0.28	31	1.13	31	2.55	31	4.54	31	7.11	31	10.25	31	13.97	31	18.28	31	23.19	31	28.69
32	32	0.29	32	1.17	32	2.63	32	4.69	32	7.33	32	10.58	32	14.42	32	18.87	32	23.93	32	29.62
33	33	0.30	33	1.21	33	2.72	33	4.84	33	7.56	33	10.91	33	14.87	33	19.46	33	24.68	33	30.55
34	34	0.31	34	1.25	34	2.80	34	4.98	34	7.79	34	11.24	34	15.32	34	20.05	34	25.43	34	31.47
35	35	0.32	35	1.28	35	2.88	35	5.13	35	8.02	35	11.57	35	15.77	35	20.64	35	26.18	35	32.40
36	36	0.33	36	1.32	36	2.96	36	5.27	36	8.25	36	11.90	36	16.22	36	21.23	36	26.93	36	33.32
37	37	0.34	37	1.35	37	3.05	37	5.42	37	8.48	37	12.23	37	16.67	37	21.82	37	27.67	37	34.25
38	38	0.35	38	1.39	38	3.13	38	5.57	38	8.71	38	12.56	38	17.12	38	22.41	38	28.42	38	35.17
39	39	0.36	39	1.43	39	3.21	39	5.71	39	8.94	39	12.89	39	17.57	39	23.00	39	29.17	39	36.10
40	40	0.37	40	1.46	40	3.29	40	5.86	40	9.17	40	13.22	40	18.02	40	23.59	40	29.92	40	37.02
41	41	0.38	41	1.50	41	3.38	41	6.01	41	9.40	41	13.55	41	18.47	41	24.18	41	30.66	41	37.95
42	42	0.38	42	1.54	42	3.46	42	6.15	42	9.63	42	13.88	42	18.92	42	24.77	42	31.41	42	38.88
43	43	0.39	43	1.57	43	3.54	43	6.30	43	9.86	43	14.21	43	19.38	43	25.36	43	32.16	43	39.80
44	44	0.40	44	1.61	44	3.62	44	6.45	44	10.08	44	14.54	44	19.83	44	25.95	44	32.91	44	40.73
45	45	0.41	45	1.65	45	3.71	45	6.59	45	10.31	45	14.87	45	20.28	45	26.54	45	33.66	45	41.65
46	46	0.42	46	1.68	46	3.79	46	6.74	46	10.54	46	15.20	46	20.73	46	27.13	46	34.40	46	42.58
47	47	0.43	47	1.72	47	3.87	47	6.89	47	10.77	47	15.54	47	21.18	47	27.72	47	35.15	47	43.51
48	48	0.44	48	1.76	48	3.95	48	7.03	48	11.00	48	15.87	48	21.63	48	28.30	48	35.90	48	44.43
49	49	0.45	49	1.79	49	4.03	49	7.18	49	11.23	49	16.20	49	22.08	49	28.89	49	36.65	49	45.36
50	50	0.46	50	1.83	50	4.12	50	7.33	50	11.46	50	16.53	50	22.53	50	29.48	50	37.40	50	46.28
51	51	0.47	51	1.87	51	4.20	51	7.47	51	11.69	51	16.86	51	22.98	51	30.07	51	38.15	51	47.21
52	52	0.48	52	1.90	52	4.28	52	7.62	52	11.92	52	17.19	52	23.43	52	30.66	52	38.90	52	48.13
53	53	0.48	53	1.94	53	4.36	53	7.77	53	12.15	53	17.52	53	23.88	53	31.25	53	39.64	53	49.06
54	54	0.49	54	1.98	54	4.45	54	7.91	54	12.38	54	17.85	54	24.33	54	31.84	54	40.39	54	49.98

The approximate Parallax in *Right Ascension* is obtained from TABLE 13., or Table of the "☾'s APPROXIMATE PARALLAX in LONGITUDE," by substituting the ☾'s *Co-Latitude,* for the " Alt. of the Nonag." at the Side, and her *Horary Angle from the Meridian;* for " the ☾'s Distance from the Nonagesimal," at the Head, as the proper Arguments to be used for this purpose.

When the ☽ is in the eastern hemisphere, her Right Ascension is *diminished* by Parallax, but when in the western, it is *increased.* On the Meridian, there is no Parallax in Right Ascension.

TABLE 18.

The Moon's Parallax in Right Ascension, continued.

Approx. Par. in R. A.	The ☾'s Declination. 11°	12°	13°	14°	15°	16°	17°	18°	19°	20°
1′	1′ 1″.12	1′ 1″.34	1′ 1″.56	1′ 1″.84	1′ 2″.12	1′ 2″.42	1′ 2″.74	1′ 3″.09	1′ 3″.46	1′ 3″.85
2	2 2.25	2 2.68	2 3.16	2 3.67	2 4.23	2 4.84	2 5.48	2 6.18	2 6.91	2 7.70
3	3 3.37	3 4.02	3 4.73	3 5.51	3 6.35	3 7.25	3 8.22	3 9.26	3 10.37	3 11.55
4	4 4.49	4 5.36	4 6.31	4 7.35	4 8.47	4 9.67	4 10.97	4 12.35	4 13.83	4 15.40
5	5 5.62	5 6.70	5 7.89	5 9.18	5 10.58	5 12.09	5 13.71	5 15.44	5 17.29	5 19.25
6	6 6.74	6 8.04	6 9.47	6 11.02	6 12.70	6 14.51	6 16.45	6 18.53	6 20.74	6 23.10
7	7 7.86	7 9.38	7 11.05	7 12.86	7 14.82	7 16.93	7 19.19	7 21.61	7 24.20	7 26.95
8	8 8.98	8 10.72	8 12.63	8 14.69	8 16.93	8 19.34	8 21.93	8 24.70	8 27.66	8 30.80
9	9 10.11	9 12.06	9 14.20	9 16.53	9 19.05	9 21.76	9 24.67	9 27.79	9 31.12	9 34.66
10	10 11.23	10 13.40	10 15.78	10 18.37	10 21.17	10 24.18	10 27.42	10 30.88	10 34.57	10 38.51
11	11 12.35	11 14.74	11 17.36	11 20.21	11 23.28	11 26.60	11 30.16	11 33.97	11 38.03	11 42.36
12	12 13.48	12 16.09	12 18.94	12 22.04	12 25.40	12 29.02	12 32.90	12 37.05	12 41.49	12 46.21
13	13 14.60	13 17.43	13 20.52	13 23.88	13 27.52	13 31.43	13 35.64	13 40.14	13 44.94	13 50.06
14	14 15.72	14 18.77	14 22.10	14 25.72	14 29.63	14 33.85	14 38.38	14 43.23	14 48.40	14 53.91
15	15 16.85	15 20.11	15 23.67	15 27.55	15 31.75	15 36.27	15 41.12	15 46.32	15 51.86	15 57.76
16	16 17.97	16 21.45	16 25.25	16 29.39	16 33.87	16 38.69	16 43.86	16 49.40	16 55.32	17 1.61
17	17 19.09	17 22.79	17 26.83	17 31.23	17 35.98	17 41.11	17 46.61	17 52.49	17 58.77	18 5.46
18	18 20.21	18 24.13	18 28.41	18 33.06	18 38.10	18 43.52	18 49.35	18 55.58	19 2.23	19 9.31
19	19 21.34	19 25.47	19 29.99	19 34.90	19 40.22	19 45.94	19 52.09	19 58.67	20 5.69	20 13.16
20	20 22.46	20 26.81	20 31.56	20 36.74	20 42.33	20 48.36	20 54.83	21 1.76	21 9.14	21 17.01
21	21 23.58	21 28.15	21 33.14	21 38.57	21 44.45	21 50.78	21 57.57	22 4.84	22 12.60	22 20.86
22	22 24.71	22 29.49	22 34.72	22 40.41	22 46.56	22 53.20	23 0.31	23 7.93	23 16.06	23 24.71
23	23 25.83	23 30.83	23 36.30	23 42.25	23 48.68	23 55.61	24 3.05	24 11.02	24 19.51	24 28.57
24	24 26.95	24 32.17	24 37.88	24 44.08	24 50.80	24 58.03	25 5.80	25 14.11	25 22.97	25 32.42
25	25 28.08	25 33.51	25 39.46	25 45.92	25 52.91	26 0.45	26 8.54	26 17.19	26 26.43	26 36.27
26	26 29.20	26 34.85	26 41.03	26 47.76	26 55.03	27 2.87	27 11.28	27 20.28	27 29.89	27 40.12
27	27 30.32	27 36.19	27 42.61	27 49.59	27 57.15	28 5.29	28 14.02	28 23.37	28 33.34	28 43.97
28	28 31.44	28 37.53	28 44.19	28 51.43	28 59.26	29 7.70	29 16.76	29 26.46	29 36.80	29 47.82
29	29 32.57	29 38.87	29 45.77	29 53.27	30 1.38	30 10.12	30 19.50	30 29.54	30 40.26	30 51.67
30	39 33.69	30 40.21	30 47.35	30 55.10	31 3.50	31 12.54	31 22.25	31 32.63	31 43.72	31 55.52
31	31 34.81	31 41.55	31 48.93	31 56.94	32 5.61	32 14.96	32 24.99	32 35.72	32 47.17	32 59.37
32	32 35.94	32 42.89	32 50.50	32 58.78	33 7.73	33 17.38	33 27.73	33 38.81	33 50.63	34 3.22
33	33 37.06	33 44.23	33 52.08	34 0.61	34 9.85	34 19.79	34 30.47	34 41.91	34 54.09	35 7.07
34	34 38.18	34 45.58	34 53.66	35 2.45	35 11.96	35 22.21	35 33.21	35 44.98	35 57.54	36 10.92
35	35 39.31	35 46.92	35 55.24	36 4.29	36 14.08	36 24.63	36 35.95	36 48.07	37 1.00	37 14.77
36	36 40.43	36 48.26	36 56.82	37 6.13	37 16.20	37 27.05	37 38.69	37 51.16	38 4.46	38 18.62
37	37 41.55	37 49.60	37 58.39	38 7.96	38 18.31	38 29.47	38 41.44	38 54.25	39 7.92	39 22.47
38	38 42.67	38 50.94	38 59.97	39 9.80	39 20.43	39 31.88	39 44.18	39 57.33	40 11.37	40 26.33
39	39 43.80	39 52.28	40 1.55	40 11.64	40 22.55	40 34.30	40 46.92	41 0.42	41 14.83	41 30.18
40	40 44.92	40 53.62	41 3.13	41 13.47	41 24.66	41 36.72	41 49.66	42 3.51	42 18.29	42 34.03
41	41 46.04	41 54.96	42 4.71	42 15.31	42 26.78	42 39.14	42 52.40	43 6.60	43 21.74	43 37.88
42	42 47.17	42 56.30	43 6.29	43 17.15	43 28.90	43 41.56	43 55.14	44 9.68	44 25.20	44 41.73
43	43 48.29	43 57.64	44 7.86	44 18.98	44 31.01	44 43.97	44 57.88	45 12.77	45 28.66	45 45.58
44	44 49.41	44 58.98	45 9.44	45 20.82	45 33.13	45 46.39	46 0.63	46 15.86	46 32.12	46 49.43
45	45 50.54	46 0.32	46 11.02	46 22.66	46 35.25	46 48.81	47 3.37	47 18.95	47 35.57	47 53.28
46	46 51.66	47 1.66	47 12.60	47 24.49	47 37.36	47 51.23	48 6.11	48 22.04	48 39.03	48 57.13
47	47 52.78	48 3.00	48 14.18	48 26.33	48 39.48	48 53.65	49 8.85	49 25.12	49 42.49	50 0.98
48	48 53.90	49 4.34	49 15.76	49 28.17	49 41.60	49 56.06	50 11.59	50 28.21	50 45.94	51 4.83
49	49 55.03	50 5.68	50 17.33	50 30.00	50 43.71	50 58.48	51 14.33	51 31.30	51 49.40	52 8.68
50	50 56.15	51 7.02	51 18.91	51 31.84	51 45.83	52 0.90	52 17.08	52 34.39	52 52.86	53 12.53
51	51 57.27	52 8.36	52 20.49	52 33.68	52 47.95	53 3.32	53 19.82	53 37.47	53 56.32	54 16.38
52	52 58.40	53 9.70	53 22.07	53 35.51	53 50.06	54 5.73	54 22.56	54 40.56	54 59.77	55 20.23
53	53 59.52	54 11.04	54 23.65	54 37.35	54 52.18	55 8.15	55 25.30	55 43.65	56 3.23	56 24.09
54	55 0.64	55 12.38	55 25.23	55 39.19	55 54.30	56 10.57	56 28.04	56 46.74	57 6.69	57 27.94

The approximate Parallax in *Right Ascension* is obtained from Table 13, or Table of the "☾'s approximate Parallax in Longitude," by substituting the ☾'s *Co-Latitude* for the "Alt. of the Nonag." at the Side, and her *Horary Angle from the Meridian*, for "the ☾'s Distance from the Nonagesimal," at the Head, as the proper Arguments to be used for this purpose.

When the ☾ is in the eastern hemisphere, her Right Ascension is *diminished* by Parallax, but when in the western, it is *increased*. On the Meridian, there is no Parallax in Right Ascension.

TABLE 18.

The Moon's Parallax in Right Ascension, concluded.

Approx. Par. in R. A.	The ☾'s Declination. 21°	22°	23°	24°	25°	26°	27°	28°	29°	30°
1′	1′ 4″.27	1′ 4″.71	1′ 5.18	1′ 5″.68	1′ 6″.20	1′ 6″.76	1′ 7″.34	1′ 7″,95	1′ 8″.60	1′ 9″.28
2	2 8.54	2 9.42	2 10.36	2 11.36	2 12.41	2 13.51	2 14.68	2 15.91	2 17.20	2 18.56
3	3 12.81	3 14.14	3 15.54	3 17.03	3 18.61	3 20.27	3 22.02	3 23.86	3 25.80	3 27.85
4	4 17.07	4 18.85	4 20.73	4 22.71	4 24.81	4 27.02	4 29.36	4 31.82	4 34.40	4 37.13
5	5 21.34	5 23.56	5 25.91	5 28.39	5 31.01	5 33.78	5 36.70	5 39.77	5 43.01	5 46.41
6	6 25.61	6 28.27	6 31.09	6 34.07	6 37.22	6 40.54	6 44.04	6 47.73	6 51.61	6 55.69
7	7 29.88	7 32.98	7 36.27	7 39.75	7 43.42	7 47.29	7 51.38	7 55.68	8 0.21	8 4.97
8	8 34.15	8 37.70	8 41.45	8 45.43	8 49.62	8 54.05	8 58.72	9 3.63	9 8.81	9 14.26
9	9 38.42	9 42.41	9 46.63	9 51.10	9 55.82	10 0.80	16 6.06	10 11.59	10 17.41	10 23.54
10	10 42.69	10 47.12	10 51.82	10 56.78	11 2.03	11 7.56	11 13.40	11 19.54	11 26.01	11 32.82
11	11 46.96	11 51.83	11 57.00	12 2.46	12 8.23	12 14.32	12 20.74	12 27.50	12 34.61	12 42.10
12	12 51.22	12 56.55	13 2.18	13 8.14	13 14.43	13 21.07	13 28.08	13 35.45	13 43.21	13 51.38
13	13 55.49	14 1.26	14 7.36	14 13.82	14 20.63	14 27.83	14 35.41	14 43.40	14 51.81	15 0.67
14	14 59.76	15 5.97	15 12.54	15 19.49	15 26.84	15 34.59	15 42.75	15 51.36	16 0.41	16 9.95
15	16 4.03	16 10.68	16 17.72	16 25.17	16 33.04	16 41.34	16 50.09	16 59.31	17 9.02	17 19.23
16	17 8.30	17 15.39	17 22.91	17 30.85	17 39.24	17 48.10	17 57.43	18 7.27	18 17.62	18 28.51
17	18 12.57	18 20.11	18 28.09	18 36.53	18 45.45	18 54.85	19 4.77	19 15.22	19 26.22	19 37.79
18	19 16.84	19 24.82	19 33.27	19 42.21	19 51.65	20 1.61	20 12.11	20 23.18	20 34.82	20 47.08
19	20 21.11	20 29.53	20 38.45	20 47.89	20 57.85	21 8.37	21 19.45	21 31.13	21 43.42	21 56.36
20	21 25.37	21 34.24	21 43.63	21 53.56	22 4.05	22 15.12	22 26.79	22 39.08	22 52.02	23 5.64
21	22 29.64	22 38.95	22 48.81	22 59.24	23 10.26	23 21.88	23 34.13	23 47.04	24 0.63	24 14.92
22	23 33.91	23 43.67	23 54.00	24 4.92	24 16.46	24 28.63	24 41.47	24 54.99	25 9.23	25 24.20
23	24 38.18	24 48.38	24 59.18	25 10.60	25 22.66	25 35.39	25 48.81	26 2.95	26 17.83	26 33.49
24	25 42.45	25 53.09	26 4.36	26 16.28	26 28.86	26 42.15	26 56.15	27 10.90	27 26.43	27 42.77
25	26 46.72	26 57.80	27 9.54	27 21.96	27 35.07	27 48.90	28 3.49	28 18.86	28 35.03	28 52.05
26	27 50.99	28 2.51	28 14.72	28 27.63	28 41.27	28 55.66	29 10.83	29 26.81	29 43.63	30 1.33
27	28 55.25	29 7.23	29 19.90	29 33.31	29 47.47	30 2.42	30 18.17	30 34.76	30 52.23	31 10.61
28	29 59.52	30 11.94	30 25.09	30 38.99	30 53.68	31 9.17	31 25.51	31 42.72	32 0.84	32 19.90
29	31 3.79	31 16.65	31 30.27	31 44.67	31 59.88	32 15.93	32 32.85	32 50.67	33 9.44	33 29.18
30	32 8.06	32 21.36	32 35.45	32 50.35	33 6.08	33 22.68	33 40.19	33 58.63	34 18.04	34 38.46
31	33 12.33	33 26.07	33 40.63	33 56.02	34 12.28	34 29.44	34 47.53	35 6.58	35 26.64	35 47.74
32	34 16.60	34 30.79	34 45.81	35 1.70	35 18.49	35 36.20	35 54.87	36 14.53	36 35.24	36 57.02
33	35 20.87	35 35.50	35 50.99	36 7.38	36 24.69	36 42.95	37 2.21	37 22.49	37 43.84	38 6.31
34	36 25.14	36 40.21	36 56.18	37 13.06	37 30.89	37 49.71	38 9.55	38 30.44	38 52.44	39 15.59
35	37 29.40	37 44.92	38 1.36	38 18.74	38 37.09	38 56.46	39 16.89	39 38.40	40 1.04	40 24.87
36	38 33.67	38 49.63	39 6.54	39 24.42	39 43.30	40 3.22	40 24.23	40 46.35	41 9.64	41 34.15
37	39 37.94	39 54.35	40 11.72	40 30.09	40 49.50	41 9.98	41 31.57	41 54.31	42 18.25	42 43.43
38	40 42.21	40 59.06	41 16.90	41 35.77	41 55.70	42 16.73	42 38.90	43 2.26	43 26.85	43 52.72
39	41 46.48	42 3.77	42 22.08	42 41.45	43 1.90	43 23.49	43 46.24	44 10.21	44 35.45	45 2.00
40	42 50.75	43 8.48	43 27.26	43 47.13	44 8.11	44 30.24	44 53.58	45 18.17	45 44.05	46 11.28
41	43 55.02	44 13.20	44 32.45	44 52.81	45 14.31	45 37.00	46 0.92	46 26.12	46 52.65	47 20.56
42	44 59.29	45 17.91	45 37.63	45 58.48	46 20.51	46 43.76	47 8.26	47 34.08	48 1.25	48 29.84
43	46 3.55	46 22.62	46 42.81	47 4.16	47 26.72	47 50.51	48 15.60	48 42.03	49 9.85	49 39.13
44	47 7.82	47 27.33	47 47.99	48 9.84	48 32.92	48 57.27	49 22.94	49 49.98	50 18.45	50 48.41
45	48 12.09	48 32.04	48 53.17	49 15.52	49 39.12	50 4.03	50 30.28	50 57.94	51 27.06	51 57.69
46	49 16.36	49 36.76	49 58.35	50 21.20	50 45.32	51 10.78	51 37.62	52 5.89	52 35.66	53 6.97
47	50 20.63	50 41.47	51 3.54	51 26.88	51 51.53	52 17.54	52 44.96	53 13.85	53 44.26	54 16.25
48	51 24.90	51 46.18	52 8.72	52 32.55	52 57.73	53 24.29	53 52.30	54 21.80	54 52.86	55 25.54
49	52 29.17	52 50.89	53 13.90	53 38.23	54 3.93	54 31.05	54 59.64	55 29.76	56 1.46	56 34.82
50	53 33.44	53 55.60	54 19.08	54 43.91	55 10.13	55 37.81	56 6.98	56 37.71	57 10.06	57 44.10
51	54 37.70	55 0.32	55 24.26	55 49.59	56 16.34	56 44.56	57 14.32	57 45.66	58 18.66	58 53.38
52	55 41.97	56 5.03	56 29.44	56 55.27	57 22.54	57 51.32	58 21.66	58 53.62	59 27.27	60 2.66
53	56 46.24	57 9.74	57 34.63	58 0.94	58 28.74	58 58.07	59 29.00	60 1.57	60 35.87	61 11.95
54	57 50.51	58 14.45	58 39.81	59 6.62	59 34.94	60 4.83	60 36.34	61 9.53	61 44.47	62 21.23

The approximate Parallax in *Right Ascension* is obtained from Table 13., or, Table of " the ☾'s approximate Parallax in Longitude," by substituting the ☾'s *Co-Latitude,* for the " Alt. of the Nonag." at the Side, and her *Horary Angle from the Meridian,* for " the ☾'s Distance from the Nonagesimal" at the Head, as the proper Arguments to be used for this purpose.

When the ☾ is in the eastern hemisphere, her Right Ascension is *diminished* by Parallax, but when in the western, it is *increased.* On the Meridian, there is no Parallax in Right Ascension.

TABLE 19. The *Second Part* of the Moon's Parallax in Declination, for every Degree of her Declination; and for every Minute of her approx. Par. in Decl. Args. = the ☾'s Declin. at the Head, and her approx. Paral. in Declin. at the Side. (Mr. J. Utting.)

Approx. Par. in Declin.	The ☾'s Declination. 1°	2°	3°	4°	5°	6°	7°	8°	9°	10°
1′	0′ 1″.05	0′ 2″.09	0′ 3″.14	0′ 4″.19	0′ 5″.23	0′ 6″.27	0′ 7″.31	0′ 8″.35	0′ 9″.39	0′ 10″.42
2	0 2.09	0 4.19	0 6.28	0 8.37	0 10.46	0 12.54	0 14.62	0 16.70	0 18.77	0 20.84
3	0 3.14	0 6.28	0 9.42	0 12.56	0 15.69	0 18.82	0 21.94	0 25.05	0 28.16	0 31.26
4	0 4.19	0 8.38	0 12.56	0 16.74	0 20.92	0 25.09	0 29.25	0 33.40	0 37.54	0 41.68
5	0 5.24	0 10.47	0 15.70	0 20.93	0 26.15	0 31.36	0 36.56	0 41.75	0 46.93	0 52.09
6	0 6.28	0 12.56	0 18.84	0 25.11	0 31.38	0 37.63	0 43.87	0 50.10	0 56.32	1 2.51
7	0 7.33	0 14.66	0 21.98	0 29.30	0 36.61	0 43.90	0 51.19	0 58.45	1 5.70	1 12.93
8	0 8.38	0 16.75	0 25.12	0 33.48	0 41.83	0 50.18	0 58.50	1 6.80	1 15.09	1 23.35
9	0 9.42	0 18.85	0 28.26	0 37.67	0 47.06	0 56.45	1 5.81	1 15.15	1 24.47	1 33.77
10	0 10.47	0 20.94	0 31.40	0 41.85	0 52.29	1 2.72	1 13.12	1 23.50	1 33.86	1 44.19
11	0 11.51	0 23.03	0 34.54	0 46.04	0 57.52	1 8.99	1 20.43	1 31.85	1 43.24	1 54.61
12	0 12.57	0 25.13	0 37.68	0 50.22	1 2.75	1 15.26	1 27.75	1 40.20	1 52.63	2 5.03
13	0 13.61	0 27.22	0 40.82	0 54.41	1 7.98	1 21.53	1 35.06	1 48.56	2 2.02	2 15.45
14	0 14.66	0 29.32	0 43.96	0 58.59	1 13.21	1 27.81	1 42.37	1 56.91	2 11.41	2 25.86
15	0 15.71	0 31.41	0 47.10	1 2.78	1 18.44	1 34.08	1 49.68	2 5.26	2 20.79	2 36.28
16	0 16.75	0 33.50	0 50.24	1 6.96	1 23.67	1 40.35	1 56.99	2 13.61	2 30.18	2 46.70
17	0 17.80	0 35.60	0 53.38	1 11.15	1 28.90	1 46.62	2 4.31	2 21.96	2 39.56	2 57.12
18	0 18.85	0 37.69	0 56.52	1 15.33	1 34.13	1 52.89	2 11.62	2 30.31	2 48.95	3 7.54
19	0 19.89	0 39.79	0 59.66	1 19.52	1 39.35	1 59.17	2 18.93	2 38.66	2 58.34	3 17.96
20	0 20.94	0 41.88	1 2.80	1 23.71	1 44.59	2 5.43	2 26.24	2 47.01	3 7.72	3 28.38
21	0 21.99	0 43.97	1 5.94	1 27.89	1 49.82	2 11.71	2 33.56	2 55.36	3 17.11	3 38.80
22	0 23.04	0 46.07	1 9.08	1 32.08	1 55.05	2 17.98	2 40.87	3 3.71	3 26.49	3 49.22
23	0 24.08	0 48.16	1 12.22	1 36.26	2 0.27	2 24.25	2 48.18	3 12.06	3 35.88	3 59.64
24	0 25.13	0 50.26	1 15.36	1 40.45	2 5.50	2 30.52	2 55.49	3 20.41	3 45.27	4 10.05
25	0 26.18	0 52.35	1 18.50	1 44.64	2 10.73	2 36.79	3 2.80	3 28.76	3 54.65	4 20.47
26	0 27.23	0 54.44	1 21.64	1 48.82	2 15.96	2 43.07	3 10.12	3 37.11	4 4.04	4 30.89
27	0 28.27	0 56.54	1 24.78	1 53.01	2 21.19	2 49.34	3 17.43	3 45.46	4 13.42	4 41.31
28	0 29.32	0 58.63	1 27.92	1 57.19	2 26.42	2 55.61	3 24.74	3 53.81	4 22.81	4 51.73
29	0 30.37	1 0.73	1 31.06	2 1.38	2 31.65	3 1.88	3 32.05	4 2.16	4 32.20	5 2.15
30	0 31.41	1 2.82	1 34.21	2 5.56	2 36.88	3 8.15	3 39.37	4 10.51	4 41.58	5 12.57
31	0 32.46	1 4.91	1 37.35	2 9.75	2 42.11	3 14.42	3 46.68	4 18.86	4 50.97	5 22.99
32	0 33.51	1 7.01	1 40.49	2 13.93	2 47.34	3 20.69	3 53.99	4 27.21	5 0.35	5 33.41
33	0 34.56	1 9.10	1 43.63	2 18.12	2 52.57	3 26.97	4 1.30	4 35.56	5 9.74	5 43.82
34	0 35.60	1 11.20	1 46.77	2 22.30	2 57.80	3 33.24	4 8.61	4 43.91	5 19.13	5 54.24
35	0 36.65	1 13.29	1 49.91	2 26.49	3 3.03	3 39.51	4 15.93	4 52.26	5 28.51	6 4.66
36	0 37.70	1 15.38	1 53.05	2 30.67	3 8.25	3 45.78	4 23.24	5 0.61	5 37.90	6 15.08
37	0 38.74	1 17.48	1 56.19	2 34.86	3 13.48	3 52.05	4 30.55	5 8.96	5 47.28	6 25.50
38	0 39.79	1 19.57	1 59.33	2 39.04	3 18.71	3 58.33	4 37.86	5 17.31	5 56.67	6 35.92
39	0 40.84	1 21.67	2 2.47	2 43.23	3 23.94	4 4.60	4 45.17	5 25.66	6 6.06	6 46.34
40	0 41.89	1 23.76	2 5.61	2 47.42	3 29.17	4 10.87	4 52.49	5 34.02	6 15.44	6 56.76
41	0 42.93	1 25.85	2 8.75	2 51.60	3 34.40	4 17.14	4 59.80	5 42.37	6 24.83	7 7.18
42	0 43.98	1 27.95	2 11.89	2 55.79	3 39.63	4 23.41	5 7.11	5 50.72	6 34.22	7 17.59
43	0 45.03	1 30.04	2 15.03	2 59.97	3 44.86	4 29.68	5 14.42	5 59.07	6 43.60	7 28.01
44	0 46.07	1 32.14	2 18.17	3 4.16	3 50.09	4 35.96	5 21.73	6 7.42	6 52.99	7 38.43
45	0 47.12	1 34.23	2 21.31	3 8.34	3 55.32	4 42.23	5 29.05	6 15.77	7 2.37	7 48.85
46	0 48.17	1 36.32	2 24.45	3 12.53	4 0.55	4 48.50	5 36.36	6 24.12	7 11.76	7 59.27
47	0 49.22	1 38.42	2 27.59	3 16.71	4 5.78	4 54.77	5 43.67	6 32.47	7 21.15	8 9.69
48	0 50.26	1 40.51	2 30.73	3 20.90	4 11.01	5 1.04	5 50.98	6 40.82	7 30.53	8 20.11
49	0 51.31	1 42.61	2 33.87	3 25.08	4 16.24	5 7.32	5 58.29	6 49.17	7 39.92	8 30.53
50	0 52.36	1 44.70	2 37.01	3 29.27	4 21.47	5 13.59	6 5.61	6 57.52	7 49.30	8 40.95
51	0 53.40	1 46.79	2 40.15	3 33.46	4 26.70	5 19.86	6 12.92	7 5.87	7 58.69	8 51.36
52	0 54.45	1 48.89	2 43.29	3 37.64	4 31.93	5 26.13	6 20.23	7 14.22	8 8.08	9 1.78
53	0 55.50	1 50.98	2 46.43	3 41.83	4 37.15	5 32.40	6 27.54	7 22.57	8 17.46	9 12.20
54	0 56.55	1 53.08	2 49.57	3 46.01	4 42.38	5 38.67	6 34.86	7 30.92	8 26.85	9 22.02

The Moon's *approximate Parallax* in *Declination* is obtained from Table 13, by substituting *her Latitude* as the proper Argument at the Side, instead of the "Alt. of the Nonag.;" and *her Polar Distance,* for "the Distance from the Nonagesimal," as the Argument at the Head, for the *positive Portion;* and the *negative Port on* is obtained from the same Table (13.) by substituting the ☾'s *Co-Latitude* at the Side, and *Complement of the Horary Angle* at the Head, as the proper Arguments.

ZODIACAL TABLES.

TABLE 19.

The *Second Part* of the Moon's Parallax in Declination, continued.

Approx. Par. in Declin.	The ☽'s Declination.									
	11°	12°	13°	14°	15°	16°	17°	18°	19°	20°
1′	0′ 11″.45	0′ 12″.47	0′ 13″.50	0′ 14″.52	0′ 15″.53	0′ 16″.54	0′ 17″.54	0′ 18″.54	0′ 19″.53	0′ 20″.52
2	0 22 .90	0 24 .95	0 26 .99	0 29 .03	0 31 .06	0 33 .08	0 35 .09	0 37 .08	0 39 .07	0 41 .04
3	0 34 .35	0 37 .42	0 40 .49	0 43 .55	0 46 59	0 49 .62	0 52 .63	0 55 .62	0 58 .60	1 1 .56
4	0 45 .79	0 49 .90	0 53 .99	0 58 .06	1 2 .12	1 6 .15	1 10 .17	1 14 .16	1 18 .14	1 22 .08
5	0 57 .24	1 2 .37	1 7 .49	1 12 .58	1 17 .65	1 22 .69	1 27 .71	1 32 .71	1 37 .67	1 42 .61
6	1 8 .69	1 14 .85	1 20 .99	1 27 .09	1 33 .18	1 39 .23	1 45 .25	1 51 .25	1 57 .20	2 3 .13
7	1 20 .14	1 27 .32	1 34 .48	1 41 .61	1 48 .70	1 55 .77	2 2 .80	2 9 .79	2 16 .74	2 23 .65
8	1 31 .59	1 39 .80	1 47 .98	1 56 .12	2 4 .23	2 12 .31	2 20 .34	2 28 .33	2 36 .27	3 44 .17
9	1 43 .04	1 52 .27	2 1 .48	2 10 .64	2 19 .76	2 28 .84	2 37 .88	2 46 .87	2 55 .81	3 4 .69
10	1 54 .49	2 4 .75	2 14 .97	2 25 .15	2 35 .29	2 45 .38	2 55 .42	3 5 .41	3 15 .34	3 25 .21
11	2 5 .93	2 17 .22	2 28 .47	2 39 .67	2 50 .82	3 1 .92	3 12 .97	3 23 .95	3 34 .88	3 45 .73
12	2 17 .38	2 29 .70	2 41 .97	2 54 .18	3 6 .35	3 18 .46	3 30 .51	3 42 .49	3 54 .41	4 6 .25
13	2 28 .83	2 42 .17	2 55 .46	3 8 .70	3 21 .88	3 35 .00	3 48 .05	4 1 .03	4 13 .94	4 26 .78
14	2 40 .28	2 54 .65	3 8 .96	3 23 .21	3 37 .41	3 51 .53	4 5 .59	4 19 .57	4 33 .48	4 47 .30
15	2 51 .73	3 7 .12	3 22 .46	3 37 .73	3 52 .94	4 8 .07	4 23 .13	4 38 .12	4 53 .01	5 7 .82
16	3 3 .18	3 19 .60	3 35 .95	3 52 .24	4 8 .47	4 24 .61	4 40 .68	4 56 .66	5 12 .55	5 28 .34
17	3 14 .62	3 32 .07	3 49 .45	4 6 .76	4 23 .99	4 41 .15	4 58 .22	5 15 .20	5 32 .08	5 48 .86
18	3 26 .07	3 44 .55	4 2 .95	4 21 .27	4 39 .52	4 57 .69	5 15 .76	5 33 .74	5 51 .61	6 9 .38
19	3 37 .52	3 57 .02	4 16 .44	4 35 .79	4 55 .05	5 14 .22	5 33 .30	5 52 .28	6 11 .15	6 29 .90
20	3 48 .97	4 9 .49	4 29 .94	4 50 .31	5 10 .58	5 30 .77	5 50 .85	6 10 .82	6 30 .68	6 50 .42
21	4 0 .42	4 21 .97	4 43 .44	5 4 .82	5 26 .11	5 47 .30	6 8 .39	6 29 .36	6 50 .22	7 10 .95
22	4 11 .87	4 34 .44	4 56 .94	5 19 .34	5 41 .64	6 3 .84	6 25 .93	6 47 .90	7 9 .75	7 31 .47
23	4 23 .32	4 46 .92	5 10 .43	5 33 .85	5 57 .17	6 20 .38	6 43 .47	7 6 .44	7 29 .28	7 51 .99
24	4 34 .77	4 59 .39	5 23 .93	5 48 .37	6 12 .70	6 36 .92	7 1 .01	7 24 .98	7 48 .82	8 12 .51
25	4 46 .21	5 11 .87	5 37 .43	6 2 .88	6 28 .23	6 53 .46	7 18 .56	7 43 .53	8 8 .35	8 33 .03
26	4 57 .66	5 24 .34	5 50 .92	6 17 .40	6 43 .76	7 9 .99	7 36 .10	8 2 .07	8 27 .89	8 53 .55
27	5 9 .11	5 36 .82	6 4 .42	6 31 .91	6 59 .29	7 26 .53	7 53 .64	8 20 .61	8 47 .42	9 14 .07
28	5 20 .56	5 49 .29	6 17 .92	6 46 .43	7 14 .82	7 43 .07	8 11 .18	8 39 .15	9 6 .95	9 34 .59
29	5 32 .01	6 1 .77	6 31 .42	7 0 .94	7 30 .34	7 59 .61	8 28 .72	8 57 .69	9 26 .49	9 55 .11
30	5 43 .46	6 14 .24	6 44 .91	7 15 .46	7 45 .87	8 16 .15	8 46 .27	9 16 .23	9 46 .02	10 15 .64
31	5 54 .90	6 26 .72	6 58 .41	7 29 .97	8 1 .40	8 32 .69	9 3 .81	9 34 .77	10 5 .56	10 36 .16
32	6 6 .35	6 39 .19	7 11 .91	7 44 .49	8 16 .93	8 49 .22	9 21 .35	9 53 .31	10 25 .09	10 56 .68
33	6 17 .80	6 51 .67	7 25 .41	7 59 .01	8 32 .46	9 5 .76	9 38 .90	10 11 .85	10 44 .63	11 17 .20
34	6 29 .25	7 4 .14	7 38 .90	8 13 .52	8 47 .99	9 22 .30	9 56 .44	10 30 .40	11 4 .16	11 37 .72
35	6 40 .70	7 16 .62	7 52 .40	8 28 .04	9 3 .52	9 38 .84	10 13 .98	10 48 .94	11 23 .69	11 58 .24
36	6 52 .15	7 29 .09	8 5 .90	8 42 .55	9 19 .05	9 55 .38	10 31 .52	11 7 .48	11 43 .23	12 18 .76
37	7 3 .59	7 41 .57	8 19 .39	8 57 .07	9 34 .58	10 11 .91	10 49 .06	11 26 .02	12 2 .76	12 39 .28
38	7 15 .04	7 54 .04	8 32 .89	9 11 .58	9 50 .11	10 28 .45	11 6 .61	11 44 .56	12 22 .30	12 59 .80
39	7 26 .49	8 6 .51	8 46 .39	9 26 .10	10 5 .64	10 44 .99	11 24 .15	12 3 .10	12 41 .83	13 20 .33
40	7 37 .94	8 18 .99	8 59 .88	9 40 .61	10 21 .17	11 1 .53	11 41 .69	12 21 .64	13 1 .36	13 40 .85
41	7 49 .39	8 31 .46	9 13 .38	9 55 .13	10 36 .70	11 18 .07	11 59 .23	12 40 .18	13 20 .90	14 1 .37
42	8 0 .84	8 43 .94	9 26 .88	10 9 .64	10 52 .22	11 34 .61	12 16 .78	12 58 .72	13 40 .43	14 21 .89
43	8 12 .29	8 56 .41	9 40 .37	10 24 .16	11 7 .75	11 51 .14	12 34 .32	13 17 .26	13 59 .97	14 42 .41
44	8 23 .74	9 8 .89	9 53 .87	10 38 .67	11 23 .28	12 7 .68	12 51 .86	13 35 .81	14 19 .50	15 2 .93
45	8 35 .18	9 21 .36	10 7 .37	10 53 .19	11 38 .81	12 24 .22	13 9 .40	13 54 .35	14 39 .03	15 23 .45
46	8 46 .63	9 33 .84	10 20 .86	11 7 .70	11 54 .34	12 40 .76	13 26 .94	14 12 .89	14 58 .57	15 43 .97
47	8 58 .08	9 46 .31	10 34 .36	11 22 .22	12 9 .87	12 57 .30	13 44 .49	14 31 .43	15 18 .10	16 4 .50
48	9 9 .53	9 58 .79	10 47 .86	11 36 .73	12 25 .40	13 13 .83	14 2 .03	14 49 .97	15 37 .64	16 25 .02
49	9 20 .98	10 11 .26	11 1 .35	11 51 .25	12 40 .93	13 30 .37	14 19 .57	15 8 .51	15 57 .17	16 45 .54
50	9 32 .43	10 23 .74	11 14 .85	12 5 .77	12 56 .46	13 46 .91	14 37 .12	15 27 .05	16 16 .71	17 6 .06
51	9 43 .88	10 36 .21	11 28 .35	12 20 .28	13 11 .99	14 3 .45	14 54 .66	15 45 .59	16 36 .24	17 26 .58
52	9 55 .32	10 48 .69	11 41 .84	12 34 .80	13 27 .52	14 19 .99	15 12 .20	16 4 .13	16 55 .77	17 47 .10
53	10 6 .77	11 1 .16	11 55 .34	12 49 .31	13 43 .04	14 36 .53	15 29 .74	16 22 .67	17 15 .31	18 7 .62
54	10 18 .22	11 13 .64	12 8 .84	13 3 .83	13 58 .57	14 53 .06	15 47 .28	16 41 .22	17 34 .84	18 28 .14

The Moon's *approximate Parallax* in *Declination* is obtained from Table 13, by substituting *her Latitude* as the proper Argument at the Side, instead of the "Alt. of the Nonag.;" and her *Polar Distance,* for "the Distance from the Nonagesimal," as the Argument at the Head, for the *positive Portion;* and the *negative Portion* is obtained from the same Table (13.) by substituting the ☾'s *Co-Latitude* at the Side, and *Complement of the Horary Angle* at the Head, as the proper Arguments.

TABLE 19.

The *Second Part* of the MOON'S PARALLAX in DECLINATION, concluded.

Approx. Par. in Declin.	The ☽'s Declination. 21°	22°	23°	24°	25°	26°	27°	28°	29°	30°
1′	0′ 21″.50	0′ 22″.48	0′ 23″.44	0′ 24″.40	0′ 25″.36	0′ 26″.30	0′ 27″.24	0′ 28″.17	0′ 29″.09	0′ 30″.00
2	0 43.00	0 44.95	0 46.89	0 48.81	0 50.71	0 52.60	0 54.48	0 56.34	0 58.18	1 0.00
3	1 4.51	1 7.43	1 10.33	1 13.21	1 16.07	1 18.91	1 21.72	1 24.50	1 27.27	1 30.00
4	1 26.01	1 29.91	1 33.78	1 37.62	1 41.43	1 45.21	1 48.96	1 52.67	1 56.36	2 0.00
5	1 47.51	1 52.38	1 57.22	2 2.02	2 6.79	2 11.51	2 16.20	2 20.84	2 25.44	2 30.00
6	2 9.01	2 14.86	2 20.66	2 26.42	2 32.14	2 37.81	2 43.44	2 49.01	2 54.53	3 0.00
7	2 30.51	2 37.33	2 44.11	2 50.83	2 57.50	3 4.12	3 10.68	3 17.18	3 23.62	3 30.00
8	2 52.02	2 59.81	3 7.55	3 15.23	3 22.86	3 30.42	3 37.92	3 45.34	3 52.71	4 0.00
9	3 13.52	3 22.29	3 31.00	3 39.64	3 48.21	3 56.72	4 5.16	4 13.51	4 21.80	4 30.00
10	3 35.02	3 44.76	3 54.44	4 4.04	4 13.57	4 23.02	4 32.39	4 41.68	4 50.89	5 0.00
11	3 56.52	4 7.24	4 17.88	4 28.45	4 38.93	4 49.33	4 59.63	5 9.85	5 19.98	5 30.00
12	4 18.03	4 29.72	4 41.33	4 52.85	5 4.29	5 15.63	5 26.87	5 38.02	5 49.06	6 0.00
13	4 39.53	4 52.19	5 4.77	5 17.25	5 29.64	5 41.93	5 54.11	6 6.19	6 18.15	6 30.00
14	5 1.03	5 14.67	5 28.22	5 41.66	5 55.00	6 8.23	6 21.35	6 34.36	6 47.24	7 0.00
15	5 22.53	5 37.14	5 51.66	6 6.06	6 20.36	6 34.53	6 48.59	7 2.52	7 16.33	7 30.00
16	5 44.03	5 59.62	6 15.10	6 30.47	6 45.71	7 0.84	7 15.83	7 30.69	7 45.42	8 0.00
17	6 5.54	6 22.10	6 38.55	6 54.87	7 11.07	7 27.14	7 43.07	7 58.86	8 14.51	8 30.00
18	6 27.04	6 44.57	7 1.99	7 19.27	7 36.43	7 53.44	8 10.31	8 27.03	8 43.60	9 0.00
19	6 48.54	7 7.05	7 25.44	7 43.68	8 1.78	8 19.74	8 37.55	8 55.20	9 12.69	9 30.00
20	7 10.04	7 29.53	7 48.88	8 8.08	8 27.14	8 46.05	9 4.79	9 23.37	9 41.77	10 0.00
21	7 31.54	7 52.00	8 12.32	8 32.49	8 52.50	9 12.35	9 32.03	9 51.53	10 10.86	10 30.00
22	7 53.05	8 14.48	8 35.77	8 56.89	9 17.86	9 38.65	9 59.27	10 19.70	10 39.95	11 0.00
23	8 14.55	8 36.96	8 59.21	9 21.30	9 43.21	10 4.95	10 26.51	10 47.87	11 9.04	11 30.00
24	8 36.05	8 59.43	9 22.65	9 45.70	10 8.57	10 31.25	10 53.75	11 16.04	11 38.13	12 0.00
25	8 57.55	9 21.91	9 46.10	10 10.10	10 33.93	10 57.56	11 20.99	11 44.21	12 7.22	12 30.00
26	9 19.05	9 44.38	10 9.54	10 34.51	10 59.28	11 23.86	11 48.23	12 12.38	12 36.30	13 0.00
27	9 40.56	10 6.86	10 32.99	10 58.91	11 24.64	11 50.16	12 15.47	12 40.54	13 5.39	13 30.00
28	10 2.06	10 29.34	10 56.43	11 23.32	11 50.00	12 16.46	12 42.71	13 8.71	13 34.48	14 0.00
29	10 23.56	10 51.81	11 19.87	11 47.72	12 15.36	12 42.77	13 9.95	13 36.88	14 3.57	14 30.00
30	10 45.06	11 14.29	11 43.32	12 12.13	12 40.71	13 9.07	13 37.18	14 5.05	14 32.66	15 0.00
31	11 6.56	11 36.77	12 6.76	12 36.53	13 6.07	13 35.37	14 4.42	14 33.22	15 1.75	15 30.00
32	11 28.07	11 59.24	12 30.20	13 0.93	13 31.43	14 1.67	14 31.66	15 1.39	15 30.83	16 0.00
33	11 49.57	12 21.72	12 53.65	13 25.34	13 56.78	14 27.98	14 58.90	15 29.55	15 59.92	16 30.00
34	12 11.07	12 44.20	13 17.09	13 49.74	14 22.14	14 54.28	15 26.14	15 57.72	16 29.01	17 0.00
35	12 32.57	13 6.67	13 40.54	14 14.15	14 47.50	15 20.58	15 53.38	16 25.89	16 58.10	17 30.00
36	12 54.07	13 29.15	14 3.98	14 38.55	15 12.86	15 46.88	16 20.62	16 54.06	17 27.19	18 0.00
37	13 15.58	13 51.63	14 27.42	15 2.95	15 38.21	16 13.18	16 47.86	17 22.23	17 56.28	18 30.00
38	13 37.08	14 14.10	14 50.87	15 27.36	16 3.57	16 39.49	17 15.10	17 50.39	18 25.37	19 0.00
39	13 58.58	14 36.58	15 14.31	15 51.76	16 28.93	17 5.79	17 42.34	18 18.56	18 54.46	19 30.00
40	14 20.08	14 59.06	15 37.76	16 16.17	16 54.28	17 32.09	18 9.58	18 46.73	19 23.54	20 0.00
41	14 41.59	15 21.53	16 1.20	16 40.57	17 19.64	17 58.39	18 36.82	19 14.90	19 52.63	20 30.00
42	15 3.09	15 44.01	16 24.64	17 4.98	17 45.00	18 24.70	19 4.06	19 43.07	20 21.72	21 0.00
43	15 24.59	16 6.48	16 48.09	17 29.38	18 10.36	18 51.00	19 31.30	20 11.24	20 50.81	21 30.00
44	15 46.09	16 28.96	17 11.53	17 53.78	18 35.71	19 17.30	19 58.53	20 39.41	21 19.90	22 0.00
45	16 7.59	16 51.44	17 34.98	18 18.19	19 1.07	19 43.60	20 25.77	21 7.57	21 48.99	22 30.00
46	16 29.10	17 13.91	17 58.42	18 42.59	19 26.43	20 9.90	20 53.01	21 35.74	22 18.08	23 0.00
47	16 50.60	17 36.39	18 21.86	19 7.00	19 51.78	20 36.21	21 29.25	23 3.91	22 47.16	23 30.00
48	17 12.10	17 58.87	18 45.31	19 31.40	20 17.14	21 2.51	21 47.49	22 32.08	23 16.25	24 0.00
49	17 33.60	18 21.34	19 8.75	19 55.80	20 42.50	21 28.81	22 14.73	23 0.25	23 45.34	24 30.00
50	17 55.10	18 43.82	19 32.19	20 20.21	21 7.86	21 55.11	22 41.97	23 28.42	24 14.43	25 0.00
51	18 16.61	19 6.30	19 55.64	20 44.61	21 33.21	22 21.42	23 9.21	23 56.58	24 43.52	25 30.00
52	18 38.11	19 28.77	20 19.08	21 9.02	21 58.57	22 47.72	23 36.45	24 24.75	25 12.61	26 0.00
53	18 59.61	19 51.25	20 42.53	21 33.42	22 23.93	23 14.02	24 3.60	24 52.92	25 41.70	26 30.00
54	19 21.11	20 13.73	21 5.97	21 57.83	22 49.28	23 40.32	24 30.93	25 21.09	26 10.78	27 0.00

The Moon's *approximate Parallax* in *Declination* is obtained from TABLE 13, by substituting *her Latitude* as the proper Argument at the Side, instead of the "Alt. of the Nonag.;" and her *Polar Distance* for "the Distance from the Nonagesimal," as the Argument at the Head, for the *positive Portion;* and the *negative Portion* is obtained from the same Table (13.) by substituting the ☾'s *Co-Latitude* at the Side, and *Complement of the Horary Angle* at the Head, as the proper Arguments.

The Reduction of the Moon's Parallaxes, in Altitude, Longitude, Latitude, Right Ascension, Declination, and Azimuth, from the Sphere to the Spheroid, for 60′ of Horizontal Parallax, and for the Latitude of Greenwich, with the Variation of Reduction for 100′ of Latitude, North of Greenwich. (Compression $\frac{1}{300}$.) (Mr. J. Utting.)

TABLE 20. Reduction of the Moon's Parallax in Altitude.

Arg. the Moon's Azimuth.

Arg. ☾'s Alt.	0°	5°	10°	15°	20°	25°	30°	35°	40°	45°	50°	55°	60°	65°	70°	75°	80°	85°	90°
0°	0″.00	0″.00	0″.00	0″.00	0″.00	0″.00	0″.00	0″.00	0″.00	0″.00	0″.00	0″.00	0″.00	0″.00	0″.00	0″.00	0″.00	0″.00	0″.00
5	0 .99	0 .99	0 .98	0 .96	0 .93	0 .90	0 .86	0 .81	0 .76	0 .70	0 .64	0 .57	0 .50	0 .42	0 .34	0 .26	0 .17	0 .08	0 .00
10	1 .98	1 .97	1 .95	1 .91	1 .86	1 .79	1 .71	1 .62	1 .51	1 .40	1 .27	1 .13	0 .99	0 .84	0 .68	0 .51	0 .34	0 .17	0 .00
15	2 .95	2 .93	2 .90	2 .85	2 .77	2 .67	2 .55	2 .41	2 .26	2 .08	1 .89	1 .69	1 .47	1 .24	1 .01	0 .76	0 .51	0 .26	0 .00
20	3 .89	3 .88	3 .83	3 .76	3 .66	3 .53	3 .37	3 .19	2 .98	2 .75	2 .50	2 .23	1 .95	1 .64	1 .33	1 .01	0 .68	0 .34	0 .00
25	4 .81	4 .79	4 .74	4 .65	4 .52	4 .36	4 .17	3 .94	3 .68	3 .40	3 .09	2 .76	2 .40	2 .03	1 .64	1 .24	0 .84	0 .42	0 .00
30	5 .69	5 .67	5 .60	5 .50	5 .35	5 .16	4 .93	4 .66	4 .36	4 .02	3 .66	3 .26	2 .85	2 .40	1 .95	1 .47	0 .99	0 .50	0 .00
35	6 .53	6 .50	6 .43	6 .31	6 .13	5 .92	5 .65	5 .35	5 .00	4 .62	4 .20	3 .74	3 .26	2 .76	2 .23	1 .69	1 .13	0 .57	0 .00
40	7 .32	7 .29	7 .20	7 .07	6 .87	6 .63	6 .34	5 .99	5 .60	5 .17	4 .70	4 .20	3 .66	3 .09	2 .50	1 .89	1 .27	0 .64	0 .00
45	8 .05	8 .02	7 .93	7 .77	7 .56	7 .29	6 .97	6 .58	6 .16	5 .69	5 .17	4 .62	4 .02	3 .40	2 .75	2 .08	1 .40	0 .70	0 .00
50	8 .72	8 .68	8 .59	8 .42	8 .19	7 .90	7 .55	7 .14	6 .68	6 .16	5 .60	5 .00	4 .36	3 .68	2 .98	2 .26	1 .51	0 .76	0 .00
55	9 .32	9 .29	9 .18	9 .00	8 .76	8 .45	8 .07	7 .64	7 .14	6 .59	5 .99	5 .35	4 .66	3 .94	3 .19	2 .41	1 .62	0 .81	0 .00
60	9 .86	9 .82	9 .71	9 .52	9 .26	8 .93	8 .54	8 .07	7 .55	6 .97	6 .34	5 .65	4 .93	4 .17	3 .37	2 .55	1 .71	0 .86	0 .00
65	10 .31	10 .27	10 .16	9 .96	9 .69	9 .35	8 .93	8 .45	7 .90	7 .29	6 .63	5 .92	5 .16	4 .36	3 .53	2 .67	1 .79	0 .90	0 .00
70	10 .69	10 .65	10 .53	10 .33	10 .05	9 .69	9 .26	8 .76	8 .19	7 .56	6 .87	6 .13	5 .35	4 .52	3 .66	2 .77	1 .86	0 .93	0 .00
75	10 .99	10 .95	10 .83	10 .62	10 .33	9 .96	9 .52	9 .00	8 .42	7 .77	7 .07	6 .31	5 .50	4 .65	3 .76	2 .85	1 .91	0 .96	0 .00
80	11 .21	11 .17	11 .04	10 .83	10 .53	10 .16	9 .71	9 .18	8 .59	7 .93	7 .20	6 .43	5 .60	4 .74	3 .83	2 .90	1 .95	0 .98	0 .00
85	11 .34	11 .29	11 .17	10 .95	10 .65	10 .27	9 .82	9 .29	8 .68	8 .02	7 .29	6 .50	5 .67	4 .79	3 .88	2 .93	1 .97	0 .99	0 .00
90	11 .38	11 .34	11 .21	10 .99	10 .69	10 .31	9 .86	9 .32	8 .72	8 .05	7 .32	6 .53	5 .69	4 .81	3 .89	2 .95	1 .98	0 .99	0 .00

Note.—For 100′ of Latitude, North of Greenwich, the Reduction must be diminished by 0″.14873 for each 10″ of the Reduction, in the above Table; and in proportion for any other Latitude.

If the Latitude is South of Greenwich, the above Correction must be *added* to the *Reduction*, before it is applied to the Moon's Parallax in Altitude.

TABLE 21. Reduction of the Moon's Parallax in Longitude.

☾'s Long.	Reduction.	Var. 100′. +
0°	7″.28	0″.167
3	7 .27	0 .166
6	7 .24	0 .166
9	7 .19	0 .165
12	7 .12	0 .163
15	7 .03	0 .161
18	6 .92	0 .158
21	6 .79	0 .155
24	6 .65	0 .152
27	6 .48	0 .149
30	6 .30	0 .145
33	6 .10	0 .140
36	5 .89	0 .135
39	5 .65	0 .129
42	5 .41	0 .124
45	5 .14	0 .118
48	4 .87	0 .111
51	4 .58	0 .105
54	4 .28	0 .098
57	4 .96	0 .091
60	3 .64	0 .084
63	3 .30	0 .076
66	2 .96	0 .068
69	2 .61	0 .060
72	2 .25	0 .052
75	1 .88	0 .043
78	1 .51	0 .034
81	1 .14	0 .026
84	0 .76	0 .018
87	0 .38	0 .009
[illegible]	0 .00	0 .000

TABLE 22. Reduction of the Moon's Parallax in Latitude.

☾'s Lat.	Reduction.	Var. 100′. +
0° 0	16″.76	0″.385
20	16 .76	0 .385
40	16 .76	0 .385
1 0	16 .76	0 .385
10	16 .76	0 .385
20	16 .76	0 .385
30	16 .76	0 .384
40	16 .75	0 .384
50	16 .75	0 .384
2 0	16 .75	0 .384
10	16 .75	0 .384
20	16 .75	0 .384
30	16 .75	0 .384
40	16 .74	0 .381
50	16 .74	0 .384
3 0	16 .74	0 .384
10	16 .74	0 .384
20	16 .73	0 .384
30	16 .73	0 .383
40	16 .73	0 .383
50	16 .72	0 .383
4 0	16 .72	0 .383
10	16 .72	0 .383
20	16 .71	0 .383
30	16 .71	0 .383
40	16 .71	0 .383
50	16 .70	0 .383
5 0	16 .70	0 .383
10	16 .69	0 .383
20	16 .69	0 .383
30	16 .69	0 .383

TABLE 23. Reduction of the Moon's Parallax in Right Ascen.

☾'s Dec.	Reduction.	Var. 100′. +
0°	0″.00	0″.000
1	0 .32	0 .007
2	0 .64	0 .015
3	0 .96	0 .022
4	1 .28	0 .029
5	1 .59	0 .036
6	1 .91	0 .044
7	2 .23	0 .051
8	2 .54	0 .058
9	2 .86	0 .066
10	3 .17	0 .073
11	3 .49	0 .080
12	3 .80	0 .087
13	4 .11	0 .094
14	4 .42	0 .101
15	4 .73	0 .108
16	5 .04	0 .115
17	5 .34	0 .122
18	5 .65	0 .129
19	5 .95	0 .136
20	6 .25	0 .143
21	6 .55	0 .150
22	6 .85	0 .157
23	7 .14	0 .164
24	7 .43	0 .171
25	7 .72	0 .178
26	8 .01	0 .184
27	8 .30	0 .190
28	8 .58	0 .196
29	8 .86	0 .203
30	9 .14	0 .210

TABLE 24. Reduction of the Moon's Parallax in Declin.

☾'s Dec.	Reduction.	Var. 100′. +
0°	18″.27	0″.419
1	18 .27	0 .419
2	18 .26	0 .419
3	18 .25	0 .419
4	18 .23	0 .418
5	18 .20	0 .417
6	18 .17	0 .417
7	18 .14	0 .416
8	18 .10	0 .415
9	18 .05	0 .414
10	18 .00	0 .413
11	17 .94	0 .411
12	17 .87	0 .410
13	17 .80	0 .409
14	17 .73	0 .407
15	17 .65	0 .405
16	17 .57	0 .403
17	17 .47	0 .401
18	17 .38	0 .399
19	17 .28	0 .396
20	17 .17	0 .393
21	17 .06	0 .391
22	16 .94	0 .388
23	16 .82	0 .386
24	16 .69	0 .383
25	16 .56	0 .380
26	16 .42	0 .377
27	16 .28	0 .373
28	16 .13	0 .370
29	15 .98	0 .366
30	15 .83	0 .363

TABLE 25. Reduction of the Moon's Parallax in Azimuth.

☾'s Azim.	Reduction.	Var. 100′. —
0°	0″.00	0″.000
3	0 .60	0 .009
6	1 .19	0 .018
9	1 .78	0 .026
12	2 .37	0 .035
15	2 .95	0 .044
18	3 .52	0 .053
21	4 .08	0 .061
24	4 .63	0 .069
27	5 .17	0 .077
30	5 .69	0 .084
33	6 .20	0 .092
36	6 .69	0 .099
39	7 .16	0 .106
42	7 .62	0 .113
45	8 .05	0 .119
48	8 .46	0 .125
51	8 .84	0 .131
54	9 .21	0 .137
57	9 .54	0 .142
60	9 .86	0 .147
63	10 .14	0 .151
66	10 .40	0 .155
69	10 .63	0 .159
72	10 .82	0 .161
75	10 .99	0 .164
78	11 .13	0 .166
81	11 .24	0 .167
84	11 .32	0 .168
87	11 .37	0 .169
90	11 .38	0 .169

PENDIX TO TABLE 13. The MOON'S APPROXIMATE PARALLAX in Longitude, &c. completed.—See Page 238, &c. Args. The Alt. of the Nonag., the ☾ Lat. or Co-Lat. at the *Side;* and the ☾'s Dist. from the Nonag., the ☾'s Hor. Ang. or the ☾'s Polar Dist. at the *Head* of the Columns. (MR. J. UTTING.)

Alt. of Non. or ☾'s Lat.	The ☾'s distance from the Nonagesimal; the ☾'s Horary Angle; or the ☾'s Polar Distance, as the case may be.											
	1°	2°	3°	4°	5°	6°	7°	8°	9°	10°	11°	12°
1°	0′ 1″.10	0′ 2″.20	0′ 3″.29	0′ 4″.39	0′ 5″.48	0′ 6″.57	0′ 7″.66	0′ 8″.75	0′ 9″.83	0′ 10″.91	0′ 11″.98	0′ 13″.06
2	0 2.20	0 4.39	0 6.58	0 8.77	0 10.95	0 13.13	0 15.31	0 17.49	0 19.65	0 21.82	0 23.97	0 26.12
3	0 3.29	0 6.58	0 9.86	0 13.14	0 16.42	0 19.69	0 22.96	0 26.22	0 29.47	0 32.72	0 35.95	0 39.17
4	0 4.39	0 8.77	0 13.14	0 17.51	0 21.88	0 26.24	0 30.60	0 34.94	0 39.28	0 43.61	0 47.91	0 52.20
5	0 5.48	0 10.95	0 16.42	0 21.88	0 27.34	0 32.79	0 38.23	0 43.66	0 49.08	0 54.48	0 59.87	1 5.24
6	0 6.57	0 13.13	0 19.69	0 26.24	0 32.79	0 39.33	0 45.86	0 52.37	0 58.87	1 5.34	1 11.80	1 18.24
7	0 7.66	0 15.31	0 22.96	0 30.60	0 38.23	0 45.86	0 53.45	1 1.04	1 8.63	1 16.18	1 23.71	1 31.22
8	0 8.75	0 17.49	0 26.22	0 34.94	0 43.66	0 52.37	1 1.04	1 9.72	1 18.37	1 27.00	1 35.60	1 44.17
9	0 9.83	0 19.65	0 29.47	0 39.28	0 49.08	0 58.87	1 8.63	1 18.37	1 28.10	1 37.79	1 47.46	1 57.09
10	0 10.91	0 21.82	0 32.72	0 43.61	0 54.48	1 5.34	1 16.18	1 27.00	1 37.79	1 48.55	1 59.28	2 9.97
☾'s Co-Lat.												
70°	0 59.04	1 58.06	2 57.05	3 55.98	4 54.84	5 53.61	6 52.27	7 50.81	8 49.20	9 47.43	10 45.49	11 43.34
71	0 59.41	1 58.79	2 58.14	3 57.44	4 56.66	5 55.80	6 54.82	7 53.71	8 52.48	9 51.08	10 49.49	11 47.70
72	0 59.76	1 59.49	2 59.19	3 58.83	4 58.39	5 57.88	6 57.25	7 56.49	8 55.60	9 54.54	10 53.29	11 51.85
73	1 0.08	2 0.15	3 0.18	4 0.15	5 0.05	5 59.86	6 59.55	7 59.12	8 58.56	9 57.82	10 56.90	11 55.78
74	1 0.39	2 0.77	3 1.11	4 1.40	5 1.61	6 1.73	7 1.73	8 1.61	9 1.35	10 0.92	11 0.30	11 59.49
75	1 0.68	2 1.36	3 1.99	4 2.57	5 3.07	6 3.48	7 3.78	8 3.95	9 3.98	10 3.83	11 3.50	12 2.98
76	1 0.96	2 1.91	3 2.81	4 3.66	5 4.44	6 5.12	7 5.70	8 6.14	9 6.44	10 6.56	11 6.50	12 6.25
77	1 1.22	2 2.42	3 3.58	4 4.69	5 5.72	6 6.66	7 7.49	8 8.18	9 8.73	10 9.11	11 9.31	12 9.30
78	1 1.46	2 2.89	3 4.29	4 5.64	5 6.90	6 8.08	7 9.14	8 10.07	9 10.86	10 11.45	11 11.90	12 12.13
79	1 1.67	2 3.33	3 4.95	4 6.51	5 7.99	6 9.39	7 10.67	8 11.82	9 12.82	10 13.65	11 14.29	12 14.73
80	1 1.87	2 3.73	3 5.55	4 7.31	5 8.99	6 10.59	7 12.06	8 13.41	9 14.61	10 15.64	11 16.47	12 17.11
81	1 2.05	2 4.09	3 6.09	4 8.03	5 9.90	6 11.67	7 13.33	8 14.86	9 16.23	10 17.44	11 18.45	12 19.27
82	1 2.22	2 4.42	3 6.58	4 8.68	5 10.71	6 12.64	7 14.46	8 16.15	9 17.68	10 19.05	11 20.22	12 21.20
83	1 2.36	2 4.70	3 7.01	4 9.25	5 11.42	6 13.50	7 15.46	8 17.29	9 18.97	10 20.47	11 21.78	12 22.90
84	1 2.48	2 4.95	3 7.38	4 9.75	5 12.04	6 14.24	7 16.33	9 18.28	9 20.08	10 21.71	11 23.14	12 24.38
85	1 2.59	2 5.16	3 7.69	4 10.17	5 12.57	6 14.87	7 17.06	8 19.12	9 21.02	10 22.75	11 24.39	12 25.63
86	1 2.68	2 5.33	3 7.95	4 10.51	5 13.00	6 15.39	7 17.66	8 19.80	9 21.79	10 23.61	11 25.23	12 26.66
87	1 2.74	2 5.46	3 8.15	4 10.78	5 13.34	6 15.79	7 18.13	8 20.34	9 22.39	10 24.28	11 25.97	12 27.46
88	1 2.79	2 5.56	3 8.29	4 10.97	5 13.57	6 16.07	7 18.46	8 20.72	9 22.82	10 24.75	11 26.49	12 28.03
89	1 2.82	2 5.62	3 8.38	4 11.08	5 13.71	6 16.25	7 18.66	8 20.95	9 23.08	10 25.04	11 26.80	12 28.37
90	1 2.83	2 5.64	3 8.41	4 11.12	5 13.76	6 16.30	7 18.73	8 21.02	9 23.16	10 25.13	11 26.91	12 28.48

Alt. of Non. or ☾'s Lat.	The ☾'s distance from the Nonagesimal; the ☾'s Horary Angle; or the ☾'s Polar Distance, as the case may be.											
	13°	14°	15°	16°	17°	18°	19°	20°	21°	22°	23°	24°
1°	0′ 14″.13	0′ 15″.20	0′ 16″.26	0′ 17″.32	0′ 18″.37	0′ 19″.42	0′ 20″.46	0′ 21″.49	0′ 22″.52	0′ 23″.54	0′ 24″.55	0′ 25″.56
2	0 28.26	0 30.39	0 32.51	0 34.63	0 36.73	0 38.82	0 40.90	0 42.97	0 45.03	0 47.07	0 49.09	0 51.10
3	0 42.38	0 45.58	0 48.76	0 51.93	0 55.09	0 58.23	1 1.34	1 4.44	1 7.52	1 10.58	1 13.62	1 16.64
4	0 56.49	1 0.75	1 4.99	1 9.22	1 13.42	1 17.60	1 21.76	1 25.89	1 29.99	1 34.07	1 38.12	1 42.14
5	1 10.58	1 15.91	1 21.21	1 26 49	1 31.74	1 36.96	1 42.16	1 47.32	1 52.45	1 57.54	2 2.60	2 7.62
6	1 24.65	1 31.04	1 37.39	1 43.72	1 50.02	1 56.28	2 2.51	2 8.70	2 14.85	2 20.97	2 27.03	2 33.06
7	1 38.69	1 46.14	1 53.55	2 0.93	2 8.27	2 15.57	2 29.57	2 30.05	2 37.23	2 44.35	2 51.82	2 58.45
8	1 52.71	2 1.21	2 9.67	2 18.10	2 26.42	2 34.82	2 43.12	2 51.36	2 59.55	3 7.69	3 15.77	3 23.78
9	2 6.68	2 16.24	2 25.76	2 35.23	2 44.65	2 54.03	3 3.35	3 12.61	3 21.82	3 30.96	3 40.04	3 49.05
10	2 20.62	2 31.23	2 41.80	2 52.31	3 2.77	3 13.18	3 23.52	3 33.81	3 44.03	3 54.18	4 4.26	4 14.27
☾'s Co-Lat.												
70°	12 40.99	13 38.40	14 35.56	15 32.45	16 29.06	17 25.37	18 21.36	19 17.02	20 12.32	21 7.25	22 1.80	22 55.95
71	12 45.70	13 43.47	14 40.98	15 38.23	16 35.18	17 31.85	18 28.19	19 24.19	20 19.83	21 15.11	22 9.99	23 4.48
72	12 50.18	13 48.29	14 46.14	15 43.73	16 41.02	17 38.01	18 34.68	19 31.01	20 26.97	21 22.58	22 17.78	23 12.59
73	12 54.44	13 52.86	14 51.03	15 48.94	16 46.53	17 43.85	18 40.83	19 37.47	20 33.76	21 29.66	22 25.17	23 20.27
74	12 58.45	13 57.18	14 55.65	15 53.85	16 51.76	17 49.37	18 46.61	19 43.57	20 40.14	21 36.34	22 32.14	23 27.58
75	13 2.22	14 1.24	15 0.00	15 58.48	16 56.68	17 54.56	18 52.11	19 49.32	20 46.16	21 42.63	22 38.70	23 34.36
76	13 5.77	14 5.05	15 4.07	16 2.82	17 1.28	17 59.42	18 57.23	19 54.70	20 51.80	21 48.53	22 44.85	23 40.76
77	13 9.07	14 8.60	15 7.87	16 6.86	17 5.56	18 3.95	19 2.00	19 59.71	20 57.05	21 54.02	22 50.58	23 46.72
78	13 12.13	14 11.89	15 11.39	16 10.61	17 9.54	18 8.15	19 6.43	20 4.37	21 1.93	21 59.11	22 55.89	23 52.25
79	13 14.94	14 14.92	15 14.63	16 14.06	17 13.20	18 12.02	19 10.51	20 8.65	21 6.42	22 3.81	23 0.79	23 57.35
80	13 17.52	14 17.69	15 17.59	16 17.22	17 16.55	18 15.56	19 14.24	20 12.57	21 10.52	22 8.10	23 5.26	24 2.01
81	13 19.85	14 20.20	15 20.28	16 20.08	17 19.58	18 18.76	19 17.61	20 16.11	21 14.24	22 11.98	23 9.31	24 6.22
82	13 21.94	14 22.44	15 22.68	16 22.64	17 22.30	18 21.64	19 20.64	20 19.29	21 17.57	22 15.46	23 12.94	24 10.00
83	13 23.78	14 24.43	15 24.80	16 24.90	17 24.69	18 24.17	19 23.31	20 22.10	21 20.51	22 18.53	23 16.14	24 13.34
84	13 25.36	14 26.15	15 26.64	16 26.86	17 26.77	18 26.37	19 25.63	20 24.53	21 23.06	22 21.20	23 18.92	24 16.23
85	13 26.73	14 27.60	15 28.20	16 28.52	17 28.53	18 28.23	19 27.59	20 26.59	21 25.22	22 23.45	23 21.27	24 18.68
86	13 27.85	14 28.80	15 29.40	16 29.88	17 29.97	18 29.75	19 29.19	20 28.27	21 26.98	22 25.30	23 23.20	24 20.69
87	13 28.71	14 29.73	15 30.47	16 30.93	17 31.09	18 30.94	19 30.44	20 29.59	21 28.36	22 26.74	23 24.70	24 22.25
88	13 29.33	14 30.39	15 31.18	16 31.69	17 31.90	18 31.79	19 31.33	20 30.52	21 29.34	22 27.76	23 25.77	24 23.36
89	13 29.70	14 30.79	15 31.60	16 32.14	17 32.37	18 32.29	19 31.86	20 31.08	21 29.93	22 28.38	23 26.41	24 24.03
90	13 29.82	14 30.92	15 31.75	16 32.29	17 32.54	18 32.46	19 32.05	20 31.27	21 30.12	22 28.58	23 26.63	24 24.29

APPENDIX TO TABLE 13. The MOON'S APPROXIMATE PARALLAX in LONGITUDE, &c. completed.—See Page 238, &c.

Alt. of Non. or ☾'s Lat.	The ☾'s distance from the Nonagesimal; the ☾'s Horary Angle; or the ☾'s Polar Distance, as the case may be.										
	25°	26°	27°	28°	29°	30°	31°	32°	33°	34°	35°
1°	0′ 26″.55	0′ 27″.54	0′ 28″.52	0′ 29″.49	0′ 30″.46	0′ 31″.41	0′ 32″.36	0′ 33″.29	0′ 34″.22	0′ 35″.13	0′ 36″.03
2	0 53.10	0 55.08	0 57.04	0 58.98	1 0.91	1 2.82	1 4.71	1 6.58	1 8.43	1 10.26	1 12.07
3	1 19.63	1 22.60	1 25.54	1 28.45	1 31.31	1 34.20	1 37.04	1 39.84	1 42.61	1 45.36	1 48.07
4	1 46.13	1 50.08	1 54.00	1 57.90	2 1.75	2 5.56	2 9.34	2 13.08	2 16.78	2 20.43	2 24.04
5	2 12.60	2 17.54	2 22.44	2 27.30	2 32.11	2 36.88	2 41.60	2 46.27	2 50.88	2 55.45	2 59.97
6	2 39.03	2 44.96	2 50.84	2 56.66	3 2.44	3 8.15	3 13.81	3 19.41	3 24.95	3 30.43	3 35.84
7	3 5.42	3 12.33	3 19.18	3 26.46	3 32.70	3 39.36	3 45.96	3 52.49	3 58.95	4 5.33	4 11.64
8	3 31.74	3 39.64	3 47.46	3 55.22	4 2.90	4 10.51	4 18.05	4 25.50	4 32.88	4 40.17	4 47.38
9	3 58.00	4 6.87	4 15.67	4 24.39	4 33.02	4 41.58	4 50.05	4 58.43	5 6.72	5 14.92	5 23.04
10	4 24.19	4 34.04	4 43.80	4 53.48	5 3.07	5 12.57	5 21.97	5 31.27	5 40.47	5 49.57	5 58.56
☾'s Co-Lat.											
70°	23 49.67	24 42.96	25 35.80	26 28.17	27 20.06	28 11.45	29 2.32	29 52.66	30 42.46	31 31.68	32 20.35
71	23 58.54	24 52.16	25 45.32	26 38.02	27 30.22	28 21.93	29 13.12	30 3.77	30 53.87	31 43.42	32 32.38
72	24 6.96	25 0.90	25 54.38	26 47.38	27 39.89	28 31.90	29 23.39	30 14.34	31 4.74	31 54.57	32 43.81
73	24 14.94	25 9.18	26 2.95	26 56.25	27 49.05	28 41.35	29 33.12	30 24.35	31 15.02	32 5.18	32 54.65
74	24 22.48	25 17.00	26 11.05	27 4.63	27 57.70	28 50.27	29 42.31	30 33.81	31 24.75	32 15.11	33 4.88
75	24 29.58	25 24.36	26 18.67	27 12.51	28 5.85	28 58.67	29 50.96	30 42.71	31 33.90	32 24.50	33 14.51
76	24 36.23	25 31.26	26 25.81	27 19.89	28 13.47	29 6.53	29 59.05	30 51.04	31 42.46	32 33.30	33 23.53
77	24 42.43	25 37.69	26 32.47	27 26.78	28 20.58	29 13.87	30 6.62	30 58.81	31 50.44	32 41.50	33 31.95
78	24 48.18	25 43.65	26 38.65	27 33.17	28 27.18	29 20.67	30 13.62	31 6.02	31 57.85	32 49.10	33 39.75
79	24 53.47	25 49.14	26 44.34	27 39.05	28 33.25	29 26.93	30 20.07	31 12.66	32 4.67	32 56.11	33 46.94
80	24 58.31	25 54.16	26 49.54	27 44.42	28 38.79	29 32.65	30 25.97	31 18.73	32 10.91	33 2.51	33 53.50
81	25 2.69	25 58.71	26 54.25	27 49.29	28 43.82	29 37.84	30 31.31	31 24.22	32 16.55	33 8.31	33 59.45
82	25 6.62	26 2.78	26 58.46	27 53.65	28 48.32	29 42.48	30 36.09	31 29.14	32 21.61	33 13.50	34 4.77
83	25 10.08	26 6.37	27 2.18	28 57.50	28 52.30	29 46.58	30 40.31	31 33.49	32 26.09	33 18.09	34 9.48
84	25 13.09	26 9.49	27 5.41	27 0.84	28 55.75	29 50.14	30 43.98	31 37.26	32 29.96	33 22.07	34 13.56
85	25 15.63	26 12.13	27 8.15	28 3.67	28 58.68	28 53.15	30 47.08	31 40.45	32 33.24	33 25.43	34 17.01
86	25 17.72	26 14.29	27 10.38	28 5.98	29 1.07	29 55.62	30 49.62	31 43.06	32 35.92	33 28.19	34 19.84
87	25 19.34	26 15.97	27 12.12	28 7.78	29 2.92	29 57.53	30 51.59	31 45.09	32 38.01	33 30.34	34 22.04
88	25 20.50	26 17.18	27 13.37	28 9.07	29 4.25	29 58.90	30 53.00	31 46.55	32 39.51	33 31.87	34 23.61
89	25 21.19	26 17.90	27 14.12	28 9.84	29 5.05	29 59.73	30 53.85	31 47.42	32 40.40	33 32.79	34 24.55
90	25 21.43	26 18.14	27 14.37	28 10.10	29 5.31	30 0.00	30 54.14	31 47.71	32 40.70	33 33.09	34 24.87

Alt. of Non. or ☾'s Lat.	The ☾'s distance from the Nonagesimal; the ☾'s Horary Angle; or the ☾'s Polar Distance, as the case may be.										
	36°	37°	38°	39°	40°	41°	42°	43°	44°	45°	46°
1°	0′ 36″.93	0′ 37″.81	0′ 38″.68	0′ 39″.54	0′ 40′.38	0′ 41″.22	0′ 42″.04	0′ 42″.85	0′ 43″.65	0′ 44″.43	0′ 45″.20
2	1 13.85	1 15.61	1 17.35	1 19.07	1 20.76	1 22.43	1 24.07	1 25.69	1 27.28	1 28.84	1 30.38
3	1 50.75	1 53.39	1 56.00	1 58.57	2 1.11	2 3.61	2 6.07	2 8.50	2 8.88	2 13.22	2 15.53
4	2 27.61	2 31.13	2 34.61	2 38.04	2 41.42	2 44.75	2 48.03	2 51.27	2 54.44	2 57.56	3 0.64
5	3 4.42	3 8.83	3 13.17	3 17.46	3 21.68	3 25.85	3 29.95	3 33.99	3 37.96	3 41.86	3 45.70
6	3 41.19	3 46.46	3 51.67	3 56.81	4 1.88	4 6.88	4 11.80	4 16.64	4 21.40	4 26.09	4 30.69
7	4 17.88	4 24.03	4 30.11	4 36.10	4 42.01	4 47.83	4 53.57	4 59.21	5 4.77	5 10.23	5 15.60
8	4 54.49	5 1.52	5 8.46	5 15.30	5 22.05	5 28.70	5 35.25	5 41.70	5 48.04	5 54.28	6 0.41
9	5 31.02	5 38.94	5 46.72	5 54.41	6 2.00	6 9.47	6 16.83	6 24.08	6 31.21	6 38.22	6 45.11
10	6 7.44	6 16.21	6 24.87	6 33.41	6 41.83	6 50.12	6 58.30	7 6.34	7 14.25	7 22.04	7 29.68
☾'s Co-Lat.											
70°	33 8.41	33 55.88	34 42.72	35 28.92	36 14.48	36 59.38	37 43.60	38 27.13	39 9.96	39 52.07	40 33.45
71	33 20.74	34 8.49	34 55.63	35 42.12	36 27.96	37 13.13	37 57.63	38 41.43	39 24.53	40 6.90	40 48.54
72	33 32.46	34 20.49	35 7.90	35 54.67	36 40.78	37 26.22	38 10.97	38 55.02	39 38.37	40 20.99	41 2.88
73	33 43.57	34 26.87	35 19.54	36 1.56	36 52.92	37 38.60	38 23.61	39 7.91	39 51.50	40 34.35	41 16.47
74	33 54.06	34 42.61	35 30.52	36 17.78	37 4.38	37 50.31	38 35.55	39 20.08	40 3.89	40 46.97	41 29.31
75	34 3.93	34 52.71	35 40.86	36 28.35	37 15.19	38 1.34	38 46.79	39 31.53	40 15.56	40 58.85	41 41.38
76	34 13.17	35 2.18	35 50.55	36 38.26	37 25.30	38 11.66	38 57.32	39 42.26	40 26.49	41 9.97	41 52.70
77	34 21.79	35 11.00	35 59.58	36 47.49	37 34.73	38 21.28	39 7.13	39 52.26	40 36.68	41 20.34	42 3.25
78	34 29.79	35 19.19	36 7.95	36 56.05	37 43.47	38 30.20	39 16.23	40 1.54	40 46.12	41 29.95	42 13.03
79	34 37.15	35 26.72	36 15.66	37 3.93	37 51.52	38 38.41	39 24.61	40 10.18	40 54.82	41 38.81	42 22.05
80	34 43.88	35 33.61	36 22.71	37 11.13	37 58.88	38 45.92	39 32.27	40 17.89	41 2.78	41 26.91	42 30.28
81	34 49.98	35 39.86	36 29.09	37 17.66	38 5.55	38 52.74	39 39.21	40 24.96	41 9.98	41 54.24	42 37.74
82	34 55.43	35 45.45	36 34.81	37 23.50	38 11.52	38 58.83	39 45.43	40 31.30	41 16.43	42 0.81	42 44.42
83	35 0.25	35 50.38	36 39.86	37 28.65	38 16.79	39 4.21	39 50.92	40 36.89	41 22.13	42 6.61	42 50.32
84	35 4.43	35 54.66	36 44.24	37 33.14	38 21.36	39 8.87	39 55.67	40 41.74	41 27.07	42 11.62	42 55.44
85	35 7.97	35 58.29	36 47.95	37 36.94	38 25.23	39 12.82	39 59.70	40 45.84	41 31.25	42 15.90	42 59.77
86	35 10.87	36 1.25	36 50.98	37 40.04	38 28.40	39 16.06	40 3.00	40 49.21	41 34.68	42 19.39	43 3.32
87	35 13.13	36 3.56	36 53.34	37 42.45	38 30.86	39 18.57	40 5.57	40 51.82	41 37.34	42 22.09	43 6.07
88	35 14.74	36 5.21	36 55.03	37 44.18	38 32.63	39 20.37	40 7.40	40 53.69	41 39.25	42 24.04	43 8.05
89	35 15.70	36 6.20	36 56.04	37 45.21	38 33.68	39 21.45	40 8.50	40 54.81	41 40.39	42 25.20	43 9.23
90	35 16.03	36 6.53	36 56.38	37 45.55	38 34.04	39 21.81	40 8.87	40 55.20	41 40.77	42 25.58	43 9.62

APPENDIX TO TABLE 13. The MOON'S APPROXIMATE PARALLAX in Longitude, &c. completed.—See page 238, &c.

The ☾'s distance from the Nonagesimal; the ☾'s Horary Angle; or the ☾'s Polar Distance, as the case may be.

Alt. of Non. or ☾'s Lat.	47°	48°	49°	50°	51°	52°	53°	54°	55°	56°	57°
1°	0′ 45″.95	0′ 46″.69	0′ 47″.42	0′ 48″.13	0′ 48″.83	0′ 49″.51	0′ 50″.18	0′ 50″.83	0′ 51″.47	0′ 52″.09	0′ 52″.69
2	1 31.89	1 33.37	1 34.82	1 36.24	1 37.63	1 39.00	1 40.34	1 41.64	1 42.93	1 44.20	1 45.39
3	2 17.80	2 20.02	2 22.21	2 24.34	2 26.42	2 28.47	2 30.47	2 32.42	2 34.34	2 36.20	2 38.01
4	3 3.66	3 6.62	3 9.53	3 12.38	3 15.17	3 17.89	3 20.56	3 23.17	3 25.71	3 28.19	3 30.61
5	3 49.47	3 53.17	3 56.79	4 0.35	4 3.84	4 7.25	4 10.58	4 13.84	4 17.02	4 20.12	4 23.14
6	4 35.21	4 39.65	4 44.00	4 48.26	4 52.44	4 56.53	5 0.53	5 4.44	5 8.25	5 11.97	5 15.59
7	5 20.87	5 26.04	5 31.11	5 36.09	5 40.96	5 45.72	5 50.39	5 54.94	5 59.39	6 3.72	6 7.95
8	6 6.43	6 12.33	6 18.13	6 23.81	6 29.37	6 34.81	6 40.13	6 45.34	6 50.41	6 55.37	7 0.19
9	6 51.88	6 58.52	7 5.03	7 11.41	7 17.66	7 23.78	7 29.76	7 35.61	7 41.32	7 46.88	7 52.31
10	7 37.30	7 44.56	7 51.79	7 58.88	8 5.82	8 12.61	8 19.25	8 25.74	8 32.08	8 38.26	8 44.28
☾'s Co-Lat.											
70°	41 14.09	41 53.98	42 33.10	43 11.45	43 49.00	44 25.76	45 1.70	45 36.82	46 11.10	46 44.55	47 17.13
71	41 29.43	42 9.57	42 48.93	43 27.51	44 5.29	44 42.28	45 18.44	45 53.79	46 28.28	47 1.93	47 34.71
72	41 44.01	42 24.38	43 3.97	43 42.79	44 20.80	44 57.99	45 34.37	46 9.91	46 44.61	47 18.46	47 51.44
73	41 57.82	42 38.42	43 18.23	43 57.26	44 35.48	45 12.88	45 49.46	46 25.20	47 0.09	47 34.13	48 7.29
74	42 10.88	42 51.68	43 31.70	44 10.93	44 49.34	45 26.94	46 3.71	46 39.64	47 14.71	47 48.92	48 22.25
75	42 23.16	43 4.16	43 44.37	44 23.79	45 2.39	45 40.18	46 17.12	46 53.22	47 28.46	48 2.84	48 36.33
76	42 34.66	43 15.85	43 56.25	44 35.84	45 14.61	45 52.57	46 29.69	47 5.95	47 41.35	48 15.88	48 49.53
77	42 45.39	43 26.75	44 7.31	44 47.08	45 26.02	46 4.13	46 41.40	47 17.81	47 53.36	48 28.04	49 1.83
78	42 55.33	43 36.86	44 17.59	44 57.50	45 36.59	46 14.85	46 52.26	47 28.82	48 4.51	48 39.02	49 13.24
79	43 4.50	43 46.17	44 27.04	45 7.09	45 46.32	46 24.72	47 2.26	47 38.95	48 14.76	48 49.70	49 23.74
80	43 12.87	43 54.68	44 35.68	45 15.86	45 55.22	46 33.74	47 11.40	47 48.21	48 24.14	48 59.19	49 33.34
81	43 20.46	44 2.38	44 43.50	45 23.81	46 3.28	46 41.91	47 19.68	47 56.60	48 32.63	49 7.79	49 42.05
82	43 27.25	44 49.29	45 30.51	45 30.92	46 10.49	46 49.23	47 27.10	48 4.12	48 40.24	49 15.49	49 49.82
83	43 33.25	44 15.38	45 56.70	45 37.20	46 16.86	46 55.69	47 33.66	48 10.75	48 46.97	49 22.29	49 56.71
84	43 38.45	44 20.67	45 2.07	45 42.65	46 22.39	47 1.30	47 39.34	48 16.51	48 52.80	49 28.19	50 2.68
85	43 42.85	44 25.14	45 6.61	45 47.27	46 27.08	47 6.04	47 44.15	48 21.38	48 57.73	49 33.18	50 7.72
86	43 46.46	44 28.81	45 10.33	45 51.04	46 30.91	47 9.93	47 48.09	48 25.37	49 1.77	49 37.27	50 11.86
87	43 49.26	44 31.66	45 13.23	45 53.98	46 33.79	47 12.95	47 51.15	48 28.47	49 4.90	49 40.44	50 15.07
88	43 51.27	44 33.69	45 15.29	45 56.08	46 36.02	47 15.11	47 53.34	48 30.69	49 7.15	49 42.72	50 17.37
89	43 52.47	44 34.91	45 16.53	45 57.34	46 37.30	47 16.41	47 54.65	48 32.02	49 8.50	49 44.08	50 18.75
90	43 52.87	44 35.32	45 16.95	45 57.76	46 37.73	47 16.84	47 55.09	48 32.46	49 8.95	49 44.54	50 19.21

The ☾'s distance from the Nonagesimal; the ☾'s Horary Angle; or the ☾'s Polar Distance, as the case may be.

Alt. of Non. or ☾'s Lat.	58°	59°	60°	61°	62°	63°	64°	65°	66°	67°	68°
1°	0′ 53″.28	0′ 53″.85	0′ 54″.41	0′ 54″.95	0′ 55″.47	0′ 55″.98	0′ 56″.47	0′ 56″.94	0′ 57″.40	0′ 57″.84	0′ 58″.26
2	1 46.55	1 47.69	1 48.81	1 49.89	1 50.93	1 51.94	1 52.92	1 53.86	1 54.78	1 55.65	1 56.49
3	2 39.78	2 41.50	2 43.17	2 44.79	2 46.36	2 47.88	2 49.34	2 50.76	2 52.12	2 53.43	2 54.69
4	3 32.96	3 35.26	3 37.49	3 39.64	3 41.73	3 43.75	3 45.71	3 47.60	3 49.42	3 51.16	3 52.84
5	4 26.08	4 28.95	4 31.73	4 34.43	4 37.05	4 39.58	4 42.01	4 44.37	4 46.63	4 48.81	4 50.91
6	5 19.12	5 22.55	5 25.89	5 29.12	5 32.26	5 35.29	5 38.22	5 41.05	5 43.77	5 46.39	5 48.90
7	6 12.06	6 16.06	6 19.95	6 23.72	6 27.38	6 30.91	6 34.33	6 37.62	6 40.80	6 43.85	6 46.78
8	7 4.89	7 9.46	7 13.90	7 18.20	7 22.38	7 26.42	7 30.32	7 34.08	7 37.71	7 41.19	7 44.54
9	7 57.59	8 2.73	8 7.71	8 12.55	8 17.24	8 21.78	8 26.17	8 30.40	8 34.48	8 38.40	8 42.16
10	8 50.14	8 55.84	9 1.38	9 6.75	9 11.96	9 17.00	9 21.87	9 26.56	9 31.09	9 35.44	9 39.62
☾'s Co-Lat.											
70°	47 48.86	48 19.71	48 49.67	49 18.75	49 46.92	50 14.18	50 40.52	51 5.94	51 30.43	51 53.96	52 16.56
71	48 6.64	48 37.69	49 7.84	49 37.09	50 5.44	50 32.87	50 59.38	51 24.95	51 49.59	52 13.28	52 36.01
72	48 23.55	48 54.77	49 25.10	49 54.53	50 23.04	50 50.62	51 17.29	51 43.01	52 7.80	52 31.63	52 54.50
73	48 39.57	49 10.96	49 41.46	50 11.05	50 39.72	51 7.47	51 34.28	52 0.15	52 25.06	52 49.01	53 12.01
74	48 54.71	49 26.26	49 56.92	50 26.66	50 35.48	51 23.36	51 50.31	52 16.31	52 41.36	53 5.44	53 28.56
75	49 8.95	49 40.65	50 11.46	50 41.34	51 10.30	51 38.32	52 5.41	52 31.53	52 56.70	53 20.90	53 44.13
76	49 22.29	49 54.14	50 25.08	50 55.09	51 24.19	51 52.34	52 19.55	52 45.79	53 11.07	53 35.38	53 58.71
77	49 34.73	50 6.71	50 37.78	51 7.93	51 37.14	52 5.41	52 32.73	52 59.09	53 24.47	53 48.88	54 12.31
78	49 46.26	50 18.37	50 49.56	51 19.82	51 49.15	52 17.53	52 44.95	53 11.41	53 36.90	54 1.40	54 24.92
79	49 56.88	50 29.10	51 0.41	51 30.78	52 0.21	52 28.69	52 56.21	53 22.76	53 48.34	54 12.93	54 36.54
80	50 6.59	50 38.92	51 10.33	51 40.80	52 10.32	52 38.89	53 6.50	53 33.14	53 58.80	54 23.47	54 47.15
81	50 15.39	50 47.82	51 19.31	51 49.86	52 19.48	52 48.13	53 15.82	53 42.53	54 8.27	54 33.02	54 56.77
82	50 23.26	50 55.76	51 27.35	51 57.98	52 27.68	52 56.40	53 24.17	53 50.95	54 16.76	54 41.57	55 5.38
83	50 30.22	51 2.80	51 34.45	52 5.16	52 34.92	53 3.72	53 31.54	53 58.38	54 24.25	54 49.11	55 12.98
84	50 36.25	51 8.90	51 40.61	52 11.38	52 41.20	53 10.05	53 37.93	53 54.83	54 30.75	54 55.66	55 19.58
85	50 41.35	51 14.06	51 45.83	52 16.65	52 46.52	53 15.42	53 43.35	54 10.30	54 36.25	55 1.20	55 25.16
86	50 45.54	51 18.29	51 50.10	52 20.96	52 50.87	53 19.81	53 47.78	54 14.76	54 40.75	55 5.74	55 29.73
87	50 48.79	51 21.57	51 53.42	52 24.31	52 54.25	53 23.22	53 51.22	54 18.24	54 44.26	55 9.27	55 33.29
88	50 51.11	51 23.92	51 55.79	53 26.71	52 56.68	53 25.67	53 53.69	54 20.72	54 46.76	55 11.79	55 35.83
89	50 52.51	51 25.33	51 57.22	52 28.15	52 58.13	53 27.14	53 55.17	54 22.21	54 48.26	55 13.30	55 37.35
90	50 52.97	51 25.80	51 57.69	52 28.63	52 58.61	53 27.62	53 55.66	54 22.71	84 48.76	55 13.82	55 37.86

APPENDIX TO TABLE 13. The Moon's Approximate Parallax in Longitude, &c. completed.—See page 238, &c.

The ☾'s distance from the Nonagesimal; the ☾'s Horary Angle; or the ☾'s Polar Distance, as the case may be.

Alt. of Non. or ☾'s Lat.	69°	70°	71°	72°	73°	74°	75°	76°	77°	78°	79°
1°	0′ 58″.66	0′ 59″.04	0′ 59″.41	0′ 59″.76	1′ 0″.08	1′ 0″.39	1′ 0″.68	1′ 0″.96	1′ 1″.22	1′ 1″.46	1′ 1″.67
2	1 57.29	1 58.06	1 58.79	1 59.49	2 0.15	2 0.77	2 1.36	2 1.91	2 2.42	2 2.89	2 3.33
3	2 55.90	2 57.05	2 58.14	2 59.19	3 0.18	3 1.11	3 1.99	3 2.81	3 3.58	3 4.29	3 4.95
4	3 54.45	3 55.98	3 57.44	3 58.83	4 0.15	4 1.40	4 2.57	4 3.66	4 4.69	4 5.64	4 6.51
5	4 52.92	4 54.84	4 56.66	4 58.39	5 0.05	5 1.61	5 3.07	5 4.44	5 5.72	5 6.90	5 7.99
6	5 51.31	5 53.61	5 55.80	5 57.88	5 59.86	6 1.73	6 3.48	6 5.12	6 6.66	6 8.08	6 9.39
7	6 49.59	6 52.27	6 54.82	6 57.25	6 59.55	7 1.73	7 3.78	7 5.70	7 7.49	7 9.14	7 10.67
8	7 47.75	7 50.81	7 53.71	7 56.49	7 59.12	8 1.61	8 3.95	8 6.14	8 8.18	8 10.07	8 11.82
9	8 45.76	8 49.20	8 52.48	8 55.60	8 58.56	9 1.35	9 3.98	9 6.44	9 8.73	9 10.86	9 12.82
10	9 43.61	9 47.43	9 51.08	9 54.54	9 57.82	10 0.92	10 3.83	10 6.56	10 9.11	10 11.45	10 13.65
☾'s Co-Lat.											
70°	52 38.20	52 58.88	53 18.59	53 37.32	53 55.08	54 11.85	54 27.62	54 42.41	54 56.19	55 8.97	55 20.74
71	52 57.78	53 18.59	53 38.42	53 57.27	54 15.13	54 32.01	54 47.88	55 2.76	55 16.63	55 29.48	55 41.33
72	53 16.40	53 37.32	53 57.27	54 16.23	54 34.20	54 51.17	55 7.14	55 22.10	55 36.05	55 48.98	56 0.90
73	53 34.03	53 55.08	54 15.13	54 34.20	54 52.27	55 9.33	55 25.39	55 40.43	55 54.46	56 7.47	56 19.44
74	53 50.69	54 11.85	54 32.01	54 51.17	55 9.33	55 26.49	55 42.63	55 57.75	56 11.85	56 24.92	56 36.96
75	54 6.38	54 27.62	54 47.88	55 7.14	55 25.39	55 42.63	55 58.85	56 14.04	56 28.21	56 41.34	56 53.44
76	54 21.06	54 42.41	55 2.76	55 22.10	55 40.43	55 57.75	56 14.04	56 29.31	56 43.54	56 56.73	57 8.89
77	54 34.75	54 56.19	55 16.63	55 36.05	55 54.46	56 11.85	56 28.21	56 43.54	56 57.83	57 11.08	57 23.29
78	54 47.44	55 8.97	55 29.48	55 48.98	56 7.47	56 24.92	56 41.34	56 56.73	57 11.08	57 24.38	57 36.64
79	54 59.14	55 20.74	55 41.33	56 0.90	56 19.44	56 36.96	56 53.44	57 8.89	57 23.29	57 36.64	57 48.93
80	55 9.82	55 31.50	55 52.16	56 11.79	56 30.40	56 47.97	57 4.50	57 20.00	57 34.44	57 47.84	58 0.17
81	55 19.50	55 41.24	56 1.96	56 21.65	56 40.31	56 57.94	57 14.52	57 30.06	57 44.55	57 57.98	58 10.35
82	55 28.17	55 49.97	56 10.74	56 30.48	56 49.19	57 6.86	57 23.49	57 39.07	57 53.59	58 7.06	58 19.47
83	55 35.84	55 57.68	56 18.50	56 38.28	56 57.04	57 14.75	57 31.41	57 47.03	58 1.59	58 15.08	58 27.52
84	55 42.48	56 4.36	56 25.22	56 45.05	57 3.84	57 21.58	57 38.28	57 53.93	58 8.52	58 22.04	58 34.50
85	55 48.10	56 10.02	56 30.91	56 50.77	57 9.60	57 27.37	57 44.10	57 59.77	58 14 38	58 27.93	58 40.41
86	55 52.70	56 14.65	56 35.58	56 55.46	57 14.31	57 32.11	57 48.86	58 4.56	58 19.19	58 32.75	58 45.25
87	55 56.28	56 18.26	56 39.20	56 59.11	57 17.98	67 35.80	57 52.57	58 8.28	58 22.93	58 36.51	58 49.02
88	55 58.84	56 20.83	56 41.79	57 1.72	57 20.60	57 38.43	57 55.22	58 10.94	58 25.60	58 39.19	58 51.71
89	56 0.37	56 22.38	56 43.35	57 3.28	57 22.17	57 40.01	57 56.80	58 12.53	58 27.20	58 40.79	58 53.32
90	56 0.89	56 22.89	56 43.87	57 3.80	57 22.70	57 40.54	57 57.33	58 13.06	58 27.73	58 41.33	58 53.86

The ☾'s distance from the Nonagesimal; the ☾'s Horary Angle; or the ☾'s Polar Distance, as the case may be.

Alt. of Non. or ☾'s Lat.	80°	81°	82°	83°	84°	85°	86°	87°	88°	89°	90°
1°	1′ 1″.87	1′ 2″.05	1′ 2″.22	1′ 2″.36	1′ 2″.48	1′ 2″.59	1′ 2″.68	1′ 2″.74	1′ 2″.79	1′ 2″.82	1′ 2″.83
2	2 3.73	2 4.09	2 4.42	2 4.70	2 4.95	2 5.16	2 5.33	2 5.46	2 5.56	2 5.62	2 5.64
3	3 5.55	3 6.09	3 6.58	3 7.01	3 7.38	3 7.69	3 7.95	3 8.15	3 8.29	3 8.38	3 8.41
4	4 7.31	4 8.03	4 8.68	4 9.25	4 9.75	4 10.17	4 10.51	4 10.78	4 10.97	4 11.08	4 11.12
5	5 8.99	5 9.90	5 10.71	5 11.42	5 12.04	5 12.57	5 13.00	5 13.31	5 13.57	5 13.71	5 13.76
6	6 10.59	6 11.67	6 12.64	6 13.50	6 14.24	6 14.87	6 15.39	6 15.79	6 16.07	6 16.25	6 16.30
7	7 12.06	7 13.33	7 14.46	7 15.46	7 16.33	7 17.06	7 17.66	7 18.13	7 18.46	7 18.66	7 18.73
8	8 13.41	8 14.86	8 16.15	8 17.29	8 18.28	8 19.12	8 19.80	8 20.34	8 20.72	8 20.95	8 21.02
9	9 14.61	9 16.23	9 17.68	9 18.97	9 20.08	9 21.02	9 21.79	9 22.39	9 22.82	9 23.08	9 23.16
10	10 15.64	10 17.44	10 19.05	10 20.47	10 21.71	10 22.75	10 23.61	10 24.28	10 24.75	10 25.04	10 25.13
☾'s Co-Lat.											
70°	55 31.50	55 41.24	55 49.97	55 57.68	56 4.36	56 10.02	56 14.65	56 18.26	56 20.83	56 22.38	56 22.89
71	55 52.16	56 1.96	56 10.74	56 18.50	56 25.22	56 30.91	56 35.58	56 39.20	56 41.79	56 43.35	56 43.87
72	56 11.79	56 21.65	56 30.48	56 38.28	56 45.04	56 50.77	56 55.46	56 59.11	57 1.72	57 3.28	57 3.80
73	56 30.40	56 40.31	56 49.19	56 57.04	57 3.84	57 9.60	57 14.31	57 17.98	57 20.60	57 22.17	57 22.70
74	56 47.97	56 57.94	57 6.86	57 14.75	57 21.58	57 27.37	57 32.11	57 35.80	57 38.43	57 40.01	57 40.54
75	57 4.50	57 14.52	57 23.49	57 31.41	57 38.28	57 44.10	57 48.86	57 52.57	57 55.22	57 56.80	57 57.33
76	57 20.00	57 30.06	57 39.07	57 47.03	57 53.93	57 59.77	58 4.56	58 8.28	58 10.94	58 12.53	58 13.06
77	57 34.44	57 44.55	57 53.59	58 1.59	58 8.42	58 14.38	58 19.19	58 22.93	58 25.60	58 27.20	58 27.73
78	57 47.84	57 57.98	58 7.06	58 15.08	58 22.04	58 27.93	58 32.75	58 36.51	58 39.19	58 40.79	58 51.33
79	58 0.17	58 10.35	58 19.47	58 27.52	58 34.50	58 40.41	58 45.25	58 49.02	58 51.71	58 53.32	58 53.86
80	58 11.45	58 21.66	58 30.81	58 38.88	58 45.89	58 51.82	58 56.67	59 0.45	59 3.15	59 4.77	59 5.31
81	58 21.66	58 31.90	58 41.07	58 49.17	58 56.20	59 2.15	59 7.02	59 10.80	59 13.51	59 15.14	59 15.68
82	58 30.81	58 41.07	58 50.27	58 58.39	59 5.44	59 11.40	59 16.28	59 20.08	59 22.79	59 24.42	59 24.94
83	58 38.88	58 49.17	58 58.39	59 6.53	59 13.59	59 19.57	59 24.46	59 28.27	59 30.99	59 32.62	59 33.17
84	58 45.89	58 56.20	59 5.44	59 13.59	59 20.67	59 26.65	59 31.56	59 35.37	59 38.10	59 39.73	59 40.28
85	58 51.82	59 2.15	59 11.40	59 19.57	59 26.65	59 32.65	59 37.57	59 41.41	59 44.12	59 45.75	59 46.30
86	58 56.67	59 7.02	59 16.28	59 24.46	59 31.56	59 37.57	59 42 48	59 46.31	59 49.04	59 50.68	59 51.23
87	59 0.45	59 10.80	59 20.08	59 28.27	59 35.37	59 41.41	59 46.31	59 50.14	59 52.88	59 54.52	59 55.07
88	59 3.15	59 13.51	59 22.79	59 30.99	59 38.10	59 44.12	59 49.04	59 52.88	59 55.62	59 57.26	59 57.81
89	59 4.77	59 15.14	59 24.42	59 32.62	59 39.73	59 45.75	59 50.68	59 54.52	59 57.26	59 58.90	59 59.45
90	59 5.31	59 15.68	59 24.94	59 33.17	59 40.28	59 46.30	59 51.23	59 55.07	59 57.81	59 59.45	60 0.00

AS A TABLE OF DECIMAL ARGUMENTS IN SPACE.

Deg.	Parts.	Deg.	Parts.	Deg.	Parts.	Deg.	Parts.	Deg.	Parts.	Deg.	Parts.	Min.	Parts.
1°	28	61°	1694	121°	3361	181°	5028	241°	6694	301°	8361	1°	0 . 5
2	55	62	1722	122	3388	182	5055	242	6722	302	9388	2	0 . 9
3	83	63	1750	123	3416	183	5083	243	6749	303	8416	3	1 . 4
4	111	64	1778	124	3444	184	5111	244	6777	304	8444	4	1 . 9
5	139	65	1806	125	3474	185	5139	245	6805	305	8472	5	2 . 3
6	166	66	1833	126	3500	186	5166	246	6833	306	8500	6	2 . 8
7	194	67	1861	127	3528	187	5194	247	6861	307	8528	7	3 . 3
8	222	68	1888	128	3555	188	5222	248	6888	308	8555	8	3 . 7
9	250	69	1916	129	3583	189	5250	249	6916	309	8583	9	4 . 2
10	277	70	1944	130	3611	190	5278	250	6944	310	8611	10	4 . 6
11	305	71	1972	131	3639	191	5306	251	6972	311	8638	11	5 . 1
12	333	72	2000	132	3666	192	5333	252	7000	312	8666	12	5 . 6
13	361	73	2028	133	3694	193	5361	253	7028	313	8694	13	6 . 0
14	389	74	2056	134	3722	194	5388	254	7055	314	8722	14	6 . 5
15	417	75	2083	135	3750	195	5416	255	7083	315	8749	15	6 . 9
16	444	76	2111	136	3777	196	5444	256	7111	316	8777	16	7 . 4
17	472	77	2139	137	3805	197	5472	257	7139	317	8805	17	7 . 9
18	500	78	2167	138	3833	198	5500	258	7166	318	8833	18	8 . 4
19	528	79	2194	139	3861	199	5528	259	7194	319	8861	19	8 . 8
20	555	80	2222	140	3888	200	5555	260	7222	320	8888	20	9 . 2
21	583	81	2250	141	3916	201	5583	261	7250	321	8916	21	9 . 7
22	611	82	2277	142	3944	202	5611	262	7278	322	8944	22	10 . 2
23	639	83	2305	143	3972	203	5638	263	7306	323	8972	23	10 . 7
24	666	84	2333	144	4000	204	5666	264	7333	324	9000	24	11 . 2
25	694	85	2361	145	4028	205	5694	265	7361	325	9028	25	11 . 6
26	722	86	2388	146	4056	206	5722	266	7388	326	9055	26	12 . 1
27	750	87	2416	147	4083	207	5749	267	7416	327	9083	27	12 . 5
28	778	88	2444	148	4111	208	5777	268	7444	328	9111	28	13 . 0
29	805	89	2472	149	4139	209	5805	269	7474	329	9139	29	13 . 5
30	833	90	2500	150	4166	210	5833	270	7500	330	9166	30	13 . 9
31	861	91	2528	151	4194	211	5861	271	7528	331	9194	31	14 . 4
32	888	92	2556	152	4222	212	5888	272	7555	332	9222	32	14 . 9
33	916	93	2583	153	4250	213	5916	273	7583	333	9250	33	15 . 3
34	944	94	2611	154	4278	214	5944	274	7611	334	9277	34	15 . 8
35	972	95	2638	155	4306	215	5972	275	7639	335	9305	35	16 . 3
36	1000	96	2666	156	4333	216	6000	276	7666	336	9333	36	16 . 8
37	1028	97	2694	157	4361	217	6028	277	7694	337	9361	37	17 . 2
38	1056	98	2722	158	4388	218	6055	278	7722	338	9388	38	17 . 7
39	1083	99	2749	159	4416	219	6083	279	7750	339	9416	39	18 . 1
40	1111	100	2777	160	4444	220	6111	280	7777	340	9444	40	18 . 5
41	1139	101	2805	161	4472	221	6139	281	7805	341	9472	41	19 . 0
42	1166	102	2833	162	4500	222	6166	282	7833	342	9500	42	19 . 4
43	1194	103	2861	163	4528	223	6194	283	7861	343	9528	43	19 . 9
44	1222	104	2888	164	4556	224	6222	284	7888	344	9556	44	20 . 3
45	1250	105	2916	165	4583	225	6250	285	7916	345	9583	45	20 . 8
46	1277	106	2944	166	4611	226	6277	286	7944	346	9611	46	21 . 3
47	1305	107	2972	167	4638	227	6305	287	7972	347	9638	47	21 . 7
48	1333	108	3000	168	4666	228	6333	288	8000	348	9666	48	22 . 2
49	1361	109	3028	169	4694	229	6361	289	8028	349	9694	49	22 . 7
50	1388	110	3055	170	4722	230	6388	290	8055	350	9722	50	23 . 1
51	1416	111	3083	171	4749	231	6416	291	8083	351	9750	51	23 . 6
52	1444	112	3111	172	4777	232	6444	292	8111	352	9777	52	24 . 0
53	1472	113	3139	173	4805	233	6472	293	8139	353	9805	53	24 . 5
54	1500	114	3166	174	4833	234	6500	294	8166	354	9833	54	25 . 0
55	1528	115	3194	175	4861	235	6528	295	8194	355	9861	55	25 . 4
56	1555	116	3222	176	4888	236	6556	296	8222	356	9888	56	25 . 9
57	1583	117	3250	177	4916	237	6583	297	8250	357	9916	57	26 . 4
58	1611	118	3278	178	4944	238	6611	298	8278	358	9944	58	26 . 8
59	1639	119	3306	179	4972	239	6638	299	8306	359	9972	59	27 . 3
60	1666	120	3333	180	5000	240	6666	300	8333	360	10000	60	27 . 8

A TABLE of the LENGTHS of CIRCULAR ARCS, for Degrees, Minutes, and Seconds, calculated to RADIUS UNITY. (MR. J. UTTING.)

Deg.	Arc.	Deg.	Arc.	Deg.	Arc.	Min.	Arc.	Sec.	Arc.
1°	.01745329	61°	1 .06465084	121°	2 .11184840	1′	.00029089	1″	.00000485
2	.03490658	62	1 .08210414	122	2 .12930169	2	.00058178	2	.00000970
3	.05235988	63	1 .09955743	123	2 .14675498	3	.00087266	3	.00001454
4	.06981317	64	1 .11701072	124	2 .16420827	4	.00116355	4	.00001939
5	.08726646	65	1 .13446401	125	2 .18166156	5	.00145444	5	.00002424
6	.10471976	66	1 .15191731	126	2 .19911486	6	.00174533	6	.00002909
7	.12217305	67	1 .16937060	127	2 .21656815	7	.00203622	7	.00003394
8	.13962634	68	1 .18682389	128	2 .23402144	8	.00232711	8	.00003879
9	.15707963	69	1 .20427718	129	2 .25147474	9	.00261799	9	.00004363
10	.17453293	70	1 .22173048	130	2 .26892803	10	.00290888	10	.00004848
11	.19198622	71	1 .23918377	131	2 .28638132	11	.00319977	11	.00005333
12	.20943951	72	1 .25663706	132	2 .30383461	12	.00349066	12	.00005818
13	.22689280	73	1 .27409035	133	2 .32128791	13	.00378155	13	.00006303
14	.24434610	74	1 .29154365	134	2 .33874120	14	.00407243	14	.00006787
15	.26179939	75	1 .30899694	135	2 .35619449	15	.00436332	15	.00007272
16	.27925268	76	1 .32645023	136	2 .37364778	16	.00465421	16	.00007757
17	.29670597	77	1 .34390352	137	2 .39110108	17	.00494510	17	.00008242
18	.31415927	78	1 .36135682	138	2 .40855437	18	.00523599	18	.00008727
19	.33161256	79	1 .37881011	139	2 .42600766	19	.00552688	19	.00009211
20	.34906585	80	1 .39626340	140	2 .44346095	20	.00581776	20	.00009696
21	.36651914	81	1 .41371669	141	2 .46091425	21	.00610865	21	.00010181
22	.38397244	82	1 .43116999	142	2 .47836754	22	.00639954	22	.00010666
23	.40142573	83	1 .44862328	143	2 .49582083	23	.00669043	23	.00011151
24	.41887902	84	1 .46607657	144	2 .51327412	24	.00698132	24	.00011636
25	.43633231	85	1 .48352986	145	2 .53072742	25	.00727221	25	.00012120
26	.45378561	86	1 .50098316	146	2 .54818071	26	.00756309	26	.00012605
27	.47123890	87	1 .51843645	147	2 .56563400	27	.00785398	27	.00013090
28	.48869219	88	1 .53588974	148	2 .58308729	28	.00814487	28	.00013575
29	.50614548	89	1 .55334303	149	2 .60054059	29	.00843576	29	.00014060
30	.52359878	90	1 .57079633	150	2 .61799388	30	.00872665	30	.00014544
31	.54105207	91	1 .58824962	151	2 .63544717	31	.00901753	31	.00015029
32	.55850536	92	1 .60570291	152	2 .65290046	32	.00930842	32	.00015514
33	.57595865	93	1 .62315620	153	2 .67035376	33	.00959931	33	.00015999
34	.59341195	94	1 .64060950	154	2 .68780705	34	.00989020	34	.00016484
35	.61086524	95	1 .65806279	155	2 .70526034	35	.01018109	35	.00016968
36	.62831853	96	1 .67551608	156	2 .72271363	36	.01047198	36	.00017453
37	.64577182	97	1 .69296937	157	2 .74016693	37	.01076286	37	.00017938
38	.66322512	98	1 .71042267	158	2 .75762022	38	.01105375	38	.00018423
39	.68067841	99	1 .72787596	159	2 .77507351	39	.01134464	39	.00018908
40	.69813170	100	1 .74532925	160	2 .79252680	40	.01163553	40	.00019393
41	.71558499	101	1 .76278254	161	2 .80998010	41	.01192642	41	.00019877
42	.73303829	102	1 .78023584	162	2 .82743339	42	.01221730	42	.00020362
43	.75049158	103	1 .79768913	163	2 .84488668	43	.01250819	43	.00020847
44	.76794487	104	1 .81514242	164	2 .86233997	44	.01279908	44	.00021332
45	.78539816	105	1 .83259571	165	2 .87979327	45	.01308997	45	.00021817
46	.80285146	106	1 .85004901	166	2 .89724656	46	.01338086	46	.00022301
47	.82030475	107	1 .86750230	167	2 .91469985	47	.01367175	47	.00022786
48	.83775804	108	1 .88495559	168	2 .93215314	48	.01396263	48	.00023271
49	.85521133	109	1 .90240888	169	2 .94960644	49	.01425352	49	.00023756
50	.87266463	110	1 .91986218	170	2 .96705973	50	.01454441	50	.00024241
51	.89011792	111	1 .93731547	171	2 .98451302	51	.01483530	51	.00024725
52	.90757121	112	1 .95476876	172	3 .00196631	52	.01512619	52	.00025210
53	.92502450	113	1 .97222205	173	3 .01941961	53	.01541707	53	.00025695
54	.94247780	114	1 .98967535	174	3 .03687290	54	.01570796	54	.00026180
55	.95993109	115	2 .00712864	175	3 .05432619	55	.01599885	55	.00026665
56	.97738438	116	2 .02458193	176	3 .07177948	56	.01628974	56	.00027150
57	.99483767	117	2 .04203522	177	3 .08923278	57	.01658063	57	.00027634
58	1 .01229097	118	2 .05948852	178	3 .10668607	58	.01687152	58	.00028119
59	1 .02974426	119	2 .07694181	179	3 .12413936	59	.01716240	59	.00028604
60	1 .04719755	120	2 .09439510	180	3 .14159265	60	.01745329	60	.00029089

The Parallaxes of the Planets Venus, Mars, Jupiter and Saturn, in Altitude, Longitude, Latitude, Right Ascension, and Declination. (Mr. J. Utting.)

TABLE 1.

Common to the Parallaxes, in Long. Lat. R. A. and Declin.

Arg. A.	Argument B. 5°	10°	15°	20°	25°	30°	35°	40°	45°	50°	55°	60°	65°	70°	75°	80°	85°	90°	Arg. A.
10°	0″.15	0″.30	0″.45	0″.59	0″.73	0″.87	1″.00	1″.12	1″.23	1″.33	1″.42	1″.50	1″.57	1″.63	1″.68	1″.71	1″.73	1″.74	10°
15	0 .23	0 .45	0 .67	0 .89	1 .09	1 .29	1 .48	1 .66	1 .83	1 .98	2 .12	2 .24	2 .35	2 .43	2 .50	2 .55	2 .58	2 .59	15
20	0 .30	0 .59	0 .89	1 .17	1 .45	1 .71	1 .96	2 .20	2 .42	2 .62	2 .80	2 .96	3 .10	3 .21	3 .30	3 .37	3 .41	3 .42	20
25	0 .37	0 .73	1 .09	1 .45	1 .79	2 .11	2 .42	2 .72	2 .99	3 .24	3 .46	3 .66	3 .83	3 .97	4 .08	4 .16	4 .21	4 .23	25
30	0 .44	0 .87	1 .29	1 .71	2 .11	2 .50	2 .87	3 .21	3 .54	3 .83	4 .10	4 .33	4 .53	4 .70	4 .83	4 .92	4 .98	5 .00	30
35	0 .50	1 .00	1 .48	1 .96	2 .42	2 .87	3 .29	3 .69	4 .06	4 .39	4 .70	4 .97	5 .20	5 .39	5 .54	5 .65	5 .72	5 .74	35
40	0 .56	1 .12	1 .66	2 .20	2 .72	3 .21	3 .69	4 .13	4 .55	4 .92	5 .27	5 .57	5 .83	6 .04	6 .21	6 .33	6 .40	6 .43	40
45	0 .62	1 .23	1 .83	2 .42	2 .99	3 .54	4 .06	4 .55	5 .00	5 .42	5 .79	6 .12	6 .41	6 .64	6 .83	6 .96	7 .04	7 .07	45
50	0 .67	1 .33	1 .98	2 .62	3 .24	3 .83	4 .39	4 .92	5 .42	5 .87	6 .28	6 .63	6 .94	7 .20	7 .40	7 .54	7 .63	7 .66	50
55	0 .71	1 .42	2 .12	2 .80	3 .46	4 .10	4 .70	5 .27	5 .79	6 .28	6 .71	7 .09	7 .42	7 .70	7 .91	8 .07	8 .16	8 .19	55
60	0 .75	1 .50	2 .24	2 .96	3 .66	4 .33	4 .97	5 .57	6 .12	6 .63	7 .09	7 .50	7 .85	8 .14	8 .37	8 .53	8 .63	8 .66	60
65	0 .79	1 .57	2 .35	3 .10	3 .83	4 .53	5 .20	5 .83	6 .41	6 .94	7 .42	7 .85	8 .21	8 .52	8 .75	8 .93	9 .03	9 .06	65
70	0 .82	1 .63	2 .43	3 .21	3 .97	4 .70	5 .39	6 .04	6 .64	7 .20	7 .70	8 .14	8 .52	8 .83	9 .08	9 .25	9 .36	9 .40	70

TABLE 2.

Parallax in Altitude.

Arg. Planet's Altitude.	Argument = Horizontal Parallax. 10″	20″	30″
0°	10″.00	20″.00	30″.00
3	9 .99	19 .97	29 .96
6	9 .95	19 .89	29 .84
9	9 .88	19 .75	29 .63
12	9 .78	19 .56	29 .34
15	9 .66	19 .32	28 .98
18	9 .51	19 .02	28 .53
21	9 .34	18 .67	28 .01
24	9 .14	18 .27	27 .41
27	8 .91	17 .82	26 .73
30	8 .66	17 .32	25 .98
33	8 .39	16 .77	25 .16
36	8 .09	16 .18	24 .27
39	7 .77	16 .54	23 .31
42	7 .43	14 .86	22 .29
45	7 .07	14 .14	21 .21
48	6 .69	13 .38	20 .07
51	6 .29	12 .59	18 .88
54	5 .88	11 .76	17 .63
57	5 .45	10 .89	16 .34
60	5 .00	10 .00	15 .00
65	4 .23	8 .45	12 .68
70	3 .42	6 .84	10 .26
75	2 .59	5 .18	7 .76
80	1 .74	3 .47	5 .21
85	0 .87	1 .74	2 .61
90	0 .00	0 .00	0 .00

TABLE 3.

Parallax in Longitude.

Arg. Planet's Latitude.	Argument = Approx. Par. in Long. 10″	20″	30″
0° 10′	10″.00	20″.00	30″.00
20	10 .00	20 .00	30 .00
30	10 .00	20 .00	30 .00
40	10 .00	20 .00	30 .00
50	10 .00	20 .00	30 .00
1 0	10 .00	20 .00	30 .00
10	10 .00	20 .00	30 .01
20	10 .00	20 .01	30 .01
30	10 .00	20 .01	30 .01
40	10 .00	20 .01	30 .01
50	10 .01	20 .01	30 .02
2 0	10 .01	20 .01	30 .02
10	10 .01	20 .01	30 .02
20	10 .01	20 .02	30 .02
30	10 .01	20 .02	30 .03
40	10 .01	20 .02	30 .03
50	10 .01	20 .02	30 .04
3 0	10 .01	20 .03	30 .04
10	10 .02	20 .03	30 .05
20	10 .02	20 .03	30 .05
30	10 .02	20 .04	30 .06

TABLE 4.

Parallax in Latitude.

Arg. Planet's Latitude.	Argument = Approx. Par. in Lat. 10″	20″	30″
0° 10′	0″.03	0″.06	0″.09
20	0 .06	0 .12	0 .17
30	0 .08	0 .17	0 .26
40	0 .12	0 .23	0 .35
50	0 .15	0 .29	0 .44
1 0	0 .17	0 .35	0 .52
10	0 .20	0 .41	0 .61
20	0 .23	0 .47	0 .70
30	0 .26	0 .52	0 .79
40	0 .29	0 .58	0 .87
50	0 .32	0 .64	0 .96
2 0	0 .35	0 .70	1 .05
10	0 .38	0 .76	1 .13
20	0 .41	0 .81	1 .22
30	0 .44	0 .87	1 .31
40	0 .47	0 .93	1 .40
50	0 .49	0 .99	1 .48
3 0	0 .52	1 .05	1 .57
10	0 .55	1 .10	1 .66
20	0 .58	1 .16	1 .74
30	0 .61	1 .22	1 .83

Table 1. is computed for an Horizontal Parallax of 10″; proportion must therefore be made for any other Parallax.

All the above Tables are entered with the same Arguments as those for the Moon, substituting the Planet in place of the Moon.

TABLE 5.

Parallax in R. A.

Arg. Planet's Declin.	Argument = Approx. Par. in R. A. 10″	20″	30″
1°	10″.00	20″.00	30″.00
2	10 .01	20 .01	30 .02
3	10 .01	20 .03	30 .04
4	10 .02	20 .05	30 .07
5	10 .04	20 .08	30 .11
6	10 .06	20 .11	30 .17
7	10 .08	20 .15	30 .23
8	10 .10	20 .20	30 .29
9	10 .12	20 .24	30 .37
10	10 .15	20 .31	30 .46
11	10 .19	20 .37	30 .56
12	10 .22	20 .45	30 .67
13	10 .26	20 .53	30 .79
14	10 .30	20 .61	30 .92
15	10 .35	20 .71	31 .06
16	10 .40	20 .81	31 .21
17	10 .46	20 .91	31 .37
18	10 .51	21 .03	31 .54
19	10 .58	21 .15	31 .73
20	10 .64	21 .28	31 .93
21	10 .71	21 .42	32 .13
22	10 .79	21 .57	32 .36
23	10 .86	21 .73	32 .59
24	10 .95	21 .89	32 .84
25	11 .03	22 .07	33 .10
26	11 .13	22 .25	33 .38
27	11 .22	22 .45	33 .67

TABLE 6.

Parallax in Declination.

Arg. Planet's Declin.	Argument = Approx. Par. in Decl. 10″	20″	30″
1°	0″.17	0″.35	0″.52
2	0 .35	0 .70	1 .05
3	0 .52	1 .05	1 .57
4	0 .70	1 .40	2 .09
5	0 .87	1 .74	2 .61
6	1 .05	2 .09	3 .14
7	1 .22	2 .44	3 .66
8	1 .39	2 .78	4 .18
9	1 .56	3 .13	4 .69
10	1 .74	3 .47	5 .21
11	1 .91	3 .82	5 .72
12	2 .08	4 .16	6 .24
13	2 .25	4 .50	6 .75
14	2 .42	4 .84	7 .26
15	2 .59	5 .18	7 .76
16	2 .76	5 .51	8 .27
17	2 .92	5 .85	8 .77
18	3 .09	6 .18	9 .27
19	3 .26	6 .51	9 .77
20	3 .42	6 .84	10 .26
21	3 .58	7 .17	10 .75
22	3 .75	7 .49	11 .24
23	3 .91	7 .81	11 .72
24	4 .07	8 .13	12 .20
25	4 .23	8 .45	12 .68
26	4 .38	8 .77	13 .15
27	4 .54	9 .08	13 .62

TABLE 7. CONVERSION of the LONGITUDE of a PLANET, into RIGHT ASCENSION in TIME.

PART I.

Argument = Geocentric Longitude.

	O VI	I VII	II VIII	
	–	–	–	
	m.	m.	m.	
0°	0 .0	8 .4	8 .9	30°
1	0 .3	8 .6	8 .8	29
2	0 .6	8 .8	8 .6	28
3	1 .0	9 .0	8 .4	27
4	1 .3	9 .1	8 .1	26
5	1 .6	9 .2	7 .8	25
6	2 .0	9 .3	7 .6	24
7	2 .3	9 .4	7 .3	23
8	2 .6	9 .5	7 .1	22
9	2 .9	9 .6	6 .8	21
10	3 .2	9 .7	6 .6	20
11	3 .5	9 .7	6 .3	19
12	3 .9	9 .8	6 .0	18
13	4 .2	9 .8	5 .7	17
14	4 .5	9 .8	5 .4	16
15	4 .8	9 .9	5 .1	15
16	5 .1	9 .9	4 .8	14
17	5 .4	9 .9	4 .5	13
18	5 .6	9 .9	4 .2	12
19	5 .9	9 .8	3 .9	11
20	6 .1	9 .8	3 .5	10
21	6 .4	9 .7	3 .2	9
22	6 .6	9 .7	2 .8	8
23	6 .9	9 .6	2 .4	7
24	7 .1	9 .5	2 .1	6
25	7 .3	9 .4	1 .7	5
26	7 .6	9 .3	1 .3	4
27	7 .8	9 .2	1 .0	3
28	8 .0	9 .1	0 .6	2
29	8 .2	9 .0	0 .3	1
30	8 .4	8 .9	0 .0	0
	+	+	+	
	XI V	X IV	IX III	

PART II.

Argument = Longitude in time corrected.

	h. h. 0 12	h. h. 1 13	h. h. 2 14	h. h. 3 15	h. h. 4 16	h. h. 5 17	
	– +	– +	– +	– +	– +	– +	
m.	m.	m.	m.	m.	m.	m.	m.
0	1 .60	1 .54	1 .39	1 .13	0 .80	0 .41	60
2	1 .60	1 .54	1 .39	1 .12	0 .79	0 .40	58
4	1 .60	1 .54	1 .38	1 .11	0 .78	0 .39	56
6	1 .60	1 .54	1 .37	1 .10	0 .77	0 .38	54
8	1 .60	1 .53	1 .36	1 .09	0 .76	0 .37	52
10	1 .60	1 .53	1 .35	1 .08	0 .74	0 .35	50
12	1 .60	1 .53	1 .34	1 .07	0 .73	0 .34	48
14	1 .60	1 .52	1 .33	1 .06	0 .72	0 .33	46
16	1 .60	1 .52	1 .32	1 .05	0 .71	0 .31	44
18	1 .60	1 .51	1 .31	1 .04	0 .69	0 .30	42
20	1 .60	1 .51	1 .30	1 .03	0 .68	0 .28	40
22	1 .59	1 .50	1 .30	1 .02	0 .67	0 .27	38
24	1 .59	1 .50	1 .29	1 .01	0 .66	0 .26	36
26	1 .59	1 .49	1 .28	1 .00	0 .64	0 .24	34
28	1 .59	1 .49	1 .27	0 .99	0 .63	0 .23	32
30	1 .59	1 .48	1 .26	0 .97	0 .61	0 .21	30
32	1 .58	1 .48	1 .25	0 .96	0 .60	0 .20	28
34	1 .58	1 .47	1 .24	0 .95	0 .59	0 .19	26
36	1 .58	1 .47	1 .23	0 .94	0 .57	0 .17	24
38	1 .58	1 .46	1 .22	0 .93	0 .56	0 .16	22
40	1 .57	1 .45	1 .21	0 .92	0 .54	0 .14	20
42	1 .57	1 .45	1 .21	0 .91	0 .53	0 .13	18
44	1 .57	1 .44	1 .20	0 .90	0 .52	0 .12	16
46	1 .57	1 .44	1 .19	0 .89	0 .51	0 .10	14
48	1 .56	1 .43	1 .18	0 .88	0 .50	0 .09	12
50	1 .56	1 .42	1 .17	0 .86	0 .48	0 .07	10
52	1 .56	1 .42	1 .17	0 .85	0 .47	0 .06	8
54	1 .55	1 .41	1 .16	0 .84	0 .46	0 .04	6
56	1 .55	1 .40	1 .15	0 .83	0 .44	0 .03	4
58	1 .55	1 .40	1 .14	0 .82	0 .43	0 .01	2
60	1 .54	1 .39	1 .13	0 .80	0 .41	0 .00	0
	– +	– +	– +	– +	– +	– +	
	h. h. 23 11	h. h. 22 10	h. h. 21 9	h. h. 20 8	h. h. 19 7	h. h. 18 6	

PART III.

Arg. = Declin.

	m.
1°	0 .00
2	0 .00
3	0 .01
4	0 .01
5	0 .01
6	0 .01
7	0 .01
8	0 .01
9	0 .02
10	0 .02
11	0 .02
12	0 .02
13	0 .03
14	0 .03
15	0 .03
16	0 .04
17	0 .04
18	0 .05
19	0 .06
20	0 .06
21	0 .07
22	0 .08
23	0 .09
24	0 .09
25	0 .10
26	0 .11
27	0 .12
28	0 .13
29	0 .14
30	0 .15
31	0 .17
32	0 .18
33	0 .19
34	0 .20
35	0 .21

TABLE 8. The DIFFERENCE between the MERIDIAN ALTITUDE of a PLANET, or other body, and its ALTITUDE one Minute before or after the Time of its Transit. Arguments = the Declination of the Body, and Lat. of the Place.

Lat.	Dec. of the same Name with the Latitude.							Dec. of a contrary Name.					
	24°	20°	16°	12°	8°	4°	0°	4°	8°	12°	16°	20°	24°
0°	4″.4	5″.4	6″.9	9″.2	14″.0	28″.1		28″.1	14″.1	9″.2	6″.9	5″.4	4″.4
2	4 .8	5 .9	7 .8	11 .1	18 .6	56 .1	56″.2	18 .7	11 .2	7 .9	6 .2	5 .0	4 .1
4	5 .2	6 .7	9 .1	13 .8	27 .8		28 .1	14 .0	9 .3	7 .0	5 .5	4 .6	3 .8
6	5 .8	7 .6	10 .8	18 .3	55 .4	58 .8	18 .7	11 .2	8 .0	6 .1	4 .9	4 .2	3 .6
8	6 .4	8 .8	13 .4	27 .3		27 .8	14 .0	9 .3	7 .0	5 .6	4 .6	3 .9	3 .4
10	7 .3	10 .5	17 .8	54 .2	54 .9	18 .5	11 .1	8 .0	6 .2	5 .0	4 .2	3 .6	3 .2
12	8 .4	13 .0	26 .5		27 .3	13 .8	9 .2	7 .0	5 .6	4 .6	3 .9	3 .4	3 .0
14	9 .8	17 .1	52 .5	53 .4	18 .1	11 .0	7 .9	6 .1	5 .1	4 .2	3 .7	3 .2	2 .8
16	12 .4	25 .4		26 .5	13 .4	9 .1	6 .9	5 .5	4 .6	3 .9	3 .4	3 .0	2 .7
18	16 .4	50 .3	51 .4	17 .5	10 .7	7 .7	6 .0	4 .9	4 .2	3 .6	3 .2	2 .9	2 .6
20	24 .2		25 .4	13 .0	8 .8	6 .7	5 .4	4 .5	3 .9	3 .4	3 .0	2 .7	2 .4
22	47 .7	49 .0	16 .8	10 .3	7 .5	5 .9	4 .9	4 .2	3 .6	3 .2	2 .9	2 .6	2 .3
24		24 .2	12 .4	8 .4	6 .4	5 .2	4 .4	3 .8	3 .4	3 .0	2 .7	2 .4	2 .2
26	46 .2	15 .9	9 .8	7 .1	5 .7	4 .7	4 .0	3 .5	3 .2	2 .8	2 .5	2 .3	2 .1
28	22 .7	11 .7	8 .0	6 .2	5 .0	4 .3	3 .7	3 .3	2 .9	2 .6	2 .4	2 .2	2 .0
30	14 .9	9 .2	6 .8	5 .3	4 .5	3 .9	3 .4	3 .0	2 .7	2 .4	2 .2	2 .1	1 .9
35	8 .9	6 .5	5 .1	4 .2	3 .6	3 .2	2 .8	2 .6	2 .3	2 .2	2 .0	1 .8	1 .7
40	5 .0	4 .1	3 .6	3 .1	2 .8	2 .6	2 .3	2 .2	2 .0	1 .9	1 .8	1 .6	1 .5
45	3 .9	3 .2	2 .8	2 .5	2 .3	2 .2	2 .0	1 .9	1 .7	1 .6	1 .5	1 .4	1 .3
50	2 .9	2 .4	2 .2	2 .0	1 .9	1 .8	1 .7	1 .6	1 .5	1 .4	1 .3	1 .3	1 .2
55	2 .1	1 .9	1 .8	1 .6	1 .5	1 .5	1 .4	1 .4	1 .3	1 .2	1 .1	1 .1	1 .0
60	1 .5	1 .4	1 .4	1 .3	1 .2	1 .2	1 .1	1 .1	1 .1	1 .0	1 .0	0 .9	0 .9
70	0 .9	0 .8	0 .8	0 .8	0 .8	0 .7	0 .7	0 .7	0 .7	0 .7	0 .7	0 .6	0 .6

TABLE A.

The Parallaxes and Diameters of the Planets.

Argument = The Sun's Long. — the Planet's Geo. Long.

Mercury.

Arg. = Elon. Super. ☌	Parallax. ☿ in apog.	Parallax. ☿ in mean dis.	Parallax. ☿ in peri.	Diam. in mean dis.
0° 0′	5″.93	6″.27	6″.65	4″.2
10 0	6 .14	6 .54	7 .02	4 .4
17 0	6 .59	7 .19	8 .27	4 .8
17 54	. . .	. . .	9 .14	. . .
20 0	6 .92	7 .76	. . .	5 .2
22 30	7 .30	8 .86	. . .	5 .9
22 46	. . .	9 .44	. . .	6 .2
25 0	7 .88	. . .	. . .	. . .
27 0	8 .71	. . .	. . .	. . .
27 49	9 .84	. . .	. . .	. . .
27 0	11 .11	. . .	. . .	. . .
25 0	12 .28	. . .	. . .	. . .
22 30	13 .25	10 .05	. . .	6 .6
20 0	13 .98	11 .47	. . .	7 .6
17 0	14 .68	12 .38	10 .11	8 .2
10 0	15 .77	13 .62	11 .90	9 .0
0 0	16 .31	14 .20	12 .55	9 .4
Inferior ☌				

Venus.

Arg. Super. ☌	Parallax. ☉ in apog.	Parallax. ☉ in mean dis.	Parallax. ☉ in per.	Diam. in mean dis.
0° 0′	5″.00	5″.05	5″.10	9″.2
10 0	5 .11	5 .16	5 .21	9 .4
20 0	5 .47	5 .52	5 .56	10 .0
30 0	6 .24	6 .27	6 .30	11 .4
40 0	7 .97	7 .93	7 .87	14 .6
42 30	8 .86	8 .70	8 .56	16 .0
45 0	10 .90	10 .12	9 .72	18 .6
45 21	. . .	. . .	13 .06	. . .
46 40	. . .	12 .60	. . .	23 .2
47 22	12 .17	. . .	. . .	. . .
45 0	13 .60	15 .68	17 .56	28 .8
42 30	16 .80	18 .35	20 .03	33 .8
40 0	18 .55	20 .03	21 .68	36 .8
30 0	23 .77	25 .34	27 .12	46 .6
20 0	27 .08	28 .77	30 .69	53 .0
10 0	29 .02	30 .78	32 .78	56 .6
0 0	29 .65	31 .83	33 .48	58 .6
Inferior ☌				

Mars, Jupiter, Saturn, Herschel.

Arg. = Elon.	Mars. Parallax in aph.	Mars. Parallax at m. dis.	Mars. Parallax in per.	Mars. Dia. at m. dis.	Jupiter. At mean dis. Paral.	Jupiter. At mean dis. Diam.	Saturn. At mean dis. Paral.	Saturn. At mean dis. Diam.	Herschel. At mean dis. Paral.	Herschel. At mean dis. Diam.	Arg.
S 0 0°	3″.26	3″.45	3″.66	3″.6	1″.40	29″.6	0″.82	14″.4	0″.43	3″.68	S 12 0°
10	3 .29	3 .48	3 .69	3 .7	1 .41	29 .8	0 .82	14 .6	0 .43	3 .69	20
20	3 .39	3 .69	3 .82	3 .8	1 .42	30 .0	0 .84	14 .8	0 .43	3 .70	10
1 0	3 .54	3 .77	4 .04	4 .0	1 .44	30 .4	0 .84	15 .0	0 .44	3 .72	11 0
10	3 .78	4 .05	4 .37	4 .2	1 .47	31 .0	0 .85	15 .2	0 .44	3 .74	20
20	4 .10	4 .44	4 .85	4 .6	1 .50	31 .8	0 .86	15 .4	0 .44	3 .76	10
2 0	4 .52	4 .96	5 .52	5 .2	1 .54	32 .6	0 .87	15 .6	0 .44	3 .78	10 0
10	5 .07	5 .64	6 .27	5 .8	1 .59	33 .6	0 .88	15 .8	0 .45	3 .81	20
20	5 .74	6 .51	7 .61	6 .8	1 .64	34 .8	0 .90	16 .0	0 .45	3 .84	10
3 0	6 .53	7 .57	9 .12	8 .0	1 .70	36 .0	0 .92	16 .2	0 .46	3 .88	9 0
10	7 .44	8 .80	10 .93	9 .2	1 .76	37 .4	0 .94	16 .5	0 .46	3 .92	20
20	8 .42	10 .15	12 .96	10 .6	1 .82	38 .8	0 .95	16 .8	0 .46	3 .95	10
4 0	9 .45	11 .54	15 .08	12 .0	1 .88	40 .0	0 .97	17 .2	0 .47	3 .98	8 0
10	10 .41	12 .90	17 .15	13 .4	1 .93	41 .0	0 .98	17 .4	0 .47	4 .01	20
20	11 .29	14 .14	19 .03	14 .8	1 .98	42 .0	1 .00	17 .6	0 .47	4 .04	10
5 0	12 .05	15 .18	20 .61	15 .8	2 .02	42 .8	1 .01	17 .7	0 .47	4 .06	7 0
10	12 .60	15 .96	21 .80	16 .6	2 .04	43 .2	1 .01	17 .8	0 .47	4 .08	20
20	12 .95	16 .45	22 .54	17 .0	2 .06	43 .6	1 .02	18 .0	0 .48	4 .10	10
6 0	13 .07	16 .61	22 .78	17 .2	2 .08	44 .0	1 .02	18 .0	0 .48	4 .10	6 0

Parallaxes.

	At the superior Conjunction. Aphelion.	At the superior Conjunction. Mean dis.	At the superior Conjunction. Per.	At the inferior Conjunction. Aphelion.	At the inferior Conjunction. Mean dis.	At the inferior Conjunction. Per.
Pallas	2″.168	2″.307	2″.469	4″.922	4″.920	5″.054
Ceres	2 .262	2 .309	2 .358	4 .713	4 .922	5 .149
Juno	2 .216	2 .372	2 .551	4 .519	5 .215	6 .168
Vesta	2 .524	2 .591	2 .662	6 .016	6 .411	6 .860

The Parallax is always $= \frac{8''.7}{\text{distance of Planet}}$; the earth's radius vector being = 1.

TABLE B.

Aberration common to all the Planets.

FIRST PART.

Arg. in Long. = The Elongation.

Arg. in Lat. = The Elong. — $\overset{s}{\text{III}}$

	$\overset{s}{\text{O}}$ — VI +	$\overset{s}{\text{I}}$ — VII +	$\overset{s}{\text{II}}$ — VIII +	
0°	20″.25	17″.54	10″.13	30°
1	20 .25	17 .36	9 .82	29
2	20 .24	17 .18	9 .51	28
3	20 .23	16 .99	9 .20	27
4	20 .21	16 .79	8 .88	26
5	20 .18	16 .59	8 .56	25
6	20 .14	16 .39	8 .24	24
7	20 .10	16 .18	7 .91	23
8	20 .06	15 .96	7 .59	22
9	20 .01	15 .74	7 .26	21
10	19 .95	15 .52	6 .93	20
11	19 .88	15 .29	6 .59	19
12	19 .81	15 .05	6 .26	18
13	19 .74	14 .81	5 .92	17
14	19 .65	14 .57	5 .58	16
15	19 .56	14 .32	5 .24	15
16	19 .47	14 .07	4 .90	14
17	19 .37	13 .81	4 .56	13
18	19 .26	13 .55	4 .21	12
19	19 .15	13 .29	3 .86	11
20	19 .03	13 .02	3 .52	10
21	18 .91	12 .75	3 .17	9
22	18 .78	12 .47	2 .82	8
23	18 .64	12 .19	2 .47	7
24	18 .50	11 .91	2 .12	6
25	18 .36	11 .62	1 .77	5
26	18 .20	11 .33	1 .41	4
27	18 .05	11 .03	1 .06	3
28	17 .88	10 .73	0 .71	2
29	17 .71	10 .43	0 .35	1
30	17 .54	10 .13	0 .00	0
	V + XI —	IV + X —	III + IX —	

SECOND PART.

Arg. in Long. G. — π.

Arg. in Lat. G. — π + $\overset{s}{\text{III}}$.

Arg.	$\overset{s}{\text{O}}$ — VI +	$\overset{s}{\text{I}}$ — VII +	$\overset{s}{\text{II}}$ — VIII +	Arg.
0°	0″.34	0″.29	0″.17	30°
5	0 .34	0 .28	0 .14	25
10	0 .33	0 .26	0 .12	20
15	0 .33	0 .24	0 .09	15
20	0 .32	0 .22	0 .06	10
25	0 .31	0 .20	0 .03	5
30	0 .29	0 .17	0 .00	0
	V + XI —	IV + X —	III + IX —	

In this Table, G is = the Planet's geocentric Longitude, and π = the Longitude of the Sun's Perigee.

The Numbers taken out of both Parts of Table B, must be multiplied by the Secant of the Planet's Geocentric Latitude for the Aberration in Longitude; and by the Sine of the same for the Aberration in Latitude.

The *Elongation* must be taken positive when to the West of the Sun, and negative when to the East.

TABLE C.

ABERRATION in LONGITUDE and LATITUDE.

Arg. in Long. = P − ϱ. (FIRST PART.) Arg. in Lat. = P − ϱ + III.[s]

	O −		VI +			I −		VII +			II −		VIII +			
	Merc.	Venus.	Mars.	Vesta.	Juno.	Merc.	Venus.	Mars.	Vesta.	Juno.	Merc.	Venus.	Mars.	Vesta.	Juno.	
0°	33″.14	23″.79	16″.48	13″.18	12″.65	28″.70	20″.61	14″.27	11″.41	10″.96	16″.57	11″.90	8″.24	6″.59	6″.33	30°
1	33 .14	23 .79	16 .47	13 .18	12 .65	28 .41	20 .40	14 .12	11 .29	10 .85	16 .07	11 .54	7 .99	6 .39	6 .14	29
2	33 .12	23 .78	16 .46	13 .17	12 .64	28 .10	20 .18	13 .97	11 .17	10 .73	15 .56	11 .19	7 .74	6 .19	5 .94	28
3	33 .09	23 .76	16 .45	13 .16	12 .63	27 .79	19 .95	13 .82	11 .05	10 .61	15 .05	10 .80	7 .49	5 .98	5 .74	27
4	33 .06	23 .74	16 .43	13 .14	12 .62	27 .47	19 .73	13 .66	10 .92	10 .49	14 .53	10 .43	7 .23	5 .78	5 .55	26
5	33 .02	23 .70	16 .41	13 .12	12 .61	27 .15	19 .49	13 .50	10 .79	10 .36	14 .01	10 .06	6 .97	5 .57	5 .35	25
6	32 .96	23 .66	16 .38	13 .10	12 .58	26 .81	19 .25	13 .33	10 .66	10 .23	13 .48	9 .68	6 .70	5 .36	5 .15	24
7	32 .90	23 .62	16 .35	13 .08	12 .56	26 .47	19 .00	13 .16	10 .52	10 .10	12 .95	9 .30	6 .44	5 .15	4 .95	23
8	32 .82	23 .56	16 .32	13 .05	12 .53	26 .11	18 .75	12 .98	10 .38	9 .97	12 .41	8 .91	6 .17	4 .94	4 .74	22
9	32 .73	23 .50	16 .28	13 .02	12 .49	25 .75	18 .49	12 .80	10 .24	9 .83	11 .88	8 .53	5 .90	4 .72	4 .53	21
10	32 .64	23 .43	16 .24	12 .98	12 .45	25 .39	18 .23	12 .62	10 .09	9 .69	11 .33	8 .14	5 .63	4 .51	4 .33	20
11	32 .54	23 .36	16 .19	12 .94	12 .41	25 .01	17 .96	12 .43	9 .94	9 .55	10 .79	7 .75	5 .36	4 .29	4 .12	19
12	32 .42	23 .27	16 .13	12 .89	12 .37	24 .63	17 .68	12 .24	9 .79	9 .40	10 .24	7 .35	5 .09	4 .07	3 .91	18
13	32 .30	23 .18	16 .07	12 .84	12 .32	24 .24	17 .40	12 .05	9 .64	9 .25	9 .69	6 .96	4 .82	3 .85	3 .70	17
14	32 .16	23 .09	16 .0	12 .79	12 .27	23 .84	17 .12	11 .85	9 .48	9 .10	9 .12	6 .56	4 .54	3 .63	3 .49	16
15	32 .01	22 .98	15 .92	12 .73	12 .22	23 .43	16 .82	11 .65	9 .32	8 .94	8 .57	6 .16	4 .26	3 .41	3 .27	15
16	31 .86	22 .87	15 .84	12 .67	12 .16	23 .02	16 .53	11 .45	9 .16	8 .78	8 .02	5 .76	3 .98	3 .19	3 .06	14
17	31 .69	22 .75	15 .76	12 .60	12 .10	22 .60	16 .23	11 .24	8 .99	8 .62	7 .45	5 .35	3 .70	2 .97	2 .85	13
18	31 .52	22 .63	15 .67	12 .53	12 .03	22 .17	15 .92	11 .03	8 .82	8 .46	6 .89	4 .95	3 .42	2 .74	2 .63	12
19	31 .34	22 .50	15 .58	12 .46	11 .96	21 .74	15 .61	10 .81	8 .65	8 .30	6 .32	4 .54	3 .14	2 .52	2 .42	11
20	31 .14	22 .36	15 .48	12 .38	11 .89	21 .30	15 .29	10 .59	8 .47	8 .13	5 .75	4 .13	2 .86	2 .29	2 .20	10
21	30 .94	22 .21	15 .38	12 .30	11 .81	20 .86	14 .97	10 .37	8 .29	7 .96	5 .18	3 .72	2 .58	2 .06	1 .98	9
22	30 .72	22 .06	15 .27	12 .22	11 .73	20 .40	14 .65	10 .14	8 .11	7 .79	4 .61	3 .31	2 .30	1 .84	1 .76	8
23	30 .50	21 .90	15 .16	12 .13	11 .65	19 .94	14 .32	9 .91	7 .93	7 .62	4 .04	2 .90	2 .01	1 .61	1 .54	7
24	30 .27	21 .74	15 .05	12 .04	11 .56	19 .48	13 .99	9 .68	7 .75	7 .44	3 .46	2 .49	1 .72	1 .38	1 .32	6
25	30 .03	21 .57	14 .93	11 .94	11 .47	19 .01	13 .65	9 .45	7 .56	7 .26	2 .89	2 .07	1 .44	1 .15	1 .10	5
26	29 .79	21 .39	14 .81	11 .84	11 .37	18 .53	13 .31	9 .21	7 .37	7 .08	2 .31	1 .66	1 .15	0 .92	0 .88	4
27	29 .53	21 .20	14 .68	11 .74	11 .27	18 .05	12 .96	8 .97	7 .18	6 .89	1 .73	1 .24	0 .86	0 .69	0 .66	3
28	29 .26	21 .01	14 .55	11 .63	11 .17	17 .56	12 .61	8 .73	6 .99	6 .71	1 .16	0 .83	0 .58	0 .46	0 .44	2
29	28 .98	20 .81	14 .41	11 .52	11 .07	17 .07	12 .25	8 .49	6 .79	6 .52	0 .58	0 .41	0 .29	0 .23	0 .22	1
30	28 .70	20 .61	14 .27	11 .41	10 .96	16 .57	11 .90	8 .24	6 .59	6 .33	0 .00	0 .00	0 .00	0 .00	0 .00	0
	V +					IV +					III +					

Arg. in Long. = G − φ. (SECOND PART.) Arg. in Lat. = G − φ + III.[s]

	O −		VI +			I −		VII +			II −		VIII +			
	Merc.	Venus.	Mars.	Vesta.	Juno.	Merc.	Venus.	Mars.	Vesta.	Juno.	Merc.	Venus.	Mars.	Vesta.	Juno.	
0°	6″.81	0″.16	1″.54	1″.17	3″.22	5″.90	0″.14	1″.33	1″.02	2″.79	3″.40	0″.08	0″.77	0″.59	1″.61	30°
3	6 .79	0 .16	1 .54	1 .17	3 .22	5 .71	0 .13	1 .29	0 .99	2 .70	3 .09	0 .08	0 .70	0 .54	1 .46	27
6	6 .76	0 .16	1 .53	1 .16	3 .21	5 .51	0 .13	1 .25	0 .95	2 .61	2 .77	0 .07	0 .63	0 .48	1 .31	24
9	6 .71	0 .16	1 .52	1 .15	3 .19	5 .20	0 .12	1 .21	0 .91	2 .51	2 .44	0 .07	0 .56	0 .42	1 .15	21
12	6 .65	0 .16	1 .50	1 .14	3 .16	5 .06	0 .12	1 .15	0 .87	2 .40	2 .10	0 .06	0 .48	0 .36	0 .99	18
15	6 .58	0 .16	1 .48	1 .13	3 .11	4 .82	0 .11	1 .08	0 .83	2 .28	1 .76	0 .05	0 .40	0 .30	0 .83	15
18	6 .48	0 .15	1 .46	1 .11	3 .06	4 .56	0 .11	1 .02	0 .79	2 .16	1 .42	0 .04	0 .32	0 .24	0 .67	12
21	6 .35	0 .15	1 .44	1 .09	3 .00	4 .29	0 .10	0 .96	0 .74	2 .03	1 .07	0 .03	0 .24	0 .18	0 .51	9
24	6 .22	0 .15	1 .41	1 .07	2 .94	4 .00	0 .10	0 .90	0 .69	1 .90	0 .71	0 .02	0 .16	0 .12	0 .34	6
27	6 .07	0 .14	1 .37	1 .05	2 .87	3 .71	0 .09	0 .84	0 .64	1 .76	0 .36	0 .01	0 .08	0 .06	0 .17	3
30	5 .90	0 .14	1 .33	1 .02	2 .79	3 .40	0 .08	0 .77	0 .59	1 .61	0 .00	0 .00	0 .00	0 .00	0 .00	0
	V +		XI −			IV +		X −			III +		IX −			

TABLE D. Arg. in Lat. = (C + L) and (C − L)

	O −		VI +			I −		VII +			II −		VIII +			
	Merc.	Venus.	Mars.	Vesta.	Juno.	Merc.	Venus.	Mars.	Vesta.	Juno.	Merc.	Venus.	Mars.	Vesta.	Juno.	
0°	2″.03	0″.70	0″.26	0″.82	1″.45	1″.76	0″.61	0″.23	0″.71	1″.26	1″.01	0″.35	0″.13	0″.41	0″.72	30°
5	2 .02	0 .70	0 .26	0 .82	1 .44	1 .66	0 .58	0 .22	0 .67	1 .19	0 .86	0 .30	0 .11	0 .35	0 .61	25
10	2 .00	0 .69	0 .26	0 .81	1 .43	1 .55	0 .54	0 .21	0 .63	1 .11	0 .69	0 .24	0 .09	0 .28	0 .50	20
15	1 .96	0 .68	0 .26	0 .79	1 .40	1 .43	0 .50	0 .19	0 .58	1 .03	0 .52	0 .18	0 .07	0 .21	0 .38	15
20	1 .90	0 .66	0 .25	0 .77	1 .36	1 .30	0 .45	0 .17	0 .53	0 .93	0 .35	0 .12	0 .05	0 .14	0 .25	10
25	1 .84	0 .64	0 .24	0 .74	1 .31	1 .16	0 .40	0 .15	0 .47	0 .83	0 .18	0 .06	0 .02	0 .07	0 .13	5
30	1 .76	0 .61	0 .23	0 .71	1 .26	1 .01	0 .35	0 .13	0 .41	0 .72	0 .00	0 .00	0 .00	0 .00	0 .00	0
	V +		XI −			IV +		X −			III +		IX −			

P = Annual Parallax; ϱ = the Reduction to the Ecliptic; G = the Geocentric Longitude; φ = the Longitude of the perihelion in the Orbit; C = the Argument of Latitude on the Orbit; and L = the Geocentric Latitude.

The Numbers in Table C, in both the Parts, must be multiplied by the Secant of the Geocentric Latitude (L) for the Aberration in Longitude, and by the Sine of the same (L) for the Aberration in Latitude. The Annual Parallax must have the same Sign as the Elongation.

TABLE C.

Aberration in Longitude and Latitude, concluded.

(FIRST PART.)

The Arg. in Long. = P − ϱ. The Arg. in Lat. = P − ϱ + III.[s]

	O −		VI +			I −		VII +			II −		VIII +			
	Ceres.	Pallas.	Jupiter.	Saturn.	Hersch.	Ceres.	Pallas.	Jupiter.	Saturn.	Hersch.	Ceres.	Pallas.	Jupiter.	Saturn.	Hersch.	
0°	12″.11	11″.44	8″.89	6″.55	4″.63	10″.49	9″.91	7″.70	5″.68	4″.01	6″.05	5″.72	4″.44	3″.28	2″.31	30°
3	12.09	11.43	8.88	6.55	4.62	10.16	9.60	7.45	5.50	3.88	5.50	5.20	4.03	2.98	2.10	27
6	12.04	11.38	8.84	6.52	4.60	9.80	9.26	7.19	5.30	3.74	4.93	4.65	3.61	2.67	1.88	24
9	11.96	11.30	8.78	6.47	4.57	9.41	8.89	6.91	5.09	3.59	4.33	4.10	3.18	2.35	1.66	21
12	11.84	11.19	8.69	6.41	4.52	9.00	8.50	6.61	4.87	3.44	3.74	3.54	2.74	2.03	1.43	18
15	11.70	11.05	8.58	6.33	4.47	8.56	8.09	6.29	4.67	3.27	3.13	2.96	2.30	1.70	1.20	15
18	11.52	10.88	8.45	6.23	4.40	8.10	7.66	5.95	4.39	3.09	2.52	2.38	1.85	1.36	0.96	12
21	11.30	10.68	8.30	6.12	4.32	7.62	7.20	5.59	4.13	2.91	1.89	1.79	1.39	1.03	0.72	9
24	11.06	10.45	8.12	5.99	4.22	7.12	6.73	5.22	3.85	2.72	1.27	1.20	0.93	0.69	0.48	6
27	10.79	10.20	7.92	5.84	4.12	6.60	6.23	4.84	3.57	2.52	0.63	0.60	0.46	0.34	0.24	3
30	10.49	9.91	7.70	5.68	4.01	6.05	5.72	4.44	3.28	2.31	0.00	0.00	0.00	0.00	0.00	0
	V +					IV +					III +					

(SECOND PART.)

Arg. in Long. = G − φ. Arg. in Lat. = G − φ + III.[s]

	O −		VI +			I −		VII +			II −		VIII +			
0°	0″.95	2″.80	0″.43	0″.37	0″.22	0″.82	2″.43	0″.37	0″.32	0″.19	0″.47	1″.40	0″.21	0″.18	0″.11	30°
5	0.94	2.79	0.43	0.37	0.22	0.78	2.29	0.35	0.30	0.18	0.40	1.18	0.18	0.16	0.09	25
10	0.93	2.75	0.42	0.36	0.22	0.73	2.15	0.33	0.28	0.17	0.32	0.96	0.15	0.13	0.07	20
15	0.91	2.70	0.41	0.36	0.21	0.67	1.98	0.30	0.26	0.15	0.24	0.72	0.11	0.09	0.06	15
20	0.89	2.63	0.40	0.35	0.20	0.61	1.80	0.27	0.24	0.14	0.16	0.48	0.07	0.06	0.04	10
25	0.86	2.54	0.39	0.33	0.20	0.54	1.61	0.25	0.21	0.12	0.08	0.24	0.04	0.03	0.02	5
30	0.82	2.43	0.37	0.32	0.19	0.47	1.40	0.21	0.18	0.11	0.00	0.00	0.00	0.00	0.00	0
	V +		XI −			IV +		X −			III +		IX −			

TABLE D.

Arg. in Lat. = (C + L) and (C − L).

	O −		VI +			I −		VII +			II −		VIII +			
	Ceres.	Pallas.	Jupiter.	Saturn.	Hersch.	Ceres.	Pallas.	Jupiter.	Saturn.	Hersch.	Ceres.	Pallas.	Jupiter.	Saturn.	Hersch.	
0°	1″.13	3″.57	0″.10	0″.14	0″.03	0″.98	3″.09	0″.09	0″.12	0″.03	0″.56	1″.78	0″.05	0″.07	0″.02	30°
5	1.13	3.55	0.10	0.14	0.03	0.92	2.92	0.08	0.12	0.03	0.47	1.51	0.04	0.06	0.01	25
10	1.12	3.51	0.10	0.14	0.03	0.86	2.73	0.08	0.11	0.02	0.38	1.22	0.03	0.05	0.01	20
15	1.09	3.45	0.10	0.14	0.03	0.79	2.52	0.07	0.10	0.02	0.29	0.92	0.03	0.04	0.01	15
20	1.06	3.35	0.10	0.13	0.03	0.72	2.29	0.07	0.09	0.02	0.19	0.62	0.02	0.02	0.01	10
25	1.02	3.23	0.09	0.13	0.03	0.64	2.05	0.06	0.08	0.02	0.10	0.31	0.01	0.01	0.00	5
30	0.98	3.09	0.09	0.12	0.03	0.56	1.78	0.05	0.07	0.02	0.00	0.00	0.00	0.00	0.00	0
	V +		XI −			IV +		X −			III +		IX −			

(SECOND PART.)

Arg. in Lat. = (φ − ☊ + L) and (φ − ☊ − L).

	O −		VI +			I −		VII +			II −		VIII +			
	Merc.	Vesta.	Juno.	Ceres.	Pallas.	Merc.	Vesta.	Juno.	Ceres.	Pallas.	Merc.	Vesta.	Juno.	Ceres.	Pallas.	
0°	0″.42	0″.07	0″.37	0″.09	0″.87	0″.36	0″.06	0″.32	0′.08	0″.76	0″.21	0″.04	0″.18	0″.04	0″.44	30°
5	0.42	0.07	0.37	0.09	0.87	0.34	0.06	0.30	0.07	0.72	0.18	0.03	0.15	0.04	0.37	25
10	0.41	0.07	0.36	0.09	0.86	0.32	0.06	0.28	0.47	0.67	0.14	0.02	0.12	0.03	0.30	20
15	0.40	0.07	0.36	0.09	0.84	0.30	0.05	0.26	0.06	0.62	0.11	0.02	0.09	0.02	0.23	15
20	0.39	0.07	0.35	0.08	0.82	0.27	0.05	0.24	0.06	0.56	0.07	0.01	0.06	0.02	0.15	10
25	0.38	0.07	0.33	0.08	0.79	0.24	0.05	0.21	0.05	0.50	0.04	0.01	0.03	0.01	0.08	5
30	0.36	0.06	0.32	0.08	0.76	0.21	0.04	0.18	0.04	0.44	0.00	0.00	0.00	0.00	0.00	0
	V +		XI −			IV +		X −			III +		IX −			

TABLE E.

Arg. in Long. = (P + 2 C). Arg. in Lat. = (P + 2 C + III).[s]

	O +		VI −			I +		VII −			II +		VIII −			
	Merc.	Vesta.	Juno.	Ceres.	Pallas.	Merc.	Vesta.	Juno.	Ceres.	Pallas.	Merc.	Vesta.	Juno.	Ceres.	Pallas.	
0°	0″.12	0″.05	0″.17	0″.10	1″.11	0″.10	0.04	0″.14	0″.09	0″.96	0″.06	0″.03	0″.08	0″.05	0″.56	30°
5	0.12	0.05	0.17	0.10	1.10	0.10	0.04	0.13	0.08	0.91	0.05	0.02	0.07	0.04	0.47	25
10	0.12	0.05	0.16	0.10	1.09	0.09	0.04	0.12	0.08	0.85	0.04	0.02	0.06	0.04	0.38	20
15	0.12	0.05	0.16	0.10	1.07	0.09	0.04	0.11	0.07	0.79	0.03	0.01	0.04	0.03	0.29	15
20	0.11	0.05	0.15	0.10	1.04	0.08	0.03	0.10	0.07	0.71	0.02	0.01	0.03	0.02	0.19	10
25	0.11	0.05	0.15	0.09	1.01	0.07	0.03	0.09	0.06	0.64	0.01	0.00	0.01	0.01	0.10	5
30	0.10	0.04	0.14	0.09	0.96	0.06	0.03	0.08	0.05	0.56	0.00	0.00	0.00	0.00	0.00	0
	V −		XI +			IV −		X +			III −		IX +			

P = Annual Parallax; ϱ = Reduction to the Ecliptic; G = the Geocentric Longitude; φ = the Longitude of the Perihelion in the Orbit; C = the Argument of Latitude on the Orbit; and L = the Geocentric Latitude.

The Numbers in both Parts of Table C. must be multiplied by the Secant of L for the Aberration in Longitude; and by the Sine of L for the Aberration in Latitude; which also must be done with the Numbers in Table E, when this Table is used.

VELOCITY OF LIGHT.

Radius of the ⊖'s Orbit.	Velocity of Light.	Radius of the ⊖'s Orbit.	Velocity of Light.	Radius of the ⊖'s Orbit.	Velocity of Light.	Radius of the ⊖'s Orbit.	Velocity of Light.	Radius of the ⊖'s Orbit.	Velocity of Light.	Radii of the ⊖'s Orbit.	Velocity of Light.
	m. s.		m. s.		m. s.		m. s.		m. s.		h. m. s.
.0001	0 0 .05	.0042	0 2 .05	.0083	0 4 .05	.25	2 1 .88	.66	5 21 .75	1	0 8 7 .5
.0002	0 0 .10	.0043	0 2 .10	.0084	0 4 .10	.26	2 6 .75	.67	5 26 .63	2	0 16 15 .0
.0003	0 0 .15	.0044	0 2 .15	.0085	0 4 .14	.27	2 11 .63	.68	5 31 .50	3	0 24 22 .5
.0004	0 0 .20	.0045	0 2 .19	.0086	0 4 .19	.28	2 16 .50	.69	5 36 .38	4	0 32 30 .0
.0005	0 0 .24	.0046	0 2 .24	.0087	0 4 .24	.29	2 21 .38	.70	5 41 .25	5	0 40 37 .5
.0006	0 0 .29	.0047	0 2 .29	.0088	0 4 .29	.30	2 26 .25	.71	5 46 .13	6	0 48 45 .0
.0007	0 0 .34	.0048	0 2 .34	.0089	0 4 .34	.31	2 31 .13	.72	5 51 .00	7	0 56 52 .5
.0008	0 0 .39	.0049	0 2 .39	.0090	0 4 .39	.32	2 36 .00	.73	5 55 .88	8	1 5 0 .0
.0009	0 0 .44	.0050	0 2 .44	.0091	0 4 .44	.33	2 40 .88	.74	6 0 .75	9	1 13 7 .5
.0010	0 0 .49	.0051	0 2 .49	.0092	0 4 .49	.34	2 45 .75	.75	6 5 .63	10	1 21 15 .0
.0011	0 0 .54	.0052	0 2 .54	.0093	0 4 .53	.35	2 50 .63	.76	6 10 .50	11	1 29 22 .5
.0012	0 0 .59	.0053	0 2 .58	.0094	0 4 .58	.36	2 55 .50	.77	6 15 .38	12	1 37 30 .0
.0013	0 0 .63	.0054	0 2 .63	.0095	0 4 .63	.37	3 0 .38	.78	6 20 .25	13	1 45 37 .5
.0014	0 0 .68	.0055	0 2 .68	.0096	0 4 .68	.38	3 5 .25	.79	6 25 .13	14	1 53 45 .0
.0015	0 0 .73	.0056	0 2 .73	.0097	0 4 .73	.39	3 10 .13	.80	6 30 .00	15	2 1 52 .5
.0016	0 0 .78	.0057	0 2 .78	.0098	0 4 .78	.40	3 15 .00	.81	6 34 .88	16	2 10 0 .0
.0017	0 0 .83	.0058	0 2 .83	.0099	0 4 .83	.41	3 19 .88	.82	6 39 .75	17	2 18 7 .5
.0018	0 0 .88	.0059	0 2 .88	.0100	0 4 .88	.42	3 24 .75	.83	6 44 .63	18	2 26 15 .0
.0019	0 0 .93	.0060	0 2 .93	.02	0 9 .75	.43	3 29 .63	.84	6 49 .50	19	2 34 22 .5
.0020	0 0 .98	.0061	0 2 .97	.03	0 14 .63	.44	3 34 .50	.85	6 54 .38	20	2 42 30 .0
.0021	0 1 .02	.0062	0 3 .02	.04	0 19 .50	.45	3 39 .38	.86	6 59 .25	21	2 50 37 .5
.0022	0 1 .07	.0063	0 3 .07	.05	0 24 .38	.46	3 44 .25	.87	7 4 .13	22	2 58 45 .0
.0023	0 1 .12	.0064	0 3 .12	.06	0 29 .25	.47	3 49 .13	.88	7 9 .00	23	3 6 52 .5
.0024	0 1 .17	.0065	0 3 .17	.07	0 34 .13	.48	3 54 .00	.89	7 13 .88	24	3 15 0 .0
.0025	0 1 .22	.0066	0 3 .22	.08	0 39 .00	.49	3 58 .88	.90	7 18 .75	25	3 23 7 .5
.0026	0 1 .27	.0067	0 3 .27	.09	0 43 .88	.50	4 3 .75	.91	7 23 .63	26	3 31 15 .0
.0027	0 1 .32	.0068	0 3 .32	.10	0 48 .75	.51	4 8 .63	.92	7 28 .50	27	3 39 22 .5
.0028	0 1 .37	.0069	0 3 .36	.11	0 53 .63	.52	4 13 .50	.93	7 33 .38	28	3 47 30 .0
.0029	0 1 .41	.0070	0 3 .41	.12	0 58 .50	.53	4 18 .38	.94	7 38 .25	29	3 55 37 .5
.0030	0 1 .46	.0071	0 3 .46	.13	1′ 3 .38	.54	4 23 .25	.95	7 43 .13	30	4 3 45 .0
.0031	0 1 .51	.0072	0 3 .51	.14	1 8 .25	.55	4 28 .13	.96	7 48 .00	31	4 11 52 .5
.0032	0 1 .56	.0073	0 3 .56	.15	1 13 .13	.56	4 33 .00	.97	7 52 .88	32	4 20 0 .0
.0033	0 1 .61	.0074	0 3 .61	.16	1 18 .00	.57	4 37 .88	.98	7 57 .75	33	4 28 7 .5
.0034	0 1 .66	.0075	0 3 .66	.17	1 22 .88	.58	4 42 .75	.99	8 2 .63	34	4 36 15 .0
.0035	0 1 .71	.0076	0 3 .71	.18	1 27 .75	.59	4 47 .63	1 .00	8 7 .50	35	4 44 22 .5
.0036	0 1 .76	.0077	0 3 .75	.19	1 32 .63	.60	4 52 .50				
.0037	0 1 .80	.0078	0 3 .80	.20	1 37 .50	.61	4 57 .38				
.0038	0 1 .85	.0079	0 3 .85	.21	1 42 .38	.62	5 2 .25				
.0039	0 1 .99	.0080	0 3 .90	.22	1 47 .25	.63	5 7 .13				
.0040	0 1 .99	.0081	0 3 .95	.23	1 52 .13	.64	5 12 .00				
.0041	0 2 .00	.0082	0 4 .00	.24	1 57 .00	.65	5 16 .88				

POLE STAR.

TABLE 1. The MEAN PLACES of the POLE STAR for each of 21 successive Years, according to BESSEL.

Years.	R. A.	Ann.Var.	Diff.	2d Diff.	Declin.	An.Var.	Diff.
	0h	+	+	+	88°	+	−
1805	53m 31s.0860	13s.4209			16′ 2″.305	19″.521	
1806	53 44 .5069	13 .4958	.0749	6	16 21 .826	19 .517	.004
1807	53 58 .0027	13 .5713	.0755	7	16 41 .343	19 .511	.006
1808	54 11 .5740	13 .6475	.0762	7	17 0 .854	19 .507	.004
1809	54 25 .2215	13 .7244	.0769	8	17 20 .361	19 .502	.005
1810	54 38 .9459	13 .8021	.0777	7	17 39 .863	19 .498	.004
1811	54 52 .7480	13 .8805	.0784	6	17 59 .361	19 .492	.006
1812	55 6 .6285	13 .7595	.0690	8	18 18 .853	19 .487	.005
1813	55 20 .5880	13 .0393	.0798	9	18 38 .340	19 .483	.004
1814	55 34 .6273	14 .1200	.0807	6	18 57 .823	19 .477	.006
1815	55 48 .7473		.0813		19 17 .300		.005

Years.	R. A.	An. Var.	Diff.	2d Diff.	Declin.	An.Var.	Diff.
	0h	+	+	+	88°	+	−
1815	55m 48s.7473	14s.2013	.0813		19′ 17″.300	19″.472	.005
1816	56 2 .9486	14 .2835	.0822	7	19 36 .772	19 .467	.005
1817	56 17 .2321	14 .3664	.0829	7	19 56 .239	19 .462	.005
1818	56 31 .5985	14 .4500	.0836	9	20 15 .701	19 .457	.005
1819	56 46 .0485	14 .5345	.0845	8	20 35 .158	19 .451	.006
1820	57 0 .5830	14 .6198	.0853	8	20 54 .609	19 .446	.005
1821	57 15 .2028	14 .7059	.0861	9	21 14 .055	19 .440	.006
1822	57 29 .9087	14 .7929	.0870	7	21 33 .495	19 .435	.005
1823	57 44 .7016	14 .8806	.0877	10	21 52 .930	19 .430	.005
1824	57 59 .5822	14 .9693	.0887		22 12 .360	19 .424	.006
1825	58 14 .5515				22 31 .784		

TABLE 2. CORRECTIONS of the observed ZENITH DISTANCE of the POLE STAR taken out of the Meridian; for various Latitudes; and for any Horary Angle counted from the time of its Meridian Passage; its Polar Distance being = 1° 40′. (HORNER.)

Horary Angle.		LATITUDES.																					Horary Angle.	
h.	m.	40°	41°	42°	43°	44°	45°	46°	47°	48°	49°	50°	51°	52°	53°	54°	55°	56°	57°	58°	59°	60°	h.	m.
0	0	0″.0	0″.0	0″.0	0″.0	0″.0	0″.0	0″.0	0″.0	0″.0	0″.0	0″.0	0″.0	0″.0	0″.0	0″.0	0″.0	0″.0	0″.0	0″.0	0″.0	0″.0	24	0
	10	0.2	0.2	0.2	0.2	0.2	0.2	0.2	0.2	0.2	0.2	0.2	0.2	0.2	0.2	0.2	0.3	0.3	0.3	0.3	0.3	0.3		50
	20	0.6	0.6	0.6	0.6	0.6	0.7	0.7	0.7	0.8	0.8	0.8	0.9	0.9	0.9	1.0	1.0	1.0	1.1	1.1	1.1	1.2		40
	30	1.3	1.3	1.4	1.4	1.5	1.5	1.6	1.6	1.7	1.8	1.9	1.9	2.0	2.1	2.1	2.2	2.3	2.4	2.5	2.6	2.7		30
	40	2.3	2.3	2.4	2.5	2.6	2.7	2.8	2.9	3.2	3.1	3.2	3.4	3.5	3.6	3.8	3.9	4.1	4.2	4.4	4.6	4.8		20
	50	3.5	3.7	3.8	3.9	4.1	4.2	4.4	4.5	4.7	4.9	5.0	5.2	5.4	5.6	5.8	6.1	6.3	6.6	6.8	7.1	7.4		10
1	0	5.0	5.2	5.4	5.6	5.8	6.0	6.2	6.5	6.7	6.9	7.2	7.5	7.7	8.0	8.3	8.7	9.0	9.4	9.8	10.2	10.6	23	0
	10	6.8	7.0	7.3	7.5	7.8	8.1	8.4	8.8	9.1	9.4	9.7	10.0	10.4	10.8	11.2	11.7	12.1	12.7	13.2	13.7	14.3		50
	20	8.7	9.1	9.4	9.7	10.1	10.5	10.8	11.3	11.7	12.1	12.5	13.0	13.5	14.0	14.5	15.1	15.7	16.3	17.0	17.7	18.5		40
	30	10.7	11.4	11.8	12.2	12.6	13.1	13.6	14.1	14.6	15.1	15.7	16.3	16.9	17.5	18.2	18.9	19.7	20.4	21.3	22.1	23.1		30
	40	13.4	13.9	14.4	14.9	15.4	16.0	16.6	17.2	17.8	18.4	19.1	19.8	20.6	21.4	22.2	23.0	23.9	24.9	25.9	27.0	28.1		20
	50	16.0	16.5	17.1	17.8	18.4	19.1	19.8	20.5	21.3	22.0	22.8	23.7	24.5	25.5	26.4	27.5	28.5	29.7	30.9	32.2	33.5		10
2	0	18.7	19.4	20.1	20.8	21.6	22.3	23.1	24.0	24.9	25.8	26.7	27.7	28.8	29.8	31.0	32.2	33.4	34.8	36.2	37.7	39.3	22	0
	10	21.6	22.4	23.2	24.0	24.9	25.8	26.7	27.7	28.7	29.7	30.8	32.0	33.2	34.4	35.7	37.1	38.6	40.1	41.8	43.5	45.3		50
	20	24.6	25.5	26.4	27.3	28.3	29.3	30.4	31.5	32.7	33.8	35.1	36.4	37.7	39.2	40.7	42.2	43.9	45.7	47.5	49.5	51.6		40
	30	27.6	28.6	29.7	30.8	31.9	33.0	34.2	35.5	36.7	38.1	39.5	40.9	42.5	44.1	45.7	47.5	49.4	51.4	53.5	55.7	58.0		30
	40	30.8	31.9	33.1	34.3	35.9	36.8	38.1	39.5	40.9	42.4	44.0	45.6	47.3	49.1	50.9	52.9	55.0	57.2	59.5	61.9	64.5		20
	50	34.0	35.2	36.5	37.8	39.1	40.6	42.0	43.5	45.1	46.8	48.5	50.3	52.2	54.1	56.2	58.4	60.6	63.0	65.6	68.3	71.2		10
3	0	37.2	38.5	39.9	41.4	42.9	44.4	46.0	47.7	49.4	51.2	53.1	55.0	57.1	59.2	61.5	63.9	66.3	68.9	71.7	74.7	77.8	21	0
	10	40.4	41.8	43.4	44.9	46.5	48.2	49.9	51.7	53.6	55.5	57.6	59.7	61.9	64.3	66.7	69.3	72.0	74 8	77.8	81.1	84.5		50
	20	43.5	45.1	46.8	48.5	50.2	52.0	53.9	55.8	57.8	59.9	62.1	64.4	66.8	69.2	71.9	74.7	77.6	80.7	83.9	87.4	91.0		40
	30	46.6	48.3	50.0	51.9	53.7	55.7	57.7	59.7	61.9	64.2	66.5	68.9	71.5	74.2	77.0	80.0	83.1	86.4	89.8	93.5	97.4		30
	40	49.6	51.4	53.3	55.2	57.2	59.3	61.4	63.6	65.9	68.3	70.8	73.4	76.1	79.0	82.0	85.1	88.4	91.9	95.6	99.5	103.7		20
	50	52.6	54.5	56.4	58.4	60.5	62.7	65.0	67.3	69.7	72.3	74.9	77.7	80.5	83.6	86.7	90.0	93.5	97.2	101.1	105.3	109.6		10
4	0	55.3	57.4	59.4	61.5	63.7	66.0	68.4	70.9	73.4	76.1	78.9	81.8	84.8	87.9	91.3	94.8	98.4	102.3	106.4	110.8	115.4	20	0
	10	58.0	60.1	62.2	64.5	66.8	69.2	71.7	74.2	76.9	79.7	82.6	85.6	88.8	92.1	95.6	99.2	103.1	107.1	111.4	115.9	120.8		50
	20	60.4	62.6	64.9	67.2	69.6	72.1	74.7	77.4	80.2	83.1	86.1	89.2	92.5	96.0	99.6	103.4	107.4	111.6	116.0	120.8	125.8		40
	30	62.7	65.0	67.3	69.7	72.2	74.8	77.5	80.3	83.2	86.2	89.3	92.6	96.0	99.6	103.3	107.2	111.4	115.7	120.3	125.2	130.4		30
	40	64.8	67.1	69.5	72.0	74.6	77.3	80.0	82.9	85.9	89.0	92.2	95.6	99.1	102.8	106.6	110.7	115.0	119.5	124.2	129.3	134.6		20
	50	66.6	69.0	71.5	74.1	76.7	79.5	82.3	85.2	88.3	91.5	94.8	98.3	101.9	105.7	109.6	113.8	118.2	122.8	127.7	132.9	138.3		10
5	0	68.3	70.7	73.2	75.9	78.6	81.4	84.3	87.3	90.4	93.7	97.1	100.6	104.3	108.2	112.2	116.5	121.0	125.7	130.7	136.0	141.5	19	0
	10	69.6	72.1	74.7	77.4	80.1	83.0	86.0	89.1	92.3	95.6	99.0	102.6	106.4	110.3	114.4	118.7	123.3	128.1	133.2	138.6	144.3		50
	20	70.7	73.3	75.9	78.6	81.4	84.3	87.3	90.4	93.7	97.1	100.6	104.2	108.0	112.0	116.2	120.6	125.2	130.1	135.2	140.7	146.4		40
	30	71.6	74.2	76.8	79.6	82.4	85.3	88.4	91.5	94.7	98.2	101.7	105.4	109.3	113.3	117.5	122.0	126.6	131.6	136.7	142.2	148.0		30
	40	72.2	74.8	77.5	80.2	83.0	86.0	89.1	92.2	95.6	99.0	102.5	106.2	110.1	114.1	118.4	122.9	127.6	132.5	137.7	143.2	149.1		20
	50	72.5	75.1	77.8	80.5	83.4	86.4	89.4	92.6	95.9	99.3	102.9	106.6	110.5	114.5	118.8	123.3	128.0	133.0	138.2	143.7	149.6		10
6	0	72.5	75.1	77.8	80.5	13.4	86.4	89.4	92.6	95.9	99.3	102.9	106.6	110.5	114.5	118.8	123.3	128.0	132.9	138.1	143.7	149.5	18	0
Horary Angle.		40°	41°	42°	43°	44°	45°	46°	47°	48°	49°	50°	51°	52°	53°	54°	55°	56°	57°	58°	59°	60°	Horary Angle.	
		LATITUDES.																						

The Co-Latitude = the observed and corrected Zenith distance, ± P. cos. t — tabular number; P being = the Star's true apparent Polar Distance, and t = the Horary Angle from the time of its superior Meridian Passage; but the tabular quantity must be diminished $\frac{1}{100}$th part, for every minute of *diminution* of the Star's Polar Distance below 1° 40′, and in the same proportion for smaller portions.

When the Star is apparently moving in the *upper* semi-circle, the amount of (P. cosine t) will be *additive*; but when in the *lower*, *subtractive*.

POLE STAR.

TABLE 2.

CORRECTIONS of the observed ZENITH DISTANCE of the POLE STAR, &c. concluded.

LATITUDES.

Horary Angle.	40°	41°	42°	43°	44°	45°	46°	47°	48°	49°	50°	51°	52°	53°	54°	55°	56°	57°	58°	59°	60°	Horary Angle.
h. m.																						h. m.
6 0	72″.5	75″.1	77″.8	80″.5	83″.4	86″.4	89″.4	92″.6	95″.9	99″.3	102″.9	106″.6	110″.5	114″.6	118″.8	123″.3	128″.0	133″.0	138″.2	143″.7	149″.5	18 0
10	72.2	74.8	77.5	80.3	83.1	86.1	89.1	92.3	95.6	99.0	102.6	106.2	110.1	114.1	118.4	122.8	127.5	132.4	137.6	143.1	148.9	50
20	71.7	74.3	77.0	79.7	82.5	85.5	88.5	91.6	94.9	98.3	101.8	105.4	109.3	113.3	117.5	121.9	126.5	131.4	136.5	141.9	147.7	40
30	70.9	73.5	76.1	78.8	81.6	84.5	87.5	90.6	93.8	97.1	100.6	104.2	108.0	112.0	116.1	120.5	125.0	129.8	134.9	140.2	145.9	30
40	69.9	72.4	75.0	77.6	80.3	83.2	86.2	89.2	92.4	95.7	99.1	102.7	106.4	110.3	114.3	118.6	123.1	127.8	132.8	138.1	143.6	20
50	68.6	71.0	73.5	76.1	78.8	81.6	84.5	87.5	90.7	93.9	97.2	100.7	104.4	108.2	112.2	116.3	120.8	125.4	130.2	135.4	140.8	10
7 0	67.0	69.4	71.9	74.4	77.1	79.8	82.6	85.5	88.6	91.7	94.9	98.4	102.0	105.7	109.6	113.6	117.9	122.5	127.2	132.2	137.5	17 0
10	65.3	67.5	69.9	72.4	75.0	77.7	80.4	83.3	86.2	89.3	92.5	95.8	99.2	102.8	106.6	110.6	114.7	119.1	123.7	128.6	133.8	50
20	63.3	65.5	67.8	70.2	72.7	75.3	77.9	80.7	83.6	86.6	89.6	92.8	96.2	99.6	103.3	107.2	111.2	115.3	119.9	124.6	129.6	40
30	61.0	63.2	65.4	67.8	70.2	72.7	75.2	77.9	80.7	83.6	86.5	89.6	92.8	96.2	99.7	103.4	107.3	111.4	115.7	120.2	125.0	30
40	58.7	60.8	62.9	65.1	67.4	69.8	72.3	74.8	77.5	80.2	83.1	86.1	89.1	92.4	95.7	09.3	103.0	106.9	111.0	115.4	120.0	20
50	56.1	58.1	60.2	62.3	64.5	66.7	69.1	71.6	74.1	76.7	79.4	82.3	85.2	88.3	91.5	94.9	98.5	102.2	106.1	110.3	114.7	10
8 0	53.4	55.3	57.3	59.3	61.4	63.6	65.8	68.1	70.5	73.0	75.6	78.3	81.1	84.0	87.1	90.3	93.7	97.3	101.0	105.0	109.1	16 0
10	50.5	52.4	54.2	56.1	58.1	60.2	62.3	64.5	66.8	69.1	71.6	74.1	76.8	79.6	82.5	85.5	88.7	92.1	95.6	99.4	103.3	50
20	47.6	49.3	51.1	52.9	54.7	56.7	58.7	60.8	62.9	65.1	67.4	69.8	72.3	75.0	77.7	80.5	83.6	86.7	90.0	93.5	97.3	40
30	44.6	46.2	47.9	49.6	51.3	53.1	55.0	56.9	58.9	61.0	63.2	65.4	67.7	70.2	72.7	75.4	78.2	81.2	84.3	87.5	91.0	30
40	41.6	43.1	44.6	46.2	47.8	49.5	51.2	53.0	54.9	56.8	58.8	60.9	63.1	65.4	67.7	70.2	72.8	75.6	78.4	81.5	84.7	20
50	38.6	39.9	41.3	42.7	44.2	45.8	47.4	49.0	50.7	52.5	54.4	56.3	58.3	60.5	62.7	65.0	67.3	69.9	72.5	75.3	78.3	10
9 0	35.5	36.7	37.9	39.3	40.6	42.0	43.5	45.0	46.6	48.3	50.0	51.7	53.6	55.5	57.5	59.7	61.8	64.2	66.6	69.2	72.0	15 0
10	32.3	33.4	34.6	35.8	37.0	38.3	39.7	41.1	42.5	44.0	45.5	47.2	48.9	50.6	52.4	54.4	56.4	58.5	60.7	63.1	65.6	50
20	29.2	30.2	31.3	32.4	33.5	34.7	35.9	37.1	38.4	39.8	41.2	42.7	44.2	45.8	47.4	49.1	51.0	52.9	54.9	57.0	59.3	40
30	26.1	27.1	28.0	29.0	30.0	31.1	32.2	33.3	34.4	35.7	36.9	38.2	39.6	41.0	42.5	44.0	47.7	47.4	49.2	51.1	53.1	30
40	23.2	24.0	24.9	25.7	26.6	27.5	28.5	29.5	30.5	31.6	32.7	33.8	35.1	36.3	37.6	39.0	40.5	42.0	43.6	45.3	47.1	20
50	20.3	21.1	21.8	22.5	23.3	24.1	25.0	25.9	26.8	27.7	28.7	29.7	30.8	31.8	33.0	34.2	35.5	36.8	38.2	39.7	41.2	10
10 0	17.6	18.2	18.9	19.5	20.2	20.9	21.6	22.4	23.2	24.0	24.8	25.7	26.6	27.6	28.5	29.6	30.7	31.9	33.1	34.3	35.7	14 0
10	15.0	15.5	16.1	16.6	17.2	17.8	18.4	19.1	19.8	20.5	21.2	21.9	22.7	23.5	24.3	25.2	26.2	27.1	28.2	29.3	30.4	50
20	12.6	13.0	13.4	13.9	14.4	14.9	15.4	16.0	16.5	17.1	17.7	18.3	19.0	19.7	20.4	21.1	21.9	22.7	23.6	24.5	25.5	40
30	10.3	10.6	11.0	11.4	11.8	12.2	12.6	13.1	13.5	14.0	14.5	15.0	15.5	16.1	16.7	17.3	17.9	18.6	19.3	20.1	20.9	30
40	8.2	8.5	8.8	9.1	9.4	9.7	10.1	10.4	10.8	11.2	11.6	12.0	12.4	12.8	13.3	13.8	14.3	14.9	15.4	16.0	16.7	20
50	6.4	6.6	6.8	7.0	7.3	7.5	7.8	8.1	8.4	8.7	9.0	9.2	9.6	9.9	10.3	10.7	11.1	11.5	11.9	12.4	12.9	10
11 0	4.7	4.9	5.0	5.2	5.4	5.6	5.8	6.0	6.2	6.4	6.6	6.9	7.1	7.4	7.6	7.9	8.2	8.5	8.8	9.2	9.5	13 0
10	3.3	3.4	3.6	3.7	3.8	3.9	4.1	4.2	4.3	4.5	4.6	4.8	5.0	5.2	5.3	5.5	5.7	6.0	6.2	6.4	6.6	50
20	2.1	2.2	3.3	2.3	2.4	2.5	2.6	2.6	2.7	2.8	3.0	3.1	3.2	3.3	3.4	3.6	3.7	3.8	4.0	4.1	4.3	40
30	1.2	1.2	1.3	1.3	1.4	1.4	1.4	1.5	1.6	1.6	1.7	1.7	1.8	1.9	1.9	2.0	2.1	2.1	2.2	2.3	2.4	30
40	0.5	0.5	0.6	0.6	0.6	0.6	0.6	0.6	0.7	0.7	0.7	0.8	0.8	0.8	0.9	0.9	0.9	0.9	1.0	1.0	1.1	20
50	0.2	0.2	0.2	0.2	0.2	0.2	0.2	0.2	0.2	0.2	0.2	0.2	0.2	0.2	0.2	0.2	0 2	0.2	0.2	0.2	0.2	10
12 0	0.0	0.0	0.0	0.0	0.0	0.0	0.0	0.0	0.0	0.0	0.0	0.0	0.0	0.0	0.0	0.0	0.0	0.0	0.0	0.0	0.0	12 0

TABLE 3. LITTROW's small general Table for determining the LATITUDE, at *any time*, by the POLE STAR.

θ	M	N	θ	M	N	θ	M	N	θ	M	N	θ	M	N	θ	M	N
h. m.			h. m.			h. m.			h. m.			h. m.			h. m.		
0 0	0″.00	0″.00	1 0	5″.85	0″.11	2 0	21″.82	0″.37	3 0	43″.63	0″.60	4 0	65″.45	0″.63	5 0	81″.42	0″.41
10	0.17	0.00	10	7.89	0.14	10	25.19	0.41	10	47.44	0.62	10	68.66	0.61	10	83.18	0.35
20	0.66	0.01	20	10.20	0.19	20	28.71	0.46	20	51.21	0.64	20	71.68	0.59	20	84.64	0.28
30	1.49	0.03	30	12.78	0.23	30	32.34	0.50	30	54.93	0.65	30	74.49	0.55	30	85.78	0.22
40	2.63	0.05	40	15.59	0.27	40	36.06	0.53	40	58.56	0.65	40	77.06	0.51	40	86.60	0.15
50	4.09	0.08	50	18.61	0.32	50	39.83	0.57	50	62.07	0.64	50	79.38	0.46	50	87.10	0.17
1 0	5.85	0.11	2 0	21.82	0.37	3 0	43.63	0.60	4 0	65.45	0.63	5 0	81.42	0.41	6 0	87.26	0.00

$\psi = Z + p \cos. t - M \text{ cotan. } Z + N$; where ψ is = the Latitude; Z = the Zenith Distance; $p = 1° 40$, or 100′; t = the Horary Angle; $\theta = t$ in the first Quadrant; $= 12^h - t$ in the second; $= t - 12^h$ in the third; and $= 24^h - t$ in the fourth; M and N being the Tabular Quantities. The quantity M is $= \frac{1}{2} p^2 \sin.^2 t$, and is always positive, but the quantity $N = \frac{1}{3} p^3 \sin.^2 t \cos. t$, becomes negative in the second and third Quadrants of t. When p, (the Polar Distance) augments or diminishes 1′, the Tabular Quantity must also be augmented or diminished by 0.02 M; and for any other quantity of variation in the same proportion.

TABLE 4. The Correction of the Pole Star's observed Zenith Distance, taken at any time, for the Latitude of Greenwich, and Polar Distance 1° 40′, with the Var. in 100′ of Lat. and in 100″ diminution of Polar Distance. Argument = the Horary Angle.

SID. HOURS.																			
Horary Angle.	O			I			II			III			IV			V			Horary Angle.
	Cor.	Var. in 100′ Lat.	Var. in 100″ dim.	Cor.	Var. in 100′ Lat.	Var. in 100″ dim.	Cor.	Var. in 100′ Lat.	Var. in 100″ dim.	Cor.	Var. in 100′ Lat.	Var. in 100″ dim.	Cor.	Var. in 100′ Lat.	Var. in 100″ dim.	Cor.	Var. in 100′ Lat.	Var. in 100″ dim.	
m.																			m.
0	0″.00	0″.00	0″.00	7″.60	0″.50	0″.08	28″.21	1″.73	0″.28	56″.06	3″.42	0″.56	83″.26	5″.05	0″.83	102″.48	6″.15	1″.02	60
1	0.02	0.00	0.00	7.85	0.51	0.08	28.63	1.75	0.29	56.54	3.45	0.56	83.66	5.08	0.84	102.70	6.17	1.03	59
2	0.04	0.00	0.00	8.10	0.53	0.08	29.06	1.78	0.29	57.02	3.48	0.57	84.06	5.10	0.84	102.91	6.19	1.03	58
3	0.06	0.00	0.00	8.35	0.54	0.08	29.49	1.81	0.29	57.49	3.50	0.57	84.46	5.13	0.84	103.12	6.20	1.03	57
4	0.08	0.01	0.00	8.61	0.56	0.09	29.92	1.83	0.30	57.97	3.53	0.58	84.86	5.15	0.85	103.33	6.21	1.03	56
5	0.10	0.01	0.00	8.87	0.57	0.09	30.35	1.86	0.30	58.45	3.56	0.58	85.25	5.18	0.85	103.53	6.22	1.04	55
6	0.12	0.01	0.00	9.13	0.59	0.09	30.79	1.89	0.31	58.92	3.59	0.59	85.64	5.20	0.86	103.73	6.23	1.04	54
7	0.14	0.01	0.00	9.40	0.60	0.09	31.23	1.91	0.31	59.40	3.62	0.59	86.03	5.22	0.86	103.93	6.24	1.04	53
8	0.16	0.01	0.00	9.67	0.62	0.10	31.67	1.94	0.32	59.88	3.64	0.60	86.42	5.24	0.86	104.12	6.25	1.04	52
9	0.19	0.01	0.00	9.95	0.63	0.10	32.11	1.97	0.32	60.35	3.67	0.60	86.81	5.26	0.87	104.31	6.26	1.04	51
10	0.22	0.01	0.00	10.23	0.65	0.10	32.55	2.00	0.33	60.83	3.70	0.61	87.20	5.28	0.87	104.50	6.27	1.05	50
11	0.26	0.02	0.00	10.51	0.66	0.11	32.99	2.02	0.33	61.30	3.73	0.61	87.58	5.30	0.88	104.68	6.28	1.05	49
12	0.31	0.02	0.00	10.79	0.68	0.11	33.43	2.05	0.33	61.77	3.76	0.62	87.96	5.32	0.88	104.86	6.29	1.05	48
13	0.37	0.02	0.00	11.08	0.70	0.11	33.87	2.08	0.34	62.25	3.79	0.62	88.34	5.34	0.88	105.03	6.30	1.05	47
14	0.43	0.03	0.00	11.37	0.71	0.11	34.32	2.11	0.34	62.72	3.81	0.63	88.71	5.36	0.89	105.20	6.31	1.05	46
15	0.50	0.03	0.00	11.67	0.73	0.12	34.77	2.14	0.35	63.19	3.84	0.63	89.08	5.38	0.89	105.36	6.32	1.05	45
16	0.57	0.04	0.01	11.97	0.75	0.12	35.22	2.16	0.35	63.67	3.87	0.64	89.45	5.40	0.89	105.52	6.33	1.06	44
17	0.64	0.05	0.01	12.28	0.76	0.12	35.67	2.19	0.36	64.14	3.90	0.64	89.81	5.42	0.90	105.68	6.34	1.06	43
18	0.72	0.05	0.01	12.59	0.78	0.13	36.12	2.22	0.36	64 61	3.92	0.65	90.17	5.44	0.90	105.83	6.35	1.06	42
19	0.80	0.06	0.01	12.91	0.80	0.13	36.58	2.25	0.37	65.09	3.95	0.65	90.53	5.46	0.91	105.98	6.36	1.06	41
20	0.88	0.07	0.01	13.23	0.82	0.13	37.04	2.28	0.37	65.56	3.98	0.66	90.89	5.48	0.91	106.12	6.37	1.06	40
21	0.97	0.07	0.01	13.55	0.83	0.14	37.50	2.30	0.37	66.03	4.01	0.66	91.24	5.50	0.91	106.26	6.38	1.06	39
22	1.06	0.08	0.01	13.87	0.85	0.14	37.96	2.33	0.38	66.50	4.04	0.66	91.59	5.52	0.92	106.39	6.39	1.06	38
23	1.15	0.09	0.01	14.20	0.87	0.14	38.42	2.36	0.38	66.97	4.06	0.67	91.94	5.54	0.92	106.52	6.39	1.07	37
24	1.25	0.09	0.01	14.53	0.89	0.15	38.88	2.39	0.39	67.44	4.09	0.67	92.28	5.56	0.92	106.65	6.40	1.07	36
25	1.35	0.10	0.01	14.86	0.91	0.15	39.35	2.41	0.39	67.91	4.12	0.68	92.62	5.58	0.93	106.77	6.40	1.07	35
26	1.46	0.11	0.01	15.19	0.93	0.15	39.82	2.44	0.40	68.38	4.15	0.68	92.96	5.60	0.93	106.89	6.41	1.07	34
27	1.58	0.12	0.02	15.53	0.95	0.16	40.29	2.47	0.40	68.84	4.17	0.69	93.30	5.62	0.93	107.01	6.41	1.07	33
28	1.70	0.13	0.02	15.87	0.97	0.16	40.76	2.50	0.41	69.30	4.20	0.69	93.63	5.64	0.94	107.12	6.42	1.07	32
29	1.83	0.14	0.02	16.21	0.99	0.16	41.23	2.53	0.41	69.76	4.23	0.70	93.96	5.66	0.94	107.23	6.42	1.07	31
30	1.96	0.15	0.02	16.56	1.01	0.17	41.70	2.56	0.42	70.22	4.25	0.70	94.29	5.67	0.94	107.34	6.43	1.07	30
31	2.09	0.16	0.02	16.91	1.03	0.17	42.17	2.58	0.42	70.68	4.28	0.71	94.61	5.69	0.95	107.44	6.43	1.07	29
32	2.23	0.17	0.02	17.26	1.05	0.17	42.64	2.61	0.43	71.14	4.31	0.71	94.93	5.70	0.95	107.54	6.44	1.08	28
33	2.37	0.18	0.02	17.62	1.07	0.18	43.11	2.64	0.43	71.60	4.34	0.72	95.24	5.72	0.95	107.64	6.44	1.08	27
34	2.51	0.19	0.03	17.98	1.10	0.18	43.58	2.67	0.44	72.05	4.36	0.72	95.55	5.73	0.96	107.73	6.45	1.08	26
35	2.66	0.20	0.03	18.34	1.12	0.18	44.05	2.70	0.44	72.50	4.39	0.72	95.86	5.75	0.96	107.82	6.45	1.08	25
36	2.81	0.21	0.03	18.71	1.14	0.19	44.52	2.73	0.45	72.95	4.41	0.73	96.16	5.77	0.96	107.90	6.45	1.08	24
37	2.96	0.22	0.03	19.08	1.16	0.19	44.99	2.76	0.45	73.40	4.44	0.73	96.46	5.79	0.96	107.98	6.46	1.08	23
38	3.11	0.23	0.03	19.45	1.18	0.19	45.46	2.79	0.45	73.85	4.47	0.74	96.76	5.80	0.97	108.05	6.46	1.08	22
39	3.27	0.24	0.03	19.82	1.20	0.20	45.93	2.82	0.46	74.30	4.50	0.74	97.06	5.82	0.97	108.12	6.46	1.08	21
40	3.43	0.25	0.03	20.20	1.23	0.20	46.40	2.85	0.46	74.75	4.53	0.75	97.35	5.84	0.97	108.18	6.47	1.08	20
41	3.60	0.26	0.04	20.58	1.25	0.21	46.88	2.87	0.47	75.19	4.56	0.75	97.64	5.85	0.98	108.24	6.47	1.08	19
42	3.77	0.27	0.04	20.96	1.27	0.21	47.36	2.90	0.47	75.53	4.59	0.76	97.93	5.87	0.98	108.30	6.47	1.08	18
43	3.95	0.28	0.04	21.34	1.30	0.21	47.84	2.93	0.48	75.97	4.61	0.76	98.21	5.89	0.98	108.35	6.47	1.08	17
44	4.13	0.29	0.04	21.73	1.32	0.22	48.32	2.96	0.48	76.41	4.64	0.76	98.49	5.90	0.98	108.40	6.48	1.08	16
45	4.32	0.31	0.04	22.12	1.35	0.22	48.80	2.99	0.49	76.85	4.67	0.77	98.77	5.92	0.99	108.45	6.48	1.08	15
46	4.51	0.32	0.05	22.51	1.37	0.23	49.29	3.02	0.49	77.28	4.69	0.77	99.04	5.94	0.99	108.49	6.48	1.08	14
47	4.70	0.33	0.05	22.91	1.39	0.23	49.77	3.04	0.50	77.71	4.72	0.78	99.31	5.95	0.99	108.53	6.48	1.09	13
48	4.90	0.34	0.05	23.31	1.42	0.23	50.25	3.07	0.50	78.14	4.75	0.78	99.58	5.97	1.00	108.56	6.48	1.09	12
49	5.10	0.35	0.05	23.71	1.44	0.24	50.73	3.10	0.51	78.67	4.78	0.79	99.84	5.98	1.00	108.58	6.48	1.09	11
50	5.31	0.37	0.05	24.11	1.47	0.24	51.22	3.13	0.51	79.10	4.80	0.79	100.10	6.00	1.00	108.59	6.49	1.09	10
51	5.52	0.38	0.06	24.51	1.49	0.25	51.70	3.15	0.52	79.52	4.83	0.80	100.36	6.01	1.00	108.59	6.49	1.09	9
52	5.73	0.39	0.06	24.91	1.52	0.25	52.18	3.18	0.52	79.94	4.85	0.80	100.61	6.03	1.01	108.59	6.49	1.09	8
53	5.95	0.41	0.06	25.32	1.54	0.25	52.66	3.21	0.53	80.36	4.88	0.80	100.86	6.04	1.01	108.59	6.49	1.09	7
54	6.17	0.42	0.06	25.73	1.57	0.26	53.14	3.24	0.53	80.78	4.90	0.81	101.10	6.06	1.01	108.58	6.49	1.09	6
55	6.39	0.43	0.06	26.14	1.59	0.26	53.62	3.27	0.54	81.20	4.93	0.81	101.34	6.07	1.01	108.58	6.49	1.09	5
56	6.62	0.45	0.07	26.55	1.62	0.27	54.10	3.30	0.54	81.62	4.95	0.82	101.58	6.09	1.02	108.58	6.49	1.09	4
57	6.86	0.47	0.07	26.96	1.64	0.27	54.59	3.33	0.55	82.03	4.98	0.82	101.81	6.10	1.02	108.58	6.49	1.09	3
58	7.10	0.48	0.07	27.37	1.67	0.27	55.08	3.36	0.55	82.44	5.00	0.82	102.04	6.12	1.02	108.58	6.49	1.09	2
59	7.35	0.49	0.07	27.79	1.70	0.28	55.57	3.39	0.56	82.85	5.03	0.83	102.26	6.14	1.02	108.58	6.49	1.09	1
60	7.60	0.50	0.08	28.21	1.73	0.28	56.06	3.42	0.56	83.26	5.05	0.83	102.48	6.15	1.02	108.58	6.49	1.09	0
	XXIII			XXII			XXI			XX			XIX			XVIII			
SID. HOURS.																			

Co-Lat. $= Z \pm P \cos. t$ — Tab. Num. as in Tab. 2. If the Lat. be South of Greenwich, the Var. in 100′ is —; but if North +.

POLE STAR.

TABLE 4.

The Correction of the Pole Star's observed Zenith Distance, &c. concluded.

SID. HOURS.

Horary Angle.	VI Cor.	VI Var. in 100' Lat.	VI Var. in 100" dim.	VII Cor.	VII Var. in 100' Lat.	VII Var. in 100" dim.	VIII Cor.	VIII Var. in 100' Lat.	VIII Var. in 100" dim.	IX Cor.	IX Var. in 100' Lat.	IX Var. in 100" dim.	X Cor.	X Var. in 100' Lat.	X Var. in 100" dim.	XI Cor.	XI Var. in 100' Lat.	XI Var. in 100" dim.	Horary Angle.
m.																			m.
0	108".57	6".48	1".09	100".18	5".94	1".00	79".69	4".66	0".80	52".67	3".10	0".53	26".16	1".53	0".26	6".98	0".42	0".07	60
1	108.56	6.48	1.09	99.93	5.92	1.00	79.27	4.64	0.79	52.20	3.07	0.52	25.77	1.51	0.26	6.75	0.40	0.07	59
2	108.54	6.47	1.09	99.67	5.90	1.00	78.85	4.62	0.79	51.73	3.04	0.52	25.38	1.49	0.25	6.52	0.39	0.07	58
3	108.51	6.47	1.09	99.41	5.89	0.99	78.43	4.59	0.78	51.26	3.01	0.51	24.99	1.46	0.25	6.30	0.38	0.06	57
4	108.47	6.46	1.08	99.14	5.87	0.99	78.01	4.57	0.78	50.79	2.98	0.51	24.60	1.44	0.25	6.08	0.37	0.06	56
5	108.43	6.46	1.08	98.87	5.85	0.99	77.59	4.54	0.78	50.32	2.95	0.50	24.21	1.41	0.24	5.87	0.36	0.06	55
6	108.38	6.45	1.08	98.60	5.84	0.99	77.17	4.52	0.77	49.85	2.92	0.50	23.82	1.39	0.24	5.66	0.34	0.06	54
7	108.33	6.45	1.08	98.33	5.82	0.98	76.74	4.50	0.77	49.38	2.89	0.49	23.44	1.37	0.23	5.46	0.33	0.05	53
8	108.28	6.44	1.08	98.05	5.80	0.98	76.31	4.47	0.76	48.92	2.87	0.49	23.06	1.34	0.23	5.26	0.32	0.05	52
9	108.22	6.44	1.08	97.77	5.78	0.98	75.88	4.45	0.76	48.46	2.84	0.48	22.68	1.32	0.23	5.06	0.31	0.05	51
10	108.16	6.43	1.08	97.49	5.76	0.97	75.45	4.42	0.75	48.00	2.82	0.48	22.30	1.30	0.22	4.87	0.30	0.05	50
11	108.10	6.43	1.08	97.20	5.74	0.97	75.02	4.40	0.75	47.54	2.79	0.48	21.92	1.27	0.22	4.68	0.29	0.05	49
12	108.03	6.42	1.08	96.91	5.72	0.97	74.59	4.38	0.75	47.08	2.76	0.47	21.54	1.25	0.22	4.49	0.28	0.04	48
13	107.96	6.42	1.08	96.62	5.70	0.97	74.16	4.35	0.74	46.62	2.73	0.47	21.17	1.23	0.21	4.31	0.27	0.04	47
14	107.89	6.41	1.08	96.32	5.68	0.96	73.72	4.33	0.74	46.16	2.70	0.46	20.80	1.21	0.21	4.13	0.26	0.04	46
15	107.81	6.41	1.08	96.02	5.66	0.96	73.28	4.30	0.73	45.70	2.67	0.46	20.43	1.19	0.20	3.95	0.25	0.04	45
16	107.73	6.40	1.08	95.72	5.64	0.96	72.84	4.28	0.73	45.24	2.64	0.45	20.07	1.16	0.20	3.78	0.24	0.04	44
17	107.64	6.40	1.08	95.42	5.62	0.95	72.40	4.26	0.72	44.78	2.62	0.45	19.71	1.14	0.20	3.61	0.23	0.04	43
18	107.55	6.39	1.08	95.11	5.60	0.95	71.96	4.23	0.72	44.32	2.60	0.44	19.35	1.12	0.19	3.44	0.22	0.03	42
19	107.45	6.39	1.07	94.80	5.58	0.95	71.52	4.21	0.72	43.86	2.57	0.44	19.00	1.10	0.19	3.28	0.21	0.03	41
20	107.35	6.38	1.07	94.49	5.56	0.94	71.08	4.18	0.71	43.41	2.55	0.43	18.65	1.08	0.19	3.12	0.20	0.03	40
21	107.25	6.38	1.07	94.17	5.54	0.94	70.63	4.16	0.71	42.95	2.52	0.43	18.30	1.06	0.18	2.97	0.19	0.03	39
22	107.14	6.37	1.07	93.85	5.52	0.94	70.18	4.13	0.70	42.49	2.49	0.42	17.95	1.04	0.18	2.82	0.18	0.03	38
23	107.03	6.37	1.07	93.53	5.50	0.94	69.73	4.11	0.70	42.03	2.46	0.42	17.60	1.02	0.18	2.67	0.17	0.03	37
24	106.91	6.36	1.07	93.20	5.48	0.93	69.28	4.08	0.69	41.58	2.43	0.42	17.26	1.00	0.17	2.53	0.16	0.03	36
25	106.79	6.36	1.07	92.87	5.46	0.93	68.83	4.06	0.69	41.13	2.41	0.41	16.92	0.98	0.17	2.39	0.15	0.02	35
26	106.66	6.35	1.07	92.54	5.44	0.93	68.38	4.03	0.68	40.68	2.38	0.41	16.58	0.96	0.17	2.26	0.14	0.02	34
27	106.53	6.35	1.07	92.21	5.42	0.92	67.93	4.01	0.68	40.23	2.35	0.40	16.24	0.94	0.16	2.13	0.13	0.02	33
28	106.40	6.34	1.06	91.87	5.40	0.92	67.48	3.98	0.67	39.78	2.33	0.40	15.91	0.92	0.16	2.00	0.13	0.02	32
29	106.26	6.33	1.06	91.53	5.38	0.92	67.03	3.96	0.67	39.33	2.30	0.39	15.58	0.90	0.16	1.88	0.12	0.02	31
30	106.12	6.32	1.06	91.19	5.35	0.91	66.57	3.93	0.67	38.88	2.28	0.39	15.26	0.88	0.15	1.76	0.12	0.02	30
31	105.98	6.31	1.06	90.84	5.33	0.91	66.12	3.90	0.66	38.43	2.25	0.38	14.94	0.86	0.15	1.76	0.11	0.02	29
32	105.83	6.30	1.06	90.49	5.31	0.90	65.66	3.87	0.66	37.98	2.22	0.38	14.63	0.84	0.15	1.65	0.11	0.02	28
33	105.68	6.29	1.06	90.14	5.29	0.90	65.20	3.84	0.65	37.53	2.19	0.38	14.32	0.82	0.14	1.54	0.10	0.01	27
34	105.52	6.28	1.06	89.78	5.27	0.90	64.74	3.81	0.65	37.09	2.17	0.37	14.01	0.80	0.14	1.43	0.10	0.01	26
35	105.36	6.27	1.05	89.42	5.25	0.89	64.28	3.79	0.64	36.65	2.14	0.37	13.70	0.78	0.14	1.32	0.09	0.01	25
36	105.20	6.26	1.05	89.06	5.22	0.89	63.82	3.76	0.64	36.21	2.12	0.36	13.39	0.77	0.13	1.22	0.09	0.01	24
37	104.03	6.25	1.05	88.70	5.20	0.89	63.36	3.74	0.63	35.77	2.09	0.36	13.08	0.75	0.13	1.12	0.08	0.01	23
38	104.86	6.24	1.05	88.33	5.18	0.88	62.90	3.71	0.63	35.33	2.07	0.35	12.78	0.73	0.13	1.03	0.08	0.01	22
39	104.69	6.23	1.05	87.96	5.16	0.88	63.44	3.69	0.62	[illegible].89	2.04	0.35	12.48	0.72	0.12	0.94	0.07	0.01	21
40	104.51	6.22	1.05	87.59	5.13	0.88	61.98	3.66	0.62	34.45	2.02	0.34	12.18	0.70	0.12	0.86	0.07	0.01	20
41	104.33	6.21	1.04	86.21	5.11	0.87	61.52	3.64	0.62	34.02	1.99	0.34	11.89	0.68	0.12	0.78	0.06	0.01	19
42	104.15	6.20	1.04	86.83	5.09	0.87	61.05	3.61	0.61	33.59	1.97	0.34	11.60	0.66	0.12	0.70	0.06	0.01	18
43	103.96	6.19	1.04	86.45	5.06	0.86	60.59	3.58	0.61	33.16	1.94	0.33	11.31	0.65	0.11	0.63	0.05	0.01	17
44	103.77	6.18	1.04	86.07	5.04	0.86	60.12	3.55	0.60	32.73	1.92	0.33	11.03	0.63	0.11	0.56	0.05	0.00	16
45	103.58	6.17	1.04	85.69	5.02	0.86	59.66	3.52	0.60	32.31	1.89	0.32	10.75	0.62	0.11	0.49	0.04	0.00	15
46	103.38	6.15	1.03	85.30	5.00	0.85	59.19	3.49	0.59	31.89	1.87	0.32	10.47	0.60	0.10	0.43	0.04	0.00	14
47	103.18	6.14	1.03	84.91	4.97	0.85	58.73	3.46	0.59	31.47	1.84	0.31	10.20	0.59	0.10	0.37	0.03	0.00	13
48	102.97	6.12	1.03	84.52	4.95	0.85	58.26	3.44	0.58	31.05	1.82	0.31	9.93	0.58	0.10	0.32	0.03	0.00	12
49	102.76	6.11	1.03	84.13	4.93	0.84	57.80	3.41	0.58	30.63	1.79	0.31	9.67	0.56	0.10	0.27	0.02	0.00	11
50	102.54	6.09	1.03	83.74	4.90	0.84	57.33	3.38	0.57	30.21	1.77	0.30	9.41	0.55	0.09	0.19	0.02	0.00	10
51	102.32	6.08	1.02	83.34	4.88	0.83	56.87	3.36	0.57	29.80	1.75	0.30	9.15	0.53	0.09	0.17	0.02	0.00	9
52	102.09	6.06	1.02	82.94	4.86	0.83	56.40	3.33	0.56	29.30	1.72	0.29	8.90	0.52	0.09	0.13	0.01	0.00	8
53	101.86	6.05	1.02	82.54	4.83	0.83	55.94	3.30	0.56	28.99	1.70	0.29	8.65	0.51	0.09	0.11	0.01	0.00	7
54	101.63	6.03	1.02	82.14	4.81	0.82	55.47	3.27	0.55	28.58	1.67	0.29	8.40	0.50	0.08	0.09	0.01	0.00	6
55	101.40	6.02	1.01	81.74	4.78	0.82	55.01	3.24	0.55	28.17	1.65	0.28	8.16	0.48	0.08	0.07	0.01	0.00	5
56	101.16	6.00	1.01	81.33	4.76	0.81	54.54	3.22	0.55	27.76	1.62	0.28	7.92	0.47	0.08	0.05	0.01	0.00	4
57	100.92	5.99	1.01	80.92	4.74	0.81	54.08	3.19	0.54	27.36	1.60	0.27	7.68	0.46	0.08	0.03	0.01	0.00	3
58	100.68	5.97	1.01	80.51	4.71	0.81	53.61	3.16	0.54	26.96	1.57	0.27	7.44	0.44	0.07	0.02	0.00	0.00	2
59	100.43	5.96	1.00	80.10	4.69	0.80	53.14	3.13	0.53	26.56	1.55	0.27	7.21	0.43	0.07	0.01	0.00	0.00	1
60	100.18	5.94	1.00	79.69	4.66	0.80	52.67	3.10	0.53	26.16	1.53	0.26	6.98	0.42	0.07	0.00	0.00	0.00	0
	XVII			XVI			XV			XIV			XIII			XII			

SID. HOURS.

Co-Lat. = Z ± P. cos. t − Tab. Num. as in Tab. 2. If the Lat. be South of Greenwich, the Var. in 100' is −; but if North +.

Mr. Baily's general Table for determining the Latitude by the Pole Star, expanded.

TABLE 5.

Sid. Hours.

Horary Angle.	O XII		I XIII		II XIV		III XV		IV XVI		V XVII		Horary Angle.
	B	C	B	C	B	C	B	C	B	C	B	C	
m.													m.
0	0".00	0".00	5".61	0".11	20".95	0".40	41".90	0".80	62".86	1".20	78′.20	1".49	60
1	0.00	0.00	5.79	0.11	21.27	0.40	42.27	0.81	63.18	1.21	78.38	1.49	59
2	0.01	0.00	5.98	0.11	21.59	0.41	42.64	0.82	63.50	1.21	78.56	1.49	58
3	0.02	0.00	6.17	0.12	21.91	0.41	43.01	0.82	63.81	1.22	78 73	1.50	57
4	0.03	0.00	6.36	0.12	22.23	0.42	43.37	0.83	64.12	1.22	78.90	1.50	56
5	0.04	0.00	6.56	0.12	22.55	0.42	43.74	0.84	64.43	1.23	79.07	1.50	55
6	0.06	0.00	6.76	0.13	22.88	0.43	44.10	0.84	64.74	1.23	79.24	1.51	54
7	0.08	0.00	6.96	0.13	23.21	0.44	44.47	0.85	65.04	1.24	79.40	1.51	53
8	0.10	0.00	7.16	0.13	23.54	0.44	44.83	0.86	65.34	1.24	79.56	1.51	52
9	0.13	0.00	7.37	0.14	23.87	0.45	45.20	0.86	65.64	1.25	79.72	1.52	51
10	0.16	0.00	7.58	0.14	24.20	0.46	45.56	0.87	65.94	1.25	79.88	1.52	50
11	0.19	0.00	7.79	0.14	24.53	0.46	45.93	0.88	66.23	1.26	80.03	1.52	49
12	0.23	0.00	8.00	0.15	24.86	0.47	46.29	0.88	66.52	1.27	80.18	1.52	48
13	0.27	0.00	8.21	0.15	25.19	0.48	46.66	0.89	66.81	1.27	80.33	1.53	47
14	0.31	0.01	8.43	0.16	25.53	0.48	47.02	0.90	67.10	1.28	80.47	1.53	46
15	0.36	0.01	8.65	0.16	25.87	0.49	47.38	0.90	67.39	1.28	80.61	1.53	45
16	0.41	0.01	8.87	0.17	26.21	0.49	47.74	0.91	67.68	1.29	80.75	1.54	44
17	0.46	0.01	9.10	0.17	26.55	0.50	48.10	0.92	67.97	1.29	80.89	1.54	43
18	0.52	0.01	9.33	0.18	26.89	0.51	48.46	0.92	68.26	1.30	81.02	1.54	42
19	0.58	0.01	9.56	0.18	27.23	0.51	48.82	0.93	68.55	1.30	81.15	1.54	41
20	0.64	0.01	9.80	0.19	27.57	0.52	49.18	0.94	68.84	1.31	81.28	1.55	40
21	0.71	0.01	10.04	0.19	27.91	0.53	49.54	0.94	69.12	1.31	81.40	1.55	39
22	0.78	0.02	10.28	0.19	28.26	0.53	49.90	0.95	69.40	1.32	81.52	1.55	38
23	0.85	0.02	10.52	0.20	28.61	0.54	50.26	0.96	69.68	1.32	81.64	1.55	37
24	0.92	0.02	10.76	0.20	28.96	0.55	50.62	0.96	69.95	1.33	81.75	1.56	36
25	1.00	0.02	11.01	0.20	29.31	0.56	50.98	0.97	70.22	1.33	81.86	1.56	35
26	1.08	0.02	11.26	0.21	29.66	0.56	51.34	0.98	70.49	1.34	81.97	1.56	34
27	1.16	0.02	11.51	0.21	30.01	0.57	51.70	0.98	70.76	1.34	82.08	1.56	33
28	1.25	0.03	11.76	0.22	30.36	0.58	52.05	0.99	71.02	1.35	82.18	1.56	32
29	1.34	0.03	12.01	0.22	30.71	0.58	52.40	0.99	71.28	1.35	82.28	1.56	31
30	1.43	0.03	12.27	0.23	31.06	0.59	52.75	1.00	71.54	1.36	82.38	1.57	30
31	1.53	0.03	12.53	0.23	31.41	0.60	53.10	1.01	71.80	1.36	82.47	1.57	29
32	1.63	0.03	12.79	0.24	31.76	0.60	53.45	1.01	72.05	1.37	82.56	1.57	28
33	1.73	0.03	13.05	0.24	32.11	0.61	53.80	1.02	72.30	1.37	82.65	1.57	27
34	1.84	0.04	13.31	0.25	32.47	0.62	54.15	1.03	72.55	1.38	82.73	1.57	26
35	1.95	0.04	13.58	0.25	32.83	0.63	54.50	1.03	72.80	1.38	82.81	1.57	25
36	2.06	0.04	13.85	0.26	33.19	0.63	54.85	1.04	73.05	1.39	82.89	1.57	24
37	2.17	0.04	14.12	0.26	33.55	0.64	55.20	1.05	73.29	1.39	82.96	1.58	23
38	2.29	0.04	14.40	0.27	33.91	0.65	55.55	1.05	73.53	1.40	83.03	1.58	22
39	2.41	0.05	14.68	0.27	34.27	0.65	55.90	1.06	73.77	1.40	83.10	1.58	21
40	2.53	0.05	14.97	0.28	34.63	0.66	56.24	1.07	74.01	1.41	83.17	1.58	20
41	2.66	0.05	15.26	0.28	34.99	0.67	56.58	1.07	74.25	1.41	83.23	1.58	19
42	2.79	0.05	15.55	0.29	35.35	0.68	56.92	1.08	74.48	1.41	83.29	1.58	18
43	2.92	0.05	15.84	0.29	35.71	0.68	57.26	1.09	74.71	1.42	83.35	1.58	17
44	3.06	0.06	16.13	0.30	36.07	0.69	57.60	1.09	74.94	1.42	83.40	1.58	16
45	3.20	0.06	16.42	0.30	36.43	0.70	57.94	1.10	75.16	1.43	83.45	1.58	15
46	3.34	0.06	16.71	0.31	36.79	0.70	58.28	1.11	75.38	1.43	83.50	1.59	14
47	3.48	0.07	17.00	0.32	37.15	0.71	58.62	1.11	75.60	1.44	83.54	1.59	13
48	3.63	0.07	17.29	0.32	37.52	0.72	58.96	1.12	75.81	1.44	83.58	1.59	12
49	8.78	0.07	17.58	0.33	37.88	0.72	59.29	1.13	76.02	1.44	83.62	1.59	11
50	3.93	0.08	17.87	0.34	38.25	0.73	59.62	1.13	76.23	1.45	83.65	1.59	10
51	4.09	0.08	18.17	0.34	38.61	0.74	59.95	1.14	76.44	1.45	83.68	1.59	9
52	4.25	0.08	18.47	0.35	38.98	0.75	60.28	1.15	76.65	1.46	83.71	1.59	8
53	4.41	0.08	18.77	0.36	39.34	0.76	60.61	1.15	76.85	1.46	83.73	1.59	7
54	4.57	0.09	19.07	0.36	39.71	0.76	60.94	1.16	77.05	1.47	83.75	1.59	6
55	4.74	0.09	19.38	0.37	40.07	0.77	61.26	1.17	77.25	1.47	83.77	1.59	5
56	4.91	0.09	19.69	0.37	40.44	0.78	61.58	1.17	77.45	1.47	83.78	1.59	4
57	5.08	0.10	20.00	0.38	40.80	0.78	61.90	1.18	77.64	1.48	83.79	1.59	3
58	5.25	0.10	20.31	0.38	41.17	0.79	62.22	1.19	77.83	1.48	83.80	1.59	2
59	5.43	0.10	20.63	0.39	41.53	0.80	62.54	1.19	78.02	1.48	83.81	1.59	1
60	5.61	0.11	20.95	0.40	41.90	0.80	62.86	1.20	78.20	1.49	83.81	1.59	0
Horary Angle.	B	C	B	C	B	C	B	C	B	C	B	C	Horary Angle.
	XI	XIII	X	XXII	IX	XXI	VIII	XX	VII	XIX	VI	XVIII	

Sid. Hours.

TABLE 6.

Corrections of B and C.

Arg. = N.P.D. of the Pole Star.			Multiplier of B	Multiplier of C
1°	38′	0	1.000	1.000
1	37	55	.999	.998
		50	.997	.995
		45	.995	.993
		40	.993	.990
		35	.992	.988
		30	.990	.985
		25	.988	.983
		20	.986	.980
		15	.985	.978
		10	.983	.975
		5	.982	.973
1	37	0	.980	.970
1	36	55	.978	.968
		50	.976	.965
		45	.975	.963
		40	.973	.960
		35	.972	.958
		30	.970	.955
		25	.968	.953
		20	.966	.950
		15	.965	.948
		10	.963	.945
		5	.962	.943
1	36	0	.960	.940
1	35	55	.959	.938
		50	.957	.935
		45	.955	.933
		40	.953	.930
		35	.952	.928
		30	.950	.926
		25	.948	.924
		20	.946	.921
		15	.945	.919
		10	.943	.916
		5	.942	.914
1	35	0	.940	.911
1	34	55	.938	.908
		50	.936	.906
		45	.935	.903
		40	.933	.901
		35	.932	.898
		30	.930	.896
		25	.928	.893
		20	.926	.891
		15	.925	.888
		10	.923	.886
		5	.922	.883
1	34	0	.920	.881
1	33	55	.918	.878
		50	.916	.876
		45	.915	.873
		40	.913	.871
		35	.912	.868
		30	.910	.866
		25	.908	.863
		20	.906	.861
		15	.905	.858
		10	.903	.856
		5	.902	.853
1	33	0	.900	.851

♃ = z ± (P + C) cos. t − B cot z; where ♃ is the Co-Latitude, z the observed Zenith Distance of the Pole Star corrected, P the Polar Distance = 1° 38′, t the Horary Angle from the upper Culmination, and B and C the respective Tabular Quantities.

TABLE 7. The Greenwich MEAN PLACES, ANNUAL PRECESSIONS, ARGUMENTS, and MAXIMA of ABERRATION, LUNAR NUTATION, and SOLAR NUTATION of the Star, for 60 successive Years from 1820 to 1880.

RIGHT ASCENSION.

	R. A. h. 0		Annual Precess.	A. s. 8		Max.	L. s. 8		Max.	S. s. 8		Max.
	m.	s.	+	°	′	s	°	′	s	°	′	s
1820	57	1.510	14.471	14	32	43.222	16	30	22.232	13	31	1.014
1821	57	16.023	14.556	14	28	43.367	16	28	22.303	13	29	1.017
1822	57	30.622	14.642	14	24	43.512	16	26	22.374	13	26	1.020
1823	57	45.357	14.729	14	20	43.658	16	24	22.446	13	23	1.024
1824	58	0.130	14.817	14	16	43.805	16	22	22.518	13	21	1.027
1825	58	14.991	14.906	14	12	43.953	16	20	22.590	13	18	1.031
1826	58	29.942	14.997	14	8	44.102	16	17	22.663	13	15	1.034
1827	58	44.984	15.088	14	4	44.252	16	15	22.736	13	12	1.038
1828	59	0.116	15.180	14	0	44.403	16	12	22.809	13	9	1.041
1829	59	15.342	15.273	13	56	44.555	16	10	22.883	13	6	1.045
1830	59	30.662	15.367	13	51	44.708	16	7	22.957	13	3	1.048
1831	59	45.781	15.461	13	47	44.862	16	5	23.032	13	0	1.052
		Iʰ										
1832	0	1.289	15.556	13	43	45.016	16	2	23.107	12	57	1.055
1833	0	16.893	15.652	13	39	45.171	16	0	23.183	12	54	1.059
1834	0	32.594	15.749	13	35	45.327	15	57	23.259	12	51	1.062
1835	0	48.442	15.847	13	30	45.484	15	55	23.336	12	48	1.066
1836	1	4.338	15.945	13	26	45.642	15	52	23.413	12	46	1.069
1837	1	20.332	16.044	13	22	45.801	15	50	23.491	12	43	1.073
1838	1	36.426	16.144	13	18	45.961	15	47	23.569	12	40	1.076
1839	1	52.621	16.245	13	14	46.122	15	45	23.647	12	37	1.080
1840	2	8.917	16.347	13	9	46.285	15	42	23.726	12	34	1.084
1841	2	25.315	16.450	13	5	46.449	15	40	23.805	12	31	1.087
1842	2	41.817	16.554	13	0	46.614	15	37	23.885	12	28	1.091
1843	2	58.423	16.659	12	56	46.780	15	34	23.965	12	25	1.095
1844	3	15.135	16.765	12	51	46.947	15	32	24.046	12	22	1.098
1845	3	31.954	16.873	12	47	47.115	15	29	24.127	12	19	1.102
1846	3	48.881	16.982	12	42	47.284	15	26	24.209	12	16	1.106
1847	4	5.918	17.092	12	38	47.455	15	24	24.291	12	13	1.110
1848	4	23.066	17.203	12	33	47.627	15	21	24.374	12	9	1.114
1849	4	40.325	17.315	12	29	47.800	15	18	24.458	12	6	1.118
1850	4	57.697	17.428	12	24	47.974	15	15	24.542	12	2	1.122
1851	5	15.182	17.542	12	20	48.148	15	12	24.627	11	58	1.126
1852	5	32.781	17.657	12	15	48.324	15	10	24.712	11	55	1.130
1853	5	50.496	17.773	12	10	48.501	15	7	24.798	11	51	1.134
1854	6	8.328	17.891	12	5	48.680	15	4	24.884	11	48	1.138
1855	6	26.279	18.011	12	0	48.861	15	1	24.971	11	44	1.142
1856	6	44.350	18.132	11	55	49.043	14	58	25.058	11	41	1.146
1857	7	2.544	18.255	11	50	49.227	14	55	25.146	11	37	1.150
1858	7	20.861	18.379	11	45	49.412	14	52	25.234	11	33	1.154
1859	7	39.302	18.504	11	40	49.598	14	49	25.323	11	30	1.158
1860	7	57.869	18.631	11	35	49.786	14	46	25.412	11	26	1.163
1861	8	16.564	18.759	11	30	49.975	14	43	25.502	11	22	1.167
1862	8	35.387	18.888	11	25	50.165	14	40	25.593	11	18	1.171
1863	8	54.340	19.018	11	19	50.357	14	37	25.684	11	14	1.176
1864	9	13.423	19.149	11	14	50.551	14	34	25.776	11	11	1.180
1865	9	32.638	19.281	11	9	50.746	14	31	25.869	11	7	1.185
1866	9	51.985	19.414	11	3	50.943	14	28	25.962	11	3	1.189
1867	10	11.466	19.547	10	58	51.141	14	25	26.056	10	59	1.194
1868	10	31.080	19.681	10	53	51.340	14	21	26.150	10	56	1.198
1869	10	50.828	19.816	10	48	51.541	14	18	26.245	10	52	1.203
1870	11	10.712	19.970	10	43	51.743	14	14	26.340	10	48	1.207
1871	11	30.733	20.090	10	38	51.946	14	11	26.436	10	44	1.212
1872	11	50.893	20.231	10	32	52.150	14	7	26.533	10	40	1.216
1873	12	11.196	20.375	10	27	52.355	14	4	26.630	10	36	1.221
1874	12	31.644	20.522	10	21	52.562	14	0	26.728	10	32	1.225
1875	12	52.241	20.672	10	16	52.770	13	57	26.827	10	28	1.230
1876	13	12.989	20.824	10	10	52.980	13	53	26.926	10	24	1.234
1877	13	33.890	20.979	10	5	53.192	13	50	27.026	10	20	1.239
1878	13	54.948	21.137	9	59	53.405	13	46	27.127	10	16	1.244
1879	14	16.165	21.298	9	54	53.619	13	43	27.329	10	12	1.249
1880	14	37.545	21.462	9	48	53.835	13	39	27.332	10	8	1.254

NORTH POLAR DISTANCE.

	N. P. D. 1°		Annual Precess.	A′. s. 11		Max.	L′. s. 11		Max.	S′. s. 11		Max.
	′	″	−	°	′	″	°	′	″	°	′	″
1820	39	5.570	19.427	17	33	20.111	11	5	7.355	14	31	0.400
1821	38	46.145	19.422	17	30	20.110	11	1	7.356	14	27	0.400
1822	38	26.726	19.416	17	27	20.108	10	57	7.358	14	23	0.400
1823	38	7.313	19.411	17	24	20.107	10	52	7.360	14	19	0.400
1824	37	47.905	18.405	17	21	20.105	10	48	7.361	14	15	0.400
1825	37	28.503	19.400	17	17	20.104	10	43	7.363	14	11	0.400
1826	37	9.106	19.394	17	13	20.102	10	39	7.364	14	7	0.400
1827	36	49.715	19.389	17	9	20.101	10	34	7.366	14	3	0.400
1828	36	30.329	19.383	17	5	20.099	10	30	7.368	13	59	0.400
1829	36	10.949	19.378	17	1	20.098	10	26	7.369	13	55	0.400
1830	35	51.574	19.372	16	57	20.096	10	21	7.371	13	51	0.400
1831	35	32.205	19.366	16	53	20.095	10	17	7.372	13	47	0.400
1832	35	13.002	19.360	16	49	20.094	10	12	7.374	13	43	0.401
1833	34	53.645	19.354	16	45	20.092	10	7	7.376	13	38	0.401
1834	34	34.294	19.348	16	41	20.091	10	2	7.377	13	34	0.401
1835	34	14.949	19.342	16	37	20.090	9	57	7.379	13	30	0.401
1836	33	55.610	19.336	16	33	20.088	9	52	7.380	13	25	0.401
1837	33	36.277	19.330	16	29	20.087	9	47	7.382	13	21	0.401
1838	33	16.950	19.324	16	25	20.085	9	42	7.383	13	17	0.401
1839	32	57.629	19.318	16	21	20.084	9	37	7.385	13	12	0.401
1840	32	38.314	19.312	16	17	20.082	9	32	7.387	13	8	0.401
1841	32	19.005	19.306	16	13	20.081	9	27	7.388	13	4	0.401
1842	31	59.702	19.300	16	9	20.079	9	21	7.390	12	59	0.401
1843	31	40.406	19.293	16	5	20.078	9	16	7.392	12	55	0.401
1844	31	21.117	19.286	16	1	20.076	9	11	7.394	12	51	0.401
1845	31	1.835	19.279	15	57	20.074	9	5	7.395	12	46	0.401
1846	30	42.560	19.272	15	53	20.073	9	0	7.397	12	42	0.401
1847	30	23.292	19.265	15	49	20.071	8	54	7.399	12	37	0.401
1848	30	4.031	19.258	15	45	20.070	8	49	7.400	12	32	0.401
1849	29	44.777	19.251	15	41	20.068	8	43	7.402	12	27	0.401
1850	29	25.530	19.244	15	36	20.067	8	38	7.404	12	22	0.401
1851	29	6.290	19.237	15	32	20.065	8	32	7.406	12	18	0.401
1852	28	47.057	19.230	15	28	20.064	8	26	7.408	12	13	0.401
1853	28	27.831	19.223	15	24	20.062	8	21	7.410	12	8	0.401
1854	28	8.612	19.216	15	20	20.061	8	15	7.412	12	4	0.401
1855	27	49.400	19.209	15	15	20.059	8	9	7.414	11	59	0.401
1856	27	30.195	19.201	15	11	20.058	8	4	7.416	11	54	0.401
1857	27	10.998	19.193	15	6	20.056	7	58	7.418	11	49	0.401
1858	26	51.809	19.185	15	2	20.055	7	53	7.420	11	44	0.401
1859	26	32.628	19.177	14	57	20.053	7	47	7.422	11	39	0.401
1860	26	13.455	19.169	14	53	20.052	7	42	7.424	11	34	0.401
1861	25	54.290	19.161	14	48	20.050	7	36	7.426	11	29	0.401
1862	25	35.133	19.153	14	44	20.049	7	30	7.428	11	24	0.401
1863	25	15.984	19.145	14	39	20.047	7	24	7.431	11	19	0.401
1864	24	56.843	19.137	14	35	20.046	7	18	7.433	11	14	0.401
1865	24	37.710	19.129	14	30	20.044	7	12	7.435	11	9	0.401
1866	24	18.585	19.121	14	26	20.043	7	6	7.437	11	4	0.402
1867	23	59.468	19.112	14	21	20.041	7	0	7.440	10	59	0.402
1868	23	40.359	19.103	14	17	20.040	6	54	7.442	10	54	0.402
1869	23	21.259	19.094	14	12	20.038	6	48	7.444	10	49	0.402
1870	23	2.168	19.085	14	7	20.037	6	41	7.447	10	43	0.402
1871	22	43.087	19.076	14	3	20.035	6	35	7.449	10	38	0.402
1872	22	24.015	19.067	13	58	20.034	6	29	7.452	10	32	0.402
1873	22	4.952	19.058	13	53	20.032	6	22	7.454	10	27	0.402
1874	21	45.898	19.049	13	48	20.031	6	16	7.457	10	21	0.402
1875	21	26.853	19.040	13	43	20.029	6	10	7.459	10	16	0.402
1876	21	7.817	19.031	13	38	20.028	6	3	7.462	10	10	0.402
1877	20	48.791	19.021	13	33	20.026	5	57	7.464	10	5	0.402
1878	20	29.775	19.011	13	28	20.025	5	50	7.467	9	59	0.402
1879	20	10.769	19.001	13	23	20.023	5	44	7.469	9	54	0.402
1880	19	51.773	18.991	13	18	20.022	5	37	7.472	9	48	0.402

The Right Ascensions and North Polar Distances are brought forward by the respective Precessions of July 1, or *mean* between the two contiguous Tabular Precessions, which were calculated for January 0 of each year.

The proper motions of the Pole Star may be applied to the Precessions of 1820 in R. A. and Dec., when finally determined; and any Multiples thereof, accumulating from this Epoch, may be applied to the Precessions of any subsequent year.

A CATALOGUE of CIRCUMPOLAR STARS, in Pairs, that differ about 12 Hours in Right Ascension from each other respectively. The Epoch 1800.

Flam. Stars.	Bayer's.	Mag.	RIGHT ASCENSION IN TIME.												NORTH POLAR DISTANCE.											
			R. A.			Ann. Var.	A.			Log. of Max.	L.			Log. of Max.	N. P. D.			Ann. Var.	A'.			Log. of Max.	L'.			Log. of Max.
			h.	m.	s.		S.				S.								S.	°	'		S.	°	'	
69 UrsæMaj.	δ	3	12	5	27 . 1	+ 3s. 01	2	28°	31'	0 .3706	4	16°	25'	0 .1753	41°	51'	19". 8	+20". 06	5	15	32	1 .2495	5	28	5	0 .8564
15 Cassiop.	κ	4	0	21	44 . 0	3 . 31	8	24	6	0 .4192	7	15	8	0 .2253	28	10	27 . 7	−19 . 97	0	7	49	1 .2514	11	22	29	0 .8573
5 Draconis	κ	3 . 4	12	24	51 . 3	2 . 62	2	23	15	0 .5782	3	27	20	0 .3174	19	6	27 . 3	+19 . 95	5	10	36	1 .2913	5	21	41	0 .8583
17 Cassiop.	ζ	4	0	25	54 . 1	3 . 27	8	23	1	0 .3133	7	5	43	0 .1602	37	12	20 . 6	−19 . 94	0	11	43	1 .2156	11	21	34	0 .8584
18 Cassiop.	α	3	0	29	14 . 4	3 . 31	8	22	1	0 .3394	7	8	8	0 .1771	34	33	42 . 4	−19 . 90	0	9	5	1 .2242	11	21	34	0 .8591
24 Cassiop.	η	4	0	37	5 . 4	3 . 51	8	20	0	0 .3554	7	8	25	0 .1925	33	14	57 . 7	−19 . 80	0	6	54	1 .2269	11	18	26	0 .8609
27 Cassiop.	γ	3	0	44	44 . 5	3 . 51	8	18	15	0 .3843	7	10	42	0 .2205	30	22	8 . 6	−19 . 68	0	3	40	1 .2330	11	15	52	0 .8627
77 UrsæMaj.	ε	3	12	45	10 . 9	2 . 68	2	17	43	0 .3589	4	14	33	0 .1353	32	57	7 . 5	+19 . 68	5	5	58	1 .2612	5	14	59	0 .8628
46 Androm.		5	1	10	37 . 5	3 . 46	8	10	39	0 .3442	6	24	0	0 .1343	45	31	23 . 0	−19 . 12	0	7	30	1 .1345	11	6	40	0 .8716
37 Cassiop.	δ	3	1	12	50 . 9	3 . 76	8	10	16	0 .3875	7	7	7	0 .2299	30	48	33 . 8	−19 . 06	11	27	32	1 .2176	11	6	10	0 8727
48 Androm.	ω	5 . 6	1	15	45 . 2	3 . 49	8	9	38	0 .2452	6	25	23	0 .1426	45	37	53 . 8	−18 . 98	0	6	37	1 .1300	11	4	12	0 .8737
49 Androm.	ξ	5 . 6	1	18	10 . 2	3 . 53	8	8	56	0 .2558	6	26	16	0 .1494	44	1	46 . 5	−18 . 90	0	4	47	1 .1367	11	4	8	0 .8745
79 UrsæMaj.	ζ pr.	3	13	19	49 . 8	2 . 42	2	9	29	0 .3504	4	13	59	0 .1006	34	1	34 . 2	+18 . 98	4	29	22	1 .2653	5	5	24	0 .8757
54 Androm.	φ	5	1	31	12 . 2	3 . 68	8	5	31	0 .2848	6	27	46	0 .1751	40	19	30 . 0	−18 . 50	11	28	48	1 .1529	11	0	44	0 .8810
85 UrsæMaj.	η	4 . 5	13	39	38 . 3	2 . 36	2	3	24	0 .3977	4	20	44	0 .0456	39	41	0 . 8	+18 . 20	4	22	11	1 .2562	4	28	28	0 .8851
45 Cassiop.	ε	3 . 4	1	40	9 . 7	3 . 15	8	3	1	0 .4383	7	7	2	0 .2726	27	19	23 . 2	−18 . 18	11	19	11	1 .2224	10	27	53	0 .8855
50 Cassiop.	f	4 . 5	1	46	38 . 7	4 . 86	8	1	10	0 .5991	7	14	30	0 .3927	18	33	22 . 7	−17 . 93	11	12	49	1 .2519	10	26	11	0 .8885
57 Androm.	γ	3 . 4	1	51	40 . 7	3 . 60	8	0	0	0 .2257	6	21	4	0 .1439	48	38	14 . 5	−17 . 73	0	1	0	1 .0763	10	24	19	0 .8915
11 Draconis	α	3 . 4	13	58	58 . 5	1 . 61	1	28	13	0 .4828	3	25	43	0 .3280	24	39	52 . 3	+17 . 42	4	23	49	1 .2945	4	22	17	0 .8954
5 UrsæMin.	a	4	14	28	8 . 3	− 0 . 31	1	20	34	0 .7421	2	26	59	0 .3339	13	24	53 . 5	+16 . 02	4	20	58	1 .3064	4	14	36	0 .9117
13 Persei	ϑ	4	2	30	36 . 8	+ 3 . 94	7	20	0	0 .2863	6	21	51	0 .1892	41	37	39 . 3	−15 . 88	11	15	14	1 .0934	10	13	49	0 .9130
23 Persei	γ	4	2	50	22 . 4	4 . 25	7	15	16	0 .3222	6	22	26	0 .2196	37	17	21 . 0	−14 . 77	11	6	58	1 .1143	10	9	20	0 .9244
7 UrsæMin.	β	3	14	51	26 . 7	− 0 . 30	1	14	42	0 .6973	2	26	26	0 .2457	15	1	39 . 6	+14 . 71	4	15	6	1 .3065	4	8	45	0 .9250
26 Persei	β	Var.	2	55	12 . 2	+ 3 . 86	7	13	43	0 .2286	6	15	50	0 .1575	49	49	33 . 0	−14 . 48	11	17	52	0 .9829	10	7	49	0 .9271
27 Persei	κ	5	2	56	3 . 9	3 . 98	7	13	29	0 .2541	6	17	29	0 .1753	45	54	44 . 0	−14 . 43	11	12	38	1 .0306	10	7	42	0 .9274
33 Persei	α	2 . 3	3	10	6 . 8	4 . 20	7	10	23	0 .2983	6	18	21	0 .2032	40	51	47 . 0	−13 . 55	11	4	26	1 .0569	10	4	16	0 .9351
Camel. 2 H.		4	3	12	59 . 4	4 . 74	7	9	33	0 .4056	6	22	55	0 .2668	30	46	20 . 0	−13 . 36	10	26	6	1 .1449	10	3	36	0 .9362
Camel. 4 H.		5 . 6	3	14	50 . 3	4 . 49	7	8	48	0 .3536	6	20	25	0 .2355	35	15	14 . 3	−13 . 24	10	28	18	1 .1078	10	3	10	0 .9375
35 Persei	σ	5	3	16	32 . 7	4 . 16	7	8	22	0 .2839	6	16	56	0 .1941	42	42	30 . 7	−13 . 13	11	3	24	0 .9644	10	2	40	0 .9383
11 UrsæMin.	ι	5 . 6	15	17	21 . 0	− 0 . 16	1	8	17	0 .6386	2	27	47	0 .1245	17	27	40 . 0	+13 . 07	4	8	27	1 .3064	4	2	52	0 .9390
12 Draconis	ι	3	15	20	29 . 7	+ 1 . 31	1	7	30	0 .4123	4	3	56	9 .9394	30	19	46 . 0	+12 . 86	4	4	24	1 .2975	4	1	52	0 .9402
Ursæ Min.	γ	3 . 4	15	21	8 . 7	− 0 . 20	1	7	21	0 .6397	2	27	4	0 .1173	17	27	16 . 0	+12 . 82	4	7	33	1 .3062	4	2	9	0 .9407
37 Persei	ψ	5	3	22	20 . 1	+ 4 . 20	7	7	0	0 .2872	6	16	35	0 .1953	42	29	12 . 0	−12 . 74	11	1	40	1 .0302	10	1	25	0 .9413
39 Persei	δ	3 . 4	3	28	44 . 8	4 . 21	7	5	27	0 .2818	6	15	41	0 .1955	42	51	58 . 0	−12 . 30	11	0	34	1 .0175	9	0	38	0 .9442
41 Persei		4 . 5	3	31	38 . 9	4 . 03	7	4	47	0 .2439	6	13	37	0 .1730	48	3	58 . 1	−12 . 10	11	4	40	0 .9542	9	29	20	0 .9459
16 UrsæMin.	ζ	4	15	51	30 . 5	− 2 . 45	0	29	50	0 .8166	2	2	1	0 .2763	11	35	52 . 0	+10 . 67	4	1	43	1 .2995	3	24	43	0 .9549
47 Persei	λ	6	3	51	44 . 4	+ 4 . 41	7	0	0	0 .3108	6	14	22	0 .2122	40	12	26 . 2	−10 . 65	10	20	0	1 .0310	9	25	0	0 .9552
13 Draconis	ϑ	3 . 4	15	58	9 . 3	1 . 14	0	28	20	0 .4112	4	7	4	9 .8312	30	53	47 . 4	+10 . 17	3	26	11	1 .2978	3	23	32	0 .9580
51 Persei	μ	4 . 5	4	0	15 . 9	4 . 36	6	27	52	0 .2961	6	12	48	0 .2056	42	6	52 . 6	−10 . 01	10	18	49	1 .0004	9	23	19	0 .9588
22 Herculis	τ	4	16	13	44 . 1	1 . 79	0	24	43	0 .2863	5	4	27	9 .8500	43	12	21 . 0	+ 8 . 97	3	20	37	1 .2746	3	20	0	0 .9638
14 Draconis	η	3	16	21	17 . 8	0 . 79	0	22	56	0 .4532	3	29	23	0 .7624	28	1	49 . 0	+ 8 . 38	3	22	0	1 .3041	3	18	46	0 .9667
35 Herculis	σ	4	16	27	39 . 5	1 . 91	0	21	17	0 .2591	5	11	5	9 .8625	47	8	34 . 5	+ 7 . 87	3	16	53	1 .2639	3	17	19	0 .9689
Camel. 17 H.		4 . 5	4	30	16 . 2	5 . 86	6	20	52	0 .5158	6	14	46	0 .3358	24	1	11 . 0	− 7 . 33	10	2	5	0 .9996	9	17	5	0 .9698

These Stars, and similar Pairs, may be used in determining the Meridian Line, and also in examining the Refraction at different Altitudes.

A CATALOGUE of CIRCUMPOLAR STARS, in Pairs, &c. concluded.

FLAM. STARS.	Bayer's.	Mag.	R. A. (RIGHT ASCENSION IN TIME.)	Ann. Var.	A.	Log. of Max.	L.	Log. of Max.	N. P. D. (NORTH POLAR DISTANCE.)	Ann. Var.	A′.	Log. of Max.	L′.	Log. of Max.
			h. m. s.		S.		S.				S. ° ′		S. ° ′	
13 Aurigæ	α	1	5 1 56 .1	+ 4″.41	6 13° 22′	0.2847	6 5° 56′	0 .2003	44° 13′ 22″.5	— 5″.03	9 26 29	0.9104	9 10 55	0 .9782
22 Ursæ Min.	ε	4	17 6 55 .7	— 6 .61	0 12 13	1.0045	0 25 1	0 .4160	7 39 24 .6	+ 4 .60	3 13 40	1.2912	3 9 58	0 .9790
22 Draconis	ζ	3	17 8 13 .9	+ 0 .12	0 11 53	0.5217	3 9 48	0 .5175	24 2 17 .7	+ 4 .49	3 11 43	1.3086	3 9 28	0 .9796
23 Draconis	β	2	17 25 55 .1	1 .33	0 7 50	0.3448	5 15 42	9 .8009	37 32 41 .3	+ 2 .97	3 6 51	1.2891	3 6 22	0 .9823
85 Herculis	ι	4	17 33 48 .8	1 .69	0 6 0	0.2884	5 22 41	9 .7891	43 52 48 .5	+ 2 .29	3 5 9	1.2742	3 5 0	0 .9831
33 Aurigæ	δ	3 .4	5 43 3 .9	4 .92	6 3 53	0.3572	6 2 13	0 .2481	35 45 2 .9	— 1 .48	9 6 49	1.0172	9 3 9	0 .9841
34 Aurigæ	β	2	5 44 51 .5	4 .34	6 3 28	0.2800	6 1 32	0 .1986	45 5 24 .7	— 1 .32	9 7 17	0.8716	9 2 51	0 .9842
32 Draconis	ξ	3 .4	17 50 4 .1	1 .02	0 2 20	0.3927	5 23 20	9 .5698	33 5 31 .0	+ 0 .87	3 2 4	1.2993	3 1 53	0 .9845
33 Draconis	γ	2	17 51 57 .8	1 .37	0 1 47	0.3418	5 26 40	9 .7243	38 28 55 .5	+ 0 .70	3 1 31	1.2951	3 1 30	0 .9843
34 Draconis	ψ²	4	17 58 39 .3	— 1 .05	0 0 16	0.6409	6 1 58	9 .5881	17 58 45 .5	+ 0 .12	3 0 17	1.3034	3 0 0	0 .9844
2 Lyncis	b	4 .5	6 1 57 .7	+ 5 .30	5 29 34	0.4207	5 29 47	0 .2791	30 56 16 .7	— 0 .17	8 29 15	1.0719	8 29 25	0 .9844
44 Draconis	χ	4 .5	18 24 38 .6	— 1 .18	11 24 22	0.6544	11 2 29	9 .6733	17 21 26 .7	— 2 .15	2 24 26	1.3284	2 25 38	0 .9834
42 Camelop.	p	5	6 30 1 .4	+ 6 .31	5 23 2	0.5532	5 25 0	0 .3559	28 13 57 .7	+ 2 .62	8 20 0	1.1523	8 24 22	0 .9828
23 Ursæ Min.	δ	3	18 36 35 .1	—18 .98	11 21 38	1.3520	11 10 29	0 .9221	13 26 17 .5	— 3 .19	2 20 21	1.2805	2 23 14	0 .9822
47 Draconis	α	5	18 48 14 .3	+ 0 .88	11 18 58	0.4288	7 5 36	9 .6851	30 51 9 .4	— 4 .19	2 19 43	1.3013	2 21 10	0 .9802
50 Draconis	l	5 .6	18 52 43 .5	— 1 .86	11 18 1	0.7211	10 19 33	9 .9368	14 48 32 .0	— 4 .57	2 17 4	1.3021	2 20 0	0 .9794
52 Draconis	υ	5	18 56 47 .3	— 0 .70	11 16 56	0.6187	9 28 29	9 .7177	18 58 21 .0	— 4 .92	2 16 34	1.3090	2 19 22	0 .9786
1 Ursæ Maj.	ο	4 .5	8 13 31 .0	+ 5 .09	4 28 50	0.4392	5 0 58	0 .1259	38 37 46 .0	+11 .04	7 14 54	1.1405	8 3 51	0 .9529
37 Cygni	γ	3	20 15 3 .0	2 .14	10 28 38	0.2328	6 20 56	9 .9113	50 22 35 .2	—11 .15	2 6 0	1.2430	2 3 32	0 .9523
50 Cygni	α	1	20 34 36 .8	2 .03	10 23 50	0.2637	6 28 24	9 .9208	45 25 40 .2	—12 .53	2 0 53	1.2613	1 29 20	0 .9430
3 Cephei	η	3 .4	20 41 11 .4	1 .24	10 22 15	0.4323	7 29 33	9 .9402	28 56 3 .7	—12 .98	1 24 54	1.2994	1 27 51	0 .9390
58 Cygni	ν²	4	20 49 43 .1	2 .23	10 20 0	0.2321	6 24 20	9 .9435	49 35 45 .8	—13 .54	1 29 39	1.2408	1 25 44	0 .9355
62 Cygni	ξ	4	20 57 39 .3	2 .17	10 18 4	0.3184	7 11 43	9 .9452	46 51 51 .0	—14 .04	1 23 56	1.2810	1 24 1	0 .9310
5 Cephei	α	3	21 13 47 .3	1 .40	10 14 2	0.4335	8 0 22	0 .0126	28 15 31 .2	—15 .01	1 18 0	1.2975	1 20 0	0 .9220
23 Ursæ Maj.	h	4	9 15 35 .5	4 .86	4 13 33	0.4648	5 0 25	0 .3030	26 4 30 .0	+15 .12	6 28 31	1.1934	7 20 40	0 .9214
24 Ursæ Maj.	d	5	9 16 32 .1	5 .55	4 13 32	0.5908	4 25 20	0 .3862	19 18 12 .2	+15 .17	7 1 44	1.2266	7 20 40	0 .9205
25 Ursæ Maj.	θ	2	9 19 23 .5	4 .08	4 12 36	0.3257	5 6 54	0 .2141	37 23 14 .0	+15 .34	6 20 30	1.1185	7 18 32	0 .9189
8 Cephei	β	3	21 26 0 .9	0 .81	10 10 56	0.5687	8 17 38	0 .1424	20 18 57 .2	—15 .70	1 12 39	1.3043	1 16 58	0 .9149
21 Cephei	ζ	4	22 3 55 .5	2 .05	10 1 8	0.3697	7 19 40	0 .0584	32 46 52 .7	—17 .55	1 8 29	1.2792	1 6 56	0 .8937
33 Ursæ Maj.	λ	3 .4	10 4 58 .4	3 .69	4 0 52	1.2458	5 7 43	0 .1555	46 5 35 .5	+17 .59	6 1 26	1.0903	7 6 11	0 .8933
23 Cephei	ε	4 .5	22 7 41 .1	2 .13	10 0 0	0.3541	7 17 42	0 .0555	33 57 0 .5	—17 .70	1 8 11	1.2774	1 5 24	0 .8922
34 Ursæ Maj.	μ	3	10 10 21 .5	3 .61	3 29 29	0.2338	5 8 17	0 .1492	48 30 0 .0	+17 .81	5 29 31	1.0867	7 5 24	0 .8865
5 Lacertæ	e	5 .6	22 21 13 .1	2 .47	9 26 36	0.2643	7 5 0	0 .0332	43 18 47 .1	—18 .23	1 10 0	1.2381	1 2 18	0 .8848
7 Lacertæ	g	4	22 23 4 .6	2 .42	9 26 6	0.2852	7 7 53	0 .0449	40 44 31 .0	—18 .30	1 8 15	1.2512	1 2 18	0 .8838
32 Cephei	ι	4	22 42 35 .3	2 .09	9 21 2	0.4748	8 0 0	0 .1815	24 50 57 .1	—18 .93	0 27 21	1.2907	0 25 23	0 .8746
48 Ursæ Maj.	β	2	10 49 40 .6	3 .71	3 19 3	0.3659	4 24 9	0 .2149	32 32 55 .5	+19 .13	6 0 21	1.2093	6 23 3	0 .8716
50 Ursæ Maj.	α	1 .2	10 51 15 .5	3 .82	3 18 39	0.4370	4 18 55	0 .2628	27 10 21 .6	+19 .17	6 4 0	1.2355	6 23 4	0 .8712
1 Androm.	ο	4	22 52 44 .5	2 .73	9 18 22	0.2206	6 28 56	0 .0508	48 44 46 .0	—19 .21	1 7 8	1.2076	0 22 29	0 .8704
52 Ursæ Maj.	ψ	3 .4	10 58 22 .0	3 .43	3 16 44	0.2514	5 2 56	0 .1430	44 25 7 .2	+19 .34	5 21 36	1.1529	6 20 47	0 .8686
1 Draconis	λ	3 .4	11 19 21 .8	3 .72	3 11 2	0.5695	4 7 1	0 .3478	19 34 0 .3	+19 .75	6 1 15	1.2723	6 13 20	0 .8617
16 Androm.	λ	4 .5	23 27 48 .7	2 .90	9 8 53	0.2476	7 2 4	0 .0852	44 37 25 .1	—19 .87	0 27 30	1.2109	0 10 47	0 .8596
17 Androm.	ι	5	23 28 21 .9	2 .90	9 8 36	0.2238	6 28 58	0 .0764	47 50 17 .5	—19 .87	0 29 30	1.1896	0 10 33	0 .8596
19 Androm.	κ	5	23 30 35 .7	2 .91	9 8 8	0.2294	7 0 0	0 .0823	46 46 21 .3	—19 90	0 28 53	1.1937	0 10 0	0 .8592
35 Cephei	γ	3	23 31 14 .5	2 .35	9 7 49	0.7261	8 12 2	0 .4476	13 29 1 .0	—19 .91	0 12 17	1.3019	0 10 0	0 .8590
63 Ursæ Maj.	χ	4	11 35 25 .5	3 .21	3 6 40	0.2746	4 27 43	0 .1374	41 6 43 .4	+19 .95	5 15 55	1.1942	6 8 21	0 .8583
64 Ursæ Maj.	γ	2	11 43 14 .5	3 .21	3 4 38	0.3311	4 21 57	0 .1667	34 11 37 .0	+20 .01	5 18 37	1.2287	6 5 52	0 .8571
11 Cassiop.	β	2 .3	23 58 34 .9	3 .11	9 0 24	0.3696	7 13 25	0 .1797	31 57 14 .5	—20 .06	0 14 50	1.2455	0 0 47	0 .8563

These Stars, and similar Pairs, may be used for determining the Meridian Line, and also in examining the Refraction at different Altitudes.

A CATALOGUE of STARS, arranged in PAIRS, suitable for verifying the Position of the Transit Instrument. The Epoch 1800.

FLAM. STARS.	RIGHT ASCENSION IN TIME.								DECLINATION.					
	Bayer's.	Mag.	R. A.	An.Var.	A.	Log. of Max.	L.	Log. of Max.	Dec. N. + S. —	An.Var.	A′.	Log. of Max.	L′.	Log. of Max.
			h. m. s.	+	S. ° ′		S. ° ′			+	S. ° ′		S. ° ′	
Ap. Sculp.	α	5	0 48 56 .80	2ˢ. 899	8 16 39	0.1592	5 10 30	0 . 0439	—30° 26′ 23″. 2	19″. 58	4 18 13	1 . 1287	11 13 41	0 . 8640
69 Piscium	σ^1	5 . 6	0 51 53 . 75	3 . 241	8 15 51	0.1606	6 17 40	0 . 0883	+30 43 40 . 2	19 . 52	6 25 23	1 . 0495	5 12 49	0 . 8648
48 Androm.	ω	5 . 6	1. 15 44 . 87	3 . 486	8 9 38	0.2452	6 25 23	0 . 1426	+44 22 9 . 4	18 . 94	6 6 37	1 . 1300	5 11 12	0 . 8740
Phœn.	γ	3	1 19 39 . 26	2 . 617	8 8 34	0.2452	4 27 57	0 . 0467	—44 20 45 . 0	18 . 83	4 22 34	1 . 2244	11 4 3	0 . 8756
39 Arietis		4	2 36 2 . 11	3 . 516	7 18 44	0.1643	6 12 6	0 . 1118	+28 24 29 . 6	15 . 56	6 11 36	0 . 8835	4 12 50	0 . 9163
12 Eridani		3 . 4	3. 3 34 . 00	2 . 518	7 11 35	0.1743	5 14 5	9 . 9710	—29 46 56 . 0	13 . 95	3 27 18	1 . 1881	10 5 44	0 . 9316
Eridani	ϑ	2 . 3	2. 50 40 . 80	2 . 272	7 15 8	0.2339	5 3 16	9 . 9626	—41 6 37 . 4	14 . 72	4 4 19	1 . 2405	10 8 55	0 . 9245
26 Persei	β	2	2 55 12 . 65	3 . 841	7 13 43	0.2286	6 15 50	0 . 1575	+40 10 29 . 2	14 . 45	5 17 52	0 . 9829	4 7 49	0 . 9271
25 Tauri	η	3	3 35 37 . 04	3 . 530	7 3 49	0.1556	6 7 22	0 . 1678	+23 28 37 . 9	11 . 80	6 16 35	0 . 6955	3 28 23	0 . 9480
27 Eridani	m^1	5	3 38 14 . 20	2 . 581	7 3 6	0.1573	5 19 58	9 . 9755	—23 50 55 . 3	11 . 61	3 19 20	1 . 1555	9 27 53	0 . 9491
34 Eridani	γ^1	2 . 3	3. 48 41 . 68	2 . 781	7 0 37	0.1333	5 25 1	0 . 0025	—14 5 8 . 1	10 . 88	3 12 51	1 . 0729	9 25 39	0 . 9539
87 Tauri	α	1	4. 24 27 . 23	3 . 415	6 22 9	0.1422	6 3 31	0 . 0904	+16 5 44 . 7	8 . 11	7 23 26	0 . 5806	3 18 18	0 . 9678
53 Eridani		4	4 29 1 . 18	2 . 739	6 21 0	0.1397	5 26 13	9 . 9953	—14 42 11 . 3	7 . 74	3 9 16	1 . 0903	9 17 17	0 . 9695
9 Leporis	β	4	5. 19 40 . 80	2 . 557	6 9 17	0.1590	5 27 19	9 . 9655	—20 55 41 . 8	3 . 51	3 5 9	1 . 1501	9 7 32	0 . 9814
11 Leporis	α	3	5 23 54 . 35	2 . 634	6 8 17	0.1513	5 28 2	9 . 9777	—17 58 27 . 2	3 . 14	3 4 19	1 . 1270	9 6 44	0 . 9820
123 Tauri	ζ	3 . 4	5 25 41 . 35	3 . 568	6 7 53	0.1594	6 1 39	0 . 1089	+21 0 27 . 0	2 . 98	7 11 15	0 . 1608	3 6 23	0 . 9822
Columbæ	β	3	5 43 54 . 55	2 . 100	6 3 41	0.2215	5 27 32	9 . 8797	—35 51 4 . 0	1 . 42	3 3 0	1 . 2411	9 3 0	0 . 9839
37 Aurigæ	ϑ	4	5. 46 4 . 89	4 . 072	6 3. 12	0.2289	6 1 9	0 . 1660	+35 11 0 . 3	0 . 22	3 9 52	0 . 6185	3 2 35	0 . 9842
										—				
44 Aurigæ	κ	4	6 2 37 . 66	3 . 818	5 29 29	0.1914	5 29 43	0 . 1376	+29 33 23 . 5	0 . 22	2 27 22	0 . 3284	2 29 28	0 . 9844
1 Canis Maj.	ζ	3	6 12 37 . 75	2 . 292	5 27 6	0.1927	6 1 25	9 . 9175	—29 58 55 . 9	1 . 10	2 27 51	1 . 2122	8 27 39	0 . 9842
7 Gemin.	η	4 . 5	6 2 48 . 10	3 . 615	5 29 25	0.1650	5 29 9	0 . 1147	+22 33 5 . 4	0 . 24	9 15 43	9 . 5507	2 29 31	0 . 9844
13 Gemin.	μ	3	6 10 51 . 29	3 . 616	5 27 34	0.1647	5 29 8	0 . 1153	+22 36 9 . 7	0 . 94	10 18 54	9 . 6867	2 28 4	0 . 9842
2 Canis Maj.	β	2	6 13 53 . 20	2 . 632	5 26 49	0.1517	6 0 46	9 . 9771	—17 52 2 . 4	1 . 21	2 28 43	1 . 1259	8 27 25	0 . 9841
24 Gemin.	γ	3	6 29 9 . 07	3 . 455	5 24 3	0.1443	5 29 0	0 . 0951	+16 33 26 . 7	2 . 28	9 10 17	0 . 4068	2 25 11	0 . 9829
9 Canis Maj.	α	1	6 36 19 . 85	2 . 641	5 21 40	0.1479	6 1 47	9 . 9838	—16 27 4 . 1	3 . 16	2 26 0	1 . 1113	8 23 13	0 . 9820
69 Gemin.	υ	5	7 23 34 . 60	3 . 706	5 10 42	0.1772	5 24 56	0 . 1269	+27 19 41 . 2	7 . 14	0 13 23	0 . 5325	2 14 7	0 . 9720
25 Canis Maj.	δ	3	7 0 15 . 06	2 . 431	5 16 8	0.1747	6 5 20	9 . 9447	—26 5 1 . 8	5 . 20	2 21 15	1 . 1835	8 18 40	0 . 9778
55 Gemin.	δ	3 . 4	7 8 9 . 54	3 . 586	5 14 21	0.1614	5 26 34	0 . 1118	+22 20 19 . 7	5 . 87	11 12 26	0 . 5163	2 17 30	0 . 9759
21 Canis Maj.	ε	2	6 50 45 . 54	2 . 349	5 18 20	0.1857	6 5 13	9 . 9295	—28 42 29 . 7	4 . 40	2 22 21	1 . 2023	8 20 29	0 . 9797
60 Gemin.	ι	4	7 13 17 . 10	3 . 740	5 13 9	0.2005	5 25 13	0 . 1307	+28 10 59 . 3	6 . 29	0 29 13	0 . 5148	2 16 12	0 . 9747
31 Canis Maj.	η	2 . 3	7 16 10 . 40	2 . 365	5 12 26	0.1846	6 7 45	9 . 9348	—28 55 16 . 5	6 . 53	2 18 38	1 . 2012	8 15 35	0 . 9739
64 Gemin.	B1	5 . 6	7 16 51 . 40	3 . 746	5 12 16	0.1828	5 24 56	0 . 1316	+28 31 0 . 8	6 . 59	0 22 11	0 . 5333	2 15 28	0 . 9737
78 Gemin.	β	2	7 33 3 . 08	3 . 677	5 8 26	0.1811	5 24 8	0 . 1298	+28 29 48 . 5	7 . 90	0 15 9	0 . 5995	2 12 11	0 . 9688
3 Navis	τ	5	7 35 46 . 47	2 . 401	5 19 42	0.3245	6 15 46	9 . 7446	—28 29 4 . 1	8 . 13	2 21 2	1 . 2880	8 21 37	0 . 9680
77 Gemin.	κ	4	7 32 20 . 93	3 . 631	5 8 34	0.1675	5 24 53	0 . 1180	+24 51 55 . 2	7 . 85	11 28 46	0 . 5240	2 12 26	0 . 9690
7 Navis	ξ	4	7 40 53 . 00	2 . 515	5 6 45	0.1590	6 7 49	9 . 9641	—24 21 59 . 0	8 . 53	2 16 9	1 . 1614	8 10 37	0 . 9663
10 Ursæ Maj.	N	4 . 5	8 47 35 . 90	3 . 975	4 27 33	0.2472	5 14 35	0 . 1713	+42 33 53 . 0	13 . 38	0 21 6	0 . 9916	1 26 15	0 . 9363
Argus	λ	2 . 3	9 0 38 . 18	2 . 195	4 17 18	0.2456	6 27 50	9 . 9537	—42 37 44 . 8	14 . 20	1 27 58	1 . 2631	7 23 10	0 . 9294
Argus	q	4	10 6 20 . 82	2 . 507	4 0 34	0.2249	6 28 43	0 . 0121	—41 7 56 . 6	17 . 61	1 15 25	1 . 2124	7 6 0	0 . 8925
34 Ursæ Maj	μ	3	10 10 21 . 20	3 . 625	3 29 29	0.2338	5 8 17	0 . 1492	+42 30 1 . 9	17 . 77	11 29 31	1 . 0867	1 5 24	0 . 8905
7 Hyd. & Crat.	α	4	10 50 2 . 27	2 . 937	3 19 58	0.1165	6 10 11	0 . 0326	—17 14 7 . 2	19 . 10	1 28 33	0 . 0402	6 22 57	0 . 8715
70 Leonis	ϑ	3	11 3 43 . 15	3 . 159	3 15 17	0.1198	5 20 47	0 . 0616	+16 31 17 . 0	19 . 43	10 10 53	0 . 9308	0 18 40	0 . 8633

As the subsidiary Number (S) for the Solar Nutation never differs considerably from (L) the subsidiary Number for Lunar Nutation, the latter may be substituted for the former, with the constant Log. of the Max. 8 .8000 used with twice the Sun's Longitude, without any sensible error in the Reduction; but if great accuracy be required, then the Solar Nutation must be collected from the Tables adapted for this purpose, at pages 28 and 36. The same may be said with respect to the preceding Table of CIRCUMPOLAR STARS.

A Catalogue of Stars arranged in Pairs, &c. concluded.

Flam. Stars.	Right Ascension in Time.								Declination.					
	Bayer's.	Mag.	R. A.	An.Var.	A.	Log. of Max.	L.	Log. of Max.	Dec. N. + S. −	An.Var.	A′.	Log. of Max.	L′.	Log. of Max.
			h. m. s.	+	S. ° ′		S. ° ′			−	S. ° ′		S. ° ′	
1 orvi	α	4	11 58 6.96	3s.056	3 0 33	0.1309	6 14 20	0.0555	−23° 36′ 42″.9	20″.02	1 12 33	1.0421	6 0 41	0.8562
23 Berenices	K	4.5	12 24 51.72	2.999	2 23 15	0.1317	5 15 25	0.0576	+23 44 1.5	19.91	10 14 41	1.0601	11 21 41	0.8583
5 Centauri	ϑ	2	13 54 57.83	3.517	1 29 12	0.1908	6 17 34	0.1232	−35 22 34.4	17.56	0 7 47	1.0177	4 23 42	0.8932
Centauri	η	3	14 22 52.0	3.741	1 22 21	0.2305	6 18 46	0.1534	−41 16 9.0	16.26	11 24 37	1.0325	4 16 0	0.9089
27 Bootis	γ	3	14 24 0.90	2.422	1 21 37	0.2173	5 4 5	9.9873	+39 11 24.3	16.20	10 8 35	1.2229	10 15 42	0.9095
30 Bootis	ζ	3.4	14 31 35.86	2.847	1 19 41	0.1221	5 22 40	0.0147	+14 35 42.8	15.80	9 20 10	1.0581	10 13 43	0.9138
9 Libræ	α²	3	14 39 49.93	3.285	1 17 31	0.1246	6 6 27	0.0771	−15 12 5.4	15.35	1 18 52	0.7908	4 11 33	0.9186
3 Cor. Bor.	β	4	15 19 35.00	2.477	1 11 41	0.1408	5 14 35	9.9669	+29 48 15.0	12.90	9 24 45	1.1904	10 2 3	0.9399
40 Libræ		4.5	15 26 24.20	3.644	1 6 0	0.1751	6 9 37	0.1242	−29 6 23.4	12.43	0 0 0	0.7864	4 0 28	0.9432
39 Libræ		5.6	15 24 54.50	3.602	1 6 24	0.1680	6 9 11	0.1187	−27 27 35.9	12.54	0 4 42	0.7729	4 0 48	0.9427
5 Cor. Bor.	α	2	15 26 13.24	2.538	1 6 0	0.1681	5 17 10	9.9690	+27 23 49.4	12.45	9 22 46	1.1754	10 0 34	0.9435
7 Cor. Bor.	ζ	4	15 31 50.80	2.250	1 4 39	0.2201	5 10 0	9.9363	+37 17 40.6	12.05	9 25 31	1.2355	9 29 21	0.9462
Lupi	η	4	15 46 54.10	3.922	1 1 3	0.2069	6 10 58	0.1583	−37 48 34.0	10.99	11 5 13	0.8756	3 26 4	0.9531
28 Serpentis	β	4.5	15 36 57.60	2.751	1 3 27	0.1356	5 23 46	9.9985	+16 3 30.9	11.71	9 15 35	1.0933	9 28 12	0.9486
46 Libræ	ϑ	5	15 42 27.09	3.379	1 2 9	0.1369	6 4 56	0.0870	−16 7 47.0	11.31	1 17 41	0.6760	3 26 53	0.9511
11 Cor. Bor.	κ	5	15 43 41.85	2.250	1 1 40	0.2140	5 11 55	9.9342	+36 17 13.8	11.22	9 23 6	1.2287	9 26 41	0.9516
Lupi	ϑ	4	15 53 29.30	3.895	0 29 26	0.2140	6 9 59	0.1532	−36 14 32.5	10.50	11 5 24	0.8387	3 24 35	0.9560
20 Cor. Bor.	ν¹	5	16 14 49.67	2.246	0 24 23	0.2062	5 16 7	9.9072	+34 16 45.3	8.87	9 16 26	1.2516	9 20 11	9.9645
Regulæ	α	5	16 18 20.33	3.882	0 23 21	0.2068	6 7 36	0.1495	−34 15 11.3	8.60	11 0 6	0.7479	3 19 18	0.9659
21 Scorpii	α	1	16 17 9.63	3.639	0 23 49	0.1701	6 5 49	0.1209	−25 58 25.6	8.69	11 28 56	0.5853	3 19 37	0.9652
27 Herculis	β	2.3	16 21 37.37	2.574	0 22 46	0.1570	5 23 22	9.9702	+21 56 8.3	8.33	9 13 7	1.1532	9 18 49	0.9670
23 Scorpii	τ	3.4	16 23 27.00	3.704	0 22 20	0.1784	6 5 56	0.1273	−27 47 7.3	8.18	11 19 9	0.5943	3 18 25	0.9675
40 Herculis	ζ	3	16 33 44.69	2.287	0 19 58	0.2071	5 19 51	9.9234	+31 58 25.5	7.36	9 13 53	1.2166	9 19 25	0.9711
26 Scorpii	ε	3	16 37 13.58	3.900	0 19 7	0.2068	6 6 12	0.1500	−33 54 49.7	7.08	10 23 1	0.6971	3 15 43	0.9721
35 Ophiuchi	η	2.3	16 58 54.71	3.415	0 14 34	0.1441	6 2 11	0.0905	−15 27 47.9	5.28	2 4 32	0.5201	3 11 29	0.9776
64 Herculis	α	3.4	17 5 31.64	2.721	0 12 31	0.1425	5 27 42	9.9920	+14 37 48.6	4.72	9 5 36	1.0937	9 10 7	0.9790
67 Herculis	π	3	17 8 5.52	2.079	0 12 5	0.2281	5 22 7	9.8797	+37 2 49.3	4.50	9 9 0	1.2435	9 9 28	0.9796
75 Herculis	ρ	4	17 16 47.62	2.062	0 10 0	0.2287	5 22 57	9.8749	+37 20 23.9	3.75	9 7 40	1.2467	9 8 8	0.9806
34 Scorpii	υ	3.4	17 17 10.67	4.053	0 9 49	0.3174	6 3 33	0.1648	−37 7 5.4	3.72	9 26 31	0.7069	3 7 59	0.9809
35 Scorpii	λ	3	17 20 1.86	4.049	0 9 12	0.2268	6 3 18	0.1640	−36 56 24.5	3.48	9 24 59	0.6997	3 7 27	0.9815
										+				
3 Lyræ	α	1	18 30 9.89	2.028	11 23 5	0.2369	6 5 16	9.8606	+38 36 21.9	3.00	8 24 42	1.2516	8 24 23	0.9828
Cor. Australis	α	5	18 55 50.72	4.082	11 17 8	0.2330	5 24 0	0.0687	−38 11 52.6	4.83	7 27 37	0.7556	2 19 29	0.9787
10 Lyræ	β	4	18 42 41.77	2.205	11 20 11	0.2063	6 5 35	9.9026	+33 8 23.9	3.71	8 23 4	1.2231	8 22 1	0.9810
38 Sagittarii	ζ	4	18 49 52.39	3.819	11 18 32	0.1917	5 26 38	0.1390	−30 8 59.7	4.32	7 14 9	0.4865	2 20 37	0.9800
59 Sagittarii	b²	5	19 44 39.02	3.691	11 5 44	0.1755	5 23 37	0.1262	−27 41 5.8	8.84	6 8 10	0.6216	2 9 56	0.9647
15 Vulpeculæ	g	4.5	19 52 51.70	2.457	11 3 48	0.1735	6 10 0	9.9540	+27 12 37.0	9.46	8 13 23	1.1847	8 8 14	0.9616
18 Capricor.	ω	5.6	20 39 51.39	3.599	10 22 27	0.1681	5 20 32	0.1185	−27 39 17.7	12.86	5 24 51	0.7815	1 28 3	0.9402
32 Vulpeculæ	q	4.5	20 46 1.86	2.545	10 20 56	0.1667	6 13 35	9.9742	+27 18 21.6	13.27	5 8 29	1.1736	7 26 36	0.9369
66 Cygni	υ	4.5	21 9 41.40	2.453	10 15 0	0.1931	6 19 54	9.9732	+34 3 53.5	14.72	7 28 31	1.2071	7 21 2	0.9242
9 Pisc. Austr.	ι	4.5	21 32 58.49	3.599	10 15 45	0.1338	5 28 28	0.1127	−33 55 43.2	16.03	5 26 54	0.9573	1 15 8	0.9113
44 Pegasi	η	3	22 33 38.16	2.787	9 23 19	0.1893	6 22 9	9.9497	+29 10 47.6	18.62	7 18 30	1.1417	6 28 28	0.8789
24 Pis. Austr.	α	1	22 46 43.53	3.322	9 19 55	0.1623	5 13 6	0.0959	−30 40 38.5	19.01	5 8 17	1.0235	0 24 1	0.8729
53 Pegasi	β	2	22 54 4.97	2.868	9 17 52	0.1464	6 16 55	1.0334	+27 0 7.5	19.20	7 15 56	1.1167	6 21 40	0.8699
76 Aquarii	δ	3.4	22 44 0.71	3.193	9 20 29	0.1164	5 20 54	0.0535	−16 52 44.5	18.94	4 12 58	0.9112	0 24 51	0.8740
54 Pegasi	α	1	22 54 47.03	2.958	9 17 39	0.1096	6 8 18	0.0332	+14 7 58.4	19.21	8 2 40	1.0116	6 21 24	0.8693
21 Androm.	α	1	23 58 4.27	3.016	9 0 32	0.1471	6 17 6	0.0616	+27 59 11.0	20.02	7 8 7	1.0765	6 0 29	0.8563
Ap. Sculp.	κ²	5.6	0 1 3.27	3.058	8 29 38	0.1507	5 12 8	0.0637	−28 54 49.7	20.02	4 24 4	1.0827	11 29 32	0.8562

A ZONE OF STARS suitable for the ZENITH SECTOR adapted for the Year 1817.

The Mean Places are determined by S. Groombridge, Esq.; and the Subsidiary Numbers for giving the Corrections in N. P. D. by the Rev. Fearon Fallows, M.A.

Stars.	Mag.	No. of Obs.	Right Ascen.			Annual Precess.	No. of Obs.	N. P. Distance.			Annual Precess.	Aberr. in N. P. D.				☽ Nut. in N. P. D.				⊙ Nut. in N. P. D.			
												A′			Max.	L′			Max.	S′			Max.
			h.	m.	s.	+					—	S.				S.				S.			
12 Cassiopeæ	6	6	0	14	46.06	3ˢ.221	6	29°	11′	3″.96	20″.003	0	9°	19	17″.87	11	25°	3	7″.19	11	26°	0	0″.40
15 κ Cassiop.	4	6	0	22	40.58	3.320	7	28	4	47.46	19.946	0	6	59	17.91	11	22	25	7.21	11	23	49	.40
17 ζ Cassiop.	4	4	0	26	49.68	3.274	6	37	6	40.76	19.907	0	11	3	16.35	11	21	1	7.22	11	22	42	.40
18 α Cassiop.	3	5	0	30	11.18	3.321	12	34	28	3.56	19.870	0	8	46	16.75	11	19	54	7.23	11	21	47	.40
24 η Cass. **	4	5	0	48	4.21	3.406	20	33	9	25.84	19.768	0	6	17	16.82	11	17	19	7.26	11	19	38	.40
Cassiop. 18 Hev.	6	6	0	42	13.84	3.492	6	29	52	56.57	19.705	0	3	32	17.30	11	15	56	7.30	11	18	30	.40
27 γ Cassiop.	3	6	0	45	44.50	3.522	9	30	16	36.46	19.646	0	2	58	17.17	11	14	48	7.30	11	17	33	.40
33 ϑ Cassiop.	4 5	5	1	0	1.21	3.547	18	35	49	34.59	19.361	0	2	58	15.89	11	10	12	7.37	11	13	43	.40
34 φ Cassiop. pr.	5.6	4	1	8	25.81	3.677	4	32	45	28.10	19.157	11	29	11	16.28	11	7	32	7.43	11	11	28	.40
34 φ Cass. seq.		6	1	8	38.43	3.678	6	32	44	2.28	19.152	11	21	20	16.47	11	7	29	7.43	11	11	24	.40
37 δ Cassiop.	3	3	1	13	55.63	3.783	6	30	43	11.42	19.010	11	26	44	16.55	11	5	49	7.47	11	10	0	.40
39 χ Cassiop.	6	6	1	22	3.17	3.825	6	31	42	42.05	18.773	11	25	24	16.18	11	3	19	7.53	11	7	49	.40
44 Cassiop. pr.	6	2	1	30	20.65	3.941	3	30	22	52.55	18.507	11	22	40	16.26	11	0	48	7.60	11	5	37	.40
44 Cass. seq.		5	1	31	1.80	3.950	5	30	22	34.38	18.484	11	22	30	14.25	11	0	36	7.60	11	5	26	.40
54 φ Androm.	5	3	1	32	14.82	3.685	8	40	14	15.38	18.442	11	28	36	14.23	11	0	13	7.61	11	5	6	.40
1 Persei.	5.6	5	1	40	2.27	3.852	5	35	45	48.98	18.164	11	23	29	14.99	10	27	55	7.68	11	3	3	.41
45 ε Cassiop.	3.4	4	1	41	20.43	4.177	6	27	14	13.66	18.078	11	18	3	16.61	10	27	32	7.69	11	2	42	.41
53 Cassiop.	6	7	1	49	34.93	4.301	7	26	30	1.27	17.796	11	15	52	16.59	10	25	9	7.77	11	0	33	.41
4 Persei	6	5	1	50	11.03	3.905	5	36	24	11.69	17.772	11	21	37	14.60	10	24	59	7.78	11	0	23	.41
4 Persei seq.		2	1	51	16.87	3.911	2	36	11	37.36	17.727	11	21	13	14.63	10	24	40	7.79	11	0	6	.41
5 Persei	6	6	1	58	49.44	4.078	6	33	13	30.51	17.409	11	17	26	15.08	10	22	31	7.86	10	28	7	.41
55 Cassiop.	6	6	2	0	15.42	4.545	6	24	20	27.71	17.347	11	12	0	16.80	10	22	8	7.87	10	27	45	.41
Persei Hev. 4		5	2	4	4.11	4.095	5	33	49	45.37	17.178	11	16	30	14.83	10	21	4	7.91	10	26	46	.41
Do.		6	2	4	9.00	4.095	6	33	48	8.65	17.174	11	16	28	14.84	10	21	3	7.91	10	26	43	.41
7 χ Persei	6.7	6	2	5	17.88	4.223	6	33	20	20.46	17.123	11	15	53	14.90	10	20	43	7.93	10	26	28	.41
Persei Hev. 4		6	2	6	18.09	4.117	6	33	43	1.48	17.076	11	15	52	14.80	10	20	27	7.94	10	26	11	.41
Persei f		6	2	6	27.30	4.118	6	33	40	57.29	17.069	11	15	48	14.80	10	20	25	7.94	10	26	9	.41
Cassiop. Hev. 35	4.5	5	2	14	9.29	4.871	8	23	25	44.55	16.708	11	8	9	16.69	10	18	19	8.02	10	24	9	.41
13 ϑ Persei	4	7	2	31	44.87	3.994	21	41	33	12.19	15.808	11	15	2	12.37	10	13	40	8.20	10	19	39	.41
Persei Hev. 9 η		6	2	37	24.85	4.276	18	34	52	20.21	15.499	11	8	32	13.81	10	12	13	8.26	10	18	12	.41
23 γ Persei	4	6	2	51	35.98	4.246	11	37	13	8.55	14.683	11	6	16	12.91	10	8	40	8.42	10	14	38	.42
27 κ Persei	5	6	2	57	12.14	4.028	6	45	50	40.71	14.345	11	12	18	10.54	10	7	20	8.48	10	13	14	.42
Camelop. Hev. 1	6	6	3	4	0.72	5.124	6	25	1	45.35	13.923	10	25	42	15.43	10	5	43	8.55	10	11	22	.42
33 α Persei	2.3	8	3	11	18.63	4.215	18	40	48	0.59	13.457	11	3	25	11.51	10	4	0	8.62	10	9	44	.42
39 δ Persei	3.4	9	3	29	54.28	4.212	20	42	48	29.02	12.207	10	29	30	10.50	9	29	45	8.82	10	5	10	.42
Camelop. Hev. 6	5.6	5	3	30	8.21	5.126	5	27	14	46.83	12.193	10	19	19	14.45	9	29	42	8.82	10	5	7	.42
Capella	1		5	3	11.16	4.343		44	11	52.68	4.918	9	26	3	8.11	9	10	44	9.52	9	13	4	.43
* * * * *											+												
48 β Ursæ Maj.	2	14	10	50	43.22	3.695	30	32	38	20.46	19.135	6	1	6	16.29	6	2	44	7.43	6	18	47	.40
50 α Ursæ Maj.	1.2	7	10	52	20.49	3.824	28	27	15	48.39	19.117	6	3	42	17.28	6	22	14	7.42	6	18	21	.40
64 γ Ursæ Maj.	2	11	11	44	28.70	3.199	25	35	17	15.56	19.996	5	17	40	16.89	6	5	19	7.20	6	4	20	.40
69 δ Ursæ Maj.	3	11	12	6	18.66	3.012	19	31	56	59.82	20.036	5	14	42	17.81	5	27	53	7.18	5	28	17	.40
77 ε Ursæ Maj.	3	17	12	45	56.47	2.657	24	33	2	41.55	19.642	5	5	45	18.25	5	14	44	7.30	5	17	30	.40
79 ζ Ursæ Maj.	3	12	13	16	31.74	2.420	34	34	6	57.57	18.936	4	28	51	18.51	5	5	1	7.49	5	9	17	.40
85 η Ursæ Maj.	2.3	21	13	40	24.02	2.387	59	39	46	12.08	18.143	4	21	52	17.93	4	27	50	7.68	5	3	0	.40
11 α Draconis	3.4	9	13	59	26.18	1.623	14	24	44	49.05	17.383	4	23	34	19.83	4	22	21	7.87	4	27	58	.41
17 κ Bootis seq.	5.6	7	14	6	54.79	2.146	8	37	21	1.54	17.049	4	17	15	18.65	4	20	18	7.94	4	26	2	.41
12 ι Draconis	3	6	15	20	52.28	1.315	24	30	23	23.16	12.826	4	5	17	19.85	4	1	46	8.72	4	7	21	.42
13 ϑ Draconis	3.4	16	15	58	28.51	1.150	37	30	56	38.96	10.138	3	26	9	19.94	3	23	34	9.08	3	28	16	.43
14 η Draconis	3	13	16	21	31.67	0.787	30	28	4	10.32	8.349	3	21	43	20.12	3	18	50	9.27	3	22	48	.43
16 Draconis	6	6	16	31	52.28	1.406	6	36	43	42.15	7.519	3	18	7	19.59	3	16	46	9.34	3	20	22	.43
17 Draconis	6	6	16	31	54.85	1.403	6	36	42	13.50	7.515	3	18	6	19.60	3	16	45	9.34	3	20	21	.43
18 g Draconis	5	6	16	39	40.34	0.281	6	25	3	47.62	6.881	3	18	9	20.23	3	15	13	9.39	3	18	22	.43
19 h Draconis	5	6	16	55	2.38	0.264	6	24	35	6.50	5.605	3	14	44	20.25	3	12	14	9.48	3	14	57	.43
21 μ Draconis	4	8	17	1	33.03	1.242	12	35	17	6.90	5.057	3	12	10	19.78	3	10	59	9.51	3	13	42	.43
22 ζ Draconis	3	6	17	8	16.21	0.149	19	24	3	35.11	4.488	3	11	48	20.24	3	9	42	9.54	3	11	53	.43
23 β Draconis	2	11	17	26	17.97	1.348	44	37	33	31.94	2.938	3	6	53	19.63	3	6	18	9.60	3	7	44	.43
24 √¹ Draconis	5	3	17	29	51.57	2.876	6	34	41	12.82	2.654	3	6	22	19.86	3	5	41	9.61	3	7	0	.43
25 √² Draconis	5	3	17	29	56.99	2.876	6	34	41	54.84	2.648	3	6	22	19.86	3	5	40	9.61	3	6	58	.43

To obtain the Arguments in *Declination*, add six signs to A′, L′, and S′, respectively.

A ZONE OF STARS suitable for the ZENITH SECTOR adapted for the Year 1817.

The MEAN PLACES are determined by S. GROOMBRIDE, Esq.; and the SUBSIDIARY NUMBERS for giving the CORRECTIONS in N. P. D. by the REV. FEARON FALLOWS, M.A.

STARS.	Mag.	No. of Obs.	RIGHT ASCEN.	Annual Precess.	No. of Obs.	N. P. DISTANCE.	Annual Precess.	ABERR. IN N. P. D.		☽ NUT. IN N. P. D.		☉ NUT. IN N. P. D.	
								A′	Max.	L′	Max.	S′	Max.
			h. m. s.	+			+	S.		S.		S.	
26 Draconis	6	6	17 33 5.76	0s.572	6	27° 58′ 58.59	2″.347	3 5° 57′	20″.10	3 5° 1	9″.62	3 6° 10	0″.43
85 ι Herculis	4	4	17 34 18.06	1.689	9	43 53 27.49	2.243	3 4 56	18.97	3 4 48	9.62	3 5 53	.43
32 ξ Draconis	3.4	5	17 50 21.90	1.017	16	33 5 44.16	0.842	3 2 3	19.97	3 1 48	9.65	3 2 12	.43
33 γ Draconis	2	28	17 52 21.55	1.387	86	38 29 7.72	0.671	3 1 33	19.56	3 1 25	9.65	3 1 45	.43
36 Draconis	5	6	18 12 49.27	0.292	6	25 39 50.06	1.119	2 27 7	20.24	2 27 37	9.64	2 27 4	.43
39 Draconis	5	4	18 21 14.30	0.880	5	31 18 10.96	1.856	2 25 26	20.13	2 26 2	9.63	2 25 8	.43
42 Draconis	6	6	18 25 49.23	2.352	6	24 32 58.09	2.221	2 24 13	20.25	2 25 16	9.62	2 24 10	.43
45 d Draconis	5.6	5	18 29 25.09	1.034	5	33 5 30.07	2.566	2 23 45	19.96	2 24 32	9.61	2 23 15	.43
46 c Draconis	5	4	18 39 6.03	1.231	6	34 38 33.79	3.401	2 21 49	19.85	2 22 42	9.59	2 21 2	.43
47 o Draconis	5	11	18 48 29.41	0.876	15	30 49 59.01	4.210	2 19 31	20.07	2 20 56	9.55	2 18 51	.43
48 Draconis	5	6	18 53 39.03	1.024	6	32 25 30.48	4.649	2 18 48	19.96	2 19 57	9.53	2 17 40	.43
49 Draconis	6.7	6	18 57 6.00	1.190	6	34 36 1.59	4.943	2 18 2	19.83	2 19 17	9.52	2 16 52	.43
51 Draconis	6	5	19 0 48.19	1.347	7	36 52 50.61	5.255	2 17 31	19.65	2 18 33	9.50	2 16 0	.43
53 Draconis	5	4	19 8 12.52	1.134	9	33 26 58.96	5.877	2 15 35	19.89	2 17 9	9.64	2 14 17	.43
54 Draconis	5	5	19 10 38.55	1.078	7	32 36 27.12	6.082	2 14 57	19.94	2 16 39	9.45	2 13 43	.43
57 δ Draconis	3	6	19 12 28.72	0.027	6	22 39 38.61	6.233	2 13 20	20.25	2 16 18	9.44	2 13 17	.43
1 κ Cygni	4	10	19 12 52.00	1.380	38	36 57 56.63	6.266	2 15 3	19.62	2 16 14	9.44	2 13 12	.43
58 π Draconis	4	12	19 19 42.02	0.328	15	24 38 14.19	6.830	2 11 56	20.24	2 14 54	9.40	2 11 36	.43
10 ι Cygni	5	11	19 25 5.19	1.510	48	38 39 23.27	7.269	2 12 49	19.42	2 13 50	9.36	2 10 21	.43
13 ϑ Cygni	4	19	19 31 31.77	1.611	25	40 11 56.24	7.794	2 11 45	19.25	2 12 34	9.32	2 8 51	.43
16 c′ Cygni pr.	7	6	19 36 56.98	1.611	6	39 53 42.92	8.228	2 10 40	19.26	2 11 29	9.28	2 7 34	.43
Do. sequens	7	6	19 36 59.76	1.611	4	39 54 8.38	8.232	2 10 39	19.26	2 11 28	9.28	2 7 33	.43
20 Cygni	5.6	5	19 46 1.88	1.507	5	37 28 20.01	8.926	2 8 26	19.47	2 9 38	9.21	2 5 26	.43
Cephei Hev. 2	6	7	19 52 24.65	1.154	7	31 38 25.93	9.440	2 5 56	19.92	2 8 19	9.17	2 3 54	.43
66 Draconis	5	6	20 2 37.22	0.954	6	28 32 1.56	10.220	2 3 7	20.06	2 6 10	9.07	2 1 28	.43
68 Draconis	6	6	20 8 33.93	0.984	6	28 28 25.01	10.663	2 1 49	20.05	2 4 56	9.02	2 0 3	.43
71 Draconis	7.8	9	20 16 32.40	1.017	9	28 19 18.02	11.247	2 0 3	20.04	2 3 13	8.95	1 28 7	.42
43 ω′ Cygni	5.6	6	20 21 26.60	1.822	7	41 13 11.01	11.599	2 2 11	18.89	2 2 9	8.90	1 26 56	.42
2 ϑ Cephei	5	7	20 26 29.60	1.020	7	27 37 8.35	11.957	1 27 44	20.05	2 1 3	8.85	1 25 42	.42
α Cygni	1		20 35 11.53	2.040		45 21 41.24	12.559	2 0 44	18.24	1 29 6	8.76	1 23 35	.42
Cephei Hev. 6	5	5	20 40 48.27	1.499	5	33 4 23.76	12.938	1 26 5	19.63	1 27 50	8.71	1 22 13	.42
3 η Cephei	3.4	10	20 41 32.68	1.222	13	28 51 19.02	12.986	1 24 53	19.98	1 27 39	8.70	1 22 3	.42
5 α Cephei	3	13	21 14 11.85	1.414	33	28 11 15.24	15.024	1 17 36	19.84	1 19 52	8.35	1 13 55	.42
9 Cephei	5	6	21 33 0.25	1.611	6	28 44 29.20	16.060	1 13 46	19.70	1 15 6	8.15	1 9 9	.41
10 Cephei	4.5	6	21 40 10.07	1.726	6	29 43 16.33	16.427	1 12 35	19.58	1 13 12	8.08	1 7 18	.41
Cephei	6.7	6	21 46 56.66	1.947	6	34 39 0.58	16.808	1 12 57	19.03	1 11 24	8.00	1 5 35	.41
17 ξ Cephei	5	5	21 58 29.19	1.697	13	26 15 41.84	17.291	1 7 26	19.73	1 8 13	7.89	1 2 35	.41
19 Cephei	5.6	6	21 59 30.43	1.836	6	28 36 30.76	17.336	1 8 5	19.54	1 7 57	7.88	1 2 18	.41
23 ε Cephei	4.5	4	22 8 18.01	2.133	16	33 51 59.23	17.710	1 8 16	18.92	1 5 27	7.79	1 0 1	.41
25 Cephei	6	6	22 12 14.73	1.934	6	28 6 39.29	17.869	1 5 11	19.48	1 4 18	7.75	0 29 0	.41
3 Lacertæ	4	6	22 16 22.54	2.327	9	38 41 4.82	17.991	1 8 52	18.27	1 3 25	7.72	0 28 9	.41
27 δ Cephei	4.5	5	22 22 23.36	2.198	6	32 31 8.90	18.253	1 4 54	18.89	1 1 22	7.67	0 26 19	.44
7 Lacertæ	4	7	22 23 45.93	2.433	7	40 39 20.49	18.303	1 8 6	17.83	1 0 57	7.65	0 25 57	.41
9 Lacertæ	5.6	6	22 29 52.69	2.445	6	39 23 49.53	18.514	1 6 19	17.94	0 29 16	7.60	0 24 20	.40
30 Cephei	5	6	22 32 10.33	2.100	7	27 21 53.63	18.579	1 0 38	19.39	0 28 26	7.58	0 23 43	.40
32 ι Cephei	4	9	22 43 11.29	2.109	14	24 45 35.46	18.928	0 27 13	19.53	0 25 5	7.49	0 20 47	.40
1 e Cassiopeæ	6	7	22 58 54.19	2.495	6	31 34 4.79	19.335	0 26 46	18.64	0 20 11	7.38	0 16 36	.40
34 o Cephei	7	9	23 11 9.04	2.395	42	22 53 20.11	19.591	0 20 21	19.45	0 16 13	7.31	0 13 18	.40
4 Cassiopeæ	5	6	23 16 44.92	2.611	10	28 43 15.96	19.690	0 23 27	17.43	0 14 23	7.28	0 11 46	.40
Cassiop. Hev. 1	5	6	23 21 36.92	2.654	6	28 27 33.01	19.763	0 22 14	18.20	0 12 48	7.26	0 10 27	.40
18 Andromedæ	6.7	6	23 30 18.21	2.864	6	40 32 28.26	19.875	0 24 51	16.82	0 9 56	7.23	0 8 5	.40
5 τ Cassiopeæ	5	5	23 38 9.26	2.864	6	32 25 1.65	19.953	0 18 51	17.99	0 7 19	7.21	0 5 58	.40
6 Cassiopeæ	5.6	6	23 39 58.68	2.855	6	28 48 7.29	19.968	0 16 43	18.46	0 6 43	7.20	0 5 28	.40
Cassiop. Bode 16 p.		5	23 47 44.97	2.968	5	35 10 39.75	20.015	0 18 16	17.41	0 4 7	7.19	0 3 21	.40
Cassiop. Bode 16		6	23 47 57.83	2.969	6	35 18 42.06	20.016	0 18 18	17.38	0 4 3	7.19	0 3 18	.40
8 σ Cassiop. **	6	6	23 49 46.74	2.996	6	35 15 49.64	20.024	0 17 54	17.36	0 3 27	7.19	0 2 47	.40
9 Cassiopeæ	6	6	23 54 52.27	3.015	6	28 43 53.28	20.039	0 13 27	18.25	0 1 44	7.18	0 1 23	.40
11 β Cassiop.	2.3	10	23 59 7.51	3.058	32	31 51 33.83	20.044	0 14 14	17.74	0 0 11	7.18	0 0 10	.40

To obtain the Arguments in *Declination*, add six signs to A′, L′, S′, respectively.

TABLE 1. The MEAN PRECESSIONS of the EQUINOCTIAL POINTS in LONGITUDE, at the commencement of each Century, from the Christian Era to A. D. 3000; and for the beginning of each Year from 1700 to 1900 inclusive. (MR. J. UTTING.)

Years of the Christian Era.	Gen. Prec.	Luni-Sol. Prec.	Constant part of Prec. in R. A.	Variable part of Prec. in R. A. and in Dec.	Years of the Christian Era.	Gen. Prec.	Luni-Sol. Prec.	Constant part of Prec. in R. A.	Variable part of Prec. in R. A. and in Dec.
A. D.					A. D.				
B. 000	49″.74854	50″.76680	45″.45587	20″.22014	1741	50″.17387	50″.34269	45″.99315	20″.05126
100	49 .77297	50 .74244	45 .48673	20 .21044	1742	50 .17411	50 .34245	45 .99345	20 .05117
200	49 .79740	50 .71808	45 .51759	20 .20074	1743	50 .17436	50 .34220	45 .99376	20 .05107
300	49 .82183	50 .69372	45 .54845	20 .19104	B. 1744	50 .17460	50 .34196	45 .99407	20 .05097
B. 400	49 .84626	50 .66936	45 .57931	20 .18134	1745	50 .17485	50 .34172	45 .99438	20 .05087
500	49 .87069	50 .64500	45 .61017	20 .17164	1746	50 .17509	50 .34147	45 .99469	20 .05078
600	49 .89512	50 .62064	45 .64103	20 .16194	1747	50 .17534	50 .34123	45 .99500	20 .05068
700	49 .91955	50 .59628	45 .67189	20 .15224	B. 1748	50 .17558	50 .34099	45 .99530	20 .05058
B. 800	49 .94398	50 .57192	45 .70275	20 .14254	1749	50 .17582	50 .34074	45 .99561	20 .05049
900	49 .96841	50 .54756	45 .73361	20 .13284	1750	50 .17607	50 .34050	45 .99592	20 .05039
1000	49 .99284	50 .52320	45 .76447	20 .12314	1751	50 .17631	50 .34026	45 .99623	20 .05029
1100	50 .01727	50 .49884	45 .79533	20 .11344	B. 1752	50 .17656	50 .34001	45 .99654	20 .05020
B. 1200	50 .04170	50 .47448	45 .82619	20 .10374	1753	50 .17680	50 .33977	45 .99685	20 .05010
1300	50 .06613	50 .45012	45 .85705	20 .09404	1754	50 .17705	50 .33952	45 .99715	20 .05000
1400	50 .09056	50 .42576	45 .88791	20 .08434	1755	50 .17729	50 .33928	45 .99746	20 .04991
1500	50 .11499	50 .40140	45 .91877	20 .07464	B. 1756	50 .17753	50 .33904	45 .99777	20 .04981
B. 1600	50 .13942	50 .37704	45 .94963	20 .06494	1757	50 .17778	50 .33879	45 .99808	20 .04971
1700	50 .16385	50 .35268	45 .98049	20 .05524	1758	50 .17802	50 .33855	45 .99839	20 .04961
1701	50 .16410	50 .35244	45 .98080	20 .05514	1759	50 .17827	50 .33831	45 .99870	20 .04952
1702	50 .16434	50 .35219	45 .98111	20 .05505	B. 1760	50 .17851	50 .33806	45 .99901	20 .04942
1703	50 .16459	50 .35195	45 .98142	20 .05495	1761	50 .17875	50 .33782	45 .99932	20 .04932
B. 1704	50 .16483	50 .35170	45 .98173	20 .05485	1762	50 .17900	50 .33758	45 .99962	20 .04923
1705	50 .16508	50 .35146	45 .98203	20 .05476	1763	50 .17924	50 .33733	45 .99993	20 .04913
1706	50 .16532	50 .35122	45 .98234	20 .05466	B. 1764	50 .17949	50 .33709	46 .00024	20 .04903
1707	50 .16556	50 .35097	45 .98265	20 .05456	1765	50 .17973	50 .33685	46 .00055	20 .04894
B. 1708	50 .16581	50 .35073	45 .98296	20 .05446	1766	50 .17998	50 .33660	46 .00086	20 .04884
1709	50 .16605	50 .35049	45 .98327	20 .05437	1767	50 .18022	50 .33636	46 .00117	20 .04874
1710	50 .16630	50 .35024	45 .98358	20 .05427	B. 1768	50 .18046	50 .33611	46 .00148	20 .04864
1711	50 .16654	50 .35000	45 .98389	20 .05417	1769	50 .18071	50 .33587	46 .00178	20 .04855
B. 1712	50 .16678	50 .34976	45 .98419	20 .05408	1770	50 .18095	50 .33563	46 .00209	20 .04845
1713	50 .16703	50 .34951	45 .98450	20 .05398	1771	50 .18120	50 .33538	46 .00240	20 .04835
1714	50 .16727	50 .34927	45 .98481	20 .05388	B. 1772	50 .18144	50 .33514	46 .00271	20 .04826
1715	50 .16752	50 .34903	45 .98512	20 .05379	1773	50 .18169	50 .33490	46 .00302	20 .04816
B. 1716	50 .16776	50 .34878	45 .98543	20 .05369	1774	50 .18193	50 .33465	46 .00333	20 .04806
1717	50 .16801	50 .34854	45 .98574	20 .05359	1775	50 .18218	50 .33441	46 .00364	20 .04797
1718	50 .16825	50 .34829	45 .98605	20 .05349	B. 1776	50 .18242	50 .33417	46 .00395	20 .04787
1719	50 .16850	50 .34805	45 .98635	20 .05340	1777	50 .18266	50 .33392	46 .00425	20 .04777
B. 1720	50 .16874	50 .34781	45 .98666	20 .05330	1778	50 .18291	50 .33368	46 .00456	20 .04767
1721	50 .16898	50 .34756	45 .98697	20 .05320	1779	50 .18315	50 .33344	46 .00487	20 .04758
1722	50 .16923	50 .34732	45 .98728	20 .05311	B. 1780	50 .18340	50 .33319	46 .00518	20 .04748
1723	50 .16947	50 .34708	45 .98759	20 .05301	1781	50 .18364	50 .33295	46 .00549	20 .04738
B. 1724	50 .16972	50 .34683	45 .98790	20 .05291	1782	50 .18389	50 .33270	46 .00580	20 .04729
1725	50 .16996	50 .34659	45 .98821	20 .05282	1783	50 .18413	50 .33246	46 .00611	20 .04719
1726	50 .17021	50 .34635	45 .98852	20 .05272	B. 1784	50 .18437	50 .33222	46 .00641	20 .04709
1727	50 .17045	50 .34610	45 .98882	20 .05262	1785	50 .18462	50 .33197	46 .00672	20 .04699
B. 1728	50 .17069	50 .34586	45 .98913	20 .05252	1786	50 .18486	50 .33173	46 .00703	20 .04690
1729	50 .17094	50 .34561	45 .98944	20 .05243	1787	50 .18511	50 .33149	46 .00734	20 .04680
1730	50 .17118	50 .34537	45 .98975	20 .05233	B. 1788	50 .18535	50 .33124	46 .00765	20 .04670
1731	50 .17143	50 .34513	45 .99006	20 .05223	1789	50 .18560	50 .33100	46 .00796	20 .04661
B. 1732	50 .17167	50 .34488	45 .99037	20 .05214	1790	50 .18584	50 .33076	46 .00827	20 .04651
1733	50 .17192	50 .34464	45 .99068	20 .05204	1791	50 .18608	50 .33051	46 .00858	20 .04641
1734	50 .17216	50 .34440	45 .99098	20 .05194	B. 1792	50 .18633	50 .33027	46 .00888	20 .04632
1735	50 .17240	50 .34415	45 .99129	20 .05184	1793	50 .18657	50 .33002	46 .00919	20 .04622
B. 1736	50 .17265	50 .34391	45 .99160	20 .05175	1794	50 .18682	50 .32978	46 .00950	20 .04612
1737	50 .17289	50 .34367	45 .99191	20 .05165	1795	50 .18706	50 .32954	46 .00981	20 .04602
1738	50 .17314	50 .34342	45 .99222	20 .05155	B. 1796	50 .18731	50 .32929	46 .01012	20 .04593
1739	50 .17338	50 .34318	45 .99253	20 .05146	1797	50 .18755	50 .32905	46 .01043	20 .04583
B. 1740	50 .17363	50 .34294	45 .99284	20 .05136	1798	50 .18779	50 .32881	46 .01073	20 .04573
1741	50 .17387	50 .34269	45 .99315	20 .05126	1799	50 .18804	50 .32856	46 .01104	20 .04564
1742	50 .17411	50 .34245	45 .99345	20 .05117	C. 1800	50 .18828	50 .32832	46 .01135	20 .04554

The *constant* part of the Prec. in R. A. belonging to any given Epoch, in this Table, may be substituted for the constant Number given in the Notes of Table 1, pages 20 and 21, after being divided by 15, to convert the Arc into Time.

PRECESSIONS.

TABLE 1.

The MEAN PRECESSIONS, &c., of the EQUINOCTIAL POINTS for Years, &c., concluded.

Years of the Christian Era.	Gen. Prec.	Luni-Sol. Prec.	Constant part of Prec. in R. A.	Variable part of Prec. in R. A. and in Dec.
A. D.				
1801	50″. 18853	50″. 32808	46″. 01166	20″. 04544
1802	50 . 18877	50 . 32783	46 . 01197	20 . 04535
1803	50 . 18902	50 . 32759	46 . 01228	20 . 04525
B. 1804	50 . 18926	50 . 32735	46 . 01259	20 . 04515
1805	50 . 18950	50 . 32710	46 . 01290	20 . 04505
1806	50 . 18975	50 . 32686	46 . 01320	20 . 04496
1807	50 . 18999	50 . 32661	46 . 01351	20 . 04486
B. 1808	50 . 19024	50 . 32637	46 . 01382	20 . 04476
1809	50 . 19048	50 . 32613	46 . 01413	20 . 04467
1810	50 . 19073	50 . 32588	46 . 01444	20 . 04457
1811	50 . 19097	50 . 32564	46 . 01475	20 . 04447
B. 1812	50 . 19121	50 . 32540	46 . 01506	20 . 04438
1813	50 . 19146	50 . 32515	46 . 01537	20 . 04428
1814	50 . 19170	50 . 32491	46 . 01567	20 . 04418
1815	50 . 19195	50 . 32467	46 . 01598	20 . 04408
B. 1816	50 . 19219	50 . 32442	46 . 01629	20 . 04399
1817	50 . 19244	50 . 32418	46 . 01660	20 . 04389
1818	50 . 19268	50 . 32393	46 . 01691	20 . 04379
1819	50 . 19292	50 . 32369	46 . 01722	20 . 04370
B. 1820	50 . 19317	50 . 32345	46 . 01753	20 . 04360
1821	50 . 19341	50 . 32320	46 . 01783	20 . 04350
1822	50 . 19366	50 . 32296	46 . 01814	20 . 04341
1823	50 . 19390	50 . 32272	46 . 01845	20 . 04331
B. 1824	50 . 19415	50 . 32247	46 . 01876	20 . 04321
1825	50 . 19439	50 . 32223	46 . 01907	20 . 04311
1826	50 . 19464	50 . 32199	46 . 01938	20 . 04302
1827	50 . 19488	50 . 32174	46 . 01969	20 . 04292
B. 1828	50 . 19512	50 . 32150	46 . 01999	20 . 04282
1829	50 . 19537	50 . 32126	46 . 02030	20 . 04273
1830	50 . 19561	50 . 32101	46 . 02061	20 . 04263
1831	50 . 19586	50 . 32077	46 . 02092	20 . 04253
B. 1832	50 . 19610	50 . 32052	46 . 02123	20 . 04243
1833	50 . 19635	50 . 32028	46 . 02154	20 . 04234
1834	50 . 19659	50 . 32004	46 . 02185	20 . 04224
1835	50 . 19683	50 . 31979	46 . 02216	20 . 04214
B. 1836	50 . 19708	50 . 31955	46 . 02246	20 . 04205
1837	50 . 19732	50 . 31931	46 . 02277	20 . 04195
1838	50 . 19757	50 . 31906	46 . 02308	20 . 04185
1839	50 . 19781	50 . 31882	46 . 02339	20 . 04176
B. 1840	50 . 19806	50 . 31858	46 . 02370	20 . 04166
1841	50 . 19830	50 . 31833	46 . 02401	20 . 04156
1842	50 . 19854	50 . 31809	46 . 02432	20 . 04146
1843	50 . 19879	50 . 31785	46 . 02462	20 . 04137
B. 1844	50 . 19903	50 . 31760	46 . 02493	20 . 04127
1845	50 . 19928	50 . 31736	46 . 02524	20 . 04117
1846	50 . 19952	50 . 31711	46 . 02555	20 . 04108
1847	50 . 19977	50 . 31687	46 . 02586	20 . 04098
B. 1848	50 . 20001	50 . 31663	46 . 02617	20 . 04088
1849	50 . 20025	50 . 31638	46 . 02648	20 . 04079
1850	50 . 20050	50 . 31614	46 . 02678	20 . 04069
1851	50 . 20074	50 . 31590	46 . 02709	20 . 04059
B. 1852	50 . 20099	50 . 31565	46 . 02740	20 . 04050
1853	50 . 20123	50 . 31541	46 . 02771	20 . 04040
1854	50 . 20148	50 . 31517	46 . 02802	20 . 04030
1855	50 . 20172	50 . 31492	46 . 02833	20 . 04020
B. 1856	50 . 20196	50 . 31468	46 . 02864	20 . 04011
1857	50 . 20221	50 . 31443	46 . 02895	20 . 04001
1858	50 . 20245	50 . 31419	46 . 02926	20 . 03991
1859	50 . 20270	50 . 31395	46 . 02956	20 . 03982
B. 1860	50 . 20294	50 . 31370	46 . 02987	20 . 03972
A.D.				
1861	50″. 20319	50″. 31346	46″. 03018	20″. 03962
1862	50 . 20343	50 . 31322	46 . 03049	20 . 03952
1863	50 . 20367	50 . 31297	46 . 03080	20 . 03943
B. 1864	50 . 20392	50 . 31273	46 . 03111	20 . 03933
1865	50 . 20416	50 . 31249	46 . 03142	20 . 03923
1866	50 . 20441	50 . 31224	46 . 03172	20 . 03914
1867	50 . 20465	50 . 31200	46 . 03203	20 . 03904
B. 1868	50 . 20490	50 . 31175	46 . 03234	20 . 03894
1869	50 . 20514	50 . 31151	46 . 03265	20 . 03885
1870	50 . 20538	50 . 31127	46 . 03296	20 . 03875
1871	50 . 20563	50 . 31102	46 . 03327	20 . 03865
B. 1872	50 . 20587	50 . 31078	46 . 03358	20 . 03855
1873	50 . 20612	50 . 31054	46 . 03388	20 . 03846
1874	50 . 20636	50 . 31029	46 . 03419	20 . 03836
1875	50 . 20660	50 . 31005	46 . 03450	20 . 03826
B. 1876	50 . 20685	50 . 30981	46 . 03481	20 . 03817
1877	50 . 20709	50 . 30956	46 . 03512	20 . 03807
1878	50 . 20734	50 . 30932	46 . 03543	20 . 03797
1879	50 . 20758	50 . 30908	46 . 03574	20 . 03788
B. 1880	50 . 20783	50 . 30883	46 . 03604	20 . 03778
1881	50 . 20807	50 . 30859	46 . 03635	20 . 03768
1882	50 . 20831	50 . 30834	46 . 03666	20 . 03758
1883	50 . 20856	50 . 30810	46 . 03697	20 . 03749
B. 1884	50 . 20880	50 . 30786	46 . 03728	20 . 03739
1885	50 . 20905	50 . 30761	46 . 03759	20 . 03729
1886	50 . 20929	50 . 30737	46 . 03790	20 . 03720
1887	50 . 20954	50 . 30713	46 . 03821	20 . 03710
B. 1888	50 . 20978	50 . 30688	46 . 03851	20 . 03700
1889	50 . 21003	50 . 30664	46 . 03882	20 . 03690
1890	50 . 21027	50 . 30640	46 . 03913	20 . 03681
1891	50 . 21051	50 . 30615	46 . 03944	20 . 03671
B. 1892	50 . 21076	50 . 30591	46 . 03975	20 . 03661
1893	50 . 21100	50 . 30567	46 . 04006	20 . 03652
1894	50 . 21125	50 . 30542	46 . 04037	20 . 03642
1895	50 . 21149	50 . 30518	46 . 04067	20 . 03632
B. 1896	50 . 21174	50 . 30494	46 . 04098	20 . 0[illegible]623
1897	50 . 21198	50 . 30469	46 . 04129	20 . 03613
1898	50 . 21222	50 . 30445	46 . 04160	20 . 03603
1899	50 . 21247	50 . 30421	46 . 04191	20 . 03593
C. 1900	50 . 21271	50 . 30396	46 . 04222	20 . 03584
B. 2000	50 . 23714	50 . 27960	46 . 07308	20 . 02614
2100	50 . 26157	50 . 25524	46 . 10394	20 . 01644
2200	50 . 28600	50 . 23088	46 . 13480	20 . 00674
2300	50 . 31043	50 . 20652	46 . 16566	19 . 99704
B. 2400	50 . 33486	50 . 18216	46 . 19652	19 . 93734
2500	50 . 35929	50 . 15780	46 . 22738	19 . 97764
2600	50 . 38372	50 . 13344	46 . 25824	19 . 96794
2700	50 . 40815	50 . 10908	46 . 28910	19 . 95824
B. 2800	50 . 43258	50 . 0[illegible]472	46 . 31996	19 . 94854
2900	50 . 45701	50 . 06036	46 . 35082	19 . 93884
3000	50 . 48144	50 . 03600	46 . 38168	19 . 92914
B. 4000	50 . 72574	49 . 79240	46 . 69028	19 . 83214

The following Quantities must be deducted from the result as given by the TABLE, for the first, second, and third Years after Bissextile, to obtain the Precessions correctly.

Years aft. Bissext.	—	—	—	—
1	0″. 03435	0″. 03447	0″. 03149	0″. 01372
2	0 . 06870	0 . 06893	0 . 06299	0 . 02745
3	0 . 10305	0 . 10340	0 . 09448	0 . 04117

TABLE 2. The AMOUNT of the GENERAL and LUNI-SOLAR PREC. of the Equin. Points in Long., and of the CONST. and VAR. QUANTITIES of PREC. in R. A. and DECL. respectively; with the Amt. of the correspg. Var. of Prec. from A. D. 1750 to A.D. 1880. (MR. J. UTTING.)

Years of the Christian Era.	General Precession.		Luni-Solar Precession.		Constant part of Prec. in R. A.		Variable part of the Prec. in R. A. and in Declin.	
	Precession.	Variation.	Precession.	Variation.	Precession.	Variation.	Precession.	Variation.
		+		−		+		−
1750	0′ 0″.00000	0′.00000	0′ 0″.00000	0″.00000	0′ 0″.00000	0″.00000	0′ 0 .00000	0″.00000
1751	0 50 .14172	0 .00024	0 50 .30604	0 .00021	0 45 .96444	0 .00031	0 20 .03667	0 .00010
B. 1752	1 40 .28345	0 .00073	1 40 .61208	0 .00073	1 31 .92888	0 .00093	0 40 .07333	0 .00029
1753	2 30 .56255	0 .00147	2 31 .05595	0 .00146	2 18 .01924	0 .00185	1 0 .16489	0 .00058
1754	3 20 .70427	0 .00244	3 21 .36199	0 .00244	3 3 .98368	0 .00309	1 20 .20156	0 .00097
1755	4 10 .84600	0 .00366	4 11 .66803	0 .00365	3 49 .94812	0 .00463	1 40 .23823	0 .00146
B. 1756	5 0 .98772	0 .00513	5 1 .97407	0 .00512	4 35 .91256	0 .00648	2 0 .27489	0 .00204
1757	5 51 .26682	0 .00684	5 52 .41794	0 .00682	5 22 .00292	0 .00864	2 20 .36645	0 .00272
1758	6 41 .40854	0 .00879	6 42 .72398	0 .00877	6 7 .96736	0 .01111	2 40 .40312	0 .00349
1759	7 31 .55027	0 .01099	7 33 .03002	0 .01096	6 53 .93180	0 .01389	3 0 .43979	0 .00437
B. 1760	8 21 .69199	0 .01344	8 23 .33606	0 .01340	7 39 .89624	0 .01698	3 20 .47645	0 .00534
1761	9 11 .97109	0 .01612	9 13 .77993	0 .01608	8 25 .98660	0 .02037	3 40 .56801	0 .00640
1762	10 2 .11282	0 .01906	10 4 .08597	0 .01900	9 11 .95104	0 .02407	4 0 .60468	0 .00757
1763	10 52 .25454	0 .02223	10 54 .39201	0 .02217	9 57 .91548	0 .02809	4 20 .64135	0 .00883
B. 1764	11 42 .39626	0 .02565	11 44 .69805	0 .02558	10 43 .87992	0 .03241	4 40 .67801	0 .01019
1765	12 32 .67536	0 .02932	12 35 .14192	0 .02923	11 29 .97028	0 .03704	5 0 .76957	0 .01164
1766	13 22 .81709	0 .03322	13 25 .44796	0 .03313	12 15 .93472	0 .04197	5 20 .80624	0 .01319
1767	14 12 .95881	0 .03738	14 15 .75400	0 .03727	13 1 .89916	0 .04722	5 40 .84291	0 .01484
B. 1768	15 3 .10054	0 .04177	15 6 .06004	0 .04165	13 47 .86360	0 .05278	6 0 .87957	0 .01659
1779	15 53 .37964	0 .04642	15 56 .50390	0 .04628	14 33 .95396	0 .05864	6 20 .97113	0 .01843
1770	16 43 .52136	0 .05130	16 46 .80995	0 .05115	15 19 .91840	0 .06482	6 41 .00780	0 .02037
1771	17 33 .66308	0 .05643	17 37 .11599	0 .05627	16 5 .88284	0 .07130	7 1 .04447	0 .02241
B. 1772	18 23 .80481	0 .06181	18 27 .42203	0 .06163	16 51 .84728	0 .07809	7 21 .08113	0 .02455
1773	19 14 .08391	0 .06742	19 17 .86589	0 .06723	17 37 .93764	0 .08519	7 41 .17269	0 .02678
1774	20 4 .22563	0 .07329	20 8 .17193	0 .07308	18 23 .90208	0 .09259	8 1 .20936	0 .02911
1775	20 54 .36736	0 .07940	20 58 .47798	0 .07917	19 9 .86652	0 .10031	8 21 .24603	0 .03153
B. 1776	21 44 .50908	0 .08575	21 48 .78402	0 .08550	19 55 .83096	0 .10833	8 41 .28269	0 .03405
1777	22 34 .78818	0 .09234	22 39 .22788	0 .09208	20 41 .92132	0 .11667	9 1 .37425	0 .03667
1778	23 24 .92990	0 .09918	23 29 .53392	0 .09890	21 27 .88576	0 .12531	9 21 .41092	0 .03939
1779	24 15 .07163	0 .10627	24 19 .83996	0 .10596	22 13 .85020	0 .13426	9 41 .44759	0 .04220
B. 1780	25 5 .21335	0 .11360	25 10 .14601	0 .11327	22 59 .81464	0 .14352	10 1 .48425	0 .04511
1781	25 55 .49245	0 .12117	26 0 .58987	0 .12082	23 45 .90500	0 .15308	10 21 .57581	0 .04812
1782	26 45 .63418	0 .12899	26 50 .89591	0 .12861	24 31 .86944	0 .16296	10 41 .61248	0 .05123
1783	27 35 .77590	0 .13705	27 41 .20195	0 .13665	25 17 .83388	0 .17314	11 1 .64915	0 .05443
B. 1784	28 25 .91762	0 .14536	28 31 .50799	0 .14494	26 3 .79832	0 .18363	11 21 .68581	0 .05773
1785	29 16 .19672	0 .15391	29 21 .95186	0 .15346	26 49 .88868	0 .19444	11 41 .77737	0 .06112
1786	30 6 .33845	0 .16270	30 12 .25790	0 .16223	27 35 .85312	0 .20555	12 1 .81404	0 .06462
1787	30 56 .48017	0 .17174	31 2 .56394	0 .17124	28 21 .81756	0 .21396	12 21 .85071	0 .06821
B. 1788	31 46 .62190	0 .18102	31 52 .86998	0 .18050	29 7 .78200	0 .22869	12 41 .88737	0 .07189
1789	32 36 .90100	0 .19055	32 43 .31585	0 .19000	29 53 .87236	0 .24074	13 1 .97893	0 .07568
1790	33 27 .04272	0 .20032	33 33 .61989	0 .19974	30 39 .83680	0 .25309	13 22 .01560	0 .07956
1791	34 17 .18444	0 .21034	34 23 .92593	0 .20973	31 25 .80124	0 .26574	13 42 .05227	0 .08353
B. 1792	35 7 .32617	0 .22060	35 14 .23197	0 .21996	32 11 .76568	0 .27871	14 2 .08893	0 .08761
1793	35 57 .60527	0 .23110	36 4 .67584	0 .23044	32 57 .85604	0 .29198	14 22 .18049	0 .09178
1794	36 47 .74699	0 .24185	36 54 .98188	0 .24115	33 43 .82048	0 .30556	14 42 .21716	0 .09605
1795	37 37 .83872	0 .25285	37 45 .28792	0 .25211	34 29 .78492	0 .31945	15 2 .25383	0 .10042
B. 1796	38 28 .03044	0 .26408	38 35 .59396	0 .26332	35 15 .74936	0 .33364	15 22 .29049	0 .10488
1797	39 18 .30954	0 .27557	39 26 .03783	0 .27477	36 1 .83972	0 .34815	15 42 .38205	0 .10944
1798	40 8 .45126	0 .28729	40 16 .34387	0 .28646	36 47 .80416	0 .36296	16 2 .41872	0 .11410
1799	40 58 .59299	0 .29926	41 6 .64991	0 .29840	37 33 .76860	0 .37809	16 22 .45539	0 .11885
C. 1800	41 48 .73471	0 .31148	41 56 .95595	0 .31058	38 19 .73304	0 .39352	16 42 .49205	0 .12370
1801	42 38 .87644	0 .32392	42 47 .26199	0 .32298	39 5 .69747	0 .40924	17 2 .52872	0 .12864
1802	43 29 .01816	0 .33662	43 37 .56803	0 .33565	39 51 .66191	0 .42529	17 22 .56539	0 .13369
1803	44 19 .15988	0 .34957	44 27 .87407	0 .34856	40 37 .62635	0 .44165	17 42 .60205	0 .13883
B. 1804	45 9 .30161	0 .36276	45 18 .18011	0 .36171	41 23 .59079	0 .45832	18 2 .63872	0 .14407
1805	45 59 .58071	0 .37620	46 8 .62398	0 .37511	42 9 .63115	0 .47529	18 22 .73028	0 .14940
1806	46 49 .72243	0 .38988	46 58 .93002	0 .38875	42 55 .64559	0 .49257	18 42 .76695	0 .15484
1807	47 39 .86416	0 .40380	47 49 .23606	0 .40264	43 41 .61003	0 .51017	19 2 .80361	0 .16037
B. 1808	48 30 .00588	0 .41797	48 39 .54210	0 .41676	44 27 .57446	0 .52806	19 22 .84028	0 .16599
1809	49 20 .28498	0 .43239	49 29 .98597	0 .43114	45 13 .66483	0 .54628	19 42 .93184	0 .17172
1810	50 10 .42670	0 .44705	50 20 .29201	0 .44575	45 59 .62927	0 .56480	20 2 .96851	0 .17754
1811	51 0 .56843	0 .46195	51 10 .59805	0 .46061	46 45 .59371	0 .58362	20 23 .00517	0 .18346
B. 1812	51 50 .71015	0 .47709	52 0 .90409	0 .47571	47 31 .55814	0 .60276	20 43 .04184	0 .18947
1813	52 40 .98925	0 .49248	52 51 .34796	0 .49106	48 17 .64851	0 .62220	21 3 .13340	0 .19559
1814	53 31 .13097	0 .50812	53 41 .65400	0 .50665	49 3 .61295	0 .64196	21 23 .17007	0 .20180
1815	54 21 .27270	0 .52400	54 31 .96004	0 .52248	49 49 .57739	0 .66202	21 43 .20673	0 .20810
B. 1816	55 11 .41442	0 .54012	55 22 .26608	0 .53856	50 35 .54183	0 .68239	22 3 .24340	0 .21451
1817	56 1 .69352	0 .55649	56 12 .70995	0 .55488	51 21 .65219	0 .70307	22 23 .33496	0 .22101

TABLE 2.

The Amount of the General and Luni-Solar Precessions of the Equinoctial Points, &c. concluded.

Years of the Christian Era.	General Precessions.		Luni-Solar Precession.		Constant part of Prec. in R. A.		Variable part of the Prec. in R. A. and in Declin.	
	Precession.	Variation.	Precession.	Variation.	Precession.	Variation.	Precession.	Variation.
		+		−		+		−
1818	56′ 51″. 83524	0″. 57310	57′ 3″. 01599	0″. 57144	52′ 7″. 59663	0″. 72407	22′ 43″. 37163	0″. 22760
1819	57 41 . 97697	0 . 58996	57 53 . 32203	0 . 58825	52 53 . 56107	0 . 74536	23 3 . 40829	0 . 23430
B. 1820	58 32 . 11869	0 . 60706	58 43 . 62807	0 . 60530	53 39 . 52551	0 . 76697	23 23 . 44496	0 . 24109
1821	59 22 . 39779	0 . 62440	59 34 . 07194	0 . 62260	54 25 . 61587	0 . 78888	23 43 . 53652	0 . 24798
1822	60 12 . 53952	0 . 64199	60 24 . 37798	0 . 64013	55 11 . 58031	0 . 81110	24 3 . 57319	0 . 25496
1823	61 2 . 68124	0 . 65983	61 14 . 68402	0 . 65792	55 57 . 54475	0 . 83363	24 23 . 60985	0 . 26205
B. 1824	61 52 . 82296	0 . 67791	62 4 . 99006	0 . 67594	56 43 . 50918	0 . 85647	24 43 . 64652	0 . 26922
1825	62 43 . 10206	0 . 69623	62 55 . 43393	0 . 69421	57 29 . 59955	0 . 87962	25 3 . 73808	0 . 27650
1826	63 33 . 24379	0 . 71479	63 45 . 73997	0 . 71272	58 15 . 56399	0 . 90307	25 23 . 77475	0 . 28387
1827	64 23 . 38551	0 . 73361	64 36 . 04601	0 . 73148	59 1 . 52843	0 . 92684	25 43 . 81141	0 . 29135
B. 1828	65 13 . 52724	0 . 75266	65 26 . 35205	0 . 75048	59 47 . 49287	0 . 95091	26 3 . 84808	0 . 29891
1829	66 3 . 80634	0 . 77196	66 16 . 79591	0 . 76972	60 33 . 58323	0 . 97529	26 23 . 93964	0 . 30658
1830	66 53 . 94806	0 . 79150	67 7 . 10196	0 . 78921	61 19 . 54767	0 . 99999	26 43 . 97631	0 . 31434
1831	67 44 . 08978	0 . 81129	67 57 . 40800	0 . 80894	62 5 . 51211	1 . 02499	27 4 . 01297	0 . 32220
B. 1832	68 34 . 23151	0 . 83132	68 47 . 71404	0 . 82892	62 51 . 47655	1 . 05030	27 24 . 04964	0 . 33015
1833	69 24 . 51061	0 . 85160	69 38 . 15790	0 . 84913	63 37 . 56691	1 . 07592	27 44 . 14120	0 . 33821
1834	70 14 . 65233	0 . 87212	70 28 . 46394	0 . 86959	64 23 . 53135	1 . 10184	28 4 . 17787	0 . 34636
1835	71 4 . 79406	0 . 89289	71 18 . 76999	0 . 89030	65 9 . 49579	1 . 12808	28 24 . 21453	0 . 35460
B. 1836	71 54 . 93578	0 . 91390	72 9 . 07603	0 . 91125	65 55 . 46023	1 . 15462	28 44 . 25120	0 . 36295
1837	72 45 . 21488	0 . 93515	72 59 . 51989	0 . 93244	66 41 . 55059	1 . 18148	29 4 . 34276	0 . 37139
1838	73 35 . 35660	0 . 95665	73 49 . 82593	0 . 95388	67 27 . 51503	1 . 20864	29 24 . 37943	0 . 37992
1839	74 25 . 49833	0 . 97839	74 40 . 13197	0 . 97556	68 13 . 47947	1 . 23611	29 44 . 41609	0 . 38856
B. 1840	75 15 . 64005	1 . 00038	75 30 . 43802	0 . 99748	68 59 . 44391	1 . 26389	30 4 . 45276	0 . 39729
1841	76 5 . 91915	1 . 02261	76 20 . 88188	1 . 01965	69 45 . 53427	1 . 29197	30 24 . 54432	0 . 40612
1842	76 56 . 06088	1 . 04508	77 11 . 18792	1 . 04206	70 31 . 49871	1 . 32037	30 44 . 58099	0 . 41505
1843	77 46 . 20260	1 . 06780	78 1 . 49396	1 . 06471	71 17 . 46315	1 . 34907	31 4 . 61765	0 . 42407
B. 1844	78 36 . 34432	1 . 09077	78 51 . 80000	1 . 08761	72 3 . 42759	1 . 37808	31 24 . 65432	0 . 43319
1845	79 26 . 62342	1 . 11397	79 42 . 24387	1 . 11075	72 49 . 51795	1 . 40741	31 44 . 74588	0 . 44241
1846	80 16 . 76515	1 . 13743	80 32 . 54991	1 . 13413	73 35 . 48239	1 . 43704	32 4 . 78255	0 . 45172
1847	81 6 . 90687	1 . 16112	81 22 . 85595	1 . 15776	74 21 . 44683	1 . 46698	32 24 . 81921	0 . 46113
B. 1848	81 57 . 04860	1 . 18506	82 13 . 16199	1 . 18163	75 7 . 41127	1 . 49722	32 44 . 85588	0 . 47064
1849	82 47 . 32770	1 . 20925	83 3 . 60586	1 . 20575	75 53 . 50163	1 . 52778	33 4 . 94744	0 . 48024
1850	83 37 . 46942	1 . 23368	83 53 . 91190	1 . 23011	76 39 . 46607	1 . 55864	33 24 . 98411	0 . 48995
1851	84 27 . 61114	1 . 25835	84 44 . 21794	1 . 25471	77 25 . 43051	1 . 58981	33 45 . 02077	0 . 49975
B. 1852	85 17 . 75287	1 . 28327	85 34 . 52398	1 . 27956	78 11 . 39495	1 . 62129	34 5 . 05744	0 . 50964
1853	86 8 . 03197	1 . 30844	86 24 . 96785	1 . 30465	78 57 . 48531	1 . 65308	34 25 . 14900	0 . 51963
1854	86 58 . 17369	1 . 33384	87 15 . 27389	1 . 32998	79 43 . 44975	1 . 68518	34 45 . 18567	0 . 52972
1855	87 48 . 31542	1 . 35949	88 5 . 57993	1 . 35556	80 29 . 41419	1 . 71759	35 5 . 22233	0 . 53991
B. 1856	88 38 . 45714	1 . 38539	88 55 . 88597	1 . 38138	81 15 . 37863	1 . 75030	35 25 . 25900	0 . 55020
1857	89 28 . 73624	1 . 41153	89 46 . 32984	1 . 40744	82 1 . 46899	1 . 78333	35 45 . 35056	0 . 56058
1858	90 18 . 87796	1 . 43791	90 36 . 63588	1 . 43375	82 47 . 43343	1 . 81666	36 5 . 38723	0 . 57105
1859	91 9 . 01969	1 . 46454	91 26 . 94192	1 . 46030	83 33 . 39787	1 . 85030	36 25 . 42389	0 . 58163
B. 1860	91 59 . 16141	1 . 49141	92 17 . 24796	1 . 48709	84 19 . 36230	1 . 88426	36 45 . 46056	0 . 59230
1861	92 49 . 44051	1 . 51853	93 7 . 69183	1 . 51413	85 5 . 45267	1 . 91851	37 5 . 55212	0 . 60307
1862	93 39 . 58224	1 . 54589	93 57 . 99787	1 . 54141	85 51 . 41711	1 . 95308	37 25 . 58879	0 . 61394
1863	94 29 . 72396	1 . 57350	94 48 . 30391	1 . 56894	86 37 . 38155	1 . 98796	37 45 . 62545	0 . 62490
B. 1864	95 19 . 86568	1 . 60135	95 38 . 60995	1 . 59671	87 23 . 34599	2 . 02314	38 5 . 66212	0 . 63596
1865	96 10 . 14478	1 . 62944	96 29 . 05382	1 . 62472	88 9 . 43635	2 . 05864	38 25 . 75368	0 . 64712
1866	97 0 . 28651	1 . 65778	97 19 . 35986	1 . 65298	88 55 . 40079	2 . 09444	38 45 . 79035	0 . 65837
1867	97 50 . 42823	1 . 68636	98 9 . 66590	1 . 68148	89 41 . 36523	2 . 13055	39 5 . 82701	0 . 66972
B. 1868	98 40 . 56996	1 . 71519	98 59 . 97194	1 . 71022	90 27 . 32967	2 . 16697	39 25 . 86368	0 . 68117
1869	99 30 . 84906	1 . 74426	99 50 . 41580	1 . 73921	91 13 . 42003	2 . 20370	39 45 . 95524	0 . 69271
1870	100 20 . 99078	1 . 77358	100 40 . 72185	1 . 76844	91 59 . 38447	2 . 24074	40 5 . 99191	0 . 70436
1871	101 11 . 13250	1 . 80314	101 31 . 02789	1 . 79791	92 45 . 34891	2 . 27809	40 26 . 02857	0 . 71610
B. 1872	102 1 . 27423	1 . 83294	102 21 . 33393	1 . 82763	93 31 . 31335	2 . 31574	40 46 . 06524	0 . 72793
1873	102 51 . 55333	1 . 86299	103 11 . 77779	1 . 85759	94 17 . 40371	2 . 35370	41 6 . 15680	0 . 73986
1874	103 41 . 69505	1 . 89328	104 2 . 08383	1 . 88780	95 3 . 36815	2 . 39198	41 26 . 19347	0 . 75189
1875	104 31 . 83678	1 . 92382	104 52 . 38988	1 . 91825	95 49 . 33259	2 . 43056	41 46 . 23013	0 . 76403
B. 1876	105 21 . 97850	1 . 95460	105 42 . 69592	1 . 94894	96 35 . 29703	2 . 46945	42 6 . 26680	0 . 77625
1877	106 12 . 25760	1 . 98563	106 33 . 13978	1 . 97987	97 21 . 38739	2 . 50864	42 26 . 35836	0 . 78858
1878	107 2 . 39932	2 . 01690	107 23 . 44582	2 . 01105	98 7 . 35183	2 . 54815	42 46 . 39503	0 . 80099
1879	107 52 . 54105	2 . 04841	108 13 . 75186	2 . 04248	98 53 . 31627	2 . 58796	43 6 . 43169	0 . 81351
B. 1880	108 42 . 68277	2 . 08017	109 4 . 05791	2 . 07414	99 39 . 28071	2 . 62809	43 26 . 46836	0 . 82612

TABLE 3. The General and Luni-Solar Precession of the Equinoctial Points: with the Constant and Variable Quantities of the Precession in Right Ascension and Declination respectively, for Months and Days.

Common Years *.	Bissextile Years.	January.				February.				March.			
		General.	Luni-Solar.	R. A.	Declin.	General.	Luni-Sol.	R. A.	Declin.	General.	Luni-Solar.	R. A.	Declin.
1	0	0″.137	0″.138	0″.126	0″.055	4″.397	4″.412	4″.031	1″.757	8″.245	8.272	7″.558	3″.294
2	1	0.275	0.276	0.252	0.110	4.535	4.549	4.157	1.811	8.382	8.409	7.684	3.348
3	2	0.412	0.414	0.378	0.165	4.672	4.687	4.283	1.866	8.519	8.547	7.810	3.403
4	3	0.550	0.551	0.504	0.220	4.809	4.825	4.409	1.921	8.657	8.685	7.936	3.458
5	4	0.687	0.689	0.630	0.274	4.947	4.963	4.533	1.976	8.794	8.823	8.062	3.513
6	5	0.824	0.827	0.756	0.329	5.084	5.101	4.661	2.031	8.932	8.961	8.188	3.568
7	6	0.962	0.965	0.882	0.384	5.222	5.239	4.787	2.086	9.069	9.099	8.314	3.623
8	7	1.099	1.103	1.008	0.439	5.359	5.377	4.913	2.141	9.206	9.237	8.440	3.678
9	8	1.237	1.241	1.134	0.494	5.496	5.514	5.039	2.196	9.344	9.374	8.566	3.733
10	9	1.374	1.379	1.260	0.549	5.634	5.652	5.165	2.251	9.481	9.512	8.692	3.788
11	10	1.512	1.516	1.386	0.604	5.771	5.790	5.291	2.306	9.619	9.650	8.818	3.843
12	11	1.649	1.654	1.512	0.659	5.909	5.928	5.417	2.360	9.756	9.788	8.944	3.897
13	12	1.766	1.792	1.638	0.714	6.046	6.066	5.543	2.415	9.894	9.926	9.070	3.952
14	13	1.924	1.930	1.764	0.769	6.183	6.204	5.669	2.470	10.031	10.064	9.196	4.007
15	14	2.061	2.068	1.890	0.823	6.321	6.342	5.795	2.525	10.168	10.202	9.322	4.062
16	15	2.199	2.206	2.016	0.878	6.458	6.479	5.921	2.580	10.306	10.340	9.448	4.117
17	16	2.336	2.344	2.141	0.933	6.596	6.617	6.047	2.635	10.443	10.477	9.574	4.172
18	17	2.473	2.481	2.267	0.988	6.733	6.755	6.173	2.690	10.581	10.615	9.700	4.227
19	18	2.611	2.619	2.393	1.043	6.871	6.893	6.299	2.745	10.718	10.753	9.826	4.282
20	19	2.748	2.757	2.519	1.098	7.008	7.031	6.424	2.800	10.855	10.891	9.952	4.337
21	20	2.886	2.895	2.645	1.153	7.145	7.169	6.550	2.854	10.993	11.029	10.078	4.391
22	21	3.023	3.033	2.771	1.208	7.283	7.307	6.676	2.909	11.130	11.167	10.204	4.446
23	22	3.160	3.171	2.897	1.263	7.420	7.444	6.802	2.964	11.268	11.305	10.330	4.501
24	23	3.298	3.309	3.023	1.317	7.558	7.582	6.928	3.019	11.405	11.442	10.456	4.556
25	24	3.435	3.447	3.149	1.372	7.695	7.720	7.054	3.074	11.542	11.580	10.581	4.611
26	25	3.573	3.584	3.275	1.427	7.832	7.858	7.180	3.129	11.680	11.718	10.707	4.666
27	26	3.710	3.722	3.401	1.482	7.970	7.996	7.306	3.184	11.817	11.856	10.833	4.721
28	27	3.847	3.860	3.527	1.537	8.107	8.134	7.432	3.239	11.955	11.994	10.959	4.776
29	28	3.985	3.998	3.653	1.592	8.245	8.272	7.558	3.294	12.092	12.132	11.085	4.831
30	29	4.122	4.136	3.779	1.647					12.229	12.270	11.211	4.885
31	30	4.260	4.274	3.905	1.702					12.367	12.407	11.337	4.940
	31									12.504	12.545	11.463	4.995

* Note.—In the Months of January and February, the Day of the Month must *always* be taken from the *first* Column.

Common Years.	Bissextile Years.	April.				May.				June.			
		General.	Luni-Solar.	R. A.	Declin.	General.	Luni-Sol.	R. A.	Declin.	General.	Luni-Solar.	R. A.	Declin.
1	0	12″.504	12″.545	11″.463	4″.995	16″.627	16″.681	15″.242	6″.642	20″.886	20″.955	19″.147	8″.344
2	1	12.642	12.683	11.589	5.050	16.764	16.819	15.368	6.697	21.024	21.093	19.273	8.399
3	2	12.779	12.821	11.715	5.105	16.901	16.957	15.494	6.752	21.161	21.230	19.399	8.454
4	3	12.827	12.959	11.841	5.160	17.039	17.095	15.620	6.807	21.299	21.368	19.525	8.508
5	4	12.964	13.097	11.967	5.215	17.176	17.233	15.746	6.862	21.436	21.506	19.651	8.563
6	5	13.101	13.235	12.093	5.270	17.314	17.370	15.872	6.917	21.573	21.644	19.777	8.618
7	6	13.329	13.372	12.219	5.325	17.451	17.508	15.998	6.971	21.711	21.782	19.903	8.673
8	7	13.466	13.510	12.345	5.380	17.588	17.646	16.124	7.026	21.848	21.920	20.029	8.728
9	8	13.604	13.648	12.471	5.434	17.726	17.784	16.250	7.081	21.986	22.058	20.155	8.783
10	9	13.741	13.786	12.597	5.489	17.863	17.922	16.376	7.136	22.123	22.195	20.281	8.838
11	10	13.878	13.924	12.723	5.544	18.001	18.060	16.502	7.191	22.260	22.333	20.407	8.893
12	11	14.016	14.062	12.849	5.599	18.138	18.198	16.628	7.246	22.398	22.471	20.533	8.948
13	12	14.153	14.200	12.975	5.654	18.276	18.335	16.754	7.301	22.535	22.609	20.659	9.002
14	13	14.291	14.337	13.101	5.709	18.413	18.473	16.880	7.356	22.673	22.747	20.785	9.057
15	14	14.428	14.475	13.227	5.764	18.550	18.611	17.006	7.411	22.810	22.885	20.911	9.112
16	15	14.565	14.613	13.353	5.819	18.688	18.749	17.132	7.465	22.947	23.023	21.037	9.167
17	16	14.703	14.751	13.479	5.874	18.825	18.887	17.258	7.520	23.085	23.160	21.163	9.222
18	17	14.840	14.889	13.605	5.928	18.963	19.025	17.384	7.575	23.222	23.298	21.289	9.277
19	18	14.978	15.027	13.731	5.983	19.100	19.163	17.510	7.630	23.360	23.436	21.415	9.332
20	19	15.115	15.165	13.857	6.038	19.237	19.300	17.636	7.685	23.497	23.574	21.541	9.387
21	20	15.253	15.302	13.983	6.093	19.375	19.438	17.762	7.740	23.635	23.712	21.667	9.442
22	21	15.390	15.440	14.109	6.148	19.512	19.576	17.888	7.795	23.772	23.850	21.793	9.496
23	22	15.527	15.578	14.235	6.203	19.650	19.714	18.014	7.850	23.909	23.988	21.919	9.551
24	23	15.665	15.716	14.361	6.258	19.787	19.852	18.140	7.905	24.047	24.126	22.045	9.606
25	24	15.802	15.854	14.487	6.313	19.924	19.990	18.266	7.959	24.184	24.263	22.171	9.661
26	25	15.940	15.992	14.613	6.368	20.062	20.128	18.392	8.014	24.322	24.401	22.297	9.716
27	26	16.077	16.130	14.738	6.422	20.199	20.265	18.518	8.069	24.459	24.539	22.423	9.771
28	27	16.214	16.267	14.864	6.477	20.337	20.403	18.644	8.124	24.596	24.677	22.549	9.826
29	28	16.352	16.405	14.990	6.532	20.474	20.541	18.770	8.179	24.734	24.815	22.675	9.881
30	29	16.489	16.543	15.116	6.587	20.612	20.679	18.896	8.234	24.871	24.953	22.801	9.936
31	30	16.627	16.681	15.242	6.642	20.749	20.817	19.021	8.289	25.009	25.091	22.927	9.991
	31					20.886	20.955	19.147	8.344				

TABLE 3.

The GENERAL and LUNI-SOLAR PRECESSIONS, &c., for Months and Days, concluded.

Common Years.	Bissextile Years.	July.				August.				September.			
		General.	Luni-Solar.	R. A.	Declin.	General.	Luni-Sol.	R. A.	Declin.	General.	Luni-Solar.	R. A.	Declin.
1	0	25″. 009	25″. 091	22″. 927	9″. 991	29″. 268	29″. 364	26″. 832	11″. 692	33″. 528	33″. 638	30″. 737	13″. 394
2	1	25 . 146	25 . 228	23 . 053	10 . 045	29 . 406	29 . 502	26 . 958	11 . 747	33 . 665	33 . 776	30 . 863	13 . 449
3	2	25 . 283	25 . 366	23 . 178	10 . 100	29 . 543	29 . 640	27 . 084	11 . 802	33 . 803	33 . 914	30 . 989	13 . 504
4	3	25 . 420	25 . 504	23 . 304	10 . 155	29 . 681	29 . 778	27 . 210	11 . 857	33 . 940	34 . 051	31 . 115	13 . 559
5	4	25 . 558	25 . 642	23 . 430	10 . 210	29 . 818	29 . 916	27 . 335	11 . 912	34 . 078	34 . 189	31 . 241	13 . 613
6	5	25 . 696	25 . 780	23 . 556	10 . 265	29 . 955	30 . 053	27 . 461	11 . 967	34 . 215	34 . 327	31 . 367	13 . 668
7	6	25 . 833	25 . 918	23 . 682	10 . 320	30 . 093	30 . 191	27 . 587	12 . 022	34 . 353	34 . 465	31 . 493	13 . 723
8	7	25 . 970	26 . 056	23 . 808	10 . 375	30 . 230	30 . 329	27 . 713	12 . 076	34 . 490	34 . 603	31 . 618	13 . 778
9	8	26 . 108	26 . 193	23 . 934	10 . 430	30 . 368	30 . 467	27 . 839	12 . 131	34 . 627	34 . 741	31 . 744	13 . 833
10	9	26 . 245	26 . 331	24 . 060	10 . 485	30 . 505	30 . 605	27 . 965	12 . 186	34 . 765	34 . 879	31 . 870	13 . 888
11	10	26 . 383	26 . 469	24 . 186	10 . 539	30 . 642	30 . 743	28 . 091	12 . 241	34 . 902	35 . 016	31 . 996	13 . 943
12	11	26 . 520	26 . 607	24 . 312	10 . 594	30 . 780	30 . 881	28 . 217	12 . 296	35 . 040	35 . 154	32 . 122	13 . 998
13	12	26 . 658	26 . 745	24 . 438	10 . 649	30 . 917	31 . 019	28 . 343	12 . 351	35 . 177	35 . 292	32 . 248	14 . 053
14	13	26 . 795	26 . 883	24 . 564	10 . 704	31 . 055	31 . 156	28 . 469	12 . 406	35 . 314	35 . 430	32 . 374	14 . 108
15	14	26 . 932	27 . 021	24 . 690	10 . 759	31 . 192	31 . 294	28 . 595	12 . 461	35 . 452	35 . 568	32 . 500	14 . 162
16	15	27 . 070	27 . 158	24 . 816	10 . 814	31 . 329	31 . 432	28 . 721	12 . 516	35 . 589	35 . 706	32 . 626	14 . 217
17	16	27 . 207	27 . 296	24 . 942	10 . 869	31 . 467	31 . 570	28 . 847	12 . 570	35 . 727	35 . 844	32 . 752	14 . 272
18	17	27 . 345	27 . 434	25 . 068	10 . 924	31 . 604	31 . 708	28 . 973	12 . 625	35 . 864	35 . 981	32 . 878	14 . 327
19	18	27 . 482	27 . 572	25 . 194	10 . 979	31 . 742	31 . 846	29 . 099	12 . 680	36 . 001	36 . 119	33 . 004	14 . 382
20	19	27 . 619	27 . 710	25 . 320	11 . 033	31 . 879	31 . 984	29 . 225	12 . 735	36 . 139	36 . 257	33 . 130	14 . 437
21	20	27 . 757	27 . 848	25 . 446	11 . 088	32 . 017	32 . 121	29 . 351	12 . 790	36 . 276	36 . 395	33 . 256	14 . 492
22	21	27 . 894	27 . 986	25 . 572	11 . 143	32 . 154	32 . 259	29 . 477	12 . 845	36 . 414	36 . 533	33 . 382	14 . 547
23	22	28 . 032	28 . 123	25 . 698	11 . 198	32 . 291	32 . 397	29 . 603	12 . 900	36 . 551	36 . 671	33 . 508	14 . 602
24	23	28 . 169	28 . 261	25 . 824	11 . 253	32 . 429	32 . 535	29 . 729	12 . 955	36 . 688	36 . 809	33 . 634	14 . 656
25	24	28 . 306	28 . 399	25 . 950	11 . 308	32 . 566	32 . 673	29 . 455	13 . 010	36 . 826	36 . 946	33 . 760	14 . 711
26	25	28 . 444	28 . 537	26 . 076	11 . 363	32 . 704	32 . 811	29 . 981	13 . 065	36 . 963	37 . 084	33 . 886	14 . 766
27	26	28 . 581	28 . 675	26 . 202	11 . 418	32 . 841	32 . 949	30 . 107	13 . 119	37 . 101	37 . 222	34 . 012	14 . 821
28	27	28 . 719	28 . 813	26 . 328	11 . 473	32 . 978	33 . 086	30 . 233	13 . 174	37 . 238	37 . 360	34 . 138	14 . 876
29	28	28 . 856	28 . 951	26 . 454	11 . 528	33 . 116	33 . 224	30 . 359	13 . 229	37 . 376	37 . 498	34 . 264	14 . 931
30	29	28 . 994	29 . 088	26 . 580	11 . 582	33 . 253	33 . 362	30 . 485	13 . 284	37 . 513	37 . 636	34 . 390	14 . 986
31	30	29 . 131	29 . 226	26 . 706	11 . 637	33 . 391	33 . 500	30 . 611	13 . 339	37 . 650	37 . 774	34 . 516	15 . 041
	31	29 . 268	29 . 364	26 . 832	11 . 692	33 . 528	33 . 638	30 . 737	13 . 394				

Common Years.	Bissextile Years.	October.				November.				December.			
		General.	Luni-Solar.	R. A.	Declin.	General.	Luni-Sol.	R. A.	Declin.	General.	Luni-Solar.	R. A.	Declin.
1	0	37″. 650	37″. 774	34″. 516	15″. 041	41″. 910	42″. 047	38″. 421	16″. 742	46″. 032	46″. 183	42″. 200	18″. 389
2	1	37 . 788	37 . 912	34 . 642	15 . 096	42 . 048	42 . 185	38 . 547	16 . 797	46 . 170	46 . 321	42 . 326	18 . 444
3	2	37 . 925	38 . 049	34 . 768	15 . 151	42 . 185	42 . 323	38 . 673	16 . 852	46 . 307	46 . 459	42 . 452	18 . 499
4	3	38 . 063	38 . 187	34 . 894	15 . 205	42 . 322	42 . 461	38 . 799	16 . 907	46 . 445	46 . 597	42 . 578	18 . 554
5	4	38 . 200	38 . 325	35 . 020	15 . 260	42 . 460	42 . 599	38 . 925	16 . 962	46 . 582	46 . 735	42 . 704	18 . 609
6	5	38 . 337	38 . 463	35 . 146	15 . 315	42 . 597	42 . 737	39 . 051	17 . 017	46 . 719	46 . 872	42 . 830	18 . 664
7	6	38 . 475	38 . 601	35 . 272	15 . 370	42 . 735	42 . 874	39 . 177	17 . 072	46 . 857	47 . 010	42 . 956	18 . 719
8	7	38 . 612	38 . 739	35 . 398	15 . 425	42 . 872	43 . 012	39 . 303	17 . 127	46 . 994	47 . 148	43 . 082	18 . 773
9	8	39 . 750	38 . 877	35 . 524	15 . 480	43 . 009	43 . 150	39 . 429	17 . 182	47 . 132	47 . 286	43 . 208	18 . 828
10	9	38 . 887	39 . 014	35 . 650	15 . 535	43 . 147	43 . 288	39 . 555	17 . 236	47 . 269	47 . 424	43 . 334	18 . 883
11	10	39 . 024	39 . 152	35 . 775	15 . 590	43 . 284	43 . 426	39 . 681	17 . 291	47 . 406	47 . 562	43 . 460	18 . 938
12	11	39 . 162	39 . 290	35 . 901	15 . 645	43 . 422	43 . 564	39 . 807	17 . 346	47 . 544	47 . 700	43 . 586	18 . 993
13	12	39 . 299	39 . 428	36 . 027	15 . 699	43 . 559	43 . 702	39 . 932	17 . 401	47 . 631	47 . 837	43 . 712	19 . 048
14	13	39 . 437	39 . 566	36 . 153	15 . 754	43 . 696	43 . 839	40 . 058	17 . 456	47 . 819	47 . 975	43 . 838	19 . 103
15	14	39 . 574	39 . 704	36 . 279	15 . 809	43 . 834	43 . 977	40 . 184	17 . 511	47 . 956	48 . 113	43 . 964	19 . 158
16	15	39 . 711	39 . 842	36 . 405	15 . 864	43 . 971	44 . 115	40 . 310	17 . 566	48 . 094	48 . 251	44 . 090	19 . 213
17	16	39 . 849	39 . 979	36 . 531	15 . 919	44 . 109	44 . 253	40 . 436	17 . 621	48 . 231	48 . 389	44 . 215	19 . 267
18	17	39 . 986	40 . 117	36 . 657	15 . 974	44 . 246	44 . 391	40 . 562	17 . 676	48 . 368	48 . 527	44 . 341	19 . 322
19	18	40 . 124	40 . 255	36 . 783	16 . 029	44 . 383	44 . 529	40 . 688	17 . 730	48 . 506	48 . 665	44 . 467	19 . 377
20	19	40 . 261	40 . 393	36 . 909	16 . 084	44 . 520	44 . 667	40 . 814	17 . 785	48 . 643	48 . 802	44 . 593	19 . 432
21	20	40 . 399	40 . 531	37 . 035	16 . 139	44 . 658	44 . 805	40 . 940	17 . 840	48 . 781	48 . 940	44 . 719	19 . 487
22	21	40 . 536	40 . 669	37 161	16 . 193	44 . 796	44 . 942	41 . 066	17 . 895	48 . 918	49 . 078	44 . 845	19 . 542
23	22	40 . 673	40 . 807	37 . 287	16 . 248	44 . 933	45 . 080	41 . 192	17 . 950	49 . 055	49 . 216	44 . 971	19 . 597
24	23	40 . 811	40 . 944	37 . 413	16 . 303	45 . 070	45 . 218	41 . 318	18 . 005	49 . 193	49 . 354	45 . 097	19 . 652
25	24	40 . 948	41 . 082	37 . 539	16 . 358	45 . 208	45 . 356	41 . 444	18 . 060	49 . 330	49 . 492	45 . 223	19 . 707
26	25	41 . 086	41 . 220	37 . 665	16 . 413	45 . 345	45 . 494	41 . 570	18 . 115	49 . 467	49 . 630	45 . 349	19 . 761
27	26	41 . 223	41 . 358	37 . 791	16 . 468	45 . 483	45 . 632	41 . 696	18 . 170	49 . 605	49 . 767	45 . 475	19 . 816
28	27	41 . 360	41 . 496	37 . 917	16 . 523	45 . 620	45 . 770	41 . 822	18 . 224	49 . 742	49 . 905	45 . 601	19 . 871
29	28	41 . 498	41 . 634	38 . 043	16 . 578	45 . 758	45 . 907	41 . 948	18 . 279	49 . 880	50 . 043	45 . 727	19 . 926
30	29	41 . 635	41 . 772	38 . 169	16 . 633	45 . 895	46 . 045	42 . 074	18 . 334	50 . 017	50 . 181	45 . 853	19 . 981
31	30	41 . 773	41 . 909	38 . 295	16 . 687	16 . 032	46 . 183	42 . 200	18 . 389	50 . 156	50 . 319	45 . 979	20 . 036
	31	41 . 910	42 . 047	38 . 421	16 . 742					50 . 293	50 . 457	46 . 105	20 . 091

TABLE 1. The AUGMENTATION of the MOON'S SEMI-DIAMETER determined from the ALTITUDE of the NONAGESIMAL DEGREE, and the ☾'s apparent DISTANCE therefrom.

PART I.

Args. = Alt. Nonag. + ap. dis. ☽ fr. nonag.
Alt. Nonag. — ap. dis. ☽ fr. nonag.

	O VI + —	I VII + —	II VIII + —	
0°	0″.00	4″.10	7″.10	30°
1	0 .14	4 .22	7 .16	29
2	0 .29	4 .34	7 .23	28
3	0 .43	4 .46	7 .30	27
4	0 .57	4 .58	7 .36	26
5	0 .72	4 .70	7 .42	25
6	0 .86	4 .82	7 .48	24
7	1 .00	4 .93	7 .54	23
8	1 .14	5 .04	7 .59	22
9	1 .28	5 .16	7 .65	21
10	1 .42	5 .27	7 .70	20
11	1 .56	5 .38	7 .74	19
12	1 .70	5 .49	7 .79	18
13	1 .84	5 .59	7 .83	17
14	1 .98	5 .70	7 .87	16
15	2 .12	5 .80	7 .92	15
16	2 .25	5 .89	7 .95	14
17	2 .39	5 .99	7 .98	13
18	2 .53	6 .09	8 .02	12
19	2 .67	6 .18	8 .04	11
20	2 .80	6 .27	8 .06	10
21	2 .94	6 .36	8 .09	9
22	3 .07	6 .45	8 .11	8
23	3 .20	6 .54	8 .13	7
24	3 .33	6 .63	8 .15	6
25	3 .46	6 .71	8 .16	5
26	3 .59	6 .79	8 .17	4
27	3 .72	6 .87	8 .18	3
28	3 .85	6 .95	8 .19	2
29	3 .97	7 .02	8 .19	1
30	4 .10	7 .10	8 .19	0
	+ — XI V	+ — X IV	+ — IX III	

PART II.

Arg. = Aggreg. of Numbers fr. Part I.	Cor. +
1″.0	0″.00
2 .0	0 .00
3 .0	0 .01
4 .0	0 .02
5 .0	0 .03
6 .0	0 .04
7 .0	0 .05
8 .0	0 .06
8 .5	0 .07
9 .0	0 .08
9 .5	0 .09
10 .0	0 .10
10 .5	0 .11
11 .0	0 .12
11 .5	0 .13
12 .0	0 .14
12 .3	0 .15
12 .7	0 .16
13 .0	0 .17
13 .3	0 .18
13 .7	0 .19
14 .0	0 .20
14 .3	0 .21
14 .7	0 .22
15 .0	0 .23
15 .3	0 .24
15 .7	0 .25
16 .0	0 .26
16 .3	0 .27
16 .7	0 .28
17 .0	0 .29

PART III.

Arg. = Parallax of the Moon in Lat.

Arg. = True Lat. of the Moon.	0′	10′ —	20′ —	30′ —	40′ —	50′ —	60′ —
SouthLat.							
6° 0	0″.00	0″.29	0″.59	0″.90	1″.22	1″.54	1″.88
5 40	0 .00	0 .27	0 .56	0 .85	1 .16	1 .46	1 .78
5 20	0 .00	0 .26	0 .53	0 .80	1 .09	1 .38	1 .69
5 0	0 .00	0 .24	0 .49	0 .76	1 .02	1 .30	1 .59
4 40	0 .00	0 .23	0 .46	0 .71	0 .96	1 .22	1 .49
4 20	0 .00	0 .21	0 .43	0 .66	0 .90	1 .14	1 .40
4 0	0 .00	0 .19	0 .40	0 .61	0 .83	1 .06	1 .30
3 40	0 .00	0 .18	0 .37	0 .56	0 .77	0 .98	1 .20
3 20	0 .00	0 .16	0 .33	0 .52	0 .70	0 .90	1 .11
3 0	0 .00	0 .14	0 .30	0 .47	0 .64	0 .82	1 .01
2 40	0 .00	0 .13	0 .27	0 .42	0 .58	0 .74	0 .91
2 20	0 .00	0 .11	0 .24	0 .37	0 .51	0 .66	0 .82
2 0	0 .00	0 .10	0 .21	0 .33	0 .45	0 .58	0 .72
1 40	0 .00	0 .08	0 .17	0 .28	0 .39	0 .50	0 .63
1 20	0 .00	0 .06	0 .14	0 .23	0 .32	0 .42	0 .53
1 0	0 .00	0 .05	0 .11	0 .18	0 .26	0 .34	0 .43
0 50	0 .00	0 .04	0 .09	0 .16	0 .23	0 .30	0 .39
0 40	0 .00	0 .03	0 .08	0 .13	0 .19	0 .26	0 .34
0 30	0 .09	0 .02	0 .06	0 .11	0 .16	0 .22	0 .29
0 20	0 .00	0 .02	0 .05	0 .08	0 .13	0 .18	0 .24
0 10	0 .00	0 .01	0 .03	0 .06	0 .10	0 .14	0 .19
0 00	0 .00	+	0 .02	0 .04	0 .06	0 .10	0 .14
N. Lat.							
0 10	0 .00	0 .01	+	0 .01 +	0 .03	0 .06	0 .10
0 20	0 .00	0 .02	0 .02	0 .01	+	0 .02 +	0 .05
0 30	0 .00	0 .02	0 .03	0 .04	0 .03	0 .02	+
0 40	0 .00	0 .03	0 .05	0 .06	0 .06	0 .06	0 .05
0 50	0 .00	0 .04	0 .07	0 .09	0 .10	0 .10	0 .10
1 00	0 .00	0 .05	0 .08	0 .11	0 .13	0 .14	0 .15
1 20	0 .00	0 .06	0 .11	0 .15	0 .19	0 .22	0 .24
1 40	0 .00	0 .08	0 .14	0 .20	0 .26	0 .30	0 .34
2 00	0 .00	0 .10	0 .18	0 .25	0 .32	0 .38	0 .44
2 20	0 .00	0 .11	0 .21	0 .30	0 .39	0 .46	0 .53
2 40	0 .00	0 .13	0 .24	0 .35	0 .45	0 .54	0 .63
3 0	0 .00	0 .14	0 .27	0 .39	0 .51	0 .62	0 .72
3 20	0 .00	0 .16	0 .30	0 .44	0 .58	0 .70	0 .82
3 40	0 .00	0 .18	0 .34	0 .49	0 .64	0 .78	0 .92
4 0	0 .00	0 .19	0 .37	0 .54	0 .70	0 .86	1 .01
4 20	0 .00	0 .21	0 .40	0 .59	0 .77	0 .94	1 .11
4 40	0 .00	0 .23	0 .43	0 .64	0 .83	1 .02	1 .20
5 0	0 .00	0 .24	0 .46	0 .68	0 .90	1 .10	1 .30
5 20	0 .00	0 .26	0 .50	0 .73	0 .96	1 .18	1 .40
5 40	0 .00	0 .27	0 .53	0 .78	1 .02	1 .26	1 .49
6 0	0 .00	0 .29	0 .56	0 .83	1 .09	1 .34	1 .59
	0′	+ 10′	+ 20′	+ 30′	+ 40′	+ 50′	+ 60′

Arg. = Parallax of the Moon in Lat.

PART IV.

Arg. = Semi-diameter of the Moon.

Arg. = Sum of preceding Equations.	14′		15′						16′
	40″ —	50″ —	0″ —	10″ —	20″ —	30″ —	40″ —	50″ —	0″
1″	0″.17	0″.15	0″.13	0″.10	0″.08	0″.06	0″.04	0″.02	0″
2	0 .38	0 .29	0 .25	0 .21	0 .17	0 .12	0 .08	0 .04	0
3	0 .50	0 .44	0 .37	0 .31	0 .25	0 .19	0 .12	0 .06	0
4	0 .97	0 .58	0 .50	0 .42	0 .33	0 .25	0 .17	0 .08	0
5	0 .83	0 .73	0 .62	0 .52	0 .42	0 .31	0 .21	0 .10	0
6	1 .00	0 .88	0 .75	0 .62	0 .50	0 .37	0 .25	0 .12	0
7	1 .16	1 .02	0 .87	0 .73	0 .58	0 .44	0 .29	0 .1[illegible]	0
8	1 .33	1 .16	1 .00	0 .83	0 .67	0 .50	0 .33	0 .17	0
9	1 .50	1 .31	1 .13	0 .94	0 .75	0 .56	0 .37	0 .19	0
10	1 .67	1 .46	1 .25	1 .04	0 .83	0 .62	0 .42	0 .2[illegible]	0
11	1 .83	1 .60	1 .37	1 .15	0 .92	0 .69	0 .46	0 .23	0
12	2 .00	1 .75	1 .50	1 .25	1 .00	0 .75	0 .50	0 .25	0
13	2 .16	1 .89	1 .62	1 .35	1 .08	0 .81	0 .54	0 .27	0
14	2 .33	2 .04	1 .75	1 .46	1 .17	0 .87	0 .58	0 .29	0
15	2 .50	2 .18	1 .87	1 .56	1 .25	0 .94	0 .62	0 .31	0
	+ 20″	+ 10″	+ 0″	+ 50″	+ 40″	+ 30″	+ 20″	+ 10″	0″
	17′			16′					

Arg. = Semi-diameter of the Moon.

TABLE 2. DIAMETER of the MOON'S ENLIGHTENED DISC, for Semi-diameter 16′ 40″, or 1000″. With the Differences for 100′ Angular Distance of the ☽ from the ☉, and Var. for 100″ Var. of Semi-diam. Arg. = the Distance of the ☽ from the ☉. (Mr. J. UTTING.)

Dist. of the ☽ from ☉	Diameter.	Diff. for 100′.	Var. for 100″.	Dist. of the ☽ from ☉	Dist. of the ☽ from ☉	Diameter.	Diff. for 100′.	Var. for 100″.	Dist. of the ☽ from ☉	Dist. of the ☽ from ☉	Diameter.	Diff. for 100′.	Var. for 100″.	Dist. of the ☽ from ☉
0°	0′ 0″.0	0″.25	0″.00	360°	60°	8′ 20′.0	25″.32	50″.00	300°	120°	25′ 0″.0	25″.06	150″.00	240°
1	0 0.2	0.76	0.02	359	61	8 35.2	25.56	51.52	299	121	25 15.0	24.80	151.50	239
2	0 0.6	1.27	0.06	358	62	8 50.5	25.80	53.05	298	122	25 29.9	24.53	152.99	238
3	0 1.4	1.78	0.14	357	63	9 6.0	26.03	54.60	297	123	25 44.6	24.26	154.46	237
4	0 2.4	2.28	0.24	356	64	9 21.6	26.26	56.16	296	124	25 59.2	23.97	155.92	236
5	0 3.8	2.79	0.38	355	65	9 37.4	26.47	57.74	295	125	26 13.6	23.68	157.36	235
6	0 5.5	3.29	0.55	354	66	9 53.3	26.68	59.33	294	126	26 27.8	23.39	158.78	234
7	0 7.5	3.80	0.75	353	67	10 9.3	26.87	60.93	293	127	26 41.8	23.08	160.18	233
8	0 9.7	4.30	0.97	352	68	10 25.4	27.07	62.54	292	128	26 55.7	22.77	161.57	232
9	0 12.3	4.80	1.23	351	69	10 41.6	27.25	64.16	291	129	27 9.3	22.45	162.93	231
10	0 15.2	5.30	1.52	350	70	10 58.0	27.42	65.80	290	130	27 22.8	22.12	164.28	230
11	0 18.4	5.80	1.84	349	71	11 14.4	27.59	67.44	289	131	27 36.1	21.79	165.61	229
12	0 21.9	6.30	2.19	348	72	11 31.0	27.74	69.10	288	132	27 49.1	21.45	166.92	228
13	0 25.6	6.79	2.56	347	73	11 47.6	27.89	70.76	287	133	28 2.0	21.10	168.20	227
14	0 29.7	7.28	2.97	346	74	12 4.4	28.03	72.44	286	134	28 14.7	20.75	169.47	226
15	0 34.1	7.77	3.41	345	75	12 21.2	28.16	74.12	285	135	28 27.1	20.39	170.71	225
16	0 38.7	8.26	3.87	344	76	12 38.1	28.29	75.81	284	136	28 39.3	20.02	171.93	224
17	0 43.7	8.75	4.37	343	77	12 55.0	28.40	77.50	283	137	28 51.4	19.65	173.14	223
18	0 48.9	9.23	4.89	342	78	13 12.1	28.51	79.21	282	138	29 3.1	19.28	174.31	222
19	0 54.5	9.71	5.45	341	79	13 29.2	28.60	80.92	281	139	29 14.7	18.89	175.47	221
20	1 0.3	10.19	6.03	340	80	13 46.4	28.69	82.64	280	140	29 26.0	18.50	176.60	220
21	1 6.4	10.66	6.64	339	81	14 3.6	28.77	84.36	279	141	29 37.1	18.11	177.71	219
22	1 12.8	11.13	7.28	338	82	14 20.8	28.84	86.08	278	142	29 48.0	17.71	178.80	218
23	1 19.5	11.60	7.95	337	83	14 38.1	28.90	87.81	277	143	29 58.6	17.30	179.86	217
24	1 26.5	12.06	8.65	336	84	14 55.5	28.95	89.55	276	144	30 9.0	16.89	180.90	216
25	1 33.7	12.52	9.37	335	85	15 12.8	29.00	91.28	275	145	30 19.2	16.48	181.92	215
26	1 41.2	12.98	10.12	334	86	15 30.2	29.03	93.02	274	146	30 29.0	16.06	182.90	214
27	1 49.0	13.43	10.90	333	87	15 47.7	29.06	94.77	273	147	30 38.7	15.63	183.87	213
28	1 57.1	13.88	11.71	332	88	16 5.1	29.08	96.51	272	148	30 48.0	15.20	184.80	212
29	2 5.4	14.33	12.54	331	89	16 22.5	29.09	98.25	271	149	30 57.2	14.76	185.72	211
30	2 14.0	14.76	13.40	330	90	16 40.0	29.09	100.00	270	150	31 6.0	14.33	186.60	210
31	2 22.8	15.20	14.28	329	91	16 57.5	29.08	101.75	269	151	31 14.6	13.88	187.46	209
32	2 32.0	15.63	15.20	328	92	17 14.9	29.06	103.49	268	152	31 22.9	13.43	188.29	208
33	2 41.3	16.06	16.13	327	93	17 32.3	29.03	105.23	267	153	31 31.0	12.98	189.10	207
34	2 51.0	16.48	17.10	326	94	17 49.8	29.00	106.98	266	154	31 38.8	12.52	189.88	206
35	3 0.8	16.89	18.08	325	95	18 7.2	28.95	108.72	265	155	31 46.3	12.06	190.63	205
36	3 11.0	17.30	19.10	324	96	18 24.5	28.90	110.45	264	156	31 53.5	11.60	191.35	204
37	3 21.4	17.71	20.14	323	97	18 41.9	28.84	112.19	263	157	32 0.5	11.13	192.05	203
38	3 32.0	18.11	21.20	322	98	18 59.2	28.77	113.92	262	158	32 7.2	10.66	192.72	202
39	3 42.9	18.50	22.29	321	99	19 16.4	28.69	115.64	261	159	32 13.6	10.19	193.36	201
40	3 54.0	18.89	23.40	320	100	19 33.6	28.60	117.36	260	160	32 19.7	9.71	193.97	200
41	4 5.3	19.28	24.53	319	101	19 50.8	28.51	119.08	259	161	32 25.5	9.23	194.55	199
42	4 16.9	19.65	25.69	318	102	20 7.9	28.40	120.79	258	162	32 31.1	8.75	195.11	198
43	4 28.6	20.02	26.86	317	103	20 25.0	28.29	122.50	257	163	32 36.3	8.26	195.63	197
44	4 40.7	20.39	28.07	316	104	20 41.9	28.16	124.19	256	164	32 41.3	7.77	196.13	196
45	4 52.9	20.75	29.29	315	105	20 58.8	28.03	125.88	255	165	32 45.9	7.28	196.59	195
46	5 5.3	21.10	30.53	314	106	21 15.6	27.89	127.56	254	166	32 50.3	6.79	197.03	194
47	5 18.0	21.45	31.80	313	107	21 32.4	27.74	129.24	253	167	32 54.4	6.30	197.44	193
48	5 30.9	21.79	33.08	312	108	21 49.0	27.59	130.90	252	168	32 58.1	5.80	197.81	192
49	5 43.9	22.12	34.39	311	199	22 5.6	27.42	132.56	251	169	33 1.6	5.30	198.16	191
50	5 57.2	22.45	35.72	310	110	22 22.0	27.25	134.20	250	170	33 4.8	4.80	198.48	190
51	6 10.7	22.77	37.07	309	111	22 38.4	27.07	135.84	249	171	33 7.7	4.30	198.77	189
52	6 24.3	23.08	38.43	308	112	22 54.6	26.87	137.46	248	172	33 10.3	3.80	199.03	188
53	6 38.2	23.39	39.82	307	113	23 10.7	26.68	139.07	247	173	33 12.5	3.29	199.25	187
54	6 52.2	23.68	41.22	306	114	23 26.7	26.47	140.67	246	174	33 14.5	2.79	199.45	186
55	7 6.4	23.97	42.64	305	115	23 42.6	26.26	142.26	245	175	33 16.2	2.28	199.62	185
56	7 20.8	24.26	44.08	304	116	23 58.4	26.05	143.84	244	176	33 17.6	1.78	199.76	184
57	7 35.4	24.53	45.54	303	117	24 14.0	25.80	145.40	243	177	33 18.6	1.27	199.86	183
58	7 50.1	24.80	47.01	302	118	24 29.5	25.56	146.95	242	178	33 19.4	0.76	199.94	182
59	8 5.0	25.06	48.50	301	119	24 44.8	25.32	148.48	241	179	33 19.8	0.25	199.98	181
60	8 20.0		50.00	300	120	25 0.0		150.00	240	180	33 20.0		200.00	180

TABLE 3. To be used with TABLE 10 of the LUNAR TABLES, for obtaining by inspection the Time of the MOON'S SEMI-DIAMETER'S PASSAGE over the MERIDIAN. (MR. J. UTTING.) Arguments = the Tabular Number, and the ☾'s Apparent Semi-diameter.

Tabular Number, from TABLE 10.	14′ 40″	14′ 50″	15′ 0″	15′ 10″	15′ 20″	15′ 30″	15′ 40″	15′ 50″	16′ 0″	16′ 10″	16′ 20″	16′ 30″	16′ 40″
12ˢ.4	70ˢ.97	71ˢ.77	72ˢ.58	73ˢ.39	74ˢ.19	75ˢ.00	75ˢ.81	76ˢ.71	77ˢ.42	78ˢ.23	79ˢ.03	79ˢ.84	80ˢ.65
12 .5	70 .40	71 .20	72 .00	72 .80	73 .60	74 .40	75 .20	76 .00	76 .80	77 .60	78 .40	79 .20	80 .00
12 .6	69 .84	70 .63	71 .43	72 .22	73 .02	73 .81	74 .60	75 .40	76 .19	76 .98	77 .78	78 .57	79 .37
12 .7	69 .29	70 .08	70 .87	71 .65	72 .44	73 .23	74 .02	74 .80	75 .59	76 .38	77 .17	77 .95	78 .74
12 .8	68 .75	69 .53	70 .31	71 .09	71 .88	72 .66	73 .44	74 .22	75 .00	75 .78	76 .56	77 .34	78 .13
12 .9	68 .22	68 .99	69 .77	70 .54	71 .32	72 .09	72 .87	73 .64	74 .42	75 .19	75 .97	76 .74	77 .52
13 .0	67 .69	68 .46	69 .23	70 .00	70 .77	71 .54	72 .31	73 .08	73 .85	74 .62	75 .38	76 .15	76 .92
13 .1	67 .18	67 .94	68 .70	69 .47	70 .23	70 .99	71 .76	72 .52	73 .28	74 .05	74 .81	75 .57	76 .34
13 .2	66 .67	67 .42	68 .18	68 .94	69 .70	70 .45	71 .21	71 .97	72 .73	73 .48	74 .24	75 .00	75 .76
13 .3	66 .17	66 .92	67 .67	68 .42	69 .17	69 .92	70 .68	71 .43	72 .18	72 .93	73 .68	74 .44	75 .19
13 .4	65 .67	66 .42	67 .16	67 .91	68 .66	69 .40	70 .15	70 .90	71 .64	72 .39	73 .13	73 .88	74 .63
13 .5	65 .19	65 .93	66 .67	67 .41	68 .15	68 .89	69 .63	70 .37	71 .11	71 .85	72 .59	73 .33	74 .07
13 .6	64 .71	65 .44	66 .18	66 .91	67 .65	68 .38	69 .12	69 .85	70 .59	71 .32	72 .06	72 .79	73 .53
13 .7	64 .23	64 .96	65 .69	66 .42	67 .15	67 .88	68 .61	69 .34	70 .07	70 .80	71 .53	72 .26	72 .99
13 .8	63 .77	64 .49	65 .22	65 .94	66 .67	67 .39	68 .12	68 .84	69 .57	70 .29	71 .01	71 .74	72 .46
13 .9	63 .31	64 .03	64 .75	65 .47	66 .19	66 .91	67 .63	68 .35	69 .06	69 .78	70 .50	71 .22	71 .94
14 .0	62 .86	63 .57	64 .29	65 .00	65 .71	66 .43	67 .14	67 .86	68 .57	69 .29	70 .00	70 .71	71 .43
14 .1	62 .41	63 .12	63 .83	64 .54	65 .25	65 .96	66 .67	67 .38	68 .09	68 .79	69 .50	70 .21	70 .92
14 .2	61 .97	62 .68	63 .38	64 .08	64 .79	65 .49	66 .20	66 .90	67 .61	68 .31	69 .01	69 .72	70 .42
14 .3	61 .54	62 .24	62 .94	63 .64	64 .34	65 .04	65 .73	66 .43	67 .13	67 .83	68 .53	69 .23	69 .93
14 .4	61 .11	61 .81	62 .50	63 .19	63 .89	64 .58	65 .28	65 .97	66 .67	67 .36	68 .06	68 .75	69 .44
14 .5	60 .69	61 .38	62 .07	62 .76	63 .45	64 .14	64 .83	65 .52	66 .21	66 .90	67 .59	68 .28	68 .97
14 .6	60 .27	60 .96	61 .64	62 .33	63 .01	63 .70	64 .38	65 .07	65 .75	66 .44	67 .12	67 .81	68 .49

MEAN REFRACTIONS.

The MEAN REFRACTIONS of CASSINI, LA CAILLE, and BOUGUER, to each Degree of Zenith Distance.

App. Zenith Dist.	Cassini.	La Caille.	Bouguer.	App. Alt.	App. Zenith Dist.	Cassini.	La Caille.	Bouguer	App. Alt.	App. Zenith Dist.	Cassini.	La Caille.	Bouguer.	App. Alt.
1°	1″	1″.1	0″.5	89°	31°	35″	40″.0	27″	59°	61°	1′ 46″	1′ 59″.1	1′ 20″	29°
2	2	2 .3	1 .0	88	32	37	41 .6	28	58	62	1 51	2 4 .0	1 24	28
3	3	3 .5	1 .5	87	33	38	43 .2	29	57	63	1 55	2 9 .2	1 27	27
4	4	4 .6	2	86	34	40	44 .9	30	56	64	2 0	2 14 .7	1 31	26
5	5	5 .8	3	85	35	41	46 .6	31	55	65	2 6	2 20 .5	1 36	25
6	6	7 .0	4	84	36	43	48 .3	32	54	66	2 12	2 26 .6	1 41	24
7	7	8 .2	5	83	37	45	50 .1	33	53	67	2 18	2 33 .0	1 45	23
8	8	9 .3	6	82	38	47	51 .9	35	52	68	2 25	2 39 .8	1 51	22
9	9	10 .5	7	81	39	49	53 .8	36	51	69	2 31	2 47 .0	1 57	21
10	10	11 .7	8	80	40	50	55 .8	38	50	70	2 39	2 54 .7	2 3	20
11	11	12 .9	9	79	41	52	57 .9	39	49	71	2 49	3 3 .0	2 10	19
12	12	14 .1	10	78	42	54	60 .0	40	48	72	3 0	3 12 .0	2 17	18
13	13	15 .4	11	77	43	56	62 .1	41	47	73	3 11	3 23 .0	2 26	17
14	14	16 .6	12	76	44	58	64 .3	42	46	74	3 24	3 35 .0	2 36	16
15	16	17 .8	12	75	45	59	66 .5	44	45	75	3 38	3 49 .0	2 48	15
16	17	19 .1	13	74	46	61	68 .8	46	44	76	3 54	4 5 .0	2 58	14
17	18	20 .3	14	73	47	63	71 .2	48	43	77	4 12	4 24 .0	3 14	13
18	19	21 .6	15	72	48	65	73 .7	49	42	78	4 32	4 45 .0	3 31	12
19	20	22 .9	16	71	49	67	76 .3	51	41	79	4 58	5 9 .0	3 54	11
20	21	24 .2	17	70	50	70	79 .0	53	40	80	5 28	5 37 .0	4 20	10
21	22	25 .5	18	69	51	72	81 .9	55	39	81	6 4	6 10 .0	4 50	9
22	24	26 .8	18	68	52	75	84 .9	58	38	82	6 47	6 51 .0	5 21	8
23	25	28 .2	19	67	53	78	88 .0	60	37	83	7 44	7 41 .0	6 5	7
24	26	29 .6	20	66	54	80	91 .2	62	36	84	8 55	8 42 .0	7 4	6
25	27	31 .0	21	65	55	83	94 .6	64	35	85	10 32	10 9 .6	8 18	5
26	28	32 .4	22	64	56	87	98 .1	66	34	86	12 48	12 9 .3	10 5	4
27	30	33 .9	23	63	57	90	101 .8	68	33	87	16 6	14 57 .3	12 25	3
28	31	35 .4	24	62	58	94	105 .8	70	32	88	21 4	19 0 .5	15 53	2
29	33	36 .9	25	61	59	98	110 .0	74	31	89	27 56	24 57 .4	20 31	1
30	34	38 .4	26	60	60	102	114 .4	77	30	90	32 20	33 30 .0	27 0	0

Bouguer's Refractions were derived from Observations made at the level of the Sea in the Torrid Zone.

TABLE 1. A Table of Mean Refractions with their respective Zenith Distances. Barom. 28 Paris Inches; and + 10° of Réaumur's Thermometer; or 29 .851 English; and 54 .5 of Fahrenheit's Thermometer.

App. Zenith Dist.	Refrac.	Diff.
1°	1″.0	1″.0
2	2 .0	1 .0
3	3 .0	1 .1
4	4 .1	1 .0
5	5 .1	1 .0
6	6 .1	1 .0
7	7 .1	1 .0
8	8 .1	1 .1
9	9 .2	1 .0
10	10 .2	1 .0
11	11 .2	1 .1
12	12 .3	1 .1
13	13 .4	1 .0
14	14 .4	1 .1
15	15 .5	1 .1
16	16 .6	1 .1
17	17 .7	1 .1
18	18 .8	1 .1
19	19 .9	1 .2
20	21 .1	1 .1
21	22 .2	1 .2
22	23 .4	1 .2
23	24 .6	1 .2
24	25 .8	1 .2
25	27 .0	1 .3
26	28 .3	1 .2
27	29 .5	1 .3
28	30 .8	1 .3
29	32 .1	1 .3
30	33 .4	

App. Zenith Dist.	Refrac.	Diff.
31°	34″.8	1″.4
32	36 .2	1 .4
33	37 .6	1 .5
34	39 .1	1 .5
35	40 .6	1 .5
36	42 .1	1 .5
37	43 .6	1 .6
38	45 .2	1 .7
39	46 .9	1 .7
40	48 .6	1 .7
41	50 .3	1 .8
42	52 .1	1 .9
43	54 .0	1 .9
44	55 .9	2 .0
45	57 .9	2 .0
46	59 .9	2 .2
47	62 .1	2 .2
48	64 .3	2 .3
49	66 .6	2 .3
50	68 .9	2 .5
51	71 .4	2 .6
52	74 .0	2 .7
53	76 .7	2 .9
54	79 .6	3 .0
55	82 .6	3 .1
56	85 .7	3 .3
57	89 .0	3 .5
58	92 .5	3 .6
59	96 .1	3 .9
60	100 .0	

App. Zenith Distance.	Refrac.	Diff.	Log.	Diff.
60° 30′	1′ 42″.1	2″.0	2 .0088	88
61 0	1 44 .1	2 .2	2 .0176	90
61 30	1 46 .3	2 .2	2 .0266	90
62 0	1 48 .5	2 .3	2 .0356	91
62 30	1 50 .8	2 .4	2 .0447	92
63 0	1 53 .2	2 .5	2 .0539	94
63 30	1 55 .7	2 .5	2 .0633	95
64 0	1 58 .2	2 .7	2 .0728	96
64 30	2 0 .9	2 .7	2 .0824	97
65 0	2 3 .6	2 .9	2 .0921	98
65 30	2 6 .5	2 .9	2 .1019	101
66 0	2 9 .4	3 .1	2 .1120	101
66 30	2 12 .5	3 .2	2 .1221	103
67 0	2 15 .7	3 .3	2 .1324	105
67 30	2 19 .0	3 .4	2 .1429	107
68 0	2 22 .4	3 .6	2 .1536	109
68 30	2 26 .0	3 .8	2 .1645	110
69 0	2 29 .8	3 .9	2 .1755	113
69 30	2 33 .7	4 .2	2 .1868	115
70 0	2 37 .9	4 .3	2 .1983	117
70 30	2 42 .2	4 .5	2 .2100	119
71 0	2 46 .7	4 .8	2 .2219	123
71 30	2 51 .5	5 .0	2 .2342	124
72 0	3 56 .5	5 .2	2 .2466	128
72 30	3 1 .7	5 .6	2 .2594	131
73 0	3 7 .3	5 .8	2 .2725	134
73 30	3 13 .1	6 .3	2 .2859	137
74 0	3 19 .4	6 .5	2 .2996	141
74 30	3 25 .9	7 .0	2 .3137	145
75 0	3 32 .9		2 .3282	

App. Zenith Distance.	Refrac.	Diff.	Log.	Diff.
75° 0′	3′ 32″.9	5″.1	2 .3282	102
75 20	3 38 .0	5 .1	2 .3384	101
75 40	3 43 .1	5 .3	2 .3485	103
76 0	3 48 .4	5 .6	2 .3588	105
76 20	3 54 .0	5 .9	2 .3693	107
76 40	3 59 .9	6 .1	2 .3800	110
77 0	4 6 .0	6 .5	2 .3910	112
77 20	4 12 .5	6 .7	2 .4022	115
77 40	4 19 .2	7 .1	2 .4137	117
78 0	4 26 .3	7 .5	2 .4254	120
78 20	4 33 .8	7 .9	2 .4374	123
78 40	4 41 .7	8 .3	2 .4497	127
79 0	4 50 .0	8 .8	2 .4624	130
79 20	4 58 .8	9 .3	2 .4754	133
79 40	5 8 .1	9 .8	2 .4887	136
80 0	5 17 .9	10 .5	2 .5023	141
80 20	5 28 .4	11 .1	2 .5164	144
80 40	5 39 .5	11 .8	2 .5308	149
81 0	5 51 .3	12 .7	2 .5457	154
81 20	6 4 .0	13 .5	2 .5611	158
81 40	6 17 .5	14 .5	2 .5769	164
82 0	6 32 .0	15 .6	2 .5933	169
82 20	6 47 .6	16 .8	2 .6102	176
82 40	7 4 .4	18 .2	2 .6278	182
83 0	7 22 .6	19 .6	2 .6460	188
83 20	7 42 .2	21 .3	2 .6648	196
83 40	8 3 .5	23 .2	2 .6844	203
84 0	8 26 .7	25 .3	2 .7047	212
84 20	8 52 .0	27 .8	2 .7259	221
84 40	9 19 .8	30 .4	2 .7480	231
85 0	9 50 .2	16 .4	2 .7711	119
85 10	10 6 .6	17 .3	2 .7830	121
85 20	10 23 .9	18 .2	2 .7951	125
85 30	10 42 .1	19 .1	2 .8076	127
85 40	11 1 .2	20 .2	2 .8203	131
85 50	11 21 .4	21 .2	2 .8334	133
86 0	11 42 .6	22 .5	2 .8467	137
86 10	12 5 .1	23 .7	2 .8604	140
86 20	12 28 .8	25 .2	2 .8744	143
86 30	12 54 .0	26 .6	2 .8887	147
86 40	13 20 .6	28 .2	2 .9034	151
86 50	13 48 .8	30 .0	2 .9185	154
87 0	14 18 .8	31 .8	2 .9339	158
87 10	14 50 .6	33 .9	2 .9497	162
87 20	15 24 .5	36 .0	2 .9659	166
87 3)	16 0 .5	38 .3	2 .9825	170
87 40	16 38 .8	40 .8	2 .9995	174
87 50	17 19 .6	43 .5	3 .0169	178
88 0	18 3 .1	46 .4	3 .0347	182
88 10	18 49 .5	49 .4	3 .0529	186
88 20	19 38 .9	52 .6	3 .0715	189
88 30	20 31 .5	56 .0	3 .0904	193
88 40	21 27 .5	59 .4	3 .1097	196
88 50	22 26 .9	63 .0	3 .1293	199
89 0	23 29 .9	66 .4	3 .1492	200
89 10	24 36 .3	69 .8	3 .1692	200
89 20	25 46 .1	72 .6	3 .1892	200
89 30	26 58 .7	74 .7	3 .2092	197
89 40	28 13 .4	76 .6	3 .2289	191
89 50	29 30 .0	75 .7	3 .2480	182
90 0	30 45 .7		3 .2662	

TABLE 2.

Par. Bar.	No. A.	Log. (1 + A.)
In. Lines.		
26. 0	−0 . 0714	9 . 9678
26. 1	−0 . 0685	9 . 9692
26. 2	−0 . 0655	9 . 9706
26. 3	−0 . 0625	9 . 9720
26. 4	−0 . 0595	9 . 9733
26. 5	−0 . 0565	9 . 9747
26. 6	−0 . 0536	9 . 9761
26. 7	−0 . 0506	9 . 9775
26. 8	−0 . 0476	9 . 9788
26. 9	−0 . 0446	9 . 9802
26 .10	−0 . 0417	9 . 9815
26 .11	−0 . 0387	9 . 9829
27. 0	−0 . 0357	9 . 9842
27. 1	−0 . 0327	9 . 9855
27. 2	−0 . 0298	9 . 9869
27. 3	−0 . 0268	9 . 9882
27. 4	−0 . 0238	9 . 9895
27. 5	−0 . 0208	9 . 9909
27. 6	−0 . 0179	9 . 9922
27. 7	−0 . 0149	9 . 9935
27. 8	−0 . 0119	9 . 9948
27. 9	−0 . 0089	9 . 9961
27 .10	−0 . 0060	9 . 9974
27 .11	−0 . 0030	9 . 9987
28. 0	−0 . 0000	0 . 0000
28. 1	+0 . 0030	0 . 0013
28. 2	+0 . 0060	0 . 0026
28. 3	+0 . 0089	0 . 0039
28. 4	+0 . 0119	0 . 0051
28. 5	+0 . 0149	0 . 0064
28. 6	+0 . 0179	0 . 0077

TABLE 3.

Réaum. Therm.	No. B.	Log. (1 + B.)
−10	+0 . 1040	0 . 0429
− 9	+0 . 0983	0 . 0407
− 8	+0 . 0926	0 . 0385
− 7	+0 . 0870	0 . 0362
− 6	+0 . 0815	0 . 0340
− 5	+0 . 0760	0 . 0318
− 4	+0 . 0706	0 . 0296
− 3	+0 . 0652	0 . 0274
− 2	+0 . 0599	0 . 0253
− 1	+0 . 0546	0 . 0231
− 0	+0 . 0494	0 . 0209
+ 1	+0 . 0443	0 . 0188
+ 2	+0 . 0391	0 . 0167
+ 3	+0 . 0341	0 . 0145
+ 4	+0 . 0291	0 . 0124
+ 5	+0 . 0241	0 . 0103
+ 6	+0 . 0192	0 . 0083
+ 7	+0 . 0143	0 . 0062
+ 8	+0 . 0095	0 . 0041
+ 9	+0 . 0047	0 . 0020
+10	+0 . 0000	0 . 0000
+11	−0 . 0047	9 . 9980
+12	−0 . 0093	9 . 9959
+13	−0 . 0139	9 . 9939
+14	−0 . 0185	9 . 9919
+15	−0 . 0230	9 . 9899
+16	−0 . 0275	9 . 9879
+17	−0 . 0319	9 . 9859
+18	−0 . 0363	9 . 9839
+19	−0 . 0406	9 . 9820
+20	−0 . 0450	9 . 9800
+21	−0 . 0492	9 . 9781
+22	−0 . 0535	9 . 9761
+23	−0 . 0577	9 . 9742
+24	−0 . 0618	9 . 9723
+25	−0 . 0660	9 . 9704
+30	−0 . 0861	9 . 9609

TABLE 4.

App. Zenith Distance.	No. C.
80° 0'	− 0". 05
81 0	− 0 . 07
82 0	− 0 . 10
83 0	− 0 . 14
84 0	− 0 . 21
85 0	− 0 . 33
86 0	− 0 . 55
86 10	− 0 . 60
86 20	− 0 . 66
86 30	− 0 . 73
86 40	− 0 . 81
86 50	− 0 . 90
87 0	− 0 . 99
87 10	− 1 . 10
87 20	− 1 . 23
87 30	− 1 . 39
87 40	− 1 . 57
87 50	− 1 . 77
88 0	− 2 . 00
88 10	− 2 . 27
88 20	− 2 . 59
88 30	− 2 . 97
88 40	− 3 . 42
88 50	− 3 . 95
89 0	− 4 . 58
89 10	− 5 . 35
89 20	− 6 . 27
89 30	− 7 . 39
89 40	− 8 . 75
89 50	−10 . 44
90 0	−12 . 49

TABLE 5.

English Barom.	No. A.	Log. (1+A.)
28 . 0	−0 . 0620	9 . 9722
28 . 1	−0 . 0587	9 . 9737
28 . 2	−0 . 0553	9 . 9753
28 . 3	−0 . 0519	9 . 9768
28 . 4	−0 . 0486	9 . 9784
28 . 5	−0 . 0453	9 . 9799
28 . 6	−0 . 0419	9 . 9814
28 . 7	−0 . 0386	9 . 9829
28 . 8	−0 . 0352	9 . 9844
28 . 9	−0 . 0319	9 . 9859
29 . 0	−0 . 0285	9 . 9874
29 . 1	−0 . 0252	9 . 9889
29 . 2	−0 . 0218	9 . 9904
29 . 3	−0 . 0185	9 . 9919
29 . 4	−0 . 0151	9 . 9934
29 . 5	−0 . 0118	9 . 9949
29 . 6	−0 . 0084	9 . 9963
29 . 7	−0 . 0050	9 . 9978
29 . 8	−0 . 0017	9 . 9993
29 . 9	+0 . 0017	0 . 0007
30 . 0	+0 . 0050	0 . 0022
30 . 1	+0 . 0083	0 . 0036
30 . 2	+0 . 0116	0 . 0050
30 . 3	+0 . 0150	0 . 0065
30 . 4	+0 . 0184	0 . 0079
30 . 5	+0 . 0217	0 . 0093
30 . 6	+0 . 0251	0 . 0108

TABLE 6.

Fahr. Ther.	No. B.	Log. (1+B.)	Fahr. Ther.	No. B.	Log. (1+B.)
10°	+0 . 1027	0 . 0425	46°	+0 . 0181	0 . 0078
12	+0 . 0976	0 . 0405	48	+0 . 0138	0 . 0060
14	+0 . 0926	0 . 0385	50	+0 . 0095	0 . 0041
16	+0 . 0876	0 . 0365	52	+0 . 0053	0 . 0023
18	+0 . 0827	0 . 0345	54	+0 . 0011	0 . 0005
20	+0 . 0778	0 . 0325	56	−0 . 0031	9 . 9986
22	+0 . 0730	0 . 0306	58	−0 . 0073	9 . 9968
24	+0 . 0682	0 . 0286	60	−0 . 0114	9 . 9950
26	+0 . 0634	0 . 0267	62	−0 . 0155	9 . 9932
28	+0 . 0587	0 . 0248	64	−0 . 0195	9 . 9914
30	+0 . 0540	0 . 0228	66	−0 . 0235	9 . 9897
32	+0 . 0494	0 . 0209	68	−0 . 0275	9 . 9879
34	+0 . 0448	0 . 0190	70	−0 . 0314	9 . 9861
36	+0 . 0403	0 . 0171	72	−0 . 0353	9 . 9844
38	+0 . 0358	0 . 0152	74	−0 . 0392	9 . 9826
40	+0 . 0313	0 . 0134	76	−0 . 0430	9 . 9809
42	+0 . 0269	0 . 0115	78	−0 . 0469	9 . 9791
44	+0 . 0224	0 . 0096	80	−0 . 0507	9 . 9774
46	+0 . 0181	0 . 0078	90	−0 . 0691	9 . 9688

TABLE 7.

App. Zen. Distance.	No. C.
80° 0'	−0 . 02
81 0	−0 . 03
82 0	−0 . 04
83 0	−0 . 06
84 0	−0 . 09
85 0	−0 . 15
86 0	−0 . 24
86 10	−0 . 27
86 20	−0 . 29
86 30	−0 . 32
86 40	−0 . 36
86 50	−0 . 40
87 0	−0 . 44
87 10	−0 . 49
87 20	−0 . 55
87 30	−0 . 62
87 40	−0 . 70
87 50	−0 . 79
88 0	−0 . 88
88 10	−1 . 01
88 20	−1 . 16
88 30	−1 . 32
88 40	−1 . 52
88 50	−1 . 76
89 0	−2 . 04
89 10	−2 . 38
89 20	−2 . 78
89 30	−3 . 28
89 40	−3 . 89
89 50	−4 . 64
90 0	−5 . 55

TABLE 1. A TABLE of MEAN REFRACTIONS for every 20′ of ZENITH DISTANCE to 78°, and for every 10′ from thence to 90° 30′. The Barom. at 29.60 English Inches; Bradley's Thermom. 50°, or Fahrenheit's 48°.75.

Zenith Distance.	$\delta\theta$	Diff.	$\frac{d\delta\theta}{d\tau}$	$\frac{d^2\delta\theta}{2d\tau^2}$	Log. $\delta\theta$.	Diff.	Zenith Distance.	$\delta\theta$	Diff.	$\frac{d\delta\theta}{d\tau}$	$\frac{d^2\delta\theta}{2d\tau^2}$	Log. $\delta\theta$.	Diff.
			−	+						−	+		
0° 0′	0″.00	0″.33	0″.0000	0″.00000	− ∞	∞	21° 0′	22″.09	0″.39	0″.0460	0″.00010	1.3442	75
20	0.33	0.34	0.0007	0.00000	9.5250	3011	20	22.48	0.38	0.0468	0.00010	1.3517	75
40	0.67	0.33	0.0014	0.00000	9.8261	1761	40	22.86	0.39	0.0476	0.00010	1.3592	73
1 0	1.00	0.34	0.0021	0.00000	0.0022	1249	22 0	23.25	0.39	0.0484	0.00010	1.3665	72
20	1.34	0.34	0.0028	0.00001	0.1271	972	20	23.64	0.40	0.0492	0.00010	1.3737	71
40	1.68	0.33	0.0035	0.00001	0.2243	791	40	24.04	0.39	0.0501	0.00010	1.3808	71
2 0	2.01	0.34	0.0042	0.00001	0.3034	669	23 0	24.43	0.39	0.0509	0.00011	1.3879	70
20	2.35	0.33	0.0049	0.00001	0.3703	580	20	24.82	0.40	0.0517	0.00011	1.3949	69
40	2.68	0.34	0.0056	0.00001	0.4283	513	40	25.22	0.40	0.0525	0.00011	1.4018	68
3 0	3.02	0.33	0.0063	0.00001	0.4796	458	24 0	25.62	0.40	0.0534	0.00011	1.4086	68
20	3.35	0.34	0.0070	0.00001	0.5254	415	20	26.02	0.41	0.0542	0.00011	1.4154	67
40	3.69	0.34	0.0077	0.00002	0.5669	379	40	26.43	0.40	0.0551	0.00011	1.4221	66
4 0	4.03	0.33	0.0084	0.00002	0.6048	349	25 0	26.83	0.41	0.0558	0.00012	1.4287	66
20	4.36	0.34	0.0091	0.00002	0.6397	323	20	27.24	0.41	0.0567	0.00012	1.4353	64
40	4.70	0.34	0.0098	0.00002	0.6720	301	40	27.65	0.42	0.0576	0.00012	1.4417	65
5 0	5.04	0.33	0.0105	0.00002	0.7021	282	26 0	28.07	0.41	0.0585	0.00012	1.4482	64
20	5.37	0.34	0.0112	0.00002	0.7303	265	20	28.48	0.42	0.0593	0.00012	1.4546	63
40	5.71	0.34	0.0119	0.00002	0.7568	250	40	28.90	0.42	0.0602	0.00013	1.4609	62
6 0	6.05	0.34	0.0126	0.00003	0.7818	236	27 0	29.32	0.42	0.0611	0.00013	1.4671	63
20	6.39	0.34	0.0133	0.00003	0.8054	225	20	29.74	0.43	0.0620	0.00013	1.4734	62
40	6.73	0.34	0.0140	0.00003	0.8279	214	40	30.17	0.43	0.0629	0.00013	1.4796	61
7 0	7.07	0.34	0.0147	0.00003	0.8493	205	28 0	30.60	0.43	0.0638	0.00013	1.4857	60
20	7.41	0.34	0.0154	0.00003	0.8698	195	20	31.03	0.43	0.0647	0.00013	1.4917	61
40	7.75	0.34	0.0161	0.00003	0.8893	187	40	31.46	0.44	0.0656	0.00014	1.4978	59
8 0	8.09	0.34	0.0169	0.00003	0.9080	179	29 0	31.90	0.43	0.0665	0.00014	1.5037	60
20	8.43	0.34	0.0176	0.00004	0.9259	173	20	32.33	0.44	0.0674	0.00014	1.5097	59
40	8.77	0.35	0.0183	0.00004	0.9432	166	40	32.77	0.45	0.0683	0.00014	1.5156	58
9 0	9.12	0.34	0.0190	0.00004	0.9598	161	30 0	33.22	0.45	0.0693	0.00014	1.5214	58
20	9.46	0.34	0.0197	0.00004	0.9759	155	20	33.67	0.45	0.0702	0.00015	1.5272	58
40	9.80	0.35	0.0204	0.00004	0.9914	150	40	34.12	0.45	0.0712	0.00015	1.5330	57
10 0	10.15	0.34	0.0211	0.00004	1.0064	145	31 0	34.57	0.46	0.0721	0.00015	1.5387	57
20	10.49	0.35	0.0218	0.00004	1.0209	141	20	35.03	0.46	0.0731	0.00015	1.5444	57
40	10.84	0.35	0.0226	0.00005	1.0350	137	40	35.49	0.46	0.0740	0.00015	1.5501	56
11 0	11.19	0.35	0.0233	0.00005	1.0487	133	32 0	35.95	0.47	0.0750	0.00016	1.5557	56
20	11.54	0.34	0.0240	0.00005	1.0620	130	20	36.42	0.47	0.0760	0.00016	1.5613	56
40	11.88	0.35	0.0248	0.00005	1.0750	126	40	36.89	0.47	0.0770	0.00016	1.5669	55
12 0	12.23	0.35	0.0255	0.00005	1.0876	122	33 0	37.36	0.48	0.0780	0.00016	1.5724	56
20	12.58	0.36	0.0262	0.00005	1.0998	120	20	37.84	0.48	0.0790	0.00016	1.5780	54
40	12.94	0.35	0.0269	0.00006	1.1118	117	40	38.32	0.49	0.0800	0.00017	1.5834	55
13 0	13.29	0.35	0.0277	0.00006	1.1235	114	34 0	38.81	0.48	0.0810	0.00017	1.5889	54
20	13.64	0.36	0.0284	0.00006	1.1349	111	20	39.29	0.50	0.0820	0.00017	1.5943	54
40	14.00	0.35	0.0291	0.00006	1.1460	109	40	39.79	0.49	0.0830	0.00017	1.5997	54
14 0	14.35	0.36	0.0299	0.00006	1.1569	106	35 0	40.28	0.50	0.0840	0.00018	1.6051	54
20	14.71	0.35	0.0306	0.00006	1.1675	104	20	40.78	0.51	0.0851	0.00018	1.6105	53
40	15.06	0.36	0.0314	0.00006	1.1779	102	40	41.29	0.51	0.0861	0.00018	1.6158	53
15 0	15.42	0.36	0.0321	0.00007	1.1881	101	36 0	41.80	0.51	0.0872	0.00018	1.6211	53
20	15.78	0.36	0.0329	0.00007	1.1982	98	20	42.31	0.52	0.0882	0.00018	1.6264	53
40	16.14	0.36	0.0336	0.00007	1.2080	96	40	42.83	0.52	0.0893	0.00019	1.6317	53
16 0	16.50	0.37	0.0344	0.00007	1.2176	94	37 0	43.35	0.52	0.0904	0.00019	1.6370	52
20	16.87	0.36	0.0351	0.00007	1.2270	93	20	43.87	0.53	0.0915	0.00019	1.6422	52
40	17.23	0.37	0.0359	0.00007	1.2363	92	40	44.40	0.54	0.0926	0.00019	1.6474	52
17 0	17.60	0.36	0.0366	0.00008	1.2455	89	38 0	44.94	0.54	0.0937	0.00020	1.6526	52
20	17.96	0.37	0.0374	0.00008	1.2544	88	20	45.48	0.55	0.0948	0.00020	1.6578	52
40	18.33	0.37	0.0381	0.00008	1.2632	87	40	46.03	0.55	0.0960	0.00020	1.6630	51
18 0	18.70	0.37	0.0389	0.00008	1.2719	85	39 0	46.58	0.55	0.0971	0.00020	1.6681	52
20	19.07	0.38	0.0397	0.00008	1.2804	84	20	47.13	0.56	0.0983	0.00020	1.6733	51
40	19.45	0.37	0.0405	0.00008	1.2888	83	40	47.69	0.57	0.0995	0.00021	1.6784	52
19 0	19.82	0.37	0.0412	0.00009	1.2971	81	40 0	48.26	0.57	0.1007	0.00021	1.6836	51
20	20.19	0.38	0.0420	0.00009	1.3052	80	20	48.83	0.58	0.1019	0.00021	1.6887	51
40	20.57	0.38	0.0428	0.00009	1.3132	79	40	49.41	0.58	0.1031	0.00021	1.6938	52
20 0	20.95	0.38	0.0436	0.00009	1.3211	78	41 0	49.99	0.59	0.1044	0.00022	1.6990	50
20	21.33	0.38	0.0444	0.00009	1.3289	77	20	50.58	0.60	0.1056	0.00022	1.7040	51
40	21.71	0.38	0.0452	0.00009	1.3366	76	40	51.18	0.60	0.1068	0.00022	1.7091	51

TABLE 1. A TABLE of MEAN REFRACTIONS for every 20′ of Zenith Distance to 78°, and for every 10′ from thence to 90° 30′. The Barom. at 29.60 English inches; Bradley's Therm. 50°, or Fahrenheit's 48° 75′.

Zenith Distance.	$\delta\theta$	Diff.	$\frac{d\delta\theta}{d\tau}$	$\frac{d^2\delta\theta}{2d\tau^2}$	Log. $\delta\theta$	Diff.	λ
			−	+			
42° 0′	0′ 51″.78	0″.61	0″.1081	0″.00023	1 .7142	50	
20	0 52 .39	0 .61	0 .1093	0 .00023	1 .7192	51	
40	0 53 .00	0 .62	0 .1106	0 .00023	1 .7243	50	
43 0	0 53 .62	0 .63	0 .1119	0 .00023	1 .7293	51	
20	0 54 .25	0 .63	0 .1132	0 .00023	1 .7344	50	
40	0 54 .88	0 .64	0 .1145	0 .00024	1 .7394	51	
44 0	0 55 .52	0 .65	0 .1159	0 .00024	1 .7445	50	
20	0 56 .17	0 .66	0 .1172	.00024	1 .7495	51	
40	0 56 .83	0 .66	0 .1186	0 .00025	1 .7546	50	
45 0	0 57 .49	0 .67	0 .1200	0 .00025	1 .75961	503	1 .0018
20	0 58 .16	0 .68	0 .1214	0 .00025	1 .76464	505	1 .0018
40	0 58 .84	0 .69	0 .1228	0 .00026	1 .76969	504	1 .0018
46 0	0 59 .53	0 .70	0 .1243	0 .00026	1 .77473	506	1 .0019
20	1 0 .23	0 .70	0 .1257	0 .00026	1 .77979	503	1 .0019
40	1 0 .93	0 .71	0 .1272	0 .00027	1 .78482	505	1 .0019
47 0	1 1 .64	0 .72	0 .1287	0 .00027	1 .78987	506	1 .0019
20	1 2 .36	0 .73	0 .1302	0 .00027	1 .79493	506	1 .0019
40	1 3 .09	0 .74	0 .1317	0 .00028	1 .79999	506	1 .0020
48 0	1 3 .83	0 .75	0 .1333	0 .00028	1 .80505	508	1 .0020
20	1 4 .58	0 .76	0 .1349	0 .00028	1 .81013	507	1 .0020
40	1 5 .34	0 .77	0 .1365	0 .00028	1 .81520	508	1 .0021
49 0	1 6 .11	0 .78	0 .1381	0 .00029	1 .82028	509	1 .0021
20	1 6 .89	0 .79	0 .1397	0 .00029	1 .82537	510	1 .0022
40	1 7 .68	0 .80	0 .1413	0 .00029	1 .83047	511	1 .0022
50 0	1 8 .48	0 .82	0 .1430	0 .00030	1 .83558	512	1 .0023
20	1 9 .30	0 .82	0 .1447	0 .00030	1 .84070	513	1 .0023
40	1 10 .12	0 .83	0 .1464	0 .00030	1 .84583	514	1 .0024
51 0	1 10 .95	0 .85	0 .1482	0 .00031	1 .85097	516	1 .0025
20	1 11 .80	0 .86	0 .1500	0 .00031	1 .85613	518	1 .0025
40	1 12 .66	0 .87	0 .1518	0 .00032	1 .86131	517	1 .0026
52 0	1 13 .53	0 .89	0 .1537	0 .00032	1 .86648	520	1 .0026
20	1 14 .42	0 .90	0 .1556	0 .00032	1 .87168	521	1 .0026
40	1 15 .32	0 .91	0 .1576	0 .00033	1 .87689	524	1 .0027
53 0	1 16 .23	0 .93	0 .1595	0 .00033	1 .88213	524	1 .0027
20	1 17 .16	0 .94	0 .1615	0 .00034	1 .88737	526	1 .0028
40	1 18 .10	0 .95	0 .1635	0 .00034	1 .89263	527	1 .0029
54 0	1 19 .05	0 .97	0 .1655	0 .00034	1 .89790	530	1 .0029
20	1 20 .02	0 .99	0 .1675	0 .00035	1 .90320	532	1 .0030
40	1 21 .01	1 .00	0 .1695	0 .00035	1 .90852	533	1 .0031
55 0	1 22 .01	1 .02	0 .1716	0 .00036	1 .91385	537	1 .0032
20	1 23 .03	1 .03	0 .1737	0 .00036	1 .91922	538	1 .0032
40	1 24 .06	1 .06	0 .1758	0 .00037	1 .92460	540	1 .0033
56 0	1 25 .12	1 .07	0 .1780	0 .00037	1 .93000	544	1 .0034
20	1 26 .19	1 .09	0 .1803	0 .00038	1 .93544	545	1 .0035
40	1 27 .28	1 .11	0 .1826	0 .00038	1 .94089	549	1 .0036
57 0	1 28 .39	1 .12	0 .1849	0 .00038	1 .94638	551	1 .0037
20	1 29 .51	1 .16	0 .1873	0 .00039	1 .95189	555	1 .0038
40	1 30 .67	1 .17	0 .1897	0 .00039	1 .95744	558	1 .0039
58 0	1 31 .84	1 .19	0 .1922	0 .00040	1 .96302	560	1 .0040
20	1 33 .03	1 .22	0 .1947	0 .00041	1 .96862	564	1 .0041
40	1 34 .25	1 .23	0 .1972	0 .00041	1 .97426	567	1 .0042
59 0	1 35 .48	1 .27	0 .1998	0 .00042	1 .97993	570	1 .0043
20	1 36 .75	1 .28	0 2025	0 .00042	1 .98563	574	1 .0044
40	1 38 .03	1 .31	0 .2052	0 .00043	1 .99137	577	1 .0045
60 0	1 39 .34	1 .34	0 .2079	0 .00043	1 .99714	582	1 .0046
20	1 40 .68	1 .37	0 .2107	0 .00044	2 .00296	585	1 .0047
40	1 42 .05	1 .39	0 .2136	0 .00044	2 .00881	590	1 .0048
61 0	1 43 .44	1 .43	0 .2166	0 .00045	2 .01471	593	1 .0049
20	1 44 .87	1 .45	0 .2196	0 .00046	2 .02064	597	1 .0051
40	1 46 .32	1 .48	0 .2227	0 .00047	2 .02661	603	1 .0052

Zenith Distance.	$\delta\theta$	Diff.	$\frac{d\delta\theta}{d\tau}$	$\frac{d^2\delta\theta}{2d\tau^2}$	Log. $\delta\theta$	Diff.	λ
			−	+			
62° 0′	1′ 47″.80	1″.52	0″.2258	0″.00047	2 .03264	606	1 .0054
20	1 49 .32	1 .55	0 .2290	0 .00048	2 .03870	612	1 .0055
40	1 50 .87	1 .59	0 .2323	0 .00048	2 .04482	616	1 .0057
63 0	1 52 .46	1 .62	0 .2357	0 .00049	2 .05098	622	1 .0058
20	1 54 .08	1 .66	0 .2391	0 .00050	2 .05720	627	1 .0060
40	1 55 .74	1 .69	0 .2426	0 .00051	2 .06347	632	1 .0061
64 0	1 57 .43	1 .74	0 .2462	0 .00051	2 .06979	638	1 .0063
20	1 59 .17	1 .78	0 .2499	0 .00052	2 .07617	644	1 .0064
40	2 0 .95	1 .82	0 .2637	0 .00053	2 .08261	649	1 .0066
65 0	2 2 .77	1 .87	0 .2576	0 .00054	2 .08910	656	1 .0068
20	2 4 .64	1 .91	0 .2615	0 .00055	2 .09566	661	1 .0070
40	2 6 .55	1 .97	0 .2656	0 .00055	2 .10227	669	1 .0072
66 0	2 8 .52	2 .01	0 .2698	0 .00056	2 .10896	676	1 .0075
20	2 10 .53	2 .07	0 .2741	0 .00057	2 .11572	682	1 .0078
40	2 12 .60	2 .12	0 .2785	0 .00058	2 .12254	690	1 .0081
67 0	2 14 .72	2 .18	0 .2830	0 .00059	2 .12944	696	1 .0083
20	2 16 .90	2 .24	0 .2877	0 .00060	2 .13640	706	1 .0086
40	2 19 .14	2 .31	0 .2925	0 .00061	2 .14346	713	1 .0089
68 0	2 21 .45	2 .36	0 .2974	0 .00062	2 .15059	721	1 .0092
20	2 23 .81	2 .44	0 .3025	0 .00063	2 .15780	729	1 .0095
40	2 26 .25	2 .51	0 .3077	0 .00064	2 .16509	739	1 .0098
69 0	2 28 .76	2 .59	0 .3131	0 .00065	2 .17248	749	1 .0101
20	2 31 .35	2 .66	0 .3186	0 .00067	2 .17997	757	1 .0104
40	2 34 .01	2 .74	0 .3243	0 .00068	2 .18754	768	1 .0107
70 0	2 36 .75	2 .84	0 .3302	0 .00069	2 .19522	777	1 .0111
20	2 39 .59	2 .92	0 .3363	0 .00070	2 .20299	789	1 .0115
40	2 42 .51	3 .02	0 .3426	0 .00071	2 .21088	799	1 .0119
71 0	2 45 .53	3 .12	0 .3491	0 .00072	2 .21887	811	1 .0124
20	2 48 .65	3 .22	0 .3559	0 .00074	2 .22698	823	1 .0129
40	2 51 .87	3 .34	0 .3629	0 .00075	2 .23521	835	1 .0134
72 0	2 55 .21	3 .45	0 .3701	0 .00077	2 .24356	847	1 .0139
20	2 58 .66	3 .58	0 .3776	0 .00078	2 .25203	861	1 .0144
40	3 2 .24	3 .71	0 .3853	0 .00080	2 .26064	875	1 .0150
73 0	3 5 .95	3 .84	0 .3934	0 .00082	2 .26939	890	1 .0156
20	3 9 .79	3 .99	0 .4018	0 .00084	2 .27829	902	1 .0162
40	3 13 .78	4 .15	0 .4106	0 .00086	2 .28731	920	1 .0168
74 0	3 17 .93	4 .31	0 .4197	0 .00088	2 .29651	936	1 .0175
20	3 22 .24	4 .49	0 .4292	0 .00090	2 .30587	954	1 .0182
40	3 26 .73	4 .67	0 .4390	0 .00092	2 .31541	969	1 .0190
75 0	3 31 .40	4 .86	0 .4492	0 .00094	2 .32510	989	1 .0197
20	3 36 .26	5 .09	0 .4598	0 .00096	2 .33499	1008	1 .0204
40	3 41 .35	5 .30	0 .4710	0 .00098	2 .34507	1029	1 .0212
76 0	3 46 .65	5 .54	0 .4826	0 .00100	2 .35536	1049	1 .0220
20	3 52 .19	5 .80	0 .4948	0 .00103	2 .36585	1072	1 .0230
40	3 57 .99	6 .08	0 .5076	0 .00105	2 .37657	1094	1 .0241
77 0	4 4 .07	6 .36	0 .5212	0 .00108	2 .38751	1118	1 .0252
20	4 10 .43	6 .68	0 .5355	0 .00111	2 .39869	1143	1 .0264
40	4 17 .11	7 .03	0 .5506	0 .00114	2 .41012	1171	1 .0281
78 0	4 24 .14	3 .65	0 .5667	0 .00118	2 .42183	597	1 .0299
10	4 27 .79	3 .73	0 .5751	0 .00120	2 .42780	604	1 .0308
20	4 31 .52	3 .86	0 .5837	0 .00122	2 .43384	609	1 .0318
30	4 35 .38	3 .94	0 .5925	0 .00124	2 .43993	617	1 .0328
40	4 39 .32	4 .05	0 .6016	0 .00127	2 .44610	625	1 .0338
50	4 43 .37	4 .16	0 .6109	0 .00129	2 .45235	634	1 .0347
79 0	4 47 .53	4 .28	0 .6204	0 .00131	2 .45869	642	1 .0357
10	4 51 .81	4 .40	0 .6302	0 .00133	2 .46511	650	1 .0367
20	4 56 .21	4 .53	0 .6403	0 .00135	2 .47161	658	1 .0377
30	5 0 .74	4 .65	0 .6507	0 .00137	2 .47819	667	1 .0387
40	5 5 .39	4 .80	0 .6615	0 .00139	2 .48486	677	1 .0398
50	5 10 .19	4 .94	0 .6726	0 .00141	2 .49163	686	1 .0409

TABLE 1. A TABLE of MEAN REFRACTIONS for every 20′ of Zenith Distance to 78°, and for every 10′ from thence to 90° 30′. The Barom. at 29.60 English inches; Bradley's Therm. 50°, or Fahrenheit's 48° 75′.

Zenith Dist.	δθ	Diff.	$\frac{d\delta\theta}{d\tau}$	$\frac{d^2\delta\theta}{2d\tau^2}$	A	Log. δθ	Diff.	λ	Coeffic. $(\tau - 50)^2$
			−	+					
80° 0′	5′ 15″.13		0″.6841	0″.00143	1.0041	2.49849		1.0420	−0″.00003
		5″.08					695		
10	5 20.21		0.6958	0.00145	1.0042	2.50544		1.0431	−0.00003
		5.25					705		
20	5 25.46		0.7079	0.00147	1.0043	2.51249		1.0442	−0.00003
		5.40					716		
30	5 30.86		0.7204	0.00149	1.0045	2.51965		1.0454	−0.00003
		5.58					727		
40	5 36.44		0.7333	0.00152	1.0046	2.52692		1.0466	−0.00003
		5.76					736		
50	5 42.20		0.7468	0.00154	1.0047	2.53428		1.0479	−0.00004
		5.94					747		
81 0	5 48.14		0.7608	0.00157	1.0049	2.54175		1.0493	−0.00004
		6.14					760		
10	5 54.28		0.7754	0.00160	1.0050	2.54935		1.0508	−0.00004
		6.36					772		
20	6 0.64		0.7906	0.00163	1.0052	2.55707		1.0523	−0.00004
		6.57					784		
30	6 7.21		0.8063	0.00166	1.0054	2.56491		1.0540	−0.00004
		6.80					797		
40	6 14.01		0.8227	0.00170	1.0056	2.57288		1.0559	−0.00004
		7.05					811		
50	6 21.06		0.8398	0.00174	1.0058	2.58099		1.0579	−0.00004
		7.30					824		
82 0	6 28.36		0.8576	0.00179	1.0060	2.58923		1.0600	−0.00004
		7.55					838		
10	6 35.91		0.8762	0.00184	1.0062	2.59761		1.0622	−0.00004
		7.84					851		
20	6 43.75		0.8957	0.00189	1.0065	2.60612		1.0646	−0.00005
		8.14					866		
30	6 51.89		0.9160	0.00194	1.0067	2.61478		1.0671	−0.00005
		8.46					883		
40	7 0.35		0.9372	0.00199	1.0070	2.62361		1.0697	−0.00005
		8.78					898		
50	7 9.13		0.9592	0.00204	1.0073	2.63259		1.0725	−0.00005
		9.14					915		
83 0	7 18.27		0.9821	0.00209	1.0075	2.64174		1.0754	−0.00005
		9.51					933		
10	7 27.78		1.0061	0.00214	1.0078	2.65107		1.0784	−0.00005
		9.91					950		
20	7 37.69		1.0312	0.00219	1.0081	2.66057		1.0815	−0.00005
		10.31					968		
30	7 48.00		1.0574	0.00225	1.0084	2.67025		1.0846	−0.00005
		10.75					986		
40	7 58.75		1.0849	0.00231	1.0088	2.68011		1.0879	−0.00006
		11.22					1006		
50	8 9.97		1.1139	0.00237	1.0092	2.69017		1.0914	−0.00006
		11.71					1025		
84 0	8 21.68		1.1445	0.00244	1.0096	2.70042		1.0951	−0.00006
		12.23					1047		
10	8 33.91		1.1768	0.00251	1.0100	2.71089		1.0992	−0.00006
		12.81					1068		
20	8 46.72		1.2109	0.00259	1.0105	2.72157		1.1036	−0.00007
		13.41					1092		
30	9 0.13		1.2470	0.00267	1.0110	2.73249		1.1082	−0.00007
		14.07					1118		
40	9 14.20		1.2851	0.00276	1.0115	2.74367		1.1130	−0.00007
		14.81					1144		
50	9 29.01		1.3253	0.00285	1.0121	2.75511		1.1178	−0.00007
		15.56					1172		
85 0	9 44.57		1.3678	0.00295	1.0127	2.76683		1.1229	−0.00008
		16.36					1199		
10	10 0.93		1.4127	0.00306	1.0133	2.77882		1.1283	−0.00008
		17.22					1228		
20	10 18.15		1.4607	0.00317	1.0140	2.79110		1.1342	−0.00008
		18.14					1255		
30	10 36.29		1.5121	0.00329	1.0147	2.80365		1.1408	−0.00008
		19.12					1287		
40	10 55.41		1.5673	0.00342	1.0155	2.81652		1.1478	−0.00009
		20.19					1317		
50	11 15.60		1.6254	0.00357	1.0163	2.82969		1.1549	−0.00009
		21.37					1352		
86 0	11 36.97		1.6878	0.00372	1.0172	2.84321		1.1624	−0.00009
		22.63					1388		
10	11 59.60		1.7549	0.00389	1.0182	2.85709		1.1706	−0.00009
		24.01					1426		
20	12 23.61		1.8271	0.00406	1.0192	2.87135		1.1794	−0.00009
		25.49					1463		
30	12 49.10		1.9048	0.00426	1.0204	2.88598		1.1888	−0.00009
		27.12					1505		
40	13 16.22		1.9888	0.00447	1.0216	2.90103		1.1989	−0.00009
		28.91					1549		
50	13 45.13		2.0797	0.00471	1.0230	2.91652		1.2098	−0.00009
		30.84					1594		
87 0	14 15.97		2.1783	0.00496	1.0244	2.93246		1.2215	−0.00009
		32.95					1640		
10	14 48.92		2.2854	0.00525	1.0261	2.94886		1.2341	−0.00008
		35.31					1692		
20	15 24.23		2.4024	0.00556	1.0278	2.96578		1.2477	−0.00007
		37.89					1745		
30	16 2.12		2.5304	0.00590	1.0298	2.98323		1.2624	−0.00006
		40.72					1800		
40	16 42.84		2.6706	0.00629	1.0318	3.00123		1.2783	−0.00005
		43.85					1859		
50	17 26.69		2.8245	0.00673	1.0342	3.01982		1.2955	−0.00003
		47.34					1921		
88 0	18 14.03		2.9944	0.00722	1.0368	3.03903		1.3141	−0.00000
		51.23					1988		
10	19 5.26		3.1829	0.00774	1.0397	3.05891		1.3342	+0.00003
		55.58					2057		
20	20 0.84		3.3924	0.00840	1.0429	3.07948		1.3560	+0.00007
		60.42					2133		
30	21 1.26		3.6256	0.00912	1.0465	3.10081		1.3797	+0.00013
		65.88					2210		
40	22 7.14		3.8866	0.00994	1.0504	3.12291		1.4057	+0.00020
		72.05					2297		
50	23 19.19		4.1804	0.01087	1.0546	3.14588		1.4341	+0.00027
		79.01					2385		
89 0	24 38.20		4.5124	0.01197	1.0593	3.16973		1.4653	+0.00038
		86.94					2482		
10	26 5.14		4.8896	0.01323	1.0648	3.19455		1.4995	+0.00054
		96.00					2586		
20	27 41.14		5.3202	0.01483	1.0710	3.22041		1.5373	+0.00077
		106.40					2696		
30	29 27.54		5.8142	0.01659	1.0780	3.24737		1.5789	+0.00103
		118.39					2815		
40	31 25.93		6.3854	0.01883	1.0860	3.27552		1.6252	+0.00137
		132.32					2946		
50	33 33.25		7.0525	0.02150	1.0952	3.30498		1.6767	+0.00182
		148.61					3085		
90 0	36 6.86		7.8298	0.02474	1.1059	3.33583		1.7344	+0.00243
		167.75					3239		
10	38 54.61		8.7542	0.02878	1.1183	3.36822		1.7999	+0.00325
		190.48					3406		
20	42 5.09		9.8487	0.03380	1.1325	3.40228		1.8721	+0.00433
		217.68					3591		
30	45 42.77		11.1748	0.04032	1.1495	3.43819		1.9556	+0.00591

TABLE 2. LOGARITHMIC CORRECTIONS depending on the state of the BAROMETER.

PARTS OF THE ENGLISH FOOT.		PARTS OF THE FRENCH FOOT.				PARTS OF THE FRENCH METRE.			
Inches.	Log.	Lines.	In.	Lin.	Log.	Met.	Log.	Met.	Log.
27.5	−0.03196	315	26	3	−0.02450	0.725	−0.01578	0.760	+0.00469
27.6	−0.03038	316	26	4	−0.02312	0.726	−0.01518	0.761	+0.00526
27.7	−0.02881	317	26	5	−0.02175	0.727	−0.01459	0.762	+0.00583
27.8	−0.02725	318	26	6	−0.02038	0.728	−0.01399	0.763	+0.00640
27.9	−0.02569	319	26	7	−0.01902	0.729	−0.01339	0.764	+0.00697
28.0	−0.02413	320	26	8	−0.01766	0.730	−0.01280	0.765	+0.00754
28.1	−0.02258	321	26	9	−0.01631	0.731	−0.01220	0.766	+0.00811
28.2	−0.02104	322	26	10	−0.01495	0.732	−0.01161	0.767	+0.00868
28.3	−0.01950	323	26	11	−0.01361	0.733	−0.01102	0.768	+0.00925
28.4	−0.01797	324	27	0	−0.01226	0.734	−0.01042	0.769	+0.00981
28.5	−0.01645	325	27	1	−0.01093	0.735	−0.00983	0.770	+0.01037
28.6	−0.01492	326	27	2	−0.00959	0.736	−0.00924	0.771	+0.01093
28.7	−0.01341	327	27	3	−0.00826	0.737	−0.00865	0.772	+0.01150
28.8	−0.01190	328	27	4	−0.00694	0.738	−0.00806	0.773	+0.01206
28.9	−0.01039	329	27	5	−0.00561	0.739	−0.00748	0.774	+0.01262
29.0	−0.00889	330	27	6	−0.00430	0.740	−0.00689	0.775	+0.01318
29.1	−0.00740	331	27	7	−0.00298	0.741	−0.00630	0.776	+0.01374
29.2	−0.00591	332	27	8	−0.00167	0.742	−0.00572	0.777	+0.01430
29.3	−0.00442	333	27	9	−0.00037	0.743	−0.00513	0.778	+0.01486
29.4	−0.00294	334	27	10	+0.00094	0.744	−0.00455	0.779	+0.01542
29.5	−0.00147	335	27	11	+0.00223	0.745	−0.00396	0.780	+0.01597
29.6	−0.00000	336	28	0	+0.00353	0.746	−0.00338	0.781	+0.01653
29.7	+0.00147	337	28	1	+0.00482	0.747	−0.00280	0.782	+0.01709
29.8	+0.00293	338	28	2	+0.00611	0.748	−0.00222	0.783	+0.01764
29.9	+0.00438	339	28	3	+0.00739	0.749	−0.00164	0.784	+0.01820
30.0	+0.00583	340	28	4	+0.00867	0.750	−0.00106	0.785	+0.01875
30.1	+0.00728	341	28	5	+0.00994	0.751	−0.00048	0.786	+0.01930
30.2	+0.00872	342	28	6	+0.01122	0.752	+0.00010	0.787	+0.01985
30.3	+0.01015	343	28	7	+0.01248	0.753	+0.00067	0.788	+0.02041
30.4	+0.01158	344	28	8	+0.01375	0.754	+0.00125	0.789	+0.02096
30.5	+0.01301	345	28	9	+0.01501	0.755	+0.00183	0.790	+0.02151
30.6	+0.01443	346	28	10	+0.01627	0.756	+0.00240	0.791	+0.02206
30.7	+0.01585	347	28	11	+0.01752	0.757	+0.00298	0.792	+0.02261
30.8	+0.01726	348	29	0	+0.01877	0.758	+0.00355	0.793	+0.02315
30.9	+0.01867	349	29	1	+0.02002	0.759	+0.00412	0.794	+0.02370
31.0	+0.02007	350	29	2	+0.02126	0.760	+0.00469	0.795	+0.02425

TABLE 3. LOGARITHMIC CORRECTIONS depending on the state of the INTERIOR THERMOMETER.

FAHRENHEIT'S SCALE.		RÉAUMUR'S SCALE.		CENTESIMAL SCALE.	
Ther.	Log.	Ther.	Log.	Ther.	Log.
−40°	+0.00402	−35°	+0.00432	−35°	+0.00362
−30	+0.00358	−30	+0.00382	−30	+0.00322
−20	+0.00313	−25	+0.00332	−25	+0.00281
−10	+0.00268	−20	+0.00281	−20	+0.00241
0	+0.00223	−15	+0.00231	−15	+0.00200
+10	+0.00178	−10	+0.00180	−10	+0.00160
+20	+0.00134	−5	+0.00130	−5	+0.00120
+30	+0.00088	0	+0.00080	0	+0.00080
+40	+0.00044	+5	+0.00030	+5	+0.00040
+50	+0.00000	+10	−0.00020	+10	+0.00000
+60	−0.00044	+15	−0.00070	+15	−0.00040
+70	−0.00089	+20	−0.00120	+20	−0.00080
+80	−0.00133	+25	−0.00170	+25	−0.00120
+90	−0.00177	+30	−0.00220	+30	−0.00160
+100	−0.00222	+35	−0.00270	+35	−0.00200

TABLE 4.

LOGARITHMIC CORRECTIONS depending on the State of the EXTERIOR THERMOMETER.

BRADLEY'S THERMOMETER.				FAHRENHEIT'S SCALE.				RÉAUMUR'S SCALE.		CENTESIMAL SCALE.	
Ther.	Log.	Ther.	Log.	Ther.	Log.	Ther.	Log.	Ther.	Log.	Ther.	Log.
−25°	+0 .07379	+35°	+0 .01379	−25°	+0 .07245	+35°	+0 .01262	−30°	+0 .08382	−30°	+0 .06925
−24	+0 .07272	+36	+0 .01285	−24	+0 .07138	+36	+0 .01169	−29	+0 .08136	−29	+0 .06735
−23	+0 .07166	+37	+0 .01192	−23	+0 .07031	+37	+0 .01076	−28	+0 .07891	−28	+0 .06545
−22	+0 .07059	+38	+0 .01099	−22	+0 .06925	+38	+0 .00983	−27	+0 .07648	−27	+0 .06357
−21	+0 .06952	+39	+0 .01006	−21	+0 .06819	+39	+0 .00891	−26	+0 .07406	−26	+0 .06168
−20	+0 .06846	+40	+0 .00914	−20	+0 .06713	+40	+0 .00799	−25	+0 .07166	−25	+0 .05980
−19	+0 .06740	+41	+0 .00822	−19	+0 .06607	+41	+0 .00707	−24	+0 .06925	−24	+0 .05794
−18	+0 .06634	+42	+0 .00730	−18	+0 .06502	+42	+0 .00615	−23	+0 .06688	−23	+0 .05609
−17	+0 .06529	+43	+0 .00638	−17	+0 .06397	+43	+0 .00523	−22	+0 .06450	−22	+0 .05424
−16	+0 .06424	+44	+0 .00546	−16	+0 .06293	+44	+0 .00432	−21	+0 .06215	−21	+0 .05240
−15	+0 .06319	+45	+0 .00455	−15	+0 .06188	+45	+0 .00340	−20	+0 .05980	−20	+0 .05056
−14	+0 .06214	+46	+0 .00363	−14	+0 .06084	+46	+0 .00249	−19	+0 .05748	−19	+0 .04874
−13	+0 .06109	+47	+0 .00272	−13	+0 .05980	+47	+0 .00158	−18	+0 .05516	−18	+0 .04692
−12	+0 .06005	+48	+0 .00181	−12	+0 .05877	+48	+0 .00068	−17	+0 .05284	−17	+0 .04511
−11	+0 .05902	+49	+0 .00090	−11	+0 .05773	+49	−0 .00023	−16	+0 .05056	−16	+0 .04331
−10	+0 .05799	+50	−0 .00000	−10	+0 .05670	+50	−0 .00113	−15	+0 .04828	−15	+0 .04152
− 9	+0 .05696	+51	−0 .00090	− 9	+0 .05567	+51	−0 .00203	−14	+0 .04601	−14	+0 .03973
− 8	+0 .05593	+52	−0 .00180	− 8	+0 .05465	+52	−0 .00293	−13	+0 .04376	−13	+0 .03795
− 7	+0 .05490	+53	−0 .00270	− 7	+0 .05362	+53	−0 .00383	−12	+0 .04152	−12	+0 .03617
− 6	+0 .05387	+54	−0 .00360	− 6	+0 .05260	+54	−0 .00473	−11	+0 .03927	−11	+0 .03440
− 5	+0 .05285	+55	−0 .00450	− 5	+0 .05158	+55	−0 .00562	−10	+0 .03705	−10	+0 .03264
− 4	+0 .05183	+56	−0 .00540	− 4	+0 .05056	+56	−0 .00651	− 9	+0 .03484	− 9	+0 .03089
− 3	+0 .05081	+57	−0 .00629	− 3	+0 .04954	+57	−0 .00740	− 8	+0 .03264	− 8	+0 .02914
− 2	+0 .04979	+58	−0 .00718	− 2	+0 .04853	+58	−0 .00829	− 7	+0 .03045	− 7	+0 .02740
− 1	+0 .04878	+59	−0 .00807	− 1	+0 .04752	+59	−0 .00918	− 6	+0 .02827	− 6	+0 .02567
0	+0 .04777	+60	−0 .00896	0	+0 .04651	+60	−0 .01006	− 5	+0 .02610	− 5	+0 .02395
+ 1	+0 .04676	+61	−0 .00985	+ 1	+0 .04550	+61	−0 .01094	− 4	+0 .02395	− 4	+0 .02223
+ 2	+0 .04576	+62	−0 .01073	+ 2	+0 .04459	+62	−0 .01182	− 3	+0 .02180	− 3	+0 .02052
+ 3	+0 .04476	+63	−0 .01161	+ 3	+0 .04350	+63	−0 .01270	− 2	+0 .01967	− 2	+0 .01881
+ 4	+0 .04376	+64	−0 .01219	+ 4	+0 .04251	+64	−0 .01358	− 1	+0 .01754	− 1	+0 .01711
+ 5	+0 .04276	+65	−0 .01337	+ 5	+0 .04152	+65	−0 .01446	0	+0 .01542	0	+0 .01542
+ 5	+0 .04176	+66	−0 .01425	+ 6	+0 .04053	+66	−0 .01533	+ 1	+0 .01333	+ 1	+0 .01374
+ 7	+0 .04076	+67	−0 .01512	+ 7	+0 .03953	+67	−0 .01620	+ 2	+0 .01123	+ 2	+0 .01206
+ 8	+0 .03977	+68	−0 .01599	+ 8	+0 .03854	+68	−0 .01707	+ 3	+0 .00914	+ 3	+0 .01038
+ 9	+0 .03878	+69	−0 .01686	+ 9	+0 .03755	+69	−0 .01794	+ 4	+0 .00707	+ 4	+0 .00873
+10	+0 .03779	+70	−0 .01773	+10	+0 .03656	+70	−0 .01881	+ 5	+0 .00500	+ 5	+0 .00707
+11	+0 .03680	+71	−0 .01860	+11	+0 .03558	+71	−0 .01968	+ 6	+0 .00295	+ 6	+0 .00542
+12	+0 .03582	+72	−0 .01946	+12	+0 .03460	+72	−0 .02054	+ 7	+0 .00091	+ 7	+0 .00377
+13	+0 .03484	+73	−0 .02033	+13	+0 .03362	+73	−0 .02141	+ 8	−0 .00113	+ 8	+0 .00213
+14	+0 .03386	+74	−0 .02119	+14	+0 .03264	+74	−0 .02227	+ 9	−0 .00315	+ 9	+0 .00050
+15	+0 .03288	+75	−0 .02205	+15	+0 .03167	+75	−0 .02313	+10	−0 .00517	+10	−0 .00113
+16	+0 .03191	+76	−0 .02291	+16	+0 .03070	+76	−0 .02398	+11	−0 .00718	+11	−0 .00275
+17	+0 .03094	+77	−0 .02377	+17	+0 .02973	+77	−0 .02484	+12	−0 .00918	+12	−0 .00436
+18	+0 .02997	+78	−0 .02462	+18	+0 .02876	+78	−0 .02569	+13	−0 .01117	+13	−0 .00597
+19	+0 .02900	+79	−0 .02548	+19	+0 .02779	+79	−0 .02654	+14	−0 .01314	+14	−0 .00757
+20	+0 .02803	+80	−0 .02633	+20	+0 .02683	+80	−0 .02739	+15	−0 .01511	+15	−0 .00918
+21	+0 .02706	+81	−0 .02718	+21	+0 .02587	+81	−0 .02824	+16	−0 .01707	+16	−0 .01077
+22	+0 .02609	+82	−0 .02803	+22	+0 .02491	+82	−0 .02909	+17	−0 .01902	+17	−0 .01235
+23	+0 .02513	+83	−0 .02888	+23	+0 .02395	+83	−0 .02994	+18	−0 .02097	+18	−0 .01393
+24	+0 .02418	+84	−0 .02973	+24	+0 .02299	+84	−0 .03078	+19	−0 .02290	+19	−0 .01550
+25	+0 .02323	+85	−0 .03057	+25	+0 .02203	+85	−0 .03161	+20	−0 .02484	+20	−0 .01707
+26	+0 .02227	+86	−0 .03141	+26	+0 .02108	+86	−0 .03246	+21	−0 .02674	+21	−0 .01863
+27	+0 .02132	+87	−0 .03225	+27	+0 .02013	+87	−0 .03330	+22	−0 .02865	+22	−0 .02019
+28	+0 .02037	+88	−0 .03309	+28	+0 .01919	+88	−0 .03414	+23	−0 .03055	+23	−0 .02174
+29	+0 .01942	+89	−0 .03393	+29	+0 .01824	+89	−0 .03498	+24	−0 .03246	+24	−0 .02329
+30	+0 .01848	+90	−0 .03476	+30	+0 .01730	+90	−0 .03581	+25	−0 .03435	+25	−0 .02484
+31	+0 .01754	+91	−0 .03559	+31	+0 .01636	+91	−0 .03664	+26	−0 .03623	+26	−0 .02636
+32	+0 .01660	+92	−0 .03643	+32	+0 .01542	+92	−0 .03747	+27	−0 .03810	+27	−0 .02789
+33	+0 .01566	+93	−0 .03726	+33	+0 .01448	+93	−0 .03830	+28	−0 .03996	+28	−0 .02941
+34	+0 .01472	+94	−0 .03809	+34	+0 .01355	+94	−0 .03913	+29	−0 .04180	+29	−0 .03094
+35	+0 .01379	+95	−0 .03892	+35	+0 .01262	+95	−0 .03996	+30	−0 .04364	+30	−0 .03246

THE

EXPLANATION AND USE

OF THE

TABLES.

THE TABLES OF REFRACTION.

WHEN a spectator, situated on the earth's surface, views a star or other celestial body, either by the unassisted eye, or through a common astronomical telescope, he sees it in that direction, in which a ray of light coming from it enters his eye; and if such direction coincided with a straight line connecting the star with his eye, the apparent and the correct astronomical places of such star would be the same; but as every ray of light, on coming out of a less into a more dense medium, is bent or refracted towards the denser, as may be explained on optical principles, the spectator refers the body viewed to a higher position in a celestial arc than is real, and a correction, to counteract the effect of this illusory appearance, is of the first importance in practical astronomy.

Not only is the amount of this correction greater than that of any other, but its uncertainty, at low altitudes, arising out of physical causes not yet completely developed, renders all calculation from assumed theories inadequate for fixing a star in its true place, under all the variations of altitude of the object, and changes of weight and temperature of the atmosphere. Dr. Bradley was the first astronomer, who made any successful attempt to establish a formula that embraced the three conditions of altitude, weight, and temperature, in assigning the proper quantity of refraction due to a given observation; but it does not appear quite certain, whether his formula was deduced from theory alone, or, what is more probable, from an union of theory with practical observations. Whenever a star is elevated by refraction, the elevation takes place only in a vertical line, the refraction being least, or rather being nothing, in the zenith of any place, and greatest at the horizon, where the stratum of air is usually of the greatest density, and the refracted line of the greatest length.

DR. BRADLEY'S REFRACTIONS.

THE basis of Dr. Bradley's law of refraction was, that, without reference to the changeable state of the atmosphere, the mean quantity of correction for the simple effect of refraction, would vary with, and in a certain ratio to the variation of the zenith distance of the body observed; but the correction depending on this law, was found to require the aid of other auxiliary quantities depending on variations in the atmosphere itself. At length, the subjoined *formula* was adopted by the ingenious and indefatigable observer, viz.:

$$\text{Refraction} = \frac{a}{29.6} \times \text{tang.}\ (z - 3r) \times 57'' \times \frac{400}{350 + h}$$

In which, a = barometer in inches: z = zenith distance of object observed: $r = 57''$ tang. z: h = height of Fahrenheit's thermometer: 29 .6 = the mean standard height of the barometer: $57''$ = the mean refraction at 45°.

The observations that finally led Dr. Bradley to adopt this formula, were partly made on circumpolar stars, during the three years that the brass mural quadrant of Greenwich was turned, that the telescope might view the north; and partly on the sun and southern stars, on the following years, when the same instrument was turned to view the south. At that period, the sun's parallax was taken at $10\frac{1}{3}''$, and the latitude of the Observatory, derived from these series of observations, was understood to be $51^\circ\ 28'\ 39\frac{1}{2}''$; "but," says Dr. Maskelyne (Explanation and Use of the Tables, page v.) "had he made use of the true parallax, $8''.8$ or $8\frac{3}{4}''$, as found by the two late transits of Venus over the Sun, he would have made the refraction, at the altitude of 45°, to be $56\frac{1}{2}''$, instead of $57''$, and the latitude of the Observatory, $51^\circ\ 28'\ 40''$, instead of $51^\circ\ 28'\ 39\frac{1}{2}''$. But," continues the Doctor, "his rule for refractions cannot be corrected for all altitudes, without examining his observations of refractions made at various times."

Dr. Bradley's Table of Mean Refractions for the Barometer at 29 .6 inches, and Thermometer at 50°, and also his Table of Corrections for the Variations of barometrical and thermometrical Measurements, are founded on the latter part of the preceding formula, and have been in constant use at the Royal Observatory of Greenwich for a long succession of years. Experience, however, has proved that the corrected refraction is about 5″ in defect at 80° of zenith distance; and that, at lower altitudes, the scale of corrections becomes too erroneous to be admissible in reducing nice observations. It may, notwithstanding, be necessary to use Dr. Bradley's refractions sometimes, particularly in cases where modern determinations of the declinations or North Polar distances of stars, are compared with the same as determined by him, for the purpose of ascertaining the proper motions; Bradley's being the only remote observations that can be relied on.

The mean refractions are contained in Table 1. for every 10′ of zenith distance, and the decimal factors for computing the corrections arising out of the state of the atmosphere, are given in Table 2.; the logarithms of both which are respectively contained in Tables 3. and 4., which may be used instead of the two former. The mode of ascertaining the amount of the corrected refractions in every case will be sufficiently understood, without formal rules, from an inspection of the subjoined examples. When the *zenith distance* is given, the correction is additive; but when the *altitude* is observed, the correction corresponding to its complement (or zen. dist.) must be subtracted from the observation.

Example 1.——In a certain northern latitude, where the southern zenith distance of α Lyræ was observed to be 13° 48′ 39″, while the barometer indicated 29 .65 inches, and Fahrenheit's thermometer 40°, let it be required to find the proper correction for the effect of refraction on this zenith distance?

Table 1. (page 1) at 13° 40′ gives	13″.83	mean refrac. at bar. 29 .60, ther. 50°
Proportional part 8′ 39″	0 .16	
Table 2. (page 3) dec. factor for bar. and ther. +	0 .38	(viz. 13″.99 × .027)
Whole correction to be added............ =	14 .37	

By logarithms thus:— Table 3. (page 5)	13° 40′ 0″	log.	1 .14082
p. p.	8 39	.....	513
Table 4. (page 8) bar. and ther. ...			10 .01173
Log. of whole correction			1 .15768 = 14″.37 as before.

Example 2.——The altitude of α Cygni was observed below the North Pole, on a clear evening, to be 6° 14′ 30″, when the barometer stood at 30 .5, and thermometer at 65°, what will be the proper correction for refraction under these circumstances?

First 90° − 6° 14′ 30″ = 83° 45′ 30″ = zenith distance.

Table 1. at 83° 40′ 0″ gives	8′ 2″.51
p.p. 5 30	0 9 .92
Table 2. dec. factor .007 × 8′ 12″.43... −	0 3 .45
Whole cor. to be subtracted from the alt. ..	8 8 .98

By logarithms thus:—

Table 3. 83° 40′ 0″.log.	2 .68351
p. p. ... 5 30	564
Table 4. bar. and ther. ..	9 .99702
Whole logarithmic cor. .	2 .68617 = 485″.48 = 8′ 5″.48

The difference in the corrections of this example, as given by the two methods, arises from the second differences being neglected in taking out the proportional part of the logarithmic difference between 83° 40′ and 83° 50′, which is here great.

THE FRENCH TABLES OF REFRACTION.

THE *French* Tables are considered as giving the mean refractions very correctly as far as to 80° of zenith distance; but cannot be implicitly depended on further, particularly near the horizon. These Tables are founded on the profound investigations of La Place (see Mec. Celeste, IV., p. 27.), and on numerous observations made by De Lambre and Piazzi. De Lambre had previously formed a table of mean refractions, as deduced from his own observations, but the application of La Place's formula introduced some slight differences, which have modified it. The tables, thus corrected, were published in 1806, by the French Board of Longitude, and are those which are called the French Tables.

Formula.—If θ be called the apparent zenith distance of a star, metre 0 .76 $(1 + y)$ the observed height of the barometer, x the number of degrees of the centigrade thermometer, the zero of which is at the temperature of thawing ice, and $a = 60''.616$, a constant quantity determined by a multiplicity of observations by Piazzi and De Lambre, the true refraction, or $\delta\theta$, will be obtained from the following expression:

$$\textit{Formula.}\text{—}\delta\theta = \frac{a(1+y)\tan.\theta}{(1+0.00375x)\left(1+\frac{x}{5412}\right)} + \frac{\frac{1}{2}a^2(1+y)}{(1+0.00375x)\,1+\frac{x}{5412}} \cdot \frac{(1+2\cos.^2\theta)\tan.\theta}{\cos.^2\theta} -$$

$$- \frac{a(1+y)}{(1+0.00375x)\,1+\frac{x}{5412}} \times 0.00125254\frac{\tan.\theta}{\cos.^2\theta} - a.0.00375x.0.0125254\frac{\tan.\theta}{\cos.^2\theta}.$$

This formula, however, is not made to extend further than 74° of zenith distance, and the following formula of sexagesimal seconds is substituted for the zenith distances greater than 74°; at which point both the formulæ give the same result.

$\delta\theta = 2790''.2\ (0.75479 - 0.049042T^2)\sin\theta\,\frac{2\Psi T}{\sqrt{\pi}} + 10021''.4\ \sin.2\theta$; in which formula $T = 25.961924\cos.\theta$, and

$\Psi T = c^{T^2} \int dt\, c^{-t^2}$, the integral being taken from $t = T$ to infinity, c being the number, the hyperbolic logarithm of which is unity, and π the ratio of the semi-circumference to radius. In the calculated Table, the logarithm of each refraction was given to four places, and the refraction is thence deduced in natural numbers.

The French Refractions were originally published in eight successive Tables; the first Table converted the measure of the old French inches into English inches, as a scale for the barometer; the second converted the degrees of Réaumur's thermometer into the centigrade division; the third converted the degrees of Fahrenheit's thermometer into centesimal degrees; the fourth gave the logarithms of the mean refractions for whole degrees as far as 80°, and for every 10′ from thence to the horizon, on a supposition of y being $= 0$, and $x = 10°$ of the centigrade thermometer; the fifth contained the fourth term of the first expression of $\delta\theta$, in supposing $x = 10$. This term was calculated for the arc lying between 66° and 85°, it being insensible near the zenith, and the refractions themselves being cousidered too uncertain near the horizon; the sixth contained the logarithms of the factor $(1 + y)$; the seventh comprised the logarithms of the co-efficient $\frac{1}{(1+0.00375x)\left(1+\frac{x}{5412}\right)}$, and Table eight contained the logarithms, used as arguments, for finding the natural numbers of the whole refraction as finally corrected. In the *Connoissance des Tems*, a selection is made from these Tables, containing only the mean refractions themselves, and the factors for the barometrical and thermometrical corrections, whereas, in the Greenwich Tables, the logarithms with their corresponding mean refractions for every 10′ and the logarithmic factors for the corrections depending on the English barometer and Fahrenheit's thermometer, are all that are deemed necessary to be given. Tables 1 and 2 are here given according to the Greenwich arrangement, from page 10 to page 13, both inclusive, and the comparative scales to three different barometers, and as many thermometers, are inserted in page 18, which may be referred to whenever occasion may require, whichever of the Tables of Refraction be used with a Foreign barometer and thermometer. The small French original Table 5 is omitted in both these selections, and also in the present collection, on which account it is here subjoined as Table 3 for the use of those astronomers who may have occasion to observe low stars.

TABLE 3. (or French 5.)

Zenith Distance.	—
67	0″. 0
68	0 . 1
74	0 . 1
75	0 . 2
77	0 . 2
78	0 . 3
79	0 . 4
80	0 . 5
81	0 . 7
82	1 . 0
83	1 . 6
84	2 . 5
85	4 . 3

When the logarithms taken out of Tables 1 and 2 are added together, and their corresponding refractions determined from any common table of logarithms, the small quantity contained in Table 3 must be subtracted therefrom, and the remainder will be the correct refraction, when the zenith distance requires Table 3 to be used; but in all smaller zenith distances, Tables 1 and 2 will be sufficient. The same examples are worked by all the different Tables of Refraction, in order that a comparative view may be taken of their relative advantages, as to facility in the operation, and accuracy in the results.

Example 1. (Zenith distance 13° 48′ 39″. Barom. 29 .65. Fahr. ther. 40°.)

Table 1. (page 10) zenith distance 13° 40′ 0″ log. 1 .1511
Proportional part.............. 8 39 .... .0048

Log. of mean refraction .... 1 .1559 for bar. 29 .94, and ther. 50°.

Table 2. log. of barom. 29 .65.................. 9 .9958
Do. log. of therm. 40°.................. 0 .0092

Log. of whole correction, or refrac. .... 1 .1609

True refraction, 14″.49, to be subtracted from the apparent zenith distance, or added to the apparent altitude.

Example 2. (Zenith distance 83° 45′ 30″. Barom. 30 .50. Ther. 65° Fahr.)

Table 1.........83° 40′ 0″ log. 2 .6872 at bar. 29 .94, ther. 50°.
p. p. .... 5 30 .... .0055
Table 2. log. bar. 30 50 .... .0081
Do. log. ther. 65° .... 9 .9864

Log. of the refrac.............. 2 .6872 = 485″.52
Table 3. deduct in this case p. p. .... 2 .29

483 .23

Hence 8′ 3″.23 is the corrected refraction.

MR. GROOMBRIDGE'S REFRACTIONS.

In the Philosophical Transactions of London, 1810, Dr. Brinkley proposed the following formula to be substituted for Dr. Bradley's, viz:

$$R = 56''.9 \times \text{tang.}\ (z - 3.2\,r) \times \frac{a}{29.6} \times \frac{500}{450+h},$$

which formula arose out of various observations made on certain circumpolar stars: and, about the same time, or soon after, Mr. Groombridge suggested a further alteration, derived from a series of careful observations made with his four foot mural circle (by Troughton) on fifty circumpolar stars of different zenith distances, down to within 2° of the horizon of Blackheath; his latitude deduced from these observations was 51° 28′ 2″.103; the mean refraction at 45° = 58″.1192; and the co-efficient of r = 3 .3625, as determined by Polaris. The formula arising out of these determinations will stand thus, without reference to the state of the atmosphere, viz.: tang. $(z - 3.3625r)$. 58″.1192 = the mean refraction. Then, from a mean of 210 observations on the effects produced by changes of atmospheric temperature, the following formulæ were proposed for producing the requisite corrections of the mean refractions, viz.: $(49° - h)$ .0024, when h denotes the degree *below* the mean of Farenheit's thermometer; $(49° - h)$ .0023, when the degree is *above* the mean; and, in both cases, the thermometer is supposed to be placed *within* the Observatory: but when the position is *external*, the correction must be $(45° - h)$ .0021; the mean state of the barometer being considered at 29 .6 English inches.

In the year 1814, the same Astronomer published still further alterations in his co-efficients, which he was induced to do from his numerous observations on stars exceeding 80° of zenith distance, and his formula, as finally settled, for mean refractions, now stands thus, viz.: tang. $(z - 3.63429r)$ 58″.132967; the corrections for temperature remain as before, when the thermometer is *internal*, but when external he has substituted $(45° - h)$ .0020.

The co-efficient of r, as thus finally determined, will not, however, be applicable further than to the zenith distance 87°; at 88°, the co-efficient becomes 3 .35709; at 89°, 3 .07989; and at 90°, only 2 .80269. From these respective co-efficients of r, Table 1 was computed, and the change of the co-efficients at the points 87°, 88°, 89°, and 90°, will account for the want of uniformity in the second differences, if examined at those points in the Table.

The corrections of the mean refractions, for the state of the barometer and thermometer, are so contrived, that the *sum* of the two tabular numbers taken from Table 2, becomes the proper factor for the correction of the mean refraction, without the use of logarithms.

Example 1. (Zenith distance 13° 48′ 39″. Barom. 29 .65. Ther. within 40°.)

Table 1. (page 14) zen. dist...............	13° 40′ 0″ gives	0′ 14″. 12
p. p..................................	0 8 39	0 0 . 17
Mean refrac. at bar. 29 .60. Ther. within 49°..............		0 14 . 29
Table 2. (page 16) bar. 29 .65 No. + .0017		
Ditto .... ther. 40 within + .0207		
Factor or sum.................. + .0224 × 14″. 29 =		+ 0 0 . 32
The corrected refraction		0 14 . 61

Example 2. (Zenith distance 83° 45′ 30″. Bar. 30 .50. Ther. within 65°.)

Table 1. (page 15.) zen. dist...............	83° 45′ 0″ gives	8′ 11″. 45
p. p..................................	0 0 30	0 0 . 57
Mean refract. at mean temp...............................		8 12 . 02
Table 2. bar................. + .0304		
Ditto... ther................ — .0352		
Sum for factor............... — .0048 × 8′ 12″. 02 =		— 0 2 . 36
Corrected refraction......................................		8 9 . 66

PIAZZI'S REFRACTIONS.

This celebrated Astronomer, regardless of theoretic formulæ, constructed a Table of Mean Refractions from his own observations alone, partly from circumpolar stars observed above and below the pole by his circular instrument, by Ramsden, that reverses in azimuth as well as in altitude; and partly by other stars, the observed zenith distances of which he compared with computed zenith distances, in given azimuths, as the stars ascended and descended, and the differences in the results indicated the corresponding elevations by refraction, while the variations of those differences pointed out the effects produced by changes of atmospheric pressure and temperature. The stars which were most frequently observed in this way were α Canis Minoris, α Lyræ, and α Tauri. The first of these was observed in 32 different zenith distances, between 38° and 50° both inclusive; the second was observed at 41 zenith distances, beginning with 61° and ending with 73° 48′; and the third had 46 different distances measured from 70° to 89° 30′. The intermediate points of zenith distance were taken by means of other stars, or were interpolated; and, what is very remarkable, the scale of refractions, thus practically determined, agrees with Bradley's formula, in a manner that is surprising. I have given Piazzi's Table in its original state, in entire degrees, and therefore shall omit exemplifying its use; particularly as it is not accompanied by the requisite corrections for the weight and temperature of the atmosphere. For the same reasons, also, the small Tables of Cassini, La Caille, and Bouguer, in page 289, are inserted merely by way of contrast with one another, and to show what necessity there existed of adapting Tables of Refractions to different climates, and of improving them from time to time, as the progressive improvements in astronomical instruments demanded.

DR. BRINKLEY'S REFRACTIONS.

In the 12th Volume of the Transactions of the Royal Irish Academy, Dr. Brinkley has written two elaborate papers on Astronomical Refractions, in the latter of which he has shown that, whatever law of variation in the density of the atmosphere be assumed, a formula, in the shape of a fluxional expression, may be determined, that will give the mean refractions correctly as far as to 74° of zenith distance. La Place had done this in his Mec. Celeste, and had integrated the expression by approximation; but Dr. Brinkley has done the same thing in a more simple way, by using the common principle of the given ratio of the sines of incidence and of refraction, which principle, besides it simplicity, has the advantage of avoiding hypothetic principles respecting the rays of light. For the investigation of the fluxional equation for refraction, and also for the approximate integration thereof, proposed by Dr. Brinkley, the reader may consult the original paper both with advantage and pleasure; it being too long to be quoted, and too interesting to be abridged. The resulting refractions accord with those of De Lambre, given in the French Tables, as low as to 80° of zenith distance, but the form of Dr. Brinkley's Tables, derived from his investigations and observations, is quite different, and nearly as convenient in the application, except at low altitudes. From a comparison of the co-lat. of Dublin Observatory determined by stars near the Pole, with the same determined by stars more remote, this eminent Astronomer found, by 525 observations of circumpolar stars, that when the barometer is $= 29.60$, and thermometer $= 50^\circ$, the mean refraction at 45° will be $= 57''.42$, the same being by the French Tables $= 57''.57$; and, what will appear very extraordinary, a general expression for refraction at any zenith distance, has been derived from experiments, independently of astronomical observations, that gives the refraction at the same point $= 57''.67$. The experiments were those of MM. Biot and Arago on the refractive force of air, and of Mr. Dalton and M. Gay-Lussac on the effects of the change of temperature on the density of air. Dr. Brinkley conceives that, from the nature of the *direct* experiments here alluded to, the result is entitled to a preference over the result derived from astronomical observations, where the principle of differences may be considered as somewhat vague. Dr. Brinkley puts, as La Place did before him, $\theta =$ zenith distance; $m : 1 =$ the ratio of the sine of incidence to the sine of refraction; $a =$ the radius of the earth; $l =$ the height of an uniform atmosphere; and $R =$ the refraction determined by the *formula*, which is this:

$$R = \frac{(m-1)\tan.\theta}{\sin. 1''} - \frac{(m-1)\, l \tan.\theta}{a \cos.^2\theta \sin. 1''} + \frac{(m-1)^2 \tan. 3\theta}{2 \sin. 1''}.$$

If the refractive force of air be assumed proportional to its density, the value of m will vary, and its changes will be known by the variations of the barometer and thermometer. Let m' be the value of m, when the height of the barometer $= 29.60$ inches, and the height of Fahrenheit's thermometer $= 50^\circ$; let also b represent the height of the barometer, and t the height of the thermometer corresponding to m. It appears, by the results of the experiments of Dalton and Gay-Lussac, that a column of air denoted by unity at the temperature of 32° of Fahrenheit becomes 1.375 at the temperature of boiling water; it becomes, therefore, for t degrees of the thermometer $= 1.00283\,(t-32)$. The increase of height in the barometer from the expansion of mercury by increase of temperature, may be considered $\frac{1}{10000}$ for every degree of the thermometer. Hence $m-1 : m'-1 ::$ density of air barometer b, and thermometer t : density barometer 29.60, and thermometer $50^\circ :: b \times \left(1-(t-50)\times,0001\right) \times 1.0375 : 29.6 \times \left(1+.002083\,(t-32)\right)$.

Therefore $m-1 = (m'-1 \times \frac{b}{29.60} \times \left(1-(t-50)\,0001\right) \times \frac{1.0375}{1+002083\,(t-32)}$. The quantity $(m'-1)$ may be deduced from the experiments of Biot and Arago. When the metrical barometer is at 0.76 and centesimal thermometer at 0, viz. English barometer $= 29.93$ inches, and Fahrenheit's thermometer $= 32^\circ$, these philosophers have found $m-1 = .0002946$. Hence $m'-1 = \frac{.0002946 \times 29.60}{1.0081 \times 1.0375 \times 29.93} = .0002803$; and $\frac{m'-1}{\sin. 1''} = 57''.82$.

The height of an uniform atmosphere is not affected by the variation of the barometer, and therefore if l' represent the height of an uniform atmosphere, the thermometer being $= 50^\circ$ $l = l' \times \frac{1+00208\,(t-32)}{1.0375}$; a very accurate value of $\frac{l'}{a}$ is not required. If we take $l' = 5.095$ miles, and the semi-diameter of the earth $= 3979$ miles, then $\frac{l'}{a} = \frac{5.095}{3979} = .00128$. It is evident that the third and fourth terms of the value of the refraction in the foregoing equation, cannot be sensibly affected by the variation of m, and therefore its mean value may be used as to these terms.

Hence substituting for $\frac{m-1}{\sin. 1''}$ and $\frac{l'}{a}$ in the equation, we have

$$\text{Refraction} = \frac{1.0375}{1+,002083\,(t-32)} \times \left(1-,0001\,(t-50)\right) \frac{b}{29.60} \times 57''.82 \text{ tang. } \theta - \frac{b}{29.60} \times 0''.0739 \frac{\tan.\theta}{\cos.^2\theta} +$$
$$+\ 0''.000238 \frac{\tan.^3\theta}{\cos.^2\theta} + 0,0080 \tan.^3\theta.$$

[See also the Appendix to Dr. Brinkley's Elements of Astronomy, where this law of atmospheric refraction is investigated.]

Dr. Brinkley computed two sets of Tables of Refractions agreeably to the preceding formula, one beginning with 1° of zenith distance, and ending with 80°, and the other beginning with 80°, and ending with 90°. The principal Table in the first set contains the logarithmic tangents of the zenith distances as far as from 1° to 80°, but from 80° to 90° the calculations, according to the formula, depend on a series of recent observations, made at low altitudes. The second, or thermometrical Table, is the same in both sets; the third Table is exclusively appropriated to zenith distances less than 80°; and the fourth to zenith distances greater than 80°; besides which the logarithm of the barometer, as taken out of an ordinary Table of logarithms, is necessary, to complete the calculation of the correct refractions. These Tables, calculated at different times, and disunited, have been here consolidated thus: Table 1. now contains the logarithm of the English barometer to every hundredth part of an inch; Table 2. contains the logarithm of the correction for Fahrenheit's thermometer; the logarithmic tangent of the zenith distance will be most conveniently taken out of a common book of logarithmic sines, tangents, &c. to one minute, or even less, by inspection, and therefore is here omitted. Table 3. contains a barometrical correction depending on the zenith distance, when 80°, or less; Table 4. contains the correction for low altitudes; and Table 5. gives the logarithm of a correction to be used when the thermometer is placed near the barometer. This arrangement reduces the two sets of Tables given by the author, into one regular series, which must be used according to the subjoined rule: viz.

Log. tang. of zen. dist. + log. of barom. Table 1. + log. ther. Table 2. = log. of apparent refraction; then the apparent refraction + cor. in Table 3. = mean refraction for zenith distances not exceeding 80°.

But for zenith distances exceeding 80°, the rule is more complex, thus:

1. Log. Table 4. + Arith. co. Table 2. + Table 5. = log. A.
2. Log. Table 4. + Table 2. + log. bar. + 7 .2773 = log. B.
3. Log. Table 2. + log. bar. + log. tang. (zen. dist. − A + B) = log. of refrac. in seconds.

Example 1. (Zenith distance 13° 48′ 39″. Bar. 29 .65. Ther. 40°.)

Log. tang. zenith distance.... 13° 48′ 39	....	9 .3906
Table 1. log. barom. (page 17)	....	1 .4721
Table 2. log. therm.	....	0 .2992
Mean refraction...........	14″. 52 corresponding to	1 .1619
Sub. No. in Table 3.	0 . 00 (page 18.)	
Correct refraction.........	14 . 52	

Example 2. (Zenith distance 83° 45′ 30″. Bar. 30 .50. Ther. 65°.)

Table 4. zenith distance.............	83° 0′ 0″.... log.	1 .4981
p. p..................................	45 30 .... log.	0 .0413
		1 .5394
Table 2. ar. co. log.	....	9 .7236
Table 5. log.	....	0 .2932
Minutes in log. A 35′.99 or 35′ 59″	....	1 .5562 = log. A
Also Table 4. log.	....	1 .5394
Table 2. log.	....	0 .2764
Log. barom.	....	1 .4843
Const. log.	....	7 .2773
Minutes for log B 3′.78 or 3′ 47″	....	0 .5774 = log. B

Hence......Zen. dist. − A + B

	83°	45′	30″
−	0	35	59
+	0	3	47

83 13 18 log. tang....	10 .9249
Table 2................	0 .2764
Log. barom. 30 .50	1 .4843
Log. of correct refraction .	2 .6856 = 484″. 85 or 8′ 4″. 85.

DR. YOUNG'S REFRACTIONS.

Dr. Young's Table of Refractions is computed on principles explained by him in the Philosophical Transactions of London for 1819, and is said to agree more perfectly with the latest observations than any other Table that has been published.

$$\textit{Formula.}\quad .0002825 = v\frac{r}{s} + (2 .47 + .5\,v^2)\,\frac{r^2}{s^2} + 3600\ v\,\frac{r^3}{s^3} + 3600\,(1 .235 + 25\,v^2)\,\frac{r^4}{s^4};$$

r being the refraction, v the sine of the altitude, and s the cosine.

The apparent altitude being found in the first column, the second shows the refraction when the barometer stands at 30 inches, which is its mean height on the level of the sea, and the thermometer at 50° of Fahrenheit. The third column contains the difference to be subtracted or added for every minute of altitude, reckoned from the nearest number in the first column. The fourth shows the number of seconds to be added for every inch that the height of the barometer exceeds 30, or to be subtracted for each inch that it wants of 30; and the last column contains the number of seconds to be subtracted for each degree that the thermometer stands above 50°, or to be added for each degree that its height wants of 50°.

If great accuracy be required, we must also deduct, from the observed height of the barometer, .003 for each degree that the thermometer near it is above 50°, and add an equal quantity for an equal depression. In fact, however, the Table, as it now stands, is found to require the temperature to be estimated from the height of the thermometer within; and if we employed the height of the thermometer without, which would be more consistent with the theory, it would probably be necessary to suppose the standard temperature of the Table 48° only, instead of 50°.

Example 1.

(Zen. dist. 13° 48′ 39″. Barom. 29.65. Ther. 40°.)

90° − 13° 48′ 39″ = 76° 11′ 21″ = the altitude.

Then alt. 76° 0′ 0″ .... gives	0′ 14″.4 (p. 19.)
11′.35 × .018 (diff. for 1′)	−0 .2
Mean refraction	0 14 .2
Diff. of barom. 0 .35 × 0 .48	−0 .17
Diff. of ther. 10 × .029	+0 .29
True refraction	0 14 .32

Example 2.

(Zen. dist. 83° 45′ 30″. Barom. 30 .50. Ther. 65°.)

90° − 83° 45′ 30″ = 6° 14′ 30″ = the altitude.

Alt. 6° 10′ 0″	8′ 20″.0
p. p. 4 30	−5 .4
Mean refraction	8 14 .6
Diff. of barom. 16″.7 × .5	+8 .35
Diff. of ther. 1 .10 × 15	−16 .50
True refraction ....=	8 6 .45

CARLINI'S REFRACTIONS.

THE Refractions of Carlini, according to the Milan Almanac, are calculated from the following

Formula. $R = 1624'' \sin. \Theta \{(1.2824065 - 1.4351870T^2)\Psi + 0.7175935T\}$

in which Θ is the apparent zenith distance.

$$T = 28 \cos. \Theta,$$

$$\Psi = e^{TT}\int e^{-tt}\,dt, \text{ the integral being taken from } t = T \text{ to } t = \infty.$$

In zenith distances not greater than 80°, the value of R, converted into a series, has been used thus:

$$R = 58'' \text{tang. } \Theta\ 1 - 1.7175935\left(\tfrac{1}{2}T^2 - \frac{2.3}{4T^4} + \frac{3.3.5}{4T^6} - \&c. - \frac{1.3}{4T^4} - \frac{2.3.5}{8T^6} + \frac{3.3.5.7}{16T^8} - \&c.\right)$$

The refraction for the altitude of 28 inch. + x lines. of the Parisian barometer, and $10 + y$ degrees of Réaumur's thermometer will be had by multiplying R by $\left(1 + \frac{x}{28 \times 12}\right)\frac{1}{1 - 0.0047086 \times y}$.

Let $1 + \frac{x}{28 \times 12} = 1 + A$, $\frac{1}{1 - 0.0047086 \times y} = 1 + B$, then the refraction sought will be $= R + R(A + B + AB)$ and its logarithm $= \log. R + \log. (1 + A) + \log. (1 + B)$.

Table 1. contains the mean refraction in minutes and seconds for the height of 28 inches of the Parisian barometer, and for the temperature of 10° of Réaumur's thermometer; or for 29.851 of the English barometer, and 54 .5 of Fahrenheit's thermometer. From 60° of zenith distance downwards, there is placed, in a parallel column by the side of the refraction, the logarithm of the refraction reduced to seconds.

Table 2. contains A, and log. (1 + A) for the Parisian barometer; Table 3., in like manner, gives B, and log. (1 + B) for Réaumur's thermometer; and the mean refraction obtained from these three tables only, will suffice where the zenith distance is less than 80°, and when the barometer and thermometer above specified are noted: but in observations nearer the horizon it is necessary to apply to the mean refraction thus found, another correction C taken from Table 4., which is found by multiplying the tabular quantity by y, *i. e.* by the degree of the thermometer above 10°. The number C results from the formula $- 14''.093 \sin. \Theta \{(1 + 2T^2)\Psi + T\}$.

Tables 5, 6, and 7, give the values of the same quantities, viz.: A, log. (1 + A), B, log. (1 + B), and C, corresponding to the height of the English barometer and degree indicated by Fahrenheit's thermometer; but, in this case, the number C taken from Table 7. must be multiplied by the observed degree of the thermometer lessened by 54°.5, which is the temperature of the first Table, according to Fahrenheit's thermometer.

Example 1.

(Zen. dist. 13° 48′ 39″. Eng. barom. 29 .65. Fahr. therm. 40°.)

Table 1. zen. dist. (p. 290) ..	13° 0′ 0″ ..	13″. 40
p. p.	48 39 ..	0 . 81
Mean refraction		14 . 21=R
A Table 5. −.0067 (p. 291.)		
B Table 6. +.0313		
Hence AB −.0002		
A+B+AB +.0244 × R		0 . 33
Refraction corrected..................		=14″. 54

Otherwise,

By Logarithms.

Log. R..............................	1 .1526
Log. (1 + A) Table 5.	9 .9970
Log. (1 + B) Table 6.	0 .0134
Log. of correct refraction...............	1 .1630

Correct refraction 14″.55

Example 2.

(Zen. dist. 83° 45′ 30″. Eng. barom. 30 .50. Fahr. therm. 65°.)

Table 1. zen. dist.	83° 40′ 0″..	8′ 3″. 50
p. p.	5 30 ..	12 . 76
Mean refraction		8 16 .25=R
A Table 5. +0 .0217		
B Table 6. −0 .0215		
Hence AB −0 .0005		
A+B+AB −0 .0003 × R		−0 . 15
C Ta. 7. (z. d. 83° 45′ 30″) .08 × 10 .5 (65° − 54°.5)		−0 . 84
Correct refraction.........................		8 15 . 26

Otherwise,

By Logarithms.

Log. 496″.25	2 .6957
Log. (1 + A).........................	0 .0093
Log. (1 + B)	9 .9905
Log.	2 .6955

Refraction..............	8′ 16″. 03
Table 7. C ....−	0 . 84
	8 15 . 19

BESSEL'S REFRACTIONS.

No modern astronomer has been so assiduous, perhaps, or so successful, in promoting both the theory and practice of astronomy, as Bessel. In Section IV. of his "Fundamenta Astronomiæ," he has investigated the subject of refraction, and, retaining the characters of La Place, has deduced the following formula, on which he calculated a set of Tables that comprehend quantities even too minute to be generally serviceable in practical astronomy.

$$\textit{Formula.}\quad \left\{\delta\theta + \frac{d\delta\theta}{d\tau}(\tau - 50) + \frac{d^2\delta\theta}{1.2.d\tau^2}(\tau - 50)^2\right\} \times \left\{1 + \frac{d\delta\theta}{\delta\theta.db}\left(\frac{b : 29.6}{1 + (\tau' - 50)\,0.0001025} - 1\right)\right\}$$

where τ = the degree of the external thermometer; b = the English barometer; and τ' = the degree of the thermometer attached to the barometer. The different portions composing the formula are placed over the columns in the Table that contain the corresponding quantities. The exponents, λ and A, in the Table, were so determined, that $-\lambda.\ 0.00208333\delta\theta = \frac{d\delta\theta}{d\tau}$; and $A.\delta\theta = \frac{d\delta\theta}{db}$; but the quantities represented by them may be dispensed with in high altitudes; λ commences at zenith distance 45°, and, in the original Table, A at 77°; but it is made here to commence at 80°, for the sake of arranging the columns so as to complete the page. The subsidiary Tables, for the use of the different barometers and thermometers, are derived from the subjoined formulæ, viz.:

Barometer in English inches and parts		log. b − 1 .47129
	Paris inches and lines	log. b − 2 .52281
	Metres	log. b + 0 .12388
Interior Thermometer (τ')	of Fahrenheit 0 .00000	− log. [1 + (τ' − 50) 0 .0001025]
	of Réaumur .. 0 .00080	− log. (1 + τ' 0 .00023105)
	Centesimal ... 0 .00080	− log. (1 + τ' 0 .00018484)
Exterior Thermometer (τ)	of Bradley ... 0 .00000	− log. [1 + (τ − 50°) 0 .0020833]
	Fahrenheit . −0 .00113	− log. [1 + (τ − 50°) 0 .0020779]
	Réaumur .. +0 .01543	− log. (1 + τ. 0 .0048570)
	Centesimal +0 .01543	− log. (1 + τ. 0 .0038856)

In using Bessel's Tables, the quantities, log. $\delta\theta$; A; and λ; must be taken out, in the first place, from Table 1. (p. 292 to 294); then the log. tabular quantities for the barometer and *interior* thermometer from Tables 2. and 3. respectively (p. 295), the sum of which, according to their signs, must be multiplied by A; afterwards the log. tabular quantity belonging to the argument for the *exterior* thermometer must be taken from Table 4., and multiplied by the quantity λ. The two products, thus obtained, added to the log. $\delta\theta$, give the logarithm of the true refraction; and to the true refraction thus obtained, may, lastly, be added, if deemed necessary, the small quantity out of the last column, viz.: $\frac{co\text{-}effic.}{(\tau-50)^2}$, which is of the second order.

If, instead of the logarithmic calculation, the refractions themselves be taken out of the Table, using also the fourth and fifth columns, for the small quantities, the logarithm of the amount may be taken out of a Table of common logarithms, and substituted for the log. $\delta\theta$, or log. of mean refraction, as given in Table 1. When the zenith distance of the object is greater than 45°, both λ and A may be omitted, as in the first example; if greater than 45°, and less than 80°, then λ may be used without A; but if greater than 80°, then both λ and A may be used, as in the second example.

Example 1.

(Zen. dist. 13° 48′ 39″. Eng. barom. 29.65. Fahr. therm. 40°.)

Table 1. zen. dist.	13° 40′ 0″	log. 1.1460
p. p.	8 39	0.0047
		1.1507
Table 2. Eng. barom. 29.65		+0.0073
Table 3. Fahr. therm. 40° (interior)..		+0.0004
Log. of correct refraction		1.1584 = 14″.40

Or thus, by taking in minor quantities,

Table ...	13° 40′ 0″ ..	14″.0	
p. p......	8 39 ..	0.15	
$\frac{d\delta\theta}{\delta\tau}$		−0.03	
$\frac{d^2\delta\theta}{2d\tau^2}$		+0.00	
		14.12 ..	log. 1.1498
Table 2. Eng. barom.			log. 0.0073
Table 3. Fahr. therm. (interior).....			log. 0.0004
Log. of correct refraction..............			1.1575 = 14″.37

Example 2. (Zenith distance 83° 45′ 30″. English barom. 30.50. Fahr. therm. interior 65°.)

Table 1. .... 83° 40′ 0″	log. 2.68011 ..	A 1.0090 ..	λ 1.0896
p. p......... 5 30	log. 0.00553		
Log. of mean refraction	2.68566	2.68566	
Table 2. English barometer 30.50	+0.01301		
Table 3. Fahrenheit's thermometer interior .	+0.00066		
	.01367 × 1.0090 =	0.01379	
Log. of whole correction, using the internal thermometer alone		2.69945 = 500″.55	or 8′ 20″.55

If the *exterior* thermometer were also given, say = 60°, then the further correction for λ would be in Table 4. − 0.01006 × 1.0896 = −.01096, and consequently the log. = 2.68849 .... = 488″.08

Also (65° − 50°)² × −0.0006 from the last column, gives.................... −0.013

488.067 or 8′ 8″.067

Hence it appears important, according to Bessel's method, that both the internal and external thermometers should be noted.

Gauss has lately transformed Bessel's Refractions by a convenient logarithmic arrangement, which gives the correction as far as 79° of zenith distance, but no further. He puts ζ = tang. of zenith distance, a = the log. for the barometer of the old French measure in Table I.; b = the logarithmic number for Réaumur's thermometer in Table II., and ζ, c, and λ, in Table III.; the logarithms being composed of five places, under the units' places of which, the numbers in column C must be placed: then a + log. tang. $\zeta + \lambda b - c - 10t$ = the correction, or proper refraction. But as this arrangement is peculiarly adapted to the use of the old French barometer, and Réaumur's thermometer, it would be of little use in an English observatory. Indeed Dr. Brinkley's Refractions are taken out in a manner very similar, and with quite as much convenience, as far as Gauss's Tables extend. [Vide Schumacher's "Sammlung von Hülfstafeln, 1822."]

In the third cahier of Baron Zach's Correspondence, Vol. VII., Bessel has announced his intention of publishing, in his seventh *Recueil*, a new set of Tables of Refraction, which he has lately calculated, and which he considers as more correct than those which he gave in his "Fundamenta Astronomiæ." They are founded on observations of circumpolar and other stars, recently made with his new circular instrument by Reichenbach, after he had tabulated the errors of division, and of flexure, and had made his new Catalogue of the Declinations of Dr. Maskelyne's 36 Stars. With these refractions he has made the summer and winter solstices produce nearly the same obliquity of the ecliptic; and the latitude determined by stars in every possible position of the heavens, will now come out as nearly the same as the errors of observation will admit. The declinations of Messieurs Piazzi, Oriani, Pond, and Brinkley, are, however, opposed to those of Bessel, and the differences are, several of them, so considerable, that amateurs possessing inferior instruments will wait with anxious suspense the developement of this apparent mystery; while the astronomers conducting the public institutions will not fail to re-examine their respective instruments, for the purpose of detecting where the cause of this discrepancy lies. In the mean time, it may be acceptable to the reader to be informed, that Bessel's last Tables, not yet published, are thus determined: Let ϱ be the refraction of the Tables of the *Fundamenta Astronomiæ*; A and λ, the exponents; h, the height of the barometer in lines of the Parisian foot; τ', the degrees of the centesimal thermometer attached to the barometer; and f, the degrees of Fahrenheit's thermometer exposed to the north: then the true refraction is thus found, viz.:

$$= \varrho . 1 .003282 \left\{ \frac{h}{232,28} \cdot \frac{5550 + 10}{5550 + \tau'} \cdot \frac{53700 + \tau'}{53700 + 10} \right\}^{A} \left\{ \frac{180 + 16,75 . 036438}{180 + f - 32) . 036438} \right\}^{\lambda}$$

Comparative Refractions resulting from the two Examples, as worked from the various Tables.

	First Example.	Second Example.
Dr. Bradley's Tables	14″. 37	8′ 8″. 98
The French	14 . 49	8 3 . 23
Mr. Groombridge's	14 . 61	8 9 . 66
Dr. Brinkley's	14 . 32	8 4 . 85
Dr. Young's	14 . 32	8 6 . 46
Carlini's	14 . 54	8 15 . 26
Bessel's	14 . 37	8 8 . 07
Average....................	14 . 43	8 8 . 07

ANNUAL PRECESSION, ABERRATION, LUNAR NUTATION, AND SOLAR NUTATION.

When a star has been observed, and its apparent place corrected for refraction, its mean place for a given time cannot be known, until the respective corrections for precession, aberration, lunar nutation, and solar nutation, have been ascertained and applied, with their proper signs, to its right ascension and polar distance, or declination, as the case may be. Various tables have been constructed for determining these corrections; and it is of the utmost importance that the practical astronomer should avail himself of such as are, at the same time, both convenient and correct. While the Rev. Fearon Fallows, M.A. was waiting for a vessel to convey him to the Cape of Good Hope, after his appointment to the situation of *Astronomer* there under the English government, he prepared for himself a series of tables for this purpose, on the plan of Baron Zach's "Nouvelles Tables," published at Marseilles in 1812 and 1813, but with such modifications and alterations as, on mature deliberation, appeared to render them more perfect, as well as more commodious, to the practical astronomer. These tables were just finished before this eminent mathematician sailed, when he favoured me with a copy, with permission to have them printed in my intended collection. Of course I was glad to avail myself of such an offer, and have accordingly adopted the tables in question, as the best I could select. They form the series contained in pages 20—38, and are numbered I. to XIV. inclusive, in due succession. The principles of their construction, and mode of using, are given in the author's own language, to which I propose to add such Examples as may render their use familiar, together with such further explanation as may be serviceable to young astronomers.

ANNUAL PRECESSION.

In consequence of the earth being a spheroid, with its equatorial exceeding its polar diameter, the action of the sun and moon, when out of the equator, on the prominent parts of the earth, is considered as the physical cause of a regressive motion of that point of the ecliptic which intersects the equator, called the vernal equinox, or first point of Aries; and successive determinations of a star's longitude, made in different years, establish the astronomical fact, that this motion amounts to upwards of 50″ in each year. Hence arises an *annual variation* both in the mean right ascension and declination, or polar distance, of every star; but each star, from its peculiar situation in the heavens, has its appropriate variation. Strictly speaking, however, it is not the *place* of the star that varies, but the points, with which it is compared, that change; for not only have the equinoctial points an annual regressive motion, but the angle included between the equator and ecliptic, called the obliquity of the ecliptic, has also an annual change, that tends to diminish its quantity. A star's right ascension, therefore, will be altered by the regressive displacement of the equinoctial points; and its declination will be increased or diminished, according to its position, by the gradual approach that the ecliptic makes towards the equator. The actual quantity of the regression, or precession, of the equinoctial points, must be determined by practical astronomy, which affords observations made at distant periods: for the aggregate quantity of regressive motion divided by the intervening number of years, will give a mean annual precession. In coming to nice decisions of this nature, it is obvious, that the exactness of the determined mean quantity will depend partly on the correctness of the observations chosen for the purpose, and partly on the length of the interval between the selected observations; and hence, it may be inferred, that future astronomers will have an advantage over astronomers of the present day; but astronomers of our own time must never forget, that *good* observations only will, in this respect, benefit succeeding ages.

The retrograde motion of the equinoctial points is not, however, perfectly uniform during the year, because the motions of the sun and moon are not uniform; neither is the change in the obliquity of the ecliptic, quite uniform. The irregular effect produced on the point, called the pole of the ecliptic, by the sun's place, as to declination, is called the *solar inequality* of *precession*, or ☉ *nutation;* and that produced by the moon's place, with reference to the *node*, is called *lunar nutation.* These two causes of *inequality*, in the precession and obliquity, require separate corrections, and will be considered hereafter.

When the epoch of a star is carried forwards a given number of years, by means of its annual precession in right ascension, and north polar distance, those precessions must be calculated for the mean period between the old and new epochs, otherwise they will not be the *mean precessions;* and if the observations made near the new epoch give the mean place of the star exactly the same as the mean precessions have made it, the star is said to have no *proper motion;* but if the calculated and observed mean places do not accord, the difference either in right ascension, or north polar distance, divided by the number of years between the respective epochs, will give the annual proper motion of the same denomination, and will be − or + accordingly as the calculated place is in excess or defect, when compared with the observed place reduced to its mean place by the tables.

When the annual precession and the annual proper motion are added together algebraically, the sum is called the *annual variation*, and is substituted for the precession in determining the mean place for a given time.

ANNUAL PRECESSION IN RIGHT ASCENSION.

Construction of Table 1.

The formulæ from which the annual precessions are calculated, were taken from Bessel's "Fundamenta Astronomiæ," page 297, but the co-efficients are adapted to the year 1820.

$$\text{The precession in R. A.} = 46''.0175 + 20''.0436 \text{ sin. R. A. tang. dec.}$$

Table 1. (page 20) contains the factor 20″.0436 sin. R. A., divided by 15, in order that the arguments and tabular quantities might be given *in time;* which is much more convenient to the practical astronomer, than when the terms are *in space.* Hence the constant portion of the precession in R. A. is $\frac{46''.0175}{15} = 3^s.0678$. The same plan is adopted in the subsequent tables, which are all newly, and, it is to be hoped, correctly calculated.

When the log. of the table is taken out, the log. tang. of the star's declination remains to be added to it, and then the natural number, corresponding to the sum of the two logarithms, is the *variable* quantity of the precession in time, and must be applied, with its proper sign, + or −, to the *constant* quantity 3ˢ.0678, to produce the whole precession in right ascension.

Table I. (Nat. Num.) page 21, contains the natural numbers belonging to the logarithms in the preceding Table, and therefore retains the same numeral prefixed with the addition of (Nat. Num.) The quantities given in this Table must be multiplied by the natural tang. of the star's declination, when the *variable* quantity of the precession is obtained without the assistance of logarithms; this must be applied with its proper sign, as before directed, to the *constant* quantity 3ˢ.0678.

Example 1.—In the year 1830, the approximate right ascension of Aldebaran will be $4^h\ 26^m\ 11^s$, and its approximate north polar distance 16° 9′ 37″ N.; let it be required to determine its annual precession in right ascension?

(+) Table I. log.	0 .08841	
Tang. dec. .	+9 .46208	
	9 .55049	$+0^s.3551$
Constant		+3 .0678
Annual precession		+3 .4229

Otherwise, by Nat. Num., thus:

Tab. No. $+1^s.226$ × .2897 (Nat. tang. dec.) .. =	$+ 0^s.3351$
Constant	+ 3 .0678
Annual precession as before	+ 3 .4229

Example 2.—At the same time the approximate right ascension of Rigel will be $5^h\ 6^m\ 22^s$ and its approximate declination 8° 24′ 11″ S.; what will be its annual precession in right ascension?

(−) Table I. log.	0 .11390	
Tang. dec... +	9 .16944	
	9 .28334	$- 0^s.192$
Constant............		+ 3 .068
Annual precession		+ 2 .876

By Nat. Num.

Tab. No. − $1^s.299$ × .1477 (nat. tang. dec.)... =	$- 0^s.192$
Constant..................................	+ 3 .068
Annual precession	+ 2 .876

Example 3.—Suppose the approximate right ascension of γ Draconis at any period to be $17^h\ 52^m\ 40^s$, and its declination 51° 30′ 45″ N.; let its annual precession in right ascension be required?

(−) Table I. log......	0 .12566	
Tang. dec. log. +	10 .09957	
	10 .22523	$- 1^s.679$
Constant.................		+ 3 .068
Annual precession..........		+ 1 .389

By Nat. Num.

Tab. No. − $1^s.336$ × 1 .257 (nat. tang. dec.).. =	$- 1^s.679$
Constant..................................	+ 3 .068
Annual precession	+ 1 .389

ABERRATION.

If the earth were not in motion, or if, the earth being in motion, light were *instantaneous*, the *apparent* and *true* places of a star, as seen by an inhabitant of the earth, would be the same; but as neither of these assumptions is correct, it becomes necessary that the astronomer should be able to ascertain and allow for the difference of the apparent and true right ascensions and declinations of the stars as determined instrumentally; in order that he may be able to correct the effect produced by the aberration of light on the place of each individual star. As the earth is changing her position with respect to the sun and stars every day of the year, the quantity of aberration both in right ascension and declination, or north polar distance, is constantly changing throughout the year, and to avoid going through a long and tedious calculation for every separate day for any given star, or number of stars, the Tables before us were so contrived, as to afford the easy means of determining certain *subsidiary numbers* peculiar to each star, be its position what it may, by the application of which, in conjunction with the sun's longitude, the exact aberration, either in right ascension or north polar distance, may readily be determined for any day, or number of successive days, and that both with ease and certainty, provided the numbers in question be correctly deduced from the Tables. [See pages 113—144.]

As the right ascension that forms the argument of the following Tables is always given in *time*, in hours at the top and minutes at the side, the observer has no occasion to convert time into degrees and minutes, &c. of *space*, but can enter the Tables with the quantities determined by the clock and transit instrument.

In the Tables for correcting the effects of lunar and solar nutation, *subsidiary numbers* are also obtained in a similar way, as will hereafter appear, so that all the corrections are found and applied according to one uniform system.

ABERRATION IN RIGHT ASCENSION.

CONSTRUCTION OF TABLES II. AND III.

THE aberration in right ascension $= \left.\begin{matrix} -18''.580 \cos. R. A. \cos. \odot \\ -20''.255 \sin. R. A. \sin. \odot \end{matrix}\right\} \text{Sec. Decl.}$

Let m be a positive quantity, and A an arc such, that

$$m \sin. A = -18''.580 \cos. R. A. \sec. dec. \ (1)$$
$$m \cos. A = -20''.255 \sin. R. A. \sec. dec. \ (2).$$

Hence $\tang. A = \frac{18''.580}{20''.255}.$ cotang. R. A, and log. tang. A = log. cotang. R. A —. 0374865;

Consequently A is deduced immediately from the logarithmic Tables, and registered in Table II., in the columns marked A, which call *tabular* A, being necessarily less than 90° or $\overset{s}{\text{III}}$.

Now in order to find the *true* value of A, which will satisfy equations (1) and (2), we know that tang. A = + tang. (6^s + A) = — tang. (6^s — A) = — tang. (12^s — A);

consequently the *annual numbers* or true values of A will be = ($\overset{s}{6}$ + tab. A) if the R. A. be greater than $\overset{s}{\text{O}}$ or less than $\overset{s}{\text{III}}$.

= ($\overset{s}{6}$ — tab. A) .... $\overset{s}{\text{III}}$ .... $\overset{s}{\text{VI}}$.

= (tab. A) .... $\overset{s}{\text{VI}}$ .... $\overset{s}{\text{IX}}$.

= ($\overset{s}{12}$ — tab. A) .... $\overset{s}{\text{IX}}$ .... $\overset{s}{\text{XII}}$.

Hence the formation of the precepts at the top and bottom of the Table.

From equations (1) and (2) $m = \frac{18''.580. \cos. R. A. \sec. dec.}{\sin. tab. A}$, or $= \frac{20''.255 \sin. R. A. \sec. dec.}{\cos. tab. A}$.

Table III. contains the factor $\frac{18''.580 . \cos. R. A.}{15 \sin. tab. A}$, or $\frac{20''.255 . \sin. R. A.}{15 \cos. tab. A}$. in order that the results may be given in *time*.

By substitution, the aberration in right ascension will be $= m \sin. A \cos. \odot + m \cos A. \sin. \odot = m \sin. (A + \odot)$.

Now since the greatest value of the sine of an arc is radius or unity, m = maximum aberration.

In Table II. the value of the subsidiary number A, is taken out by inspection for hours, minutes, and seconds of the star's right ascension, the column p. p. giving the proportional part to be added or subtracted, when the seconds in the argument are read in the first or last column, called "min.", as the case may be.

	s	°	′	
Thus for Aldebaran (1830).. 4^h 26^m R. A. =	0	21	45	from Table II.
11^s p. p.....sub.			3	
	0	21	42	= tab. A
Precept 6 +	6	0	0	
	6	21	42	= the true number A.

	s	°	′	
For Rigel.................. 5^h 6^m........	0	12	25	Table II.
22 p. p.			— 5	
	0	12	20	= tab. A
Precept 6 +	6	0	0	
	6	12	20	= the true A.

	s	°	′	
For γ Draconis .... 17^h 52^m......	0	1	50	Table II.
p. p. 40^s			— 9	
Precept $\overset{s}{0}$ +	0	1	41	= the true A.

Table III. gives a quantity, which, multiplied by the secant of the star's declination, produces the maximum aberration in right ascension. Either the logarithms or natural numbers may be used from the Table.

Thus for Aldebaran the log. of $4^h\ 26^m$ . . . = 0 .12491
($\frac{11}{60}$ of dif. 11) for 11^s. . + 2

0 .12493
Sec. dec. 16° 9′ 39″ 10 .01751

Max. ab. 1^s.388 . . . . 0 .14244

By natural numbers thus:

Tab.No. Nat. sec.
1^s.33 × 1 .0411 = 1^s.388 = max. ab.

For Rigel. . . . . $5^h\ 6^m\ 22^s$. . . . . log. . . . 0 .12859
Sec. dec. 8° 24′ 11″ log. . 10 .00469

Log. max. . . . . 0 .13328

Maximum ab. 1^s.359

By natural numbers:

1^s.345 × 1 .0108 = 1^s359 = max. ab.

For γ Draconis . . . . $17^h\ 52^m\ 40^s$. . . . log. . . . 0 .13043
Sec. dec. 51° 30′ 45″ log. 10 .20595

Log. max. . . . . 0 .33638

Maximum ab. 2^s.169

By natural numbers:

1^s.350 × 1 .6067 = 2 .169 = max. ab.

LUNAR NUTATION IN RIGHT ASCENSION.

CONSTRUCTION OF TABLES IV. AND V.

THE lunar nutation in R. A. = − 9″. 648 cos. R. A. cos. ☊ } tang. dec.
+ 7 . 1822 sin. R. A. sin. ☊ }
− 16 . 54414 sin. ☊

Let m be a positive quantity as before, and L an arc such, that m sin. L = − 9″.648 cos. R. A. tang. dec. = p
m cos. L = + 7 .1822 sin. R. A. tang. dec. − 16″.54414 = q

Then, by a similar process as before, the *true* values of L may be obtained; also the same precepts at the top of the columns, and the maximum m.

The lunar nutation will be found = m sin. (L + ☊)

Table IV. (nat. num.) page 25, contains the factor + 7″ .1822 sin. R. A.
Table V. (nat. num.) page 27, ... − 9 .648 cos. R. A.
Table IV. page 24 . . . . log. sin. R. A. + .8562514
Table V. page 27 . . . . log. cos. R. A. + .9844373

The mode of performing the operations is sufficiently explained in the margin of the Tables, both for the logarithmic and natural numbers; but as the former is generally the more convenient Table, the Examples here are worked by logarithmic operations.

Example 1.——ALDEBARAN. [R. A. = $4^h\ 26^m\ 11^s$. Dec. 16° 9′ 37″. N.]

q − Table 4, log. . . . . 0 .81878
+ tang. dec. log. . . . . 9 .46208

Nat. num. − 1 .909 0 .28086
Constant − 16 .544

q − 18 .453

Log. p . . . . . . − 0 .04802 }
+ cosec. tab. L 11 .20817 }
Constant. . . . . . 8 .82391 }

Log max. . . = 0 .08010 and max. ab. = 1^s.203

p − Table V. log. . . . . 0 .58594
+ tang. dec. log. . . . . 9 .46208

Log. p. . . . . 0 .04802
Sub. log. q. . . . . 1 .26507

Log. tang. L′ . . . . =8 .78295

	s		
Tabular L′	0	3°	28
p − } q − } Precept	6 + L′		
True L	=6	3	28

Example 2.——Rigel. [R. A. = 5^h 6^m 22^s. Dec. = 8^0 24′ 11″. S.]

q + Table IV. log. 0 .84425
+ tang. dec. log. 9 .16944

Nat. num. + 1 .032 0 .01369
Constant − 16 .544

p + Table V. log. 0 .34967
+ tang. dec. log. 9 .16944

Log. p .... +9 .51911

q = − 15 .512 .... Sub. log. q ... 1 .19067

Tab. L. = 0 s 1° 13′ .... Log. tang. L. 1 .32844

q − } p + } Precept.... 6 − L.

True L... = 5 28 47

Log. p .... = 9 .51911
+ cosec. tab. L. = 11 .67298
Constant..... + 8 .82391

Log. max. .... 0 .01600 and max. nut. = 1^s.038

Example 3.——γ Draconis. [R. A. = 17^h 52^m 40^s. Dec. = 51^0 30′ 45″. N.]

q + Table IV. log. 0 .85603
+ tang. dec. log. 10 .09957

Nat. num. + 9 .029 0 .95560
Constant − 16 .544

p + Table V. log. 9 .48996
+ tang. dec. log. 10 .09957

Log. p .... +9 .58953

q = − 7 .515 .... Sub. log. q ... 0 .87593

Tab. L..... 0 s 2° 58′ .... Log. tang. L. . 8 .71360

+ p } − q } Precept.... 6 − L.

True L... = 5 27 2

Log. p .... 9 .58953
+ cosec. tab. L. 11 .28604
+ constant... 8 .82391

Log. max.... = 9 .69948 and max. nut. = 0^s.501

N.B. By an error in the calculation, true L′ of Aldebaran is printed 6 s 3° 33′ in the argument at page 117, and the maximum of aberration is consequently put 1^s.231 instead of 1^s.203, which vitiates the column " ☾ nut." in the second or hundredth place of decimals.

SOLAR NUTATION IN RIGHT ASCENSION.

Construction of Tables VI. and VII.

Solar nutation in R. A. is = − 0″. 3982 sin. R. A. 2 ⊙ } tang. dec.
− 0 . 4340 cos. R. A. 2 ⊙ }
− 0 . 9173 sin. 2 ⊙. Where ⊙ = the sun's longitude.

Let m be a positive quantity as before, and S an arc such, that m sin. S = − 0 .4340 cos. R. A. tang. dec. = p
m cos. S = − 0 .3982 sin. R. A. tang. dec. = 0 .9173 = q

Then, by a process similar to what has been explained, the true values of S will be obtained: also the same precepts will apply above and below the respective columns of the Table, and likewise the maximum of solar nutation.

Table VI. contains the factor − 0 .4340 cos. R. A.
Table VII. .... − 0 .3982 sin. R. A.

The method of taking the numbers from the respective Tables of ⊙ nutation will appear from the subjoined Examples.

Example 1.——ALDEBARAN. [R. A. = $4^h\ 26^m\ 11^s$. Dec. = 16° 9′ 40″ N.]

q − Table VI. log...9 .56270
+ tang. dec. log...9 .46208

Nat. num. − 0 .106 9 .02478
Constant. − 0 .916

q = − 1 .023

p − Table VII. log...9 .23730
+ tang. dec. log...9 .46208

Log. p −........ 8 .69938
Log. q − sub..... 0 .00987

Log. tang. tab. S = 8 .68951

Tabular S .... = 0s 2° 49′
p −, q − } Precept.... 6 + S

True S.......... 6 2 49

Log. p .... 8 .69938
+ cosec. S log...11 .30856
Constant..... + 8 .82391

Log. max. s. nut. = 8 .83185
Max. sol. nut.. = 0s.068

Example 2.——RIGEL. [R. A. = $5^h\ 6^m$ 22. Dec. = 8° 24′ 11″ S.]

q + Table VI. log...9 .58810
+ tang. dec..log...9 .16944

Nat. num. + 0 .057 8 .757554
Constant. − 0 .917

q = − 0 .860

p + Table VII. log. 9 .00570
p. p. .... − 290
+ tang. dec. log...9 .16944

Log. p + ...8 .17224
Log. q − sub......9 .93449

Log. tang. S... = 8 .23775

Tabular S .... = 0s 1° 0′
p +, q − } Precept.....6 − S

True S ....= 5 29 0

Log. p .... 8 .17224
+ Log. cosec. S..11 .75815
+ constant...... 8 .82391

Log. max. s. nut. = 8 .75430
Max. sol. nut... = 0 .057

Example 3.——γ DRACONIS. [R. A. = $17^h\ 52^m\ 40^s$. Dec. 51° 30′ 45″ N.]

q + Table VI. log.. 9 .59980
+ tang. dec. log. .10 .09957

Nat. num. + 0 .5004 9 .69937
Constant. − 0 .9170

q = − 0 .4166

p + Table VII. log. 8 .14160
+ tang. dec. log...10 .09957

Log. p + ... 8 .24117
Log. q − sub...... 9 .61972

Log. tang. tab. S.. =8 .62145

Tabular S....... =0s 2° 24′
p +, q − } Precept....6 − S

True S......... = 6 27 36

Log. p .... 8 .24117
+ cosec. S.....11 .37803
+ constant..... 8 .82391

Log. max. s. nut. 8 .44311
Max. solar nut. =0s.028

N.B. In page 135 the maximum of solar nutation in the right ascension of γ Draconis, is, by erroneous calculation, tabulated 0s.070; this will be rectified hereafter.

ANNUAL PRECESSION IN NORTH POLAR DISTANCE.

Construction of Table VIII.

The precession in north polar distance is $= 20''.0436$ Cos. R. A.; which product is so easily tabulated, that it requires no explanation, and the quantities may always be taken out by simple inspection.

Let it be required to find the annual precession in north polar distance of the three stars, specified in the preceding Examples respectively, according to the data there given?

Example 1. Table VIII. gives (for *Aldebaran*) .... $4^h\ 26^m$ R. A. $= -7''.992$
p. p. for 11^s ($\frac{11}{60}$ of dif. 80) sub. 15
Annual precession in north polar distance required .... -7.977

Example 2. Table VIII. (for *Rigel*) gives .... $5^h\ 6^m$.......... $-4''.679$
p. p. for 22^s ($\frac{22}{60}$ of 85) sub. 31
Annual precession in north polar distance .... -4.648

Example 3. Table VIII. (γ *Draconis*) .... $17^h\ 52^m$........ $+0''.700$
p. p. for 40^s ($\frac{2}{3}$ of 88) sub. 59
Annual precession in north polar distance .... $+0.641$

In this series of Tables, the north polar distance has been substituted for the *declination*, to avoid all ambiguity respecting the application of the positive and negative signs, after the quantities are ascertained.

ABERRATION IN NORTH POLAR DISTANCE.

Construction of Tables IX. and X.

$$\text{The aberration in north polar distance} = \left.\begin{matrix} +20''.255 \text{ cos. R. A. sin. } \odot \\ -18\ .580 \text{ sin. R. A. cos. } \odot \end{matrix}\right\} \text{sin. declin.} \\ +8\ .0659 \text{ cos. } \odot \text{ cos. dec.}$$

Let m be a positive quantity, and A$'$ an arc such, that m sin. A$' = 18''.580$ sin. R. A. sin. dec. $+ 8''.0659$ cos. dec. (1)

m cos. A$' = 20''.255$ cos. R. A. sin. dec. (2)

Hence tang. $A' = \dfrac{-18''.580 \text{ sin. R. A. sin. dec.} + 8''.0659 \text{ cos. dec.}}{+20''.255 \text{ cos. R. A. sin. dec.}}$, by dividing the numerator and denominator by cos. dec.

$= \dfrac{-18''.580 \text{ sin. R. A. tang. dec.} + 8''.0659}{+20''.255 \text{ cos. R. A. tang. dec.}} = \dfrac{p}{q}$: where $p = -18''.580$ sin. R. A. tang. dec. $+ 8''.0659$,

and $q = +20''.255$ cos. R. A. tang. dec.

Hence log. tang. A$' =$ log. $p -$ log. q.

Table IX. contains the factor $20''.255$ cos. R. A.
Table X. .... $18''.580$ sin. R. A.

Consequently A$'$ is deduced immediately from the logarithmic Tables, which, being less than 90^0 or $\overset{s}{\text{III}}$ is called tabular A$'$.

Now in order to find the *true* value of A$'$, which will satisfy the equations (1) and (2), since tang. A$' = +$ tang. $(\overset{s}{6} + A') =$ $-$ tang. $(\overset{s}{6} - A') = -$ tang. $(\overset{s}{12} - A')$ the *annual numbers*, or true values of A$'$, will be

$\overset{s}{0}$ + tab. A$'$ if p be plus and q plus
6 + tab. A$'$ if p be minus and q minus
6 − tab. A$'$ if p be plus and q minus
12 − tab. A$'$ if p be minus and q plus

Hence the formation of the precepts, or four cases contained in the small Table denominated "Positive Argument, in several of the pages.

Since m sin. tab. A′ $= p$ cos. dec.
and m cos. tab. A′ $= q$ cos. dec.

$$m = \frac{p \text{ cos. dec.}}{\text{sin. tab. A}'} = p \text{ cosec. tab. A}'.\text{cos. dec. or } = q \text{ sec. tab. A}' \text{ cos. dec.}$$

Consequently log. m = log. p + log. cosec. tab. A′ + log. cos. dec.
− log. q + log. sec. tab. A′ + log. cos. dec.

By substitution, the aberration in north polar distance will be $= m$ sin. A′ cos. ☉ + m cos. A′ sin. ☉ $= m$ sin. (A′ + ☉). Hence m is the maximum of aberration.

The aberration in north polar distance requires more calculation than the aberration in right ascension. The annual arc A is here distinguished by an accent, thus A, and its value depends on two quantities, q and p, that must be previously found. Table IX. (page 31) gives a logarithm which, added to the logarithmic tangent of the star's declination, makes logarithm q; [or the number in Table IX. (page 32) multiplied by the natural tangent of the star's declination, produces q]. In like manner, the logarithm in Table X. (page 33) added to the logarithmic tangent of the star's declination will give the logarithm of a number to be added to, or subtracted from the constant quantity 8″.066, according to the signs, which, in these Tables, must be particularly attended to, then the resulting number will be p; [which also may be had from Table X. (page 34) without logarithms, though not so conveniently.] Then subtract the logarithm of q from the logarithm of p, and the remainder is the logarithmic tangent of A′ (corresponding to tabular A in Table II.) before the precept, depending on one of the four cases in the small corner Table, is applied. Lastly, add together the logarithm of p, the logarithm cosin. declination, and the logarithm cosec. of A′, and the sum will be the logarithm of the maximum of aberration in north polar distance.

Example 1.——ALDEBARAN. [R. A. = 4^h 26^m 11^s. Dec. 16° 9′ 40″ N.]

q + Table IX. 4^h 26^m.... log. .... 0 .90723
p. p... 11^s.................. −99
Log. tang. dec... 18° 9′ 37″... 9 .46208

Log. q .... 10 .36832

p − Table X... 4^h 26^m.... log. .... 1 .23145
p. p... 11^s.................. +15
Log. tang. dec. .... 9 .46208

Nat. num. .. − 4 .940 .... 10 .69368
+ 8 .066

p = + 3 .123 .... log. p 10 .49499 +
− log. q 10 .36832 +

By precept {p + / q +} True A′ = 1s 23° 15′...............Tang. A′ 0 .12667

Lastly,......log. p ....10 .49499
+ log. cos. dec.... 9 .98249
+ log. cosec. A′ ..10 .09623

Log. max.......= 0 .57371
Maximum 3″.748

Example 2.——RIGEL. [R. A. = 5^h 6^m 22^s. Dec. = 8° 24′ 11″ S.]

q − Table IX.. 5^h 6^m... log. .... 0 .67472
p. p. for 22^s .... −292
Tang. dec. .... 9 .16944

Log. q .... 9 .84124

p + Table X.................. log. 1 .25688
p. p.......................... +17
Tang. dec.....................9 .16944

Nat. num. .. + 2 .670 .... 0 .42649
+ 8 .066

p = + 10 .736 ...... log. p 1 .03084
Sub. log. q 9 .84124

A′ = 2s 26° 18′...............................Tang. A′ = 1 .18960
{p + / q −} Precept 6 − A′

True A′ = 3 3 42

Log. p .... 1 .03084
+ cos. dec..... 9 .99531
+ cosec. A′ ...10 .00090

Log. max. ..= 1 .02705
Hence max. of ab. = 10″.643

Example 3.——γ DRACONIS. [R. A. = 17^h 52^m 40^s. Dec. 51° 30′ 45″ N.]

q – Table IX. .... log. 9 .80956
Tang. dec. log. .... 10 .09957

Log. q .... 9 .90913

p + Table X. .... log. 1 .26882
Tang. dec. log. .... 10 .09957

Nat. num... + 23 .356 .... 1 .36839
Constant ... + 8 .066

p = + 31 .422 ...... log. p 1 .49723
Sub. log. q 9 .90913

Tab. A′ = 2^s 28° 31′ .... Log. tang. A′ = 1 .58810
p +, q – } Precept 6 – A′

True A′ = 3 1 29

Log. p .... 1 .49723
+ cos. dec.... 9 .79404
+ cosec A′ .. 10 .00015

Log. max. .. = 1 .29142 and max. of ab. = 19″.562

LUNAR NUTATION IN NORTH POLAR DISTANCE.

CONSTRUCTION OF TABLES XI. AND XII.

THE lunar nutation in north polar distance is taken = – 9″.648 sin. R. A. cos. ☊
+ 7 .1822 cos. R. A. sin. ☊

Let m be a positive quantity as before, and L′ an arc such, that m sin. L′ = – 9″.648 sin. R. A.
m cos. L′ = + 7 .1822 cos. R. A.

Then the true values of L′, with the formation of the precepts at the top and bottom of the Table XI., may be determined precisely in the same manner as in the preceding determinations, and are registered in the columns L′ of the Table.

The lunar nutation in north polar distance will thus be found always = m sin. (L′ + ☊), in which expression m is = the maximum of nutation.

The numbers in Tables XI. and XII. are taken out by simple inspection, and require no further explanation.

Example 1.——ALDEBARAN. [R. A. = 4^h 26^m 11^s. Dec. = 16° 9′ 40″ N.]

Table XI. .... = 2^s 12° 4′ for 4^h 26^m
p. p. for 11^s .... +2

2 12 6
Precept .... 12 – L′

True L′ .... = 9 17 54

Table XII.. 4^h 26^m ... log. 0 .96849
p. p. ($\frac{11}{60}$ of 33) .... +6

Log. max. .... 0 .96855

Max. of lunar nut. = 9″.302

Example 2.——RIGEL. [R. A. = 5^h 6^m 22^s. Dec. = 8° 24′ 11″ S]

Table XI. .... = 2^s 19° 56
Precept .... 12 – L′

True L′ 9 10 4

Table XII. .. log. max..... 0 .97917

Max. of lunar nut. 9″.532

Example 3.——γ DRACONIS. [R. A. = 17^h 52^m 40^s. Dec. 51° 30′ 45″ N.]

Table XI. .... = 2^s 28° 39
Precept .... 6 – L′

True L′ .... = 3 1 21

Table XII. .. log. max..... 0 .98436

Max. of lunar nut. 9″.647

SOLAR NUTATION IN NORTH POLAR DISTANCE.

CONSTRUCTION OF TABLES XIII. and XIV.

SOLAR nutation in north polar distance $= +\ 0''.3982$ cos. R. A. sin. 2⊙
$-\ 0\ .4340$ sin. R. A. cos. 2⊙

Let m be a positive quantity as before, and S′ an arc such, that m sin. S′ $= -\ 0''.4340$ sin. R. A.
m cos. S′ $= -\ 0\ .3982$ cos. R. A.

The true values of S′ and m, as well as the formation of the precepts above and below the proper columns, are determined in the same manner as has been before described, and as they are registered in the columns S′ of the Table. The solar nutation will be found $= m$ sin. (S + 2 ⊙) where m is $=$ the maximum of solar nutation.

The numbers for solar nutation in north polar distance are taken by inspection from Tables XIII. and XIV. in the same manner as those for lunar nutation are from Tables XI. and XII.

Example 1.——ALDEBARAN. R. A. $= 4^h\ 26^m\ 11^s$. Dec. $= 16°\ 9'\ 40''$ N.]

	S		Table XIV...log. max. ....9 .632
Table XIII. tab. S′	= 2	8° 18′	Max. of solar nut. 0″.429
Precept............	12	— S′	
True S′...........	=9	21 42	

Example 2.——RIGEL. [R. A. $= 5^h\ 6^m\ 22^s$. Dec. $= 8°\ 24'\ 11'$ S.]

	S		Table XIV..log. max...9 .639
Table XIII...tab. S′..	2	17° 40′	Max. = 0″.433
Precept.............	12	— S′	
True S′.............	9	12 20	

Example 3.——γ DRACONIS. [R. A. $= 17^h\ 52^m\ 40^s$. Dec. 51° 30′ 45″ N.]

	S		Table XIV..log. max...9 .637
Table XIII. .. tab. S′.	2	28° 19	Max. of solar nut. 0″.434
Precept.............	6	— S′	
True S′...........	=3	1 41	

It is presumed that the preceding Examples will suffice for illustrating the different computations to be derived from the foregoing series of Tables, from I. to XIV.; for which purpose each Table has been singly exemplified: but in computing the subsidiary numbers for all the separate corrections of any given star, it will generally be convenient to adopt some *type* for facilitating the calculations, such as the subjoined form modified by Mr. Fallows, which I propose to fill up with the Examples already worked; from which it will appear, that all the logarithmic numbers required for the declination of any star, in finding the subsidiary numbers, may be taken out at one opening of the Book of Logarithms, and inserted at once; and that all the other blanks may be filled up from the Tables with little trouble.

Should any computer be disposed to adopt this method of obtaining the subsidiary numbers, for giving a succession of corrections for other stars, it would be worth his while to follow the example of the Cape astronomer, who has had some Volumes of blank types printed, for the purpose of being filled up with figures from the respective Tables, as occasion may require; which plan conduces both to accuracy and expedition.

1830. SUBSIDIARY NUMBERS FOR RIGHT ASCENSION.

ALDEBARAN. Approx. R. A. = $4^h\ 26^m\ 11^s$. Approx. Dec. = $16^d\ 9^m\ 40^s$ N.

PRECESSION.		LUNAR NUTATION.			SOLAR NUTATION.		
+ Tab. I.= 0 .08841 + tan. Dec.= 9 .46208		(q−) Tab. IV.= 0 .81878 + tan. Dec.= 9 .46208	(p−) Tab. V. = 0 .58594 + tan. Dec. = 9 .46208		(q−) Tab. VI.=9 .56270 + tan. Dec.=9 .46208	(p−) Tab. VII.=9 .23738 + tan. Dec.=9 .46208	
N°= 0 .355 + 3 .068	l.= 9 .55049	N°=− 1 .909 } l.= 0 .28086 −16 .544 } q.=−18 .453	− l. p. = 0 .04802 sub. − l. q. = 1 .26507		N°=− 0 .106 } l.= 9 .02478 − 0 .917 } q.=− 1 .023	− l. p.=8 .69938 sub. − l. p.=0 .00987	
An. Prec.=+3s.423			L.......=0S 3° 28′ Pr.......=6 + L.	tan.L.= 8 .78295		S........=0S 2° 49′ Pr.......=6 + S.	tan. S. = 8 .68951
ABERRATION.			True L.=6S 3° 28′	l. p.= 0 .05802 +Cosec.L.=11 .20817 + 8 .82391		True S.=6S 2° 49′	l. p.......= 8 .6994 +Cosec. S.=11 .3086 + 8 .8239
Tab. II...=0S 21° 42′ = A. Pr.........=6 + A.	Tab. III.= 0 .12493 + Sec. Dec.=10 .01751						
True A...=6 21 42	Log. Max.= 0 .14244 Max.= 1s.388		Max.=1s.203	Log. Max.= 0 .08010		Max.=0s.068	Log. Max.= 8 .8319

RIGEL. Approx. R. A. = $5^h\ 6^m\ 22$. Approx. Dec. = $8^d\ 24^m\ 11^s$ S.

PRECESSION.		LUNAR NUTATION.			SOLAR NUTATION.		
− Tab. I.= 0 .11390 + tan. Dec.= 9 .16944		(q+) Tab. IV.=0 .84425 + tan. Dec.=9 .16944	(p+) Tab. V.=0 .34967 + tan. Dec.=9 .16944		(q+) Tab. VI.=9 .58810 + tan. Dec.=9 .16944	(p+) Tab. VII.=9 .00280 + tan. Dec.=9 .16944	
N°=−0 .192 +3 .068	l.= 9 .28334	N°=+ 1 .032 } l.= 0 .01369 −16 .544 } q.=−15 .512	+ l. p.=9 .51911 sub. − l. q.=1 .19067		N°=+ 0 .057 } l.= 8 .75754 − 0 .917 } q.=− 0 .860	+ l. p.=8 .17224 sub. − l. q.=9 .93449	
An. Prec.=+2s.876			L........=0S 1° 13′ Pr. .....=6 − L.	tan. L.=8 .32844		S. ....=0S 1° 0′ Pr.......=6 − S.	tan. S.=8 .23775
ABERRATION.			True L.=5S 28° 47′	l. p.= 0 .51911 +Cosec. L.=11 .67298 + 8 .82391		True S.=5S 29° 0′	l. p.......= 8 .1722 +Cosec. S.=11 .7581 + 8 .8239
Tab. II...=0S 12° 20′ = A. Pr. ....=6 + A.	Tab. III.= 0 .12859 + Sec. Dec.=10 .00469						
True A...=6 12 20	Log. Max.= 0 .13328 Max.=1s.359		Max.=1s.038	Log. Max.= 8 .82391		Max.=0s.057	Log. Max.=8 .7542

γ DRACONIS. Approx. R. A. = $17^h\ 52^m\ 40^s$. Approx. Dec. = $51^d\ 30^m\ 45^s$ N.

PRECESSION.		LUNAR NUTATION.			SOLAR NUTATION.		
− Tab. I.= 0 .12566 + tan. Dec.=10 .09957		(q+) Tab. IV.= 0 .85603 + tan. Dec.=10 .09957	(p+) Tab. V.= 9 .48996 + tan. Dec.=10 .09957		(q+) Tab. VI.= 9 .59980 + tan. Dec.=10 .09957	(p+) Tab. VII.= 8 .14160 + tan. Dec.=10 .09957	
N°=−1 .679 +3 .068	l.=10 .22523	N°=+ 9 .029 } l.= 0 .95560 −16 .544 } q.= − 7 .515	+ l. p.= 9 .58953 sub. − l. q.= 0 .87593		N°=+0 .5004 } l.= 9 .69937 −0 .917 } q.=−0 .4156	+ l. p.= 8 .24117 sub. − l. q.= 9 .61972	
An. Prec.=+1s.389			L.......=0S 2° 58′ Pr......=6 − L.	tan L.= 8 .71360		S. ......=0S 2° 24′ Pr......=6 − S.	tan. S.= 8 .62145
ABERRATION.			True L.=5S 27° 2′	l. p.= 9 .58953 +Cosec.L.=11 .28604 + 8 .82391		True S.=5S 27° 36′	l. p.......... 8 .2412 +Cosec. S.=11 .3780 + 8 .8239
Tab. II...=0S 1° 41′ = A. Pr.........=0 + A.	Tab. III.= 0 .13043 + Sec. Dec.=10 .20595						
True A...=0 1 41	Log. Max.= 0 .33638 Max.=2s.169		Max.=0s.501	Log. Max.= 9 .69948		Max.=0s.0277	Log. Max.= 8 .4431

1830. SUBSIDIARY NUMBERS FOR NORTH POLAR DISTANCE.

ALDEBARAN.

ABERRATION.			LUNAR NUTATION.	SOLAR NUTATION.
(q+) Tab. IX.= 0.90624 + tan. Dec.= 9.46208	(p−) Tab. X.=1.23160 + tan. Dec.=9.46208		Tab. XI. L′......= 2^{S} 12° 6′ Pr. ....=12 − L′.	Tab. XIII. S′. = 2^{S} 8° 18′ Pr...............=12 − S′.
l. q.=10.36832	Num.=−4.940 +8.066	l.=0.69368	True L′.........= 9 17 54	True S′. ....= 9 21 42
PRECESSION.	p.=+3.126	− l. p.=10.49499 sub. + l. q.=10.36832	Tab. XII. Log. Max.= 0.96852 Max.= 9″.301	Tab. XIV. Log. Max.=9.632 Max.=0″.429
Tab. VIII.= An. Prec.=−7″.978	A′. ......= 1^{S} 23° 15′ Pr. ......= 0 + A′.	tan. A′......= 0.12667		
	True A′.= 1 23 15	l. p. ....=10.49499 + Cos. Dec.= 9.98249 +Cosec. A′.=10.09623		
	Max.=3″.748	Log. Max...= 0.57311		

RIGEL.

ABERRATION.			LUNAR NUTATION.	SOLAR NUTATION.
(q−) Tab. IX.= 0.67180 + tan. Dec.= 9.16944	(p+) Tab. X.=1.25705 + tan. Dec.=9.16944		Tab. XI. L′.......= 2^{S} 19° 56′ Pr. ....=12 − L′.	Tab. XIII. S′ = 2^{S} 17° 40′ Pr...............=12 − S′.
l. q.= 9.84124	Num.=+ 2.670 + 8.066	l.=0.42649	True L′...........= 9 10 4	True S′.........= 9 12 20
PRECESSION.	p.=+10.736	l. p.= 1.03084 sub. − l. q.= 9.84124	Tab. XII. Log. Max. = 0.97917 Max. = 9″.532	Tab. XIV. Log. Max.=9.636 Max.= 0″.433
Tab. VIII.= An. Prec.=−4″.651	A′. ......= 2^{S} 26° 18′ Pr. ......= 6 − A′.	tan. A′.......= 1.18960		
	True A′.= 3 3 42	l. p. ....= 1.03084 + Cos. Dec.= 9.99531 +Cosec. A′.=10.00090		
	Max.=10″.643	Log. Max...= 1.02705		

γ DRACONIS.

ABERRATION.			LUNAR NUTATION.	SOLAR NUTATION.
(q−) Tab. IX.= 9.80956 + tan. Dec.=10.09957	(p+) Tab. X.= 1.26882 + tan. Dec.=10.09957		Tab. XI. L′.........=22^{S} 8° 39′ Pr. ....= 6 − L′.	Tab. XIII. S′. ...=2^{S} 28° 19′ Pr.. ....=6 − S′.
l. q.= 9.90913	Num.=+23.356 + 8.066	l.= 1.36839	True L′. ....=3 1 21	True S′. ....=3 1 41
PRECESSION.	p.=+31.422	l. p.= 1.49723 − l. q.= 9.90913	Tab. XII. Log. Max. = 0.98436 Max.=9″.647	Tab. XIV. Log. Max.=9.637 Max.=0″.434
Tab. VIII.= An. Prec.=+0″.639	A′. ......= 2^{S} 28° 31′ Pr. ......= 6 − A′.	tan. A′.......= 1.58810		
	True A′.= 3 1 29	l. p. ....= 1.49723 +Cos. Dec.= 9.79404 +Cosec. A′.=10.00015		
	Max.=19″.562	Log. Max...= 1.29142		

When the subsidiary numbers of stars have been determined in the manner above described, they may be arranged in a Catalogue of reference according to the subjoined plan, and will serve for determining the corrections for many years preceding and following the epoch which is assumed for the mean period.

Specimen of a Catalogue of SUBSIDIARY NUMBERS for 1830.

	IN RIGHT ASCENSION.							IN NORTH POLAR DISTANCE.						
Stars.	Ann. Prec.	Aberration.		☾ Nutation.		⊙ Nutation.		Ann. Prec.	Aberration.		☾ Nutation.		⊙ Nutation.	
		A.	Max.	L.	Max.	S.	Max.		A′.	Max.	L′.	Max.	S′.	Max.
		s		s		s			s		s		s	
Aldebaran	+3s.423	6 21° 42′	1s.388	6 3° 28′	1s.203	6 2° 49′	0s.068	−7″.978	1 23° 15	3″.748	9 17° 54′	9″.301	9 21° 42	0″.429
Rigel........	+2 .876	6 12 20	1 .359	5 28 47	1 .038	5 29 0	0 .057	−4 .651	3 3 42	10 .643	9 10 4	9 .532	9 12 20	0 .433
γ Draconis	+1 .389	0 1 41	2 .169	5 27 2	0 .501	5 27 36	0 .028	+0 .639	3 1 29	19 .562	3 1 21	9 .647	3 1 41	0 .434

Table XV.

When the annual precession, either in right ascension or north polar distance, is determined, as above explained and registered, Table XV. is so contrived, as to give by inspection the precession belonging to any particular day in the year, let the annual precession be what it may, by which means the use of a factor for the given day is avoided. In this Table, the star's mean place is supposed to be adapted to midnight, between December of the preceding, and January of the following year, as is the case in all the foreign catalogues; hence when the epoch is noon of January 1st, one-half of the daily difference, viz. .0014, must be subtracted from the precession given by the Table, when great accuracy is wanted. The first column, headed (1), contains the factors, or simple fractions of a year, corresponding to each successive day; the next column (2) contains *double* the same quantities; and the column (3) contains *three times* that quantity; and so of the rest, up to the twentieth column; these successive columns therefore give the *products* of the first column multiplied by the numbers 1, 2, 3, 4, 5, &c. respectively standing over them; and the product of any factor, belonging to a given day, multiplied by the numeral at the top of any column, is given at the intersection of the horizontal with the vertical lines. Thus the factor for June 1 is 0 .416, and under 10 at top, or over 10 at the bottom, in the same horizontal line, is the product 4 .165, and in the column 20, stands 8 .330; so that if 10″ or 20″ were the annual precession, the same would respectively be, for June 1, 4″.165 or 8″.33. The first column was originally calculated for five decimals, but the nearest figure was substituted in the third or thousandth place, when the rest were cut off, and therefore the products derived from the five figures are more correct than if they had been derived from the *three* only, as they now stand in column (1). The numerals at the top and bottom of the columns may be used as integers or decimals, or both successively, as the case may demand, and the successive products added together, when properly placed under one another, will give the whole product required. For instance, let 15″.43 be the annual precession; then, if the precession were required for the 10th of May, there will be this short operation: viz.

15″.	......5″.343	
.4	...... .142	
.03	 10	but 15″.43 × 0 .356 = 5″.493 only, because the factor 0 .356 is somewhat too small; and greater
Whole amount	= 5″.495	

accuracy is obtained from the Table, than from the product given by the factor itself in the usual way. It will not be difficult to recollect, that the tabular numbers belonging to the *units*, if under 20, are placed as in the Table with respect to the *decimal point;* if it is a decimal in the first place, as .4, its product, or tabular quantity, must stand next to the *right* of the *point;* but if the figure belongs to the *second*, or hundredth's place of decimals, as .03, then the product belonging to it will be removed two places to the right of the point; and, in like manner, a figure in the *thousandth's* place, will require its product to be moved three figures to the right of the point. Also if the numeral consists of several integers, the point itself must be moved to where the *unit* places it to the right. Thus if 154″.3 had been the annual quantity, the result would have stood thus:

15″.	......53″.43
4 .	...... 1 .42
.3	...... .10
Whole amount	= 54 .95

In like manner, if on October 1, the annual precession were 99″.0, or 9″.9, or 0″.99, the corresponding precessions for that day would be

6757 6 .757 74 .327	or........	6 .757 6757 7 .4327	or........	.6757 6757 .74327	according to the nature of the data.

Example.—Let it be required to determine the precession, both in right ascension and north polar distance, of the stars Aldebaran, Rigel, and γ Draconis, on the 10th of August, 1830, supposing their annual precessions as registered in the small preceding Table?

Star	Annual precession	R. A.		N. P. D.	
ALDEBARAN.——Prec. in R. A... = +3^s.423 in N. P. D. = −7 .978		3. ..	1^s.825	7″. ..	4″.258
		.4 ..	243	.9 ..	547
		.02 ..	12	.07 ..	42
		.003..	2	.008..	5
		Prec.	+2 .082	Prec.	−4 .852
RIGEL.——Prec. in R. A... = +2 .876 in N. P. D. = −4 .651		2. ..	1^s.217	4. ..	4″.258
		.8 ..	487	.6 ..	465
		.07 ..	42	.05 ..	30
		006..	4	001..	0
		Prec.	+1 .750	Prec.	−2 .828
γ DRACONIS.——Prec. in R. A... = +1^s.389 in N. P. D. = +0 .639		1. ..	0^s.608	0 .6 ..	0″.365
		.3 ..	183	.03 ..	18
		.08 ..	48	.009..	5
		009..	5	Prec.	+0 .388
		Prec.	+0 .844		

TABLE XVI.

It has been before explained, that on any given day of the year, the *maximum* of aberration multiplied by the natural sine of the quantity A, added to the sun's longitude, will give the aberration belonging to that day in right ascension, or in north polar distance if A′ be substituted for A; but it must not be the sun's longitude at *noon*, as given in the Nautical Almanac, or as collected from the Solar Tables, but the sun's longitude at the moment when the star in question is on the meridian; that being the time when its right ascension or north polar distance is usually observed. Table XVI. was therefore constructed for giving the reduction of the sun's longitude at noon to the time of any star's culmination. It is calculated for every ten days of the year, and for every hour of a star's right ascension in time. For instance, if it were required to know what number of minutes are to be added to the sun's longitude on the 19th of June, for a star that has 14^h 30^m of right ascension, or that culminates at this sidereal time, the number in the Table is 20′, so that the sun's longitude at noon must be increased by 20 to give the proper argument denoted by (☉), when we say that the aberration $= m$ sin. (A + ☉) in all cases. And the same is equally true of (☉) when we say that the solar nutation $= m$ sin. (A + 2☉).

TABLE XVII.

If greater accuracy in the sun's longitude, used as an argument, should be required, than the preceding Table gives, this Table may be entered with the difference produced by subtracting the sun's right ascension from the star's at the time of its culmination, and the addition to be made will be given in minutes and seconds. Thus, on the 19th of June, 1823, the sun's right ascension at noon is given (Nautical Almanac) = 5^h 48^m 22, which, subtracted from 14^h 30^m 0^s, the star's right ascension in the preceding example, the difference 8^h 41^m 38^s, used as an argument, gives 21′ 25″, as the addition to be made, instead of 20′, given in Table XVI.; which is copied from the Greenwich Collection of Tables, and which was computed probably for a mean between the leap years.

When the sun's longitude for *Greenwich* is taken as an argument at another Observatory, it becomes necessary to make a further allowance + or −, accordingly as that observatory lies west or east of Greenwich, for the meridian of which the ☉ is calculated.

This Table was calculated to serve one of these purposes as well as the other, and it is presumed will be found at the same time convenient and correct, provided that the change of longitude be assumed as equable for short intervals. Should it be required to ascertain how much will require to be added to the sun's longitude, as given in the Nautical Almanac, at an observatory 1^h 12^m 30^s

west of Greenwich, $+2' 58''.6$ would be the tabular quantity; but if the observatory were situated so much to the east, the correction would be $-2' 58''.6$. In 4^m or 1^0 difference of longitude, the correction would be $\pm 9''.9$; but so small a quantity in an *argument* may safely be neglected in ordinary operations.

Table XVIII.

When the *subsidiary numbers* for a given star are prepared and tabulated, and the sun's longitude for the time of the star's culmination is ascertained on a given day at a given place, by means of the foregoing Tables, it remains that the maximum of the aberration, of the lunar nutation and solar nutation, be multiplied each by the natural sine of its own argument, and that the proper signs, + or −, be prefixed to each product according to circumstances, which would not be difficult to an expert mathematician; but as celestial observations may be, and often are made by persons who are not skilled in trigonometrical calculations, and who therefore might often be at a loss to find the natural sine belonging to each degree of the four quadrants of the circle, and also to fix the right positive and negative signs in their proper places, it seemed necessary to introduce a Table, such as XVIII., in which all the requisites might be presented to the eye, and the products obtained without the trouble of multiplication. The arguments for each correction are equally applicable, and extend to every second minute of the circle, and also to all the factors up to 20, integral or decimal, as in the general Table of daily Precessions (Table XV.), and forty-five pages were found necessary to contain all the products required; these pages have been calculated at the expense of considerable labour, and it is trusted, that the practical astronomer will find, in the use of this Table, considerable mitigation of his trouble in the final determination of his corrections, without the use of logarithms, or the tedious work of taking proportional parts. The first column in each page (from 47 to 91) contains the argument increasing in degrees and minutes downwards, and the last column gives the degrees and minutes of the argument increasing upwards, in their respective quarters of the circle, which always bear the proper positive or negative signs to be prefixed to the resulting numbers. These numbers, or products of 1, 2, 3, 4, &c. into the natural sines of the proper arcs, are taken out exactly as in Table XV. already explained; the figures constituting the maxima of aberration, ☾ nutation, and ☉ nutation, being those that stand above and below their respective columns in the Table. A few Examples will render further explanation unnecessary.

Example 1.—Let it be required to determine the apparent right ascension and north polar distance of Aldebaran on the 10th of October, 1823?

Table XVIII.

Aber. in R. A. } A $6^s\ 21^0\ 42'$ + ☉ $6^s\ 16^0\ 22'$..... = $1^s\ 8^0\ 4'$ } Arg. $1^s\ 8^0\ 39'$ } 1 .3 ..0 .8120
Table XVI. for 10th Oct. $4^h\ 25^m$ R. A. of star gives +35 } Max. $1^s.388$ } .08 .. 499
.008.. 49

Cor..... + $0^s.8668$

☾ Nut. in R. A. } ...... L $6^s\ 3^0\ 28'$ + ☊ $9^s\ 23^0\ 28'$ .... = Arg. $3^s\ 26^0\ 56'$ } 1 .2 ..1 .0699
Max of ☾ Nut. $1^s.203$ } .003.. 26

Cor..... + $1^s.0725$

* ☉ Nut. in R. A. } ...... S $6^s\ 2^0\ 49'$ + 2 ☉ $1^s\ 3^0\ 54'$ .... = Arg. $7^s\ 6^0\ 43'$ } .06 0359
Max. of ☉ Nut. .068 } .008... 47

Cor..... − .0406

Sum of the three corrections in R. A. .. + $1^s.8990$

Star's mean R. A. 1820 .. (Naut. Alm. 1823) .... $4^h\ 25^m$ 36 . 28 }
Ann. variation 3 .43 × 3 years .... + 10 . 29 } $4^h\ 25^m\ 49^s.229$
Do. for 10th Oct. Table XV. (as explained above) . + 2 . 659 }

The apparent R. A. of Aldebaran .... = 4 25 51 .128

According to Dr. Maskelyne's Table in Naut. Alm, 1823........... = 4 25 50 .86

* The *mean longitude* of the ☾'s ☊ is here taken from the Lunar Tables 1 and 2, page 181, and not the equated longitude, as given in the Nautical Almanac, which for this purpose is not so proper. The reduction of the ☊'s epoch (− $1' 35''.3$) to noon of January 1, is here disregarded.

Table XVIII.

Aber. in N. P. D. } A′ 1^s 23° 15′ + ⊙ 6^s 16° 22′ ... = 8^s 9° 37′ } Arg. 8^s 10° 12′ } 3″. .. 2″.823
Table XVI. gives as before .... + 35 } Max. 3″.748 } .7 .. .659
.04 .. 37
.008 .. 7

Cor. for aberration − 3″.526

☾ Nut. in N. P. D. } .... L′ 9^s 17° 54′ + ☊ 9^s 23° 28′ = Arg. 7^s 11° 22′ } 9. .. 5″.948
Max. ☾ Nut. 9″.301 } .3 .. 198
.001 .. 6

Cor. for lunar nutation − 6″.152

⊙ Nut. in N. P. D. } .... S′ 9^s 21° 42′ + 2 ⊙ 1^s 3° 54′ = Arg. 10^s 25° 36′ } .4 .. .226
Max. ⊙ Nut. 0″.429 } .02 .. 11
.009 .. 5

Cor. for solar nutation − .242

Sum of three corrections − 9″.920

Mean N. P. D. 1820 (Naut. Alm. 1823)	73° 51′ 39″.67
Ann. Var. − 7″.95 × 3 years	−23 .850
October 10, by Table XV.	− 6 .164
The apparent N. P. D. of Aldebaran	73° 50′ 59″.736
According to the Naut. Alm. 1823	73 50 59 .7

If the apparent or observed north polar distance, corrected for refraction, had been given 73° 50′ 59″.736, the precessions (or annual variations), aberration, as well as ☾ and ⊙ nutations, as above determined, must have been applied with contrary signs, or their amount with a contrary sign, to obtain the *mean* north polar distance for the beginning of 1820; or for 1823, by omitting three years' annual variation. Though the Tables are constructed for determining the *apparent* places from the *mean*, and have their signs arranged accordingly, yet they are equally convenient for reversing the operation by a mere change of the signs. And this is equally true with respect to right ascension, as well as north polar distance.

When the observatory is in a southern latitude, north polar distance then becomes south polar distance, and all the signs relating thereto must be changed accordingly.

Example 2.—Let the observed or apparent right ascension and north polar distance of γ Draconis be $17^h\ 52^m\ 34^s.05$ and 38° 29′ 25″.8, on the 10th of May, as given in the Nautical Almanac of 1824, and let it be required to determine its mean place for the beginning of the said year?

Table XVIII.

Aber. in R. A. } A 0^s 1° 41′ + ⊙ 1^s 19° 43′ + ...(Table XVI.)... 36′ = Arg. 1^s 22° 0′ } 2. .. 1 .576
Max. 2 .169 } .1 .. .079
.06 .. 47
.009 .. 7

Cor. for aberration + 1″.709

☾ Nut. in R. A. } L 5^s 27° 2′ + ☾ ☊ 9^s 12° 15′ ..(Lunar Tables).. = Arg. 3^s 9° 17′ } .5 .. .4934
Max. 0 .501 } .001 .. 9

Cor. for ☾ Nutation + $0^s.4943$

⊙ Nut. in R. A. } S $\overset{s}{5}$ 27° 36′ + 2 ⊙ $\overset{s}{3}$ 10° 38′................= Arg. $\overset{s}{9}$ 8° 14′ } .02 .. .0198
Max. .028 } .008.. 8

Cor. for ⊙ Nutation − .0206

Precession on May 10 (Table XV.) An. Pr. = + 1 .389..........................1. ..0 .356
.3 .. 107
.08 .. 28
.009.. 3

+ 0 .494

Amount of three corrections and precession (or annual variation) + 2 .676

Hence apparent R. A...........17^h 52^m 34^s.05
Corrections with sign changed ... − 2 .676

Mean R. A. for 1823.........=17 52 31 .374

Aber. in N. P. D. } A′ $\overset{s}{3}$ 1° 29′ + ⊙ $\overset{s}{1}$ 20° 19′............... = Arg. $\overset{s}{4}$ 21° 48′ } 19. ..11″.751
Max. 19 .562 } .5 .. 309
.06 .. 37
.002.. 1

Cor. for aberration + 12 .098

☾ Nut. in N. P. D. } L′ $\overset{s}{3}$ 1° 21′ + ☾ ☊ $\overset{s}{9}$ 12° 15′............ = Arg. $\overset{s}{0}$ 13° 36′ } 9. ..2 .116
Max. 9″.647 } .6 .. 141
.04 .. 94
.007.. 2

Cor. for ☾ Nutation + 2 .353

⊙ Nut. in N. P. D. } S′ $\overset{s}{3}$ 1° 41′ + 2 ⊙ $\overset{s}{3}$ 10° 38′............= Arg. $\overset{s}{6}$ 12° 19′ } .4 .. .0853
Max. 0 .434 } .03 .. 64
.004.. 8

Cor. for ⊙ Nutation − .0925

Sum of the three corrections = + 14 .359

Apparent N. P. Distance of γ Draconis, Naut. Alm. 1824..=38° 29′ 25″.8
Prec. for 10th May from Ann. Var. + 0 .70 (Table XV.) .... + .249 } with sign changed − 14 .608
Amount of Cor...+14 .359 }

Mean N. P. Distance for Jan. 0, 1824...38 29 11 .192

If the mean north polar distance had been required for May 10, 1824, the precession (or annual variation) must have been omitted, and the sum of the corrections for aberration, ☾ nutation, and ⊙ nutation, with the contrary sign, must have been applied thus...38° 29′ 25″.8
− 14 .359

Mean N. P. D. 10th May, 1824 ....38 29 11 .441

In like manner the same subsidiary arcs and maxima will give the corrections for any number of successive days, from this Table, provided that the variation in the sun's longitude, or double the variation, as the case may be, be daily added to the next preceding solar arguments respectively; and also that the daily variation of the mean longitude of the moon's node be also subtracted from day to day from the lunar arguments; or the same may be done for every *ten days*, and the intermediate days filled up by interpolation. The corrections of 48 principal stars, contained in pages 113—144, for every degree of the sun's longitude, and of the moon's node, were calculated in this manner, and are so arranged as to afford great facility to the astronomer who may have occasion to observe those stars, or any of them, or who may be disposed to compute a Table of daily corrections for any particular year; the principal part of the work being already computed.

DELAMBRE'S UNIVERSAL TABLES OF ABERRATION AND NUTATION, in R. A. in *space*, and in N. P. D. (pages 92—94.)

Though the preceding series of Tables are peculiarly adapted for giving the corrections for a succession of days, by means of the subsidiary arcs and maxima, yet it will frequently happen that the corrections for a single observation, or for a mean of a small number of observations made near each other, may be required, either for determining the latitude, or, where the latitude is known, for finding the mean place at a given epoch; in either of these cases, the Universal Tables of Delambre may be found convenient, and will save the trouble of computing the subsidiary numbers.

Example 1.—Let it be required to determine the aberration and ☾ nutation for the R. A. and N. P. D. of Aldebaran on the 10th of October, 1823, as before, the R. A. of ∗ being = $4^h\ 25^m\ 49^s.25$ or $\overset{s}{2}\ 6^\circ\ 27'\ 15''$, and long. of ⊙ = $\overset{s}{6}\ 16^\circ\ 22'\ 19''$?

Aberration.—Tab. A. Arg. = $\overset{s}{2}\ 6^\circ\ 27'\ 15'' - \overset{s}{6}\ 16^\circ\ 22'\ 19'' = \overset{s}{7}\ 20^\circ\ 4'\ 56''$....... $+12''.46$
Tab. B. Arg. = 2 6 27 15 + 6 16 22 19 = 8 22 49 34 .... − 0 .10

Sum.. +12 .36

∗'s An. Var. in N. P. D. 1820 (Naut. Alm. 1823) −7″.95 × 3 years. −23 .85
Oct. 10 (Table XV.) An. Var.................................. − 6 .157

Amount of An. Var. to Oct. 10, 1823............................ −30 .007
∗'s N. P. D. (Naut. Alm. 1823) for 1820 .... 73 51 39 .67

Do. Oct. 10, 1823.. 73 51 9 .663
Sub. from.. 90

∗'s Declination Oct. 10, 1823............................ = 16 8 50 .337

Nat. Secant of Dec. 16° 8′ 50″....1 .0410 × + 12″.36 = + 12″.97086 in space,
And, dividing by 15, cor. for aber. in R. A. in time... = + $0^s.865$, nearly as before.

Tab. C. Arg. = $\overset{s}{7}\ 20^\circ\ 4'\ 56''$ (as in Tab. A)....... + 14″.90
Tab. D. Arg. = 8 22 49 34 (as in Tab. B)....... − 0 .83

Sum.. + 14 .07

Nat. sine of dec. (16° 8′ 50″) .2780 × + 14″.07 .... = + 3″.91 1st part.

Tab. E. Arg. ⊙ Long. $\overset{s}{6}\ 16^\circ\ 22'\ 19''$ + ∗'s Dec. 16° 8′ 50″ − $\overset{s}{7}\ 3^\circ\ 1'\ 9''$ − 3″.38 }
Do. Arg. ⊙ Long. 6 16 22 19 − ∗'s Dec. 16 8 50 = 6 0 13 29 − 4 .03 } − 7 .41 2nd part.

Cor. for aberration in N. P. D. = − 3 .50

Lunar Nutation.—Tab. F. R. A. of ∗ $\overset{s}{2}\ 6^\circ\ 27'$ − Long. ☾'s ☊ $\overset{s}{9}\ 23^\circ\ 28'$ = Arg. $\overset{s}{4}\ 12^\circ\ 59'$ + 5″.98
Tab. G. R. A. of ∗ 2 6 27 + Long. ☾'s ☊ 9 23 28 = Arg. 11 29 55 − 1 .29

+ 4 .69

Tang. dec. 16° 8′ 50″.2894 × + 4″.69 .. = + 1″.357 1st part.
Tab. H. Long. ☾'s ☊ $\overset{s}{9}\ 23^\circ\ 28'$......... + 15 .810 2nd part.

17 .167 sum in space.
Divide by 15 .... + $1^s.14$ Cor. for lunar nutation in R. A.

Tab. I. Arg. as in F. = $\overset{s}{4}\ 12^\circ\ 59'$ ... − 6″.41 1st part.
Tab. K. Arg. as in G. = 11 29 55 ... 0 .00 2nd part.

− 6 .41 Cor. for lunar nutation in N. P. D.

Solar Nutation.—Tab. F. (R. A.) of ∗ $\overset{s}{2}$ 6° 27′ − 2 ⊙ $\overset{s}{1}$ 3° 54′ = Arg. $\overset{s}{1}$ 2° 33′ − 7″.40 } − 7 .16
Tab. G. (R. A.) of ∗ 2 6 27 + 2 ⊙ 1 3 54 = Arg. 3 10 21 + 0 .24 }

Tang. dec. 16° 8′ 50″ N. .2894 × 7″.16 − 2″.072 1st part.
Tab. H. 2 ⊙ $\overset{s}{1}$ 3° 54′.............. − 9 .63 2nd part.

Sum − 11 .702 in space.
Divide by 15...................... 0ˢ.7801 in time.

Then − 0 .7801 × .06221 = − 0ˢ.048 Cor. for the Solar Nutation in R. A.

Tab. I. Arg. as in Tab. F. = $\overset{s}{1}$ 2° 33′ − 4″.72
Tab. K. Arg. as in Tab. G. = 3 10 21 − 1 .27

Sum − 5 .99 × .06221 = − .372 Cor. for the ⊙ Nut. in N. P. D.

Hence it appears that Delambre's Universal Tables give the aberration very nearly the same as the Tables before us, but that they give the lunar nutation a trifle greater, and consequently the solar nutation derived therefrom proportionably great. The difference, however, lies in the hundredth's place of the second, and will generally be less than the error of observation. This small discrepancy arises out of the difference in the co-efficients for the nutation respectively adopted. In Bessel's last determination of these co-efficients, he has diminished the quantities still further, but not so as materially to affect the results.

The factor .06221 for converting lunar into solar nutation was determined by Zach. [See page 120 of Vol. I. Tabulæ Speciales.]

Differential Tables.

When a star, whose mean place is known, has other stars in its vicinity, it is sometimes convenient to ascertain the places of those other stars by direct measurement of the differences of their respective right ascensions and declinations, as compared with the known right ascension and declination of the principal star; which may be done very correctly by a telescope, having a spider's line micrometer well fixed in the meridian.

After observations of this sort have been made, the series of eleven small differential Tables, contained in pages 96 and 97, will show whether the variations in aberration and lunar nutation, corresponding to the measured differences of right ascension and declination, will amount to a quantity considerable enough to render a direct calculation of them necessary from the preceding general Tables. The field of view of an astronomical telescope will seldom take in a measure of more than 30′, and when the power is considerable, not more than 10′, more or less according to the change of power; and an inspection of the Tables will point out what corresponding change will take place in the corrections, + or −, as the case may be, and where such change is inconsiderable, the micrometrical measurements of the differences of right ascension and declination (or north polar distance) may be at once applied to the known mean or apparent place of the principal star, with which the small stars are compared: but if the amount appears from the Tables to be considerable, then the corrections belonging to them must be had by direct calculation. Had the arguments of the Tables been extended to every single degree, the corrections themselves might indeed have been deduced from them, but the work of double proportions would even then have become tedious.

If, in the first Table, the argument (R. A. ∗ − ⊙'s long.) were $\overset{s}{\text{O}}$ or $\overset{s}{\text{VI}}$, the change of aberration in right ascension, + or −, is nothing, but at $\overset{s}{\text{III}}$ it is + 0″.527, and at $\overset{s}{\text{IX}}$ − 0″.527, where the aberration is a maximum, supposing the star's declination to be 50°; but, in Table 2, the maximum takes place at $\overset{s}{\text{O}}$ and $\overset{s}{\text{VI}}$ of the argument, and the minimum at III and IX, and the sum of the two tabular quantities, taken from Tables 1 and 2, will give the whole variation of aberration in right ascension, corresponding to *one degree* in the change of right ascension and declination respectively; and when proportional parts are taken for quantities of change more or less than one degree, proportional variations in the correction will result. In like manner, proportional variations of the other corrections, for aberration in declination, and for lunar nutation in right ascension and declination, may be had from Tables 3, 4, 5, 6, and 7, in succession. The remaining Tables, from 8 to 11 inclusively, will save the trouble of proportional calculations, by exhibiting the *differences* that may be allowed in the right ascension and declination of a star's place, before the respective corrections will individually amount to *one second*. If, for instance, in Table VIII., the star's declination be 30°, the change in right ascension may be 2°, and in declination 3°, before the aberration in right ascension be changed 1″; or the same effect will be produced if the change in right ascension be only 1°, with a change in declination of 4°.

Camerarius contrived these Tables, or at least some of them, about the year 1797 (See the Acts of the Academy of Sciences at Erfurt), and the general formulæ as altered by him, according to the differential method, stand thus, where d denotes difference, viz.:

Table 1. d aber....... = [19''.41734 sin. (R. A. − ☉)] d R. A. sec. dec. ∗
Table 2. ....[−19 .41734 cos. (R. A. − ☉) tang. dec. ∗] d dec. sec dec. ∗
Table 3. d aber. in dec... = 19 .41734 sin. dec. ∗ cos. (R. A. − ☉) d A. R.
Table 4. .... 19 .41734 cos. dec. ∗ sin. (R. A. − ☉) d dec.
Table 5. d nut. R. A. ... = 8 .415 [sin. (R. A. ∗ − ☊ ☾)] tang. dec. ∗
Table 6. d nut. R. A. − $\frac{\text{cos. (R. A.} - ☊ ☾) \, d \text{ dec.}}{\text{cos. dec. } ∗}$
Table 7. d nut. dec... = 8''.415 cos. (R. A. ∗ − ☊ ☾) d R. A.

ANNUAL AND DIURNAL CORRECTIONS OF ABERRATION.

In the present improved and improving state of practical astronomy, it might be deemed an omission to pass in silence over any correction, however minute, which has occupied the attention of eminent astronomers. Laplace and Delambre, on taking into consideration the effect that the eccentricity of the earth's orbit must have on the annual aberration of light, have determined formulæ by which this effect may be computed, thus:

For R. A..... $\frac{0''.33531}{\text{cos. dec. } ∗}$ sin. (R. A. ∗ − 8° 42′ 40''.)

And for dec... −0''.022 cos. dec. ∗ + 0''.33531 cos. (R. A. ∗ −8° 42′ 40'').

[Vide *Conn. des tems* anneé IX. p. 471, and anneé XI. p. 344.]

The corrections resulting from these formulæ may be considered as constant for many years, and a small Table may easily be constructed for a given number of stars, that shall exhibit these corrections. If it be required to ascertain what are the corrections of the annual aberration in right ascension and declination of Polaris, when this star's right ascension is $0^h\ 57^m$ or 14° 15′ in arc, and its declination 88° 21′, the process will stand thus, viz.:

In R. A.

Log.....0''.33531	9 .52543
Sub. cos. 88° 21′.................	8 .45930
	1 .06613
Add sin. 5° 32′ (14° 15′ − 8° 43′).....	8 .98419
Cor. = +1''.123 (or 0''.075)........	0 .05032

In Declin.

Log..... −0''.022		8 .34243
Cos. dec. 88° 21′.................		8 .45930
−0''.0006		6 .80173
Log. + 0 .33531		9 .52543
Cos. 5° 32		9 .99797
	+0 .3338	9 .52340
	−0 .0006	
Cor. =	+0 .3332	

These two corrections being both positive, show, according to Zach [Tab. Spec., Vol. I. p. 104], that if they were *omitted*, the aberration of both kinds would be *too great*. The same author has given an approximate Table of these corrections for 38 principal stars, in the page referred to, by means of taking the whole aberration given in his Special Tables when the sun's perigee is used as the argument, and then taking $\frac{1}{60}$ part of it for the tabular quantity; but as the formulæ give results a little different, Zach's Table has been omitted.

From analogy a *diurnal aberration* of light has been deduced by Zach from the preceding annual aberration, and the following formulæ were determined for calculating the tabular quantities contained in the Table of Diurnal Aberration at page 98; viz.:

For R. A....... $\frac{-0''.304 \text{ cos. lat. cos. hor. } \angle ∗}{\text{cos. dec. } ∗}$

And for dec. ... −0 .304 cos. lat. sin. dec. ∗ sin. horary ∠ ∗,

but in meridian observations, the horary ∠ = 0, therefore the diurnal aberration in declination becomes nothing; and in right ascension it becomes −0''.304 cos. lat. sec. dec. ∗ : and from this the Table before us is calculated, though it does not appear that any astronomer has hitherto availed himself of its use.

CORRECTIONS OF LUNAR NUTATION.

It may be proper to remark here that the celebrated Bessel has lately introduced some additional quantities, as co-efficients of the lunar nutation, which render the computations more intricate without adding materially to their accuracy. They are as follow, viz.:

In R. A. .... + 0″.0877 cos. R. A. cos. 2 ☊ tang. dec.
− 0 .1846 sin. 2 ☽
−(0 .0874 cos. R. A. cos. 2 ☽ + .0801 sin. R. A. sin. 2 ☽) tan. dec.

In Declin. ... −0 .0877 sin. R. A. cos. 2 ☊
+0 .0874 sin. R. A. cos. 2 ☽ −0″.0801 cos. R. A. sin. 2 ☽

In which additional formulæ ☽ denotes the moon's true longitude.

The corrections arising out of these formulæ may be used when the utmost precision is required, as in determining the disputed question of parallax of the stars.

A GENERAL TABLE OF THE ABERRATION OF LIGHT, ADAPTED FOR THE PLANETS OR COMETS.

Light will pass through the mean distance from the sun to the earth in $8^m\ 13^s.2$, and therefore through any other comparative distance D in $8^m\ 13^s.2 \times$ D. The quantity of aberration of light in any case must be ascertained from the combined consideration of the velocity of light and of the diurnal geocentric motion of the body observed. Let g represent the diurnal geocentric motion of any moving body, and D its distance from the earth at the time, then this motion in a given time will be $= \frac{8^m\ 13^s.2}{24^h}.D.g. = \frac{493^s.2}{1440^m}.D.g$, if the geocentric motion be estimated in minutes; or $\frac{493^s.2}{86400}.D.g$, if the said motion be given in seconds: hence the logarithms of the constant factors for the diurnal geocentric motion of a celestial body, will be

for minutes = log. 9 .5346606
for seconds = log. 7 .7565094,

and for the horary geocentric motion,

for minutes = log. 0 .9148718
for seconds = log. 9 .1367206.

For instance, if the sun's aberration were required, his distance from the earth being = 1, and his mean daily motion = 59′ 8″.3, or 3548″.3 = g, the logarithm of this quantity added to the constant logarithm 7 .7565094 will give the logarithm of 20″.255 for the effect of aberration; and thus the effect of aberration for any other distance and daily motion will be $= \frac{20''.255}{3548''.3}$ D.g. In the case of the moon, if her distance from the earth be taken at 0 .00253 = D, and her mean diurnal motion = 13° 10′ 35′ = 47435″.g, then the sum of the logarithms of these two quantities added to the constant logarithm before used, will give the effect of aberration = 0″.685 only, a quantity too minute to be generally noticed.

Lambert published in the Berlin Collection of Astronomical Tables, a General Table of the Aberration of Light derived from the expression $\frac{493''.2}{1440'}$. D . g, which was free from the inconvenience attending Tables of double entry, and which required both the geocentric motion of the body, and its distance from the earth to be expressed in minutes of arc, the sun's distance from the earth being assumed = 10′. The same Table was afterwards re-formed by Zach, who introduced more modern data, and is that which now forms the subject of explanation. The use of this Table (p. 98) may be exemplified by taking the diurnal geocentric motion and distance of the comet of 1769 as given by Pingré, viz. D = 6′.67, g in long. = 1° 3′, and g in lat. = 25′, if the distance of the sun be put = 10′.

Then Arg. I. = D + g = 1° 9′.67 produces from the Table 41″.57
Arg. II. = D − g = 0 56 .33 .... 27 .17

The difference, or aberration in longitude .... 14 .40
Pingré by trig. calc. found it........................ 14 .20

Again Arg. I. = D + g = 31′.67 gives from the Table..... 8″.59
Arg. II. = D − g = 18 .33 .... 2 .88

Diff. or aberration in latitude......................... 5 .71
Pingré's calculation .... 5 .6

When a planet or comet is stationary, or $g = 0$, then the aberration is also nothing. And in cases where $D + g$ exceeds $2^{\circ} 40'$ the limit of the Table, one-half of the argument may be substituted, and then the resulting aberration must be multiplied by 4; when the argument is $1^{\circ} 20'$ the aberration is $54''.80$, but when double, or $2^{\circ} 40'$, it is $219''.20 = 4 \times 54''.80$.

When the aberration is applied to the *mean*, or *calculated* place of a planet or comet, and its motion is *direct*, the sign $-$ must be prefixed, but when the motion is *retrograde*, the prefixed sign must be $+$, to give the *apparent* or *observed* place; but on the contrary, if the aberration is applied to the *apparent*, to give the *mean* place, these signs must be reversed; *i. e.* in a direct motion $+$ must be used, and $-$ when retrograde.

As the distance of any planet from the earth varies inversely as its apparent disc, the measurement of the disc by means of a good micrometer will, at any time, give the distance of any of the larger planets with sufficient accuracy for the argument, provided its mean distance and mean diameter be previously known.

REDUCTION TO THE MERIDIAN BY DELAMBRE'S METHOD.

SINCE the repeating and altitude and azimuth circles were made useful in observatories, it has been found convenient to construct Tables for the purpose of reducing altitudes or zenith distances taken at a short distance out of the meridian, to what they would have been if taken exactly in the meridian. When a good chronometer or clock is at hand, to indicate the hour angles at the different instants of observation, the measures may be taken with the face of the circle placed alternately east and west, and the facility with which the reduction may be made renders this method of observing in many cases peculiarly desirable.

Part I. of the Table (pages 99—101) was originally computed by DELAMBRE, from the formula $\frac{2''. \sin^2 \frac{1}{2} p'}{\sin 1''}$; where p′ is the hour angle in sidereal time: but Schumacher having discovered some errors in it, re-computed it, and extended the argument from 15^{m} to 36^{m}; and also added Part II., computed under his own direction, as a second correction, particularly when the hour angle is more than 10^{m}.

If we call the tabular quantity in Part I. m, and that in Part II. n, the altitude of the pole (or the lat.) L, the declination of the star D, the zenith distance out of the culmination, z (the dec. and zen. dist. being $+$ if north, and $-$ if south), and the zenith distance at the culmination Z, we have $Z = z - \left(\frac{\cos L \cos D}{\sin Z}\right) . m + \left(\frac{\cos L \cos D}{\sin Z}\right)^2 . \text{cotang } Z . n$.

Example.

Horary Angle.	Part I. Reduction.	Part II. Reduction.
11^{m} 30^{s}	259″.6	0″.16
10 20	209 .6	0 .11
7 45	117 .9	0 .03
6 30	83 .0	0 .02
3 0	17 .7	0 .00
1 40	5 .4	0 .00
1 30	4 .4	0 .00
4 20	36 .9	0 .00
7 0	96 .2	0 .02
8 40	147 .5	0 .05
12 0	282 .7	0 .19
16 30	534 .3	0 .69
	12) 1795 .2	12) 1 .27
Mean	149 .6	0 .106

POLE STAR.

Log. cos L (51° 30′) .... 9 .7941496
Log. cos D (88 20) .... 8 .4636649
Log. sin Z (36 50) ar. co. .... 0 .2222185

Const. log. .... 8 .4800330
Log. 149″.6 (m) .... 2 .1749316

First part.............. = 4″.5182....... 0 .6549646

Const. log. .... 8 .4800330
Log. 0″.106 (n)......................... 0 .0253059
Cotang Z (36° 50′)..................... 10 .1255162

Second part .... 0″.0013....... 7 .1108881

Whole reduction .... 4 .5195 to be added to the mean altitude gained out of the meridian near the upper culmination, or to be subtracted from z.

In the preceding Example, the latitude of the place and declination of the star, and consequently its zenith distance, are supposed to be previously known; but in all cases where the latitude alone, or the declination alone, is known, z must be substituted for Z in the formula, and then the resulting reduction, which will not materially differ from the true reduction, applied to z, will give Z, and also L, very near the truth; after which, the operation pointed out by the formula must be repeated with Z and L, as though they had been previously known. This repetition will give the reduction near enough for all observations made near the meridian, but if the horary angle be great, the last reduction may be considered as only approximate, and the operation may be repeated a second time, to give the reduction perfectly correct.

When observations of this kind are made on stars below the pole, the altitude is the least, or zenith distance the greatest, on the meridian, therefore the reduction will require a *contrary sign.*

REDUCTION TO THE MERIDIAN BY DR. YOUNG'S METHOD.

DR. YOUNG has simplified the preceding formula by omitting its second term, and has so transformed it, that a Table of Natural Versed Sines to seconds of time, is all that is required besides a Table of Logarithms, for determining the reduction. His rule is this:

"Add together the logarithmic cosines of the latitude and declination, the secant of the altitude, and the arithmetical complement of the sine of 1″ for a *constant logarithm:*

Collect into one sum the versed sines of the horary angle of the several observations, divide by their number, and add the logarithm of the mean thus found to the constant logarithm, and the natural number, corresponding to the sum, will be the correction in seconds, which must be subtracted from the observed altitude, when the observation is immediately under the pole, but in all other cases added to it."

Let the same Example of the Pole Star be taken, as was given to illustrate the preceding method.

Horary Angles.	V. S.
11ᵐ 30ˢ	1259
10 20	1016
7 45	572
6 30	402
3 0	86
1 40	26
1 30	21
4 20	179
7 0	466
8 40	715
12 0	1370
16 30	2590
	12) 8702
Mean	725

Latitude ... 51° 30′ } log. cos.	9 .7941496
Declination . 88 20 }	8 .4636649
Altitude.... 53 10.. log. sec.	10 .2222185
Log. sine 1″ ar. co.	5 .3144251
Constant logarithm	3 .7944581
Log. 725 (+ 4 to Index)	6 .8603380
4″.5164 reduction	0 .6547961

The numbers of the Table are millionths of the radius; if they are considered as integers 4 must be added to the index of the logarithm. When a solar clock is used for observations of the star's altitude, it will be more accurate to add log. 5 .3168000 instead of 5 .3144251. Dr. Young's DEMONSTRATION is given in a small 8vo pamphlet published by order of the Commissioners of the Board of Longitude, 1819.

TERRESTRIAL GRADUATION.

THE accuracy with which the trigonometrical survey has been executed in every part of England, and in many parts of Scotland, gives the English and Scotch astronomers of the present day the advantage of obtaining their longitudes and latitudes with comparatively little trouble. The exact position of many of the church steeples, and of other conspicuous marks, has been recorded in the volumes of the Trigonometrical Survey already published, and the remainder will no doubt be contained in the succeeding volumes. The advantage thus afforded will be embraced with avidity by such amateurs, as have furnished themselves with astronomical instruments for the purpose of making observations on celestial phenomena, or who may hereafter be induced to do so. The series of Tables under the head "TERRESTRIAL GRADUATION," (pages 104—108) will be found useful in affording the requisite data for effecting the calculations in all cases where the latitude is nearly known; and though the skilful astronomer may not think it right to depend entirely on his latitude and longitude derived from a trigonometrical station or mark in his neighbourhood, yet such deduction, corroborating his latitude deduced from celestial observations, will give him confidence also as to his longitude, which is not so easily obtained. Triangulation is now encouraged in almost every part of the civilized world, and already several measures have been taken showing the length of a degree both of longitude and latitude, in various parallels, from which Tables have been constructed; but it remains to be ascertained whether the globe itself is so regular a body, as to admit of correct Tables to be founded on the same quantity of compression. On this account the Tables derived from distant measurements have been collated, including No. 1 to 8, and the astronomer may select that compression which, from further investigation, may turn out to be most eligible in his particular situation.

When any of the first four Tables is used, the number of yards or fathoms constituting a degree of longitude at the equator, must be assumed as equal to 60 geographical miles, and then the number of yards composing a degree of latitude or longitude in any given parallel may be had by direct proportion, from the geographical miles given in the Tables, according to the compression deemed most eligible. According to Puissant and Svanberg, the equatorial degree, reduced into English measure, is taken at 60847 fathoms, viz. $\frac{100149}{9} \cdot \frac{39\ .3702}{72}$; Cagnoli has assumed it = 60893; Lieut. Col. W. Lampton = 60857; and the Author of the 8th Table = 60845; the determination of General Mudge in the Trigonometrical Survey being = 60823 in latitude 52° 34′; or, which amounts to the same thing, = 60845 in latitude 54° 44′ 8″, where a degree of latitude in the spheroid has been demonstrated to be equal to the equatorial degree, viz. where the square of the sine is equal $\frac{2}{3}$ of radius. [Vide James Adams on the Ellipse.]

The tabular numbers of Lieut. Col. W. Lampton's Table were computed from the formula given in the Appendix to Vol. XII. of the ASIATIC RESEARCHES, in page 90, where

m = the degree in latitude l.
p = the perpendicular degree } on the equator, where $p = d$.
d = the degree of longitude }
m = the degree on the meridian }
p' = the perpendicular degree } in any other latitude $'l$.
d' = the degree of longitude }
1 = the polar axis.
$1 + e$ = the equatorial diameter.

$$\text{Then } m' = m \frac{\sqrt{\cos^2 l.(1+e)^2 + \sin^2 l}^{\,3}}{\sqrt{\cos^2 {'l}.(1+e)^2 + \sin^2 {'l}}^{\,3}}$$

$$p' = \frac{p(1+e)}{\sqrt{\cos^2 {'l}.(1+e)^2 + \sin^2 {'l}}}$$

$$d' = \frac{d\,(1+e)}{\sqrt{(1+e^2) + \tan^2 {'l}}}$$

In which m = 60491 .4; $'l$ = 13° 34′ 44″; p' or d' = 60857.05 fathoms; and $1+e$ = 1 .0032896. [Vide Vol. XIII. ASIATIC RESEARCHES, p. 122. 1820.] The construction of the other Tables may be understood without further explanation; but such readers as wish for particular information respecting the Tables of Svanberg and Cagnoli, may consult "*Exposition des Operations faites en Lapponie,*" &c. *Stockholm*, 1805; and "*Trigonometrie rectiligne et spherique,*" *par Antoine Cagnoli, traduite de l'Italien par N. M. Champre. Second edition.* 1808. *A Paris.*

The following Examples, it is presumed, will sufficiently illustrate the use of these Tables of TERRESTRIAL GRADUATION, as they have reference to the determination of both the longitude and latitude of a given place in the neighbourhood of a trigonometrical station or other ascertained mark.

Example 1.—The longitude of the Pagoda, in Kew Gardens, is given in the Trigonometrical Survey = 1^m 10^s.4 west of Greenwich, and its latitude = 51° 28′ 16″ N., and when, in the year 1811, I erected a private observatory at East Sheen, in the next parish, I was naturally desirous of deriving the position of my station therefrom, and made use of this simple process with success. After having established a small circular instrument on my pillar, I found that the Pagoda bore 56° 58′ 40″ to the west of my meridian line, and when I had placed a staff at 8 .50 chains (of Gunter) from the pillar, in a line directed towards my distant meridian mark, I took this as a station from which to measure the angle included between the two lines, drawn or pointing thereto from my pillar and from the Pagoda respectively; which angle, by a good Troughton's reflecting circle, I found to be 118° 20′. Hence the angle, subtended at the Pagoda by 8 .50 chains, was found to be 4° 42′ 20″, viz. the supplement of the sum of the other two; and as the Pagoda was not distant, a small triangle was deemed sufficient for this purpose. The operation performed by plane trigonometry stands thus:

As the sine of 4° 41′ 20″ ar. co. ....log. 1 .087538
Is to the opposite side 8 .50 chains .... 0 .929419
So is the sine of 61° 40′ (Sup. 118° 20′) .... 9 .944582

To the distance 91 .525 chains.................. 1 .961539 = 2013 .55 yards.

As radius ....10 .000000

91 .525 chains distance........................ 1 .961539
Sine 56° 58′ 40″ .... 9 .923482

76 .74 chains dif. of longitude .... 1 .885021 = 1688 .28 yards.

As radius ....10 .000000

91 .525 chains distance........................ 1 .961539
Sine 33° 1′ 20″ (or cos 56° 58′ 40″) .... 9 .736367

49 .878 chains dif. of latitude.................... 1 .697906 = 1097 .316 yards.

If we take the length of a degree of longitude at the equator at 60845, as determined in the Trigonometrical Survey, or at 121690 yards, Table 1 (or Table 4, col. $\frac{1}{321}$) gives 37 .4 geographical miles for a degree in latitude 51° 27′ 44″, and, consequently, $\frac{37.4}{60}$ of 121690 = 75853 .4 yards will be a degree of longitude in this parallel of latitude; but the difference of latitude between the Pagoda and Observatory in question, came out = 1688 .28 yards, and 1° is equal 4^m or 240^s in time: hence $\frac{1688.28}{75853.4}$ of $240^s = 5^s.34$ the difference of longitude in time; and as the Pagoda is more to the west than the Observatory, this difference must be subtracted;

Longitude of the Pagoda	$=1^m\ 10^s.4$ W.
Diff.	-5.34
Longitude of the Observatory	$=1\ \ 5.06$ W.

On the 30th of April, 1822, the difference between Greenwich and East Sheen solar time was taken, by means of a good Hardy's Chronometer, and, allowing for rate, was found to be $1^m\ 4^s.8$ W.

In Table 2 (or Table 3, col. $\frac{1}{321}$) a degree of the meridian is given = 59 .97 geographical miles, in the parallel of latitude from 51° to 52°, and the difference of latitude has been calculated = 1097 .316 yards, therefore assuming a degree at the equator = 121690 miles as before, we shall have $\frac{1097.316}{121690}$ of 59′.97 = 32″.45 for the difference of latitude, to be subtracted, as the Pagoda is to the north of the Observatory; hence 51° 28′ 16″ − 32″.45 = 51° 27′ 43″.55 is the latitude required.

With a new circle of 15 inches diameter by Fayrer, the latitude was determined by the pole star thus, with Dr. Brinkley's Refractions.

Face east, April 27, 1821	51° 27′ 41″.62
Face west, May 9, 1821	51 27 44 .46
Average	51 27 43 .04

Example 2.—On removing my astronomical instruments from East Sheen to South Kilworth in Leicestershire, it became a matter of interest to ascertain the latitude and longitude of the Rectory-house. In the Trigonometrical Survey it appeared that the position of either North Kilworth or South Kilworth church steeple is erroneously given, on account of the *southern* church having the greater latitude assigned. On communicating this apparent discrepancy to Major (then Captain) Coleby, he was so obliging as to examine the original calculations, and found that, by some mistake, a wrong triangle had been taken in determining the position of South Kilworth, but that North Kilworth was placed right. He was moreover so obliging as to re-calculate the position of South Kilworth steeple from the proper data, and found its longitude $4^m\ 25^s.66$ west, and its latitude 52° 25′ 56″ north; the position of North Kilworth steeple being, in longitude, $4^m\ 21^s.53$ west, and latitude 52° 26′ 39″ north. Now the Observatory at the centre of the Rectory-house is just 30 yards south, and 22 yards east, of the centre of the steeple; and, as 33 .8 yards is 1″ of latitude, and 21 yards 1″ or $0^s.066$ of longitude, in this parallel, the latitude of the Observatory derived from South Kilworth steeple is 52° 25′ 55″.93, and its longitude $4^m\ 25^s.59$. But to ascertain the position of the Observatory as derived by triangulation from North Kilworth steeple (both steeples having fine pointed spires), I took the angle of bearing of North Kilworth steeple, and found it 41° 29′ to the east of my meridian mark, by a Hadley's sextant; then placing a staff in the meridian line at 233 yards, from the Observatory, I measured the angle included by the two lines coming from it, and from North Kilworth steeple, and made it 133° 39′ 30″, and consequently the small complemental angle subtended by 233 yards = 4° 51′ 50″, and the whole distance = 1990 .3 yards from the Observatory to North Kilworth steeple.

Hence by means of the sine of the bearing 41° 29′, of its cosine, and the distance 1990 .3 above given, the difference of longitude comes out, by plane trigonometry, 1318 .4 yards, and the difference of latitude 1491.

According to Lampton, Table 7 gives a degree of latitude in parallel 52° 26′ = 60833 fathoms, or 121666 yards, and a degree of longitude in the same parallel 37178 fathoms, or 74356 yards: hence $\frac{1491}{121666}$ of 60′ = 44″.1 is the difference of latitude, and $\frac{1318.4}{74356}$ of 60′ = 1′ 3″.8, or in time $0^m\ 4^s.25$, is the difference of longitude; consequently, the longitude of the Observatory derived from North Kilworth steeple is $4^m\ 21^s.53 + 4^s.25 = 4^m\ 25.78$, instead of $4^m\ 25^s.59$, as determined from South Kilworth steeple, and the latitude comes out 52° 26′ 39″ − 44″.1 = 52° 25′ 54″.9, instead of 52° 25′ 55″.93; and the difference of 1″ in the two determinations of the latitude has arisen most probably from some inaccuracy in taking the large angle of more than 133° at twice, by means of a staff dividing the large angle into two, the index error being not precisely the same at the different distances from the respective objects.

SPACE AND TIME.

THE conversion of space into time, and the contrary, is an operation of frequent occurrence in astronomical calculations, and is rendered easy by the two small Tables given in page 109. The mode of using these Tables will be evident from a simple inspection of their construction.

Example 1.—Let it be required to turn an arc of 109° 20′ 35″ of the equator, or of terrestrial longitude, into its corresponding time?

In the first vertical column of "SPACE INTO TIME."	100°	$= 7^h\ 20^m\ 0^s$
	9	= 0 36 0
In the third column..............................	20′	= 0 1 20
In the fifth column	35″	= 0 0 2 .33
	Answer	7 57 22 .33

Let it be required to turn $7^h\ 57^m\ 22^s.33$ into degrees and parts of the equator, or of terrestrial longitude?

Column 1. "TIME INTO SPACE."..........	7^h	= 105° 0′ 0″
Column 3.	57^m	= 14 15 0
	22^s	= 5 30
	0 .33 or $\frac{1}{3}$	= 5
	Answer	109 20 35

Example 2.

Space into Time.		*Time into Space.*	
200°	........$13^h\ 20^m\ 0^s$	13^h	........195° 0′ 0″
8	.... 0 32 0	55^m	.... 13 45 0
45′	.... 0 3 0	2^s	.... 0 0 30
36″	.... 0 0 2 .4	0 .4	.... 0 0 6
	13 55 2 .4		208 45 36

SIDEREAL AND SOLAR TIME.

As the use of a sidereal clock is indispensable in every observatory, and as several phenomena have their occurrence calculated and noted down in solar time, such as occultations, eclipses, &c., it becomes necessary to convert solar into sidereal time, and the reverse: for which purpose the two subsidiary Tables in page 110 will be found convenient. The first Table contains the positive equation, or quantity to be added to solar time to produce its equivalent in sidereal time; and the second Table contains the negative equation, or quantity to be subtracted from sidereal time to give its equivalent in the denomination of solar time. The former equation is usually called *acceleration* of sidereal on solar time, and the latter *retardation* of solar as compared with sidereal. A solar day is measured by the earth's rotation on its axis, as it regards the *sun*, and is a synodic period; but a sidereal day is measured by the absolute rotation of the earth on its axis, as it regards a *star*, or fixed point, and is the only uniform standard measure of time that we know of in nature. The daily difference of these periods is commensurate with the earth's daily motion in her orbit; viz. on an average 59′ 8″.33 = $3^m\ 56^s.55$; and $\frac{24^h}{3^m\ 56^s.55} = 365\frac{1}{4}$ solar days nearly, within which time the acceleration will amount to another day; also $\frac{24^h}{3^m\ 55^s.908} = 366\frac{1}{4}$ sidereal days nearly; so that in every year there is a sidereal day more than there are solar days in the same time; and on these data the Tables are respectively founded. Their uses will appear from the work of the subjoined example.

Example. What portion of sidereal time is equivalent to $15^h\ 32^m\ 18^s$ of solar time, and the reverse?

15^h	......$2^m\ 27^s.847$	15^h	......$2^m\ 27^s.442$
32^m	...... 5 .257	34^m	...... 5 .570
18^s......	0 .050	$51^s.154$	0 .142
Acceleration	+2 33 .154	Retardation	−2 33 .154
Solar time	15 32 18 .000	Sidereal time	15 34 51 .154
Sidereal time	15 34 51 .154	Solar time	15 32 18 .0

To ascertain the solar time that corresponds to the indicated sidereal time, and the contrary, at any moment on a given day, it is necessary to know the sun's right ascension or distance in time from the vernal equinox, which might be calculated from the solar Tables after the plan of Zach, but may be had by inspection from the Nautical Almanac. The rule is this When the sidereal time is given to find the corresponding mean solar time, subtract the sun's right ascension, as given in the Nautical Almanac, for the specified day, from 24 hours, which call the *Complement to the Equinox*, or distance of the equinox to the sun (as given in the *Connoissance des Tems)*, and add it to the given sidereal time; then take out the retardation corresponding to this sum, and subtract it therefrom, and the remainder will be the *apparent* solar time, to which apply the equation of time, with its proper sign from the Nautical Almanac, and the sum or difference will be the mean solar time.

Example. Let it be required to calculate what the solar time was when the sidereal clock indicated $23^h\ 50^m\ 30^s$ on the 7th of April, 1823.

Sidereal time indicated		$23^h\ 50^m\ 30^s$
Complem. to equinox ($24^h - 1^h\ 2^m\ 8^s.6$)		22 57 51 .4
		22 48 21 .4 sum.
Then 22^h equation	$-3^m\ 36^s.419$	−3 44 .34
48^m	−0 7 .864	
21 .4	−0 0 .058	22 44 37 .06 ap. ⊙ time.
Equation of time at noon		+ 2 19 .30
Mean solar time		22 46 56 .36

When the solar time is given, and the sidereal is required, the operation and signs must all be reversed.

Thus,.... Mean solar time, 7th April, 1823		$22^h\ 46^m\ 56^s.36$
Equation of time subtract here		−2 19 .30
Apparent solar time		22 44 37 .06
Acceleration 22^h	$3^m\ 36$ .84	+3 44 .17
.... 44^m	0 7 .23	
.... 37^s	0 0 .10	22 48 21 .23 sum.
Complement to the equinox		−22 57 51 .40
Sidereal time required		23 50 29 .83

As the astronomical solar day begins at noon, 10 o'clock A.M. must be substituted for 22 solar hours; and in the last subtraction, the 24 borrowed must be added to $22^h\ 48^m\ 21^s.23$, to make 46^h, from which $22^h\ 57^m\ 51^s.4$ is to be subtracted.

The column of *solar days*, at the beginning of the page 110, will be serviceable, as well as the column of *solar hours* under it, in comparing the rates of two clocks, one solar and the other sidereal; or in bringing the latter to keep true time, when the rate of the former is known.

By ZACH's method, from the SOLAR TABLES.—Given, as before, the sidereal time $= 23^h\ 50^m\ 30^s$ on the 7th of April, 1823, to determine the corresponding solar?

Solar Table 3, 1823		$0^h\ 0^m\ 36$ .64	(p. 156)
Table 4, 7th April		0 59 11 .40	
Sum		0 59 48 .04	
SOL. TAB. 18. ☾ equat. of equin.	☊ $^s10\ 3^0$	+0 .94	(p. 175)
SOL. TAB. 19. ⊙ equat. long. ..	0 17	−0 .04	
Sun's mean R. A.		0 59 48 .94	

22^h	$3^m\ 36^s.419$
50^m	8 .191
41^s	0 .112
Equation	−3 44 .722

Sidereal time given	$23^h\ 50^m\ 30^s.0$
Sun's mean R. A. sub.	0 59 48 .94
Diff. in sidereal time	22 50 41 .06
Acceleration of ditto	−3 44 .72
Mean solar time	=22 46 56 .34

If the equation of time be applied to the mean solar time with a contrary sign, it will give the *apparent* time.

Given the mean solar time on April 7, 1823, = 22^h 46^m 56 .34, to find the sidereal?

22^h	3 36 .84
46^m	7 .56
56^s.34....................	0 .16
	3 44 .56

Mean solar time........	22^h	46^m	56^s.34
Acceleration .... +		3	44 .56
	22	50	40 .90
Sun's mean R. A...... +	0	59	48 .94
Sidereal time..........	23	50	29 .84 as before.

The sun's mean right ascension is here taken from the preceding operation, as given by the SOLAR TABLES. The fraction of a second has been lost by limiting the decimals to two figures each.

THE CLOCK'S RATE.

THERE perhaps never was a clock or chronometer constructed, that did not gain or lose some portion of time every day, however nicely regulated; and astronomers seem to have agreed, that it is better to take an account of a clock's *rate* or daily deviation from the exact truth, than attempt to perfect it by regulation, when its amount is not considerable. Hence a proportional part of the daily rate, whether it be gain or loss, may frequently be required to be taken into the account, when nice observations are made, and particularly when their object is to determine right ascension in time. The Table contained in pages 111 and 112 has been calculated to answer the purpose of indicating the due proportion of a clock's daily rate, at any given interval of time, counted from the moment at which the rate was assigned. It is constructed on the same principle as the General Tables of Precession and Aberration before explained. The whole daily rate may consist of any number of units, or decimals, or both, which are found at the top or bottom of the Table successively, and in the line denoting the interval, at either side, the quantity will always be found. Suppose the daily rate of any clock to be + 2^s.57, and that after an interval of 4^h 36^m, an allowance is required to be made for it:

In the line 4^h 36^m, and in columns 2^s ......0^s.333
.5 ...... .096
.07...... .013 } + 0 .442 will be the allowance required. If the rate had been − 9″.32, which is too much to be allowed, and the interval 11^h 24^m, the proportion would have been thus:

9^s ...... 4 .275
.3 ...... .142
.02...... .099
− 4 .426

but if the rate had been .93 the proportion would have been .... 0^s.9 ...... .427
.03...... 14
− .441

and in like manner a proper proportion of any other rate may be obtained.

THE CORRECTIONS OF 48 PRINCIPAL STARS.

WHEN the reader has made himself master of the XVIII general Tables, extending from page 20 to page 91, which, it is presumed, have been sufficiently explained and exemplified, he will be prepared to comprehend the convenient arrangement of the arguments, mean places, and maxima of the aberration, ☾ nutation, and ☉ nutation of 48 principal stars, which have been separately calculated for the year 1830, and which are contained in pages 113 to 144. According to this arrangement all the corrections of any of the given stars may be readily obtained for any given day, for many years to come, when the correct annual variation is taken from Table XV., by simple inspection for the given day, and increased by as many years' variation as have elapsed since 1820, which is the epoch of the mean places. Each star has its corrections running through columns that occupy two successive pages, and the sun's true longitude and the mean place of the moon's ascending node, for the given day, are all the data that the Table requires to be found elsewhere. The former of these is found in the Nautical Almanac of the given year, or, if that be not at hand, it may be had from an almanac of *another year*, for any year from 1750 to 1899, by means of Table 3. of the SOLAR TABLES (p. 156), as will be explained hereafter; and the latter must be taken from Tables 1 and 2. of the LUNAR TABLES at page 181. The sun's longitude thus obtained will, however, be for *noon at Greenwich*, and it ought to be for the moment of the star's culmination, which will require the further use of Table XVI. (p. 45), constructed for this purpose, and also of Table XVII. (p. 46), if the longitude of the observatory be considerable, as it regards Greenwich.

The list of 48 principal stars contains all the Greenwich enlarged list of 46, except α¹ Capricorni (which has the same corrections as α² Capricorni), with the addition of three zenith stars, for the three new observatories, viz.: θ URSÆ MAJORIS for Cambridge, ζ URSÆ MAJORIS for Edinburgh, and ε SAGITTARII for the Cape of Good Hope; all which stars will be also as useful as the other Greenwich stars, whenever they are observed elsewhere.

The utility as well as convenience of the Table of CORRECTIONS before us, considered as a GENERAL TABLE, will appear from the examples that are subjoined.

Example 1.

What will be the mean place, the amount of the corrections, and the true apparent place of α Orionis when on the meridian of Greenwich on April 10, 1824?

R. A.	N. P. D.	Arguments in R. A.	Arguments in N. P. D.
1820. R. A. 5h 45m 25s.92	1820 .... = 82° 38′ 6″.04	A ...... = 6s 3° 13′	A′.... = 2s 28° 23′
An. Var. + 3s.25 × 4y + 13.00	An. Var. − 1″.37 × 4. − 5.48	☉ ..... = 0 20 33	☉.... = 0 20 33
1824 .... = 5 45 38.920	1824 .... = 82 38 0.56	Tab. XVI. 0 0 12	Tab. XVI. 0 0 12
April 10. Tab. XV. 3s .... + 0.822	April 10. Tab. XV. 1″ .... − 0.274	Arg.... = 6 23 58	Arg... = 4 9 8
.2 .... + .055	.3 .... − 82		
.05 .... + .13	.07 .... − 2	L...... = 6 0 15	L′.... = 9 2 36
Mean R. A. ...... = 5 45 39.810	Mean N. P. D. 82 38 0.20	☊ 1824 = 9 19 5	☊.... = 9 19 4
		Ap. 10. B.. − 5 18	Ap. 10... − 5 15
Arg. 6s 23° 58′ Aber.. − 0.553	Arg. 4s 9° 8′ Aber. + 4.36	Arg.... = 3 14 2	Arg... = 6 16 25
Arg. 3 14 2 ☾ nut.. + 1.132	Arg. 6 16 25 ☾ nut. − 2.81		
Arg. 7 11 42 ☉ nut. − 0.044	Arg. 10 14 43 ☉ nut. − 0.31	S...... = 6 0 12	S′.... = 9 3 13
Correction in R. A. = + 0.535	Cor. in N. P. D... = + 1.24	☉ 2.... = 1 11 30	☉ 2 .. = 1 11 30
True apparent R. A. = 5 45 40.345	True appar. N. P. D.. 82 38 1.44	Arg.... = 7 11 42	Arg.. = 10 14 43

In this Example Mr. Pond's mean place and annual variations have been taken.

Example 2.

Let it be required to determine the true apparent place of θ URSÆ MAJORIS on the 1st of July, 1829?

As the Nautical Almanac for this year is not yet published, the Almanac for 1823 may be used with Table 3 of the SOLAR TABLES (p. 156) instead, thus: take out the tabular numbers in the column "Long." for the respective years, and apply their *difference* to the sun's longitude of the 1st of July, 1823, and the amount will be very nearly the proper longitude for July 1, 1829; when the tabular number of the required year is *greater* than that of the year for which the Almanac is calculated, the sign + must be applied to their *difference*; but if it be *smaller*, then the sign −.

	Years.		
Thus, SOLAR TABLE 3,	1823, in col. "long." stand..	19′.7	
	1829, ...	52.0	
	Dif.	+32.3	or 32′ 18″
	☉ long. July 1, 1823... =	3s 8° 46.21	
	Do. July 1, 1829... =	3 9 18.39	

In this case, both the years are common years; but when the sun's longitude is required for any day in a *leap year*, and the almanac at hand is for a *common* year, *add one* to the given day in the common year, and take from the almanac the longitude for it, instead of the longitude of the given day, and apply the tabular quantity of minutes thereto, with the proper sign, and the result will be the longitude for the given day of the required year, whether before or after. On the contrary, if the sun's longitude be required for a given day of a *common year*, and an almanac for some *bissextile* be at hand, one day must be *subtracted* from the said day in the bissextile, and the longitude belonging to such day must be taken, instead of that of the given day, and the tabular quantity applied thereto will be the proper longitude for the required day of the distant year. This contrivance was first published by ZACH, but as the sun's longitude is given in the *Nautical Almanac*, and also now in the *Connoissance des Tems* for the *noon* of each day, Zach's numbers do not answer their intended purpose, and others more suitable have been here substituted, which may be depended on for determining the arguments; for the greatest error will never exceed a very few minutes, nor the corresponding correction be affected, at the most, above one-tenth of a second.

R. A.	N. P. D.	Arguments in R. A.	Arguments in N. P. D.
1820.............. 9^h 20^m $45^s.21$	1820............... 37° 30′ 30″.40	A...... = 4s 12° 5′	A′.... = 6s 19° 55′
An. Var. + 4.07 × 9^y 36.63	An. Var. + 15″.34 × 9 = 2 18.06	⊙..... = 3 9 19	⊙.... = 3 9 19
July 1. Tab. XV. { 4. ... 1.995	July 1. Tab. XV. { 15″. .. 7.48	Tab. XVI. + 5	Tab. XVI. + 5
.07 ... 35	.3 .. 0.15	Arg.... = 7 21 29	Arg... = 9 29 19
Mean R. A. 1829, July 1. } .. = 9 21 23.870	.04.. 2	L..... = 5 6 45	L′.... = 7 18 3
Arg. 7s 21° 29′ Aber. − 1.649	Mean N. P. D. 1829, July 1. } = 37 32 56.11	☊....... 6 2 45	☊....... 6 2 45
Arg. 11 9 30 ☾ nut. − 0.571	Arg. 9s 20° 19′ Aber. − 11.47	Arg... = 11 9 30	Arg... = 1 20 48
Arg. 11 29 35 ⊙nut. − 0.001	Arg. 1 20 48 ☾ nut. + 6.41	S..... = 5 10 47	S′..... = 7 12 5
Amount of correct. = − 2.221	Arg. 2 0 53 ⊙ nut. + 0.36	⊙ 2... = 6 18 48	⊙ 2... = 6 18 48
Apparent R. A... = 9 21 21.649	Apparent N. P. D. = 37 32 51.41	Arg.... ≐ 11 29 35	Arg... = 2 0 53

If this Example had been for Edinburgh, where the longitude is about 12^m 41^s west of Greenwich, the addition to the sun's longitude would have been about 31″, according to Table XVII. (p. 46), and consequently the argument for aberration must have been increased 31″, and that for solar nutation upwards of 1′, but the resulting corrections would not have been materially different, particularly in right ascension.

If, however, the corrections had been required for the Cape of Good Hope, where the longitude is 1^h 13^m 36^s east, according to the same Table about 3′ must have been *subtracted* from each of the two arguments for aberration, and twice this quantity from each of the arguments for solar nutation, which diminution would have rendered these corrections sensibly different. The same Table of Corrections for the 48 Stars is equally applicable to all places, when the sun's longitude is altered accordingly, in those arguments of which it forms a constituent part. The change in the place of the ☾'s ☊ amounts to little more than 3′ in a whole day, and therefore may be disregarded in every part of Europe. If the difference of longitude be great, as at Madras, New Holland, &c., a proportional part of the tabular differences may be added or subtracted, accordingly as the sun's longitude is less or greater at the place in question than at Greenwich, for as the sun's daily motion is very nearly a degree, those tabular differences of single degrees may be substituted for daily differences, without any sensible error in the corrections.

The Table before us is the only one published that has given the corrections to *single degrees*, and as the form is equally applicable to all years within certain limits, it will be found useful, not only for giving the corrections due to any individual observation, but for tabulating a succession of monthly, weekly, or daily corrections, as may best suit the purpose of the practical astronomer. In the first volume of the *Tabulæ speciales*, Zach has published two general Tables, one in right ascension and the other in declination, for giving the corrections of the 36 original Greenwich stars for every ten degrees of the argument, for the year 1802, but without the solar nutation; and has contrived to make all the corrections *additive* by applying − 4ˢ to the Table of right ascension, and − 40″ to that of declination; which method of tabulation may be convenient for the use of the amateur astronomer, but disguises the quantities, and leaves him in the dark with respect to the due application of the signs. Besides, the co-efficient of the aberration is taken = 20″.0 only, which is now generally allowed to be = 20″.255.

Before Mr. Fallows left England, he so arranged the signs and degrees of the circle, with their corresponding + and − signs, on an enlarged trigonometrical sliding rule, made by Mr. Bates of the Poultry, London, that, when the arguments are prepared, the corrections may be read at one set of the slide, for all the four quadrants, and with a degree of accuracy that renders the rule truly valuable; the integer with the decimals of the tenth and hundredth places, may always be obtained with certainty, and the third decimal figure may also be had by *estimation*, and in many cases with correctness, particularly where the integral portion consists of an unit, as is almost always the case in right ascension ; and in north polar distance where the integral part is often composed of two figures, the third decimal is not wanted, on account of the correction being in *space*, or fifteen times lower in denomination than the correction in *time*. A description of this rule will probably be given in its place in the 2nd Volume.

Notwithstanding the great care that was taken in calculating the subsidiary numbers A, A′, L, L′, S, and S′, for supplying the arguments of these 48 stars, and for obtaining their maxima of aberration, ☾ nutation, and ☉ nutation, in right ascension and north polar distance respectively, there are some slight errors which I have detected since the Table was printed, and which will be pointed out and remedied hereafter.

The mean places also of these stars having been given differently by the various eminent astronomers of the present time, the computer is puzzled in his determination what numbers to adopt, as well with respect to the mean places, as to the annual variations. For these reasons I have composed four additional small Tables, which are subjoined to the Table of Corrections, for the purpose of gratifying each computer with the choice of his own elements of computation: and though much time and labour have been expended in the composition of these auxiliary Tables, yet the assistance they afford to the computer, who is anxious to avail himself of improved Arguments, will not, it is hoped, fail to be acceptable. The Table of the Comparative Mean Places, page 145, was carefully extracted from the best authorities, and arranged in such way, that the column of hours and minutes of right ascension, as well as the column of degrees and minutes of north polar distance, are *common* to all the authors, and the columns of *seconds*, both of time and space, belong to the names under which they stand, and any one of the mean places, for the epoch 1820, may be taken and substituted for the Greenwich place given at the head of the large Table of Corrections: or an average of the whole may be substituted, if thought necessary, while discrepancies continue to exist.

Also in the following "Table of Comparative Annual Precessions, Secular Changes of Precession, Proper Motions, and Annual Variations, of different eminent Astronomers for the Epoch 1800" (p. 146), the annual variations of Maskelyne, Piazzi, or Bessel may any of them be taken, or an average of the whole; but the Secular Changes could not be included in the same Table, and therefore those by Schumacher, contained in the preceding Table, may be adopted, together with his annual variations, for the epoch 1820, which are founded on the results given in Bessel's Fundamenta Astronomiæ. Still it will remain matter of interest to see, in the present Table, how the three authors have differed in their respective deductions of the Proper Motions, and consequently in their Annual Variations, which are compounded of the annual precessions and proper motions, when regard is had to their algebraic signs.

The third auxiliary Table, which gives the Changes that take place in the Arguments in 30 years (p. 147) will be found particularly useful, inasmuch as a proper application of it to the arguments computed for 1830, will adapt those arguments to *any year* for a century or more back or to come, and will thus render the Table of Corrections perpetual. This observation is equally true with regard to the fourth auxiliary Table (p. 148), which gives the annual precessions, and maxima of aberration, ☾ nutation, and ☉ nutation for two epochs, at an interval of thirty years from each other, so that proportional parts of the respective differences may be applied to bring them to any other epoch. The computation of the columns of these two Tables for the epoch 1800, to contrast with those already computed for the epoch 1830, operated fortunately as a check on the former computations, and afforded the certain means of detecting some errors, which otherwise might have remained concealed, in the Table of Corrections for 1830, which was printed at the time those errors were detected. Except in one solitary instance of α Tauri, in each of these (third and fourth) auxiliary Tables, I have reason to believe that all the errors alluded to are now corrected, as well in the subsidiary numbers as in the maxima; and therefore the computer is strongly recommended to take his Arguments, adapted to his own epoch, from the Table in page 147, in preference to those subjoined to the Table of Corrections itself. If, however, he prefers *correcting* the errors in the numbers given in the Table of Corrections, he may avail himself of the Table of *Corrigenda*, which for this purpose is here inserted.

CORRIGENDA.

Stars.		Instead of s	°	′	Insert s	°	′	Stars.		Instead of s	°	′	Insert s	°	′
α Cassiopeæ......	S	7	2°	52′	7	2°	16′	α Scorpii........	S′	3	24°	23′	3	23°	23′
α Ursæ Minoris ..	A′	11	16	58	11	16	55	α Herculis.......	S′	3	12	26	3	12	13
α Ceti...........	S′	10	14	56	10	14	11	β Draconis.......	S	5	20	52	5	18	33
α Tauri.........	L	6	3	33	6	3	28	γ Draconis... ...	L	5	27	8	5	27	2
α Orionis........	A	6	13	12	6	3	13	ε Sagittarii.......	S	5	29	15	5	29	12
α Orionis........	A′	2	29	50	2	28	23	α Lyræ..........	S	6	4	8	6	4	28
α Geminorum	L′	8	14	20	8	14	6	α Lyræ..........	S′	2	22	5	2	22	50
θ Ursæ Majoris...	S′	7	11	50	7	12	5	β Aquilæ........	S	6	1	23	6	1	20
β Leonis.........	S′	6	5	5	6	5	21	α Cephei.........	L′	1	19	33	1	19	48
β Virginis........	S′	6	5	58	6	4	58	β Cephei.........	L	8	17	42	8	17	54
α Virginis	L	6	5	24	6	5	33	β Cephei.........	S	8	15	13	8	15	11
α Virginis	A′	2	3	35	2	3	32	α Piscis Austrini..	S	5	8	41	5	16	10
α Virginis	S′	5	5	16	5	5	6	α Pegasi.........	S	6	7	9	6	6	49
ζ Ursæ Majoris...	S	4	13	40	4	20	12	α Pegasi.........	A′	2	1	49	2	2	5
ζ Ursæ Majoris...	L′	5	4	41	5	4	51	α Andromedæ....	A′	1	6	48	1	6	43
η Ursæ Majoris...	A	2	3	49	2	2	49								

The errors that have been detected in the subsidiary numbers registered in the preceding Table, do not fortunately affect the corrections to be taken out of the corresponding columns of the Table of Corrections in any material degree, except in three instances; viz. α Orionis, ζ Ursæ Majoris, and α Piscis Austrini; but unless the corrected numbers contained in the Table at page 147, be always used in preference to those subjoined to the "*Corrections* of the 48 principal Stars," it will be necessary to make the alterations here pointed out with a pen before the corrections are taken out of the Table.

The maximum of lunar nutation in the right ascension of α Lyræ is found to be 0^s.727 instead of 0^s.969, as it was computed at first, and is therefore recomputed in the subsequent page; as are also the lunar nutations in right ascension of α Virginis and α Aquilæ, and the solar nutation of β and γ Draconis, all which were found erroneous; the true quantities are now given in one Table, which must be referred to when any of those quantities are sought for.

On recomputation, the minor errors, chiefly found in the *maxima* of solar nutation, affect only the thousandth, or third figure of the decimal numbers, and therefore might be disregarded, particularly as they gradually diminish till they became evanescent in descending the columns; but as we wish the reader to be in possession of the full extent of the errors we have discovered, we will specify them here, and point out the ready means of their rectification.

For ζ Ursæ Majoris.... (☉ nut. in R. A.) instead of 0 .070 use 0^s.063 from the same column of α Virginis.
For ε Sagittarii........ (☉ nut. in R. A.) instead of 0 .085 use 0 .079 from the same column of α Cassiopeæ.
For β Aquilæ......... (☉ nut. in R. A.) instead of 0 .057 use 0 .059 from the same column of α Aquilæ.
For α Piscis Austrini... (☉ nut. in R. A.) instead of 0 .071 use 0 .068 from the same column of α Tauri.
For α Pegasi.......... (☉ nut. in R. A.) instead of 0 .057 use 0 .060 from the same column of β Virginis.

The last column of the Table of Corrections of 48 principal Stars, in page 144, has the "Common Argument" numbered improperly, but may be readily corrected with a pen, within the blank portion of the column; it now begins with III^s 0 IX^s and ends with I^s 15° VII^s, like the last column in the preceding page (143), whereas it ought to begin I^s 15° VII^s and to end O^s 0° VI^s, like the same column in all the preceding *even* pages of the Table.

THE LUNAR NUTATION IN R. A. OF α URSÆ MAJORIS, α VIRGINIS, AND α LYRÆ; ALSO THE SOLAR NUTATION OF β AND γ DRACONIS, IN THE SAME, RE-COMPUTED.

Arg.	☾ Nut. α Ursæ Majoris.	☾ Nut. α Virginis.	☾ Nut. α Lyræ.	⊙ Nut. β and γ Drac.	Arg.	☾ Nut. α Ursæ Majoris.	☾ Nut. α Virginis.	☾ Nut. α Lyræ.	⊙ Nut. β and γ Drac.	Arg.	☾ Nut. α Ursæ Majoris.	☾ Nut. α Virginis.	☾ Nut. α Lyræ.	⊙ Nut. β and γ Drac.
+ −					+ −					+ −				
S S					S S					S S				
III 0° IX	1^s.815	1^s.137	0^s.727	0^s.028	IV 0° X	1^s.572	0^s.984	0^s.629	0^s.024	V 0° XI	0^s.907	0^s.568	0^s.363	0^s.014
1	1.815	1.137	.727	.028	1	1.556	.974	.623	.024	1	.879	.551	.352	.014
2	1.814	1.136	.726	.028	2	1.540	.963	.616	.024	2	.851	.533	.341	.013
3	1.813	1.135	.725	.028	3	1.523	.953	.609	.023	3	.823	.516	.330	.013
4	1.811	1.134	.724	.028	4	1.505	.942	.602	.023	4	.795	.498	.318	.012
5	1.809	1.133	.723	.028	5	1.486	.931	.595	.023	5	.766	.480	.307	.012
6	1.807	1.131	.722	.028	6	1.468	.919	.587	.023	6	.737	.462	.295	.011
7	1.804	1.129	.721	.028	7	1.449	.907	.580	.022	7	.708	.444	.284	.011
8	1.800	1.126	.720	.028	8	1.430	.895	.572	.022	8	.679	.425	.272	.010
9	1.795	1.123	.719	.028	9	1.410	.883	.564	.022	9	.650	.407	.260	.010
10	1.789	1.120	.717	.028	10	1.389	.869	.556	.021	10	.620	.388	.248	.010
11	1.782	1.116	.714	.028	11	1.369	.857	.548	.021	11	.590	.370	.236	.009
12	1.775	1.112	.711	.027	12	1.348	.844	.540	.021	12	.560	.351	.224	.009
13	1.768	1.107	.708	.027	13	1.326	.831	.532	.020	13	.530	.332	.212	.008
14	1.760	1.102	.705	.027	14	1.305	.817	.523	.020	14	.500	.313	.200	.008
15	1.752	1.097	.702	.027	15	1.283	.803	.514	.020	15	.469	.294	.188	.007
16	1.744	1.092	.699	.027	16	1.260	.789	.505	.019	16	.438	.275	.175	.007
17	1.735	1.087	.695	.027	17	1.237	.775	.496	.019	17	.407	.255	.163	.006
18	1.726	1.081	.691	.027	18	1.214	.760	.487	.019	18	.377	.236	.151	.006
19	1.716	1.075	.687	.026	19	1.190	.745	.477	.018	19	.346	.217	.138	.005
20	1.705	1.068	.682	.026	20	1.166	.730	.467	.018	20	.315	.198	.126	.005
21	1.694	1.061	.678	.026	21	1.192	.715	.457	.018	21	.283	.178	.113	.004
22	1.682	1.054	.673	.026	22	1.117	.700	.447	.017	22	.252	.158	.101	.004
23	1.670	1.046	.669	.026	23	1.091	.684	.437	.017	23	.221	.138	.089	.003
24	1.657	1.038	.664	.026	24	1.066	.668	.427	.016	24	.190	.119	.076	.003
25	1.644	1.030	.658	.025	25	1.041	.652	.417	.016	25	.157	.099	.063	.002
26	1.630	1.021	.653	.025	26	1.015	.636	.407	.016	26	.127	.079	.051	.002
27	1.616	1.012	.647	.025	27	0.988	.619	.396	.015	27	.095	.060	.038	.001
28	1.602	1.003	.641	.025	28	0.962	.602	.385	.015	28	.063	.040	.025	.001
29	1.587	0.994	.635	.025	29	0.934	.585	.374	.015	29	.032	.020	.013	.000
30	1.572	0.984	.629	.024	30	0.907	.568	.363	.014	30	.000	.000	.000	.000
− +					− +					− +				
S S					S S					S S				
VI 0° O	0^s.000	0^s.000	0^s.000	0^s.000	VII 0° I	0^s.907	0^s.568	0^s.363	0^s.014	VIII 0° II	1^s.572	0^s.984	0^s.629	.024
1	.032	.020	.013	.000	1	0.934	.585	.374	.015	1	1.587	0.994	.635	.025
2	.063	.040	.025	.001	2	0.962	.602	.385	.015	2	1.602	1.003	.641	.025
3	.095	.060	.038	.001	3	0.988	.619	.396	.015	3	1.616	1.012	.647	.025
4	.127	.079	.051	.002	4	1.015	.636	.407	.016	4	1.630	1.021	.653	.025
5	.157	.099	.063	.002	5	1.041	.652	.417	.016	5	1.644	1.030	.658	.025
6	.190	.119	.076	.003	6	1.066	.668	.427	.016	6	1.657	1.038	.664	.026
7	.221	.138	.089	.003	7	1.091	.684	.437	.017	7	1.670	1.046	.669	.026
8	.252	.158	.101	.004	8	1.117	.700	.447	.017	8	1.682	1.054	.673	.026
9	.283	.178	.113	.004	9	1.192	.715	.457	.018	9	1.694	1.061	.678	.026
10	.315	.198	.126	.005	10	1.166	.730	.467	.018	10	1.705	1.068	.682	.026
11	.346	.217	.138	.006	11	1.190	.745	.477	.018	11	1.716	1.075	.687	.026
12	.377	.236	.151	.006	12	1.214	.760	.487	.019	12	1.726	1.081	.691	.027
13	.407	.255	.163	.006	13	1.237	.775	.496	.019	13	1.735	1.087	.695	.027
14	.438	.275	.175	.007	14	1.260	.789	.505	.019	14	1.744	1.092	.699	.027
15	.469	.294	.188	.007	15	1.283	.803	.514	.020	15	1.752	1.097	.702	.027
16	.500	.313	.200	.008	16	1.305	.817	.523	.020	16	1.760	1.102	.705	.027
17	.530	.332	.212	.008	17	1.326	.831	.532	.020	17	1.768	1.107	.708	.027
18	.560	.351	.224	.009	18	1.348	.844	.540	.021	18	1.775	1.112	.711	.027
19	.590	.370	.236	.009	19	1.369	.857	.548	.021	19	1.782	1.116	.714	.028
20	.620	.388	.248	.010	20	1.389	.869	.556	.021	20	1.789	1.120	.717	.028
21	.650	.407	.260	.010	21	1.410	.883	.564	.022	21	1.795	1.123	.719	.028
22	.679	.425	.272	.010	22	1.430	.895	.572	.022	22	1.800	1.126	.720	.028
23	.708	.444	.284	.011	23	1.449	.907	.580	.022	23	1.804	1.129	.721	.028
24	.737	.462	.295	.011	24	1.468	.919	.587	.023	24	1.807	1.131	.722	.028
25	.766	.480	.307	.012	25	1.486	.931	.595	.023	25	1.809	1.133	.723	.028
26	.795	.498	.318	.012	26	1.505	.942	.602	.023	26	1.811	1.134	.724	.028
27	.823	.516	.330	.013	27	1.523	.953	.609	.023	27	1.813	1.135	.725	.028
28	.851	.533	.341	.013	28	1.540	.963	.616	.024	28	1.814	1.136	.726	.028
29	.879	.551	.352	.014	29	1.556	.974	.623	.024	29	1.815	1.137	.727	.028
30	.907	.568	.363	.014	30	1.572	.984	.629	.024	30	1.815	1.137	.727	.028

Example 3.

Let it be required to determine the true apparent place of γ Draconis, on the 16th of August, 1824, taking the mean places of Piazzi, the annual variations of Schumacher, the arguments from the Table in page 147, the maximum of ⊙ nutation from the re-computed Table in the last page, and the rest from the TABLE OF CORRECTIONS of 48 Stars?

R. A.	N. P. D.	Arguments in R. A.	Arguments in N. P. D.
1820 …………… 17h 52m 25s.510	1820 …………… 38° 29′ 10″.240	A…… = 0s 1° 42′	A′…. = 3s 1° 30′
An. Var. 1.388 × 4 + 5.552	An. Var. 0″.714 × 4 + 2.856	⊙ ….. = 4 23 29	⊙ …. = 4 23 29
Aug. 16. Tab. XV. { 1. … 0.625	Aug. 16. Tab. XV. { .7 … 0.437	Tab. XVI. 19	Tab. XVI. 19
.3 … .187	.014… 87		
.09… 56		Arg…. = 4 25 30	Arg… = 7 25 18
	Mean N. P. D…. = 38 22 13.62		
Mean R. A. …. = 17 52 31.930		L…… = 5 27 10	L′…. = 3 1 22
		☊ ….. = 9 7 4	☊ …. = 9 7 4
Arg. 4s 25° 30′ Aber. − 0.782	Arg. 7s 25° 18′…… −16.02	Arg…. = 3 4 14	Arg… = 0 8 26
Arg. 3 4 14 ☾ nut. + 0.998	Arg. 0 8 26 …… + 1.46		
Arg. 3 15 10 ⊙ nut. + 0.027	Arg. 0 19 18 …… + 0.14	S…… = 5 27 34	S′…. = 3 1 42
		⊙ 2 … = 9 17 36	⊙ 2 .. = 9 17 36
True appar. R. A. = 17 52 32.173	True app. N. P. D. = 38 28 59.20		
		Arg…. = 3 15 10	Arg… = 0 19 18

Example 4.

What will be the apparent right ascension and declination of α ANDROMEDÆ at the beginning of the year 1860, if the mean places and annual variations be taken on an average of all those determined by Maskelyne, Zach, Piazzi, Brinkley, and Pond?

In this case the two epochs are 1820 and 1860, and the annual variation must be first obtained for 1840, the mean period, by applying a portion of the secular variation equal to $\frac{20}{100}$ or $\frac{1}{5}$ of the quantity tabulated for this star, in page 145, first for right ascension, and then for north polar distance, and the mean right ascension and north polar distance thus brought up will be those which belong to the star at the remote epoch. The mean right ascension for 1820, on an average of five observers, comes out 23h 59m 6s 06, and the annual variation, on an average of Maskelyne, Piazzi, and Bessel, is +3s.058, but as Maskelyne determined a *positive* proper motion, and each of the two others a *negative* one, all varying in amount, it will be better to substitute the mean precession taken on an average, which is 3s.063, and consider it as the annual variation, till the proper motion is better determined. The secular change of the annual variation in right ascension of this star is tabulated (p. 145) +0s.0176, $\frac{1}{5}$th of which is 0s.0035; therefore 3s.0665 may be taken as the precession due to the middle epoch 1840: which is somewhat smaller than Schumacher has given the annual variation. Again the average of four north polar distances for 1820 comes out 61° 54′ 10″.205, and the average of three annual variations −19″.918, almost the same as Schumacher has given it, the three proper motions being found here all positive: then $\frac{1}{5}$ of the secular change of the annual variation is $\frac{1}{5}$ of − .0044 or − .0009, and consequently −19″.9189 will be the annual variation in north polar distance due to 1840; and the work will stand thus:

R. A.	N. P. D.	Arguments in R. A.	Arguments in N. P. D.
1820…………… 23h 59m 6s.06	1820 …………… 61° 54′ 10″.205	A ….. = 8s 29° 40′	A′…. = 1s 7° 3′
3s.0665 × 40 years.. + 2 2.66	An. var. × 40 …… − 13 16.756	⊙ ….. = 9 11 24	⊙ …. = 9 11 24
Mean R. A…… = 0 1 8.72	Mean N.P.D.1860 = 61 40 53.449	Tab. XVI. 13	Tab. XVI. 13
		Arg…. = 6 11 17	Arg… = 10 18 40
Arg. 6s 11° 17′ Aber. − 0.274	Arg. 10s 18° 40′ Aber. − 7.880	L ….. = 6 17 26	L′…. = 11 29 37
Arg. 5 0 12 ☾ nut. + 0.573	Arg. 10 12 23 ☾ nut. − 5.330	☊ ….. = 10 12 46	☊ … = 10 12 46
Arg. 1 7 32 ⊙ nut. + 0.038	Arg. 6 22 56 ⊙ nut. − 0.156		
Apparent R. A. … = 0 1 9s.057	Apparent N. P. D. = 61 40 40.083	Arg…. = 5 0 12	Arg… = 10 12 23
		S…… = 6 14 18	S′…. = 11 29 42
		⊙ 2…. = 6 23 14	⊙ 2 .. = 6 23 14
		Arg…. = 1 7 32	Arg… = 6 22 56

In this example the sun's longitude (☉) was thus obtained;

Solar Table 3. 1823 .... 19 .7	Jan. 2. (viz. + 1)	$\overset{s}{9}$ 11° 22′	= ☉ 1823
1860 B .... 21 .8	Diff. add	2	
Diff. .... + 2 .1		9 11 24	= ☉ 1860

The annual variation in right ascension of β Ursæ Minoris is −, a contrary sign (+) must therefore be applied to its secular change, the annual variation being diminishing.

As α Andromedæ wanted only 54 of the vernal equinox in January, 1820, and as its annual variation may be taken at $+ 3^s.07$, $\frac{54^s}{3^s.07} = 17.6$ years will elapse from this epoch, before it will require to be transferred from the end to the beginning of the catalogue, which will happen near the beginning of August of the year 1837.

When a star is among those given in the *Fundamenta Astronomiæ*, its secular change of annual variation, either in right ascension or north polar distance may be found from the annual variations given for the years 1755 and 1800; thus for γ Pegasi, annual variation in right ascension for $\left.\begin{array}{l}1755 = 46''.011\\ 1800 = 46.076\end{array}\right\}$ diff. in 45 years = 0″.065 increasing; then as 45 : 0″.065 :: 100 : 0″.1444 and $\frac{0''.1444}{15} = 0^s.0096$. But if the secular change is not known, the precession for the middle period may be found with sufficient accuracy by the approximate right ascension and declination, obtained from the annual variation known to belong to the first epoch, and then the annual variation so gained may be applied as a factor for the whole interval, to bring up the mean place to the last epoch: thus, suppose the mean right ascension of α Andromedæ at 1820 = $23^h\ 59^m\ 6^s.06$, and corresponding annual variation = $3^s.063$; then $20 \times 3.063 = 1^m\ 1^s.26$; and this added to the said mean right ascension, gives for the approximate mean right ascension of 1840, $0^h\ 0^m\ 7^s.32$, where the tabular quantity (p. 21) is $0^s.000$; hence the constant quantity in the margin, viz. $3^s.0678$, must be taken for the annual variation, which is nearly as before. But it has been said, that the annual variation for a *mean epoch* must be used, which in this case is $\frac{3^s.063 + 3^s.067}{2} = 3^s.065$ for 1830, and when this is substituted for bringing up the mean right ascension from 1820 to 1840, the calculated annual variation resulting therefrom will be more correct, for carrying up the right ascension from 1820 to 1860.

The annual variation for a given year may also be thus computed: let P denote the position of a star as given in a modern catalogue, Piazzi's for instance, and B that of Bradley for 1755, as calculated by Bessel, p and p' the precessions of the *Fundamenta Astronomiæ* for 1755 and 1800, and μ the proper motion taken from the same work, then the positions for any other year, say 1821, may be calculated by the following formula; $B + (P - B)\,\frac{21+45}{45} + (p' - p)\,.\,\frac{21+45}{90}\,.\,21$; or the position for 1821 may be put $= P + (P - B)\,.\,\frac{7}{15} + (p' - p)\,.\,15.4$. But the annual variation in the *Fundamenta* contains the proper motion, so that for 1821 it will be $p + (p' - p)\,.\,\frac{7}{15} + \frac{\mu}{45}$.

[Vide Bessel's *Fundamenta Astronomiæ*, p. 136, and Schumacher's *Astronomische Hülfstafeln* for 1821.]

I subjoin three Tables, which have been printed at a lithographic press and circulated privately among practical astronomers, since the preceding Tables were printed, in order to show the true spirit of competition which prevails in some of the principal observatories; and leave the reader to amuse himself with a critical examination of their contents. Every new instrument of competent magnitude, will in future probably produce some inferences of a similar nature, to establish its credit; but whether the discrepancies we perceive arise out of the observations themselves, or out of the annual variations and corrections applied to them, is a matter that remains to be ascertained. From the contents of the last column of Table 2, a conclusion, it is understood, has been (rather hastily) drawn, that there is a general motion of all the southern stars towards the south, and of the northern stars towards the north!

(TABLE 1.)

BESSEL'S

RECENT DETERMINATION OF THE DECLINATION

OF

THE 36 STARS.

Name of Star.	Declination for 1815, by Cary's Circle.	Probable error.	Declination for 1820, by Reichenbach's Circle.	Probable error.
α Aurigæ	45° 47′ 44″. 93	0″. 57	45° 48′ 9″. 12	0″. 18
α Cygni	44 37 26 . 45	0 . 58	44 38 28 . 47	0 . 18
α Lyræ	38 37 4 . 01	0 . 75	38 37 17 . 77	0 . 24
α Geminorum	32 16 54 . 66	0 . 75	32 16 21 . 05	0 . 23
β Geminorum	28 27 44 . 14	0 . 75	28 27 5 . 54	0 . 22
β Tauri	28 26 20 . 09	0 . 76	28 26 40 . 40	0 . 23
α Andromedæ	28 4 3 . 28	0 . 76	28 5 46 . 59	0 . 22
α Coronæ	27 20 34 . 61	0 . 76	27 19 34 . 44	0 . 22
α Arietis	22 34 56 . 11	0 . 80	22 36 22 . 32	0 . 23
α Bootis	20 9 0 . 75	0 . 75	20 7 25 . 43	0 . 21
α Tauri	16 7 37 . 29	0 . 77	16 8 17 . 16	0 . 22
β Leonis	15 36 20 . 92	0 . 79	15 34 40 . 04	0 . 24
α Herculis	14 36 32 . 52	0 . 76	14 36 10 . 45	0 . 24
α Pegasi	14 12 41 . 94	0 . 77	14 14 19 . 05	0 . 24
γ Pegasi	14 9 13 . 02	0 . 75	14 10 56 . 22	0 . 23
α Leonis	12 51 59 . 71	0 . 75	12 50 33 . 58	0 . 22
α Ophiuchi	12 42 10 . 37	0 . 77	12 41 55 . 66	0 . 24
γ Aquilæ	10 10 12 . 28	0 . 76	10 10 53 . 97	0 . 23
α Aquilæ	8 23 14 . 89	0 . 74	8 24 0 . 69	0 . 21
α Orionis	7 21 45 . 21	0 . 76	7 21 50 . 69	0 . 22
α Serpentis	7 0 53 . 37	0 . 77	6 59 54 . 84	0 . 23
β Aquilæ	5 57 9 . 29	0 . 75	5 57 50 . 84	0 . 23
α Canis Min	5 41 23 . 13	0 . 75	5 40 40 . 32	0 . 21
α Ceti	3 21 23 . 46	0 . 89	3 22 37 . 67	0 . 24
β Virginis	2 48 23 . 20	0 . 77	2 46 42 . 81	0 . 29
α Aquarii	− 1 12 48 . 98	0 . 89	− 1 11 25 . 48	0 . 23
α Hydræ	− 7 51 44 . 39	0 . 83	− 7 53 1 . 68	0 . 23
β Orionis	− 8 25 27 . 36	0 . 80	− 8 25 4 . 22	0 . 24
α Virginis	−10 11 33 . 88	0 . 73	−10 13 7 . 69	0 . 22
α[1] Capricorni	−13 4 20 . 91	0 . 78	−13 3 25 . 59	0 . 35
α[2] Capricorni	−13 6 40 . 64	0 . 89	−13 5 43 . 49	0 . 35
α[1] Libræ	−15 13 14 . 64	0 . 84	−15 14 33 . 27	0 . 25
α[2] Libræ	−15 15 58 . 17	0 . 82	−15 17 15 . 05	0 . 25
α Canis Maj	−16 28 14 . 68	0 . 73	−16 28 37 . 15	0 . 23
α Scorpii	−26 0 39 . 17	0 . 85	−26 1 23 . 00	0 . 26
α Piscis Aust	−30 36 2 . 81	0 . 91	−30 34 28 . 68	0 . 37

(TABLE 2.)

GREENWICH NORTH POLAR DISTANCES OF 44 STARS IN 1822.

Annual Variation, 1818.	Approximate Right Ascension.	Name of Star.	Predicted North Polar Distance, 1822.	Stars south, by observation.
	h. m.			
−20″. 20	0″. 4	γ Pegasi	75° 48′ 20″. 20	
−19 . 89	0 . 30	α Cassiop	34 26 24 . 12	+1″. 7 or 1″.5
−17 . 40	1 . 57	α Arietis	67 23 00 . 60	+2 . 0
−14 . 59	2 . 53	α Ceti	86 36 49 . 45	+2 . 0
−13 . 46	3 . 11	α Persei	40 46 51 . 65	+1 . 0
− 7 . 92	4 . 25	Aldebaran	73 51 24 . 09	+1 . 5
− 4 . 54	5 . 3	Capella	44 11 39 . 65	+1 . 8
− 4 . 74	5 . 6	Rigel	98 24 51 . 18	+2 . 0
− 3 . 80	5 . 15	β Tauri	61 33 9 . 54	+1 . 2
− 1 . 36	5 . 45	α Orionis	82 38 3 . 47	+2 . 0
+ 4 . 41	6 . 37	Sirius	106 28 40 . 94	+3 . 5
+ 7 . 12	7 . 23	Castor	57 43 50 . 96	+1 . 3
+ 8 . 63	7 . 30	Procyon	84 19 32 . 12	+2 . 7
+ 8 . 02	7 . 34	Pollux	61 33 8 . 74	+0 . 8 or 1. 0
+15 . 19	9 . 19	α Hydræ	97 53 28 . 41	+1 . 0
+17 . 23	9 . 59	Regulus	77 9 57 . 77	+0 . 4
+19 . 26	10 . 52	α Ursæ Maj	27 17 24 . 89	−0 . 6
+20 . 04	11 . 40	β Leonis	74 25 57 . 69	+0 . 3
+19 . 98	11 . 44	γ Ursæ Maj	35 18 55 . 17	−0 . 6
+18 . 94	13 . 16	Spica Virg	100 13 41 . 29	+0 . 5
+18 . 15	13 . 40	η Ursæ Maj.	39 47 41 . 45	+0 . 2
+18 . 97	14 . 7	Arcturus	69 53 9 . 86	+0 . 3
+15 . 30	14 . 41	α^1 Libræ	105 17 40 . 37	+0 . 6
+15 . 32	14 . 41	α^2 Libræ	105 14 56 . 59	+0 . 6
+14 . 74	14 . 51	β Ursæ Min	15 7 1 . 64	−0 . 6
+12 . 45	15 . 27	α Cor. Bor	62 40 17 . 62	+0 . 5
+11 . 72	15 . 35	α Serpentis	83 00 24 . 80	+0 . 2
+ 8 . 59	16 . 18	Antares	116 1 33 . 91	+1 . 5
+ 4 . 57	17 . 6	α Herculis	75 23 55 . 13	+0 . 5
+ 3 . 08	17 . 26	α Ophiuchi	77 18 6 . 66	+0 . 8
+ 0 . 67	17 . 52	γ Draconis	38 29 9 . 65	+0 . 2
− 3 . 02	18 . 31	α Lyræ	51 22 33 . 39	+0 . 8
− 8 . 34	19 . 37	γ Aquilæ	79 48	
− 9 . 06	19 . 42	α Aquilæ	81 35 37 . 37	+1 . 0
− 8 . 56	19 . 46	β Aquilæ	84 1	
	20 . 8	α^1 Capricorni	103 5 16 . 13	
−10 . 68	20 . 8	α^2 Capricorni	103 5 16 . 13	
−12 . 63	20 . 35	α Cygni	45 21 3 . 43	+1 . 5
−15 . 07	21 . 14	α Cephei	28 9 57 . 21	+0 . 6
−15 . 68	21 . 26	β Cephei	20 13 9 . 43	+0 . 3
−17 . 27	21 . 56	α Aquarii	91 10 46 . 12	+2 . 3
−18 . 86	22 . 48	Fomalhaut		
−19 . 32	22 . 56	α Pegasi	75 44 57 . 84	+3 . 2 : :
−19 . 95	23 . 59	α Andromedæ	61 53 30 . 70	+1 . 1

(TABLE 3.)

COMPARATIVE DIFFERENCES OF NORTH POLAR DISTANCE

OF THE

GREENWICH STARS,

AT THREE OBSERVATORIES.

Name of Stars.	Difference between Greenwich and Dublin, 1822.	Difference between Greenwich and Bessel, with Bradley's Refractions, 1820.	Difference between Dr. Brinkley and Bessel.	Supposed error of Bessel's Catalogue for 1815.
Polaris	−0″. 2			
β Ursæ Min.	+1 . 0			
β Cephei	0 . 0			
α Ursæ Maj.	−0 . 2			
α Cephei	−0 . 8			
α Cassiop	−2 . 5			
γ Ursæ Maj	−0 . 1			
γ Draconis	−0 . 9			
η Ursæ Maj	−0 . 6			
α Persei	−1 . 1			
Capella	−1 . 6	+0″. 5	2″. 1	3″. 27
α Cygni	−1 . 3	+1 . 8	3 . 1	1 . 45
α Lyræ	−1 . 3	+2 . 2	3 . 5	1 . 19
Castor	−1 . 3	+1 . 3	2 . 6	4 . 00
Pollux	−0 . 8	+1 . 5	2 . 3	3 . 16
β Tauri	−1 . 0	+1 . 7	2 . 7	3 . 51
α Androm	−1 . 6	+1 . 0	2 . 6	6 . 12
α Cor. Bor.	−1 . 4	+2 . 3	3 . 7	0 . 79
α Arietis	−1 . 5	+1 . 1	2 . 6	1 . 69
Arcturus	−0 . 7	+2 . 3	3 . 0	2 . 08
Aldebaran	−1 . 4	+1 . 4	2 . 8	2 . 91
β Leonis	−1 . 5	+2 . 0	3 . 5	1 . 58
α Herculis		+3 . 3		4 . 28
α Pegasi		+1 . 3		4 . 36
Regulus	−0 . 1	+2 . 8	2 . 9	2 . 99
α Ophiuchi	−1 . 4	+3 . 2	4 . 6	4 . 39
α Aquilæ	−1 . 9	+2 . 7	4 . 6	4 . 70
α Orionis	−1 . 6	+1 . 0	2 . 6	1 . 33
α Serpentis	−1 . 4	+3 . 7	5 . 1	3 . 83
Procyon	−1 . 7	+2 . 3	4 . 0	4 . 67
α Ceti	−1 . 9	+1 . 7	3 . 6	4 . 64
α Aquarii	−2 . 9	+2 . 3	5 . 2	1 . 48
α Hydræ		+2 . 8		2 . 42
Rigel		+1 . 5		2 . 56
Spica Virg.	−0 . 9	+3 . 8	4 . 7	4 . 58
Sirius	−2 . 0	+1 . 7	3 . 7	3 . 88

THE LAST GREENWICH

CATALOGUE OF 45 PRINCIPAL FIXED STARS.

1756 and 1813. An. Var. 1818.	No.	Names of Stars.	1813 and 1823. Obs. An. Var. 1818.	Right Ascen. 1823.	Predicted North Polar Distance, 1823.	Observed N. P. D. 1823. Bradley's Refrac. interior Ther.	Stars south of predicted place.	No. of Observations. 1756.	1812, 1813.	1822, 1823.	N. P. D. 1818. viz. (1813 + 1823)/2	Total No. of Obs.
−20″.09	1	γ Pegasi.......	19″.86	0h 4′ 8″.09	75° 48′ 0″.11	75° 48′ 2″.4	2″.3	14	25	13	75° 49′ 41″.7	45
−19 .85	2	α Cassiop.......	19 .70	0 30 31 .29	34 26 4 .23	34 26 5 .7	1 .5	5	77	70	34 27 44 .3	140
......	3	Polaris.........	......	0 57 46 .38	....	1 38 7 .5	...	...	200	250	1 39 44 .6	450
−17 .40	4	α Arietis........	17 .22	1 57 13 .09	67 22 42 .60	67 22 44 .4	1 .8	10	77	90	67 24 10 .6	167
−14 .59	5	α Ceti..........	14 .41	2 53 2 .29	86 36 34 .86	86 36 36 .8	2 .0	7	18	46	86 37 48 .7	64
−13 .41	6	α Persei........	13 .34	3 11 44 .48	40 46 38 .34	40 46 39 .2	0 .9	10	66	70	40 47 45 .7	136
− 7 .92	7	Aldebaran......	7 .77	4 25 46 .58	73 51 16 .17	73 51 17 .7	1 .5	60	69	85	73 51 56 .4	154
− 4 .54	8	Capella.........	4 .38	5 3 37 .83	44 11 35 .11	44 11 36 .8	1 .7	59	108	95	44 11 58 .8	203
− 4 .74	9	Rigel...........	4 .53	5 6 2 .21	98 24 46 .44	98 24 48 .4	2 .0	37	30	11	98 25 11 .2	41
− 3 .80	10	β Tauri.........	3 .72	5 15 6 .76	61 33 5 .74	61 33 6 .5	0 .8	14	70	68	61 33 25 .0	138
− 1 .36	11	α Orionis.......	1 .15	5 45 35 .66	82 38 2 .11	82 38 4 .2	2 .1	65	66	63	82 38 9 .9	129
+ 4 .41	12	Sirius...........	4 .73	6 37 20 .85	106 28 45 .35	106 28 48 .7	3 .4	84	30	82	106 28 25 .1	112
+ 7 .12	13	Castor..........	7 .22	7 23 17 .62	57 43 58 .08	57 43 59 .0	0 .9	19	47	77	57 43 23 .0	124
+ 8 .63	14	Procyon........	8 .92	7 30 2 .19	84 19 40 .37	84 19 43 .3	2 .9	64	40	84	84 18 58 .7	124
+ 8 .02	15	Pollux..........	8 .06	7 34 28 .52	61 33 16 .76	61 33 17 .0	0 .2	38	58	74	61 32 36 .8	132
+15 .19	16	α Hydræ.......	15 .32	9 18 53 .50	97 53 43 .20	97 53 44 .5	1 .3	5	10	35	97 52 27 .9	45
+17 .23	17	Regulus.........	17 .28	9 53 56 .30	77 10 15 .00	77 10 15 .6	0 .6	25	68	77	77 8 49 .3	145
+19 .26	18	α Urs. Maj.....	19 .22	10 52 43 .41	27 17 44 .16	27 17 43 .7	−0 .5	7	70	80	27 16 7 .5	150
+20 .04	19	β Leonis........	20 .08	11 40 1 .66	74 26 17 .73	74 26 18 .1	0 .4	11	22	32	74 24 37 .7	54
+19 .98	20	γ Urs. Maj.....	19 .95	11 44 28 .63	35 19 15 .18	35 19 14 .8	−0 .4	5	60	45	35 17 34 .9	105
+18 .94	21	Spica Virg.....	18 .94	13 15 52 .91	100 14 0 .73	100 14 0 .7	0 .0	19	20	46	100 12 26 .0	66
+18 .15	22	η Urs. Maj.....	18 .16	13 40 33 .54	39 47 59 .60	39 47 59 .4	−0 .2	Z.Sec.	100	95	39 46 28 .9	195
+18 .97	23	Arcturus......	19 .01	14 7 35 .61	69 53 28 .83	69 53 29 .2	0 .4	106	120	106	69 51 54 .1	226
+15 .30	24	α^1 Libræ.......	15 .36	14 40 54 .93	105 15 11 .91	105 15 12 .5	0 .6	4	12	14	105 13 56 .8	26
+15 .32	25	α^2 Libræ.......	15 .38	14 41 6 .36	105 17 55 .67	105 17 56 .3		11	15	20	105 16 39 .5	35
+14 .74	26	β Urs. Min.....	14 .67	14 51 19 .53	15 7 16 .38	15 7 15 .7	−0 .7	1	110	109	15 6 2 .3	219
+12 .45	27	α Cor. bor.......	12 .51	15 27 11 .95	62 41 0 .07	62 41 0 .6	0 .5	7	90	42	62 39 58 .1	132
+11 .72	28	α Serpentis.....	11 .73	15 35 33 .53	83 0 36 .52	83 0 36 .6	0 .1	5	77	44	82 59 38 .0	121
+ 8 .59	29	Antares.........	8 .75	16 18 34 .23	116 1 42 .50	116 1 44 .1	1 .6	21	36	10	116 28 25 .1	46
+ 4 .57	30	α Herculis......	4 .61	17 6 35 .00	75 23 59 .70	75 24 0 .4	0 .7	7	55	30	75 23 37 .0	85
+ 3 .08	31	α Ophiuchi ...	3 .16	17 26 43 .49	77 18 9 .74	77 18 11 .0	1 .3	3	83	32	77 17 54 .7	115
+ 0 .67	32	γ Draconis.....	0 .69	17 52 30 .14	38 29 10 .31	38 29 10 .5	0 .2	Z.Sec.	140	130	38 29 7 .1	270
− 3 .02	33	α Lyræ.........	2 .94	18 30 56 .98	51 22 30 .37	51 22 31 .2	0 .8	89	170	150	51 22 45 .9	320
− 8 .34	34	γ Aquilæ......		19 37 50 .82	79 48 35 .58	....	...	...	...	...	....	...
− 9 .06	35	α Aquilæ......	8 .93	19 42 8 .94	81 35 28″.31	81 35 29 .5	1 .2	104	140	112	81 36 14 .2	252
− 8 .56	36	β Aquilæ......		19 46 37 .27	84 1 37 .94	....	...	...	...	...	....	...
−10 .66	37	α^1 Capricor....	10 .75	20 7 49 .91	103 2 48 .83	103 2 49 .6	1 .0	6	35	22	103 3 42 .5	57
−10 .68	38	α^2 Capricor....	10 .76	20 8 13 .69	103 5 5 .45	103 5 6 .6		10	20	22	103 5 59 .4	50
−12 .63	39	α Cygni........	12 .47	20 35 24 .19	45 20 50 .80	45 20 52 .4	1 .6	55	120	120	45 21 54 .5	240
−15 .07	40	α Cephei.......	14 .99	21 14 21 .05	28 9 42 .20	28 9 42 .8	0 .6	5	70	57	28 10 57 .7	127
−15 .68	41	β Cephei........	15 .66	21 26 20 .44	20 12 53 .78	20 12 54 .0	0 .2	5	70	62	20 14 12 .5	132
−17 .27	42	α Aquarii......	17 .00	21 56 41 .55	91 10 28 .85	91 10 31 .4	2 .5	9	20	55	91 11 56 .4	75
......	43	Fomalhaut.....	......	22 47 50 .97	....	....	...	...	...	...	....	...
−19 .32	44	α Pegasi........	19 .00	22 55 57 .29	75 44 38 .52	75 44 41 .8	3 .3	6	30	50	75 46 16 .7	80
−19 .95	45	α Andromed...	19 .78	23 59 15 .61	61 53 10 .75	61 53 12 .5	1 .8	11	40	40	61 54 .51 .0	80

The observed North Polar Distances of 1823 were taken by direct measurement from the Pole, with the Circle in good order, and 27 of them also by reflection from quicksilver, and their exact accordance renders it probable, that the results are very correct. The mean of about 1300 observations of the Pole Star during the last ten years is 1° 39′ 44″.5 for the North Polar Distance for Jan. 1, 1818, and the mean of all the annual variations 19″.425.

BESSEL'S TABLE OF 1818.

So long ago as the year 1770, Dr. Maskelyne gave a Table of Corrections in *Space* of the right ascensions of 31 principal stars for every tenth day in the year; and in 1802 extended it to 36 stars, with the corrections given in *Time*; in both of which the correction for the moon's node was given for every tenth degree of its longitude. The latter of these was re-published by Mr. F. Wollaston in his *Fasciculus Astronomicus*, and a similar Table was computed and published by him for finding the corrections in north polar distance; those Tables were found very convenient, but as the coefficients have been improved since they were computed, they may now be considered as nearly obsolete. Bessel has not only improved the coefficients from time to time, but has constructed a Table, according to an arrangement somewhat different, which has lately been preferred by computers, who have undertaken to give the apparent right ascensions of the Greenwich stars for each successive day of the year. A *particular* Table of this sort requires many previous calculations, and must be renewed from time to time; but when correctly constructed, it affords the ready means of obtaining the daily apparent right ascensions, and consequently of watching the rate of the clock made use of. Our preceding "TABLE OF CORRECTIONS" will at all times supply the means of renewing such parts of the *particular* Table alluded to, as may require it; and also of extending it, so as to give the corrections for north polar distance also, according to the plan lately adopted by Dr. Young in the Nautical Almanac. Nothing is more easy than to substitute the day of the month for the sun's longitude on that day, as part of the argument; which may also be done with respect to the longitude of the ☾'s ☊.

The arrangement of Bessel's Table, as here printed, will require some explanation. The quantities which vary from year to year are very judiciously separated from those which recur every year, with but little alteration. The upper portion of the Table contains the *mean* right ascension for the beginning of each year, combined with the lunar nutation for successive intervals of 100 days, with an adjoining column of centesimal differences (improperly headed "☾ nut.") which differences, in this denomination, greatly facilitate interpolation for the intermediate days. The lower portion of the Table contains in one sum, in the column "Aber." the amount of the precession, aberration, and solar nutation, for the moment of the star's culmination, and the contiguous smaller column, improperly marked "☉ nut.", is the column of decimal differences. When the upper part of the Table, which extends only to the year 1825 inclusive, is out of date, it may very readily be extended by the aid of the preceding unlimited Table of "CORRECTIONS OF 48 STARS."

Example 1.

Let it be required to find the apparent right ascension of α ORIONIS at Greenwich on April 10, 1824?

Preparatory Tables.

Tab. 1. 1824 B.	− 0 .92
Tab. 2. Greenwich	+ 0 .01
Tab. 3. before June 17	+ 1 .00
	+ 0 .09

10 April, 1824, upper part		5^h	45^m	39^s.719
.09 (as 100 : 19 :: .09 : ,002 nearly)			+	.002
April 10, lower part			+	.298
As 100 : 133 :: 09 : ,012			−	.012
Apparent R. A.	=	5	45	40 .007
By the Nautical Almanac, 1824,	=	5	45	40 .33
By the Table of 48 Stars	=	5	45	40 .345

If we were to judge from the *primâ facie* evidence of this comparison, we should say that the corrections given by Bessel's Table differ from the corrections obtained from the two other sources as much as one-third of a second; but, on looking into the mean places and annual variations respectively made use of, we shall find that this is not the case. Pond's mean right ascension on the 10th of April, 1824, comes out, as we have seen before (p. 336), 5^h 45^m 39^s.810 before the corrections are applied, and this right ascension has been used in both the latter examples; but Bessel's, according to the right ascension given at p. 145, and annual variation given at p. 146, will come out 5^h 45^m 39^s.565, taking the seconds of right ascension (1820) at 25 .70 and annual variation = 3^s.244, instead of 25^s.92 and 3^s.25, as given by Pond; this allowance would reduce the difference to one-tenth of a second.

In this example the 10th of April is itself found in the Table, and a very small portion only of the corrections arise out of the tabular differences, that were taken without regard being had to the second differences, which, however, should not be neglected where extreme correctness is the object.

Example 2.

Let the apparent right ascension of γ Pegasi be required for the time of its culmination at Greenwich, on the 1st of March, 1823?

Preparatory Tables.

Tab. 1. 1823	− 0 .67	January 0, 1823, upper part	0h 4m 8s.771
Tab. 2. Greenwich	+ 0 .01	As 100d : 66 diff. :: 60d.34 : 39	+ 0 .039
Tab. 3. before March 21	+ 1 .00	March 1, lower part	− 0 .649
	+ 0 .34		0 4 8 .161
January	31	In Tilloch's Journal, January, 1823	0 4 8 .16
February	28		
March	1		
	60 .34		

In this example the difference in the lower part of the Table was only 2 or .002 in 100 days, the proportional part for .34 of a day amounts to nothing worth notice; but in the upper part $\frac{66}{100}$ of 60 .34 days was a quantity not to be omitted, and was therefore taken into the account without regard to the second difference arising from the gradual diminution of the tabular differences. The difference arising from January 0 to April 10, in 100 days, is tabulated 66 opposite January 0, but as the preceding difference in 100 days is 72, and the following difference in the same space of time 60, it is clear that the *second difference* in 100 days at this period is = 6, or .6 in ten days, or .06 in one day; now according to the most convenient mode of distributing second differences, the mean first difference must stand opposite the mean time, viz. the 50th day from January 0, in this instance, and if + .6 be applied successively to each 10 days upwards, and − .6 to each 10 downwards, the column of differences, supposing the intervals each 10 days, would be duly proportioned, and at 60 days would stand 65 .4 (66 − .6) to be substituted for 66 above used. This suggestion may be useful to the computer who undertakes to prepare the daily apparent right ascensions, or north polar distances of any number of stars. In the last example, the substitution of 65 .4 for 66 as the difference, would make no sensible alteration in the result, in consequence of 60d.34 being nearly in the middle of the interval of 100 days, and the whole difference itself being small. When 66 is used, the correction is + 0s.0398344, and when 65 .4 is substituted it is + 0s.03946236, but in some cases an attention to the second differences, and a consequent proper tabulation of first differences become important.

In the first example, the differences in the upper part of the Table run 29, 19, and 11, and 19 stands opposite April 10; here the second differences are respectively 10 and 8, and if 9 be taken as the mean second difference, the tabular differences to be used at the beginning, middle, and end of the interval, become 23 .5, 19, and 14 .5, viz. 19 + (.9 × 5), 19, and 19 − (.9 × 5) respectively; and $\frac{23.5}{100}$ of .09 produces .00115, instead of .0171 which has been put down .002. Also in the lower part of the Table, the tabular differences run 157, 133, 104, the middle one being opposite April 10, and belonging to April 10 + 5, or April 15; and the respective second differences are 24 and 29, say 26 as a mean, in ten days, or 2 .6 in one, then the proper equated differences will be 133 + (2 .6 × 5) = 146 at April 10, or beginning of the interval, 133 at the middle, viz. April 15, and 133 − (2 .6 × 5) = 120 at the end of the interval, or beginning of the next following, which in the Table stands 104. Then the correction arising from $\frac{146}{100}$ of .09 will be found .01314 instead of .012, which latter is the correction belonging to April 15. But the omission will seldom affect the result, except in the *third* decimal figure, when the second differences are disregarded, and in general a simple inspection of the tabular differences will show what allowance may safely be made + or −, when it is considered that the tabular number opposite any one day belongs to the *middle of the interval* from which it is derived, viz. 50, or 5 days counted forwards, according to the part of the Table (upper or lower) to which it belongs.

Example 3.—Let the apparent right ascension of γ Pegasi be required for February 1, 1823?

Prep. Tab. = +0 .34 as before.

January 0, 1823, as before		0h 4m 8s.771
As 100d : 87 :: 32d.34 : .021		+ 0 .028
January 30 (below)		− 0 .473
$\frac{87}{100}$ of 2d [81 + (3d × 2)]		− 0 .018
Apparent R. A.	=	0 4 8 .308
In Tilloch's Journal, January, 1823	=	0 4 8 .30

Hence during the interval, from February 1 to March 1, 1823, 0s.14 is the whole difference in the right ascension of γ Pegasi, which continues to diminish, and if the apparent right ascensions were computed in the same manner for the 10th and 20th of February, it would be found that, out of the whole difference 0s.14, 0s.07 would belong to the interval of the first ten days, 0s.04 to the middle interval, and 0s.03 to the last, which differences should be distributed accordingly, as seen in the Journal already referred to. And in like manner the daily apparent right ascensions may be calculated and arranged for any other month, due attention being paid to the day in which an increase in the differences changes into a decrease, and *vice versâ*. When the right ascension of the star has been observed by a transit instrument on any evening, and is found the same as the computation here exemplified gives it, the clock shows true sidereal time, but otherwise the difference, + or −, is its error at the moment of the star's culmination.

Let it be required to carry forward the upper portion of the Table, for γ Pegasi, through the year 1826, in order to show how the deficiency may at any time be supplied by the preceding Table of "Corrections of 48 Stars," as the years advance?

			Column supplied for 1826.	
1820. Bessel's R. A. of γ Pegasi (p. 145)		0^h 3^m $58^s.59$		
An. var. + $3^s.075$ × 6 years.....................		+ 18 .450		
Right ascension on January 0, 1826..............		0 4 17 .040	0^h 4^m	
	S			Diff.
L of γ Pegasi, 1830..........	6 8° 24′			
☾ ☊ 1826 (p. 181)...........	8 10 22 14″			
January 0. Arg.	2 18 46 14 (p. 113)	+ 1 .095	$18^s.135$	23
April 10. ☊ deduct...........	− 5 14 33			
April 10. Arg.	2 13 31 41.......	+ 1 .072	18 .112	35
July 19. (☊ 100 days)	− 5 17 44			
July 19. Arg.	2 8 13 57	+ 1 .037	18 .077	43
October 27. (☊ 100 days)	− 5 17 44			
October 27. Arg.	2 2 56 13	+ 0 .994	18 .034	51
Dec. 66. (☊ 100 days)........	− 5 17 44			
Dec. 66. Arg.................	1 27 38 29	+ 0 .943	17 .983	

The respective lunar nutations are here added in succession to the mean right ascension, and the sums tabulated with their proper differences; and in like manner, any other column of the upper portion of the Table, which is never published far in advance, may be supplied for any year forward or back that may be wanted; or, what may be sometimes desirable, the amount of the mean right ascension and correction for lunar nutation may be readily had in this way for any *particular day*, to be united with the other three corrections, taken from the more permanent portion of the Table.

SOLAR TABLES.

TABLES 1 AND 2. SUN'S MEAN LONGITUDE AND PERIGEE.

Though the Nautical Almanac supplies the practical Astronomer with the sun's true apparent longitude, right ascension in time, and declination for every day of the year, together with the equation of time, the sun's semi-diameter in arc, the time of his passing the meridian, his hourly motion, and logarithm of his distance, all computed with great care, and conveniently arranged; yet other particulars relating to the sun are frequently required to be known in the reduction of solar observations, particularly as the computations in the Almanac are exclusively for noon at Greenwich. The Solar Tables, which have been here selected, and printed in a series under one title, are regularly numbered in Arabic figures from 1 to 25, and may be considered as supplementary to the computations published under the direction of the Honourable Board of Longitude.

Table 1, which contains the mean longitude of the sun, and of his perigee for 150 years from the time of Bradley, is adapted to the Royal Observatory of Greenwich, and furnishes the means of obtaining his mean anomaly, which is sometimes wanted as an argument, particularly for Table 9 of this series. Formerly the mean anomaly of the sun, or of a planet, was its angular distance from the apogee or aphelion, and Dr. Halley's Tables were constructed on this mode of calculation; but now the mean anomaly means the angular distance from the perigee or perihelion, as the case may be: and therefore if the tabular mean longitude of the sun's perigee be subtracted from his own mean longitude for the same time, the difference will always be his *mean anomaly*, at the beginning of the solar year, which the Table supposes to be at noon of January 1 in every year, the solar year being more advanced than the civil year by twelve hours.

When the numbers have been taken out of TABLE 1 for the given year, the quantities contained in Table 2, opposite the given day of the required month, must be added respectively thereto, which will give the mean longitude of the sun, and of his perigee corresponding to the period in question. For instance, if the sun's mean anomaly were required for June 1, 1825, or for April 20, 1826, the numbers would stand thus:

	S					S			
Table 1........1825........☉ =	9	10°	50′	9″.0	perigee =	9	9°	54′	51″
Table 2........June 1......☉ =	4	28	49	57 .8	...... =				25 .6
☉'s long........................		29	40	6 .8		9	9	55	16 .6
☉'s perigee	9	9	55	16 .6					
Mean anomaly	4	29	44	50 .2					

	S								
Table 1........1826........☉ =	9	10°	35′	49″.4	perigee =	9	9	55	52
Table 2........April 20 ....☉ =	3	17	26	8 .0	...... =				18 .5
☉'s long........................	0	28	1	57 .4		9	9	56	10 .5
☉'s perigee.....................	9	9	56	10 .5					
Mean anomaly...................	3	18	5	46 .9					

In the two additional small columns in Table 2, the number of days from the beginning of common years, and the fraction of a year that commences at noon of January 1, are introduced to fill up the page. These factors are placed one day more *forward* than they are printed in the first column of Table XV. (page 39), for the daily precessions of the stars.

TABLES 3 AND 4. COMPARATIVE EPOCHS.

As it will frequently happen that the sun's TRUE LONGITUDE and MEAN RIGHT ASCENSION may require to be known, the former as part of an *argument*, and the latter for the purpose of converting solar into sidereal time, when the Nautical Almanac for the given time is not at hand, the Tables 3 and 4 have been so arranged, as to satisfy the *desiderata* with sufficient accuracy for the respective purposes, for a period of 150 years. We have had occasion to explain and exemplify the uses of these two Tables at pages 334 and 336, it will not, therefore, be necessary to dwell further on them in this place.

TABLE 5. CONVERSION OF THE SUN'S TRUE LONGITUDE INTO APPARENT RIGHT ASCENSION IN TIME.

THOUGH the Nautical Almanac gives the right ascension in time corresponding to the sun's apparent longitude at Greenwich every day at noon, yet, at another place, situated either east or west of the prime meridian, neither the sun's longitude nor the corresponding right ascension will be that which is given in the said Almanac. Table 5 therefore, extending from page 157 to 159, is computed for the purpose of converting the sun's longitude in *space*, at any time or place, into his corresponding right ascension in *time*. This Table was computed by a very accurate and industrious computer, from the subjoined *formula*, viz. $\frac{\text{Rad} \times \text{cosin obliq}}{\text{co-tang ☉'s long}}$ = tang. ☉'s R. A. where the obliquity of the ecliptic was taken, according to Delambre's determination for 1800, = 23° 27′ 57″; and the columns of differences in 10′, and variation in 100″ change of the obliquity, render the Table not only convenient to the astronomer, but of unlimited duration in its application. Several subsequent Tables have also been computed for the same obliquity, and, after a similar plan for the columns of difference and variation, by the same gentleman, whose name is therefore prefixed, and to whose obliging promptitude in affording his assistance, I cannot but feel greatly indebted.

From $\overset{s}{\text{O}}$ to $\overset{s}{\text{III}}$, and from $\overset{s}{\text{VI}}$ to $\overset{s}{\text{IX}}$, signs, the ☉'s R. A. is less than his longitude;

From $\overset{s}{\text{III}}$ to $\overset{s}{\text{VI}}$, and from $\overset{s}{\text{IX}}$ to $\overset{s}{\text{XII}}$, the ☉'s R. A. is greater than his longitude.

The variation of the right ascension for a diminution of the obliquity is *additive*, when the sun's longitude exceeds his right ascension; but when the right ascension, on the contrary, exceeds the longitude, the variation is *subtractive*. The time contained in the principal columns is so arranged, that in the three first signs of the ☉'s longitude it is to be taken without any addition or diminution; in the next three, the tabular quantity must be subtracted from 12^h; in the third quadrant, it must be added to 12^h: and in the last three signs it must be subtracted from 24^h, as directed above and below the Table.

Example 1.—Given the sun's longitude $\overset{s}{3}$ 15° 11′ 20″.76, and the obliquity of the ecliptic 23° 27′ 49″.76, to determine his right ascension in time?

Solar Table 5....$\overset{s}{3}$ 15° 10′.................... $12^h - 4^h\, 54^m\, 8^s.91$
As 10′ : diff. 43 .04 :: 1′ 20″.76...... −5 .79
As 100″ : var. 0 .78 :: 7 .24 dim... +0 .06

Sun's R. A. in time .... $\begin{cases} 12^h - 4\ 54\ \ 3 .18 \\ \text{or} ....7\ \ 5\ 56 .82 \end{cases}$

By the formula, thus:

Tang. ⊙ long. (sup. to 180°)...74° 48 39″. 24....log. 10 .5662466
Cosin. obliq. eclip. ....23 27 49 . 76....log. 9 .9625169

Tang. ⊙'s R. A. (sup. to 80°)..73 30 47 . 62....log. 10 .5287635
Subtract from ....180 0 0 . 00

Sun's R. A. in space.........106 29 12 . 38

Then by Table (p. 109) for converting space into time, we have,

100°	....	$6^h 40^m\ 0^s.000$
6	....	0 24 0 .000
29′	....	0 1 56 .000
12″	....	0 .799
0 . 38	....	0 .025

Sun's R. A. in time....7 5 56 .824 as before.

Example 2.—Given the ⊙'s longitude in the Nautical Almanac on the 24th of April, 1823, = $\overset{s}{1}$ 3° 28′ 30″, required his right ascension at the time?

Solar Table 5....$\overset{s}{1}$ 3° 20′ .... $0^h + 2^h\, 4^m 24^s.770$
8 .5 (.85 × diff. 38 .56) .... +32 .776
5″.3 diminution of obliq. (.053 × 1 .28) .... + .057

Sun's R. A. by Table ....= 2 4 57 .603

By Nautical Almanac = 2 4 57 .6

Example 3.—Given the sun's longitude on January 1, 1824, $\overset{s}{9}$ 10° 6′ 13″, what will be the right ascension, disregarding the small diminution of the obliquity?

Solar Table 5....$\overset{s}{9}$ 10° 0′ .... $24^h - 5^h\, 16^m 28^s.58$
6 .2 (.62 × 43 .36)................ −26 .88

24^h − 5 16 1 .70

R. A. by Solar Table .... 18 43 58 .3
By Nautical Almanac .... 18 43 58 .3

TABLE 6. THE SUN'S DECLINATION TO EVERY 10′ OF HIS LONGITUDE.

THE formula from which this Table was computed, is this, viz. $\frac{\text{sin. ☉'s long.} \times \text{sin. obliq.}}{\text{radius}} = \text{sin. ☉'s declination.}$

When the sun's longitutide is between O^{s} and III^{s}, the declination is N. increasing;
between III and VI, ... N. decreasing;
between VI and IX, ... S. increasing;
between IX and XII,.............. S. decreasing.
The variation for a diminution of the obliquity of the ecliptic is always subtractive.

Example 1.—Suppose the ☉'s longitude to be 3^{s} 15° 11′ 20″.76, and the obliquity of the ecliptic 23° 27′ 49″.76, what will be his declination?

Tab. 6....... 3^{s} 15° 10′..................	22° 36′ 8″.26
(As 10′ : Diff. 68″.06 :: 1′ 20″.76 :).......	− 9 .16
(As 100″ : Var. 95″.9 :: 7″.24 :)..........	− 6 .94
Sun's declin. N.	= 22 35 52 .16

By the Formula.

		log.
Sin. ☉'s long. (sup. to 180°)	74° 48′ 39″.24	9 .9845571
Sin. obliq. eclip...........	23 27 49 .76	9 .6000684
Sin. ☉'s declin. N........	22 35 52 .18	9 .5846255

Example 2.—The sun's longitude at noon, on the 1st of March, 1824, is given in the Nautical Almanac 11^{s} 10° 54′ 55″, and obliquity 23° 27′ 48″.3, let the corresponding declination be found from the Table?

Tab. 6. ...... 11^{s} 10° 50′	7° 30′ 43″.78	Dec. by Table 7° 28′ 49″.55 south.
(As 10′ : 227″.74 diff. :: 4′.9 :) ... −	1 51 .59	By Naut. Alm. 7 28 50
(As 100″ : 30 .38 var. :: 8″.7 :) .. −	2 .64	

TABLE 7. PROPORTIONAL PARTS OF THE SUN'S DIURNAL CHANGES.

THIS Table is founded on the same principle as the general Tables XV. and XVIII., at pages 39 and 47, &c. which have been already explained. When the daily variation of the sun's right ascension or declination, as given in the Nautical Almanac, is made the argument at the top or bottom of the Table, or both if necessary; in the horizontal line of the interval, or hour angle from noon, at the side, will be found the proportional part of the said daily variation. Or if the difference of longitude between Greenwich and any other meridian be taken at the side, with the daily variation as before, the proportional part of the right ascension or declination to be added or subtracted as the case may be, will be found in like manner. The advantage of this Table is, that the results are taken out of one denomination, in integers and decimals, and the constant factor 60, will readily turn the decimal portion into integers of the next lower denomination, whether they may be in time or space; and in this manner the working of double proportions, as well as of logarithmic calculations is avoided.

The same Table may also be advantageously used in lunar computations, in any of the daily changes, where the second difference can be disregarded; such as in finding the ☾'s horizontal semi-diameter, or parallax, for a given time, under any given meridian, provided the difference in 24 hours be the head argument. The equation of time may likewise be found for a given hour and minute under any given meridian. It was considered that 10^{m} would be too long an interval, and that multiples of 4 minutes in the side argument, would give the most convenient intervals, or differences of longitude, in case the tabular proportions for single minutes should in any case be required.

Example 1.—On the 1st of March, 1824, the sun's right ascension is given in the Nautical Almanac 22^{h} 49^{m} $34^{s}.1$, and on the 2nd, 22^{h} 53^{m} $18^{s}.3$; what will be his right ascension at 5 o'clock in the evening of the 1st, at Greenwich?

March 2, 1824. ☉'s R. A...........	22^{h} 53^{m} $18^{s}.3$
1	22 49 34 .1
Diff.	3 44 .2
Or.........................	224 .2

At 5^{h} (p. 164.) Tab. 7.	200^{s}	$41^{s}.670$
	20	4 .167
	4	0 .833
	.2..........	41
Increase in 5^{h}..............	=	46 .711

☉'s R. A. March 1, at noon.....................	22^{h} 49^{m} $34^{s}.1$
Increase in 5^{h}	46 .7
☉'s R. A. at the time required....................	22 50 20 .8

Example 2.—On the 1st of March, 1824, the sun's declination is given in the Nautical Almanac 7° 28′ 50″ S. decreasing by the daily difference 22′ 53″, required the declination at 5 o'clock at Greenwich, or at noon at a place 75° west of Greenwich?

Tab. 7. 22′ 53″ = 1373″....	1300.″	270″.8
	70.	14 .58
	3.	.625
Decrease in 5ʰ..........................		286 .005

March 1. Dec...............	7° 28′ 50″ S
Decrease....................	— 4 46 . (or 286″)
Dec. required	= 7 24 4

Or otherwise in minutes and decimals, thus:

22′ 53″ = 22′.88 nearly......	20′.	4′.167 by Tab. 7.
	2.	.417
	.8	.169
	.08........	16
		4 .769
		60
		— 4′ 46″.140 = decrease nearly as before.

In the first example 200ˢ was taken from the column headed 20, and in the second 1300 from the column 13, but as ciphers were added to the arguments, the decimal point was moved in each case, from 4 .167 to 41 .67, and from 2 .708 to 270 .8; viz. as many figures to the right as there were ciphers added to the integral quantities respectively, which must always be attended to in similar cases. In like manner, 70. was substituted for 7, and consequently 14 .58 for 1 .458.

Example 3.—In the Nautical Almanac of 1824, the sun's right ascension and declination south, are given on the 17th of October 13ʰ 29ᵐ 5ˢ.3, and 9° 20′ 31″; and on the following day, 13ʰ 32ᵐ 50ˢ.4, and 9° 42′ 25″; let it be required to determine his meridian right ascension and declination on the 18th at Madras, supposing its longitude east to be 5ʰ 21ᵐ 56ˢ?

	Right ascension.	Declination.
	13ʰ 32ᵐ 50ˢ.4	9° 42′ 25″
	13 29 5 .3	9 20 31
Difference	3 45 .1	21′ 54″

Use the arguments at side, 5ʰ 20ᵐ and 5ʰ 24ᵐ successively, and then apply a due proportion of the difference to the first result, thus:

Tab. 7. 5ʰ 20ᵐ} R. A....... 3ᵐ	0ᵐ.667		5ʰ 24ᵐ} 3.	0ᵐ.675
.7	.155		.7	.157
.05........	11		.05..........	11
	0 .833			0 .843
	60			60
	49ˢ.98			50ˢ.58
Increase for 1ᵐ 56ˢ......................	+ .29			49 .98
Proportional part for 5ʰ 21ᵐ 56ˢ	50 .27		As 4ᵐ : 0ˢ.60 :: 1ᵐ 56ˢ : 0ˢ.29	

Then 13ʰ 32ᵐ 50ˢ.4 — 50ˢ.27 (being east) = 13ʰ 32ᵐ 0ˢ.13 = right ascension at noon at Madras.

Again, Tab. 7. 5ʰ 20ᵐ} Dec. 20′.	4′.445		5ʰ 24ᵐ......... 20′.	4′.500
1.	.223		1.	.225
.9.......	20		.9.....	20
	4 .688			4 .745
Increase for 1ᵐ 56ˢ	0 .027			4 .688
Proportion for 5ʰ 21ᵐ 56ˢ..............	4 .715 or 4′ 42″.9		As 4ᵐ : 0 .057 :: 1ᵐ 56ˢ : 0.27	

Then 9° 42′ 25″ — 4′ 42″.9 = 9° 37′ 43″.1 = the declination at Madras.

Though this example, which was chosen on purpose, required two operations, partly in consequence of the great difference of longitude, and partly because that longitude, used as an argument, fell in the middle of a four minute interval, yet one operation will in general, particularly in Europe, be sufficient for all common purposes. The error in right ascension would here have been 0ˢ.29, but in declination it would not have been more than 0″.027, if the first operation only, from the argument $5^h\ 20^m$, had been resorted to. Indeed even in this example one operation would have given very nearly a true result, if a *mean* between the tabular quantities belonging to the two arguments $5^h\ 20^m$ and $5^h\ 24^m$ had been taken for $5^h\ 21^m\ 56^s$ thus,

	Right ascension.		Declination.
$5^h\ 20^m$ } 3.	.... .671	20.	..........4′.472
5 24 } .7	.... .156	1.	.... .224
.05	 11	.9	 20
	0′.838 = 50ˢ.28		4 .716 or 4′ 42″.9

Example 4.—Required the time of the moon's passage over the meridian, in longitude 155° east, on the 1st of March 1818?

Tab. 7. Arg. 155°, or $10^h\ 20^m$, and 62^m diff. of meridian passages 60^m..........		25^m. 83
2		0 . 86
Diff. of meridians............................ (East)		− 26 . 69
Time of ☾'s meridian passage at Greenwich, 1st of March	20	0 . 0
Time required..	19^h	33^m. 31
	or 19^h	33^m 19ˢ

[See the work of this example in p. 205 of PROFESSOR LAX's TABLES, to be used with the Nautical Almanac, 1821.]

Example 5.—On the 12th of October, 1824, the ☾'s semi-diameter at noon is given in the Nautical Almanac 15′ 30″, and on the following day 15′ 38″ at Greenwich, what will be the semi-diameter at $9^h\ 56^m$?

First 15′ 38″ − 15′ 30″.................... =	8″	= diff. in 24^h
Tab. 7. with $9^h\ 56^m$ at side, and 8″ at top	3″.311	= the addition
Semi-diameter at noon.......................	15′ 30	
Semi-diameter required.................. =	15 33 .311	

Example 6.—On the 12th of October, 1824, the ☾'s horizontal parallax at noon is given 56′ 51″ in the Nautical Almanac, and at noon of the following day 57′ 23″, what will be her horizontal parallax at $9^h\ 56^m$?

First........................ 57′ 23″ − 56′ 51″ = 32″ = diff. in 24^h		
Tab. 7. with Args. $9^h\ 56^m$ and 32″..................	{ 30″.......	12″.42
	{ 2	0 .83
Diff. in $9^h\ 56^m$		+ 13 .25
☾'s hor par. at noon......................................		56 51 .0
Ditto at $9^h\ 56^m$..		57 4 .25

Example 7.—On the 12th of October, 1824, the equation of time in the Nautical Almanac is given $13^m\ 29^s.9$, and on the following day, $13^m\ 44^s.2$, making a difference = $14^s.3$, what will be the equation at $9^h\ 56^m$ in the evening?

Tab. 7. with Args. $9^h\ 56^m$ and $14^s.3$	{ 14ˢ.	5 .795
	{ .3......	.137
Increase in $9^h\ 56^m$		+ 5 .932
Equation at noon of the 12th............................		13 29 .9
Equation at $9^h\ 56^m$		$13^m\ 35^s.832$

Example 8.—Let it be required to know the meridian equation of time at Moscow in longitude 2^h 30^m 12^s east, and at Monte Video Lighthouse, in longitude 3^h 44^m 52 west, on the 1st of June, 1824?

May 31. Equation	2^m $42^s.0$		
		diff. $8^s.8$	
June 1. Do.	2 33 .2		
		diff. 9 .3	
2. Do.	2 23 .9		

Tab. 7. Arg. 4^h 30^m	8^s.	.833	Arg. 3^h 45^m	9.	1 .406
	.8....	83		.3	47
Proportion for 2^h 30^m		+ 0 .916	Proportion for 3^h 45^m		− 1 .453
June 1. Equation		2 33 .2			2 33 .2
Equation at Moscow		= 2^m $34^s.116$	At Monte Video		= 2^m $31^s.746$

In like manner the proportional changes in the right ascension and declination of the moon may be obtained as *approximate* quantities, to be corrected by the Table *of second differences* hereafter explained.

In all cases where the side argument exceeds 12 hours, one half of it may be substituted as the argument, and then the resulting amount of the tabular quantities must be *doubled*.

TABLE 8. REDUCTION OF THE ECLIPTIC TO THE EQUATOR.

THIS Table differs from Table 5. in this respect, that the sun's place in the ecliptic reduced to the equator, or his longitude reduced into right ascension is given in *space*, whereas in Table 5. the latter is given in time. The construction of this Table depends on the same formula, and the tabular quantities are taken out in a similar manner, except that in the Table before us the variation of the reduction is always *subtractive* for a diminution of the obliquity of the ecliptic, when it does not exceed 23° 27′ 57″; otherwise in going back beyond 1800, when the obliquity exceeds this quantity, the variation becomes *additive*.

Example 1.—When the sun's longitude is $\overset{s}{3}$ 15° 11′ 20″.76, what is his right ascension in arc, the obliquity at the time being 23° 27′ 49″.76?

☉'s long.	$\overset{s}{3}$ 15° 10′	Reduction	+ 1° 17′ 46″.31
As 10′ : diff. 45″.68 :: 1′ 20″.76 :		+	6 .15
As. 100″ var. : 11″.79 :: 7 .24 :		−	0 .85
Reduction of the ecliptic to the equator		+	1 17 51 .61
Sun's longitude given			105 11 20 .76
Sun's right ascension in arc			106 29 12 .37 (= 17^h 5^m $56^s.824$)

This is the same result as was determined in the first example of Table 5. and either one or the other of these two Tables may be used, accordingly as the right ascension is required to be in time or arc; or in cases where great care is necessary, they may both be used as a check upon each other. More examples are here unnecessary, as the mode of using this Table is obvious.

The converse of this Table is given as Table 1. of the ZODIACAL TABLES.

TABLE 9. THE SUN'S SEMI-DIAMETER, HORARY MOTION, AND HORIZONTAL PARALLAX.

As the sun's semi-diameter, horary motion, and horizontal parallax, vary inversely as his distance from the earth, and as that distance depends on his angular distance from the perigee, the columns in the Nautical Almanac will apply only to Greenwich, and such other places as are on the same meridian, or nearly so; hence it has been thought desirable to introduce a Table that will exhibit the three requisites for any given mean anomaly of the sun, which is constantly varying with the lapse of time, or change of longitude. The mean anomaly may be obtained from Tables 1. and 2. as above directed, and its signs and degrees, used as arguments, must be taken, the former at the top, middle, or bottom of the Table, and the latter at the proper side, under and opposite which respectively will be found by inspection the three quantities required, or the nearest thereto without taking proportions, which the small differences will supply almost at sight. This Table will save the trouble arising from working the proportions according to Table 7.; or otherwise will operate as a check on the computations given in the Nautical Almanac. It was first published by Delambre, under the sanction of the Board of Longitude in Paris, 1806.

Example 1.—Let the sun's semi-diameter, horary motion, and horizontal parallax be required for the 1st of May, 1824, at Greenwich?

	S	°	′	″
Tab. 1. 1824 B., mean long.........	9	10°	5′	20″.3
Tab. 2. May 1. (Bis.) add	4	0	14	56 .2
	1	10	19	16 .5
Subtract..........................	9	9	54	9 .7
Mean anomaly......................	4	0	25	6 .8

	S	°	′	″
Mean long. of perigee	9	9°	53′	49″.
Tab. 2. add				20 .7
	9	9	54	9 .7

	Semi-diameter	Horary motion	Horizontal parallax
Tab. 9.	= 15′ 53″.1	= 2′ 25″.30	= 8″.72
Nautical Almanac ...	= 15 53 .3	2 25 .3	* * *
Conn. des Tems.	= 15 52 .3	2 25 .3	* * *
Jahrbuch...........	= 15 53 .3	2 25 .24	* * *

Doubts have been entertained as to the exact mean diameter of the sun, owing to the supposed inflection of light by the lines of the micrometer.

TABLE 10. REDUCTION TO EITHER SOLSTICE.

In almost every computation that has reference to the longitude, latitude, right ascension, or declination of any heavenly body, but particularly of the sun, the obliquity of the ecliptic is an essential element: hence its determination by direct observation is one of the first operations that the practical astronomer imposes upon himself, when he has been so fortunate as to obtain an instrument competent to the purpose. Indeed the determination of the obliquity puts to the test, at the same time, the powers of his instrument, the accuracy of his Table of Refractions, his own skill in observing and computing, and the correctness of his adopted latitude. As the obliquity is equal to the sun's greatest declination properly corrected, at either solstice, its determination depends on solstitial observations of the sun's meridian altitude, and provided the sun's longitude were three or nine signs exactly at *noon* on the respective days of observation, the operation would be as simple as that by which the latitude of a place is determined; but as it will rarely happen that the sun's declination is a maximum at noon, on either of two given days, it becomes necessary to reduce the observations of the sun's meridian altitude, made before and after his passage over either solstitial point, to what an observation would be, if made in one of those two situations; and if, when the proper reductions are made, the obliquity determined near the winter solstice accords with that determined near the summer solstice, and both with the average of other accredited determinations, the observer will have great reason to be satisfied with his instrument, as well as with his latitude, which is involved in the computation. The difference between the sun's greatest declination, or declination at one of the solstitial points, and his observed declination on a preceding or following day is called the *reduction*, which is always a quantity to be added to the observed declination; this quantity depends on the sun's longitude at the moment of making the observation; and the formula, which has been used by the French* astronomers for obtaining it, is this; $\sin D' = 2 \operatorname{tang} \omega \sin^2 \frac{1}{2} L' - 2 \operatorname{tang}^3 \omega \sin^4 \frac{1}{2} L'$: where D′ is the reduction, ω the obliquity, and L′ the sun's distance in longitude from one of the solstices. In Delambre's Astronomie, Tome II. p. 269, is a Table for giving the reduction corresponding to various distances of the sun from either solstice, when counted on the ecliptic; but as the application of this Table requires a previous calculation from the Solar Tables, *the Conn. des Tems*, or the *Nautical Almanac*, of the sun's longitude for the instant of observation, we have preferred making the sun's *right ascension* the argument, and have adapted the reduction thereto, as being an argument determined by the observation itself, when a sidereal clock is used, as is usual in all established observatories, public or private. In this Table the obliquity is put = 23° 27′ 40″, and the reduction is the *difference* between this and the sun's declination at the various points of the ecliptic corresponding to the observed right ascensions, and when the variation in 100″ change of the obliquity is applied, the quantities will thus be all additive until the year 1835.

* Vide Traité Elementaire d'Astronomie Physique, Tome II. p. 336. par J. B. Biot.

Example 1.—On a certain day in June, when the sun's right ascension at noon was $5^h\ 33^m\ 18^s$, and the obliquity $23^\circ\ 27'\ 43''.37$, the sun's declination was determined by observation; required the reduction to the solstice?

$6^h - 5^h\ 33^m\ 18^s = 26^m\ 42^s$ dist. from solstice.	Tab. 10. $26^m\ 40^s$ gives .... $8'\ 29''.90$
$23^\circ\ 27'\ 43''.37 - 23^\circ\ 27'\ 40'' =$ change $3''.37$	As 10^s : $6''.4$ :: 2^s : .... + 1 .28
	As $100''$: 0 .46 : $3''.37$: .... + .01

Reduction to be applied to the observed declination in order to give the apparent obliquity. + 8 31 .19

Before the *mean* obliquity is obtained, the sun's latitude (or deviation from the ecliptic), and lunar nutation of the obliquity must be applied, as will be explained presently.

According to the *formula*, as given by M. Biot, L′ being $= 6^\circ\ 7'\ 37''$.

Log. 2	0 .3010300	Log.	0 .3010300
Log. tang ω	9 .6375143	Log. $\text{tang}^3\ \omega$	8 .9125429
Log. $\sin^2 \frac{1}{2}$ L′	7 .4557688	Log. $\sin^4 \frac{1}{2}$ L′	4 .9115376
Log. of 1st part	7 .3943131	Log. of 2nd part	4 .1251105
First part	+ 24792097	Second part	− 23339
Second part	− 23339		
Sin D′	24778758		
Log. sin D′	7 .3940795		
D′ = reduction	+ $8'\ 31''.10$		

According to Delambre's Table.

6°	$8'\ 10''.28$
As $10'$: $27''.57$ :: $7'\ 37''$:	+ 20 .95
As $100''$: 0 .69 :: 13 .63 :	− 0 .09
Reduction	8 31 .14

In this way the reduction may be applied to any number of observations, and the average of all the results may be adopted as the true one.

This example is taken from an observation of the sun's apparent declination on the 15th of June, 1809, by Messrs. Biot and Mathieu at Paris, which gave the obliquity $23^\circ\ 27'\ 44''.56$, when $+0''.85$ was applied for the sun's latitude.

TABLE 11. THE SUN'S LATITUDE.

According to the theory of La Place, the perturbations experienced by the earth, as occasioned by the respective attractions of the Moon, Venus, and Jupiter, are sufficient to occasion slight deviations in her motion from the plane of the ecliptic, sometimes to the amount of 1″; and Delambre calculated a Table of the equations due to these perturbations, which was published in the "Tables Astronomiques" before referred to, and of which the Table before us is a copy. According to this author, the sun's apparent deviation from the ecliptic, arising from the earth's oscillations is called his *latitude;* and in cases where great precision is required, such as in observations of the equinoxes and solstices, or for the purpose of comparison with the Solar Tables themselves, this Table may be found useful. The arguments and application of the equations belonging to them, have been explained in the note subjoined to the Tables in page 172.

Example 1.—Let it be required to know the sun's latitude at noon on the 15th of June, 1824?

As the arguments in this Table are not given in signs and degrees, but in the thousandth parts of the circle, the Table given in page 261, must be used for converting the quantities out of one denomination into the other, before the tabular corrections can be taken out; and because the Table denominated "The Circle divided into 10000 parts," to be used on this occasion, subdivides the circle lower than is here wanted, the last figure must be omitted in every instance, and when it exceeds

5, the preceding figure must be increased by unity; thus in taking the millesimal argument corresponding to 1^s 25°, or 55°, which in the Table (p. 261) is 1528, (or $\frac{1528}{10000}$), the 8 may be omitted, and the 2 preceding increased to 3, so that 153 (or $\frac{153}{1000}$) will be the proper argument required.

Instead of the sun's and moon's *mean* longitudes, the true longitudes given in the Nautical Almanac may be substituted, when the Lunar Tables of Bürg are not at hand.

	S		
☉'s long. 15 June, 1824	2	24°	14′
☾'s long. ditto	10	7	58
Diff. = A	3	16	16
Heliocentric long. of ⊖ = B	8	24	14
Heliocentric long. of Venus = C	1	25	0
Heliocentric long. of Jupiter = E	3	20	10
Sup. long. ☾ ☊ = N	2	19	5

	S				Arg.	Corr.
A + B + N	2	29°	35′	=	250...	−0″.68
2 B − C	3	23	28	=	314...	+0 .07
3 C − 4 B	6	8	4	=	522...	−0 .25
B − 2 E	1	13	54	=	122...	+0 .11
Sun's latitude required						−0 .75

Example 2.—Let the sun's latitude be required on the 15th of December, in the year 1824?

	S		
☉'s long. 15 Dec., 1824	8	23°	33′
☾'s ditto	6	21	58
Diff. or A	2	1	35
B	2	23	33
C	11	18	26
E	4	5	3
N	2	29	31

	S				Arg.	Corr.
A + B + N	7	24°	39′	=	653...	−0″.56
2 B − C	5	28	40	=	497...	−0 .10
3 C − 4 B	11	21	6	=	975...	+0 .23
B − 2 E	6	13	27	=	536...	−0 .16
Sun's latitude						−0 .59

TABLE 12. CORRECTIONS OF THE SUN'S LONGITUDE, &C. DEPENDING ON HIS LATITUDE.

THE directions subjoined to this Table sufficiently explain its use, as well as its connection with the preceding Table; it will therefore be a sufficient exemplification, if the corrections of the sun's longitude, right ascension, and declination, be ascertained for the 15th of June and December, 1824, agreeably to the two preceding examples.

Example 1.

☉'s lat. − 0″.75 × − 0″.04 = + 0″.03 cor. in long.
− 0 .75 × − 0 .04 = + 0 .03 cor. in R. A.
− 0 .75 × + 1 .00 = − 0 .75 cor. in dec.

Example 2.

☉'s lat. − 0″.59 × − 0″.067 = + 0″.04 cor. in long.
− 0 .59 × − 0 .067 = + 0 .04 cor. in R. A.
− 0 .59 × + 1 .00 = − 0 .59 cor. in dec.

As the sun's latitude is *negative* in both these examples, the signs of the quantities, taken from this Table, are all changed agreeably to the directions there subjoined. In applying the correction for declination, it must always be understood that north declination is designated by +, and south by −; hence the negative correction will diminish the declination when north, but increase it when south; and the reverse will be the case, when the correction for declination has the sign +.

TABLE 13. THE SUN'S HORARY MOTIONS IN SIDEREAL TIME.

IN many computations where the sun's place for a given time is required to be known, as is the case in eclipses and occultations, it is convenient to know what are the hourly changes that take place, under different circumstances, in his longitude, right ascension, declination, and semi-diameter. These horary changes were calculated by DELAMBRE and published in the "TABLES ASTRONOMIQUES" by the French Board of Longitude, from which the TABLE now before us was transcribed. When great precision is not required, it may be sufficient to take the day of the month as the argument; but it will be more correct to take the sun's true longitude for the noon of the given day out of the Nautical Almanac, and to proportion for the time elapsed, according to the diurnal change, by means of our SOLAR TABLE 7, already explained, and then with this argument to use the column of differences in Table 13.

Example.—Let it be required to know the sun's horary motion in longitude, right ascension, and declination, and also his semi-diameter in sidereal time, at 6 o'clock P. M. on July 10, 1824, at Greenwich; his longitude at noon being $3^{s}\ 18^{\circ}\ 4'\ 46''$, and on the following day $3^{s}\ 19^{\circ}\ 1'\ 58''$?

	s			
Sun's long. 11 July	3	19°	1′	58″
10 do.	3	18	4	46
Daily difference ...			57	12

Tab. 7. ⎱ 6 hours. ⎰	50.	12′.50
	7.	1 .75
	.2.....	0 .05
Change in 6^{h}......		14 .30 or 18″

	s			
Sun's long. at noon	3	18°	4′	46″
Change in 6^{h}.....		+14	18	
Long. at 6^{h}	3	18	19	4

Tab. 13. Arg. $3^{s}\ 18^{\circ}\ 19'$. { Hor. mot. in long. = 143″.04; hor. mot. in R. A. = 153″.07
Hor. mot. in dec. = 19″.29, and semi-diameter in time = $1^{m}\ 8^{s}.3$ }

When the horary motions are required for a given time, at any place east or west of Greenwich, the time given must be increased or diminished, according to the difference of longitude, before the proportion of the sun's increase of longitude since the preceding noon is applied.

TABLE 14. THE MEAN OBLIQUITY OF THE ECLIPTIC.

In consequence of the modern improvements made in astronomical instruments, and in the Tables for giving the corrections due to celestial observations, it has become necessary to reform the Table for exhibiting the mean obliquity of the ecliptic, by making the most approved determinations of the secular variation the basis of the Table, in conjunction with the mean obliquity obtained at various recent epochs. The secular diminution for many years used at Greenwich is now found to be too great, and the Table carried forwards with that diminution is consequently no longer considered a proper standard, with which to compare the deductions of modern astronomers. Indeed it is no easy matter confidently to give a preference to the conclusions of any individual astronomer, while others may have an equal title to that distinction: I have therefore selected Delambre's and Dr. Brinkley's obliquities and secular diminutions, and have carried them forwards to the end of the century, as containing within their limits the deductions of almost all the most eminent observers, such as Messrs. Pond, Groombridge, Oriani, Bessel, Carlini, Conti, Arago, &c. &c., and in the course of a few years it will appear probably, which of the two columns gives the obliquity nearer to the truth. The obliquity standing opposite each year must be taken as the *mean* obliquity belonging to the beginning of such year. Thus at the beginning of 1800, Delambre's mean obliquity is 23° 27′ 57″.0, and Dr. Brinkley's 23° 27′ 56″.04, and in the year 1819 they are the same, within the $\frac{1}{1000}$ of a second.

TABLE 15. THE MONTHLY OBLIQUITY OF THE ECLIPTIC.

This Table gives the diminution due to the first day of each month, to be deducted from the obliquity of any given year, according to the secular variation of both Delambre and Dr. Brinkley, and the authority of either one author or the other may be adopted, as may hereafter appear to be more conformable to the truth.

TABLE 16. LUNAR EQUATION OF THE OBLIQUITY. TABLE 17. SOLAR EQUATION OF THE OBLIQUITY.

When the sun's greatest apparent declination has been determined by a series of observations near either solstice, the equations contained in these two small Tables must be applied (together with the sun's latitude) according to their signs respectively, to the observed obliquity (or greatest declination): and the quantity so corrected will be the *mean* obliquity at the solstice in question, and may be reduced to the beginning of the year, by applying a proportional part of the annual diminution with its proper sign.

TABLE 18. LUNAR NUTATION OF THE EQUINOXES IN R. A. IN TIME.——TABLE 19. SOLAR NUTATION OF THE EQUINOXES IN R. A. IN TIME.

These two small Tables are used in determining the mean distance of the sun from the equinoctial point; the quantities contained therein must be applied with their respective signs to the observed right ascension of the sun, together with the equation of time applied with its contrary sign, to obtain his *mean** right ascension; with which the right ascension of a star may be compared, by the aid of a good sidereal clock, with an established rate. Such comparison may affect the right ascension of a whole catalogue of stars. [See Nautical Almanac, 1823, p. 179.]

TABLE 20. GREENWICH OBSERVATIONS OF THE OBLIQUITY.

It has always been understood, that, with Bradley's Refractions, the summer and winter determinations of the obliquity have seldom agreed at Greenwich Observatory, and that this was the principal reason why other formulæ have been contrived for giving a different scale of refractions at low altitudes. The Table now under our notice was copied from a manuscript at Greenwich Observatory, which is considered as an authentic document, and therefore will be considered, probably, as an object of some curiosity among astronomers, who may not have had an opportunity of referring to it. It will, at any rate, afford the means of tracing the comparative deviations from the truth of the instruments made use of at the different periods; and, what will appear remarkable, the discrepancies between the summer and winter observations of the obliquity were smaller in the first four or five years, with the quadrant, than at any subsequent period, and probably with the same refractions. In the year 1760 the difference became so great, that a blank was left unfilled by the winter observation; and, on that account, the observations were omitted for the next four years; and Dr. Maskelyne entertained an opinion, for a time, that the summer and winter obliquities were not subject to the same diminution. The earliest and latest determinations made at Greenwich, accord, in a very satisfactory manner, with the column of Dr. Brinkley's Obliquity, as do the *summer* observations of Greenwich, with but few exceptions, throughout.

TABLES 21 AND 22. GENERAL TABLES OF THE CORRECTIONS OF NOON, &c., BY DELAMBRE.

Before the transit instrument was introduced into general use in practical astronomy, the right ascension of the stars was determined by observing the interval of time elapsed between the moments when they were observed to have equal altitudes at each side of the meridian, the middle point being the time of the meridian passage. Le Monnier and La Caille both practised this method, which requires no meridian line to be previously known, and which is capable of considerable accuracy when the corresponding altitudes are taken with a good instrument, and in those parts of the heavens where the change of altitude is the most rapid. If the sun had no daily variation of declination, the time of exact noon might be ascertained in the same way, with equal ease; but as that is not the case, the noon thus determined will require a correction depending, partly on the diurnal, or horary, change of declination, partly on the interval elapsed between the two corresponding equal altitudes, and partly on the latitude of the place of observation. Delambre published a Table (xxxv of "Tables Astronomiques par le bureau des longitudes de France") for giving generally the equation for noon and midnight, but as the quantities were required to be taken out by double entry, in which various proportions were to be worked, the same author afterwards contrived to simplify his Table, without sacrificing its accuracy (Tome I. Astronomie Theoretique et Pratique, p. 576. Paris, 1814.), which is the one now before us, and which was constructed from the following *formula*, viz.:

$\text{Cor.} = \frac{d\,D}{360^0} \cdot \frac{t}{\sin(15^0.\,t)} \tan\text{g}\,L + \left(\frac{dD \tan D}{360^0}\right) \frac{t}{\tan(15^0.\,t)}$; in which dD is put for the diurnal change of declination, D the sun's declination at noon, L the latitude of the place, and t one half of the interval elapsed between the two observations. The two factors $\frac{dD}{360^0}$ and $\frac{dD \tan D}{360^0}$, which depend on the longitude of the sun, are contained in Table 21, and the factors $\frac{t}{\sin(15^0.\,t)}$ and $\frac{t}{\tan(15^0.\,t)}$ constitute Table 22., the tangent of the latitude being taken out of an ordinary Table of sines, tangents, &c. In using the Tables, the logarithms of the two first columns of each, $\left(\frac{\log dD}{360^0}\right.$ and $\left.\frac{\log t}{\sin(15^0.\,t)}\right)$ must be added to the log tangent of the latitude, for one portion, and the logarithms from the second columns $\left(\frac{\log dD \tan D}{360^0}\right.$ and $\left.\frac{\log t}{\tan(15^0.\,t)}\right)$ must be added together for the second portion, regard being had to the signs; then the natural quantities respectively belonging to these two logarithms, must be algebraically applied together, and will be the proper correction in seconds of time, to be added to, or subtracted from, the time of noon depending on half the elapsed interval.

* The time determined by the sun will not accord exactly with the time determined by a star, if these two equations are omitted.

Example 1.—Suppose that equal altitudes of the sun were taken on the 20th of April, 1823, the sun's longitude at noon being $\overset{s}{0}$ 29° 34′ 52″, and that the interval between the observations was 8^h 12^m, in latitude 51° 30′; required the correction for noon?

	Col. 1.	Col. 2.	
Table 21. ⊙'s long. 29°...........	log. 0 .5407 −	.... log. 9 .8345 +	
35′...........	−24	.... +62	
Table 22. t = 4^h 5^m.............	log. 0 .6681 +	.... log. 0 .3502 +	
1^m.............	+ 7	.... −28	
		Sum of log. 0 .1881 +	1^s.542...the natural product for part 2.
Log Tab. Tan. 51° 30′...............	0 .0994 +		
	Sum of log. 1 .3065......................		− 20 .254...the natural product for part 1.
Whole correction			− 18 .712

In this example, the time of each observation is omitted to be given, since the *correction* for noon is all that is required; but when the beginning and end of the interval are also given, as in the next example, it may be ascertained how much the clock or chronometer used is fast or slow, particularly if its *rate* be known. The signs preceding the logarithmic numbers have reference to their *addition*, but the signs following them have reference to their natural *products*, where like signs produce +, and unlike −.

Example 2.—On the 1st of May, 1823, equal altitudes of the sun were taken, for the purpose of ascertaining the true time by a chronometer, that had been set by guess, in latitude 52° 26′ N.; let it be required to know the error of the chronometer, from the subjoined observations?

A. M.		P. M.
9^h 29^m 25^s alt. of sun's upper limb 42° 22′	....	2^h 25^m 19^s lower limb.
9 33 9 .... lower limb 42 22	....	2 29 43 upper limb.
9 31 17 ...... center therefore. 42 22	....	2 27 31 .5 = the mean.
The whole interval .... 2^h 28^m 43^s + 2^h 27^m 31^s .5		= 4 56 14 .5
The half interval		= 2 28 7 .25
And noon uncorrected 9^h 31^m 17^s + 2^h 28^m 7 .25		= 11 59 24 .25 by chronom.

Hence the correction and true time may be had thus:

	Col. 1.	Col. 2.	
Table 21. ⊙'s long....... $\overset{s}{1}$ 10° or 40°	0 .4860 −	.... 9 .9088 +	
16′	−17	.... +9	
Table 22. t.............2^h 25^m	0 .6114 +	.... 0 .5180 +	
3 . 1	+12	.... −30	
Log. tan. of lat.. 52° 26′	0 .11397 + ...log.	0 .4247 +	2^s.659 = part 2.
First part ....log.	1 .21087	....	− 16 .250 = part. 1.
			− 13 .591 = correction.

Noon uncorrected..................	11^h 59^m 24^s.25
Correction	−13 .59
Apparent noon by chronom..........	11 59 10 .66 (the rate not known.)
Noon, by Naut. Alm. 12^h − equat. 3^m 1^s.5	11 56 58 .50
Error of chronom. for app. time	+2 12 .16

This example has been worked on a supposition, as is usual, that the altitude of the sun varies by quantities proportional to the time, which is not strictly the case, except when the declination happens to be equal to the latitude, and of the same denomination, north or south. It will be more correct to obtain the half interval (t) from the time of the upper limb's contact with the middle line in the morning, compared with the time of the lower limb's contact in the evening; or from the time of the lower limb's contact in the morning, compared with the time of the upper limb's contact in the evening; or, what is still better, from an average of both, thus:

Upper limb's contact A. M. $9^h 29^m 25^s$, viz. from 12^h........$2^h 30^m 35^s$
Lower limb's contact P. M. 2 25 19
Half the sum = t $2^h 27^m 57^s$
Again, lower limb A. M....9 33 9 from 12^h.............2 26 51
Upper limb P. M.................................2 29 43
Half the sum..........= t' $2^h 28^m$ 7^s

The mean or $\frac{t+t'}{2}$ ....= 2 28 2 = the true half interval.

Time of sun's centre add .... 9 31 17 as before.

Noon uncorrected....... = 11 59 19
Correction .... −13 .59 as before.

Noon, by chronom...... = 11 59 5 .41
Noon, by Naut. Alm....... 11 56 58 .50

Chronom. fast at noon ..... 2 6 .91 by equal altitudes.
Do. by sun's transit at noon, taken the same day ..... } 2 6 .95 by $2\frac{1}{2}$ feet transit instrument.

Diff. of the two methods ... 0 0 .04

The equal-altitude instrument, used on this occasion, had an index error of + 1′ 16″ in altitude; but as its vertical axis of motion was watched by a good level, and as the altitude remained *unaltered*, its exact amount was of no importance, in regard to the determination of the correction for noon.

The first Table (21) is useful only when the sun is the object observed; but the second (22) may be used with the planets, if it should be thought desirable to take equal altitudes of any of them, which will seldom be the case where there is a transit instrument.

TABLES 23 AND 24. CORRECTIONS FOR EQUAL ALTITUDES, ETC., BY ZACH.

ANTECEDENTLY to the publication of DELAMBRE'S ASTRONOMIE, viz. in 1809, Baron Zach published his "*Tables abregées et portatives du Soleil,*" in which were the corrections for equal altitudes computed from a formula similar to Delambre's, and which were re-published in an extended form, at Marseilles, in 1812. [NOUVELLES TABLES.] These corrections are contained in Tables 23 and 24 of our present SOLAR TABLES, which have been printed in addition to Delambre's, that the reader may take his choice of the two forms; and, that he may judge the better which mode is preferable, the same examples will be here repeated. According to Zach $\frac{dD}{360 \sin 15^0} = a$, $\frac{-dD \tan D}{36 \tan 150^0} = b$, $\frac{t. \sin 15^0}{\sin (15^0. t)} = \tan \alpha$, and $\frac{t. \tan 150''}{10 \tan (15^0. t)} = \beta$: then log a + log tan α + log tan lat. = log of first part; and log b + log tan β = log of second part, and the sum taken algebraically is the whole correction.

Example 1.

Tab. 24. Long. ☉ 29° 35′ .... a − 13″.30, and log a.............. 1 .12385 −
Tab. 23. Angle α (for $t = 4^h 6^m$) 50° 22′ 7″ and log tan ....10 .08184 +
Log tan of latitude 51° 30′.......................................10 .09939 +

1 .30508 − 20^s.188 = 1st part.

Tab. 24. b − 11^s.942 .... log b............................ 1 .07708 −
β − 7° 19′ 23″ log tan β .... 9 .10856 −

0 .18564 + 1 .533 = 2d part.

Whole correction... −18 .655

Example 2.

Tab. 24. ☉'s long. 1ˢ 10°......... a — 11ˢ.775 and log a .... 1.07096 —
Tab. 23 (t = 2ʰ 28ᵐ.1) ∠ α + 46° 41′ 38″ and log α .... 10.02570 +
Log tan of lat 52° 26′.. 10.11397 +

First part of cor.. log. 1.21063 ...... —15ˢ.706
Tab. 24. b — 14.052, log b.................................. 1.14774 —
Tab. 23. β (2ʰ 28ᵐ.1) ∠ β — 10° 50′ 55″, log tan β .... 9.28254 —

Second part of cor... log 0.43028 ...... + 2.693

Whole correction.. —13.013

The corrections derived from these different Tables are not precisely the same, though nearly so. The place of the sun's apogee and the obliquity of the ecliptic may not have been taken the same by both authors, and the sun's declination at noon has been assumed by Zach as equal to half the sum of the two declinations belonging to the respective times of observation, which will produce a small error when t is a long interval, which will happen when the correction is computed for midnight. To avoid the occurrence of these small errors Zach has shown how a and b may be omitted by substituting the sun's (or planet's) daily change of declination with its proper sign for a, and the tangent of $\frac{1}{2}$ (δ + D) the two declinations, for b; and the work may be abridged by the adoption of *constant* logarithms A and B. The formula for the first part of the correction is $\frac{dD \tan \alpha \tan L}{360 \sin 15^0}$, and for the second part $\frac{dD \tan \beta . \tan \frac{1}{2}(\delta + D)}{36 \tan 150^0}$; hence for the term $\frac{1}{360 \sin 15^0}$ the constant log is = 8.0307013, and for the term $\frac{1}{36 \tan 30^0}$ the constant log is = 8.6822581. For example, MARS was in opposition to the sun on the 8th of April, 1809, and, that the time of his meridian passage might be known, equal altitudes were taken of him at Florence, at an interval of 8ʰ 20ᵐ, in latitude 43° 46′ 40″ N., when his declination was 5° 9′ 40″ S., and his change of declination in 24 hours + 6′ 38″, and the work was performed thus;

Log 398″ (6′ 38″) .... 2.59988 +
Log A (constant).......................... 8.03070 +
Log tan α (+ 50° 33′ 40″) Tab. 23........... 0.08484 +
Log tan lat 43° 46′ 40″ .... 9.98146 +

Log of the first part .... 0.69688 = + 4ˢ.97

Log 398″............................... 2.59988 +
Log B (constant) .... 8.68226 +
Log tan β (— 7° 8′ 16″) Tab. 23............. 9.09769 —
Log tan dec. (5° 9′ 40″)...................... 8.95579 +

Log of second part .... 9.33562 = — 0.22

Whole correction for the time of culmination............. + 4.75

In Table 24, the columns a and b have their seconds marked (″) instead of (ˢ) for *time*, according to the old plan, after Zach's example, which mode of marking seconds of time has in general been avoided in these Tables, to prevent confusion.

It may afford the computer some service to notice here, that the tangent of the latitude and of the declination are *positive* when north, and *negative* when south: and that the daily motion in declination of the sun, or of a planet, is + when approaching the north pole, and — when receding from it: hence the sun's dD is + in the ascending six signs from the winter to the summer solstice, and *vice versâ*.

The signs for the sines and tangents of the half interval, or horary ∠ t, depend on the number of degrees contained in the respective angles. The sign of the quantity a must always be changed in north latitudes when the correction is midnight, and also in south latitudes, when the correction is for noon; but the sign of b never varies.

If any astronomer should wish to construct abridged Tables for a given latitude L, he may make a tan L = c, and then his correction will be c. tang α + b. tang β.

Example for the correction at *midnight.*

Suppose that equal altitudes be taken of the sun on two successive days at an interval of eighteen hours and twenty minutes, with the sun's longitude at midnight $\overset{s}{11}$ 25°, and in latitude 43° 7′ N.; let it be required to determine the correction due to midnight?

Log a (− 15ˢ.251)	1 .183298 +	(+ for −, being midnight)
Log tan α	0 .545524 +	
Log tan lat.	9 .971429 +	
	1 .700251 =	+ 50ˢ.148 first part.
Log b (+ 2ˢ.374)	0 .375481 +	
Log tan β	9 .761600 +	
	0 .137081 =	+ 1 .371 second part.
The whole correction		51 .519

TABLE 25. SUN'S PARALLAX IN ALTITUDE.

THE difference of the sun's apparent altitudes as seen at the same instant from the centre and surface of the earth, or the angle subtended at the sun by the earth's semi-diameter, is called his *parallax* (change of place), and its effect is to depress or diminish his altitude inversely as its sine; the greatest effect being produced at the horizon, and the least at the zenith, where it becomes 0. The quantity of the horizontal parallax being inversely as the sun's distance from the earth is constantly changing, and therefore a Table becomes necessary to exhibit the proper quantity to be added to the apparent altitude, under all the combined variations of distance and altitude. With the horizontal parallax, taken from the SOLAR TABLE 9, enter the proper column, and descend till the line containing the altitude shows the correction to be applied. Thus, when the horizontal parallax is 8″.6 and altitude 24°, the correction is + 5″.05.

LUNAR TABLES.

THE comparatively small distance of the moon from our planet, and the irregularity of her motion in her varying orbit have made her an object of frequent observation; and the perfection which the lunar theory has now attained, has rendered the Tables of her motions so correct, that both Astronomy and Navigation are greatly indebted to the opportunities she affords, by her rapid change of place among the heavenly bodies, of ascertaining the longitude of any given station on the surface of the earth, or of the sea. The influence also which she constantly exerts, in common with the sun and larger planets, on the position of the earth itself, very sensibly affects the apparent places of the fixed stars, as we have seen from the Tables of lunar nutation. Hence auxiliary Tables for converting the apparent into the true place of the moon, and *vice versâ,* are as necessary in practical astronomy as stellar, solar, or planetary Tables.

TABLE 1. EPOCHS OF THE MEAN LONGITUDE OF THE MOON'S ASCENDING NODE.

WE have already seen that the argument, on which the tabular quantity of the moon's nutation depends, both in right ascension and north polar distance, contains the mean longitude of the moon's ascending node, as one of its constituent parts; and this Table exhibits that longitude for the beginning of each of 120 successive years, by simple inspection, that can require no particular explanation.

TABLE 2. RETROGRADE MOTION OF THE MOON'S ASCENDING NODE FOR EACH DAY OF THE YEAR.

WHEN the longitude of the moon's ascending node is required for a given time, the tabular quantity due to the YEAR must first be taken from Table 1, and then the quantity standing opposite the DAY in Table 2, must be subtracted therefrom, and the remainder will be the required longitude. Thus,

	s
For 1825, mean long. of ☾'s ☊ =	8 29° 41′ 57″.8
July 1. (com. year)...............	−9 35 6
Mean long. ☊ July 1, 1825.......	8 19 6 51 .8

TABLE 3. THE MOON'S SEMI-DIAMETER FOR EVERY SECOND OF HER EQUATORIAL HORIZONTAL PARALLAX.

LIKE the parallax of the sun, the parallax of the moon varies inversely as her distance from the earth; and as the earth is a spheroid, the moon's horizontal parallax is not the same in all latitudes: hence, when we speak of the moon's horizontal parallax generally, we mean as seen from the equator, which is that given by the Nautical Almanac, Connoissance des Tems, and Jahrbuch. Her apparent diameter also varies inversely as her distance, or directly as her equatorial horizontal parallax, and the Table before us gives the semi-diameter corresponding to each second of parallax, which will be found very useful in eclipses, occultations, and on various other occasions where the diameter, or semi-diameter, of the moon is one of the elements of computation. The mean semi-diameter of the moon seems not to be finally settled among astronomers and computers; while Burg gives it 1″.15 more than the present Table, Burckhardt gives it 0″.35 less.

Example.—When the moon's equatorial horizontal parallax is 58′ 42″, what will be her semi-diameter by the Table? Answer 16′ 0″.10.

TABLE 4. REDUCTION OF THE MOON'S PARALLAX IN THE SPHEROID, ETC.

THIS Table contains the reduction, or differences between the moon's horizontal parallax in the sphere and in the spheroid, in the different parallels of latitude, which differences are always to be subtracted from the parallax in the sphere; and, as the compression has not yet been finally ascertained, the reduction has been computed for two different compressions, agreeably to recent determinations, and either one column or the other may be used, or a mean between both, as future determinations may direct. If we put c for the compression, p for the horizontal parallax, and L for the latitude, then, omitting the second term of the series, the *formula* for computing the reduction in the Table is, $- c\,p\,\sin.^2 L$ = the reduction.

Example 1.—Let the reduced horizontal parallax of the moon be required for the latitude 51° 30′, on the 24th of May, 1823, at noon?

Nautical Almanac..........	54′ 41″	....	54′ 41″
Compression $\frac{1}{309}$ (Table 4)..	−6 .41	Compression $\frac{1}{317}$..	−6 .24
Reduced hor par...........	54 34 .59	....	54 34 .76

Example 2.—Let the mean reduced horizontal parallax of the moon be required for ten o'clock, P.M., at Edinburgh, on the said day?

Naut. Alm... at noon ....54′ 41″, at midnight 54 31 } As 12ʰ : − 10″ :: 10ʰ 12ᵐ 41ˢ : − 8″.5

Then hor. par. 54′ 41″ − 8″.5 = 54′ 32″.50 ... and lat. 56° gives reduc. { −7″.21 for $\frac{1}{309}$, −7 .01 for $\frac{1}{317}$ }

Then hor. par. 54′ 41″ − 8″.5 =	54′ 32″.50
Mean reduction	−7 .11
Reduced hor. par..............	54 25 .39

As Edinburgh lies 12ᵐ 41ˢ (more or less) *west* of Greenwich, this quantity has been *added* to 10ʰ, the Edinburgh time, to give the time at Greenwich, for which the corresponding variation of the horizontal parallax, since Greenwich noon, must be proportioned. If the longitude had been *east*, it must have been *subtracted* from the hour given at the place of observation.

In the SOLAR TABLE 7., with 20″ at the top, and 10ʰ 12ᵐ at the side, the proportional part 8″.5 is seen by inspection: the variation in the hor. par. being 20″ in 24ʰ.

TABLE 5. LOGARITHMS OF THE EARTH'S RADII, WITH THREE DIFFERENT COMPRESSIONS.

It has lately been a fashionable experiment to try what number of vibrations a pendulum, of a given length, will make in a minute in different latitudes, in order to determine the compression of the earth, or the different lengths of his radii at the respective places of trial; it being known that the number of vibrations in a given time will always be in direct proportion to the square root of the *impelling force*, which, in a detached pendulum, is *gravity*. This Table, therefore, which gives the logarithms of the earth's radii in each parallel of latitude, agreeably to three different compressions (which are supposed to embrace the two extremes, and nearly the mean), it is presumed, will afford considerable assistance to such persons as are engaged in this scientific pursuit. Indeed some of the nicer corrections, particularly of the moon's place, are so connected with the true figure of the earth, that he who finally determines this figure, will be considered as a great friend to lunar computations, whether the result be obtained by means of the pendulum, by triangulation, or by Cagnoli's proposed method of observing the duration of a stellar occultation.

The logarithms of the earth's radii may also be used for reducing the equatorial horizontal parallaxes to any given latitude in the spheroid; but not quite so conveniently, as may be done by the preceding Table of the simple reductions.

Example. As radius .... 10 .00000
Sine 54′ 41″ (as before)...... 8 .20156
Log. of 51° 30′ (Table 5.) .... 9 .99916 in col. $\frac{1}{317}$

Sine 54′ 34″.6 .... 8 .20072 for the reduced paral.

TABLE 6. ANGLES OF THE VERTICAL WITH THE RADIUS; OR REDUCTION OF THE LATITUDE IN EACH PARALLEL.

If we consider the earth as a spheroid, its radii can no longer be considered as equal to one another, and consequently the parallaxes subtended thereby will necessarily vary, even in the same degree of altitude, in different latitudes: neither will the vertical line exactly coincide with the earth's radius. The object of this Table is, to give a measure of the horizontal parallax in every degree of latitude such, that, if subtracted from the said latitude, the remainder will be the latitude reduced to the centre; and an object seen from such reduced latitude will appear in the same point, as if it were actually seen from the centre of the earth. When this reduced latitude is used with the reduced parallax, as given by Table 4., then the same rules will apply to computations depending on them, as usually apply to computations on the sphere. But the maximum of the reduction will depend on the compression assumed; if we take a degree contiguous to the horizon, and take the $\frac{1}{309}$th part thereof, as the excess of the equatorial over the polar parallax, the result will be 11″.6, for the maximum of the reduction, and, in like manner for any other compression. If we call the two semi-axes m and n respectively, the tabular angles of the vertical with radius, are derived from the *formula*

$$\left(\frac{m^2-n^2}{m^2+n^2}\right)\frac{\sin 2\,L}{\sin 1''}+\left(\frac{m^2-n^2}{m^2-n^2}\right)^2\frac{\sin 4\,L}{\sin 2''}+\text{\&c. but } m=n+1\text{, therefore }\frac{m^2-n^2}{m^2+n^2}=\frac{2\,n+1}{2\,n^2+2\,n+1}.$$

Example.—Let it be required to know the latitude of Greenwich reduced to the centre, with the compression $\frac{1}{317}$?

Latitude .... 51° 25′ 40″
Reduction .... − 10 35 .6

Reduced lat......... 51 18 4 .4

TABLE 7. AUGMENTATION OF THE MOON'S SEMI-DIAMETER IN ALTITUDE.

When the moon is at the horizon of any place she is more remote than when she approaches the meridian, for the rotation of the earth gradually brings an inhabitant nearer her, as she appears to ascend in the heavens; consequently her apparent diameter increases with her altitude, and in eclipses and occultations this increase, and subsequent decrease in her descent, are found sufficient to affect the computations of those phenomena. This Table contains the quantities, or corrections, to be added to the horizontal semi-diameter, given in Table 3. or in the Nautical Almanac, agreeably to any given altitude, and the mode of taking it out is so evident, that no particular directions are wanted. For instance, if the moon's semi-diameter be required at 32° 30′ of altitude, when her semi-diameter at the horizon is 16′ 20″, the correction is + 9″.20, and her semi-diameter consequently 16′ 29″.2.

TABLE 8. THE MOON'S PARALLAX IN ALTITUDE.

It has been already said, that the moon's parallax is greatest at the horizon, and that its sine diminishes as the sine of the moon's altitude increases; the Table before us exhibits in each column of horizontal parallax, the diminished parallax corresponding to all the respective altitudes that increase successively by the common difference 10′; in all cases therefore $\frac{1}{10}$ of the difference of any two contiguous parallaxes in the same column, will be the diminution belonging to 1′, which greatly facilitates the use of the Table; also the last column gives the differences, at any altitude, corresponding to the difference of 100″ at the horizon, betwixt two horizontal parallaxes, standing in different columns, which differences may be used as factors for any number of additional seconds in the horizontal parallax, and will thus, in a certain degree obviate the inconvenience of double entry. An example or two will render the use of this additional column familiar.

Example 1.—When the moon's horizontal parallax is 59′ 20″, and her altitude 50° 30′, what will be her diminished parallax?

At 50° 30′.... col. 59′ paral........ 37′ 31″.72
20″........... + 12 .72 = $\frac{20}{100}$ or .2 of diff. 63″.61

Diminished parallax................. 37 44 .44

Example 2.—Let the altitude be 46° 25′, and the horizontal parallax 57′ 33″?

46° 20′.......... 39′ 21″.38 } dif. .721 × 5′ = 3″.605
46 30.......... 39 14 .17 }

46 25.......... 39 17 .78 (viz. 39′ 21″.38 − 3″.60)
$\frac{33}{100}$ of 68″.95.... + 22 .98

Dim. paral ..... 39 41 .06

In this example the factor for 46° 20′ is 69″.05, and for 46° 30′ it is 68″.84; the mean (68″.95) is therefore the proper factor for 46° 25′.

At 60° of altitude the diminished parallax is just one-half of the horizontal parallax; below 60° it is more than one half, and above less.

TABLE 9. TIME OF THE MOON'S SEMI-DIAMETER PASSING THE MERIDIAN.

It seldom happens that the moon is full at the moment of her meridian passage, and, as the dark edge can seldom be distinctly seen in a transit-instrument, it is of importance in all lunar transits, that the time of the moon's semi-diameter's passage over a line fixed in the meridian should be readily obtained, and particularly, when it is considered, that her diameter and daily motion are both constantly varying. The theorem for determining T, or time in seconds of the semi-diameter's passage, is this,

$$T = \frac{\left(43200^{s} + \frac{\text{diff. R. A.}''}{15}\right) ☾ \text{ semi-diam.}'' \times \text{sec. } ☾\text{'s dec.}}{648000''}$$

: where 43200^{s} are the seconds in twelve hours, and 648000″ the seconds in 180°. Tables for giving the time of the moon's semi-diameter's passage have been computed by Wollaston, Mackay, Burg, and, recently, by Lambert, of Washington; but as they are not extended far enough to supersede the necessity of working proportions, and as most of them are Tables of double entry, we have selected those which are capable of being simplified, and of having their principal argument subdivided into a denomination low enough for giving the required tabular quantity, with as little trouble as possible. The Table numbered 9. is the first of two Tables transmitted by Lambert to the Astronomical Society of London, in 1822; but it was found necessary to extend the side argument to single seconds, and the top one to each 10′; and the natural secant of the ☾'s declination is proposed to be used as a factor, instead of the second Table, which is a troublesome Table of double entry.

Example 1.—When the moon's change in right ascension in 12 solar hours is 7° 15′, her apparent semi-diameter 15′ 45, and declination 24° 46′ N., let the time of her meridian passage be required?

7° 10′ in 12^{h} ...... and 15′ 45″ ...... $65^{s}.50$ } diff. = .07
7 20 .... do. ...... 65 .57 }
7 15 .... 65 .50 + .035 65 .535
Nat. secant 24° 46′, 1 .1013 × $65^{s}.535$ = $72^{s}.17$, or 1^{m} $12^{s}.17$ the time required.

Example 2.—On the 26th of May, 1823, according to the Nautical Almanac, the moon passed the meridian of Greenwich at $13^h\ 44^m$; her right ascension at noon being given 261° 25′ 6″, at midnight 268° 3′ 6″, and at noon, on the 27th, 274° 37′ 21″; her declination at noon 26° 17′ 13″ S., and at midnight 26° 3′ 49″ S.; her semi-diameter at noon 14′ 45″, and at midnight 14′ 44″; and her horizontal parallax at the passage 54′ 5″; let the time of her semi-diameter's meridian passage be required?

From these data the diff. of R. A. in the preceding 12 hours, will be............... 6° 38′ 0″
in the following 12 hours,..................... 6 34 15
The declination on the meridian will be..26 1 55 S.
☾'s alt. on the meridian 38° 31′ 20″ − 26° 1′ 55″ 12 29 25
Hor. semi-diam. ☾ at $13^h\ 44^m$ = 14′ 44″ + 3″ aug. (at 12°½ alt.).................. 14 47

Then Table 9, with 14′ 47″, and the mean diff. of R. A. 6° 36′, gives $61^s.30$ (diff. in 10′ = .05)
Nat. secant of dec. (26° 2′) 1 .1129 × $61^s.30$ = 68 .22, or $1^m\ 8^s.22$, the time of passage.

By the Theorem.

Log. $43200^s + 1584\left(\frac{6^\circ\ 36'}{15}\right)$ = log. 44784...........4 .651123
Log. ☾'s semi-diam. 14′ 47″, or 887″ ....2 .947924
Log. sec. 26° 2′..............................10 .046463
Ar. co. log. 648000″ (constant log.) .... 4 .188425
Sum (rejecting 20 in Index) = log. $68^s.22$ as before .. 1 .833935

TABLE 10. SECONDS OF THE LUNAR DISC, THAT PASS THE MERIDIAN IN ONE SECOND OF SIDEREAL TIME, BY BURG.

THIS Table is calculated to give the same result as the preceding one, but in a different way. It was published among the *Tables Astronomiques*, of the French Board of Longitude, but with the arguments both at top and side in entire degrees only, so that it was encumbered with all the inconveniences of double entry, with a paucity of arguments. It is not, however, introduced here solely for the purpose of giving the time of the meridian passage of the ☾'s semi-diameter, but furnishes one of the arguments for TABLE 12., as will be seen presently, and for which the extended form that has been here given it, renders it more convenient than it was in its original state. The tabular number is used as a divisor of the apparent semi-diameter, and has, from inadvertence, the signature (*s*) instead of (″), as has also Table 3. of the PROMISCUOUS TABLES (page 289), which is made an appendage to this Table, to obviate the operation of dividing when great accuracy is not wanted. The same examples given under the last Table, will serve to show how far this method accords with the former one, and also with the *Theorem*.

Example 1. (as before).
Table 10. .... Dec. 24° 46′ and 7° 15′ (12^h motion in R. A.) .... 13″.07 = the divisor,

and $\frac{15^\circ\ 45''}{13''.07}$ or $\frac{945''}{13''.07}$ = $72^s.22$ or $1^m\ 12^s.22$ for the required time.

[Table 3. (PROMISCUOUS) with 13″.1 and 15′ 40″ gives $72^s.31$, by inspection.]

Example 2. (as before).
Table 10. .... Dec. 26° 2′ and 6° 36′ gives 12″.98,

and $\frac{14'\ .47''}{12''.98}$ or $\frac{887''}{12''.98}$ = $68^s.33$ or $1^m\ 8^s.33$ for the required time.

[Table 3. (PROMISCUOUS) by inspection, with 13″.0 and 14′ 50″ gives $68^s.46$.]

This Table was constructed from the *formula* (15 − 0 .04155 d A) cos. (L − z), where 0 .01455 = $\frac{1^h}{24^h\ 4^m}$, d A = the ☾'s motion in R. A. in 24^h, and L − z = the latitude − zen. dist. viz. = the declination.

TABLE 11. INTERPOLATION BY SECOND DIFFERENCES.

THE moon's changes of longitude, latitude, right ascension, or declination, in twelve hours, are seldom so uniform as to admit of proportional parts thereof being taken, for a shorter period, without reference to the second differences. The Nautical Almanac and Connoissance des Tems contain each a Table of the equation of second differences, which may be referred to when either of them is at hand; but as this collection would not be complete without such Table, one has been introduced among the LUNAR TABLES, in a form that is at the same time concise and comprehensive. When the longitude or latitude, &c. of the moon is required for a given time, the approximate quantity must first be obtained by direct proportion, on a supposition of uniform variation, from the current 12 hours, and then the equation for second differences, taken out of this Table and duly applied, will give the correction. The ample directions given at pages 171 to 173 of the Nautical Almanac for the proper mode of using the equation of second differences, will supersede the necessity of repeating them here, it being presumed that every practical astronomer has that useful publication on his table; but an additional example inserted here may tend to facilitate the operation.

Example 1.—According to the Nautical Almanac, the moon will pass the meridian of Greenwich at $10^h\ 21^m$ on the 5th of October, 1824; let it be required to ascertain what will be her longitude, latitude, right ascension, and declination at the time?

☾'s LONGITUDE.		S	First Diff.	Second Diff.	Mean of second Diff.
	1824, Oct. 5, noon	11 12° 35′ 54″			
	Midnight ..	11 18 34 30	5° 58′ 36″		
	Oct. 6, Noon......	11 24 34 51	6 0 21	+ 1′ 45″	+ 1′ 50″
	Midnight ..	0 0 37 7	6 2 16	+ 1 55	

As 12^h : 5° 58′ 36″ :: $10^h\ 21^m$: 5° 9′ 18″, the proportional part to be added.
Long. at noon, 5th Oct. 11 12 35 54

Approx. long. at $10^h\ 21^m$...... 11 17 45 12 }
Equat. at 1^m .... 3″.6 } .... − 6.6 } True Longitude = S 11 17° 45′ 5″.4
50ˢ .... 3 .0 }

☾'s LATITUDE.			First Diff.	Second Diff.	Mean of second Diff.
	1824, Oct. 5. Noon.......	4° 38′ 59″ N.			
	Midnight ...	4 48 59	10′ 0″		
	Oct. 6. Noon	4 55 49	6 50	− 3′ 10″	− 3′ 13″
	Midnight ...	4 59 23	3 34	− 3 16	

As 12^h : 10′ :: $10^h\ 21^m$: .... 8′ 37″.5 the proportional part to be added.
Latitude at noon, Oct. 5,............. 4 38 59

Approx. lat. at $10^h\ 21^m$.............. 4 47 36 .5 }
Equat. for 3′ ......10″.8 } 11″.5 ..... + 11 .5 } True latitude = 4° 47′ 48″.0
13″ ..... 0 .7 }

☾'s RIGHT ASC.			First Diff.	Second Diff.	Mean of second Diff.
	1824, Oct. 5. Noon.....	342° 10′ 51″			
	Midnight..	347 36 51	5° 26′ 0″		
	Oct. 6. Noon.....	353 4 16	5 27 25	+ 1′ 25″	+ 2′ 8″
	Midnight..	358 34 32	5 30 16	+ 2 51	

As 12^h : 5° 26′ :: $10^h\ 21^m$: .... 4° 41′ 10″.5 proportional part to be added.
Right ascension at noon 342 10 51

Approximate R. A. .... 346 52 1 .5 }
Equat. for 2′ ...... 7″.2 } 7″.7....... − 7 .7 } True right ascension 346° 51′ 53″.8
8″ ...... 0 .5 } In time............ $23^h\ 7^m\ 30^s.6$

☾'s DECLINATION.			First Diff.	Second Diff.	Mean of second Diff.
1824, Oct. 5.	Noon....	$2^{0}\ 32'\ 26''$ S.			
			$2^{0}\ 27'\ 2''$		
	Midnight	0 5 24 S.		+ 0′ 39″	
			2 27 41		+ 1″.5
Oct. 6.	Noon....	2 22 17 N.		− 0 36	
			2 27 5		
	Midnight	4 49 22 N.			

As 12^{h} : $2^{0}\ 27'\ 2''$ :: $10^{h}\ 21^{m}$ :: .... $2^{0}\ 6'\ 49''$ proportion to be subtracted.
Declination at noon .... 2 32 26 S.

Approximate dec...................... 0 25 37 S.
Equation for 1″.5 .... − 0 .1
} True declination $0^{0}\ 25'\ 36''.9$ S.

This example affords all the variety that can well occur with respect to the application of the signs with the mean of the second differences, in effecting which there can be no difficulty, if care be taken to put + to the second difference, where the first difference is increasing, and the contrary, and if the *mean* be put equal to half the sum of the second differences *algebraically*, and the equation with a contrary sign to that of the mean.

TABLE 12. FACTORS BY WHICH THE EQUATORIAL INTERVAL IN A TRANSIT INSTRUMENT MUST BE MULTIPLIED FOR THE TIME OF THE MOON'S MERIDIAN PASSAGE OVER ONE OF ITS SPACES.

THIS Table is one of those which were published by the French Board of Longitude; it depends on Table 10 for one of its arguments, and on the moon's horizontal parallax for the other. It is useful in those cases where the moon's enlightened limb has not had its passage over the middle line observed, but has been noticed at the line preceding or following. The factors in the Table are derived from the *formula* $\frac{15\ F\ (1 - \sin p \cos L)}{(15 - 0\ .04155\ dA)\cos D}$, which are respectively multiplied by F, the equatorial interval in sidereal time; p being the horizontal parallax.

Example.—In a transit instrument that has its equatorial interval = 25^{s}, the moon's enlightened limb was observed in contact with the line next following the middle line (or wire) when her equatorial horizontal parallax was 58′ 30″, her declination $8^{0}\ 27'$ N., and her motion in right ascension (dA) $6^{0}\ 15'$ in 12 hours; what allowance must be made for reducing the observation to the meridian wire?

Table 10., with the motion in R. A. at top, and dec. at side gives 14″.31 for the side-argument;

Table 12, with 14″.31, and horiz. par. 58′ 30″, gives the factor 1 .037; and $25^{s} \times 1\ .037 = 25^{s}.92$ is the allowance, or time of passing over one space, included by two contiguous lines in the focus of the eye-piece, which must be subtracted in this case from the time noted at the passage over the next following wire. This Table was computed for Paris, but will do for Greenwich if 3′ be subtracted from the horizontal parallax. The corrections have different signs to the east and west of the meridian, which therefore balance one another when the observations are made at both sides.

TABLE 13. CORRECTION OF THE MOON'S OBSERVED ZENITH DISTANCE.

TABLE 13. was also taken from the *Tables Astronomiques*, and is intended to remedy the inconvenience experienced in taking the moon's meridian altitude, or zenith distance, either before or after the passage of the enlightened limb has been observed. The correction contained in the Table is very small, never exceeding 1″.4, and is derived from the expression $-\frac{1}{2} N^{2} \sin 1'' \tan (L - z)$, where $N = (15 - 0\ .04155\ dA) \times (b - c) \cos (L - z)$, and $c = b \pm \frac{\text{semi-diam. ☾}}{(15 - 0\ .04155\ dA)\cos D}$ = the moment when the centre passed the meridian.

Example.—On the 21st of June, 1823, the zenith distance of the moon's upper limb was taken at 15^{s} after her first or enlightened limb was observed to be on the middle wire of a transit-instrument, her right ascension and declination at noon being $244^{0}\ 44'\ 43''$ and $25^{0}\ 31'\ 51''$ S., and at midnight $251^{0}\ 22'\ 15''$ and $26^{0}\ 3'\ 35''$ S. respectively, her semi-diameter 14′ 46″, and the time of the meridian passage $10^{h}\ 44^{m}$, what correction must be applied to the observed zenith distance of the limb, exclusively of the effect of parallax and refraction?

Table 10, with mot. in R. A. 6° 37′, and dec. 26°, gives .... + 12″.99 (first limb)
Semi-diameter of the moon in seconds .... 886 .00
+ 12″.99 × − 15ˢ (the interval)........................... − 194 .85

The top argument to be used for Tab. 13......................... 691 .15
Tab. 13., with 691″ and dec. 26° gives .... + 0 .6 for the correction.

This correction is positive because the declination, being south, is negative, and is thereby diminished; or the north polar distance augmented.

TABLE 14. DECREASE OF THE MOON'S OR SUN'S DIAMETER WHEN INCLINED TO THE HORIZON.

When the diameter of the sun or moon is measured in a horizontal direction, it is not affected by refraction; but when it is measured vertically, and particularly at a low altitude, the lower limb is more elevated by refraction than the upper one, and therefore the apparent vertical diameter is diminished in proportion to the difference of the two refractions. The same effect is produced in the measure of an oblique diameter, but more or less, accordingly as the diameter measured is more or less inclined to the horizon. This Table is computed to give the decrease of the diameter, at every 3° of inclination, on a supposition that the mean diameter is = 30′.

Example 1.—When the line connecting the centre of the moon and of a contiguous star makes an angle of 45° with the horizon, what is the diminution of her diameter in that direction, her altitude being 37°?

With the inclin. 45° at the side, and 37° alt. at the bottom, the decrease is − 0″.7.

Example 2.—At what altitude will the sun's vertical diameter be decreased just three seconds, when his diameter is 30′?
Answer. At 24° alt. or 66° zen. dist. in 90° of inclination.

The same may be ascertained from a Table of Refractions, thus:

Bradley's Tab. (page 2.) Zen. dist 66° .... Refrac. 2′ 7″.40
66° 30′..................... 2 10 .44

Diff. in 30′ .... 3 .04

This correction must be subtracted from the horizontal diameter of the sun, when the measure of his vertical diameter is used, at a given altitude, for graduating the scale of a micrometer: the horizontal diameter being taken from No. 9 of the Solar Tables, or from the Nautical Almanac for the given day.

TABLE 15. THE LUMINOUS PORTION OF THE MOON IN DIGITS.

This Table shows how many digits, or twelfth parts of the moon's diameter, are luminous in any part of her lunation, or orbit as she regards the sun.

Example 1.—On the 8th of June, 1823, the sun's longitude is given in the Nautical Almanac 2ˢ 18° 44′, and the moon's at noon 2ˢ 9 52′; what portion of the moon's disc was then visible?

☉'s Long....... 2ˢ 18° 44′ } Diff. = 8° 52′ = Argument { Tab. 15. 8°........ 0 .059 } 60′ : .015 :: 52′ : .013
☾'s Long....... 2 9 52 } { 9°........ 0 .074 }
8° 52′ = 0 .072 digits. (viz. 0 .059 + .013)

At noon the ☾'s diameter is given 33′ 12″, one-twelfth of which is 2′ 46″.33 or 166″⅓, the value of one digit at the time: hence 166″.33 × .072 = 11″.98 is the luminous portion. As the sun's longitude exceeded the moon's, or was east, the new moon had not taken place.

The same result, or nearly so, may be had from Table 2. of the PROMISCUOUS TABLES, page 288, without regarding the digits, thus;

Arg. 8°................ 9″.7 } Diff. in 100′ : 4″.3 :: 52′ : 2″.236
9°................12 .3 }

Then 9″.7 + 2″.236 = ... 11″.93 }
As 100″ var. : 0″.97 :: 4″ : .... — .03 }

Luminous portion .... = 11 .90 supposing the ☾'s semi-diameter = 1000″, which is only 996″.

These two Tables are however more calculated to gratify curiosity, than to promote any useful purpose.

ZODIACAL TABLES.

THERE are several Tables necessary for giving the elements of computation of eclipses, occultations, &c. which could not be classed under any one of the foregoing series of Tables, since the sun, moon, zodiacal stars, and planets, may all be included in the respective phenomena. We have therefore chosen an epithet of comprehensive signification, ZODIACAL, to prefix to these Tables, as being indicative of that portion of the heavens, where the phenomena in question occur. Many of these Tables, particularly of the moon's parallaxes, are new, and the others are chiefly re-calculated, with a view to reconciling the different methods of computation, and to removing some of the obstacles that may have prevented the more frequent notices of approaching occultations.

TABLE 1. REDUCTION OF THE EQUATOR TO THE ECLIPTIC.

THE uses of this Table are explained in the margin of the Table itself It differs from No. 8. of the SOLAR TABLES only in the disposition of the arguments, and is computed from the following *formula*, viz. co-tang ☉'s long $= \frac{\text{rad. cosin obliq.}}{\text{tang ☉'s R. A.}}$

Example 1.—Let the sun's longitude be required when his right ascension was observed to be 106° 29′ 12″.38, the obliquity of the ecliptic at the time being 23° 27′ 49″.76 ?

	S	Reduction.	
Arguments	3 16° 20′	1° 17′ 13″.23	
	3 16 30	1 17 55 .78	.... 1° 17′ 55″.78

As 10′ : 42 .55 :: 47″.62 : (co. to 30′)................................. — 3 .38
As 100″ : 10″.97 var. :: 7″.24 : (dim. obliq.)............................ — 0 .79

Reduction of the equator to the ecliptic................................ — 1 17 51 .61
Sun's right ascension given.. 106 29 12 .38

Sun's longitude required .... 105 11 20 .77

By the *formula*.

Cosin obliq. eclip. .... 23° 27′ 49″.76 .... log 9 .9625169
Tang ☉'s R. A. (comp. to 180°) .... 73 30 47 .62 Ar. Co.......... log 9 .4712365

Cotan ☉'s longitude (comp. to 180°)..... 74 48 39 .24 .... log 9 .4337534
Subtract from 180 0 0

Sun's longitude required...............105 11 20 .76 as before.

In this as in the SOLAR TABLE 8, when the obliquity is less than 23° 27′ 57″, the variation is subtractive, but when more, it is additive, as it regards the reduction.

Example 2.—In the Nautical Almanac of 1823, the sun's right ascension on the 24th of April, is given $2^h\ 4^m\ 57^s.6$, what was his longitude at the time?

⊙'s R. A. .... $2^h\ 4^m\ 57^s.6$ = 1 s 1° 14′ 24″ (by Tab. at page 109)
Zod. Tab. 1. 1 s 1° 10′ .... +2 13 56 .91
As 10′ : diff. 22″.70 :: 4′ 24″ : + 9″.99 }
As 100″ : 19″.92 var. :: 5″.3 dim. − 1 .05 } .... + 8 .94

The sun's longitude .... 1 3 28 29 .85
By the Nautical Almanac, 1823 .... 1 3 28 30

Example 3.—The sun's right ascension on the 1st of January, 1824, is given in the Nautical Almanac $18^h\ 43^m\ 58^s.3$; let his longitude on that day be required?

⊙'s R. A. .... $18^h\ 43^m\ 58^s.3$ .... = 9 s 10° 59′ 34″.5
Zod. Tab. 1. .... 9 s 10° 50′ .... 0° 52′ 38″.04 }
As 10 : 46″.48 diff. :: 9′ 34″.5 : .... + 44 .50 } .... − 0 53 21 .85
As 100″ : 7″.5 var. :: 8″.7 dim. : ...... − 0 .65 }
True longitude .... 9 10 6 12 .65
By the Nautical Almanac, 1824 .... 9 10 6 13

Example 4.—Let the ecliptic (or approximate) longitude of Pollux be required for the 9th of July, 1824, when his apparent right ascension, as given in the Nautical Almanac is $7^h\ 34^m\ 34^s$, or 3 s 23° 38′ 30″?

Arg. 3 s 20° gives reduc. .... 1° 44′ 43″.57
As 10° : diff. 35″.50 :: 3° 38′.5 : .... +12 .92
As 100″ : var. 14″.87 :: 10″.5 dim. obliq. .... − 1 .56

The whole reduction subtract .... 1 44 54 .93
Right ascension given .... 3 23 38 30 .00

The star's ecliptic longitude .... 3 21 53 35 .07

TABLE 2. CORRECTION OF THE APPROXIMATE (OR ECLIPTIC) LONGITUDE OF A ZODIACAL STAR OR PLANET.

THIS Table of DR. MASKELYNE was found too contracted, and abounding with errors, probably typographical, which circumstances led to a minute examination and correction of its tabular quantities, as well as to an extension of the side argument, for the purpose of diminishing the corrections depending on the proportional parts; and it is presumed, that the Table in its present form will be found greatly improved; particularly as the right ascension is now given in both the denominations of *time* and *space*. Before this Table can be used, however, the ecliptic declination corresponding to the star's approximate longitude must be found from SOLAR TABLE 6, page 160, &c., and this declination ± the given declination of the star, accordingly as they are, one or both, N. or S., will give the approximate latitude for the head argument; the right ascension, either in time or space, being the side argument: then the tabular correction, applied with its proper sign to the approximate longitude, will give the true longitude of the star in question. As the star's latitude is measured from the ecliptic, it will require some caution to place the N. or S. accordingly.

Example.—In Piazzi's Catalogue the right ascension and declination of φ Ophiuchi are given 244° 55′ 37″ and 16° 9′ 47″.6 S. for the epoch 1800, let it be required to determine what the corresponding longitude of this star was at the time, the obliquity being then, according to DELAMBRE, 23° 27′ 57″?

Zod. Tab. 1. R. A. .... 8 s 4° 50′ .... 1° 51′ 2″.97
As 10′ : diff. 33″.27 :: 5′ 37″ : .... −18 .69

Reduction .... + 1 50 44 .28
Right ascension given .... 8 4 55 37 .0

Ecliptic longitude .... 8 6 46 21 .28

Solar Tab. 6. Approx. longitude φ Ophiuchi 8s 6° 40′ gives declin. 21° 26′ 48″.13 S.
As 10′ : 102″ :: 6′ 21″.3 : .... −1 4 .8

Ecliptic declin. φ Ophiuchi...................... 21 27 52 .93 S.
Declination of do. given........................ 16 9 47 .6 S.

Approximate latitude.......................... 5 18 5 .33 N.

Zodiacal Table 2. R. A. φ Oph. 8s 4° 30′......55′ 0″.7 with 5° 20′ approx. lat. N.
8 5 0......54 0 .3 with do.

Diff. in 30′ R. A. 1 0 .4

Approx. lat. .... 5° 20′ with 8s 4° 30′ R. A. 55′ 0″.7
5 10 with do. ....53 17 .1

Diff. in 10′ change of lat. 1 43 .6

As 30′ : 60″.4 :: 25′ 37″ : − 51 .58 proportional part for diff. in R. A.
As 10′ : 1′ 43″.6 :: 1′ 55″ : − 19 .86 proportional part for diff. in approx. lat.

−1 11 .44 } True correction −53′ 49″.26 (lat. N.)
Tab. correc. for approx. long. 55 0 .7 }
Approximate longitude of φ Ophiuchi ....246 46 21 .28

True longitude of do. by the Tables..................245° 52 32

Table 4. with long. 8s 5° and lat. 5° 18′ gives a further correction for the diminution of obliquity, +0″.1, which may be here disregarded.

By Dr. Maskelyne's logarithmic method.

Log tang dec...... 16° 9′ 47″.6 S........9 .462164
Log sine R. A.....244 55 37 (sub.)....9 .957020 tang............10 .329884

Tang A (aux. ∠)... 17 44 40 S..........9 .505144 Ar. Co. cos...... 0 .021169
Obliquity .... 23 27 57 N.

Log cos B .... 5 43 15 N. .... 9 .997831

Tang true longitude 65° 52′ 32″, or 245° 52′ 32″.... 0 .348884

[See Mackay's Longitude, Vol. I., p. 44.]

By Dr. O. Gregory's method.

Log sin R. A. ..244° 55′ 37″ .....9 .957020 } Tang R. A....244° 55′ 37″....10 .329884
Log tang dec ... 16 9 47 .6.....9 .462164 } Sin (φ + obl.) 84 16 43 .... 9 997831
Tang ∠ φ...... 72 15 20 .....0 .494856 } Cosec. φ...... 72 15 20 ....10 .021169

Tang sum.... 65 52 32 .... 0 .348884
+180

True longitude 245 52 32

[See Dr. O. Gregory's Elements of Trigonometry, p. 156.]

In cases where great precision is wanted, the logarithmic calculus is shorter than the tabular; but in many instances, the proportions may be omitted, which may generally be done, when the object of the computation is only to ascertain whether or not an occultation of an individual star will take place in a given lunation: and what is here said of the longitude of a star, is equally true of the latitude.

[On this subject the reader may consult with advantage the first volume of Professor Woodhouse's Treatise on Astronomy, Theoretical and Practical, page 157, et seq.]

TABLE 3. CORRECTION OF THE APPROXIMATE LATITUDE OF A STAR OR PLANET.

TABLE 3. has also been extended from Dr. Maskelyne's Table, and gives nearly the same trouble as its predecessor. When the ecliptic declination and approximate latitude are obtained from the preceding Tables, by the process which has just been explained, they are used as the proper arguments for this Table, and the approximate latitude is always diminished by the tabular correction. But as the obliquity was taken at 23° 28' a further correction for the seconds of diminution must be had from Table 5.

Example.—Let the latitude of φ Ophiuchi be determined for the epoch 1800, its right ascension and declination being, as before, 244° 55' 37" and 16° 9' 47".6 respectively, and the obliquity 23° 27' 57" ?

With the ecliptic dec. 21° 27' 53", and approx. lat. 5° 18' 5".33 N., already determined, enter Zodiacal Table 3., and the tabular correction is had from the nearest arguments, thus:

Dec. 21° 0', and approx. lat. 5° 20', give 5' 36".0 }
21 30 .... 5 20 ....4 31 .6 } diff. 1' 4".4 for dec.
21 0 .... 5 10 ....5 25 .4 }
21 30 .... 5 10 ....4 23 .0 } diff. 1' 2".4 for lat.

As 10' : 64".4 :: diff. 7' 53" : .... −0' 50".77 }
As 10' : 62".4 :: diff. 1' 55" : .... −0 11 .96 }
Approximate Correction .... 5 36 .0 }

True correction for obliquity 23° 28'................ 4 33 .27 to be subtracted.
Approximate latitude of φ Oph...................... 5 18 5 .33

True latitude for obl. 23° 28' .... 5 13 32 .06
Table 5. for 3" dim. obl. (☉'s long. 8ˢ 6°)........... −2 .74 (viz. $\frac{91''.4 \times 3''}{100''}$)

True latitude for obliq. 23° 27' 57".................. 5 13 29 .32 N.

By Dr. Maskelyne's logarithmic method.

Log tang declin .... 16° 9' 47".6 S........ 9 .462164
Sin R. A. (244° 55' 37") ..... 64 55 37 .. (sub.) 9 .957020

Tang A (aux. ∠) .... 17 44 40 S. .... 9 .505144
Obliquity 1800............. 23 27 57 N.

Tang B................... 5 43 15 .... 9 .000784
Sine true longitude .... 245 52 32 .... 9 .960309

Tang true latitude......... 5 13 29 N. .... 8 .961093

By Dr. O. Gregory's method.

Log sine R. A. (244° 55' 37").. 64° 55' 37" .... 9 .957020
Log tang declin .... 16 9 47 .6..... (sub.) 9 .462164

Tang (aux. ∠) φ .... 72 15 20 .... 0 .494856
Obliquity 1800 .... 23 27 57

(φ + obl.).................. 95 43 17
Supplement (φ + obl.) .... 84 16 43 log cos.... 8 .998786
Sine dec .. 9 .444647
Sec φ..... 10 .515893

The sine true lat. of φ Oph. ... 5 13 29 N. .... 8 .959326

TABLES 4 AND 5. CORRECTIONS OF THE LONGITUDE AND LATITUDE OF A STAR OR PLANET, ARISING OUT OF 100″ DIMINUTION OF THE OBLIQUITY.

THE quantities taken out of these two Tables are small at present, but will increase annually, as the angle of the ecliptic continues to diminish. The mode of applying these secondary corrections has been explained in the two last examples, under the Tables 2. and 3. respectively.

TABLE 6. THE PRINCIPAL ZODIACAL STARS THAT MAY BE OCCULTED BY THE MOON, ETC.

THIS Table contains a methodical arrangement of the Zodiacal Stars, given by Dr. Young in one of the numbers of Professor Brande's Journal, and might have been extended, so as to contain the original larger number, from whence the extract was made, if the Catalogue had been at hand. The apparent path of the moon among the Zodiacal Stars, in any lunation, depends on the place of her ascending node, which lies in that path, and which therefore becomes the argument of the Table. Such stars as lie within the moon's diameter, when she passes before them, would always be occulted provided she had no parallax: but as the quantity of her parallax depends on different circumstances, it cannot with certainty be affirmed whether an individual star, lying apparently in the moon's path, will or will not suffer an occultation, until a rough computation has been made, or until the Tables exhibiting the parallaxes in question have been inspected. The principal stars arranged opposite the moon's node for the time, in horizontal lines in this Table, are those that *may* be occulted in any place, and *will be* occulted somewhere in each lunation, while the moon passes over or near the place of the node.

Since this Table was printed, Dr. Young has given the elements for computing the occultations of the stars contained therein, in the Nautical Almanac of 1824, which Elements it is hoped he will continue annually.

For instance, in the month of July, 1824, the time of the ☾'s conjunction in R. A., and the other Elements of Computation, are given for σ ♏, ϑ Ophiuchi, B Ophiuchi, ν ♐, ο ♐, π ♐, η ♓, η ♉, 132 ♉, δ ♊, ξ ♌, ο ♌, and π ♌; all which, except η ♓, are found in the horizontal line of the Table indicated by $\overset{s}{\text{IX}}$ 9°, the place of the moon's node, as given for the time in the Nautical Almanac, and by our Lunar Tables 1. and 2.

TABLE 7. THE LONGITUDE AND LATITUDE OF THE NONAGESIMAL DEGREE, ETC.

IN all eclipses and occultations, the moon's parallax is so essential an element of computation, that various methods of determining it have been devised by mathematicians. Cassini's method, however, seems to have afforded the easiest and surest means of obtaining the proposed end, and has, therefore, been adopted the most generally. As the ecliptic is a great circle of the sphere, one half of it is always above the horizon, and the middle, or most elevated point of it, is 90° from the intersection of the horizon, which point was therefore denominated, the *nonagesimal;* its elevation above the horizon is called its *altitude,* and its distance from the equinoctial point, at the beginning of the sign Aries, is called its *longitude.* These are the names that have been long in use, and are, therefore, retained in the Table now under our consideration, though Monsieur Biot proposes to substitute the terms *longitude* and *latitude* of the *zenith;* and for R. A. of the *mid-heaven* (or *medium cœli),* he uses the *right ascension* of the *zenith;* which nomenclature is equally applicable, and may hereafter become equally familiar.

In the computation of this Table, the obliquity of the ecliptic is taken at 23° 28′, and the reduced latitude of Greenwich at 51° 17′ 48″, from the compression $\frac{1}{309}$; and the columns of differences, and variations both for latitude and obliquity, are so contrived, as to render the Table at the same time both comprehensive and permanent. The different terms of the *formula* from which the computations were effected may be thus familiarly explained.

$$\begin{cases} \text{Cos mid-heaven (or R. A. zen.)} \times \text{cos reduced lat.} = \text{cos arc I.} \\ \text{Co-tang reduced lat.} \times \text{sin mid-heaven (or R. A. zen.)} = \text{co-tang arc II.} \\ \text{Arc II} \pm \text{the obliq.} = \text{arc III.} \\ \text{Sin arc III} \times \text{sin arc I} = \text{cos altitude nonagesimal (or lat. zen.)} \\ \text{Cos arc III} \times \text{tang arc I} = \text{tang long. nonagesimal (or long. zen.)} \end{cases}$$

In the application of this formula it will be necessary to take notice, that arc I is greater than a quadrant in the second and third quadrants of the mid-heaven (or of the R. A. zen.); that arc II is always less than a quadrant; that in arc III the obliquity has the sign + when Aries is east of the meridian, and − when west; and that whenever arc III is less than a quadrant, the longitude of the nonagesimal (or zen.) is of the same affection as the R. A. of the mid-heaven (or zen.), but when greater than a quadrant, then of the same affection as arc I. The R. A. of the mid-heaven is the sun's right ascension added to the hour angle, when both are turned into space.

Example 1.—At $17^h\ 27^m\ 15^s.7$ apparent solar time at Greenwich on the 7th of July, 1823 (viz. July 8th at $27^m\ 15^s.7$ past 5 o'clock A.M.), when the sun's right ascension in arc was $106^\circ\ 29'\ 12''.38$; and the obliquity of the ecliptic $23^\circ\ 27'\ 49''.76$; what were the longitude and altitude of the nonagesimal degree (or of the zenith), supposing the latitude of Greenwich to be $51^\circ\ 28'\ 40''$, and the compression $\frac{1}{314.3}$?

By the formula.

Longitude of Greenwich	51° 28′ 40″
Reduction of lat. (Lunar Tab. 6.)	− 10 41
Latitude reduced	51 17 59

Sun's right ascension given	106° 29′ 12″.38	
Time $17^h\ 27^m\ 15^s.7$ in arc (p. 109.)	261 48 55 .50	
Cos. R. A. of mid-heaven (or of zen.)	8 18 7 .88	log. 9 .9954247
Cos. reduced latitude	51 17 59 .00	9 .7960512
Cos. arc I	51 46 45 .04	9 .7914759
Co-tang. reduced latitude	51 17 59 .00	9 .9037187
Sin. R. A. mid-heaven	8 18 7 .88	9 .1595492
Co-tang. arc II	83 24 4 .35	9 .0632679
Obliq. eclip.	− 23 27 49 .76	
Sin. arc III	59 56 14 .59	9 .9372563
Sin. arc I	51 46 45 .04	9 .8952193
Cos. *altitude of the nonagesimal*	47 9 37 .57	9 .8324756
Cos. arc III	59 56 14 .59	9 .6997910
Tang. arc I	51 46 45 .04	10 .1037433
Tang. *longitude of the nonagesimal*:	32 27 39 .50	9 .8035343

By Zodiacal Table 7.

Arg. R. A. of the mid-heaven	Long. of Nonag.	Alt. of Nonag.
8° 0′ 0″.00	32° 14′ 37″.00	47° 3′ 47″.00
18 7 .88	+ 13 2 .00	+ 6 4 .80
Cor. for 11″ diff. lat.	+ 0 8 .25	− 0 9 .23
Cor. for 10″.24 dim. obliq.	− 0 8 .02	− 0 5 .49
Long. of nonagesimal	32 27 39 .23	Alt. of nonag. 47 9 37 .08

Example 2.—The ecliptic conjunction of the sun and moon on the 10th of July, 1824, is given in the Nautical Almanac $16^h\ 20^m\frac{1}{4}$, what will be the longitude and altitude of the *nonagesimal* at the time?

Sun's R. A. on the 11th	$7^h\ 22^m\ 26$ .3 or 110° 36′ 34″.5	diff. 1° 1′ 12″
on the 10th	7 18 21 .5 or 109 35 22 .5	
As 24^h : diff. 61′ 12″ :: $16^h\ 20^m\ 15^s$: ...	+ 41 39 .6	
Sun's right ascension at the time	110 17 2 .1	
Hour angle $16^h\ 20^m\ 15^s$	245 3 45	
R. A. of mid-heaven	355 20 47 .1	

Argument 355°	22° 42′ 58″	42° 28′ 37″
0 20′ 47″	+ 15 32 .5	+ 7 36
Cor. for 10″.4 dim. obliq.	− 0 10 .5	− 0 4
Long. of nonagesimal	22 58 20	Alt. non. 42 36 9

TABLE 8. THE ANGLE OF POSITION OF EACH POINT OF THE ECLIPTIC, ETC.

TABLE 9. AUGMENTATION OF THE ECLIPTIC ANGLE, ETC.

IF a portion of a large circle be drawn from the pole of the ecliptic to a given star, and another portion from the pole of the equator to the same star, the angle contained between the two curves will be the *angle of position* of the said star, and the arcs will be respectively the complements of the star's latitude and declination. As Tables have been given for computing the longitude and latitude of a star, to facilitate the computation of occultations, and as one of the methods of obtaining a star's aberration in right ascension and north polar distance, employs the longitude, latitude, angle of position, and north polar distance of a star, it has been deemed proper to insert the present Tables, as a suitable appendage to Tables 2, 3, 4, and 5, of this series; in order that the computer may have his option of the different methods. Our original plan was to admit such Tables only as are calculated for giving the *corrections* in *right ascension* and *north polar distance*, and to recommend the conversion of *corrected* or apparent right ascensions and north polar distances into *apparent longitudes* and *latitudes*, in preference to computing the corrections for mean longitudes and latitudes; but as various Tables of the *mean* longitudes and latitudes, with the secular variations of stars, have been published, and may hereafter be extended, the Tables before us may have their value in those cases where the corrections are required to be computed from the longitudes and latitudes. Professor Woodhouse has given so full an explanation, with suitable examples, of the uses of the angle of position in determining aberrations, that the reader will best consult his own benefit by referring to it. [See a TREATISE on ASTRONOMY, THEORETICAL and PRACTICAL, p. 275, &c. Part I. Vol. I. new edit.]

Though Dr. Maskelyne published a Table for giving the angle of position of each degree of the ecliptic for the obliquity 23° 28′, which might have been adopted here, yet a new Table has been computed for every 10′ of the ecliptic's longitude, and for the obliquity 23° 27′, as more suitable for a future epoch. If P be put for the angle of position, L for the longitude of the star, N for its north polar distance, and O for the obliquity of the ecliptic, then the *formula* will be $P = \frac{\cos L . \sin O}{\sin N}$ for the star's angle of position, of which the quantity depending on (cos L. sin O) is contained in Table 8, and the other portion, or modification, in Table 9, which follows.

Example.—Let it be required to determine the angle of position of Aldebaran for July 1, 1820, supposing its longitude at the time to be 2ˢ 7° 16′ 23″, its latitude 5° 28′ 44″ S., and its right ascension in arc 66° 24′ 30″?

By Zod. Tab. 1. R. A. 2ˢ 6° 20′ gives the reduction	+	1° 45′ 54″.4
As 10′ : 35″.29 diff. :: 4′.5 :		−16 .9
As 100″ : 21 .15 var. :: 7″ :		− 1 .5
Whole reduction	+	1 45 36
Right ascension		66 24 30
Ecliptic longitude (or arg.)		68 10 6

Tab. 8. Ecliptic longitude 2ˢ 8° 10′.1	9° 9′ 48″.96
As 100″ : 46″.30 var. ; increase of obliq. 50″ :	+23 .15
Tab. 9. Arg. 9° with 5° 30′ lat. gives augmentation	2 31 .10
0 10′ do.	+ 2 .83
True angle of position	9 12 46 .04

By the formula.

Log. cos. of longitude 67° 16′ 23″	9 .586969
Log. sin. obliq. 23° 27′ 50″	9 .600069
	19 .187038
Log. sin. N. P. D. 73° 51′ 40″ .... (sub.)	9 .982538
Log. sin. of the angle of position 9° 12′ 50″	9 .204500

By another *formula*, [viz. $P = \frac{\cos \text{R. A.} \sin \text{obliq.}}{\sin * \text{'s co-lat.}}$

Log. sin. R. A. 66° 24′ 30″	9 .602294
Log. sin. obliq. 23 27 50	9 .600069
	19 .202363
Log. sin. co-lat * 84° 31′ 16″ .... (sub.)....	9 .998011
Log. sin. angle of position 9° 12′ 43″	9 .204352

Either of these formulæ will give the angle of position with less trouble than the Tables give it, provided that Taylor's logarithms are used. The difference between the results of the two formulæ, arises probably from a want of an exact accordance between the longitude and right ascension of Aldebaran, both which were taken from pages 178 and 179 of the Nautical Almanac of 1823.

When the angle of position of any star is a right angle, the precession in right ascension is 0; when this angle is acute, the precession is +, but when obtuse, it becomes —, as is the case, for instance, with β Ursæ Minoris.

The angle of position is also an element in the computation of parallaxes, in eclipses and occultations, when the parallactic method is used.

TABLE 10. THE ANGLE BETWEEN THE ECLIPTIC AND A LINE PARALLEL TO THE EQUATOR.

THIS Table may be used, in conjunction with the two preceding Tables, for determining the parallaxes in eclipses and occultations by the parallactic method, if it should be deemed desirable. Dr. Maskelyne computed the original Table for the obliquity 23° 28′, with a column of variation for 10″ variation of the obliquity, but the present Table is adapted for the obliquity 23° 27′ 57″, and has its variation, without the decimal point, for 100″ variation of the obliquity. The sun's declination must be made the argument when solar eclipses are computed; but for occultations, the declination of that point of the ecliptic to which the star's place is referred must be used.

Example 1.—Let it be required to determine the angle that a parallel of the sun's declination made with the ecliptic in 1800, when that declination was 14° 15′ south or north?

Dec. 14°...................... gives	19° 1′ 30″
One half of the diff. in 30′............	— 11 10
The angle required...............	= 18 50 20

Example 2.—What will be the said angle on the 10th of July, 1824, at 15h 25m 30s, Greenwich time?

Sun's dec. 10th July, 1824.............	22° 14′ 28″ N.
11th do.	22 6 38 N.
As 24h : .... diff.	7 50 :: 15h 25m.5 : 5′ 2″
Dec. at 15h 25m.5............	22 14 28 — 5′ 2″ = 22° 9′ 26″
Then dec. 22° 5′ gives	8 8 40
As 5′ : diff. 14′ 25″ :: 4′ 26″ (or 4′.42)......................	—12 44
The angle required.......................................	7 55 56

TABLE 11. THE ECLIPTIC'S LONGITUDE CORRESPONDING TO EVERY 10′ OF ITS DECLINATION.

TABLE 11, of the ZODIACAL series is the converse of the SOLAR TABLE 6, and is computed from the simple formula ⊙'s long. $= \frac{\text{radius. sin ⊙'s dec.}}{\text{sin obliquity.}}$

When the declination is N increasing, the longitude will be in the first quadrant, and will be indicated by the tabular quantity; when it is north decreasing, the longitude will be in the second quadrant, and will be = 180° diminished by the tabular quantity; when south increasing, the longitude will be in the third quadrant, and the tabular quantity must be increased by 180°; and when south decreasing, the longitude will be in the fourth quadrant, and will equal 360° diminished by the tabular quantity.

When the sun is near either of the equinoxes, where his declination varies quickly, his longitude may be deduced from his observed declination, or meridian altitude in a known latitude, with considerable precision.

Example 1.—In the Nautical Almanac of 1824, the sun's declination for the 1st of March is given 7° 28′ 50″ south decreasing, let his corresponding longitude be determined?

Zod. Tab. 11	7° 20′	18° 41′ 45″.2
	As 10′ : diff. 1579″.4 :: 8′ 50″ : ..	+ 23 15 .1
	As 100″ : 78″.1 :: dim. ob. 8″.7 :	+ 0 5 .4
	Tabular quantity .... (sub.)	19 5 5 .7
	From	360 0 0
	Sun's longitude by Table	340 54 54 .3 or 11^s 10° 54′ 54″.3
	Sun's longitude by Nautical Almanac	11^s 10 54 55

If we substitute 7° 28′ 49″.55 for the sun's declination, as determined by our Solar Table 6. (Example 2. p. 352) instead of 7° 28′ 50″, the longitude will come out 11^s 10° 55′ 55″.38; hence, in this instance, an error of half a second in the measure of the declination will produce a corresponding error in the determination of the longitude of upwards of a second.

By the formula.

Radius	10 .0000000
Sine 7° 28′ 50″	9 .1145767
	19 .1145767
Sine 23° 27′ 48″.3	9 .6000615
Sine Tab. Num. 19° 5′ 7″	9 .5145152

The same may be done more expeditiously by adding the log. cosec. of the obliquity to the log. sine of the dec.

Hence the longitude is only 11^s 10° 54′ 53″ by the formula, which, indeed, is most to be depended on with Taylor's logarithms.

TABLE 12. THE ECLIPTIC'S RIGHT ASCENSION IN SPACE TO EVERY 10′ OF ITS DECLINATION.

When the sun's longitude corresponding to his observed declination has been determined by the preceding Table, his right ascension may be had from our Solar Table 5; but as this proceeding would require two processes, we have computed the present Table, to give the right ascension in arc, corresponding to the sun's declination at one opening. The precepts with respect to the tabular quantities in the four quadrants, are the same as in Table 11.

Formula.—Cos. longitude × sec. declination = cos. right ascension.

Example.—Required the sun's right ascension on the 1st of March, 1824, his declination being 7° 28′ 50″ south decreasing?

Zod. Tab. 12.	7° 20′	17° 14′ 42″.2
	As 10′ : 1473″.4 :: 8′ 50″ :	+ 21 41 .5
	As 100″ : 85 .1 diff. :: 8″.7 dim. obliq....	+ 7 .4
	Subtract tab. num.	17 36 31 .1
	From	360 0 0
	R. A. in arc	242 23 28 .9 = 22^h 49^m $33^s.9$
	By the Nautical Almanac	22 49 34 .1

By the formula.

Cos.... 19° 5′ 7″ (tab. num. for long.) .... log. 9 .9754470
Sec. dec. 7 28 50 (ar. co. cos.) .... log. 0 .0037121

Cos. R. A. 17° 36′ 31″ or 342° 23′ 29″ .... log. 9 .9791591
In time 22ʰ 49ᵐ 33 .9

TABLE 13. THE MOON'S APPROXIMATE PARALLAX IN LONGITUDE, ETC.

When the reader was instructed how to find the Longitude and Altitude of the Nonagesimal degree at any given time, by means of Zodiacal Table 7, he was left in suspense as to their utility, which will now appear. The moon's parallax, both in longitude and latitude, depends on the place of the nonagesimal degree, as it regards the moon's place at any specified time, and the present Table is so constructed, as to require the Altitude of the Nonagesimal point to be one of its Arguments, and the moon's Distance in Longitude from the said point to be the other, in order that the tabular quantity may give the parallax in longitude *nearly*, which it would do at the first operation, provided the moon's *apparent* distance from the nonagesimal point were previously known; but as the *true* calculated distance, supposed to be seen from the centre of the earth, is not that which appears to an inhabitant of the earth, an *approximate* parallax must be computed and applied to the moon's calculated distance from the nonagesimal point, before correct *data* can be had, for computing the true quantity of parallax, and by means of it the moon's *apparent* longitude; which is that required to be known in eclipses and occultations.

According to La Lande and Biot, the moon's parallax in long. $= \frac{\text{hor. par.} \times \text{sin alt. nonag.} \times \text{sin ☾'s dist. from nonag.}}{\text{cos ☾'s latitude.}}$

The Table before us (13) contains the product of the sines of the altitude and distance of the nonagesimal, on a supposition that the horizontal parallax is 60′; Table 15, gives the augmentation for the moon's latitude; and Table 17, gives the correction for an increase or decrease in the horizontal parallax from 60′. By means of these three Tables, the *approximate* parallax in longitude is obtained, and applied to the moon's *true* distance, to obtain nearly the *apparent* distance; and with this apparent distance, as a new argument, the whole operation is repeated, to procure the *correct* parallax.

The parallax in longitude *increases* the longitude of the moon, when she is to the *east* of the nonagesimal point, but decreases it, when she is to the west; and must be applied accordingly. The horizontal parallax of the moon, which must be used with Table 17, is the equatorial horizontal parallax reduced to the latitude of the place of observation by the aid of Lunar Table 4; because the nonagesimal point given in Zodiacal Table 7, is for the *reduced* latitude of Greenwich, when allowance is made for the earth's compression; but if the computations were made on a supposition that the earth is a perfect sphere, then the reduction of the latitude should be omitted, and the parallax in the sphere only would be had by the same process.

When Table 13 was constructed, it was intended to apply only to the determination of the parallax in longitude, in which *sines* only are concerned; and the numbers in the side argument from 10° to 70° were limited by the length of the page to 60, in which those were omitted which were the least likely to occur; but, on examining the formulæ for the computation of parallaxes in latitude, right ascension, and declination, it was discovered that the Table already computed (of sines only) might be equally useful for determining these also, provided such arguments only were adopted, as consist of *sines:* and in consequence of this suggestion Table 13 was completed from 1° to 10°, and from 70° to 90° inclusive, in the form of an Appendix, at pages 257—260; and the two parts must consequently be considered as one general Table for different purposes, when used with appropriate arguments. But as it is a matter of indifference, which argument applies at the head, and which at the side of the Table, it will seldom happen that both arguments cannot be used without turning to the appendix,

Example.—Let it be required to find the moon's parallax in longitude, at Greenwich, when her longitude was 104° 28′ 0″.33, her latitude 1° 12′ 50″.355 N., her equatorial horizontal parallax 61′ 21″.15, longitude of the nonagesimal point of the ecliptic 32° 27′ 39″.5, altitude of do. 47° 9′ 37″.57, and the compression $\frac{1}{314\cdot 3}$?

☾'s longitude .... 104° 28′ 0″.33
Longitude of nonages.............. 32 27 39 .50

☾'s dist. from nonages............ 72 0 20 .83

☾'s equat. hor. paral.................. 61′ 21″.15
Reduction for $\frac{1}{314\cdot 3}$ (Lunar Tab. 4) ... — 7 .15

Hor. par. for Greenwich.............. 61 14 .0

	72°	73°	Diff.	
Zod. Tab. 13....... 47°	41′ 44″.01	41′ 57″.82	13″.81	As 60′ : 13″.81 :: 20″.83 : 0″.08
48	42 24 .38	42 38 .42	14 .04	As 60 : 14 .04 :: 20 .83 : 0 .08

Then ...... 41′ 44″.01 + 0″.08 = 41′ 44″.09 } diff. = 40″.37. As 60′ : 40″.37 :: 9′ 37″.57 : 6″.48
42 24 .38 + 0 .08 = 42 24 .46 }

Hence...... 41 44 .01 + 0 .08 + 6 48 = 41′ 50″.57 = approx. paral. in long. for 60′ hor. par.
Tab. 17. Augmentation for 1′ 14″ excess..... + 51 .61
Tab. 15. Augmentation for ☾'s true latitude.. + 0 .58

Approximate parallax in longitude.......... 42 42 .76 to be added to ☾'s dist. from nonages.
☾'s calculated distance from nonages. .... 72° 0 20 .83

☾'s corrected distance from nonages.... = 72 43 3 .59 to be used as the new argument.

Repetition of the process with the corrected argument.

As 60′ : 13″.81 :: 43′ 3″.59 : 9″.91	41′ 44″.01 + 9″.91 = 41′ 53″.92 } diff. = 40″.54
As 60 : 14 .04 :: 43 3 .59 : 10″.08	42 24 .38 + 10 .08 = 42 34 .46 }
As 60 : 40 .54 :: 9 37 .57 : 6 .50	41 53 .92 + 6 .50 = 42 0 .42 = par. for 60′ hor. par.

Tab. 17. Augmentation for excess in hor. par. 1′ 14″............ + 51 .81
Tab. 15. Augmentation for apparent latitude of moon .... + 0 .11

Moon's corrected parallax in longitude..................... + 42 52 .34
Moon's true longitude given .... 104° 28 0 .33

Moon's apparent longitude .... 105 10 52 .67

As this corrected parallax does not exceed the approximate parallax more than 10″, it may be taken as the true parallax, without a second repetition. For the mere purpose of getting the approximate parallax in longitude *nearly*, it would have been sufficient to have taken what the Tables 13 and 17 offer from simple inspection, viz. 42′ and +50″, which would have pointed out the difference, in round numbers, between the moon's real and apparent longitude, and consequently would have furnished a ready estimate of the *probability* of an eclipse or occultation taking place, which is the first object of research. In this case, the moon's longitude exceeds the longitude of the nonagesimal, or is to the east; therefore the parallax in longitude must be *added* to her true longitude, to produce her *apparent* longitude; and consequently the time of an apparent conjunction will happen before the true calculated time, as many hours and parts nearly, as the moon's hourly motion at the time is contained in the true parallax. If the parallax is to be taken corresponding to the *observed* longitude, with a view to correcting the Lunar Tables, the parallax in longitude must be applied with a contrary sign, to gain the true longitude.

In this example the *apparent* latitude of the moon could not be had, till the moon's approximate parallax in latitude had been computed, as in the following example; but as the parallaxes, both in longitude and latitude, are equally concerned in an occultation, no inconvenience will arise from both being computed conjointly.

By the formula.

Hor. par. of moon............................	3674″.0	.... log. 3 .5651392
Sin. alt. of nonagesimal	47° 9′ 37″.57	.... 9 .8652583
Sin. moon's distance from nonagesimal	72 0 20 .83	.... 9 .9782206
Sec. moon's *true* latitude..................	1 12 50 .355	.... 10 .0000975
Moon's *approx. paral. in longitude*..........	+ 42 42 .81	.... 3 .4087156
Moon's apparent distance from nonagesimal..	72 43 3 .62	

Repetition with more correct data.

Hor. par. of moon	3674″.0	.... log. 3 .5651392
Sin. alt. of nonagesimal....................	47° 9′ 37″.57	.... 9 .8652583
Sin. apparent distance from nonagesimal.....	72 43 3 .64	.... 9 .9799363
Sec. moon's *apparent* latitude	31 12 .78	.... 10 .0000179
Moon's TRUE PARALLAX in LONGITUDE	+ 42 52 .48	.... 3 .4103517
Do. by the TABLES	+ 42 52 .34	

TABLE 14. THE MOON'S APPROXIMATE PARALLAX IN LATITUDE, ETC.

THE moon's parallax in latitude consists of two parts;

Part I. = ☾'s hor. par. × cos. alt. of nonages. × cos ☾'s apparent lat.

Part II. = ☾'s hor. par. × sin ☾'s appar. lat. × sin alt. of nonages. × cos ☾'s dist. from nonages.

This Table (14) comprehends the moon's hor. par. × cos. alt. of nonages. for the horizontal parallax 60′. The following Table (15) contains the diminution arising from cos. moon's latitude, and Table 17 gives the correction for the excess or defect of horizontal parallax compared with 60′: the amount of these three tabular quantities, composing Part I., or principal part of the parallax in latitude, applied to the moon's true latitude, will give her *approximate* latitude, or apparent latitude *nearly*, with which the process must be repeated to get Part I. correctly. With respect to Part II. of the parallax in latitude, which is comparatively small, it is contained in Table 16, one of the arguments being the approximate tabular number taken from Table 13, and the other the moon's approximate latitude just determined. Then the sum or difference, according to circumstances, of the *two parts* above specified, will be the moon's parallax in latitude; and this parallax applied to the moon's true latitude, with the proper sign, will give the apparent latitude wanted.

The reason why the tabular number, from Table 13. is made one of the arguments for Table 16, for the second portion of the moon's parallax in latitude, is, that the formula for Table 13, viz. "hor. par. × sin alt. nonag. × sin ☾'s dist. from the nonages." is also the formula for Part II. of the parallax in latitude, when the term "× sin ☾'s lat." is taken away, which therefore will make the other argument, provided that the sine of the complement of the moon's distance be used as an argument in Table 13, instead of the sine of the distance itself, for gaining the tabular number as an argument for Table 16. Hence the tabular quantity of Table 16 comprises the product of the entire formula for Part II., without any additional Table. With respect to the application of the signs to Part II. of the parallax in latitude, if the distances of the moon from the nonagesimal and from the elevated pole of the ecliptic, be both above or both under 90°, the sign − must be applied, but if one distance be above and the other below 90°, then + will be the proper sign.

The parallax thus determined will increase the moon's latitude when it is south, at a place in northern latitude, but will decrease it when north; its effect being to depress the true place in altitude.

Example.—Let the parallax in latitude be determined under the same circumstances as the parallax in longitude has been determined in the preceding example?

ZOD. TAB. 14.....47° 0′	40′ 55″.19	
47 10	40 47 .53	40′ 47″.53
	As 10′ : diff. 7 .66 :: 22″.43 :	+ 0 .29
	Tabular approx. paral. in lat. for 60′ hor. par.	40 47 .82
TAB. 17..........	Augmentation for excess of hor. par. 1′ 14″	+ 50 .32
	Approx. parallax for 61′ 14″ hor. par.	41 38 .14
TAB. 15..........	Diminution for latitude 1° 12′ 50″..................................	− 0 .56
	Approximate parallax in latitude (Part I.).........................	41 37 .58

Moon's true lat. 1° 12′ 50″.35 − 41′ 37″.58 = 31′ 12″.77 = apparent latitude, to be used in the repeated operation for par. in longitude.

Repetition for Part I.

Moon's approximate parallax in latitude for 61′ 14″ hor. par....................	= 41′ 38″.14
Diminution for apparent latitude 31′ 12″.77..................................	− 0 .10
Part I. of the corrected parallax in latitude....................................	41 38 .04

The head argument for entering Table 16, to obtain Part II. of the parallax in latitude, must now be thus obtained;

Moon's correct distance from the nonagesimal..........................	= 72° 43′ 3″.59
The *sine* of its complement is the sine of................................	17 16 56 .41

which must therefore be taken as the head argument, in entering Table 13, to obtain the Tabular number for the head argument of Table 16, which in fact is the approximate parallax in longitude procured by the complement of the moon's distance from the nonagesimal, instead of the distance itself.

Operation for obtaining Part II.

		17°	18°	diff.	
Zod. Tab. 13...	47°	12′ 49″.78	13′ 33″.60	43 .82	As 60′ : 43″.82 :: 16′ 56″.41 : 12″.37
	48	13 2 .19	13 46 .72	44 .53	As 60 : 44 .53 :: 16 56 .41 : 12 .57

Then....... 12′ 49″.78 + 12″.37 = 13′ 2″.15 } diff. = 12″.61. As 60′ : 12″.61 :: 9′ 37″.57 : 2″.02
13 2 .19 + 12 .57 = 13 14 .76 }
Hence...... 13 2 .15 + 2 .02 = 13 4 .17 = tabular number for 60′ of hor. parallax.

Tab. 17. Augmentation for 1′ 14″ of excess = 16″.08 to be added to 13′ 4″.17, consequently 13′ 20″.25 is the head argument for Table 16, with which and the moon's apparent latitude *nearly* 31′ 12″.77 we obtain

From Table 16, Part II................................	— 0′	7″.27
Part I. before determined	41	38 .04
The true parallax in latitude..............................	41	30 .77

The second part is here subtracted from the first, because the moon's distance from the nonagesimal, and also from the north pole of the ecliptic, are both acute angles: and the parallax here determined must apply to the moon's true latitude with a *negative* sine, because that latitude is north.

Thus, moon's true latitude	1° 12′ 50″.35
Parallax in latitude	— 0 41 30 .77
Moon's true apparent latitude	0 31 19 .58

This example has been worked with a degree of minuteness that will seldom be necessary; for if 13′ had been taken for the tabular number (in Table 13), for the head argument of Table 16, as being the nearest minute, the second part of the parallax in latitude would have come out the same within a *second*. Indeed, in cases where the latitude is very small, and particularly in eclipses, where the latitude *must be* small, the second part of the parallax may be disregarded, as of little or no amount.

By the formula. *Previous operation with the moon's true latitude.*

Horizontal parallax of moon	3674″.0	log. 3 .5651392
Cos. altitude of nonagesimal....................	47° 9′ 37″.57	9 .8324756
Cos. moon's true latitude.......................	1 12 50 .35	9 .9999025
Moon's approximate parallax in latitude	— 41 37 .57 or 2497″.57....	3 .3975173
Moon's approximate latitude....................	0 31 12 .78	

Operation repeated with the apparent latitude nearly.

Horizontal parallax of moon.........................	3674″.0	log. 3 .5651392
Cos. altitude of nonagesimal	47° 9′ 37″.57	9 .8324756
Cos. appar. lat. nearly............................	31 12 .78	9 .9999821
Moon's parallax in lat.........................	41 38 .03 or 2498″.03	3 .3975969 = Part I.

Horizontal parallax of moon	3674″.0	log. 3 .5651392
Sin. altitude of nonagesimal	47° 9′ 37″.57	9 .8652583
Cos. moon's apparent distance	72 43 3 .64	9 .4728738
Sin. moon's appar. lat. nearly....................	31 12 .78	7 .9580568
Moon's parallax in latitude......................	7 .27	0 .8613281 = Part II.

Part I. of moon's parallax in latitude	0° 41′ 38″.03
Part II. of do.	− 7 .27
Moon's whole parallax in latitude subtract	0 41 30 .76
Moon's true latitude given	1 12 50 .35
Moon's apparent latitude	31 19 .59
Do. by the Tables	31 19 .58

In the preceding examples for gaining the parallaxes in longitude and latitude, the results have been obtained by working three successive proportions, which may in general be done by two only, with nearly equal accuracy; and frequently the process may be otherwise simplified, by taking the *approximate* parallaxes from the Tables, within a few seconds, by simple inspection; as will be shown hereafter. For if the true parallaxes do not differ from the approximate more than 10″ or 12″, the operation need not be repeated.

Example 2.—On July 10, 1824, Greenwich time, the moon will be at her ecliptic conjunction with the sun at $16^h\ 20\frac{1}{4}^m$, when the longitude of the nonagesimal degree will be 22° 58′ 20″, and its altitude 42° 36′ 9″ (see the bottom of page 377); let it be required to determine her apparent longitude and latitude at the time, from the data given in the Nautical Almanac for that day?

Preparation of the data.

		S	First diff.	Second diff.	Mean of second diff.
☾'s long. July 10	Noon	9 10° 28′ 58″			
	Midnight	9 16 32 44	6° 3′ 46″		
11.	Noon	9 22 34 29	6 1 45	− 2′ 1″	− 1′ 54″
	Midnight	9 28 34 27	5 59 58	− 1 47	

As $12^h : 6° 1' 45'' :: 4^h 20^m.5$:	2° 10′ 37″ proportional part.
Equation for second diff.	+ 13 (p. 201.)
Longitude July 10, midnight	S 9 16 32 44
Longitude at $16^h 20^m 15^s$	9 18 43 34 = 288° 43′ 34″.

		North.	First diff.	Second diff.	Mean.
☾'s lat. July 10.	Noon	0° 5′ 12″			
	Midnight	0 38 43	33′ 31″		
11.	Noon	1 11 36	32 53	− 38″	− 43″.5
	Midnight	1 43 30	31 54	− 59	

As $12^h : 4^h 20 .25 :: 32' 53''$:	11′ 53″ prop. part.
Equation of second diff.	+ 5
Latitude July 10, midnight	0° 38 43
Latitude at $16^h 20^m 15^s$	0 50 41 N.

☾'s longitude	288° 43′ 34″
Long. of nonages.	22 58 20
☾'s dist. from nonages.	265 45 14

Hence the moon must be in the east.

☾'s equatorial hor. par. at midnight	54′ 35″
Do. at $16^h 20^m 15^s$	54 32
Reduction for $\frac{1}{309}$ (LUNAR TAB. 4.)	− 6 .4
Hor. paral. for Greenwich	54 25 .6

Operation for the parallaxes.

Longitude.

Tab. 13. (42° 36′ and 85° 45′) .... + 40′ 30″
(Supplement of) ☾'s distance .... 85° 45 14

Apparent distance *nearly* 86 25 44

Tab. 13.	86°	87°	Diff.	Mean diff.
42°	40′ 3″.00	40′ 5″.57	2″.57	2″.59
43	40 49 .21	40 51 .82	2 .61	
Diff.	46 .21	46 .25		46 .23

As 60′ : 46″.23 :: 36′ 9″ : .... 27″.85
As 60 : 2 .59 :: 25 44 : .... 1 .12
First tabular number .... 40 3 .00

For 60′ hor. par. .... 40 31 .97
Tab. 17. for diff. 5′ 34″.4 .... − 3 45 .96
Tab. 15. augmentation for lat. .... + 0 0 .22

True parallax in long. .... 36 46 .23

By the formula.

Hor. par. 3265″.6 .... log. 3 .513963
Sin. alt. nonag. .... 42° 36′ 9″ .... 9 .830520
Sin. appar. dist. .... 68 25 44 .... 9 .999156
Sec. ☾'s appar. lat. .... 0 10 37 ....10 .000002

2206″.2 = 36′ 46″.2 .... 3 .343641

True long. .... S 9 18° 43′ 34″
Parallax .... + 36 46 .2

Apparent longitude .... = 9 19 20 20 .2

True lat. 0 50 41
Parallax .... − 40 3 .2

Apparent latitude .... = 10 37 .8 N.

Latitude.

Comp. of ☾'s dist. = 3° 34′ 16″, Alt. nonag. = 42° 36′ 9″
Tab. 18. Alt. nonag. 42° .... 44′ 35″.32
43 .... 43 52 .87

As 60 : diff. 42″.45 :: 36′ 9″ : .... 25 .57
44′ 35″.32 − 25″.57 .... = 44 9 .75
Tab. 17. diff. 5′ 34″.4 hor. par. .... − 4 6 .09

Part I. approx. par. for 54′ 25″.6 .... 40 3 .66
Tab. 15. dim. for 10′ 37″ lat. .... − .01

Part I. of paral. in dec. .... 40 3 .65
☾'s appar. lat. nearly 50′ 41″ − 40′ 4″ .... = 10 37

Tab. 13. with 3° 34′ and 42° 36′ gives arg. ...... 2 .32
Tab. 16. with 2′ 32″ and 10′ 37″ Part II. .. − 0 .40

Part. I. .... 40 3 .65
Part II. .... − 0 40

Paral. in lat. .... = 40 3 .25.

By the formula.

Hor. par. 3265″.6 .... log. 3 .513963
Cos. alt. nonag. 42° 36′ 9″ .... 9 .866900
Cos. ☾'s appar. lat. 10′ 53″ .... 9 .999998

Part I. 2403″.6 = 40′ 3″.6 .... 3 .380861

Hor. par. 3265″.6 .... 3 .513963
Sin. alt. .... 42° 36′ 9″ .... 9 .830520
Cos. ☾'s appar. dist. ... 86 25 44 .... 8 .794366
Sin. ☾'s appar. lat. .... 10 53 .... 7 .458187

Part II. .... − 0″.39 .... 9 .597036
Part I. .... 40 3 .6

40 3 .21 = paral. in latitude.

TABLES 15, 16, and 17. CORRECTIONS OF THE MOON'S PARALLAX IN LONG. AND LAT.

These Tables are so connected with Tables 13 and 14, that their respective constructions and uses will be best understood by a reference to the examples which have been given for determining the moon's parallaxes in longitude and latitude, in which examples their respective applications have been sufficiently explained and exemplified.

The *formulæ* for constituting the four successive Tables, that are used in connexion, may be thus recapitulated; viz.

Table 13. = hor. par. × alt. nonages. × sin ☾'s distance from nonages.
14. = hor. par. × cos alt. nonages.
15. = cos ☾'s apparent latitude.
16. = sin ☾'s apparent latitude.
17. reduces the parallaxes derived from hor. par. 60′ to their proper quantities, corresponding to other horizontal parallaxes.

TABLE 18. THE MOON'S PARALLAX IN RIGHT ASCENSION, ETC.

We have already seen that Zodiacal Table 13, consisting of *sines* in both of its arguments, may be applied to various purposes, provided that the arguments consist of no other denomination; the moon's approximate parallax in right ascension and in declination may therefore be deduced from the same Table, by using the sines of different appropriate arcs as the arguments, which may be taken indifferently at the head or side.

The formula for giving the approximate parallax in right ascension for the sphere may be reduced into this expression, viz. par. in R. A. = hor. paral. × sin co-lat. × sin ☾'s horary ∠ × sec. dec. ☾. Table 13, will give (sin co-lat. × sin true horary ∠) for 60′ of horizontal parallax, and the product arising from the remaining term (sec. declin. ☾), will be had from Table 18, which contains the tabular quantities arising from the whole. The parallax thus procured must be modified by Table 17, to suit the given horizontal parallax, accordingly as it is more or less than 60′. In the eastern hemisphere its sign is −, but in the western +.

Example.—Let it be required to determine the moon's parallax in right ascension on the 10th of July, 1824, at $16^h\ 9^m\ 30^s$ Greenwich time, according to the data given in the Nautical Almanac?

Previous determination of the arguments.

			First Diff.	Second Diff.	Mean of second Diff.
☾'s R. A. July 10, 1824.	Noon	281° 23′ 47″			
	Midnight	287 51 35	6° 27′ 48″		
				− 9′ 5″	
July 11.	Noon	294 10 18	6 18 43		− 9′ 12″
				− 9 19	
	Midnight	300 19 42	6 9 24		

As 12^h : 6° 18′ 43″ :: $4^h\ 9^m\ 30^s$: ... 2° 11′ 14″ the prop. part to be added.

R. A. at midnight on the 10th	287 51 35
Approximate R. A.	290 2 49
Lunar Tab. 11. Mean of second diff.	+ 1 1 .17
☾'s true R. A. at $16^h\ 9^m\ 30^s$	290 3 50 .17
Right ascension of mid-heaven	352 38 34 .50
☾'s *true horary angle*	62 34 44 .33

The moon will therefore be in the *western* hemisphere, with her equat. hor. paral. = 54′ 28″.

☉'s R. A. 10th	$7^h\ 18^m\ 21^s.5$
11th	7 22 26 .3
Diff. in 24^h	4 4 .8
As 24^h : $4^m\ 4^s.8$ :: $16^h\ 9^m.5$:	2 42 .8
☉'s R. A.	= 7 21 4 .3
Hour angle from noon	16 9 30
R. A. of mid-heaven	23 30 34 .3
Do. in arc	= 352° 38′ 34″.50

			First Diff.	Second Diff.	Mean.
☾'s declin. July 10, 1824.	Noon	22° 57′ 45″ S.			
	Midnight	21 47 50	1° 9′ 55″		
				+ 14′ 17″	
July 11.	Noon	20 23 38	1 24 12		+ 13′ 32″
				+ 12 47	
	Midnight	18 46 39	1 36 59		

As 12^h : 1° 24′ 12″ :: $4^h\ 9^m\ 30^s$:	− 25′ 49″.7
Tab. of second diff.	− 1 32 .8
☾'s dec. at midnight	21 47 50
☾'s true declination S.	21 23 33 .1

☉'s dec. July 10. Noon	22° 14′ 28″ N.
11. Do.	22 6 38 N.
As 24^h : −7′ 50″ :: $16^h\ 9^m\ 30^s$:	− 5 16 .4
☉'s true declination	22 9 11 .6 N.

Tabular operation (for the sphere).

Arguments for Tab. 13.		62°	63°	diff.	mean diff.
Co-lat. = 38° 31′ 20″	38°	32′ 36″.95	32′ 54″.81	17″.86	18″.06
Horary ∠ ☾ 62 34 44	39	33 20 .37	33 38 .63	18 .26	
	diff.	43 .42	43 .82		43 .6

As 60′ : 43″.6 :: 31′ 20″ : .... 22″.74 proportion for co-lat.
As 60 : 18 .06 :: 34 44 : .... 10 .44 proportion for hor. ∠
First tabular quantity .... 32′ 36 .95 to be added.

Approximate paral...................... 33 10 .13 for the argument of Tab. 18.

Arguments for Tab. 18.			21°	22°	diff.	
Approx. paral....	33′ 10″.13	33′	35′ 20″.87	35′ 35 .50	14″.63	14″.85 mean diff.
☾'s declin......	21° 23 33	34	36 25 .14	36 40 .21	15 .07	
			1 4 .27	1 4 .71		64 49 mean diff.

As 60″ : 64″.69 :: 10″.13 : .... 10″.89 proportion for approx. paral.
As 60′ : 14″.85 :: 23′ 33″ : ...... 5 .83 proportion for ☾'s declin.
First tabular quantity .... 35′ 20 .87 to be added.

Parallax in R. A. .... 35 37 .59 for 60′ of horiz. paral.
Tab. 17. With 60′ − 54′ 28″ = 5′ 32″ gives − 3 17 .07 to be subtracted.

Parallax in the sphere by the Tables 32 20 .52 for 54′ 28″ horiz. paral.

Instead of using Table 17, the same result may be obtained by direct proportion thus, As 60′ : 54′ 28″ :: 35′ 37″.59 : 32′ 20″.52.

By the formula (with the same arguments) for the sphere.

Horiz. paral. 54′ 28″ = 3268″.......................... log. 3 .5142820
Sine 38° 31′ 20″ (co-lat.)................................ 9 .7943612
Sine 62 34 44 (horary ∠ ☾)........................... 9 .9482397
Secant 21 23 33 .1 (declination)...........................10 .0310020

Parallax in R. A. 1940″.4 = 32′ 20″.4....................... 3 .2878849
By the Tables............ 32 20 .5

The parallax for the sphere, determined by either the Tables, or by the formula, must now be applied to the moon's *true* horary angle, to obtain the *apparent* horary angle; thus 62° 34′ 44″ + 32′ 20″.5 = 63° 7′ 4″.5; because the moon is to the west of the meridian.

Also the moon's *true* declination may be converted into the *apparent*, by applying the parallax in declination (obtained as in the subsequent example) thus 21° 23′ 33″ + 33′ 9″ = 21° 56′ 42″ S. The latitude of the place, and the moon's equatorial horizontal parallax must now be *reduced*, according to the Lunar Tables 4 and 6, and then, according to Mayer's plan, the parallax in the spheroid may be had from the usual formula for the sphere.

For the spheroid (according to Mayer).

Sin. hor. par. reduced 3261″.6.......................... log. 3 .5134307
Sin. co-lat. (from reduced lat.) 38° 42′ 12″..................... 9 .7960801
Sin. apparent horary angle... 63 7 4 .5..................... 9 .9503351
Sec. true dec. .... 21 23 33 .... 0 .0310020

Paral. in R. A. 1953″.7 = 32′ 33″.7.......................... 3 .2908479

The *true* declination is here used agreeably to Biot's formula [Astron. I. 268.].

Instead of reducing the latitude and equatorial horizontal parallax La Lande proposed to increase the latter according to this formula*, viz. Augmentation = hor. par. × sine α × tang. latitude, where α is the angle of the vertical with radius; and then to apply the usual formula for the sphere thus;

For the augmentation, hor. par. 60′ = 3600′	log.	3.5563025
Sin. α 10′ 52″		7.4998217
Tang. lat. 51° 28′ 40″		10.0990491
Augmentation ... 14″.295 for 60′ hor. par.		1.1551733

For the spheroid (According to La Lande).

Augmented parallax 3261″.6 + 14″.3 = 3275″.9		log. 3.5153306
Sin. co-lat. of Greenwich	38° 31′ 20″	9.7943612
Sin. apparent horary angle	63 7 4.5	9.9503351
Sec. true dec.	21 23 33	0.0310020
Parallax in R. A. 1954″.4 or...	32′ 34″.4	3.2910289

For the spheroid by the Tables.

MAYER's method.

Tab. 13, with 38° and 63° gives	32′ 54″.81
p. p. for diff. of co-lat. 42′ 12″	+ 30.95
p. p. for diff. of hor. ∠ 7′ 4″.5	+ 2.05
Argument for 60′ hor. par.	33 27.81
Tab. 17. diff. hor. par. 5′ 38″.4	− 3 8.58
Argument for Tab. 18	30 19.23
Tab. 18, with 33′ and 21° gives	32′ 8″.06
p. p. for diff. 19″.23	+ 20.67
p. p. for diff. 23′ 33″	+ 5.31
Paral. with reduced hor. par.	32 34.04

LA LANDE's method.

Tab. 13, with 38° and 63° gives	32′ 54″.81
p. p. for diff. of co-lat. 31′ 20″	+ 22.98
p. p. for diff. of hor. ∠ 7′ 4″.5	+ 2.05
Argument for Tab. 18	33 19.84
Tab. 18, with 33′ and 21° gives	35′ 20″.87
p. p. for diff. 19″.84	+ 21.32
p. p. for diff. 23′ 33″ of dec.	+ 5.31
Paral. for hor. par. 60′	35 47.50
Tab. 17. for diff. 5′ 24″.1 may come here	− 3 13.38
Parallax in R. A.	32 34.12

* According to La Lande's method (first proposed by Chairaut), the moon's place, as seen from the centre of the spheroid, is obtained by two operations. If a vertical line demitted from any point O, on the earth's surface, be prolonged so as to meet the polar axis in a point K, the centre of the earth being C, then the parallax due to the point O, is transferred by the augmentation N K, to the point K, by the formula here given; and is afterwards modified for the centre C, by means of the small Tables of REDUCTION inserted at page 256.

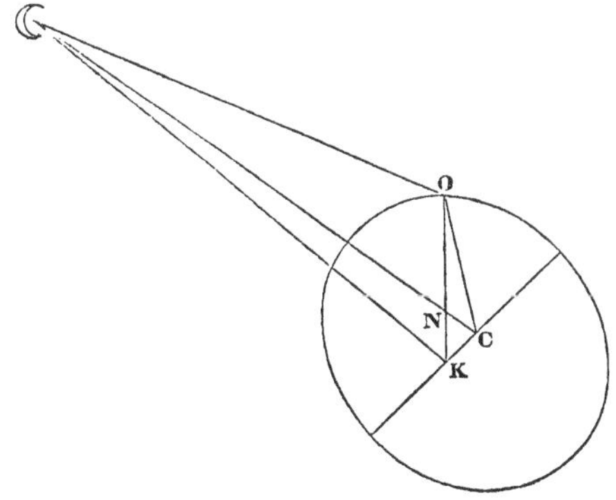

Moon's true right ascension given................	290° 3′ 50″
Parallax in R. A. add (☾ being west)	0 32 34
Moon's apparent R. A.	290 36 24 = $19^h\ 22^m\ 25^s.6$

If the *apparent* declination had been used, the parallax in R. A. would have been 32′ 42″ by either of the two methods.

The moon's parallax in right ascension and in declination will be found useful on various occasions, but particularly when the apparent difference of right ascension and declination of the moon and of a star, or planet, has been measured at the time of its nearest appulse, by a micrometer attached to an equatorial telescope.

In the notes subjoined to Tables 18 and 19, and also in the appendix to Table 13, the *moon's* lat. and co-lat. are by mistake printed for the lat. and co-lat. of the *place*.

TABLE 19. THE MOON'S PARALLAX IN DECLINATION, ETC.

The moon's parallax in declination is composed of two parts, one positive and the other negative, viz. for the spheroid, according to Mayer,

1. *Hor. paral.* × *sine reduced latitude of place* × *sine* ☾*'s polar distance.*
2. − *hor. par.* × *sine* ☾*'s dec.* × *sine reduced co-lat. of place* × *sine comp. of horary* ∠.

The former of these may be obtained from Table 13, by using one of the terms as one argument, and the other term as the other. The latter, or negative portion, is obtained partly from Table 13, and partly from Table 19, at present under consideration; the sine of the reduced latitude, and the sine of the complement of the moon's horary angle must be the arguments for Table 13, and the resulting *tabular quantity* (called in this Table *approx. par. in declin.*) must be used as the side argument of the Table before us, while the moon's declination is used as the head argument. The difference between the two portions thus obtained will be the moon's parallax in declination in the *spheroid*. But if the latitude be not previously reduced (by Lunar Table 6.) the parallax will be for the *sphere*. It may also be proper to remark here, that the argument "approx. par. in declin." at the head of the first column, is that which arises from Table 13, by using two terms of the *negative* portion only.

If La Lande's plan of using the *augmented* horizontal parallax, without reducing the latitude, be preferred, Table 13 will be equally serviceable, and then the final correction, taken from the Table of reduction (23), must be applied, to give the parallax in the spheroid, agreeably to the directions hereafter given for the use of the small Tables of reduction.

Example.—Let it be required to determine the moon's parallax in declination on the 10th of July, 1824, at $16^h\ 9^m\ 30^s$ Greenwich time, according to the data given in the Nautical Almanac?

The arguments have been determined for the last example; viz. The ☾'s dec. = 21° 23′ 33″ S., the reduced lat. = 51° 17′ 48″, ☾'s true horary ∠ = 62° 34′ 44″, ☾'s polar distance at $16^h\ 9^m\ 30^s$ = 68° 36′ 27″, and equatorial horiz. paral. = 54′ 28″, do. reduced 54′ 21″.6.

According to Mayer.

By the Tables.

First part.

	68°	69°	Diff.	Mean diff.
Tab. 13. 51	43′ 14″.01	43′ 31″.91	17″.90	18″
52	43 50 .27	44 8 .42	18 .15	
	36 .26	36 .51		36 .38

As 60′ : 36″.38 :: 17′ 48″ :	+ 10″.8
As 60 : 18″ :: 36′ 27″ :	+ 11 .0
First tabular quantity	43′ 14 .0
For 60′ hor. par.............................	43 35 .8
Tab. 17. for 5′ 38″.4 diff. of hor. par...........	− 4 5 .9
First portion	+39 29 .9

By the formula, for the spheroid.

First Part.

Hor. par. 3261″.6	log. 3 .5134307
Sine reduced lat. .... 51° 17′ 48″	9 .8923139
Sine true polar dist... 68 36 27.............	9 .9689980
The first portion 2370″......................	3 .3747426

Or + 39′ 30″, as by the Tables nearly.

Second Part (approx.) *by the Tables.*

	27°	28°	Diff.	Mean.
Tab. 13. 38°	16′ 46″.22	17′ 20″.53	34 .31	34 .69
39	17 8 .54	17 43 .61	35 .07	
	22 .32	23 .08		22 .7

As 60′ : 22″.7 :: 42′ 12″ : .... 15″.96
As 60 : 34 .69 :: 25 16 : .... 14 .63
First tabular number .... 16′ 46 .22

17 16 .81
Tab. 17. for 5′ 38″.4 diff. of hor. par. .... —1 37 .49

Argument for Tab. 19 .... 15 39 .32

	21°	22°	Mean diff.
Tab. 19..... 15′	5′ 22″.53	5′ 37″.14	15 .10
16	5 44 .03	5 59 .62	
	21 .50	22 .48	21 .99

As 60″ : 21″.99 :: 39″.32 : .... 14″.41
As 60′ : 15″.10 :: 23′ 33″ : .... 5 .93
First tabular number .... 5 22 .53

Second portion .... — 5 42 .87
First portion .... 39 29 .90

Approx. paral. .... 33 47 .03

Second part, repeated.

	26°	27°	Mean diff.
Tab. 13..... 38°	16′ 11″.60	16′ 46″.22	35 .5
39	16 33 .15	17 8 .54	

Mean diff. .... 21 .9

As 60′ : 21″.9 :: 42′ 12″ : .... 15″.4
As 60 : 35 .5 :: 52 34 : .... 31 .1
First tabular number .... 6 11 .6

Arg. for Tab. 19. .... 16 58 .1

	21°	22°	Mean diff.
Tab. 19..... 16	5′ 44″.03	5′ 59″.62	16″.07
17	6 5 .54	6 22 .10	

Mean diff. .... 21″.99

As 60″ : 21″.99 :: 58″.1 .... 0′ 21″.29
As 60 : 16 .07 :: 52′ 34″ : .... 0 14 .08
First tabular number .... 5 44 .03

The sum .... 6 19 .4
Tab. 17. diff. 5′ 38″.4 .... — 35 .6

First part .... 39 29 .9

Second part .... — 5 43 .8
True par. in dec. .... 33 46 .1

Second Part (approx.) *by the formula.*

Hor. par. 3261″.6 .... log. 3 .5134307
Sine ☾'s true dec. .... 21° 23′ 33″ .... 9 .5620011
Sine red. co-lat. .... 38 42 12 .... 9 .7960801
Sine compl. true hor. ∠ 27 25 16 .... 9 .6632550

Second portion 342″.58 .... 2 .5347669

First portion .... 0° 39′ 30″
Second do. .... 0 5 42 .6

Approx. paral. .... 0 33 47 .4
True declination 21 23 33

Apparent declination .... 21 57 20 .4

Second part repeated with appar. dec.

Hor. par. 3261″.6 .... log. 3 .5134307
Sine apparent dec. ...... 21° 57′ 20″ .... 9 .5727410
Sine reduced co-lat. .... 38 42 12 .... 9 .7960801
Sine comp. appar. hor. ∠ 26 52 34 .... 9 .6551984

344″.5 .... = — 5′ 44″.5 2 .5371870
First portion .... 39 30 .0

True par. in dec. .... 33 45 .5
True dec. .... 21 23 33 .0

Appar. dec. .... = 21 57 18 5 S.

If the *first part* should also be repeated with the *apparent* polar distance 68° 2′ 40″ substituted for the true polar distance, it would be found = 39′ 21″, and the paral. in dec. thus determined would be 39′ 21″ — 5′ 44″ = 33′ 37″.

According to La Lande.

By the formula, with the augmented parallax.

Augmented hor. par. 3261″.6 + 14″.3 = 3275″.9		log. 3 .5153306
Sine latitude Greenwich	51° 28′ 40″	9 .8934133
Sine *apparent* polar distance ☾	68 3 18	9 .9673350
First part 2377″.3 = 39′ 37″.3		3 .3760789
Aug. hor. par. 3275″.9		3 .5153306
Sine apparent dec. ☾	21° 57′ 20″.................	9 .5727410
Sine co-lat. Greenwich	38 31 20	9 .7943612
Sine comp. apparent horary ∠ ☾	26 52 34	9 .6551984
Second part 344″.7 .... = − 5′ 44″.7		2 .5374480
First part ....39 37 .3		
Parallax in dec. ....33 52 .6 as seen from K.		

The Tables will give the same, according to this method.

The equation (or reduction) +16″.9, taken from Table 24., must yet be applied to the last determined parallax, according to La Lande, to reduce it from the point K to the centre C; which will greatly increase its difference from the result derived according to Mayer's method: and this difference would have been still greater, if the *true* polar distance and declination had been used for the *apparent*.

Dr. T. Young, when speaking of Mayer's method, observes, that "supposing the moon, as may easily happen, to be considerably to the north of the east or west, this correction (arising from the *reduction of the latitude*) will be not merely superfluous, but absolutely erroneous, since in truth a smaller correction of an opposite nature is required. When the moon is due east, or due west, her altitude is not affected by the obliquity of the surface; since the perpendicular to the meridian is obviously parallel to the surface of the sphere: and when she passes beyond these points, her apparent altitude, instead of being diminished by the ellipticity, is actually increased by it."* On this account, and because it has long been a complaint, that the different methods do not bring out the same result, we have contrasted the two methods of Mayer and La Lande in this Example; which contrast, we trust, may induce the skilful theorist to reconcile the discrepancies. At the same time, it appears that these Tables will apply to any of the methods, provided that the arguments for Table 13 are always composed of *sines*. In *right ascension* no reduction (or equation) is required for the earth's compression.

In the work of the preceding Examples, both in right ascension and declination, we have presented to the eye all the minutiæ for explaining the triple and double proportions, and have repeated the operations with the corrected arguments, to obtain the true results; which process appears formidable; but the repetitions may be avoided, by correcting the arguments in the first instance, by the approximate parallaxes taken by simple inspection from the Tables, and applied to the *true* places, given as the data of the Example; a few seconds more or less, in the *arguments*, being of little importance, as they affect the parallaxes to be determined therefrom in a very small degree. The subjoined Example will explain this proposed abridgment of the work, both in right ascension and declination, the parallaxes in which are best determined conjointly.

Example 2.—Given the moon's true horary angle 97° 32′ 59″.44 east of the meridian; the moon's true declination 23° 53′ 3″.95 north; her horizontal parallax adapted for the latitude 61′ 14″; and the reduced latitude of the place 51° 17′ 59″; to determine the moon's parallax in right ascension and declination?

* *The Quarterly Journal of Science, Literature, and the Arts,* Vol. X., p. 412.

Paral. in R. A. according to Mayer.

Reduced hor. par. 3674" .... log.	3 .5651392
Sin. co-lat. reduced .... 38° 42′ 1″. ...	9 .7960512
Sec. true dec. .... 23 53 3 .95...	10 .0388809
Sum of three logarithms	3 .4000713
Sin. true horary ∠ .... 97 32 59 .44...	9 .9962187
Approx. paral. ...... 2490″.52 = 41 30 .52...	3 .3962900
Sin. appar. horary ∠ .... 98 14 29 .96...	9 .9954914
Sum of three logarithms	3 .4000713
True paral. .... 2486″.35 = 41 26 .35...	3 .3955627

Paral. in Dec. according to Mayer.

Reduced hor. par. 3674″ .... log.	3 .5651392
Sin. red. lat. .... 51° 17′ 59″. ...	9 .8923325
Sin. true polar dis. .... 66 6 56 .05...	9 .9611191
Part. I. .... 2621″.75 = 43 41 .75...	3 .4185908
Red. hor. par. 3674″	3 .5651392
Sin. true dec. ☾ .... 23 53 3 .95...	9 .6073405
Sin. red. co-lat. .... 38 42 1 .0 ...	9 .7960512
Sin. comp. true horary ∠ ..... 7 32 59 .44...	9 .1185579
	2 .0870888
Part II. 122″.20... = − 2′ 2″.20	
Part I. .... 43 41 .75	
True paral. .... 41 39 .55	

Paral. in R. A. according to La Lande.

Augmentation for Greenwich.	Hor. par. 3674″	3 .5651392
	Sin. α 10′ 52″	7 .4998217
	Tang. lat. .... 51° 28′ 40″. ...	10 .0990491
Augmentation..............	14 .59...	1 .1640100
Augmented par. 3688″.59		3 .5668601
Sin. co-lat. .... 38° 31′ 20″. ...		9 .7943612
Sin. appar. horary ∠ .... 98 14 29 .96...		9 .9954914
Sec. true dec. .... 23 53 3 .95...		10 .0388809
True paral. .... 2486″.53 = 41 26 .53...		3 .3955936

Paral. in Dec. according to La Lande.

☾'s augmented par. 3688″.59	3 .5668601
Sin. latitude .... 51° 28′ 40″. ...	9 .8934133
Sin. true polar dist. .... 66 6 56 .05...	9 .9611191
Part I. 2638″.72 .... = 43 58 .72...	3 .4213925
☾'s augmented par. 3688″.59	3 .5668601
Sin. ☾'s true dec. .... 23 53 3 .95...	9 .6073405
Sin. co-lat. .... 38 31 20 ...	9 .7943612
Sin. comp. true horary ∠ ..... 7 32 59 .44...	9 .1185579
Part II. 122″.21 = − 2′ 2″.21	2 .0871197
Part I. .... 43 58 .72	
41 56 .51	
Reduction Table 24.. −17 .04	
True paral. .... 41 39 .47	

In the work of this example the arguments consist of the *true* quantities, except in the instance of the horary angle in right ascension, which should always be the *apparent* horary angle. The two methods appear thus to bring out the same results.

TABLES 20 TO 25. REDUCTION OF THE MOON'S PARALLAXES.

It has been stated in the note at page 389, that La Lande's method of determining the parallaxes of the moon by means of her augmented horizontal parallax, requires two operations; first, it gives the parallax as seen from a point K in the polar axis intersected by a vertical, and then reduces such parallax to C, the centre of the earth. The small Tables now presenting themselves, and numbered 20 to 25, with the exception of 23, were computed from La Lande's formulæ, for the purpose of giving those reductions, together with an azimuthal correction, when the parallactic method of computing eclipses and occultations is adopted.

The formulæ from which the Tables are computed are as follow, viz.:

Altitude..........	Tab. 20	= hor. par. × sin. α × cos. azim. × sin. alt. ☾.
Longitude........	21	= hor. par. × sin. α × sec. lat. of place × sin. obliq. × cos. ☾'s long.
Latitude	22	= hor. par. × sin. α × sec. lat. of place × cos. obliq. × sec. ☾'s lat.
[Right Ascension...	23	= hor. par. × sin. α × sec. lat. of place × sin. ☽'s dec.]
Declination.......	24	= hor. par. × sin. α × sec. lat. of place × cos. ☾'s dec.
Azimuth.........	25	= hor. par. × sin. α × sin. azimuth.

The correction contained in Table 20 is occasioned by the moon's parallax in azimuth arising from the earth's compression, which affects the moon's altitude more or less in certain situations. When the moon is situated between the prime vertical and the elevated pole, the tabular correction for altitude in the spheroid is additive; but in all other cases, subtractive.

The most simple application of Tables 21, 22, and 24, is, to consider them as *diminishing* the parallaxes due to the point K; but, as La Lande applies them to the longitude, latitude, and declination of the moon, under limited circumstances, we will describe his modes of application.

The correction for longitude in Table 21, in all northern latitudes, must be subtracted from the true longitude of the moon, as seen from the earth's centre, whenever the north declination is diminishing, or the south declination increasing; but otherwise must be added. In southern latitudes, the application is just the reverse.

In northern latitudes, the correction in Table 22 must be subtracted, when the moon's latitude is north, unless the moon be between the zenith and elevated pole, in which particular case, it must be added; but when the moon's latitude is south, the correction is additive. In southern latitudes, the contrary. In the computation of this Table, the cosine was accidentally used as a factor, instead of the secant, but if the tabular number be taken = 16″.8, as a *constant quantity*, the difference in the results will be imperceptible.

Table 23 was computed from its supposed analogy to Table 21, but its tabular corrections are superfluous; which circumstance was not observed in time to prevent the Table being printed.

The correction given in Table 24 must be added to the moon's declination if south, but subtracted if north, provided the observer is in a northern latitude; but if he is in a southern latitude, the reverse must take place.

Table 25 is useful when the parallactic method of computation is used. The correction applies to the true difference of the azimuths of the moon and other body, as observed in a given latitude. When the moon is to the north of the sun's or star's vertical, the correction is additive; but if to the south of the said line, it is subtractive; supposing the observer in a northern latitude. [Astronomie par M. De La Lande, Tome II. p. 509.]

We have already given an instance of the application of the reduction from Table 24 to the parallax of the moon in declination; and the following additional example for determining her parallax in longitude and latitude, will afford the means of applying the reductions from Tables 21 and 22, as well as of comparing the results of Mayer's and La Lande's methods of determining the parallaxes of these denominations. When Mayer's method is used, the altitude of the nonagesimal and moon's distance therefrom, must be derived from the reduced latitude according to Zodiacal Table 7, and the reduced horizontal parallax must be applied, as before directed; but when La Lande's method is adopted, the altitude of the nonagesimal and moon's distance must be computed from the apparent or unreduced latitude, and the augmented horizontal parallax must be substituted for the reduced horizontal parallax, as was the case when the parallax in right ascension and declination were determined by La Lande's process; after which, the reductions from Tables 21 and 22 must be respectively applied.

		°	′	″
Given....	The reduced horizontal parallax.................	0	61	14.
	Altitude of the nonagesimal.....................	47	9	37 .57
	Moon's true distance from nonagesimal	72	0	20 .83
	Moon's true latitude...........................	1	12	50 .35
	Reduced latitude of the place...................	51	17	59

Required.. The parallax in longitude and latitude, according to Mayer's method?

Paral. in Longitude according to Mayer.

Reduced hor. par. 3674″ ....log.		3 .5651392
Sin. true alt. nonag...........	47° 9′ 37″.57...	9 .8652583
Sec. moon's true lat...........	1 12 50 .35...	10 .0000975
Sum of three logs.		3 .4304950
Sin. ☾'s true dist. from nonag.	72 0 20 .83...	9 .9782206
Approx. par. 2562″.81 ......=	+42 42 .81...	3 .4087156
Sin. app. dist. from nonag.	72 43 3 .64...	9 .9799363
Sum of three logs.		3 .4304950
Moon's true paral.....2572″.95 =	42 52 .95...	3 .4104313

Paral. in Latitude according to Mayer.

Reduced hor. par. 3674″ ....log.		3 .5651392
Cosin. true alt. nonag.	47° 9′ 37″.57...	9 .8324756
Cosin. ☾'s true lat.	1 12 50 .35...	9 .9999025
Part I. .... 2497″.57 =	41 37 .57...	3 .3975173
Reduced hor. par. 3674″		3 .5651392
Sin. true alt. nonag...........	47 9 37 .57...	9 .8652583
Cos. true dist. from nonag.....	72 0 20 .83...	9 .4898474
Sin. ☾'s true lat............	1 12 50 .35...	8 .3260591
Part II. of par.	−17 .63..	1 .2463040
Part I.	41 37 .57	
True paral..................	41 19 .94	

Given......The ☾'s augmented parallax......3688″.59

True alt. of nonagesimal	47° 0′ 37″.30
True dist. of ☾ from nonagesimal	71 52 29 .96
☾'s true latitude	1 12 50 .35
Latitude of place (apparent)	51 28 40

Required...The parallax in longitude and latitude according to La Lande's method?

Paral. in Longitude according to La Lande.

		log.
Augmented paral. 3688″.59	log.	3 .5668601
Sin. true alt. nonag.	47° 0′ 37″.3	9 .8642007
Sec. ☾'s true lat.	1 12 50 .35	10 .0000975
Sum of three logs.		3 .4311583
Sin. ☾'s true dist. from non.	71 52 29 .96	9 .9778973
Approx. par...... 2564″.81 =	+42 44 .81	3 .4090556
Sin. appar. dist. nonag.	72 35 14 .77	9 .9796279
Sum of three logs.		3 .4311583
Paral..... 2575″.05 =	42′ 55″.05	3 .4107862
Reduction (Tab. 21.)	−1 .83	
True paral.	42 53 .22	

Paral. in Latitude according to La Lande.

		log.
Augmented paral. 3688″.59	log.	3 .5668601
Cos. true alt. nonag.	47° 0′ 37″.3	9 .8336991
Cos. ☾'s true latitude	1 12 50 .35	9 .9999025
Part I...... 2514″.56 =	41 54 .56	3 .4004617
Augmented paral. 3688″.59		3 .5668601
Sin. true alt. nonag.	47° 0′ 37″.3	9 .8642007
Cos. ☾'s true dist. nonag.	71 52 29 .96	9 .4928879
Sin. ☾'s true latitude	1 12 50 .35	8 .3260591
Part II	−17″.78	1 .2500078
Part I	41 54 .56	
Paral.	41 36 .78	
Reduction	−16 .84 (Tab. 22.)	
True paral.	41 19 .94	

Comparison of the Parallaxes.

	Longitude.	Latitude.	Right. Ascen.	Declination.
Mayer	42′ 52″. 95	41′ 19″. 94	41′ 26″. 35	41′ 39″. 55
La Lande	42 53 . 22	41 19 . 94	41 26 . 53	41 39 . 47
Difference	+0 0 . 27	0 0 . 00	+0 0 . 18	−0 0 . 08

RECAPITULATION of the *formulæ* constituting the TABLES 13, 17, 18, and 19, when applied to determine the moon's parallaxes in right ascension and declination.

For parallax in right ascension.

Table 13 = hor. par. × sin. co-lat. of place × sin. ☾'s horary ∠ = A.
18 = A × secant ☾'s declination.

For parallax in declination.

13 = hor. par. × sin. lat. of place × sin. ☾'s polar dist. = Part I.
13 = hor. par. × sin. co-lat. of place × sin. comp. ☾'s horary ∠ = B.
19 = B × sin. ☾'s declination = Part II.
17 reduces the parallaxes derived from hor. par. 60′ to their proper quantities, corresponding to other horizontal parallaxes, as before explained in page 386.

As the longitude and altitude of the nonagesimal made use of in the preceding example, when worked according to La Lande's method, are those that belong to the sphere, it may be acceptable to the reader to see the operation here, by which they were ascertained; especially as ZODIACAL TABLE 7, from which the data for Mayer's method are derived, was computed for the spheroid. For this purpose we have as data the right ascension of the zenith = 8° 18′ 7″.88, the apparent latitude of the place = 51° 28′ 40″, and the obliquity of the ecliptic = 23° 27′ 49″.76, as given in Example 1., at page 377.

COMPUTATION OF THE LONGITUDE AND ALTITUDE OF THE NONAGESIMAL FOR THE SPHERE.

Cosin. R. A. of the zenith	8° 18′ 7″.88	log. 9.9954247
Cosin. latitude	51 28 40 .00	9.7943612
Cosin Arc I.	51 57 15 .15	9.7897859
Co-tang. lat.	51 28 40 .00	9.9009509
Sin. R. A. zenith	8 18 7 .88	9.1595492
Co-tang. Arc. II.	83 26 33 .95	9.0605001
Obliq. of ecliptic (sub.)	23 27 49 .76	
Sin. Arc. III.	59 58 44 .19	9.9374384
Sin. Arc. I.	51 57 15 .15	9.8962607
Cosin. ALT. OF NONAGES.	47 0 37 .30	9.8336991
Cosin. Arc. III.	59 58 44 .19	9.6992463
Tang. Arc. I.	51 57 15 .15	10.1064749
Tang. LONG. OF NONAGES.	32 35 30 .37	9.8057217

Moon's longitude given	104° 28′ 0″.33
Longitude of the nonagesimal	32 35 30 .37
Moon's true distance from the nonages.	71 52 29 .96

In page 392, we had occasion to notice Dr. Thomas Young's remark on the incompetency of Mayer's method of deducing the lunar parallaxes truly in certain situations of the moon, and have therefore computed a couple of Tables for giving the azimuthal correction of altitudes taken on the spheroid, agreeably to the formula suggested by this ingenious and learned author; which Tables, together with a Table of the mean longitudes and latitudes of Dr. Maskelyne's 35 Stars, and five small Tables for giving the corrections in longitude and latitude, we propose to introduce in this place, as being connected with the subject before us. The application of the different parallaxes to the computation of eclipses, occultations, &c., will be reserved until the subject is treated of practically in our second volume.

TABLE A. Azimuthal Correction of Parallax for the Spheroid. (Mr. J. Utting.)
Compression $\frac{1}{300}$.

Azimuth of the object.	Argument. The latitude of the place.									
	45° 0′	46° 40′	48° 20′	50° 0′	51° 40′	53° 20′	55° 0′	56° 40′	58° 20′	60° 0′
	Or, the angle of the vertical with radius.									
	11′ 8″.62	11′ 7″.62	11′ 4″.33	10′ 58″.83	10′ 51″.09	10′ 41″.12	10′ 29″.00	10′ 14″.73	9′ 58″.36	9′ 39″.98
0° 0′	11′ 8″.62	11′ 6″.49	10′ 59″.84	10′ 48″.82	10′ 33″.54	10′ 14″.19	9′ 51″.07	9′ 24″.46	8′ 54″.71	8′ 22″.28
1 40	11 8.34	11 6.21	10 59.56	10 48.85	10 33.27	10 13.93	9 50.82	9 24.22	8 54.49	8 22.07
3 20	11 7.49	11 5.36	10 58.72	10 47.72	10 32.47	10 13.15	9 50.07	9 23.50	8 53.81	8 21.43
5 0	11 6.08	11 3.95	10 57.33	10 46.35	10 31.13	10 11.85	9 48.82	9 22.31	8 52.68	8 20.37
6 40	11 4.10	11 1.98	10 55.38	10 44.43	10 29.26	10 10.03	9 47.07	9 20.64	8 51.10	8 18.88
8 20	11 1.56	10 59.45	10 52.87	10 41.97	10 26.85	10 7.70	9 44.83	9 18.49	8 49.07	8 16.97
10 0	10 58.47	10 56.37	10 49.81	10 38.96	10 23.92	10 4.86	9 42.09	9 15.88	8 46.59	8 14.65
11 40	10 54.81	10 52.72	10 46.21	10 35.42	10 20.45	10 1.50	9 38.86	9 12.79	8 43.67	8 11.90
13 20	10 50.60	10 48.53	10 42.05	10 31.33	10 16.46	9 57.63	9 35.13	9 4.24	8 40.30	8 8.70
15 0	10 45.84	10 43.78	10 37.35	10 26.71	10 11.95	9 53.26	9 30.93	9 5.22	8 36.49	8 5.16
16 40	10 40.53	10 38.49	10 32.12	10 21.56	10 6.93	9 48.38	9 26.24	9 0.74	8 32.25	8 1.18
18 20	10 34.68	10 32.66	10 26.35	10 15.89	10 1.38	9 43.01	9 21.06	8 55.80	8 27.57	7 56.78
20 0	10 28.30	10 26.30	10 20.04	10 9.69	9 55.33	9 37.15	9 15.42	8 50.41	8 22.47	7 51.99
21 40	10 21.38	10 19.40	10 13.22	10 2.98	9 48.78	9 30.79	9 9.31	8 44.57	8 16.94	7 46.79
23 20	10 13.94	10 11.98	10 5.87	9 55.76	9 41.73	9 23.96	9 2.73	8 38.29	8 10.98	7 41.20
25 0	10 5.98	10 4.05	9 58.02	9 48.03	9 34.18	9 16.64	8 55.69	8 31.57	8 4.62	7 35.22
26 40	9 57.50	9 55.60	9 49.65	9 39.81	9 26.15	9 8.86	8 48.20	8 24.42	7 57.84	7 28.85
28 20	9 48.52	9 46.65	9 40.79	9 31.09	9 17.64	9 0.61	8 40.26	8 16.83	7 50.66	7 22.11
30 0	9 39.04	9 37.20	9 31.44	9 21.90	9 8.66	8 51.90	8 31.88	8 8.83	7 43.08	7 14.99
31 40	9 29.07	9 27.26	9 21.60	9 12.22	8 59.22	8 42.73	8 23.07	8 0.42	7 35.10	7 7.50
33 20	9 18.62	9 16.84	9 11.29	9 2.08	8 49.31	8 33.15	8 13.83	7 51.59	7 26.75	6 59.65
35 0	9 7.70	9 5.96	9 0.51	8 51.48	8 38.97	8 23.11	8 4.17	7 42.37	7 18.01	6 51.44
36 40	8 56.32	8 54.61	8 49.27	8 40.43	8 28.18	8 12.65	7 54.11	7 32.76	7 8.91	6 42.89
38 20	8 44.48	8 42.81	8 37.59	8 28.95	8 16.96	8 1.78	7 43.64	7 22.77	6 59.44	6 33.99
40 0	8 32.19	8 30.56	8 25.46	8 17.03	8 5.32	7 50.49	7 32.78	7 12.40	6 49.62	6 24.77
41 40	8 19.48	8 17.89	8 12.92	8 4.69	7 53.27	7 38.81	7 21.54	7 1.66	6 39.45	6 15.21
43 20	8 6.34	8 4.79	7 59.95	7 51.94	7 40.82	7 26.74	7 9.93	6 50.57	6 28.94	6 5.34
45 0	7 52.79	7 51.28	7 46.58	7 38.79	7 27.98	7 14.30	6 57.95	6 39.13	6 18.10	5 55.16
46 40	7 38.83	7 37.37	7 32.81	7 25.25	7 14.76	7 1.48	6 45.61	6 27.35	6 6.94	5 44.68
48 20	7 24.50	7 23.08	7 18.66	7 11.33	7 1.17	6 48.31	6 32.94	6 15.25	5 55.48	5 33.70
50 0	7 9.78	7 8.41	7 4.14	6 57.05	6 47.23	6 34.79	6 19.93	6 2.82	5 43.71	5 22.86
51 40	6 54.70	6 53.38	6 49.26	6 42.42	6 32.94	6 20.94	6 6.60	5 50.10	5 31.65	5 11.53
53 20	6 39.27	6 38.00	6 34.03	6 27.45	6 18.32	6 6.77	5 52.96	5 37.07	5 19.31	4 59.94
55 0	6 23.50	6 22.28	6 18.47	6 12.15	6 3.38	5 52.28	5 39.02	5 23.76	5 6.70	4 48.09
56 40	6 7.41	6 6.24	6 2.59	5 56.53	5 48.14	5 37.50	5 24.80	5 10.17	4 53.83	4 36.01
58 20	5 51.01	5 49.89	5 46.40	5 40.62	5 32.59	5 22.43	5 10.30	4 56.33	4 40.71	4 23.68
60 0	5 34.31	5 33.25	5 29.92	5 24.41	5 16.77	5 7.09	4 55.53	4 42.23	4 27.36	4 11.14
61 40	5 17.33	5 16.32	5 13.16	5 7.93	5 0.68	4 51.49	4 40.52	4 27.89	4 13.78	3 58.38
63 20	5 0.08	4 59.12	4 56.13	4 51.19	4 44.33	4 35.65	4 25.27	4 13.33	3 59.98	3 45.42
65 0	4 42.57	4 41.67	4 38.66	4 34.20	4 27.75	4 19.57	4 9.80	3 58.55	3 45.98	3 32.27
66 40	4 24.83	4 23.98	4 21.35	4 16.98	4 10.93	4 3.27	3 54.11	3 43.57	3 31.79	3 18.94
68 20	4 6.86	4 6.07	4 3.62	3 59.55	3 53.91	3 46.76	3 38.23	3 28.40	3 17.42	3 5.44
70 0	3 48.68	3 47.95	3 45.68	3 41.91	3 36.68	3 30.06	3 22.16	3 13.06	3 2.88	2 51.79
71 40	3 30.31	3 29.64	3 27.55	3 24.07	3 19.28	3 13.19	3 5.92	2 57.55	2 48.19	2 37.99
73 20	3 11.76	3 11.15	3 9.24	3 6.08	3 1.70	2 56.15	2 49.52	2 41.89	2 33.01	2 24.05
75 0	2 53.05	2 52.50	2 50.78	2 47.93	2 43.97	2 38.96	2 32.98	2 26.09	2 18.39	2 10.00
76 40	2 34.19	2 33.70	2 32.17	2 29.63	2 26.10	2 21.64	2 16.31	2 10.17	2 3.31	1 55.83
78 20	2 15.21	2 14.78	2 13.43	2 11.20	2 8.11	2 4.20	1 59.52	1 54.14	1 48.13	1 41.57
80 0	1 56.10	1 55.73	1 54.58	1 52.67	1 50.01	1 46.65	1 42.64	1 38.02	1 32.85	1 27.22
81 40	1 36.90	1 36.60	1 35.63	1 34.03	1 31.82	1 29.02	1 25.66	1 21.81	1 17.50	1 12.80
83 20	1 17.62	1 17.37	1 16.60	1 15.32	1 13.55	1 11.30	1 8.62	1 5.53	1 2.08	0 58.31
85 0	0 58.27	0 58.09	0 57.51	0 56.55	0 55.22	0 53.53	0 51.51	0 49.20	0 46.90	0 43.78
86 40	0 38.88	0 38.75	0 38.37	0 37.73	0 36.84	0 35.71	0 34.37	0 32.82	0 31.09	0 29.20
88 20	0 19.45	0 19.38	0 19.19	0 18.87	0 18.43	0 17.86	0 17.19	0 16.42	0 15.55	0 14.61
90 0	0 0.00	0 0.00	0 0.00	0 0.00	0 0.00	0 0.00	0 0.00	0 0.00	0 0.00	0 0.00
	Arg. Latitude of the place.									
	45° 0′	43° 20′	41° 40′	40° 0′	38° 20′	36° 40′	35° 0′	33° 20′	31° 40′	30° 0′

TABLE B. Proportional Logarithms of the Azimuthal Correction of Parallax for the Spheroid. (Mr. J. Utting.)

Compression $\frac{1}{300}$.

Azimuth.	Latitude of place.										Azimuth.
	45° 0′	46° 40′	48° 20′	50° 0′	51° 40′	53° 20′	55° 0′	56° 40′	58° 20′	60° 0′	
0° 0′	1 .2048	1 .2055	1 .2078	1 .2115	1 .2167	1 .2234	1 .2318	1 .2419	1 .2536	1 .2673	0° 0′
1 40	1 .2050	1 .2057	1 .2079	1 .2116	1 .2169	1 .2236	1 .2320	1 .2420	1 .2538	1 .2675	1 40
3 20	1 .2055	1 .2063	1 .2085	1 .2122	1 .2174	1 .2242	1 .2326	1 .2426	1 .2544	1 .2680	3 20
5 0	1 .2065	1 .2072	1 .2094	1 .2131	1 .2183	1 .2251	1 .2335	1 .2435	1 .2553	1 .2689	5 0
6 40	1 .2078	1 .2085	1 .2107	1 .2144	1 .2196	1 .2264	1 .2348	1 .2448	1 .2566	1 .2702	6 40
8 20	1 .2094	1 .2102	1 .2124	1 .2161	1 .2213	1 .2281	1 .2364	1 .2465	1 .2583	1 .2719	8 20
10 0	1 .2115	1 .2122	1 .2144	1 .2181	1 .2233	1 .2301	1 .2385	1 .2485	1 .2603	1 .2739	10 0
11 40	1 .2139	1 .2146	1 .2168	1 .2205	1 .2257	1 .2325	1 .2409	1 .2509	1 .2627	1 .2763	11 40
13 20	1 .2167	1 .2174	1 .2196	1 .2233	1 .2285	1 .2353	1 .2437	1 .2537	1 .2655	1 .2791	13 20
15 0	1 .2199	1 .2206	1 .2228	1 .2265	1 .2317	1 .2385	1 .2469	1 .2569	1 .2687	1 .2823	15 0
16 40	1 .2234	1 .2242	1 .2264	1 .2301	1 .2353	1 .2421	1 .2505	1 .2605	1 .2723	1 .2859	16 40
18 20	1 .2274	1 .2282	1 .2304	1 .2341	1 .2393	1 .2461	1 .2544	1 .2645	1 .2763	1 .2899	18 20
20 0	1 .2318	1 .2326	1 .2348	1 .2385	1 .2437	1 .2505	1 .2588	1 .2689	1 .2807	1 .2943	20 0
21 40	1 .2366	1 .2374	1 .2396	1 .2433	1 .2485	1 .2553	1 .2636	1 .2737	1 .2855	1 .2991	21 40
23 20	1 .2419	1 .2426	1 .2448	1 .2485	1 .2537	1 .2605	1 .2689	1 .2789	1 .2907	1 .3043	23 20
25 0	1 .2475	1 .2483	1 .2505	1 .2542	1 .2594	1 .2662	1 .2745	1 .2846	1 .2964	1 .3100	25 0
26 40	1 .2536	1 .2544	1 .2566	1 .2603	1 .2655	1 .2723	1 .2807	1 .2907	1 .3025	1 .3161	26 40
28 20	1 .2602	1 .2610	1 .2632	1 .2669	1 .2721	1 .2789	1 .2872	1 .2973	1 .3091	1 .3227	28 20
30 0	1 .2673	1 .2680	1 .2702	1 .2739	1 .2791	1 .2859	1 .2943	1 .3043	1 .3161	1 .3297	30 0
31 40	1 .2748	1 .2756	1 .2778	1 .2815	1 .2867	1 .2935	1 .3018	1 .3119	1 .3237	1 .3373	31 40
33 20	1 .2829	1 .2836	1 .2858	1 .2895	1 .2948	1 .3015	1 .3099	1 .3199	1 .3317	1 .3453	33 20
35 0	1 .2914	1 .2922	1 .2944	1 .2981	1 .3033	1 .3101	1 .3185	1 .3285	1 .3403	1 .3539	35 0
36 40	1 .3006	1 .3013	1 .3035	1 .3072	1 .3124	1 .3192	1 .3276	1 .3376	1 .3494	1 .3630	36 40
38 20	1 .3103	1 .3110	1 .3132	1 .3169	1 .3221	1 .3289	1 .3373	1 .3473	1 .3591	1 .3727	38 20
40 0	1 .3206	1 .3213	1 .3235	1 .3272	1 .3324	1 .3392	1 .3476	1 .3576	1 .3694	1 .3830	40 0
41 40	1 .3315	1 .3322	1 .3344	1 .3381	1 .3433	1 .3501	1 .3585	1 .3685	1 .3803	1 .3939	41 40
43 20	1 .3430	1 .3438	1 .3460	1 .3497	1 .3549	1 .3617	1 .3701	1 .3801	1 .3919	1 .4055	43 20
45 0	1 .3553	1 .3561	1 .3583	1 .3620	1 .3672	1 .3740	1 .3823	1 .3924	1 .4042	1 .4178	45 0
46 40	1 .3683	1 .3691	1 .3713	1 .3750	1 .3802	1 .3870	1 .3953	1 .4054	1 .4172	1 .4308	46 40
48 20	1 .3821	1 .3829	1 .3851	1 .3888	1 .3940	1 .4008	1 .4091	1 .4192	1 .4310	1 .4446	48 20
50 0	1 .3967	1 .3975	1 .3997	1 .4034	1 .4086	1 .4154	1 .4237	1 .4338	1 .4456	1 .4592	50 0
51 40	1 .4122	1 .4130	1 .4152	1 .4189	1 .4241	1 .4309	1 .4393	1 .4493	1 .4611	1 .4747	51 40
53 20	1 .4287	1 .4295	1 .4317	1 .4354	1 .4406	1 .4474	1 .4557	1 .4658	1 .4776	1 .4912	53 20
55 0	1 .4462	1 .4470	1 .4492	1 .4529	1 .4581	1 .4649	1 .4732	1 .4833	1 .4951	1 .5087	55 0
56 40	1 .4648	1 .4656	1 .4678	1 .4715	1 .4767	1 .4835	1 .4918	1 .5019	1 .5137	1 .5273	56 40
58 20	1 .4847	1 .4854	1 .4876	1 .4913	1 .4965	1 .5033	1 .5117	1 .5217	1 .5335	1 .5471	58 20
60 0	1 .5058	1 .5066	1 .5088	1 .5125	1 .5177	1 .5245	1 .5328	1 .5429	1 .5547	1 .5683	60 0
61 40	1 .5285	1 .5292	1 .5314	1 .5351	1 .5403	1 .5471	1 .5555	1 .5655	1 .5773	1 .5909	61 40
63 20	1 .5528	1 .5535	1 .5557	1 .5594	1 .5646	1 .5711	1 .5798	1 .5898	1 .6016	1 .6152	63 20
65 0	1 .5789	1 .5796	1 .5818	1 .5855	1 .5907	1 .5975	1 .6059	1 .6159	1 .6277	1 .6413	65 0
66 40	1 .6070	1 .6078	1 .6100	1 .6137	1 .6189	1 .6257	1 .6340	1 .6441	1 .6559	1 .6695	66 40
68 20	1 .6375	1 .6383	1 .6405	1 .6442	1 .6494	1 .6562	1 .6645	1 .6746	1 .6864	1 .7000	68 20
70 0	1 .6708	1 .6715	1 .6737	1 .6774	1 .6826	1 .6894	1 .6978	1 .7078	1 .7196	1 .7332	70 0
71 40	1 .7071	1 .7079	1 .7101	1 .7138	1 .7190	1 .7258	1 .7341	1 .7442	1 .7560	1 .7696	71 40
73 20	1 .7472	1 .7480	1 .7502	1 .7539	1 .7591	1 .7659	1 .7742	1 .7843	1 .7961	1 .8097	73 20
75 0	1 .7918	1 .7925	1 .7948	1 .7985	1 .8037	1 .8104	1 .8188	1 .8289	1 .8406	1 .8543	75 0
76 40	1 .8419	1 .8427	1 .8449	1 .8486	1 .8538	1 .8606	1 .8689	1 .8790	1 .8908	1 .9044	76 40
78 20	1 .8990	1 .8997	1 .9019	1 .9056	1 .9109	1 .9176	1 .9260	1 .9360	1 .9478	1 .9615	78 20
80 0	1 .9651	1 .9659	1 .9681	1 .9718	1 .9770	1 .9838	1 .9921	2 .0022	2 .0140	2 .0276	80 0
81 40	2 .0436	2 .0444	2 .0466	2 .0503	2 .0555	2 .0623	2 .0707	2 .0807	2 .0925	2 .1061	81 40
83 20	2 .1400	2 .1407	2 .1429	2 .1466	2 .1519	2 .1586	2 .1670	2 .1771	2 .1888	2 .2025	83 20
85 0	2 .2645	2 .2652	2 .2675	2 .2712	2 .2764	2 .2831	2 .2915	2 .3016	2 .3133	2 .3270	85 0
86 40	2 .4403	2 .4410	2 .4432	2 .4469	2 .4522	2 .4589	2 .4673	2 .4774	2 .4891	2 .5028	86 40
88 20	2 .7411	2 .7419	2 .7441	2 .7478	2 .7530	2 .7598	2 .7682	2 .7782	2 .7900	2 .8036	88 20
	Latitude of place.										
	45° 0′	43° 20′	41° 40′	40° 0′	38° 20′	36° 40′	35° 0′	33° 20′	31° 40′	30° 0′	

TABLE 1.

The MEAN LONGITUDES and LATITUDES of DR. MASKELYNE'S 35 STARS for 1756 and 1805.

Stars.	Longitude in 1756.	Longitude in 1805.	An. Precession in Longitude.	Var. of Long. for 10″ change of obliq.	Latitude in 1756.	Latitude in 1805.	An. Var. in Lat. (since / 1805 / before)	Var. of Lat. for 10″ change of obliq.
γ Pegasi........	5° 45′ 22″.14	6° 26′ 19″.03	50″.1406	+ 2″.23	12° 35′ 36″.79 N.	12° 35′ 41″.04	+0″.0867−	− 5″.31
α Arietis........	34 15 5.15	34 56 8.95	50.2816	+ 1.45	9 57 28.84 N.	9 57 34.38	+0.1131−	− 6.08
α Ceti...........	40 54 43.62	41 35 42.28	50.1767	− 1.67	12 36 18.62 S.	12 36 0.49	+0.3700−	− 7.04
α Tauri.........	66 22 46.46	67 3 49.57	50.2676	− 0.37	5 29 3.14 S.	5 28 49.51	+0.2782−	− 9.21
α Aurigæ......	78 26 57.47	79 7 59.37	50.2429	+ 0.80	22 51 46.23 N.	22 51 44.45	−0.0363+	− 9.87
β Orionis.......	73 25 11.99	74 6 15.03	50.2700	− 1.66	31 9 13.04 S.	31 8 49.64	+0.4776−	− 9.81
β Tauri.........	79 10 3.33	79 51 1.99	50.1767	+ 0.16	5 21 57.14 N.	5 22 10.26	+0.2678−	− 9.85
α Orionis.......	85 20 47.62	86 1 50.30	50.2588	− 0.20	16 3 33.68 S.	16 3 8.85	+0.5067−	− 9.98
α Canis Maj....	100 43 10.79	101 23 38.02	49.5353	+ 1.63	39 32 52.89 S.	39 33 39.62	−0.9537+	− 9.96
α Geminorum.	106 50 33.74	107 31 29.20	50.1114	− 0.55	10 4 32.20 N.	10 4 49.19	+0.3467−	− 9.57
α Canis Min...	112 25 10.40	113 5 43.68	49.6588	+ 1.13	15 58 6.66 S.	15 58 45.21	−0.7868+	− 9.33
β Geminorum.	109 50 55.84	110 31 28.32	49.6424	− 0.41	6 40 1.75 N.	6 40 13.47	+0.2392−	− 9.37
α Hydræ	143 53 8.05	144 34 0.76	50.0553	+ 3.36	22 23 55.23 S.	22 23 43.56	+0.2382−	− 7.46
α Leonis........	146 26 16.35	147 7 6.67	50.0065	− 0.18	0 27 25.94 N.	0 27 31.95	+0.1227−	− 5.43
β Leonis........	168 13 55.62	168 54 41.61	49.9182	− 2.13	12 17 9.02 N.	12 16 57.11	−0.2431+	− 4.20
β Virginis......	173 42 11.93	174 23 51.73	51.0163	− 0.11	0 41 36.81 N.	0 41 33.68	−0.0639+	− 1.00
α Virginis......	200 26 16.73	201 7 14.20	50.1524	+ 0.32	2 2 9.93 S.	2 2 22.26	−0.2516+	+ 3.63
α Boötis........	200 49 45.13	201 30 44.99	50.2012	− 5.55	30 54 19.87 N.	30 52 17.58	−2.4957+	+ 9.89
α² Libræ	221 40 54.93	222 21 50.47	50.1131	− 0.05	0 21 51.69 N.	0 21 28.44	−0.4745+	+ 6.74
α Coronæ......	218 51 18.47	219 32 41.38	50.6635	− 7.57	44 21 0.00 N.	44 20 41.23	−0.3831+	+10.93
α Serpentis.....	228 39 7.64	229 20 17.49	50.4051	− 3.13	25 31 44.61 N.	25 31 30.00	−0.2982+	+ 8.62
α Scorpii	246 21 23.63	247 2 22.92	50.1896	+ 0.32	4 32 12.92 S.	4 32 38.26	−0.5180+	+ 9.23
α Herculis	252 44 38.01	253 25 36.58	50.1749	− 2.18	37 18 53.78 N.	37 18 34.09	−0.4018+	+11.70
α Ophiuchi ...	259 1 50.78	259 42 56.39	50.3186	− 1.30	35 52 53.38 N.	35 52 23.11	−0.6178+	+ 9.96
α Lyræ.........	281 53 42.97	282 34 55.72	50.4643	+ 4.05	61 44 44.24 N.	61 44 34.45	−0.1998+	+10.14
γ Aquilæ......	297 32 34.89	298 13 26.52	50.0333	+ 2.87	31 16 9.58 N.	31 15 53.73	−0.3235+	+ 9.51
α Aquilæ	298 19 59.67	299 1 27.09	50.7637	+ 2.73	29 18 41.19 N.	29 18 37.77	−0.0698+	+ 9.40
β Aquilæ	299 1 51.55	299 42 43.07	50.1310	+ 2.49	26 43 4.08 N.	26 42 21.05	−0.8782+	+ 9.24
α¹ Capricor....	300 21 53.07	301 2 50.90	50.1598	+ 0.63	7 0 56.94 N.	7 0 32.71	−0.4945+	+ 8.73
α² Capricor....	300 27 0.25	301 8 4.37	50.2882	+ 0.63	6 57 9.25 N.	6 57 10.61	+0.0278−	+ 8.60
α Cygni	331 58 25.07	332 38 47.62	49.4398	+15.33	59 55 3.15 N.	59 54 54.87	−0.1690+	+17.45
α Aquarii......	329 57 0.49	330 37 59.52	50.1843	+ 1.64	10 40 28.07 N.	10 40 20.33	−0.1580+	+ 5.44
α Piscis Aust..	330 25 32.69	331 6 52.49	50.6082	− 1.59	21 6 15.02 S.	21 6 39.06	−0.4906+	+ 6.89
α Pegasi	350 5 12.30	350 46 9.68	50.1506	+ 3.48	19 24 39.49 N.	19 24 38.92	−0.0116+	+ 8.33
α Andromedæ	10 54 45.40	11 35 37.68	50.0465	+ 4.72	25 41 4.40 N.	25 41 0.04	−0.0890+	−10.99

This Table was taken from Zach's SUPPLEMENT AUX NOUVELLES TABLES, 1813. The Longitudes and Latitudes for 1756 were computed from Bradley's Observations; the Longitudes for 1805 from Maskelyne's Right Ascensions; and the Latitudes for 1805 from Piazzi's Declinations.

The Longitude and Latitude of any of these Stars may be brought up to any given time, by applying the Annual Precession to the Longitude of 1805, when increased in proportion to the time elapsed, and the Annual Variation in Latitude, increased in the same proportion, to the Latitude of 1805, in the same way that the Right Ascensions and Declinations, or North Polar Distances of Stars, are brought up from the common catalogues, by means of the GENERAL TABLE of DAILY PRECESSIONS.

TABLE 2.

ABERRATION of the STARS in LONGITUDE and LATITUDE.

Arg. for the Long. = ⊙'s long. − ∗'s long.
Arg. for the lat. = ⊙'s long. − ∗'s long. − 3 signs.

	O − VI +	I − VII +	II − VIII +	
0°	20".26	17".54	10".13	30°
1	20.25	17.36	9.82	29
2	20.24	17.18	9.51	28
3	20.23	16.99	9.20	27
4	20.21	16.79	8.88	26
5	20.18	16.59	8.56	25
6	20.14	16.39	8.24	24
7	20.10	16.18	7.91	23
8	20.06	15.96	7.59	22
9	20.01	15.74	7.26	21
10	19.95	15.52	6.93	20
11	19.88	15.29	6.59	19
12	19.81	15.05	6.26	18
13	19.74	14.81	5.92	17
14	19.65	14.57	5.58	16
15	19.56	14.32	5.24	15
16	19.47	14.07	4.90	14
17	19.37	13.81	4.56	13
18	19.26	13.55	4.21	12
19	19.15	13.29	3.87	11
20	19.03	13.02	3.52	10
21	18.91	12.75	3.17	9
22	18.78	12.47	2.82	8
23	18.64	12.19	2.47	7
24	18.50	11.91	2.12	6
25	18.36	11.62	1.77	5
26	18.21	11.33	1.41	4
27	18.05	11.03	1.06	3
28	17.88	10.73	0.71	2
29	17.71	10.43	0.35	1
30	17.54	10.13	0.00	0
	XI − V +	X − IV +	IX − III +	

To obtain the star's aberration in longitude, multiply the tabular number by the secant of the ∗'s latitude.
To obtain the star's aberration in latitude, multiply the tabular number by the sine of the ∗'s latitude.

TABLE 3.

LUNAR EQUATION of the EQUINOXES in LONGITUDE.

Arg. = Long. ☾'s ☊.

	O − VI +	I − VII +	II − VIII +	
0°	0".0	8".9	15".5	30°
1	0.3	9.2	15.6	29
2	0.6	9.5	15.8	28
3	0.9	9.7	15.9	27
4	1.2	10.0	16.1	26
5	1.6	10.2	16.2	25
6	1.9	10.5	16.3	24
7	2.2	10.8	16.4	23
8	2.5	11.0	16.6	22
9	2.8	11.2	16.7	21
10	3.1	11.5	16.8	20
11	3.4	11.7	16.9	19
12	3.7	12.0	17.0	18
13	4.0	12.2	17.1	17
14	4.3	12.4	17.2	16
15	4.6	12.6	17.3	15
16	4.9	12.8	17.3	14
17	5.2	13.1	17.4	13
18	5.5	13.3	17.5	12
19	5.8	13.5	17.5	11
20	6.1	13.7	17.6	10
21	6.4	13.9	17.6	9
22	6.7	14.1	17.7	8
23	7.0	14.3	17.7	7
24	7.3	14.5	17.8	6
25	7.5	14.6	17.8	5
26	7.8	14.8	17.9	4
27	8.1	15.0	17.9	3
28	8.4	15.1	17.9	2
29	8.7	15.3	18.0	1
30	8.9	15.5	18.0	0
	XI + V −	X + IV −	IX + III −	

TABLE 4.

SOLAR EQUATION of the EQUINOXES in LONGITUDE.

Arg. = ⊙'s long., or day of the month.

Day.	⊙'s long.	Equat.	⊙'s long.	Day.
	s	−	s	
Mar. 21	O 0°	−0".0	VI 0°	Sep. 23
26	5	0.2	5	28
31	10	0.3	10	Oct. 3
Apr. 5	15	0.5	15	9
10	20	0.6	20	14
15	25	0.7	25	19
20	I 0	0.8	VII 0	24
25	5	0.9	5	29
May 1	10	1.0	10	Nov. 3
6	15	1.0	15	8
11	20	1.0	20	13
16	25	0.9	25	18
21	II 0	0.9	VIII 0	22
27	5	0.8	5	27
June 1	10	0.7	10	Dec. 2
6	15	0.5	15	7
11	20	0.3	20	12
16	25	0.2	25	17
		+		
22	III 0	0.0	IX 0	22
27	5	0.2	5	27
July 2	10	0.3	10	Jan. 1
7	15	0.5	15	6
13	20	0.7	20	11
18	25	0.8	25	16
23	IV 0	0.9	X 0	20
28	5	0.9	5	25
Aug. 3	10	1.0	10	30
8	15	1.0	15	Feb. 4
13	20	1.0	20	9
18	25	0.9	25	14
23	V 0	0.9	XI 0	19
29	5	0.8	5	24
Sep. 3	10	0.7	10	Mar. 1
8	15	0.5	15	6
13	20	0.3	20	11
18	25	0.2	25	16
23	VI 0	0.0	XII 0	21

TABLE 5.

SECULAR VARIATION of the NORTHERN STARS in LONGITUDE.

	O − VI +	I − VII +	II − VIII +	III + IX −	IV + X −	V + XI −
0°	47".5	37".8	18".0	6".7	29".5	44".5
3	47.1	36.2	15.6	9.2	31.5	45.4
6	46.6	34.5	13.2	11.6	33.4	46.1
9	45.9	32.8	10.8	14.0	35.1	46.8
12	45.1	30.8	8.4	16.4	36.8	47.3
15	44.2	28.8	5.9	18.8	38.3	47.6
18	43.2	26.8	3.4	21.0	39.8	47.9
21	42.0	24.7	0.9	23.3	41.1	48.0
			±			
24	40.7	22.5	1.7	25.4	42.4	48.0
27	39.3	20.2	4.2	27.5	43.5	47.8
30	37.8	18.0	6.7	29.5	44.5	47.5

Multiply the tabular number by the tangent of the star's latitude, and, if it be *south*, change the signs; the product will be the secular variation.

TABLE 6.

SECULAR VARIATION of the NORTHERN STARS in LATITUDE.

	O + VI −	I + VII −	II + VIII −	III + IX −	IV + X −	V + XI −
0°	6".7	29".5	44".5	47".5	37".8	18".0
3	9.2	31.5	45.4	47.1	36.2	15.6
6	11.6	33.4	46.1	46.6	34.5	13.2
9	14.0	35.1	46.8	45.9	32.8	10.8
12	16.4	36.8	47.3	45.1	30.8	8.4
15	18.8	38.3	47.6	44.2	28.8	5.9
18	21.0	39.8	47.9	43.2	26.8	3.4
21	23.3	41.1	48.0	42.0	24.7	0.9
						∓
24	25.4	42.4	48.0	40.7	22.5	1.7
27	27.5	43.5	47.8	39.3	20.2	4.2
30	29.5	44.5	47.5	37.8	18.0	6.7

If the latitude of the star be south, change the sign in the Table; the tabular number is the secular variation.

TABLES A. AND B.

THESE two Tables are constructed from the formulæ suggested by Dr. Thomas Young in Brande's Journal above referred to, and produce the same correction of the moon's altitude in different degrees of azimuth, arising out of the earth's ellipticity, except that Table A. is computed for the compression $\frac{1}{309}$, and Table B for $\frac{1}{306}$. The arguments vary by 100′.

The formula for A. is, *correction* = α × *sin.* 2 *lat.* × *cos. azimuth;* where α denotes the angle contained between the vertical and radius taken from Lunar Table 6. (p. 185.)

The formula for B. is, *correction* = *const. log.* 1 .2048 × *co-sec.* 2 *lat.* × *sec. azimuth.*

Example for Table A.—Given the latitude of the place 48° 20′, the angle of the vertical with radius 11′ 4″.33, and azimuth of the moon 23° 20′ to determine the correction for her altitude?

With lat. 48° 20′ (or α = 11′ 4″.33) at the head of the Table, and azimuth 23° 20′ at the side, the cor. is 10′ 5″.87

By the formula.

Angle α (reduction of lat.) 664″.33........................ log. 2 .822384
Sin. 83° 20′, supplem. of 48° 20′ × 2.......................... 9 .997054
Cos. azimuth 23° 20′... 9 .962945

Correction of altitude 605″.87 = 10′ 5″.87 .... 2 .782383

Example for Table B.—Given the latitude of the place 51° 40′, the moon's azimuth 63° 20′, and the constant logarithm 1 .2048, to determine the proportional logarithm of the correction?

With lat. 51° 40′ at the head of the Table, and 63° 20′ at the side, the proper log. is 1 .5646.

By the formula.

Cosec. 76° 40′ (Sup. 51° 40′ × 2)........................ log. 10 .01187
Sec. azimuth 63° 20′... 10 .34795
Const. log... 1 .20480

Proportional logarithm required .... 1 .56462
Corresponding numbers (from a Tab. of prop. log.).... 4′ 54″ = correction.

The correction taken from either of these Tables supposes the place of the apparent zenith always brought nearer to the Pole than the true zenith, and in the case of lunar distances must be applied to the altitudes of both the moon and star. When the moon or star is north of the east or west in a northern hemisphere, the correction must be added to the altitude, but subtracted when south; and the reverse in a southern hemisphere. The altitudes thus corrected will give the true parallax, from the *reduced* equatorial parallax, by the ordinary rules, and also the refraction obtained therefrom will be without sensible error. The maximum of this correction takes place in latitude 45°, when the object has 0° of azimuth; from whence it decreases both towards the Equator and the Pole, but the scale of diminution is not the same on each side of the 45°, on account of the difference in the reduction of the latitude (LUN. TAB. 6.), which is less from 0° to 45° of lat., than it is from 45° to 90°, and in some cases an error of about 2″ may arise from this cause. The Tables were computed from 45° to 90° of latitude, as most suitable for Europe.

TABLE 1. THE MEAN LONGITUDES AND LATITUDES OF 35 STARS.

THE use of this Table, and method of applying the tabular quantities, have been explained in the notes subjoined to the Table itself, and the example given for the explanation of the five following Tables, will at the same time render the application of this sufficiently intelligible.

GENERAL TABLES 2 TO 6. CORRECTIONS OF THE MEAN LONGITUDES AND LATITUDES.

THE notes accompanying these five small Tables will explain the manner of using them, without further directions. They become necessary when the longitude and latitude of a star are derived from the *mean* right ascension and declination; but when the *apparent* right ascension and declination furnish the data, the longitudes and latitudes will not require the corrections in question.

Example.—Let it be required to determine the mean and apparent longitudes and latitudes of α Scorpii for the 1st of May, 1824, and also its secular variations in longitude and latitude, by the aid of the six preceding Tables?

LONGITUDE.

Table 1, 1805			247° 2′ 22″.92
An. preces. 50″.1896 × 19 years		+	15 53 .60
May 1 (Table XV. page 41)		+	16 .65
Mean longitude			247 18 33 .17
⊙'s longitude	S 1 11° 0′ 50″		
*'s longitude	−8 7 18 33		
Arg. Tab. 2....	6 26 17 43 Aberration	+	18 .08
☾'s ☊ for Tab. 3.	9 10 54 0 ☾ nutation	+	17 .50
May 1. Tab. 4	⊙ nutation	−	1 .00
*'s *apparent* longitude			247 19 7 .75

Tab. 5. with Arg. S 8 7° 18′ gives .... 12″.0

Nat. Tang. Lat. 4° 33′ = .0794 × 12″ = Sec. Var. 0″.9528

LATITUDE.

Tab. 1. 1805		4° 32′ 38″.26
An. Var. 0″.5180 × 19 years		− 0 9 .84
May (Tab. XV.)		− 0 0 .17
Mean latitude		4 32 28 .25
Arg. Tab. 2..... S 3 26° 17′ 43″	Aberration...	+ 8 .99
Apparent latitude		4 32 37 .24

Tab. 6. with Arg. S 8 7° 18′ gives sec. var. .... 46″.4

THE CIRCLE DIVIDED INTO 10000 PARTS.

IN many Tables, particularly of the sun and moon, it has been found convenient to suppose the circle divided into 1000, or 10000 parts, instead of signs, degrees, and minutes, &c., for the sake of expressing the arguments in figures of one common denomination, and a Table showing the value of the circular parts, in degrees, minutes, &c., is become necessary for converting the numbers given in one denomination, into those of the other, as occasion may require. Such Table is that which next presents itself for explanation. The construction of this Table is so simple, that it can hardly be misunderstood. We have had occasion to introduce its application in determining the sun's latitude from the Table in page 172, and the Examples given to explain that application in pages 357 and 358, where the common and millesimal arguments stand side by side, will render further exemplification unnecessary.

CIRCULAR ARCS.

THIS Table supposes the circle, of which radius is unity, and the circumference consequently 6 .28318530, divided into equal parts, and shows, by inspection, how many of those parts are due to any given number of degrees, minutes, and seconds: for instance, in 33° 34′ 35″, we have, ... 33° = .57595865

34′ = .00989020

35″ = .00016968

The decimal portion of the circle .... = .58601853

If the radius be of any other denomination, the corresponding parts may be had by direct proportion, thus

As 1 : number for radius :: tabular number : number required.

By reversing the operation, it may also become useful on certain occasions where astronomy is concerned; for instance, if the number of fathoms in a given arc on the surface of the earth be accurately measured, the radius of that arc may be inferred, when the corresponding celestial arc is determined.

PLANETARY TABLES.

TABLES 1 TO 6. PARALLAXES.

As the planets, like the moon, are constantly changing their respective distances from the earth, their parallaxes, like the moon's, are constantly varying; and it is of importance to be able to ascertain the parallax of a planet, for the purpose of ascertaining its true place in the heavens, as compared with that of some other body. Indeed, Tables have been lately published at Copenhagen, that exhibit the relative positions of the moon and brighter planets, Venus, Jupiter, and Saturn, from day to day, with a view of substituting these planets for the less bright stars, in taking lunar distances under circumstances when the stars cannot be seen. In operations of this nature, and also in comparing a planet's place with a contiguous star, the planetary parallaxes, as well as planetary aberrations, become essential subjects of computation. We have, therefore, introduced, after the lunar parallaxes, six small Tables, for giving the planetary parallaxes in longitude, latitude, right ascension, and declination, and have employed the same formulæ relatively, as were used in the construction of the Tables for giving the moon's parallaxes; but, on a supposition that the horizontal parallax in each case is 10″. Hence the resulting tabular parallax will require to be augmented or diminished, in the same proportion as 10″ are to the given horizontal parallax of the planet in question; and by this contrivance, the Tables become general for all the planets, notwithstanding four only of the most conspicuous are specified in the page (263) containing the Tables.

The arguments A and B, in Table 1, which contains quantities common to all the parallaxes of the different denominations, are not always in the same terms, though always *sines* of some given arc or other, that form constituent parts of each formula, as was the case with ZODIACAL TABLE 13, which, it has been seen, was rendered useful in giving the products of *sines*, used as the arguments or factors for different purposes, where sines only were employed. Indeed PLANETARY TABLE 1 gives $\frac{1}{360}$th part of the tabular quantities contained in ZODIACAL TABLE 13; or is as 10″ . 60′; but, as the quantities are small, the Planetary Table is compressed into a small compass, by making the arguments vary by 5° from one another. Whenever the sine of a degree represented by A, is not contained in the side column, it may be taken for the head argument, and B for the side, and *vice versâ*, they being of the same denomination, always sines.

The following scheme will show what constitutes the arguments A and B in each different case.

TAB. 1. In *longitude*, A = sin. alt. nonages., and B = sin. dist. from nonages.
In *latitude*, A = sin. co-alt. nonag., and B = sin. co-lat. of planet, for the first part.
A = ± sin. alt. nonag., and B = sin. co-dist. nonages, for the second part.
In *right ascension*, A = sin. co-lat. of place, and B = sin. horary angle.
In *declination*, A = sin. lat. of place, and B = sin. polar distance, for the first part.
A = − sin. co-lat. of place, and B = sin. comp. horary angle, for the second part.
TAB. 2. gives the paral. in alt. from the common *formula* (hor. par. × cos. alt.).
TAB. 3. completes the paral. in long. and is = sec. lat. of planet × approx. paral.
TAB. 4. gives the correction of Part II. of the paral. in lat. and is = sin. lat. of planet × appr. par. Part II.
TAB. 5. completes the paral. in R. A. and is = sec. of planet's dec. × approx. par.
TAB. 6. gives the correction of Part II. of the paral. in dec. and is = sin. dec. of planet × appr. par. Part II.

The quantity taken from Table 1 is called the approximate parallax in longitude, latitude, right ascension, or declination, as the case may be, and would be the true parallax respectively, if the planet had no latitude or no declination. These tabular quantities become the respective head arguments for Tables 3, 4, 5, and 6.

The horizontal parallaxes of the planets, to be used for the conversion of the tabular parallaxes into the parallaxes peculiar to each planet, must be obtained from Table A in page 265, after the tabular operations are performed, that suppose the horizontal parallax to be always 10″. Or, when the geocentric distances of the planets are known at any time, their corresponding horizontal parallaxes may be obtained by dividing 8″.7, the sun's parallax, by such distances respectively; provided that they are expressed in terms that suppose the radius of the earth's orbit to be unity.

Example.—On the 1st of November, 1824, the geocentric longitude of Saturn, according to the Nautical Almanac, will be $\overset{s}{2}$ 6° 17′, his geocentric latitude 2° 5′ S., his right ascension in time 4^h 19^m, his declination 19° 20′ N., and the time of his culmination 13^h 50^m solar time; let it be required to determine his parallax in longitude, latitude, right ascension, and declination, at 8 o'clock in the evening at Greenwich?

For the parallax in longitude.

Sun's R. A. Nov. 1, 1824	14^h	26^m	$33^s.8$	diff. = 3^m $55^s.8$
Nov. 2	14	30	29 .6	
$\frac{1}{3}$ of diff.	+	1	18 .6	
Sun's R. A. at 8 hours	14	27	52 .4	
Add	8			
R. A. of mid-heaven	22	27	52 .4	= 336° 58′

Arg. 336° 58′ (ZOD. TAB. 7) gives long. nonag. = 8° 32′ 0″.7 and alt. nonag. = 35° 34′ 58″.7
Longitude of Saturn at 8^h 66 17

Distance of Saturn from nonag................57 45

With Arg. A (alt. nonag.) 35° 35′, and Arg. B 57° 45′ (distance), TABLE 1 gives the tabular quantity for 10″ of hor. paral. = 4″.85 for approx. paral. in longitude. TABLE 3, for lat. 2° 5′, converts 10″ approx. par., into 10″.01.

Hence as 10″ : 10″.01 :: 4″.85 : 4″.855. We have, therefore, 4″.855 for 10″ hor. par.

As Saturn passes the meridian at 13^h 50^m solar time, his elongation will be nearly 5 signs, and his horizontal parallax by Table A, page 265, = 1″.01. Hence as 10″ : 4″ 855 :: 1″.01 : 0″.49.

The true parallax in longitude will, therefore, be 0″.49, which must be added to the longitude of the planet, because its position is to the east of the nonagesimal degree. (Vide page 382.)

For the parallax in latitude.

With Arg. A (co-alt. nonag.) 54° 25′, and Arg. B (co-lat. of planet) 87° 55′, TABLE 1 gives 8″.12 as the tabular quantity for Part I.

Again, with Arg. A (alt. nonag.) 35° 35′, and Arg. B (co-dist.) 32° 15′, the same Table gives 3″.08, for Part II.

TABLE 4 gives 0″.365 for 10″; and as 10″ : 0″.365 :: 3″.08 : 0″.112 for Part II.

Then, because the planet's distance from the nonagesimal is less than 90°, and its distance from the elevated pole above 90°, its latitude being south, the sum of the two parts, viz. 8″.12 + 0″.112 = 8″.232, will be the approximate parallax for 10″ of horizontal parallax, seeing the proportional part of the correction for latitude from TABLE 4 applies always to the second part.

Lastly, as 10″ : 8″.232 :: 1″.01 : 0″.8314; hence 0″.8314 is the true parallax in latitude, and must be added to the planet's true latitude, because it is south, and Greenwich is in a northern latitude. (Vide page 383.)

For the parallax in right ascension.

With Arg. A (reduced co-lat.) 38° 42′ 12″, and Arg. B (horary angle) 1^h 50^m (13^h 50^m) = 42° 30′, TABLE 1 gives 4″.11; and as 10″ : 4″.11 :: 1 .01 : 0″.41 for the approximate parallax. But Table 5 with 19° dec. gives 10″.85; therefore, as 10″ : 10″.85 :: 0″.41 : 0″.44. Hence the true parallax in *right ascension* is 0″.44, which must be subtracted, as the planet is in the eastern hemisphere. (Vide page 387.)

For the parallax in declination.

With Arg. A (reduced lat.) 51° 17′ 48″, and Arg. B. (polar dist.) 70° 40′, TABLE 1. gives 7″.30 for Part I.; also with Arg. A (co-lat. red.) 38° 42′ 12″, and Arg. B (comp. horary angle) 47° 30′, the same Table gives − 4″.58 for Part II. of the approximate parallax, for 10″ of hor. paral. Then Table 6. gives 3″.26 with dec. 19°, in column 10″, and as 10″ : 4″.58 :: 3″.26 : 1 .493; then 7″.30 − 1″.493 will be the parallax in dec. for 10″ of hor. paral., viz. 5″.80, or .580 for 1″; and consequently .580 × 1″.01 = .590 will be the true parallax in *declination*, to be subtracted from the true declination, which in this case is north.

According to the formulæ.

Longitude.

Horizontal paral. of Saturn 1″.01 ....log.	0 .004321
Sin. alt. of nonagesimal ..35° 35′	9 .764838
Sin. distance from nonag. 57 45	9 .927231
Sec. planet's lat......... 2 5	10 .000287
Planet's true *parallax in longitude* 0″.497	9 .696677

Latitude.

Horizontal parallax 1″.01	0 .004321
Cos. alt. of nonagesimal	9 .910235
Cos. latitude of Saturn	9 .999713
First part of parallax in lat. 0″.821	9 .914269

Hor. paral. 1″.01	0 .004321
Sin. alt. of nonagesimal	9 .764838
Cos. Saturn's dist. from nonag.	9 .727228
Sin. Saturn's lat.	8 .560540
Second part of paral. in lat. 0″.011	8 .056927
First part ....0 .821	
Paral. in latitude........0 .832	

Right ascension.

Horizontal paral. 1″.01	0 .004321
Sin. co-lat. (reduced).....38° 42′	9 .796049
Sin. horary angle 42 30	9 .829683
Sec. Saturn's dec.........19 20	10 .025208
True parallax in *right ascension* 0″.45	9 .655261

Declination.

Horizontal paral. 1″.01	0 .004321
Sin. reduced lat.....51° 18′	9 .892334
Sin. Polar dist......70 40	9 .974792
First part of paral. in dec. ..... 0″.744	9 .871447
Horizontal paral. 1″.01	0 .004321
Sin. dec........19° 20′	9 .519911
Cos. lat. (red)	9 .796049
Cos. horary ∠ 42 30	9 .867631
Second part of paral. in dec. −0″.154	9 .187912
Parallax in *declination*....... 0 .590	

TABLE 7. CONVERSION OF THE GEOCENTRIC LONGITUDE OF A PLANET INTO RIGHT ASCENSION IN TIME.

THIS Table consists of three parts, and was originally computed by Delambre, probably for computing the right ascension of the planets, given in the *Connoisance des Tems*, and, as these are not given for the four small planets in any of the Almanacs, except the *Milan Effemeridi*, it may be found useful for supplying the omission, provided their geocentric longitudes and latitudes have been previously computed. When the geocentric longitude of the planet is reduced to time, the equation from Part I. must be applied to it, to obtain the longitude in time corrected. The geocentric latitude of the planet, with + prefixed if north, but − if south, must, in the next place, be multiplied by the equation taken from Part II., and the product added to the longitude in time corrected, if the signs are alike, but if unlike, subtracted. Lastly, the said product multiplied by the equation taken from Part III. will give a second product, which must also be applied to the former quantity, and the sum will be the planet's right ascension in time.

Example.—On the 1st of November, 1824, the geocentric longitude of Saturn is given in the Nautical Almanac 2^S 6° 17′, his geocentric latitude 2° 5′ south, and his declination 19° 20′ north, let it be required to determine his right ascension in time, and also the time of his meridian passage at Greenwich, the sun's right ascension on the said day being 14^h 26^m 34^s?

Geocentric longitude of Saturn 2^S 6° 17′ in time	4^h 25^m.13
Equation to geoc. long. from Part I.	−7 .60
Geocentric longitude in time corrected	4 17 .53
Equation from Part II. $-0^m.70 \times -2°.083$ (latitude)	+1 .46
Equation from Part III. 0 .06 × 1^m.46	+0 .08
Right ascension of Saturn by the Table	4 19 .07
Ditto, by the Nautical Almanac	4 19
Sun's right ascension given (subtract)	14 26
Saturn's meridian passage by Table, and by Naut. Alm.	13 53

TABLE 8. THE DIFFERENCE BETWEEN THE MERIDIAN ALTITUDE OF A PLANET, OR OTHER BODY, AND ITS ALTITUDE, ONE MINUTE BEFORE OR AFTER ITS TRANSIT.

THOUGH the Tables of "REDUCTION to the MERIDIAN," given in pages 99—103, and explained in pages 329 and 330, give the correction accurately for altitudes taken out of the meridian, yet cases may occur, when the computation, necessary to be employed in the use of those Tables, may be dispensed with, and when the correction may be had near enough from an approximate Table by inspection. This purpose is intended to be answered by the Table now presenting itself, which is so simple in its use, that no further direction is required than the work of an example. The tabular quantity is that which is due to *one minute* of time before or after culmination, but may be extended to several minutes, by multiplying it by the *square* of the minutes and parts in question; and though the result will not be strictly accurate, yet, at short intervals from the meridian, provided the object be not near the zenith, it will be very nearly so.

Example.—In latitude 29° N., a star which is 14° south of the equator, had its altitude taken $5^m\ 30^s$, after its culmination, let it be required to determine its correction for altitude?

Answer $2''.4 \times 5.5 \times 5.5 = 72''.6$, or $+1'\ 12''.6$.

TABLE A. THE PARALLAXES AND DIAMETERS OF THE PLANETS. (FROM ZACH.)

THE parallaxes given in this Table are the horizontal parallaxes of each of the ten planets; which are obtained by dividing the sun's mean parallax 8''.7 by the several respective distances from the earth, depending on their elongations from the sun, which elongations are obtained by subtracting the planet's geocentric longitude at the given time, from the sun's longitude at the same time. The distances of the seven nearest planets, however, are not always the same at the same elongation, and therefore each of these planets has three columns of horizontal parallax, adapted for their nearest, mean, and greatest distances, which will be a guide for the horizontal parallax to be taken at any intermediate point of their orbits. But the variation of distance in the orbits of Jupiter, Saturn, and Herschel, are not perceptible enough, to require more than one column each. Like the horizontal parallaxes, the apparent diameters also of the planets vary inversely as their distances, and therefore the same argument will serve also for obtaining the respective diameters of the different planets, which diameters are contained in as many parallel columns: hence, if the diameter of any planet be measured at any time by a micrometer, the Table will show by inspection the corresponding horizontal parallax of that planet, except with respect to the four last discovered planets, which are too small to allow of the variations in their diameters to be accurately determined by the nicest micrometer hitherto constructed.

Example.—Let the horizontal parallax, and diameter of Saturn, be required for November 1, 1824?

	S			
Sun's longitude Nov. 1, 1824.......	7	9°	3'	
Saturn's geocentric longitude	2	6	17	
Argument for Table A............	5	2	46	gives 1''.01 for the hor. par.
				17 .7 for the diameter.

TABLES B, C, D, AND E. ABERRATION OF THE PLANETS IN LONGITUDE AND LATITUDE.

WHEN the orbits of the planets are considered as concentric circles described round the sun, the aberration of a planet may be easily computed in any point of such orbit; but, in the present improved state of practical astronomy, the aberrations so determined would not be considered as sufficiently correct. So long ago as 1794, DELAMBRE computed Tables of aberration for the planets then known, from formulæ that included all the essential terms for ensuring accuracy, which Tables were published in the *Connoissance des Tems* of that year; but, as the sun's mean aberration was taken at 20''.0, instead of 20''.25, Baron Zach, in 1806 [*Tabulæ Speciales*, Vol. I. p. 195 *et seq.*], composed four Tables on Delambre's plan, in which he took the sun's aberration at 20''.255, and neglected some of the minor quantities, to render the computations less complex. In the *Connoissance des Tems* of the year 1818, PUISSANT computed Tables of aberration for all the planets, including the four recently discovered ones, from formulæ, which he had previously demonstrated in a Memoir presented to the Institute of France in 1814, to which, the reader, who wishes for further information on this head, is referred. These are the Tables which have found a place in our volume, and which are, in a certain degree, explained by marginal notes in the pages that contain them respectively. The arguments are numerous, and the computation, even with the help of the Tables, is consequently complex. The Tables B and C are used with their proper arguments for the aberration both in longitude and latitude; but Table D for the aberration in latitude only. In the last Table (E), the quantities for Venus, Mars, Jupiter, Saturn, and Herschel, are so small, that they are omitted, as is also the second part for all the planets. The co-efficients of the various formulæ are detailed in the *Connoissance des Tems* above referred to, and the subjoined example, it is presumed, will render the use of these Tables intelligible.

Example.—On the 1st of October, 1823, let the aberration of Saturn, both in longitude and latitude, be determined for midnight from the following data, viz:

		s	°	′
The sun's longitude at the time	⊙ =	6	7°	59′
The planet's geocentric longitude	G =	1	23	19
The planet's elongation east (G – ⊙)	T =	–7	15	20
The planet's geocentric latitude south	L =		–2	25
Heliocentric longitude of ditto	H =	1	18	49
The annual parallax (G – H)	P =		–4	30
Longitude of the planet's perihelion	φ =	3	0	0
Longitude of the planet's node	☊ =	3	22	9
Argument of the planet's latitude (☊ – G)	C =	1	28	50
Reduction to the ecliptic	ϱ =		0	0.5
Longitude of the sun's perigee	π =	9	9	53
Distance from perihelion to node (φ – ☊)	=	11	7	51

According to the preceding data, the arguments and their corresponding aberrations in longitude, will stand thus:

		Arguments.		Tabular quantities.
			s	
Table B	First part	T =	7 15°.3	+ 14″.32
	Second part	G – π =	3 13 .4	+ 0 .07
Table C	First part	P – ϱ =	0 4 .5	– 6 .54
	Second part	G – φ =	10 23 .3	– 0 .29

Sum..................+7 .56...log. 0 .87852
Sec. L 2° 25′..........................10 .00038

Aber. in long. +7″.57............... 0 .87890

The arguments and their corresponding aberrations in latitude, will be as follow:

		Arguments.		Aber. in lat.
			s s	
Table B	First part	T – III =	10 15°.3	–14″.40
	Second part	G – π + III =	6 13 .4	+ 0 .33
Table C	First part	P – ϱ + III =	3 4 .5	+ 0 .50
	Second part	G – φ + III =	1 23 .3	– 0 .23

Sum ....–13 .80...log. 1 .13988
Sine 2° 25′ .... 8 .62496

			s	
Table D	First part	C + L =	1 26°.4	– 0 .08
	Ditto	C – L =	2 1 .3	– 0 .07

Sum multiplied by sine of lat.................. – 0 .58...... 9 .76484

Aberration in latitude........................ – 0 .73

Neither the second part of Table D, nor Table E, is wanted for the aberration of Saturn, but, when necessary for any of the nearer planets, may be applied in a similar manner.

Bouvard has given Tables of the aberration in longitude of Jupiter, Saturn, and Uranus (Herschel), which were published by the French Board of Longitude in 1821, along with the other Tables of those planets, but as the circle is divided into 400°, and subdivided into minutes and seconds centesimally, they are not calculated for general use.

According to this author the general formulæ for the lunar and solar nutations of the planets in longitude, are equivalent to

Lunar nutation = −18″.0144 sin. supplement of the node.
Solar nutation = + 1 .0692 sin. 2 ☉'s long.

The quantities arising from these formulæ may either be tabulated, or applied by logarithms in the usual way, along with the aberration. The lunar nutation must have the negative sign in the first semi-circle, and the positive in the second; and the solar nutation must have the positive sign in the first and fourth quadrants, but the negative in the second and third. The centesimal seconds of Bouvard may be converted into the sexagesimal denomination by multiplying the tabular quantity by .324. Thus, in the example before us, the supplement of the moon's node is $2^s\ 7^\circ\ 51'$, or 188 .5, as the argument, in circular parts, of Bouvard's Table XLIII. (p. 38 of Tables Astronomiques, 1821, Paris), and the corresponding centesimal seconds of ☾ nutation is −51″.5, which, multiplied by .324, give 16″.68 sexagesimal, for the proper nutation, which is agreeable to the formula before given, viz.

18″.01.......log. 1 .255272 }
Sin. 67° 51′...... 9 .966705 } Sum = log. 1 .221977 = −16″.68.

In like manner, for the solar nutation, twice the sun's longitude in this example is 31° 56′, or 88 circular parts by our Table in page 261, and according to Bouvard's Table, with arg. 88, the centesimal nutation is +1″.3, which, multiplied by .324, produces 0″.42 as the sexagesimal quantity; and according to the formula, we have

1″.069.......log. 0 .028977 }
Sin. 31° 56′...... 9 .723400 } Sum of logs. 9 .752377 = +0″.56.

VELOCITY OF LIGHT.

Various instances may occur in practical astronomy, in which the velocity of light becomes one of the elements of computation, as in the eclipses of Jupiter's satellites, the occultation of stars by the planets, &c.; in which cases, when the distance of the planet from the earth is known, an allowance can be made for the time elapsed since the moment of the phenomenon's actual occurrence. The Table before us considers the mean radius of the earth's orbit as *unity*, and the distance of the heavenly body from which light is transmitted to the earth, must be computed in terms of the same denomination, in order to obtain the time which will be taken up by the passage of light from such body moving in its relative orbit.

Example.—According to Delambre the mean distance of Jupiter from the earth is 5 .2027 radii of the earth's orbit, his greatest distance 6 .4703, his least distance 3 .9351, and the distance of the fourth satellite from Jupiter 0 .0125; let it be required to determine from these data, how soon, after the computed time, an emersion of the fourth satellite of Jupiter will be visible, at each of those distances respectively, supposing that an emersion could be seen in all those three situations?

For the mean distance of the satellite, we have 5 .2027 + 0 .0125 = 5 .2152.

	h	m	s
Then 5 radii give	0	40	37 .50
0 .20		1	37 .50
.0027			1 .32
Time of the passage of light from the mean distance	0	42	16 .32

For the greatest distance, we have 6 .4703 + 0 .0125 = 6 .4828.

	h	m	s
6 radii give	0	48	45 .00
0 .68		3	54 .00
.0028			1 .37
Time of the passage from the greatest dist.	0	52	40 .37

For the smallest distance, we have 3 .9351 + 0 .0125 = 3 .9476.

3 radii give..	0^h	24^m	22^s .50
0 .94	7	38	.25
.0076..		3	.71
Time of the passage at the least dist.	0	32	4 .46

Hence, it appears that upwards of ten minutes may be allowed before and after the mean calculated time of an emersion of Jupiter's fourth satellite, at the two extreme distances from the earth, and that the greatest difference arising out of those extreme positions is 20^m 36^s very nearly.

POLE STAR.

TABLE 1.

THERE is no star in the northern hemisphere, which is so frequently observed as the Pole Star, and yet no star of its magnitude which changes its apparent place so much in every successive year. For this reason, that eminent German Astronomer, Bessel, has computed the mean place of this star, together with the annual variations, for twenty-one successive years, from 1805 to 1825 inclusive, agreeably to the results of his own observations. This Table is No. 1 of our series of Tables that respect the POLE STAR, and the simple inspection of its contents will show what the *mean* right ascension in time, and mean declination have been at the commencement of each of the years specified, together with the annual variations, and their first and second differences.

TABLES 2, 3, 4, 5, and 6.

THE Pole Star had long been a guide to mariners, to indicate generally the course by which their vessels might be steered, even before the compass was introduced in navigation, but its altitude taken out of the meridian does not appear to have been systematically applied for determining the latitude till the year 1800, when Mr. John Stevens, an English marine officer, in the service of the East India Company, and now a member of the Astronomical Society of London, constructed a Table for the reduction of this star, observed in any position, to the meridian, which Table gives its meridian altitude, or latitude of the place, sufficiently near for marine purposes. In the year 1816, however, an eminent astronomer of Vienna, Littrow, proposed, in Lindenau's Astronomical Journal (Vol. III. p. 208), a method of rendering the Pole Star serviceable at all times, for determining the latitude of any place with great precision. The same subject was introduced into Zach's *Correspondance Astronomique*, in November, 1820, and excited the attention of HORNER, of Zurich, who, in August, 1821, published a Table, with subsequent corrections, in the same work, for the purpose of reducing the observed zenith distance of the Pole Star, taken at any time, to the true zenith distance taken on the meridian. In January, 1822, Littrow published, in the same periodical work, a shorter General Table than Horner's, containing two different columns of tabular quantities; which Table is applicable to any latitude, but is not so simple in its application as Horner's. In these Tables, the differences in the argument, depending on the star's horary angle, are not less than 10 minutes of sidereal time; for which reason, we have constructed a Table, adapted for the latitude of Greenwich, for every minute of the horary angle, with columns of variation for every 100′ difference of latitude, and for 100″ diminution of the ecliptic's obliquity: which Table will be found very convenient, and applicable in every part of England, Ireland, and even in the southern parts of Scotland. Lastly, Mr. F. Baily, whose zeal for the success of practical astronomy is well known, having observed that, in Horner's Table, the latitude is supposed to be nearly known (though the latitude is the object sought for), computed a General Table that does not require the latitude to be previously known, still making the horary angle the principal argument, but adding an auxiliary Table of Corrections, for the tabular quantities B and C, which corrections depend on the north polar distance of the Pole Star at the time of observation. These several Tables constitute Numbers 2 to 6, under the common head POLE STAR; and the practical Astronomer is at liberty to prefer any one which may appear best suited for his purpose. We have taken the liberty to expand Mr. Baily's Table to every minute of the argument, as well as to extend the Table of Corrections to a more remote period than it was originally computed for. The formulæ contained in the marginal notes will supply the place of formal rules, for the application of these Tables respectively, and the subjoined examples will illustrate such application. Similar Tables, by Schumacher, are annexed to an "Ephemeris of the

Distances of the four Planets, Venus, Mars, Jupiter, and Saturn, from the Moon's Center, for 1822," printed (in English) for the royal Danish Sea-Chart-Office, at Copenhagen, and Don Philippe Banzà has introduced Horner's Table into his Spanish Nautical Almanac for 1826; which circumstances are here mentioned to show the opinion entertained of the utility of such Tables on the Continent.

Example.—Suppose in latitude about 50° N., the zenith distance *(z)*, of the Pole Star, to have been determined by observation to be 39° 12′ 16″.4, when its apparent polar distance (P), by a good Table, was 1° 38′, and the star just four sidereal hours from the moment of its upper culmination, let the exact co-latitude of the place be determined from these data, according to each of the Tables 2, 3, 4, and 5, respectively?

According to HORNER's Table 2. (Lat. 50°).

4h 0m horary ∠ gives a tabular quantity (in col. 50°)............= 1′ 18″.9
Subtract for 2′ diff. of P, $\frac{2}{100}$ of ditto .... −1 .6

−1 17 .3 = correct tab. No.

Cos. 4h or 60° log..... 9 .69897 } Sum 1 .69019............= 0° 49 0 .0
Minutes of P 98′..... 1 .99122 }
Measured zenith distance *(z)* corrected.......................39 12 16 .4

The true co-latitude ....39 59 59 .1

According to LITTROW's Table 3. ($t = \theta$).

4h 0m in col. θ gives M = 65″.45, and N...............= +0′ 0″.63
Then 65″.45 × 0.04 (for −2′ of P) = 62″.83 = M′
M′ 62″.83 ....log. 1 .79817 } Sum 1 .88663 = 77″.02 −1 17 .02
Co-tang. *z* 39° 12′ 16″.4 .10 .08846 }
Cos. *t* = cos. 60° ... log. 9 .69897 } Sum 1 .69019........ + 0° 49 0 .0
P 98′.................. 1 .99122 }
z corrected .... 39 12 16 .4

The true co-latitude (ψ).............................. 40 0 0 .01

According to the GREENWICH Table 4.

IV Sid. hours 0 min....... +83″.26 }
Diff. of lat. (5″ 05) in 88′ var. −4 .44 } Tabular number ....− 0° 1′ 17″.1
Dim. of P 2′ (2 × 0″.83) .. −1 .66 }
Cos. *t* 60°............ log. 9 .69897 } Sum 1 .69019 ....+ 0 49 0 0
Minutes of P 98′ .... 1 .99122 }
z corrected... 39 12 16 .4

39 59 59 .3

In Table 4, for " Var. in 100″ dim." put " $\frac{1}{100}$ of Cor." at the head of each third column.

According to Mr. BAILY's Tables 5 and 6.

Sid. hour IV 0m ...... C = 1″.20, and P + C .... = 1° 38′ 1″.2
Log. 5881″.2... 3 .76946 } Sum .. 3 .46843 = 2940″.60
Cos. *t* 60° ..... 9 .69897 }
B 62″.86 ...... 1 .79837 } Sum .. 1 .88683 = −77 .06
Co-tang. *z*.....10 .08846 }

2863 .54 = 0° 47′ 43″.54
z corrected ....39 12 16 .40

Co-latitude ....39 59 59 .94

As the north polar distance (or P) is just 1° 38′, Table 6 does not alter the quantity derived from Table 5.

By a rigid trigonometrical calculus, the co-latitude comes out, from the preceding data, = 40° 0′ 0″.0.

TABLE 7. THE GREENWICH MEAN PLACES, ETC., OF THE POLE STAR, FOR SIXTY SUCCESSIVE YEARS, FROM 1820 TO 1880.

As Bessel's Table of the mean places of the Pole Star terminates with the year 1825, it has been deemed advisable to give a Table that may be referred to for many years to come, and that contains not only the mean places as determined at Greenwich, but also the subsidiary numbers for computing the corrections, adapted to the years specified in the Table, which numbers, it is presumed, will be found not only correct, but conducive to the purpose of the computer, who may be disposed to give a specific Table of the daily places of this interesting body, from year to year. The construction of this Table has been effected with considerable labour and great care, and contains much elementary matter in a small compass. Its application will be intelligible by any person who has made himself acquainted with the mode of applying the corrections of 48 principal stars, explained and exemplified in pages 336—342.

[See the note to the LAST GREENWICH CATALOGUE, at page 346.]

CIRCUMPOLAR STARS.

IN taking the right ascensions of the stars, it is of the utmost importance that the transit instrument be properly fixed in the true meridian of the place of observation. When the Pole Star can be observed at two successive passages over the meridian, above and below the pole, the error of the instrument's position may be easily detected, but in skies that are liable to be obscured by clouds, almost every day in the year, the practical Astronomer will be glad to avail himself of other circumpolar stars that may be seen passing the meridian, when the Pole Star is at a distance from either point of its passage.

When the middle wire of a transit instrument, is made to bisect a true meridian mark, it will also bisect the polar point, when properly elevated, provided all its adjustments are perfect; and in this case, any circumpolar star, as well as the Pole Star, will descend through the western semi-circle, in precisely the same time that it will take in ascending through the eastern semi-circle of its apparent diurnal path, namely, in twelve sidereal hours; the corrections for aberration, &c., being considered as varying insensibly in 24 hours. Hence if a star is found from observation to remain longer in the western semi-circle than in the eastern, by a second or any given number of seconds of time, one half of this difference will be the error of position of the instrument, and the meridian mark must be moved, as well as the object-end of the telescope, towards the west, to meet the star at the lower extremity of its western semi-circle; but if the eastern semi-circle be found the larger, the mark must be moved towards the east, in order that the line of collimation of the telescope, may bisect both the meridian mark and the polar point. The absolute quantity of space, at the meridian mark, that shall correspond to the error in time, will always depend on the star's polar distance, and may be computed for a given distance of the said mark from the Observatory; but experience of the time during which the star in question passes from one wire to another, will always enable an observer to estimate the fractional portion due to any small number of seconds, without rigid computation. If the clock's rate is not previously known, the star's transit should again be observed, at the end of twenty-four hours, to gain what has been the daily error of the clock, which, of course, must be allowed for in noting the times of the respective transits.

But the most accurate, as well as most convenient way of using circumpolar stars, for effecting the purpose in question, is, to select them in pairs of nearly equal polar distances, and at nearly twelve hours of right ascension from each other, so that one star may be observed at the upper culmination, and the other at the lower, and then again, after an interval of twelve sidereal hours, when their positions will be reversed with respect to altitude; for in this mode of observing, first suggested by Mr. Butt, it will not be necessary to depend on the going of the clock made use of for a longer period than the short time that intervenes between the upper culmination of one star, and the lower culmination of the other, at either of the two times of making the observations. The absolute times between the periods of observing, nay even the right ascensions of the stars themselves, are not required, in this method, to be known or determined by the transit instrument, but only the *differences* of the times of the two passages at each respective period. If any star α, precedes another star β, in both its superior and inferior meridian passages, by the same number of minutes and seconds precisely, it is evident that the two semi-circles round the polar point, have been passed through in equal times by both stars, and that, therefore, the instrument has been directed to bisect the two concentric circles passed over in the polar point. But if the difference of the times of the passages is not the same in the reversed positions of the two stars, one-fourth of the difference of these differences will be the error, in time, of the transit instrument's position, which may be accordingly rectified, by moving the side-screw at the pivot of the axis, more or less, in the required direction, accordingly as the stars are remote from or near the pole; which a little experience will effect, better than any calculation, without the aid of a graduated meridian mark. If, however, a considerable interval of time elapses before the reversed position of the two stars furnishes the second observation, the apparent places of the stars themselves will have altered, for which an allowance in right ascension must be made; but this will not be necessary after the short interval of only twelve hours. The Table before us contains a selection of stars arranged in pairs, differing from each other about twelve hours, more or less, together with their right ascensions, north polar distances, annual variations, and subsidiary numbers, according to Zach, for the epoch 1800: so that the true places may be also had from the same observations, if circumstances, or the object of the observations, should require it. When a graduated circle forms a part of the instrument, and it is found to be nearly in the meridian, the true time and latitude of the place may both be thus known, as well as the error of the instrument's position, provided that the altitudes and times indicated by the clock, be noted at the respective transits.

There are three advantages in the method of verifying the position of a transit instrument by a pair of circumpolar stars; first, the stars are moving in contrary directions when observed, and it will appear from the passages, on which side of the meridian-mark the greater semi-circle lies, *i. e.* which semi-circle is passed over in the longer time; secondly, it is an independent method, that has no reference to right ascensions determined under any other meridian, particularly if the successive observations are made in the reversed positions after an interval of one-half of a sidereal day; and thirdly, when this plan of observing is adopted, the computations of the corrections, for obtaining the apparent right ascension of each star, will be avoided.

Example.—On the 27th of October, 1823, β Cephei was observed, in the evening, to pass the transit instrument's middle wire, above the pole, at a certain observatory, at $21^h\ 26^m\ 22^s.74$; and θ Ursæ Majoris had been previously observed to pass below the pole, at $21^h 21^m\ 1^s.5$ by the sidereal clock, adjusted to true sidereal time, agreeably to a meridian mark then fixed; and, on the morning of the 28th, θ Ursæ Majoris was observed to pass above at $9^h\ 21^m\ 3^s.5$, and β Cephei below at $9^h\ 26^m\ 20^s.7$: let it be required to ascertain what was the deviation of the transit telescope, or error of the meridian-mark, according to these observations, by each of the two methods above explained?

First method.

β Cephei above	$21^h\ 26^m\ 22^s.7$	θ Ursæ Majoris below	$21^h\ 21^m\ 1^s.5$
Ditto... below	9 26 20 .7	Ditto........ above	9 21 3 .5
Western semi-circle smaller by......	2 .0	Eastern semi-circle larger by.........	2 .0

Hence, by each of the two stars, the meridian-mark and telescope require to be carried towards the east, a quantity of space equal to the motion in one second of time, or one-half of the differences respectively.

Second method.

Again β Cephei above	$21^h\ 26^m\ 22^s.7$	θ Ursæ Majoris above	$9^h\ 21^m\ 3^s.5$
θ Ursæ Majoris below	21 21 1 .5	β Cephei below	9 26 20 .7
Diff. at first position	5 21 .2	Diff. at second position .... =	5 17 .2
Ditto at second ditto	5 17 .2		
Diff. of two differences .... =	0 4 .0; one-fourth of which, as before, = 1^s.		

Now let us suppose, that the absolute times of the passages of the two stars, were not the true times, and that the differences only were truly taken, then, because the difference at the first position exceeds the difference at the second, the following star (β Cephei) came too late to the mark at the first position, and too soon at the second; but, in coming to the upper culmination, or first position, it passed through the eastern semi-circle, which, therefore, was too large, and required the meridian-mark to be moved towards the east, that the two semi-circles might be equalized. It may not appear why one-fourth of the difference of the two differences is the error of position by the second method, since one-half of the difference is the same error by the first method. This will be comprehended, if we consider that, when the mark is out of position, the line of collimation of the telescope cuts off a small portion of the semi-circle, both above and below, which small arcs are equal to each other, and the measure of one of them is the error of position, or half the difference of the two semi-circles. Now according to the first method, the two stars are considered as separately passing through different semi-circles, and each affording the means of determining the excess or defect of its own semi-circle; but, by the second method, the whole circle is considered as passed over in twelve hours, by one star or the other, and the two small arcs cut off from one semi-circle, are two similar arcs added to the other, and consequently, the small arc which is the measure of the error of position, is to be allowed for *four times*.

If we suppose the clock indicating nearly true sidereal time, its error may be determined by computing the apparent right ascension of either, or both, of the two stars, thus:

For β Cephei.

Mean R. A. 1800	21h 26m 0s .90
23 years precess. 0s.81 ...	18 .63
October 27, ditto	0 .67
Aber. (Tab. XVIII.) ..	+0 .96
Lunar nutation	−0 .24
Solar nutation	−0 .04
Star's app. R. A.	21 26 20 .97
Time by clock at the transit above	21 26 21 .74
Error of clock	+0 .91

Arguments.

	s		
☉ long. 27 Oct.	7	3° 16′	
Tab. XVI. ∗'s R. A.		+18	
	7	3	34
A	10	10	56
Arg. for Aberr.	5	14	30
Log. max. 0 .5687			
Max. = 3s.705			

	s		
☾'s ☊	9	22° 35′	
L	8	17	38
Arg. for ☾ nut.	6	10	13
Log. max. 0 .1424			
Max. 1s.388			

Arg. ☉ nut. 10s 25°.
Max. 0s .06

For θ Ursæ Majoris.

Mean R. A. 1800	9h 19m 23s .50
23 years prec. 4s.08	1 33 .84
October 27	3 .33
Aber. (Tab. XVIII.).	−0 .50
Lunar nutation......	+1 .63
Solar nutation.......	−0 .07
Star's appar. R. A.	9 21 2s .71
Time by clock.........	9 21 1 .50
Clock's error	−0 .23

Arguments.

	s		
☉ long. as before..........	7	3° 16′	
For ∗'s R. A.		+18	
	7	3	34
A	4	12	36
Argument for aber..........	11	16	10
Log. max. 0 .3257			
Max. 2s.117			

	s		
☾'s ☊	9	22° 35′	
L	5	6	54
Arg. for ☾ nut.	2	29	29
Log. max. 0 .2141			
Max. 1s.637			

Arg. ☉ nut. 7s 14°.
Max. 0s .09

From the preceding computations it appears, as might be expected, that, as the meridian-mark was out of adjustment, one semi-circle, *i. e.* one star's passage, gives the error of the clock +, and the other −; but, as the north polar distances of the two stars are nearly double, the one of the other, the quantities of the opposite errors are not the same.

The corrections of these two stars might have been taken more readily from the Table of Corrections of 48 Principal Stars, pages 123 and 141.

CORRESPONDING HIGH AND LOW STARS.

Another method of verifying the position of a transit-instrument is, by means of a pair of stars having nearly the same right ascension, but differing considerably in declination, or having nearly the same declination of opposite denominations, one north and the other south. If the clock's time, deduced from both stars in succession, be the same, the instrument may be considered as well placed in the meridian; otherwise, half the difference of the times will be very nearly the error of position, and the object-end of the telescope must be turned, by the proper adjusting-screw, to or from the lower star, according to the quantity and denomination of the error. If the second star comes too soon into the field, the object-end of the telescope must be turned from it, and vice versâ. For this purpose, it is necessary that the *apparent* right ascensions of both stars, be known at the time of observation; otherwise, they cannot be truly compared with the observed times of their transits, which are always the apparent times, when the rate of the clock is known, and its error in time applied with a contrary sign. Hence the right ascensions, declinations, annual variations in both, and subsidiary numbers for the corrections, adapted to the epoch 1800, are all given in this Table, and will be found serviceable for their respective purposes. When any pair of the 36 stars, frequently observed by the Astronomer Royal, are among the observations of the evening, the Tables of their daily places, now regularly published in one of the monthly journals, will afford the ready means of comparing their observed with their computed right ascensions; and, provided their declinations differ sufficiently, the transit-instrument's position may be inferred from thence without further trouble. Of these stars, 24 have their apparent places, given also for every tenth day, in the Nautical Almanac. But if a pair of corresponding high and low stars, taken from the present Table, be found more convenient, which they generally will be, as there is a choice for every hour of right ascension, the corrections deduced from the subsidiary numbers, as already explained, must be applied to the mean right ascensions, when brought up by the annual variations, with their respective proper signs. This method of examining the position of a transit-instrument, and of adjusting its meridian mark, is not so independent as the one preceding, inasmuch as it is taken for granted that the right ascensions of the stars employed are truly determined by the Astronomer, on whose catalogue reliance is and must necessarily be placed, till the circumpolar method has confirmed the correctness of this assumption.

Example.—On the 11th of October, 1823, α Piscis Austrini culminated at 22ʰ 47ᵐ 51ˢ.5, and β Pegasi at 22ʰ 54ᵐ 17ˢ.8; what were the errors of the clock, and of the transit instrument's position?

α Piscis Austrini.

		h	m	s
Mean R. A. 1800		22	46	34 .53
23 years precession			1	16 .41
11 October, Tab. XV.				2 .58
☉ + A .. = 4ˢ 7° 42′ and max. aber. 1ˢ.453	Tab. XVIII.			+1 .16
☊ + L .. = 3 6 24 and max. ☾ nut. 1 .247	Tab. XVIII.			+1 .24
2 ☉ + S = 6 19 0 and ☉ nut.... 0 .06	Tab. XVIII.			−0 .02
Star's apparent R. A.		22	47	55 .90
Time by clock		22	47	51 .50
Clock slow				4 .40

β Pegasi.

	h	m	s
Mean R. A.	22	54	4 .97
23 years precession		1	5 .92
11 October			2 .24
☉ + A .. = 4ˢ 5° 39′, max aber..... 1 .14..			+0 .93
☊ + L .. = 4 10 13, max. ☾ nut. .10 .80..			+8 .25
2 ☉ + S = 7 22 0, max. ☉ nut.. 0 .06..			−0 .05
Star's app. R. A.	22	54	22 .26
Time by clock	22	54	17 .80
Clock slow			4 .46

Hence the mean error of the clock is $-4^s.43$, and the deviation of the transit telescope $\frac{0^s.06}{2} = 0.03$.

	h	m	s
According to the Tab. of Cor. of 48 principal stars, the app. R. A. of α Pis. Aust. is	22	47	55 .95
According to Schumacher's *Astronomische Hulfstafeln*, 1823	22	47	55 .98
According to Tilloch's Journal, Sept. 1823	22	47	56 .01

A ZONE OF STARS SUITABLE FOR THE ZENITH SECTOR, ETC.

The obstacles to practical astronomy, that arise out of the variable states of the atmosphere, so far as *refraction* is concerned, are best obviated by taking observations in the zenith, as often as the objects of the observer will admit of deductions therefrom applicable to his purpose. The conductors of trigonometrical surveys are sensible of the superior advantage of zenith observations in determining a celestial arc that shall correspond to a measured terrestrial line; and also in ascertaining the latitude of a given station from a zenith star, which has had its declination, or its polar distance, correctly measured in an established Observatory. The zone of stars that were used by Major Colby, had the north polar distances determined, as we understand, by Stephen Groombridge, Esq., of Blackheath, to whose successful labours the astronomical world is much indebted; and the subsidiary numbers, for giving the corrections, were computed from the series of Tables contained in the former part of this Volume, by Mr. Fallows, before he sailed from England. Hence it is presumed, that any observer, who has a good zenith sector, zenith micrometer, or altitude and azimuth circle of considerable dimensions, may, by the help of these stars, deduce the latitude of any given place in England, Ireland, or Scotland, both with ease and accuracy; for as the distance from the equator to the zenith is always equal to the pole's elevation, or latitude of the place, there can be no difficulty in ascertaining that distance when the apparent declination of the observed star is known, and also its measured distance from the zenith of the place of observation, whether north or south; then the apparent declination increased by the zenith distance, when the star lies to the south of the zenith point, but diminished thereby, when it is on the northern side, will be the latitude, without any correction for refraction, provided the star and zenith point are both in the field of view of the telescope, which they must be when the zenith micrometer is used. The only requisite correction will be, the conversion of the true declination of the star into the apparent by using the subsidiary numbers A′, L′, and S′, to obtain the arguments for the general Table XVIII. (p. 47, &c.), from which the quantities are respectively taken by inspection.

Example.—On the 20th of July, 1818, the zenith distance of γ Draconis was measured by a zenith instrument, on an average of the reversed positions, and was found 2′ 52″, towards the north, at a given station; let it be required to ascertain the latitude of that station?

γ Drac. 1817, N. P. D.	38° 29′ 7″.72
One year's precession	0 .67
July 20	0 .37
☉ + A′.. 6ˢ 29° 2′,..max. aber......19″.56	−9 .47
☊ + L′. 4 6 16,..max. ☾ nut.... 9 .65	+7 .79
2 ☉ + S′ 10 26 43,..max. ☉ nut.... 0 .43	−0 .24
Apparent N. P. D. of star	38 29 24 .84
Apparent declination....................................	51 30 35 .16
Measured Z. D. towards north....................................	−2 52 .0
True latitude of the station....................................	51 27 43 .16

PRECESSIONS.

TABLES 1, 2, AND 3.

We have already said, in page 308, that an annual precession, or retrocession of the equinoctial point, is occasioned by the attraction of the sun and moon (when out of the equator), on the prominent parts of the earth, considered as a spheroid, but no distinction was made, in that explanation, between the *luni-solar* and the *general* precession. The former is that which is occasioned by the joint actions of the moon and sun only, and its quantity is measured along the ecliptic circle from some point where the intersection took place at a given time, as 1750, and is considered proportional to the time elapsed, without reference to the diminution of the obliquity, or angle contained between the equator and ecliptic; the latter precession has reference, moreover, to the perturbations of the larger planets, and to the change taking place in the obliquity, which alters the attraction of the sun and moon a little, and which, consequently, alters the plane of the equator itself, by two small quantities called the solar and lunar nutations. The laws by which the changes in question take place, have been developed by La Place, and various astronomers of eminence, such as Delambre, Zach, and Bessel have deduced the respective quantities of both denominations of precession of the equinoctial points, by laborious comparisons of distant authentic observations of the same stars. Zach made use of the Catalogues of Bradley, Barry, and Mayer, besides his own, and, from an average of four series of comparisons, concluded that, for 1750, the mean general precession is 50″.0540, which Delambre had previously determined to be 50″.25; but Bessel, depending on comparisons between the observations of Bradley and Piazzi, made the general precession for the same period, 50″.176068. The luni-solar precession, corresponding to the same period, was made by Zach 50″.2390, and by Bessel 50″.340499. The constant part of the precession in right ascension, and also the variable part in right ascension, as well as in declination, are derived from the luni-solar precession in longitude.

The formulæ from which the variations of the precession in right ascension and declination, are computed, are as follow, viz.:

$$\text{The variation of pr. in long.} = \frac{\text{prec.} \times \text{cos. lat.} \times \text{cos. angle of position}}{\text{cos. declination.}}$$

$$\text{The variation of pr. in dec.} = \text{variat. in R. A.} \times \text{cos. dec.} \times \text{tang.} \angle \text{ position.}$$

Or otherwise, according to La Place,

$$\begin{aligned} \text{Variat. in R. A.} &= \text{prec. in long.} \times (\text{cos. obliq.} \times \text{sin. R. A.} \times \text{tang. dec.}) \\ &\quad - \text{variat. obliq.} \times \text{cos. R. A.} \times \text{tang. dec.} - 0''201683. \\ \text{And variat. in dec.} &= \text{prec. in long.} \times \text{sin. obliq.} \times \text{cos. R. A.} \\ &\quad + \text{variat.} \times \text{obliq.} \times \text{sin. R. A.} \end{aligned}$$

Agreeably to the latter of these methods, Zach computed the constant part of the precession in R. A., for 1750, to be

$$50''.2390 \times \text{cos. } 23^\circ\ 28'\ 23'' - 0''.20168 \ \ldots\ldots = 45''.87992$$

and the variable part in R. A. and Dec.

$$50''.2390 \times \text{sin. } 23^\circ\ 28'\ 23'' \times \text{sin. R. A. } 0^\circ\ 0' \times \text{tang. dec. } 0^\circ\ 0' \ldots\ldots = +\ 20\ .01109$$

But Bessel's constant and variable quantities come out differently; viz. the constant = 45″.99592, and variable = 20″.05039. Hence the annual precessions in right ascension and declination, or north polar distances, derived from these respective numbers, differ from each other and produce not the same proper motions; which circumstance points out the necessity of establishing a scale of precessions in longitude from which astronomers in general may take the co-efficients of their formulæ for deducing the annual precessions. With this view, we have computed the three Tables now to be explained, which admit of the quantities corresponding to any given time to be taken out by inspection. The epoch from which the different precessions are made to increase or diminish, is 1750, and the numbers determined by Bessel are assumed as the foundation, thus, viz.

General precession....... $= 50''.17607 + 0''.0002442966\ (t - 1750)$
Luni-solar precession .... $= 50\ .34050 - 0\ .0002435890\ (t - 1750)$
Constant in R. A. .... $= 45\ .99592 + 0\ .0003086450\ (t - 1750)$
Variable in R. A. and Dec. $= 20\ .05039 - 0\ .0000970204\ (t - 1750)$

When the time for which any of the precessions may be wanted, is antecedent to 1750, the sign of the variations must be changed respectively.

Table 1 is constructed for solar years, and shows the mean precessions of the four different denominations for the beginning of each of 17 centuries from the Christian era, for the commencement of each year, from 1700 to 1900, and for each century again from 1900 to 3000. The columns for the constant and variable parts of the luni-solar precession, are particularly useful for showing, by inspection, the co-efficients for determining the precession of a star both in right ascension and declination, for any given year since 1700; and may be used also for computing the *amount* of the respective precessions accumulating in any given number of years, by multiplying the whole interval by the tabular numbers belonging to the *middle* of the two extreme periods. This result will be correct, if the whole interval in years, divided by 4, give a quotient equal to the number of *leap years* indicated by the Table within the given period, without a remainder : but if the *common* year 1800 be included in the interval, there may be one leap year more in the quotient than in the Table, in which case one day's precession must be subtracted from the amount. Otherwise, if there be a remainder of 1, 2, or 3, after the division by 4, the quantities contained in the small subjoined Table, agreeably to such remainder, must be subtracted, in place of a whole day's motion.

But to avoid a liability to error in obtaining the true amount of any of the precessions from Table 1, Table 2 was computed for the sole purpose of giving the amounts accumulating from the year 1750, to any subsequent year, as far as 1880, without any further trouble than applying the variation belonging to any given year to the amount of the precession annexed to the same year, by paying a due regard to the signs of the variation in question; for if the amounts of the two extreme periods, including the interval, be separately taken from the Table, the *difference* of these two amounts will be the amount belonging to the difference of the times, or interval given. In this way the middle of the interval, and also the tabular numbers belonging to it, are altogether disregarded; and, if the interval consisted of years and days, the application of the latter may be readily made from the corresponding numbers contained in Table 3, which may be considered as supplementary. In tabulating the four different kinds of mean precessions in Table 2, the year following leap year has an increase or decrease of one day's motion, and therefore requires no subsequent correction, on account of the leap years; and the columns of variation increase in the following ratio, if we put V.I for the annual variation of 1750—51, the variation for the following year will be $=$ V. $(1 + 2)$; for the next $=$ V. $(1 + 2 + 3)$; for the next $=$ V. $(1 + 2 + 3 + 4)$, &c., &c.; and in this way, the amount of the variations is given for any number of years after 1750, corresponding to the amount of the mean precessions for the same elapsed interval, and the sum or difference of the two tabular quantities so taken, will be the true amount. This Table is particularly adapted to the important purpose of determining the *proper motion* of a star, and certain results connected with, and partly dependent on it; and, in the present state of practical astronomy, we beg leave to call the attention of astronomers to the utility of computations, thus rendered easy, which may reconcile deductions of a practical nature, that at present are at variance with each other, and which will continue so, until the *proper motions* of different astronomers are derived by a systematic process, founded on true principles, from authentic mean places of the stars, separated from each other by a long interval.

When the amounts of the constant and variable parts of the tabular precessions are obtained, the corresponding amounts of the precessions in right ascension and declination of any star may be easily obtained from the common formulæ, viz.

Precession in R. A. .... = constant part + (variable part sin. R. A. tang. dec.)
Precession in Declination = variable part cos. R. A.

The right ascension and declination to be used, are those belonging to the middle of the given interval.

When these last amounts are compared with the differences arising out of the mean places given at the respective epochs, it will appear whether the aggregate obtained by computation, exceeds or falls short of the aggregate obtained by distant observations; and, consequently, whether the star in question has an annual motion not accounted for by the computation, which may be called its *proper motion*.

Example.—In Bessel's FUNDAMENTA ASTRONOMIÆ, the mean right ascension of α Lyræ is given $277^\circ\ 9'\ 41''.6$ for the year 1755, according to Bradley's observations, and its mean declination $38^\circ\ 34'\ 11''.4$ N.; which quantities, in Mr. Pond's Catalogue of 1823, are increased to $277^\circ\ 44'\ 15''$ and $38^\circ\ 37'\ 28''.8$ respectively. Let it be required to determine from the Tables, the amount of the annual precessions of both kinds for the 68 years' interval, and also the proper motions arising from a comparison of the computed with the observed mean places?

By Table 1.

1789.....constant 46″.00796 × 68 years ....52′	8″.54
Deduct one day for the common year 1800...	.12
Amount of the constant part...............52	8 .42
1789.....variable part 20″.04661 × 68 years 22	43 .169
Deduct one day's motion as above..........	.055
Amount of the variable part...............22	43 .114

$$\frac{277^\circ\,9'\,41''.6 + 277^\circ\,44'\,15''}{2} = \text{mean R. A. } 277^\circ\,26'\,58''.3$$

$$\frac{38^\circ\,34'\,11''.4 + 38^\circ\,37'\,28''.8}{2} = \text{mean Dec. } 38\ \ 35\ \ 50\ .3\,\text{N.}$$

By Table 2.

1823..Constant part	55′ 57″ .54475
Variation	+.83363
Amount in 73 years	55 58 .37838
1755..Constant part................	3 42 .94812
Variation	+0 .00463
Amount in 5 years (sub.)	3 49 .95275
Amount in 68 years	52 8 .42653
1823..Variable part	24 23 .60985
Variation..................	−0 .26205
Amount in 73 years................	24 23 .34780
1755..Variable part	1 40 .23823
Variation	−0 .00146
Amount in 5 years (sub.)............	1 40 .23677
Amount in 68 years................	22 43 .11103

Then according to the formulæ.

Precession in Right Ascension.

Log. of 1363″.11 tab. amount of variable part..	3 .134531
Log. sin. R. A. 277° 26′ 58″.3 (supplement)...	9 .996318 −
Log. tang. dec. 38° 35′ 50 .3 N.	9 .902117
True amount 1078″.86.... = −17′ 58″.86..	3 .032966
Constant part (as above) ....52 8 .42	
Whole amount of prec..........34 9 .56	
R. A. in 1755 ....277 9 41 .60	
Computed R. A. 1824 .....277 43 51 .16	
Mr. Pond's R. A. ....277 44 15 .00	
Difference in 68 years ...... +23 .84	

$$\frac{23''.84}{68} = 0''.0506 \text{ or } 0^s.0233 \text{ for the proper motion.}$$

Precession in Declination.

Log. of 1363″.11 tab. amount of var. part.....	3 .134531
Log. cos. R. A. 277° 26′ 58″.3 (suppl.)	9 .112837 +
True amount 176″.754 ..... = + 2′ 56″.75..	2 .247368
Dec. in 1755 ....38 34 11 .40	
Computed dec. in 1823......38 37 8 .15	
Mr. Pond's dec. ....38 37 28 .80	
Difference in 68 years ...... +20 .65	

$$\frac{20''.65}{68} = +\ 0''.3037 \text{ for the annual proper motion.}$$

The sign for the sines of the first and second quadrants is +, and for the third and fourth −: but the sign for the cosines is + in the first and fourth quadrants, and − in the second and third.

PROMISCUOUS TABLES.

TABLE 1. AUGMENTATION OF THE MOON'S SEMI-DIAMETER, ETC.

We have already said that the moon's semi-diameter increases with her altitude, and our Lunar Table 7 gives the augmentation corresponding to any given altitude; but, at the time of an eclipse or occultation, there may be reasons why the proper altitude cannot be conveniently measured, and a subsequent computation of it may be troublesome. On this account Delambre contrived a Table consisting of four parts, which will give the augmentation from some of the computed elements of the eclipse or occultation, used as arguments, without reference to the altitude; which Table was adopted and extended by

Dr. Mackay, and in that state now presents itself for explanation. When the altitude of the nonagesimal, and the moon's apparent distance therefrom, are put into signs, degrees, &c., take their sum as one argument, and their difference, when the said distance is subtracted from the altitude, as another; with these, obtain the two corresponding equations from Part I., and add them, according to their signs, into one sum; then, with this sum as an argument, enter Part II., and add the equation therefrom, to the said sum, with its proper sign. In the next place, if it be an occultation of a star that forms the observation, the equation from Part III. must, in like manner, be applied, by using as arguments the true latitude and parallax in latitude of the moon; but in a solar eclipse, this equation vanishes. Lastly, with the sum of the foregoing equations as one argument, and the horizontal semi-diameter of the moon as the other, the equation from Part IV. must be taken, and added, in the same way, to the aggregate of the former equations, and the whole sum will be the proper augmentation.

Example.—Let the altitude of the nonagesimal be 55° 17′, the apparent distance of the moon from the nonagesimal 14° 43′, the moon's true latitude 24′ 2″ south, her parallax in latitude 35′ 40″, and her horizontal semi-diameter 15′ 30″; then from these data, let the augmentation of her semi-diameter be determined?

	S		
Altitude of the nonagesimal	1 25° 17′		
Moon's distance from do.	0 14 43		
Sum......Arg. 1	2 10 0	Part I.	+ 7″.70
Difference Arg. 2	1 10 34		+ 5 .33
Amount of the equations of Part I.			+13 .03
Equation to 13″.03 ...... Part II.			+ 0 .17
☾'s lat. 24′ 2″ S. and paral. in lat. 35′ 40″ Part III.			− 0 .12
Sum of the preceding equations			+13 .08
☾'s semi-diam. 15′ 30″ and sum 13 .08 Part IV.			− 0 .81
Augmentation required			+12 .27
Moon's horizontal semi-diameter			15 30 .0
Moon's augmented semi-diameter			15 42 .27

TABLE 2 answers the same purpose as LUNAR TABLE 15, and its mode of application has been explained in page 372; also TABLE 3, to be used with LUNAR TABLE 10, if thought desirable, has already been applied in page 368.

MEAN REFRACTIONS.

THE mean refractions of Cassini, La Caille, and Bouguer have already been noticed in page 301; Carlini's have been explained and exemplified in pages 304, 305; and Bessel's, in pages 305—307.

HAVING now given, what we trust will be considered as a satisfactory explanation of all the Tables, originally collected or computed, we might here conclude our first volume; but, as some foreign Tables for giving the aberrations and nutations, solar and lunar, in right ascension and declination, have lately fallen in our way, which have been used on the Continent by certain eminent astronomers, for the reduction of their observations of the stars, we have thought it not inconsistent with our plan to add these to our collection, for the gratification of our English promoters of astronomy. The Tables of Gauss, which we shall first notice, like those of Delambre, are *general*, and may be advantageously employed for reducing any individual star from its apparent to its mean place, and the contrary, for the purpose of rendering detached observations useful in the construction of a Catalogue.

GENERAL TABLES by PROFESSOR GAUSS.

The General Tables of Gauss are taken from the *Effemeridi Astronomiche di Milano, per l'anno* 1823, but the formulæ from which they are computed, are not given along with the explanation of their use. The reader may however consult on this subject Professor Littrow's work on theoretic and practical astronomy, lately published in the German language.

Table 1 gives the aberration in right ascension in arc, by using the sun's longitude as the argument, and taking the quantities agreeably to the formula $-$ a. sec δ. cos $(\odot + A - \alpha)$; in which expression, α represents the star's right ascension in arc, δ its declination, $\odot$ the longitude of the sun, and a, A, respectively the tabular quantities; the latter of which is evidently the reduction of the ecliptic to the equator.

The aberration in declination consists of three parts, the first of which is taken from Table 1 as before, according to this formula, $-$ a sin δ. sin (⊙ + A $-$ α); the second and third are taken from Table II., with the arguments ⊙ + δ and ⊙ $-$ δ, where the south declination is considered negative.

The lunar nutation in right ascension is taken from Table III., agreeably to the formula $-$ b tang δ. cos. (☊ + B $-$ α) + c, the tabular quantities b, B, and c, being taken out of the Table with the longitude of the moon's node, as the argument.

The lunar nutation in declination is taken also from Table III., according to the formula $-$ b sin (☊ + B $-$ α).

The solar nutation, both in right ascension and declination, is taken from Table IV. The nutation in right ascension is composed of two parts, depending respectively on the arguments 2 ⊙, and 2 ⊙ $-$ α; the latter of which tabular quantities must be multiplied by the tangent of the star's declination, taken negatively if the declination be south. The nutation in declination is taken out with the argument 2 ⊙ $-$ α, according to the signs of the column, if the declination be north, but if south, the contrary.

Example.—Let it be required to determine the aberration and nutation, lunar and solar, of β Cephei, both in right ascension and declination, at the time of culmination on the 27th of October, 1823, according to the General Tables of Professor Gauss?

In Right Ascension.

	S		
⊙	7 3° 38′	..log. a	1.2812−
Sec. δ.....	69 47	....	10.4615+
A........	+2 19 }		
α	10 21 35 }		
⊙ + A − α	8 14 22	..cos...	9.4305−
Aber. 14″.90 = +0ˢ.99.....			1.1732+

	S		
☊	9 22° 35′	log. b	0.8803−
Tang. δ..	69 47	......	10.4338+
B	+6 42 }		
α........	10 21 35 }		
☊ + B − α	11 7 42	..cos..	9.9662+
Nat. No. −19″.07		....	1.2803−
c......	+15 .29		
☾ Nut. in arc −3″.78 = −0ˢ.25			

	S	
2 ⊙.........	2 7° 16′	−0″.94
2 ⊙ − α.....	3 15 41 }	
+ 0″.12..tang. δ.....	}	+0 .32
⊙ nut. in arc.		−0 .62
Do. in time		−0ˢ.04 }
Aberration		+0 .99 }
☾ nut.		−0 .25 }
Whole correction..........		+0 .70

In Declination.

	S		
⊙	7 3° 38′	log. a....	1.2812−
δ........	69 47	sin.	9.9724+
⊙ + A − α	74 22	(sup.) sin	9.9836−
First part......	+17″.27....		1.2372
⊙ + δ second ..−	0 .93		
⊙ − δ third....+	3 .25		
Aberration	+19 .59		

	S		
☊	9 22° 35′	log. b....	0.8803−
☊ + B − α	22 18	(sup.) sin.	9.5792−
☾ nut. + 2″.88		....	0.4595+

	S	
2 ⊙ − α	3 15° 41′	− 0″.45
Solar nutation		− 0 .45
Aberration		+19 .59
☾ Nutation..............		+ 2 .88
Correction in dec..........		+22 .02
Do. in N. P. D...........		−22 .02

Professor Gauss contrived also another method of obtaining the corrections of a star due to any given day, including also the precessions, by separating the variable quantities from the constant parts of the formulæ, and by making Tables of the variable parts common to all the stars. These Tables, denominated A, B, C, D, E, and F, are published in Schumacher's *Hülfstafeln* for the year 1820, which we might have copied here, but that Bessel has made some alterations in the arrangement of the quantities, that render them more convenient for computations, and that, consequently, have gained them a preference in the estimation of several astronomers.

TABLE 1. (ABERRATION.)

	OS Log. a	VIS A +	IS Log. a	VIIS A +	IIS Log. a	VIIIS A +	
0°	1.2690	0° 0′	1.2790	2° 11′	1.2977	2° 6′	30°
1	1.2690	0 5	1.2796	2 14	1.2983	2 3	29
2	1.2691	0 11	1.2802	2 16	1.2988	2 0	28
3	1.2692	0 16	1.2808	2 18	1.2993	1 57	27
4	1.2692	0 22	1.2815	2 20	1.2998	1 54	26
5	1.2693	0 27	1.2821	2 21	1.3003	1 51	25
6	1.2695	0 32	1.2827	2 23	1.3008	1 47	24
7	1.2696	0 37	1.2834	2 24	1.3012	1 44	23
8	1.2698	0 43	1.2840	2 25	1.3017	1 40	22
9	1.2700	0 48	1.2847	2 26	1.3021	1 36	21
10	1.2703	0 53	1.2853	2 27	1.3025	1 32	20
11	1.2705	0 58	1.2860	2 28	1.3028	1 28	19
12	1.2708	1 3	1.2866	2 28	1.3032	1 24	18
13	1.2711	1 8	1.2873	2 28	1.3036	1 20	17
14	1.2714	1 12	1.2879	2 28	1.3039	1 16	16
15	1.2718	1 17	1.2886	2 28	1.3042	1 11	15
16	1.2721	1 22	1.2892	2 28	1.3045	1 7	14
17	1.2725	1 26	1.2899	2 27	1.3048	1 3	13
18	1.2729	1 30	1.2905	2 27	1.3050	0 58	12
19	1.2733	1 34	1.2912	2 26	1.3053	0 53	11
20	1.2738	1 39	1.2918	2 25	1.3055	0 49	10
21	1.2742	1 42	1.2924	2 24	1.3057	0 44	9
22	1.2747	1 46	1.2931	2 22	1.3059	0 39	8
23	1.2752	1 50	1.2938	2 21	1.3060	0 34	7
24	1.2757	1 53	1.2944	2 19	1.3061	0 30	6
25	1.2762	1 57	1.2949	2 17	1.3063	0 25	5
26	1.2768	2 0	1.2956	2 15	1.3064	0 20	4
27	1.2773	2 3	1.2961	2 13	1.3064	0 15	3
28	1.2779	2 6	1.2966	2 11	1.3065	0 10	2
29	1.2785	2 9	1.2972	2 8	1.3065	0 5	1
30	1.2790	2 11	1.2977	2 6	1.3065	0 0	0
	Log. a	A −	Log. a	A −	Log. a	A −	
	VS	XIS	IVS	XS	IIIS	IXS	

TABLE 2. (ABERRATION.)

	OS VIS − +	IS VIIS − +	IIS VIIIS − +	
0°	4″.03	3″.49	2″.02	30°
1	4.03	3.46	1.96	29
2	4.03	3.42	1.89	28
3	4.03	3.38	1.83	27
4	4.02	3.34	1.77	26
5	4.02	3.30	1.70	25
6	4.01	3.26	1.64	24
7	4.00	3.22	1.58	23
8	3.99	3.18	1.51	22
9	3.98	3.13	1.45	21
10	3.97	3.09	1.38	20
11	3.96	3.04	1.31	19
12	3.95	3.00	1.25	18
13	3.93	2.95	1.18	17
14	3.91	2.90	1.11	16
15	3.90	2.85	1.04	15
16	3.88	2.80	0.98	14
17	3.86	2.75	0.91	13
18	3.84	2.70	0.84	12
19	3.81	2.65	0.77	11
20	3.79	2.59	0.70	10
21	3.77	2.54	0.63	9
22	3.74	2.48	0.56	8
23	3.71	2.43	0.49	7
24	3.68	2.37	0.42	6
25	3.66	2.31	0.35	5
26	3.63	2.26	0.28	4
27	3.59	2.20	0.21	3
28	3.56	2.14	0.14	2
29	3.53	2.08	0.07	1
30	3.49	2.02	0.00	0
	+ − VS XIS	+ − IVS XS	+ − IIIS IXS	

TABLE 3. (LUNAR NUTATION.)

	OS Log. b	B −	VIS c − +	IS Log. b	B −	VIIS c − +	IIS Log. b	B −	VIIIS − + c	
0°	0.9844	6° 0′	0′.00	0.9588	6° 45′	8″.27	0.8960	7° 48′	14″.33	30°
1	0.9844	0 15	0.29	0.9571	6 54	8.52	0.8939	7 40	14.47	29
2	0.9843	0 31	0.58	0.9554	7 3	8.77	0.8917	7 32	14.61	28
3	0.9842	0 46	0.87	0.9536	7 12	9.01	0.8896	7 23	14.74	27
4	0.9840	1 1	1.15	0.9518	7 20	9.25	0.8875	7 14	14.87	26
5	0.9837	1 16	1.44	0.9500	7 28	9.49	0.8854	7 4	14.99	25
6	0.9834	1 32	1.73	0.9481	7 36	9.72	0.8834	6 53	15.11	24
7	0.9830	1 47	2.02	0.9462	7 43	9.96	0.8814	6 42	15.23	23
8	0.9825	2 2	2.30	0.9442	7 49	10.19	0.8795	6 29	15.34	22
9	0.9821	2 17	2.59	0.9422	7 55	10.41	0.8776	6 17	15.45	21
10	0.9815	2 31	2.87	0.9402	8 1	10.63	0.8758	6 3	15.55	20
11	0.9809	2 46	3.16	0.9382	8 6	10.85	0.8740	5 49	15.64	19
12	0.9802	3 1	3.44	0.9361	8 10	11.07	0.8723	5 35	15.73	18
13	0.9795	3 15	3.72	0.9340	8 14	11.28	0.8707	5 20	15.82	17
14	0.9787	3 29	4.00	0.9318	8 17	11.49	0.8691	5 4	15.90	16
15	0.9779	3 43	4.28	0.9297	8 20	11.70	0.8677	4 48	15.98	15
16	0.9770	3 57	4.56	0.9275	8 23	11.90	0.8663	4 31	16.05	14
17	0.9760	4 11	4.84	0.9253	8 24	12.10	0.8649	4 14	16.12	13
18	0.9750	4 24	5.11	0.9231	8 25	12.30	0.8637	3 56	16.18	12
19	0.9739	4 37	5.39	0.9208	8 25	12.49	0.8625	3 38	16.24	11
20	0.9728	4 50	5.66	0.9186	8 25	12.67	0.8615	3 20	16.29	10
21	0.9716	5 3	5.93	0.9163	8 24	12.86	0.8605	3 1	16.34	9
22	0.9704	5 16	6.20	0.9140	8 23	13.04	0.8596	2 41	16.38	8
23	0.9691	5 28	6.46	0.9118	8 21	13.21	0.8588	2 22	16.42	7
24	0.9678	5 40	6.73	0.9095	8 18	13.38	0.8582	2 2	16.45	6
25	0.9664	5 51	6.99	0.9072	8 15	13.55	0.8576	1 42	16.48	5
26	0.9650	6 3	7.25	0.9050	8 11	13.72	0.8571	1 22	16.50	4
27	0.9635	6 14	7.51	0.9027	8 6	13.88	0.8568	1 2	16.52	3
28	0.9620	6 24	7.77	0.9005	8 1	14.03	0.8565	0 41	16.53	2
29	0.9604	6 35	8.02	0.8983	7 55	14.18	0.8563	0 21	16.54	1
30	0.9588	6 45	8.27	0.8960	7 48	14.33	0.8563	0 0	16.54	0
	Log. b	+ B	− + c	Log. b	+ B	− + c	Log. b	+ B	− + c	
	VS		XIS	IVS		XS	IIIS		IXS	

TABLE 4. (SOLAR NUTATION.)

	Right Ascension. Part I. Argument. 2 ☉	Right Ascension. Part II. Argument. 2 ☉ − α.	Declin. Argument. 2 ☉ − α.	
0°	−0″.00+	−0″.47−	−0″.00+	360°
10	0.18	0.46	0.08	350
20	0.35	0.44	0.16	340
30	0.51	0.41	0.24	330
40	0.66	0.36	0.30	320
50	0.79	0.30	0.36	310
60	0.89	0.24	0.41	300
70	0.96	0.16	0.44	290
80	1.01	0.08	0.46	280
90	1.03	+0.00+	0.47	270
100	1.01	0.08	0.46	260
110	0.96	0.16	0.44	250
120	0.89	0.24	0.41	240
130	0.79	0.30	0.36	230
140	0.66	0.36	0.30	220
150	0.51	0.41	0.24	210
160	0.35	0.44	0.16	200
170	0.18	0.46	0.08	190
180	−0.00+	+0.47+	−0.00+	180

BESSEL'S MODIFICATION OF GAUSS'S TABLES.

In the *Hülfstafeln* of the year 1821, and also in No. 4 of *Astronomische Nachrichten*, Schumacher has published an account of Bessel's modification of Gauss's Tables already mentioned, together with the Tables themselves, which may be useful for reducing a number of observations of the same stars made within a comparatively short space of time, and which Bessel affirms he has himself found very convenient for this purpose. We propose, therefore, to detail the principles on which these new Tables are constructed, and to illustrate their use by a few examples; after which, we propose to insert Schumacher's Catalogue of 500 Stars, including the constants of each star, as computed from Bessel's formulæ, under Schumacher's own direction.

The formulæ of Bessel's corrections, in their unaltered state, are these:

Right Ascension.

+(46″.0175 + 20″.0436 tang. dec. sin. R. A.) *t*.
−(15″.3958 + 6″.6821 tang. dec. sin. R. A.) sin. ☊.
−(1″.2255 + 0″.5319 tang. dec. sin. R. A.) sin. 2 ⊙.
−(8″.9771 cos. ☊ − .08768 . cos. 2 ☊) tang. dec. cos. R. A.
−(0″.5799 cos. 2 ⊙) tang. dec. cos. R. A.
−(18″.5837 cos. ⊙) cos. R. A. sec. dec.
−(20″.2550 sin. ⊙) sin. R. A. sec. dec.

Declination.

+20″.0436 cos. R. A. *t*.
−6″.6821 sin. ☊ cos. R. A.
−0″.5319 sin. 2 ⊙ cos. R. A.
+(8″.9771 cos. ☊ − 0 .08768 cos. 2 ☊) sin. R. A.
−(0″.5799 cos. 2 ⊙ sin. R. A.
−(18″.5837 cos. ⊙) × (tang. ω cos. dec. − sin. R. A. sin. dec.
−(20″.2550 sin. ⊙) cos. R. A. sin. dec.

The co-efficients for the precession are those belonging to the year 1820; and are contained in the first lines respectively, where *t* denotes the time elapsed since the commencement of the year, which is assumed to be always when the sun's *mean* right ascension is $18^h\,40^m$ or $\overset{s}{9}\;10^\circ\,0'$.

For the sake of facilitating the computations, Bessel reduced some of the co-efficients into other forms, thus,

$$\frac{6''.6821}{20.0436} = 0.3334, \text{ and } \frac{0''.5319}{20.0436} = 0.0265.$$

Whence the following numbers are obtained, viz.:

$$46''.0175 \times 0''.3334 = 15''.3412 = (15.3958 - .0546)$$
$$46.0175 \times 0.0265 = 1.2212 = (1.2255 - .0043);$$

And, by substitution, the formulæ may be reduced into the subjoined form.

Right Ascension.

+(*t* − .3334 sin. ☊ − 0 .0265 sin. 2 ⊙) ×
(46″.0175 + 20″.0436 . tang. dec. sin. R. A.
−(8″.9771 cos. ☊ + 0 .5799 cos. 2 ⊙) tang. dec. cos. R. A.
−18″.5837 cos. ⊙ cos. R. A. sec. dec.
−20″.255 sin. ⊙ sin. R. A. sec. dec.
−0 .0546 sin. ☊ − 0 .0043 sin. 2 ⊙.

Declination.

+(*t* − .3334 sin. ☊ − 0 .0265 sin. 2 ⊙) 20″.0436 cos. R. A.
+(8″.9771 cos. ☊ + .5799 cos. 2 ⊙) sin. R. A.
−18″.5837 cos. ⊙ (tang. ω cos. dec. − sin. R. A. sin. dec.
−20″.255 sin. ⊙ cos. R. A. sin. dec.

The Author, in the next place, proceeds to subdivide the expressions in such way, that the constant parts of the formulæ are kept distinct from the variable parts, and also, those that are peculiar to each individual star are separated from those that are common to all the stars. The quantities that are common to both right ascension and declination, are respectively designated A, B, C, and D, and formed into Tables; the two first, which include the ☾'s ☊, are variable, and require to be computed every year, but are common to all the stars; as are also C, D, which are rendered constant by the peculiar manner in which the arguments are obtained, in sidereal, instead of solar time. Besides these four quantities, which are computed and tabulated, the two first for every ten days of the sidereal year, and the two last for every day, there are four constant quantities peculiar to each star, both in right ascension and declination; those in right ascension being expressed by the letters *a*, *b*, *c*, and *d*, and those in declination by *a′*, *b′*, *c′*, and *d′*, which are separate quantities. The various products may be tabulated according to the subjoined expressions, viz.

Right Ascension.

$A = +t - .3334 \sin. \Omega - .0265 \sin. 2 \odot + 0.0032 \sin. 2 \Omega$

$B = -8''.9771 \cos. \Omega - .5799 \cos. 2 \odot + .0877 \cos. 2 \Omega$

$C = -18''.5837 \cos. \odot$

$D = -20''.255 \sin. \odot$

$a = +46''.0175 + 20''.0436$ tang. dec. sin. R. A.

$b = +$ tang. dec. cos. R. A.

$c = +$ cos. R. A. sec. dec.

$d = +$ sin. R. A. sec. dec.

Declination.

$A = +t - .3334 \sin. \Omega - .0265 \sin. 2 \odot$

$B = -8''.9771 \cos. \Omega - .5799 \cos. 2 \odot$

$C = -18''.5837 \cos. \odot$

$D = -20''.255 \sin. \odot$

$a' = 20''.0436$ cos. R. A.

$b' = -$ sin. R. A.

$c' = +$ tang. ω. cos. dec. $-$ sin. R. A. sin. dec.

$d' = +$ cos. R. A. sin. dec.

Then, when the argument is properly prepared, the whole correction will be obtained, either from the formulæ or the Tables, thus:

$$\text{Cor. in R. A.} = \begin{cases} Aa + Bb + Cc + Dd + t.m \\ -0.0521 \sin. \Omega - 0.0004 \sin. 2\,\Omega - 0.0042 \sin. 2 \odot \end{cases}$$

$$\text{Cor. in dec.} \ldots = Aa' + Bb' + c' + Dd' + t.m'$$

Where m and m' are respectively the proper motions in right ascension and declination.

But before the tabular logarithmic quantities can be taken correctly from the Tables A, B, C, and D, the common arguments peculiar to the time of culmination of any given star, must be ascertained for the proposed day. For this purpose, the year is divided into $366\frac{1}{4}$ sidereal days, the first of which called Jan. 0, commences at noon on the 31st of December preceding: but the solar year takes its date from the moment when the sun's mean right ascension is $18^h\ 40^m$, to avoid that annual variation in the sun's longitude which amounts to an arc equal to his motion in $5^h\ 48^m\ 48^s$, the fractional part of the solar year. By this contrivance of substituting the sun's mean right ascension for his longitude, in the argument, the numbers in the Tables C and D, may remain the same in each successive year, which would not have been the case if the solar days had been used in those Tables, because the annual variation of the sun's longitude is almost 15', and in four years, amounts to 59' 8''.33. Another reason for the substitution of sidereal for solar days, is, that on certain solar days, some stars will transit the meridian twice, which cannot be the case in a sidereal day; and, thirdly, as the star's right ascension is always given in sidereal time, it may form a part of the argument, so as to adapt it to the moment of culmination, without altering the denomination of time.

In every part of the civil year, the length of the solar day exceeds that of the sidereal day, by $3^m\ 56^s.556$ of sidereal time, or $3^m\ 55^s.908$ of solar; and if the modes of reckoning time commenced at the beginning of the year, it would be an easy matter to substitute one denomination for the other by a simple Table, but the sidereal year begins on the 21st of March, and the same integers will not, therefore, express the successive days of both a solar and sidereal year. Hence Professor Gauss, and after him, Bessel, adopted a plan of allowing 1, and sometimes 2, as the solar year begins on the noon of December 31, to the numeral of the civil day, in order to have the day for which the Tables are computed, thus, if we call f the civil day, g the right ascension of a given star, and we want the numeral i to be applied to the civil date, to correspond with the Tables, then when g is less than $18^h\ 40^m$, $i = 1$, from Jan. 0 to R. A. $\odot = g$; later $i = +2$; when g exceeds $18^h\ 40^m$, $i = 0$ from Jan. 0 to R. A. $\odot = g$; later $= +1$. Again, as the sun's mean right ascension is not the same in each successive year at the noon of December 31, but differs from $18^h\ 40^m$, more or less as the sun's longitude varies, a correction of the argument, denominated k, becomes necessary, the annual variation of which is $0^d.24226$, except in a leap year, when it must be augmented by 1 in January and February. The value of k is found from the following formula; viz. $k = -0.87942 - \frac{1}{4}(T - 4\,M) + T \times 0.0077446$; where T denotes the number of years elapsed since 1800, and M the number of leap-years in the same time. For instance, for the year 1824 M is $= 5$ before Feb. 29, but $= 6$ after; hence $-\dfrac{T - 4\,M}{4} = \dfrac{24 - 20}{4} = -1$, and $24 \times 0.0077446 = +0.18587$; therefore $k = -0.87942 - 1 + 0.18587 = -1.6935$ before 29th Feb., but after, $k = -0.87942 + 0.18587 = -0.6935$, because $-\dfrac{T - 4\,M}{4} = \dfrac{24 - 24}{4} = 0$. Also, as the Tables are computed for the longitude of Paris, there requires a further correction of the argument, to reduce the time to the given meridian, which is called h. Then, if we wish to obtain the correct argument, we must take $f + g + h + i + k$; the amount of which will give the date, together with the united reductions, with which the four Tables, A, B, C, and D, must all be successively entered, for gaining the tabular quantities. This corrected argument will generally consist of a fractional portion of a day added to the integral numeral; and hence, when accuracy is the object to be attained, interpolation will become necessary in taking out the tabular quantities: but, in ordinary cases, the numeral expressing the day, after the four reductions have been applied, may be taken as the argument of the Tables, without affecting the results in any considerable degree.

As the constant numbers in the Tables and Catalogue of 500 Stars are all logarithmic, and are applied in pairs, to obtain the separate natural numbers, it must be remembered that when the two logarithms have *contrary* signs, the natural product will be *negative;* but when *like* signs, the same will be *positive.* The sum of all the four products taken algebraically will be the correction in arc, for the given day, in precession, aberration, lunar nutation, and solar nutation, both in right ascension and declination respectively; but the amount of the corrections in right ascension must be subsequently divided by 15, to convert space into time, or otherwise, the constant log. of $\frac{1}{15}$ 8.8239 may be included with each pair, in which case, the resulting corrections will be in time. The proper motion (*t.m*) for a few months, and also the minor quantities in the second line of "Cor. in R. A.," may, in ordinary cases, be omitted, as amounting to a very small aggregate.

In No. 34 of Schumacher's *Astronomiche Nachrichten*, Mr. Bessel has given an AUXILIARY Table for assisting to compute the SUPPLEMENTARY Table A and B, which was suggested to him by Mr. F. Baily, and which we will add to the Tables previously given, after having exemplified the use of these Tables, and given a Table containing the decimal fractions of a day for pointing out the numbers *g* and *h*.

Example 1.—Let it be required to determine the corrections of α Orionis, when passing the meridian of Greenwich, on April 10, 1824?

	In Right Ascension.				Sum of Cor.
f 10.000	Log. A +9.7639	Log. B −0.4245	Log. C −1.2400	Log. D −0.8554	+28″.223
g 0.240	*a* +1.6867	*b* +7.9128	*c* +8.8050	*d* +0.0027	− 0.022
h 0.006					
i .....+ 1.000	Log... +1.4506	Log... −8.3373	Log... −0.0450	Log... −0.8581	− 1.111
k .....− 0.693	Nat.No. +28.223	Nat.No. −0.022	Nat.No. −1.111	Nat.No. −7.213	− 7.213
Arg. 10.553	The whole correction in R. A.				+19″.877 ÷ 15 = +1ˢ.325

In Declination.				Sum of Cor.
Log. A +9.7639	Log. B −0.4245	Log. C −1.2400	Log. D −0.8554	+0″.739
a′ +0.1052	*b′* −9.9991	*c′* +9.4808	*d′* +7.9092	+2.653
Log. .. +9.8691	Log.... +0.4236	Log.... −0.7208	Log. .. −8.7646	−5.258
Nat. No. +0.739	Nat. No. +2.653	Nat. No. −5.258	Nat. No. −0.058	−0.058
	The whole correction in Declination..............			−1.924

Example 2.—Let it be required to determine the corrections in right ascension and declination of β Cephei on the 27th of October, 1823, when passing the meridian of Greenwich?

	In Right Ascension.				Sum of Cor.
f Oct. ... 27.000	Log. *a* +1.0869	Log. *b* +0.3277	Log. *c* +0.3553	Log. *d* −0.2547	+13″.505
g 0.893	A +0.0436	B −0.5715	C +1.1898	D +1.0493	− 7.927
h 0.006					
i ...... +1.000	Log... +1.1305	Log... −0.8992	Log... +1.5451	Log... −1.3040	+35.090
k ...... −1.451	Nat.No. +13″.505	Nat. No. −7″.929	Nat.No. +35″.090	Nat.No. −20″.140	−20.140
Arg. 27.448	Whole correction in time $\frac{20''.526}{15}$ = + 1ˢ.368.				+20″.526 = Cor. in arc.

In Declination.				Sum of Cor.
Log. *a′* +1.1962	Log. *b′* +9.7934	Log. *c′* +9.8651	Log. *d′* +9.8664	+17″.371
A +0.0436	B −0.5715	C +1.1898	D +1.0493	− 2.316
Log. .. +1.2398	Log.... −0.3649	Log.... +1.0549	Log.... +0.9157	+12.750
Nat. No. +17″.371	Nat. No. +2″.316	Nat. No. +12″.750	Nat. No. +8.235	+ 8.235
	The whole correction in declination			+36.040

DECIMAL FRACTIONS OF A DAY.

TABLE 1.

Min.	O	I	II	III	IV	V	VI	VII	VIII	IX	X	XI	Min.
0	.0000	.0417	.0833	.1250	.1666	.2083	.2500	.2917	.3333	.3749	.4166	.4583	0
1	.0007	.0424	.0840	.1257	.1673	.2090	.2507	.2924	.3340	.3756	.4173	.4590	1
2	.0014	.0431	.0847	.1264	.1680	.2097	.2514	.2931	.3347	.3763	.4180	.4597	2
3	.0021	.0438	.0854	.1271	.1687	.2104	.2521	.2938	.3354	.3770	.4187	.4604	3
4	.0028	.0445	.0861	.1278	.1694	.2111	.2528	.2945	.3361	.3777	.4194	.4611	4
5	.0035	.0452	.0868	.1285	.1701	.2118	.2535	.2952	.3368	.3784	.4201	.4618	5
6	.0042	.0459	.0875	.1292	.1708	.2125	.2542	.2959	.3375	.3791	.4208	.4625	6
7	.0049	.0466	.0882	.1299	.1715	.2132	.2549	.2966	.3382	.3798	.4215	.4632	7
8	.0056	.0473	.0889	.1306	.1722	.2139	.2556	.2973	.3389	.3805	.4222	.4639	8
9	.0063	.0480	.0896	.1313	.1729	.2146	.2563	.2980	.3396	.3812	.4229	.4646	9
10	.0069	.0486	.0902	.1319	.1736	.2152	.2569	.2986	.3402	.3818	.4235	.4652	10
11	.0076	.0493	.0909	.1326	.1743	.2159	.2576	.2993	.3409	.3825	.4242	.4659	11
12	.0083	.0500	.0916	.1333	.1750	.2166	.2583	.3000	.3416	.3832	.4249	.4666	12
13	.0090	.0507	.0923	.1340	.1757	.2173	.2590	.3007	.3423	.3839	.4256	.4673	13
14	.0097	.0514	.0930	.1347	.1764	.2180	.2597	.3014	.3430	.3846	.4263	.4680	14
15	.0104	.0521	.0937	.1354	.1771	.2187	.2604	.3021	.3437	.3853	.4270	.4687	15
16	.0111	.0528	.0944	.1361	.1778	.2194	.2611	.3028	.3444	.3860	.4277	.4694	16
17	.0118	.0535	.0951	.1368	.1785	.2201	.2618	.3035	.3451	.3867	.4284	.4701	17
18	.0125	.0542	.0958	.1375	.1792	.2208	.2625	.3042	.3458	.3874	.4291	.4708	18
19	.0132	.0549	.0965	.1382	.1799	.2215	.2632	.3049	.3465	.3881	.4298	.4715	19
20	.0139	.0556	.0972	.1389	.1806	.2222	.2639	.3056	.3472	.3888	.4305	.4722	20
21	.0146	.0563	.0979	.1396	.1813	.2229	.2646	.3063	.3479	.3895	.4312	.4729	21
22	.0153	.0570	.0986	.1403	.1820	.2236	.2653	.3070	.3486	.3902	.4319	.4736	22
23	.0160	.0577	.0993	.1410	.1827	.2243	.2660	.3077	.3493	.3909	.4326	.4743	23
24	.0167	.0584	.1000	.1417	.1834	.2250	.2667	.3084	.3500	.3916	.4333	.4750	24
25	.0174	.0591	.1007	.1424	.1841	.2257	.2674	.3091	.3507	.3923	.4340	.4757	25
26	.0181	.0598	.1014	.1431	.1848	.2264	.2681	.3098	.3514	.3930	.4347	.4764	26
27	.0188	.0605	.1021	.1438	.1855	.2271	.2688	.3105	.3521	.3937	.4354	.4771	27
28	.0195	.0612	.1028	.1445	.1862	.2278	.2695	.3112	.3528	.3944	.4361	.4778	28
29	.0202	.0619	.1035	.1452	.1869	.2285	.2702	.3119	.3535	.3951	.4368	.4785	29
30	.0208	.0625	.1042	.1458	.1875	.2291	.2708	.3125	.3542	.3957	.4374	.4791	30
31	.0215	.0632	.1049	.1465	.1882	.2298	.2715	.3132	.3549	.3964	.4381	.4798	31
32	.0222	.0639	.1056	.1472	.1889	.2305	.2722	.3139	.3556	.3971	.4388	.4805	32
33	.0229	.0646	.1063	.1479	.1896	.2312	.2729	.3146	.3563	.3978	.4395	.4812	33
34	.0236	.0653	.1069	.1486	.1903	.2319	.2736	.3153	.3569	.3985	.4402	.4819	34
35	.0243	.0660	.1076	.1493	.1910	.2326	.2743	.3160	.3576	.3992	.4409	.4826	35
36	.0250	.0667	.1083	.1500	.1917	.2333	.2750	.3167	.3583	.3999	.4416	.4833	36
37	.0257	.0674	.1090	.1507	.1924	.2340	.2757	.3174	.3590	.4006	.4423	.4840	37
38	.0264	.0681	.1097	.1514	.1931	.2347	.2764	.3181	.3597	.4013	.4430	.4847	38
39	.0271	.0688	.1104	.1521	.1938	.2354	.2771	.3188	.3604	.4020	.4437	.4854	39
40	.0278	.0695	.1111	.1528	.1945	.2361	.2778	.3195	.3611	.4027	.4444	.4861	40
41	.0285	.0702	.1118	.1535	.1952	.2368	.2785	.3202	.3618	.4034	.4451	.4868	41
42	.0292	.0709	.1125	.1542	.1959	.2375	.2792	.3209	.3625	.4041	.4458	.4875	42
43	.0299	.0716	.1132	.1549	.1966	.2382	.2799	.3216	.3632	.4048	.4465	.4882	43
44	.0306	.0723	.1139	.1556	.1973	.2389	.2806	.3223	.3639	.4055	.4472	.4889	44
45	.0313	.0730	.1146	.1563	.1980	.2396	.2813	.3230	.3646	.4062	.4479	.4896	45
46	.0320	.0737	.1153	.1570	.1987	.2403	.2820	.3237	.3653	.4069	.4486	.4903	46
47	.0327	.0744	.1160	.1577	.1994	.2410	.2827	.3244	.3660	.4076	.4493	.4910	47
48	.0334	.0751	.1167	.1584	.2001	.2417	.2834	.3251	.3667	.4083	.4500	.4917	48
49	.0341	.0758	.1174	.1591	.2008	.2424	.2841	.3258	.3674	.4090	.4507	.4924	49
50	.0347	.0764	.1180	.1597	.2014	.2430	.2847	.3264	.3680	.4096	.4513	.4930	50
51	.0354	.0771	.1187	.1604	.2021	.2437	.2854	.3271	.3687	.4103	.4520	.4937	51
52	.0361	.0777	.1194	.1611	.2028	.2444	.2861	.3277	.3694	.4110	.4527	.4944	52
53	.0368	.0784	.1201	.1618	.2035	.2451	.2868	.3284	.3701	.4117	.4534	.4951	53
54	.0375	.0791	.1208	.1625	.2042	.2458	.2875	.3291	.3707	.4124	.4541	.4958	54
55	.0382	.0798	.1215	.1632	.2049	.2465	.2882	.3298	.3714	.4131	.4548	.4965	55
56	.0389	.0805	.1222	.1639	.2056	.2472	.2889	.3305	.3721	4138	.4555	.4972	56
57	.0396	.0812	.1229	.1646	.2063	.2479	.2896	.3312	.3728	.4145	.4562	.4979	57
58	.0403	.0819	.1236	.1653	.2070	.2486	.2903	.3319	.3735	.4152	.4569	.4986	58
59	.0410	.0826	.1243	.1660	.2077	.2493	.2910	.3326	.3742	.4159	.4576	.4993	59
60	.0417	.0833	.1250	.1666	.2083	.2500	.2917	.3333	.3749	.4166	.4583	.5000	60
	.5 + XII	.5 + XIII	.5 + XIV	.5 + XV	.5 + XVI	.5 + XVII	.5 + XVIII	.5 + XIX	.5 + XX	.5 + XXI	.5 + XXII	.5 + XXIII	

TABLE 2.

AUXILIARY TABLE

For constructing the SUPPLEMENTARY TABLE.

Days.	−3′ 10″.1188 .τ	t−0.02652 Sin 2 ⊙	− 0″.57992 Cos 2 ⊙
Jan. 0	0° 0′ 0″.0	0.00908	+0″.5449
10	0 31 41.2	0.04449	+0.4419
20	1 3 22.4	0.07774	+0.2842
30	1 35 3.6	0.10812	+0.0916
Feb. 9	2 6 44.8	0.13526	−0.1119
19	2 38 25.9	0.15921	−0.2996
March 1	3 10 7.1	0.18043	−0.4525
11	3 41 48.3	0.19967	−0.5491
21	4 13 29.5	0.21794	−0.5798
31	4 45 10.7	0.23633	−0.5422
April 10	5 16 51.9	0.25585	−0.4417
20	5 48 33.1	0.27739	−0.2910
30	6 20 14.3	0.30158	−0.1081
May 10	6 51 55.4	0.32871	+0.0864
20	7 23 36.6	0.35878	+0.2702
30	7 55 17.8	0.39144	+0.4235
June 9	8 26 59.0	0.42607	+0.5298
19	8 58 40.2	0.46188	+0.5778
29	9 30 21.4	0.49792	+0.5626
July 9	10 2 2.6	0.53326	+0.4860
19	10 33 43.8	0.56703	+0.3563
29	11 5 24.9	0.59851	+0.1874
August 8	11 37 6.1	0.62724	−0.0026
18	12 8 47.3	0.65303	−0.1928
28	12 40 28.5	0.67603	−0.3623
Sept. 7	13 12 9.7	0.69667	−0.4918
17	13 43 50.9	0.71569	−0.5660
27	14 15 32.1	0.73400	−0.5756
Octob. 7	14 47 13.3	0.75263	−0.5185
17	15 18 54.5	0.77261	−0.4000
27	15 50 35.6	0.79484	−0.2335
Nov. 6	16 22 16.8	0.81995	−0.0380
16	16 53 58.0	0.84827	+0.1629
26	17 25 39.2	0.87970	+0.3448
Dec. 6	17 57 20.4	0.91378	+0.4848
16	18 29 1.6	0.94968	+0.5651
26	19 0 42.8	0.98634	+0.5752
36	19 32 23.9	1.02258	+0.5137

TABLE 3.

SUPPLEMENTARY TABLE for PRECESSION and ABERRATION.

(According to BESSEL.)

Days.	1821.		1822.		1823.		1824.	
	Log. A	Log. B —	Log. A	Log. B —	Log. A	Log. B —	Log. A	Log. B
Jan. 0	8.9191	0.9101	9.2705	0.8457	9.4314	0.7042	9.5104	0.3924—
10	9.0822	0.9143	9.3500	0.8494	9.4876	0.7075	9.5567	0.3969—
20	9.1950	0.9217	9.4138	0.8563	9.5331	0.7152	9.5958	0.4103—
30	9.2784	0.9304	9.4654	0.8652	9.5735	0.7257	9.6290	0.4292—
Feb. 9	9.3423	0.9400	9.5075	0.8745	9.6057	0.7368	9.6568	0.4489—
19	9.3929	0.9479	9.5421	0.8827	9.6326	0.7465	9.6797	0.4657—
March 1	9.4340	0.9542	9.5712	0.8887	9.6553	0.7532	9.6994	0.4765—
11	9.4690	0.9581	9.5964	0.8915	9.6750	0.7555	9.7165	0.4791—
21	9.5000	0.9586	9.6193	0.8905	9.6931	0.7529	9.7322	0.4722—
31	9.5291	0.9557	9.6411	0.8857	9.7105	0.7449	9.7474	0.4549—
April 10	9.5577	0.9504	9.6629	0.8772	9.7278	0.7318	9.7630	0.4267—
20	9.5866	0.9415	9.6853	0.8656	2.7466	0.7136	9.7793	0.3875—
30	9.6162	0.9299	9.7087	0.8516	3.7670	0.6919	9.7969	0.3378—
May 10	9.6467	0.9186	9.7331	0.8363	9.7878	0.6672	9.8156	0.3043—
20	9.6777	0.9074	9.7582	0.8212	9.8096	0.6436	9.8356	0.2142—
30	9.7086	0.8976	9.7837	0.8076	9.8322	0.6209	9.8561	0.1472—
June 9	9.7390	0.8904	9.8090	0.7967	9.8547	0.6022	9.8770	0.0852—
19	9.7682	0.8859	9.8337	0.7897	9.8768	0.5893	9.8975	0.0367—
29	9.7956	0.8848	9.8571	0.7870	9.8980	0.5831	9.9171	0.0101—
July 9	9.8210	0.8876	9.8789	0.7886	9.9178	0.5839	9.9356	0.0081—
19	9.8441	0.8932	9.8988	0.7939	9.9353	0.5906	9.9526	0.0286—
29	9.8648	0.9004	9.9168	0.8018	9.9515	0.6011	9.9677	0.0626—
Aug. 8	9.8830	0.9090	9.9327	0.8109	9.9658	0.6136	9.9811	0.1014—
18	9.8990	0.9175	9.9468	0.8198	9.9784	0.6257	9.9928	0.1371—
28	9.9131	0.9243	9.9591	0.8272	9.9893	0.6354	0.0029	0.1454—
Sept. 7	9.9256	0.9294	9.9700	0.8319	9.9985	0.6416	0.0119	0.1794—
17	9.9369	0.9315	9.9800	0.8329	0.0077	0.6412	0.0200	0.1757—
27	9.9477	0.9299	9.9895	0.8298	0.0187	0.6348	0.0303	0.1542—
Oct. 7	9.9583	0.9253	9.9989	0.8223	0.0243	0.6214	0.0353	0.1096—
17	9.9692	0.9170	0.0086	0.8106	0.0329	0.6010	0.0434	0.0362—
27	9.9809	0.9063	0.0190	0.7951	0.0424	0.5741	0.0522	9.9165—
Nov. 6	9.9934	0.8932	0.0302	0.7770	0.0525	0.5414	0.0617	9.7485—
16	0.0068	0.8797	0.0423	0.7575	0.0637	0.5053	0.0723	9.4422—
26	0.0210	0.8669	0.0551	0.7386	0.0757	0.4687	0.0837	9.0864—
Dec. 6	0.0358	0.8561	0.0685	0.7223	0.0883	0.4354	0.0957	9.3233
16	0.0507	0.8488	0.0822	0.7105	0.1012	0.4096	0.1067	9.6661
26	0.0654	0.8457	0.0956	0.7043	0.1133	0.3940	0.1196	9.6726
36	0.0767	0.8457	0.1085	0.7038	0.1263	0.3920	0.1321	9.6882

$$A = t - 0.02652 \times \sin 2 \odot - 0.33342 \times \sin \Omega + 0.00320 \times \sin 2 \Omega.$$
$$B = - 0''.57992 \times \cos 2 \odot - 8''.97707 \times \cos \Omega + 0''.08763 \times \cos 2 \Omega.$$

TABLE 4.

Days.	January.		February.		March.		April.		May.		June.		Days.
	Log. C	Log. D	Log. C	Log. D	Log. C	Log. D	Log. C	Log. D	Log. C	Log. D	Log. C	Log. D	
0	0.5091—	1.2999	1.0901—	1.1812	1.2410—	0.8479	1.2619—	0.5625—	1.1556—	1.1113—	0.8152—	1.2779—	0
1	0.5507—	1.2985	1.0986—	1.1743	1.2438—	0.8270	1.2605—	0.6012—	1.1494—	1.1200—	0.7954—	1.2805—	1
2	0.5885—	1.2969	1.1068—	1.1672	1.2464—	0.8048	1.2589—	0.6367—	1.1431—	1.1284—	0.7746—	1.2830—	2
3	0.6231—	1.2952	1.1147—	1.1597	1.2488—	0.7813	1.2572—	0.6694—	1.1365—	1.1365—	0.7526—	1.2854—	3
4	0.6550—	1.2933	1.1224—	1.1521	1.2511—	0.7563	1.2553—	0.6997—	1.1297—	1.1443—	0.7293—	1.2876—	4
5	0.6848—	1.2913	1.1298—	1.1442	1.2533—	0.7297	1.2534—	0.7279—	1.1226—	1.1519—	0.7046—	1.2898—	5
6	0.7123—	1.2891	1.1369—	1.1360	1.2553—	0.7012	1.2513—	0.7541—	1.1153—	1.1592—	0.6782—	1.2917—	6
7	0.7381—	1.2868	1.1438—	1.1274	1.2571—	0.6707	1.2491—	0.7788—	1.1079—	1.1663—	0.6500—	1.2936—	7
8	0.7625—	1.2844	1.1504—	1.1186	1.2589—	0.6376	1.2467—	0.8020—	1.0999—	1.1732—	0.6197—	1.2953—	8
9	0.7853—	1.2818	1.1569—	1.1095	1.2604—	0.6017	1.2442—	0.8238—	1.0919—	1.1798—	0.5870—	1.2969—	9
10	0.8068—	1.2790	1.1630—	1.1000	1.2619—	0.5624	1.2415—	0.8445—	1.0835—	1.1861—	0.5516—	1.2984—	10
11	0.8272—	1.2761	1.1690—	1.0902	1.2632—	0.5190	1.2387—	0.8642—	1.0749—	1.1923—	0.5130—	1.2997—	11
12	0.8465—	1.2731	1.1747—	1.0800	1.2644—	0.4707	1.2358—	0.8828—	1.0660—	1.1983—	0.4703—	1.3010—	12
13	0.8649—	1.2699	1.1802—	1.0695	1.2654—	0.4163	1.2327—	0.9006—	1.0568—	1.2040—	0.4229—	1.3021—	13
14	0.8824—	1.2665	1.1856—	1.0585	1.2663—	0.3541	1.2295—	0.9175—	1.0472—	1.2096—	0.3696—	1.3030—	14
15	0.8991—	1.2629	1.1907—	1.0472	1.2671—	0.2815	1.2261—	0.9336—	1.0373—	1.2149—	0.3088—	1.3039—	15
16	0.9150—	1.2592	1.1957—	1.0353	1.2678—	0.1940	1.2225—	0.9490—	1.0271—	1.2202—	0.2378—	1.3046—	16
17	0.9303—	1.2553	1.2004—	1.0231	1.2683—	0.0842	1.2189—	0.9638—	1.0165—	1.2252—	0.1526—	1.3053—	17
18	0.9449—	1.2513	1.2050—	1.0103	1.2687—	9.9365	1.2150—	0.9780—	1.0055—	1.2300—	0.0465—	1.3058—	18
19	0.9589—	1.2470	1.2093—	0.9971	1.2689—	9.7101	1.2110—	0.9916—	0.9941—	1.2346—	9.9058—	1.3061—	19
20	0.9722—	1.2426	1.2135—	0.9834	1.2690—	9.2022	1.2069—	1.0047—	0.9824—	1.2391—	9.6964—	1.3064—	20
21	0.9852—	1.2380	1.2175—	0.9690	1.2690—	9.2718—	1.2025—	1.0172—	0.9701—	1.2434—	9.2741—	1.3065—	21
22	0.9975—	1.2333	1.2214—	0.9540	1.2689—	9.7292—	1.1980—	1.0293—	0.9573—	1.2475—	9.0828	1.3065—	22
23	1.0094—	1.2283	1.2251—	0.9383	1.2686—	9.9465—	1.1934—	1.0409—	0.9441—	1.2515—	9.6325	1.3064—	23
24	1.0211—	1.2231	1.2286—	0.9219	1.2683—	0.0906—	1.1885—	1.0521—	0.9303—	1.2553—	9.8675	1.3062—	24
25	1.0319—	1.2178	1.2319—	0.9047	1.2677—	0.1987—	1.1835—	1.0628—	0.9159—	1.2590—	0.0187	1.3058—	25
26	1.0425—	1.2122	1.2351—	0.8867	1.2671—	0.2851—	1.1783—	1.0732—	0.9010—	1.2625—	0.1313	1.3054—	26
27	1.0527—	1.2065	1.2381—	0.8678	1.2663—	0.3570—	1.1730—	1.0833—	0.8854—	1.2659—	0.2201	1.3048—	27
28	1.0625—	1.2005	1.2410—	0.8479	1.2654—	0.4186—	1.1674—	1.0930—	0.8690—	1.2691—	0.2938	1.3041—	28
29	1.0721—	1.1943	1.2438—	0.8270	1.2643—	0.4723—	1.1616—	1.1023—	0.8519—	1.2721—	0.3566	1.3033—	29
30	1.0812—	1.1879			1.2632—	0.5198—	1.1556—	1.1113—	0.8340—	1.2751—	0.4115	1.3023—	30
31	1.0901—	1.1812			1.2619—	0.5625—	1.1494—	1.1200—	0.8152—	1.2779—	0.4600	1.3012—	31
32	1.0986—	1.1743			1.2605—	0.6012—			0.7954—	1.2805—			32

Days.	July.		August.		September.		October.		November.		December.		Days.
0	0.4115	1.3023—	1.0533	1.2061—	1.2338	0.8942—	1.2663	0.3555	1.1707	1.0872	0.8558	1.2715	0
1	0.4600	1.3012—	1.0626	1.2004—	1.2368	0.8763—	1.2654	0.4170	1.1649	1.0970	0.8371	1.2746	1
2	0.5035	1.3001—	1.0716	1.1946—	1.2396	0.8576—	1.2644	0.4707	1.1588	1.1065	0.8173	1.2776	2
3	0.5429	1.2987—	1.0802	1.1886—	1.2424	0.8378—	1.2632	0.5185	1.1525	1.1157	0.7964	1.2804	3
4	0.5791	1.2973—	1.0886	1.1823—	1.2450	0.8170—	1.2619	0.5615	1.1460	1.1245	0.7743	1.2831	4
5	0.6123	1.2957—	1.0967	1.1758—	1.2474	0.7950—	1.2605	0.6003	1.1393	1.1331	0.7509	1.2856	5
6	0.6430	1.2940—	1.1046	1.1691—	1.2497	0.7717—	1.2589	0.6359	1.1323	1.1413	0.7260	1.2880	6
7	0.6715	1.2922—	1.1122	1.1622—	1.2519	0.7469—	1.2572	0.6687	1.1251	1.1493	0.6985	1.2902	7
8	0.6983	1.2903—	1.1195	1.1550—	1.2539	0.7204—	1.2554	0.6991	1.1176	1.1570	0.6711	1.2922	8
9	0.7233	1.2882—	1.1266	1.1476—	1.2559	0.6920—	1.2534	0.7275	1.1098	1.1644	0.6405	1.2942	9
10	0.7469	1.2860—	1.1335	1.1400—	1.2576	0.6616—	1.2513	0.7540	1.1017	1.1716	0.6073	1.2960	10
11	0.7691	1.2837—	1.1401	1.1321—	1.2593	0.6287—	1.2491	0.7789	1.0934	1.1785	0.5715	1.2976	11
12	0.7901	1.2812—	1.1465	1.1239—	1.2608	0.5930—	1.2467	0.8022	1.0848	1.1852	0.5321	1.2991	12
13	0.8101	1.2786—	1.1527	1.1154—	1.2622	0.5539—	1.2441	0.8243	1.0758	1.1917	0.4886	1.3005	13
14	0.8290	1.2758—	1.1588	1.1066—	1.2634	0.5107—	1.2414	0.8452	1.0665	1.1979	0.4401	1.3017	14
15	0.8471	1.2730—	1.1646	1.0975—	1.2646	0.4625—	1.2386	0.8650	1.0569	1.2040	0.3854	1.3028	15
16	0.8644	1.2699—	1.1702	1.0881—	1.2656	0.4081—	1.2356	0.8838	1.0469	1.2098	0.3228	1.3037	16
17	0.8808	1.2668—	1.1757	1.0783—	1.2664	0.3459—	1.2325	0.9018	1.0365	1.2154	0.2494	1.3046	17
18	0.8966	1.2635—	1.1809	1.0682—	1.2672	0.2730—	1.2291	0.9190	1.0258	1.2208	0.1608	1.3052	18
19	0.9116	1.2600—	1.1859	1.0578—	1.2678	0.1850—	1.2257	0.9354	1.0146	1.2260	0.0492	1.3058	19
20	0.9261	1.2564—	1.1908	1.0469—	1.2683	0.0745—	1.2221	0.9510	1.0029	1.2311	9.8982	1.3061	20
21	0.9400	1.2527—	1.1955	1.0357—	1.2687	9.9258—	1.2183	0.9660	0.9909	1.2359	9.6637	1.3064	21
22	0.9533	1.2488—	1.2001	1.0240—	1.2689	9.6964—	1.2144	0.9803	0.9783	1.2405	9.1173	1.3065	22
23	0.9661	1.2447—	1.2045	1.0118—	1.2690	9.1790—	1.2103	0.9941	0.9652	1.2450	9.3000—	1.3065	23
24	0.9785	1.2405—	1.2087	0.9991—	1.2690	9.2900	1.2060	1.0074	0.9516	1.2493	9.7226—	1.3064	24
25	0.9903	1.2361—	1.2127	0.9860—	1.2689	9.7332	1.2015	1.0201	0.9374	1.2534	9.9330—	1.3061	25
26	1.0018	1.2315—	1.2166	0.9723—	1.2686	9.9474	1.1968	1.0324	0.9225	1.2573	0.0741—	1.3057	26
27	1.0128	1.2268—	1.2203	0.9581—	1.2682	0.0903	1.1920	1.0442	0.9069	1.2611	0.1804—	1.3051	27
28	1.0235	1.2219—	1.2239	0.9432—	1.2677	0.1976	1.1870	1.0555	0.8907	1.2647	0.2655—	1.3044	28
29	1.0338	1.2168—	1.2274	0.9276—	1.2671	0.2835	1.1817	1.0665	0.8737	1.2682	0.3367—	1.3035	29
30	1.0438	1.2115—	1.2307	0.9113—	1.2663	0.3555	1.1763	1.0770	0.8558	1.2715	0.3976—	1.3026	30
31	1.0533	1.2061—	1.2338	0.8942—	1.2654	0.4170	1.1707	1.0872	0.8371	1.2746	0.4510—	1.3015	31
32	1.0626	1.2004—	1.2368	0.8763—			1.1649	1.0970			0.4982—	1.3002	32
	Log. C	Log. D	Log. C	Log. D	Log. C	Log. D	Log. C	Log. D	Log. C	Log. D	Log. C	Log. D	

Stars 1821.	Mag.	R.A. in Time.	Ann. Var.	R. A. in Arc.	Ann. Var.	Log. a	Log. b	Log. c	Log. d	Declination.	Ann. Var.	Log. a′	Log. b′	Log. c′	Log. d′
		h. m. s.	+		+			+	+			+	–		
γ Pegasi	2.3	0 4 1.92	3ˢ.077	1° 0′ 24″.95	46″.160	1.6637	9.4028	0.0134	8.2583	+14° 11′ 19″.90	+20.050	1.3019	8.2448	9.6196	9.3893
ι Ceti	4	0 10 18.16	3.053	2 34 32.40	45.795	1.6614	9.2376–	0.0060	8.6591	– 9 48 57.50	20.045	1.3016	8.6526	9.6388	9.2312–
κ Cassiopeæ	4	0 22 53.35	3.308	5 43 20.18	49.623	1.6968	0.2711	0.3254	9.3263	+61 56 31.00	19.933	1.2998	8.9987	9.0651	9.9435
ζ Cassiopeæ	4	0 27 2.78	3.278	6 45 41.73	49.170	1.6913	0.1184	0.2166	9.2906	+52 54 37.96	19.925	1.2990	9.0709	9.2248	9.8988
π Androm.	4.5	0 27 20.84	3.176	6 50 12.53	47.633	1.6771	9.8050	0.0720	9.1508	+32 43 58.89	19.949	1.2989	9.0757	9.4782	9.7299
ε Androm.	4	0 29 6.66	3.138	7 16 39.91	47.075	1.6756	9.7283	0.0519	9.1582	+28 20 18.30	19.649	1.2985	9.1027	9.5079	9.6729
δ Androm.	3	0 29 46.44	3.173	7 26 36.18	47.599	1.6767	9.7557	0.0583	9.1744	+29 52 50.44	19.813	1.2983	9.1124	9.4939	9.6937
α Cassiopeæ	3	0 30 24.32	3.327	7 36 1.98	49.902	1.6977	0.1599	0.2436	9.3689	+55 33 15.76	19.836	1.2982	9.1215	9.1349	9.9124
β Ceti	2.3	0 34 36.09	3.016	8 39 1.31	45.244	1.6530	9.5313–	0.0193	9.2015	–18 58 13.61	19.870	1.2970	9.1773	9.6622	9.5070–
ζ Androm.	4	0 37 52.22	3.162	9 28 3.32	47.423	1.6761	9.6280	0.0310	9.2531	+23 17 30.46	19.689	1.2961	9.2161	9.5233	9.5911
η Cassiopeæ	4	0 38 19.40	3.530	9 34 51.07	52.954	1.7086	9.1791	0.2562	9.4836	+56 51 47.64	19.290	1.2959	9.2213	8.9907	9.9168
ν Androm.	4	0 39 58.36	3.260	9 59 35.41	48.907	1.6897	9.9187	0.1097	9.3558	+40 6 5.89	19.710	1.2954	9.2394	9.3428	9.8024
γ Cassiopeæ	3	0 45 58.10	3.511	11 29 31.44	52.670	1.7230	0.2253	0.2889	9.5971	+59 44 44.80	19.722	1.2932	9.2994	8.6682	9.9276
μ Androm.	4	0 46 50.88	3.284	11 42 43.20	49.261	1.6914	0.8763	0.0916	9.4082	+37 31 34.55	19.682	1.2929	9.3075	9.3436	9.7756
ε Piscium	4	0 53 39.60	3.109	13 24 56.98	46.641	1.6682	9.0724	9.9912	9.3687	+ 6 55 30.52	19.518	1.2900	9.3655	9.6052	9.0692
η Ceti	3.4	0 59 35.44	3.021	14 53 51.55	45.311	1.6532	9.2792–	9.9934	9.4183	–11 7 57.38	19.293	1.2872	9.4104	9.6772	9.2709–
β Androm.	2	0 59 43.82	3.316	14 55 57.31	49.740	1.6953	9.8250	0.0700	9.4960	+34 40 9.74	19.331	1.2871	9.4111	9.3230	9.7401
θ Cassiopeæ	4.5	1 0 15.91	3.573	15 3 58.71	53.600	1.7262	0.1266	0.2176	9.6477	+54 11 39.07	19.323	1.2867	9.4149	8.6346	9.8938
ψ Cassiopeæ	4.5	1 13 24.60	4.069	18 21 9.00	61.035	1.7853	0.3535	0.3889	9.9097	+67 11 28.49	19.032	1.2794	9.4981	9.0863–	9.9420
δ Cassiopeæ	3	1 14 11.01	3.822	18 32 45.19	57.327	1.7539	0.2033	0.2688	9.7945	+59 18 5.19	18.992	1.2789	9.5025	8.7153–	9.9113
θ Ceti	3	1 15 4.60	2.994	18 46 8.95	44.915	1.6530	9.1813–	9.9818	9.5130	– 9 6 34.35	18.801	1.2783	9.5075	9.6808	9.1758–
η Piscium	4	1 21 54.99	3.192	20 28 47.35	47.876	1.6796	9.3818	9.9856	9.5578	+14 25 16.12	18.825	1.2737	9.5439	9.5227	9.3679
51 Androm.	3.4	1 27 3.12	3.618	21 45 46.78	54.273	1.7339	0.0091	0.1400	9.7412	+47 43 2.13	18.490	1.2700	9.5691	8.2482	9.8370
τ Ceti	3.4	1 35 45.03	2.783	23 56 15.49	41.739	1.6390	0.4431–	9.9801	9.6274	+16 52 56.14	19.238	1.2630	9.6083	9.7269	9.4239–
ε Cassiopeæ	3.4	1 41 37.36	4.187	25 24 20.47	62.811	1.7974	0.2446	0.2956	9.9722	+62 46 57.00	18.068	1.2579	9.6325	9.2625–	9.9049
ζ Ceti	3	1 42 37.35	2.950	25 39 20.29	44.255	1.6463	9.2526–	9.9633	9.6448	–11 13 26.48	17.886	1.2570	9.6364	9.7076	9.2442–
α Triang.	3.4	1 42 53.96	3.386	25 43 29.36	50.795	1.7057	9.6931	0.0116	9.6945	+28 42 6.88	17.814	1.2567	9.6375	9.2362	9.6361
γ Arietis ½	4.5	1 43 43.54	3.270	25 55 53.16	49.044	1.6896	9.4763	9.9767	9.6636	+18 24 {55.86 / 46.91}	18.000	1.2560	9.6408	9.4372	9.4534
β Arietis	3	1 44 45.68	3.280	26 11 25.14	49.198	1.6922	9.5124	9.9798	9.6716	+19 55 45.99	17.929	1.2550	9.6448	9.4109	9.4855
f Cassiopeæ	4.5	1 48 20.53	4.867	27 5 7.93	73.010	1.8653	9.4262	0.4492	0.1579	+71 32 53.19	17.867	1.2516	9.6583	9.4691–	9.9266
υ Ceti	4.5	1 51 33.98	2.820	27 53 29.73	42.299	1.6257	9.5516–	9.9790	9.7027	–21 56 53.36	17.771	1.2484	9.6701	9.7615	9.5190–
γ Androm.	3.4	1 52 56.81	3.625	28 14 12.15	54.377	1.7355	9.8913	0.0703	9.8003	+41 27 55.68	17.601	1.2470	9.6750	8.0777	9.7659
α Arietis	3	1 57 6.36	3.352	29 16 29.21	50.279	1.6998	9.5603	9.9754	9.7241	+22 36 42.69	17.383	1.2427	9.6893	9.3277	9.5255
β Triang.	4	1 58 55.61	3.537	29 43 54.12	53.054	1.7222	9.7699	0.0208	9.7776	+34 8 11.29	17.435	1.2408	9.6954	8.9082	9.6878
ο Ceti	Var.	2 10 18.52	3.025	32 34 37.86	45.378	1.6575	8.7473–	9.9266	9.7321	– 3 47 39.36	16.729	1.2277	9.7311	9.6708	8.7463–
35 Cas. Hev.	4.5	2 14 27.88	4.774	33 36 58.27	71.612	1.8551	0.2841	0.3214	0.1441	+66 35 20.34	16.642	1.2226	9.7432	9.5259–	9.8832
ν Ceti	4.5	2 26 29.14	3.124	36 37 17.13	46.857	1.6723	8.8293	9.9061	9.7772	+ 4 48 26.80	16.105	1.2066	9.7756	9.5826	8.5288
δ Ceti	4	2 30 19.03	3.069	37 34 45.45	46.035	1.6620	7.7935–	9.8991	9.7852	– 0 26 57.70	15.837	1.2011	9.7852	9.6423	7.7935–
ε Ceti	4 5	2 30 54.59	2.894	37 43 38.89	43.410	1.6362	9.2488–	9.9088	9.7973	–12 38 11.81	15.647	1.2002	9.7867	9.7461	9.2381–
θ Persei	4	2 32 1.28	4.028	38 0 19.20	60.422	1.7777	9.9491	0.0749	9.9677	+48 27 51.22	15.702	1.1986	9.7894	9.2384–	9.7707
35 Arietis	4	2 32 57.79	3.482	38 14 26.83	52.228	1.7187	9.6011	9.9449	9.8416	+26 56 21.57	15.757	1.1972	9.7917	9.0273	9.5512
γ Ceti	3	2 34 1.92	3.094	38 30 28.82	46.415	1.6680	8.5294	9.8938	9.7916	+ 2 28 32.83	15.510	1.1957	9.7942	9.6093	9.5289
μ Ceti	4	2 35 16.32	3.219	38 49 4.04	48.283	1.6820	9.1084	9.8974	9.8027	+ 9 21 11.98	15.679	1.1945	9.7972	9.5137	9.1025
π Ceti	4	2 35 36.17	2.843	38 54 2.56	42.641	1.6308	9.3075–	9.9054	9.8122	–14 37 9.99	15.735	1.1932	9.7979	9.7623	9.2932–
39 Arietis	4	2 37 16.35	3.538	39 19 5.22	53.073	1.7235	9.6232	9.9446	9.8579	+28 29 50.26	15.406	1.1907	9.8018	8.8984	9.5672
p Persei	4.5	2 39 19.26	3.749	39 49 48.88	56.238	1.7473	9.7715	9.9863	9.9075	+37 34 30.81	15.353	1.1874	9.8065	8.6684–	9.6705
41 Arietis	3	2 39 28.13	3.499	39 52 2.01	52.489	1.7195	9.5831	9.9333	9.8551	+26 30 59.77	15.287	1.1872	9.8069	9.0094	9.5349
τ Eridani	4.5	2 42 55.15	2.714	40 43 47.27	40.710	1.6107	9.4804–	9.9116	9.8466	–21 44 44.06	15.245	1.1816	9.8146	9.8095	9.4483–
η Eridani	3	2 47 41.43	2.927	41 55 21.52	43.904	1.6410	9.1005–	9.8777	9.8310	– 9 33 53.24	14.768	1.1737	9.8249	9.7320	9.0944–
γ Persei	3.4	2 51 53.27	4.270	42 58 19.09	64.044	1.8062	9.9840	0.0828	0.0521	+52 47 47.48	14.645	1.1665	9.8336	9.4479–	9.7654
α Ceti	2	2 52 56.01	3.121	43 13 57.12	46.817	1.6705	8.6340	9.8632	9.8364	+ 3 22 56.33	14.524	1.1646	9.8357	9.5942	9.6332
ρ Persei	3.4	2 53 41.01	3.796	43 26 0.14	56.945	1.7546	9.7560	9.9653	9.9416	+38 8 21.24	14.476	1.1632	9.8373	8.9203–	9.6517
11 Eridani	4	2 54 30.09	2.641	43 37 31.34	39.616	1.5995	9.5150–	9.9000	9.8792	–24 19 49.38	14.496	1.1617	9.8388	9.8323–	9.4745
β Persei	Var.	2 56 33.03	3.851	44 8 15.50	57.765	1.7622	9.7835	9.9732	9.9602	+40 15 30.80	14.426	1.1581	9.8429	9.0748–	9.6663
δ Arietis	4	3 1 24.30	3.410	45 21 7.93	51.145	1.7070	9.3848	9.8711	9.8764	+19 2 36.83	14.129	1.1480	9.8521	9.2508	9.3604
12 Eridani	3.4	3 4 28.52	2.551	46 7 7.78	38.259	1.5772	9.5969–	9.9019	9.9189	–29 41 48.91	14.594	1.1429	9.8578	9.8658	9.5358–
ζ Eridani	4	3 7 8.53	2.903	46 47 8.00	43.546	1.6392	9.0587–	9.8415	9.8686	– 9 29 24.44	13.781	1.1376	9.8626	9.7390	9.0527–
16 Eridani	3.4	3 11 33.27	2.658	47 53 19.07	39.869	1.6008	9.4419–	9.8606	9.9044	–22 24 55.51	13.470	1.1285	9.8703	9.8351	9.4077–
α Persei	2.3	3 11 35.47	4.220	47 53 49.98	63.294	1.8010	9.8904	0.0113	0.0553	+49 12 55.56	13.435	1.1285	9.8704	9.4444–	9.7056
ο Tauri	4.5	3 15 11.32	3.212	48 47 49.73	48.187	1.6834	8.9876	9.8234	9.8811	+ 8 23 33.34	13.163	1.1208	9.8764	9.5046	8.9829
ξ Tauri	4	3 17 28.59	3.233	49 22 8.81	48.495	1.6853	9.0184	9.8192	9.8857	+ 9 6 8.24	13.045	1.1158	9.8802	9.4893	9.0129
17 Eridani	4.5	3 21 44.54	2.967	50 26 8.11	44.511	1.6481	9.8029–	9.8068	9.8801	– 5 41 43.21	12.765	1.1065	9.8870	9.7062	8.8008–
ε Eridani	4	3 24 30.08	2.818	51 7 31.24	42.265	1.6359	9.0471–	9.8044	9.8980	–10 4 13.25	12.524	1.0998	9.8913	9.7509	9.0404–

Stars 1821.	Mag.	R. A. in Time.	Ann. Var.	R. A. in Arc.	Ann. Var.	Log. a	Log. b	Log. c	Log. d	Declination.	Ann. Var.	Log. a′	Log. b′	Log. c′	Log. d′
		h. m. s.	+		+			+	+		+	+	—		
19 Eridani..	4	3 25 52.96	2s.637	51° 28′ 14″.43	39″.549	1.5978	9.4060—	9.8280	9.9269	−22° 14′ 20″.06	12″.466	1.0966	9.8934	9.8437	9.3725—
δ Persei......	3.4	3 30 13.35	4.219	52 33 20.21	63.285	1.8007	9.8174	9.9518	0.0677	+47 12 20.65	12.112	1.0861	9.8998	9.4590—	9.6495
ν Persei......	4.5	3 33 3.23	4.019	53 15 46.49	60.292	1.7816	9.7313	9.9058	0.0328	+42 0 14.71	11.990	1.0790	9.9038	9.3299—	9.6023
b Pleiadum	4.5	3 34 15.67	3.538	53 33 55.03	53.064	1.7246	9.4129	9.8115	9.9433	+23 32 37.70	11.942	1.0759	9.9055	8.8839	9.3752
δ Eridani ...	3.4	3 34 40.88	2.870	53 40 13.13	43.053	1.6344	9.0353—	9.7798	9.9133	−10 22 29.52	12.662	1.0748	9.9061	9.7574	9.0281—
η Tauri......	3	3 36 50.97	3.538	54 12 50.18	53.071	1.7251	9.4062	9.8047	9.9469	+23 32 38.98	11.681	1.0692	9.9091	8.8684	9.3684
ζ Persei......	3.4	3 42 53.90	3.742	55 43 28.43	56.133	1.7490	9.5353	9.8192	9.9857	+31 20 34.01	11.214	1.0527	9.9172	8.7719—	9.4668
ε Persei......	3.4	3 45 52.15	3.988	56 28 2.28	59.815	1.7769	9.6581	9.8547	0.0334	+39 28 59.09	11.058	1.0445	9.9209	9.2901—	9.5456
γ Eridani ...	2.3	3 49 40.84	2.791	57 25 12.58	41.863	1.6212	9.1287—	9.7443	9.9388	−14 1 25.37	10.715	1.0333	9.9257	9.7961	9.1156—
λ Tauri......	4	3 50 46.01	3.302	57 41 30.15	49.535	1.6956	9.0546	9.7375	9.9365	+11 58 41.86	10.793	1.0301	9.9270	9.3965	9.0450
μ Persei......	4.5	4 1 47.39	4.357	60 26 50.80	65.362	1.8152	9.7377	9.8670	0.0135	+47 56 34.09	9.785	0.9953	9.9395	9.5504	9.5637
ο Eridani....	4.5	4 3 7.57	2.910	60 46 53.50	43.653	1.6412	9.7968—	9.6921	9.9444	−7 18 40.97	9.867	0.9907	9.9409	9.7337	8.7933—
γ Tauri......	3.4	4 9 36.91	3.398	62 24 12.88	50.975	1.7062	9.0995	9.6812	9.9630	+15 11 16.28	9.279	0.9680	9.9476	9.2712	9.0841
*41 Eridani	3.4	4 11 6.86	2.249	62 46 42.93	33.736	1.5300	9.4931—	9.7429	9.9316	−34 14 26.04	9.260	0.9624	9.9490	9.9340	9.4105—
δ1 Tauri	4	4 12 37.31	3.438	63 9 15.90	51.576	1.7120	9.1432	9.6744	9.9702	+17 6 54.42	9.118	0.9569	9.9505	9.1824	9.1235
δ2 Tauri	4.5	4 13 47.17	3.444	63 26 48.57	51.661	1.7119	9.1363	9.6698	9.9710	+17 1 18.00	8.996	0.9525	9.9516	9.1851	9.1168
ε Tauri......	4	4 18 10.49	3.485	64 32 34.42	52.273	1.7174	9.1647	9.6570	9.9794	+18 46 30.23	8.628	0.9356	9.9556	9.0804	9.1410
α Tauri......	1	4 25 39.70	3.425	66 24 48.91	51.375	1.7104	9.0637	9.6197	9.9796	+16 8 28.25	7.916	0.9045	9.9621	9.2099	9.0463
ν Eridani....	4	4 27 22.64	2.986	66 50 39.63	44.783	1.6514	8.4083—	9.5956	9.9644	−3 43 33.11	7.885	0.8969	9.9635	9.6927	8.4074—
υ Eridani....	3	4 28 35.79	2.330	67 8 56.85	34.951	1.5434	9.3669—	9.6558	0.0311	−30 56 6.52	7.750	0.8914	9.9645	9.7274	9.3002—
53 Eridani...	4	4 29 59.02	2.738	67 29 45.29	41.074	1.6146	9.0005—	9.5973	9.9800	−14 39 35.41	7.565	0.8851	9.9656	9.8154	8.9862—
54 Eridani..	4	4 32 36.82	2.617	68 9 12.33	39.251	1.5937	9.1322—	9.5978	9.9847	−20 1 15.86	7.408	0.8729	9.9676	9.8607	9.1052—
1 Orionis ...	4	4 40 7.98	3.256	70 1 59.70	48.840	1.6831	8.5994	9.5363	9.9760	+6 38 23.78	6.823	0.8357	9.9731	9.5085	8.5964
3 Orionis ...	4	4 41 41.01	3.193	70 25 15.11	47.897	1.6791	8.4920	9.5270	9.9760	+5 17 28.64	6.736	0.8274	9.9741	9.5382	8.4901
z Orionis....	4	4 44 55.35	3.107	71 13 50.24	46.600	1.6695	8.0800	9.5078	9.9766	+2 8 23.98	6.479	0.8098	9.9763	9.6003	8.0797
ι Aurigæ ...	4	4 45 18.71	3.880	71 19 40.61	58.207	1.7656	9.3157	9.5811	0.0523	+32 52 22.65	6.450	0.8075	9.9765	9.1751—	9.2400
10 Camelop.	4.5	4 47 32.31	5.288	71 53 4.62	79.322	1.8989	9.7331	9.7959	0.2812	+60 9 57.73	6.228	0.7952	9.9779	9.7843—	9.4309
ε Aurigæ....	4	4 49 8.26	4.277	72 17 3.92	64.149	1.8073	9.4613	9.6231	0.1187	+43 32 48.90	6.099	0.7857	9.9789	9.5336—	9.3215
ζ Aurigæ ...	4	4 49 58.95	4.171	72 29 44.23	62.565	1.7960	9.4144	9.3992	0.1003	+40 48 10.70	6.016	0.7807	9.9794	9.4693—	9.2935
ι Tauri	4.5	4 52 24.11	3.570	73 6 0.26	53.556	1.7284	9.0549	9.4942	0.0116	+21 19 31.90	5.803	0.7658	9.9808	8.7511	9.0241
η Aurigæ ...	4	4 53 58.45	4.184	73 29 36.81	62.754	1.7973	9.3924	9.5756	0.1038	+40 58 50.84	5.624	0.7560	9.9817	9.4787—	9.2703
ε Leporis	3.4	4 57 52.62	2.525	74 28 9.23	37.874	1.5795	9.0475—	9.4625	0.0186	−22 37 1.26	5.378	0.7301	9.9838	9.8872	9.0127—
β Eridani ...	3	4 59 3.18	2.943	74 45 47.77	44.148	1.6456	8.3891—	9.4215	9.9863	−5 19 31.29	5.228	0.7220	9.9845	9.7174	8.3872—
λ Eridani ...	4	5 0 34.88	2.866	75 8 43.18	42.991	1.6330	8.6081—	9.4142	9.9906	−8 59 28.00	5.124	0.7112	9.9852	9.7633	8.6028—
α Aurigæ ...	1	5 3 29.00	4.409	75 52 9.25	66.140	1.8196	9.3998	9.5443	0.1433	+45 48 15.97	4.457	0.6903	9.9866	9.5940	9.2431
ι Leporis ...	4.5	5 3 56.80	2.791	75 59 11.96	41.863	1.6217	8.7149—	9.3938	9.9966	−12 5 27.19	4.902	0.6866	9.9869	9.7977	8.7052—
β Orionis ...	1	5 5 56.44	2.877	76 29 3.55	43.157	1.6348	8.5388—	9.3734	9.9925	−8 24 55.15	4.699	0.6710	9.9878	9.7574	8.5341—
τ Orionis ...	4	5 8 55.17	2.913	77 13 47.59	43.700	1.6395	8.4365—	9.3478	9.9924	−7 2 44.43	4.412	0.6468	9.9891	9.7407	8.4332—
λ Leporis ...	4.5	5 11 19.64	2.754	77 49 54.62	41.317	1.6166	8.6998—	9.3358	0.0021	−13 22 7.27	4.223	0.6262	9.9901	9.8118	8.6878—
β Tauri......	2	5 14 59.19	3.785	78 44 43.66	56.771	1.7533	9.0242	9.3463	0.0474	+28 26 46.83	3.717	0.5931	9.9916	8.9321—	8.9683
η Orionis ...	4.5	5 15 28.77	3.008	78 52 11.50	45.123	1.6544	7.9377—	9.2861	9.9922	−2 34 12.09	3.849	0.5882	9.9917	9.6791	7.9373—
γ Orionis ...	2	5 15 32.06	3.215	78 53 0.86	48.230	1.6826	8.3195	9.2876	9.9943	+6 10 42.51	3.833	0.5876	9.9918	9.5131	8.3170
β Leporis...	4	5 20 34.72	2.568	80 8 40.76	30.527	1.5852	8.8155—	9.2630	0.0231	−20 54 30.96	3.368	0.5358	9.9935	9.8791	8.7859—
δ Orionis ...	2	5 22 51.87	3.062	80 42 58.03	45.926	1.6615	7.0932—	9.2077	9.9943	−0 26 22.97	3.186	0.5103	9.9943	9.6450	7.0932—
α Leporis....	3.4	5 24 50.14	2.643	81 12 32.11	39.643	1.5977	8.6949—	9.2059	0.0166	−17 57 27.82	3.080	0.4879	9.9949	9.8559	8.6732—
φ Orionis ...	4.5	5 24 59.56	3.282	81 14 53.36	49.230	1.6927	8.3993	9.1881	0.0007	+9 21 33.79	3.034	0.4847	9.9949	9.4274	8.3935
λ Orionis ...	4	5 25 16.76	3.294	81 19 11.44	49.409	1.6941	8.4164	9.1851	0.0014	+9 48 22.60	3.065	0.4814	9.9950	9.4139	8.4100
ι Orionis....	3.4	5 26 41.17	2.939	81 40 17.55	44.079	1.6427	8.1850—	9.1633	9.9978	−6 2 2.18	2.975	0.4636	9.9954	9.7289	8.1826—
ζ Tauri......	3.4	5 26 56.93	3.572	81 44 7.69	53.576	1.7295	8.7423	9.1875	0.0254	+21 1 26.99	2.888	0.4618	9.9955	8.7000	8.7124
ε Orionis	2.3	5 27 7.93	3.038	81 46 58.93	45.570	1.6586	7.5191—	9.1552	9.9956	−1 19 25.44	2.865	0.4577	9.9955	9.6597	7.5190
σ Orionis....	4	5 29 45.54	3.002	82 26 23.04	45.031	1.6540	7.7946—	9.1196	9.9967	−2 42 38.40	2.696	0.4218	9.9962	9.6816	7.7941—
ζ Orionis ...	3	5 31 43.84	3.027	82 55 57.54	45.412	1.6562	7.6429—	9.0903	9.9970	−2 2 41.84	2.477	0.3927	9.9967	9.6713	7.6426—
γ Leporis ...	4	5 36 59.86	2.492	84 14 57.89	37.375	1.5769	8.6184—	9.0353	0.0322	−22 30 49.04	1.628	0.3036	9.9978	9.8932	8.5839—
ζ Leporis ...	4.5	5 38 50.43	2.709	84 42 36.42	40.638	1.6097	8.3896—	8.9796	0.0130	−14 53 43.98	1.897	0.2676	9.9981	9.8296	8.3748—
κ Orionis ...	3	5 39 15.96	2.840	84 48 59.40	42.601	1.6293	8.1906—	8.9622	0.0045	−9 44 24.29	1.847	0.2591	9.9982	9.7754	8.1843—
136 Tauri....	4.5	5 42 4.61	3.763	85 31 9.22	56.444	1.7516	8.6104	8.9451	0.0510	+27 33 32.72	1.482	0.1965	9.9987	8.8833—	8.5580
*δ Aurigæ..	3.4	5 44 45.83	4.922	86 11 27.43	73.830	1.8681	8.9645	9.0551	0.2325	+54 15 23.64	1.190	0.1261	9.9990	9.7453—	8.7310
α Orionis ...	1	5 45 29.17	3.245	86 22 14.61	48.677	1.6867	7.9128	8.8050	0.0027	+7 21 55.24	1.275	0.1052	9.9991	9.4808	7.9092
*β Aurigæ..	2	5 46 23.93	4.400	86 35 58.9	66.000	1.8193	8.7743	8.9254	0.1491	+44 55 1.15	1.164	0.0774	9.9992	9.5993—	8.6244
θ Aurigæ ...	4	5 47 30.71	4.084	86 52 40.69	61.255	1.7868	8.6162	8.8349	0.0981	+37 11 22.25	0.997	0.0410	9.9994	9.4112—	8.5175
η Leporis....	4	5 48 15.05	2.728	87 3 45.80	40.913	1.6123	8.1130—	8.7231	0.0129	−14 12 23.33	1.252	0.0137	9.9994	9.8234	8.0996—
ν Orionis....	4.5	5 57 20.82	3.420	89 20 12.37	51.307	1.7102	7.4850	8.0782	0.0146	+14 46 54.33	0.297	9.3748	0.0000	9.2163	7.4703
θ Leporis ...	4.5	6 58 3.42	2.716	89 30 51.24	40.743	1.6093	7.3542—	7.9432	0.0149	−14 55 40.72	0.114	9.2406	0.0000	9.8306	7.3393—

Stars 1821	Mag.	R. A. in Time.	Ann. Var.	R. A. in Arc.	Ann. Var.	Log. a	Log. b	Log. c	Log. d	Declination.	Ann. Var.	Log. a′	Log. b′	Log. c′	Log. d′
		h. m. s.	+		+			—	+		—	—	—		
2 Lyncis....	4.5	6 3 49.00	5ˢ.301	90° 57′ 15″.06	79″.518	1.9001	8.4437—	8.5104	0.2889	+59° 3′ 37″.99	0ˢ.334	9.5146	9.9990	9.8024—	8.1548—
κ Aurigæ ...	4	6 3 58.06	3.816	90 59 24.97	57.244	1.7588	7.9912—	8.2981	0.0605	+29 33 15.73	0.672	9.5340	9.9999	9.0629—	7.9306—
η Geminor...	4.5	6 4 4.26	3.616	91 1 0.32	57.247	1.7351	7.8673—	8.2836	0.0345	+22 32 57.45	0.351	9.5428	9.9999	8.2420	7.8327—
*5 Monosc.	4.5	6 6 7.36	2.922	91 31 50.33	43.828	1.6418	7.4644	8.4292	0.0024	— 6 13 36.45	0.450	9.7251	9.9998	9.7323	7.4618
μ Geminor.	3	6 12 6.72	3.629	93 1 52.90	54.431	1.7352	8.3426—	8.7580	0.0341	+22 35 45.04	1.156	0.0232	9.9994	8.2312	8.3079—
ζ Canis Maj.	3	6 13 26.64	2.302	93 21 39.53	34.528	1.5375	8.5294	8.8306	0.0617	—29 59 30.13	1.280	0.0689	9.9992	9.9420	8.4670
β Canis Maj.	2.3	6 14 48.92	2.637	93 42 13.86	39.561	1.5973	8.3188	8.8317	0.0206	—17 52 33.13	1.379	0.1106	9.9991	9.8570	8.2973
λ Canis Maj.	4	6 15 34.19	2.190	93 53 32.87	32.855	1.5166	8.6501	8.9099	0.0772	—33 21 9.51	1.510	0.1329	9.9990	9.9596	8.5719
γ Geminor.	3	6 27 21.85	3.463	96 50 27.70	51.941	1.7154	8.5488—	9.0943	0.0153	+16 32 35.16	2.400	0.3769	9.9969	9.1250	8.5304—
ε Geminor...	3	6 32 54.76	3.693	98 13 39.81	55.402	1.7435	8.8302—	9.1994	0.0393	+25 17 55.84	2.856	0.4570	9.9955	8.4843—	8.7864—
ξ² Geminor.	4	6 35 14.48	3.371	98 48 37.15	50.566	1.7043	8.5513—	9.1966	0.0063	+13 4 50.10	3.212	0.4864	9.9948	9.2991	8.5399—
α Canis Maj.	1	6 37 15.54	2.644	99 18 52.06	39.661	1.6039	8.6801	9.2273	0.0125	—16 28 35.11	4.440	0.5107	9.9942	9.8427	8.6619
κ¹ Canis Maj.	4	6 43 9.29	2.238	100 47 19.37	33.569	1.5259	9.0732	9.3453	0.0653	—32 18 26.99	3.723	0.5739	9.9922	9.9503	9.0002
o¹ Canis Maj.	4	6 46 42.29	2.488	101 40 34.35	37.321	1.5718	8.9541	9.3453	0.0301	—23 57 59.68	3.953	0.5936	9.9909	9.9000	8.9149
ι Canis Maj.	4.5	6 48 9.07	2.670	102 2 15.99	40.048	1.6030	8.7999	9.3382	0.0094	—16 49 42.51	4.112	0.6209	9.9903	9.8442	8.7809
ε Canis Maj.	2.3	6 51 35.38	2.355	102 53 50.74	35.328	1.5478	9.0877	9.4058	0.0460	—28 44 2.72	4.436	0.6504	9.9889	9.9290	9.0306
ζ Geminor.	4	6 53 29.12	3.565	103 22 16.95	53.471	1.7278	8.9443—	9.3934	0.0174	+20 49 26.83	4.672	0.6657	9.9881	8.7776	8.9148—
σ Canis Maj.	3.4	6 54 35.29	2.385	103 38 49.40	35.779	1.5539	9.0927	9.4256	0.0404	—27 41 6.99	4.779	0.6746	9.9876	9.9221	9.0399
o² Canis Maj.	4	6 55 32.87	2.501	103 53 13.04	37.508	1.5743	9.0201	9.4181	0.0250	—23 34 42.77	4.836	0.6820	9.9871	9.8955	8.9823
γ Canis Maj.	4	6 55 39.38	2.712	103 54 50.66	40.684	1.6092	8.8204	9.3969	0.0029	—15 22 30.78	4.792	0.6828	9.9871	9.8299	8.8045
δ Canis Maj.	3.4	7 1 6.72	2.436	105 16 40.84	36.541	1.5628	9.1112	9.4675	0.0311	—26 6 54.61	5.246	0.7225	9.9844	9.9108	9.0644
22 Monosc.	4.5	7 2 43.38	3.068	105 40 50.65	46.020	1.6623	6.9834	9.4318	9.9835	— 0 12 15.41	5.397	0.7334	9.9835	9.6410	6.9834
27 Canis Ma.	4.5	7 6 57.45	2.443	106 44 21.75	36.648	1.5639	9.1485	9.5059	0.0277	—26 2 58.57	5.707	0.7612	9.9812	9.9088	9.1020
λ Geminor.	4.5	7 7 48.10	3.458	106 57 1.44	51.875	1.7146	8.9461—	9.4838	9.9998	+16 51 17.17	5.851	0.7663	9.9807	9.1399	8.9270—
δ Geminor.	3.4	7 9 25.24	3.593	107 21 19.62	53.896	1.7313	9.0876—	9.5084	0.0135	+22 18 11.52	5.982	0.7763	9.9798	9.5950	9.0539—
ι Geminor...	4	7 14 35.82	3.741	108 38 56.73	56.115	1.7496	9.2332—	9.5595	0.0312	+28 8 41.06	6.436	0.8064	9.9766	8.8076—	9.1785—
η Canis Maj.	3	7 17 0.39	2.360	109 15 5.81	35.393	1.5508	9.2612	9.5762	0.0330	—28 57 35.44	6.605	0.8199	9.9750	9.9227	9.2032
β Canis Min.	3	7 17 25.94	3.253	109 21 29.05	48.788	1.6892	8.7023—	9.5254	9.9797	+ 8 38 33.35	6.648	0.8221	9.9747	9.4584	8.6973—
α Geminor.	3.4	7 23 9.95	3.842	110 47 20.82	57.634	1.7623	9.3305—	9.6230	0.0436	+32 16 16.52	7.170	0.8519	9.9708	9.1212—	9.2776—
α Canis Min.	1.2	7 29 55.82	3.146	112 28 52.97	47.197	1.6808	8.5799—	9.5847	9.9678	+ 5 40 35.66	8.647	0.8842	9.9657	9.5321	8.5778—
26 Monosc.	4.5	7 32 41.52	2.863	113 10 22.74	42.938	1.6340	8.8015	9.6005	9.9690	— 9 8 24.24	7.875	0.8968	9.9635	9.7593	8.7959
κ Geminor.	4	7 33 37.66	3.625	113 24 17.35	54.381	1.7366	9.2641—	9.6411	0.0048	+24 49 7.89	8.009	9.9008	9.9627	7.9412	9.2220—
β Geminor.	2	7 34 21.15	3.682	113 35 9.70	55.224	1.7479	9.3360—	9.6581	0.0180	+28 26 59.47	8.078	0.9039	9.9621	8.7400—	9.2801—
ξ Navis.....	4	7 41 45.98	2.523	115 26 29.75	37.846	1.5775	9.2901	9.6737	9.9964	—24 24 59.02	8.560	0.9349	9.9557	9.8856	9.2494
ι Navis.....	3.4	7 59 54.89	2.548	119 58 43.29	38.216	1.5839	9.3430	9.7373	9.9762	—23 47 35.58	9.885	1.0006	9.9376	9.8731	9.3045
β Cancri.....	4	8 6 47.73	3.258	121 41 55.93	48.863	1.6897	8.9548—	9.7268	9.9361	+ 9 43 49.23	10.554	1.0223	9.9298	9.4533	8.9485—
o Ursæ Maj	4.5	8 15 17.67	5.072	123 49 25.08	76.077	1.8834	0.0073—	0.0642	0.2381	+61 18 17.39	11.332	1.0474	9.9195	9.7163—	9.6887—
δ Hydræ.....	4	8 28 10.04	3.177	127 2 30.60	47.662	1.6794	8.8244—	9.7825	9.9048	+ 6 19 20.19	12.015	1.0817	9.9021	9.5359	8.8217—
δ Cancri.....	4.5	8 34 29.84	3.425	128 37 25.72	51.378	1.7105	9.3275—	9.8191	9.9166	+18 48 21.06	12.738	1.0972	9.8928	9.2015	9.3036—
ε Hydræ.....	4	8 37 17.20	3.185	129 19 17.95	47.773	1.6809	8.8953—	9.8052	9.8918	+ 7 4 6.86	12.757	1.1035	9.8885	9.5258	8.8919—
ζ Hydræ.....	4	8 45 55.91	3.189	131 28 58.60	47.837	1.6791	8.8859—	9.8240	9.8775	+ 6 37 16.32	13.264	1.1230	9.8746	9.5375	8.8830—
ι Ursæ Maj.	3.4	8 46 53.90	4.157	131 43 28.44	62.362	1.7999	9.8800—	0.0039	0.0537	+48 44 12.17	13.635	1.1251	9.8729	9.4389—	9.6992—
κ Ursæ Maj.	4.5	8 51 21.66	4.151	132 50 24.91	62.261	1.7943	9.8758—	0.0057	0.0385	+47 51 20.23	13.797	1.1344	9.8653	9.4022—	9.7026—
θ Hydræ ...	4.5	9 5 2.65	3.128	136 15 39.70	46.919	1.6609	8.5874—	9.8595	9.8403	+ 3 3 50.19	14.809	1.1608	9.8397	9.5982	8.5858—
38 Lyncis...	4	9 7 40.09	3.760	136 55 1.41	56.401	1.7524	9.7494—	9.9644	9.9353	+37 33 16.93	14.605	1.1654	9.8345	8.8589—	9.6485—
40 Lyncis...	4.5	9 10 7.34	3.685	137 31 50.08	55.271	1.7447	9.7154—	9.9552	9.9168	+35 8 37.47	14.744	1.1697	9.8294	8.5280—	9.6280—
h Ursæ Maj.	4	9 17 17.54	4.848	139 19 23.16	72.723	1.8611	0.1886—	0.2355	0.1697	+63 50 13.73	15.110	1.1818	9.8141	9.5951—	9.8329—
α Hydræ ...	2	9 18 47.55	2.945	139 41 47.87	44.172	1.6456	9.0238	9.8864	9.8149	— 7 53 12.42	15.226	1.1843	9.8108	9.7149	9.0197
θ Ursæ Maj.	3	9 20 49.16	4.072	140 12 17.42	61.083	1.7975	0.0003—	0.1010	0.0216	+52 29 10.02	16.011	1.1875	9.8062	9.3859—	9.7849—
λ Leonis ...	4.5	9 20 49.44	3.443	140 12 21.53	51.641	1.7130	9.5291—	9.9240	9.8446	+23 45 9.32	15.423	1.1885	9.8062	9.1445	9.4906—
o Leonis.....	4	9 31 35.11	3.205	142 53 43.56	48.074	1.6840	9.1782—	9.9094	9.7881	+10 42 8.80	16.000	1.2037	9.7805	9.4976	9.1706—
ε Leonis.....	3	9 35 40.08	3.423	143 55 1.17	51.351	1.7112	9.5681—	9.9488	9.8114	+24 35 39.32	16.184	1.2094	0.7701	9.1748	9.5268—
υ Ursæ Maj.	4.5	9 38 10.95	4.382	144 32 44.22	65.730	1.8200	0.1473—	0.2103	0.0628	+59 52 24.45	16.555	1.2129	9.7635	9.4531	9.8479—
μ Leonis.....	3	9 42 33.33	3.422	145 38 19.96	51.333	1.7139	9.7210—	9.9662	9.8011	+26 50 43.43	16.599	1.2187	9.7516	9.1219	9.5715—
π Leonis	4.5	9 50 44.54	3.180	147 41 10.66	47.704	1.6785	9.1217—	9.9322	9.7333	+ 8 53 58.45	16.953	1.2289	9.7280	9.5392	9.1164—
η Leonis.....	3.4	9 57 33.57	3.281	149 23 21.38	49.213	1.6925	9.4370—	9.9557	9.7278	+17 37 54.75	17.251	1.2368	9.7069	9.4140	9.4161—
α Leonis.....	1	9 58 49.94	3.202	149 42 22.25	48.024	1.6842	9.2940—	9.9472	9.7138	+12 50 19.13	17.252	1.2382	9.7028	9.4929	9.2830—
λ Hydræ ...	4.5	10 1 51.47	2.918	150 27 52.04	43.766	1.6436	9.2469	9.9483	9.7016	—11 28 21.24	17.508	1.2415	9.6928	9.7188	9.2381
λ Ursæ Maj.	3.4	10 6 15.59	3.665	151 33 53.87	54.975	1.7417	9.9260—	0.0858	9.8194	+43 48 13.35	17.701	1.2461	9.6777	8.2131—	9.7844—
ζ Leonis.....	4.4	10 6 42.79	3.353	151 40 41.88	50.294	1.7017	9.5994—	9.9849	9.7165	+24 18 22.37	17.550	1.2466	9.6762	9.3017	9.5591—
γ Leonis.....	2	10 10 5.13	3.320	152 31 17.01	49.800	1.6948	9.5263—	9.9771	9.6932	+20 44 36.11	17.908	1.2500	9.6641	9.3846	9.4974—
μ Ursæ Maj.	3	10 11 37.61	3.621	152 54 24.19	54.319	1.7353	9.9100—	0.0812	9.7901	+42 23 46.98	17.787	1.2515	9.6584	8.1287	9.7783—
30 Leon. Mi.	4.5	10 15 36.80	3.462	153 54 12.04	51.925	1.7171	9.7937—	0.0384	9.7284	+34 42 15.63	18.087	1.2552	9.6433	9.0269	9.7087—

Stars 1821.	Mag.	R. A. in Time.	Ann. Var.	R. A. in Arc.	Ann. Var.	Log. a	Log. b	Log. c	Log. d	Declination.	Ann. Var.	Log. a'	Log. b'	Log. c'	Log. d'
		h. m. s.	+		+			—	+		—	—	—		
μ Hydræ........	4	10 17 25 .85	2ˢ.889	154° 21′ 27″.81	43″.330	1.6389	9.4103	9.9720	9.6532	−15° 55′ 30″.53	18″.177	1.2569	9.6362	9.7293	9.3933
31 Leonis Min.	4.5	10 17 28 .93	3.489	154 22 13 .93	52.330	1.7219	9.8419—	0.0562	9.7373	+37 37 16 .09	18.162	1.2570	9.6360	8.9017	9.7406—
ϱ Leonis Min....	4	10 23 22 .49	3.169	155 50 39 .34	47.529	1.6765	9.2164—	9.9672	9.6189	+10 13 31 .72	18.282	1.2621	9.6119	9.5496	9.2094—
37 Leonis Min.	4	10 28 37 .15	3.405	157 9 17 .28	51.070	1.7080	9.7754—	0.0405	9.6650	+32 54 12 .48	18.441	1.2665	9.5891	9.1861	9.6995—
42 Leonis Min.	4.5	10 35 52 .73	3.355	158 58 10 .93	50.326	1.7029	9.7595—	0.0399	9.6247	+31 37 22 .42	18.711	1.2720	9.5549	9.2587	9.6896—
ν Hydr. & Crat.	4	10 40 47 .61	2.945	160 11 54 .15	44.172	1.6451	9.4093	9.9801	9.5455	−15 15 30 .73	18.631	1.2755	9.5299	9.7058	9.3938
46 Leonis Min.	4.5	10 43 16 .30	3.381	160 49 4 .46	50.718	1.7047	9.8233—	0.0628	9.6042	+35 10 39 .34	19.158	1.2772	9.5166	9.2187	9.7357—
54 Leonis Min.	4.5	10 45 54 .15	3.275	161 28 32 .26	49.122	1.6909	9.6593—	0.0221	9.5473	+25 42 10 .89	18.968	1.2788	9.5020	9.4036	9.6141—
β Ursæ Maj.....	2	10 50 58 .50	3.703	162 44 37 .52	55.540	1.7427	0.1731—	0.2479	9.7401	+57 20 22 .42	19.115	1.2823	9.4722	8.1908—	9.9052—
α Hydr. & Crat.	4	10 51 3 .25	2.908	162 45 48 .73	43.617	1.6450	9.4747	0.0002	9.4920	−17 20 49 .72	18.990	1.2820	9.4718	9.7013	9.4545
α Ursæ Maj.....	1.2	10 52 35 .79	3.804	163 8 53 .35	57.062	1.7581	0.2684—	0.3197	9.8010	+62 42 54 .11	19.291	1.2829	9.4622	8.7684—	9.9297—
χ Leonis........	4.5	10 55 46 .38	3.097	163 56 35 .75	46.451	1.6705	9.1468—	9.9873	9.4464	+ 8 18 8 .90	19.302	1.2847	9.4418	9.5906	9.1423—
ψ Ursæ Majoris	3.4	10 59 33 .96	3.423	164 53 29 .36	51.333	1.7104	9.9918—	0.1388	9.5701	+45 28 4 .90	19.438	1.2867	9.4160	9.0740	9.8377—
β Hydr. & Crat.	4	11 2 52 .11	2.944	165 43 1 .62	44.163	1.6438	9.5895	0.0187	9.4246	−21 50 58 .51	19.512	1.2884	9.3922	9.6943	9.5571
δ Leonis........	3	11 4 34 .26	3.206	166 8 33 .91	48.093	1.6804	9.5826—	0.0185	9.4107	+21 30 12 .16	19.571	1.2891	9.3793	9.4997	9.5513—
θ Leonis........	3	11 4 49 .89	3.155	166 12 28 .39	47.331	1.6760	9.4563—	0.0053	9.3954	+16 24 26 .16	19.481	1.2893	9.3773	9.5429	9.4382—
ξ Ursæ Majoris	4	11 8 36 .28	3.217	167 9 4 .21	48.261	1.6890	9.7937—	0.0631	9.4212	+32 32 4 .29	20.163	1.2910	9.3471	9.3915	9.7196—
υ Ursæ Majoris	4	11 8 47 .13	3.270	167 11 46 .96	49.052	1.6904	9.8192—	0.0709	9.4274	+34 4 13 .71	19.453	1.2910	9.3456	9.3718	9.7374—
δ Hydr. & Crat.	3.4	11 10 23 .80	2.991	167 35 57 .04	44.869	1.6528	9.3803	0.0025	9.3447	−13 48 34 .33	19.330	1.2917	9.3319	9.6746	9.3676
σ Leonis........	4	11 11 54 .19	3.113	167 58 28 .99	46.688	1.6677	9.0800—	9.9936	9.3220	+ 7 0 34 .76	19.636	1.2923	9.3188	9.6079	9.0768—
ι Leonis	4	11 14 34 .13	3.131	168 38 44 .51	46.965	1.6704	9.3005—	0.0002	9.3030	+11 30 58 .32	19.685	1.2934	9.2942	9.5866	9.2916—
γ Hydr. & Crat.	4	11 15 56 .48	2.978	168 59 7 .18	44.664	1.6519	9.4691	0.0106	9.2999	−16 42 3 .80	19.594	1.2939	9.2812	9.6727	9.4504
τ Leonis........	4	11 18 43 .69	3.083	169 40 53 .39	46.247	1.6652	8.8199—	9.9939	9.2541	+ 3 50 29 .99	19.707	1.2949	9.2531	9.6243	8.8190—
λ Draconis......	3.4	11 20 39 .88	3.705	170 9 58 .16	55.579	1.7450	0.4400—	0.4662	9.7051	+70 19 2 .90	19.857	1.2955	9.2325	8.1650—	9.9674—
e Leonis........	3.5	11 21 10 .28	3.064	170 17 34 .15	45.956	1.6618	8.5404	9.9940	9.2272	− 2 0 59 .20	19.779	1.2957	9.2269	9.6432	8.5401
ξ Hydr. & Crat.	4	11 24 12 .46	2.923	171 3 6 .83	43.849	1.6449	9.7712	0.0610	9.2582	−30 52 5 .29	19.854	1.2967	9.1918	9.6555	9.7048
θ Hydr. & Crat.	4	11 27 36 .11	3.033	171 54 1 .61	45.497	1.6588	9.1860	0.0008	9.1541	− 8 48 42 .11	19.773	1.2976	9.1489	9.6537	9.1809
υ Leonis........	4.5	11 27 46 .94	3.068	171 56 43 .47	46.022	1.6630	7.4544—	9.9957	9.1465	+ 0 9 52 .10	19.802	1.2977	9.1465	9.6371	7.4544—
ζ Hydr. & Crat.	4	11 35 41 .74	3.020	173 55 26 .13	45.302	1.6566	9.4924	0.0178	9.0450	−17 21 20 .16	19.941	1.2995	9.0247	9.6492	9.4722
χ Ursæ Majoris	4	11 36 32 .92	3.207	174 8 13 .82	48.099	1.6845	0.0551—	0.1788	9.1903	+48 46 18 .30	19.925	1.2997	9.0093	9.3206	9.8740—
ν Virginis	4.5	11 36 39 .22	3.089	174 9 48 .35	46.311	1.6655	9.1191—	0.0015	9.0110	+ 7 31 56 .28	20.133	1.2995	9.0073	9.6201	9.1153—
93 Leonis	4	11 38 44 .24	3.103	174 41 3 .56	46.539	1.6697	9.5871—	0.0286	8.9973	+21 12 50 .15	19.949	1.3001	8.9668	9.5695	9.5567—
β Leonis........	2.3	11 39 55 .51	3.065	174 58 47 .74	45.976	1.6675	9.4436—	0.0146	8.9583	+15 34 22 .06	20.015	1.3003	8.9420	9.5962	9.4272—
β Virginis	3.4	11 41 22 .47	3.122	175 20 31 .22	46.829	1.6637	8.6838—	9.9991	8.9101	+ 2 46 25 .08	20.257	1.3006	8.9096	9.6331	8.6833—
β Hydr. & Crat.	4	11 43 53 .43	3.005	175 58 21 .47	45.077	1.6542	9.8100	0.0749	8.9226	−32 54 44 .65	20.033	1.3009	8.8466	9.6048	9.7340
γ Ursæ Majoris	2	11 44 22 .34	3.209	176 5 29 .56	48.135	1.6808	0.1487—	0.2370	9.0716	+54 41 24 .71	20.045	1.3010	8.8335	9.2906	9.9107—
ο Virginis	4.5	11 56 5 .17	3.060	179 1 17 .59	45.902	1.6635	9.2340—	0.0062	8.2387	+ 9 43 39 .83	20.033	1.3019	8.2324	9.6283	9.2277—
α Corvi..........	4.5	11 59 11 .86	3.074	179 47 57 .83	46.104	1.6626	9.6430	0.0384	7.5824	−23 43 46 .68	20.093	1.3020	7.5441	9.6007	9.6047
									—				+		
ε Corvi..........	4	12 0 55 .89	3.068	180 13 58 .33	46.014	1.6632	9.5981	0.0317	7.6405	−21 37 26 .48	20.070	1.3020	7.6088	9.6042	9.5664
δ Ursæ Majoris	3	12 6 30 .47	3.014	181 37 37 .04	45.216	1.6543	0.2045—	0.2759	8.7293	+58 1 39 .50	20.296	1.3018	8.4532	9.4047	9.9284—
γ Corvi..........	3	12 6 36 .61	3.067	181 39 9 .09	46.006	1.6645	9.4727	0.0182	8.4783	−16 32 46 .03	19.954	1.3018	8.4599	9.6105	9.4543
η Virginis.......	3.4	12 10 44 .88	3.065	182 41 15 .59	45.974	1.6629	7.7591—	9.9995	8.6710	+ 0 19 47 .28	20.021	1.3016	8.6711	9.6378	7.7591—
a Com. Beren.	4.5	12 18 1 .03	3.010	184 30 15 .38	45.155	1.6549	9.7210—	0.0520	8.9484	+27 49 7 .82	19.956	1.3007	8.8950	9.6238	9.6677—
δ Corvi	3	12 20 37 .11	3.104	185 9 16 .65	46.554	1.6676	9.4417	0.0144	8.9696	−15 31 0 .52	20.085	1.3002	8.9535	9.5957	9.4256
η Corvi	4.5	12 22 51 .57	3.076	185 42 53 .57	46.139	1.6680	9.4320	0.0133	9.0136	−15 12 8 .24	19.958	1.2998	8.9981	9.5941	9.4165
β Corvi..........	2 3	12 24 59 .74	3.116	186 14 56 .09	46.739	1.6713	9.6126	0.0315	9.0709	−22 24 19 .60	19.999	1.2994	9.0368	9.5560	9.5785
8 Canum Ven.	4.5	12 25 13 .30	2.869	186 18 19 .48	43.037	1.6436	9.9569—	0.1286	9.1719	+42 19 58 .11	19.585	1.2994	9.0407	9.5963	9.8257—
κ Draconis......	3.4	12 25 46 .24	2.612	186 26 33 .55	39.181	1.5974	0.4548—	0.4797	9.5325	+70 46 34 .84	19.980	1.2992	9.0500	9.3960	9.9723—
γ Virginis.......	4	12 32 35 .55	3.035	188 8 52 .87	45.525	1.6631	7.9054	9.9956	9.1515	− 0 27 55 .79	19.893	1.2976	9.1515	9.6364	7.9054
... Comes.......	4	12 32 35 .66	3.035	188 8 54 .97	45.525	1.6631	7.9060	9.9956	9.1515	− 0 27 58 .49	19.826	1.2976	9.1515	9.6364	7.9060
ε Ursæ Majoris	3	12 46 6 .94	2.667	191 31 44 .13	39.998	1.6006	0.1775—	0.2542	9.5638	+56 56 0 .48	19.753	1.2931	9.3007	9.6067	9.9144—
δ Virginis.......	3.4	12 46 35 .21	3.017	191 38 48 .22	45.257	1.6600	8.8744—	9.9922	9.3063	+ 4 22 23 .84	19.691	1.2930	9.3051	9.6515	8.8732—
12 Canum Ven.	2.3	12 47 38 .06	2.818	191 54 30 .96	42.266	1.6298	9.9034—	0.1018	9.4259	+39 17 14 .02	19.562	1.2926	9.3146	9.6689	9.7921—
36 Com. Beren.	4.5	12 50 3 .67	2.967	192 30 55 .12	44.502	1.6491	9.5110—	0.0123	9.3586	+18 22 43 .26	19.415	1.2916	9.3358	9.6814	9.4883—
Virginis.......	3.4	12 53 15 .89	2.987	193 18 58 .37	44.807	1.6536	9.3128—	9.9976	9.3718	+11 55 28 .16	19.444	1.2902	9.3623	9.6742	9.3033—
41 Com Beren.	4	12 58 34 .47	2.881	194 38 37 .06	43.217	1.6360	9.7220—	0.0421	9.4592	+28 35 17 .57	19.485	1.2877	9.4028	9.7008	9.6655—
ψ Hydræ	4.5	12 59 25 .64	3.203	194 51 24 .53	48.052	1.6822	9.5951	0.0186	9.4122	−22 9 28 .84	19.445	1.2872	9.4089	9.4847	9.5618
θ Virginis.......	4.5	13 0 41 .27	3.097	195 10 22 .22	46.451	1.6669	8.8883	9.9360	9.4192	− 4 34 45 .88	19.382	1.2866	9.4178	9.6147	8.8869
42 Com. Beren.	4.5	13 1 16 .33	2.917	195 19 4 .93	43.752	1.6458	9.5082—	0.0073	9.4448	+18 28 44 .39	19.192	1.2863	9.4219	9.6949	9.4853—
61 Virginis......	4.5	13 9 3 .11	3.116	197 15 46 .64	46.744	1.6801	9.4737	0.0001	9.4925	−17 18 43 .62	20.158	1.2820	9.4724	9.5133	9.4536
γ Hydræ	4.5	13 9 12 .54	3.237	197 18 8 .04	48.555	1.6853	9.5911—	0.0134	9.5069	+22 13 22 .27	19.093	1.2819	9.4733	9.7112	9.5576—
α Virginis	1	13 15 46 .60	3.144	198 56 36 .13	47.153	1.6738	9.2319	9.9828	9.5183	−10 13 22 .24	18.963	1.2778	9.5114	9.5677	9.2249
ζ Ursæ Majoris	3	13 16 40 .70	2.422	199 10 10 .53	36.331	1.5600	0.1440—	0.2261	9.7672	+55 51 47 .81	18.959	1.2772	9.5163	9.7121	9.8931—

Stars 1821.	Mag.	R. A. in Time.	Ann. Var.	R. A. in Arc.	Ann. Var.	Log. a	Log. b	Log. c	Log. d	Declination.	Ann. Var.	Log. a'	Log. b'	Log. c'	Log. d'
		h. m. s.	+		+			—	—		—	—	+		
g Urs. Maj.	5	13 18 2.08	2″.426	199° 30′ 31″.18	36″.386	1.5578	0.1440—	0.2259	9.7752	+55° 55′ 22″.40	18″.968	1.2763	9.5237	9.7158	9.8925—
ζ Virginis..	4	13 25 31.80	3.051	201 23 41.99	45.767	1.6625	7.7211—	9.9690	9.5621	— 0 19 26.00	18.536	1.2710	9.5621	9.6396	7.7211—
η Urs. Maj.	2.3	13 40 28.80	2.370	205 7 0.63	35.550	1.5539	0.0363—	0.1507	9.8217	+50 12 36.64	18.151	1.2589	9.6278	9.7810	9.8424—
υ Bootis ...	4	13 40 50.25	2.886	205 12 33.68	43.288	1.6381	9 4334—	9.9752	9.6480	+16 41 27.19	18.029	1.2586	9.6293	9.7309	9.4147—
k Centauri	4.5	13 41 32.34	3.443	205 23 5.09	51.642	1.7110	9.7534	0.0280	9.7042	—32 6 4.18	18.134	1.2567	9.6322	9.1457	9.6813
η Bootis ...	3	13 46 9.48	2.856	206 32 22.15	42.839	1.6323	9.4960—	9.9768	9.6752	+19 17 58.14	18.258	1.2537	9.6501	9.7461	9.4708—
i Draconis	4.5	13 46 11.09	1.738	206 32 46.42	26.068	1.4193	0.2051—	0.3354	0.0340	+65 36 33.29	18.022	1.2536	9.6502	9.7682	9.9110—
τ Virginis	4.5	13 52 32.59	3.044	208 8 8.81	45.667	1.6591	8.5705—	9.9458	9.6739	+ 2 24 56.57	17.745	1.2475	9.6735	9.6566	8.5702—
π Hydræ ..	4.5	13 56 12.15	3.387	209 3 2.23	50.808	1.7052	9.6262	9.9873	9.7319	—25 48 52.47	17.663	1.2437	9.6862	9.2536	9.5805
α Draconis	3.4	13 59 32.13	1.605	209 53 1.89	24.069	1.3868	0.2740—	0.3159	0.0753	+65 14 2.66	17.388	1.2401	9.6974	9.8022	9.8961—
κ Virginis	4	14 3 21.37	3.189	210 50 24.88	47.832	1.6787	9.1544	9.9397	9.7157	— 9 26 14.26	17.263	1.2358	9.7098	9.5367	9.1485
ι Virginis	4	14 6 38.34	3.136	211 39 35.16	47.033	1.6718	8.8841	9.9318	9.7218	— 5 8 26.43	17.424	1.2321	9.7201	9.5858•	8.8823
α Bootis ...	1	14 7 30.17	2.728	211 52 26.01	40.927	1.6247	9.4929—	9.9564	9.7500	+20 7 9.15	18.957	1.2311	9.7227	9.7702	9.4655—
λ Virginis	4	14 9 26.11	3.228	212 21 32.99	48.417	1.6849	9.2739	9.9372	9.7390	—12 32 27.26	16.864	1.2288	9.7285	9.4878	9.2634
λ Bootis ...	4	14 9 33.98	2.281	212 23 29.64	34.213	1.5383	9.9556—	0.0921	9.8945	+46 54 51.36	16.774	1.2286	9.7289	9.8374	9.7901—
ι Bootis	4.5	14 9 48.90	2.122	212 27 13.43	31.833	1.5072	0.0365—	0.1388	9.9422	+52 11 46.79	16.849	1.2283	9.7297	9.8389	9.8239—
θ Bootis ...	4	14 19 5.43	2.032	214 46 21.42	30.473	1.4917	0.0324—	0.1319	9.9735	+52 40 55.01	16.904	1.2166	9.7561	9.8553	9.8151—
ρ Bootis ...	4	14 24 6.82	2.588	216 1 42.32	38.822	1.5898	9.6893—	9.9755	9.8372	+31 9 41.60	16.090	1.2099	9.7695	9.8298	9.6217—
γ Bootis ...	3.4	14 24 51.90	2.426	216 12 58.53	36.386	1.5610	9.8166—	0.0169	9.8816	+39 5 46.15	15.986	1.2088	9.7715	9.8509	8.7065—
			—		—										
a Urs. Min.	4	14 28 2.35	0.270	217 0 35.20	4.045	0.6244—	0.5216—	0.5338	0.4111	+76 29 29.23	16.062	1.2043	9.7796	9.8368	8.8901—
			+		+										
π Bootis ...	3.4	14 32 18.37	2.804	218 4 35.57	42.058	1.6252	9.3866—	9.9159	9.8099	+17 11 27.99	15.766	1.1982	9.7901	9.7759	9.3667—
ζ Bootis ...	3.4	14 32 35.81	2.850	218 8 57.22	42.757	1.6316	9.3084—	9.9097	9.8048	+14 30 9.82	15.697	1.1977	9.7908	9.7596	9.2943—
μ Virginis	4.5	14 33 38.26	3.149	218 24 33.83	47.242	1.6728	8 8249	9.8957	9.7948	— 4 52 24.70	16.004	1.1962	9.7933	9.5794	8.8233
34 Bootis...	4.5	14 35 32.81	2 632	218 53 12.10	39.487	1.5969	9.6039—	9.9425	9.8491	+27 17 38.91	15.574	1.1933	9.7978	9.8285	9.5526—
ο Bootis ...	4.5	14 36 53.11	2.791	219 13 16.62	41.866	1.6229	9.3938—	9.9103	9.8221	+17 43 40.13	15.552	1.1912	9.8009	9.7824	9.3727—
ε Bootis	3	14 37 9.76	2.613	219 17 26.34	39.198	1.5946	9.6113—	9.9421	9.8550	+27 50 3.39	15.516	1.1908	9.8016	9.8322	9.5579—
109 Virgin.	4	14 37 12.04	3.019	219 18 0 56	45.278	1.6574	8.5548—	9.8891	9.8021	+ 2 39 14.52	15.489	1.1907	9.8017	9.6655	8.5543—
α^1 Libræ...	6	14 40 48.33	3.298	220 12 1.31	49.473	1.6950	9.3184	9.8985	9.8254	—15 14 40.94	15.344	1.1851	9.8099	9.3963	9.3029
α^2 Libræ...	3	14 40 59.77	3.302	220 14 53.63	49.530	1.6951	9.3194	9.8983	9.8260	—15 17 25.04	15.306	1.1847	9.8103	9.3950	9.3037
ξ Bootis ...	3.4	14 43 7.85	2.762	220 46 57.80	41.423	1.6159	9.4367—	9.9058	9.8416	+19 50 55.86	15.294	1.1813	9.8150	9.7994	9.4101—
			—		—										
β Urs. Min.	3	14 51 20.06	0.305	222 50 0.09	4.581	0.6479—	0.4338—	0.4491	0.4162	+74 53 13.35	14.820	1.1673	9.8324	9.8862	9.8500—
			+		+										
δ Libræ ...	4.5	14 51 25.12	3.188	222 51 16.85	47.821	1.6802	9.0019	9 8692	9.8366	— 7 48 4.91	14.635	1.1672	9.8326	9.5285	8.9979
γ Scorp. / 20 Libr.	3.4	14 53 36.80	3.483	223 24 11.96	52.246	1.7186	9.5214	9.9025	9.8783	—24 34 12.24	14.522	1.1633	9.8370	9.0375	9.4801
β Bootis...	3	14 55 11.92	2.256	223 47 58.79	33.842	1.5304	9.7991—	9.9813	9.9631	+41 6 5.58	14.520	1.1605	9.8402	9.8933	9.6762—
δ Lupi	4.5	15 6 57.92	3 618	226 44 28.83	54.268	1.7345	9.5882	9.8961	9.9225	—29 28 57.23	13.804	1.1380	9.8623	9.2880	9.5280
β Libræ ...	2.3	15 7 23.08	3.211	226 50 46.19	48.170	1.6836	9.0205	9.8401	9.8681	— 8 42 52.94	13.679	1.1372	9.8630	9.5032	9.0155
δ Bootis ...	3.4	15 8 16.85	2.414	227 4 12.74	36.208	1.5578	9.6620—	9.9146	9.9460	+33 59 19.74	13.699	1.1353	9.8646	8.8860	9.5806—
μ Bootis....	4	15 17 43.34	2.257	229 25 50.17	33.862	1.5330	9.7061—	9.9167	9.9841	+38 0 40.26	12.875	1.1186	9.8806	9.9083	9.6026—
β Cor. Bor.	4	15 20 26.93	2.475	230 6 44.70	37.132	1.5709	9.5637—	9.8683	9.9463	+29 43 44.80	12.743	1.1095	9.8850	9.8793	9.5021—
*ι Draconis	3	15 20 57.27	1.314	230 14 19.12	19.720	1.4736	0.0374—	0.1017	0.1815	+59 35 44.86	12.800	1.1079	9.8858	9.9458	9.7416—
			—		—										
*γ Ur. M.	3.4	15 21 4.45	0.203	230 16 6.75	3.050	0.4302—	0.3061—	0.3266	0.4069	+72 28 16.77	12.727	1.1076	9.8860	9.9366	9.7849—
			+		+										
37 Libræ ..	4	15 24 24.21	3.259	231 6 3.09	48.879	1.6867	9.0189	9.8039	9.8970	— 9 26 35.37	12.813	1.1001	9.8911	9.4778	9.0130
γ Libræ ...	4.5	15 25 31.65	3.341	231 22 57.54	50.119	1.6987	9.1974	9.8087	9.9063	—14 11 0.39	12.389	1.0974	9.8928	9.3610	9.1840
*θ Cor. Bo.	4.5	15 25 42.44	2.404	231 25 36.63	36.061	1.5592	9.5901—	9.8663	9.9645	+31 58 9.03	12.570	1.0969	9.8931	9.8933	9.5187—
δ Serpentis	3	15 26 15.17	2.856	231 33 47.57	42.844	1.6327	9.0880—	9.8018	9.9022	+11 8 41.92	12.350	1.0957	9.8939	9.7614	9.0798—
α Cor. Bor.	2	15 27 6.87	2.531	231 46 31.87	37.965	1.5784	9.5047—	9.8426	9.9466	+27 19 24.81	12.408	1.0936	9.8952	9.8729	9.4533—
40 Libræ. .	4.5	15 27 41.04	3.649	231 55 15.61	54.731	1.7390	9.5371	9.8491	9.9550	—29 10 45 97	12.364	1.0922	9.8961	7.6826—	9.4781
η Libræ ...	4.5	15 34 0 85	3.361	233 30 12.81	50.417	1.7021	9.2052	9.7896	9.9205	—15 5 36.02	11.922	1.0765	9.9052	9.3217	9.1908
α Serpentis	2.3	15 35 27.62	2.948	233 51 50.50	44.214	1.6138	8.3595—	9.7739	9.9105	+ 6 59 46.94	11.734	1.0728	9.9072	9.7236	8.8563—
λ Serpentis	4.5	15 37 45.64	2 905	234 26 21.53	43.574	1.6110	8.9080—	9.7687	9.9145	+ 7 55 15.02	11.687	1.0667	9.9104	9.7340	8.9038—
β Serpentis	3.4	15 37 55.51	2.760	234 28 52.69	41.400	1.6164	9.2214	9.7813	9.9277	—15 59 25.90	11.551	1.0664	9.9106	9.2856	9.2042
μ Serpentis	3.4	15 40 17.27	3.121	235 4 18.98	46.818	1.6706	8.4585	9.7584	9.9143	— 2 52 24.68	11.416	1.0559	9.9137	9.5937	8.4579
κ Serpentis	4	15 40 40.83	2.689	235 10 12.40	40.340	1 6069	9.2863—	9.7803	9.9378	+18 42 6.90	11.447	1.0588	9.9143	9.8289	9.2628—
ε Serpentis	3	15 41 53.86	2.983	235 28 27.95	44.742	1.6490	8.6975—	9.7551	9.9175	+ 5 1 27.17	11.259	1.0556	9.9159	9.7029	8.6958—
δ Cor. Bor.	4.5	15 42 4.99	2.502	235 31 14.90	37.524	1.5767	9.4530—	9.8016	9.9648	+26 37 25.20	11.329	1.0550	9.9161	9.8793	9.4043—
θ Libræ ...	4.5	15 43 38.58	3.408	235 54 46.95	51.115	1.7062	9.2116	9.7661	9.9357	—16 11 40.65	11.055	1.0507	9.9181	9.2691	9.1940
ρ Scorpii...	4	15 45 51.19	3.678	236 27 47.79	55.166	1.7416	9.4803	9.7992	9.9778	—28 40 53.86	11.056	1.0445	9.9209	8.2847—	9.4235
π Scorpii...	3.4	15 48 2.27	3.608	237 0 38.34	54.113	1.7330	9.4162	9.7808	9.9685	—25 35 18.93	10.933	1.0381	9.9236	8.4655	9.3714
γ Serpentis	3	15 48 11.41	2.766	237 2 51.21	41.489	1.6140	9.2002—	9.7533	9.9416	+16 15 12.31	12.134	1.0383	9.9238	9.8139	9.1825—
δ Scorpii...	3	15 49 45.79	3.529	237 26 30.48	52.933	1.7233	9.3395	9.7641	9.9589	—22 6 6.30	10.704	1.0331	9.9257	8.9296	9.3064
ε Cor. Bor.	4.5	15 50 10.76	2.484	237 32 41.33	37.257	1.5710	9.4443—	9.7814	9.9779	+27 24 10.12	10.771	1.0318	9.9262	9.8886	9.3927—
			—		—										
*ζ Ur. Min.	4	15 50 39.00	2.251	237 39 45.04	36.760	1.5572—	0.4136—	0.4226	0.6213	+78 20 22.96	10.746	1.0301	9.9268	9.9615	9.7191—

Stars 1821.	Mag.	R. A. in Time.	Ann. Var.	R. A. in Arc.	Ann. Var.	Log. a	Log. b	Log. c	Log. d	Declination.	Ann. Var.	Log. a'	Log. b'	Log. c'	Log. d'
		h. m. s.	+		+			—	—		—	—	+		
ξ Libræ	4.5	15 54 32.10	3ˢ.282	238° 38′ 1″.47	49″.231	1.6929	8.9998	9.7243	9.9392	—10° 52′ 12″.01	10.392	1.0186	9.9314	9.4204	8.9919
π Serpentis...	4.5	15 54 34.90	2.572	238 38 43.43	38.574	1.5871	9.3505—	9.7532	9.9684	+23 18 31.94	10.332	1.0192	9.9314	9.8672	9.3136—
β Scorpii	2	15 55 2.58	3.473	238 45 40.98	52.089	1.7162	9.2593	9.7400	9.9571	—19 18 19.06	10.358	1.0170	9.9320	9.1037	9.2341
ω¹ Scorpii	4.5	15 56 21.28	3.497	239 5 19.21	52.462	1.7188	9.2759	9.7382	9.9610	—20 10 27.56	10.247	1.0129	9.9335	9.0474	9.2484
ω² Scorpii	4.5	15 56 55.19	3.498	239 13 47.83	52.465	1.7194	9.2787	9.7370	9.9622	—20 22 27.54	10.194	1.0111	9.9341	9.0324	9.2507
*θ Draconis..	3.4	15 58 33.27	1.142	239 38 18.94	17.140	1.2596	9.9257—	9.9924	0.2247	+59 2 46.42	9.804	1.0059	9.9359	9.9837	9.6369—
ν Scorpii......	4	16 1 36.10	3.473	240 24 7.95	52.100	1.7161	9.2303	9.7180	9.9636	—18 59 7.20	9.898	0.9958	9.9393	9.1056	9.2060
δ Ophiuchi...	3	16 4 58.17	3.128	241 14 32.58	46.921	1.6722	8.4330	9.6829	9.9435	— 3 13 27.42	9.686	0.9844	9.9428	9.5844	8.4323
ε Ophiuchi...	3	16 8 51.40	3.161	242 12 50.94	47.410	1.6752	8.5393	9.6697	9.9480	— 4 14 48.25	9.231	0.9707	9.9468	9.5651	8.5381
σ Scorpii	4	16 10 19.36	3.625	242 34 51.29	54.381	1.7354	9.3349	9.7065	9.9915	—25 9 7.13	9.135	0.9629	9.9482	8.1929	9.5416
γ Herculis...	3.4	16 14 1.46	2.638	243 30 21.83	39.569	1.5981	9.2005—	9.6753	9.9777	+19 34 53.84	8.804	0.9516	9.9518	9.8506	9.1747—
*τ Herculis...	4	16 14 21.80	1.796	243 35 27.84	26.940	1.4304	9.6746—	9.8123	0.1162	+46 44 28.05	9.068	0.9502	9.9521	9.9776	9.5104—
α Scorpii	1	16 18 26.89	3.664	244 36 45.75	54.955	1.7392	9.3208	9.6786	0.0023	—26 1 24.67	8.575	0.9344	9.9559	7.8041—	9.2744
φ Ophiuchi...	4.5	16 20 54.09	3.425	245 13 35.55	51.377	1.7102	9.0858	9.6399	9.9757	—16 12 43.88	8.380	0.9245	9.9581	9.2129	9.0682
λ Ophiuchi...	4	16 21 53.24	3.017	245 28 18.61	45.254	1.6557	8.2378—	9.6186	9.9593	+ 2 23 6.04	8.304	0.9204	9.9589	9.6735	8.2374—
*η Draconis...	3	16 21 34.32	0.733	245 23 34.80	11.800	1.0739	9.8924—	9.9467	0.2859	+61 55 17.11	8.269	0.9216	9.9587	0.0028	9.5651—
β Herculis ...	2.3	16 22 31.19	2.563	245 37 47.92	38.445	1.5875	9.2195—	9.6480	9.9920	+21 53 12.38	8.256	0.9178	9.9595	9.8706	9.1870—
h Herculis ...	4.5	16 24 14.01	2.801	246 3 30.10	42.014	1.6351	8.9313—	9.6177	9.9703	+11 52 50.10	8.165	0.9105	9.9609	9.7874	8.9219—
τ Scorpii......	3.4	16 24 45.13	3.718	246 11 21.09	55.772	1.7459	9.3287	9.6595	0.0148	—27 49 58.13	8.033	0.9084	9.9614	8.6368—	9.2753
ζ Ophiuchi...	3.4	16 27 18.65	3.293	246 49 39.70	49.396	1.6931	8.8498	9.6019	9.9704	—10 11 40.99	7.796	0.8971	9.9635	9.4224	8.8429
σ Herculis ...	4	16 28 19.89	1.925	247 4 58.38	28.882	1.4611	9.5572—	9.7249	0.0988	+42 48 42.72	7.723	0.8925	9.9643	9.9751	9.4226—
			—		—										
*A Draconis	4.5	16 28 22.33	0.174	247 5 34.98	2.620	0.3946—	0.0095—	0.0389	0.4131	+69 9 20.35	7.747	0.8922	9.9643	0.0066	9.5608—
			+		+										
ζ Herculis ...	3	16 34 32.07	2.256	248 38 1.00	33.837	1.5363	9.3562—	9.6328	0.0403	+31 56 0.97	6.779	0.8637	9.9691	9.9350	9.2849—
η Herculis ...	3	16 36 45.13	2.042	249 11 17.02	30.626	1.4871	9.4631—	9.6617	0.0818	+39 16 7.48	7.162	0.8528	9.9707	9.9674	9.3520—
ε Scorpii......	3	16 38 35.29	3.866	249 38 49.29	57.984	1.7684	9.3696	9.6225	0.0532	—33 57 27.11	7.283	0.8437	9.9720	9.2140—	9.2884
ι Ophiuchi ...	4	16 45 32.46	2.829	251 23 6.86	42.428	1.6385	8.7707—	9.5114	9.9840	+10 28 11.35	6.323	0.8064	9.9767	9.7774	8.7635—
κ Ophiuchi...	4	16 49 11.87	2.829	252 17 58.10	42.439	1.6311	8.7140—	9.4891	9.9852	+ 9 39 44.67	5.983	0.7852	9.9789	9.7692	8.7078—
ε Herculis.....	3	16 53 26.26	2.284	253 21 33.93	34.264	1.5364	9.2391—	9.5248	0.0493	+31 11 49.16	5.644	0.7591	9.9814	9.9383	9.1713—
η Ophiuchi...	2.3	17 0 7.27	3.431	255 1 49.02	51.461	1.7108	8.8549	9.4282	0.0011	—15 29 31.47	4.971	0.7145	9.9850	9.2048	8.8388
μ Draconis...	4	17 1 37.76	1.226	255 24 26.42	18.392	1.2698	9.5514—	9.6396	0.2240	+54 42 34.05	4.988	0.7035	9.9858	0.0173	9.3131—
A Ophiuchi ..	4.5	17 4 21.11	3.681	256 5 22.22	55.218	1.7454	9.0754	9.4285	0.0346	—26 19 42.10	5.894	0.6834	9.9872	8.6179—	9.0279
			—		—										
ε Ursæ Min.	4	17 4 36.63	6.589	256 9 9.48	98.840	1.9926—	0.2489—	0.2528	0.8610	+82 18 57.42	4.766	0.6805	9.9872	0.0087	9.3751—
			+		+										
α Herculis ...	3.4	17 6 29.55	2.725	256 37 15.50	40.882	1.6121	8.7802—	9.3786	0.0023	+14 36 9.89	4.517	0.6668	9.9880	9.8230	8.7660—
41 Ophiuchi	4.5	17 7 25.80	3.077	256 51 27.04	46.153	1.6637	6.9681	9.3567	9.9885	— 0 14 3.19	4.593	0.6592	9.9885	9.6335	6.9681
δ Herculis	4	17 7 40.16	2.442	256 55 2.38	36.634	1.5669	9.0246—	9.3977	0.0315	+25 3 29.90	4.606	0.6571	9.9886	9.9062	8.9817—
ζ Draconis ...	3	17 8 16.86	0.144	257 4 12.87	2.167	0.3562	9.6999—	9.7394	0.3784	+65 56 8.36	4.470	0.6519	9.9888	0.0281	9.3133—
π Herculis ...	3.4	17 8 48.83	2.083	257 12 12.43	31.241	1.4952	9.2227—	9.4431	0.0868	+37 1 1.59	4.388	0.6476	9.9891	9.9702	9.1250—
ϱ Ophiuchi...	4.5	17 10 16.79	3.582	257 34 10.10	53.733	1.7283	8.9151	9.3625	0.0193	—20 54 31.89	4.492	0.6355	9.9897	8.7553	8.8855
u Herculis ...	4	17 10 43.08	2.209	257 40 46.19	33.134	1.5205	9.1466—	9.4071	0.0678	+33 17 58.11	4.248	0.6314	9.9899	9.9538	9.0688—
ν Serpentis ...	4.5	17 10 46.19	3.369	257 41 32.78	50.542	1.7026	8.6800	9.3394	0.0006	—12 39 15.93	4.184	0.6312	9.9899	9.3211	8.6693
θ Ophiuchi..	3.4	17 11 1.40	3.673	257 45 22.21	55.100	1.7409	8.9814	9.3686	0.0321	—24 48 32.03	4.229	0.6290	9.9900	8.2057—	8.9493
e Herculis ...	4.5	17 11 29.68	2.058	257 52 25.27	30.875	1.4912	9.2071—	9.4228	0.0907	+37 29 12.98	3.994	0.6246	9.9902	9.9728	9.1067—
ϱ Herculis ...	4	17 17 30.31	2.006	259 22 34.61	30.088	1.4914	9.1478—	9.3652	0.0920	+37 19 5.43	3.582	0.5680	9.9925	9.9736	9.0483—
σ Ophiuchi...	4.5	17 17 38.19	2.973	259 24 32.87	44.599	1.6487	8.1410—	9.2656	9.9938	+ 4 18 18.89	3.602	0.5669	9.9925	9.7047	8.1398—
λ Herculis....	4.5	17 23 30.35	2.417	260 52 35.22	36.253	1.5594	8.8932—	9.2475	0.0417	+26 15 9.22	3.119	0.5028	9.9945	9.9170	8.8459—
β Draconis...	2	17 26 23.22	1.343	261 35 48.35	20.138	1.3061	9.2788—	9.3797	0.2103	+52 26 17.21	2.909	0.4673	9.9953	0.0207	9.0639—
α Ophiuchi...	2	17 26 37.93	2.773	261 39 22.08	41.600	1.6185	8.5146—	9.1725	0.0061	+12 41 56.42	3.031	0.4643	9.9954	9.8068	8.5038—
o Serpentis...	4.5	17 31 21.38	3.361	262 50 20.64	50.420	1.7035	8.4510	9.1066	0.0075	—12 46 7.66	2.465	0.3985	9.9966	9.3097	8.4401
*ι Herculis...	4	17 34 24. 2	1.687	263 36 3.51	25.310	1.4033	9.0638—	9.2061	0.1564	+46 6 25.13	2.183	0.3495	9.9973	0.0073	8.9047—
β Ophiuchi...	3	17 34 37.80	2.957	263 39 26.98	44.362	1.6474	7.9537—	9.0447	9.9988	+ 4 39 5.81	1.969	0.3460	9.9973	9.7103	7.9522—
γ Ophiuchi...	4	17 38 55.03	2.997	264 43 45.44	44.954	1.6537	7.6500—	8.9637	9.9987	+ 2 47 2.95	1.862	0.2662	9.9982	9.6829	7.6495—
μ Herculis ...	4	17 39 26.85	2.331	264 51 42.75	34.970	1.5500	8.6747—	9.0055	0.0516	+27 49 55.30	2.512	0.2548	9.9983	9.9288	8.8213—
ν Ophiuchi...	4	17 49 10.30	3.297	267 17 34.47	49.456	1.6942	7.9089	8.6805	0.0058	— 9 44 23.97	0.925	9.9786	9.9995	9.4129	7.9026
θ Herculis....	4	17 50 6.50	2.043	267 31 37.48	30.647	1.4882	8.5165—	8.7342	0.0989	+37 16 51.31	0.769	9.9380	9.9996	9.9780	8.4172—
*ξ Herculis ..	4	17 50 48.43	2.320	267 42 6.30	34.800	1.5415	8.3520—	8.6627	0.0590	+29 16 30.14	0.753	9.9069	9.9996	9.9381	8.2926—
ξ Draconis ...	3.4	17 50 25.55	1.024	267 36 23.27	15.354	1.1846	8.8067—	8.8836	0.2624	+56 54 12.17	0.785	9.9232	9.9996	0.0309	8.5439—
o Ophiuchi...	4	17 51 40.75	3.001	267 55 11.26	45.013	1.6531	7.2720—	8.5605	0.0003	+ 2 57 0.97	0.669	9.8639	9.9997	9.6857	7.2714—
γ Draconis...	2	17 52 27.28	1.388	268 6 43.48	20.819	1.3185	8.6174—	8.7238	0.2057	+51 30 50.84	0.714	9.8215	9.9998	0.0222	8.4114—
γ Sagittarii...	4	17 54 10.9[illegible]	3.856	268 34 44.64	57.844	1.7618	8.1631	8.4587	0.0642	—30 24 51.34	0.695	9.7016	9.9999	9.1198—	8.0988
p Ophiuchi...	4.5	17 56 24.70	3.024	269 6 10.46	45.361	1.6544	6.8434—	8.1952	0.0004	+ 2 33 0.40	0.435	9.5024	9.9999	9.6797	6.8430—
ϑ Ophiuchi...	4	17 58 51.63	2.836	269 42 54.43	42.547	1.6299	6.9225—	7.7027	0.0061	+ 9 32 52.69	0.063	9.0128	0.0000	9.7737	6.9164—
								+			+	+			
o Herculis ...	4	18 0 33.56	2.334	270 8 23.35	35.015	1.5444	7.1256	7.4435	0.0571	+28 44 43.02	0.104	8.6532	0.0000	9.9352	7.0685

STARS 1821.	Mag.	R. A. in Time.	Ann. Var.	R. A. in Arc.	Ann. Var.	Log. a	Log. b	Log. c	Log. d	Declination.	Ann. Var.	Log. a′	Log. b′	Log. c′	Log. d′
		h. m. s.	+		+			+	—		+	+	+		
μ Sagittarii	3.4	18 3 3.44	3ˢ.581	270° 45′ 50″.94	53″.708	1.7304	7.7110—	8.1548	0.0301	—21° 5′ 37″.08	0″.317	0.4183	0.0000	8.6541	7.6809—
ϑ Sagittarii	3.4	18 9 31.86	3.836	272 22 58.68	57.537	1.7599	8.3784—	8.6808	0.0616	—29 53 29.19	0.809	9.9180	9.9996	9.0850—	8.3164—
η Serpentis	4	18 12 2.88	3.096	273 0 43.19	46.440	1.6725	7.4305—	8.7211	0.0000	— 2 56 8.01	0.452	0.0208	9.9994	9.5824	7.4299—
κ Lyræ....	4.5	18 13 35.44	2.009	273 23 51.57	31.488	1.4981	8.6339	8.8648	0.0912	+35 59 24.42	1.228	0.0752	9.9992	9.9721	8.5419
λ Sagittarii	4	18 16 55.23	3.704	274 13 50.90	55.553	1.7447	8.5466—	8.9125	0.0434	—25 30 30.34	1.259	0.1685	9.9988	8.5771—	8.5020—
			—		—										
χ Draconis	4.5	18 24 15.98	1.078	276 3 59.73	16.176	1.2502—	9.5294	9.5496	0.5231	+72 39 10.94	1.776	0.3261	9.9976	0.0328	9.0038
			+		+										
α Lyræ....	1	18 30 52.92	2.027	277 43 7.88	30.406	1.4793	9.0306	9.2353	0.1032	+38 37 23.55	3.003	0.4296	9.9960	9.9812	8.9234
φ Sagittarii	4.5	18 34 28.55	3.761	278 37 8.32	56.421	1.7496	8.8859—	9.2265	0.0458	—27 9 48 32	3.017	0.4768	9.9951	9.9230	8.8352—
β Lyræ....	3	18 43 27.94	2.205	280 51 59.12	33.075	1.5205	9.0905	9.3526	0.0694	+33 9 42.25	3.797	0.5770	9.9921	9.9545	9 0133
σ Sagittarii	3	18 44 9.61	3.727	281 2 26.25	55.910	1.7469	8.9801—	9.3304	0.0401	—26 30 27.72	3.777	0.5837	9.9919	2.6960—	8.9318—
θ Serp. pr.	4.5	18 47 19.16	2.977	281 49 47.44	44.659	1.6498	8.1543	9.3128	9.9917	+ 3 58 50.10	4.311	0.6133	9.9907	9.6998	8.1532
θ Serp. seq.	5	18 47 20.58	2.977	281 50 8.75	44.654	1.6498	8.1544	9.3130	9.9917	+ 3 58 47.80	4.368	0.6135	9.9907	9 6998	8.1534
ζ Sagittarii	3.4	18 51 12.95	3.824	282 48 14.30	57.357	1.7586	9.1092—	9.4086	0.0521	—30 7 29.69	4.433	0.6471	9.9891	9.0569—	9.0462—
ε Aquilæ...	3.4	18 51 29.38	2.710	282 52 20.66	40.651	1.6111	8.7708	9.3626	0.0037	+14 50 1.38	4.426	0.6495	9.9889	9.8255	8.7561
γ Lyræ....	3	18 52 14.44	2.233	283 3 36.57	33.500	1.5264	9.1574	9.4278	0 0624	+32 27 3.27	4.580	0.6557	9.9886	9.9489	9.0837
ο Sagittarii	4.5	18 53 56.89	3.599	283 29 16.65	53.985	1.7315	8.9740—	9.4006	0.0206	—21 59 30.98	4.719	0 6693	9.9878	8.5835	8.9412—
τ Sagittarii	4	18 55 45.64	3.749	283 56 13.19	56 233	1.7507	9.1060—	9.4356	0.0408	—27 55 11.19	4.613	0.6833	9.9870	8.8510—	9.0523—
λ Aquilæ...	3	18 56 44.51	3.179	284 11 7.70	47.678	1.6791	8.3428—	9.3905	9.9883	— 5 8 26 96	4.943	0.6969	9.9866	9.5383	8.3411—
ζ Aquilæ...	3	18 57 10.65	2.745	284 17 39.79	41.181	1.6161	8.7764	9.4049	9.9987	+13 36 21.06	5.023	0.6942	9.9863	9.8128	8.7641
π Sagittarii	4.5	18 59 6.78	3.572	284 46 41.70	53.587	1.7290	8.9976—	9 4374	0.0161	—21 17 50.87	5.155	0.7082	9.9854	8.7259	8.9668—
δ Draconis	3	19 12 28.75	0.040	288 7 11.30	0.598	9.5763	9.8723	9.9071	9.3923	+67 20 49.88	6.298	0.7948	9.9779	0.0188	9.4579
κ Cygni...	4	19 12 57.37	1.382	288 14 20.52	20.729	1.3163	9.6191	9.7165	0.1986	+53 2 30.56	6.380	0.7974	9.9776	0.0086	9.3981
δ Aquilæ...	3.4	19 16 28.12	3.022	289 7 1.85	45.323	1.6542	8.1995	9.5157	9 9759	+ 2 46 0.50	6.754	0.8170	9.9754	9.6805	8.1990
			—		—										
τ Draconis	4.5	19 18 55.76	1.089	289 43 56.35	16.335	1.1976—	0.0436	0.0630	0.5083	+73 1 12.47	6.863	0.8305	9.9737	0.0116	9.5091
			+		+										
*π Dracon.	4	19 19 43.47	0.331	289 55 52.26	4.960	0.6918	9.8713	0.9127	0.3533	+65 22 15.50	6.875	0.8345	9.9732	0.0151	9.4912
6 Vulpecu.	4	19 21 14.95	2.483	290 18 44.24	37.251	1.5743	9.1954	9.5408	0.0124	+24 18 35.03	6.909	0.8422	9.9721	9.8930	9.1550
β Cygn. pr.	3	19 23 29.94	2.412	290 52 29 08	36.173	1.5591	9.2700	9.6043	0.0230	+27 35 26 73	7.198	0.8536	9 9705	9.9125	9.2176
μ Aquilæ..	4.5	19 25 20.37	2.920	291 20 5 48	43.796	1.6407	8.6505	9.5641	9.9724	+ 7 0 27.92	7 243	0.8626	9.9692	9.7360	8.6473
h Sagittarii	4.5	19 25 48.11	3.660	291 27 1.70	54.894	1.7390	9.2371—	9.6068	0.0125	—25 16 4.94	7.397	0.8648	9 9388	7.6789—	9.1934—
κ Aquilæ..	4	19 27 15.44	3.232	291 48 51.63	48.476	1.6852	8.6846—	9.5737	9.9714	— 7 24 57.77	7.517	0.8718	9 9677	9.4922	8.6809—
θ Cygni....	4	19 31 37.99	1.603	292 34 29.8[illegible]	24.043	1.3831	9.6635	9.7805	0.1546	+49 48 40.91	8.069	0.8922	9 [illegible]643	9.9929	9.4733
α Sagittæ	4	19 32 5.46	2.6 7	2[illegible]3 1 21.85	40.156	1.6038	9.0939	9.6[illegible]31	9.9348	+17 36 38.63	7.9[illegible]3	0.8940	9.9639	9.8402	9.0731
φ Cygni...	4	19 32 18.54	2.367	2[illegible]3 4 38.0[illegible]	3[illegible].50[illegible]	1.5500	9.3503	9.6546	0.0251	+29 44 52.74	8.0[illegible]	0.895[illegible]	9.9638	9.9208	9.2889
γ Aquilæ	3	19 37 45.13	2.854	294 26 13.48	42 812	1.6308	8.8711	9.6236	9.9661	+10 11 6.12	8.343	0 9284	9.9592	9.7695	8 8642
δ Cygni...	3.4	19 39 22.59	1.873	294 50 38.91	28.083	1.4475	9.6189	9.7717	0.1061	+44 41 57.85	8.466	0 9253	9.9578	9.9763	9.4706
δ Sagittæ...	4	19 39 24.18	2.678	294 51 2.68	40.171	1 6029	9.1379	9.6456	9.9798	+18 5 59.38	8.49[illegible]	0.9254	9.9578	9.8417	9.1158
α Aquilæ	1.2	19 42 3 07	2.927	295 30 42.71	43.912	1.6369	8.8037	9.6388	9.9601	+ 8 24 13.70	9.089	9.9360	9.9554	9.7492	8.7990
η Aquilæ...	4	19 43 20.67	3 051	295 50 10.11	45.762	1.6613	7.6254	9.6393	9.9543	+ 0 33 18.23	8 758	0.9411	9.9543	9.6461	7.6254
β Aquilæ	3.4	19 46 31.36	2.949	296 37 46.85	44.230	1 6449	8.6708	9.6539	9.9537	+ 5 58 4.44	8.567	0 9529	9.9513	9.7199	8.6684
γ Sagittæ	4.5	19 50 47.47	2.661	297 41 52.12	39.911	1.6010	9.2046	9.6916	9.9715	+19 0 54.57	9.539	0 9691	9.9471	9.8444	9.1802
c Sagittæ..	4.5	19 51 38.09	3.704	297 54 31.32	55.564	1.7444	9.3996—	9.7252	0.0012	—28 11 48.39	9.497	9.9721	9.9463	8.5443—	9.3447—
θ Aquilæ...	3.4	20 2 3.89	3.100	300 30 58.39	46 503	1.6667	8.0760—	9.7058	9.9354	— 1 20 38.56	10.239	1.0075	9.9352	9.6167	8.0759—
α¹ Caprico.	4	20 7 43.23	3.331	301 55 45.61	49.959	1.6986	9.0885—	9 7347	9.9401	—13 3 9.91	10.695	1.0252	9.9288	9.3639	9.0771—
ο Cygni....	4	20 7 59.50	1.884	301 59 52 45	28.256	1.4516	9.7424	9.8840	0.0883	+46 12 12.96	10.728	1 0261	9.9284	9.9602	9.5826
α² Caprico.	3	20 8 6.99	3.337	302 1 44.18	50.049	1.6987	9.0910—	9.7360	9.9397	—13 2 26.63	10.716	1.026[illegible]	9.9233	9.3631	9.0796—
23 Vulpec.	4.5	20 8 20.98	2.478	302 5 14.69	37.172	1.5713	9.4375	9.7764	9.9792	+27 16 19.16	10.754	1.0271	9.9280	9.8887	9.3863
*33 Cygni	4.5	20 9 13.08	1.389	302 18 16.24	20.837	1.3197	9.8994	9.9806	0.1797	+56 1 22.29	10.690	1.0298	9.9270	9.9747	9.6466
32 Cygni...	4.5	20 9 56.42	1.855	302 29 6.25	27.822	1.4437	9.7629	9.8976	0.0937	+47 10 6.34	10.777	1.0320	9.9261	9.9608	9.5953
β Caprico.	3.4	20 10 56.61	3.379	302 44 9.54	50.679	1.7045	9.1712—	9.7488	9.9406	—15 20 12.93	10.975	1.0349	9.9249	9.2924	9.1554—
			—		—										
κ Cephei...	4.5	20 14 40.49	1.849	303 40 7.41	27.734	1.4343—	0.3863	0.3973	0.5737	+77 10 1.10	11.076	1.0459	9.9203	9.9580	9.7328
			+		+										
γ Cygni...	3	20 15 47.99	2.144	303 56 59.92	32.167	1.5081	9.6660	9.8608	0.0326	+39 41 20.70	11.259	1.0489	9.9188	9.9364	9.5522
ι Cygni....	4.5	20 22 4.93	2.446	305 31 13 93	36.690	1.5645	9.5217	9.8257	9.9721	+29 46 39.39	11.725	1.0669	9.9106	9.8926	9.4602
ε Delphini	4	20 24 39.40	2.862	303 9 50.98	42.926	1.6330	9.0474	9.7785	9.9147	+10 42 10.18	11.901	1.0728	9.9070	9.7607	9.0398
β Delphini	4	20 29 9.04	2.807	307 17 15.60	42.099	1.6237	9.1785	9.7954	9.9[illegible]	+13 58 53.02	12.273	1.0842	9.9007	9.7877	9.1654
α Delphini	3.4	20 31 19.39	2.789	307 49 50.84	41.830	1.6200	9.2244	9.8034	9.9132	+15 17 15.32	12.416	1.0896	9.8975	9.7972	9.2088
α Cygni...	1	20 35 20.10	2.036	308 49 53.45	30.534	1.4856	9.7919	9.9451	0.0394	+44 33 44.02	12.611	1.0992	9.8915	9.9326	9.6441
ψ Caprico.	4.5	20 35 29.25	3.578	308 52 18.77	53.665	1.7292	9.4840	9.8437	9.9373	+25 54 19.67	12.491	1.0995	9.8913	9.8637	9.4380
ι Aquarii...	4.5	20 37 58.71	3.256	309 29 40.70	48.838	1.6883	9.0560—	9.8103	9.8946	—10 8 32.27	12.857	1.1053	9.8877	9.4643	9.0492—
3 Aquarii..	4	20 38 16.86	3.166	309 34 12.08	47.488	1.6772	8.8014	9.8063	9.8891	— 5 40 30.2[illegible]	12.826	1.1061	9.8870	9.5511	8.7993—
γ Delph. se.	4	20 38 22.01	2.793	309 35 30.22	41.896	1.6205	9.2469	9.8204	9.9029	+15 29 11.67	12.683	1.1062	9.8868	9.7952	9.2309
ε Cygni....	3	20 38 57.96	2.419	309 44 29.38	36.290	1.5550	9.6233	9.8837	9.9638	+33 18 23.60	13.246	1.1070	9.8859	9.8949	9.5454
η Cephei...	3.4	20 41 37.05	1.220	310 24 15 76	18.303	1.2628	0.0706	9.1281	0.1981	+61 8 45.97	13.808	1.1136	9.8817	9.9427	9.7541

Stars 1821.	Mag.	R. A. in Time.	Ann. Var.	R. A. in Arc.	Ann. Var.	Log. a	Log. b	Log. c	Log. d	Declination.	Ann. Var.	Log. a'	Log. b'	Log. c'	Log. d'
		h. m. s.	+		+			+	−		+	+	+		
μ Aquarii	4.5	20 42 59.41	3ˢ.242	310° 44′ 51″.12	48″.635	1.6866	9.0451−	9.8209	9.8856	− 9° 38′ 46″.76	13″.173	1.1166	9.8794	9.4785	9.0389−
9 Vulpeculæ	4.5	20 46 55.70	2.548	311 43 55.55	38.221	1.5829	9.5375	9.8749	9.9245	+27 23 0.35	13.445	1.1252	9.8729	9.8625	9.4859
ν Cygni	4	20 50 29.87	2.226	312 37 28.02	33.393	1.5241	9.7619	9.9496	9.9856	+40 29 0.96	13.681	1.1326	9.8668	9.9073	9.6431
ξ Cygni	4	20 58 24.86	2.168	314 36 12.88	32.520	1.5133	9.8194	9.9839	9.9899	+43 13 4.90	14.113	1.1484	9.8525	9.9052	9.6820
ζ Cygni	3	21 5 18.87	2.539	316 19 43.10	38.081	1.5819	9.6119	9.9196	9.8995	+29 29 52.88	14.473	1.1612	9.8392	9.8560	9.5516
δ Equulei	4.5	21 5 45.66	2.923	316 26 24.89	43.849	1.6411	9.0738	9.8659	9.8440	+ 9 17 18.46	14.324	1.1620	9.8383	9.7320	9.0680
α Equulei	4.5	21 6 52.30	3.000	316 43 4.43	45.002	1.6526	8.7595	9.8635	9.8374	+ 4 30 51.43	14.572	1.1640	9.8361	9.6872	8.7581
σ Cygni	4.5	21 10 22.53	2.336	317 35 38.02	35.034	1.5467	9.7712	9.9756	9.9363	+38 38 52.24	14.740	1.1702	9.8289	9.8809	9.6638
υ Cygni	4.5	21 10 32.84	2.445	317 38 12.65	36.682	1.5665	9.7000	9.9508	9.9107	+34 9 1.85	14.855	1.1705	9.8285	9.8677	9.6178
e Pegasi	4	21 13 48.44	2.771	318 27 6.64	41.571	1.6173	9.4122	9.8986	9.8461	+19 2 40.14	15.144	1.1761	9.8217	9.7971	9.3877
α Cephei	3	21 14 18.14	1.431	318 34 20.49	21.459	1.3275	0.1461	0.2009	0.1466	+61 49 46.73	14.977	1.1769	9.8206	9.8966	9.8202
ζ Capricorni	4	21 16 25.56	3.440	319 6 23.35	51.610	1.7130	9.5101−	9.9150	9.8526	−23 10 44.58	15.269	1.1804	9.8160	9.1502	9.4735−
β Aquarii	3	21 22 7.67	3.165	320 31 55.02	47.481	1.6761	8.9341−	9.8903	9.8059	− 6 21 7.36	15.580	1.1895	9.8032	9.5576	8.9315−
β Cephei	3	21 26 18.76	0.802	321 34 27.96	12.035	1.0869	0.3277	0.3553	0.2547	+69 46 34.95	15.593	1.1962	9.7934	9.8652	9.8664
γ Capricorni	4	21 30 9.42	3.340	322 32 23.49	50.096	1.6977	9.3975−	9.9202	9.8046	−17 27 50.94	15.965	1.2016	9.7841	9.3645	9.3770−
ι Piscis Australis	4.5	21 34 15.32	3.600	323 33 49.82	53.997	1.7324	9.7318−	9.9861	9.8543	−33 50 8.41	16.099	1.2074	9.7737	9.4746	9.6512−
ε Pegasi	2.3	21 35 23.57	2.948	323 50 53.54	44.216	1.6447	9.1098	9.9126	9.7763	+ 9 3 36.31	16.266	1.2091	9.7708	9.7173	9.1043
π Cygni	4.5	21 35 44.48	2.111	323 56 7.27	31.671	1.5020	9.9896	0.1030	9.9652	+50 22 33.90	16.218	1.2095	9.7699	9.8635	9.7942
g Pegasi	4.5	21 36 2.37	2.843	324 0 35.59	42.650	1.6286	9.3806	9.9264	9.7875	+16 32 3.96	16.310	1.2099	9.7691	9.7659	9.3622
*κ Pegasi	4	21 36 32.62	2.706	324 8 9.39	40.590	1.6084	9.5739	9.9508	9.8099	+24 49 40.28	16.357	1.2106	9.7678	9.8061	9.5318
δ Capricorni	3.4	21 37 8.75	3.323	324 17 13.64	49.844	1.6953	9.3930−	9.9288	9.7855	−16 55 57.01	16.087	1.2115	9.7662	9.3895	9.3738−
11 Cephei	4.5	21 39 14.26	0.895	324 48 33.91	13.424	1.1280	0.3629	0.3886	0.2369	+70 29 14.87	16.400	1.2143	9.7606	9.8377	9.8867
10 Cephei	4.5	21 40 16.20	1.711	325 4 3.04	25.671	1.4133	0.1575	0.2187	0.0628	+60 17 47.81	16.377	1.2157	9.7579	9.8528	9.8526
α Aquarii	3	21 56 35.34	3.082	329 8 47.06	46.235	1.6649	8.2492−	9.9338	9.7101	− 1 11 3.29	17.261	1.2357	9.7100	9.6267	8.2491−
ι Aquarii	4.5	21 56 45.71	3.254	329 11 25.69	48.806	1.6877	9.3538−	9.9485	9.7240	−14 43 51.95	17.314	1.2359	9.7094	9.4617	9.3392−
ι Pegasi	4	21 58 41.09	2.786	329 40 16.37	41.788	1.6171	9.5943	9.9770	9.7442	+24 28 33.34	17.419	1.2407	9.7033	9.7812	9.5534
θ Pegasi	4	22 1 10.31	3.034	330 17 34.72	45.509	1.6541	8.9080	9.9407	9.6970	+ 5 19 20.72	17.544	1.2407	9.6951	9.6799	8.9062
π Pegasi	4	22 2 3.04	2.658	330 30 45.56	39.868	1.5996	9.7407	0.0128	9.7652	+32 18 15.31	17.525	1.2417	9.6922	9.7993	9.6676
ζ Cephei	4	22 4 38.59	2.054	331 9 38.79	30.816	1.4906	0.1353	0.2102	9.9510	+57 19 14.73	17.511	1.2444	9.6834	9.8064	9.8677
θ Aquarii	4.5	22 7 22.53	3.172	331 50 40.95	47.581	1.6763	9.1285−	9.9503	9.6788	− 8 40 10.10	17.702	1.2472	9.6738	9.5538	9.1235−
ε Cephei	4.5	22 8 27.07	2.188	332 6 46.01	32.826	1.5056	0.1199	0.2005	9.9242	+56 9 11.36	17.722	1.2484	9.6700	9.7994	9.8657
γ Aquarii	4	22 12 24.40	3.103	333 6 5.96	46.547	1.6663	8.5511−	9.9506	9.6559	− 2 17 2.77	18.003	1.2522	9.6555	9.5187	8.5509−
31 Pegasi	4.5	22 12 42.58	2.952	333 10 38.68	44.274	1.6455	9.2515	9.9591	9.6629	+11 13 29.79	17.973	1.2525	9.6544	9.7110	9.2430
3 Lacertæ	4	22 16 31.52	2.328	334 7 52.81	34.915	1.5450	0.0510	0.1584	9.8441	+51 20 4.73	17.831	1.2561	9.6398	9.7866	9.8467
ζ Aquarii	4	22 19 36.59	3.088	334 54 8.87	46.316	1.6642	8.1682−	9.9570	9.6276	− 0 55 55.30	18.224	1.2589	9.6275	9.6306	8.1682−
β Piscis Australis	4	22 21 18.41	3.442	335 19 36.17	51.634	1.7119	9.7752−	0.0361	9.6983	−33 15 34.64	18.214	1.2602	9.6206	9.1271	9.6975−
δ Cephei	4.5	22 22 31.70	2.194	335 37 55.47	32.909	1.5189	0.1553	0.2293	9.8853	+57 30 4.72	18.251	1.2614	9.6155	9.7643	9.8855
7 Lacertæ	4	22 23 55.97	2.449	335 58 59.52	36.731	1.5624	0.0271	0.1466	9.7959	+49 21 53.20	18.309	1.2626	9.6096	9.7720	9.8408
η Aquarii	4	22 26 9.24	3.082	336 32 18.56	46.235	1.6643	8.2195−	9.9626	9.6001	− 1 2 6.88	18.443	1.2645	9.6000	9.6302	8.2194−
ε Piscis Australis	4	22 30 44.17	3.339	337 41 2.48	50.086	1.6995	9.6914−	0.0201	9.6334	−27 58 21.82	18.626	1.2687	9.5795	9.3122	9.6374−
ζ Pegasi	3	22 32 32.09	2.982	338 8 1.39	44.727	1.6504	9.2095	9.9741	9.5776	+ 9 54 4.49	18.705	1.2696	9.5711	9.6916	9.2030
η Pegasi	3	22 34 37.12	2.795	338 39 16.79	41.926	1.6225	9.7180	0.0235	9.6205	+29 17 18.88	18.700	1.2711	9.5611	9.7456	9.6586
λ Pegasi	4.5	22 37 55.52	2.885	339 28 52.77	43.268	1.6343	9.5015	0.0063	9.5795	+22 37 42.43	18.902	1.2735	9.5447	9.7287	9.5567
μ Pegasi	4	22 41 21.80	2.873	340 20 26.96	43.089	1.6341	9.6155	0.0121	9.5650	+23 39 32.70	18.858	1.2759	9.5269	9.7264	9.5774
λ Aquarii	4	22 43 16.22	3.127	340 48 59.78	46.907	1.6722	9.1512−	9.9800	9.5215	− 8 31 41.48	19.016	1.2771	9.5167	9.5804	9.1464−
ι Cephei	4	22 43 19.19	2.092	340 49 47.85	31.380	1.5014	0.3117	0.3535	9.8947	+65 15 37.18	18.786	1.2772	9.5164	9.6811	9.9334
δ Aquarii	3	22 45 8.39	3.195	341 17 5.81	47.926	1.6809	9.4554−	9.9953	9.5252	−16 46 7.21	19.087	1.2784	9.5063	9.5092	9.4365−
α Piscis Australis	1	22 47 44.30	3.339	341 56 4.55	50.082	1.6963	9.7494−	0.0430	9.5565	−30 34 6.88	18.922	1.2800	9.4915	9.3340	9.6844−
ο Andromedæ	4	22 53 41.75	2.728	343 25 26.29	40.923	1.6125	9.9263	0.1062	9.5800	+41 21 57.11	19.207	1.2836	9.4553	9.7112	9.8017
β Pegasi	2	22 55 6.34	2.891	343 46 35.17	43.360	1.6349	9.6917	0.0329	9.4968	+27 6 53.84	19.470	1.2843	9.4462	9.7107	9.6411
α Pegasi	2	22 55 51.22	2.979	343 57 43.52	44.680	1.6494	9.3874	9.9963	9.4549	+14 14 43.37	19.327	1.2847	9.4413	9.6890	9.3738
56 Pegasi	4.5	22 58 24.25	2.908	344 36 3.78	43.624	1.6394	9.6429	0.0251	9.4651	+24 30 19.71	19.407	1.2861	9.4241	9.7034	9.6019
c Aquarii	4.5	22 59 53.43	3.214	344 58 21.49	48.216	1.6825	9.5943−	0.0181	9.4470	−22 8 24.80	19.528	1.2869	9.4138	9.4833	9.5611−
γ Piscium	4.5	23 7 53.10	3.105	346 58 16.51	46.571	1.6612	8.5937	9.9890	9.3534	+ 2 18 23.26	19.555	1.2907	9.3530	9.6462	8.5934
λ Andromedæ	4.5	23 28 49.45	2.894	352 12 21.79	43.407	1.6359	0.0034	0.1502	9.2866	+45 29 24.52	19.512	1.2979	9.1323	9.6031	9.8491
ι Piscium	4.5	23 30 44.31	3.083	352 41 12.13	46.249	1.6610	8.9074	9.9979	9.1062	+ 4 39 26.44	19.465	1.2984	9.1048	9.6463	8.9060
γ Cephei	3	23 32 3.66	2.346	353 0 54.95	35.186	1.5532	0.6209	0.6328	9.7210	+76 37 59.67	20.035	1.2987	9.0850	9.3398	9.9848
ω Piscium	4.5	23 50 7.11	3.077	357 31 52.91	46.155	1.6621	9.0118	0.0019	8.6365	+ 5 52 24.84	19.940	1.3016	8.6342	9.6397	9.0096
30 Piscium	4.5	23 52 46.45	3.075	358 11 36.69	46.123	1.6637	9.0894−	0.0030	8.5019	− 7 0 29.38	20.053	1.3017	8.4986	9.6304	9.0862−
ζ Ceti	4	23 54 33.83	3.082	358 38 27.49	46.223	1.6644	9.5201−	0.0225	8.3978	−18 19 51.33	20.142	1.3019	8.3751	9.6070	9.4975−
α Andromedæ	1	23 59 9.43	3.073	359 47 13.97	46.093	1.6625	9.7275	0.0545	7.6240	+28 6 9.61	19.917	1.3020	7.5696	9.5850	9.6731
β Cassiopeæ	2.3	23 59 40.36	3.122	359 55 5.41	46.823	1.6624	0.2069	0.2778	7.4319	+58 9 41.97	19.831	1.3020	7.1541	9.3621	9.9292

ANNUAL VARIATIONS

OF

SCHUMACHER'S CONSTANTS FOR 67 NORTHERN STARS,

RECKONED FROM 1820.

STARS.	Log. a	Log. b	Log. c	Log. d	Log. a'	Log. b'	Log. c'	Log. d'	STARS.	Log. a	Log. b	Log. c	Log. d	Log. a'	Log. b'	Log. c'	Log. d'
κ Cassiopeæ	+ 0.9	+ 0.9	+ 0.7	+11.1	−0.1	+10.3	− 9.6	+ 0.1	β Ursæ Mi.	−16.7	− 1.2	− 1.1	−1.3	+0.1	−0.1	+0.0	− 0.0
ζ Cassiopeæ	0.7	0.8	0.4	9.0	0.2	8.7	5.9	0.2	ι Draconis	+ 0.3	1.1	1.0	0.1	−0.5	+0.4	0.2	0.6
α Cassiopeæ	0.6	0.8	0.5	8.3	0.2	7.7	7.6	0.2	γ Ursæ Mi.	−28.1	0.8	0.6	0.7	0.1	−0.1	0.0	+ 0.1
η Cassiopeæ	0.7	0.7	0.4	7.1	0.2	6.5	11.4	0.1	ζ Ursæ Mi.	1.4	6.0	+ 0.2	1.6	+1.2	0.4	−0.3	1.2
γ Cassiopeæ	0.9	0.8	0.5	6.1	0.3	5.4	25.2	0.0	θ Draconis	+ 0.5	1.1	− 1.0	0.1	−0.6	+0.3	+0.2	− 0.7
θ Cassiopeæ	0.7	0.6	0.3	4.7	0.2	4.1	26.1	0.0	η Draconis	1.1	1.0	0.8	0.1	0.6	0.0	0.1	0.6
ψ Cassiopeæ	1.2	0.7	0.5	4.7	0.4	3.9	+10.8	− 0.3	A Draconis	−11.1	0.3	0.2	0.5	+0.2	0.1	0.0	+ 0.1
δ Cassiopeæ	0.8	0.5	0.3	4.2	0.4	3.6	22.5	0.3	μ Draconis	+ 0.2	1.7	1.6	0.0	−1.5	0.1	0.1	− 1.5
ε Cassiopeæ	1.0	0.3	0.1	3.5	0.7	2.8	6.9	0.5	ε Ursæ Mi.	− 1.8	+ 7.6	+ 7.7	1.2	+8.5	−0.5	−0.5	+ 8.3
f Cassiopeæ	1.5	0.5	0.3	4.2	0.7	3.0	5.2	0.6	ζ Draconis	+ 5.7	− 0.3	− 0.3	0.0	−0.[illegible]	+0.1	0.0	− 0.2
35 Cassiop.	1.1	− 0.1	− 0.2	3.0	1.0	2.3	4.0	0.8	β Draconis	0.1	3.0	2.9	0.0	2.8	0.0	0.0	3.0
γ Persei	0.6	0.6	0.8	1.8	1.3	1.4	3.6	1.0	ξ Draconis	0.2	7.9	7.9	0.0	7.8	0.0	+0.1	7.8
10 Camelop.	0.4	3.7	4.9	0.8	5.1	0.6	1.0	5.0	γ Draconis	0.1	13.6	13.6	0.0	14.0	0.0	0.0	13.5
δ Aurigæ	0.1	23.7	23.6	0.2	+9.0	0.1	0.2	23.7	χ Draconis	0.4	3.1	4.1	+0.2	3.1	0.0	0.1	3.2
2 Lyncis	0.0	+90.3	+90.4	− 0.1	9.8	0.0	− 0.1	+90.4	δ Draconis	−41.0	+ 0.4	+ 0.4	0.3	0.0	0.0	0.0	+ 0.1
ο Ursæ Maj.	− 0.7	1.8	2.0	1.4	2.4	− 1.1	1.2	2.2	κ Cygni	0.1	1.6	1.5	0.0	+1.3	0.0	−0.1	1.4
h Ursæ Maj.	0.9	0.5	0.7	2.4	1.3	1.8	3.2	1.2	τ Draconis	+ 2.3	− 0.4	− 0.5	0.6	−0.9	0.1	0.0	− 1.0
θ Ursæ Maj.	0.6	0.4	0.6	2.0	1.1	1.5	3.9	0.9	π Draconis	− 2.4	+ 0.7	+ 0.6	0.3	+0.4	−0.1	0.0	+ 0.3
υ Ursæ Maj.	0.9	0.1	0.4	2.4	0.9	1.9	4.2	0.8	33 Cygni	0.2	1.1	1.0	0.0	0.7	0.3	0.2	0.8
β Ursæ Maj	0.7	− 0.5	− 0.3	4.4	0.2	3.7	80.6	0.2	κ Cephei	+ 3.8	0.1	0.1	1.4	−0.9	+0.3	+0.3	− 0.8
α Ursæ Maj.	1.0	0.6	0.4	4.8	0.4	3.9	21.4	0.2	η Cephei	− 0.4	1.1	1.0	0.2	+0.5	−0.4	−0.3	+ 0.6
λ Draconis	1.4	1.1	1.0	8.0	0.2	6.8	97.1	0.1	α Cephei	0.1	1.2	1.0	0.1	0.3	0.5	0.4	0.6
γ Ursæ Maj.	0.6	0.8	0.5	15.6	0.1	15.0	+ 5.0	− 0.2	β Cephei	1.7	1.2	1.1	0.6	−0.1	0.3	0.3	0.3
δ Ursæ Maj	0.7	1.0	0.7	+31.5	0.0	+32.2	3.8	0.3	π Cygni	+ 0.1	1.2	0.8	−0.5	+0.5	0.9	0.6	0.8
κ Draconis	1.0	1.5	1.3	6.0	−0.1	7.3	3.7	0.2	11 Cephei	− 1.5	1.3	1.2	+0.6	0.2	0.3	0.4	0.3
ε Ursæ Maj.	0.5	1.1	0.8	3.5	0.2	4.1	2.0	0.4	10 Cephei	0.0	1.2	1.0	−0.2	−0.1	0.8	0.7	0.5
ζ Ursæ Maj.	0.3	1.1	0.8	1.7	0.3	2.3	1.3	0.5	ζ Cephei	+ 0.2	1.2	0.9	0.6	0.4	1.2	0.8	0.5
g Ursæ Maj.	0.3	1.0	0.9	1.6	0.3	2.1	1.3	0.5	ε Cephei	0.3	1.2	1.0	0.8	0.3	1.3	0.8	0.6
η Ursæ Maj.	0.2	1.1	0.8	1.1	0.4	1.6	0.9	0.6	3 Lacertæ	0.2	1.1	0.9	1.1	.0.4	1.5	0.8	0.6
i Draconis	0.0	1.2	0.8	0.6	0.3	1.1	0.8	0.4	δ Cephei	0.4	1.2	0.9	0.9	0.3	1.5	1.0	0.6
α Draconis	+ 0.2	1.1	1.1	0.1	0.3	0.9	0.8	0.4	ι Cephei	0.4	1.3	1.1	1.1	0.2	1.9	1.5	0.4
ι Bootis	− 0.1	1.1	0.8	0.6	0.5	1.0	0.6	0.6	γ Cephei	1.2	2.0	1.9	4.3	+0.1	6.1	4.1	0.2
θ Bootis	0.1	1.3	1.0	0.3	0.5	0.9	0.6	0.8	β Cassiopeæ	0.7	1.0	0.6	0.0	0.0	0.0	4.4	0.2
a Ursæ Min.	21.0	1.4	1.3	− 1.5	+0.1	− 0.2	0.0	0.0									

In the preceding Catalogue of 500 Stars, the right ascensions and declinations, together with the annual variations, were taken from Schumacher's *Astronomische Hulfstafeln* for the year 1821, and in all cases where Pond's determination is annexed, we have preferred its adoption. The particular stars which have an * prefixed, had the blanks, left by Schumacher, filled up from Piazzi's Catalogue of 1814; these stars not having been observed by both Bradley and Piazzi; or at least not having been observed both in right ascension and declination. The right ascensions *in arc* have however no reference to Pond's observations.

The logarithms or constants a, b, c, d, and a', b', c', d', for computing the reductions, in conjunction with Tables A, B, C, and D, were taken from Schumacher's *Astronomische Hulfstafeln* of the year 1822, after the errata had been carefully corrected; and it is presumed they will afford the ready means of ascertaining the time, or of determining the latitude when a Table containing the diurnal places of the 46 Greenwich stars is not at hand, or when no observation could be taken of any of them.

The value of k for 1821 is −0 .967, for 1822 −1 .209, for 1823 −1 .451, and for 1824 −1 .693, and −0 .307 after February, the numbers for any subsequent year may be had by adding 0 .242 to those of the preceding.

As the *constants* are subject to greater variations annually, the nearer the places of the stars are to the north pole, the computers, Rosenberg and Scherke, have very properly added a Table of the annual variations of the constants of all such northern stars as are likely to have their places most affected by changes of right ascension and declination, which Table we have here given, for the purpose of rendering the constants applicable for a longer period than they would have been without attention being paid to their annual changes. These variations, multiplied by the number of years elapsed since 1820, and applied with their proper signs to their respective constants, will render them as correct as if they had been computed for the latter epoch.

TABLE 1. MEAN OBLIQUITY.

Years.	Mean Obliquity.	D	k	Years.	Mean Obliquity.	D	k
1750	23° 23′ 17″.65	1505	+0.02	1813	23° 27′ 48′.86	4120	−0.25
1760	13 .08	5157	0.60	1814	48 .40	4485	0.49
1770	8 .51	2012	0.17	1815	47 .95	4850	0.74
1780	3 .94	5664	0.74	1816	47 .49	5215	+0.02
1790	27 59 .37	2519	0.32	1817	47 .03	5581	−0.22
1800	54 .80	6171	−0.10	1818	46 .57	5946	0.46
1801	54 .34	6535	0.34	1819	46 .12	6311	0.70
1802	53 .89	102	0.59	1820	45 .66	6676	+0.05
1803	53 .43	467	0.83	1821	45 .20	244	−0.19
1804	52 .97	832	0.07	1822	44 .75	609	0.43
1805	52 .52	1198	0.31	1823	44 .29	974	0.67
1806	52 .06	1563	0.56	1824	43 .83	1339	+0.08
1807	51 .60	1928	0.80	1825	43 .38	1705	−0.16
1808	51 .14	2293	0.04	1826	42 .92	2070	0.40
1809	50 .69	2659	0.28	1827	42 .46	2435	0 64
1810	50 .23	3024	0.52	1828	42 .00	2800	+0.12
1811	49 .77	3389	0.77	1829	41 .55	3166	−0.13
1812	49 .32	3754	0.01	1830	41 .09	3531	0.37

TABLE 2. LUNAR NUTATION of the EQUINOXES, and EQUATION of the OBLIQUITY.

Arg.	Nut. in Long.	Equat. of Obliq.	Arg.	Nut. in Long.	Equat. of Obliq.
0	+0″.00	+8″.89	3800	−6″.19	−8″.43
100	1 .52	8 .85	3900	7 .62	8 .09
200	3 .03	8 .74	4000	8 .99	7 .67
300	4 .51	8 .56	4100	10 .28	7 .18
400	5 .96	8 .31	4200	11 .48	6 .64
500	7 .36	7 .98	4300	12 .57	6 .03
600	8 .69	7 .59	4400	13 .56	5 .38
700	9 .96	7 .14	4500	14 .42	4 .68
800	11 .15	6 .63	4600	15 .16	3 .94
900	12 .25	6 .05	4700	15 .76	3 .17
1000	13 .24	5 .43	4800	16 .23	2 .37
1100	14 .13	4 .76	4900	16 .56	1 .56
1200	14 .90	4 .05	5000	16 .74	0 .73
1300	15 .54	3 .31	5100	16 .78	+0 .10
1400	16 .06	2 .53	5200	16 .68	0 .93
1500	16 .44	1 .73	5300	16 .43	1 .74
1600	16 .68	0 .91	5400	16 .05	2 .54
1700	16 .78	0 .08	5500	15 .53	3 .32
1800	16 .74	−0 .74	5600	14 .89	4 .07
1900	16 .55	1 .57	5700	14 .11	4 .78
2000	16 .23	2 .39	5800	13 .23	5 .44
2100	15 .76	3 .18	5900	12 .23	6 .07
2200	15 .15	3 .95	6000	11 .13	6 .64
2300	14 .41	4 .69	6100	9 .94	7 .15
2400	13 .54	5 .39	6200	8 .68	7 .60
2500	12 .56	6 .04	6300	7 .33	7 .99
2600	11 .46	6 .65	6400	5 .93	8 .31
2700	10 .26	7 .19	6500	4 .48	8 .56
2800	8 .97	7 .68	6600	3 .00	8 .74
2900	7 .60	8 .09	6700	1 .49	8 .85
3000	6 .16	8 .44	6800	+0 .03	8 .89
3100	4 .66	8 .71	6900	1 .55	8 .85
3200	3 .13	8 .91	7000	3 .05	8 .74
3300	1 .56	9 .03	7100	4 .54	8 .56
3400	−0 .01	9 .07	7200	5 .98	8 .30
3500	1 .59	9 .03	7300	7 .38	7 .98
3600	3 .16	8 .91	7400	8 .71	7 .58
3700	4 .69	8 .71			

In Table 1. the mean obliquity for 1800 is taken at 23° 27′ 54″.8, and the annual variation — 0.457.

In Table 2. the formula for the ☾ nutation is — 16″.78332 sin ☊ + 0″.16094 sin 2 ☊. and for the equation of the obliquity it is + 8″.97709 cos ☊ — 0″.08768 cos 2 ☊.

In Table 3. the ⊙ equation of the equinoxes in longitude is derived from — 1″.33594 sin 2 ⊙. and the ⊙ equation of the obliquity from + 0″.57992 cos 2 ⊙.

TABLE 3. SOLAR NUTATION of the EQUINOXES, and SOLAR EQUATION of the OBLIQUITY.

Date		Nut.	Obliq.	Date		Nut.	Obliq.
January	0	+0″.46	−0″.55	July	9	+0″.75	−0″.72
	10	0 .87	0 .45		19	1 .07	0 .60
	20	1 .16	0 .31		29	1 .27	0 .44
	30	1 .32	0 .13	August	8	1 .34	0 .26
February	9	1 .31	+0 .06		18	1 .26	0 .04
	19	1 .14	0 .24		28	1 .04	+0 .14
March	1	0 .83	0 .38	Septemb.	7	0 .68	0 .22
	11	0 .42	0 .46		17	0 .26	0 .24
	21	−0 .03	0 .48		27	−0 .20	0 .23
	31	0 .48	0 .43	October	7	0 .63	0 .16
April	10	0 .87	0 .31		17	0 .99	0 .02
	20	1 .16	0 .15		27	1 .24	−0 .16
	30	1 .32	−0 .05	Novemb.	6	1 .34	0 .37
May	10	1 .32	0 .25		16	1 .27	0 .58
	20	1 .17	0 .45		26	1 .05	0 .77
	30	0 .90	0 .61	Decemb.	6	0 .69	0 .93
June	9	0 .53	0 .73		16	0 .26	1 .01
	19	0 .09	0 .79		26	+0 .22	1 .02
	29	+0 .34	0 .79		36	0 .66	0 .97
July	9	0 .75	0 .72				

TABLE 4. OBSERVED OBLIQUITY.

		By observation.	By Table 1.	Difference.
1814.	Summer......	23° 27′ 43″.34	23° 27′ 44″.11	+0″.77
	Winter	44 .60	45 .34	+0 .74
1815.	Summer.......	45 .78	46 .62	+0 .84
	Winter.......	47 .75	47 .91	+0 .16
1816.	Summer......	48 .98	49 .15	+0 .17
	Winter	50 .22	50 .33	+0 .11
1817.	Summer......	51 .24	51 .37	+0 .13
	Winter	52 .86	52 .29	−0 .57
1818.	Summer......	53 .04	53 .01	−0 .03
	Winter	51 .91	53 .55	+1 .64
1820.	Summer......	53 .96	53 .75	−0 .21
	Winter	53 .53	53 .33	−0 .20
1821.	Summer......	51 .79	52 .67	+0 .88
	Winter	52 .72	51 .77	−0 .95
1822.	Summer......	50 .47	50 .69	+0 .22
	Winter	50 .84	49 .38	−1 .46

In the 34th number of Schumacher's *Astronomische Nachrichten*, Bessel has given the Tables which are here copied, the use of which will be understood from the two subjoined examples.

Example 1.—Let the nutation of the equinoxes in longitude be required for Feb. 13, 1824?

Table 2. The date + D = 1383..... + 15″.98 }
Table 3. The date + k = Feb. 12 .08 + 1 .27 } + 17″.25 = nut.

In January and February of every Leap-Year the date must be diminished by 1.

Example 2.—Let it be required to determine the apparent obliquity of the ecliptic on June 21 .05 1824?

Table 1. 1824.................. 23° 27′ 43″.83 }
Table 2. Date + D = 1513 +1 .62 } 23° 27′ 44″.66 = ap. obl.
Table 3. Date + k June 21 .13 −0 79 }

The numbers in the column D of Table 1. are so many days' motion from the node at the beginning of each year, to which must be added the number of days since the year commenced to gain the argument for Table 2.; the whole period of the retrograde motion of the ☊ being 6798 days. The quantity k has reference to the sun's longitude at the beginning of each year and is applied to the civil date, in order to give the argument for Table 3. The first ten obliquities in Table 4. were determined by *Cary*'s Circle, and the last six by *Reichenbach*'s. [Vide Table 14. p. 174.]

TABLE 1. MEAN REFRACTIONS.

Zen. Dist.	δθ	Log. δθ	Diff.
0°	0″.00	0.0000	
1	1 .02	0.0085	
2	2 .04	0.3097	3012
3	3 .06	0.4860	1763
4	4 .08	0.6112	1252
5	5 .11	0.7086	974
6	6 .14	0.7882	796
7	7 .17	0.8557	675
8	8 .21	0.9144	587
9	9 .25	0.9663	519
10	10 .30	1.0129	466
11	11 .35	1.0553	424
12	12 .42	1.0941	388
13	13 .49	1.1300	359
14	14 .56	1.1634	334
15	15 .66	1.1947	313
16	16 .75	1.2241	294
17	17 .86	1.2519	278
18	18 .98	1.2784	265
19	20 .11	1.3036	252
20	21 .26	1.3277	241
21	22 .42	1.3507	230
22	23 .60	1.3729	222
23	24 .80	1.3944	215
24	26 .01	1.4151	207
25	27 .24	1.4352	201
26	28 .49	1.4547	195
27	29 .76	1.4736	189
28	31 .05	1.4921	185
29	32 .38	1.5102	181
30	33 .72	1.5279	177
31	35 .09	1.5452	173
32	36 .49	1.5622	170
33	37 .93	1.5790	168
34	39 .39	1.5954	164
35	40 .89	1.6116	162
36	42 .42	1.6276	160
37	44 .00	1.6435	159
38	45 .61	1.6591	156
39	47 .27	1.6746	155
40	48 .99	1.6901	155
41	50 .75	1.7055	154
42	52 .57	1.7207	152
43	54 .43	1.7358	151
44	56 .35	1.7510	152
45	58 .36	1.76111	151
46	1 0 .43	1.78123	1512
47	1 2 .57	1.79637	1514
48	1 4 .80	1.81155	1518
49	1 7 .11	1.82678	1523
50	1 9 .52	1.84208	1530
51	1 12 .02	1.85747	1539
52	1 14 .64	1.87298	1551
53	1 17 .38	1.88863	1565
54	1 20 .24	1.90440	1577
55	1 23 .25	1.92036	1596
56	1 26 .41	1.93653	1617
57	1 29 .73	1.95291	1638
58	1 33 .23	1.96955	1664
59	1 36 .93	1.98646	1691
60	1 40 .85	2.00368	1722
61	1 45 .01	2.02124	1756
62	1 49 .44	2.03918	1794
63	1 54 .17	2.05754	1836
64	1 59 .22	2.07635	1881

Zen. Dist.	δθ	Log. δθ	Diff.	dδθ/dτ
65°	2′ 4″.65	2.09567		
66	2 10 .48	2.11555	1988	
67	2 16 .78	2.13603	2048	
68	2 23 .61	2.15719	2116	
69	2 31 .04	2.17910	2191	
70 00	2 39 .16	2.20185	2275	
10	2 40 .59	2.20573	388	
20	2 42 .04	2.20963	390	
30	2 43 .52	2.21356	393	
40	2 45 .02	2.21752	396	
50	2 46 .53	2.22150	398	
71 00	2 48 .08	2.22552	402	
10	2 49 .65	2.22956	404	
20	2 51 .25	2.23363	407	
30	2 52 .87	2.23773	410	
40	2 54 .53	2.24186	413	
50	2 56 .21	2.24603	417	
72 00	2 57 .92	2.25022	419	
10	2 59 .66	2.25445	423	
20	3 1 .43	2.25870	425	
30	3 3 .23	2.26299	429	
40	3 5 .06	2.26732	433	
50	3 6 .93	2.27168	436	
73 00	3 8 .83	2.27608	440	
10	3 10 .77	2.28051	443	
20	3 12 .74	2.28498	447	
30	3 14 .75	2.28948	450	
40	3 16 .80	2.29402	454	
50	3 18 .88	2.29860	458	
74 00	3 21 .01	2.30322	462	
10	3 23 .18	2.30789	467	
20	3 25 .39	2.31259	470	
30	3 27 .66	2.31734	475	
40	3 29 .95	2.32213	479	
50	3 32 .30	2.32696	483	—
75 00	3 34 .70	2.33184	488	0.009
10	3 37 .16	2.33677	493	0.009
20	3 39 .65	2.34174	497	0.010
30	3 42 .21	2.34676	502	0.010
40	3 44 .82	2.35183	507	0.011
50	3 47 .48	2.35695	512	0.011
76 00	3 50 .21	2.36212	517	0.012
10	3 53 .00	2.36735	523	0.012
20	3 55 .85	2.37263	528	0.013
30	3 58 .76	2.37796	533	0.013
40	4 1 .74	2.38334	538	0.014
50	4 4 .79	2.38879	545	0.014
77 00	4 7 .91	2.39430	551	0.015
10	4 11 .11	2.39987	557	0.015
20	4 14 .39	2.40550	563	0.016
30	4 17 .74	2.41119	569	0.016
40	4 21 .19	2.41695	576	0.017
50	4 24 .72	2.42278	583	0.017
78 00	4 28 .33	2.42867	589	0.018
10	4 32 .04	2.43463	596	0.018
20	4 35 .84	2.44066	603	0.019
30	4 39 .75	2.44677	611	0.020
40	4 43 .76	2.45295	618	0.021
50	4 47 .88	2.45921	626	0.022
79 00	4 52 .12	2.46556	635	0.023
10	4 46 .47	2.47198	642	0.024
20	5 0 .94	2.47848	650	0.025
30	5 5 .54	2.48507	659	0.026
40	5 10 .28	2.49176	669	0.027
50	5 15 .16	2.49853	677	0.028
80 00	5 20 .19	2.50541	688	0.030

Zen. Dist.	δθ	Log. δθ	Diff.	dδθ/dτ	dδθ/dp
80° 00	5′ 20″.19	2.50541		0.030	0.04
10	5 25 .36	2.51237	696	0.031	0.04
20	5 30 .70	2.51944	707	0.033	0.04
30	5 36 .20	2.52660	716	0.034	0.04
40	5 41 .88	2.53387	727	0.036	0.05
50	5 47 .74	2.54125	738	0.038	0.05
81 00	5 53 .79	2.54874	749	0.040	0.05
10	6 0 .04	2.55635	759	0.042	0.06
20	6 6 .50	2.56407	772	0.044	0.06
30	6 13 .18	2.57192	785	0.046	0.07
40	6 20 .09	2.57989	797	0.049	0.07
50	6 27 .26	2.58800	811	0.051	0.08
82 00	6 34 .68	2.59624	824	0.053	0.08
10	6 42 .37	2.60462	838	0.057	0.09
20	6 50 .33	2.61313	851	0.060	0.09
30	6 58 .59	2.62179	866	0.063	0.10
40	7 7 .19	2.63062	883	0.067	0.10
50	7 16 .13	2.63961	899	0.071	0.11
83 00	7 25 .40	2.64875	914	0.074	0.11
10	7 35 .05	2.65806	931	0.079	0.12
20	7 45 .10	2.66755	949	0.084	0.12
30	7 55 .58	2.67722	967	0.089	0.13
40	8 6 .50	2.68708	986	0 095	0.14
50	8 17 .90	2.69714	1006	0.101	0.15
84 00	8 29 .80	2.70740	1026	0.107	0.16
10	8 42 .24	2.71787	1047	0.114	0.17
20	8 55 .25	2.72856	1069	0.122	0.18
30	9 8 .88	2.73948	1092	0.130	0.20
40	9 23 .16	2.75063	1115	0.139	0.21
50	9 38 .12	2.76202	1139	0.149	0.23
85 00	9 53 .84	2.77367	1165	0.159	0.25
10	10 10 .35	2.78558	1191	0.171	0.26
20	10 27 .73	2.79777	1219	0.184	0.28
30	10 46 .03	2.81025	1248	0.198	0.31
40	11 5 .30	2.82302	1277	0.213	0.33
50	11 25 .66	2.83611	1309	0.229	0.36
86 00	11 47 .15	2.84951	1340	0.248	0.39
10	12 9 .88	2.86325	1374	0.269	0.43
20	12 33 .97	2.87735	1410	0.292	0.47
30	12 59 .51	2.89182	1447	0.317	0.51
40	13 26 .61	2.90666	1484	0.345	0.56
50	13 55 .40	2.92189	1523	0.376	0.62
87 00	14 26 .04	2.93754	1565	0.410	0.68
10	14 58 .71	2.95362	1608	0.448	0.75
20	15 33 .60	2.97016	1654	0.490	0.83
30	16 10 .89	2.98717	1701	0.538	0.91
40	16 50 .8	3.00466	1749	0.593	1.01
50	17 33 .6	3.02267	1801	0.654	1.13
88 00	18 19 .6	3.04122	1855	0.722	1.26
10	19 9 .0	3.06031	1909	0.799	1.41
20	20 2 .2	3.07998	1967	0.887	1.59
30	20 59 .6	3.10024	2026	0.987	1.79
40	22 1 .7	3.12113	2089	1.101	2.02
50	23 8 .9	3.14268	2155	1.231	2.29
89 00	24 21 .8	3.16489	2221	1.380	2.61
10	25 40 .9	3.18779	2290	1.551	2.98
20	27 7 .1	3.21140	2361	1.749	3.41
30	28 40 .8	3.23574	2434	1.977	3.93
40	30 23 .2	3.26083	2509	2.241	4.54
50	32 15 .0	3.28667	2584	2.549	5.26
90 00	34 17 .5	3.31334	2667	2.909	6.12

TABLE 2. THERMOMETER.

	Log.	Diff.		Log.	Diff.
50°	0 .00000		50	0 .00000	
49	0 .00090	91	51	9 .99910	
48	0 .00181		52	9 .99820	
47	0 .00272		53	9 .99730	
46	0 .00363	92	54	9 .99640	
45	0 .00455		55	9 .99550	
44	0 .00546		56	9 .99460	90
43	0 .00638		57	9 .99371	
42	0 .00730		58	9 .99282	
41	0 .00822		59	9 .99193	
40	0 .00914		60	9 .99104	
39	0 .01006	93	61	9 .99016	89
38	0 .01099		62	9 .98927	
37	0 .01192		63	9 .98839	
36	0 .01285	94	64	9 .98751	
35	0 .01379		65	9 .98663	
34	0 .01472		66	9 .98575	88
33	0 .01566		67	9 .98488	
32	0 .01660		68	9 .98401	
31	0 .01754		69	9 .98314	
30	0 .01848		70	9 .98227	
29	0 .01942	95	71	9 .98140	87
28	0 .02037		72	9 .98054	
27	0 .02132		73	9 .97967	
26	0 .02227		74	9 .97881	
25	0 .02323	96	75	9 .97795	
24	0 .02418		76	9 .97709	
23	0 .02513		77	9 .97623	
22	0 .02609		78	9 .97537	86
21	0 .02706	97	79	9 .97452	
20	0 .02803		80	9 .97367	
19	0 .02900		81	9 .97282	
18	0 .02997		82	9 .97197	
17	0 .03094		83	9 .97112	
16	0 .03191		84	9 .97027	85
15	0 .03288	98	85	9 .96943	
14	0 .03386		86	9 96859	
13	0 .03484		87	9 .96775	
12	0 .03582		88	9 .96691	
11	0 .03680	99	89	9 .96607	84
10	0 .03779		90	9 .96524	

TABLE 3. BAROMETER.

Inches.	Log.	Diff.
31 .0	0 .01424	
30 .9	0 .01284	140
30 .8	0 .01143	141
30 .7	0 .01002	141
30 .6	0 .00860	142
30 .5	0 .00718	142
30 .4	0 .00575	143
30 .3	0 .00432	143
30 .2	0 .00289	143
30 .1	0 .00145	144
30 .0	0 .00000	145
29 .9	9 .99855	145
29 .8	9 .99709	146
29 .7	9 .99563	146
29 .6	9 .99417	146
29 .5	9 .99270	147
29 .4	9 .99123	147
29 .3	9 .98975	148
29 .2	9 .98826	149
29 .1	9 .98677	149
29 .0	9 .98528	149
28 .9	9 .98378	150
28 .8	9 .98227	151
28 .7	9 .98076	151
28 .6	9 .97924	152
28 .5	9 .97772	152
28 .4	9 .97620	152
28 .3	9 .97466	154
28 .2	9 .97313	153
28 .1	9 .97158	155
28 .0	9 .97004	154
27 .9	9 .96848	156
27 .8	9 .96692	156
27 .7	9 .96536	156
27 .6	9 .96379	157
27 .5	9 .96221	158

TABLE 4.

	+	Diff.
10°	0 .00173	
20	0 .00130	43
30	0 .00087	43
40	0 .00043	44
	—	
50	0 .00000	43
60	0 .00043	43
70	0 .00087	44
80	0 .00130	43
90	0 .00173	43

MR. IVORY's Tables of Refraction were published in Part II. of the Philosophical Transactions of London for 1823, and appeared just in time to allow of their being added to our collection. The reader who wishes to examine the principles on which Mr. Ivory constructed these Tables, is referred to his valuable paper "On the Astronomical Refractions," which gives an ample detail of all the particulars.

In Table 1., the first column contains the zenith distances; the second gives the corresponding values of $\delta\theta$, or the mean refractions at the temperature of 50° of FAHRENHEIT, and the barometric pressure 30 English inches; the third comprises the logarithms of these mean refractions: and when the zenith distance exceeds 75° the values of $\frac{d\delta\theta}{d\tau}$ and $\frac{d\delta\theta}{dp}$ are added in two other columns, as the factors for determining the effects of any changes in the temperature and barometric pressure respectively. The height of the barometer in English inches is supposed to be expressed by p, and the temperature as shewn by Fahrenheit's thermometer by τ.

The true refraction is computed by the following formula, viz.

$$r = \frac{1}{1+\beta\,(\tau-50)} \times \frac{p}{30} \times \delta\theta + \frac{d\delta\theta}{d\tau}(\tau - 50°) - \frac{d\delta\theta}{dp}(30 - p).$$

The first term is the mean refraction corrected for the observed temperature and pressure in the manner usually practised by astronomers. When the zenith distance does not exceed 75°, the two remaining terms are to be accounted evanescent; and even when the zenith distance is 80° or a little more, the same terms may on many occasions be omitted: otherwise the two terms, amounting generally to some seconds, must be added to the first term with their proper signs. Three subsidiary Tables are added for facilitating the corrections for the barometer and thermometer. Table 2. contains the logarithms of $\frac{1}{1+\beta\,(\tau-50°)} = \frac{1}{1+\frac{\tau-50}{480}}$, for 40° on each side of the mean temperature 50°; negative indices being avoided by substituting the arithmetical complements. Table 3. contains the logarithms, or the arithmetical complements of $\frac{p}{30}$ for the values of p between 31 and $27\frac{1}{2}$. Table 4. contains the small corrections, positive or negative, to be applied to the numbers in Table 3., in order to reduce the observed length of the barometric column to the mean temperature of 50°. The numbers of this Table are the logarithms of $\frac{1}{1+\frac{\tau-50}{10000}}$ equal to $-\frac{\tau-50}{10000} \times .434$.

Example 1. (Zen. dist. 13° 48′ 39″. Barom. 29 .65. Therm. 40°.)

Zen. dist. 13°	log. 1 .13000
0 48′ 39″	+ .02790
Thermometer 40°	0 .00914
Barometer 29 .65	9 .99490
Tab. 4.	+ .00043
True refraction 11″.534	1 .16237

Example 2. (Zen. Dist. 83° 45′ 30″. Barom. 30 .5. Therm. 65°.)

Zen. Dist. 83° 40′	log. 2 .68708	
5 30″	+ 553	
Thermom. 65°	9 .98663	
Barom. ...30 .5	0 .00718	
Tab. 4 .....65°	— 0 .00065	
Nat. Num. 485″.04	2 .68577 = 8′ 5″.04 = r	
$\frac{d\delta\theta}{d\tau}$ — .098 × 15		—1 .47
$\frac{d\delta\theta}{dp}$ — 0 .14 × — 0 .5		+0 .07
True refraction		8 3 .64

If we turn to page 307 we shall perceive, on examining the small Table of *Comparative Refractions*, that the true refraction in Example 1., as here given, is nearly the same as Carlini's; but that the French refraction agrees most closely with Mr. Ivory's in the second Example. At 86° of zenith distance, the difference between the French refractions and Mr. Ivory's amounts to about 2″, and at greater zenith distances it will be still more, as it ought to be, to accord with the best observations.

TABLE I. (Barometer 28 French inches. Thermometer 0° Réaumur.)

App. Zen.	Log. R	Diff. for 1′.	App. Zen.	Log. R	Diff. in 1′.	App. Zen.	Log. R	Diff. in 1′.	n	App. Zen.	Log. R	Diff. in 1′.	n	App. Zen.	Log. R	Diff. in 1′.	n
0° 0′	— ∞		21° 0′	1.3645	37	42° 0′	1.7343	25		55° 30′	1.9421	27	3	66° 0′	2.1292	34	7
20	9.5451	0.01505	20	1.3718	37	20	1.7393	25		40	1.9448	27	3	10	2.1326	34	8
40	9.8462	0.00880	40	1.3793	36	40	1.7444	25		50	1.9475	27	3	20	2.1360	34	8
1 0	0.0223	0.00624	22 0	1.3866	36	43 0	1.7494	25		56 0	1.9502	27	3	30	2.1394	34	8
20	0.1472	0.00486	20	1.3938	35	20	1.7545	25		10	1.9529	27	3	40	2.1428	34	8
40	0.2444	0.00395	40	1.4009	35	40	1.7595	25		20	1.9556	27	3	50	2.1462	34	8
2 0	0.3235	.00334	23 0	1.4080	35	44 0	1.7646	25		30	1.9583	27	3	67 0	2.1497	35	8
20	0.3904	290	20	1.4150	34	20	1.7696	25		40	1.9611	27	4	10	2.1532	35	8
40	0.4484	256	40	1.4219	34	40	1.7747	25		50	1.9638	27	4	20	2.1567	35	9
3 0	0.4997	229	24 0	1.4287	34	45 0	1.7797	25		57 0	1.9666	27	4	30	2.1602	35	9
20	0.5455	207	20	1.4355	33	20	1.7848	25		10	1.9693	28	4	40	2.1637	36	9
40	0.5870	189	40	1.4422	33	40	1.7898	25		20	1.9721	28	4	50	2.1672	36	9
4 0	0.6249	.00174	25 0	1.4488	33	46 0	1.7948	25		30	1.9748	28	4	68 0	2.1708	36	9
20	0.6598	161	20	1.4554	32	20	1.7999	25		40	1.9776	28	4	10	2.1744	36	9
40	0.6921	150	40	1.4618	32	40	1.8049	25		50	1.9804	28	4	20	2.1781	36	9
5 0	0.7222	141	26 0	1.4683	32	47 0	1.8100	25		58 0	1.9832	28	4	30	2.1817	36	10
20	0.7504	132	20	1.4747	31	20	1.8150	25		10	1.9860	28	4	40	2.1854	36	10
40	0.7769	125	40	1.4810	31	40	1.8201	25		20	1.9888	28	4	50	2.1890	37	10
6 0	0.8019	.00118	27 0	1.4872	31	48 0	1.8252	25	2	30	1.9916	28	4	69 0	2.1927	37	10
20	0.8255	112	20	1.4935	31	10	1.8277	25	2	40	1.9944	28	4	10	2.1965	37	10
40	0.8480	107	40	1.4997	30	20	1.8303	25	2	50	1.9973	28	4	20	2.2002	38	10
7 0	0.8694	102	28 0	1.5058	30	30	1.8328	25	2	59 0	2.0001	28	4	30	2.2040	38	10
20	0.8899	097	20	1.5118	30	40	1.8354	25	2	10	2.0030	28	4	40	2.2078	38	11
40	0.9094	093	40	1.5178	30	50	1.8379	25	2	20	2.0058	29	4	50	2.2117	38	11
8 0	0.9281	.00089	29 0	1.5238	30	49 0	1.8404	25	2	30	2.0087	29	4	70 0	2.2155	39	11
20	0.9460	86	20	1.5298	29	10	1.8430	25	2	40	2.0116	29	4	10	2.2194	39	11
40	0.9633	83	40	1.5357	29	20	1.8455	25	2	50	2.0144	29	5	20	2.2233	39	11
9 0	0.9799	80	30 0	1.5415	29	30	1.8481	25	2	60 0	2.0173	29	5	30	2.2272	40	11
20	0.9960	77	20	1.5473	29	40	1.8506	25	2	10	2.0202	29	5	40	2.2312	40	12
40	1.0115	75	40	1.5531	28	50	1.8532	26	2	20	2.0232	29	5	50	2.2352	40	12
10 0	1.0265	00072	31 0	1.5588	28	50 0	1.8557	26	2	30	2.0261	29	5	71 0	2.2392	40	12
20	1.0410	70	20	1.5645	28	10	1.8583	26	2	40	2.0290	29	5	10	2.2432	41	12
40	1.0551	68	40	1.5702	28	20	1.8609	26	2	50	2.0320	29	5	20	2.2473	41	13
11 0	1.0688	66	32 0	1.5758	28	30	1.8634	26	2	61 0	2.0349	30	5	30	2.2514	41	13
20	1.0821	65	20	1.5814	28	40	1.8660	26	2	10	2.0379	30	5	40	2.2555	42	13
40	1.0951	63	40	1.5870	27	50	1.8686	26	2	20	2.0408	30	5	50	2.2597	42	13
12 0	1.1077	.00061	33 0	1.5925	27	51 0	1.8711	26	2	30	2.0438	30	5	72 0	2.2639	42	14
20	1.1199	60	20	1.5980	27	10	1.8737	26	2	40	2.0468	30	5	10	2.2681	42	14
40	1.1319	58	40	1.6035	27	20	1.8763	26	2	50	2.0498	30	5	20	2.2724	43	14
13 0	1.1436	57	34 0	1.6090	27	30	1.8789	26	2	62 0	2.0528	30	5	30	2.2767	43	14
20	1.1550	55	20	1.6144	27	40	1.8815	26	3	10	2.0559	30	5	40	2.2810	44	15
40	1.1661	54	40	1.6198	27	50	1.8840	26	3	20	2.0589	31	5	50	2.2854	44	15
14 0	1.1770	.00053	35 0	1.6252	27	52 0	1.8866	26	3	30	2.0619	31	6	73 0	2.2897	44	16
20	1.1876	52	20	1.6306	26	10	1.8892	26	3	40	2.0650	31	6	10	2.2942	44	16
40	1.1980	51	40	1.6359	26	20	1.8918	26	3	50	2.0681	31	6	20	2.2986	45	16
15 0	1.2082	50	36 0	1.6412	26	30	1.8945	26	3	63 0	2.0712	31	6	30	2.3032	45	16
20	1.2183	49	20	1.6465	26	40	1.8971	26	3	10	2.0743	31	6	40	2.3077	46	17
40	1.2281	48	40	1.6518	26	50	1.8997	26	3	20	2.0774	31	6	50	2.3123	46	17
16 0	1.2377	.00047	37 0	1.6571	26	53 0	1.9023	26	3	30	2.0805	31	6	74 0	2.3169	47	17
20	1.2471	46	20	1.6623	26	10	1.9049	26	3	40	2.0837	32	6	10	2.3216	47	18
40	1.2564	46	40	1.6675	26	20	1.9075	26	3	50	2.0868	32	6	20	2.3263	48	18
17 0	1.2656	45	38 0	1.6727	26	30	1.9102	26	3	64 0	2.0900	32	6	30	2.3310	48	19
20	1.2745	44	20	1.6779	26	40	1.9128	26	3	10	2.0932	32	6	40	2.3358	48	19
40	1.2833	43	40	1.6831	25	50	1.9154	26	3	20	2.0964	32	6	50	2.3407	49	19
18 0	1.2920	.00042	39 0	1.6882	25	54 0	1.9181	26	3	30	2.0996	32	6	75 0	2.3455	49	20
20	1.3005	42	20	1.6934	25	10	1.9207	26	3	40	2.1028	32	7	10	2.3505	49	20
40	1.3089	41	40	1.6985	25	20	1.9234	27	3	50	2.1061	32	7	20	2.3554	50	20
19 0	1.3172	40	40 0	1.7037	25	30	1.9260	27	3	65 0	2.1093	33	7	30	2.3605	50	20
20	1.3253	40	20	1.7088	25	40	1.9287	27	3	10	2.1126	33	7	40	2.3655	51	21
40	1.3333	39	40	1.7139	25	50	1.9313	27	3	20	2.1159	33	7	50	2.3707	51	21
20 0	1.3412	.00039	41 0	1.7191	25	55 0	1.9340	27	3	30	2.1192	33	7	76 0	2.3758	52	22
20	1.3490	38	20	1.7241	25	10	1.9367	27	3	40	2.1225	33	7	10	2.3810	53	22
40	1.3567	38	40	1.7292	25	20	1.9394	27	3	50	2.1258	33	7	20	2.3863	53	23

TABLE 1. concluded. (Barometer 28 French inches. Thermometer 0° Réaumur.

App. Zen.	Log. R.	Diff. for 1′.	m.	n.	App. Zen.	Log. R	Diff. for 1′.	m.	n.	App. Zen.	Log. R	Diff. for 1′.	m.	n.	App. Zen.	Log. R	Diff. for 1′.	m.	n.
76° 30′	2.3917	54	2	24	80° 0′	2.5192	70	4	42	83° 30′	2.6916	99	8	85	87° 0′	2.9559	166	24	221
40	2.3970	55	2	24	10	2.5262	71	4	43	40	2.7015	101	9	88	10	2.9725	171	26	234
50	2.4025	55	2	24	20	2.5333	72	4	44	50	2.7117	103	9	91	20	2.9896	177	28	248
77 0	2.4080	56	3	25	30	2.5404	73	4	45	84 0	2.7220	105	10	95	30	3.0073	182	30	262
10	2.4136	56	3	25	40	2.5477	74	5	47	10	2.7325	107	10	99	40	3.0255	188	32	278
20	2.4192	57	3	26	50	2.5551	75	5	48	20	2.7432	110	10	104	50	3.0444	195	34	295
30	2.4249	57	3	27	81 0	2.5626	76	5	49	30	2.7542	112	11	108	88 0	3.0639	202	37	314
40	2.4306	59	3	28	10	2.5702	77	5	51	40	2.7655	115	11	113	10	3.0840	209	40	334
50	2.4365	59	3	29	20	2.5780	79	5	52	50	2.7770	118	12	118	20	3.1049	217	43	356
78 0	2.4424	60	3	30	30	2.5858	80	5	54	85 0	2.7888	121	13	123	30	3.1266	225	46	380
10	2.4484	61	3	31	40	2.5938	81	6	56	10	2.8009	124	13	128	40	3.1491	234	50	406
20	2.4544	61	3	32	50	2.6020	83	6	58	20	2.8132	126	14	134	50	3.1725	242	55	434
30	2.4605	62	3	33	82 0	2.6102	84	6	60	30	2.8259	130	15	141	89 0	3.1968	253	59	485
40	2.4667	63	3	34	10	2.6187	85	6	62	40	2.8388	133	15	148	10	3.2222	264	65	499
50	2.4730	64	3	35	20	2.6272	87	6	65	50	2.8521	136	16	155	20	3.2486	276	71	537
79 0	2.4793	64	3	36	30	2.6359	89	7	67	86 0	2.8658	140	17	162	30	3.2762	288	78	579
10	2.4858	65	4	37	40	2.6448	90	7	70	10	2.8789	144	18	171	40	3.3051	302	86	625
20	2.4923	66	4	38	50	2.6538	92	7	72	20	2.8942	148	19	179	50	3.3353	317	95	677
30	2.4989	67	4	39	83 0	2.6630	94	7	75	30	2 9089	152	20	189	90 0	3.3670	333	106	734
40	2.5056	68	4	40	10	2.6724	95	8	78	40	2.9241	157	22	199					
50	2.5124	69	4	41	20	2.6819	97	8	81	50	2.9398	161	23	210					

TABLE 2.

BAROMETER, in Paris Inches.

B	b	B	b
	−		+
25.0	0.0492	28.0	0.0000
.1	0.0475	.1	0.0015
.2	0.0458	.2	0.0031
.3	0.0440	.3	0.0046
.4	0.0423	.4	0.0062
.5	0.0406	.5	0.0077
.6	0.0306	.6	0.0092
.7	0.0372	.7	0.0107
.8	0.0355	.8	0.0122
25.9	0.0339	28.9	0.0137
26.0	0.0322	29.0	0.0152
.1	0.0305	.1	0.0167
.2	0.0289	.2	0.0182
.3	0.0272	.3	0.0197
.4	0.0255	.4	0.0212
.5	0.0239	.5	0.0227
.6	0.0223	.6	0.0241
.7	0.0206	.7	0.0256
.8	0.0190	.8	0.0271
26.9	0.0174	29.9	0.0285
27.0	0.0158	30.0	0.0300
.1	0.0142	.1	0.0314
.2	0.0126	.2	0.0328
.3	0.0110	.3	0.0343
.4	0.0094	.4	0.0357
.5	0.0078	.5	0.0371
.6	0.0062	.6	0.0386
.7	0.0047	.7	0.0400
.8	0.0031	.8	0.0414
27.9	0.0015	30.9	0.0428
28.0	0.0000	31.0	0.0442

TABLE 3.

EXTERIOR THERMOMETER of Réaumur.

T	t	T	t
−	+	+	−
0°	0.0000	0°	0.0000
1	0.0020	1	0.0020
2	0.0040	2	0.0039
3	0.0060	3	0.0059
4	0.0080	4	0.0078
5	0.0100	5	0.0098
6	0.0120	6	0.0117
7	0.0141	7	0.0136
8	0.0161	8	0.0155
9	0.0182	9	0.0174
10	0.0202	10	0.0193
11	0.0223	11	0.0212
12	0.0244	12	0.0231
13	0.0265	13	0.0250
14	0.0286	14	0.0268
15	0.0307	15	0.0287
16	0.0329	16	0.0305
17	0.0350	17	0.0324
18	0.0372	18	0.0342
19	0.0393	19	0.0360
20	0.0415	20	0.0379
21	0.0437	21	0.0397
22	0.0459	22	0.0415
23	0.0481	23	0.0433
24	0.0503	24	0.0451
25	0.0525	25	0.0468
26	0.0547	26	0.0486
27	0.0570	27	0.0504
28	0.0593	28	0.0521
29	0.0615	29	0.0539
30	0.0638	30	0.0556

TABLE 4.

INTERIOR THERMOMETER of Réaumur.

T′	t′	T′	t′
−	+	+	−
0°	0.0000	0°	0.0000
5	0.0005	5	0.0005
10	0.0010	10	0.0010
15	0.0015	15	0.0015
20	0.0020	20	0.0020
25	0.0024	25	0.0024
30	0.0029	30	0.0029

Comparison of the English and French Barometers when both are decimally divided.

ENGLISH.	FRENCH.	ENGLISH.	FRENCH.
Inches.	Inches.	Inches.	Inches.
27.5	25.80	29.3	27.49
.6	.89	.4	.58
.7	.99	.5	.68
.8	26.08	.6	.77
.9	.18	.7	.87
28.0	.27	.8	.96
.1	.36	.9	28.05
.2	.46	30.0	.15
.3	.55	.1	.24
.4	.65	.2	.34
.5	.74	.3	.43
.6	.83	.4	.52
.7	.93	.5	.61
.8	27.02	.6	.70
.9	.12	.7	.80
29.0	.21	.8	.89
.1	.30	.9	.99
.2	.39	31.0	29.08

TABLE 1. (For Barom. 28 French inches, and Réaumur's Therm. 0°.)

App. Alt.	Refraction.	App. Alt.	Refraction.	App. Alt.	Refraction.	App. Alt.	Refraction.	App. Alt.	Refraction.	App. Alt.	Refrac.	App. Alt.	Refrac.	App. Alt.	Refra:
0° 00′	38′ 48″.2	8° 20	6′ 32″.5	16° 40′	3′ 18″.9	25° 00′	2′ 8″.6	33° 20′	1′ 31″.4	41° 40	1′ 7″.7	58° 00′	0′ 37″.7	74° 00′	0′ 17″.3
10	36 4.3	30	6 25.3	50	3 16.9	10	2 7.7	30	1 30.9	50	1 7.3	20	37.2	20	16.9
20	33 39.0	40	6 18.4	17 00	3 14.9	20	2 6.7	40	1 30.3	42 00	1 6.9	40	36.7	40	16.5
30	31 29.0	50	6 11.7	10	3 12.9	30	2 5.8	50	1 29.7	20	1 6.1	59 00	36.2	75 00	16.1
40	29 32.7	9 00	6 5.3	20	3 11.0	40	2 4.9	34 00	1 29.2	40	1 5.3	20	35.7	20	15.8
50	27 48.2	10	5 59.1	30	3 9.1	50	2 3.9	10	1 28.6	43 00	1 4.6	40	35.3	40	15.4
1 00	26 13.4	20	5 53.0	40	3 7.3	26 00	2 3.0	20	1 28.1	20	1 3.8	60 00	34.8	76 00	15.0
10	24 47.7	30	5 47.1	50	3 5.4	10	2 2.1	30	1 27.5	40	1 3.1	20	34.3	20	14.7
20	23 29.7	40	5 41.4	18 00	3 3.6	20	2 1.3	40	1 27.0	44 00	1 2.4	40	33.9	40	14.3
30	22 18.5	50	5 35.9	10	3 1.8	30	2 0.4	50	1 26.5	20	1 1.7	61 00	33.4	77 00	13.9
40	21 13.3	10 00	5 30.6	20	3 0.1	40	1 59.5	35 00	1 25.9	40	1 0.9	20	33.0	20	13.5
50	20 13.4	10	5 25.4	30	2 58.4	50	1 58.7	10	1 25.4	45 00	1 0.2	40	32.5	40	13.1
2 00	19 18.6	20	5 20.3	40	2 56.7	27 00	1 57.8	20	1 24.9	20	59.1	62 00	32.1	78 00	12.8
10	18 27.7	30	5 15.4	50	2 55.1	10	1 57.0	30	1 24.3	40	58.8	20	31.6	20	12.4
20	17 40.4	40	5 10.7	19 00	2 53.5	20	1 56.2	40	1 23.8	46 00	58.2	40	31.2	40	12.0
30	16 56.9	50	5 6.1	10	2 51.9	30	1 55.3	50	1 23.3	20	57.5	63 00	30.7	79 00	11.7
40	16 16.4	11 00	5 1.6	20	2 50.3	40	1 54.5	36 00	1 22.8	40	56.8	20	30.3	20	11.4
50	15 38.7	10	4 57.2	30	2 48.8	50	1 53.7	10	1 22.3	47 00	56.2	40	29.8	40	11.0
3 00	15 3.5	20	4 53.0	40	2 47.2	28 00	1 52.9	20	1 21.8	20	55.5	64 00	29.4	80 00	10.6
10	14 30.6	30	4 48.8	50	2 45.7	10	1 52.2	30	1 21.3	40	54.9	20	29.0	20	10.3
20	13 59.7	40	4 44.7	20 00	2 44.3	20	1 51.4	40	1 20.8	48 00	54.3	40	28.5	40	9.9
30	13 30.8	50	4 40.8	10	2 42.8	30	1 50.6	50	1 20.3	20	53.6	65 00	28.1	81 00	9.5
40	13 3.8	12 00	4 37.0	20	2 41.4	40	1 49.9	37 00	1 19.9	40	53.0	20	27.7	20	9.2
50	12 36.7	10	4 33.2	30	2 40.0	50	1 49.1	10	1 19.4	49 00	52.4	40	27.3	40	8.8
4 00	12 14.2	20	4 29.6	40	2 38.6	29 00	1 48.4	20	1 18.9	20	51.8	66 00	26.8	82 00	8.5
10	11 50.4	30	4 26.0	50	2 37.2	10	1 47.6	30	1 18.5	40	51.2	20	26.4	20	8.1
20	11 30.0	40	4 22.6	21 00	2 35.9	20	1 46.9	40	1 18.0	50 00	50.6	40	26.0	40	7.8
30	11 9.8	50	4 19.2	10	2 34.6	30	1 46.2	50	1 17.5	20	50.0	67 00	25 6	83 00	7.4
40	10 50.5	13 00	4 15.9	20	2 33.3	40	1 45.5	38 00	1 17.0	40	49.4	20	25.2	20	7.0
50	10 32.3	10	4 12.6	30	2 32.0	50	1 44.8	10	1 16.5	51 00	48.8	40	24.8	40	6.7
5 00	10 14.9	20	4 9.5	40	2 30.7	30 00	1 44.1	20	1 16.1	20	48.2	68 00	24.4	84 00	6.3
10	9 58.5	30	4 6.4	50	2 29.4	10	1 43.4	30	1 15.7	40	47.6	20	23.9	20	6.0
20	9 42.8	40	4 3.4	22 00	2 28.2	20	1 42.7	40	1 15.2	52 00	47.1	40	23.5	40	5.6
30	9 27.8	50	4 0.5	10	2 27.0	30	1 42.0	50	1 14.8	20	46.5	69 00	23.1	85 00	5.3
40	9 13.6	14 00	3 57.6	20	2 25.8	40	1 41.4	39 00	1 14.3	40	46.0	20	22.7	20	4.9
50	9 0.1	10	3 54.8	30	2 24.6	50	1 40.7	10	1 13.9	53 00	45.4	40	22.3	40	4.6
6 00	8 47.2	20	3 52.0	40	2 23.4	31 00	1 40.0	20	1 13.5	20	44.9	70 00	21.9	86 00	4.2
10	8 34.9	30	3 49.7	50	2 22.3	10	1 39.4	30	1 13.0	40	44.3	20	21.5	20	3.9
20	8 22.9	40	3 46.7	23 00	2 21.2	20	1 38.7	40	1 12.6	54 00	43.8	40	21.1	40	3.5
30	8 11.6	50	3 44.1	10	2 20.0	30	1 38.1	50	1 12.2	20	43.2	71 00	20.8	87 00	3.2
40	8 0.7	15 00	3 41.6	20	2 18.9	40	1 37.5	40 00	1 11.7	40	42.7	20	20.4	20	2.8
50	7 50.3	10	3 39.1	30	2 17.9	50	1 36.8	10	1 11.3	55 00	42.2	40	20.0	40	2.5
7 00	7 40.3	20	3 36.7	40	2 16.8	32 00	1 36.2	20	1 10.9	20	41.7	72 00	19.6	88 00	2.1
10	7 30.6	30	3 34.3	50	2 15.7	10	1 35.6	30	1 10.5	40	41.1	20	19.2	20	1.8
20	7 21.4	40	3 32.0	24 00	2 14.7	20	1 35.0	40	1 10.1	56 00	40.6	40	18.8	40	1.4
30	7 12.5	50	3 29.7	10	2 13.6	30	1 34.4	50	1 9.7	20	40.1	73 00	18.4	89 00	1.1
40	7 3.8	16 00	3 27.5	20	2 12.6	40	1 33.8	41 00	1 9.3	40	39.6	20	18.0	20	0.7
50	6 55.6	10	3 25.3	30	2 11.6	50	1 33.2	10	1 8.9	57 00	39.1	40	17.7	40	0.4
8 00	6 47.6	20	3 23.1	40	2 10.6	33 00	1 32.6	20	1 8.5	20	38.6	74 00	17.3	90 00	0.0
10	6 40.0	30	3 21.0	50	2 9.6	10	1 32.0	30	1 8.1	40	38.2				

TABLE 4. (Barom. 28 French inches. Therm. +8 Réaumur.)

True Alt.	Refraction.	True Alt.	Refraction.	True Alt.	Refraction.	True Alt.	Refraction.	True Alt.	Refraction.	True Alt.	Refraction.	True Alt.	Refraction.	True Alt.	Refraction.
−0° 30′	33′ 11″.3	2° 10	16′ 27″.2	4° 50′	9′ 54″.5	10°	5′ 17″.3	26°	1′ 58″.8	42°	1′ 4″.9	58°	0′ 36″.1	74°	0′ 16″.5
20	31 37.3	20	15 51.3	5 00	9 38.9	11	4 49.7	27	1 53.7	43	1 2.6	59	34.8	75	15.4
10	30 8.2	30	15 17.3	20	9 10.2	12	4 26.3	28	1 49.0	44	1 0.4	69	33.6	76	14.3
0 00	28 44.0	40	14 45.2	40	8 41.0	13	4 6.3	29	1 44.7	45	0 58.2	61	32.3	77	13.1
+ 10	27 24.0	50	14 14.9	6 00	8 20.1	14	3 48.8	30	1 40.4	46	56.3	62	31.0	78	12.0
20	26 9.4	3 00	13 46.3	20	7 57.8	15	3 33.5	31	1 36.1	47	54.4	63	29.7	79	11.0
30	24 59.4	10	13 19.3	40	7 37.1	16	3 19.9	32	1 32.9	48	52.6	64	28.4	80	9.9
40	23 53.0	20	12 53.6	7 00	7 17.8	17	3 7.9	33	1 29.4	49	50.7	65	27.2	81	8.8
50	22 50.3	30	12 29.4	20	7 0.3	18	2 57.1	34	1 26.1	50	48.8	66	26.0	82	7.8
1 00	21 51.6	40	12 6.3	40	6 46.1	19	2 47.3	35	1 22.9	51	47.2	67	24.8	83	6.7
10	20 56.7	50	11 44.6	8 00	6 29.6	20	2 38.5	36	1 19.9	52	45.6	68	23.6	84	5.6
20	20 4.8	4 00	11 23.8	20	6 15.4	21	2 30.4	37	1 17.1	53	44.0	69	22.4	85	4.6
30	19 15.4	10	11 3.9	40	6 2.1	22	2 23.1	38	1 14.4	54	42.4	70	21.2	86	3.5
40	18 29.8	20	10 45.4	9 00	5 49.9	23	2 16.2	39	1 11.8	55	40.8	71	20.0	87	2.5
50	17 46.4	30	10 27.7	20	5 38.3	24	2 10.0	40	1 9.3	56	38.6	72	18.8	88	1.4
2 00	17 5.5	40	10 10.4	40	5 27.6	25	2 4.1	41	1 7.1	57	37.3	73	17.1	89	0.4

TABLE 2. Barometrical Corrections for French Inches and Decimal Parts.

App. Alt.	25.0	25.2	25.4	25.6	25.8	26.0	26.2	26.4	26.6	26.8	27.0	27.2	27.4	27.6	27.8	28.0
	—	—	—	—	—	—	—	—	—	—	—	—	—	—	—	—
0° 00	4' 39".4	4' 18".4	3' 59".8	3' 41".2	3' 20".2	3' 1".6	2' 43".0	2' 22".0	2' 3".4	1' 47".1	1' 28".5	1' 9".8	0' 53".6	0' 34".9	0' 16".3	0' 0".0
10	4 19.7	4 0.2	3 42.9	3 25.6	3 6.1	2 58.9	2 31.5	2 12.0	1 54.7	1 39.6	1 22.3	1 4.9	49.8	32.5	15.2	0.0
20	4 2.3	3 44.1	3 27.9	3 11.8	2 53.6	2 37.5	2 21.3	2 3.1	1 47.0	1 32.9	1 16.7	1 0.6	46.4	30.3	14.1	0.0
30	3 46.7	3 29.7	3 14.5	2 59.4	2 42.5	2 27.3	2 9.2	1 55.2	1 40.1	1 26.9	1 11.8	0 56.7	43.4	28.3	13.2	0.0
40	3 32.7	3 16.8	3 2.6	2 48.4	2 32.5	2 18.3	2 4.1	1 48.1	1 34.0	1 21.6	1 7.4	53.2	40.8	26.6	12.4	0.0
50	3 20.2	3 5.1	2 51.8	2 38.5	2 23.5	2 10.1	1 56.8	1 41.8	1 28.4	1 16.7	1 3.4	50.0	38.4	25.0	11.7	0.0
1 00	3 8.8	2 54.6	2 42.0	2 29.5	2 15.3	2 2.6	1 50.1	1 36.0	1 23.4	1 12.4	0 59.8	47.2	36.2	23.6	11.0	0.0
10	2 58.5	2 45.1	2 33.2	2 21.3	2 8.0	1 56.0	1 44.1	1 30.8	1 18.9	1 8.4	56.5	44.6	34.2	22.3	10.4	0.0
20	2 49.2	2 36.1	2 25.2	2 13.9	2 1.3	1 50.0	1 38.7	1 26.0	1 14.7	1 4.9	53.6	42.3	32.4	21.1	9.9	0.0
30	2 40.6	2 28.6	2 17.8	2 7.2	1 55.1	1 44.4	1 33.7	1 21.6	1 10.9	1 11.6	50.9	40.2	30.8	20.1	9.4	0.0
40	2 32.8	2 21.3	2 11.1	2 1.0	1 49.5	1 39.3	1 29.1	1 17.7	1 7.5	0 58.6	48.4	38.2	29.3	19.1	8.9	0.0
50	2 25.6	2 14.7	2 5.0	1 55.3	1 44.4	1 34.7	1 24.9	1 14.2	1 4.3	55.8	46.1	36.4	27.9	18.2	8.5	0.0
2 00	2 19.0	2 8.6	1 59.3	1 50.1	1 39.6	1 30.4	1 21.1	1 10.7	1 1.4	53.2	44.0	34.8	26.7	17.4	8.1	0.0
10	2 12.9	2 3.0	1 54.1	1 45.2	1 35.3	1 26.4	1 17.5	1 7.6	0 58.7	51.0	42.1	33.2	25.5	16.6	7.8	0.0
20	2 7.3	1 57.7	1 49.2	1 40.7	1 31.2	1 22.7	1 14.2	1 4.7	56.2	48.8	40.3	31.8	24.4	15.9	7.4	0.0
30	2 2.0	1 52.9	1 44.7	1 36.6	1 27.5	1 19.3	1 11.9	1 2.0	53.9	46.8	38.7	30.5	23.4	15.3	7.1	0.0
40	1 57.2	1 48.4	1 40.6	1 32.8	1 24.0	1 16.2	1 8.4	0 59.6	51.8	44.9	37.1	29.3	22.5	14.7	6.8	0.0
50	1 52.7	1 44.1	1 36.7	1 29.2	1 20.7	1 13.2	1 5.7	57.3	49.8	43.2	35.7	28.2	21.6	14.1	6.6	0.0
3 00	1 48.4	1 40.3	1 33.1	1 25.8	1 17.7	1 10.5	1 3.3	55.1	47.9	41.6	34.3	27.1	20.8	13.5	6.3	0.0
20	1 40.8	1 33.2	1 26.5	1 19.8	1 12.2	1 5.5	0 58.8	51.2	44.5	38.6	31.9	25.2	19.3	12.6	5.9	0.0
40	1 34.1	1 27.0	1 20.7	1 14.5	1 7.4	1 1.1	54.9	47.8	41.5	36.1	29.8	23.5	18.0	11.8	5.5	0.0
4 00	1 28.1	1 21.5	1 15.6	1 9.8	1 3.1	0 57.3	51.4	44.8	38.9	33.8	27.9	22.0	16.9	11.0	5.1	0.0
20	1 22.8	1 16.6	1 11.1	1 5.5	0 59.3	53.8	48.3	42.1	36.6	31.7	26.2	20.7	15.9	10.4	4.7	0.0
40	1 18.1	1 12.2	1 7.0	1 1.8	55.9	50.7	45.5	39.7	34.5	29.9	24.7	19.5	15.0	9.8	4.3	0.0
5	1 12.8	1 8.3	1 3.3	0 58.4	52.9	48.0	43.1	37.5	32.6	28.3	23.4	18.4	14.1	9.2	3.9	0.0
6	1 3.3	0 58.5	0 54.3	50.1	45.4	41.1	36.9	32.1	28.0	24.3	20.0	15.8	12.1	7.9	3.5	0.0
7	0 55.2	51.1	47.4	43.7	39.6	35.9	32.2	28.1	24.4	21.2	17.5	13.8	10.6	6.9	3.2	0.0
8	48.9	45.2	42.0	38.7	35.1	31.8	28.5	24.9	21.6	18.8	15.5	12.2	9.4	6.1	2.9	0.0
9	43.8	40.6	37.6	34.7	31.4	28.5	25.6	22.4	19.4	16.8	13.9	11.0	8.4	5.5	2.6	0.0
10	39.7	36.7	34.1	31.4	28.4	25.8	23.1	20.2	17.5	15.2	12.6	9.9	7.6	5.0	2.3	0.0
12	33.3	30.8	28.5	25.8	23.8	21.8	19.4	16.9	14.7	12.8	10.5	8.3	6.4	4.2	1.9	0.0
14	28.5	26.4	24.5	22.1	20.4	18.5	16.3	14.5	12.6	10.9	9.0	7.1	5.5	3.6	1.7	0.0
16	24.9	23.0	21.4	19.3	17.8	16.2	14.5	12.7	11.0	9.5	7.9	6.2	4.8	3.1	1.5	0.0
18	22.0	20.4	18.9	17.1	15.8	14.3	12.9	11.2	9.7	8.4	7.0	5.5	4.2	2.8	1.3	0.0
20	19.7	18.2	16.9	15.3	14.1	12.8	11.5	10.0	8.7	7.6	6.2	4.9	3.8	2.5	1.2	0.0
25	15.4	14.3	13.3	12.0	11.0	10.0	9.0	7.8	6.8	5.9	4.9	3.9	3.0	1.9	0.9	0.0
30	12.5	11.5	10.7	9.7	9.0	8.1	7.3	6.3	5.5	4.8	4.0	3.1	2.4	1.6	0.7	0.0
35	10.3	9.5	8.8	8.0	7.4	6.7	6.0	5.2	4.6	4.0	3.3	2.6	2.0	1.3	0.6	0.0
40	8.6	7.9	7.4	6.7	6.2	5.6	5.0	4.4	3.8	3.3	2.7	2.2	1.7	1.1	0.5	0.0
45	7.2	6.7	6.2	5.6	5.2	4.7	4.2	3.7	3.2	2.8	2.3	1.8	1.4	0.9	0.4	0.0
50	6.1	5.6	5.2	4.7	4.3	3.9	3.5	3.1	2.7	2.3	1.9	1.5	1.2	0.8	0.3	0.0
60	4.2	3.9	3.6	3.2	3.0	2.7	2.4	2.1	1.8	1.6	1.3	1.0	0.8	0.5	0.2	0.0
70	2.6	2.4	2.3	2.0	1.9	1.7	1.5	1.3	1.2	1.0	0.8	0.7	0.5	0.3	0.2	0.0
80	1.3	1.2	1.1	1.0	0.9	0.8	0.7	0.6	0.6	0.5	0.4	0.3	0.2	0.2	0.1	0.0
90	0.0	0.0	0.0	0.0	0.0	0.0	0.0	0.0	0.0	0.0	0.0	0.0	0.0	0.0	0.0	0.0
	+ 31.3	+ 31.1	+ 30.9	+ 30.6	+ 30.4	+ 30.2	+ 30.0	+ 29.7	+ 29.5	+ 29.3	+ 29.0	+ 28.8	+ 28.6	+ 28.4	+ 28.2	+ 28.0

TABLE 3. Thermometrical Corrections, additive under, and subtractive above 0° of Réaumur's Thermometer.

App. Alt.	0°	1°	2°	3°	4°	5°	6°	7°	8°	9°	10°	11°	12°	13°	14°	15°
0° 00	0".0	18".9	37".9	56".8	1' 15".8	1' 34".7	1' 54".9	2' 15".0	2' 35".2	2' 55".3	3' 15".5	3' 37".1	3' 58".7	4' 20".3	4' 41".9	5' 3".5
10	0.0	17.1	34.2	51.2	1 8.3	1 25.4	1 43.5	2 1.6	2 19.6	2 37.7	2 55.8	3 15.2	3 34.5	3 53.9	4 13.2	4 32.6
20	0.0	15.4	30.9	46.3	1 1.8	1 17.2	1 33.4	1 49.7	2 5.9	2 22.2	2 38.4	2 55.9	3 13.4	3 30.9	3 48.4	4 5.9
30	0.0	14.0	28.0	42.0	0 56.0	1 10.0	1 24.8	1 39.6	1 54.4	2 9.2	2 24.0	2 39.8	2 55.7	3 11.5	3 27.4	3 43.2
40	0.0	12.8	25.6	38.4	51.2	1 4.0	1 17.4	1 30.8	1 44.3	1 57.7	2 11.1	2 25.6	2 40.1	2 54.5	3 9.0	3 23.5
50	0.0	11.7	23.4	35.2	46.9	0 58.6	1 11.0	1 23.3	1 35.7	1 48.0	2 0.4	2 13.6	2 26.8	2 39.9	2 53.1	3 6.3
1 00	0.0	11.0	22.0	32.9	43.9	54.9	1 6.0	1 17.2	1 28.3	1 39.5	1 50.6	2 2.8	2 15.1	2 27.3	2 39.6	2 51.8
10	0.0	10.0	20.0	29.9	39.9	49.9	1 0.3	1 10.6	1 21.0	1 31.3	1 41.7	1 53.1	2 4.5	2 15.9	2 27.3	2 38.7
20	0.0	9.3	18.6	27.9	37.2	46.5	0 56.3	1 6.0	1 15.8	1 25.5	1 35.3	1 45.7	1 56.2	2 6.6	2 17.1	2 27.5
30	0.0	8.6	17.3	25.9	34.6	43.2	52.3	1 1.4	1 10.6	1 19.7	1 28.8	1 38.5	1 48.2	1 57.9	2 7.6	2 17.3
40	0.0	8.1	16.2	24.3	32.4	40.5	49.0	0 57.5	1 5.9	1 14.4	1 22.9	1 31.9	1 40.9	1 50.0	1 59.0	2 8.0
50	0.0	7.6	15.1	22.7	30.2	37.8	45.8	53.7	1 1.7	1 9.6	1 17.6	1 26.1	1 34.5	1 42.9	1 51.4	1 59.9
2 00	0.0	7.1	14.2	21.3	28.4	35.5	43.0	50.5	57.9	1 5.4	1 12.9	1 20.8	1 28.8	1 36.7	1 44.7	1 52.6
20	0.0	6.4	12.8	19.1	25.5	31.9	38.5	45.1	51.8	58.4	1 5.0	1 12.1	1 19.1	1 26.2	1 33.2	1 40.3
40	0.0	5.7	11.4	17.0	22.7	28.4	34.4	40.4	46.3	52.3	58.3	1 4.6	1 11.0	1 17.3	1 23.7	1 30.0
3 00	0.0	5.1	10.3	15.4	20.6	25.7	31.1	36.6	42.0	47.5	52.9	58.6	1 4.3	1 10.0	1 15.7	1 21.4
20	0.0	4.7	9.4	14.1	18.8	23.5	28.4	33.3	38.3	43.2	48.1	53.3	0 58.5	1 3.7	1 8.8	1 14.0
40	0.0	4.3	8.6	13.0	17.3	21.6	26.1	30.6	35.2	39.7	44.2	49.0	0 53.8	0 58.6	1 3.4	1 8.2

TABLE 3. concluded. THERMOMETRICAL CORRECTIONS additive under, and subtractive above, 0° of Réaumur's Thermometer.

App. Alt.	0°	1°	2°	3°	4°	5°	6°	7°	8°	9°	10°	11°	12°	13°	14°	15°
4° 00′	0″.0	4″.0	8″.0	11″.9	15″.9	19″.9	24″.1	28″.3	32″.5	36″.7	40″.9	45″.3	49″.7	54″.1	58″.5	62″.9
4 20	0.0	3.7	7.4	11.1	14.8	18.5	22.4	26.2	30.1	33.9	37.8	41.9	46.0	50.1	54.2	58.2
4 40	0.0	3.4	6.8	10.3	13.7	17.1	20.7	24.3	27.8	31.4	35.0	38.8	42.7	46.5	50.4	54.2
5 00	0.0	3.2	6.4	9.7	12.9	16.1	19.5	22.9	26.2	29.6	33.0	36.5	40.1	43.6	47.2	50.7
6	0.0	2.7	5.5	8.2	10.9	13.6	16.4	19.2	22.0	24.8	27.6	30.6	33.5	36.5	39.5	42.5
7	0.0	2.4	4.7	7.0	9.3	11.6	14.0	16.4	18.7	21.1	23.5	26.1	28.6	31.2	33.7	36.3
8	0.0	2.0	4.0	6.0	8.0	10.1	12.2	14.3	16.4	18.5	20.6	22.8	25.0	27.3	29.5	31.7
9	0.0	1.8	3.6	5.4	7.1	8.9	10.8	12.7	14.5	16.4	18.3	20.3	22.2	24.2	26.1	28.1
10	0.0	1.6	3.3	4.8	6.4	8.0	9.7	11.3	13.0	14.6	16.3	18.0	19.8	21.6	23.4	25.2
12	0.0	1.3	2.7	4.0	5.4	6.6	8.0	9.4	10.8	12.2	13.6	15.1	16.6	18.1	19.6	21.1
14	0.0	1.1	2.3	3.4	4.5	5.6	6.8	8.0	9.2	10.4	11.6	12.9	14.1	15.4	16.6	17.9
16	0.0	1.0	2.0	2.9	3.9	4.9	5.9	7.0	8.0	9.1	10.1	11.2	12.3	13.4	14.5	15.6
18	0.0	0.9	1.7	2.6	3.4	4.3	5.2	6.1	7.1	8.0	8.9	9.9	10.8	11.8	12.7	13.7
20	0.0	0.8	1.5	2.3	3.0	3.8	4.6	5.4	6.2	7.0	7.8	8.7	9.5	10.4	11.2	12.1
25	0.0	0.6	1.2	1.8	2.4	3.0	3.6	4.3	4.9	5.6	6.2	6.9	7.5	8.2	8.8	9.5
30	0.0	0.5	1.0	1.4	1.9	2.4	2.9	3.4	4.0	4.5	5.0	5.6	6.1	6.6	7.1	7.6
35	0.0	0.4	0.8	1.2	1.6	2.0	2.4	2.8	3.3	3.7	4.1	4.5	5 0	5.4	5.9	6.3
40	0.0	0.3	0.7	1.0	1.4	1.7	2.0	2.4	2.7	3.1	3.4	3.8	4.2	4.5	4.9	5.3
45	0.0	0.3	0.5	0.8	1.1	1.4	1.7	2.0	2.2	2.5	2.8	3.2	3.6	4.0	4.2	4.4
50	0.0	0.2	0.4	0.7	0.9	1.1	1.3	1.6	1.8	2.1	2.3	2.6	2.9	3.2	3.4	3.7
60	0.0	0.2	0.3	0.5	0.6	0.8	1.0	1.1	1.3	1.5	1.7	1.9	2.1	2.2	2.4	2.6
70	0.0	0.1	0.2	0.4	0.5	0.6	0.7	0.8	0.9	1.0	1.1	1.2	1.3	1.4	1.5	1.6
80	0.0	0.1	0.1	0.2	0.2	0.3	0.3	0.4	0.4	0.5	0.5	0.5	0.6	0.6	0.7	0.8
90	0.0	0.0	0.0	0.0	0.0	0.0	0.0	0.0	0.0	0.0	0.0	0 0	0.0	0.0	0.0	0.0

	15°	16°	17°	18°	19°	20°	21°	22°	23°	24°	25°	26°	27°	28°	29°	30°
0° 00	5′ 3″.5	5′ 26″.7	5′ 49″.8	6′ 13″.0	6′ 36″.7	6′ 59″.3	7′ 19″.1	7′ 39″.0	7′ 58″.8	8′ 18″.7	8′ 38″.5	8′ 54″.2	9′ 29″.7	10′ 5″.1	10′ 40″.6	11′ 16″.1
10	4 32.6	4 53.2	5 13.8	5 34.5	5 55.1	6 15.7	6 33.4	6 51.2	7 8.9	7 26.7	7 44.4	7 58.1	8 29.4	9 0.8	9 32.1	10 3.5
20	4 5.9	4 24.5	4 43.1	5 1.8	5 20.4	5 39.0	5 54.8	6 10.5	6 26.3	6 42.0	6 57.8	7 10.5	7 39.0	8 7.5	8 36.0	9 4.5
30	3 43.2	4 0.1	4 16.9	4 33.8	4 50.6	5 7.5	5 21.8	5 36.1	5 50.4	6 4.7	6 19.0	6 30.5	6 56.2	7 22.0	7 47.7	8 13.5
40	3 23.5	3 38.8	3 54.1	4 9.4	4 24.7	4 40.0	4 53.1	5 6.1	5 19.2	5 32.2	5 45.3	5 55.6	6 19.0	6 42.5	7 5.9	7 29.3
50	3 6.3	3 20.4	3 34.5	3 48.6	4 2.7	4 16.8	4 28.7	4 40.6	4 52.4	5 4.3	5 16.2	5 25.7	5 47.1	6 8.4	6 29.8	6 51.2
1 00	2 51.8	3 4.7	3 17.6	3 30.6	3 43.5	3 56.4	4 7.3	4 18.2	4 29.2	4 40.1	4 51.0	4 59.7	5 19.2	5 38.8	5 58.3	6 17.9
10	2 38.7	2 50.7	3 2.6	3 14.6	3 26.5	3 38.5	3 48.6	3 58.6	4 8.7	4 18.7	4 28.8	4 36.8	4 54.8	5 12.9	5 30.9	5 49.0
20	2 27.5	2 38.5	2 49.5	3 0.6	3 11.6	3 22.6	3 31.9	3 41.3	3 50.6	4 0.0	4 9.3	4 16.7	4 33 4	4 50.0	5 6.7	5 23.4
30	2 17.3	2 27.5	2 37.8	2 48.0	2 58.3	3 8.5	3 17.2	3 25.9	3 34.6	3 43.3	3 52.0	3 58.8	4 14.3	4 29.7	4 45.2	5 0.7
40	2 8.0	2 17.6	2 27.2	2 36.7	2 46.3	2 55.9	3 4.0	3 12.1	3 20.3	3 28.4	3 36.5	3 42.8	3 57.3	4 11.7	4 26.2	4 40.6
50	1 59.9	2 8.9	2 17.9	2 26.9	2 35.9	2 44.9	2 52.5	3 0.1	3 7.6	3 15.2	3 22.8	3 28.7	3 42.2	3 55.8	4 9.3	4 22.8
2 00	1 52.6	2 1.1	2 9.5	2 18.0	2 26.4	2 34.9	2 42.0	2 49.1	2 56.2	3 3.3	3 10.4	3 16.0	3 28.6	3 41.3	3 53.9	4 6.6
10	1 46.3	1 54.2	2 2.1	2 10.0	2 17.9	2 25.8	2 32.5	2 39.2	2 45.8	2 52.5	2 59.2	3 3.4	3 15.3	3 27.3	3 39.2	3 52.1
20	1 40.3	1 47.8	1 55.3	2 2.8	2 10.3	2 17.8	2 24.1	2 30.3	2 36.6	2 42.8	2 49.1	2 54.1	3 5.3	3 16.6	3 27.8	3 39.1
30	1 34.9	1 42.0	1 49.1	1 56.3	2 3.4	2 10.5	2 16.4	2 22.3	2 28.2	2 34.1	2 40.0	2 44.7	2 55.3	3 6.0	3 16.6	3 27.2
40	1 30.0	1 36.7	1 43.5	1 50.2	1 57.0	2 3.7	2 9.3	2 14.9	2 20.5	2 26.1	2 31.7	2 36.2	2 46.2	2 56.3	3 6.3	3 16.4
50	1 25.7	1 32.1	1 38.5	1 44.9	1 51.3	1 57.7	2 3.1	2 8.4	2 13.8	2 19.1	2 24.5	2 28.6	2 38.1	2 47.7	2 57.2	3 6.7
3 00	1 21.4	1 27.3	1 33.2	1 39.0	1 44.9	1 50.8	1 56.1	2 1.4	2 6.6	2 11.9	2 17.2	2 21.0	2 30.1	2 39.3	2 48.4	2 57.5
20	1 14.0	1 19.6	1 25.1	1 30.7	1 36.2	1 41.8	1 46.5	1 51.1	1 55.8	2 0.4	2 5.1	2 8.7	2 16.9	2 25.2	2 33.4	2 41.7
40	1 8.2	1 13.3	1 18.3	1 23.4	1 28.4	1 33 5	1 37.7	1 42.0	1 46.2	1 50.5	1 54.7	1 58.0	2 5.6	2 13.1	2 20.7	2 28.2
4 00	1 2.9	1 7.6	1 12.2	1 16.9	1 21.5	1 26.2	1 30.1	1 34.0	1 38.0	1 41.9	1 45.8	1 48.8	1 55.8	2 2.7	2 9.7	2 16.6
20	0 58.2	1 2.5	1 6.9	1 11.2	1 15.6	1 19.9	1 23.5	1 27.1	1 30.8	1 34.4	1 38.0	1 40.8	1 47.3	1 53.7	2 0.2	2 6.6
40	54.2	0 58.3	1 2.3	1 6.4	1 10.4	1 14.5	1 18.0	1 21.6	1 25.1	1 28.7	1 31.2	1 34.6	1 40.4	1 46.3	1 52.1	1 58.0
5	50.7	54.5	0 58.3	1 2.1	1 5.9	1 9.7	1 12.8	1 15.9	1 19.1	1 22.2	1 25.3	1 27.8	1 33.4	1 39.1	1 44.7	1 50.3
6	42.5	45.6	48.8	0 52.0	0 55.1	0 58.2	1 1.5	1 4.8	1 8.1	1 11.4	1 14.7	1 18.2	1 21.7	1 25.2	1 28.7	1 32.2
7	36.3	39.0	41.7	44.4	47.1	49.8	0 52.6	0 55.4	0 58 2	1 1.0	1 3.8	1 6.8	1 9.8	1 12.9	1 15.9	1 18.9
8	31.7	34.0	36.4	38.7	41.1	43.4	45.9	48.4	50.8	0 53.3	0 55.8	0 58.4	1 1.0	1 3.5	1 6.1	1 8.7
9	28.1	30.2	32.3	34.3	36.4	38.5	40.7	42.9	45.0	47.2	49.4	51.7	0 54.0	0 56.3	0 58.6	1 0.9
10	25.2	27.1	28.9	30.1	32.6	34.5	36.5	38.4	40.4	42.3	44.3	46.4	48.5	50.5	52.6	0 54.7
12	21.1	22.7	24.2	25.8	27.3	28.9	30.5	32.2	33.8	35.5	37.1	38.8	40.5	42.2	43.9	45.6
14	17.9	19.2	20.5	21.7	23.0	24.3	25.7	27.1	28.5	29.9	31.3	32.7	34.2	35.6	37.1	38.5
16	15.6	16.7	17.8	18.9	20.0	21.1	22.3	23.5	24.7	25.9	27.1	28.4	29.6	30.9	32.1	33.4
18	13.7	14.7	15.7	16.7	17.7	18.7	19.8	20.8	21.9	22.9	24.0	25.1	26.2	27.4	28.5	29.6
20	12.1	13.0	13.9	14.8	15.7	16.6	17.5	18.5	19.4	20.4	21.3	22.3	23.3	24.3	25.3	26.3
25	9.5	10.2	10.9	11.6	12.3	13.0	13.7	14.5	15.0	16.0	16.7	17.5	18.3	19.0	19.8	20.6
30	7.6	8.2	8.8	9.3	9.9	10.5	11.1	11.7	12.2	12.8	13.4	14.0	14.7	15.3	16.0	16.6
35	6.3	6.8	7.2	7.7	8.1	8.6	9.1	9.6	10.0	10.5	11.0	11.5	12.1	12.6	13.2	13.7
40	5.3	5.7	6.1	6.4	6.8	7.2	7.6	8.0	8.5	8.9	9.3	9.7	10.1	10.6	11.0	11.4
45	4.4	4.7	5.1	5.4	5.8	6.1	6.2	6.4	6.5	6.7	6.8	7.4	7.9	8.5	9.0	9.6
50	3.7	4.1	4.5	4.8	5.2	5.6	5.9	6.1	6.2	6.3	6.5	6.8	7.1	7.4	7.7	8.0
60	2.6	2.8	3.0	3.1	3.3	3.5	3.7	3.9	4.0	4.2	4.4	4.6	4.8	5.0	5.3	5.5
70	1.6	1.8	1.9	2.0	2.1	2.2	2.4	2.6	2.7	2.8	2.9	3.1	3.2	3.3	3.4	3.5
80	0.8	0.8	0.9	0.9	1.0	1.1	1.2	1.3	1.3	1.4	1.4	1.4	1.5	1.5	1.6	1.7
90	0.0	0.0	0.0	0.0	0.0	0.0	0.0	0.0	0.0	0.0	0.0	0.0	0.0	0.0	0.0	0.0

In the eighth volume of Baron Zach's *Correspondance Astronomique* (No. V.), Professor Littrow, of Vienna, informs us that Bessel has recently been applying his three feet circle by Reichenbach in renewing his investigations of the subject of celestial refractions, and that he has found, from a new series of observations, that the errors in his old Tables derived from Bradley's observations are very small, particularly in the higher altitudes. The inference that has been drawn from the new observations is, that the refraction given by the Tables contained in the *Fundamenta Astronomiæ* (see pages 294—296 of this volume), ought to be augmented by the 0 .003282th part, in order to satisfy these observations. Mr. Littrow therefore recomputed the Tables of refraction, which are contained in our pages 439 and 440, and which include all the corrections indicated by Bessel in *Recueil* VII. of his observations, lately published. In the preface to these Tables the author has developed the principles on which his computations are founded; to which the reader may refer for the particulars. The mean refractions computed by Littrow are adapted for twenty-eight Paris inches of the barometer, and zero of Réaumur's thermometer, which may suit the continental astronomers, but is not convenient for reducing English observations; we have consequently given as an appendage a small Table for converting the degrees of the English barometer into corresponding parts of Paris decimal inches: the conversion of Fahrenheit's scale into Réaumur's, for using the thermometer, being given in page 18 of the present volume. By the aid of these small comparative Tables, observations taken and registered with the English barometer and Fahrenheit's thermometer, may be reduced by the Tables before us, as well as if the barometer with Paris decimal inches and Réaumur's thermometer had been the instruments attended to, at the time of making the observations. The author has preferred computing his refractions in the logarithmic form, and uses both the interior and exterior thermometers; the zenith distance is the argument, which is given for every 10′; and two small auxiliary quantities m and n are introduced in the lower altitudes, or greater zenith distances. If we put r' = the true refraction, we shall have $\log. r' = \log. R + (b + t') + t + m(b + t') + nt$ as will be seen in the subjoined example. Or when the zenith distance is not too great, the computation may be rendered more simple thus; $\log. r' = \log. R + b + (n + 1)t$.

In the ninth volume of the *Correspondance* (No. II.), Baron Zach has made a new arrangement of the refractions re-computed by Littrow, by leaving out the logarithms, and giving only the natural numbers, both in the Table of mean refractions, and also in the Tables of barometric and thermometric corrections, so that the true refractions may be obtained by simple addition of the tabular quantities, and with nearly as great accuracy as by Littrow's method, when proportional parts are correctly taken. In all ordinary cases, and particularly for nautical purposes, Zach's arrangement will be found sufficiently correct, and very convenient in use. He has also given an additional Table IV. to be used when the refraction corresponding to the *true* altitudes are wanted for particular occasions; this Table is constructed for twenty-eight Paris inches, and + 8° of Réaumur's thermometer, as being most agreeable to an average pressure and temperature of the air at the surface of the sea. We will give the same example for both arrangements of the refractions by Littrow and Zach respectively.

Example.—Let the observed zenith distance of a star be 88° 43′ 24″, when the barometer in Paris inches indicated 28 .574, the interior thermometer + 3°.4, and the exterior — 10°.5, according to Réaumur's scale, and let the true refraction be required?

According to Littrow.

$\log. R. = 3.1570,\ m = 0.052,\ n = 0.415$

$b + t' = 0.0085$, B gives b .... 0 .0088, T gives $t = 0.0212$

$t = 0.0212$, T′ gives t' .. −0 .0003

$m(b + t') = 0.0004$ $\quad b + t' = 0.0085$

$nt = 0.0088$

$\log. r' = 3.1959$ and r' (true refrac.) = 1570″.0 = 26′ 10″.0

(Or otherwise more simply thus:)

Log. R.... = 3 .1570
Log. b .... 0 .0088
Log. $(n + 1)t$ 0 .0300 } = 3 .1958 = 26′ 9″.8 = r'

According to Zach.

Table 1. gives mean refraction		23′	56″. 2
Table 2. for barometer	+	0 +	30 . 6
Table 3. for thermometer	+	1	42 . 9
True refraction (r')		26	9 . 7

The corresponding heights of the English barometer and Fahrenheit's thermometer, in this example, by the small Tables will be found respectively 30 .46 and 32°.

(Argument = R. A. ∗) TABLE 1. (Tab. No. = $-1^{s}.3362$ sin. R. A.)

Multiply the Tab. No. by the tang. of the star's dec., and add to it $3^{s}.0678$.

S.	−	+	−	+	−	+	−	+	−	+	−	+	S.
	O	XII	I	XIII	II	XIV	III	XV	IV	XVI	V	XVII	
N.	+	−	+	−	+	−	+	−	+	−	+	−	N.
Min.	Nat. No.	Log.	Nat. No.	Log.	Nat. No.	Log.	Nat. No.	Log.	Nat. No.	Log.	Nat. No.	Log.	Min.
0	$0^{s}.000$	− ∞	$0^{s}.346$	9.5389	$0^{s}.668$	9.8249	$0^{s}.945$	9.9754	$1^{s}.157$	0.0634	$1^{s}.291$	0.1108	60
1	0.006	7.7657	0.352	9.5459	0.673	9.8281	0.949	9.9773	1.160	0.0645	1.292	0.1113	59
2	0.012	8.0667	0.358	9.5528	0.678	9.8314	0.953	9.9791	1.163	0.0656	1.293	0.1118	58
3	0.017	8.2428	0.363	9.5596	0.683	9.8346	0.957	9.9810	1.166	0.0667	1.295	0.1123	57
4	0.023	8.3677	0.369	9.5662	0.688	9.8377	0.961	9.9828	1.169	0.0677	1.296	0.1128	56
5	0.029	8.4646	0.375	9.5728	0.693	9.8409	0.965	9.9846	1.172	0.0688	1.298	0.1133	55
6	0.035	8.5438	0.380	9.5792	0.698	9.8440	0.969	9.9865	1.174	0.0698	1.299	0.1137	54
7	0.041	8.6107	0.386	9.5856	0.703	9.8471	0.973	9.9882	1.177	0.0708	1.301	0.1142	53
8	0.047	8.6687	0.392	9.5918	0.708	9.8501	0.977	9.9900	1.180	0.0718	1.302	0.1146	52
9	0.053	8.7198	0.397	9.5980	0.713	9.8531	0.981	9.9918	1.182	0.0728	1.304	0.1150	51
10	0.058	8.7656	0.402	9.6040	0.718	9.8561	0.985	9.9935	1.185	0.0738	1.305	0.1155	50
11	0.064	8.8069	0.408	9.6100	0.723	9.8591	0.989	9.9953	1.188	0.0748	1.307	0.1159	49
12	0.070	8.8447	0.414	9.6159	0.728	9.8620	0.993	9.9970	1.191	0.0758	1.308	0.1163	48
13	0.076	8.8794	0.419	9.6217	0.733	9.8649	0.997	9.9987	1.193	0.0767	1.309	0.1167	47
14	0.082	8.9116	0.425	9.6274	0.738	9.8678	1.001	0.0003	1.196	0.0777	1.310	0.1171	46
15	0.087	8.9415	0.430	9.6330	0.743	9.8706	1.005	0.0020	1.198	0.0786	1.311	0.1175	45
16	0.093	8.9695	0.436	9.6385	0.748	9.8735	1.009	0.0037	1.201	0.0795	1.312	0.1178	44
17	0.099	8.9958	0.442	9.6440	0.752	9.8762	1.013	0.0053	1.203	0.0805	1.313	0.1182	43
18	0.105	9.0205	0.447	9.6494	0.757	9.8790	1.017	0.0069	1.206	0.0814	1.314	0.1186	42
19	0.111	9.0440	0.452	9.6547	0.761	9.8818	1.021	0.0085	1.208	0.0823	1.315	0.1189	41
20	0.117	9.0662	0.457	9.6599	0.766	9.8845	1.024	0.0101	1.211	0.0832	1.316	0.1192	40
21	0.123	9.0873	0.463	9.6651	0.771	9.8872	1.028	0.0117	1.213	0.0840	1.317	0.1196	39
22	0.129	9.1075	0.468	9.6702	0.776	9.8898	1.031	0.0133	1.216	0.0849	1.318	0.1199	38
23	0.134	9.1267	0.474	9.6752	0.780	9.8925	1.035	0.0149	1.218	0.0858	1.319	0.1202	37
24	0.140	9.1451	0.479	9.6802	0.785	9.8951	1.038	0.0164	1.221	0.0866	1.320	0.1205	36
25	0.146	9.1628	0.485	9.6851	0.790	9.8977	1.042	0.0179	1.223	0.0871	1.321	0.1208	35
26	0.152	9.1797	0.490	9.6900	0.795	9.9003	1.046	0.0194	1.226	0.0883	1.322	0.1211	34
27	0.158	9.1961	0.496	9.6947	0.799	9.9028	1.050	0.0209	1.228	0.0891	1.323	0.1214	33
28	0.163	9.2118	0.501	9.6995	0.804	9.9054	1.054	0.0224	1.230	0.0899	1.324	0.1216	32
29	0.169	9.2269	0.506	9.7041	0.808	9.9079	1.057	0.0239	1.233	0.0907	1.324	0.1219	31
30	0.174	9.2416	0.511	9.7087	0.813	9.9103	1.060	0.0254	1.235	0.0915	1.325	0.1222	30
31	0.180	9.2557	0.517	9.7133	0.818	9.9128	1.064	0.0268	1.238	0.0923	1.326	0.1224	29
32	0.186	9.2694	0.522	9.7178	0.823	9.9152	1.067	0.0282	1.240	0.0931	1.326	0.1226	28
33	0.192	9.2827	0.528	9.7222	0.827	9.9176	1.071	0.0297	1.242	0.0938	1.327	0.1229	27
34	0.198	9.2956	0.533	9.7266	0.832	9.9200	1.074	0.0311	1.244	0.0946	1.327	0.1221	26
35	0.203	9.3081	0.539	9.7309	0.837	9.9224	1.078	0.0325	1.246	0.0953	1.328	0.1233	25
36	0.209	9.3202	0.544	9.7352	0.841	9.9248	1.081	0.0338	1.248	0.0960	1.328	0.1235	24
37	0.215	9.3320	0.550	9.7394	0.846	9.9271	1.085	0.0352	1.250	0.0968	1.329	0.1237	23
38	0.221	9.3435	0.555	9.7436	0.850	9.9294	1.088	0.0366	1.252	0.0975	1.329	0.1239	22
39	0.227	9.3547	0.560	9.7477	0.855	9.9317	1.092	0.0379	1.254	0.0982	1.330	0.1241	21
40	0.232	9.3656	0.565	9.7518	0.859	9.9340	1.095	0.0393	1.256	0.0989	1.331	0.1242	20
41	0.238	9.3762	0.570	9.7559	0.864	9.9362	1.099	0.0406	1.258	0.0996	1.331	0.1244	19
42	0.244	9.3865	0.576	9.7599	0.868	9.9384	1.102	0.0419	1.260	0.1002	1.332	0.1245	18
43	0.250	9.3966	0.581	9.7638	0.873	9.9406	1.105	0.0432	1.262	0.1009	1.332	0.1247	17
44	0.255	9.4065	0.586	9.7677	0.877	9.9428	1.108	0.0445	1.264	0.1016	1.333	0.1248	16
45	0.261	9.4161	0.592	9.7716	0.881	9.9450	1.112	0.0457	1.266	0.1022	1.333	0.1249	15
46	0.267	9.4255	0.597	9.7754	0.886	9.9472	1.115	0.0470	1.268	0.1028	1.334	0.1251	14
47	0.272	9.4348	0.602	9.7792	0.890	9.9493	1.118	0.0482	1.270	0.1035	1.334	0.1252	13
48	0.278	9.4438	0.607	9.7829	0.894	9.9514	1.121	0.0495	1.271	0.1041	1.334	0.1253	12
49	0.284	9.4526	0.612	9.7866	0.899	9.9535	1.124	0.0507	1.273	0.1047	1.335	0.1254	11
50	0.289	9.4612	0.617	9.7903	0.903	9.9556	1.127	0.0519	1.274	0.1053	1.335	0.1255	10
51	0.295	9.4697	0.623	9.7939	0.907	9.9576	1.130	0.0531	1.276	0.1059	1.336	0.1256	9
52	0.301	9.4780	0.628	9.7975	0.911	9.9597	1.133	0.0543	1.278	0.1065	1.336	0.1256	8
53	0.306	9.4861	0.633	9.8010	0.916	9.9617	1.136	0.0555	1.279	0.1071	1.336	0.1257	7
54	0.312	9.4941	0.638	9.8046	0.920	9.9637	1.139	0.0567	1.281	0.1076	1.336	0.1257	6
55	0.318	9.5019	0.643	9.8080	0.924	9.9657	1.142	0.0578	1.283	0.1082	1.336	0.1258	5
56	0.324	9.5096	0.648	9.8115	0.929	9.9677	1.145	0.0590	1.285	0.1087	1.336	0.1258	4
57	0.329	9.5171	0.653	9.8149	0.933	9.9696	1.148	0.0601	1.286	0.1093	1.336	0.1259	3
58	0.335	9.5245	0.658	9.8182	0.937	9.9716	1.151	0.0612	1.288	0.1098	1.336	0.1259	2
59	0.341	9.5318	0.663	9.8216	0.941	9.9735	1.154	0.0623	1.290	0.1103	1.336	0.1259	1
60	0.346	9.5389	0.668	9.8249	0.945	9.9754	1.157	0.0634	1.291	0.1108	1.336	0.1259	0
	Nat. No.	Log.	Nat. No.	Log.	Nat. No.	Log.	Nat. No.	Log.	Nat. No.	Log.	Nat. No.	Log.	
N.	+	−	+	−	+	−	+	−	+	−	+	−	N.
	XI	XXIII	X	XXII	IX	XXI	VIII	XX	VII	XIX	VI	XVIII	
S.	−	+	−	+	−	+	−	+	−	+	−	+	S.

(Argument = R. A. ∗) TABLE 2. (Tab. No. = − 1^s.350 sin R. A.)

Min.	− O	+ XII	− I	+ XIII	− II	+ XIV	− III	+ XV	− IV	+ XVI	− V	+ XVII	Min.
	Nat. No.	Log.	Nat. No.	Log.	Nat. No.	Log.	Nat. No.	Log.	Nat. No.	Log.	Nat. No.	Log.	
0	0^s.000	0.0000	0^s.349	9.5428	0^s.675	9.8293	0^s.954	9.9795	1^s.168	0.0674	1^s.304	0.1153	60
1	0.006	7.7782	0.355	9.5502	0.680	9.8325	0.958	9.9814	1.171	0.0686	1.305	0.1156	59
2	0.012	8.0792	0.361	9.5575	0.685	9.8357	0.963	9.9836	1.174	0.0697	1.306	0.1159	58
3	0.017	8.2304	0.367	9.5647	0.690	9.8388	0.967	9.9854	1.177	0.0708	1.308	0.1166	57
4	0.023	8.3617	0.373	9.5717	0.695	9.8420	0.971	9.9872	1.180	0.0719	1.309	0.1169	56
5	0.029	8.4624	0.379	9.5786	0.700	9.8451	0.975	9.9890	1.183	0.0730	1.311	0.1176	55
6	0.035	8.5441	0.384	9.5843	0.705	9.8482	0.979	9.9908	1.186	0.0741	1.312	0.1179	54
7	0.041	8.6128	0.390	9.5911	0.710	9.8513	0.983	9.9926	1.189	0.0752	1.314	0.1186	53
8	0.047	8.6721	0.396	9.5977	0.715	9.8543	0.987	9.9943	1.192	0.0763	1.315	0.1189	52
9	0.053	8.7243	0.401	9.6031	0.720	9.8573	0.991	9.9961	1.194	0.0770	1.317	0.1196	51
10	0.059	8.7709	0.406	9.6085	0.725	9.8603	0.995	9.9978	1.197	0.0781	1.318	0.1199	50
11	0.065	8.8129	0.412	9.6149	0.730	9.8633	0.999	9.9996	1.200	0.0792	1.320	0.1206	49
12	0.071	8.8513	0.418	9.6212	0.735	9.8663	1.003	0.0013	1.203	0.0803	1.321	0.1209	48
13	0.077	8.8865	0.423	9.6263	0.740	9.8692	1.007	0.0030	1.205	0.0810	1.322	0.1212	47
14	0.083	8.9191	0.429	9.6325	0.745	9.8722	1.011	0.0048	1.208	0.0821	1.323	0.1216	46
15	0.088	8.9445	0.434	9.6375	0.750	9.8751	1.015	0.0065	1.210	0.0828	1.324	0.1219	45
16	0.094	8.9731	0.440	9.6435	0.755	9.8779	1.019	0.0082	1.213	0.0839	1.325	0.1222	44
17	0.100	9.0000	0.446	9.6493	0.760	9.8808	1.023	0.0099	1.215	0.0846	1.326	0.1225	43
18	0.106	9.0253	0.451	9.6542	0.765	9.8837	1.027	0.0116	1.218	0.0856	1.327	0.1229	42
19	0.112	9.0492	0.457	9.6599	0.769	9.8859	1.031	0.0133	1.220	0.0864	1.328	0.1232	41
20	0.118	9.0719	0.462	9.6646	0.774	9.8887	1.034	0.0145	1.222	0.0871	1.329	0.1235	40
21	0.124	9.0934	0.468	9.6702	0.779	9.8915	1.038	0.0162	1.225	0.0881	1.330	0.1239	39
22	0.130	9.1139	0.473	9.6749	0.784	9.8943	1.041	0.0175	1.228	0.0892	1.331	0.1242	38
23	0.136	9.1335	0.479	9.6803	0.788	9.8965	1.045	0.0191	1.230	0.0899	1.332	0.1245	37
24	0.142	9.1523	0.484	9.6848	0.793	9.8993	1.048	0.0204	1.233	0.0910	1.333	0.1248	36
25	0.148	9.1703	0.490	9.6902	0.798	9.9020	1.052	0.0220	1.235	0.0917	1.334	0.1252	35
26	0.154	9.1875	0.495	9.6946	0.803	9.9047	1.056	0.0237	1.238	0.0927	1.335	0.1255	34
27	0.160	9.2041	0.501	9.6998	0.807	9.9069	1.060	0.0253	1.240	0.0934	1.336	0.1258	33
28	0.165	9.2175	0.506	9.7042	0.812	9.9096	1.064	0.0269	1.242	0.0941	1.337	0.1261	32
29	0.171	9.2330	0.511	9.7084	0.816	9.9117	1.068	0.0286	1.245	0.0952	1.338	0.1265	31
30	0.176	9.2455	0.516	9.7126	0.821	9.9143	1.071	0.0298	1.247	0.0959	1.339	0.1268	30
31	0.182	9.2601	0.522	9.7177	0.826	9.9170	1.075	0.0314	1.250	0.0969	1.340	0.1270	29
32	0.188	9.2742	0.527	9.7218	0.831	9.9196	1.078	0.0326	1.252	0.0976	1.340	0.1272	28
33	0.194	9.2878	0.533	9.7267	0.835	9.9217	1.082	0.0342	1.254	0.0983	1.341	0.1274	27
34	0.200	9.3010	0.538	9.7308	0.840	9.9243	1.085	0.0354	1.256	0.0990	1.341	0.1275	26
35	0.205	9.3118	0.544	9.7356	0.845	9.9269	1.089	0.0370	1.258	0.0997	1.342	0.1276	25
36	0.211	9.3243	0.549	9.7396	0.849	9.9289	1.092	0.0382	1.260	0.1004	1.342	0.1278	24
37	0.217	9.3365	0.555	9.7443	0.854	9.9315	1.096	0.0398	1.263	0.1014	1.343	0.1280	23
38	0.223	9.3483	0.561	9.7490	0.859	9.9340	1.099	0.0410	1.265	0.1021	1.343	0.1282	22
39	0.229	9.3598	0.566	9.7528	0.864	9.9365	1.103	0.0426	1.267	0.1028	1.344	0.1284	21
40	0.234	9.3692	0.571	9.7566	0.868	9.9385	1.106	0.0438	1.269	0.1035	1.345	0.1286	20
41	0.240	9.3802	0.576	9.7604	0.873	9.9410	1.110	0.0453	1.271	0.1041	1.345	0.1287	19
42	0.246	9.3909	0.582	9.7649	0.877	9.9430	1.113	0.0465	1.273	0.1048	1.346	0.1290	18
43	0.252	9.4014	0.587	9.7686	0.882	9.9455	1.116	0.0477	1.275	0.1055	1.346	0.1292	17
44	0.258	9.4116	0.592	9.7723	0.886	9.9474	1.119	0.0488	1.277	0.1062	1.347	0.1294	16
45	0.264	9.4216	0.598	9.7767	0.890	9.9494	1.123	0.0504	1.279	0.1068	1.347	0.1295	15
46	0.270	9.4314	0.603	9.7803	0.895	9.9518	1.126	0.0515	1.281	0.1075	1.348	0.1296	14
47	0.275	9.4393	0.608	9.7839	0.899	9.9538	1.129	0.0527	1.283	0.1082	1.348	0.1297	13
48	0.281	9.4487	0.613	9.7875	0.903	9.9557	1.132	0.0538	1.284	0.1086	1.348	0.1298	12
49	0.287	9.4579	0.618	9.7910	0.908	9.9581	1.135	0.0550	1.286	0.1092	1.349	0.1300	11
50	0.292	9.4654	0.623	9.7945	0.912	9.9600	1.138	0.0561	1.287	0.1096	1.349	0.1302	10
51	0.298	9.4742	0.629	9.7987	0.916	9.9619	1.141	0.0573	1.289	0.1103	1.350	0.1303	9
52	0.304	9.4829	0.634	9.8021	0.920	9.9638	1.144	0.0584	1.291	0.1109	1.350	0.1303	8
53	0.309	9.4900	0.639	9.8055	0.925	9.9661	1.147	0.0596	1.292	0.1113	1.350	0.1303	7
54	0.315	9.4983	0.644	9.8089	0.929	9.9680	1.150	0.0607	1.294	0.1119	1.350	0.1303	6
55	0.321	9.5065	0.649	9.8122	0.933	9.9699	1.153	0.0618	1.296	0.1126	1.350	0.1303	5
56	0.327	9.5147	0.654	9.8156	0.938	9.9722	1.156	0.0630	1.298	0.1133	1.350	0.1303	4
57	0.332	9.5211	0.660	9.8195	0.942	9.9741	1.159	0.0641	1.299	0.1136	1.350	0.1303	3
58	0.338	9.5289	0.665	9.8228	0.946	9.9759	1.162	0.0652	1.301	0.1143	1.350	0.1303	2
59	0.344	9.5366	0.670	9.8261	0.950	9.9777	1.165	0.0663	1.303	0.1149	1.350	0.1303	1
60	0.349	9.5428	0.675	9.8293	0.954	9.9795	1.168	0.0674	1.304	0.1153	1.350	0.1303	0
	Nat. No.	Log.	Nat. No.	Log.	Nat. No.	Log.	Nat. No.	Log.	Nat. No.	Log.	Nat. No.	Log.	
	XI −	XXIII +	X −	XXII +	IX −	XXI +	VIII −	XX +	VII −	XIX +	VI −	XVIII +	

(Argument = R. A. ∗) TABLE 3. (Tab. No. = − $1^s.239$ Cos R. A.)

Min.	− O	+ XII	− I	+ XIII	− II	+ XIV	− III	+ XV	− IV	+ XVI	− V	+ XVII	Min.
	Nat. No.	Log.	Nat. No.	Log.	Nat. No.	Log.	Nat. No.	Log.	Nat. No.	Log.	Nat. No.	Log.	
0	$1^s.239$	0.0931	$1^s.196$	0.0777	$1^s.073$	0.0306	$0^s.876$	9.9425	$0^s.619$	9.7917	$0^s.321$	9.5065	60
1	1.239	0.0931	1.195	0.0774	1.070	0.0294	0.872	9.9405	0.615	9.7889	0.315	9.4983	59
2	1.239	0.0931	1.194	0.0770	1.067	0.0282	0.868	9.9385	0.610	9.7853	0.310	9.4914	58
3	1.239	0.0931	1.192	0.0763	1.065	0.0273	0.864	9.9365	0.605	9.7818	0.305	9.4843	57
4	1.239	0.0931	1.191	0.0759	1.062	0.0261	0.860	9.9345	0.601	9.7789	0.300	9.4771	56
5	1.238	0.0930	1.189	0.0752	1.059	0.0249	0.857	9.9330	0.596	9.7752	0.294	9.4683	55
6	1.238	0.0930	1.188	0.0748	1.056	0.0237	0.853	9.9309	0.591	9.7716	0.289	9.4609	54
7	1.238	0.0929	1.186	0.0741	1.053	0.0224	0.849	9.9289	0.586	9.7679	0.284	9.4533	53
8	1.238	0.0928	1.185	0.0737	1.050	0.0212	0.845	9.9269	0.582	9.7649	0.279	9.4456	52
9	1.238	0.0927	1.183	0.0730	1.048	0.0204	0.841	9.9248	0.577	9.7612	0.273	9.4362	51
10	1.237	0.0927	1.181	0.0722	1.045	0.0191	0.837	9.9227	0.572	9.7574	0.268	9.4281	50
11	1.237	0.0926	1.180	0.0719	1.042	0.0179	0.833	9.9206	0.567	9.7536	0.263	9.4200	49
12	1.237	0.0925	1.178	0.0711	1.039	0.0166	0.829	9.9186	0.562	9.7497	0.258	9.4116	48
13	1.237	0.0924	1.176	0.0704	1.036	0.0154	0.825	9.9165	0.558	9.7466	0.252	9.4014	47
14	1.236	0.0923	1.175	0.0700	1.033	0.0141	0.821	9.9143	0.553	9.7427	0.247	9.3926	46
15	1.236	0.0921	1.173	0.0691	1.030	0.0128	0.817	9.9122	0.548	9.7388	0.242	9.3838	45
16	1.236	0.0920	1.171	0.0682	1.027	0.0116	0.813	9.9101	0.543	9.7348	0.236	9.3729	44
17	1.235	0.0919	1.169	0.0678	1.024	0.0103	0.809	9.9079	0.538	9.7308	0.231	9.3636	43
18	1.235	0.0917	1.168	0.0674	1.021	0.0090	0.804	9.9053	0.533	9.7267	0.226	9.3541	42
19	1.234	0.0915	1.166	0.0667	1.018	0.0077	0.800	9.9031	0.528	9.7226	0.220	9.3424	41
20	1.234	0.0913	1.164	0.0660	1.015	0.0065	0.796	9.9009	0.523	9.7185	0.215	9.3324	40
21	1.233	0.0911	1.162	0.0652	1.012	0.0052	0.792	9.8987	0.519	9.7152	0.210	9.3222	39
22	1.233	0.0910	1.160	0.0645	1.008	0.0035	0.788	9.8965	0.514	9.7110	0.204	9.3096	38
23	1.232	0.0908	1.158	0.0637	1.005	0.0022	0.784	9.8943	0.509	9.7067	0.199	9.2989	37
24	1.232	0.0906	1.156	0.0630	1.002	0.0009	0.780	9.8921	0.504	9.7024	0.194	9.2878	36
25	1.231	0.0904	1.154	0.0622	0.999	9.9996	0.775	9.8893	0.499	9.6981	0.188	9.2742	35
26	1.231	0.0903	1.152	0.0615	0.996	9.9983	0.771	9.8870	0.494	9.6937	0.183	9.2625	34
27	1.230	0.0899	1.150	0.0607	0.993	9.9969	0.767	9.8848	0.489	9.6893	0.178	9.2504	33
28	1.229	0.0896	1.148	0.0599	0.989	9.9952	0.763	9.8825	0.484	9.6848	0.172	9.2355	32
29	1.229	0.0894	1.146	0.0592	0.986	9.9939	0.758	9.8797	0.479	9.6803	0.167	9.2227	31
30	1.228	0.0892	1.144	0.0584	0.983	9.9926	0.754	9.8774	0.474	9.6758	0.162	9.2095	30
31	1.227	0.0890	1.142	0.0577	0.979	9.9908	0.750	9.8751	0.469	9.6712	0.156	9.1931	29
32	1.227	0.0888	1.140	0.0569	0.976	9.9894	0.745	9.8722	0.464	9.6665	0.151	9.1790	28
33	1.226	0.0885	1.138	0.0561	0.973	9.9881	0.741	9.8698	0.459	9.6618	0.146	9.1644	27
34	1.225	0.0881	1.136	0.0554	0.969	9.9863	0.737	9.8675	0.454	9.6571	0.140	9.1461	26
35	1.224	0.0878	1.134	0.0546	0.966	9.9850	0.732	9.8645	0.449	9.6522	0.135	9.1303	25
36	1.223	0.0876	1.132	0.0538	0.963	9.9836	0.728	9.8621	0.444	9.6474	0.129	9.1106	24
37	1.223	0.0874	1.129	0.0527	0.959	9.9818	0.724	9.8597	0.439	9.6425	0.124	9.0934	23
38	1.222	0.0871	1.127	0.0519	0.956	9.9805	0.719	9.8567	0.434	9.6375	0.119	9.0755	22
39	1.221	0.0867	1.125	0.0512	0.952	9.9786	0.715	9.8543	0.429	9.6325	0.113	9.0531	21
40	1.220	0.0864	1.123	0.0504	0.949	9.9773	0.710	9.8513	0.424	9.6274	0.108	9.0334	20
41	1.219	0.0860	1.120	0.0492	0.945	9.9754	0.706	9.8488	0.419	9.6222	0.103	9.0128	19
42	1.218	0.0856	1.118	0.0484	0.942	9.9741	0.702	9.8463	0.414	9.6170	0.097	8.9868	18
43	1.217	0.0853	1.116	0.0477	0.938	9.9722	0.697	9.8432	0.408	9.6107	0.092	8.9638	17
44	1.216	0.0849	1.113	0.0465	0.935	9.9708	0.693	9.8407	0.403	9.6053	0.086	8.9345	16
45	1.215	0.0846	1.111	0.0457	0.931	9.9689	0.688	9.8376	0.398	9.5999	0.081	8.9085	15
46	1.214	0.0842	1.109	0.0449	0.928	9.9675	0.684	9.8351	0.393	9.5944	0.076	8.8808	14
47	1.213	0.0839	1.106	0.0438	0.924	9.9657	0.679	9.8319	0.388	9.5888	0.070	8.8451	13
48	1.212	0.0835	1.104	0.0430	0.921	9.9643	0.675	9.8293	0.383	9.5832	0.065	8.8129	12
49	1.210	0.0828	1.101	0.0418	0.917	9.9624	0.670	9.8261	0.378	9.5775	0.059	8.7709	11
50	1.209	0.0824	1.099	0.0410	0.913	9.9609	0.666	9.8235	0.373	9.5717	0.054	8.7324	10
51	1.208	0.0821	1.096	0.0398	0.910	9.9590	0.661	9.8202	0.366	9.5635	0.049	8.6902	9
52	1.207	0.0817	1.094	0.0390	0.906	9.9571	0.656	9.8169	0.362	9.5587	0.043	8.6335	8
53	1.206	0.0813	1.091	0.0378	0.902	9.9552	0.652	9.8142	0.357	9.5527	0.038	8.5798	7
54	1.204	0.0806	1.089	0.0370	0.899	9.9538	0.647	9.8109	0.352	9.5465	0.032	8.5052	6
55	1.203	0.0803	1.086	0.0358	0.895	9.9518	0.643	9.8082	0.347	9.5403	0.027	8.4314	5
56	1.202	0.0799	1.083	0.0346	0.891	9.9499	0.638	9.8048	0.341	9.5327	0.022	8.3424	4
57	1.201	0.0795	1.081	0.0338	0.887	9.9479	0.633	9.8014	0.336	9.5263	0.016	8.2041	3
58	1.199	0.0788	1.078	0.0326	0.883	9.9460	0.629	9.7987	0.331	9.5198	0.011	8.0414	2
59	1.198	0.0785	1.075	0.0314	0.880	9.9445	0.624	9.7952	0.326	9.5132	0.005	7.6990	1
60	1.196	0.0777	1.073	0.0306	0.876	9.9425	0.619	9.7917	0.321	9.5065	0.000	0.0000	0
	Nat. No.	Log.	Nat. No.	Log.	Nat. No.	Log.	Nat. No.	Log.	Nat. No.	Log.	Nat. No.	Log.	
	XI +	XXIII −	X +	XXII −	IX +	XXI −	VIII +	XX −	VII +	XIX −	VI +	XVIII −	

(Argument = R. A ✱.) TABLE 4. (Tab. No. = + 0 .648 Cos R. A.)

S. N.	+ 0 −	− XII +	+ I −	− XIII +	+ II −	− XIV +	+ III −	− XV +	+ IV −	− XVI +	+ V −	− XVII +	S. N.
Min.	Nat. No.	Log.	Nat. No.	Log.	Nat. No.	Log.	Nat. No.	Log.	Nat. No.	Log.	Nat. No.	Log.	Min.
0	0*.643	9.8082	0.621	9.7931	0*.557	9.7459	0*.455	9.6580	0*.322	9.5079	0*.166	9.2201	60
1	0.643	9.8082	0.621	9.7928	0.556	9.7451	0.453	9.6561	0.319	9.5038	0.164	9.2148	59
2	0.643	9.8082	0.620	9.7924	0.554	9.7435	0.451	9.6542	0.317	9.5011	0.161	9.2068	58
3	0.643	9.8082	0.619	9.7917	0.553	9.7427	0.449	9.6522	0.314	9.4969	0.158	9.1987	57
4	0.643	9.8082	0.618	9.7910	0.551	9.7412	0.447	9.6503	0.312	9.4942	0.156	9.1931	56
5	0.643	9.8082	0.617	9.7903	0.550	9.7404	0.445	9.6484	0.309	9.4900	0.153	9.1847	55
6	0.643	9.8082	0.617	9.7907	0.548	9.7388	0.443	9.6464	0.307	9.4871	0.150	9.1761	54
7	0.643	9 8082	0.616	9.7896	0.547	9.7380	0.441	9.6444	0.304	9.4829	0.147	9.1673	53
8	0.643	9.8082	0.615	9.7889	0.545	9.7364	0.439	9.6425	0.302	9.4800	0.145	9.1614	52
9	0.643	9.8082	0.614	9.7882	0.544	9.7356	0.437	9.6405	0 299	9.4757	0.142	9.1523	51
10	0.643	9.8081	0.613	9.7878	0.542	9.7340	0.435	9.6385	0.297	9.4728	0.139	9.1430	50
11	0.643	9.8081	0.613	9.7875	0.541	9.7332	0.432	9.6355	0.294	9.4683	0.136	9.1335	49
12	0.643	9.8081	0.612	9.7868	0.539	9.7316	0.430	9.6335	0.292	9.4654	0.134	9.1271	48
13	0.643	9.8081	0.611	9.7860	0.538	9.7308	0.428	9.6314	0.289	9.4609	0.131	9.1173	47
14	0.643	9.8080	0.610	9.7853	0.536	9.7292	0.426	9.6294	0.287	9.4579	0.128	9.1072	46
15	0.642	9.8078	0.609	9.7846	0.535	9.7284	0.424	9.6274	0.284	9.4533	0.126	9.1004	45
16	0.642	9.8075	0.608	9.7839	0.533	9.7267	0.422	9.6253	0.282	9.4502	0.123	9.0899	44
17	0.641	9.8072	0.607	9.7832	0.532	9.7259	0.420	9.6232	0.279	9.4456	0.120	9.0792	43
18	0.641	9.8070	0.606	9.7825	0.530	9.7243	0.418	9.6212	0.277	9.4425	0.117	9.0682	42
19	0.641	9.8069	0.605	9.7818	0.528	9.7226	0.416	9.6191	0.274	9.4378	0.114	9.0569	41
20	0.641	9.8067	0.604	9.7810	0.527	9.7218	0.413	9.6160	0.272	9.4346	0.112	9.0492	40
21	0.640	9.8065	0.603	9.7803	0.525	9.7202	0.411	9.6138	0.269	9.4298	0.109	9.0374	39
22	0.640	9.8063	0.602	9.7796	0.524	9.7193	0.409	9.6117	0.267	9.4265	0.106	9.0253	38
23	0.640	9.8062	0.601	9.7789	0.522	9.7177	0.407	9.6096	0.264	9.4216	0.103	9.0128	37
24	0.640	9.8061	0.600	9.7782	0.520	9.7160	0.405	9.6075	0.262	9.4183	0.101	9.0043	36
25	0.639	9.8058	0.599	9.7774	0.519	9.7152	0.403	9.6053	0.259	9.4133	0.098	8.9912	35
26	0.639	9.8056	0.598	9.7767	0.517	9.7135	0.400	9.6021	0.256	9.4082	0.095	8.9777	34
27	0.639	9.8054	0.597	9.7760	0.515	9.7118	0.398	9.5999	0.254	9.4048	0.092	8.9638	33
28	0.638	9.8052	0 596	9.7752	0.514	9.7110	0.396	9.5977	0.251	9.3997	0.090	8.9542	32
29	0.638	9.8050	0.595	9.7745	0.512	9.7093	0.394	9.5955	0.249	9.3962	0.087	8.9395	31
30	0.638	9.8048	0.594	9.7738	0.510	9.7080	0.392	9.5933	0.246	9.3909	0.084	8.9243	30
31	0.637	9.8046	0.593	9.7731	0.509	9.7067	0.389	9.5899	0.244	9.3874	0.081	8.9085	29
32	0.637	9.8044	0.592	9.7723	0.507	9.7050	0.387	9.5877	0.241	9.3820	0.078	8.8921	28
33	0.637	9.8041	0.591	9.7716	0.505	9.7033	0.385	9.5855	0.238	9.3766	0.076	8.8808	27
34	0.636	9.8038	0.590	9.7709	0.503	9.7016	0.383	9.5832	0.236	9.3729	0.073	8.8633	26
35	0.636	9.8035	0.589	9.7701	0.502	9.7007	0.380	9.5798	0.233	9.3674	0.070	8.8451	25
36	0.635	9.8030	0.588	9.7694	0.500	9.6990	0.378	9.5775	0.231	9.3636	0.067	8.8261	24
37	0.635	9.8028	0.586	9.7679	0.498	9.6972	0.376	9.5752	0.228	9.3579	0.064	8.8062	23
38	0.634	9.8025	0.585	9.7672	0.496	9.6955	0.374	9.5729	0.225	9.3522	0.062	8.7924	22
39	0.634	9.8021	0.584	9.7664	0.495	9.6946	0.371	9.5694	0.223	9.3483	0.059	8.7709	21
40	0.633	9.8018	0.583	9.7657	0.493	9.6928	0.369	9.5670	0.220	9.3424	0.056	8.7482	20
41	0.633	9.8014	0.582	9.7649	0.491	9.6911	0.367	9.5647	0.217	9.3365	0.053	8.7243	19
42	0.632	9.8010	0.581	9.7642	0.489	9.6893	0.364	9.5611	0.215	9.3324	0.050	8.6990	18
43	0.632	9.8007	0.579	9.7627	0.487	9.6875	0.362	9.5587	0.212	9.3263	0.048	8.6812	17
44	0.631	9.8003	0.578	9.7619	0.485	9.6857	0.360	9.5563	0.209	9.3201	0.045	8.6532	16
45	0.631	9.8000	0.577	9.7612	0.484	9.6848	0.357	9.5527	0.207	9.3160	0.042	8.6232	15
46	0.630	9.7997	0.576	9.7604	0.482	9.6830	0.355	9.5502	0.204	9.3096	0.039	8.5911	14
47	0.630	9.7993	0.574	9.7589	0.480	9.6812	0 353	9.5478	0.201	9.3032	0.036	8.5563	13
48	0.629	9.7990	0.573	9.7582	0.478	9.6794	0 350	9.5441	0.199	9.2989	0.034	8.5315	12
49	0.629	9.7987	0.572	9.7574	0.476	9.6776	0.348	9.5416	0.196	9.2923	0.031	8.4914	11
50	0.628	9.7980	0.571	9.7566	0.474	9.6758	0.346	9.5391	0.193	9.2856	0.028	8.4472	10
51	0.627	9.7976	0.569	9.7551	0.472	9.6739	0.343	9.5353	0.191	9.2810	0.025	8.3979	9
52	0.627	9.7973	0.568	9.7543	0.470	9.6721	0.341	9.5328	0.188	9.2742	0.022	8.3424	8
53	0.626	9.7966	0.567	9.7536	0.469	9.6712	0.338	9.5289	0.185	9.2672	0.020	8.3010	7
54	0.625	9.7962	0.565	9.7520	0.467	9.6693	0.336	9.5263	0.183	9.2625	0.017	8.2304	6
55	0.625	9.7959	0.564	9.7513	0.465	9.6675	0.334	9.5237	0.180	9.2553	0.014	8.1461	5
56	0.624	9.7954	0.563	9.7505	0.463	9 6656	0.331	9.5198	0.177	9.2480	0.011	8.0414	4
57	0.623	9.7950	0.561	9.7490	0.461	9.6637	0.329	9.5172	0.175	9.2430	0.008	7.9031	3
58	0.623	9.7945	0.560	9.7482	0.459	9.6618	0.326	9.5132	0.172	9.2355	0.006	7.7782	2
59	0.622	9.7938	0.558	9.7466	0.457	9.6599	0.324	9.5105	0.169	9.2279	0.003	7.4771	1
60	0.621	9.7931	0.557	9.7459	0.455	9.6580	0.322	9.5079	0.166	9.2201	0.000	0.0000	0
	Nat. No.	Log.	Nat. No.	Log.	Nat. No.	Log.	Nat. No.	Log.	Nat. No.	Log.	Nat. No.	Log.	
N. S.	+ XI −	− XXIII +	+ X −	− XXII +	+ IX −	− XXI +	+ VIII −	− XX +	+ VII −	− XIX +	+ VI −	− XVIII +	N. S.

(Argument = R. A. *) TABLE 5. (Tab. No. = − 20″.0436 cos R. A.)

S. N.	− 0 +	+ XII −	− I +	+ XIII −	− II +	+ XIV −	− III +	+ XV −	− IV +	+ XVI −	− V +	+ XVII −	S. N.
Min.	Nat. No.	Log.	Nat. No.	Log.	Nat. No.	Log.	Nat. No.	Log.	Nat. No.	Log.	Nat. No.	Log.	Min.
0	20″.044	1.3020	19″.361	1.2869	17″.358	1.2395	14″.173	1.1515	10″.022	1.0010	5″.188	0.7150	60
1	20.044	1.3020	19.338	1.2864	17.314	1.2384	14.111	1.1496	9.946	0.9976	5.103	0.7078	59
2	20.043	1.3020	19.315	1.2859	17.270	1.2373	14.049	1.1476	9.870	0.9943	5.019	0.7006	58
3	20.042	1.3019	19.291	1.2854	17.226	1.2362	13.986	1.1457	9.794	0.9910	4.934	0.6932	57
4	20.041	1.3019	19.267	1.2848	17.181	1.2350	13.923	1.1437	9.717	0.9875	4.849	0.6857	56
5	20.039	1.3019	19.243	1.2843	17.136	1.2339	13.860	1.1418	9.641	0.9841	4.764	0.6780	55
6	20.037	1.3018	19.218	1.2837	17.090	1.2327	13.797	1.1398	9.564	0.9806	4.679	0.6702	54
7	20.034	1.3018	19.193	1.2831	17.044	1.2316	13.734	1.1378	9.487	0.9771	4.594	0.7622	53
8	20.031	1.3017	19.168	1.2826	16.998	1.2304	13.670	1.1358	9.409	0.9735	4.509	0.6541	52
9	20.028	1.3016	19.142	1.2820	16.951	1.2292	13.606	1.1337	9.322	0.9695	4.424	0.6458	51
10	20.025	1.3016	19.116	1.2814	16.904	1.2280	13.542	1.1317	9.255	0.9664	4.338	0.6373	50
11	20.021	1.3015	19.089	1.2808	16.857	1.2268	13.477	1.1296	9.177	0.9627	4.253	0.6287	49
12	20.016	1.3014	19.062	1.2802	16.810	1.2256	13.412	1.1275	9.100	0.9590	4.167	0.6198	48
13	20.011	1.3013	19.035	1.2796	16.762	1.2243	13.347	1.1254	9.022	0.9553	4.082	0.6109	47
14	20.006	1.3012	19.008	1.2789	16.714	1.2231	13.281	1.1232	8.943	0.9515	3.996	0.6016	46
15	20.001	1.3011	18.980	1.2783	16.666	1.2218	13.215	1.1211	8.865	0.9477	3.910	0.5922	45
16	19.995	1.3009	18.952	1.2777	16.617	1.2206	13.249	1.1189	8.787	0.9438	3.825	0.5826	44
17	19.989	1.3008	18.923	1.2770	16.568	1.2193	13.083	1.1167	8.708	0.9399	3.739	0.5728	43
18	19.982	1.3006	18.894	1.2763	16.519	1.2180	13.017	1.1145	8.629	0.9360	3.643	0.5615	42
19	19.975	1.3005	18.865	1.2757	16.469	1.2167	12.951	1.1123	8.550	0.9294	3.567	0.5523	41
20	19.968	1.3003	18.835	1.2750	16.419	1.2153	12.884	1.1100	8.471	0.9279	3.481	0.5417	40
21	19.960	1.3002	18.805	1.2743	16.369	1.2140	12.817	1.1078	8.391	0.9238	3.394	0.5307	39
22	19.952	1.3000	18.774	1.2736	16.318	1.2127	12.749	1.1055	8.312	0.9197	3.308	0.5196	38
23	19.943	1.2998	18.743	1.2728	16.267	1.2113	12.681	1.1032	8.232	0.9155	3.222	0.5081	37
24	19.934	1.2996	18.712	1.2721	16.216	1.2099	12.613	1.1008	8.152	0.9113	3.136	0.4964	36
25	19.925	1.2994	18.681	1.2714	16.164	1.2084	12.545	1.0985	8.072	0.9070	3.049	0.4842	35
26	19.915	1.2992	18.649	1.2707	16.112	1.2071	12.477	1.0961	7.992	0.9027	2.963	0.4717	34
27	19.905	1.2990	18.617	1.2699	16.060	1.2057	12.409	1.0937	7.912	0.8983	2.876	0.4588	33
28	19.894	1.2987	18.584	1.2691	16.008	1.2043	12.340	1.0913	7.831	0.8938	2.790	0.4456	32
29	19.883	1.2985	18.551	1.2684	15.955	1.2029	12.271	1.0889	7.751	0.8894	2.703	0.4318	31
30	19.872	1.2982	18.518	1.2676	15.902	1.2015	12.202	1.0864	7.670	0.8848	2.616	0.4176	30
31	19.860	1.2980	18.485	1.2668	15.848	1.2000	12.133	1.0839	7.589	0.8802	2.530	0.4031	29
32	19.848	1.2977	18.451	1.2660	15.794	1.1985	12.063	1.0815	7.508	0.8755	2.443	0.3879	28
33	19.836	1.2975	18.416	1.2652	15.740	1.1970	11.993	1.0789	7.427	0.8708	2.356	0.3722	27
34	19.823	1.2972	18.381	1.2644	15.687	1.1955	11.923	1.0764	7.346	0.8661	2.269	0.3558	26
35	19.810	1.2969	18.346	1.2635	15.632	1.1940	11.852	1.0738	7.264	0.8612	2.182	0.3389	25
36	19.797	1.2966	18.311	1.2627	15.577	1.1925	11.781	1.0712	7.183	0.8563	2.095	0.3212	24
37	19.783	1.2963	18.275	1.2619	15.522	1.1909	11.710	1.0686	7.101	0.8513	2.008	0.3028	23
38	19.769	1.2960	18.239	1.2610	15.466	1.1894	11.639	1.0659	7.019	0.8463	1.921	0.2835	22
39	19.754	1.2957	18.202	1.2601	15.410	1.1878	11.568	1.0633	6.937	0.8412	1.834	0.2634	21
40	19.739	1.2953	18.165	1.2592	15.354	1.1862	11.497	1.0606	6.855	0.8360	1.747	0.2423	20
41	19.724	1.2950	18.128	1.2583	15.298	1.1846	11.425	1.0579	6.773	0.8308	1.660	0.2201	19
42	19.708	1.2946	18.091	1.2575	15.241	1.1830	11.353	1.0551	6.691	0.8255	1.573	0.1967	18
43	19.692	1.2943	18.053	1.2565	15.184	1.1814	11.281	1.0523	6.608	0.8201	1.485	0.1717	17
44	19.675	1.2939	18.015	1.2556	15.127	1.1798	11.208	1.0495	6.526	0.8146	1.398	0.1455	16
45	19.658	1.2935	17.977	1.2547	15.070	1.1781	11.135	1.0467	6.443	0.8091	1.311	0.1176	15
46	19.641	1.2932	17.938	1.2538	15.012	1.1764	11.062	1.0438	6.360	0.8035	1.224	0.0878	14
47	19.624	1.2928	17.899	1.2528	14.954	1.1748	10.989	1.0410	6.277	0.7978	1.136	0.0554	13
48	19.606	1.2924	17.859	1.2519	14.896	1.1731	10.916	1.0381	6.194	0.7920	1.049	0.0208	12
49	19.588	1.2920	17.819	1.2509	14.837	1.1713	10.843	1.0351	6.111	0.7861	0.960	9.9823	11
50	19.569	1.2916	17.779	1.2499	14.778	1.1696	10.769	1.0322	6.027	0.7801	0.874	9.9415	10
51	19.550	1.2911	17.738	1.2489	14.719	1.1679	10.695	1.0292	5.944	0.7741	0.788	9.8965	9
52	19.530	1.2907	17.697	1.2479	14.659	1.1661	10.621	1.0262	5.860	0.7679	0.700	9.8451	8
53	19.510	1.2903	17.656	1.2469	14.599	1.1643	10.547	1.0231	5.777	0.7617	0.612	9.7868	7
54	19.490	1.2898	17.615	1.2459	14.539	1.1625	10.473	1.0201	5.693	0.7553	0.525	9.7202	6
55	19.469	1.2893	17.573	1.2448	14.479	1.1607	10.398	1.0169	5.609	0.7489	0.437	9.6405	5
56	19.448	1.2889	17.531	1.2438	14.418	1.1589	10.323	1.0138	5.525	0.7423	0.350	9.5441	4
57	19.427	1.2884	17.488	1.2427	14.357	1.1561	10.248	1.0106	5.441	0.7357	0.262	9.4183	3
58	19.405	1.2879	17.445	1.2417	14.296	1.1552	10.173	1.0074	5.356	0.7288	0.175	9.2430	2
59	19.383	1.2874	17.402	1.2406	14.235	1.1534	10.098	1.0042	5.272	0.7220	0.087	8.9395	1
60	19.361	1.2869	17.358	1.2395	14.173	1.1515	10.022	1.0010	5.188	0.7150	0.000	− ∞	0
	Nat. No.	Log.	Nat. No.	Log.	Nat. No.	Log.	Nat. No.	Log.	Nat. No.	Log.	Nat. No.	Log.	
N. S.	− XI +	+ XXIII −	− X +	+ XXII −	− IX +	+ XXI −	− VIII +	+ XX −	− VII +	+ XIX −	− VI +	+ XVIII −	N. S.

(Argument = R. A. ⁎) TABLE 6. (Tab. No. = − 20″.255 cos R. A.)

Min.	O −	XII +	I −	XIII +	II −	XIV +	III −	XV +	IV −	XVI +	V −	XVII +	Min.
	Nat. No.	Log.	Nat. No.	Log.	Nat. No.	Log.	Nat. No.	Log.	Nat. No.	Log.	Nat. No.	Log.	
0	20″.255	1.3065	19″.565	1.2915	17″.541	1.2441	14″.322	1.1560	10″.128	1.0055	5″.243	0.7195	60
1	20.255	1.3065	19.542	1.2910	17.497	1.2430	14.259	1.1541	10.051	1.0022	5.158	0.7124	59
2	20.254	1.3065	19.518	1.2904	17.452	1.2419	14.196	1.1522	9.974	0.9989	5.072	0.7051	58
3	20.254	1.3065	19.494	1.2899	17.407	1.2407	14.133	1.1503	9.897	0.9955	4.986	0.6977	57
4	20.252	1.3065	19.470	1.2894	17.362	1.2396	14.070	1.1483	9.820	0.9921	4.900	0.6902	56
5	20.250	1.3064	19.445	1.2888	17.316	1.2385	14.006	1.1463	9.743	0.9887	4.814	0.6825	55
6	20.248	1.3064	19.420	1.2883	17.270	1.2373	13.942	1.1443	9.665	0.9852	4.728	0.6747	54
7	20.246	1.3063	19.395	1.2877	17.224	1.2361	13.878	1.1423	9.587	0.9817	4.642	0.6668	53
8	20.243	1.3063	19.370	1.2871	17.177	1.2350	13.814	1.1403	9.509	0.9781	4.556	0.6586	52
9	20.240	1.3062	19.344	1.2866	17.130	1.2338	13.749	1.1383	9.431	0.9746	4.470	0.6503	51
10	20.236	1.3061	19.318	1.2860	17.083	1.2326	13.684	1.1362	9.353	0.9709	4.384	0.6419	50
11	20.232	1.3060	19.291	1.2854	17.035	1.2314	13.619	1.1341	9.275	0.9673	4.298	0.6332	49
12	20.227	1.3059	19.264	1.2847	16.987	1.2301	13.553	1.1320	9.196	0.9636	4.212	0.6244	46
13	20.222	1.3058	19.236	1.2841	16.939	1.2289	13.487	1.1299	9.117	0.9598	4.126	0.6154	47
14	20.217	1.3057	19.208	1.2835	16.890	1.2276	13.421	1.1278	9.038	0.9561	4.039	0.6062	46
15	20.211	1.3056	19.180	1.2829	16.841	1.2264	13.355	1.1257	8.959	0.9522	3.952	0.5968	45
16	20.206	1.3055	19.152	1.2822	16.792	1.2251	13.288	1.1235	8.880	0.9484	3.865	0.5871	44
17	20.199	1.3053	19.123	1.2815	16.742	1.2238	13.221	1.1213	8.800	0.9445	3.778	0.5773	43
18	20.193	1.3052	19.093	1.2809	16.692	1.2225	13.154	1.1191	8.720	0.9405	3.691	0.5672	42
19	20.186	1.3050	19.063	1.2802	16.642	1.2212	13.087	1.1169	8.640	0.9365	3.604	0.5568	41
20	20.178	1.3049	19.033	1.2795	16.592	1.2199	13.019	1.1146	8.560	0.9325	3.517	0.5462	40
21	20.170	1.3047	19.002	1.2788	16.541	1.2186	12.951	1.1123	8.480	0.9284	3.430	0.5353	39
22	20.162	1.3045	18.971	1.2781	16.490	1.2172	12.883	1.1100	8.400	0.9243	3.343	0.5241	38
23	20.153	1.3043	18.940	1.2774	16.439	1.2159	12.815	1.1077	8.320	0.9201	3.256	0.5127	37
24	20.144	1.3042	18.909	1.2767	16.387	1.2145	12.746	1.1054	8.239	0.9159	3.169	0.5009	36
25	20.134	1.3039	18.877	1.2760	16.335	1.2131	12.677	1.1030	8.158	0.9116	3.082	0.4887	35
26	20.125	1.3037	18.845	1.2752	16.283	1.2117	12.608	1.1007	8.077	0.9072	2.995	0.4762	34
27	20.115	1.3035	18.813	1.2745	16.230	1.2103	12.539	1.0983	7.996	0.9029	2.908	0.4634	33
28	20.105	1.3033	18.780	1.2737	16.177	1.2089	12.470	1.0959	7.915	0.8984	2.820	0.4501	32
29	20.094	1.3031	18.747	1.2729	16.123	1.2075	12.400	1.0934	7.833	0.8939	2.732	0.4364	31
30	20.082	1.3028	18.713	1.2722	16.069	1.2060	12.330	1.0910	7.751	0.8894	2.644	0.4222	30
31	20.070	1.3026	18.679	1.2714	16.015	1.2045	12.260	1.0885	7.669	0.8848	2.556	0.4076	29
32	20.058	1.3023	18.645	1.2706	15.961	1.2031	12.190	1.0860	7.587	0.8801	2.468	0.3924	28
33	20.046	1.3020	18.610	1.2698	15.906	1.2016	12.119	1.0835	7.505	0.8754	2.380	0.3767	27
34	20.033	1.3017	18.575	1.2689	15.851	1.2001	12.048	1.0809	7.423	0.8706	2.292	0.3604	26
35	20.020	1.3015	18.540	1.2681	15.796	1.1986	11.977	1.0784	7.341	0.8658	2.204	0.3434	25
36	20.006	1.3012	18.504	1.2673	15.741	1.1970	11.906	1.0758	7.259	0.8609	2.116	0.3258	24
37	19.992	1.3008	18.468	1.2664	15.685	1.1955	11.834	1.0731	7.177	0.8559	2.028	0.3074	23
38	19.977	1.3005	18.431	1.2656	15.629	1.1939	11.762	1.0705	7.094	0.8509	1.940	0.2881	22
39	19.962	1.3002	18.394	1.2647	15.573	1.1924	11.690	1.0678	7.011	0.8458	1.852	0.2680	21
40	19.947	1.2999	18.357	1.2638	15.516	1.1908	11.618	1.0651	6.928	0.8406	1.764	0.2468	20
41	19.932	1.2996	18.319	1.2629	15.459	1.1892	11.545	1.0624	6.845	0.8353	1.676	0.2246	19
42	19.916	1.2992	18.281	1.2620	15.402	1.1876	11.472	1.0597	6.762	0.8300	1.588	0.2012	18
43	19.899	1.2988	18.243	1.2611	15.345	1.1860	11.399	1.0569	6.679	0.8246	1.500	0.1764	17
44	19.883	1.2984	18.205	1.2602	15.287	1.1843	11.326	1.0541	6.595	0.8192	1.412	0.1501	16
45	19.866	1.2981	18.166	1.2593	15.229	1.1827	11.253	1.0513	6.511	0.8136	1.324	0.1221	15
46	19.849	1.2977	18.127	1.2583	15.171	1.1810	11.180	1.0484	6.427	0.8080	1.236	0.0922	14
47	19.831	1.2973	18.087	1.2574	15.112	1.1793	11.106	1.0456	6.343	0.8023	1.148	0.0601	13
48	19.813	1.2969	18.047	1.2564	15.053	1.1776	11.032	1.0426	6.259	0.7965	1.060	0.0253	12
49	19.794	1.2965	18.007	1.2554	14.994	1.1759	10.958	1.0397	6.175	0.7906	0.971	9.9876	11
50	19.775	1.2961	17.966	1.2545	14.934	1.1742	10.884	1.0368	6.091	0.7847	0.883	9.9462	10
51	19.756	1.2957	17.925	1.2535	14.874	1.1724	10.809	1.0338	6.007	0.7786	0.795	9.9005	9
52	19.736	1.2953	17.834	1.2525	14.814	1.1707	10.734	1.0307	5.923	0.7725	0.707	9.8494	8
53	19.716	1.2948	17.842	1.2515	14.753	1.1689	10.659	1.0277	5.838	0.7662	0.618	9.7914	7
54	19.696	1.2944	17.800	1.2504	14.692	1.1671	10.584	1.0246	5.753	0.7599	0.530	9.7245	6
55	19.675	1.2939	17.758	1.2494	14.631	1.1653	10.508	1.0215	5.668	0.7534	0.442	9.6453	5
56	19.654	1.2934	17.715	1.2484	14.570	1.1635	10.432	1.0184	5.583	0.7469	0.354	9.5484	4
57	19.632	1.2930	17.672	1.2473	14.508	1.1616	10.356	1.0152	5.498	0.7402	0.265	9.4235	3
58	19.610	1.2925	17.629	1.2462	14.446	1.1598	10.280	1.0120	5.413	0.7334	0.177	9.2474	2
59	19.588	1.2920	17.585	1.2452	14.384	1.1579	10.204	1.0088	5.328	0.7265	0.089	8.9464	1
60	19.565	1.2915	17.541	1.2441	14.322	1.1560	10.128	1.0055	5.243	0.7195	0.000	− ∞	0
	Nat. No.	Log.	Nat. No.	Log.	Nat. No.	Log.	Nat. No.	Log.	Nat. No.	Log.	Nat. No.	Log.	
	+ XI	− XXIII	+ X	− XXII	+ IX	− XXI	+ VIII	− XX	+ VII	− XIX	+ VI	− XVIII	

(Argument = R. A. ✱) TABLE 7. (Tab. No. +18″.580 sin R. A.)

Min.	O +	XII −	I +	XIII −	II +	XIV −	III +	XV −	IV +	XVI −	V +	XVII −	Min.
	Nat. No.	Log.	Nat. No.	Log.	Nat. No.	Log.	Nat. No.	Log.	Nat. No.	Log.	Nat. No.	Log.	
0	0″.000	− ∞	4″.809	0.6820	9″.290	0.9680	13″.138	1.1185	16″.090	1.2066	17″.947	1.2540	60
1	0.081	8.9089	4.887	0.6891	9.361	0.9713	13.195	1.1204	16.130	1.2077	17.968	1.2545	59
2	0.162	9.2099	4.965	0.6959	9.431	0.9745	13.252	1.1223	16.170	1.2087	17.989	1.2550	58
3	0.243	9.3860	5.043	0.7027	9.500	0.9777	13.309	1.1241	16.210	1.2098	18.009	1.2555	57
4	0.324	9.5109	5.121	0.7094	9.569	0.9809	13.365	1.1260	16.250	1.2109	18.028	1.2560	56
5	0.405	9.6078	5.199	0.7159	9.639	0.9840	13.422	1.1278	16.289	1.2119	18.048	1.2564	55
6	0.486	9.6870	5.277	0.7224	9.708	0.9871	13.478	1.1296	16.328	1.2129	18.067	1.2569	54
7	0.567	9.7539	5.355	0.7287	9.777	0.9902	13.534	1.1314	16.367	1.2140	18.086	1.2573	53
8	0.648	9.8119	5.432	0.7350	9.846	0.9933	13.589	1.1332	16.405	1.2150	18.104	1.2578	52
9	0.729	9.8630	5.510	0.7411	9.915	0.9963	13.645	1.1349	16.443	1.2160	18.122	1.2582	51
10	0.810	9.9087	5.588	0.7472	9.984	0.9993	13.700	1.1367	16.480	1.2170	18.140	1.2586	50
11	0.891	9.9501	5.665	0.7532	10.052	1.0022	13.754	1.1384	16.518	1.2180	18.157	1.2590	49
12	0.972	9.9879	5.742	0.7590	10.119	1.0052	13.808	1.1401	16.555	1.2189	18.174	1.2595	48
13	1.053	0.0226	5.820	0.7648	10.187	1.0081	13.852	1.1418	16.592	1.2199	18.191	1.2599	47
14	1.134	0.0547	5.897	0.7705	10.254	1.0109	13.916	1.1435	16.628	1.2208	18.207	1.2602	46
15	1.215	0.0846	5.974	0.7762	10.322	1.0138	13.969	1.1452	16.664	1.2218	18.223	1.2606	45
16	1.296	0.1126	6.050	0.7817	10.389	1.0166	14.022	1.1468	16.700	1.2227	18.238	1.2610	44
17	1.377	0.1389	6.127	0.7872	10.457	1.0194	14.075	1.1485	16.735	1.2236	18.253	1.2614	43
18	1.458	0.1637	6.203	0.7925	10.524	1.0222	14.128	1.1501	16.770	1.2245	18.268	1.2617	42
19	1.539	0.1871	6.279	0.7979	10.591	1.0249	14.181	1.1517	16.805	1.2254	18.283	1.2621	41
20	1.619	0.2093	6.355	0.8031	10.657	1.0276	14.233	1.1533	16.839	1.2263	18.298	1.2624	40
21	1.700	0.2305	6.431	0.8083	10.723	1.0303	14.286	1.1549	16.873	1.2272	18.312	1.2627	39
22	1.781	0.2506	6.507	0.8134	10.789	1.0330	14.337	1.1565	16.907	1.2281	18.325	1.2631	38
23	1.862	0.2699	6.583	0.8184	10.855	1.0356	14.388	1.1580	16.940	1.2289	18.339	1.2634	37
24	1.942	0.2883	6.658	0.8234	10.921	1.0383	14.439	1.1596	16.973	1.2298	18.352	1.2637	36
25	2.023	0.3059	6.734	0.8283	10.987	1.0409	14.490	1.1611	17.006	1.2306	18.364	1.2640	35
26	2.104	0.3229	6.810	0.8331	11.052	1.0434	14.541	1.1626	17.039	1.2315	18.376	1.2643	34
27	2.184	0.3392	6.885	0.8379	11.117	1.0460	14.591	1.1641	17.071	1.2323	18.388	1.2645	33
28	2.264	0.3549	6.960	0.8426	11.182	1.0485	14.641	1.1656	17.103	1.2331	18.399	1.2648	32
29	2.345	0.3701	7.035	0.8473	11.247	1.0510	14.691	1.1670	17.135	1.2339	18.410	1.2651	31
30	2.426	0.3847	7.110	0.8519	11.311	1.0535	14.740	1.1685	17.166	1.2347	18.420	1.2653	30
31	2.506	0.3989	7.185	0.8564	11.375	1.0560	14.790	1.1700	17.197	1.2354	18.431	1.2656	29
32	2.586	0.4126	7.260	0.8609	11.439	1.0584	14.839	1.1714	17.227	1.2362	18.441	1.2658	28
33	2.667	0.4259	7.335	0.8654	11.503	1.0608	14.888	1.1728	17.257	1.2370	18.450	1.2660	27
34	2.747	0.4388	7.410	0.8698	11.567	1.0632	14.936	1.1742	17.287	1.2377	18.459	1.2663	26
35	2.827	0.4512	7.484	0.8741	11.630	1.0656	14.984	1.1756	17.317	1.2385	18.468	1.2665	25
36	2.907	0.4634	7.557	0.8784	11.693	1.0679	15.032	1.1770	17.346	1.2392	18.477	1.2667	24
37	2.987	0.4752	7.631	0.8826	11.756	1.0703	15.080	1.1784	17.375	1.2399	18.486	1.2669	23
38	3.067	0.4867	7.705	0.8868	11.819	1.0726	15.127	1.1797	17.403	1.2406	18.494	1.2671	22
39	3.147	0.4978	7.779	0.8909	11.881	1.0748	15.174	1.1811	17.432	1.2413	18.502	1.2672	21
40	3.226	0.5087	7.852	0.8950	11.943	1.0771	15.220	1.1824	17.460	1.2420	18.509	1.2674	20
41	3.306	0.5193	7.926	0.8990	12.005	1.0794	15.267	1.1837	17.487	1.2427	18.516	1.2676	19
42	3.386	0.5297	7.999	0.9030	12.066	1.0816	15.313	1.1850	17.514	1.2434	18.523	1.2677	18
43	3.465	0.5398	8.073	0.9070	12.128	1.0838	15.359	1.1863	17.541	1.2441	18.529	1.2679	17
44	3.545	0.5496	8.146	0.9109	12.189	1.0860	15.404	1.1876	17.568	1.2447	18.535	1.2680	16
45	3.624	0.5593	8.218	0.9148	12.251	1.0882	15.449	1.1889	17.594	1.2454	18.540	1.2681	15
46	3.704	0.5687	8.291	0.9186	12.312	1.0903	15.494	1.1902	17.620	1.2460	18.545	1.2682	14
47	3.783	0.5779	8.363	0.9224	12.372	1.0924	15.538	1.1914	17.646	1.2466	18.550	1.2684	13
48	3.863	0.5869	8.435	0.9261	12.432	1.0946	15.582	1.1926	17.671	1.2473	18.554	1.2685	12
49	3.942	0.5958	8.507	0.9298	12.492	1.0967	15.626	1.1939	17.696	1.2479	18.558	1.2686	11
50	4.022	0.6044	8.579	0.9335	12.552	1.0987	15.669	1.1951	17.720	1.2485	18.562	1.2686	10
51	4.101	0.6128	8.651	0.9371	12.611	1.1008	15.713	1.1963	17.744	1.2491	18.566	1.2687	9
52	4.181	0.6211	8.723	0.9407	12.671	1.1028	15.756	1.1975	17.768	1.2496	18.569	1.2688	8
53	4.260	0.6294	8.795	0.9442	12.730	1.1049	15.799	1.1987	17.792	1.2502	18.572	1.2688	7
54	4.339	0.6372	8.867	0.9477	12.789	1.1069	15.842	1.1998	17.815	1.2508	18.574	1.2689	6
55	4.417	0.6451	8.938	0.9512	12.848	1.1089	15.884	1.2010	17.838	1.2514	18.576	1.2689	5
56	4.495	0.6527	9.008	0.9546	12.906	1.1108	15.926	1.2021	17.860	1.2519	18.578	1.2690	4
57	4.574	0.6603	9.079	0.9580	12.965	1.1128	15.968	1.2032	17.883	1.2524	18.579	1.2690	3
58	4.653	0.6677	9.149	0.9614	13.023	1.1147	16.010	1.2044	17.905	1.2530	18.579	1.2690	2
59	4.731	0.6749	9.220	0.9647	13.081	1.1166	16.050	1.2055	17.926	1.2535	18.580	1.2691	1
60	4.809	0.6820	9.290	0.9680	13.138	1.1185	16.090	1.2066	17.947	1.2540	18.580	1.2691	0
	Nat. No.	Log.	Nat. No.	Log.	Nat. No.	Log.	Nat. No.	Log.	Nat. No.	Log.	Nat. No.	Log.	
	+ XI	− XXIII	+ X	− XXII	+ IX	− XXI	+ VIII	− XX	+ VII	− XIX	+ VI	− XVIII	

(Argument = Dec. of ✱.) TABLE 8. (Tab. No. = + 8″.0659 cos dec.)

NORTH — SOUTH +

Dec.	Nat. No.	Log.	Dec.	Nat. No.	Log.	Dec.	Nat. No.	Log.	Dec.	Nat. No.	Log.	Dec.	Nat. No.	Log.
0°	8″.066	0 .9067	19°	7″.626	0 .8823	37°	6″.442	0 .8090	55°	4″.626	0 .6652	73°	2″.358	0 .3725
1	8 .065	0 .9066	20	7 .579	0 .8796	38	6 .356	0 .8032	56	4 .510	0 .6542	74	2 .223	0 .3469
2	8 .061	0 .9064	21	7 .530	0 .8768	39	6 .268	0 .7971	57	4 .393	0 .6428	75	2 .088	0 .3197
3	8 .055	0 .9061	22	7 .479	0 .8738	40	6 .179	0 .7909	58	4 .274	0 .6383	76	1 .951	0 .2903
4	8 .046	0 .9056	23	7 .425	0 .8707	41	6 .087	0 .7844	59	4 .154	0 .6185	77	1 .814	0 .2586
5	8 .035	0 .9050	24	7 .369	0 .8674	42	5 .994	0 .7777	60	4 .033	0 .6056	78	1 .677	0 .2245
6	8 .021	0 .9042	25	7 .311	0 .8640	43	5 .900	0 .7709	61	3 .910	0 .5922	79	1 .539	0 .1872
7	8 .005	0 .9034	26	7 .250	0 .8603	44	5 .803	0 .7637	62	3 .787	0 ·5783	80	1 .401	0 .1464
8	7 .987	0 .9024	27	7 .187	0 .8565	45	5 .703	0 .7561	63	3 .662	0 .5637	81	1 .262	0 .1011
9	7 .967	0 .9013	28	7 .122	0 .8526	46	5 .603	0 .7484	64	3 .536	0 .5485	82	1 .123	0 .0504
10	7 .944	0 .9000	29	7 .055	0 .8485	47	5 .501	0 .7404	65	3 .409	0 .5326	83	0 .983	9 .9926
11	7 .918	0 .8986	30	6 .985	0 .8442	48	5 .397	0 .7322	66	3 .281	0 .5160	84	0 .843	9 .9258
12	7 .890	0 .8971	31	6 .913	0 .8397	49	5 .292	0 .7236	67	3 .152	0 .4986	85	0 .703	9 .8470
13	7 .859	0 .8954	32	6 .840	0 .8351	50	5 .185	0 .7147	68	3 .022	0 .4803	86	0 .563	9 .7505
14	7 .826	0 .8935	33	6 .765	0 .8303	51	5 .076	0 .7055	69	2 .891	0 .4610	87	0 .423	9 .6263
15	7 .791	0 .8916	34	6 .687	0 .8252	52	4 .966	0 .6960	70	2 .759	0 .4408	88	0 .282	9 .4502
16	7 .753	0 .8895	35	6 .607	0 .8200	53	4 .854	0 .6861	71	2 .626	0 .4193	89	0 .141	9 .1492
17	7 .713	0 .8872	36	6 .526	0 .8146	54	4 .741	0 .6759	72	2 .493	0 .3967	90	0 .000	0 .0000
18	7 .671	0 .8849												

WHEN Mr. Fallows transmitted to the Admiralty his approximate Catalogue of Stars, he sent also a copy of the ten Tables by which his observations were reduced; consisting entirely of natural numbers, and arranged so as to admit of the numbers being taken in pairs to be multiplied by each other, in a manner similar to Bessel's, and subsequently by a natural sine, tangent, or secant, as the case may be, in order to convert the tabular products into the requisite corrections. Of these Tables, five are nearly the same as our Tables I, V, VIII, IX, X, in Mr. Fallows's first series for obtaining ZACH'S CONSTANTS, but as they have not the same signs, we have given the author's second series entire, for finding at once the amount of the corrections in precession, aberration, and nutation, both lunar and solar, for any given observation, or number of observations, of the same star, both in right ascension and in declination; and have also shewn how the tabular quantities may be converted into *constants* for any given star, or number of stars.

The formulæ from which Tables of this kind are usually derived, are so ingeniously separated into detached parts by Mr. Fallows, that the right ascension of the star in time is the common argument of eight out of the ten Tables; but the tenth requires to be annually computed, as depending partly on the sun's longitude, or 2 ☉, and partly on the moon's ascending node, or 2 ☊. It would occupy too much space in this stage of our work, to copy the author's analysis and transposition of the formulæ, which we understand he contrived before he had seen Bessel's new method of obtaining the corrections by a similar device; we will therefore give such a synopsis of the Tables, and of their tabular values, as we trust will enable the reader, who has examined the construction and use of Bessel's Constants, before explained, to understand the nature and application of their contents. And that the application of the quantities may assimilate with Bessel's, we propose to adopt his nomenclature; for though the same letters do not denote the same values, according to both authors, yet the aggregate of the results and mode of applying them to the mean, or observed places of the stars, will be the same; particularly as we have added columns of logarithms, corresponding to the natural numbers, computed by Mr. Fallows, for the use of those computers who prefer the logarithmic numbers.

SYNOPSIS.

Constants.

TAB. 1. = − 1ˢ.3362 sin R. A. tang dec + 3ˢ.0678 = precession in R. A. .... = a
2. = − 1 .350 sin R. A. = p, and p. sec dec.. = b
3. = − 1 .239 cos R. A. = q, and q. sec dec. .... = c
4. = + 0 .643 cos R. A. = s, and s. tang dec. .... = d
5. = − 20″.0436 cos R. A. = the annual precession .... = a'
6. = − 20 .255 cos R. A. = p', and p' sin dec. .... = b'
7. = + 18 .580 sin R. A. = q', and (q' sin dec) + r' .... = c'
8. = + 8 .0659 cos dec = r'
9. = − 9 .648 sin R. A. = s' .... = d'
10. $t - \frac{\sin ☊}{3} - \frac{\sin 2 ☉}{40}$ .... = A

$0.93046\ (\cos ☊ - \cos 2\ ☊ + \frac{2}{31} \cos 2\ ☉)$ .... = D

Tab. of natural or logarithmic sines, &c. (sin ☉) for the time of culmination............... = B
Ditto .... (cos ☉) ditto .... = C

Then the whole correction in R. A. = A a + B b + C c + D d.
And in declination .... = A a' + B b' + C c' + D d'.

A SECOND SERIES OF TABLES BY THE REV. F. FALLOWS, M.A.

(Argument = R. A. ✱) TABLE 9. (Tab. No. = −9″.648 sin. R. A.)

S. N. Min.	− O +	+ XII −	− I +	+ XIII −	− II +	+ XIV −	− III +	+ XV −	− IV +	+ XVI −	− V +	+ XVII −	S. N. Min.
	Nat. No.	Log.	Nat. No.	Log.	Nat. No.	Log.	Nat. No.	Log.	Nat. No.	Log.	Nat. No.	Log.	
0	0″.000	− ∞	2″.497	0.3974	4″.824	0.6834	6″.823	0.8339	8″.355	0.9220	9″.319	0.9694	60
1	0.042	8.6243	2.538	0.4044	4.861	0.6867	6.853	0.8358	8.376	0.9231	9.330	0.9699	59
2	0.085	8.9253	2.578	0.4113	4.897	0.6899	6.883	0.8377	8.397	0.9241	9.341	0.9704	58
3	0.127	9.1014	2.619	0.4181	4.933	0.6931	6.912	0.8395	8.418	0.9250	9.351	0.9709	57
4	0.169	9.2263	2.659	0.4248	4.969	0.6963	6.941	0.8414	8.438	0.9263	9.361	0.9713	56
5	0.211	9.3232	2.700	0.4313	5.005	0.6994	6.970	0.8432	8.459	0.9273	9.371	0.9718	55
6	0.253	9.4024	2.740	0.4378	5.041	0.7025	6.999	0.8450	8.479	0.9283	9.381	0.9723	54
7	0.295	9.4693	2.781	0.4442	5.077	0.7056	7.028	0.8468	8.499	0.9294	9.391	0.9727	53
8	0.337	9.5273	2.821	0.4504	5.113	0.7087	7.056	0.8486	8.519	0.9304	9.401	0.9732	52
9	0.379	9.5784	2.861	0.4565	5.149	0.7117	7.085	0.8503	8.539	0.9314	9.410	0.9736	51
10	0.421	9.6241	2.901	0.4626	5.185	0.7147	7.113	0.8521	8.558	0.9324	9.419	0.9740	50
11	0.463	9.6655	2.941	0.4685	5.220	0.7176	7.141	0.8538	8.577	0.9334	9.428	0.9744	49
12	0.505	9.7032	2.981	0.4744	5.255	0.7205	7.169	0.8555	8.596	0.9343	9.437	0.9748	48
13	0.547	9.7380	3.021	0.4802	5.290	0.7235	7.197	0.8572	8.615	0.9353	9.446	0.9752	47
14	0.589	9.7701	3.061	0.4859	5.325	0.7263	7.225	0.8589	8.634	0.9362	9.454	0.9756	46
15	0.631	9.8000	3.101	0.4915	5.360	0.7292	7.253	0.8606	8.653	0.9372	9.462	0.9760	45
16	0.673	9.8280	3.141	0.4971	5.395	0.7320	7.281	0.8622	8.671	0.9381	9.470	0.9764	44
17	0.715	9.8543	3.181	0.5025	5.430	0.7348	7.309	0.8639	8.690	0.9390	9.478	0.9767	43
18	0.757	9.8791	3.221	0.5079	5.465	0.7376	7.337	0.8655	8.708	0.9399	9.486	0.9771	42
19	0.799	9.9025	3.261	0.5133	5.500	0.7403	7.364	0.8671	8.726	0.9408	9.494	0.9775	41
20	0.841	9.9247	3.300	0.5185	5.534	0.7430	7.391	0.8687	8.744	0.9417	9.501	0.9778	40
21	0.883	9.9459	3.340	0.5237	5.569	0.7457	7.418	0.8703	8.762	0.9426	9.508	0.9781	39
22	0.925	9.9660	3.380	0.5288	5.603	0.7484	7.445	0.8718	8.780	0.9435	9.515	0.9784	38
23	0.967	9.9853	3.419	0.5338	5.637	0.7510	7.472	0.8734	8.797	0.9443	9.522	0.9788	37
24	1.009	0.0037	3.458	0.5388	5.671	0.7537	7.498	0.8749	8.814	0.9452	9.529	0.9791	36
25	1.051	0.0213	3.497	0.5437	5.705	0.7563	7.525	0.8765	8.831	0.9460	9.536	0.9794	35
26	1.093	0.0383	3.536	0.5485	5.739	0.7588	7.551	0.8780	8.848	0.9468	9.542	0.9796	34
27	1.135	0.0546	3.575	0.5533	5.773	0.7614	7.577	0.8795	8.865	0.9477	9.548	0.9799	33
28	1.176	0.0703	3.614	0.5580	5.806	0.7639	7.603	0.8810	8.881	0.9485	9.554	0.9802	32
29	1.218	0.0855	3.653	0.5627	5.840	0.7654	7.629	0.8824	8.898	0.9493	9.560	0.9805	31
30	1.260	0.1001	3.692	0.5673	5.874	0.7689	7.655	0.8839	8.914	0.9501	9.566	0.9807	30
31	1.301	0.1143	3.731	0.5718	5.907	0.7713	7.680	0.8854	8.930	0.9508	9.571	0.9810	29
32	1.343	0.1280	3.770	0.5763	5.940	0.7738	7.705	0.8868	8.946	0.9516	9.576	0.9812	28
33	1.385	0.1413	3.809	0.5807	5.973	0.7762	7.730	0.8882	8.962	0.9524	9.581	0.9814	27
34	1.426	0.1541	3.847	0.5851	6.006	0.7786	7.755	0.8896	8.977	0.9531	9.586	0.9816	26
35	1.468	0.1666	3.886	0.5895	6.039	0.7810	7.780	0.8910	8.992	0.9539	9.591	0.9819	25
36	1.509	0.1788	3.924	0.5938	6.072	0.7833	7.805	0.8924	9.007	0.9546	9.595	0.9821	24
37	1.551	0.1906	3.963	0.5980	6.105	0.7856	7.830	0.8938	9.022	0.9553	9.599	0.9822	23
38	1.592	0.2020	4.001	0.6022	6.138	0.7880	7.855	0.8951	9.037	0.9560	9.603	0.9824	22
39	1.634	0.2132	4.039	0.6063	6.170	0.7902	7.879	0.8965	9.052	0.9567	9.607	0.9826	21
40	1.675	0.2241	4.077	0.6104	6.202	0.7925	7.903	0.8978	9.066	0.9574	9.611	0.9828	20
41	1.717	0.2347	4.115	0.6144	6.234	0.7948	7.927	0.8991	9.081	0.9581	9.615	0.9829	19
42	1.758	0.2451	4.153	0.6184	6.266	0.7970	7.951	0.9004	9.095	0.9588	9.619	0.9831	18
43	1.800	0.2552	4.191	0.6224	6.298	0.7992	7.975	0.9017	9.109	0.9595	9.622	0.9832	17
44	1.841	0.2650	4.229	0.6263	6.330	0.8014	7.998	0.9030	9.123	0.9601	9.625	0.9834	16
45	1.883	0.2747	4.267	0.6301	6.362	0.8036	8.022	0.9043	9.137	0.9608	9.628	0.9835	15
46	1.924	0.2841	4.305	0.6340	6.394	0.8057	8.045	0.9055	9.150	0.9614	9.631	0.9836	14
47	1.965	0.2933	4.343	0.6377	6.425	0.8078	8.068	0.9068	9.163	0.9620	9.633	0.9837	13
48	2.006	0.3023	4.380	0.6415	6.456	0.8100	8.091	0.9080	9.176	0.9626	9.635	0.9838	12
49	2.047	0.3111	4.418	0.6452	6.487	0.8120	8.114	0.9093	9.189	0.9633	9.637	0.9839	11
50	2.088	0.3198	4.455	0.6488	6.518	0.8141	8.137	0.9105	9.202	0.9639	9.639	0.9840	10
51	2.129	0.3282	4.492	0.6525	6.549	0.8162	8.160	0.9117	9.214	0.9645	9.641	0.9841	9
52	2.170	0.3365	4.529	0.6560	6.580	0.8182	8.182	0.9129	9.226	0.9650	9.642	0.9842	8
53	2.211	0.3447	4.566	0.6596	6.611	0.8202	8.204	0.9140	9.238	0.9656	9.643	0.9842	7
54	2.252	0.3526	4.603	0.6631	6.642	0.8223	8.226	0.9152	9.250	0.9662	9.644	0.9843	6
55	2.293	0.3604	4.640	0.6666	6.672	0.8242	8.248	0.9164	9.262	0.9667	9.645	0.9843	5
56	2.334	0.3681	4.677	0.6700	6.702	0.8262	8.270	0.9175	9.274	0.9673	9.646	0.9844	4
57	2.375	0.3756	4.714	0.6734	6.733	0.8282	8.292	0.9186	9.286	0.9678	9.647	0.9844	3
58	2.416	0.3830	4.751	0.6768	6.763	0.8301	8.313	0.9198	9.297	0.9684	9.647	0.9844	2
59	2.457	0.3903	4.788	0.6801	6.793	0.8320	8.334	0.9209	9.308	0.9689	9.648	0.9844	1
60	2.497	0.3974	4.824	0.6834	6.823	0.8339	8.355	0.9220	9.319	0.9694	9.648	0.9844	0
	Nat. No.	Log.	Nat. No.	Log.	Nat. No.	Log.	Nat. No.	Log.	Nat. No.	Log.	Nat. No.	Log.	
N. S.	+ XI −	− XXIII +	+ X −	− XXII +	+ IX −	− XXI +	+ VIII −	− XX +	+ VII −	− XIX +	+ VI −	− XVIII +	N. S.

TABLE 10. (Arg. = day of the month.) $A = t - \frac{\sin \Omega}{3} - \frac{\sin 2 \odot}{40}$. $D = 0.93046 (\cos \Omega - \cos 2 \Omega + \frac{2}{31} \cos 2 \odot)$.

1824.	A	Log. A	D	Log. D	1824.	A	Log. A	D	Log. D	1824.	A	Log. A	D	Log. D
	+	+	+	+		+	+	+	+		+	+	+	+
Jan. 1	.321	9.5065	.255	9.4065	Mar. 1	.498	9.6972	.311	9.4928	May 1	.629	9.7987	.221	9.3444
2	.324	.5105	.255	.4067	2	.500	.6990	.311	.4932	2	.632	.8007	.218	.3385
3	.327	.5145	.255	.4069	3	.502	.7007	.312	.4938	3	.635	.8028	.216	.3345
4	.331	.5198	.255	.4073	4	.504	.7024	.312	.4942	4	.637	.8041	.213	.3284
5	.334	.5237	.256	.4082	5	.506	.7042	.312	.4946	5	.640	.8062	.210	.3222
6	.338	.5289	.256	.4084	6	.508	.7059	.313	.4950	6	.643	.8082	.207	.3160
7	.342	.5340	.256	.4088	7	.511	.7084	.313	.4955	7	.646	.8102	.204	.3096
8	.345	.5378	.257	.4095	8	.513	.7101	.313	.4955	8	.649	.8122	.201	.3032
9	.348	.5416	.257	.4099	9	.515	.7118	.313	.4950	9	.652	.8142	.198	.2967
10	.352	.5465	.258	.4110	10	.517	.7135	.312	.4946	10	.655	.8162	.195	.2900
11	.356	.5515	.258	.4116	11	.519	.7152	.312	.4942	11	.658	.8182	.192	.2833
12	.360	.5563	.259	.4124	12	.521	.7168	.312	.4938	12	.661	.8202	.189	.2765
13	.364	.5611	.259	.4133	13	.523	.7185	.312	.4934	13	.664	.8222	.186	.2695
14	.367	.5647	.260	.4150	14	.524	.7193	.311	.4928	14	.667	.8241	.184	.2648
15	.370	.5682	.261	.4166	15	.526	.7210	.311	.4922	15	.670	.8261	.181	.2577
16	.374	.5729	.261	.4174	16	.528	.7226	.310	.4914	16	.673	.8280	.178	.2504
17	.377	.5763	.262	.4183	17	.530	.7243	.309	.4900	17	.676	.8299	.175	.2430
18	.381	.5809	.263	.4200	18	.532	.7259	.308	.4886	18	.679	.8319	.172	.2355
19	.385	.5855	.264	.4216	19	.534	.7275	.308	.4882	19	.682	.8338	.169	.2279
20	.388	.5888	.265	.4232	20	.536	.7292	.307	.4871	20	.686	.8363	.167	.2227
21	.391	.5899	.266	.4249	21	.538	.7308	.306	.4857	21	.689	.8382	.165	.2175
22	.394	.5955	.267	.4265	22	.540	.7324	.306	.4850	22	.693	.8407	.162	.2095
23	.397	.5988	.268	.4281	23	.542	.7340	.305	.4843	23	.695	.8420	.160	.2041
24	.400	.6021	.270	.4314	24	.544	.7356	.304	.4829	24	.699	.8445	.158	.1987
25	.404	.6064	.271	.4330	25	.546	.7372	.303	.4814	25	.702	.8463	.156	.1931
26	.407	.6096	.272	.4346	26	.548	.7388	.302	.4800	26	.706	.8488	.153	.1847
27	.410	.6128	.273	.4362	27	.550	.7404	.300	.4771	27	.709	.8506	.151	.1790
28	.413	.6160	.274	.4378	28	.552	.7419	.299	.4757	28	.713	.8531	.148	.1703
29	.416	.6191	.276	.4409	29	.554	.7435	.298	.4742	29	.716	.8549	.146	.1644
30	.419	.6222	.277	.4425	30	.556	.7451	.296	.4713	30	.719	.8567	.143	.1553
31	.422	.6253	.278	.4434	31	.558	.7466	.294	.4683	31	.723	.8591	.141	.1492
Feb. 1	.425	.6284	.279	.4456	Apr. 1	.560	.7482	.292	.4654	June 1	.726	.8609	.139	.1430
2	.428	.6314	.281	.4487	2	.562	.7497	.291	.4639	2	.730	.8633	.137	.1367
3	.431	.6345	.282	.4502	3	.564	.7513	.289	.4609	3	.733	.8651	.135	.1303
4	.434	.6375	.284	.4533	4	.566	.7528	.287	.4579	4	.737	.8675	.133	.1239
5	.437	.6405	.285	.4548	5	.569	.7551	.286	.4564	5	.740	.8692	.131	.1173
6	.440	.6435	.287	.4579	6	.571	.7566	.284	.4533	6	.744	.8716	.129	.1106
7	.443	.6464	.288	.4594	7	.573	.7582	.282	.4502	7	.747	.8733	.128	.1072
8	.445	.6484	.289	.4609	8	.575	.7600	.280	.4472	8	.751	.8756	.126	.1004
9	.448	.6513	.290	.4624	9	.577	.7612	.278	.4440	9	.754	.8774	.125	.0969
10	.454	.6542	.292	.4654	10	.579	.7627	.275	.4393	10	.758	.8797	.123	.0899
11	.453	.6561	.293	.4669	11	.581	.7642	.273	.4362	11	.761	.8814	.121	.0828
12	.456	.6590	.294	.4683	12	.583	.7657	.270	.4314	12	.765	.8837	.120	.0792
13	.458	.6609	.295	.4698	13	.585	.7672	.268	.4281	13	.768	.8854	.118	.0719
14	.461	.6637	.296	.4713	14	.588	.7694	.265	.4232	14	.772	.8876	.117	.0682
15	.463	.6656	.297	.4728	15	.590	.7709	.262	.4183	15	.775	.8893	.116	.0645
16	.466	.6684	.298	.4742	16	.592	.7723	.260	.4150	16	.779	.8915	.115	.0607
17	.468	.6702	.299	.4757	17	.595	.7745	.258	.4116	17	.783	.8938	.114	.0569
18	.471	.6730	.300	.4771	18	.597	.7760	.255	.4065	18	.786	.8954	.113	.0531
19	.474	.6758	.301	.4786	19	.599	.7774	.253	.4031	19	.790	.8976	.112	.0492
20	.476	.6776	.302	.4800	20	.602	.7796	.250	.3979	20	.793	.8993	.111	.0453
21	.478	.6794	.303	.4814	21	.604	.7810	.248	.3945	21	.797	.9015	.110	.0414
22	.480	.6812	.304	.4829	22	.607	.7832	.245	.3892	22	.801	.9036	.109	.0374
23	.483	.6839	.305	.4843	23	.609	.7846	.243	.3856	23	.804	.9053	.108	.0334
24	.485	.6857	.306	.4857	24	.612	.7868	.240	.3802	24	.808	.9074	.107	.0294
25	.487	.6875	.307	.4871	25	.614	.7882	.238	.3766	25	.811	.9090	.106	.0253
26	.489	.6893	.308	.4886	26	.617	.7903	.235	.3711	26	.815	.9112	.106	.0233
27	.491	.6911	.309	.4900	27	.619	.7917	.232	.3655	27	.818	.9128	.105	.0212
28	.493	.6928	.309	.4907	28	.622	.7938	.230	.3617	28	.822	.9149	.105	.0191
29	.495	.6946	.310	.4914	29	.624	.7952	.227	.3560	29	.826	.9170	.104	.0180
					30	.627	.7973	.224	.3502	30	.829	.9186	.104	.0170

TABLE 10. (Arg. = day of the month.) Col. A. $= t - \frac{\sin \Omega}{3} - \frac{\sin 2 \odot}{40}$. Col. D. $= 0.93046 \left(\cos \Omega - \cos 2\Omega + \frac{2}{31} \cos. 2 \odot\right)$.

	A	Log. A	D	Log. D
	+	+	+	+
July 1	.833	9.9206	.104	9.0170
2	.837	.9227	.104	.0176
3	.840	.9243	.104	.0182
4	.844	.9263	.104	.0188
5	.847	.9279	.104	.0194
6	.851	.9299	.104	.0200
7	.854	.9315	.104	.0260
8	.858	.9335	.105	.0212
9	.861	.9350	.105	.0232
10	.865	.9370	.106	.0253
11	.868	.9385	.106	.0269
12	.872	.9405	.107	.0285
13	.875	.9420	.107	.0301
14	.878	.9435	.107	.0317
15	.882	.9455	.108	.0334
16	.885	.9469	.108	.0334
17	.889	.9489	.109	.0374
18	.892	.9504	.109	.0394
19	.895	.9518	.110	.0414
20	.899	.9538	.111	.0453
21	.902	.9552	.112	.0492
22	.906	.9571	.113	.0531
23	.909	.9586	.114	.0569
24	.912	.9600	.115	.0607
25	.915	.9614	.116	.0645
26	.918	.9628	.117	.0682
27	.921	.9643	.119	.0755
28	.924	.9657	.120	.0792
29	.927	.9671	.121	.0828
30	.930	.9685	.122	.0864
31	.933	.9699	.123	.0899
Aug. 1	.936	.9713	.124	.0934
2	.939	.9727	.125	.0969
3	.942	.9741	.126	.1004
4	.945	.9754	.127	.1038
5	.948	.9768	.129	.1106
6	.950	.9777	.130	.1139
7	.953	.9791	.131	.1173
8	.956	.9805	.132	.1206
9	.958	.9814	.133	.1239
10	.961	.9827	.134	.1271
11	.964	.9841	.135	.1303
12	.966	.9850	.136	.1335
13	.969	.9863	.137	.1367
14	.971	.9872	.138	.1399
15	.974	.9886	.139	.1430
16	.976	.9894	.140	.1461
17	.979	.9908	.141	.1492
18	.981	.9917	.142	.1523
19	.984	.9930	.143	.1553
20	.986	.9939	.144	.1584
21	.989	.9952	.145	.1614
22	.991	.9961	.146	.1644
23	.994	.9974	.147	.1673
24	.996	.9983	.148	.1703
25	.999	.9996	.149	.1732
26	1.001	0.0004	.150	.1761
27	1.003	0.0013	.150	.1775
28	1.005	0.0022	.151	.1790
29	1.008	0.0035	.152	.1818
30	1.010	0.0043	.153	.1847
31	1.012	0.0052	.154	.1875

	A	Log. A	D	Log. D
	+	+	+	+
Sept. 1	1.014	0.0060	.154	9.1875
2	1.016	.0069	.154	.1882
3	1.019	.0082	.154	.1889
4	1.021	.0090	.154	.1896
5	1.023	.0099	.155	.1903
6	1.025	.0107	.155	.1908
7	1.027	.0116	.155	.1914
8	1.029	.0124	.155	.1919
9	1.031	.0133	.155	.1925
10	1.033	.0141	.156	.1931
11	1.035	.0149	.156	.1925
12	1.037	.0158	.156	.1919
13	1.039	.0166	.156	.1914
14	1.041	.0175	.156	.1908
15	1.043	.0183	.155	.1903
16	1.045	.0191	.155	.1889
17	1.047	.0199	.154	.1875
18	1.048	.0204	.154	.1861
19	1.050	.0212	.153	.1847
20	1.052	.0220	.153	.1832
21	1.054	.0228	.152	.1818
22	1.055	.0233	.151	.1790
23	1.057	.0241	.150	.1761
24	1.059	.0249	.149	.1732
25	1.061	.0257	.148	.1703
26	1.063	.0265	.147	.1673
27	1.064	.0269	.146	.1644
28	1.066	.0278	.145	.1614
29	1.068	.0286	.143	.1553
30	1.070	.0294	.142	.1523
Oct. 1	1.072	.0302	.141	.1492
2	1.074	.0310	.140	.1461
3	1.075	.0314	.138	.1399
4	1.077	.0322	.137	.1367
5	1.079	.0330	.135	.1303
6	1.081	.0338	.133	.1239
7	1.083	.0346	.131	.1173
8	1.085	.0354	.129	.1106
9	1.087	.0362	.127	.1038
10	1.090	.0374	.125	.0969
11	1.092	.0382	.123	.0899
12	1.094	.0390	.121	.0828
13	1.096	.0398	.119	.0755
14	1.098	.0406	.117	.0682
15	1.100	.0414	.114	.0569
16	1.103	.0426	.112	.0492
17	1.105	.0434	.110	.0414
18	1.107	.0441	.107	.0294
19	1.109	.0449	.105	.0212
20	1.111	.0457	.102	.0086
21	1.114	.0469	.100	9.0000
22	1.116	.0477	.097	8.9868
23	1.118	.0484	.095	8.9777
24	1.121	.0496	.092	8.9638
25	1.123	.0504	.090	8.9542
26	1.126	.0515	.087	8.9395
27	1.128	.0523	.084	8.9243
28	1.131	.0535	.081	8.9085
29	1.133	.0542	.078	8.8921
30	1.136	.0554	.075	8.8751
31	1.138	.0561	.072	8.8573

	A	Log. A	D	Log. D
	+	+	+	+
Nov. 1	1.141	0.0573	.069	8.8388
2	1.144	.0584	.066	.8195
3	1.146	.0592	.063	.7993
4	1.149	.0603	.060	.7782
5	1.152	.0615	.057	.7559
6	1.154	.0622	.054	.7324
7	1.157	.0633	.051	.7076
8	1.160	.0645	.048	.6812
9	1.162	.0652	.044	.6435
10	1.165	.0663	.041	.6128
11	1.168	.0674	.038	.5798
12	1.171	.0686	.035	.5441
13	1.174	.0697	.032	.5052
14	1.177	.0708	.029	.4624
15	1.180	.0719	.027	.4314
16	1.183	.0730	.024	.3802
17	1.186	.0741	.022	.3424
18	1.189	.0752	.019	.2788
19	1.192	.0763	.016	.2041
20	1.195	.0774	.014	.1461
21	1.198	.0785	.011	8.0414
22	1.202	.0799	.009	7.9542
23	1.205	.0810	.006	7.7782
24	1.209	.0824	.004	7.6021
25	1.212	.0835	.001	7.0000
			—	
26	1.216	.0849	.002	7.3010
27	1.219	.0860	.005	7.6990
28	1.222	.0871	.007	7.8451
29	1.226	.0885	.010	8.0000
30	1.229	.0896	.013	.1139
Dec. 1	1.232	.0906	.016	.2041
2	1.236	.0920	.018	.2553
3	1.239	.0931	.020	.3010
4	1.243	.0945	.022	.3424
5	1.246	.0955	.024	.3802
6	1.250	.0969	.026	.4150
7	1.253	.0980	.028	.4472
8	1.257	.0993	.030	.4771
9	1.260	.1004	.032	.5052
10	1.264	.1017	.03	.5315
11	1.267	.1028	.036	.5563
12	1.271	.1041	.038	.5798
13	1.274	.1052	.040	.6021
14	1.278	.1065	.041	.6128
15	1.281	.1075	.043	.6335
16	1.285	.1089	.044	.6435
17	1.288	.1099	.045	.6532
18	1.292	.1113	.046	.6628
19	1.295	.1123	.047	.6721
20	1.299	.1136	.048	.6812
21	1.302	.1146	.049	.6902
22	1.306	.1159	.050	.6990
23	1.309	.1169	.051	.7076
24	1.313	.1183	.052	.7160
25	1.317	.1196	.053	.7243
26	1.320	.1206	.053	.7283
27	1.324	.1219	.054	.7324
28	1.327	.1229	.054	.7364
29	1.331	.1242	.055	.7404
30	1.334	.1252	.055	.7404
31	1.338	.1265	.055	.7404

THE SECOND SERIES of TABLES by Mr. FALLOWS may be used for giving the corrections of a star, either with or without reference to the *constants*, both which methods will best be understood from the work of an example, without formal rules, by attending to the subjoined explanation.

Let it be required to find the sum of the corrections of Fomalhaut, in right ascension and declination, for July 20, 1824, at the time of its passing the meridian of Greenwich, its right ascension being taken at $22^h\ 48^m$, and its declination 30° 33′ south?

For right ascension.

The sun's longitude at Greenwich noon is $\overset{s}{3}$ 27° 37′ 9″, and Table XVI., page 45, gives 35′ to be added for the time of culmination, and the natural sine and cosine of 118° 12′ are .881 and .473 for B and C respectively, to stand at the head of the two first columns of the subjoined *type;* then A and D taken from the ANNUAL TABLE (10) must stand at the head of the third and fourth columns, which four numbers will fill the first horizontal line. Then under A put the star's precession in time, as obtained from Table 1.; under B put the tabular number from Table 2. $= p$; and under C put the tabular number from Table 3 $= q$; next multiply those numbers respectively together, and the sum of their products multiplied by the secant of the declination will give a portion of the corrections called X. Again under D put the number from Table 4. $= s$; the product of the numbers under A will be the portion Y of the correction; also the product of the numbers under D multiplied by the tangent of the declination will be the last portion Z of the correction. Lastly X + Y + Z, when their signs are carefully attended to, will be the amount of the corrections in right ascension in time. The work when performed by both sets of columns, will stand thus, both in natural numbers, and by logarithms.

	By natural numbers.				*By logarithms.*				
	B	C	A	D	B	C	A	D	
Table I. .414+ 590	.881+ .418+	.473− 1 .178−	.901+ 3 .312+	.112+ .616+	9.9449+ 9.6212+	9 .6749− 0 .0712−	9 .9547+ 0 .5201+	9 .0492+ 9 .7896+	
207 37 .244+ 3 .068	.352 9 6	471 82 3	2 .981 3	.061 6 1	9.5661+ 0.368 + 0.557 +	9 .7461+ 0 .557 +	0 .4748+ 2 .984 +	8 .8388+ 9 .7710+ 8 .6098 0 .040 +	log. log.tan. dec. log. nat. num.
Prec. in R. A. = 3 .312+	.367+ .556+	.556+	2 .984+ Tang. dec.	.068+ .590	0.925 + nat. sum.				
Nat. sec. of dec.	923+ 1 .161			034 6	9.9661 log. of do. 0.0649 log. sec. of dec.				
	.923 92 55 1			.040+	0.0310 log. X				
	1 .071+ = X 2 .984+ = Y 0 .040+ = Z				1.074 += X 2.984 += Y 0.040 += Z				
	4 .095+ = sum of corrections.				4ˢ.098 += sum of corrections.				

For declination.

The *type* for giving the corrections in DECLINATION contains five columns which are filled up in the same way as the type for Right Ascension, except that C stands over each of two separate columns. The precession from Table 5 stands under A, the number from Table 6. under B, from Table 7. under the first C, from Table 8. under the second C, and from Table 9. under D, and the products are treated as appears in the subjoined type.

Table VIII. at XXII^h 48^m Precession = 19″.062 −

By natural numbers.

	B	C	C	A	D
	.881 + 19 .264 −	.473 − 5 .742 −	.473 − 6 .945 +	.901 + 19 .062 −	.112 + 2 .981 +
	15 .411 1 .541 .019	2 .297 .402 17	2 .778 .486 21	17 .156 19	0 .298 30 6
	16 .971 − 2 .716 +	2 .716 +	3 .285 −	17 .175 −	0 .334 +
Sin dec. =	14 .255 − .508 +				
	7 .127 .114				
	7 .241 − = X′ 3 .285 − 17 .175 − 0 .334 +				
	27″.366 − = the sum of the corrections.				

By logarithms.

B	C	C	A	D
9 .9449 + 1 .2847 −	9 .6749 − 0 .7591 −	9 .6749 − 0 .8417 +	9 .9547 + 1 .2802 −	9 .0492 + 0 .4744 +
1 .2296 −	0 .4340 +	0 .5166 −	1 .2349 −	9 .5236 +
16 .967 − 2 .717 +	2 .717 +	3 .286 −	17 .176 −	0 .334 +
14 .250 − nat. sum.				
1 .1538 − log. 9 .7061 log. sin dec.				
0 .8599				
7 .242 − = X′ 3 .286 − 17 .176 − 0 .334 +				
27″.370 − = the sum of the corrections.				

When a series of observations are made on a star at any number of equal intervals, the reduction may be obtained by taking a *mean* of all the numbers, standing in the first horizontal line of the type, when taken out for the respective days, as though the observation had been a single one made on the mean day; which is an advantage peculiar to these Tables.

But it will be found generally most convenient to determine the *constants* for any given number of stars that are frequently observed, and to register them as applicable to all observations of the same stars, in the way Schumacher has done with the constants derived from Bessel's formulæ.

Method of computing the constants.

Table 1. gives − 1^s.3362 sin. R. A. × tang. dec. + 3^s.0678 = ann. prec. .... = a
2. .... − 1 .350 sin. R. A. = p; and p sec. dec. .... = b
3. .... − 1 .239 cos. R. A. = q; and q sec. dec. .... = c
4. .... + 0 .643 cos. R. A. = s; and s tang. dec. .... = d
5. .... −20″.0436 cos. R. A. = annual precession .... = a'
6. .... −20 .255 cos. R. A. = p'; and p' sin. dec. .... = b'
7. .... +18 .580 sin. R. A. = q'; and q' (sin. dec.) + r' .... = c'
8. .... + 8 .0659 cos. dec. = r'
9. .... − 9 .648 sin. R. A. = s' .... = d'
10. .... Column A. $= t - \frac{\text{sin. ☽ ☊}}{3} - \frac{\text{sin. 2 ☉}}{40}$ .... = A
Column D. 0 .93046 (cos. ☊ − cos. 2 ☊ + $\frac{2}{31}$ cos. 2 ☉) .... = D
From a Table of sines, tangents, &c. sin. ☉ at culmination .... = B
Do. .... cos. ☉ at do. C

Constants for Fomalhaut.

In Right Ascension. According to the formula .... (a) 3 .312 +
+ .418 × 1 .161 .... (b) .485 +
−1 .178 × 1 .161 .... (c) 1 .367 −
+ .612 × .590 .... (d) 0 .361 +

In Declination.			
	According to the formula......................	(a')	19 .062 −
	− 19″.264 × .508	(b')	9 .786 −
	(− 5 .742 × .508) + 6 .945	(c')	4 .028 +
	+ 2 .981	(d')	2 .981 +
	July 20⅝	(A)	.901 +
	July 20⅝	(D)	.112 +
	Nat. sin. 117° 37′ (☉ July 20) + 35′ (Tab. XVI.) ..	(B)	.881 +
	Nat cos. do. do.	(C)	.473 −

When the *constants* are thus prepared, either in natural numbers or logarithms, the corrections both in right ascension and declination are obtained (like Bessel's) thus;

Cor. in R. A......... $= A a + B b + C c + D d.$
Cor. in Declin........ $= A a' + B b' + C c' + D d'.$

THE EXAMPLE OF FOMALHAUT WORKED BY THE CONSTANTS.

In Right Ascension.

	A a	B b	C c	D d
	.901 + 3 .312 +	.881 + .485 +	.473 − 1 .367 −	.112 + 0 .361 +
Products	2 .984 + .427 + .646 + .040 +	.427 +	.646 +	.040 +
	4 .097 +	sum of corrections.		

In Declination.

	A a'	B b'	C c'	D d'
	.901 + 19 .062 −	.881 + 9 .786 −	.473 − 4 .028 +	.112 + 2 .981 +
Products	17 .175 − 8 .621 − 1 .905 − 0 .334 +	− 8 .621 −	1 .905 −	0 .334 +
	27 .367 −	= sum of corrections.		

By Logarithms.

	A a	B b	C c	D d
	9 .9547 + 0 .5201 +	9 .9449 + 9 .6857 +	9 .6749 − 0 .1358 −	9 .0492 + 9 .5575 +
Log.	0 .4748 +	9 .6306 +	9 .8107 +	8 .6067 +
Nat. No.	2 .984 + .427 + .646 + .040 +	.427 +	.646 +	.040 +
	4″.097	sum as before.		

By Logarithms.

	A a'	B b'	C c'	D d'
	9 .9547 + 3 .2802 −	9 .9449 + 0 .9906 −	9 .6749 − 0 .6051 +	9 .0492 + 0 .4744 +
Log.	3 .2349 −	0 .9355 −	0 .2800 −	9 .5236 +
Nat. No.	17 .175 − 8 .620 − 1 .906 − 0 .334 +	8 .620 −	1 .905 −	0 .334 +
	27″.367 −	= sum as before.		

This second series of Tables by Mr. Fallows was originally composed of natural numbers only, on a supposition that the corrections might be obtained more readily by short multiplication than by logarithms; but as many computers may prefer the logarithmic calculus, we have substituted columns of logarithms for columns of differences, which being small may be dispensed with; thus the Tables are rendered capable of giving the quantities both in logarithms and natural numbers, and we have consequently given examples of both modes of computation. The AUXILIARY or ANNUAL Table 10, given for the year 1824, originally contained four columns for B, C, A, and D, but without any designation, the first two of which gave the nat. sines and cosines of the sun's longitude for the noon of Greenwich on each day; and thus, like the last two, required to be re-computed every successive year: but we propose to take the quantities B and C from any common Table of sines and cosines, either in natural numbers or logarithms, for the sun's longitude as given in the Nautical Almanac, when increased by the quantity of minutes pointed out by Table XVI., page 45, by which means the sun's longitude will correspond with the time of the star's culmination. Thus the two columns which we have designated, after Bessel, B and C, are dispensed with in Table 10, and the columns A and D alone will require to be annually computed.

THE END.

G. Woodfall, Printer,
Angel Court, Skinner Street, London.

APPENDIX TO VOLUME I.

OF

AN INTRODUCTION

TO

PRACTICAL ASTRONOMY.

APPENDIX TO VOLUME I.

OF

AN INTRODUCTION

TO

PRACTICAL ASTRONOMY.

WHEN the Preface to this volume was sent to the press an opinion was expressed, that, " when the deductions of several astronomers concur in pointing out defects in the corrections (then) adopted, the co-efficients themselves may be varied, until it is found that the Tables, arising out of such variations, may constitute a general and permanent code of corrections." The Astronomical Society of London, always alive to every suggestion that has for its object the promotion of their favourite science, has already effected to a certain extent the anticipated desideratum; a general catalogue of 2881 principal stars has been reduced to Jan. 1, 1830, with much care; and such *constants* have been annexed to each star, as greatly abbreviate the computations, in reducing the apparent to the mean places due to individual observations, when any of those stars have been observed. The selection of the co-efficients was judiciously left to the discretion of the president for the time, Mr. Baily, whose introductory paper " On the Construction and Use of some New Tables," &c. has been universally admired, and generally approved. The numbers from which the annual precessions are computed are the same as those given in our Table 1, at page 282, for the year 1830; the constant of aberration is taken at 20".36, according to a mean of the deductions of Messrs. Struve, Bessel, and Brinkley; the constant of lunar nutation, agreeably to Dr. Brinkley's recent determination, is assumed equal 9".25; and instead of Bessel's solar numbers + 0 .3982, − 0 .434, and − 0 .9173, adopted in this volume, he has used + 0".500, − 0".545, and − 1".151, for giving the solar nutation; which numbers are nearly so many means between Bessel's and Mr. Herschel's, used in the construction of his special Tables, published in the first volume of the MEMOIRS; as are also his constants of aberration, and of lunar nutation very nearly. The author has also added a minor correction depending on twice the mean longitude of the moon's node; but whether those adopted will ultimately prove the true numbers, when deductions from future observations shall have been multiplied, time alone can determine. The constant of aberration resulting from 1022 observations taken at Greenwich by Troughton's circle in the years 1825 and 1826, and computed by W. Richardson, one of the assistant observers, is 20".411; and that derived from 1011 observations made with Jones's Circle is 20".454; the mean of which is 20".4325, and also very nearly the mean between Mr. Herschel's 20".5 and Mr. Baily's 20".36. If however it should turn out, that any one of the data from which the constants of the catalogues in question are computed should be in any appreciable degree erroneous, the quantities are so intermixed in the computation of these constants, that many of them must be vitiated by such intimate connexion, and incapable of rectification without a recomputation. The plan and arrangement of the new catalogue are similar to that which we have given under the title of " SCHUMACHER'S CATALOGUE AND CONSTANTS OF 500 STARS ", contained in pages 427—434 of this volume, except that the arguments in the Astronomical Society's catalogue are given in *solar* time.

From the sanction that has been very justly given to the catalogue and annexed constants of the Astronomical Society, it has been inferred, that they must eventually supersede all other works of the same kind, so far as occasional reference is had to the observed places of the stars selected for the ordinary purposes of astronomy; but it must be recollected, that, though the catalogue is denominated GENERAL, it contains only a limited number of stars, with constants appropriated to those individual stars; and therefore does not supply all the wants of practical astronomers, whose attention may be directed to stars not there included. Tables, which are not thus confined in their application, are therefore still necessary to be occasionally referred to, and we are prepared to show, that the Tables on Baron Zach's principle, as modified in our collection, are capable of satisfying all the demands of practical astronomy whatever the maxima of aberration, and of nutation, lunar and solar, may hereafter be fixed upon as the *most correct:* and when a number of constant subsidiary numbers are computed, such an arrangement as we have given in a tabulated form, for 48 principal stars, will be peculiarly adapted to facilitate the formation of the monthly, weekly, or daily apparent places of any number of stars that may be fixed upon, as points of reference in the heavens, or as arbiters of the clock's rate in any observatory; and as the Nautical Almanac now contains the apparent places of 60 such stars for every tenth day of the year, the author has been induced to add this Appendix to his Volume of Tables, explanatory of the *extended use* that may be very conveniently made of them, whenever the maxima of aberration and of lunar and solar nutation are assumed to be *different* from those of Bessel, on which the fourteen consecutive Tables were constructed, from page 20 to 38, both inclusive. The explanations follow one another in the order according to which the Tables are numbered, and when our readers have perused them, they will perceive that, whilst the *proportions* that guide the construction of the respective Tables remain unaltered, the numbers resulting from the use of those Tables, and depending on given maxima, may be made the same that would be derived from any other maxima that may be substituted; for a single *factor*, expressive of the required ratio, will convert one into the other result by the most simple operation: and thus the utility of the Tables is rendered *permanent* under all changes of the co-efficients, for which they were originally computed.

TABLE I. Precession in Right Ascension. (P. 20, 21.)

Let c represent the constant and v the variable part of the precession in right ascension, then the precession of a star in right ascension becomes $= c + v.$ sin. R. A. tang. dec.

The numbers employed in Table I. are those derived from Bessel's formula, and reduced to 1820, as seen in the Table at page 282; where $c = \left(\frac{46''.01753}{15}\right)\ 3^s.0678$ and $v = \left(\frac{20''.04360}{15}\right)\ 1^s.33624$. Since our Table I. contains the part $v.$ sin. R. A. it is evident that, if all the natural numbers be divided by v, the sine of right ascension will remain. Now if c' be the constant and v' the variable part of the precession in right ascension, different from that employed in Table I., this Table may still be employed, provided all the natural numbers be multiplied by a constant quantity: for $\frac{v'.\ \text{sin. R. A. tang. dec.}}{v.\ \text{sin. R. A. tang. dec.}} = \frac{v'}{v}$. Then to reduce Table I. to $v'.$ sin. R. A., there will arise an equation where $v'.$ sin. R. A. $= v.$ sin. R. A. $\frac{v'}{v}$.

By way of example, let Table I. be employed in computing the precession agreeably to the constant and variable parts of precession registered at page 282, for the year 2600. Here for 2600, $c' = \left(\frac{46''.25824}{15}\right)\ 3^s.0839$ and $v = \left(\frac{19''.96794}{15}\right)\ 1^s.33119$.

Variable part to be employed, or v' for 2600 .... $= 1^s.33119$ .... log. 0 .12424
Variable employed in Table, or v for 1820 ..... $= 1\ .33624$ .... log. 0 .12594
Constant factor for variable part, or $\frac{v'}{v}$ = Nat. No. 0 .99610...... log. 9 .99830

Therefore to the logarithmic numbers add 9 .99830, or multiply the natural numbers by .9961; and in either case use the constant $+\ 3^s.0839$ for the year in question.

Example.——In the year 1830, the approximate right ascension of Aldebaran will be $4^h\ 26^m\ 11^s$ and its declination $16^\circ\ 9'\ 37''$ N.; let its annual precession in right ascension be determined, by using the co-efficients of Bessel for the year 2600?

Table I.	log. 0 .08843 +	
Tang. dec..............	+ 9 .46206	
Constant	+ 9 .99830	
	9 .54879	$+\ 0^s.3538$
Constant		+ 3 .0839
Annual precession		+ 3 .4377

Otherwise, by Nat. Numb. thus:

Tab. No.	Tang. dec.	Constant.		
$+\ 1^s.226$	$\times$ .2898	$\times$ .9961 =	..	$+\ 0^s.3539$
Constant				+ 3 .0839
Annual precession..............				+ 3 .4378

TABLE II. Annual Number in Right Ascension. (P. 22.)

The annual number in right ascension is the longitude of the ecliptic, corresponding to the star's right ascension, subtracted from nine signs, to make it a positive argument: therefore this Table will always remain the same, with the exception of the slight alteration caused by the change in the obliquity of the ecliptic. The annual number in right ascension may also be found by first reducing the star's right ascension to the ecliptic, by Table 1 of our Zodiacal series, page 204, and then subtracting the result from nine signs, thus:

Example.——Let the true A of Aldebaran for 1830 be found?

First method.

			s		
R. A.	4^h 26^m	=	0	21°	45′
p. p.	11^s	=			− 3
R. A............	4 26 11	=	0	21	42, which

s added to 6, as directed by the precept, becomes s 6 21° 42′ for the true value of A, as given in page 117.

Second method.

	s			
4^h 26^m 11^s.............. =	2	6°	32′	45″
Reduction by Zod. Tab. 1...	+ 1	45	6	
	2	8	17	51
Taken from..............	9			
A..................... =	6	21	42	9

TABLE III. Logarithm of the Maximum, and the Maximum of Aberration. (P. 23.)

Since A is not dependent on any co-efficient of aberration, this Table may be used for any different co-efficient, by applying a constant factor. Let m represent the co-efficient employed in this Table, and m' the co-efficient to be substituted: then since the values in this Table = $(m \div 15)$. sin. R. A. sec. Tab. A; and as $\frac{(m' \div 15) . \text{sin. R. A. sec. Tab. A}}{(m \div 15) . \text{sin. R. A. sec. Tab. A}} = \frac{m'}{m}$, the required tabular values for the new co-efficient = values in the Table × $\frac{m'}{m}$: therefore if the natural values in this Table be multiplied by the constant factor, or if the logarithmic numbers have the constant log. number added to them, they will thereby be reduced to the same values, as if the new co-efficient had been used in the construction of the Table itself.

Example.——Let there be found a constant factor, which shall reduce the values in Table III. to the values belonging to the co-efficient 20″.5 instead of 20″.255?

Log. of required co-efficient	m' = 20″.5	1 .31175
of the tabular co-efficient	m = 20 .255	1 .30653
The factor, or $\frac{m'}{m}$	1 .0121	0 .00522

Then let the maximum of aberration in right ascension of Aldebaran be found for 1830; agreeably to the principal co-efficient 20″.5?

[Approx. R. A. 4^h 26^m 11^s, and Dec. 16° 9′ 37″.]

To R. A. 4^h 26^m 11^s	log. 0 .12493
Log. sec. dec. 16° 9′ 37″	10 .01751
Constant............................	0 .00522
Max. Ab. = 1^s.405...................	0 .14766

By Natural Numbers, thus:

Tab. No.		Nat. sec.		Constant.		
1^s.333	×	1 .0412	×	1 .0121	=	1^s.404 max. ab.

By Mr. Herschel's Table II. = 1 .4049

TABLES IV. and V. Argument L and Maximum of Lunar Nutation in Right Ascension. (P. 24—27.)

Agreeably to the method of notation given at page 311 of this volume, the

$$\text{Lunar nutation in R. A.} = \left.\begin{matrix} -\ 9''.648\ \ \text{cos. R. A. cos. } \Omega \\ -\ 7\ .1822\ \ \text{sin. R. A. sin. } \Omega \end{matrix}\right\} \text{tang. dec.}$$
$$-16\ .54414\ \text{sin. } \Omega$$

The co-efficients consisting of the second and third terms of this formula are dependent on the first: for — 7″.1822 = — 9″.648 cot. 2 obliq. 2 sin. obliq. and — 16″.54414 = — 7″.1822 cot. obliq. Now if the co-efficient or first term be altered, the remaining two must be altered in the same ratio, so long as the obliquity of the ecliptic remains unaltered; but for many years, the decrease in the obliquity of the ecliptic will not sensibly affect the value of L; and this value will always remain the same, so long as the several terms of the formula from which it is deduced, continue in the same ratio with respect to each other, as those of any other formula to be employed. At page 311 it is stated that "*m*. sin. L = — 9 .648 cos. R. A. tang. dec." In this equation it is evident that as sin. L, and cos. R. A. tang. dec. remain unaltered by any change in the co-efficient, *m* alone must vary with the new co-efficient, or first term.

From what has been said, it appears that these two Tables may be employed in finding L in ordinary cases, and thence by a factor the maximum may be obtained as correctly as from other co-efficients. But in cases of great delicacy, it is advisable to reduce the terms to the precise year, and thence to compute L and its maximum.

Example.——Let it be required to find a factor, which shall reduce the maximum found by these Tables to the maximum found from the formula, where the co-efficient, or the first term, is — 8″.977?

Log. of co-efficient to be substituted		— 8″.977	0 .95313
of co-efficient employed in Tables..........		— 9 .648	0 .98444
of $\frac{-8''.977}{-9''.648}$, or factor...........	.9304	....	9 .96869

Example.——Let L and the maximum of ☾ nut. in R. A. of Aldebaran for 1830 be found; agreeably to the formula, where the co-efficient of the first term is — 8″.977?

[Approx. R. A. $4^h\ 26^m\ 11^s$, and Dec. 16° 9′ 37″.]

q—Table IV..........	log. 0 .81880	*p*—Table V.	log. 0 .58434		
Tang. dec. 16° 9′ 37″ ...	log. 9 .46206	Tang dec.	log. 9 .46206		
Nat. numb. — 1″.909...	log. 0 .28086	Log. *p*....................	0 .04640	Constant log.	9 .96869
Constant..	—16 .544	Sub. log. *q*	1 .26607	....	1 .26607
q	—18 .453	Log. tang. L 0^s 3° 27′......	8 .78033	Log. sec. L................	0 .00079
		p—, *q*— ...... 6 + L		Log. for time	8 .82391
		True L.. = 6 3 27		Max. 1^s.1468 and log. max. =	0 .05946

* According to Mr. Herschel's Table III. N′.. = 0 3 27 .... Max. 1 .1469

Without the constant, the maximum is 1^s.2325; and 1^s.2325 × .9304 = 1 .1467 max. ☾ nut.

In finding the maximum in time, the two constant logarithms may be added together, to make one constant 8 .79260 = 9 .96869 + 8 .82391.

TABLES VI. and VII.—Argument S, and Maximum of Solar Nutation in Right Ascension. (P. 28, 29.)

According to the method of notation adopted at page 312, the

$$\text{Solar nutation in R. A.} = \left.\begin{array}{l} -0''.4340 \text{ cos. R. A. cos. } 2\odot \\ -0\ .3982 \text{ sin. R. A. sin. } 2\odot \end{array}\right\} \text{tang. dec.}$$
$$-0\ .9173 \text{ sin. } 2\odot.$$

In this formula the second and third terms depend on the first co-efficient: for —.3982 = —0″.434 cos. obl. and —0″.9173 = 0″.3982 co-tang. obliquity. Therefore, according to what has been said on the Lunar nutation, S will not vary by changing the co-efficients, so long as they be in the same ratio with those employed in constructing the Tables.

* Our subsidiary arc L differs from Mr. Herschel's N′ by 6 signs, owing to our precept for regulating the arguments; but as the signs of the corrections given in his Tables are the reverse of ours, the resulting corrections are the same in both.

Example.——Let a factor be determined, which shall reduce the maximum of solar nutation in R. A. found by these Tables, to the maximum found from the formula, where the principal co-efficient is −0″.580?

Log. of co-efficient proposed to be used...	−0″.580	9.7634
of co-efficient employed in the Tables	−0 .434	9.6375
then $\frac{-0''.580}{-0\ .434}$, or factor......	=1.336, or log.	0.1259

Then let S and the maximum of ⊙ nut. in R. A. for Aldebaran for 1830 be found, such as will arise from the formula, which has its principal co-efficient −0″.580?

[Approx. R. A. 4h 26m 11s, and decl. 16° 9′ 37″ N.]

q−Table VI.........	log. q .5627	p−Table VII............	log. 9.2373		
Tang. decl. 16° 9′ 37...	log. 9.4621	Tang. decl................	log. 9.4621		
Nat. numb. −0″.106	log. 9.0248	Log. p −	8.6994	Constant log..............	0.1259
Constant... −0 .917		Log. q − sub.................	0.0099	....	0.0099
q−1 .023		Log. tang. S 0 2° 48′.........	8.6895	Log. sec. S	0.0005
		p− q− .... 6 + S		Log. for time.............	8.8239
		True S =6 2 48		Max. 0s.091 and log. max. =	8.9602
		Mr. Herschel's N″=0 2 48	0.0912		

Without the constant, the maximum in time is 0s .068; and 0s .068 × 1 .338=0 .091

In finding the maximum in time, the two constant logarithms may be added together to make one constant, viz. 8 .9498=0 .1259+8 .8239.

TABLE VIII.—Annual Precession in North Polar Distance. (P. 30.)

The co-efficient of the annual precession in north polar distance is that of the variable part of the precession in right ascension in space, with the sign changed. This being the case, the factor used for reducing Table I. in natural numbers, will serve equally for this Table.

Example.——Let the annual precession of Aldebaran, in north polar distance, be found for 1830, using Bessel's co-efficient for the year 2600, viz. −19″.96794?

[Approx. R. A. 4h 26m 11s.]

Table VIII. R. A. 4h 26m		=−7″.992
p.p. for	11s	= 15
R. A. 4 26 11		=−7 .977, which multiplied by ·9961 due to Table I., becomes precession =−7″.946 for the given year.

TABLES IX. and X.—Argument A′ and Maximum of Aberration in North Polar Distance. (P. 31—34.)

The annual number in north polar distance is the longitude of the ecliptic, pointed out by a circle passing through the star and drawn perpendicularly to its circle of declination; and this longitude is subtracted from nine signs, to make A′ positive.

Thus it is seen that A′ does not depend on any co-efficient, consequently an alteration in the principal co-efficient produces an alteration in the maximum only.

The co-efficient employed in these two Tables and in Table III., being necessarily alike, the factor, or constant logarithm, used for reducing Table III., will serve equally for reducing the maximum found from these Tables.

Example.——Let the argument A and maximum of aberration in north polar distance of Aldebaran for 1830, be found; by substituting the principal co-efficient 20″.5 for 20″.255?

[Approx. R. A. $4^h\ 26^m\ 11^s$ and decl. 16° 9′ 37″ N.]

$q+$Table IX.	log. 0.90643	$p-$Table X.		log. 1.23160		
Tang. decl.........	log. 9.46206	Tang. decl.	16° 9′ 37	log. 9.46206	Log. cos. decl............	9.98249
$q+$	log. 0.36849	Nat. numb.....	−4.939	log. 0.69366	Constant	0.00522
		Constant.......	+8.066			
		p............	+3.127	log. 0.49513	....	0.49513
		$q+$sub.................		log. 0.36849		
		Log. tang. A′..	1^s 23° 14	0.12664	Cosec. A′...........	0.09632
		$p+$ } $q+$ }	0+A′		Max. 3″.7946 and log. max.	=0.57916
		True A.......	=1 23 14			
		Mr. Herschel's Tab. VI. gives N.	=1 23 13	....	Max. 3.7929	

Without the constant logarithm, the maximum is 3″.748; and 3″.748 multiplied by 1.0121, the factor for Table III. becomes 3″.795 for the maximum, nearly as before.

TABLES XI. and XII.—Argument L′ and Maximum of Lunar Nutation in North Polar Distance. (P. 35, 36.)

At page 316, the Lunar nutation in N. P. D., by the formula, $= -9''.648$ sin. R. A. cos. ☊ $+7''.1822$ cos. R. A. sin. ☊. As the co-efficient of the second term of this formula depends on that of the first, the value of L is not altered by any change of co-efficient.

Then we have $\text{tang. L}' = \dfrac{-9''.648 \text{ sin. R. A.}}{+7''.1822 \text{ cos. R. A.}} = \dfrac{-9''.648}{+7.1822}$ tang. R. A.; but since $\dfrac{-9''.648}{+7''.1822} = \dfrac{1}{-\text{cot. 2obliq. 2 sin. obliq.}}$, the true value of L does not depend on any co-efficient; but on the obliquity of the ecliptic and right ascension of the star.

The same principal co-efficient may be used for the maximum of nutation in north polar distance, as is used for the corresponding maximum in right ascension; and in Table XII. we have maximum $=9''.648$ sin. R. A. co-sec. L: but here the co-secant of L and sine of right ascension remain unaltered by a change of the co-efficient; therefore the same constant logarithm and factor employed in reducing the maximum obtained from Tables IV. and V. will also be required here.

Example.——Let L′ and the maximum of Lunar nutation in north polar distance of Aldebaran be found for 1830; using the formula which has the co-efficient of its principal term $-8''.977$?

[Approx. R. A. $4^h\ 26^m\ 11^s$.]

Argument L′.	S	Log. Maximum and Maximum.	
Table XI. $4^h\ 26^m\ 11^s$ Tab. L..............	=2 12° 6′	Table XII. $4^h\ 26^m\ 11^s$ log. max...............	0.96855
Precept................................	12 — L	Constant logarithm..........................	9.96869
True value of L′..........................	=9 17 54	Max. 8″654, and log. max.....................	=0.93724
		Or,	
		Table XII. $4^h\ 26^m\ 11^s$ gives 9″.301; and 9″.301 × ·9304=8″.654 for max. ☾ nut.	

TABLES XIII. and XIV.—Argument S′ and Maximum of Solar Nutation in North Polar Distance. (P. 37, 38.)

According to the formula given at page 317, the Solar nutation in N. P. D. $= -0''.4340$ sin. R. A. cos. 2☉ $+0''.3982$ cos. R. A. sin. 2☉.

Then, $\text{tang. S}' = \dfrac{-0''.4340 \text{ sin R. A.}}{+0''.3982 \text{ cos R. A.}} = \dfrac{-0''.4340}{+0''.3982}$ tang. R. A.; but since $\dfrac{-0''.4340}{+0''.3982} = -$secant of obliquity, it follows that S′ does not depend on any co-efficient, but on the obliquity of the ecliptic and right ascension of the star. Indeed, the argument S′ is twelve signs lessened by the ecliptic longitude of the star's right ascension: and will, for north polar distance, be three signs greater than A, on the computation of which it is an admirable check.

In Table XIV. the maximum is $= -0''.4340$ sin R. A. cosec S'. Now as sine R. A. cosec S' remains unaltered by a change of the co-efficient, and as the formula in this case is part of the formula for right ascension, the same constant logarithm and factor determined for reducing the maximum found from Tables VI. and VII. will be required here.

Example.—— Let S' and the maximum of solar nutation in north polar distance of Aldebaran for 1830, be found; using the formula, which has the principal co-efficient $-0''.588$?

[Approx. R. A. $4^h\ 26^m\ 11^s$.]

Argument S'.		Log. Maximum and Maximum.	
Table XIII. $4^h\ 26^m\ 11^s$ tab. S'=	$2^s\ 8'\ 18'$	Table XIV. $4^h\ 26^m\ 11^s$ log. max.......	9 .632
Precept	12 —— S'	Constant logarithm	0 .126
True S'......................	=9 21' 42	Max. 0''.573; and log. max............	9 .758

Or:

Table XIV. $4^h\ 26^m\ 11^s$ gives 0''.429; and 0''.429 × 1·336=0''.573 for max. ☉ nut.

The principal object of the preceding pages has been to show, that the arguments derived from the Series of Fourteen General Tables, will not be altered by any co-efficient, different from the corresponding one employed in constructing the respective Tables; but only the maxima, which may be deduced by means of constant factors, found by dividing the co-efficient to be substituted by the corresponding one already employed.

Although these general Tables do not give some recently employed minute quantities depending on twice the mean longitude of the Moon's node, and on twice the true longitude of the Moon; yet they may become serviceable in these respects. It is pleasing to observe, that the arguments and maxima already found for the Lunar and Solar nutation, will become useful for finding these minute corrections; for the arguments will require little or no alteration; and the maxima may be deduced by using constant factors, in the following manner:—

Corrections depending on twice the ☊ and twice the ☾'s longitude.

Bessel's formulæ for the corrections depending on twice the mean longitude of the Moon's ascending node; and on twice the Moon's true longitude, will, in part for the former, and wholly for the latter, be found at page 328. In cases where the utmost accuracy is to be observed, it becomes necessary to take these minute quantities into account.

As Bessel's formula for the corrections due to twice the Moon's node is but in part given at page 328, the deficiency will be supplied by Mr. Herschel's Paper, at page 426 of Vol. I. of *The Memoirs of the Astronomical Society of London.*

Lunar nutation on 2 ☊, in R. A.... = +0''.0877 cos R. A. cos 2 ☊ } tang. decl.
+0 .0641 sin R. A. sin 2 ☊ }
+0 .1476 sin 2 ☊.

Lunar nutation on 2 ☊, in N. P. D.. = +0''.0877 sin R. A. cos 2 ☊
−0 .0641 cos R. A. sin 2 ☊.

These formulæ are so nearly allied to the formulæ from which L and L' with their maxima are gained, that if six signs be added to L and L', they will always be within thirty-two minutes of what the preceding formulæ would produce; and if the maximum of L and L', found from our series of General Tables, be multiplied by 00901, or if the logarithmic maxima have 7 .9546 added to each of them, they will be reduced nearly to the same maxima that would be gained from the preceding formulæ; and would not err more than $0^s.002$ in right ascension, even in the case of the pole star half a century hence, and by a quantity quite insensible in north polar distance.

For the sake of distinction, let the argument for 2 ☊ be called N for right ascension, and N' for north polar distance, and the following example will explain our proposed method of gaining the quantity required.

Example.——Let N and N' with their maxima for Aldebaran, be found for 1830; having L = $\overset{s}{6}$ 3° 27', and the maximum $1^s.232$; also L' = $\overset{s}{9}$ 17° 56', and the maximum 9''.301?

Argument N.		Maximum of N.
	s	
L..........	=6 3° 27′	
Constant.....	6 add	Max. L−1^s.232; and 1^s.232 × ·00901=0^s.011 max. N.
N..........	=0 3 27	

Argument N′.		Maximum of N′.
	s	
L′..........	=9 17° 54′	
Constant.....	6 add	Max. L′=9″.301; and 9″.301 × ·00901=0″.084 max. N′.
N′..........	=3 17 54	

By a strict calculation of those values from the formulæ, there is obtained, N = $\overset{s}{0}$ 3° 31′ and max. 0^s.011; also, N′ = $\overset{s}{3}$ 17° 36′ and max. 0″.084.

The agreement between L and N may be confirmed, thus:—

$$\text{Tang. N} = \text{tang. L} \times 1\ .019;\ \text{and}\ 1.019 \ldots\ldots\ldots\ldots = \frac{+0''.0877 \times -7''.1822}{-9''.6480 \times +0''.0641};$$

$$\text{Likewise, with sufficient accuracy, the factor } .00901 = \sqrt{\frac{+0''.0877 \times +0''.0641}{-7''.1822 \times -9''.6480}}.$$

[Correction depending on 2 ☾.]

According to Bessel's formulæ given at page 328, we shall have

Lunar nutation for 2 ☾ in R. A. $= -0''.0874$ cos. R. A. cos. 2 ☾ } tang. dec.
−0 .0801 sin. R. A. sin. 2 ☾ }
−0 .1846 sin. 2 ☽.

Lunar nutation for 2 ☾ in N. P. D. $= -0''.0874$ sin. R. A. cos. 2 ☾
+0 .0801 cos. R. A. sin. 2 ☾

In these formulæ, the signs of the co-efficients are like those of the Solar nutation; and as the several terms of the preceding formulæ are in the same ratio with those belonging to the Solar nutation, S and S′ already computed will require no alteration: for if the arguments from these formulæ be called M and M′, then tang. M = tang. S $\frac{-0''.0874 \times -0''.3982}{-0''.4340 \times -0''.0801}$ = tang. S × 1 .001; but this latter factor may be disregarded, as it never can cause an error of more than two minutes in the argument. Since the two formulæ have their corresponding co-efficients in the same ratio; and S remains thereby unaltered by any change in the co-efficient, the maximum due to 2 ☽ will be = max. 2 ☉ $\frac{-0''.0874}{-0''.4340}$; and this last fraction is equal to .2014: therefore, to find the arguments M and M′, take S and S′ without alteration; and to find the maxima of M and M′, multiply the maxima of S and S′ by .2014, and the results will be the maxima of M and M′.

Example.——Let M and M′ with their maxima for Aldebaran be found for 1830; having S = $\overset{s}{6}$ 2° 48′, and the maximum 0^s.068; also S′ = $\overset{s}{9}$ 21° 42′ and the maximum 0″.429?

Argument M.		Maximum of M.
	s	
S............	6 2° 48′	
Constant.......	0. add.	Max. S = 0^s.068; and 0^s.068 × ·2014=0^s.014 max. M.
M...........	6 2 48	

Argument M′.		Maximum of M′.
	s	
S′............	9 21° 42′	
Constant.......	0. add.	Max. S′ = 0″.429; and 0″.429 × ·2014=0″.086 max. M′.
M′...........	9 21 42	

By a strict calculation of these values from the formulæ, there is obtained: M $\overset{s}{6}$ 2° 48′ and max. 0^s.014; also, M′ $\overset{s}{9}$ 21° 41′ and max. 0″.086. With these constants the minor corrections may be obtained from Table XVIII. when the arguments are completed by the addition of 2 ☊ and 2 ☾ respectively. We have taken the example of Aldebaran according to the data given in pages 311—318, because some errors of computation vitiated the results there given, which are here corrected.

In the second part of the "Memoirs of the Astronomical Society of London," are published Subsidiary Tables for facilitating the computation of annual Tables of the apparent places of forty-six principal fixed stars, which were computed under the able direction of the present President, Mr. Herschel, and which, therefore, we will denominate Mr. Herschel's Tables: as the co-efficients employed in the preceding Example are the same as those adopted in the construction of the said Tables, after allowing for the negative maxima, by adding six signs to convert declination into north polar distance; the comparison of the respective computations will be seen in the subjoined Table.

SUBSIDIARY NUMBERS FOR ALDEBARAN FOR 1830.

	OUR TABLES.					MR. HERSCHEL'S.			
	R. A.		N. P. D.			R. A.		N. P. D.	
	Argument.	Max.	Argument.	Max.		Argument.	Max.	Argument.	Max.
	S		S			S		S	
A	6 21° 42′	1s.405	1 23° 14′	3″.795	A′	6 21° 42′	1s.405	1 23° 13′	3″.793
L	6 3 27	1 .147	9 17 54	8 .654	L′	6 3 27	1 .147	9 17 54	8 .654
S	6 2 48	0 .091	9 21 42	0 .573	S′	6 2 48	0 .091	9 21 42	0 .556
N	0 3 27	0 .011	3 17 54	0 .084	N′	0 3 27	0 .011	3 17 54	0 .087
M	6 2 48	0 .014	9 21 42	0 .086	M′	6 2 48	0 .014	9 21 42	0 .084

Considering the small variation in the maximum of S′ and M′, Mr. Herschel has employed for all the stars, the mean maximum of S′ = 0″.556, and of M′ = 0″.084; and he likewise takes $\frac{1}{100}$ of the maximum of L′ for that of N′. The adoption of those mean maxima will never cause an error of more than 0″.03 in practice; but the circumstance is named merely to reconcile the discrepancy in the maxima of S′, N′ and M′. Mr. Herschel has not employed the mean annual precessions; but the annual variations, taken from the catalogue, affixed to the "Nautical Almanac" for 1824.

Table XV. (Pages 39 and 320.)

In our explanation of Table XV. as a General Table of Precession, we inadvertently stated, that "the star's mean place is supposed to be adapted to midnight, between December of the preceding, and January of the following year"; but this is not the case; the epoch of the Table is January 0, or December 31 *at noon*, agreeably to the reckoning of time (t) required in Bessel's formulæ, given at pages 421 and 422, for finding which by inspection it is very convenient. The foreign and English catalogues have not the same epoch with respect to the commencement of the year, and it is only when the stars are reduced to January 0, that the tabular numbers can be taken in our Table opposite the days of the month in common years, according to civil reckoning, and in leap years after February opposite the next *following* days (not *preceding* as stated in the note there subjoined): but when, as at Greenwich, the epoch is January 1 at noon, the next preceding day must be taken instead of the day of the date in all common years, and in leap years after February, the tabular date must be taken as it stands opposite the proper numbers. With this attention to the precept the Table will answer for either of the two epochs, as the case may demand. As the days of observation are usually put down according to civil reckoning in all observatories, the Table was computed from the formula $\frac{pt}{365}$ instead of $\frac{pt}{365.25}$, that the figures might terminate with the year; but the tabular quantities would be very nearly the same, whichever were adopted; $\frac{1}{365}$ being $= .0027397$ and $\frac{1}{365.25}$ $= .0027378$ of a year, for one day's factor, and the difference when a maximum at the end of the year, even in the column 20″, will amount to only 0″.013.

When however great accuracy is required in the reduction of a star's place, an error of greater magnitude will frequently arise from the use of this Table, as well as from ordinary daily factors, by reason of the tabular quantities being adapted to

3 P 2

the *noon* of each day, instead of the hour of the star's meridian passage, which is a variable time, and yet the time on which the accuracy of the precession depends. The following Table has been computed to remedy this source of inaccuracy, by giving an additional tabular quantity proportional to the time elapsed from the preceding noon to the hour of the meridian passage, and also to the annual variation, or precession, in columns corresponding to the principal Table under our consideration. The arguments of the supplemental Table are, at the side the solar hours elapsed from the preceding noon to the time of observation, which will always be indicated near enough by a solar clock, or ordinary watch; and at the head those columns must be used that contain the figures of the annual variation, or precession, and the numbers must be taken out and arranged in the same manner as the explanation of the larger Table (XV.) fully directs: then the sum of the two quantities derived from the two Tables will be the amount of precession accruing from the epoch to the hour of observation. Suppose, for example, that α Persei was observed passing the meridian of Greenwich on December 11, 1827, at 9^h 45^m by the solar clock, since a few minutes more or less will not sensibly affect the argument, 10 hours may be taken as one of the arguments, and precession in N. P. D. 13″.3, taken from the Nautical Almanac, as the other; then the quantity given by the large Table will be

13″.	12″.253 }		12″.536
0 .3	.283 }		
		Supplemental Table	.015
		Correct precession	12 .551

SUPPLEMENTAL TABLE for the Reduction of a Star's Precession from Noon to the Time of its Meridian Passage.
(As an Appendix to Table XV. p. 39.)
Arguments, the time elapsed at the side: precession at the head.

Hours elapsed.	1″	2″	3″	4″	5″	6″	7″	8″	9″	10″	11″	12″	13″	14″	15″	16″	17″	18″	19″	20″
1	.00011	.0002	.0003	.0004	.0006	.0007	.0008	.0009	.0010	.0011	.0012	.0013	.0014	.0015	.0017	.0018	.0019	.0020	.0021	.0022
2	.00023	.0005	.0007	.0009	.0011	.0013	.0016	.0018	.0021	.0023	.0025	.0026	.0028	.0032	.0034	.0036	.0038	.0042	.0044	.0046
3	.00034	.0007	.0010	.0014	.0017	.0020	.0024	.0028	.0031	.0034	.0037	.0040	.0040	.0048	.0051	.0056	.0059	.0062	.0065	.0068
4	.00046	.0009	.0014	.0019	.0023	.0028	.0032	.0038	.0043	.0046	.0051	.0056	.0061	.0064	.0069	.0076	.0081	.0086	.0091	.0092
5	.00057	.0011	.0017	.0023	.0029	.0035	.0040	.0046	.0052	.0057	.0063	.0070	.0076	.0080	.0086	.0092	.0098	.0104	.0110	.0114
6	.00068	.0014	.0020	.0027	.0034	.0041	.0048	.0054	.0061	.0068	.0075	.0082	.0089	.0096	.0103	.0108	.0115	.0122	.0129	.0136
7	.00079	.0016	.0024	.0032	.0040	.0048	.0055	.0064	.0072	.0079	.0087	.0096	.0104	.0110	.0118	.0128	.0136	.0144	.0152	.0158
8	.00091	.0018	.0027	.0036	.0045	.0054	.0064	.0072	.0081	.0091	.0100	.0108	.0117	.0128	.0137	.0144	.0153	.0162	.0172	.0182
9	.00102	.0020	.0031	.0041	.0051	.0061	.0071	.0082	.0092	.0102	.0112	.0122	.0132	.0142	.0152	.0164	.0174	.0184	.0194	.0204
10	.00114	.0023	.0034	.0046	.0057	.0068	.0079	.0092	.0103	.0114	.0123	.0136	.0147	.0158	.0169	.0184	.0195	.0206	.0217	.0228
11	.00125	.0025	.0038	.0050	.0063	.0075	.0088	.0100	.0116	.0125	.0138	.0150	.0163	.0176	.0189	.0200	.0216	.0232	.0241	.0250
12	.00137	.0027	.0041	.0055	.0069	.0082	.0096	.0110	.0124	.0137	.0151	.0164	.0178	.0192	.0207	.0220	.0234	.0248	.0261	.0274
13	.00148	.0030	.0044	.0059	.0074	.0089	.0104	.0118	.0133	.0148	.0163	.0178	.0193	.0208	.0222	.0236	.0251	.0266	.0281	.0296
14	.00159	.0032	.0048	.0064	.0080	.0096	.0112	.0128	.0143	.0159	.0175	.0192	.0208	.0224	.0240	.0256	.0271	.0286	.0302	.0318
15	.00171	.0034	.0051	.0068	.0085	.0102	.0120	.0136	.0154	.0171	.0188	.0204	.0221	.0240	.0255	.0272	.0290	.0308	.0325	.0342
16	.00182	.0036	.0055	.0073	.0091	.0109	.0127	.0146	.0164	.0182	.0200	.0218	.0236	.0254	.0273	.0292	.0310	.0328	.0346	.0364
17	.00194	.0039	.0058	.0078	.0097	.0116	.0136	.0156	.0175	.0194	.0213	.0232	.0251	.0272	.0291	.0312	.0331	.0350	.0369	.0388
18	.00205	.0041	.0062	.0082	.0103	.0123	.0144	.0164	.0185	.0205	.0226	.0246	.0267	.0288	.0309	.0328	.0349	.0370	.0390	.0410
19	.00217	.0043	.0065	.0087	.0109	.0130	.0152	.0174	.0195	.0217	.0239	.0260	.0282	.0304	.0327	.0348	.0369	.0390	0412	.0434
20	.00228	.0046	.0068	.0091	.0114	.0137	.0160	.0182	.0205	.0228	.0251	.0274	.0297	.0320	.0342	.0364	.0387	.0410	.0433	.0456
21	.00239	.0048	.0072	.0096	.0120	.0143	.0167	.0192	.0215	.0239	.0263	.0286	.0310	.0334	.0360	.0384	.0407	.0430	.0454	.0478
22	.00251	.0050	.0075	.0100	.0125	.0151	.0176	.0200	.0226	.0251	.0276	.0302	.0327	.0352	.0375	.0400	0426	.0452	.0477	.0502
23	.00262	.0052	.0079	.0105	.0131	.0157	.0183	.0210	.0236	.0262	.0288	.0314	.0340	.0366	.0393	.0420	.0446	.0472	.0498	.0524
24	.00274	.0055	.0082	.0110	.0137	.0164	.0192	.0220	.0247	.0274	.0301	.0328	.0355	.0384	.0411	.0440	.0467	.0494	.0521	.0548

Having now shown that the arguments and maxima, computed from our Series of Fourteen General Tables, may be subservient in producing the same results, that the computations from other more recent Tables afford; the following Table of comparative co-efficients and factors is subjoined, to afford the ready means of reconciling discrepancies in the computed apparent places of the stars, when the Tables from which they were computed are not at hand.

A TABLE OF FACTORS FOR REDUCING THE MAXIMA OF ABERRATION, AND OF LUNAR AND SOLAR NUTATION, computed from our Series of Fourteen General Tables, to Maxima of the same value that would be gained from the various Formulæ, or Tables of other Authors.

DESCRIPTION.	Precessions.				Aberration.			Lunar Nutation.			Solar Nutation.		
	Con-stant in R. A. in Time.	Co-effic. of variab. part in R. A. and of preces. in Declin.	Factors for var. part.		Constant.	Factors.		Co-effic. of Obliq.	Factors.		Prin-cipal Co-effic.	Factors.	
			Loga-rithm.	Nat. Numb.		Loga-rithm.	Nat. Numb.		Loga-rithm.	Nat. Numb.		Loga-rithm.	Nat. Numb.
BAILY'S Tables, in Append. to Vol. II. of Mem. of Astron. Soc. of London	+ 3^s.0680	+ 20''.0426	9 .99998	.99995	20''.360	0 .00225	1 .0052	9''.250	9 .98170	.9587	0''.545	0 .0989	1 .256
BESSEL'S formula, as given at page 421 ...	3 .0678	20 .0436	0 .00000	1 .00000	20 .255	0 .00000	1 .0000	8 .977	9 .96869	.9304	0 .5799	0 .1259	1 .336
DELAMBRE'S Tables, at page 92	....	....	....	....	20 .255	0 .00000	1 .0000	10 .055	0 .01795	1 .0422			
GROOMBRIDGE'S Tables, in Vol. I. of Mem. of Astron. Soc. of London	3 .0599	20 .0095	9 .99926	.99829	20 .255	0 .00000	1 .0000	9 .630	9 .99919	.9981	Involved in the precession.		
HERSCHEL'S Tables, at page 421 of Vol. I. of Mem. of Astron. Soc. of London	Ann. Var. from N. Alm. of 1824.				20 .500	0 .00522	1 .0121	8 .977	9 .96869	.9304	0 .5799	0 .1259	1 .336
MASKELYNE'S Tables, in his first volume of Observations, p. 177	3 .0680	20 .0500	0 .00014	1 .00033	20 .000	9 .99450	.9874						
Recent Observations at Greenwich, reduced by RICHARDSON......	....	....	....	....	20 .4325	0 .00380	1 .0088						

FACTORS FOR DETERMINING CORRECTIONS DEPENDING ON 2 ☊ AND 2 ☾, EMPLOYED BY SOME AUTHORS.

DESCRIPTION.	Lunar Nutation for 2 ☊.			Lunar Nutation for 2 ☾.		
	Principal Co-effic.	Factors.		Principal Co-effic.	Factors.	
		Loga-rithm.	Nat. Numb.		Loga-rithm.	Nat. Numb.
BAILY'S Tables............	0''.090	9 .3167	.2074	0 .090	9 .3167	.2074
BESSEL'S formula.........	0 .088	7 .9546	.0090	0 .087	9 .3040	.2014
HERSCHEL'S Tables	0 .088	7 .9546	.0090	0 .087	9 .3040	.2014

USE OF THE ADJOINING TABLE.

[Positive arguments and maxima for 2 ☊.]

For BAILY, add six signs to S and S′, and taking their maxima, as found from the Fourteen General Tables, reduce them to the maxima of N and N′; but, for BESSEL and HERSCHEL, add six signs to L and L′, originally found, and reduce their maxima to those of N and N′.

[Positive arguments and maxima for 2 ☾.]

For all the three authors, retain S and S′, and reduce their original maxima to those of M and M′.

In these directions the character ☊ denotes the mean longitude of the moon's ascending node; and ☾ is put for the moon's true longitude.

SUPPLEMENTAL PRINCIPAL STARS.

AT the time when the subsidiary numbers of forty-eight principal stars were printed at page 147, twenty-four of them only had their apparent places computed in the Nautical Almanac, but lately the list has been increased to sixty, and we have been requested to give in this Appendix the subsidiary numbers of the sixteen additional stars not before computed, which we have great pleasure in doing, for the use of those computers who prefer using our numbers, and method of gaining the successive apparent places for each ten days of the year. As the mean places of these stars are given annually in the Nautical Almanac, it will not be necessary to give the right ascensions and declinations due to any epoch, but only the subsidiary numbers, which, by the use of our General Table XVIII. with the sun's longitude, and place of the moon's node for the time, to complete the arguments, will give the corrections by inspection for any day, or successive number of days, as has been explained at page 322. The annexed Tables may be considered as a continuation of the Tables for the subsidiary numbers of the forty-eight stars contained in pages 147 and 148. Indeed the corrections corresponding to every single degree of each argument might have been given, as at pages 113, et seq. and for any maxima, if such addition would not have increased the size of the Appendix too much.

Subsidiary Numbers of a Supplemental List of Principal Stars.

RIGHT ASCENSION.

STARS.	A		Change in 30 years.	L		Change in 30 years.	S		Change in 30 years.	Precession.		Max. of Aber.		Max. ☾ Nut.		Max. ⊙ Nut.	
	1800.	1830.		1800.	1830.		1800.	1830.		1800.	1830.	1800.	1830.	1800.	1830.	1800.	1830.
	s ° ′	s ° ′	′	s ° ′	s ° ′	′	s ° ′	s ° ′	′								
γ Orionis	6 10 28	6 10 7	−21	6 0 41	6 0 40	− 1	6 0 31	6 0 32	+ 1	+ 3s.209	+ 3s.210	1s.354	1s.354	1s.145	1s.154	0s.070	0s.064
δ Orionis	6 8 46	6 8 25	−21	5 29 57	5 29 58	+ 1	5 29 58	5 29 58	0	3.057	3.058	1.347	1.348	1.099	1.100	.073	.060
ε Orionis	6 7 47	6 7 27	−20	5 29 53	5 29 53	0	5 29 55	5 29 55	0	3.037	3.038	1.348	1.349	1.092	1.092	.073	.061
ζ Orionis	6 6 44	6 6 23	−21	5 29.51	5 29 51	0	5 29 54	5 29 53	− 1	3.020	3.021	1.349	1.350	1.086	1.086	.074	.061
β Aurigæ......	6 3 32	6 2 58	−34	6 1 30	6 1 19	−11	6 1 15	6 1 4	−11	4.397	4.398	1.906	1.907	1.580	1.580	.088	.087
δ Ursæ Major.	2 28 32	2 28 6	−26	4 17 21	4 16 24	−57	4 22 34	4 22 19	−15	3.017	3.003	2.347	2.336	1.499	1.491	.077	.076
α Draconis	1 28 4	1 27 51	−13	3 25 41	3 25 55	+14	4 0 40	4 0 55	+15	1.624	1.625	3.037	3.021	1.350	1.340	.064	.063
ε Boötis........	1 18 30	1 18 10	−20	5 14 19	5 14 28	+ 9	5 17 10	5 17 17	+ 7	2.622	2.621	1.453	1.452	0.979	0.979	.054	.054
δ Ophiuchi.....	0 26 59	0 26 37	−22	6 0 53	6 0 53	0	6 0 43	6 0 43	0	3.133	3.135	1.327	1.328	1.126	1.127	.062	.063
δ Ursæ Minor.	11 21 36	11 23 47	+131	11 15 53	11 19 31	+218	11 18 28	11 22 28	+180	−18.889	−19.168	22.471	22.645	6.976	6.984	.383	.385
β Lyræ.........	11 20 11	11 19 56	−15	6 5 38	6 5 44	+ 6	6 4 32	6 4 39	+ 7	+ 2.211	+ 2.210	1.608	1.609	0.793	0.799	.044	.044
ζ Aquilæ	11 17 4	11 16 45	−19	6 2 11	6 2 14	+ 3	6 1 46	6 1 49	+ 3	2.755	2.754	1.383	1.383	0.992	0.989	.055	.055
δ Draconis......	11 13 17	11 13 17	− 0	8 28 16	8 28 36	+20	8 27 52	8 28 16	+24	0.030	0.023	3.474	3.481	0.479	0.480	.022	.022
δ Aquilæ	11 12 36	11 12 15	−21	6 0 31	6 0 33	+ 2	6 0 26	6 0 26	0	3.008	3.007	1.341	1.340	1.081	1.081	.060	.059
1st 61 Cygni. .	10 17 59	10 17 38	−21	6 22 37	6 22 51	+14	6 18 41	6 18 52	+11	2.329	2.330	1.640	1.643	0.908	0.910	.049	.049
β Aquarii......	10 12 12	10 11 48	−24	5 27 0	5 27 14	+14	5 27 34	5 27 46	+12	3.164	3.162	1.294	1.292	1.069	1.140	.059	.063

NORTH POLAR DISTANCE.

STARS.	A′		Change in 30 years.	L′		Change in 30 years.	S′		Change in 30 years.	Precession.		Max. of Aber.		Max. ☾ Nut.		Max. ⊙ Nut.	
	1800.	1830.		1800.	1830.		1800.	1830.		1800.	1830.	1800.	1830.	1800.	1830.	1800.	1830.
	s ° ′	s ° ′	′	s ° ′	s ° ′	′	s ° ′	s ° ′	′	″	″	″	″	″	″	″	″
γ Orionis	2 25 57	2 26 4	+ 7	9 8 33	9 8 14	−19	9 10 28	9 10 7	−21	− 3.960	− 3.824	6.082	6.067	9.564	9.570	0.433	0.433
δ Orionis	3 0 11	3 0 10	− 1	9 7 8	9 6 50	−18	9 8 46	9 8 24	−22	3.325	3.193	8.213	8.204	9.589	9.594	.433	.434
ε Orionis	3 0 28	3 0 27	− 1	9 6 20	9 6 3	−17	9 7 47	9 7 26	−21	2.957	2.824	8.494	8.487	9.602	9.606	.434	.434
ζ Orionis	3 0 37	3 0 34	− 3	9 5 29	9 5 11	−18	9 6 44	9 6 23	−21	2.559	2.427	8.723	8.718	9.613	9.617	.434	.434
β Aurigæ......	9 7 17	9 6 14	−63	9 2 50	9 2 24	−26	9 3 28	9 2 57	−31	1.323	1.130	7.436	7.431	9.638	9.642	.434	.434
δ Ursæ Major.	5 14 56	5 14 31	−25	5 28 10	5 27 40	−30	5 28 31	5 28 7	−24	+20.038	+20.034	17.818	17.811	7.183	7.184	.398	.398
α Draconis.	4 23 44	4 23 28	−16	4 22 29	4 22 15	−14	4 28 5	4 27 53	−12	17.403	17.368	19.848	19.829	7.862	7.870	.408	.408
ε Boötis........	4 0 17	4 0 1	−16	4 12 31	4 12 11	−20	4 18 30	4 18 9	−21	15.563	15.491	14.603	14.591	8.250	8.264	.413	.413
δ Ophiuchi. ...	2 25 40	2 25 35	− 5	3 22 26	3 22 7	−19	3 26 58	3 26 37	−21	9.726	9.606	7.177	7.150	9.135	9.141	.426	.426
δ Ursæ Minor.	2 20 18	2 22 49	+151	2 23 11	2 24 57	+106	2 21 36	2 23 48	+132	− 3.186	− 2.363	19.067	19.048	9.594	9.619	.434	.434
β Lyræ.........	2 23 1	2 22 50	−11	2 22 2	2 21 49	−13	2 20 11	2 19 56	−15	3.703	3.808	16.861	16.864	9.574	9.570	.433	.433
ζ Aquilæ.......	2 24 32	2 24 23	− 9	2 19 27	2 19 12	−15	2 17 4	2 16 45	−19	4.868	4.985	12.127	12.134	9.521	9.514	.432	.432
δ Draconis......	2 13 20	2 13 19	− 1	2 16 18	2 16 18	0	2 13 17	2 13 17	0	6.233	6.236	20.255	20.255	9.438	9.438	.431	.431
δ Aquilæ	2 28 0	2 27 55	− 5	2 15 44	2 15 26	−18	2 12 36	2 12 15	−21	6.478	6.603	8.899	8.914	9.421	9.412	.431	.431
1st 61 Cygni...	1 29 3	1 28 45	−18	1 23 50	1 23 31	−19	1 17 59	1 17 39	−20	14.045	14.130	16.903	16.916	8.527	8.512	.417	.417
β Aquarii......	3 14 39	3 14 23	−16	1 18 10	1 17 45	−25	1 12 12	1 11 48	−24	15.411	15.500	6.907	6.938	8.280	8.263	.413	.413

Mr. Baily's Tables, published as an Appendix to Vol. II. of the Memoirs of the Astronomical Society of London, and also separately.

The Tables here referred to profess to give the corrections for precession, aberration, and nutations, including the corrections depending on twice the mean longitude of the moon's node. These Tables are constructed to the mean places of the stars for the beginning of 1830; and unless the several constants be reduced to any other year, the corrections obtained from them will be similar to corrections obtained from subsidiary numbers and maxima calculated for the year 1830. Indeed there are but few of the constants intended to have their variation assigned to them; and unless the proper motion of a star is set down in Piazzi's catalogue for 1800, as amounting to half a second in arc, and determined from the observations of Bradley or Meyer, it is not included in the precession.

The author has adopted Greenwich mean solar time instead of Paris sidereal time, and has supposed the fictitious year to commence at mean noon of January the 1st, when the sun's mean longitude is exactly 281°, instead of mean noon at Paris, when it is 280° on January 0, or December the 31st of the preceding year. Likewise a, b, c, d and a', b', c', d' are calculated upon the same principles as M. Schumacher's c, d, a, b and c', d', a', b', except that Mr. Baily's constants (a, b, c, d) in right ascension have the constant logarithm 8 .8239 added to them, to reduce the corrections into arcs of time. By the use of known factors our Tables will supply the results arising from his constants, with the advantage of keeping every step of the process separate.

Example.——Let the apparent right ascension and north polar distance of α Persei be found on March 21, 1829, at Greenwich apparent noon; using the factors for the values employed by Mr. Baily, and supposing his Tables not to be at hand?

[1830, Approx. R. A. $3^h\ 12^m\ 13^s.4$, and Dec. 49° 14′ 56″ N.]

R. A. $3^h\ 12^m\ 13^s.4$ Tab. 1 .... log.	9 .99734 +		
Dec. 49° 14′ 56″ .... log. tang.	10 .06465		
Constant from Table .... log.	9 .99997	Constant number	+ 3^s.0680
log.	0 .06196	Of var. part	+ 1 .1533
		Ann. precess.	+ 4 .2213

RIGHT ASCENSION.							NORTH POLAR DISTANCE.						
Arg.	Arg. and Max. from pages 147 and 148 of Volume I.				Factors from preceding Table.	Reduced Max.	Arg.	Arg. and Max. from pages 147 and 148 of Volume I.				Factors from preceding Table.	Reduced Max.
	S	°	′					S	°	′			
Pr.	......			......	......	$+4^s.221$	Pr.	......			−13″.398 ×	.99999	−13 .397
A	7	9	30	$1^s.994$ ×	1 .0052	2 .004	A′	11	3	37	11 .501 ×	1 .0052	11 .561
L	6	18	13	1 .596 ×	.9587	1 .530	L′	10	3	47	8 .634 ×	.9587	8 .278
S	6	14	57	0 .087 ×	1 .2256	0 .109	S′	10	9	30	0 .418 ×	1 .2256	0 .525
N	0	14	57	0 .087 ×	.2078	0 .018	N′	4	9	30	0 .418 ×	.2078	0 .087

	S			
Naut. Alm. 1825, ☉'s long............	0	0°	36′	4″
Sol. Tab. 3. {1829 .. + 52′ 0″; 1825 .. + 50 2} diff. . +			1	58
Sun's true long. Mar. 21st, 1829	0	0	38	2

	S			
Lun. Tab. 1, 1829 ☾'s node........	6	12°	19′	54″
2, Mar. 21 .. −4° 11′ 0″; Constant.. − 1 35 } −		4	12	35
Long. ☾'s ☊ Mar. 21st, 1829	6	8	7	19

Right Ascension.

	s	°	′		
Table XV. Mar. 21. and + 4ˢ.221, precess......					+ 0ˢ.925
⊙ ..... =	0	0°	38′		
A ..... =	7	9	30		
Arg.......	7	10	8	and max 2ˢ.004, aber. ...	− 1 .292
☾'s ☊ .. =	6	8°	8′		
L...... =	6	18	13		
Arg.......	0	26	21	and max. 1 .530, ☾′ nut..	+ 0 .679
2 ⊙.... =	0	1°	16′		
S...... =	6	14	57		
Arg.......	6	16	13	and max. 0 .109, ⊙ nut..	− 0 .031
2 ☾'s ☊ =	0	16°	16′		
N ..... =	0	14	57		
Arg.......	1	1	13	and max. 0 .018, ☾″ nut.	+ 0 .009
Sum..................					+ 0 .290
Jan. 0, 1829, mean R. A.					3^h 12^m 9 .620
Mar. 21, 1829, app. R. A.................					3 12 9 .910

North Polar Distance.

	s	°	′		
Table XV. Mar. 21 and prec. − 13″.397, precess..					− 2″.94
⊙ ..... =	0	0°	38′		
A′..... =	11	3	7		
Arg.......	11	3	45	and max. 11 .561, aber. ..	− 5 .02
☾'s ☊ .. =	6	8°	8′		
L′ ..... =	10	3	47		
Arg.......	4	11	55	and max. 8 .278, ☾′ nut.	+ 6 .16
2 ⊙.... =	0	1°	16′		
S′...... =	10	9	30		
Arg.......	10	10	46	and max. 0 .525, ⊙ nut.	− 0 .40
2 ☾'s ☊ =	0	16°	16′		
N′..... =	4	9	30		
Arg.......	4	25	46	and max. 0 .087, ☾″ nut.	+ 0 .05
Sum...............					− 2 .24
Jan. 0, 1829, mean N. P. D..............					40° 45′ 18 .50
Mar. 21, 1829, app. N. P. D.............					40 45 16 .26

Mr. Baily, in his "Memoir", has omitted to observe that the mean place of the star should be reduced to the instant when the sun's mean longitude is exactly 281° at Greenwich. When the sun's mean longitude is exactly 281°, neither the civil nor the mean solar year commences universally; but Mr. Baily has supposed the mean fictitious year to commence at this instant; and as far as regards the longitudes of the sun and of the moon's node, his reduction x sufficiently allows for the deviation. But in the calculation, t, or time, is made to commence as above stated, and on fictitious January the 1st, to be equal to nothing; whereas in the year 1819 the fictitious year commences at 17^h 7^m past true mean noon. In this case, supposing the required time to be January the 1st at 17^h 7^m mean solar time, it is obvious that the desired time will be at the precise commencement of the fictitious year, so that the tabular date will be January the 1st, and will properly allow for the sun and node; but on January the 1st (tabular date) t, or time, is nothing, whereas from noon to 17^h 7^m, a star by precession will have moved + 0″.039, supposing its annual motion to be 20″. The error from this source would have been less if the fictitious year had been made to commence with the true mean solar year 1830; or perhaps, for the convenience of computation, 280° 40′ might have been used instead of 281°. Considering indeed the variation in the constants and other changes, the numbers adopted may be quite as near as needful; but without examining the matter, discrepancies in two calculations could not readily be reconciled: for in the year 1804, the value of x is 25^h 54^m from January the 1st in that year.

Take the difference between $\overset{s}{9}$ 11° and the sun's mean longitude for the given year, in Table 1 of Solar Tables (at page 113), which call plus when $\overset{s}{9}$ 11° is the greater, and minus when the less; then if the star's mean place be given to January 0 at noon, to the difference just found apply + 3548″, and to the logarithm of the sum add the constant logarithm 3 .887, and the logarithm of the annual precession; the sum will be the logarithm of a number to be applied to the star's place on January 0, to reduce the same to the instant when the sun's mean longitude was exactly 281°. This correction may however be obtained in a much more concise way by Mr. Baily's Tables, but in the present case they are supposed not to be at hand.

Example.——For January 0, 1829, given the mean R. A. of α Persei 3^h 12^m $9^s.62$, and N. P. D. 40° 45′ 18″.50; required its mean place, when the sun's mean longitude is exactly 281°, in the same year, at the meridian of Greenwich?

	s			
☉'s mean long. requir..........	+ 9	11°		
Sol. Tab. 1. 1829 ☉'s long.	− 9	10	51′	59″
Difference....................	+ 0	0	8	1 = + 481″
One day from Jan. 0 to Jan. 1...................				= + 3548

Sum = + 4029″, seconds of sun's longitude from January 0 to time when sun's mean longitude is 281°.

Log. of + 4029″............................	3 .605	....	3 .605 +
Constant............................	3 .887	....	3 .887
Ann. prec. in R. A. + 4ˢ.221............	0 .625	Ann. prec. in N. P. D. − 13″.397..	1 .127 −
Correction in R. A. ....+ 0ˢ.013 ..	8 .117	Corr. in N. P. D..... − 0″.04....	8 .619 −
1829, Jan. 0, ∗'s mean R. A....... 3ʰ 12ᵐ 9 .620		N. P. D. .... 40° 45′ 18 .50	
1829, fictitious Jan. 1, ∗'s m. R. A. 3 12 9 .633		N. P. D..... 40 45 18 .46	

Result of the Calculation from Mr. Baily's Tables of the Apparent Place of α Persei on March 21, 1829.

R. A.			
a A	− 1ˢ.275	a' A...........	+ 5″.23
b B........................	− 0 .017	b' B...........	− 0 .11
c C........................	+ 1 .118	c' C...........	+ 3 .55
d D	+ 0 .449	d' D	− 6 .46
Sum	+ 0 .275	Sum .... + 2 .21; with sign changed ...	− 2″.21
Mean R. A. as above.......	3ʰ 12ᵐ 9 .633	N. P. D.	40° 45′ 18 .46
App. R. A. on Mar. 21.....	3 12 9 .908	N. P. D.	40 45 16 .25
Bypreceding calculation R. A.	3 12 9 .910	N. P. D.	40 45 16 .26

For the use of those computers who prefer the constants given in the Catalogue of the Astronomical Society of London, we subjoin some useful Tables, from which the numbers A, B, C, and D may be taken by inspection; the two former for an unlimited time, and the two latter till 1830 inclusive, and also C′ and D′ for a long period. The Table, which contains the logarithms of A and B, has for its argument the true longitude of the Sun at the time of the observation, which may be taken from the Nautical Almanac, or other Ephemeris, with scarcely any trouble; and therefore the Table is not only conducive to accuracy and convenience, but is equally applicable to any year, month, day, and hour, and gives the exact numbers at once, by the aid of the differences for each 10′ of the Sun's longitude, exhibited in a separate column. The Tables containing C and D for six successive years are consequently computed for *apparent noon*, to correspond to A and B; and the columns, being completed for every five days in succession, and comprising the daily differences besides, afford the easy means of adapting those numbers to the time of the star's meridian passage, or to the longitude of any observatory east or west of Greenwich. The Tables exhibiting the values of C′ and D′ for Leap Year, and for each of the three following years, are also computed to every five days in order that they may be more readily united with C″ and D″, given in Table VII of the Astronomical Society's Catalogue, to obtain C and D with more facility for any future year; it being of great importance that the annual numbers A, B, C and D, should be easily computed, to unite with the constants a, b, c, d, and a', b', c' and d' for determining the corrections, in converting the apparent into the mean places, and vice versâ, agreeably to the precepts. The uses of these respective Tables are fully explained by their skilful Author in the excellent document that precedes the Catalogue. The method of finding the four numbers from our Tables will be explained by the subjoined Example.

Required A, B, C, and D on the 30th of August, 1829, at the time when α Cygni is on the meridian of Greenwich?

	s	°	′		
Sun's longitude August 30, by Naut. Alm.	5	6	47	Then 5ˢ 7° gives + 1 .2353 and	− 0 .9007
Table XVI. (p. 45) gives..............			+23	0 10′ .. +5 and	−30
Sun's longitude at the star's passage	5	7	10	A...... = +1 .2358	B = −0 .8977

August 29, 1829, gives................	+9 .8283	....	+0 .9532
For 1 day add daily difference	12	...... sub.	6
10 hours difference of R. A..........	5		2
C.......	= +9 .8300	D.......	= +0 .9524

A TABLE, containing the Logarithms of A. and B. to be used in lieu of the Tables A. and B. prefixed to the Catalogue of the Astronomical Society. Argument = Sun's Long. at the Time of Observation.

	S 0−		S VI+		S I−		S VII+		S II−		S VIII+		
	Log. A.	Diff. 10′	Log. B.	Diff. 10′	Log. A.	Diff. 10′	Log. B.	Diff. 10′	Log. A.	Diff. 10′	Log. B.	Diff. 10′	
0° 0′	1.2713			3010.3	1.2088	7.3	1.0077	21.7	0.9703	22.0	1.2463	7.2	30° 0′
0 20	1.2713	.1	9.0735	1505.1	1.2074	7.4	1.0121	21.4	0.9659	22.3	1.2478	7.1	29 40
0 40	1.2713	.2	9.3746	880.4	1.2059	7.5	1.0164	21.2	0.9614	22.6	1.2492	7.0	29 20
1 0	1.2712	.3	9.5506	586.9	1.2044	7.7	1.0206	20.8	0.9569	23.0	1.2506	6.9	29 0
1 30	1.2712	.4	9.7267	416.3	1.2021	7.8	1.0269	20.4	0.9500	23.5	1.2527	6.8	28 30
2 0	1.2710	.5	9.8516	322.9	1.1997	8.0	1.0330	20.0	0.9429	24.0	1.2547	6.6	28 0
2 30	1.2709	.6	9.9485	263.7	1.1973	8.1	1.0390	19.6	0.9357	24.5	1.2567	6.5	27 30
3 0	1.2707	.7	0.0276	222.6	1.1949	8.3	1.0449	19.3	0.9284	25.1	1.2587	6.4	27 0
3 30	1.2705	.8	0.0945	193.0	1.1924	8.4	1.0507	18.9	0.9208	25.6	1.2606	6.2	26 30
4 0	1.2702	.9	0.1524	170.2	1.1899	8.6	1.0563	18.6	0.9131	26.2	1.2624	6.1	26 0
4 30	1.2700	1.1	0.2034	152.2	1.1873	8.8	1.0619	18.2	0.9052	26.8	1.2643	6.0	25 30
5 0	1.2697	1.2	0.2491	137.6	1.1847	8.9	1.0674	17.9	0.8973	27.4	1.2661	5.8	25 0
5 30	1.2693	1.3	0.2904	125.5	1.1820	9.1	1.0727	17.6	0.8890	28.1	1.2678	5.7	24 30
6 0	1.2689	1.4	0.3280	115.4	1.1793	9.3	1.0780	17.2	0.8806	28.7	1.2695	5.6	24 0
6 30	1.2685	1.5	0.3626	106.8	1.1765	9.4	1.0832	16.9	0.8720	29.4	1.2712	5.4	23 30
7 0	1.2680	1.6	0.3947	99.3	1.1737	9.6	1.0882	16.6	0.8632	30.1	1.2728	5.3	23 0
7 30	1.2676	1.7	0.4245	92.9	1.1708	9.8	1.0932	16.3	0.8541	30.9	1.2744	5.2	22 30
8 0	1.2671	1.8	0.4523	87.2	1.1678	10.0	1.0981	16.0	0.8449	31.7	1.2759	5.0	22 0
8 30	1.2666	1.9	0.4785	82.1	1.1649	10.1	1.1029	15.7	0.8354	32.5	1.2775	4.9	21 30
9 0	1.2659	2.1	0.5031	77.6	1.1618	10.3	1.1077	15.5	0.8256	33.4	1.2789	4.8	21 0
9 30	1.2653	2.2	0.5264	73.5	1.1587	10.5	1.1123	15.2	0.8156	34.2	1.2804	4.7	20 30
10 0	1.2647	2.3	0.5484	69.9	1.1556	10.7	1.1168	14.9	0.8054	35.2	1.2818	4.5	20 0
10 30	1.2640	2.4	0.5694	66.6	1.1524	10.9	1.1213	14.7	0.7948	36.2	1.2831	4.4	19 30
11 0	1.2633	2.5	0.5894	63.5	1.1491	11.1	1.1257	14.4	0.7839	37.2	1.2844	4.3	19 0
11 30	1.2625	2.6	0.6084	60.7	1.1458	11.3	1.1300	14.2	0.7728	38.3	1.2857	4.2	18 30
12 0	1.2617	2.7	0.6267	58.2	1.1424	11.5	1.1343	13.9	0.7613	39.5	1.2870	4.0	18 0
12 30	1.2609	2.9	0.6441	55.8	1.1389	11.7	1.1385	13.7	0.7494	40.7	1.2882	3.9	17 30
13 0	1.2600	3.0	0.6609	53.7	1.1354	11.9	1.1426	13.4	0.7372	42.0	1.2894	3.8	17 0
13 30	1.2591	3.1	0.6770	51.6	1.1319	12.1	1.1466	13.2	0.7246	43.4	1.2905	3.7	16 30
14 0	1.2582	3.2	0.6925	49.8	1.1282	12.3	1.1505	13.0	0.7116	44.8	1.2916	3.6	16 0
14 30	1.2572	3.3	0.7074	48.0	1.1245	12.5	1.1544	12.7	0.6982	46.3	1.2927	3.4	15 30
15 0	1.2563	3.4	0.7218	46.3	1.1208	12.7	1.1583	12.5	0.6843	48.0	1.2937	3.3	15 0
15 30	1.2552	3.6	0.7357	44.8	1.1170	13.0	1.1620	12.3	0.6699	49.8	1.2947	3.2	14 30
16 0	1.2541	3.7	0.7491	43.4	1.1131	13.2	1.1657	12.1	0.6550	51.6	1.2957	3.1	14 0
16 30	1.2530	3.8	0.7621	42.0	1.1091	13.4	1.1693	11.9	0.6395	53.7	1.2966	3.0	13 30
17 0	1.2519	3.9	0.7747	40.7	1.1051	13.7	1.1729	11.7	0.6234	55.8	1.2975	2.9	13 0
17 30	1.2507	4.1	0.7869	39.5	1.1010	13.9	1.1764	11.5	0.6066	58.2	1.2984	2.7	12 30
18 0	1.2495	4.2	0.7988	38.3	1.0968	14.2	1.1799	11.3	0.5892	60.7	1.2992	2.6	12 0
18 30	1.2483	4.3	0.8103	37.2	1.0926	14.4	1.1832	11.1	0.5710	63.5	1.3000	2.5	11 30
19 0	1.2470	4.4	0.8214	36.2	1.0883	14.7	1.1866	10.9	0.5519	66.6	1.3007	2.4	11 0
19 30	1.2457	4.5	0.8323	35.2	1.0839	14.9	1.1898	10.7	0.5319	69.9	1.3014	2.3	10 30
20 0	1.2443	4.7	0.8428	34.2	1.0794	15.2	1.1930	10.5	0.5110	73.5	1.3021	2.2	10 0
20 30	1.2429	4.8	0.8531	33.4	1.0748	15.5	1.1962	10.3	0.4889	77.6	1.3028	2.1	9 30
21 0	1.2415	4.9	0.8631	32.5	1.0702	15.7	1.1993	10.1	0.4656	82.1	1.3034	1.9	9 0
21 30	1.2400	5.0	0.8729	31.7	1.0655	16.0	1.2023	10.0	0.4410	87.2	1.3040	1.8	8 30
22 0	1.2385	5.2	0.8825	30.9	1.0607	16.3	1.2053	9.8	0.4149	92.9	1.3045	1.7	8 0
22 30	1.2370	5.3	0.8916	30.1	1.0558	16.6	1.2082	9.6	0.3870	99.3	1.3050	1.6	7 30
23 0	1.2353	5.4	0.9007	29.4	1.0508	16.9	1.2111	9.4	0.3572	106.8	1.3055	1.5	7 0
23 30	1.2337	5.6	0.9095	28.7	1.0457	17.2	1.2140	9.3	0.3252	115.4	1.3060	1.4	6 30
24 0	1.2320	5.7	0.9181	28.1	1.0405	17.6	1.2167	9.1	0.2905	125.5	1.3064	1.3	6 0
24 30	1.2303	5.8	0.9265	27.4	1.0353	17.9	1.2195	8.9	0.2529	137.6	1.3068	1.2	5 30
25 0	1.2286	6.0	0.9347	26.8	1.0299	18.2	1.2221	8.8	0.2116	152.2	1.3071	1.1	5 0
25 30	1.2268	6.1	0.9428	26.2	1.0244	18.6	1.2248	8.6	0.1660	170.2	1.3074	.9	4 30
26 0	1.2250	6.2	0.9506	25.6	1.0189	18.9	1.2274	8.4	0.1149	193.0	1.3077	.8	4 0
26 30	1.2231	6.4	0.9583	25.1	1.0132	19.3	1.2299	8.3	0.0570	222.6	1.3080	.7	3 30
27 0	1.2212	6.5	0.9658	24.5	1.0074	19.6	1.2324	8.1	9.9901	263.7	1.3082	.6	3 0
27 30	1.2192	6.7	0.9732	24.0	1.0015	20.0	1.2348	8.0	9.9110	322.9	1.3084	.5	2 30
28 0	1.2172	6.8	0.9804	23.5	0.9955	20.4	1.2372	7.8	9.8141	416.3	1.3085	.4	2 0
28 30	1.2152	6.9	0.9874	23.0	0.9894	20.8	1.2395	7.7	9.6892	586.9	1.3086	.3	1 30
29 0	1.2131	7.0	0.9944	22.6	0.9832	21.2	1.2418	7.5	9.5132	880.4	1.3087	.2	1 0
29 20	1.2117	7.1	0.9989	22.3	0.9789	21.4	1.2434	7.4	9.3371	1505.1	1.3088	.1	0 40
29 40	1.2103	7.2	1.0033	22.0	0.9746	21.7	1.2448	7.3	9.0361	3010.3	1.3088		0 20
30 0	1.2088		1.0077		0.9703		1.2463				1.3088		0 0
	Log. A.	Diff. 10′	Log. B.	Diff. 10′	Log. A.	Diff. 10′	Log. B.	Diff. 10′	Log. A.	Diff. 10′	Log. B.	Diff. 10′	
	S V+ XI−		S V− XI+		S IV+ X−		S IV− X+		S III+ IX−		S III− IX+		

A TABLE containing the Logarithms of C and D for every Fifth Day of 1825.
(Adapted to Apparent Noon.)

1825.	Log. C.	Daily Diff.	Log. D.	Daily Diff.	1825.	Log. C.	Daily Diff.	Log. D.	Daily Diff.
Jan. 1	+9.5475	42	+9.6674	+ 5	July 5	+9.9331	17	+0.3093	− 2
6	9.5687	39	9.6698	− 28	10	9.9416	16	0.3082	7
11	9.5884	37	9.6560	60	15	9.9497	16	0.3045	12
16	9.6067	34	9.6261	82	20	9.9575	15	0.2984	16
21	9.6236	31	9.5853	110	25	9.9648	14	0.2903	19
26	9.6391	29	9.5304	140	30	9.9718	13	0.2806	22
31	9.6534	26	9.4603	175	Aug. 4	9.9783	12	0.2696	23
Feb. 5	9.6666	24	9.3729	216	9	9.9844	11	0.2581	23
10	9.6786	22	9.2647	263	14	9.9901	11	0.2464	22
15	9.6897	20	9.1331	315	19	9.9955	10	0.2352	20
20	9.6998	19	8.9758	358	24	0.0005	9	0.2252	16
25	9.7092	18	8.7967	338	29	0.0052	9	0.2171	11
Mar. 2	9.7180	16	8.6279	−149	Sept. 3	0.0096	8	0.2115	− 5
7	9.7262	16	8.5532	+179	8	0.0138	8	0.2091	+ 2
12	9.7340	15	8.6426	378	13	0.0178	8	0.2102	10
17	9.7415	15	8.8315	405	18	0.0216	8	0.2152	18
22	9.7488	14	9.0340	364	23	0.0254	7	0.2242	26
27	9.7560	14	9.2160	316	28	0.0291	7	0.2371	32
April 1	9.7632	14	9.3739	271	Oct. 3	0.0328	8	0.2532	38
6	9.7704	14	9.5093	234	8	0.0366	8	0.2723	43
11	9.7778	15	9.6261	203	13	0.0405	8	0.2938	46
16	9.7855	16	9.7276	177	18	0.0445	8	0.3168	48
21	9.7934	16	9.8162	155	23	0.0487	9	0.3409	49
26	9.8016	17	9.8936	136	28	0.0531	9	0.3653	48
May 1	9.8101	18	9.9615	119	Nov. 2	0.0577	10	0.3895	47
6	9.8190	18	0.0210	105	7	0.0625	10	0.4130	45
11	9.8281	19	0.0733	91	12	0.0676	11	0.4354	42
16	9.8374	19	0.1186	79	17	0.0729	11	0.4562	38
21	9.8470	19	0.1581	68	22	0.0785	11	0.4753	34
26	9.8567	20	0.1920	58	27	0.0841	12	0.4925	30
31	9.8665	20	0.2209	49	Dec. 2	0.0899	12	0.5075	26
June 5	9.8764	20	0.2453	40	7	0.0959	12	0.5203	21
10	9.8862	20	0.2653	32	12	0.1018	12	0.5309	17
15	9.8960	19	0.2813	24	17	0.1078	12	0.5392	12
20	9.9056	19	0.2934	17	22	0.1137	12	0.5452	7
25	9.9150	18	0.3020	10	27	0.1196	11	0.5489	+ 3
30	+9.9242	+18	+0.3072	+ 4	32	+0.1253		+0.5506	

A TABLE containing the Logarithms of C and D for every Fifth Day of 1826.
(Adapted to Apparent Noon.)

1826.	Log. C.	Daily Diff.	Log. D.	Daily Diff.	1826.	Log. C.	Daily Diff.	Log. D.	Daily Diff.
Jan. 1	+9.5256	+43	+0.5504	− 1	July 5	+9.9139	+17	+0.6982	− 2
6	9.5471	40	0.5500	5	10	9.9225	17	0.6974	4
11	9.5672	37	0.5478	8	15	9.9308	16	0.6955	6
16	9.5858	34	0.5440	10	20	9.9386	15	0.6927	7
21	9.6029	32	0.5389	12	25	9.9461	14	0.6891	9
26	9.6187	29	0.5328	14	30	9.9531	13	0.6848	9
31	9.6331	27	0.5259	15	Aug. 4	9.9597	12	0.6801	10
Feb. 5	9.6464	24	0.5186	15	9	9.9658	12	0.6751	10
10	9.6585	22	0.5113	14	14	9.9716	11	0.6701	10
15	9.6695	20	0.5044	12	19	9.9769	10	0.6653	9
20	9.6797	19	0.4983	10	24	9.9819	9	0.6610	7
25	9.6890	17	0.4934	7	29	9.9866	9	0.6574	5
Mar. 2	9.6977	16	0.4901	− 3	Sept. 3	9.9910	8	0.6547	− 3
7	9.7058	15	0.4887	+ 1	8	9.9951	8	0.6534	0
12	9.7135	15	0.4894	6	13	9.9990	8	0.6534	+ 3
17	9.7209	14	0.4923	10	18	0.0028	7	0.6548	6
22	9.7281	14	0.4974	15	23	0.0065	7	0.6577	9
27	9.7353	14	0.5047	19	28	0.0101	7	0.6621	12
April 1	9.7424	14	0.5140	22	Oct. 3	0.0138	7	0.6679	14
6	9.7496	15	0.5248	25	8	0.0175	8	0.6749	16
11	9.7570	15	0.5371	27	13	0.0214	8	0.6830	18
16	9.7647	16	0.5504	28	18	0.0254	8	0.6921	20
21	9.7726	17	0.5645	29	23	0.0296	9	0.7019	21
26	9.7809	17	0.5789	29	28	0.0340	9	0.7122	21
May 1	9.7895	18	0.5934	28	Nov. 2	0.0387	10	0.7228	21
6	9.7984	18	0.6076	27	7	0.0435	10	0.7333	21
11	9.8075	19	0.6212	26	12	0.0486	11	0.7436	20
16	9.8170	19	0.6342	24	17	0.0540	11	0.7535	18
21	9.8267	20	0.6462	22	22	0.0595	12	0.7627	17
26	9.8365	20	0.6572	19	27	0.0653	12	0.7711	15
31	9.8464	20	0.6669	17	Dec. 2	0.0712	12	0.7786	13
June 5	9.8564	20	0.6754	14	7	0.0771	12	0.7850	11
10	9.8664	20	0.6826	12	12	0.0832	12	0.7903	8
15	9.8763	19	0.6884	9	17	0.0893	12	0.7945	6
20	9.8860	19	0.6929	6	22	0.0953	12	0.7974	3
25	9.8956	19	0.6960	3	27	0.1012	+12	0.7991	+ 1
30	+9.9049	+18	+0.6977	+ 1	32	+0.1070		+0.7996	

A TABLE containing the Logarithms of C and D for every Fifth Day of 1827.
(Adapted to Apparent Noon.)

1827.	Log. C.	Daily Diff.	Log. D.	Daily Diff.	1827.	Log. C.	Daily Diff.	Log. D.	Daily Diff.
Jan. 1	+9.4469	+51	+0.7999	−1	July 5	+9.8737	+18	+0.8727	−2
6	9.4720	47	0.7993	3	10	9.8829	18	0.8718	3
11	9.4954	43	0.7976	5	15	9.8918	17	0.8701	5
16	9.5167	39	0.7951	7	20	9.9002	16	0.8677	6
21	9.5364	36	0.7918	8	25	9.9081	15	0.8648	7
26	9.5543	33	0.7879	9	30	9.9156	14	0.8615	7
31	9.5707	30	0.7836	9	Aug. 4	9.9225	13	0.8579	8
Feb. 5	9.5856	27	0.7791	9	9	9.9291	12	0.8541	8
10	9.5992	24	0.7745	9	14	9.9351	11	0.8502	7
15	9.6113	22	0.7702	8	19	9.9408	10	0.8465	7
20	9.6222	20	0.7664	6	24	9.9460	10	0.8431	6
25	9.6324	18	0.7632	5	29	9.9508	9	0.8402	5
Mar. 2	9.6416	18	0.7608	3	Sept. 3	9.9555	8	0.8379	3
7	9.6504	17	0.7594	−1	8	9.9597	8	0.8363	−2
12	9.6589	16	0.7591	+2	13	9.9638	8	0.8355	0
17	9.6667	15	0.7600	4	18	9.9677	8	0.8357	+2
22	9.6744	15	0.7621	6	23	9.9715	8	0.8367	4
27	9.6821	15	0.7652	9	28	9.9753	8	0.8388	6
April 1	9.6897	15	0.7698	11	Oct. 3	9.9791	8	0.8419	8
6	9.6974	16	0.7752	13	8	9.9829	8	0.8459	10
11	9.7053	17	0.7815	14	13	9.9869	8	0.8507	11
16	9.7136	17	0.7885	15	18	9.9911	9	0.8562	12
21	9.7221	18	0.7961	16	23	9.9954	9	0.8624	13
26	9.7309	19	0.8040	16	28	0.0000	10	0.8689	14
May 1	9.7402	19	0.8122	16	Nov. 2	0.0048	10	0.8757	14
6	9.7497	20	0.8203	16	7	0.0098	11	0.8826	14
11	9.7595	20	0.8282	15	12	0.0152	11	0.8894	13
16	9.7697	21	0.8358	14	17	0.0208	12	0.8960	12
21	9.7800	21	0.8429	13	22	0.0266	12	0.9021	11
26	9.7906	21	0.8495	12	27	0.0327	12	0.9077	10
31	9.8013	21	0.8553	10	Dec. 2	0.0388	13	0.9127	9
June 5	9.8120	21	0.8604	9	7	0.0451	13	0.9170	7
10	9.8227	21	0.8647	7	12	0.0515	13	0.9203	5
15	9.8333	21	0.8681	5	17	0.0579	13	0.9229	3
20	9.8439	20	0.8706	3	22	0.0642	12	0.9245	+2
25	9.8541	20	0.8722	+1	27	0.0704	12	0.9253	0
30	+9.8640	19	+0.8729	0	32	+0.0765		+0.9253	

A TABLE containing the Logarithms of C and D for every Fifth Day of 1828.
(Adapted to Apparent Noon.)

1828.	Log. C.	Daily Diff.	Log. D.	Daily Diff.	1828.	Log. C.	Daily Diff.	Log. D.	Daily Diff.
Jan. 1	+9.2853	+71	+0.9254	−2	July 4	+9.8116	+21	+0.9610	−3
6	9.3208	64	0.9244	4	9	9.8220	20	0.9596	4
11	9.3526	58	0.9226	5	14	9.8320	19	0.9577	5
16	9.3814	52	0.9202	6	19	9.8415	18	0.9552	6
21	9.4072	47	0.9172	7	24	9.8503	17	0.9523	7
26	9.4306	42	0.9137	8	29	9.8587	16	0.9489	7
31	9.4516	38	0.9099	8	Aug. 3	9.8665	15	0.9453	7
Feb. 5	9.4704	34	0.9060	8	8	9.8738	13	0.9416	8
10	9.4873	31	0.9020	8	13	9.8805	12	0.9378	7
15	9.5026	28	0.8982	7	18	9.8868	11	0.9341	7
20	9.5167	25	0.8946	6	23	9.8925	11	0.9307	6
25	9.5292	23	0.8916	5	28	9.8980	10	0.9277	5
Mar. 1	9.5407	22	0.8891	3	Sept. 2	9.9030	9	0.9251	4
6	9.5515	20	0.8874	−2	7	9.9077	9	0.9230	3
11	9.5613	19	0.8864	0	12	9.9122	9	0.9216	−1
16	9.5709	18	0.8864	+2	17	9.9165	8	0.9211	0
21	9.5801	18	0.8873	3	22	9.9205	8	0.9212	+2
26	9.5892	18	0.8890	5	27	9.9247	8	0.9222	4
31	9.5982	18	0.8916	7	Oct. 2	9.9288	8	0.9241	5
April 5	9.6073	19	0.8950	8	7	9.9330	9	0.9267	7
10	9.6167	19	0.8992	9	12	9.9373	9	0.9300	8
15	9.6262	20	0.9039	10	17	9.9418	9	0.9339	9
20	9.6363	21	0.9091	11	22	9.9465	10	0.9383	10
25	9.6467	21	0.9146	11	27	9.9515	10	0.9431	10
30	9.6574	23	0.9203	11	Nov. 1	9.9567	11	0.9482	10
May 5	9.6687	23	0.9260	11	6	9.9622	12	0.9533	10
10	9.6802	24	0.9316	11	11	9.9680	12	0.9585	10
15	9.6920	24	0.9370	10	16	9.9741	13	0.9635	9
20	9.7042	24	0.9421	9	21	9.9805	13	0.9681	9
25	9.7164	25	0.9467	8	26	9.9870	13	0.9724	7
30	9.7288	25	0.9509	7	Dec. 1	9.9937	14	0.9761	6
June 4	9.7412	25	0.9544	6	6	0.0005	14	0.9792	5
9	9.7535	25	0.9573	4	11	0.0074	14	0.9816	3
14	9.7657	24	0.9594	3	16	0.0144	14	0.9832	+2
19	9.7777	24	0.9609	+1	21	0.0213	14	0.9841	0
24	9.7894	23	0.9616	0	26	0.0281	13	0.9842	−1
29	+9.8007	+22	+0.9617	−1	31	+0.0347		+0.9835	

A TABLE containing the Logarithms of C and D for every Fifth Day of 1829.
(Adapted to Apparent Noon.)

1829.	Log. C.	Daily Diff.	Log. D.	Daily Diff.
Jan. 1	+8.9263		+0.9834	
		+152		−3
6	9.0022		0.9818	
		125		4
11	9.0648		0.9796	
		106		6
16	9.1176		0.9767	
		91		7
21	9.1632		0.9734	
		78		8
26	9.2022		0.9696	
		69		8
31	9.2365		0.9656	
		60		8
Feb. 5	9.2665		0.9615	
		52		8
10	9.2927		0.9573	
		46		8
15	9.3158		0.9533	
		41		7
20	9.3363		0.9496	
		37		7
25	9.3547		0.9463	
		33		6
Mar. 2	9.3711		0.9435	
		31		4
7	9.3865		0.9414	
		28		3
12	9.4007		0.9400	
		27		−1
17	9.4141		0.9393	
		25		0
22	9.4268		0.9395	
		25		+2
27	9.4395		0.9405	
		25		3
April 1	9.4519		0.9422	
		25		5
6	9.4643		0.9447	
		26		6
11	9.4771		0.9478	
		27		7
16	9.4904		0.9513	
		27		8
21	9.5037		0.9554	
		28		9
26	9.5177		0.9597	
		29		9
May 1	9.5321		0.9642	
		30		9
6	9.5469		0.9687	
		31		9
11	9.5622		0.9731	
		31		8
16	9.5776		0.9773	
		31		8
21	9.5933		0.9813	
		31		7
26	9.6090		0.9850	
		32		6
31	9.6248		0.9879	
		31		5
June 5	9.6403		0.9904	
		31		4
10	9.6557		0.9923	
		30		3
15	9.6709		0.9936	
		30		+1
20	9.6857		0.9942	
		28		0
25	9.6999		0.9941	
		28		−1
30	+9.7137		+0.9934	
		+26		3

1829.	Log. C.	Daily Diff.	Log. D.	Daily Diff.
July 5	+9.7269		+0.9921	
		+25		−4
10	9.7393		0.9901	
		24		5
15	9.7513		0.9875	
		22		6
20	9.7624		0.9845	
		21		7
25	9.7729		0.9810	
		20		8
30	9.7828		0.9772	
		18		8
Aug. 4	9.7919		0.9731	
		17		8
9	9.8004		0.9689	
		16		8
14	9.8082		0.9647	
		14		8
19	9.8154		0.9606	
		13		8
24	9.8221		0.9567	
		12		7
29	9.8283		0.9532	
		12		6
Sept. 3	9.8341		0.9501	
		11		5
8	9.8394		0.9475	
		10		4
13	9.8446		0.9456	
		10		2
18	9.8495		0.9445	
		10		−1
23	9.8543		0.9440	
		9		+1
28	9.8590		0.9443	
		9		2
Oct. 3	9.8637		0.9455	
		10		4
8	9.8685		0.9473	
		10		5
13	9.8736		0.9498	
		10		6
18	9.8787		0.9529	
		11		7
23	9.8842		0.9565	
		12		8
28	9.8899		0.9605	
		12		8
Nov. 2	9.8960		0.9647	
		13		9
7	9.9023		0.9690	
		14		9
12	9.9091		0.9733	
		14		8
17	9.9161		0.9774	
		15		8
22	9.9234		0.9812	
		15		7
27	9.9309		0.9846	
		15		6
Dec. 2	9.9385		0.9874	
		16		5
7	9.9463		0.9897	
		16		3
12	9.9542		0.9913	
		16		+2
17	9.9620		0.9921	
		15		0
22	9.9697		0.9922	
		15		−1
27	9.9773		0.9916	
		+15		3
32	+9.9847		+0.9902	

A TABLE containing the Logarithms of C and D, for every Fifth Day of 1830.
(Adapted to Apparent Noon.)

1830.	Log. C.	Daily Diff.	Log. D.	Daily Diff.
Jan. 1	−8.5302		+0.9902	
				−4
6	8.2504		0.9880	
				6
11	−7.3424		0.9851	
				7
16	+8.1106		0.9817	
		+654		8
21	8.4378		0.9777	
		354		9
26	8.6149		0.9734	
		238		9
31	8.7340		0.9687	
		179		10
Feb. 5	8.8235		0.9639	
		139		10
10	8.8932		0.9590	
		111		9
15	8.9489		0.9543	
		92		9
20	8.9948		0.9498	
		79		8
25	9.0342		0.9457	
		69		7
Mar. 2	9.0686		0.9422	
		60		6
7	9.0986		0.9393	
		55		4
12	9.1261		0.9371	
		49		3
17	9.1508		0.9357	
		46		−1
22	9.1742		0.9351	
		44		0
27	9.1962		0.9352	
		43		+2
April 1	9.2178		0.9362	
		43		3
6	9.2393		0.9379	
		43		5
11	9.2606		0.9403	
		43		6
16	9.2819		0.9432	
		43		7
21	9.3034		0.9465	
		44		7
26	9.3255		0.9501	
		44		8
May 1	9.3477		0.9540	
		45		8
6	9.3700		0.9579	
		45		8
11	9.3927		0.9617	
		46		7
16	9.4155		0.9654	
		45		7
21	9.4379		0.9688	
		45		6
26	9.4602		0.9718	
		44		5
31	9.4822		0.9743	
		43		4
June 5	9.5038		0.9762	
		42		3
10	9.5247		0.9775	
		41		+1
15	9.5452		0.9782	
		39		0
20	9.5649		0.9782	
		38		−1
25	9.5838		0.9775	
		36		3
30	+9.6018		+0.9762	
		+34		−4

1830.	Log. C.	Daily Diff.	Log. D.	Daily Diff.
July 5	+9.6188		+0.9741	
		+32		−6
10	9.6350		0.9713	
		30		7
15	9.6501		0.9680	
		28		8
20	9.6643		0.9642	
		26		9
25	9.6775		0.9598	
		25		9
30	9.6898		0.9551	
		23		10
Aug. 4	9.7012		0.9502	
		21		10
9	9.7117		0.9450	
		20		10
14	9.7215		0.9398	
		18		10
19	9.7304		0.9347	
		16		10
24	9.7385		0.9298	
		15		9
29	9.7462		0.9252	
		14		8
Sept. 3	9.7533		0.9211	
		13		7
8	9.7597		0.9176	
		13		6
13	9.7660		0.9148	
		12		4
18	9.7720		0.9127	
		12		3
23	9.7778		0.9114	
		12		−1
28	9.7836		0.9110	
		11		+1
Oct. 3	9.7892		0.9114	
		12		2
8	9.7951		0.9126	
		12		4
13	9.8010		0.9145	
		12		5
18	9.8072		0.9170	
		13		6
23	9.8137		0.9202	
		14		7
28	9.8205		0.9237	
		14		8
Nov. 2	9.8276		0.9275	
		15		8
7	9.8352		0.9315	
		16		8
12	9.8431		0.9354	
		16		8
17	9.8513		0.9392	
		17		7
22	9.8597		0.9426	
		18		6
27	9.8685		0.9457	
		18		5
Dec. 2	9.8774		0.9481	
		18		4
7	9.8864		0.9499	
		18		2
12	9.8955		0.9510	
		18		+1
17	9.9045		0.9513	
		18		−1
22	9.9134		0.9507	
		18		3
27	9.9222		0.9494	
		+17		−5
32	+9.9307		+0.9471	

A TABLE of the Values of C′ and D′ for Leap Year. (Adapted to Apparent Noon.)

Argument.	C′ = t − .0249 sin 2 ☉	D′ = − .5147 cos 2 ☉	Argument.	C′ = t − .0249 sin 2 ☉	D′ = − .5447 cos 2 ☉
January 1	+0.0086	+0.5110	July 4	+0.5169	+0.4946
6	0.0263	0.4696	9	0.5342	0.4500
11	0.0436	0.4133	14	0.5511	0.3928
16	0.0604	0.3440	19	0.5676	0.3248
21	0.0765	0.2639	24	0.5834	0.2483
26	0.0920	0.1755	29	0.5987	0.1638
31	0.1068	+0.0819	August...... 3	0.6133	+0.0752
February ... 5	0.1207	−0.0141	8	0.6272	−0.0157
10	0.1339	0.1095	13	0.6404	0.1064
15	0.1463	0.2016	18	0.6530	0.1942
20	0.1581	0.2870	23	0.6648	0.2766
25	0.1691	0.3634	28	0.6761	0.3513
March 1	0.1796	0.4286	September.. 2	0.6869	0.4157
6	0.1897	0.4804	7	0.6971	0.4696
11	0.1994	0.5177	12	0.7070	0.5095
16	0.2089	0.5392	17	0.7166	0.5348
21	0.2183	0.5444	22	0.7260	0.5446
26	0.2277	0.5335	27	0.7355	0.5385
31	0.2373	0.5068	October..... 2	0.7450	0.5166
April......... 5	0.2472	0.4652	7	0.7548	0.4794
10	0.2574	0.4101	12	0.7649	0.4278
15	0.2681	0.3432	17	0.7754	0.3632
20	0.2794	0.2664	22	0.7865	0.2876
25	0.2914	0.1823	27	0.7983	0.2032
30	0.3040	0.0932	November... 1	0.8107	0.1122
May......... 5	0.3173	−0.0016	6	0.8238	−0.0176
10	0.3314	+0.0897	11	0.8378	+0.0779
15	0.3461	0.1785	16	0.8525	0.1710
20	0.3615	0.2620	21	0.8679	0.2591
25	0.3775	0.3380	26	0.8840	0.3393
30	0.3940	0.4044	December... 1	0.9007	0.4090
June......... 4	0.4110	0.4595	6	0.9180	0.4661
9	0.4284	0.5017	11	0.9356	0.5085
14	0.4460	0.5299	16	0.9536	0.5350
19	0.4638	0.5435	21	0.9717	0.5447
24	0.4816	0.5420	26	0.9898	0.5371
29	+0.4994	+0.5256	31	+1.0077	+0.5126

A TABLE of the Values of C′ and D′ for the first after Leap Year. (Adapted to Apparent Noon.)

Argument.	C′ = t − 0.0249 sin 2 ☉	D′ = − 0.5447 cos 2 ☉	Argument.	C′ = t − 0.0249 sin 2 ☉	D′ = − 0.5447 cos 2 ☉
January..... 1	+0.0092	+0.5059	July 5	+0.5175	+0.4890
6	0.0269	0.4622	10	0.5347	0.4424
11	0.0441	0.4039	15	0.5516	0.3836
16	0.0608	0.3329	20	0.5679	0.3141
21	0.0769	0.2514	25	0.5837	0.2359
26	0.0922	0.1621	30	0.5989	0.1512
31	0.1068	+0.0680	August...... 4	0.6134	+0.0620
February.... 5	0.1207	−0.0282	9	0.6272	−0.0290
10	0.1338	0.1233	14	0.6403	0.1193
15	0.1461	0.2144	19	0.6528	0.2065
20	0.1577	0.2988	24	0.6645	0.2880
25	0.1687	0.3738	29	0.6757	0.3615
March....... 2	0.1791	0.4370	September.. 3	0.6864	0.4249
7	0.1891	0.4868	8	0.6965	0.4763
12	0.1988	0.5218	13	0.7064	0.5141
17	0.2083	0.5409	18	0.7160	0.5372
22	0.2176	0.5439	23	0.7254	0.5447
27	0.2271	0.5306	28	0.7349	0.5363
April......... 1	0.2367	0.5016	October..... 3	0.7444	0.5121
6	0.2466	0.4579	8	0.7542	0.4727
11	0.2569	0.4011	13	0.7644	0.4191
16	0.2678	0.3326	18	0.7750	0.3528
21	0.2791	0.2546	23	0.7862	0.2756
26	0.2912	0.1697	28	0.7980	0.1900
May 1	0.3039	−0.0800	November... 2	0.8106	0.0985
6	0.3173	+0.0117	7	0.8238	−0.0035
11	0.3315	0.1029	12	0.8379	+0.0918
16	0.3463	0.1909	17	0.8527	0.1844
21	0.3618	0.2736	22	0.8682	0.2715
26	0.3778	0.3483	27	0.8844	0.3503
31	0.3945	0.4131	December... 2	0.9012	0.4183
June......... 5	0.4115	0.4664	7	0.9185	0.4732
10	0.4289	0.5066	12	0.9363	0.5135
15	0.4466	0.5328	17	0.9542	0.5375
20	0.4644	0.5442	22	0.9723	0.5446
25	0.4822	0.5405	27	0.9904	0.5346
30	+0.4999	+0.5220	32	+1.0083	+0.5077

A TABLE of the Values of C′ and D′ for the second after Leap Year.
(Adapted to Apparent Noon.)

Argument.	$C' = t - .0249 \sin 2\odot$	$D' = -0.5447 \cos 2\odot$	Argument.	$C' = t - .0249 \sin 2\odot$	$D' = -0.5447 \cos 2\odot$
January..... 1	+0.0090	+0.5076	July 5	+0.5173	+0.4909
6	0.0267	0.4647	10	0.5346	0.4450
11	0.0440	0.4071	15	0.5515	0.3867
16	0.0607	0.3366	20	0.5678	0.3177
21	0.0768	0.2556	25	0.5836	0.2399
26	0.0922	0.1667	30	0.5989	0.1554
31	0.1068	+0.0726	August...... 4	0.6134	+0.0663
February.... 5	0.1207	−0.0236	9	0 6272	−0.0247
10	0.1339	0.1188	14	0.6404	0.1150
15	0.1462	0.2101	19	0.6529	0.2024
20	0.1578	0.2949	24	0.6646	0.2842
25	0.1688	0.3703	29	0.6759	0.3582
March....... 2	0.1793	0.4342	September.. 3	0.6866	0.4221
7	0.1893	0.4848	8	0.6967	0.4741
12	0.1991	0.5205	13	0.7066	0.5126
17	0.2085	0.5404	18	0.7162	0.5364
22	0.2179	0.5441	23	0.7256	0.5447
27	0.2273	0.5329	28	0.7351	0.5371
April 1	0.2369	0.5034	October. 3	0.7446	0.5136
6	0.2468	0.4604	8	0.7544	0.4750
11	0.2571	0.4041	13	0.7646	0.4220
16	0.2679	0.3362	18	0.7752	0.3563
21	0.2792	0.2586	23	0.7863	0.2796
26	0.2913	0.1739	28	0.7981	0.1944
May......... 1	0.3040	−0.0844	November... 2	0.8106	0.1030
6	0.3173	+0.0073	7	0.8238	−0.0082
11	0.3315	0.0986	12	0.8379	+0.0871
16	0.3463	0.1868	17	0.8526	0.1799
21	0.3617	0.2698	22	0.8681	0.2674
26	0.3777	0.3449	27	0.8843	0.3467
31	0.3943	0.4102	December.. 2	0.9011	0.4152
June......... 5	0.4114	0.4642	7	0.9184	0.4709
10	0.4287	0.5050	12	0.9361	0.5119
15	0.4464	0.5319	17	0.9540	0.5367
20	0.4642	0.5440	22	0.9721	0.5447
25	0.4820	0.5411	27	0.9902	0.5355
30	+0.4998	+0.5232	32	+1.0082	+0.5094

A TABLE of the Values of C′ and D′ for the third after Leap Year.
(Adapted to Apparent Noon.)

Argument.	$C' = t - 0.0249 \sin 2\odot$	$D' = -0.5447 \cos 2\odot$	Argument.	$C' = t - 0.0249 \sin 2\odot$	$D' = -0.5447 \cos 2\odot$
January. 1	+0.0088	+0.5093	July......... 5	+0.5171	+0.4927
6	0.0265	0.4671	10	0.5344	0.4475
11	0.0438	0.4102	15	0.5513	0.3897
16	0.0605	0.3403	20	0.5677	0.3213
21	0.0767	0.2598	25	0.5835	0.2438
26	0.0921	0.1712	30	0.5988	0.1596
31	0.1068	+0.0772	August...... 4	0.6133	+0.0706
February.... 5	0.1207	−0.0189	9	0.6272	−0.0203
10	0.1339	0.1142	14	0.6404	0.1106
15	0.1463	0.2058	19	0.6529	0.1983
20	0.1579	0.2909	24	0.6647	0.2804
25	0.1690	0.3668	29	0.6760	0.3548
March........ 2	0.1795	0.4314	September.. 3	0.6867	0.4193
7	0.1895	0.4827	8	0.6969	0.4719
12	0.1994	0.5191	13	0.7068	0.5111
17	0.2087	0.5398	18	0.7164	0.5356
22	0.2181	0.5443	23	0.7258	0.5447
27	0.2275	0.5352	28	0.7353	0.5378
April......... 1	0.2371	0.5051	October...... 3	0.7448	0.5151
6	0.2470	0.4628	8	0.7546	0.4772
11	0.2572	0.4070	13	0.7647	0.4249
16	0.2680	0.3397	18	0.7753	0.3597
21	0.2793	0.2626	23	0.7864	0.2835
26	0.2913	0.1781	28	0.7982	0.1988
May......... 1	0.3040	−0.0888	November... 2	0.8106	0.1075
6	0.3173	+0.0029	7	0.8238	−0.0128
11	0.3314	0.0943	12	0.8378	+0.0824
16	0.3462	0.1827	17	0.8525	0.1754
21	0.3616	0.2659	22	0.8680	0.2633
26	0.3776	0.3414	27	0.8842	0.3430
31	0.3941	0.4073	December... 2	0.9009	0.4121
June......... 5	0.4112	0.4619	7	0.9182	0.4685
10	0.4285	0.5034	12	0.9358	0.5102
15	0.4462	0.5309	17	0.9538	0.5359
20	0.4640	0.5438	22	0.9719	0.5447
25	0.4818	0.5416	27	0.9900	0.5363
30	+0.4996	+0.5243	32	+1.0080	+0.5110

BESSEL'S FORMULÆ. (P. 421, 422.)

AT page 422, in the formula for determining the value of k, the quantities are calculated from the sun's mean motion in a mean solar day, but as the intervals, in the Annual Tables, are those for the mean solar time of sidereal days, the values in the formula should be found by employing the sun's mean motion in a sidereal day. Formidable as this discrepancy may appear, the error in the right ascension of Polaris will not amount to $0^s.01$, from this source, during a century.

It is much to be wished that astronomers would come to a decision, in fixing some point of the sun's mean longitude for an universal stellar epoch. Bessel has indeed adopted 280° as the sun's mean longitude for January 0, and Mr. Baily 281° for January 1, which assumptions agree sufficiently near, allowing for the difference of a day; but still there is no need of this. Yet as many Catalogues already exist in which January 0 or January 1 is adopted as the epoch, it would perhaps be no easy matter to introduce the vernal equinox as an universal meridian epoch. Since this is the case, there should be chosen some point in the sun's longitude, which nearly accords with January 1 at noon, for the greater part of Europe, and perhaps for this purpose 280° 30′ might be chosen.

When the sun's mean longitude is exactly 280°, or at fictitious noon of January 0, 1800, the mean longitude of the moon's node, by the Tables of Burckhardt, was $\overset{s}{1}$ 3° 16′ 4″, or $\overset{s}{1}$ 3°.26788; and as the node, in a mean solar year, is 19° 20′ 28″.4 retrograde; for any other year, when the sun's mean longitude is exactly 280°, we have

$$\text{Mean long. } ☾\text{'s } ☊ = \overset{s}{1}\ 3^\circ.26788 - 19^\circ.3412\,27\,(t - 1800).$$

If the apparent place of a star is to be calculated by Schumacher's constants from its mean place, then the place of the star should be reduced to the commencement of the year in which the apparent place is required, and to the instant when the sun's mean longitude is exactly 280°. This may be done by comparing the sun's mean longitude, at the star's epoch, with 280°, and thence finding the effect of the precession, by a process similar to that shown for correcting the star's epoch for Mr. Baily's Tables. But it will be more convenient to unite with A a quantity which shall at once give the reduction from true January 0, in any year, to the time for which the calculation is made. For instance, let this quantity be called A‴, then for the meridian of Greenwich, and in any year,

$$A''' = +\,0.00026 - .00002120\,(t - 1800) + \begin{cases} +\ .000683; & \text{if } t \text{ be 1st after Bissextile.} \\ +\ .001365; & \text{if } t \text{ be 2d after Bissextile.} \\ +\ .002048; & \text{if } t \text{ be 3d after Bissextile.} \\ +\ .002730; & \text{if } t \text{ be a Bissextile year.} \end{cases}$$

For the meridian of Greenwich, and in portions of sidereal time, for any year, we shall have

$$k = -\,0.873352 + .0077638\,(t - 1800) + \begin{cases} -\ 0.250685; & \text{if } t \text{ be 1st after Bissextile.} \\ -\ 0.501370; & \text{if } t \text{ be 2d after Bissextile.} \\ -\ 0.752055; & \text{if } t \text{ be 3d after Bissextile.} \\ -\ 1.002740; & \text{if } t \text{ be a Bissextile year,} \end{cases}$$

provided the calculation be made before March the 1st in this last year; but if after March 0, then the two first terms of the formula will give the true value of k without any addition.

Since the mean place of the moon's node and A‴ and k will remain constant quantities, so long as 280° of the sun's mean longitude is fixed as the stellar epoch; those quantities are calculated for the nineteenth century, and may for any year therein be found in Table I.: and, for the eighteenth and twentieth centuries, by applying the quantities in Table II. to those for the corresponding year in the nineteenth.

It must be observed, here, that Tables B and C (in Vol. I.), for nice calculations, should not be used for periods very remote from 1821, but must, as well as the constants, be from time to time recalculated.

The formulæ for determining the values of A and B may be found at page 425, and an auxiliary Table containing a part of the same. By way of simplifying the computations, let us put

$$A' = t - 0.02652 \times \sin 2 \odot \text{; and } A'' = -0.33342 \times \sin ☊ + 0.00320 \times \sin 2 ☊\text{; also}$$
$$B' = -0''.57792 \times \cos 2 \odot\text{; and } B'' = -8''.97707 \times \cos ☊ + 0''.08763 \times \cos 2 ☊.$$

Now the values of A′ and B′ are already computed, and may be found in auxiliary Table 2, at page 425, but as the place of the moon's node is differently divided, for the sake of tabular calculation, a part of the Table is here repeated, and forms Table III. in a decimal form.

Using $+h$ to express the western, and $-h$ to express the eastern longitude of a place from Greenwich, and following the notation at page 422, we have

$$\text{Arg.} = f + g + h + i + k.$$

Having the longitude of a place, we may find the value of h by Table 1, at page 424, though in strictness h should be expressed in sidereal time; but if mean solar time be used, no practical error can arise therefrom.

The values of A″ and B″ before alluded to, will be found in Tables IV. and V., and may be taken from thence with the mean longitude of the moon's node, found by Tables I. and III.

When the correction of the star's place is required from true January 0, then

$$A = A' + A'' + A''';$$

but when the same is required from fictitious January 0, then

$$A = A' + A''\text{; and in either case, } B = B' + B''.$$

Explanation and Use of the following Tables I. and II.

Table I. The mean longitude of the moon's node is common to all years. When the correction is required to be applied to a star's mean place at the preceding true Greenwich January 0, then A‴ must be united with A′+ A″; or the star's annual precession multiplied by A‴, will produce a correction to be applied to the star's mean place or to the sum of the four corrections. In taking out the value of k, care must be taken to take the second column of k, after the 29th of February in a Bissextile year.

Table II. This Table is intended for reducing the values given in Table I. from 1800 to 1900, to those belonging to the corresponding year at a century preceding and following.

Example 1.——Let the moon's node, A‴, and k, be found for 1755?

		☾'s ☊ (s)	A‴	k
Table I.....	1855,	☾'s ☊ = 1 19°.5004,	A‴= + .00114,	k = − 1 .198
II.....	−100,	+ 4 14 .1227	− .00061	+ 0 .227
	1755,	☾'s ☊ = 6 3 .6231,	A‴= + .00053,	k = − 0 .971

Example 2.——Let the moon's node, A‴, and k, be found for 1784?

		☾'s ☊ (s)	A‴	k	
Table I.....	1884 B.,	☾'s ☊ = 6 28°.6048,	A‴= + .00121,	k = − 1 .224 and	− 0 .221
II....	−100,	+ 4 14 .1227	− .00061	+ 0 .227	+ 0 .227
	1784 B.,	☾'s ☊ =11 12 .7275,	A′= + .00060,	k = − 0 .997 and	+ 0 .006 after Feb.

TABLE I.

Showing the Mean Longitude of the Moon's Node, at the Fictitious Noon of January 0, and the Values of A‴ and of k, adapted to Sidereal Time for the Meridian of Greenwich.

Years.	☾'s Node to Mean Solar Time.	Values in Sidereal Time.			Years.	☾'s Node to Mean Solar Time.	Values in Sidereal Time.		
		A‴	k				A‴	k	
	S			(Aft. Feb.)		S			(Aft. Feb.)
1800 C.	1 3°.2679	$+^{d}.00026$	$-\ 0^{d}.873$						
1801	0 13.9267	+ .00092	− 1.116		1853	2 28°.1829	$-^{d}.00018$	−0.713	
1802	11 24.5854	+ .00158	− 1.359		1854	2 8.8416	+ .00048	−0.955	
1803	11 5.2442	+ .00224	− 1.602		1855	1 19.5004	+ .00114	−1.198	
1804 B.	10 15.9030	+ .00291	− 1.845	− 0.842	1856 B.	1 0.1592	+ .00180	−1.441	−0.439
1805	9 26.5618	+ .00084	− 1.085		1857	0 10.8180	− .00026	−0.681	
1806	9 7.2205	+ .00150	− 1.328		1858	11 21.4767	+ .00040	−0.924	
1807	8 17.8793	+ .00216	− 1.571		1859	11 2.1355	+ .00106	−1.167	
1808 B.	7 28.5381	+ .00282	− 1.814	− 0.811	1860 B.	10 12.7943	+ .00172	−1.410	−0.408
1809	7 9.1968	+ .00075	− 1.054		1861	9 23.4530	− .00035	−0.650	
1810	6 19.8556	+ .00141	− 1.297		1862	9 4.1118	+ .00031	−0.893	
1811	6 0.5144	+ .00207	− 1.540		1863	8 14.7706	+ .00097	−1.136	
1812 B.	5 11.1732	+ .00274	− 1.783	− 0.780	1864 B.	7 25.4294	+ .00163	−1.379	−0.376
1813	4 21.8319	+ .00067	− 1.023		1865	7 6.0881	− .00043	−0.619	
1814	4 2.4907	+ .00133	− 1.266		1866	6 16.7469	+ .00023	−0.862	
1815	3 13.1495	+ .00199	− 1.509		1867	5 27.4054	+ .00089	−1.105	
1816 B.	2 23.8083	+ .00265	− 1.752	− 0.749	1868 B.	5 8.0644	+ .00155	−1.348	−0.345
1817	2 4.4670	+ .00058	− 0.992		1869	4 18.7232	− .00052	−0.588	
1818	1 15.1258	+ .00124	− 1.235		1870	3 29.3820	+ .00014	−0.831	
1819	0 25.7846	+ .00191	− 1.478		1871	3 10.0408	+ .00080	−1.074	
1820 B.	0 6.4433	+ .00257	− 1.721	− 0.718	1872 B.	2 20.6995	+ .00146	−1.317	−0.314
1821	11 17.1021	+ .00050	− 0.961		1873	2 1.3583	− .00060	−0.557	
1822	10 27.7609	+ .00116	− 1.204		1874	1 12.0171	+ .00006	−0.800	
1823	10 8.4197	+ .00183	− 1.447		1875	0 22.6759	+ .00072	−1.043	
1824 B.	9 19.0784	+ .00248	− 1.690	− 0.687	1876 B.	0 3.3346	+ .00138	−1.285	−0.283
1825	8 29.7372	+ .00041	− 0.930		1877	11 13.9934	− .00069	−0.526	
1826	8 10.3960	+ .00108	− 1.173		1878	10 24.6522	− .00003	−0.769	
1827	7 21.0548	+ .00174	− 1.416		1879	10 5.3110	+ .00063	−1.012	
1828 B.	7 1.7135	+ .00240	− 1.659	− 0.656	1880 B.	9 15.9697	+ .00130	−1.255	−0.252
1829	6 12.3723	+ .00033	− 0.899		1881	8 26.6285	− .00077	−0.495	
1830	5 23.0311	+ .00099	− 1.142		1882	8 7.2873	− .00011	−0.738	
1831	5 3.6899	+ .00165	− 1.385		1883	7 17.9460	+ .00055	−0.981	
1832 B.	4 14.3486	+ .00231	− 1.628	− 0.625	1884 B.	6 28.6048	+ .00121	−1.224	−0.221
1833	3 25.0074	+ .00024	− 0.868		1885	6 9.2636	− .00086	−0.464	
1834	3 5.6662	+ .00091	− 1.111		1886	5 19.9224	− .00020	−0.707	
1835	2 16.3249	+ .00157	− 1.354		1887	5 0.5881	+ .00046	−0.750	
1836 B.	1 26.9837	+ .00223	− 1.597	− 0.594	1888 B.	4 11.2399	+ .00112	−1.193	−0.190
1837	1 7.6425	+ .00016	− 0.837		1889	3 21.8987	− .00094	−0.433	
1838	0 18.3012	+ .00082	− 1.080		1890	3 2.5575	− .00028	−0.676	
1839	11 28.9600	+ .00148	− 1.323		1891	2 13.2162	+ .00038	−0.919	
1840 B.	11 9.6188	+ .00214	− 1.566	− 0.563	1892 B.	1 23.8750	+ .00104	−1.162	−0.159
1841	10 20.2776	+ .00007	− 0.806		1893	1 4.5338	− .00103	−1.402	
1842	10 0.9364	+ .00073	− 1.049		1894	0 15.1926	− .00037	−0.645	
1843	9 11.5951	+ .00139	− 1.292		1895	11 25.8513	+ .00029	−0.888	
1844 B.	8 22.2539	+ .00205	− 1.534	− 0.532	1896 B.	11 6.5101	+ .00095	−1.131	−0.128
1845	8 2.9127	− .00001	− 0.775		1897	10 17.1689	− .00111	−0.371	
1846	7 13.5715	+ .00065	− 1.018		1898	9 27.8276	− .00045	−0.614	
1847	6 24.2302	+ .00131	− 1.261		1899	9 8.4864	+ .00021	−0.857	
1848 B.	6 4.8890	+ .00197	− 1.503	− 0.501	1900 C.	8 19.1452	+ .00087	−1.100	−0.097
1849	5 15.5478	− .00010	− 0.744		1901	7 29.8040	+ .00153	−1.343	
1850	4 26.2065	+ .00056	− 0.987		1902	7 10.4627	+ .00219	−1.586	
1851	4 6.8653	+ .00122	− 1.229		1903	6 21.1215	+ .00285	−1.828	
1852 B.	3 17.5241	+ .00188	− 1.472	− 0.470	1904 B.	6 1.7803	+ .00351	−2.071	−1.066

TABLE II.—Secular Variations to be applied to the Values in any Year in the 19th Century, to produce those for the same Year in other Centuries.

Years.	☾'s Node to Mean Solar Time.	Variations in Sidereal Time.		
		A‴	k	
			All Years.	B. Years.
	s			
− 100	+ 4 14°.1227	− .00061	+ 0 .227	+ 0 .227
+ 100	+ 7 15 .8773	+ .00061	− 0 .227	− 0 .227

TABLE III.—Motion of the Moon's Node in intervals of Ten Sidereal Days.

Days.	☾'s ☊	Days.	☾'s ☊
Jan. 0	0°.0000	July 9	10 .0339
10	0 .5281	19	10 .5620
20	1 .0562	29	11 .0901
30	1 .5843	Aug. 8	11 .6182
Feb. 9	2 .1124	18	12 .1463
19	2 .6405	28	12 .6744
Mar. 1	3 .1686	Sept. 7	13 .2025
11	3 .6967	17	13 .7306
21	4 .2248	27	14 .2587
31	4 .7529	Oct. 7	14 .7868
April 10	5 .2810	17	15 .3149
20	5 .8091	27	15 .8430
30	6 .3372	Nov. 6	16 .3711
May 10	6 .8653	16	16 .8992
20	7 .3934	26	17 .4273
30	7 .9215	Dec. 6	17 .9554
June 9	8 .4496	16	18 .4835
19	8 .9777	26	19 .0116
29	9 .5058	36	19 .5397

TABLE IV. Arg. = Mean Long. ☾'s ☊ (A″ = −0″.33342 sin. ☊ + 0″.0032 sin. 2 ☊)

Deg.	s O. — A″	Diff.	s I. — A″	Diff.	s II. — A″	Diff.	s III. — A″	Diff.	s IV. — A″	Diff.	s V. — A″	Diff.	Deg.
		+		+		+		+		—		—	
0	0 .00000	570	0 .16394	496	0 .28599	292	0 .33342	8	0 .29153	289	0 .16948	512	30
1	0 .00570	572	0 .16890	491	0 .28891	283	0 .33350	—6	0 .28864	300	0 .16436	518	29
2	0 .01142	570	0 .17381	487	0 .29174	275	0 .33344	14	0 .28564	309	0 .15918	522	28
3	0 .01712	569	0 .17868	480	0 .29449	267	0 .33330	24	0 .28255	316	0 .15396	528	27
4	0 .02281	569	0 .18348	476	0 .29716	258	0 .33306	35	0 .27939	326	0 .14868	532	26
5	0 .02850	568	0 .18824	470	0 .29974	248	0 .33271	44	0 .27613	334	0 .14336	536	25
6	0 .03418	568	0 .19294	464	0 .30222	240	0 .33227	56	0 .27279	342	0 .13800	542	24
7	0 .03986	566	0 .19758	460	0 .30462	231	0 .33171	65	0 .26937	353	0 .13258	545	23
8	0 .04552	565	0 .20218	452	0 .30693	221	0 .33106	75	0 .26584	359	0 .12713	550	22
9	0 .05117	564	0 .20670	447	0 .30914	212	0 .33031	86	0 .26225	368	0 .12163	553	21
10	0 .05681	561	0 .21117	441	0 .31126	203	0 .32945	95	0 .25857	376	0 .11610	558	20
11	0 .06242	560	0 .21558	435	0 .31329	194	0 .32850	106	0 .25481	385	0 .11052	561	19
12	0 .06802	558	0 .21993	427	0 .31523	184	0 .32744	116	0 .25096	392	0 .10491	563	18
13	0 .07360	556	0 .22420	422	0 .31707	174	0 .32628	126	0 .24704	400	0 .09928	567	17
14	0 .07916	554	0 .22842	415	0 .31881	165	0 .32502	136	0 .24304	407	0 .09361	571	16
15	0 .08470	551	0 .23257	407	0 .32046	156	0 .32366	145	0 .23897	415	0 .08790	574	15
16	0 .09021	549	0 .23664	402	0 .32202	146	0 .32221	156	0 .23482	424	0 .08216	576	14
17	0 .09570	545	0 .24066	394	0 .32348	136	0 .32065	166	0 .23058	429	0 .07640	578	13
18	0 .10115	543	0 .24460	387	0 .32484	126	0 .31899	176	0 .22629	437	0 .07062	580	12
19	0 .10658	540	0 .24847	380	0 .32610	117	0 .31723	185	0 .22192	445	0 .06482	583	11
20	0 .11198	537	0 .25227	372	0 .32727	106	0 .31538	196	0 .21747	451	0 .05898	584	10
21	0 .11735	534	0 .25599	365	0 .32833	97	0 .31342	205	0 .21296	458	0 .05315	587	9
22	0 .12269	529	0 .25964	357	0 .32930	87	0 .31137	215	0 .20838	464	0 .04728	588	8
23	0 .12798	526	0 .26321	350	0 .33017	76	0 .30922	224	0 .20374	472	0 .04140	588	7
24	0 .13324	522	0 .26671	340	0 .33093	66	0 .30698	234	0 .19902	476	0 .03552	590	6
25	0 .13846	518	0 .27011	334	0 .33159	57	0 .30464	244	0 .19426	484	0 .02962	591	5
26	0 .14364	514	0 .27345	326	0 .33216	48	0 .30220	253	0 .18942	490	0 .02371	593	4
27	0 .14878	510	0 .27671	317	0 .33264	37	0 .29967	263	0 .18452	495	0 .01778	592	3
28	0 .15388	506	0 .27988	310	0 .33301	25	0 .29704	271	0 .17957	501	0 .01186	593	2
29	0 .15894	500	0 .28298	301	0 .33326	16	0 .29433	280	0 .17456	508	0 .00593	593	1
30	0 .16394		0 .28599		0 .33342		0 .29153		0 .16948		0 .00000		0
		—		—		—		+		+		+	
Deg.	s XI. +		s X. +		s IX. +		s VIII. +		s VII. +		s VI. +		Deg.

TABLE V. Arg. = Mean Longitude ☾'s ☊ (B' = −8".97707 cos. ☊ + 0".08763 cos. 2 ☊.)

Deg.	S 0. —		S I. —		S II. —		S III. ∓		S IV. +		S V. +		Deg.
	B"	Diff.	B"	Diff.	B"	Diff.	B"	Diff.	B"	Diff.	B"	Diff.	
		—		—		—		∓		+		+	
0	8".8894	13	7".7305	768	4".5324	1338	0".0877	1567	4".4447	1377	7".8182	798	30
1	8.8881	40	7.6537	791	4.3986	1351	0.0690	1568	4.5824	1363	7.8980	773	29
2	8.8841	65	7.5745	814	4.2635	1365	0.2258	1568	4.7187	1349	7.9753	749	28
3	8.8776	92	7.4931	836	4.1270	1377	0.3826	1568	4.8536	1335	8.0502	723	27
4	8.8684	118	7.4095	859	3.9893	1391	0.5394	1566	4.9871	1319	8.1225	697	26
5	8.8566	145	7.3236	881	3.8502	1402	0.6960	1566	5.1190	1305	8.1923	673	25
6	8.8421	170	7.2355	903	3.7100	1415	0.8526	1564	5.2495	1289	8.2596	647	24
7	8.8251	197	7.1452	924	3.5685	1426	1.0090	1561	5.3784	1272	8.3243	620	23
8	8.8054	222	7.0528	945	3.4259	1437	1.1651	1558	5.5056	1256	8.3865	595	22
9	8.7832	249	6.9583	967	3.2822	1447	1.3209	1556	5.6312	1239	8.4460	569	21
10	8.7583	275	6.8616	987	3.1375	1458	1.4765	1551	5.7551	1222	8.5029	542	20
11	8.7308	300	6.7629	1008	2.9917	1467	1.6316	1547	5.8773	1204	8.5571	515	19
12	8.7008	326	6.6621	1028	2.8450	1477	1.7863	1543	5.9977	1185	8.6086	489	18
13	8.6682	352	6.5593	1048	2.6973	1485	1.9406	1537	6.1162	1167	8'.6575	462	17
14	8.6330	377	6.4545	1068	2.5488	1494	2.0943	1532	6.2329	1148	8.7037	434	16
15	8.5953	404	6.3477	1086	2.3994	1502	2.2475	1526	6.3477	1129	8.7471	407	15
16	8.5549	428	6.2391	1106	2.2492	1510	2.4001	1519	6.4606	1109	8.7878	380	14
17	8.5121	453	6.1285	1125	2.0982	1517	2.5520	1511	6.5715	1089	8.8258	352	13
18	8.4668	479	6.0160	1143	1.9465	1523	2.7031	1505	6.6804	1069	8.8610	324	12
19	8.4189	504	5.9017	1161	1.7942	1530	2.8536	1496	6.7873	1048	8.8934	297	11
20	8.3685	528	5.7856	1179	1.6412	1535	3.0032	1487	6.8921	1026	8.9231	268	10
21	8.3157	554	5.6677	1197	1.4877	1541	3.1519	1479	6.9947	1005	8.9499	241	9
22	8.2603	578	5.5480	1213	1.3336	1545	3.2998	1469	7.0952	984	8.9740	212	8
23	8.2025	602	5.4267	1230	1.1791	1550	3.4467	1459	7.1936	961	8.9952	184	7
24	8.1423	627	5.3037	1247	1.0241	1554	3.5926	1449	7.2897	939	9.0136	157	6
25	8.0796	650	5.1790	1262	0.8687	1557	3.7375	1438	7.3836	916	9.0293	127	5
26	8.0146	675	5.0528	1279	0.7130	1560	3.8813	1427	7.4752	893	9.0420	100	4
27	7.9471	699	4.9249	1293	0.5570	1562	4.0240	1414	7.5645	869	9.0520	71	3
28	7.8772	721	4.7956	1309	0.4008	1565	4.1654	1403	7.6514	846	9.0591	42	2
29	7.8051	746	4.6647	1323	0.2443	1566	4.3057	1390	7.7360	822	9.0633	14	1
30	7.7305		4.5324		0.0877		4.4447		7.8182		9.0647		0
		+		+		+		±		—		—	
Deg.	S XI. —		S X. —		S IX. —		S VIII. ∓		S VII. +		S VI. +		Deg.

EXPLANATION AND USE OF THE PRECEDING TABLES III. IV. AND V.

TABLE III. is useful for finding the place of the moon's node, when A and B are not required for the whole year.

Example.——Let it be required to find the place of the moon's node, for calculating the corrections due to α Persei at its transit over the meridian of Königsburg, on March 21st, 1829; supposing the logarithms of A and B not to be at hand?

Given civil date .... $f = +21^{d}.000$

Right ascension of star... $3^{h}\ 12^{m}$ .... } Table I. p. 424, { $g = +\ 0.133$

Longitude of Königsburg 1 22 E. of Greenwich........ } Table I. p. 424, { $h = -\ 0.057$

As g is less than $18^{h}\ 40^{m}$, and ☉'s R. A. not equal to that of the star........ $i = +\ 1.000$

By Table 1 of this series, for 1829........ $k = -\ 0.899$

Argument to the Tables that give A, B, C, D........ March = 21.177,

which date falls between March 21st and 31st; then

Table I... 1829, ☾'s ☊........ = S 6 12°.3723

III., March 21........ = − 4.2248

1829, March 21, ☾'s ☊........ = 6 8.1475

TABLES IV. AND V. can require no more than an example to explain them.

Example.——Required the logarithms of A and B, on March 21st, 1829?

☾'s ☊ 6^s 8°.1475 Table IV......... $A'' = +0.04815$; and with ☾'s ☊, in Table V......... $B'' = +8''.9704$
Auxiliary Table (p. 425), Mar. 21 ... $A' = +0.21794$; and in the same Table .... $B' = -0.5798$

Natural number $A = A' + A''$ .... $= +0.26609$; and natural namber $B = B' + B''$ ...... $= +8.3906$
Logarithm of A..................... $= +9.4250$; and logarithm of B .... $= +0.9238$

In the same way on March 31st the moon's node may be found to be 6^s 7°.6194; and thence $A = +0.28137$ and $B = +8.4399$; whence log. $A = +9.4493$ and of $B = +0.9263$.

These two logarithms of A are such as require the star's place to be found for the instant when the sun's mean longitude is exactly 280°. By means of $A''' = +.00033$, the place of the star may be brought from true Greenwich January 0, 1829, to this instant; or A''' may be united with A.

March 21, nat. numb. for A	$= +0.26609$	March 31, nat. numb. for A	$= +0.28137$
Table I. 1829 .... A'''	$= +0.00033$	A'''	$= +0.00033$
Nat. numb. for $A = A' + A'' + A'''$..	$= +0.26642$	Nat. numb. for $A = A' + A'' + A'''$..	$= +0.28170$
Logarithm of A..................	$= +9.4256$	Logarithm of A	$= +9.4498$

For expediting the calculation of an Annual Table of A and B, it is not necessary to find the moon's node on every tenth day, and thence A'' and B''. The motion of the moon's node in 100 sidereal days is 5°.2810 retrograde, so that if 5°.2810 be four times deducted in succession from the place of the moon's node on January 0, its place will remain for April 10, July 19, Oct. 27, and fictitious Dec. 66, which numbers, with that on January 0, will be arguments for finding A'' and B'' from Tables IV. and V.

Example.——Let the values of A'' and B'' be found for the radical dates of 1829?

Date.	☾'s ☊.	A''	Diff.	B''	Diff.
1829.	S				
January..... 0	6 12°.3723	+0.07277		+8″.8479	
			−3083		+1454
April 10	6 7.0913	+0.04194		+8.9933	
			−3120		+ 666
July 19	6 1.8103	+0.01074		+9.0599	
			−3131		− 126
October 27	5 26.5293	−0.02057		+9.0473	
			−3112		− 914
December... 66	5 21.2483	−0.05169		+8.9559	

From these values of A'' and B'', those for the intermediate intervals of ten days may be found by simple proportion *.

By way of example, let the logarithms of A and B for 1829 be found.

1829. Jan. 0	A''............	$= +0.07277$	and B''	$= +8''.8479$
	A'............	$= +0.00908$	B'	$= +0.5449$
	A	$= +0.08185$	B	$= +9.3928$
	Log.	8.9130	Log.................	0.9728

* This method of interpolating B'' will (at its maximum) cause an error of 0ˢ.023 in the R. A. of Polaris; and of A'', an error of 0ˢ.006 in the R. A. of Polaris and of δ Ursæ Minoris. From either of these sources the greatest error in the declination of any star will be 0″.01.

Jan. 0	A″............ = + 0.07277	and	B″	= + 8″.8479
	$\frac{1}{10}$th diff........ − 308.3			+ 145.4
Jan. 10......	A′............ = + 0.069687		B″	= + 8.86244
	A′............ = + 0.04449		B′	= + 0.4419
	A .... = + 0.11418		B	= + 9.3043
	Log. 9.0576		Log.	0.9687

Jan. 10......	A″............ = + 0.069687	and	B″	= + 8″.86244
	$\frac{1}{10}$th diff........ − 308.3			+ 145.4
Jan. 20......	A″............ = + 0.066604		B″	= + 8.87698
	A′............ = + 0.07774		B′	= + 0.2842
	A .... = + 0.14434		B	= + 9.1612
	Log. 9.1594		Log.	0.9620

And, in like manner, the process may be carried on until April 10th, when the difference will be, for A″, −312.0 and for B″, +66.6; and the same order may be observed until fictitious December 36.

The subjoined Supplementary Table is similar to those published at page 425, without the correction A‴, which, in cases of extreme accuracy, should be united with A″ at each radical date; and this may be done without altering the column of differences.

Supplementary Table for 1829.

1829.	Log. of A.	Log. of B.	1829.	Log. of A.	Log. of B.	1829.	Log. of A.	Log. of B.
Jan. 0	8.9130	0.9728	May 10	9.5579	0.9590	Sept. 7	9.8399	0.9326
10	9.0576	0.9687	20	9.5891	0.9680	17	9.8498	0.9287
20	9.1594	0.9620	30	9.6210	0.9754	27	9.8590	0.9282
30	9.2346	0.9534	June 9	9.6525	0.9806	Oct. 7	9.8682	0.9310
Feb. 9	9.2916	0.9442	19	9.6830	0.9831	17	9.8780	0.9369
19	9.3356	0.9356	29	9.7117	0.9827	27	9.8889	0.9452
March 1	9.3705	0.9285	July 9	9.7381	0.9795	Nov. 6	9.9011	0.9543
11	9.3994	0.9243	19	9.7618	0.9739	16	9.9146	0.9634
21	9.4250	0.9235	29	9.7826	0.9660	26	9.9293	0.9715
31	9.4492	0.9262	Aug. 8	9.8005	0.9569	Dec. 6	9.9449	0.9775
April 10	9.4739	0.9320	18	9.8157	0.9471	16	9.9607	0.9808
20	9.5000	0.9400	28	9.8288	0.9391	26	9.9764	0.9808
30	9.5280	0.9493				36	9.9913	0.9776

For the Meridian of Greenwich $k = -0.899$ according to Table I.

To employ the Sidereal Tables for Apparent or Mean Solar Time*.

It is obvious that on any day, if the right ascension of a star coincide with that of the sun at noon, the star's transit will be at noon. Then if it be required to calculate the apparent place of a star, when it is not in the meridian, put g' for the sun's right ascension on the given day at noon, expressed in the decimal fraction of a day, and h' for the apparent hour required, expressed in the same measure. Then i will always be 0 when g' is more, and 1 when less than $18^h\ 43^m\ 31^s$, corresponding to ⊙ $\overset{S}{9}$ 10°, and if we retain the same method of notation as at page 424 of volume I. we shall have

$$\text{Arg.} = f + g' + h + h' + i + k$$ for the tabular date, to enter the Tables with.

* When mean solar time is given, it is easily turned into apparent time, by applying the equation of time with its proper sign.

Example.——Let the argument for calculating the quantities from the Tables A, B, C and D, on March 21st, 1829, at Greenwich apparent noon, be found?

Civil date given		$f=+$	$21^d.000$
Sun's R. A. by N. Almanac.. 0^h 2^m	Table at page 424.	$g'=+$	0 .001
Longitude from Greenwich .. 0 0		$h=$	0 .000
Apparent time............ 0 0		$h'=+$	0 .000
As g' is less than 18^h 43^m 30^s		$i=+$	1 .000
Table I, 1829		$k=-$	0 .899
Argument to Tables A, B, C and D		1829, Mar.	21 .102

This argument is that required for calculating the corrections due to α Persei, on March 21st, 1829, at Greenwich apparent noon. By way of a continuation of the example of α Persei, the apparent place of this star will now be found thus, supposing as before its mean place to be given for true January 0, 1829.

R. A. in time.		In N. P. D.	
A a	$+1^s.123$	A a'	$+3''.58$
B b	$+0$ .434	B b'	-6 .22
C c	-1 .271	C c'	$+5$ .17
D d	-0 .016	D d'	-0 .11
A''' × ann. pr. (+4''.220 × +.00033).	+0 .001	A''' × ann. pr. (+13''.435 × +0.00033)	+0 .00
Corr. in R. A. in time	+0 .271	Corr. in declination	+2.42 in N. P. D. − 2''.42
Jan. 0, 1829, mean R. A.	3^h 12^m 9 .620	Jan. 0, 1829, mean N. P. D.	40° 45' 18 .50
March 21, 1829, app. R. A.	3 12 9 .891	March 21, 1829, app. N. P. D.	40 45 16 .08

If the logarithm of A, containing A''', had been employed, then the last correction would have been included in A a and A a'.

This mode of using the sidereal Tables, in gaining the corrections for *apparent* time, is not intended to supersede the Tables constructed for mean solar time; but merely to show that they may be readily used for calculating the apparent place of a star, when it is not in the meridian.

It might easily be shown, that the subsidiary arguments (A, L, S and A', L', S') may be used, and their maxima reduced so as to give the same corrections here obtained; but as the whole of both series of Tables are contained in this volume, such process will be unnecessary.

THE REV. F. FALLOWS'S SECOND SERIES OF GENERAL TABLES at Pages 445—458.

IN these Tables, the co-efficients employed are nearly the same as those employed in the construction of the series of fourteen general Tables near the beginning of the volume. For the purpose of giving the same results by this second series, as by subsidiary numbers gained from the first series, Mr. FALLOWS adopted the following formulæ;

$$\left.\begin{aligned} A &= t - 0.3583 \sin \Omega - 0.0199 \sin 2 \odot. \\ D &= \cos \Omega + 0.04498 \cos 2 \odot. \end{aligned}\right\} A$$

After the author had constructed his Tables, and seen BESSEL'S new formulæ, for the purpose of gaining the same results by this second series as those deduced from BESSEL'S new formulæ, he has shown that the quantities B and C will require no alteration, and that the change in the nutations may be effected by making

$$\left.\begin{aligned} A &= t - \frac{\sin. \Omega}{3} - \frac{\sin. 2 \odot}{40} \\ D &= 0.93046 \left(\cos \Omega - \frac{\cos. 2 \Omega}{100} + \frac{2}{31} \cos. 2 \odot\right) \end{aligned}\right\} B$$

From these formulæ Table 10 was computed to Greenwich apparent noon, supposing t to commence at noon on January 1st, in common years: but by an omission cos 2 ☊ was put for $\frac{\cos 2 ☊}{100}$ at the head.

In formula B, the quantity depending on twice the node, and a slight error in the solar nutation in A being rectified, the formulæ for producing Bessel's new results by this second series, will require

$$\left.\begin{aligned} A &= t - \frac{\sin. ☊}{3} + \frac{\sin. 2\,☊}{313} - \frac{\sin. 2\,\odot}{38} \\ D &= 0.93046\left(\cos. ☊ - \frac{\cos. 2\,☊}{102} + \frac{2}{31}\cos. 2\,\odot\right) \end{aligned}\right\} C$$

For facilitating the calculation of the annual Table, the same plan may be adopted here as has been used for computing Bessel's A and B at pages 487 and 488; and, indeed, since Bessel's A and Fallows's A (in formula C) are almost the same, Table IV containing the values of A″, at page 485, becomes useful for both purposes. The values of D′, of A′, and of D″ will be given in the three following Tables.

Table of the Values of D.

Arg. = Long. of ☊. (D′ = +0.9305 cos ☊ −0.0091 cos 2☊.)

Deg.	S O. + D′	Diff.	S I. + D′	Diff.	S II. + D′	Diff.	S III. ± D′	Diff.	S IV. − D′	Diff.	S V. − D′	Diff.	Deg.
0	0.9214	—	0.8013	—	0.4698	—	0.0091	—	0.4607	+	0.8104	+	30
1	0.9212	2	0.7933	80	0.4559	139	$\overline{0.0072}$	163	0.4750	143	0.8186	82	29
2	0.9208	4	0.7851	82	0.4419	140	0.0234	+162	0.4891	141	0.8266	80	28
3	0.9202	6	0.7767	84	0.4278	141	0.0397	163	0.5031	140	0.8344	78	27
4	0.9192	10	0.7680	87	0.4135	143	0.0559	162	0.5169	138	0.8419	75	26
5	0.9180	12	0.7591	89	0.3991	144	0.0721	162	0.5306	137	0.8491	72	25
6	0.9165	15	0.7500	91	0.3845	146	0.0884	163	0.5441	135	0.8561	70	24
7	0.9147	18	0.7406	94	0.3699	146	0.1046	162	0.5575	134	0.8628	67	23
8	0.9127	20	0.7310	96	0.3551	148	0.1208	162	0.5707	132	0.8693	65	22
9	0.9104	23	0.7212	98	0.3402	149	0.1369	161	0.5837	130	0.8754	61	21
10	0.9078	26	0.7112	100	0.3252	150	0.1530	161	0.5965	128	0.8813	59	20
11	0.9049	29	0.7010	102	0.3101	151	0.1691	161	0.6092	127	0.8869	56	19
12	0.9018	31	0.6905	105	0.2949	152	0.1852	161	0.6217	125	0.8923	54	18
13	0.8985	33	0.6799	106	0.2796	153	0.2012	160	0.6340	123	0.8974	51	17
14	0.8948	37	0.6690	109	0.2642	154	0.2171	159	0.6461	121	0.9021	47	16
15	0.8909	39	0.6579	111	0.2487	155	0.2330	159	0.6580	119	0.9066	45	15
16	0.8867	42	0.6467	112	0.2331	156	0.2488	158	0.6696	116	0.9108	42	14
17	0.8823	44	0.6352	115	0.2175	156	0.2645	157	0.6811	115	0.9148	40	13
18	0.8776	47	0.6235	117	0.2018	157	0.2802	157	0.6924	113	0.9185	37	12
19	0.8726	51	0.6117	118	0.1860	158	0.2958	156	0.7035	111	0.9219	34	11
20	0.8674	52	0.5997	120	0.1701	159	0.3113	155	0.7144	109	0.9249	30	10
21	0.8619	55	0.5875	122	0.1542	159	0.3267	154	0.7250	106	0.9277	28	9
22	0.8562	57	0.5751	124	0.1382	160	0.3420	153	0.7354	104	0.9302	25	8
23	0.8502	60	0.5625	126	0.1222	160	0.3573	153	0.7456	102	0.9323	21	7
24	0.8439	63	0.5497	128	0.1061	161	0.3724	151	0.7556	100	0.9342	19	6
25	0.8374	65	0.5368	129	0.0900	161	0.3874	150	0.7653	97	0.9359	17	5
26	0.8307	67	0.5237	131	0.0739	161	0.4023	149	0.7748	95	0.9373	14	4
27	0.8237	70	0.5105	132	0.0577	162	0.4171	148	0.7841	93	0.9383	10	3
28	0.8165	72	0.4971	134	0.0415	162	0.4317	146	0.7931	90	0.9390	7	2
29	0.8090	75	0.4835	136	0.0253	162	0.4463	146	0.8019	88	0.9394	4	1
30	0.8013	77	0.4698	137	0.0091	162	0.4607	144	0.8104	85	0.9396	2	0
		+		+		+		—		—		—	
Deg.	S XI. +		S X. +		S IX. +		S VIII. ±		S VII. −		S VI. −		Deg.

TABLE of the VALUES of A'.

Arg. = ☉'s Long. (A'= −0 .0265 sin 2 ☉.)

Deg.	S O. VI. } − A'	Diff.	S I. VII. } − A	Diff.	S II. VIII. } − A	Diff.	Deg.
0	0 .0000	9	0 .0230	4	0 .0230	5	30
1	0 .0009	10	0 .0234	5	0 .0225	5	29
2	0 .0019	9	0 .0239	4	0 .0220	5	28
3	0 .0028	9	0 .0243	3	0 .0215	6	27
4	0 .0037	9	0 .0246	3	0 .0209	6	26
5	0 .0046	9	0 .0249	3	0 .0203	6	25
6	0 .0055	9	0 .0252	3	0 .0197	6	24
7	0 .0064	9	0 .0255	3	0 .0191	7	23
8	0 .0073	9	0 .0258	2	0 .0184	6	22
9	0 .0082	9	0 .0260	2	0 .0178	7	21
10	0 .0091	8	0 .0262	1	0 .0171	8	20
11	0 .0099	9	0 .0263	1	0 .0163	7	19
12	0 .0108	9	0 .0264	1	0 .0156	8	18
13	0 .0117	8	0 .0265	0	0 .0148	7	17
14	0 .0125	8	0 .0265	0	0 .0141	8	16
15	0 .0133	8	0 .0265	0	0 .0133	8	15
16	0 .0141	7	0 .0265	0	0 .0125	8	14
17	0 .0148	8	0 .0265	1	0 .0117	9	13
18	0 .0156	7	0 .0264	1	0 .0108	9	12
19	0 .0163	8	0 .0263	1	0 .0099	8	11
20	0 .0171	7	0 .0262	2	0 .0091	9	10
21	0 .0178	6	0 .0260	2	0 .0082	9	9
22	0 .0184	7	0 .0258	3	0 .0073	9	8
23	0 .0191	6	0 .0255	3	0 .0064	9	7
24	0 .0197	6	0 .0252	3	0 .0055	9	6
25	0 .0203	6	0 .0249	3	0 .0046	9	5
26	0 .0209	6	0 .0246	3	0 .0037	9	4
27	0 .0215	5	0 .0243	4	0 .0028	9	3
28	0 .0220	5	0 .0239	5	0 .0019	10	2
29	0 .0225	5	0 .0234	4	0 .0009	9	1
30	0 .0230		0 .0230		0 .0000		0
Deg.	S V. XI. } +		S IV. X. } +		S III. IX. } +		Deg.

TABLE of the VALUES of D''.

Arg. = ☉'s Long. (D''= +0 .0601 cos 2 ☉.)

Deg.	S O. VI. } + D''	Diff.	S I. VII. } ± D''	Diff.	S II. VIII. } − D''	Diff.	Deg.
0	0 .0601	0	0 .0302	19	0 .0302	17	30
1	0 .0601	1	0 .0283	19	0 .0319	17	29
2	0 .0600	2	0 .0264	19	0 .0336	17	28
3	0 .0598	3	0 .0245	20	0 .0353	17	27
4	0 .0595	3	0 .0225	19	0 .0370	16	26
5	0 .0592	4	0 .0206	20	0 .0386	16	25
6	0 .0588	5	0 .0186	20	0 .0402	15	24
7	0 .0583	5	0 .0166	20	0 .0417	15	23
8	0 .0578	6	0 .0146	21	0 .0432	15	22
9	0 .0572	7	0 .0125	21	0 .0447	14	21
10	0 .0565	8	0 .0104	20	0 .0461	13	20
11	0 .0557	8	0 .0084	21	0 .0474	12	19
12	0 .0549	9	0 .0063	21	0 .0486	12	18
13	0 .0540	10	0 .0042	21	0 .0498	12	17
14	0 .0530	9	0 .0021	21	0 .0510	11	16
15	0 .0521	11	0 .0000	21	0 .0521	9	15
16	0 .0510	12	$\overline{0.0021}$	21	0 .0530	10	14
17	0 .0498	12	0 .0042	21	0 .0540	9	13
18	0 .0486	12	0 .0063	21	0 .0549	8	12
19	0 .0474	13	0 .0084	20	0 .0557	8	11
20	0 .0461	14	0 .0104	21	0 .0565	7	10
21	0 .0447	15	0 .0125	21	0 .0572	6	9
22	0 .0432	15	0 .0146	20	0 .0578	5	8
23	0 .0417	15	0 .0166	20	0 .0583	5	7
24	0 .0402	16	0 .0186	20	0 .0588	4	6
25	0 .0386	16	0 .0206	19	0 .0592	3	5
26	0 .0370	17	0 .0225	20	0 .0595	3	4
27	0 .0353	17	0 .0245	19	0 .0598	2	3
28	0 .0336	17	0 .0264	19	0 .0600	1	2
29	0 .0319	17	0 .0283	19	0 .0601	0	1
30	0 .0302		0 .0302		0 .0601		0
Deg.	S V. XI. } +		S IV. X. } ±		S III. IX. } −		Deg.

If the annual Table is to be computed for mean solar time, and the year to commence on January 0 at noon in common years, the value of t may be taken at sight by entering column 10 of Table XV. at pages 39—44 of this volume, and depressing the numbers there taken out, by placing the decimal point one figure to the left: for then it will be the same as from column one, but to four places of decimals. If t be made to commence on January the first at noon, then enter Table XV. as before; but take out for the *preceding* day. Using TABLE IV for A'' at page 485, and the three preceding TABLES for the values of D', A', and D'', we shall have, in either case,

$$A = t + A' + A''$$
$$D = D' + D''$$

Example.—Let A and D, (in formula C) of Mr. FALLOWS's second series, be found for July the 19th, 1824, at Greenwich noon; supposing the year to commence on January the first in common years?

By Nautical Almanac of 1824, for July 19th.... ☾'s ☊ = 9ˢ 8° 29' and ☉'s longitude = 3ˢ 26° 40'.

With preceding day, July 18, enter column 10 in Table XV. and depressing one figure.... } t = + 0 .5453

With ☾'s ☊.... 9ˢ 8° 29' in Table IV...A''= +0 .3288; and D'.............. = +0 .1459

With ☉'s long... 3 26 40..............A' = +0 .0213; and D'' .... = −0 .0359

A = t + A' + A''........................ = +0 .8954; and D = D' + D''....... = +0 .1100

At page 455 of Vol. I. July 19.............A = + .895 ; and D .... = + .110

3 S

In the late new Catalogue, published by the Astronomical Society of London, the constant of aberration is somewhat increased, and the co-efficients of the nutations differ from those employed by Mr. Fallows; but since the co-efficients of precession are not sensibly altered, the second series of Mr. Fallows may be still employed by making a slight alteration in the formulæ of A, B, C and D. Nearly the same results may be obtained by Mr. Fallows's second series, as by the Catalogue of the Astronomical Society, by making

$$\left.\begin{array}{l} A = t - 0.3436 \sin \text{☊} + 0.0041 \sin 2\,\text{☊} - \dfrac{\sin 2\odot}{40} \\ B = +\,1.0052 \sin \odot \text{ at the time for which the calculation is made} \\ C = +\,1.0052 \cos \odot \text{ at} \ldots\ldots\ldots\ldots\ldots\text{ditto}\ldots\ldots\ldots\ldots \\ D = \ 0.95875 \left(\cos \text{☊} - \dfrac{\cos 2\,\text{☊}}{103} + 0.05892 \cos 2\odot\right) \end{array}\right\} D$$

By changing the formulæ of the variable annual quantities, the second series, and constants deduced therefrom, may be employed,

To obtain the sum of the corrections the same as by the Table of 48 Stars at pages 113—148, and including the precession.... } by using formula A.

To obtain the sum of the corrections the same as by Bessel's formula at page 422.................................... } by using formula C.

To obtain the sum of the corrections the same as by New Catalogue of the Astronomical Society of London .... } by using formula D.

Should it be found desirable to construct annual tables of A and D, to every tenth day of the year, then it will expedite the calculation if the values of $t + A'$, and of D'' be made out for every tenth day of four successive years: for by these means a great deal of time may be saved, and the true values thus found would not in fifty years vary more than 0 .0004 from the true value of $t + A'$; nor more than 0 .0008 from the true value of D''. It is almost unnecessary to add, that the auxiliary Tables here given may be employed in calculating the annual values for sidereal time; and that if the annual Table is to be computed for sidereal time, the value of t and of the Sun's longitude and Moon's node must be found in sidereal time.

If constants be employed, it must be observed that those constants will vary considerably when the star is near the pole; but the variation in natural numbers or in logarithms may be found, by computing the constants of the desired star to two intervals of time, and dividing their difference by the number of years intervening, to give the annual variation.

Arguments for the Sun's Latitude.

In the note of explanation subjoined to Solar Table 11, in page 172, the argument A is put $= \odot - ☾$, instead of $= ☾ - \odot$, where the character ☾ denotes the moon's mean longitude, and $\odot$ the sun's mean longitude; and in consequence of this erroneous precept for the formation of argument A, the two examples worked at page 358 are incorrect, so far as this argument is concerned, and ought to be performed in the following manner:

Example 1.——

	s	°	′
☾ 15 June, 1824....................	10	7	58
⊙ Ditto..........................	2	24	14
A = ☾ − ⊙........................	7	13	44
Helioc. long. of ⊖ = B.............	8	24	14
Ditto of Venus = C..............	1	25	0
Ditto of Jupiter = E..............	3	20	10
Supp. long. ☾'s ☊ = N..............	2	19	44

	s	°	′		Solar Tab. 11.
A+B+N....	6	27	42	= 577 ..	− 0″.31
2 B−C......	3	23	28	= 314 ..	− 0 .06 (+ before)
3 C−4 B....	6	8	4	= 522 ..	− 0 .25
B−2 E......	1	13	54	= 122 ..	+ 0 .11
Sun's latitude required (instead of −0″.75, as before given)......					− 0 .51

Example 2.

	s	°	′
☾ 15 December, 1824..............	6	21	58
⊙ Ditto..........................	8	23	33
A = ☾ − ⊙.......................	9	28	25
B...............................	2	23	33
C...............................	11	18	26
E...............................	4	5	3
N...............................	2	29	31

	s	°	′		Solar Tab. 11.
A+B+N....	3	21	29	= 309	− 0 .64
2 B−C......	5	28	40	= 497	− 0 .10
3 C−4 B.....	11	21	6	= 975	+ 0 .23
B−2 E......	6	13	27	= 536	− 0 .16
Sun's latitude required (instead of −0″.59 before given)............					− 0 .67

The numbers here computed must consequently be substituted for those before determined, as the proper factors to be used in the two examples given in the same page (358), to illustrate the use of SOLAR TABLE 12.

In these examples the data were taken from the Nautical Almanac of 1824, where the true longitudes are given instead of the mean longitudes of the sun and moon; but the error arising from this substitution is scarcely perceptible. To avoid the trouble of computing the arguments from quantities to be deduced from some ephemeris of the given year, or from the original Tables of the solar, lunar, and planetary motions computed by Delambre, Carlini, Burg, Burckhardt, Damoiseau, Bouvard, &c. we have arranged a series of short Tables on the most simple plan, that will give the respective arguments, already computed, by inspection; in the formation of which the mean motions only are adopted, and united agreeably to the proper precepts; and as the greatest error, that can arise from the use of these convenient Tables, will never exceed 0″.09 in the sun's latitude, they will be found very useful in all cases where nice computations relating to the sun's declination are required, by saving much time, and by preventing a liability to error from a complicated union of the quantities constituting the arguments.

The titles of the five successive Tables sufficiently explain their use; but to show the extent of their accuracy, we will give the arguments due to the two preceding examples and corresponding corrections taken from TABLE 11; from the arrangement of which the advantage of the method will be obvious.

Example 1.					*Example* 2.				
Date.	A+B+N	2 B—C	3 C—4 B	B—2 E	Date.	A+B+N	2 B—C	3 C—4 B	B—2 E
Tab. 1. 1824..	512	141	138	776	Tab. 1. 1824..	512	141	138	776
3. June..	549	155	362	344	3. Dec...	274	342	802	760
4. 15...	514	014	034	032	4. 15...	514	014	034	032
Arguments. ...	575	310	534	152	Arguments	300	497	974	568
Corrections....	—0 .30	—0 .06	—0 .25	+0 .08	Corrections	—0 .65	—0 .10	+0 .23	—0 .14
Sum of the corrections or sun's latitude......=—0″.53					Sum of the corrections or sun's latitude......=—0″.66				

When the year is in the 19th century, the numbers taken from Tables 1, 3, 4, and 5, when added together will produce the four arguments; but if it be in the preceding or following century, the numbers from Table 2 must also be applied. If the given place be east or west of Greenwich, its time must be reduced to Greenwich time, by subtraction in the former case and addition in the latter, at the rate of one hour for 15°.

Example 3.—Required the sun's latitude at New York, on June 27th, 1766, at $19^h\ 44^m$ mean time?

Mean time at place...... (1766, June 27^d) $17^h\ 44^m$
Longitude of New York, west .... 4 56 (add)

Greenwich Time........ (1766, June 27^d) 22 40

	A+B+N	2 B—C	3 C—4 B	B—2 E
Table 1, 1866...............	229	870	951	691
2, —100..................	809	550	351	868
3, June	549	155	362	344
4, 27^d.......	955	027	062	059
5, 23^h...	035	001	002	002
1766, June 27^d 23^h	577	603	728	964

The sun's latitude.

A+B+N = 577	Table 11, page 172.	+0″.31
2 B — C = 603		—0 .07
3 C — 4 B= 728		—0 .09
B — 2 E= 964		+0 .16
Sum = sun's latitude........		+0 .31

Example 4.—Required the sun's latitude at Paramatta, on January 12th, 1924, at $6^h\ 24^m$ mean time?

Mean time at Paramatta...... (1924, Jan. 12^d) $6^h\ 24^m$
Longitude of Paramatta, east................ 10 4 (sub.)

Greenwich time............. (1924, Jan. 11^d) 20 20

	A+B+N	2 B—C	3 C—4 B	B—2 E
Table 1, 1824 B.............	512	141	138	776
2, +100...............	191	450	649	132
3, Jan. B.........	963	999	998	998
4, 11^d....	367	010	024	023
5, 20^h	031	001	002	002
1924, Jan. 11^d 20^h	064	601	811	931

The sun's latitude.

A+B+N = 064	Page 172.	—0″.26
2 B — C = 601		—0 .07
3 C —4 B= 811		+0 .03
B —2 E= 931		+0 .16
Sum = sun's latitude.......		—0 .14

THE SUN'S LATITUDE.

TABLES for DETERMINING the ARGUMENTS for TABLE 11, at page 172.

TABLE 1.

EPOCHS for the NINETEENTH CENTURY.

Years.	A+B+N	2 B — C	3 C — 4 B	B — 2 E	Years.	A+B+N	2 B — C	3 C — 4 B	B — 2 E
1801	790	527	978	654	1851	886	252	803	220
1802	203	901	854	485	1852 B.	336	627	681	053
1803	616	276	730	316	1853	749	002	557	884
1804 B.	066	651	608	149	1854	162	376	433	715
1805	479	025	484	980	1855	575	750	309	546
1806	892	399	360	811	1856 B.	025	125	187	379
1807	305	774	236	641	1857	438	500	063	209
1808 B.	755	149	114	474	1858	851	874	939	040
1809	168	523	990	305	1859	264	248	815	871
1810	581	897	866	136	1860 B.	714	623	693	704
1811	994	272	742	967	1861	127	998	569	535
1812 B.	444	647	620	800	1862	540	372	445	365
1813	857	021	496	630	1863	953	746	321	196
1814	270	395	372	461	1864 B.	403	121	199	029
1815	683	770	248	292	1865	816	496	075	860
1816 B.	133	145	126	125	1866	229	870	951	691
1817	546	519	002	956	1867	642	244	827	521
1818	959	893	878	787	1868 B.	092	619	705	354
1819	372	268	754	617	1869	505	994	581	185
1820 B.	822	643	632	450	1870	918	368	457	016
1821	236	017	508	281	1871	331	742	333	847
1822	649	392	384	112	1872 B.	781	117	211	680
1823	062	766	260	943	1873	194	492	087	511
1824 B.	512	141	138	776	1874	607	866	963	341
1825	925	515	014	607	1875	020	240	839	172
1826	338	890	890	437	1876 B.	470	615	717	005
1827	751	264	766	268	1877	883	990	593	836
1828 B.	201	639	644	101	1878	296	364	469	667
1829	614	013	520	932	1879	709	738	345	498
1830	027	388	396	763	1880 B.	159	113	223	331
1831	440	762	272	593	1881	572	488	099	161
1832 B.	890	137	150	426	1882	985	863	975	992
1833	303	511	026	257	1883	398	237	851	823
1834	716	886	902	088	1884 B.	848	612	729	656
1835	129	260	778	919	1885	261	986	605	487
1836 B.	579	635	657	752	1886	674	360	481	318
1837	992	009	532	583	1887	087	734	357	148
1838	405	384	408	413	1888 B.	537	109	235	981
1839	818	758	284	244	1889	950	484	111	812
1840 B.	268	133	162	077	1890	364	858	987	643
1841	681	507	038	908	1891	777	232	863	474
1842	094	882	914	739	1892 B.	227	608	741	307
1843	507	256	790	569	1893	640	982	617	137
1844 B.	957	631	668	402	1894	053	356	493	968
1845	370	005	544	233	1895	466	730	369	799
1846	783	380	420	064	1896 B.	916	105	247	632
1847	196	754	296	895	1897	329	480	123	463
1848 B.	646	129	174	728	1898	742	854	999	294
1849	059	503	050	559	1899	155	228	875	124
1850	472	878	926	389	1900 C.	568	603	751	955

TABLE 2.

MOTIONS to be ADDED to the EPOCHS in Table 1, to obtain the EPOCHS for other CENTURIES.

Centuries.		A+B+N	2 B — C	3 C — 4 B	B — 2 E
Past...	—200	618	100	701	737
	—100	809	550	351	868
Future	+100	191	450	649	132
	+200	419	901	301	265

TABLE 3.

MOTIONS for MONTHS.

Months.		A+B+N	2 B — C	3 C — 4 B	B — 2 E
January ...	C.	000	000	000	000
	B.	963	999	998	998
February..	C.	139	032	074	071
	B.	102	031	072	068
March		168	060	142	134
April		307	092	216	205
May		410	123	288	273
June		549	155	362	344
July		651	186	434	412
August		791	217	509	483
September......		930	249	583	553
October		032	280	655	621
November		171	312	730	692
December		274	342	802	760

In the months January and February of *Bissextile* years, use the line *marked* B; but in all *other* years use the line C.

TABLE 4.

MOTIONS for DAYS.

Days.	A+B+N	2 B — C	3 C — 4 B	B — 2 E
1	000	000	000	000
2	037	001	002	002
3	073	002	005	005
4	110	003	007	007
5	147	004	010	009
6	184	005	012	011
7	220	006	014	014
8	257	007	017	016
9	294	008	019	018
10	331	009	022	020
11	367	010	024	023
12	404	011	026	025
13	441	012	029	027
14	478	013	031	030
15	514	014	034	032
16	551	015	036	034
17	588	016	038	036
18	625	017	041	039
19	661	018	043	041
20	698	019	046	043
21	735	021	048	046
22	772	022	050	048
23	808	023	053	050
24	845	024	055	052
25	882	025	058	055
26	919	026	060	057
27	955	027	062	059
28	992	028	065	061
29	029	029	067	064
30	066	030	070	066
31	102	031	072	068

TABLE 5.

MOTIONS for HOURS.

Hours.	A+B+N	2 B — C	3 C — 4 B	B — 2 E
0	000	000	000	000
1	002	000	000	000
2	003	000	000	000
3	005	000	000	000
4	006	000	000	000
5	008	000	001	000
6	009	000	001	001
7	011	000	001	001
8	012	000	001	001
9	014	000	001	001
10	015	000	001	001
11	017	000	001	001
12	018	001	001	001
13	020	001	001	001
14	021	001	001	001
15	023	001	002	001
16	024	001	002	002
17	026	001	002	002
18	028	001	002	002
19	029	001	002	002
20	031	001	002	002
21	032	001	002	002
22	034	001	002	002
23	035	001	002	002
24	037	001	002	002

CATALOGUE OF 520 ZODIACAL STARS LIABLE TO OCCULTATION.

ON a re-examination of *the principal stars that may be occulted by the moon*, given in our ZODIACAL TABLE 6 (pp. 223—224), as extracted from the *Jahrbuch* of 1780, the arrangement, being in the order of longitude instead of right ascension, may frequently disappoint the observer, who looks out for a regular succession of occultations on any given evening, since the occurrence of the phenomena may not take place in the order pointed out by the Table. On this account, and because the subjects of occultations, and of the transits of moon-culminating stars, have lately engaged the attention of computers and observers in various parts of the globe, we have composed a catalogue of all the *occultable* stars given by Zach and Piazzi down to the sixth magnitude inclusive, viz. 520; which also include all that are found above the magnitude 6.7 in the Catalogue of the Astronomical Society, except *four*, which, having been observed only once each, are omitted. By reason of the extreme difficulty of observing a star smaller than one of the sixth magnitude in contact with the enlightened limb of the moon, with an ordinary telescope, it is presumed that the selection that has been made will be considered sufficiently extensive. In consequence of the various methods that have been devised of computing the occultation of a star, which will be found in the second volume, the Catalogue contains, in so many separate columns, the longitudes and latitudes of each star, in the order of their respective longitudes, as well as their corresponding right ascensions and declinations, together with the annual variations of the former, and annual precessions of the latter computed for the year 1830; the mean places being given for 1820. The same page also contains the north-east angle that the star's meridian makes with the ecliptic, and the annual variation of this angle, the principal columns of which have been taken chiefly from the *Bononiæ Ephemerides* for 1817 to 1822, computed by *Pietro Caturegli;* which computations have greatly facilitated our labours. In the opposite adjoining pages we have given the *subsidiary numbers* for converting the mean into the apparent places of each star, which have been carefully computed, with much labour and expense, and which may be all taken out at one opening of the book, and applied in the way that has been directed at pages 309—324. The method of finding the aberration in longitude and latitude by arguments, agreeably to Zach's plan for aberration in right ascension and declination, is we apprehend new, and therefore may require explanation. To obtain the argument in longitude, the star's longitude is only subtracted from *nine* signs; but the maximum must be found by multiplying the constant 20″.255 by the secant of the star's latitude: and for the argument in latitude the star's longitude is subtracted from *six* signs; the maximum in this case being obtained by multiplying 20″.255 by the sine of the star's latitude. When the star's place is given in this denomination, the nutation may be supplied by using Tables 3 and 4 at page 400; and the precession in longitude is given in Table 3 at page 285. The arguments in longitude and latitude may be adapted to any year by subtracting 50″.3 from each of the arguments of the epoch, for every following year: but if this adaptation should be neglected, the error arising from the omission will not amount to one-tenth of a second in longitude in twenty years; nor will it amount to the same quantity in the maximum of latitude in an entire century. The sum of the annual variation and aberration in latitude must be applied according to its sign, without regarding whether the latitude be north or south: likewise, the sum of the precession, aberration and nutations in declination must be applied according to its sign, without regarding whether the declination be north or south.

To this Catalogue is subjoined, in one page, the numbers of the Astronomical Society's Catalogue, that correspond to the figures indicating the numerical succession of the Catalogue we have thus arranged, in order that the constants of that valuable work may be substituted by such persons as prefer the method of gaining the corrections in right ascension and declination by their means.

After having concluded our Catalogue, the *numerals* by which they are there denoted, afforded the means of a convenient classification of such stars as may be expected to suffer occultation at certain times, depending on the longitudes of the moon and of its node; and in justice we are bound to acknowledge here our great obligation to Mr. Henry Jenkins of Isleworth, for the sole computation and arrangement of the fourteen pages that follow the Catalogue, and that exhibit a distinction between those stars that may be expected to be occulted when viewed by a spectator in *England,* or in places having the same parallel of latitude, and those that will be seen occulted in other countries: the former are everywhere followed by a *round point* or period, and the latter by a *comma* or tailed point. The method of computing the elements for effecting such ingenious arrangement, and in the order of right ascension too, will be explained hereafter.

A CATALOGUE OF 520 ZODIACAL STARS FOR 1820. (Pietro Caturegli.)

No.	Mag.	Stars.	Longitude.	Ann. Var.	Latitude.	Ann. Var.	North East Angle of the *'s Meridian with the Ecliptic.	Ann. Var. of the ∠.	R. Ascension in Time.	Ann. Precess. 1830.	Declination.	Ann. Precess. 1830.
			S ° ′ ″	+	° ′ ″		° ′ ″	+	h. m. s.	+	° ′ ″	
1	6	12 Ceti.	0 2 48 25 .1	50″.12	6 37 5″.0 S	+0″.02	66 28 25″.7	2″.23	0 20 51	3s.057	4 57 10″ S	−19″.957
2	6	10 Ceti.	0 3 34 18 .9	49 .98	2 41 23 .3 S	−0 .17	66 34 52 .6	1 .92	0 17 23	3 .066	1 2 48 S	−19 .982
3	6	44 Piscium...	0 4 5 4 .9	50 .16	0 44 37 .2 S	−0 .10	66 35 48 .8	1 .81	0 16 10	3 .069	0 56 34 N	+19 .992
4	6	13 Ceti.	0 4 8 3 .2	50 .45	6 47 21 .0 S	+0 .08	66 31 19 .7	2 .67	0 25 58	3 .056	4 35 4 S	−19 .911
5	6	35 Piscium...	0 4 26 8 .2	50 .13	6 36 14 .6 N	+0 .09	66 22 43 .1	0 .90	0 5 43	3 .072	7 49 18 N	+20 .037
6	5 .6	d Piscium. ...	0 5 28 5 .1	50 .08	5 27 58 .2 N	+0 .02	66 27 14 .5	1 .40	0 11 21	3 .077	7 11 24 N	+20 .017
7	6	45 Piscium...	0 6 25 44 .8	49 .99	4 30 40 .7 N	+0 .13	66 31 25 .7	1 .84	0 16 25	3 .080	6 41 46 N	+19 .990
8	5	20 Ceti.........	0 9 13 42 .7	49 .97	6 17 34 .0 S	−0 .01	66 50 30 .7	4 .20	0 43 49	3 .058	2 7 25 S	−19 .670
9	6	Pisc. 20 May.	0 10 39 8 .1	49 .74	0 8 37 .7 N	−1 .02	66 53 40 .0	3 .78	0 38 56	3 .085	4 21 36 N	+19 .748
10	6	60 Piscium...	0 11 0 34 .2	50 .12	1 31 5 .5 N	+0 .10	66 52 14 .2	3 .72	0 38 5	3 .092	5 45 28 N	+19 .759
11	6	62 Piscium...	0 11 25 39 .4	50 .10	1 56 47 .1 N	+0 .13	66 52 51 .1	3 .79	0 38 57	3 .094	6 18 56 N	+19 .748
12	5	δ Piscium.....	0 11 37 52 .9	50 .07	2 10 25 .0 N	+0 .14	66 53 4 .7	3 .83	0 39 21	3 .094	6 36 16 N	+19 .742
13	6	58 Piscium...	0 12 58 26 .5	50 .13	6 22 21 .0 N	+0 .18	66 43 8 .4	3 .71	0 37 38	3 .110	10 59 27 N	+19 .767
14	6	33 Ceti.	0 14 41 8 .8	50 .19	4 40 12 .8 S	−0 .14	67 20 20 .6	5 .68	1 1 18	3 .078	1 29 8 N	+19 .320
15	4	ε Piscium.....	0 15 0 54 .7	50 .06	1 4 58 .1 N	+0 .30	67 12 30 .1	5 .05	0 53 37	3 .106	6 55 8 N	+19 .488
16	5	e Piscium.....	0 15 25 2 .5	49 .74	1 29 50 .0 S	−0 .11	67 20 56 .7	5 .50	0 59 7	3 .097	4 41 46 N	+19 .370
17	6	F Piscium....	0 16 48 39 .3	50 .20	4 16 20 .3 S	−0 .19	67 34 12 .4	6 .28	1 8 31	3 .086	2 39 54 N	+19 .141
18	6	ζ¹ Piscium...	0 17 21 30 .1	50 .18	0 13 1 .7 S	−0 .23	67 30 24 .4	5 .96	1 4 20	3 .112	6 37 18 N	+19 .247
19	5	μ Piscium....	0 20 36 26 .0	50 .36	3 4 1 .7 S	−0 .04	68 1 23 .8	7 .30	1 20 45	3 .112	5 12 48 N	+18 .796
20	5	ν Piscium.....	0 22 59 28 .9	50 .02	4 42 23 .5 S	−0 .20	68 25 35 .2	8 .21	1 32 4	3 .110	4 34 24 N	+18 .431
21	4	η Piscium. ...	0 24 18 12 .7	50 .18	5 21 57 .3 N	+0 .17	67 59 46 .8	7 .57	1 21 52	3 .189	14 24 54 N	+18 .763
22	6	π Piscium. ...	0 24 24 13 .5	50 .08	1 52 56 .5 N	+0 .41	68 18 22 .9	7 .95	1 27 34	3 .168	11 13 4 N	+18 .581
23	6	101 Piscium .	0 25 1 8 .8	50 .15	4 21 1 .0 N	+0 .26	68 11 48 .1	7 .91	1 26 9	3 .188	13 44 18 N	+18 .628
24	5	ο Piscium	0 25 13 28 .3	50 .28	1 38 3 .7 S	−0 .08	68 39 24 .1	8 .56	1 35 54	3 .147	8 14 54 N	+18 .295
25	6	105 Piscium .	0 26 32 0 .3	50 .14	5 38 1 .9 N	+0 .27	68 18 22 .5	8 .28	1 29 59	3 .211	15 29 22 N	+18 .500
26	6	54 Ceti........	0 27 9 30 .2	49 .85	0 20 49 .4 S	−0 .21	68 54 25 .2	9 .04	1 41 19	3 .171	10 8 54 N	+18 .096
27	6	ι Arietis.......	1 1 0 16 .4	50 .21	5 26 56 .5 N	+0 .20	69 5 59 .5	9 .77	1 47 32	3 .253	16 56 5 N	+17 .857
28	5	ξ¹ Ceti........	1 1 31 37 .8	50 .16	4 16 53 .3 S	−0 .25	69 57 26 .9	10 .69	2 3 28	3 .164	7 59 53 N	+17 .181
29	6	A Arietis.....	1 2 34 55 .3	50 .15	5 20 6 .6 N	+0 .31	69 25 3 .7	10 .30	1 53 51	3 .269	17 23 3 N	+17 .598
30	6	15 Arietis. ...	1 4 32 54 .9	50 .13	5 57 11 .8 N	+0 .18	69 45 2 .2	10 .91	2 0 39	3 .296	18 38 49 N	+17 .304
31	6	ξ¹ Arietis.....	1 4 51 44 .0	50 .09	3 34 7 .7 S	−0 .33	70 38 19 .3	11 .60	2 15 11	3 .197	9 47 27 N	+16 .631
32	5	ξ² Ceti........	1 4 57 10 .0	50 .21	5 52 14 .3 S	−0 .28	70 46 32 .6	11 .79	2 18 36	3 .171	7 38 53 N	+16 .463
33	6	85 Arietis.....	1 5 18 2 .6	50 .21	3 45 53 .2 S	−0 .33	70 44 55 .9	11 .74	2 17 7	3 .199	9 45 1 N	+16 .536
34	6	ϑ¹ Arietis	1 6 21 41 .3	50 .22	5 44 23 .9 N	+0 .31	70 10 10 .1	11 .52	2 8 8	3 .315	19 3 50 N	+16 .966
35	6	υ Arietis......	1 8 10 52 .9	50 .20	2 43 48 .4 S	−0 .36	71 21 45 .1	12 .41	2 26 49	3 .235	11 39 46 N	+16 .040
36	6	27 Arietis.....	1 8 31 14 .9	50 .12	2 41 23 .3 N	+0 .21	70 59 58 .5	12 .35	2 20 56	3 .304	16 54 13 N	+16 .342
37	6	85 Ceti........	1 9 2 23 .6	50 .21	4 48 3 .2 S	−0 .36	71 42 0 .1	12 .85	2 32 48	3 .214	9 58 12 N	+15 .722
38	4	μ Ceti.........	1 9 24 43 .7	50 .34	5 34 30 .6 S	−0 .53	71 50 5 .0	13 .00	2 35 13	3 .207	9 20 58 N	+15 .590
39	5 .6	38 Arietis	1 10 7 39 .8	50 .33	3 21 15 .3 S	−0 .24	71 53 15 .3	13 .08	2 35 9	3 .242	11 41 1 N	+15 .594
40	5 .6	ν Arietis	1 11 37 29 .9	50 .14	6 9 3 .7 N	+0 .27	71 23 12 .0	13 .22	2 28 37	3 .382	21 10 38 N	+15 .946
41	6	μ Arietis......	1 11 48 59 .3	50 .16	4 2 39 .1 N	+0 .22	71 40 54 .8	13 .33	2 32 14	3 .357	19 14 22 N	+15 .753
42	6	σ Arietis......	1 12 25 40 .8	50 .17	1 18 35 .3 S	−0 .31	72 20 31 .7	13 .65	2 47 34	3 .290	14 20 6 N	+15 .236
43	5	π Arietis......	1 12 37 20 .1	50 .17	1 7 16 .6 N	+0 .26	72 11 17 .2	13 .64	2 39 15	3 .325	16 42 36 N	+15 .365
44	6	40 Arietis. ...	1 12 41 29 .7	50 .25	1 51 36 .6 N	+0 .37	72 7 38 .2	13 .64	2 38 27	3 .339	17 31 44 N	+15 .410
45	6	ρ² Arietis.	1 14 21 32 .2	50 .10	1 30 1 .6 N	+0 .15	72 37 25 .9	14 .14	2 45 42	3 .350	17 35 49 N	+14 .996
46	6	ρ³ Arietis.....	1 14 24 10 .4	50 .22	1 10 31 .7 N	+0 .07	72 39 58 .6	14 .15	2 46 17	3 .346	17 18 1 N	+14 .962
47	6	47 Arietis.....	1 15 31 42 .0	50 .16	3 35 35 .8 N	+0 .39	72 44 21 .2	14 .47	2 47 47	3 .395	19 56 29 N	+14 .876
48	5	ε Arietis	1 15 59 10 .7	50 .16	4 9 23 .5 N	+0 .39	72 48 28 .4	14 .61	2 48 56	3 .408	20 36 53 N	+14 .807
49	6	53 Arietis.....	1 16 53 7 .2	50 .18	0 17 37 .8 N	+0 .40	73 27 4 .7	14 .87	2 57 18	3 .358	17 10 43 N	+14 .304
50	4	57 Arietis.....	1 18 19 58 .7	50 .39	1 48 26 .2 N	+0 .34	73 44 19 .3	15 .28	3 1 21	3 .397	19 2 21 N	+14 .054
51	5	ζ Arietis......	1 19 25 50 .0	50 .04	2 52 31 .6 N	+0 .25	73 57 54 .6	15 .59	3 4 34	3 .426	20 22 16 N	+13 .852
52	6	τ¹ Arietis.....	1 20 52 58 .4	50 .33	2 35 24 .4 N	+0 .16	74 26 39 .8	16 .01	3 10 51	3 .439	20 29 30 N	+13 .449
53	5 .6	f Tauri	1 21 4 35 .7	50 .22	5 56 4 .8 S	−0 .41	75 9 52 .4	15 .97	3 20 57	3 .293	12 18 48 N	+12 .784
54	6	65 Arietis	1 21 31 12 .0	50 .13	2 4 12 .0 N	+0 .49	74 41 51 .2	16 .18	3 14 4	3 .437	20 9 30 N	+13 .240
55	5 .6	64 Arietis.....	1 22 28 44 .0	50 .19	5 52 27 .9 N	+0 .40	74 35 47 .3	16 .59	3 13 42	3 .517	24 4 50 N	+13 .263
56	6	7 Tauri	1 24 39 25 .9	50 .16	5 3 30 .9 N	+0 .44	75 24 50 .9	17 .17	3 23 47	3 .529	23 51 12 N	+12 .589
57	6	11 Tauri......	1 26 15 43 .0	50 .16	5 33 52 .6 N	+0 .44	75 54 28 .4	17 .65	3 30 2	3 .557	24 44 22 N	+12 .159
58	4 .5	Electra	1 26 53 55 .2	50 .15	4 10 20 .6 N	+0 .43	76 16 47 .6	17 .73	3 34 12	3 .538	23 32 26 N	+11 .866

SUBSIDIARY NUMBERS OF THE ZODIACAL STARS FOR 1830.

No.	Longitude. Aberration. A	Max.	Latitude. Aberration. A′	Max	Right Ascension in Time. Aberration. A	Max.	⊙ Nutation. S	Max.	☾ Nutation. L	Max.	Declination. Aberration. A′	Max.	⊙ Nutation. S′	Max.	☾ Nutation. L′	Max.
	s ° ′	″	s ° ′	″	s ° ′	s	s ° ′	s	s ° ′	s	s ° ′	″	s ° ′	″	s ° ′	″
1	8 27 3	20.39	5 27 3	2.34	8 24 11	1.244	5 27 41	0.061	5 27 9	1.101	3 11 53	8.363	11 24 11	0.398	11 22 51	7.207
2	8 26 17	20.27	5 26 17	0.95	8 25 8	1.240	5 29 32	0.061	5 29 25	1.102	3 2 28	8.089	11 25 8	0.398	11 24 0	7.199
3	8 25 46	20.26	5 25 46	0.27	8 25 27	1.240	6 0 28	0.061	6 0 35	1.105	8 27 30	8.052	5 25 27	0.398	5 24 24	7.196
4	8 25 44	20.39	5 25 44	2.39	8 22 47	1.244	5 27 52	0.061	5 27 22	1.099	3 10 57	8.363	11 22 47	0.399	11 21 8	7.221
5	8 25 26	20.39	5 25 26	2.33	8 28 18	1.250	6 3 44	0.062	6 4 36	1.109	8 10 41	8.387	5 28 18	0.398	5 27 55	7.184
6	8 24 24	20.34	5 24 24	1.93	8 26 47	1.249	6 3 26	0.061	6 4 13	1.109	8 12 4	8.280	5 26 47	0.398	5 26 2	7.190
7	8 23 26	20.31	5 23 26	1.59	8 25 23	1.248	6 3 11	0.061	6 3 55	1.110	8 13 9	8.193	5 25 23	0.398	5 24 19	7.197
8	8 20 38	20.37	5 20 38	2.22	8 17 57	1.244	5 29 2	0.061	5 28 49	1.100	3 5 0	8.224	11 17 57	0.399	5 15 15	7.288
9	8 19 13	20.26	5 19 13	0.05	8 19 17	1.246	6 2 3	0.062	6 2 31	1.111	8 18 52	7.952	5 19 16	0.399	5 16 51	7.267
10	8 18 51	20.26	5 18 51	0.54	8 19 30	1.248	6 2 43	0.062	6 3 20	1.113	8 15 18	7.971	5 19 29	0.399	5 17 8	7.263
11	8 18 26	20.26	5 18 26	0.69	8 19 16	1.250	6 2 57	0.062	6 3 38	1.115	8 13 52	7.969	5 19 15	0.399	5 16 51	7.267
12	8 18 14	20.26	5 18 14	0.77	8 19 10	1.250	6 3 5	0.062	6 3 48	1.115	8 13 9	7.978	5 19 9	0.399	5 16 43	7.268
13	8 16 53	20.38	5 16 53	2.25	8 19 38	1.265	6 5 8	0.062	6 6 19	1.125	8 2 24	8.261	5 19 37	0.399	5 17 17	7.262
14	8 15 10	20.32	5 15 10	1.65	8 13 14	1.248	6 0 42	0.061	6 0 52	1.107	8 26 13	7.950	5 13 13	0.401	5 9 38	7.385
15	8 14 51	20.26	5 14 51	0.38	8 15 18	1.254	6 3 11	0.062	6 3 55	1.119	8 12 16	7.853	5 15 17	0.400	5 12 4	7.338
16	8 14 27	20.26	5 14 27	0.53	8 13 49	1.251	6 2 9	0.062	6 2 39	1.114	8 18 3	7.824	5 13 49	0.400	5 10 19	7.372
17	8 13 3	20.31	5 13 3	1.51	8 11 17	1.250	6 1 14	0.061	6 1 30	1.110	8 23 18	7.856	5 11 17	0.401	5 7 20	7.432
18	8 12 30	20.26	5 12 30	0.08	8 12 25	1.256	6 3 0	0.062	6 3 42	1.121	8 13 2	7.744	5 12 25	0.401	5 8 40	7.405
19	8 9 15	20.28	5 9 15	1.08	8 8 1	1.258	6 2 19	0.062	6 2 51	1.120	8 16 48	7.633	5 8 2	0.403	5 3 33	7.523
20	8 6 52	20.32	5 6 52	1.66	8 5 1	1.261	6 1 59	0.062	6 2 27	1.119	8 18 36	7.600	5 5 1	0.403	5 0 7	7.617
21	8 5 33	20.34	5 5 33	1.89	8 7 43	1.294	6 6 16	0.064	6 7 43	1.157	7 22 31	7.784	5 7 44	0.402	5 3 13	7.532
22	8 5 27	20.26	5 5 27	0.67	8 6 12	1.280	6 4 50	0.063	6 5 57	1.145	8 0 47	7.514	5 6 12	0.404	5 1 29	7.578
23	8 4 50	20.31	5 4 50	1.54	8 6 35	1.292	6 5 56	0.064	6 7 17	1.156	7 24 3	7.642	5 6 35	0.404	5 2 54	7.567
24	8 4 38	20.26	5 4 38	0.58	8 4 0	1.272	6 3 30	0.063	6 4 19	1.135	8 8 49	7.384	5 4 0	0.405	4 28 59	7.649
25	8 3 20	20.35	5 3 20	1.99	8 5 34	1.303	6 6 37	0.064	6 8 8	1.166	7 19 27	7.702	5 5 34	0.404	5 0 45	7.598
26	8 2 42	20.26	5 2 42	0.12	8 2 33	1.280	6 4 15	0.063	6 5 14	1.145	8 3 36	7.280	5 2 35	0.406	4 27 23	7.697
27	7 28 51	20.34	4 28 51	1.92	8 0 56	1.320	6 6 55	0.065	6 8 31	1.183	7 14 50	7.432	5 0 56	0.407	4 25 36	7.756
28	7 28 20	20.31	4 28 20	1.51	7 26 46	1.282	6 3 11	0.063	6 3 55	1.141	8 9 56	7.082	4 26 47	0.408	4 21 5	7.912
29	7 27 17	20.34	4 27 17	1.88	7 29 17	1.326	6 6 59	0.066	6 8 35	1.188	7 13 31	7.346	4 29 17	0.407	4 23 47	7.817
30	7 25 19	20.36	4 25 19	2.10	7 27 30	1.339	6 7 20	0.066	6 9 1	1.200	7 9 32	7.268	4 27 31	0.408	4 21 51	7.884
31	7 25 0	20.29	4 25 0	1.26	7 23 45	1.293	6 3 44	0.064	6 4 36	1.153	8 5 6	6.820	4 23 45	0.409	4 18 53	8.032
32	7 24 55	20.36	4 24 55	2.07	7 22 52	1.288	6 2 55	0.063	6 3 35	1.142	8 11 17	6.942	4 22 52	0.409	4 16 59	8.069
33	7 24 34	20.29	4 24 34	1.33	7 23 15	1.294	6 3 42	0.064	6 4 33	1.153	8 5 15	6.793	4 23 15	0.409	4 17 23	8.053
34	7 23 30	20.29	4 23 30	1.32	7 25 34	1.346	6 7 19	0.067	6 9 0	1.206	7 8 0	7.123	4 25 34	0.409	4 19 48	7.960
35	7 21 41	20.27	4 21 41	0.97	7 20 45	1.307	6 4 15	0.065	6 5 15	1.171	7 29 42	6.520	4 20 45	0.410	4 14 48	8.155
36	7 21 20	20.27	4 21 20	0.95	7 22 17	1.335	6 6 14	0.066	6 7 40	1.198	7 13 31	6.639	4 22 16	0.410	4 16 21	8.093
37	7 20 49	20.32	4 20 49	1.70	7 19 15	1.303	6 3 35	0.064	6 4 25	1.159	8 5 0	6.539	4 19 14	0.412	4 13 16	8.219
38	7 20 26	20.35	4 20 26	1.97	7 18 38	1.301	6 3 20	0.064	6 4 6	1.156	8 6 59	6.575	4 18 38	0.413	4 12 38	8.245
39	7 19 44	20.28	4 19 44	1.18	7 18 39	1.311	6 4 8	0.065	6 5 6	1.170	7 29 52	6.379	4 18 39	0.413	4 12 39	8.245
40	7 18 14	20.37	4 18 14	2.17	7 20 18	1.374	6 7 33	0.068	6 9 17	1.232	7 0 39	6.780	4 20 17	0.411	4 14 29	8.175
41	7 18 3	20.30	4 18 3	1.43	7 19 23	1.359	6 6 47	0.067	6 8 20	1.220	7 6 4	6.502	4 19 23	0.412	4 13 24	8.214
42	7 17 26	20.26	4 17 26	0.46	7 17 1	1.329	6 4 55	0.066	6 6 3	1.189	7 21 32	6.146	4 17 1	0.414	4 11 2	8.314
43	7 17 14	20.26	4 17 14	0.39	7 17 36	1.343	6 5 45	0.067	6 7 5	1.205	7 13 48	6.200	4 17 36	0.413	4 11 37	8.288
44	7 17 10	20.26	4 17 10	0.70	7 17 48	1.348	6 6 3	0.067	6 7 26	1.211	7 11 8	6.241	4 17 48	0.413	4 11 50	8.280
45	7 15 30	20.26	4 15 30	0.53	7 15 58	1.353	6 5 53	0.067	6 7 15	1.214	7 10 45	6.062	4 15 58	0.415	4 10 1	8.358
46	7 15 27	20.26	4 15 27	0.42	7 15 49	1.351	6 5 46	0.067	6 7 6	1.212	7 11 41	6.034	4 15 49	0.415	4 9 51	8.365
47	7 14 20	20.29	4 14 20	1.27	7 15 27	1.373	6 6 35	0.068	6 8 6	1.233	7 2 55	6.119	4 15 27	0.415	4 9 29	8.381
48	7 13 52	20.30	4 13 52	1.47	7 15 9	1.379	6 6 46	0.069	6 8 19	1.238	7 0 36	6.132	4 15 9	0.415	4 9 12	8.393
49	7 12 59	20.26	4 12 59	0.11	7 13 4	1.355	6 5 28	0.067	6 6 43	1.216	7 12 6	5.766	4 13 4	0.417	4 7 9	8.483
50	7 11 32	20.26	4 11 32	0.64	7 12 3	1.372	6 5 55	0.068	6 7 17	1.231	7 5 21	5.691	4 12 3	0.417	4 6 12	8.525
51	7 10 26	20.28	4 10 26	1.02	7 11 15	1.385	6 6 13	0.069	6 7 39	1.243	7 0 32	5.667	4 11 15	0.417	4 5 26	8.560
52	7 8 59	20.27	4 8 59	0.91	7 9 41	1.390	6 6 3	0.069	6 7 27	1.247	6 29 59	5.484	4 9 42	0.418	4 4 58	8.625
53	7 8 47	20.36	4 8 47	2.09	7 7 12	1.337	6 3 31	0.066	6 4 20	1.187	8 0 10	5.555	4 7 13	0.420	4 1 38	8.727
54	7 8 20	20.26	4 8 20	0.73	7 8 54	1.388	6 5 51	0.069	6 7 13	1.246	7 0 46	5.373	4 8 53	0.419	4 3 13	8.659
55	7 7 23	20.36	4 7 23	2.07	7 9 0	1.427	6 6 58	0.071	6 8 34	1.279	6 16 59	5.726	4 8 59	0.419	4 3 18	8.656
56	7 5 12	20.33	4 5 12	1.79	7 6 30	1.430	6 6 32	0.071	6 8 2	1.281	6 16 29	5.372	4 6 32	0.420	4 1 0	8.759
57	7 3 26	20.35	4 3 26	1.96	7 4 58	1.443	6 6 31	0.072	6 8 1	1.292	6 12 9	5.265	4 5 0	0.422	3 29 35	8.822
58	7 2 58	20.30	4 2 58	1.47	7 3 58	1.432	6 6 3	0.071	6 7 27	1.284	6 16 24	4.992	4 3 59	0.422	3 28 39	8.863

A CATALOGUE OF 520 ZODIACAL STARS FOR 1820. (CONTINUED).

No.	Mag.	Stars.	Longitude. S	°	′	″	Ann. Var. +	Latitude. °	′	″	Ann. Var.	North East Angle of the ✱'s Meridian with the Ecliptic. °	′	″	Ann. Var. of the ∠. +	R. Ascension in Time. h	m	s.	Ann. Preces. 1830. +	Declination. °	′	″	Ann. Preces. 1830. +
59	5.6	Celeno	1	26	55	15.0	50.08	4	20	49.0 N	+0.36	76	16	10.4	17.75	3	34	7	3s.542	23	42	57 N	11.873
60	5	Taygeta	1	27	3	4.4	50.18	4	30	0.9 N	+0.37	76	17	56.4	17.79	3	34	30	3.546	23	53	43 N	11.847
61	5	Maja	1	27	10	0.9	50.21	4	22	18.0 N	+0.34	76	21	10.7	17.81	3	35	8	3.545	23	47	53 N	11.802
62	5	Merope	1	27	11	12.5	50.26	3	56	22.6 N	+0.51	76	24	15.8	17.79	3	35	39	3.537	23	22	54 N	11.763
63	3	η Tauri	1	27	28	41.7	50.11	4	1	56.7 N	+0.37	76	29	51.2	17.87	3	36	48	3.542	23	32	27 N	11.684
64	5	Atlas	1	27	50	32.0	50.19	3	53	58.5 N	+0.47	76	38	21.9	17.95	3	38	28	3.542	23	29	44 N	11.563
65	5.6	Pleione	1	27	51	54.7	50.17	3	58	49.9 N	+0.58	76	38	22.2	17.97	3	38	29	3.545	23	34	44 N	11.562
66	6	32 Tauri	1	29	14	29.6	50.09	1	59	30.9 N	+0.47	77	19	5.0	18.16	3	46	14	3.518	21	57	7 N	11.002
67	6	Tauri	2	0	48	13.4	50.19	3	33	49.7 S	−0.46	78	17	28.9	18.16	3	57	41	3.418	16	51	3 N	10.153
68	5	A′ Tauri	2	0	56	17.3	50.32	1	14	28.2 N	+0.28	77	59	43.0	18.50	3	54	4	3.520	21	34	54 N	10.424
69	6	ω^1 Tauri	2	1	29	54.6	50.51	1	22	59.4 S	−0.42	78	23	54.3	18.43	3	58	41	3.468	19	7	36 N	10.077
70	6	48 Tauri	2	2	17	15.4	50.41	5	48	24.6 S	−0.46	78	57	4.6	18.33	4	5	33	3.382	14	56	36 N	9.534
71	6	41 Tauri	2	2	25	30.7	50.13	6	34	27.0 N	+0.30	78	3	1.6	19.38	3	55	35	3.655	27	6	24 N	10.308
72	3.4	Hyadum prim.	2	3	17	8.0	50.31	5	45	8.6 S	−0.35	79	18	41.7	18.52	4	9	33	3.390	15	11	5 N	9.246
73	6	h^2 Tauri	2	3	23	17.8	50.24	6	18	30.9 S	−0.46	79	22	36.2	18.51	4	10	24	3.379	14	39	23 N	9.181
74	5.6	ω^2 Tauri	2	3	32	50.9	50.16	0	46	12.0 S	−0.46	79	6	41.9	18.90	4	6	43	3.500	20	7	40 N	9.463
75	6	63 Tauri	2	4	20	25.9	50.37	4	45	38.5 S	−0.70	79	38	55.6	18.76	4	13	6	3.419	16	20	59 N	8.970
76	4	Hyadum sec.	2	4	21	4.3	50.28	3	59	17.9 S	−0.42	79	36	39.7	18.81	4	12	34	3.436	17	6	47 N	9.011
77	4.5	δ^2 Tauri	2	4	36	34.5	50.30	4	7	50.5 S	−0.40	79	42	53.7	18.85	4	13	44	3.435	17	1	9 N	8.920
78	5	π Tauri	2	4	46	25.4	50.33	6	55	26.4 S	−0.41	79	54	51.5	18.71	4	16	27	3.375	14	17	55 N	8.707
79	5.6	71 Tauri	2	4	50	50.9	50.31	6	1	13.8 S	−0.32	79	53	59.7	18.78	4	16	5	3.395	15	12	6 N	8.735
80	5	δ^3 Tauri	2	5	0	49.8	50.17	3	42	26.8 S	−0.33	79	50	34.2	18.95	4	15	5	3.447	17	30	28 N	8.813
81	6	φ Tauri	2	5	24	3.6	50.22	5	46	57.1 N	+0.36	79	17	17.7	19.99	4	9	18	3.670	26	54	41 N	9.261
82	5	ϑ^1 Tauri	2	5	26	10.6	50.30	5	45	53.3 S	−0.35	80	6	24.5	18.89	4	18	18	3.405	15	33	17 N	8.560
83	5.6	ϑ^2 Tauri	2	5	26	33.9	50.28	5	51	27.9 S	−0.41	80	6	49.4	18.88	4	18	23	3.403	15	27	51 N	8.552
84	6	75 Tauri	2	5	28	15.1	50.20	5	22	14.1 S	−0.47	80	6	3.2	18.92	4	18	10	3.414	15	56	55 N	8.571
85	6	χ Tauri	2	5	36	4.4	50.09	3	59	58.2 N	+0.26	79	31	39.1	19.81	4	11	38	3.629	25	11	46 N	9.080
86	5.6	κ^1 Tauri	2	5	41	6.9	50.15	0	36	29.3 N	+0.43	79	49	29.6	19.45	4	14	39	3.550	21	52	25 N	8.846
87	6	80 Tauri	2	5	45	36.1	50.20	6	8	34.5 S	−0.47	80	14	44.8	18.92	4	19	53	3.399	15	14	12 N	8.434
88	5.6	81 Tauri	2	5	53	20.8	50.22	6	6	32.6 S	−0.35	80	17	32.8	18.94	4	20	23	3.400	15	17	33 N	8.394
89	4	ε Tauri	2	5	56	45.5	50.33	2	35	13.9 S	−0.32	80	7	48.4	19.21	4	18	7	3.479	18	46	22 N	8.574
90	5.6	Tauri 160 May	2	5	56	46.6	50.30	5	36	33.0 S	−0.39	80	17	24.9	18.98	4	20	16	3.412	15	47	40 N	8.404
91	5	υ^1 Tauri	2	5	58	52.2	50.27	1	5	7.0 N	+0.15	79	54	16.2	19.55	4	15	32	3.564	22	23	48 N	8.775
92	6	85 Tauri	2	6	12	13.8	50.20	5	59	38.4 S	−0.48	80	24	18.9	19.00	4	21	35	3.405	15	27	27 N	8.300
93	6	υ^2 Tauri	2	6	14	21.2	50.15	1	13	42.0 N	+0.34	79	59	39.9	19.62	4	16	32	3.569	22	34	54 N	8.697
94	5	ρ Tauri	2	6	31	44.3	50.14	7	3	40.8 S	−0.43	80	34	28.3	18.99	4	23	38	3.383	14	27	31 N	8.136
95	1	Aldebaran	2	7	16	21.1	50.22	5	28	46.4 S	−0.26	80	47	6.2	19.21	4	25	36	3.423	16	8	24 N	7.979
96	5.6	σ^1 Tauri	2	7	56	44.6	50.14	6	17	52.2 S	−0.39	81	4	39.7	19.26	4	28	53	3.409	15	26	13 N	7.714
97	5.6	σ^2 Tauri	2	7	59	22.5	50.29	6	11	15.4 S	−0.36	81	5	21.7	19.27	4	28	59	3.412	15	33	9 N	7.705
98	5	τ Tauri seq.	2	9	38	25.4	50.18	0	41	30.5 N	+0.38	81	22	13.1	20.17	4	31	27	3.584	22	36	10 N	7.504
99	6	Tauri 172 M.	2	10	1	35.9	50.30	3	36	41.7 S	−0.60	81	45	32.6	19.78	4	35	46	3.484	18	24	4 N	7.154
100	6	96 Tauri	2	10	29	55.5	50.20	6	31	36.8 S	−0.54	82	4	7.2	19.60	4	39	26	3.419	15	34	53 N	6.854
101	5.6	i Tauri	2	11	14	12.8	50.24	3	39	25.4 S	−0.43	82	14	14.4	19.95	4	40	51	3.490	18	31	51 N	6.737
102	6	k Tauri	2	13	29	46.9	50.16	2	20	17.5 N	+0.42	82	50	40.9	21.00	4	47	8	3.654	24	45	47 N	6.215
103	4.5	ι Tauri	2	14	16	13.6	50.19	1	13	10.5 S	−0.43	83	20	46.0	20.62	4	52	20	3.568	21	19	26 N	5.782
104	5	m Tauri	2	14	59	21.9	50.84	4	14	53.6 S	−0.47	83	45	34.7	20.36	4	56	48	3.496	18	23	40 N	5.406
105	5.6	l^1 Tauri	2	15	15	38.8	50.06	2	29	26.5 S	−0.40	83	48	17.8	20.59	4	57	10	3.541	20	10	19 N	5.377
106	6	105 Tauri	2	15	24	13.3	50.23	1	12	43.7 S	−0.55	83	48	38.8	20.76	4	57	10	3.574	21	27	27 N	5.375
107	6	Tauri 152 Caille	2	15	40	26.5	50.17	1	20	0.1 N	+0.49	83	48	29.6	21.15	4	57	9	3.642	24	1	3 N	5.377
108	5.6	n Tauri	2	18	3	21.8	50.05	1	1	49.3 S	−0.30	84	51	18.7	21.07	5	8	27	3.592	21	54	2 N	4.420
109	6	111 Tauri	2	18	56	33.9	50.19	5	48	52.8 S	−0.50	85	24	54.6	20.57	5	13	55	3.473	17	12	24 N	3.953
110	5.6	115 Tauri	2	19	38	54.2	50.28	5	16	35.0 S	−0.51	85	41	25.9	20.68	5	16	40	3.490	17	47	52 N	3.717
111	6	117 Tauri	2	19	48	41.0	50.19	6	0	27.3 S	−0.50	85	46	28.8	20.61	5	17	35	3.472	17	4	51 N	3.639
112	5	o Tauri	2	19	58	55.5	50.15	1	18	50.4 S	−0.49	85	43	22.1	21.20	5	16	50	3.593	21	46	26 N	3.703
113	2	Nath.	2	20	3	36.9	50.14	5	22	14.4 N	+0.33	85	31	1.4	22.35	5	14	55	3.778	28	26	45 N	3.863
114	5.6	119 Tauri	2	20	52	51.5	50.20	4	42	31.6 S	−0.50	86	11	8.7	20.83	5	21	40	3.508	18	27	3 N	3.287
115	6	120 Tauri	2	21	11	26.3	50.20	4	46	38.8 S	−0.50	86	18	55.6	20.84	5	22	59	3.507	18	24	9 N	3.174
116	6	121 Tauri	2	21	52	56.9	50.23	0	41	49.1 N	+0.46	86	28	26.6	21.65	5	24	28	3.654	23	54	35 N	3.043

SUBSIDIARY NUMBERS OF THE ZODIACAL STARS FOR 1830 (CONTINUED).

No.	Longitude. Aberration. A (s ° ′)	Max.	Latitude. Aberration. A′ (s ° ′)	Max.	Right Ascension in Time. Aberration. A (s ° ′)	Max.	⊙ Nutation. S (s ° ′)	Max.	☾ Nutation. L (s ° ′)	Max.	Declination. Aberration. A′ (s ° ′)	Max.	⊙ Nutation. S′ (s ° ′)	Max.	☾ Nutation. L′ (s ° ′)	Max.
59	7 2 56	20″.31	4 2 56	1″.54	7 3 59	1s.433	6 6 6	0s.071	6 7 30	1s.284	6 15 41	5″.019	4 4 1	0″.422	3 28 41	8″.862
60	7 2 49	20.31	4 2 49	1.59	7 3 54	1.436	6 6 8	0.071	6 7 32	1.286	6 15 0	5.027	4 3 54	0.422	3 28 36	8.866
61	7 2 42	20.31	4 2 42	1.54	7 3 45	1.435	6 6 5	0.071	6 7 29	1.285	6 15 6	4.991	4 3 45	0.422	3 28 26	8.871
62	7 2 40	20.30	4 2 40	1.39	7 3 37	1.431	6 5 57	0.071	6 7 20	1.282	6 16 38	4.931	4 3 38	0.422	3 28 21	8.876
63	7 2 23	20.30	4 2 23	1.42	7 3 21	1.433	6 5 56	0.071	6 7 19	1.283	6 16 1	4.912	4 3 21	0.423	3 28 6	8.887
64	7 2 1	20.30	4 2 1	1.38	7 2 57	1.433	6 5 53	0.071	6 7 15	1.284	6 16 0	4.853	4 2 57	0.423	3 27 44	8.904
65	7 2 0	20.30	4 2 0	1.41	7 2 57	1.434	6 5 53	0.071	6 7 15	1.285	6 15 45	4.862	4 2 57	0.423	3 27 44	8.904
66	7 0 37	20.26	4 0 37	0.71	7 1 3	1.421	6 5 13	0.071	6 6 26	1.278	6 21 40	4.478	4 1 4	0.424	3 26 2	8.977
67	6 29 3	20.29	3 29 3	1.26	6 28 19	1.382	6 3 44	0.068	6 4 36	1.233	7 15 49	4.273	3 28 19	0.425	3 23 37	9.080
68	6 28 55	20.26	3 28 55	0.44	6 29 10	1.421	6 4 51	0.070	6 5 59	1.272	6 23 5	4.218	3 29 11	0.425	3 24 22	9.049
69	6 28 22	20.26	3 28 22	0.49	6 28 5	1.401	6 4 11	0.069	6 5 9	1.252	7 5 1	4.080	3 28 5	0.426	3 23 24	9.089
70	6 27 34	20.35	3 27 34	2.05	6 26 27	1.372	6 3 8	0.067	6 3 51	1.219	7 25 2	4.353	3 26 26	0.426	3 21 59	9.152
71	6 27 26	20.38	3 27 26	2.32	6 28 49	1.485	6 5 59	0.073	6 7 21	1.325	5 28 53	4.752	3 28 49	0.425	3 24 3	9.062
72	6 26 34	20.35	3 26 34	2.03	6 25 30	1.376	6 3 5	0.068	6 3 47	1.221	7 24 40	4.238	3 25 29	0.427	3 21 9	9.181
73	6 26 28	20 37	3 26 28	2.23	6 25 18	1.373	6 2 57	0.067	6 3 38	1.217	7 27 1	4.318	3 25 17	0.427	3 20 58	9.186
74	6 26 19	20.26	3 26 19	0.27	6 26 10	1.413	6 4 7	0.072	6 5 4	1 263	7 0 9	3.603	3 26 10	0.427	3 21 44	9.156
75	6 25 31	20.32	3 25 31	1.65	6 24 40	1.385	6 3 12	0.068	6 3 56	1.232	7 20 1	3.975	3 24 39	0.427	3 20 26	9.208
76	6 25 31	20.30	3 25 31	1.41	6 24 47	1.390	6 3 21	0.069	6 4 8	1.238	7 16 24	3.891	3 24 46	0.427	3 20 32	9.205
77	6 25 15	20.30	3 25 15	1.46	6 24 31	1.390	6 3 18	0.069	6 4 4	1.238	7 17 1	3.875	3 24 30	0.427	3 20 18	9.212
78	6 25 5	20.40	3 25 5	2.44	6 23 52	1.373	6 2 44	0.067	6 3 22	1.216	7 29 23	4.275	3 23 51	0.428	3 19 45	9.234
79	6 25 0	20.36	3 25 0	2.12	6 23 57	1.378	6 2 54	0.068	6 3 35	1.223	7 25 43	4.115	3 23 57	0.428	3 19 49	9.231
80	6 24 50	20.29	3 24 50	1.31	6 24 11	1.394	6 3 21	0.069	6 4 7	1.243	7 14 43	3.777	3 24 11	0.428	3 20 1	9.223
81	6 24 28	20.35	3 24 28	2.04	6 25 33	1.489	6 5 18	0.074	6 6 32	1.332	5 26 14	4.248	3 25 32	0.427	3 21 12	9.177
82	6 24 25	20.35	3 24 25	2.04	6 23 26	1.381	6 2 54	0.068	6 3 35	1.227	7 24 27	3.996	3 23 25	0.428	3 19 23	9.248
83	6 24 25	20.36	3 24 25	2.06	6 23 24	1.381	6 2 53	0.068	6 3 35	1.226	7 24 47	4.002	3 23 23	0.428	3 19 21	9.249
84	6 24 23	20.34	3 24 23	1.89	6 23 28	1.384	6 2 59	0.068	6 3 40	1.230	7 22 38	3.927	3 23 27	0.428	3 19 24	9.247
85	6 24 16	20.30	3 24 16	1.41	6 24 59	1.468	6 4 53	0.073	6 6 0	1.312	6 3 32	3.917	3 24 59	0.427	3 20 43	9.196
86	6 24 11	20.26	3 24 11	0.21	6 24 17	1.433	6 4 9	0.071	6 5 7	1.281	6 20 45	3.565	3 24 17	0.428	3 20 6	9.220
87	6 24 7	20.37	3 24 7	2.17	6 23 4	1.380	6 2 48	0.068	6 3 27	1.224	7 26 9	4.027	3 23 3	0.428	3 19 4	9.260
88	6 23 58	20.37	3 23 58	2.16	6 22 56	1.381	6 2 48	0.068	6 3 27	1.225	7 26 2	4.010	3 22 56	0.428	3 18 57	9.264
89	6 23 55	20.27	3 23 55	0.91	6 23 28	1.406	6 3 29	0.070	6 4 17	1.254	7 8 33	3.570	3 23 27	0.428	3 19 25	9.247
90	6 23 55	20.35	3 23 55	1.98	6 22 58	1.384	6 2 54	0.068	6 3 34	1.229	7 23 56	3.930	3 22 57	0.428	3 18 59	9.263
91	6 23 53	20.26	3 23 53	0.38	6 24 4	1.438	6 4 12	0.071	6 5 11	1.286	6 17 40	3.550	3 24 4	0.428	3 19 55	9.227
92	6 23 39	20.36	3 23 39	2.12	6 22 39	1.382	6 2 48	0.068	6 3 27	1.226	7 25 27	3.947	3 22 39	0.428	3 18 42	9.272
93	6 23 37	20.26	3 23 37	0.44	6 23 50	1.441	6 4 12	0.071	6 5 11	1.289	6 16 50	3.529	3 23 49	0.428	3 19 43	9.235
94	6 23 20	20.41	3 23 20	2.49	6 22 10	1.376	6 2 35	0.068	6 3 10	1.219	7 29 55	4.100	3 22 10	0.429	3 18 18	9.287
95	6 22 35	20.35	3 22 35	1.93	6 21 43	1.388	6 2 48	0.068	6 3 27	1.232	7 23 14	3.749	3 21 43	0.429	3 17 54	9.300
96	6 21 55	20.37	3 21 55	2.22	6 20 56	1.384	6 2 36	0.068	6 3 12	1.228	7 26 58	3.812	3 20 56	0.429	3 17 15	9.324
97	6 21 52	20.37	3 21 52	2.18	6 20 54	1.385	6 2 37	0.068	6 3 13	1.229	7 26 30	3.787	3 20 54	0.429	3 17 13	9.325
98	6 20 13	20.26	3 20 13	0.25	6 20 20	1.447	6 3 37	0.072	6 4 27	1.299	6 15 39	3.029	3 20 20	0.430	3 16 44	9.342
99	6 19 50	20.29	3 19 50	1.28	6 19 19	1.409	6 2 50	0.070	6 3 30	1.255	7 13 55	3.154	3 19 19	0.430	3 15 43	9.370
100	6 19 22	20.38	3 19 22	2.30	6 18 28	1.389	6 2 19	0.068	6 2 52	1.230	7 28 43	3.587	3 18 28	0.430	3 15 10	9.393
101	6 18 37	20.29	3 18 37	1.29	6 18 8	1.412	6 2 41	0.070	6 3 19	1.259	7 13 36	2.990	3 18 8	0.431	3 14 53	9.402
102	6 16 22	20.27	3 16 22	0.83	6 16 40	1.476	6 3 15	0.073	6 4 1	1.317	5 28 23	2.633	3 16 39	0.431	3 13 38	9.442
103	6 15 35	20.26	3 15 35	0.43	6 15 27	1.440	6 2 38	0.071	6 3 14	1.284	6 26 3	2.367	3 15 27	0.432	3 12 38	9.470
104	6 14 52	20.31	3 14 52	1.50	6 14 24	1.415	6 2 8	0.070	6 2 38	1.258	7 19 18	2.646	3 14 24	0.432	3 11 46	9.490
105	6 14 36	20.27	3 14 36	0.88	6 14 20	1.430	6 2 19	0.071	6 2 51	1.275	7 6 37	2.336	3 14 19	0.432	3 11 42	9.493
106	6 14 27	20.26	3 14 27	0.43	6 14 19	1.443	6 2 27	0.071	6 3 1	1.287	6 25 29	2.203	3 14 19	0.432	3 11 42	9.493
107	6 14 11	20.26	3 14 11	0.47	6 14 20	1.470	6 2 44	0.073	6 3 22	1.312	6 1 55	2.214	3 14 19	0.432	3 11 42	9.493
108	6 11 48	20.26	3 11 48	0.37	6 11 42	1.450	6 2 3	0.072	6 2 32	1.293	6 23 18	1.815	3 11 42	0.432	3 9 33	9.543
109	6 10 55	20.35	3 10 55	0.25	6 10 27	1.409	6 1 28	0.069	6 1 48	1.250	8 2 54	2.593	3 10 27	0.432	3 8 31	9.565
110	6 10 13	20.34	3 10 13	1.87	6 9 49	1.414	6 1 25	0.070	6 1 45	1.255	8 1 6	2.377	3 9 49	0.432	3 7 59	9.574
111	6 10 3	20.36	3 10 3	2.12	6 9 37	1.409	6 1 20	0.069	6 1 39	1.249	8 5 13	2.579	3 9 37	0.432	3 7 49	9.578
112	6 9 53	20.26	3 9 53	0.46	6 9 47	1.450	6 1 43	0.072	6 2 6	1.292	6 26 57	1.558	3 9 47	0.433	3 7 57	9.575
113	6 9 48	20.34	3 9 48	1.89	6 10 13	1.532	6 2 18	0.075	6 2 50	1.359	4 19 26	2.448	3 10 12	0.433	3 8 19	9.568
114	6 8 59	20.32	3 8 59	1.67	6 8 40	1.421	6 1 13	0.070	6 1 36	1.262	8 0 29	2.135	3 8 39	0.433	3 7 3	9.591
115	6 8 40	20.32	3 8 40	1.69	6 8 22	1.420	6 1 15	0.070	6 1 32	1.262	8 1 24	2.116	3 8 21	0.434	3 6 48	9.595
116	6 7 59	20.26	3 7 59	0.25	6 8 1	1.474	6 1 32	0.073	6 1 53	1.314	5 26 35	1.249	3 8 1	0.434	3 6 31	9.599

A CATALOGUE OF 520 ZODIACAL STARS FOR 1820 (CONTINUED).

No.	Mag.	Stars.	Longitude.	Ann. Var.	Latitude.	Ann. Var.	North East Angle of the ✱'s Meridian with the Ecliptic.	Ann. Var. of the ∠.	R. Ascension in Time.	Ann. Precess. 1830.	Declination.	Ann. Precess. 1830.
			S ° ′ ″	+	° ′ ″	″	° ′ ″	+	h m ″	+	° ′ ″	″
117	6	122 Tauri	2 21 58 6.7	50.18	6 18 38.7 S	−0.50	86 39 59.8	20.72	5 26 37	3.471	16 55 11 N	+2.859
118	3.4	ζ Tauri	2 22 16 8.9	50.09	2 13 3.0 S	−0.46	86 42 39.0	21.23	5 26 53	3.576	21 1 25 N	2.834
119	6	125 Tauri	2 22 55 25.7	50.17	2 31 0.0 N	+0.52	86 52 38.7	22.03	5 28 35	3.708	25 47 12 N	2.686
120	6	130 Tauri	2 24 28 53.1	50.30	5 41 59.8 S	−0.47	87 41 48.6	20.90	5 36 57	3.491	17 39 10 N	1.962
121	5	132 Tauri	2 24 59 12.2	50.06	1 7 49.7 N	+0.52	87 48 32.5	21.89	5 37 58	3.674	24 29 51 N	1.871
122	4.5	136 Tauri	2 26 0 14.6	50.19	4 9 31.2 N	+0.46	88 12 23.5	22.50	5 42 1	3.763	27 33 34 N	1.517
123	5	χ¹ Orionis	2 26 10 35.0	50.03	3 10 32.5 S	−0.39	88 22 43.9	21.27	5 43 44	3.559	20 14 2 N	1.371
124	6	χ² Orionis	2 26 17 46.4	50.31	3 42 24.0 S	−0.53	88 26 3.1	21.20	5 44 17	3.546	19 42 20 N	1.321
125	5.6	139 Tauri	2 27 2 6.5	50.21	2 29 31.4 N	+0.46	88 41 16.5	22.20	5 46 49	3.717	25 55 17 N	1.097
126	6	141 Tauri	2 27 52 40.7	50.16	1 3 33.1 S	−0.52	89 5 11.3	21.61	5 50 49	3.618	22 23 12 N	0.750
127	5.6	χ⁴ Orionis	2 28 18 16.2	50.44	3 46 5.4 S	−0.44	89 16 57.0	21.22	5 52 48	3.546	19 41 4 N	0.577
128	5	χ³ Orionis	2 28 24 32.2	50.28	3 19 16.2 S	−0.40	89 19 30.5	21.29	5 53 14	3.558	20 7 59 N	0.540
129	5	Propus	2 28 25 58.0	50.14	0 11 26.6 S	−0.34	89 19 15.4	21.75	5 53 11	3.642	23 15 50 N	0.543
130	6	3 Gemin	2 29 43 24.6	50.13	0 19 47.4 S	−0.43	89 52 49.6	21.73	5 58 48	3.639	23 7 59 N	+0.052
131	6	68 Orionis	3 0 19 10.6	50.32	3 38 28.6 S	−0.48	90 8 7.0	21.60	6 1 21	3.550	19 49 17 N	−0.170
132	4	κ Aurigæ	3 0 51 5.7	49.93	6 5 34.2 N	+0.14	90 23 25.6	22.97	6 3 54	3.825	29 33 18 N	0.397
133	4.5	Tejat prior	3 0 55 30.4	50.08	0 54 38.3 S	−0.45	90 23 56.6	21.63	6 4 0	3.623	22 32 58 N	0.404
134	5.6	71 Orionis	3 1 0 29.9	50.18	4 15 2.6 S	−0.36	90 25 30.4	21.15	6 4 16	3.533	19 12 33 N	0.424
135	3	Tejat post	3 2 47 6.1	50.30	0 50 15.1 S	−0.34	91 12 1.5	21.60	6 12 4	3.623	22 35 49 N	1.106
136	6	15 Gemin. seq.	3 3 59 1.8	50.12	2 30 53.3 S	−0.27	91 41 48.5	21.31	6 17 3	3.576	20 53 25 N	1.541
137	6	16 Gemin.	3 4 2 15.0	50.14	2 48 25.9 S	−0.50	91 42 58.5	21.26	6 17 14	3.569	20 35 43 N	1.558
138	5	ν Gemin.	3 4 17 17.9	50.11	3 4 41.9 S	−0.43	91 49 10.1	21.22	6 18 16	3.561	20 19 1 N	1.648
139	6	49 Aurigæ	3 5 16 35.7	50.19	4 47 54.6 N	+0.49	92 22 48.5	22.52	6 23 52	3.779	28 9 8 N	2.138
140	6	54 Aurigæ	3 6 13 14.9	50.19	5 6 3.8 N	+0.48	92 48 42.0	22.52	6 28 12	3.785	28 24 42 N	2.514
141	3	Alhena	3 6 35 20.2	50.22	6 45 47.1 S	−0.40	92 43 53.9	20.65	6 27 18	3.462	16 32 39 N	2.432
142	6	28 Gemin.	3 7 18 44.8	50.11	5 53 30.4 N	+0.49	93 19 35.6	22.60	6 33 20	3.805	29 8 35 N	2.960
143	3	Mebsuta	3 7 25 28.4	50.16	2 2 54.0 N	+0.48	93 15 44.8	21.84	6 32 51	3.693	25 17 59 N	2.917
144	5.6	26 Gemin.	3 7 37 48.4	50.27	5 26 23.5 S	−0.35	93 10 58.5	20.74	6 31 55	3.493	17 48 45 N	2.833
145	6	37 Gemin.	3 9 58 16.3	50.19	2 30 37.9 N	+0.48	94 23 2.6	21.69	6 44 14	3.695	25 35 28 N	3.897
146	6	ω¹ Gemin.	3 11 41 29.3	50.17	1 31 14.2 N	+0.43	95 5 7.6	21.33	6 51 26	3.660	24 27 47 N	4.512
147	4	Mekbuda	3 12 28 38.8	50.18	2 3 40.2 S	−0.42	95 16 50.8	20.72	6 53 25	3.562	20 49 29 N	4.681
148	6	47 Gemin.	3 13 24 6.0	50.20	4 22 41.6 N	+0.47	95 57 9.0	21.61	7 0 13	3.729	27 8 34 N	5.257
149	6	M Gemin.	3 13 58 28.6	50.25	1 42 0.4 N	+0.38	96 3 41.5	21.08	7 1 29	3.652	24 25 13 N	5.364
150	6	o Gemin.	3 14 0 50.1	50.17	6 33 0.5 S	−0.27	95 45 45.6	20.05	6 58 2	3.444	16 12 36 N	5.070
151	6	53 Gemin.	3 14 16 42.4	50.11	5 32 13.8 N	+0.45	96 23 52.0	21.70	7 4 42	3.755	28 12 4 N	5.635
152	5	51 Gemin.	3 15 11 9.1	50.15	6 10 46.6 S	−0.42	96 14 36.4	19.95	7 3 2	3.447	16 27 21 N	5.491
153	3.4	Wasat.	3 16 0 22.7	50.13	0 11 57.4 S	−0.40	96 48 55.3	20.52	7 9 22	3.590	22 18 17 N	6.024
154	4.5	λ Gemin.	3 16 16 6.1	50.07	5 39 15.1 S	−0.41	96 41 34.9	19.87	7 7 45	3.455	16 51 24 N	5.887
155	6	A Gemin.	3 16 20 33.8	50.09	2 56 50.7 N	+0.38	97 7 25.9	20.93	7 12 29	3.670	25 23 15 N	6.285
156	4	ι Gemin.	3 16 26 49.7	49.96	5 44 33.4 N	+0.35	97 20 45.6	21.38	7 14 32	3.744	28 8 50 N	6.456
157	5.6	q Gemin.	3 16 38 33.1	50.18	1 39 48.1 S	−0.44	97 0 19.2	20.24	7 11 19	3.550	20 46 28 N	6.186
158	5.6	B¹ Gemin.	3 17 11 17.1	50.03	6 10 31.0 N	+0.35	97 41 35.0	21.33	7 18 7	3.750	28 28 48 N	6.751
159	5.6	B² Gemin.	3 17 19 22.8	50.13	5 59 21.8 N	+0.27	97 44 12.0	21.23	7 18 36	3.744	28 16 44 N	6.791
160	6	P Gemin.	3 17 50 10.3	50.25	0 28 21.7 S	−0.33	97 32 51.3	20.21	7 17 3	3.572	21 48 18 N	6.661
161	5	υ Gemin.	3 18 49 51.3	50.21	5 12 5.2 N	+0.38	98 18 52.6	20.86	7 24 49	3.709	27 17 16 N	7.300
162	5	K Gemin.	3 20 4 31.9	50.13	5 48 55.3 S	−0.47	98 10 57.2	19.32	7 23 20	3.430	16 12 21 N	7.174
163	2	Pollux.	3 20 43 51.6	49.60	6 40 15.6 N	+0.23	99 13 33.4	20.72	7 34 18	3.731	28 27 8 N	8.064
164	6	c Gemin.	3 20 50 1.5	50.18	4 24 40.4 N	+0.41	99 4 51.4	20.34	7 33 7	3.671	26 12 16 N	7.970
165	4	κ Gemin.	3 21 9 3.1	50.05	3 3 38.4 N	+0.39	99 6 28.2	20.08	7 33 34	3.634	24 49 15 N	8.006
166	6	F Gemin.	3 21 9 46.8	50.20	3 45 53.4 S	−0.34	98 41 48.4	19.32	7 29 4	3.471	18 4 33 N	7.642
167	6	G Gemin.	3 22 34 42.5	50.02	2 39 54.0 S	−0.32	99 18 5.3	19.17	7 35 41	3.486	18 56 28 N	8.174
168	5	φ Gemin.	3 22 43 49.0	50.16	5 45 36.9 N	+0.37	99 57 46.5	20.13	7 42 28	3.686	27 13 22 N	8.713
169	6	ω¹ Cancri	3 24 38 30.5	50.17	4 44 20.0 N	+0.45	100 37 57.5	19.56	7 50 1	3.641	25 52 37 N	9.303
170	6	1 Cancri	3 25 38 27.1	50.16	4 51 9.7 S	−0.42	100 20 21.3	18.43	7 46 45	3.415	16 15 50 N	9.047
171	6	3 Cancri	3 26 14 3.0	50.17	3 11 11.5 S	−0.42	100 39 7.5	18.43	7 50 28	3.447	17 47 39 N	9.335
172	6	μ¹ Cancri	3 26 24 45.5	50.06	2 17 34.8 N	+0.40	101 6 19.9	18.87	7 55 38	3.567	23 8 30 N	9.734
173	6	5 Cancri	3 26 34 22.4	50.16	3 59 17.0 S	−0.42	100 43 50.4	18.31	7 51 14	3.427	16 56 38 N	9.384
174	6	ζ Cancri	3 28 49 27.7	50.23	2 16 48.7 S	−0.31	101 39 24.7	17.95	8 1 52	3.445	18 11 0 N	10.206

SUBSIDIARY NUMBERS OF THE ZODIACAL STARS FOR 1830 (continued).

No.	Longitude. Aberration. A	Max.	Latitude. Aberration. A′	Max.	Right Ascension in Time. Aberration. A	Max.	⊙ Nutation. S	Max.	☾ Nutation. L	Max.	Declination. Aberration. A′	Max.	⊙ Nutation. S′	Max.	☾ Nutation. L′	Max.
	s ° ′	″	s ° ′	″	s ° ′	s	s ° ′	s	s ° ′	s	s ° ′	″	s ° ′	″	s ° ′	″
117	6 7 54	20.37	3 7 54	2.23	6 7 32	1.410	6 1 2	0.069	6 1 17	1.248	8 10 23	2.505	3 7 31	0.434	3 6 7	9.604
118	6 7 36	20.27	3 7 36	0.78	6 7 28	1.444	6 1 16	0.071	6 1 33	1.286	7 12 4	1.384	3 7 27	0.434	3 6 4	9.604
119	6 6 57	20.27	3 6 57	0.89	6 7 4	1.498	6 1 27	0.074	6 1 47	1.334	4 27 37	1.399	3 7 4	0.434	3 5 45	9.610
120	6 5 23	20.35	3 5 23	2.01	6 5 9	1.416	6 0 45	0.070	6 0 55	1.282	8 13 53	2.165	3 5 9	0.434	3 4 12	9.628
121	6 4 52	20.26	3 4 52	0.40	6 4 55	1.483	6 0 58	0.074	6 1 11	1.320	5 7 1	0.852	3 4 54	0.434	3 4 0	9.630
122	6 3 51	20.30	3 3 51	1.47	6 3 59	1.523	6 0 52	0.075	6 1 5	1.353	3 26 30	1.589	3 3 58	0.434	3 3 14	9.636
123	6 3 41	20.28	3 3 41	1.12	6 3 36	1.439	6 0 35	0.071	6 0 44	1.280	8 7 25	1.248	3 3 36	0.434	3 2 52	9.639
124	6 3 34	20.29	3 3 34	1.31	6 3 28	1.434	6 0 33	0.071	6 0 41	1.275	8 11 24	1.411	3 3 28	0.434	3 2 49	9.640
125	6 2 50	20.27	3 2 50	0.88	6 2 53	1.501	6 0 36	0.074	6 0 44	1.336	3 29 4	0.999	3 2 52	0.434	3 2 20	9.642
126	6 1 59	20.26	3 1 59	0.38	6 1 58	1.460	6 0 22	0.072	6 0 26	1.301	7 22 10	0.470	3 1 57	0.434	3 1 36	9.645
127	6 1 33	20.29	3 1 33	1.33	6 1 30	1.434	6 0 15	0.071	6 0 18	1.275	8 21 38	1.352	3 1 30	0.434	3 1 14	9.647
128	6 1 27	20.28	3 1 27	1.17	6 1 24	1.438	6 0 14	0.071	6 0 17	1.279	8 20 53	1.185	3 1 24	0.434	3 1 9	9.647
129	6 1 26	20.26	3 1 26	0.06	6 1 25	1.470	6 0 16	0.073	6 0 20	1.310	6 18 23	0.229	3 1 25	0.434	3 1 10	9.647
130	6 0 8	20.26	3 0 8	0.12	6 0 8	1.468	6 0 2	0.073	6 0 2	1.309	8 19 55	0.113	3 0 7	0.434	3 0 7	9.648
131	5 29 32	20.29	2 29 32	1.28	5 29 33	1.435	5 29 56	0.071	5 29 55	1.276	9 2 36	1.283	2 29 34	0.434	2 29 39	9.648
132	5 29 1	20.37	2 29 1	2.15	5 28 58	1.552	5 29 45	0.076	5 29 42	1.374	2 24 43	2.155	2 28 57	0.434	2 29 10	9.648
133	5 28 56	20.26	2 28 56	0.32	5 28 56	1.462	5 29 48	0.072	5 29 46	1.302	9 25 38	0.361	2 28 56	0.434	2 29 9	9.648
134	5 28 51	20.31	2 28 51	1.50	5 28 53	1.430	5 29 49	0.070	5 29 47	1.270	9 5 39	1.512	2 28 53	0.434	2 29 6	9.648
135	5 27 5	20.26	2 27 5	0.30	5 27 6	1.462	5 29 28	0.072	5 29 21	1.302	10 23 20	0.535	2 27 6	0.434	2 27 39	9.643
136	5 25 53	20.27	2 25 53	0.89	5 25 57	1.445	5 29 19	0.071	5 29 9	1.286	10 0 51	1.082	2 25 58	0.434	2 26 43	9.636
137	5 25 51	20.27	2 25 51	0.99	5 25 54	1.442	5 29 19	0.071	5 29 10	1.283	9 26 26	1.162	2 25 55	0.434	2 26 41	9.635
138	5 25 34	20.28	2 25 34	1.09	5 25 40	1.439	5 29 17	0.071	5 29 7	1.280	9 26 58	1.275	2 25 41	0.434	2 26 31	9.633
139	5 24 35	20.32	2 24 35	1.70	5 24 23	1.530	5 28 45	0.075	5 28 27	1.359	1 27 33	1.903	2 24 23	0.434	2 25 27	9.624
140	5 23 38	20.33	2 23 38	1.80	5 23 23	1.533	5 28 31	0.076	5 28 10	1.362	1 24 8	2.063	2 23 24	0.434	2 24 38	9.615
141	5 23 16	20.39	2 23 16	2.39	5 23 36	1.407	5 29 8	0.069	5 28 56	1.245	9 15 48	2.568	2 23 37	0.434	2 24 48	9.617
142	5 22 33	20.36	2 22 33	2.08	5 22 12	1.543	5 28 12	0.076	5 27 47	1.369	1 22 43	2.405	2 22 12	0.434	2 23 39	9.602
143	5 22 26	20.26	2 22 26	0.72	5 22 18	1.491	5 28 27	0.074	5 28 5	1.330	0 24 1	1.379	2 22 19	0.434	2 23 46	9.602
144	5 22 14	20.34	2 22 14	1.92	5 22 32	1.416	5 28 55	0.070	5 28 40	1.256	9 23 11	2.224	2 22 32	0.434	2 23 56	9.605
145	5 19 53	20.27	2 19 53	0.89	5 19 41	1.493	5 27 54	0.074	5 27 25	1.330	0 19 50	1.808	2 19 42	0.433	2 21 37	9.566
146	5 18 10	20.26	2 18 10	0.54	5 18 2	1.477	5 27 41	0.074	5 27 8	1.317	0 4 30	1.888	2 18 2	0.433	2 20 14	9.539
147	5 17 23	20.26	2 17 23	0.73	5 17 35	1.439	5 27 56	0.071	5 27 27	1.281	10 26 19	2.020	2 17 35	0.433	2 19 52	9.530
148	5 16 28	20.31	2 16 28	1.55	5 15 59	1.509	5 27 0	0.074	5 26 19	1.343	0 27 30	2.624	2 15 59	0.432	2 18 34	9.500
149	5 15 53	20.26	2 15 53	0.60	5 15 43	1.474	5 27 14	0.073	5 26 36	1.315	0 1 25	2.240	2 15 43	0.432	2 18 19	9.493
150	5 15 51	20.38	2 15 51	2.31	5 16 30	1.399	5 28 14	0.069	5 27 49	1.239	9 27 37	3.082	2 16 30	0.432	2 19 0	9.510
151	5 15 35	20.35	2 15 35	1.95	5 14 58	1.522	5 26 40	0.075	5 25 54	1.353	0 25 48	2.987	2 14 58	0.432	2 17 42	9.477
152	5 14 40	20.37	2 14 40	2.18	5 15 22	1.400	5 28 3	0.069	5 27 36	1.241	10 0 24	3.102	2 15 22	0.432	2 18 2	9.486
153	5 13 51	20.26	2 13 51	0.07	5 13 52	1.449	5 27 9	0.072	5 26 29	1.293	11 12 10	2.426	2 13 52	0.431	2 16 47	9.452
154	5 13 36	20.35	2 13 36	1.99	5 14 16	1.401	5 27 52	0.069	5 27 22	1.244	10 3 54	3.089	2 14 16	0.431	2 17 7	9.461
155	5 13 31	20.28	2 13 31	1.04	5 13 9	1.483	5 26 38	0.073	5 25 51	1.323	0 5 51	2.736	2 13 9	0.431	2 16 12	9.435
156	5 13 25	20.35	2 13 25	2.03	5 12 40	1.519	5 26 11	0.075	5 25 18	1.350	0 20 59	3.295	2 12 40	0.431	2 15 47	9.402
157	5 13 13	20.26	2 13 13	0.59	5 13 25	1.433	5 27 16	0.071	5 26 38	1.279	10 29 57	2.560	2 13 25	0.431	2 16 25	9.442
158	5 12 40	20.37	2 12 40	2.18	5 11 50	1.522	5 25 58	0.075	5 25 1	1.353	0 20 57	3.481	2 11 50	0.431	2 15 5	9.401
159	5 12 32	20.36	2 12 32	2.11	5 11 43	1.519	5 25 58	0.075	5 25 1	1.351	0 19 48	3.453	2 11 43	0.431	2 14 59	9.398
160	5 12 1	20.26	2 12 1	0.17	5 12 5	1.441	5 26 55	0.071	5 26 12	1.287	11 8 29	2.685	2 12 5	0.431	2 15 18	9.408
161	5 11 2	20.33	2 11 2	1.84	5 10 16	1.503	5 25 48	0.074	5 24 49	1.338	0 12 38	3.463	2 10 16	0.430	2 13 46	9.359
162	5 9 47	20.35	2 9 47	2.05	5 10 37	1.392	5 27 29	0.069	5 26 54	1.235	10 4 43	3.548	2 10 37	0.430	2 14 4	9.369
163	5 9 8	20.39	2 9 8	2.35	5 8 2	1.516	5 25 9	0.075	5 24 2	1.348	0 14 32	4.006	2 8 2	0.429	2 11 53	9.294
164	5 9 1	20.31	2 9 1	1.56	5 8 20	1.486	5 25 35	0.073	5 24 33	1.325	0 4 47	3.554	2 8 19	0.429	2 12 7	9.302
165	5 8 43	20.28	2 8 43	1.08	5 8 13	1.409	5 25 47	0.073	5 24 48	1.311	11 27 0	3.399	2 8 13	0.429	2 12 2	9.301
166	5 8 42	20.29	2 8 42	1.33	5 9 17	1.404	5 27 1	0.069	5 26 20	1.250	10 15 38	3.348	2 9 17	0.429	2 12 56	9.330
167	5 7 17	20.27	2 7 17	0.94	5 7 43	1.408	5 26 40	0.070	5 25 54	1.256	10 21 25	3.426	2 7 43	0.429	2 11 37	9.284
168	5 7 8	20.35	2 7 8	2.04	5 6 7	1.496	5 24 58	0.074	5 23 49	1.333	0 6 51	4.051	2 6 8	0.428	2 10 12	9.233
169	5 5 13	20.32	2 5 13	1.67	5 4 19	1.475	5 24 53	0.073	5 23 42	1.317	11 28 42	4.100	2 4 20	0.428	2 8 41	9.173
170	5 4 13	20.32	2 4 13	1.71	5 5 6	1.383	5 26 48	0.068	5 26 4	1.231	10 9 29	4.020	2 5 7	0.427	2 9 22	9.200
171	5 3 38	20.28	2 3 38	1.12	5 4 14	1.394	5 26 24	0.069	5 25 34	1.243	10 17 18	3.917	2 4 14	0.427	2 8 37	9.170
172	5 3 27	20.27	2 3 27	0.81	5 3 0	1.441	5 25 11	0.071	5 24 4	1.289	11 14 54	3.999	2 3 0	0.426	2 7 32	9.127
173	5 3 17	20.30	2 3 17	1.41	5 4 3	1.387	5 26 33	0.068	5 25 45	1.235	10 13 14	3.792	2 4 3	0.427	2 8 27	9.164
174	5 1 2	20.27	2 1 2	0.78	5 1 30	1.392	5 25 59	0.069	5 25 3	1.243	10 20 9	4.186	2 1 30	0.425	2 6 13	9.074

A CATALOGUE OF 520 ZODIACAL STARS FOR 1820 (CONTINUED).

No.	Mag.	Stars.	Longitude.				Ann. Var.	Latitude.			Ann. Var.	North East Angle of the ✱'s Meridian with the Ecliptic.			Ann. Var. of the ∠.	Right Ascension in Time.			Ann. Prec. 1830.	Declination.			Ann. Precess. 1830.
			S	°	′	″	+	°	′	″		°	′	″	+	h.	m.	s.	+	°	′	″	−
175	6	12 Cancri	3	28	52	6″.5	50″.15	6	22	45″.8 S	−0″.40	101	26	5″.1	17″.72	7	58	38	3ˢ.360	14	9	28 N	9″.964
176	6	λ Cancri	3	29	18	4.5	50.19	4	21	43.7 N	+0.33	102	22	20.7	18.23	8	9	49	3.582	24	34	53 N	10.800
177	6	D^1 Cancri	4	1	15	50.6	50.26	1	1	12.7 S	−0.31	102	36	56.9	17.47	8	13	3	3.449	18	54	9 N	11.035
178	6	D^2 Cancri	4	2	8	25.6	49.99	2	7	27.5 S	−0.21	102	50	31.9	17.20	8	15	38	3.419	17	37	57 N	11.223
179	6	η Cancri	4	2	53	43.3	50.17	1	33	22.5 N	+0.33	103	23	49.3	17.22	8	22	17	3.484	21	2	44 N	11.700
180	5.6	ϑ Cancri	4	3	12	56.6	50.05	0	46	56.5 S	−0.30	103	18	44.9	17.01	8	21	19	3.435	18	41	46 N	11.631
181	6	29 Cancri	4	3	27	59.7	50.16	4	43	30.5 S	−0.38	103	7	34.6	16.79	8	18	34	3.357	14	48	1 N	11.434
182	6	39 Cancri	4	4	41	5.5	50.12	1	34	15.9 N	+0.28	104	0	40.7	16.74	8	29	44	3.466	20	38	10 N	12.223
183	6	40 Cancri	4	4	42	49.2	50.11	1	32	26.8 N	+0.30	104	1	5.7	16.74	8	29	50	3.465	20	35	59 N	12.230
184	5	Asell. Bor.	4	5	1	40.2	50.02	3	10	35.3 N	+0.29	104	16	51.2	16.74	8	32	51	3.493	22	6	32 N	12.439
185	4.5	Asell. Austr.	4	6	12	19.2	50.14	0	4	17.9 N	+0.14	104	23	9.9	16.27	8	34	26	3.422	18	48	38 N	12.546
186	6	A^2 Cancri	4	8	19	54.0	49.95	5	37	18.6 S	−0.35	104	40	2.3	15.62	8	37	3	3.301	12	45	55 N	12.723
187	6	ο^1 Cancri	4	9	52	1.8	50.26	1	51	19.7 S	−0.31	105	23	50.7	15.25	8	47	11	3.352	16	0	22 N	13.395
188	5	ο^2 Cancri	4	9	52	21.5	50.15	1	35	6.4 S	−0.31	105	25	12.4	15.25	8	47	31	3.357	16	15	54 N	13.417
189	6	α^1 Cancri	4	10	35	25.4	50.11	5	29	32.7 S	−0.33	105	22	34.1	15.06	8	46	5	3.285	12	18	28 N	13.323
190	5.6	ξ Cancri	4	10	41	44.5	50.22	5	24	30.2 N	+0.28	106	21	10.5	15.15	8	59	0	3.465	22	46	4 N	14.147
191	6	79 Cancri	4	10	55	45.0	50.15	5	25	44.3 N	+0.33	106	25	36.7	15.08	8	59	59	3.462	22	43	18 N	14.208
192	5	Sertan	4	11	7	38.4	50.15	5	5	41.1 S	−0.33	105	33	45.1	14.92	8	48	38	3.287	12	32	54 N	13.488
193	5.6	κ Cancri	4	13	39	37.7	50.26	5	35	1.4 S	−0.32	106	17	1.8	14.27	8	58	0	3.258	11	23	11 N	14.083
194	6	82 Cancri	4	14	8	14.3	50.15	0	57	41.3 S	−0.30	106	44	15.5	14.07	9	5	17	3.326	15	40	58 N	14.529
195	6	83 Cancri	4	14	9	46.3	50.09	1	57	7.5 N	+0.18	107	0	17.7	14.04	9	8	55	3.369	18	27	51 N	14.747
196	5	ξ Leonis	4	19	8	16.8	50.07	3	9	42.7 S	−0.25	107	56	10.0	12.68	9	22	14	3.249	12	5	34 N	15.509
197	6	h Leonis	4	19	38	28.2	49.96	4	40	6.0 S	−0.20	107	58	23.7	12.60	9	22	17	3.224	10	30	17 N	15.512
198	6	ψ Leonis	4	20	58	22.5	50.24	0	20	16.2 N	+0.20	108	39	41.9	12.00	9	33	55	3.277	14	50	25 N	16.136
199	4	ο Leonis	4	21	44	24.0	49.96	3	45	54.2 S	−0.15	108	33	10.3	11.96	9	31	32	3.219	10	42	25 N	16.011
200	6	18 Leonis	4	22	19	1.9	50.20	1	32	3.0 S	−0.28	108	50	23.2	11.68	9	36	41	3.242	12	38	6 N	16.278
201	5.6	ν Leonis	4	24	49	26.8	50.07	0	2	33.1 N	+0.24	109	32	14.1	10.82	9	48	32	3.238	13	18	0 N	16.860
202	3.4	η Leonis	4	25	23	22.4	50.18	4	51	17.8 N	+0.20	110	6	43.6	10.36	9	57	30	3.283	17	38	11 N	17.273
203	6	10 Sextantis	4	25	37	9.4	50.02	3	24	27.9 S	−0.22	109	28	42.0	10.80	9	46	53	3.193	9	46	56 N	16.782
204	6	11 Sextantis	4	26	13	12.9	50.14	3	50	45.9 S	−0.13	109	35	10.0	10.65	9	48	35	3.184	9	10	9 N	16.862
205	4.5	π Leonis	4	26	48	5.6	50.05	3	55	13.1 S	−0.16	109	42	30.6	10.48	9	50	41	3.179	8	54	14 N	16.962
206	1	Regulus	4	27	19	35.5	49.94	0	27	33.2 N	+0.12	110	6	21.6	10.01	9	58	47	3.220	12	50	37 N	17.328
207	6	34 Leonis	4	27	34	4.2	50.35	2	2	7.7 N	+0.14	110	17	9.2	9.82	10	1	56	3.233	14	14	25 N	17.465
208	5	a Leonis	4	27	54	10.6	50.01	1	25	28.5 S	−0.11	110	5	18.5	9.96	9	58	21	3.197	10	52	35 N	17.309
209	6	37 Leonis	4	28	35	3.6	50.25	2	49	18.6 N	+0.14	110	33	30.5	9.44	10	7	0	3.231	14	37	21 N	17.678
210	6	14 Sextantis	4	29	11	20.3	50.14	5	37	48.7 S	−0.20	110	7	50.7	9.91	9	57	22	3.144	6	29	9 N	17.266
211	6	42 Leonis	4	29	18	3.4	50.33	4	26	15.2 N	+0.21	110	51	4.2	9.08	10	12	9	3.240	15	52	51 N	17.887
212	6	16 Sextantis	4	29	33	46.2	50.02	4	53	30.0 S	−0.22	110	14	12.7	9.74	9	59	48	3.150	7	3	1 N	17.373
213	6	i Leonis	5	1	56	57.1	50.16	4	34	31.3 N	+0.26	111	20	19.5	8.18	10	22	35	3.215	15	3	30 N	18.279
214	6	44 Leonis	5	2	19	48.3	50.17	1	1	33.7 S	−0.18	110	57	39.7	8.55	10	15	46	3.168	9	41	48 N	18.026
215	6	45 Leonis	5	2	31	27.6	50.21	0	5	53.3 N	+0.13	111	4	3.4	8.39	10	18	8	3.175	10	40	37 N	18.115
216	6	43 Leonis	5	2	37	44.4	50.16	3	18	47.6 S	−0.17	110	53	28.3	8.67	10	13	35	3.145	7	27	14 N	17.941
217	4	ρ Leonis	5	3	52	29.0	50.06	0	8	34.1 N	+0.11	111	18	0.8	7.95	10	23	19	3.166	10	13	49 N	18.306
218	6	49 Leonis	5	4	37	50.4	50.07	0	15	41.8 S	−0.17	111	23	53.4	7.75	10	25	35	3.157	9	34	41 N	18.385
219	6	k Leonis	5	5	8	17.1	50.04	5	55	39.4 N	+0.04	111	58	42.0	6.98	10	36	53	3.195	15	8	36 N	18.756
220	5.6	48 Leonis	5	5	12	31.8	49.95	1	51	42.8 S	−0.10	111	24	7.2	7.73	10	25	24	3.141	7	52	39 N	18.373
221	6	l Leonis	5	7	9	58.4	50.18	2	48	37.6 N	+0.10	111	59	28.9	6.62	10	39	47	3.160	11	29	46 N	18.845
222	6	37 Sextantis	5	8	1	18.4	50.16	1	20	49.7 S	−0.09	111	51	16.8	6.79	10	36	43	3.128	7	19	11 N	18.750
223	6	34 Sextantis	5	8	16	31.2	49.77	4	15	34.6 S	−0.00	111	46	46.3	7.03	10	33	19	3.106	4	31	17 N	18.644
224	6	36 Sextantis	5	9	16	41.0	49.88	5	1	54.8 S	−0.03	111	54	18.6	6.80	10	35	53	3.096	3	25	58 N	18.724
225	5.6	c Leonis	5	11	29	38.2	49.98	0	12	32.0 S	−0.02	112	21	43.8	5.55	10	51	25	3.116	7	4	1 N	19.166
226	4.5	χ Leonis	5	12	0	28.9	49.71	1	20	45.5 N	−0.10	112	30	5.5	5.20	10	55	44	3.121	8	18	29 N	19.273
227	6	55 Leonis	5	12	23	25.1	50.11	5	38	53.0 S	−0.15	112	18	46.4	5.92	10	46	26	3.080	1	41	42 N	19.034
228	5	d Leonis	5	12	24	17.3	49.98	2	31	13.3 S	−0.03	112	22	48.0	5.54	10	51	15	3.099	4	34	58 N	19.162
229	6	g Leonis	5	14	30	59.4	50.14	5	33	43.1 S	−0.07	112	34	0.4	5.25	10	54	24	3.074	0	58	2 N	19.241
230	5.6	65 Leonis	5	14	31	41.8	49.72	3	25	45.5 S	+0.10	112	35	46.5	4.98	10	57	43	3.086	2	55	55 N	19.320
231	4	ι Leonis	5	15	2	22.0	50.24	6	6	8.8 N	+0.04	113	6	52.0	3.63	11	14	32	3.121	11	31	14 N	19.660
232	4	σ Leonis	5	16	11	42.4	50.07	1	41	46.8 N	−0.04	112	55	40.2	3.80	11	11	51	3.102	7	0	52 N	19.613

SUBSIDIARY NUMBERS OF THE ZODIACAL STARS FOR 1830 (CONTINUED).

No.	Longitude. Aberration. A	Max.	Latitude. Aberration. A′	Max.	Right Ascension in Time. Aberration. A	Max.	☉ Nutation. S	Max.	☾ Nutation. L	Max.	Declination. Aberration. A′	Max.	☉ Nutation. S′	Max.	☾ Nutation. L′	Max.
	S ° ′		S ° ′		S ° ′		S ° ′		S ° ′		S ° ′		S ° ′		S ° ′	
175	5 1 0	20s.38	2 1 0	2s.25	5 2 17	1s.365	5 26 54	0s.067	5 26 11	1s.211	10 2 18	4s.600	2 2 17	0s.426	2 6 56	9s.102
176	5 0 34	20 .31	2 0 34	1 .54	4 29 36	1 .450	5 24 18	0 .072	5 22 59	1 .298	11 19 35	4 .611	1 29 36	0 .424	2 4 32	9 .004
177	4 28 36	20 .26	1 28 36	0 .36	4 28 49	1 .392	5 25 28	0 .069	5 24 26	1 .246	10 24 23	4 .456	1 28 49	0 .424	2 3 50	8 .974
178	4 27 43	20 .26	1 27 43	0 .75	4 28 12	1 .381	5 25 41	0 .068	5 24 41	1 .234	10 18 31	4 .578	1 28 12	0 .423	2 3 17	8 .949
179	4 26 58	20 .26	1 26 58	0 .54	4 26 35	1 .407	5 24 40	0 .070	5 23 26	1 .260	11 3 23	4 .720	1 26 35	0 .423	2 1 50	8 .885
180	4 26 39	20 .26	1 26 39	0 .27	4 26 50	1 .387	5 25 16	0 .069	5 24 10	1 .243	10 23 21	4 .689	1 26 49	0 .423	2 2 3	8 .895
181	4 26 24	20 .32	1 26 24	1 .67	4 27 29	1:.360	5 26 17	0 .067	5 25 25	1 .211	10 7 0	4 .895	1 27 29	0 .423	2 2 39	8 .922
182	4 25 11	20 .26	1 25 11	0 .55	4 24 47	1 .399	5 24 31	0 .069	5 23 15	1 .255	11 1 25	4 .951	1 24 46	0 .421	2 0 11	8 .813
183	4 25 9	20 .26	1 25 9	0 .54	4 24 45	1 .399	5 24 31	0 .069	5 23 16	1 .254	11 1 17	4 .951	1 24 45	0 .421	2 0 10	8 .812
184	4 24 50	20 .28	1 24 50	1 .12	4 24 0	1 .412	5 24 1	0 .070	5 22 39	1 .266	11 7 4	5 .130	1 24 1	0 .421	1 29 30	8 .782
185	4 23 39	20 .26	1 23 39	0 .02	4 23 39	1 .382	5 24 50	0 .069	5 23 39	1 .237	10 23 55	5 .049	1 23 38	0 .421	1 29 9	8 .765
186	4 21 32	20 .35	1 21 32	1 .98	4 23 0	1 .340	5 27 19	0 .066	5 26 28	1 .189	9 25 34	5 .214	1 23 0	0 .420	1 28 33	8 .739
187	4 20 0	20 .26	1 20 0	0 .65	4 20 31	1 .354	5 25 16	0 .067	5 24 10	1 .212	10 13 18	5 .431	1 20 31	0 .418	1 26 13	8 .634
188	4 19 59	20 .26	1 19 59	0 .56	4 20 26	1 .355	5 25 11	0 .067	5 24 4	1 .213	10 14 16	5 .426	1 20 26	0 .418	1 26 9	8 .631
189	4 19 16	20 .34	1 19 16	1 .94	4 20 46	1 .333	5 26 21	0 .066	5 25 29	1 .185	10 0 7	5 .702	1 20 47	0 .419	1 26 28	8 .646
190	4 19 10	20 .34	1 19 10	1 .91	4 17 34	1 .405	5 22 56	0 .070	5 21 19	1 .260	11 6 58	6 .006	1 17 34	0 .417	1 23 26	8 .509
191	4 18 56	20 .34	1 18 56	1 .92	4 17 19	1 .404	5 22 55	0 .070	5 21 18	1 .259	11 6 40	6 .030	1 17 19	0 .417	1 23 11	8 .499
192	4 18 44	20 .33	1 18 44	1 .80	4 20 9	1 .333	5 26 14	0 .066	5 25 21	1 .186	10 1 5	5 .719	1 20 9	0 .418	1 25 53	8 .619
193	4 16 12	20 .35	1 16 12	1 .97	4 17 50	1 .322	5 26 25	0 .065	5 25 35	1 .175	9 27 47	6 .001	1 17 50	0 .417	1 23 41	8 .521
194	4 15 33	20 .26	1 15 33	0 .34	4 16 0	1 .343	5 24 56	0 .067	5 23 46	1 .203	10 12 32	5 .856	1 16 0	0 .416	1 21 54	8 .443
195	4 15 32	20 .26	1 15 32	0 .69	4 15 6	1 .361	5 23 58	0 .067	5 22 35	1 .222	10 22 1	5 .974	1 15 6	0 .415	1 21 2	8 .404
196	4 10 43	20 .28	1 10 43	1 .12	4 11 45	1 .314	5 25 47	0 .065	5 24 48	1 .173	10 1 4	6 .341	1 11 45	0 .413	1 17 43	8 .261
197	4 10 13	20 .32	1 10 13	1 .65	4 11 45	1 .307	5 26 20	0 .064	5 25 29	1 .163	9 26 2	6 .450	1 11 45	0 .413	1 17 42	8 .260
198	4 8 53	20 .26	1 8 53	0 .12	4 8 47	1 .323	5 24 37	0 .066	5 23 23	1 .186	10 9 54	6 .495	1 8 47	0 .410	1 14 42	8 .136
199	4 8 7	20 .29	1 8 7	1 .33	4 9 23	1 .303	5 26 7	0 .064	5 25 13	1 .161	9 27 4	6 .580	1 9 23	0 .410	1 15 20	8 .161
200	4 7 33	20 .26	1 7 33	0 .54	4 8 4	1 .309	5 25 21	0 .065	5 24 17	1 .171	10 3 3	6 .574	1 8 5	0 .410	1 13 59	8 .107
201	4 5 2	20 .26	1 5 2	0 .02	4 5 1	1 .307	5 24 56	0 .065	5 23 46	1 .171	10 5 6	6 .789	1 5 1	0 .409	1 10 49	7 .982
202	4 4 28	20 .32	1 4 28	1 .71	4 2 42	1 .331	5 23 7	0 .066	5 21 32	1 .193	10 17 27	7 .160	1 2 41	0 .408	1 8 20	7 .891
203	4 4 25	20 .29	1 4 15	1 .20	4 5 27	1 .292	5 26 16	0 .064	5 25 24	1 .152	9 24 42	6 .860	1 5 27	0 .409	1 11 16	8 .000
204	4 3 38	20 .30	1 3 38	1 .36	4 5 0	1 .289	5 26 28	0 .063	5 25 39	1 .149	9 22 2	6 .872	1 5 0	0 .409	1 10 48	7 .982
205	4 3 4	20 .30	1 3 4	1 .38	4 4 28	1 .287	5 26 33	0 .064	5 25 45	1 .146	9 22 16	6 .964	1 4 28	0 .409	1 10 13	7 .962
206	4 2 33	20 .26	1 2 33	0 .17	4 2 22	1 .300	5 24 57	0 .064	5 23 36	1 .165	10 3 47	6 .974	1 2 21	0 .408	1 8 0	7 .879
207	4 2 18	20 .26	1 2 18	0 .72	4 1 32	1 .306	5 24 21	0 .065	5 23 3	1 .171	10 7 45	7 .067	1 1 32	0 .408	1 7 6	7 .848
208	4 1 57	20 .26	1 1 57	0 .50	4 2 29	1 .292	5 25 43	0 .064	5 24 43	1 .154	9 28 4	6 .979	1 2 28	0 .408	1 8 7	7 .883
209	4 1 17	20 .27	1 1 17	1 .00	4 0 13	1 .306	5 24 7	0 .065	5 22 46	1 .171	10 8 44	7 .185	1 0 12	0 .407	1 5 39	7 .798
210	4 0 40	20 .35	1 0 40	1 .99	4 2 30	1 .277	5 27 26	0 .063	5 26 50	1 .132	9 15 42	7 .228	1 2 29	0 .408	1 8 7	7 .883
211	4 0 34	20 .31	1 0 34	1 .57	3 28 52	1 .312	5 23 32	0 .065	5 22 3	1 .176	10 12 8	7 .352	0 28 52	0 .406	1 4 11	7 .749
212	4 0 18	20 .32	1 0 18	1 .73	4 2 6	1 .277	5 27 12	0 .063	5 26 33	1 .135	9 17 15	7 .200	1 2 6	0 .408	1 7 42	7 .869
213	3 27 55	20 .32	0 27 55	1 .62	3 26 7	1 .303	5 23 42	0 .064	5 22 15	1 .167	10 9 25	7 .530	0 26 7	0 .405	1 1 9	7 .653
214	3 27 32	20 .26	0 27 32	0 .37	3 27 55	1 .279	5 25 55	0 .063	5 25 4	1 .143	9 24 52	7 .258	0 27 55	0 .406	1 3 9	7 .715
215	3 27 20	20 .26	0 27 20	0 .04	3 27 17	1 .282	5 25 34	0 .063	5 24 32	1 .146	9 27 35	7 .290	0 27 17	0 .405	1 2 27	7 .693
216	3 27 14	20 .28	0 27 14	1 .17	3 28 31	1 .272	5 26 55	0 .063	5 26 13	1 .133	9 18 37	7 .315	0 28 30	0 .406	1 3 48	7 .735
217	3 25 59	20 .26	0 25 59	0 .05	3 25 55	1 .278	5 25 42	0 .063	5 24 42	1 .143	9 26 21	7 .367	0 25 55	0 .405	1 0 56	7 .648
218	3 25 14	20 .26	0 25 14	0 .09	3 25 20	1 .275	5 25 57	0 .063	5 25 1	1 .139	9 24 32	7 .400	0 25 20	0 .404	1 0 15	7 .627
219	3 24 43	20 .36	0 24 43	2 .09	3 22 20	1 .298	5 23 28	0 .064	5 21 57	1 .160	10 9 3	7 .832	0 22 19	0 .403	0 26 51	7 .533
220	3 24 39	20 .26	0 24 39	0 .66	3 25 23	1 .269	5 26 40	0 .063	5 25 54	1 .132	9 19 54	7 .422	0 25 24	0 .404	1 0 18	7 .629
221	3 22 42	20 .27	0 22 42	1 .00	3 21 34	1 .278	5 25 0	0 .063	5 23 59	1 .142	9 29 35	7 .649	0 21 33	0 .403	0 25 57	7 .511
222	3 21 50	20 .26	0 21 50	0 .48	3 22 23	1 .263	5 26 50	0 .062	5 26 6	1 .127	9 18 29	7 .562	0 22 22	0 .403	0 26 55	7 .535
223	3 21 35	20 .31	0 21 35	1 .51	3 23 16	1 .258	5 28 3	0 .062	5 27 36	1 .118	9 11 3	7 .653	0 23 17	0 .403	0 27 56	7-.563
224	3 20 35	20 .33	0 20 35	1 .78	3 22 36	1 .256	5 28 31	0 .062	5 28 10	1 .113	9 8 17	7 .741	0 22 36	0 .403	0 27 10	7 .541
225	3 18 22	20 .26	0 18 22	0 .08	3 18 27	1 .258	5 26 51	0 .062	5 26 8	1 .123	9 17 52	7 .713	0 18 27	0 .401	0 22 21	7 .426
226	3 17 51	20 .26	0 17 51	0 .48	3 17 17	1 .261	5 26 17	0 .062	5 25 26	1 .126	9 21 5	7 .772	0 17 16	0 .401	0 20 59	7 .397
227	3 17 28	20 .35	0 17 28	1 .99	3 19 47	1 .251	5 29 16	0 .061	5 29 6	1 .107	9 4 0	7 .916	0 19 47	0 .402	0 23 55	7 .460
228	3 17 27	20 .27	0 17 27	0 .89	3 18 30	1 .253	5 27 58	0 .062	5 27 30	1 .115	9 11 21	7 .763	0 18 30	0 .401	0 22 25	7 .426
229	3 15 21	20 .35	0 15 21	1 .96	3 17 38	1 .201	5 29 35	0 .061	5 29 29	1 .105	9 2 14	7 .988	0 17 39	0 .401	0 21 24	7 .406
230	3 15 20	20 29	0 15 20	1 .21	3 16 45	1 .249	5 28 41	0 .062	5 28 23	1 .110	9 7 12	7 .867	0 16 45	0 .401	0 20 22	7 .384
231	3 14 49	20 .37	0 14 49	2 .15	3 12 13	1 .269	5 24 43	0 .063	5 23 30	1 .129	9 28 48	8 .198	0 12 14	0 .399	0 14 57	7 .291
232	3 13 40	20 .26	0 13 40	0 .60	3 12 57	1 .253	5 26 48	0 .062	5 26 3	1 .118	9 17 40	7 .914	0 12 57	0 .400	0 15 49	7 .304

A CATALOGUE OF 520 ZODIACAL STARS FOR 1820 (CONTINUED).

No.	Mag.	Stars.	Longitude.	Ann. Var.	Latitude.	Ann. Var.	North East Angle of the ✱'s Meridian with the Ecliptic.	Ann. Var. of the ∠.	Right Ascension in Time.	Ann. Prec. 1830.	Declination.	Ann. Precess. 1830.
			S ° ′ ″	+	° ′ ″		° ′ ″	+	h. m. s.	+	° ′ ″	
233	5.6	75 Leonis	5 16 52 41.7	50″.22	2 22 24.3 S	+0″.10	112 50 53.6	4″.10	11 8 1	3s.083	3 0 0 N	−19″.541
234	5.6	69 Leonis	5 16 53 19.0	50.14	4 38 12.9 S	−0.05	112 49 9.1	4.39	11 4 32	3.073	0 54 32 N	−19.470
235	6	76 Leonis	5 17 23 53.8	49.96	2 32 50.6 S	+0.01	112 53 27.2	3.96	11 9 40	3.081	2 38 13 N	−19.572
236	5.6	79 Leonis	5 18 40 20.6	49.98	2 16 12.1 S	−0.01	113 0 2.5	3.52	11 14 48	3.079	2 23 45 N	−19.663
237	4	τ Leonis	5 18 59 47.7	50.16	0 33 20.0 S	0.00	113 3 41.2	3.19	11 18 40	3.084	3 50 50 N	−19.727
238	6	89 Leonis	5 20 23 55.8	50.18	0 16 29.9 N	+0.02	113 10 35.6	2.63	11 25 9	3.082	4 3 37 N	−19.819
239	5	ξ¹ Virginis	5 20 48 6.7	50.04	6 6 52.7 N	+0.04	113 28 5.2	1.71	11 36 0	3.090	9 15 34 N	−19.939
240	5.6	ξ² Virginis	5 21 25 28.8	49.99	6 21 49.4 N	−0.01	113 30 31.3	1.48	11 38 39	3.087	9 14 45 N	−19.961
241	4.5	ν Virginis	5 21 38 24.4	50.22	4 35 40.7 N	−0.18	113 24 47.6	1.05	11 36 36	3.085	7 32 20 N	−19.944
242	4.5	E Leonis	5 21 51 52.2	50.12	5 42 10.6 S	+0.01	113 13 40.9	2.97	11 21 7	3.060	2 0 39 S	+19.763
243	4.5	υ Leonis	5 22 31 28.3	50.09	3 2 49.9 S	−0.03	113 15 6.6	2.40	11 27 44	3.068	0 10 10 N	−19.852
244	3.4	Zavijava	5 24 36 38.6	51.01	0 41 34.4 N	+0.02	113 22 55.2	1.22	11 41 18	3.072	2 46 50 N	−19.981
245	5	π Virginis	5 25 2 35.7	50.36	6 9 23.4 N	+0.04	113 35 26.2	0.33	11 51 39	3.074	7 37 7 N	−20.031
246	5.6	B Virginis	5 26 1 0.4	49.97	3 20 59.9 N	−0.14	113 29 6.4	0.40	11 50 44	3.072	4 39 30 N	−20.029
247	6	Virginis	5 27 33 35.6	50.13	5 46 50.6 S	+0.03	113 30 41.6	1.18	11 41 50	3.060	4 19 54 S	+19.984
248	6	r Virginis	5 28 56 42.6	49.99	2 42 52.9 N	−0.14	113 29 26.9	0.45	12 0 28	3.068	2 54 37 N	−20.044
249	5.6	c Virginis	6 0 50 44.5	49.74	5 4 35.0 N	+0.20	113 31 51.4	1.39	12 11 12	3.063	4 19 0 N	−20.017
250	6	n Virginis	6 2 4 47.4	50.07	1 8 11.0 N	−0.18	113 26 48.1	1.23	12 9 27	3.068	0 12 53 N	−20.021
251	3.4	η Virginis	6 2 19 14.3	50.16	1 22 19.1 N	−0.13	113 26 34.6	1.34	12 10 42	3.068	0 20 6 N	−20.020
252	4	γ¹ Virginis	6 7 39 4.3	49.51	2 48 26.7 N	−0.31	113 14 32.2	3.23	12 32 33	3.069	0 27 37 S	+19.835
253	4	γ² Virginis	6 7 39 6.6	49.46	2 48 24.2 N	−0.34	113 14 31.9	3.23	12 32 33	3.070	0 27 40 S	+19.836
254	5.6	q Virginis	6 8 58 30.7	49.86	5 19 50.1 S	+0.25	113 25 42.2	2.55	12 24 30	3.090	8 27 26 S	+19.925
255	6	χ Virginis	6 9 38 33.6	49.86	3 27 30.5 S	+0.26	113 17 44.8	3.02	12 29 58	3.090	7 0 9 S	+19.866
256	6	28 Virginis	6 10 3 54.1	50.16	2 44 20.0 S	+0.14	113 14 25.1	3.25	12 32 39	3.090	6 30 28 S	+19.834
257	6	38 Virginis	6 11 6 44.2	50.00	1 59 16.0 N	−0.24	113 1 18.8	4.21	12 43 59	3.080	2 34 19 S	+19.666
258	6	k Virginis	6 12 41 27.2	49.86	2 22 0.6 N	−0.37	112 53 10.3	4.76	12 50 23	3.083	2 50 16 S	+19.550
259	6	48 Virginis	6 13 36 48.8	50.08	2 54 50.0 N	−0.29	112 47 33.1	5.12	12 54 38	3.083	2 41 29 S	+19.465
260	5.6	ψ Virginis	6 13 41 17.7	50.05	3 25 22.7 S	+0.25	113 1 45.6	4.33	12 45 0	3.108	8 33 31 S	+19.649
261	4.5	ϑ Virginis	6 15 43 25.5	50.08	1 45 15.4 N	−0.29	112 36 41.3	5.63	13 0 38	3.096	4 34 30 S	+19.335
262	5.6	g Virginis	6 17 13 53.0	50.08	3 15 2.3 S	+0.25	112 41 55.4	5.50	12 58 29	3.127	9 46 29 S	+19.382
263	6	50 Virginis	6 17 29 58.6	50.16	2 41 50.9 S	+0.20	112 38 6.3	5.65	13 0 20	3.126	9 22 3 S	+19.340
264	6	65 Virginis	6 18 34 44.9	50.21	3 34 34.6 N	−0.21	112 13 46.1	6.74	13 14 0	3.098	3 58 45 S	+18.992
265	6	66 Virginis	6 18 56 40.6	50.21	3 27 56.6 N	−0.21	112 11 8.9	6.84	13 15 11	3.100	4 13 10 S	+18.959
266	6	58 Virginis	6 19 20 23.3	50.19	2 10 51.7 S	+0.26	112 23 48.2	6.30	13 8 2	3.135	9 35 39 S	+19.153
267	6	l² Virginis	6 21 4 33.1	49.95	3 8 15.1 N	−0.38	111 54 30.7	7.45	13 22 37	3.113	5 19 20 S	+18.739
268	1	Spica	6 21 19 45.8	50.10	2 2 25.9 S	+0.29	112 8 21.4	6.96	13 15 43	3.147	10 13 4 S	+18.944
269	6	l³ Virginis	6 21 35 1.8	50.20	4 15 12.8 N	−0.17	111 48 0.3	7.73	13 26 10	3.107	4 28 32 S	+18.626
270	5	i Virginis	6 22 15 12.7	49.87	3 20 17.4 S	+0.41	112 6 40.5	7.11	13 17 13	3.161	11 46 3 S	+18.901
271	6	h Virginis	6 22 44 12.5	50.02	0 24 39.7 S	+0.39	111 50 27.8	7.58	13 23 30	3.146	9 14 1 S	+18.711
272	5.6	69 Virginis	6 23 38 14.0	50.03	6 18 21.5 S	+0.30	112 11 24.4	7.25	13 17 52	3.189	15 2 10 S	+18.882
273	5.6	m Virginis	6 24 11 52.4	49.89	1 43 23.9 N	−0.29	111 30 10.7	8.26	13 32 11	3.140	7 47 25 S	+18.425
274	6	75 Virginis	6 24 37 22.6	50.05	5 15 26.5 S	+0.29	111 56 48.4	7.69	13 23 15	3.191	14 25 59 S	+18.718
275	6	86 Virginis	6 26 31 13.4	50.15	1 22 23.4 S	+0.32	111 19 13.7	8.68	13 36 21	3.180	11 31 12 S	+18.278
276	6	83 Virginis	6 27 32 1.8	50.05	5 0 11.6 S	+0.38	111 28 3.8	8.67	13 34 47	3.216	15 16 9 S	+18.334
277	6	85 Virginis	6 27 38 5.2	50.15	4 31 19.4 S	+0.28	111 24 13.4	8.74	13 35 54	3.213	14 51 30 S	+18.294
278	6	87 Virginis	6 28 47 36.3	50.09	6 19 14.1 S	+0.36	111 23 38.6	8.97	13 37 39	3.238	16 57 12 S	+18.231
279	5.6	89 Virginis	6 29 26 49.1	50.02	6 21 57.8 S	+0.40	111 17 6.7	9.18	13 40 6	3.245	17 13 55 S	+18.141
280	6	94 Virginis	6 29 58 40.1	50.03	3 40 51.8 N	−0.36	110 22 59.8	10.19	13 56 46	3.161	8 1 38 S	+17.474
281	6	95 Virginis	7 0 13 30.8	50.02	3 19 21.0 N	−0.37	110 21 9.9	10.23	13 57 12	3.166	8 26 57 S	+17.456
282	4	κ Virginis	7 1 58 49.3	49.99	2 55 15.8 N	−0.25	110 1 12.6	10.72	14 3 18	3.182	9 25 54 S	+17.188
283	6	Virg. 562 M.	7 3 31 14.0	49.97	2 55 15.4 S	+0.44	110 8 36.4	10.77	14 1 1	3.255	15 26 46 S	+17.289
284	4	λ Virginis	7 4 26 19.9	50.09	0 30 21.0 N	−0.29	109 39 25.4	11.27	14 9 23	3.228	12 32 13 S	+16.909
285	6	2 Libræ	7 4 53 35.8	50.04	2 25 5.3 N	−0.51	109 25 26.2	11.53	14 9 45	3.211	10 53 8 S	+16.700
286	6	Virginis	7 5 11 15.4	50.15	4 20 46.6 S	+0.33	109 55 59.0	11.21	14 5 30	3.287	17 21 20 S	+17.087
287	6	5 Libræ	7 11 16 7.5	50.19	0 33 31.9 N	−0.37	108 1 34.7	13.29	14 36 3	3.290	14 41 39 S	+15.543
288	5.6	μ Libræ	7 11 39 18.5	50.04	2 3 6.3 N	−0.48	107 48 22.0	13.46	14 39 28	3.273	13 23 32 S	+15.353
289	6	ξ¹ Libræ	7 12 10 43.0	50.16	4 33 53.3 N	−0.39	107 30 6.8	13.69	14 44 37	3.243	11 9 23 S	+15.060
290	6	8 Libræ	7 12 30 56.1	50.11	0 23 5.9 N	−0.54	107 42 30.8	13.65	14 40 45	3.304	15 14 30 S	+15.280

SUBSIDIARY NUMBERS OF THE ZODIACAL STARS FOR 1830 (CONTINUED).

No.	Longitude. Aberration. A	Max.	Latitude. Aberration. A′	Max.	Right Ascension in Time. Aberration. A	Max.	☉ Nutation. S	Max.	☾ Nutation. L	Max.	Declination. Aberration. A′	Max.	☉ Nutation. S′	Max.	☾ Nutation. L′	Max.
	S ° ′	″	S ° ′	″	S ° ′	″	S ° ′	″	S ° ′	″	S ° ′	″	S ° ′	″	S ° ′	″
233	3 12 59	20.27	0 12 59	0.84	3 13 59	1.246	5 28 39	0.062	5 28 20	1.109	9 7 22	7.908	0 13 59	0.400	0 17 3	7.324
234	3 12 58	20.32	0 12 58	1.64	3 14 55	1.246	5 29 36	0.061	5 29 31	1.105	9 2 5	8.003	0 14 56	0.400	0 18 11	7.344
235	3 12 28	20.27	0 12 28	0.90	3 13 32	1.245	5 28 49	0.063	5 28 32	1.108	9 6 27	7.927	0 13 33	0.400	0 16 31	7.315
236	3 11 11	20.27	0 11 11	0.80	3 12 9	1.244	5 28 55	0.062	5 28 39	1.107	9 5 54	7.953	0 12 10	0.399	0 14 52	7.290
237	3 10 52	20.26	0 10 52	0.19	3 11 6	1.245	5 28 14	0.062	5 27 50	1.109	9 9 34	7.941	0 11 7	0.399	0 13 36	7.272
238	3 9 28	20.26	0 9 28	0.09	3 9 20	1.244	5 28 8	0.062	5 27 42	1.109	9 10 6	7.976	0 9 21	0.399	0 11 28	7.246
239	3 9 4	20.37	0 9 4	2.16	3 6 24	1.256	5 25 40	0.062	5 24 40	1.116	9 22 49	8.310	0 6 24	0.399	0 7 52	7.212
240	3 8 26	20.38	0 8 26	2.25	3 5 40	1.256	5 25 40	0.062	5 24 40	1.115	9 22 43	8.341	0 5 40	0.398	0 6 59	7.205
241	3 8 13	20.32	0 8 13	1.62	3 6 14	1.250	5 26 31	0.061	5 25 40	1.112	9 18 42	8.191	0 6 14	0.398	0 7 40	7.211
242	3 8 0	20.35	0 8 0	2.01	3 10 26	1.243	6 0 58	0.061	6 1 12	1.100	2 24 57	8.205	6 10 27	0.399	6 12 47	7.263
243	3 7 20	20.28	0 7 20	1.08	3 8 39	1.241	5 29 57	0.061	5 29 56	1.103	9 0 17	8.061	0 8 39	0.399	0 10 37	7.237
244	3 5 15	20.26	0 5 15	0.25	3 4 56	1.241	5 28 43	0.061	5 28 25	1.105	9 6 51	8.044	0 4 58	0.398	0 6 5	7.200
245	3 4 49	20.37	0 4 49	2.17	3 2 8	1.250	5 26 24	0.062	5 25 35	1.108	9 18 37	8.348	0 2 8	0.398	0 2 38	7.185
246	3 3 51	20.28	0 3 51	1.18	3 2 23	1.243	5 27 49	0.061	5 27 19	1.106	9 11 30	8.147	0 2 23	0.398	0 2 56	7.186
247	3 2 18	20.35	0 2 18	2.04	3 4 48	1.243	6 2 5	0.061	6 2 34	1.102	2 19 16	8.297	6 4 49	0.398	6 5 55	7.198
248	3 0 55	20.27	0 0 55	0.96	2 29 44	1.240	5 28 39	0.061	5 28 20	1.103	9 7 8	8.125	11 29 43	0.398	11 29 40	7.183
249	2 29 1	20.33	11 29 1	1.79	2 26 49	1.242	5 27 59	0.061	5 27 31	1.104	9 10 30	8.251	11 26 49	0.398	11 26 5	7.190
250	2 27 47	20.26	11 27 47	0.40	2 27 17	1.239	5 29 56	0.061	5 29 54	1.103	9 0 24	8.068	11 27 18	0.398	11 26 40	7.188
251	2 27 32	20.26	11 27 32	0.48	2 26 57	1.239	5 29 52	0.061	5 29 50	1.103	9 0 42	8.070	11 26 57	0.398	11 26 15	7.189
252	2 22 13	20.27	11 22 13	0.99	2 21 0	1.241	6 0 14	0.061	6 0 18	1.104	2 28 43	8.043	5 21 0	0.399	5 18 58	7.242
253	2 22 13	20.27	11 22 13	0.99	2 21 0	1.241	6 0 13	0.061	6 0 18	1.104	2 28 43	8.045	5 21 0	0.399	5 18 58	7.242
254	2 20 53	20.34	11 20 53	1.88	2 23 12	1.254	6 4 0	0.062	6 4 55	1.115	2 8 47	8.236	5 23 12	0.399	5 21 38	7.216
255	2 20 13	20.29	11 20 13	1.22	2 21 42	1.250	6 3 18	0.062	6 4 4	1.114	2 12 14	8.086	5 21 42	0.399	5 19 48	7.233
256	2 19 48	20.27	11 19 48	0.97	2 20 58	1.249	6 3 4	0.062	6 3 46	1.113	2 13 27	8.043	5 20 58	0.399	5 18 56	7.243
257	2 18 45	20.26	11 18 45	0.70	2 17 54	1.244	6 1 13	0.061	6 1 30	1.108	2 23 25	7.946	5 17 54	0.399	5 15 12	7.289
258	2 17 10	20.27	11 17 10	0.84	2 16 10	1.246	6 1 20	0.061	6 1 38	1.109	2 22 46	7.912	5 16 10	0.400	5 13 7	7.321
259	2 16 15	20.28	11 16 15	1.03	2 15 2	1.247	6 1 15	0.062	6 1 33	1.109	2 23 9	7.901	5 15 1	0.400	5 11 45	7.345
260	2 16 10	20.29	11 16 10	1.21	2 17 38	1.258	6 3 58	0.062	6 4 53	1.121	2 4 54	7.999	5 17 37	0.399	5 14 52	7.294
261	2 14 8	20.26	11 14 8	0.62	2 13 25	1.251	6 2 6	0.062	6 2 35	1.114	2 18 22	7.805	5 13 24	0.401	5 9 50	7.381
262	2 12 38	20.28	11 12 38	1.15	2 13 59	1.265	6 4 27	0.062	6 5 29	1.129	2 4 54	7.885	5 13 59	0.400	5 10 31	7.368
263	2 12 22	20.27	11 12 22	0.95	2 13 29	1.264	6 4 15	0.063	6 5 14	1.129	2 5 55	7.842	5 13 29	0.401	5 9 56	7.379
264	2 11 17	20.29	11 11 17	1.27	2 9 50	1.254	6 1 48	0.062	6 2 13	1.115	2 19 58	7.747	5 9 50	0.402	5 5 39	7.471
265	2 10 55	20.29	11 10 55	1.22	2 9 30	1.254	6 1 54	0.062	6 2 20	1.113	2 16 21	7.728	5 9 31	0.402	5 5 16	7.480
266	2 10 31	20.27	11 10 31	0.77	2 11 25	1.267	6 4 18	0.063	6 5 18	1.132	2 5 15	7.746	5 11 25	0.401	5 7 30	7.429
267	2 8 47	20.28	11 8 47	1.11	2 7 31	1.259	6 2 21	0.062	6 2 54	1.120	2 16 33	7.622	5 7 32	0.403	5 2 59	7.537
268	2 8 32	20.26	11 8 32	0.72	2 9 22	1.271	6 4 31	0.063	6 5 34	1.137	2 3 32	7.657	5 9 22	0.402	5 5 6	7.484
269	2 8 17	20.31	11 8 17	1.50	2 6 35	1.259	6 1 58	0.062	6 2 26	1.110	2 18 48	7.645	5 6 35	0.404	5 1 55	7.566
270	2 7 36	20.28	11 7 36	1.18	2 8 58	1.279	6 5 11	0.063	6 6 23	1.143	1 29 27	7.697	5 8 58	0.402	5 4 38	7.495
271	2 7 7	20.26	11 7 7	0.15	2 7 18	1.270	6 4 2	0.063	6 4 58	1.135	2 6 6	7.531	5 7 18	0.403	5 2 43	7.544
272	2 6 13	20.37	11 6 13	2.22	2 8 47	1.296	6 6 36	0.064	6 8 7	1.155	1 21 9	7.917	5 8 48	0.402	5 4 27	7.500
273	2 5 40	20.26	11 5 40	0.61	2 4 59	1.269	6 3 21	0.063	6 4 8	1.132	2 10 2	7.440	5 4 59	0.404	5 0 5	7.616
274	2 5 14	20.34	11 5 14	1.85	2 7 20	1.295	6 6 16	0.064	6 7 42	1.157	1 22 25	7.758	5 7 22	0.403	5 2 47	7.543
275	2 3 20	20.26	11 3 20	0.48	2 3 53	1.284	6 4 53	0.063	6 6 0	1.149	1 29 50	7.371	5 3 53	0.405	4 28 51	7.653
276	2 2 19	20.33	11 2 19	1.77	2 4 17	1.304	6 6 27	0.065	6 7 56	1.167	1 19 49	7.587	5 4 17	0.405	4 29 19	7.639
277	2 2 14	20.31	11 2 14	1.59	2 4 0	1.302	6 6 16	0.064	6 7 43	1.166	1 20 51	7.533	5 4 0	0.405	4 28 59	7.649
278	2 1 4	20.37	11 1 4	2.23	2 3 32	1.316	6 7 7	0.065	6 8 44	1.177	1 15 22	7.668	5 3 32	0.405	4 28 28	7.665
279	2 0 25	20.38	11 0 25	2.25	2 2 52	1.319	6 7 11	0.065	6 8 50	1.179	1 14 31	7.638	5 2 54	0.405	4 27 44	7.687
280	1 29 53	20.29	10 29 53	1.30	1 28 30	1.279	6 3 15	0.063	6 4 0	1.142	2 9 42	7.152	4 28 31	0.408	4 22 58	7.845
281	1 29 38	20.28	10 29 38	1.17	1 28 23	1.281	6 3 25	0.063	6 4 12	1.141	2 8 32	7.122	4 28 24	0.408	4 22 50	7.849
282	1 27 53	20.28	10 27 53	1.03	1 26 48	1.287	6 3 44	0.064	6 4 36	1.148	2 5 54	6.995	4 26 50	0.408	4 21 8	7.910
283	1 26 20	20.28	10 26 20	1.03	1 27 24	1.316	6 6 5	0.065	6 7 29	1.180	1 18 26	7.032	4 27 25	0.408	4 21 46	7.888
284	1 25 25	20.26	10 25 25	0.18	1 25 15	1.303	6 4 50	0.064	6 5 58	1.167	1 26 51	6.806	4 25 15	0.409	4 19 27	7.972
285	1 24 58	20.27	10 24 58	0.85	1 24 7	1.297	6 4 10	0.064	6 5 8	1.159	2 1 48	6.775	4 24 7	0.409	4 18 16	8.017
286	1 24 40	20.31	10 24 40	1.54	1 26 14	1.331	6 6 44	0.066	6 8 17	1.194	1 12 51	7.047	4 26 15	0.408	4 20 31	7.932
287	1 18 36	20.26	10 18 36	0.20	1 18 25	1.328	6 5 9	0.066	6 6 20	1.190	1 20 19	6.258	4 18 25	0.413	4 12 25	8.254
288	1 18 12	20.26	10 18 12	0.72	1 17 33	1.322	6 4 39	0.065	6 5 43	1.182	1 24 35	6.222	4 17 33	0.413	4 11 35	8.291
289	1 17 41	20.31	10 17 41	1.61	1 16 14	1.313	6 3 49	0.065	6 4 42	1.169	2 1 53	6.272	4 16 14	0.414	4 10 8	8.374
290	1 17 21	20.26	10 17 21	0.14	1 17 13	1.334	6 5 14	0.066	6 6 26	1.195	1 18 34	6.150	4 17 13	0.414	4 11 14	8.305

A CATALOGUE OF 520 ZODIACAL STARS FOR 1820 (CONTINUED).

No.	Mag.	Stars.	Longitude.	Ann. Var.	Latitude.	Ann. Var.	North East Angle of the ✱'s Meridian with the Ecliptic.	Ann. Var. of the ∠.	Right Ascension in Time.	Ann. Prec. 1830.	Declination.	Ann. Precess. 1830.
			S ° ′ ″	+	° ′ ″		° ′ ″	−	h. m. s.	+	° ′ ″	+
291	3	Kiffa Austr.	7 12 34 20″.9	50″.02	0 21 22″.5 N	−0″.51	107 41 44″.3	13″.66	14 40 56	3″.305	15 17 12 S	15″.271
292	5	ξ^2 Libræ.........	7 12 35 39.7	49.99	5 11 53.3 N	−0.45	107 21 12.3	13.83	14 47 0	3.237	10 40 34 S	14.921
293	6	Libræ............	7 13 14 23.2	50.18	1 15 52.4 S	+0.38	107 39 34.7	13.82	14 41 32	3.335	17 2 2 S	15.236
294	6	Libræ............	7 15 32 1.7	50.16	4 16 3.8 S	+0.39	107 29 59.8	14.47	14 46 57	3.404	20 35 13 S	14.923
295	6	ν^1 Libræ.........	7 16 15 26.3	50.00	1 13 0.2 N	−0.52	106 36 13.1	14.71	14 56 36	3.328	15 32 5 S	14.347
296	5.6	ι^1 Libræ.........	7 18 29 34.7	50.12	1 49 40.3 S	+0.58	106 12 53.0	15.33	15 1 59	3.400	19 6 10 S	14.014
297	6	Solitarii	7 19 5 26.9	50.21	6 0 54.3 S	+0.41	106 29 28.4	15.58	14 59 22	3.475	23 17 22 S	14.175
298	6	Lib. 600 May..	7 19 16 27.5	50.24	6 3 54.3 N	−0.53	105 23 4.6	15.54	15 14 0	3.277	11 43 10 S	13.244
299	6	o^2 Libræ.........	7 19 46 27.0	50.20	3 20 16.2 N	−0.41	105 24 2.4	15.65	15 13 0	3.327	14 28 54 S	13.309
300	6	28 Libræ	7 20 2 44.1	50.20	0 17 5.5 N	−0.48	105 33 2.2	15.73	15 10 46	3.381	17 29 46 S	13.458
301	6	ζ^1 Libræ.........	7 21 23 9.0	50.31	2 7 7.7 N	−0.42	104 59 13.1	16.06	15 18 7	3.362	16 4 47 S	12.971
302	6	ζ^3 Libræ.........	7 21 55 15.2	50.30	2 21 35.5 N	−0.54	104 47 58.9	16.19	15 20 32	3.363	15 59 1 S	12.810
303	6	ζ^4 Libræ.........	7 22 30 5.9	50.04	2 15 15.9 N	−0.47	104 37 17.5	16.33	15 22 46	3.370	16 14 2 S	12.660
304	4.5	γ Libræ	7 22 37 17.1	50.33	4 24 23.8 N	−0.36	104 26 12.6	16.33	15 25 28	3.333	14 10 50 S	12.476
305	6	41 Libræ........	7 24 27 21.8	50.39	0 12 28.1 N	−0.64	104 8 39.6	16.87	15 28 33	3.427	18 42 1 S	12.261
306	4.5	η Libræ	7 24 50 16.2	50.01	4 1 21.1 N	−0.62	103 44 15.8	16.85	15 33 57	3.359	15 5 25 S	11.885
307	5	κ Libræ	7 25 14 35.7	50.00	0 0 28.2 N	−0.60	103 53 50.4	17.08	15 31 35	3.438	19 5 9 S	12.052
308	5.6	42 Libræ	7 25 48 23.6	50.02	4 6 56.8 S	+0.51	104 5 33.0	17.38	15 29 39	3.524	23 13 25 S	12.184
309	4.5	ϑ Libræ	7 27 21 19.0	50.27	3 29 10.0 N	−0.32	102 55 24.9	17.43	15 43 35	3.390	16 11 34 S	11.198
310	5	ψ Libræ.........	7 27 53 4.0	50.05	6 6 28.2 N	−0.54	102 35 11.1	17.46	15 48 7	3.343	13 45 5 S	10.867
311	5	λ Libræ	7 27 57 42.2	50.11	0 6 29.2 N	−0.56	102 57 27.8	17.73	15 42 54	3.463	19 37 11 S	11.247
312	5	b Scorpii	7 28 36 44.2	50.15	5 27 55.5 S	+0.62	103 14 54.1	18.28	15 40 10	3.585	25 11 43 S	11.441
313	5.6	49 Libræ	7 28 52 19.3	49.54	4 1 45.2 N	−0.81	102 21 44.0	17.74	15 50 15	3.342	15 59 33 S	10.710
314	6	f^2 Scorpii	7 28 53 39.8	50.17	3 35 21.7 S	+0.45	102 57 17.5	18.20	15 43 14	3.549	23 25 57 S	11.220
315	6	f^1 Scorpii	7 29 0 11.0	50.17	4 8 2.2 S	+0.45	102 58 9.9	18.27	15 43 10	3.561	23 59 15 S	11.225
316	5	a^1 Scorpii	7 29 6 16.1	50.07	4 55 29.8 S	+0.57	103 0 47.2	18.36	15 42 49	3.579	24 46 50 S	11.250
317	6	a^2 Scorpii	7 29 19 11.0	50.11	4 47 36.8 S	+0.60	102 55 17.4	18.41	15 43 52	3.579	24 42 2 S	11.174
318	3	Icklarkrau	8 0 3 30.6	50.20	1 57 44.2 S	+0.56	102 23 7.0	18.35	15 49 42	3.527	22 5 58 S	10.748
319	3.4	π Scorpii	8 0 25 41.7	50.29	5 27 9.7 S	+0.67	102 35 1.1	18.75	15 47 59	3.606	25 35 11 S	10.874
320	2	Acrab. præc....	8 0 40 36.9	50.24	1 1 52.1 N	−0.55	101 55 25.0	18.28	15 54 59	3.469	19 18 12 S	10.360
321	5.6	Acrab. seq.	8 0 40 43.2	50.25	1 2 6.5 N	−0.55	101 55 21.8	18.28	15 55 0	3.469	19 18 0 S	10.355
322	4.5	ω^1 Scorpii	8 1 9 22.3	50.15	0 14 38.1 N	−0.60	101 48 28.3	18.44	15 56 18	3.490	20 10 19 S	10.257
323	4.5	ω^2 Scorpii	8 1 19 36.3	50.08	0 4 31.2 N	−0.58	101 45 29.9	18.49	15 56 52	3.496	20 22 20 S	10.214
324	6	Scorpii	8 1 22 25.0	50.11	5 0 26.9 S	+0.65	102 11 9.4	18.95	15 52 29	3.607	25 21 14 S	10.540
325	4	ν Scorpii.........	8 2 7 49.8	50.15	1 39 23.7 N	−0.55	101 21 5.0	18.55	16 1 33	3.469	18 59 2 S	9.861
326	6	m Scorpii	8 2 30 43.5	50.17	5 15 27.3 S	+0.46	101 46 53.6	19.25	15 57 10	3.626	25 50 6 S	10.189
327	6	Scorpii	8 2 52 15.9	50.18	0 10 51.7 S	+0.47	101 12 31.5	18.84	16 3 6	3.515	20 55 53 S	9.740
328	5	c^2 Scorpii	8 3 44 5.5	50.30	6 39 19.1 S	+0.34	101 27 8.4	19.71	16 1 14	3.673	27 26 58 S	9.880
329	5.6	o Scorpii	8 4 55 27.9	50.10	2 37 51.9 S	+0.56	100 37 17.2	19.50	16 9 49	3.590	23 43 31 S	9.222
330	5	ψ Ophiuchi	8 5 2 14.8	50.12	1 34 40.6 N	−0.64	100 16 29.8	19.12	16 13 35	3.495	19 36 22 S	8.930
331	4	σ Scorpii	8 5 17 11.6	50.13	4 0 47.9 S	+0.52	100 35 48.0	19.74	16 10 16	3.626	25 9 3 S	9.188
332	5	χ Ophiuchi......	8 5 27 46.6	50.01	3 14 50.7 N	−0.56	100 0 49.9	19.07	16 16 36	3.461	18 2 17 S	8.694
333	5	g Scorp. (Ophi.)	8 5 55 24.8	50.09	1 44 0.1 S	+0.61	100 9 53.9	19.61	16 14 48	3.578	23 1 20 S	8.833
334	4.5	φ Ophiuchi	8 6 9 16.7	50.25	5 13 18.4 N	−0.54	99 38 59.9	19.04	16 20 51	3.421	16 12 38 S	8.359
335	5	ω Ophiuchi	8 7 6 47.3	50.13	0 27 13.8 N	−0.55	99 33 7.0	19.61	16 21 29	3.537	21 4 17 S	8.306
336	6	i Scorpii.........	8 7 13 32.0	50.23	3 12 59.7 S	+0.48	99 46 5.7	20.04	16 19 17	3.626	24 42 32 S	8.479
337	1	Antares	8 7 14 59.1	50.15	4 32 45.0 S	+0.58	99 52 7.2	20.22	16 18 23	3.658	26 1 20 S	8.550
338	5	m Scorpii	8 8 47 24.6	50.13	4 27 24.1 N	−0.58	98 40 52.8	19.52	16 31 10	3.456	17 23 3 S	7.528
339	3.4	τ Scorpii	8 8 56 39.7	50.16	6 5 47.2 S	+0.66	99 18 32.1	20.81	16 24 42	3.714	27 49 54 S	8.048
340	6	25 Scorpii	8 11 0 55.0	50.18	3 6 14.3 S	+0.49	98 13 46.2	20.73	16 35 51	3.656	25 11 23 S	7.145
341	6	18 Ophiuchi ...	8 11 33 23.2	50.38	2 8 36.0 S	+0.63	97 56 44.2	20.68	16 38 48	3.635	24 18 50 S	6.904
342	6	Ophiuchi	8 12 6 34.1	50.19	5 48 15.1 N	−0.49	97 19 45.8	19.87	16 45 39	3.444	16 30 38 S	6.341
343	6	29 Ophiuchi ...	8 13 43 5.5	50.06	3 53 13.7 N	−0.45	96 45 50.7	20.25	16 51 20	3.499	18 36 37 S	5.866
344	6	Ophi. 606 May.	8 13 54 34.3	50.11	2 20 1.5 S	+0.63	96 58 57.1	21.06	16 48 57	3.657	24 48 30 S	6.063
345	6	26 Ophiuchi ...	8 13 56 22.8	50.18	2 13 27.8 S	+0.49	96 57 50.5	21.05	16 49 8	3.655	24 42 15 S	6.048
346	6	28 Scorpii	8 14 59 28.4	50.18	1 19 13.2 N	−0.66	96 21 14.6	20.70	16 55 28	3.569	21 18 15 S	5.519
347	4.5	A Ophiuchi	8 17 31 9.4	49.78	3 28 1.1 S	+1.79	95 30 28.3	21.72	17 4 18	3.710	26 19 16 S	4.772
348	6	Ophi. 679 May.	8 17 55 43.1	50.25	0 57 11.6 S	+0.65	95 13 26.9	21.35	17 7 8	3.644	23 51 45 S	4.531

SUBSIDIARY NUMBERS OF THE ZODIACAL STARS FOR 1830 (CONTINUED).

No.	Longitude.		Latitude.		Right Ascension in Time.						Declination.					
	Aberration.		Aberration.		Aberration.		☉ Nutation.		☾ Nutation.		Aberration.		☉ Nutation.		☾ Nutation.	
	A	Max.	A′	Max.	A	Max.	S	Max.	L	Max.	A′	Max.	S′	Max.	L′	Max.
	s ° ′	″	s ° ′	″	s ° ′	s	s ° ′	s	s ° ′		s ° ′ ″	″	s ° ′ ″	″	s ° ′	″
291	1 17 17	20.26	10 17 17	0.12	1 17 11	1.334	6 5 14	0.066	6 6 27	1.196	1 18 25	6.146	4 17 11	0.414	4 11 11	8.307
292	1 17 16	20.33	10 17 16	1.84	1 15 39	1.313	6 3 37	0.065	6 4 27	1.167	2 3 29	6.279	4 15 39	0.415	4 9 41	8.372
293	1 16 37	20.26	10 16 37	0.45	1 17 2	1.346	6 5 49	0.067	6 7 9	1.208	1 12 39	6.147	4 17 2	0.414	4 11 3	8.314
294	1 14 20	20.31	10 14 20	1.51	1 15 39	1.378	6 6 49	0.068	6 8 23	1.237	1 0 54	6.191	4 15 39	0.415	4 9 42	8.372
295	1 13 36	20.26	10 13 36	0.43	1 13 15	1.344	6 4 59	0.066	6 6 8	1.203	1 17 42	5.790	4 13 15	0.416	4 7 21	8.475
296	1 11 22	20.26	10 11 22	0.65	1 11 54	1.373	6 5 55	0.068	6 7 17	1.232	1 5 6	5.676	4 11 54	0.417	4 6 3	8.533
297	1 10 46	20.36	10 10 46	2.12	1 12 33	1.408	6 7 11	0.070	6 8 50	1.262	0 23 4	6.056	4 12 33	0.417	4 6 39	8.505
298	1 10 35	20.36	10 10 35	2.14	1 8 56	1.331	6 3 29	0.065	6 4 18	1.180	2 1 39	5.745	4 8 55	0.419	4 3 14	8.659
299	1 10 5	20.28	10 10 5	1.18	1 9 9	1.345	6 4 18	0.066	6 5 17	1.201	1 22 4	5.485	4 9 10	0.419	4 3 28	8.648
300	1 9 49	20.26	10 9 49	0.10	1 9 44	1.365	6 5 13	0.068	6 6 24	1.223	1 10 51	5.417	4 9 44	0.418	4 4 0	8.624
301	1 8 28	20.26	10 8 28	0.75	1 7 55	1.358	6 4 37	0.067	6 5 41	1.214	1 16 23	5.274	4 7 54	0.420	4 2 17	8.701
302	1 7 56	20.27	10 7 56	0.84	1 7 19	1.359	6 4 32	0.067	6 5 35	1.214	1 16 50	5.223	4 7 19	0.420	4 1 44	8.726
303	1 7 22	20.27	10 7 22	0.80	1 6 46	1.362	6 4 32	0.067	6 5 35	1.217	1 15 58	5.157	4 6 46	0.420	4 1 13	8.748
304	1 7 14	20.31	10 7 14	1.55	1 6 7	1.351	6 3 56	0.067	6 4 50	1.202	1 23 55	5.259	4 6 7	0.421	4 0 38	8.775
305	1 5 24	20.26	10 5 24	0.07	1 5 21	1.383	6 5 1	0.069	6 6 11	1.238	1 6 15	4.934	4 5 22	0.421	3 29 55	8.807
306	1 5 1	20.30	10 5 1	1.42	1 4 2	1.359	6 3 58	0.067	6 4 53	1.212	1 21 7	4.989	4 4 3	0.422	3 28 44	8.860
307	1 4 37	20.26	10 4 37	0.00	1 4 36	1.388	6 5 2	0.069	6 6 12	1.243	1 4 40	4.849	4 4 37	0.422	3 29 15	8.837
308	1 4 3	20.30	10 4 3	1.45	1 5 4	1.426	6 6 9	0.071	6 7 34	1.277	0 18 4	5.113	4 5 5	0.421	3 29 40	8.818
309	1 2 30	20.29	10 2 30	1.23	1 1 43	1.371	6 3 59	0.068	6 4 54	1.223	1 17 25	4.671	4 1 43	0.423	3 26 38	8.952
310	1 1 59	20.37	10 1 59	2.15	1 0 38	1.358	6 3 18	0.067	6 4 4	1.205	1 27 33	4.876	4 0 37	0.424	3 25 38	8.994
311	1 1 54	20.26	10 1 54	0.04	1 1 53	1.398	6 4 48	0.069	6 5 55	1.251	1 2 23	4.527	4 1 53	0.423	3 26 46	8.946
312	1 1 15	20.34	10 1 15	1.93	1 2 31	1.454	6 6 13	0.072	6 7 39	1.300	0 9 10	4.991	4 2 32	0.423	3 27 21	8.920
313	1 0 59	20.30	10 0 59	1.42	1 0 6	1.373	6 3 45	0.068	6 4 38	1.223	1 18 50	4.539	4 0 6	0.424	3 25 10	9.013
314	1 0 58	20.29	10 0 58	1.27	1 1 47	1.435	6 5 41	0.071	6 6 59	1.285	0 15 42	4.688	4 1 48	0.423	3 26 41	8.949
315	1 0 51	20.30	10 0 51	1.46	1 1 48	1.441	6 5 49	0.071	6 7 9	1.290	0 13 26	4.746	4 1 49	0.423	3 26 42	8.949
316	1 0 45	20.32	10 0 45	1.74	1 1 54	1.450	6 6 1	0.072	6 7 24	1.297	0 10 19	4.850	4 1 54	0.423	3 26 47	8.945
317	1 0 32	20.32	10 0 32	1.70	1 1 38	1.450	6 5 57	0.072	6 7 9	1.297	0 10 28	4.805	4 1 38	0.423	3 26 33	8.955
318	0 29 48	20.26	9 29 48	0.70	1 0 14	1.424	6 5 8	0.071	6 6 19	1.275	0 20 55	4.380	4 0 14	0.424	3 25 18	9.008
319	0 29 26	20.34	9 29 26	1.92	1 0 39	1.462	6 5 59	0.072	6 7 21	1.306	0 6 19	4.780	4 0 38	0.424	3 25 39	8.993
320	0 29 11	20.26	9 29 11	0.37	0 28 59	1.401	6 4 20	0.069	6 5 21	1.252	1 4 6	4.185	3 28 59	0.425	3 24 12	9.056
321	0 29 11	20.26	9 29 11	0.37	0 28 58	1.401	6 4 20	0.069	6 5 20	1.252	1 4 7	4.183	3 28 58	0.425	3 24 11	9.056
322	0 28 42	20.26	9 28 42	0.09	0 28 39	1.409	6 4 29	0.070	6 5 31	1.260	0 29 54	4.129	3 28 39	0.425	3 23 53	9.068
323	0 28 32	20.26	9 28 32	0.03	0 28 31	1.411	6 4 30	0.070	6 5 33	1.262	0 28 55	4.111	3 28 30	0.425	3 23 47	9.073
324	0 28 29	20.33	9 28 29	1.77	0 29 34	1.462	6 5 44	0.072	6 7 4	1.306	0 6 25	4.595	3 29 34	0.425	3 24 43	9.034
325	0 27 44	20.26	9 27 44	0.58	0 27 24	1.401	6 4 4	0.069	6 5 0	1.251	1 5 59	4.011	3 27 24	0.426	3 22 48	9.113
326	0 27 21	20.34	9 27 21	1.85	0 29 26	1.470	6 5 38	0.073	6 6 56	1.314	0 3 33	4.501	3 28 26	0.425	3 23 42	9.075
327	0 26 59	20.26	9 26 59	0.06	0 27 1	1.419	6 4 24	0.070	6 5 25	1.269	0 26 5	3.921	3 26 59	0.426	3 22 29	9.127
328	0 26 7	20.39	9 26 7	2.35	0 27 28	1.492	6 5 48	0.073	6 7 7	1.330	11 26 15	4.617	3 27 28	0.426	3 22 52	9.111
329	0 24 56	20.27	9 24 56	0.93	0 25 26	1.450	6 4 41	0.072	6 5 46	1.297	0 11 8	3.825	3 25 25	0.427	3 21 6	9.182
330	0 24 49	20.26	9 24 49	0.56	0 24 33	1.411	6 3 47	0.070	6 4 39	1.260	1 3 32	3.637	3 24 32	0.427	3 20 20	9.211
331	0 24 34	20.30	9 24 34	1.42	0 25 19	1.467	6 4 56	0.072	6 6 4	1.310	0 3 59	3.959	3 25 19	0.427	3 21 0	9.184
332	0 24 24	20.28	9 24 24	1.15	0 23 50	1.399	6 3 24	0.069	6 4 11	1.247	1 12 18	3.683	3 23 49	0.428	3 19 43	9.235
333	0 23 56	20.26	9 23 56	0.61	0 24 14	1.445	6 4 21	0.072	6 5 21	1.293	0 14 21	3.607	3 24 14	0.428	3 20 4	9.221
334	0 23 42	20.31	9 23 42	1.84	0 22 49	1.387	6 2 57	0.068	6 3 38	1.232	1 22 1	3.836	3 22 49	0.428	3 18 51	9.267
335	0 22 45	20.26	9 22 45	0.16	0 22 40	1.428	6 3 45	0.071	6 4 38	1.275	0 25 29	3.346	3 22 40	0.428	3 18 43	9.272
336	0 22 38	20.28	9 22 38	1.14	0 23 12	1.465	6 4 28	0.072	6 5 30	1.309	0 4 30	3.596	3 23 10	0.428	3 19 10	9.256
337	0 22 37	20.31	9 22 37	1.61	0 23 24	1.481	6 4 44	0.073	6 5 49	1.321	11 28 1	3.797	3 23 23	0.428	3 19 21	9.249
338	0 21 4	20.31	9 21 4	1.59	0 20 24	1.399	6 2 50	0.069	6 3 29	1.244	1 18 13	3.414	3 20 23	0.429	3 16 47	9.340
339	0 20 55	20.37	9 20 55	2.15	0 21 55	1.508	6 4 45	0.074	6 5 50	1.341	11 17 50	3.887	3 21 55	0.429	3 18 4	9.295
340	0 18 51	20.28	9 18 51	1.10	0 19 17	1.477	6 3 49	0.073	6 4 42	1.318	11 18 13	3.077	3 19 17	0.430	3 15 51	9.371
341	0 18 18	20.26	9 18 18	0.76	0 18 36	1.468	6 3 34	0.072	6 4 23	1.310	0 3 14	2.879	3 18 36	0.430	3 15 16	9.390
342	0 17 45	20.35	9 17 45	2.05	0 17 1	1.397	6 2 16	0.069	6 2 48	1.239	1 26 9	3.274	3 17 1	0.431	3 13 56	9.431
343	0 16 9	20.30	9 16 9	1.37	0 15 41	1.415	6 2 21	0.070	6 2 53	1.259	1 16 7	2.732	3 15 41	0.431	3 12 50	9.436
344	0 15 57	20.27	9 15 57	0.83	0 16 14	1.477	6 3 11	0.073	6 3 55	1.317	11 27 27	2.566	3 16 14	0.431	3 13 17	9.449
345	0 15 55	20.27	9 15 55	0.78	0 16 12	1.476	6 3 10	0.073	6 3 54	1.319	11 28 11	2.557	3 16 11	0.431	3 13 15	9.451
346	0 14 52	20.26	9 14 52	0.46	0 14 43	1.441	6 2 30	0.071	6 3 5	1.286	0 26 42	2.270	3 14 43	0.432	3 12 2	9.480
347	0 12 21	20.29	9 12 21	1.22	0 12 40	1.500	6 2 38	0.074	6 3 15	1.335	11 10 2	2.276	3 12 41	0.433	3 10 20	9.526
348	0 11 56	20.26	9 11 56	0.33	0 12 1	1.471	6 2 17	0.073	6 2 49	1.311	0 1 33	1.854	3 12 1	0.433	3 9 48	9.538

A CATALOGUE OF 520 ZODIACAL STARS FOR 1820 (CONTINUED).

No.	Mag.	Stars.	Longitude.	Ann. Var.	Latitude.	Ann. Var.	North East Angle of the ✱'s Meridian with the Ecliptic.	Ann. Var. of the ∠.	Right Ascension in Time.	Ann. Precess. 1830.	Declination.	Ann. Precess. 1830.
			S ° ′ ″	+	° ′ ″		° ′ ″	−	h. m. s.	+	° ′ ″	+
349	5 .6	39 Oph. seq....	8 17 55 44″.1	50″.18	1 10 12″.6 S	+0″.50	95 13 57″.6	21″.38	17 7 8	3ˢ.650	24 4 46 S	4ˢ.538
350	4 .5	ϱ Ophiuchi......	8 18 22 40 .0	50 .32	2 3 11 .3 N	−0 .56	94 55 34 .0	20 .96	17 10 13	3 .567	20 54 30 S	4 .269
351	3 .4	ϑ Ophiuchi	8 18 52 52 .2	50 .12	1 49 6 .6 S	+0 .57	94 51 8 .0	21 .58	17 10 58	3 .672	24 48 31 S	4 .204
352	6	Scorp. 191 Caille	8 19 16 21 .4	50 .01	1 46 13 .9 N	−0 .71	94 33 40 .0	21 .08	17 13 56	3 .578	21 15 47 S	3 .950
353	6	43 Ophiuchi ...	8 19 23 10 .9	50 .07	4 56 5 .6 S	+0 .55	94 45 43 .5	22 .19	17 12 2	3 .762	27 57 23 S	4 .110
354	5 .6	B Ophiuchi	8 19 49 11 .0	50 .16	0 55 45 .1 S	+0 .63	94 25 4 .3	21 .52	17 15 23	3 .652	23 59 52 S	3 .826
355	5	45 Ophiuchi ...	8 20 22 0 .0	50 .18	6 35 52 .4 S	+0 .50	94 23 56 .5	22 .64	17 15 52	3 .817	29 41 32 S	3 .781
356	5	e² Ophiuchi	8 20 57 33 .7	50 .24	0 39 35 .5 S	+0 .59	93 55 17 .1	21 .57	17 20 26	3 .649	23 48 44 S	3 .389
357	6	2 Sagittarii......	8 22 33 30 .4	50 .18	1 27 39 .4 N	−0 .50	93 11 1 .2	21 .36	17 27 56	3 .597	21 47 42 S	2 .743
358	5	d Ophiuchi	8 23 38 29 .0	50 .09	1 43 37 .6 N	−0 .54	92 43 4 .3	21 .38	17 32 39	3 .593	21 35 5 S	2 .334
359	5	p Sagittarii......	8 24 43 41 .6	50 .13	4 23 46 .7 S	+0 .51	92 22 8 .3	22 .40	17 36 14	3 .768	27 45 2 S	2 .020
360	5	4 Sagittarii......	8 27 26 20 .7	50 .12	0 21 1 .5 S	+0 .58	91 6 49 .9	21 .83	17 48 48	3 .656	23 47 17 S	0 .925
361	5	Sagit. 709 May.	8 27 55 34 .1	50 .20	0 40 54 .9 N	−0 .57	90 53 43 .3	21 .67	17 51 0	3 .628	22 45 52 S	0 .734
362	6	Sagittarii.........	8 28 6 3 .7	50 .18	2 43 29 .4 N	−0 .50	90 48 29 .8	21 .37	17 51 53	3 .573	20 43 29 S	0 .658
363	6	a Sagittarii......	8 28 7 58 .9	49 .96	0 49 13 .6 S	+0 .56	90 48 53 .1	21 .92	17 51 49	3 .670	24 16 11 S	0 .662
364	5	γ^1 Sagittarii.....	8 28 34 56 .0	50 .06	6 7 17 .3 S	+0 .47	90 38 55 .3	22 .98	17 53 31	3 .826	29 34 36 S	0 .511
365	5	Sagittarii.........	8 29 16 6 .8	50 .24	5 0 21 .9 S	+0 .70	90 19 53 .0	22 .74	17 56 41	3 .792	28 27 57 S	0 .234
												−
366	6	Sagittarii.........	9 0 10 11 .2	50 .30	0 15 51 .0 S	+0 .61	89 55 35 .2	21 .83	18 0 44	3 .655	23 43 35 S	0 .117
367	3 .4	μ^1 Sagittarii.....	9 0 41 57 .8	50 .11	2 21 55 .8 N	−0 .58	89 42 4 .8	21 .41	18 3 0	3 .583	21 5 42 S	0 .314
368	6	14 Sagittarii....	9 0 48 5 .3	50 .15	1 42 32 .9 N	−0 .68	89 39 22 .9	21 .51	18 3 27	3 .601	21 45 1 S	0 .354
369	6	μ^2 Sagittarii.....	9 1 2 50 .0	50 .10	2 41 15 .5 N	−0 .55	89 33 13 .9	21 .36	18 4 29	3 .574	20 46 15 S	0 .444
370	6	16 Sagittarii....	9 1 3 18 .9	50 .08	3 1 37 .9 N	−0 .61	89 33 5 .1	21 .31	18 4 30	3 .565	20 25 51 S	0 .445
371	5 .6	Sagittarii........	9 1 30 47 .3	50 .19	3 38 33 .8 S	+0 .49	89 19 23 .9	22 .43	18 6 47	3 .751	27 5 48 S	0 .648
372	3 .4	Kaus Med.......	9 2 3 55 .5	50 .28	6 26 52 .6 S	+0 .59	89 3 6 .5	23 .02	18 9 28	3 .835	29 53 36 S	0 .884
373	6	21 Sagittarii....	9 3 25 32 .6	50 .08	2 47 28 .9 N	−0 .63	88 32 35 .0	21 .29	18 14 38	3 .568	20 37 38 S	1 .330
374	4	Kaus Bor........	9 3 48 20 .9	50 .08	2 6 12 .6 S	+0 .73	88 19 18 .4	22 .06	18 16 52	3 .703	25 30 34 S	1 .527
375	6	Sagittarii.........	9 4 9 56 .4	50 .03	5 30 3 .9 N	−0 .50	88 15 32 .9	20 .72	18 17 26	3 .495	17 53 56 S	1 .574
376	6	Sag. 728 May..	9 4 39 17 .4	50 .13	4 33 2 .2 N	−0 .64	88 2 36 .5	21 .01	18 19 37	3 .522	18 49 56 S	1 .765
377	6	Sag. 730 May..	9 4 58 7 .0	50 .09	4 51 27 .7 N	−0 .55	87 54 56 .4	20 .95	18 20 53	3 .513	18 30 53 S	1 .875
378	6	Sag. 740 May..	9 6 18 32 .7	50 .13	0 20 7 .1 S	+0 .52	87 15 43 .5	21 .62	18 27 34	3 .648	23 38 50 S	2 .459
379	6	26 Sagittarii.....	9 7 2 57 .8	50 .18	0 42 56 .9 S	+0 .48	86 56 2 .8	21 .63	18 30 53	3 .657	23 59 23 S	2 .746
380	6	Sagittarii.........	9 7 38 11 .7	50 .19	1 56 26 .3 S	+0 .48	86 38 53 .5	21 .80	18 33 45	3 .689	25 10 48 S	2 .995
381	4 .5	φ Sagittarii......	9 7 39 57 .1	50 .30	3 55 51 .0 S	+0 .54	86 34 40 .0	22 .17	18 34 25	3 .745	27 9 53 S	3 .052
382	6	28 Sagittarii....	9 8 11 16 .8	50 .21	0 38 27 .3 N	−0 .51	86 28 45 .6	21 .33	18 35 29	3 .616	22 34 11 S	3 .144
383	6	29 Sagittarii....	9 9 7 54 .3	50 .16	2 38 6 .5 N	−0 .51	86 7 52 .4	20 .97	18 38 59	3 .560	20 31 8 S	3 .445
384	6	30 Sagittarii....	9 9 14 43 .7	50 .02	0 47 3 .2 N	−0 .55	86 2 0 .0	21 .22	18 40 1	3 .609	22 21 27 S	3 .534
385	6	31 Sagittarii....	9 9 33 43 .9	49 .99	0 59 53 .1 N	−0 .59	85 54 19 .1	21 .16	18 41 19	3 .602	22 7 18 S	3 .646
386	3	σ Sagittarii......	9 9 52 14 .7	50 .13	3 25 26 .6 S	+0 .58	85 37 31 .6	21 .86	18 44 6	3 .721	26 30 32 S	3 .886
387	5	ν^1 Sagittarii......	9 9 57 26 .3	50 .07	0 7 59 .8 N	−0 .53	85 42 42 .5	21 .26	18 43 18	3 .623	22 57 20 S	3 .816
388	6	33 Sagittarii.....	9 10 2 53 .0	50 .12	1 30 59 .5 N	−0 .51	85 42 57 .3	21 .05	18 43 14	3 .586	21 34 10 S	3 .810
389	5	ν^2 Sagittarii......	9 10 10 42 .5	50 .29	0 11 13 .2 N	−0 .57	85 37 12 .8	21 .23	18 44 14	3 .621	22 53 7 S	3 .896
390	6	Sagittarii.........	9 10 20 21 .9	50 .18	0 20 0 .0 S	+0 .48	85 32 3 .9	21 .29	18 45 6	3 .634	23 23 31 S	3 .971
391	6	ξ^1 Sagittarii.....	9 10 53 28 .3	50 .06	2 8 31 .2 N	−0 .46	85 22 53 .0	20 .88	18 46 38	3 .567	20 52 51 S	4 .102
392	5	ξ^2 Sagittarii.....	9 10 56 7 .5	50 .10	1 41 4 .4 N	−0 .57	85 20 55 .5	20 .94	18 46 59	3 .578	21 19 57 S	4 .131
393	4	τ Sagittarii......	9 12 19 24 .2	50 .02	5 3 15 .0 S	+0 .74	84 28 50 .4	21 .88	18 55 42	3 .754	27 55 17 S	4 .876
394	4 .5	ο Sagittarii......	9 12 28 36 .7	50 .15	0 53 12 .0 N	−0 .49	84 40 36 .6	20 .88	18 53 53	3 .592	21 59 38 S	4 .720
395	4 .5	π Sagittarii......	9 13 44 18 .3	50 .09	1 27 38 .8 N	−0 .60	84 10 29 .3	20 .66	18 59 3	3 .571	21 17 58 S	5 .158
396	6	Sagit. 767 May.	9 14 17 2 .9	50 .08	0 45 13 .6 N	−0 .53	83 55 12 .1	20 .68	19 1 42	3 .586	21 56 47 S	5 .381
397	6	ψ Sagittarii......	9 14 31 45 .7	50 .19	2 54 15 .4 S	+0 .55	83 38 35 .3	21 .19	19 4 30	3 .681	25 33 23 S	5 .618
398	6	Sagit. 771 May.	9 14 39 58 .6	50 .17	1 49 54 .1 S	+0 .65	83 38 26 .6	21 .00	19 4 34	3 .651	24 28 28 S	5 .624
399	5	d Sagittarii......	9 15 50 6 .9	50 .04	3 16 52 .4 N	−0 .49	83 23 27 .2	20 .16	19 7 6	3 .514	19 15 49 S	5 .834
400	6	Sagit. 776 May.	9 16 3 57 .0	50 .09	0 14 1 .4 S	+0 .48	83 8 19 .3	20 .57	19 9 50	3 .601	22 43 37 S	6 .063
401	6	p Sagittarii	9 16 9 48 .6	50 .20	5 45 49 .0 S	+0 .46	82 46 28 .9	21 .44	19 13 16	3 .747	28 12 13 S	6 .351
402	6	Sagit. 775 May.	9 16 37 32 .7	49 .93	6 38 1 .0 N	−0 .97	83 11 55 .6	19 .74	19 8 44	3 .429	15 50 21 S	5 .970
403	6	χ^1 Sagittarii	9 16 49 13 .4	50 .18	2 27 52 .1 S	+0 .58	82 42 19 .7	20 .78	19 14 19	3 .653	24 50 50 S	6 .437
404	5 .6	ϱ^2 Sagittarii	9 16 54 43 .3	50 .32	3 47 19 .0 N	−0 .60	82 58 46 .1	19 .97	19 11 21	3 .496	18 37 53 S	6 .188
405	5	ϱ^1 Sagittarii	9 16 56 18 .8	50 .07	4 14 37 .9 N	−0 .55	82 59 12 .6	19 .92	19 11 14	3 .485	18 10 35 S	6 .178
406	6	χ^3 Sagittarii	9 16 57 10 .8	50 .22	1 56 4 .6 S	+0 .56	82 40 53 .1	20 .68	19 14 36	3 .639	24 18 18 S	6 .460

SUBSIDIARY NUMBERS OF THE ZODIACAL STARS FOR 1830 (CONTINUED).

No.	Longitude. Aberration.		Latitude. Aberration.		Right Ascension in Time. Aberration.		☉ Nutation.		☾ Nutation.		Declination. Aberration.		☉ Nutation.		☾ Nutation.	
	a	Max.	a′	Max.	A	Max.	S	Max.	L	Max.	A′	Max.	S′	Max.	L′	Max.
	s ° ′	″	s ° ′	″	s ° ′	s	s ° ′	s	s ° ′	s	s ° ′	″	s ° ′	″	s ° ′	″
349	0 11 56	20.26	9 11 56	0.41	0 12 2	1.473	6 2 19	0.072	6 2 51	1.313	11 29 15	1.872	3 12 2	0.433	3 9 49	9.538
350	0 11 29	20.26	9 11 29	0.72	0 11 19	1.440	6 1 54	0.071	6 2 20	1.283	1 4 21	1.866	3 11 18	0.433	3 9 12	9.550
351	0 10 59	20.26	9 10 59	0.64	0 11 8	1.482	6 2 12	0.073	6 2 42	1.321	11 20 18	1.809	3 11 7	0.433	3 9 4	9.553
352	0 10 35	20.26	9 10 35	0.62	0 10 27	1.445	6 1 47	0.071	6 2 12	1.287	1 2 4	1.710	3 10 27	0.433	3 8 32	9.564
353	0 10 28	20.33	9 10 28	1.74	0 10 53	1.524	6 2 24	0.075	6 2 58	1.353	10 24 13	2.402	3 10 53	0.433	3 8 52	9.558
354	0 10 2	20.26	9 10 2	0.33	0 10 7	1.474	6 1 56	0.073	6 2 23	1.313	11 28 5	1.574	3 10 7	0.433	3 8 14	9.570
355	0 9 30	20.39	9 9 30	2.33	0 10 0	1.550	6 2 21	0.076	6 2 53	1.373	10 12 56	2.780	3 10 0	0.433	3 8 9	9.572
356	0 8 54	20.26	9 8 54	0.24	0 8 56	1.473	6 1 42	0.073	6 2 6	1.312	11 29 17	1.384	3 8 56	0.433	3 7 17	9.587
357	0 7 18	20.26	9 7 18	0.54	0 7 14	1.452	6 1 16	0.072	6 1 34	1.293	1 2 25	1.219	3 7 14	0.434	3 5 52	9.608
358	0 6 13	20.26	9 6 13	0.61	0 6 8	1.451	6 1 4	0.072	6 1 19	1.292	1 9 16	1.121	3 6 7	0.434	3 4 59	9.620
359	0 5 8	20.31	9 5 8	1.55	0 5 18	1.525	6 1 10	0.075	6 1 27	1.354	10 2 52	1.751	3 5 18	0.434	3 4 19	9.627
360	0 2 25	20.26	9 2 25	0.12	0 2 25	1.476	6 0 27	0.073	6 0 34	1.314	11 14 9	0.394	3 2 25	0.434	3 1 59	9.644
361	0 1 56	20.26	9 1 56	0.24	0 1 56	1.464	6 0 21	0.072	6 0 26	1.304	1 11 18	0.382	3 1 55	0.434	3 1 34	9.646
362	0 1 46	20.27	9 1 46	0.96	0 1 43	1.444	6 0 18	0.071	6 0 21	1.285	2 16 24	1.000	3 1 43	0.434	3 1 24	9.646
363	0 1 44	20.26	9 1 44	0.29	0 1 44	1.481	6 0 20	0.073	6 0 25	1.320	10 14 30	0.420	3 1 44	0.434	3 1 24	9.646
364	0 1 17	20.37	9 1 12	2.16	0 1 21	1.553	6 0 19	0.076	6 0 24	1.375	9 6 45	2.168	3 1 20	0.434	3 1 5	9.647
365	0 0 36	20.33	9 0 36	1.77	0 0 39	1.536	6 0 8	0.076	6 0 10	1.363	9 3 40	1.769	3 0 39	0.434	3 0 31	9.648
366	11 29 41	20.26	8 29 41	0.09	11 29 41	1.475	5 29 57	0.073	5 29 56	1.314	8 27 2	0.917	2 29 42	0.434	2 29 45	9.648
367	11 29 11	20.27	8 29 11	0.84	11 29 11	1.447	5 29 52	0.072	5 29 50	1.288	3 7 45	0.847	2 29 11	0.434	2 29 21	9.648
368	11 29 4	20.26	8 29 4	0.61	11 29 4	1.454	5 29 50	0.072	5 29 48	1.295	3 12 18	0.622	2 29 4	0.434	2 29 15	9.648
369	11 28 49	20.27	8 28 49	0.95	11 28 50	1.444	5 29 48	0.071	5 29 45	1.285	3 9 27	0.968	2 28 51	0.434	2 29 3	9.648
370	11 28 48	20.28	8 28 48	1.07	11 28 50	1.441	5 29 49	0.071	5 29 46	1.282	3 8 19	1.086	2 28 51	0.434	2 29 3	9.648
371	11 28 21	20.29	8 28 21	1.29	11 28 18	1.517	5 29 36	0.075	5 29 33	1.349	8 16 49	1.307	2 28 18	0.434	2 28 38	9.647
372	11 27 48	20.38	8 27 48	2.28	11 27 41	1.557	5 29 27	0.076	5 29 19	1.378	8 18 51	2.301	2 27 42	0.434	2 28 7	9.645
373	11 26 26	20.27	8 26 26	0.98	11 26 31	1.442	5 29 25	0.071	5 29 17	1.283	3 25 1	1.121	2 26 31	0.434	2 27 11	9.640
374	11 26 3	20.26	8 26 3	0.74	11 25 19	1.496	5 29 11	0.074	5 29 0	1.332	7 16 23	0.963	2 26 0	0.434	2 26 45	9.636
375	11 25 42	20.34	8 25 42	1.94	11 25 52	1.418	5 29 24	0.070	5 29 14	1.257	3 13 46	2.053	2 25 53	0.434	2 26 39	9.635
376	11 25 12	20.31	8 25 12	1.61	11 25 22	1.426	5 29 17	0.070	5 29 7	1.267	3 19 7	1.757	2 25 22	0.434	2 26 14	9.632
377	11 24 54	20.32	8 24 54	1.71	11 25 4	1.423	5 29 15	0.070	5 29 5	1.263	3 19 15	1.824	2 25 5	0.434	2 26 0	9.630
378	11 23 33	20.26	8 23 33	0.12	11 23 32	1.472	5 28 46	0.073	5 28 30	1.312	6 0 21	0.996	2 23 33	0.434	2 24 45	9.616
379	11 22 49	20.26	8 22 49	0.25	11 24 46	1.476	5 28 37	0.073	5 28 18	1.315	6 5 39	1.134	2 22 46	0.434	2 24 8	9.608
380	11 22 13	20.26	8 22 13	0.68	11 22 7	1.490	5 28 25	0.074	5 28 3	1.327	6 21 47	1.385	2 22 7	0.434	2 23 35	9.601
381	11 22 12	20.30	8 22 12	1.39	11 21 58	1.515	5 28 16	0.075	5 27 52	1.347	7 10 37	1.851	2 21 58	0.434	2 23 28	9.599
382	11 21 40	20.26	8 21 40	0.22	11 21 43	1.459	5 28 30	0.072	5 28 9	1.301	5 11 29	1.285	2 21 43	0.434	2 23 16	9.596
383	11 20 44	20.27	8 20 44	0.93	11 20 55	1.438	5 28 30	0.071	5 28 9	1.281	4 16 55	1.670	2 20 55	0.433	2 22 37	9.585
384	11 20 37	20.26	8 20 37	0.28	11 20 40	1.456	5 28 20	0.072	5 27 57	1.298	5 9 34	1.449	2 20 40	0.433	2 22 25	9.582
385	11 20 18	20.26	8 20 18	0.35	11 20 22	1.454	5 28 18	0.072	5 27 54	1.296	5 6 47	1.509	2 20 22	0.433	2 22 10	9.577
386	11 19 59	20.29	8 19 59	1.21	11 19 43	1.504	5 27 50	0.074	5 27 20	1.338	6 27 31	1.976	2 19 43	0.433	2 21 38	9.567
387	11 19 54	20.26	8 19 54	0.05	11 19 54	1.462	5 28 9	0.072	5 27 43	1.303	5 18 6	1.537	2 19 55	0.433	2 21 47	9.570
388	11 19 49	20.26	8 19 49	0.54	11 19 55	1.448	5 28 15	0.072	5 27 51	1.290	5 0 33	1.625	2 19 55	0.433	2 21 48	9.571
389	11 19 41	20.26	8 19 41	0.06	11 19 41	1.461	5 28 7	0.072	5 27 41	1.303	5 17 15	1.569	2 19 41	0.433	2 21 37	9.567
390	11 19 31	20.26	8 19 31	0.12	11 19 30	1.466	5 28 2	0.073	5 27 35	1.308	5 23 41	1.603	2 19 30	0.433	2 21 28	9.564
391	11 18 58	20.26	8 18 58	0.76	11 19 9	1.440	5 28 9	0.071	5 27 46	1.283	4 24 18	1.818	2 19 9	0.433	2 21 10	9.558
392	11 18 56	20.26	8 18 56	0.59	11 19 4	1.445	5 28 10	0.071	5 27 46	1.287	4 28 39	1.735	2 19 4	0.433	2 21 6	9.557
393	11 17 32	20.33	8 17 32	1.78	11 17 3	1.521	5 27 9	0.075	5 26 29	1.352	6 29 33	2.652	2 17 3	0.432	2 19 26	9.520
394	11 17 23	20.26	8 17 23	0.31	11 17 28	1.450	5 27 48	0.072	5 27 17	1.292	5 8 1	1.925	2 17 28	0.433	2 19 47	9.529
395	11 16 7	20.26	8 16 7	0.52	11 16 16	1.442	5 27 40	0.071	5 27 7	1.285	5 2 11	2.140	2 16 16	0.432	2 18 47	9.506
396	11 15 35	20.26	8 15 35	0.27	11 15 40	1.447	5 27 29	0.072	5 26 54	1.291	5 8 36	2.184	2 15 40	0.432	2 18 16	9.492
397	11 15 20	20.28	8 15 20	1.02	11 15 0	1.488	5 26 59	0.074	5 26 16	1.325	6 9 33	2.482	2 15 1	0.432	2 17 45	9.478
398	11 15 12	20.26	8 15 12	0.65	11 14 59	1.474	5 27 7	0.073	5 26 25	1.315	6 1 1	2.354	2 14 59	0.432	2 17 44	9.478
399	11 14 2	20.28	8 14 2	1.16	11 14 25	1.421	5 27 36	0.070	5 27 2	1.265	4 17 54	2.619	2 14 24	0.431	2 17 15	9.464
400	11 13 48	20.26	8 13 48	0.08	11 13 46	1.453	5 27 5	0.072	5 26 24	1.297	5 15 41	2.441	2 13 46	0.431	2 16 42	9.449
401	11 13 42	20.35	8 13 42	2.04	11 12 58	1.520	5 26 14	0.075	5 25 22	1.351	6 21 50	3.266	2 12 57	0.431	2 16 2	9.431
402	11 13 14	20.39	8 13 14	2.34	11 14 2	1.394	5 27 57	0.069	5 27 29	1.234	3 29 22	3.354	2 14 2	0.431	2 16 55	9.456
403	11 13 2	20.27	8 13 2	0.87	11 12 43	1.469	5 26 37	0.073	5 25 50	1.317	6 1 27	2.746	2 12 43	0.431	2 15 50	9.425
404	11 12 57	20.29	8 12 57	1.34	11 13 25	1.414	5 27 32	0.070	5 26 57	1.259	4 14 54	2.828	2 13 25	0.431	2 16 25	9.442
405	11 12 55	20.30	8 12 55	1.50	11 13 27	1.410	5 27 36	0.070	5 27 2	1.254	4 12 4	2.904	2 13 26	0.431	2 16 26	9.442
406	11 12 54	20.26	8 12 54	0.68	11 12 39	1.469	5 26 41	0.073	5 25 55	1.314	5 27 30	2.688	2 12 39	0.431	2 15 47	9.423

A CATALOGUE OF 520 ZODIACAL STARS FOR 1820 (CONTINUED).

No.	Mag.	Stars.	Longitude.				Ann. Var.	Latitude.			Ann. Var.	North East Angle of the ✱'s Meridian with the Ecliptic.			Ann. Var. of the ∠.	R. Ascension in Time.			Ann. Precess. 1830.	Declination.			Ann. Precess. 1830.
			S	°	′	″	+ ″	°	′	″	″	°	′	″	−	h.	m.	s.	+	°	′	″	−
407	5.6	υ Sagittarii	9	17	12	51.5	50.00	6	7	6.0 N	−0.48	82	56	54.2	19.70	19	11	25	3.439	16	16	58 S	6.193
408	6	Sag. 788 May.	9	18	34	47.1	50.00	0	30	2.1 N	−0.57	82	9	11.8	20.09	19	20	12	3.567	21	40	35 S	6.920
409	6	h^1 Sagittarii	9	19	12	24.9	50.07	3	2	59.2 S	+0.57	81	40	57.8	20.45	19	25	5	3.650	25	6	12 S	7.321
410	4.5	h^2 Sagittarii	9	19	19	46.2	50.22	3	14	10.1 S	+0.45	81	37	12.3	20.46	19	25	44	3.654	25	16	15 S	7.374
411	5.6	e^1 Sagittarii	9	21	41	17.0	50.15	5	4	45.0 N	−0.55	81	9	48.3	19.12	19	30	24	3.437	16	41	46 S	7.749
412	5	e^2 Sagittarii	9	22	8	36.2	50.18	5	10	13.3 N	−0.48	80	59	35.9	19.04	19	32	13	3.432	16	32	10 S	7.895
413	6	f Sagittarii	9	22	24	58.1	50.01	1	25	48.9 N	−0.61	80	41	25.0	19.33	19	35	51	3.516	20	11	3 S	8.187
414	6	Sag. 799 May.	9	22	30	7.7	50.20	5	46	46.4 N	−0.69	80	53	4.4	18.93	19	33	17	3.416	15	52	37 S	7.980
415	6	ω Sagittarii	9	23	19	30.9	50.40	5	24	12.2 S	+0.44	79	49	49.1	19.95	19	44	48	3.671	26	46	3 S	8.897
416	5	b Sagittarii	9	23	24	32.5	50.11	6	18	7.9 S	+0.51	79	42	50.5	20.06	19	45	53	3.693	27	38	12 S	8.983
417	5.6	57 Sagittarii	9	23	53	40.7	50.18	1	53	3.8 N	−0.40	80	8	57.7	18.99	19	41	44	3.494	19	29	30 S	8.654
418	5.6	a Sagittarii	9	24	2	32.9	50.13	5	26	17.6 S	+0.51	79	32	26.5	19.79	19	47	58	3.665	26	40	24 S	9.145
419	6	Sagittarii	9	25	16	58.3	50.19	2	9	26.9 S	+0.42	79	20	37.6	17.81	19	50	41	3.574	23	13	19 S	9.354
420	6	g Sagittarii	9	25	55	42.5	50.17	5	6	36.4 N	−0.53	79	34	4.2	18.36	19	47	44	3.407	15	57	30 S	9.123
421	6	63 Sagittarii	9	27	15	24.5	50.16	6	43	34.4 N	−0.37	79	9	42.9	18.02	19	51	53	3.365	14	7	37 S	9.443
422	6	4 Capricor.	9	29	14	25.9	50.19	2	4	19.8 S	+0.40	77	51	36.9	18.18	20	7	26	3.533	22	21	22 S	10.623
423	5.6	σ Capricor.	10	0	9	46.8	50.05	0	28	26.3 N	−0.50	77	43	59.8	17.76	20	9	0	3.470	19	40	18 S	10.738
424	4	Prima Giedi	10	1	15	23.7	50.08	7	0	27.0 N	−0.46	77	45	21.4	17.21	20	7	40	3.330	13	3	21 S	10.638
425	3	Secun. Giedi	10	1	20	39.5	50.24	6	57	4.3 N	−0.16	77	43	22.6	17.19	20	8	3	3.330	13	5	39 S	10.666
426	3.4	Dabih Major	10	1	31	57.6	50.11	4	36	28.2 N	−0.46	77	31	48.6	17.23	20	10	53	3.375	15	20	28 S	10.877
427	6	Capricor.	10	1	41	53.4	50.20	3	15	41.8 S	+0.39	76	51	54.7	17.65	20	18	57	3.532	22	58	46 S	11.464
428	6	Capric. 840 M.	10	1	48	55.3	50.02	5	55	23.3 S	+0.42	76	32	49.5	17.84	20	22	8	3.586	25	32	33 S	11.691
429	5	ν Capricor.	10	1	55	16.8	50.13	6	35	43.2 N	−0.46	77	30	20.0	17.08	20	10	40	3.333	13	19	1 S	10.859
430	5	π Capricor.	10	2	12	1.7	50.08	0	55	6.4 N	−0.47	77	2	55.3	17.25	20	17	0	3.442	18	47	36 S	11.322
431	5	ϱ Capricor.	10	2	39	5.6	49.97	1	13	3.7 N	−0.43	76	54	53.5	17.12	20	18	35	3.432	18	24	1 S	11.436
432	6	ο Capricor. seq.	10	2	42	22.3	50.18	0	24	54.0 N	−0.48	76	50	3.9	17.15	20	19	34	3.447	19	10	11 S	11.507
433	4.5	ψ Capricor.	10	4	38	58.5	50.18	7	0	16.2 S	+0.63	75	25	24.9	17.10	20	35	26	3.571	25	54	31 S	12.615
434	5	υ Capricor.	10	5	9	2.4	50.24	0	14	33.2 N	−0.31	75	59	23.1	16.54	20	29	47	3.427	18	45	50 S	12.225
435	6	τ^1 Capricor.	10	5	17	0.2	50.22	3	18	4.8 N	−0.38	76	10	27.7	16.40	20	27	15	3.368	15	45	45 S	12.050
436	6	17 Capricor.	10	5	39	29.4	50.13	3	23	30.7 S	+0.45	75	29	8.8	16.57	20	35	43	3.490	22	9	37 S	12.633
437	6	τ^2 Capricor.	10	5	47	0.9	50.12	3	22	5.4 N	−0.39	76	0	49.0	16.28	20	29	12	3.363	15	34	39 S	12.185
438	6	Capricor.	10	5	49	31.6	50.21	4	42	0.8 S	+0.37	75	17	28.7	16.60	20	37	51	3.512	23	23	6 S	12.780
439	6	Capricor.	10	6	34	12.2	50.22	5	59	56.8 S	+0.36	74	53	41.3	16.45	20	42	26	3.527	24	26	58 S	13.084
440	6	19 Capricor.	10	8	35	15.4	49.96	0	28	56.8 S	+0.35	74	48	35.2	15.64	20	44	37	3.406	18	35	53 S	13.227
441	6	20 Capricor.	10	9	22	15.8	50.11	1	52	0.3 S	+0.36	74	26	9.8	15.45	20	49	22	3.420	19	43	31 S	13.538
442	6	Aquarii 858 M.	10	9	55	52.0	50.10	5	44	2.4 N	−0.37	74	50	23.8	15.23	20	43	14	3.286	12	14	50 S	13.135
443	6	21 Capricor.	10	10	5	2.1	50.07	0	30	32.9 S	+0.42	74	20	25.5	15.22	20	50	43	3.391	18	13	32 S	13.624
444	5	η Capricor.	10	10	13	36.0	50.08	2	58	28.4 S	+0.41	74	3	36.4	15.23	20	54	9	3.430	20	33	32 S	13.842
445	5.6	χ^1 Capricor.	10	10	46	4.3	50.05	4	32	10.5 S	+0.48	73	43	31.8	15.11	20	58	14	3.449	21	54	32 S	14.099
446	6	8 Aquarii	10	11	7	54.8	50.15	3	51	8.0 N	−0.33	74	21	28.9	14.91	20	50	1	3.307	13	44	38 S	13.580
447	6	χ^3 Capricor.	10	11	10	26.3	50.28	3	59	23.0 S	+0.52	73	39	46.8	14.97	20	59	14	3.435	21	16	16 S	14.161
448	6	9 Aquarii	10	11	16	41.6	50.00	3	18	34.9 N	−0.34	74	16	37.5	14.87	20	51	12	3.315	14	13	35 S	13.654
449	5.6	ϑ Capricor.	10	11	19	33.6	50.11	0	33	59.5 S	+0.46	73	57	24.8	14.87	20	55	49	3.378	17	56	28 S	13.946
450	6	φ Capricor.	10	12	30	37.7	50.09	4	30	43.3 S	+0.40	73	12	16.6	14.56	21	5	22	3.427	21	23	31 S	14.535
451	6	30 Capricor.	10	13	50	30.4	50.18	2	8	6.6 S	+0.35	73	4	9.0	14.14	21	7	51	3.376	18	43	55 S	14.684
452	5	ν Aquarii	10	13	52	45.7	50.11	4	46	56.8 N	−0.36	73	36	21.0	14.19	20	59	46	3.269	12	5	36 S	14.193
453	5	29 Capricor.	10	14	11	8.7	50.24	0	42	27.7 N	−0.25	73	13	40.5	14.05	21	5	46	3.329	15	54	44 S	14.558
454	6	33 Capricor.	10	14	21	28.0	50.17	5	18	25.4 S	+0.50	72	34	39.8	13.99	21	13	56	3.417	21	36	35 S	15.041
455	4	ζ Capricor.	10	14	25	6.6	49.91	6	58	39.9 S	+0.53	72	21	9.5	13.98	21	16	22	3.441	23	11	2 S	15.181
456	6	35 Capricor.	10	14	56	10.9	50.28	5	51	48.0 S	+0.31	72	20	51.1	13.30	21	17	1	3.418	21	58	4 S	15.218
457	5.6	36 Capricor.	10	15	3	38.0	50.16	6	32	58.8 S	+0.33	72	13	37.5	13.76	21	18	26	3.426	22	35	1 S	15.298
458	5	Capricor.	10	15	10	0.4	50.11	1	21	6.1 S	+0.36	72	46	10.4	13.74	21	12	13	3.350	17	35	38 S	14.941
459	6	18 Aquarii	10	16	49	56.6	50.20	2	16	3.4 N	−0.28	72	36	45.8	13.32	21	14	21	3.281	13	33	37 S	15.064
460	6	17 Aquarii	10	17	39	6.8	50.14	5	44	46.1 N	−0.29	72	36	35.5	13.20	21	13	17	3.225	10	4	47 S	15.002
461	5	ι Capricor.	10	17	40	58.5	50.09	4	57	44.1 S	+0.36	71	42	32.8	12.91	21	26	59	3.372	20	15	58 S	15.770
462	6	19 Aquarii	10	18	3	11.4	50.15	5	10	19.8 N	−0.29	72	28	16.3	13.07	21	15	32	3.230	10	30	27 S	15.131
463	6	Capric. 898 M.	10	19	0	3.8	50.06	5	34	51.8 S	+0.38	71	17	44.2	12.46	21	33	8	3.364	20	26	11 S	16.096

SUBSIDIARY NUMBERS OF THE ZODIACAL STARS FOR 1830 (CONTINUED).

No.	Longitude.		Latitude.		Right Ascension in Time.						Declination.					
	Aberration.		Aberration.		Aberration.		☉ Nutation.		☾ Nutation.		Aberration.		☉ Nutation.		☾ Nutation.	
	A	Max.	A′	Max.	A	Max.	S	Max.	L	Max.	A′	Max.	S′	Max.	L′	Max.
	s ° ′	″	s ° ′	″	s ° ′	s	s ° ′	s	s ° ′	s	s ° ′	″	s ° ′	″	s ° ′	″
407	11 12 39	20.37	8 12 39	2.16	11 13 24	1.396	5 27 49	0.069	5 27 19	1.238	4 2 6	3.579	2 13 24	0.431	2 16 24	9.441
408	11 11 17	20.26	8 11 17	0.18	11 11 21	1.439	5 26 48	0.071	5 26 4	1.285	5 7 53	2.786	2 11 21	0.430	2 14 40	9.388
409	11 10 39	20.28	8 10 39	1.08	11 10 12	1.475	5 26 7	0.073	5 25 13	1.317	6 0 30	3.136	2 10 12	0.430	2 13 43	9.357
410	11 10 32	20.28	8 10 32	1.14	11 10 3	1.477	5 26 3	0.073	5 25 9	1.319	6 1 20	3.179	2 10 3	0.430	2 13 35	9.353
411	11 8 10	20.33	8 8 10	1.79	11 8 58	1.393	5 27 12	0.069	5 26 33	1.238	4 8 40	3.598	2 8 58	0.429	2 12 40	9.321
412	11 7 43	20.33	8 7 43	1.82	11 8 33	1.391	5 27 10	0.069	5 26 31	1.236	4 8 16	3.663	2 8 32	0.429	2 12 19	9.308
413	11 7 27	20.26	8 7 27	0.51	11 7 41	1.419	5 26 28	0.070	5 25 38	1.261	4 28 49	3.334	2 7 41	0.429	2 11 35	9.283
414	11 7 22	20.35	8 7 22	2.04	11 8 18	1.386	5 27 15	0.068	5 26 36	1.230	4 5 23	3.805	2 8 17	0.429	2 12 6	9.301
415	11 6 32	20.34	8 6 32	1.91	11 5 34	1.488	5 24 57	0.074	5 23 47	1.328	6 4 6	4.056	2 5 34	0.428	2 9 46	9.215
416	11 6 27	20.37	8 6 27	2.22	11 5 18	1.499	5 24 44	0.074	5 23 31	1.336	6 7 28	4.243	2 5 19	0.427	2 9 32	9.206
417	11 5 58	20.26	8 5 58	0.67	11 6 18	1.411	5 26 22	0.070	5 25 31	1.260	4 25 18	3.545	2 6 19	0.428	2 10 23	9.239
418	11 5 49	20.34	8 5 49	1.92	11 4 49	1.486	5 24 49	0.074	5 23 37	1.326	6 2 55	4.155	2 4 49	0.427	2 9 7	9.189
419	11 4 35	20.26	8 4 35	0.76	11 4 10	1.443	5 25 23	0.072	5 24 17	1.292	5 15 49	3.841	2 4 11	0.427	2 8 34	9.168
420	11 3 56	20.33	8 3 56	1.81	11 4 53	1.381	5 26 50	0.068	5 26 6	1.230	4 8 13	4.091	2 4 53	0.427	2 9 10	9.192
421	11 2 36	20.39	8 2 36	2.37	11 3 54	1.368	5 27 5	0.067	5 26 24	1.212	4 1 19	4.472	2 3 54	0.427	2 8 19	9.159
422	11 0 37	20.26	8 0 37	0.73	11 0 11	1.427	5 24 16	0.071	5 22 57	1.280	5 12 8	4.808	2 0 11	0.425	2 5 3	9.024
423	10 29 42	20.26	7 29 42	0.17	10 29 47	1.401	5 25 26	0.070	5 24 22	1.254	4 27 30	4.324	1 29 47	0.424	2 4 42	9.010
424	10 28 36	20.40	7 28 36	2.47	11 0 8	1.355	5 26 56	0.066	5 26 13	1.200	3 29 22	4.945	2 0 8	0.425	2 5 0	9.022
425	10 28 31	20.40	7 28 31	2.45	11 0 2	1.355	5 26 55	0.066	5 26 12	1.200	3 29 32	4.943	2 0 2	0.424	2 4 54	9.019
426	10 28 20	20.32	7 28 20	1.62	10 29 20	1.367	5 26 21	0.067	5 25 30	1.217	4 8 27	4.670	1 29 20	0.424	2 4 19	8.993
427	10 28 10	20.28	7 28 10	1.15	10 27 23	1.428	5 24 19	0.071	5 23 0	1.279	5 11 49	4.756	1 27 23	0.423	2 2 33	8.918
428	10 28 3	20.36	7 28 3	2.09	10 26 37	1.455	5 23 34	0.072	5 22 6	1.301	5 21 17	5.147	1 26 37	0.423	2 1 51	8.887
429	10 27 56	20.39	7 27 56	2.33	10 29 24	1.355	5 26 48	0.067	5 26 4	1.201	4 0 40	4.949	1 29 24	0.424	2 4 22	8.995
430	10 27 40	20.26	7 27 40	0.32	10 27 51	1.390	5 25 22	0.069	5 24 18	1.244	4 23 43	4.567	1 27 52	0.423	2 2 59	8.936
431	10 27 13	20.26	7 27 13	0.43	10 27 29	1.386	5 25 25	0.069	5 24 22	1.240	4 21 59	4.623	1 27 29	0.423	2 2 39	8.921
432	10 27 10	20.26	7 27 10	0.15	10 27 15	1.392	5 25 12	0.069	5 24 6	1.246	4 25 22	4.632	1 27 15	0.423	2 2 26	8.912
433	10 25 13	20.40	7 25 13	2.47	10 23 23	1.453	5 22 56	0.072	5 21 18	1.299	5 20 15	5.644	1 23 23	0.420	1 28 55	8.755
434	10 24 43	20.26	7 24 43	0.09	10 24 47	1.383	5 25 0	0.069	5 23 51	1.239	4 23 15	4.917	1 24 46	0.421	2 0 11	8.812
435	10 24 35	20.28	7 24 35	1.17	10 25 23	1.362	5 25 50	0.067	5 24 53	1.216	4 11 20	4.989	1 25 23	0.422	2 0 44	8.837
436	10 24 12	20.29	7 24 12	1.20	10 23 19	1.411	5 23 55	0.070	5 22 31	1.265	5 7 1	5.223	1 23 19	0.420	1 28 51	8.752
437	10 24 5	20.29	7 24 5	1.19	10 24 55	1.360	5 25 50	0.067	5 24 52	1.214	4 10 48	5.045	1 24 54	0.421	2 0 19	8.818
438	10 24 2	20.32	7 24 2	1.66	10 22 47	1.423	5 23 31	0.070	5 22 1	1.275	5 11 19	5.402	1 22 48	0.420	1 28 22	8.731
439	10 23 17	20.36	7 23 17	2.12	10 21 41	1.432	5 23 3	0.071	5 21 27	1.282	5 14 24	5.676	1 21 41	0.419	1 27 20	8.684
440	10 21 16	20.26	7 21 16	0.17	10 21 8	1.374	5 24 36	0.069	5 23 21	1.232	4 23 6	5.322	1 21 9	0.419	1 28 49	8.661
441	10 20 29	20.26	7 20 29	0.66	10 19 58	1.382	5 24 9	0.069	5 22 48	1.239	4 27 7	5.488	1 19 58	0.418	1 25 43	8.612
442	10 19 56	20.35	7 19 56	2.02	10 21 29	1.334	5 26 25	0.066	5 25 35	1.185	3 29 44	5.660	1 21 29	0.419	1 27 8	8.676
443	10 19 47	20.26	7 19 47	0.18	10 19 38	1.369	5 24 32	0.068	5 23 17	1.228	4 21 35	5.485	1 19 38	0.418	1 25 24	8.598
444	10 19 38	20.28	7 19 38	1.05	10 18 47	1.387	5 23 46	0.069	5 22 19	1.244	4 29 53	5.668	1 18 47	0.418	1 24 35	8.562
445	10 19 6	20.31	7 19 6	1.60	10 17 46	1.397	5 23 13	0.069	5 21 40	1.254	5 4 11	5.889	1 17 46	0.417	1 23 38	8.518
446	10 18 44	20.30	7 18 44	1.36	10 19 48	1.338	5 25 51	0.066	5 24 53	1.193	4 5 15	5.633	1 19 49	0.418	1 25 34	8.605
447	10 18 41	20.30	7 18 41	1.41	10 17 31	1.390	5 23 23	0.069	5 21 52	1.247	5 2 0	5.870	1 17 31	0.417	1 23 23	8.507
448	10 18 35	20.28	7 18 35	1.17	10 19 31	1.340	5 25 41	0.066	5 24 41	1.197	4 7 1	5.618	1 19 31	0.418	1 25 17	8.593
449	10 18 32	20.26	7 18 32	0.20	10 18 22	1.364	5 24 29	0.068	5 23 13	1.223	4 20 29	5.616	1 18 23	0.417	1 24 11	8.544
450	10 17 21	20.31	7 17 21	1.59	10 15 59	1.388	5 23 9	0.069	5 21 35	1.245	5 1 54	6.062	1 15 59	0.416	1 21 53	8.442
451	10 16 1	20.26	7 16 1	0.75	10 15 22	1.364	5 23 55	0.068	5 22 31	1.224	4 22 59	5.957	1 15 22	0.416	1 21 17	8.416
452	10 15 59	20.32	7 15 59	1.69	10 17 23	1.325	5 26 10	0.065	5 25 16	1.179	4 0 10	5.955	1 17 23	0.417	1 23 15	8.502
453	10 15 40	20.26	7 15 40	0.25	10 15 53	1.344	5 24 51	0.067	5 23 40	1.204	4 13 19	5.862	1 15 53	0.416	1 21 48	8.438
454	10 15 30	20.34	7 15 30	1.87	10 13 50	1.385	5 22 49	0.069	5 21 10	1.244	5 1 53	6.334	1 13 50	0.414	1 19 47	8.350
455	10 15 27	20.40	7 15 27	2.46	10 13 14	1.400	5 22 13	0.069	5 20 27	1.255	5 6 17	6.584	1 13 13	0.414	1 19 12	8.325
456	10 14 55	20.36	7 14 55	2.07	10 13 3	1.387	5 22 36	0.069	5 20 54	1.244	5 2 43	6.466	1 13 3	0.414	1 19 2	8.317
457	10 14 48	20.38	7 14 48	2.31	10 12 42	1.393	5 22 21	0.069	5 20 36	1.248	5 4 20	6.577	1 12 42	0.414	1 18 41	8.302
458	10 14 42	20.26	7 14 42	0.48	10 14 16	1.353	5 24 10	0.067	5 22 50	1.214	4 19 1	6.031	1 14 16	0.415	1 20 12	8.368
459	10 13 2	20.27	7 13 2	0.80	10 13 44	1.326	5 25 24	0.066	5 24 20	1.185	4 6 50	6.115	1 13 44	0.414	1 19 42	8.346
460	10 12 12	20.35	7 12 12	2.03	10 14 1	1.310	5 26 35	0.064	5 25 48	1.161	3 24 30	6.369	1 14 1	0.415	1 19 57	8.357
461	10 12 11	20.33	7 12 11	1.75	10 10 32	1.366	5 22 53	0.068	5 21 15	1.226	4 26 48	6.582	1 10 33	0.412	1 16 30	8.210
462	10 11 48	20.33	7 11 48	1.82	10 13 27	1.310	5 26 24	0.064	5 25 34	1.162	3 25 54	6.357	1 13 26	0.414	1 19 25	8.333
463	10 10 52	20.35	7 10 52	1.97	10 8 58	1.365	5 22 42	0.068	5 21 1	1.225	4 26 43	6.746	1 8 58	0.410	1 14 55	8.144

A CATALOGUE OF 520 ZODIACAL STARS FOR 1820. (CONCLUDED).

No.	Mag.	Stars.	Longitude.	Ann. Var.	Latitude.	Ann. Var.	North East Angle of the ✱'s Meridian with the Ecliptic.	Ann. Var. of the ∠.	Right Ascension in Time.	Ann. Prec. 1830.	Declination.	Ann. Precess. 1830.
			S ° ′ ″	+	° ′ ″		° ′ ″		h. m. s.	+	° ′ ″	
464	5	κ Capric......	10 19 7 21 .8	50″. 24	4 49 23″ .1 S	+0″.26	71 21 12″ .8	−12″.44	21 32 35	3ˢ.353	19 40 50 S	−16ˢ.067
465	4	Nashira pri...	10 19 16 2 .5	50 .31	2 32 23 .4 S	+0 .42	71 33 39 .9	−12 .45	21 30 6	3 .322	17 28 9 S	−15 .937
466	6	d^1 Capricor ..	10 20 28 28 .6	49 .90	0 10 32 .3 S	+0 .57	71 28 28 .8	−12 .16	21 31 45	3 .280	14 50 34 S	−16 .024
467	6	d^2 Capricor...	10 20 41 50 .7	50 .19	0 38 33 .8 S	+0 .27	71 22 47 .6	−12 .07	21 33 14	3 .284	15 13 0 S	−16 .101
468	6	45 Capric. d^3	10 20 47 57 .9	50 .24	1 2 44 .9 S	−0 .05	71 19 9 .1	−12 .02	21 34 11	3 .288	15 34 4 S	−16 .150
469	3 .4	Nashira post.	10 21 1 15 .3	50 .34	2 34 13 .3 S	+0 .58	71 7 20 .8	−11 .90	21 37 5	3 .304	16 56 12 S	−16 .300
470	5	ξ Aquarii.....	10 21 36 8 .5	50 .15	5 58 25 .1 N	−0 .30	71 36 2 .7	−12 .12	21 28 10	3 .192	8 39 18 S	−15 .832
471	5 .6	λ Capricor....	10 22 29 40 .3	50 .00	1 56 38 .5 N	−0 .13	71 8 48 .6	−11 .62	21 36 44	3 .236	12 11 25 S	−16 .281
472	6	c^1 Capricor...	10 22 53 57 .1	49 .92	4 13 20 .9 N	−0 .18	71 11 37 .7	−11 .65	21 35 23	3 .204	9 54 10 S	−16 .214
473	5	μ Capricor...	10 23 18 6 .8	50 .30	0 39 55 .4 S	+0 .25	70 45 23 .8	−11 .26	21 43 28	3 .259	14 23 36 S	−16 .617
474	6	29 Aquarii...	10 24 13 22 .0	50 .48	4 38 2 .6 S	+0 .26	70 9 55 .2	−10 .76	21 52 35	3 .293	17 49 34 S	−17 .050
475	5 .6	35 Aquarii...	10 25 8 8 .9	50 .20	6 38 8 .5 S	+0 .15	69 44 9 .5	−10 .35	21 59 6	3 .303	19 23 42 S	−17 .343
476	4 .5	ι Aquarii.....	10 26 12 20 .3	50 .16	2 3 57 .1 S	+0 .29	69 59 34 .1	−10 .26	21 56 42	3 .247	14 44 11 S	−17 .237
477	6	e Aquarii	10 27 58 15 .6	50 .03	0 16 18 .4 S	+0 .23	69 16 37 .4	− 9 .82	22 0 59	3 .213	12 26 42 S	−17 .424
478	5 .6	30 Aquarii...	10 28 2 33 .5	50 .14	5 5 12 .1 N	−0 .21	70 5 2 .9	−10 .20	21 53 48	3 .158	7 23 14 S	−17 .105
479	6	37 Aquarii...	10 28 12 41 .0	50 .11	0 25 52 .2 N	−0 .32	69 46 50 .1	− 9 .79	22 0 55	3 .204	11 42 7 S	−17 .421
480	6	42 Aquarii...	10 28 55 52 .2	50 .20	1 59 34 .2 S	+0 .20	69 26 51 .6	− 9 .39	22 7 9	3 .221	13 43 27 S	−17 .684
481	6	45 Aquarii...	10 29 15 41 .3	50 .21	2 37 36 .3 S	+0 .20	69 19 43 .3	− 9 .23	22 9 20	3 .226	14 12 5 S	−17 .774
482	6	Aqua. 919 M.	11 0 18 40 .1	49 .96	1 32 18 .2 N	−0 .25	69 26 31 .0	− 9 .23	22 7 22	3 .177	9 55 57 S	−17 .693
483	6	50 Aquarii...	11 0 24 56 .2	50 .21	3 19 2 .8 S	+0 .19	69 3 3 .1	− 8 .80	22 14 48	3 .219	14 26 15 S	−17 .990
484	4 .5	Ancha	11 0 44 40 .6	50 .06	2 43 7 .5 N	−0 .34	69 25 38 .8	− 9 .19	22 7 19	3 .163	8 40 30 S	−17 .692
485	6	56 Aquarii...	11 1 20 54 .5	50 .22	4 49 1 .2 S	+0 .18	68 44 25 .8	− 8 .36	22 20 38	3 .222	15 30 5 S	−18 .208
486	6	ρ Aquarii	11 1 31 1 .0	50 .20	2 22 54 .6 N	−0 .09	69 15 52 .3	− 8 .92	22 10 43	3 .161	8 43 11 S	−17 .829
487	5	σ Aquarii	11 2 52 31 .3	50 .19	1 13 4 .7 S	+0 .17	68 47 32 .9	− 8 .18	22 21 7	3 .181	11 35 40 S	−18 .226
488	6	58 Aquarii...	11 3 1 34 .6	50 .28	1 31 18 .7 S	+0 .21	68 44 39 .2	− 8 .10	22 22 8	3 .183	11 49 24 S	−18 .263
489	6	51 Aquarii...	11 3 30 27 .7	50 .09	4 48 30 .4 N	−0 .31	69 0 49 .1	− 8 .54	22 14 44	3 .127	5 45 36 S	−17 .987
490	6	τ^1 Aquarii ...	11 5 28 36 .7	50 .14	5 54 41 .8 S	+0 .16	67 58 27 .3	− 6 .87	22 38 9	3 .192	15 0 7 S	−18 .795
491	5 .6	τ^2 Aquarii ...	11 6 4 53 .5	50 .11	5 39 23 .4 S	+0 .25	67 54 48 .5	− 6 .69	22 40 3	3 .185	14 32 21 S	−18 .852
492	6	63 Aquarii...	11 6 54 37 .3	50 .01	4 7 18 .8 N	−0 .16	68 25 23 .5	− 7 .43	22 28 26	3 .114	5 9 7 S	−18 .481
493	6	70 Aquarii...	11 6 59 29 .8	50 .21	2 44 49 .3 S	+0 .14	68 2 16 .6	− 6 .69	22 39 2	3 .161	11 30 8 S	−18 .822
494	6	67 Aquarii...	11 7 8 35 .6	50 .08	1 4 18 .0 N	−0 .14	68 15 32 .8	− 7 .04	22 33 50	3 .135	7 54 4 S	−18 .661
495	6	74 Aquarii...	11 7 42 55 .1	50 .22	4 11 33 .1 S	+0 .13	67 49 23 .4	− 6 .30	22 43 59	3 .164	12 34 14 S	−18 .966
496	4	λ Aquarii	11 9 3 37 .5	50 .04	0 22 56 .2 S	+0 .10	67 54 44 .1	− 6 .27	22 43 13	3 .133	8 32 4 S	−18 .945
497	6	78 Aquarii...	11 9 39 10 .3	49 .71	0 13 4 .1 S	+0 .16	67 50 37 .3	− 6 .09	22 45 11	3 .128	8 9 29 S	−19 .000
498	6	81 Aquarii...	11 11 16 25 .2	50 .25	0 44 5 .3 S	+0 .04	67 36 59 .0	− 5 .51	22 52 2	3 .122	8 1 26 S	−19 .182
499	6	82 Aquarii...	11 11 43 30 .2	50 .19	0 23 39 .4 S	+0 .10	67 34 53 5	− 5 .41	22 53 12	3 .118	7 32 12 S	−19 .211
500	6	h^1 Aquarii ...	11 11 52 59 .4	50 .14	1 40 41 .3 S	+0 .15	67 29 37 .6	− 5 .21	22 55 46	3 .123	8 39 44 S	−19 .274
501	5 .6	ψ^1 Aquarii...	11 13 46 37 .9	50 .46	3 59 25 .1 S	+0 .40	67 9 9 .9	− 4 .31	23 6 27	3 .122	10 3 56 S	−19 .508
502	6	3 Piscium ...	11 13 53 39 .5	50 .13	6 1 26 .0 N	−0 .08	67 30 26 .7	− 5 .50	22 51 24	3 .073	0 46 40 S	−19 .166
503	5	ψ^2 Aquarii...	11 14 12 54 .2	50 .28	4 16 42 .0 S	+0 .06	67 5 32 .4	− 4 .13	23 8 33	3 .120	10 9 45 S	−19 .551
504	5	ψ^3 Aquarii...	11 14 17 5 .4	50 .30	4 46 27 .3 S	+0 .07	67 3 2 .6	− 4 .05	23 9 35	3 .122	10 35 32 S	−19 .571
505	5 .6	χ Aquarii	11 14 32 52 .0	50 .11	2 50 13 .3 S	+0 .06	67 9 18 .5	− 4 .20	23 7 31	3 .114	8 42 19 S	−19 .531
506	5	φ Aquarii	11 14 37 33 .3	50 .02	1 2 20 .5 S	+0 .28	67 14 38 .7	− 4 .39	23 5 0	3 .106	7 0 58 S	−19 .480
507	6	96 Aquarii...	11 16 8 24 .7	50 .36	0 41 13 .1 S	+0 .11	67 7 19 .9	− 3 .95	23 10 3	3 .098	6 6 22 S	−19 .580
508	5 .6	κ^1 Piscium ...	11 20 23 11 .7	49 .88	4 26 32 .2 N	+0 .04	66 53 7 .7	− 3 .27	23 17 42	3 .066	0 16 23 N	+19 .711
509	6	κ^2 Piscium...	11 20 24 31 .1	49 .85	4 16 58 .0 N	−0 .22	66 53 2 .7	− 3 .24	23 18 1	3 .067	0 8 11 N	+19 .717
510	6	16 Piscium...	11 22 54 29 .2	50 .04	4 16 15 .4 N	+0 .14	66 43 21 .3	− 2 .45	23 27 12	3 .064	1 6 19 N	+19 .845
511	5 .6	20 Piscium...	11 23 37 3 .9	50 .19	1 19 59 .7 S	0 .00	66 38 16 .2	− 1 .45	23 38 41	3 .076	3 45 38 S	−19 .962
512	5	λ Piscium	11 24 5 3 .1	49 .79	3 25 19 .1 N	−0 .01	66 40 1 .3	− 1 .96	23 32 52	3 .066	0 47 28 N	+19 .909
513	6	21 Piscium ...	11 25 29 47 .2	49 .88	2 2 8 .6 N	+0 .01	66 36 49 .6	− 1 .31	23 40 15	3 .068	0 4 40 N	+19 .973
514	4 .5	30 Piscium ..	11 25 31 50 .6	50 .10	5 42 31 .5 S	+0 .03	66 25 32 .5	− 0 .23	23 52 43	3 .073	7 0 49 S	−20 .035
515	6	19 Piscium ..	11 25 45 24 .8	49 .95	4 33 8 .8 N	+0 .13	66 34 54 .4	− 1 .58	23 37 12	3 .062	2 29 24 N	+19 .949
516	5	27 Piscium ..	11 25 46 0 .2	49 .88	3 7 41 .1 S	−0 .02	66 31 34 .9	− 0 .51	23 49 27	3 .073	4 33 12 S	−20 .025
517	5	33 Piscium ..	11 26 25 38 .4	50 .09	5 46 13 .0 S	−0 .19	66 24 50 .1	+ 0 .07	23 56 7	3 .070	6 42 52 S	−20 .044
518	5	29 Piscium ..	11 26 41 44 .0	50 .01	2 57 27 .5 S	−0 .18	66 31 0 .8	− 0 .24	23 52 35	3 .071	4 1 43 S	−20 .034
519	6	22 Piscium ..	11 26 48 39 .0	50 .15	3 29 13 .2 N	+0 .03	66 33 41 .5	− 1 .10	23 42 45	3 .064	1 55 50 N	+19 .991
520	4 .5	ν Piscium	0 0 4 3 .5	50 .11	6 22 5 .3 N	−0 .13	66 24 21 .8	− 0 .46	23 50 4	3 .062	5 52 4 N	+20 .028

SUBSIDIARY NUMBERS OF THE ZODIACAL STARS FOR 1830 (CONCLUDED).

No.	Longitude. Aberration. A	Max.	Latitude. Aberration. A′	Max.	Right Ascension in Time. Aberration. A	Max.	⊙ Nutation. S	Max.	☾ Nutation. L	Max.	Declination. Aberration. A′	Max.	⊙ Nutation. S′	Max.	☾ Nutation. L′	Max.
	S ° ′	″	S ° ′	″	S ° ′	s	S ° ′	s	S ° ′	s	S ° ′	″	S ° ′	″	S ° ′	″
464	10 10 44	20.32	7 10 44	1.70	10 9 7	1.359	5 22 56	0.067	5 21 19	1.219	4 24 41	6.686	1 9 7	0.410	1 15 4	8.149
465	10 10 36	20.27	7 10 36	0.89	10 9 45	1.343	5 23 47	0.067	5 22 21	1.205	4 18 8	6.476	1 9 45	0.411	1 15 42	8.177
466	10 9 23	20.26	7 9 23	0.06	10 9 19	1.324	5 24 39	0.066	5 23 25	1.187	4 9 57	6.449	1 9 19	0.410	1 15 16	8.159
467	10 9 10	20.26	7 9 10	0.23	10 8 57	1.326	5 24 25	0.065	5 23 8	1.172	4 7 35	6.981	1 8 57	0.410	1 14 53	8.143
468	10 9 4	20.26	7 9 4	0.37	10 8 43	1.327	5 24 21	0.066	5 23 49	1.191	4 12 8	6.510	1 8 43	0.410	1 14 38	8.133
469	10 8 50	20.27	7 8 50	0.91	10 7 57	1.335	5 23 48	0.066	5 22 23	1.198	4 16 17	6.622	1 7 58	0.410	1 13 52	8.102
470	10 8 16	20.36	7 8 16	2.11	10 10 14	1.296	5 26 53	0.064	5 26 10	1.150	3 20 55	6.711	1 10 14	0.411	1 16 12	8.197
471	10 7 22	20.26	7 7 22	0.69	10 8 3	1.307	5 25 31	0.065	5 24 28	1.169	4 1 41	6.587	1 8 4	0.410	1 13 59	8.106
472	10 6 58	20.31	7 6 58	1.49	10 8 24	1.300	5 26 21	0.064	5 25 31	1.156	3 24 47	6.692	1 8 24	0.410	1 14 19	8.121
473	10 6 34	20.26	7 6 34	0.24	10 6 20	1.316	5 24 35	0.065	5 23 20	1.180	4 8 34	6.692	1 6 20	0.409	1 12 11	8.035
474	10 5 38	20.32	7 5 38	1.64	10 3 58	1.334	5 23 8	0.066	5 21 35	1.197	4 18 14	7.054	1 3 58	0.408	1 9 43	7.941
475	10 4 43	20.39	7 4 43	2.34	10 2 17	1.344	5 22 24	0.066	5 20 39	1.203	4 22 4	7.361	1 2 16	0.408	1 7 53	7.875
476	10 3 39	20.26	7 3 39	0.73	10 2 55	1.312	5 24 15	0.065	5 22 55	1.176	4 9 17	6.972	1 2 53	0.408	1 8 34	7.900
477	10 1 53	20.26	7 1 53	0.09	10 1 48	1.297	5 26 4	0.064	5 25 10	1.159	4 2 37	7.013	1 1 47	0.408	1 7 22	7.857
478	10 1 49	20.33	7 1 49	1.79	10 3 40	1.281	5 27 6	0.063	5 26 26	1.137	3 18 5	7.113	1 3 39	0.408	1 9 23	7.929
479	10 1 39	20.26	7 1 39	0.15	10 1 49	1.294	5 25 22	0.064	5 24 17	1.158	4 0 28	7.011	1 1 48	0.408	1 7 23	7.857
480	10 0 56	20.26	7 0 56	0.71	10 0 11	1.301	5 24 28	0.064	5 23 12	1.167	4 6 13	7.188	1 0 10	0.407	1 5 37	7.796
481	10 0 36	20.27	7 0 36	0.93	9 29 36	1.303	5 24 15	0.064	5 22 55	1.168	4 7 31	7.212	0 29 36	0.407	1 3 59	7.775
482	9 29 33	20.26	6 29 33	0.54	10 0 7	1.283	5 25 59	0.063	5 25 3	1.146	3 25 27	7.141	1 0 7	0.407	1 5 33	7.794
483	9 29 27	20.28	6 29 27	1.17	9 28 11	1.302	5 24 4	0.065	5 22 42	1.166	4 8 2	7.333	0 28 10	0.406	1 3 26	7.726
484	9 29 7	20.27	6 29 7	0.96	10 0 8	1.279	5 26 29	0.063	5 25 40	1.140	3 21 55	7.185	1 0 8	0.407	1 5 34	7.794
485	9 28 31	20.32	6 28 31	1.70	9 26 38	1.306	5 23 33	0.065	5 22 4	1.169	4 10 40	7.523	0 26 38	0.405	1 1 44	7.670
486	9 28 21	20.27	6 28 21	0.84	9 29 15	1.278	5 26 26	0.063	5 25 37	1.140	3 22 5	7.225	0 29 14	0.407	1 4 36	7.762
487	9 26 59	20.26	6 26 59	0.43	9 26 30	1.285	5 25 9	0.064	5 24 2	1.150	4 0 7	7.347	0 26 31	0.405	1 1 35	7.666
488	9 26 50	20.26	6 26 50	0.54	9 26 14	1.285	5 25 3	0.064	5 23 54	1.151	4 0 44	7.371	0 26 14	0.405	1 1 17	7.657
489	9 26 21	20.32	6 26 21	1.70	9 28 12	1.268	5 27 37	0.062	5 27 4	1.126	3 14 5	7.434	0 28 11	0.406	1 3 27	7.725
490	9 24 23	20.36	6 24 23	2.09	9 22 0	1.297	5 23 22	0.064	5 21 50	1.160	4 9 28	7.885	0 21 59	0.403	0 26 28	7.524
491	9 23 47	20.35	6 23 47	1.99	9 21 30	1.293	5 23 41	0.064	5 22 13	1.156	4 7 26	7.845	0 21 29	0.402	0 25 52	7.508
492	9 22 57	20.30	6 22 57	1.45	9 24 36	1.261	5 27 48	0.062	5 27 18	1.121	3 12 39	7.579	0 24 36	0.404	0 29 26	7.603
493	9 22 52	20.27	6 22 52	0.97	9 21 46	1.278	5 25 0	0.063	5 23 51	1.143	3 29 38	7.637	0 21 45	0.403	0 26 12	7.517
494	9 22 43	20.26	6 22 43	0.38	9 23 8	1.266	5 26 36	0.063	5 25 49	1.130	3 20 2	7.520	0 23 8	0.403	0 27 47	7.558
495	9 22 9	20.30	6 22 9	1.48	9 20 26	1.281	5 24 30	0.063	5 23 14	1.146	4 2 18	7.774	0 20 26	0.402	0 24 40	7.478
496	9 20 48	20.26	6 20 48	0.14	9 20 38	1.265	5 26 16	0.063	5 25 24	1.130	3 21 44	7.624	0 20 38	0.402	0 24 55	7.484
497	9 20 13	20.26	6 20 13	0.08	9 20 7	1.263	5 26 26	0.063	5 25 36	1.128	3 20 41	7.644	0 20 7	0.402	0 24 18	7.470
498	9 18 35	20.26	6 18 35	0.26	9 18 17	1.261	5 26 26	0.062	5 25 37	1.126	3 20 22	7.724	0 18 17	0.401	0 22 10	7.421
499	9 18 8	20.26	6 18 8	0.14	9 17 58	1.259	5 26 39	0.062	5 25 52	1.124	3 19 5	7.734	0 17 59	0.401	0 21 48	7.414
500	9 17 59	20.26	6 17 59	0.59	9 17 16	1.262	5 26 8	0.062	5 25 14	1.127	3 22 0	7.779	0 17 17	0.401	0 20 59	7.397
501	9 16 5	20.30	6 16 5	1.41	9 14 25	1.264	5 25 26	0.062	5 24 23	1.128	3 25 27	7.976	0 14 26	0.400	0 17 34	7.332
502	9 15 58	20.36	6 15 58	2.12	9 18 27	1.249	5 29 40	0.061	5 29 35	1.103	3 1 48	8.000	0 18 28	0.401	0 22 23	7.426
503	9 15 39	20.31	6 15 39	1.51	9 13 50	1.264	5 25 22	0.062	5 24 18	1.127	3 25 39	8.013	0 13 51	0.400	0 16 54	7.321
504	9 15 35	20.32	6 15 35	1.68	9 13 33	1.266	5 25 10	0.063	5 24 4	1.129	3 26 41	8.054	0 13 34	0.400	0 16 33	7.315
505	9 15 19	20.27	6 15 19	1.00	9 14 7	1.259	5 26 3	0.062	5 25 8	1.108	3 22 1	7.924	0 14 7	0.400	0 17 13	7.327
506	9 15 14	20.26	6 15 14	0.37	9 14 48	1.254	5 26 50	0.062	5 26 5	1.119	3 17 42	7.849	0 14 49	0.400	0 18 2	7.341
507	9 13 43	20.26	6 13 43	0.24	9 13 26	1.251	5 27 13	0.062	5 26 35	1.115	3 15 20	7.883	0 13 25	0.400	0 16 24	7.313
508	9 9 29	20.31	6 9 29	1.57	9 11 21	1.243	6 0 9	0.061	6 0 11	1.102	8 29 12	8.085	6 11 22	0.399	6 13 55	7.277
509	9 9 27	20.31	6 9 27	1.51	9 11 16	1.242	6 0 5	0.061	6 0 7	1.103	8 29 31	8.077	6 11 15	0.399	6 13 48	7.275
510	9 6 57	20.31	6 6 57	1.51	9 8 47	1.241	6 0 31	0.061	6 0 38	1.102	8 27 16	8.123	6 8 47	0.399	6 10 47	7.239
511	9 6 15	20.26	6 6 15	0.47	9 5 40	1.242	5 28 15	0.061	5 27 51	1.106	3 9 20	8.048	0 5 40	0.398	0 6 58	7.206
512	9 5 47	20.29	6 5 47	1.21	9 7 15	1.240	6 0 24	0.061	6 0 30	1.102	8 27 54	8.103	6 7 15	0.399	6 8 54	7.221
513	9 4 22	20.26	6 4 22	0.72	9 5 15	1.240	6 0 4	0.061	6 0 5	1.103	8 29 40	8.070	6 5 15	0.398	6 6 27	7.203
514	9 4 20	20.35	6 4 20	2.02	9 1 50	1.248	5 26 42	0.061	5 25 56	1.108	3 17 10	8.310	0 1 50	0.398	0 2 17	7.184
515	9 4 6	20.31	6 4 6	1.61	9 6 4	1.241	6 1 12	0.061	6 1 29	1.102	8 23 36	8.028	6 6 4	0.398	6 7 28	7.209
516	9 4 6	20.33	6 4 6	1.81	9 2 43	1.243	5 27 43	0.061	5 27 11	1.106	3 12 4	8.147	0 2 44	0.398	0 3 22	7.188
517	9 3 26	20.35	6 3 26	2.04	9 0 55	1.247	5 26 51	0.061	5 26 6	1.106	3 16 24	8.318	0 0 55	0.398	0 1 9	7.183
518	9 3 10	20.28	6 3 10	1.04	9 1 52	1.242	5 28 7	0.061	5 27 41	1.105	3 9 56	8.132	0 1 52	0.398	0 2 19	7.184
519	9 3 3	20.29	6 3 3	1.23	9 4 34	1.240	6 0 56	0.061	6 1 10	1.102	8 25 4	8.138	6 4 34	0.398	6 5 37	7.196
520	8 29 48	20.38	5 29 48	2.25	9 2 34	1.245	6 2 49	0.061	6 3 28	1.103	8 15 33	8.366	6 2 34	0.398	6 3 9	7.187

A TABLE exhibiting the NUMERALS of the ASTRONOMICAL SOCIETY'S CATALOGUE, which correspond with the Stars given in our preceding ZODIACAL CATALOGUE.

Zod. Cat.	Ast. Soc.	Zod. Cat.	Ast. Soc.	Zod. Cat.	Ast. Soc.	Zod. Cat.	Ast. Soc.	Zod. Cat.	Ast. Soc.	Zod. Cat.	Ast. Soc.	Zod. Cat.	Ast. Soc.	Zod. Cat.	Ast. Soc.	Zod. Cat.	Ast. Soc.
1	37	59	402	117	683	175	993	233	1328	291	1681	349	1973	407	2251	465	2566
2	32	60	405	118	684	176	1012	234	1321	292	1691	350	1981	408	2276	466	2568
3	28	61	409	119	688	177	1016	235	1332	293	1682	351	1986	409	2287	467	2575
4	50	62	410	120	706	178	1023	236	1339	294	1690	352	1990	410	2290	468	2576
5	16	63	414	121	709	179	1037	237	1347	295	1711	353	1988	411	2302	469	2586
6	23	64	418	122	722	180	1035	238	1357	296	1721	354	1993	412	2308	470	2562
7	29	65	419	123	729	181	1033	239	1369	297	1716	355	1994	413	2313	471	2585
8	86	66	435	124	731	182	1050	240	1372	298	1746	356	2005	414	2310	472	2579
9	79	67	454	125	737	183	1051	241	1371	299	1744	357	2020	415	2335	473	2600
10	77	68	448	126	745	184	1060	242	1373	300	1740	358	2032	416	2339	474	2613
11	80	69	457	127	748	185	1066	243	1362	301	1750	359	2039	417	2328	475	2627
12	81	70	468	128	750	186	1072	244	1376	302	1755	360	2054	418	2344	476	2622
13	74	71	451	129	749	187	1093	245	1386	303	1759	361	2063	419	2349	477	2632
14	125	72	478	130	760	188	1094	246	1385	304	1764	362	2069	420	2343	478	2614
15	103	73	480	131	768	189	1090	247	1377	305	1778	363	2067	421	2354	479	2630
16	116	74	471	132	774	190	1109	248	1397	306	1787	364	2076	422	2384	480	2653
17	139	75	490	133	775	191	1111	249	1417	307	1781	365	2085	423	2393	481	2660
18	132	76	488	134	777	192	1097	250	1412	308	1780	366	2093	424	2386	482	2656
19	162	77	492	135	790	193	1106	251	1415	309	1811	367	2096	425	2388	483	2672
20	184	78	504	136	799	194	1122	252	1465	310	1820	368	2097	426	2397	484	2655
21	166	79	503	137	801	195	1130	253	1466	311	1807	369	2098	427	2406	485	2686
22	178	80	499	138	804	196	1154	254	1446	312	1800	370	2099	428	2412	486	2661
23	172	81	477	139	817	197	1155	255	1460	313	1826	371	2103	429	2396	487	2688
24	189	82	510	140	822	198	1174	256	1467	314	1810	372	2105	430	2403	488	2690
25	181	83	511	141	820	199	1171	257	1484	315	1808	373	2118	431	2404	489	2671
26	195	84	508	142	832	200	1177	258	1495	316	1806	374	2122	432	2408	490	2724
27	205	85	484	143	831	201	1193	259	1500	317	1812	375	2123	433	2445	491	2726
28	237	86	496	144	828	202	1206	260	1488	318	1823	376	2128	434	2432	492	2701
29	222	87	515	145	850	203	1192	261	1511	319	1818	377	2132	435	2424	493	2725
30	231	88	517	146	870	204	1194	262	1506	320	1836	378	2152	436	2446	494	2714
31	249	89	507	147	872	205	1197	263	1510	321	β^2 Scorp.	379	2157	437	2430	495	2732
32	255	90	516	148	882	206	1209	264	1531	322	1837	380	2162	438	2449	496	2730
33	251	91	502	149	885	207	1214	265	1532	323	1838	381	2163	439	2468	497	2736
34	239	92	520	150	879	208	1207	266	1521	324	1829	382	2164	440	2474	498	2748
35	272	93	505	151	891	209	1222	267	1545	325	1851	383	2174	441	2484	499	2750
36	259	94	523	152	888	210	1205	268	1533	326	1840	384	2175	442	2471	500	2754
37	286	95	528	153	900	211	1232	269	1551	327	1854	385	2176	443	2490	501	2773
38	293	96	536	154	898	212	1210	270	1535	328	1850	386	2180	444	2497	502	2746
39	292	97	537	155	907	213	1252	271	1547	329	1871	387	2179	445	2506	503	2778
40	276	98	543	156	911	214	1237	272	1536	330	1874	388	2178	446	2487	504	2781
41	283	99	550	157	906	215	1242	273	1556	331	1872	389	2181	447	2507	505	2776
42	307	100	555	158	919	216	1233	274	1546	332	1882	390	2183	448	2494	506	2772
43	302	101	559	159	922	217	1254	275	1565	333	1877	391	2185	449	2501	507	2783
44	301	102	575	160	916	218	1259	276	1560	334	1891	392	2187	450	2514	508	2798
45	313	103	585	161	933	219	1279	277	1564	335	1893	393	2208	451	2520	509	2799
46	314	104	592	162	929	220	1256	278	1567	336	1888	394	2205	452	2508	510	2817
47	317	105	593	163	948	221	1284	279	1574	337	1885	395	2220	453	2516	511	2841
48	319	106	595	164	946	222	1277	280	1605	338	1907	396	2224	454	2537	512	2829
49	342	107	594	165	947	223	1270	281	1606	339	1900	397	2227	455	2543	513	2845
50	348	108	623	166	940	224	1275	282	1615	340	1911	398	2229	456	2545	514	2870
51	354	109	640	167	951	225	1303	283	1610	341	1917	399	2230	457	2546	515	2838
52	362	110	651	168	963	226	1310	284	1624	342	1934	400	2241	458	2528	516	2864
53	379	111	655	169	976	227	1295	285	1631	343	1944	401	2255	459	2538	517	2877
54	370	112	652	170	970	228	1302	286	1617	344	1939	402	2236	460	2531	518	2869
55	369	113	642	171	978	229	1309	287	1669	345	1941	403	2256	461	2560	519	2852
56	383	114	663	172	989	230	1313	288	1677	346	1954	404	2250	462	2540	520	2865
57	392	115	667	173	981	231	1338	289	1688	347	1967	405	2248	463	2573		
58	403	116	671	174	998	232	1334	290	1680	348	1974	406	2258	464	2570		

THE PRINCIPAL FIXED STARS THAT MAY BE OCCULTED BY THE MOON,

With the corresponding Places of the ☾'s ☊, when the OCCULTATIONS may be expected.

☾'s ☊		RIGHT ASCENSION.			
		0°.................. 15°	15°........... 30°	30°................ 45°	45°.. 60°
S	°				
XI	27	3. 10, 9. 11, 12, 15.	22.	36. 41, 47. 48,	51. 56, 57, 59. 58. 60. 61. 62. 63. 64. 65.
	24	2. 10. 9. 11, 12, 15.	22.	36. 41, 47. 48.	51. 56, 57, 59. 58. 60. 61. 62. 63. 64. 65.
	21	2. 10. 11, 12, 15.	22.	36. 41. 47. 48.	56, 57, 59. 58. 60. 61. 62. 63. 64. 65.
	18	10. 11. 12. 15.	23, 22.	36. 41. 47. 48.	55, 56, 57, 59. 58. 60. 61. 62. 63. 64. 65.
	15	10. 11. 12.	23,	41. 47. 48.	55, 56, 57, 59. 58. 60. 61. 62. 63. 64. 65.
	12	10. 11. 12.	23,	41. 47. 48.	55, 56, 57, 59. 58. 60. 61. 62. 63. 64. 65.
	9	10. 11. 12.	23, 29,	41. 47. 48.	55, 56, 57, 59. 58. 60. 61. 62. 63. 64. 65.
	6	11. 12.	23, 27, 29,	34, 41. 47. 48.	55, 56. 57, 59. 58. 60. 61. 62. 63. 64. 65. 71,
	3	11. 12.	21, 23, 27, 29,	34, 41. 48.	55, 56. 57, 59. 58. 60. 61. 62. 63. 64. 65. 71,
	0	12.	21, 23, 27, 29,	34, 40, 41. 48.	55, 56. 57, 59. 58. 60. 61. 62. 63. 64. 65. 71,
X	27	7,	21, 23. 25, 27, 29,	30, 34, 40, 41. 48.	55, 56. 57, 59. 58. 60. 61. 62. 63. 64. 65.
	24	7,	21, 23. 25, 27, 29,	30, 34, 40, 41. 48.	55, 56. 57, 59. 58. 60. 61. 62. 63. 64. 65.
	21	7,	21, 23. 25, 27, 29,	30, 34, 40, 41. 48.	55, 56. 57, 59. 58. 60. 61. 62. 63. 64. 65.
	18	7,	21, 23. 25, 27, 29,	30, 34, 40, 41. 48.	55, 56. 57, 59. 58. 60. 61. 62. 63. 64. 65.
	15	7,	21, 23. 25, 27, 29,	30, 34, 40, 41. 48.	55, 56. 57, 59. 58. 60. 61. 62. 63. 64. 65.
	12	7,	21, 23. 25, 27, 29,	30, 34, 40, 41. 48.	55, 56, 57, 59. 58. 60. 61. 62. 63. 64. 65.
	9	6, 7,	21, 23. 25, 27, 29,	30, 34, 40, 41. 48.	55, 56, 57, 59. 58. 60. 61. 62. 63. 64. 65.
	6	6, 7.	21, 23. 25, 27, 29,	30, 34, 40, 41. 48.	55, 56, 57, 59. 58. 60. 61. 62. 63. 64. 65.
	3	6, 7.	21, 23. 25, 27, 29,	30, 34, 40, 41. 48.	55, 56, 57, 59. 58. 60. 61. 62. 63. 64. 65.
	0	6, 7.	21, 23. 25, 27, 29,	30, 34, 40, 41. 48.	55, 56, 57, 59. 58. 60. 61. 62. 63. 64. 65.
IX	27	6, 7.	21, 23. 25, 27, 29,	30, 34, 40, 41. 47. 48.	55, 56, 57, 59. 58. 60. 61. 62. 63. 64. 65.
	24	6, 7.	21, 23. 25, 27, 29,	30, 34, 40, 41. 47. 48.	56, 57, 59. 58. 60, 61. 62. 63. 64. 65.
	21	6, 7.	21, 23. 25, 27, 29,	30, 34, 41. 47. 48.	56, 59, 58. 60, 61, 62. 63. 64. 65.
	18	6, 7. 13,	21, 23. 25, 27, 29,	30, 34, 41. 47. 48.	56, 59, 58, 60, 61, 62. 63. 64. 65.
	15	6, 7. 13,	21, 23. 25, 27, 29,	30, 34, 41. 47. 48.	56, 59, 58, 60, 61, 62, 63, 64, 65,
	12	6, 7. 13,	21, 23. 25, 27, 29,	30, 34, 41. 47. 48.	51. 56, 59, 58, 60, 61, 62, 63, 64, 65,
	9	6, 7. 13,	21, 23. 25, 27, 29,	34, 41. 47. 48.	51. 52. 59, 58, 60, 61, 62, 63, 64, 65, 66.
	6	5, 6, 7.	21, 23. 25, 27, 29,	34, 41. 47. 48,	51. 52. 59, 58, 60, 61, 62, 63, 64, 65, 66.
	3	5, 6, 7.	21, 23. 25, 27, 29,	41. 47. 48,	51. 52. 54. 59, 58, 61, 62, 63, 64, 65, 66.
	0	6, 7.	21, 23. 25, 27, 29,	36. 41, 47. 48,	51. 52. 54. 58, 62, 63, 64, 65, 66. 68.
VIII	27	6, 7.	21, 23. 25, 27, 29,	36. 41, 47, 48,	51. 52. 54. 62, 66. 68.
	24	6, 7.	21, 23. 25, 27, 29,	36. 41, 47, 48,	50. 51. 52. 54. 66. 68.
	21	6, 7.	21, 23.	36. 41, 44. 47, 48,	50. 51, 52. 54. 66. 68.
	18	6, 7.	21, 23,	36. 41, 44. 45. 47,	50. 51, 52, 54. 66, 68.
	15	6, 7.	23,	36. 41, 44. 45. 47,	50. 51, 52, 54. 66, 68.
	12	6, 7.	23,	36. 44. 45. 46. 47,	50. 51, 52, 54, 66, 68,
	9	6, 7.	23,	36, 44. 43. 45. 46.	50. 51, 52, 54, 66, 68,
	6	6, 7.	23,	36, 44. 43. 45. 46. 49.	50, 52, 54, 66, 68,
	3	6, 7,	23, 22.	36, 44, 43. 45. 46. 49.	50, 54, 68,
	0	6, 7,	22.	36, 44, 43. 45. 46. 49.	50, 54, 68, 69.
VII	27	7,	22.	36, 44, 43. 45, 46. 49.	50, 69.
	24	7, 12.	22.	36, 44, 43. 45, 46, 49.	50, 69.
	21	7, 11. 12.	22.	44, 43, 45, 46, 49.	69.
	18	7, 11. 12.	22.	43, 45, 46, 49,	69.
	15	7, 10. 11. 12.	22.	43, 46, 49,	69.
	12	10. 11. 12. 15.	22,	43, 42. 46, 49,	69,
	9	10. 11. 12. 15.	22, 26.	42. 49,	69,
	6	10. 11. 12. 15.	22, 26.	42. 49,	67. 69,
	3	10. 11. 12, 15.	22, 26.	42.	67. 69,
	0	10. 11, 12, 15.	18. 22, 26.	42.	67. 69,
VI	27	10. 9. 11, 12, 15.	18. 26.	42.	67.
	24	10, 9. 11, 12, 15,	18. 26.	42,	67.
	21	10, 9. 11, 12, 15,	18. 24. 26,	35. 42,	67.
	18	10, 9. 11, 15,	18. 24. 26,	35. 39. 42,	67.
	15	10, 9. 15,	18. 24. 26,	35. 39. 42,	67,
	12	9. 15, 16.	18, 24. 26,	35. 39. 42,	67,
	9	3. 9, 16.	18, 24. 26,	31. 35. 39.	67,
	6	3. 9, 16.	18, 24.	31. 33. 35. 39.	67,
	3	3. 9, 16.	18, 24,	31. 33. 35, 39.	67,
	0	3. 9, 16.	18, 19. 24,	31. 33. 35, 39.	67,
☾'s ☊		0h..................... 1h	1h............... 2h	2h..................... 3h	3h.. 4h
		RIGHT ASCENSION.			

The numbers in this Table point out the Stars in the preceding Catalogue.

☾'s ☊		RIGHT ASCENSION.			
S	°	0°.................... 15°	15°......... 30°	30°...................... 45°	45°.. 60°
V	27	3. 9, 16.	19. 24,	28. 31. 33. 35, 37. 39,	67,
	24	3. 16,	19. 24,	28. 31. 33. 35, 37. 39,	67,
	21	3, 16,	19. 24,	28. 31. 33. 35, 37. 39,	67,
	18	3, 2. 16,	19.	28. 31, 33. 35, 37. 39,	67,
	15	3, 2. 16,	19.	28. 31, 33, 37. 39,	53.
	12	3, 2. 16,	17. 19. 20.	28. 31, 33, 37. 39, 38.	53.
	9	2.	17. 19, 20.	28. 31, 33, 37. 39, 38.	53.
	6	2.	17. 19, 20.	28. 31, 33, 37. 39, 38.	53.
	3	2.	14. 17. 19, 20.	28. 31, 33, 37. 38.	53.
	0	2,	14. 17. 19, 20.	28, 31, 33, 32. 37. 38.	53.
IV	27	2,	14. 17. 19, 20.	28, 31, 33, 32. 37, 38.	53.
	24	2,	14. 17. 19, 20.	28, 33, 32. 37, 38.	53.
	21	2,	14. 17. 20.	28, 33, 32. 37, 38.	53.
	18	2,	14. 17, 20.	28, 33, 32. 37, 38.	53.
	15	2,	14. 17, 20,	28, 33, 32. 37, 38.	53.
	12		14. 17, 20,	28, 33, 32. 37, 38.	53.
	9		14. 17, 20,	28, 33, 32. 37, 38.	53. 67,
	6		14, 17, 20,	28, 33, 32. 37, 38.	53. 67,
	3		14, 17, 20,	28, 33, 32. 37, 38.	53. 67,
	0		14, 17, 20,	28, 33, 32. 37, 38.	53. 67,
III	27		14, 17, 20,	28, 33, 32. 37, 38.	53. 67,
	24		14, 17, 20,	28, 33, 32. 37, 38.	67,
	21	8.	14, 17, 20,	28, 33, 32. 37, 38.	67,
	18	8.	14, 17, 20,	28, 33, 32. 37. 38.	67,
	15	8.	14, 17, 20,	28, 31, 33, 32. 37. 38.	67.
	12	8.	14, 17, 20,	28, 31, 33, 32. 37. 39, 38.	67.
	9	8.	14, 17, 20,	28, 31, 33, 32. 37. 39, 38.	67.
	6	1. 8.	14, 17, 20,	28, 31, 33, 37. 39, 38.	67.
	3	1. 8.	14, 17, 20,	28, 31, 33, 37. 39,	67. 69,
	0	1. 8.	14, 17, 20.	28, 31, 33, 35, 37. 39,	67. 69,
II	27		14, 17, 20.	28. 31, 33, 35, 37. 39,	67. 69,
	24		14, 17, 20.	28. 31, 33. 35, 37. 39,	69,
	21		14. 17, 20.	28. 31, 33. 35, 37. 39.	69,
	18		14. 17, 19, 20.	28. 31. 33. 35, 39.	69.
	15		14. 17. 19, 20.	28. 31. 33, 35, 39.	69.
	12		14. 17. 19, 20.	28. 31. 33. 35. 39. 42,	69.
	9		14. 17. 19, 20.	28. 31. 33. 35. 39. 42,	69.
	6		14. 17. 19, 20.	28. 31. 33. 35. 39. 42,	69.
	3		14. 17. 19, 20.	31. 33. 35. 39. 42,	68,
	0		14. 17. 19, 24,	31. 35. 42,	68,
I	27		14. 17. 19. 24,	35. 42. 49,	68,
	24	2,	17. 19. 24,	35. 42. 49,	68,
	21	2,	19. 24,	42. 49,	66, 68,
	18	2, 16,	19. 24,	42. 49,	66, 68.
	15	2, 16,	19. 24. 26,	43, 42. 46, 49.	66, 68.
	12	2, 16,	19. 24. 26,	43, 42. 45, 46, 49.	50, 54, 66, 68.
	9	2, 16,	24. 26,	43, 45, 46, 49.	50, 54, 66, 68.
	6	2. 16,	24. 26,	44, 43, 45, 46, 49.	50, 52, 54, 66. 68.
	3	2. 16.	18, 24. 26.	44, 43, 45, 46, 49.	50, 51, 52, 54, 66. 68.
	0	2. 16.	18, 24. 26.	44, 43. 45, 46. 49.	50, 51, 52, 54, 66.
O	27	3, 2. 16.	18, 26.	44, 43. 45. 46.	50. 51, 52, 54. 62, 64, 65, 66.
	24	3, 2. 9, 16.	18, 26.	36, 44, 43. 45. 46.	50. 51, 52, 54. 59, 58, 62, 63, 64, 65, 66.
	21	3, 2. 9, 16.	18. 26.	36, 44. 43. 45. 46.	50. 51, 52. 54. 59, 58, 61, 62, 63, 64, 65, 66.
	18	3, 2. 9, 15, 16.	18. 22,	36, 44. 43. 45. 46. 47,	50. 51, 52. 54. 59, 58, 60, 61, 62, 63, 64, 65,
	15	3, 9, 15,	18. 22,	36, 44. 43. 45. 46. 47,	50. 51. 52. 54. 59, 58, 60, 61, 62, 63, 64, 65,
	12	3. 9, 15,	18. 22,	36, 44. 45. 47, 48,	50. 51. 52. 54. 59, 58, 60, 61, 62, 63, 64, 65,
	9	3. 10, 9. 15,	18. 22,	36. 41, 44. 47, 48,	51. 52. 59, 58, 60, 61, 62. 63, 64. 65.
	6	3. 10, 9. 15,	18. 22,	36. 41, 44. 47, 48,	51. 52. 56, 59, 58, 60, 61, 62. 63. 64. 65.
	3	3. 10, 9. 11, 12, 15.	22.	36. 41, 44. 47, 48,	51. 52. 56, 59, 58: 60, 61, 62. 63. 64. 65.
	0	3. 10, 9. 11, 12, 15.	22.	36. 41, 47. 48,	51. 56, 57, 59. 58. 60, 61. 62. 63. 64. 65.
☾'s ☊		0h..................... 1h	1h........... 2h	2h......................... 3h	3h.. 4h
		RIGHT ASCENSION.			

The numbers in this Table point out the Stars in the preceding Catalogue.

☾'s ☊	RIGHT ASCENSION.	
	60°.......... 75°	75°.......... 90°
S °		
XI 27	81, 85.	113, 122.
24	81, 85.	113, 122.
21	81, 85.	113, 122.
18	81, 85.	113, 122.
15	81, 85.	113, 122.
12	81, 85.	113, 122.
9	81, 85.	113, 122.
6	81, 85.	113, 122.
3	81, 85.	113, 122.
0	81, 85.	113, 122.
X 27	81, 85.	113, 122.
24	81, 85.	113, 122.
21	81, 85.	113, 122.
18	81, 85.	113, 122,
15	81, 85.	113, 122, 125.
12	81, 85.	113, 122, 125.
9	81, 85.	119. 122, 125.
6	81, 85.	119. 122, 125.
3	85.	119. 122, 125.
0	85.	119. 122, 125.
IX 27	85. 102.	119. 125.
24	85, 102.	119. 121. 125,
21	85, 102.	119. 121. 125,
18	85, 102.	119, 121. 125,
15	85, 102. 107.	116. 119, 121. 125,
12	85, 102. 107.	116. 119, 121. 125, 129. 130.
9	85, 102. 107.	116. 119, 121. 129. 130.
6	93. 102, 107.	116. 119, 121. 129. 130.
3	91. 93. 98. 102, 107.	116. 121, 129. 130.
0	91. 93. 98. 102, 107.	116. 121, 126. 129. 130.
VIII 27	86. 91. 93. 98. 102, 107,	116, 121, 126. 129. 130.
24	86. 91. 93. 98. 102, 107,	116, 121, 126. 129, 130,
21	86. 91. 93. 98. 107,	108. 112. 116, 126. 129, 130,
18	86. 91. 93, 98. 107,	108. 112. 116, 126. 129, 130,
15	86. 91, 93, 98, 103. 107, 106.	108. 112. 116, 126. 129, 130,
12	86. 91, 93, 98, 103. 106.	108. 112. 118. 126, 129, 130,
9	74. 86, 91, 93, 98, 103. 106.	108. 112. 118. 126,
6	74. 86, 91, 93, 98, 103. 106.	108. 112. 118. 123. 126, 128.
3	74. 86, 91, 98, 103. 105. 106.	108, 112, 118. 123. 126, 128.
0	74. 86, 103. 105. 106,	108, 112, 118. 123. 126, 127. 128.
VII 27	74. 86, 103, 105. 106,	108, 112, 118. 123. 124. 127. 128.
24	74. 103, 105. 106,	108, 112, 118, 123. 124. 127. 128.
21	74, 89. 103, 105. 106,	108, 112, 118, 123. 124. 127. 128.
18	74, 89. 103, 105. 106,	118, 123. 124. 127. 128.
15	74, 89. 101. 103, 105.	118, 123, 124. 127. 128,
12	74, 89. 99. 101. 105,	118, 123, 124. 127. 128,
9	74, 89. 99. 101. 104. 105,	114. 115. 118, 123, 124. 127, 128,
6	80. 89. 99. 101. 104. 105,	114. 115. 123, 124, 127, 128,
3	76. 80. 89, 99. 101. 104. 105,	114. 115. 123, 124, 127, 128,
0	76. 77. 80. 89, 99. 101. 104. 105,	114. 115. 123, 124, 127, 128,
VI 27	76. 77. 80. 89, 99. 101. 104. 105,	110. 114. 115. 123, 124, 127, 128,
24	76. 77. 80. 89, 99. 101. 104.	110. 114. 115. 120. 124, 127, 128,
21	76. 75. 77. 80. 89, 99, 101, 104.	110. 114. 115. 120. 124, 127,
18	76. 75. 77. 80. 89, 99, 101, 104.	109. 110. 114. 115. 120. 124, 127,
15	76. 75. 77. 80, 99, 101, 104,	109. 110. 114. 115. 120. 124, 127,
12	76. 75. 77. 80, 84. 95. 99, 101, 104,	109. 110. 111. 114, 115. 120. 124, 127,
9	76, 75. 77, 80, 84. 90. 95. 99, 101, 104,	109. 110. 111. 114, 115, 120. 124, 127,
6	76, 75. 77, 80, 84. 90. 95. 99, 101, 104,	109. 110. 111. 114, 115, 120. 124, 127,
3	72. 76, 75. 77, 80, 84. 82. 83. 90. 95. 99, 101, 104,	109. 110. 111. 114, 115, 117. 120. 124, 127,
0	70. 72. 76, 75. 77, 80, 84. 82. 83. 90. 92. 95. 99, 101, 104,	109. 110. 111. 114, 115, 117. 120. 124, 127,
	4h.......... 5h	5h.......... 6h
☾'s ☊	RIGHT ASCENSION.	

The numbers in this Table point out the Stars in the preceding Catalogue.

☾'s ☊	RIGHT ASCENSION.	
	60°.......... 75°	75°.......... 90°
S V 27°	70. 72. 76, 75. 77, 80, 79. 84. 82. 83. 90. 92. 95. 99, 101, 104,	109. 110. 111. 114, 115, 117. 120. 124, 127,
24	70. 72. 76, 75, 77, 80, 79. 84. 82. 83. 87. 90. 88. 92. 95. 97. 101, 104,	109. 110. 111. 114, 115, 117. 120. 124, 127,
21	70. 72. 76, 75, 77, 80, 79. 84. 82. 83. 87. 90. 88. 92. 95. 97. 101, 104,	109. 110. 111. 114, 115, 117. 120. 124, 127,
18	70. 72. 76, 75, 77, 80, 79. 84. 82. 83. 87. 90. 88. 92. 95. 96. 97. 101, 104,	109. 110. 111. 114, 115, 117. 120. 124, 127,
15	70. 72. 76, 75, 77, 80, 79. 84. 82. 83. 87. 90. 88. 92. 95. 96. 97. 101, 104,	109. 110. 111. 114, 115, 117. 120. 124, 127,
12	70. 72. 73. 76, 75, 77, 80, 79. 84. 82. 83. 87. 90. 88. 92. 95. 96. 97. 100. 101, 104,	109. 110. 111. 114, 115, 117. 120. 124, 127,
9	70. 72. 73. 76, 75, 77, 80, 79. 84. 82. 83. 87. 90. 88. 92. 95. 96. 97. 100. 101, 104,	109. 110. 111. 114, 115, 120. 124, 127,
6	70. 72. 73. 76, 75, 77, 80, 79. 84. 82. 83. 87. 90. 88. 92. 95. 96. 97. 100. 101, 104,	109. 110. 111. 114, 115, 120. 124, 127,
3	70. 72. 73. 76, 75, 77, 80, 79. 84. 82. 83. 87. 90. 88. 92. 95. 96. 97. 101, 104,	109. 110. 111. 114, 115, 120. 124, 127, 128,
0	70. 72. 73. 76, 75, 77, 80, 79. 84. 82. 83. 87. 90. 88. 92. 95. 96. 97. 101, 104,	109. 110. 111. 114, 115. 120. 124, 127, 128,
IV 27	70. 72. 73. 76, 75, 77, 80, 79. 84. 82. 83. 87. 90. 88. 92. 95. 96. 97. 101, 104,	109. 110. 111. 114. 115. 120. 123, 124, 127, 128,
24	70. 72. 73. 76, 75, 77, 80, 79. 84. 82. 83. 87. 90. 88. 92. 95. 97. 99, 101, 104,	109. 110. 114. 115. 120. 123, 124, 127, 128,
21	70. 72. 76, 75, 77, 80, 79. 84. 82. 83. 87. 90. 88. 92. 95. 99, 101, 104,	109. 110. 114. 115. 123, 124, 127, 128,
18	70. 72. 76, 75, 77, 80, 79. 84. 82. 83. 87. 90. 88. 92. 95. 99, 101, 104,	110. 114. 115. 123, 124, 127, 128,
15	70. 72. 76, 75, 77, 80, 79. 84. 82. 83. 90. 92. 95. 99, 101, 104,	110. 114. 115. 123, 124, 127. 128,
12	70. 72. 76, 75. 77, 80, 79. 84. 82. 83. 90, 95. 99, 101, 104.	110. 114. 115. 123, 124. 127. 128,
9	70. 72. 76, 75. 77, 80, 84. 82. 83. 90. 95. 99, 101, 104.	114. 115. 123, 124. 127. 128.
6	70. 72. 76, 75. 77, 80, 84. 82. 90. 95. 99, 101, 104.	114. 115. 118, 123. 124. 127. 128.
3	70. 76, 75. 77, 80, 84. 90. 95. 99, 101, 104. 105,	114. 118, 123. 124. 127. 128.
0	76, 75. 77. 80, 84. 99, 101, 104. 105,	118, 123. 124. 127. 128.
III 27	76. 75. 77. 80, 84. 99, 101, 104. 105,	118, 123. 124. 127. 128.
24	76. 75. 77. 80, 89, 99. 101. 104. 105,	118, 123. 124. 126, 128.
21	76. 75. 77. 80. 89, 99. 101. 104. 105,	118, 123. 126, 128.
18	76. 75. 77. 80. 89, 99. 101. 105,	112, 118. 123. 126, 130,
15	76. 77. 80. 89, 99. 101. 103, 105. 106,	108, 112, 118. 126, 129, 130,
12	76. 77. 80. 89, 99. 101. 103, 105. 106,	108, 112, 118. 126, 129, 130,
9	76. 77. 80. 89, 99. 101. 103, 105. 106,	108, 112, 118. 126. 129, 130,
6	76. 80. 89. 99. 103, 105. 106,	108, 112, 118. 126. 129, 130,
3	80. 89. 103, 105. 106,	108, 112. 118. 126. 129, 130.
0	89. 103. 105. 106.	108. 112. 126. 129. 130.
II 27	74, 89. 103. 106.	108. 112. 116, 121, 126. 129. 130.
24	74, 89. 103. 106.	108. 112. 116, 121, 126. 129. 130.
21	74, 89. 103. 106.	108. 112. 116, 121, 129. 130.
18	74, 103. 106.	108. 116, 121, 129. 130.
15	74, 98, 103. 107, 106.	108. 116. 121, 125, 129.
12	74. 86, 98, 107,	116. 121. 125,
9	74. 86, 91, 98, 107,	116. 119, 121. 125,
6	74. 86, 91, 93, 98, 107,	116. 119, 121. 125,
3	74. 86, 91, 93, 98. 102, 107,	116. 119, 121. 125,
0	74. 86. 91, 93, 98. 102, 107.	116. 119, 121. 125,
I 27	74. 86. 91, 93, 98. 102, 107.	119, 121. 125.
24	86. 91. 93. 98. 102, 107.	119. 125.
21	86. 91. 93. 98. 102, 107.	119. 122, 125.
18	86. 91. 93. 98. 102. 107.	119. 122, 125.
15	86. 91. 93. 102. 107.	119. 122, 125.
12	91. 93. 102.	119. 122, 125.
9	91. 93. 102.	119. 122, 125.
6	93. 102.	119. 122,
3	85, 102.	122.
0	85, 102.	122.
O 27	85,	113, 122.
24	85,	113, 122.
21	85,	113, 122.
18	85,	113, 122.
15	85.	113, 122.
12	85.	113, 122.
9	85.	113, 122.
6	85.	113, 122.
3	81, 85.	113, 122.
0	81, 85.	113, 122.
	4h.......... 5h	5h.......... 6h
☾'s ☊	RIGHT ASCENSION.	

The numbers in this Table point out the Stars in the preceding Catalogue.

☾'s ☊	RIGHT ASCENSION.			
	90°......................... 105°	105°................................ 120°	120°...................... 135°	135°.......................... 150°
S °				
XI 27	132, 139. 140. 142,	148. 151, 156, 159, 161, 164. 168, 169,	176. 184.	
24	132, 139. 140, 142,	148. 151, 156, 159, 161, 164. 168, 169,	176, 184.	195.
21	132, 139. 140, 142,	148. 151, 156, 161, 164. 169,	176, 184.	195.
18	132, 139. 140, 142,	148. 151, 156, 161, 164, 165. 169,	176, 184.	195. 206.
15	132, 139. 140, 142,	148. 151, 161, 164, 165. 169,	176, 184.	195. 206.
12	132, 139, 140, 142,	148. 151, 155. 161, 164, 165. 169,	176, 184.	195. 206.
9	139, 140,	148, 155. 161, 164, 165. 169, 172.	176, 182. 183. 184,	195. 198. 201. 206.
6	139, 140,	148, 155. 164, 165. 169, 172.	176, 179. 182. 183. 184,	195, 198. 201. 206.
3	139, 140,	148, 155. 164, 165. 172.	179. 182. 183. 184,	195, 198. 201. 206.
0	139, 140,	148, 155. 164, 165. 172.	179. 182. 183. 184,	195, 198. 201. 206,
X 27	139, 140, 145.	148, 155. 165. 172.	179. 182. 183. 184,	195, 198. 201. 208. 206,
24	139, 140, 145.	148, 155. 165, 172.	179. 182. 183.	195, 198. 201. 208. 206,
21	139, 145.	148, 155. 165, 172,	179. 182. 183. 185.	195, 198, 200. 201, 208. 206,
18	139, 143. 145.	149. 155, 165, 172,	179, 182, 183, 185.	194. 198, 200. 201, 208. 206,
15	143. 145. 146.	149. 155, 165, 172,	179, 182, 183, 185.	194. 198, 200. 201, 208.
12	143. 145. 146.	149. 155, 165, 172,	179, 182, 183, 185.	194. 198, 200. 201, 208.
9	143. 145. 146.	149. 155, 172,	180. 179, 182, 183, 185.	194. 200. 208,
6	143. 145, 146.	149. 155, 172,	180. 179, 185. 188.	194. 200. 208,
3	143. 145, 146.	149.	177. 180. 185, 187. 188.	194. 200, 208,
0	143. 145, 146.	149,	177. 180. 185, 187. 188.	194, 200, 203. 208,
IX 27	143, 145, 146,	149, 153. 160.	177. 180. 185, 187. 188.	194, 196. 200, 203. 204. 205. 208,
24	143, 145, 146,	149, 153. 160.	177. 178. 180. 185, 187. 188.	194, 196. 199. 200, 203. 204. 205.
21	143, 146,	149, 153. 160.	177. 178. 180, 187. 188.	194, 196. 199. 200, 203. 204. 205.
18	143, 146,	149, 153. 160.	174. 177, 178. 180, 187. 188,	194, 196. 199. 200, 203. 204. 205.
15	143, 146,	153. 160.	174. 177, 178. 180, 187, 188,	196. 199. 203. 204. 205.
12	146,	153, 157. 160.	174. 177, 178. 180, 187, 188,	196. 199. 203, 204. 205.
9	135.	153, 157. 160,	174. 177, 178. 180, 187, 188,	196, 197. 199. 203, 204. 205.
6	133. 135.	153, 157. 160, 167. 171.	174. 177, 178, 187, 188,	196, 197. 199. 203, 204. 205.
3	133. 135. 147.	153, 157. 160, 167. 171.	174. 178, 187,	196, 197. 199, 203, 204, 205,
0	133. 135. 147.	153, 157. 160, 167. 171.	174, 178,	196, 197. 199, 203, 204, 205, 210.
VIII 27	133. 135. 147.	157. 160, 167. 171.	174, 178,	196, 197. 199, 203, 204, 205, 210.
24	133. 135. 147.	157, 167. 171. 173.	174, 178, 192.	196, 197. 199, 203, 204, 205, 210.
21	133. 135, 136. 147.	157, 166. 167. 171. 173.	174, 178, 181. 192.	196, 197. 199, 203, 204, 205, 210.
18	133, 135, 136. 137. 147.	157, 166. 167, 171. 173.	174, 181. 192. 193.	197. 199, 204, 205, 210.
15	133, 135, 136. 137. 138. 147,	157, 166. 167, 171, 173.	174, 181. 189. 192. 193.	197. 199, 204, 205, 210.
12	133, 135, 136. 137. 138. 147,	157, 166. 167, 170. 171, 173.	181. 189. 192. 193.	197, 199, 204, 205, 210.
9	133, 135, 136. 137. 138. 147,	166. 167, 170. 171, 173.	181. 186. 189. 192. 193.	197, 199, 204, 205, 210.
6	133, 136. 137. 138. 147,	166. 167, 170. 171, 173.	181. 186. 189. 192. 193.	197, 199, 204, 205, 210.
3	131. 136, 137. 138. 147,	166. 167, 170. 171, 173,	181. 186. 189. 192. 193.	197, 199, 204, 205, 210.
0	131. 136, 137. 138.	166, 170. 171, 173,	181. 186. 189. 192. 193.	197, 199, 204, 205, 210.
VII 27	131. 134. 136, 137, 138.	166, 170. 171, 173,	181. 186. 189. 192. 193.	197, 199, 204, 205, 210.
24	131. 134. 136, 137, 138,	166, 170. 173,	181, 186. 189. 192. 193.	197, 199, 204, 205, 210.
21	131. 134. 136, 137, 138,	166, 170. 173,	181, 186. 189. 192. 193.	197, 199, 204, 205, 210.
18	131. 134. 136, 137, 138,	154. 162. 166, 170. 173,	181, 186. 189. 192. 193.	197, 199, 204, 205, 210.
15	131. 134. 137, 138, 144.	154. 162. 166, 170. 173,	181, 186. 189. 192. 193.	197, 199, 204, 205, 210.
12	131, 134. 137, 138, 144.	154. 162. 166, 170, 173,	181, 186. 189. 192. 193.	197, 199, 204, 205, 210.
9	131, 134. 138, 144.	154. 162. 166, 170, 173,	181, 186. 189. 192. 193.	197, 199, 204, 205, 210.
6	131, 134. 138, 144.	154. 162. 166, 170, 173,	181, 186. 189. 192. 193.	197, 199, 204, 205, 210.
3	131, 134, 144.	152. 154. 162. 166, 170, 173, 175.	181, 186. 189. 192. 193.	197, 199, 204, 205, 210.
0	131, 134, 144.	152. 154. 162. 166, 170, 173, 175.	181, 186. 189. 192. 193.	197, 199, 203, 204, 205, 210.
VI 27	131, 134, 144.	152. 154. 162. 166, 170, 173, 175.	181, 186. 189. 192. 193.	197, 199, 203, 204, 205, 210.
24	131, 134, 144.	152. 154. 162. 166, 170, 173, 175.	181, 186. 189. 192. 193.	197. 199, 203, 204, 205,
21	131, 134, 144.	152. 154. 162. 166, 170, 173,	181, 186. 189. 192. 193.	197. 199, 203, 204, 205,
18	131, 134, 144. 150.	152. 154. 162. 166, 170, 173,	181, 186. 189. 192. 193.	196, 197. 199, 203, 204, 205.
15	131, 134, 144. 150.	152. 154. 162. 166, 170, 173,	181, 186. 189. 192. 193.	196, 197. 199, 203, 204. 205.
12	131, 134, 144. 150.	152. 154. 162. 166, 170, 173,	181, 186. 189. 192. 193.	196, 197. 199, 203, 204. 205.
9	131, 134, 144. 150.	152. 154. 162. 166, 170, 173,	181. 186. 189. 192.	196, 197. 199. 203. 204. 205.
6	131, 134, 144.	152. 154. 162. 166, 170. 173,	181. 186. 189. 192.	196, 197. 199. 203. 204. 205.
3	134, 144.	152. 154. 162. 166, 170. 173,	181. 192.	196, 197. 199. 203. 204. 205.
0	134, 144.	152. 154. 162. 166, 170. 173,	181. 192.	196, 197. 199. 203. 204. 205. 208,
	6^h........................... 7^h	7^h.................................... 8^h	8^h......................... 9^h	9^h .... 10^h
☾'s ☊	RIGHT ASCENSION.			

The numbers in this Table point out the Stars in the preceding Catalogue.

☾'s ☊		RIGHT ASCENSION.			
		90°............................ 105°	105°...................................... 120°	120°.......................... 135°	135°...................... 150°
S	°				
V	27	134, 144.	154. 162. 166, 170. 171, 173,	181.	196. 199. 200, 203. 204. 208,
	24	131, 134, 144.	154. 162. 166, 170. 171, 173,	181.	196. 199. 200, 203. 208,
	21	131, 134, 144.	154. 166, 170. 171, 173,	181.	196. 199. 200, 203. 208,
	18	131, 134, 144.	154. 166, 170. 171, 173.	181. 187,	196. 200, 208,
	15	131, 134, 144.	154. 166, 170. 171, 173.	181. 187, 188,	196. 200, 208.
	12	131, 134, 144.	166, 167, 170. 171, 173.	174, 178, 187, 188,	196. 200, 208.
	9	131, 134, 144.	166. 167, 170. 171, 173.	174, 178, 187, 188,	194, 200. 201, 208.
	6	131, 134, 144.	166. 167, 171. 173.	174, 178, 187, 188,	194, 200. 201, 208. 206,
	3	131, 134, 138, 144.	166. 167, 171. 173.	174, 178, 187. 188,	194, 198, 200. 201, 208. 206,
	0	131, 134, 138,	166. 167, 171. 173.	174, 178, 187. 188.	194, 198, 200. 201, 208. 206,
IV	27	131, 134. 137, 138,	166. 167, 171. 173.	174. 177, 178. 180, 187. 188.	194. 198, 200. 201, 206,
	24	131, 134. 137, 138, 147,	166. 167. 171.	174. 177, 178. 180, 187. 188.	194. 198, 200. 201. 206,
	21	131, 134. 136, 137, 138, 147,	157, 166. 167. 171.	174. 177, 178. 180, 185, 187. 188.	194. 198, 201. 206.
	18	131, 134. 136, 137, 138, 147,	157, 167. 171.	174. 177, 178. 180, 185, 187. 188.	194. 198. 201. 206.
	15	131. 134. 136, 137, 138, 147,	157, 167.	174. 177, 178. 180, 185, 188.	194. 198. 201. 206.
	12	131. 134. 136, 137, 138. 147,	157, 167.	174. 177. 178. 180. 185,	194. 198. 201. 206.
	9	131. 134. 136, 137. 138. 147,	157, 160, 167.	174. 177. 180. 185,	198. 201. 206.
	6	131. 134. 136, 137. 138. 147.	157. 160,	177. 180. 185.	195, 198. 206.
	3	131. 136. 137. 138. 147.	153, 157. 160,	177. 180. 182, 183, 185.	195, 198.
	0	131. 136. 137. 138. 147.	153, 157. 160,	177. 180. 179, 182, 183, 185.	195,
III	27	131. 133, 135, 136. 137. 138. 147.	153, 157. 160,	177. 180. 179, 182, 183, 185.	195,
	24	133, 135, 136. 137. 138. 147.	153, 157. 160.	179, 182, 183, 185.	195.
	21	133, 135, 136. 137. 147.	153, 157. 160.	179, 182, 183, 185.	195.
	18	133, 135, 136.	153. 160.	179, 182. 183.	195.
	15	133, 135,	153. 160. 172,	179. 182. 183. 184,	195.
	12	133. 135.	153. 160. 172,	179. 182. 183. 184,	195. 202,
	9	133. 135. 146,	149, 153. 160. 172,	179. 182. 183. 184,	195. 202,
	6	133. 135. 146,	149, 153. 172,	179. 182. 183. 184,	195. 202,
	3	133. 135. 146,	149, 153. 172,	179. 182. 183. 184,	202,
	0	133. 135. 143, 146,	149, 165, 172.	179. 182. 184,	202,
II	27	133. 135. 143, 145, 146,	149, 155, 165, 172.	184.	202,
	24	143, 145, 146.	149. 155, 165, 172.	176, 184.	202,
	21	143, 145, 146.	149. 155, 165, 172.	176, 184.	202,
	18	143, 145, 146.	149. 155, 165, 172.	176, 184. 190,	191, 202,
	15	143. 145, 146.	149. 155, 165. 172.	176, 184. 190,	191, 202.
	12	143. 145, 146.	149. 155. 164, 165. 169, 172.	176, 184. 190,	191, 202.
	9	143. 145. 146.	149. 155. 164, 165. 169,	176, 184. 190,	191, 202.
	6	143. 145.	148, 155. 164, 165. 169,	176, 184. 190,	191, 202.
	3	143. 145.	148, 155. 164, 165. 169,	176. 190,	191, 202.
	0	143. 145.	148, 155. 161, 164, 165. 169,	176. 190,	191, 202.
I	27	143. 145.	148, 155. 161, 164, 165. 169,	176. 190,	191, 202.
	24	145.	148, 155. 161, 164, 165. 169,	176. 190,	191, 202.
	21	139,	148, 155. 161, 164. 168, 169,	176. 190,	191, 202.
	18	139, 140,	148, 151, 161, 164. 168, 169.	176. 190,	191, 202.
	15	139, 140,	148. 151, 156, 161, 164. 168, 169.	176. 190,	191, 202.
	12	139, 140,	148. 151, 156, 159, 161, 164. 168, 169.	176. 190,	191, 202.
	9	139, 140,	148. 151, 156, 159, 161, 164. 168, 169.	176. 190,	191, 202.
	6	139, 140,	148. 151, 156, 159, 161, 164. 168, 169.	176. 190,	191, 202,
	3	139, 140, 142,	148. 151, 156, 158, 159, 161, 164. 168, 169.	176. 190,	191, 202,
	0	139, 140, 142,	148. 151, 156, 158, 159, 161, 164. 168, 169.	176. 190,	191, 202,
O	27	139. 140, 142,	148. 151, 156, 158, 159, 161, 164. 168, 169.	176. 190,	191, 202,
	24	139. 140, 142,	148. 151, 156, 158, 159, 161, 164. 168, 169.	176. 190,	191, 202,
	21	132, 139. 140, 142,	148. 151, 156, 158, 159, 161. 164. 168, 169.	176. 190,	191, 202,
	18	132, 139. 140, 142,	148. 151, 156, 158, 159, 161. 164. 168, 169.	176. 190,	191, 202,
	15	132, 139. 140. 142,	148. 151, 156, 158, 159, 161, 164. 168, 169.	176. 190,	191, 202,
	12	132, 139. 140. 142,	148. 151, 156, 158, 159, 161, 164. 168, 169.	176. 190,	191, 202,
	9	132, 139. 140. 142,	148. 151, 156, 158, 159, 161, 164. 168, 169.	176. 190,	191, 202,
	6	132, 139. 140. 142,	148. 151, 156, 158, 159, 161, 164. 168, 169.	176. 184. 190,	191,
	3	132, 139. 140. 142,	148. 151, 156, 158, 159, 161, 164. 168, 169.	176. 184.	
	0	132, 139. 140. 142,	148. 151, 156, 158, 159, 161, 164. 168, 169.	176. 184.	
☾'s ☊		6h................................ 7h	7h 8h	8h 9h	9h 10h
		RIGHT ASCENSION.			

The numbers in this Table point out the Stars in the preceding Catalogue.

RIGHT ASCENSION.

☾'s ☊		150°.........165°	165°.........180°	180°.........195°	195°.........210°	210°.........225°
S	°					
V	27	216. 214, 220. 222, 228. 225,	237, 238, 244,	250, 251, 257, 258,	261, 267, 273. 280, 281,	282. 285. 289,
	24	214, 220. 222. 225,	237. 238, 244,	250, 251, 257, 258, 259,	261. 265, 267, 273. 280, 281,	282. 285. 289, 292,
	21	214, 220. 218, 222. 225,	237. 238, 244,	250. 251, 252, 253, 257, 258, 259,	261. 264, 265, 267, 273. 280, 281,	282. 285. 289, 292,
	18	214, 217, 220. 218, 222. 225,	237. 238. 244.	250. 251. 252, 253, 257. 258, 259,	261. 264, 265, 267, 273. 280, 281.	282. 285. 289, 292,
	15	214. 215, 217, 220. 218, 222. 225,	237. 238. 244.	250. 251. 252, 253, 257. 258. 259,	261. 264, 265, 267, 269, 280. 281.	282. 289, 292,
	12	214. 215, 217, 220. 218, 222. 225. 226,	232, 237. 238. 244.	248, 250. 251. 252, 253, 257. 258. 259,	261. 264, 265, 267. 269, 280. 281.	282. 289, 292,
	9	214. 215, 217, 218. 222. 225. 226,	232, 238. 244.	248, 250. 251. 252, 253, 257. 258. 259.	261. 264, 265, 267. 269, 280. 281.	282. 289, 292,
	6	214. 215, 217, 218. 225. 226,	232, 238. 244.	248, 250. 251. 252. 253. 257. 258. 259.	264. 265. 267. 269, 280. 281.	289. 292,
	3	214. 215. 217. 218. 225. 226,	232, 238. 244. 246,	248, 251. 252. 253. 257. 258. 259.	264. 265. 267. 269, 280. 281.	289. 292,
	0	214. 215. 217. 218. 225. 226,	232, 246,	248, 252. 253. 258. 259.	264. 265. 267. 269, 280. 281.	289. 292,
IV	27	215. 217. 218. 226.	232. 246,	248, 252. 253. 258. 259.	264. 265. 267. 269. 280. 281.	289. 292,
	24	215. 217. 218. 226.	232. 246,	248. 252. 253. 259.	264. 265. 267. 269. 280.	289. 292,
	21	215. 217. 221, 226.	232. 246,	248. 252. 253. 259.	264. 265. 269. 280.	289. 292,
	18	207, 215. 217. 221, 226.	232. 246,	248. 252. 253.	264. 265. 269. 280.	289. 292,
	15	207, 221, 226.	232. 246.	248. 252. 253.	264. 265. 269. 280.	289. 292,
	12	207, 209, 221, 226.	232. 241, 246.	248. 249,	264. 265. 269. 280.	289. 292.
	9	207, 209, 221,	241, 246.	248. 249,	264. 269. 280.	289. 292,
	6	207, 209, 221.	241, 246.	249,	269. 280.	289. 292,
	3	207. 209, 221.	241, 246.	249,	269. 280.	289. 292,
	0	207. 209, 221.	241, 246.	249,	269. 280.	289. 292,
III	27	207. 209. 221.	241, 246.	249,	269. 280.	289. 292,
	24	207. 209. 221.	241, 246.	249,	269. 280.	289. 292,
	21	207. 209. 211, 213, 221..	241,	249,	269. 280.	289. 292,
	18	207. 209. 211, 213, 221.	241.	249,	269. 280.	289. 292,
	15	209. 211, 213, 221.	241.	249,	269. 280.	289, 292,
	12	209. 211, 213,	241. 245,	249,	269. 280.	289, 292,
	9	209. 211, 213,	239, 241. 245,	249.	269. 280.	289, 292,
	6	209. 211, 213,	231, 239, 241. 245,	249.	269. 280.	289, 292,
	3	211, 213,	231, 239, 241. 245,	249.	269. 280. 281.	289, 292,
	0	211. 213. 219,	231, 239, 241. 245,	249.	264. 269. 280. 281.	289,
II	27	211. 213. 219,	231, 239, 241. 240, 245,	249.	264. 265. 269. 280. 281.	282. 289,
	24	211. 213. 219,	231, 239, 241. 240, 245,	249.	264. 265. 269. 280. 281.	282. 289,
	21	211. 213. 219,	231, 239, 241. 240, 245,	249,	264. 265. 267. 269. 280. 281.	282. 285. 288. 289,
	18	211. 213. 219,	231, 239, 241. 240, 245,	249,	264. 265. 267. 269. 280. 281.	282. 285. 288.
	15	211. 213. 219,	231, 239, 241. 245,	249,	264. 265. 267. 269, 280. 281.	282. 285. 288. 295.
	12	211. 213. 219,	231, 239, 241. 245,	249,	264. 265. 267. 269, 280, 281.	282. 285. 288. 295.
	9	211. 213. 219,	231, 239, 241. 245,	249, 259.	264. 265. 267. 269, 280, 281,	282. 285. 288. 295.
	6	211. 213. 219,	231, 239, 241.	249, 259.	264. 265. 267. 269, 280, 281,	282. 285, 288. 295.
	3	211 213. 219,	231, 239, 241.	249, 259.	264. 265. 267. 269, 280, 281,	282, 285, 287. 288, 295.
	0	211. 213. 219,	231, 241. 246.	249, 252. 253. 259.	264, 265, 267. 269, 273. 280, 281,	282, 285, 287. 288, 290. 291. 295.
I	27	211. 213. 219,	231, 241. 246.	249, 252. 253. 258. 259.	264, 265, 267, 269, 273. 280, 281,	282, 285, 287. 288, 290. 291. 295,
	24	211. 213. 219,	241. 246.	249, 252. 253. 258. 259.	264, 265, 267, 273. 281,	282, 284. 285, 287. 288, 290. 291. 295,
	21	211. 213. 219,	241, 246.	248. 249, 252. 253. 257. 258. 259.	261. 264, 265, 267, 273.	282, 284. 285, 287. 288, 290. 291. 295,
	18	211. 213. 219,	241, 246.	248. 249, 252. 253. 257. 258. 259.	261. 264, 265, 267, 273.	284. 285, 287. 290. 291. 295,
	15	211. 213. 219,	241, 246.	248. 252. 253. 257. 258. 259,	261. 265, 267, 273.	284. 287, 290, 291, 293. 295,
	12	211. 213. 219,	241, 246.	248. 252. 253. 257. 258. 259,	261. 267, 273,	284. 287, 290, 291, 293. 295,
	9	211. 213.	241, 246.	248. 252, 253, 257. 258. 259,	261. 273,	284, 287, 290, 291, 293.
	6	211. 213.	241, 246.	248. 252, 253, 257. 258, 259,	261. 273,	284, 287, 290, 291, 293.
	3	211. 213,	241, 246,	248. 251. 252, 253, 257, 258, 259,	261, 271. 273,	284, 287, 290, 291, 293.
	0	211. 213, 221.	246,	248. 250. 251. 252, 253, 257, 258,	261, 271. 273,	284, 293.
O	27	211, 213, 221.	246,	248, 250. 251. 252, 253, 257, 258,	261, 271. 275.	284, 293,
	24	211, 213, 221.	246,	248, 250. 251. 252, 253, 257, 258,	261, 271. 275.	293,
	21	209. 211, 213, 221.	232. 216,	248, 250. 251. 257,	261, 271. 275.	293,
	18	209. 211, 213, 221.	232. 244.	248, 250. 251. 257,	271. 275.	293,
	15	209. 211, 213, 221.	232. 244.	248, 250. 251,	271, 275.	283. 293,
	12	209. 211, 213, 221. 226.	232. 244.	250, 251,	268. 271, 275.	283.
	9	207. 209. 211, 213, 221. 226.	232. 238. 244.	250, 251,	266. 268. 271, 275,	283. 294.
	6	207. 209. 213, 221, 226.	232. 238. 244.	250, 251,	266. 268. 271, 275,	283. 294.
	3	207. 209. 221, 226.	232, 238. 244.	250, 251,	263. 266. 268. 271, 275,	283. 294.
	0	207. 209, 221, 226.	232, 238. 244,	250,	263. 266. 268. 270. 275,	283. 286. 294.
☾'s ☊		10^h.........11^h	11^h.........12^h	12^h.........13^h	13^h.........14^h	14^h.........15^h

RIGHT ASCENSION.

The numbers in this Table point out the Stars in the preceding Catalogue.

RIGHT ASCENSION.

☾'s ☊		150°........................165°	165°...........180°	180°.........................195°	195°.................... 210°	210°.........................225
S	°					
V	27	216. 214, 220. 222, 228. 225,	237, 238, 244,	250, 251, 257, 258,	261, 267, 273. 280, 281,	282. 285. 289,
	24	214, 220. 222. 225,	237. 238, 244,	250, 251, 257, 258, 259,	261. 265, 267, 273. 280, 281,	282. 285. 289, 292,
	21	214, 220. 218, 222. 225,	237. 238, 244,	250. 251, 252, 253, 257, 258, 259,	261. 264, 265, 267, 273. 280, 281,	282. 285. 289, 292,
	18	214, 217, 220. 218, 222. 225,	237. 238. 244.	250. 251. 252, 253, 257. 258, 259,	261. 264, 265, 267, 273. 280, 281.	282. 285. 289, 292,
	15	214. 215, 217, 220. 218, 222. 225,	237. 238. 244.	250. 251. 252, 253, 257. 258. 259,	261. 264, 265, 267, 269, 280. 281.	282. 289, 292,
	12	214. 215, 217, 220. 218, 222. 225. 226,	232, 237. 238. 244.	248, 250. 251. 252, 253, 257. 258. 259,	261. 264, 265, 267. 269, 280. 281.	282. 289, 292,
	9	214. 215, 217, 218. 222. 225. 226,	232, 238. 244.	248, 250. 251. 252, 253, 257. 258. 259.	261. 264, 265, 267. 269, 280. 281.	282. 289, 292,
	6	214. 215, 217, 218. 225. 226,	232, 238. 244.	248, 250. 251. 252. 253. 257. 258. 259.	264. 265. 267. 269, 280. 281.	289. 292,
	3	214. 215. 217. 218. 225. 226,	232, 238. 244. 246,	248, 251. 252. 253. 257. 258. 259.	264. 265. 267. 269, 280. 281.	289. 292,
	0	214. 215. 217. 218. 225. 226,	232, 246,	248, 252. 253. 258. 259.	264. 265. 267. 269, 280. 281.	289. 292,
IV	27	215. 217. 218. 226.	232. 246,	248, 252. 253. 258. 259.	264. 265. 267. 269. 280. 281.	289. 292,
	24	215. 217. 218. 226.	232. 246,	248. 252. 253. 259.	264. 265. 267. 269. 280.	289. 292,
	21	215. 217. 221, 226.	232. 246,	248. 252. 253. 259.	264. 265. 269. 280.	289. 292,
	18	207, 215. 217. 221, 226.	232. 246,	248. 252. 253.	264. 265. 269. 280.	289. 292,
	15	207, 221, 226.	232. 246.	248. 252. 253.	264. 265. 269. 280.	289. 292,
	12	207, 209, 221, 226.	232. 241, 246.	248. 249,	264. 265. 269. 280.	289. 292.
	9	207, 209, 221,	241, 246.	248. 249,	264. 269. 280.	289. 292,
	6	207, 209, 221.	241, 246.	249,	269. 280.	289. 292,
	3	207. 209, 221.	241, 246.	249,	269. 280.	289. 292,
	0	207. 209, 221.	241, 246.	249,	269. 280.	289. 292,
III	27	207. 209. 221.	241, 246.	249,	269. 280.	289. 292,
	24	207. 209. 221.	241, 246.	249,	269. 280.	289. 292,
	21	207. 209. 211, 213, 221..	241,	249,	269. 280.	289. 292,
	18	207. 209. 211, 213, 221.	241.	249,	269. 280.	289. 292,
	15	209. 211, 213, 221.	241.	249,	269. 280.	289, 292,
	12	209. 211, 213,	241. 245,	249,	269. 280.	289, 292,
	9	209. 211, 213,	239, 241. 245,	249.	269. 280.	289, 292,
	6	209. 211, 213,	231, 239, 241. 245,	249.	269. 280.	289, 292,
	3	211, 213,	231, 239, 241. 245,	249.	269. 280. 281.	289, 292,
	0	211. 213. 219,	231, 239, 241. 245,	249.	264. 269. 280. 281.	289,
II	27	211. 213. 219,	231, 239, 241. 240, 245,	249.	264. 265. 269. 280. 281.	282. 289,
	24	211. 213. 219,	231, 239, 241. 240, 245,	249.	264. 265. 269. 280. 281.	282. 289,
	21	211. 213. 219,	231, 239, 241. 240, 245,	249,	264. 265. 267. 269. 280. 281.	282. 285. 288. 289,
	18	211. 213. 219,	231, 239, 241. 240, 245,	249,	264. 265. 267. 269. 280. 281.	282. 285. 288.
	15	211. 213. 219,	231, 239, 241. 245,	249,	264. 265. 267. 269, 280. 281.	282. 285. 288. 295.
	12	211. 213. 219,	231, 239, 241. 245,	249,	264. 265. 267. 269, 280, 281.	282. 285. 288. 295.
	9	211. 213. 219,	231, 239, 241. 245,	249, 259.	264. 265. 267. 269, 280, 281,	282. 285. 288. 295.
	6	211. 213. 219,	231, 239, 241.	249, 259.	264. 265. 267. 269, 280, 281,	282. 285, 288. 295.
	3	211 213. 219,	231, 239, 241.	249, 259.	264. 265. 267. 269, 280, 281,	282, 285, 287. 288, 295.
	0	211. 213. 219,	231, 241. 246.	249, 252. 253. 259.	264, 265, 267. 269, 273. 280, 281,	282, 285, 287. 288, 290. 291. 295.
I	27	211. 213. 219,	231, 241. 246.	249, 252. 253. 258. 259.	264, 265, 267, 269, 273. 280, 281,	282, 285, 287. 288, 290. 291. 295,
	24	211. 213. 219,	241. 246.	249, 252. 253. 258. 259.	264, 265, 267, 273. 281,	282, 284. 285, 287. 288, 290. 291. 2
	21	211. 213. 219,	241, 246.	248. 249, 252. 253. 257. 258. 259.	261. 264, 265, 267, 273.	282, 284. 285, 287. 288, 290. 291. 2
	18	211. 213. 219,	241, 246.	248. 249, 252. 253. 257. 258. 259.	261. 264, 265, 267, 273.	284. 285, 287. 290. 291. 295,
	15	211. 213. 219,	241, 246.	248. 252. 253. 257. 258. 259,	261. 265, 267, 273.	284. 287, 290, 291, 293. 295,
	12	211. 213. 219,	241, 246.	248. 252. 253. 257. 258. 259,	261. 267, 273,	284. 287, 290, 291, 293. 295,
	9	211. 213.	241, 246.	248. 252, 253, 257. 258. 259,	261. 273,	284, 287, 290, 291, 293.
	6	211. 213.	241, 246.	248. 252, 253, 257. 258, 259,	261. 273,	284, 287, 290, 291, 293.
	3	211. 213,	241, 246,	248. 251. 252, 253, 257, 258, 259,	261, 271. 273,	284, 287, 290, 291, 293.
	0	211. 213, 221.	246,	248. 250. 251. 252, 253, 257, 258,	261, 271. 273,	284, 293.
O	27	211, 213, 221.	246,	248, 250. 251. 252, 253, 257, 258,	261, 271. 275.	284, 293,
	24	211, 213, 221.	246,	248, 250. 251. 252, 253, 257, 258,	261, 271. 275.	293,
	21	209. 211, 213, 221.	232. 246,	248, 250. 251. 257,	261, 271. 275.	293,
	18	209. 211, 213, 221.	232. 244.	248, 250. 251. 257,	271. 275.	293,
	15	209. 211, 213, 221.	232. 244.	248, 250. 251,	271, 275.	283. 293,
	12	209. 211, 213, 221. 226.	232. 244.	250, 251,	268. 271, 275.	283.
	9	207. 209. 211, 213, 221. 226.	232. 238. 244.	250, 251,	266. 268. 271, 275,	283. 294.
	6	207. 209. 213, 221, 226.	232. 238. 244.	250, 251,	266. 268. 271, 275,	283. 294.
	3	207. 209. 221, 226.	232, 238. 244.	250, 251,	263. 266. 268. 271, 275,	283. 294.
	0	207. 209, 221, 226.	232, 238. 244,	250,	263. 266. 268. 270. 275,	283. 286. 294.
☾'s ☊		10h.............................11h	11h..............12h	12h............................13h	13h...................... 14h	14h.............................15

RIGHT ASCENSION.

The numbers in this Table point out the Stars in the preceding Catalogue.

☾'s ☊	RIGHT ASCENSION.		
	225°.. 240°	240°.. 255°	255°.. 270°
S °			
XI 27	308, 312. 316. 315, 314, 317. 319. 324. 326.	331, 337, 339.	353, 359, 364. 365,
24	308, 312. 316. 315, 314, 317, 319. 324. 326.	331, 337, 339.	353, 359, 364. 365,
21	308, 312. 316, 315, 314, 317, 319. 324, 326.	331, 337, 339.	353, 355. 359, 364. 365,
18	308, 312. 316, 315, 314, 317, 319. 324, 326.	331, 337, 339.	353, 355. 359, 364. 365,
15	308, 312. 316, 315, 314, 317, 319. 324, 326.	331, 337, 339.	353, 359, 364. 365,
12	308, 312. 316, 315, 317, 319. 324, 326,	331, 337, 339.	353, 359, 364. 365,
9	308, 312. 316, 315, 317, 319. 324, 326,	331, 337, 339.	353, 359, 364. 365,
6	308, 312. 316, 315, 317, 319. 324, 326,	331, 337, 339.	353, 359, 365,
3	308, 312. 316, 315, 317, 319. 324, 326,	328. 331, 337, 339.	353, 359, 365.
0	308, 312. 316, 315, 317, 319. 324, 326,	331, 337, 339.	353, 359, 365.
X 27	308, 312. 316, 315, 317, 319. 324, 326,	331, 337, 339.	353, 359, 365.
24	308, 312. 316, 315, 317, 319. 324, 326,	331, 337, 339.	347, 353. 359, 365.
21	308, 312. 316, 315, 317, 319. 324, 326.	331, 337, 339.	347, 353. 359, 365.
18	308, 312. 316, 315, 317, 319. 324, 326.	331, 337, 339.	347, 353. 359. 365.
15	308, 312. 316, 315, 317, 319. 324, 326.	331, 337,	347, 353. 359. 365.
12	308, 312. 316, 315, 314, 317, 319. 324, 326.	331, 337,	347, 353. 359.
9	308, 312. 316, 315, 314, 317, 319. 324. 326.	331, 337, 336, 340,	347, 353. 359.
6	308, 312. 316. 315, 314, 317, 319. 324. 326.	331, 337. 336, 340,	347, 353. 359.
3	308, 312. 316. 315, 314, 317. 319. 324. 326.	331, 337. 336, 340,	347, 359.
0	308, 312. 316. 315, 314, 317. 319. 324. 326.	331, 337. 336, 340,	347,
IX 27	308, 312. 316. 315, 314, 317. 319. 324. 326.	331, 337. 336, 340, 344, 345,	347, 351,
24	308, 312. 316. 315, 314, 317. 319. 324. 326.	329, 331. 337. 336, 340, 341, 344, 345,	347. 351,
21	308, 316. 315, 314, 317. 324.	329, 331. 337. 336, 340, 341, 344, 345,	347. 351, 363,
18	308, 316. 315. 314, 317. 324.	329, 331. 337. 336, 340, 341, 344, 345,	347. 351, 363,
15	308. 316. 315. 314, 317.	329, 331. 336, 340. 341, 344, 345,	347. 349, 348, 351, 354, 360, 363,
12	308. 315. 314, 317.	329, 331. 333, 336. 340. 341, 344, 345,	347. 349, 348, 351, 354, 356, 360, 363,
9	308. 315. 314. 318,	329, 331. 333, 336. 340. 341, 344. 345,	349, 348, 351, 354, 356, 360, 363,
6	308. 315. 314. 318,	329, 333, 336. 340, 341, 344. 345.	349, 348, 351. 354, 356, 360, 363,
3	308. 315. 314. 318,	329. 333, 336. 340. 341. 344. 345.	349, 348, 351. 354, 356, 360, 361, 363.
0	308. 314. 318,	329. 333, 336. 341. 344. 345.	349, 348, 351. 354, 356, 360, 361, 363.
VIII 27	314. 318,	329. 333, 341. 344. 345.	349. 348. 351. 354. 356, 360. 361, 363.
24	296, 318,	329. 333. 341. 345.	349. 348. 351. 354. 356. 360. 361, 363.
21	296, 318.	329. 333. 341.	349. 348. 354. 356. 357, 360. 361,
18	296, 318.	327, 333.	349. 348. 354. 356. 357, 358, 360. 361,
15	296, 318. 323,	327, 333. 335, 346,	348. 352, 354. 356. 357, 358, 361.
12	296, 311, 318. 322, 323,	327, 333. 335, 346,	352, 357, 358, 361. 362,
9	296. 307, 311, 318. 322, 323,	327, 335, 346,	350, 352, 357, 358, 361. 362,
6	296. 305, 307, 311, 322, 323,	327, 335, 346,	350, 352, 357, 358, 361. 362,
3	296. 300, 305, 307, 311, 320, 321, 322, 323,	327. 330, 335, 346,	350, 352, 357. 358, 361. 362,
0	296. 300, 305, 307, 311, 320, 321, 322, 323,	327. 330, 335. 346,	350, 352, 357. 358. 362,
VII 27	296. 300, 305, 307, 311, 320, 321, 322. 323.	325, 327. 330, 335. 346.	350, 352. 357. 358. 362,
24	296. 300, 305, 307. 311. 320, 321, 322. 323.	325, 327. 330, 335. 346.	350, 352. 357. 358. 362,
21	300, 305. 307. 311. 320, 321, 322. 323.	325, 327. 330, 335. 346.	350. 352. 357. 358. 362.
18	300, 305. 307. 311. 320. 321. 322. 323.	325, 330, 335. 346.	350. 352. 358. 362.
15	300. 305. 307. 311. 320. 321. 322. 323.	325, 330. 346.	350. 352. 358. 362.
12	300. 301, 303, 305. 307. 311. 320. 321.	325. 330. 332, 343,	350. 352. 358. 362.
9	300. 301, 302, 303, 305. 320. 321.	325. 330. 332, 343,	350. 362.
6	300. 301, 302, 303, 320. 321.	325. 330. 332, 343,	362.
3	300. 301, 302, 303, 309, 320. 321.	325. 330. 332, 343,	
0	301, 302, 303, 309,	325. 330. 332, 338, 343,	
VI 27	299, 301. 302, 303, 309, 313,	325. 332, 338, 343,	
24	299, 301. 302. 303. 306, 309, 313,	332. 338, 343,	
21	299, 301. 302. 303. 306, 309, 313,	332. 338, 343.	
18	299, 301. 302. 303. 306, 309, 313,	332. 338, 343.	
15	299, 301. 302. 303. 304, 306, 309, 313,	332. 334, 338, 343.	
12	299, 301. 302. 303. 304, 306, 309. 313,	332. 334, 338, 343.	
9	299. 301. 302. 303. 304, 306, 309. 313,	332. 334, 338, 342, 343.	
6	299. 302. 304, 306, 309. 313,	332. 334, 338, 342, 343.	
3	299. 304, 306, 309. 313.	334, 338. 342, 343.	
0	299. 304, 306, 309. 313.	334, 338. 342, 343.	
☾'s ☊	15h.. 16h	16h.. 17h	17h.. 18h
	RIGHT ASCENSION.		

The numbers in this Table point out the Stars in the preceding Catalogue.

☾'s ☊	RIGHT ASCENSION.		
S °	225°…………………… 240°	240°…………………… 255°	255°…………………… 270°
V 27	299. 304, 306. 309. 313.	334, 338. 342, 343.	
24	299. 304, 306. 309. 313.	334, 338. 342, 343.	
21	299. 304. 306. 309. 313.	334, 338. 342, 343.	
18	299. 304. 306. 310, 313.	334, 338. 342, 343.	
15	304. 306. 310, 313.	334, 338. 342, 343.	
12	304. 306. 310, 313.	334, 338. 342, 343.	
9	298, 304. 306. 310, 313.	334, 338. 342, 343.	
6	298, 304. 306. 310, 313.	334, 338. 342, 343.	
3	298, 304. 306. 310, 313.	334, 338. 342, 343.	
0	298, 304. 306. 310, 313.	334, 338. 342, 343.	
IV 27	298, 304. 306. 310, 313.	334, 338. 342, 343.	
24	298, 304. 306. 310, 313.	334, 338. 342, 343.	
21	298, 304. 306. 310, 313.	334, 338. 342, 343.	
18	298, 304. 306. 310, 313.	334, 338. 342, 343.	362.
15	298, 304. 306. 310, 313.	334, 338. 342, 343.	362.
12	298, 304. 306. 310, 313.	334, 338, 343.	362.
9	298, 304. 306. 310, 313.	334, 338, 343.	362.
6	298, 304. 306. 313.	332. 334, 338, 343,	362.
3	298, 304. 306. 309. 313.	332. 334, 338, 343,	362.
0	298, 304. 306. 309. 313.	332. 334, 338, 343,	350. 358. 362,
III 27	304. 306. 309. 313.	332. 334, 338, 343,	350. 358. 362,
24	299. 304. 306. 309. 313.	332. 338, 343,	350. 352. 357. 358. 362,
21	299. 304, 306. 309. 313,	332. 338, 343,	350. 352. 357. 358. 361. 362,
18	299. 304, 306, 309. 313,	332. 338, 343,	350. 352. 357. 358. 361. 362,
15	299. 304, 306, 309. 313,	332, 343, 346.	350. 352. 357. 358, 361. 362,
12	299. 304, 306, 309. 313,	332, 346.	350, 352. 357. 358, 361. 362,
9	299. 304, 306, 309, 313,	330. 332, 346.	350, 352, 357, 358, 360. 361,
6	299. 302. 303. 304, 306, 309, 313,	325. 330. 332, 346.	350, 352, 357, 358, 360. 361,
3	299. 301. 302. 303. 304, 306, 309, 313,	325. 330. 332, 346.	350, 352, 357, 358, 360. 361, 363.
0	299, 301. 302. 303. 304, 306, 309, 313,	325. 330. 332, 346,	350, 352, 357, 358, 360. 361, 363.
II 27	299, 301. 302. 303. 306, 309, 320. 321.	325. 330. 332, 335. 346,	350, 352, 356. 357, 360. 361, 363.
24	299, 301. 302. 303. 309, 320. 321.	325. 330, 335. 346,	352, 354. 356. 357, 360, 361, 363.
21	299, 301. 302. 303. 320. 321.	325, 330, 335. 346,	348. 354. 356. 360, 363,
18	299, 301. 302, 303, 320. 321. 322. 323.	325, 330, 335. 346,	349. 348. 354. 356. 360, 363,
15	299, 301, 302, 303, 311. 320. 321. 322. 323.	325, 327. 330, 335. 346,	349. 348. 354. 356, 360, 363,
12	299, 301, 302, 303, 305. 311. 320, 321, 322. 323.	325, 327. 330, 335,	349. 348. 351. 354. 356, 360, 363,
9	300. 301, 302, 303, 305. 307. 311. 320, 321, 322. 323.	325, 327. 330, 335,	349. 348, 351. 354, 356, 360, 363,
6	300. 301, 302, 303, 305. 307. 311. 320, 321, 322. 323.	325, 327. 335,	349, 348, 351. 354, 356, 363,
3	300. 301, 303, 305. 307. 311. 320, 321, 322, 323,	327. 335, 341. 344. 345.	349, 348, 351. 354, 356, 363,
0	300. 301, 305. 307. 311. 320, 321, 322, 323,	327, 333. 335, 341. 344. 345.	349, 348, 351, 354, 356,
I 27	300. 305. 307. 311, 320, 321, 322, 323,	327, 333. 335, 341. 344. 345.	349, 348, 351, 354,
24	300, 305, 307, 311, 318. 322, 323,	327, 333. 341. 344. 345.	349, 348, 351, 354,
21	300, 305, 307, 311, 318. 322, 323,	327, 329. 333. 340. 341. 344. 345.	347. 349, 351,
18	300, 305, 307, 311, 318. 323,	327, 329. 333. 340. 341, 344, 345,	347. 349, 351, 359.
15	300, 305, 307, 311, 318.	329. 333, 336. 340. 341, 344, 345,	347. 351, 359.
12	296. 300, 305, 307, 318.	329. 333, 336. 340. 341, 344, 345,	347. 351, 359. 365.
9	296. 300, 318.	329. 333, 336. 340. 341, 344, 345,	347. 359. 365.
6	296. 318,	329, 333, 336. 340, 341, 344, 345,	347, 359. 365.
3	296. 314. 318,	329, 331. 333, 336. 340, 341, 344, 345.	347, 353. 359. 365.
0	296. 314. 318,	329, 331. 333, 336, 340, 341, 344,	347, 353. 359, 365.
O 27	296, 314. 318,	329, 331. 337. 336, 340,	347, 353. 359, 365.
24	296, 315. 314. 313,	329, 331. 337. 336, 340,	347, 353. 359, 365.
21	296, 308. 315. 314. 318,	329, 331. 337. 336, 340,	347, 353. 359, 365,
18	296, 308. 315. 314.	329, 331. 337. 336, 340,	347, 353. 359, 364. 365,
15	296, 308. 316. 315. 314, 317. 324.	329, 331, 337. 336, 340,	347, 353. 359, 364. 365,
12	296, 308. 316. 315. 314, 317. 324. 326.	331, 337. 336, 340,	347, 353, 359, 364. 365,
9	308. 316. 315. 314, 317. 324. 326.	331, 337. 336,	353, 359, 364. 365,
6	308. 316. 315, 314, 317. 319. 324. 326.	331, 337, 336,	353, 359, 364. 365,
3	308, 312. 316. 315, 314, 317. 319. 324. 326.	331, 337,	353, 359, 364. 365,
0	308, 312. 316. 315, 314, 317. 319. 324. 326.	331, 337, 339.	353, 359, 364. 365,
☾'s ☊	15h…………………… 16h	16h…………………… 17h	17h…………………… 18h
	RIGHT ASCENSION.		

The numbers in this Table point out the Stars in the preceding Catalogue.

☾'s ☊		RIGHT ASCENSION.		
		270° .... 285°	285° .... 300°	300° .... 315°
S	°			
XI	27	371, 381, 393,	401. 415. 418.	427, 436, 438. 444, 445. 447.
	24	371, 381, 393.	401. 415. 418.	427, 436, 438. 444, 445. 447.
	21	371, 381, 393.	401. 410, 415. 418.	427, 436, 438. 444, 445. 447.
	18	371, 381, 393.	401. 410, 415. 418.	427, 436, 438. 441, 444, 447.
	15	371, 381, 386, 393.	409, 410,	427, 436, 441, 444. 447.
	12	371, 381, 386, 393.	409, 410,	427, 436. 441, 444. 447.
	9	371, 381, 386, 393.	397, 409, 410,	422, 427. 436. 441, 444.
	6	371, 381, 386, 393.	397, 409, 410, 419,	422, 427. 436. 441, 444.
	3	371, 381, 386, 393.	397, 403, 409, 410, 419,	422, 427. 436. 441, 444. 449,
	0	371, 381, 386, 393.	397, 403, 409, 410, 419,	422, 427. 436. 440, 441. 443, 444. 449,
X	27	371, 381, 386,	397, 403, 409, 410, 419,	422, 427. 440, 441. 443, 449,
	24	371, 381. 386,	397, 403, 406, 409, 410, 419,	422, 427. 440, 441. 443, 449,
	21	371, 381. 386,	397, 398, 403, 406, 409. 410. 419,	422. 440, 441. 443, 449,
	18	371, 381. 386.	397, 398, 403, 406, 409. 410. 419.]	422. 434, 440, 441. 443, 449.
	15	371. 374, 380, 381. 386.	397. 398, 403, 406, 409. 410. 419.	422. 434, 440, 441. 443. 449.
	12	371. 374, 380, 381. 386.	397. 398, 403. 406, 409. 410. 419.	422. 432, 434, 440. 443. 449.
	9	371. 374, 380, 381. 386.	397. 398, 403. 406, 409. 419.	422. 423, 432, 434, 440. 443. 449.
	6	371. 374, 380, 386.	397. 398, 403. 406. 419.	423, 430, 432, 434, 440. 443. 449.
	3	371. 374, 380, 386.	397. 398. 400, 403. 406.	423, 430, 431, 432, 434, 440. 443. 449.
	0	371. 374, 379, 380, 390,	398. 400, 403. 406.	423, 430, 431, 432, 434. 440.
IX	27	374, 379, 380. 390,	398. 400, 406. 408,	423, 430, 431, 432. 434.
	24	374. 378, 379, 380. 387, 390,	398. 400, 406. 408,	423. 430, 431, 432. 434.
	21	374. 378, 379, 380. 387, 389, 390,	396, 400, 408, 413,	423. 430. 431, 432. 434.
	18	366, 374. 378, 379, 380. 387, 389, 390, 394,	396, 400, 408, 413, 417,	423. 430. 431. 432. 434. 448,
	15	366, 374. 378, 379, 380. 382, 384, 387, 389, 390, 394,	396, 400. 408, 413, 417,	423. 430. 431. 432. 434. 448,
	12	366, 378, 379. 382, 384, 385, 387, 389, 390. 394, 395,	396, 400. 408, 413, 417,	423. 430. 431. 435, 437, 446, 448,
	9	366, 378, 379. 382, 384, 385, 388, 387, 389, 390. 394, 395,	396, 400. 408. 413, 417,	430. 431. 435, 437, 446, 448,
	6	366, 378. 379. 382, 384, 385, 388, 387. 389. 390. 392, 394, 395,	396, 400. 408. 413, 417,	430. 431. 435, 437, 446, 448,
	3	366, 378. 379. 382, 384, 385, 388, 387. 389. 390. 392, 394, 395,	396. 400. 408. 413. 417,	431. 435, 437, 446, 448,
	0	366, 378. 379. 382, 384, 385, 388, 387. 389. 390. 391, 392, 394. 395,	396. 408. 413. 417.	435, 437, 446, 448.
VIII	27	366. 368, 378. 382. 384. 385, 388, 387. 389. 391, 392, 394. 395,	396. 413. 417.	435, 437, 446, 448.
	24	366. 368, 382. 383, 384. 385. 388, 387. 389. 391, 392, 394. 395.	396. 399, 413. 417.	435. 437. 446, 448.
	21	366. 368, 382. 383, 384. 385. 388, 391, 392, 394. 395.	396. 399, 413. 417.	426, 435. 437. 446. 448.
	18	366. 367, 368, 373, 382. 383, 384. 385. 388. 391, 392. 394. 395.	399, 404, 417.	426, 435. 437. 446. 448.
	15	367, 368, 369, 373, 382. 383, 384. 385. 388. 391, 392. 395.	399, 404, 417.	426, 435. 437. 446. 448.
	12	367, 368, 369, 370, 373, 383, 385. 388. 391. 392. 395.	399, 405, 404,	426, 435. 437. 446. 448.
	9	367, 368. 369, 370, 373, 383, 388. 391. 392.	399, 405, 404,	426, 435. 437. 442, 446. 448.
	6	367, 368. 369, 370, 373, 383, 388. 391. 392.	399, 405, 404,	426, 435. 437. 442, 446.
	3	367, 368. 369, 370, 373, 383. 391.	399. 405, 404, 411, 412,	426, 435. 437. 442, 446.
	0	367. 368. 369, 370, 373, 383. 391.	399. 405, 404, 411, 412,	426, 442, 446.
VII	27	367. 368. 369, 370, 373. 383.	399. 405, 404, 411, 412,	426, 442, 446.
	24	367. 368. 369. 370, 373. 376, 383.	399. 405, 404. 411, 412,	426. 442, 446.
	21	367. 369. 370. 373. 376, 377, 383.	399. 405, 404. 411, 412, 414,	426. 442, 446.
	18	367. 369. 370. 373. 376, 377, 383.	399. 405, 404. 411, 412, 414,	426. 442, 446.
	15	367. 369. 370. 373. 376, 377,	399. 405. 404. 411, 412, 414,	426. 442, 446.
	12	369. 370. 373. 376, 377,	405. 404. 411, 412, 414,	426. 442, 446.
	9	370. 375, 376, 377,	405. 404. 411, 412, 414, 420,	426. 442, 446.
	6	370. 375, 376, 377,	405. 404. 407, 411, 412, 414, 420,	429, 426. 442, 446.
	3	375, 376, 377,	405. 404. 407, 411, 412, 414, 420,	429, 426. 442, 446.
	0	375, 376, 377,	405. 404. 407, 411, 412, 414, 420,	429, 426. 442, 446.
VI	27	375, 376, 377,	405. 404. 407, 411, 412, 414, 420,	426. 442, 446.
	24	375, 376. 377,	405. 404. 407, 411, 412, 414, 420,	426. 442, 446.
	21	375, 376. 377,	405. 404. 407, 411, 412, 414, 420,	426. 442, 446.
	18	375, 376. 377,	402, 405. 404. 407, 411, 412, 414, 420,	426. 442, 446.
	15	375, 376. 377,	402, 405. 404. 407, 411, 412, 414, 420,	426. 442, 446. 448.
	12	375, 376. 377,	405. 404. 407, 411, 412, 414, 420,	426. 442, 446. 448.
	9	375, 376. 377,	405. 404. 407, 411, 412, 414, 420,	426. 435. 437. 442, 446. 448.
	6	375, 376. 377.	405. 404. 407, 411, 412, 414, 420,	426, 435. 437. 446. 448.
	3	375, 376. 377,	405. 404. 407, 411, 412, 414, 420,	426, 435. 437. 446. 448.
	0	375, 376. 377,	405. 404. 407, 411, 412, 414, 420,	426, 435. 437. 446, 448.
☾'s ☊		18h 19h	19h 20h	20h 21h
		RIGHT ASCENSION.		

The numbers in this Table point out the Stars in the preceding Catalogue.

☾'s ☊		RIGHT ASCENSION. 270° ... 285° (18h ... 19h)	285° ... 300° (19h ... 20h)	300° ... 315° (20h ... 21h)
S V	27°	375, 376. 377,	405. 404. 411, 412, 414, 420,	426, 435. 437. 446, 448.
	24	375, 376. 377,	405. 404. 411, 412, 414, 420,	426, 435. 437. 446, 448.
	21	375, 376. 377,	405. 404. 411, 412, 420,	426, 435. 437. 446, 448,
	18	375, 376. 377,	399. 405. 404. 411, 412, 420,	426, 435. 437. 446, 448,
	15	375, 376. 377,	399. 405, 404. 411, 412, 420,	426, 435, 437, 446, 448,
	12	375, 376, 377,	399. 405, 404. 411, 412,	436, 435, 437, 446, 448,
	9	375, 376, 377,	399. 405, 404. 411,	435, 437, 448,
	6	375, 376, 377,	399. 405, 404,	435, 437, 448,
	3	375, 376, 377,	399. 405, 404, 417.	431. 435, 437,
	0	375, 376, 377,	399. 405, 404, 417.	431. 435, 437,
IV	27	370. 376, 377, 383.	399, 405, 404, 417.	430. 431.
	24	370. 373. 376, 377, 383. 391.	399, 405, 404, 413. 417.	430. 431. 434.
	21	369. 370. 373. 376, 377, 383. 391.	399, 405, 404, 413. 417.	423. 430. 431. 432. 434. 449.
	18	369. 370. 373. 376, 377, 383. 391.	399, 404, 413. 417.	423. 430. 431. 432. 434. 440. 443. 449.
	15	367. 369. 370. 373. 376, 383. 391. 392. 395.	399, 413. 417,	423. 430. 431, 432. 434. 440. 443. 449.
	12	367. 369. 370. 373. 383, 388. 391. 392. 395.	399, 413. 417,	423. 430, 431, 432. 434. 440. 443. 449.
	9	367. 369. 370, 373, 383, 388. 391. 392. 395.	396. 399, 408. 413, 417,	423. 430, 431, 432. 434. 440. 443. 449.
	6	367. 368. 369, 370, 373, 383, 385. 388. 391, 392. 394. 395.	396. 408. 413, 417,	423, 430, 431, 432, 434, 440. 443. 449.
	3	367. 368. 369, 370, 373, 383, 384. 385. 388. 391, 392. 394. 395.	396. 408. 413, 417,	423, 430, 431, 432, 434, 440, 441. 443, 449.
	0	367, 368. 369, 370, 373, 382. 383, 384. 385. 388. 391, 392, 394. 395,	396. 408. 413, 417,	423, 430, 432, 434, 440, 441. 443, 449,
III	27	367, 368. 369, 370, 373, 382. 383, 384. 385. 388, 387. 389. 391, 392, 394. 395,	396. 400. 408. 413,	423, 432, 434, 440, 441. 443, 449,
	24	367, 368. 369, 370, 373, 382. 383, 384. 385. 388, 387. 369. 391, 392, 394, 395,	396, 400. 408, 413,	423, 432, 434, 440, 441. 443, 449,
	21	367, 368, 369, 370, 373, 382. 384. 385, 388, 387. 389. 390. 391, 392, 394, 395,	396, 400. 408,	422. 423, 440, 441. 443, 444. 449,
	18	367, 368, 369, 378. 382. 384, 385, 388, 387. 389. 390. 392, 394, 395,	396, 400. 408,	422. 440, 441, 444.
	15	367, 368, 378. 382, 384, 385, 388, 387. 389. 390. 392, 394, 395,	396, 400, 408, 419.	422. 441, 444.
	12	366. 367, 368, 378. 379. 382, 384, 385, 388, 387, 389, 390. 394,	396, 400, 408, 419.	422. 436. 441, 444.
	9	366. 368, 378. 379. 382, 384, 385, 387, 389, 390. 394,	396, 398. 400, 406. 408, 419.	422. 427. 436. 441, 444. 447.
	6	366. 368, 378. 379. 382, 384, 385, 387, 389, 390,	398. 400, 406. 419.	422, 427. 436. 441, 444. 447.
	3	366. 378, 379. 382, 384, 387, 389, 390,	398. 400, 403. 406. 419.	422, 427. 436. 441, 444, 445. 447.
	0	366. 378, 379, 380. 382, 387, 389, 390,	398. 400, 403. 406. 409. 419,	422, 427. 436. 444, 445. 447.
II	27	366, 378, 379, 380. 387, 389, 390,	397. 398. 403. 406. 409. 410. 419,	422, 427. 436. 444, 445. 447.
	24	366, 374. 378, 379, 380. 390,	397. 398. 403. 406, 409. 410. 419,	422, 427. 436, 438. 444, 445. 447.
	21	366, 374. 378, 379, 380. 390,	397. 398, 403. 406, 409. 410. 419,	422, 427, 436, 438. 444, 445. 447,
	18	366, 374. 378, 379, 380.	397. 398, 403, 406, 409. 410. 419,	422, 427, 436, 438. 444, 445. 447,
	15	366, 374. 379, 380, 386.	397. 398, 403, 406, 409, 410. 419,	427, 436, 438. 444, 445. 447,
	12	366, 374, 380, 386.	397, 398, 403, 406, 409, 410,	427, 436, 438. 445. 447,
	9	374, 380, 386.	397, 398, 403, 406, 409, 410,	427, 436, 438. 445, 447,
	6	374, 380, 381. 386.	397, 403, 409, 410,	427, 436, 438. 445, 447,
	3	371. 374, 380, 381. 386.	397, 403, 409, 410,	427, 436, 438. 445, 447,
	0	371. 374, 380, 381. 386,	397, 403, 409, 410, 415. 418.	427, 436, 438, 439. 445, 447,
I	27	371. 374, 381. 386,	397, 409, 410, 415. 418.	428. 438, 439. 445, 447,
	24	371. 374, 381. 386, 393.	397, 409, 410, 415. 418.	428. 438, 439. 445, 447,
	21	371. 381, 386, 393.	397, 410, 415. 418.	428. 438, 439. 445, 447,
	18	371, 381, 386, 393.	410, 415. 418.	428. 438, 439. 445, 447,
	15	371, 381, 386, 393.	415. 418.	428. 438, 439. 445, 447,
	12	371, 381, 386, 393.	415. 418.	428. 438, 439. 445, 447,
	9	371, 381, 386, 393.	415. 418.	428. 438, 439. 445, 447,
	6	371, 381, 386, 393.	415. 418.	428. 438, 439. 445, 447,
	3	371, 381, 393.	415. 418.	428. 438, 439. 445, 447,
	0	371, 381, 393,	415. 416. 418.	428. 438, 439. 445, 447,
O	27	371, 381, 393,	415. 416. 418.	428. 438, 439. 445, 447,
	24	371, 381, 393,	415. 416. 418.	428. 438, 439. 445, 447,
	21	371, 381, 393,	415. 416. 418.	428. 438, 439. 445, 447,
	18	371, 381, 393,	415. 416. 418.	428. 438, 439. 445, 447,
	15	371, 381, 393,	401. 415. 416. 418.	428. 438, 439. 445, 447,
	12	371, 381, 393,	401. 415. 416. 418.	428. 438, 445, 447,
	9	371, 381, 393,	401. 415. 418.	428. 436, 438. 445. 447,
	6	371, 372. 381, 393,	401. 415. 418.	428. 436, 438. 444, 445. 447,
	3	371, 372. 381, 393,	401. 415. 418.	427, 436, 438. 444, 445. 447,
	0	371, 372. 381, 393,	401. 415. 418.	427, 436, 438. 444, 445. 447.

The numbers in this Table point out the Stars in the preceding Catalogue.

☾'s ☊		Right Ascension.		
		315°.......330°	330°.......345°	345°.......360°
S	°			
XI	27	450. 465, 469, 476,	480, 481. 483. 487, 488, 493. 496, 497, 498, 499, 500.	506. 507, 511.
	24	450. 451, 465, 469, 476,	480, 481. 483. 487, 488, 493. 496, 497, 498, 499, 500.	506. 507. 511.
	21	450. 451, 465, 469, 476,	480. 481. 487, 488. 496, 497, 498, 499, 500.	506. 507.
	18	451, 465. 468, 469. 476.	480. 481. 487, 488. 496, 497, 498. 499, 500.	506. 507. 513,
	15	451, 458, 465. 468, 469. 473, 476.	477, 480. 481. 487. 488. 496. 497, 498. 499.	506. 507. 513,
	12	451, 458, 465. 467, 468, 469. 473, 476.	479, 477, 480. 487. 488. 496. 497. 498. 499.	506. 507. 513,
	9	451, 458, 465. 467, 468, 469. 473, 476.	479, 477, 480. 487. 488. 494, 496. 497. 498. 499.	507. 513,
	6	451, 458, 465. 466, 467, 468, 469, 473, 476.	479, 477, 480. 487. 488. 494, 496. 497. 498. 499.	513,
	3	451. 458, 465. 466, 467, 468. 473,	479, 477, 487. 494, 496. 497. 499.	513. 519,
	0	451. 458. 466, 467, 468. 473.	479, 477. 494, 496. 497.	512, 513. 519,
X	27	451. 458. 466, 467. 468. 473.	479, 477. 482, 494, 497.	512, 513. 519,
	24	451. 458. 466, 467. 468. 473.	479. 477. 482, 494.	512, 513. 519,
	21	453, 458. 466. 467. 468. 473.	479. 477. 482, 494.	512, 513. 519,
	18	453, 458. 466. 467. 468. 473.	479. 477. 482, 486, 494.	510, 512, 513. 519,
	15	453, 458. 466. 467. 471,	479. 484, 482, 486, 494.	509, 510, 512, 515, 513. 519.
	12	453, 466. 471,	479. 484, 482. 486, 494.	508, 509, 510, 512. 515, 519.
	9	453, 466. 471,	479. 484, 482. 486, 494.	508, 509, 510, 512. 515, 519.
	6	453. 459, 471,	484, 482. 486,	508, 509, 510, 512. 515, 519.
	3	453. 459, 471,	484, 482. 486. 492,	508, 509, 510, 512. 515, 519.
	0	453. 459, 471.	484. 482. 486. 492,	508, 509, 510, 512. 515, 519.
IX	27	453. 459, 471.	484. 482. 486. 492,	508, 509, 510. 512. 515, 519.
	24	453. 459, 471.	484. 486. 492,	508, 509. 510. 512. 515. 519.
	21	453. 459. 471.	484. 486. 489, 492,	508. 509. 510. 512. 515. 519.
	18	459. 472, 471.	484. 486. 489, 492,	508. 509. 510. 512. 515.
	15	459. 472, 471.	484. 489, 492.	508. 509. 510. 515.
	12	459. 472,	484. 489, 492.	508. 509. 510. 515.
	9	459. 472, 478,	484. 489, 492.	508. 509. 510. 515.
	6	459. 472, 478,	489, 492. 502,	508. 509. 510. 515. 520,
	3	459. 472, 478,	489, 492. 502,	508. 509. 510. 515. 520,
	0	452, 462, 472, 478,	489, 492. 502,	508. 509. 510. 515. 520,
VIII	27	452, 462, 472. 478,	489, 492. 502,	508. 509. 510. 515. 520,
	24	452, 462, 472. 478,	489. 492. 502,	508. 509. 510. 515. 520,
	21	452, 462, 472. 478,	489. 492. 502,	508. 509. 510. 515.
	18	452, 460, 462, 472. 478,	489. 492. 502,	508. 509. 510. 515.
	15	452, 460, 462, 470, 472. 478,	489. 492. 502	508. 509. 510. 515.
	12	452, 460, 462, 470, 472. 478,	489. 492. 502,	508. 509. 510. 515.
	9	452, 460, 462, 470, 472. 478,	489. 492. 502,	508. 509. 510. 515.
	6	452, 460, 462, 470, 472. 478,	489. 492. 502,	508. 509. 510. 515.
	3	452. 460, 462, 470, 472. 478,	489. 492. 502,	508. 509. 510. 515. 519.
	0	452. 460, 462, 470, 472. 478.	489. 492. 502,	508. 509. 510. 512. 515. 519.
VII	27	452. 460, 462, 470, 472. 478.	489. 492. 502,	508. 509. 510. 512. 515. 519.
	24	452. 460, 462, 470, 472. 478,	489. 492. 502,	508. 509. 510. 512. 515, 519.
	21	452. 460, 462, 470, 472. 478,	489. 492. 502,	508. 509. 510. 512. 515, 519.
	18	452. 460, 462. 470, 472. 478,	489. 492.	508, 500. 510. 512. 515, 519.
	15	452. 460, 462, 470, 472. 478,	489. 492.	508, 509, 510, 512. 515, 519.
	12	452. 460, 462, 470, 472. 478,	489, 492.	508, 509, 510, 512. 515, 519.
	9	452. 460, 462, 470, 472. 478,	489, 492.	508, 509, 510, 512. 515, 519.
	6	452. 460, 462, 470, 472. 478,	489, 492.	508, 509, 510, 512. 515, 513. 519,
	3	452. 460, 462, 470, 472. 478,	489, 492.	508, 509, 510, 512, 513. 519,
	0	452. 460, 462, 470, 472. 478,	489, 492.	508, 509, 510, 512, 513. 519,
VI	27	452. 460, 462, 472. 478,	489, 492,	508, 509, 512, 513. 519,
	24	452, 460, 462, 472. 478,	489, 492,	512, 513. 519,
	21	452, 460, 462, 472. 478,	484. 489, 492,	512, 513. 519,
	18	452, 460, 462, 472. 478,	484. 486. 489, 492,	512, 513.
	15	452, 462, 472, 478,	484. 486. 492,	513,
	12	452, 462, 472,	484. 486. 492,	513,
	9	452, 462, 472,	484. 486.	513,
	6	452, 462, 472,	484. 486.	513,
	3	452, 472,	484. 482. 486. 494.	513,
	0	452, 459. 472, 471.	484, 482. 486. 494.	
		21^h.......22^h	22^h.......23^h	23^h.......24^h
☾'s ☊		Right Ascension.		

The numbers in this Table point out the Stars in the preceding Catalogue.

☾'s ☊		RIGHT ASCENSION. 315°.......... 330°	330°.......... 345°	345°.......... 360°
S V	27°	452, 459. 472, 471.	484, 482. 486, 494.	
	24	459. 471.	484, 482. 486, 494.	507. 511.
	21	459. 471.	484, 482. 486, 494. 497. 499.	507. 511.
	18	459. 471.	479. 484, 482. 486, 494, 496. 497. 498. 499.	506. 507. 511.
	15	459. 471.	479. 482, 486, 494, 496. 497. 498. 499.	506. 507. 511.
	12	459, 471,	479. 482, 494, 496. 497. 498. 499.	506. 507. 511.
	9	459, 471,	479. 477. 482, 494, 496. 497. 498. 499. 500.	506. 507. 511. 518.
	6	453. 459, 471,	479. 477. 482, 494, 496. 497. 498. 499, 500.	506. 507, 511, 516. 518.
	3	453. 459, 466. 471,	479. 477. 482, 487. 496, 497, 498. 499, 500.	506. 507, 511, 516. 518.
	0	453. 459, 466. 471, 473.	479, 477. 487. 488. 496, 497, 498, 499, 500.	506, 507, 511, 516. 518.
IV	27	453. 459, 466. 467. 473.	479, 477. 487. 488. 496, 497, 498, 499, 500.	506, 505. 507, 511, 516. 518.
	24	453. 466. 467. 468. 473.	479, 477. 487. 488. 496, 497, 498, 499, 500.	506, 505. 507, 511, 516. 518.
	21	453, 466. 467. 468. 473.	479, 477, 480. 487. 488. 493. 496, 498, 500,	506, 505. 516. 518.
	18	453, 466. 467. 468. 473. 476.	479, 477, 480. 487. 488. 493. 498, 500,	506, 505. 516. 518,
	15	453, 458. 466, 467. 468. 473, 476.	477, 480. 481. 487, 488, 493. 500,	505. 516, 518,
	12	453, 458. 466, 467, 468. 473, 476.	477, 480. 481. 487, 488, 493. 500,	501. 505. 516, 518,
	9	453, 458. 466, 467, 468. 473, 476.	480. 481. 487, 488, 493. 500,	501. 505, 503. 516, 518,
	6	458. 465. 466, 467, 468, 469. 473, 476.	480. 481. 483. 487, 488, 493. 500,	501. 505, 503. 516, 518,
	3	451. 458. 465. 466, 467, 468, 469. 473, 476.	480, 481. 483. 487, 488, 493, 495.	501. 505, 503. 504. 516, 518,
	0	451. 458. 465. 467, 468, 469. 473, 476,	480, 481. 483. 488, 493, 495.	501. 505, 503. 504. 516,
III	27	451. 458, 465. 468, 469. 476,	480, 481, 483. 493, 495.	501. 505, 503. 504. 516, 514.
	24	451. 458, 465. 468, 469. 476,	480, 481, 483. 493, 495.	501. 505, 503. 504. 514. 517.
	21	451. 458, 465. 469. 476,	480, 481, 483. 493, 495.	501, 505, 503. 504. 514. 517.
	18	451, 458, 465, 469, 476,	481, 483, 485. 493, 495.	501, 503. 504. 514. 517.
	15	451, 458, 465, 469, 474. 476,	481, 483, 485. 493, 495.	501, 503, 504. 514. 517.
	12	451, 465, 469, 474.	481, 483, 485. 495,	501, 503, 504. 514. 517.
	9	451, 465, 469, 474.	481, 483, 485. 491. 495,	501, 503, 504. 514. 517.
	6	451, 465, 464. 469, 474.	483, 485. 491. 495,	501, 503, 504. 514. 517.
	3	450. 451, 461. 465, 464. 469, 474.	483, 485. 491. 495,	501, 503, 504, 514. 517.
	0	450. 461. 464. 474.	483, 485. 490. 491. 495,	501, 503, 504, 514. 517.
II	27	450. 461. 464. 474.	483, 485. 490. 491. 495,	501, 503, 504, 514. 517.
	24	450. 454. 461. 464. 474.	485. 490. 491. 495,	501, 503, 504, 514. 517.
	21	450. 454. 461. 464. 463. 474.	485, 490. 491. 495,	501, 503, 504, 514. 517.
	18	450. 454. 461. 464. 463. 474,	485, 490. 491. 495,	501, 503, 504, 514. 517.
	15	450. 454. 461. 464. 463. 474,	485, 490. 491. 495,	501, 503, 504, 514. 517.
	12	450, 454. 456. 461. 464. 463. 474,	485, 490. 491. 495,	501, 503, 504, 514. 517.
	9	450, 454. 456. 461. 464, 463. 474,	485, 490. 491. 495,	501, 503, 504, 514. 517.
	6	450, 454. 456. 461. 464, 463. 474,	485, 490. 491. 495,	501, 503, 504, 514. 517.
	3	450, 454. 456. 461, 464, 463. 474,	485, 490. 491. 495,	501, 503, 504, 514. 517.
	0	450, 454. 456. 461, 464, 463. 474,	485, 490. 491. 495,	501, 503, 504, 514. 517.
I	27	450, 454. 456. 461, 464, 463. 474, 475.	485, 490. 491. 495,	501, 503, 504, 514. 517.
	24	450, 454. 456. 461, 464, 463. 474, 475.	485, 490. 491. 495,	501, 503, 504, 516,
	21	450, 454. 456. 461, 464, 463. 474,	485, 490. 491. 495,	501, 503, 504. 516, 518,
	18	450, 454. 456. 457. 461, 464, 463. 474,	485, 490. 491. 495,	501, 503, 504. 516, 518,
	15	450, 454. 456. 457. 461, 464, 463. 474,	485, 490. 491. 495,	501, 503, 504. 516, 518,
	12	450, 454. 456. 457. 461, 464, 463. 474,	485, 490. 491. 495,	501, 503. 504. 516, 518,
	9	450, 454. 456. 461, 464, 463. 474,	485. 491. 495,	501, 503. 504. 516, 518,
	6	450, 454. 456. 461, 464, 463. 474,	485. 491. 495,	501. 505, 503. 504. 516, 518,
	3	450, 454. 456. 461, 464, 463. 474,	483, 485. 495.	501. 505, 503. 504. 516. 518.
	0	450, 454. 456. 461. 464, 463. 474,	483, 485. 495.	501. 505, 503. 504. 516. 518.
O	27	450, 454. 456. 461. 464. 463. 474.	483, 485. 493, 495.	501. 505, 503. 504. 516. 518.
	24	450, 454. 456. 461. 464. 463. 474.	483, 485. 493, 495.	501. 505, 503. 511, 516. 518.
	21	450, 454. 456. 461. 464. 463. 474.	483, 485. 493, 495.	501. 505, 503. 511, 516. 518.
	18	450, 454. 456. 461. 464. 463. 474.	481, 483, 485. 493, 495. 500,	501. 505. 511, 516. 518.
	15	450, 454. 461. 464. 463. 474.	481, 483, 485. 493, 495. 500,	501. 505. 511, 518.
	12	450. 454. 461. 464. 474.	481, 483, 493, 495. 500,	506, 505. 511,
	9	450. 454. 461. 464. 469, 474.	480, 481, 483. 493. 500,	506, 505. 507, 511.
	6	450. 454. 461. 465, 464. 469, 474. 476,	480, 481, 483. 488, 493. 498, 500,	506, 505. 507, 511.
	3	450. 461. 465, 464. 469, 476,	480, 481, 483. 487, 488, 493. 498, 500,	506, 505. 507, 511.
	0	450. 465, 469, 476,	480, 481. 483. 487, 488, 493. 498, 499, 500.	506, 507, 511.
☾'s ☊		21h.......... 22h RIGHT ASCENSION.	22h.......... 23h	23h.......... 24h

The numbers in this Table point out the Stars in the preceding Catalogue.

A TABLE OF PROPORTIONAL LOGARITHMS TO TWENTY-FOUR HOURS.

M. or S.	H. or M. 0	1	2	3	4	5	6	7	8	9	10	11	12	13	14	15	16	17	18	19	20	21	22	23
0		1.3802	1.0792	9031	7782	6812	6021	5351	4771	4260	3802	3388	3010	2663	2341	2041	1761	1498	1249	1015	792	580	378	185
1	3.1584	1.3730	1.0756	9007	7764	6798	6009	5341	4762	4252	3795	3382	3004	2657	2336	2036	1756	1493	1245	1011	788	577	375	182
2	2.8573	1.3660	1.0720	8983	7745	6784	5997	5331	4753	4244	3788	3375	2998	2652	2331	2031	1752	1489	1241	1007	785	573	371	179
3	2.6812	1.3590	1.0685	8959	7728	6769	5985	5320	4744	4236	3781	3369	2992	2646	2325	2027	1747	1485	1237	1003	781	570	368	175
4	2.5563	1.3522	1.0649	8936	7710	6755	5973	5310	4735	4228	3773	3362	2986	2641	2320	2022	1743	1481	1233	999	777	566	365	172
5	2.4594	1.3455	1.0615	8912	7692	6741	5961	5300	4726	4220	3766	3355	2980	2635	2315	2017	1738	1476	1229	996	774	563	361	169
6	2.3802	1.3388	1.0580	8889	7674	6726	5949	5290	4717	4212	3759	3349	2974	2629	2310	2012	1734	1472	1225	992	770	559	358	166
7	2.3133	1.3323	1.0546	8865	7657	6712	5937	5279	4708	4204	3752	3342	2968	2624	2305	2008	1729	1468	1221	988	767	556	355	163
8	2.2553	1.3259	1.0512	8842	7639	6698	5925	5269	4699	4196	3745	3336	2962	2618	2300	2003	1725	1464	1217	984	763	552	352	160
9	2.2041	1.3195	1.0478	8819	7622	6684	5913	5259	4691	4188	3738	3329	2956	2613	2295	1998	1720	1460	1213	980	759	549	348	157
10	2.1584	1.3133	1.0444	8796	7604	6670	5902	5249	4682	4180	3730	3323	2950	2607	2289	1993	1716	1455	1209	977	756	546	345	153
11	2.1170	1.3071	1.0411	8773	7587	6656	5890	5239	4673	4172	3723	3316	2945	2602	2284	1989	1711	1451	1205	973	752	542	342	150
12	2.0792	1.3010	1.0378	8751	7570	6642	5878	5229	4664	4164	3716	3310	2939	2596	2279	1984	1707	1447	1201	969	749	539	339	146
13	2.0444	1.2950	1.0345	8728	7552	6628	5867	5219	4655	4156	3709	3304	2933	2591	2274	1979	1703	1443	1197	965	745	535	335	144
14	2.0122	1.2891	1.0313	8706	7535	6614	5855	5209	4646	4149	3702	3297	2927	2585	2269	1974	1698	1438	1193	962	741	532	332	141
15	1.9823	1.2833	1.0280	8683	7518	6601	5843	5199	4638	4141	3695	3291	2921	2580	2264	1969	1694	1434	1190	958	738	528	329	138
16	1.9542	1.2776	1.0248	8661	7501	6587	5832	5189	4629	4133	3688	3284	2915	2575	2259	1965	1689	1430	1186	954	734	525	326	135
17	1.9279	1.2719	1.0216	8639	7484	6573	5821	5179	4620	4125	3681	3278	2909	2569	2255	1960	1685	1426	1182	950	731	522	322	132
18	1.9031	1.2663	1.0185	8617	7467	6559	5809	5169	4611	4117	3674	3271	2903	2564	2249	1955	1680	1422	1178	947	727	518	319	129
19	1.8796	1.2607	1.0154	8595	7451	6546	5797	5159	4603	4110	3667	3265	2897	2558	2244	1951	1676	1417	1174	943	724	515	316	125
20	1.8573	1.2553	1.0122	8573	7434	6532	5786	5149	4594	4102	3660	3258	2891	2553	2239	1946	1671	1413	1170	939	720	511	313	122
21	1.8361	1.2499	1.0091	8552	7417	6519	5774	5139	4585	4094	3653	3252	2886	2547	2234	1941	1667	1409	1166	935	716	508	309	119
22	1.8159	1.2446	1.0061	8530	7401	6505	5763	5129	4577	4086	3646	3246	2880	2542	2229	1936	1663	1405	1162	932	713	505	306	116
23	1.7966	1.2393	1.0030	8509	7384	6492	5752	5120	4568	4079	3639	3239	2874	2537	2224	1932	1658	1401	1158	928	709	501	303	113
24	1.7782	1.2341	1.0000	8487	7368	6478	5740	5110	4559	4071	3632	3233	2868	2531	2219	1927	1654	1397	1154	924	706	498	300	110
25	1.7604	1.2289	9970	8466	7351	6465	5729	5100	4551	4063	3625	3227	2862	2526	2214	1922	1649	1393	1150	920	702	495	296	107
26	1.7434	1.2239	9940	8445	7335	6451	5718	5090	4542	4056	3618	3220	2856	2520	2209	1918	1645	1388	1146	917	699	491	293	104
27	1.7270	1.2188	9911	8424	7319	6438	5707	5081	4534	4048	3611	3214	2850	2515	2203	1913	1641	1384	1142	913	695	488	290	101
28	1.7112	1.2139	9881	8403	7302	6426	5695	5071	4525	4040	3604	3208	2845	2510	2198	1908	1636	1380	1138	909	692	484	287	98
29	1.6960	1.2090	9852	8382	7286	6413	5684	5061	4517	4033	3597	3201	2839	2504	2193	1904	1632	1376	1134	905	688	481	283	94
30	1.6812	1.2041	9823	8361	7270	6399	5673	5052	4508	4025	3590	3195	2833	2499	2188	1899	1627	1372	1130	902	685	478	280	91
31	1.6670	1.1993	9794	8341	7254	6385	5662	5042	4499	4017	3583	3189	2827	2493	2183	1894	1623	1368	1126	898	681	474	277	88
32	1.6532	1.1946	9765	8320	7238	6372	5651	5032	4491	4010	3577	3183	2822	2488	2178	1890	1619	1364	1123	895	677	471	274	85
33	1.6399	1.1900	9737	8300	7222	6359	5640	5023	4483	4002	3570	3176	2816	2483	2173	1885	1614	1359	1119	891	674	468	271	82
34	1.6269	1.1852	9708	8280	7206	6346	5629	5013	4474	3995	3563	3170	2810	2477	2169	1880	1610	1355	1115	887	670	464	267	79
35	1.6143	1.1806	9680	8259	7190	6333	5618	5004	4466	3987	3556	3164	2804	2472	2164	1876	1605	1351	1111	883	667	461	264	76
36	1.6021	1.1761	9652	8239	7175	6320	5607	4994	4457	3979	3549	3158	2798	2467	2159	1871	1601	1347	1107	880	663	458	261	73
37	1.5902	1.1716	9625	8219	7159	6307	5596	4985	4449	3972	3542	3151	2793	2461	2154	1866	1597	1343	1103	876	660	454	258	70
38	1.5786	1.1671	9597	8199	7143	6295	5585	4975	4440	3964	3535	3145	2787	2456	2149	1862	1592	1339	1099	872	656	451	255	67
39	1.5673	1.1627	9570	8179	7128	6282	5574	4966	4432	3957	3529	3139	2781	2451	2144	1857	1588	1335	1095	868	653	447	251	64
40	1.5563	1.1584	9542	8159	7112	6269	5563	4956	4424	3949	3522	3133	2776	2446	2139	1852	1584	1331	1092	865	649	444	248	61
41	1.5456	1.1540	9515	8140	7097	6256	5552	4947	4415	3942	3515	3127	2770	2440	2134	1848	1579	1327	1088	861	646	441	245	58
42	1.5351	1.1498	9489	8120	7081	6243	5541	4937	4407	3934	3508	3120	2764	2435	2129	1843	1575	1322	1084	858	642	438	242	55
43	1.5249	1.1455	9462	8101	7066	6231	5531	4928	4399	3927	3502	3114	2758	2430	2124	1839	1571	1318	1080	854	639	434	239	52
44	1.5149	1.1413	9435	8081	7051	6218	5520	4919	4390	3920	3495	3108	2753	2424	2119	1834	1566	1314	1076	850	635	431	235	48
45	1.5052	1.1372	9409	8062	7035	6205	5509	4909	4382	3912	3488	3102	2746	2419	2114	1829	1562	1310	1072	847	632	427	232	45
46	1.4956	1.1331	9383	8043	7020	6193	5498	4900	4374	3905	3481	3096	2741	2414	2109	1825	1558	1306	1068	843	628	424	229	42
47	1.4863	1.1290	9357	8023	7005	6180	5488	4891	4366	3897	3475	3089	2736	2409	2104	1820	1553	1302	1064	839	625	421	226	39
48	1.4771	1.1249	9331	8004	6990	6168	5477	4881	4357	3890	3468	3083	2730	2403	2099	1816	1549	1298	1061	836	621	418	223	36
49	1.4682	1.1209	9305	7985	6975	6155	5466	4872	4349	3883	3461	3077	2724	2398	2095	1811	1545	1294	1057	832	618	414	220	33
50	1.4594	1.1170	9279	7966	6960	6143	5456	4863	4341	3875	3455	3071	2719	2393	2090	1806	1540	1290	1053	828	614	411	216	30
51	1.4508	1.1130	9254	7948	6945	6131	5445	4853	4333	3868	3448	3065	2713	2388	2085	1802	1536	1286	1049	824	611	408	213	27
52	1.4424	1.1092	9228	7929	6930	6118	5435	4844	4325	3860	3441	3059	2708	2382	2080	1797	1532	1282	1045	821	608	404	210	24
53	1.4341	1.1053	9203	7910	6915	6106	5424	4835	4316	3853	3435	3053	2702	2377	2075	1793	1528	1278	1041	817	604	401	207	21
54	1.4260	1.1015	9178	7892	6900	6094	5414	4826	4308	3846	3428	3047	2696	2372	2070	1788	1523	1274	1037	814	601	398	204	18
55	1.4180	1.0977	9153	7873	6885	6081	5403	4817	4300	3839	3421	3041	2691	2367	2065	1784	1519	1270	1034	810	597	394	201	15
56	1.4102	1.0939	9129	7855	6871	6069	5393	4808	4292	3831	3415	3035	2685	2362	2061	1779	1515	1265	1030	806	594	391	197	12
57	1.4025	1.0902	9104	7836	6856	6057	5382	4799	4284	3824	3408	3028	2679	2356	2056	1775	1510	1261	1026	803	590	388	194	9
58	1.3949	1.0865	9079	7818	6842	6045	5372	4789	4276	3817	3401	3022	2674	2351	2051	1770	1506	1257	1022	799	587	384	191	6
59	1.3875	1.0828	9055	7800	6827	6033	5362	4780	4268	3809	3395	3016	2668	2346	2046	1766	1502	1253	1018	795	583	381	188	3
60	1.3802	1.0792	9031	7782	6812	6021	5351	4771	4260	3802	3388	3010	2663	2341	2041	1761	1498	1249	1015	792	580	378	185	0

A TABLE of Hourly Proportional Parts of the Moon's Changes in Twelve Hours.

Moon in 12 hours.	Hours after Noon or Midnight.											Moon in 12 hours.
	11h	10h	9h	8h	7h	6h	5h	4h	3h	2h	1h	
°	° ′	° ′	° ′	° ′	° ′	° ′	° ′	° ′	° ′	° ′	° ′	°
′	′ ″	′ ″	′ ″	′ ″	′ ″	′ ″	′ ″	′ ″	′ ″	′ ″	′ ″	
″	″ ‴	″ ‴	″ ‴	″ ‴	″ ‴	″ ‴	″ ‴	″ ‴	″ ‴	″ ‴	″ ‴	″
0	0 . 0	0 . 0	0 . 0	0 . 0	0 . 0	0 . 0	0 . 0	0 . 0	0 . 0	0 . 0	0 . 0	0
1	0 .55	0 .50	0 .45	0 .40	0 .35	0 .30	0 .25	0 .20	0 .15	0 .10	0 . 5	1
2	1 .50	1 .40	1 .30	1 .20	1 .10	1 . 0	0 .50	0 .40	0 .30	0 .20	0 .10	2
3	2 .45	2 .30	2 .15	2 . 0	1 .45	1 .30	1 .15	1 . 0	0 .45	0 .30	0 .15	3
4	3 .40	3 .20	3 . 0	2 .40	2 .20	2 . 0	1 .40	1 .20	1 . 0	0 .40	0 .20	4
5	4 .35	4 .10	3 .45	3 :20	2 .55	2 .30	2 . 5	1 .40	1 .15	0 .50	0 .25	5
6	5 .30	5 . 0	4 .30	4 . 0	3 .30	3 . 0	2 .30	2 . 0	1 .30	1 . 0	0 .30	6
7	6 .25	5 .50	5 .15	4 .40	4 . 5	3 .30	2 .55	2 .20	1 .45	1 .10	0 .35	7
8	7 .20	6 .40	6 . 0	5 .20	4 .40	4 . 0	3 .20	2 .40	2 . 0	1 .20	0 .40	8
9	8 .15	7 .30	6 .45	6 . 0	5 .15	4 .30	3 .45	3 . 0	2 .15	1 .30	0 .45	9
10	9 .10	8 .20	7 .30	6 .40	5 .50	5 . 0	4 .10	3 .20	2 .30	1 .40	0 .50	10
11	10 . 5	9 .10	8 .15	7 .20	6 .25	5 .30	4 .35	3 .40	2 .45	1 .50	0 .55	11
12	11 . 0	10 . 0	9 . 0	8 . 0	7 . 0	6 . 0	5 . 0	4 . 0	3 . 0	2 . 0	1 . 0	12
13	11 .55	10 .50	9 .45	8 .40	7 .35	6 .30	5 .25	4 .20	3 .15	2 .10	1 . 5	13
14	12 .50	11 .40	10 .30	9 .20	8 .10	7 . 0	5 .50	4 .40	3 .30	2 .20	1 .10	14
15	13 .45	12 .30	11 .15	10 . 0	8 .45	7 .30	6 .15	5 . 0	3 .45	2 .30	1 .15	15
16	14 .40	13 .20	12 . 0	10 .40	9 .20	8 . 0	6 .40	5 .20	4 . 0	2 .40	1 .20	16
17	15 .35	14 .10	12 .45	11 .20	9 .55	8 .30	7 . 5	5 .40	4 .15	2 .50	1 .25	17
18	16 .30	15 . 0	13 .30	12 . 0	10 .30	9 . 0	7 .30	6 . 0	4 .30	3 . 0	1 .30	18
19	17 .25	15 .50	14 .15	12 .40	11 . 5	9 .30	7 .55	6 .20	4 .45	3 .10	1 .35	19
20	18 .20	16 .40	15 . 0	13 .20	11 .40	10 . 0	8 .20	6 .40	5 . 0	3 .20	1 .40	20
21	19 .15	17 .30	15 .45	14 . 0	12 .15	10 .30	8 .45	7 . 0	5 .15	3 .30	1 .45	21
22	20 .10	18 .20	16 .30	14 .40	12 .50	11 . 0	9 .10	7 .20	5 .30	3 .40	1 .50	22
23	21 . 5	19 .10	17 .15	15 .20	13 .25	11 .30	9 .35	7 .40	5 .45	3 .50	1 .55	23
24	22 . 0	20 . 0	18 . 0	16 . 0	14 . 0	12 . 0	10 . 0	8 . 0	6 . 0	4 . 0	2 . 0	24
25	22 .55	20 .50	18 .45	16 .40	14 .35	12 .30	10 .25	8 .20	6 .15	4 .10	2 . 5	25
26	23 .50	21 .40	19 .30	17 .20	15 .10	13 . 0	10 .50	8 .40	6 .30	4 .20	2 .10	26
27	24 .45	22 .30	20 .15	18 . 0	15 .45	13 .30	11 .15	9 . 0	6 .45	4 .30	2 .15	27
28	25 .40	23 .20	21 . 0	18 .40	16 .20	14 . 0	11 .40	9 .20	7 . 0	4 .40	2 .20	28
29	26 .35	24 .10	21 .45	19 .20	16 .55	14 .30	12 . 5	9 .40	7 .15	4 .50	2 .25	29
30	27 .30	25 . 0	22 .30	20 . 0	17 .30	15 . 0	12 .30	10 . 0	7 .30	5 . 0	2 .30	30
31	28 .25	25 .50	23 .15	20 .40	18 . 5	15 .30	12 .55	10 .20	7 .45	5 .10	2 .35	31
32	29 .20	26 .40	24 . 0	21 .20	18 .40	16 . 0	13 .20	10 .40	8 . 0	5 .20	2 .40	32
33	30 .15	27 .30	24 .45	22 . 0	19 .15	16 .30	13 .45	11 . 0	8 .15	5 .30	2 .45	33
34	31 .10	28 .20	25 .30	22 .40	19 .50	17 . 0	14 .10	11 .20	8 .30	5 .40	2 .50	34
35	32 . 5	29 .10	26 .15	23 .20	20 .25	17 .30	14 .35	11 .40	8 .45	5 .50	2 .55	35
36	33 . 0	30 . 0	27 . 0	24 . 0	21 . 0	18 . 0	15 . 0	12 . 0	9 . 0	6 . 0	3 . 0	36
37	33 .55	30 .50	27 .45	24 .40	21 .35	18 .30	15 .25	12 .20	9 .15	6 .10	3 . 5	37
38	34 .50	31 .40	28 .30	25 .20	22 .10	19 . 0	15 .50	12 .40	9 .30	6 .20	3 .10	38
39	35 .45	32 .30	29 .15	26 . 0	22 .45	19 .30	16 .15	13 . 0	9 .45	6 .30	3 .15	39
40	36 .40	33 .20	30 . 0	26 .40	23 .20	20 . 0	16 .40	13 .20	10 . 0	6 .40	3 .20	40
41	37 .35	34 .10	30 .45	27 .20	23 .55	20 .30	17 . 5	13 .40	10 .15	6 .50	3 .25	41
42	38 .30	35 . 0	31 .30	28 . 0	24 .30	21 . 0	17 .30	14 . 0	10 .30	7 . 0	3 .30	42
43	39 .25	35 .50	32 .15	28 .40	25 . 5	21 .30	17 .55	14 .20	10 .45	7 .10	3 .35	43
44	40 .20	36 .40	33 . 0	29 .20	25 .40	22 . 0	18 .20	14 .40	11 . 0	7 .20	3 .40	44
45	41 .15	37 .30	33 .45	30 . 0	26 .15	22 .30	18 .45	15 . 0	11 .15	7 .30	3 .45	45
46	42 .10	38 .20	34 .30	30 .40	26 .50	23 . 0	19 .10	15 .20	11 .30	7 .40	3 .50	46
47	43 . 5	39 .10	35 .15	31 .20	27 .25	23 .30	19 .35	15 .40	11 .45	7 .50	3 .55	47
48	44 . 0	40 . 0	36 . 0	32 . 0	28 . 0	24 . 0	20 . 0	16 . 0	12 . 0	8 . 0	4 . 0	48
49	44 .55	40 .50	36 .45	32 .40	28 .35	24 .30	20 .25	16 .20	12 .15	8 .10	4 . 5	49
50	45 .50	41 .40	37 .30	33 .20	29 .10	25 . 0	20 .50	16 .40	12 .30	8 .20	4 .10	50
51	46 .45	42 .30	38 .15	34 . 0	29 .45	25 .30	21 .15	17 . 0	12 .45	8 .30	4 .15	51
52	47 .40	43 .20	39 . 0	34 .40	30 .20	26 . 0	21 .40	17 .20	13 . 0	8 .40	4 .20	52
53	48 .35	44 .10	39 .45	35 .20	30 .55	26 .30	22 . 5	17 .40	13 .15	8 .50	4 .25	53
54	49 .30	45 . 0	40 .30	36 . 0	31 .30	27 . 0	22 .30	18 . 0	13 .30	9 . 0	4 .30	54
55	50 .25	45 .50	41 .15	36 .40	32 . 5	27 .30	22 .55	18 .20	13 .45	9 .10	4 .35	55
56	51 .20	46 .40	42 . 0	37 .20	32 .40	28 . 0	23 .20	18 .40	14 . 0	9 .20	4 .40	56
57	52 .15	47 .30	42 .45	38 . 0	33 .15	28 .30	23 .45	19 . 0	14 .15	9 .30	4 .45	57
58	53 .10	48 .20	43 .30	38 .40	33 .50	29 . 0	24 .10	19 .20	14 .30	9 .40	4 .50	58
59	54 . 5	49 .10	44 .15	39 .20	34 .25	29 .30	24 .35	19 .40	14 .45	9 .50	4 .55	59
Arg.	11h	10h	9h	8h	7h	6h	5h	4h	3h	2h	1h	Arg.

In making use of this Table, it must be noticed that the argument on the side is the moon's change in *twelve* hours, and may be read degrees, minutes or seconds, so that the numbers taken out be read agreeably to their title in the same line with the argument. When the argument is seconds, the *thirds* may be divided by sixty, and more conveniently read thus: 0″, 5″, 10″, 15″, 20″, 25″, 30″, 35″, 40″, 45″, 50″ or 55″
″.0 ″.1 ″.2 ″.3 ″.3 ″.4 ″.5 ″.6 ″.7 ″.8 ″.8 or ″.9

In taking out a number, it will be best to write down the argument first, and then casting the eye to the given hour, the proportional parts for that hour and for one hour may be taken by inspection.

North-East Angle of the ⁎'s Meridian with the Ecliptic.

In our explanation of the Catalogue of 520 Stars, at page 495 of this Appendix, we were limited to a single page, that all the columns, relating to the same star, might be seen at one opening of the volume; it will consequently be necessary to give some additional explanation here of the uses to which *the north-east angle of a star's meridian with the ecliptic* may be applied, which was before purposely omitted. Before the moon's right ascension was given in the Nautical Almanac to the accuracy of *seconds*, this angle was of great use in computing visible occultations by the parallactic method, and is still useful in the *inverse* problem of the *third method* given in our second volume; it is also convenient in reducing the moon's parallaxes in longitude and latitude into corresponding parallaxes in right ascension and declination, and *vice versâ:* it may likewise be found useful in the reduction of a star's aberration, nutation, and precession in longitude and latitude into similar corrections in right ascension and declination, and the converse. We will take the example of the star Aldebaran, given at page 378, to show how its angle of position may be determined for July 1, 1820, from this column, more readily than by any of the common methods, thus:

Zod. Cat. No. 95, Aldebaran's north-east angle, 1820	80° 47′ 6″.2
Table XV. July 1, with +19″.21 variation	+9 .6
	80 47 15 .8
Take this angle from	90 0 0
True angle of position by the Catalogue..................................	9 12 44 .2
Page 378 by the Tables..	9 12 46 .0
By the formula at page 378, using the longitude	9 12 50
By another formula at page 379, using the R. A.	9 12 43

The Principal Fixed Stars that may be occulted by the Moon.

For the same reason that we postponed pointing out the uses to which the north-east angle of a star's meridian with the ecliptic may be conveniently put, we deferred explaining the method of computing the elements for effecting the preceding arrangement of stars that may be occulted, in the order of right ascension, which subject we will now resume. A *general* list of stars liable to lunar occultations may be constructed variously according to the limits of *latitude* and *parallax* which we assign to the moon, which are constantly varying; hence it will be convenient to speak of three denominations, according to which an arrangement may be made; the *external* extreme, the *mean*, and the *internal* extreme. The *external* extreme would afford such an extensive list, that not a single star selected could undergo occultation without being pointed out; according to the *mean* there would be a few stars superfluous in the list, and a few deficient; and the *internal* extreme arrangement would contain only such as *must* suffer occultation in some part of the world; but several useful stars would be omitted. The list given in the Jahrbuch of 1780, was computed from the *mean* number 1° 12′, and the moon's *greatest* inclination 5° 19′; but in our list all stars of the given magnitude are employed, the latitudes of which are less than 6° 39′; the *mean* arrangement is employed, with the addition of 2′ as a compensation for the intervals, in the side argument of the ☾'s ☊, having differences of 3°; and the mean inclination, 5° 8′ 44″ is adopted. As there appears to be some ambiguity in the method of determining the proper successive places in the list given in the Jahrbuch, it may be proper to explain the process by which our list was arranged, which was according to the following rule; viz.

Rule.—Write down the latitude of the star twice over with the sign + if north, and − if south, after it; under these place 1° 14′+ and 1° 14′−, and call their sums respectively a and b; then from a and b determine the corresponding arcs a' and b'; and also a'' and b'' by the annexed formulæ; viz.

$$\text{Sin. } a' = \cot - 5°\ 8'\ 44'' \times \text{tang } a \quad (1)$$
$$\text{Sin. } b' = \cot - 5\ \ 8\ \ 44 \times \text{tang } b \quad (2)$$
$$\text{Sin. } a'' = \text{cotan } 5\ \ 8\ \ 44 \times \text{tang } a \quad (3)$$
$$\text{Sin. } b'' = \text{cotan } 5\ \ 8\ \ 44 \times \text{tang } b \quad (4)$$

To the star's longitude apply a' and b', and to its opposite longitude a'' and b''; and call the sums respectively c, d, and c', d': then the space from d to c will be that passed over by the moon's north node, while the star in question is liable to be occulted with reference to that node; and the space included between d' and c' will be that travelled over by the south node, while the star is subject to occultation. An example will render the rule perfectly intelligible.

Example.——Let it be required to determine the space in the ecliptic over which each of the moon's nodes must pass to produce an occultation in No. 15 (ε Piscium), supposing the latitude of the star to be 1° 5′, and the longitude 0^{s} 15° 10′ at the time?

Star's latitude..........	1° 5′+	.... and	1° 5′+
Mean limit	1 14 +	.... and	1 14 −
	a=2 19		b=0 9
Log cot 5° 8′ 44″...	11.04552−	.... and	11.04552+
Tang a 2 19 0 ...	8.60698+	.... and	8.60698+
Sin a' −26 41 47 ...	9.65250−	.. Sin a''+ 26° 41′ 47″	9.65250+
Log cot 5° 8′ 44″...	11.04552−	.... and	11.04552+
Tang b−0 9 0 ...	7.41797−	.... and	7.41797−
Sin b' +1 39 58 ...	8.46349+	.. Sin b''−1° 39′ 58″...	8.46349−

Long. of star........	0^{s} 15° 10′	.. Opposite longitude..	6^{s} 15° 10′
a'........ say ...... −	0 26 42	.. a''.... say .... +	26 42
	c=11 18 28		c'=7 11 52
b' .... say...... +	1 40	.. b''.... say .... −	1 40
	d= 0 16 50		d'=6 13 30

Hence it appears that ε Piscium may be occulted by the moon while her north node is passing over 28° 22′, in a retrograde direction, viz. from d=0^{s} 16° 50′ to c=11^{s} 18° 28′, and again while her south node is passing over the same arc with her north node passing from c'=7^{s} 11° 52′ to d'=6^{s} 13° 30′. In our ZODIACAL TABLE 6, copied from the Jahrbuch, the limits are given 0^{s} 15° to 11^{s} 18°, and 7^{s} 12° to 6^{s} 15° inclusive; but no notice is there taken of the moon's right ascension, which regulates the time of the phenomenon.

We have given the rule and example at full length for the satisfaction of those readers, who may wish to examine the accuracy of the arrangement in our Table: for instance, in our example, under the column of the RIGHT ASCENSION 0°————15°, at the head, and 0^{h}————1^{h} at the foot, we find the number 15, (representing ε Piscium) occurring with a *comma* after it, from ☾'s ☊ 0^{s} 18° to 0^{s} 6′ inclusive, and from 0^{s} 3′ to XI^{s} 18′ with a *point* annexed, (thus, 15.) showing that in the former interval the star will be occulted *somewhere* on the globe, but that the phenomenon will be visible in *England* during the latter intercepted arc. These have reference to the north node; and with respect to the south node, in the same vertical column we find 15. commencing at VII^{s} 12° and ending with VI^{s} 27°; likewise 15, beginning with VI^{s} 24° and ceasing with VI^{s} 12°. This distinction of the annexed marks will save the observer as well as computer much trouble, in pointing out at once what he has to expect, without previous computation.

It would have been tedious to have determined the place of each individual star, by working the operation at full length, as we have given it, and therefore a subsidiary Table was composed of cotan 5° 8′ 44″ × tan a, or b; from which the numbers were taken out by simple inspection, which reduced the example to the subjoined work;

		s					s				
Star's longitude............		0	15°	10′	Opposite longitude.....		6	15°	10′	∗'s lat. =1°	5′+
Always with	sign of a changed	−	26	42	Always with	same sign	+	26	42	limit =1	14 ±
	c=	11	18	28		c'=	7	11	52	Sum a =2	19+
	sign of b changed	+	1	40		same sign	−	1	40	Sum b =0	9−
	d=	0	16	50		d'=	6	13	30		

Table of Proportional Logarithms to twenty-four Hours.

This Table is useful for finding the proportional parts of the sun's daily variation in right ascension and declination, and of the equation of time for any longitude, or hour after noon. The proportional logarithms are found by subtracting the logarithms of the minutes in the several arguments from 3 .15836, the logarithm of twenty-four hours, when reduced to minutes, so that the prop. log. of 24 becomes = 0. The value of the resulting term, given by the Table, depends on the denomination of the argument used, and may be found thus:

Rule.——To the proportional logarithm of the daily variation add the proportional logarithm of the time from noon, or of the difference of longitude, or of their amount, and the sum will be the proportional logarithm of the correction due to that interval, which must be added, or subtracted, accordingly as the quantity in question is increasing or decreasing.

Example.——Required the sun's right ascension and declination, and also the equation of time, at 3^h 30^m P. M. at Edinburgh, on the 10th of July, 1828?

	R. A.		Declination		Equation	
⊙ R. A. at noon 10th.......	= 7^h 18^m $28^s.6$		22° 14′ 1″		$+4^m$ 56^s 5	
at Greenwich .. 11th.......	= 7 22 33 .5		22 6 11		+5 4 .8	
Daily variation.................	4 4 .9	p. l. .. 7694 ..	7 50	p. l. 4863......	8 .3	p. l. 2 .2400
Diff. long. $12^m.8$ + time 3^h 30^m= 3^h $42^m.8$ at Gr..		... 8105		8105		8105
Correction	$+0^m$ 38^s.......	1 .5799 ..	−1′ 12″.7	1.2968	$+1^s.28$	3 .0505
July 10, at Greenwich..........	7 18 28 .6		22 14 1		4 56 .50	
at Edinburgh..........	7 19 6 .6 = R. A.		22 12 48 .3 = Dec. N.		+4 57 .78 = Equa.	

When one or both of the arguments are small quantities, falling near the beginning of the Table, as in the case of the equation in this example, it will be convenient to diminish the index of the resulting proportional logarithm, and then the answer divided by 10, 100, or 1000, accordingly as 1, 2, or 3 is the index subtracted, will be the true quantity: if, for instance, we substitute 1 .0505 or 0 .0505 for 3 .0505, in the former case we get 2^m 8^s = 128^s, or, dividing by 100, $1^s.28$; and in the latter we have 21^m 22^s = 1282^s, which, divided by 1000, will become $1^s.282$.

The Table of hourly proportional parts of the moon's changes, in twelve hours, requires no further explanation than is contained in the margin.

On the Parallaxes.

Though we have published a variety of new Tables relating to the different parallaxes of the moon, computed with great care, and have given several examples to illustrate their application in the various computations, yet we regret to find that in some instances in pages 386—393 the *second part* of parallax, both in latitude and declination, has inadvertently been applied with a wrong sign, which improper application has vitiated the resulting parallaxes, by rendering some part of each operation erroneous; and as the errors are transmitted through different successive lines, they could not be corrected without too great an increase of our tabulated errata: we propose, therefore, to repeat the work of the first example, given in page 381, that we may have an opportunity of correcting some of the conclusions that are incorrectly deduced. We are the more disposed to adopt this measure, because different authors have embarrassed the subject, not only by proposing different formulæ as the groundwork of the operations, but also by not attending sufficiently to the *quality* of the terms introduced; the terms *true*, *apparent*, and *mean*, are frequently either omitted, or confounded with one another, and the computer is at a loss to know which is most proper to apply, particularly in lunar computations, when these epithets are not specified in the data. On this account some of the examples were worked by the data differently modified, to show the resulting differences. We have lately had occasion to consult a practised and skilful computer on this subject, who has investigated and varied the common formulæ, with a view of rendering the moon's parallax in altitude convertible into the proper parallaxes in latitude and declination, and the converse; which conversion is not feasible by Lalande's and Mayer's formulæ, as the operations are usually performed. We will first repeat the operations by Mayer's method, as being upon the whole more eligible in practice. This author gives his work of finding the parallaxes in longitude and latitude, in his Lunar Tables of 1770 (p. 40), in which he does not use the *true* polar distance, or angle, but the *visible* or *apparent* one, thus:

$$\text{Parallax in longitude} = \frac{\text{hor par} \times \text{sin alt nonag} \times \text{sin ☾'s } \textit{appar.} \text{ dist. from nonages}}{\text{cos ☾'s } \textit{true} \text{ latitude}}$$

Parallax in latitude.. { first part... = hor par × cos alt nonages. × cos ☾'s *visible* latitude
second part = hor par × sin alt nonages. × cos ☾'s *appar.* dist. from nonages × sin ☾'s *visible* latitude.

The latter formula turns the parallax in altitude into that of declination, or of latitude by means of the *visible* angle P ☾ Z, or p ☾ Z, where P is the pole of the equator, p the pole of the ecliptic, ☾ the moon, and Z the zenith point: whereas, in effect, we have obtained the parallax in R. A. from the *true* parallax in altitude by considering the visible or apparent angle $= P$ ☾ Z; and yet the parallax in declination has been computed from the *approximate* parallax in altitude supposing the *true* angle $= P$ ☾ Z; hence arises a cause of discrepancy in the conclusions. The lunar parallax in altitude may be thus expressed; viz.

Parallax in alt. = hor par × sin (true Z ☾ + par in alt), and the operations performed in the following manner will then accord with one another.

Example.——Let it be required to find, by Mayer's method improved, the moon's parallax in longitude, latitude, right ascension, and declination, at Greenwich, when her true longitude was 104° 28′ 0″.33, her true latitude 1° 12′ 50″.355 N., her equatorial horizontal parallax, 61′ 21″.15, longitude of the nonagesimal point of the ecliptic, 32° 27′ 39″.5, altitude of ditto, 47° 9′ 37″.57, and the compression $\frac{1}{314}$, or reduced parallax for Greenwich, 61′ 14″?

Operation for Longitude and Latitude, by Mayer's Method.

Preparation.					
☾'s true longitude	104° 28′ 0″.33	....		latitude........	1° 12′ 50″.36 N
long. nonag.	32 27 39 .50				
☾'s true dist. from nonag.	72 0 20 .83	....	72° 0′ 20″.83		
approx. par. in longitude	+ 42 52 .92	.... half.......	+ 21 26 .46	approx. par. in lat.	− 41 30 .54
visible distance	= 72 43 13 .75	.... mean dist...	72 21 47 .29	visible latitude	= 0 31 19 .82 N

Red. hor par.......	3674″........ log.	3 .5651392	....		log. 3 .5651392	log	3 .5651392
Sin alt nonages.......	47 9 37 .57 ...	9 .8652583	cos true alt non..	47° 9′ 37″.57	.. 9 .8324756	sin	9 .8652583
Sin ☾'s vis. dist. non..	72 43 13 .75 ...	9 .9799429	cos ☾'s vis. lat..	0 31 19 .82	.. 9 .9999820	sin	7 .9596848
Sec. ☾'s true lat.....	1 12 50 .36 ...	10 .0000975	.... first part..	41′ 38″.025	..... 3 .3975968	cos m.d.	9 .4814184
Parallax in long. = 42′ 52″.99.........		3 .4104379	.... second part	− 7 .439	....	log	0 .715007
			parallax in lat...	= 41 30 .59			

By the rule at p. 383, the second part of parallax is here subtracted from the first, because the moon's *visible* distance from the nonagesimal, and also from the elevated or north pole of the ecliptic are both *acute* angles.

OPERATION for RIGHT ASCENSION and DECLINATION, by MAYER'S METHOD.

Preparation.							
☾'s true R. A....	105° 51′ 7″.33				dec.......	23° 53′ 3″.95
R.A. of Midheaven	−8 18 7 .88					
True horary angle	97 32 59 .45	....	97 32 59 .45			
Approx. par. in R.A.	+ 41 26 .30	.......half +	20 43 .15	....	approx. par in dec	− 46 0 .90
Visib. hor angle...	98 14 25 .75	mean hor ∠	97 53 42 .60	....	visib. dec..	23 7 3 .05

Red. hor par. 3674″.............	log. 3 .5651392	.................log	3 .5651392	.... log	3 .5651392
Sin colat red...... 38° 42′ 1″	... 9 .7960512	..... cos colat red.....	9 .8923325	.... sin	9 .7960512
Sin vis. hor ∠.... 98 14 25 .75	... 9 .9954927	..... cos ☾'s vis. dec..	9 .9636470	.... sin	9 .5939705
Sec ☾'s true dec... 23 53 3 .95	... 0 .0388809				
		first part.. 43′ 57″.025	3 .4211187		
Par in R. A. 41′ 26″.36.............	3 .3955640				
			mean hor ∠ 97° 53′ 42″.60..	cos	9 .1378634
		second part +2 3 .887	....		2 .0930243
Par in R. A. 41 26 .36	par in dec............	=46 0 .94			

The second part of parallax is here added to the first, because the moon's *visible* distance from the meridian, and also from the *elevated* pole of the equator are unlike angles: that is, one is obtuse and the other acute.

If we take the reduced latitude of Greenwich =51° 17′ 59″, the R. A. of the zenith 8° 18′ 7″.88; the obliquity 23 27 49 .76; the ☾'s true longitude 104° 28′ 0″.33; and her latitude 1° 12′ 50″.36; by spherical trigonometry we shall obtain the following numbers; viz.

	☾'s long.	☾'s lat.	R. A.	Declin.	Altitude.
True......	104 28 0 .33	1° 12′ 50″.36 N	105° 51′ 7″.33	23° 53′ 3″.95 N	
Apparent..	105 10 53 .25	0 31 19 .82 N	106 32 33 .63	23 7 3 .05 N	
Parallaxes..	+42 52 .92	−41 30 .54	+41 26 .30	−46 0 .90	59′ 40″.65

In order to see how far our computation of parallaxes by the method proposed by La Lande agrees with the parallaxes above deduced, we will now repeat the computation of the examples by La Lande's method, given in pages 393 and 395; and in doing this we will take the correct parallax as the approximate, that we may give to all parts of the computation its exact value.

OPERATION for LONGITUDE and LATITUDE, by LA LANDE'S METHOD.

Preparation.					
☾'s true longitude.........	104° 28′ 0″.33	....		true latitude......	1° 12′ 50″.36 N.
Long. of nonag............	32 35 30 .37 (p. 396)				
☾'s true dist. from nonag. ..	71 52 29 .96	....	7ˢ 52′ 29″.96		
Approx. par. in long........	+42 52 .92	...half......	+21 26 .46	approx. par. in lat.	−41 30 .54
Vis. dist. from nonag.	72 35 22 .88	....mean dist..	72 13 56 42	visible latitude...	0 31 19 .82 N.

Equatorial horizontal parallax reduced (as at p. 381) ...	= 3674″..............	log. 3 .5651392
Sin. 10′ 41″ (red. of lat. as at p. 377)		7 .4924322
Tang. of app. lat. 51° 28′ 40″		10 .0990491
Augmentation..................................	= 14″.343	1 .1566205
Augmented horizontal parallax......................	= 3688 .343	

Augmented hor. par. 3688″.343....	log. 3 .5668313	....	log. 3 .5668313	log... 3 .5668313
Sin. alt. nonag. ..... 47° 0′ 37″.30	9 .8642007	cos. tr. alt. nonag. 47° 0′ 37″.30	9 .8336991	sin... 9 .8642007
Sin. ☾ 's vis. dist. non. 72 35 22 .85	9 .9796332	cos. ☾ 's vis. lat.. 0 31 19 .32	9 .9999820	sin... 7 .9596848
Sec. ☾ 's tr. lat. ..... 1 12 50 .36	10 .0000975	first part 2514.85= 41 54 .85	3 .4005124	m. dist. 9 .4845245
Paral. .....2574″.91 = 42 54 .91	3 .4107627	second part ....= −7 .50		0 .8752413
Reduction (Table 21.) −1 .83				
		parallax .... 41 47 .35		
True parallax......... 42 53 .08		reduction ...... −16 .84 (Table 22.)		
		True parallax .. 41 30 .51		

The parallax in longitude is to be added to the ☾ 's true longitude, as the moon is to the east, of the nonagesimal; likewise, the parallax in latitude is to be subtracted from the true latitude, as the true latitude is less than 90° from the elevated pole of the ecliptic.

The second part of parallax is here subtracted from the first; because the moon's visible distance from the nonagesimal, and also from the elevated pole of the ecliptic are like angles; that is, both less than 90°.

Operation for Right Ascension and Declination, by La Lande's Method.

Preparation.

☾ 's true R. A...... 105° 51′ 7″.33	....	true declin. 23° 53′ 3″.95 N.
R. A. of midheaven.. −8 18 7 .88		
True hor. angle..... 97 32 59 .45	.... 97° 32′ 59″.45	
Approx. par. in R. A. + 41 26 .30	half....... + 20 43 .15	...... approx. par. in dec.. − 46 0 .90
Visible hor. angle ... 98 14 25 .75	m. hor. ang. 97 53 42 .60	...... visible declin....... 23 7 3 .05 N.

Aug. hor. par. 3688″.343........	log. 3 .5668313	....	log. 3 .5668313	... log. 3 .5668313
Sin. app. co-lat....... 38° 31′ 20″	9 .7943612....	cos. app. co-lat. 38° 31′ 20″	..... 9 .8934133	sin. ... 9 .7943612
Sin. vis. hor. ang..... 98 14 25 .75	9 .9954927....	cos. ☾ 's vis. dec. 23 7 3 .05	..... 9 .9636470	sin. ... 9 .5939705
Sec. ☾ 's true dec..... 23 53 3 .95	10 .0388809	first part 2653″.94 = 44′ 13″.94....	3 .6238916	
True paral. 2486″.37 = 41′ 26″.37 ..	3 .3955661....	second part....... +2 3 .89	hor. ∠ 97° 53′ 42″.60	9 .1378634
		46 17 .83		2 .0930264
		reduction .... − 16 .81 (Table 24.)		
		true parallax....... 46 1 .02		

The parallax in right ascension is to be added to the moon's true right ascension, as the moon is to the east of the meridian. The parallax in declination is to be subtracted from the moon's true declination, as the true declination is less than 90° from the elevated pole of the equator. The second part of parallax in declination is to be added to the first, as the moon's visible distance from the meridian and also from the elevated pole of the equator are unlike angles; that is, one is more than 90°, and the other less.

Comparison of the Parallaxes.

	Longitude.	Latitude.	Right Ascension.	Declination.
By spherical trig.	+42′ 52″.92	−41′ 30″.54	+41′ 26″.30	−46′ 0″.90
By Mayer's formula......	+42 52 .99	−41 30 .59	+41 26 .36	−46 0 .94
By La Lande's formula..	+42 53 .08	−41 30 .51	+41 26 .37	−46 1 .02

It should be here remarked, that whenever the latitude of the place is considerably different from that of Greenwich, the corrections required from Tables 20, 21, 22, 24, and 25 should be calculated from the formulæ at page 393.

Enlarged Trigonometrical Sliding Rule.

We have stated at pages 337 and 338, that Mr. Bates, mathematical instrument-maker, of the Poultry, London, constructed an enlarged sliding rule for Mr. Fallows, before he left England, which we proposed to describe in our second volume; but as its application is connected with the tabulated subsidiary numbers given in this volume, it may be introduced more properly in this Appendix. When the Cape astronomer had examined a sliding rule for working trigonometrical proportions logarithmically, which we had put into his hands, and which he had not before seen, after making himself acquainted with the method of reading the divisions and subdivisions, agreeably to any assumed value of the integer representing unity, he was struck with the facility by which any proportion might be worked, in which sines, tangents, and natural numbers were mutually concerned; and immediately saw, that an enlargement of the logarithmic divisions on the scales would render the instrument exceedingly useful in all computations, where a constant quantity is required to be multiplied by a succession of sines or tangents; inasmuch as that the products might be obtained by inspection at one position of the slider. He therefore determined to have a brass sliding rule made from an enlarged radius, that should contain a Gunter's logarithmic line of numbers at each edge of the slider, and the semicircle of sines from $\overset{\text{s}}{\text{O}}$ to $\overset{\text{s}}{\text{VI}}$ having the word *plus* stamped on it on one side of the rule, and the semicircle, from $\overset{\text{s}}{\text{VI}}$ to $\overset{\text{s}}{\text{XII}}$, stamped with the word *minus*, on the other side, and in such way that the dividing strokes of the slider might be in the same plane with, and contiguous to, the lines on the two sides of the rule. By such arrangement, one face of the rule was sufficient for giving all the successive products that are contained in our long Table XVIII, which occupies forty-five pages. The length of the scale fixed upon is just three feet, allowing 34 inches of graduations in each line; the sines at one side of the rule begin at 0° 35′ and ascend to III plus at the other end of the rule, where the same divisions are numbered back again to $\overset{\text{s}}{\text{V}}$ 29° 25′, at which point the scale becomes evanescent. At the other side of the rule the same divisions are numbered from $\overset{\text{s}}{\text{V}}$ 0° 35′ to IX minus at the opposite end and back again, in the inverse order, to $\overset{\text{s}}{\text{XI}}$ 29° 25′; so that two lines of similar graduations contain all the sines of the circle that can be included. Through the space of 20°, at the beginning of the first and third quadrants, and also at the end of the second and fourth, the degree is subdivided into 5′ spaces; from $\overset{\text{s}}{\text{O}}$ 20° to $\overset{\text{s}}{\text{I}}$ 10° forwards, and from $\overset{\text{s}}{\text{IV}}$ 20° to $\overset{\text{s}}{\text{V}}$ 10° backwards, the degree is subdivided into spaces of 10′ each; and again from $\overset{\text{s}}{\text{VI}}$ 20° to $\overset{\text{s}}{\text{VII}}$ 10° forwards, and from $\overset{\text{s}}{\text{X}}$ 20° to $\overset{\text{s}}{\text{XI}}$ 10° backwards: from $\overset{\text{s}}{\text{I}}$ 10° to $\overset{\text{s}}{\text{II}}$ forwards, and from $\overset{\text{s}}{\text{IV}}$ to $\overset{\text{s}}{\text{IV}}$ 20° backwards, the value of a division is 15′; which is also the case from $\overset{\text{s}}{\text{VII}}$ 10° to $\overset{\text{s}}{\text{VIII}}$ forwards, and from $\overset{\text{s}}{\text{X}}$ to $\overset{\text{s}}{\text{X}}$ 20° back: in the remainder of the scale, from $\overset{\text{s}}{\text{II}}$ to $\overset{\text{s}}{\text{IV}}$ at one side, and from $\overset{\text{s}}{\text{VII}}$ to $\overset{\text{s}}{\text{X}}$ at the other, the values of the divisions are 30′, or entire degrees, accordingly as the degree is divided or not. In all cases, as the degrees are numbered by small figures separately stamped, it is easily seen whether the degree is divided into 10, 6, 4, or 2 parts, which must be read accordingly 6′, 10′, 15′, or 30′, as being $\frac{1}{10}$th, $\frac{1}{6}$th, $\frac{1}{4}$th, or $\frac{1}{2}$ of a degree. The further subdivisions must be read by estimation, for which purpose a sliding quadrangular frame of brass, carrying a fine silver wire stretched across it, has been fitted to embrace the rule, and when the fractional part of the division is cut off by it, it will lie across the fractional part of the corresponding division on the logarithmic line of numbers marked on the slider, which will always of itself show the integer and two decimal figures with certainty, and the third, or thousandth numeral, by estimation, in using the indicating wire. On the back of the rule, two similar logarithmic lines of numbers are inserted on the sides, and two trigonometrical lines, one of sines and the other of tangents, both logarithmic, are marked and numbered in degrees and parts on the slider, by means of which trigonometrical operations may be performed; so that the rule is not only adapted to give by inspection the correction in aberration, and nutation, solar and lunar, due to any maximum quantity multiplied by the sine of its proper argument, at the same time pointing out the proper sign, + or −, due to the correction; but will also give an approximate value in any simple trigonometrical operation at one position of the slider on the posterior face. We shall however confine our examples, of the use of the first face, to operations for which the instrument was principally intended.

General Rule.——Bring the maximum, or constant quantity found on the slider, whether it be for A, A′, L, L′, S, or S′, to the extreme line of the rule denoting $\overset{\text{s}}{\text{III}}$ or $\overset{\text{s}}{\text{IX}}$, and let it remain; then in coincidence with the *argument*, found on one of the sides of the rule, will be the *product* of the maximum by the sine found; and if the argument should have a fractional part of a degree not given on the rule, the wire, laid over that estimated fractional part, will point out on the slider the integer, if any, and decimal portion, constituting the product required as the correction sought.

Example 1.——Let it be required to determine the corrections in aberration, and in nutation of both kinds, in R. A. for ε Boötis, at its meridian passage on June 2, 1827?

This is one of the 16 supplemental principal stars, now given in the Nautical Almanac, for which we have computed the subsidiary numbers in page 472, which, for the day in question, will be these; viz.

Aberration.	s		
A	1	18°	12′
☉	2	11	7
Table XVI. (15h.6) ...			23
Arg. for ab. =	3	29	42
Max.		1s.452	

☾ Nutation.	s		
☊	7	12°	59′
L	5	14	27
Arg. ..	0	27	26
Max...		0s.979	

☉ Nutation.	s		
2 ☉.........	4	23°	0′
S...........	5	17	16
Arg. ..	10	10	16
Max...		0s.054	

Then by the rule,

				s			s
with 14 .52	for	1 .452	at III we have, opposite	3	29°.7 + 12 .61	for	+ 1 .261 ab.
with 9 .79	for	.979	at do. we have, opposite	0	27 .26 + 4 .51	for	+ .451 ☾ nut.
with 5 .4	for	.054	at do. we have, opposite	10	10 .16 − 4 .5	for	− .045 ☉ nut.

Sum of the required corrections................... = + 1 .667
With precession + 2 .621 June 1, in Table XV. gives .. + 1 .156

Sum of four corrections............................ + 2 .823

Example 2.——Let it be required to determine the corrections in north polar distance, of the same star, for the same day?

In this case we shall have the arguments, maxima, and corrections from the rule, thus:

	s				
Arg. in aberration =	6	11°	33′,	max. 14″.592, and correction by rule	−2″.922
Arg. in ☾ nut....	11	25	12,	max. 8 .263, and ditto...........	−0 .692
Arg. in ☉ nut....	9	11	11,	max. 0 .413, and ditto...........	−0 .402

Sum of the required corrections.............. −4 .016
With + 15 .498 Table XV. gives the precession + 6 .450

Sum of the four corrections .... + 2 .434

When the arguments are previously prepared, the corrections can be ascertained in this manner, almost as fast as an amanuensis can commit them to paper. But the chief advantage of the method is, that when the slider has been once set to its position, proper for a given maximum, a succession of corrections can be read off due to any number of intervals that depend on the changes in the sun's longitude, or moon's node respectively, whether these intervals be daily, weekly, or monthly. When the arguments are prepared for each ten days, taken in succession, interpolation for intermediate days affords an easy and ready method of obtaining the daily corrections throughout any year for which such arguments are tabulated. Nor is it necessary that the maxima made use of should be those immediately derived from our Tables, since the operation will be precisely the same, if our maxima be previously modified by the factors we have given, for converting their values into such as depend on any other constants of aberration and nutation. For instance, if we wish to identify the corrections, given by the rule, with those that would arise from the constants adopted in the Catalogue of the Astronomical Society, the maxima arising from our factors (p. 471) would stand thus:

			Max.
Max. of aber. in R. A.....	1s.452 × 1 .0052	....	= 1 .4595
Max. of ☾ nut. in R. A. ...	0 .979 × 0 .9587	....	= 0 .9385
Max. of ☉ nut. in R. A. ..	0 .054 × 1 .256	....	= 0 .0678
Max. of aber. in N. P. D..	14″.592 × 1 .0052	....	= 15 .0297
Max. of ☾ nut. in do......	8 .263 × 0 .9587	....	= 7 .9217
Max. of ☉ nut. in do......	0 .413 × 1 .256	....	= 0 .5187

If we substitute these numbers for the maxima derived from our Tables, we shall have from the rule, or from Table XVIII, the following corrections; viz.

Corr. in aber. in R. A.	$+1^s.267$,	in N. P. D.	$-3''.064$
in ☾ nut. in do..................	$+0\ .441$,	in do......	$-0\ .663$
in ⊙ nut. in do..................	$-0\ .052$,	in do......	$-0\ .501$
Sum of the three	$+1\ .656$	....	$-4\ .228$
Precession as above.........	$+1\ .156$	....	$+6\ .450$
Sum of four corrections......	$+2\ .812$	....	$+2\ .222$
Do. without the factors	$+2\ .823$	....	$+2\ .434$
Difference.................	$0\ .011$	....	$0\ .212$

It has been strongly recommended to astronomers, in some of the annual addresses to the Astronomical Society, to reduce all their observations by means of the constants adopted in their catalogue, in preference to all others. If this recommendation is intended to apply, exclusively, to the reduction of future observations, for obtaining new catalogues of the *mean* places, it will have a good effect in rendering such catalogues comparative among themselves; but if the new constants are intended to be employed in computing the *apparent* places from the *mean* places already determined in existing catalogues, an alteration in the constants will produce a difference between the *observed* and *computed apparent* places of every star, which ought to be identical, when both the observations and computations are correct. If we call the sum of the corrections depending on old constants c, and the sum of those derived from new constants c', then the error, introduced into the computed apparent places, will be $c \sim c'$; which, in the preceding example, is $0^s.011$ in R. A. and $0''.212$ in N. P. D. Hence the same constants must necessarily be employed in computing the *apparent* places, that are involved in the *mean* places of the catalogue that is consulted. It will make no difference, however, in the use of our sliding rule, what the constants are, from which the maxima are obtained; for, as we have said, when the slider is set to the required position, it will always give, by inspection, the whole series of products, that arise successively from all the sines, without altering its position; and when a sliding index is used, the readings may be obtained almost as fast as they can be transcribed. The following Table contains the corrections of ε Boötis for twenty successive days in the month of June, 1827, as given by the slider, after the arguments had been prepared in their proper columns, to be read in succession on the rule, as the index was carried, by short steps, from one argument to another. The mean place for Jan. 0, 1827, as given in the Nautical Almanac of that year, was previously changed to January 1, at $14^h\ 37^m$, the time of culmination, by adding one day's precession, and as much more from our Supplemental Table in page 470, as brought up the place to the said date: and in taking out the daily precessions from Table XV, the quantities were taken from the preceding day, as directed in page 469: otherwise, if the epoch had remained Jan. 0, 1827, the days given in the Table would have been proper. The results, depending on the mechanical operation, accord with the apparent places given in the Nautical Almanac, which probably were computed with our constants.

DIURNAL CHANGES

IN THE APPARENT RIGHT ASCENSION OF ε BOÖTIS.

1827. June.	Arguments for Aberration. ☉ + 1s 18° 12′	Corr. in Aber.	Arguments for ☾ nut. ☊ + 7s 12° 59′	Corr. in ☾ nut.	Arguments for ☉ nut. 2☉ + 4s 23° 0′	Corr. in ☉ nut.	Precession in R. A.	Sum of 4 corr.	Apparent R. A.
	s ° ′	+	s ° ′	+	s ° ′	−	+	+	
2	3 29 42	1s.26	0 27 26	0s.45	10 10 16	0s.04	1s.16	2.83	14h 37m 28s.96
3	4 0 40	1.25	0 27 23	0.45	10 12 11	0.04	1.17	2.83	28.96
4	4 1 37	1.23	0 27 20	0.45	10 14 6	0.04	1.18	2.82	28.95
5	4 2 34	1.22	0 27 16	0.45	10 16 1	0.04	1.19	2.82	28.95
6	4 3 31	1.21	0 27 13	0.44	10 17 56	0.04	1.20	2.81	28.94
7	4 4 28	1.20	0 27 10	0.44	10 19 51	0.03	1.20	2.81	28.94
8	4 5 25	1.18	0 27 7	0.44	10 21 46	0.03	1.21	2.80	28.93
9	4 6 22	1.17	0 27 4	0.44	10 23 41	0.03	1.22	2.80	28.93
10	4 7 20	1.15	0 27 1	0.44	10 25 36	0.03	1.23	2.79	28.92
11	4 8 17	1.14	0 26 57	0.44	10 27 30	0.03	1.24	2.79	28.92
12	4 9 14	1.13	0 26 54	0.44	10 29 24	0.03	1.24	2.78	28.91
13	4 10 11	1.11	0 26 51	0.44	11 1 18	0.02	1.25	2.78	28.91
14	4 11 8	1.09	0 26 48	0.44	11 3 12	0.02	1.26	2.77	28.90
15	4 12 4	1.08	0 26 45	0.44	11 5 6	0.02	1.27	2.77	28.90
16	4 13 1	1.06	0 26 42	0.44	11 7 0	0.02	1.28	2.76	28.89
17	4 13 59	1.04	0 26 38	0.44	11 8 54	0.02	1.29	2.75	28.88
18	4 14 56	1.03	0 26 35	0.44	11 10 48	0.02	1.29	2.74	28.87
19	4 15 53	1.01	0 26 32	0.44	11 12 42	0.02	1.30	2.73	28.86
20	4 16 50	0.99	0 26 29	0.44	11 14 35	0.01	1.30	2.72	28.85
21	4 17 47	0.97	0 26 26	0.44	11 16 29	0.01	1.31	2.70	28.83
22	4 18 44	0.95	0 26 23	0.43	11 18 23	0.01	1.32	2.69	28.82

Mean right ascension on Jan. 1, 1827, at the time of the meridian passage.......... 14 37 26.13

IN THE APPARENT NORTH POLAR DISTANCE OF ε BOÖTIS.

1827. June.	Argts. for Aber. ☉ + 4s 0° 2′	Corr. in Aber.	Argts. for ☾ nut. ☊ + 4s 12° 13′	Corr. in ☾ nut.	Argts. for ☉ nut. 2☉ + 4s 18° 11′	Corr. in ☉ nut.	Precession in N. P. D.	Sum of 4 corr.	Apparent N. P. D.
	s ° ′	−	s ° ′	−	s ° ′	−	+		
2	6 11 33	2″.92	11 25 12	0″.69	9 11 11	0″.40	6.45	+2.44	62h 11m 32s.70
3	6 12 30	3.15	11 25 9	0.70	9 13 5	0.40	6.50	+2.25	32.53
4	6 13 27	3.38	11 25 6	0.71	9 14 59	0.40	6.54	+2.05	32.31
5	6 14 24	3.62	11 25 3	0.72	9 16 54	0.40	6.59	+1.85	32.11
6	6 15 21	3.86	11 25 0	0.72	9 18 48	0.39	6.64	+1.67	31.93
7	6 16 18	4.09	11 24 56	0.73	9 20 42	0.39	6.68	+1.47	31.73
8	6 17 15	4.32	11 24 53	0.74	9 22 37	0.38	6.73	+1.29	31.55
9	6 18 12	4.56	11 24 50	0.74	9 24 31	0.38	6.78	+1.10	31.36
10	6 19 9	4.78	11 24 47	0.75	9 26 25	0.37	6.83	+0.93	31.19
11	6 20 6	5.01	11 24 43	0.76	9 28 20	0.37	6.87	+0.73	30.99
12	6 21 3	5.23	11 24 40	0.77	10 0 14	0.36	6.92	+0.56	30.82
13	6 22 0	5.45	11 24 37	0.78	10 2 8	0.35	6.97	+0.39	30.65
14	6 22 57	5.69	11 24 34	0.78	10 4 2	0.34	7.01	+0.20	30.46
15	6 23 55	5.90	11 24 30	0.79	10 5 56	0.33	7.06	+0.04	30.30
16	6 24 52	6.13	11 24 27	0.80	10 7 51	0.33	7.11	−0.14	30.12
17	6 25 50	6.35	11 24 24	0.81	10 9 45	0.32	7.16	−0.32	29.94
18	6 26 47	6.56	11 24 21	0.81	10 11 39	0.31	7.20	−0.48	29.78
19	6 27 45	6.78	11 24 17	0.82	10 13 34	0.30	7.25	−0.65	29.61
20	6 28 42	7.00	11 24 14	0.83	10 15 28	0.29	7.30	−0.82	29.44
21	6 29 39	7.21	11 24 11	0.84	10 17 22	0.28	7.34	−0.99	29.27
22	7 0 36	7.42	11 24 8	0.85	10 19 17	0.27	7.39	−1.15	29.11

Mean N. P. D. on Jan. 1, 1827, at the meridian passage.......... 62 11 30.26

ADDITIONAL ERRATA IN VOLUME I.

Page	Argument.	Column.	Instead of	Insert
xiv	74° 50′	In Errata	....	p. 195
	84 0	....	....	p. 196
5	App. Zen. Dis. 27° 0′	Log. Refrac.	1 .46204	1 .46240
10, 11, 12	at the head	....	29 .94	29 .922
13	Barom. 29 .75	Log.	9 .6973	9 .9973
17	Fah. Ther. 40	Log.	0 .2292	0 .2992
21	III 52^{m}	....	1 .143	1 .133
			s	s
22	II | XIV 4^{m}	....	1 26° 40′	1 26° 46′
23	II^{h} 30^{m}	....	.10975	.10875
	III 17	....	.11356	.11366
30	I 27	4	18″.574	18″.584
	head	....	I XII	I XIII
			s	s
35	V | XVII 9^{m}	....	2 20° 37′	2 20° 27′
36	V 15^{m}	....	6 .98073	0 .98073
37	IV 44	....	4 12 .28	2 12 .28
38	XV 60	....	0 .436	0 .426
61	— V 28° 0′ to 29° 58′	18	Add 0 .2 to each number down the Column, except to the last.	
90	VII 27	18	17 .276	17 .976
	VII 27 10	18	17 .278	17 .978
97	....	....	page 99	97
107	at the bottom	....	.3702	.3708
110	Solar Days 17	Add seconds	2 .893	2 .793
	18	Add seconds	3 .057	2 .957
	19	Add seconds	3 .221	3 .121
111	6^{h} 48^{m}	9^{s}	3 .550	2 .550
			s	s
113, 114	Schedir	☉ Nut. R. A.	7 2° 52′	7 2° 16′
	Polaris	Aber. N. P. D.	11 16 58	11 16 55
114	α Cassiop. V 27	Ditto	0 .78	0 .88
	V 28	Ditto	0 .48	0 .58
	V 29	Ditto	0 .23	0 .29
115	α Arietis III 27	Ditto	7 .69	7 .00
	β Orionis V 7	☾ Nut. R. A.	.415	.405
115, 116	α Ceti	☉ Nut. N. P. D.	10 14 56	10 14 11
117, 118	α Tauri	☾ Nut. R. A.	6 3 33	6 3 28
118	α Aurigæ	V 27 ☾ Nut. R. A.	0^{s}.073	0 .083
			s	
119, 120	α Orionis	Aber. R. A.	6 13° 12′	6 3 13
		Aber. N. P. D.	2 29 50	2 28 23
121	α Can. Min. IV 3	☉ Nut. R. A.	.953	.053
		☾ Nut. N. P. D.	.78	.73
121, 122	α Geminorum	☾ Nut. N. P. D.	8 14 20	8 14 6
122	....	☉ Nut. R. A.	9^{s}.054	0 .054
	α Gemin. V 4	Aber. R. A.	.792	.692
	β Gemin. V 10	☾ Nut. R. A.	.60	.460
			s	
123, 124	θ Ursæ Maj.	☉ Nut. N. P. D.	7 11° 50′	7 12 5
124	....	Ann. Var. N. P. D.	+15 .01	+16 .01
125, 126	β Leonis	☉ Nut. N. P. D.	6 5 5	6 5 21
	β Virginis	☉ Nut. N. P. D.	6 5 58	6 4 58
126	β Leonis IV 23	Aber. R. A.	.764	.774
	IV 26	Aber. N. P. D.	6 .23	6 .33
127	ζ Urs. Maj. IV 11	Aber. N. P. D.	13 .88	13 .98
127, 128	α Virginis	☾ Nut. R. A.	6 5 24	6 5 33
		Aber. N. P. D.	2 3 35	2 3 32
		☾ Nut. N. P. D.	5 5 16	5 5 6
	ζ Ursæ Maj.	☉ Nut. R. A.	4 13 40	4 20 12
		☾ Nut. N. P. D.	5 4 41	5 40 51
128	γ Urs. Maj. IV 15	☉ Nut. N. P. D.	3 .28	0 .28
	α Virginis V 1	Aber. R. A.	.517	.617
	ζ Urs. Maj. IV 16	Aber. R. A.	1 .539	1 .549
	V 25	☾ Nut. R. A.	.119	.109
	IV 16	☾ Nut. N. P. D.	12 .65	12 .85
129	α² Libræ IV 5	Aber. R. A.	1 .192	1 .092
	III 26	Aber. N. P. D.	4 .52	5 .52
129, 130	η Ursæ Maj.	Aber. R. A.	2 3 49	2 2 49
130	α Boötis V 6	Aber. N. P. D.	5 .16	5 .06
	V 11	Aber. N. P. D.	4 .15	4 .05
132	α Cor. Bor. V 28	Aber. N. P. D.	0 .43	0 .53
133	α Scorpii III 27	Aber. R. A.	1 .329	1 .320
	α Herculis IV 14	Aber. N. P. D.	9 .93	8 .93
	IV 15	Aber. N. P. D.	9 .78	8 .78
	β Dracon. III 12	☾ Nut. R. A.	.391	.491

Page	Argument.	Column.	Instead of	Insert
			s	s
133, 134	α Scorpii	☉ Nut. N. P. D.	3 24 23	3 23 23
	α Herculis	☉ Nut. N. P. D.	3 12 26	3 12 13
	β Draconis	☉ Nut. R. A.	5 20 52	5 18 33
134	α Herculis V 1	☾ Nut. N. P. D.	4 .52	4 .62
135, 136	γ Draconis	☾ Nut. R. A.	5 27 8	5 27 2
	ι Sagittarii	☉ Nut. R. A.	5 29 15	5 29 12
137	α Aquilæ IV 7	Aber. R. A.	1 .063	1 .073
137, 138	α Lyræ	☉ Nut. R. A.	6 4 8	6 4 28
	s	☉ Nut. N. P. D.	2 22 5	2 22 50
	α Lyræ III 0°	☾ Nut. R. A.	0^{s}.969	0 .727 vide p. 340
138	γ Aquilæ V 8	Aber. R. A.	.517	.507
	α² Capric. III 29	☉ Nut. N. P. D.	.471	.371
	α Cygni IV 15	☉ Nut. N. P. D.	.397	.297
139, 140	β Aquilæ	☉ Nut. R. A.	6 1° 23′	6 1° 20′
	α² Capricorni	Increase the Column of Aber. in R. A. by a 68th part.		
140	α Cygni V 14	☉ Nut. R. A.	.612	.012
141	α Cephei IV 2	Aber. N. P. D.	17 .83	16 .83
			s	s
141, 142	α Cephei	☾ Nut. N. P. D.	1 19° 33′	1 19° 48′
	β Cephei	☾ Nut. R. A.	8 17 42	8 17 54
		☉ Nut. R. A.	8 15 13	8 15 11
142	α Cephei V 7	Aber. R. A.	1 .167	1 .067
	V 8	Aber. R. A.	1 .123	1 .023
	β Cephei V 12	Aber. N. P. D.	6 .13	6 .23
143, 144	α Piscis Austr.	☉ Nut. R. A.	5 8 41	5 16 10
	α Pegasi	☉ Nut. R. A.	6 7 9	6 6 49
		Aber. N. P. D.	2 1 49	2 2 5
	α Andromedæ	Aber. N. P. D.	1 6 48	1 6 43
144	α Piscis Austr.	Ann. Var. N. P. D.	+19″.10	— 19″.10
	α Pegasi	Ann. Var. N. P. D.	+19. 43	— 19 .43
	Common Arg.	Right hand.	Read as in preceding alternate Pages.	
	α Pegasi V 13	Aber. N. P. D.	3 .30	3 00
146	α Lyræ	Maskelyne Pr. R. A.	3^{s}.013	2 .013
	α Ursæ Maj. Piazzi	Ann. Var. N. P. D.	....	19 .17
			s	
147	α Canis Maj.	A′ 1800	2 16° 0′	2 26 0
	α Geminorum	S 1800	11 25 15	5 25 15
	α Cygni	S′ 1800	1 23 29	1 23 44
		1830	1 23 14	1 23 29
	α Aurigæ 1830	A	6 12 56	6 12 51
		Change in 30 Years	— 26	— 31
	α Geminorum 1800	S	11 25 11	5 25 15
	α Cor. Bor. 1830	A′	5 22 28	3 22 28
148	α Ceti	Prec. N. P. D. 1830	14″.640	14 .575
	α Ursæ Min. 1800	Aber. R. A.	40 .916	40 .525
	1800	Aber. N. P. D.	19 .920	20 .112
	1830	Aber. N. P. D.	20 .096	20 .089
	α Orionis 1800	☾ Nut. R. A.	1 .179	1 .165
	β Virginis 1830	☾ Nut. N. P. D.	7 .290	7 .200
	α² Libræ 1830	☾ Nut. N. P. D.	8 .397	8 .307
	α Ophiuchi 1800	☾ Nut. N. P. D.	9 .620	9 .602
	ι Sagitt. 1800, 1830	☉ Nut. N. P. D.	0 .343	0 .434
149	α Can. Maj. July 9	Aber.	— 0 .012	+ 0 .012
	γ Pegasi April 10	H. m.	6 .484	5 .484
150	α Leonis Jan. 0	Aber.	1 .889	0 .889
152	α Aquarii April 10	Aber.	— 0 .037	+ 0 .037
170	12^{m} 40^{s}	Diff.	2 .01	3 .01
	30 0	Reduction.	10 43 .34	10 45 .34
172	Explanation	At bottom	☉ — ☾	☾ — ☉
	Tab. 2. 330	2 B — C	blank	0 .02
		3 C — 4 B	blank	0 .17
181	Tab. 1. 1820	☾'s ☊	25′ 55″.9	26 55 .9
195	74′ 50″	3	14 23 .67	14 23 .37
201	2^{h} 50^{m}	2	18″.8	10 .8
	0 20	13	11 .6	10 .5
	1 30	13	41 .7	42 .7
	Seconds 0^{h} 0^{m}	Col. 18	12^{h} 15^{m}	0 0
	0 10	Ditto	12 50	11 .50
	bottom	....	X — III —	IX — III —
239	Alt. Nonag. 37	17°	10′ 13″.43	10 33 .43
	42	15	10 33 .46	10 23 .46
270	Tab. 3 1^{h} 0^{m}	last column.	0″.17	0 .07
275	79 Ursæ Maj.	R. A.	13^{h} 19^{m} 49^{s}.8	13 15 49 .8
287	Part I.	under the Table	+ —	— + always.
	Part IV. 4″	40″	0″.97	0 .67

ADDITIONAL ERRATA.

Page	Argument.	Column.	Instead of	Insert
298	Example 2.	pp. 5′ 30″	0′ 9″.92	0 6.42
		whole correc.	8 8.98	8 5.48
300	Example 2.	Log. of the Refrac.	485″.52	486.64
	Ditto	Sum of Refrac.	8′ 3″.23	8 4.35
	20th line	from bottom	foot	feet
301	last line but 7	....	it	its
302	In the formula	at middle of page	0001 and 002083	.0001 and .002083
305	Example 2.	pp. 5′ 30″	12.76	5.64
		Mean Refrac.	8 16.25	8 9.14
		Log. of gives	496″.25 and log.	489″.14 and log.
		Correct Refrac.	8′ 15″.26	8′ 8″.15
307	Second Example in the Table.	Bradley's	8 8.98	8 5.48
		French	8 3.23	8 4.35
		Carlini's	8 15.26	8 8.15
		Average	8 8.07	8 6.71
309	Example 1.	first line	North Polar Dist.	Declination.
			$+0^s.3351$	$+0^s.3551$
310	Formula	line 8 from top	Cotang. A.	Cotang. R. A. twice
311	Exam. 1. Aldebaran	See p. 464 in the Appendix.		
	Lunar Nutation.	Formula in R. A.	+7.1822	—7.1822
		Log. cos. R. A.	Tab. V. p. 27	Tab. V. p. 26
312	....	....	Dele the note N.B.	
	Solar Nut. R. A.	second line /	— 0.3982	+0.3982
			sin R. A 2 ⊙	sin R. A. sin 2 ⊙
		third line	cos R. A. 2 ⊙	cos R. A. cos 2 ⊙
		sixth line	= 0.9173 = q	— 0.9173 = q
		ninth line	Table VI.	VII.
		last line	Table VII.	VI.
	Line 8 from top	Log. tang. L	1.32844	8.32844
314	Aber. in N. P. D.	line 5	= 18.580	= — 18.580
315	From the head	line 5	— log. q	= log. q
318	Aldebaran	Lunar Nut.	See Appendix, p. 464.	
320	Tab. XV.	Epoch	See Appendix, p. 469.	
			s	
325	Tab. E.	line 13 from bottom	—7 3° 1′ 9″	=7 3 1 9
326	Tang. dec.	line 3 from top	× 7″.16 — 2″.072	×—7.16=—2.072
329	Pole Star.	line 7	const. log 8.4800330	8.4800330 × 2
		8	0.0253059	9.0253059
		10	0″.0013	0.0001
			7.1108881	6.1108881
		11	4.5195	4.5183
331	From the head	1	60843	60849
		10	m	m'
		33	4° 42′ 20″	4 41 20
332	From the head	14	miles	yards
333	Space and Time.	line 6	$7^h\ 20^m\ 0^s$	6 40 0
	From the head	lines 10 and 11	7 57 22.33	7 17 22.33
		13	57^m = 14° 15′	17^m = 4° 15′
334	From the head	line 5	specified day	preceding noon
	Equation.	lines 14 and 35	3 36.419	3 36.249
			— 3 44.34	— 3 44.17
		line 16	22 44 37.06	22 44 37.23
		lines 18 and 20	22 46 56.36	22 46 56.53
		line 22	22 44 37.06	22 44 37.23
		24	22 48 21.23	22 48 21.40
		26	23 50 29.83	23 50 30
335	Clock's Rate.	12	$0^s.335$	$0^s.383$
		13	+0.442	+0.492
		18	.099	.009
			s	s
336	Arguments	in N. P. D.	4 9° 8′	3 19° 8′
	N. P. D.	Arg.	Ditto	Ditto
		Arg. 3 19 8	+4″.36	+5.30
		Cor. in N. P. D.	+1.24	+2.18

Page	Argument.	Column.	Instead of	Insert
336	....	True App. N. P. D.	82° 38′ 1″.44	82° 38′ 2″.38
338	line 5	from head	2d volume	Appendix
339	α Virginis	Corrigenda.	S′	L′
341	N. P. D.	Mean N. P. D.	38 22 13.62	38 29 13.62
				s
350	⊙'s Long.	line 7 from head	29° 40′ 6″.8	2 9° 40′ 6″.8
354	From bottom	line 4	.137	.124
		3	+5.932	+5.919
		1	$13^m\ 35^s.832$	13 35.819
355	From the head	6	Arg. $4^h\ 30^m$	Arg. $2^h\ 30^m$
	From the foot	13	$17^h\ 5^m$ 56.824	$7^h\ 5^m$ 56.824
356	From the head	lines 5 and 7	1st of May	2d of May
		line 11	15′ 53″.1	15′ 53″.3
		13	15 52.3	15 53.3
358	Examples 1 and 2.	Arg. erroneous.	⊙—☾ put for ☾—⊙	Vide App^x. p. 492.
	Table 12.	In Exam. 1 and 2.	Corrections wrong.	Vide App^x. p. 493.
363	From foot	line 4	is midnight	is for midnight
			s	
365	From head	3	8 19° 6′ 51″.8	8 20 6 51.8
366		3	his	its
		30	$\frac{\sin 2 L}{\sin 1''}$ +	$\frac{\sin 2 L}{\sin 1''}$ —
		32	51 25 40	51 28 40
368	Table 10.	Example 1.	15° 45′	15′ 45″
369	Table 11.	Example 1.	Oct. 5, noon	Oct. 4, midnight, and re-compute.
	From head	lines 22 and 23.	1^m and 50^s	1′ and 50″
375	From foot	line 13	5° 43′ 15″	5° 43′ 17″
	Tab. 5.	21	⊙'s long.	✱'s long.
377	From head	6	Longitude	Latitude
377, 378	Example 2.	in both pages	Conjunction.	Opposition.
380	From foot	line 2	242° &c.	342° &c.
385	From foot	10	— 43.5	— 48.5
		last line	east	west
386	line 11	....	dec.	lat.
	20	....	68°	86°
386. 391. 392. 393. 395.	The second part of Parallax is applied with a wrong sign, and therefore vitiates the results. Vide Appendix, p. 533, &c.			
400	head of Tab. 2.	Arg. for Lat.	— 3 signs	— 3 signs N. lat. + 3 signs S. lat.
402	Longitude.	line from head 10	Aber. +18″.08	+ 18.08 × sec ✱'s lat. = 18.13
	Latitude.	8	Aber. + 8.99	+ 8.99 × sin ✱'s lat. = 0.71
	App. Longitude.	13	247 19 7.75	247 19 7.8
	App. Latitude.	9	4 32 37.24	4 32 28.96
412	Circumpol^r. Stars	Methods given here not correct. Vide Vol. II. pp. 327—332.		
414	High & Low Stars	Deviation of Instrum^t. not correct. Vide Vol. II. pp. 332-334.		
415	Precessions.	line from foot 14	Precess. in Long.	Precess. in R. A.
417	Precess. in R. A.	10	= 0.0506	= 0.3506
421	line 2	....	1821	1822
431	β Serpentis	Declination.	—15 59 25.9	+15 59 25.9
437	Zen. Dis. 45°	Log. δθ	1.76111	1.76611
	59	δθ	4′ 46″.47	4 56.47
	64 65	Diff.	blank	1932
452	Synopsis	from foot line 5	— cos 2 ☊	— $\frac{\cos 2 ☊}{100}$
454, 455	at the head	formula for D	— cos 2 ☊	— $\frac{\cos 2 ☊}{100}$
466	line 20	from head	value of L	value of L′

CONTENTS OF THE APPENDIX.

G. WOODFALL, Printer, Angel Court, Skinner Street, London.

For EU product safety concerns, contact us at Calle de José Abascal, 56–1°,
28003 Madrid, Spain or eugpsr@cambridge.org.

www.ingramcontent.com/pod-product-compliance
Ingram Content Group UK Ltd.
Pitfield, Milton Keynes, MK11 3LW, UK
UKHW050227090726
473066UK00011B/1027

* 9 7 8 1 1 0 8 0 6 4 0 5 7 *